INTEGRATED MATH 1

John A. Carter
Gilbert Cuevas
Roger Day
Carol Malloy
Berchie Holliday
Beatrice Moore Luchin
Jerry Cummins
Ruth Casey
Viken Hovsepian
Dinah Zike

Mc Graw Hill Education

mheonline.com

Send all inquiries to:
McGraw-Hill Education
8787 Orion Place
Columbus, OH 43240

ISBN: 978-0-07-663858-1
MHID: 0-07-663858-8

Printed in the United States of America.

5 6 7 8 9 QVS 17 16

Contents in Brief

Authors

Our lead authors ensure that the Macmillan/McGraw-Hill and Glencoe/McGraw-Hill mathematics programs are truly vertically aligned by beginning with the end in mind—success in Integrated Math I and beyond. By "backmapping" the content from the high school programs, all of our mathematics programs are well articulated in their scope and sequence.

Lead Authors

John A. Carter, Ph.D.
Principal
Adlai E. Stevenson High School
Lincolnshire, Illinois

Areas of Expertise: Using technology and manipulatives to visualize concepts; mathematics achievement of English-language learners

Gilbert J. Cuevas, Ph.D.
Professor of Mathematics Education
Texas State University—San Marcos
San Marcos, Texas

Areas of Expertise: Applying concepts and skills in mathematically rich contexts; mathematical representations

Roger Day, Ph.D., NBCT
Mathematics Department Chairperson
Pontiac Township High School
Pontiac, Illinois

Areas of Expertise: Understanding and applying probability and statistics; mathematics teacher education

Carol Malloy, Ph.D.
Associate Professor
University of North Carolina at Chapel Hill
Chapel Hill, North Carolina

Areas of Expertise: Representations and critical thinking; student success in Algebra 1

Program Authors

Ruth Casey
Mathematics Consultant
Regional Teacher Partner
University of Kentucky
Lexington, Kentucky

Areas of Expertise: Graphing technology and mathematics

Jerry Cummins
Mathematics Consultant
Former President, National Council of Supervisors of Mathematics
Western Springs, Illinois

Areas of Expertise: Graphing technology and mathematics

Dr. Berchie Holliday, Ed.D.
National Mathematics Consultant
Silver Spring, Maryland

Areas of Expertise: Using mathematics to model and understand real-world data; the effect of graphics on mathematical understanding

Beatrice Moore Luchin
Mathematics Consultant
Houston, Texas

Areas of Expertise: Mathematical literacy; working with English language learners

Contributing Author

Dinah Zike FOLDABLES
Educational Consultant
Dinah-Might Activities, Inc.
San Antonio, Texas

These professionals were instrumental in providing valuable input and suggestions for improving the effectiveness of the mathematics instruction.

Lead Consultants

Viken Hovsepian
Professor of Mathematics
Rio Hondo College
Whittier, California

Jay McTiche
Educational Author and Consultant
Columbia, Maryland

Consultants

Mathematical Content

Grant A. Fraser, Ph.D.
Professor of Mathematics
California State University, Los Angeles
Los Angeles, California

Arthur K. Wayman, Ph.D.
Professor of Mathematics Emeritus
California State University, Long Beach
Long Beach, California

Gifted and Talented

Shelbi K. Cole
Research Assistant
University of Connecticut
Storrs, Connecticut

College Readiness

Robert Lee Kimball, Jr.
Department Head, Math and Physics
Wake Technical Community College
Raleigh, North Carolina

Differentiation for English-Language Learners

Susana Davidenko
State University of New York
Cortland, New York

Alfredo Gómez
Mathematics/ESL Teacher
George W. Fowler High School
Syracuse, New York

Graphing Calculator

Ruth M. Casey
T^3 National Instructor
Frankfort, Kentucky

Jerry Cummins
Former President
National Council of Supervisors of Mathematics
Western Springs, Illinois

Mathematical Fluency

Robert M. Capraro
Associate Professor
Texas A&M University
College Station, Texas

Pre-AP

Dixie Ross
Lead Teacher for Advanced Placement Mathematics
Pflugerville High School
Pflugerville, Texas

Reading and Writing

ReLeah Cossett Lent
Author and Educational Consultant
Morgantown, GA

Lynn T. Havens
Director of Project CRISS
Kalispell, Montana

Online Guide

connectED.mcgraw-hill.com

The eStudentEdition allows you to access your math curriculum anytime, anywhere.

The icons found throughout your textbook provide you with the opportunity to connect the print textbook with online interactive learning.

Investigate

Animations illustrate math concepts through movement and audio.

Vocabulary tools include fun Vocabulary Review Games.

Multilingual eGlossary presents key vocabulary in 13 languages.

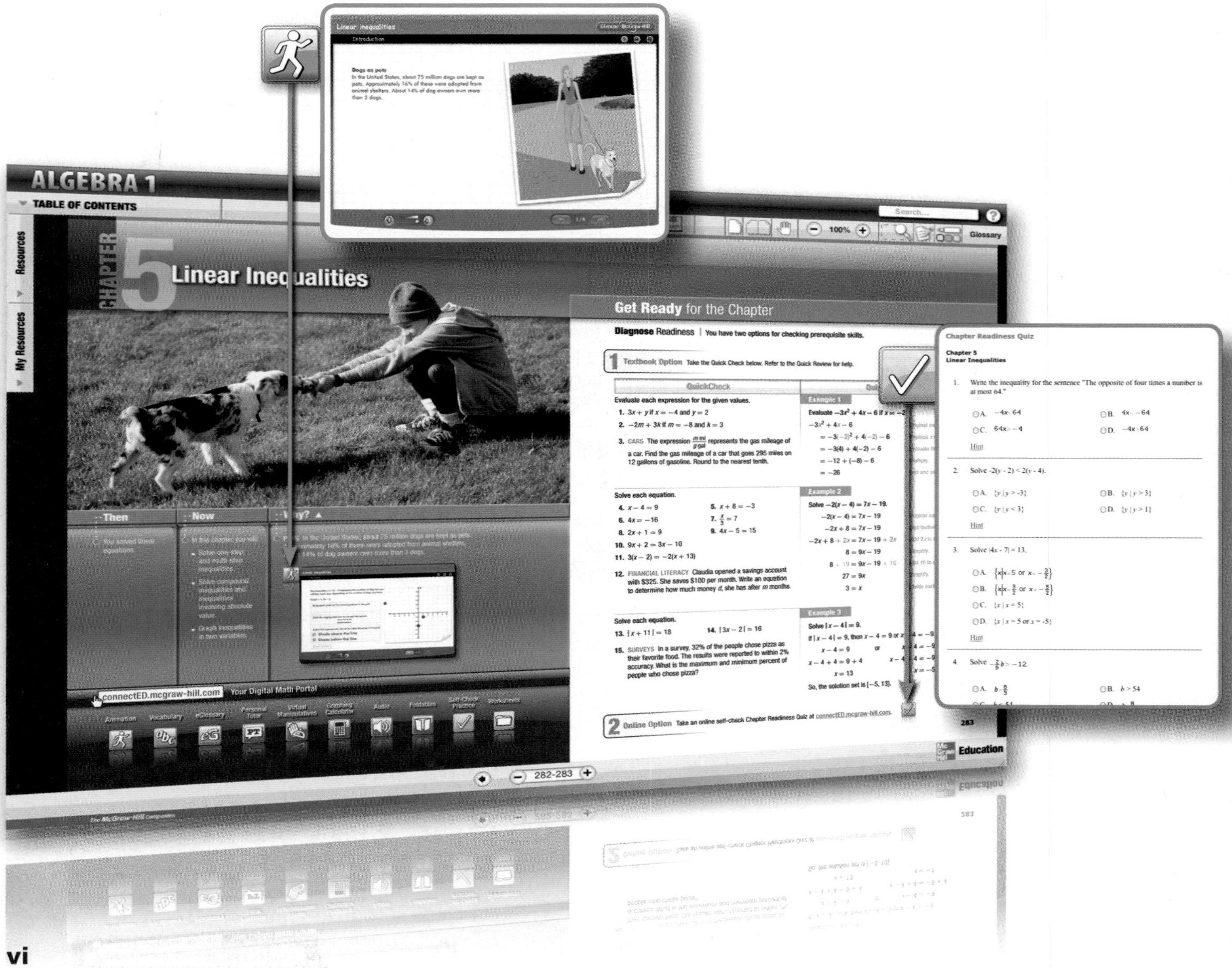

Learn

Personal Tutor presents an experienced educator explaining step-by-step solutions to problems.

Virtual Manipulatives are outstanding tools for enhancing understanding.

Graphing Calculator provides other calculator keystrokes for each Graphing Technology Lab.

Audio is provided to enhance accessibility.

Foldables provide a unique way to enhance study skills.

Practice

Self-Check Practice allows you to check your understanding and send results to your teacher.

Worksheets provide additional practice.

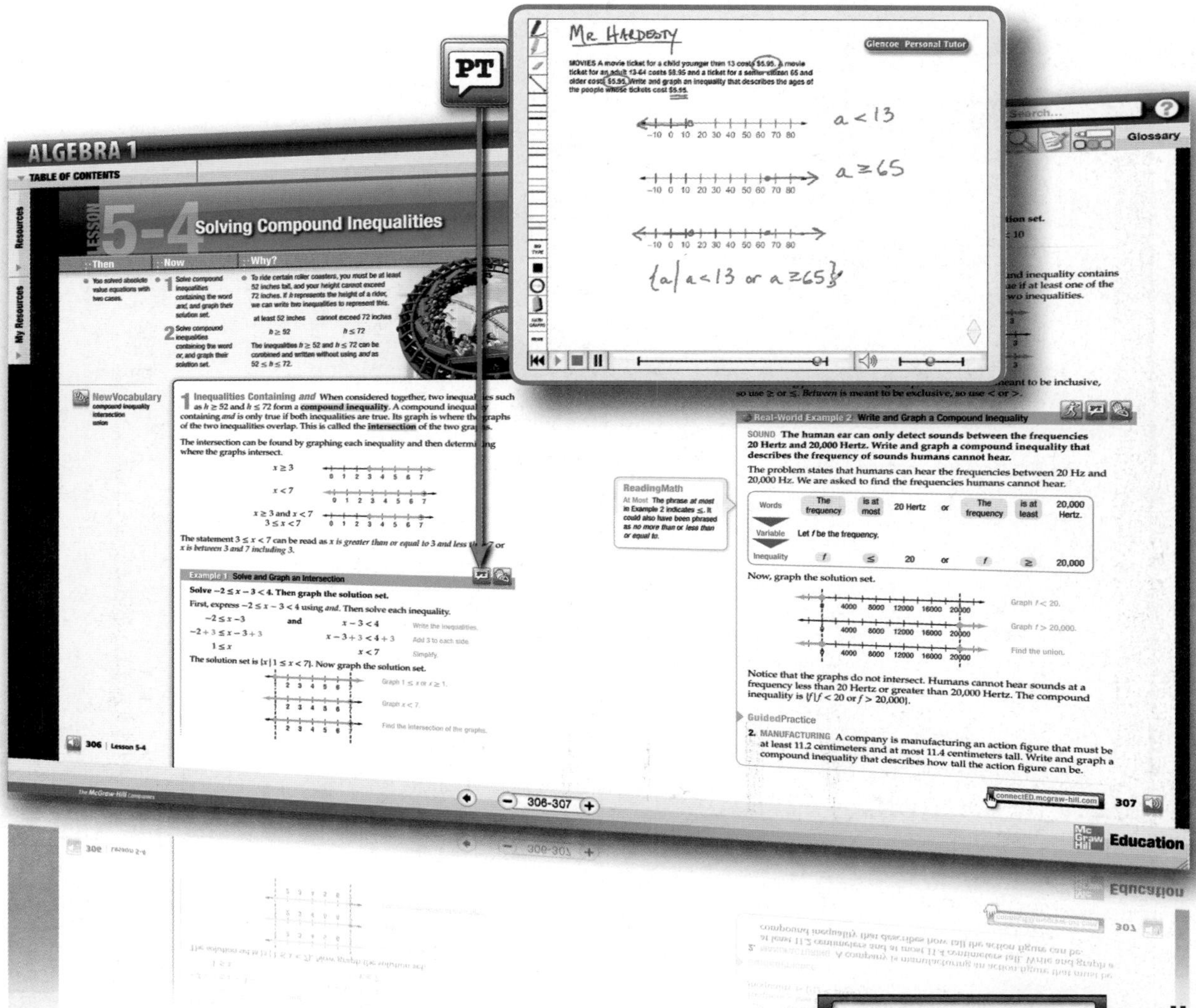

CHAPTER 0 Preparing for Integrated Math I

connectED.mcgraw-hill.com **Your Digital Math Portal**

Vocabulary p. P2

Multilingual eGlossary p. P2

Personal Tutor p. P18

Foldables p. P2

CHAPTER 1

Expressions, Equations, and Functions

JupiterImages/Comstock Images/Alamy

connectED.mcgraw-hill.com **Your Digital Math Portal**

Animation pp. 18, 72

Vocabulary pp. 4, 139

Multilingual eGlossary pp. 4, 74

Personal Tutor pp. 6, 99

CHAPTER 2

Linear Equations

Virtual Manipulatives pp. 23, 90

Graphing Calculator pp. 55, 118

Foldables pp. 4, 74

Self-Check Practice pp. 3, 73

CHAPTER 3

Linear Functions

connectED.mcgraw-hill.com **Your Digital Math Portal**

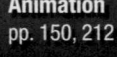

Animation pp. 150, 212

Vocabulary pp. 152, 272

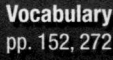

Multilingual eGlossary pp. 152, 214

Personal Tutor pp. 158, 248

Flame/Alamy

CHAPTER 4 Equations of Linear Functions

Virtual Manipulatives pp. 158, 248

Graphing Calculator pp. 169, 215

Foldables pp. 152, 214

Self-Check Practice pp. 151, 213

CHAPTER 5

Linear Inequalities

connectED.mcgraw-hill.com **Your Digital Math Portal**

Animation pp. 282, 332

Vocabulary pp. 284, 378

Multilingual eGlossary pp. 284, 334

Personal Tutor pp. 293, 342

CHAPTER 6

Systems of Linear Equations and Inequalities

Virtual Manipulatives pp. 291, 336

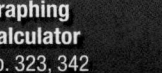

Graphing Calculator pp. 323, 342

Foldables pp. 284, 334

Self-Check Practice pp. 283, 383

CHAPTER 7

Exponents and Exponential Functions

connectED.mcgraw-hill.com **Your Digital Math Portal**

Michael Dunning/Photographer's Choice/Getty Images

CHAPTER 8

Radical Functions, Rational Functions and Geometry

CHAPTER 9 Statistics and Probability

connectED.mcgraw-hill.com **Your Digital Math Portal**

Animation pp. 520, 558

Vocabulary pp. 522, 642

Multilingual eGlossary pp. 522, 642

Personal Tutor pp. 531, 572

CHAPTER 10 Tools of Geometry

LatinStock Collection/Alamy

Virtual Manipulatives pp. 533, 604

Graphing Calculator pp. 548

Foldables pp. 522, 642

Self-Check Practice pp. 521, 559

CHAPTER 11

Parallel and Perpendicular Lines

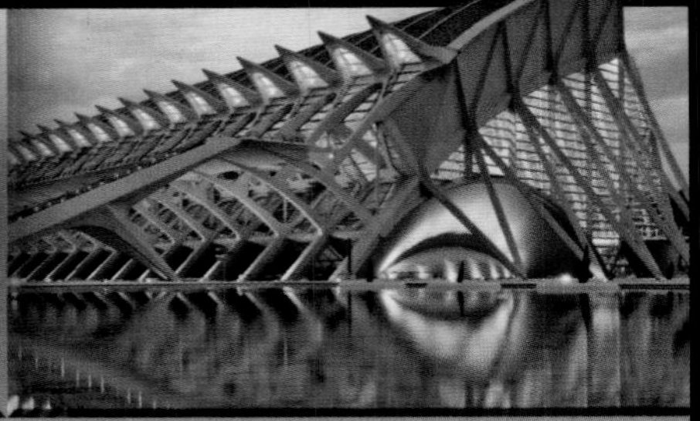

connectED.mcgraw-hill.com **Your Digital Math Portal**

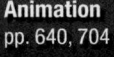

Animation pp. 640, 704

Vocabulary pp. 642, 706

Multilingual eGlossary pp. 642, 706

Personal Tutor pp. 688, 747

Jeremy Woodhouse/Masterfile

CHAPTER 12

Congruent Triangles

Scott Markewitz/Getty Images

Virtual Manipulatives pp. 688, 753

Graphing Calculator pp. 657, 764

Foldables pp. 642, 706

Self-Check Practice pp. 641, 705

CHAPTER 13

Quadrilaterals

connectED.mcgraw-hill.com **Your Digital Math Portal**

Animation pp. 800, 858

Vocabulary pp. 802, 860

Multilingual eGlossary pp. 802, 860

Personal Tutor pp. 804, 861

CHAPTER 14

Similarity, Transformations and Symmetry

Barry Rosenthal/The Image Bank/Getty Images

Virtual Manipulatives pp. 806, 884

Graphing Calculator pp. 812, 939

Foldables pp. 802, 860

Self-Check Practice pp. 801, 859

CHAPTER 15

Circles

connectED.mcgraw-hill.com **Your Digital Math Portal**

Animation pp. 672, 962

Vocabulary pp. 674, 964

Multilingual eGlossary pp. 674, 964

Personal Tutor pp. 697, 941

Student Handbook

Reference

Rubberball/Getty Images

CHAPTER 0

Preparing for Integrated Math I

Now

Chapter 0 contains lessons on topics from previous courses. You can use this chapter in various ways.

- Begin the school year by taking the Pretest. If you need additional review, complete the lessons in this chapter. To verify that you have successfully reviewed the topics, take the Posttest.
- As you work through the text, you may find that there are topics you need to review. When this happens, complete the individual lessons that you need.
- Use this chapter for reference. When you have questions about any of these topics, flip back to this chapter to review definitions or key concepts.

connectED.mcgraw-hill.com Your Digital Math Portal

Animation | Vocabulary | eGlossary | Personal Tutor | Virtual Manipulatives | Graphing Calculator | Audio | Foldables | Self-Check Practice | Worksheets

Get Started on the Chapter

You will review several concepts, skills, and vocabulary terms as you study Chapter 0. To get ready, identify important terms and organize your resources.

FOLDABLES® Study Organizer

Throughout this text, you will be invited to use Foldables to organize your notes.

Why should you use them?

- They help you organize, display, and arrange information.
- They make great study guides, specifically designed for you.
- You can use them as your math journal for recording main ideas, problem-solving strategies, examples, or questions you may have.
- They give you a chance to improve your math vocabulary.

How should you use them?

- Write general information — titles, vocabulary terms, concepts, questions, and main ideas – on the front tabs of your Foldable.
- Write specific information — ideas, your thoughts, answers to questions, steps, notes, and definitions — under the tabs.
- Use the tabs for:
  - math concepts in parts, like types of triangles,
  - steps to follow, or
  - parts of a problem, like *compare* and *contrast* (2 parts) or *what*, *where*, *when*, *why*, and *how* (5 parts).
- You may want to store your Foldables in a plastic zipper bag that you have three-hole punched to fit in your notebook.

When should you use them?

- Set up your Foldable as you begin a chapter, or when you start learning a new concept.
- Write in your Foldable every day.
- Use your Foldable to review for homework, quizzes, and tests.

Review Vocabulary

English		Español
integer	p. P7	entero
absolute value	p. P11	valor absolute
opposites	p. P11	opuestos
reciprocal	p. P18	recíproco
perimeter	p. P23	perímetro
circle	p. P24	círculo
diameter	p. P24	diámetro
center	p. P24	centro
circumference	p. P24	circunferencia
radius	p. P24	radio
area	p. P26	area
volume	p. P29	volumen
surface area	p. P31	area de superficie
probability	p. P33	probabilidad
sample space	p. P33	espacio muestral
complements	p. P33	complementos
tree diagram	p. P34	diagrama de árbol
odds	p. P35	probabilidades
mean	p. P37	media
median	p. P37	mediana
mode	p. P37	moda
range	p. P38	rango
quartile	p. P38	cuartil
interquartile range	p. P38	amplitud intercuartílica
outliers	p. P39	valores atípicos
bar graph	p. P41	gráfica de barras
histogram	p. P41	histograma
line graph	p. P42	gráfica lineal
circle graph	p. P42	gráfica circular
box-and-whisker plot	p. P43	diagrama de caja y patillas

CHAPTER 0 Pretest

Determine whether you need an estimate or an exact answer. Then solve.

1. **SHOPPING** Addison paid \$1.29 for gum and \$0.89 for a package of notebook paper. She gave the cashier a \$5 bill. If the tax was \$0.14, how much change should Addison receive?

2. **DISTANCE** Luis rode his bike 1.2 miles to his friend's house, then 0.7 mile to the video store, then 1.9 miles to the library. If he rode the same route back home, about how far did he travel in all?

Find each sum or difference.

3. $20 + (-7)$
4. $-15 + 6$
5. $-9 - 22$
6. $18.4 - (-3.2)$
7. $23.1 + (-9.81)$
8. $-5.6 + (-30.7)$

Find each product or quotient.

9. $11(-8)$
10. $-15(-2)$
11. $63 \div (-9)$
12. $-22 \div 11$

Replace each ● with <, >, or = to make a true sentence.

13. $\frac{7}{20}$ ● $\frac{2}{5}$
14. 0.15 ● $\frac{1}{8}$
15. Order 0.5, $-\frac{1}{7}$, -0.2, and $\frac{1}{3}$ from least to greatest.

Find each sum or difference. Write in simplest form.

16. $\frac{5}{6} + \frac{2}{3}$
17. $\frac{11}{12} - \frac{3}{4}$
18. $\frac{1}{2} + \frac{4}{9}$
19. $-\frac{3}{5} + \left(-\frac{1}{5}\right)$

Find each product or quotient.

20. $2.4(-0.7)$
21. $-40.5 \div (-8.1)$

Name the reciprocal of each number.

22. $\frac{4}{11}$
23. $-\frac{3}{7}$

Find each product or quotient. Write in simplest form.

24. $\frac{2}{21} \div \frac{1}{3}$
25. $\frac{1}{5} \cdot \frac{3}{20}$
26. $\frac{6}{25} \div \left(-\frac{3}{5}\right)$
27. $\frac{1}{9} \cdot \frac{3}{4}$
28. $-\frac{2}{21} \div \left(-\frac{2}{15}\right)$
29. $2\frac{1}{2} \cdot \frac{2}{15}$

Express each percent as a fraction in simplest form.

30. 20%
31. 7.5%

Use the percent proportion to find each number.

32. 18 is what percent of 72?
33. 35 is what percent of 200?
34. 24 is 60% of what number?

35. **TEST SCORES** James answered 14 items correctly on a 16-item quiz. What percent did he answer correctly?

36. **BASKETBALL** Emily made 75% of the baskets that she attempted. If she made 9 baskets, how many attempts did she make?

Find the perimeter and area of each figure.

37.

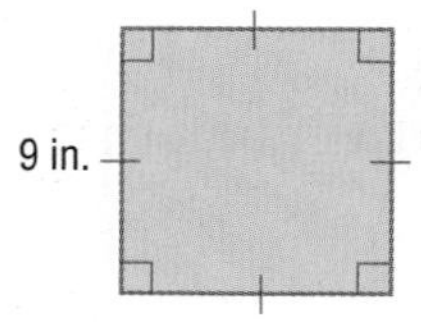

38.

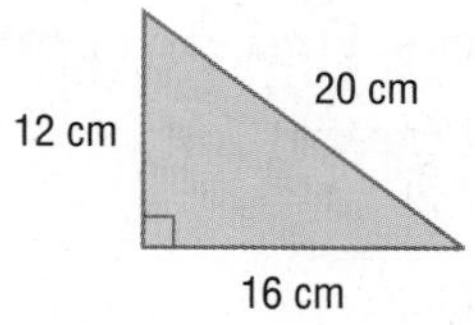

39. A parallelogram has side lengths of 7 inches and 11 inches. Find the perimeter.

40. **GARDENS** Find the perimeter of the garden.

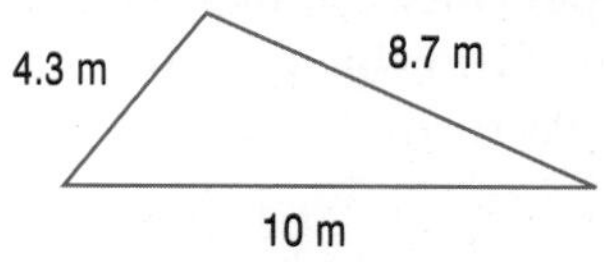

Find the circumference and area of each circle. Round to the nearest tenth.

41.

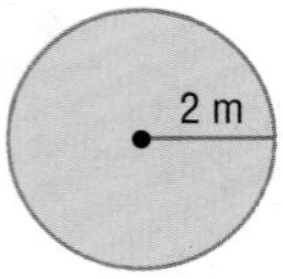

42.

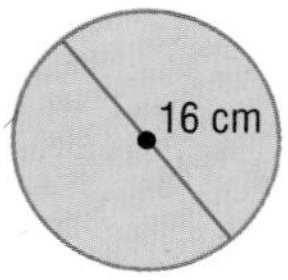

43. **BIRDS** The floor of a birdcage is a circle with a circumference of about 47.1 inches. What is the diameter of the birdcage floor? Round to the nearest inch.

Find the volume and surface area of each rectangular prism given the measurements below.

44. $\ell = 3$ cm, $w = 1$ cm, $h = 3$ cm

45. $\ell = 6$ ft, $w = 2$ ft, $h = 5$ ft

46. Find the volume and surface area of the rectangular prism.

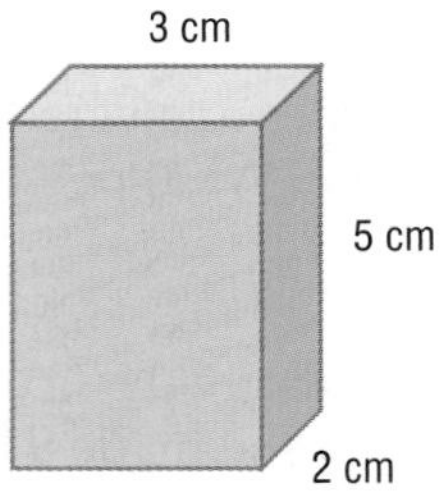

One pencil is randomly selected from a case containing 3 red, 4 green, 2 black, and 6 blue pencils. Find each probability.

47. P(green)

48. P(red or blue)

49. Use a tree diagram to find the sample space for the event *a die is rolled, and a coin is tossed*. State the number of possible outcomes.

One coin is randomly selected from a jar containing 20 pennies, 15 nickels, 3 dimes, and 12 quarters. Find the odds of each outcome. Write in simplest form.

50. a penny

51. a penny or nickel

52. A coin is tossed 50 times. The results are shown in the table. Find the experimental probability of heads. Write as a fraction in simplest form.

Lands Face-Up	Number of Times
heads	22
tails	28

Find the mean, median, and mode for each set of data.

53. {10, 11, 18, 24, 30}

54. {4, 8, 9, 9, 10, 14, 16}

55. Find the range, median, lower quartile, and upper quartile for {16, 19, 21, 24, 25, 31, 35}.

56. **SCHOOL** Devonte's scores on his first four Spanish tests are 92, 85, 90, and 92. What test score must Devonte earn on the fifth test so that the mean will be exactly 90?

57. **MUSIC** The table shows the results of a survey in which students were asked to choose which of four instruments they would like to learn. Make a bar graph of the data.

Favorite Instrument	
Instrument	**Number of Students**
drums	8
guitar	12
piano	5
trumpet	7

58. Make a double box-and-whisker plot of the data.
A: 42, 50, 38, 59, 50, 44, 46, 62, 47, 35, 55, 56
B: 47, 49, 48, 49, 40, 54, 56, 42, 57, 45, 45, 46

59. **EXPENSES** The table shows how Dylan spent his money at the fair. What type of graph is the best way to display these data? Explain your reasoning and make a graph of the data.

Money Spent at the Fair	
How Spent	**Amount ($)**
rides	6
food	10
games	4

LESSON 0-1 Plan for Problem Solving

Objective

- Use the four-step problem-solving plan.

NewVocabulary

four-step problem-solving plan
defining a variable

Common Core State Standards

Mathematical Practices

1 Make sense of problems and persevere in solving them.

Using the **four-step problem-solving plan** can help you solve any word problem.

KeyConcept Four-Step Problem Solving Plan

Step 1 **Understand** the Problem.

Step 2 **Plan** the Solution.

Step 3 **Solve** the Problem.

Step 4 **Check** the Solution.

Each step of the plan is important.

Step 1 Understand the Problem

To solve a verbal problem, first read the problem carefully and explore what the problem is about.

- Identify what information is given.
- Identify what you need to find.

Step 2 Plan the Solution

One strategy you can use is to write an equation. Choose a variable to represent one of the unspecified numbers in the problem. This is called **defining a variable**. Then use the variable to write expressions for the other unspecified numbers in the problem.

Step 3 Solve the Problem

Use the strategy you chose in Step 2 to solve the problem.

Step 4 Check the Solution

Check your answer in the context of the original problem.

- Does your answer make sense?
- Does it fit the information in the problem?

Example 1 Use the Four-Step Plan

FLOORS Ling's hallway is 10 feet long and 4 feet wide. He paid $200 to tile his hallway floor. How much did Ling pay per square foot for the tile?

Understand We are given the measurements of the hallway and the total cost of the tile. We are asked to find the cost of each square foot of tile.

Plan Write an equation. Let f represent the cost of each square foot of tile. The area of the hallway is 10×4 or 40 ft^2.

40	times	the cost per square foot	equals	200.
40	$\cdot$	f	$=$	200

Solve $40 \cdot f = 200$. Find f mentally by asking, "What number times 40 is 200?"

$f = 5$

The tile cost $5 per square foot.

Check If the tile costs $5 per square foot, then 40 square feet of tile costs $5 \cdot 40$ or $200. The answer makes sense.

When an exact value is needed, you can use estimation to check your answer.

PT

Example 2 Use the Four-Step Plan

TRAVEL Emily's family drove 254.6 miles. Their car used 19 gallons of gasoline. Describe the car's gas mileage.

Understand We are given the total miles driven and how much gasoline was used. We are asked to find the gas mileage of the car.

Plan Write an equation. Let G represent the car's gas mileage.

gas mileage = number of miles ÷ number of gallons used

$G = 254.6 \div 19$

Solve $G = 254.6 \div 19$

$= 13.4$ mi/gal

The car's gas mileage is 13.4 miles per gallon.

Check Use estimation to check your solution.

260 mi ÷ 20 gal = 13 mi/gal

Since the solution 13.4 is close to the estimate, the answer is reasonable.

Exercises

Determine whether you need an estimate or an exact answer. Then use the four step problem-solving plan to solve.

1. **DRIVING** While on vacation, the Jacobson family drove 312.8 miles the first day, 177.2 miles the second day, and 209 miles the third day. About how many miles did they travel in all?

2. **PETS** Ms. Hernandez boarded her dog at a kennel for 4 days. It cost $18.90 per day, and she had a coupon for $5 off. What was the final cost for boarding her dog?

3. **MEASUREMENT** William is using a 1.75-liter container to fill a 14-liter container of water. About how many times will he need to fill the smaller container?

4. **SEWING** Fabric costs $5.15 per yard. The drama department needs 18 yards of the fabric for their new play. About how much should they expect to pay?

5. **FINANCIAL LITERACY** The table shows donations to help purchase a new tree for the school. How much money did the students donate in all?

Number of Students	Amount of Each Donation
20	$2.50
15	$3.25

6. **SHOPPING** Is $12 enough to buy a half gallon of milk for $2.30, a bag of apples for $3.99, and four cups of yogurt that cost $0.79 each? Explain.

LESSON 0-2

Real Numbers

Objective

- Classify and use real numbers.

NewVocabulary

positive number
negative number
natural number
whole number
integer
rational number
square root
principal square root
perfect square
irrational number
real number
graph
coordinate

A number line can be used to show the sets of natural numbers, whole numbers, integers, and rational numbers. Values greater than 0, or **positive numbers**, are listed to the right of 0, and values less than 0, or **negative numbers**, are listed to the left of 0.

natural numbers: 1, 2, 3, …

whole numbers: 0, 1, 2, 3, …

integers: …, −3, −2, −1, 0, 1, 2, 3, …

rational numbers: numbers that can be expressed in the form $\frac{a}{b}$, where a and b are integers and $b \neq 0$

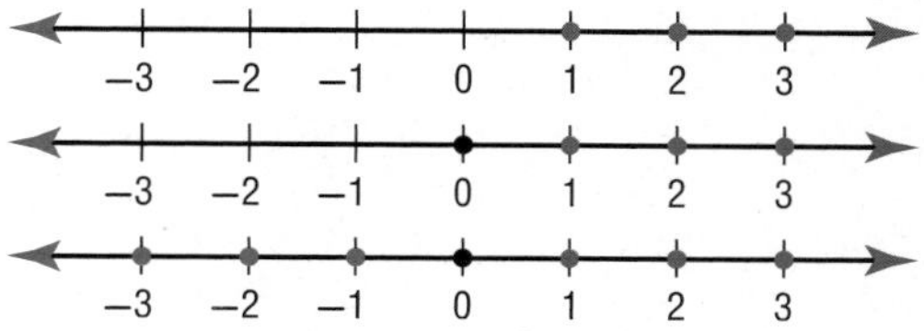

A **square root** is one of two equal factors of a number. For example, one square root of 64, written as $\sqrt{64}$, is 8 since $8 \cdot 8$ or 8^2 is 64. The nonnegative square root of a number is the **principal square root**. Another square root of 64 is −8 since $(-8) \cdot (-8)$ or $(-8)^2$ is also 64. A number like 64, with a square root that is a rational number, is called a **perfect square**. The square roots of a perfect square are rational numbers.

A number such as $\sqrt{3}$ is the square root of a number that is not a perfect square. It cannot be expressed as a terminating or repeating decimal; $\sqrt{3} \approx 1.73205\ldots$. Numbers that cannot be expressed as terminating or repeating decimals, or in the form $\frac{a}{b}$, where a and b are integers and $b \neq 0$, are called **irrational numbers**. Irrational numbers and rational numbers together form the set of **real numbers**.

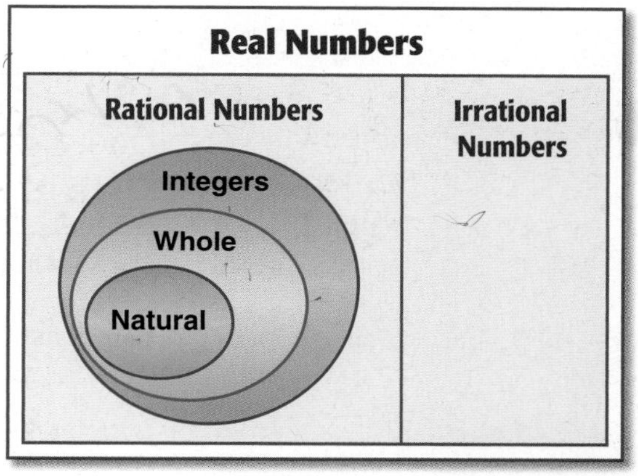

Example 1 Classify Real Numbers

Name the set or sets of numbers to which each real number belongs.

a. $\frac{5}{22}$

Because 5 and 22 are integers and $5 \div 22 = 0.2272727\ldots$ or $0.2\overline{27}$, which is a repeating decimal, this number is a rational number.

b. $\sqrt{81}$

Because $\sqrt{81} = 9$, this number is a natural number, a whole number, an integer, and a rational number.

c. $\sqrt{56}$

Because $\sqrt{56} = 7.48331477\ldots$, which is not a repeating or terminating decimal, this number is irrational.

To **graph** a set of numbers means to draw, or plot, the points named by those numbers on a number line. The number that corresponds to a point on a number line is called the **coordinate** of that point. The rational numbers and the irrational numbers complete the number line.

Example 2 Graph and Order Real Numbers

Graph each set of numbers on a number line. Then order the numbers from least to greatest.

a. $\left\{\frac{5}{3}, -\frac{4}{3}, \frac{2}{3}, -\frac{1}{3}\right\}$

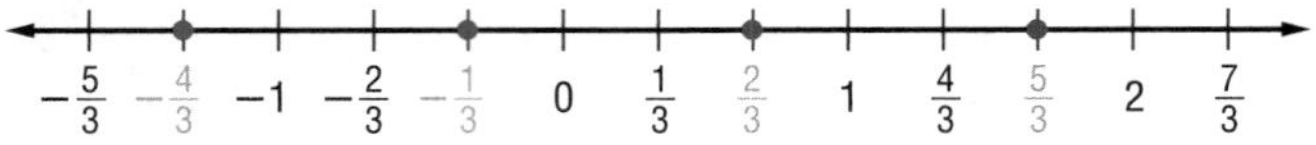

From least to greatest, the order is $-\frac{4}{3}$, $-\frac{1}{2}$, $\frac{2}{3}$, and $\frac{5}{3}$.

b. $\left\{6\frac{4}{5}, \sqrt{49}, 6.\overline{3}, \sqrt{57}\right\}$

Express each number as a decimal. Then order the decimals.

$6\frac{4}{5} = 6.8$ $\quad \sqrt{49} = 7$ $\quad 6.\overline{3} = 6.33333333\ldots$ $\quad \sqrt{57} = 7.5468344\ldots$

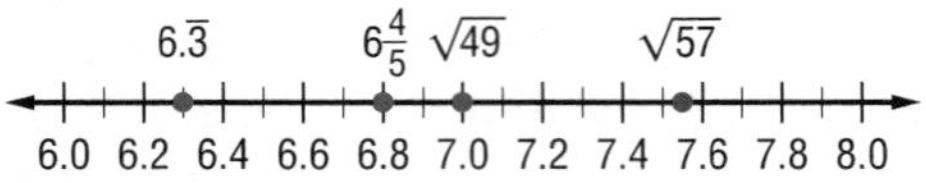

From least to greatest, the order is $6.\overline{3}$, $6\frac{4}{5}$, $\sqrt{49}$, and $\sqrt{57}$.

c. $\left\{\sqrt{20}, 4.7, \frac{12}{3}, 4\frac{1}{3}\right\}$

$\sqrt{20} = 4.47213595\ldots$ $\quad 4.7 = 4.7$ $\quad \frac{12}{3} = 4.0$ $\quad 4\frac{1}{3} = 4.33333333\ldots$

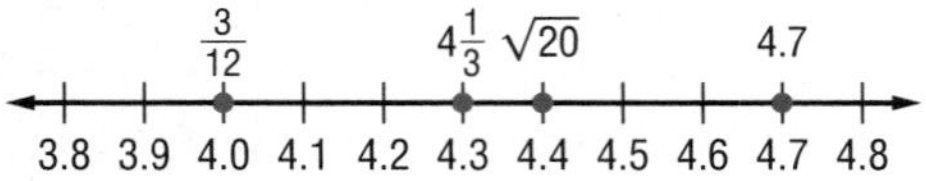

From least to greatest, the order is $\frac{12}{3}$, $4\frac{1}{3}$, $\sqrt{20}$, and 4.7.

Any repeating decimal can be written as a fraction.

Example 3 Write Repeating Decimals as Fractions

Write $0.\overline{7}$ as a fraction in simplest form.

Step 1

$N = 0.777\ldots$	Let N represent the repeating decimal.
$10N = 10(0.777\ldots)$	Since only one digit repeats, multiply each side by 10.
$10N = 7.777\ldots$	Simplify.

Step 2 Subtract N from $10N$ to eliminate the part of the number that repeats.

$$10N = 7.777\ldots$$
$$\underline{-(N = 0.777\ldots)}$$

$9N = 7$	Subtract.
$\frac{9N}{9} = \frac{7}{9}$	Divide each side by 9.
$N = \frac{7}{9}$	Simplify.

> **StudyTip**
> Perfect Squares Keep a list of perfect squares in your notebook. Refer to it when you need to simplify a square root.

Perfect squares can be used to simplify square roots of rational numbers.

KeyConcept Perfect Square

Words Rational numbers with square roots that are rational numbers.

Examples 25 is a perfect square since $\sqrt{25} = 5$.

144 is a perfect square since $\sqrt{144} = 12$.

Example 4 Simplify Roots

Simplify each square root.

a. $\sqrt{\frac{4}{121}}$

$$\sqrt{\frac{4}{121}} = \sqrt{\left(\frac{2}{11}\right)^2} \quad 2^2 = 4 \text{ and } 11^2 = 121$$

$$= \frac{2}{11} \quad \text{Simplify.}$$

b. $-\sqrt{\frac{49}{256}}$

$$-\sqrt{\frac{49}{256}} = -\sqrt{\left(\frac{7}{16}\right)^2} \quad 7^2 = 49 \text{ and } 16^2 = 256$$

$$= -\frac{7}{16}$$

You can estimate roots that are not perfect squares.

Example 5 Estimate Roots

Estimate each square root to the nearest whole number.

a. $\sqrt{15}$

Find the two perfect squares closest to 15. List some perfect squares.

1, 4, 9, 16, 25, 36, …

15 is between 9 and 16.

$9 < 15 < 16$ Write an inequality.

$\sqrt{9} < \sqrt{15} < \sqrt{16}$ Take the square root of each number.

$3 < \sqrt{15} < 4$ Simplify.

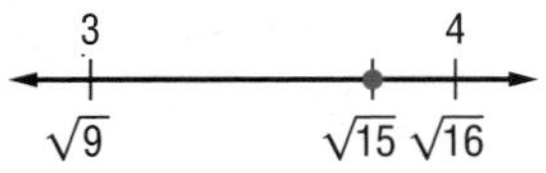

Since 15 is closer to 16 than 9, the best whole-number estimate for $\sqrt{15}$ is 4.

b. $\sqrt{130}$

Find the two perfect squares closest to 130. List some perfect squares.

81, 100, 121, 144

130 is between 121 and 144.

$121 < 130 < 144$ — Write an inequality.

$\sqrt{121} < \sqrt{130} < \sqrt{144}$ — Take the square root of each number.

$11 < \sqrt{130} < 12$ — Simplify.

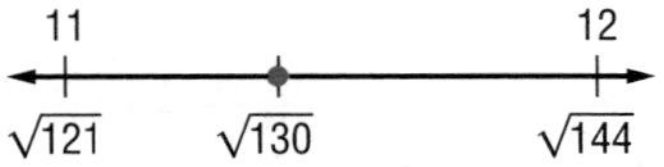

Since 130 is closer to 121 than to 144, the best whole number estimate for $\sqrt{130}$ is 11.

CHECK $\sqrt{130} \approx 11.4018$ — Use a calculator.

Rounded to the nearest whole number, $\sqrt{130}$ is 11. So the estimate is valid.

StudyTip

Draw a Diagram
Graphing points on a number line can help you analyze your estimate for accuracy.

Exercises

Name the set or sets of numbers to which each real number belongs.

1. $-\sqrt{64}$
2. $\frac{8}{3}$
3. $\sqrt{28}$
4. $\frac{56}{7}$
5. $-\sqrt{22}$
6. $\frac{36}{6}$
7. $-\frac{5}{12}$
8. $\frac{18}{3}$
9. $\sqrt{10.24}$
10. $\frac{-54}{19}$
11. $\sqrt{\frac{82}{20}}$
12. $-\frac{72}{8}$

Graph each set of numbers on a number line. Then order the numbers from least to greatest.

13. $\left\{\frac{7}{5}, -\frac{3}{5}, \frac{3}{4}, -\frac{6}{5}\right\}$
14. $\left\{\frac{1}{2}, -\frac{7}{9}, \frac{1}{9}, -\frac{4}{9}\right\}$
15. $\left\{2\frac{1}{4}, \sqrt{7}, 2.\overline{3}, \sqrt{8}\right\}$
16. $\left\{\frac{4}{5}, \sqrt{2}, 0.\overline{1}, \sqrt{3}\right\}$
17. $\left\{-3.5, -\frac{15}{5}, -\sqrt{10}, -3\frac{3}{4}\right\}$
18. $\left\{\sqrt{64}, 8.8, \frac{26}{3}, 8\frac{2}{7}\right\}$

Write each repeating decimal as a fraction in simplest form.

19. $0.\overline{5}$
20. $0.\overline{4}$
21. $0.\overline{13}$
22. $0.\overline{21}$

Simplify each square root.

23. $-\sqrt{25}$
24. $\sqrt{361}$
25. $\pm\sqrt{36}$
26. $\sqrt{0.64}$
27. $\pm\sqrt{1.44}$
28. $-\sqrt{6.25}$
29. $\sqrt{\frac{16}{49}}$
30. $\sqrt{\frac{169}{196}}$
31. $\sqrt{\frac{25}{324}}$

Estimate each root to the nearest whole number.

32. $\sqrt{112}$
33. $\sqrt{252}$
34. $\sqrt{415}$
35. $\sqrt{670}$

LESSON 0-3

Operations with Integers

Objective

- Add, subtract, multiply, and divide integers.

NewVocabulary

absolute value
opposites
additive inverses

An integer is any number from the set {..., −3, −2, −1, 0, 1, 2, 3, ...}. You can use a number line to add integers.

Example 1 Add Integers with the Same Sign

Use a number line to find −3 + (−4).

Step 1 Draw an arrow from 0 to −3.

Step 2 Draw a second arrow 4 units to the left to represent adding −4.

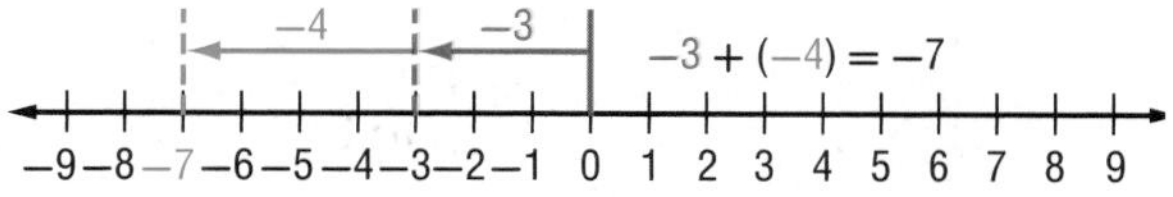

The second arrow ends at −7. So, −3 + (−4) = −7.

You can also use absolute value to add integers. The **absolute value** of a number is its distance from 0 on the number line.

Same Signs (+ + or − −)		Different Signs (+ − or − +)	
3 + 5 = 8	3 and 5 are positive. Their sum is positive.	3 + (−5) = −2	−5 has the greater absolute value. Their sum is negative.
−3 + (−5) = −8	−3 and −5 are negative. Their sum is negative.	−3 + 5 = 2	5 has the greater absolute value. Their sum is positive.

Example 2 Add Integers Using Absolute Value

Find −11 + (−7).

$-11 + (-7) = -(|-11| + |-7|)$ Add the absolute values. Both numbers are negative, so the sum is negative.

$= -(11 + 7)$ Absolute values of nonzero numbers are always positive.

$= -18$ Simplify.

Every positive integer can be paired with a negative integer. These pairs are called **opposites**. A number and its opposite are **additive inverses**. Additive inverses can be used when you subtract integers.

Example 3 Subtract Positive Integers

Find 18 − 23.

$18 - 23 = 18 + (-23)$ To subtract 23, add its inverse.

$= -(|-23| - |18|)$ Subtract the absolute values. Because $|-23|$ is greater than $|18|$, the result is negative.

$= -(23 - 18)$ Absolute values of nonzero numbers are always positive.

$= -5$ Simplify.

StudyTip

Products and Quotients The product or quotient of two numbers having the *same sign* is positive. The product or quotient of two numbers having *different signs* is negative.

Same Signs (+ + or − −)		Different Signs (+ − or − +)	
3(5) = 15	3 and 5 are positive. Their product is positive.	3(−5) = −15	3 and −5 have different signs. Their product is negative.
−3(−5) = 15	−3 and −5 are negative. Their product is positive.	−3(5) = −15	−3 and 5 have different signs. Their product is negative.

PT

Example 4 Multiply and Divide Integers

Find each product or quotient.

a. 4(−5)

$4(-5) = -20$ different signs → negative product

b. −51 ÷ (−3)

$-51 \div (-3) = 17$ same sign → positive quotient

c. −12(−14)

$-12(-14) = 168$ same sign → positive product

d. −63 ÷ 7

$-63 \div 7 = -9$ different signs → negative quotient

Exercises

Find each sum or difference.

1. $-8 + 13$ **2.** $11 + (-19)$ **3.** $-19 - 8$

4. $-77 + (-46)$ **5.** $12 - 34$ **6.** $41 + (-56)$

7. $50 - 82$ **8.** $-47 - 13$ **9.** $-80 + 102$

Find each product or quotient.

10. $5(18)$ **11.** $60 \div 12$ **12.** $-12(15)$

13. $-64 \div (-8)$ **14.** $8(-22)$ **15.** $54 \div (-6)$

16. $30(14)$ **17.** $-23(5)$ **18.** $-200 \div 2$

19. WEATHER The outside temperature was −4°F in the morning and 13°F in the afternoon. By how much did the temperature increase?

20. DOLPHINS A dolphin swimming 24 feet below the ocean's surface dives 18 feet straight down. How many feet below the ocean's surface is the dolphin now?

21. MOVIES A movie theater gave out 50 coupons for $3 off each movie. What is the total amount of discounts provided by the theater?

22. WAGES Emilio earns $11 per hour. He works 14 hours a week. His employer withholds $32 from each paycheck for taxes. If he is paid weekly, what is the amount of his paycheck?

23. FINANCIAL LITERACY Talia is working on a monthly budget. Her monthly income is $500. She has allocated $200 for savings, $100 for vehicle expenses, and $75 for clothing. How much is available to spend on entertainment?

LESSON 0-4 Adding and Subtracting Rational Numbers

Objective

- Compare and order; add and subtract rational numbers.

You can use different methods to compare rational numbers. One way is to compare two fractions with common denominators. Another way is to compare decimals.

PT

Example 1 Compare Rational Numbers

Replace ● with <, >, or = to make $\frac{2}{3}$ ● $\frac{5}{6}$ a true sentence.

Method 1 Write the fractions with the same denominator.

The least common denominator of $\frac{2}{3}$ and $\frac{5}{6}$ is 6.

$\frac{2}{3} = \frac{4}{6}$

$\frac{5}{6} = \frac{5}{6}$

Since $\frac{4}{6} < \frac{5}{6}, \frac{2}{3} < \frac{5}{6}$.

Method 2 Write as decimals.

Write $\frac{2}{3}$ and $\frac{5}{6}$ as decimals. You may want to use a calculator.

2 [÷] 3 [ENTER] .6666666667

So, $\frac{2}{3} = 0.\overline{6}$.

5 [÷] 6 [ENTER] .8333333333

So, $\frac{5}{6} = 0.8\overline{3}$.

Since $0.\overline{6} < 0.8\overline{3}, \frac{2}{3} < \frac{5}{6}$.

You can order rational numbers by writing all of the fractions as decimals.

PT

Example 2 Order Rational Numbers

Order $5\frac{2}{9}$, $5\frac{3}{8}$, 4.9, and $-5\frac{3}{5}$ from least to greatest.

$5\frac{2}{9} = 5.\overline{2}$ $\qquad$ $5\frac{3}{8} = 5.375$

$4.9 = 4.9$ $\qquad$ $-5\frac{3}{5} = -5.6$

$-5.6 < 4.9 < 5.\overline{2} < 5.375$. So, from least to greatest, the numbers are $-5\frac{3}{5}$, 4.9, $5\frac{2}{9}$, and $5\frac{3}{8}$.

To add or subtract fractions with the same denominator, add or subtract the numerators and write the sum or difference over the denominator.

Example 3 Add and Subtract Like Fractions

Find each sum or difference. Write in simplest form.

a. $\frac{3}{5} + \frac{1}{5}$

$\frac{3}{5} + \frac{1}{5} = \frac{3+1}{5}$ — The denominators are the same. Add the numerators.

$= \frac{4}{5}$ — Simplify.

StudyTip

Mental Math If the denominators of the fractions are the same, you can use mental math to determine the sum or difference.

b. $\frac{7}{16} - \frac{1}{16}$

$\frac{7}{16} - \frac{1}{16} = \frac{7-1}{16}$ — The denominators are the same. Subtract the numerators.

$= \frac{6}{16}$ — Simplify.

$= \frac{3}{8}$ — Rename the fraction.

c. $\frac{4}{9} - \frac{7}{9}$

$\frac{4}{9} - \frac{7}{9} = \frac{4-7}{9}$ — The denominators are the same. Subtract the numerators.

$= -\frac{3}{9}$ — Simplify.

$= -\frac{1}{3}$ — Rename the fraction.

To add or subtract fractions with unlike denominators, first find the least common denominator (LCD). Rename each fraction with the LCD, and then add or subtract. Simplify if possible.

Example 4 Add and Subtract Unlike Fractions

Find each sum or difference. Write in simplest form.

a. $\frac{1}{2} + \frac{2}{3}$

$\frac{1}{2} + \frac{2}{3} = \frac{3}{6} + \frac{4}{6}$ — The LCD for 2 and 3 is 6. Rename $\frac{1}{2}$ as $\frac{3}{6}$ and $\frac{2}{3}$ as $\frac{4}{6}$.

$= \frac{3+4}{6}$ — Add the numerators.

$= \frac{7}{6}$ or $1\frac{1}{6}$ — Simplify.

b. $\frac{3}{8} - \frac{1}{3}$

$\frac{3}{8} - \frac{1}{3} = \frac{9}{24} - \frac{8}{24}$ — The LCD for 8 and 3 is 24. Rename $\frac{3}{8}$ as $\frac{9}{24}$ and $\frac{1}{3}$ as $\frac{8}{24}$.

$= \frac{9-8}{24}$ — Subtract the numerators.

$= \frac{1}{24}$ — Simplify.

c. $\frac{2}{5} - \frac{3}{4}$

$\frac{2}{5} - \frac{3}{4} = \frac{8}{20} - \frac{15}{20}$ — The LCD for 5 and 4 is 20. Rename $\frac{2}{5}$ as $\frac{8}{20}$ and $\frac{3}{4}$ as $\frac{15}{20}$.

$= \frac{8-15}{20}$ — Subtract the numerators.

$= -\frac{7}{20}$ — Simplify.

StudyTip

Number Line To use a number line, put your pencil at the first number. If you are adding or subtracting a positive number, then move left to find the difference. To find the sum, move your pencil to the right.

You can use a number line to add rational numbers.

Example 5 Add Decimals

Use a number line to find 2.5 + (−3.5).

Step 1 Draw an arrow from 0 to 2.5.

Step 2 Draw a second arrow 3.5 units to the left.

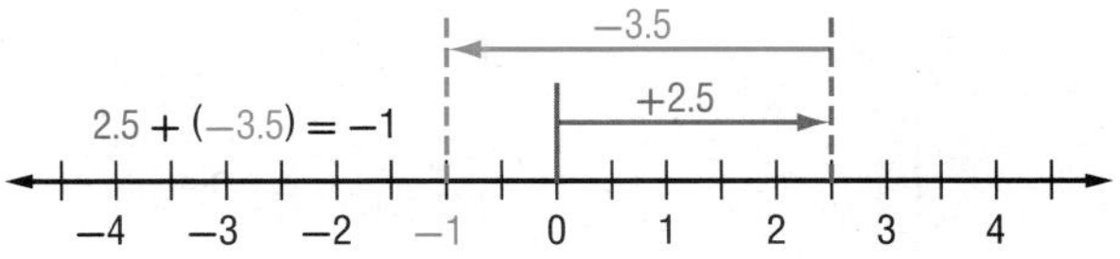

The second arrow ends at −1.

So, 2.5 + (−3.5) = −1.

You can also use absolute value to add rational numbers.

Same Signs (+ + or − −)		Different Signs (+ − or − +)	
3.1 + 2.5 = 5.6	3.1 and 2.5 are positive, so the sum is positive.	3.1 + (−2.5) = 0.6	3.1 has the greater absolute value, so the sum is positive.
−3.1 + (−2.5) = −5.6	−3.1 and −2.5 are negative, so the sum is negative.	−3.1 + 2.5 = −0.6	−3.1 has the greater absolute value, so the sum is negative.

Example 6 Use Absolute Value to Add Rational Numbers

Find each sum.

a. $-13.12 + (-8.6)$

$-13.12 + (-8.6) = -(|-13.12| + |-8.6|)$ Both numbers are negative, so the sum is negative.

$= -(13.12 + 8.6)$ Absolute values of nonzero numbers are always positive.

$= -21.72$ Simplify.

b. $\frac{7}{16} + \left(-\frac{3}{8}\right)$

$\frac{7}{16} + \left(-\frac{3}{8}\right) = \frac{7}{16} + \left(-\frac{6}{16}\right)$ The LCD is 16. Replace $-\frac{3}{8}$ with $-\frac{6}{16}$.

$= \left(\left|\frac{7}{16}\right| - \left|-\frac{6}{16}\right|\right)$ Subtract the absolute values. Because $\left|\frac{7}{16}\right|$ is greater than $\left|-\frac{6}{16}\right|$, the result is positive.

$= \frac{7}{16} - \frac{6}{16}$ Absolute values of nonzero numbers are always positive.

$= \frac{1}{16}$ Simplify.

To subtract a negative rational number, add its inverse.

Example 7 Subtract Decimals

Find −32.25 − (−42.5).

$-32.25 - (-42.5) = -32.25 + 42.5$	To subtract −42.5, add its inverse.
$= \lvert 42.5 \rvert - \lvert -32.25 \rvert$	Subtract the absolute values. Because $\lvert 42.5 \rvert$ is greater than $\lvert -32.25 \rvert$, the result is positive.
$= 42.5 - 32.25$	Absolute values of nonzero numbers are always positive.
$= 10.25$	Simplify.

Exercises

Replace each ● with <, >, or = to make a true sentence.

1. $-\frac{5}{8}$ ● $\frac{3}{8}$
2. $\frac{4}{5}$ ● 0.71
3. $\frac{5}{6}$ ● 0.875
4. 1.2 ● $1\frac{2}{9}$
5. $\frac{8}{15}$ ● $0.5\overline{3}$
6. $-\frac{7}{11}$ ● $-\frac{2}{3}$

Order each set of rational numbers from least to greatest.

7. 3.8, 3.06, $3\frac{1}{6}$, $3\frac{3}{4}$
8. $2\frac{1}{4}$, $1\frac{7}{8}$, 1.75, 2.4
9. 0.11, $-\frac{1}{9}$, −0.5, $\frac{1}{10}$
10. $-4\frac{3}{5}$, $-3\frac{2}{5}$, −4.65, −4.09

Find each sum or difference. Write in simplest form.

11. $\frac{2}{5} + \frac{1}{5}$
12. $\frac{3}{9} + \frac{4}{9}$
13. $\frac{5}{16} - \frac{4}{16}$
14. $\frac{6}{7} - \frac{3}{7}$
15. $\frac{2}{3} + \frac{1}{3}$
16. $\frac{5}{8} + \frac{7}{8}$
17. $\frac{4}{3} + \frac{4}{3}$
18. $\frac{7}{15} - \frac{2}{15}$
19. $\frac{1}{3} - \frac{2}{9}$
20. $\frac{1}{2} + \frac{1}{4}$
21. $\frac{1}{2} - \frac{1}{3}$
22. $\frac{3}{7} + \frac{5}{14}$
23. $\frac{7}{10} - \frac{2}{15}$
24. $\frac{3}{8} + \frac{1}{6}$
25. $\frac{13}{20} - \frac{2}{5}$

Find each sum or difference. Write in simplest form if necessary.

26. −1.6 + (−3.8)
27. −32.4 + (−4.5)
28. −38.9 + 24.2
29. −9.16 − 10.17
30. 26.37 + (−61.1)
31. 72.5 − (−81.3)
32. 43.2 + (−27.9)
33. 79.3 − (−14)
34. 1.34 − (−0.458)
35. $-\frac{1}{6} - \frac{2}{3}$
36. $\frac{1}{2} - \frac{4}{5}$
37. $-\frac{2}{5} + \frac{17}{20}$
38. $-\frac{4}{5} + \left(-\frac{1}{3}\right)$
39. $-\frac{1}{12} - \left(-\frac{3}{4}\right)$
40. $-\frac{7}{8} - \left(-\frac{3}{16}\right)$

41. **GEOGRAPHY** About $\frac{7}{10}$ of the surface of Earth is covered by water. The rest of the surface is covered by land. How much of Earth's surface is covered by land?

LESSON 0-5

Multiplying and Dividing Rational Numbers

Objective

- Multiply and divide rational numbers.

NewVocabulary
multiplicative inverses
reciprocals

The product or quotient of two rational numbers having the *same sign* is positive. The product or quotient of two rational numbers having *different signs* is negative.

PT

Example 1 Multiply and Divide Decimals

Find each product or quotient.

a. $7.2(-0.2)$

different signs ⟶ negative product

$7.2(-0.2) = -1.44$

b. $-23.94 \div (-10.5)$

same sign ⟶ positive quotient

$-23.94 \div (-10.5) = 2.28$

To multiply fractions, multiply the numerators and multiply the denominators. If the numerators and denominators have common factors, you can simplify before you multiply by canceling.

PT

Example 2 Multiply Fractions

Find each product.

a. $\frac{2}{5} \cdot \frac{1}{3}$

$\frac{2}{5} \cdot \frac{1}{3} = \frac{2 \cdot 1}{5 \cdot 3}$ — Multiply the numerators. Multiply the denominators.

$= \frac{2}{15}$ — Simplify.

b. $\frac{3}{5} \cdot 1\frac{1}{2}$

$\frac{3}{5} \cdot 1\frac{1}{2} = \frac{3}{5} \cdot \frac{3}{2}$ — Write $1\frac{1}{2}$ as an improper fraction.

$= \frac{3 \cdot 3}{5 \cdot 2}$ — Multiply the numerators. Multiply the denominators.

$= \frac{9}{10}$ — Simplify.

c. $\frac{1}{4} \cdot \frac{2}{9}$

$\frac{1}{4} \cdot \frac{2}{9} = \frac{1}{\cancel{4}_2} \cdot \frac{\cancel{2}^1}{9}$ — Divide by the GCF, 2.

$= \frac{1 \cdot 1}{2 \cdot 9}$ or $\frac{1}{18}$ — Multiply the numerators. Multiply the denominators and simplify.

PT

Example 3 Multiply Fractions with Different Signs

Find $\left(-\frac{3}{4}\right)\left(\frac{3}{8}\right)$.

$\left(-\frac{3}{4}\right)\left(\frac{3}{8}\right) = -\left(\frac{3}{4} \cdot \frac{3}{8}\right)$ — different signs ⟶ negative product

$= -\left(\frac{3 \cdot 3}{4 \cdot 8}\right)$ or $\frac{9}{32}$ — Multiply the numerators. Multiply the denominators and simplify.

Two numbers whose product is 1 are called **multiplicative inverses** or **reciprocals**.

Example 4 Find the Reciprocal

Name the reciprocal of each number.

a. $\frac{3}{8}$

$\frac{3}{8} \cdot \frac{8}{3} = 1$ — The product is 1.

The reciprocal of $\frac{3}{8}$ is $\frac{8}{3}$.

b. $2\frac{4}{5}$

$2\frac{4}{5} = \frac{14}{5}$ — Write $2\frac{4}{5}$ as $\frac{14}{5}$.

$\frac{14}{5} \cdot \frac{5}{14} = 1$ — The product is 1.

The reciprocal of $2\frac{4}{5}$ is $\frac{5}{14}$.

To divide one fraction by another fraction, multiply the dividend by the reciprocal of the divisor.

Example 5 Divide Fractions

Find each quotient.

a. $\frac{1}{3} \div \frac{1}{2}$

$\frac{1}{3} \div \frac{1}{2} = \frac{1}{3} \cdot \frac{2}{1}$ — Multiply $\frac{1}{3}$ by $\frac{2}{1}$, the reciprocal of $\frac{1}{2}$.

$= \frac{2}{3}$ — Simplify.

b. $\frac{3}{8} \div \frac{2}{3}$

$\frac{3}{8} \div \frac{2}{3} = \frac{3}{8} \cdot \frac{3}{2}$ — Multiply $\frac{3}{8}$ by $\frac{3}{2}$, the reciprocal of $\frac{2}{3}$.

$= \frac{9}{16}$ — Simplify.

StudyTip

Use Estimation You can justify your answer by using estimation. $\frac{3}{8}$ is close to $\frac{1}{2}$ and $\frac{2}{3}$ is close to 1. So, the quotient is close to $\frac{1}{2}$ divided by 1 or $\frac{1}{2}$.

c. $\frac{3}{4} \div 2\frac{1}{2}$

$\frac{3}{4} \div 2\frac{1}{2} = \frac{3}{4} \div \frac{5}{2}$ — Write $2\frac{1}{2}$ as an improper fraction

$= \frac{3}{4} \cdot \frac{2}{5}$ — Multiply $\frac{3}{4}$ by $\frac{2}{5}$, the reciprocal of $2\frac{1}{2}$.

$= \frac{6}{20}$ or $\frac{3}{10}$ — Simplify.

d. $-\frac{1}{5} \div \left(-\frac{3}{10}\right)$

$-\frac{1}{5} \div \left(-\frac{3}{10}\right) = -\frac{1}{5} \cdot \left(-\frac{10}{3}\right)$ — Multiply $-\frac{1}{5}$ by $-\frac{10}{3}$, the reciprocal of $-\frac{3}{10}$.

$= \frac{10}{15}$ or $\frac{2}{3}$ — Same sign ⟶ positive quotient; simplify.

Exercises

Find each product or quotient. Round to the nearest hundredth if necessary.

1. $6.5(0.13)$
2. $-5.8(2.3)$
3. $42.3 \div (-6)$
4. $-14.1(-2.9)$
5. $-78 \div (-1.3)$
6. $108 \div (-0.9)$
7. $0.75(-6.4)$
8. $-23.94 \div 10.5$
9. $-32.4 \div 21.3$

Find each product. Simplify before multiplying if possible.

10. $\frac{3}{4} \cdot \frac{1}{5}$
11. $\frac{2}{5} \cdot \frac{3}{7}$
12. $-\frac{1}{3} \cdot \frac{2}{5}$
13. $-\frac{2}{3} \cdot \left(-\frac{1}{11}\right)$
14. $2\frac{1}{2} \cdot \left(-\frac{1}{4}\right)$
15. $3\frac{1}{2} \cdot 1\frac{1}{2}$
16. $\frac{2}{9} \cdot \frac{1}{2}$
17. $\frac{3}{2} \cdot \left(-\frac{1}{3}\right)$
18. $\frac{1}{3} \cdot \frac{6}{5}$
19. $-\frac{9}{4} \cdot \frac{1}{18}$
20. $\frac{11}{3} \cdot \frac{9}{44}$
21. $\left(-\frac{30}{11}\right) \cdot \left(-\frac{1}{3}\right)$
22. $-\frac{3}{5} \cdot \frac{5}{6}$
23. $\left(-\frac{1}{3}\right)\left(-7\frac{1}{2}\right)$
24. $\frac{2}{7} \cdot 4\frac{2}{3}$

Name the reciprocal of each number.

25. $\frac{6}{7}$
26. $\frac{1}{22}$
27. $-\frac{14}{23}$
28. $2\frac{3}{4}$
29. $-5\frac{1}{3}$
30. $3\frac{3}{4}$

Find each quotient.

31. $\frac{2}{3} \div \frac{1}{3}$
32. $\frac{16}{9} \div \frac{4}{9}$
33. $\frac{3}{2} \div \frac{1}{2}$
34. $\frac{3}{7} \div \left(-\frac{1}{5}\right)$
35. $-\frac{9}{10} \div 3$
36. $\frac{1}{2} \div \frac{3}{5}$
37. $2\frac{1}{4} \div \frac{1}{2}$
38. $-1\frac{1}{3} \div \frac{2}{3}$
39. $\frac{11}{12} \div 1\frac{2}{3}$
40. $4 \div \left(-\frac{2}{7}\right)$
41. $-\frac{1}{3} \div \left(-1\frac{1}{5}\right)$
42. $\frac{3}{25} \div \frac{2}{15}$

43. **PIZZA** A large pizza at Pizza Shack has 12 slices. If Bobby ate $\frac{1}{4}$ of the pizza, how many slices of pizza did he eat?

44. **MUSIC** Samantha practices the flute for $4\frac{1}{2}$ hours each week. How many hours does she practice in a month?

45. **BAND** How many band uniforms can be made with $131\frac{3}{4}$ yards of fabric if each uniform requires $3\frac{7}{8}$ yards?

46. **CARPENTRY** How many boards, each 2 feet 8 inches long, can be cut from a board 16 feet long if there is no waste?

47. **SEWING** How many 9-inch ribbons can be cut from $1\frac{1}{2}$ yards of ribbon?

LESSON 0-6

The Percent Proportion

Objective

- Use and apply the percent proportion.

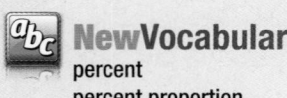

NewVocabulary

percent
percent proportion

A **percent** is a ratio that compares a number to 100. To write a percent as a fraction, express the ratio as a fraction with a denominator of 100. Fractions should be expressed in simplest form.

Example 1 Percents as Fractions

Express each percent as a fraction or mixed number.

a. 79%

$79\% = \frac{79}{100}$ Definition of percent

b. 107%

$107\% = \frac{107}{100}$ Definition of percent

$= 1\frac{7}{100}$ Simplify.

c. 0.5%

$0.5\% = \frac{0.5}{100}$ Definition of percent

$= \frac{5}{1000}$ Multiply the numerator and denominator by 10 to eliminate the decimal.

$= \frac{1}{200}$ Simplify.

In the **percent proportion**, the ratio of a part of something to the whole (base) is equal to the percent written as a fraction.

part → $\frac{a}{b} = \frac{p}{100}$ ← percent
whole →

Example: 25% (percent) of 40 (whole) is 10 (part).

You can use the percent proportion to find the part.

Example 2 Find the Part

40% of 30 is what number?

$\frac{a}{b} = \frac{p}{100}$ The percent is 40, and the base is 30. Let *a* represent the part.

$\frac{a}{30} = \frac{40}{100}$ Replace *b* with 30 and *p* with 40.

$100a = 30(40)$ Find the cross products.

$100a = 1200$ Simplify.

$\frac{100a}{100} = \frac{1200}{100}$ Divide each side by 100.

$a = 12$ Simplify.

The part is 12. So, 40% of 30 is 12.

You can also use the percent proportion to find the percent of the base.

PT

Example 3 Find the Percent

SURVEYS Kelsey took a survey of students in her lunch period. 42 out of the 70 students Kelsey surveyed said their family had a pet. What percent of the students had pets?

$\frac{a}{b} = \frac{p}{100}$ — The part is 42, and the base is 70. Let p represent the percent.

$\frac{42}{70} = \frac{p}{100}$ — Replace a with 42 and b with 70.

$4200 = 70p$ — Find the cross products.

$\frac{4200}{70} = \frac{70p}{70}$ — Divide each side by 70.

$60 = p$ — Simplify.

The percent is 60, so $\frac{60}{100}$ or 60% of the students had pets.

PT

StudyTip

Percent Proportion In percent problems, the whole, or base usually follows the word *of*.

Example 4 Find the Whole

67.5 is 75% of what number?

$\frac{a}{b} = \frac{p}{100}$ — The percent is 75, and the part is 67.5. Let b represent the base.

$\frac{67.5}{b} = \frac{75}{100}$ — Replace a with 67.5 and p with 75.

$6750 = 75b$ — Find the cross products.

$\frac{6750}{75} = \frac{75b}{75}$ — Divide each side by 75.

$90 = b$ — Simplify.

The base is 90, so 67.5 is 75% of 90.

Exercises

Express each percent as a fraction or mixed number in simplest form.

1. 5% **2.** 60% **3.** 11%

4. 120% **5.** 78% **6.** 2.5%

7. 0.6% **8.** 0.4% **9.** 1400%

Use the percent proportion to find each number.

10. 25 is what percent of 125?
11. 16 is what percent of 40?
12. 14 is 20% of what number?
13. 50% of what number is 80?
14. What number is 25% of 18?
15. Find 10% of 95.
16. What percent of 48 is 30?
17. What number is 150% of 32?
18. 5% of what number is 3.5?
19. 1 is what percent of 400?
20. Find 0.5% of 250.
21. 49 is 200% of what number?
22. 15 is what percent of 12?
23. 36 is what percent of 24?

24. **BASKETBALL** Madeline usually makes 85% of her shots in basketball. If she attempts 20, how many will she likely make?

25. **TEST SCORES** Brian answered 36 items correctly on a 40-item test. What percent did he answer correctly?

26. **CARD GAMES** Juanita told her dad that she won 80% of the card games she played yesterday. If she won 4 games, how many games did she play?

27. **SOLUTIONS** A glucose solution is prepared by dissolving 6 milliliters of glucose in 120 milliliters of pure solution. What is the percent of glucose in the resulting solution?

28. **DRIVER'S ED** Kara needs to get a 75% on her driving education test in order to get her license. If there are 35 questions on the test, how many does she need to answer correctly?

29. **HEALTH** The U.S. Food and Drug Administration requires food manufacturers to label their products with a nutritional label. The label shows the information from a package of macaroni and cheese.

Nutrition Facts		
Serving Size 1 cup (228g)		
Servings per container 2		
Amount per serving		
Calories 250	Calories from Fat 110	
		%Daily value*
Total Fat 12g		18%
	Saturated Fat 3g	15%
Cholesterol 30mg		10%
Sodium 470mg		20%
Total Carbohydrate 31g		10%
	Dietary Fiber 0g	0%
	Sugars 5g	
Protein 5g		
Vitamin A 4%	•	Vitamin C 2%
Calcium 20%	•	Iron 4%

a. The label states that a serving contains 3 grams of saturated fat, which is 15% of the daily value recommended for a 2000-Calorie diet. How many grams of saturated fat are recommended for a 2000-Calorie diet?

b. The 470 milligrams of sodium (salt) in the macaroni and cheese is 20% of the recommended daily value. What is the recommended daily value of sodium?

c. For a healthy diet, the National Research Council recommends that no more than 30 percent of the total Calories come from fat. What percent of the Calories in a serving of this macaroni and cheese come from fat?

30. **TEST SCORES** The table shows the number of points each student in Will's study group earned on a recent math test. There were 88 points possible on the test. Express all answers to the nearest tenth of a percent.

Name	Will	Penny	Cheng	Minowa	Rob
Score	72	68	81	87	75

a. Find Will's percent correct on the test.

b. Find Cheng's percent correct on the test.

c. Find Rob's percent correct on the test.

d. What was the highest percentage? The lowest?

31. **PET STORE** In a pet store, 15% of the animals are hamsters. If the store has 40 animals, how many of them are hamsters?

LESSON 0-7 Perimeter

Objective

- Find the perimeter of two-dimensional figures.

New Vocabulary

perimeter
circle
diameter
circumference
center
radius

Perimeter is the distance around a figure. Perimeter is measured in linear units.

Rectangle

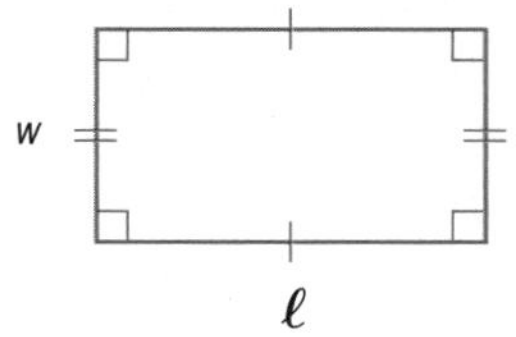

$P = 2(\ell + w)$ or
$P = 2\ell + 2w$

Parallelogram

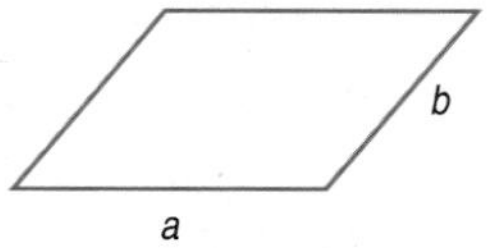

$P = 2(a + b)$ or
$P = 2a + 2b$

Square

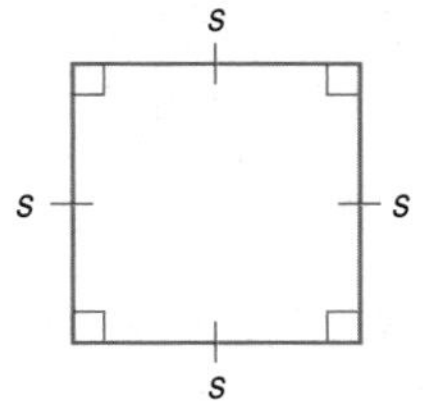

$P = 4s$

Triangle

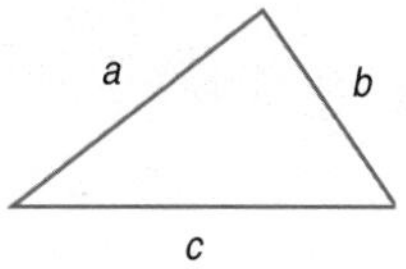

$P = a + b + c$

Example 1 Perimeters of Rectangles and Squares

Find the perimeter of each figure.

a. a rectangle with a length of 5 inches and a width of 1 inch

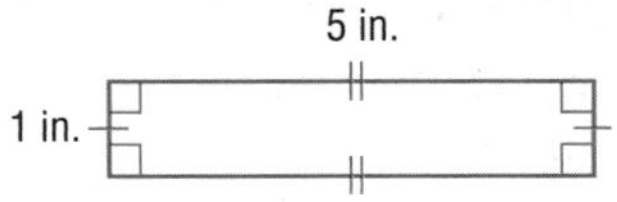

$P = 2(\ell + w)$	Perimeter formula
$= 2(5 + 1)$	$\ell = 5$, $w = 1$
$= 2(6)$	Add.
$= 12$	The perimeter is 12 inches.

b. a square with a side length of 7 centimeters

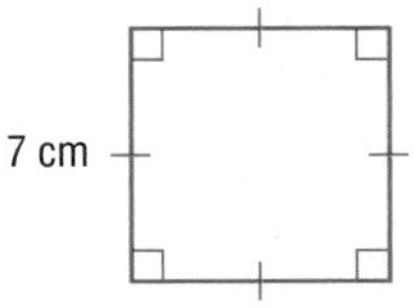

$P = 4s$	Perimeter formula
$= 4(7)$	Replace s with 7.
$= 28$	The perimeter is 28 centimeters.

Example 2 Perimeters of Parallelograms and Triangles

Find the perimeter of each figure.

StudyTip

Congruent Marks
The hash marks on the figures indicate sides that have the same length.

a.

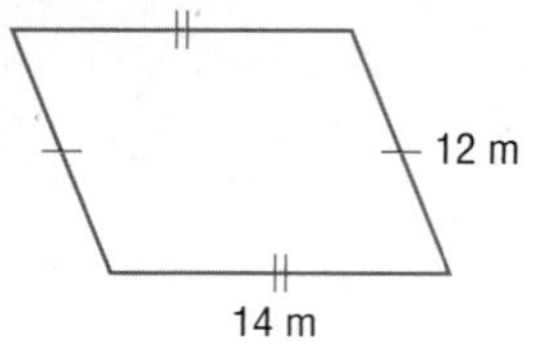
12 m
14 m

$P = 2(a + b)$	Perimeter formula
$= 2(14 + 12)$	$a = 14, b = 12$
$= 2(26)$	Add.
$= 52$	Multiply.

The perimeter of the parallelogram is 52 meters.

b.

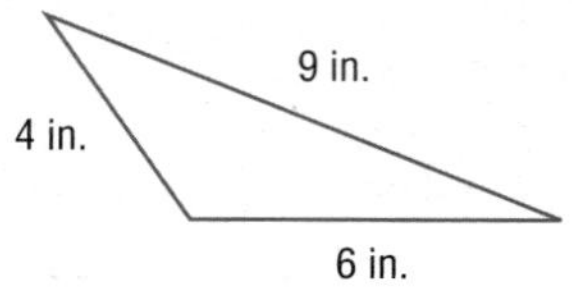
9 in.
4 in.
6 in.

$P = a + b + c$	Perimeter formula
$= 4 + 6 + 9$	$a = 4, b = 6, c = 9$
$= 19$	Add.

The perimeter of the triangle is 19 inches.

A **circle** is the set of all points in a plane that are the same distance from a given point.

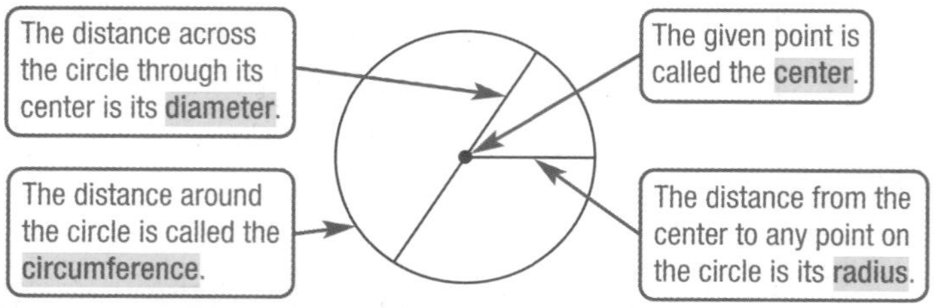
The distance across the circle through its center is its diameter.
The given point is called the center.
The distance around the circle is called the circumference.
The distance from the center to any point on the circle is its radius.

The formula for the circumference of a circle is $C = \pi d$ or $C = 2\pi r$.

Example 3 Circumference

Find each circumference to the nearest tenth.

StudyTip

Pi To perform a calculation that involves π, use a calculator.

a. The radius is 4 feet.

$C = 2\pi r$	Circumference formula
$= 2\pi(4)$	Replace r with 4.
$= 8\pi$	Simplify.

The exact circumference is 8π feet.

8 [π] [ENTER] **25.13274123**

The circumference is about 25.1 feet.

b. The diameter is 15 centimeters.

$C = \pi d$	Circumference formula
$= \pi(15)$	Replace d with 15.
$= 15\pi$	Simplify.
≈ 47.1	Use a calculator to evaluate 15π.

The circumference is about 47.1 centimeters.

c.

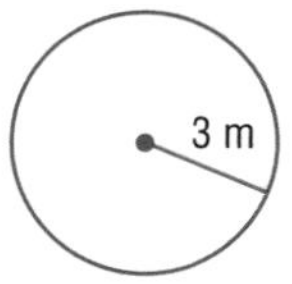
3 m

$C = 2\pi r$	Circumference formula
$= 2\pi(3)$	Replace r with 3.
$= 6\pi$	Simplify.
≈ 18.8	Use a calculator to evaluate 6π.

The circumference is about 18.8 meters.

Exercises

Find the perimeter of each figure.

1. 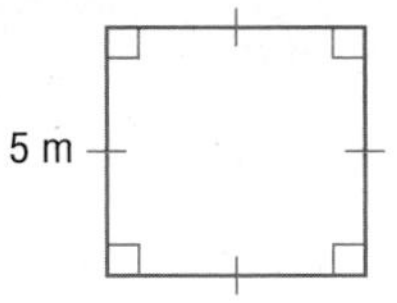

2. 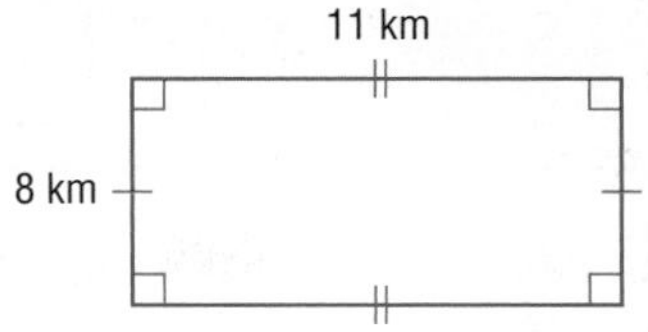

3. 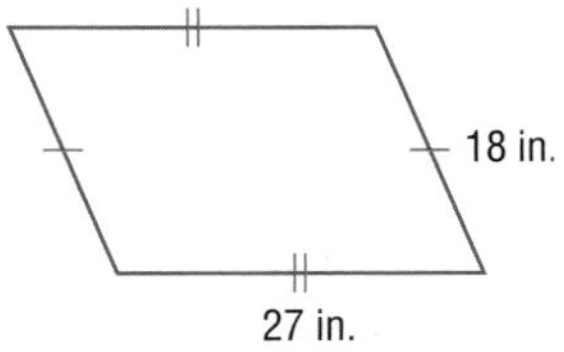

4. 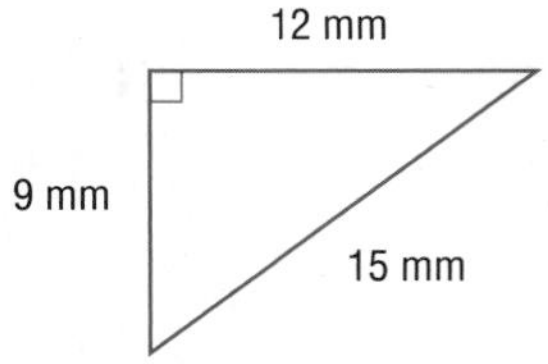

5. a square with side length 8 inches
6. a rectangle with length 9 centimeters and width 3 centimeters
7. a triangle with sides 4 feet, 13 feet, and 12 feet
8. a parallelogram with side lengths $6\frac{1}{4}$ inches and 5 inches
9. a quarter-circle with a radius of 7 inches

Find the circumference of each circle. Round to the nearest tenth.

10. 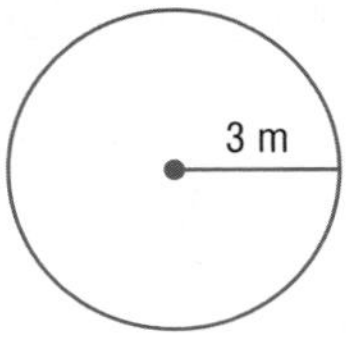

11. 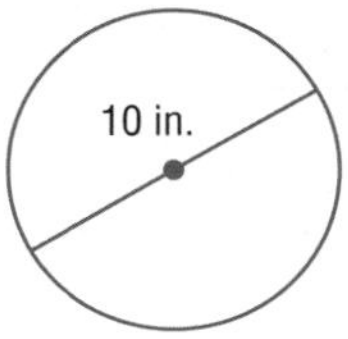

12. 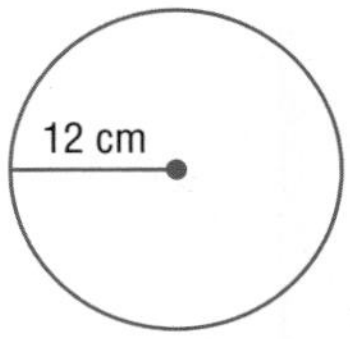

13. **GARDENS** A square garden has a side length of 5.8 meters. What is the perimeter of the garden?

14. **ROOMS** A rectangular room is $12\frac{1}{2}$ feet wide and 14 feet long. What is the perimeter of the room?

15. **CYCLING** The tire for a 10-speed bicycle has a diameter of 27 inches. Find the distance traveled in 10 rotations of the tire. Round to the nearest tenth.

16. **GEOGRAPHY** Earth's circumference is approximately 25,000 miles. If you could dig a tunnel to the center of the Earth, how long would the tunnel be? Round to the nearest tenth mile.

Find the perimeter of each figure. Round to the nearest tenth.

17. 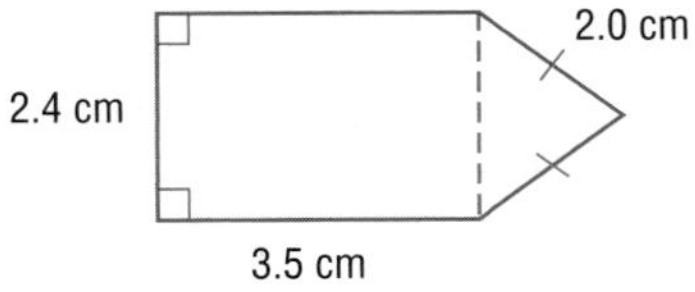

18. 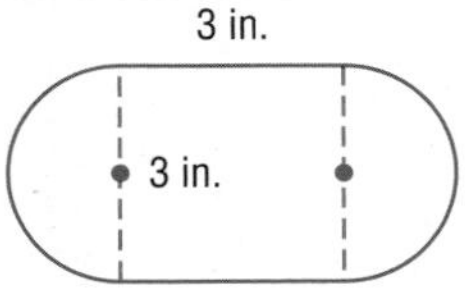

19. 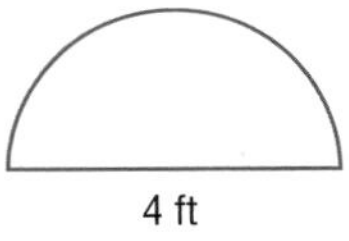

20. 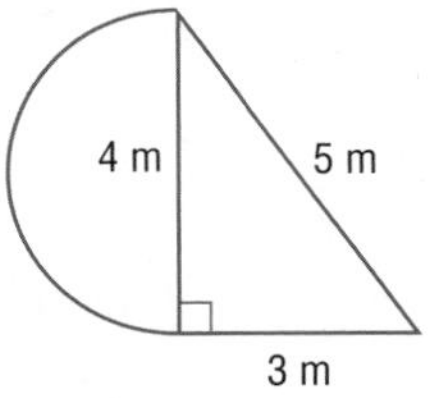

LESSON 0-8 Area

Objective

- Find the area of two-dimensional figures.

NewVocabulary
area

Area is the number of square units needed to cover a surface. Area is measured in square units.

Rectangle

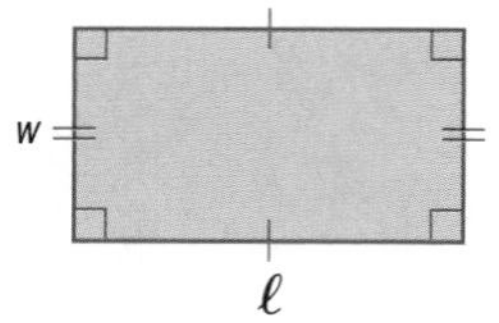

$A = \ell w$

Parallelogram

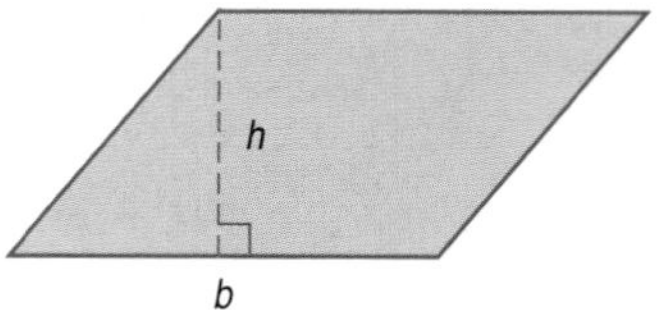

$A = bh$

Square

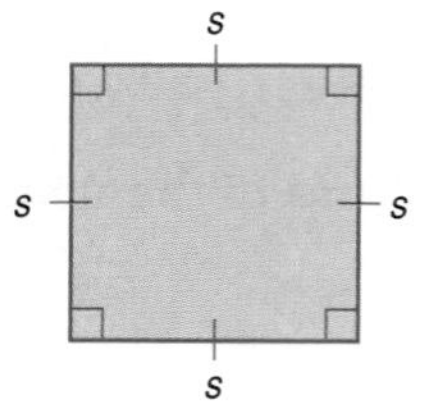

$A = s^2$

Triangle

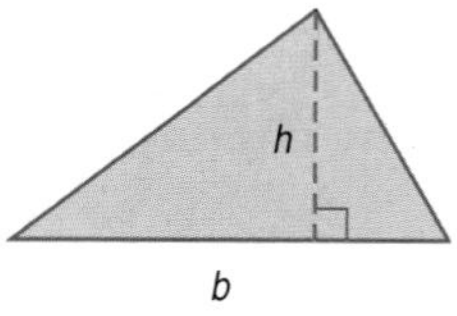

$A = \frac{1}{2}bh$

Example 1 Areas of Rectangles and Squares

Find the area of each figure.

a. a rectangle with a length of 7 yards and a width of 1 yard

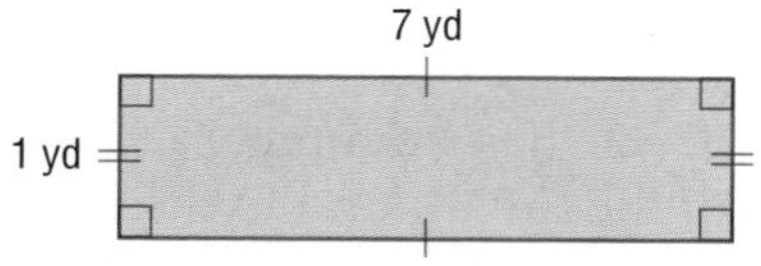

$A = \ell w$ — Area formula

$= 7(1)$ — $\ell = 7$, $w = 1$

$= 7$ — The area of the rectangle is 7 square yards.

b. a square with a side length of 2 meters

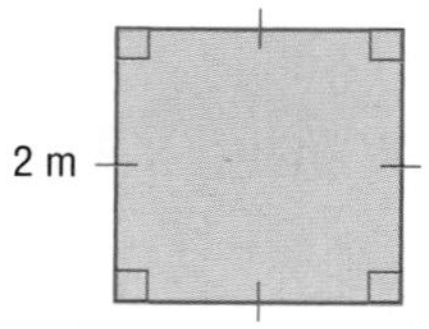

$A = s^2$ — Area formula

$= 2^2$ — $s = 2$

$= 4$ — The area is 4 square meters.

Example 2 Areas of Parallelograms and Triangles

Find the area of each figure.

a. **a parallelogram with a base of 11 feet and a height of 9 feet**

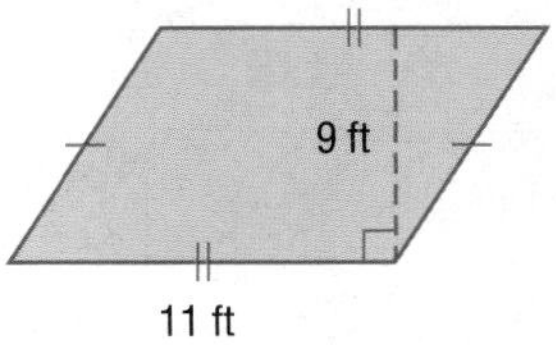

$A = bh$ Area formula

$= 11(9)$ $b = 11, h = 9$

$= 99$ Multiply.

The area is 99 square feet.

b. **a triangle with a base of 12 millimeters and a height of 5 millimeters**

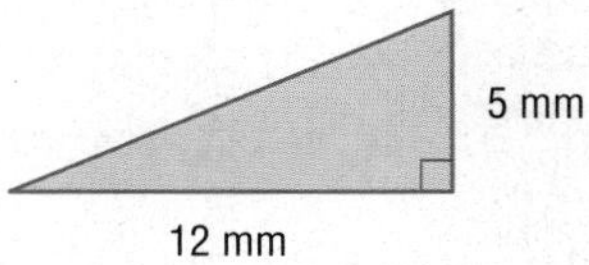

$A = \frac{1}{2}bh$ Area formula

$= \frac{1}{2}(12)(5)$ $b = 12, h = 5$

$= 30$ Multiply.

The area is 30 square millimeters.

The formula for the area of a circle is $A = \pi r^2$.

Example 3 Areas of Circles

Find the area of each circle to the nearest tenth.

a. **a radius of 3 centimeters**

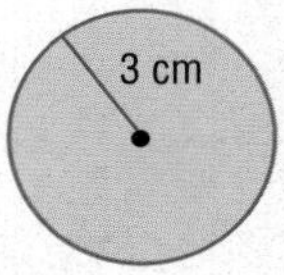

$A = \pi r^2$ Area formula

$= \pi(3)^2$ Replace r with 3.

$= 9\pi$ Simplify.

≈ 28.3 Use a calculator to evaluate 9π.

The area is about 28.3 square centimeters.

b. **a diameter of 21 meters**

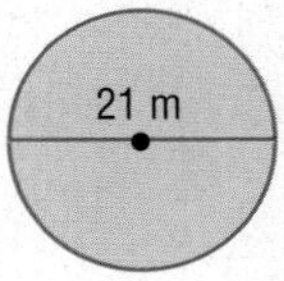

$A = \pi r^2$ Area formula

$= \pi(10.5)^2$ Replace r with 10.5.

$= 110.25\pi$ Simplify.

≈ 346.4 Use a calculator to evaluate 110.25π.

The area is about 346.4 square meters.

StudyTip

Mental Math You can use mental math to check your solutions. Square the radius and then multiply by 3.

Example 4 Estimate Area

Estimate the area of the polygon if each square represents 1 square mile.

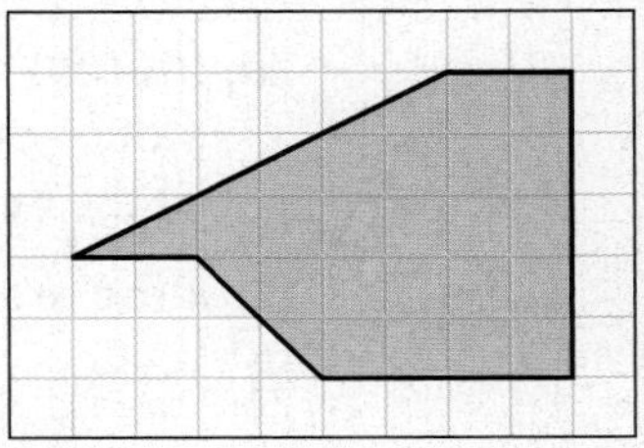

One way to estimate the area is to count each square as one unit and each partial square as a half unit, no matter how large or small.

$A \approx$ squares + partial squares

$\approx 21(1) + 8(0.5)$ 21 whole squares and 8 partial squares

$\approx 21 + 4$ or 25

The area of the polygon is about 25 square miles.

Exercises

Find the area of each figure.

1. 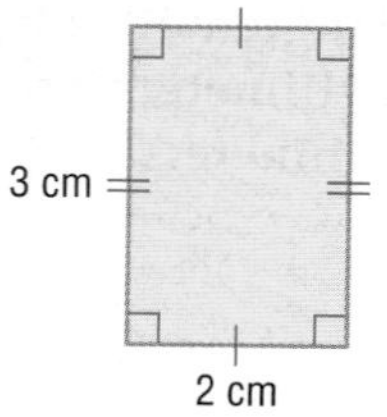

2. 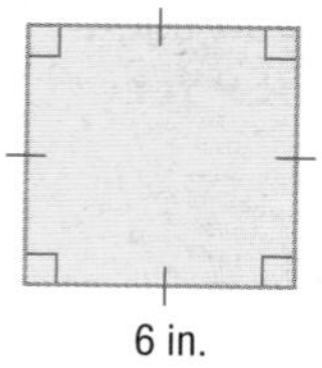

3. 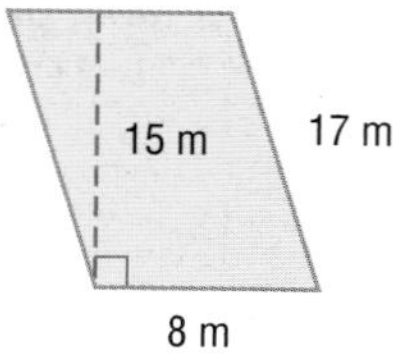

Find the area of each figure. Round to the nearest tenth if necessary.

4. a triangle with a base 12 millimeters and height 11 millimeters
5. a square with side length 9 feet
6. a rectangle with length 8 centimeters and width 2 centimeters
7. a triangle with a base 6 feet and height 3 feet
8. a quarter-circle with a diameter of 4 meters
9. a semi-circle with a radius of 3 inches

Find the area of each circle. Round to the nearest tenth.

10. 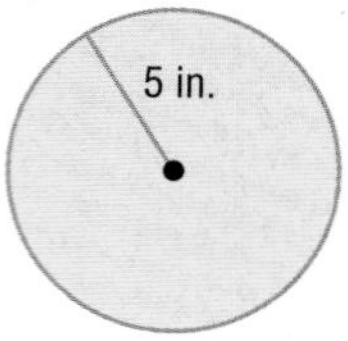

11. 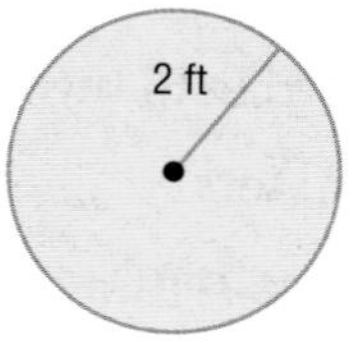

12. 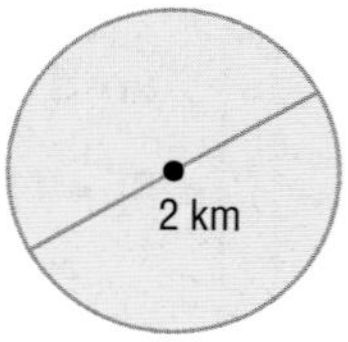

13. The radius is 4 centimeters.
14. The radius is 7.2 millimeters.
15. The diameter is 16 inches.
16. The diameter is 25 feet.
17. **CAMPING** The square floor of a tent has an area of 49 square feet. What is the side length of the tent?

Estimate the area of each polygon in square units.

18. 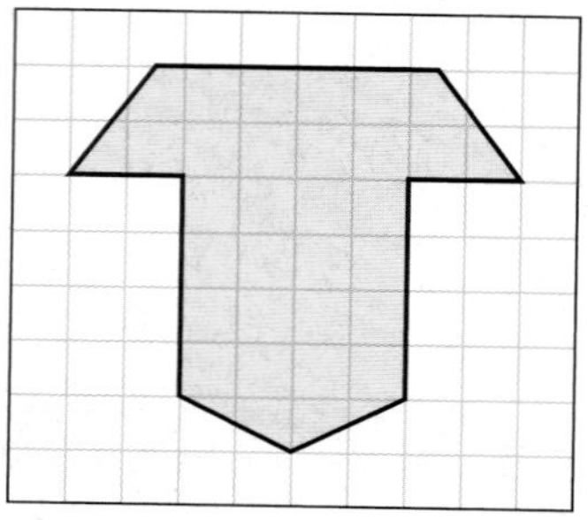

19. 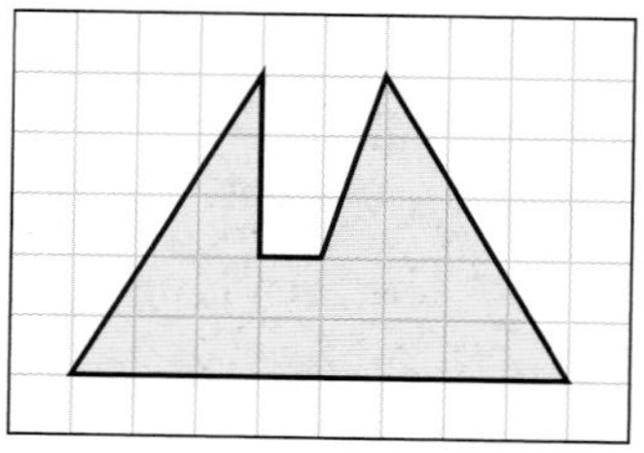

20. **HISTORY** Stonehenge is an ancient monument in Wiltshire, England. The giant stones of Stonehenge are arranged in a circle 30 meters in diameter. Find the area of the circle. Round to the nearest tenth square meter.

Find the area of each figure. Round to the nearest tenth.

21. 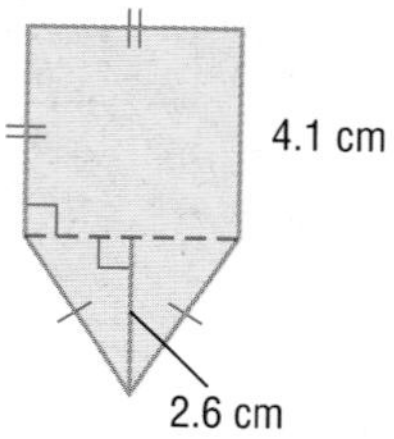

22. 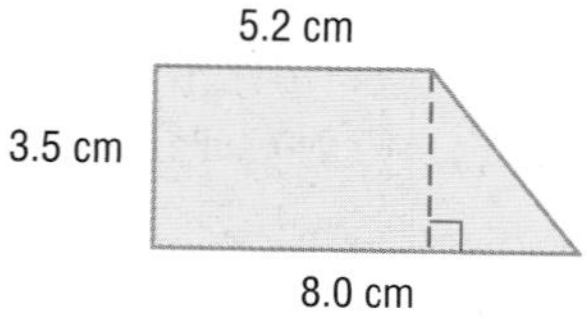

23. 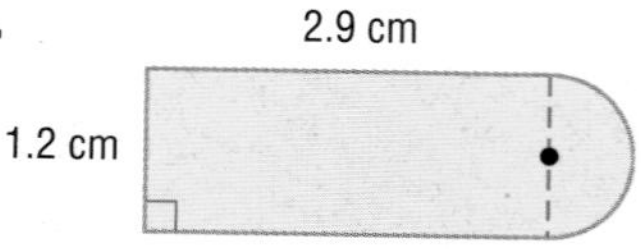

LESSON 0-9 Volume

Objective

- Find the volumes of rectangular prisms and cylinders.

NewVocabulary
volume

Volume is the measure of space occupied by a solid. Volume is measured in cubic units.

To find the volume of a rectangular prism, multiply the length times the width times the height. The formula for the volume of a rectangular prism is shown below.

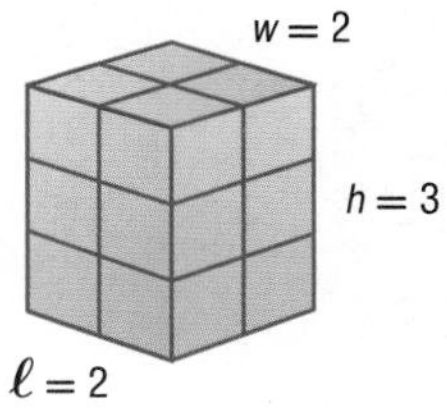

$$V = \ell \cdot w \cdot h$$

The prism at the right has a volume of $2 \cdot 2 \cdot 3$ or 12 cubic units.

Example 1 Volumes of Rectangular Prisms

Find the volume of each rectangular prism.

a. The length is 8 centimeters, the width is 1 centimeter, and the height is 5 centimeters.

$V = \ell \cdot w \cdot h$	Volume formula
$= 8 \cdot 1 \cdot 5$	Replace ℓ with 8, w with 1, and h with 5.
$= 40$	Simplify.

The volume is 40 cubic centimeters.

b.

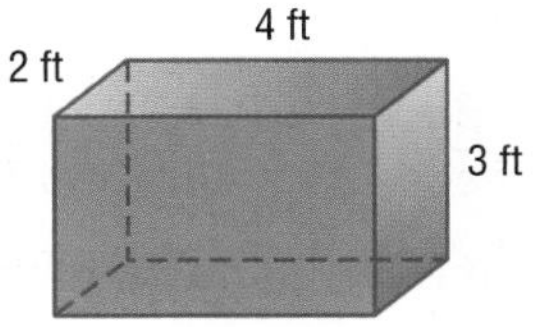

The prism has a length of 4 feet, width of 2 feet, and height of 3 feet.

$V = \ell \cdot w \cdot h$	Volume formula
$= 4 \cdot 2 \cdot 3$	Replace ℓ with 4, w with 2, and h with 3.
$= 24$	Simplify.

The volume is 24 cubic feet.

The volume of a solid is the product of the area of the base and the height of the solid. For a cylinder, the area of the base is πr^2. So the volume is $V = \pi r^2 h$.

Example 2 Volume of a Cylinder

Find the volume of the cylinder.

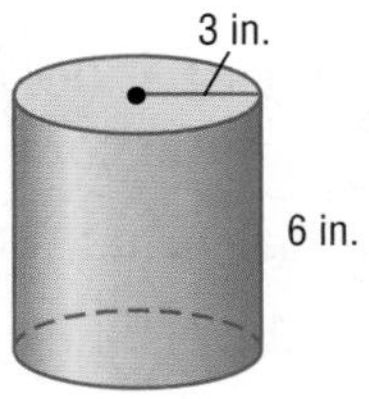

$V = \pi r^2 h$	Volume of a cylinder
$= \pi(3^2)6$	$r = 3$, $h = 6$
$= 54\pi$	Simplify.
≈ 169.6	Use a calculator.

The volume is about 169.6 cubic inches.

Exercises

Find the volume of each rectangular prism given the length, width, and height.

1. $\ell = 5$ cm, $w = 3$ cm, $h = 2$ cm
2. $\ell = 10$ m, $w = 10$ m, $h = 1$ m
3. $\ell = 6$ yd, $w = 2$ yd, $h = 4$ yd
4. $\ell = 2$ in., $w = 5$ in., $h = 12$ in.
5. $\ell = 13$ ft, $w = 9$ ft, $h = 12$ ft
6. $\ell = 7.8$ mm, $w = 0.6$ mm, $h = 8$ mm

Find the volume of each rectangular prism.

7. 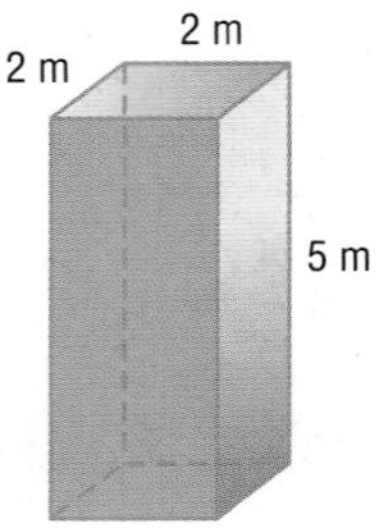

8. 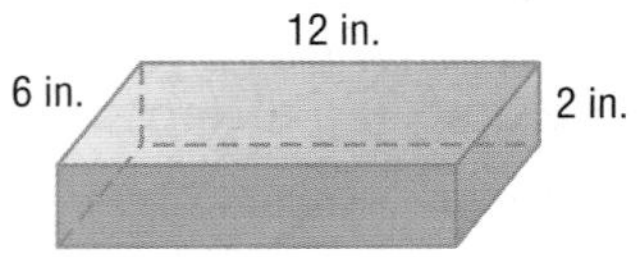

9. **GEOMETRY** A cube measures 3 meters on a side. What is its volume?

10. **AQUARIUMS** An aquarium is 8 feet long, 5 feet wide, and 5.5 feet deep. What is the volume of the aquarium?

11. **COOKING** What is the volume of a microwave oven that is 18 inches wide by 10 inches long with a depth of $11\frac{1}{2}$ inches?

12. **BOXES** A cardboard box is 32 inches long, 22 inches wide, and 16 inches tall. What is the volume of the box?

13. **SWIMMING POOLS** A children's rectangular pool holds 480 cubic feet of water. What is the depth of the pool if its length is 30 feet and its width is 16 feet?

14. **BAKING** A rectangular cake pan has a volume of 234 cubic inches. If the length of the pan is 9 inches and the width is 13 inches, what is the height of the pan?

15. **GEOMETRY** The volume of the rectangular prism at the right is 440 cubic centimeters. What is the width?

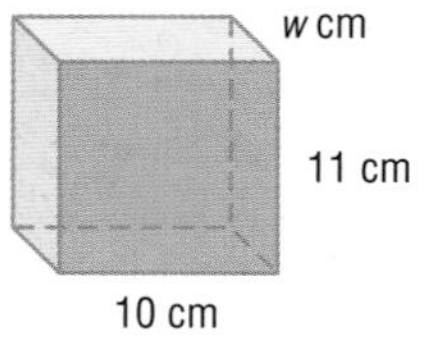

Find the volume of each cylinder. Round to the nearest tenth.

16. 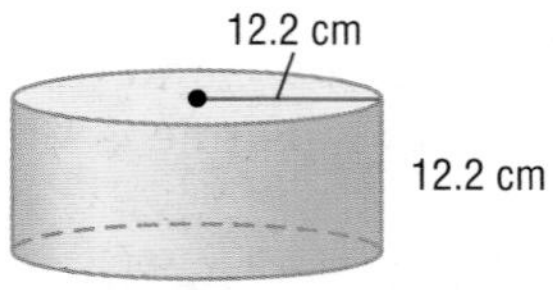

17. 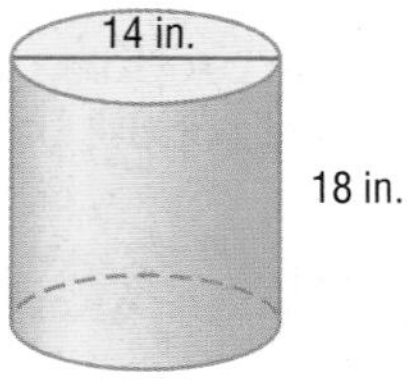

18. 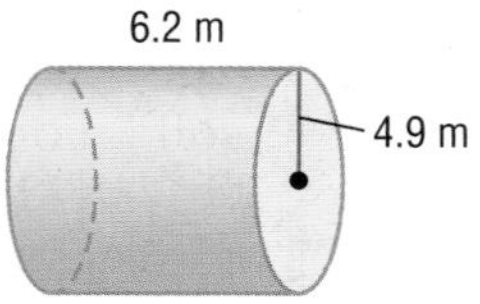

19. **FIREWOOD** Firewood is usually sold by a measure known as a *cord*. A full cord may be a stack 8 × 4 × 4 feet or a stack 8 × 8 × 2 feet.
    - **a.** What is the volume of a full cord of firewood?
    - **b.** A "short cord" of wood is 8 × 4 × the length of the logs. What is the volume of a short cord of $2\frac{1}{2}$-foot logs?
    - **c.** If you have an area that is 12 feet long and 2 feet wide in which to store your firewood, how high will the stack be if it is a full cord of wood?

LESSON 0-10

Surface Area

Objective

- Find the surface areas of rectangular prisms and cylinders.

NewVocabulary
surface area

Surface area is the sum of the areas of all the surfaces, or faces, of a solid. Surface area is measured in square units.

KeyConcept Surface Area

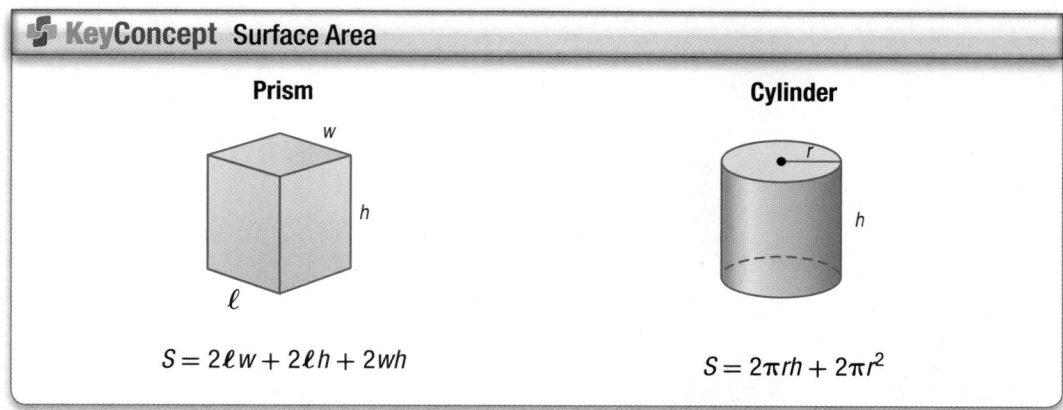

Prism

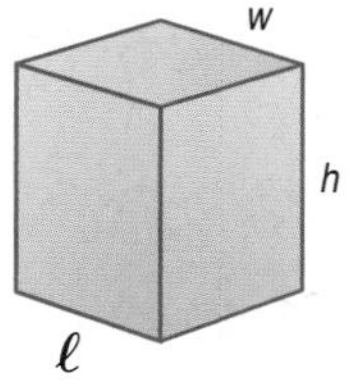

$S = 2\ell w + 2\ell h + 2wh$

Cylinder

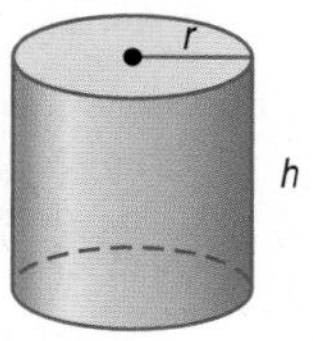

$S = 2\pi rh + 2\pi r^2$

Example 1 Find Surface Areas

PT

Find the surface area of each solid. Round to the nearest tenth if necessary.

a.

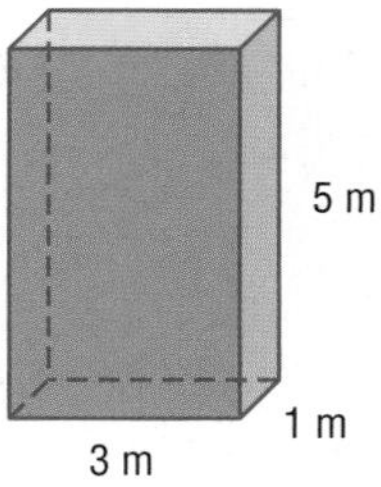

The prism has a length of 3 meters, width of 1 meter, and height of 5 meters.

$S = 2\ell w + 2\ell h + 2wh$	Surface area formula
$= 2(3)(1) + 2(3)(5) + 2(1)(5)$	$\ell = 3, w = 1, h = 5$
$= 6 + 30 + 10$	Multiply.
$= 46$	Add.

The surface area is 46 square meters.

b.

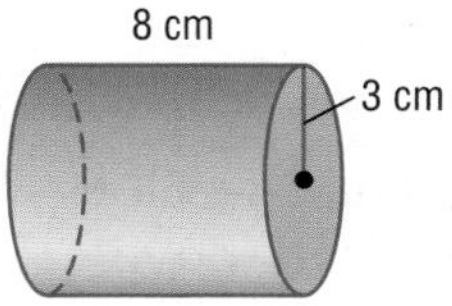

The height is 8 centimeters and the radius of the base is 3 centimeters. The surface area is the sum of the area of each base, $2\pi r^2$, and the area of the side, given by the circumference of the base times the height or $2\pi rh$.

$S = 2\pi rh + 2\pi r^2$	Formula for surface area of a cylinder.
$= 2\pi(3)(8) + 2\pi(3^2)$	$r = 3, h = 8$
$= 48\pi + 18\pi$	Simplify.
$\approx 207.3 \text{ cm}^2$	Use a calculator.

Exercises

Find the surface area of each rectangular prism given the measurements below.

1. $\ell = 6$ in., $w = 1$ in., $h = 4$ in

2. $\ell = 8$ m, $w = 2$ m, $h = 2$ m

3. $\ell = 10$ mm, $w = 4$ mm, $h = 5$ mm

4. $\ell = 6.2$ cm, $w = 1$ cm, $h = 3$ cm

5. $\ell = 7$ ft, $w = 2$ ft, $h = \frac{1}{2}$ ft

6. $\ell = 7.8$ m, $w = 3.4$ m, $h = 9$ m

Find the surface area of each solid.

7.

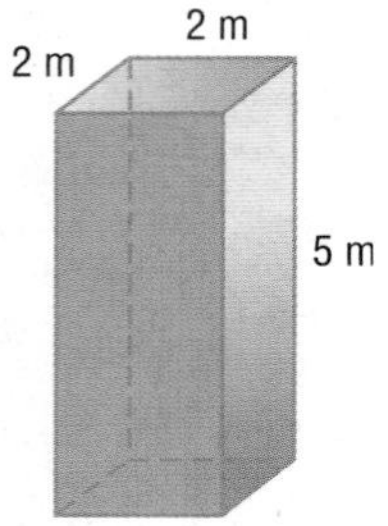

8.

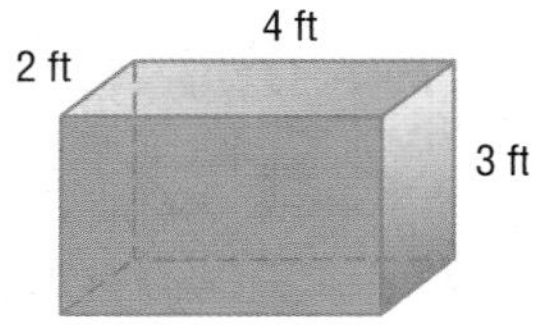

9.

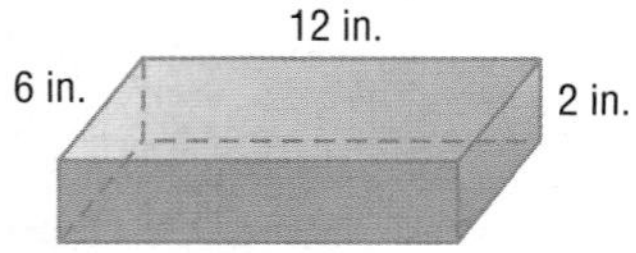

10.

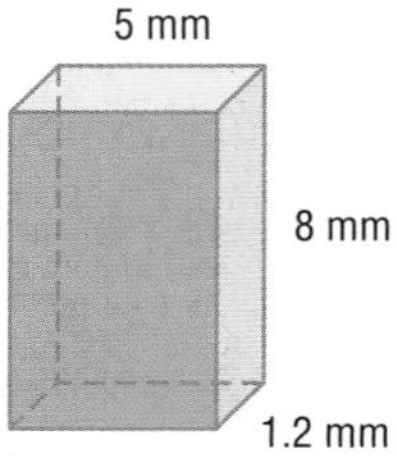

11.

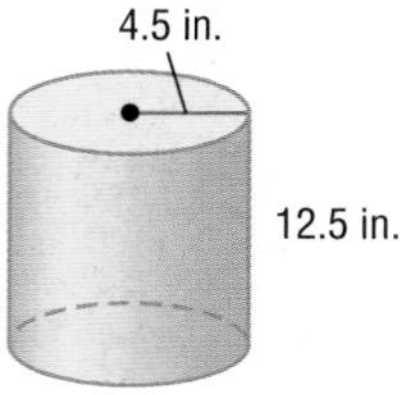

12.

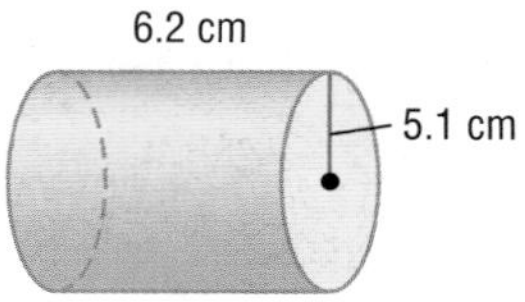

13. **GEOMETRY** What is the surface area of a cube with a side length of 2 meters?

14. **GIFTS** A gift box is a rectangular prism 14 inches long, 5 inches wide, and 4 inches high. If the box is to be covered in fabric, how much fabric is needed if there is no overlap?

15. **BOXES** A new refrigerator is shipped in a box 34 inches deep, 66 inches high, and $33\frac{1}{4}$ inches wide. What is the surface area of the box in square feet? Round to the nearest square foot. (*Hint:* 1 ft^2 = 144 in^2)

16. **PAINTING** A cabinet is 6 feet high, 3 feet wide, and 2 feet long. The entire outside surface of the cabinet is being painted except for the bottom. What is the surface area of the cabinet that is being painted?

17. **SOUP** A soup can is 4 inches tall and has a diameter of $3\frac{1}{4}$ inches. How much paper is needed for the label on the can? Round your answer to the nearest tenth.

18. **CRAFTS** For a craft project, Sarah is covering all the sides of a box with stickers. The length of the box is 8 inches, the width is 6 inches, and the height is 4 inches. If each sticker has a length of 2 inches and a width of 4 inches, how many stickers does she need to cover the box?

LESSON 0-11 Simple Probability and Odds

Objective

- Find the probability and odds of simple events.

NewVocabulary

probability
sample space
equally likely
complements
tree diagram
odds

The **probability** of an event is the ratio of the number of favorable outcomes for the event to the total number of possible outcomes. When you roll a die, there are six possible outcomes: 1, 2, 3, 4, 5, or 6. This list of all possible outcomes is called the **sample space**.

When there are n outcomes and the probability of each one is $\frac{1}{n}$, we say that the outcomes are **equally likely**. For example, when you roll a die, the 6 possible outcomes are equally likely because each outcome has a probability of $\frac{1}{6}$. The probability of an event is always between 0 and 1, inclusive. The closer a probability is to 1, the more likely it is to occur.

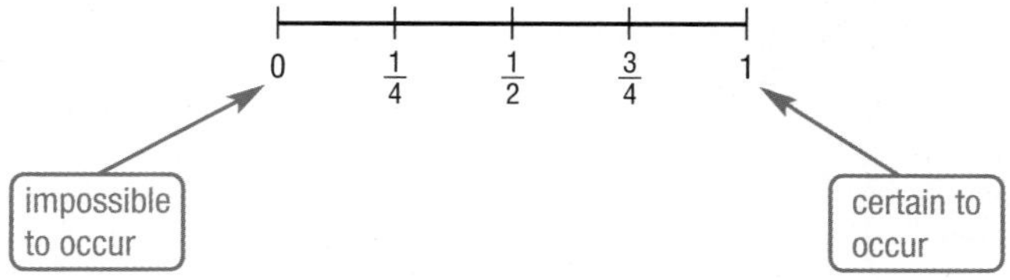

Example 1 Find Probabilities

A die is rolled. Find each probability.

a. rolling a 1 or 5

There are six possible outcomes. There are two favorable outcomes, 1 and 5.

$$\text{probability} = \frac{\text{number of favorable outcomes}}{\text{total number of possible outcomes}} = \frac{2}{6}$$

So, $P(1 \text{ or } 5) = \frac{2}{6}$ or $\frac{1}{3}$.

b. rolling an even number

Three of the six outcomes are even numbers. So, there are three favorable outcomes.

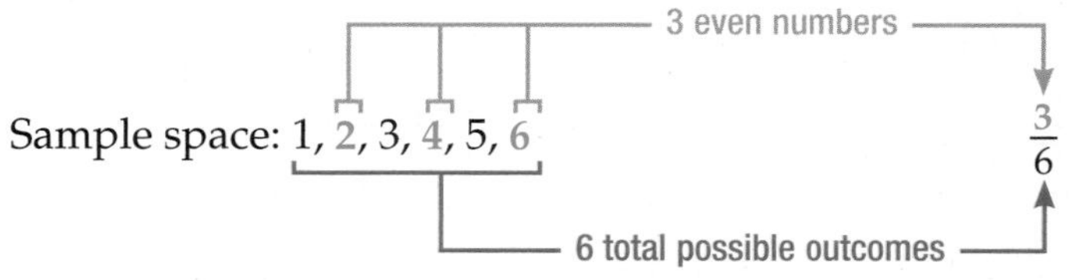

So, $P(\text{even number}) = \frac{3}{6}$ or $\frac{1}{2}$.

The events for rolling a 1 and for *not* rolling a 1 are called **complements**.

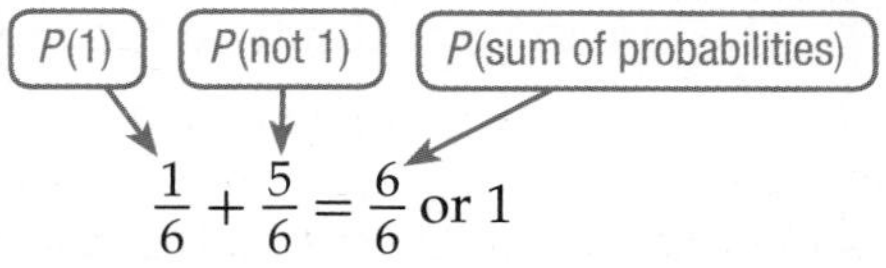

The sum of the probabilities for any two complementary events is always 1.

Example 2 Find Probabilities

A bowl contains 5 red chips, 7 blue chips, 6 yellow chips, and 10 green chips. One chip is randomly drawn. Find each probability.

a. blue

There are 7 blue chips and 28 total chips.

$$P(\text{blue chip}) = \frac{7}{28} \quad \begin{matrix} \leftarrow \text{number of favorable outcomes} \\ \leftarrow \text{number of possible outcomes} \end{matrix}$$

$$= \frac{1}{4}$$

The probability can be stated as $\frac{1}{4}$, 0.25, or 25%.

b. red or yellow

There are 5 + 6 or 11 chips that are red or yellow.

$$P(\text{red or yellow}) = \frac{11}{28} \quad \begin{matrix} \leftarrow \text{number of favorable outcomes} \\ \leftarrow \text{number of possible outcomes} \end{matrix}$$

$$\approx 0.39$$

The probability can be stated as $\frac{11}{28}$, about 0.39, or about 39%.

StudyTip

Alternate Method A chip drawn will either be green or not green. So, another method for finding *P*(not green) is to find *P*(green) and subtract that probability from 1.

c. not green

There are 5 + 7 + 6 or 18 chips that are not green.

$$P(\text{not green}) = \frac{18}{28} \quad \begin{matrix} \leftarrow \text{number of favorable outcomes} \\ \leftarrow \text{number of possible outcomes} \end{matrix}$$

$$= \frac{9}{14} \text{ or about } 0.64$$

The probability can be stated as $\frac{9}{14}$, about 0.64, or about 64%.

One method used for counting the number of possible outcomes is to draw a **tree diagram**. The last column of a tree diagram shows all of the possible outcomes.

Example 3 Use a Tree Diagram to Count Outcomes

School baseball caps come in blue, yellow, or white. The caps have either the school mascot or the school's initials. Use a tree diagram to determine the number of different caps possible.

StudyTip

Counting Outcomes When counting possible outcomes, make a column in your tree diagram for each part of the event.

Color	Design	Outcomes
blue	mascot	blue, mascot
	initials	blue, initials
yellow	mascot	yellow, mascot
	initials	yellow, initials
white	mascot	white, mascot
	initials	white, initials

The tree diagram shows that there are 6 different caps possible.

This example is an illustration of the **Fundamental Counting Principle**, which relates the number of outcomes to the number of choices.

KeyConcept Fundamental Counting Principle

Words	If event M can occur in m ways and is followed by event N that can occur in n ways, then the event M followed by N can occur in $m \cdot n$ ways.
Example	If there are 4 possible sizes for fish tanks and 3 possible shapes, then there are $4 \cdot 3$ or 12 possible fish tanks.

PT

Example 4 Use the Fundamental Counting Principle

a. An ice cream shop offers one, two, or three scoops of ice cream from among 12 different flavors. The ice cream can be served in a wafer cone, a sugar cone, or in a cup. Use the Fundamental Counting Principle to determine the number of choices possible.

There are 3 ways the ice cream is served, 3 different servings, and there are 12 different flavors of ice cream.

Use the Fundamental Counting Principle to find the number of possible choices.

number of scoops		number of flavors		number of serving options		number of choices of ordering ice cream
3	$\cdot$	12	$\cdot$	3	=	108

So, there are 108 different ways to order ice cream.

b. Jimmy needs to make a 3-digit password for his log-on name on a Web site. The password can include any digit from 0-9, but the digits may not repeat. How many possible 3-digit passwords are there?

If the first digit is a 4, then the next digit cannot be a 4.

We can use the Fundamental Counting Principle to find the number of possible passwords.

1st digit		2nd digit		3rd digit		number of passwords
10	$\cdot$	9	$\cdot$	8	=	720

So, there are 720 possible 3-digit passwords.

StudyTip

Odds The sum of the number of successes and the number of failures equals the size of the sample space, or the number of possible outcomes.

The **odds** of an event occurring is the ratio that compares the number of ways an event can occur (successes) to the number of ways it cannot occur (failures).

PT

Example 5 Find the Odds

Find the odds of rolling a number less than 3.

There are six possible outcomes; 2 are successes and 4 are failures.

So, the odds of rolling a number less than 3 are $\frac{1}{2}$ or 1:2.

Exercises

One coin is randomly selected from a jar containing 70 nickels, 100 dimes, 80 quarters, and 50 one-dollar coins. Find each probability.

1. P(quarter)

2. P(dime)

3. P(quarter or nickel)

4. P(value greater than \$0.10)

5. P(value less than \$1)

6. P(value at most \$1)

One of the polygons below is chosen at random. Find each probability.

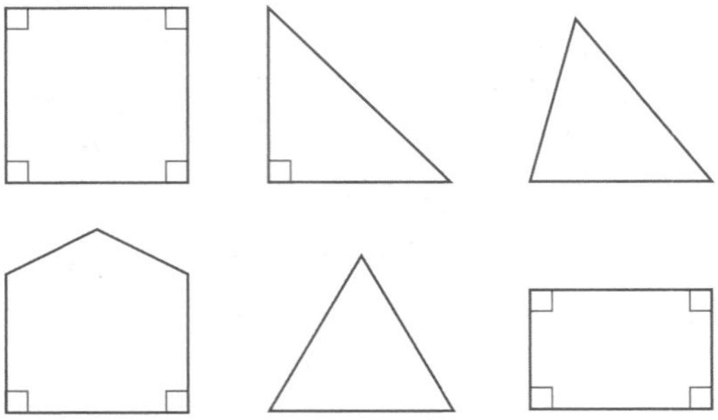

7. P(triangle)

8. P(pentagon)

9. P(not a quadrilateral)

10. P(more than 2 right angles)

Use a tree diagram to find the sample space for each event. State the number of possible outcomes.

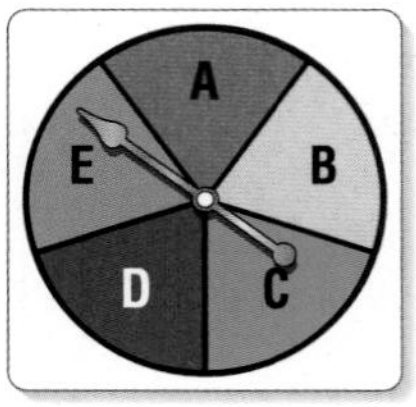

11. The spinner at the right is spun and two coins are tossed.

12. At a restaurant, you choose two sides to have with breakfast. You can choose white or whole wheat toast. You can choose sausage links, sausage patties, or bacon.

13. How many different 3-character codes are there using A, B, or C for the first character, 8 or 9 for the second character, and 0 or 1 for the third character?

A bag is full of different colored marbles. The probability of randomly selecting a red marble from the bag is $\frac{1}{8}$. The probability of selecting a blue marble is $\frac{13}{24}$. Find each probability.

14. P(not red)

15. P(not blue)

Find the odds of each outcome if a computer randomly picks a letter in the name THE UNITED STATES OF AMERICA.

16. the letter A

17. the letter T

18. a vowel

19. a consonant

Margaret wants to order a sub at the local deli.

20. Find the number of possible orders of a sub with one topping and one dressing option.

21. Find the number of possible ham subs with mayonnaise, any combination of toppings or no toppings at all.

22. Find the number of possible orders of a sub with any combination of dressing and/or toppings.

Subs	
ham, salami, roast beef, turkey, bologna, pepperoni	
Dressing	**Toppings**
mayonnaise, mustard, vinegar, oil	lettuce, onions, peppers, olives

LESSON 0-12 Measures of Center, Variation, and Position

Objective

- Find measures of central tendency, variation, and position.

NewVocabulary

variable
data
measurement or quantitative data
categorical or qualitative data
univariate data
measures of center or central tendency
mean
median
mode
measures of spread or variation
range
quartile
measures of position
lower quartile
upper quartile
five-number summary
interquartile range
outlier

A **variable** is a characteristic of a group of people or objects that can assume different values called **data**. Data that have units and can be measured are called **measurement** or **quantitative data**. Data that can be organized into different categories are called **categorical** or **qualitative data**. Some examples of both types of data are listed below.

Measurement Data	Categorical Data
Times: 15 s, 20 s, 45 s, 19 s **Ages:** 10 yr, 15 yr, 14 yr, 16 yr **Distance:** 5 mi, 30 mi, 18 mi	**Favorite color:** blue, red, purple, green **Hair color:** black, blonde, brown **Phone Numbers:** 555-1234, 555-5678

Measurement data in one variable, called **univariate data**, are often summarized using a single number to represent what is average or typical. Measures of what is average are also called **measures of center** or **central tendency**. The most common measures of center are mean, median, and mode.

KeyConcept Measures of Center

- The **mean** is the sum of the values in a data set divided by the total number of values in the set.
- The **median** is the middle value or the mean of the two middle values in a set of data when the data are arranged in numerical order.
- The **mode** is the value or values that appear most often in a set of data. A set of data can have no mode, one mode, or more than one mode.

PT

Example 1 Measures of Center

BASEBALL The table shows the number of hits Marcus made for his team. Find the mean, median, and mode.

Team Played	Hits
Badgers	3
Hornets	6
Bulldogs	5
Vikings	2
Rangers	3
Panthers	7

Mean: To find the mean, find the sum of all the hits and divide by the number of games in which he made these hits.

$$\text{mean} = \frac{3+6+5+2+3+7}{6} = \frac{26}{6} \text{ or about 4 hits}$$

Median: To find the median, order the numbers from least to greatest and find the middle value or values.

2, 3, 3, 5, 6, 7

$\frac{3+5}{2}$ or 4 hits — Since there is an even number of values, find the mean of the middle two.

Mode: From the arrangement of the data values, we can see that the value that occurs most often in the set is 3, so the mode of the data set is 3 hits.

Marcus's mean and median number of hits for these games was 4, and his mode was 3 hits.

Two very different data sets can have the same mean, so statisticians also use **measures of spread** or **variation** to describe how widely the data values vary. One such measure is the **range**, which is the difference between the greatest and least values in a set of data.

PT

Example 2 Range

WALKING The times in minutes it took Olivia to walk to school each day this week are 18, 15, 15, 12, and 14. Find the range.

range = greatest value − least value — Definition of range

= 18 − 12 or 6 — The greatest value is 18, and the least value is 12.

The range of the times is 6 minutes.

Statisticians often talk about the position of a value relative to other values in a set. **Quartiles** are common **measures of position** that divide a data set arranged in ascending order into four groups, each containing about one fourth or 25% of the data. The median marks the second quartile Q_2 and separates the data into upper and lower halves. The first or **lower quartile** Q_1 is the median of the lower half, while the third or **upper quartile** Q_3 is the median of the upper half.

> **StudyTip**
> **Calculating Quartiles** When the number of values in a set of data is odd, the median is not included in either half of the data when calculating Q_1 or Q_3.

25 31 36 39 40 41 44 45 49 50 54

25 ↑ minimum; 36 ↑ Q_1 lower quartile; 41 ↑ Q_2 median; 49 ↑ Q_3 upper quartile; 54 ↑ maximum

|← 25% →|← 25% →|← 25% →|← 25% →|

The three quartiles, along with the minimum and maximum values, are called a **five- number summary** of a data set.

PT

Example 3 Five-Number Summary

FUNDRAISER The number of boxes of donuts Aang sold for a fundraiser each day for the last 11 days were 22, 16, 35, 26, 14, 17, 28, 29, 21, 17, and 20. Find the minimum, lower quartile, median, upper quartile, and maximum of the data set. Then interpret this five-number summary.

Order the data from least to greatest. Use the list to determine the quartiles.

14, 16, 17, 17, 20, 21, 22, 26, 28, 29, 35

14 ↑ Min.; 17 ↑ Q_1; 21 ↑ Q_2; 28 ↑ Q_3; 35 ↑ Max.

The minimum is 14, the lower quartile is 17, the median is 21, the upper quartile is 28, and the maximum is 35. Over the last 11 days, Aang sold a minimum of 14 boxes and a maximum of 35 boxes. He sold fewer than 17 boxes 25% of the time, fewer than 21 boxes 50% of the time, and fewer than 28 boxes 75% of the time.

The difference between the upper and lower quartiles is called the **interquartile range**. The interquartile range, or IQR, contains about 50% of the values.

14, 16, 17, 17, 20, 21, 22, 26, 28, 29, 35

17 ↑ Q_1; 28 ↑ Q_3

|← $IQR = Q_1 - Q_3$ or 11 →|

Before deciding on which measure of center best describes a data set, check for outliers. An **outlier** is an extremely high or extremely low value when compared with the rest of the values in the set. To check for outliers, look for data values that are beyond the upper or lower quartiles by more than 1.5 times the interquartile range.

Example 4 Effect of Outliers

PT

TEST SCORES Students taking a make-up test received the following scores: 88, 79, 94, 90, 45, 71, 82, and 88.

a. Identify any outliers in the data.

First determine the median and upper and lower quartiles of the data.

45, 71, 79, 82, 88, 88, 90, 94

$Q_1 = \frac{71 + 79}{2}$ or 75 $\quad Q_2 = \frac{82 + 88}{2}$ or 85 $\quad Q_3 = \frac{88 + 90}{2}$ or 89

Find the interquartile range.

$IQR = Q_3 - Q_1 = 89 - 75$ or 14

Use the interquartile range to find the values beyond which any outliers would lie.

$Q_1 - 1.5(IQR)$	and	$Q_3 + 1.5(IQR)$	Values beyond which outliers lie
$75 - 1.5(14)$		$89 + 1.5(14)$	$Q_1 = 75$, $Q_3 = 89$, and $IQR = 14$
54		110	Simplify.

There are no scores greater than 110, but there is one score less than 54. The score of 45 can be considered an outlier for this data set.

b. Find the mean and median of the data set with and without the outlier. Describe what happens.

Data Set	Mean	Median
with outlier	$\frac{88 + 79 + 94 + 90 + 45 + 71 + 82 + 88}{8}$ or about 79.6	85
without outlier	$\frac{88 + 79 + 94 + 90 + 71 + 82 + 88}{7}$ or about 84.6	88

Removal of the outlier causes the mean and median to increase, but notice that the mean is affected more by the removal of the outlier than the median.

StudyTip

Interquartile Range When the interquartile range is a small value, the data in the set are close together. A large interquartile range means that the data are spread out.

Exercises

Find the mean, median, mode, and range for each data set.

1. number of students helping at the cookie booth each hour: 3, 5, 8, 1, 4, 11, 3
2. weight in pounds of boxes loaded onto a semi truck: 201, 201, 200, 199, 199
3. car speeds in miles per hour observed by a highway patrol officer: 60, 53, 53, 52, 53, 55, 55, 57
4. number of songs downloaded by students last week in Ms. Turner's class: 3, 7, 21, 23, 63, 27, 29, 95, 23
5. ratings of an online video: 2, 5, 3.5, 4, 4.5, 1, 1, 4, 2, 1.5, 2.5, 2, 3, 3.5

6. **SCHOOL SUPPLIES** The table shows the cost of school supplies. Find the mean, median, mode and range of the costs.

Cost of School Supplies	
Supply	**Cost**
pencils	$0.50
pens	$2.00
paper	$2.00
pocket folder	$1.25
calculator	$5.25
notebook	$3.00
erasers	$2.50
markers	$3.50

7. **BOWLING** Sue's average for 9 games of bowling is 108. What is the lowest score she can receive for the tenth game to have an mean of 110?

8. **LAUNDRY** Two brands of laundry detergents were tested to determine how many times a shirt could be washed before it faded. The results for 6 shirts in number of washes follow.

   Brand A: 16, 15, 13, 14, 16, 16

   Brand B: 11, 16, 18, 12, 15, 18

   a. Find the mean and range for each brand.

   b. Which brand performed more consistently? Explain.

Find the minimum, lower quartile, median, upper quartile, and maximum values for each data set.

9. prices in dollars of smartphones: 311, 309, 312, 314, 399, 312

10. attendance at an event for the last nine years: 68, 99, 73, 65, 67, 62, 80, 81, 83

11. books a student checks out of the library: 17, 9, 10, 17, 18, 5, 2

12. ounces of soda dispensed into 36-ounce cups:
    36.1, 35.8, 35.2, 36.5, 36.0, 36.2, 35.7, 35.8, 35.9, 36.4, 35.6

13. ages of riders on a roller coaster:
    45, 17, 16, 22, 25, 19, 20, 21, 32, 37, 19, 21, 24, 20, 18, 22, 23, 19

14. **NUTRITION** The table shows the number of servings of fruit and vegetables that Cole eats one week. Find the minimum, median, lower quartile, upper quartile, and maximum number of servings. Then interpret this five-number summary.

Fruits and Vegetables	
Day	**Number of Servings**
Monday	5
Tuesday	7
Wednesday	5
Thursday	4
Friday	3
Saturday	3
Sunday	8

Find the mean and median of the data set, and then identify any outliers. If the set has an outlier, find the mean and median without the outlier, and state which measure is affected more by the removal of this value.

15. distance traveled in miles to visit relatives during winter break:
    210, 45, 10, 108, 452, 225, 35, 95, 140, 25, 65, 250

16. time spent on social networking Web sites in minutes per day:
    25, 35, 45, 30, 65, 50, 25, 100, 45, 35, 5, 105, 110, 190, 40, 30, 80

17. batting averages for the last 10 seasons: 0.267, 0.305, 0.304, 0.201, 0.284, 0.302, 0.311, 0.289, 0.300, 0.292

18. **CHALLENGE** The cost of 8 different pairs of pants at a department store are $39.99, $31.99, $19.99, $14.99, $19.99, $23.99, $36.99, and $26.99.

    a. Find the mean, median, mode, and range of the pants prices.

    b. Suppose each pair of pants needs to be hemmed at an additional cost of $8 per pair. Including these alteration costs, what are the mean, median, mode, and range of the pant prices?

    c. Suppose the original price of each pair of pants is discounted by 25%. Find the mean, median, mode, and range of the discounted pant prices.

    d. Make a conjecture as to the effect on the mean, median, mode, and range of a data set if the same value n is added to each value in the data set. What is the effect on these same measures if each item in a data set is multiplied by the same value n?

LESSON 0-13 Representing Data

Objective

- Represent sets of data using different visual displays.

NewVocabulary
frequency table
bar graph
cumulative frequency
histogram
line graph
stem-and-leaf plot
circle graph
box-and-whisker plot

Common Core State Standards

Content Standards
S.ID.1 Represent data with plots on the real number line (dot plots, histograms, and box plots).

A **frequency table** uses tally marks to record and display frequencies of events. A **bar graph** compares categories of data with bars representing the frequencies.

Example 1 Make a Bar Graph

Make a bar graph to display the data.

Sport	Tally	Frequency
basketball	𝍸 𝍸 𝍸	15
football	𝍸 𝍸 𝍸 𝍸 𝍸	25
soccer	𝍸 𝍸 𝍸 III	18
baseball	𝍸 𝍸 𝍸 𝍸 I	21

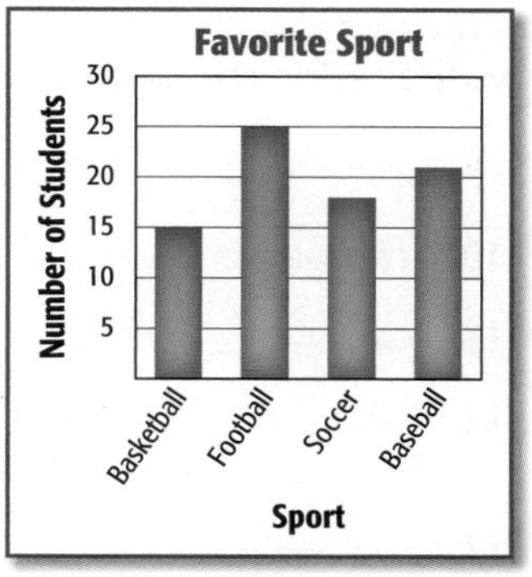

Step 1 Draw a horizontal axis and a vertical axis. Label the axes as shown. Add a title.

Step 2 Draw a bar to represent each sport. The vertical scale is the number of students who chose each sport. The horizontal scale identifies the sport.

The **cumulative frequency** for each event is the sum of its frequency and the frequencies of all preceding events. A **histogram** is a type of bar graph used to display numerical data that have been organized into equal intervals.

Example 2 Make a Histogram and a Cumulative Frequency Histogram

Make histograms of the frequency and the cumulative frequency.

Age at Inauguration	40–44	45–49	50–54	55–59	60–64	65–69
U.S. Presidents	2	7	13	12	7	3

Find the cumulative frequency for each interval.

Age	< 45	< 50	< 55	< 60	< 65	< 70
Presidents	2	2 + 7 = 9	9 + 13 = 22	22 + 12 = 34	34 + 7 = 41	41 + 3 = 44

Make each histogram like a bar graph but with no space between the bars.

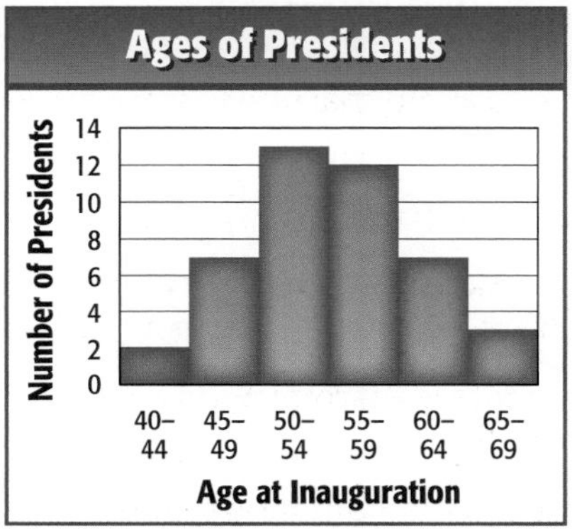

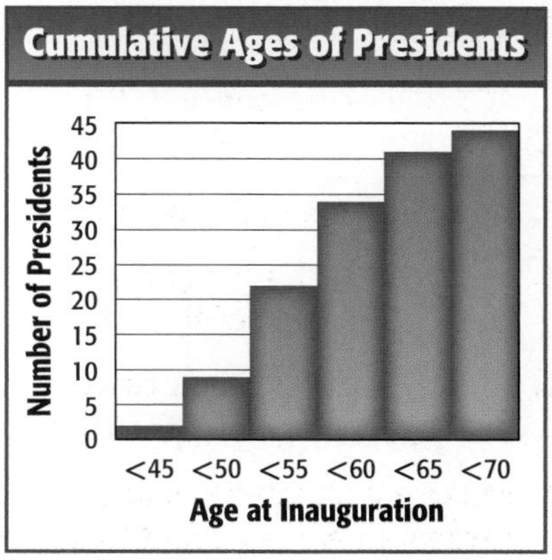

Another way to represent data is by using a line graph. A **line graph** usually shows how data change over a period of time.

Example 3 Make a Line Graph

Sales at the Marshall High School Store are shown in the table. Make a line graph of the data.

School Store Sales Amounts					
September	$670	December	$168	March	$412
October	$229	January	$290	April	$309
November	$300	February	$388	May	$198

Step 1 Draw a horizontal axis and a vertical axis and label them as shown. Include a title.

Step 2 Plot the points.

Step 3 Draw a line connecting each pair of consecutive points.

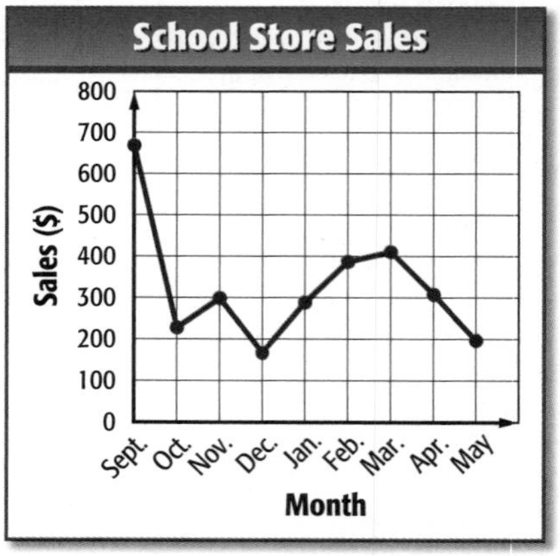

Data can also be organized and displayed by using a stem-and-leaf plot. In a **stem-and-leaf plot**, the digits of the least place value usually form the *leaves*, and the rest of the digits form the *stems*.

Real-WorldLink

The fastest animal on land is the cheetah. Cheetahs can run at speeds up to 60 miles per hour.

Source: Infoplease

Real-World Example 4 Make a Stem-and-Leaf Plot

ANIMALS **The speeds (mph) of 20 of the fastest land animals are listed at the right. Use the data to make a stem-and-leaf plot.**

42	40	40	35	50
32	50	36	50	40
45	70	43	45	32
40	35	61	48	35

Source: *The World Almanac*

The least place value is ones. So, 32 miles per hour would have a stem of 3 and a leaf of 2.

Stem	Leaf
3	2 2 5 5 5 6
4	0 0 0 0 2 3 5 5 8
5	0 0 0
6	1
7	0

Key: 3|2 = 32

Martin Harvey/Gallo Images/Getty Images

A **circle graph** is a graph that shows the relationship between parts of the data and the whole. The circle represents all of the data.

Example 5 Make a Circle Graph

The table shows how Lily spent 8 hours of one day at summer camp. Make a circle graph of the data.

Summer Camp	
Activity	Hours
canoeing	3
crafts	1
eating	2
hiking	2

First, find the ratio that compares the number of hours for each activity to 8. Then multiply each ratio by 360° to find the number of degrees for each section of the graph.

Canoeing: $\frac{3}{8} \cdot 360° = 135°$

Crafts: $\frac{1}{8} \cdot 360° = 45°$

Eating: $\frac{2}{8} \cdot 360° = 90°$

Hiking: $\frac{2}{8} \cdot 360° = 90°$

Summer Camp

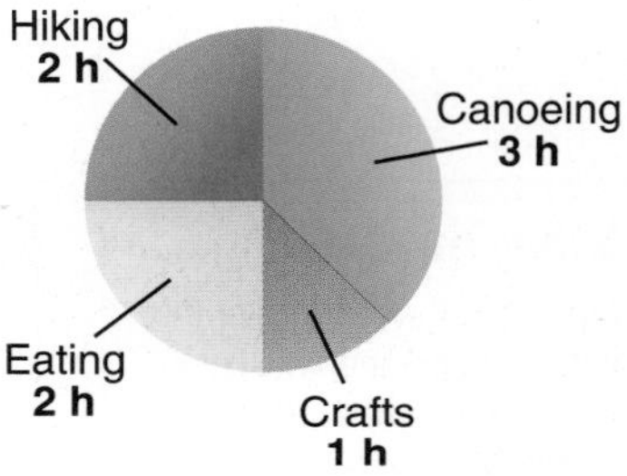

WatchOut!

Circle Graphs The sum of the measures of each section of a circle graph should be 360°.

A **box-and-whisker plot** is a graphical representation of the five-number summary of a data set. The box in a box-and-whisker plot represents the interquartile range.

Example 6 Make a Box-and-Whisker Plot

Draw a box-and-whisker plot for these data. Describe how the outlier affects the quartile points.

14, 30, 16, 20, 18, 16, 20, 18, 22, 13, 8

Step 1 Order the data from least to greatest. Then determine the maximum, minimum and the quartiles.

8, 13, **14**, 16, 16, **18**, 18, 20, **20**, 22, 30

min. ↑ 8, Q_1 ↑ 13, Q_2 ↑ 18, Q_3 ↑ 20, max. ↑ 30

Determine the interquartile range.

$$IQR = Q_3 - Q_1$$
$$= 20 - 14 \text{ or } 6$$

Check to see if there are any outliers.

$14 - 1.5(6) = 5$ $\quad$ $20 + 1.5(6) = 29$

Numbers less than 5 or greater than 29 are outliers.

The only outlier is 30.

Step 2 Draw a number line that includes the minimum and maximum values in the data. Place dots above the number line to represent the three quartile points, any outliers, the minimum value that is not an outlier, and the maximum value that is not an outlier.

Step 3 Draw the box and the whiskers. The vertical rules go through the quartiles. The outliers are not connected to the whiskers.

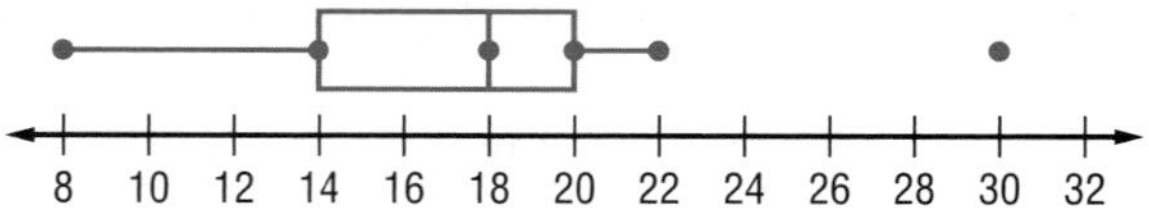

Step 4 Omit 30 from the data. Repeat Step 1 to determine Q_1, Q_2, and Q_3.

8, 13, **14,** 16, **16, 18,** 18, **20,** 20, 22

↑ Q_1 ↑ $Q_2 = 17$ ↑ Q_3

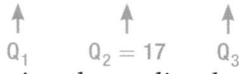

Removing the outlier does not affect Q_1 or Q_2 and thus does not affect the interquartile range. The value of Q_2 changes from 18 to 17.

StudyTip

Parallel Box-and-Whisker Plot A double box-and-whisker plot is sometimes called a *parallel box-and-whisker plot.*

Example 7 Compare Data

CLIMATE Lucas is going to go to college in either Dallas or Nashville. He wants to live in a place that does not get too cold. So he decides to compare the average monthly low temperatures of each city.

Average Monthly Low Temperatures (°F)		
Month	Dallas	Nashville
Jan.	36	28
Feb.	41	31
Mar.	49	39
Apr.	56	47
May	65	57
June	73	65
July	77	70
Aug.	76	68
Sept.	69	61
Oct.	58	49
Nov.	47	40
Dec.	39	32

Source: weather.com

a. Draw a double box-and-whisker plot for the data.

Determine the quartiles and outliers for each city.

Dallas

36, 39, 41, 47, 49, 56, 58, 65, 69, 73, 76, 77

↑ $Q_1 = 44$ ↑ $Q_2 = 57$ ↑ $Q_3 = 71$

Nashville

28, 31, 32, 39, 40, 47, 49, 57, 61, 65, 68, 70

↑ $Q_1 = 35.5$ ↑ $Q_2 = 48$ ↑ $Q_3 = 63$

There are no outliers. Draw the plots using the same number line.

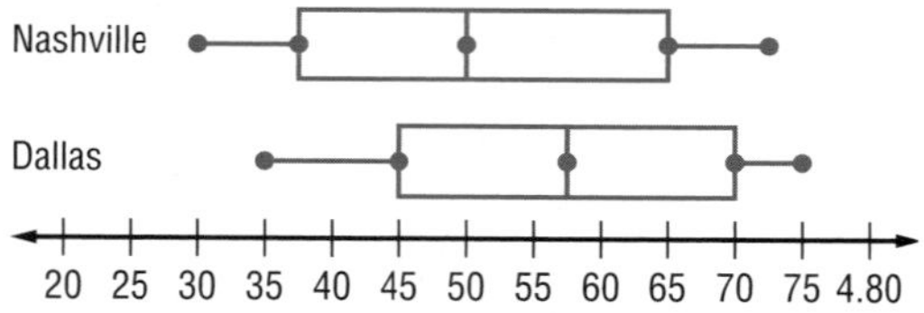

b. Use the double box-and-whisker plot to compare the data.

The interquartile range of temperatures for both cities is about the same. However, all quartiles of the Dallas temperatures are shifted to the right of those of Nashville, meaning Dallas has higher average low temperatures.

c. One night in August, a weather reporter stated the low for Nashville as being "only 65." Is it appropriate for the weather reporter to use the word only in the statement? Is 65 an unusually low temperature for Nashville in August? Explain your answer.

No, 65 is not an unusually low temperature for August in Nashville. It is lower than the average, but not by much.

When displaying data, some graphs are better choices than others.

Example 8 Select a Display

Which type of graph is the best way to display each set of data? Explain.

a. the results of the women's Olympic High Jump event from 1972 to 2008

Since the data would show change over time, a line graph would give the reader a clear picture of changes in height.

b. the percent of students in class who have 0, 1, 2, 3, or more than 3 pets

Since the data would show how parts are related to the whole, a circle graph would give the reader a clear picture of how different segments of the class relate to the whole class.

Exercises

1. SURVEYS Alana surveyed several students to find the number of hours of sleep they typically get each night. The results are shown in the table. Make a bar graph of the data.

Hours of Sleep					
Alana	8	Kwam	7.5	Tomas	7.75
Nick	8.25	Kate	7.25	Sharla	8.5

2. PLAYS The frequency table at the right shows the ages of people attending a high school play.

a. Make a histogram to display the data.

b. Make a cumulative frequency histogram showing the number of people attending who were less than 20, 40, 60, or 80 years old.

Age	Tally	Frequency
0–19	𝍸 𝍸 𝍸 𝍸 𝍸 𝍸 𝍸 𝍸 𝍸 II	47
20–39	𝍸 𝍸 𝍸 𝍸 𝍸 𝍸 𝍸 𝍸 III	43
40–59	𝍸 𝍸 𝍸 𝍸 𝍸 𝍸 I	31
60–79	𝍸 III	8

3. LAWN CARE Marcus started a lawn care service. The chart shows how much money he made over summer break. Make a line graph of the data.

Lawn Care Profits ($)								
Week	1	2	3	4	5	6	7	8
Profit	25	40	45	50	75	85	95	95

Use each set of data to make a stem-and-leaf plot and a box-and-whisker plot. Describe how the outliers affect the quartile points.

4. {65, 63, 69, 71, 73, 59, 60, 70, 72, 66, 71, 58}

5. {31, 30, 28, 26, 22, 34, 26, 31, 47, 32, 18, 33, 26, 23, 18}

6. FINANCIAL LITERACY The table shows how Ping spent his allowance of $40. Make a circle graph of the data.

Allowance	
How Spent	**Amount ($)**
savings	15
downloaded music	8
snacks	5
T-shirt	12

7. JOGGING The table shows the number of miles Hannah jogged each day for 10 days. Make a line graph of the data.

Day	1	2	3	4	5	6	7	8	9	10
Miles Jogged	2	2	3	3.5	4	4.5	2.5	3	4	5

8. **BASKETBALL** Two basketball teams are analyzing the number of points they scored in each game this season.

Lions: 48, 52, 55, 49, 53, 55, 51, 50, 46, 53, 47, 55, 50, 51, 60, 52, 57, 56, 58, 55
Eagles: 35, 39, 37, 40, 44, 42, 53, 42, 40, 44, 48, 46, 43, 47, 45, 41, 45, 43, 47, 48

a. Make a double box-and-whisker plot to display the data.
b. How does the number of points scored by the Lions compare to the number of points scored by the Eagles?
c. In the first game of the post season, a sports announcer reported the Lions scored a whopping 60 points. Is it appropriate for the announcer to use the word whopping in the statement? Is 60 an unusually high number of points for the Lions to score? Explain your answer.

9. **TESTS** Mr. O'Neil teaches two algebra classes. The test scores for the two classes are shown.

Third Period											
77	98	85	79	76	86	84	91	67	88	93	87
99	78	81	80	82	84	83	85	84	95	90	88
Sixth Period											
91	93	88	75	80	78	81	90	82	95	76	88
89	79	93	88	85	94	83	88	91	72	88	70

a. Make a double box-and-whisker plot to display the data.
b. Write a brief description of each data distribution.
c. How do the scores from the third period class compare to the scores from the sixth period class?

Which type of graph is the best way to display each set of data? Explain.

10. an organization's dollar contributions to 4 different charities

11. the prices of a college football ticket from 1990 to the present

12. the percent of glass, plastic, paper, steel, and aluminum in a recycling center

13. **DISCUS** The winning distances for the girls' discus throw at an annual track meet are shown below.

Year	1999	2000	2001	2002	2003	2004	2005	2006	2007	2008	2009	2010
Distance (m)	119	124	126	129	130	130	133	135	136	137	138	140

a. Make a stem-and-leaf plot to display the winning distances.
b. Make a histogram to display the winning distances.
c. What does the stem-and-leaf plot show you that the histogram does not?
d. If this trend continues, what would you expect the winning distance to be in 2030? Is your answer reasonable? Explain.

14. **DRINKS** Tate is buying drinks for a party. He is comparing 2-liter bottles to 12-packs of 12-ounce cans. The prices of 2-liter bottles are $0.99, $1.99, $1.87, $1.79, $1.29, $1.43, and $1.15. The prices of 12-packs are $2.50, $4.25, $3.34, $2.65, $3.19, $3.89, and 2.99.

a. Make a double box-and-whisker plot to display the data.
b. Notice that instead of comparing price per item it would be more beneficial to compare price per ounce. What is the price per ounce of each item if a 2-liter is approximately 67 ounces and a 12-pack is 144 ounces? Round to the nearest cent.
c. Make a new double box-and-whisker plot from the data obtained in part **b**.
d. Which is the better deal, the 12-packs of cans or the 2-liter bottles? Explain.

CHAPTER 0 Posttest

Determine whether you need an estimate or an exact answer. Then use the four-step problem-solving plan to solve.

1. **DISTANCE** Fabio rode his scooter 2.3 miles to his friend's house, then 0.7 mile to the grocery store, then 2.1 miles to the library. If he rode the same route back home, about how far did he travel in all?

2. **SHOPPING** The regular price of a T-shirt is \$9.99. It is on sale for 15% off. Sales tax is 6%. If you give the cashier a \$10 bill, how much change will you receive?

Find each sum or difference.

3. $-31 + (-4)$
4. $48 - 55$
5. $-71 - (-10)$
6. $31 - 42.9$
7. $-11.5 + 8.1$
8. $-0.38 - (-1.06)$

Find each product or quotient.

9. $-21(-5)$
10. $-81 \div (-3)$
11. $-120 \div 8$
12. $-39 \div -3$

Replace each ● with <, >, or = to make a true sentence.

13. $-0.62 \bullet -\frac{6}{7}$
14. $\frac{12}{44} \bullet \frac{8}{11}$
15. Order $4\frac{4}{5}$, 4.85, $2\frac{5}{8}$, and 2.6 from least to greatest.

Find each sum or difference. Write in simplest form.

16. $\frac{1}{7} + \frac{5}{7}$
17. $\frac{7}{8} - \frac{1}{8}$
18. $\frac{1}{6} + \left(-\frac{1}{2}\right)$
19. $-\frac{1}{12} - \left(-\frac{3}{4}\right)$

Find each product or quotient.

20. $-1.2(9.3)$
21. $-20.93 \div (-2.3)$
22. $10.5 \div (-1.2)$
23. $(-3.4)(-2.8)$

Name the reciprocal of each number.

24. 6
25. $1\frac{2}{5}$
26. $-2\frac{3}{7}$
27. $-\frac{1}{2}$
28. $\frac{4}{3}$
29. $5\frac{1}{3}$

Find each product or quotient. Write in simplest form.

30. $\frac{2}{5} \cdot \frac{5}{9}$
31. $\frac{4}{5} \div \frac{1}{5}$
32. $-\frac{7}{8} \cdot 2$
33. $\frac{1}{3} \div 2\frac{1}{4}$
34. $-6 \cdot \left(-\frac{3}{4}\right)$
35. $\frac{7}{18} \div \left(-\frac{14}{15}\right)$

36. **PICNIC** Joseph is mixing $5\frac{1}{2}$ gallons of orange drink for his class picnic. Every $\frac{1}{2}$ gallon requires 1 packet of orange drink mix. How many packets of orange drink mix does Joseph need?

Express each percent as a fraction in simplest form.

37. 6%
38. 140%

Use the percent proportion to find each number.

39. 50% of what number is 31?
40. What number is 110% of 51?
41. Find 8% of 95.

42. **SOLUTIONS** A solution is prepared by dissolving 24 milliliters of saline in 150 milliliters of pure solution. What is the percent of saline in the pure solution?

43. **SHOPPING** Marta got 60% off a pair of shoes. If the shoes cost \$9.75 (before sales tax), what was the original price of the shoes?

Find the perimeter and area of each figure.

44.

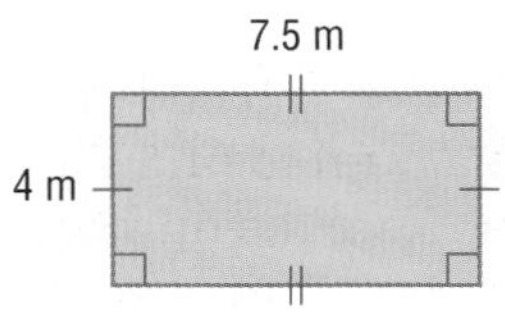

45.

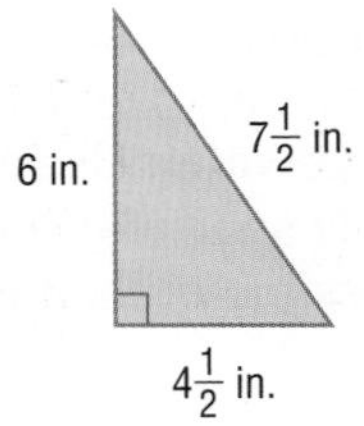

46. A parallelogram has a base of 20 millimeters and a height of 6 millimeters. Find the area.

47. **GARDENS** Find the perimeter of the garden.

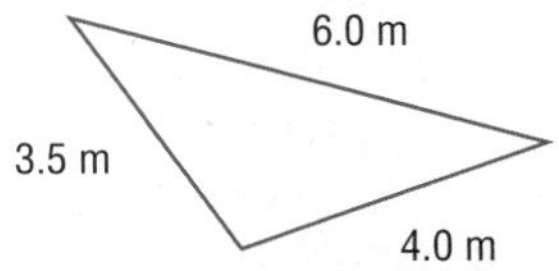

Find the circumference and area of each circle. Round to the nearest tenth.

48.

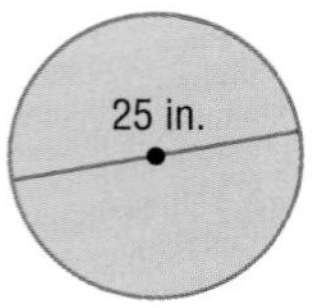

49.

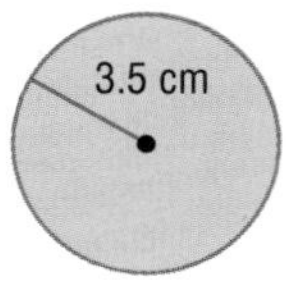

50. **PARKS** A park has a circular area for a fountain that has a circumference of about 16 feet. What is the radius of the circular area? Round to the nearest tenth.

Find the volume and surface area of each rectangular prism given the measurements below.

51. $\ell = 1.5$ m, $w = 3$ m, $h = 2$ m

52. $\ell = 4$ in., $w = 1$ in., $h = \frac{1}{2}$ in.

53. Find the volume and surface area of the rectangular prism.

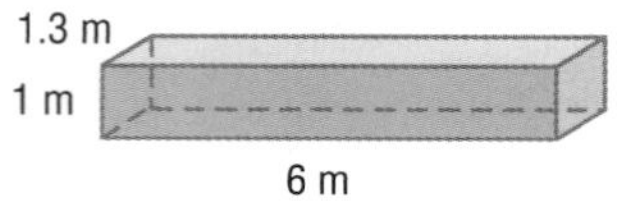

One marble is randomly selected from a jar containing 3 red, 4 green, 2 black, and 6 blue marbles. Find each probability.

54. P(red or blue)

55. P(green or red)

56. P(not black)

57. P(not blue)

58. A movie theater is offering snack specials. You can choose a small, medium, large, or jumbo popcorn with or without butter, and soda or bottled water. Use a tree diagram to find the sample space for the event. State the number of possible outcomes.

One coin is randomly selected from a jar containing 20 pennies, 15 nickels, 3 dimes, and 12 quarters. Find the odds of each outcome. Write in simplest form.

59. a dime

60. a value less than $0.25

61. a value greater than $0.10

62. a value less than $0.05

63. **SCHOOL** In a science class, each student must choose a lab project from a list of 15, write a paper on one of 6 topics, and give a presentation about one of 8 subjects. How many ways can students choose to do their assignments?

64. **GAMES** Marcos has been dealt seven different cards. How many different ways can he play his cards if he is required to play one card at a time?

Find the mean, median, and mode for each set of data.

65. {99, 88, 88, 92, 100}

66. {30, 22, 38, 41, 33, 41, 30, 24}

67. Find the range, median, lower quartile, and upper quartile for {77, 75, 72, 70, 79, 77, 70, 76}.

68. **TESTS** Kevin's scores on the first four science tests are 88, 92, 82, and 94. What score must he earn on the fifth test so that the mean will be 90?

69. **FOOD** The table shows the results of a survey in which students were asked to choose their favorite food. Make a bar graph of the data.

Favorite Foods

Food	Number of Students
pizza	15
chicken nuggets	10
cheesy potatoes	8
ice cream	5

70. Make a double box-and-whisker plot of the data.
A: 26, 18, 26, 29, 18, 20, 35, 32, 31, 24, 26, 22
B: 16, 20, 16, 19, 21, 30, 25, 22, 21, 19, 16, 17

71. **BUDGET** The table shows how Kat spends her allowance. Which graph is the best way to display these data? Explain your reasoning and make a graph of the data.

Category	Amount ($)
Savings	25
Clothes	10
Entertainment	15

CHAPTER 1

Expressions, Equations, and Functions

Then

You have learned how to perform operations on whole numbers.

Now

In this chapter, you will:

- Write algebraic expressions.
- Use the order of operations.
- Solve equations.
- Represent and interpret relations and functions.
- Use function notation.
- Interpret the graphs of functions.

Why? ▲

SCUBA DIVING A scuba diving store rents air tanks and wet suits. An algebraic expression can be written to represent the total cost to rent this equipment. This expression can be evaluated to determine the total cost for a group of people to rent the equipment.

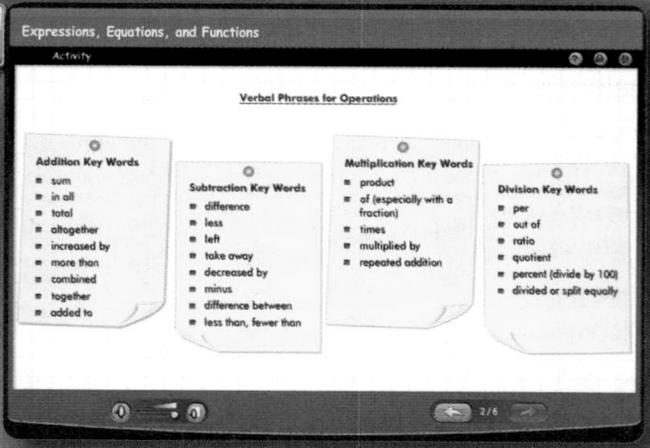

connectED.mcgraw-hill.com **Your Digital Math Portal**

Animation | Vocabulary | eGlossary | Personal Tutor | Virtual Manipulatives | Graphing Calculator | Audio | Foldables | Self-Check Practice | Worksheets

Jupiterimages/Comstock Images/Al

Get Ready for the Chapter

Diagnose Readiness | You have two options for checking prerequisite skills.

1 Textbook Option

Take the Quick Check below. Refer to the Quick Review for help.

QuickCheck

Write each fraction in simplest form. If the fraction is already in simplest form, write *simplest form.*

1. $\frac{24}{36}$ **2.** $\frac{34}{85}$ **3.** $\frac{36}{12}$

4. $\frac{27}{45}$ **5.** $\frac{11}{18}$ **6.** $\frac{5}{65}$

7. $\frac{19}{1}$ **8.** $\frac{16}{44}$ **9.** $\frac{64}{88}$

10. **ICE CREAM** Fifty-four out of 180 customers said that cookie dough ice cream was their favorite flavor. What fraction of customers was this?

QuickReview

Example 1

Write $\frac{24}{40}$ in simplest form.

Find the greatest common factor (GCF) of 24 and 40.

factors of 24: 1, 2, 3, 4, 6, 8, 12, 24
factors of 40: 1, 2, 4, 5, 8, 10, 20, 40

The GCF of 24 and 40 is 8.

$\frac{24 \div 8}{40 \div 8} = \frac{3}{5}$ Divide the numerator and denominator by their GCF, 8.

Find the perimeter of each figure.

11.

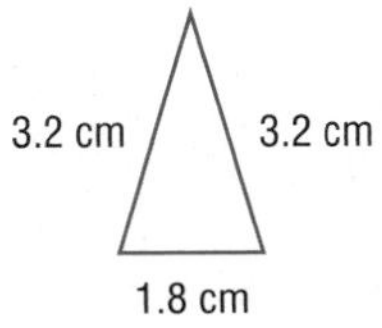

12.

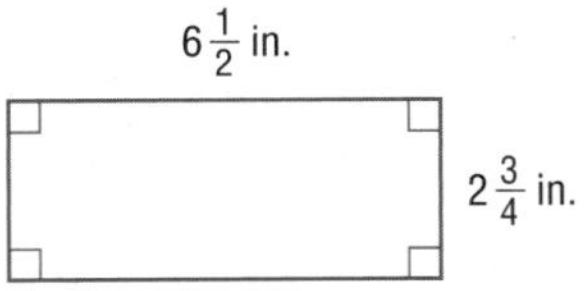

13. **FENCING** Jolon needs to fence a garden. The dimensions of the garden are 6 meters by 4 meters. How much fencing does Jolon need to purchase?

Example 2

Find the perimeter.

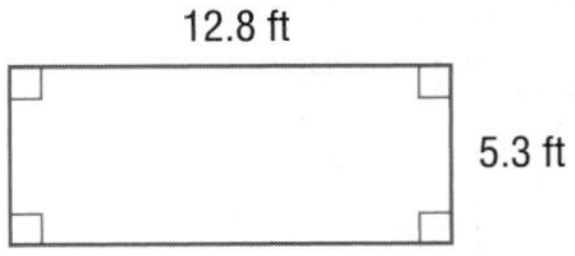

$P = 2\ell + 2w$

$= 2(12.8) + 2(5.3)$ $\ell = 12.8$ and $w = 5.3$

$= 25.6 + 10.6$ or 36.2 Simplify.

The perimeter is 36.2 feet.

Evaluate.

14. $6 \cdot \frac{2}{3}$ **15.** $4.2 \cdot 8.1$ **16.** $\frac{3}{8} \div \frac{1}{4}$

17. $5.13 \div 2.7$ **18.** $3\frac{1}{5} \cdot \frac{3}{4}$ **19.** $2.8 \cdot 0.2$

20. **CONSTRUCTION** A board measuring 7.2 feet must be cut into three equal pieces. Find the length of each piece.

Example 3

Find $2\frac{1}{4} \div 1\frac{1}{2}$.

$2\frac{1}{4} \div 1\frac{1}{2} = \frac{9}{4} \div \frac{3}{2}$ Write mixed numbers as improper fractions.

$= \frac{9}{4}\left(\frac{2}{3}\right)$ Multiply by the reciprocal.

$= \frac{18}{12}$ or $1\frac{1}{2}$ Simplify.

2 Online Option

Take an online self-check Chapter Readiness Quiz at connectED.mcgraw-hill.com.

Get Started on the Chapter

You will learn several new concepts, skills, and vocabulary terms as you study Chapter 1. To get ready, identify important terms and organize your resources. You may wish to refer to Chapter 0 to review prerequisite skills.

FOLDABLES® StudyOrganizer

Expressions, Equations, and Functions Make this Foldable to help you organize your Chapter 1 notes about expressions, equations, and functions. Begin with five sheets of plain paper.

1 **Fold** the sheets of paper in half along the width. Then cut along the crease.

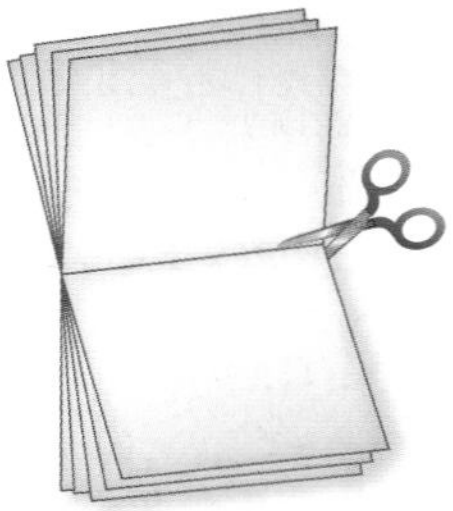

2 **Staple** the ten half-sheets together to form a booklet.

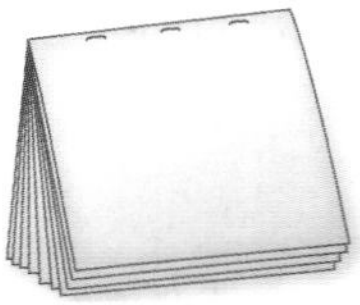

3 **Cut** nine centimeters from the bottom of the top sheet, eight centimeters from the second sheet, and so on.

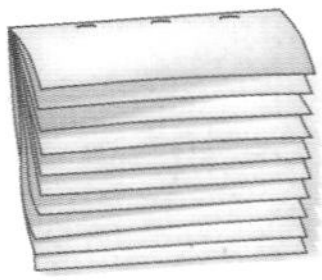

4 **Label** each of the tabs with a lesson number. The ninth tab is for Properties and the last tab is for Vocabulary.

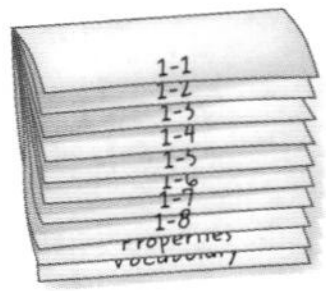

NewVocabulary

English		Español
algebraic expression	p. 5	expression algebraica
variable	p. 5	variable
term	p. 5	término
power	p. 5	potencia
coefficient	p. 28	coeficiente
equation	p. 33	ecuación
solution	p. 33	solución
identity	p. 35	identidad
relation	p. 40	relacíon
domain	p. 40	domino
range	p. 40	rango
independent variable	p. 42	variable independiente
dependent variable	p. 42	variable dependiente
function	p. 47	función
intercept	p. 56	intersección
line symmetry	p. 57	simetría
end behavior	p. 57	comportamiento final

ReviewVocabulary

additive inverse inverso aditivo a number and its opposite

multiplicative inverse inverso multiplicativo two numbers with a product of 1

perimeter perímetro the distance around a geometric figure

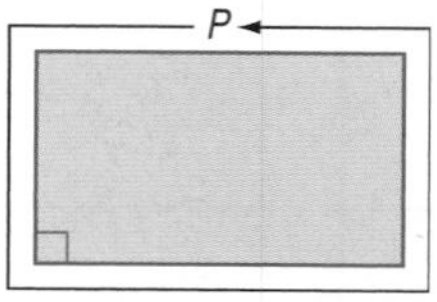

LESSON 1-1

Variables and Expressions

Then

- You performed operations on integers.

Now

1. Write verbal expressions for algebraic expressions.
2. Write algebraic expressions for verbal expressions.

Why?

- Cassie and her friends are at a baseball game. The stadium is running a promotion where hot dogs are \$0.10 each. Suppose *d* represents the number of hot dogs Cassie and her friends eat. Then 0.10*d* represents the cost of the hot dogs they eat.

NewVocabulary

algebraic expression
variable
term
factor
product
power
exponent
base

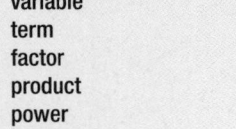

Common Core State Standards

Content Standards

A.SSE.1a Interpret parts of an expression, such as terms, factors, and coefficients.

A.SSE.2 Use the structure of an expression to identify ways to rewrite it.

Mathematical Practices

4 Model with mathematics.

1 Write Verbal Expressions

An **algebraic expression** consists of sums and/or products of numbers and variables. In the algebraic expression 0.10*d*, the letter *d* is called a variable. In algebra, **variables** are symbols used to represent unspecified numbers or values. Any letter may be used as a variable.

$$0.10d \qquad 2x + 4 \qquad 3 + \frac{z}{6} \qquad p \cdot q \qquad 4cd \div 3mn$$

A **term** of an expression may be a number, a variable, or a product or quotient of numbers and variables. For example, 0.10*d*, 2*x* and 4 are each terms.

The term that contains *x* or other letters is sometimes referred to as the *variable term.* → $2x + 4$ ← A term that does not have a variable is a *constant term.*

In a multiplication expression, the quantities being multiplied are **factors**, and the result is the **product**. A raised dot or set of parentheses are often used to indicate a product. Here are several ways to represent the product of *x* and *y*.

$$xy \qquad x \cdot y \qquad x(y) \qquad (x)y \qquad (x)(y)$$

An expression like x^n is called a **power**. The word *power* can also refer to the exponent. The **exponent** indicates the number of times the base is used as a factor. In an expression of the form x^n, the **base** is *x*. The expression x^n is read "*x* to the *n*th power." When no exponent is shown, it is understood to be 1. For example, $a = a^1$.

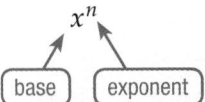

x^n — base (*x*), exponent (*n*)

PT

Example 1 Write Verbal Expressions

Write a verbal expression for each algebraic expression.

a. $3x^4$

three times *x* to the fourth power

b. $5z^2 + 16$

5 times *z* to the second power plus sixteen

GuidedPractice

1A. $16u^2 - 3$

1B. $\frac{1}{2}a + \frac{6b}{7}$

StudyTip

CCSS Modeling When writing an expression to model a situation, begin by identifying the important quantities and their relationships.

2 Write Algebraic Expressions

Another important skill is translating verbal expressions into algebraic expressions.

KeyConcept Translating Verbal to Algebraic Expressions

Operation	Verbal Phrases
Addition	more than, sum, plus, increased by, added to
Subtraction	less than, subtracted from, difference, decreased by, minus
Multiplication	product of, multiplied by, times, of
Division	quotient of, divided by

Example 2 Write Algebraic Expressions

Write an algebraic expression for each verbal expression.

a. a number t more than 6

The words *more than* suggest addition.
Thus, the algebraic expression is $6 + t$ or $t + 6$.

b. 10 less than the product of 7 and f

Less than implies subtraction, and *product* suggests multiplication.
So the expression is written as $7f - 10$.

c. two thirds of the volume v

The word *of* with a fraction implies that you should multiply.
The expression could be written as $\frac{2}{3}v$ or $\frac{2v}{3}$.

GuidedPractice

2A. the product of p and 6

2B. one third of the area a

Variables can represent quantities that are known and quantities that are unknown. They are also used in formulas, expressions, and equations.

Real-World Example 3 Write an Expression

SPORTS MARKETING Mr. Martinez orders 250 key chains printed with his athletic team's logo and 500 pencils printed with their Web address. Write an algebraic expression that represents the cost of the order.

Let k be the cost of each key chain and p be the cost of each pencil. Then the cost of the key chains is $250k$ and the cost of the pencils is $500p$. The cost of the order is represented by $250k + 500p$.

GuidedPractice

3. COFFEE SHOP Katie estimates that $\frac{1}{8}$ of the people who order beverages also order pastries. Write an algebraic expression to represent this situation.

Masterfile

Real-WorldCareer

Sports Marketing
Sports marketers promote and manage athletes, teams, facilities and sports-related businesses and organizations. A minimum of a bachelor's degree in sports management or business administration is preferred.

Check Your Understanding

Example 1 **Write a verbal expression for each algebraic expression.**

1. $2m$ **2.** $\frac{2}{3}r^4$ **3.** $a^2 - 18b$

Example 2 **Write an algebraic expression for each verbal expression.**

4. the sum of a number and 14

5. 6 less a number t

6. 7 more than 11 times a number

7. 1 minus the quotient of r and 7

8. two fifths of the square of a number j

9. n cubed increased by 5

Example 3 **10.** **GROCERIES** Mr. Bailey purchased some groceries that cost d dollars. He paid with a $50 bill. Write an expression for the amount of change he will receive.

Practice and Problem Solving

Example 1 **Write a verbal expression for each algebraic expression.**

11. $4q$ **12.** $\frac{1}{8}y$ **13.** $15 + r$ **14.** $w - 24$

15. $3x^2$ **16.** $\frac{r^4}{9}$ **17.** $2a + 6$ **18.** $r^4 \cdot t^3$

Example 2 **Write an algebraic expression for each verbal expression.**

19. x more than 7

20. a number less 35

21. 5 times a number

22. one third of a number

23. f divided by 10

24. the quotient of 45 and r

25. three times a number plus 16

26. 18 decreased by 3 times d

27. k squared minus 11

28. 20 divided by t to the fifth power

Example 3 **29.** **GEOMETRY** The volume of a cylinder is π times the radius r squared multiplied by the height h. Write an expression for the volume.

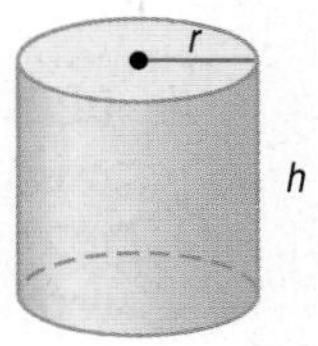

30. **FINANCIAL LITERACY** Jocelyn makes x dollars per hour working at the grocery store and n dollars per hour babysitting. Write an expression that describes her earnings if she babysat for 25 hours and worked at the grocery store for 15 hours.

Write a verbal expression for each algebraic expression.

31. $25 + 6x^2$ **32.** $6f^2 + 5f$ **33.** $\frac{3a^5}{2}$

34. CCSS **SENSE-MAKING** A certain smartphone family plan costs $55 per month plus additional usage costs. If x is the number of cell phone minutes used above the plan amount and y is the number of megabytes of data used above the plan amount, interpret the following expressions.

a. $0.25x$

b. $2y$

c. $0.25x + 2y + 55$

35. **DREAMS** It is believed that about $\frac{3}{4}$ of our dreams involve people that we know.

a. Write an expression to describe the number of dreams that feature people you know if you have d dreams.

b. Use the expression you wrote to predict the number of dreams that include people you know out of 28 dreams.

36. **SPORTS** In football, a touchdown is awarded 6 points and the team can then try for a point after a touchdown.

a. Write an expression that describes the number of points scored on touchdowns T and points after touchdowns p by one team in a game.

b. If a team wins a football game 27-0, write an equation to represent the possible number of touchdowns and points after touchdowns by the winning team.

c. If a team wins a football game 21-7, how many possible number of touchdowns and points after touchdowns were scored during the game by both teams?

37. **MULTIPLE REPRESENTATIONS** In this problem, you will explore the multiplication of powers with like bases.

a. Tabular Copy and complete the table.

10^2	×	10^1	=	10 × 10 × 10	=	10^3
10^2	×	10^2	=	10 × 10 × 10 × 10	=	10^4
10^2	×	10^3	=	10 × 10 × 10 × 10 × 10	=	?
10^2	×	10^4	=	?	=	?

b. Algebraic Write an equation for the pattern in the table.

c. Verbal Make a conjecture about the exponent of the product of two powers with like bases.

H.O.T. Problems Use Higher-Order Thinking Skills

38. **REASONING** Explain the differences between an algebraic expression and a verbal expression.

39. **OPEN ENDED** Define a variable to represent a real-life quantity, such as time in minutes or distance in feet. Then use the variable to write an algebraic expression to represent one of your daily activities. Describe in words what your expression represents, and explain your reasoning.

40. **CCSS CRITIQUE** Consuelo and James are writing an algebraic expression for *three times the sum of n squared and 3*. Is either of them correct? Explain your reasoning.

Consuelo	James
$3(n^2 + 3)$	$3n^2 + 3$

41. **CHALLENGE** For the cube, x represents a positive whole number. Find the value of x such that the volume of the cube and 6 times the area of one of its faces have the same value.

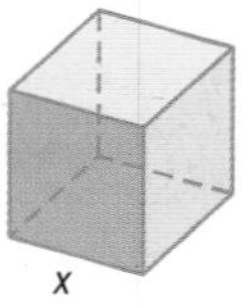

42. **WRITING IN MATH** Describe how to write an algebraic expression from a real-world situation. Include a definition of algebraic expression in your own words.

Standardized Test Practice

43. Which expression best represents the volume of the cube?

A the product of three and five

B three to the fifth power

C three squared

D three cubed

44. Which expression best represents the perimeter of the rectangle?

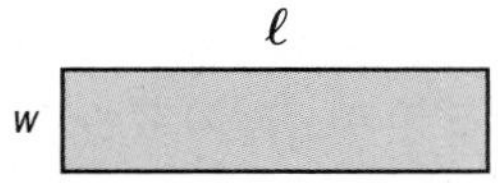

F $2\ell w$

G $\ell + w$

H $2\ell + 2w$

J $4(\ell + w)$

45. SHORT RESPONSE The yards of fabric needed to make curtains is 3 times the length of a window in inches, divided by 36. Write an expression that represents the yards of fabric needed in terms of the length of the window ℓ.

46. GEOMETRY Find the area of the rectangle.

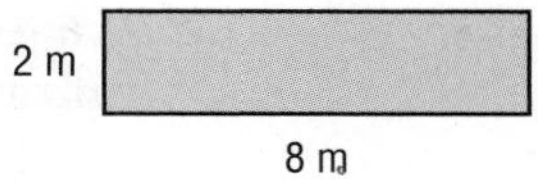

A 14 square meters

B 16 square meters

C 50 square meters

D 60 square meters

Spiral Review

47. AMUSEMENT PARKS A roller coaster enthusiast club took a poll to see what each member's favorite ride was. Make a bar graph of the results. (Lesson 0-13)

Our Favorite Rides							
Ride	Big Plunge	Twisting Time	The Shiner	Raging Bull	The Bat	Teaser	The Adventure
Number of Votes	5	22	16	9	25	6	12

48. SPORTS The results for an annual 5K race are shown at the right. Make a box-and-whisker plot for the data. Write a sentence describing what the length of the box-and-whisker plot tells about the times for the race. (Lesson 0–13)

Annual 5K Race Results			
Joe	14:48	Carissa	19:58
Jessica	19:27	Jordan	14:58
Lupe	15:06	Taylor	20:47
Dante	20:39	Mi-Ling	15:48
Tia	15:54	Winona	21:35
Amber	20:49	Angel	16:10
Amanda	16:30	Catalina	20:21

Find the mean, median, and mode for each set of data. (Lesson 0–12)

49. {7, 6, 5, 7, 4, 8, 2, 2, 7, 8}

50. {−1, 0, 5, 2, −2, 0 ,−1, 2, −1, 0}

51. {17, 24, 16, 3, 12, 11, 24, 15}

52. SPORTS Lisa has a rectangular trampoline that is 6 feet long and 12 feet wide. What is the area of her trampoline in square feet? (Lesson 0–8)

Find each product or quotient. (Lesson 0–5)

53. $\frac{3}{5} \cdot \frac{7}{11}$

54. $\frac{4}{3} \div \frac{7}{6}$

55. $\frac{5}{6} \cdot \frac{8}{3}$

Skills Review

Evaluate each expression.

56. $\frac{3}{5} + \frac{4}{9}$

57. $5.67 - 4.21$

58. $\frac{5}{6} - \frac{8}{3}$

59. $10.34 + 14.27$

60. $\frac{11}{12} + \frac{5}{36}$

61. $37.02 - 15.86$

LESSON 1-2 Order of Operations

Then	Now	Why?
You expressed algebraic expressions verbally.	**1** Evaluate numerical expressions by using the order of operations. **2** Evaluate algebraic expressions by using the order of operations.	The admission prices for SeaWorld Adventure Park in Orlando, Florida, are shown in the table. If four adults and three children go to the park, the expression below represents the cost of admission for the group. 4(78.95) + 3(68.95)

Ticket	Price ($)
Adult	78.95
Child	68.95

Ken Biggs/Stone/Getty Images

NewVocabulary
evaluate
order of operations

Common Core State Standards

Content Standards
A.SSE.1b Interpret complicated expressions by viewing one or more of their parts as a single entity.
A.SSE.2 Use the structure of an expression to identify ways to rewrite it.

Mathematical Practices
7 Look for and make use of structure.

1 Evaluate Numerical Expressions

To find the cost of admission, the expression 4(78.95) + 3(68.95) must be evaluated. To **evaluate** an expression means to find its value.

Example 1 Evaluate Expressions

Evaluate 3^5.

$3^5 = 3 \cdot 3 \cdot 3 \cdot 3 \cdot 3$ — Use 3 as a factor 5 times.
$= 243$ — Multiply.

GuidedPractice

1A. 2^4 **1B.** 4^5 **1C.** 7^3

The numerical expression that represents the cost of admission contains more than one operation. The rule that lets you know which operation to perform first is called the **order of operations**.

KeyConcept Order of Operations

Step 1 Evaluate expressions inside grouping symbols.

Step 2 Evaluate all powers.

Step 3 Multiply and/or divide from left to right.

Step 4 Add and/or subtract from left to right.

Example 2 Order of Operations

Evaluate $16 - 8 \div 2^2 + 14$.

$16 - 8 \div 2^2 + 14 = 16 - 8 \div 4 + 14$ — Evaluate powers.
$= 16 - 2 + 14$ — Divide 8 by 4.
$= 14 + 14$ — Subtract 2 from 16.
$= 28$ — Add 14 and 14.

GuidedPractice

2A. $3 + 42 \cdot 2 - 5$ **2B.** $20 - 7 + 8^2 - 7 \cdot 11$

StudyTip

Grouping Symbols Grouping symbols such as parentheses ( ), brackets [ ], and braces { } are used to clarify or change the order of operations.

When one or more grouping symbols are used, evaluate within the innermost grouping symbols first.

PT

Example 3 Expressions with Grouping Symbols

Evaluate each expression.

a. $4 \div 2 + 5(10 - 6)$

$4 \div 2 + 5(10 - 6) = 4 \div 2 + 5(4)$ — Evaluate inside parentheses.

$= 2 + 5(4)$ — Divide 4 by 2.

$= 2 + 20$ — Multiply 5 by 4.

$= 22$ — Add 2 to 20.

b. $6\left[32 - (2 + 3)^2\right]$

$6\left[32 - (2 + 3)^2\right] = 6\left[32 - (5)^2\right]$ — Evaluate innermost expression first.

$= 6[32 - 25]$ — Evaluate power.

$= 6[7]$ — Subtract 25 from 32.

$= 42$ — Multiply.

StudyTip

Grouping Symbols A fraction bar is considered a grouping symbol. So, evaluate expressions in the numerator and denominator before completing the division.

c. $\dfrac{2^3 - 5}{15 + 9}$

$\dfrac{2^3 - 5}{15 + 9} = \dfrac{8 - 5}{15 + 9}$ — Evaluate the power in the numerator.

$= \dfrac{3}{15 + 9}$ — Subtract 5 from 8 in the numerator.

$= \dfrac{3}{24}$ or $\dfrac{1}{8}$ — Add 15 and 9 in denominator, and simplify.

GuidedPractice

3A. $5 \cdot 4(10 - 8) + 20$ **3B.** $15 - \left[10 + (3 - 2)^2\right] + 6$ **3C.** $\dfrac{(4 + 5)^2}{3(7 - 4)}$

2 Evaluate Algebraic Expressions

To evaluate an algebraic expression, replace the variables with their values. Then find the value of the numerical expression using the order of operations.

PT

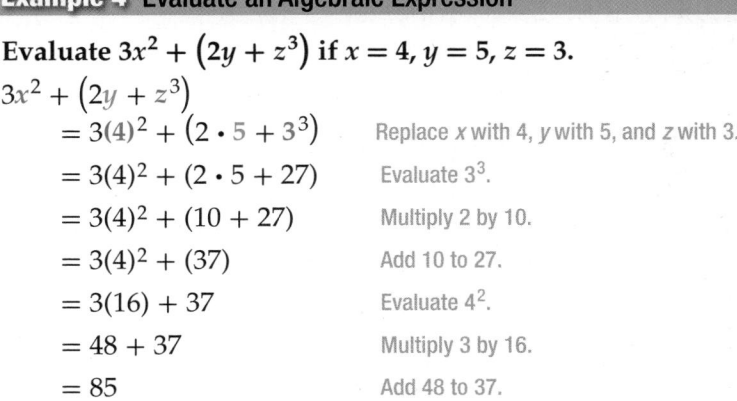

Example 4 Evaluate an Algebraic Expression

Evaluate $3x^2 + \left(2y + z^3\right)$ if $x = 4$, $y = 5$, $z = 3$.

$3x^2 + \left(2y + z^3\right)$

$= 3(4)^2 + \left(2 \cdot 5 + 3^3\right)$ — Replace x with 4, y with 5, and z with 3.

$= 3(4)^2 + (2 \cdot 5 + 27)$ — Evaluate 3^3.

$= 3(4)^2 + (10 + 27)$ — Multiply 2 by 10.

$= 3(4)^2 + (37)$ — Add 10 to 27.

$= 3(16) + 37$ — Evaluate 4^2.

$= 48 + 37$ — Multiply 3 by 16.

$= 85$ — Add 48 to 37.

GuidedPractice

Evaluate each expression.

4A. $a^2(3b + 5) \div c$ if $a = 2$, $b = 6$, $c = 4$ **4B.** $5d + (6f - g)$ if $d = 4$, $f = 3$, $g = 12$

Real-WorldLink

The National Oceanic & Atmospheric Administration (NOAA) developed the Science on a Sphere system to educate people about Earth's processes. There are five computers and four video projectors that power the sphere.

Source: NOAA

Real-World Example 5 Write and Evaluate an Expression

ENVIRONMENTAL STUDIES **Science on a Sphere (SOS)® demonstrates the effects of atmospheric storms, climate changes, and ocean temperature on the environment. The volume of a sphere is four thirds of π multiplied by the radius r to the third power.**

a. Write an expression that represents the volume of a sphere.

Words	four thirds	of	π multiplied by radius to the third power
Variable	Let r = radius.		
Equation	$\frac{4}{3}$	$\times$	πr^3 or $\frac{4}{3}\pi r^3$

b. Find the volume of the 3-foot radius sphere used for SOS.

$V = \frac{4}{3}\pi r^3$	Volume of a sphere
$= \frac{4}{3}\pi(3)^3$	Replace r with 3.
$= \left(\frac{4}{3}\right)\pi(27)$	Evaluate $3^3 = 27$.
$= 36\pi$	Multiply $\frac{4}{3}$ by 27.

The volume of the sphere is 36π cubic feet.

GuidedPractice

5. **FOREST FIRES** According to the California Department of Forestry, an average of 539.2 fires each year are started by burning debris, while campfires are responsible for an average of 129.1 each year.

A. Write an algebraic expression that represents the number of fires, on average, in d years of debris burning and c years of campfires.

B. How many fires would there be in 5 years?

Check Your Understanding

Examples 1–3 **Evaluate each expression.**

1. 9^2 **2.** 4^4 **3.** 3^5

4. $30 - 14 \div 2$ **5.** $5 \cdot 5 - 1 \cdot 3$ **6.** $(2 + 5)4$

7. $[8(2) - 4^2] + 7(4)$ **8.** $\frac{11 - 8}{1 + 7 \cdot 2}$ **9.** $\frac{(4 \cdot 3)^2}{9 + 3}$

Example 4 **Evaluate each expression if $a = 4$, $b = 6$, and $c = 8$.**

10. $8b - a$ **11.** $2a + (b^2 \div 3)$ **12.** $\frac{b(9 - c)}{a^2}$

Example 5

13. **BOOKS** Akira bought one new book for \$20 and three used books for \$4.95 each. Write and evaluate an expression to find how much money the books cost.

14. **CCSS REASONING** Koto purchased food for herself and her friends. She bought 4 cheeseburgers for \$2.25 each, 3 French fries for \$1.25 each, and 4 drinks for \$4.00. Write and evaluate an expression to find how much the food cost.

Practice and Problem Solving

Examples 1–3 **Evaluate each expression.**

15. 7^2 **16.** 14^3 **17.** 2^6

18. $35 - 3 \cdot 8$ **19.** $18 \div 9 + 2 \cdot 6$ **20.** $10 + 8^3 \div 16$

21. $24 \div 6 + 2^3 \cdot 4$ **22.** $(11 \cdot 7) - 9 \cdot 8$ **23.** $29 - 3(9 - 4)$

24. $(12 - 6) \cdot 5^2$ **25.** $3^5 - (1 + 10^2)$ **26.** $108 \div [3(9 + 3^2)]$

27. $[(6^3 - 9) \div 23]4$ **28.** $\frac{8 + 3^3}{12 - 7}$ **29.** $\frac{(1 + 6)9}{5^2 - 4}$

Example 4 **Evaluate each expression if $g = 2$, $r = 3$, and $t = 11$.**

30. $g + 6t$ **31.** $7 - gr$ **32.** $r^2 + (g^3 - 8)^5$

33. $(2t + 3g) \div 4$ **34.** $t^2 + 8rt + r^2$ **35.** $3g(g + r)^2 - 1$

Example 5

36. **GEOMETRY** Write an algebraic expression to represent the area of the triangle. Then evaluate it to find the area when $h = 12$ inches.

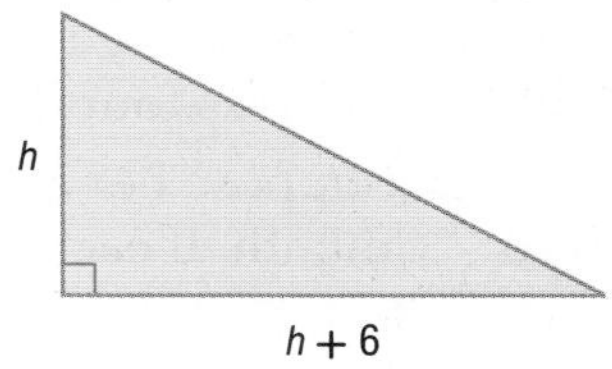

37. **AMUSEMENT PARKS** In 2004, there were 3344 amusement parks and arcades. This decreased by 148 by 2009. Write and evaluate an expression to find the number of amusement parks and arcades in 2009.

38. CCSS **STRUCTURE** Kamilah sells tickets at Duke University's athletic ticket office. If p represents a preferred season ticket, b represents a blue zone ticket, and g represents a general admission ticket, interpret and then evaluate the following expressions.

a. $45b$ **b.** $15p + 35g$ **c.** $6p + 11b + 22g$

Duke University Football Ticket Prices	
Preferred Season Ticket	$100
Blue Zone	$80
General Admission	$70

Source: Duke University

Evaluate each expression.

39. 4^2 **40.** 12^3 **41.** 3^6

42. 11^5 **43.** $(3 - 4^2)^2 + 8$ **44.** $23 - 2(17 + 3^3)$

45. $3[4 - 8 + 4^2(2 + 5)]$ **46.** $\frac{2 \cdot 8^2 - 2^2 \cdot 8}{2 \cdot 8}$

47. $25 + \left[(16 - 3 \cdot 5) + \frac{12 + 3}{5}\right]$ **48.** $7^3 - \frac{2}{3}(13 \cdot 6 + 9)4$

Evaluate each expression if $a = 8$, $b = 4$, and $c = 16$.

49. $a^2bc - b^2$ **50.** $\frac{c^2}{b^2} + \frac{b^2}{a^2}$ **51.** $\frac{2b + 3c^2}{4a^2 - 2b}$

52. $\frac{3ab + c^2}{a}$ **53.** $\left(\frac{a}{b}\right)^2 - \frac{c}{a - b}$ **54.** $\frac{2a - b^2}{ab} + \frac{c - a}{b^2}$

55. **SALES** One day, 28 small and 12 large merchant spaces were rented. Another day, 30 small and 15 large spaces were rented. Write and evaluate an expression to show the total rent collected.

THE FLEA MARKET
MERCHANT SPACE RENTALS
Small space $7.00/day
Large space $9.75/day
Open Daily from 9:00–6:00

56. SHOPPING Evelina is shopping for back-to-school clothes. She bought 3 skirts, 2 pairs of jeans, and 4 sweaters. Write and evaluate an expression to find how much she spent, without including sales tax.

Clothing	
skirt	\$25.99
jeans	\$39.99
sweater	\$22.99

57. PYRAMIDS The pyramid at the Louvre has a square base with a side of 35.42 meters and a height of 21.64 meters. The Great Pyramid in Egypt has a square base with a side of 230 meters and a height of 146.5 meters. The expression for the volume of a pyramid is $\frac{1}{3}Bh$, where B is the area of the base and h is the height.

a. Draw both pyramids and label the dimensions.

b. Write a verbal expression for the difference in volume of the two pyramids.

c. Write an algebraic expression for the difference in volume of the two pyramids. Find the difference in volume.

58. FINANCIAL LITERACY A sales representative receives an annual salary s, an average commission each month c, and a bonus b for each sales goal that she reaches.

a. Write an algebraic expression to represent her total earnings in one year if she receives four equal bonuses.

b. Suppose her annual salary is \$52,000 and her average commission is \$1225 per month. If each of the four bonuses equals \$1150, what does she earn annually?

H.O.T. Problems Use Higher-Order Thinking Skills

59. ERROR ANALYSIS Tara and Curtis are simplifying $[4(10) - 3^2] + 6(4)$. Is either of them correct? Explain your reasoning.

Tara

$[4(10) - 3^2] + 6(4)$
$= [4(10) - 9] + 6(4)$
$= 4(1) + 6(4)$
$= 4 + 6(4)$
$= 4 + 24$
$= 28$

Curtis

$[4(10) - 3^2] + 6(4)$
$= [4(10) - 9] + 6(4)$
$= (40 - 9) + 6(4)$
$= 31 + 6(4)$
$= 31 + 24$
$= 55$

60. REASONING Explain how to evaluate $a[(b - c) \div d] - f$ if you were given values for a, b, c, d, and f. How would you evaluate the expression differently if the expression was $a \cdot b - c \div d - f$?

61. CCSS PERSEVERANCE Write an expression using the whole numbers 1 to 5 using all five digits and addition and/or subtraction to create a numeric expression with a value of 3.

62. OPEN ENDED Write an expression that uses exponents, at least three different operations, and two sets of parentheses. Explain the steps you would take to evaluate the expression.

63. WRITING IN MATH Choose a geometric formula and explain how the order of operations applies when using the formula.

64. WRITING IN MATH Equivalent expressions have the same value. Are the expressions $(30 + 17) \times 10$ and $10 \times 30 + 10 \times 17$ equivalent? Explain why or why not.

Standardized Test Practice

65. Let m represent the number of miles. Which algebraic expression represents the number of feet in m miles?

A $5280m$

B $\frac{5280}{m}$

C $m + 5280$

D $5280 - m$

66. SHORT RESPONSE

Simplify: $[10 + 15(2^3)] \div [7(2^2) - 2]$

Step 1 $[10 + 15(8)] \div [7(4) - 2]$

Step 2 $[10 + 120] \div [28 - 2]$

Step 3 $130 \div 26$

Step 4 $\frac{1}{5}$

Which is the first *incorrect* step? Explain the error.

67. EXTENDED RESPONSE Consider the rectangle below.

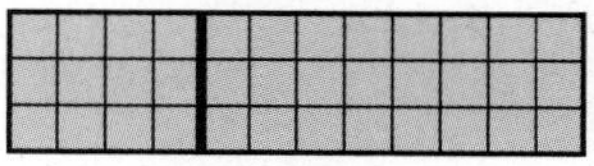

Part A Which expression models the area of the rectangle?

F $4 + 3 \times 8$ **H** $3 \times 4 + 8$

G $3 \times (4 + 8)$ **J** $3^2 + 8^2$

Part B Draw one or more rectangles to model each other expression.

68. GEOMETRY What is the perimeter of the triangle if $a = 9$ and $b = 10$?

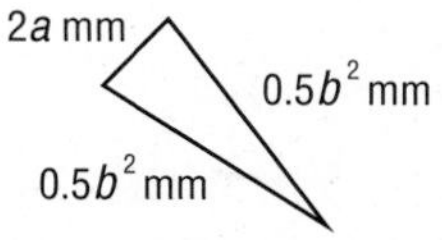

A 164 mm **C** 28 mm

B 118 mm **D** 4 mm

Spiral Review

Write a verbal expression for each algebraic expression. (Lesson 1-1)

69. $14 - 9c$

70. $k^3 + 13$

71. $\frac{4 - v}{w}$

72. MONEY Destiny earns \$8 per hour babysitting and \$15 for each lawn she mows. Write an expression to show the amount of money she earns babysitting h hours and mowing m lawns. (Lesson 1-1)

Find the area of each figure. (Lesson 0-8)

73.

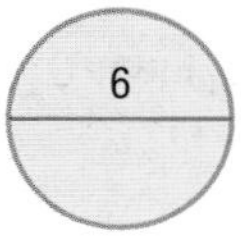

74.

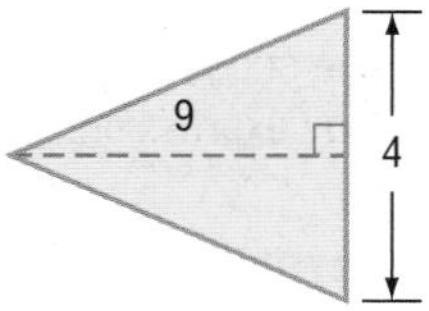

75.

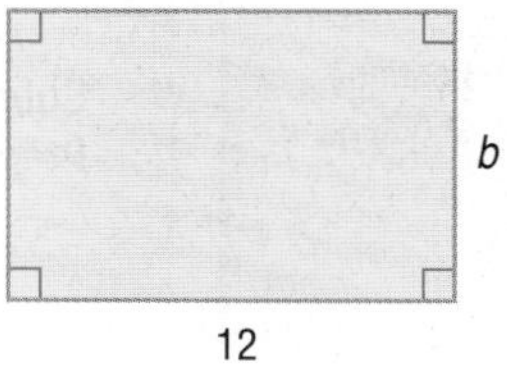

76. SCHOOL Aaron correctly answered 27 out of 30 questions on his last biology test. What percent of the questions did he answer correctly? (Lesson 0-6)

Skills Review

Find the value of each expression.

77. $5.65 - 3.08$

78. $6 \div \frac{4}{5}$

79. $4.85(2.72)$

80. $1\frac{1}{12} + 3\frac{2}{3}$

81. $\frac{4}{9} \cdot \frac{3}{2}$

82. $7\frac{3}{4} - 4\frac{7}{10}$

LESSON 1-3 Properties of Numbers

Then

- You used the order of operations to simplify expressions.

Now

1. Recognize the properties of equality and identity.
2. Recognize the Commutative and Associative Properties.

Why?

- Natalie lives 32 miles away from the mall. The distance from her house to the mall is the same as the distance from the mall to her house. This is an example of the Reflexive Property.

NewVocabulary
equivalent expressions
additive identity
multiplicative identity
multiplicative inverse
reciprocal

Common Core State Standards

Content Standards
A.SSE.1b Interpret complicated expressions by viewing one or more of their parts as a single entity.
A.SSE.2 Use the structure of an expression to identify ways to rewrite it.

Mathematical Practices
2 Reason abstractly and quantitatively.
3 Construct viable arguments and critique the reasoning of others.

1 Properties of Equality and Identity

The expressions $4k + 8k$ and $12k$ are called **equivalent expressions** because they represent the same number. The properties below allow you to write an equivalent expression for a given expression.

KeyConcept Properties of Equality

Property	Words	Symbols	Examples
Reflexive Property	Any quantity is equal to itself.	For any number a, $a = a$.	$5 = 5$ $4 + 7 = 4 + 7$
Symmetric Property	If one quantity equals a second quantity, then the second quantity equals the first.	For any numbers a and b, if $a = b$, then $b = a$.	If $8 = 2 + 6$, then $2 + 6 = 8$.
Transitive Property	If one quantity equals a second quantity and the second quantity equals a third quantity, then the first quantity equals the third quantity.	For any numbers a, b, and c, if $a = b$ and $b = c$, then $a = c$.	If $6 + 9 = 3 + 12$ and $3 + 12 = 15$, then $6 + 9 = 15$.
Substitution Property	A quantity may be substituted for its equal in any expression.	If $a = b$, then a may be replaced by b in any expression.	If $n = 11$, then $4n = 4 \cdot 11$

The sum of any number and 0 is equal to the number. Thus, 0 is called the **additive identity**.

KeyConcept Addition Properties

Property	Words	Symbols	Examples
Additive Identity	For any number a, the sum of a and 0 is a.	$a + 0 = 0 + a = a$	$2 + 0 = 2$ $0 + 2 = 2$
Additive Inverse	A number and its opposite are additive inverses of each other.	$a + (-a) = 0$	$3 + (-3) = 0$ $4 - 4 = 0$

There are also special properties associated with multiplication. Consider the following equations.

$4 \cdot n = 4$

The solution of the equation is 1. Since the product of any number and 1 is equal to the number, 1 is called the **multiplicative identity**.

$6 \cdot m = 0$

The solution of the equation is 0. The product of any number and 0 is equal to 0. This is called the **Multiplicative Property of Zero**.

Two numbers whose product is 1 are called **multiplicative inverses** or **reciprocals**. Zero has no reciprocal because any number times 0 is 0.

KeyConcept Multiplication Properties

Property	Words	Symbols	Examples
Multiplicative Identity	For any number *a*, the product of *a* and 1 is *a*.	$a \cdot 1 = a$ $1 \cdot a = a$	$14 \cdot 1 = 14$ $1 \cdot 14 = 14$
Multiplicative Property of Zero	For any number *a*, the product of *a* and 0 is 0.	$a \cdot 0 = 0$ $0 \cdot a = 0$	$9 \cdot 0 = 0$ $0 \cdot 9 = 0$
Multiplicative Inverse	For every number $\frac{a}{b}$, where $a, b \neq 0$, there is exactly one number $\frac{b}{a}$ such that the product of $\frac{a}{b}$ and $\frac{b}{a}$ is 1.	$\frac{a}{b} \cdot \frac{b}{a} = 1$ $\frac{b}{a} \cdot \frac{a}{b} = 1$	$\frac{4}{5} \cdot \frac{5}{4} = \frac{20}{20}$ or 1 $\frac{5}{4} \cdot \frac{4}{5} = \frac{20}{20}$ or 1

PT

Example 1 Evaluate Using Properties

Evaluate $7(4 - 3) - 1 + 5 \cdot \frac{1}{5}$. Name the property used in each step.

$7(4 - 3) - 1 + 5 \cdot \frac{1}{5} = 7(1) - 1 + 5 \cdot \frac{1}{5}$ Substitution: $4 - 3 = 1$

$= 7 - 1 + 5 \cdot \frac{1}{5}$ Multiplicative Identity: $7 \cdot 1 = 7$

$= 7 - 1 + 1$ Multiplicative Inverse: $5 \cdot \frac{1}{5} = 1$

$= 6 + 1$ Substitution: $7 - 1 = 6$

$= 7$ Substitution: $6 + 1 = 7$

StudyTip

CCSS Arguments As you evaluate an expression, you are constructing an argument using stated assumptions, definitions, and previously established results. The properties of numbers are valid reasons for steps in the argument.

GuidedPractice

Name the property used in each step.

1A. $2 \cdot 3 + (4 \cdot 2 - 8)$

$= 2 \cdot 3 + (8 - 8)$?

$= 2 \cdot 3 + (0)$?

$= 6 + 0$?

$= 6$?

1B. $7 \cdot \frac{1}{7} + 6(15 \div 3 - 5)$

$= 7 \cdot \frac{1}{7} + 6(5 - 5)$?

$= 7 \cdot \frac{1}{7} + 6(0)$?

$= 1 + 6(0)$?

$= 1 + 0$?

$= 1$?

2 Use Commutative and Associative Properties

Nikki walks 2 blocks to her friend Sierra's house. They walk another 4 blocks to school. At the end of the day, Nikki and Sierra walk back to Sierra's house, and then Nikki walks home.

The distance from Nikki's house to school	equals	the distance from the school to Nikki's house.
$2 + 4$	$=$	$4 + 2$

This is an example of the **Commutative Property** for addition.

KeyConcept Commutative Property

Words	The order in which you add or multiply numbers does not change their sum or product.
Symbols	For any numbers a and b, $a + b = b + a$ and $a \cdot b = b \cdot a$.
Examples	$4 + 8 = 8 + 4$ $\quad$ $7 \cdot 11 = 11 \cdot 7$

An easy way to find the sum or product of numbers is to group, or associate, the numbers using the **Associative Property**.

KeyConcept Associative Property

Words	The way you group three or more numbers when adding or multiplying does not change their sum or product.
Symbols	For any numbers a, b, and c, $(a + b) + c = a + (b + c)$ and $(ab)c = a(bc)$.
Examples	$(3 + 5) + 7 = 3 + (5 + 7)$ $\quad$ $(2 \cdot 6) \cdot 9 = 2 \cdot (6 \cdot 9)$

Tanya Constantine/Digital Vision/Getty Images

Real-WorldLink

A child's birthday party may cost about $200 depending on the number of children invited.

Source: Family Corner

Real-World Example 2 Apply Properties of Numbers

PARTY PLANNING Eric makes a list of items that he needs to buy for a party and their costs. Find the total cost of these items.

Party Supplies	
Item	**Cost ($)**
balloons	6.75
decorations	14.00
food	23.25
beverages	20.50

Balloons		Decorations		Food		Beverages
6.75	+	14.00	+	23.25	+	20.50

$= 6.75 + 23.25 + 14.00 + 20.50$ Commutative (+)

$= (6.75 + 23.25) + (14.00 + 20.50)$ Associative (+)

$= 30.00 + 34.50$ Substitution

$= 64.50$ Substitution

The total cost is $64.50.

GuidedPractice

2. FURNITURE Rafael is buying furnishings for his first apartment. He buys a couch for $300, lamps for $30.50, a rug for $25.50, and a table for $50. Find the total cost of these items.

Example 3 Use Multiplication Properties

Evaluate $5 \cdot 7 \cdot 4 \cdot 2$ using the properties of numbers. Name the property used in each step.

$5 \cdot 7 \cdot 4 \cdot 2$	$= 5 \cdot 2 \cdot 7 \cdot 4$	Commutative (×)
	$= (5 \cdot 2) \cdot (7 \cdot 4)$	Associative (×)
	$= 10 \cdot 28$	Substitution
	$= 280$	Substitution

GuidedPractice

Evaluate each expression using the properties of numbers. Name the property used in each step.

3A. $2.9 \cdot 4 \cdot 10$

3B. $\frac{5}{3} \cdot 25 \cdot 3 \cdot 2$

Check Your Understanding

Example 1 **Evaluate each expression. Name the property used in each step.**

1. $(1 \div 5)5 \cdot 14$

2. $6 + 4(19 - 15)$

3. $5(14 - 5) + 6(3 + 7)$

4. **FINANCIAL LITERACY** Carolyn has 9 quarters, 4 dimes, 7 nickels, and 2 pennies, which can be represented as $9(25) + 4(10) + 7(5) + 2$. Evaluate the expression to find how much money she has. Name the property used in each step.

Examples 2–3 **Evaluate each expression using the properties of numbers. Name the property used in each step.**

5. $23 + 42 + 37$

6. $2.75 + 3.5 + 4.25 + 1.5$

7. $3 \cdot 7 \cdot 10 \cdot 2$

8. $\frac{1}{4} \cdot 24 \cdot \frac{2}{3}$

Practice and Problem Solving

Example 1 **Evaluate each expression. Name the property used in each step.**

9. $3(22 - 3 \cdot 7)$

10. $7 + (9 - 3^2)$

11. $\frac{3}{4}[4 \div (7 - 4)]$

12. $[3 \div (2 \cdot 1)]\frac{2}{3}$

13. $2(3 \cdot 2 - 5) + 3 \cdot \frac{1}{3}$

14. $6 \cdot \frac{1}{6} + 5(12 \div 4 - 3)$

Example 2

15. **GEOMETRY** The expression $2 \cdot \frac{22}{7} \cdot 14^2 + 2 \cdot \frac{22}{7} \cdot 14 \cdot 7$ represents the approximate surface area of the cylinder at the right. Evaluate this expression to find the approximate surface area. Name the property used in each step.

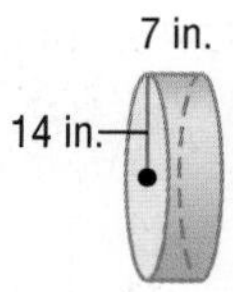

16. **CCSS REASONING** A traveler checks into a hotel on Friday and checks out the following Tuesday morning. Use the table to find the total cost of the room including tax.

Hotel Rates Per Day		
Day	**Room Charge**	**Sales Tax**
Monday–Friday	$72	$5.40
Saturday–Sunday	$63	$5.10

Examples 2–3 **Evaluate each expression using properties of numbers. Name the property used in each step.**

17. $25 + 14 + 15 + 36$

18. $11 + 7 + 5 + 13$

19. $3\frac{2}{3} + 4 + 5\frac{1}{3}$

20. $4\frac{4}{9} + 7\frac{2}{9}$

21. $4.3 + 2.4 + 3.6 + 9.7$

22. $3.25 + 2.2 + 5.4 + 10.75$

23. $12 \cdot 2 \cdot 6 \cdot 5$

24. $2 \cdot 8 \cdot 10 \cdot 2$

25. $0.2 \cdot 4.6 \cdot 5$

26. $3.5 \cdot 3 \cdot 6$

27. $1\frac{5}{6} \cdot 24 \cdot 3\frac{1}{11}$

28. $2\frac{3}{4} \cdot 1\frac{1}{8} \cdot 32$

29. **SCUBA DIVING** The sign shows the equipment rented or sold by a scuba diving store.

a. Write two expressions to represent the total sales to rent 2 wet suits, 3 air tanks, 2 dive flags, and selling 5 underwater cameras.

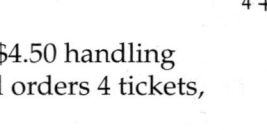

b. What are the total sales?

30. **COOKIES** Bobby baked 2 dozen chocolate chip cookies, 3 dozen sugar cookies, and a dozen oatmeal raisin cookies. How many total cookies did he bake?

Evaluate each expression if $a = -1$, $b = 4$, and $c = 6$.

31. $4a + 9b - 2c$

32. $-10c + 3a + a$

33. $a - b + 5a - 2b$

34. $8a + 5b - 11a - 7b$

35. $3c^2 + 2c + 2c^2$

36. $3a - 4a^2 + 2a$

37. **FOOTBALL** A football team is on the 35-yard line. The quarterback is sacked at the line of scrimmage. The team gains 0 yards, so they are still at the 35-yard line. Which identity or property does this represent? Explain.

Find the value of x. Then name the property used.

38. $8 = 8 + x$

39. $3.2 + x = 3.2$

40. $10x = 10$

41. $\frac{1}{2} \cdot x = \frac{1}{2} \cdot 7$

42. $x + 0 = 5$

43. $1 \cdot x = 3$

44. $5 \cdot \frac{1}{5} = x$

45. $2 + 8 = 8 + x$

46. $x + \frac{3}{4} = 3 + \frac{3}{4}$

47. $\frac{1}{3} \cdot x = 1$

48. **GEOMETRY** Write an expression to represent the perimeter of the triangle. Then find the perimeter if $x = 2$ and $y = 7$.

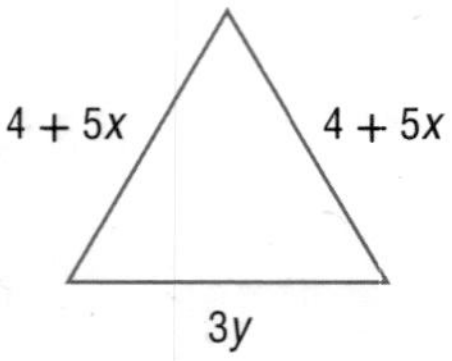

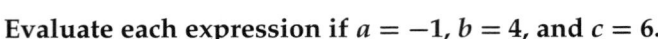

49. **SPORTS** Tickets to a baseball game cost $25 each plus a $4.50 handling charge per ticket. If Sharon has a coupon for $10 off and orders 4 tickets, how much will she be charged?

50. CCSS **PRECISION** The table shows prices on children's clothing.

a. Interpret the expression $5(8.99) + 2(2.99) + 7(5.99)$.

b. Write and evaluate three different expressions that represent 8 pairs of shorts and 8 tops.

c. If you buy 8 shorts and 8 tops, you receive a discount of 15%. Find the greatest and least amount of money you can spend on the 16 items at the sale.

Shorts	Shirts	Tank Tops
$7.99	$8.99	$6.99
$5.99	$4.99	$2.99

51. GEOMETRY A regular octagon measures $(3x + 5)$ units on each side. What is the perimeter if $x = 2$?

52. MULTIPLE REPRESENTATIONS You can use *algebra tiles* to model and explore algebraic expressions. The rectangular tile has an area of x, with dimensions 1 by x. The small square tile has an area of 1, with dimensions 1 by 1.

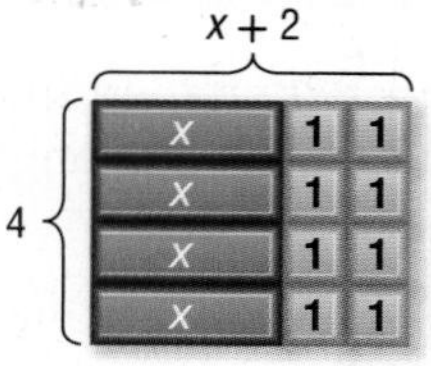

a. Concrete Make a rectangle with algebra tiles to model the expression $4(x + 2)$ as shown above. What are the dimensions of this rectangle? What is its area?

b. Analytical What are the areas of the green region and of the yellow region?

c. Verbal Complete this statement: $4(x + 2) =$ _?_. Write a convincing argument to justify your statement.

53. GEOMETRY A **proof** is a logical argument in which each statement you make is supported by a statement that is accepted as true. It is given that $\overline{AB} \cong \overline{CD}$, $\overline{AB} \cong \overline{BD}$, and $\overline{AB} \cong \overline{AC}$. Pedro wants to prove $\triangle ADB \cong \triangle ADC$. To do this, he must show that $\overline{AD} \cong \overline{AD}$, $\overline{AB} \cong \overline{DC}$ and $\overline{BD} \cong \overline{AC}$.

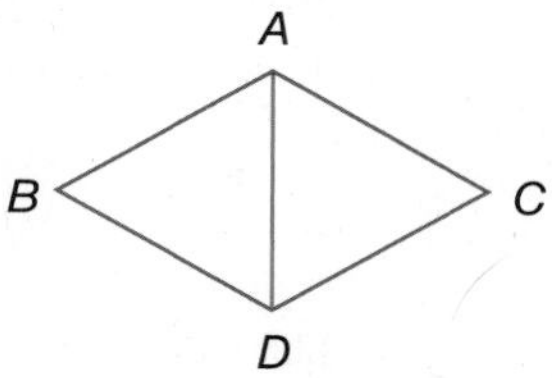

a. Copy the figure and label $\overline{AB} \cong \overline{CD}$, $\overline{AB} \cong \overline{BD}$, and $\overline{AB} \cong \overline{AC}$.

b. Explain how he can use the Reflexive and Transitive Properties to prove $\triangle ADB \cong \triangle ADC$.

c. If AC is x centimeters, write an equation for the perimeter of $ACDB$.

H.O.T. Problems Use Higher-Order Thinking Skills

54. OPEN ENDED Write two equations showing the Transitive Property of Equality. Justify your reasoning.

55. CCSS ARGUMENTS Explain why 0 has no multiplicative inverse.

56. REASONING The sum of any two whole numbers is always a whole number. So, the set of whole numbers {0, 1, 2, 3, 4, … } is said to be closed under addition. This is an example of the **Closure Property**. State whether each statement is *true* or *false*. If false, justify your reasoning.

a. The set of whole numbers is closed under subtraction.

b. The set of whole numbers is closed under multiplication.

c. The set of whole numbers is closed under division.

57. CHALLENGE Does the Commutative Property *sometimes*, *always* or *never* hold for subtraction? Explain your reasoning.

58. REASONING Explain whether 1 can be an additive identity. Give an example to justify your answer.

59. WHICH ONE DOESN'T BELONG? Identify the equation that does not belong with the other three. Explain your reasoning.

$x + 12 = 12 + x$	$7h = h \cdot 7$	$1 + a = a + 1$	$(2j)k = 2(jk)$

60. WRITING IN MATH Determine whether the Commutative Property applies to division. Justify your answer.

Standardized Test Practice

61. A deck is shaped like a rectangle with a width of 12 feet and a length of 15 feet. What is the area of the deck?

A 3 ft^2
B 27 ft^2
C 108 ft^2
D 180 ft^2

62. GEOMETRY A box in the shape of a rectangular prism has a volume of 56 cubic inches. If the length of each side is multiplied by 2, what will be the approximate volume of the box?

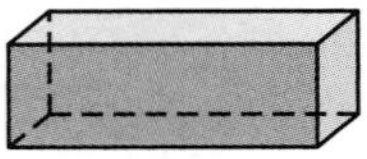

F 112 in^3 **H** 336 in^3
G 224 in^3 **J** 448 in^3

63. $27 \div 3 + (12 - 4) =$

A $\frac{-11}{5}$ **C** 17
B $\frac{27}{11}$ **D** 25

64. GRIDDED RESPONSE Ms. Beal had 1 bran muffin, 16 ounces of orange juice, 3 ounces of sunflower seeds, 2 slices of turkey, and half a cup of spinach. Find the total number of grams of protein she consumed.

Protein Content

Food	Protein (g)
bran muffin (1)	3
orange juice (8 oz)	2
sunflower seeds (1 oz)	2
turkey (1 slice)	12
spinach (1 c)	5

Spiral Review

Evaluate each expression. (Lesson 1-2)

65. $3 \cdot 5 + 1 - 2$ **66.** $14 \div 2 \cdot 6 - 5^2$ **67.** $\frac{3 \cdot 9^2 - 3^2 \cdot 9}{3 \cdot 9}$

68. GEOMETRY Write an expression for the perimeter of the figure. (Lesson 1-1)

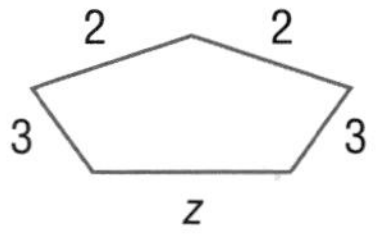

Find the perimeter and area of each figure. (Lessons 0-7 and 0-8)

69. a rectangle with length 5 feet and width 8 feet

70. a square with length 4.5 inches

71. SURVEY Andrew took a survey of his friends to find out their favorite type of music. Of the 34 friends surveyed, 22 said they liked rock music the best. What percent like rock music the best? (Lesson 0-6)

Name the reciprocal of each number. (Lesson 0-5)

72. $\frac{6}{17}$ **73.** $\frac{2}{23}$ **74.** $3\frac{4}{5}$

Skills Review

Find each product. Express in simplest form.

75. $\frac{12}{15} \cdot \frac{3}{14}$ **76.** $\frac{5}{7} \cdot \left(-\frac{4}{5}\right)$ **77.** $\frac{10}{11} \cdot \frac{21}{35}$

78. $\frac{63}{65} \cdot \frac{120}{126}$ **79.** $-\frac{4}{3} \cdot \left(-\frac{9}{2}\right)$ **80.** $\frac{1}{3} \cdot \frac{2}{5}$

EXTEND 1-3

Algebra Lab
Accuracy

All measurements taken in the real world are approximations. The greater the care with which a measurement is taken, the more accurate it will be. **Accuracy** refers to how close a measured value comes to the actual or desired value. For example, a fraction is more accurate than a rounded decimal.

CCSS Common Core State Standards
Content Standards
N.Q.3 Choose a level of accuracy appropriate to limitations on measurement when reporting quantities.
Mathematical Practices
6 Attend to precision.

Activity 1 When Is Close Good Enough?

Measure the length of your desktop. Record your results in centimeters, in meters, and in millimeters.

Analyze the Results

1. Did you round to the nearest whole measure? If so, when?
2. Did you round to the nearest half, tenth, or smaller? If so, when?
3. Which unit of measure was the most appropriate for this task?
4. Which unit of measure was the most accurate?

Deciding where to round a measurement depends on how the measurement will be used. But calculations should not be carried out to greater accuracy than that of the original data.

Activity 2 Decide Where to Round

a. Elan has \$13 that he wants to divide among his 6 nephews. When he types 13 ÷ 6 into his calculator, the number that appears is 2.166666667. Where should Elan round?

Since Elan is rounding money, the smallest increment is a penny, so round to the hundredths place. This will give him 2.17, and $\$2.17 \times 6 = \$13.02$. Elan will be two pennies short, so round to \$2.16. Since $\$2.16 \times 6 = \$12.96$, Elan can give each of his nephews \$2.16.

b. Dante's mother brings him a dozen cookies, but before she leaves she eats one and tells Dante he has to share with his two sisters. Dante types 11 ÷ 3 into his calculator and gets 3.666666667. Where should Dante round?

After each sibling receives 3 cookies, there are two cookies left. In this case, it is more accurate to convert the decimal portion to a fraction and give each sibling $\frac{2}{3}$ of a cookie.

c. Eva measures the dimensions of a box as 8.7, 9.52, and 3.16 inches. She multiplies these three numbers to find the measure of the volume. The result shown on her calculator is 261.72384. Where should Eva round?

Eva should round to the tenths place, 261.7, because she was only accurate to the tenths place with one of her measures.

Exercises

5. Jessica wants to divide \$23 six ways. Her calculator shows 3.833333333. Where should she round?

6. Ms. Harris wants to share 2 pizzas among 6 people. Her calculator shows 0.3333333333. Where should she round?

7. The measurements of an aquarium are 12.9, 7.67, and 4.11 inches. The measure of the volume is given by the product 406.65573. Where should the number be rounded?

Algebra Lab

Accuracy *Continued*

For most real-world measurements, a decision must be made on the level of accuracy needed or desired.

Activity 3 Find an Appropriate Level of Accuracy

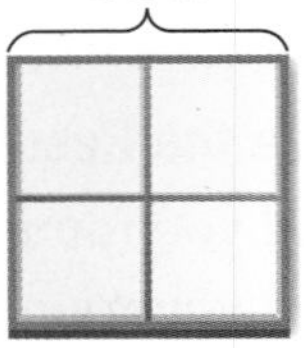

a. Jon needs to buy a shade for the window opening shown, but the shades are only available in whole inch increments. What size shade should he buy?

He should buy the 27-inch shade because it will be enough to cover the glass.

b. Tom is buying flea medicine for his dog. The amount of medicine depends on the dog's weight. The medicine is available in packages that vary by 10 dog pounds. How accurate does Tom need to be to buy the correct medicine?

He needs to be accurate to within 10 pounds.

c. Tyrone is building a jet engine. How accurate do you think he needs to be with his measurements?

He needs to be very accurate, perhaps to the thousandth of an inch.

Exercises

8. Matt's table is missing a leg. He wants to cut a piece of wood to replace the leg. How accurate do you think he needs to be with his measurements?

For each situation, determine where the rounding should occur and give the rounded answer.

9. Sam wants to divide \$111 seven ways. His calculator shows 15.85714286.

10. Kiri wants to share 3 pies among 11 people. Her calculator shows 0.2727272727.

11. Evan's calculator gives him the volume of his soccer ball as 137.2582774. Evan measured the radius of the ball to be 3.2 inches.

For each situation, determine the level of accuracy needed. Explain.

12. You are estimating the length of your school's basketball court. Which unit of measure should you use: 1 foot, 1 inch, or $\frac{1}{16}$ inch?

13. You are estimating the height of a small child. Which unit of measure should you use: 1 foot, 1 inch, or $\frac{1}{16}$ inch?

14. **TRAVEL** Curt is measuring the driving distance from one city to another. How accurate do you think he needs to be with his measurement?

15. **MEDICINE** A nurse is administering medicine to a patient based on his weight. How accurate do you think she needs to be with her measurements?

LESSON 1-4

The Distributive Property

:: Then

- You explored Associative and Commutative Properties.

:: Now

1. Use the Distributive Property to evaluate expressions.
2. Use the Distributive Property to simplify expressions.

:: Why?

- John burns approximately 420 Calories per hour by inline skating. The chart below shows the time he spent inline skating in one week.

Day	Mon	Tue	Wed	Thu	Fri	Sat	Sun
Time (h)	1	$\frac{1}{2}$	0	1	0	2	$2\frac{1}{2}$

To determine the total number of Calories that he burned inline skating that week, you can use the Distributive Property.

NewVocabulary
like terms
simplest form
coefficient

Common Core State Standards

Content Standards
A.SSE.1a Interpret parts of an expression, such as terms, factors, and coefficients.
A.SSE.2 Use the structure of an expression to identify ways to rewrite it.

Mathematical Practices
1 Make sense of problems and persevere in solving them.
8 Look for and express regularity in repeated reasoning.

1 Evaluate Expressions

There are two methods you could use to calculate the number of Calories John burned inline skating. You could find the total time spent inline skating and then multiply by the Calories burned per hour. Or you could find the number of Calories burned each day and then add to find the total.

Method 1 Rate Times Total Time

$$420\left(1 + \frac{1}{2} + 1 + 2 + 2\frac{1}{2}\right)$$
$$= 420(7)$$
$$= 2940$$

Method 2 Sum of Daily Calories Burned

$$420(1) + 420\left(\frac{1}{2}\right) + 420(1) + 420(2) + 420\left(2\frac{1}{2}\right)$$
$$= 420 + 210 + 420 + 840 + 1050$$
$$= 2940$$

Either method gives the same total of 2940 Calories burned. This is an example of the **Distributive Property**.

KeyConcept Distributive Property

Symbol For any numbers a, b, and c,
$a(b + c) = ab + ac$ and $(b + c)a = ba + ca$ and
$a(b - c) = ab - ac$ and $(b - c)a = ba - ca$.

Examples

$3(2 + 5) = 3 \cdot 2 + 3 \cdot 5$
$3(7) = 6 + 15$
$21 = 21$

$4(9 - 7) = 4 \cdot 9 - 4 \cdot 7$
$4(2) = 36 - 28$
$8 = 8$

The Symmetric Property of Equality allows the Distributive Property to be written as follows.

If $a(b + c) = ab + ac$, then $ab + ac = a(b + c)$.

SuperStock/age fotostock

Real-WorldLink

The record attendance for a single baseball game was set in 1959. There were 92,706 spectators at a game between the Los Angeles Dodgers and the Chicago White Sox.

Source: Baseball Almanac

Real-World Example 1 Distribute Over Addition

PT

SPORTS A group of 7 adults and 6 children are going to a University of South Florida Bulls baseball game. Use the Distributive Property to write and evaluate an expression for the total ticket cost.

USF Bulls Baseball Tickets	
Ticket	**Cost ($)**
Adult Single Game	5
Children Single Game (12 and under)	3
Groups of 10 or more Single Game	2
Senior Single Game (65 and over)	3

Source: USF

Understand You need to find the cost of each ticket and then find the total cost.

Plan 7 + 6 or 13 people are going to the game, so the tickets are $2 each.

Solve Write an expression that shows the product of the cost of each ticket and the sum of adult tickets and children's tickets.

$2(7 + 6) = 2(7) + 2(6)$ Distributive Property

$= 14 + 12$ Multiply.

$= 26$ Add.

The total cost is $26.

Check The total number of tickets needed is 13 and they cost $2 each. Multiply 13 by 2 to get 26. Therefore, the total cost of tickets is $26.

StudyTip

CCSS Sense-Making and Perseverance The four-step problem solving plan is a tool for making sense of any problem. When making and executing your plan, continually ask yourself, "Does this make sense?" Monitor and evaluate your progress and change course if necessary.

GuidedPractice

1. SPORTS A group of 3 adults, an 11-year old, and 2 children under 10 years old are going to a baseball game. Write and evaluate an expression to determine the cost of tickets for the group.

You can use the Distributive Property to make mental math easier.

Example 2 Mental Math

PT

Use the Distributive Property to rewrite 7 • 49. Then evaluate.

$7 \cdot 49 = 7(50 - 1)$ Think: $49 = 50 - 1$

$= 7(50) - 7(1)$ Distributive Property

$= 350 - 7$ Multiply.

$= 343$ Subtract.

GuidedPractice

Use the Distributive Property to rewrite each expression. Then evaluate.

2A. 304(15)

2B. $44 \cdot 2\frac{1}{2}$

2C. 210(5)

2D. 52(17)

2 Simplify Expressions

You can use algebra tiles to investigate how the Distributive Property relates to algebraic expressions.

Gail Mooney/Masterfile

Problem-SolvingTip

Make a Model It can be helpful to visualize a problem using algebra tiles or folded paper.

The rectangle at the right has 3 x-tiles and 6 1-tiles. The area of the rectangle is $x + 1 + 1 + x + 1 + 1 + x + 1 + 1$ or $3x + 6$. Therefore, $3(x + 2) = 3x + 6$.

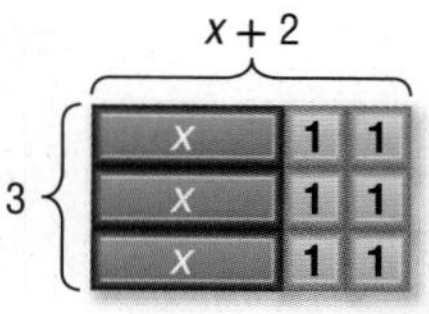

Example 3 Algebraic Expressions

Rewrite each expression using the Distributive Property. Then simplify.

a. $\mathbf{7(3w - 5)}$

$7(3w - 5) = 7 \cdot 3w - 7 \cdot 5$ Distributive Property

$= 21w - 35$ Multiply.

b. $\mathbf{(6v^2 + v - 3)4}$

$(6v^2 + v - 3)4 = 6v^2(4) + v(4) - 3(4)$ Distributive Property

$= 24v^2 + 4v - 12$ Multiply.

GuidedPractice

3A. $(8 + 4n)2$

3B. $-6(r + 3g - t)$

3C. $(2 - 5q)(-3)$

3D. $-4(-8 - 3m)$

ReviewVocabulary

term a number, a variable, or a product or quotient of numbers and variables

Like terms are terms that contain the same variables, with corresponding variables having the same power.

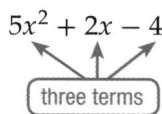

$5x^2 + 2x - 4$ — three terms

$6a^2 + a^2 + 2a$ — like terms ($6a^2$, a^2); unlike terms (a^2, $2a$)

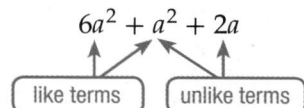

The Distributive Property and the properties of equality can be used to show that $4k + 8k = 12k$. In this expression, $4k$ and $8k$ are like terms.

$4k + 8k = (4 + 8)k$ Distributive Property

$= 12k$ Substitution

An expression is in **simplest form** when it contains no like terms or parentheses.

Example 4 Combine Like Terms

a. Simplify $\mathbf{17u + 25u}$**.**

$17u + 25u = (17 + 25)u$ Distributive Property

$= 42u$ Substitution

b. Simplify $\mathbf{6t^2 + 3t - t}$**.**

$6t^2 + 3t - t = 6t^2 + (3 - 1)t$ Distributive Property

$= 6t^2 + 2t$ Substitution

GuidedPractice

Simplify each expression. If not possible, write *simplified*.

4A. $6n - 4n$

4B. $b^2 + 13b + 13$

4C. $4y^3 + 2y - 8y + 5$

4D. $7a + 4 - 6a^2 - 2a$

Math HistoryLink

Kambei Mori
(c. 1600–1628)
Kambei Mori was a Japanese scholar who popularized the abacus. He changed the focus of mathematics from philosophy to computation.

Example 5 Write and Simplify Expressions

Use the expression *twice the difference of 3x and y increased by five times the sum of x and 2y.*

a. Write an algebraic expression for the verbal expression.

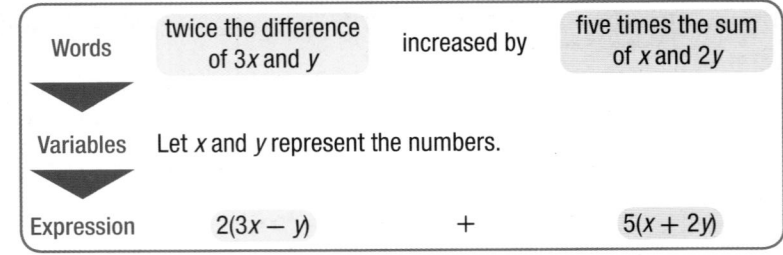

Words	twice the difference of $3x$ and y	increased by	five times the sum of x and $2y$
Variables	Let x and y represent the numbers.		
Expression	$2(3x - y)$	$+$	$5(x + 2y)$

b. Simplify the expression, and indicate the properties used.

$2(3x - y) + 5(x + 2y) = 2(3x) - 2(y) + 5(x) + 5(2y)$	Distributive Property
$= 6x - 2y + 5x + 10y$	Multiply.
$= 6x + 5x - 2y + 10y$	Commutative (+)
$= (6 + 5)x + (-2 + 10)y$	Distributive Property
$= 11x + 8y$	Substitution

GuidedPractice

5. Use the expression *5 times the difference of q squared and r plus 8 times the sum of 3q and 2r.*

A. Write an algebraic expression for the verbal expression.

B. Simplify the expression, and indicate the properties used.

The **coefficient** of a term is the numerical factor. For example, in $6ab$, the coefficient is 6, and in $\frac{x^2}{3}$, the coefficient is $\frac{1}{3}$. In the term y, the coefficient is 1 since $1 \cdot y = y$ by the Multiplicative Identity Property.

ConceptSummary Properties of Numbers

The following properties are true for any numbers a, b, and c.

Properties	Addition	Multiplication
Commutative	$a + b = b + a$	$ab = ba$
Associative	$(a + b) + c = a + (b + c)$	$(ab)c = a(bc)$
Identity	0 is the identity. $a + 0 = 0 + a = a$	1 is the identity. $a \cdot 1 = 1 \cdot a = a$
Zero	—	$a \cdot 0 = 0 \cdot a = 0$
Distributive	$a(b + c) = ab + ac$ and $(b + c)a = ba + ca$	
Substitution	If $a = b$, then a may be substituted for b.	

age fotostock/SuperStock

Check Your Understanding

Example 1 **1. PILOT** A pilot at an air show charges \$25 per passenger for rides. If 12 adults and 15 children ride in one day, write and evaluate an expression to describe the situation.

Example 2 **Use the Distributive Property to rewrite each expression. Then evaluate.**

2. $14(51)$ **3.** $6\frac{1}{9}(9)$

Example 3 **Use the Distributive Property to rewrite each expression. Then simplify.**

4. $2(4 + t)$ **5.** $(g - 9)5$

Example 4 **Simplify each expression. If not possible, write *simplified*.**

6. $15m + m$ **7.** $3x^3 + 5y^3 + 14$ **8.** $(5m + 2m)10$

Example 5 **Write an algebraic expression for each verbal expression. Then simplify, indicating the properties used.**

9. 4 times the sum of 2 times x and six

10. one half of 4 times y plus the quantity of y and 3

Practice and Problem Solving

Example 1 **11. TIME MANAGEMENT** Margo uses dots to track her activities on a calendar. Red dots represent homework, yellow dots represent work, and green dots represent track practice. In a typical week, she uses 5 red dots, 3 yellow dots, and 4 green dots. How many activities does Margo do in 4 weeks?

12. CCSS REASONING The Red Cross is holding blood drives in two locations. In one day, Center 1 collected 715 pints and Center 2 collected 1035 pints. Write and evaluate an expression to estimate the total number of pints of blood donated over a 3-day period.

Example 2 **Use the Distributive Property to rewrite each expression. Then evaluate.**

13. $(4 + 5)6$ **14.** $7(13 + 12)$ **15.** $6(6 - 1)$

16. $(3 + 8)15$ **17.** $14(8 - 5)$ **18.** $(9 - 4)19$

19. $4(7 - 2)$ **20.** $7(2 + 1)$ **21.** $7 \cdot 497$

22. $6(525)$ **23.** $36 \cdot 3\frac{1}{4}$ **24.** $\left(4\frac{2}{7}\right)21$

Example 3 **Use the Distributive Property to rewrite each expression. Then simplify.**

25. $2(x + 4)$ **26.** $(5 + n)3$

27. $(4 - 3m)8$ **28.** $-3(2x - 6)$

Example 4 **Simplify each expression. If not possible, write *simplified*.**

29. $13r + 5r$ **30.** $3x^3 - 2x^2$ **31.** $7m + 7 - 5m$

32. $5z^2 + 3z + 8z^2$ **33.** $(2 - 4n)17$ **34.** $11(4d + 6)$

35. $7m + 2m + 5p + 4m$ **36.** $3x + 7(3x + 4)$ **37.** $4(fg + 3g) + 5g$

Example 5 **Write an algebraic expression for each verbal expression. Then simplify, indicating the properties used.**

38. the product of 5 and m squared, increased by the sum of the square of m and 5

39. 7 times the sum of a squared and b minus 4 times the sum of a squared and b

40. **GEOMETRY** Find the perimeter of an isosceles triangle with side lengths of $5 + x$, $5 + x$, and xy. Write in simplest form.

41. **GEOMETRY** A regular hexagon measures $3x + 5$ units on each side. What is the perimeter in simplest form?

Simplify each expression.

42. $6x + 4y + 5x$
43. $3m + 5g + 6g + 11m$
44. $4a + 5a^2 + 2a^2 + a^2$
45. $5k + 3k^3 + 7k + 9k^3$
46. $6d + 4(3d + 5)$
47. $2(6x + 4) + 7x$

48. **FOOD** Kenji is picking up take-out food for his study group.
   a. Interpret the expression $4(2.49) + 3(1.29) + 3(0.99) + 5(1.49)$.
   b. How much would it cost if Kenji bought four of each item on the menu?

Menu

Item	Cost ($)
sandwich	2.49
cup of soup	1.29
side salad	0.99
drink	1.49

Use the Distributive Property to rewrite each expression. Then simplify.

49. $\left(\frac{1}{3} - 2b\right)27$
50. $4(8p + 4q - 7r)$
51. $6(2c - cd^2 + d)$

Simplify each expression. If not possible, write *simplified*.

52. $6x^2 + 14x - 9x$
53. $4y^3 + 3y^3 + y^4$
54. $a + \frac{a}{5} + \frac{2}{5}a$

55. **MULTIPLE REPRESENTATIONS** The area of the model is $2(x - 4)$ or $2x - 8$. The expression $2(x - 4)$ is in *factored form.*

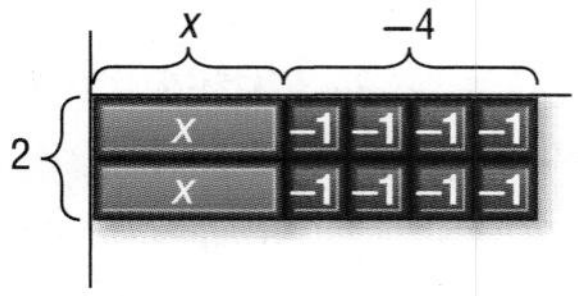

   a. **Geometric** Use algebra tiles to form a rectangle with area $2x + 6$. Use the result to write $2x + 6$ in factored form.
   b. **Tabular** Use algebra tiles to form rectangles to represent each area in the table. Record the factored form of each expression.
   c. **Verbal** Explain how you could find the factored form of an expression.

Area	Factored Form
$2x + 6$	
$3x + 3$	
$3x - 12$	
$5x + 10$	

H.O.T. Problems Use Higher-Order Thinking Skills

56. **CCSS PERSEVERANCE** Use the Distributive Property to simplify $6x^2[(3x - 4) + (4x + 2)]$.

57. **REASONING** Should the Distributive Property be a property of multiplication, addition, or both? Explain your answer.

58. **WRITING IN MATH** Why is it helpful to represent verbal expressions algebraically?

59. **WRITING IN MATH** Use the data about skating on page 25 to explain how the Distributive Property can be used to calculate quickly. Also, compare the two methods of finding the total Calories burned.

Standardized Test Practice

60. Which illustrates the Symmetric Property of Equality?

A If $a = b$, then $b = a$.

B If $a = b$, and $b = c$, then $a = c$.

C If $a = b$, then $b = c$.

D If $a = a$, then $a + 0 = a$.

61. Anna is three years younger than her sister Emily. Which expression represents Anna's age if we express Emily's age as y years?

F $y + 3$ **H** $3y$

G $y - 3$ **J** $\frac{3}{y}$

62. Which property is used below?
If $4xy^2 = 8y^2$ and $8y^2 = 72$, then $4xy^2 = 72$.

A Reflexive Property

B Substitution Property

C Symmetric Property

D Transitive Property

63. SHORT RESPONSE A drawer contains the socks in the chart. What is the probability that a randomly chosen sock is blue?

Color	Number
white	16
blue	12
black	8

Spiral Review

Evaluate each expression. Name the property used in each step. (Lesson 1-3)

64. $14 + 23 + 8 + 15$

65. $0.24 \cdot 8 \cdot 7.05$

66. $1\frac{1}{4} \cdot 9 \cdot \frac{5}{6}$

67. SPORTS Braden runs 6 times a week for 30 minutes and lifts weights 3 times a week for 20 minutes. Write and evaluate an expression for the number of hours Braden works out in 4 weeks. (Lesson 1-2)

SPORTS Refer to the table showing Blanca's cross-country times for the first 8 meets of the season. Round answers to the nearest second. (Lesson 0-12)

Cross Country	
Meet	**Time**
1	22:31
2	22:21
3	21:48
4	22:01
5	21:48
6	20:56
7	20:34
8	20:15

68. Find the mean of the data.

69. Find the median of the data.

70. Find the mode of the data.

71. SURFACE AREA What is the surface area of the cube? (Lesson 0-10)

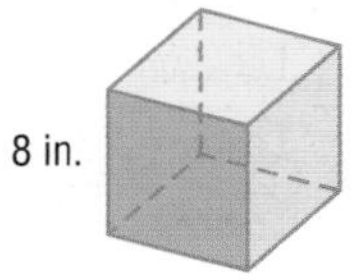

Skills Review

Evaluate each expression.

72. $12(7 + 2)$

73. $11(5) - 8(5)$

74. $(13 - 9) \cdot 4$

75. $3(6) + 7(6)$

76. $(1 + 19) \cdot 8$

77. $16(5 + 7)$

CHAPTER 1

Mid-Chapter Quiz

Lessons 1-1 through 1-4

Write a verbal expression for each algebraic expression. (Lesson 1-1)

1. $21 - x^3$

2. $3m^5 + 9$

Write an algebraic expression for each verbal expression. (Lesson 1-1)

3. five more than *s* squared

4. four times *y* to the fourth power

5. **CAR RENTAL** The XYZ Car Rental Agency charges a flat rate of \$29 per day plus \$0.32 per mile driven. Write an algebraic expression for the rental cost of a car for *x* days that is driven *y* miles. (Lesson 1-1)

Evaluate each expression. (Lesson 1-2)

6. $24 \div 3 - 2 \cdot 3$

7. $5 + 2^2$

8. $4(3 + 9)$

9. $36 - 2(1 + 3)^2$

10. $\dfrac{40 - 2^3}{4 + 3(2^2)}$

11. **AMUSEMENT PARK** The costs of tickets to a local amusement park are shown. Write and evaluate an expression to find the total cost for 5 adults and 8 children. (Lesson 1-2)

12. **MULTIPLE CHOICE** Write an algebraic expression to represent the perimeter of the rectangle shown below. Then evaluate it to find the perimeter when $w = 8$ cm. (Lesson 1-2)

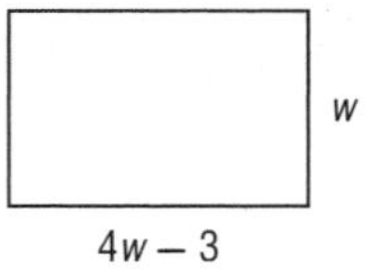

A 37 cm

B 232 cm

C 74 cm

D 45 cm

Evaluate each expression. Name the property used in each step. (Lesson 1-3)

13. $(8 - 2^3) + 21$

14. $3(1 \div 3) \cdot 9$

15. $[5 \div (3 \cdot 1)]\frac{3}{5}$

16. $18 + 35 + 32 + 15$

17. $0.25 \cdot 7 \cdot 4$

Use the Distributive Property to rewrite each expression. Then evaluate. (Lesson 1-4)

18. $3(5 + 2)$

19. $(9 - 6)12$

20. $8(7 - 4)$

Use the Distributive Property to rewrite each expression. Then simplify. (Lesson 1-4)

21. $4(x + 3)$

22. $(6 - 2y)7$

23. $-5(3m - 2)$

24. **DVD SALES** A video store chain has three locations. Use the information in the table below to write and evaluate an expression to estimate the total number of DVDs sold over a 4-day period. (Lesson 1-4)

Location	Daily Sales Numbers
Location 1	145
Location 2	211
Location 3	184

25. **MULTIPLE CHOICE** Rewrite the expression $(8 - 3p)(-2)$ using the Distributive Property. (Lesson 1-4)

F $16 - 6p$

G $-10p$

H $-16 + 6p$

J $10p$

LESSON 1-5

Equations

∴ Then	∴ Now	∴ Why?
● You simplified expressions.	● **1** Solve equations with one variable. **2** Solve equations with two variables.	● Mark's baseball team scored 3 runs in the first inning. At the top of the third inning, their score was 4. The open sentence below represents the change in their score. $3 + r = 4$ The solution is 1. The team got 1 run in the second inning.

NewVocabulary

open sentence
equation
solving
solution
replacement set
set
element
solution set
identity

Common Core State Standards

Content Standards

A.CED.1 Create equations and inequalities in one variable and use them to solve problems.

A.REI.3 Solve linear equations and inequalities in one variable, including equations with coefficients represented by letters.

Mathematical Practices

3 Construct viable arguments and critique the reasoning of others.

1 Solve Equations

A mathematical statement that contains algebraic expressions and symbols is an **open sentence**. A sentence that contains an equals sign, =, is an **equation**.

expression → $3x + 7$ $3x + 7 = 13$ ← equation

Finding a value for a variable that makes a sentence true is called **solving** the open sentence. This replacement value is a **solution**.

A set of numbers from which replacements for a variable may be chosen is called a **replacement set**. A **set** is a collection of objects or numbers that is often shown using braces. Each object or number in the set is called an **element**, or member. A **solution set** is the set of elements from the replacement set that make an open sentence true.

Example 1 Use a Replacement Set

PT

Find the solution set of the equation $2q + 5 = 13$ if the replacement set is {2, 3, 4, 5, 6}.

Use a table to solve. Replace q in $2q + 5 = 13$ with each value in the replacement set.

q	$2q + 5 = 13$	True or False?
2	$2(2) + 5 = 13$	false
3	$2(3) + 5 = 13$	false
4	$2(4) + 5 = 13$	true
5	$2(5) + 5 = 13$	false
6	$2(6) + 5 = 13$	false

Since the equation is true when $q = 4$, the solution of $2q + 5 = 13$ is $q = 4$.

The solution set is {4}.

GuidedPractice

Find the solution set for each equation if the replacement set is {0, 1, 2, 3}.

1A. $8m - 7 = 17$

1B. $28 = 4(1 + 3d)$

You can often solve an equation by applying the order of operations.

PT

Standardized Test Example 2 Apply the Order of Operations

Solve $6 + (5^2 - 5) \div 2 = p$.

A 3 **B** 6 **C** 13 **D** 16

Read the Test Item

You need to apply the order of operations to the expression in order to solve for p.

Solve the Test Item

$6 + (5^2 - 5) \div 2 = p$	Original equation
$6 + (25 - 5) \div 2 = p$	Evaluate powers.
$6 + 20 \div 2 = p$	Subtract 5 from 25.
$6 + 10 = p$	Divide 20 by 2.
$16 = p$	Add.

The correct answer is D.

Test-Taking Tip

Rewrite the Equation If you are allowed to write in your testing booklet, it can be helpful to rewrite the equation with simplified terms.

Guided Practice

2. Solve $t = 9^2 \div (5 - 2)$.

F 3 **G** 6 **H** 14.2 **J** 27

Some equations have a unique solution. Other equations do not have a solution.

PT

Example 3 Solutions of Equations

Solve each equation.

a. $\mathbf{7 - (4^2 - 10) + n = 10}$

Simplify the equation first and then look for a solution.

$7 - (4^2 - 10) + n = 10$	Original equation
$7 - (16 - 10) + n = 10$	Evaluate powers.
$7 - 6 + n = 10$	Subtract 10 from 16.
$1 + n = 10$	Subtract 6 from 7.

The only value for n that makes the equation true is 9. Therefore, this equation has a unique solution of 9.

b. $\mathbf{n(3 + 2) + 6 = 5n + (10 - 3)}$

$n(3 + 2) + 6 = 5n + (10 - 3)$	Original equation
$n(5) + 6 = 5n + (10 - 3)$	Add 3 + 2.
$n(5) + 6 = 5n + 7$	Subtract 3 from 10.
$5n + 6 = 5n + 7$	Commutative (×)

No matter what real value is substituted for n, the left side of the equation will always be one less than the right side. So, the equation will never be true. Therefore, there is no solution of this equation.

Study Tip

Guess and Check When the solution to an equation is not easy to see, substitute values for x and test the equation. Continue to test values until you get a true statement. For example, if $3x + 16 = 73$, test values for x.

$3(10) + 16 = 48$ too low
$3(20) + 16 = 76$ too high
$3(19) + 16 = 73$ ✓

Guided Practice

3A. $(18 + 4) + m = (5 - 3)m$

3B. $8 \cdot 4 \cdot k + 9 \cdot 5 = (36 - 4)k - (2 \cdot 5)$

ReadingMath

Identities An identity is an equation that shows that a number or expression is equivalent to itself.

An equation that is true for every value of the variable is called an **identity**.

PT

Example 4 Identities

Solve $(2 \cdot 5 - 8)(3h + 6) = [(2h + h) + 6]2$.

$(2 \cdot 5 - 8)(3h + 6) = [(2h + h) + 6]2$	Original Equation
$(10 - 8)(3h + 6) = [(2h + h) + 6]2$	Multiply $2 \cdot 5$.
$2(3h + 6) = [(2h + h) + 6]2$	Subtract 8 from 10.
$6h + 12 = [(2h + h) + 6]2$	Distributive Property
$6h + 12 = [3h + 6]2$	Add $2h + h$.
$6h + 12 = 6h + 12$	Distributive Property

No matter what value is substituted for h, the left side of the equation will always be equal to the right side. So, the equation will always be true. Therefore, the solution of this equation could be any real number.

GuidedPractice

Solve each equation.

4A. $12(10 - 7) + 9g = g(2^2 + 5) + 36$

4B. $2d + (2^3 - 5) = 10(5 - 2) + d(12 \div 6)$

4C. $3(b + 1) - 5 = 3b - 2$

4D. $5 - \frac{1}{2}(c - 6) = 4$

2 Solve Equations with Two Variables

Some equations contain two variables. It is often useful to make a table of values and use substitution to find the corresponding values of the second variable.

PT

Example 5 Equations Involving Two Variables

MOVIE RENTALS Mr. Hernandez pays \$10 each month for movies delivered by mail. He can also rent movies in the store for \$1.50 per title. Write and solve an equation to find the total amount Mr. Hernandez spends this month if he rents 3 movies from the store.

The cost of the movie plan is a flat rate. The variable is the number of movies he rents from the store. The total cost is the price of the plan plus \$1.50 times the number of movies from the store. Let C be the total cost and m be the number of movies.

$C = 1.50m + 10$	Original equation
$= 1.50(3) + 10$	Substitute 3 for m.
$= 4.50 + 10$	Multiply.
$= 14.50$	

Mr. Hernandez spends \$14.50 on movie rentals in one month.

GuidedPractice

5. TRAVEL Amelia drives an average of 65 miles per hour. Write and solve an equation to find the time it will take her to drive 36 miles.

Check Your Understanding

Example 1 **Find the solution set of each equation if the replacement set is {11, 12, 13, 14, 15}.**

1. $n + 10 = 23$
2. $7 = \frac{c}{2}$
3. $29 = 3x - 7$
4. $(k - 8)12 = 84$

Example 2 5. **MULTIPLE CHOICE** Solve $\frac{d+5}{10} = 2$.

A 10 B 15 C 20 D 25

Examples 3–4 **Solve each equation.**

6. $x = 4(6) + 3$
7. $14 - 82 = w$
8. $5 + 22a = 2 + 10 \div 2$
9. $(2 \cdot 5) + \frac{c^3}{3} = c^3 \div (1^5 + 2) + 10$

Example 5 10. **RECYCLING** San Francisco has a recycling facility that accepts unused paint. Volunteers blend and mix the paint and give it away in 5-gallon buckets. Write and solve an equation to find the number of buckets of paint given away from the 30,000 gallons that are donated.

Practice and Problem Solving

Example 1 **Find the solution set of each equation if the replacement sets are *y*: {1, 3, 5, 7, 9} and *z*: {10, 12, 14, 16, 18}.**

11. $z + 10 = 22$
12. $52 = 4z$
13. $\frac{15}{y} = 3$
14. $17 = 24 - y$
15. $2z - 5 = 27$
16. $4(y + 1) = 40$
17. $22 = \frac{60}{y} + 2$
18. $111 = z^2 + 11$

Examples 2–4 **Solve each equation.**

19. $a = 32 - 9(2)$
20. $w = 56 \div (2^2 + 3)$
21. $\frac{27+5}{16} = g$
22. $\frac{12 \cdot 5}{15 - 3} = y$
23. $r = \frac{9(6)}{(8+1)3}$
24. $a = \frac{4(14-1)}{3(6)-5} + 7$
25. $(4 - 2^2 + 5)w = 25$
26. $7 + x - (3 + 32 \div 8) = 3$
27. $3^2 - 2 \cdot 3 + u = (3^3 - 3 \cdot 8)(2) + u$
28. $(3 \cdot 6 \div 2)v + 10 = 3^2 v + 9$
29. $6k + (3 \cdot 10 - 8) = (2 \cdot 3)k + 22$
30. $(3 \cdot 5)t + (21 - 12) = 15t + 3^2$
31. $(2^4 - 3 \cdot 5)q + 13 = (2 \cdot 9 - 4^2)q + \left(\frac{3 \cdot 4}{12} - 1\right)$
32. $\frac{3 \cdot 22}{18+4}r - \left(\frac{4^2}{9+7} - 1\right) = r + \left(\frac{8 \cdot 9}{3} \div 3\right)$
33. **SCHOOL** A conference room can seat a maximum of 85 people. The principal and two counselors need to meet with the school's juniors to discuss college admissions. If each student must bring a parent with them, how many students can attend each meeting? Assume that each student has a unique set of parents.
34. **CCSS MODELING** The perimeter of a regular octagon is 128 inches. Find the length of each side.

Example 5 **35.** **SPORTS** A 200-pound athlete who trains for four hours per day requires 2836 Calories for basic energy requirements. During training, the same athlete requires an additional 3091 Calories for extra energy requirements. Write an equation to find C, the total daily Calorie requirement for this athlete. Then solve the equation.

36. **ENERGY** An electric generator can power 3550 watts of electricity. Write and solve an equation to find how many 75-watt light bulbs a generator could power.

Make a table of values for each equation if the replacement set is {−2, −1, 0, 1, 2}.

37. $y = 3x - 2$

38. $3.25x + 0.75 = y$

Solve each equation using the given replacement set.

39. $t - 13 = 7$; {10, 13, 17, 20}

40. $14(x + 5) = 126$; {3, 4, 5, 6, 7}

41. $22 = \frac{n}{3}$; {62, 64, 66, 68, 70}

42. $35 = \frac{g - 8}{2}$; {78, 79, 80, 81}

Solve each equation.

43. $\frac{3(9) - 2}{1 + 4} = d$

44. $j = 15 \div 3 \cdot 5 - 4^2$

45. $c + (3^2 - 3) = 21$

46. $(3^3 - 3 \cdot 9) + (7 - 2^2)b = 24b$

47. CCSS **SENSE-MAKING** Blood flow rate can be expressed as $F = \frac{p_1 - p_2}{r}$, where F is the flow rate, p_1 and p_2 are the initial and final pressure exerted against the blood vessel's walls, respectively, and r is the resistance created by the size of the vessel.

a. Write and solve an equation to determine the resistance of the blood vessel for an initial pressure of 100 millimeters of mercury, a final pressure of 0 millimeters of mercury, and a flow rate of 5 liters per minute.

b. Use the equation to complete the table below.

Initial Pressure p_1 (mm Hg)	Final Pressure p_2 (mm Hg)	Resistance r (mm Hg/L/min)	Blood Flow Rate F (L/min)
100	0		5
100	0	30	
	5	40	4
90		10	6

Determine whether the given number is a solution of the equation.

48. $x + 6 = 15$; 9

49. $12 + y = 26$; 14

50. $2t - 10 = 4$; 3

51. $3r + 7 = -5$; 2

52. $6 + 4m = 18$; 3

53. $-5 + 2p = -11$; -3

54. $\frac{q}{2} = 20$; 10

55. $\frac{w - 4}{5} = -3$; -11

56. $\frac{g}{3} - 4 = 12$; 48

Make a table of values for each equation if the replacement set is {−2, −1, 0, 1, 2}.

57. $y = 3x + 5$

58. $-2x - 3 = y$

59. $y = \frac{1}{2}x + 2$

60. $4.2x - 1.6 = y$

61. **GEOMETRY** The length of a rectangle is 2 inches greater than the width. The length of the base of an isosceles triangle is 12 inches, and the lengths of the other two sides are 1 inch greater than the width of the rectangle.

a. Draw a picture of each figure and label the dimensions.

b. Write two expressions to find the perimeters of the rectangle and triangle.

c. Find the width of the rectangle if the perimeters of the figures are equal.

62. **CONSTRUCTION** The construction of a building requires 10 tons of steel per story.

a. Define a variable and write an equation for the number of tons of steel required if the building has 15 stories.

b. How many tons of steel are needed?

63. **MULTIPLE REPRESENTATIONS** In this problem, you will further explore writing equations.

a. Concrete Use centimeter cubes to build a tower similar to the one shown at the right.

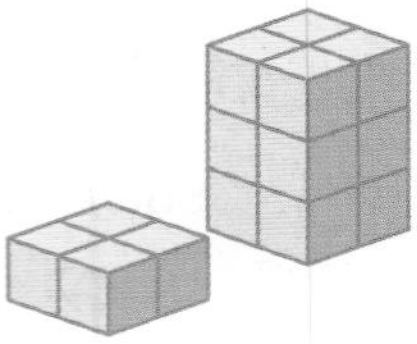

b. Tabular Copy and complete the table shown below. Record the number of layers in the tower and the number of cubes used in the table.

Layers	1	2	3	4	5	6	7
Cubes	?	?	?	?	?	?	?

c. Analytical As the number of layers in the tower increases, how does the number of cubes in the tower change?

d. Algebraic Write a rule that gives the number of cubes in terms of the number of layers in the tower.

H.O.T. Problems Use Higher-Order Thinking Skills

64. **REASONING** Compare and contrast an expression and an equation.

65. **OPEN ENDED** Write an equation that is an identity.

66. **REASONING** Explain why an open sentence always has at least one variable.

67. CCSS **CRITIQUE** Tom and Li-Cheng are solving the equation $x = 4(3 - 2) + 6 \div 8$. Is either of them correct? Explain your reasoning.

Tom

$x = 4(3 - 2) + 6 \div 8$
$= 4(1) + 6 \div 8$
$= 4 + 6 \div 8$
$= 4 + \frac{6}{8}$
$= 4\frac{3}{4}$

Li-Cheng

$x = 4(3 - 2) + 6 \div 8$
$= 4(1) + 6 \div 8$
$= 4 + 6 \div 8$
$= 10 \div 8$
$= \frac{5}{4}$

68. **CHALLENGE** Find all of the solutions of $x^2 + 5 = 30$.

69. **OPEN ENDED** Write an equation that involves two or more operations with a solution of -7.

70. **WRITING IN MATH** Explain how you can determine that an equation has no real numbers as a solution. How can you determine that an equation has all real numbers as solutions?

Standardized Test Practice

71. Which of the following is *not* an equation?

A $y = 6x - 4$

B $\frac{a+4}{2} = \frac{1}{4}$

C $(4 \cdot 3b) + (8 \div 2c)$

D $55 = 6 + d^2$

72. SHORT RESPONSE The expected attendance for the Drama Club production is 65% of the student body. If the student body consists of 300 students, how many students are expected to attend?

73. GEOMETRY A speedboat and a sailboat take off from the same port. The diagram shows their travel. What is the distance between the boats?

F 12 mi

G 15 mi

H 18 mi

J 24 mi

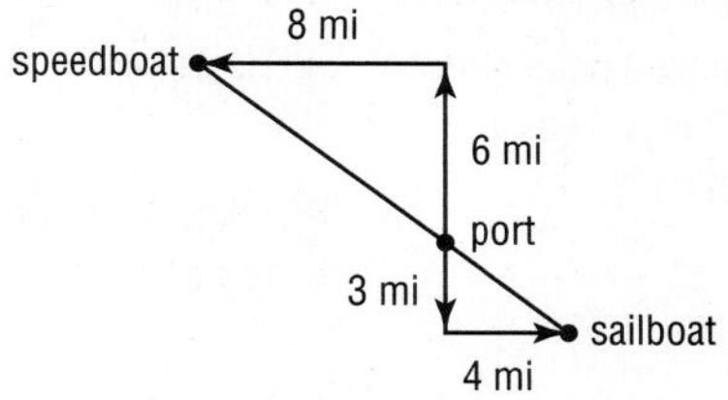

74. Michelle can read 1.5 pages per minute. How many pages can she read in two hours?

A 90 pages
B 150 pages
C 120 pages
D 180 pages

Spiral Review

75. ZOO A zoo has about 500 children and 750 adults visit each day. Write an expression to represent about how many visitors the zoo will have over a month. (Lesson 1-4)

Find the value of p in each equation. Then name the property that is used. (Lesson 1-3)

76. $7.3 + p = 7.3$

77. $12p = 1$

78. $1p = 4$

79. MOVING BOXES The figure shows the dimensions of the boxes Steve uses to pack. How many cubic inches can each box hold? (Lesson 0-9)

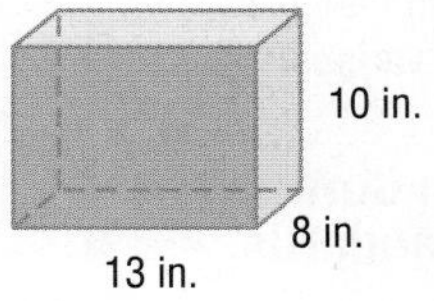

Express each percent as a fraction. (Lesson 0-6)

80. 35%

81. 15%

82. 28%

For each problem, determine whether you need an estimate or an exact answer. Then solve. (Lessons 0-6 and 0-1)

83. TRAVEL The distance from Raleigh, North Carolina, to Philadelphia, Pennsylvania, is approximately 428 miles. The average gas mileage of José's car is 45 miles per gallon. About how many gallons of gas will be needed to make the trip?

84. PART-TIME JOB An employer pays $8.50 per hour. If 20% of pay is withheld for taxes, what are the take-home earnings from 28 hours of work?

Skills Review

Find each sum or difference.

85. $1.14 + 5.6$

86. $4.28 - 2.4$

87. $8 - 6.35$

88. $\frac{4}{5} + \frac{1}{6}$

89. $\frac{2}{7} + \frac{3}{4}$

90. $\frac{6}{8} - \frac{1}{2}$

LESSON 1-6 Relations

Then

- You solved equations with one or two variables.

Now

1. Represent relations.
2. Interpret graphs of relations.

Why?

- The deeper in the ocean you are, the greater pressure is on your body. This is because there is more water over you. The force of gravity pulls the water weight down, creating a greater pressure.

The equation that relates the total pressure of the water to the depth is $P = rgh$, where

$P =$ the pressure,
$r =$ the density of water,
$g =$ the acceleration due to gravity, and
$h =$ the height of water above you.

NewVocabulary
coordinate system
coordinate plane
x- and *y*-axes
origin
ordered pair
x- and *y*-coordinates
relation
mapping
domain
range
independent variable
dependent variable

Common Core State Standards

Content Standards

A.REI.10 Understand that the graph of an equation in two variables is the set of all its solutions plotted in the coordinate plane, often forming a curve (which could be a line).

F.IF.1 Understand that a function from one set (called the domain) to another set (called the range) assigns to each element of the domain exactly one element of the range. If *f* is a function and *x* is an element of its domain, then $f(x)$ denotes the output of *f* corresponding to the input *x*. The graph of *f* is the graph of the equation $y = f(x)$.

Mathematical Practices

1 Make sense of problems and persevere in solving them.

1 Represent a Relation

This relationship between the depth and the pressure exerted can be represented by a line on a coordinate grid.

A **coordinate system** is formed by the intersection of two number lines, the *horizontal axis* and the *vertical axis*.

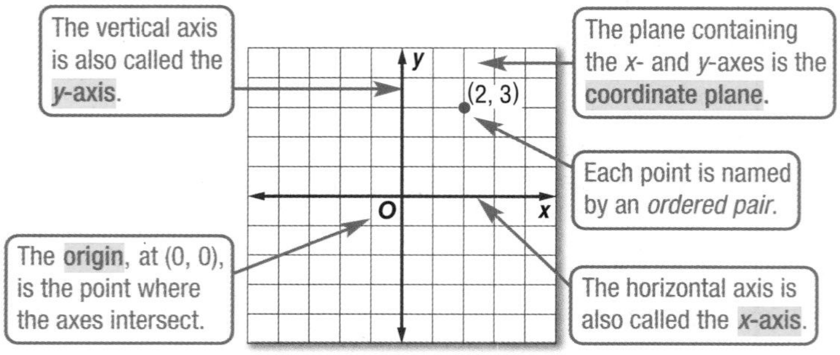

The vertical axis is also called the y-axis.
The plane containing the x- and y-axes is the coordinate plane.
y
(2, 3)
Each point is named by an ordered pair.
O
x
The origin, at (0, 0), is the point where the axes intersect.
The horizontal axis is also called the x-axis.

A point is represented on a graph using ordered pairs.

- An **ordered pair** is a set of numbers, or *coordinates*, written in the form (x, y).
- The x-value, called the **x-coordinate**, represents the horizontal placement of the point.
- The y-value, or **y-coordinate**, represents the vertical placement of the point.

A set of ordered pairs is called a **relation**. A relation can be represented in several different ways: as an equation, in a graph, with a table, or with a mapping.

A **mapping** illustrates how each element of the *domain* is paired with an element in the *range*. The set of the first numbers of the ordered pairs is the **domain**. The set of second numbers of the ordered pairs is the **range** of the relation. This mapping represents the ordered pairs $(-2, 4)$, $(-1, 4)$, $(0, 6)$ $(1, 8)$, and $(2, 8)$.

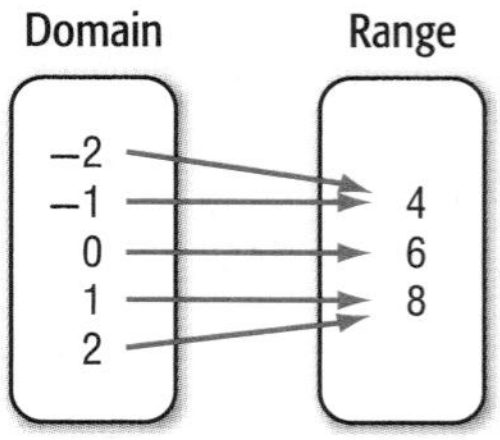

Domain
Range
−2
−1
0
1
2
4
6
8

Jeff Hunter/Photographer's Choice RF/Getty Images

StudyTip

CCSS Sense-Making Each representation of the same relation serves a different purpose. Graphing the points can show the pattern between the points. A mapping shows you at a glance if elements are paired with the same element.

Study the different representations of the same relation below.

Ordered Pairs

(1, 2)
(−2, 4)
(0, −3)

Table

x	y
1	2
−2	4
0	−3

Graph

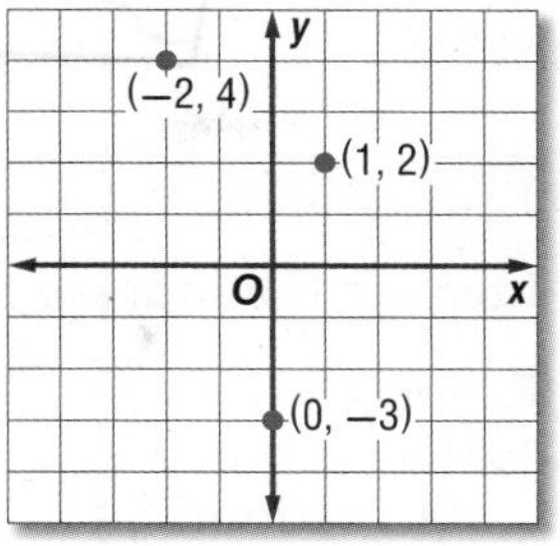

Mapping

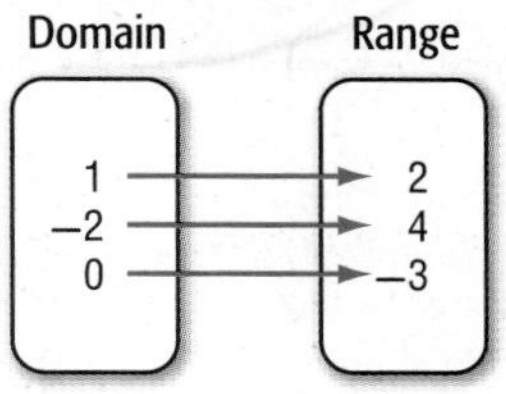

The x-values of a relation are members of the domain and the y-values of a relation are members of the range. In the relation above, the domain is {−2, 1, 0} and the range is {−3, 2, 4}.

Example 1 Representations of a Relation

a. Express {(2, 5), (−2, 3), (5, −2), (−1, −2)} as a table, a graph, and a mapping.

Table
Place the x-coordinates into the first column of the table. Place the corresponding y-coordinates in the second column of the table.

x	y
2	5
−2	3
5	−2
−1	−2

Graph
Graph each ordered pair on a coordinate plane.

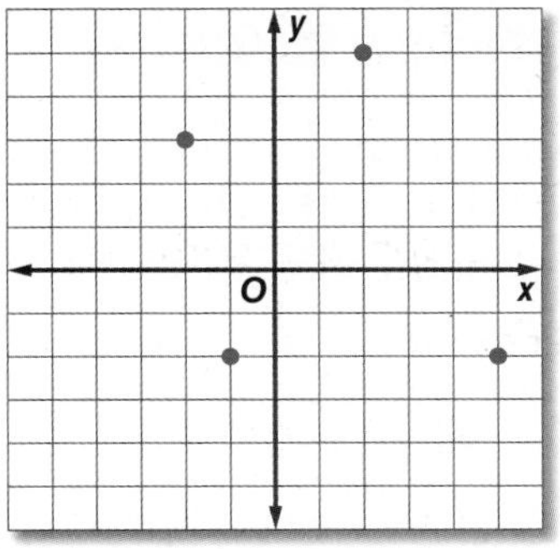

Mapping
List the x-values in the domain and the y-values in the range. Draw arrows from the x-values in the domain to the corresponding y-values in the range.

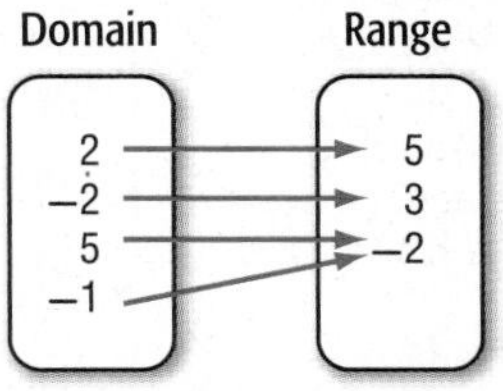

b. Determine the domain and the range of the relation.

The domain of the relation is {2, −2, 5, −1}. The range of the relation is {5, 3, −2}.

GuidedPractice

1A. Express {(4, −3), (3, 2), (−4, 1), (0, −3)} as a table, graph, and mapping.

1B. Determine the domain and range.

In a relation, the value of the variable that determines the output is called the **independent variable**. The variable with a value that is dependent on the value of the independent variable is called the **dependent variable**. The domain contains values of the independent variable. The range contains the values of the dependent variable.

Real-World Example 2 Independent and Dependent Variables

Identify the independent and dependent variables for each relation.

a. DANCE **The dance committee is selling tickets to the Fall Ball. The more tickets that they sell, the greater the amount of money they can spend for decorations.**

The number of tickets sold is the independent variable because it is unaffected by the money spent on decorations. The money spent on decorations is the dependent variable because it depends on the number of tickets sold.

b. MOVIES **Generally, the average price of going to the movies has steadily increased over time.**

Time is the independent variable because it is unaffected by the cost of attending the movies. The price of going to the movies is the dependent variable because it is affected by time.

Real-WorldLink

In 1948, a movie ticket cost \$0.36. In 2008, the average ticket price in the United States was \$7.18.

Source: National Association of Theatre Owners

GuidedPractice

Identify the independent and dependent variables for each relation.

2A. The air pressure inside a tire increases with the temperature.

2B. As the amount of rain decreases, so does the water level of the river.

2 Graphs of a Relation

A relation can be graphed without a scale on either axis. These graphs can be interpreted by analyzing their shape.

Example 3 Analyze Graphs

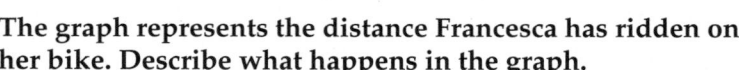

The graph represents the distance Francesca has ridden on her bike. Describe what happens in the graph.

Bike Ride

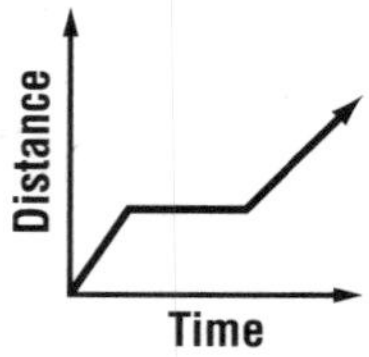

As time increases, the distance increases until the graph becomes a horizontal line.

So, time is increasing but the distance remains constant. At this section Francesca stopped. Then she continued to ride her bike.

GuidedPractice

Describe what is happening in each graph.

3A. Driving to School

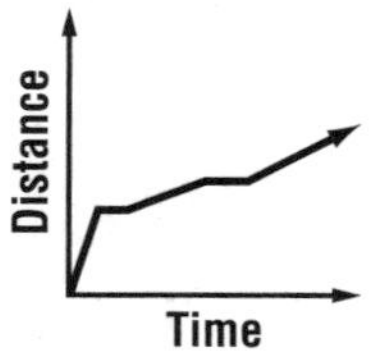

3B. Change in Income

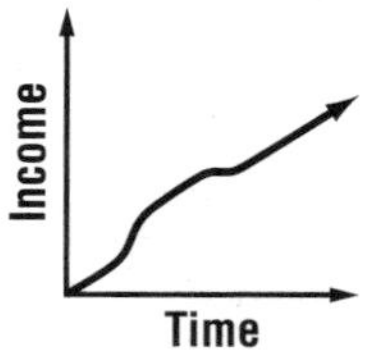

Erik Dreyer/Taxi/Getty Images

Check Your Understanding

Example 1 **Express each relation as a table, a graph, and a mapping. Then determine the domain and range.**

1. $\{(4, 3), (-2, 2), (5, -6)\}$
2. $\{(5, -7), (-1, 4), (0, -5), (-2, 3)\}$

Example 2 **Identify the independent and dependent variables for each relation.**

3. Increasing the temperature of a compound inside a sealed container increases the pressure inside a sealed container.
4. Mike's cell phone is part of a family plan. If he uses more minutes than his share, then there are fewer minutes available for the rest of his family.
5. Julian is buying concert tickets for himself and his friends. The more concert tickets he buys the greater the cost.
6. A store is having a sale over Labor Day weekend. The more purchases, the greater the profits.

Example 3 CCSS **MODELING** **Describe what is happening in each graph.**

7. The graph represents the distance the track team runs during a practice.

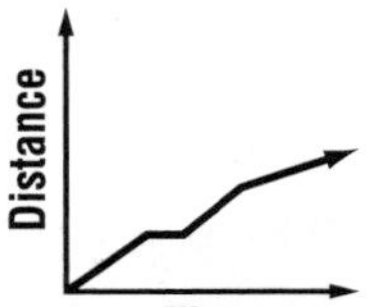

8. The graph represents revenues generated through an online store.

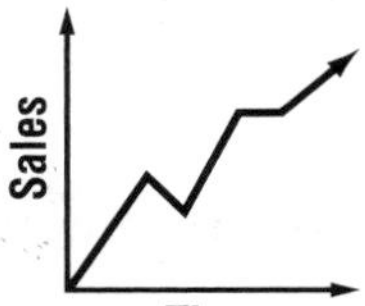

Practice and Problem Solving

Example 1 **Express each relation as a table, a graph, and a mapping. Then determine the domain and range.**

9. $\{(0, 0), (-3, 2), (6, 4), (-1, 1)\}$
10. $\{(5, 2), (5, 6), (3, -2), (0, -2)\}$
11. $\{(6, 1), (4, -3), (3, 2), (-1, -3)\}$
12. $\{(-1, 3), (3, -6), (-1, -8), (-3, -7)\}$
13. $\{(6, 7), (3, -2), (8, 8), (-6, 2), (2, -6)\}$
14. $\{(4, -3), (1, 3), (7, -2), (2, -2), (1, 5)\}$

Example 2 **Identify the independent and dependent variables for each relation.**

15. The Spanish classes are having a fiesta lunch. Each student that attends is to bring a Spanish side dish or dessert. The more students that attend, the more food there will be.
16. The faster you drive your car, the longer it will take to come to a complete stop.

Example 3 CCSS **MODELING** **Describe what is happening in each graph.**

17. The graph represents the height of a bungee jumper.

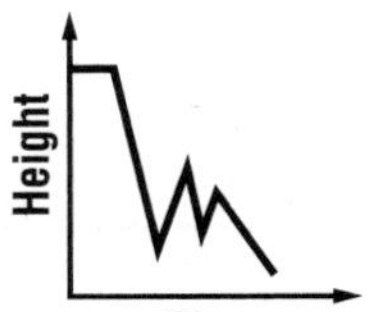

18. The graph represents the sales of lawn mowers.

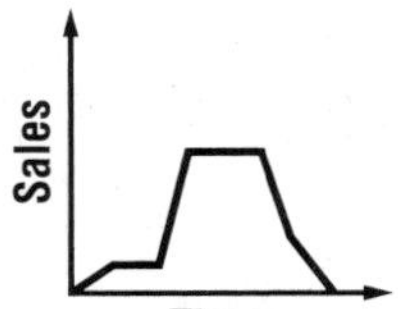

CCSS MODELING Describe what is happening in each graph.

19. The graph represents the value of a rare baseball card.

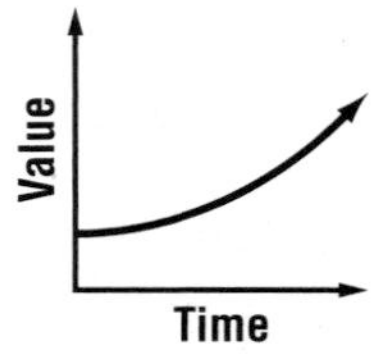

20. The graph represents the distance covered on an extended car ride.

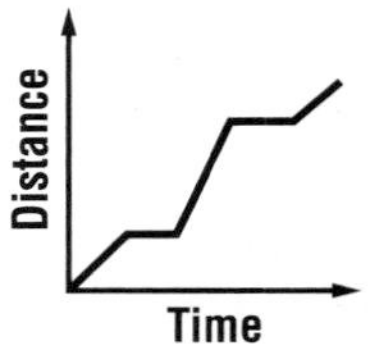

For Exercises 21–23, use the graph at the right.

21. Name the ordered pair at point *A* and explain what it represents.

22. Name the ordered pair at point *B* and explain what it represents.

23. Identify the independent and dependent variables for the relation.

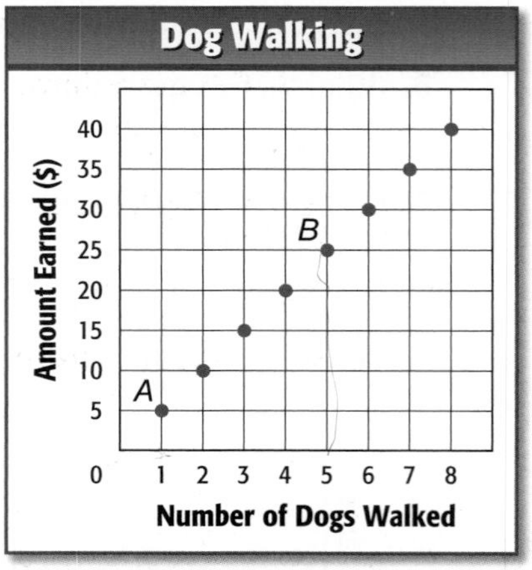

For Exercises 24–26, use the graph at the right.

24. Name the ordered pair at point *C* and explain what it represents.

25. Name the ordered pair at point *D* and explain what it represents.

26. Identify the independent and dependent variables.

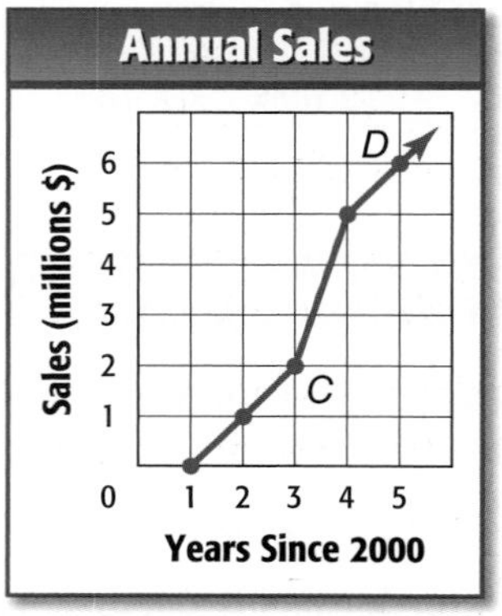

Express each relation as a set of ordered pairs. Describe the domain and range.

27.

Buying Aquarium Fish	
Number of Fish	**Total Cost**
1	$2.50
2	$4.50
5	$10.50
8	$16.50

28.

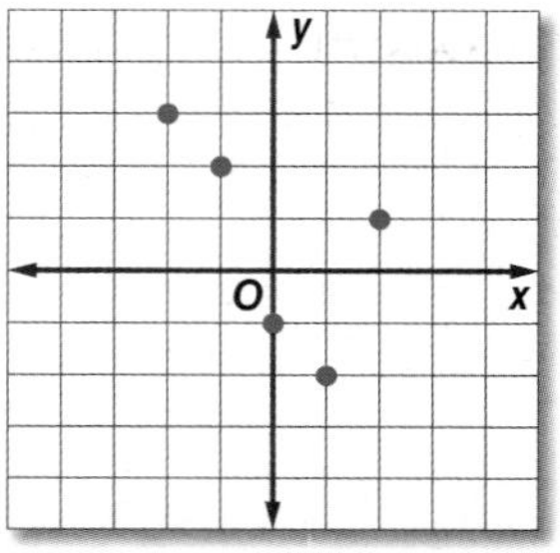

Express the relation in each table, mapping, or graph as a set of ordered pairs.

29.

x	y
4	−1
8	9
−2	−6
7	−3

30.

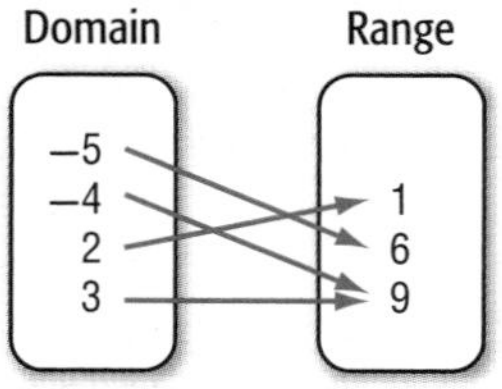

31.

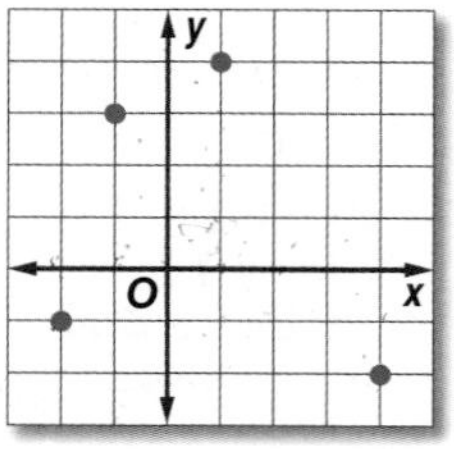

32. **SPORTS** In a triathlon, athletes swim 2.4 miles, bicycle 112 miles, and run 26.2 miles. Their total time includes transition time from one activity to the next. Which graph best represents a participant in a triathlon? Explain.

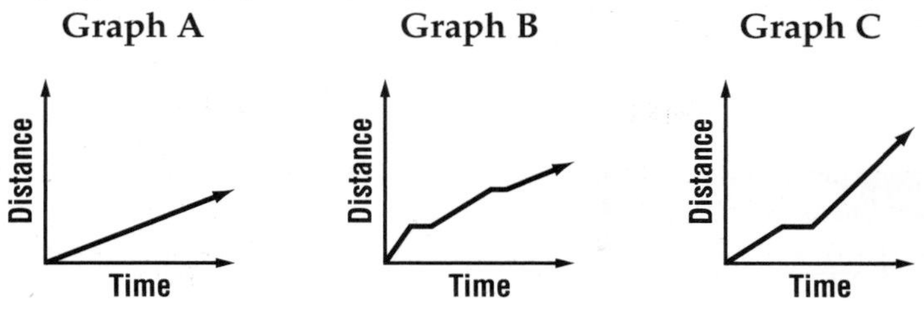

Draw a graph to represent each situation.

33. **ANTIQUES** A grandfather clock that is over 100 years old has increased in value from when it was first purchased.

34. **CAR** A car depreciates in value. The value decreases quickly in the first few years.

35. **REAL ESTATE** A house typically increases in value over time.

36. **EXERCISE** An athlete alternates between running and walking during a workout.

37. **PHYSIOLOGY** A typical adult has about 2 pounds of water for every 3 pounds of body weight. This can be represented by the equation $w = 2\left(\frac{b}{3}\right)$, where w is the weight of water in pounds and b is the body weight in pounds.

a. Make a table to show the relation between body and water weight for people weighing 100, 105, 110, 115, 120, 125, and 130 pounds. Round to the nearest tenth if necessary.

b. What are the independent and dependent variables?

c. State the domain and range, and then graph the relation.

d. Reverse the independent and dependent variables. Graph this relation. Explain what the graph indicates in this circumstance.

H.O.T. Problems Use Higher-Order Thinking Skills

38. **OPEN ENDED** Describe a real-life situation that can be represented using a relation and discuss how one of the quantities in the relation depends on the other. Then represent the relation in three different ways.

39. **CHALLENGE** Describe a real-world situation where it is reasonable to have a negative number included in the domain or range.

40. CCSS **PRECISION** Compare and contrast dependent and independent variables.

41. **CHALLENGE** The table presents a relation. Graph the ordered pairs. Then reverse the y-coordinate and the x-coordinate in each ordered pair. Graph these ordered pairs on the same coordinate plane. Graph the line $y = x$. Describe the relationship between the two sets of ordered pairs.

x	y
0	1
1	3
2	5
3	7

42. **WRITING IN MATH** Use the data about the pressure of water on page 40 to explain the difference between dependent and independent variables.

Standardized Test Practice

43. A school's cafeteria employees surveyed 250 students asking what beverage they drank with lunch. They used the data to create the table below.

Beverage	Number of Students
milk	38
chocolate milk	112
juice	75
water	25

What percent of the students surveyed preferred drinking juice with lunch?

A 25% **C** 35%

B 30% **D** 40%

44. Which of the following is equivalent to $6(3 - g) + 2(11 - g)$?

F $2(20 - g)$ **H** $8(5 - g)$

G $8(14 - g)$ **J** $40 - g$

45. SHORT RESPONSE Grant and Hector want to build a clubhouse at the midpoint between their houses. If Grant's house is at point G and Hector's house is at point H, what will be the coordinates of the clubhouse?

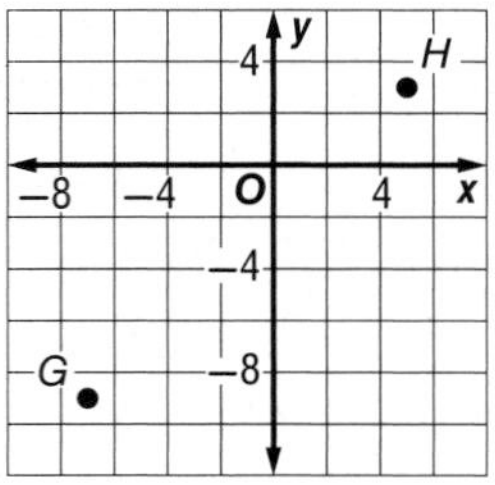

46. If $3b = 2b$, which of the following is true?

A $b = 0$

B $b = \frac{2}{3}$

C $b = 1$

D $b = \frac{3}{2}$

Spiral Review

Solve each equation. (Lesson 1-5)

47. $6(a + 5) = 42$

48. $92 = k + 11$

49. $17 = \frac{45}{w} + 2$

50. HOT-AIR BALLOON A hot-air balloon owner charges $150 for a one-hour ride. If he gave 6 rides on Saturday and 5 rides on Sunday, write and evaluate an expression to describe his total income for the weekend. (Lesson 1-4)

51. LOLLIPOPS A bag of lollipops contains 19 cherry, 13 grape, 8 sour apple, 15 strawberry, and 9 orange flavored lollipops. What is the probability of drawing a sour apple flavored lollipop? (Lesson 0-11)

Find the perimeter of each figure. (Lesson 0-7)

52.

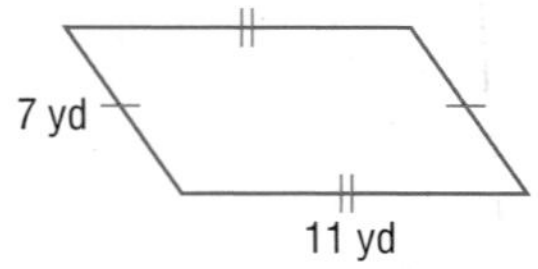

53.

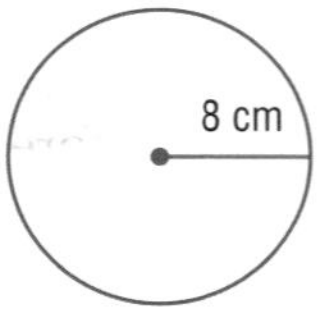

54.

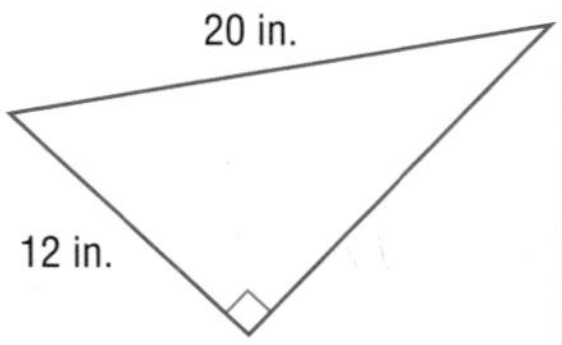

Skills Review

Evaluate each expression.

55. 8^2

56. $(-6)^2$

57. $(2.5)^2$

58. $(-1.8)^2$

59. $(3 + 4)^2$

60. $(1 - 4)^2$

LESSON

1-7 Functions

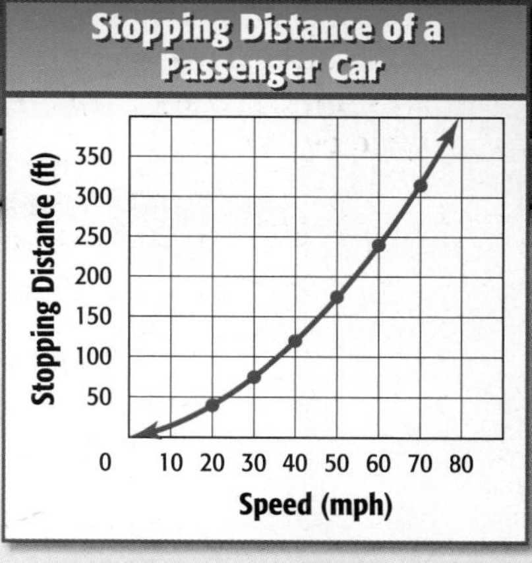

Then

- You solved equations with elements from a replacement set.

Now

1. Determine whether a relation is a function.
2. Find function values.

Why?

- The distance a car travels from when the brakes are applied to the car's complete stop is the stopping distance. This includes time for the driver to react. The faster a car is traveling, the longer the stopping distance. The stopping distance is a function of the speed of the car.

NewVocabulary

function
discrete function
continuous function
vertical line test
function notation
nonlinear function

CCSS Common Core State Standards

Content Standards

F.IF.1 Understand that a function from one set (called the domain) to another set (called the range) assigns to each element of the domain exactly one element of the range. If f is a function and x is an element of its domain, then $f(x)$ denotes the output of f corresponding to the input x. The graph of f is the graph of the equation $y = f(x)$.

F.IF.2 Use function notation, evaluate functions for inputs in their domains, and interpret statements that use function notation in terms of a context.

Mathematical Practices

3 Construct viable arguments and critique the reasoning of others.

1 Identify Functions

A **function** is a relationship between input and output. In a function, there is exactly one output for each input.

KeyConcept Function

Words A function is a relation in which each element of the domain is paired with *exactly* one element of the range.

Examples

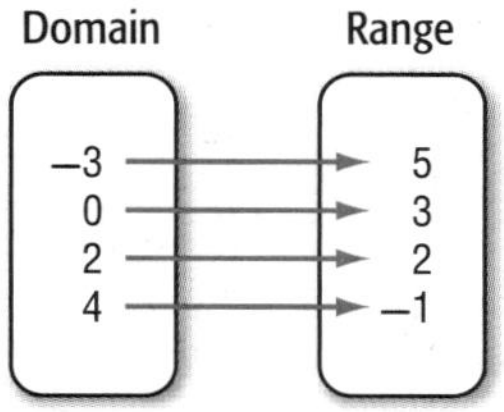

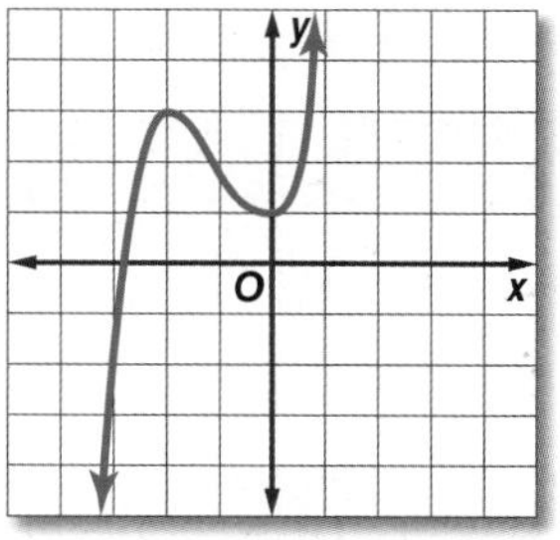

Example 1 Identify Functions

Determine whether each relation is a function. Explain.

a.

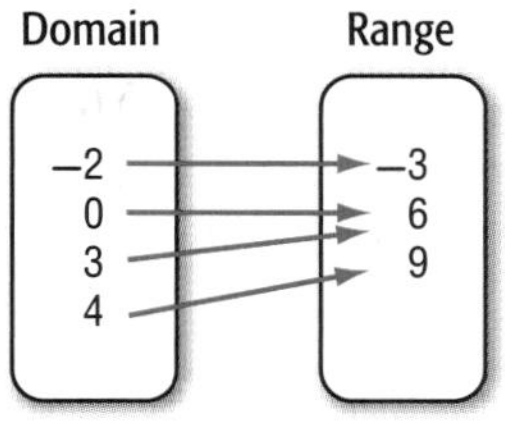

For each member of the domain, there is only one member of the range. So this mapping represents a function. It does not matter if more than one element of the domain is paired with one element of the range.

b.

Domain	1	3	5	1
Range	4	2	4	−4

The element 1 in the domain is paired with both 4 and −4 in the range. So, when x equals 1 there is more than one possible value for y. This relation is not a function.

GuidedPractice

1. $\{(2, 1), (3, -2), (3, 1), (2, -2)\}$

A graph that consists of points that are not connected is a **discrete function**. A function graphed with a line or smooth curve is a **continuous function**.

Real-World Example 2 Draw Graphs

ICE SCULPTING At an ice sculpting competition, each sculpture's height was measured to make sure that it was within the regulated height range of 0 to 6 feet. The measurements were as follows: Team 1, 4 feet; Team 2, 4.5 feet; Team 3, 3.2 feet; Team 4, 5.1 feet; Team 5, 4.8 feet.

a. Make a table of values showing the relation between the ice sculpting team and the height of their sculpture.

Team Number	1	2	3	4	5
Height (ft)	4	4.5	3.2	5.1	4.8

b. Determine the domain and range of the function.

The domain of the function is {1, 2, 3, 4, 5} because this set represents values of the independent variable. It is unaffected by the heights.

The range of the function is {4, 4.5, 3.2, 5.1, 4.8} because this set represents values of the dependent variable. This value depends on the team number.

c. Write the data as a set of ordered pairs. Then graph the data.

Use the table. The team number is the independent variable and the height of the sculpture is the dependent variable. Therefore, the ordered pairs are (1, 4), (2, 4.5), (3, 3.2), (4, 5.1), and (5, 4.8).

Because the team numbers and their corresponding heights cannot be between the points given, the points should not be connected.

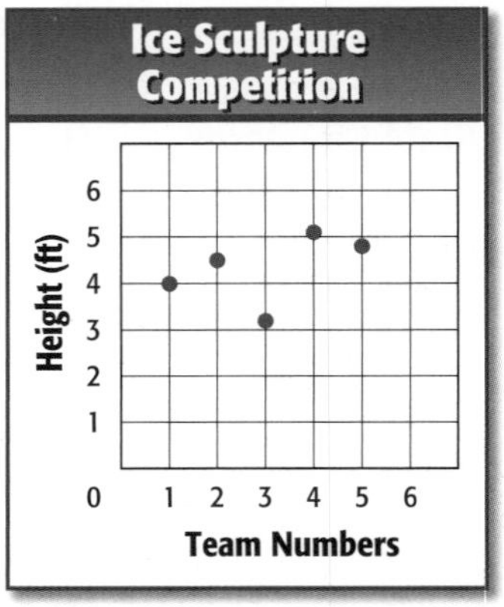

d. State whether the function is *discrete* or *continuous*. Explain your reasoning.

Because the points are not connected, the function is discrete.

Real-WorldLink

The Icehotel, located in the Arctic Circle in Sweden, is a hotel made out of ice. The ice insulates the igloo-like hotel so the temperature is at least −8°C.

Source: Icehotel

GuidedPractice

2. A bird feeder will hold up to 3 quarts of seed. The feeder weighs 2.3 pounds when empty and 13.4 pounds when full.

A. Make a table that shows the bird feeder with 0, 1, 2, and 3 quarts of seed in it weighing 2.3, 6, 9.7, 13.4 pounds respectively.

B. Determine the domain and range of the function.

C. Write the data as a set of ordered pairs. Then graph the data.

D. State whether the function is *discrete* or *continuous*. Explain your reasoning.

> **StudyTip**
> **Vertical Line Test** One way to perform the vertical line test is to use a pencil. Place your pencil vertically on the graph and move from left to right. If the pencil passes over the graph in only one place, then the graph represents a function.

You can use the **vertical line test** to see if a graph represents a function. If a vertical line intersects the graph more than once, then the graph is not a function. Otherwise, the relation is a function.

Function

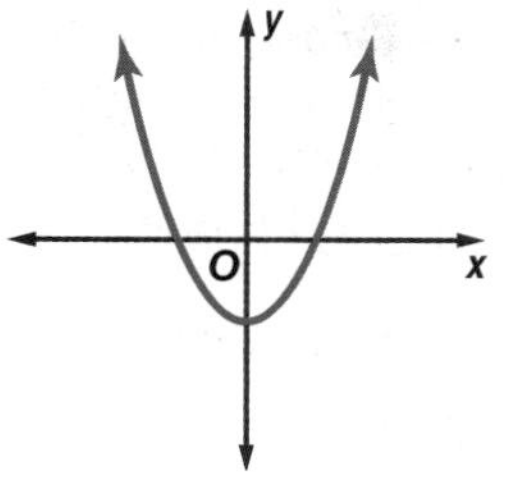

Not a Function

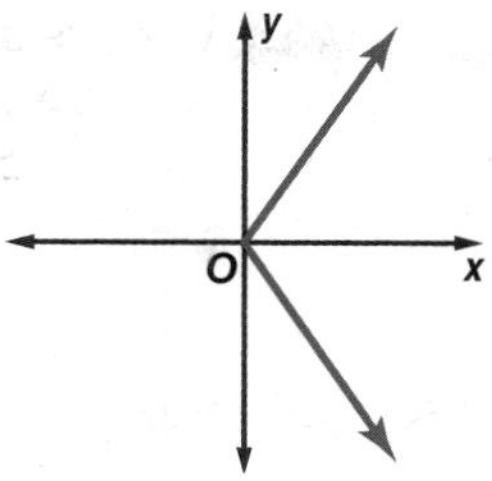

Function

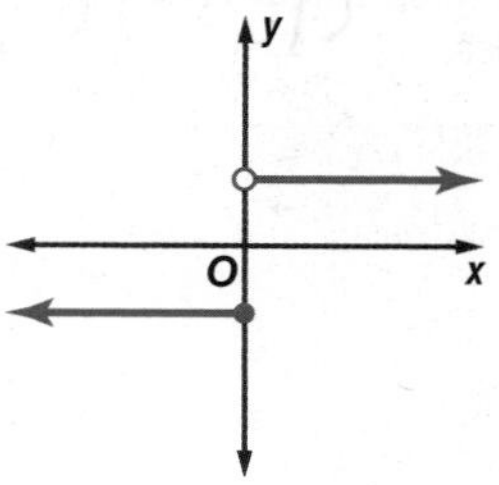

Recall that an equation is a representation of a relation. Equations can also represent functions. Every solution of the equation is represented by a point on a graph. The graph of an equation is the set of all its solutions, which often forms a curve or a line.

Example 3 Equations as Functions

Determine whether $-3x + y = 8$ is a function.

First make a table of values. Then graph the equation.

x	-1	0	1	2
y	5	4.5	11	14

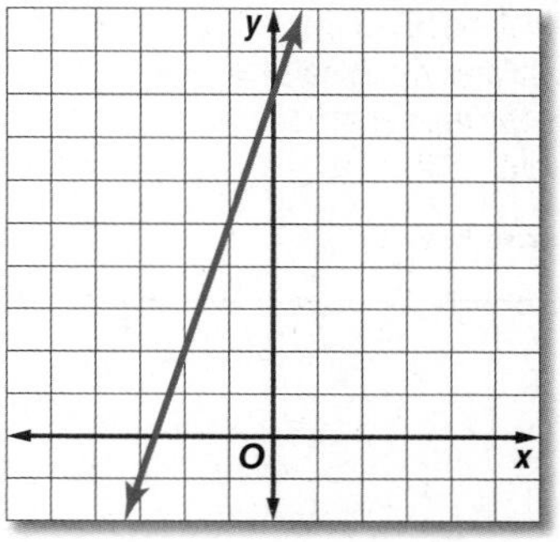

Connect the points with a smooth graph to represent all of the solutions of the equation. The graph is a line. To use the vertical line test, place a pencil at the left of the graph to represent a vertical line. Slowly move the pencil across the graph.

For any value of x, the vertical line passes through no more than one point on the graph. So, the graph and the equation represent a function.

GuidedPractice **Determine whether each relation is a function.**

3A. $4x = 8$

3B. $4x = y + 8$

A function can be represented in different ways.

ConceptSummary Representations of a Function

Table

x	y
-2	1
0	-1
2	1

Mapping

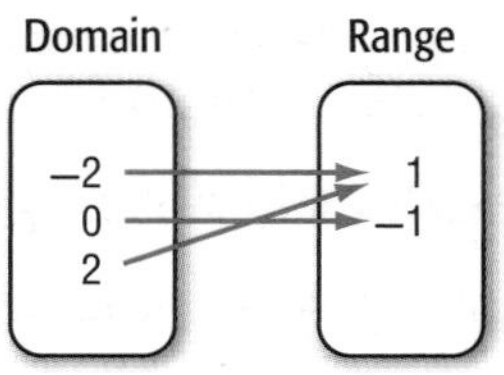

Equation

$$f(x) = \frac{1}{2}x^2 - 1$$

Graph

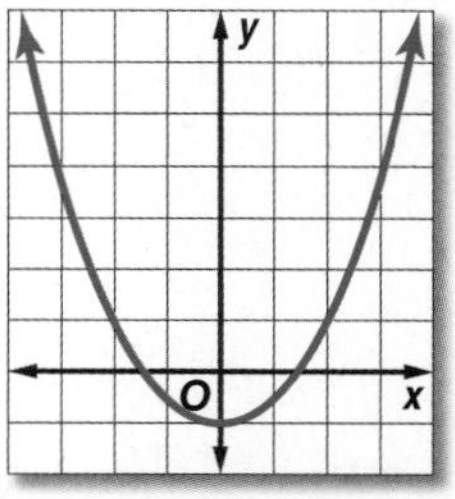

StudyTip

Function Notation Functions are indicated by the symbol $f(x)$. This is read *f of x*. Other letters, such as *g* or *h*, can be used to represent functions.

2 Find Function Values

Equations that are functions can be written in a form called **function notation**. For example, consider $y = 3x - 8$.

Equation	Function Notation
$y = 3x - 8$	$f(x) = 3x - 8$

In a function, x represents the elements of the domain, and $f(x)$ represents the elements of the range. The graph of $f(x)$ is the graph of the equation $y = f(x)$. Suppose you want to find the value in the range that corresponds to the element 5 in the domain. This is written $f(5)$ and is read *f of 5*. The value $f(5)$ is found by substituting 5 for x in the equation.

PT

Example 4 Function Values

For $f(x) = -4x + 7$, find each value.

a. $f(2)$

$f(2) = -4(2) + 7$ — $x = 2$

$= -8 + 7$ — Multiply.

$= -1$ — Add.

b. $f(-3) + 1$

$f(-3) + 1 = [-4(-3) + 7] + 1$ — $x = -3$

$= 19 + 1$ — Simplify.

$= 20$ — Add.

GuidedPractice

For $f(x) = 2x - 3$, find each value.

4A. $f(1)$

4B. $6 - f(5)$

4C. $f(-2)$

4D. $f(-1) + f(2)$

A function with a graph that is not a straight line is a **nonlinear function**.

PT

Example 5 Nonlinear Function Values

If $h(t) = -16t^2 + 68t + 2$, find each value.

a. $h(4)$

$h(4) = -16(4)^2 + 68(4) + 2$ — Replace t with 4.

$= -256 + 272 + 2$ — Multiply.

$= 18$ — Add.

b. $2[h(g)]$

$2[h(g)] = 2[-16(g)^2 + 68(g) + 2]$ — Replace t with g.

$= 2(-16g^2 + 68g + 2)$ — Simplify.

$= -32g^2 + 136g + 4$ — Distributive Property

GuidedPractice

If $f(t) = 2t^3$, find each value.

5A. $f(4)$

5B. $3[f(t)] + 2$

5C. $f(-5)$

5D. $f(-3) - f(1)$

Check Your Understanding

Examples 1, 3 **Determine whether each relation is a function. Explain.**

1.

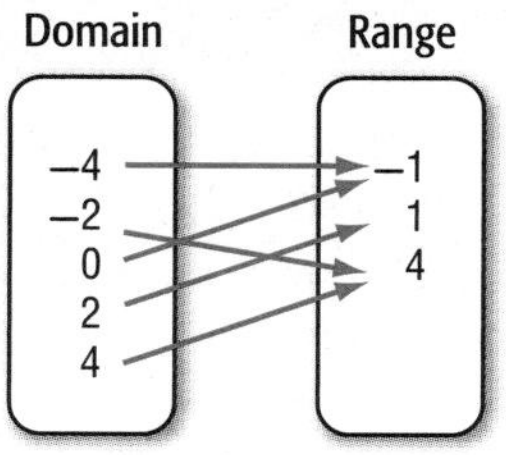

2.

Domain	Range
2	6
5	7
6	9
6	10

3. {(2, 2), (−1, 5), (5, 2), (2, −4)}

4. $y = \frac{1}{2}x - 6$

5.

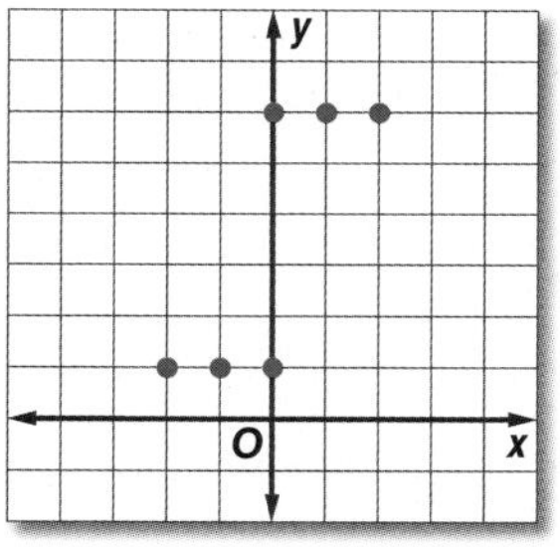

6.

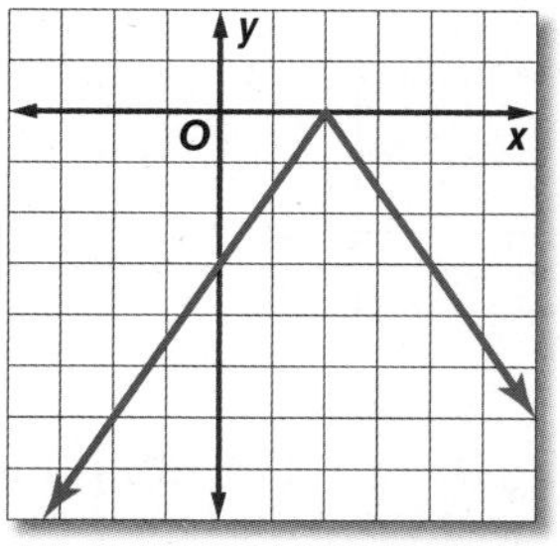

7.

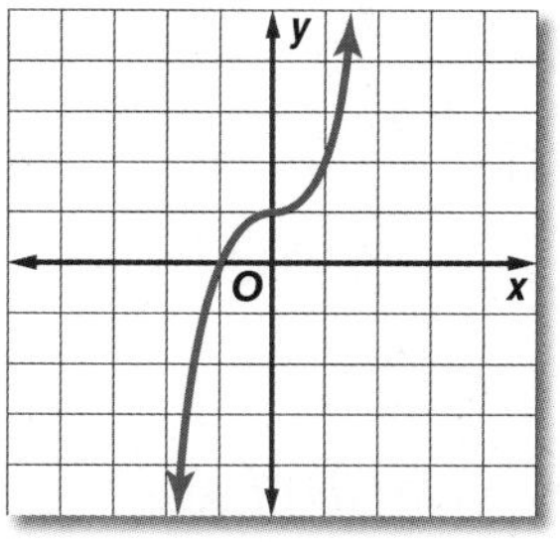

8.

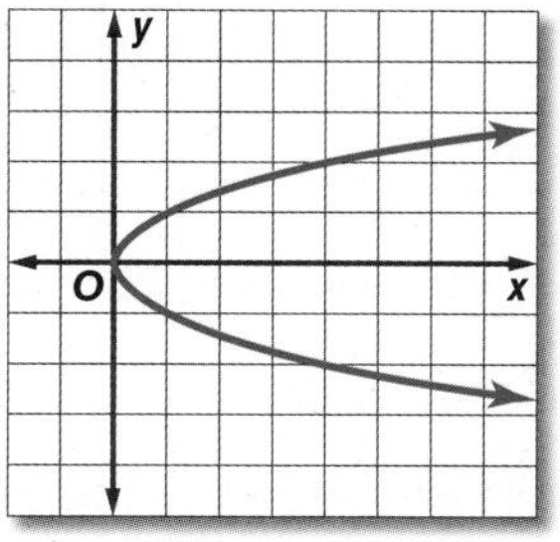

Example 2

9. SCHOOL ENROLLMENT The table shows the total enrollment in U.S. public schools.

School Year	2004–05	2005–06	2006–07	2007–08
Enrollment (in thousands)	48,560	48,710	48,948	49,091

Source: *The World Almanac*

a. Write a set of ordered pairs representing the data in the table if x is the number of school years since 2004–2005.

b. Draw a graph showing the relationship between the year and enrollment.

c. Describe the domain and range of the data.

10. CCSS **REASONING** The cost of sending cell phone pictures is given by $y = 0.25x$, where x is the number of pictures that you send and y is the cost in dollars.

a. Write the equation in function notation. Interpret the function in terms of the context.

b. Find $f(5)$ and $f(12)$. What do these values represent?

c. Determine the domain and range of this function.

Examples 4–5 **If $f(x) = 6x + 7$ and $g(x) = x^2 - 4$, find each value.**

11. $f(-3)$

12. $f(m)$

13. $f(r - 2)$

14. $g(5)$

15. $g(a) + 9$

16. $g(-4t)$

17. $f(q + 1)$

18. $f(2) + g(2)$

19. $g(-b)$

Practice and Problem Solving

Example 1 **Determine whether each relation is a function. Explain.**

20. 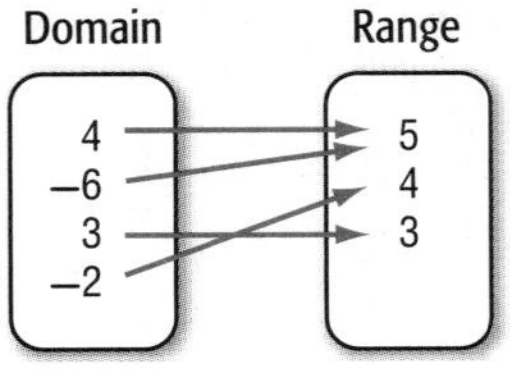

21. 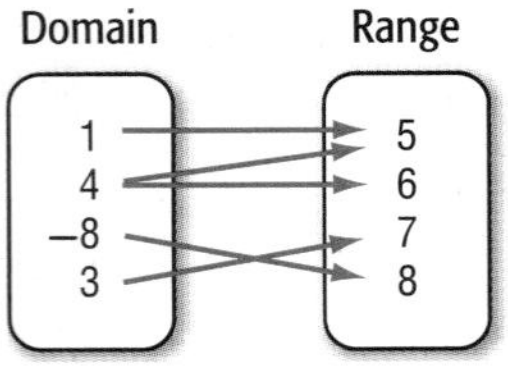

22.

Domain	Range
4	6
−5	3
6	−3
−5	5

23.

Domain	Range
−4	2
3	−5
4	2
9	−7
−3	−5

24. 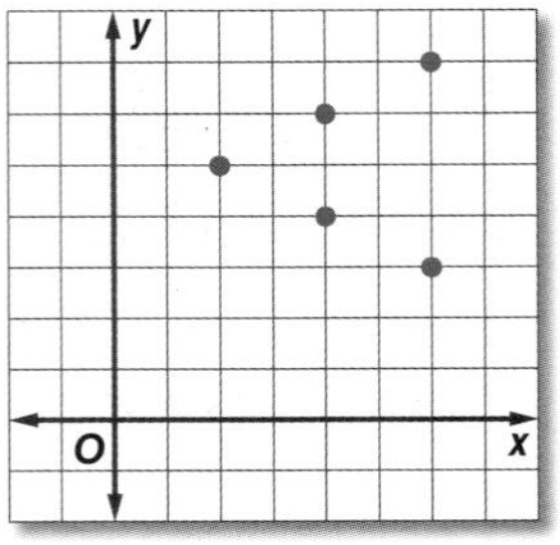

25. 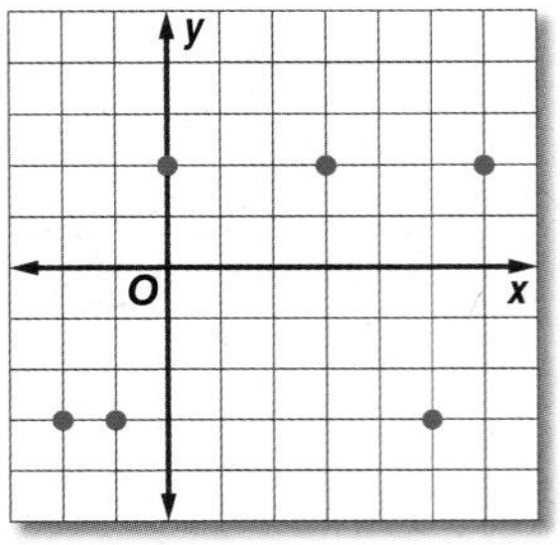

Example 2 **26.** CCSS **SENSE-MAKING** The table shows the median home prices in the United States, from 2007 to 2009.

Year	Median Home Price ($)
2007	234,300
2008	213,200
2009	212,200

a. Write a set of ordered pairs representing the data in the table.

b. Draw a graph showing the relationship between the year and price.

c. What is the domain and range for this data?

Example 3 **Determine whether each relation is a function.**

27. $\{(5, -7), (6, -7), (-8, -1), (0, -1)\}$

28. $\{(4, 5), (3, -2), (-2, 5), (4, 7)\}$

29. $y = -8$

30. $x = 15$

31. $y = 3x - 2$

32. $y = 3x + 2y$

Examples 4–5 **If $f(x) = -2x - 3$ and $g(x) = x^2 + 5x$, find each value.**

33. $f(-1)$

34. $f(6)$

35. $g(2)$

36. $g(-3)$

37. $g(-2) + 2$

38. $f(0) - 7$

39. $f(4y)$

40. $g(-6m)$

41. $f(c - 5)$

42. $f(r + 2)$

43. $5[f(d)]$

44. $3[g(n)]$

45 **EDUCATION** The average national math test scores $f(t)$ for 17-year-olds can be represented as a function of the national science scores t by $f(t) = 0.8t + 72$.

a. Graph this function. Interpret the function in terms of the context.

b. What is the science score that corresponds to a math score of 308?

c. What is the domain and range of this function?

Determine whether each relation is a function.

46. 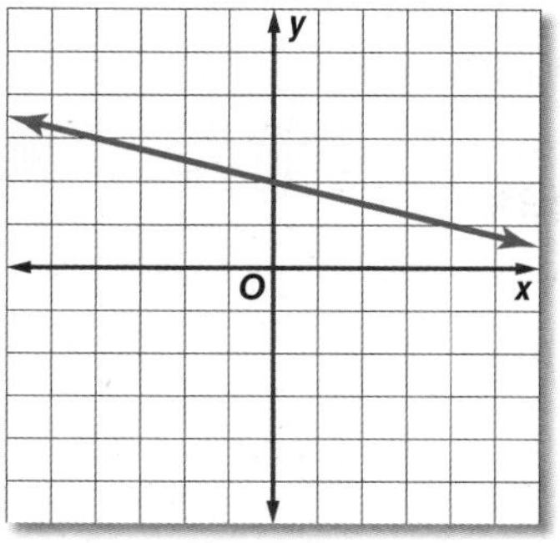

47. 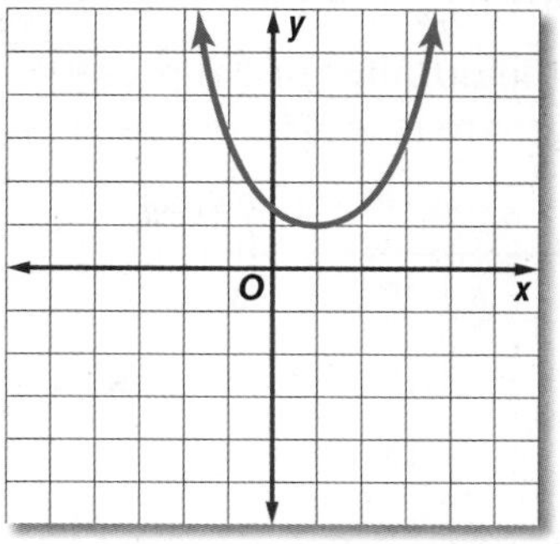

48. BABYSITTING Christina earns $7.50 an hour babysitting.

a. Write an algebraic expression to represent the money Christina will earn if she works h hours.

b. Choose five values for the number of hours Christina can babysit. Create a table with h and the amount of money she will make during that time.

c. Use the values in your table to create a graph.

d. Does it make sense to connect the points in your graph with a line? Why or why not?

H.O.T. Problems Use Higher-Order Thinking Skills

49. OPEN ENDED Write a set of three ordered pairs that represent a function. Choose another display that represents this function.

50. REASONING The set of ordered pairs $\{(0, 1), (3, 2), (3, -5), (5, 4)\}$ represents a relation between x and y. Graph the set of ordered pairs. Determine whether the relation is a function. Explain.

51. CHALLENGE Consider $f(x) = -4.3x - 2$. Write $f(g + 3.5)$ and simplify by combining like terms.

52. WRITE A QUESTION A classmate graphed a set of ordered pairs and used the vertical line test to determine whether it was a function. Write a question to help her decide if the same strategy can be applied to a mapping.

53. CCSS PERSEVERANCE If $f(3b - 1) = 9b - 1$, find one possible expression for $f(x)$.

54. ERROR ANALYSIS Corazon thinks $f(x)$ and $g(x)$ are representations of the same function. Maggie disagrees. Who is correct? Explain your reasoning.

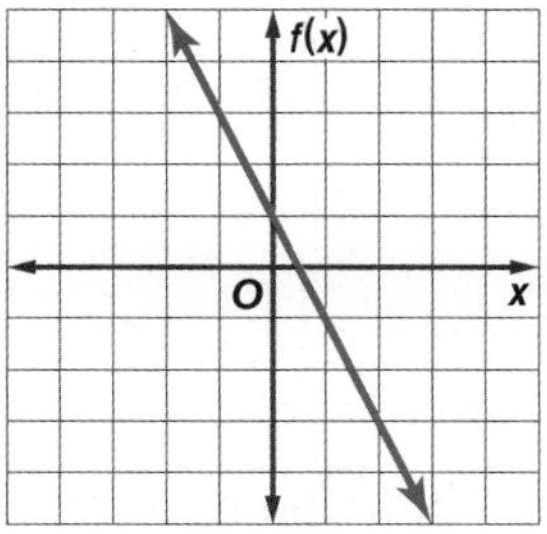

x	$g(x)$
−1	1
0	−1
1	−3
2	−5
3	−7

55. WRITING IN MATH How can you determine whether a relation represents a function?

Standardized Test Practice

56. Which point on the number line represents a number whose square is less than itself?

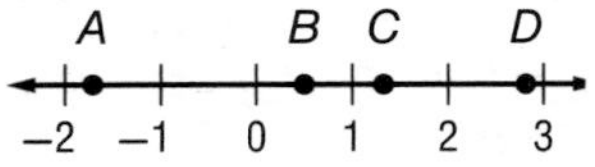

A *A* **C** *C*
B *B* **D** *D*

57. Determine which of the following relations is a function.

F {(–3, 2), (4, 1), (–3, 5)}
G {(2, –1), (4, –1), (2, 6)}
H {(–3, –4), (–3, 6), (8, –2)}
J {(5, –1), (3, –2), (–2, –2)}

58. GEOMETRY What is the value of x?

A 3 in.
B 4 in.
C 5 in.
D 6 in.

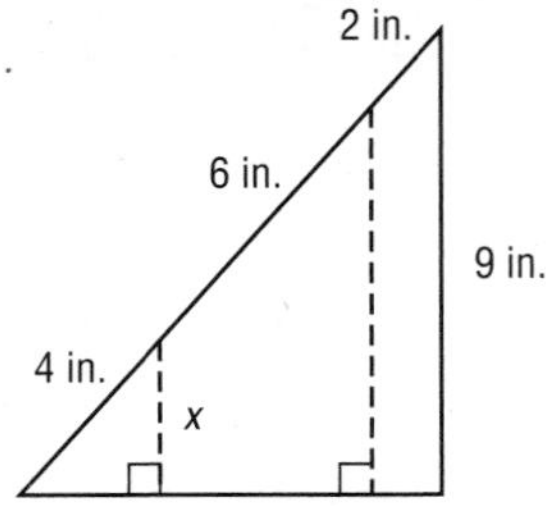

59. SHORT RESPONSE Camille made 16 out of 19 of her serves during her first volleyball game. She made 13 out of 16 of her serves during her second game. During which game did she make a greater percent of her serves?

Spiral Review

Solve each equation. (Lesson 1-5)

60. $x = \frac{27 + 3}{10}$

61. $m = \frac{3^2 + 4}{7 - 5}$

62. $z = 32 + 4(-3)$

63. SCHOOL SUPPLIES The table shows the prices of some items Tom needs. If he needs 4 glue sticks, 10 pencils, and 4 notebooks, write and evaluate an expression to determine Tom's cost. (Lesson 1-4)

School Supplies Prices	
glue stick	\$1.99
pencil	\$0.25
notebook	\$1.85

Write a verbal expression for each algebraic expression. (Lesson 1-1)

64. $4y + 2$

65. $\frac{2}{3}x$

66. $a^2b + 5$

Find the volume of each rectangular prism. (Lesson 0-9)

67.

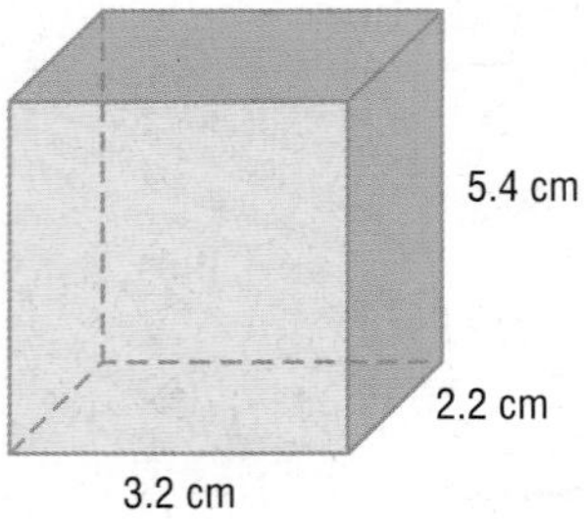

68.

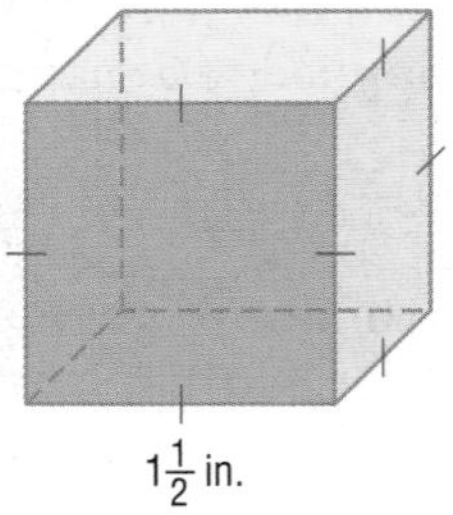

69.

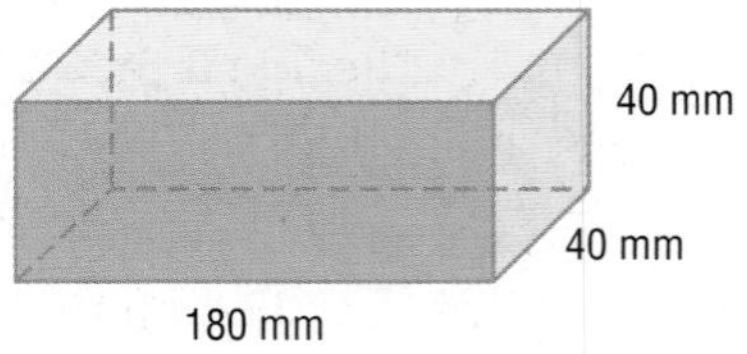

Skills Review

Evaluate each expression.

70. If $x = 3$, then $6x - 5 =$ ___?___.

71. If $n = -1$, then $2n + 1 =$ ___?___.

72. If $p = 4$, then $3p + 4 =$ ___?___.

73. If $q = 7$, then $7q - 9 =$ ___?___

74. If $k = -11$, then $4k + 6 =$ ___?___

75. If $y = 10$, then $8y - 15 =$ ___?___

Graphing Technology Lab
Representing Functions

You can use TI-Nspire Technology to explore the different ways to represent a function.

CCSS Common Core State Standards
Content Standards
A.CED.2 Create equations in two or more variables to represent relationships between quantities; graph equations on coordinate axes with labels and scales.
Mathematical Practices
5 Use appropriate tools strategically.

Activity

Graph $f(x) = 2x + 3$ on the TI-Nspire graphing calculator.

Step 1 Add a new **Graphs** page.

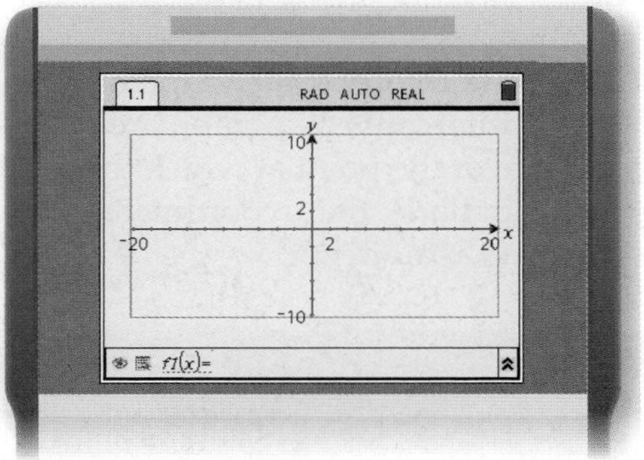

Step 2 Enter $2x + 3$ in the entry line.

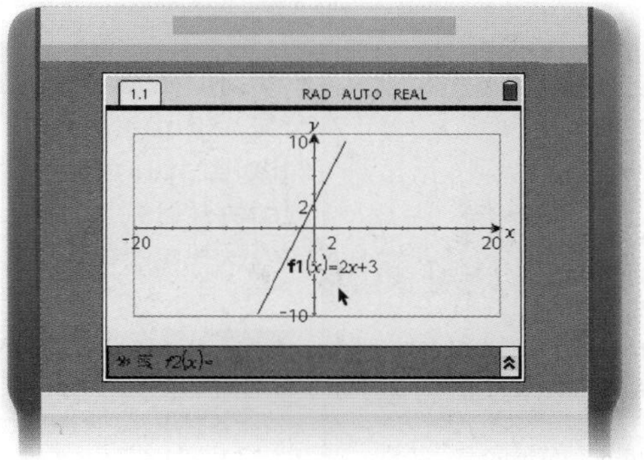

Represent the function as a table.

Step 3 Select the **Show Table** option from the **View** menu to add a table of values on the same display.

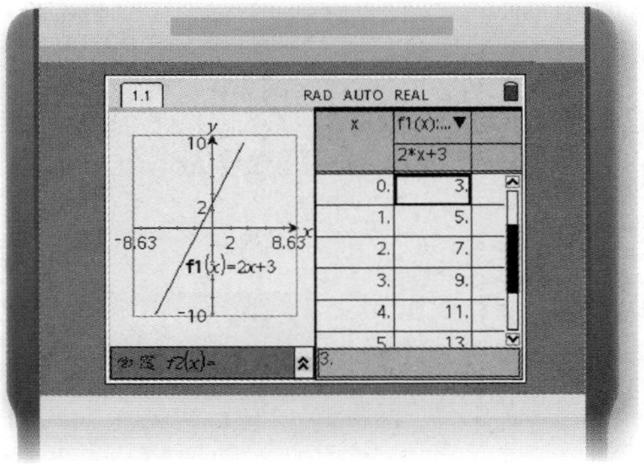

Step 4 Press **ctrl** and **tab** to toggle from the table to the graph. On the graph side, select the line and move it. Notice how the values in the table change.

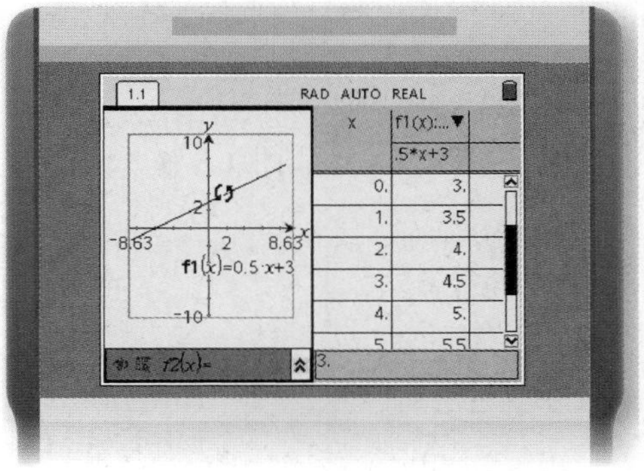

Analyze the Results

CCSS TOOLS Graph each function. Make a table of five ordered pairs that also represents the function.

1. $g(x) = -x - 3$

2. $h(x) = \frac{1}{3}x + 3$

3. $f(x) = -\frac{1}{2}x - 5$

4. $f(x) = 3x - \frac{1}{2}$

5. $g(x) = -2x + 5$

6. $h(x) = \frac{1}{5}x + 4$

LESSON 1-8 Interpreting Graphs of Functions

Then	Now	Why?
You identified functions and found function values.	**1** Interpret intercepts, and symmetry of graphs of functions. **2** Interpret positive, negative, increasing, and decreasing behavior, extrema, and end behavior of graphs of functions.	Sales of video games, including hardware, software, and accessories, have increased at times and decreased at other times over the years. Annual retail video game sales in the U.S. from 2000 to 2009 can be modeled by the graph of a nonlinear function.

NewVocabulary
- intercept
- *x*-intercept
- *y*-intercept
- line symmetry
- positive
- negative
- increasing
- decreasing
- extrema
- relative maximum
- relative minimum
- end behavior

Common Core State Standards

Content Standards

F.IF.4 For a function that models a relationship between two quantities, interpret key features of graphs and tables in terms of the quantities, and sketch graphs showing key features given a verbal description of the relationship.

Mathematical Practices

1 Make sense of problems and persevere in solving them.

1 Interpret Intercepts and Symmetry

To interpret the graph of a function, estimate and interpret key features. The **intercepts** of a graph are points where the graph intersects an axis. The y-coordinate of the point at which the graph intersects the y-axis is called a **y-intercept**. Similarly, the x-coordinate of the point at which a graph intersects the x-axis is called an **x-intercept**.

Real-World Example 1 Interpret Intercepts

PT

PHYSICS The graph shows the height y of an object as a function of time x. Identify the function as *linear* or *nonlinear*. Then estimate and interpret the intercepts.

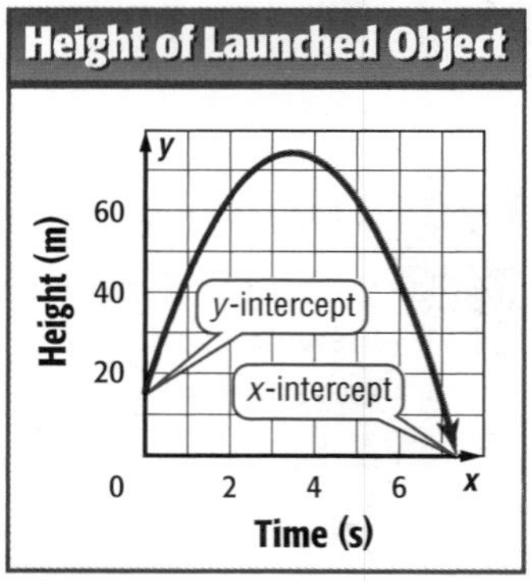

Linear or Nonlinear: Since the graph is a curve and not a line, the graph is nonlinear.

y-Intercept: The graph intersects the y-axis at about (0, 15), so the y-intercept of the graph is about 15. This means that the object started at an initial height of about 15 meters above the ground.

x-Intercept(s): The graph intersects the x-axis at about (7.4, 0), so the x-intercept is about 7.4. This means that the object struck the ground after about 7.4 seconds.

GuidedPractice

1. The graph shows the temperature y of a medical sample thawed at a controlled rate. Identify the function as *linear* or *nonlinear*. Then estimate and interpret the intercepts.

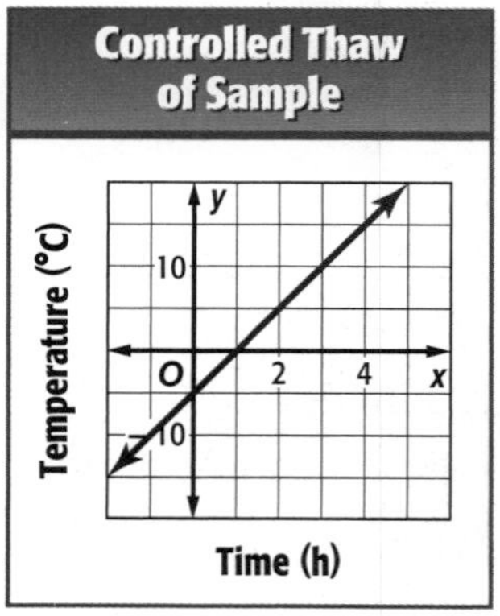

Kiyoshi Ota/Bloomberg via Getty Images

The graphs of some functions exhibit another key feature: symmetry. A graph possesses **line symmetry** in the y-axis or some other vertical line if each half of the graph on either side of the line matches exactly.

PT

Real-World Example 2 Interpret Symmetry

PHYSICS An object is launched. The graph shows the height y of the object as a function of time x. Describe and interpret any symmetry.

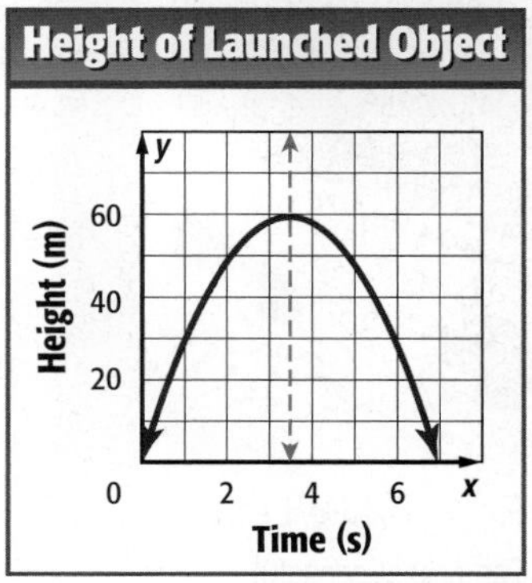

The right half of the graph is the mirror image of the left half in approximately the line $x = 3.5$ between approximately $x = 0$ and $x = 7$.

In the context of the situation, the symmetry of the graph tells you that the time it took the object to go up is equal to the time it took to come down.

> **StudyTip**
> **Symmetry** The graphs of most real-world functions do not exhibit symmetry over the entire domain. However, many have symmetry over smaller portions of the domain that are worth analyzing.

GuidedPractice

2. Describe and interpret any symmetry exhibited by the graph in Guided Practice 1.

2 Interpret Extrema and End Behavior

Interpreting a graph also involves estimating and interpreting where the function is increasing, decreasing, positive, or negative, and where the function has any extreme values, either high or low.

KeyConcepts Positive, Negative, Increasing, Decreasing, Extrema, and End Behavior

A function is **positive** where its graph lies *above* the x-axis, and **negative** where its graph lies *below* the x-axis.

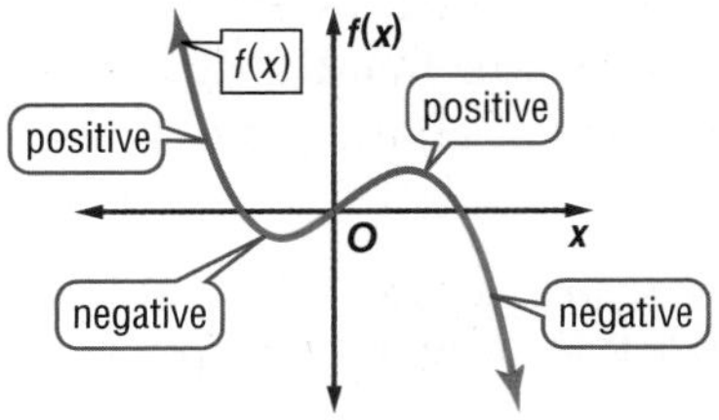

A function is **increasing** where the graph goes *up* and **decreasing** where the graph goes *down* when viewed from left to right.

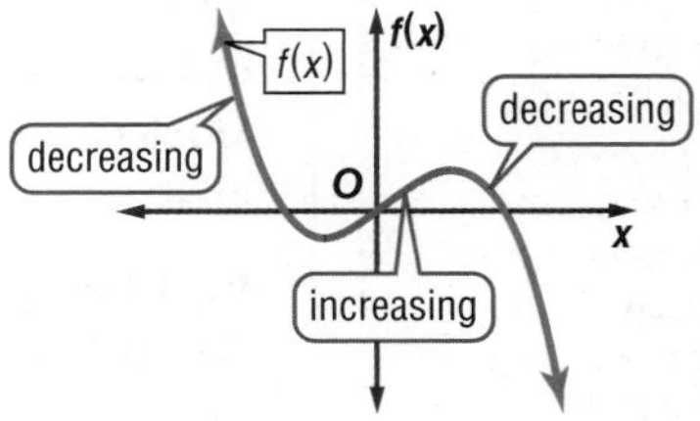

The points shown are the locations of relatively high or low function values called **extrema**. Point A is a **relative minimum**, since no other nearby points have a lesser y-coordinate. Point B is a **relative maximum**, since no other nearby points have a greater y-coordinate.

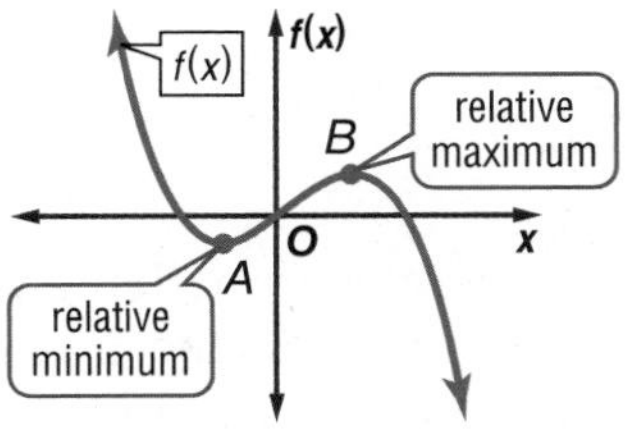

End behavior describes the values of a function at the positive and negative extremes in its domain.

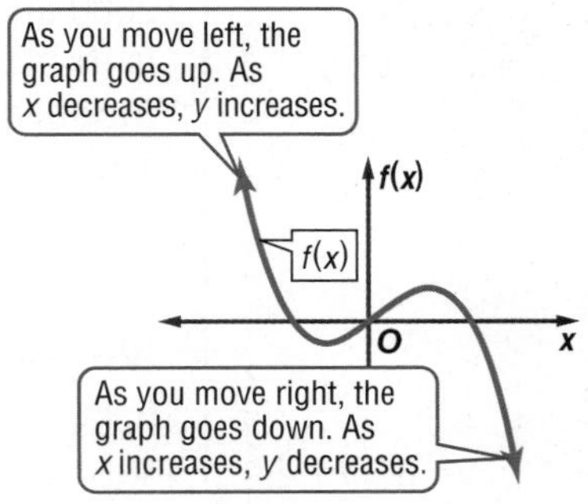

> **StudyTip**
> **End Behavior** The end behavior of some graphs can be described as approaching a specific y-value. In this case, a portion of the graph looks like a horizontal line.

Real-WorldLink

The first successful commercially sold portable video game system was released in 1989 and sold for $120.

Source: *PCWorld*

StudyTip

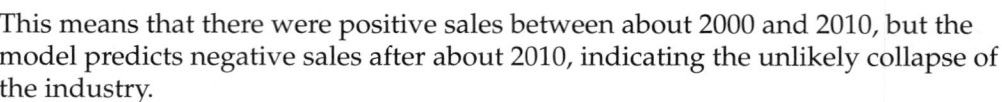

Constant A function is *constant* where the graph does not go up or down as the graph is viewed from left to right.

Real-World Example 3 Interpret Extrema and End Behavior

VIDEO GAMES U.S. retail sales of video games from 2000 to 2009 can be modeled by the function graphed at the right. Estimate and interpret where the function is positive, negative, increasing, and decreasing, the x-coordinates of any relative extrema, and the end behavior of the graph.

Positive: between about $x = -0.6$ and $x = 10.4$

Negative: for about $x < -0.6$ and $x > 10.4$

This means that there were positive sales between about 2000 and 2010, but the model predicts negative sales after about 2010, indicating the unlikely collapse of the industry.

Increasing: for about $x < 1.5$ and between about $x = 3$ and $x = 8$

Decreasing: between about $x = 2$ and $x = 3$ and for about $x > 8$

This means that sales increased from about 2000 to 2002, decreased from 2002 to 2003, increased from 2003 to 2008, and have been decreasing since 2008.

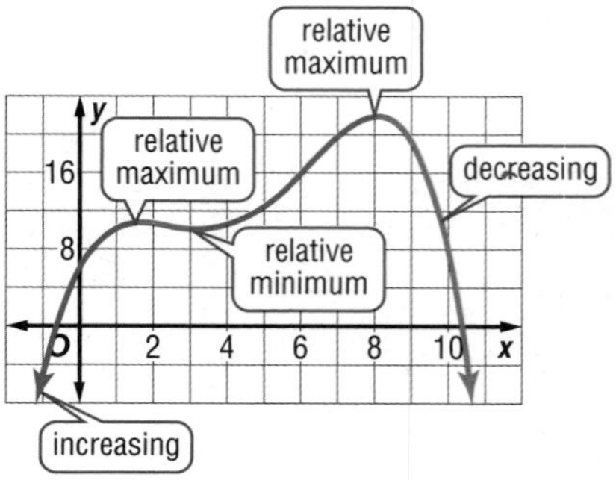

Relative Maximums: at about $x = 1.5$ and $x = 8$

Relative Minimum: at about $x = 3$

The extrema of the graph indicate that the industry experienced two relative peaks in sales during this period: one around 2002 of approximately $10.5 billion and another around 2008 of approximately $22 billion. A relative low of $10 billion in sales came in about 2003.

End Behavior:
As x increases or decreases, the value of y decreases.

The end behavior of the graph indicates negative sales several years prior to 2000 and several years after 2009, which is unlikely. This graph appears to only model sales well between 2000 and 2009 and can only be used to predict sales in 2010.

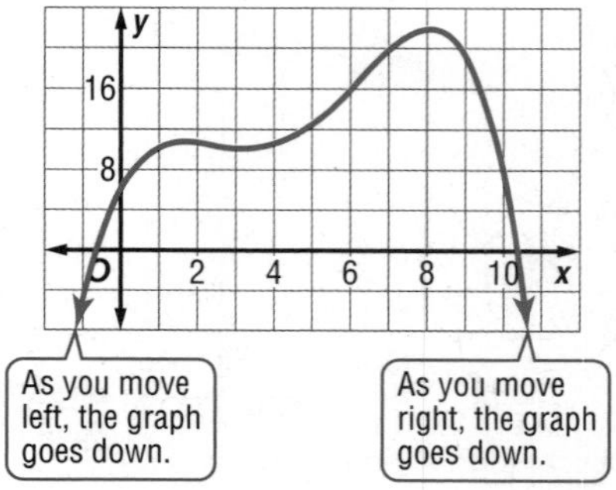

GuidedPractice

3. Estimate and interpret where the function graphed in Guided Practice 1 is positive, negative, increasing, or decreasing, the x-coordinate of any relative extrema, and the end behavior of the graph.

Check Your Understanding

Examples 1–3 **CCSS SENSE-MAKING** Identify the function graphed as *linear* or *nonlinear*. Then estimate and interpret the intercepts of the graph, any symmetry, where the function is positive, negative, increasing, and decreasing, the x-coordinate of any relative extrema, and the end behavior of the graph.

1.
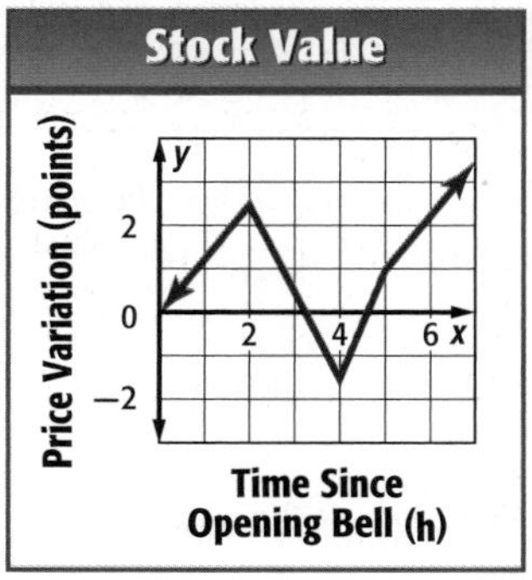

2.
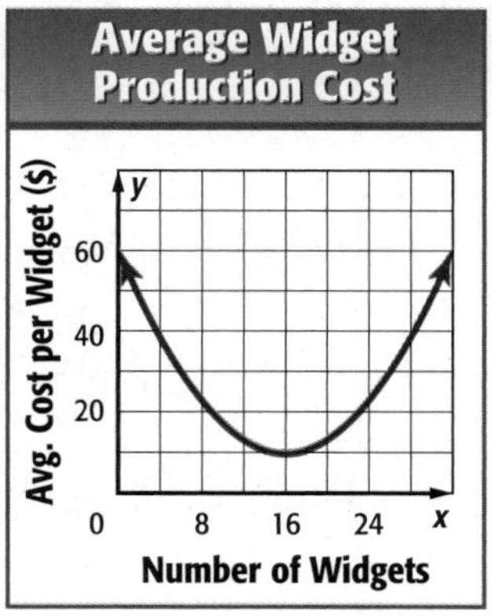

3.
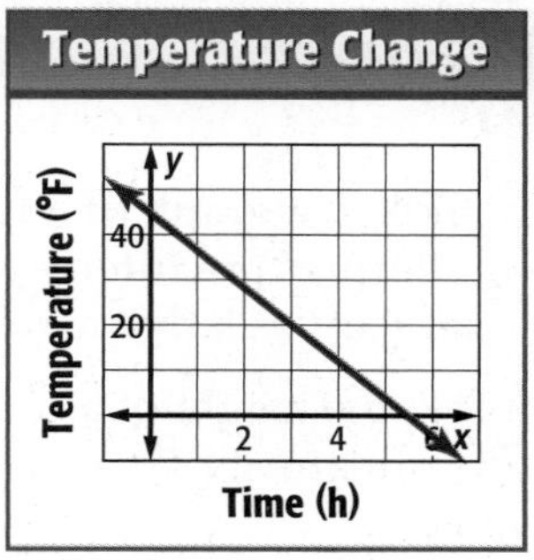

Practice and Problem Solving

Examples 1–3 **CCSS SENSE-MAKING** Identify the function graphed as *linear* or *nonlinear*. Then estimate and interpret the intercepts of the graph, any symmetry, where the function is positive, negative, increasing, and decreasing, the x-coordinate of any relative extrema, and the end behavior of the graph.

4.
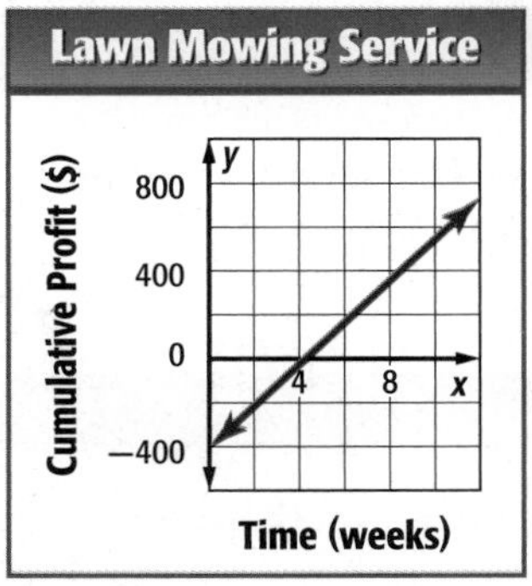

5.
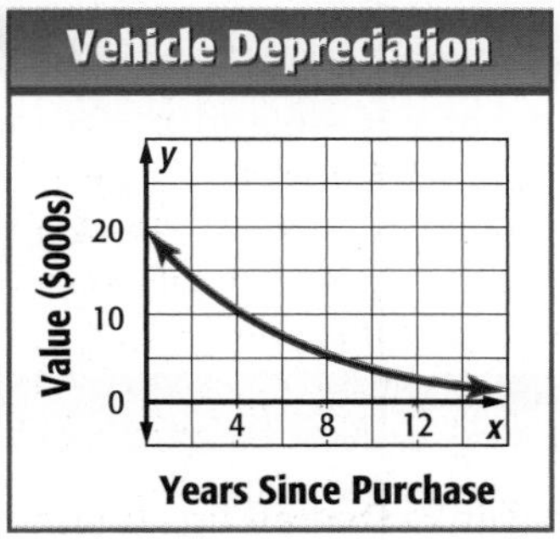

6.
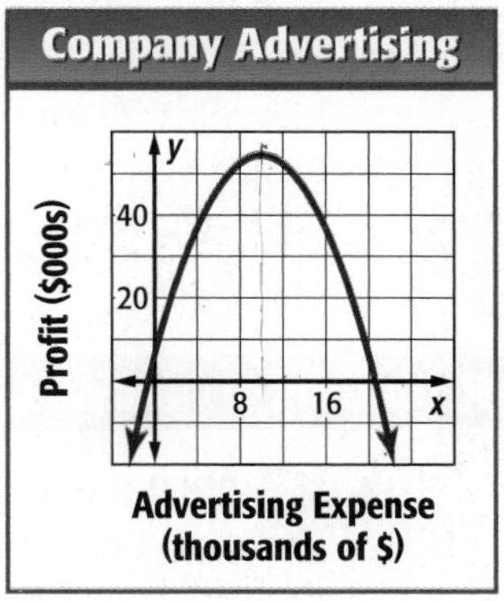

7.
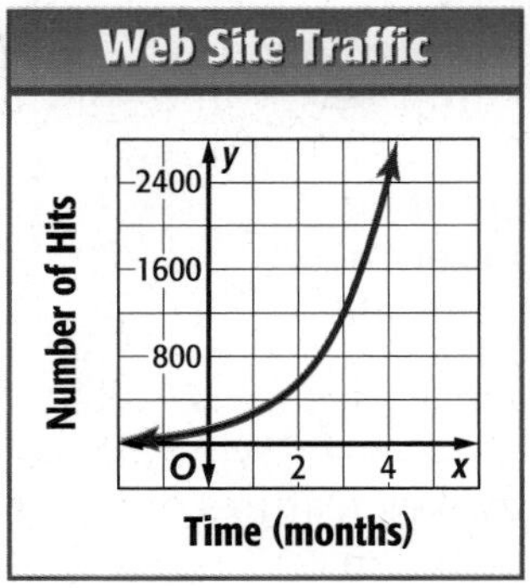

8.
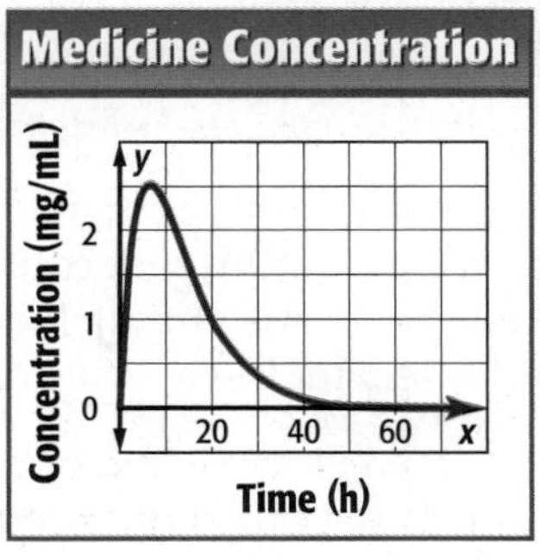

9.
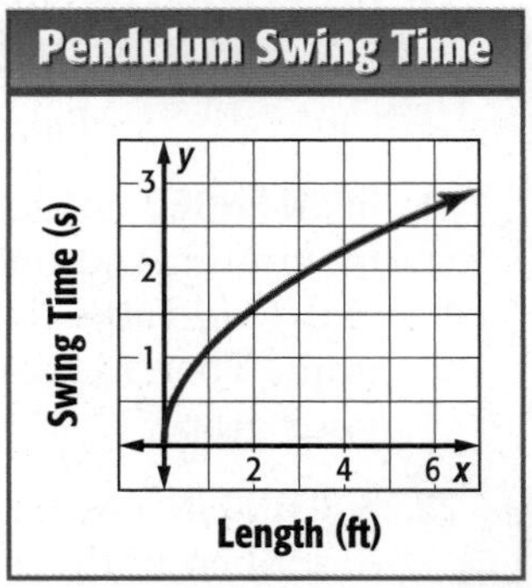

10. **FERRIS WHEEL** At the beginning of a Ferris wheel ride, a passenger cart is located at the same height as the center of the wheel. The position y in feet of this cart relative to the center t seconds after the ride starts is given by the function graphed at the right. Identify and interpret the key features of the graph. (*Hint:* Look for a pattern in the graph to help you describe its end behavior.)

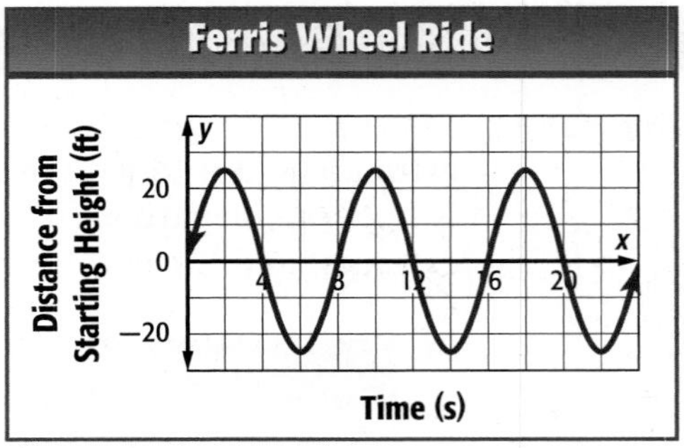

Sketch a graph of a function that could represent each situation. Identify and interpret the intercepts of the graph, where the graph is increasing and decreasing, and any relative extrema.

11. the height of a corn plant from the time the seed is planted until it reaches maturity 120 days later

12. the height of a football from the time it is punted until it reaches the ground 2.8 seconds later

13. the balance due on a car loan from the date the car was purchased until it was sold 4 years later

Sketch graphs of functions with the following characteristics.

14. The graph is linear with an x-intercept at -2. The graph is positive for $x < -2$, and negative for $x > -2$.

15. A nonlinear graph has x-intercepts at -2 and 2 and a y-intercept at -4. The graph has a relative minimum of -4 at $x = 0$. The graph is decreasing for $x < 0$ and increasing for $x > 0$.

16. A nonlinear graph has a y-intercept at 2, but no x-intercepts. The graph is positive and increasing for all values of x.

17. A nonlinear graph has x-intercepts at -8 and -2 and a y-intercept at 3. The graph has relative minimums at $x = -6$ and $x = 6$ and a relative maximum at $x = 2$. The graph is positive for $x < -8$ and $x > -2$ and negative between $x = -8$ and $x = -2$. As x decreases, y increases and as x increases, y increases.

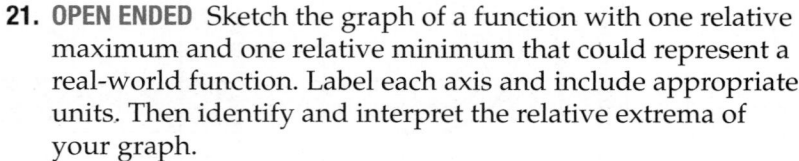

H.O.T. Problems Use Higher-Order Thinking Skills

18. CCSS **CRITIQUE** Katara thinks that all linear functions have exactly one x-intercept. Desmond thinks that a linear function can have at most one x-intercept. Is either of them correct? Explain your reasoning.

19. **CHALLENGE** Describe the end behavior of the graph shown.

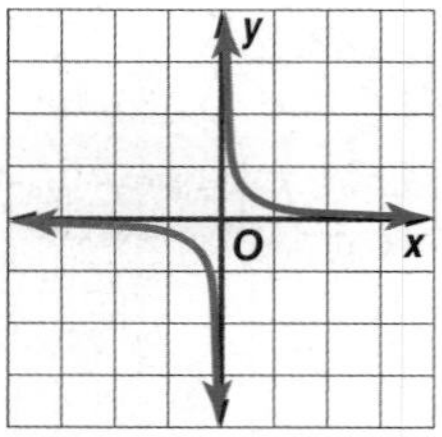

20. **REASONING** Determine whether the following statement is *true* or *false*. Explain.

Functions have at most one y-intercept.

21. **OPEN ENDED** Sketch the graph of a function with one relative maximum and one relative minimum that could represent a real-world function. Label each axis and include appropriate units. Then identify and interpret the relative extrema of your graph.

22. **WRITING IN MATH** Describe how you would identify the key features of a graph described in this lesson using a table of values for a function.

Standardized Test Practice

23. Which sentence best describes the end behavior of the function shown?

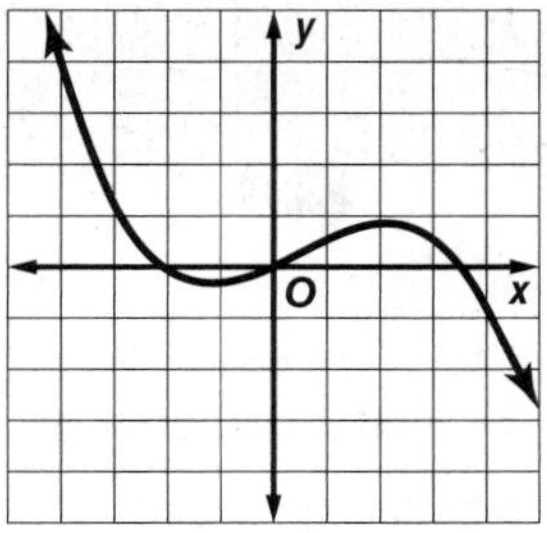

A As x increases, y increases, and as x decreases, y increases.

B As x increases, y increases, and as x decreases, y decreases.

C As x increases, y decreases, and as x decreases, y increases.

D As x increases, y decreases, and as x decreases, y decreases.

24. Which illustrates the Transitive Property of Equality?

F If $c = 1$, then $c \cdot \frac{1}{c} = 1$.

G If $c = d$ and $d = f$, then $c = f$.

H If $c = d$, then $d = c$.

J If $c = d$ and $d = c$, then $c = 1$.

25. Simplify the expression $5d(7 - 3) - 16d + 3 \cdot 2d$.

A $10d$ **C** $21d$

B $14d$ **D** $25d$

26. What is the probability of selecting a red card or an ace from a standard deck of cards?

F $\frac{1}{26}$ **G** $\frac{1}{2}$ **H** $\frac{7}{13}$ **J** $\frac{15}{26}$

Spiral Review

Determine whether each relation is a function. (Lesson 1-7)

27.

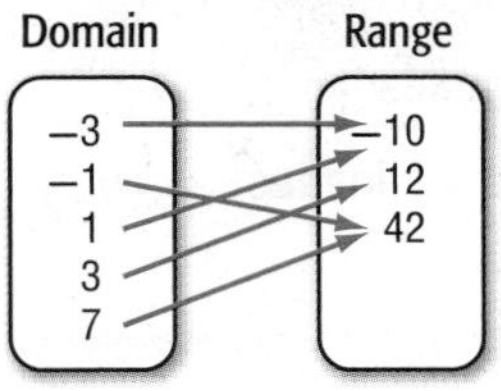

28. $\{(0, 2), (3, 5), (0, -1), (-2, 4)\}$

29.

x	y
17	6
18	6
19	5
20	4

30. GEOMETRY Express the relation in the graph at the right as a set of ordered pairs. Describe the domain and range. (Lesson 1-6)

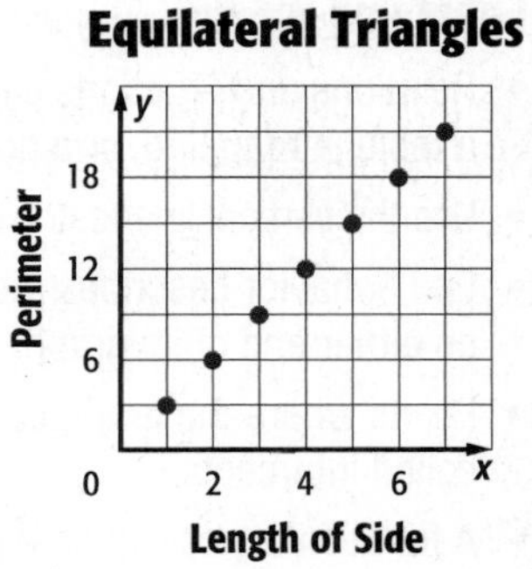

Use the Distributive Property to rewrite each expression. (Lesson 1-4)

31. $\frac{1}{2}d(2d + 6)$ **32.** $-h(6h - 1)$ **33.** $3z - 6x$

34. CLOTHING Robert has 30 socks in his sock drawer. 16 of the socks are white, 6 are black, 2 are red, and 6 are yellow. What is the probability that he randomly pulls out a black sock? (Lesson 0-11)

Skills Review

Evaluate each expression.

35. $(-7)^2$ **36.** 3.2^2 **37.** $(-4.2)^2$ **38.** $\left(\frac{1}{4}\right)^2$

CHAPTER 1 Study Guide and Review

Study Guide

KeyConcepts

Order of Operations (Lesson 1-2)

- Evalute expressions inside grouping symbols.
- Evaluate all powers.
- Multiply and/or divide in order from left to right.
- Add or subtract in order from left to right.

Properties of Equality (Lessons 1-3 and 1-4)

- For any numbers a, b, and c:

Reflexive: $a = a$

Symmetric: If $a = b$, then $b = a$.

Transitive: If $a = b$ and $b = c$, then $a = c$.

Substitution: If $a = b$, then a may be replaced by b in any expression.

Distributive: $a(b + c) = ab + ac$ and $a(b - c) = ab - ac$

Commutative: $a + b = b + a$ and $ab = ba$

Associative: $(a + b) + c = a + (b + c)$ and $(ab)c = a(bc)$

Solving Equations (Lesson 1-5)

- Apply order of operations and the properties of real numbers to solve equations.

Relations, Functions, and Interpreting Graphs of Functions (Lessons 1-6 through 1-8)

- Relations and functions can be represented by ordered pairs, a table, a mapping, or a graph.
- Use the vertical line test to determine if a relation is a function.
- End behavior describes the long-term behavior of a function on either end of its graph.
- Points where the graph of a function crosses an axis are called intercepts.
- A function is positive on a portion of its domain where its graph lies above the x-axis, and negative on a portion where its graph lies below the x-axis.

FOLDABLES StudyOrganizer

Be sure the Key Concepts are noted in your Foldable.

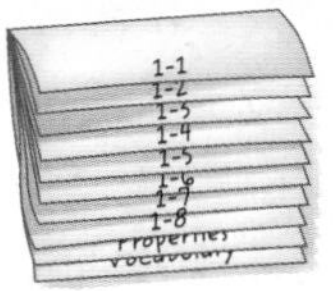

KeyVocabulary

algebraic expression (p. 5)
base (p. 5)
coefficient (p. 28)
coordinate system (p. 40)
dependent variable (p. 42)
domain (p. 40)
end behavior (p. 57)
equation (p. 33)
exponent (p. 5)
function (p. 47)
independent variable (p. 42)
intercept (p. 56)
like terms (p. 27)
line symmetry (p. 57)
mapping (p. 40)
ordered pair (p. 40)
order of operations (p. 10)
origin (p. 40)
power (p. 5)
range (p. 40)
reciprocal (p. 17)
relation (p. 40)
relative maximum (p. 57)
relative minimum (p. 57)
replacement set (p. 33)
simplest form (p. 27)
solution (p. 33)
term (p. 5)
variables (p. 5)
vertical line test (p. 49)

VocabularyCheck

State whether each sentence is *true* or *false*. If *false*, replace the underlined term to make a true sentence.

1. A <u>coordinate system</u> is formed by two intersecting number lines.
2. An <u>exponent</u> indicates the number of times the base is to be used as a factor.
3. An expression is <u>in simplest form</u> when it contains like terms and parentheses.
4. In an expression involving multiplication, the quantities being multiplied are called <u>factors</u>.
5. In a <u>function</u>, there is exactly one output for each input.
6. <u>Order of operations</u> tells us to perform multiplication before subtraction.
7. Since the product of any number and 1 is equal to the number, 1 is called the <u>multiplicative inverse</u>.

Lesson-by-Lesson Review

1-1 Variables and Expressions

Write a verbal expression for each algebraic expression.

8. $h - 7$ **9.** $3x^2$ **10.** $5 + 6m^3$

Write an algebraic expression for each verbal expression.

11. a number increased by 9

12. two thirds of a number d to the third power

13. 5 less than four times a number

Evaluate each expression.

14. 2^5 **15.** 6^3 **16.** 4^4

17. BOWLING Fantastic Pins Bowling Alley charges $2.50 for shoe rental plus $3.25 for each game. Write an expression representing the cost to rent shoes and bowl g games.

Example 1

Write a verbal expression for $4x + 9$.

nine more than four times a number x

Example 2

Write an algebraic expression for *the difference of twelve and two times a number cubed.*

Variable Let x represent the number.

Expression $12 - 2x^3$

Example 3

Evaluate 3^4.

The base is 3 and the exponent is 4.

$3^4 = 3 \cdot 3 \cdot 3 \cdot 3$ Use 3 as a factor 4 times.

$= 81$ Multiply.

1-2 Order of Operations

Evaluate each expression.

18. $24 - 4 \cdot 5$ **19.** $15 + 3^2 - 6$

20. $7 + 2(9 - 3)$ **21.** $8 \cdot 4 - 6 \cdot 5$

22. $[(2^5 - 5) \div 9]11$ **23.** $\frac{11 + 4^2}{5^2 - 4^2}$

Evaluate each expression if $a = 4$, $b = 3$, and $c = 9$.

24. $c + 3a$

25. $5b^2 \div c$

26. $(a^2 + 2bc) \div 7$

27. ICE CREAM The cost of a one-scoop sundae is $2.75, and the cost of a two-scoop sundae is $4.25. Write and evaluate an expression to find the total cost of 3 one-scoop sundaes and 2 two-scoop sundaes.

Example 4

Evaluate the expression $3(9 - 5)^2 \div 8$.

$3(9 - 5)^2 \div 8 = 3(4)^2 \div 8$ Work inside parentheses.

$= 3(16) \div 8$ Evaluate 4^2.

$= 48 \div 8$ Multiply.

$= 6$ Divide.

Example 5

Evaluate the expression $(5m - 2n) \div p^2$ if $m = 8$, $n = 4$, $p = 2$.

$(5m - 2n) \div p^2$

$= (5 \cdot 8 - 2 \cdot 4) \div 2^2$ Replace m with 8, n with 4, and p with 2.

$= (40 - 8) \div 2^2$ Multiply.

$= 32 \div 2^2$ Subtract.

$= 32 \div 4$ Evaluate 2^2.

$= 8$ Divide.

Study Guide and Review *Continued*

1-3 Properties of Numbers

Evaluate each expression using properties of numbers. Name the property used in each step.

28. $18 \cdot 3(1 \div 3)$

29. $[5 \div (8 - 6)]\frac{2}{5}$

30. $(16 - 4^2) + 9$

31. $2 \cdot \frac{1}{2} + 4(4 \cdot 2 - 7)$

32. $18 + 41 + 32 + 9$

33. $7\frac{2}{5} + 5 + 2\frac{3}{5}$

34. $8 \cdot 0.5 \cdot 5$

35. $5.3 + 2.8 + 3.7 + 6.2$

36. **SCHOOL SUPPLIES** Monica needs to purchase a binder, a textbook, a calculator, and a workbook for her algebra class. The binder costs \$9.25, the textbook \$32.50, the calculator \$18.75, and the workbook \$15.00. Find the total cost for Monica's algebra supplies.

Example 6

Evaluate $6(4 \cdot 2 - 7) + 5 \cdot \frac{1}{5}$. Name the property used in each step.

$6(4 \cdot 2 - 7) + 5 \cdot \frac{1}{5}$

$= 6(8 - 7) + 5 \cdot \frac{1}{5}$ Substitution

$= 6(1) + 5 \cdot \frac{1}{5}$ Substitution

$= 6 + 5 \cdot \frac{1}{5}$ Multiplicative Identity

$= 6 + 1$ Multiplicative Inverse

$= 7$ Substitution

1-4 The Distributive Property

Use the Distributive Property to rewrite each expression. Then evaluate.

37. $(2 + 3)6$

38. $5(18 + 12)$

39. $8(6 - 2)$

40. $(11 - 4)3$

41. $-2(5 - 3)$

42. $(8 - 3)4$

Rewrite each expression using the Distributive Property. Then simplify.

43. $3(x + 2)$

44. $(m + 8)4$

45. $6(d - 3)$

46. $-4(5 - 2t)$

47. $(9y - 6)(-3)$

48. $-6(4z + 3)$

49. **TUTORING** Write and evaluate an expression for the number of tutoring lessons Mrs. Green gives in 4 weeks.

Tutoring Schedule

Day	Students
Monday	3
Tuesday	5
Wednesday	4

Example 7

Use the Distributive Property to rewrite the expression 5(3 + 8). Then evaluate.

$5(3 + 8) = 5(3) + 5(8)$ Distributive Property

$= 15 + 40$ Multiply.

$= 55$ Simplify.

Example 8

Rewrite the expression $6(x + 4)$ using the Distributive Property. Then simplify.

$6(x + 4) = 6 \cdot x + 6 \cdot 4$ Distributive Property

$= 6x + 24$ Simplify.

Example 9

Rewrite the expression $(3x - 2)(-5)$ using the Distributive Property. Then simplify.

$(3x - 2)(-5)$

$= (3x)(-5) - (2)(-5)$ Distributive Property

$= -15x + 10$ Simplify.

1-5 Equations

Find the solution set of each equation if the replacement sets are *x*: {1, 3, 5, 7, 9} and *y*: {6, 8, 10, 12, 14}.

50. $y - 9 = 3$

51. $14 + x = 21$

52. $4y = 32$

53. $3x - 11 = 16$

54. $\frac{42}{y} = 7$

55. $2(x - 1) = 8$

Solve each equation.

56. $a = 24 - 7(3)$

57. $z = 63 \div (3^2 - 2)$

58. **AGE** Shandra's age is four more than three times Sherita's age. Write an equation for Shandra's age. Solve if Sherita is 3 years old

Example 10

Solve the equation $5w - 19 = 11$ if the replacement set is *w*: {2, 4, 6, 8, 10}.

Replace *w* in $5w - 19 = 11$ with each value in the replacement set.

w	$5w - 19 = 11$	True or False?
2	$5(2) - 19 = 11$	false
4	$5(4) - 19 = 11$	false
6	$5(6) - 19 = 11$	true
8	$5(8) - 19 = 11$	false
10	$5(10) - 19 = 11$	false

Since the equation is true when $w = 6$, the solution of $5w - 19 = 11$ is $w = 6$.

1-6 Relations

Express each relation as a table, a graph, and a mapping. Then determine the domain and range.

59. {(1, 3), (2, 4), (3, 5), (4, 6)}

60. {(−1, 1), (0, −2), (3, 1), (4, −1)}

61. {(−2, 4), (−1, 3), (0, 2), (−1, 2)}

Express the relation shown in each table, mapping, or graph as a set of ordered pairs.

62.

x	*y*
5	3
3	−1
1	2
−1	0

63.

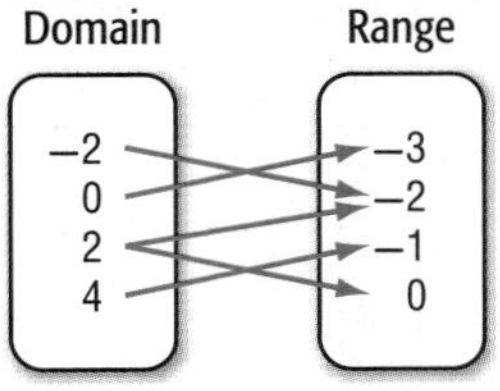

64. **GARDENING** On average, 7 plants grow for every 10 seeds of a certain type planted. Make a table to show the relation between seeds planted and plants growing for 50, 100, 150, and 200 seeds. Then state the domain and range and graph the relation.

Example 11

Express the relation {(−3, 4), (1, −2), (0, 1), (3, −1)} as a table, a graph, and a mapping.

Table

Place the *x*-coordinates into the first column. Place the corresponding *y*-coordinates in the second column.

x	*y*
−3	4
1	−2
0	1
3	−1

Graph

Graph each ordered pair on a coordinate plane.

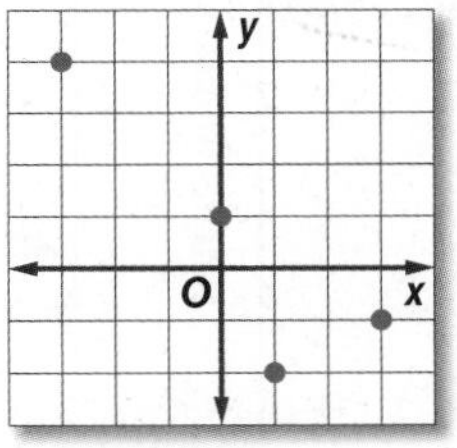

Mapping

List the *x*-values in the domain and the *y*-values in the range. Draw arrows from the *x*-values in set *X* to the corresponding *y*-values in set *Y*.

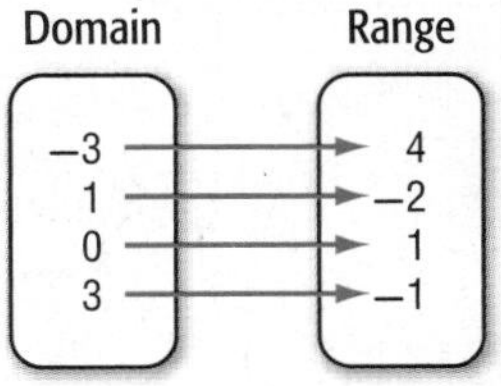

Study Guide and Review *Continued*

1-7 Functions

Determine whether each relation is a function.

65. 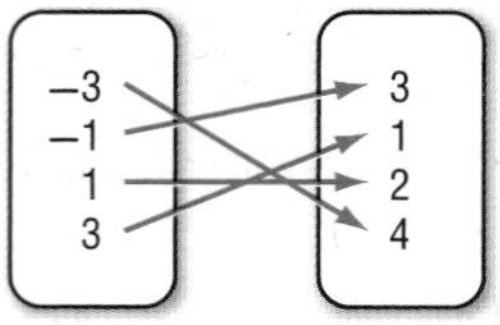

66. 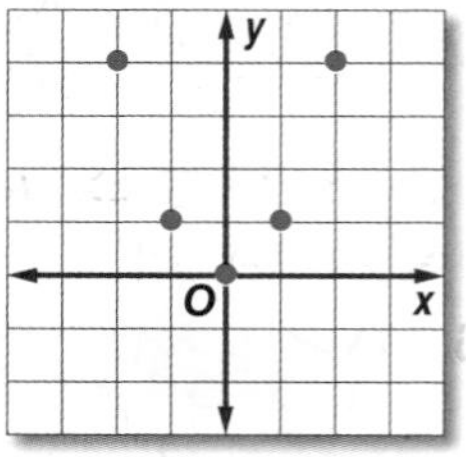

67. {(8, 4), (6, 3), (4, 2), (2, 1), (6, 0)}

If $f(x) = 2x + 4$ and $g(x) = x^2 - 3$, find each value.

68. $f(-3)$ **69.** $g(2)$ **70.** $f(0)$

71. $g(-4)$ **72.** $f(m + 2)$ **73.** $g(3p)$

74. **GRADES** A teacher claims that the relationship between number of hours studied for a test and test score can be described by $g(x) = 45 + 9x$, where x represents the number of hours studied. Graph this function.

Example 12

Determine whether $2x - y = 1$ represents a function.

First make a table of values. Then graph the equation.

x	y
−1	−3
0	−1
1	1
2	3
3	5

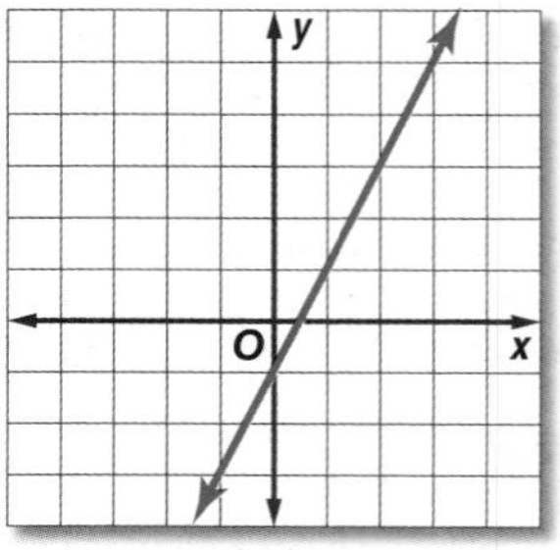

Using the vertical line test, it can be shown that $2x - y = 1$ does represent a function.

1-8 Interpreting Graphs of Functions

75. Identify the function graphed as *linear* or *nonlinear*. Then estimate and interpret the intercepts of the graph, any symmetry, where the function is positive, negative, increasing, and decreasing, the x-coordinate of any relative extrema, and the end behavior of the graph.

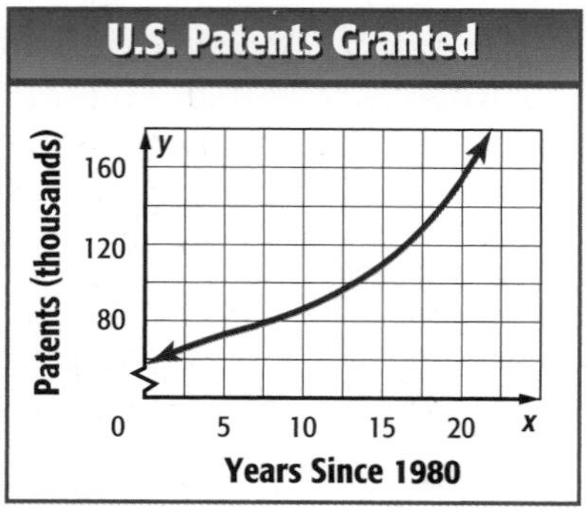

Example 13

POPULATION **The population of Haiti from 1994 to 2010 can be modeled by the function graphed below. Estimate and interpret where the function is increasing, and decreasing, the x-coordinates of any relative extrema, and the end behavior of the graph.**

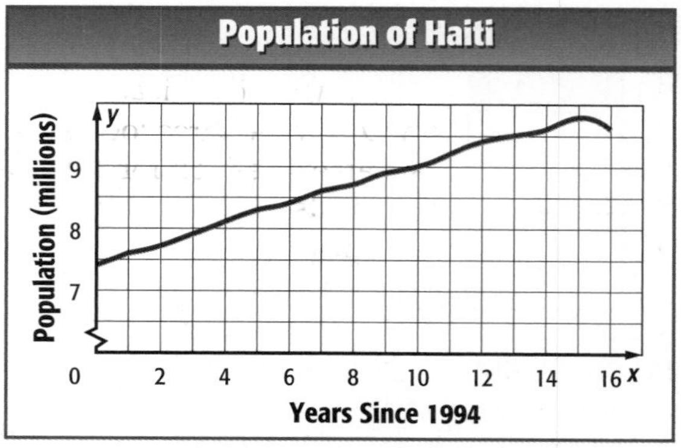

The population increased from 1994 to 2009 and decreased from 2009 to 2010. The relative maximum of the graph indicates that the population peaked in 2009.

As x increases or decreases, y decreases. The end behavior indicates a decline in population from 2009 to 2010.

CHAPTER 1 Practice Test

Write an algebraic expression for each verbal expression.

1. six more than a number
2. twelve less than the product of three and a number
3. four divided by the difference between a number and seven

Evaluate each expression.

4. $32 \div 4 + 2^3 - 3$

5. $\dfrac{(2 \cdot 4)^2}{7 + 3^2}$

6. **MULTIPLE CHOICE** Find the value of the expression $a^2 + 2ab + b^2$ if $a = 6$ and $b = 4$.

   A 68

   B 92

   C 100

   D 121

Evaluate each expression. Name the property used in each step.

7. $13 + (16 - 4^2)$

8. $\frac{2}{9}[9 \div (7 - 5)]$

9. $37 + 29 + 13 + 21$

Rewrite each expression using the Distributive Property. Then simplify.

10. $4(x + 3)$

11. $(5p - 2)(-3)$

12. **MOVIE TICKETS** A company operates three movie theaters. The chart shows the typical number of tickets sold each week at the three locations. Write and evaluate an expression for the total typical number of tickets sold by all three locations in four weeks.

Location	Tickets Sold
A	438
B	374
C	512

Find the solution of each equation if the replacement sets are x: {1, 3, 5, 7, 9} and y: {2, 4, 6, 8, 10}.

13. $3x - 9 = 12$

14. $y^2 - 5y - 11 = 13$

15. **CELL PHONES** The ABC Cell Phone Company offers a plan that includes a flat fee of \$29 per month plus a \$0.12 charge per minute. Write an equation to find C, the total monthly cost for m minutes. Then solve the equation for $m = 50$.

Express the relation shown in each table, mapping, or graph as a set of ordered pairs.

16.

x	y
−2	4
1	2
3	0
4	−2

17.

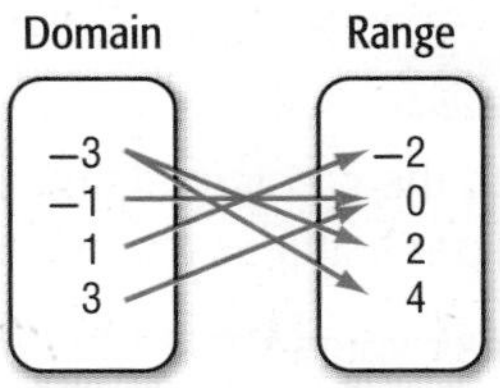

18. **MULTIPLE CHOICE** Determine the domain and range for the relation {(2, 5), (−1, 3), (0, −1), (3, 3), (−4, −2)}.

   F D: {2, −1, 0, 3, −4}, R: {5, 3, −1, 3, −2}

   G D: {5, 3, −1, 3, −2}, R: {2, −1, 0, 3, 4}

   H D: {0, 1, 2, 3, 4}, R: {−4, −3, −2, −1, 0}

   J D: {2, −1, 0, 3, −4}, R: {2, −1, 0, 3, 4}

19. Determine whether the relation {(2, 3), (−1, 3), (0, 4), (3, 2), (−2, 3)} is a function.

If $f(x) = 5 - 2x$ and $g(x) = x^2 + 7x$, find each value.

20. $g(3)$

21. $f(-6y)$

22. Identify the function graphed as *linear* or *nonlinear*. Then estimate and interpret the intercepts of the graph, any symmetry, where the function is positive, negative, increasing, and decreasing, the x-coordinate of any relative extrema, and the end behavior of the graph.

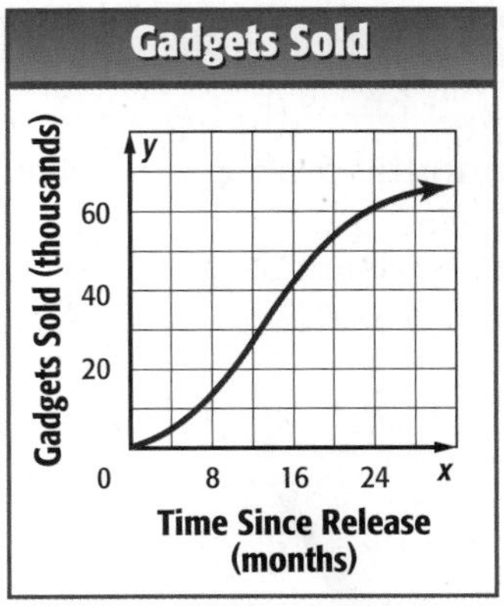

CHAPTER 1

Preparing for Standardized Tests

Eliminate Unreasonable Answers

You can eliminate unreasonable answers to help you find the correct one when solving multiple choice test items. Doing so will save you time by narrowing down the list of possible correct answers.

Strategies for Eliminating Unreasonable Answers

fStop/Alamy

Step 1

Read the problem statement carefully to determine exactly what you are being asked to find.

Ask yourself:

- What am I being asked to solve?
- What format (i.e., fraction, number, decimal, percent, type of graph) will the correct answer be?
- What units (if any) will the correct answer have?

Step 2

Carefully look over each possible answer choice and evaluate for reasonableness.

- Identify any answer choices that are clearly incorrect and eliminate them.
- Eliminate any answer choices that are not in the proper format.
- Eliminate any answer choices that do not have the correct units.

Step 3

Solve the problem and choose the correct answer from those remaining. Check your answer.

Standardized Test Example

Read each problem. Eliminate any unreasonable answers. Then use the information in the problem to solve.

Jason earns 8.5% commission on his weekly sales at an electronics retail store. Last week he had \$4200 in sales. What was his commission for the week?

A \$332 **C** \$425

B \$357 **D** \$441

Using mental math, you know that 10% of $4200 is $420. Since 8.5% is less than 10%, you know that Jason earned less than $420 in commission for his weekly sales. So, answer choices C and D can be eliminated because they are greater than $420. The correct answer is either A or B.

$\$4200 \times 0.085 = \$357$

So, the correct answer is B.

Exercises

Read each problem. Eliminate any unreasonable answers. Then use the information in the problem to solve.

1. Coach Roberts expects 35% of the student body to turn out for a pep rally. If there are 560 students, how many does Coach Roberts expect to attend the pep rally?

A 184

B 196

C 214

D 390

2. Jorge and Sally leave school at the same time. Jorge walks 300 yards north and then 400 yards east. Sally rides her bike 600 yards south and then 800 yards west. What is the distance between the two students?

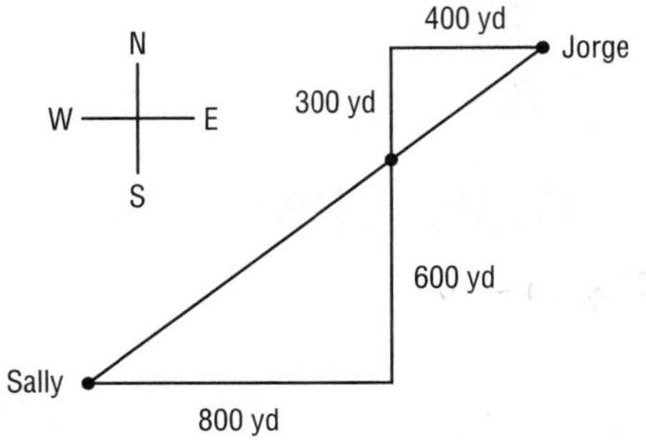

F 500 yd

G 750 yd

H 1,200 yd

J 1,500 yd

3. What is the range of the relation below?

$\{(1, 2), (3, 4), (5, 6), (7, 8)\}$

A all real numbers

B all even numbers

C $\{2, 4, 6, 8\}$

D $\{1, 3, 5, 7\}$

4. The expression $3n + 1$ gives the total number of squares needed to make each figure of the pattern where n is the figure number. How many squares will be needed to make Figure 9?

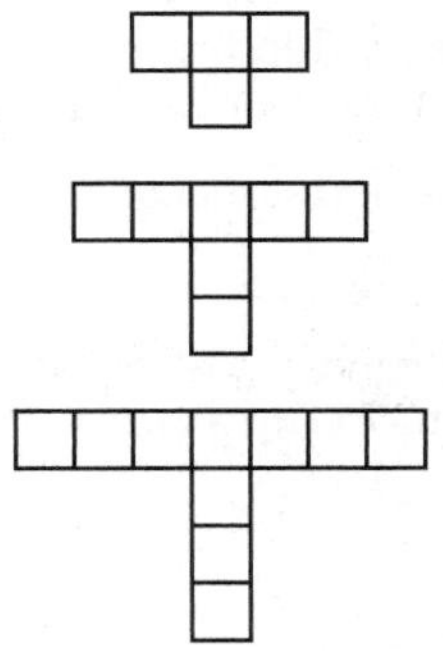

F 28 squares

G 32.5 squares

H 56 squares

J 88.5 squares

5. The expression $3x - (2x + 4x - 6)$ is equivalent to

A $-3x - 6$

B $-3x + 6$

C $3x + 6$

D $3x - 6$

Standardized Test Practice

Chapter 1

Multiple Choice

Read each question. Then fill in the correct answer on the answer document provided by your teacher or on a sheet of paper.

1. Evaluate the expression 2^6.

A 12

B 32

C 64

D 128

2. Which sentence best describes the end behavior of the function shown?

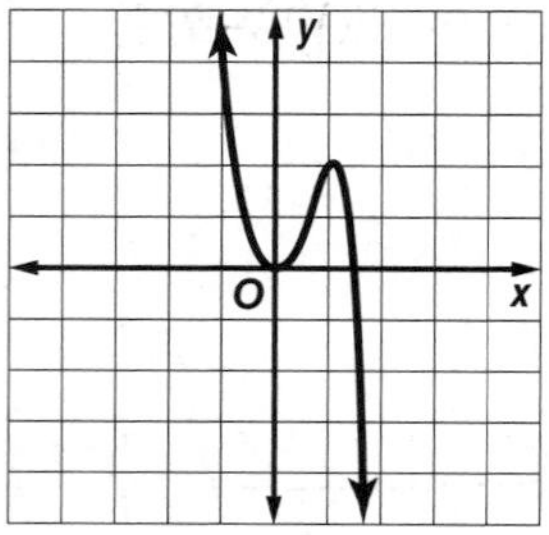

F As x increases, y increases, and as x decreases, y increases.

G As x increases, y increases, and as x decreases, y decreases.

H As x increases, y decreases, and as x decreases, y increases.

J As x increases, y decreases, and as x decreases, y decreases.

3. Let y represent the number of yards. Which algebraic expression represents the number of feet in y?

A $y - 3$

B $y + 3$

C $3y$

D $\frac{3}{y}$

4. What is the domain of the following relation?

$$\{(1, 3), (-6, 4), (8, 5)\}$$

F {3, 4, 5}

G {−6, 1, 8}

H {−6, 1, 3, 4, 5, 8}

J {1, 3, 4, 5, 8}

5. The table shows the number of some of the items sold at the concession stand at the first day of a soccer tournament. Estimate how many items were sold from the concession stand throughout the four days of the tournament.

Concession Sales Day 1 Results	
Item	Number Sold
Popcorn	78
Hot Dogs	80
Chip	48
Sodas	51
Bottled Water	92

A 1350 items

B 1400 items

C 1450 items

D 1500 items

6. There are 24 more cars than twice the number of trucks for sale at a dealership. If there are 100 cars for sale, how many trucks are there for sale at the dealership?

F 28

G 32

H 34

J 38

7. Refer to the relation in the table below. Which of the following values would result in the relation *not* being a function?

x	−6	−2	0	?	3	5
y	−1	8	3	−3	4	0

A −1

B 3

C 7

D 8

Test-Taking Tip

Question 7 A function is a relation in which each element of the domain is paired with *exactly* one element of the range.

Short Response/Gridded Response

Record your answers on the answer sheet provided by your teacher or on a sheet of paper.

8. The edge of each box below is 1 unit long.

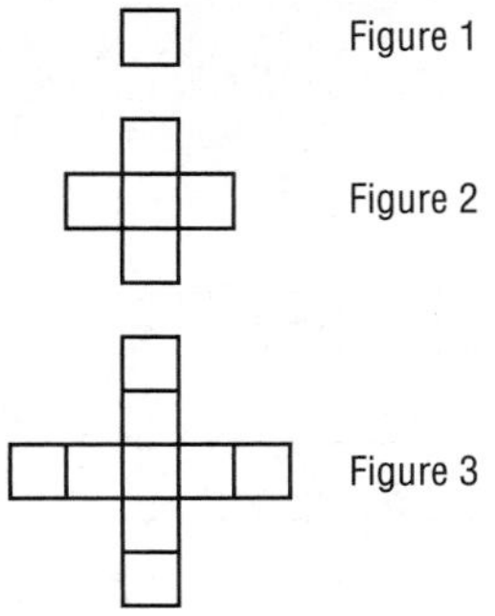

a. Make a table showing the perimeters of the first 3 figures in the pattern.

b. Look for a pattern in the perimeters of the shapes. Write an algebraic expression for the perimeter of Figure n.

c. What would be the perimeter of Figure 10 in the pattern?

9. The table shows the costs of certain items at a corner hardware store.

Item	Cost
box of nails	\$3.80
box of screws	\$5.25
claw hammer	\$12.95
electric drill	\$42.50

a. Write two expressions to represent the total cost of 3 boxes of nails, 2 boxes of screws, 2 hammers, and 1 electric drill.

b. What is the total cost of the items purchased?

10. GRIDDED RESPONSE Evaluate the expression below.

$$\frac{5^3 \cdot 4^2 - 5^2 \cdot 4^3}{5 \cdot 4}$$

11. Use the equation $y = 2(4 + x)$ to answer each question.

a. Complete the table for each value of x.

b. Plot the points from the table on a coordinate grid. What do you notice about the points?

c. Make a conjecture about the relationship between the change in x and the change in y.

x	y
1	
2	
3	
4	
5	
6	

Extended Response

Record your answers on a sheet of paper. Show your work.

12. The volume of a sphere is four-thirds the product of π and the radius cubed.

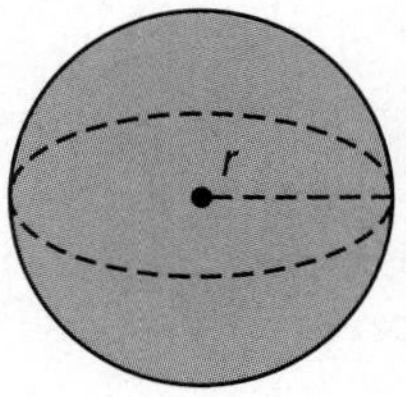

a. Write an expression for the volume of a sphere with radius r.

b. Find the volume of a sphere with a radius of 6 centimeters. Describe how you found your answer.

Need ExtraHelp?

If you missed Question...	1	2	3	4	5	6	7	8	9	10	11	12
Go to Lesson...	1-2	1-8	1-1	1-6	1-4	1-5	1-7	1-5	1-3	1-2	1-4	1-1

CHAPTER 2

Linear Equations

Then

You learned to simplify algebraic expressions.

Now

In this chapter, you will:

- Create equations that describe relationships.
- Solve linear equations in one variable.
- Solve proportions.
- Use formulas to solve real-world problems.

Why?

SHOPPING In recent years, the percent of change in sales per year at shopping malls in the U.S. averaged 5%. A store manager can use this data to set a sales goal for the upcoming year.

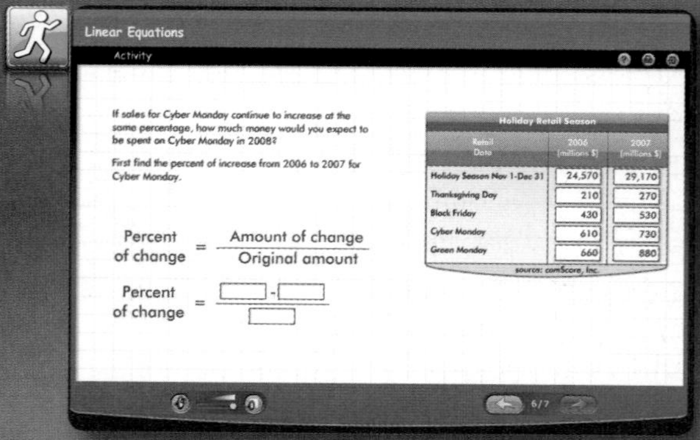

connectED.mcgraw-hill.com Your Digital Math Portal

Animation | Vocabulary | eGlossary | Personal Tutor | Virtual Manipulatives | Graphing Calculator | Audio | Foldables | Self-Check Practice | Worksheets

Get Ready for the Chapter

Diagnose Readiness | **You have two options for checking prerequisite skills.**

1 Textbook Option Take the Quick Check below. Refer to the Quick Review for help.

QuickCheck

Write an algebraic expression for each verbal expression.

1. four less than three times a number n

2. a number d cubed less seven

3. the difference between two times b and eleven

Evaluate each expression.

4. $(9 - 4)^2 + 3$

5. $\frac{3 \cdot 8 - 12 \div 2}{3^2}$

6. $5(8 - 2) \div 3$

7. $\frac{1}{3}(21) + \frac{1}{8}(32)$

8. $72 \div 9 + 3 \cdot 2^3$

9. $\frac{11 - 3}{2} + 7$

10. $2[(5 - 3)^2 + 8] + (3 - 1) \div 2$

11. BAKERY Sue buys 1 carrot cake for \$14, 6 large chocolate chip cookies for \$1.50 each, and a dozen doughnuts for \$0.45 each. How much money did Sue spend at the bakery?

Find each percent.

12. What percent of 400 is 260?

13. Twelve is what percent of 60?

14. What percent of 25 is 75?

15. ICE CREAM What percent of the people surveyed prefer strawberry ice cream?

Favorite Flavor	Number of Responses
vanilla	82
chocolate	76
strawberry	42

QuickReview

Example 1

Write an algebraic expression for the phrase *the product of eight and w increased by nine.*

the product of eight and w increased by nine

$8 \cdot w + 9$

The expression is $8w + 9$.

Example 2

Evaluate $9 - \left[\frac{8 + 2^2}{2} - 2(5 \times 2 - 8)\right]$.

$9 - \left[\frac{8 + 2^2}{2} - 2(5 \times 2 - 8)\right]$ Original expression

$= 9 - \left[\frac{8 + 2^2}{2} - 2(2)\right]$ Evaluate inside the parentheses.

$= 9 - \left(\frac{8 + 2^2}{2} - 4\right)$ Multiply.

$= 9 - \left(\frac{8 + 4}{2} - 4\right)$ Evaluate the power.

$= 9 - (6 - 4)$ Add and then divide.

$= 7$ Simplify.

Example 3

32 is what percent of 40?

$\frac{a}{b} = \frac{p}{100}$ Use the percent proportion.

$\frac{32}{40} = \frac{p}{100}$ Replace a with 32 and b with 40.

$32(100) = 40p$ Find the cross products.

$3200 = 40p$ Multiply.

$80 = p$ Divide each side by 40.

32 is 80% of 40.

2 Online Option Take an online self-check Chapter Readiness Quiz at connectED.mcgraw-hill.com.

Get Started on the Chapter

You will learn several new concepts, skills, and vocabulary terms as you study Chapter 2. To get ready, identify important terms and organize your resources. You may wish to refer to Chapter 0 to review prerequisite skills.

FOLDABLES StudyOrganizer

Linear Functions Make this Foldable to help you organize your Chapter 2 notes about linear equations. Begin with 5 sheets of grid paper.

1 **Fold** each sheet in half along the width.

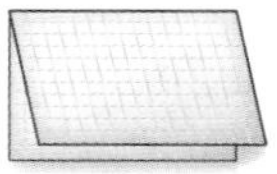

2 **Unfold** each sheet and tape to form one long piece.

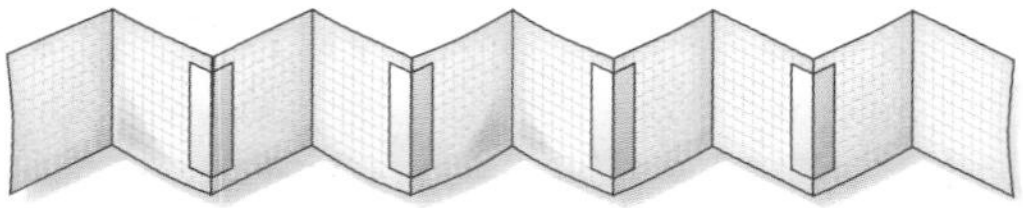

3 **Label** each page with the lesson number as shown. Refold to form a booklet.

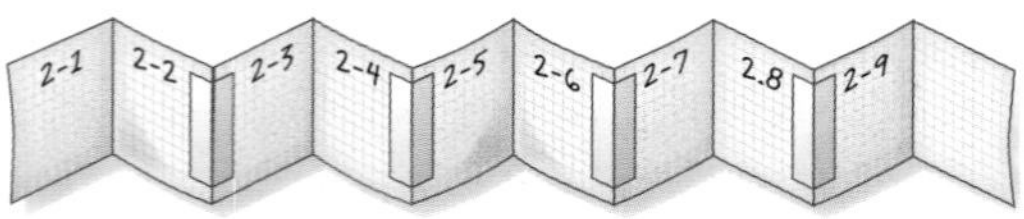

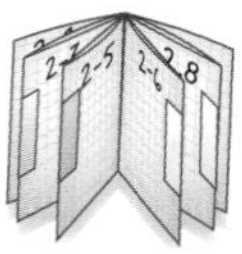

NewVocabulary

English		Español
formula	p. 76	fórmula
solve an equation	p. 83	resolver una ecuación
equivalent equations	p. 83	ecuaciones equivalentes
multi-step equation	p. 91	ecuación de varios pasos
identity	p. 98	identidad
ratio	p. 111	razón
proportion	p. 111	proporción
rate	p. 113	tasa
unit rate	p. 113	tasa unitaria
scale model	p. 114	modelo de escala
percent of change	p. 119	porcentaje de cambio
literal equation	p. 127	ecuación literal
dimensional analysis	p. 128	análisis dimensional
weighted average	p. 132	promedio ponderado

ReviewVocabulary

algebraic expression expresion algebraica an expression consisting of one or more numbers and variables along with one or more arithmetic operations

coordinate system
sistema de coordenedas
the grid formed by the intersection of two number lines, the horizontal axis and the vertical axis

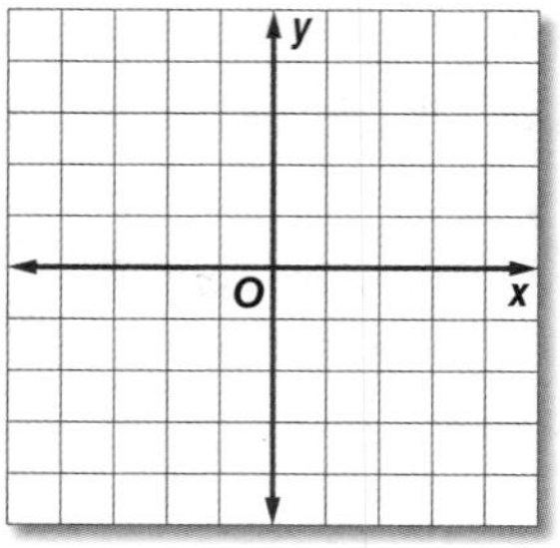

function función a relation in which each element of the domain is paired with exactly one element of the range

LESSON 2-1 Writing Equations

Then

- You evaluated and simplified algebraic expressions.

Now

1. Translate sentences into equations.
2. Translate equations into sentences.

Why?

- The Daytona 500 is widely considered to be the most important event of the NASCAR circuit. The distance around the track is 2.5 miles, and the race is a total of 500 miles. We can write an equation to determine how many laps it takes to finish the race.

NewVocabulary
formula

Common Core State Standards

Content Standards
A.CED.1 Create equations and inequalities in one variable and use them to solve problems.

Mathematical Practices
2 Reason abstractly and quantitatively.

1 Write Verbal Expressions

To write an equation, identify the unknown for which you are looking and assign a variable to it. Then, write the sentence as an equation. Look for key words such as *is*, *is as much as*, *is the same as*, or *is identical to* that indicate where you should place the equals sign.

Consider the Daytona 500 example above.

Words	The length of each lap	times the	number of laps	is	the length of the race.
Variable	Let ℓ represent the number of laps in the race.				
Equation	2.5	$\times$	ℓ	$=$	500

Example 1 Translate Sentences into Equations

PT

Translate each sentence into an equation.

a. Seven times a number squared is five times the difference of *k* and *m*.

Seven	times	*n* squared	is	five	times	the difference of *k* and *m*.
7	$\cdot$	n^2	$=$	5	$\cdot$	$(k - m)$

The equation is $7n^2 = 5(k - m)$.

b. Fifteen times a number subtracted from 80 is 25.

You can rewrite the verbal sentence so it is easier to translate. *Fifteen times a number subtracted from 80 is 25* is the same as *80 minus 15 times a number is 25*. Let *n* represent the number.

80	minus	15	times	a number	is	25.
80	$-$	15	$\cdot$	n	$=$	25

The equation is $80 - 15n = 25$.

GuidedPractice

1A. Two plus the quotient of a number and 8 is the same as 16.

1B. Twenty-seven times *k* is *h* squared decreased by 9.

Translating sentences to algebraic expressions and equations is a valuable skill in solving real-world problems.

Real-WorldLink

In 1919, Britain and France offered a flight that carried two passengers at a time. Now there are more than 87,000 flights each day in the U.S.

Source: NATCA

PT

Real-World Example 2 Use the Four-Step Problem-Solving Plan

AIR TRAVEL Refer to the information at the left. In how many days will 261,000 flights have occurred in the United States?

Understand The information given in the problem is that there are approximately 87,000 flights per day in the United States. We are asked to find how many days it will take for 261,000 flights to have occurred.

Plan Write an equation. Let d represent the number of days needed.

87,000	times	the number of days	equals	261,000.
87,000	$\cdot$	d	$=$	261,000

Solve $87{,}000\,d = 261{,}000$ Find d by asking, "What number times 87,000 is 261,000?"

$d = 3$

Check Check your answer by substituting 3 for d in the equation.

$87{,}000(3) \stackrel{?}{=} 261{,}000$ Substitute 3 for d.

$261{,}000 = 261{,}000$ ✓ Multiply.

The answer makes sense and works for the original problem.

GuidedPractice

2. GOVERNMENT There are 50 members in the North Carolina Senate. This is 70 fewer than the number in the North Carolina House of Representatives. How many members are in the North Carolina House of Representatives?

A rule for the relationship between certain quantities is called a **formula**. These equations use variables to represent numbers and form general rules.

PT

Example 3 Write a Formula

GEOMETRY Translate the sentence into a formula.

The area of a triangle equals the product of $\frac{1}{2}$ the length of the base and the height.

Words	The area of a triangle	equals	the product of $\frac{1}{2}$ the length of the base and the height.
Variables	Let A = area, b = base, and h = height.		
Equation	A	$=$	$\frac{1}{2}bh$

The formula for the area of a triangle is $A = \frac{1}{2}bh$.

GuidedPractice

3. GEOMETRY Translate the sentence into a formula.
In a right triangle, the square of the measure of the hypotenuse c is equal to the sum of the squares of the measures of the legs, a and b.

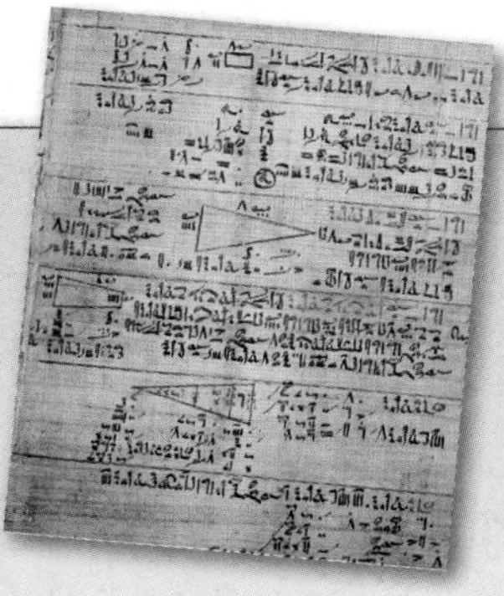

Math HistoryLink

Ahmes (about 1680–1620 B.C.) Ahmes was the Egyptian mathematician and scribe who copied the Rhind Mathematical Papyrus. The papyrus contains 87 algebra problems of the same type. The first set of problems asks how to divide n loaves of bread among 10 people.

2 Write Sentences from Equations

If you are given an equation, you can write a sentence or create your own word problem.

Example 4 Translate Equations into Sentences

PT

Translate each equation into a sentence.

a. $6z - 15 = 45$

$6z$	$-$	15	$=$	45
Six times z	minus	fifteen	equals	forty-five.

b. $y^2 + 3x = w$

y^2	$+$	$3x$	$=$	w
The sum of y squared	and	three times x	is	w.

GuidedPractice

4A. $15 = 25u^2 + 2$

4B. $\frac{3}{2}r - t^3 = 132$

When given a set of information, you can create a problem that relates a story.

Example 5 Write a Problem

PT

Write a problem based on the given information.

t = the time that Maxine drove in each turn; $t + 4$ = the time that Tia drove in each turn; $2t + (t + 4) = 28$

Sample problem:

Maxine and Tia went on a trip, and they took turns driving. During her turn, Tia drove 4 hours more than Maxine. Maxine took 2 turns, and Tia took 1 turn. Together they drove for 28 hours. How many hours did Maxine drive?

GuidedPractice

5. p = Beth's salary; $0.1p$ = bonus; $p + 0.1p = 525$

Check Your Understanding

Example 1 **Translate each sentence into an equation.**

1. Three times r less than 15 equals 6.
2. The sum of q and four times t is equal to 29.
3. A number n squared plus 12 is the same as the quotient of p and 4.
4. Half of j minus 5 is the sum of k and 13.
5. The sum of 8 and three times k equals the difference of 5 times k and 3.
6. Three fourths of w plus 5 is one half of w increased by nine.
7. The quotient of 25 and t plus 6 is the same as twice t plus 1.
8. Thirty-two divided by y is equal to the product of three and y minus four.

Example 2

9. **FINANCIAL LITERACY** Samuel has \$1900 in the bank. He wishes to increase his account to a total of \$2500 by depositing \$30 per week from his paycheck. Write and solve an equation to find how many weeks he needs to reach his goal.

10. **CCSS MODELING** Miguel is earning extra money by painting houses. He charges a \$200 fee plus \$12 per can of paint needed to complete the job. Write and use an equation to find how many cans of paint he needs for a \$260 job.

Translate each sentence into a formula.

Example 3

11. The perimeter of a regular pentagon is 5 times the length of each side.
12. The area of a circle is the product of π and the radius r squared.
13. Four times π times the radius squared is the surface area of a sphere.
14. One third the product of the length of the side squared and the height is the volume of a pyramid with a square base.

Example 4

Translate each equation into a sentence.

15. $7m - q = 23$
16. $6 + 9k + 5j = 54$
17. $3(g + 8) = 4h - 10$
18. $6d^2 - 7f = 8d + f^2$

Example 5

Write a problem based on the given information.

19. g = gymnasts on a team; $3g = 45$
20. c = cost of a notebook; $0.25c$ = markup; $c + 0.25c = 3.75$

Practice and Problem Solving

Example 1

Translate each sentence into an equation.

21. The difference of f and five times g is the same as 25 minus f.
22. Three times b less than 100 is equal to the product of 6 and b.
23. Four times the sum of 14 and c is a squared.

Example 2

24. **MUSIC** A piano has 52 white keys. Write and use an equation to find the number of octaves on a piano keyboard.

25. **GARDENING** A flat of plants contains 12 plants. Yoshi wants a garden that has three rows with 10 plants per row. Write and solve an equation for the number of flats Yoshi should buy.

Example 3

Translate each sentence into a formula.

26. The perimeter of a rectangle is equal to 2 times the length plus twice the width.
27. Celsius temperature C is five ninths times the difference of the Fahrenheit temperature F and 32.
28. The density of an object is the quotient of its mass and its volume.
29. Simple interest is computed by finding the product of the principal amount p, the interest rate r, and the time t.

Example 4

Translate each equation into a sentence.

30. $j + 16 = 35$
31. $4m = 52$
32. $7(p + 23) = 102$
33. $r^2 - 15 = t + 19$
34. $\frac{2}{5}v + \frac{3}{4} = \frac{2}{3}x^2$
35. $\frac{1}{3} - \frac{4}{5}z = \frac{4}{3}y^3$

C Squared Studios/Photodisc/Getty Images

Example 5 **Write a problem based on the given information.**

36. q = quarts of strawberries; $2.50q = 10$

37. p = the principal amount; $0.12p$ = the interest charged; $p + 0.12p = 224$

38. m = number of movies rented; $10 + 1.50m = 14.50$

39. p = the number of players in the game; $5p + 7$ = number of cards in a deck

For Exercises 40–43, match each sentence with an equation.

A. $g^2 = 2(g - 10)$

B. $\frac{1}{2}g + 32 = 15 + 6g$

C. $g^3 = 24g + 4$

D. $3g^2 = 30 + 9g$

40. One half of g plus thirty-two is as much as the sum of fifteen and six times g.

41. A number g to the third power is the same as the product of 24 and g plus 4.

42. The square of g is the same as two times the difference of g and 10.

43. The product of 3 and the square of g equals the sum of thirty and the product of nine and g.

44. **FINANCIAL LITERACY** Tim's bank contains quarters, dimes, and nickels. He has three more dimes than quarters and 6 fewer nickels than quarters. If he has 63 coins, write and solve an equation to find how many quarters Tim has.

45. **SHOPPING** Pilar bought 17 items for her camping trip, including tent stakes, packets of drink mix, and bottles of water. She bought 3 times as many packets of drink mix as tent stakes. She also bought 2 more bottles of water than tent stakes. Write and solve an equation to discover how many tent stakes she bought.

46. **MULTIPLE REPRESENTATIONS** In this problem, you will explore how to translate relations with powers.

x	2	3	4	5	6
y	5	10	17	26	37

a. **Verbal** Write a sentence to describe the relationship between x and y in the table.

b. **Algebraic** Write an equation that represents the data in the table.

c. **Graphical** Graph each ordered pair and draw the function. Describe the graph as discrete or continuous.

H.O.T. Problems Use Higher-Order Thinking Skills

47. **OPEN ENDED** Write a problem about your favorite television show that uses the equation $x + 8 = 30$.

48. **CCSS REASONING** The surface area of a three-dimensional object is the sum of the areas of the faces. If ℓ represents the length of the side of a cube, write a formula for the surface area of the cube.

49. **CHALLENGE** Given the perimeter P and width w of a rectangle, write a formula to find the length ℓ.

50. **WRITING IN MATH** How can you translate a verbal sentence into an algebraic equation?

Standardized Test Practice

51. Which equation *best* represents the relationship between the number of hours an electrician works h and the total charges c?

Cost of Electrician	
Emergency House Call	\$30 one time fee
Rate	\$55/hour

A $c = 30 + 55$
B $c = 30h + 55$
C $c = 30 + 55h$
D $c = 30h + 55h$

52. A car traveled at 55 miles per hour for 2.5 hours and then at 65 miles per hour for 3 hours. How far did the car travel in all?

F 300.5 mi
G 305 mi
H 330 mi
J 332.5 mi

53. SHORT RESPONSE Suppose each dimension of rectangle $ABCD$ is doubled. What is the perimeter of the new $ABCD$?

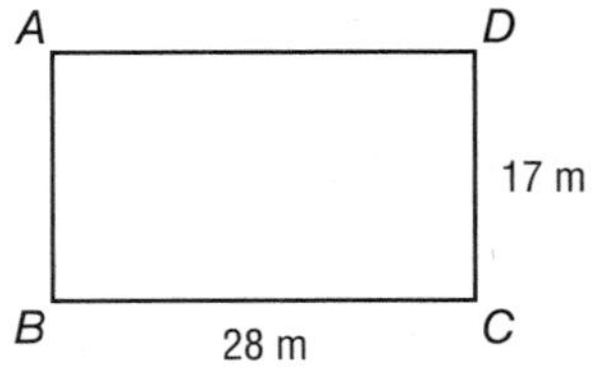

54. STATISTICS Stacy's first five science test scores were 95, 86, 83, 95, and 99. Which of the following is a true statement?

A The mode is the same as the median.
B The median is the same as the mean.
C The range is the same as the mode.
D The mode is the same as the mean.

Spiral Review

55. POPULATION Identify the function graphed as *linear* or *nonlinear.* Then estimate and interpret the intercepts of the graph, any symmetry, where the function is positive, negative, increasing, and decreasing, the x-coordinate of any relative extrema, and the end behavior of the graph. (Lesson 1-8)

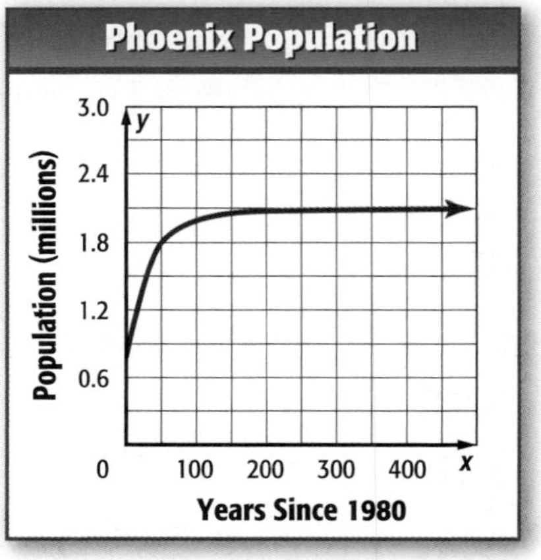

56. SHOPPING Cuties is having a sale on earrings that are regularly \$29 for each pair. If you buy 2 pairs, you get 1 pair free. (Lesson 1-7)

a. Make a table that shows the cost of buying 1 to 5 pairs of earrings.

b. Write the data as a set of ordered pairs.

c. Graph the data.

57. GEOMETRY Refer to the table below. (Lesson 1-6)

Polygon	triangle	quadrilateral	pentagon	hexagon	heptagon
Number of Sides	3	4	5	6	7
Interior Angle Sum	180	360	540	720	900

a. Identify the independent and dependent variables.

b. Identify the domain and range for this situation.

c. State whether the function is *discrete* or *continuous.* Explain.

Skills Review

Evaluate each expression.

58. 9^2 **59.** 10^6 **60.** 3^5 **61.** 5^3

EXPLORE 2-2

Algebra Lab
Solving Equations

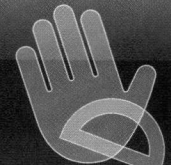

You can use **algebra tiles** to model solving equations. To **solve an equation** means to find the value of the variable that makes the equation true. An x tile represents the variable x. The 1 tile represents a positive 1. The −1 tile represents a negative 1. And, the −x tile represents the variable negative x. The goal is to get the x-tile by itself on one side of the mat by using the rules stated below.

CCSS Common Core State Standards
Content Standards
A.REI.3 Solve linear equations and inequalities in one variable, including equations with coefficients represented by letters.
Mathematical Practices
8 Look for and express regularity in repeated reasoning.

Rules for Equation Models When Adding or Subtracting:

- You can remove or add the same number of identical algebra tiles to each side of the mat without changing the equation.

- One positive tile and one negative tile of the same unit are called a zero pair. Since $1 + (-1) = 0$, you can remove or add zero pairs to either side of the equation mat without changing the equation.

Activity 1 Addition Equation

Use an equation model to solve $x + 3 = -4$.

Step 1 Model the equation. Place 1 x-tile and 3 positive 1-tiles on one side of the mat. Place 4 negative 1-tiles on the other side of the mat.

Step 2 Isolate the x-term. Add 3 negative 1-tiles to each side. The resulting equation is $x = -7$.

$x + 3 = -4$
$x + 3 + (-3) = -4 + (-3)$
$x = -7$

Activity 2 Subtraction Equation

Use an equation model to solve $x - 2 = 1$.

Step 1

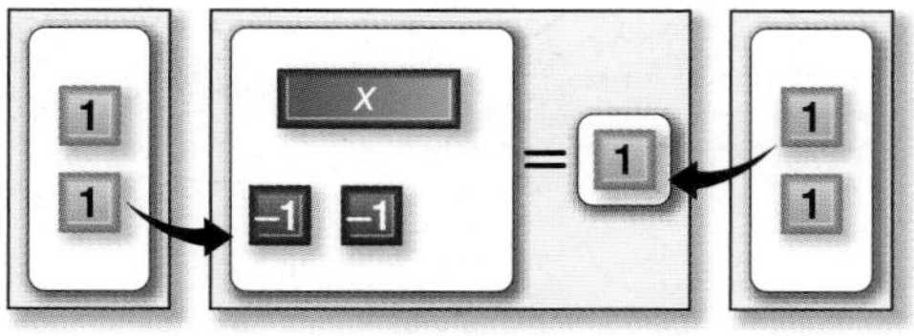

$x - 2 = 1$
$x - 2 + 2 = 1 + 2$

Place 1 x-tile and 2 negative 1-tiles on one side of the mat. Place 1 positive 1-tile on the other side of the mat. Then add 2 positive 1-tiles to each side.

Step 2

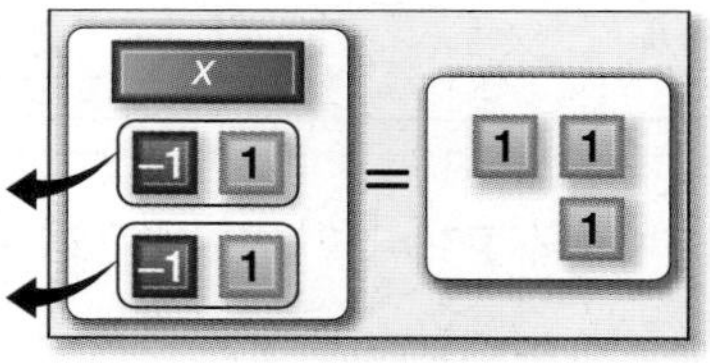

$x = 3$

Group the tiles to form zero pairs. Then remove all the zero pairs. The resulting equation is $x = 3$.

(continued on the next page)

Algebra Lab

Solving Equations *Continued*

Model and Analyze

Use algebra tiles to solve each equation.

1. $x + 4 = 9$ **2.** $x + (-3) = -4$ **3.** $x + 7 = -2$ **4.** $x + (-2) = 11$

5. **WRITING IN MATH** If $a = b$, what can you say about $a + c$ and $b + c$? about $a - c$ and $b - c$?

When solving multiplication equations, the goal is still to get the *x*-tile by itself on one side of the mat by using the rules for dividing.

Rules for the Equation Models When Dividing:

- You can group the tiles on each side of the equation mat into an equal number of groups without changing the equation.

- You can place an equal grouping on each side of the equation mat without changing the equation.

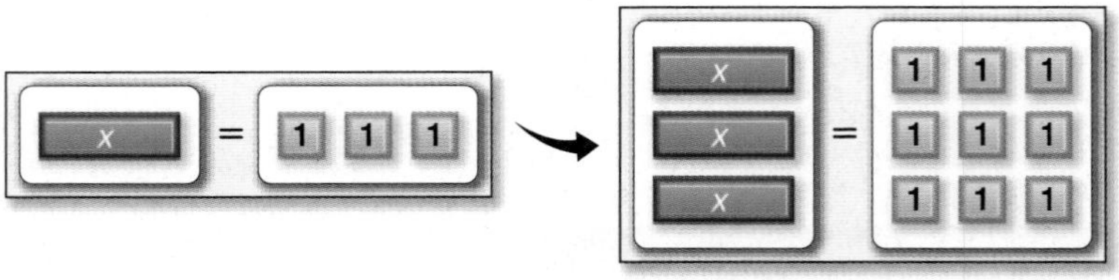

Activity 3 Multiplication Equations

Use an equation model to solve $3x = 12$.

Step 1 Model the equation. Place 3 *x*-tiles on one side of the mat. Place 12 positive 1-tiles on the other side of the mat.

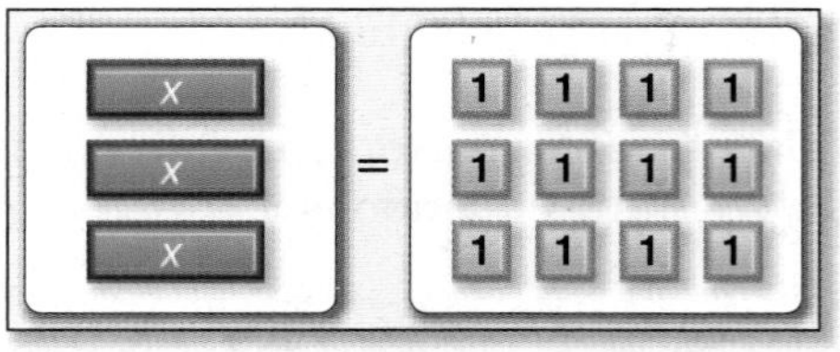

Step 2 Isolate the *x*-term. Separate the tiles into 3 equal groups to match the 3 *x*-tiles. Each *x*-tile is paired with 4 positive 1-tiles. The resulting equation is $x = 4$.

$$3x = 12$$
$$\frac{3x}{3} = \frac{12}{3}$$
$$x = 4$$

Model and Analyze

Use algebra tiles to solve each equation.

6. $5x = -15$ **7.** $-3x = -9$ **8.** $4x = 8$ **9.** $-6x = 18$

10. **MAKE A CONJECTURE** How would you use algebra tiles to solve $\frac{x}{4} = 5$? Discuss the steps you would take to solve this equation algebraically.

LESSON 2-2 Solving One-Step Equations

Then

- You translated sentences into equations.

Now

1. Solve equations by using addition and subtraction.
2. Solve equations by using multiplication and division.

Why?

- A record for the most snow angels made at one time was set in Michigan when 3784 people participated. North Dakota had 8910 people register to break the record. To determine how many more people North Dakota had than Michigan, solve the equation $3784 + x = 8910$.

NewVocabulary
solve an equation
equivalent equations

Common Core State Standards

Content Standards

A.REI.1 Explain each step in solving a simple equation as following from the equality of numbers asserted at the previous step, starting from the assumption that the original equation has a solution. Construct a viable argument to justify a solution method.

A.REI.3 Solve linear equations and inequalities in one variable, including equations with coefficients represented by letters.

Mathematical Practices

6 Attend to precision.

1 Solve Equations Using Addition or Subtraction

In an equation, the variable represents the number that satisfies the equation. To **solve an equation** means to find the value of the variable that makes the equation true.

The process of solving an equation requires assuming that the original equation has a solution and isolating the variable (with a coefficient of 1) on one side of the equation. Each step in this process results in equivalent equations. **Equivalent equations** have the same solution.

KeyConcept Addition Property of Equality

Words If an equation is true and the same number is added to each side of the equation, the resulting equivalent equation is also true.

Symbols For any real numbers a, b, and c, if $a = b$, then $a + c = b + c$.

Examples

$$14 = 14 \qquad\qquad -3 = -3$$
$$14 + 3 = 14 + 3 \qquad\qquad +9 = +9$$
$$17 = 17 \qquad\qquad 6 = 6$$

Example 1 Solve by Adding

Solve $c - 22 = 54$.

Horizontal Method

$c - 22 = 54$	Original equation
$c - 22 + 22 = 54 + 22$	Add 22 to each side.
$c = 76$	Simplify.

Vertical Method

$$c - 22 = 54$$
$$+22 = +22$$
$$c = 76$$

To check that 76 is the solution, substitute 76 for c in the original equation.

CHECK	$c - 22 = 54$	Original equation
	$76 - 22 \stackrel{?}{=} 54$	Substitute 76 for c.
	$54 = 54$ ✓	Subtract.

GuidedPractice

1A. $113 = g - 25$

1B. $j - 87 = -3$

Similar to the Addition Property of Equality, the **Subtraction Property of Equality** can also be used to solve equations.

StudyTip

Subtraction Subtracting a value is equivalent to adding the opposite of the value.

KeyConcept Subtraction Property of Equality

Words If an equation is true and the same number is subtracted from each side of the equation, the resulting equivalent equation is also true.

Symbols For any real numbers a, b, and c, if $a = b$, then $a - c = b - c$.

Examples

$$87 = 87$$
$$87 - 17 = 87 - 17$$
$$70 = 70$$

$$13 = 13$$
$$-28 = -28$$
$$-15 = -15$$

StudyTip

Solving an Equation When solving equations you can use either the horizontal method or the vertical method. Both methods will produce the same result.

Example 2 Solve by Subtracting

Solve $63 + m = 79$.

Horizontal Method

$63 + m = 79$	Original equation
$63 - \mathbf{63} + m = 79 - \mathbf{63}$	Subtract 63 from each side.
$m = 16$	Simplify.

Vertical Method

$$63 + m = 79$$
$$\underline{-63 \qquad = -63}$$
$$m = 16$$

To check that 16 is the solution, replace m with 16 in the original equation.

CHECK

$63 + m = 79$	Original equation
$63 + \mathbf{16} \stackrel{?}{=} 79$	Substitution, $m = 16$
$79 = 79$ ✓	Simplify.

GuidedPractice

2A. $27 + k = 30$

2B. $-12 = p + 16$

2 Solve Equations Using Multiplication or Division

In the equation $\frac{x}{3} = 9$, the variable x is divided by 3. To solve for x, undo the division by multiplying each side by 3. This is an example of the **Multiplication Property of Equality**.

KeyConcept Multiplication Property of Equality

Words If an equation is true and each side is multiplied by the same nonzero number, the resulting equation is equivalent.

Symbols For any real numbers a, b, and c, $c \neq 0$, if $a = b$, then $ac = bc$.

Example If $x = 5$, then $3x = 15$.

Division Property of Equality

Words If an equation is true and each side is divided by the same nonzero number, the resulting equation is equivalent.

Symbols For any real numbers a, b, and c, $c \neq 0$, if $a = b$, then $\frac{a}{c} = \frac{b}{c}$.

Example If $x = -20$, then $\frac{x}{5} = \frac{-20}{5}$ or -4.

The reciprocal of a number can be used to solve equations.

PT

Example 3 Solve by Multiplying or Dividing

Solve each equation.

a. $\frac{2}{3}q = \frac{1}{2}$

$\frac{2}{3}q = \frac{1}{2}$ — Original equation

$\frac{3}{2}\left(\frac{2}{3}\right)q = \frac{3}{2}\left(\frac{1}{2}\right)$ — Multiply each side by $\frac{3}{2}$, the reciprocal of $\frac{2}{3}$.

$q = \frac{3}{4}$ — Check the result.

b. $39 = -3r$

$39 = -3r$ — Original equation

$\frac{39}{-3} = \frac{-3r}{-3}$ — Divide each side by −3.

$-13 = r$ — Check the result.

Guided Practice

3A. $\frac{3}{5}k = 6$

3B. $-\frac{1}{4} = \frac{2}{3}b$

Review Vocabulary

reciprocal the multiplicative inverse of a number

We can also use reciprocals and properties of equality to solve real-world problems.

PT

Real-World Example 4 Solve by Multiplying

SURVEYS Of a group of 13- to 15-year-old girls surveyed, 225, or about $\frac{9}{20}$ said they talk on the telephone while they watch television. About how many girls were surveyed?

Words	Nine twentieths times those surveyed	is	225.
Variable	Let g = the number of girls surveyed.		
Equation	$\frac{9}{20}g$	=	225

$\frac{9}{20}g = 225$ — Original equation

$\left(\frac{20}{9}\right)\frac{9}{20}g = \left(\frac{20}{9}\right)225$ — Multiply each side by $\frac{20}{9}$.

$g = \frac{4500}{9}$ — $\left(\frac{20}{9}\right)\left(\frac{9}{20}\right) = 1$

$g = 500$ — Simplify.

About 500 girls were surveyed.

Guided Practice

4. STAINED GLASS Allison is making a stained glass window. Her pattern requires that one fifth of the glass should be blue. She has 288 square inches of blue glass. If she intends to use all of her blue glass, how much glass will she need for the entire project?

Ronnie Kaufman/Blend Images/Photolibrary/Index Stock

Real-World Link

Almost half of 10- to 18-year-olds in the U.S. use a cell phone. Of those, 53% play games on their phones, more than 33% download games, 52% use the calendar/organizer, and nearly all teens with camera phones snap pictures.

Source: Lexdon Business Library

Check Your Understanding

Examples 1–3 **Solve each equation. Check your solution.**

1. $g + 5 = 33$

2. $104 = y - 67$

3. $\frac{2}{3} + w = 1\frac{1}{2}$

4. $-4 + t = -7$

5. $a + 26 = 35$

6. $-6 + c = 32$

7. $1.5 = y - (-5.6)$

8. $3 + g = \frac{1}{4}$

9. $x + 4 = \frac{3}{4}$

10. $\frac{t}{7} = -5$

11. $\frac{a}{36} = \frac{4}{9}$

12. $\frac{2}{3}n = 10$

13. $\frac{8}{9} = \frac{4}{5}k$

14. $12 = \frac{x}{-3}$

15. $-\frac{r}{4} = \frac{1}{7}$

Example 4

16. **FUNDRAISING** The television show "Idol Gives Back" raised money for relief organizations. During this show, viewers could call in and vote for their favorite performer. The parent company contributed \$5 million for the 50 million votes cast. What did they pay for each vote?

17. CCSS **REASONING** Hana decides to buy her cat a bed from an online fund that gives $\frac{7}{8}$ of her purchase to shelters that care for animals. How much of Hana's money went to the animal shelter?

Practice and Problem Solving

Examples 1–3 **Solve each equation. Check your solution.**

18. $v - 9 = 14$

19. $44 = t - 72$

20. $-61 = d + (-18)$

21. $18 + z = 40$

22. $-4a = 48$

23. $12t = -132$

24. $18 - (-f) = 91$

25. $-16 - (-t) = -45$

26. $\frac{1}{3}v = -5$

27. $\frac{u}{8} = -4$

28. $\frac{a}{6} = -9$

29. $-\frac{k}{5} = \frac{7}{5}$

30. $\frac{3}{4} = w + \frac{2}{5}$

31. $-\frac{1}{2} + a = \frac{5}{8}$

32. $-\frac{t}{7} = \frac{1}{15}$

33. $-\frac{5}{7} = y - 2$

34. $v + 914 = -23$

35. $447 + x = -261$

36. $-\frac{1}{7}c = 21$

37. $-\frac{2}{3}h = -22$

38. $\frac{3}{5}q = -15$

39. $\frac{n}{8} = -\frac{1}{4}$

40. $\frac{c}{4} = -\frac{9}{8}$

41. $\frac{2}{3} + r = -\frac{4}{9}$

Example 4

42. **CATS** A domestic cat can run at speeds of 27.5 miles per hour when chasing prey. A cheetah can run 42.5 miles per hour faster when chasing prey. How fast can the cheetah go?

43. **CARS** The average time t it takes to manufacture a car in the United States is 24.9 hours. This is 8.1 hours longer than the average time it takes to manufacture a car in Japan. Write and solve an equation to find the average time in Japan.

Solve each equation. Check your solution.

44. $\frac{x}{9} = 10$ **45.** $\frac{b}{7} = -11$ **46.** $\frac{3}{4} = \frac{c}{24}$

47. $\frac{2}{3} = \frac{1}{8}y$ **48.** $\frac{2}{3}n = 14$ **49.** $\frac{3}{5}g = -6$

50. $4\frac{1}{5} = 3p$ **51.** $-5 = 3\frac{1}{2}x$ **52.** $6 = -\frac{1}{2}n$

53. $-\frac{2}{5} = -\frac{z}{45}$ **54.** $-\frac{g}{24} = \frac{5}{12}$ **55.** $-\frac{v}{5} = -45$

Write an equation for each sentence. Then solve the equation.

56. Six times a number is 132.

57. Two thirds equals negative eight times a number.

58. Five elevenths times a number is 55.

59. Four fifths is equal to ten sixteenths of a number.

60. Three and two thirds times a number equals two ninths.

61 Four and four fifths times a number is one and one fifth.

62. CCSS **PRECISION** Adelina is comparing prices for two brands of health and energy bars at the local grocery store. She wants to get the best price for each bar.

a. Write an equation to find the price for each bar of the Feel Great brand.

b. Write an equation to find the price of each bar for the Super Power brand.

c. Which bar should Adelina buy? Explain.

63. **MEDIA** The world's largest passenger plane, the Airbus A380, was first used by Singapore Airlines in 2005. The following description appeared on a news Web site after the plane was introduced.

"That airline will see the A380 transporting some 555 passengers, 139 more than a similarly set-up 747."

How many passengers will a similarly set-up 747 transport?

64. **FUEL** In 2004, approximately 5 million cars and trucks were classified as flex-fuel, which means they could run on gasoline or ethanol. In 2009, that number increased to about 8 million. How many more cars and trucks were flex-fuel in 2009?

65. **CHEERLEADING** At a certain cheerleading competition, the maximum time per team, including the set up, is 3 minutes. The Ridgeview High School squad's performance time is 2 minutes and 34 seconds. How much time does the squad have left for their set up?

66. **COMIC BOOKS** An X-Men #1 comic book in mint condition recently sold for $45,000. An Action Comics #63 (Mile High), also in mint condition, sold for $15,000. How much more did the X-Men comic book sell for than the Action Comics book?

67. **MOVIES** A certain movie made $1.6 million in ticket sales. Its sequel made $0.8 million in ticket sales. How much more did the first movie make than the sequel?

68. **CAMERAS** An electronics store sells a certain digital camera for $126. This is $\frac{2}{3}$ of the price that a photography store charges. What is the cost of the camera at the photography store?

69 **BLOGS** In 2006, 57 million American adults read online blogs. However, 45 million fewer American adults say that they maintain their own blog. How many American adults maintain a blog?

70. **SCIENCE CAREERS** According to the Bureau of Labor and Statistics, approximately 140,000,000 people were employed in the United States in 2009.

a. The number of people in production occupations times 20 is the number of working people. Write an equation to represent the number of people employed in production occupations in 2009. Then solve the equation.

b. The number of people in repair occupations is 2,300,000 less than the number of people in production occupations. How many people are in repair occupations?

71. **DANCES** Student Council has a budget of $1000 for the homecoming dance. So far, they have spent $350 dollars for music.

a. Write an equation to represent the amount of money left to spend. Then solve the equation.

b. They then spent $225 on decorations. Write an equation to represent the amount of money left.

c. If the Student Council spent their entire budget, write an equation to represent how many $6 tickets they must sell to make a profit.

H.O.T. Problems Use Higher-Order Thinking Skills

72. **WHICH ONE DOESN'T BELONG?** Identify the equation that does not belong with the other three. Explain your reasoning.

$n + 14 = 27$	$12 + n = 25$	$n - 16 = 29$	$n - 4 = 9$

73. **OPEN ENDED** Write an equation involving addition and demonstrate two ways to solve it.

74. **REASONING** For which triangle is the height not $4\frac{1}{2}b$, where b is the length of the base?

Triangle	Base (cm)	Height (cm)
$\triangle ABC$	3.8	17.1
$\triangle MQP$	5.4	24.3
$\triangle RST$	6.3	28.5
$\triangle TRW$	1.6	7.2

75. CCSS **STRUCTURE** Determine whether each sentence is *sometimes*, *always*, or *never true*. Explain your reasoning.

a. $x + x = x$ **b.** $x + 0 = x$

76. **REASONING** Determine the value for each statement below.

a. If $x - 7 = 14$, what is the value of $x - 2$?

b. If $t + 8 = -12$, what is the value of $t + 1$?

77. **CHALLENGE** Solve each equation for x. Assume that $a \neq 0$.

a. $ax = 12$ **b.** $x + a = 15$ **c.** $-5 = x - a$ **d.** $\frac{1}{a}x = 10$

78. **WRITING IN MATH** Consider the Multiplication Property of Equality and the Division Property of Equality. Explain why they can be considered the same property. Which one do you think is easier to use?

Standardized Test Practice

79. Which of the following best represents the equation $w - 15 = 33$?

A Jake added w ounces of water to his bottle, which originally contained 33 ounces of water. How much water did he add?

B Jake added 15 ounces of water to his bottle, for a total of 33 ounces. How much water w was originally in the bottle?

C Jake drank 15 ounces of water from his bottle and 33 ounces were left. How much water w was originally in the bottle?

D Jake drank 15 ounces of water from his water bottle, which originally contained 33 ounces. How much water w was left?

80. SHORT RESPONSE Charlie's company pays him for every mile that he drives on his trip. When he drives 50 miles, he is paid \$30. To the nearest tenth, how many miles did he drive if he was paid \$275?

81. The table shows the results of a survey given to 500 international travelers. Based on the data, which statement is true?

Vacation Plans

Destination	Percent
The Tropics	37
Europe	19
Asia	17
Other	17
No Vacation	10

F Fifty have no vacation plans.

G Fifteen are going to Asia.

H One third are going to the tropics.

J One hundred are going to Europe.

82. GEOMETRY The amount of water needed to fill a pool represents the pool's ____.

A volume

B surface area

C circumference

D perimeter

Spiral Review

Translate each sentence into an equation. (Lesson 2-1)

83. The sum of twice r and three times k is identical to thirteen.

84. The quotient of t and forty is the same as twelve minus half of u.

85. The square of m minus the cube of p is sixteen.

86. TOYS Identify the function graphed as *linear* or *nonlinear.* Then estimate and interpret the intercepts of the graph, any symmetry, where the function is positive, negative, increasing, and decreasing, the x-coordinate of any relative extrema, and the end behavior of the graph. (Lesson 1-8)

Distance from Toy Car

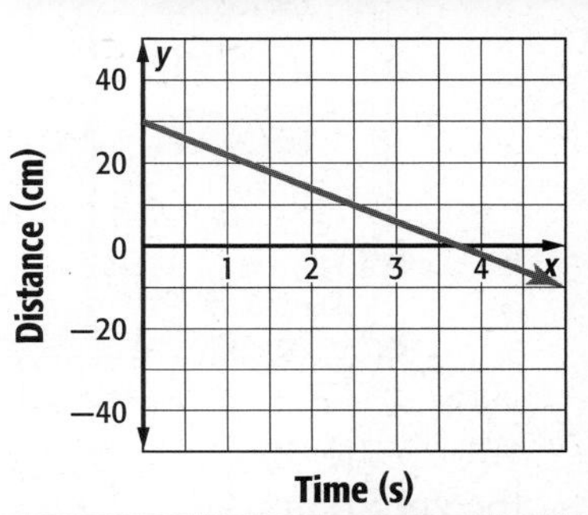

Skills Review

87. COMMUNICATION Sato communicates with friends for a project. He averages 5 hours using email, 8 hours on the phone, and 2 hours with them in person the first week. If this trend continues, write and evaluate an expression to predict how many hours he will spend communicating with friends over the next 12 weeks.

88. PETS The Poochie Pet supply store has the following items on sale. Write and evaluate an expression to find the total cost of purchasing 1 collar, 2 T-shirts, 3 kerchiefs, 1 leash, and 4 flying disks.

Item	Cost (\$)
studded collar	4.50
kerchief	3.00
doggy T-shirt	6.25
leash	5.50
flying disk	3.25

EXPLORE

2-3 Algebra Lab
Solving Multi-Step Equations

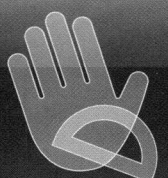

You can use algebra tiles to model solving multi-step equations.

CCSS Common Core State Standards
Content Standards
A.REI.3 Solve linear equations and inequalities in one variable, including equations with coefficients represented by letters.

Activity

Use an equation model to solve $4x + 3 = -5$.

Step 1 Model the equation.

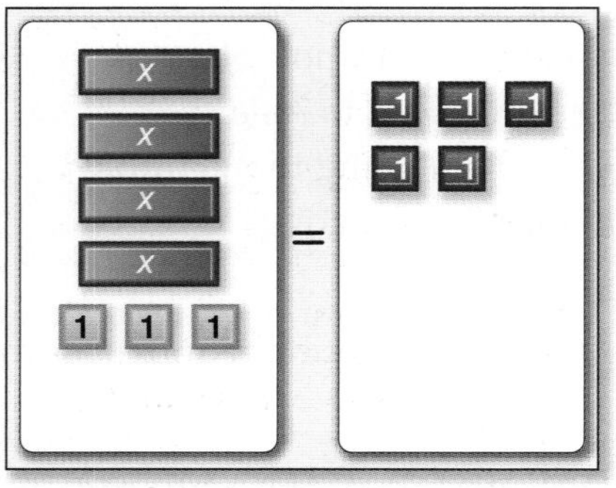

$4x + 3 = -5$

Place 4 x-tiles and 3 positive 1-tiles on one side of the mat. Place 5 negative 1-tiles on the other side.

Step 2 Isolate the x-term.

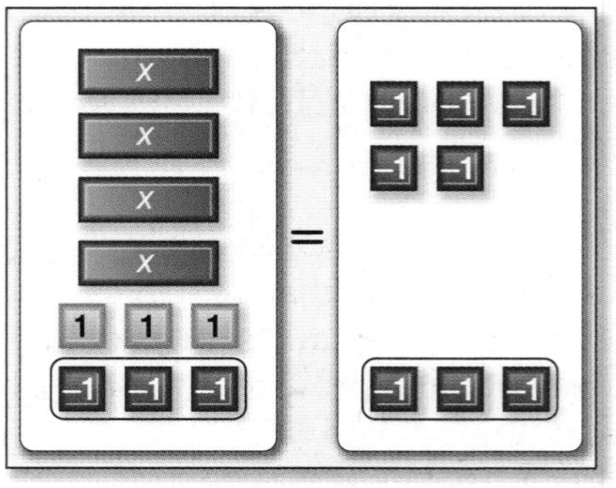

$4x + 3 - 3 = -5 - 3$

Since there are 3 positive 1-tiles with the x-tiles, add 3 negative 1-tiles to each side to form zero pairs.

Step 3 Remove zero pairs.

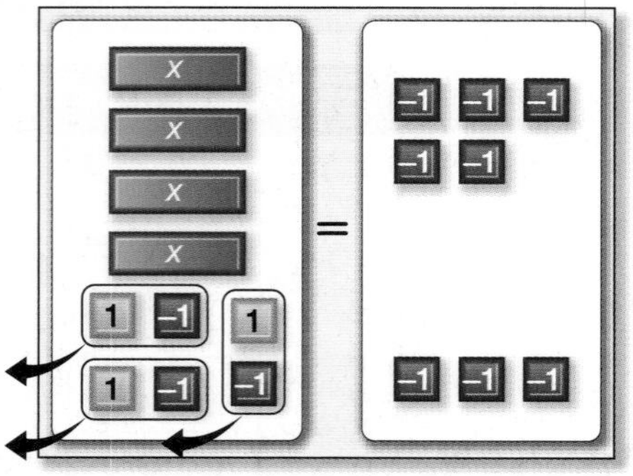

$4x = -8$

Group the tiles to form zero pairs and remove the zero pairs.

Step 4 Group the tiles.

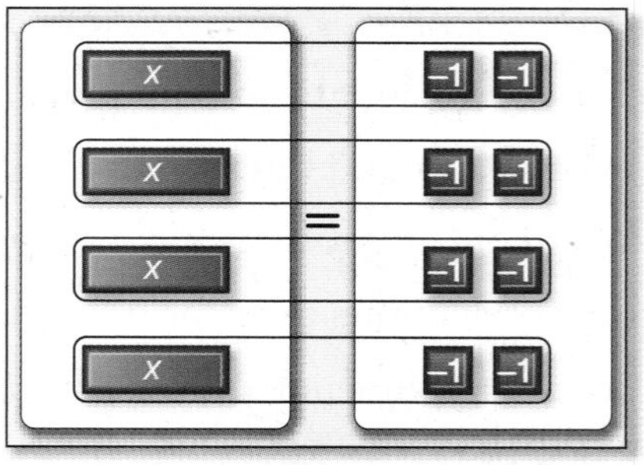

$\frac{4x}{4} = \frac{-8}{4}$

$x = -2$

Separate the remaining tiles into 4 equal groups to match the 4 x-tiles. Each x-tile is paired with 2 negative 1-tiles. The resulting equation is $x = -2$.

Model

Use algebra tiles to solve each equation.

1. $3x - 7 = -10$

2. $2x + 5 = 9$

3. $5x - 7 = 8$

4. $-7 = 3x + 8$

5. $5 + 4x = -11$

6. $3x + 1 = 7$

7. $11 = 2x - 5$

8. $7 + 6x = -11$

9. What would be your first step in solving $8x - 29 = 67$?

10. What steps would you use to solve $9x + 14 = -49$?

LESSON 2-3

Solving Multi-Step Equations

Then

- You solved one-step equations.

Now

1. Solve equations involving more than one operation.
2. Solve equations involving consecutive integers.

Why?

- The Tour de France is the premier cycling event in the world. The map shows the 2007 Tour de France course. If the length of the shortest portion of the race can be represented by k, the expression $4k + 20$ is the length of the longest stage or 236 kilometers. This can be described by the equation $4k + 20 = 236$.

NewVocabulary
multi-step equation
consecutive integers
number theory

Common Core State Standards

Content Standards

A.REI.1 Explain each step in solving a simple equation as following from the equality of numbers asserted at the previous step, starting from the assumption that the original equation has a solution. Construct a viable argument to justify a solution method.

A.REI.3 Solve linear equations and inequalities in one variable, including equations with coefficients represented by letters.

Mathematical Practices

8 Look for and express regularity in repeated reasoning.

1 Solve Multi-Step Equations

Since the above equation requires more than one step to solve, it is called a **multi-step equation**. To solve this equation, we must undo each operation by working backward.

Example 1 Solve Multi-Step Equations

Solve each equation. Check your solution.

a. $11x - 4 = 29$

$11x - 4 = 29$	Original equation
$11x - 4 + 4 = 29 + 4$	Add 4 to each side.
$11x = 33$	Simplify.
$\frac{11x}{11} = \frac{33}{11}$	Divide each side by 11.
$x = 3$	Simplify.

b. $\frac{a + 7}{8} = 5$

$\frac{a + 7}{8} = 5$	Original equation
$8\left(\frac{a + 7}{8}\right) = 8(5)$	Multiply each side by 8.
$a + 7 = 40$	Simplify.
$-7 = -7$	Subtract 7 from each side.
$a = 33$	Simplify.

You can check your solutions by substituting the results back into the original equations.

GuidedPractice

Solve each equation. Check your solution.

1A. $2a - 6 = 4$

1B. $\frac{n + 1}{-2} = 15$

Real-WorldLink

Shoppers in Shanghai, China, can pay for purchased items at a terminal that can match the buyers' fingerprints with their bank accounts.

Source: *Shanghai Daily*

Real-World Example 2 Write and Solve a Multi-Step Equation

SHOPPING Hiroshi is buying a pair of water skis that are on sale for $\frac{2}{3}$ of the original price. After he uses a \$25 gift certificate, the total cost before taxes is \$115. What was the original price of the skis? Write an equation for the problem. Then solve the equation.

Words	Two thirds	of	the price	minus	25	is	115.
Variable	Let $p =$ original price of the skis.						
Equation	$\frac{2}{3}$	$\cdot$	p	$-$	25	$=$	115

$$\frac{2}{3}p - 25 = 115 \quad \text{Original equation}$$

$$\frac{2}{3}p - 25 + 25 = 115 + 25 \quad \text{Add 25 to each side.}$$

$$\frac{2}{3}p = 140 \quad \text{Simplify.}$$

$$\frac{3}{2}\left(\frac{2}{3}p\right) = \frac{3}{2}(140) \quad \text{Multiply each side by } \frac{3}{2}.$$

$$p = 210 \quad \text{Simplify.}$$

The original price of the skis was \$210.

Guided Practice

2A. **RETAIL** A music store has sold $\frac{3}{5}$ of their hip-hop CDs, but 10 were returned. Now the store has 62 hip-hop CDs. How many were there originally?

2B. **READING** Len read $\frac{3}{4}$ of a graphic novel over the weekend. Monday, he read 22 more pages. If he has read 220 pages, how many pages does the book have?

2 Solve Consecutive Integer Problems

Consecutive integers are integers in counting order, such as 4, 5, and 6 or n, $n + 1$, and $n + 2$. Counting by two will result in *consecutive even integers* if the starting integer n is even and *consecutive odd integers* if the starting integer n is odd.

Concept Summary Consecutive Integers

Type	Words	Symbols	Example
Consecutive Integers	Integers that come in counting order.	$n, n + 1, n + 2, \dots$	$\dots, -2, -1, 0, 1, 2, \dots$
Consecutive Even Integers	Even integer followed by the next even integer.	$n, n + 2, n + 4, \dots$	$\dots, -2, 0, 2, 4, \dots$
Consecutive Odd Integers	Odd integer followed by the next odd integer.	$n, n + 2, n + 4, \dots$	$\dots, -1, 1, -3, 5, \dots$

Number theory is the study of numbers and the relationships between them.

William Thomas Cain/Stringer/Getty Images News/Getty Images

Example 3 Solve a Consecutive Integer Problem

NUMBER THEORY Write an equation for the following problem. Then solve the equation and answer the problem.

Find three consecutive odd integers with a sum of −51.

Let n = the least odd integer.

Then $n + 2$ = the next greater odd integer, and $n + 4$ = the greatest of the three integers.

Words	The sum of three consecutive odd integers	is	−51.
Equation	$n + (n + 2) + (n + 4)$	=	−51.

$n + (n + 2) + (n + 4) = -51$ Original equation

$3n + 6 = -51$ Simplify.

$-6 = -6$ Subtract 6 from each side.

$3n = -57$ Simplify.

$\frac{3n}{3} = \frac{-57}{3}$ Divide each side by 3.

$n = -19$ Simplify.

$n + 2 = -19 + 2$ or -17 $n + 4 = -19 + 4$ or -15

The consecutive odd integers are −19, −17, and −15.

CHECK −19, −17, and −15 are consecutive odd integers.
$-19 + (-17) + (-15) = -51$ ✓

Study Tip

CCSS Regularity You can use the same expressions to represent either consecutive even integers or consecutive odd integers. It is the value of n (odd or even) that differs between the two expressions.

Guided Practice

3. Write an equation for the following problem. Then solve the equation and answer the problem.

Find three consecutive integers with a sum of 21.

Check Your Understanding

Example 1 **Solve each equation. Check your solution.**

1. $3m + 4 = -11$

2. $12 = -7f - 9$

3. $-3 = 2 + \frac{a}{11}$

4. $\frac{3}{2}a - 8 = 11$

5. $8 = \frac{x - 5}{7}$

6. $\frac{c + 1}{-3} = -21$

Example 2

7. NUMBER THEORY Twelve decreased by twice a number equals −34. Write an equation for this situation and then find the number.

8. BASEBALL Among the career home run leaders for Major League Baseball, Hank Aaron has 175 fewer than twice the number that Dave Winfield has. Hank Aaron hit 755 home runs. Write an equation for this situation. How many home runs did Dave Winfield hit in his career?

Example 3 **Write an equation and solve each problem.**

9. Find three consecutive odd integers with a sum of 75.

10. Find three consecutive integers with a sum of −36.

Practice and Problem Solving

Example 1 **Solve each equation. Check your solution.**

11. $3t + 7 = -8$

12. $8 = 16 + 8n$

13. $-34 = 6m - 4$

14. $9x + 27 = -72$

15. $\frac{y}{5} - 6 = 8$

16. $\frac{f}{-7} - 8 = 2$

17. $1 + \frac{r}{9} = 4$

18. $\frac{k}{3} + 4 = -16$

19. $\frac{n-2}{7} = 2$

20. $14 = \frac{6+z}{-2}$

21. $-11 = \frac{a-5}{6}$

22. $\frac{22-w}{3} = -7$

Example 2 **23** **FINANCIAL LITERACY** The Cell+ Cellular Phone store offers the plans shown in the table. Raul chose the business plan and has budgeted $100 per month. Write an equation for this situation, and determine how many minutes per month he can use the phone and stay within budget.

Plan	Flat Monthly Fee	Anytime Minutes	Cost per Minute After Anytime Minutes
personal	$29.99	250	$0.20
business	$49.99	650	$0.15
executive	$59.99	1200	$0.10

Example 3 **Write an equation and solve each problem.**

24. Fourteen less than three fourths of a number is negative eight. Find the number.

25. Seventeen is thirteen subtracted from six times a number. What is the number?

26. Find three consecutive even integers with the sum of −84.

27. Find three consecutive odd integers with the sum of 141.

28. Find four consecutive integers with the sum of 54.

29. Find four consecutive integers with the sum of −142.

Solve each equation. Check your solution.

30. $-6m - 8 = 24$

31. $45 = 7 - 5n$

32. $\frac{2b}{3} + 6 = 24$

33. $\frac{5x}{9} - 11 = -51$

34. $65 = \frac{3}{4}c - 7$

35. $9 + \frac{2}{3}x = 81$

36. $-\frac{5}{2} = \frac{3}{4}z + \frac{1}{2}$

37. $\frac{5}{6}k + \frac{2}{3} = \frac{4}{3}$

38. $-\frac{1}{5} - \frac{4}{9}a = \frac{2}{15}$

39. $-\frac{3}{7} = \frac{3}{4} - \frac{b}{2}$

Write an equation and solve each problem.

40. **CCSS REASONING** The ages of three brothers are consecutive integers with the sum of 96. How old are the brothers?

41. **VOLCANOES** Moving lava can build up and form beaches at the coast of an island. The growth of an island in a seaward direction may be modeled as $8y + 2$ centimeters, where y represents the number of years that the lava flows. An island has expanded 60 centimeters seaward. How long has the lava flowed?

Solve each equation. Check your solution.

42. $-5x - 4.8 = 6.7$

43. $3.7q + 26.2 = 111.67$

44. $0.6a + 9 = 14.4$

45. $\frac{c}{2} - 4.3 = 11.5$

46. $9 = \frac{-6p - (-3)}{-8}$

47. $3.6 - 2.4m = 12$

48. If $7m - 3 = 53$, what is the value of $11m + 2$?

49. If $13y + 25 = 64$, what is the value of $4y - 7$?

50. If $-5c + 6 = -69$, what is the value of $6c - 15$?

51. **AMUSEMENT PARKS** An amusement park offers a yearly membership of $275 that allows for free parking and admission to the park. Members can also use the water park for an additional $5 per day. Nonmembers pay $6 for parking, $15 for admission, and $9 for the water park.

a. Write and solve an equation to find the number of visits it would take for the total cost to be the same for a member and a nonmember if they both use the water park at each visit.

b. Make a table for the costs of members and nonmembers after 3, 6, 9, 12, and 15 visits to the park.

c. Plot these points on a coordinate graph and describe what you see.

52. **SHOPPING** At The Family Farm, you can pick your own fruits and vegetables.

a. The cost of a bag of potatoes is $1.50 less than $\frac{1}{2}$ of the price of apples. Write and solve an equation to find the cost of potatoes.

b. The price of each zucchini is 3 times the price of winter squash minus $7. Write and solve an equation to find the cost of zucchini.

c. Write an equation to represent the cost of a pumpkin using the cost of the blueberries.

The Family Farm	
Fruit	**Price ($)**
Apples	6.99/bag
Pumpkins	5.00 each
Blueberries	2.99/qt
Winter squash	2.99 each

H.O.T. Problems Use Higher-Order Thinking Skills

53. **OPEN ENDED** Write a problem that can be modeled by the equation $2x + 40 = 60$. Then solve the equation and explain the solution in the context of the problem.

54. **CHALLENGE** Solve each equation for x. Assume that $a \neq 0$.

a. $ax + 7 = 5$

b. $\frac{1}{a}x - 4 = 9$

c. $2 - ax = -8$

55. **REASONING** Determine whether each equation has a solution. Justify your answer.

a. $\frac{a + 4}{5 + a} = 1$

b. $\frac{1 + b}{1 - b} = 1$

c. $\frac{c - 5}{5 - c} = 1$

56. CCSS **REGULARITY** Determine whether the following statement is *sometimes, always,* or *never* true. Explain your reasoning.

The sum of three consecutive odd integers equals an even integer.

57. **WRITING IN MATH** Write a paragraph explaining the order of the steps that you would take to solve a multi-step equation.

Standardized Test Practice

58. Which is the best estimate for the number of minutes on the calling card advertised below?

A 10 min
B 20 min
C 50 min
D 200 min

59. GRIDDED RESPONSE The scale factor for two similar triangles is 2 : 3. The perimeter of the smaller triangle is 56 cm. What is the perimeter of the larger triangle in centimeters?

60. Mr. Morrison is draining his cylindrical pool. The pool has a radius of 10 feet and a standard height of 4.5 feet. If the pool water is pumped out at a constant rate of 5 gallons per minute, about how long will it take to drain the pool? ($1 \text{ ft}^3 = 7.5$ gal)

F 37.8 min
G 7 h
H 25.4 h
J 35.3 h

61. STATISTICS Look at the golf scores for the five players in the table.

Player	1	2	3	4	5
Score	80	91	103	79	78

Which of these is the range of the golf scores?

A 10
B 25
C 35
D 40

Spiral Review

62. GAS MILEAGE A midsize car with a 4-cylinder engine travels 34 miles on a gallon of gas. This is 10 miles more than a luxury car with an 8-cylinder engine travels on a gallon of gas. How many miles does a luxury car travel on a gallon of gas? (Lesson 2-2)

63. DEER In a recent year, 1286 female deer were born in Clark County. That is 93 fewer than the number of male deer born. How many male deer were born that year? (Lesson 2-2)

Translate each equation into a verbal sentence. (Lesson 2-1)

64. $f - 15 = 6$

65. $3h + 7 = 20$

66. $k^2 + 18 = 54 - m$

67. $3p = 8p - r$

68. $\frac{3}{5}t + \frac{1}{3} = t$

69. $\frac{1}{2}v = \frac{2}{3}v + 4$

70. GEOGRAPHY The Pacific Ocean covers about 46% of Earth. If P represents the surface area of the Pacific Ocean and E represents the surface area of Earth, write an equation for this situation. (Lesson 2-1)

Find the value of n in each equation. Then name the property that is used. (Lesson 1-3)

71. $1.5 + n = 1.5$

72. $8n = 1$

73. $4 - n = 0$

74. $1 = 2n$

Skills Review

Evaluate each expression.

75. $5 + 3(4^2)$

76. $\frac{38 - 12}{2 \cdot 13}$

77. $[5(1 + 1)]^3$

78. $[8(2) - 4^2] + 7(4)$

LESSON 2-4

Solving Equations with the Variable on Each Side

Then

- You solved multi-step equations.

Now

1. Solve equations with the variable on each side.
2. Solve equations involving grouping symbols.

Why?

- The equation $y = 1.3x + 19$ represents the number of times Americans eat in their cars each year, where x is the number of years since 1985, and y is the number of times that they eat in their car. The equation $y = -1.3x + 93$ represents the number of times Americans eat in restaurants each year, where x is the number of years since 1985, and y is the number of times that they eat in a restaurant.

  The equation $1.3x + 19 = -1.3x + 93$ represents the year when the number of times Americans eat in their cars will equal the number of times Americans eat in restaurants.

NewVocabulary
identity

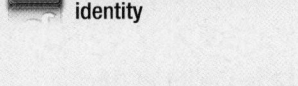

Common Core State Standards

Content Standards

A.REI.1 Explain each step in solving a simple equation as following from the equality of numbers asserted at the previous step, starting from the assumption that the original equation has a solution. Construct a viable argument to justify a solution method.

A.REI.3 Solve linear equations and inequalities in one variable, including equations with coefficients represented by letters.

Mathematical Practices

1. Make sense of problems and persevere in solving them.
5. Use appropriate tools strategically.

Gen Nishino/Taxi/Getty Images

1 Variables on Each Side

To solve an equation that has variables on each side, use the Addition or Subtraction Property of Equality to write an equivalent equation with the variable terms on one side.

Example 1 Solve an Equation with Variables on Each Side

Solve $2 + 5k = 3k - 6$. Check your solution.

$2 + 5k = 3k - 6$	Original equation
$\underline{-3k = -3k}$	Subtract $3k$ from each side.
$2 + 2k = -6$	Simplify.
$\underline{-2 \quad = -2}$	Subtract 2 from each side.
$2k = -8$	Simplify.
$\frac{2k}{2} = \frac{-8}{2}$	Divide each side by 2.
$k = -4$	Simplify.

CHECK $2 + 5k = 3k - 6$	Original equation
$2 + 5(-4) \stackrel{?}{=} 3(-4) - 6$	Substitution, $k = -4$
$2 + -20 \stackrel{?}{=} -12 - 6$	Multiply.
$-18 = -18$ ✓	Simplify.

GuidedPractice

Solve each equation. Check your solution.

1A. $3w + 2 = 7w$

1B. $5a + 2 = 6 - 7a$

1C. $\frac{x}{2} + 1 = \frac{1}{4}x - 6$

1D. $1.3c = 3.3c + 2.8$

2 Grouping Symbols

If equations contain grouping symbols such as parentheses or brackets, use the Distributive Property first to remove the grouping symbols.

Example 2 Solve an Equation with Grouping Symbols

Solve $6(5m - 3) = \frac{1}{3}(24m + 12)$.

$6(5m - 3) = \frac{1}{3}(24m + 12)$	Original equation
$30m - 18 = 8m + 4$	Distributive Property
$30m - 18 - 8m = 8m + 4 - 8m$	Subtract $8m$ from each side.
$22m - 18 = 4$	Simplify.
$22m - 18 + 18 = 4 + 18$	Add 18 to each side.
$22m = 22$	Simplify.
$\frac{22m}{22} = \frac{22}{22}$	Divide each side by 22.
$m = 1$	Simplify.

StudyTip

Solving an Equation You may want to eliminate the terms with a variable from one side before eliminating a constant.

GuidedPractice

Solve each equation. Check your solution.

2A. $8s - 10 = 3(6 - 2s)$

2B. $7(n - 1) = -2(3 + n)$

Some equations may have no solution. That is, there is no value of the variable that will result in a true equation. Some equations are true for all values of the variables. These are called **identities**.

Example 3 Find Special Solutions

Solve each equation.

a. $5x + 5 = 3(5x - 4) - 10x$

$5x + 5 = 3(5x - 4) - 10x$	Original equation
$5x + 5 = 15x - 12 - 10x$	Distributive Property
$5x + 5 = 5x - 12$	Simplify.
$-5x \quad = -5x$	Subtract $5x$ from each side.
$5 \neq -12$	

Since $5 \neq -12$, this equation has no solution.

ReadingMath

No Solution The symbol that represents no solution is ∅.

b. $3(2b - 1) - 7 = 6b - 10$

$3(2b - 1) - 7 = 6b - 10$	Original equation
$6b - 3 - 7 = 6b - 10$	Distributive Property
$6b - 10 = 6b - 10$	Simplify.
$0 = 0$	Subtract $6b - 10$ from each side.

Since the expressions on each side of the equation are the same, this equation is an identity. It is true for all values of b.

GuidedPractice

3A. $7x + 5(x - 1) = -5 + 12x$

3B. $6(y - 5) = 2(10 + 3y)$

The steps for solving an equation can be summarized as follows.

ConceptSummary Steps for Solving Equations

Step 1 Simplify the expressions on each side. Use the Distributive Property as needed.

Step 2 Use the Addition and/or Subtraction Properties of Equality to get the variables on one side and the numbers without variables on the other side. Simplify.

Step 3 Use the Multiplication or Division Property of Equality to solve.

There are many situations in which you must simplify expressions with grouping symbols in order to solve an equation.

Standardized Test Example 4 Write an Equation

Find the value of x so that the figures have the same area.

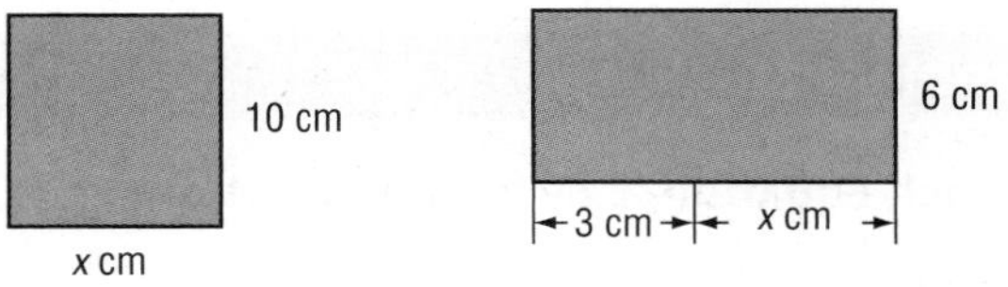

A 3 **C** 6.5

B 4.5 **D** 7

Test-TakingTip

CCSS Tools There is often more than one way to solve a problem. In this example, you can write an algebraic equation and solve for x. Or you can substitute each answer choice into the formulas to find the correct answer.

Read the Test Item

The area of the first rectangle is $10x$, and the area of the second is $6(3 + x)$. The equation $10x = 6(3 + x)$ represents this situation.

Solve the Test Item

A
$$10x = 6(3 + x)$$
$$10(3) \stackrel{?}{=} 6(3 + 3)$$
$$30 \stackrel{?}{=} 6(6)$$
$$30 \neq 36 \text{ ✗}$$

B
$$10x = 6(3 + x)$$
$$10(4.5) \stackrel{?}{=} 6(3 + 4.5)$$
$$45 \stackrel{?}{=} 6(7.5)$$
$$45 = 45 \text{ ✓}$$

Since the value 4.5 results in a true statement, you do not need to check 6.5 and 7. The answer is B.

GuidedPractice

4. Find the value of x so that the figures have the same perimeter.

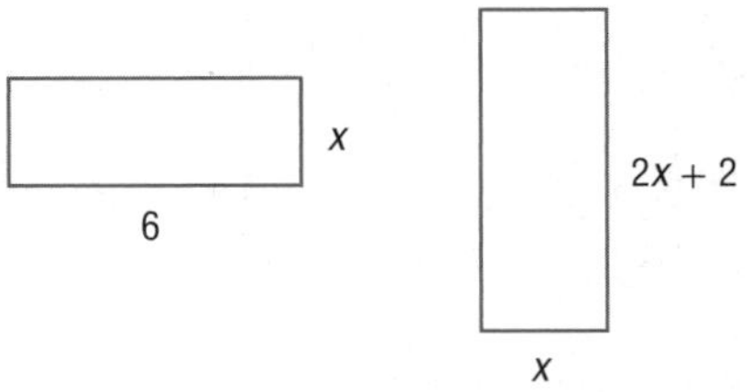

F 1.5 **G** 2 **H** 3.2 **J** 4

Check Your Understanding

Examples 1–3 **Solve each equation. Check your solution.**

1. $13x + 2 = 4x + 38$

2. $\frac{2}{3} + \frac{1}{6}q = \frac{5}{6}q + \frac{1}{3}$

3. $6(n + 4) = -18$

4. $7 = -11 + 3(b + 5)$

5. $5 + 2(n + 1) = 2n$

6. $7 - 3r = r - 4(2 + r)$

7. $14v + 6 = 2(5 + 7v) - 4$

8. $5h - 7 = 5(h - 2) + 3$

Example 4

9. **MULTIPLE CHOICE** Find the value of x so that the figures have the same perimeter.

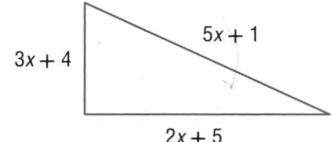

A 4 **B** 5 **C** 6 **D** 7

Practice and Problem Solving

Examples 1–3 **Solve each equation. Check your solution.**

10. $7c + 12 = -4c + 78$

11. $2m - 13 = -8m + 27$

12. $9x - 4 = 2x + 3$

13. $6 + 3t = 8t - 14$

14. $\frac{b - 4}{6} = \frac{b}{2}$

15. $\frac{5v - 4}{10} = \frac{4}{5}$

16. $8 = 4(r + 4)$

17. $6(n + 5) = 66$

18. $5(g + 8) - 7 = 103$

19. $12 - \frac{4}{5}(x + 15) = 4$

20. $3(3m - 2) = 2(3m + 3)$

21. $6(3a + 1) - 30 = 3(2a - 4)$

Example 4

22. **GEOMETRY** Find the value of x so the rectangles have the same area.

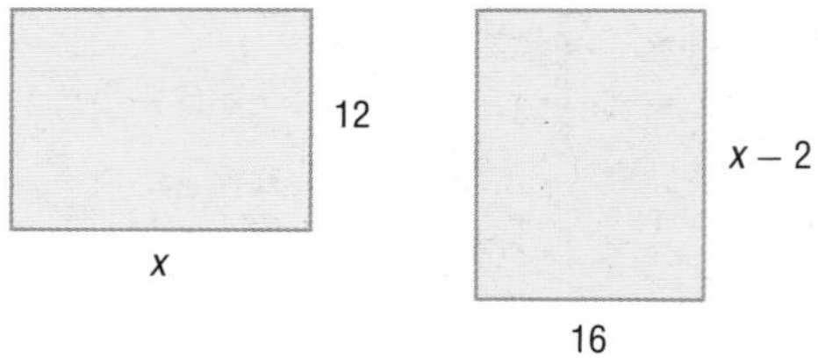

23. **NUMBER THEORY** Four times the lesser of two consecutive even integers is 12 less than twice the greater number. Find the integers.

24. CCSS **SENSE-MAKING** Two times the least of three consecutive odd integers exceeds three times the greatest by 15. What are the integers?

Solve each equation. Check your solution.

25. $2x = 2(x - 3)$

26. $\frac{2}{5}h - 7 = \frac{12}{5}h - 2h + 3$

27. $-5(3 - q) + 4 = 5q - 11$

28. $2(4r + 6) = \frac{2}{3}(12r + 18)$

29. $\frac{3}{5}f + 24 = 4 - \frac{1}{5}f$

30. $\frac{1}{12} + \frac{3}{8}y = \frac{5}{12} + \frac{5}{8}y$

31. $\frac{2m}{5} = \frac{1}{3}(2m - 12)$

32. $\frac{1}{8}(3d - 2) = \frac{1}{4}(d + 5)$

33. $6.78j - 5.2 = 4.33j + 2.15$

34. $14.2t - 25.2 = 3.8t + 26.8$

35. $3.2k - 4.3 = 12.6k + 14.5$

36. $5[2p - 4(p + 5)] = 25$

37. NUMBER THEORY Three times the lesser of two consecutive even integers is 6 less than six times the greater number. Find the integers.

38. MONEY Chris has saved twice the number of quarters that Nora saved plus 6. The number of quarters Chris saved is also five times the difference of the number of quarters and 3 that Nora has saved. Write and solve an equation to find the number of quarters they each have saved.

39. DVD A company that replicates DVDs spends \$1500 per day in building overhead plus \$0.80 per DVD in supplies and labor. If the DVDs sell for \$1.59 per disk, how many DVDs must the company sell each day before it makes a profit?

40. MOBILE PHONES The table shows the number of mobile phone subscribers for two states for a recent year. How long will it take for the numbers of subscribers to be the same?

State	Mobile Phone Subscribers (thousands)	New Subscribers Each Year (thousands)
Alabama	3765	325
Wisconsin	3842	292

41. MULTIPLE REPRESENTATIONS In this problem, you will explore $2x + 4 = -x - 2$.

a. Graphical Make a table of values with five points for $y = 2x + 4$ and $y = -x - 2$. Graph the points from the tables.

b. Algebraic Solve $2x + 4 = -x - 2$.

c. Verbal Explain how the solution you found in part **b** is related to the intersection point of the graphs in part **a**.

H.O.T. Problems Use Higher-Order Thinking Skills

42. REASONING Solve $5x + 2 = ax - 1$ for x. Assume that $a \neq 0$. Describe each step.

43. CHALLENGE Write an equation with the variable on each side of the equals sign, at least one fractional coefficient, and a solution of −6. Discuss the steps you used.

44. OPEN ENDED Create an equation with at least two grouping symbols for which there is no solution.

45. CCSS CRITIQUE Determine whether each solution is correct. If the solution is not correct, describe the error and give the correct solution.

a.
$$\begin{aligned} 2(g + 5) &= 22 \\ 2g + 5 &= 22 \\ 2g + 5 - 5 &= 22 \\ 2g &= 17 \\ g &= 8.5 \end{aligned}$$

b.
$$\begin{aligned} 5d &= 2d - 18 \\ 5d - 2d &= 2d - 18 - 2d \\ 3d &= -18 \\ d &= -6 \end{aligned}$$

c.
$$\begin{aligned} -6z + 13 &= 7z \\ -6z + 13 - 6z &= 7z - 6z \\ 13 &= z \end{aligned}$$

46. CHALLENGE Find the value of k for which each equation is an identity.

a. $k(3x - 2) = 4 - 6x$

b. $15y - 10 + k = 2(ky - 1) - y$

47. WRITING IN MATH Compare and contrast solving equations with variables on both sides of the equation to solving one-step or multi-step equations with a variable on one side of the equation.

Standardized Test Practice

48. A hang glider, 25 meters above the ground, starts to descend at a constant rate of 2 meters per second. Which equation shows the height h after t seconds of descent?

A $h = 25t + 2t$

B $h = -25t + 2$

C $h = 2t + 25$

D $h = -2t + 25$

49. GEOMETRY Two rectangular walls each with a length of 12 feet and a width of 23 feet need to be painted. It costs \$0.08 per square foot for paint. How much will it cost to paint the walls?

F \$22.08   **H** \$34.50

G \$23.04   **J** \$44.16

50. SHORT RESPONSE Maddie works at Game Exchange. They are having a sale as shown.

Item	Price	Special
video games	\$20	Buy 2 get 1 Free
DVDs	\$15	Buy 1 get 1 Free

Her employee discount is 15%. If sales tax is 7.25%, how much does she spend for a total of 4 video games?

51. Solve $\frac{4}{5}x + 7 = \frac{3}{15}x - 3$.

A $-16\frac{2}{3}$ **C** -10

B $-14\frac{4}{9}$ **D** $-6\frac{2}{3}$

Spiral Review

Solve each equation. Check your solution. (Lesson 2-3)

52. $5n + 6 = -4$

53. $-1 = 7 + 3c$

54. $\frac{1}{2}z + 7 = 16 - \frac{3}{5}z$

55. $\frac{2}{5}x + 6 = \frac{2}{3}x + 10$

56. $\frac{a}{7} - 3 = -2$

57. $9 + \frac{y}{5} = 6$

58. WORLD RECORDS In 1998, Winchell's House of Donuts in Pasadena, California, made the world's largest donut. It weighed 5000 pounds and had a circumference of 298.3 feet. What was the donut's diameter to the nearest tenth? (*Hint:* $C = \pi d$) (Lesson 2-2)

59. ZOO At a zoo, the cost of admission is posted on the sign. Find the cost of admission for two adults and two children. (Lesson 1-4)

Find the value of n. Then name the property used in each step. (Lesson 1-3)

60. $25n = 25$

61. $n \cdot 1 = 2$

62. $12 \cdot n = 12 \cdot 6$

63. $n + 0 = \frac{2}{3}$

64. $4 \cdot \frac{1}{4} = n$

65. $(10 - 8)(7) = 2(n)$

Skills Review

Translate each sentence into an equation.

66. Twice a number t decreased by eight equals seventy.

67. Five times the sum of m and k is the same as seven times k.

68. Half of p is the same as p minus 3.

Evaluate each expression.

69. $-9 - (-14)$

70. $-10 + (20)$

71. $-15 - 9$

72. $5(14)$

73. $-55 \div (-5)$

74. $-25(-5)$

LESSON 2-5

Solving Equations Involving Absolute Value

Then

- You solved equations with the variable on each side.

Now

1 Evaluate absolute value expressions.

2 Solve absolute value equations.

Why?

- In 2007, a telephone poll was conducted to determine the reading habits of people in the U.S. People in this survey were allowed to select more than one type of book.

The survey had a margin of error of $\pm 3\%$. This means that the results could be three points higher or lower. So, the percent of people who read religious material could be as high as 69% or as low as 63%.

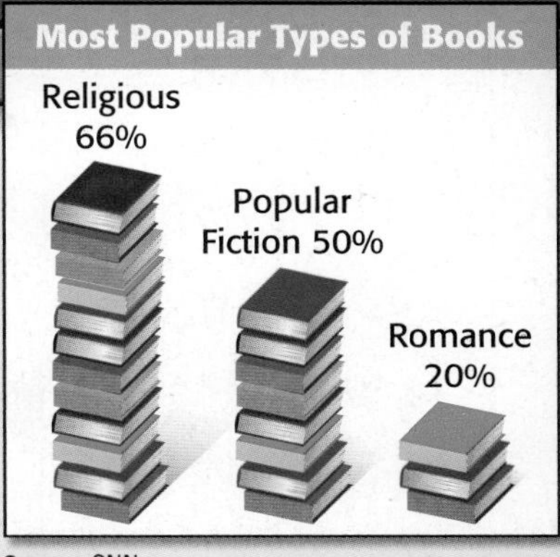

Source: CNN

CCSS Common Core State Standards

Content Standards

A.REI.1 Explain each step in solving a simple equation as following from the equality of numbers asserted at the previous step, starting from the assumption that the original equation has a solution. Construct a viable argument to justify a solution method.

A.REI.3 Solve linear equations and inequalities in one variable, including equations with coefficients represented by letters.

Mathematical Practices

3 Construct viable arguments and critique the reasoning of others.

7 Look for and make use of structure.

1 Absolute Value Expressions

Expressions with absolute values define an upper and lower range in which a value must lie. Expressions involving absolute value can be evaluated using the given value for the variable.

PT

Example 1 Expressions with Absolute Value

Evaluate $|m + 6| - 14$ if $m = 4$.

$|m + 6| - 14 = |4 + 6| - 14$ Replace m with 4.

$= |10| - 14$ $4 + 6 = 10$

$= 10 - 14$ $|10| = 10$

$= -4$ Simplify.

GuidedPractice

1. Evaluate $23 - |3 - 4x|$ if $x = 2$.

2 Absolute Value Equations

The margin of error in the example at the top of the page is an example of absolute value. The distance between 66 and 69 on a number line is the same as the distance between 63 and 66.

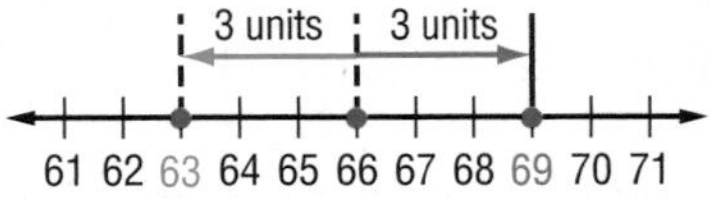

There are three types of open sentences involving absolute value, $|x| = n$, $|x| < n$, and $|x| > n$. In this lesson, we will consider only the first type. Look at the equation $|x| = 4$. This means that the distance between 0 and x is 4.

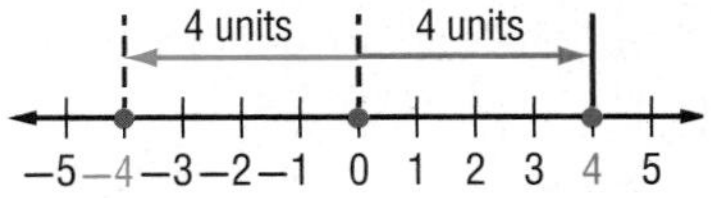

If $|x| = 4$, then $x = -4$ or $x = 4$. Thus, the solution set is $\{-4, 4\}$.

For each absolute value equation, we must consider both cases. To solve an absolute value equation, first isolate the absolute value on one side of the equals sign if it is not already by itself.

ReadingMath

Absolute Values The expression $|f + 5|$ is read the *absolute value of the quantity f plus 5.*

KeyConcept Absolute Value Equations

Words When solving equations that involve absolute values, there are two cases to consider.

Case 1 The expression inside the absolute value symbol is positive or zero.

Case 2 The expression inside the absolute value symbol is negative.

Symbols For any real numbers a and b, if $|a| = b$ and $b \geq 0$, then $a = b$ or $a = -b$.

Example $|d| = 10$, so $d = 10$ or $d = -10$.

Example 2 Solve Absolute Value Equations

Solve each equation. Then graph the solution set.

a. $|f + 5| = 17$

$|f + 5| = 17$ Original equation

Case 1

$f + 5 = 17$

$f + 5 - 5 = 17 - 5$ Subtract 5 from each side.

$f = 12$ Simplify.

Case 2

$f + 5 = -17$

$f + 5 - 5 = -17 - 5$

$f = -22$

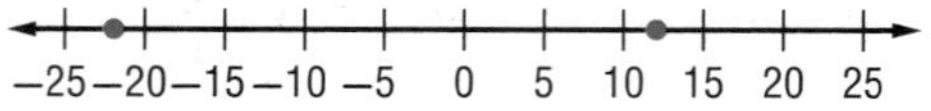

b. $|b - 1| = -3$

$|b - 1| = -3$ means the distance between b and 1 is -3. Since distance cannot be negative, the solution is the empty set $\varnothing$.

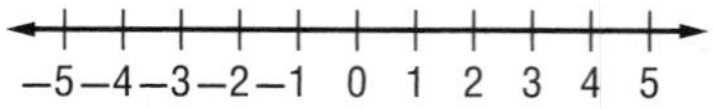

GuidedPractice

2A. $|y + 2| = 4$

2B. $|3n - 4| = -1$

Absolute value equations occur in real-world situations that describe a range within which a value must lie.

Blend Images/Photoshot

Real-WorldLink

In 2001, the number of households in the U.S. that had either a turtle, snake, lizard, or other reptile as a pet was 1,678,000.

Source: American Veterinary Medical Association

Real-World Example 3 Solve an Absolute Value Equation

SNAKES **The temperature of an enclosure for a pet snake should be about 80°F, give or take 5°. Find the maximum and minimum temperatures.**

You can use a number line to solve.

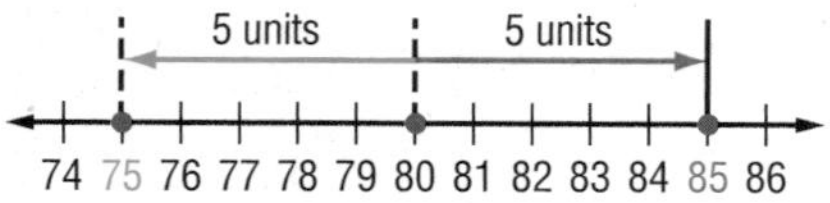

The distance from 80 to 75 is 5 units.
The distance from 80 to 85 is 5 units.

The solution set is {75, 85}. The maximum and minimum temperatures are 85° and 75°.

GuidedPractice

3. ICE CREAM Ice cream should be stored at 5°F with an allowance for 5°. Write and solve an equation to find the maximum and minimum temperatures at which the ice cream should be stored.

When given two points on a graph, you can write an absolute value equation for the graph.

> **StudyTip**
> **Find the Midpoint** To find the point midway between two points, add the values together and divide by 2. For Example 4, $11 + 19 = 30$, $30 \div 2 = 15$. So 15 is the point halfway between 11 and 19.

Example 4 Write an Absolute Value Equation

Write an equation involving absolute value for the graph.

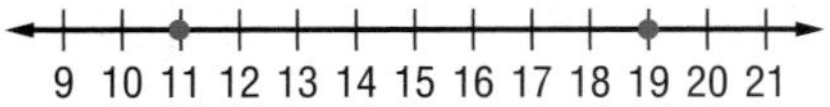

Find the point that is the same distance from 11 and from 19. This is the midpoint between 11 and 19, which is 15.

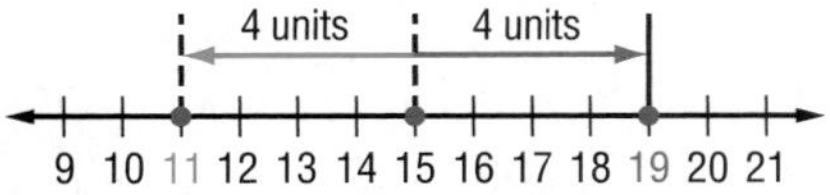

The distance from 15 to 11 is 4 units.
The distance from 15 to 19 is 4 units.

So an equation is $|x - 15| = 4$.

GuidedPractice

4. Write an equation involving absolute value for the graph.

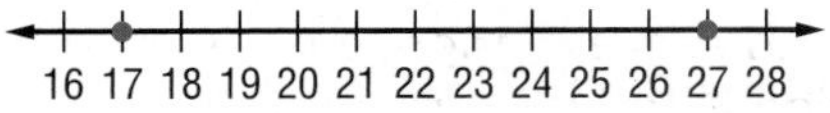

Check Your Understanding

Example 1 **Evaluate each expression if $f = 3$, $g = -4$, and $h = 5$.**

1. $|3 - h| + 13$ **2.** $16 - |g + 9|$ **3.** $|f + g| - h$

Example 2 **Solve each equation. Then graph the solution set.**

4. $|n + 7| = 5$ **5.** $|3z - 3| = 9$ **6.** $|4n - 1| = -6$

7. $|b + 4| = 2$ **8.** $|2t - 4| = 8$ **9.** $|5h + 2| = -8$

Example 3 **10. FINANCIAL LITERACY** For a company to invest in a product, they must believe they will receive a 12% return on investment (ROI) plus or minus 3%. Write an equation to find the least and the greatest ROI they believe they will receive.

Example 4 **Write an equation involving absolute value for each graph.**

11.
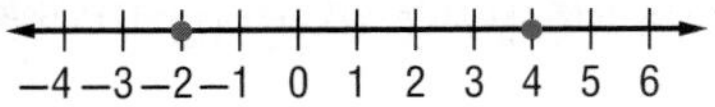

12.
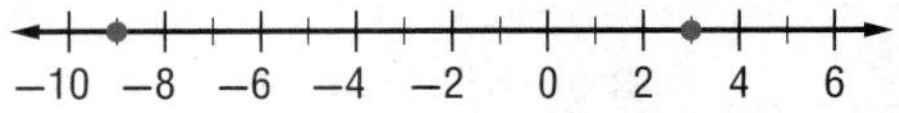

Practice and Problem Solving

Example 1 **Evaluate each expression if $a = -2$, $b = -3$, $c = 2$, $x = 2.1$, $y = 3$, and $z = -4.2$.**

13. $|2x + z| + 2y$ **14.** $4a - |3b + 2c|$ **15.** $-|5a + c| + |3y + 2z|$

16. $-a + |2x - a|$ **17.** $|y - 2z| - 3$ **18.** $3|3b - 8c| - 3$

19. $|2x - z| + 6b$ **20.** $-3|z| + 2(a + y)$ **21.** $-4|c - 3| + 2|z - a|$

Example 2 **Solve each equation. Then graph the solution set.**

22. $|n - 3| = 5$ **23.** $|f + 10| = 1$ **24.** $|v - 2| = -5$

25. $|4t - 8| = 20$ **26.** $|8w + 5| = 21$ **27.** $|6y - 7| = -1$

28. $\left|\frac{1}{2}x + 5\right| = -3$ **29.** $|-2y + 6| = 6$ **30.** $\left|\frac{3}{4}a - 3\right| = 9$

Example 3

31. SURVEY The circle graph at the right shows the results of a survey that asked, "How likely is it that you will be rich some day?" If the margin of error is ±4%, what is the range of the percent of teens who say it is very likely that they will be rich?

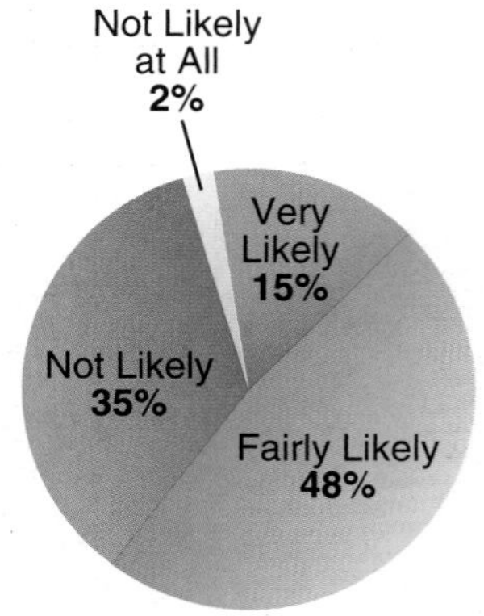

32. CHEERLEADING For competition, the cheerleading team is preparing a dance routine that must last 4 minutes, with a variation of ±5 seconds.

a. Find the least and greatest possible times for the routine in minutes and seconds.

b. Find the least and greatest possible times in seconds.

Example 4 **Write an equation involving absolute value for each graph.**

33. 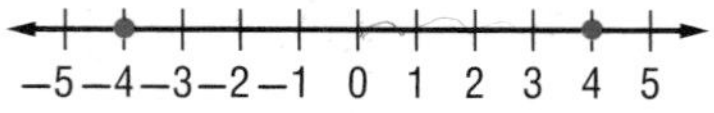

34. 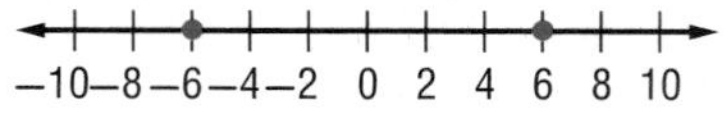

35. 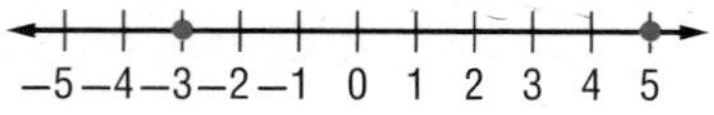

36. 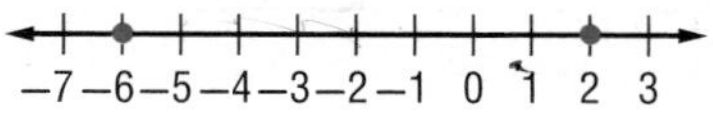

Solve each equation. Then graph the solution set.

37. $\left|-\frac{1}{2}b - 2\right| = 10$ **38.** $|-4d + 6| = 12$ **39.** $|5f - 3| = 12$

40. $2|h| - 3 = 8$ **41.** $4 - 3|q| = 10$ **42.** $\frac{4}{|p|} + 12 = 14$

43. CCSS **SENSE-MAKING** The 4 × 400 relay is a race where 4 runners take turns running 400 meters, or one lap around the track.

a. If a runner runs the first leg in 52 seconds plus or minus 2 seconds, write an equation to find the fastest and slowest times.

b. If the runners of the second and third legs run their laps in 53 seconds plus or minus 1 second, write an equation to find the fastest and slowest times.

c. Suppose the runner of the fourth leg is the fastest on the team. If he runs an average of 50.5 seconds plus or minus 1.5 seconds, what are the team's fastest and slowest times?

44. FASHION To allow for a model's height, a designer is willing to use models that require him to change hems either up or down 2 inches. The length of the skirts is 20 inches.

a. Write an absolute value equation that represents the length of the skirts.

b. What is the range of the lengths of the skirts?

c. If a 20-inch skirt was fitted for a model that is 5 feet 9 inches tall, will the designer use a 6-foot-tall model?

45. CCSS PRECISION Speedometer accuracy can be affected by many details such as tire diameter and axle ratio. For example, there is variation of ±3 miles per hour when calibrated at 50 miles per hour.

a. What is the range of actual speeds of the car if calibrated at 50 miles per hour?

b. A speedometer calibrated at 45 miles per hour has an accepted variation of ±1 mile per hour. What can we conclude from this?

Write an equation involving absolute value for each graph.

46.

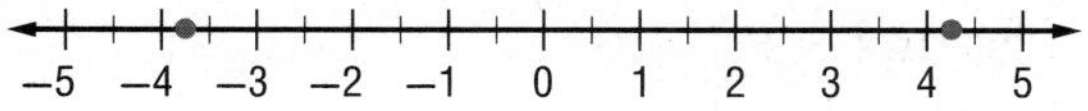

47.

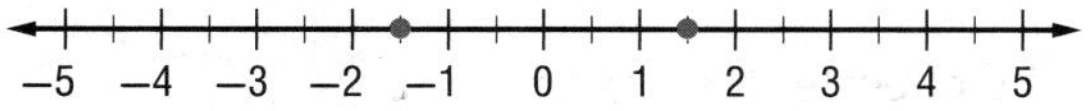

48.

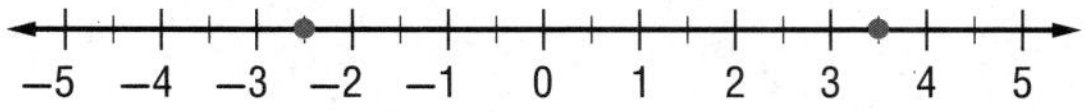

49.

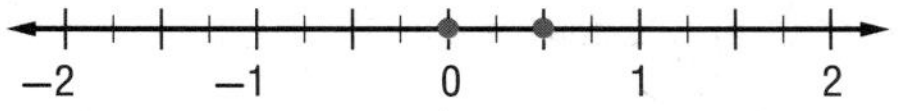

50.

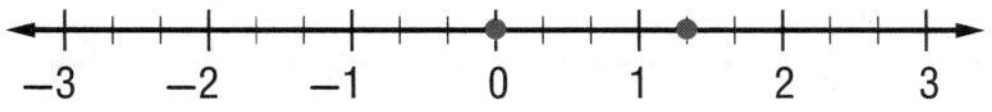

51.

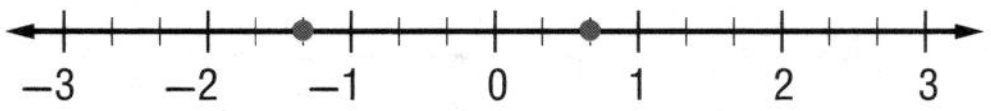

52. MUSIC A CD will record an hour and a half of music plus or minus 3 minutes for time between tracks.

a. Write an absolute value equation that represents the recording time.

b. What is the range of time in minutes that the CD could run?

c. Graph the possible times on a number line.

53. ACOUSTICS The Red Rocks Amphitheater located in the Red Rock Park near Denver, Colorado, is the only naturally occurring amphitheater. The acoustic qualities here are such that a maximum of 20,000 people, plus or minus 1000, can hear natural voices clearly.

a. Write an equation involving an absolute value that represents the number of people that can hear natural voices at Red Rocks Amphitheater.

b. Find the maximum and minimum number of people that can hear natural voices clearly in the amphitheater.

c. What is the range of people in part **b**?

54. BOOK CLUB The members of a book club agree to read within ten pages of the last page of the chapter. The chapter ends on page 203.

a. Write an absolute value equation that represents the pages where club members could stop reading.

b. Write the range of the pages where the club members could stop reading.

55. SCHOOL Teams from Washington and McKinley High Schools are competing in an academic challenge. A correct response on a question earns 10 points and an incorrect response loses 10 points. A team earns 0 points on an unattempted question. There are 5 questions in the math section.

a. What are the maximum and minimum scores a team can earn on the math section?

b. Suppose the McKinley team has 160 points at the start of the math section. Write and solve an equation that represents the maximum and minimum scores the team could have at the end of the math section.

c. What are all of the possible scores that a school can earn on the math section?

H.O.T. Problems Use Higher-Order Thinking Skills

56. OPEN ENDED Describe a real-world situation that could be represented by the absolute value equation $|x - 4| = 10$.

CCSS STRUCTURE Determine whether the following statements are *sometimes*, *always*, or *never* true, if c is an integer. Explain your reasoning.

57. The value of $|x + 1|$ is greater than zero.

58. The solution of $|x + c| = 0$ is greater than 0.

59. The inequality $|x| + c < 0$ has no solution.

60. The value of $|x + c| + c$ is greater than zero.

61. REASONING Explain why an absolute value can never be negative.

62. CHALLENGE Use the sentence $x = 7 \pm 4.6$.

a. Describe the values of x that make the sentence true.

b. Translate the sentence into an equation involving absolute value.

63. ERROR ANALYSIS Alex and Wesley are solving $|x + 5| = -3$. Is either of them correct? Explain your reasoning.

Alex

$|x + 5| = 3$ or $|x + 5| = -3$

$x + 5 = 3$ $\quad$ $x + 5 = -3$

$-5 \; -5$ $\quad$ $-5 \; -5$

$x = -2$ $\quad$ $x = -8$

Wesley

$|x + 5| = -3$

The solution is $\emptyset$.

64. WRITING IN MATH Explain why there are either two, one, or no solutions for absolute value equations. Demonstrate an example of each possibility.

Standardized Test Practice

65. Which equation represents the second step of the solution process?

Step 1: $4(2x + 7) - 6 = 3x$

Step 2: ____________

Step 3: $5x + 28 - 6 = 0$

Step 4: $5x = -22$

Step 5: $x = -4.4$

A $4(2x - 6) + 7 = 3x$
B $4(2x + 1) = 3x$
C $8x + 7 - 6 = 3x$
D $8x + 28 - 6 = 3x$

66. GEOMETRY The area of a circle is 25π square centimeters. What is the circumference?

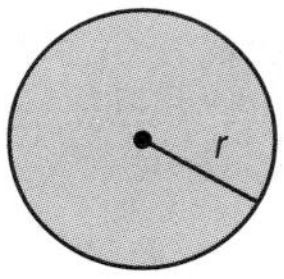

F 625π cm
G 50π cm
H 25π cm
J 10π cm

67. Tanya makes \$5 an hour and 15% commission of the total dollar value on cosmetics she sells. Suppose Tanya's commission is increased to 17%. How much money will she make if she sells \$300 worth of product and works 30 hours?

A \$201 **C** \$255
B \$226 **D** \$283

68. EXTENDED RESPONSE John's mother has agreed to take him driving every day for two weeks. On the first day, John drives for 20 minutes. Each day after that, John drives 5 minutes more than the day before.

a. Write an expression for the minutes John drives on the nth day. Explain.

b. For how many minutes will John drive on the last day? Show your work.

c. John's driver's education teacher requires that each student drive for 30 hours with an adult outside of class. Will John's sessions with his mother fulfill this requirement?

Spiral Review

Write and solve an equation for each sentence. (Lesson 2-4)

69. One half of a number increased by 16 is four less than two thirds of the number.

70. The sum of one half of a number and 6 equals one third of the number.

71. SHOE If ℓ represents the length of a man's foot in inches, the expression $2\ell - 12$ can be used to estimate his shoe size. What is the approximate length of a man's foot if he wears a size 8? (Lesson 2-3)

Skills Review

Write an equation for each problem. Then solve the equation.

72. Seven times a number equals -84. What is the number?

73. Two fifths of a number equals -24. Find the number.

74. Negative 117 is nine times a number. Find the number.

75. Twelve is one fifth of a number. What is the number?

CHAPTER 2 Mid-Chapter Quiz

Lessons 2-1 through 2-5

Translate each sentence into an equation. (Lesson 2-1)

1. The sum of three times *a* and four is the same as five times *a*.
2. One fourth of *m* minus six is equal to two times the sum of *m* and 9.
3. The product of five and *w* is the same as *w* to the third power.

4. **MARBLES** Drew has 50 red, green, and blue marbles. He has six more red marbles than blue marbles and four fewer green marbles than blue marbles. Write and solve an equation to determine how many blue marbles Drew has. (Lesson 2-2)

Solve each equation. Check your solution. (Lesson 2-2)

5. $p + 8 = 13$
6. $-26 = b - 3$
7. $\frac{t}{6} = 3$

8. **MULTIPLE CHOICE** Solve the equation $\frac{3}{5}a = \frac{1}{4}$. (Lesson 2-2)

   A -3

   B $\frac{3}{20}$

   C $\frac{5}{12}$

   D 2

Solve each equation. Check your solution. (Lesson 2-3)

9. $2x + 5 = 13$
10. $-21 = 7 - 4y$
11. $\frac{m}{6} - 3 = 8$
12. $-4 = \frac{d + 3}{5}$

13. **FISH** The average length of a yellow-banded angelfish is 12 inches. This is 4.8 times as long as an average common goldfish. (Lesson 2-3)

    a. Write an equation you could use to find the length of the average common goldfish.

    b. What is the length of an average common goldfish?

Write an equation and solve each problem. (Lesson 2-3)

14. Three less than three fourths of a number is negative 9. Find the number.
15. Thirty is twelve added to six times a number. What is the number?
16. Find four consecutive integers with a sum of 106.

Solve each equation. Check your solution. (Lesson 2-4)

17. $8p + 3 = 5p + 9$
18. $\frac{3}{4}w + 6 = 9 - \frac{1}{4}w$
19. $\frac{z + 6}{3} = \frac{2z}{4}$

20. **PERIMETER** Find the value of *x* so that the triangles have the same perimeter. (Lesson 2-4)

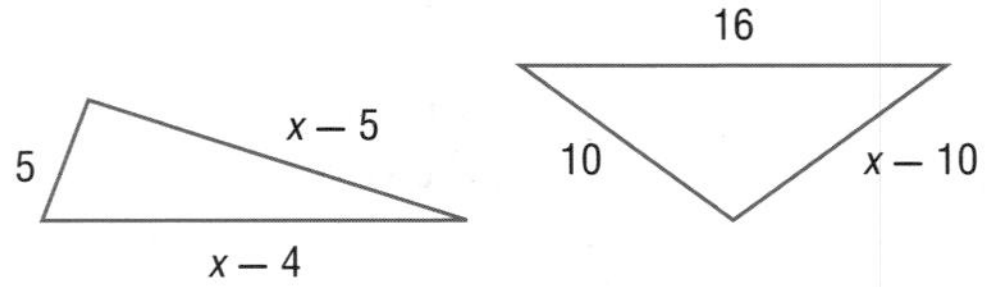

21. **PRODUCTION** ABC Sporting Goods Company produces baseball gloves. Their fixed monthly production cost is \$8000 with a per glove cost of \$5. XYZ Sporting Goods Company also produces baseball gloves. Their fixed monthly production cost is \$10,000 with a per glove cost of \$3. Find the value of *x*, the number of gloves produced monthly, so that the total monthly production cost is the same for both companies. (Lesson 2-4)

Evaluate each expression if $x = -4$, $y = 7$, and $z = -9$. (Lesson 2-5)

22. $|3x - 2| + 2y$
23. $|-4y + 2z| - 7z$

24. **MULTIPLE CHOICE** Solve $|6m - 3| = 9$. (Lesson 2-5)

    F $\{2\}$

    G $\{-1, 2\}$

    H $\{-3, 6\}$

    J $\{-3, 3\}$

25. **COFFEE** Some say to brew an excellent cup of coffee, you must have a brewing temperature of 200° F, plus or minus 5 degrees. Write and solve an equation describing the maximum and minimum brewing temperatures for an excellent cup of coffee.

LESSON 2-6

Ratios and Proportions

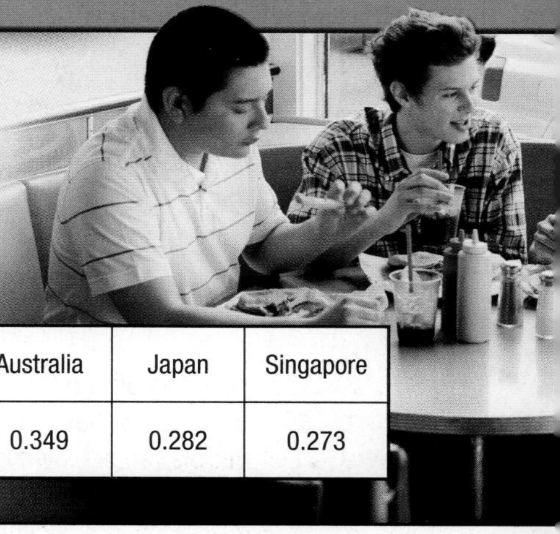

Then

- You evaluated percents by using a proportion.

Now

1. Compare ratios.
2. Solve proportions.

Why?

- Ratios allow us to compare many items by using a common reference. The table below shows the number of restaurants a certain popular fast food chain has per 10,000 people in the United States as well as other countries. This allows us to compare the number of these restaurants using an equal reference.

Countries	United States	New Zealand	Canada	Australia	Japan	Singapore
Number of Restaurants per 10,000 People	0.433	0.369	0.352	0.349	0.282	0.273

NewVocabulary

ratio
proportion
means
extremes
rate
unit rate
scale
scale model

Common Core State Standards

Content Standards

A.REI.1 Explain each step in solving a simple equation as following from the equality of numbers asserted at the previous step, starting from the assumption that the original equation has a solution. Construct a viable argument to justify a solution method.

A.REI.3 Solve linear equations and inequalities in one variable, including equations with coefficients represented by letters.

Mathematical Practices

6 Attend to precision.

1 Ratios and Proportions

The comparison between the number of restaurants and the number of people is a ratio. A **ratio** is a comparison of two numbers by division. The ratio of x to y can be expressed in the following ways.

$$x \text{ to } y \qquad x : y \qquad \frac{x}{y}$$

Suppose you wanted to determine the number of restaurants per 100,000 people in Australia. Notice that this ratio is equal to the original ratio.

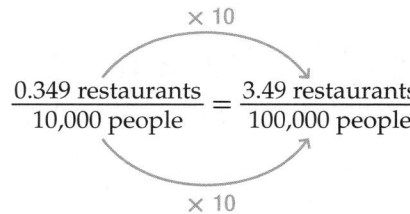

$$\frac{0.349 \text{ restaurants}}{10{,}000 \text{ people}} \overset{\times 10}{=} \frac{3.49 \text{ restaurants}}{100{,}000 \text{ people}}$$

× 10 (numerator), × 10 (denominator)

An equation stating that two ratios are equal is called a **proportion**. So, we can state that $\frac{0.349}{10{,}000} = \frac{3.49}{100{,}000}$ is a proportion.

Example 1 Determine Whether Ratios Are Equivalent

Determine whether $\frac{2}{3}$ and $\frac{16}{24}$ are equivalent ratios. Write *yes* or *no*. Justify your answer.

$$\frac{2}{3} = \frac{2}{3} \quad (\div 1 \text{ numerator and denominator}) \qquad \frac{16}{24} = \frac{2}{3} \quad (\div 8 \text{ numerator and denominator})$$

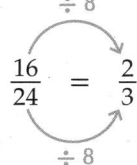

When expressed in simplest form, the ratios are equivalent.

GuidedPractice

Determine whether each pair of ratios are equivalent ratios. Write *yes* or *no*. Justify your answer.

1A. $\frac{6}{10}, \frac{2}{5}$

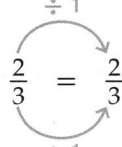

1B. $\frac{1}{6}, \frac{5}{30}$

StudyTip

Means and Extremes To solve a proportion using cross products, write an equation that sets the product of the extremes equal to the product of the means.

There are special names for the terms in a proportion.

1.5 and 1.2 are called the **means**. They are the middle terms of the proportion.

$$0.2 : 1.5 \quad = \quad 1.2 : 9.0$$

0.2 and 9.0 are called the **extremes**. They are the first and last terms of the proportion.

KeyConcept Means-Extremes Property of Proportion

Words	In a proportion, the product of the extremes is equal to the product of the means.
Symbols	If $\frac{a}{b} = \frac{c}{d}$ and $b, d \neq 0$, then $ad = bc$.
Examples	Since $\frac{2}{4} = \frac{1}{2}$, $2(2) = 4(1)$ or $4 = 4$.

Another way to determine whether two ratios form a proportion is to use cross products. If the cross products are equal, then the ratios form a proportion.

This is the same as multiplying the means, and multiplying the extremes.

Example 2 Cross Products

Use cross products to determine whether each pair of ratios forms a proportion.

a. $\frac{2}{3.5}, \frac{8}{14}$

$\frac{2}{3.5} \stackrel{?}{=} \frac{8}{14}$ Original proportion

$2(14) \stackrel{?}{=} 3.5(8)$ Cross products

$28 = 28$ ✓ Simplify.

The cross products are equal, so the ratios form a proportion.

b. $\frac{0.3}{1.5}, \frac{0.5}{2.0}$

$\frac{0.3}{1.5} \stackrel{?}{=} \frac{0.5}{2.0}$ Original proportion

$0.3(2.0) \stackrel{?}{=} 1.5(0.5)$ Cross products

$0.6 \neq 0.75$ ✗ Simplify.

The cross products are not equal, so the ratios do not form a proportion.

GuidedPractice

2A. $\frac{0.2}{1.8}, \frac{1}{0.9}$

2B. $\frac{15}{36}, \frac{35}{42}$

StudyTip

Cross Products When you find cross products, you are said to be *cross multiplying*.

2 Solve Proportions

To solve proportions, use cross products.

Example 3 Solve a Proportion

Solve each proportion. If necessary, round to the nearest hundredth.

a. $\frac{x}{10} = \frac{3}{5}$

$\frac{x}{10} = \frac{3}{5}$	Original proportion
$x(5) = 10(3)$	Find the cross products.
$5x = 30$	Simplify.
$\frac{5x}{5} = \frac{30}{5}$	Divide each side by 5.
$x = 6$	Simplify.

b. $\frac{x-2}{14} = \frac{2}{7}$

$\frac{x-2}{14} = \frac{2}{7}$	Original proportion
$(x-2)7 = 14(2)$	Find the cross products.
$7x - 14 = 28$	Simplify.
$7x = 42$	Add 14 to each side.
$x = 6$	Divide each side by 7.

GuidedPractice

3A. $\frac{r}{8} = \frac{25}{40}$

3B. $\frac{x+4}{5} = \frac{3}{8}$

The ratio of two measurements having different units of measure is called a **rate**. For example, a price of $9.99 per 10 songs is a rate. A rate that tells how many of one item is being compared to 1 of another item is called a **unit rate**.

Blend Images/SuperStock

Real-WorldCareer

Retail Buyer A retail buyer purchases goods for stores, primarily from wholesalers, for resale to the general public. Buyers use math to determine the amount of each product to order. A bachelor's degree with an emphasis on business studies is usually required.

Real-World Example 4 Rate of Growth

RETAIL In the past two years, a retailer has opened 232 stores. If the rate of growth remains constant, how many stores will the retailer open in the next 3 years?

Understand Let r represent the number of retail stores.

Plan Write a proportion for the problem.

$$\frac{232 \text{ retail stores}}{2 \text{ years}} = \frac{r \text{ retail stores}}{3 \text{ years}}$$

Solve

$\frac{232}{2} = \frac{r}{3}$	Original proportion
$232(3) = 2r$	Find the cross products.
$696 = 2r$	Simplify.
$\frac{696}{2} = \frac{2r}{2}$	Divide each side by 2.
$348 = r$	Simplify.

The retailer will open 348 stores in 3 years.

Check If the clothing retailer continues to open 232 stores every 2 years, then in the next 3 years, it will open 348 stores.

GuidedPractice

4. **EXERCISE** It takes 7 minutes for Isabella to walk around the gym track twice. At this rate, how many times can she walk around the track in a half hour?

A rate called a **scale** is used to make a **scale model** of something too large or too small to be convenient at actual size.

Real-WorldLink

The Great Smoky Mountains National Park in Tennessee is home to several waterfalls. The Ramsey Cascades is 100 feet tall. It is the tallest in the park.

Source: National Park Service

Real-World Example 5 Scale and Scale Models

MOUNTAIN TRAIL The Ramsey Cascades Trail is about $1\frac{1}{8}$ inches long on a map with scale 3 inches = 10 miles. What is the actual length of the trail?

Let ℓ represent the actual length.

$$\begin{array}{l} \text{scale} \rightarrow \\ \text{actual} \rightarrow \end{array} \frac{3}{10} = \frac{1\frac{1}{8}}{\ell} \begin{array}{l} \leftarrow \text{scale} \\ \leftarrow \text{actual} \end{array}$$

$3(\ell) = 1\frac{1}{8}(10)$ — Find the cross products.

$3\ell = \frac{45}{4}$ — Simplify.

$3\ell \div 3 = \frac{45}{4} \div 3$ — Divide each side by 3.

$\ell = \frac{15}{4}$ or $3\frac{3}{4}$ — Simplify.

The actual length is about $3\frac{3}{4}$ miles.

GuidedPractice

5. **AIRPLANES** On a model airplane, the scale is 5 centimeters = 2 meters. If the model's wingspan is 28.5 centimeters, what is the actual wingspan?

Check Your Understanding

Examples 1–2 **Determine whether each pair of ratios are equivalent ratios. Write *yes* or *no*.**

1. $\frac{3}{7}, \frac{9}{14}$
2. $\frac{7}{8}, \frac{42}{48}$
3. $\frac{2.8}{4.4}, \frac{1.4}{2.1}$

Example 3 **Solve each proportion. If necessary, round to the nearest hundredth.**

4. $\frac{n}{9} = \frac{6}{27}$
5. $\frac{4}{u} = \frac{28}{35}$
6. $\frac{3}{8} = \frac{b}{10}$

Example 4

7. **RACE** Jennie ran the first 6 miles of a marathon in 58 minutes. If she is able to maintain the same pace, how long will it take her to finish the 26.2 miles?

Example 5

8. **CCSS PRECISION** On a map of North Carolina, Raleigh and Asheville are about 8 inches apart. If the scale is 1 inch = 12 miles, how far apart are the cities?

Practice and Problem Solving

Examples 1–2 **Determine whether each pair of ratios are equivalent ratios. Write *yes* or *no*.**

9. $\frac{9}{11}, \frac{81}{99}$

10. $\frac{3}{7}, \frac{18}{42}$

11. $\frac{8.4}{9.2}, \frac{8.8}{9.6}$

12. $\frac{4}{3}, \frac{6}{8}$

13. $\frac{29.2}{10.4}, \frac{7.3}{2.6}$

14. $\frac{39.68}{60.14}, \frac{6.4}{9.7}$

Example 3 **Solve each proportion. If necessary, round to the nearest hundredth.**

15. $\frac{3}{8} = \frac{15}{a}$

16. $\frac{t}{2} = \frac{6}{12}$

17. $\frac{4}{9} = \frac{13}{q}$

18. $\frac{15}{35} = \frac{g}{7}$

19. $\frac{7}{10} = \frac{m}{14}$

20. $\frac{8}{13} = \frac{v}{21}$

21. $\frac{w}{2} = \frac{4.5}{6.8}$

22. $\frac{1}{0.19} = \frac{12}{n}$

23. $\frac{2}{0.21} = \frac{8}{n}$

24. $\frac{2.4}{3.6} = \frac{k}{1.8}$

25. $\frac{t}{0.3} = \frac{1.7}{0.9}$

26. $\frac{7}{1.066} = \frac{z}{9.65}$

27. $\frac{x-3}{5} = \frac{6}{10}$

28. $\frac{7}{x+9} = \frac{21}{36}$

29. $\frac{10}{15} = \frac{4}{x-5}$

Example 4

30. **CAR WASH** The B-Clean Car Wash washed 128 cars in 3 hours. At that rate, how many cars can they wash in 8 hours?

Example 5

31. **GEOGRAPHY** On a map of Florida, the distance between Jacksonville and Tallahassee is 2.6 centimeters. If 2 centimeters = 120 miles, what is the distance between the two cities?

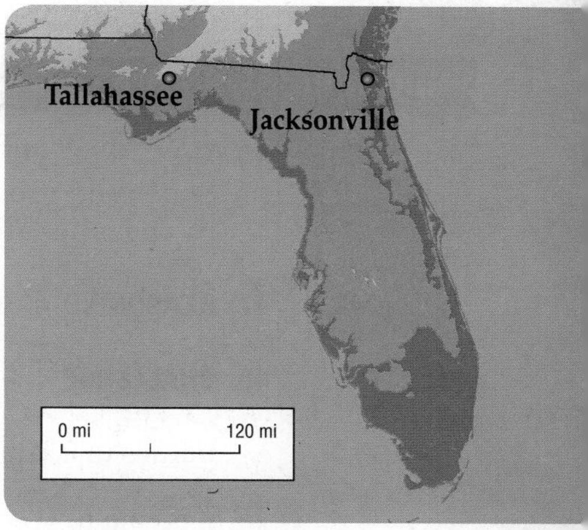

32. **CCSS PRECISION** An artist used interlocking building blocks to build a scale model of Kennedy Space Center, Florida. In the model, 1 inch equals 1.67 feet of an actual space shuttle. The model is 110.3 inches tall. How tall is the actual space shuttle? Round to the nearest tenth.

33. **MENU** On Monday, a restaurant made $545 from selling 110 hamburgers. If they sold 53 hamburgers on Tuesday, how much did they make?

Solve each proportion. If necessary, round to the nearest hundredth.

34. $\frac{6}{14} = \frac{7}{x-3}$

35. $\frac{7}{4} = \frac{f-4}{8}$

36. $\frac{3-y}{4} = \frac{1}{9}$

37. $\frac{4v+7}{15} = \frac{6v+2}{10}$

38. $\frac{9b-3}{9} = \frac{5b+5}{3}$

39. $\frac{2n-4}{5} = \frac{3n+3}{10}$

40. **ATHLETES** At Piedmont High School, 3 out of every 8 students are athletes. If there are 1280 students at the school, how many are not athletes?

41. **BRACES** Two out of five students in the ninth grade have braces. If there are 325 students in the ninth grade, how many have braces?

42. **PAINT** Joel used a half gallon of paint to cover 84 square feet of wall. He has 932 square feet of wall to paint. How many gallons of paint should he purchase?

43 **MOVIE THEATERS** Use the table at the right.

a. Write a ratio of the number of indoor theaters to the total number of theaters for each year.

b. Do any two of the ratios you wrote for part **a** form a proportion? If so, explain the real-world meaning of the proportion.

Year	Indoor	Drive-In	Total
2003	35,361	634	35,995
2004	36,012	640	36,652
2005	37,092	648	37,740
2006	37,776	649	38,425
2007	38,159	635	38,794
2008	38,201	633	38,834
2009	38,605	628	39,233

Source: North American Theater Owners

44. **DIARIES** In a survey, 36% of the students said that they kept an electronic diary. There were 900 students who kept an electronic diary. How many students were in the survey?

45. **MULTIPLE REPRESENTATIONS** In this problem, you will explore how changing the lengths of the sides of a shape by a factor changes the perimeter of that shape.

a. Geometric Draw a square *ABCD*. Draw a square *MNPQ* with sides twice as long as *ABCD*. Draw a square *FGHJ* with sides half as long as *ABCD*.

b. Tabular Complete the table below using the appropriate measures.

ABCD		*MNPQ*		*FGHJ*	
Side length		Side length		Side length	
Perimeter		Perimeter		Perimeter	

c. Verbal Make a conjecture about the change in the perimeter of a square if the side length is increased or decreased by a factor.

H.O.T. Problems Use Higher-Order Thinking Skills

46. CCSS **STRUCTURE** In 2007, organic farms occupied 2.6 million acres in the United States and produced goods worth about $1.7 billion. Divide one of these numbers by the other and explain the meaning of the result.

47. **REASONING** Compare and contrast ratios and rates.

48. **CHALLENGE** If $\frac{a+1}{b-1} = \frac{5}{1}$ and $\frac{a-1}{b+1} = \frac{1}{1}$, find the value of $\frac{b}{a}$. (*Hint:* Choose values of a and b for which the proportions are true and evaluate $\frac{b}{a}$.)

49. **WRITING IN MATH** On a road trip, Marcus reads a highway sign and then looks at his gas gauge.

Marcus's gas tank holds 10 gallons and his car gets 32 miles per gallon at his current speed of 65 miles per hour. If he maintains this speed, will he make it to Atlanta without having to stop and get gas? Explain your reasoning.

50. **WRITING IN MATH** Describe how businesses can use ratios. Write about a real-world situation in which a business would use a ratio.

Standardized Test Practice

51. In the figure, $x : y = 2 : 3$ and $y : z = 3 : 5$. If $x = 10$, find the value of z.

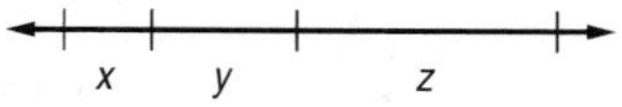

A 15
B 20
C 25
D 30

52. GRIDDED RESPONSE A race car driver records the finishing times for recent practice trials.

Trial	Time (seconds)
1	5.09
2	5.10
3	4.95
4	4.91
5	5.05

What is the mean time, in seconds, for the trials?

53. GEOMETRY If $\triangle LMN$ is similar to $\triangle LPO$, what is z?

F 240
G 140
H 120
J 70

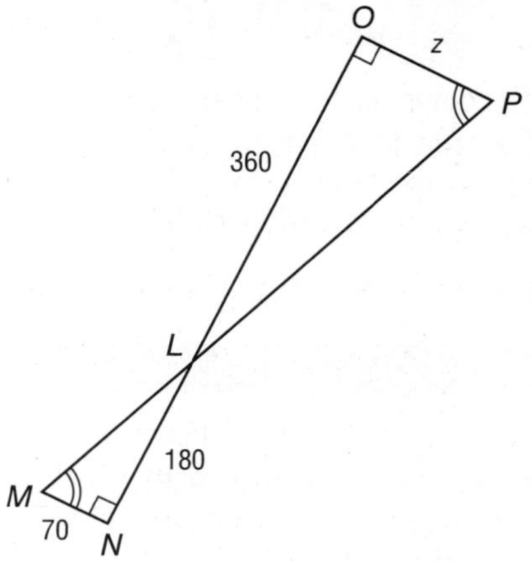

54. Which equation below illustrates the Commutative Property?

A $(3x + 4y) + 2z = 3x + (4y + 2z)$
B $7(x + y) = 7x + 7y$
C $xyz = yxz$
D $x + 0 = x$

Spiral Review

Solve each equation. (Lesson 2-5)

55. $|x + 5| = -8$

56. $|b + 9| = 2$

57. $|2p - 3| = 17$

58. $|5c - 8| = 12$

59. HEALTH When exercising, a person's pulse rate should not exceed a certain limit. This maximum rate is represented by the expression $0.8(220 - a)$, where a is age in years. Find the age of a person whose maximum pulse rate is 122 more than their age. (Lesson 2-4)

Solve each equation. Check your solution. (Lesson 2-3)

60. $15 = 4a - 5$

61. $7g - 14 = -63$

62. $9 + \frac{y}{5} = 6$

63. $\frac{t}{8} - 6 = -12$

64. GEOMETRY Find the area of $\triangle ABC$ if each small triangle has a base of 5.2 inches and a height of 4.5 inches. (Lesson 1-4)

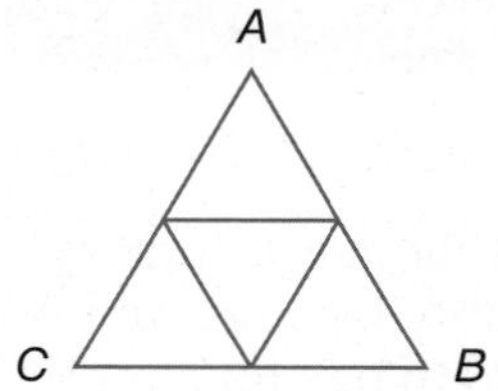

Evaluate each expression. (Lesson 1-2)

65. $3 + 16 \div 8 \cdot 5$

66. $4^2 \cdot 3 - 5(6 + 3)$

Skills Review

Solve each equation.

67. $4p = 22$

68. $5h = 33$

69. $1.25y = 4.375$

70. $9.8m = 30.87$

EXTEND 2-6

Spreadsheet Lab: Descriptive Modeling

When using numbers to model a real-world situation, it is often helpful to have a metric. A **metric** is a rule for assigning a number to some characteristic or attribute. For example, teachers use metrics to determine grades. Each teacher determines an appropriate metric for assessing a student's performance and assigning a grade.

You can use a spreadsheet to calculate different metrics.

CCSS Common Core State Standards
Content Standards
N.Q.2 Define appropriate quantities for the purpose of descriptive modeling.

Activity

Dorrie wants to buy a house. She has the following expenses: rent of $650, credit card monthly bills of $320, a car payment of $410, and a student loan payment of $115. Dorrie has a yearly salary of $46,500. Use a spreadsheet to find Dorrie's debt-to-income ratio.

Step 1 Enter Dorrie's debts in column B.

Step 2 Add her debts using a function in cell B6. Go to Insert and then Function. Then choose Sum. The sum of 1495 appears in B6.

Step 3 Now insert Dorrie's salary in column C. Remember to find her monthly salary by dividing the yearly salary by 12.

A mortgage company will use the debt-to-income ratio as a metric to determine if Dorrie qualifies for a loan. The **debt-to-income ratio** is calculated as *how much she owes per month* divided by *how much she earns each month.*

Step 4 Enter a formula to find the debt-to-income ratio in cell C6. In the formula bar, enter =B6/C2.

Lab 2-6 B Spreadsheet.xls

◇	A	B	C
1	Type of Debt	Expenses	Salary
2	Rent	650	3875
3	Credit Cards	320	
4	Car Payment	410	
5	Student Loan	115	
6		1495	0.385806
7			

Sheet 1 / Sheet 2 / Sheet 3

The ratio of about 0.39 appears. An ideal ratio would be 0.36 or less. A ratio higher than 0.36 would cause an increased interest rate or may require a higher down payment.

The spreadsheet shows a debt-to-income ratio of about 0.39. Dorrie should try to eliminate or reduce some debts or try to earn more money in order to lower her debt-to-income ratio.

Exercises

1. How could Dorrie improve her debt-to-income ratio?

2. Another metric mortgage companies use is the ratio of monthly mortgage to total monthly income. An ideal ratio is 0.28. Using this metric, how much could Dorrie afford to pay for a mortgage each month?

3. How effective are each of these metrics as measures of whether Dorrie can afford to buy a house? Explain your reasoning.

4. **CCSS MODELING** Metrics are used to compare athletes. For example, ERAs are used to compare pitchers. Find a metric and evaluate its effectiveness for modeling. Compare it to other metrics, and then define your own metric.

LESSON 2-7

Percent of Change

Then	Now	Why?
You solved proportions.	**1** Find the percent of change. **2** Solve problems involving percent of change.	Every year, millions of people volunteer their time to improve their community. The difference in the number of volunteers from one year to the next can be used to determine a percent to represent the increase or decrease in volunteers.

Jim West/imagebroker/age fotostock

NewVocabulary
percent of change
percent of increase
percent of decrease

Common Core State Standards

Content Standards

N.Q.1 Use units as a way to understand problems and to guide the solution of multi-step problems; choose and interpret units consistently in formulas; choose and interpret the scale and the origin in graphs and data displays.

A.REI.3 Solve linear equations and inequalities in one variable, including equations with coefficients represented by letters.

Mathematical Practices

8 Look for and express regularity in repeated reasoning.

1 Percent of Change

Percent of change is the ratio of the change in an amount to the original amount expressed as a percent. If the new number is greater than the original number, the percent of change is a **percent of increase.** If the new number is less than the original number, the percent of change is a **percent of decrease.**

Example 1 Percent of Change

Determine whether each percent of change is a percent of *increase* or a percent of *decrease*. Then find the percent of change.

a. original: 20
final: 23

Subtract the original amount from the final amount to find the amount of change: $23 - 20 = 3$.

Since the new amount is greater than the original, this is a percent of increase.

Use the original number, 20, as the base.

change → / original amount →

$$\frac{3}{20} = \frac{r}{100}$$

$$3(100) = r(20)$$

$$300 = 20r$$

$$\frac{300}{20} = \frac{20r}{20}$$

$$15 = r$$

The percent of increase is 15%.

b. original: 25
final: 17

Subtract the original amount from the final amount to find the amount of change: $17 - 25 = -8$.

Since the new amount is less than the original, this is a percent of decrease.

Use the original number, 25, as the base.

change → / original amount →

$$\frac{-8}{25} = \frac{r}{100}$$

$$-8(100) = r(25)$$

$$-800 = 25r$$

$$\frac{-800}{25} = \frac{25r}{25}$$

$$-32 = r$$

The percent of decrease is 32%.

GuidedPractice

1A. original: 66
new: 30

1B. original: 9.8
new: 12.1

1C. original: 24
new: 40

1D. original: 500
new: 131

Real-WorldLink

In 2009, the total revenue earned by the North American cruise industry was more than $15.16 billion.

Source: Cruise Market Watch

Real-World Example 2 Percent of Change

CRUISE **The total number of passengers on cruise ships increased 10% from 2007 to 2009. If there were 17.22 million passengers in 2009, how many were there in 2007?**

Let f = the number of passengers in 2009. Since 10% is a percent of increase, the number of passengers in 2007 is less than the number of passengers in 2009.

change → original amount →	$\frac{17.22 - f}{f} = \frac{10}{100}$	Percent proportion
	$(1722 - f)100 = 10f$	Find the cross products.
	$1722 - 100f = 10f$	Distributive Property
	$1722 - 100f + \mathbf{100f} = 10f + \mathbf{100f}$	Add $100f$ to each side.
	$1722 = 110f$	Simplify.
	$\frac{1722}{110} = \frac{110f}{110}$	Divide each side by 110.
	$15.65 \approx f$	Simplify.

There were approximately 15.65 million passengers in 2007.

GuidedPractice

2. **TUITION** A recent percent of increase in tuition at Northwestern University, in Evanston, Illinois, was 5.4%. If the new cost is $33,408 per year, find the original cost per year.

2 Solve Problems

Two applications of percent of change are sales tax and discounts. Sales tax is an example of a percent of increase. Discount is an example of a percent of decrease.

Example 3 Sales Tax

SHOPPING **Marta is purchasing wire and beads to make jewelry. Her merchandise is $28.62 before tax. If the tax is 7.25% of the total sales, what is the final cost?**

Step 1 Find the tax.

The tax is 7.25% of the price of the merchandise.

7.25% of $28.62 = 0.0725 × 28.62	7.25% = 0.0725
= 2.07495	Use a calculator.

Step 2 Find the cost with tax.

Round $2.07495 to $2.07 since tax is always rounded to the nearest cent. Add this amount to the original price: $28.62 + $2.07 = $30.69.

The total cost of Marta's jewelry supplies is $30.69.

GuidedPractice

3. **SHOPPING** A new DVD costs $24.99. If the sales tax is 6.85%, what is the total cost?

To find a discounted amount, you will follow similar steps to those for sales tax.

Example 4 Discounts

DISCOUNT Since Tyrell has earned good grades in school, he qualifies for the Good Student Discount on his car insurance. His monthly payment without the discount is $85. If the discount is 20%, what will he pay each month?

Step 1 Find the discount.

The discount is 20% of the original payment.

$20\% \text{ of } \$85 = 0.20 \times 85$ 20% = 0.20

$= 17$ Use a calculator.

Step 2 Find the cost after discount.

Subtract $17 from the original payment: $85 − $17 = $68.

With the Good Student Discount, Tyrell will pay $68 per month.

StudyTip

CCSS Regularity When translating a problem from word sentences to math sentences, the word "is" translates to =, and the word "of" translates to ×.

GuidedPractice

4. **SALES** A picture frame originally priced at $14.89 is on sale for 40% off. What is the discounted price?

Check Your Understanding

Example 1

State whether each percent of change is a percent of *increase* or a percent of *decrease*. Then find the percent of change. Round to the nearest whole percent.

1. original: 78
   new: 125
2. original: 41
   new: 24
3. original: 6 candles
   new: 8 candles
4. original: 35 computers
   new: 32 computers

Example 2

5. **GEOGRAPHY** The distance from Phoenix to Tucson is 120 miles. The distance from Phoenix to Flagstaff is about 21.7% longer. To the nearest mile, what is the distance from Phoenix to Flagstaff?

Example 3

Find the total price of each item.

6. dress: $22.50
   sales tax: 7.5%
7. video game: $35.99
   sales tax: 6.75%
8. **PROM** A limo costs $85 to rent for 3 hours plus a 7% sales tax. What is the total cost to rent a limo for 6 hours?
9. **GAMES** A computer game costs $49.95 plus a 6.25% sales tax. What is the total cost of the game?

Example 4

Find the discounted price of each item.

10. guitar: $95.00
    discount: 15%
11. DVD: $22.95
    discount: 25%
12. **SKATEBOARD** A skateboard costs $99.99. If you have a coupon for 20% off, how much will you save?
13. **CCSS MODELING** Tickets to the county fair are $8 for an adult and $5 for a child. If you have a 15% discount card, how much will 2 adult tickets and 2 child tickets cost?

Practice and Problem Solving

Example 1 **State whether each percent of change is a percent of *increase* or a percent of *decrease*. Then find the percent of change. Round to the nearest whole percent.**

14. original: 35
new: 40

15. original: 16
new: 10

16. original: 27
new: 73

17. original: 92
new: 21

18. original: 21.2 grams
new: 10.8 grams

19. original: 11 feet
new: 25 feet

20. original: $68
new: $76

21. original: 21 hours
new: 40 hours

Example 2 **22.** **GASOLINE** The average cost of regular gasoline in North Carolina increased by 73% from 2006 to 2007. If the average cost of a gallon of gas in 2006 was $2.069, what was the average cost in 2007? Round to the nearest cent.

23. **CARS** Beng is shopping for a car. The cost of a new car is $15,500. This is 25% greater than the cost of a used car. What is the cost of the used car?

Example 3 **Find the total price of each item.**

24. messenger bag: $28.00
tax: 7.25%

25. software: $45.00
tax: 5.5%

26. vase: $5.50
tax: 6.25%

27. book: $25.95
tax: 5.25%

28. magazine: $3.50
tax: 5.75%

29. pillow: $9.99
tax: 6.75%

Example 4 **Find the discounted price of each item.**

30. computer: $1099.00
discount: 25%

31. CD player: $89.99
discount: 15%

32. athletic shoes: $59.99
discount: 40%

33. jeans: $24.50
discount: 33%

34. jacket: $125.00
discount: 25%

35. belt: $14.99
discount: 20%

Find the final price of each item.

36. sweater: $14.99
discount: 12%
tax: 6.25%

37. printer: $60.00
discount: 25%
tax: 6.75%

38. board game: $25.00
discount: 15%
tax: 7.5%

39. **CONSUMER PRICE INDEX** An *index* measures the percent change of a value from a base year. An index of 115 means that there was a 15% increase from the base year. In 2000, the consumer price index of dairy products was 160.7. In 2007, it was 194.0. Determine the percent of change.

40. **FINANCIAL LITERACY** The current price of each share of a technology company is $135. If this represents a 16.2% increase over the past year, what was the price per share a year ago?

41. CCSS **MODELING** A group of girls are shopping for dresses to wear to the spring dance. One finds a dress priced $75 with a 20% discount. A second girl finds a dress priced $85 with a 30% discount.

a. Find the amount of discount for each dress.

b. Which girl is getting the better price for the dress?

42. **RECREATIONAL SPORTS** In 1995, there were 73,567 youth softball teams. By 2007, there were 86,049. Determine the percent of increase.

43 **CCSS TOOLS** Which grocery item had the greatest percent increase in cost from 2000 to 2007?

Average Retail Prices of Selected Grocery Items

Grocery Item	Cost in 2000 ($ per pound)	Cost in 2007 ($ per pound)
milk (gallon)	2.79	3.87
turkey (whole)	0.99	1.01
chicken (whole)	1.08	1.17
ground beef	1.63	2.23
apples	0.82	1.12
iceberg lettuce	0.85	0.95
peanut butter	1.96	1.88

Source: Statistical Abstract of the United States

44. **MULTIPLE REPRESENTATIONS** In this problem, you will explore patterns in percentages.

a. Tabular Copy and complete the following table.

1% of	500	is 5.	100% of		is 20.		% of 80 is 20.
2% of		is 5.	50% of		is 20.		% of 40 is 20.
4% of		is 5.	25% of		is 20.		% of 20 is 20.
8% of		is 5.	12.5% of		is 20.		% of 10 is 20.

b. Verbal Describe the patterns in the second and fifth columns.

c. Analytical Use the patterns to write the fifth row of the table.

H.O.T. Problems Use Higher-Order Thinking Skills

45. **OPEN ENDED** Write a real-world problem to find the total price of an item including sales tax.

46. **REASONING** If you have 75% of a number n, what percent of decrease is it from the number n? If you have 40% of a number a, what percent of decrease do you have from the number a? What pattern do you notice? Is this always true?

47. **ERROR ANALYSIS** Maddie and Xavier are solving for the percent change if the original amount was $25 and the new amount is $28. Is either of them correct? Explain your reasoning.

Maddie

$$\frac{3}{28} = \frac{r}{100}$$
$$3(100) = 28r$$
$$300 = 28r$$
$$10.7 = r$$

Xavier

$$\frac{3}{25} = \frac{r}{100}$$
$$3(100) = 25r$$
$$300 = 25r$$
$$12 = r$$

48. **CHALLENGE** Determine whether the following statement is *sometimes*, *always*, or *never* true. *The percent of change is less than 100%.*

49. **WRITING IN MATH** When is percent of change used in the real world? Explain how to find a percent of change between two values.

Standardized Test Practice

50. GEOMETRY The rectangle has a perimeter of P centimeters. Which equation could be used to find the length ℓ of the rectangle?

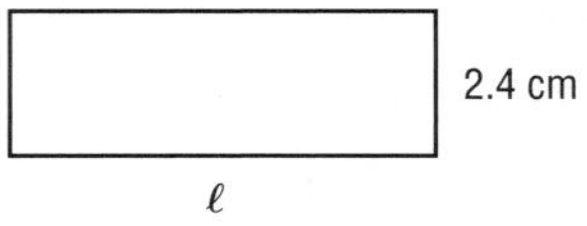

A $P = 2.4\ell$

B $P = 4.8 + \ell$

C $P = 2.4 + 2\ell$

D $P = 4.8 + 2\ell$

51. SHORT RESPONSE Henry is painting a room with four walls that are 12 feet by 14 feet. A gallon of paint costs \$18 and covers 350 square feet. If he uses two coats of paint, how much will it cost him to paint the room?

52. The number of students at Franklin High School increased from 840 to 910 over a 5-year period. What was the percent of increase?

F 8.3%

G 14.0%

H 18.5%

J 92.3%

53. PROBABILITY Two dice are rolled. What is the probability that the sum is 10?

A $\frac{1}{3}$ **B** $\frac{1}{6}$ **C** $\frac{1}{12}$ **D** $\frac{1}{36}$

Spiral Review

54. TRAVEL The Chan's minivan requires 5 gallons of gasoline to travel 120 miles. How many gallons of gasoline will they need to travel 360 miles? (Lesson 2-6)

Evaluate each expression if $x = -2$, $y = 6$, and $z = 4$. (Lesson 2-5)

55. $|3 - x| + 7$

56. $12 - |z + 9|$

57. $|y + x| - z + 4$

Solve each equation. Round to the nearest hundredth. Check your solution. (Lesson 2-4)

58. $1.03p - 4 = -2.15p + 8.72$

59. $18 - 3.8t = 7.36 - 1.9t$

60. $5.4w + 8.2 = 9.8w - 2.8$

61. $2[d + 3(d - 1)] = 18$

Solve each equation. Check your solution. (Lesson 2-3)

62. $5n + 6 = -4$

63. $-11 = 7 + 3c$

64. $15 = 4a - 5$

65. $-14 + 7g = -63$

66. RIVERS The Congo River in Africa is 2900 miles long. That is 310 miles longer than the Niger River, which is also in Africa. (Lesson 2-2)

a. Write an equation you could use to find the length of the Niger River.

b. What is the length of the Niger River?

67. FOOD Cameron purchased x pounds of apples for \$0.99 per pound and y pounds of oranges for \$1.29 per pound. Write an algebraic expression that represents the cost of the purchase. (Lesson 1-1)

Skills Review

Translate each equation into a sentence.

68. $d - 14 = 5$

69. $2f + 6 = 19$

70. $y - 12 = y + 8$

71. $3a + 5 = 27 - 2a$

72. $-6c^2 - 4c = 25$

73. $d^4 + 64 = 3d^3 + 77$

EXTEND 2-7

Algebra Lab: Percentiles

A **percentile** is a measure that is often used to report test data, such as standardized test scores. It tells us what percent of the total scores were below a given score.

- Percentiles measure rank from the bottom.
- There is no 0 percentile rank. The lowest score is at the 1st percentile.
- There is no 100th percentile rank. The highest score is at the 99th percentile.

Activity

A talent show was held for the twenty finalists in the Teen Idol contest. Each performer received a score from 0 through 30 with 30 being the highest. What is Victor's percentile rank?

Name	Score	Name	Score
Arnold	17	Ishi	27
Benito	9	James	20
Brooke	25	Kat	16
Carmen	21	Malik	10
Daniel	14	Natalie	26
Delia	29	Pearl	4
Fernando	15	Twyla	6
Heather	12	Victor	28
Horatio	5	Warren	22
Ingrid	11	Yolanda	18

Step 1 Write one score on each of 20 slips of paper.

Step 2 Arrange the slips vertically from greatest to least score.

Step 3 Find Victor's percentile rank.

Victor had a score of 28. There are 18 scores below his score. To find his percentile rank, use the following formula.

$$\frac{\text{number of scores below 28}}{\text{total number of scores}} \cdot 100 = \frac{18}{20} \cdot 100 \text{ or } 90$$

Victor scored at the 90th percentile in the contest.

Analyze the Results

1. Find the median, lower quartile, and upper quartile of the scores.
2. Which performer was at the 50th percentile? the 25th percentile? the 75th percentile?
3. Compare and contrast the values for the median, lower quartile, and upper quartile and the scores for the 25th, 50th, and 75th percentiles.
4. While Victor scored at the 90th percentile, what percent of the 30 possible points did he score?
5. CCSS **ARGUMENTS** Compare and contrast the percentile rank and the percent score.
6. Are there any outliers in the data that could alter the results of our computations?
7. **Deciles** are values that divide a set of data into ten equal-sized parts. The 1st decile contains data up to but not including the 10th percentile; the 2nd decile contains data from the 10th percentile up to but not including the 20th percentile, and so on.
   a. Which contestants' scores fall in the 6th decile?
   b. In which decile are Heather and Daniel?

LESSON 2-8 Literal Equations and Dimensional Analysis

Then

- You solved equations with variables on each side.

Now

1. Solve equations for given variables.
2. Use formulas to solve real-world problems.

Why?

- Each year, more people use credit cards to make everyday purchases. If the entire balance is not paid by the due date, compound interest is applied. The formula for computing the balance of an account with compound interest added annually is $A = P(1 + r)^t$.
  - A represents the amount of money in the account including the interest,
  - P is the amount in the account before interest is added,
  - r is the interest rate written as a decimal,
  - t is the time in years.

NewVocabulary
literal equation
dimensional analysis
unit analysis

Common Core State Standards

Content Standards
A.CED.4 Rearrange formulas to highlight a quantity of interest, using the same reasoning as in solving equations.
A.REI.3 Solve linear equations and inequalities in one variable, including equations with coefficients represented by letters.

Mathematical Practices
6 Attend to precision.

1 Solve for a Specific Variable

Some equations such as the one above contain more than one variable. At times, you will need to solve these equations for one of the variables.

Example 1 Solve for a Specific Variable

Solve $4m - 3n = 8$ for m.

$4m - 3n = 8$ — Original equation

$4m - 3n + 3n = 8 + 3n$ — Add $3n$ to each side.

$4m = 8 + 3n$ — Simplify.

$\frac{4m}{4} = \frac{8 + 3n}{4}$ — Divide each side by 4.

$m = \frac{8}{4} + \frac{3}{4}n$ — Simplify.

$m = 2 + \frac{3}{4}n$ — Simplify.

GuidedPractice

Solve each equation for the variable indicated.

1A. $15 = 3n + 6p$, for n

1B. $\frac{k - 2}{5} = 11j$, for k

1C. $28 = t(r + 4)$, for t

1D. $a(q - 8) = 23$, for q

Sometimes we need to solve equations for a variable that is on both sides of the equation. When this happens, you must get all terms with that variable onto one side of the equation. It is then helpful to use the Distributive Property to isolate the variable for which you are solving.

StudyTip

Solving for a Specific Variable When an equation has more than one variable, it can be helpful to highlight the variable for which you are solving on your paper.

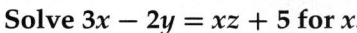

Example 2 Solve for a Specific Variable

Solve $3x - 2y = xz + 5$ for x.

$3x - 2y = xz + 5$	Original equation
$3x - 2y + 2y = xz + 5 + 2y$	Add $2y$ to each side.
$3x - xz = xz - xz + 5 + 2y$	Subtract xz from each side.
$3x - xz = 5 + 2y$	Simplify.
$x(3 - z) = 5 + 2y$	Distributive Property
$\frac{x(3-z)}{3-z} = \frac{5+2y}{3-z}$	Divide each side by $3 - z$.
$x = \frac{5+2y}{3-z}$	Simplify.

Since division by 0 is undefined, $3 - z \neq 0$ so $z \neq 3$.

GuidedPractice

Solve each equation for the variable indicated.

2A. $d + 5c = 3d - 1$, for d

2B. $6q - 18 = qr + t$, for q

2 Use Formulas

An equation that involves several variables is called a formula or **literal equation**. To solve a literal equation, apply the process of solving for a specific variable.

Real-WorldLink

The largest yo-yo in the world is 32.7 feet in circumference. It was launched by crane from a height of 189 feet.

Source: *Guinness Book of World Records*

Real-World Example 3 Use Literal Equations

YO-YOS Use the information about the largest yo-yo at the left. The formula for the circumference of a circle is $C = 2\pi r$, where C represents circumference and r represents radius.

a. Solve the formula for r.

$C = 2\pi r$	Formula for circumference
$\frac{C}{2\pi} = \frac{2\pi r}{2\pi}$	Divide each side by 2π.
$\frac{C}{2\pi} = r$	Simplify.

b. Find the radius of the yo-yo.

$\frac{C}{2\pi} = r$	Formula for radius
$\frac{32.7}{2\pi} = r$	$C = 32.7$
$5.2 \approx r$	Use a calculator.

The yo-yo has a radius of about 5.2 feet.

GuidedPractice

3. GEOMETRY The formula for the volume of a rectangular prism is $V = \ell wh$, where ℓ is the length, w is the width, and h is the height.

A. Solve the formula for w.

B. Find the width of a rectangular prism that has a volume of 79.04 cubic centimeters, a length of 5.2 centimeters, and a height of 4 centimeters.

When using formulas, you may want to use dimensional analysis. **Dimensional analysis** or **unit analysis** is the process of carrying units throughout a computation.

PT

Example 4 Use Dimensional Analysis

RUNNING A 10K run is 10 kilometers long. If 1 meter = 1.094 yards, use dimensional analysis to find the length of the race in miles. (*Hint*: 1 mi = 1760 yd)

Since the given conversion relates meters to yards, first convert 10 kilometers to meters. Then multiply by the conversion factor such that the unit meters are divided out. To convert from yards to miles, multiply by $\frac{1\text{ mi}}{1760\text{ yd}}$.

length of run		kilometers to meters		meters to yards		yards to miles
10 km	×	$\frac{1000\text{ m}}{1\text{ km}}$	×	$\frac{1.094\text{ yd}}{1\text{ m}}$	×	$\frac{1\text{ mi}}{1760\text{ yd}}$

Notice how the units cancel, leaving the unit to which you are converting.

$$10\ \cancel{\text{km}} \times \frac{1000\ \cancel{\text{m}}}{1\ \cancel{\text{km}}} \times \frac{1.094\ \cancel{\text{yd}}}{1\ \cancel{\text{m}}} \times \frac{1\text{ mi}}{1760\ \cancel{\text{yd}}} = \frac{10{,}940\text{ mi}}{1760}$$

$$\approx 6.2\text{ mi}$$

A 10K race is approximately 6.2 miles.

StudyTip

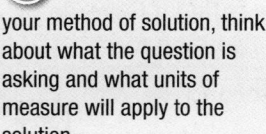

CCSS Precision As you plan your method of solution, think about what the question is asking and what units of measure will apply to the solution.

GuidedPractice

4. A car travels a distance of 100 feet in about 2.8 seconds. What is the velocity of the car in miles per hour? Round to the nearest whole number.

Check Your Understanding

Examples 1–2 **Solve each equation or formula for the variable indicated.**

1 $5a + c = -8a$, for a

2. $7h + f = 2h + g$, for g

3. $\frac{k + m}{-7} = n$, for k

4. $q = p(r + s)$, for p

Example 3

5. PACKAGING A soap company wants to use a cylindrical container to hold their new liquid soap.

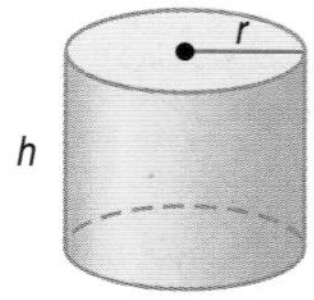

$V = \pi r^2 h$

a. Solve the formula for h.

b. What is the height of a container if the volume is 56.52 cubic inches and the radius is 1.5 inches? Round to the nearest tenth.

Example 4

6. SHOPPING Scott found a rare video game on an online auction site priced at 35 Australian dollars. If the exchange rate is \$1 U.S. = \$1.24 Australian, find the cost of the game in United States dollars. Round to the nearest cent.

7. CCSS PRECISION A fisheye lens has a minimum focus range of 13.5 centimeters. If 1 centimeter is equal in length to about 0.39 inch, what is the minimum focus range of the lens in feet?

Practice and Problem Solving

Examples 1–2 **Solve each equation or formula for the variable indicated.**

8. $u = vw + z$, for v

9. $x = b - cd$, for c

10. $fg - 9h = 10j$, for g

11. $10m - p = -n$, for m

12. $r = \frac{2}{3}t + v$, for t

13. $\frac{5}{9}v + w = z$, for v

14. $\frac{10ac - x}{11} = -3$, for a

15. $\frac{df + 10}{6} = g$, for f

Example 3

16. FITNESS The formula to compute a person's body mass index is $B = 703 \cdot \frac{w}{h^2}$. B represents the body mass index, w is the person's weight in pounds, and h represents the person's height in inches.

a. Solve the formula for w.

b. What is the weight to the nearest pound of a person who is 64 inches tall and has a body mass index of 21.45?

17. PHYSICS Acceleration is the measure of how fast a velocity is changing. The formula for acceleration is $a = \frac{v_f - v_i}{t}$. a represents the acceleration rate, v_f is the final velocity, v_i is the initial velocity, and t represents the time in seconds.

a. Solve the formula for v_f.

b. What is the final velocity of a runner who is accelerating at 2 feet per second squared for 3 seconds with an initial velocity of 4 feet per second?

Example 4

18. SWIMMING If each lap in a pool is 100 meters long, how many laps equal one mile? Round to the nearest tenth. (*Hint*: 1 foot ≈ 0.3048 meter)

19. CCSS PRECISION How many liters of gasoline are needed to fill a 13.2-gallon tank? There are about 1.06 quarts per 1 liter. Round to the nearest tenth.

Solve each equation or formula for the variable indicated.

20. $-14n + q = rt - 4n$, for n

21. $18t + 11v = w - 13t$, for t

22. $ax + z = aw - y$, for a

23. $10c - f = -13 + cd$, for c

Select an appropriate unit from the choices below and convert the rate to that unit.

ft/s	mph	mm/s	km/s

24. a car traveling at 36 ft/s

25. a snail moving at 3.6 m/h

26. a person walking at 3.4 mph

27. a satellite moving at 234,000 m/min

28. DANCING The formula $P = \frac{1.2W}{H^2}$ represents the amount of pressure exerted on the floor by a ballroom dancer's heel. In this formula, P is the pressure in pounds per square inch, W is the weight of a person wearing the shoe in pounds, and H is the width of the heel of the shoe in inches.

a. Solve the formula for W.

b. Find the weight of the dancer if the heel is 3 inches wide and the pressure exerted is 30 pounds per square inch.

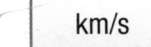

Write an equation and solve for the variable indicated.

29. Seven less than a number t equals another number r plus 6. Solve for t.

30. Ten plus eight times a number a equals eleven times number d minus six. Solve for a.

31. Nine tenths of a number g is the same as seven plus two thirds of another number k. Solve for k.

32. Three fourths of a number p less two is five sixths of another number r plus five. Solve for r.

33. **GIFTS** Ashley has 214 square inches of paper to wrap a gift box. The surface area S of the box can be found by using the formula $S = 2w(\ell + h) + 2\ell h$, where w is the width of the box, ℓ is the length of the box, and h is the height. If the length of the box is 7 inches and the width is 6 inches, how tall can Ashley's box be?

34. **DRIVING** A car is driven x miles a year and averages m miles per gallon.

a. Write a formula for g, the number of gallons used in a year.

b. If the average price of gas is p dollars per gallon, write a formula for the total gas cost c in dollars for driving this car each year.

c. Car A averages 15 miles per gallon on the highway, while Car B averages 35 miles per gallon on the highway. If you average 15,000 miles each year, how much money would you save on gas per week by using Car B instead of Car A if the cost of gas averages $3 per gallon? Explain.

H.O.T. Problems Use Higher-Order Thinking Skills

35. **CHALLENGE** The circumference of an NCAA women's basketball is 29 inches, and the rubber coating is $\frac{3}{16}$ inch thick. Use the formula $v = \frac{4}{3}\pi r^3$, where v represents the volume and r is the radius of the inside of the ball, to determine the volume of the air inside the ball. Round to the nearest whole number.

36. **REASONING** Select an appropriate unit to describe the highway speed of a car and the speed of a crawling caterpillar. Can the same unit be used for both? Explain.

37. **ERROR ANALYSIS** Sandrea and Fernando are solving $4a - 5b = 7$ for b. Is either of them correct? Explain.

Sandrea

$4a - 5b = 7$

$-5b = 7 - 4a$

$\frac{-5b}{-5} = \frac{7 - 4a}{-5}$

$b = \frac{7 - 4a}{-5}$

Fernando

$4a - 5b = 7$

$5b = 7 - 4a$

$\frac{5b}{5} = \frac{7 - 4a}{5}$

$b = \frac{7 - 4a}{5}$

38. **OPEN ENDED** Write a formula for A, the area of a geometric figure such as a triangle or rectangle. Then solve the formula for a variable other than A.

39. CCSS **PERSEVERANCE** Solve each equation or formula for the variable indicated.

a. $n = \frac{x + y - 1}{xy}$ for x

b. $\frac{x + y}{x - y} = \frac{1}{2}$ for y

40. **WRITING IN MATH** Why is it helpful to be able to represent a literal equation in different ways?

Standardized Test Practice

41. Eula is investing \$6000, part at 4.5% interest and the rest at 6% interest. If d represents the amount invested at 4.5%, which expression represents the amount of interest earned in one year by the amount paying 6%?

A $0.06d$ **C** $0.06(d + 6000)$
B $0.06(d - 6000)$ **D** $0.06(6000 - d)$

42. Todd drove from Boston to Cleveland, a distance of 616 miles. His breaks, gasoline, and food stops took 2 hours. If his trip took 16 hours altogether, what was Todd's average speed?

F 38.5 mph **H** 44 mph
G 40 mph **J** 47.5 mph

43. SHORT RESPONSE Brian has 3 more books than Erika. Jasmine has triple the number of books that Brian has. Altogether Brian, Erika, and Jasmine have 22 books. How many books does Jasmine have?

44. GEOMETRY Which of the following best describes a plane?

A a location having neither size nor shape
B a flat surface made up of points having no depth
C made up of points and has no thickness or width
D a boundless, three-dimensional set of all points

Spiral Review

Find the final price of each item. (Lesson 2-7)

45. lamp: \$120.00
discount: 20%
tax: 6%

46. dress: \$70.00
discount: 30%
tax: 7%

47. camera: \$58.00
discount: 25%
tax: 6.5%

48. jacket: \$82.00
discount: 15%
tax: 6%

49. comforter: \$67.00
discount: 20%
tax: 6.25%

50. lawnmower: \$720.00
discount: 35%
tax: 7%

Solve each proportion. If necessary, round to the nearest hundredth. (Lesson 2-6)

51. $\frac{3}{4.5} = \frac{x}{2.5}$

52. $\frac{2}{0.36} = \frac{7}{p}$

53. $\frac{m}{9} = \frac{2.8}{4.9}$

54. JOBS Laurie mows lawns to earn extra money. She can mow at most 30 lawns in one week. She profits \$15 on each lawn she mows. Identify a reasonable domain and range for this situation and draw a graph. (Lesson 1-6)

55. ENTERTAINMENT Each member of the pit orchestra is selling tickets for the school musical. The trombone section sold 50 floor tickets and 90 balcony tickets. Write and evaluate an expression to find how much money the trombone section collected. (Lesson 1-4)

Skills Review

Solve each equation.

56. $8k + 9 = 7k + 6$

57. $3 - 4q = 10q + 10$

58. $\frac{3}{4}n + 16 = 2 - \frac{1}{8}n$

59. $\frac{1}{4} - \frac{2}{3}y = \frac{3}{4} - \frac{1}{3}y$

60. $4(2a - 1) = -10(a - 5)$

61. $2(w - 3) + 5 = 3(w - 1)$

LESSON 2-9 Weighted Averages

:: Then	:: Now	:: Why?
You translated sentences into equations.	**1** Solve mixture problems. **2** Solve uniform motion problems.	Baseball players' performance is measured in large part by statistics. Slugging average (SLG) is a weighted average that measures the power of a hitter. The slugging average is calculated by using the following formula. $\text{SLG} = \frac{1\text{B} + (2 \times 2\text{B}) + (3 \times 3\text{B}) + (4 \times \text{HR})}{\text{at bats}}$

NewVocabulary
weighted average
mixture problem
uniform motion problem
rate problem

Common Core State Standards

Content Standards

A.REI.1 Explain each step in solving a simple equation as following from the equality of numbers asserted at the previous step, starting from the assumption that the original equation has a solution. Construct a viable argument to justify a solution method.

A.REI.3 Solve linear equations and inequalities in one variable, including equations with coefficients represented by letters.

Mathematical Practices

1 Make sense of problems and persevere in solving them.

4 Model with mathematics.

1 Weighted Averages

The batter's slugging percentage is an example of a weighted average. The **weighted average** M of a set of data is found by multiplying each data value by its weight and then finding the mean of the new data set.

Mixture problems are problems in which two or more parts are combined into a whole. They are solved using weighted averages. In a mixture problem, the units are usually the number of gallons or pounds and the value is the cost, value, or concentration per unit.

Real-World Example 1 Mixture Problem

RETAIL **A tea company sells blended tea for $25 per pound. To make blackberry tea, dried blackberries that cost $10.50 per pound are blended with black tea that costs $35 per pound. How many pounds of black tea should be added to 5 pounds of dried blackberries to make blackberry tea?**

Step 1 Let w be the weight of the black tea. Make a table to organize the information.

	Number of Units (lb)	Price per Unit ($)	Total Price (price)(units)
Dried Blackberries	5	10.50	10.50(5)
Black Tea	w	35	$35w$
Blackberry Tea	$5 + w$	25	$25(5 + w)$

Write an equation using the information in the table.

Price of blackberries	plus	price of tea	equals	price of blackberry tea.
10.50(5)	+	$35w$	=	$25(5 + w)$

Step 2 Solve the equation.

$10.50(5) + 35w = 25(5 + w)$	Original equation
$52.5 + 35w = 125 + 25w$	Distributive Property
$52.5 + 35w - 25w = 125 + 25w - 25w$	Subtract $25w$ from each side.
$52.5 + 10w = 125$	Simplify.
$52.5 - 52.5 + 10w = 125 - 52.5$	Subtract 52.5 from each side.
$10w = 72.5$	Simplify.
$w = 7.25$	Divide each side by 10.

To make the blackberry tea, 7.25 pounds of black tea will need to be added to the dried blackberries.

StudyTip

Mixture Problems When you organize the information in mixture problems, remember that the final mixture must contain the sum of the parts in the correct quantities and at the correct percents.

GuidedPractice

1. COFFEE How many pounds of Premium coffee beans should be mixed with 2 pounds of Supreme coffee to make the Blend coffee?

Sometimes mixture problems are expressed in terms of percents.

Real-World Example 2 Percent Mixture Problem

FRUIT PUNCH Mrs. Matthews has 16 cups of punch that is 3% pineapple juice. She also has a punch that is 33% pineapple juice. How many cups of the 33% punch will she need to add to the 3% punch to obtain a punch that is 20% pineapple juice?

Problem-SolvingTip

Make a Table Using a table is a great way to organize the given information. It also helps you understand how to write an equation to solve for the missing value.

Step 1 Let x = the amount of 33% solution to be added. Make a table.

	Amount of Punch (cups)	Amount of Pineapple Juice
3% Punch	16	0.03(16)
33% Punch	x	$0.33x$
20% Punch	$16 + x$	$0.20(16 + x)$

Write an equation using the information in the table.

Amount of pineapple juice in 3% punch	plus	amount of pineapple juice in 33% punch	equals	amount of pineapple juice in 20% punch.
$0.03(16)$	$+$	$0.33x$	$=$	$0.20(16 + x)$

Step 2 Solve the equation.

$0.03(16) + 0.33x = 0.20(16 + x)$	Original equation
$0.48 + 0.33x = 3.2 + 0.20x$	Simplify.
$0.48 + 0.33x - \mathbf{0.20x} = 3.2 + 0.20x - \mathbf{0.20x}$	Subtract $0.20x$ from each side.
$0.48 + 0.13x = 3.2$	Simplify.
$0.48 - \mathbf{0.48} + 0.13x = 3.2 - \mathbf{0.48}$	Subtract 0.48 from each side.
$0.13x = 2.72$	Simplify.
$\frac{0.13x}{0.13} = \frac{2.72}{0.13}$	Divide each side by 0.13.
$x \approx 20.9$	Round to the nearest tenth.

Mrs. Matthews should add about 20.9 cups of the 33% punch to the 16 cups of the 3% punch.

GuidedPractice

2. ANTIFREEZE One type of antifreeze is 40% glycol, and another type of antifreeze is 60% glycol. How much of each kind should be used to make 100 gallons of antifreeze that is 48% glycol?

2 Uniform Motion Problems

Uniform motion problems or **rate problems** are problems in which an object moves at a certain speed or rate. The formula $d = rt$ is used to solve these problems. In the formula, d represents distance, r represents rate, and t represents time.

Real-WorldLink

Inline skating is the fourth most popular recreational activity in the U.S.

Source: Statistical Abstract of the United States

Real-World Example 3 Speed of One Vehicle

INLINE SKATING **It took Travis and Tony 40 minutes to skate 5 miles. The return trip took them 30 minutes. What was their average speed for the trip?**

Understand We know that the boys did not travel the same amount of time on each portion of their trip. So, we will need to find the weighted average of their speeds. We are asked to find their average speed for both portions of the trip.

Plan First find the rate of the going portion, and then the return portion of the trip. Because the rate is in miles per hour we convert 40 minutes to about 0.667 hours and 30 minutes to 0.5 hours.

Going

$r = \frac{d}{t}$ Formula for rate

$\approx \frac{5 \text{ miles}}{0.667 \text{ hour}}$ or about 7.5 miles per hour Substitution $d = 5$ mi, $t = 0.667$ h

Return

$r = \frac{d}{t}$ Formula for rate

$= \frac{5 \text{ miles}}{0.5 \text{ hour}}$ or 10 miles per hour Substitution $d = 5$ mi, $t = 0.5$ h

Because we are looking for a weighted average we cannot just average their speeds. We need to find the weighted average for the round trip.

Solve

$$M = \frac{(\text{rate of going})(\text{time of going}) + (\text{rate of return})(\text{time of return})}{\text{time of going} + \text{time of return}}$$

$$\approx \frac{(7.5)(0.667) + (10)(0.5)}{0.667 + 0.5}$$ Substitution

$$\approx \frac{10.0025}{1.167} \text{ or about } 8.6$$ Simplify.

Their average speed was about 8.6 miles per hour.

Check Our solution of 8.6 miles per hour is between the going portion rate, 7.5 miles per hour, and the return rate, 10 miles per hour. So, we know that our answer is reasonable.

GuidedPractice

3. EXERCISE Austin jogged 2.5 miles in 16 minutes and then walked 1 mile in 10 minutes. What was his average speed?

The formula $d = rt$ can also be used to solve real-world problems involving two vehicles in motion.

Real-World Example 4 Speeds of Two Vehicles

FREIGHT TRAINS Two trains are 550 miles apart heading toward each other on parallel tracks. Train A is traveling east at 35 miles per hour, while Train B travels west at 45 miles per hour. When will the trains pass each other?

> **StudyTip**
>
> **CCSS Sense-Making** Drawing a diagram is not just for geometry problems. You can use diagrams to visualize many problem situations that can be represented by equations.

Step 1 Draw a diagram.

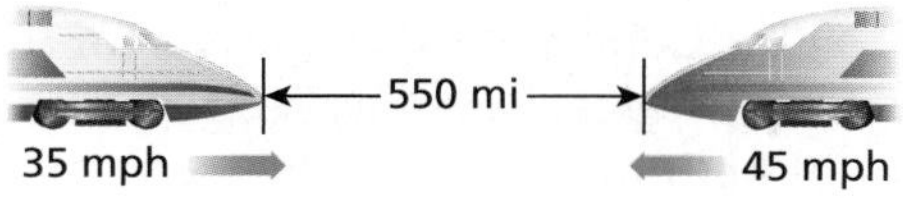

Step 2 Let t = the number of hours until the trains pass each other. Make a table.

	r	t	$d = rt$
Train A	35	t	$35t$
Train B	45	t	$45t$

Step 3 Write and solve an equation.

Distance traveled by Train A	plus	distance traveled by Train B	equals	550 miles.
$35t$	$+$	$45t$	$=$	550

$35t + 45t = 550$ Original equation

$80t = 550$ Simplify.

$\frac{80t}{80} = \frac{550}{80}$ Divide each side by 80.

$t = 6.875$ Simplify.

The trains will pass each other in about 6.875 hours.

GuidedPractice

4. CYCLING Two cyclists begin traveling in opposite directions on a circular bike trail that is 5 miles long. One cyclist travels 12 miles per hour, and the other travels 18 miles per hour. How long will it be before they meet?

Check Your Understanding

Example 1

1. FOOD Tasha ordered soup and salad for lunch. If Tasha ordered 10 ounces of soup for lunch and the total cost was $3.30, how many ounces of salad did Tasha order?

Example 2

2. CHEMISTRY Margo has 40 milliliters of 25% solution. How many milliliters of 60% solution should she add to obtain the required 30% solution?

Example 3

3. TRAVEL A boat travels 16 miles due north in 2 hours and 24 miles due west in 2 hours. What is the average speed of the boat?

4. EXERCISE Felisa jogged 3 miles in 25 minutes and then jogged 3 more miles in 30 minutes. What was her average speed in miles per minute?

Example 4

5. CYCLING A cyclist begins traveling 18 miles per hour. At the same time and at the same starting point, an inline skater follows the cyclist's path and begins traveling 6 miles per hour. After how much time will they be 24 miles apart?

Practice and Problem Solving

Example 1

6. **CANDY** A candy store wants to create a mix using two hard candies. One is priced at $5.45 per pound, and the other is priced at $7.33 per pound. How many pounds of the $7.33 candy should be mixed with 11 pounds of the $5.45 candy to sell the mixture for $6.14 per pound?

7. **BUSINESS** Party Supplies Inc. sells metallic balloons for $2 each and helium balloons for $3.50 per bunch. Yesterday, they sold 36 more metallic balloons than the number of bunches of helium balloons. The total sales for both types of balloons were $281. Let b represent the number of metallic balloons sold.

   a. Copy and complete the table representing the problem.

	Number	Price	Total Price
Metallic Balloons	b		
Bunches of Helium Balloons	$b - 36$		

   b. Write an equation to represent the problem.

   c. How many metallic balloons were sold?

   d. How many bunches of helium balloons were sold?

8. **FINANCIAL LITERACY** Lakeisha spent $4.57 on color and black-and-white copies for her project. She made 7 more black-and-white copies than color copies. How many color copies did she make?

Type of Copy	Cost per Page
color	$0.44
black-and-white	$0.07

Example 2

9. **FISH** Rosamaria is setting up a 20-gallon saltwater fish tank that needs to have a salt content of 3.5%. If Rosamaria has water that has 2.5% salt and water that has 3.7% salt, how many gallons of the water with 3.7% salt content should Rosamaria use?

10. **CHEMISTRY** Hector is performing a chemistry experiment that requires 160 milliliters of 40% sulfuric acid solution. He has a 25% sulfuric acid solution and a 50% sulfuric acid solution. How many milliliters of each solution should he mix to obtain the needed solution?

Example 3

11. **TRAVEL** A boat travels 36 miles in 1.5 hours and then 14 miles in 0.75 hour. What is the average speed of the boat?

12. CCSS **MODELING** A person walked 1.5 miles in 28 minutes and then jogged 1.2 more miles in 10 minutes. What was the average speed in miles per minute?

Example 4

13. **AIRLINERS** Two airliners are 1600 miles apart and heading toward each other at different altitudes. The first plane is traveling north at 620 miles per hour, while the second is traveling south at 780 miles per hour. When will the planes pass each other?

14. **SAILING** A ship is sailing due east at 20 miles per hour when it passes the lighthouse. At the same time a ship is sailing due west at 15 miles per hour when it passes a point. The point is 175 miles east of the lighthouse. When will these ships pass each other?

15. **CHEMISTRY** A lab technician has 40 gallons of a 15% iodine solution. How many gallons of a 40% iodine solution must he add to make a 20% iodine solution?

16. GRADES At Westbridge High School, a student's grade point average (GPA) is based on the student's grade and the class credit rating. Brittany's grades for this quarter are shown. Find Brittany's GPA if a grade of A equals 4 and a B equals 3.

Class	Credit Rating	Grade
Algebra 1	1	A
Science	1	A
English	1	B
Spanish	1	A
Music	$\frac{1}{2}$	B

17. SPORTS In a triathlon, Steve swam 0.5 mile in 15 minutes, biked 20 miles in 90 minutes, and ran 4 miles in 30 minutes. What was Steve's average speed for the triathlon in miles per hour?

18. MUSIC Amalia has 10 songs on her digital media player. If 3 songs are 5 minutes long, 3 are 4 minutes long, 2 are 2 minutes long, and 2 are 3.5 minutes long, what is the average length of the songs?

19. DISTANCE Garcia is driving to Florida for vacation. The trip is a total of 625 miles.

a. How far can he drive in 6 hours at 65 miles per hour?

b. If Garcia maintains a speed of 65 miles per hour, how long will it take him to drive to Florida?

20. TRAVEL Two buses leave Smithville at the same time, one traveling north and the other traveling south. The northbound bus travels at 50 miles per hour, and the southbound bus travels at 65 miles per hour. Let t represent the amount of time since their departure.

a. Copy and complete the table representing the situation.

	r	t	$d = rt$
Northbound bus	?	?	?
Southbound bus	?	?	?

b. Write an equation to find when the buses will be 345 miles apart.

c. Solve the equation. Explain how you found your answer.

21. TRAVEL A subway travels 60 miles per hour from Glendale to Midtown. Another subway, traveling at 45 miles per hour, takes 11 minutes longer for the same trip. How far apart are Glendale and Midtown?

H.O.T. Problems Use Higher-Order Thinking Skills

22. OPEN ENDED Write a problem that depicts motion in opposite directions.

23. CCSS ARGUMENTS Describe the conditions so that adding a 50% solution to a 100% solution would produce a 75% solution.

24. CHALLENGE Find five consecutive odd integers from least to greatest in which the sum of the first and the fifth is one less than three times the fourth.

25. CHALLENGE Describe a situation involving mixtures that could be represented by $1.00x + 0.15(36) = 0.50(x + 36)$.

26. WRITING IN MATH Describe how a gallon of 25% solution is added to an unknown amount of 10% solution to get a 15% solution.

Standardized Test Practice

27. If $2x + y = 5$, what is the value of $4x$?

A $10 - y$

B $10 - 2y$

C $\frac{5 - y}{2}$

D $\frac{10 - y}{2}$

28. Which expression is equivalent to $7x^2 3x^{-4}$?

F $21x^{-8}$

G $21x^2$

H $21x^{-6}$

J $21x^{-2}$

29. GEOMETRY What is the base of the triangle if the area is 56 square meters?

A 4 m

B 8 m

C 16 m

D 28 m

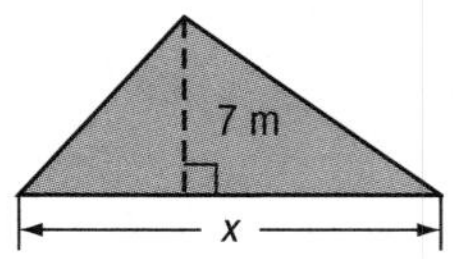

30. SHORT RESPONSE Brianne makes blankets for a baby store. She works on the blankets 30 hours per week. The store pays her \$9.50 per hour plus 30% of the profit. If her hourly rate is increased by \$0.75 and her commission is raised to 40%, how much will she earn for a week in which there was a \$300 profit?

Spiral Review

Solve each equation or formula for *x*. (Lesson 2-8)

31. $2bx - b = -5$

32. $3x - r = r(-3 + x)$

33. $A = 2\pi r^2 + 2\pi rx$

34. SKIING Yuji is registering for ski camp. The cost of the camp is \$1254, but there is a sales tax of 7%. What is the total cost of the camp including tax? (Lesson 2-7)

Translate each equation into a sentence. (Lesson 2-1)

35. $\frac{n}{-6} = 2n + 1$

36. $18 - 5h = 13h$

37. $2x^2 + 3 = 21$

Refer to the graph.

38. Name the ordered pair at point *A* and explain what it represents. (Lesson 1-6)

39. Name the ordered pair at point *B* and explain what it represents. (Lesson 1-6)

40. Identify the independent and dependent variables for the function. (Lesson 1-6)

41. BASEBALL Tickets to a baseball game cost \$18.95, \$12.95, or \$9.95. A hot dog and soda combo costs \$5.50. The Madison family is having a reunion. They buy 10 tickets in each price category and plan to buy 30 combos. What is the total cost for the tickets and meals? (Lesson 1-4)

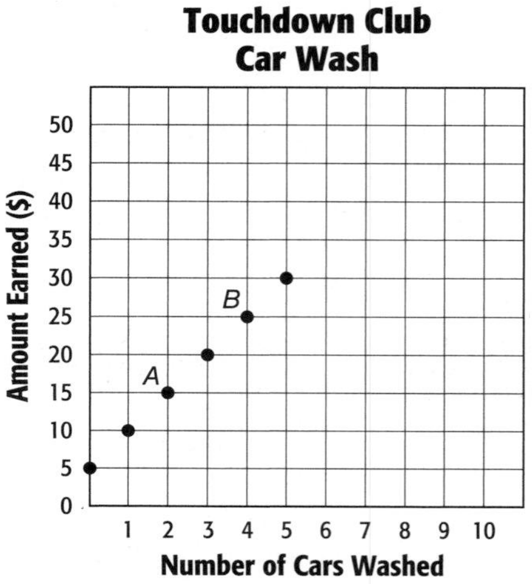

Skills Review

Solve each equation.

42. $a - 8 = 15$

43. $9m - 11 = -29$

44. $18 - 2k = 24$

45. $5 - 8y = 61$

46. $7 = \frac{h}{2} + 3$

47. $\frac{n}{6} + 1 = 5$

CHAPTER 2

Study Guide and Review

Study Guide

KeyConcepts

Writing Equations (Lesson 2-1)

- Identify the unknown you are looking for and assign a variable to it. Then, write the sentence as an equation.

Solving Equations (Lessons 2-2 to 2-4)

- Addition and Subtraction Properties of Equality: If an equation is true and the same number is added to or subtracted from each side, the resulting equation is true.
- Multiplication and Division Properties of Equality: If an equation is true and each side is multiplied or divided by the same nonzero number, the resulting equation is true.
- Steps for Solving Equations:
  - **Step 1** Simplify the expression on each side. Use the Distributive Property as needed.
  - **Step 2** Use the Addition and/or Subtraction Properties of Equality to get the variables on one side and the numbers without variables on the other side.
  - **Step 3** Use the Multiplication or Division Property of Equality to solve.

Absolute Value Equations (Lesson 2-5)

- For any real numbers a and b, if $|a| = b$ and $b \geq 0$, then $a = b$ or $a = -b$.

Ratios and Proportions (Lesson 2-6)

- The Means-Extremes Property of Proportion states that in a proportion, the product of the extremes is equal to the product of the means.

Percent of Change (Lesson 2-7)

- percent of change $= \dfrac{\text{the change in an amount}}{\text{the original amount}}$ expressed as a percent

Weighted Averages (Lesson 2-9)

- the weighted average M of a set of data $= \dfrac{\text{sum of (units} \times \text{the value per unit)}}{\text{the total number of units}}$

FOLDABLES® StudyOrganizer

Be sure the Key Concepts are noted in your Foldable.

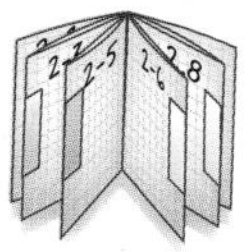

KeyVocabulary

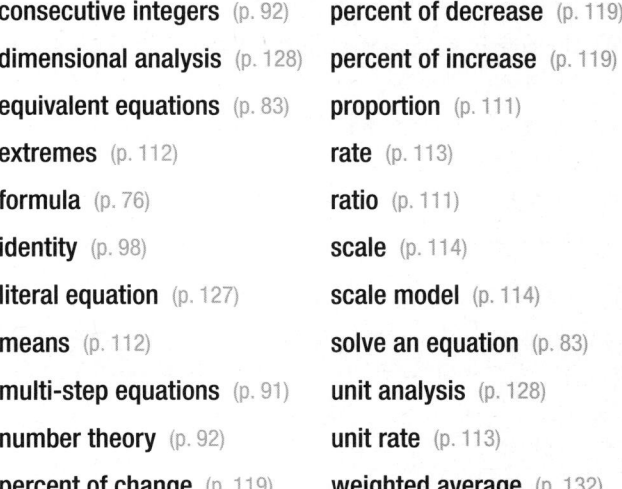

consecutive integers (p. 92)
dimensional analysis (p. 128)
equivalent equations (p. 83)
extremes (p. 112)
formula (p. 76)
identity (p. 98)
literal equation (p. 127)
means (p. 112)
multi-step equations (p. 91)
number theory (p. 92)
percent of change (p. 119)
percent of decrease (p. 119)
percent of increase (p. 119)
proportion (p. 111)
rate (p. 113)
ratio (p. 111)
scale (p. 114)
scale model (p. 114)
solve an equation (p. 83)
unit analysis (p. 128)
unit rate (p. 113)
weighted average (p. 132)

VocabularyCheck

State whether each sentence is *true* or *false*. If *false*, replace the underlined term to make a true sentence.

1. In order to write an equation to solve a problem, identify the unknown for which you are looking and assign a(n) number to it.
2. To solve an equation means to find the value of the variable that makes the equation true.
3. The numbers 10, 12, and 14 are an example of consecutive even integers.
4. The absolute value of any number is simply the distance the number is away from zero on a number line.
5. A(n) equation is a comparison of two numbers by division.
6. An equation stating that two ratios are equal is called a(n) proportion.
7. If the new number is less than the original number, the percent of change is a percent of increase.
8. The weighted average of a set of data is the sum of the product of the number of units and the value per unit divided by the sum of the number of units.

Study Guide and Review *Continued*

Lesson-by-Lesson Review

2-1 Writing Equations

Translate each sentence into an equation.

9. The sum of five times a number x and three is the same as fifteen.

10. Four times the difference of b and six is equal to b squared.

11. One half of m cubed is the same as four times m minus nine.

Translate each equation into a sentence.

12. $3p + 8 = 20$

13. $h^2 - 5h + 6 = 0$

14. $\frac{3}{4}w^2 + \frac{2}{3}w - \frac{1}{5} = 2$

15. **FENCING** Adrianne wants to create an outdoor rectangular kennel. The length will be three feet more than twice the width. Write and use an equation to find the length and the width of the kennel if Adrianne has 54 feet of fencing.

Example 1

Translate the following sentence into an equation.

Six times the sum of a number n and four is the same as the difference between two times n to the second power and ten.

$6(n + 4) = 2n^2 - 10$

Example 2

Translate $3d^2 - 9d + 8 = 4(d + 2)$ into a sentence.

Three times a number d squared minus nine times d increased by eight is equal to four times the sum of d and two.

2-2 Solving One-Step Equations

Solve each equation. Check your solution.

16. $x - 9 = 4$

17. $-6 + g = -11$

18. $\frac{5}{9} + w = \frac{7}{9}$

19. $3.8 = m + 1.7$

20. $\frac{a}{12} = 5$

21. $8y = 48$

22. $\frac{2}{5}b = -4$

23. $-\frac{t}{16} = -\frac{7}{8}$

24. **AGE** Max is four years younger than his sister Brenda. Max is 16 years old. Write and solve an equation to find Brenda's age.

Example 3

Solve $x - 13 = 9$. Check your solution.

$x - 13 = 9$	Original equation
$x - 13 + 13 = 9 + 13$	Add 13 to each side.
$x = 22$	$-13 + 13 = 0$ and $9 + 13 = 22$

To check that 22 is the solution, substitute 22 for x in the original equation.

CHECK	$x - 13 = 9$	Original equation
	$22 - 13 \stackrel{?}{=} 9$	Substitute 22 for x.
	$9 = 9$ ✓	Subtract.

2-3 Solving Multi-Step Equations

Solve each equation. Check your solution.

25. $2d - 4 = 8$

26. $-9 = 3t + 6$

27. $14 = -8 - 2k$

28. $\frac{n}{4} - 7 = -2$

29. $\frac{r + 4}{3} = 7$

30. $-18 = \frac{9 - a}{2}$

31. $6g - 3.5 = 8.5$

32. $0.2c + 4 = 6$

33. $\frac{f}{3} - 9.2 = 3.5$

34. $4 = \frac{-3u - (-7)}{-8}$

35. **CONSECUTIVE INTEGERS** Find three consecutive odd integers with a sum of 63.

36. **CONSECUTIVE INTEGERS** Find three consecutive integers with a sum of −39.

Example 4

Solve $7y - 9 = 33$. Check your solution.

$7y - 9 = 33$	Original equation
$7y - 9 + 9 = 33 + 9$	Add 9 to each side.
$7y = 42$	Simplify.
$\frac{7y}{7} = \frac{42}{7}$	Divide each side by 7.
$y = 6$	Simplify.

CHECK $7y - 9 = 33$	Original equation
$7(6) - 9 \stackrel{?}{=} 33$	Substitute 6 for y.
$42 - 9 \stackrel{?}{=} 33$	Multiply.
$33 = 33$ ✓	Subtract.

2-4 Solving Equations with the Variable on Each Side

Solve each equation. Check your solution.

37. $8m + 7 = 5m + 16$

38. $2h - 14 = -5h$

39. $21 + 3j = 9 - 3j$

40. $\frac{x - 3}{4} = \frac{x}{2}$

41. $\frac{6r - 7}{10} = \frac{r}{4}$

42. $3(p + 4) = 33$

43. $-2(b - 3) - 4 = 18$

44. $4(3w - 2) = 8(2w + 3)$

Write an equation and solve each problem.

45. Find the sum of three consecutive odd integers if the sum of the first two integers is equal to twenty-four less than four times the third integer.

46. **TRAVEL** Mr. Jones drove 480 miles to a business meeting. His travel time to the meeting was 8 hours and from the meeting was 7.5 hours. Find his rate of travel for each leg of the trip.

Example 5

Solve $9w - 24 = 6w + 18$.

$9w - 24 = 6w + 18$	Original equation
$9w - 24 - 6w = 6w + 18 - 6w$	Subtract $6w$ from each side.
$3w - 24 = 18$	Simplify.
$3w - 24 + 24 = 18 + 24$	Add 24 to each side.
$3w = 42$	Simplify.
$\frac{3w}{3} = \frac{42}{3}$	Divide each side by 3.
$w = 14$	Simplify.

Example 6

Write an equation to find three consecutive integers such that three times the sum of the first two integers is the same as thirteen more than four times the third integer.

Let x, $x + 1$, and $x + 2$ represent the three consecutive integers.

$3(x + x + 1) = 4(x + 2) + 13$

CHAPTER 2 Study Guide and Review *Continued*

2-5 Solving Equations Involving Absolute Value

Evaluate each expression if $m = -8$, $n = 4$, and $p = -12$.

47. $|3m - n|$

48. $|-2p + m| - 3n$

49. $-3|6n - 2p|$

50. $4|7m + 3p| + 4n$

Solve each equation. Then graph the solution set.

51. $|x - 6| = 11$

52. $|-4w + 2| = 14$

53. $\left|\frac{1}{3}d - 6\right| = 15$

54. $\left|\frac{2b}{3} + 8\right| = 20$

Example 7

Solve $|y - 9| = 16$. Then graph the solution set.

Case 1

$y - 9 = 16$	Original equation
$y - 9 + 9 = 16 + 9$	Add 9 to each side.
$y = 25$	Simplify.

Case 2

$y - 9 = -16$	Original equation
$y - 9 + 9 = -16 + 9$	Add 9 to each side.
$y = -7$	Simplify.

The solution set is $\{-7, 25\}$.

Graph the points on a number line.

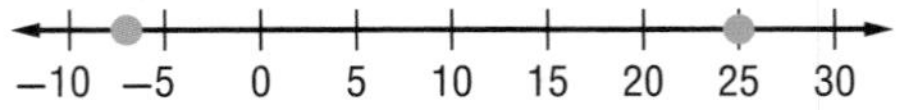

2-6 Ratios and Proportions

Determine whether each pair of ratios are equivalent ratios. Write *yes* or *no*.

55. $\frac{27}{45}, \frac{3}{5}$

56. $\frac{18}{32}, \frac{3}{4}$

Solve each proportion. If necessary, round to the nearest hundredth.

57. $\frac{4}{9} = \frac{a}{45}$

58. $\frac{3}{8} = \frac{21}{t}$

59. $\frac{9}{12} = \frac{g}{16}$

60. **CONSTRUCTION** A new gym is being built at Greenfield Middle School. The length of the gym as shown on the builder's blueprints is 12 inches. Find the actual length of the new gym.

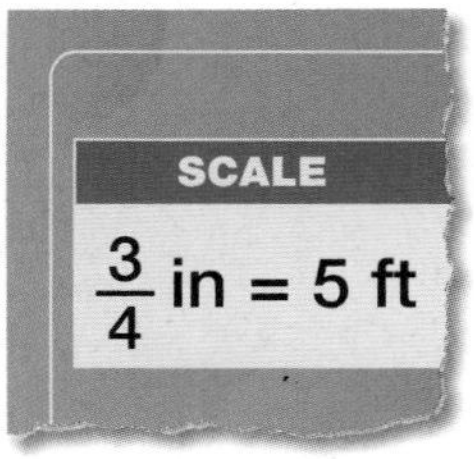

Example 8

Determine whether $\frac{7}{9}$ and $\frac{42}{54}$ are equivalent ratios. Write *yes* or *no*. Justify your answer.

First, simplify each ratio. $\frac{7}{9}$ is already in simplest form.

$$\frac{42}{54} = \frac{42 \div 6}{54 \div 6} = \frac{7}{9}$$

When expressed in simplest form, the ratios are equivalent. The answer is yes.

Example 9

Solve $\frac{r}{8} = \frac{3}{4}$. If necessary, round to the nearest hundredth.

$\frac{r}{8} = \frac{3}{4}$	Original equation
$r(4) = 3(8)$	Find the cross products.
$4r = 24$	Simplify.
$\frac{4r}{4} = \frac{24}{4}$	Divide each side by 4.
$r = 6$	Simplify.

2-7 Percent of Change

State whether each percent of change is a percent of *increase* or a percent of *decrease*. Then find the percent of change. Round to the nearest whole percent.

61. original: 40, new: 50

62. original: 36, new: 24

63. original: $72, new: $60

Find the total price of each item.

64. boots: $64, tax: 7%

65. video game: $49, tax: 6.5%

66. hockey skates: $199, tax: 5.25%

Find the discounted price of each item.

67. digital media player: $69.00, discount: 20%

68. jacket: $129, discount: 15%

69. backpack: $45, discount: 25%

70. **ATTENDANCE** An amusement park recorded attendance of 825,000 one year. The next year, the attendance increased to 975,000. Determine the percent of increase in attendance.

Example 10

State whether the percent of change is a percent of *increase* or a percent of *decrease*. Then find the percent of change. Round to the nearest whole percent.

original: 80
final: 60

Subtract the original amount from the final amount to find the amount of change. $60 - 80 = -20$. Since the new amount is less than the original, this is a percent of decrease.

Use the original number, 80, as the base.

change → / original amount → $\frac{20}{80} = \frac{r}{100}$	Percent proportion
$20(100) = r(80)$	Find cross products.
$2000 = 80r$	Simplify.
$\frac{2000}{80} = \frac{80r}{80}$	Divide each side by 80.
$25 = r$	Simplify.

The percent of decrease is 25%.

2-8 Literal Equations and Dimensional Analysis

Solve each equation or formula for the variable indicated.

71. $3x + 2y = 9$, for y

72. $P = 2\ell + 2w$, for ℓ

73. $-5m + 9n = 15$, for m

74. $14w + 15x = y - 21w$, for w

75. $m = \frac{2}{5}y + n$, for y

76. $7d - 3c = f + 2d$, for d

77. **GEOMETRY** The formula for the area of a trapezoid is $A = \frac{1}{2}h(a + b)$, where h represents the height and a and b represent the lengths of the bases. Solve for h.

Example 11

Solve $6p - 8n = 12$ for p.

$6p - 8n = 12$	Original equation
$6p - 8n + 8n = 12 + 8n$	Add $8n$ to each side.
$6p = 12 + 8n$	Simplify.
$\frac{6p}{6} = \frac{12 + 8n}{6}$	Divide each side by 6.
$\frac{6p}{6} = \frac{12}{6} + \frac{8}{6}n$	Simplify.
$p = 2 + \frac{4}{3}n$	Simplify.

Study Guide and Review *Continued*

2-9 Weighted Averages

78. **CANDY** Michael is mixing two types of candy for a party. The chocolate pieces cost $0.40 per ounce, and the hard candy costs $0.20 per ounce. Michael purchases 20 ounces of the chocolate pieces, and the total cost of his candy was $11. How many ounces of hard candy did he purchase?

79. **TRAVEL** A car travels 100 miles east in 2 hours and 30 miles north in half an hour. What is the average speed of the car?

80. **FINANCIAL LITERACY** A candle supply store sells votive wax and low-shrink wax. How many pounds of low-shrink wax should be mixed with 8 pounds of votive wax to obtain a blend that sells for $0.98 a pound?

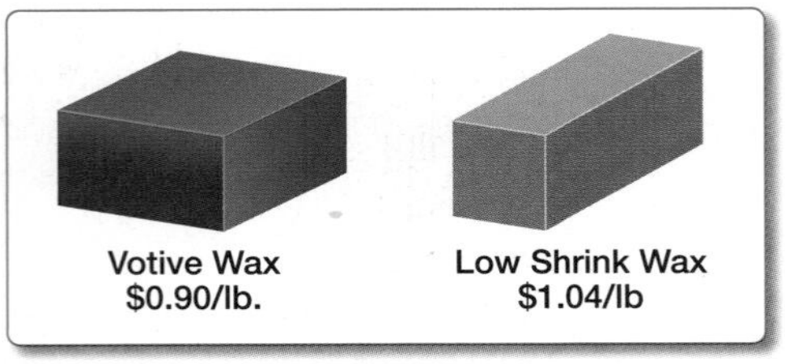

Example 12

METALS An alloy of metals is 25% copper. Another alloy is 50% copper. How much of each should be used to make 1000 grams of an alloy that is 45% copper?

Let $x =$ the amount of the 25% copper alloy. Write and solve an equation.

$0.25x + 0.50(1000 - x) = 0.45(1000)$	Original Equation
$0.25x + 500 - 0.50x = 450$	Distributive Property
$-0.25x + 500 = 450$	Simplify.
$-0.25x + 500 - 500 = 450 - 500$	Subtract 500 from each side.
$-0.25x = -50$	Simplify.
$\frac{-0.25x}{-0.25} = \frac{-50}{-0.25}$	Divide each side by –0.25.
$x = 200$	Simplify.

200 grams of the 25% alloy and 800 grams of the 50% alloy should be used.

CHAPTER 2 Practice Test

Translate each sentence into an equation.

1. The sum of six and four times d is the same as d minus nine.
2. Three times the difference of two times m and five is equal to eight times m to the second power increased by four.

Solve each equation. Check your solutions.

3. $x - 5 = -11$
4. $\frac{2}{3} = w + \frac{1}{4}$
5. $\frac{t}{6} = -3$

Solve each equation. Check your solution.

6. $2a - 5 = 13$
7. $\frac{p}{4} - 3 = 9$

8. **MULTIPLE CHOICE** At Mama Mia Pizza, the price of a large pizza is determined by $P = 9 + 1.5x$, where x represents the number of toppings added to a cheese pizza. Daniel spent \$13.50 on a large pizza. How many toppings did he get?

   A 0

   B 1

   C 3

   D 5

Solve each equation. Check your solution.

9. $5y - 4 = 9y + 8$
10. $3(2k - 2) = -2(4k - 11)$
11. **GEOMETRY** Find the value of x so that the figures have the same perimeter.

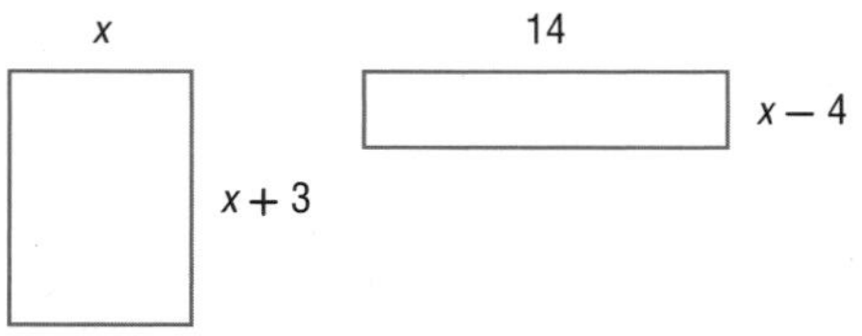

12. Evaluate the expression $|3t - 2u| + 5v$ if $t = 2$, $u = -5$, and $v = -3$.

Solve each equation. Then graph the solution set.

13. $|p - 4| = 6$
14. $|2b + 5| = 9$

Solve each proportion. If necessary, round to the nearest hundredth.

15. $\frac{a}{3} = \frac{16}{24}$
16. $\frac{9}{k + 3} = \frac{3}{5}$
17. **MULTIPLE CHOICE** Akiko uses 2 feet of thread for every three squares that she sews for her quilt. How many squares can she sew if she has 38 feet of thread?

   F 19

   G 57

   H 76

   J 228

18. State whether the percent of change is a percent of *increase* or a percent of *decrease*. Then find the percent of change. Round to the nearest whole percent.

    original: 54 new: 45

19. Find the total price of a sweatshirt that is priced at \$48 and taxed at 6.5%.
20. **SHOPPING** Kirk wants to purchase a wide-screen TV. He sees an advertisement for a TV that was originally priced at \$3200 and is 20% off. Find the discounted price of the TV.
21. Solve $5x - 3y = 9$ for y.
22. Solve $A = \frac{1}{2}bh$ for h.
23. **CHEMISTRY** Deon has 12 milliliters of a 5% solution. He also has a solution that has a concentration of 30%. How many milliliters of the 30% solution does Deon need to add to the 5% solution to obtain a 20% solution?
24. **BICYCLING** Shanee bikes 5 miles to the park in 30 minutes and 3 miles to the library in 45 minutes. What was her average speed?
25. **MAPS** On a map of North Carolina, the distance between Charlotte and Wilmington is 14.75 inches. If 2 inches equals 24 miles, what is the approximate distance between the two cities?

CHAPTER 2

Preparing for Standardized Tests

Gridded Response Questions

In addition to multiple-choice, short-answer, and extended-response questions, you will likely encounter gridded-response questions on standardized tests. For gridded-response questions, you must print your answer on an answer sheet and mark in the correct circles on the grid to match your answer.

Strategies for Solving Gridded Response Questions

Step 1

Read the problem carefully.

- **Ask yourself:** "What information is given?" "What do I need to find?" "How do I solve this type of problem?"
- **Solve the Problem:** Use the information given in the problem to solve.
- **Check your answer:** If time permits, check your answer to make sure you have solved the problem correctly.

Step 2

Write your answer in the answer boxes.

- Print only one digit or symbol in each answer box.
- Do not write any digits or symbols outside the answer boxes.
- You may write your answer with the first digit in the left answer box, or with the last digit in the right answer box. You may leave blank any boxes you do not need on the right or the left side of your answer.

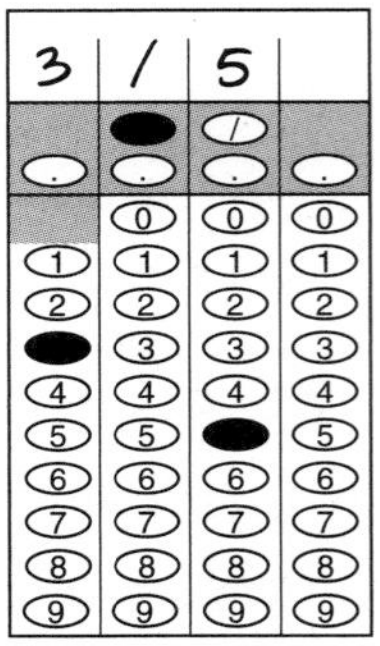

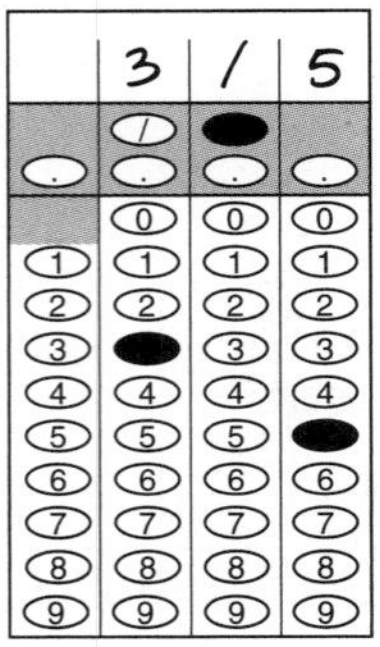

Step 3

Fill in the grid.

- Fill in only one bubble for every answer box that you have written in. Be sure not to fill in a bubble under a blank answer box.
- Fill in each bubble completely and clearly.

Standardized Test Example

Read the problem. Identify what you need to know. Then use the information in the problem to solve.

GRIDDED RESPONSE Ashley is 3 years older than her sister, Tina. Combined, the sum of their ages is 27 years. How old is Ashley?

Read the problem carefully. You are told that Ashley is 3 years older than her sister and that their ages combined equal 27 years. You need to find Ashley's age.

Solve the Problem

Words Ashley's age plus Tina's age is equal to 27 years.

Variable Let a represent Ashley's age. Then Tina's age is $a - 3$, since she is 3 years younger than Ashley.

Equation $a \quad + \quad (a - 3) \quad = \quad 27$

Solve the equation for a.

$a + (a - 3) = 27$	Original equation.
$2a - 3 = 27$	Add like terms.
$2a = 30$	Add 3 to each side.
$a = 15$	Divide each side by 2.

Since we let a represent Ashley's age, we know that she is 15 years old.

Fill in the Grid

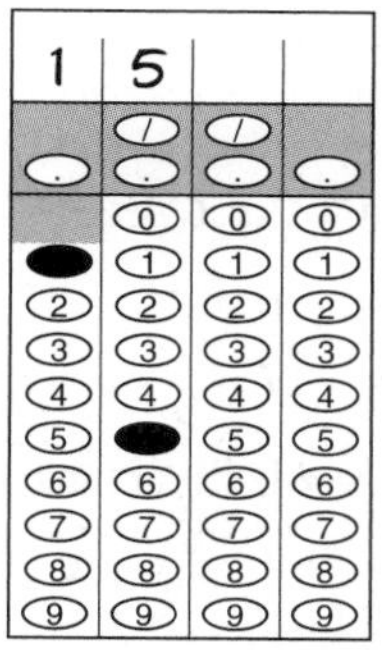

Exercises

Read each problem. Identify what you need to know. Then use the information in the problem to solve. Copy and complete an answer grid on your paper.

1. Orlando has \$1350 in the bank. He wants to increase his balance to a total of \$2550 by depositing \$40 each week from his paycheck. How many weeks will he need to save in order to reach his goal?

2. Fourteen less than three times a number is equal to 40. Find the number.

3. The table shows the regular prices and sale prices of certain items at a department store this week. What is the percent of discount during the sale?

Item	Regular Price (\$)	Sale Price (\$)
pillows	25	20
sweaters	30	24
entertainment center	125	100

4. Maureen is driving from Raleigh, North Carolina, to Charlotte, North Carolina, to visit her brother at college. If she averages 65 miles per hour on the trip, then the equation $\frac{d}{2.65} = 65$ can be solved for the distance d. What is the distance to the nearest mile from Raleigh to Charlotte?

5. Find the value of x so that the figures below have the same area.

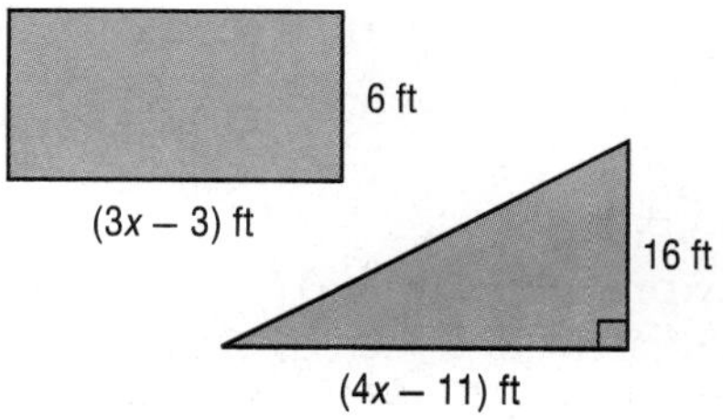

6. The sum of three consecutive whole numbers is 18. What is the greatest of the numbers?

CHAPTER 2

Standardized Test Practice

Cumulative, Chapters 1 and 2

Multiple Choice

Read each question. Then fill in the correct answer on the answer document provided by your teacher or on a sheet of paper.

1. Which point on the number line best represents the position of $\sqrt{8}$?

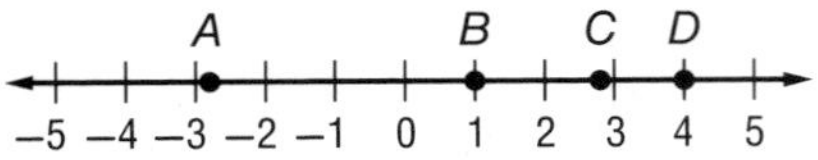

A -2.8 C 2.8

B 1 D 4

2. Find the value of x so that the figures have the same area.

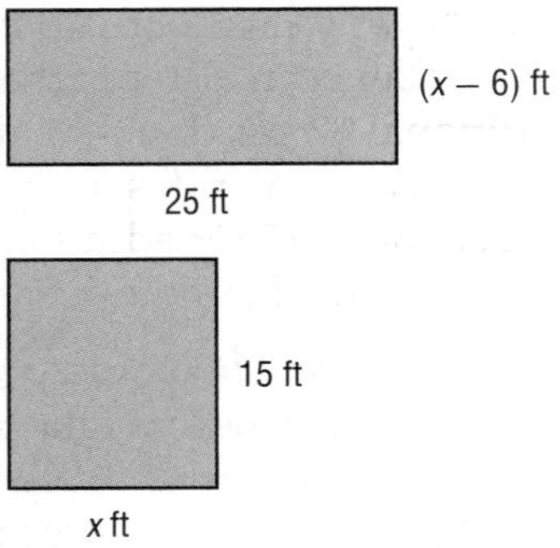

F 10 H 13

G 12 J 15

3. The elevation of Black Mountain is 27 feet more than 16 times the lowest point in the state. If the elevation of the lowest point in the state is 257 feet, what is the elevation of Black Mountain?

A 4,085 feet C 4,139 feet

B 4,103 feet D 4,215 feet

4. The expression $(3x^2 + 5x - 12) - 2(x^2 + 4x + 9)$ is equivalent to which of the following?

F $x^2 - 3x - 30$

G $x^2 + 13x + 6$

H $5x^2 + x - 18$

J $x^2 + 3x - 21$

5. The amount of soda, in fluid ounces, dispensed from a machine must satisfy the equation $|a - 0.4| = 20$. Which of the following graphs shows the acceptable minimum and maximum amounts that can be dispensed from the machine?

A 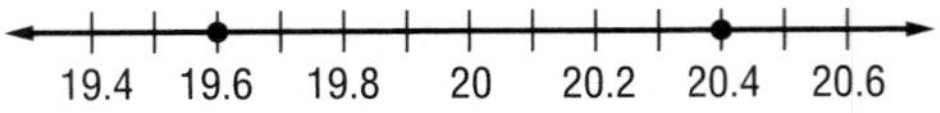

B 

C 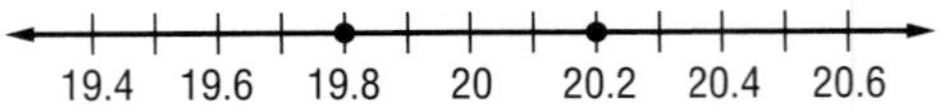

D 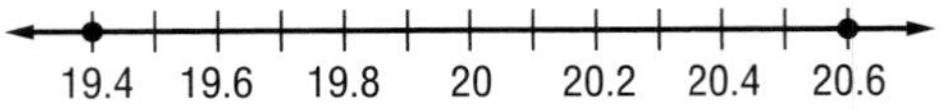

6. If a and b represent integers, $ab = ba$ is an example of which property?

F Associative Property

G Commutative Property

H Distributive Property

J Closure Property

7. The sum of one fifth of a number and three is equal to half of the number. What is the number?

A 5 C 15

B 10 D 20

8. Aaron charges \$15 to mow the lawn and \$10 per hour for other gardening work. Which expression represents his earnings?

F $10h$

G $15h$

H $15h + 10$

J $15 + 10h$

Test-Taking Tip

Question 2 Use the figures and the formula for area to set up an equation. The product of the length and width of each figure should be equal.

Short Response/Gridded Response

Record your answers on the answer sheet provided by your teacher or on a sheet of paper.

9. The formula for the lateral area of a cylinder is $A = 2\pi rh$, where r is the radius and h is the height. Solve the equation for h.

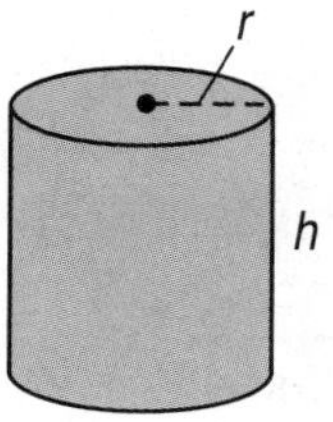

10. GRIDDED RESPONSE Solve the proportion $\frac{x}{18} = \frac{7}{21}$.

11. GRIDDED RESPONSE The table shows the cost of renting a moving van. If Miguel budgeted $75, how many miles could he drive the van and maintain his budget?

Moving Van Rentals	
Flat Fee	$50 for up to 300 miles
Variable Fee	$0.20 per mile over 300

12. Find the height of a soup can if the area of the label is 302 square centimeters and the radius of the can is 4 centimeters. Round to the nearest whole number.

13. GRIDDED RESPONSE Lara's car needed a particular part that costs $75. The mechanic charges $50 per hour to install the part. If the total cost was $350, how many hours did it take to install the part?

14. Lucinda is buying a set of patio furniture that is on sale for $\frac{4}{5}$ of the original price. After she uses a $50 gift certificate, the total cost before sales tax is $222. What was the original price of the patio furniture?

Extended Response

Record your answers on a sheet of paper. Show your work.

15. The city zoo offers a yearly membership that costs $120. A yearly membership includes free parking. Members can also purchase a ride pass for an additional $2 per day that allows them unlimited access to the rides in the park. Nonmembers pay $12 for admission to the park, $5 for parking, and $5 for a ride pass.

a. Write an equation that could be solved for the number of visits it would take for the total cost to be the same for a member and a nonmember if they both purchase a ride pass each day. Solve the equation.

b. What would the total cost be for members and nonmembers after this number of visits?

c. Georgena is deciding whether or not to purchase a yearly membership. Explain how she could use the results above to help make her decision.

Need ExtraHelp?

If you missed Question...	1	2	3	4	5	6	7	8	9	10	11	12	13	14	15
Go to Lesson...	0-2	2-4	2-3	1-4	2-5	1-3	2-4	1-1	2-8	2-6	2-3	2-8	2-3	2-3	2-4

CHAPTER 3

Linear Functions

Then

You solved linear equations algebraically.

Now

In this chapter you will:

- Identify linear equations, intercepts, and zeros.
- Graph and write linear equations.
- Use rate of change to solve problems.

Why? ▲

AMUSEMENT PARKS The Magic Kingdom in Orlando, Florida, is one of the most popular amusement parks in the world. Yearly attendance figures increase steadily each year. Quantities like populations that change with respect to time can be described using rate of change. Often you can represent these situations with linear functions.

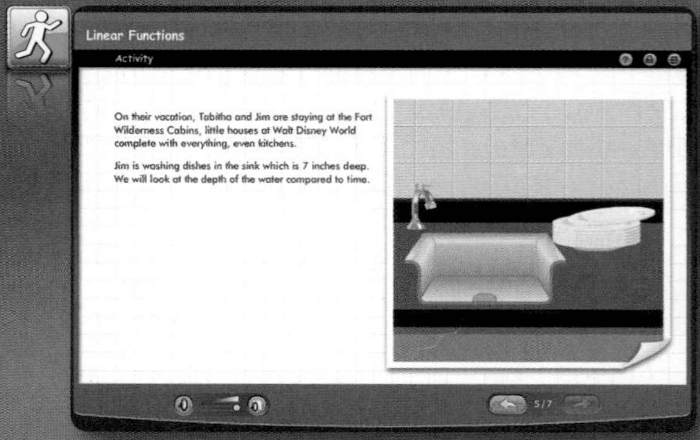

connectED.mcgraw-hill.com **Your Digital Math Portal**

Animation | Vocabulary | eGlossary | Personal Tutor | Virtual Manipulatives | Graphing Calculator | Audio | Foldables | Self-Check Practice | Worksheets

Flame/Al

Get Ready for the Chapter

Diagnose Readiness | You have two options for checking prerequisite skills.

1 Textbook Option

Take the Quick Check below. Refer to the Quick Review for help.

QuickCheck

Graph each ordered pair on a coordinate grid.

1. $(-3, 3)$ **2.** $(-2, 1)$ **3.** $(3, 0)$

4. $(-5, 5)$ **5.** $(0, 6)$ **6.** $(2, -1)$

Write the ordered pair for each point.

7. *A* **8.** *B*

9. *C* **10.** *D*

11. *F* **12.** *G*

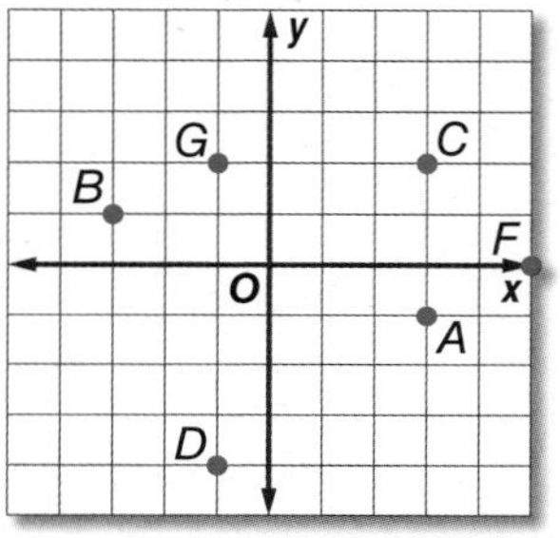

Solve each equation for *y*.

13. $3x + y = 1$ **14.** $8 - y = x$

15. $5x - 2y = 12$ **16.** $3x + 4y = 10$

17. $3 - \frac{1}{2}y = 5x$ **18.** $\frac{y + 1}{3} = x + 2$

Evaluate $\frac{a - b}{c - d}$ for each set of values.

19. $a = 7, b = 6, c = 9, d = 5$

20. $a = -3, b = 0, c = 3, d = -1$

21. $a = -5, b = -5, c = 5, d = 8$

22. $a = -6, b = 3, c = 8, d = 2$

23. **MOVIES** A movie made $297.2 million in 22 weeks. How much did the movie make on average each week?

QuickReview

Example 1

Graph (3, −2) on a coordinate grid.

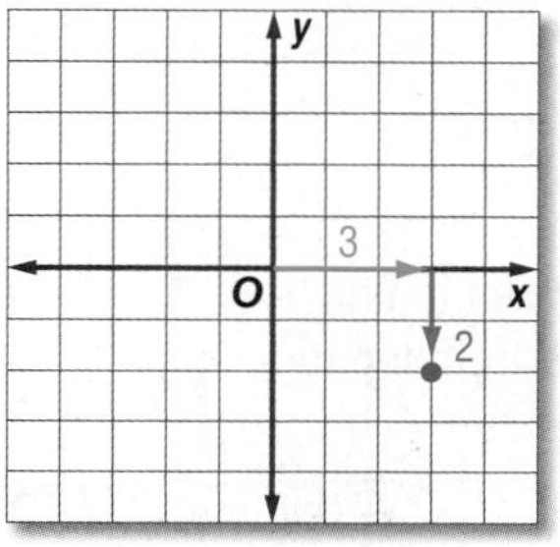

Example 2

Solve $x - 2y = 8$ for *y*.

$x - 2y = 8$	Original equation
$x - x - 2y = 8 - x$	Subtract *x* from each side.
$-2y = 8 - x$	Simplify.
$\frac{-2y}{-2} = \frac{8 - x}{-2}$	Divide each side by −2.
$y = \frac{1}{2}x - 4$	Simplify.

Example 3

Evaluate $\frac{a - b}{c - d}$ for $a = 3$, $b = 5$, $c = -2$, and $d = -6$.

$\frac{a - b}{c - d}$	Original expression
$= \frac{3 - 5}{-2 - (-6)}$	Substitute 3 for *a*, 5 for *b*, −2 for *c*, and −6 for *d*.
$= \frac{-2}{4}$	Simplify.
$= \frac{-2 \div 2}{4 \div 2}$	Divide −2 and 4 by their GCF, 2.
$= \frac{-1}{2}$ or $-\frac{1}{2}$	Simplify. The signs are different so the quotient is negative.

2 Online Option

Take an online self-check Chapter Readiness Quiz at connectED.mcgraw-hill.com.

Get Started on the Chapter

You will learn several new concepts, skills, and vocabulary terms as you study Chapter 3. To get ready, identify important terms and organize your resources. You may wish to refer to Chapter 0 to review prerequisite skills.

FOLDABLES® StudyOrganizer

Linear Functions Make this Foldable to help you organize your Chapter 3 notes about graphing relations and functions. Begin with four sheets of grid paper.

1 **Fold** each sheet of grid paper in half from top to bottom.

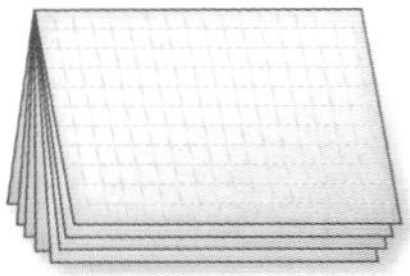

2 **Cut** along fold. Staple the eight half-sheets together to form a booklet.

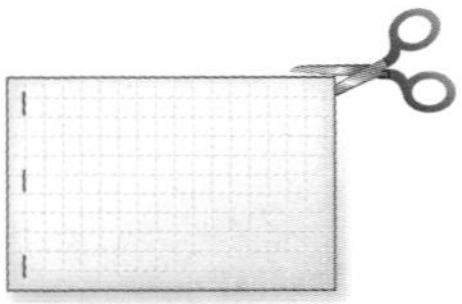

3 **Cut** tabs into margin. The top tab is 4 lines wide, the next tab is 8 lines wide, and so on. When you reach the bottom of a sheet, start the next tab at the top of the page.

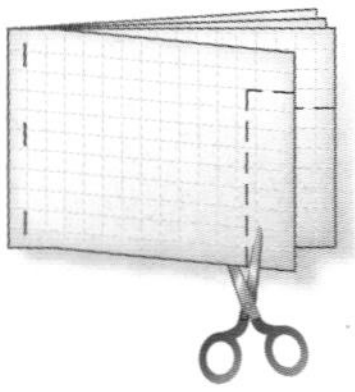

4 **Label** each of the tabs with a lesson number. Use the extra pages for vocabulary.

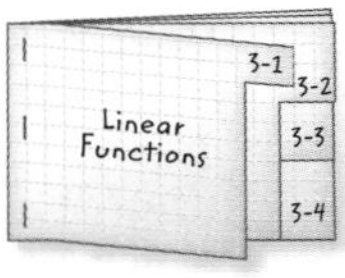

NewVocabulary

English		Español
linear equation	p. 155	ecuación lineal
standard form	p. 155	forma estándar
constant	p. 155	constante
x-intercept	p. 156	intersección *x*
y-intercept	p. 156	intersección *y*
linear function	p. 163	función lineal
parent function	p. 163	críe la función
family of graphs	p. 163	la familia de gráficas
root	p. 163	raiz
rate of change	p. 172	tasa de cambio
slope	p. 174	pendiente
direct variation	p. 182	variación directa
constant of variation	p. 182	constante de variación
arithmetic sequence	p. 189	sucesión arithmética
inductive reasoning	p. 196	razonamiento inductivo
deductive reasoning	p. 196	razonamiento deductivo

ReviewVocabulary

origin origen
the point where the two axes in a coordinate plane intersect with coordinates (0, 0)

x-axis eje *x* the horizontal number line on a coordinate plane

y-axis eje *y* the vertical number line on a coordinate plane

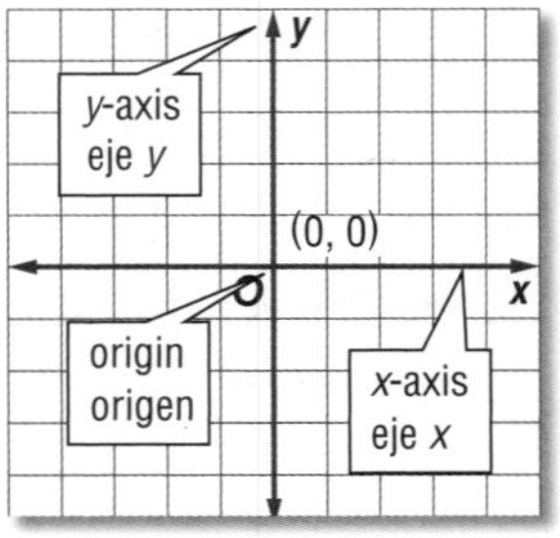

EXPLORE 3-1

Algebra Lab
Analyzing Linear Graphs

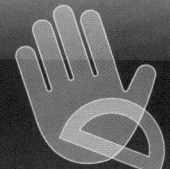

Analyzing a graph can help you learn about the relationship between two quantities. A **linear function** is a function for which the graph is a line. There are four types of linear graphs. Let's analyze each type.

CCSS Common Core State Standards
Content Standards
F.IF.4 For a function that models a relationship between two quantities, interpret key features of graphs and tables in terms of the quantities, and sketch graphs showing key features given a verbal description of the relationship.

Activity 1 Line that Slants Up

Analyze the function graphed at the right.

a. Describe the domain, range, and end behavior.

b. Describe the intercepts and any maximum or minimum points.

c. Identify where the function is positive, negative, increasing, and decreasing.

d. Describe any symmetry.

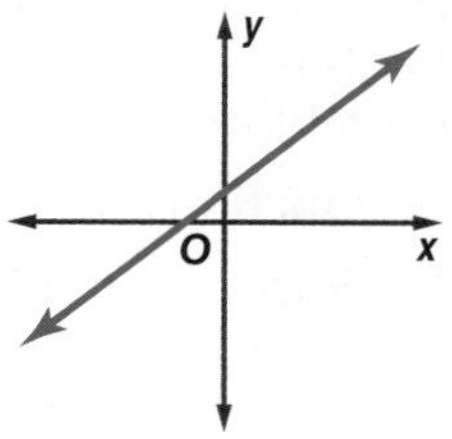

a. The domain and range are all real numbers. As you move left, the graph goes down. So as x decreases, y decreases. As you move right, the graph goes up. So as x increases, y increases.

b. There is one x-intercept and one y-intercept. There are no maximum or minimum points.

c. The function value is 0 at the x-intercept. The function values are negative to the left of the x-intercept and positive to the right. The function goes up from left to right, so it is increasing on the entire domain.

d. The graph has no symmetry.

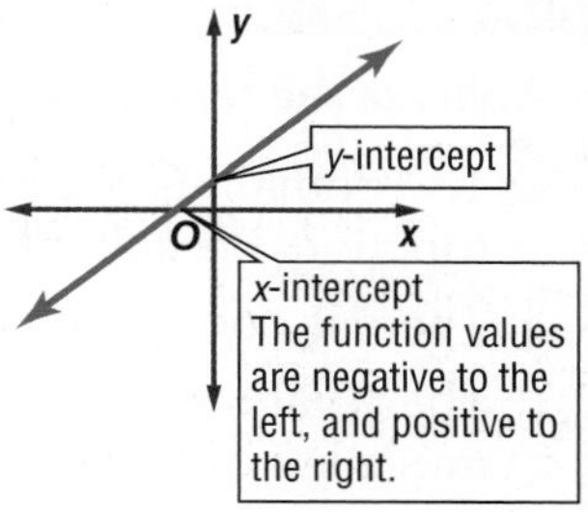

Lines that slant down from left to right have some different key features.

Activity 2 Line that Slants Down

Analyze the function graphed at the right.

a. Describe the domain, range, and end behavior.

b. Describe the intercepts and any maximum or minimum points.

c. Identify where the function is positive, negative, increasing, and decreasing.

d. Describe any symmetry.

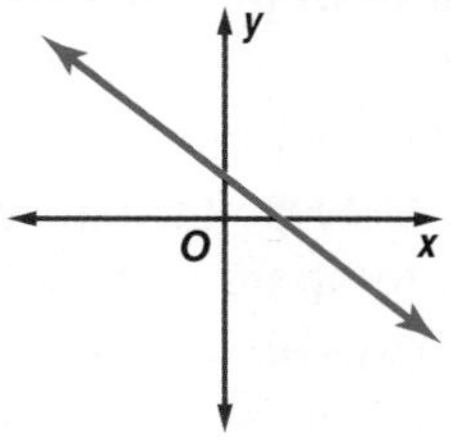

a. The domain and range are all real numbers. As you move left, the graph goes up. So as x decreases, y increases. As you move right, the graph goes down. So as x increases, y decreases.

b. There is one x-intercept and one y-intercept. There are no maximum or minimum points.

c. The function values are positive to the left of the x-intercept and negative to the right.
The function goes down from left to right, so it is decreasing on the entire domain.

d. The graph has no symmetry.

Analyzing Linear Graphs *Continued*

Horizontal lines represent special functions called **constant functions**.

Activity 3 Horizontal Line

Analyze the function graphed at the right.

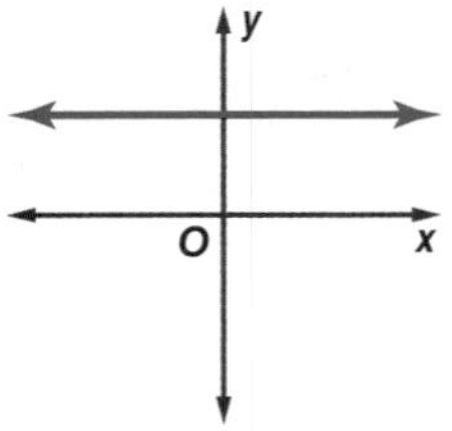

a. The domain is all real numbers, and the range is one value. As you move left or right, the graph stays constant. So as x decreases or increases, y is constant.

b. The graph does not intersect the x-axis, so there is no x-intercept. The graph has one y-intercept. There are no maximum or minimum points.

c. The function values are all positive. The function is constant on the entire domain.

d. The graph is symmetric about any vertical line.

Vertical lines represent linear relations that are *not* functions.

Activity 4 Vertical Line

Analyze the relation graphed at the right.

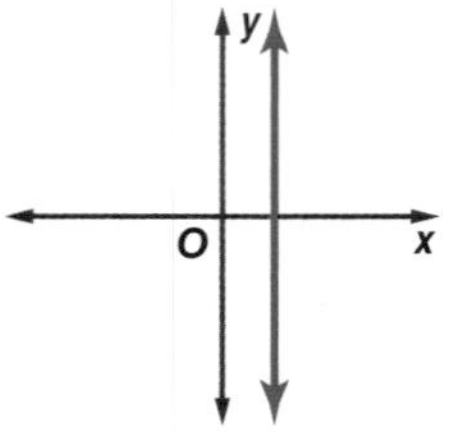

a. The domain is one value, and the range is all real numbers. This relation is not a function. Because you cannot move left or right on the graph, there is no end behavior.

b. There is one x-intercept and no y-intercept. There are no maximum or minimum points.

c. The y-values are positive above the x-axis and negative below. Because you cannot move left or right on the graph, the relation is neither increasing nor decreasing.

d. The graph is symmetric about itself.

Analyze the Results

1. Compare and contrast the key features of lines that slant up and lines that slant down.

2. How would the key features of a horizontal line below the x-axis differ from the features of a line above the x-axis?

3. Consider lines that pass through the origin.

a. How do the key features of a line that slants up and passes through the origin compare to the key features of the line in Activity 1?

b. Compare the key features of a line that slants down and passes through the origin to the key features of the line in Activity 2.

c. Describe a horizontal line that passes through the origin and a vertical line that passes through the origin. Compare their key features to those of the lines in Activities 3 and 4.

4. **CCSS TOOLS** Place a pencil on a coordinate plane to represent a line. Move the pencil to represent different lines and evaluate each conjecture.

a. *True* or *false*: A line can have more than one x-intercept.

b. *True* or *false*: If the end behavior of a line is that as x increases, y increases, then the function values are increasing over the entire domain.

c. *True* or *false*: Two different lines can have the same x- and y-intercepts.

Sketch a linear graph that fits each description.

5. as x increases, y decreases

6. one x-intercept and one y-intercept

7. has symmetry

8. is not a function

LESSON 3-1

Graphing Linear Equations

Then

- You represented relationships among quantities using equations.

Now

1. Identify linear equations, intercepts, and zeros.
2. Graph linear equations.

Why?

- Recycling one ton of waste paper saves an average of 17 trees, 7000 gallons of water, 3 barrels of oil, and about 3.3 cubic yards of landfill space.

The relationship between the amount of paper recycled and the number of trees saved can be expressed with the equation $y = 17x$, where y represents the number of trees and x represents the tons of paper recycled.

NewVocabulary
linear equation
standard form
constant
x-intercept
y-intercept

Common Core State Standards

Content Standards
F.IF.4 For a function that models a relationship between two quantities, interpret key features of graphs and tables in terms of the quantities, and sketch graphs showing key features given a verbal description of the relationship.
F.IF.7a Graph linear and quadratic functions and show intercepts, maxima, and minima.

Mathematical Practices
8 Look for and express regularity in repeated reasoning.

1 Linear Equations and Intercepts

A **linear equation** is an equation that forms a line when it is graphed. Linear equations are often written in the form $Ax + By = C$. This is called the **standard form** of a linear equation. In this equation, C is called a **constant**, or a number. Ax and By are variable terms.

KeyConcept Standard Form of a Linear Equation

Words	The standard form of a linear equation is $Ax + By = C$, where $A \geq 0$, A and B are not both zero, and A, B, and C are integers with a greatest common factor of 1.
Examples	In $3x + 2y = 5$, $A = 3$, $B = 2$, and $C = 5$. In $x = -7$, $A = 1$, $B = 0$, and $C = -7$.

PT

Example 1 Identify Linear Equations

Determine whether each equation is a linear equation. Write the equation in standard form.

a. $y = 4 - 3x$

Rewrite the equation so that it appears in standard form.

$y = 4 - 3x$ — Original equation
$y + 3x = 4 - 3x + 3x$ — Add $3x$ to each side.
$3x + y = 4$ — Simplify.

The equation is now in standard form where $A = 3$, $B = 1$, and $C = 4$. This is a linear equation.

b. $6x - xy = 4$

Since the term xy has two variables, the equation cannot be written in the form $Ax + By = C$. Therefore, this is not a linear equation.

GuidedPractice

1A. $\frac{1}{3}y = -1$

1B. $y = x^2 - 4$

A linear equation can be represented on a coordinate graph. The x-coordinate of the point at which the graph of an equation crosses the x-axis is an **x-intercept**. The y-coordinate of the point at which the graph crosses the y-axis is called a **y-intercept**.

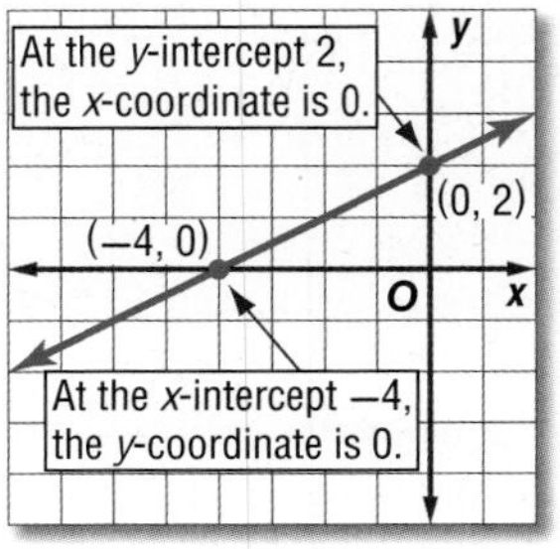

The graph of a linear equation has at most one x-intercept and one y-intercept, unless it is the equation $x = 0$ or $y = 0$, in which case every number is a y-intercept or an x-intercept, respectively.

Standardized Test Example 2 Find Intercepts from a Graph

Find the x- and y-intercepts of the line graphed at the right.

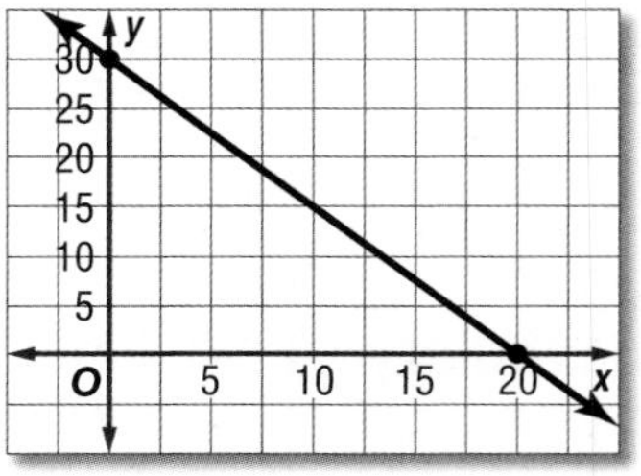

A x-intercept is 0; y-intercept is 30.

B x-intercept is 20; y-intercept is 30.

C x-intercept is 20; y-intercept is 0.

D x-intercept is 30; y-intercept is 20.

Read the Test Item

We need to determine the x- and y-intercepts of the line in the graph.

Solve the Test Item

Step 1 Find the x-intercept. Look for the point where the line crosses the x-axis.

The line crosses at (20, 0). The x-intercept is 20 because it is the x-coordinate of the point where the line crosses the x-axis.

Step 2 Find the y-intercept. Look for the point where the line crosses the y-axis.

The line crosses at (0, 30). The y-intercept is 30 because it is the y-coordinate of the point where the line crosses the y-axis.

Thus, the answer is B.

ReadingMath

Intercepts Usually, the individual coordinates are called the x-intercept and the y-intercept. The x-intercept 20 is located at (20, 0).The y-intercept 30 is located at (0, 30).

GuidedPractice

2. HEALTH Find the x- and y-intercepts of the graph.

F x-intercept is 0; y-intercept is 150.

G x-intercept is 150; y-intercept is 0.

H x-intercept is 150; no y-intercept.

J No x-intercept; y-intercept is 150.

Gym Membership

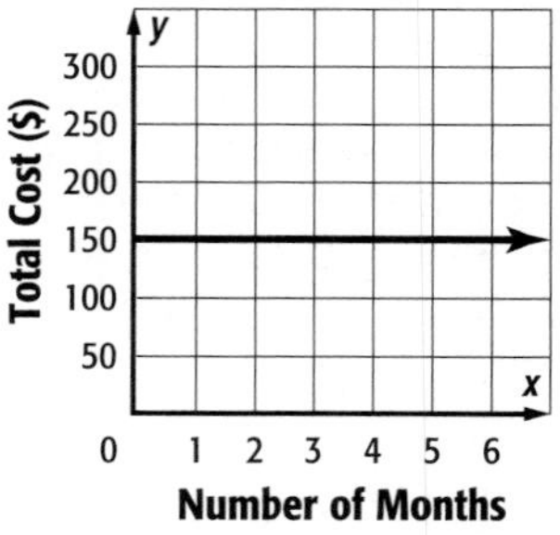

StudyTip

Defining Variables In Example 3, time is the *independent* variable, and volume of water is the *dependent* variable.

Real-World Example 3 Find Intercepts from a Table

SWIMMING POOL A swimming pool is being drained at a rate of 720 gallons per hour. The table shows the function relating the volume of water in a pool and the time in hours that the pool has been draining.

Draining a Pool	
Time (h)	Volume (gal)
x	y
0	10,080
2	8640
6	5760
10	2880
12	1440
14	0

a. Find the x- and y-intercepts of the graph of the function.

x-intercept = 14 — 14 is the value of x when $y = 0$.

y-intercept = 10,080 — 10,080 is the value of y when $x = 0$.

b. Describe what the intercepts mean in this situation.

The x-intercept 14 means that after 14 hours, the water has a volume of 0 gallons, or the pool is completely drained.

The y-intercept 10,080 means that the pool contained 10,080 gallons of water at time 0, or before it started to drain. This is shown in the graph.

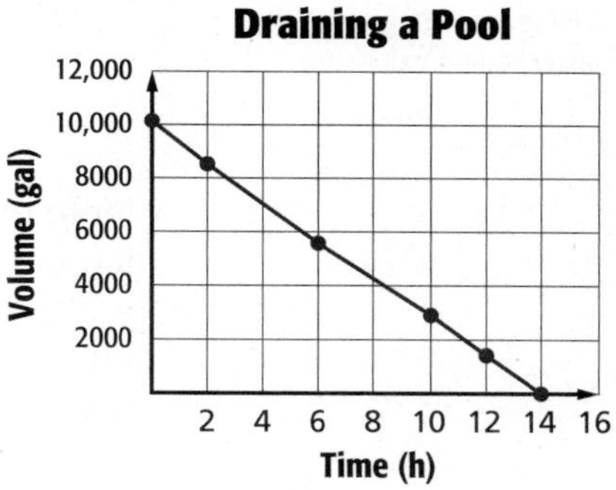

GuidedPractice

3. DRIVING The table shows the function relating the distance to an amusement park in miles and the time in hours the Torres family has driven. Find the x- and y-intercepts. Describe what the intercepts mean in this situation.

Time (h)	Distance (mi)
0	248
1	186
2	124
3	62
4	0

2 Graph Linear Equations

By first finding the x- and y-intercepts, you have the ordered pairs of two points through which the graph of the linear equation passes. This information can be used to graph the line because only two points are needed to graph a line.

StudyTip

Intercepts The x-intercept is where the graph crosses the x-axis. So the y-value is always 0. The y-intercept is where the graph crosses the y-axis. So, the x-value is always 0.

Example 4 Graph by Using Intercepts

Graph $2x + 4y = 16$ by using the x- and y-intercepts.

To find the x-intercept, let $y = 0$.

$2x + 4y = 16$	Original equation
$2x + 4(0) = 16$	Replace y with 0.
$2x = 16$	Simplify.
$x = 8$	Divide each side by 2.

The x-intercept is 8. This means that the graph intersects the x-axis at (8, 0).

To find the y-intercept, let $x = 0$.

$2x + 4y = 16$	Original equation
$2(0) + 4y = 16$	Replace x with 0.
$4y = 16$	Simplify.
$y = 4$	Divide each side by 4.

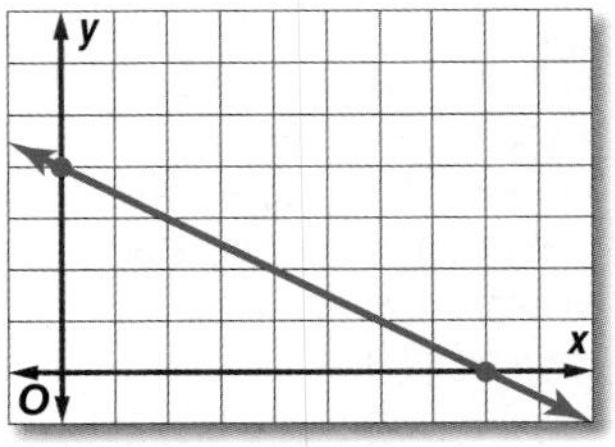

The y-intercept is 4. This means the graph intersects the y-axis at (0, 4).

Plot these two points and then draw a line through them.

StudyTip

Equivalent Equations Rewriting equations by solving for y may make it easier to find values for y.

$-x + 2y = 3 \rightarrow y = \frac{x + 3}{2}$

GuidedPractice

Graph each equation by using the x- and y-intercepts.

4A. $-x + 2y = 3$

4B. $y = -x - 5$

Note that the graph in Example 4 has both an x- and a y-intercept. Some lines have an x-intercept and no y-intercept or vice versa. The graph of $y = b$ is a horizontal line that only has a y-intercept (unless $b = 0$). The intercept occurs at $(0, b)$. The graph of $x = a$ is a vertical line that only has an x-intercept (unless $a = 0$). The intercept occurs at $(a, 0)$.

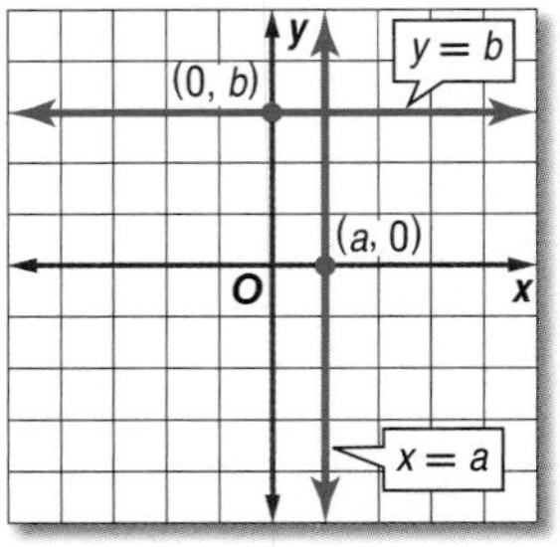

Every ordered pair that makes an equation true represents a point on the graph. So, the graph of an equation represents all of its solutions. Any ordered pair that does not make the equation true represents a point that is not on the line.

Example 5 Graph by Making a Table

Graph $y = \frac{1}{3}x + 2$.

The domain is all real numbers. Select values from the domain and make a table. When the x-coefficient is a fraction, select a number from the domain that is a multiple of the denominator. Create ordered pairs and graph them.

x	$\frac{1}{3}x + 2$	y	(x, y)
−3	$\frac{1}{3}(-3) + 2$	1	(−3, 1)
0	$\frac{1}{3}(0) + 2$	2	(0, 2)
3	$\frac{1}{3}(3) + 2$	3	(3, 3)
6	$\frac{1}{3}(6) + 2$	4	(6, 4)

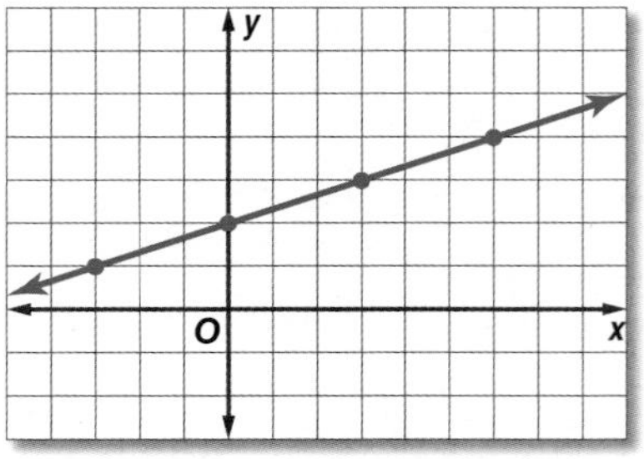

GuidedPractice

Graph each equation by making a table.

5A. $2x - y = 2$

5B. $x = 3$

5C. $y = -2$

Check Your Understanding

Example 1 **Determine whether each equation is a linear equation. Write *yes* or *no*. If yes, write the equation in standard form.**

1. $x = y - 5$ **2.** $-2x - 3 = y$ **3.** $-4y + 6 = 2$ **4.** $\frac{2}{3}x - \frac{1}{3}y = 2$

Examples 2–3 **Find the x- and y-intercepts of the graph of each linear function. Describe what the intercepts mean.**

5. **Increasing Temperature**

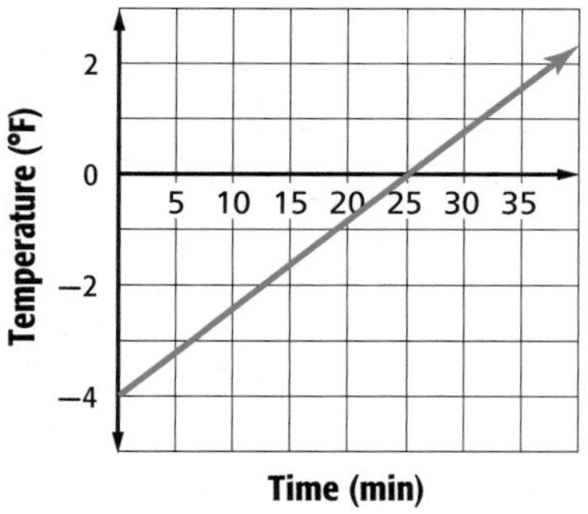

6.

Position of Scuba Diver	
Time (s)	Depth (m)
x	y
0	−24
3	−18
6	−12
9	−6
12	0

Example 4 **Graph each equation by using the x- and y-intercepts.**

7. $y = 4 + x$ **8.** $2x - 5y = 1$

Example 5 **Graph each equation by making a table.**

9. $x + 2y = 4$ **10.** $-3 + 2y = -5$ **11.** $y = 3$

12. CCSS **REASONING** The equation $5x + 10y = 60$ represents the number of children x and adults y who can attend the rodeo for $60.

a. Use the x- and y-intercepts to graph the equation.

b. Describe what these values mean.

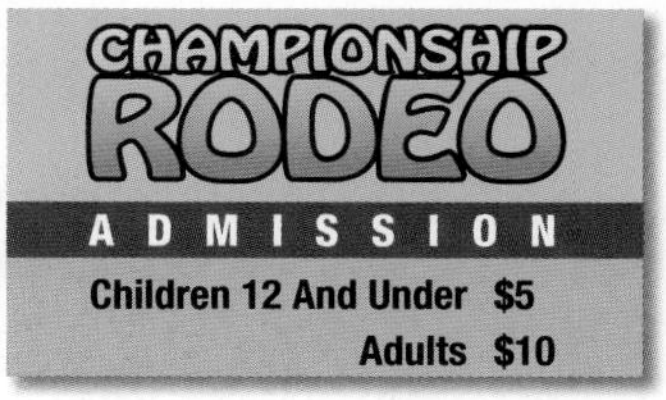

Practice and Problem Solving

Example 1 **Determine whether each equation is a linear equation. Write *yes* or *no*. If yes, write the equation in standard form.**

13. $5x + y^2 = 25$ **14.** $8 + y = 4x$ **15.** $9xy - 6x = 7$

16. $4y^2 + 9 = -4$ **17.** $12x = 7y - 10y$ **18.** $y = 4x + x$

Example 2 **Find the x- and y-intercepts of the graph of each linear function.**

19.

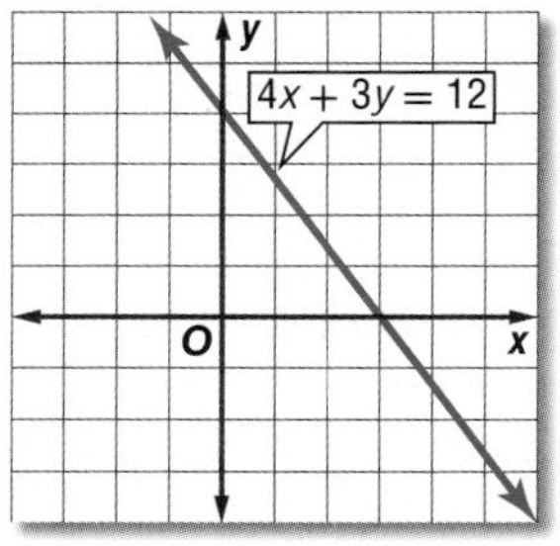

20.

x	y
−3	−1
−2	0
−1	1
0	2
1	3

Example 3 **Find the x- and y-intercepts of each linear function. Describe what the intercepts mean.**

21. **Descent of Eagle**

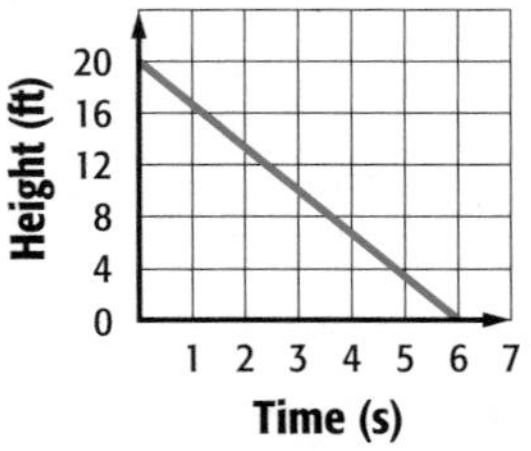

22.

Eva's Distance from Home	
Time (min)	Distance (mi)
x	y
0	4
2	3
4	2
6	1
8	0

Example 4 **Graph each equation by using the x- and y-intercepts.**

23. $y = 4 + 2x$ **24.** $5 - y = -3x$ **25.** $x = 5y + 5$

26. $x + y = 4$ **27.** $x - y = -3$ **28.** $y = 8 - 6x$

Example 5 **Graph each equation by making a table.**

29. $x = -2$ **30.** $y = -4$ **31.** $y = -8x$

32. $3x = y$ **33.** $y - 8 = -x$ **34.** $x = 10 - y$

35 **TV RATINGS** The number of people who watch a singing competition can be given by $p = 0.15v$, where p represents the number of people in millions who saw the show and v is the number of potential viewers in millions.

a. Make a table of values for the points (v, p).

b. Graph the equation.

c. Use the graph to estimate the number of people who saw the show if there are 14 million potential viewers.

d. Explain why it would not make sense for v to be a negative number.

Determine whether each equation is a linear equation. Write *yes* or *no*. If yes, write the equation in standard form.

36. $x + \frac{1}{y} = 7$ **37.** $\frac{x}{2} = 10 + \frac{2y}{3}$

38. $7n - 8m = 4 - 2m$ **39.** $3a + b - 2 = b$

40. $2r - 3rt + 5t = 1$ **41.** $\frac{3m}{4} = \frac{2n}{3} - 5$

42. **FINANCIAL LITERACY** James earns a monthly salary of $1200 and a commission of $125 for each car he sells.

a. Graph an equation that represents how much James earns in a month in which he sells x cars.

b. Use the graph to estimate the number of cars James needs to sell in order to earn $5000.

Graph each equation.

43. $2.5x - 4 = y$ **44.** $1.25x + 7.5 = y$ **45.** $y + \frac{1}{5}x = 3$

46. $\frac{2}{3}x + y = -7$ **47.** $2x - 3 = 4y + 6$ **48.** $3y - 7 = 4x + 1$

49. CCSS **REASONING** Mrs. Johnson is renting a car for vacation and plans to drive a total of 800 miles. A rental car company charges $153 for the week including 700 miles and $0.23 for each additional mile. If Mrs. Johnson has only $160 to spend on the rental car, can she afford to rent a car? Explain your reasoning.

50. **AMUSEMENT PARKS** An amusement park charges \$50 for admission before 6 P.M. and \$20 for admission after 6 P.M. On Saturday, the park took in a total of \$20,000.

a. Write an equation that represents the number of admissions that may have been sold. Let x represent the admissions sold before 6 P.M., and let y represent the admissions sold after 6 P.M.

b. Graph the equation.

c. Find the x- and y-intercepts of the graph. What does each intercept represent?

Find the x-intercept and y-intercept of the graph of each equation.

51. $5x + 3y = 15$
52. $2x - 7y = 14$
53. $2x - 3y = 5$
54. $6x + 2y = 8$
55. $y = \frac{1}{4}x - 3$
56. $y = \frac{2}{3}x + 1$

57. **ONLINE GAMES** The percent of teens who play online games can be modeled by $p = \frac{15}{4}t + 66$. p is the percent of students, and t represents time in years since 2000.

a. Graph the equation.

b. Use the graph to estimate the percent of students playing the games in 2008.

58. **MULTIPLE REPRESENTATIONS** In this problem, you will explore x- and y-intercepts of graphs of linear equations.

a. **Graphical** If possible, use a straightedge to draw a line on a coordinate plane with each of the following characteristics.

x- and y-intercept	x-intercept, no y-intercept	exactly 2 x-intercepts	no x-intercept, y-intercept	exactly 2 y-intercepts

b. **Analytical** For which characteristics were you able to create a line and for which characteristics were you unable to create a line? Explain.

c. **Verbal** What must be true of the x- and y-intercepts of a line?

H.O.T. Problems Use Higher-Order Thinking Skills

59. **CCSS REGULARITY** Copy and complete each table. State whether any of the tables show a linear relationship. Explain.

Perimeter of a Square	
Side Length	Perimeter
1	
2	
3	
4	

Area of a Square	
Side Length	Area
1	
2	
3	
4	

Volume of a Cube	
Side Length	Volume
1	
2	
3	
4	

60. **REASONING** Compare and contrast the graphs of $y = 2x + 1$ with the domain $\{1, 2, 3, 4\}$ and $y = 2x + 1$ with the domain of all real numbers.

OPEN ENDED Give an example of a linear equation of the form $Ax + By = C$ for each condition. Then describe the graph of the equation.

61. $A = 0$
62. $B = 0$
63. $C = 0$

64. **WRITING IN MATH** Explain how to find the x-intercept and y-intercept of a graph and summarize how to graph a linear equation.

Standardized Test Practice

65. Sancho can ride 8 miles on his bicycle in 30 minutes. At this rate, about how long would it take him to ride 30 miles?

A 8 hours

B 6 hours 32 minutes

C 2 hours

D 1 hour 53 minutes

66. GEOMETRY Which is a true statement about the relation graphed?

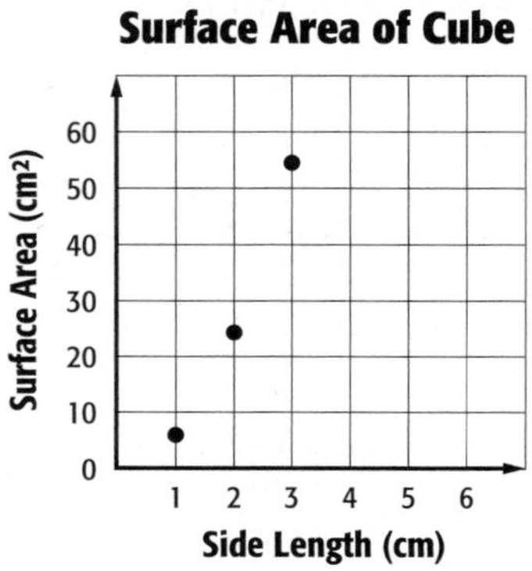

F The relation is not a function.

G Surface area is the independent quantity.

H The surface area of a cube is a function of the side length.

J As the side length of a cube increases, the surface area decreases.

67. SHORT RESPONSE Selena deposited $2000 into a savings account that pays 1.5% interest compounded annually. If she does not deposit any more money into her account, how much will she earn in interest at the end of one year?

68. A candle burns as shown in the graph.

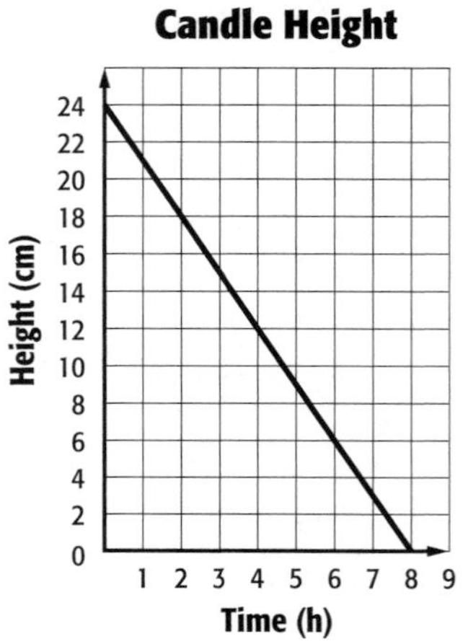

If the height of the candle is 8 centimeters, approximately how long has the candle been burning?

A 0 hours

B 24 minutes

C 64 minutes

D $5\frac{1}{2}$ hours

Spiral Review

69. FUNDRAISING The Madison High School Marching Band sold solid-color gift wrap for $4 per roll and print gift wrap for $6 per roll. The total number of rolls sold was 480, and the total amount of money collected was $2340. How many rolls of each kind of gift wrap were sold? (Lesson 2-9)

Solve each equation or formula for the variable specified. (Lesson 2-8)

70. $S = \frac{n}{2}(A + t)$, for A

71. $2g - m = 5 - gh$, for g

72. $\frac{y + a}{3} = c$, for y

73. $4z + b = 2z + c$, for z

Skills Review

Evaluate each expression if $x = 2$, $y = 5$, and $z = 7$.

74. $3x^2 - 4y$

75. $\frac{x - y^2}{2z}$

76. $\left(\frac{y}{z}\right)^2 + \frac{xy}{2}$

77. $z^2 - y^3 + 5x^2$

LESSON 3-2 Solving Linear Equations by Graphing

Then

- You graphed linear equations by using tables and finding roots, zeros, and intercepts.

Now

1. Solve linear equations by graphing.
2. Estimate solutions to a linear equation by graphing.

Why?

- The cost of braces can vary widely. The graph shows the balance of the cost of treatments as payments are made. This is modeled by the function $b = -85p + 5100$, where p represents the number of \$85 payments made, and b is the remaining balance.

Orthodontic Payments

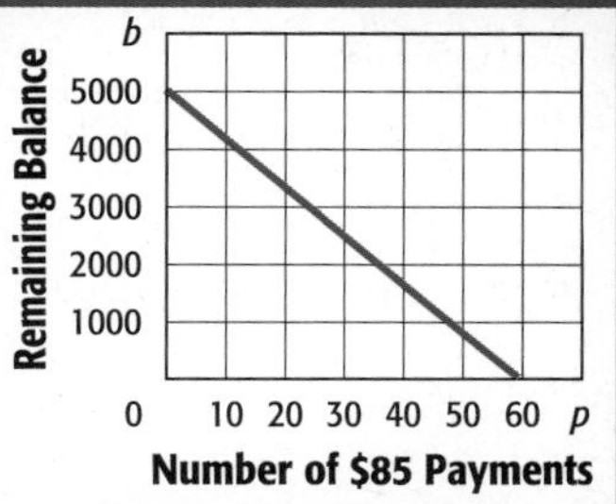

NewVocabulary
linear function
parent function
family of graphs
root
zeros

Common Core State Standards

Content Standards
A.REI.10 Understand that the graph of an equation in two variables is the set of all its solutions plotted in the coordinate plane, often forming a curve (which could be a line).

F.IF.7a Graph linear and quadratic functions and show intercepts, maxima, and minima.

Mathematical Practices
4 Model with mathematics.

1 Solve by Graphing

A **linear function** is a function for which the graph is a line. The simplest linear function is $f(x) = x$ and is called the **parent function** of the family of linear functions. A **family of graphs** is a group of graphs with one or more similar characteristics.

KeyConcept Linear Function

Parent function:	$f(x) = x$
Type of graph:	line
Domain:	all real numbers
Range:	all real numbers

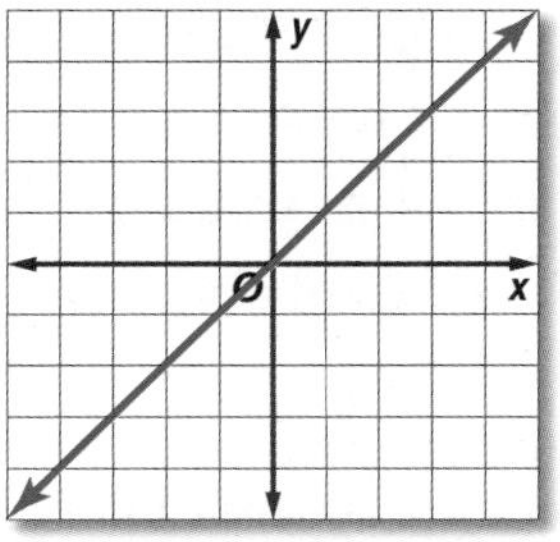

The solution or **root** of an equation is any value that makes the equation true. A linear equation has at most one root. You can find the root of an equation by graphing its related function. To write the related function for an equation, replace 0 with $f(x)$.

Linear Equation	Related Function
$2x - 8 = 0$	$f(x) = 2x - 8$ or $y = 2x - 8$

Values of x for which $f(x) = 0$ are called **zeros** of the function f. The zero of a function is located at the x-intercept of the function. The root of an equation is the value of the x-intercept. So:

- 4 is the x-intercept of $2x - 8 = 0$.
- 4 is the solution of $2x - 8 = 0$.
- 4 is the root of $2x - 8 = 0$.
- 4 is the zero of $f(x) = 2x - 8$.

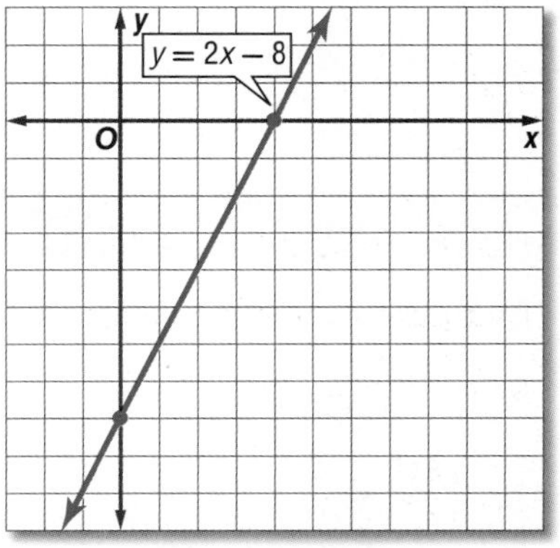

Example 1 Solve an Equation with One Root

Solve each equation.

a. $0 = \frac{1}{3}x - 2$

Method 1 Solve algebraically.

$0 = \frac{1}{3}x - 2$	Original equation
$0 + 2 = \frac{1}{3}x - 2 + 2$	Add 2 to each side.
$3(2) = 3\left(\frac{1}{3}x\right)$	Multiply each side by 3.
$6 = x$	Solve.

The solution is 6.

b. $3x + 1 = -2$

Method 2 Solve by graphing.

Find the related function. Rewrite the equation with 0 on the right side.

$3x + 1 = -2$	Original equation
$3x + 1 + 2 = -2 + 2$	Add 2 to each side.
$3x + 3 = 0$	Simplify.

The related function is $f(x) = 3x + 3$. To graph the function, make a table.

x	$f(x) = 3x + 3$	$f(x)$	$(x, f(x))$
-2	$f(-2) = 3(-2) + 3$	-3	$(-2, -3)$
-1	$f(-1) = 3(-1) + 3$	0	$(-1, 0)$

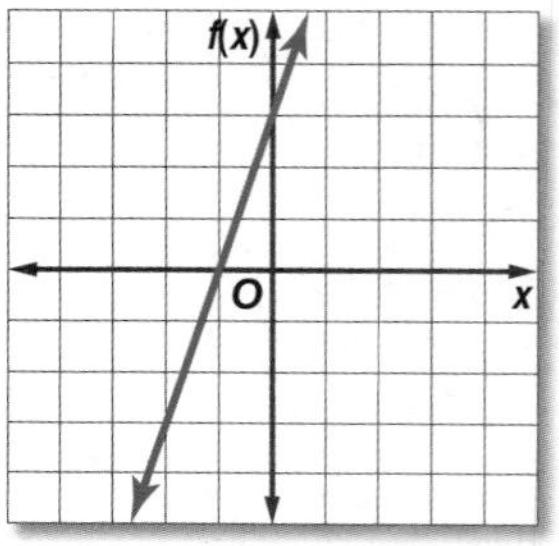

The graph intersects the x-axis at -1. So, the solution is -1.

StudyTip

Zeros from tables The zero is located at the x-intercept, so the value of y will equal 0. When looking at a table, the zero is the x-value when $y = 0$.

GuidedPractice

1A. $0 = \frac{2}{5}x + 6$

1B. $-1.25x + 3 = 0$

For equations with the same variable on each side of the equation, use addition or subtraction to get the terms with variables on one side. Then solve.

Example 2 Solve an Equation with No Solution

Solve each equation.

a. $3x + 7 = 3x + 1$

Method 1 Solve algebraically.

$3x + 7 = 3x + 1$	Original equation
$3x + 7 - 1 = 3x + 1 - 1$	Subtract 1 from each side.
$3x + 6 = 3x$	Simplify.
$3x - 3x + 6 = 3x - 3x$	Subtract $3x$ from each side.
$6 = 0$	Simplify.

The related function is $f(x) = 6$. The root of a linear equation is the value of x when $f(x) = 0$. Since $f(x)$ is always equal to 6, this equation has no solution.

b. $2x - 4 = 2x - 6$

Method 2 Solve by graphing.

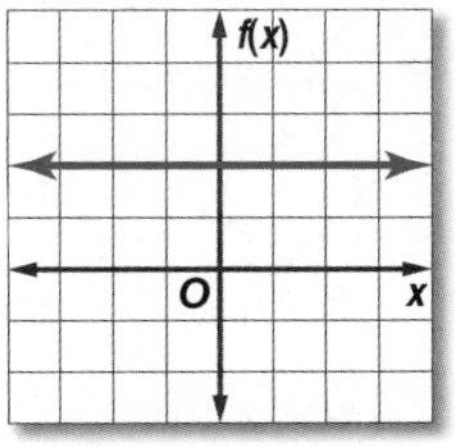

$2x - 4 = 2x - 6$	Original equation
$2x - 4 + 6 = 2x - 6 + 6$	Add 6 to each side.
$2x + 2 = 2x$	Simplify.
$2x - 2x + 2 = 2x - 2x$	Subtract $2x$ from each side.
$2 = 0$	Simplify.

Graph the related function, which is $f(x) = 2$. The graph does not intersect the x-axis. Thus, there is no solution.

Guided Practice

2A. $4x + 3 = 4x - 5$ **2B.** $2 - 3x = 6 - 3x$

2 Estimate Solutions by Graphing

Graphing may provide only an estimate. In these cases, solve algebraically to find the exact solution.

Real-World Career

Entertainment Manager

An entertainment manager supervises tech tests, calls show cues, schedules performances and performers, coaches employees and guest talent, and manages expenses. Entertainment managers need a college degree in a field such as communication or theater.

Comstock/PunchStock/Jupiter Images

Real-World Example 3 Estimate by Graphing

PT

CARNIVAL RIDES Emily is going to a local carnival. The function $m = 20 - 0.75r$ represents the amount of money m she has left after r rides. Find the zero of this function. Describe what this value means in this context.

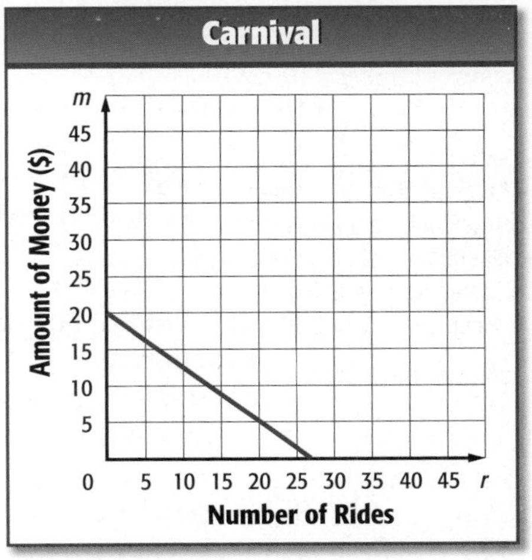

Make a table of values.

r	$m = 20 - 0.75r$	m	(r, m)
0	$m = 20 - 0.75(0)$	20	(0, 20)
5	$m = 20 - 0.75(5)$	16.25	(5, 16.25)

The graph appears to intersect the r-axis at 27.

Next, solve algebraically to check.

$m = 20 - 0.75r$	Original equation
$0 = 20 - 0.75r$	Replace m with 0.
$0 + 0.75r = 20 - 0.75r + 0.75r$	Add $0.75r$ to each side.
$0.75r = 20$	Simplify.
$\frac{0.75r}{0.75} = \frac{20}{0.75}$	Divide each side by 0.75.
$r \approx 26.67$	Simplify and round to the nearest hundredth.

The zero of this function is about 26.67. Since Emily cannot ride part of a ride, she can ride 26 rides before she will run out of money.

Guided Practice

3. FINANCIAL LITERACY Antoine's class is selling candy to raise money for a class trip. They paid \$45 for the candy, and they are selling each candy bar for \$1.50. The function $y = 1.50x - 45$ represents their profit y when they sell x candy bars. Find the zero and describe what it means in the context of this situation.

Check Your Understanding

Examples 1–2 **Solve each equation by graphing. Verify your answer algebraically.**

1. $-2x + 6 = 0$

2. $-x - 3 = 0$

3. $4x - 2 = 0$

4. $9x + 3 = 0$

5. $2x - 5 = 2x + 8$

6. $4x + 11 = 4x - 24$

7. $3x - 5 = 3x - 10$

8. $-6x + 3 = -6x + 5$

Example 3

9. NEWSPAPERS The function $w = 30 - \frac{3}{4}n$ represents the weight w in pounds of the papers in Tyrone's newspaper delivery bag after he delivers n newspapers. Find the zero and explain what it means in the context of this situation.

Practice and Problem Solving

Solve each equation by graphing. Verify your answer algebraically.

10. $0 = x - 5$

11. $0 = x + 3$

12. $5 - 8x = 16 - 8x$

13. $3x - 10 = 21 + 3x$

14. $4x - 36 = 0$

15. $0 = 7x + 10$

16. $2x + 22 = 0$

17 $5x - 5 = 5x + 2$

18. $-7x + 35 = 20 - 7x$

19. $-4x - 28 = 3 - 4x$

20. $0 = 6x - 8$

21. $12x + 132 = 12x - 100$

Example 3

22. TEXTING Sean is sending texts to his friends. The function $y = 160 - x$ represents the number of characters y the message can hold after he has typed x characters. Find the zero and explain what it means in the context of this situation.

23. GIFT CARDS For her birthday Kwan receives a \$50 gift card to download songs. The function $m = -0.50d + 50$ represents the amount of money m that remains on the card after a number of songs d are downloaded. Find the zero and explain what it means in the context of this situation.

Solve each equation by graphing. Verify your answer algebraically.

24. $-7 = 4x + 1$

25. $4 - 2x = 20$

26. $2 - 5x = -23$

27. $10 - 3x = 0$

28. $15 + 6x = 0$

29. $0 = 13x + 34$

30. $0 = 22x - 10$

31. $25x - 17 = 0$

32. $0 = \frac{1}{2} + \frac{2}{3}x$

33. $0 = \frac{3}{4} - \frac{2}{5}x$

34. $13x + 117 = 0$

35. $24x - 72 = 0$

36. SEA LEVEL Parts of New Orleans lie 0.5 meter below sea level. After d days of rain the equation $w = 0.3d - 0.5$ represents the water level w in meters. Find the zero, and explain what it means in the context of this situation.

37. CCSS **MODELING** An artist completed an ice sculpture when the temperature was $-10°C$. The equation $t = 1.25h - 10$ shows the temperature h hours after the sculpture's completion. If the artist completed the sculpture at 8:00 A.M., at what time will it begin to melt?

Solve each equation by graphing. Verify your answer algebraically.

38. $7 - 3x = 8 - 4x$

39. $19 + 3x = 13 + x$

40. $16x + 6 = 14x + 10$

41. $15x - 30 = 5x - 50$

42. $\frac{1}{2}x - 5 = 3x - 10$

43. $3x - 11 = \frac{1}{3}x - 8$

44. HAIR PRODUCTS Chemical hair straightening makes curly hair straight and smooth. The percent of the process left to complete is modeled by $p = -12.5t + 100$, where t is the time in minutes that the solution is left on the hair, and p represents the percent of the process left to complete.

a. Find the zero of this function.

b. Make a graph of this situation.

c. Explain what the zero represents in this context.

d. State the possible domain and range of this function.

45. MUSIC DOWNLOADS In this problem, you will investigate the change between two quantities.

a. Copy and complete the table.

Number of Songs Downloaded	Total Cost ($)	$\frac{\text{Total Cost}}{\text{Number of Songs Downloaded}}$
2	4	
4	8	
6	12	

b. As the number of songs downloaded increases, how does the total cost change?

c. Interpret the value of the total cost divided by the number of songs downloaded.

H.O.T. Problems Use Higher-Order Thinking Skills

46. ERROR ANALYSIS Clarissa and Koko solve $3x + 5 = 2x + 4$ by graphing the related function. Is either of them correct? Explain your reasoning.

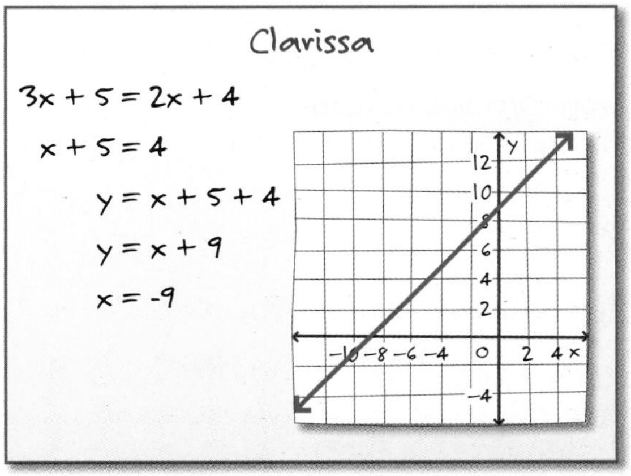

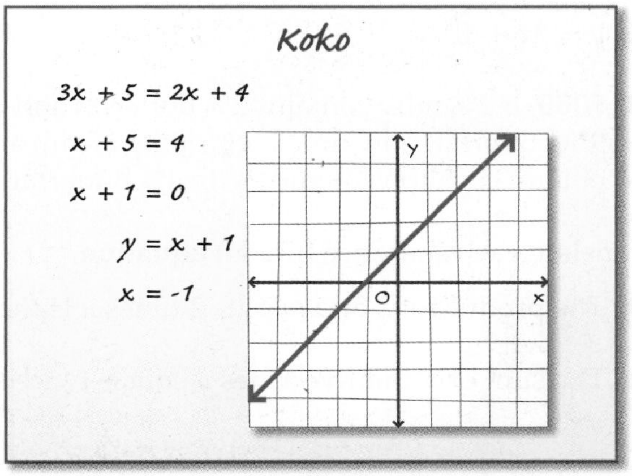

47. CHALLENGE Find the solution of $\frac{2}{3}(x + 3) = \frac{1}{2}(x + 5)$ by graphing. Verify your solution algebraically.

48. CCSS TOOLS Explain when it is better to solve an equation using algebraic methods and when it is better to solve by graphing.

49. OPEN ENDED Write a linear equation that has a root of $-\frac{3}{4}$. Write its related function.

50. WRITING IN MATH Summarize how to solve a linear equation algebraically and graphically.

Standardized Test Practice

51. What are the x- and y-intercepts of the graph of the function?

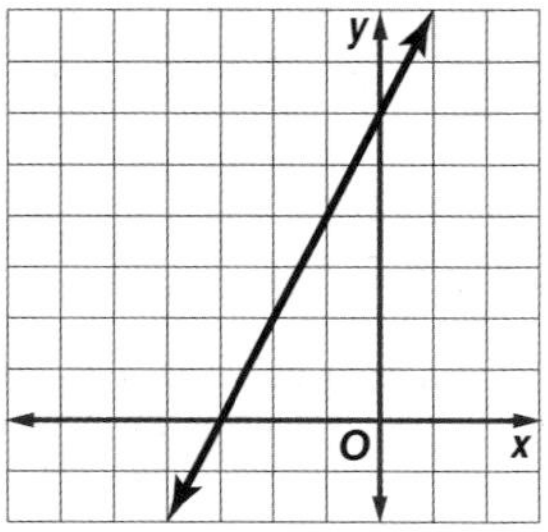

A $-3, 6$
B $6, -3$
C $3, -6$
D $-6, 3$

52. The table shows the cost C of renting a pontoon boat for h hours.

Hours	1	2	3
Cost ($)	7.25	14.5	21.75

Which equation best represents the data?

F $C = 7.25h$
G $C = h + 7.25$
H $C = 21.75 - 7.25h$
J $C = 7.25h + 21.75$

53. Which is the best estimate for the x-intercept of the graph of the linear function represented in the table?

x	y
0	5
1	3
2	1
3	−1
4	−3

A between 0 and 1
B between 2 and 3
C between 1 and 2
D between 3 and 4

54. EXTENDED RESPONSE Mr. Kauffmann has the following options for a backyard pool.

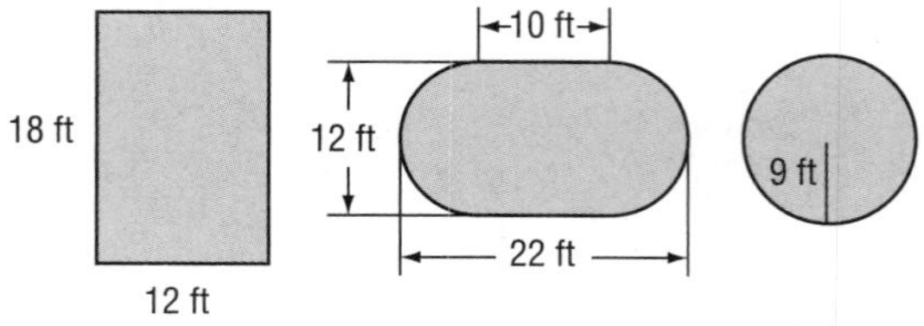

If each pool has the same depth, which pool would give the greatest area to swim? Explain your reasoning.

Spiral Review

Find the x- and y-intercepts of the graph of each linear equation. (Lesson 3-1)

55. $y = 2x + 10$

56. $3y = 6x - 9$

57. $4x - 14y = 28$

58. FOOD If 2% milk contains 2% butterfat and whipping cream contains 9% butterfat, how much whipping cream and 2% milk should be mixed to obtain 35 gallons of milk with 4% butterfat? (Lesson 2-9)

Translate each sentence into an equation. (Lesson 2-1)

59. The product of 3 and m plus 2 times n is the same as the quotient of 4 and p.

60. The sum of x and five times y equals twice z minus 7.

Skills Review

Simplify.

61. $\frac{25}{10}$

62. $\frac{-4}{-12}$

63. $\frac{6}{-12}$

64. $\frac{-36}{8}$

Evaluate $\frac{a-b}{c-d}$ for the given values.

65. $a = 6, b = 2, c = 9, d = 3$

66. $a = -8, b = 4, c = 5, d = -3$

67. $a = 4, b = -7, c = -1, d = -2$

EXTEND 3-2

Graphing Technology Lab
Graphing Linear Functions

The power of a graphing calculator is the ability to graph different types of equations accurately and quickly. By entering one or more equations in the calculator you can view features of a graph, such as the x-intercept, y-intercept, the origin, intersections, and the coordinates of specific points.

CCSS Common Core State Standards
Content Standards
N.Q.1 Use units as a way to understand problems and to guide the solution of multi-step problems; choose and interpret units consistently in formulas; choose and interpret the scale and the origin in graphs and data displays.
F.IF.7a Graph linear and quadratic functions and show intercepts, maxima, and minima.
Mathematical Practices
5 Use appropriate tools strategically.

Often linear equations are graphed in the **standard viewing window**, which is [−10, 10] by [−10, 10] with a scale of 1 on each axis. To quickly choose the standard viewing window on a TI-83/84 Plus, press [ZOOM] 6.

Activity 1 Graph a Linear Equation

Graph $3x - y = 4$.

Step 1 Enter the equation in the Y= list.

- The Y= list shows the equation or equations that you will graph.
- Equations must be entered with the y isolated on one side of the equation. Solve the equation for y, then enter it into the calculator.

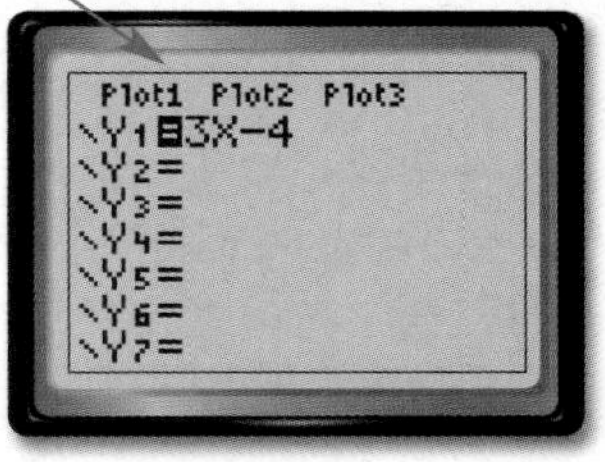

The equals sign appears shaded for graphs that are selected to be displayed.
Plot1 Plot2 Plot3
\Y1=3X-4
\Y2=
\Y3=
\Y4=
\Y5=
\Y6=
\Y7=

$3x - y = 4$	Original equation
$3x - y - 3x = 4 - 3x$	Subtract $3x$ from each side.
$-y = -3x + 4$	Simplify.
$y = 3x - 4$	Multiply each side by −1.

KEYSTROKES: [Y=] 3 [X,T,θ,n] [−] 4

Step 2 Graph the equation in the standard viewing window.

- Graph the selected equation.

KEYSTROKES: [ZOOM] 6

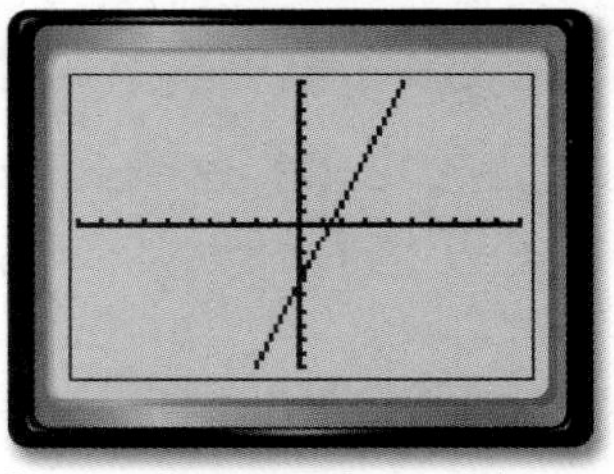

[−10, 10] scl: 1 by [−10, 10] scl: 1

Sometimes a complete graph is not displayed using the standard viewing window. A **complete graph** includes all of the important characteristics of the graph on the screen including the origin and the x- and y-intercepts. Note that the graph above is a complete graph because all of these points are visible.

When a complete graph is not displayed using the standard viewing window, you will need to change the viewing window to accommodate these important features. Use what you have learned about intercepts to help you choose an appropriate viewing window.

(continued on the next page)

Graphing Linear Functions *Continued*

Activity 2 Graph a Complete Graph

Graph $y = 5x - 14$.

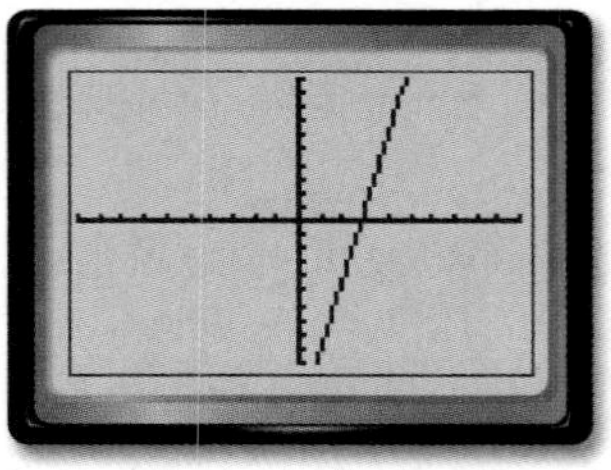

[−10, 10] scl: 1 by [−10, 10] scl: 1

Step 1 Enter the equation in the Y= list and graph in the standard viewing window.

- Clear the previous equation from the Y= list. Then enter the new equation and graph.

KEYSTROKES: [Y=] [CLEAR] 5 [X,T,θ,n] [−] 14 [ZOOM] 6

Step 2 Modify the viewing window and graph again.

- The origin and the x-intercept are displayed in the standard viewing window. But notice that the y-intercept is outside of the viewing window.

  Find the y-intercept.

$y = 5x - 14$	Original equation
$= 5(0) - 14$	Replace x with 0.
$= -14$	Simplify.

  Since the y-intercept is −14, choose a viewing window that includes a number less than −14. The window [−10, 10] by [−20, 5] with a scale of 1 on each axis is a good choice.

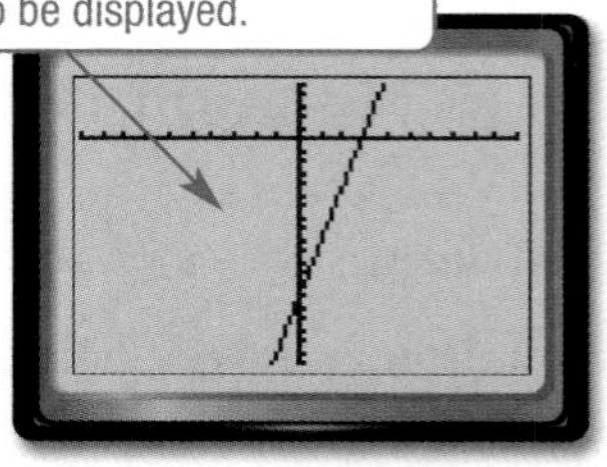
This window allows the complete graph, including the y-intercept, to be displayed.

[−10, 10] scl: 1 by [−20, 5] scl: 1

KEYSTROKES: [WINDOW] -10 [ENTER] 10 [ENTER] 1 [ENTER] -20 [ENTER] 5 [ENTER] 1 [GRAPH]

Exercises

Use a graphing calculator to graph each equation in the standard viewing window. Sketch the result.

1. $y = x + 5$

2. $y = 5x + 6$

3. $y = 9 - 4x$

4. $3x + y = 5$

5. $x + y = -4$

6. $x - 3y = 6$

CCSS **SENSE-MAKING** **Graph each equation in the standard viewing window. Determine whether the graph is complete. If the graph is not complete, adjust the viewing window and graph the equation again.**

7. $y = 4x + 7$

8. $y = 9x - 5$

9. $y = 2x - 11$

10. $4x - y = 16$

11. $6x + 2y = 23$

12. $x + 4y = -36$

Consider the linear equation $y = 3x + b$.

13. Choose several different positive and negative values for b. Graph each equation in the standard viewing window.

14. For which values of b is the complete graph in the standard viewing window?

15. How is the value of b related to the y-intercept of the graph of $y = 3x + b$?

EXPLORE 3-3

Algebra Lab
Rate of Change of a Linear Function

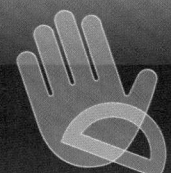

In mathematics, you can measure the steepness of a line using a ratio.

CCSS Common Core State Standards
Content Standards
F.IF.6 Calculate and interpret the average rate of change of a function (presented symbolically or as a table) over a specified interval. Estimate the rate of change from a graph.
F.LE.1a Prove that linear functions grow by equal differences over equal intervals, and that exponential functions grow by equal factors over equal intervals.

Set Up the Lab

- Stack three books on your desk.
- Lean a ruler on the books to create a ramp.
- Tape the ruler to the desk.
- Measure the rise and the run. Record your data in a table like the one at the right.
- Calculate and record the ratio $\frac{\text{rise}}{\text{run}}$.

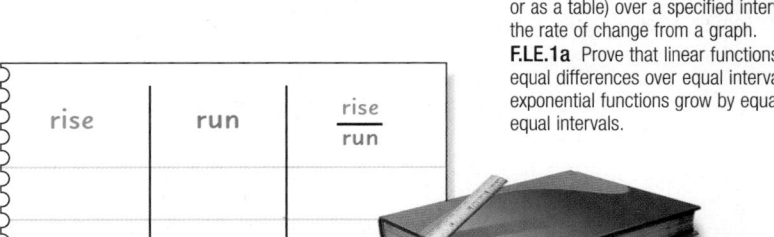

rise	run	$\frac{\text{rise}}{\text{run}}$

Activity

Step 1

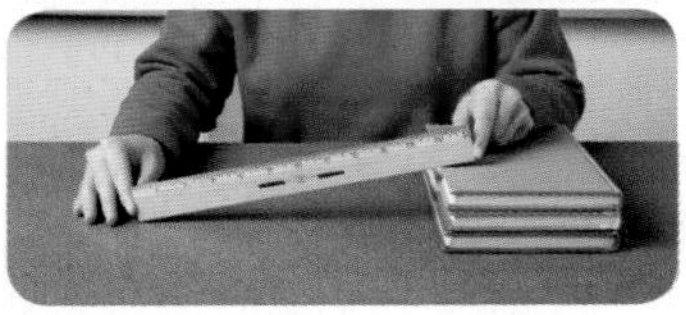

Move the books to make the ramp steeper. Measure and record the rise and the run. Calculate and record $\frac{\text{rise}}{\text{run}}$.

Step 2

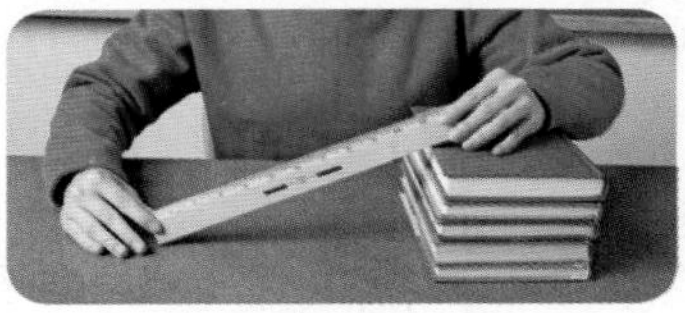

Add books to the stack to make the ramp even steeper. Measure, calculate, and record your data in the table.

Analyze the Results

1. Examine the ratios you recorded. How did they change as the ramp became steeper?

2. **MAKE A PREDICTION** Suppose you want to construct a skateboard ramp that is not as steep as the one shown at the right. List three different sets of $\frac{\text{rise}}{\text{run}}$ measurements that will result in a less steep ramp. Verify your predictions by calculating the ratio $\frac{\text{rise}}{\text{run}}$ for each ramp.

18 in.

24 in.

$m = \frac{18}{24} = \frac{3}{4}$

3. Copy the coordinate graph shown and draw a line through the origin with a $\frac{\text{rise}}{\text{run}}$ ratio greater than the original line. Then draw a line through the origin with a ratio less than that of the original line. Explain using the words *rise* and *run* why the lines you drew have a ratio greater or less than the original line.

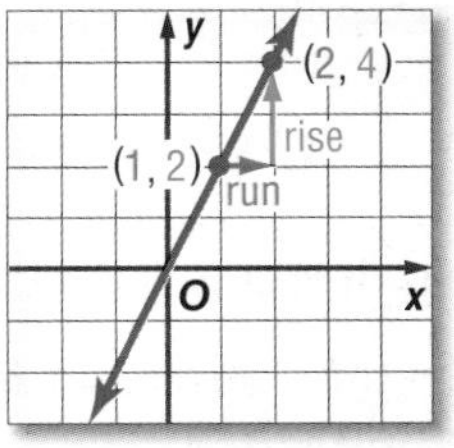

4. We have seen what happens on the graph as the $\frac{\text{rise}}{\text{run}}$ ratio gets closer to zero. What would you predict will happen when the ratio is zero? Explain your reasoning. Give an example to support your prediction.

LESSON 3-3

Rate of Change and Slope

Then

- You graphed ordered pairs in the coordinate plane.

Now

1. Use rate of change to solve problems.
2. Find the slope of a line.

Why?

- The Daredevil Drop at Wet 'n Wild Emerald Pointe in Greensboro, North Carolina, is a thrilling ride that drops you 76 feet down a steep water chute. A *rate of change* of the ride might describe the distance a rider has fallen over a length of time.

NewVocabulary

rate of change
slope

Common Core State Standards

Content Standards

F.IF.6 Calculate and interpret the average rate of change of a function (presented symbolically or as a table) over a specified interval. Estimate the rate of change from a graph.

F.LE.1a Prove that linear functions grow by equal differences over equal intervals, and that exponential functions grow by equal factors over equal intervals.

Mathematical Practices

2 Reason abstractly and quantitatively.

1 Rate of Change

Rate of change is a ratio that describes, on average, how much one quantity changes with respect to a change in another quantity.

KeyConcept Rate of Change

If x is the independent variable and y is the dependent variable, then

$$\text{rate of change} = \frac{\text{change in } y}{\text{change in } x}.$$

Real-World Example 1 Find Rate of Change

ENTERTAINMENT **Use the table to find the rate of change. Then explain its meaning.**

Number of Computer Games	Total Cost ($)
x	y
2	78
4	156
6	234

$$\text{rate of change} = \frac{\text{change in } y}{\text{change in } x} \quad \begin{matrix}\leftarrow \text{dollars} \\ \leftarrow \text{games}\end{matrix}$$

$$= \frac{\text{change in cost}}{\text{change in number of games}}$$

$$= \frac{156 - 78}{4 - 2}$$

$$= \frac{78}{2} \text{ or } \frac{39}{1}$$

The rate of change is $\frac{39}{1}$. This means that each game costs $39.

GuidedPractice

1. **REMODELING** The table shows how the tiled surface area changes with the number of floor tiles.
   A. Find the rate of change.
   B. Explain the meaning of the rate of change.

Number of Floor Tiles	Area of Tiled Surface (in^2)
x	y
3	48
6	96
9	144

J Silver/SuperStock

So far, you have seen rates of change that are *constant.* Many real-world situations involve rates of change that are not constant.

PT

Real-World Example 2 Compare Rates of Change

AMUSEMENT PARKS The graph shows the number of people who visited U.S. theme parks in recent years.

a. Find the rates of change for 2000–2002 and 2002–2004.

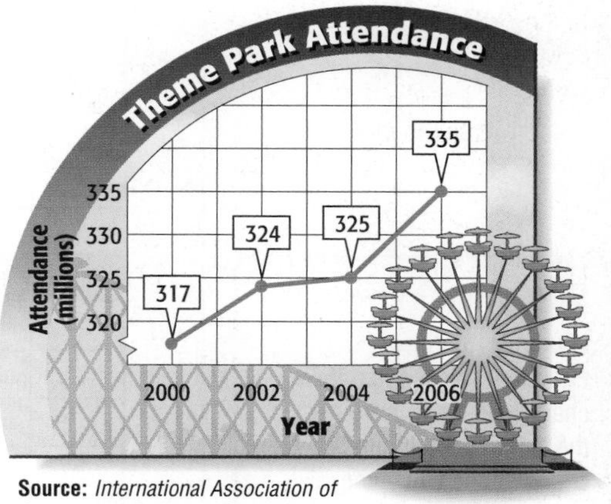

Source: *International Association of Amusement Parks and Attractions*

2000–2002:

$$\frac{\text{change in attendance}}{\text{change in time}} = \frac{324 - 317}{2002 - 2000} \quad \begin{matrix}\leftarrow \text{people} \\ \leftarrow \text{years}\end{matrix} \qquad \text{Substitute.}$$

$$= \frac{7}{2} \text{ or } 3.5 \qquad \text{Simplify.}$$

Over this 2-year period, attendance increased by 7 million, for a rate of change of 3.5 million per year.

2002–2004:

$$\frac{\text{change in attendance}}{\text{change in time}} = \frac{325 - 324}{2004 - 2002} \qquad \text{Substitute.}$$

$$= \frac{1}{2} \text{ or } 0.5 \qquad \text{Simplify.}$$

Over this 2-year period, attendance increased by 1 million, for a rate of change of 0.5 million per year.

b. Explain the meaning of the rate of change in each case.

For 2000–2002, on average, 3.5 million more people went to a theme park each year than the last.

For 2002–2004, on average, 0.5 million more people attended theme parks each year than the last.

c. How are the different rates of change shown on the graph?

There is a greater vertical change for 2000–2002 than for 2002–2004. Therefore, the section of the graph for 2000–2002 is steeper.

GuidedPractice

2. Refer to the graph above. Without calculating, find the 2-year period that has the least rate of change. Then calculate to verify your answer.

StudyTip

CCSS Reasoning A positive rate of change indicates an increase over time. A negative rate of change indicates that a quantity is decreasing.

A rate of change is constant for a function when the rate of change is the same between any pair of points on the graph of the function. Linear functions have a constant rate of change.

Example 3 Constant Rates of Change

Determine whether each function is linear. Explain.

a.

x	y
1	−6
4	−8
7	−10
10	−12
13	−14

b.

x	y
−3	10
−1	12
1	16
3	18
5	22

x	y	rate of change
1	−6	
		$\frac{-8-(-6)}{4-1}$ or $-\frac{2}{3}$
4	−8	
		$\frac{-10-(-8)}{7-4}$ or $-\frac{2}{3}$
7	−10	
		$\frac{-12-(-10)}{10-7}$ or $-\frac{2}{3}$
10	−12	
		$\frac{-14-(-12)}{13-10}$ or $-\frac{2}{3}$
13	−14	

The rate of change is constant. Thus, the function is linear.

x	y	rate of change
−3	10	
		$\frac{12-10}{-1-(-3)}$ or 1
−1	12	
		$\frac{16-12}{1-(-1)}$ or 2
1	16	
		$\frac{18-16}{3-1}$ or 1
3	18	
		$\frac{22-18}{5-3}$ or 2
5	22	

This rate of change is not constant. Thus, the function is not linear.

StudyTip

Linear or Nonlinear Function? Notice that the changes in x and y are not the same. For the rate of change to be linear, the change in x-values must be constant and the change in y-values must be constant.

GuidedPractice

3A.

x	y
−3	11
−2	15
−1	19
1	23
2	27

3B.

x	y
12	−4
9	1
6	6
3	11
0	16

2 Find Slope

The **slope** of a nonvertical line is the ratio of the change in the y-coordinates (rise) to the change in the x-coordinates (run) as you move from one point to another.

It can be used to describe a rate of change. Slope describes how steep a line is. The greater the absolute value of the slope, the steeper the line.

The graph shows a line that passes through (−1, 3) and (2, −2).

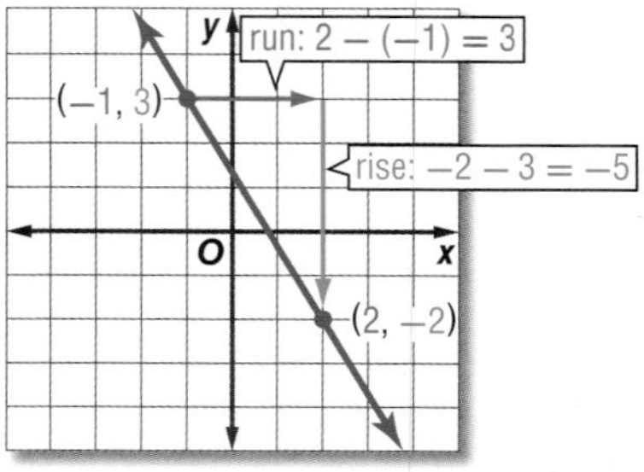

$$\textbf{slope} = \frac{\text{rise}}{\text{run}}$$

$$= \frac{\text{change in } y\text{-coordinates}}{\text{change in } x\text{-coordinates}}$$

$$= \frac{-2-3}{2-(-1)} \text{ or } -\frac{5}{3}$$

So, the slope of the line is $-\frac{5}{3}$.

Because a linear function has a constant rate of change, any two points on a nonvertical line can be used to determine its slope.

ReadingMath

Subscripts y_1 is read as *y sub one* and x_2 is read as *x sub two.* The 1 and 2 are subscripts and refer to the first and second point to which the *x*- and *y*-values correspond.

KeyConcept Slope

Words The slope of a nonvertical line is the ratio of the rise to the run.

Symbols The slope m of a nonvertical line through any two points, (x_1, y_1) and (x_2, y_2), can be found as follows.

$$m = \frac{y_2 - y_1}{x_2 - x_1} \begin{array}{l} \leftarrow \text{change in } y \\ \leftarrow \text{change in } x \end{array}$$

Graph

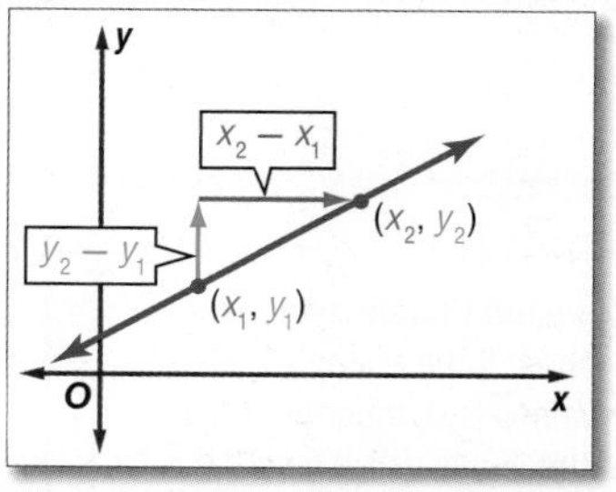

The slope of a line can be positive, negative, zero, or undefined. If the line is not horizontal or vertical, then the slope is either positive or negative.

PT

Example 4 Positive, Negative and Zero Slope

Find the slope of a line that passes through each pair of points.

a. (−2, 0) and (1, 5)

$$\begin{aligned} m &= \frac{y_2 - y_1}{x_2 - x_1} && \frac{\text{rise}}{\text{run}} \\ &= \frac{5 - 0}{1 - (-2)} && (-2, 0) = (x_1, y_1) \text{ and } (1, 5) = (x_2, y_2) \\ &= \frac{5}{3} && \text{Simplify.} \end{aligned}$$

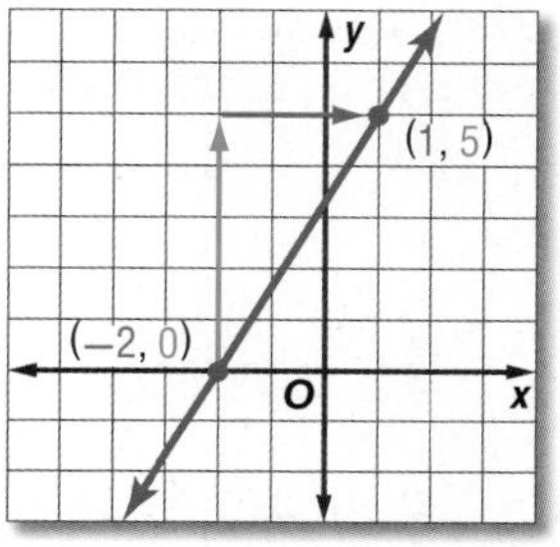

WatchOut!

Order Be careful not to transpose the order of the *x*-values or the *y*-values.

b. (−3, 4) and (2, −3)

$$\begin{aligned} m &= \frac{y_2 - y_1}{x_2 - x_1} && \frac{\text{rise}}{\text{run}} \\ &= \frac{-3 - 4}{2 - (-3)} && (-3, 4) = (x_1, y_1) \text{ and } (2, -3) = (x_2, y_2) \\ &= \frac{-7}{5} \text{ or } -\frac{7}{5} && \text{Simplify.} \end{aligned}$$

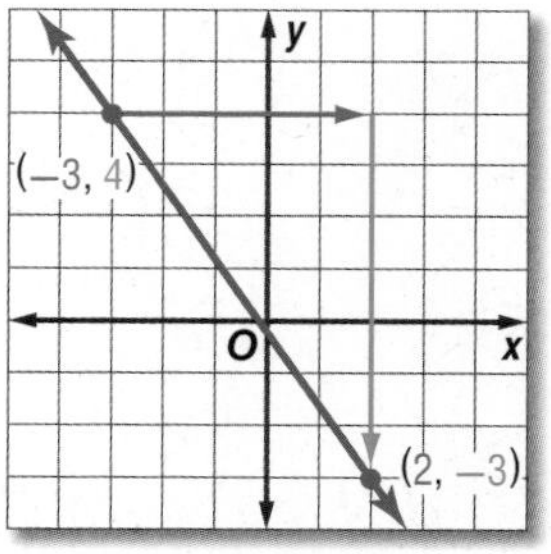

c. (−3, −1) and (2, −1)

$$\begin{aligned} m &= \frac{y_2 - y_1}{x_2 - x_1} && \frac{\text{rise}}{\text{run}} \\ &= \frac{-1 - (-1)}{2 - (-3)} && \text{Substitute.} \\ &= \frac{0}{5} \text{ or } 0 && \text{Simplify.} \end{aligned}$$

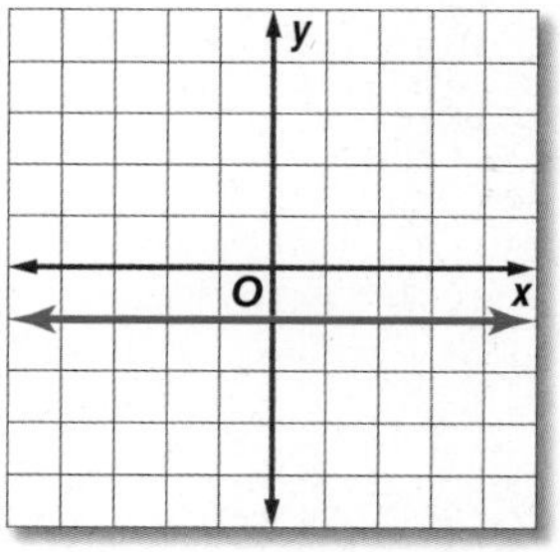

GuidedPractice

Find the slope of the line that passes through each pair of points.

4A. (3, 6), (4, 8) **4B.** (−4, −2), (0, −2) **4C.** (−4, 2), (−2, 10)

4D. (6, 7), (−2, 7) **4E.** (−2, 2), (−6, 4) **4F.** (4, 3), (−1, 11)

Example 5 Undefined Slope

Find the slope of the line that passes through (−2, 4) and (−2, −3).

$m = \frac{y_2 - y_1}{x_2 - x_1}$ $\frac{\text{rise}}{\text{run}}$

$= \frac{-3 - 4}{-2 - (-2)}$ Substitute.

$= \frac{-7}{0}$ or undefined   Simplify.

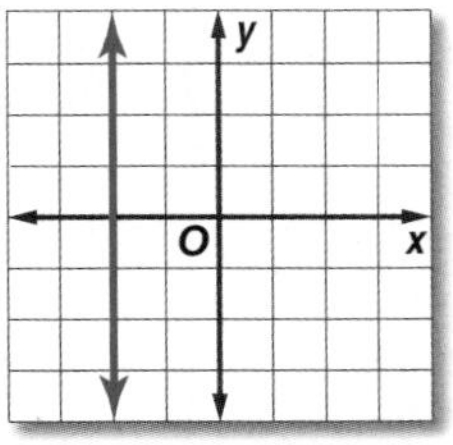

> **StudyTip**
>
> **Zero and Undefined Slopes** If the change in y-values is 0, then the graph of the line is horizontal. If the change in x-values is 0, then the slope is undefined. This graph is a vertical line.

GuidedPractice

Find the slope of the line that passes through each pair of points.

5A. (6, 3), (6, 7)   **5B.** (−3, 2), (−3, −1)

The graphs of lines with different slopes are summarized below.

ConceptSummary Slope

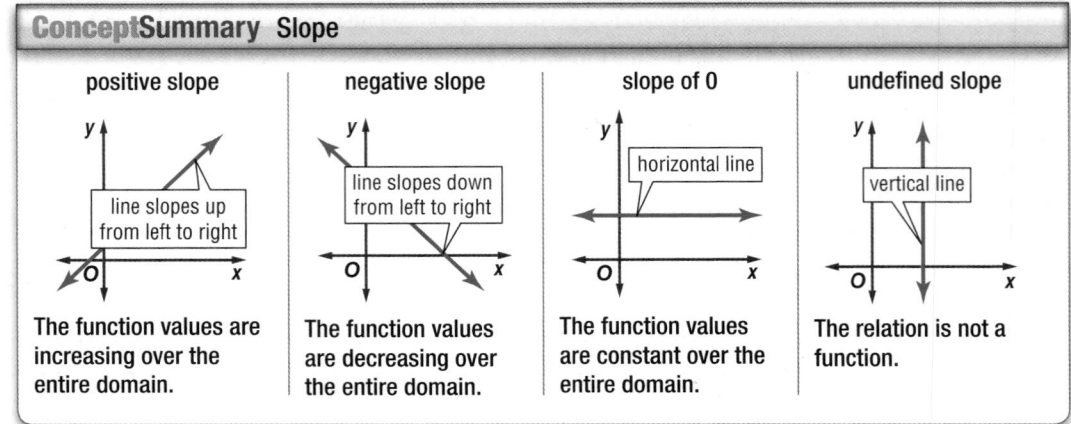

positive slope

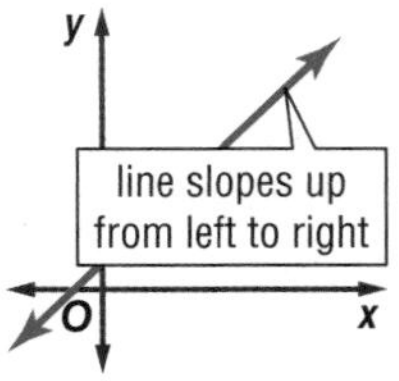

The function values are increasing over the entire domain.

negative slope

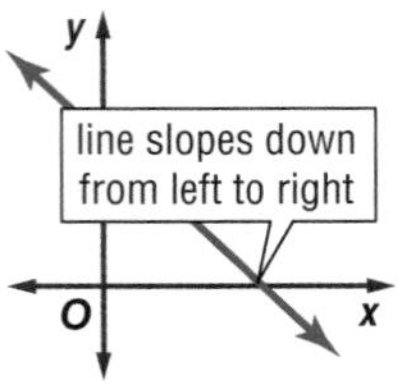

The function values are decreasing over the entire domain.

slope of 0

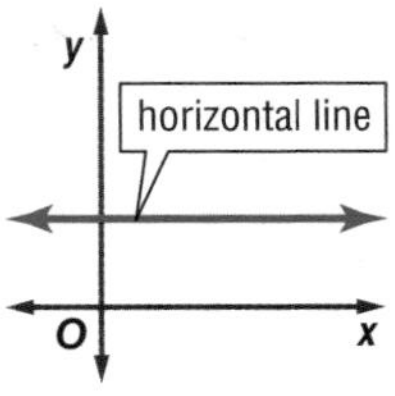

The function values are constant over the entire domain.

undefined slope

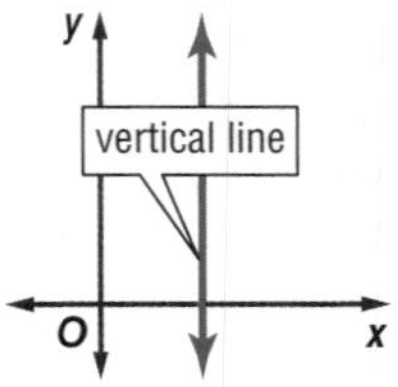

The relation is not a function.

Example 6 Find Coordinates Given the Slope

Find the value of r so that the line through (1, 4) and (−5, r) has a slope of $\frac{1}{3}$.

$m = \frac{y_2 - y_1}{x_2 - x_1}$ Slope Formula

$\frac{1}{3} = \frac{r - 4}{-5 - 1}$ Let (1, 4) = (x_1, y_1) and (−5, r) = (x_2, y_2).

$\frac{1}{3} = \frac{r - 4}{-6}$ Subtract.

$3(r - 4) = 1(-6)$ Find the cross products.

$3r - 12 = -6$ Distributive Property

$3r = 6$ Add 12 to each side and simplify.

$r = 2$ Divide each side by 3 and simplify.

So, the line goes through (−5, 2).

GuidedPractice

Find the value of r so the line that passes through each pair of points has the given slope.

6A. (−2, 6), (r, −4); $m = -5$ **6B.** (r, −6), (5, −8); $m = -8$

Check Your Understanding

Example 1 **Find the rate of change represented in each table or graph.**

1.

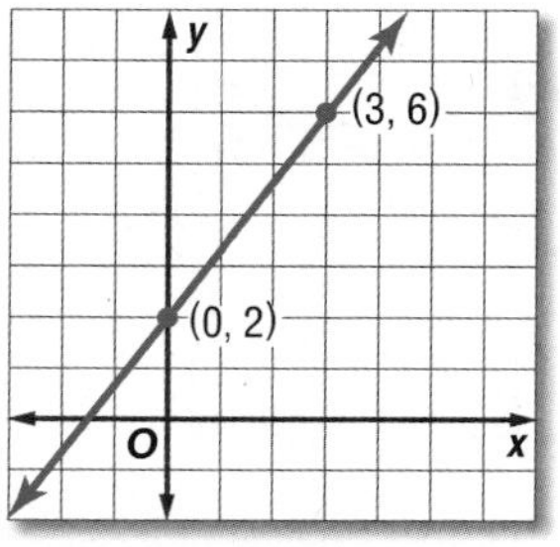

2.

x	y
3	−6
5	2
7	10
9	18
11	26

Example 2

3. CCSS **SENSE-MAKING** Refer to the graph at the right.

a. Find the rate of change of prices from 2006 to 2008. Explain the meaning of the rate of change.

b. Without calculating, find a two-year period that had a greater rate of change than 2006–2008. Explain.

c. Between which years would you guess the new stadium was built? Explain your reasoning.

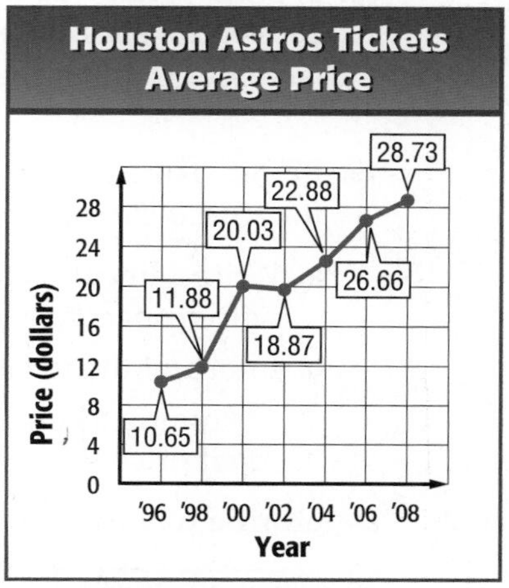

Source: *Team Marketing Report*

Example 3 **Determine whether each function is linear. Write *yes* or *no*. Explain.**

4.

x	−7	−4	−1	2	5
y	5	4	3	2	1

5.

x	8	12	16	20	24
y	7	5	3	0	−2

Examples 4–5 **Find the slope of the line that passes through each pair of points.**

6. (5, 3), (6, 9)

7. (−4, 3), (−2, 1)

8. (6, −2), (8, 3)

9. (1, 10), (−8, 3)

10. (−3, 7), (−3, 4)

11. (5, 2), (−6, 2)

Example 6 **Find the value of *r* so the line that passes through each pair of points has the given slope.**

12. $(-4, r), (-8, 3), m = -5$

13. $(5, 2), (-7, r), m = \frac{5}{6}$

Practice and Problem Solving

Example 1 **Find the rate of change represented in each table or graph.**

14.

x	y
5	2
10	3
15	4
20	5

15.

x	y
1	15
2	9
3	3
4	−3

Example 1 **Find the rate of change represented in each table or graph.**

16.

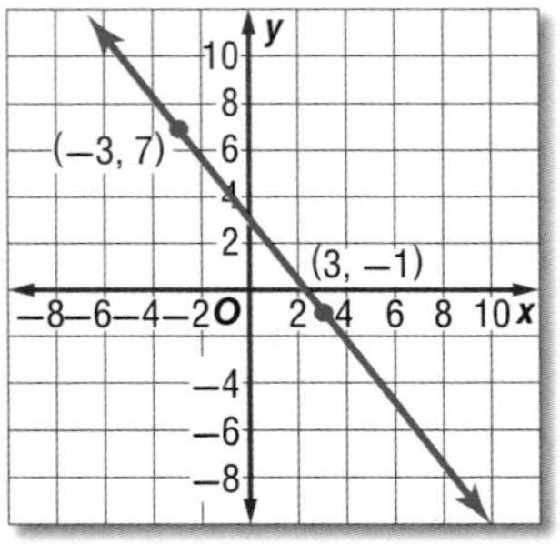

17.

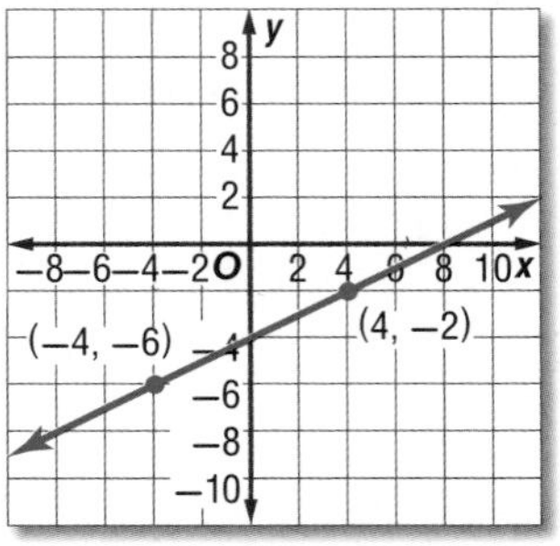

Example 2

18. SPORTS What was the annual rate of change from 2004 to 2008 for women participating in collegiate lacrosse? Explain the meaning of the rate of change.

Year	Number of Women
2004	5545
2008	6830

19. RETAIL The average retail price in the spring of 2009 for a used car is shown in the table at the right.

Age (years)	Value ($)
2	17,378
3	16,157

a. Write a linear function to model the price of the car with respect to age.

b. Interpret the meaning of the slope of the line.

c. Assuming a constant rate of change predict the average retail price for a 7-year-old car.

Example 3 **Determine whether each function is linear. Write *yes* or *no*. Explain.**

20.

x	4	2	0	−2	−4
y	−1	1	3	5	7

21.

x	−7	−5	−3	−1	0
y	11	14	17	20	23

22.

x	−0.2	0	0.2	0.4	0.6
y	0.7	0.4	0.1	0.3	0.6

23.

x	$\frac{1}{2}$	$\frac{3}{2}$	$\frac{5}{2}$	$\frac{7}{2}$	$\frac{9}{2}$
y	$\frac{1}{2}$	1	$\frac{3}{2}$	2	$\frac{5}{2}$

Examples 4–5 **Find the slope of the line that passes through each pair of points.**

24. (4, 3), (−1, 6)
25. (8, −2), (1, 1)
26. (2, 2), (−2, −2)
27. (6, −10), (6, 14)
28. (5, −4), (9, −4)
29. (11, 7), (−6, 2)
30. (−3, 5), (3, 6)
31. (−3, 2), (7, 2)
32. (8, 10), (−4, −6)
33. (−8, 6), (−8, 4)
34. (−12, 15), (18, −13)
35. (−8, −15), (−2, 5)

Example 6 **Find the value of *r* so the line that passes through each pair of points has the given slope.**

36. (12, 10), (−2, *r*), $m = -4$
37. (*r*, −5), (3, 13), $m = 8$
38. (3, 5), (−3, *r*), $m = \frac{3}{4}$
39. (−2, 8), (*r*, 4), $m = -\frac{1}{2}$

CCSS TOOLS **Use a ruler to estimate the slope of each object.**

40.

41.

42. **DRIVING** When driving up a certain hill, you rise 15 feet for every 1000 feet you drive forward. What is the slope of the road?

Find the slope of the line that passes through each pair of points.

43.

x	y
4.5	−1
5.3	2

44.

x	y
0.75	1
0.75	−1

45.

x	y
$2\frac{1}{2}$	$-1\frac{1}{2}$
$-\frac{1}{2}$	$\frac{1}{2}$

46. **MULTIPLE REPRESENTATIONS** In this problem, you will investigate why the slope of a line through any two points on that line is constant.

a. **Visual** Sketch a line ℓ that contains points A, B, A' and B' on a coordinate plane.

b. **Geometric** Add segments to form right triangles ABC and $A'B'C'$ with right angles at C and C'. Describe $\overline{AC}$ and $\overline{A'C'}$, and $\overline{BC}$ and $\overline{B'C'}$.

c. **Verbal** How are triangles ABC and $A'B'C'$ related? What does that imply for the slope between any two distinct points on line ℓ?

47. **BASKETBALL** The table shown below shows the average points per game (PPG) Michael Redd has scored in each of his first 9 seasons with the NBA's Milwaukee Bucks.

Season	1	2	3	4	5	6	7	8	9
PPG	2.2	11.4	15.1	21.7	23.0	25.4	26.7	22.7	21.2

a. Make a graph of the data. Connect each pair of adjacent points with a line.

b. Use the graph to determine in which period Michael Redd's PPG increased the fastest. Explain your reasoning.

c. Discuss the difference in the rate of change from season 1 through season 4, from season 4 through season 7, from season 7 through season 9.

H.O.T. Problems Use Higher-Order Thinking Skills

48. **REASONING** Why does the Slope Formula not work for vertical lines? Explain.

49. **OPEN ENDED** Use what you know about rate of change to describe the function represented by the table.

Time (wk)	Height of Plant (in.)
4	9.0
6	13.5
8	18.0

50. **CHALLENGE** Find the value of d so the line that passes through (a, b) and (c, d) has a slope of $\frac{1}{2}$.

51. **WRITING IN MATH** Explain how the rate of change and slope are related and how to find the slope of a line.

52. **CCSS ARGUMENTS** Kyle and Luna are finding the value of a so the line that passes through $(10, a)$ and $(-2, 8)$ has a slope of $\frac{1}{4}$. Is either of them correct? Explain.

Kyle

$$\frac{-2-10}{8-a} = \frac{1}{4}$$

$$1(8-a) = 4(-12)$$

$$8-a = -48$$

$$a = 56$$

Luna

$$\frac{8-a}{-2-10} = \frac{1}{4}$$

$$4(8-a) = 1(-12)$$

$$32-4a = -12$$

$$a = 11$$

Standardized Test Practice

53. The cost of prints from an online photo processor is given by $C(p) = 29.99 + 0.13p$. \$29.99 is the cost of the membership, and p is the number of 4-inch by 6-inch prints. What does the slope represent?

A cost per print

B cost of the membership

C cost of the membership and 1 print

D number of prints

54. Danita bought a computer for \$1200 and its value depreciated linearly. After 2 years, the value was \$250. What was the amount of yearly depreciation?

F \$950

G \$475

H \$250

J \$225

55. SHORT RESPONSE The graph represents how much the Wright Brothers National Monument charges visitors. How much does the park charge each visitor?

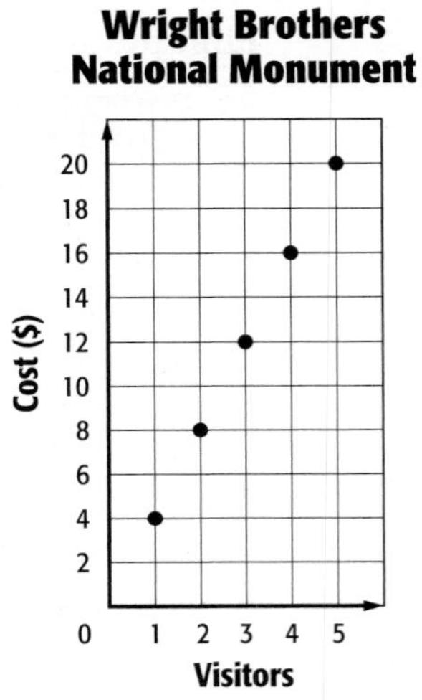

56. PROBABILITY At a gymnastics camp, 1 gymnast is chosen at random from each team. The Flipstars Gymnastics Team consists of 5 eleven-year-olds, 7 twelve-year-olds, 10 thirteen-year-olds, and 8 fourteen-year-olds. What is the probability that the age of the gymnast chosen is an odd number?

A $\frac{1}{30}$ **B** $\frac{1}{15}$ **C** $\frac{1}{2}$ **D** $\frac{3}{5}$

Spiral Review

Solve each equation by graphing. (Lesson 3-2)

57. $3x + 18 = 0$

58. $8x - 32 = 0$

59. $0 = 12x - 48$

Find the *x*- and *y*-intercepts of the graph of each linear function. (Lesson 3-1)

60.

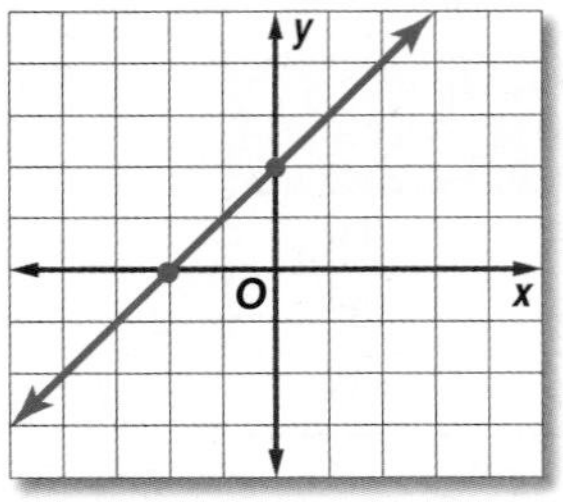

61.

x	y
−3	−4
−2	−2
−1	0
0	2
1	4

62. HOMECOMING Dance tickets are \$9 for one person and \$15 for two people. If a group of seven students wishes to go to the dance, write and solve an equation that would represent the least expensive price p of their tickets. (Lesson 1-3)

Skills Review

Find each quotient.

63. $8 \div \frac{2}{3}$

64. $\frac{3}{8} \div \frac{1}{4}$

65. $\frac{5}{8} \div 2$

66. $\frac{12 \cdot 6}{9}$

67. $\frac{2 \cdot 15}{6}$

68. $\frac{18 \cdot 5}{15}$

CHAPTER 3

Mid-Chapter Quiz

Lessons 3-1 through 3-3

Determine whether each equation is a linear equation. Write *yes* or *no*. If yes, write the equation in standard form. (Lesson 3-1)

1. $y = -4x + 3$

2. $x^2 + 3y = 8$

3. $\frac{1}{4}x - \frac{3}{4}y = -1$

Graph each equation using the *x*- and *y*-intercepts. (Lesson 3-1)

4. $y = 3x - 6$

5. $2x + 5y = 10$

Graph each equation by making a table. (Lesson 3-1)

6. $y = -2x$

7. $x = 8 - y$

8. BOOK SALES The equation $5x + 12y = 240$ describes the total amount of money collected when selling x paperback books at \$5 per book and y hardback books at \$12 per book. Graph the equation using the *x*- and *y*-intercepts. (Lesson 3-1)

Find the root of each equation. (Lesson 3-2)

9. $x + 8 = 0$

10. $4x - 24 = 0$

11. $18 + 8x = 0$

12. $\frac{3}{5}x - \frac{1}{2} = 0$

Solve each equation by graphing. (Lesson 3-2)

13. $-5x + 35 = 0$

14. $14x - 84 = 0$

15. $118 + 11x = -3$

16. MULTIPLE CHOICE The function $y = -15 + 3x$ represents the outside temperature, in degrees Fahrenheit, in a small Alaskan town where x represents the number of hours after midnight. The function is accurate for x values representing midnight through 4:00 P.M. Find the zero of this function. (Lesson 3-2)

A 0

B 3

C 5

D −15

17. Find the rate of change represented in the table. (Lesson 3-3)

x	y
1	2
4	6
7	10
10	14

Find the slope of the line that passes through each pair of points. (Lesson 3-3)

18. (2, 6), (4, 12)

19. (1, 5), (3, 8)

20. (−3, 4), (2, −6)

21. $\left(\frac{1}{3}, \frac{3}{4}\right), \left(\frac{2}{3}, \frac{1}{4}\right)$

22. MULTIPLE CHOICE Find the value of r so the line that passes through the pair of points has the given slope. (Lesson 3-3)

$$(-4, 8), (r, 12), m = \frac{4}{3}$$

F −4

G −1

H 0

J 3

23. Find the slope of the line that passes through the pair of points. (Lesson 3-3)

x	y
2.6	−2
3.1	4

24. POPULATION GROWTH The graph shows the population growth in Heckertsville since 2003. (Lesson 3-3)

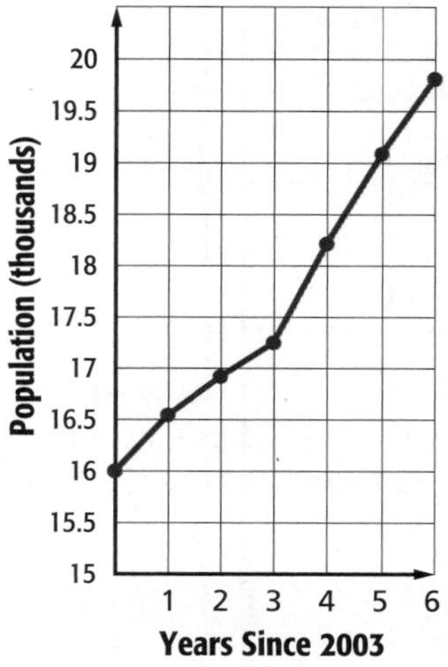

a. For which time period is the rate of change the greatest?

b. Explain the meaning of the slope from 2003 to 2009.

LESSON 3-4

Direct Variation

Then	Now	Why?
● You found rates of change of linear functions.	● **1** Write and graph direct variation equations. **2** Solve problems involving direct variation.	● Bianca is saving her money to buy a designer purse that costs \$295. To help raise the money, she charges \$8 per hour to babysit her neighbors' child. The slope of the line that represents the amount of money Bianca earns is 8, and the rate of change is constant.

NewVocabulary
direct variation
constant of variation
constant of proportionality

CCSS Common Core State Standards

Content Standards

A.REI.10 Understand that the graph of an equation in two variables is the set of all its solutions plotted in the coordinate plane, often forming a curve (which could be a line).

F.IF.7a Graph linear and quadratic functions and show intercepts, maxima, and minima.

Mathematical Practices

1 Make sense of problems and persevere in solving them.

6 Attend to precision.

1 Direct Variation Equations

A **direct variation** is described by an equation of the form $y = kx$, where $k \neq 0$. The equation $y = kx$ illustrates a constant rate of change, and k is the **constant of variation**, also called the **constant of proportionality**.

PT

Example 1 Slope and Constant of Variation

Name the constant of variation for each equation. Then find the slope of the line that passes through each pair of points.

a.

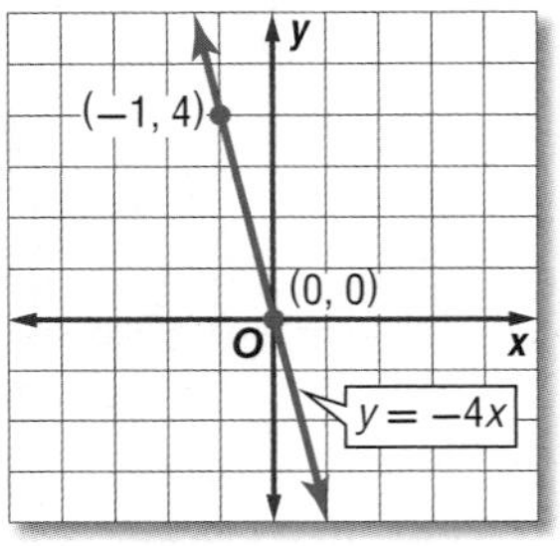

The constant of variation is −4.

$$m = \frac{y_2 - y_1}{x_2 - x_1}$$ Slope Formula

$$= \frac{4 - 0}{-1 - 0}$$ $(x_1, y_1) = (0, 0)$, $(x_2, y_2) = (-1, 4)$

$$= -4$$ The slope is −4.

b.

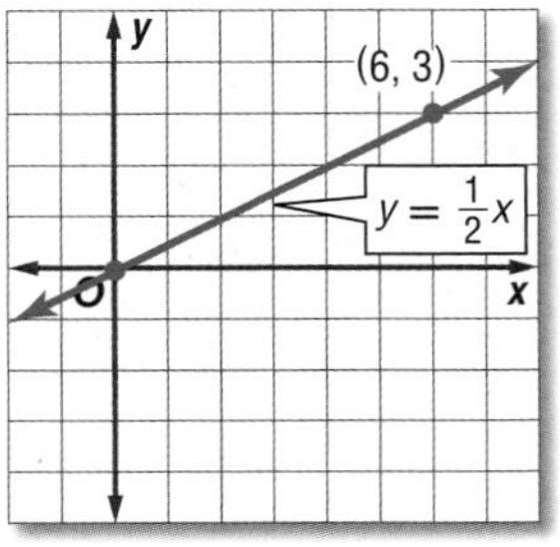

The constant of variation is $\frac{1}{2}$.

$$m = \frac{y_2 - y_1}{x_2 - x_1}$$ Slope Formula

$$= \frac{3 - 0}{6 - 0}$$ $(x_1, y_1) = (0, 0)$, $(x_2, y_2) = (6, 3)$

$$= \frac{1}{2}$$ The slope is $\frac{1}{2}$.

GuidedPractice

1A. Name the constant of variation for $y = \frac{1}{4}x$. Then find the slope of the line that passes through (0, 0) and (4, 1), two points on the line.

1B. Name the constant of variation for $y = -2x$. Then find the slope of the line that passes through (0, 0) and (1, −2), two points on the line.

The slope of the graph of $y = kx$ is k. Since $0 = k(0)$, the graph of $y = kx$ always passes through the origin. Therefore the x- and y-intercepts are zero.

StudyTip

Constant of Variation A line with a positive constant of variation will go up from left to right and a line with a negative constant of variation will go down from left to right.

Example 2 Graph a Direct Variation

Graph $y = -6x$.

Step 1 Write the slope as a ratio.

$-6 = \frac{-6}{1}$ $\frac{\text{rise}}{\text{run}}$

Step 2 Graph (0, 0).

Step 3 From the point (0, 0), move down 6 units and right 1 unit. Draw a dot.

Step 4 Draw a line containing the points.

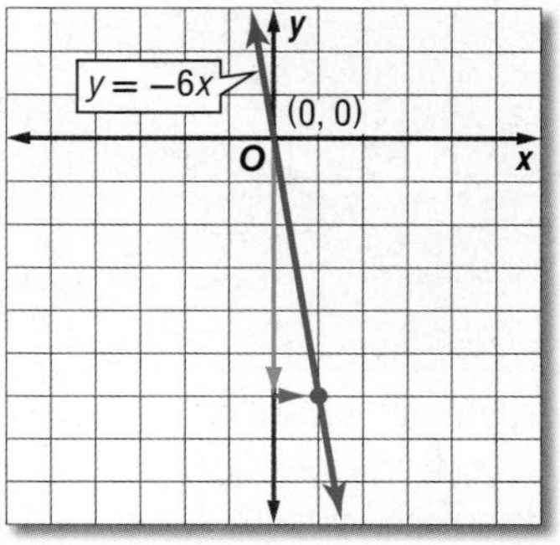

GuidedPractice

2A. $y = 6x$ **2B.** $y = \frac{2}{3}x$ **2C.** $y = -5x$ **2D.** $y = -\frac{3}{4}x$

The graphs of all direct variation equations share some common characteristics.

ConceptSummary Direct Variation Graphs

- Direct variation equations are of the form $y = kx$, where $k \neq 0$.
- The graph of $y = kx$ always passes through the origin.
- The slope is positive if $k > 0$.

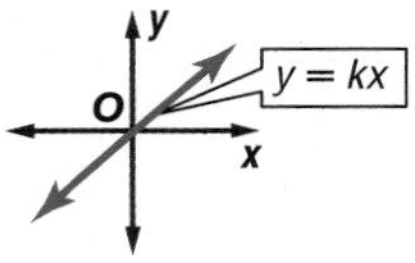

- The slope is negative if $k < 0$.

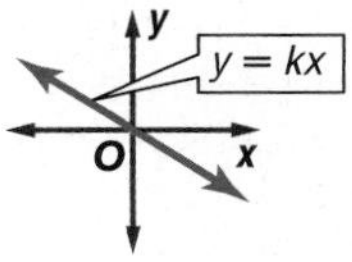

If the relationship between the values of y and x can be described by a direct variation equation, then we say that y varies directly as x.

Example 3 Write and Solve a Direct Variation Equation

Suppose y varies directly as x, and $y = 72$ when $x = 8$.

a. Write a direct variation equation that relates x and y.

$y = kx$ Direct variation formula

$72 = k(8)$ Replace y with 72 and x with 8.

$9 = k$ Divide each side by 8.

Therefore, the direct variation equation is $y = 9x$.

b. Use the direct variation equation to find x when $y = 63$.

$y = 9x$ Direct variation formula

$63 = 9x$ Replace y with 63.

$7 = x$ Divide each side by 9.

Therefore, $x = 7$ when $y = 63$.

GuidedPractice

3. Suppose y varies directly as x, and $y = 98$ when $x = 14$. Write a direct variation equation that relates x and y. Then find y when $x = -4$.

2 Direct Variation Problems

One of the most common applications of direct variation is the formula $d = rt$. Distance d varies directly as time t, and the rate r is the constant of variation.

Real-WorldLink

In 2006, domestic airlines transported over 660 million passengers an average distance of 724 miles per flight.

Source: Bureau of Transportation Statistics

Real-World Example 4 Estimate Using Direct Variation

TRAVEL The distance a jet travels varies directly as the number of hours it flies. A jet traveled 3420 miles in 6 hours.

a. Write a direct variation equation for the distance d flown in time t.

Words	Distance	equals	rate	times	time.
Variable	Let r = rate.				
Equation	3420	=	r	×	6

Solve for the rate.

$3420 = r(6)$ Original equation

$\frac{3420}{6} = \frac{r(6)}{6}$ Divide each side by 6.

$570 = r$ Simplify.

Therefore, the direct variation equation is $d = 570t$. The airliner flew at a rate of 570 miles per hour.

b. Graph the equation.

The graph of $d = 570t$ passes through the origin with slope 570.

$m = \frac{570}{1}$ $\frac{\text{rise}}{\text{run}}$

Distance Flown

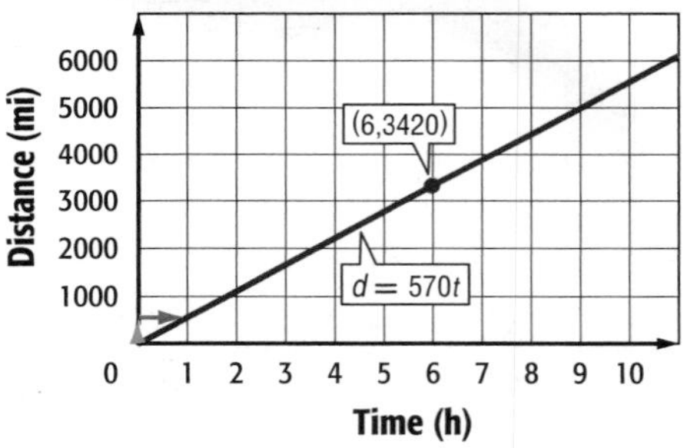

Problem-SolvingTip

CCSS Precision Notice that the question asks for an estimate, not an exact answer.

c. Estimate how many hours it will take for an airliner to fly 6500 miles.

$d = 570t$ Original equation

$6500 = 570t$ Replace d with 6500.

$\frac{6500}{570} = \frac{570t}{570}$ Divide each side by 570.

$t \approx 11.4$ Simplify.

It would take the airliner approximately 11.4 hours to fly 6500 miles.

GuidedPractice

4. HOT-AIR BALLOONS A hot-air balloon's height varies directly as the balloon's ascent time in minutes.

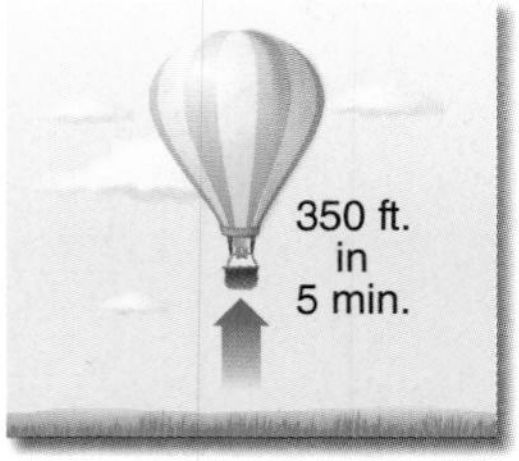

A. Write a direct variation for the distance d ascended in time t.

B. Graph the equation.

C. Estimate how many minutes it would take to ascend 2100 feet.

D. About how many minutes would it take to ascend 3500 feet?

Ed Boettcher/CORBIS Premium RF/Alamy

Check Your Understanding

Example 1 **Name the constant of variation for each equation. Then find the slope of the line that passes through each pair of points.**

1.

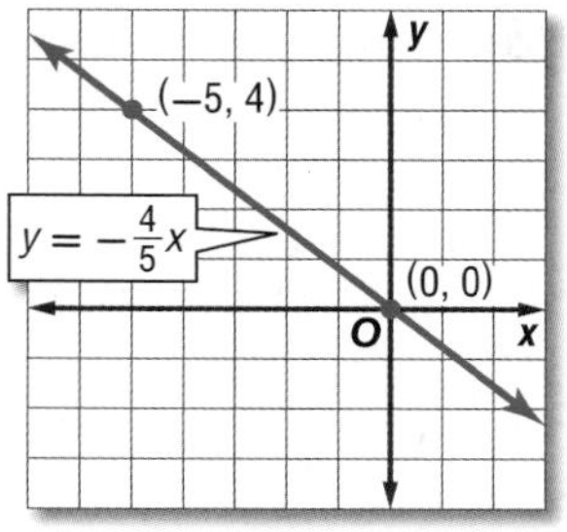

2.

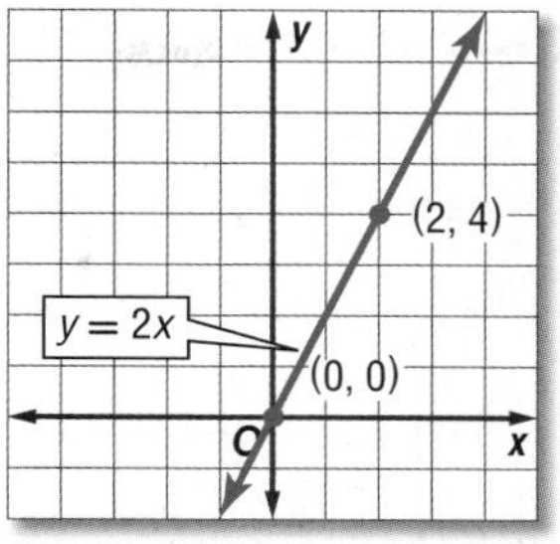

Example 2 **Graph each equation.**

3. $y = -x$ **4.** $y = \frac{3}{4}x$ **5.** $y = -8x$ **6.** $y = -\frac{8}{5}$

Example 3 **Suppose y varies directly as x. Write a direct variation equation that relates x and y. Then solve.**

7. If $y = 15$ when $x = 12$, find y when $x = 32$.

8. If $y = -11$ when $x = 6$, find x when $y = 44$.

Example 4 **9.** **CCSS REASONING** You find that the number of messages you receive on your message board varies directly as the number of messages you post. When you post 5 messages, you receive 12 messages in return.

a. Write a direct variation equation relating your posts to the messages received. Then graph the equation.

b. Find the number of messages you need to post to receive 96 messages.

Practice and Problem Solving

Example 1 **Name the constant of variation for each equation. Then find the slope of the line that passes through each pair of points.**

10.

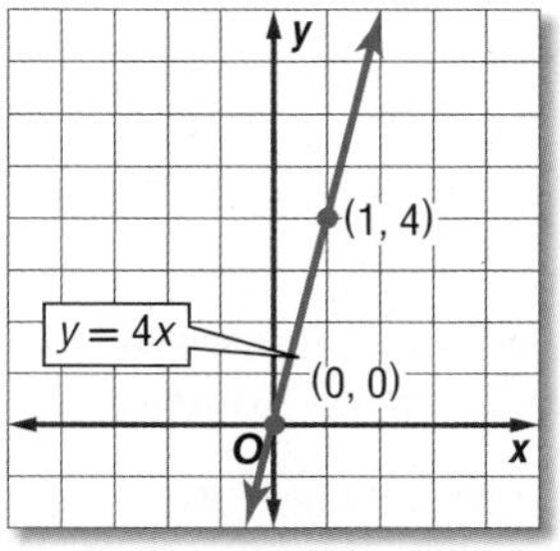

11.

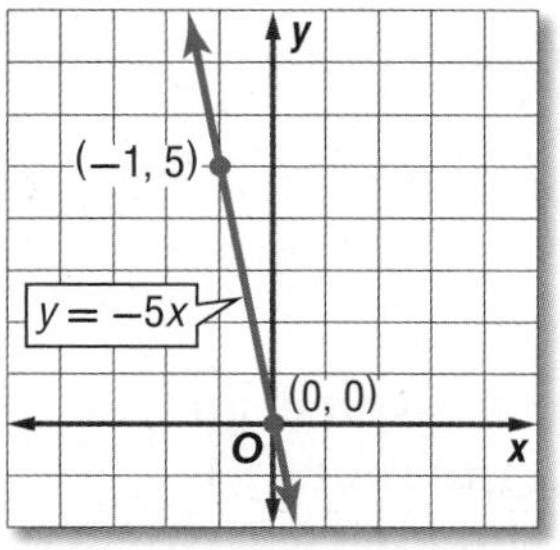

12.

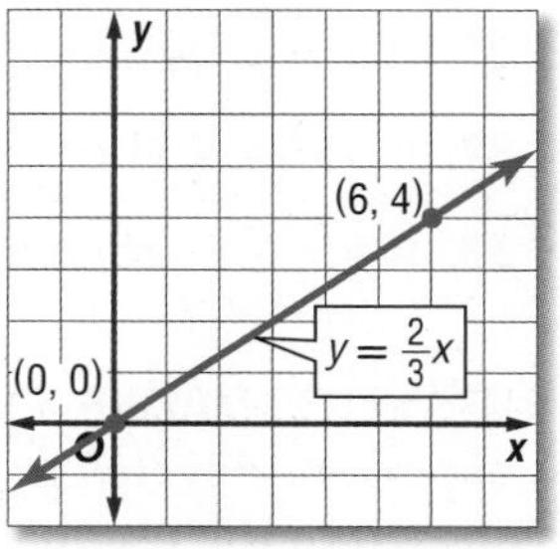

13.

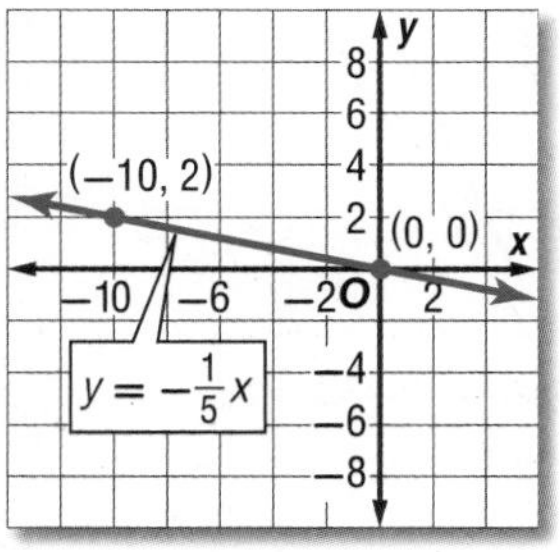

14.

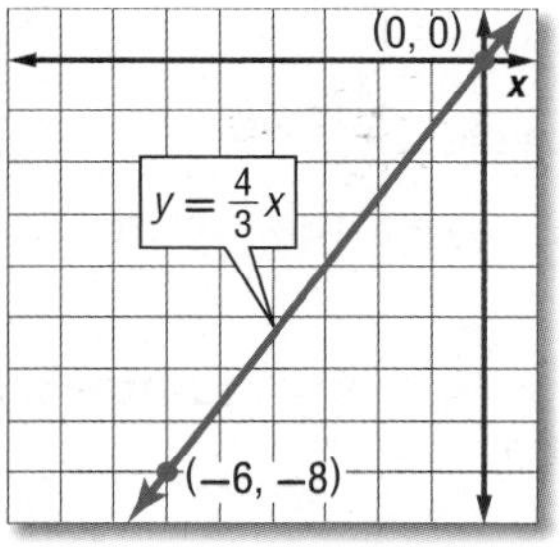

15.

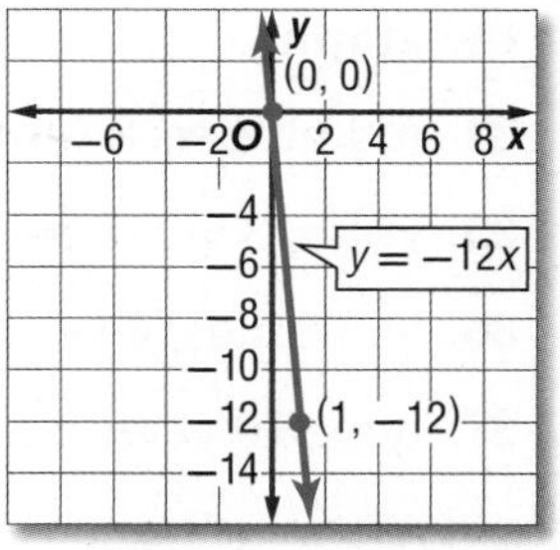

Example 2 **Graph each equation.**

16. $y = 10x$
17. $y = -7x$
18. $y = x$
19. $y = \frac{7}{6}x$
20. $y = \frac{1}{6}x$
21. $y = \frac{2}{9}x$
22. $y = \frac{6}{5}x$
23. $y = -\frac{5}{4}x$

Example 3 **Suppose y varies directly as x. Write a direct variation equation that relates x and y. Then solve.**

24. If $y = 6$ when $x = 10$, find x when $y = 18$.
25. If $y = 22$ when $x = 8$, find y when $x = -16$.
26. If $y = 4\frac{1}{4}$ when $x = \frac{3}{4}$, find y when $x = 4\frac{1}{2}$.
27. If $y = 12$ when $x = \frac{6}{7}$, find x when $y = 16$.

Example 4

28. **SPORTS** The distance a golf ball travels at an altitude of 7000 feet varies directly with the distance the ball travels at sea level, as shown.

Hitting a Golf Ball		
Altitude (ft)	0 (sea level)	7000
Distance (yd)	200	210

   a. Write and graph an equation that relates the distance a golf ball travels at an altitude of 7000 feet y with the distance at sea level x.

   b. What would be a person's average driving distance at 7000 feet if his average driving distance at sea level is 180 yards?

29. **FINANCIAL LITERACY** Depreciation is the decline in a car's value over the course of time. The table below shows the values of a car with an average depreciation.

Age of Car (years)	1	2	3	4	5
Value (dollars)	12,000	10,200	8400	6600	4800

   a. Write an equation that relates the age x of the car to the value y that it lost after each year.

   b. Find the age of the car if the value is \$300.

Suppose y varies directly as x. Write a direct variation equation that relates x and y. Then solve.

30. If $y = 3.2$ when $x = 1.6$, find y when $x = 19$.
31. If $y = 15$ when $x = \frac{3}{4}$, find x when $y = 25$.
32. If $y = 4.5$ when $x = 2.5$, find y when $x = 12$.
33. If $y = -6$ when $x = 1.6$, find y when $x = 8$.

CCSS SENSE-MAKING **Certain endangered species experience cycles in their populations as shown in the graph at the right. Match each animal below to one of the colored lines in the graph.**

34. red grouse, 8 years per cycle
35. voles, 3 years per cycle
36. lemmings, 4 years per cycle
37. lynx, 10 years per cycle

Population Cycles of Endangered Species

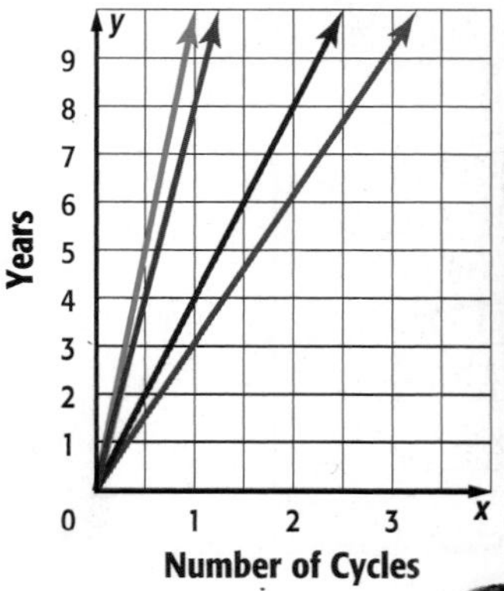

In Exercises 38–40, write and graph a direct variation equation that relates the variables.

38. PHYSICAL SCIENCE The weight W of an object is 9.8 m/s^2 times the mass of the object m.

39. MUSIC Music downloads are \$0.99 per song. The total cost of d songs is T.

40. GEOMETRY The circumference of a circle C is approximately 3.14 times the diameter d.

41. MULTIPLE REPRESENTATIONS In this problem, you will investigate the family of direct variation functions.

a. Graphical Graph $y = x$, $y = 3x$, and $y = 5x$ on the same coordinate plane.

b. Algebraic Describe the relationship among the constant of variation, the slope of the line, and the rate of change of the graph.

c. Verbal Make a conjecture about how you can determine without graphing which of two direct variation equations has the steeper graph.

42. TRAVEL A map of North Carolina is scaled so that 3 inches represents 93 miles. How far apart are Raleigh and Charlotte if they are 1.8 inches apart on the map?

43. INTERNET A company will design and maintain a Web site for your company for \$9.95 per month. Write a direct variation equation to find the total cost C for having a Web page for n months.

44. BASEBALL Before their first game, high school student Todd McCormick warmed all 5200 seats in a new minor league stadium, by literally sitting in every seat. He started at 11:50 A.M. and finished around 3 P.M.

a. Write a direct variation equation relating the number of seats to time. What is the meaning of the constant of variation in this situation?

b. About how many seats had Todd sat in by 1:00 P.M.?

c. How long would you expect it to take Todd to sit in all of the seats at a major league stadium with more than 40,000 seats?

H.O.T. Problems Use Higher-Order Thinking Skills

45. WHICH ONE DOESN'T BELONG? Identify the equation that does not belong. Explain.

$9 = rt$	$9a = 0$	$z = \frac{1}{9}x$	$w = \frac{9}{t}$

46. REASONING How are the constant of variation and the slope related in a direct variation equation? Explain your reasoning.

47. OPEN ENDED Model a real-world situation using a direct variation equation. Graph the equation and describe the rate of change.

48. CCSS STRUCTURE Suppose y varies directly as x. If the value of x is doubled, then the value of y is also *always, sometimes* or *never* doubled. Explain your reasoning.

49. ERROR ANALYSIS Eddy says the slope between any two points on the graph of a direct variation equation $y = kx$ is $\frac{1}{k}$. Adelle says the slope depends on the points chosen. Is either of them correct? Explain.

50. WRITING IN MATH How can you identify the graph of a direct variation equation?

Standardized Test Practice

51. Patricia pays $1.19 each to download songs to her digital media player. If n is the number of downloaded songs, which equation represents the cost C in dollars?

A $C = 1.19n$

B $n = 1.19C$

C $C = 1.19 \div n$

D $C = n + 1.19$

52. Suppose that y varies directly as x, and $y = 8$ when $x = 6$. What is the value of y when $x = 8$?

F 6

G 12

H $10\frac{2}{3}$

J 16

53. What is the relationship between the input (x) and output (y)?

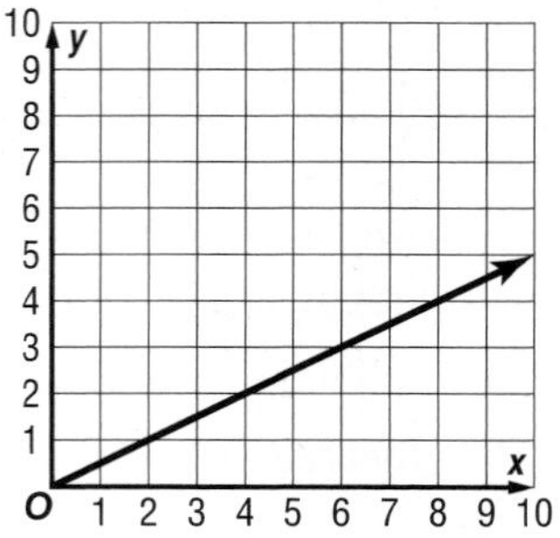

A The output is two more than the input.

B The output is two less than the input.

C The output is twice the input.

D The output is half the input.

54. SHORT RESPONSE A telephone company charges $40 per month plus $0.07 per minute. How much would a month of service cost a customer if the customer talked for 200 minutes?

Spiral Review

55. TELEVISION The graph shows the average number of television channels American households receive. What was the annual rate of change from 2004 to 2008? Explain the meaning of the rate of change. (Lesson 3-3)

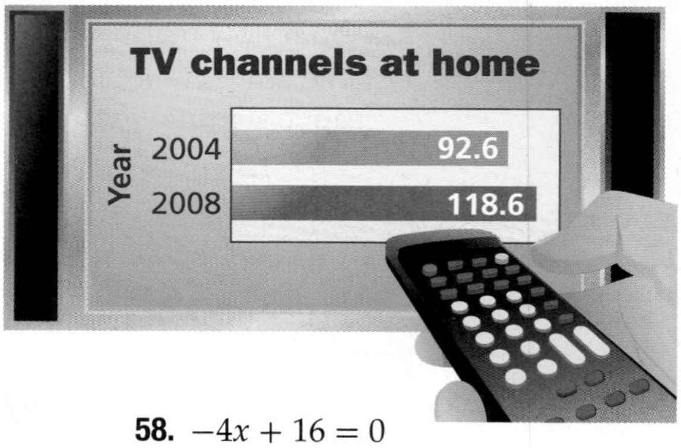

Solve each equation by graphing. (Lesson 3-2)

56. $0 = 18 - 9x$

57. $2x + 14 = 0$

58. $-4x + 16 = 0$

59. $-5x - 20 = 0$

60. $8x - 24 = 0$

61. $12x - 144 = 0$

Evaluate each expression if $a = 4$, $b = -2$, and $c = -4$. (Lesson 2-5)

62. $|2a + c| + 1$

63. $4a - |3b + 2|$

64. $-|a + 1| + |3c|$

65. $-a + |2 - a|$

66. $|c - 2b| - 3$

67. $-2|3b - 8|$

Skills Review

Find each difference.

68. $13 - (-1)$

69. $4 - 16$

70. $-3 - 3$

71. $-8 - (-2)$

72. $16 - (-10)$

73. $-8 - 4$

LESSON 3-5 Arithmetic Sequences as Linear Functions

Then

- You indentified linear functions.

Now

1. Recognize arithmetic sequences.
2. Relate arithmetic sequences to linear functions.

Why?

- During a 2000-meter race, the coach of a women's crew team recorded the team's times at several intervals.
  - At 400 meters, the time was 1 minute 32 seconds.
  - At 800 meters, it was 3 minutes 4 seconds.
  - At 1200 meters, it was 4 minutes 36 seconds.
  - At 1600 meters, it was 6 minutes 8 seconds.

  They completed the race with a time of 7 minutes 40 seconds.

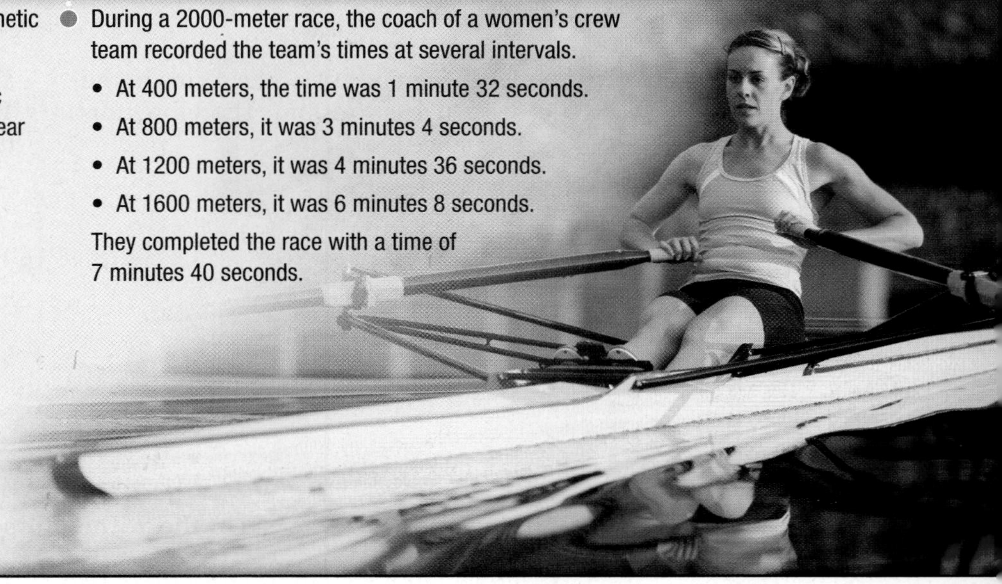

NewVocabulary

sequence
terms of the sequence
arithmetic sequence
common difference

Common Core State Standards

Content Standards

F.BF.2 Write arithmetic and geometric sequences both recursively and with an explicit formula, use them to model situations, and translate between the two forms.

F.LE.2 Construct linear and exponential functions, including arithmetic and geometric sequences, given a graph, a description of a relationship, or two input-output pairs (include reading these from a table).

Mathematical Practices

8 Look for and express regularity in repeated reasoning.

Aurora Open/Ty Milford/Getty Images

1 Recognize Arithmetic Sequences

You can relate the pattern of team times to linear functions. A **sequence** is a set of numbers, called the **terms of the sequence**, in a specific order. Look for a pattern in the information given for the women's crew team. Make a table to analyze the data.

Distance (m)	400	800	1200	1600	2000
Time (min : sec)	1:32	3:04	4:36	6:08	7:40

+ 1:32   + 1:32   + 1:32   + 1:32

As the distance increases in regular intervals, the time increases by 1 minute 32 seconds. Since the difference between successive terms is constant, this is an **arithmetic sequence**. The difference between the terms is called the **common difference** d.

KeyConcept Arithmetic Sequence

Words An arithmetic sequence is a numerical pattern that increases or decreases at a constant rate called the *common difference*.

Examples

3, 5, 7, 9, 11, . . .
+2 +2 +2 +2
$d = 2$

33, 29, 25, 21, 17, . . .
−4 −4 −4 −4
$d = -4$

The three dots used with sequences are called an *ellipsis*. The ellipsis indicates that there are more terms in the sequence that are not listed.

Math HistoryLink

Mina Rees **(1902–1997)** Rees received the first award for Distinguished Service to Mathematics from the Mathematical Association of America. She was the first president of the Graduate Center at The City University of New York. Her work in analyzing patterns is still inspiring young women to study mathematics today.

Example 1 Identify Arithmetic Sequences

Determine whether each sequence is an arithmetic sequence. Explain.

a. $-4, -2, 0, 2, \ldots$

$-4 \quad -2 \quad 0 \quad 2$

$+2 \quad +2 \quad +2$

The difference between terms in the sequence is constant. Therefore, this sequence is arithmetic.

b. $\frac{1}{2}, \frac{5}{8}, \frac{3}{4}, \frac{13}{16}, \ldots$

$\frac{1}{2} \quad \frac{5}{8} \quad \frac{3}{4} \quad \frac{13}{16}$

$+\frac{1}{8} \quad +\frac{1}{8} \quad +\frac{1}{16}$

This is not an arithmetic sequence. The difference between terms is not constant.

GuidedPractice

1A. $-26, -22, -18, -14, \ldots$

1B. $1, 4, 9, 25, \ldots$

You can use the common difference of an arithmetic sequence to find the next term.

Example 2 Find the Next Term

Find the next three terms of the arithmetic sequence 15, 9, 3, −3, … .

Step 1 Find the common difference by subtracting successive terms.

$15 \quad 9 \quad 3 \quad -3$

$-6 \quad -6 \quad -6$

The common difference is −6.

Step 2 Add −6 to the last term of the sequence to get the next term.

$-3 \quad -9 \quad -15 \quad -21$

$-6 \quad -6 \quad -6$

The next three terms in the sequence are −9, −15, and −21.

GuidedPractice

2. Find the next four terms of the arithmetic sequence 9.5, 11.0, 12.5, 14.0, … .

StudyTip

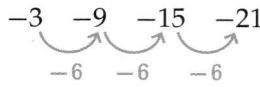

CCSS Regularity Notice the regularity in the way expressions in terms of a_1 and d change with each term of the sequence.

Each term in an arithmetic sequence can be expressed in terms of the first term a_1 and the common difference d.

Term	Symbol	In Terms of a_1 and d	Numbers
first term	a_1	a_1	8
second term	a_2	$a_1 + d$	$8 + 1(3) = 11$
third term	a_3	$a_1 + 2d$	$8 + 2(3) = 14$
fourth term	a_4	$a_1 + 3d$	$8 + 3(3) = 17$
⋮	⋮	⋮	⋮
nth term	a_n	$a_1 + (n - 1)d$	$8 + (n - 1)(3)$

KeyConcept *n*th Term of an Arithmetic Sequence

The nth term of an arithmetic sequence with first term a_1 and common difference d is given by $a_n = a_1 + (n - 1)d$, where n is a positive integer.

The Graduate Center, City University of New York

Example 3 Find the *n*th Term

a. Write an equation for the *n*th term of the arithmetic sequence −12, −8, −4, 0, … .

Step 1 Find the common difference.

$-12 \quad -8 \quad -4 \quad 0$

$+4 \quad +4 \quad +4$ — The common difference is 4.

Step 2 Write an equation.

$a_n = a_1 + (n-1)d$	Formula for the *n*th term
$= -12 + (n-1)4$	$a_1 = -12$ and $d = 4$
$= -12 + 4n - 4$	Distributive Property
$= 4n - 16$	Simplify.

StudyTip

nth Terms Since n represents the number of the term, the inputs for n are the counting numbers.

b. Find the 9th term of the sequence.

Substitute 9 for n in the formula for the nth term.

$a_n = 4n - 16$	Formula for the *n*th term
$a_9 = 4(9) - 16$	$n = 9$
$a_9 = 36 - 16$	Multiply.
$a_9 = 20$	Simplify.

c. Graph the first five terms of the sequence.

n	$4n - 16$	a_n	(n, a_n)
1	4(1) − 16	−12	(1, −12)
2	4(2) − 16	−8	(2, −8)
3	4(3) − 16	−4	(3, −4)
4	4(4) − 16	0	(4, 0)
5	4(5) − 16	4	(5, 4)

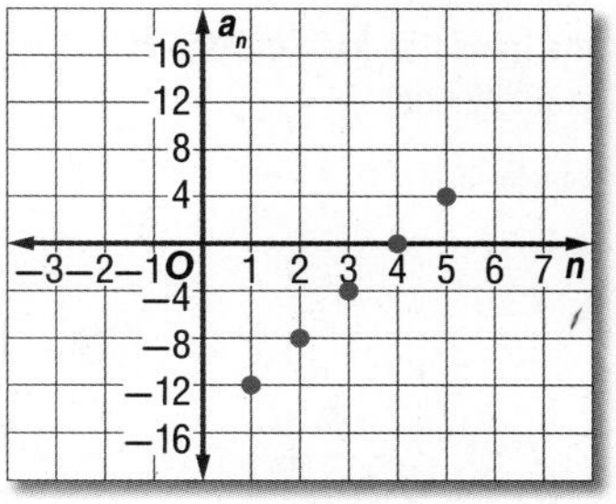

d. Which term of the sequence is 32?

In the formula for the nth term, substitute 32 for a_n.

$a_n = 4n - 16$	Formula for the *n*th term
$32 = 4n - 16$	$a_n = 32$
$32 + 16 = 4n - 16 + 16$	Add 16 to each side.
$48 = 4n$	Simplify.
$12 = n$	Divide each side by 4.

Guided Practice

Consider the arithmetic sequence 3, −10, −23, −36, … .

3A. Write an equation for the nth term of the sequence.

3B. Find the 15th term in the sequence.

3C. Graph the first five terms of the sequence.

3D. Which term of the sequence is −114?

2 Arithmetic Sequences and Functions

Arithmetic Sequences and Functions As you can see from Example 3, the graph of the first five terms of the arithmetic sequence lie on a line. An arithmetic sequence is a linear function in which n is the independent variable, a_n is the dependent variable, and d is the slope. The formula can be rewritten as the function $f(n) = (n - 1)d + a_1$, where n is a counting number.

While the domain of most linear functions are all real numbers, in Example 3 the domain of the function is the set of counting numbers and the range of the function is the set of integers on the line.

Real-WorldLink

When a Latina turns 15, her family may host a quinceañera for her birthday. The quinceañera is a traditional Hispanic ceremony and reception that signifies the transition from childhood to adulthood.

Source: Quince Girl

Real-World Example 4 Arithmetic Sequences as Functions

INVITATIONS Marisol is mailing invitations to her quinceañera. The arithmetic sequence \$0.42, \$0.84, \$1.26, \$1.68, ... represents the cost of postage.

a. Write a function to represent this sequence.

The first term, a_1, is 0.42. Find the common difference.

0.42   0.84   1.26   1.68

+0.42   +0.42   +0.42

The common difference is 0.42.

$a_n = a_1 + (n - 1)d$	Formula for the nth term
$= 0.42 + (n - 1)0.42$	$a_1 = 0.42$ and $d = 0.42$
$= 0.42 + 0.42n - 0.42$	Distributive Property
$= 0.42n$	Simplify.

The function is $f(n) = 0.42n$.

b. Graph the function and determine the domain.

The rate of change of the function is 0.42. Make a table and plot points.

n	$f(n)$
1	0.42
2	0.84
3	1.26
4	1.68
5	2.10

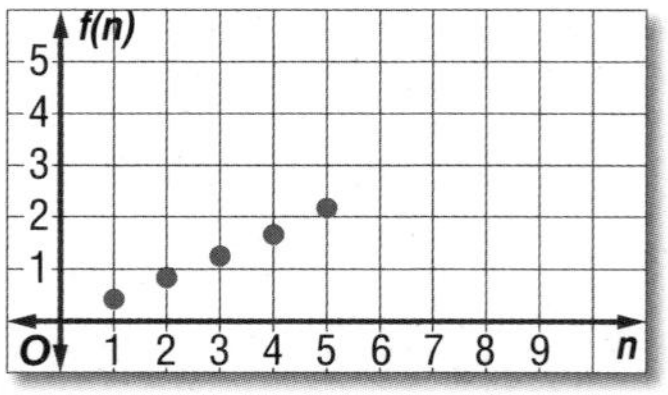

The domain of a function is the number of invitations Marisol mails. So, the domain is $\{0, 1, 2, 3, \ldots\}$.

Guided Practice

4. TRACK The chart below shows the length of Martin's long jumps.

Jump	1	2	3	4
Length (ft)	8	9.5	11	12.5

A. Write a function to represent this arithmetic sequence.

B. Then graph the function.

Check Your Understanding

Example 1 **Determine whether each sequence is an arithmetic sequence. Write *yes* or *no*. Explain.**

1. 18, 16, 15, 13, …

2. 4, 9, 14, 19, …

Example 2 **Find the next three terms of each arithmetic sequence.**

3. 12, 9, 6, 3, …

4. −2, 2, 6, 10, …

Example 3 **Write an equation for the *n*th term of each arithmetic sequence. Then graph the first five terms of the sequence.**

5. 15, 13, 11, 9, …

6. −1, −0.5, 0, 0.5, …

Example 4 **7. SAVINGS** Kaia has \$525 in a savings account. After one month she has \$580 in the account. The next month the balance is \$635. The balance after the third month is \$690. Write a function to represent the arithmetic sequence. Then graph the function.

Practice and Problem Solving

Example 1 **Determine whether each sequence is an arithmetic sequence. Write *yes* or *no*. Explain.**

8. −3, 1, 5, 9, …

9. $\frac{1}{2}, \frac{3}{4}, \frac{5}{8}, \frac{7}{16}, \ldots$

10. −10, −7, −4, 1, …

11. −12.3, −9.7, −7.1, −4.5, …

Example 2 **Find the next three terms of each arithmetic sequence.**

12. 0.02, 1.08, 2.14, 3.2, …

13. 6, 12, 18, 24, …

14. 21, 19, 17, 15, …

15. $-\frac{1}{2}, 0, \frac{1}{2}, 1, \ldots$

16. $2\frac{1}{3}, 2\frac{2}{3}, 3, 3\frac{1}{3}, \ldots$

17. $\frac{7}{12}, 1\frac{1}{3}, 2\frac{1}{12}, 2\frac{5}{6}, \ldots$

Example 3 **Write an equation for the *n*th term of the arithmetic sequence. Then graph the first five terms in the sequence.**

18. −3, −8, −13, −18, …

19. −2, 3, 8, 13, …

20. −11, −15, −19, −23, …

21. −0.75, −0.5, −0.25, 0, …

Example 4 **22. AMUSEMENT PARKS** Shiloh and her friends spent the day at an amusement park. In the first hour, they rode two rides. After 2 hours, they had ridden 4 rides. They had ridden 6 rides after 3 hours.

a. Write a function to represent the arithmetic sequence.

b. Graph the function and determine the domain.

23. CCSS **MODELING** The table shows how Ryan is paid at his lumber yard job.

Linear Feet of 2×4 Planks Cut	10	20	30	40	50	60	70
Amount Paid in Commission (\$)	8	16	24	32	40	48	56

a. Write a function to represent Ryan's commission.

b. Graph the function and determine the domain.

24. The graph is a representation of an arithmetic sequence.

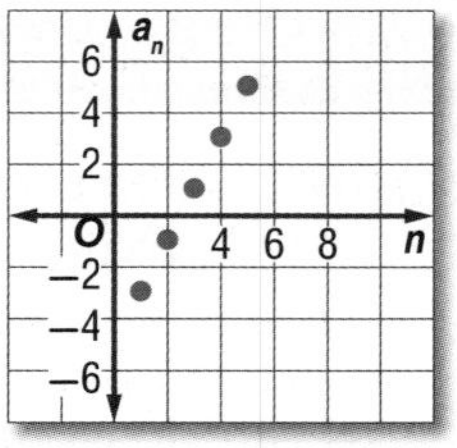

a. List the first five terms.

b. Write the formula for the nth term.

c. Write the function.

25. **NEWSPAPERS** A local newspaper charges by the number of words for advertising. Write a function to represent the advertising costs.

DAILY NEWS ADVERTISING	
10 words \$7.50	20 words \$10.00
15 words \$8.75	25 words \$11.25

26. The fourth term of an arithmetic sequence is 8. If the common difference is 2, what is the first term?

27. The common difference of an arithmetic sequence is -5. If a_{12} is 22, what is a_1?

28. The first four terms of an arithmetic sequence are 28, 20, 12, and 4. Which term of the sequence is -36?

29. **CARS** Jamal's odometer of his car reads 24,521. If Jamal drives 45 miles every day, what will the odometer reading be in 25 days?

30. **YEARBOOKS** The yearbook staff is unpacking a box of school yearbooks. The arithmetic sequence 281, 270, 259, 248 … represents the total number of ounces that the box weighs as each yearbook is taken out of the box.

a. Write a function to represent this sequence.

b. Determine the weight of each yearbook.

c. If the box weighs at least 11 ounces empty and 292 ounces when it is full, how many yearbooks were in the box?

31. **SPORTS** To train for an upcoming marathon, Olivia plans to run 3 miles per day for the first week and then increase the daily distance by a half mile each of the following weeks.

a. Write an equation to represent the nth term of the sequence.

b. If the pattern continues, during which week will she run 10 miles per day?

c. Is it reasonable to think that this pattern will continue indefinitely? Explain.

H.O.T. Problems Use Higher-Order Thinking Skills

32. **OPEN ENDED** Create an arithmetic sequence with a common difference of -10.

33. CCSS **PERSEVERANCE** Find the value of x that makes $x + 8$, $4x + 6$, and $3x$ the first three terms of an arithmetic sequence.

34. **REASONING** Compare and contrast the domain and range of the linear functions described by $Ax + By = C$ and $a_n = a_1 + (n - 1)d$.

35. **CHALLENGE** Determine whether each sequence is an arithmetic sequence. Write *yes* or *no*. Explain. If yes, find the common difference and the next three terms.

a. $2x + 1, 3x + 1, 4x + 1\ldots$

b. $2x, 4x, 8x, \ldots$

36. **WRITING IN MATH** How are graphs of arithmetic sequences and linear functions similar? different?

Standardized Test Practice

37. GRIDDED RESPONSE The population of Westerville is about 35,000. Each year the population increases by about 400. This can be represented by the following equation, where n represents the number of years from now and p represents the population.

$$p = 35{,}000 + 400n$$

In how many years will the Westerville population be about 38,200?

38. Which relation is a function?

A $\{(-5, 6), (4, -3), (2, -1), (4, 2)\}$

B $\{(3, -1), (3, -5), (3, 4), (3, 6)\}$

C $\{(-2, 3), (0, 3), (-2, -1), (-1, 2)\}$

D $\{(-5, 6), (4, -3), (2, -1), (0, 2)\}$

39. Find the formula for the nth term of the arithmetic sequence.

$$-7, -4, -1, 2, \ldots$$

F $a_n = 3n - 4$

G $a_n = -7n + 10$

H $a_n = 3n - 10$

J $a_n = -7n + 4$

40. STATISTICS A class received the following scores on the ACT. What is the difference between the median and the mode in the scores?

18, 26, 20, 30, 25, 21, 32, 19, 22, 29, 29, 27, 24

A 1 **C** 3

B 2 **D** 4

Spiral Review

Name the constant of variation for each direct variation. Then find the slope of the line that passes through each pair of points. (Lesson 3-4)

41.

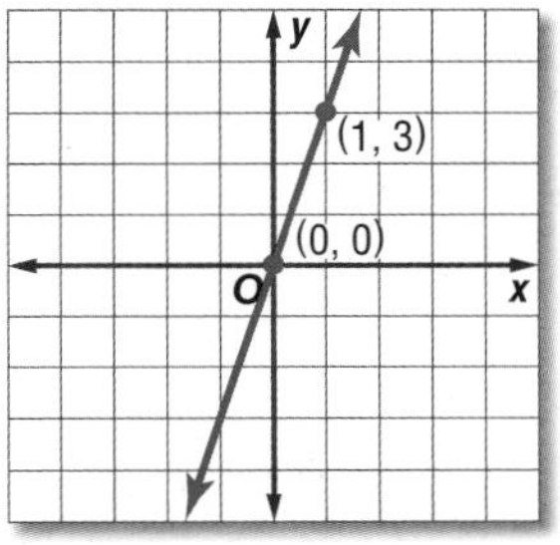

42.

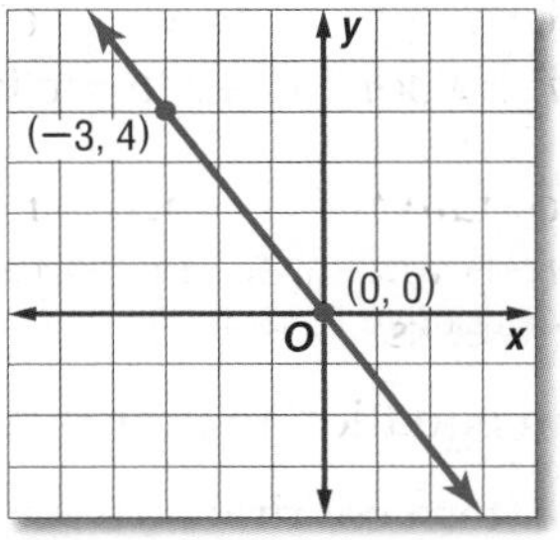

Find the slope of the line that passes through each pair of points. (Lesson 3-3)

43. $(5, 3), (-2, 6)$ **44.** $(9, 2), (-3, -1)$ **45.** $(2, 8), (-2, -4)$

Solve each equation. Check your solution. (Lesson 2-3)

46. $5x + 7 = -8$ **47.** $8 = 2 + 3n$ **48.** $12 = \frac{c - 6}{2}$

49. SPORTS The most popular sports for high school girls are basketball and softball. Write and use an equation to find how many more girls play on basketball teams than on softball teams. (Lesson 2-1)

Basketball
453,000 girls

Softball
369,000 girls

Skills Review

Graph each point on the same coordinate plane.

50. $A(2, 5)$ **51.** $B(-2, 1)$ **52.** $C(-3, -1)$

53. $D(0, 4)$ **54.** $F(5, -3)$ **55.** $G(-5, 0)$

EXTEND 3-5

Algebra Lab
Inductive and Deductive Reasoning

CCSS Common Core State Standards
Mathematical Practices
3 Construct viable arguments and critique the reasoning of others.

If Jolene is not feeling well, she may go to a doctor. The doctor will ask her questions about how she is feeling and possibly run other tests. Based on her symptoms, the doctor can diagnose Jolene's illness. This is an example of inductive reasoning. **Inductive reasoning** is used to derive a general rule after observing many events.

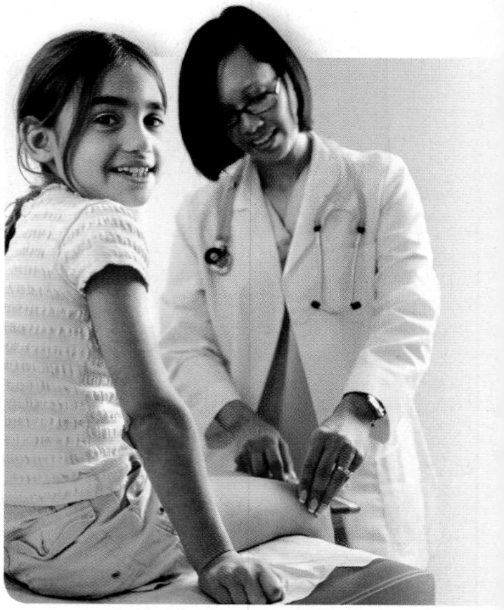

To use inductive reasoning:

Step 1 Observe many examples.

Step 2 Look for a pattern.

Step 3 Make a conjecture.

Step 4 Check the conjecture.

Step 5 Discover a likely conclusion.

With **deductive reasoning**, you come to a conclusion by accepting facts. The results of the tests ordered by the doctor may support the original diagnosis or lead to a different conclusion. This is an example of deductive reasoning. There is no conjecturing involved. Consider the two statements below.

1) If the strep test is positive, then the patient has strep throat.

2) Jolene tested positive for strep.

If these two statements are accepted as facts, then the obvious conclusion is that Jolene has strep throat. This is an example of deductive reasoning.

Exercises

1. Explain the difference between *inductive* and *deductive* reasoning. Then give an example of each.

2. When a detective reaches a conclusion about the height of a suspect from the distance between footprints, what kind of reasoning is being used? Explain.

3. When you examine a finite number of terms in a sequence of numbers and decide that it is an arithmetic sequence, what kind of reasoning are you using? Explain.

4. Suppose you have found the common difference for an arithmetic sequence based on analyzing a finite number of terms, what kind of reasoning do you use to find the 100th term in the sequence?

5. CCSS **PERSEVERANCE**

a. Copy and complete the table.

3^1	3^2	3^3	3^4	3^5	3^6	3^7	3^8	3^9
3	9	27						

b. Write the sequence of numbers representing the numbers in the ones place.

c. Find the number in the ones place for the value of 3^{100}. Explain your reasoning. State the type of reasoning that you used.

LESSON 3-6 Proportional and Nonproportional Relationships

∴ Then	∴ Now	∴ Why?

∴ Then

- You recognized arithmetic sequences and related them to linear functions.

∴ Now

1. Write an equation for a proportional relationship.
2. Write an equation for a nonproportional relationship.

∴ Why?

- Heather is planting flats of flowers. The table shows the number of flowers that she has planted and the amount of time that she has been working in the garden.

Number of flowers planted (p)	1	6	12	18
Number of minutes working (f)	5	30	60	90

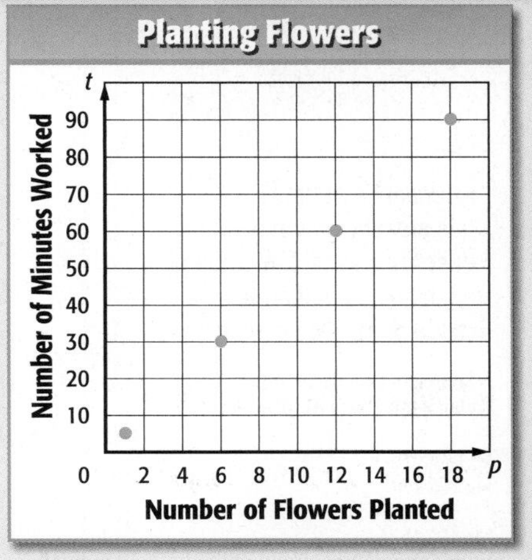

The relationship between the flowers planted and the time that Heather worked in minutes can be graphed. Let p represent the number of flowers planted. Let t represent the number of minutes that Heather has worked.

When the ordered pairs are graphed, they form a linear pattern. This pattern can be described by an equation.

CCSS Common Core State Standards

Content Standards

F.LE.1b Recognize situations in which one quantity changes at a constant rate per unit interval relative to another.

F.LE.2 Construct linear and exponential functions, including arithmetic and geometric sequences, given a graph, a description of a relationship, or two input-output pairs (include reading these from a table).

Mathematical Practices

1 Make sense of problems and persevere in solving them.

7 Look for and make use of structure.

1 Proportional Relationships

If the relationship between the domain and range of a relation is linear, the relationship can be described by a linear equation. If the equation is of the form $y = kx$, then the relationship is proportional. In a proportional relationship, the graph will pass through (0, 0). So, direct variations are proportional relationships.

KeyConcept Proportional Relationship

Words A relationship is proportional if its equation is of the form $y = kx$, $k \neq 0$. The graph passes through (0, 0).

Example $y = 3x$

x	0	1	2	3	4
y	0	3	6	9	12

The ratio of the value of x to the value of y is constant when $x \neq 0$.

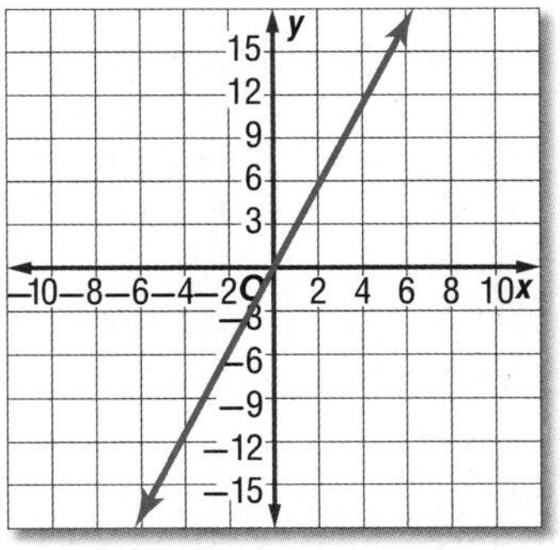

Proportional relationships are useful when analyzing real-world data. The pattern can be described using a table, a graph, and an equation.

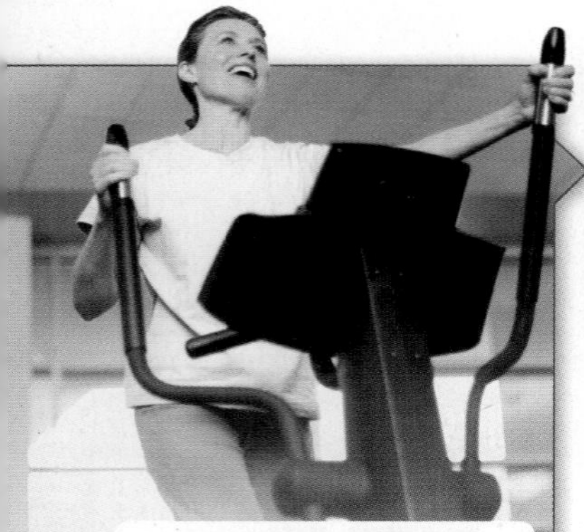

Real-World Example 1 Proportional Relationships

BONUS PAY Marcos is a personal trainer at a gym. In addition to his salary, he receives a bonus for each client he sees.

Number of Clients	1	2	3	4	5
Bonus Pay ($)	45	90	135	180	225

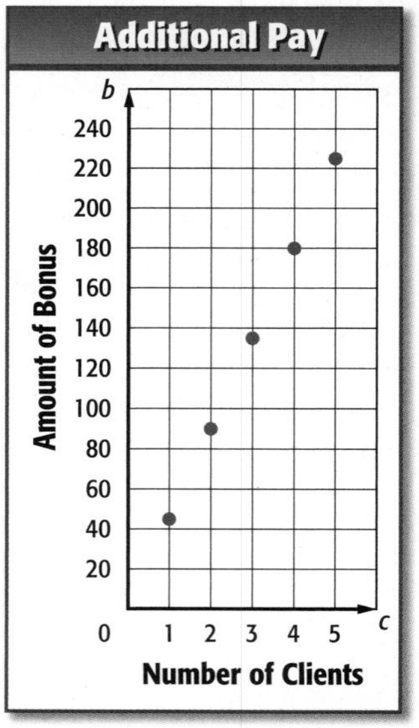

a. Graph the data. What can you deduce from the pattern about the relationship between the number of clients and the bonus pay?

The graph demonstrates a linear relationship between the number of clients and the bonus pay.

The graph also passes through the point (0, 0) because when Marcos sees 0 clients, he does not receive any bonus money. Therefore, the relationship is proportional.

b. Write an equation to describe this relationship.

Look for a pattern that can be described in an equation.

+1 +1 +1 +1

Number of Clients	1	2	3	4	5
Bonus Pay ($)	45	90	135	180	225

+45 +45 +45 +45

The difference between the values for the number of clients c is 1. The difference in the values for the bonus pay b is 45. This suggests that the k-value is $\frac{45}{1}$ or 45. So the equation is $b = 45c$. You can check this equation by substituting values for c into the equation.

CHECK If $c = 1$, then $b = 45(1)$ or 45. ✓
If $c = 5$, then $b = 45(5)$ or 225. ✓

c. Use this equation to predict the amount of Marcos's bonus if he sees 8 clients.

$b = 45c$ Original equation
$= 45(8)$ or 360 $c = 8$

Marcos will receive a bonus of $360 if he sees 8 clients.

Real-WorldLink

Attendance at fitness clubs has steadily grown over the past fifteen years. Members' ages are expanding to a range of 15–34 on average.

Source: International Health, Raquet, and Sportsclub Association

StudyTip

CCSS Structure Look for a pattern that shows a constant rate of change between the terms.

GuidedPractice

1. CHARITY A professional soccer team is donating money to a local charity for each goal they score.

Number of Goals	1	2	3	4	5
Donation ($)	75	150	225	300	375

A. Graph the data. What can you deduce from the pattern about the relationship between the number of goals and the money donated?

B. Write an equation to describe this relationship.

C. Use this equation to predict how much money will be donated for 12 goals.

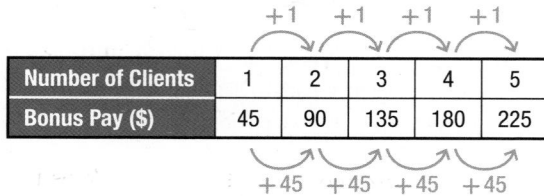

2 Nonproportional Relationships

Some linear equations can represent a nonproportional relationship. If the ratio of the value of x to the value of y is different for select ordered pairs that are on the line, the equation is nonproportional and the graph will not pass through (0, 0).

Example 2 Nonproportional Relationships

Write an equation in function notation for the graph.

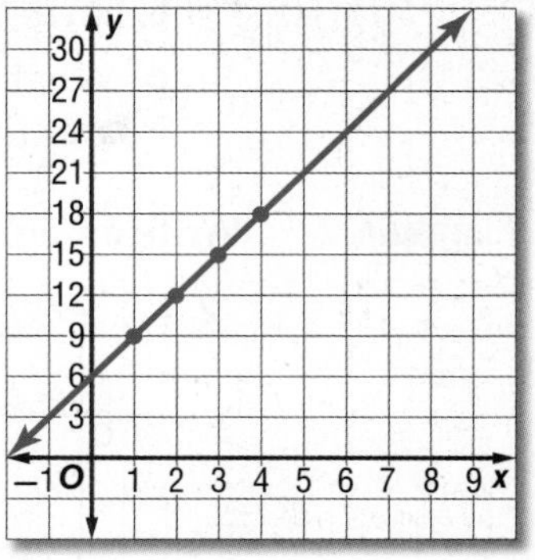

Understand You are asked to write an equation of the relation that is graphed in function notation.

Plan Find the difference between the x-values and the difference between the y-values.

Solve Select points from the graph and place them in a table.

x	1	2	3	4
y	9	12	15	18

x: +1, +1, +1; y: +3, +3, +3

Notice that $\frac{1}{9} \neq \frac{2}{12} \neq \frac{3}{15} \neq \frac{4}{18}$.

StudyTip

Graphs of Lines A value added to or subtracted from one side of the equation $y = ax$ will cause a shift along the y-axis for the graph of the line.

The difference between the x-values is 1, while the difference between the y-values is 3. This suggests that $y = 3x$ or $f(x) = 3x$.

If $x = 1$, then $y = 3(1)$ or 3. But the y-value for $x = 1$ is 9. Let's try some other values and see if we can detect a pattern.

x	1	2	3	4
$3x$	3	6	9	12
y	9	12	15	18

y is always 6 more than $3x$.

This pattern shows that 6 should be added to one side of the equation. Thus, the equation is $y = 3x + 6$ or $f(x) = 3x + 6$.

Check Compare the ordered pairs from the table to the graph. The points correspond. ✓

GuidedPractice

2. Write an equation in function notation for the relation shown in the table.

A.

x	1	2	3	4
y	3	2	1	0

B. Write an equation in function notation for the graph.

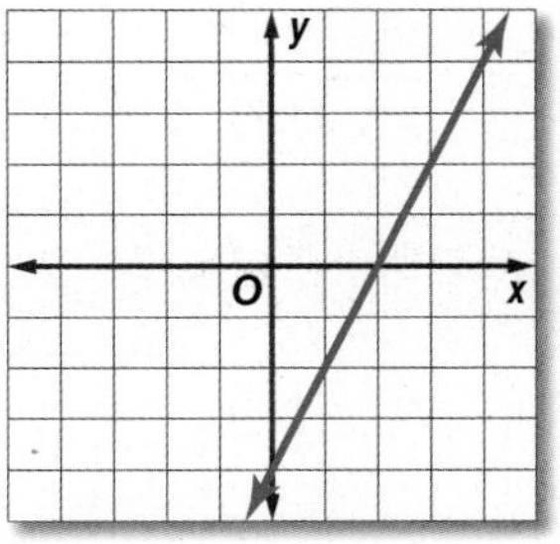

Check Your Understanding

Example 1

1. **GEOMETRY** The table shows the perimeter of a square with sides of a given length.

Side Length (in.)	1	2	3	4	5
Perimeter (in.)	4	8	12	16	20

a. Graph the data.

b. Write an equation to describe the relationship.

c. What conclusion can you make regarding the relationship between the side and the perimeter?

Example 2 **Write an equation in function notation for each relation.**

2.
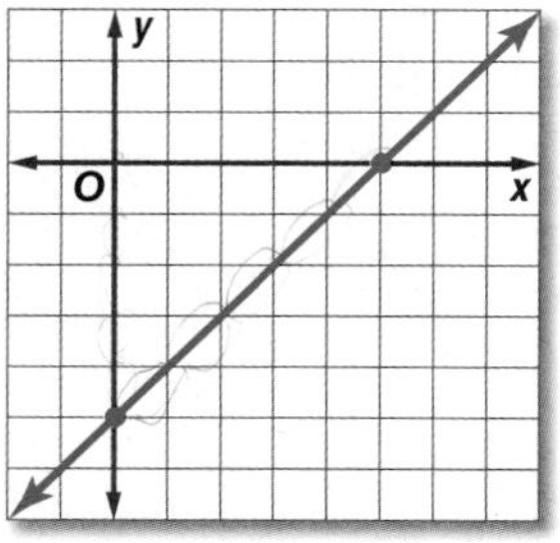

3.
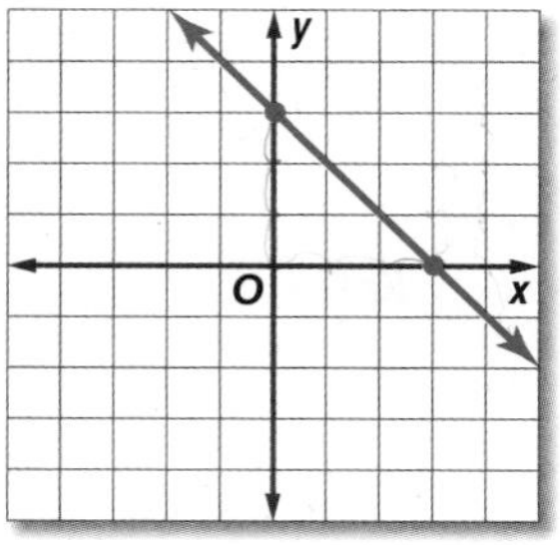

Practice and Problem Solving

Example 1

4. **CCSS STRUCTURE** The table shows the pages of comic books read.

Books Read	1	2	3	4	5
Pages Read	35	70	105	140	175

a. Graph the data.

b. Write an equation to describe the relationship.

c. Find the number of pages read if 8 comic books were read.

Example 2 **Write an equation in function notation for each relation.**

5.
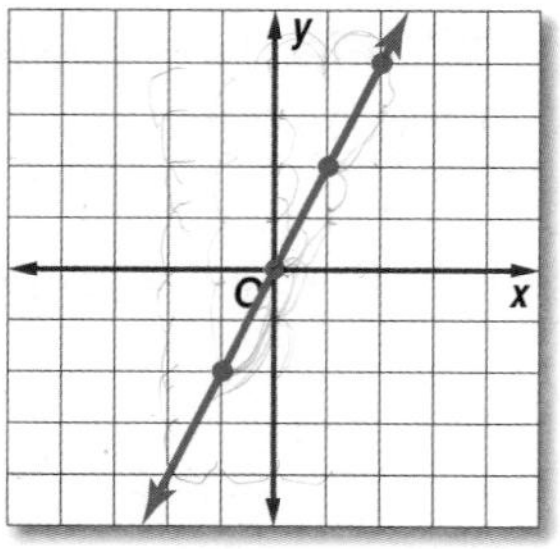

6.
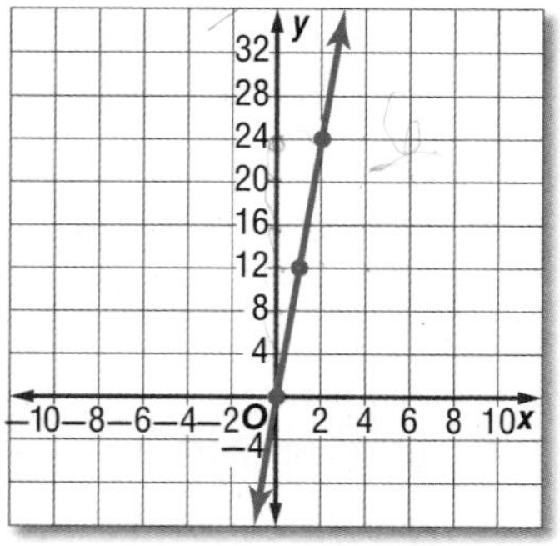

7.
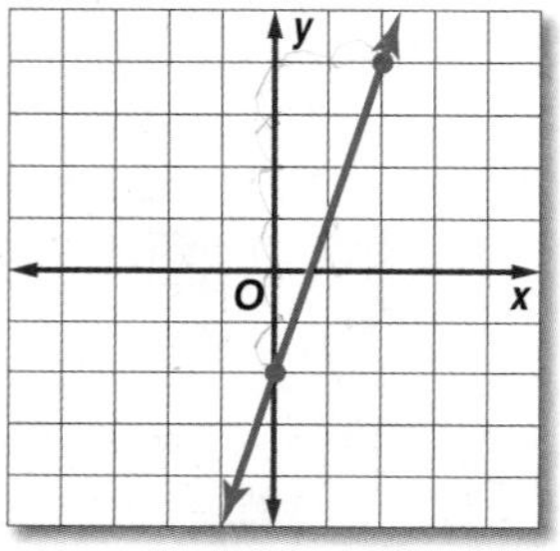

8.
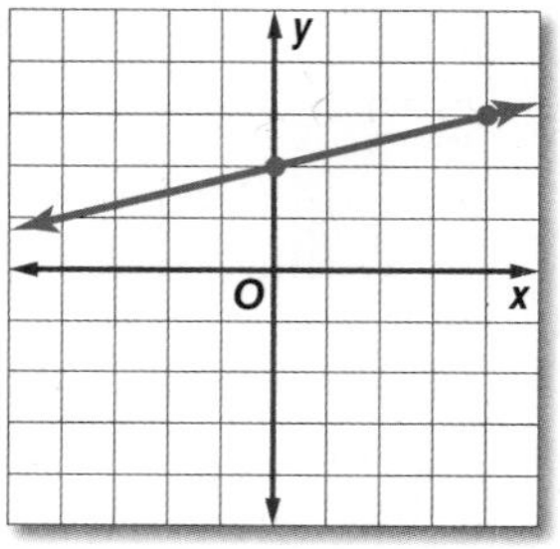

For each arithmetic sequence, determine the related function. Then determine if the function is *proportional* or *nonproportional*. Explain.

9. 0, 3, 6, ...

10. $-4, 0, 4, \ldots$

11. BOWLING Marielle is bowling with her friends. The table shows prices for renting a pair of shoes and bowling. Write an equation to represent the total price y if Marielle buys x games.

Games Bowled	Total Price ($)
2	7.00
4	11.50
6	16.00
8	20.50

12. SNOWFALL The total snowfall each hour of a winter snowstorm is shown in the table below.

Hour	1	2	3	4
Inches of Snowfall	1.65	3.30	4.95	6.60

a. Write an equation to fit the data in the table.

b. Describe the relationship between the hour and inches of snowfall.

13. FUNDRAISER The Cougar Pep Squad wants to sell T-shirts in the bookstore for the spring dance. The cost in dollars to order T-shirts in their school colors is represented by the equation $C = 2t + 3$.

a. Make a table of values that represents this relationship.

b. Rewrite the equation in function notation.

c. Graph the function.

d. Describe the relationship between the number of T-shirts and the cost.

H.O.T. Problems Use Higher-Order Thinking Skills

14. CCSS CRITIQUE Quentin thinks that $f(x)$ and $g(x)$ are both proportional. Claudia thinks they are not proportional. Is either of them correct? Explain your reasoning.

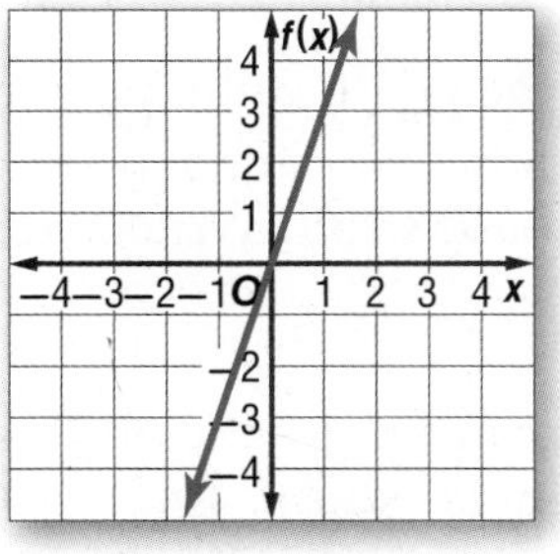

x	$g(x)$
−2	−7
−1	−4
0	−1
1	2
2	5

15. OPEN ENDED Create an arithmetic sequence in which the first term is 4. Explain the pattern that you used. Write an equation that represents your sequence.

16. CHALLENGE Describe how inductive reasoning can be used to write an equation from a pattern.

17. REASONING A **counterexample** is a specific case that shows that a statement is false. Provide a counterexample to the following statement. *The related function of an arithmetic sequence is always proportional.* Explain your reasoning.

18. WRITING IN MATH Compare and contrast proportional relationships with nonproportional relationships.

Standardized Test Practice

19. What is the slope of a line that contains the point (1, −5) and has the same y-intercept as $2x - y = 9$?

A −9 **C** 2

B −7 **D** 4

20. SHORT RESPONSE $\triangle FGR$ is an isosceles triangle. What is the measure of $\angle G$?

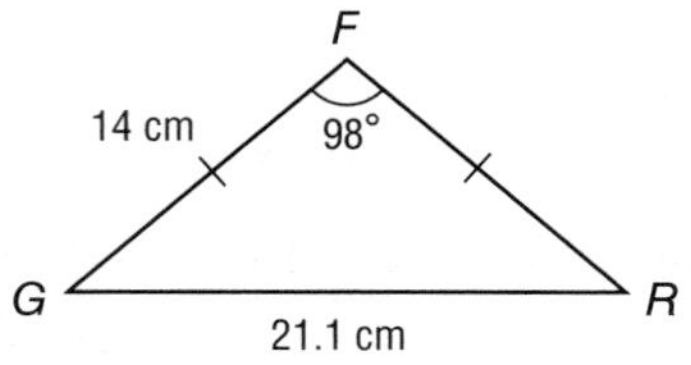

21. Luis deposits \$25 each week into a savings account from his part-time job. If he has \$350 in savings now, how much will he have in 12 weeks?

F \$600 **H** \$650

G \$625 **J** \$675

22. GEOMETRY Omar and Mackenzie want to build a pulley system by attaching one end of a rope to their 8-foot-tall tree house and anchoring the other end to the ground 28 feet away from the base of the tree house. How long, to the nearest foot, does the piece of rope need to be?

A 26 ft **C** 28 ft

B 27 ft **D** 29 ft

Spiral Review

Find the next three terms in each sequence. (Lesson 3-5)

23. 3, 13, 23, 33, …

24. −2, −1.4, −0.8, −0.2, …

25. $\frac{3}{4}, \frac{7}{8}, 1, \frac{9}{8}, \ldots$

Suppose y varies directly as x. Write a direct variation equation that relates x and y. Then solve. (Lesson 3-4)

26. If $y = 45$ when $x = 9$, find y when $x = 7$.

27. If $y = -7$ when $x = -1$, find x when $y = -84$.

28. GENETICS About $\frac{2}{25}$ of the male population in the world cannot distinguish red from green. If there are 14 boys in the ninth grade who cannot distinguish red from green, about how many ninth-grade boys are there in all? Write and solve an equation to find the answer. (Lesson 2-2)

29. GEOMETRY The volume V of a cone equals one third times the product of π, the square of the radius r of the base, and the height h. (Lesson 2-1)

a. Write the formula for the volume of a cone.

b. Find the volume of a cone if r is 10 centimeters and h is 30 centimeters.

Skills Review

Solve each equation for y.

30. $3x = y + 7$

31. $2y = 6x - 10$

32. $9y + 2x = 12$

Graph each equation.

33. $y = x - 8$

34. $x - y = -4$

35. $2x + 4y = 8$

Study Guide and Review

Study Guide

KeyConcepts

Graphing Linear Equations (Lesson 3-1)

- The standard form of a linear equation is $Ax + By = C$, where $A \geq 0$, A and B are not both zero, and A, B, and C are integers whose greatest common factor is 1.

Solving Linear Equations by Graphing (Lesson 3-2)

- Values of x for which $f(x) = 0$ are called zeros of the function f. A zero of a function is located at an x-intercept of the graph of the function.

Rate of Change and Slope (Lesson 3-3)

- If x is the independent variable and y is the dependent variable, then rate of change equals

$$\frac{\text{change in } y}{\text{change in } x}.$$

- The slope of a line is the ratio of the rise to the run.

$$m = \frac{y_2 - y_1}{x_2 - x_1}$$

Direct Variation (Lesson 3-4)

- A direct variation is described by an equation of the form $y = kx$, where $k \neq 0$.

Arithmetic Sequences (Lesson 3-5)

- The nth term a_n of an arithmetic sequence with first term a_1 and common difference d is given by $a_n = a_1 + (n - 1)d$, where n is a positive integer.

Proportional and Nonproportional Relationships (Lesson 3-6)

- In a proportional relationship, the graph will pass through (0, 0).
- In a nonproportional relationship, the graph will *not* pass through (0, 0)

FOLDABLES StudyOrganizer

Be sure the Key Concepts are noted in your Foldable.

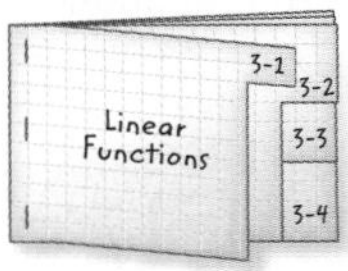

KeyVocabulary

arithmetic sequence (p. 189)
common difference (p. 189)
constant (p. 155)
constant of variation (p. 182)
deductive reasoning (p. 196)
direct variation (p. 182)
inductive reasoning (p. 196)
linear equation (p. 155)
linear function (p. 163)
rate of change (p. 172)
root (p. 163)
sequence (p. 189)
slope (p. 174)
standard form (p. 155)
terms of the sequence (p. 189)
x-intercept (p. 156)
y-intercept (p. 156)
zero of a function (p. 163)

VocabularyCheck

State whether each sentence is *true* or *false*. If *false*, replace the underlined word or number to make a true sentence.

1. The x-coordinate of the point at which the graph of an equation crosses the x-axis is an x-intercept.
2. A linear equation is an equation of a line.
3. The difference between successive terms of an arithmetic sequence is the constant of variation.
4. The regular form of a linear equation is $Ax + By = C$.
5. Values of x for which $f(x) = 0$ are called zeros of the function f.
6. Any two points on a nonvertical line can be used to determine the slope.
7. The slope of the line $y = 5$ is 5.
8. The graph of any direct variation equation passes through (0, 1).
9. A ratio that describes, on average, how much one quantity changes with respect to a change in another quantity is a rate of change.
10. In the linear equation $4x + 3y = 12$, the constant term is 12.

CHAPTER 3

Study Guide and Review *Continued*

Lesson-by-Lesson Review

3-1 Graphing Linear Equations

Find the x-intercept and y-intercept of the graph of each linear function.

11.

x	y
−8	0
−4	3
0	6
4	9
8	12

12.

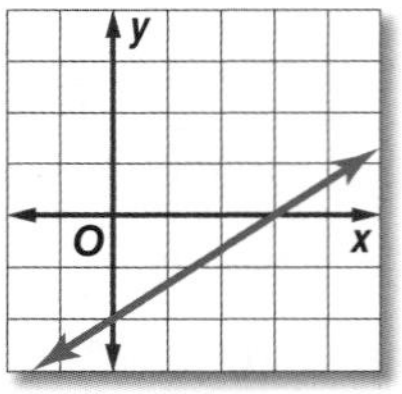

Graph each equation.

13. $y = -x + 2$

14. $x + 5y = 4$

15. $2x - 3y = 6$

16. $5x + 2y = 10$

17. **SOUND** The distance d in kilometers that sound waves travel through water is given by $d = 1.6t$, where t is the time in seconds.

a. Make a table of values and graph the equation.

b. Use the graph to estimate how far sound can travel through water in 7 seconds.

Example 1

Graph $3x - y = 4$ by using the x- and y-intercepts.

Find the x-intercept.

$3x - y = 4$

$3x - 0 = 4$ Let $y = 0$.

$3x = 4$

$x = \frac{4}{3}$

Find the y-intercept.

$3x - y = 4$

$3(0) - y = 4$ Let $x = 0$.

$-y = 4$

$y = -4$

x-intercept: $\frac{4}{3}$

y-intercept: -4

The graph intersects the x-axis at $\left(\frac{4}{3}, 0\right)$ and the y-axis at $(0, -4)$. Plot these points. Then draw the line through them.

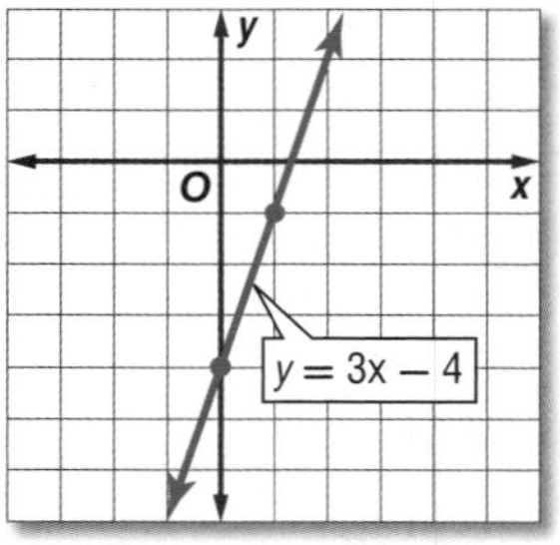

3-2 Solving Linear Equations by Graphing

Find the root of each equation.

18. $0 = 2x + 8$

19. $0 = 4x - 24$

20. $3x - 5 = 0$

21. $6x + 3 = 0$

Solve each equation by graphing.

22. $0 = 16 - 8x$

23. $0 = 21 + 3x$

24. $-4x - 28 = 0$

25. $25x - 225 = 0$

26. **FUNDRAISING** Sean's class is selling boxes of popcorn to raise money for a class trip. Sean's class paid $85 for the popcorn, and they are selling each box for $1. The function $y = x - 85$ represents their profit y for each box of popcorn sold x. Find the zero and describe what it means in this situation.

Example 2

Solve $3x + 1 = -2$ by graphing.

The first step is to find the related function.

$3x + 1 = -2$ Original equation

$3x + 1 + 2 = -2 + 2$ Add 2 to each side.

$3x + 3 = 0$ Simplify.

The related function is $y = 3x + 3$.

The graph intersects the x-axis at -1. So, the solution is -1.

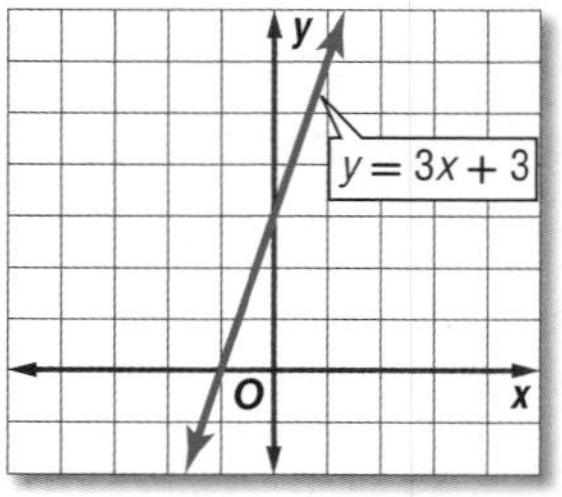

3-3 Rate of Change and Slope

Find the rate of change represented in each table or graph.

27.

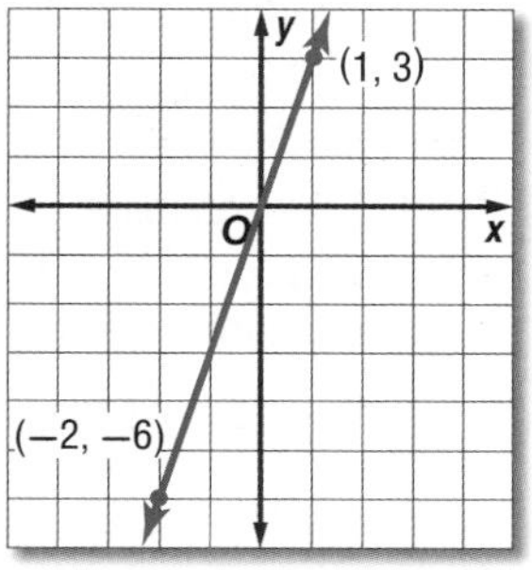

28.

x	y
−2	−3
0	−3
4	−3
12	−3

Find the slope of the line that passes through each pair of points.

29. (0, 5), (6, 2)

30. (−6, 4), (−6, −2)

31. **PHOTOS** The average cost of online photos decreased from \$0.50 per print to \$0.15 per print between 2002 and 2009. Find the average rate of change in the cost. Explain what it means.

Example 3

Find the slope of the line that passes through (0, −4) and (3, 2).

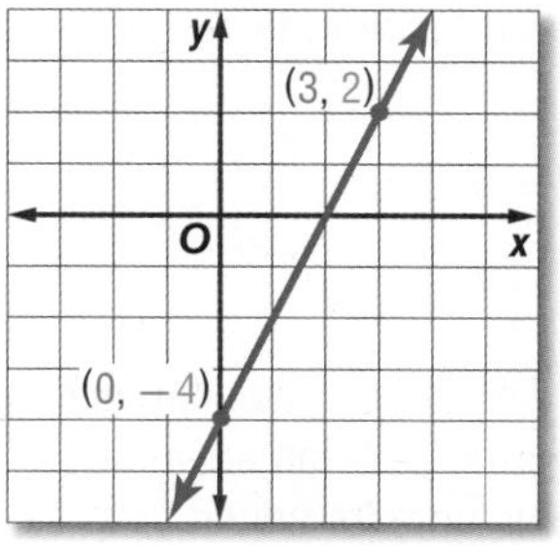

Let (0, −4) = (x_1, y_1) and (3, 2) = (x_2, y_2).

$m = \frac{y_2 - y_1}{x_2 - x_1}$ Slope formula

$= \frac{2 - (-4)}{3 - 0}$ $x_1 = 0, x_2 = 3, y_1 = -4, y_2 = 2$

$= \frac{6}{3}$ or 2 Simplify.

3-4 Direct Variation

Graph each equation.

32. $y = x$

33. $y = \frac{4}{3}x$

34. $y = -2x$

Suppose *y* varies directly as *x*. Write a direct variation equation that relates *x* and *y*. Then solve.

35. If $y = 15$ when $x = 2$, find y when $x = 8$.

36. If $y = -6$ when $x = 9$, find x when $y = -3$.

37. If $y = 4$ when $x = -4$, find y when $x = 7$.

38. **JOBS** Suppose you earn \$127 for working 20 hours.

a. Write a direct variation equation relating your earnings to the number of hours worked.

b. How much would you earn for working 35 hours?

Example 4

Suppose *y* varies directly as *x*, and $y = -24$ when $x = 8$.

a. Write a direct variation equation that relates *x* and *y*.

$y = kx$ Direct variation equation

$-24 = k(8)$ Substitute −24 for *y* and 8 for *x*.

$\frac{-24}{8} = \frac{k(8)}{8}$ Divide each side by 8.

$-3 = k$ Simplify.

So, the direct variation equation is $y = -3x$.

b. Use the direct variation equation to find *x* when $y = -18$.

$y = -3x$ Direct variation equation

$-18 = -3x$ Replace *y* with −18.

$\frac{-18}{-3} = \frac{-3x}{-3}$ Divide each side by −3.

$6 = x$ Simplify.

Therefore, $x = 6$ when $y = -18$.

Study Guide and Review *Continued*

3-5 Arithmetic Sequences as Linear Functions

Find the next three terms of each arithmetic sequence.

39. 6, 11, 16, 21, ...

40. 1.4, 1.2, 1.0, ...

Write an equation for the *n*th term of each arithmetic sequence.

41. $a_1 = 6, d = 5$

42. 28, 25, 22, 19, ...

43. SCIENCE The table shows the distance traveled by sound in water. Write an equation for this sequence. Then find the time for sound to travel 72,300 feet.

Time (s)	1	2	3	4
Distance ft)	4820	9640	14,460	19,280

Example 5

Find the next three terms of the arithmetic sequence 10, 23, 36, 49, ...

Find the common difference.

10 23 36 49
+13 +13 +13

So, $d = 13$.

Add 13 to the last term of the sequence. Continue adding 13 until the next three terms are found.

49 62 75 88
+13 +13 +13

The next three terms are 62, 75, and 88.

3-6 Proportional and Nonproportional Relationships

44. Write an equation in function notation for this relation.

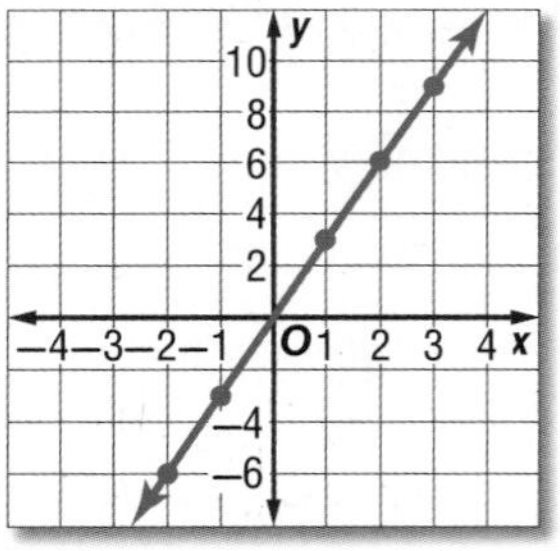

45. ANALYZE TABLES The table shows the cost of picking your own strawberries at a farm.

Number of Pounds	1	2	3	4
Total Cost ($)	1.25	2.50	3.75	5.00

a. Graph the data.

b. Write an equation in function notation to describe this relationship.

c. How much would it cost to pick 6 pounds of strawberries?

Example 6

Write an equation in function notation for this relation.

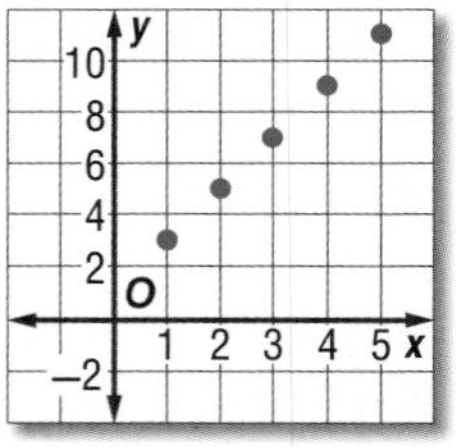

Make a table of ordered pairs for several points on the graph.

x	1	2	3	4	5
y	3	5	7	9	11

The difference in *y*-values is twice the difference of *x* values. This suggests that $y = 2x$. However, $3 \neq 2(1)$. Compare the values of *y* to the values of 2*x*.

x	1	2	3	4	5
2x	2	4	6	8	10
y	3	5	7	9	11

The difference between *y* and 2*x* is always 1. So the equation is $y = 2x + 1$. Since this relation is also a function, it can be written as $f(x) = 2x + 1$.

Practice Test

1. **TEMPERATURE** The equation to convert Celsius temperature C to Kelvin temperature K is shown.

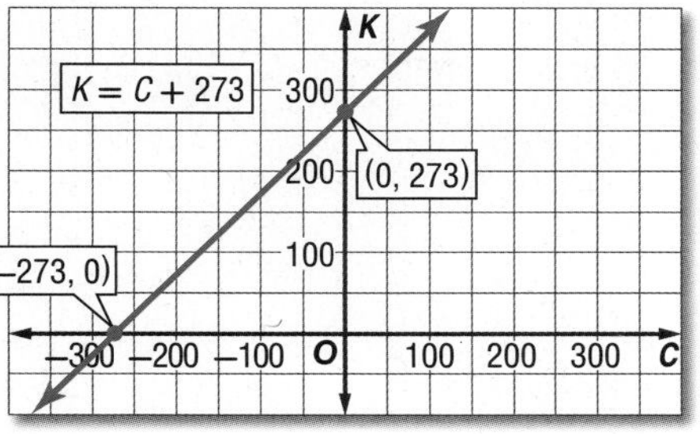

   a. State the independent and dependent variables. Explain.

   b. Determine the C- and K-intercepts and describe what the intercepts mean in this situation.

Graph each equation.

2. $y = x + 2$
3. $y = 4x$
4. $x + 2y = -1$
5. $-3x = 5 - y$

Solve each equation by graphing.

6. $4x + 2 = 0$
7. $0 = 6 - 3x$
8. $5x + 2 = -3$
9. $12x = 4x + 16$

Find the slope of the line that passes through each pair of points.

10. $(5, 8), (-3, 7)$
11. $(5, -2), (3, -2)$
12. $(-4, 7), (8, -1)$
13. $(6, -3), (6, 4)$

14. **MULTIPLE CHOICE** Which is the slope of the linear function shown in the graph?

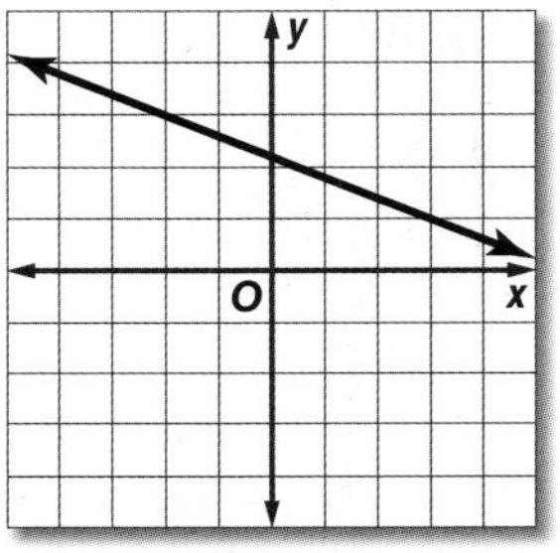

   A $-\frac{5}{2}$ C $\frac{2}{5}$

   B $-\frac{2}{5}$ D $\frac{5}{2}$

Suppose y varies directly as x. Write a direct variation equation that relates x and y. Then solve.

15. If $y = 6$ when $x = 9$, find x when $y = 12$.
16. When $y = -8$, $x = 8$. What is x when $y = -6$?
17. If $y = -5$ when $x = -2$, what is y when $x = 14$?
18. If $y = 2$ when $x = -12$, find y when $x = -4$.

19. **BIOLOGY** The number of pints of blood in a human body varies directly with the person's weight. A person who weighs 120 pounds has about 8.4 pints of blood in his or her body.

   a. Write and graph an equation relating weight and amount of blood in a person's body.

   b. Predict the weight of a person whose body holds 12 pints of blood.

Find the next three terms of each arithmetic sequence.

20. 0, −15, −30, −45, −60, …
21. 5, 8, 11, 14, …

Determine whether each sequence is an arithmetic sequence. If it is, state the common difference.

22. −40, −32, −24, −16, …
23. 0.75, 1.5, 3, 6, 12, …
24. 5, 17, 29, 41, …

25. **MULTIPLE CHOICE** In each figure, only one side of each regular pentagon is shared with another pentagon. The length of each side is 1 centimeter. If the pattern continues, what is the perimeter of a figure that has 6 pentagons?

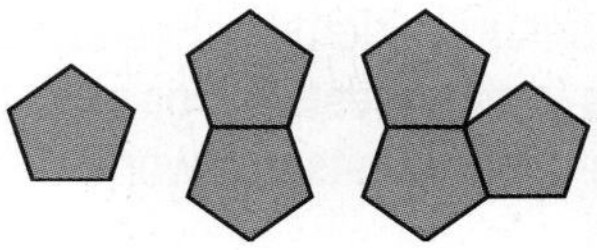

   F 30 cm

   G 25 cm

   H 20 cm

   J 15 cm

Preparing for Standardized Tests

Reading Math Problems

The first step to solving any math problem is to read the problem. When reading a math problem to get the information you need to solve, it is helpful to use special reading strategies.

Strategies for Reading Math Problems

Step 1

Read the problem quickly to gain a general understanding of it.

- **Ask yourself:** "What do I know?" "What do I need to find out?"
- **Think:** "Is there enough information to solve the problem? Is there extra information?"
- **Highlight:** If you are allowed to write in your test booklet, underline or highlight important information. Cross out any information you don't need.

Step 2

Reread the problem to identify relevant facts.

- **Analyze:** Determine how the facts are related.
- **Key Words:** Look for keywords to solve the problem.
- **Vocabulary:** Identify mathematical terms. Think about the concepts and how they are related.
- **Plan:** Make a plan to solve the problem.
- **Estimate:** Quickly estimate the answer.

Step 3

Identify any obvious wrong answers.

- **Eliminate:** Eliminate any choices that are very different from your estimate.
- **Units of Measure:** Identify choices that are possible answers based on the units of measure in the question. For example, if the question asks for area, only answers in square units will work.

Step 4

Look back after solving the problem.

Check: Make sure you have answered the question.

Ableimages/Digital Vision/Getty Images

Standardized Test Example

Read the problem. Identify what you need to know. Then use the information in the problem to solve.

Jamal, Gina, Lisa, and Renaldo are renting a car for a road trip. The cost of renting the car is given by the function $C = 12.5 + 21d$, where C is the total cost for renting the car for d days. What does the slope of the function represent?

A number of people

B cost per day

C number of days

D miles per gallon

Read the problem carefully. The number of people going on the trip is not needed information. You need to know what the slope of the function represents.

Slope is a ratio. The word "per" in answers B and D imply that they are both ratios. Since choices A and C are not ratios, eliminate them.

The problem says that C represents the cost of renting the car. So the slope cannot represent the miles per gallon of the car. The slope must represent the cost per day.

The correct answer is B.

Exercises

Read each problem. Identify what you need to know. Then use the information in the problem to solve.

1. What does the x-intercept mean in the context of the situation given below?

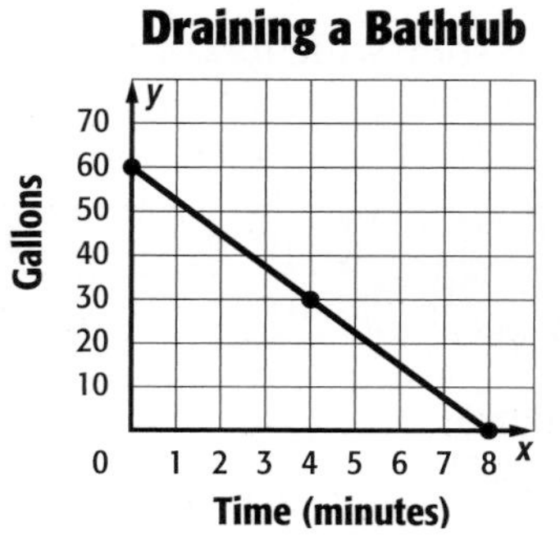

A amount of time needed to drain the bathtub

B number of gallons in the tub when the drain plug is pulled

C number of gallons in the tub after x minutes

D amount of water drained each minute

2. The amount of money raised by a charity carwash varies directly as the number of cars washed. When 11 cars are washed, $79.75 is raised. How many cars must be washed to raise $174.00?

F 10 cars

G 16 cars

H 22 cars

J 24 cars

3. The function $C = 25 + 0.45(x - 450)$ represents the cost of a monthly cell phone bill, when x minutes are used. Which statement best represents the formula for the cost of the bill?

A The cost consists of a flat fee of $0.45 and $25 for each minute used over 450.

B The cost consists of a flat fee of $450 and $0.45 for each minute used over 25.

C The cost consists of a flat fee of $25 and $0.45 for each minute used over 450.

D The cost consists of a flat fee of $25 and $0.45 for each minute used.

CHAPTER 3

Standardized Test Practice

Cumulative, Chapters 1 through 3

Multiple Choice

Read each question. Then fill in the correct answer on the answer document provided by your teacher or on a sheet of paper.

1. Horatio is purchasing a computer cable for \$15.49. If the sales tax rate in his state is 5.25%, what is the total cost of the purchase?

A \$16.42 **C** \$15.73
B \$16.30 **D** \$15.62

2. What is the value of the expression below?

$$3^2 + 5^3 - 2^5$$

F 14 **H** 102
G 34 **J** 166

3. What is the slope of the linear function graphed below?

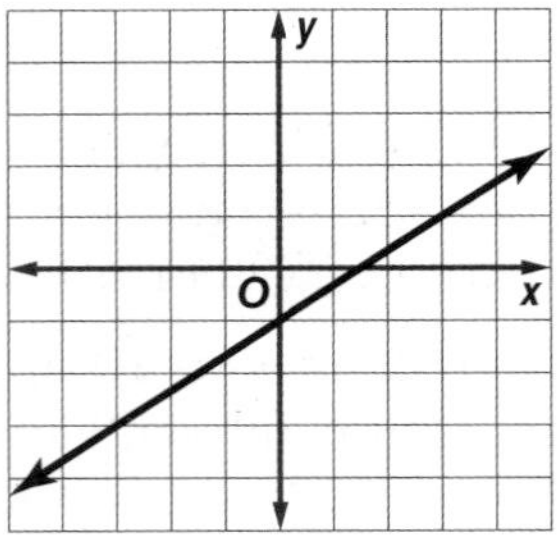

A $-\frac{1}{3}$ **C** $\frac{2}{3}$
B $\frac{1}{2}$ **D** $\frac{3}{2}$

4. Find the rate of change for the linear function represented in the table.

Hours Worked	1	2	3	4
Money Earned (\$)	5.50	11.00	16.50	22.00

F increase \$6.50/h
G increase \$5.50/h
H decrease \$5.50/h
J decrease \$6.50/h

5. Suppose that y varies directly as x, and $y = 14$ when $x = 4$. What is the value of y when $x = 9$?

A 25.5 **C** 29.5
B 27.5 **D** 31.5

6. Write an equation for the nth term of the arithmetic sequence shown below.

$$-2, 1, 4, 7, 10, 13, \ldots$$

F $a_n = 2n - 1$ **H** $a_n = 3n + 2$
G $a_n = 2n + 4$ **J** $a_n = 3n - 5$

7. The table shows the labor charges of an electrician for jobs of different lengths.

Number of Hours (n)	Labor Charges (c)
1	\$60
2	\$85
3	\$110
4	\$135

Which function represents the situation?

A $C(n) = 25n + 35$ **C** $C(n) = 35n + 25$
B $C(n) = 25n + 30$ **D** $C(n) = 35n + 40$

8. Find the value of x so that the figures have the same area.

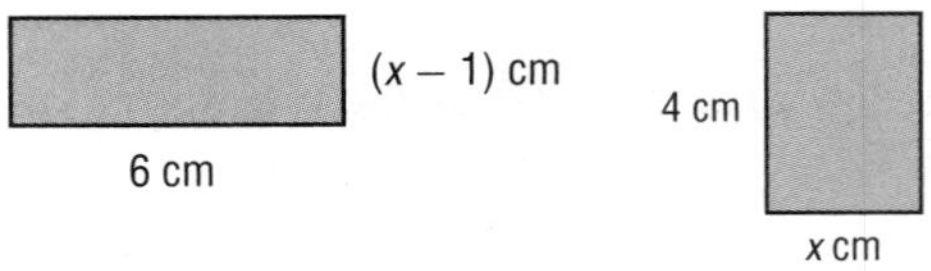

F 3 **H** 5
G 4 **J** 6

9. The table shows the total amount of rain during a storm. Write a formula to find out how much rain will fall after a given hour.

Hours (h)	1	2	3	4
Inches (n)	0.45	9.9	1.35	1.8

A $h = 0.45n$ **C** $h = 0.9n$
B $n = 0.45h$ **D** $h = 1.8n$

Test-Taking Tip

Question 3 You can *eliminate unreasonable answers* to multiple choice items. The line slopes up from left to right, so the slope is positive. Answer choice A can be eliminated.

Short Response/Gridded Response

Record your answers on the answer sheet provided by your teacher or on a sheet of paper.

10. The scale on a map is 1.5 inches = 6 miles. If two cities are 4 inches apart on the map, what is the actual distance between the cities?

11. Write a direct variation equation to represent the graph below.

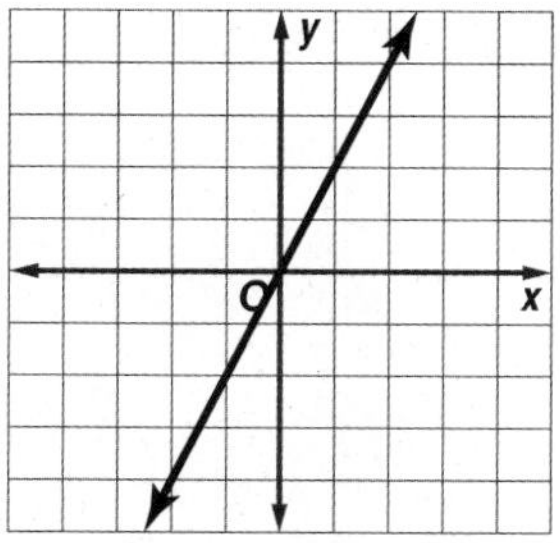

12. Justine bought a car for $18,500 and its value depreciated linearly. After 3 years, the value was $14,150. What is the amount of yearly depreciation?

13. GRIDDED RESPONSE Use the graph to determine the solution to the equation $-\frac{1}{3}x + 1 = 0$?

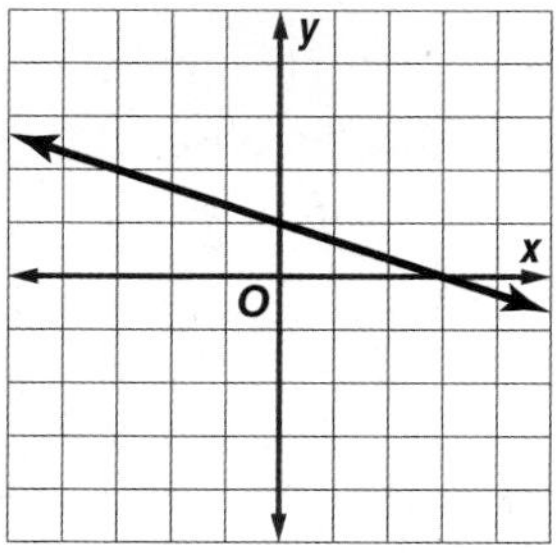

14. Write an expression that represents the total surface area (including the top and bottom) of a tower of n cubes each having a side length of s. (Do not include faces that cover each other.)

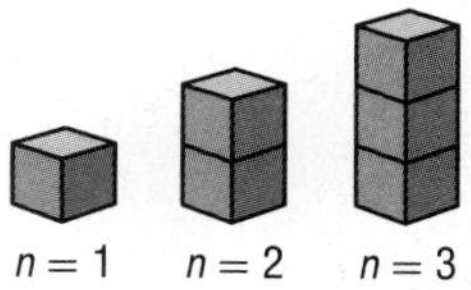

15. GRIDDED RESPONSE There are 120 members in the North Carolina House of Representatives. This is 70 more than the number of members in the North Carolina Senate. How many members are in the North Carolina Senate?

Extended Response

Record your answers on a sheet of paper. Show your work.

16. A hot air balloon was at a height of 60 feet above the ground when it began to ascend. The balloon climbed at a rate of 15 feet per minute.

a. Make a table that shows the height of the hot air balloon after climbing for 1, 2, 3, and 4 minutes.

b. Let t represent the time in minutes since the balloon began climbing. Write an algebraic equation for a sequence that can be used to find the height, h, of the balloon after t minutes.

c. Use your equation from part b to find the height, in feet, of the hot air balloon after climbing for 8 minutes.

Need ExtraHelp?

If you missed Question...	1	2	3	4	5	6	7	8	9	10	11	12	13	14	15	16
Go to Lesson...	2-7	1-2	3-3	3-3	3-4	3-5	3-6	2-4	3-4	2-6	3-4	3-3	3-2	0-10	2-1	3-5

CHAPTER 4

Equations of Linear Functions

Then

- You graphed linear functions.

Now

- In this chapter, you will:
  - Write and graph linear equations in various forms.
  - Use scatter plots and lines of fit, and write equations of best-fit lines using linear regression.
  - Find inverse linear functions.

Why?

- **TRAVEL** The number of trips people take changes from year to year. From the yearly data, patterns emerge. Rate of change can be applied to these data to determine a linear model. This can be used to predict the number of trips taken in future years.

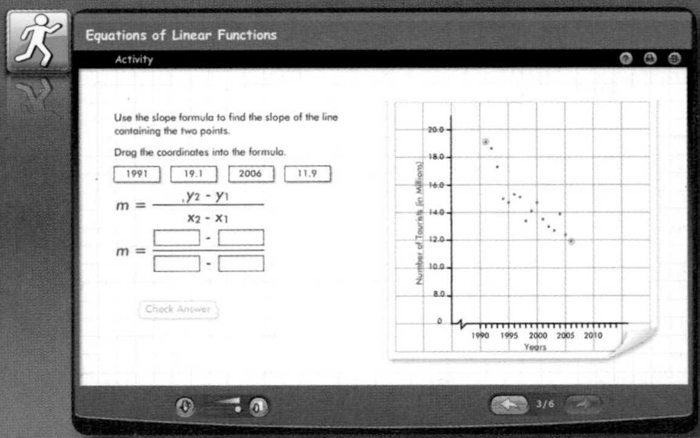

connectED.mcgraw-hill.com Your Digital Math Portal

Animation | Vocabulary | eGlossary | Personal Tutor | Virtual Manipulatives | Graphing Calculator | Audio | Foldables | Self-Check Practice | Worksheets

Get Ready for the Chapter

Diagnose Readiness | You have two options for checking prerequisite skills.

1 Textbook Option Take the Quick Check below. Refer to the Quick Review for help.

QuickCheck

Evaluate $3a^2 - 2ab + c$ for the values given.

1. $a = 2, b = 1, c = 5$
2. $a = -3, b = -2, c = 3$
3. $a = -1, b = 0, c = 11$
4. $a = 5, b = -3, c = -9$
5. **CAR RENTAL** The cost of renting a car is given by $49x + 0.3y$. Let x represent the number of days rented, and let y represent the number of miles driven. Find the cost for a five-day rental over 125 miles.

Solve each equation for the given variable.

6. $x + y = 5$ for y
7. $2x - 4y = 6$ for x
8. $y - 2 = x + 3$ for y
9. $4x - 3y = 12$ for x
10. **GEOMETRY** The formula for the perimeter of a rectangle is $P = 2w + 2\ell$, where w represents width and ℓ represents length. Solve for w.

Write the ordered pair for each point.

11. A
12. B
13. C
14. D
15. E
16. F

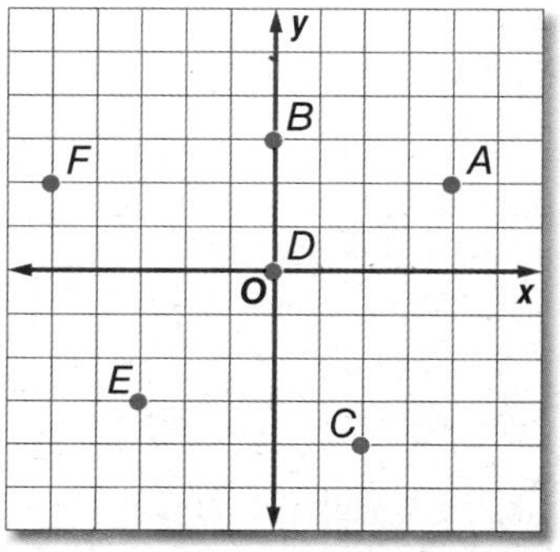

QuickReview

Example 1

Evaluate $2(m - n)^2 + 3p$ for $m = 5$, $n = 2$, and $p = -3$.

$2(m - n)^2 + 3p$	Original expression
$= 2(5 - 2)^2 + 3(-3)$	Substitute.
$= 2(3)^2 + 3(-3)$	Subtract.
$= 2(9) + 3(-3)$	Evaluate power.
$= 18 + (-9)$	Multiply.
$= 9$	Add.

Example 2

Solve $5x + 15y = 9$ for x.

$5x + 15y = 9$	Original equation
$5x + 15y - 15y = 9 - 15y$	Subtract $15y$ from each side.
$5x = 9 - 15y$	Simplify.
$\frac{5x}{5} = \frac{9 - 15y}{5}$	Divide each side by 5.
$x = \frac{9}{5} - 3y$	Simplify.

Example 3

Write the ordered pair for A.

Step 1 Begin at point A.

Step 2 Follow along a vertical line to the x-axis. The x-coordinate is -4.

Step 3 Follow along a horizontal line to the y-axis. The y-coordinate is 2.

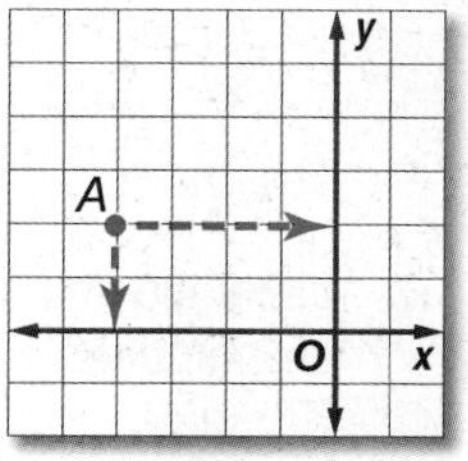

The ordered pair for point A is $(-4, 2)$.

2 Online Option Take an online self-check Chapter Readiness Quiz at connectED.mcgraw-hill.com.

Get Started on the Chapter

You will learn several new concepts, skills, and vocabulary terms as you study Chapter 4. To get ready, identify important terms and organize your resources. You may wish to refer to Chapter 0 to review prerequisite skills.

FOLDABLES® StudyOrganizer

Linear Functions Make this Foldable to help you organize your Chapter 4 notes about linear functions. Begin with one sheet of 11" by 17" paper.

1 **Fold** each end of the paper in about 2 inches.

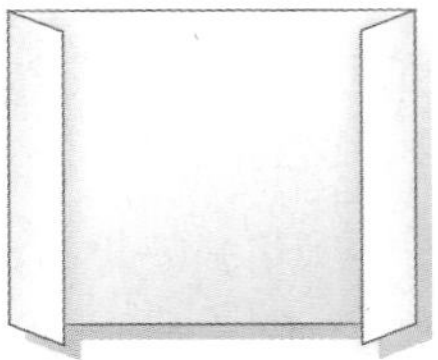

2 **Fold** along the width and the length. Unfold. Cut along the fold line from the top to the center.

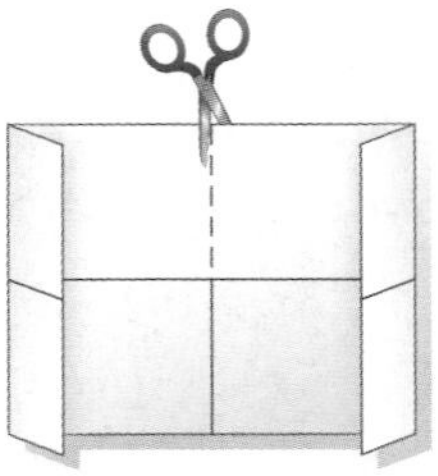

3 **Fold** the top flaps down. Then fold in half and turn to form a folder. Staple the flaps down to form pockets.

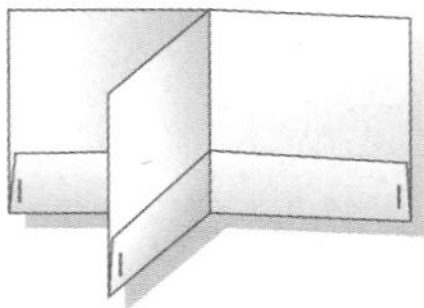

4 **Label** the front with the chapter title.

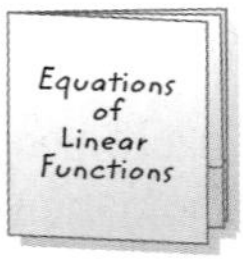

NewVocabulary

English		Español
slope-intercept form	p. 216	forma pendiente-intersección
linear extrapolation	p. 228	extrapolación lineal
point-slope form	p. 233	forma punto-pendiente
parallel lines	p. 239	rectas paralelas
perpendicular lines	p. 240	rectas perpendiculares
scatter plot	p. 247	gráfica de dispersión
line of fit	p. 248	recta de ajuste
linear interpolation	p. 249	interpolación lineal
best-fit line	p. 255	recta de ajuste óptimo
linear regression	p. 255	retroceso lineal
correlation coefficient	p. 255	coeficiente de correlación
median-fit line	p. 258	línea de mediana-ataque
inverse relation	p. 263	relación inversa
inverse function	p. 264	función inversa

ReviewVocabulary

coefficient coeficiente the numerical factor of a term

function función a relation in which each element of the domain is paired with exactly one element of the range

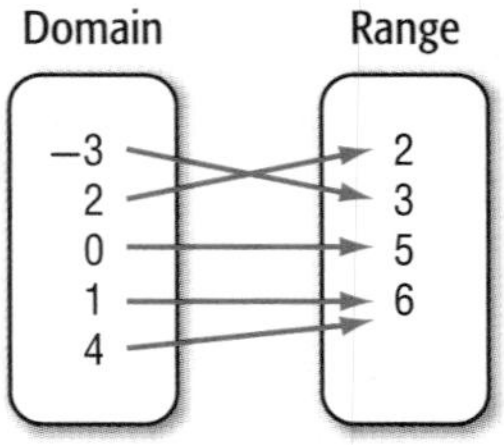

ratio razón a comparison of two numbers by division

Graphing Technology Lab
Investigating Slope-Intercept Form

Set Up the Lab

- Cut a small hole in a top corner of a plastic sandwich bag. Hang the bag from the end of the force sensor.
- Connect the force sensor to your data collection device.

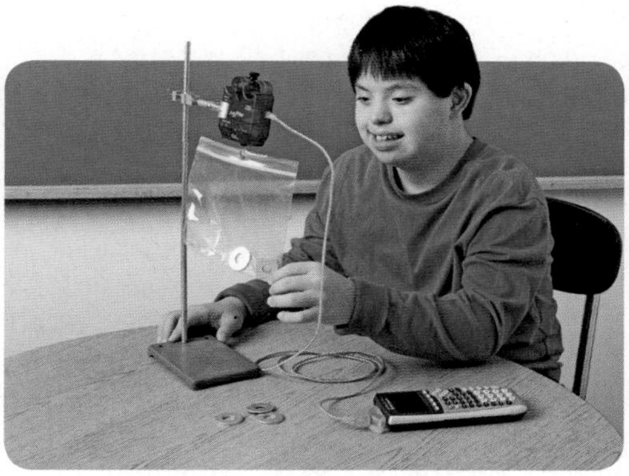

Activity Collect Data

Step 1 Use the sensor to collect the weight with 0 washers in the bag. Record the data pair in the calculator.

Step 2 Place one washer in the plastic bag. Wait for the bag to stop swinging, then measure and record the weight.

Step 3 Repeat the experiment, adding different numbers of washers to the bag. Each time, record the number of washers and the weight.

Analyze the Results

1. The domain contains values of the independent variable, number of washers. The range contains values of the dependent variable, weight. Use the graphing calculator to create a scatter plot using the ordered pairs (washers, weight).
2. Write a sentence that describes the points on the graph.
3. Describe the position of the point on the graph that represents the trial with no washers in the bag.
4. The rate of change can be found by using the formula for slope.

$$\frac{\text{rise}}{\text{run}} = \frac{\text{change in weight}}{\text{change in number of washers}}$$

   Find the rate of change in the weight as more washers are added.

5. Explain how the rate of change is shown on the graph.

Make a Conjecture

The graph shows sample data from a washer experiment. Describe the graph for each situation.

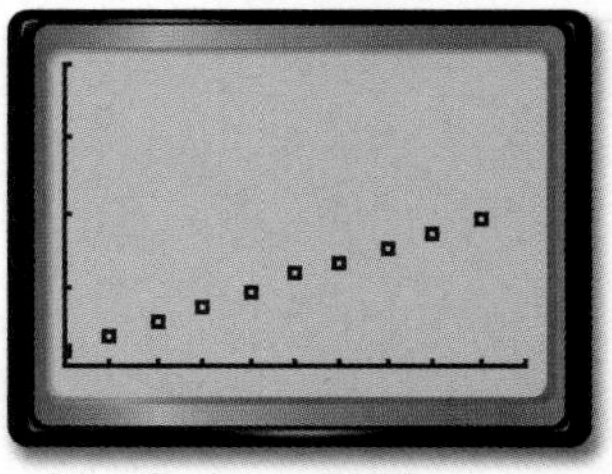

[0, 20] scl: 2 by [0, 1] scl: 0.25

6. a bag that hangs weighs 0.8 N when empty and increases in weight at the rate of the sample
7. a bag that has the same weight when empty as the sample and increases in weight at a faster rate
8. a bag that has the same weight when empty as the sample and increases in weight at a slower rate

Ed Imaging

LESSON 4-1 Graphing Equations in Slope-Intercept Form

Then

- You found rates of change and slopes.

Now

1. Write and graph linear equations in slope-intercept from.
2. Model real-world data with equations in slope-intercept form.

Why?

- Jamil has 500 songs on his digital media player. He joins a music club that lets him download 30 songs per month for a monthly fee. The number of songs that Jamil could eventually have in his player if he does not delete any songs is represented by $y = 30x + 500$.

NewVocabulary
slope-intercept form
constant function

Common Core State Standards

Content Standards
F.IF.7a Graph linear and quadratic functions and show intercepts, maxima, and minima.
S.ID.7 Interpret the slope (rate of change) and the intercept (constant term) of a linear model in the context of the data.

Mathematical Practices
2 Reason abstractly and quantitatively.
8 Look for and express regularity in repeated reasoning.

1 Slope-Intercept Form

Slope-Intercept Form An equation of the form $y = mx + b$, where m is the slope and b is the y-intercept, is in **slope-intercept form**. The variables m and b are called *parameters* of the equation. Changing either value changes the equation's graph.

KeyConcept Slope-Intercept Form

Words The slope-intercept form of a linear equation is $y = mx + b$, where m is the slope and b is the y-intercept.

Example

$$y = mx + b$$
$$y = 2x + 6$$

slope ↑ (2) ↑ y-intercept (6)

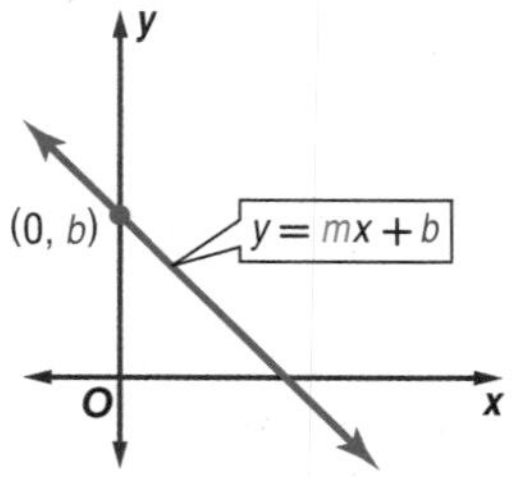

Example 1 Write and Graph an Equation

Write an equation in slope-intercept form for the line with a slope of $\frac{3}{4}$ and a y-intercept of −2. Then graph the equation.

$y = mx + b$ Slope-intercept form

$y = \frac{3}{4}x + (-2)$ Replace m with $\frac{3}{4}$ and b with −2.

$y = \frac{3}{4}x - 2$ Simplify.

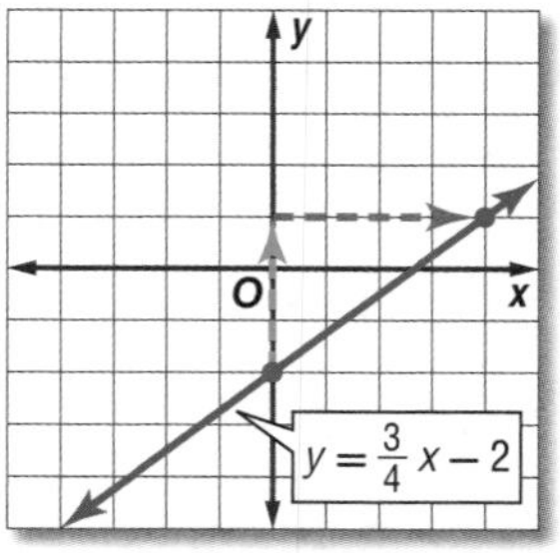

Now graph the equation.

Step 1 Plot the y-intercept (0, −2).

Step 2 The slope is $\frac{\text{rise}}{\text{run}} = \frac{3}{4}$. From (0, −2), move up 3 units and right 4 units. Plot the point.

Step 3 Draw a line through the two points.

GuidedPractice

Write an equation of a line in slope intercept form with the given slope and y-intercept. Then graph the equation.

1A. slope: $-\frac{1}{2}$, y-intercept: 3

1B. slope: −3, y-intercept: −8

When an equation is not written in slope-intercept form, it may be easier to rewrite it before graphing.

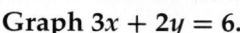

Example 2 Graph Linear Equations

Graph $3x + 2y = 6$.

Rewrite the equation in slope-intercept form.

$3x + 2y = 6$	Original equation
$3x + 2y - 3x = 6 - 3x$	Subtract $3x$ from each side.
$2y = 6 - 3x$	Simplify.
$2y = -3x + 6$	$6 - 3x = 6 + (-3x)$ or $-3x + 6$
$\frac{2y}{2} = \frac{-3x + 6}{2}$	Divide each side by 2.
$y = -\frac{3}{2}x + 3$	Slope-intercept form

Now graph the equation. The slope is $-\frac{3}{2}$, and the y-intercept is 3.

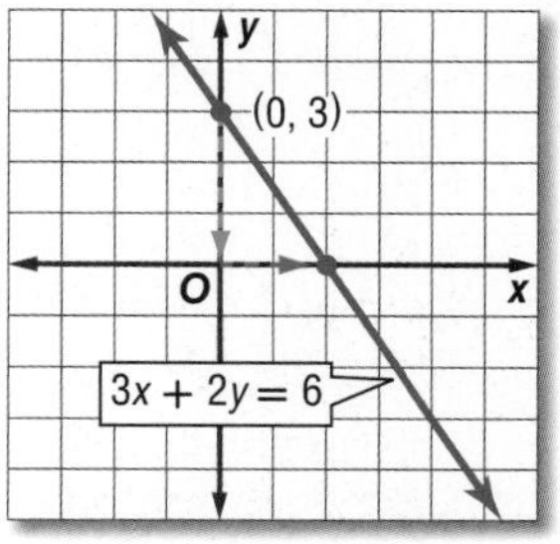

Step 1 Plot the y-intercept (0, 3).

Step 2 The slope is $\frac{\text{rise}}{\text{run}} = -\frac{3}{2}$. From (0, 3), move down 3 units and right 2 units. Plot the point.

Step 3 Draw a line through the two points.

StudyTip

Counting and Direction When counting rise and run, a negative sign may be associated with the value in the numerator or denominator. If with the numerator, begin by counting down for the rise. If with the denominator, count left when counting the run. The resulting line will be the same.

GuidedPractice

Graph each equation.

2A. $3x - 4y = 12$

2B. $-2x + 5y = 10$

Except for the graph of $y = 0$, which lies on the x-axis, horizontal lines have a slope of 0. They are graphs of **constant functions**, which can be written in slope-intercept form as $y = 0x + b$ or $y = b$, where b is any number. Constant functions do not cross the x-axis. Their domain is all real numbers, and their range is b.

Example 3 Graph Linear Equations

Graph $y = -3$.

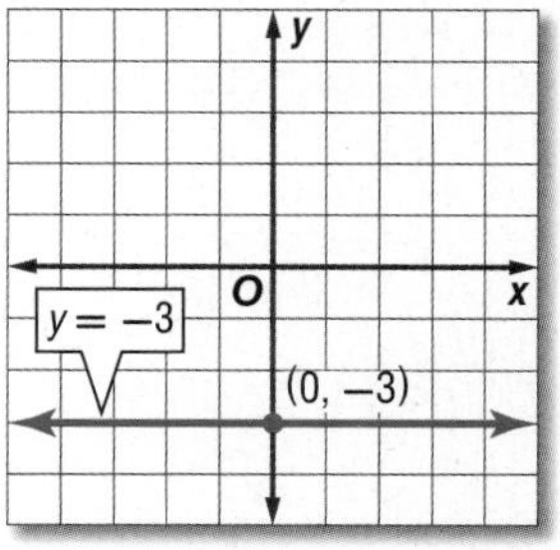

Step 1 Plot the y-intercept (0, −3).

Step 2 The slope is 0. Draw a line through the points with y-coordinate −3.

GuidedPractice

Graph each equation.

3A. $y = 5$

3B. $2y = 1$

Vertical lines have no slope. So, equations of vertical lines cannot be written in slope-intercept form.

There are times when you will need to write an equation when given a graph. To do this, locate the y-intercept and use the rise and run to find another point on the graph. Then write the equation in slope-intercept form.

Standardized Test Example 4 Write an Equation in Slope-Intercept Form

Which of the following is an equation in slope-intercept form for the line shown?

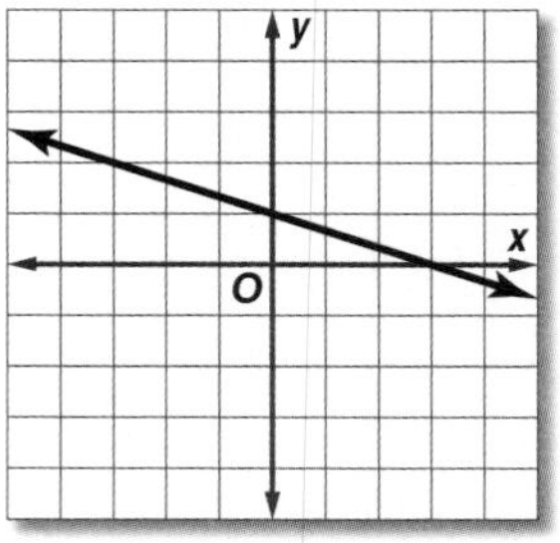

A $y = -3x + 1$

B $y = -3x + 3$

C $y = -\frac{1}{3}x + 1$

D $y = -\frac{1}{3}x + 3$

Test-Taking Tip

Eliminating Choices Analyze the graph to determine the slope and the y-intercept. Then you can save time by eliminating answer choices that do not match the graph.

Read the Test Item

You need to find the slope and y-intercept of the line to write the equation.

Solve the Test Item

Step 1 The line crosses the y-axis at (0, 1), so the y-intercept is 1. The answer is either A or C.

Step 2 To get from (0, 1) to (3, 0), go down 1 unit and 3 units to the right. The slope is $-\frac{1}{3}$.

Step 3 Write the equation.

$y = mx + b$

$y = -\frac{1}{3}x + 1$

CHECK The graph also passes through (−3, 2). If the equation is correct, this should be a solution.

$y = -\frac{1}{3}x + 1$

$2 \stackrel{?}{=} -\frac{1}{3}(-3) + 1$

$2 \stackrel{?}{=} 1 + 1$

$2 = 2$ ✓ The answer is C.

Guided Practice

4. Which of the following is an equation in slope-intercept form for the line shown?

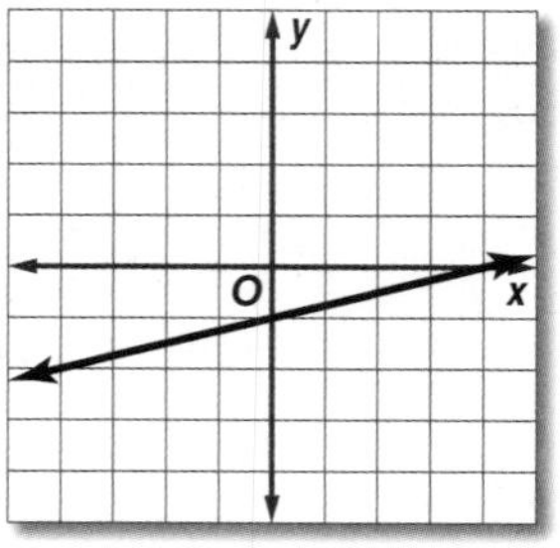

F $y = \frac{1}{4}x - 1$

G $y = \frac{1}{4}x + 4$

H $y = 4x - 1$

J $y = 4x + 4$

2 Modeling Real-World Data

Real-world data can be modeled by a linear equation if there is a constant rate of change. The rate of change represents the slope. The y-intercept is the point where the value of the independent variable is 0.

Real-WorldLink

In 1997, about 2.6 million girls competed in high school sports. The number of girls competing in high school sports has increased by an average of 0.06 million per year since 1997.

Source: National Federation of High School Associations

Real-World Example 5 Write and Graph a Linear Equation

SPORTS **Use the information at the left about high school sports.**

a. Write a linear equation to find the number of girls in high school sports after 1997.

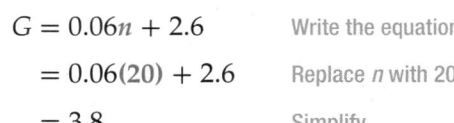

Words	Number of girls competing	equals	rate of change	times	number of years	plus	amount at start.
Variables	Let G = number of girls competing.			Let n = number of years since 1997.			
Equation	G	$=$	0.06	$\times$	n	$+$	2.6

The equation is $G = 0.06n + 2.6$.

b. Graph the equation.

The y-intercept is where the data begins. So, the graph passes through (0, 2.6).

The rate of change is the slope, so the slope is 0.06.

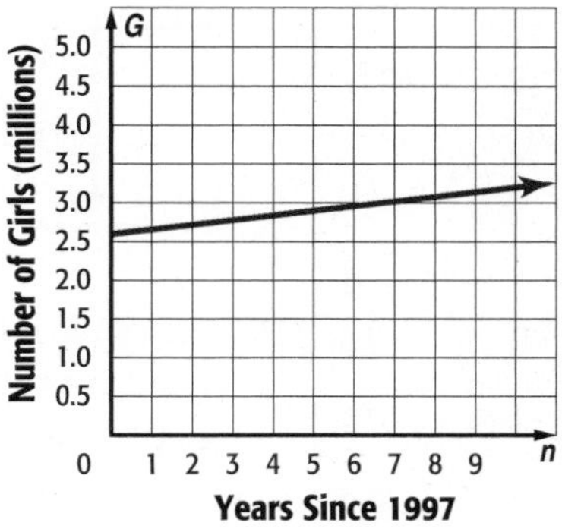

c. Estimate the number of girls competing in 2017.

The year 2017 is 20 years after 1997.

$G = 0.06n + 2.6$ — Write the equation.

$= 0.06(20) + 2.6$ — Replace n with 20.

$= 3.8$ — Simplify.

There will be about 3.8 million girls competing in high school sports in 2017.

GuidedPractice

5. FUNDRAISERS The band boosters are selling sandwiches for \$5 each. They bought \$1160 in ingredients.

A. Write an equation for the profit P made on n sandwiches.

B. Graph the equation.

C. Find the total profit if 1400 sandwiches are sold.

Check Your Understanding

Example 1 **Write an equation of a line in slope-intercept form with the given slope and y-intercept. Then graph the equation.**

1. slope: 2, y-intercept: 4

2. slope: -5, y-intercept: 3

3. slope: $\frac{3}{4}$, y-intercept: -1

4. slope: $-\frac{5}{7}$, y-intercept: $-\frac{2}{3}$

Examples 2–3 **Graph each equation.**

5. $-4x + y = 2$

6. $2x + y = -6$

7. $-3x + 7y = 21$

8. $6x - 4y = 16$

9. $y = -1$

10. $15y = 3$

Example 4 **Write an equation in slope-intercept form for each graph shown.**

11.

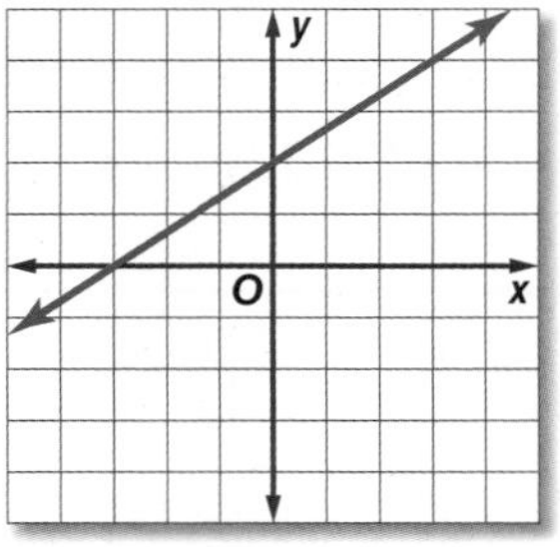

12.

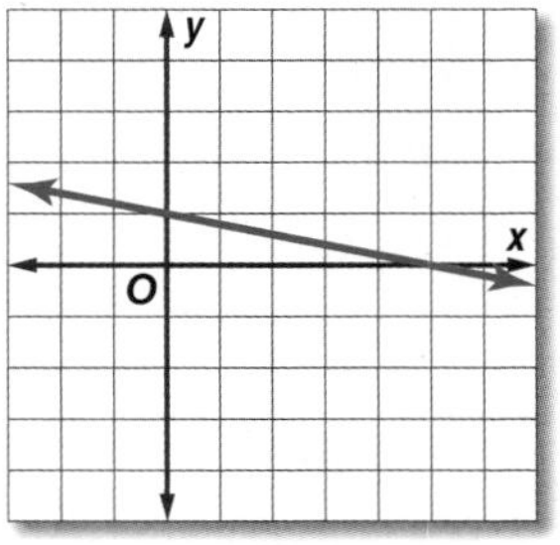

13.

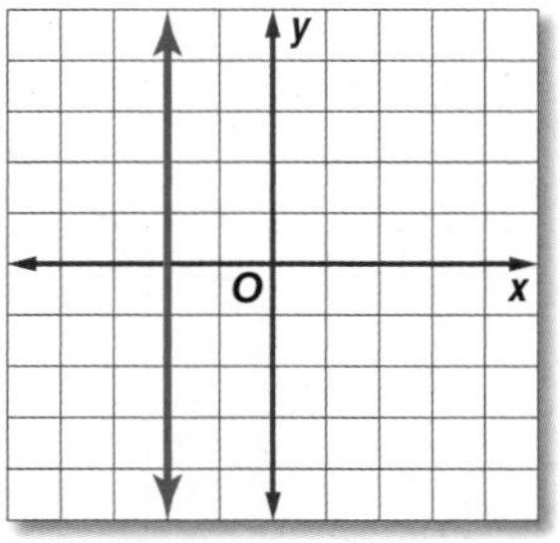

14.

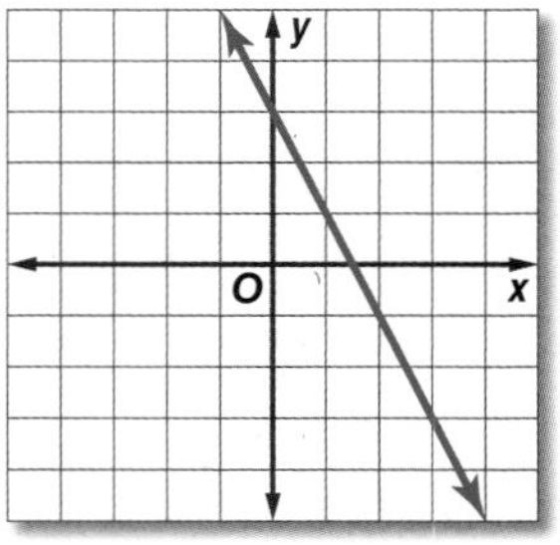

Example 5

15. **FINANCIAL LITERACY** Rondell is buying a new stereo system for his car using a layaway plan.

a. Write an equation for the total amount S that he has paid after w weeks.

b. Graph the equation.

c. Find out how much Rondell will have paid after 8 weeks.

16. **CCSS REASONING** Ana is driving from her home in Miami, Florida, to her grandmother's house in New York City. On the first day, she will travel 240 miles to Orlando, Florida, to pick up her cousin. Then they will travel 350 miles each day.

a. Write an equation that models the total number of miles m Ana has traveled, if d represents the number of days after she picks up her cousin.

b. Graph the equation.

c. How long will the drive take if the total length of the trip is 1343 miles?

Practice and Problem Solving

Example 1 **Write an equation of a line in slope-intercept form with the given slope and y-intercept. Then graph the equation.**

17 slope: 5, y-intercept: 8

18. slope: 3, y-intercept: 10

19. slope: −4, y-intercept: 6

20. slope: −2, y-intercept: 8

21. slope: 3, y-intercept: −4

22. slope: 4, y-intercept: −6

Examples 2–3 **Graph each equation.**

23. $-3x + y = 6$

24. $-5x + y = 1$

25. $-2x + y = -4$

26. $y = 7x - 7$

27. $5x + 2y = 8$

28. $4x + 9y = 27$

29. $y = 7$

30. $y = -\frac{2}{3}$

31. $21 = 7y$

32. $3y - 6 = 2x$

Example 4 **Write an equation in slope-intercept form for each graph shown.**

33.

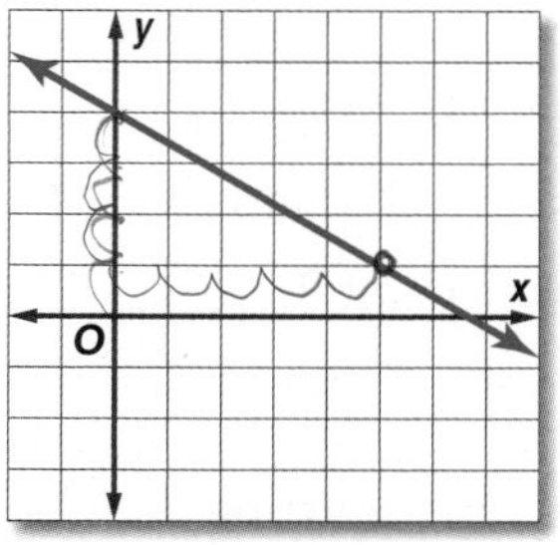

34.

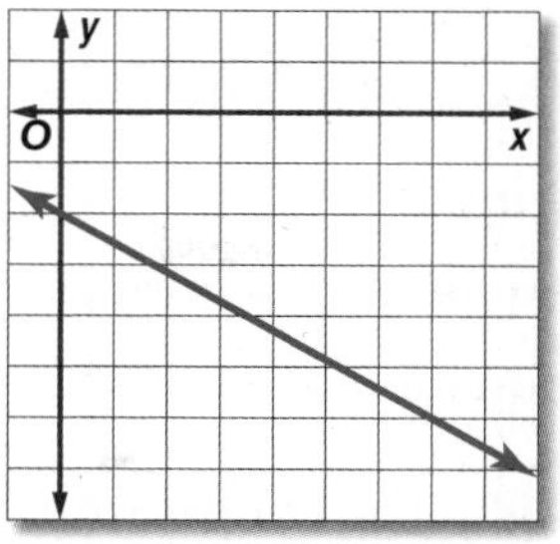

35.

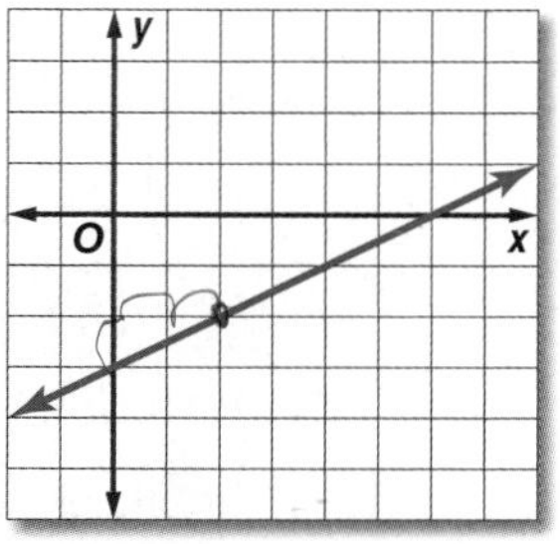

36.

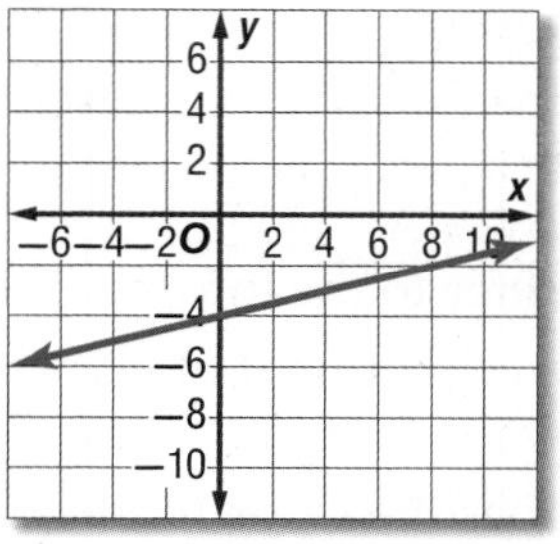

Example 5 **37** **MANATEES** In 1991, 1267 manatees inhabited Florida's waters. The manatee population has increased at a rate of 123 manatees per year.

a. Write an equation for the manatee population, P, t years since 1991.

b. Graph this equation.

c. In 2006, the manatee was removed from Florida's endangered species list. What was the manatee population in 2006?

Write an equation of a line in slope-intercept form with the given slope and y-intercept.

38. slope: $\frac{1}{2}$, y-intercept: -3

39. slope: $\frac{2}{3}$, y-intercept: -5

40. slope: $-\frac{5}{6}$, y-intercept: 5

41. slope: $-\frac{3}{7}$, y-intercept: 2

42. slope: 1, y-intercept: 4

43. slope: 0, y-intercept: 5

Graph each equation.

44. $y = \frac{3}{4}x - 2$

45. $y = \frac{5}{3}x + 4$

46. $3x + 8y = 32$

47. $5x - 6y = 36$

48. $-4x + \frac{1}{2}y = -1$

49. $3x - \frac{1}{4}y = 2$

50. **TRAVEL** A rental company charges \$8 per hour for a mountain bike plus a \$5 fee for a helmet.

a. Write an equation in slope-intercept form for the total rental cost C for a helmet and a bicycle for t hours.

b. Graph the equation.

c. What would the cost be for 2 helmets and 2 bicycles for 8 hours?

51. **CCSS REASONING** For Illinois residents, the average tuition at Chicago State University is \$157 per credit hour. Fees cost \$218 per year.

a. Write an equation in slope-intercept form for the tuition T for c credit hours.

b. Find the cost for a student who is taking 32 credit hours.

Write an equation of a line in slope-intercept form with the given slope and y-intercept.

52. slope: -1, y-intercept: 0

53. slope: 0.5, y-intercept: 7.5

54. slope: 0, y-intercept: 7

55. slope: -1.5, y-intercept: -0.25

56. Write an equation of a horizontal line that crosses the y-axis at $(0, -5)$.

57. Write an equation of a line that passes through the origin and has a slope of 3.

58. **TEMPERATURE** The temperature dropped rapidly overnight. Starting at 80°F, the temperature dropped 3° per minute.

a. Draw a graph that represents this drop from 0 to 8 minutes.

b. Write an equation that describes this situation. Describe the meaning of each variable as well as the slope and y-intercept.

59. **FITNESS** Refer to the information at the right.

a. Write an equation that represents the cost C of a membership for m months.

b. What does the slope represent?

c. What does the C-intercept represent?

d. What is the cost of a two-year membership?

60. **MAGAZINES** A teen magazine began with a circulation of 500,000 in its first year. Since then, the circulation has increased an average of 33,388 per year.

a. Write an equation that represents the circulation c after t years.

b. What does the slope represent?

c. What does the c-intercept represent?

d. If the magazine began in 1944, and this trend continues, in what year will the circulation reach 3,000,000?

61. **SMART PHONES** A telecommunications company sold 3305 smart phones in the first year of production. Suppose, on average, they expect to sell 25 phones per day.

a. Write an equation for the number of smart phones P sold t years after the first year of production, assuming 365 days per year.

b. If sales continue at this rate, how many years will it take for the company to sell 100,000 phones?

H.O.T. Problems Use Higher-Order Thinking Skills

62. **OPEN ENDED** Draw a graph representing a real-world linear function and write an equation for the graph. Describe what the graph represents.

63. **REASONING** Determine whether the equation of a vertical line can be written in slope-intercept form. Explain your reasoning.

64. **CHALLENGE** Summarize the characteristics that the graphs $y = 2x + 3$, $y = 4x + 3$, $y = -x + 3$, and $y = -10x + 3$ have in common.

65. CCSS **REGULARITY** If given an equation in standard form, explain how to determine the rate of change.

66. **WRITING IN MATH** Explain how you would use a given y-intercept and the slope to predict a y-value for a given x-value without graphing.

Standardized Test Practice

67. A music store has x CDs in stock. If 350 are sold and $3y$ are added to stock, which expression represents the number of CDs in stock?

A $350 + 3y - x$
B $x - 350 + 3y$
C $x + 350 + 3y$
D $3y - 350 - x$

68. PROBABILITY The table shows the result of a survey of favorite activities. What is the probability that a student's favorite activity is sports or drama club?

Extracurricular Activity	Students
art club	24
band	134
choir	37
drama club	46
mock trial	19
school paper	26
sports	314

F $\frac{3}{8}$
G $\frac{4}{9}$
H $\frac{3}{5}$
J $\frac{2}{3}$

69. A recipe for fruit punch calls for 2 ounces of orange juice for every 8 ounces of lemonade. If Jennifer uses 64 ounces of lemonade, which proportion can she use to find x, the number of ounces of orange juice needed?

A $\frac{2}{x} = \frac{64}{6}$
B $\frac{8}{x} = \frac{64}{2}$
C $\frac{2}{8} = \frac{x}{64}$
D $\frac{6}{2} = \frac{x}{64}$

70. EXTENDED RESPONSE The table shows the results of a canned food drive. 1225 cans were collected, and the 12th-grade class collected 55 more cans than the 10th-grade class. How many cans each did the 10th- and 12th-grade classes collect? Show your work.

Grade	Cans
9	340
10	x
11	280
12	y

Spiral Review

For each arithmetic sequence, determine the related function. Then determine if the function is *proportional* or *nonproportional*. (Lesson 3-6)

71. 3, 7, 11, …
72. 8, 6, 4, …
73. 0, 3, 6, …
74. 1, 2, 3, …

75. GAME SHOWS Contestants on a game show win money by answering 10 questions. (Lesson 3-5)

a. Find the value of the 10th question.

b. If all questions are answered correctly, how much are the winnings?

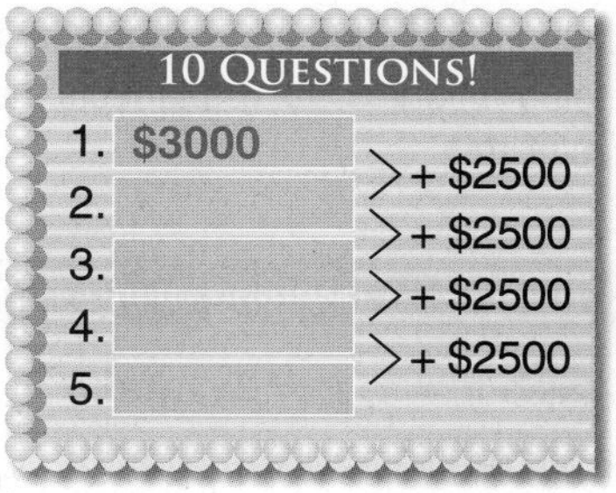

Suppose y varies directly as x. Write a direct variation equation that relates x and y. Then solve. (Lesson 3-4)

76. If $y = 10$ when $x = 5$, find y when $x = 6$.

77. If $y = -16$ when $x = 4$, find x when $y = 20$.

78. If $y = 6$ when $x = 18$, find y when $x = -12$.

79. If $y = 12$ when $x = 15$, find x when $y = -6$.

Skills Review

Find the slope of the line that passes through each pair of points.

80. (2, 3), (9, 7)
81. (−3, 6), (2, 4)
82. (2, 6), (−1, 3)
83. (−3, 3), (1, 3)

EXTEND 4-1

Graphing Technology Lab
The Family of Linear Graphs

A family of people is related by birth, marriage, or adoption. Often people in families share characteristics. The graphs in a family share at least one characteristic. Graphs in the linear family are all lines, with the simplest graph in the family being that of the parent function $y = x$. This parent function is also known as the **identity function**. Its graph contains all points with coordinates (a, a). Its domain and range are all real numbers.

You can use a graphing calculator to investigate how changing the parameters m and b in $y = mx + b$ affects the graphs in the family of linear functions.

Parent Graph
Identity Function

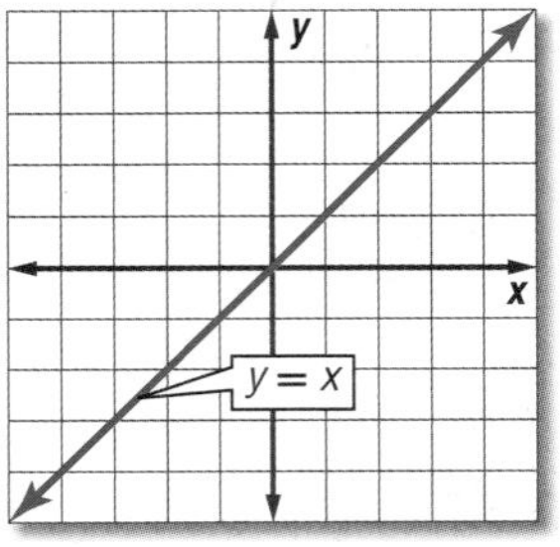

CCSS Common Core State Standards
Content Standards
F.BF.3 Identify the effect on the graph of replacing $f(x)$ by $f(x) + k$, $k\,f(x)$, $f(kx)$, and $f(x + k)$ for specific values of k (both positive and negative); find the value of k given the graphs. Experiment with cases and illustrate an explanation of the effects on the graph using technology.
S.ID.7 Interpret the slope (rate of change) and the intercept (constant term) of a linear model in the context of the data.
Mathematical Practices
7 Look for and make use of structure.

Activity 1 Changing b in $y = mx + b$

Graph $y = x$, $y = x + 4$, and $y = x - 2$ in the standard viewing window.

Enter the equations in the **Y=** list as **Y1**, **Y2**, and **Y3**. Then graph the equations.

KEYSTROKES: [Y=] [X,T,θ,n] [ENTER] [X,T,θ,n] [+] 4 [ENTER] [X,T,θ,n] [−] 2 [ENTER] [ZOOM] 6

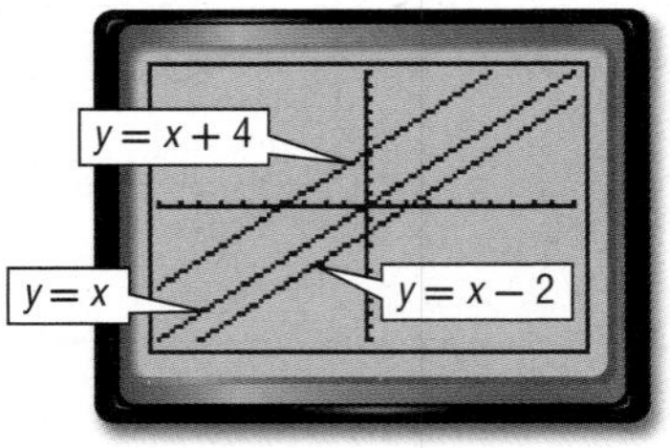

[−10, 10] scl: 1 by [−10, 10] scl: 1

1A. How do the slopes of the graphs compare?

1B. Compare the graph of $y = x + 4$ and the graph of $y = x$. How would you obtain the graph of $y = x + 4$ from the graph of $y = x$?

1C. How would you obtain the graph of $y = x - 2$ from the graph of $y = x$?

Changing the y-intercept, b, translates, or moves, a linear function up or down the y-axis. Changing m in $y = mx + b$ affects the graphs in a different way. First, investigate positive values of m.

Activity 2 Changing m in $y = mx + b$, Positive Values

Graph $y = x$, $y = 2x$, and $y = \frac{1}{3}x$ in the standard viewing window.

Enter the equations in the **Y=** list and graph.

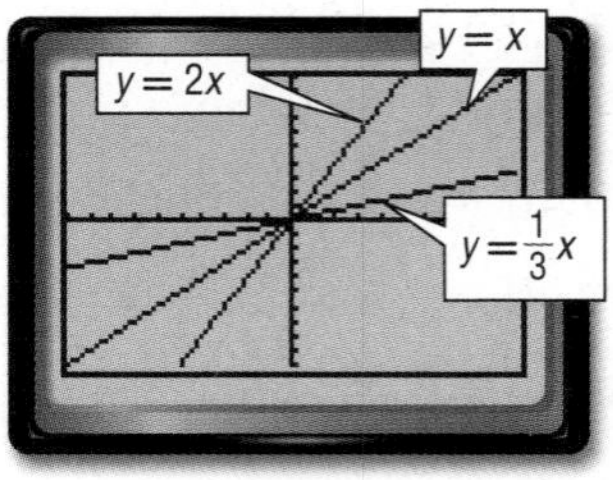

[−10, 10] scl: 1 by [−10, 10] scl: 1

2A. How do the y-intercepts of the graphs compare?

2B. Compare the graph of $y = 2x$ and the graph of $y = x$.

2C. Which is steeper, the graph of $y = \frac{1}{3}x$ or the graph of $y = x$?

Does changing m to a negative value affect the graph differently than changing it to a positive value?

Activity 3 Changing m in $y = mx + b$, Negative Values

Graph $y = x$, $y = -x$, $y = -3x$, and $y = -\frac{1}{2}x$ in the standard viewing window.

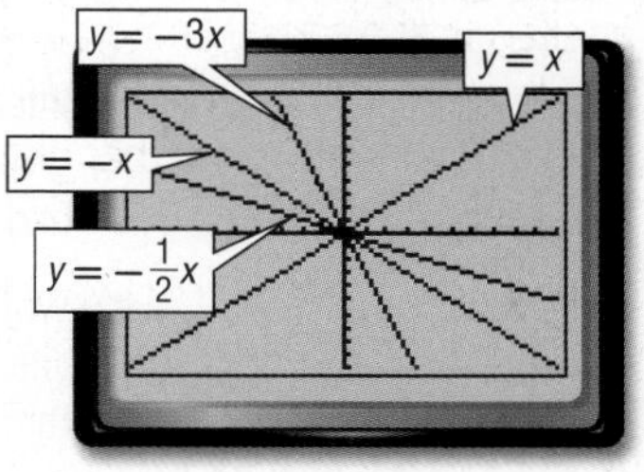

[−10, 10] scl: 1 by [−10, 10] scl: 1

Enter the equations in the **Y=** list and graph.

3A. How are the graphs with negative values of m different than graphs with a positive m?

3B. Compare the graphs of $y = -x$, $y = -3x$, and $y = -\frac{1}{2}x$. Which is steepest?

Analyze the Results

CCSS SENSE-MAKING AND PERSEVERANCE Graph each set of equations on the same screen. Describe the similarities or differences.

1. $y = 2x$
$y = 2x + 3$
$y = 2x - 7$

2. $y = x + 1$
$y = 2x + 1$
$y = \frac{1}{4}x + 1$

3. $y = x + 4$
$y = 2x + 4$
$y = \frac{3}{4}x + 4$

4. $y = 0.5x + 2$
$y = 0.5x - 5$
$y = 0.5x + 4$

5. $y = -2x - 2$
$y = -4.2x - 2$
$y = -\frac{1}{3}x - 2$

6. $y = 3x$
$y = 3x + 6$
$y = 3x - 7$

7. Families of graphs have common characteristics. What do the graphs of all equations of the form $y = mx + b$ have in common?

8. How does the value of b affect the graph of $y = mx + b$?

9. What is the result of changing the value of m on the graph of $y = mx + b$ if m is positive?

10. How can you determine which graph is steepest by examining the following equations?
$y = 3x$, $y = -4x - 7$, $y = \frac{1}{2}x + 4$

11. Explain how knowing about the effects of m and b can help you sketch the graph of an equation.

12. The equation $y = k$ can also be a parent graph. Graph $y = 5$, $y = 2$, and $y = -4$ on the same screen. Describe the similarities or differences among the graphs.

Extension

Nonlinear functions can also be defined in terms of a family of graphs. Graph each set of equations on the same screen. Describe the similarities or differences.

13. $y = x^2$
$y = -3x^2$
$y = (-3x)^2$

14. $y = x^2$
$y = x^2 + 3$
$y = (x - 2)^2$

15. $y = x^2$
$y = 2x^2 + 4$
$y = (3x)^2 - 5$

16. Describe the similarities and differences in the classes of functions $f(x) = x^2 + c$ and $f(x) = (x + c)^2$, where c is any real number.

LESSON 4-2 Writing Equations in Slope-Intercept Form

Then

- You graphed lines given the slope and the *y*-intercept.

Now

1. Write an equation of a line in slope-intercept form given the slope and one point.
2. Write an equation of a line in slope-intercept form given two points.

Why?

- In 2006, the attendance at the Columbus Zoo and Aquarium was about 1.6 million. In 2009, the zoo's attendance was about 2.2 million. You can find the average rate of change for these data. Then you can write an equation that would model the average attendance at the zoo for a given year.

NewVocabulary

constraint
linear extrapolation

Common Core State Standards

Content Standards

F.BF.1 Write a function that describes a relationship between two quantities.

a. Determine an explicit expression, a recursive process, or steps for calculation from a context.

b. Combine standard function types using arithmetic operations.

F.LE.2 Construct linear and exponential functions, including arithmetic and geometric sequences, given a graph, a description of a relationship, or two input-output pairs (include reading these from a table).

Mathematical Practices

3 Construct viable arguments and critique the reasoning of others.

6 Attend to precision.

1 Write an Equation Given the Slope and a Point

The next example shows how to write an equation of a line if you are given a slope and a point other than the *y*-intercept.

Example 1 Write an Equation Given the Slope and a Point

PT

Write an equation of the line that passes through (2, 1) with a slope of 3.

You are given the slope but not the *y*-intercept.

Step 1 Find the *y*-intercept.

$y = mx + b$	Slope-intercept form
$1 = 3(2) + b$	Replace *m* with 3, *y* with 1, and *x* with 2.
$1 = 6 + b$	Simplify.
$1 - 6 = 6 + b - 6$	Subtract 6 from each side.
$-5 = b$	Simplify.

Step 2 Write the equation in slope-intercept form.

$y = mx + b$	Slope-intercept form
$y = 3x - 5$	Replace *m* with 3 and *b* with –5.

Therefore, the equation of the line is $y = 3x - 5$.

GuidedPractice

Write an equation of a line that passes through the given point and has the given slope.

1A. (–2, 5), slope 3

1B. (4, –7), slope –1

2 Write an Equation Given Two Points

If you are given two points through which a line passes, you can use them to find the slope first. Then follow the steps in Example 1 to write the equation.

Example 2 Write an Equation Given Two Points

StudyTip

Choosing a point Given two points on a line, you may select either point to be (x_1, y_1). Be sure to remain consistent throughout the problem.

Write an equation of the line that passes through each pair of points.

a. (3, 1) and (2, 4)

Step 1 Find the slope of the line containing the given points.

$m = \frac{y_2 - y_1}{x_2 - x_1}$	Slope Formula
$= \frac{4 - 1}{2 - 3}$	$(x_1, y_1) = (3, 1)$ and $(x_2, y_2) = (2, 4)$
$= \frac{3}{-1}$ or -3	Simplify.

Step 2 Use either point to find the *y*-intercept.

$y = mx + b$	Slope-intercept form
$4 = (-3)(2) + b$	Replace *m* with −3, *x* with 2, and *y* with 4.
$4 = -6 + b$	Simplify.
$4 - (-6) = -6 + b - (-6)$	Subtract −6 from each side.
$10 = b$	Simplify.

Step 3 Write the equation in slope-intercept form.

$y = mx + \mathbf{b}$	Slope-intercept form
$y = -3x + \mathbf{10}$	Replace *m* with −3 and *b* with 10.

Therefore, the equation is $y = -3x + 10$.

b. (−4, −2) and (−5, −6)

StudyTip

Slope If the (x_1, y_1) coordinates are negative, be sure to account for both the negative signs and the subtraction symbols in the Slope Formula.

Step 1 Find the slope of the line containing the given points.

$m = \frac{y_2 - y_1}{x_2 - x_1}$	Slope Formula
$= \frac{-6 - (-2)}{-5 - (-4)}$	$(x_1, y_1) = (-4, -2)$ and $(x_2, y_2) = (-5, -6)$
$= \frac{-4}{-1}$ or 4	Simplify.

Step 2 Use either point to find the *y*-intercept.

$y = mx + b$	Slope-intercept form
$-2 = 4(-4) + b$	Replace *m* with 4, *x* with −4, and *y* with −2.
$-2 = -16 + b$	Simplify.
$-2 - (-16) = -16 + b - (-16)$	Subtract −16 from each side.
$14 = b$	Simplify.

Step 3 Write the equation in slope-intercept form.

$y = mx + \mathbf{b}$	Slope-intercept form
$y = 4x + \mathbf{14}$	Replace *m* with 4 and *b* with 14.

Therefore, the equation is $y = 4x + 14$.

GuidedPractice

Write an equation of the line that passes through each pair of points.

2A. (−1, 12), (4, −8)

2B. (5, −8), (−7, 0)

In mathematics, a **constraint** is a condition that a solution must satisfy. Equations can be viewed as constraints in a problem situation. The solutions of the equation meet the constraints of the problem.

Real-World Career

Ground Crew
Airline ground crew responsibilities include checking tickets, helping passengers with luggage, and making sure that baggage is loaded properly and secure. This job usually requires a high school diploma or GED.

Source: Airline Jobs

PT

Real-World Example 3 Use Slope-Intercept Form

FLIGHTS The table shows the number of domestic flights in the U.S. from 2004 to 2008. Write an equation that could be used to predict the number of flights if it continues to decrease at the same rate.

Year	Flights (millions)
2004	9.97
2005	10.04
2006	9.71
2007	9.84
2008	9.37

Understand You know the number of flights for 2004–2008.

Plan Let x represent the number of years since 2000, and let y represent the number of flights. Write an equation of the line that passes through (4, 9.97) and (8, 9.37).

Solve Find the slope.

$m = \frac{y_2 - y_1}{x_2 - x_1}$ — Slope formula

$= \frac{9.37 - 9.97}{8 - 4}$ — Let $(x_1, y_1) = (4, 9.97)$ and $(x_2, y_2) = (8, 9.37)$.

$= -\frac{0.6}{4}$ or -0.15 — Simplify.

Use (8, 9.37) to find the y-intercept of the line.

$y = mx + b$ — Slope-intercept form

$9.37 = -0.15(8) + b$ — Replace y with 9.37, m with −0.15, and x with 8.

$9.37 = -1.2 + b$ — Simplify.

$10.57 = b$ — Add 1.2 to each side.

Write the equation using $m = -0.15$ and $b = 10.57$.

$y = mx + b$ — Slope-intercept form

$y = -0.15x + 10.57$ — Replace m with −0.15 and b with 10.57.

Check Check your result by using the coordinates of the other point.

$y = -0.15x + 10.57$ — Original equation

$9.97 \stackrel{?}{=} -0.15(4) + 10.57$ — Replace y with 9.97 and x with 4.

$9.97 = 9.97$ ✓ — Simplify.

Guided Practice

3. FINANCIAL LITERACY In addition to his weekly salary, Ethan is paid \$16 per delivery. Last week, he made 5 deliveries, and his total pay was \$215. Write a linear equation to find Ethan's total weekly pay T if he makes d deliveries.

You can use a linear equation to make predictions about values that are beyond the range of the data. This process is called **linear extrapolation**.

Problem-Solving Tip

CCSS Precision Deciding whether an answer is reasonable is useful when an exact answer is not neccessary.

PT

Real-World Example 4 Predict from Slope-Intercept Form

FLIGHTS Estimate the number of domestic flights in 2020.

$y = -0.15x + 10.57$ — Original equation

$= -0.15(20) + 10.57$ or 7.57 million — Replace x with 20.

Guided Practice

4. MONEY Use the equation in Guided Practice 3 to predict how much money Ethan will earn in a week if he makes 8 deliveries.

Stephan Goerich/imagebroker/age fotostock

Check Your Understanding

Example 1 **Write an equation of the line that passes through the given point and has the given slope.**

1. (3, −3), slope 3
2. (2, 4), slope 2
3. (1, 5), slope −1
4. (−4, 6), slope −2

Example 2 **Write an equation of the line that passes through each pair of points.**

5. (4, −3), (2, 3)
6. (−7, −3), (−3, 5)
7. (−1, 3), (0, 8)
8. (−2, 6), (0, 0)

Examples 3, 4 **9. WHITEWATER RAFTING** Ten people from a local youth group went to Black Hills Whitewater Rafting Tour Company for a one-day rafting trip. The group paid $425.

a. Write an equation in slope-intercept form to find the total cost C for p people.

b. How much would it cost for 15 people?

Practice and Problem Solving

Example 1 **Write an equation of the line that passes through the given point and has the given slope.**

10. (3, 1), slope 2
11. (−1, 4), slope −1
12. (1, 0), slope 1
13. (7, 1), slope 8
14. (2, 5), slope −2
15. (2, 6), slope 2

Example 2 **Write an equation of the line that passes through each pair of points.**

16. (9, −2), (4, 3)
17. (−2, 5), (5, −2)
18. (−5, 3), (0, −7)
19. (3, 5), (2, −2)
20. (−1, −3), (−2, 3)
21. (−2, −4), (2, 4)

Examples 3, 4 **22.** CCSS **MODELING** Greg is driving a remote control car at a constant speed. He starts the timer when the car is 5 feet away. After 2 seconds the car is 35 feet away.

a. Write a linear equation to find the distance d of the car from Greg.

b. Estimate the distance the car has traveled after 10 seconds.

23. ZOOS Refer to the beginning of the lesson.

a. Write a linear equation to find the attendance (in millions) y after x years. Let x be the number of years since 2000.

b. Estimate the zoo's attendance in 2020.

24. BOOKS In 1904, a dictionary cost 30¢. Since then the cost of a dictionary has risen an average of 6¢ per year.

a. Write a linear equation to find the cost C of a dictionary y years after 1904.

b. If this trend continues, what will the cost of a dictionary be in 2020?

Write an equation of the line that passes through the given point and has the given slope.

25. (4, 2), slope $\frac{1}{2}$
26. (3, −2), slope $\frac{1}{3}$
27. (6, 4), slope $-\frac{3}{4}$
28. (2, −3), slope $\frac{2}{3}$
29. (2, −2), slope $\frac{2}{7}$
30. (−4, −2), slope $-\frac{3}{5}$

31. **DOGS** In 2001, there were about 56.1 thousand golden retrievers registered in the United States. In 2002, the number was 62.5 thousand.

   a. Write a linear equation to find the number of thousands of golden retrievers G that will be registered in year t, where $t = 0$ is the year 2000.

   b. Graph the equation.

   c. Estimate the number of golden retrievers that will be registered in 2017.

32. **GYM MEMBERSHIPS** A local recreation center offers a yearly membership for $265. The center offers aerobics classes for an additional $5 per class.

   a. Write an equation that represents the total cost of the membership.

   b. Carly spent $500 one year. How many aerobics classes did she take?

33. **SUBSCRIPTION** A magazine offers an online subscription that allows you to view up to 25 archived articles free. To view 30 archived articles, you pay $49.15. To view 33 archived articles, you pay $57.40.

   a. What is the cost of each archived article for which you pay a fee?

   b. What is the cost of the magazine subscription?

Write an equation of the line that passes through the given points.

34. $(5, -2), (7, 1)$ 35. $(5, -3), (2, 5)$ 36. $\left(\frac{5}{4}, 1\right), \left(-\frac{1}{4}, \frac{3}{4}\right)$ 37. $\left(\frac{5}{12}, -1\right), \left(-\frac{3}{4}, \frac{1}{6}\right)$

Determine whether the given point is on the line. Explain why or why not.

38. $(3, -1); y = \frac{1}{3}x + 5$ 39. $(6, -2); y = \frac{1}{2}x - 5$

For Exercises 40–42, determine which equation best represents each situation. Explain the meaning of each variable.

A $y = -\frac{1}{3}x + 72$	B $y = 2x + 225$	C $y = 8x + 4$

40. **CONCERTS** Tickets to a concert cost $8 each plus a processing fee of $4 per order.

41. **FUNDRAISING** The freshman class has $225. They sell raffle tickets at $2 each to raise money for a field trip.

42. **POOLS** The current water level of a swimming pool in Tucson, Arizona, is 6 feet. The rate of evaporation is $\frac{1}{3}$ inch per day.

43. CCSS **SENSE-MAKING** A manufacturer implemented a program to reduce waste. In 1998 they sent 946 tons of waste to landfills. Each year after that, they reduced their waste by an average 28.4 tons.

   a. How many tons were sent to the landfill in 2010?

   b. In what year will it become impossible for this trend to continue? Explain.

44. **COMBINING FUNCTIONS** The parents of a college student open an account for her with a deposit of $5000, and they set up automatic deposits of $100 to the account every week.

   a. Write a function $d(t)$ to express the amount of money in the account t weeks after the initial deposit.

   b. The student plans on spending $600 the first week and $250 in each of the following weeks for room and board and other expenses. Write a function $w(t)$ to express the amount of money taken out of the account each week.

   c. Find $B(t) = d(t) - w(t)$. What does this new function represent?

   d. Will the student run out of money? If so, when?

45. **CONCERT TICKETS** Jackson is ordering tickets for a concert online. There is a processing fee for each order, and the tickets are \$52 each. Jackson ordered 5 tickets and the cost was \$275.

a. Determine the processing fee. Write a linear equation to represent the total cost C for t tickets.

b. Make a table of values for at least three other numbers of tickets.

c. Graph this equation. Predict the cost of 8 tickets.

46. **MUSIC** A music store is offering a Frequent Buyers Club membership. The membership costs \$22 per year, and then a member can buy CDs at a reduced price. If a member buys 17 CDs in one year, the cost is \$111.25.

a. Determine the cost of each CD for a member.

b. Write a linear equation to represent the total cost y of a one year membership, if x CDs are purchased.

c. Graph this equation.

H.O.T. Problems Use Higher-Order Thinking Skills

47. **ERROR ANALYSIS** Tess and Jacinta are writing an equation of the line through (3, −2) and (6, 4). Is either of them correct? Explain your reasoning.

Tess

$m = \frac{4-(-2)}{6-3} = \frac{6}{3}$ or 2

$y = mx + b$

$6 = 2(4) + b$

$6 = 8 + b$

$-2 = b$

$y = 2x - 2$

Jacinta

$m = \frac{4-(-2)}{6-3} = \frac{6}{3}$ or 2

$y = mx + b$

$-2 = 2(3) + b$

$-2 = 6 + b$

$-8 = b$

$y = 2x - 8$

48. **CHALLENGE** Consider three points, (3, 7), (−6, 1) and (9, p), on the same line. Find the value of p and explain your steps.

49. **REASONING** Consider the standard form of a linear equation, $Ax + By = C$.

a. Rewrite the equation in slope-intercept form.

b. What is the slope?

c. What is the y-intercept?

d. Is this true for all real values of A, B, and C?

50. **OPEN ENDED** Create a real-world situation that fits the graph at the right. Define the two quantities and describe the functional relationship between them. Write an equation to represent this relationship and describe what the slope and y-intercept mean.

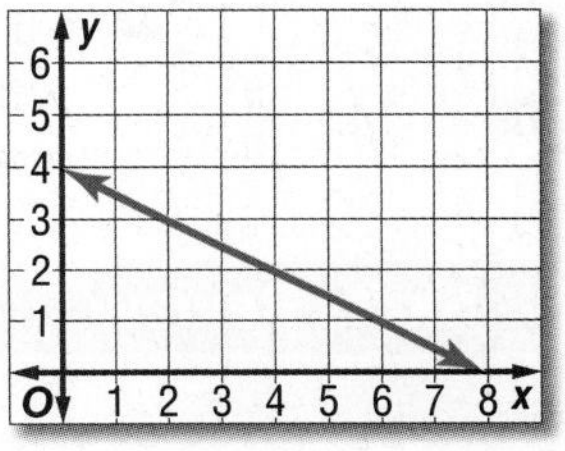

51. **WRITING IN MATH** Linear equations are useful in predicting future events. Describe some factors in real-world situations that might affect the reliability of the graph in making any predictions.

52. **CCSS ARGUMENTS** What information is needed to write the equation of a line? Explain.

Standardized Test Practice

53. Which equation *best* represents the graph?

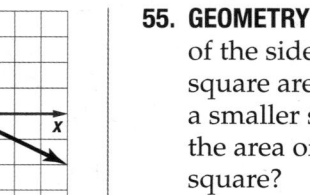

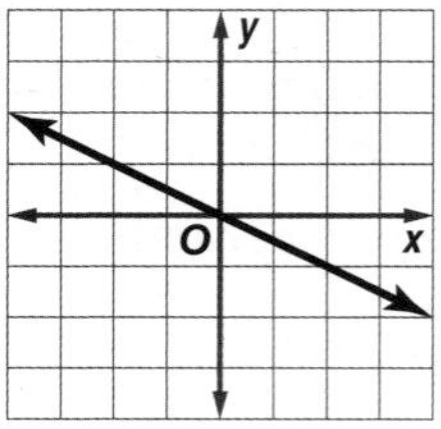

A $y = 2x$

B $y = -2x$

C $y = \frac{1}{2}x$

D $y = -\frac{1}{2}x$

54. Roberto receives an employee discount of 12%. If he buys a \$355 item at the store, what is his discount to the nearest dollar?

F \$3

G \$4

H \$30

J \$43

55. GEOMETRY The midpoints of the sides of the large square are joined to form a smaller square. What is the area of the smaller square?

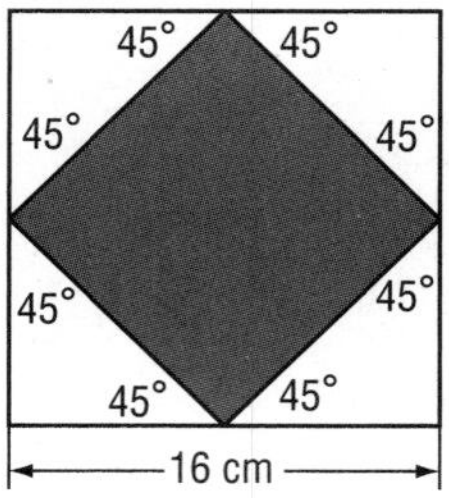

A 64 cm^2

B 128 cm^2

C 248 cm^2

D 256 cm^2

56. SHORT RESPONSE If $\frac{5(x+4)}{2} + 7 = 37$, what is the value of $3x - 9$?

Spiral Review

Graph each equation. (Lesson 4-1)

57. $y = 3x + 2$

58. $y = -4x + 2$

59. $3y = 2x + 6$

60. $y = \frac{1}{2}x + 6$

61. $3x + y = -1$

62. $2x + 3y = 6$

Write an equation in function notation for each relation. (Lesson 3-6)

63.

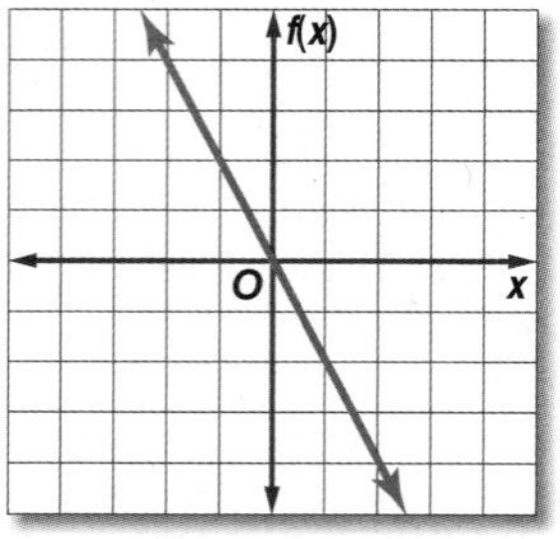

64.

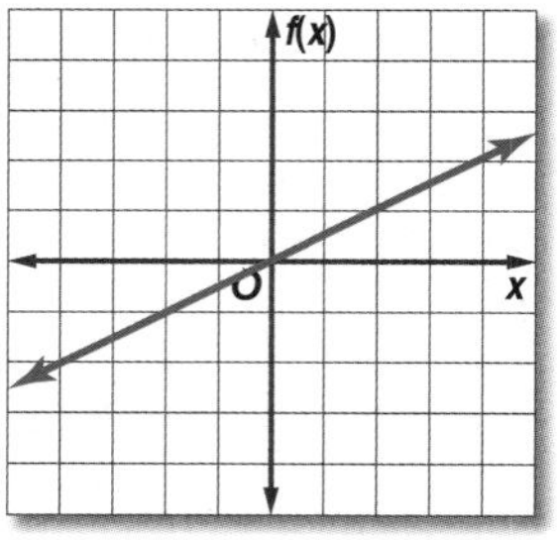

65. METEOROLOGY The distance d in miles that the sound of thunder travels in t seconds is given by the equation $d = 0.21t$. (Lesson 3-4)

a. Graph the equation.

b. Use the graph to estimate how long it will take you to hear thunder from a storm 3 miles away.

Solve each equation. Check your solution. (Lesson 2-3)

66. $-5t - 2.2 = -2.9$

67. $-5.5a - 43.9 = 77.1$

68. $4.2r + 7.14 = 12.6$

69. $-14 - \frac{n}{9} = 9$

70. $\frac{-8b - (-9)}{-10} = 17$

71. $9.5x + 11 - 7.5x = 14$

Skills Review

Find the value of r so the line through each pair of points has the given slope.

72. $(6, -2), (r, -6), m = 4$

73. $(8, 10), (r, 4), m = 6$

74. $(7, -10), (r, 4), m = -3$

75. $(6, 2), (9, r), m = -1$

76. $(9, r), (6, 3), m = -\frac{1}{3}$

77. $(5, r), (2, -3), m = \frac{4}{3}$

LESSON 4-3

Writing Equations in Point-Slope Form

Then

- You wrote linear equations given either one point and the slope or two points.

Now

1. Write equations of lines in point-slope form.
2. Write linear equations in different forms.

Why?

- Most humane societies have foster homes for newborn puppies, kittens, and injured or ill animals. During the spring and summer, a large shelter can place 3000 animals in homes each month.

If a shelter had 200 animals in foster homes at the beginning of spring, the number of animals in foster homes at the end of the summer could be represented by $y = 3000x + 200$, where x is the number of months and y is the number of animals.

NewVocabulary

point-slope form

Common Core State Standards

Content Standards

F.IF.2 Use function notation, evaluate functions for inputs in their domains, and interpret statements that use function notation in terms of a context.

F.LE.2 Construct linear and exponential functions, including arithmetic and geometric sequences, given a graph, a description of a relationship, or two input-output pairs (include reading these from a table).

Mathematical Practices

2 Reason abstractly and quantitatively.

1 Point-Slope Form

An equation of a line can be written in **point-slope form** when given the coordinates of one known point on a line and the slope of that line.

KeyConcept Point-Slope Form

Words The linear equation $y - y_1 = m(x - x_1)$ is written in point-slope form, where (x_1, y_1) is a given point on a nonvertical line and m is the slope of the line.

Symbols $y - y_1 = m(x - x_1)$

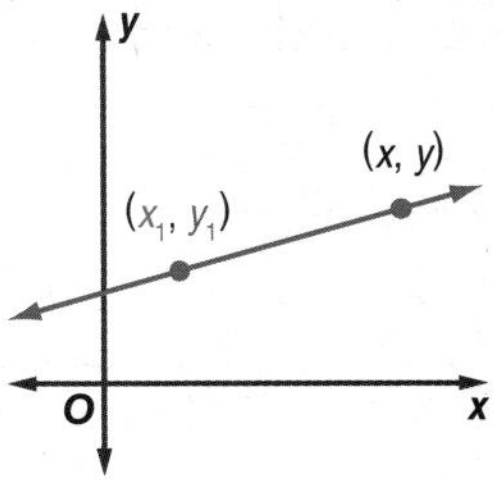

Example 1 Write and Graph an Equation in Point-Slope Form

Write an equation in point-slope form for the line that passes through (3, −2) with a slope of $\frac{1}{4}$. Then graph the equation.

$$y - y_1 = m(x - x_1) \quad \text{Point-slope form}$$

$$y - (-2) = \frac{1}{4}(x - 3) \quad (x_1, y_1) = (3, -2), m = \frac{1}{4}$$

$$y + 2 = \frac{1}{4}(x - 3) \quad \text{Simplify.}$$

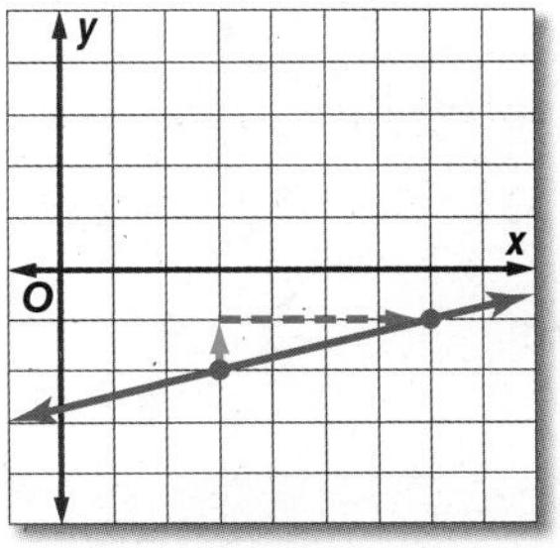

Plot the point at (3, −2) and use the slope to find another point on the line. Draw a line through the two points.

GuidedPractice

1. Write an equation in point-slope form for the line that passes through (−2, 1) with a slope of −6. Then graph the equation.

2 Forms of Linear Equations

If you are given the slope and the coordinates of one or two points, you can write the linear equation in the following ways.

StudyTip

Slope The slope of the line remains unchanged throughout the line. You can go in either direction along the line using the same rise over run and you will always end at a point on the line.

ConceptSummary Writing Equations

Given the Slope and One Point	Given Two Points
Step 1 Substitute the value of m and let the x and y coordinates be (x_1, y_1). Or, substitute the value of m, x, and y into the slope-intercept form and solve for b.	**Step 1** Find the slope.
Step 2 Rewrite the equation in the needed form.	**Step 2** Choose one of the two points to use.
	Step 3 Follow the steps for writing an equation given the slope and one point.

ReviewVocabulary

Standard form of a linear equation $Ax + By = C$, where $A \geq 0$, A and B are not both zero, and A, B, and C are integers with a greatest common factor of 1

Example 2 Standard Form

PT

Write $y - 1 = -\frac{2}{3}(x - 5)$ in standard form.

$y - 1 = -\frac{2}{3}(x - 5)$	Original equation
$3(y - 1) = 3\left(-\frac{2}{3}\right)(x - 5)$	Multiply each side by 3 to eliminate the fraction.
$3(y - 1) = -2(x - 5)$	Simplify.
$3y - 3 = -2x + 10$	Distributive Property
$3y = -2x + 13$	Add 3 to each side.
$2x + 3y = 13$	Add $2x$ to each side.

GuidedPractice

2. Write $y - 1 = 7(x + 5)$ in standard form.

To find the y-intercept of an equation, rewrite the equation in slope-intercept form.

Example 3 Slope-Intercept Form

PT

Write $y + 3 = \frac{3}{2}(x + 1)$ in slope-intercept form.

$y + 3 = \frac{3}{2}(x + 1)$	Original equation
$y + 3 = \frac{3}{2}x + \frac{3}{2}$	Distributive Property
$y = \frac{3}{2}x - \frac{3}{2}$	Subtract 3 from each side.

GuidedPractice

3. Write $y + 6 = -3(x - 4)$ in slope-intercept form.

Being able to use a variety of forms of linear equations can be useful in other subjects as well.

StudyTip

Slopes in Squares
Nonvertical opposite sides of a square have equal slopes. If the coordinates for one of the vertices are unavailable, use the slope of the opposite side.

Example 4 Point-Slope Form and Standard Form

GEOMETRY The figure shows square *RSTU*.

a. Write an equation in point-slope form for the line containing side $\overline{TU}$.

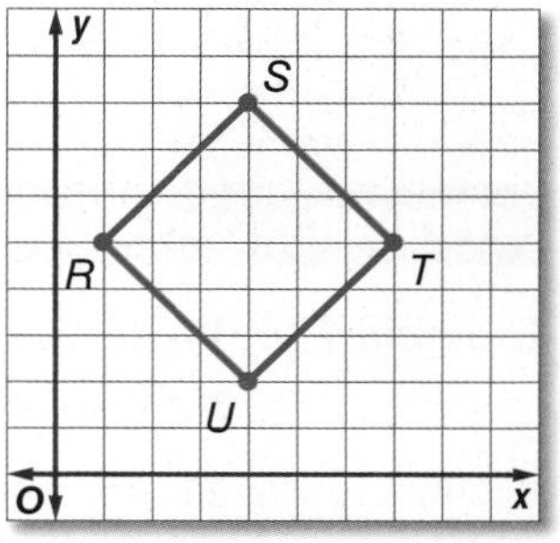

Step 1 Find the slope of $\overline{TU}$.

$m = \frac{y_2 - y_1}{x_2 - x_1}$ — Slope Formula

$= \frac{5-2}{7-4}$ or 1 — $(x_1, y_1) = (4, 2)$ and $(x_2, y_2) = (7, 5)$

Step 2 You can select either point for (x_1, y_1) in the point-slope form.

$y - y_1 = m(x - x_1)$ — Point-slope form

$y - 2 = 1(x - 4)$ — $(x_1, y_1) = (4, 2)$

$y - 5 = 1(x - 7)$ — $(x_1, y_1) = (7, 5)$

b. Write an equation in standard form for the same line.

$y - 2 = 1(x - 4)$	Original equation	$y - 5 = 1(x - 7)$
$y - 2 = 1x - 4$	Distributive Property	$y - 5 = 1x - 7$
$y = 1x - 2$	Add to each side.	$y = 1x - 2$
$-1x + y = -2$	Subtract 1*x* from each side.	$-1x + y = -2$
$x - y = 2$	Multiply each side by −1.	$x - y = 2$

GuidedPractice

4A. Write an equation in point-slope form of the line containing side $\overline{ST}$.

4B. Write an equation in standard form of the line containing $\overline{ST}$.

Check Your Understanding

Example 1 **Write an equation in point-slope form for the line that passes through the given point with the slope provided. Then graph the equation.**

1. $(-2, 5)$, slope -6 **2.** $(-2, -8)$, slope $\frac{5}{6}$ **3.** $(4, 3)$, slope $-\frac{1}{2}$

Example 2 **Write each equation in standard form.**

4. $y + 2 = \frac{7}{8}(x - 3)$ **5.** $y + 7 = -5(x + 3)$ **6.** $y + 2 = \frac{5}{3}(x + 6)$

Example 3 **Write each equation in slope-intercept form.**

7. $y - 10 = 4(x + 6)$ **8.** $y - 7 = -\frac{3}{4}(x + 5)$ **9.** $y - 9 = x + 4$

Example 4 **10. GEOMETRY** Use right triangle FGH.

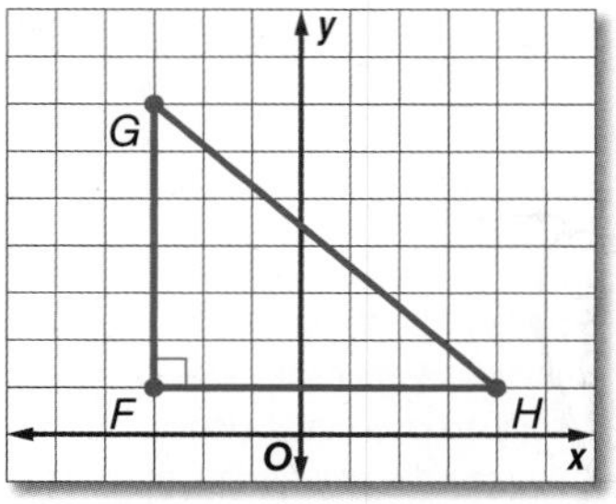

a. Write an equation in point-slope form for the line containing $\overline{GH}$.

b. Write the standard form of the line containing $\overline{GH}$.

Practice and Problem Solving

Example 1 **Write an equation in point-slope form for the line that passes through each point with the given slope. Then graph the equation.**

11. $(5, 3), m = 7$

12. $(2, -1), m = -3$

13. $(-6, -3), m = -1$

14. $(-7, 6), m = 0$

15. $(-2, 11), m = \frac{4}{3}$

16. $(-6, -8), m = -\frac{5}{8}$

17. $(-2, -9), m = -\frac{7}{5}$

18. $(-6, 0)$, horizontal line

Example 2 **Write each equation in standard form.**

19. $y - 10 = 2(x - 8)$

20. $y - 6 = -3(x + 2)$

21. $y - 9 = -6(x + 9)$

22. $y + 4 = \frac{2}{3}(x + 7)$

23. $y + 7 = \frac{9}{10}(x + 3)$

24. $y + 7 = -\frac{3}{2}(x + 1)$

25. $2y + 3 = -\frac{1}{3}(x - 2)$

26. $4y - 5x = 3(4x - 2y + 1)$

Example 3 **Write each equation in slope-intercept form.**

27. $y - 6 = -2(x - 7)$

28. $y - 11 = 3(x + 4)$

29. $y + 5 = -6(x + 7)$

30. $y - 1 = \frac{4}{5}(x + 5)$

31. $y + 2 = \frac{1}{6}(x - 4)$

32. $y + 6 = -\frac{3}{4}(x + 8)$

33. $y + 3 = -\frac{1}{3}(2x + 6)$

34. $y + 4 = 3(3x + 3)$

Example 4 **35** **MOVIE RENTALS** The number of copies of a movie rented at a video kiosk decreased at a constant rate of 5 copies per week. The 6th week after the movie was released, 4 copies were rented. How many copies were rented during the second week?

36. **CCSS REASONING** A company offers premium cable for \$39.95 per month plus a one-time setup fee. The total cost for setup and 6 months of service is \$264.70.

a. Write an equation in point-slope form to find the total price y for any number of months x. (*Hint*: The point (6, 264.70) is a solution to the equation.)

b. Write the equation in slope-intercept form.

c. What is the setup fee?

Write an equation for the line described in standard form.

37. through $(-1, 7)$ and $(8, -2)$

38. through $(-4, 3)$ with y-intercept 0

39. with x-intercept 4 and y-intercept 5

Write an equation in point-slope form for each line.

40.

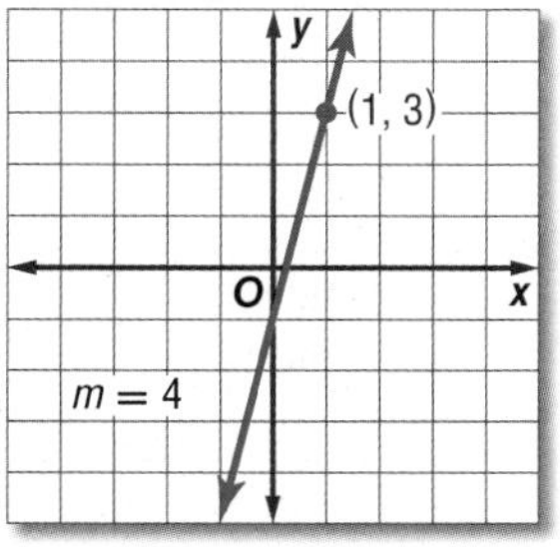

41.

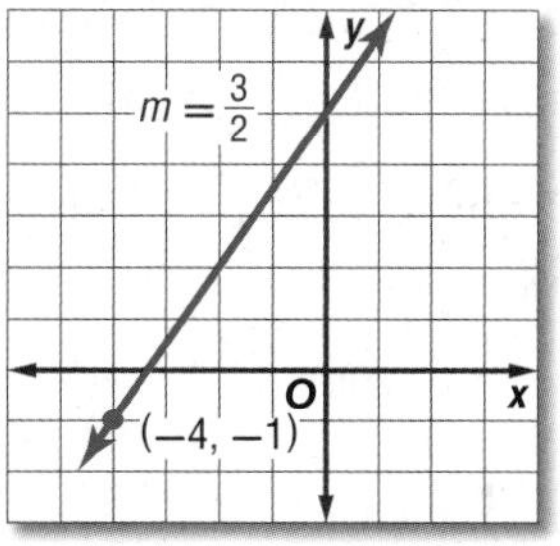

42.

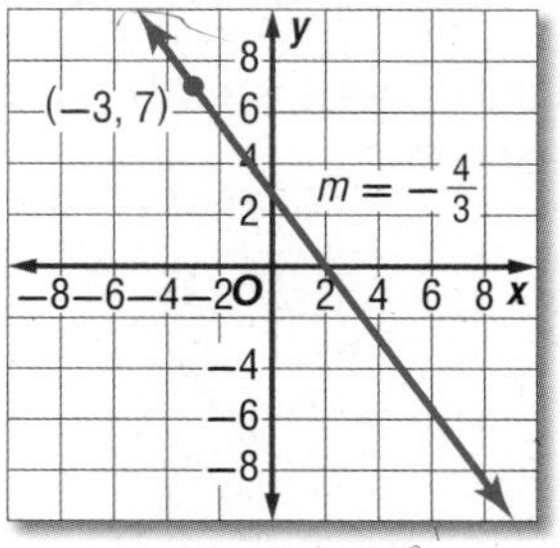

Write each equation in slope-intercept form.

43. $y + \frac{3}{5} = x - \frac{2}{5}$

44. $y - \frac{7}{2} = \frac{1}{2}(x - 4)$

45. $y + \frac{1}{3} = \frac{5}{6}\left(x + \frac{2}{5}\right)$

46. Write an equation in point-slope form, slope-intercept form, and standard form for a line that passes through $(-2, 8)$ with slope $\frac{8}{5}$.

47. Line ℓ passes through $(-9, 4)$ with slope $\frac{4}{7}$. Write an equation in point-slope form, slope-intercept form, and standard form for line ℓ.

48. **WEATHER** The barometric pressure is 598 millimeters of mercury (mmHg) at an altitude of 1.8 kilometers and 577 millimeters of mercury at 2.1 kilometers.

a. Write a formula for the barometric pressure as a function of the altitude.

b. What is the altitude if the pressure is 657 millimeters of mercury?

H.O.T. Problems Use Higher-Order Thinking Skills

49. **WHICH ONE DOESN'T BELONG?** Identify the equation that does not belong. Explain your reasoning.

$y - 5 = 3(x - 1)$	$y + 1 = 3(x + 1)$	$y + 4 = 3(x + 1)$	$y - 8 = 3(x - 2)$

50. CCSS **CRITIQUE** Juana thinks that $f(x)$ and $g(x)$ have the same slope but different intercepts. Sabrina thinks that $f(x)$ and $g(x)$ describe the same line. Is either of them correct? Explain your reasoning.

The graph of $g(x)$ is the line that passes through $(3, -7)$ and $(-6, 4)$.

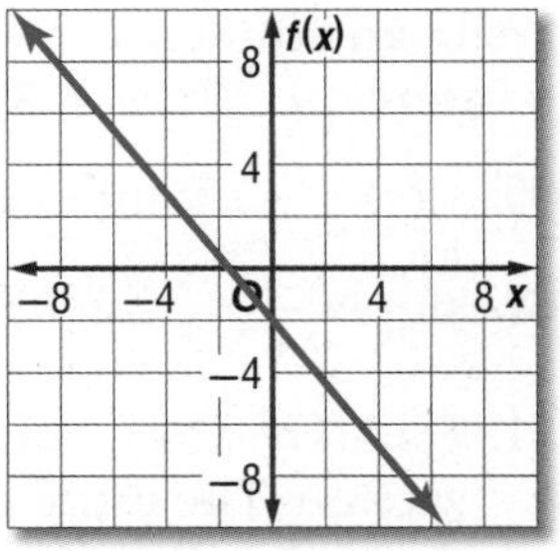

51. **OPEN ENDED** Describe a real-life scenario that has a constant rate of change and a value of y for a particular value of x. Represent this situation using an equation in point-slope form, an equation in standard form, and an equation in slope-intercept form.

52. **REASONING** Write an equation for the line that passes through $(-4, 8)$ and $(3, -7)$. What is the slope? Where does the line intersect the x-axis? the y-axis?

53. **CHALLENGE** Write an equation in point-slope form for the line that passes through the points (f, g) and (h, j).

54. **WRITING IN MATH** Why do we represent linear equations in more than one form?

Standardized Test Practice

55. Which statement is *most* strongly supported by the graph?

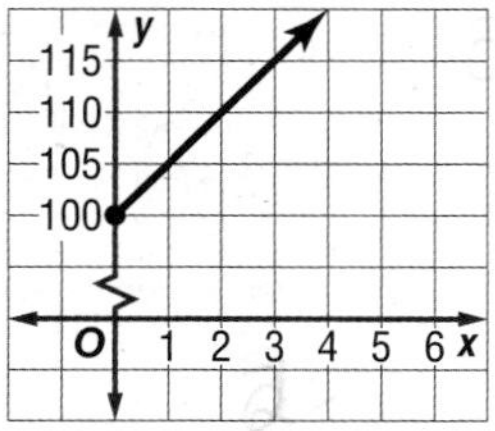

A You have $100 and spend $5 weekly.
B You have $100 and save $5 weekly.
C You need $100 for a new CD player and save $5 weekly.
D You need $100 for a new CD player and spend $5 weekly.

56. SHORT RESPONSE A store offers customers a $5 gift certificate for every $75 they spend. How much would a customer have to spend to earn $35 worth of gift certificates?

57. GEOMETRY Which triangle is similar to $\triangle ABC$?

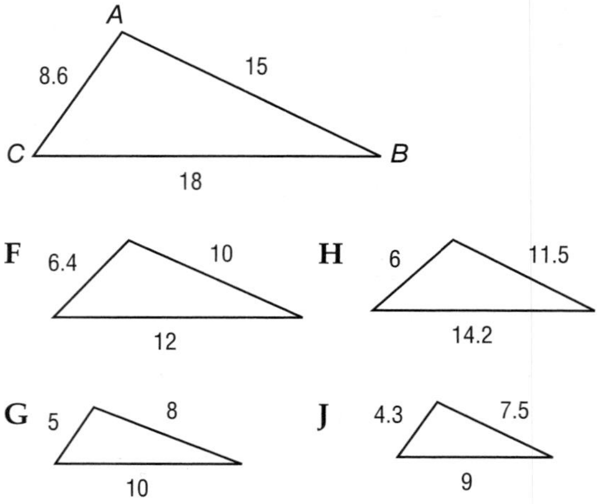

58. In a class of 25 students, 6 have blue eyes, 15 have brown hair, and 3 have blue eyes and brown hair. How many students have neither blue eyes nor brown hair?

A 4
B 7
C 10
D 22

Spiral Review

Write an equation of the line that passes through each pair of points. (Lesson 4-2)

59. (4, 2), (−2, −4)
60. (3, −2), (6, 4)
61. (−1, 3), (2, −3)
62. (2, −2), (3, 2)
63. (7, −2), (−4, −2)
64. (0, 5), (−3, 5)

Write an equation in slope-intercept form of the line with the given slope and *y*-intercept. (Lesson 4-1)

65. slope: −2, *y*-intercept: 6
66. slope: 3, *y*-intercept: −5
67. slope: $\frac{1}{2}$, *y*-intercept: 3
68. slope: $-\frac{3}{5}$, *y*-intercept: 12
69. slope: 0, *y*-intercept: 3
70. slope: −1, *y*-intercept: 0

71. THEATER The Coral Gables Actors' Playhouse has 7 rows of seats in the orchestra section. The number of seats in the rows forms an arithmetic sequence, as shown in the table. On opening night, 368 tickets were sold for the orchestra section. Was the section oversold? (Lesson 3-5)

Rows	Number of Seats
7	76
6	68
5	60

Skills Review

Solve each equation or formula for the variable specified.

72. $y = mx + b$, for m
73. $v = r + at$, for a
74. $km + 5x = 6y$, for m
75. $4b - 5 = -t$, for b

LESSON

4-4 Parallel and Perpendicular Lines

Then

- You wrote equations in point-slope form.

Now

1. Write an equation of the line that passes through a given point, parallel to a given line.
2. Write an equation of the line that passes through a given point, perpendicular to a given line.

Why?

- Notice the squares, rectangles and lines in the piece of art shown at the right. Some of the lines intersect forming right angles. Other lines do not intersect at all.

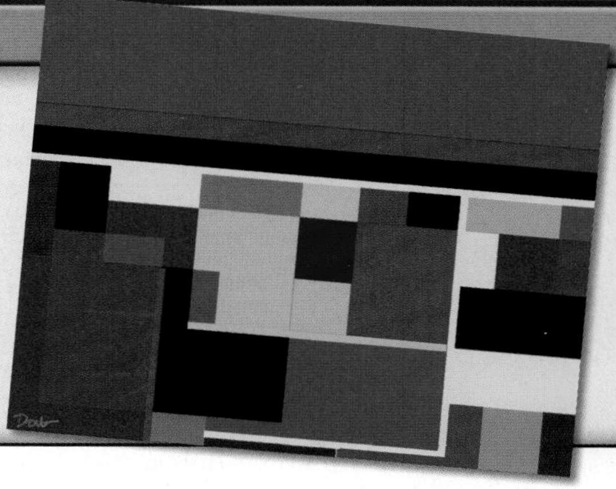

NewVocabulary
parallel lines
perpendicular lines

Common Core State Standards

Content Standards
F.LE.2 Construct linear and exponential functions, including arithmetic and geometric sequences, given a graph, a description of a relationship, or two input-output pairs (include reading these from a table).
S.ID.7 Interpret the slope (rate of change) and the intercept (constant term) of a linear model in the context of the data.

Mathematical Practices
5 Use appropriate tools strategically.

1 Parallel Lines

Lines in the same plane that do not intersect are called **parallel lines**. Nonvertical parallel lines have the same slope.

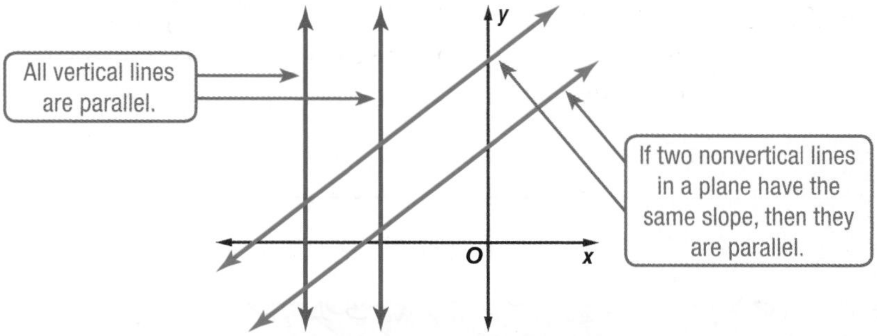
All vertical lines are parallel.
If two nonvertical lines in a plane have the same slope, then they are parallel.
y
x
O

You can write an equation of a line parallel to a given line if you know a point on the line and an equation of the given line. First find the slope of the given line. Then, substitute the point provided and the slope from the given line into the point-slope form.

PT

Example 1 Parallel Line Through a Given Point

Write an equation in slope-intercept form for the line that passes through $(-3, 5)$ and is parallel to the graph of $y = 2x - 4$.

Step 1 The slope of the line with equation $y = 2x - 4$ is 2. The line parallel to $y = 2x - 4$ has the same slope, 2.

Step 2 Find the equation in slope-intercept form.

$y - y_1 = m(x - x_1)$	Point-slope form
$y - 5 = 2[x - (-3)]$	Replace m with 2 and (x_1, y_1) with $(-3, 5)$.
$y - 5 = 2(x + 3)$	Simplify.
$y - 5 = 2x + 6$	Distributive Property
$y - 5 + 5 = 2x + 6 + 5$	Add 5 to each side.
$y = 2x + 11$	Write the equation in slope-intercept form.

GuidedPractice

1. Write an equation in point-slope form for the line that passes through $(4, -1)$ and is parallel to the graph of $y = \frac{1}{4}x + 7$.

Review Vocabulary

opposite reciprocals

The opposite reciprocal of $\frac{a}{b}$ is $-\frac{b}{a}$. Their product is -1.

2 Perpendicular Lines

Lines that intersect at right angles are called **perpendicular lines**. The slopes of nonvertical perpendicular lines are opposite reciprocals. That is, if the slope of a line is 4, the slope of the line perpendicular to it is $-\frac{1}{4}$.

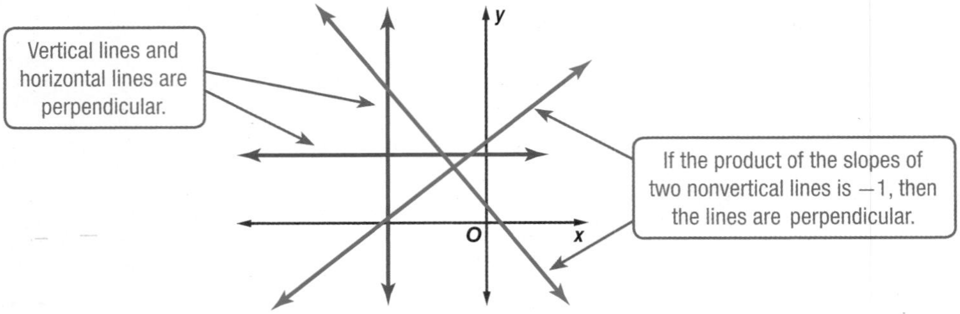

Vertical lines and horizontal lines are perpendicular.
If the product of the slopes of two nonvertical lines is −1, then the lines are perpendicular.
y
x
O

You can use slope to determine whether two lines are perpendicular.

Real-World Example 2 Slopes of Perpendicular Lines

PT

DESIGN The outline of a company's new logo is shown on a coordinate plane.

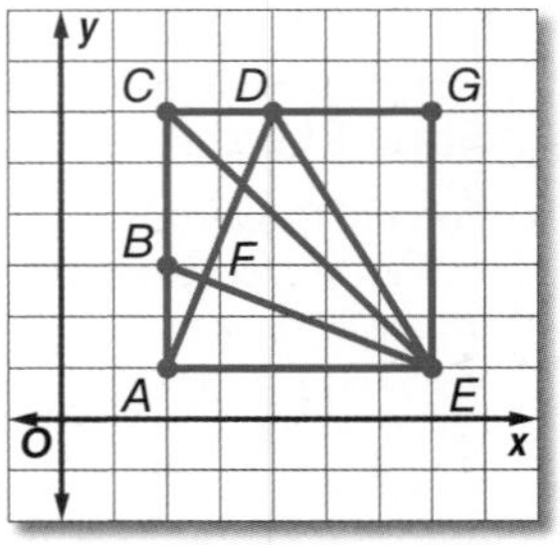

y
C
D
G
B
F
A
E
O
x

a. Is $\angle DFE$ a right angle in the logo?

If $\overline{BE}$ and $\overline{AD}$ are perpendicular, then $\angle DFE$ is a right angle. Find the slopes of $\overline{BE}$ and $\overline{AD}$.

slope of $\overline{BE}$: $m = \frac{1-3}{7-2}$ or $-\frac{2}{5}$

slope of $\overline{AD}$: $m = \frac{6-1}{4-2}$ or $\frac{5}{2}$

The line segments are perpendicular because $-\frac{2}{5} \times \frac{5}{2} = -1$. Therefore, $\angle DFE$ is a right angle.

b. Is each pair of opposite sides parallel?

If a pair of opposite sides are parallel, then they have the same slope.

slope of $\overline{AC}$: $m = \frac{6-1}{2-2}$ or undefined

Since $\overline{AC}$ and $\overline{GE}$ are both parallel to the y-axis, they are vertical and are therefore parallel.

slope of $\overline{CG}$: $m = \frac{6-6}{7-2}$ or 0

Since $\overline{CG}$ and $\overline{AE}$ are both parallel to the x-axis, they are horizontal and are therefore parallel.

Guided Practice

2. CONSTRUCTION On the plans for a treehouse, a beam represented by $\overline{QR}$ has endpoints $Q(-6, 2)$ and $R(-1, 8)$. A connecting beam represented by $\overline{ST}$ has endpoints $S(-3, 6)$ and $T(-8, 5)$. Are the beams perpendicular? Explain.

Real-World Link

Though treehouses are typically built for recreational purposes, they were originally designed as a way to be protected from wild animals, dense population, and other threats.

Source: *The Treehouse Book*

You can determine whether the graphs of two linear equations are parallel or perpendicular by comparing the slopes of the lines.

Example 3 Parallel or Perpendicular Lines

Determine whether the graphs of $y = 5$, $x = 3$, and $y = -2x + 1$ are *parallel* or *perpendicular*. Explain.

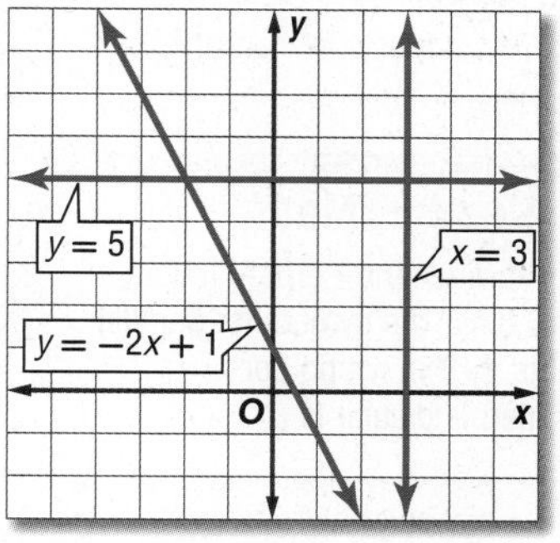

Graph each line on a coordinate plane.

From the graph, you can see that $y = 5$ is parallel to the x-axis and $x = 3$ is parallel to the y-axis. Therefore, they are perpendicular. None of the lines are parallel.

Guided Practice

3. Determine whether the graphs of $6x - 2y = -2$, $y = 3x - 4$, and $y = 4$ are *parallel* or *perpendicular*. Explain.

You can write the equation of a line perpendicular to a given line if you know a point on the line and the equation of the given line.

Study Tip

CCSS Tools Graph the given equation on a coordinate grid and plot the given point. Using a ruler, draw a line perpendicular to the given line that passes through the point.

Example 4 Perpendicular Line Through a Given Point

Write an equation in slope-intercept form for the line that passes through $(-4, 6)$ and is perpendicular to the graph of $2x + 3y = 12$.

Step 1 Find the slope of the given line by solving the equation for y.

$2x + 3y = 12$	Original equation
$2x - 2x + 3y = -2x + 12$	Subtract $2x$ from each side.
$3y = -2x + 12$	Simplify.
$\frac{3y}{3} = \frac{-2x + 12}{3}$	Divide each side by 3.
$y = -\frac{2}{3}x + 4$	Simplify.

The slope is $-\frac{2}{3}$.

Step 2 The slope of the perpendicular line is the opposite reciprocal of $-\frac{2}{3}$ or $\frac{3}{2}$. Find the equation of the perpendicular line.

$y - y_1 = m(x - x_1)$	Point-slope form
$y - 6 = \frac{3}{2}[x - (-4)]$	$(x_1, y_1) = (-4, 6)$ and $m = \frac{3}{2}$
$y - 6 = \frac{3}{2}(x + 4)$	Simplify.
$y - 6 = \frac{3}{2}x + 6$	Distributive Property
$y - 6 + 6 = \frac{3}{2}x + 6 + 6$	Add 6 to each side.
$y = \frac{3}{2}x + 12$	Simplify.

Guided Practice

4. Write an equation in slope-intercept form for the line that passes through $(4, 7)$ and is perpendicular to the graph of $y = \frac{2}{3}x - 1$.

ReadingMath

Parallel and Perpendicular Lines The symbol for parallel is ||. The symbol for perpendicular is ⊥.

ConceptSummary Parallel and Perpendicular Lines

	Parallel Lines	Perpendicular Lines
Words	Two nonvertical lines are parallel if they have the same slope.	Two nonvertical lines are perpendicular if the product of their slopes is −1.
Symbols	$\overleftrightarrow{AB} \parallel \overleftrightarrow{CD}$	$\overleftrightarrow{EF} \perp \overleftrightarrow{GH}$
Models	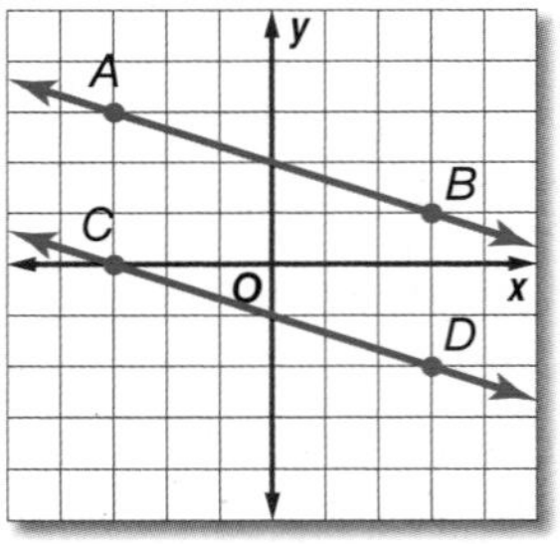	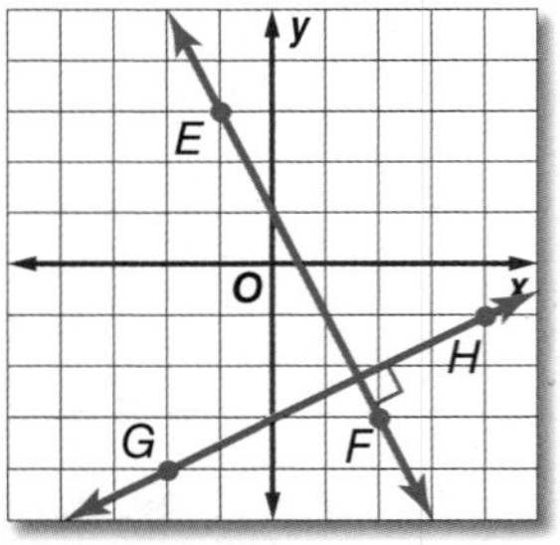

Check Your Understanding

Example 1 **Write an equation in slope-intercept form for the line that passes through the given point and is parallel to the graph of the given equation.**

1. $(-1, 2), y = \frac{1}{2}x - 3$

2. $(0, 4), y = -4x + 5$

Example 2

3. **GARDENS** A garden is in the shape of a quadrilateral with vertices $A(-2, 1)$, $B(3, -3)$, $C(5, 7)$, and $D(-3, 4)$. Two paths represented by $\overline{AC}$ and $\overline{BD}$ cut across the garden. Are the paths perpendicular? Explain.

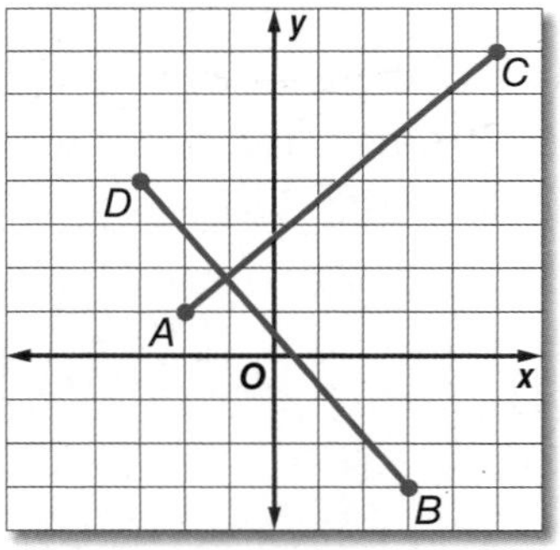

4. **CCSS PRECISION** A square is a quadrilateral that has opposite sides parallel, consecutive sides that are perpendicular, and diagonals that are perpendicular. Determine whether the quadrilateral is a square. Explain.

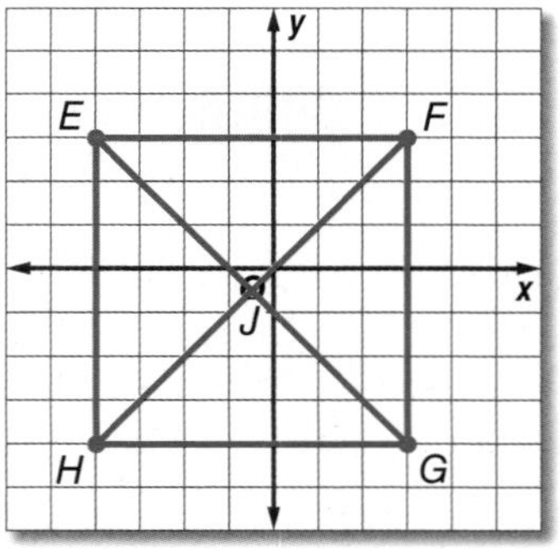

Example 3 **Determine whether the graphs of the following equations are *parallel* or *perpendicular*. Explain.**

5. $y = -2x, 2y = x, 4y = 2x + 4$

6. $y = \frac{1}{2}x, 3y = x, y = -\frac{1}{2}x$

Example 4 **Write an equation in slope-intercept form for the line that passes through the given point and is perpendicular to the graph of the equation.**

7. $(-2, 3), y = -\frac{1}{2}x - 4$

8. $(-1, 4), y = 3x + 5$

9. $(2, 3), 2x + 3y = 4$

10. $(3, 6), 3x - 4y = -2$

Practice and Problem Solving

Example 1 **Write an equation in slope-intercept form for the line that passes through the given point and is parallel to the graph of the given equation.**

11. $(3, -2), y = x + 4$ **12.** $(4, -3), y = 3x - 5$ **13.** $(0, 2), y = -5x + 8$

14. $(-4, 2), y = -\frac{1}{2}x + 6$ **15.** $(-2, 3), y = -\frac{3}{4}x + 4$ **16.** $(9, 12), y = 13x - 4$

Example 2 **17. GEOMETRY** A trapezoid is a quadrilateral that has exactly one pair of parallel opposite sides. Is *ABCD* a trapezoid? Explain your reasoning.

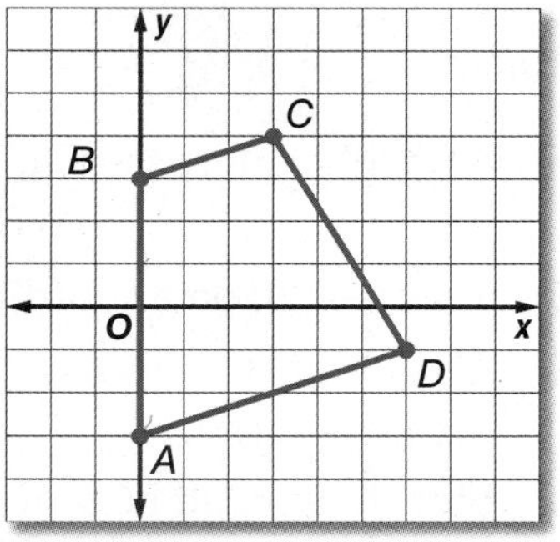

18. GEOMETRY *CDEF* is a kite. Are the diagonals of the kite perpendicular? Explain your reasoning.

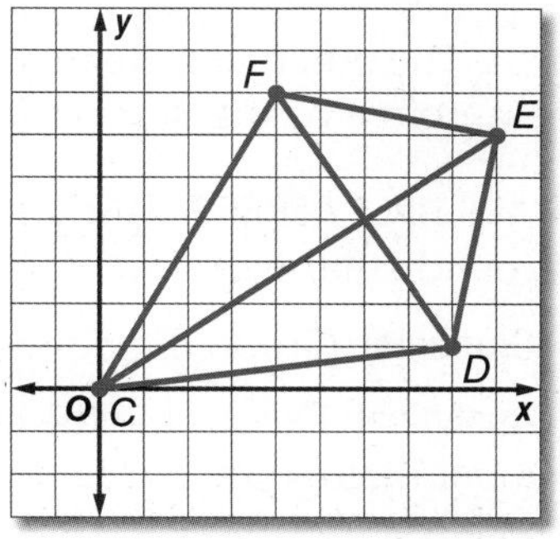

19. Determine whether the graphs of $y = -6x + 4$ and $y = \frac{1}{6}x$ are perpendicular. Explain.

20. MAPS On a map, Elmwood Drive passes through $R(4, -11)$ and $S(0, -9)$, and Taylor Road passes through $J(6, -2)$ and $K(4, -5)$. If they are straight lines, are the two streets perpendicular? Explain.

Example 3 **CCSS PERSEVERANCE** **Determine whether the graphs of the following equations are *parallel* or *perpendicular*. Explain.**

21. $2x - 8y = -24, 4x + y = -2, x - 4y = 4$

22. $3x - 9y = 9, 3y = x + 12, 2x - 6y = 12$

Example 4 **Write an equation in slope-intercept form for the line that passes through the given point and is perpendicular to the graph of the equation.**

23 $(-3, -2), y = -2x + 4$ **24.** $(-5, 2), y = \frac{1}{2}x - 3$ **25.** $(-4, 5), y = \frac{1}{3}x + 6$

26. $(2, 6), y = -\frac{1}{4}x + 3$ **27.** $(3, 8), y = 5x - 3$ **28.** $(4, -2), y = 3x + 5$

Write an equation in slope-intercept form for a line perpendicular to the graph of the equation that passes through the *x*-intercept of that line.

29. $y = -\frac{1}{2}x - 4$ **30.** $y = \frac{2}{3}x - 6$ **31.** $y = 5x + 3$

32. Write an equation in slope-intercept form for the line that is perpendicular to the graph of $3x + 2y = 8$ and passes through the *y*-intercept of that line.

Determine whether the graphs of each pair of equations are *parallel, perpendicular,* or *neither*.

33. $y = 4x + 3$
$4x + y = 3$

34. $y = -2x$
$2x + y = 3$

35. $3x + 5y = 10$
$5x - 3y = -6$

36. $-3x + 4y = 8$
$-4x + 3y = -6$

37. $2x + 5y = 15$
$3x + 5y = 15$

38. $2x + 7y = -35$
$4x + 14y = -42$

39. Write an equation of the line that is parallel to the graph of $y = 7x - 3$ and passes through the origin.

40. EXCAVATION Scientists excavating a dinosaur mapped the site on a coordinate plane. If one bone lies from $(-5, 8)$ to $(10, -1)$ and a second bone lies from $(-10, -3)$ to $(-5, -6)$, are the bones parallel? Explain.

41 ARCHAEOLOGY In the ruins of an ancient civilization, an archaeologist found pottery at $(2, 6)$ and hair accessories at $(4, -1)$. A pole is found with one end at $(7, 10)$ and the other end at $(14, 12)$. Is the pole perpendicular to the line through the pottery and the hair accessories? Explain.

42. GRAPHICS To create a design on a computer, Andeana must enter the coordinates for points on the design. One line segment she drew has endpoints of $(-2, 1)$ and $(4, 3)$. The other coordinates that Andeana entered are $(2, -7)$ and $(8, -3)$. Could these points be the vertices of a rectangle? Explain.

43. MULTIPLE REPRESENTATIONS In this problem, you will explore parallel and perpendicular lines.

a. Graphical Graph the points $A(-3, 3)$, $B(3, 5)$, and $C(-4, 0)$ on a coordinate plane.

b. Analytical Determine the coordinates of a fourth point D that would form a parallelogram. Explain your reasoning.

c. Analytical What is the minimum number of points that could be moved to make the parallelogram a rectangle? Describe which points should be moved, and explain why.

H.O.T. Problems Use Higher-Order Thinking Skills

44. CHALLENGE If the line through $(-2, 4)$ and $(5, d)$ is parallel to the graph of $y = 3x + 4$, what is the value of d?

45. REASONING Which key features of the graphs of two parallel lines are the same, and which are different? Which key features of the graphs of two perpendicular lines are the same, and which are different?

46. OPEN ENDED Graph a line that is parallel and a line that is perpendicular to $y = 2x - 1$.

Example 3

47. CCSS CRITIQUE Carmen and Chase are finding an equation of the line that is perpendicular to the graph of $y = \frac{1}{3}x + 2$ and passes through the point $(-3, 5)$. Is either of them correct? Explain your reasoning.

Carmen	Chase
$y - 5 = -3[x - (-3)]$	$y - 5 = 3[x - (-3)]$
$y - 5 = -3(x + 3)$	$y - 5 = 3(x + 3)$
$y = -3x - 9 + 5$	$y = 3x + 9 + 5$
$y = -3x - 4$	$y = 3x + 14$

48. WRITING IN MATH Illustrate how you can determine whether two lines are parallel or perpendicular. Write an equation for the graph that is parallel and an equation for the graph that is perpendicular to the line shown. Explain your reasoning.

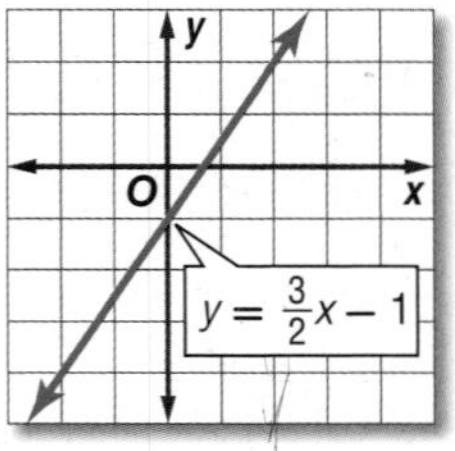

Standardized Test Practice

49. Which of the following is an algebraic translation of the following phrase?

5 less than the quotient of a number and 8

A $5 - \frac{n}{8}$

B $\frac{n}{8} - 5$

C $5 - \frac{8}{n}$

D $\frac{8}{n} - 5$

50. A line through which two points would be parallel to a line with a slope of $\frac{3}{4}$?

F (0, 5) and (−4, 2)

G (0, 2) and (−4, 1)

H (0, 0) and (0, −2)

J (0, −2) and (−4, −2)

51. Which equation best fits the data in the table?

A $y = x + 4$

B $y = 2x + 3$

C $y = 7$

D $y = 4x - 5$

x	y
1	5
2	7
3	9
4	11

52. SHORT RESPONSE Tyler is filling his 6000-gallon pool at a constant rate. After 4 hours, the pool contained 800 gallons. How many total hours will it take to completely fill the pool?

Spiral Review

Write each equation in standard form. (Lesson 4-3)

53. $y - 13 = 4(x - 2)$

54. $y - 5 = -2(x + 2)$

55. $y + 3 = -5(x + 1)$

56. $y + 7 = \frac{1}{2}(x + 2)$

57. $y - 1 = \frac{5}{6}(x - 4)$

58. $y - 2 = -\frac{2}{5}(x - 8)$

59. CANOE RENTAL Latanya and her friends rented a canoe for 3 hours and paid a total of \$45. (Lesson 4-2)

a. Write a linear equation to find the total cost C of renting the canoe for h hours.

b. How much would it cost to rent the canoe for 8 hours?

Write an equation of the line that passes through each point with the given slope. (Lesson 4-2)

60. $(5, -2), m = 3$

61. $(-5, 4), m = -5$

62. $(3, 0), m = -2$

63. $(3, 5), m = 2$

64. $(-3, -1), m = -3$

65. $(-2, 4), m = -5$

Simplify each expression. If not possible, write *simplified*. (Lesson 1-4)

66. $13m + m$

67. $14a^2 + 13b^2 + 27$

68. $3(x + 2x)$

69. FINANCIAL LITERACY At a Farmers' Market, merchants can rent a small table for \$5.00 and a large table for \$8.50. One time, 25 small and 10 large tables were rented. Another time, 35 small and 12 large were rented. (Lesson 1-2)

a. Write an algebraic expression to show the total amount of money collected.

b. Evaluate the expression.

Skills Review

Express each relation as a graph. Then determine the domain and range.

70. $\{(3, 8), (3, 7), (2, -9), (1, -9), (-5, -3)\}$

71. $\{(3, 4), (4, 3), (2, 2), (5, -4), (-4, 5)\}$

72. $\{(0, 2), (-5, 1), (0, 6), (-1, 9), (-4, -5)\}$

73. $\{(-7, 6), (-3, -4), (4, -5), (-2, 6), (-3, 2)\}$

CHAPTER 4

Mid-Chapter Quiz

Lessons 4-1 through 4-4

Write an equation in slope-intercept form for each graph shown. (Lesson 4-1)

1.

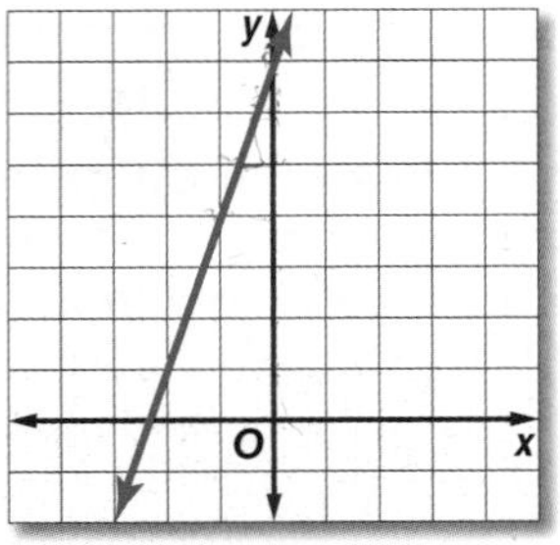

2.

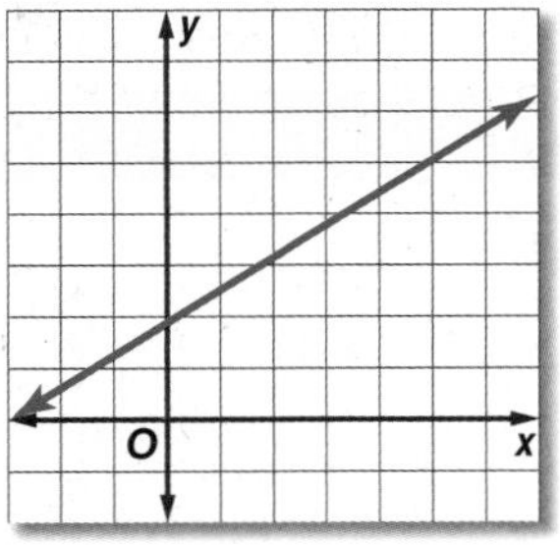

Graph each equation. (Lesson 4-1)

3. $y = 2x + 3$

4. $y = \frac{1}{3}x - 2$

5. BOATS Write an equation in slope-intercept form for the total rental cost C for a pontoon boat used for t hours. (Lesson 4-1)

Write an equation of the line with the given conditions. (Lesson 4-2)

6. (2, 5); slope 3

7. (−3, −1), slope $\frac{1}{2}$

8. (−3, 4), (1, 12)

9. (−1, 6), (2, 4)

10. (2, 1), slope 0

11. MULTIPLE CHOICE Write an equation of the line that passes through the point (0, 0) and has slope −4. (Lesson 4-2)

A $y = x - 4$

B $y = x + 4$

C $y = -4x$

D $y = 4 - x$

Write an equation in point-slope form for the line that passes through each point with the given slope. (Lesson 4-3)

12. (1, 4), $m = 6$

13. (−2, −1), $m = -3$

14. Write an equation in point-slope form for the line that passes through the point (8, 3), $m = -2$. (Lesson 4-3)

15. Write $y + 3 = \frac{1}{2}(x - 5)$ in standard form. (Lesson 4-3)

16. Write $y + 4 = -7(x - 3)$ in slope-intercept form. (Lesson 4-3)

Write each equation in standard form. (Lesson 4-3)

17. $y - 5 = -2(x - 3)$

18. $y + 4 = \frac{2}{3}(x - 3)$

Write each equation in slope-intercept form. (Lesson 4-3)

19. $y - 3 = 4(x + 3)$

20. $y + 1 = \frac{1}{2}(x - 8)$

21. MULTIPLE CHOICE Determine whether the graphs of the pair of equations are *parallel, perpendicular,* or *neither.* (Lesson 4-4)

$$y = -6x + 8$$
$$3x + \frac{1}{2}y = -3$$

F parallel

G perpendicular

H neither

J not enough information

Write an equation in slope-intercept form for the line that passes through the given point and is perpendicular to the graph of the equation. (Lesson 4-4)

22. (3, −4); $y = -\frac{1}{3}x - 5$

23. (0, −3); $y = -2x + 4$

24. (−4, −5); $-4x + 5y = -6$

25. (−1, −4); $-x - 2y = 0$

LESSON 4-5 Scatter Plots and Lines of Fit

Then

- You wrote linear equations given a point and the slope.

Now

1. Investigate relationships between quantities by using points on scatter plots.
2. Use lines of fit to make and evaluate predictions.

Why?

- The graph shows the number of people from the United States who travel to other countries. The points do not all lie on the same line; however, you may be able to draw a line that is close to all of the points. That line would show a linear relationship between the year *x* and the number of travelers each year *y*. Generally, international travel has increased.

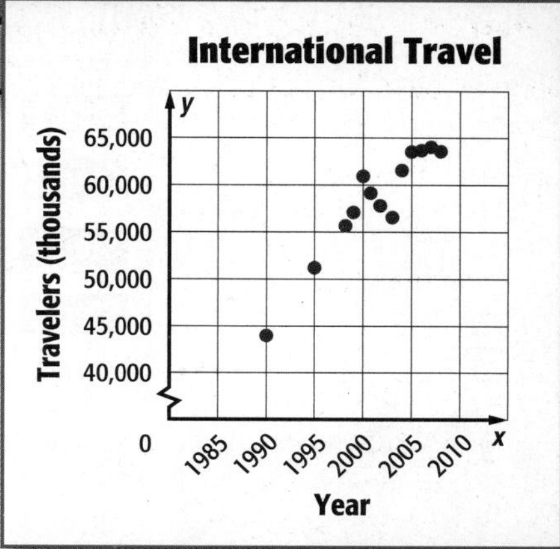

NewVocabulary
bivariate data
scatter plot
line of fit
linear interpolation

Common Core State Standards

Content Standards

S.ID.6a Fit a function to the data; use functions fitted to data to solve problems in the context of the data. Use given functions or choose a function suggested by the context. Emphasize linear, quadratic, and exponential models.

S.ID.6c Fit a linear function for a scatter plot that suggests a linear association.

Mathematical Practices

1 Make sense of problems and persevere in solving them.

4 Model with mathematics.

1 Investigate Relationships Using Scatter Plots

Data with two variables are called **bivariate data**. A **scatter plot** shows the relationship between a set of data with two variables, graphed as ordered pairs on a coordinate plane. Scatter plots are used to investigate a relationship between two quantities.

ConceptSummary Scatter Plots

Positive Correlation

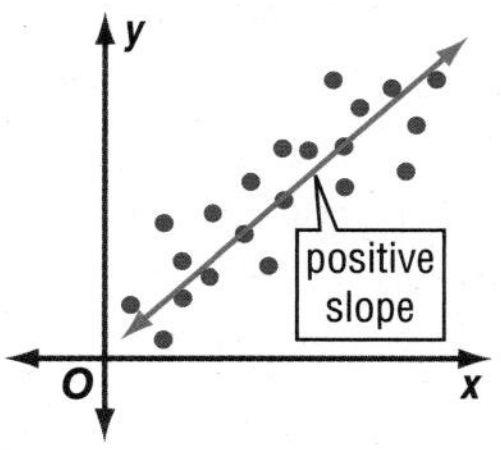

As *x* increases, *y* increases

Negative Correlation

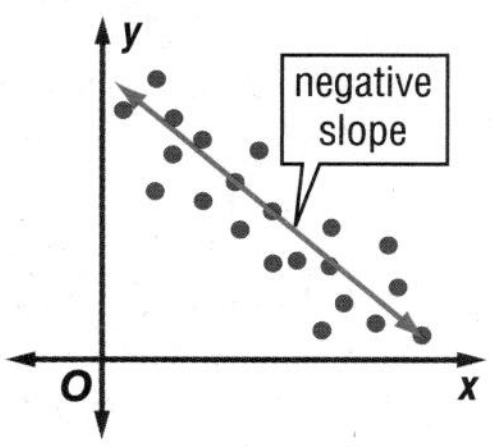

As *x* decreases, *y* decreases

No Correlation

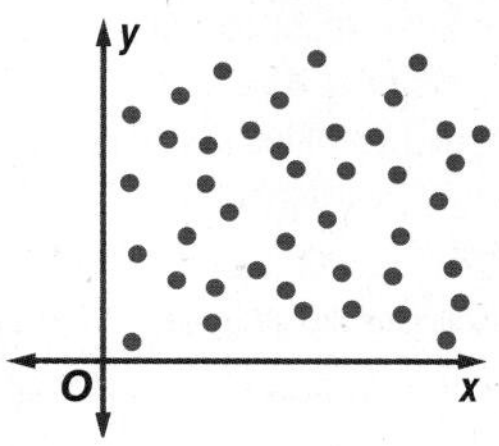

x and *y* are not related

Real-World Example 1 Evaluate a Correlation

WAGES **Determine whether the graph shows a *positive*, *negative*, or *no* correlation. If there is a positive or negative correlation, describe its meaning in the situation.**

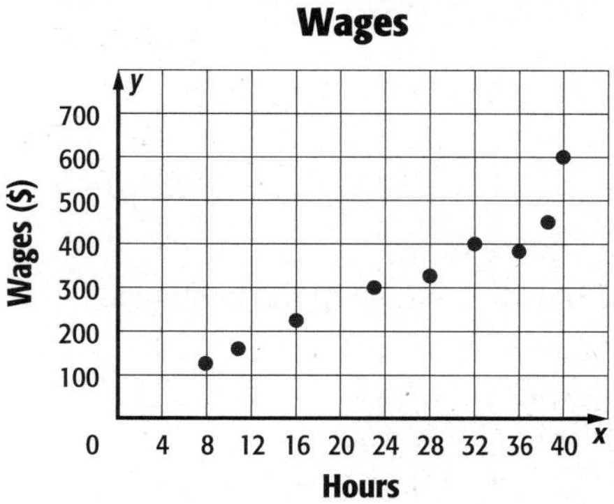

The graph shows a positive correlation. As the number of hours worked increases, the wages usually increase.

GuidedPractice

1. Refer to the graph on international travel. Determine whether the graph shows a *positive*, *negative*, or *no* correlation. If there is a positive or negative correlation, describe its meaning.

2 Use Lines of Fit

Scatter plots can show whether there is a trend in a set of data. When the data points all lie close to a line, a **line of fit** or *trend line* can model the trend.

KeyConcept Using a Linear Function to Model Data

Step 1 Make a scatter plot. Determine whether any relationship exists in the data.

Step 2 Draw a line that seems to pass close to most of the data points.

Step 3 Use two points on the line of fit to write an equation for the line.

Step 4 Use the line of fit to make predictions.

Real-WorldLink

The Kingda Ka roller coaster at Six Flags Great Adventure in Jackson, New Jersey, has broken three records: tallest roller coaster at 456 feet, fastest at 128 miles per hour, and largest vertical drop of 418 feet.

Source: Ultimate Roller Coaster

Real-World Example 2 Write a Line of Fit

ROLLER COASTERS The table shows the largest vertical drops of nine roller coasters in the United States and the number of years after 1988 that they were opened. Identify the independent and the dependent variables. Is there a relationship in the data? If so, predict the vertical drop in a roller coaster built 30 years after 1988.

Years Since 1988	1	3	5	8	12	12	12	13	15
Vertical Drop (ft)	151	155	225	230	306	300	255	255	400

Source: Ultimate Roller Coaster

Step 1 Make a scatter plot.

The independent variable is the year, and the dependent variable is the vertical drop. As the number of years increases, the vertical drop of roller coasters increases. There is a positive correlation between the two variables.

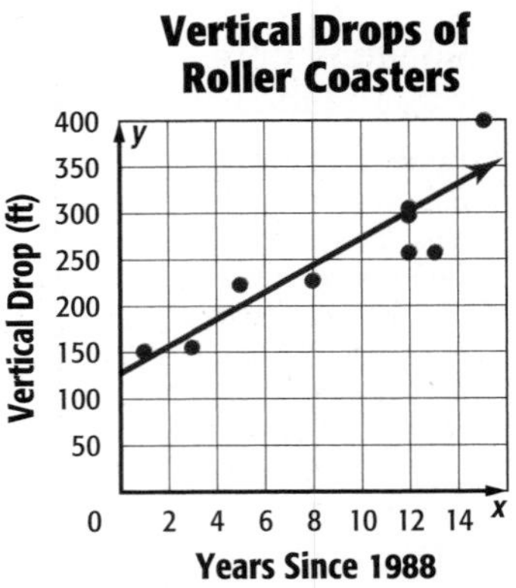

Step 2 Draw a line of fit.

No one line will pass through all of the data points. Draw a line that passes close to the points. A line of fit is shown.

Step 3 Write the slope-intercept form of an equation for the line of fit.

The line of fit passes close to (2, 150) and the data point (12, 300).

Find the slope.

$m = \frac{y_2 - y_1}{x_2 - x_1}$ $(x_1, y_1) = (2, 150)$, $(x_2, y_2) = (12, 300)$

$= \frac{300 - 150}{12 - 2}$

$= \frac{150}{10}$ or 15

Use $m = 15$ and either the point-slope form or the slope-intercept form to write the equation of the line of fit.

$y - y_1 = m(x - x_1)$

$y - 150 = 15(x - 2)$

$y - 150 = 15x - 30$

$y = 15x + 120$

A slope of 15 means that the vertical drops increased an average of 15 feet per year. To predict the vertical drop of a roller coaster built 30 years after 1988, substitute 30 for x in the equation. The vertical drop is $15(30) + 120$ or 570 feet.

Guided Practice

2. MUSIC The table shows the dollar value in millions for the sales of CDs for the year. Make a scatter plot and determine what relationship exists, if any.

Year	2000	2001	2002	2003	2004	2005	2006	2007	2008
Sales	13,215	12,909	12,044	11,233	11,447	10,520	9373	7452	5471

Reading Math

Interpolation and Extrapolation The Latin prefix *inter-* means between, and the Latin prefix *extra-* means beyond.

In Lesson 4-2, you learned that linear extrapolation is used to predict values *outside* the range of the data. You can also use a linear equation to predict values *inside* the range of the data. This is called **linear interpolation**.

Real-World Example 3 Use Interpolation or Extrapolation

TRAVEL Use the scatter plot to find the approximate number of United States travelers to international countries in 1996.

Step 1 Draw a line of fit. The line should be as close to as many points as possible.

Step 2 Write the slope-intercept form of the equation. The line of fit passes through (0, 44,623) and (18, 63,554).

International Travel

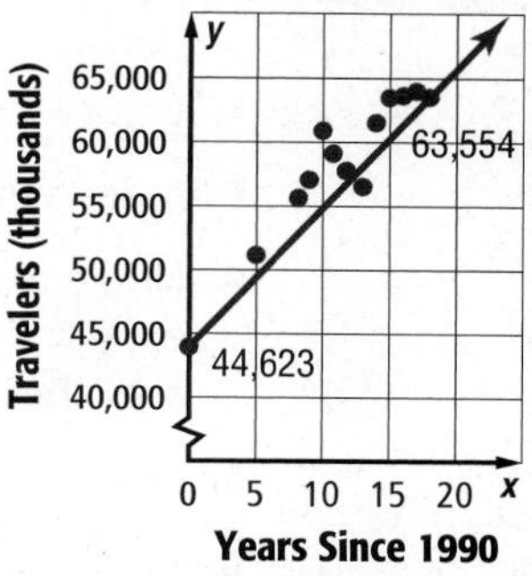

Source: *Statistical Abstract of the United States*

Find the slope.

$m = \frac{y_2 - y_1}{x_2 - x_1}$ — Slope Formula

$= \frac{63{,}554 - 44{,}623}{18 - 0}$ — $(x_1, y_1) = (0, 44{,}623)$, $(x_2, y_2) = (18, 63{,}554)$

$= \frac{18{,}931}{18}$ — Simplify.

Use $m = \frac{18{,}931}{18}$ and either the point-slope form or the slope-intercept form to write the equation of the line of fit.

$$y - y_1 = m(x - x_1)$$
$$y - 44{,}623 = \frac{18{,}931}{18}(x - 0)$$
$$y - 44{,}623 = \frac{18{,}931}{18}x$$
$$y = \frac{18{,}931}{18}x + 44{,}623$$

Step 3 Evaluate the function for $x = 1996 - 1990$ or 6.

$y = \frac{18{,}931}{18}x + 44{,}623$ — Equation of best-fit line

$= \frac{18{,}931}{18}(6) + 44{,}623$ — $x = 6$

$= 6310\frac{1}{3} + 44{,}623$ or $50{,}933\frac{1}{3}$ — Add.

In 1996, there were approximately 50,933 thousand or 50,933,000 people who traveled from the United States to international countries.

Guided Practice

3. MUSIC Use the equation for the line of fit for the data in Guided Practice 2 to estimate CD sales in 2015.

Check Your Understanding

Example 1 **Determine whether each graph shows a *positive*, *negative*, or *no* correlation. If there is a positive or negative correlation, describe its meaning in the situation.**

1. **Free Throws**

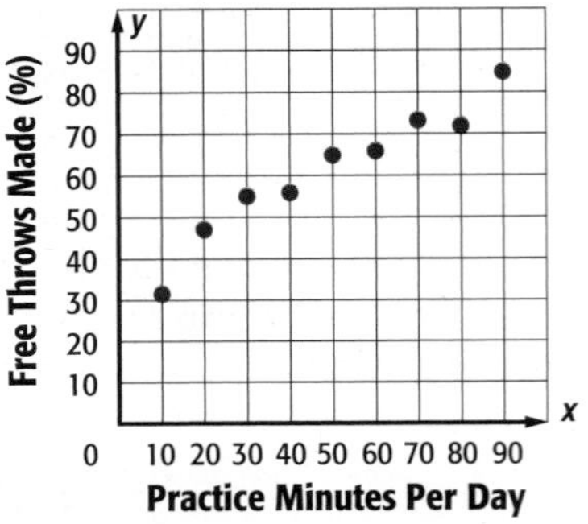

2. **Lemonade Sales**

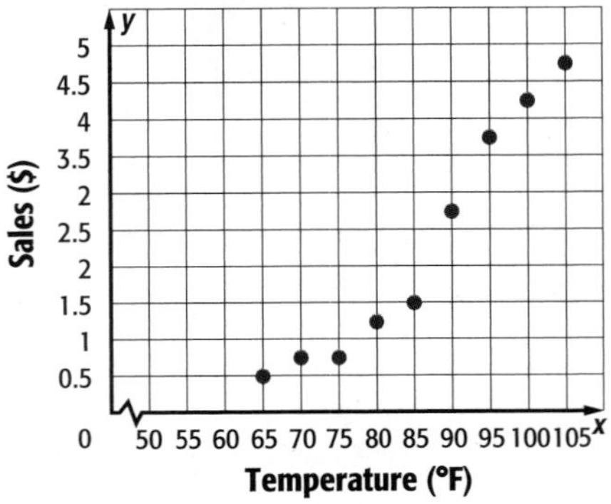

Example 2

3. CCSS **SENSE-MAKING** The table shows the median age of females when they were first married.

a. Make a scatter plot and determine what relationship exists, if any, in the data. Identify the independent and the dependent variables.

b. Draw a line of fit for the scatter plot.

c. Write an equation in slope-intercept form for the line of fit.

Example 3

d. Predict what the median age of females when they are first married will be in 2016.

e. Do you think the equation can give a reasonable estimate for the year 2056? Explain.

Year	Age
1996	24.8
1997	25.0
1998	25.0
1999	25.1
2000	25.1
2001	25.1
2002	25.3
2003	25.3
2004	25.3
2005	25.5
2006	25.9

Source: U.S. Bureau of Census

Practice and Problem Solving

Example 1 **Determine whether each graph shows a *positive*, *negative*, or *no* correlation. If there is a positive or negative correlation, describe its meaning in the situation.**

4. **Game Tickets at the Fair**

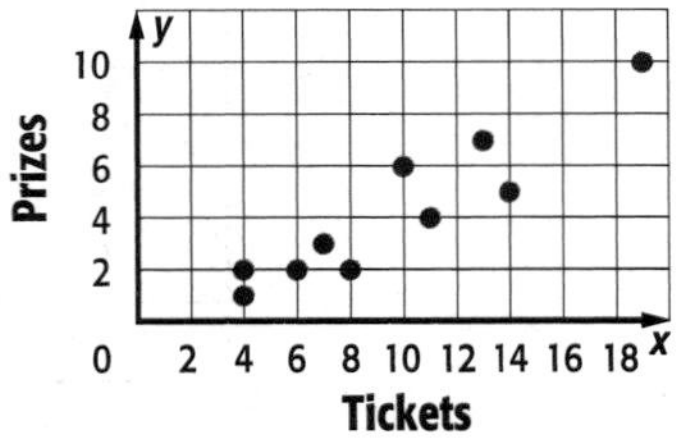

5 **NBA 3-Point Percentage**

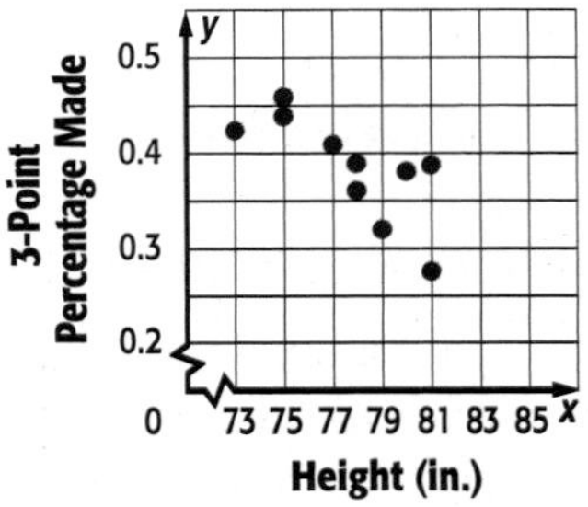

6.

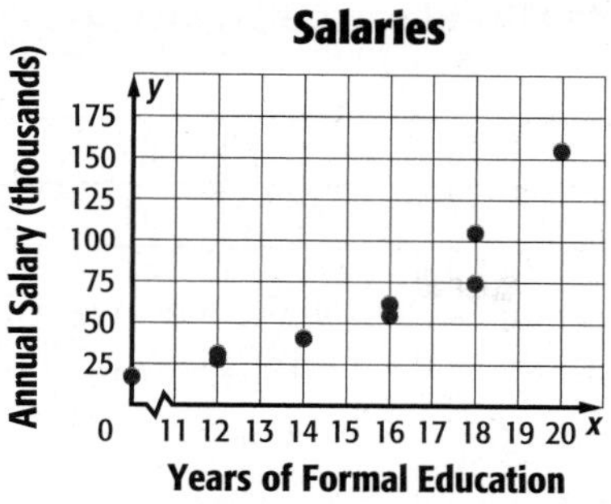

7

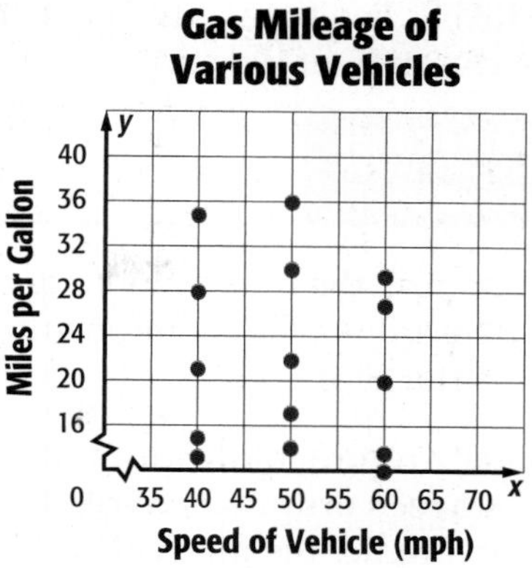

Examples 2–3 **8. MILK** Refer to the scatter plot of gallons of milk consumption per person for selected years.

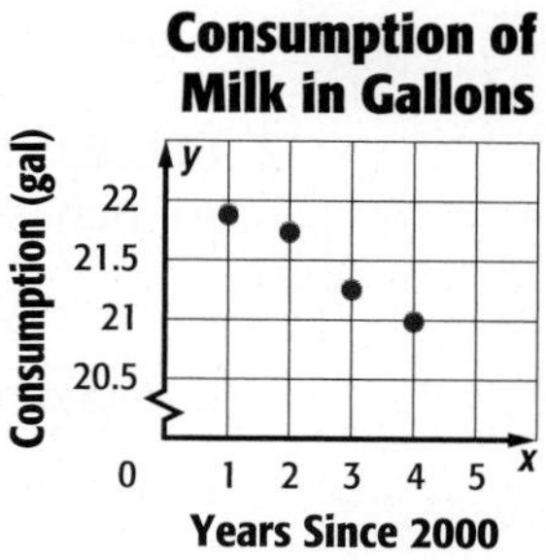

a. Use the points (2, 21.75) and (4, 21) to write the slope-intercept form of an equation for the line of fit.

b. Predict the milk consumption in 2020.

c. Predict in what year milk consumption will be 10 gallons.

d. Is it reasonable to use the equation to estimate the consumption of milk for any year? Explain.

9. FOOTBALL Use the scatter plot.

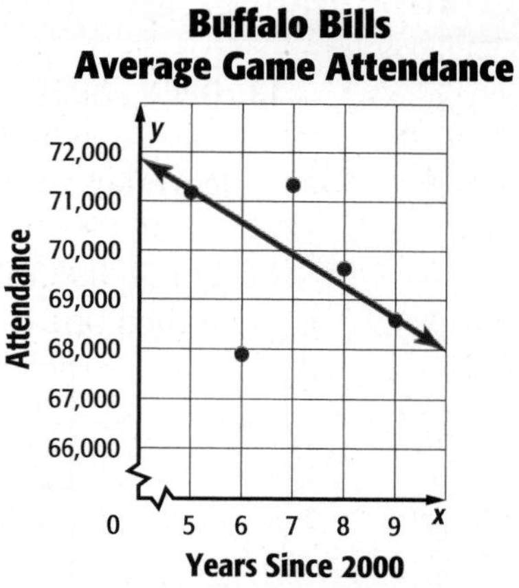

a. Use the points (5, 71,205) and (9, 68,611) to write the slope-intercept form of an equation for the line of fit shown in the scatter plot.

b. Predict the average attendance at a game in 2020.

c. Can you use the equation to make a decision about the average attendance in any given year in the future? Explain.

10. CCSS **SENSE-MAKING** The Body Mass Index (BMI) is a measure of body fat using height and weight. The heights and weights of twelve men with normal BMI are given in the table at the right.

a. Make a scatter plot comparing the height in inches to the weight in pounds.

b. Draw a line of fit for the data.

c. Write the slope-intercept form of an equation for the line of fit.

d. Predict the normal weight for a man who is 84 inches tall.

e. A man's weight is 188 pounds. Use the equation of the line of fit to predict the height of the man.

Height (in.)	Weight (lb)
62	115
63	124
65	120
67	134
67	140
68	138
68	144
68	152
69	147
72	155
73	168
73	166

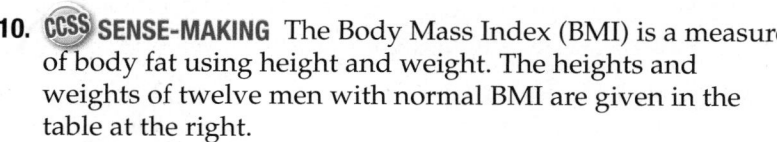

11 **GEYSERS** The time to the next eruption of Old Faithful can be predicted by using the duration of the current eruption.

Duration (min)	1.5	2	2.5	3	3.5	4	4.5	5
Interval (min)	48	55	70	72	74	82	93	100

a. Identify the independent and the dependent variables. Make a scatter plot and determine what relationship, if any, exists in the data. Draw a line of fit for the scatter plot.

b. Let x represent the duration of the previous interval. Let y represent the time between eruptions. Write the slope-intercept form of the equation for the line of fit. Predict the interval after a 7.5-minute eruption.

c. Make a critical judgment about using the equation to predict the duration of the next eruption. Would the equation be a useful model?

12. COLLECT DATA Use a tape measure to measure both the foot size and the height in inches of ten individuals.

a. Record your data in a table.

b. Make a scatter plot and draw a line of fit for the data.

c. Write an equation for the line of fit.

d. Make a conjecture about the relationship between foot size and height.

H.O.T. Problems Use Higher-Order Thinking Skills

13. OPEN ENDED Describe a real-life situation that can be modeled using a scatter plot. Decide whether there is a *positive*, *negative*, or *no* correlation. Explain what this correlation means.

14. WHICH ONE DOESN'T BELONG? Analyze the following situations and determine which one does not belong.

- hours worked and amount of money earned
- height of an athlete and favorite color
- seedlings that grow an average of 2 centimeters each week
- number of photos stored on a camera and capacity of camera

15. CCSS ARGUMENTS Determine which line of fit is better for the scatter plot. Explain your reasoning.

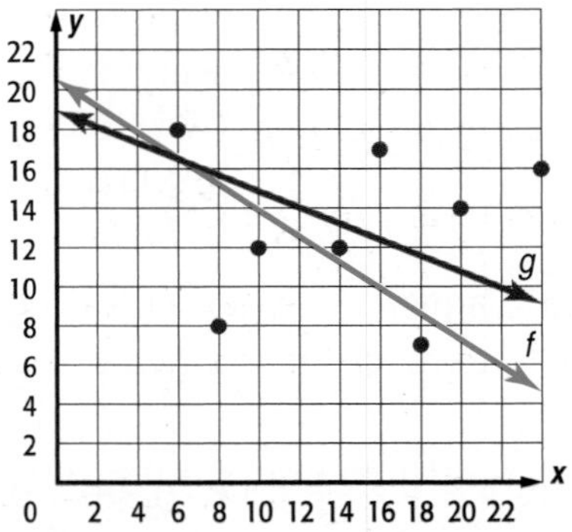

16. REASONING What can make a scatter plot and line of fit more useful for accurate predictions? Does an accurate line of fit always predict what will happen in the future? Explain.

17. WRITING IN MATH Make a scatter plot that shows the height of a person and age. Explain how you could use the scatter plot to predict the age of a person given his or her height. How can the information from a scatter plot be used to identify trends and make decisions?

Standardized Test Practice

18. Which equation best describes the relationship between the values of x and y in the table?

A $y = x - 5$
B $y = 2x - 5$
C $y = 3x - 7$
D $y = 4x - 7$

x	y
−1	−7
0	−5
2	−1
4	3

19. STATISTICS Mr. Hernandez collected data on the heights and average stride lengths of a random sample of high school students. He then made a scatter plot. What kind of correlation did he most likely see?

F positive
G constant
H negative
J no

20. GEOMETRY Mrs. Aguilar's rectangular bedroom measures 13 feet by 11 feet. She wants to purchase carpet for the bedroom that costs \$2.95 per square foot, including tax. How much will the carpet cost?

A \$70.80
B \$141.60
C \$145.95
D \$421.85

21. SHORT RESPONSE Nikia bought a one-month membership to a fitness center for \$35. Each time she goes, she rents a locker for \$0.25. If she spent \$40.50 at the fitness center last month, how many days did she go?

Spiral Review

Determine whether the graphs of each pair of equations are *parallel, perpendicular,* or *neither.* (Lesson 4-4)

22. $y = -2x + 11$
$y + 2x = 23$

23. $3y = 2x + 14$
$2x + 3y = 2$

24. $y = -5x$
$y = 5x - 18$

25. $y = 3x + 2$
$y = -\frac{1}{3}x - 2$

Write each equation in standard form. (Lesson 4-3)

26. $y - 13 = 4(x - 2)$

27. $y - 5 = -2(x + 2)$

28. $y + 3 = -5(x + 1)$

29. $y + 7 = \frac{1}{2}(x + 2)$

30. $y - 1 = \frac{5}{6}(x - 4)$

31. $y - 2 = -\frac{2}{5}(x - 8)$

Graph each equation. (Lesson 4-1)

32. $y = 2x + 3$

33. $4x + y = -1$

34. $3x + 4y = 7$

Find the slope of the line that passes through each pair of points. (Lesson 3-3)

35. (3, 4), (10, 8)

36. (−4, 7), (3, 5)

37. (3, 7), (−2, 4)

38. (−3, 2), (−3, 4)

39. (−2, −6), (−1, 10)

40. (1, −5), (−3, −5)

41. DRIVING Latisha drove 248 miles in 4 hours. At that rate, how long will it take her to drive an additional 93 miles? (Lesson 2-6)

Skills Review

Express each relation as a graph. Then determine the domain and range.

42. {(4, 5), (5, 4), (−2, −2), (4, −5), (−5, 4)}

43. {(7, 6), (3, 4), (4, 5), (−2, 6), (−3, 2)}

EXTEND 4-5

Algebra Lab

Correlation and Causation

CCSS **Common Core State Standards**
Content Standards
S.ID.9 Distinguish between correlation and causation.

You may be considering attending a college or technical school in the future. What factors cause tuition to rise—increased building costs, higher employee salaries, or the amount of bottled water consumed?

Let's see how bottled water and college tuition are related. The table shows the average college tuition and fees for public colleges and the per person U.S. consumption of bottled water per year for 2003 through 2007.

Year	2003	2004	2005	2006	2007
Water Consumed (gallons)	21.6	23.2	25.4	27.6	29.3
Tuition ($)	4645	5126	5492	5804	6191

Source: *Beverage Marketing Corporation* and *College Board*

Activity Correlation and Causation

Follow the steps to learn about correlation and causation.

Step 1 Graph the ordered pairs (gallons, tuition) to create a scatter plot. For example, one ordered pair is (21.6, 4645). Describe the graph.

Step 2 Is the correlation *positive* or *negative*? Explain.

Step 3 Do you think drinking more bottled water *causes* college tuition costs to rise? Explain.

Step 4 **Causation** occurs when a change in one variable produces a change in another variable. Correlation can be observed between many variables, but causation can only be determined from data collected from a controlled experiment. Describe an experiment that could illustrate causation.

Exercises

For each exercise, determine whether each situation illustrates *correlation* or *causation*. Explain your reasoning, including other factors that might be involved.

1. A survey showed that sleeping with the light on was positively correlated to nearsightedness.
2. A controlled experiment showed a positive correlation between the number of cigarettes smoked and the probability of developing lung cancer.
3. A random sample of students found that owning a cell phone had a negative correlation with riding the bus to school.
4. A controlled experiment showed a positive correlation between the number of hours using headphones when listening to music and the level of hearing loss.
5. DeQuan read in the newspaper that shark attacks are positively correlated with monthly ice cream sales.

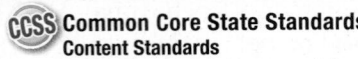

LESSON 4-6 Regression and Median-Fit Lines

:: Then	:: Now	:: Why?
You used lines of fit and scatter plots to evaluate trends and make predictions.	**1** Write equations of best-fit lines using linear regression. **2** Write equations of median-fit lines.	The table shows the total attendance, in millions of people, at the Minnesota State Fair from 2005 to 2009. You can use a graphing calculator to find the equation of a *best-fit line* and use it to make predictions about future attendance at the fair.

Year	Attendance (millions)
2005	1.633
2006	1.681
2007	1.682
2008	1.693
2009	1.790

NewVocabulary
best-fit line
linear regression
correlation coefficient
residual
median-fit line

Common Core State Standards

Content Standards

S.ID.6 Represent data on two quantitative variables on a scatter plot, and describe how the variables are related.

a. Fit a function to the data; use functions fitted to data to solve problems in the context of the data. Use given functions or choose a function suggested by the context. Emphasize linear, quadratic, and exponential models.

b. Informally assess the fit of a function by plotting and analyzing residuals.

c. Fit a linear function for a scatter plot that suggests a linear association.

S.ID.8 Compute (using technology) and interpret the correlation coefficient of a linear fit.

Mathematical Practices

5 Use appropriate tools strategically.

1 Best-Fit Lines You have learned how to find and write equations for lines of fit by hand. Many calculators use complex algorithms that find a more precise line of fit called the **best-fit line**. One algorithm is called **linear regression**.

Your calculator may also compute a number called the **correlation coefficient**. This number will tell you if your correlation is positive or negative and how closely the equation is modeling the data. The closer the correlation coefficient is to 1 or −1, the more closely the equation models the data.

Real-World Example 1 Best-Fit Line

MOVIES **The table shows the amount of money made by movies in the United States. Use a graphing calculator to write an equation for the best-fit line for that data.**

Year	2000	2001	2002	2003	2004	2005	2006	2007	2008	2009
Income ($ billion)	7.48	8.13	9.19	9.35	9.27	8.95	9.25	9.65	9.85	10.21

Before you begin, make sure that your Diagnostic setting is on. You can find this under the **CATALOG** menu. Press **D** and then scroll down and click **DiagnosticOn**. Then press ENTER.

Step 1 Enter the data by pressing STAT and selecting the **Edit** option. Let the year 2000 be represented by 0. Enter the years since 2000 into List 1 (**L1**). These will represent the *x*-values. Enter the income ($ billion) into List 2 (**L2**). These will represent the *y*-values.

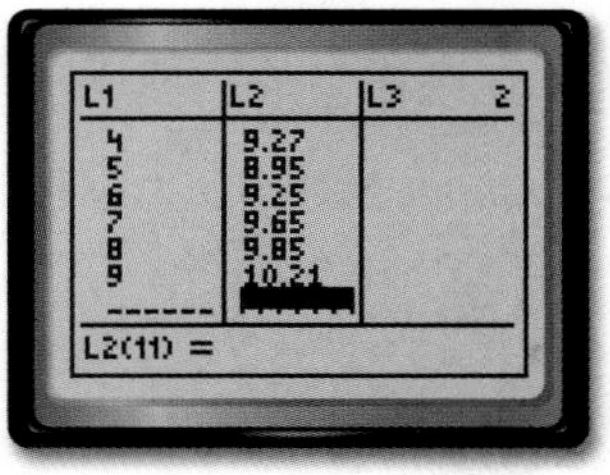

Step 2 Perform the regression by pressing STAT and selecting the **CALC** option. Scroll down to **LinReg (ax+b)** and press ENTER twice.

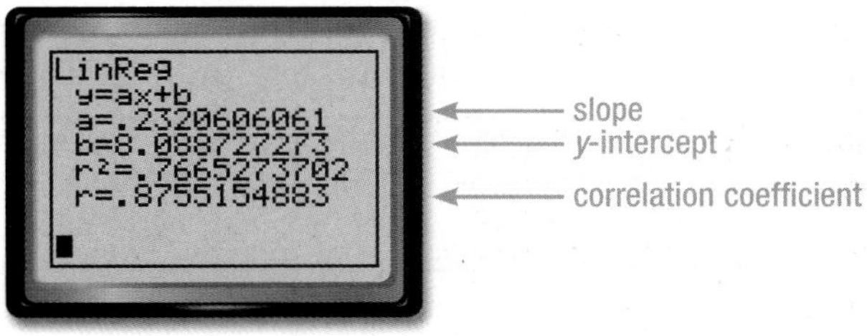

Step 3 Write the equation of the regression line by rounding the a and b values on the screen. The form that we chose for the regression was $ax + b$, so the equation is $y = 0.23x + 8.09$. The correlation coefficient is about 0.8755, which means that the equation models the data fairly well.

Real-WorldLink

In 1994, Minnesota became the first state to sanction girls' ice hockey as a high school varsity sport.

Source: ESPNET SportsZone

GuidedPractice

Write an equation of the best-fit line for the data in each table. Name the correlation coefficient. Round to the nearest ten-thousandth. Let x be the number of years since 2003.

1A. HOCKEY The table shows the number of goals of leading scorers for the Mustang Girls Hockey Team.

Year	2003	2004	2005	2006	2007	2008	2009	2010
Goals	30	23	41	35	31	43	33	45

1B. HOCKEY The table gives the number of goals scored by the team each season.

Year	2003	2004	2005	2006	2007	2008	2009	2010
Goals	63	44	55	63	81	85	93	84

We know that not all of the points will lie on the best-fit line. The difference between an observed y-value and its predicted y-value (found on the best-fit line) is called a **residual**. Residuals measure how much the data deviate from the regression line. When residuals are plotted on a scatter plot they can help to assess how well the best-fit line describes the data. If the best-fit line is a good fit, there is no pattern in the residual plot.

Real-World Example 2 Graph and Analyze a Residual Plot

HOCKEY Graph and analyze the residual plot for the data for Guided Practice 1A. Determine if the best-fit line models the data well.

After calculating the best-fit line in Guided Practice 1A, you can obtain the residual plot of the data. Turn on **Plot2** under the **STAT PLOT** menu and choose ⋰. Use **L1** for the **Xlist** and **RESID** for the **Ylist**. You can obtain **RESID** by pressing [2nd] **[STAT]** and selecting **RESID** from the list of names. Graph the scatter plot of the residuals by pressing [ZOOM] and choosing **ZoomStat**.

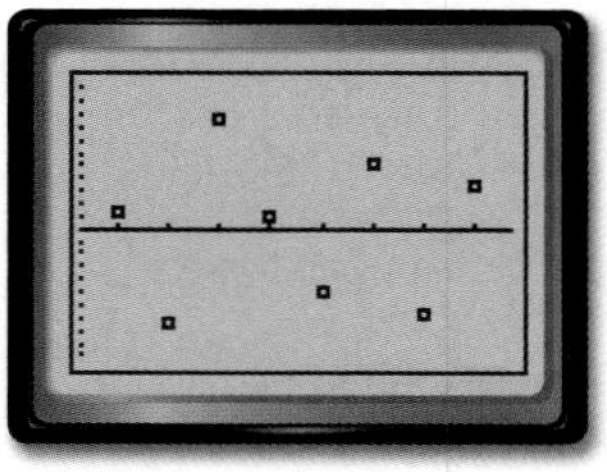

[0, 8] scl: 1 by [−10, 10] scl: 2

The residuals appear to be randomly scattered and centered about the line $y = 0$. Thus, the best-fit line seems to model the data well.

GuidedPractice

2. UNEMPLOYMENT Graph and analyze the residual plot for the following data comparing graduation rates and unemployment rates.

Graduation Rate	73	85	64	81	68	82
Unemployment Rate	6.9	4.1	3.2	5.5	4.3	5.1

A residual is positive when the observed value is above the line, negative when the observed value is below the line, and zero when it is on the line. One common measure of goodness of fit is the sum of squared vertical distances from the points to the line. The best-fit line, which is also called the *least-squares regression line*, minimizes the sum of the squares of those distances.

We can use points on the best-fit line to estimate values that are not in the data. Recall that when we estimate values that are between known values, this is called *linear interpolation*. When we estimate a number outside of the range of the data, it is called *linear extrapolation*.

Real-World Example 3 Use Interpolation and Extrapolation

PAINTBALL The table shows the points received by the top ten paintball teams at a tournament. Estimate how many points the 20th-ranked team received.

Rank	1	2	3	4	5	6	7	8	9	10
Score	100	89	96	99	97	98	78	70	64	80

Write an equation of the best-fit line for the data. Then extrapolate to find the missing value.

Step 1 Enter the data from the table into the lists. Let the ranks be the x-values and the scores be the y-values. Then graph the scatter plot.

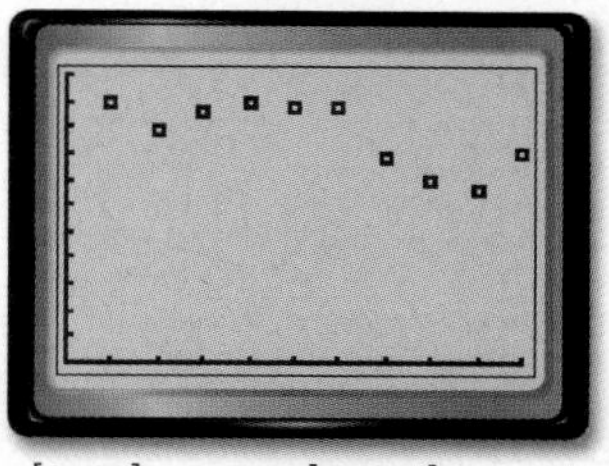

[0, 10] scl: 1 by [0, 110] scl: 10

Step 2 Perform the linear regression using the data in the lists. Find the equation of the best-fit line.

The equation is about $y = -3.32x + 105.3$.

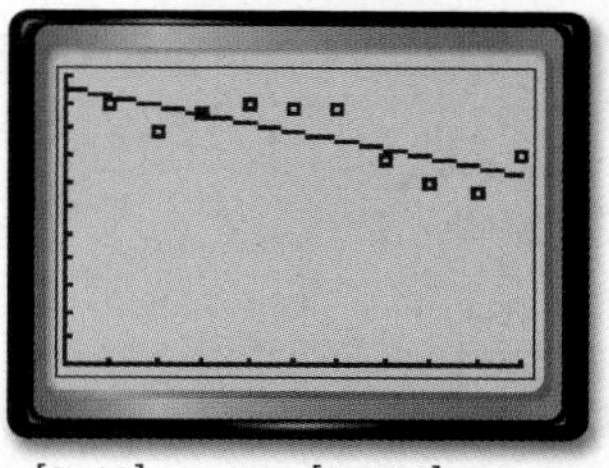

[0, 10] scl: 1 by [0, 110] scl: 10

Step 3 Graph the best-fit line. Press [Y=] [VARS] and choose **Statistics**. From the **EQ** menu, choose **RegEQ**. Then press [GRAPH].

Step 4 Use the graph to predict the points that the 20th-ranked team received. Change the viewing window to include the x-value to be evaluated. Press [2nd] **[CALC]** [ENTER] 20 [ENTER] to find that when $x = 20$, $y \approx 39$. It is estimated that the 20th ranked team received 39 points.

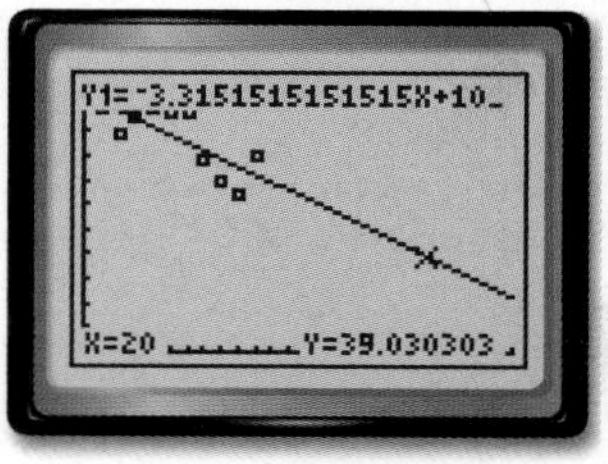

[0, 25] scl: 1 by [0, 110] scl: 1

GuidedPractice

ONLINE GAMES Use linear interpolation to estimate the percent of Americans that play online games for the following ages.

Age	15	20	30	40	50
Percent	81	54	37	29	25

Source: Pew Internet & American Life Survey

3A. 35 years

3B. 18 years

2 Median-Fit Lines

A second type of fit line that can be found using a graphing calculator is a **median-fit line**. The equation of a median-fit line is calculated using the medians of the coordinates of the data points.

Example 4 Median-Fit Line

PT

PAINTBALL Find and graph the equation of a median-fit line for the data in Example 3. Then predict the score of the 15th ranked team.

Step 1 Reenter the data if it is not in the lists. Clear the **Y=** list and graph the scatter plot.

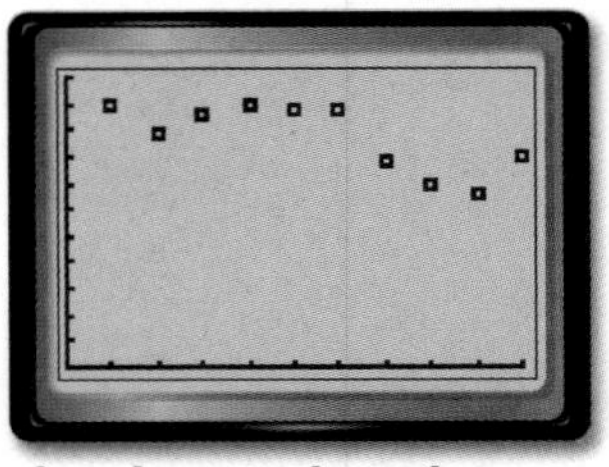

[0, 10] scl: 1 by [0, 110] scl: 10

Step 2 To find the median-fit equation, press the STAT key and select the **CALC** option. Scroll down to the **Med-Med** option and press ENTER. The value of a is the slope, and the value of b is the y-intercept.

The equation for the median-fit line is about $y = -3.71x + 108.26$.

Step 3 Copy the equation to the **Y=** list and graph. Use the **value** option to find the value of y when $x = 15$.

The 15th place team scored about 53 points.

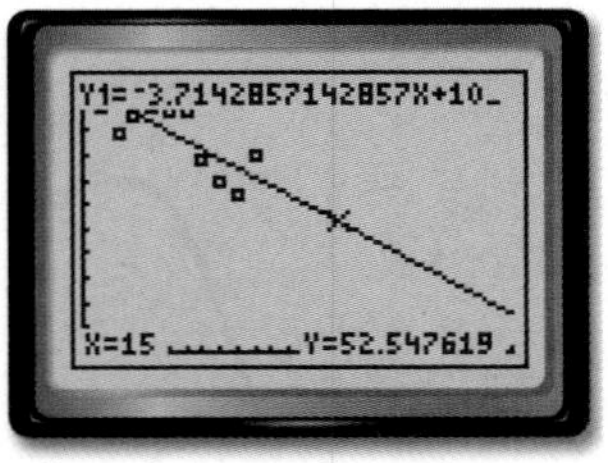

[0, 25] scl: 1 by [0, 110] scl: 1

Notice that the equations for the regression line and the median-fit line are very similar.

GuidedPractice

4. Use the data from Guided Practice 3 and a median-fit line to estimate the numbers of 18- and 35-year-olds who play online games. Compare these values with the answers from the regression line.

Real-WorldLink

Paintball is more popular with 12- to 17-year-olds than any other age group. In a recent year, 3,649,000 teens participated in paintball while 2,195,000 18- to 24-year-olds participated.

Source: *Statistical Abstract* of the *United States*

Alan Thornton/Stone/Getty Images

Check Your Understanding

Examples 1, 2

1. **POTTERY** A local university is keeping track of the number of art students who use the pottery studio each day.

Day	1	2	3	4	5	6	7
Students	10	15	18	15	13	19	20

a. Write an equation of the regression line and find the correlation coefficient.

b. Graph the residual plot and determine if the regression line models the data well.

Example 3

2. **COMPUTERS** The table below shows the percent of Americans with a broadband connection at home in a recent year. Use linear extrapolation and a regression equation to estimate the percentage of 60-year-olds with broadband at home.

Age	25	30	35	40	45	50
Percent	40	42	36	35	36	32

Example 4

3. **VACATION** The Smiths want to rent a house on the lake that sleeps eight people. The cost of the house per night is based on how close it is to the water.

Distance from Lake (mi)	0.0 (houseboat)	0.3	0.5	1.0	1.25	1.5	2.0
Price/Night ($)	785	325	250	200	150	140	100

a. Find and graph an equation for the median-fit line.

b. What would you estimate is the cost of a rental 1.75 miles from the lake?

Practice and Problem Solving

Example 1

Write an equation of the regression line for the data in each table. Then find the correlation coefficient.

4. **SKYSCRAPERS** The table ranks the ten tallest buildings in the world.

Rank	1	2	3	4	5	6	7	8	9	10
Stories	101	88	110	88	88	80	69	102	78	70

5. **MUSIC** The table gives the number of annual violin auditions held by a youth symphony each year since 2004. Let x be the number of years since 2004.

Year	2004	2005	2006	2007	2008	2009	2010
Auditions	22	19	25	37	32	35	42

Example 2

6. **RETAIL** The table gives the sales at a clothing chain since 2004. Let x be the number of years since 2004.

Year	2004	2005	2006	2007	2008	2009	2010
Sales (Millions of Dollars)	6.84	7.6	10.9	15.4	17.6	21.2	26.5

a. Write an equation of the regression line.

b. Graph and analyze the residual plot.

Examples 3, 4 **7.** **MARATHON** The number of entrants in the Boston Marathon every five years since 1975 is shown. Let x be the number of years since 1975.

Year	1975	1980	1985	1990	1995	2000	2005	2010
Entrants	2395	5417	5594	9412	9416	17,813	20,453	26,735

a. Find an equation for the median-fit line.

b. According to the equation, how many entrants were there in 2003?

8. **CAMPING** A campground keeps a record of the number of campsites rented the week of July 4 for several years. Let x be the number of years since 2000.

Year	2002	2003	2004	2005	2006	2007	2008	2009	2010
Sites Rented	34	45	42	53	58	47	57	65	59

a. Find an equation for the regression line.

b. Predict the number of campsites that will be rented in 2012.

c. Predict the number of campsites that will be rented in 2020.

9. **ICE CREAM** An ice cream company keeps a count of the tubs of chocolate ice cream delivered to each of their stores in a particular area.

a. Find an equation for the median-fit line.

b. Graph the points and the median-fit line.

c. How many tubs would be delivered to a 1500-square-foot store? a 5000-square-foot store?

Store Size (ft^2)	2100	2225	3135	3569	4587
Tubs (hundreds)	110	102	215	312	265

10. CCSS **SENSE-MAKING** The prices of the eight top-selling brands of jeans at Jeanie's Jeans are given in the table below.

Sales Rank	1	2	3	4	5	6	7	8
Price ($)	43	44	50	61	64	135	108	78

a. Find the equation for the regression line.

b. According to the equation, what would be the price of a pair of the 12th best-selling brand?

c. Is this a reasonable prediction? Explain.

11. **STATE FAIRS** Refer to the beginning of the lesson.

a. Graph a scatter plot of the data, where $x = 1$ represents 2005. Then find and graph the equation for the best-fit line.

b. Graph and analyze the residual plot.

c. Predict the total attendance in 2020.

Alex Cao/Photodisc/Getty Images

12. FIREFIGHTERS The table shows statistics from the U.S. Fire Administration.

a. Find an equation for the median-fit line.

b. Graph the points and the median-fit line.

c. Does the median-fit line give you an accurate picture of the number of firefighters? Explain.

Age	Number of Firefighters
18	40,919
25	245,516
35	330,516
45	296,665
55	167,087
65	54,559

13. ATHLETICS The table shows the number of participants in high school athletics.

Year Since 1970	1	10	20	30	35
Athletes	3,960,932	5,356,913	5,298,671	6,705,223	7,159,904

a. Find an equation for the regression line.

b. According to the equation, how many participated in 1988?

14. ART A count was kept on the number of paintings sold at an auction by the year in which they were painted. Let x be the number of years since 1950.

Year Painted	1950	1955	1960	1965	1970	1975
Paintings Solds	8	5	25	21	9	22

a. Find the equation for the linear regression line.

b. How many paintings were sold that were painted in 1961?

c. Is the linear regression equation an accurate model of the data? Explain why or why not.

H.O.T. Problems Use Higher-Order Thinking Skills

15. CCSS **ARGUMENTS** Below are the results of the World Superpipe Championships in 2008.

Men	Score	Rank	Women	Score
Shaun White	93.00	1	Torah Bright	96.67
Mason Aguirre	90.33	2	Kelly Clark	93.00
Janne Korpi	85.33	3	Soko Yamaoka	85.00
Luke Mitrani	85.00	4	Ellery Hollingsworth	79.33
Keir Dillion	81.33	5	Sophie Rodriguez	71.00

Find an equation of the regression line for each, and graph them on the same coordinate plane. Compare and contrast the men's and women's graphs.

16. REASONING For a class project, the scores that 10 randomly selected students earned on the first 8 tests of the school year are given. Explain how to find a line of best fit. Could it be used to predict the scores of other students? Explain your reasoning.

17. OPEN ENDED For 10 different people, measure their heights and the lengths of their heads from chin to top. Use these data to generate a linear regression equation and a median-fit equation. Make a prediction using both of the equations.

18. E **WRITING IN MATH** How are lines of fit and linear regression similar? different?

Standardized Test Practice

19. GEOMETRY Sam is putting a border around a poster. x represents the poster's width, and y represents the poster's length. Which equation represents how much border Sam will use if he doubles the length and the width?

A $4xy$ **C** $4(x + y)$
B $(x + y)^4$ **D** $16(x + y)$

20. SHORT RESPONSE Tatiana wants to run 5 miles at an average pace of 9 minutes per mile. After 4 miles, her average pace is 9 minutes 10 seconds. In how many minutes must she complete the final mile to reach her goal?

21. What is the slope of the line that passes through (1, 3) and (−3, 1)?

F −2 **H** $\frac{1}{2}$
G $-\frac{1}{2}$ **J** 2

22. What is an equation of the line that passes through (0, 1) and has a slope of 3?

A $y = 3x - 1$
B $y = 3x - 2$
C $y = 3x + 4$
D $y = 3x + 1$

Spiral Review

23. USED CARS Gianna wants to buy a specific make and model of a used car. She researched prices from dealers and private sellers and made the graph shown. (Lesson 4-5)

a. Describe the relationship in the data.

b. Use the line of fit to predict the price of a car that is 7 years old.

c. Is it reasonable to use this line of fit to predict the price of a 10-year-old car? Explain.

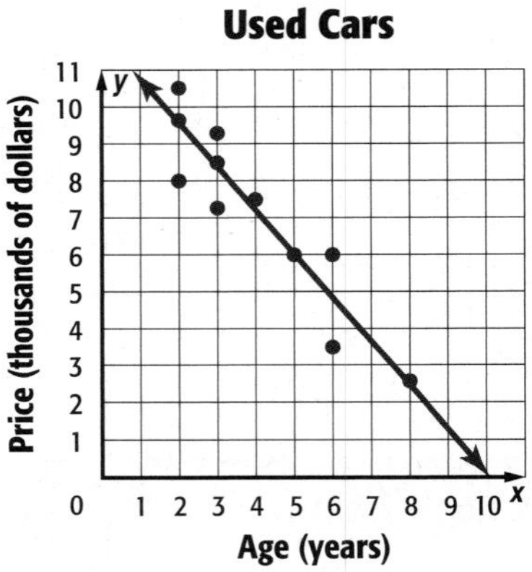

24. GEOMETRY A quadrilateral has sides with equations $y = -2x$, $2x + y = 6$, $y = \frac{1}{2}x + 6$, and $x - 2y = 9$. Is the figure a rectangle? Explain your reasoning. (Lesson 4-4)

Write each equation in standard form. (Lesson 4-3)

25. $y - 2 = 3(x - 1)$

26. $y - 5 = 6(x + 1)$

27. $y + 2 = -2(x - 5)$

28. $y + 3 = \frac{1}{2}(x + 4)$

29. $y - 1 = \frac{2}{3}(x + 9)$

30. $y + 3 = -\frac{1}{4}(x + 2)$

Find the slope of the line that passes through each pair of points. (Lesson 3-3)

31. (3, 4), (10, 8) **32.** (−4, 7), (3, 5) **33.** (3, 7), (−2, 4) **34.** (−3, 2), (−3, 4)

Skills Review

If $f(x) = x^2 - x + 1$, find each value.

35. $f(-1)$ **36.** $f(5) - 3$ **37.** $f(a)$ **38.** $f(b^2)$

Graph each equation.

39. $y = x + 2$ **40.** $x + 5y = 4$ **41.** $2x - 3y = 6$ **42.** $5x + 2y = 6$

LESSON 4-7

Inverse Linear Functions

Then

- You represented relations as tables, graphs, and mappings.

Now

1. Find the inverse of a relation.
2. Find the inverse of a linear function.

Why?

- Randall is writing a report on Santiago, Chile, and he wants to include a brief climate analysis. He found a table of temperatures recorded in degrees Celsius. He knows that a formula for converting degrees Fahrenheit to degrees Celsius is $C(x) = \frac{5}{9}(x - 32)$. He will need to find the *inverse* function to convert from degrees Celsius to degrees Fahrenheit.

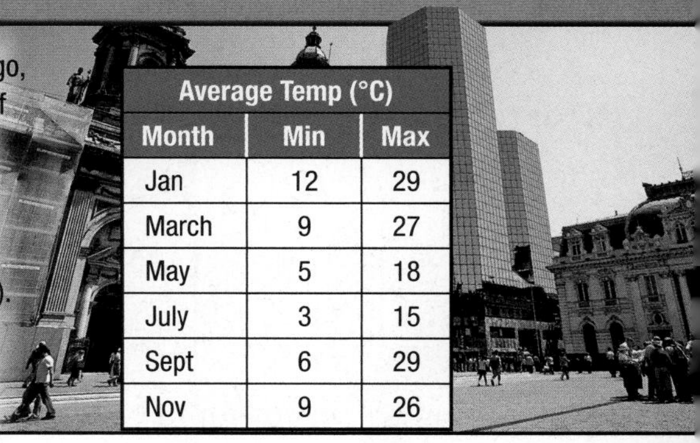

Average Temp (°C)		
Month	Min	Max
Jan	12	29
March	9	27
May	5	18
July	3	15
Sept	6	29
Nov	9	26

NewVocabulary
inverse relation
inverse function

Common Core State Standards

Content Standards
A.CED.2 Create equations in two or more variables to represent relationships between quantities; graph equations on coordinate axes with labels and scales.
F.BF.4a Solve an equation of the form $f(x) = c$ for a simple function f that has an inverse and write an expression for the inverse.

Mathematical Practices
6 Attend to precision.

1 Inverse Relations

An **inverse relation** is the set of ordered pairs obtained by exchanging the x-coordinates with the y-coordinates of each ordered pair in a relation. If (5, 3) is an ordered pair of a relation, then (3, 5) is an ordered pair of the inverse relation.

KeyConcept Inverse Relations

Words If one relation contains the element (a, b), then the inverse relation will contain the element (b, a).

Example A and B are inverse relations.

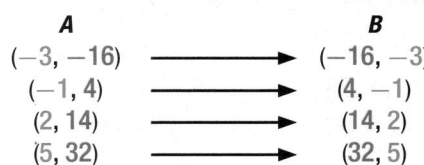

A		*B*
(−3, −16)	→	(−16, −3)
(−1, 4)	→	(4, −1)
(2, 14)	→	(14, 2)
(5, 32)	→	(32, 5)

Notice that the domain of a relation becomes the range of its inverse, and the range of the relation becomes the domain of its inverse.

PT

Example 1 Inverse Relations

Find the inverse of each relation.

a. **{(4, −10), (7, −19), (−5, 17), (−3, 11)}**

To find the inverse, exchange the coordinates of the ordered pairs.

(4, −10) → (−10, 4) (−5, 17) → (17, −5)
(7, −19) → (−19, 7) (−3, 11) → (11, −3)

The inverse is {(−10, 4), (−19, 7), (17, −5), (11, −3)}.

b.

x	−4	−1	5	9
y	−13	−8.5	0.5	6.5

Write the coordinates as ordered pairs. Then exchange the coordinates of each pair.

(−4, −13) → (−13, −4) (5, 0.5) → (0.5, 5)
(−1, −8.5) → (−8.5, −1) (9, 6.5) → (6.5, 9)

The inverse is {(−13, −4), (−8.5, −1), (0.5, 5), (6.5, 9)}.

GuidedPractice

1A. $\{(-6, 8), (-15, 11), (9, 3), (0, 6)\}$

1B.

x	-10	-4	-3	0
y	5	11	12	15

The graphs of relations can be used to find and graph inverse relations.

Example 2 Graph Inverse Relations

Graph the inverse of the relation.

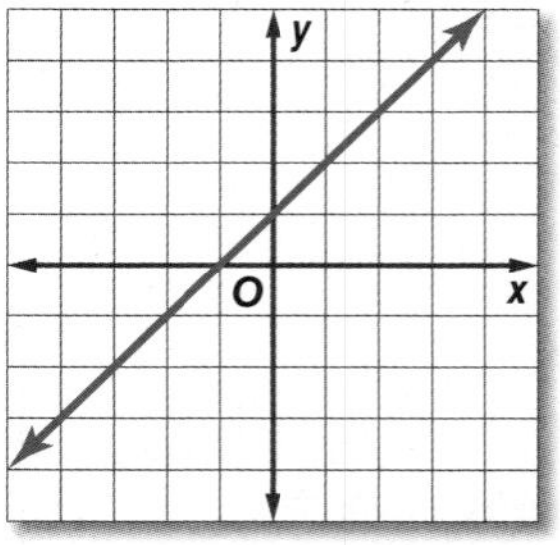

> **StudyTip**
>
> **CCSS Precision** Only two points are necessary to graph the inverse of a line, but several should be used to avoid possible error.

The graph of the relation passes through the points at $(-4, -3)$, $(-2, -1)$, $(0, 1)$, $(2, 3)$, and $(3, 4)$. To find points through which the graph of the inverse passes, exchange the coordinates of the ordered pairs. The graph of the inverse passes through the points at $(-3, -4)$, $(-1, -2)$, $(1, 0)$, $(3, 2)$, and $(4, 3)$. Graph these points and then draw the line that passes through them.

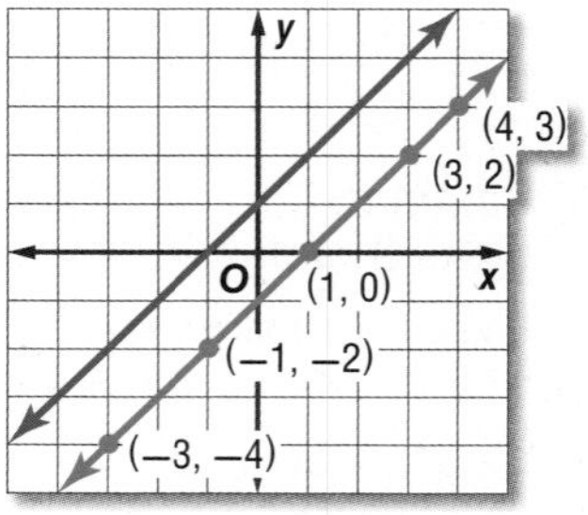

GuidedPractice

Graph the inverse of each relation.

2A.

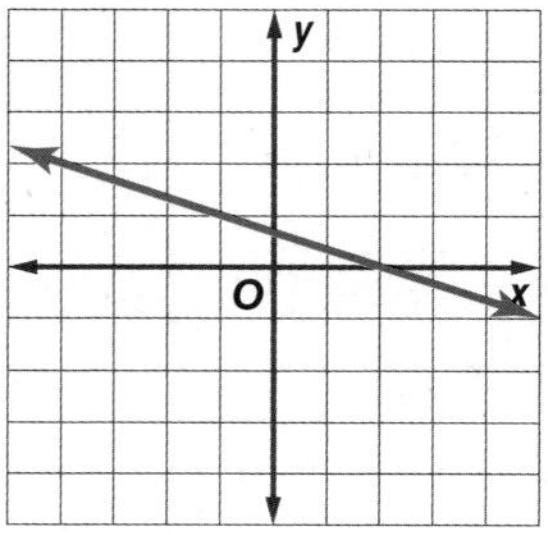

2B.

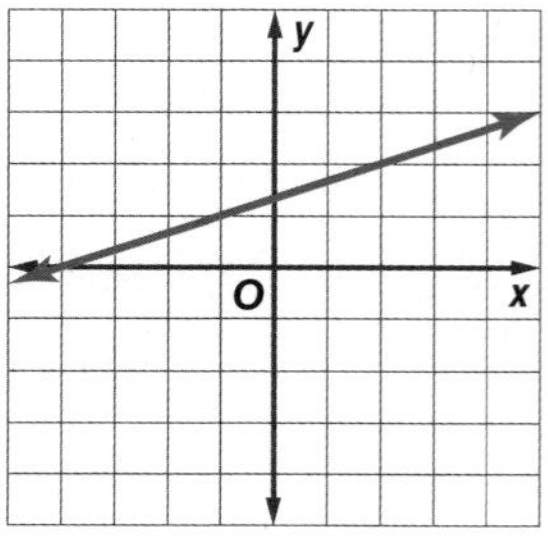

The graphs from Example 2 are graphed on the right with the line $y = x$. Notice that the graph of an inverse is the graph of the original relation reflected in the line $y = x$. For every point (x, y) on the graph of the original relation, the graph of the inverse will include the point (y, x).

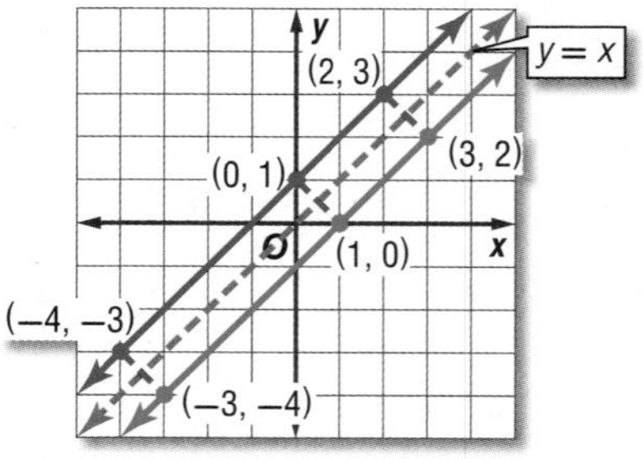

2 Inverse Functions

Inverse Functions A linear relation that is described by a function has an **inverse function** that can generate ordered pairs of the inverse relation The inverse of the linear function $f(x)$ can be written as $f^{-1}(x)$ and is read *f of x inverse* or *the inverse of f of x*.

KeyConcept Finding Inverse Functions

To find the inverse function $f^{-1}(x)$ of the linear function $f(x)$, complete the following steps.

Step 1 Replace $f(x)$ with y in the equation for $f(x)$.

Step 2 Interchange y and x in the equation.

Step 3 Solve the equation for y.

Step 4 Replace y with $f^{-1}(x)$ in the new equation.

Example 3 Find Inverse Linear Functions

Find the inverse of each function.

a. $f(x) = 4x - 8$

Step 1	$f(x) = 4x - 8$	Original equation
	$y = 4x - 8$	Replace $f(x)$ with y.
Step 2	$x = 4y - 8$	Interchange y and x.
Step 3	$x + 8 = 4y$	Add 8 to each side.
	$\frac{x+8}{4} = y$	Divide each side by 4.
Step 4	$\frac{x+8}{4} = f^{-1}(x)$	Replace y with $f^{-1}(x)$.

The inverse of $f(x) = 4x - 8$ is $f^{-1}(x) = \frac{x+8}{4}$ or $f^{-1}(x) = \frac{1}{4}x + 2$.

WatchOut!

Notation The -1 in $f^{-1}(x)$ is *not* an exponent.

CHECK Graph both functions and the line $y = x$ on the same coordinate plane. $f^{-1}(x)$ appears to be the reflection of $f(x)$ in the line $y = x$. ✓

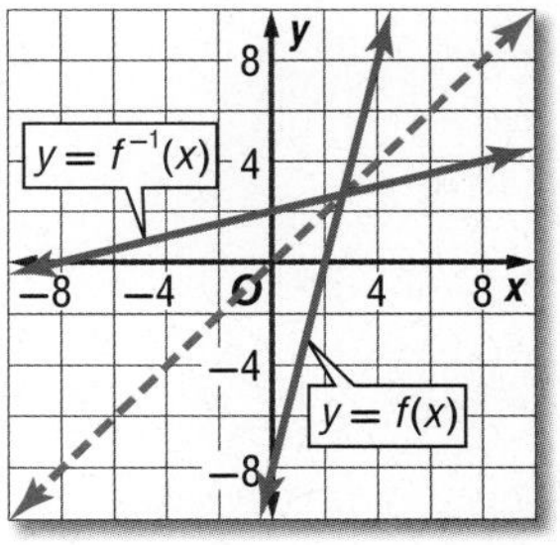

b. $f(x) = -\frac{1}{2}x + 11$

Step 1	$f(x) = -\frac{1}{2}x + 11$	Original equation
	$y = -\frac{1}{2}x + 11$	Replace $f(x)$ with y.
Step 2	$x = -\frac{1}{2}y + 11$	Interchange y and x.
Step 3	$x - 11 = -\frac{1}{2}y$	Subtract 11 from each side.
	$-2(x - 11) = y$	Multiply each side by -2.
	$-2x + 22 = y$	Distributive Property
Step 4	$-2x + 22 = f^{-1}(x)$	Replace y with $f^{-1}(x)$.

The inverse of $f(x) = -\frac{1}{2}x + 11$ is $f^{-1}(x) = -2x + 22$.

GuidedPractice

3A. $f(x) = 4x - 12$

3B. $f(x) = \frac{1}{3}x + 7$

Real-World Example 4 Use an Inverse Function

TEMPERATURE Refer to the beginning of the lesson. Randall wants to convert the temperatures from degrees Celsius to degrees Fahrenheit.

a. Find the inverse function $C^{-1}(x)$.

Step 1	$C(x) = \frac{5}{9}(x - 32)$	Original equation
	$y = \frac{5}{9}(x - 32)$	Replace $C(x)$ with y.
Step 2	$x = \frac{5}{9}(y - 32)$	Interchange y and x.
Step 3	$\frac{9}{5}x = y - 32$	Multiply each side by $\frac{9}{5}$.
	$\frac{9}{5}x + 32 = y$	Add 32 to each side.
Step 4	$\frac{9}{5}x + 32 = C^{-1}(x)$	Replace y with $C^{-1}(x)$.

The inverse function of $C(x)$ is $C^{-1}(x) = \frac{9}{5}x + 32$.

b. What do x and $C^{-1}(x)$ represent in the context of the inverse function?

x represents the temperature in degrees Celsius. $C^{-1}(x)$ represents the temperature in degrees Fahrenheit.

c. Find the average temperatures for July in degrees Fahrenheit.

The average minimum and maximum temperatures for July are 3° C and 15° C, respectively. To find the average minimum temperature, find $C^{-1}(3)$.

$C^{-1}(x) = \frac{9}{5}x + 32$	Original equation
$C^{-1}(3) = \frac{9}{5}(3) + 32$	Substitute 3 for x.
$= 37.4$	Simplify.

To find the average maximum temperature, find $C^{-1}(15)$.

$C^{-1}(x) = \frac{9}{5}x + 32$	Original equation
$C^{-1}(15) = \frac{9}{5}(15) + 32$	Substitute 15 for x.
$= 59$	Simplify.

The average minimum and maximum temperatures for July are 37.4° F and 59° F, respectively.

Real-WorldLink

The winter months in Chile occur during the summer months in the U.S. due to Chile's location in the southern hemisphere. The average daily high temperature of Santiago during its winter months is about 60° F.

Source: World Weather Information Service

GuidedPractice

4. RENTAL CAR Peggy rents a car for the day. The total cost $C(x)$ in dollars is given by $C(x) = 19.99 + 0.3x$, where x is the number of miles she drives.

A. Find the inverse function $C^{-1}(x)$.

B. What do x and $C^{-1}(x)$ represent in the context of the inverse function?

C. How many miles did Peggy drive if her total cost was $34.99?

Check Your Understanding

Example 1 **Find the inverse of each relation.**

1. {(4, −15), (−8, −18), (−2, −16.5), (3, −15.25)}

2.

x	−3	0	1	6
y	11.8	3.7	1	−12.5

Example 2 **Graph the inverse of each relation.**

3.

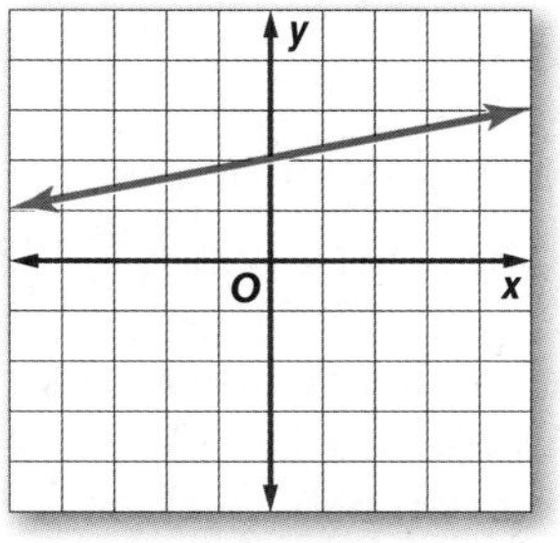

4.

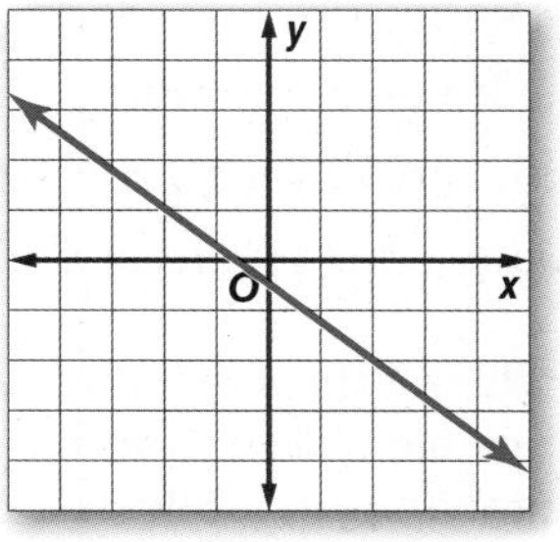

Example 3 **Find the inverse of each function.**

5. $f(x) = -2x + 7$

6. $f(x) = \frac{2}{3}x + 6$

Example 4 **7.** **CCSS REASONING** Dwayne and his brother purchase season tickets to the Cleveland Crusaders games. The ticket package requires a one-time purchase of a personal seat license costing \$1200 for two seats. A ticket to each game costs \$70. The cost $C(x)$ in dollars for Dwayne for the first season is $C(x) = 600 + 70x$, where x is the number of games Dwayne attends.

a. Find the inverse function.

b. What do x and $C^{-1}(x)$ represent in the context of the inverse function?

c. How many games did Dwayne attend if his total cost for the season was \$950?

Practice and Problem Solving

Example 1 **Find the inverse of each relation.**

8. {(−5, 13), (6, 10.8), (3, 11.4), (−10, 14)}

9. {(−4, −49), (8, 35), (−1, −28), (4, 7)}

10.

x	y
−8	−36.4
−2	−15.4
1	−4.9
5	9.1
11	30.1

11.

x	y
−3	7.4
−1	4
1	0.6
3	−2.8
5	−6.2

Example 2 **Graph the inverse of each relation.**

12.

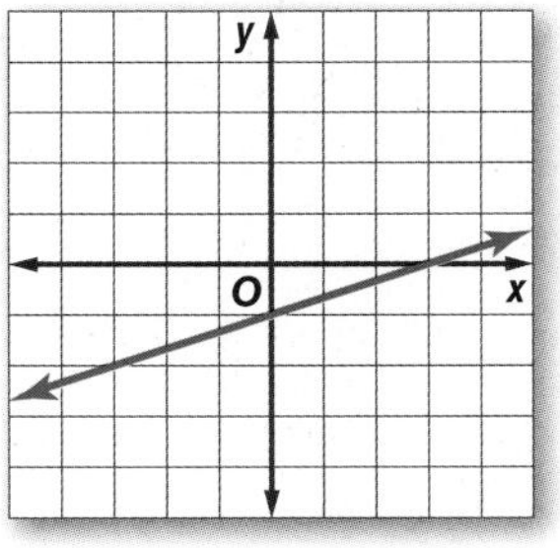

13.

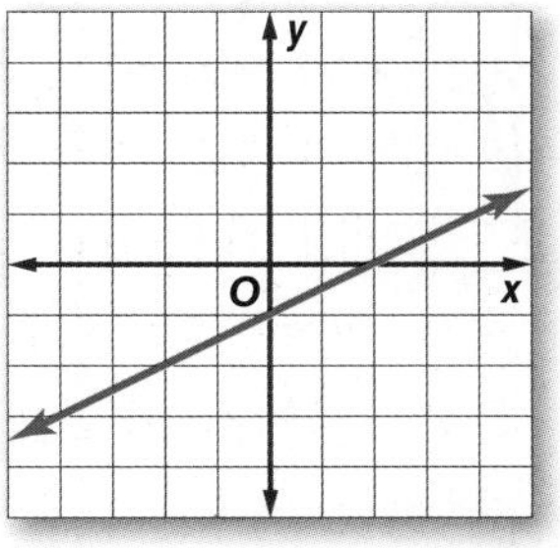

Example 3 **Find the inverse of each function.**

14. $f(x) = 25 + 4x$

15. $f(x) = 17 - \frac{1}{3}x$

16. $f(x) = 4(x + 17)$

17. $f(x) = 12 - 6x$

18. $f(x) = \frac{2}{5}x + 10$

19. $f(x) = -16 - \frac{4}{3}x$

Example 4 20. **DOWNLOADS** An online music subscription service allows members to download songs for \$0.99 each after paying a monthly service charge of \$3.99. The total monthly cost $C(x)$ of the service in dollars is $C(x) = 3.99 + 0.99x$, where x is the number of songs downloaded.

a. Find the inverse function.

b. What do x and $C^{-1}(x)$ represent in the context of the inverse function?

c. How many songs were downloaded if a member's monthly bill is \$27.75?

21. **LANDSCAPING** At the start of the mowing season, Chuck collects a one-time maintenance fee of \$10 from his customers. He charges the Fosters \$35 for each cut. The total amount collected from the Fosters in dollars for the season is $C(x) = 10 + 35x$, where x is the number of times Chuck mows the Fosters' lawn.

a. Find the inverse function.

b. What do x and $C^{-1}(x)$ represent in the context of the inverse function?

c. How many times did Chuck mow the Fosters' lawn if he collected a total of \$780 from them?

Write the inverse of each equation in $f^{-1}(x)$ notation.

22. $3y - 12x = -72$

23. $x + 5y = 15$

24. $-42 + 6y = x$

25. $3y + 24 = 2x$

26. $-7y + 2x = -28$

27. $3y - x = 3$

CCSS **TOOLS** **Match each function with the graph of its inverse.**

A.
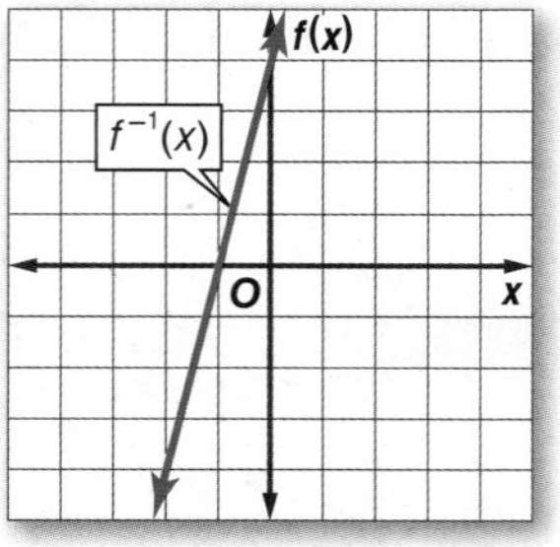

B.
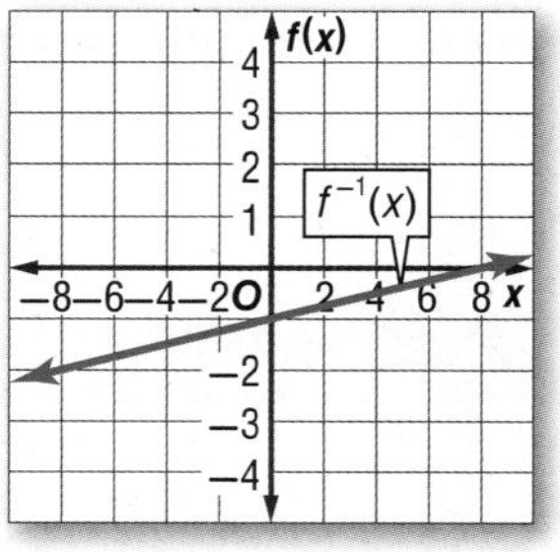

C.
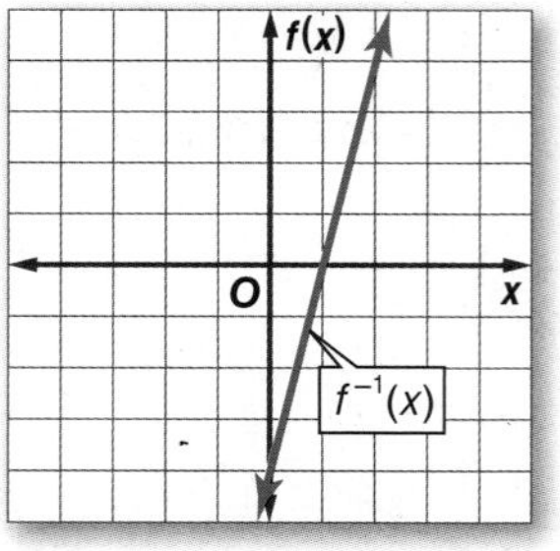

D.
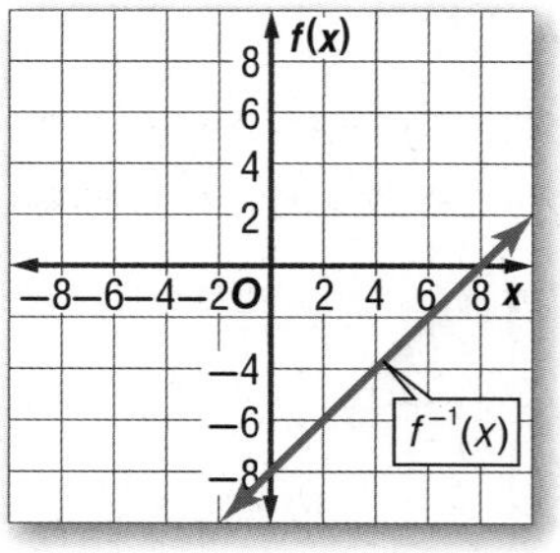

28. $f(x) = x + 4$

29. $f(x) = 4x + 4$

30. $f(x) = \frac{1}{4}x + 1$

31. $f(x) = \frac{1}{4}x - 1$

Write an equation for the inverse function $f^{-1}(x)$ that satisfies the given conditions.

32. slope of $f(x)$ is 7; graph of $f^{-1}(x)$ contains the point (13, 1)

33. graph of $f(x)$ contains the points (−3, 6) and (6, 12)

34. graph of $f(x)$ contains the point (10, 16); graph of $f^{-1}(x)$ contains the point (3, −16)

35. slope of $f(x)$ is 4; $f^{-1}(5) = 2$

36. CELL PHONES Mary Ann pays a monthly fee for her cell phone package which includes 700 minutes. She gets billed an additional charge for every minute she uses the phone past the 700 minutes. During her first month, Mary Ann used 26 additional minutes and her bill was \$37.79. During her second month, Mary Ann used 38 additional minutes and her bill was \$41.39.

a. Write a function that represents the total monthly cost $C(x)$ of Mary Ann's cell phone package, where x is the number of additional minutes used.

b. Find the inverse function.

c. What do x and $C^{-1}(x)$ represent in the context of the inverse function?

d. How many additional minutes did Mary Ann use if her bill for her third month was \$48.89?

37. MULTIPLE REPRESENTATIONS In this problem, you will explore the domain and range of inverse functions.

a. Algebraic Write a function for the area $A(x)$ of the rectangle shown.

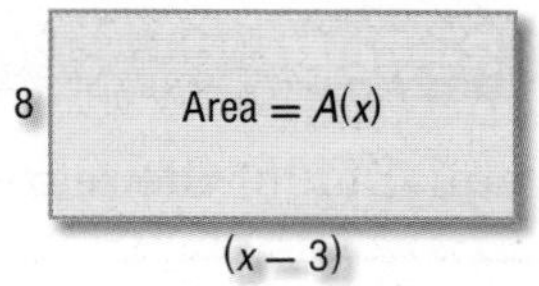

b. Graphical Graph $A(x)$. Describe the domain and range of $A(x)$ in the context of the situation.

c. Algebraic Write the inverse of $A(x)$. What do x and $A^{-1}(x)$ represent in the context of the situation?

d. Graphical Graph $A^{-1}(x)$. Describe the domain and range of $A^{-1}(x)$ in the context of the situation.

e. Logical Determine the relationship between the domains and ranges of $A(x)$ and $A^{-1}(x)$.

H.O.T. Problems Use Higher-Order Thinking Skills

38. CHALLENGE If $f(x) = 5x + a$ and $f^{-1}(10) = -1$, find a.

39. CHALLENGE If $f(x) = \frac{1}{a}x + 7$ and $f^{-1}(x) = 2x - b$, find a and b.

CCSS ARGUMENTS Determine whether the following statements are *sometimes*, *always*, or *never* true. Explain your reasoning.

40. If $f(x)$ and $g(x)$ are inverse functions, then $f(a) = b$ and $g(b) = a$.

41. If $f(a) = b$ and $g(b) = a$, then $f(x)$ and $g(x)$ are inverse functions.

42. OPEN ENDED Give an example of a function and its inverse. Verify that the two functions are inverses by graphing the functions and the line $y = x$ on the same coordinate plane.

43. WRITING IN MATH Explain why it may be helpful to find the inverse of a function.

Standardized Test Practice

44. Which equation represents a line that is perpendicular to the graph and passes through the point at (2, 0)?

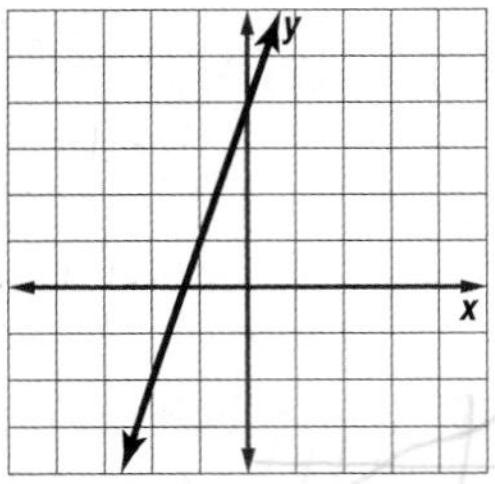

A $y = 3x - 6$

B $y = -3x + 6$

C $y = -\frac{1}{3}x + \frac{2}{3}$

D $y = \frac{1}{3}x - \frac{2}{3}$

45. A giant tortoise travels at a rate of 0.17 mile per hour. Which equation models the time t it would take the giant tortoise to travel 0.8 mile?

F $t = \frac{0.8}{0.17}$ **H** $t = \frac{0.17}{0.8}$

G $t = (0.17)(0.8)$ **J** $0.8 = \frac{0.17}{t}$

46. GEOMETRY If $\triangle JKL$ is similar to $\triangle JNM$ what is the value of a?

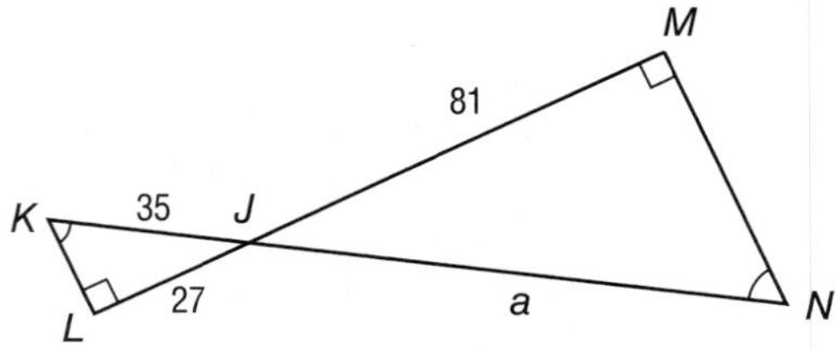

A 62.5

B 105

C 125

D 155.5

47. GRIDDED RESPONSE What is the difference in the value of $2.1(x + 3.2)$, when $x = 5$ and when $x = 3$?

Spiral Review

Write an equation of the regression line for the data in each table. (Lesson 4-6)

48.

x	1	3	5	7	9
y	3	8	15	18	21

49.

x	3	5	7	9	11
y	7.2	23.5	41.2	56.4	73.1

50.

x	1	2	3	4	5
y	21	33	39	54	64

51.

x	2	4	6	8	10
y	1.4	2.4	2.9	3.3	4.2

52. TESTS Determine whether the graph at the right shows a *positive*, *negative*, or *no* correlation. If there is a correlation, describe its meaning. (Lesson 4-5)

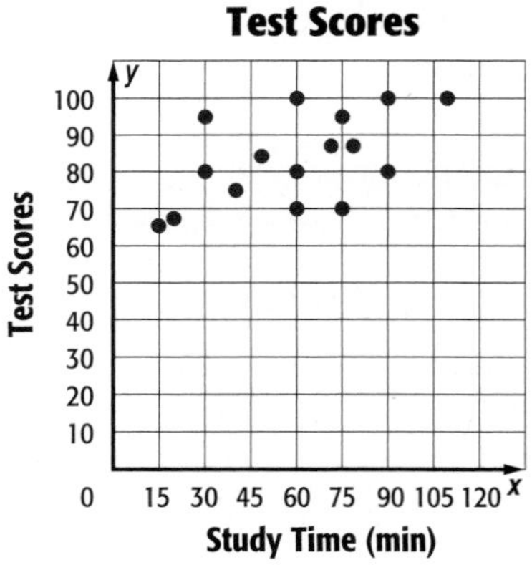

Suppose y varies directly as x. (Lesson 3-4)

53. If $y = 2.5$ when $x = 0.5$, find y when $x = 20$.

54. If $y = -6.6$ when $x = 9.9$, find y when $x = 6.6$.

55. If $y = 2.6$ when $x = 0.25$, find y when $x = 1.125$.

56. If $y = 6$ when $x = 0.6$, find x when $y = 12$.

Skills Review

Solve each equation.

57. $104 = k - 67$

58. $-4 + x = -7$

59. $\frac{m}{7} = -11$

60. $\frac{2}{3}p = 14$

61. $-82 = 18 - n$

62. $\frac{9}{t} = -27$

4-7 Algebra Lab
Drawing Inverses

EXTEND

You can use patty paper to draw the graph of an inverse relation by reflecting the original graph in the line $y = x$.

CCSS Common Core State Standards
Content Standards
F.BF.4a Solve an equation of the form $f(x) = c$ for a simple function f that has an inverse and write an expression for the inverse.

Activity Draw an Inverse

Consider the graphs shown.

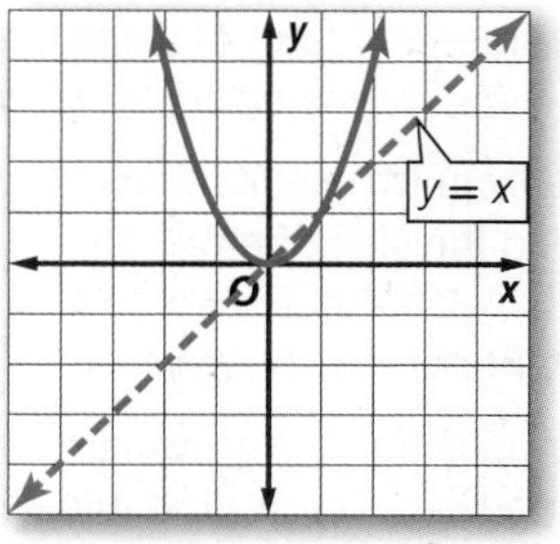

Step 1 Trace the graphs onto a square of patty paper, waxed paper, or tracing paper.

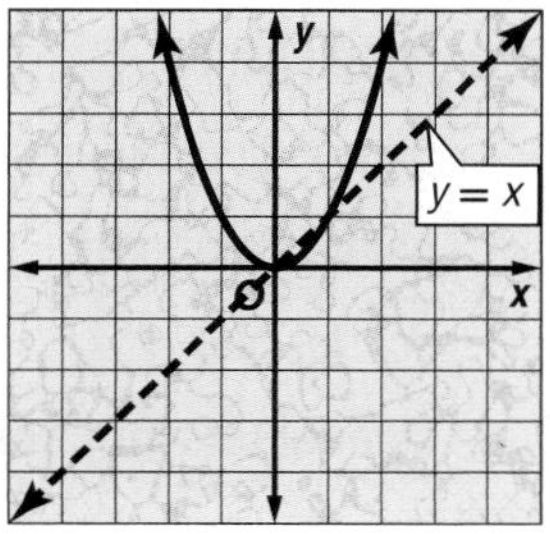

Step 2 Flip the patty paper over and lay it on the original graph so that the traced $y = x$ is on the original $y = x$.

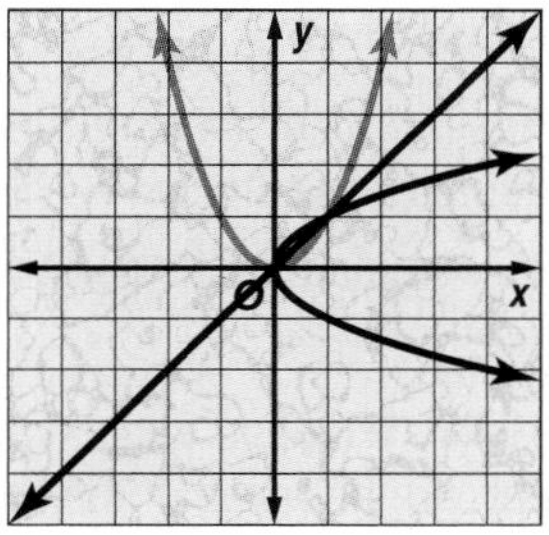

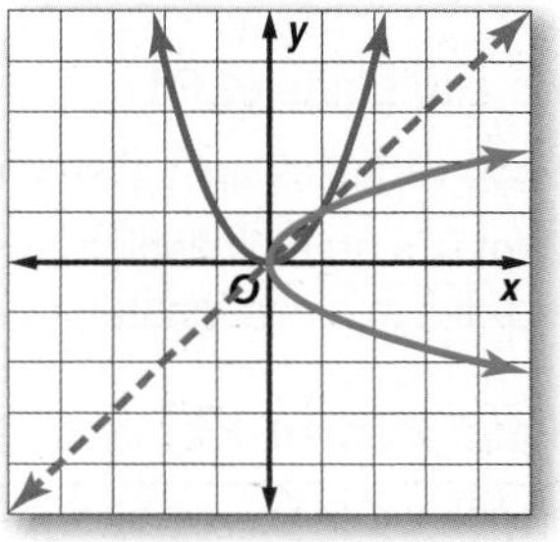

Notice that the result is the reflection of the graph in the line $y = x$ or the inverse of the graph.

Analyze The Results

1. Is the graph of the original relation a function? Explain.
2. Is the graph of the inverse relation a function? Explain.
3. What are the domain and range of the original relation? of the inverse relation?
4. If the domain of the original relation is restricted to $D = \{x \mid x \geq 0\}$, is the inverse relation a function? Explain.
5. If the graph of a relation is a function, what can you conclude about the graph of its inverse?
6. **CHALLENGE** The vertical line test can be used to determine whether a relation is a function. Write a rule that can be used to determine whether a function has an inverse that is also a function.

CHAPTER 4 Study Guide and Review

Study Guide

KeyConcepts

Slope-Intercept Form (Lessons 4-1 and 4-2)

- The slope-intercept form of a linear equation is $y = mx + b$, where m is the slope and b is the y-intercept.
- If you are given two points through which a line passes, use them to find the slope first.

Point-Slope Form (Lesson 4-3)

- The linear equation $y - y_1 = m(x - x_1)$ is written in point-slope form, where (x_1, y_1) is a given point on a nonvertical line and m is the slope of the line.

Parallel and Perpendicular Lines (Lesson 4-4)

- Nonvertical parallel lines have the same slope.
- Lines that intersect at right angles are called perpendicular lines. The slopes of perpendicular lines are opposite reciprocals.

Scatter Plots and Lines of Fit (Lesson 4-5)

- Data with two variables are called bivariate data.
- A scatter plot is a graph in which two sets of data are plotted as ordered pairs in a coordinate plane.

Regression and Median-Fit Lines (Lesson 4-6)

- A graphing calculator can be used to find regression lines and median-fit lines.

Inverse Linear Functions (Lesson 4-7)

- An inverse relation is the set of ordered pairs obtained by exchanging the x-coordinates with the y-coordinates of each ordered pair of a relation.
- A linear function $f(x)$ has an inverse function that can be written as $f^{-1}(x)$ and is read *f of x inverse* or *the inverse of f of x.*

FOLDABLES® StudyOrganizer

Be sure the Key Concepts are noted in your Foldable.

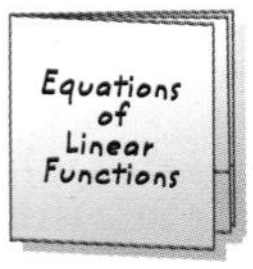

KeyVocabulary

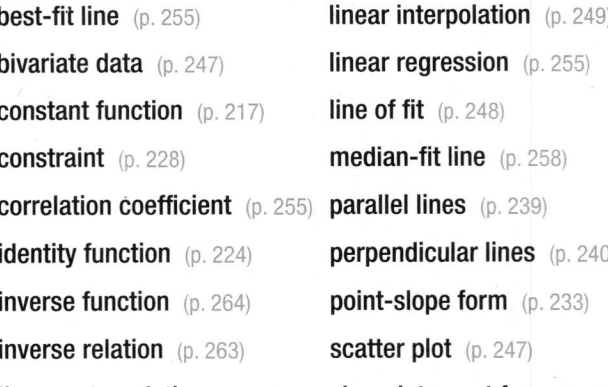

best-fit line (p. 255)	**linear interpolation** (p. 249)
bivariate data (p. 247)	**linear regression** (p. 255)
constant function (p. 217)	**line of fit** (p. 248)
constraint (p. 228)	**median-fit line** (p. 258)
correlation coefficient (p. 255)	**parallel lines** (p. 239)
identity function (p. 224)	**perpendicular lines** (p. 240)
inverse function (p. 264)	**point-slope form** (p. 233)
inverse relation (p. 263)	**scatter plot** (p. 247)
linear extrapolation (p. 228)	**slope-intercept form** (p. 216)

VocabularyCheck

State whether each sentence is *true* or *false*. If *false*, replace the underlined term to make a true sentence.

1. The y-intercept is the y-coordinate of the point where the graph crosses the y-axis.
2. The process of using a linear equation to make predictions about values that are beyond the range of the data is called linear regression.
3. An inverse relation is the set of ordered pairs obtained by exchanging the x-coordinates with the y-coordinates of each ordered pair of a relation.
4. The correlation coefficient describes whether the correlation between the variables is positive or negative and how closely the regression equation is modeling the data.
5. Lines in the same plane that do not intersect are called parallel lines.
6. Lines that intersect at acute angles are called perpendicular lines.
7. A(n) constant function can generate ordered pairs for an inverse relation.
8. The range of a relation is the range of its inverse function.
9. An equation of the form $y = mx + b$ is in point-slope form.

Lesson-by-Lesson Review

4-1 Graphing Equations in Slope-Intercept Form

Write an equation of a line in slope-intercept form with the given slope and y-intercept. Then graph the equation.

10. slope: 3, y-intercept: 5

11. slope: -2, y-intercept: -9

12. slope: $\frac{2}{3}$, y-intercept: 3

13. slope: $-\frac{5}{8}$, y-intercept: -2

Graph each equation.

14. $y = 4x - 2$

15. $y = -3x + 5$

16. $y = \frac{1}{2}x + 1$

17. $3x + 4y = 8$

18. SKI RENTAL Write an equation in slope-intercept form for the total cost of skiing for h hours with one lift ticket.

Example 1

Write an equation of a line in slope-intercept form with slope -5 and y-intercept -3. Then graph the equation.

$y = mx + b$	Slope-intercept form
$y = -5x + (-3)$	$m = -5$ and $b = -3$
$y = -5x - 3$	Simplify.

To graph the equation, plot the y-intercept $(0, -3)$.

Then move up 5 units and left 1 unit. Plot the point. Draw a line through the two points.

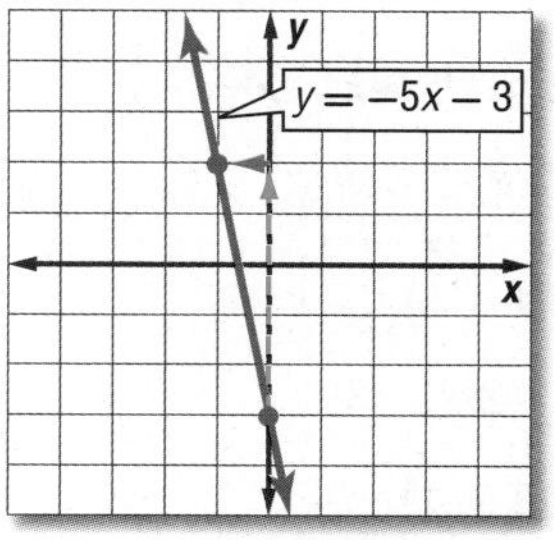

4-2 Writing Equations in Slope-Intercept Form

Write an equation of the line that passes through the given point and has the given slope.

19. (1, 2), slope 3

20. (2, -6), slope -4

21. (-3, -1), slope $\frac{2}{5}$

22. (5, -2), slope $-\frac{1}{3}$

Write an equation of the line that passes through the given points.

23. (2, -1), (5, 2)

24. (-4, 3), (1, 13)

25. (3, 5), (5, 6)

26. (2, 4), (7, 2)

27. CAMP In 2005, a camp had 450 campers. Five years later, the number of campers rose to 750. Write a linear equation that represents the number of campers that attend camp.

Example 2

Write an equation of the line that passes through (3, 2) with a slope of 5.

Step 1 Find the y-intercept.

$y = mx + b$	Slope-intercept form
$2 = 5(3) + b$	$m = 5$, $y = 2$, and $x = 3$
$2 = 15 + b$	Simplify.
$-13 = b$	Subtract 15 from each side.

Step 2 Write the equation in slope-intercept form.

$y = mx + b$	Slope-intercept form
$y = 5x - 13$	$m = 5$ and $b = -13$

Study Guide and Review *Continued*

4-3 Writing Equations in Point-Slope Form

Write an equation in point-slope form for the line that passes through the given point with the slope provided.

28. (6, 3), slope 5

29. (−2, 1), slope −3

30. (−4, 2), slope 0

Write each equation in standard form.

31. $y - 3 = 5(x - 2)$

32. $y - 7 = -3(x + 1)$

33. $y + 4 = \frac{1}{2}(x - 3)$

34. $y - 9 = -\frac{4}{5}(x + 2)$

Write each equation in slope-intercept form.

35. $y - 2 = 3(x - 5)$

36. $y - 12 = -2(x - 3)$

37. $y + 3 = 5(x + 1)$

38. $y - 4 = \frac{1}{2}(x + 2)$

Example 3

Write an equation in point-slope form for the line that passes through (3, 4) with a slope of −2.

$y - y_1 = m(x - x_1)$ Point-slope form

$y - 4 = -2(x - 3)$ Replace m with −2 and (x_1, y_1) with (3, 4).

Example 4

Write $y + 6 = -4(x - 3)$ in standard form.

$y + 6 = -4(x - 3)$ Original equation

$y + 6 = -4x + 12$ Distributive Property

$4x + y + 6 = 12$ Add $4x$ to each side.

$4x + y = 6$ Subtract 6 from each side.

4-4 Parallel and Perpendicular Lines

Write an equation in slope-intercept form for the line that passes through the given point and is parallel to the graph of each equation.

39. (2, 5), $y = x - 3$

40. (0, 3), $y = 3x + 5$

41. (−4, 1), $y = -2x - 6$

42. (−5, −2), $y = -\frac{1}{2}x + 4$

Write an equation in slope-intercept form for the line that passes through the given point and is perpendicular to the graph of the given equation.

43. (2, 4), $y = 3x + 1$

44. (1, 3), $y = -2x - 4$

45. (−5, 2), $y = \frac{1}{3}x + 4$

46. (3, 0), $y = -\frac{1}{2}x$

Example 5

Write an equation in slope-intercept form for the line that passes through (−2, 4) and is parallel to the graph of $y = 6x - 3$.

The slope of the line with equation $y = 6x - 3$ is 6. The line parallel to $y = 6x - 3$ has the same slope, 6.

$y - y_1 = m(x - x_1)$ Point-slope form

$y - 4 = 6[x - (-2)]$ Substitute.

$y - 4 = 6(x + 2)$ Simplify.

$y - 4 = 6x + 12$ Distributive Property

$y = 6x + 16$ Add 4 to each side.

4-5 Scatter Plots and Lines of Fit

47. Determine whether the graph shows a *positive*, *negative*, or *no* correlation. If there is a positive or negative correlation, describe its meaning.

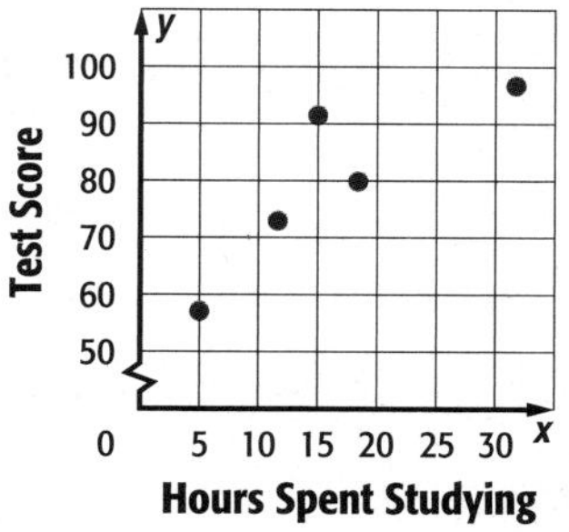

48. ATTENDANCE A scatter plot of data compares the number of years since a business has opened and its annual number of sales. It contains the ordered pairs (2, 650) and (5, 1280). Write an equation in slope-intercept form for the line of fit for this situation.

Example 6

The scatter plot displays the number of texts and the number of calls made daily. Write an equation for the line of fit.

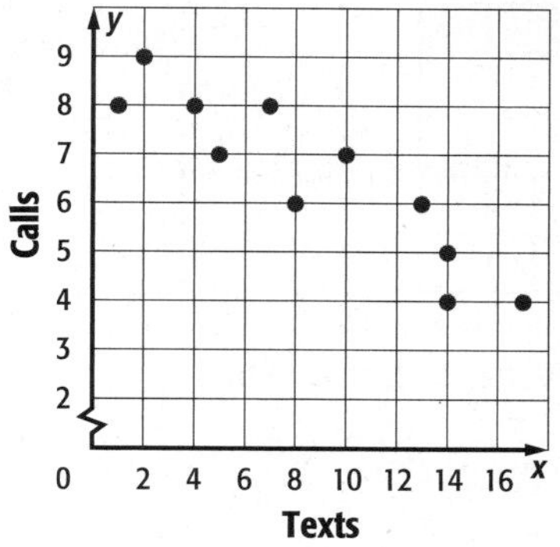

First, find the slope using (2, 9) and (17, 4).

$m = \frac{4-9}{17-2} = \frac{-5}{15}$ or $-\frac{1}{3}$ — Substitute and simplify.

Then find the y-intercept.

$9 = -\frac{1}{3}(2) + b$ — Substitute.

$9\frac{2}{3} = b$ — Add $\frac{2}{3}$ to each side.

Write the equation. $y = -\frac{1}{3}x + 9\frac{2}{3}$

4-6 Regression and Median-Fit Lines

49. SALE The table shows the number of purchases made at an outerwear store during a sale. Write an equation of the regression line. Then estimate the daily purchases on day 10 of the sale.

Days Since Sale Began	1	2	3	4	5	6	7
Daily Purchases	15	21	32	30	40	38	51

50. MOVIES The table shows ticket sales at a certain theater during the first week after a movie opened. Write an equation of the regression line. Then estimate the daily ticket sales on the 15th day.

Days Since Movie Opened	1	2	3	4	5	6	7
Daily Ticket Sales	85	92	89	78	65	68	55

Example 7

ATTENDANCE The table shows the annual attendance at an amusement park. Write an equation of the regression line for the data.

Years Since 2004	0	1	2	3	4	5	6
Attendance (thousands)	75	80	72	68	65	60	53

Step 1 Enter the data by pressing STAT and selecting the **Edit** option.

Step 2 Perform the regression by pressing STAT and selecting the **CALC** option. Scroll down to **LinReg (ax + b)** and press ENTER.

Step 3 Write the equation of the regression line by rounding the a- and b-values on the screen. $y = -4.04x + 79.68$

Study Guide and Review *Continued*

4-7 Inverse Linear Functions

Find the inverse of each relation.

51. {(7, 3.5), (6.2, 8), (−4, 2.7), (−12, 1.4)}

52. {(1, 9), (13, 26), (−3, 4), (−11, −2)}

53.

X	Y
−4	2.7
−1	3.8
0	4.1
3	7.2

54.

X	Y
−12	4
−8	0
−4	−4
0	−8

Find the inverse of each function.

55. $f(x) = \frac{5}{11}x + 10$

56. $f(x) = 3x + 8$

57. $f(x) = -4x - 12$

58. $f(x) = \frac{1}{4}x - 7$

59. $f(x) = -\frac{2}{3}x + \frac{1}{4}$

60. $f(x) = -3x + 3$

Example 8

Find the inverse of the relation.

{(5, −3), (11, 2), (−6, 12), (4, −2)}

To find the inverse, exchange the coordinates of the ordered pairs.

(5, −3) → (−3, 5) (−6, 12) → (12, −6)

(11, 2) → (2, 11) (4, −2) → (−2, 4)

The inverse is {(−3, 5), (2, 11), (12, −6), (−2, 4)}.

Example 9

Find the inverse of $f(x) = \frac{1}{4}x + 9$.

$f(x) = \frac{1}{4}x + 9$	Original equation
$y = \frac{1}{4}x + 9$	Replace $f(x)$ with y.
$x = \frac{1}{4}y + 9$	Interchange y and x.
$x - 9 = \frac{1}{4}y$	Subtract 9 from each side.
$4(x - 9) = y$	Multiply each side by 4.
$4x - 36 = y$	Distributive Property
$4x - 36 = f^{-1}(x)$	Replace y with $f^{-1}(x)$.

CHAPTER 4 Practice Test

1. Graph $y = 2x - 3$.

2. **MULTIPLE CHOICE** A popular pizza parlor charges \$12 for a large cheese pizza plus \$1.50 for each additional topping. Write an equation in slope-intercept form for the total cost C of a pizza with t toppings.

   A $C = 12t + 1.50$

   B $C = 13.50t$

   C $C = 12 + 1.50t$

   D $C = 1.50t - 12$

Write an equation of a line in slope-intercept form that passes through the given point and has the given slope.

3. $(-4, 2)$; slope -3

4. $(3, -5)$; slope $\frac{2}{3}$

Write an equation of the line in slope-intercept form that passes through the given points.

5. $(1, 4), (3, 10)$

6. $(2, 5), (-2, 8)$

7. $(0, 4), (-3, 0)$

8. $(7, -1), (9, -4)$

9. **PAINTING** The data in the table show the size of a room in square feet and the time it takes to paint the room in minutes.

Room Size	100	150	200	400	500
Painting Time	160	220	270	500	680

   a. Use the points (100, 160) and (500, 680) to write an equation in slope-intercept form.

   b. Predict the amount of time required to paint a room measuring 750 square feet.

10. **SALARY** The table shows the relationship between years of experience and teacher salary.

Years Experience	1	5	10	15	20
Salary (thousands of dollars)	28	31	42	49	64

   a. Write an equation for the best-fit line.

   b. Find the correlation coefficient and explain what it tells us about the relationship between experience and salary.

Write an equation in slope-intercept form for the line that passes through the given point and is parallel to the graph of each equation.

11. $(2, -3), y = 4x - 9$

12. $(-5, 1), y = -3x + 2$

Write an equation in slope-intercept form for the line that passes through the given point and is perpendicular to the graph of the equation.

13. $(1, 4), y = -2x + 5$

14. $(-3, 6), y = \frac{1}{4}x + 2$

15. **MULTIPLE CHOICE** The graph shows the relationship between outside temperature and daily ice cream cone sales. What type of correlation is shown?

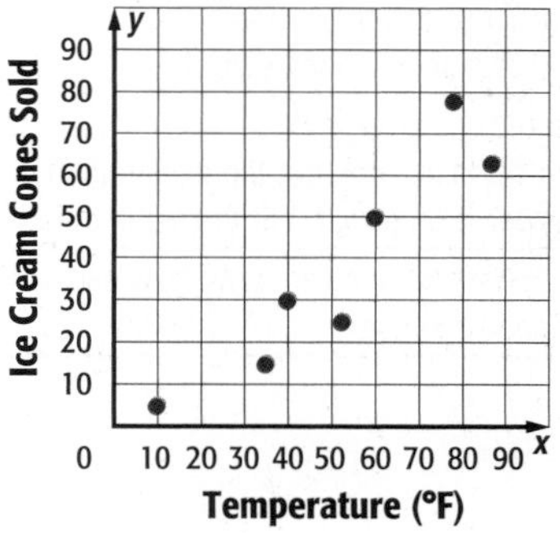

   F positive correlation

   G negative correlation

   H no correlation

   J not enough information

16. **ADOPTION** The table shows the number of children from Ethiopia adopted by U.S. citizens.

Years Since 2000	5	6	7	8	9
Number of Children	442	731	1254	1724	2277

   a. Write the slope-intercept form of the equation for the line of fit.

   b. Predict the number of children from Ethiopia who will be adopted in 2025.

Find the inverse of each function.

17. $f(x) = -5x - 30$

18. $f(x) = 4x + 10$

19. $f(x) = \frac{1}{6}x - 2$

20. $f(x) = \frac{3}{4}x + 12$

CHAPTER 4 Preparing for Standardized Tests

Short Answer Questions

Short answer questions require you to provide a solution to the problem, along with a method, explanation, and/or justification used to arrive at the solution.

Tetra Images/Photolibrary/Index Stock Imagery

Strategies for Solving Short Answer Questions

Step 1

Short answer questions are typically graded using a **rubric**, or a scoring guide. The following is an example of a short answer question scoring rubric.

Scoring Rubric	
Criteria	Score
Full Credit: The answer is correct and a full explanation is provided that shows each etep.	2
Partial Credit: • The answer is correct, but the explanation is incomplete. • The answer is incorrect, but the explanation is correct.	1
No Credit: Either an answer is not provided or the answer does not make sense.	0

Step 2

In solving short answer questions, remember to...

- explain your reasoning or state your approach to solving the problem.
- show all of your work or steps.
- check your answer if time permits.

Standardized Test Example

Read the problem. Identify what you need to know. Then use the information in the problem to solve. Show your work.

The table shows production costs for building different numbers of skateboards. Determine the missing value, x, that will result in a linear model.

Skateboards Built	Production Costs
14	$325
28	$500
x	$375
22	$425

Read the problem carefully. You are given several data points and asked to find the missing value that results in a linear model.

Example of a 2-point response:

Set up a coordinate grid and plot the three given points: (14, 325), (28, 500), (22, 425).

Then draw a straight line through them and find the x-value that produces a y-value of 375.

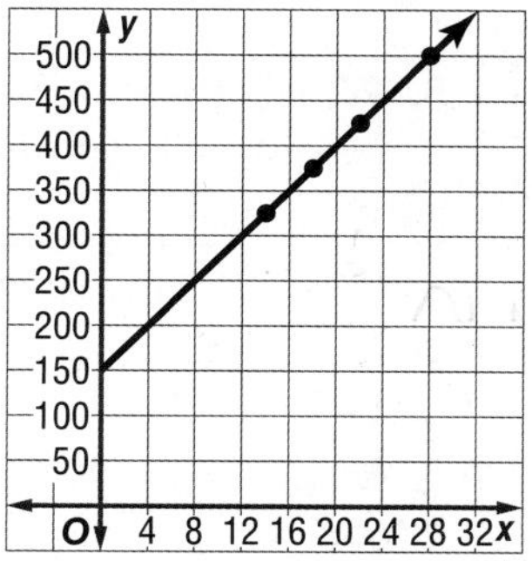

So, building 18 skateboards would result in production costs of $375. These data form a linear model.

The steps, calculations, and reasoning are clearly stated. The student also arrives at the correct answer. So, this response is worth the full 2 points.

Exercises

Read each problem. Identify what you need to know. Then use the information in the problem to solve. Show your work.

1. Given points $M(-1, 7)$, $N(3, -5)$, $O(6, 1)$, and $P(-3, -2)$, determine two segments that are perpendicular to each other.

2. Write the equation of a line that is parallel to $4x + 2y = 8$ and has a y-intercept of 5.

3. Three vertices of a quadrilateral are shown on the coordinate grid. Determine a fourth vertex that would result in a trapezoid.

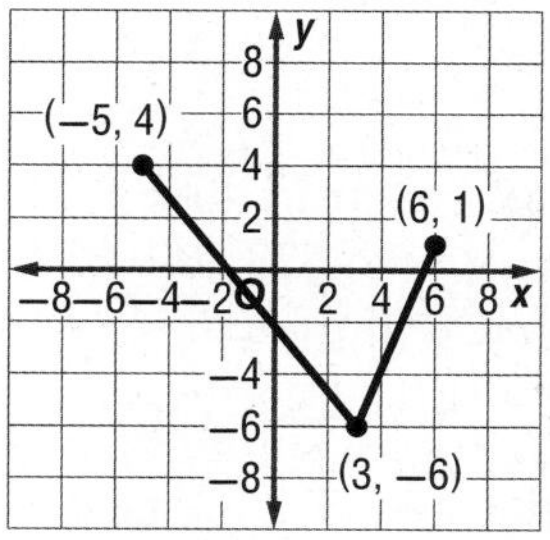

Standardized Test Practice

Cumulative, Chapters 1 through 4

Multiple Choice

Read each question. Then fill in the correct answer on the answer document provided by your teacher or on a sheet of paper.

1. What is the rate of change represented in the graph?

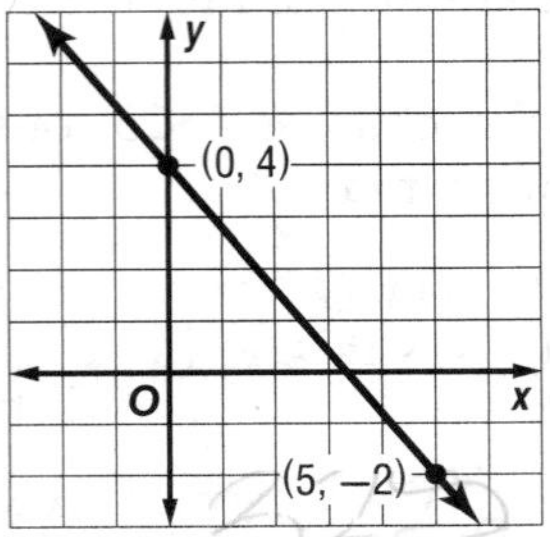

A $-\frac{2}{5}$

B $-\frac{5}{6}$

C $-\frac{6}{5}$

D $-\frac{5}{2}$

2. The table below shows the cost for renting a bicycle at a bike shop located in Venice Beach. What is a function that can represent this sequence?

Number of Hours	Cost ($)
1	10
2	14
3	18
4	22

F $f(n) = 4n + 10$

G $f(n) = 4n + 6$

H $f(n) = 10n + 4$

J $f(n) = 10n - 6$

3. Jaime bought a car in 2005 for $28,500. By 2008, the car was worth $23,700. Based on a linear model, what will the value of the car be in 2012?

A $17,300

B $17,550

C $18,100

D $18,475

4. If the graph of a line has a positive slope and a negative y-intercept, what happens to the x-intercept if the slope and the y-intercept are doubled?

F The x-intercept becomes four times as great.

G The x-intercept becomes twice as great.

H The x-intercept becomes one-fourth as great.

J The x-intercept remains the same.

5. Which absolute value equation has the graph below as its solution?

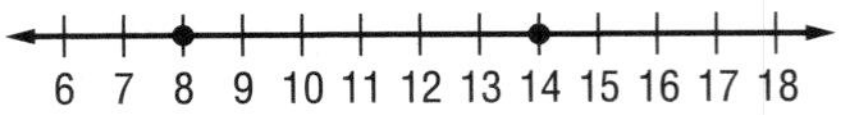

A $|x - 3| = 11$

B $|x - 4| = 12$

C $|x - 11| = 3$

D $|x - 12| = 4$

6. The table below shows the relationship between certain temperatures in degrees Fahrenheit and degrees Celsius. Which of the following linear equations correctly models this relationship?

Celsius (C)	Fahrenheit (F)
10°	50°
15°	59°
20°	68°
25°	77°
30°	86°

F $F = \frac{8}{5}C + 35$

G $F = \frac{4}{5}C + 42$

H $F = \frac{9}{5}C + 32$

J $F = \frac{12}{5}C + 26$

Test-Taking Tip

Question 3 Find the average annual depreciation between 2005 and 2008. Then extend the pattern to find the car's value in 2012.

Short Response/Gridded Response

Record your answers on the answer sheet provided by your teacher or on a sheet of paper.

7. What is the equation of the line graphed below?

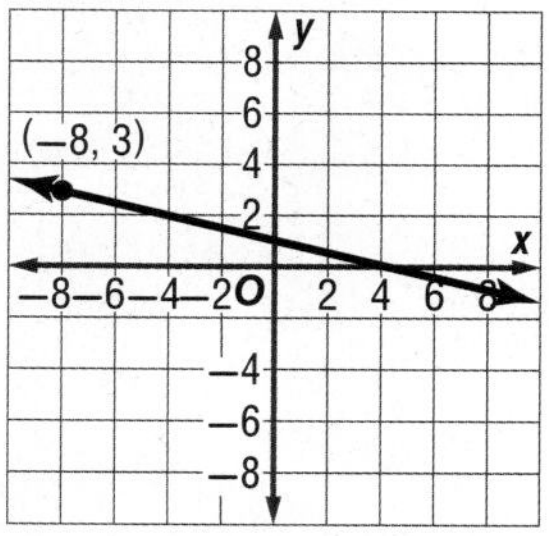

Express your answer in point slope form using the point (−8, 3).

8. GRIDDED RESPONSE The linear equation below is a best fit model for the peak depth of the Mad River when x inches of rain fall. What would you expect the peak depth of the river to be after a storm that produces $1\frac{3}{4}$ inches of rain? Round your answer to the nearest tenth of a foot if necessary.

$$y = 2.5x + 14.8$$

9. Jacob formed an advertising company in 1992. Initially, the company only had 14 employees. In 2008, the company had grown to a total of 63 employees. Find the percent of change in the number of employees working at Jacob's company. Round to the nearest tenth of a percent if necessary.

10. The table shows the total amount of rain during a storm.

Hours	1	2	3	4
Inches	0.45	0.9	1.35	1.8

a. Write an equation to fit the data in the table.

b. Describe the relationship between the hour and the amount of rain received.

11. An electrician charges a \$25 consultation fee plus \$35 per hour for labor.

a. Copy and complete the following table showing the charges for jobs that take 1, 2, 3, 4, or 5 hours.

Hours, h	Total Cost, C
1	
2	
3	
4	
5	

b. Write an equation in slope-intercept form for the total cost of a job that takes h hours.

c. If the electrician bills in quarter hours, how much would it cost for a job that takes 3 hours 15 minutes to complete?

Extended Response

Record your answer on a sheet of paper. Show your work.

12. Explain how you can determine whether two lines are parallel or perpendicular.

Need ExtraHelp?

If you missed Question...	1	2	3	4	5	6	7	8	9	10	11	12
Go to Lesson...	3-3	3-5	4-5	3-1	2-5	4-2	4-3	4-5	2-7	3-6	4-2	4-4

CHAPTER 5

Linear Inequalities

Then

You solved linear equations.

Now

In this chapter, you will:

- Solve one-step and multi-step inequalities.
- Solve compound inequalities and inequalities involving absolute value.
- Graph inequalities in two variables.

Why? ▲

PETS In the United States, about 75 million dogs are kept as pets. Approximately 16% of these were adopted from animal shelters. About 14% of dog owners own more than 3 dogs.

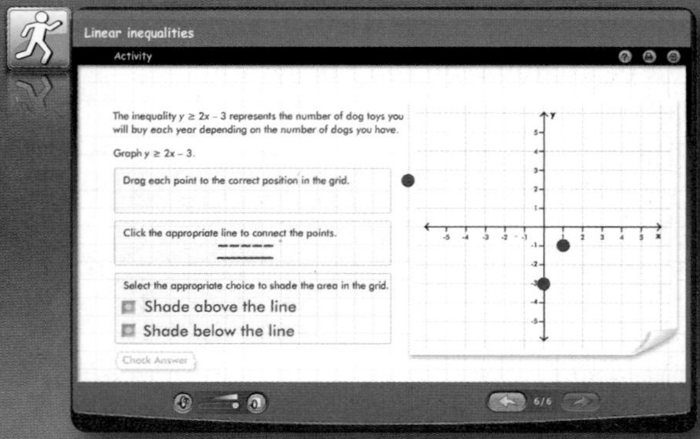

Linear inequalities
Activity
The inequality y ≥ 2x – 3 represents the number of dog toys you will buy each year depending on the number of dogs you have.
Graph y ≥ 2x – 3.
Drag each point to the correct position in the grid.
Click the appropriate line to connect the points.
Select the appropriate choice to shade the area in the grid.
Shade above the line
Shade below the line
Check Answer
6/6

connectED.mcgraw-hill.com Your Digital Math Portal

Animation | Vocabulary | eGlossary | Personal Tutor | Virtual Manipulatives | Graphing Calculator | Audio | Foldables | Self-Check Practice | Worksheets

Get Ready for the Chapter

Diagnose Readiness | **You have two options for checking prerequisite skills.**

1 Textbook Option

Take the Quick Check below. Refer to the Quick Review for help.

QuickCheck

Evaluate each expression for the given values.

1. $3x + y$ if $x = -4$ and $y = 2$
2. $-2m + 3k$ if $m = -8$ and $k = 3$
3. **CARS** The expression $\frac{m \text{ mi}}{g \text{ gal}}$ represents the gas mileage of a car. Find the gas mileage of a car that goes 295 miles on 12 gallons of gasoline. Round to the nearest tenth.

QuickReview

Example 1

Evaluate $-3x^2 + 4x - 6$ if $x = -2$.

$-3x^2 + 4x - 6$	Original expression
$= -3(-2)^2 + 4(-2) - 6$	Replace x with -2.
$= -3(4) + 4(-2) - 6$	Evaluate the power.
$= -12 + (-8) - 6$	Multiply.
$= -26$	Add and subtract.

QuickCheck

Solve each equation.

4. $x - 4 = 9$
5. $x + 8 = -3$
6. $4x = -16$
7. $\frac{x}{3} = 7$
8. $2x + 1 = 9$
9. $4x - 5 = 15$
10. $9x + 2 = 3x - 10$
11. $3(x - 2) = -2(x + 13)$
12. **FINANCIAL LITERACY** Claudia opened a savings account with $325. She saves $100 per month. Write an equation to determine how much money d, she has after m months.

QuickReview

Example 2

Solve $-2(x - 4) = 7x - 19$.

$-2(x - 4) = 7x - 19$	Original equation
$-2x + 8 = 7x - 19$	Distributive Property
$-2x + 8 + 2x = 7x - 19 + 2x$	Add $2x$ to each side.
$8 = 9x - 19$	Simplify.
$8 + 19 = 9x - 19 + 19$	Add 19 to each side.
$27 = 9x$	Simplify.
$3 = x$	Divide each side by 3.

QuickCheck

Solve each equation.

13. $|x + 11| = 18$
14. $|3x - 2| = 16$
15. **SURVEYS** In a survey, 32% of the people chose pizza as their favorite food. The results were reported to within 2% accuracy. What is the maximum and minimum percent of people who chose pizza?

QuickReview

Example 3

Solve $|x - 4| = 9$.

If $|x - 4| = 9$, then $x - 4 = 9$ or $x - 4 = -9$.

$x - 4 = 9$	or	$x - 4 = -9$
$x - 4 + 4 = 9 + 4$		$x - 4 + 4 = -9 + 4$
$x = 13$		$x = -5$

So, the solution set is $\{-5, 13\}$.

2 Online Option

Take an online self-check Chapter Readiness Quiz at connectED.mcgraw-hill.com.

Get Started on the Chapter

You will learn several new concepts, skills, and vocabulary terms as you study Chapter 5. To get ready, identify important terms and organize your resources. You may wish to refer to Chapter 0 to review prerequisite skills.

FOLDABLES® StudyOrganizer

Linear Inequalities Make this Foldable to help you organize your Chapter 5 notes about linear inequalities. Begin with a sheet of 11" by 17" paper.

1 **Fold** each side so the edges meet in the center.

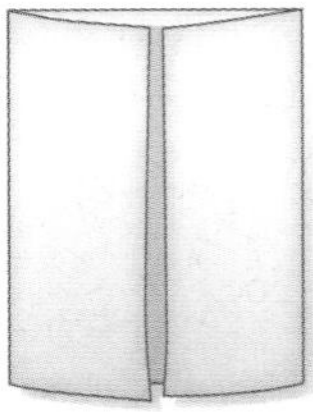

2 **Fold** in half.

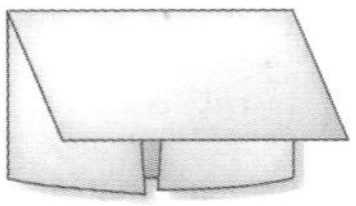

3 **Unfold** and cut from each end until you reach the vertical line.

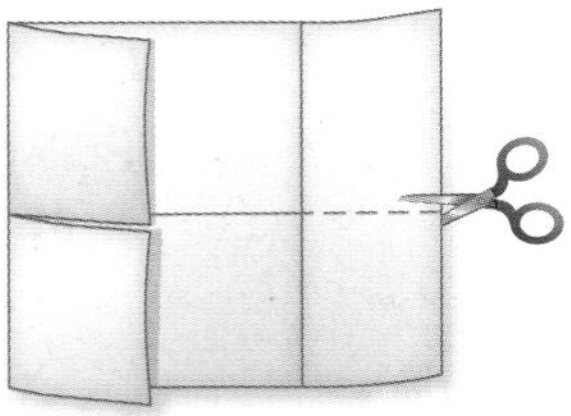

4 **Label** the front of each flap.

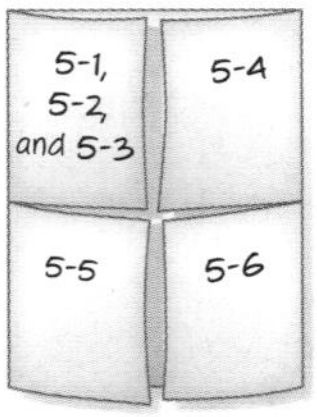

NewVocabulary

English		Español
inequality	p. 285	desigualdad
set-builder notation	p. 286	notación de construcción de conjuntos
compound inequality	p. 306	desigualdad compuesta
intersection	p. 306	intersección
union	p. 307	unión
boundary	p. 317	frontera
half-plane	p. 317	semiplano
closed half-plane	p. 317	semiplano cerrada
open half-plane	p. 317	semiplano abierto

ReviewVocabulary

equivalent equations ecuaciones equivalentes equations that have the same solution

linear equation ecuación lineal an equation in the form $Ax + By = C$, with a graph consisting of points on a straight line

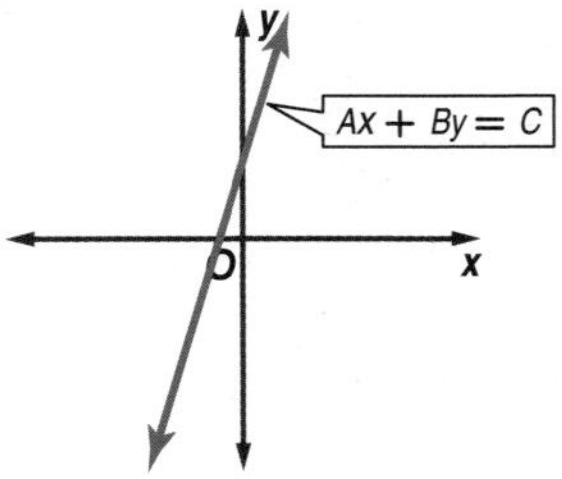

solution set conjunto solución the set of elements from the replacement set that makes an open sentence true

LESSON 5-1

Solving Inequalities by Addition and Subtraction

Then

- You solved equations by using addition and subtraction.

Now

1. Solve linear inequalities by using addition.
2. Solve linear inequalities by using subtraction.

Why?

- The data in the table show that the recommended daily allowance of Calories for girls 11–14 years old is less than that of girls between 15–18 years old.

Calories	
Girls 11–14 Years	Girls 15–18
1845	2110

Source: Vital Health Zone

$$1845 < 2110$$

If a 13-year-old girl and a 16-year-old girl each eat 150 more Calories in a day than is suggested, the 16-year-old will still eat more Calories.

$$1845 + 150 \ \underline{?}\ 2110 + 150$$

$$1995 < 2260$$

NewVocabulary

inequality
set-builder notation

Common Core State Standards

Content Standards

A.CED.1 Create equations and inequalities in one variable and use them to solve problems.

A.REI.3 Solve linear equations and inequalities in one variable, including equations with coefficients represented by letters.

Mathematical Practices

2 Reason abstractly and quantitatively.

4 Model with mathematics.

1 Solve Inequalities by Addition

An open sentence that contains $<$, $>$, $\leq$, or $\geq$ is an **inequality**. The example above illustrates the Addition Property of Inequalities.

KeyConcept Addition Property of Inequalities

Words If the same number is added to each side of a true inequality, the resulting inequality is also true.

Symbols For all numbers a, b, and c, the following are true.

1. If $a > b$, then $a + c > b + c$.
2. If $a < b$, then $a + c < b + c$.

This property is also true for $\geq$ and $\leq$.

Example 1 Solve by Adding

PT

Solve $x - 12 \geq 8$. Check your solution.

$x - 12 \geq 8$	Original inequality
$x - 12 + 12 \geq 8 + 12$	Add 12 to each side.
$x \geq 20$	Simplify.

The solution is the set {all numbers greater than or equal to 20}.

CHECK To check, substitute three different values into the original inequality: 20, a number less than 20, and a number greater than 20.

GuidedPractice

Solve each inequality. Check your solution.

1A. $22 > m - 8$

1B. $d - 14 \geq -19$

Stephen Simpson/Workbook Stock/Getty Images

ReadingMath

set-builder notation

$\{x \mid x \geq 20\}$ is read *the set of all numbers x such that x is greater than or equal to 20.*

A more concise way of writing a solution set is to use **set-builder notation**. In set-builder notation, the solution set in Example 1 is $\{x \mid x \geq 20\}$.

This solution set can be graphed on a number line. Be sure to check if the endpoint of the graph of an inequality should be a circle or a dot. If the endpoint is not included in the graph, use a circle, otherwise use a dot.

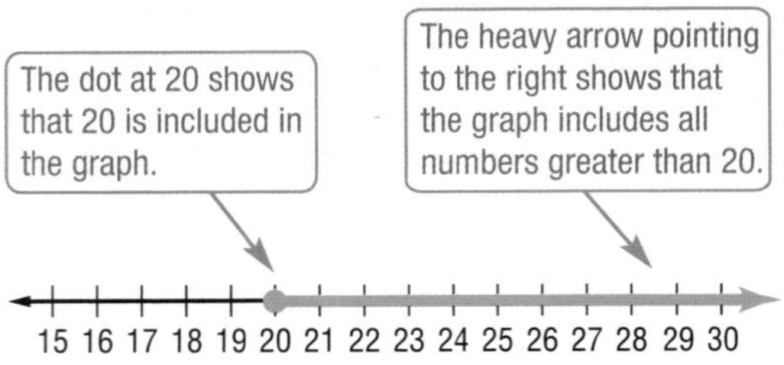

The dot at 20 shows that 20 is included in the graph.
The heavy arrow pointing to the right shows that the graph includes all numbers greater than 20.
15 16 17 18 19 20 21 22 23 24 25 26 27 28 29 30

2 Solve Inequalities by Subtraction

Subtraction can also be used to solve inequalities.

KeyConcept Subtraction Property of Inequalities

Words If the same number is subtracted from each side of a true inequality, the resulting inequality is also true.

Symbols For all numbers *a*, *b*, and *c*, the following are true.

1. If $a > b$, then $a - c > b - c$.
2. If $a < b$, then $a - c < b - c$.

This property is also true for $\geq$ and $\leq$.

Standardized Test Example 2 Solve by Subtracting

Test-TakingTip

Isolating the Variable When solving inequalities, the goal is to isolate the variable on one side of the inequality. This is the same as with solving equations.

Solve $m + 19 > 56$.

A $\{m \mid m < 41\}$ **B** $\{m \mid m < 37\}$ **C** $\{m \mid m > 37\}$ **D** $\{m \mid m > 41\}$

Read the Test Item

You need to find the solution set for the inequality.

Solve the Test Item

Step 1 Solve the inequality.

$m + 19 > 56$ Original inequality

$m + 19 - \mathbf{19} > 56 - \mathbf{19}$ Subtract 19 from each side.

$m > 37$ Simplify.

Step 2 Write in set-builder notation: $\{m \mid m > 37\}$. The answer is C.

GuidedPractice

2. Solve $p + 8 \leq 18$.

F $\{p \mid p \geq 10\}$ **G** $\{p \mid p \leq 10\}$ **H** $\{p \mid p \leq 26\}$ **J** $\{p \mid p \geq 126\}$

Terms that are constants are not the only terms that can be subtracted. Terms with variables can also be subtracted from each side to solve inequalities.

StudyTip

Writing Inequalities Simplifying the inequality so that the variable is on the left side, as in $a \geq 6$, prepares you to write the solution set in set-builder notation.

Example 3 Variables on Each Side

Solve $3a + 6 \leq 4a$. Then graph the solution set on a number line.

$3a + 6 \leq 4a$	Original inequality
$3a - 3a + 6 \leq 4a - 3a$	Subtract $3a$ from each side.
$6 \leq a$	Simplify.

Since $6 \leq a$ is the same as $a \geq 6$, the solution set is $\{a \mid a \geq 6\}$.

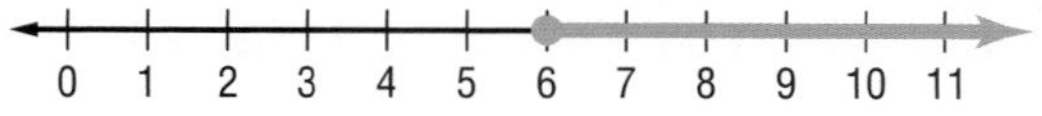

GuidedPractice

Solve each inequality. Then graph the solution set on a number line.

3A. $9n - 1 < 10n$

3B. $5h \leq 12 + 4h$

Verbal problems containing phrases like *greater than* or *less than* can be solved by using inequalities. The chart shows some other phrases that indicate inequalities.

ConceptSummary Phrases for Inequalities

<	>	≤	≥
less than fewer than	greater than more than	at most, no more than, less than or equal to	at least, no less than, greater than or equal to

Real-WorldLink

Leopard geckos are commonly yellow and white with black spots. They are nocturnal and easy to tame. They do not have toe pads like other geckos, so they do not climb.

Source: Exotic Pets

Real-World Example 4 Use an Inequality to Solve a Problem

PETS Felipe needs for the temperature of his leopard gecko's basking spot to be at least 82°F. Currently the basking spot is 62.5°F. How much warmer does the basking spot need to be?

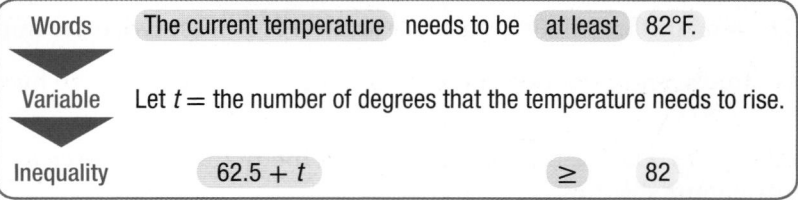

Words	The current temperature	needs to be	at least	82°F.
Variable	Let t = the number of degrees that the temperature needs to rise.			
Inequality	$62.5 + t$		$\geq$	82

$62.5 + t \geq 82$	Original inequality
$62.5 + t - 62.5 \geq 82 - 62.5$	Subtract 62.5 from each side.
$t \geq 19.5$	Simplify.

Felipe needs to raise the temperature of the basking spot 19.5°F or more.

GuidedPractice

4. SHOPPING Sanjay has $65 to spend at the mall. He bought a T-shirt for $18 and a belt for $14. If Sanjay wants a pair of jeans, how much can he spend?

Check Your Understanding

Examples 1–3 **Solve each inequality. Then graph the solution set on a number line.**

1. $x - 3 > 7$

2. $5 \geq 7 + y$

3. $g + 6 < 2$

4. $11 \leq p + 4$

5. $10 > n - 1$

6. $k + 24 > -5$

7. $8r + 6 < 9r$

8. $8n \geq 7n - 3$

Example 4 **Define a variable, write an inequality, and solve each problem. Check your solution.**

9. Twice a number increased by 4 is at least 10 more than the number.

10. Three more than a number is less than twice the number.

11. **AMUSEMENT** A thrill ride swings passengers back and forth, a little higher each time up to 137 feet. Suppose the height of the swing after 30 seconds is 45 feet. How much higher will the ride swing?

Practice and Problem Solving

Examples 1–3 **Solve each inequality. Then graph the solution set on a number line.**

12. $m - 4 < 3$

13. $p - 6 \geq 3$

14. $r - 8 \leq 7$

15. $t - 3 > -8$

16. $b + 2 \geq 4$

17. $13 > 18 + r$

18. $5 + c \leq 1$

19. $-23 \geq q - 30$

20. $11 + m \geq 15$

21. $h - 26 < 4$

22. $8 \leq r - 14$

23. $-7 > 20 + c$

24. $2a \leq -4 + a$

25. $z + 4 \geq 2z$

26. $w - 5 \leq 2w$

27. $3y + 6 \leq 2y$

28. $6x + 5 \geq 7x$

29. $-9 + 2a < 3a$

Example 4 **Define a variable, write an inequality, and solve each problem. Check your solution.**

30. Twice a number is more than the sum of that number and 9.

31. The sum of twice a number and 5 is at most 3 less than the number.

32. The sum of three times a number and −4 is at least twice the number plus 8.

33. Six times a number decreased by 8 is less than five times the number plus 21.

CCSS **MODELING** **Define a variable, write an inequality, and solve each problem. Then interpret your solution.**

34. **FINANCIAL LITERACY** Keisha is babysitting at \$8 per hour to earn money for a car. So far she has saved \$1300. The car that Keisha wants to buy costs at least \$5440. How much money does Keisha still need to earn to buy the car?

35. **TECHNOLOGY** A recent survey found that more than 21 million people between the ages of 12 and 17 use the Internet. Of those, about 16 million said they use the Internet at school. How many teens that are online do not use the Internet at school?

36. **MUSIC** A DJ added 20 more songs to his digital media player, making the total more than 61. How many songs were originally on the player?

37. TEMPERATURE The water temperature in a swimming pool increased 4°F this morning. The temperature is now less than 81°F. What was the water temperature this morning?

38. BASKETBALL A player's goal was to score at least 150 points this season. So far, she has scored 123 points. If there is one game left, how many points must she score to reach her goal?

39. SPAS Samantha received a $75 gift card for a local day spa for her birthday. She plans to get a haircut and a manicure. How much money will be left on her gift card after her visit?

Service	Cost ($)
haircut	at least 32
manicure	at least 26

40. VOLUNTEER Kono knows that he can only volunteer up to 25 hours per week. If he has volunteered for the times recorded at the right, how much more time can Kono volunteer this week?

Center	Time (h)
Shelter	3 h 15 min
Kitchen	2 h 20 min

Solve each inequality. Check your solution, and then graph it on a number line.

41. $c + (-1.4) \geq 2.3$

42. $9.1g + 4.5 < 10.1g$

43. $k + \frac{3}{4} > \frac{1}{3}$

44. $\frac{3}{2}p - \frac{2}{3} \leq \frac{4}{9} + \frac{1}{2}p$

45. MULTIPLE REPRESENTATIONS In this problem, you will explore multiplication and division in inequalities.

a. Geometric Suppose a balance has 12 pounds on the left side and 18 pounds on the right side. Draw a picture to represent this situation.

b. Numerical Write an inequality to represent the situation.

c. Tabular Create a table showing the result of doubling, tripling, or quadrupling the weight on each side of the balance. Create a second table showing the result of reducing the weight on each side of the balance by a factor of $\frac{1}{2}$, $\frac{1}{3}$, or $\frac{1}{4}$. Include a column in each table for the inequality representing each situation.

d. Verbal Describe the effect multiplying or dividing each side of an inequality by the same positive value has on the inequality.

CCSS REASONING If $m + 7 \geq 24$, then complete each inequality.

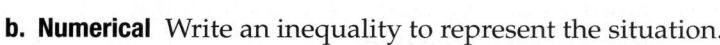

46. $m \geq$ _?_

47. $m +$ _?_ ≥ 27

48. $m - 5 \geq$ _?_

49. $m -$ _?_ ≥ 14

50. $m - 19 \geq$ _?_

51. $m +$ _?_ ≥ 43

H.O.T. Problems Use Higher-Order Thinking Skills

52. REASONING Compare and contrast the graphs of $a < 4$ and $a \leq 4$.

53. CHALLENGE Suppose $b > d + \frac{1}{3}$, $c + 1 < a - 4$, and $d + \frac{5}{8} > a + 2$. Order a, b, c, and d from least to greatest.

54. OPEN ENDED Write three linear inequalities that are equivalent to $y < -3$.

55. WRITING IN MATH Summarize the process of solving and graphing linear inequalities.

56. WRITING IN MATH Explain why $x - 2 > 5$ has the same solution set as $x > 7$.

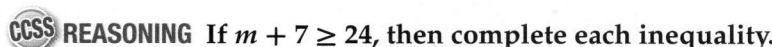

Standardized Test Practice

57. Which equation represents the relationship shown?

x	y
1	1
2	9
3	17
4	25
5	33
6	41

A $y = 7x - 8$
B $y = 7x + 8$
C $y = 8x - 7$
D $y = 8x + 7$

58. What is the solution set of the inequality $7 + x < 5$?

F $\{x \mid x < 2\}$
G $\{x \mid x > 2\}$
H $\{x \mid x < -2\}$
J $\{x \mid x > -2\}$

59. Francisco has \$3 more than $\frac{1}{4}$ the number of dollars that Kayla has. Which expression represents how much money Francisco has?

A $3\left(\frac{1}{4}k\right)$
B $\frac{1}{4}k + 3$
C $3 - \frac{1}{4}k$
D $\frac{1}{4} + 3k$

60. GRIDDED RESPONSE The mean score for 10 students on the chemistry final exam was 178. However, the teacher had made a mistake and recorded one student's score as ten points less than the actual score. What should the mean score be?

Spiral Review

Find the inverse of each function. (Lesson 4-7)

61. $f(x) = 7x - 28$

62. $f(x) = \frac{2}{5}x + 12$

63. $f(x) = -\frac{1}{3}x - 8$

64. $f(x) = 12x + 16$

Write the slope-intercept form of an equation for the line that passes through the given point and is perpendicular to the graph of each equation. (Lesson 4-4)

65. $(-2, 0)$, $y = x - 6$

66. $(-3, 1)$, $y = -3x + 7$

67. $(1, -3)$, $y = \frac{1}{2}x + 4$

68. $(-2, 7)$, $2x - 5y = 3$

69. TRAVEL On an island cruise in Hawaii, each passenger is given a lei. A crew member hands out 3 red, 3 blue, and 3 green leis in that order. If this pattern is repeated, what color lei will the 50th person receive? (Lesson 3-6)

Find the *n*th term of each arithmetic sequence described. (Lesson 3-5)

70. $a_1 = 52$, $d = 12$, $n = 102$

71. $-9, -7, -5, -3, \ldots$ for $n = 18$

72. $0.5, 1, 1.5, 2, \ldots$ for $n = 50$

73. JOBS Refer to the time card shown. Write a direct variation equation relating your pay to the hours worked and find your pay if you work 30 hours. (Lesson 3-4)

Weekly Time Card

Day	Hours
FRIDAY	2.0
SATURDAY	3.5
SUNDAY	2.0
TOTAL HOURS	7.5
PAY	\$52.50

Skills Review

Solve each equation.

74. $8y = 56$

75. $4p = -120$

76. $-3a = -21$

77. $2c = \frac{1}{5}$

78. $\frac{r}{2} = 21$

79. $-\frac{3}{4}g = -12$

80. $\frac{2}{5}w = -4$

81. $-6x = \frac{2}{3}$

EXPLORE 5-2

Algebra Lab
Solving Inequalities

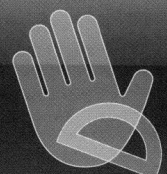

You can use algebra tiles to solve inequalities.

CCSS Common Core State Standards
Content Standards
A.REI.3 Solve linear equations and inequalities in one variable, including equations with coefficients represented by letters.

Activity Solve Inequalities

Solve $-2x \le 4$.

Step 1 Use a self-adhesive note to cover the equals sign on the equation mat. Then write a $\le$ symbol on the note. Model the inequality.

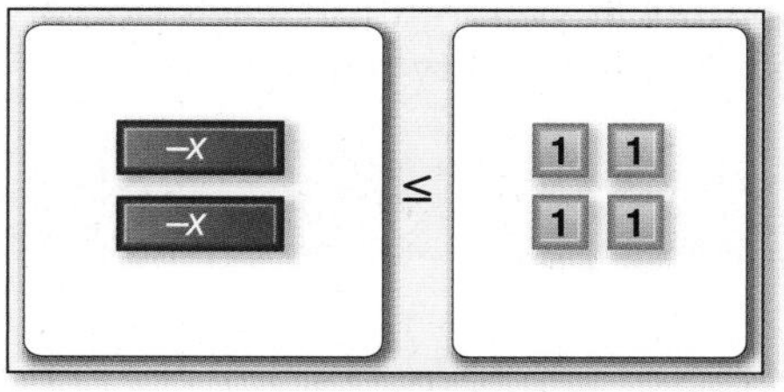

$-2x \le 4$

Step 2 Since you do not want to solve for a negative x-tile, eliminate the negative x-tiles by adding 2 positive x-tiles to each side. Remove the zero pairs.

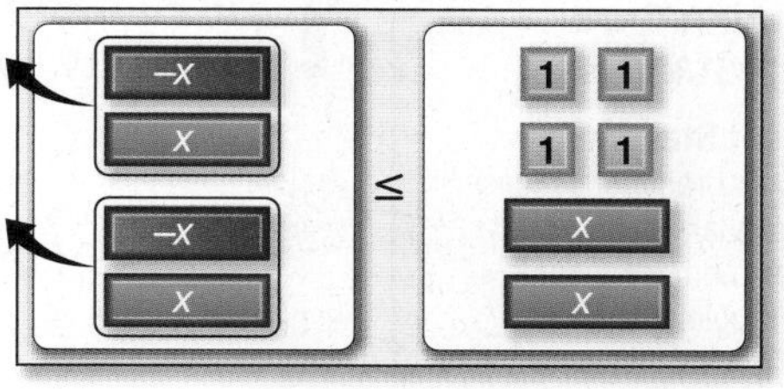

$-2x + 2x \le 4 + 2x$

Step 3 Add 4 negative 1-tiles to each side to isolate the x-tiles. Remove the zero pairs.

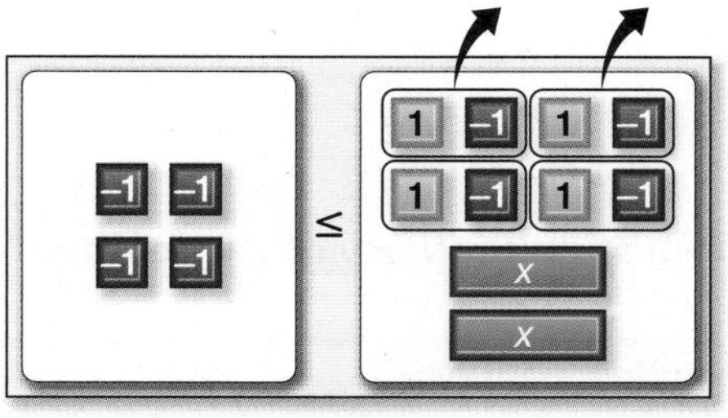

$-4 \le 2x$

Step 4 Separate the tiles into 2 groups.

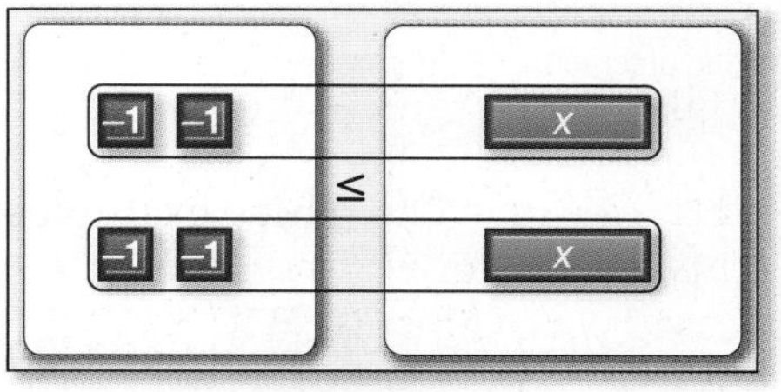

$-2 \le x$ or $x \ge -2$

Model and Analyze

Use algebra tiles to solve each inequality.

1. $-3x < 9$

2. $-4x > -4$

3. $-5x \ge 15$

4. $-6x \le -12$

5. In Exercises 1–4, is the coefficient of x in each inequality positive or negative?

6. Compare the inequality symbols and locations of the variable in Exercises 1–4 with those in their solutions. What do you find?

7. Model the solution for $3x \le 12$. How is this different from solving $-3x \le 12$?

8. Write a rule for solving inequalities involving multiplication and division. (*Hint:* Remember that dividing by a number is the same as multiplying by its reciprocal.)

LESSON 5-2

Solving Inequalities by Multiplication and Division

Then	Now	Why?
You solved equations by using multiplication and division.	**1** Solve linear inequalities by using multiplication. **2** Solve linear inequalities by using division.	Terrell received a gift card for \$20 of music downloads. If each download costs \$0.89, the number of downloads he can purchase can be represented by the inequality $0.89d \leq 20$.

CCSS Common Core State Standards

Content Standards

A.CED.1 Create equations and inequalities in one variable and use them to solve problems.

A.REI.3 Solve linear equations and inequalities in one variable, including equations with coefficients represented by letters.

Mathematical Practices

1 Make sense of problems and persevere in solving them.

6 Attend to precision.

1 Solve Inequalities by Multiplication

If you multiply each side of an inequality by a positive number, then the inequality remains true.

$4 > 2$	Original inequality
$4(3) \underset{}{\overset{?}{\_\_}} 2(3)$	Multiply each side by 3.
$12 > 6$	Simplify.

Notice that the direction of the inequality remains the same.

If you multiply each side of an inequality by a negative number, the inequality symbol changes direction.

$7 < 9$	Original inequality
$7(-2) \overset{?}{\_\_} 9(-2)$	Multiply each side by -2.
$-14 > -18$	Simplify.

These examples demonstrate the **Multiplication Property of Inequalities**.

KeyConcept Multiplication Property of Inequalities

Words	Symbols	Examples
If both sides of an inequality that is true are multiplied by a positive number, the resulting inequality is also true.	For any real numbers a and b and any positive real number c, if $a > b$, then $ac > bc$. And, if $a < b$, then $ac < bc$.	$6 > 3.5$ $6(2) > 3.5(2)$ $12 > 7$ and $2.1 < 5$ $2.1(0.5) < 5(0.5)$ $1.05 < 2.5$
If both sides of an inequality that is true are multiplied by a negative number, the direction of the inequality sign is reversed to make the resulting inequality also true.	For any real numbers a and b and any negative real number c, if $a > b$, then $ac < bc$. And, if $a < b$, then $ac > bc$.	$7 > 4.5$ $7(-3) < 4.5(-3)$ $-21 < -13.5$ and $3.1 < 5.2$ $3.1(-4) > 5.2(-4)$ $-12.4 > -20.8$

This property also holds for inequalities involving $\leq$ and $\geq$.

John Lund/Marc Romanelli/Blend Images/Getty Images

Real-World Example 1 Write and Solve an Inequality

PT

SURVEYS Of the students surveyed at Madison High School, fewer than eighty-four said they have never purchased an item online. This is about one eighth of those surveyed. How many students were surveyed?

Understand You know the number of students who have never purchased an item online and the portion this is of the number of students surveyed.

Plan Let n = the number of students surveyed. Write an open sentence that represents this situation.

Words	One eighth	times	the number of students surveyed	is less than	84.
Inequality	$\frac{1}{8}$	$\cdot$	n	$<$	84.

Solve Solve for n.

$\frac{1}{8}n < 84$ — Original inequality

$(8)\frac{1}{8}n < (8)84$ — Multiply each side by 8.

$n < 672$ — Simplify.

Check Check the endpoint with 672 and the direction of the inequality with a value less than 672.

$\frac{1}{8}(672) \stackrel{?}{=} 84$ — Check endpoint.

$84 = 84$ ✓

$\frac{1}{8}(0) \stackrel{?}{<} 84$ — Check direction.

$0 < 84$ ✓

The solution set is $\{n \mid n < 672\}$, so fewer than 672 students were surveyed.

StudyTip

CCSS Sense-Making In Example 1, you could also check the solution by substituting a number greater than 672 and verifying that the resulting inequality is false.

Real-WorldLink

More than 30,000 different orchid species flower in the wild on every continent except Antarctica.

Source: Aloha Orchid Nursery

GuidedPractice

1. **BIOLOGY** Mount Kinabalue in Malaysia has the greatest concentration of wild orchids on Earth. It contains more than 750 species, or about one fourth of all orchid species in Malaysia. How many orchid species are there in Malaysia?

You can also use multiplicative inverses with the Multiplication Property of Inequalities to solve an inequality.

Example 2 Solve by Multiplying

PT

Solve $-\frac{3}{7}r < 21$. Graph the solution on a number line.

$-\frac{3}{7}r < 21$ — Original inequality

$\left(-\frac{7}{3}\right)\left(-\frac{3}{7}r\right) > \left(-\frac{7}{3}\right)21$ — Multiply each side by $-\frac{7}{3}$. Reverse the inequality symbol.

$r > -49$ — Simplify. Check by substituting values.

The solution set is $\{r \mid r > -49\}$.

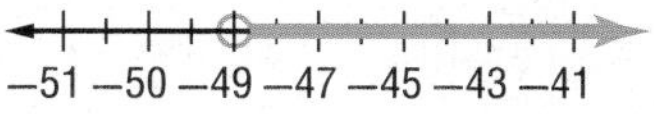

GuidedPractice

Solve each inequality. Check your solution.

2A. $-\frac{n}{6} \leq 8$ **2B.** $-\frac{4}{3}p > -10$ **2C.** $\frac{1}{5}m \geq -3$ **2D.** $\frac{3}{8}t < 5$

WatchOut!

Negatives A negative sign in an inequality does not necessarily mean that the direction of the inequality should change. For example, when solving $\frac{x}{6} > -3$, do not change the direction of the inequality.

2 Solve Inequalities by Division

If you divide each side of an inequality by a positive number, then the inequality remains true.

$-10 < -5$	Original inequality
$\frac{-10}{5} \ ? \ \frac{-5}{5}$	Divide each side by −5.
$-2 < -1$	Simplify.

Notice that the direction of the inequality remains the same. If you divide each side of an inequality by a negative number, the inequality symbol changes direction.

$15 < 18$	Original inequality
$\frac{15}{-3} \ ? \ \frac{18}{-3}$	Divide each side by −3.
$-5 > -6$	Simplify.

These examples demonstrate the **Division Property of Inequalities**.

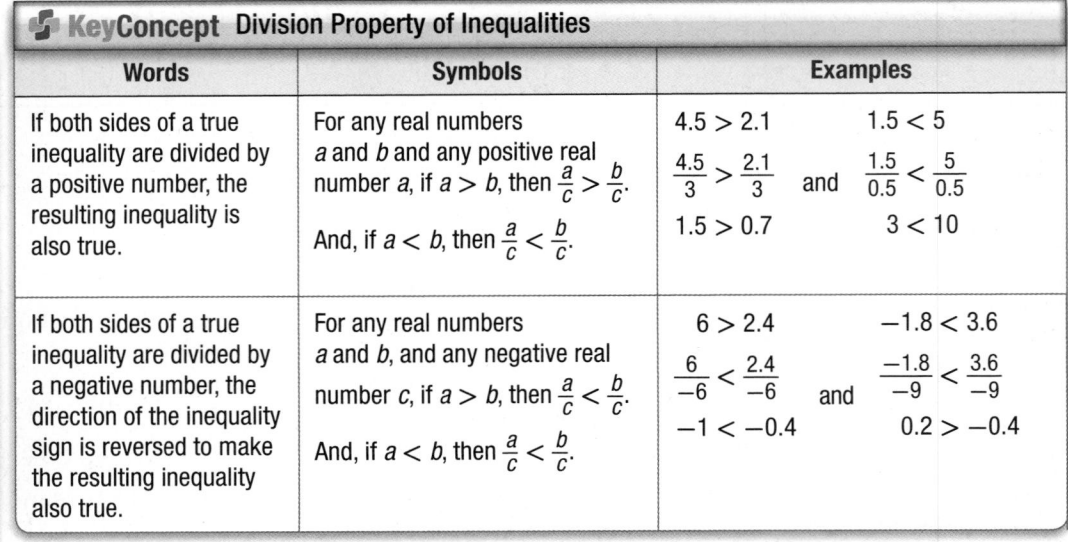

KeyConcept Division Property of Inequalities

Words	Symbols	Examples
If both sides of a true inequality are divided by a positive number, the resulting inequality is also true.	For any real numbers a and b and any positive real number a, if $a > b$, then $\frac{a}{c} > \frac{b}{c}$. And, if $a < b$, then $\frac{a}{c} < \frac{b}{c}$.	$4.5 > 2.1$, $\frac{4.5}{3} > \frac{2.1}{3}$, $1.5 > 0.7$ and $1.5 < 5$, $\frac{1.5}{0.5} < \frac{5}{0.5}$, $3 < 10$
If both sides of a true inequality are divided by a negative number, the direction of the inequality sign is reversed to make the resulting inequality also true.	For any real numbers a and b, and any negative real number c, if $a > b$, then $\frac{a}{c} < \frac{b}{c}$. And, if $a < b$, then $\frac{a}{c} < \frac{b}{c}$.	$6 > 2.4$, $\frac{6}{-6} < \frac{2.4}{-6}$, $-1 < -0.4$ and $-1.8 < 3.6$, $\frac{-1.8}{-9} < \frac{3.6}{-9}$, $0.2 > -0.4$

This property also holds true for inequalities involving ≤ and ≥.

Math HistoryLink

Thomas Harriot

(1560–1621) Harriot was a prolific astronomer. He was the first to map the Moon's surface and to see sunspots. Harriot is best known for his work in algebra.

Example 3 Divide to Solve an Inequality

Solve each inequality. Graph the solution on a number line.

a. $60t > 8$

$60t > 8$	Original inequality
$\frac{60t}{60} > \frac{8}{60}$	Divide each side by 60.
$t > \frac{2}{15}$	Simplify.

$\left\{t \mid t > \frac{2}{15}\right\}$

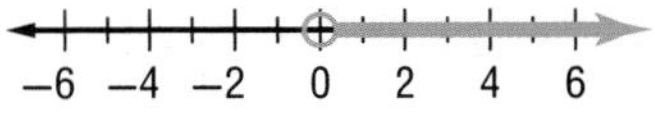

b. $-7d \leq 147$

$-7d \leq 147$	Original inequality
$\frac{-7d}{-7} \geq \frac{147}{-7}$	Divide each side by −7.
$d \geq -21$	Simplify.

$\{d \mid d \geq -21\}$

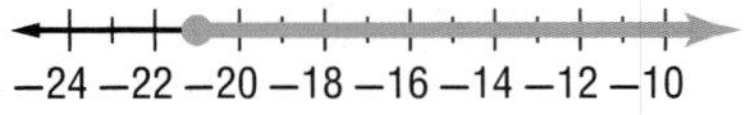

GuidedPractice

3A. $8p < 58$

3B. $-42 > 6r$

3C. $-12h > 15$

3D. $-\frac{1}{2}n < 6$

Brand X Pictures/PunchStock

Check Your Understanding

Example 1

1. **FUNDRAISING** The Jefferson Band Boosters raised more than \$5500 from sales of their \$15 band DVD. Define a variable, and write an inequality to represent the number of DVDs they sold. Solve the inequality and interpret your solution.

Examples 2–3 **Solve each inequality. Graph the solution on a number line.**

2. $30 > \frac{1}{2}n$ **3.** $-\frac{3}{4}r \leq -6$ **4.** $-\frac{c}{6} \geq 7$ **5.** $\frac{h}{2} < -5$

6. $9t > 108$ **7.** $-84 < 7v$ **8.** $-28 \leq -6x$ **9.** $40 \geq -5z$

Practice and Problem Solving

Example 1 **Define a variable, write an inequality, and solve each problem. Then interpret your solution.**

10. **CELL PHONE PLAN** Mario purchases a prepaid phone plan for \$50 at \$0.13 per minute. How many minutes can Mario talk on this plan?

11. **FINANCIAL LITERACY** Rodrigo needs at least \$560 to pay for his spring break expenses, and he is saving \$25 from each of his weekly paychecks. How long will it be before he can pay for his trip?

Examples 2–3 **Solve each inequality. Graph the solution on a number line.**

12. $\frac{1}{4}m \leq -17$ **13.** $\frac{1}{2}a < 20$ **14.** $-11 > -\frac{c}{11}$

15. $-2 \geq -\frac{d}{34}$ **16.** $-10 \leq \frac{x}{-2}$ **17.** $-72 < \frac{f}{-6}$

18. $\frac{2}{3}h > 14$ **19.** $-\frac{3}{4}j \geq 12$ **20.** $-\frac{1}{6}n \leq -18$

21. $6p \leq 96$ **22.** $4r < 64$ **23.** $32 > -2y$

24. $-26 < 26t$ **25.** $-6v > -72$ **26.** $-33 \geq -3z$

27. $4b \leq -3$ **28.** $-2d < 5$ **29.** $-7f > 5$

30. **CHEERLEADING** To remain on the cheerleading squad, Lakita must attend at least $\frac{3}{5}$ of the study table sessions offered. She attends 15 sessions. If Lakita met the requirements, what is the maximum number of study table sessions?

31. **BRACELETS** How many bracelets can Caitlin buy for herself and her friends if she wants to spend no more than \$22?

32. CCSS **PRECISION** The National Honor Society at Pleasantville High School wants to raise at least \$500 for a local charity. Each student earns \$0.50 for every quarter of a mile walked in a walk-a-thon. How many miles will the students need to walk?

33. **MUSEUM** The American history classes are planning a trip to a local museum. Admission is \$8 per person. Determine how many people can go for \$260.

34. **GASOLINE** If gasoline costs \$3.15 per gallon, how many gallons of gasoline, to the nearest tenth, can Jan buy for \$24?

Match each inequality to the graph of its solution.

35. $-\frac{2}{3}h \leq 9$ **36.** $25j \geq 8$ **37.** $3.6p < -4.5$ **38.** $2.3 < -5t$

a.
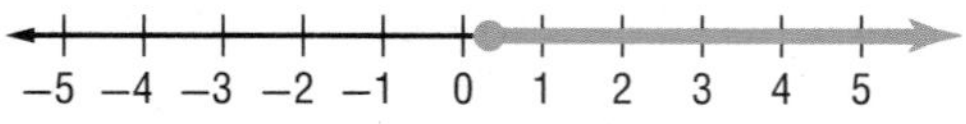

b.
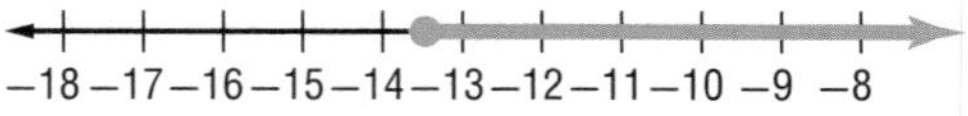

c.
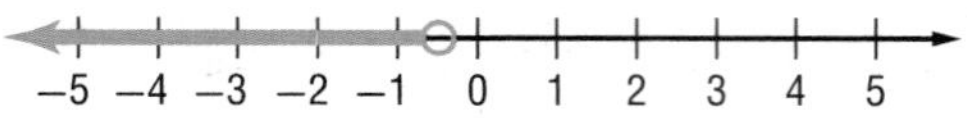

d.
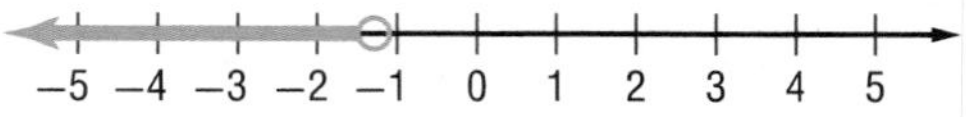

39. **CANDY** Fewer than 42 employees at a factory stated that they preferred fudge over fruit candy. This is about two thirds of the employees. How many employees are there?

40. **TRAVEL** A certain travel agency employs more than 275 people at all of its branches. Approximately three fifths of all the people are employed at the west branch. How many people work at the west branch?

41. **MULTIPLE REPRESENTATIONS** The equation for the volume of a pyramid is $\frac{1}{3}$ the area of the base times the height.

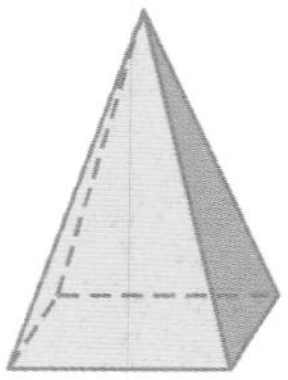

a. Geometric Draw a pyramid with a square base b cm long and a height of h cm.

b. Numerical Suppose the pyramid has a volume of 72 cm^3. Write an equation to find the height.

c. Tabular Create a table showing the value of h when $b = 1, 3, 6, 9,$ and 12.

d. Numerical Write an inequality for the possible lengths of b such that $b < h$. Write an inequality for the possible lengths of h such that $b > h$.

H.O.T. Problems Use Higher-Order Thinking Skills

42. **ERROR ANALYSIS** Taro and Jamie are solving $6d \geq -84$. Is either of them correct? Explain your reasoning.

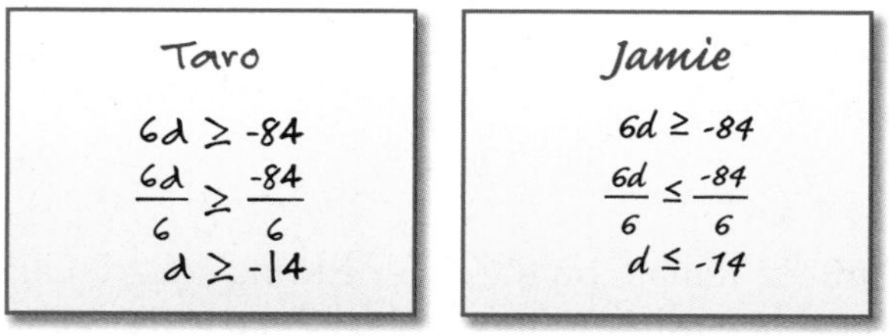

43. **CHALLENGE** Solve each inequality for x. Assume that $a > 0$.

a. $-ax < 5$ **b.** $\frac{1}{a}x \geq 8$ **c.** $-6 \geq ax$

44. **CCSS STRUCTURE** Determine whether $x^2 > 1$ and $x > 1$ are equivalent. Explain.

45. **REASONING** Explain whether the statement *If* $a > b$, *then* $\frac{1}{a} > \frac{1}{b}$ is *sometimes, always,* or *never* true.

46. **OPEN ENDED** Create a real-world situation to represent the inequality $-\frac{5}{8} \geq x$.

47. **WRITING IN MATH** How are solving linear inequalities and linear equations similar? different?

Standardized Test Practice

48. Juan's international calling card costs 9¢ for each minute. Which inequality can be used to find how long he can talk to a friend if he does not want to spend more than \$2.50 on the call?

A $0.09 \geq 2.50m$

B $0.09 \leq 2.50m$

C $0.09m \geq 2.50$

D $0.09m \leq 2.50$

49. SHORT RESPONSE Find the value of x.

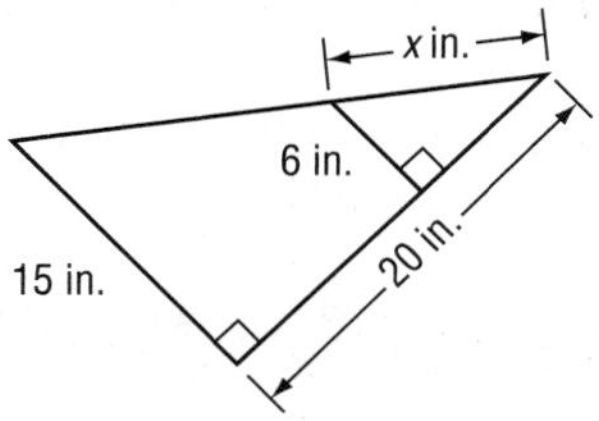

50. What is the greatest rate of decrease of this function?

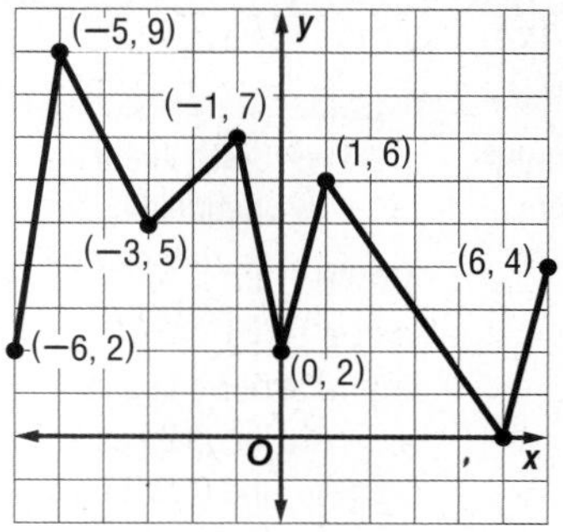

F −5 **H** −2

G −3 **J** 1

51. What is the value of x if $4x - 3 = -2x$?

A −2 **C** $\frac{1}{2}$

B $-\frac{1}{2}$ **D** 2

Spiral Review

Solve each inequality. Check your solution, and then graph it on a number line. (Lesson 5-1)

52. $-8 + 4a < 6a$

53. $2y + 11 \geq -24y$

54. $7 - 2b > 12b$

Find the inverse of each function. (Lesson 4-7)

55. $f(x) = -6x + 18$

56. $f(x) = \frac{3}{7}x + 9$

57. $f(x) = 4x - 5$

58. HOME DECOR Pam is having blinds installed at her home. The cost c of installation for any number of blinds b can be described by $c = 25 + 6.5b$. Graph the equation and determine how much it would cost if Pam has 8 blinds installed. (Lesson 3-1)

59. RESCUE A boater radioed for a helicopter to pick up a sick crew member. At that time, the boat and the helicopter were at the positions shown. How long will it take for the helicopter to reach the boat? (Lesson 2-9)

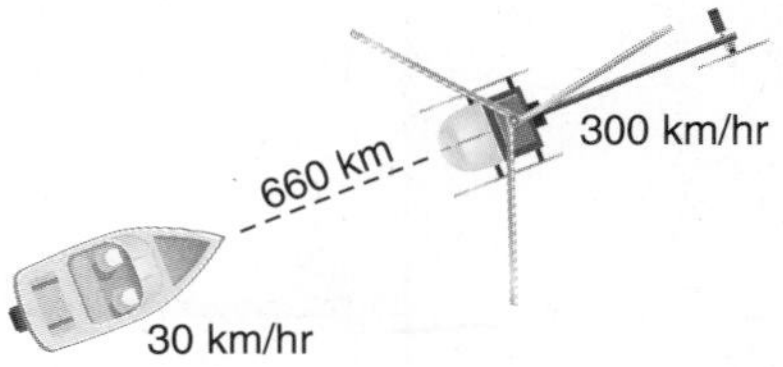

Solve each equation. (Lesson 2-5)

60. $|x + 3| = 10$

61. $|2x - 8| = 6$

62. $|3x + 1| = -2$

Skills Review

Solve each equation.

63. $4y + 11 = 19$

64. $2x - 7 = 9 + 4x$

65. $\frac{1}{4} + 2x = 4x - 8$

66. $\frac{1}{3}(6w - 3) = 3w + 12$

67. $\frac{7r + 5}{2} = 13$

68. $\frac{1}{2}a = \frac{a - 3}{4}$

LESSON

5-3 Solving Multi-Step Inequalities

Then

- You solved multi-step equations.

Now

1. Solve linear inequalities involving more than one operation.
2. Solve linear inequalities involving the Distributive Property.

Why?

- A salesperson may make a base monthly salary and earn a commission on each of her sales. To find the number of sales she needs to make to pay her monthly bills, you can use a multi-step inequality.

CCSS Common Core State Standards

Content Standards

A.CED.1 Create equations and inequalities in one variable and use them to solve problems.

A.REI.3 Solve linear equations and inequalities in one variable, including equations with coefficients represented by letters.

Mathematical Practices

7 Look for and make use of structure.

1 Solve Multi-Step Inequalities

Multi–step inequalities can be solved by undoing the operations in the same way you would solve a multi-step equation.

Real-World Example 1 Solve a Multi-Step Inequality

PT

SALES Write and solve an inequality to find the sales Mrs. Jones needs if she earns a monthly salary of \$2000 plus a 10% commission on her sales. Her goal is to make at least \$4000 per month. What sales does she need to meet her goal?

base salary + (commission × sales) ≥ income needed

$2000 + 0.10x \geq 4000$	Substitution
$0.10x \geq 2000$	Subtract 2000 from each side.
$x \geq 20{,}000$	Divide each side by 0.10.

She must make at least \$20,000 in sales to meet her monthly goal.

Guided Practice

1. FINANCIAL LITERACY The Print Shop advertises a special to print 400 flyers for less than the competition. The price includes a \$3.50 set-up fee. If the competition charges \$35.50, what does the Print Shop charge for each flyer?

When multiplying or dividing by a negative number, the direction of the inequality symbol changes. This holds true for multi-step inequalities.

Example 2 Inequality Involving a Negative Coefficient

PT

Solve $-11y - 13 > 42$. Graph the solution on a number line.

$-11y - 13 > 42$	Original inequality
$-11y > 55$	Add 13 to each side and simplify.
$\frac{-11y}{-11} < \frac{55}{-11}$	Divide each side by −11, and reverse the inequality.
$y < -5$	Simplify.

The solution set is $\{y \mid y < -5\}$.

Guided Practice **Solve each inequality.**

2A. $23 \geq 10 - 2w$

2B. $43 > -4y + 11$

Don Hammond/Design Pics Inc./Alamy

You can translate sentences into multi-step inequalities and then solve them using the Properties of Inequalities.

Real-WorldCareer

Veterinarian Veterinarians take care of sick and injured animals. Vets can work anywhere from a zoo to a research facility to owning their own practice. Vets need to earn a bachelor's degree, attend vet college for 4 years, and take a test to get licensed.

Example 3 Write and Solve an Inequality

Define a variable, write an inequality, and solve the problem.

Five minus 6 times a number is more than four times the number plus 45.

Let n be the number.

Five	minus	six times a number	is more than	four times a number	plus	forty-five.
5	$-$	$6n$	$>$	$4n$	$+$	45

$5 - 10n > 45$	Subtract $4n$ from each side and simplify.
$-10n > 40$	Subtract 5 from each side and simplify.
$\frac{-10n}{-10} < \frac{40}{-10}$	Divide each side by −10, and reverse the inequality.
$n < -4$	Simplify.

The solution set is $\{n \mid n < -4\}$.

GuidedPractice

3. *Two more than half of a number is greater than twenty-seven.*

2 Solve Inequalities Involving the Distributive Property

When solving inequalities that contain grouping symbols, use the Distributive Property to remove the grouping symbols first. Then use the order of operations to simplify the resulting inequality.

Example 4 Distributive Property

Solve $4(3t - 5) + 7 \geq 8t + 3$. Graph the solution on a number line.

$4(3t - 5) + 7 \geq 8t + 3$	Original inequality
$12t - 20 + 7 \geq 8t + 3$	Distributive Property
$12t - 13 \geq 8t + 3$	Combine like terms.
$4t - 13 \geq 3$	Subtract $8t$ from each side and simplify.
$4t \geq 16$	Add 13 to each side.
$\frac{4t}{4} \geq \frac{16}{4}$	Divide each side by 4.
$t \geq 4$	Simplify.

The solution set is $\{t \mid t \geq 4\}$.

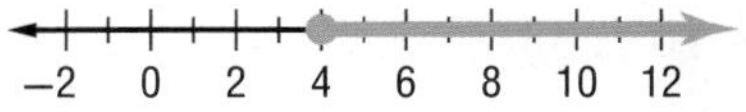

GuidedPractice

Solve each inequality. Graph the solution on a number line.

4A. $6(5z - 3) \leq 36z$

4B. $2(h + 6) > -3(8 - h)$

WatchOut!

Distributive Property If a negative number is multiplied by a sum or difference, remember to distribute the negative sign along with the number to each term inside the parentheses.

If solving an inequality results in a statement that is always true, the solution set is the set of all real numbers. This solution set is written as $\{x \mid x \text{ is a real number.}\}$. If solving an inequality results in a statement that is never true, the solution set is the empty set, which is written as the symbol $\varnothing$. The empty set has no members.

Example 5 Empty Set and All Reals

Solve each inequality. Check your solution.

a. $\mathbf{9t - 5(t - 5) \leq 4(t - 3)}$

$9t - 5(t - 5) \leq 4(t - 3)$	Original inequality
$9t - 5t + 25 \leq 4t - 12$	Distributive Property
$4t + 25 \leq 4t - 12$	Combine like terms.
$4t + 25 - \mathbf{4t} \leq 4t - 12 - \mathbf{4t}$	Subtract $4t$ from each side.
$25 \leq -12$	Simplify.

Since the inequality results in a false statement, the solution set is the empty set, ∅.

b. $\mathbf{3(4m + 6) \leq 42 + 6(2m - 4)}$

$3(4m + 6) \leq 42 + 6(2m - 4)$	Original inequality
$12m + 18 \leq 42 + 12m - 24$	Distributive Property
$12m + 18 \leq 12m + 18$	Combine like terms.
$12m + 18 - \mathbf{12m} \leq 12m + 18 - \mathbf{12m}$	Subtract $12m$ from each side.
$18 \leq 18$	Simplify.

All values of m make the inequality true. All real numbers are solutions.

StudyTip

CCSS Structure Notice that the inequality $4t + 25 < 4t - 12$ means *some number 4t plus 25 is less than or equal to that number minus 12.* No real number makes that inequality true. Observing the meaning of the expressions in each step in this way can lead you to solutions more quickly.

GuidedPractice

Solve each inequality. Check your solution.

5A. $18 - 3(8c + 4) \geq -6(4c - 1)$

5B. $46 \leq 8m - 4(2m + 5)$

Check Your Understanding

Example 1

1. CANOEING If four people plan to use the canoe with 60 pounds of supplies, write and solve an inequality to find the allowable average weight per person.

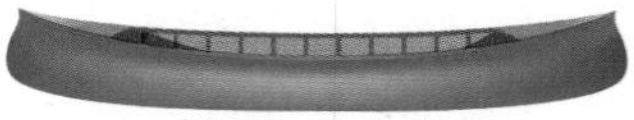

800-lb capacity

2. SHOPPING Rita is ordering a movie for \$11.95 and a few CDs. She has \$50 to spend. Shipping and sales tax will be \$10. If each CD costs \$9.99, write and solve an inequality to find the greatest number of CDs that she can buy.

Example 2

CCSS STRUCTURE Solve each inequality. Graph the solution on a number line.

3. $6h - 10 \geq 32$

4. $-3 \leq \frac{2}{3}r + 9$

5. $-3x + 7 > 43$

6. $4m - 17 < 6m + 25$

Example 3

Define a variable, write an inequality, and solve each problem. Then check your solution.

7. Four times a number minus 6 is greater than eight plus two times the number.

8. Negative three times a number plus 4 is less than five times the number plus 8.

Examples 4–5 **Solve each inequality. Graph the solution on a number line.**

9. $-6 \leq 3(5v - 2)$

10. $-5(g + 4) > 3(g - 4)$

11. $3 - 8x \geq 9 + 2(1 - 4x)$

Practice and Problem Solving

Examples 1 and 2

CCSS **STRUCTURE** **Solve each inequality. Graph the solution on a number line.**

12. $5b - 1 \geq -11$

13. $21 > 15 + 2a$

14. $-9 \geq \frac{2}{5}m + 7$

15. $\frac{w}{8} - 13 > -6$

16. $-a + 6 \leq 5$

17. $37 < 7 - 10w$

18. $8 - \frac{z}{3} \geq 11$

19. $-\frac{5}{4}p + 6 < 12$

20. $3b - 6 \geq 15 + 24b$

21. $15h + 30 < 10h - 45$

Example 3

Define a variable, write an inequality, and solve each problem. Check your solution.

22. Three fourths of a number decreased by nine is at least forty-two.

23. Two thirds of a number added to six is at least twenty-two.

24. Seven tenths of a number plus 14 is less than forty-nine.

25. Eight times a number minus twenty-seven is no more than the negative of that number plus eighteen.

26. Ten is no more than 4 times the sum of twice a number and three.

27. Three times the sum of a number and seven is greater than five times the number less thirteen.

28. The sum of nine times a number and fifteen is less than or equal to the sum of twenty-four and ten times the number.

Examples 4 and 5

CCSS **STRUCTURE** **Solve each inequality. Graph the solution on a number line.**

29. $-3(7n + 3) < 6n$

30. $21 \geq 3(a - 7) + 9$

31. $2y + 4 > 2(3 + y)$

32. $3(2 - b) < 10 - 3(b - 6)$

33. $7 + t \leq 2(t + 3) + 2$

34. $8a + 2(1 - 5a) \leq 20$

Define a variable, write an inequality, and solve each problem. Then interpret your solution.

35. **CARS** A car salesperson is paid a base salary of $35,000 a year plus 8% of sales. What are the sales needed to have an annual income greater than $65,000?

36. **ANIMALS** Keith's dog weighs 90 pounds. A healthy weight for his dog would be less than 75 pounds. If Keith's dog can lose an average of 1.25 pounds per week on a certain diet, after how long will the dog reach healthy weight?

37. Solve $6(m - 3) > 5(2m + 4)$. Show each step and justify your work.

38. Solve $8(a - 2) \leq 10(a + 2)$. Show each step and justify your work.

39. **MUSICAL** A high school drama club is performing a musical to benefit a local charity. Tickets are $5 each. They also received donations of $565. They want to raise at least $1500.

    a. Write an inequality that describes this situation. Then solve the inequality.

    b. Graph the solution.

40. **ICE CREAM** Benito has $6 to spend. A sundae costs $3.25 plus $0.65 per topping. Write and solve an inequality to find how many toppings he can order.

41 **SCIENCE** The normal body temperature of a camel is 97.7°F in the morning. If it has had no water by noon, its body temperature can be greater than 104°F.

a. Write an inequality that represents a camel's body temperature at noon if the camel had no water.

b. If C represents degrees Celsius, then $F = \frac{9}{5}C + 32$. Write and solve an inequality to find the camel's body temperature at noon in degrees Celsius.

42. **NUMBER THEORY** Find all sets of three consecutive positive even integers with a sum no greater than 36.

43. **NUMBER THEORY** Find all sets of four consecutive positive odd integers with a sum that is less than 42.

Solve each inequality. Check your solution.

44. $2(x - 4) \leq 2 + 3(x - 6)$

45. $\frac{2x - 4}{6} \geq -5x + 2$

46. $5.6z + 1.5 < 2.5z - 4.7$

47. $0.7(2m - 5) \geq 21.7$

GRAPHING CALCULATOR **Use a graphing calculator to solve each inequality.**

48. $3x + 7 > 4x + 9$

49. $13x - 11 \leq 7x + 37$

50. $2(x - 3) < 3(2x + 2)$

51. $\frac{1}{2}x - 9 < 2x$

52. $2x - \frac{2}{3} \geq x - 22$

53. $\frac{1}{3}(4x + 3) \geq \frac{2}{3}x + 2$

54. **MULTIPLE REPRESENTATIONS** In this problem, you will solve compound inequalities. A number x is greater than 4, and the same number is less than 9.

a. Numerical Write two separate inequalities for the statement.

b. Graphical Graph the solution set for the first inequality in red. Graph the solution set for the second inequality in blue. Highlight where they overlap.

c. Tabular Make a table using ten points from your number line, including points from each section. Use one column for each inequality and a third column titled "Both are True." Complete the table by writing true or false.

d. Verbal Describe the relationship between the colored regions of the graph and the chart.

e. Logical Make a prediction of what the graph of $4 < x < 9$ looks like.

H.O.T. Problems Use Higher-Order Thinking Skills

55. **CCSS REASONING** Explain how you could solve $-3p + 7 \geq -2$ without multiplying or dividing each side by a negative number.

56. **CHALLENGE** If $ax + b < ax + c$ is true for all real values of x, what will be the solution of $ax + b > ax + c$? Explain how you know.

57. **CHALLENGE** Solve each inequality for x. Assume that $a > 0$.

a. $ax + 4 \geq -ax - 5$

b. $2 - ax < x$

c. $-\frac{2}{a}x + 3 > -9$ $<$

58. **WHICH ONE DOESN'T BELONG?** Name the inequality that does not belong. Explain.

$4y + 9 > -3$	$3y - 4 > 5$	$-2y + 1 < -5$	$-5y + 2 < -13$

59. **WRITING IN MATH** Explain when the solution set of an inequality will be the empty set or the set of all real numbers. Show an example of each.

Standardized Test Practice

60. What is the solution set of the inequality $4t + 2 < 8t - (6t - 10)$?

A $\{t \mid t < -6.5\}$ **C** $\{t \mid t < 4\}$
B $\{t \mid t > -6.5\}$ **D** $\{t \mid t > 4\}$

61. GEOMETRY The section of Liberty Ave. between 5th St. and King Ave. is temporarily closed. Traffic is being detoured right on 5th St., left on King Ave. and then back on Liberty Ave. How long is the closed section of Liberty Ave.?

F 100 ft
G 120 ft
H 144 ft
J 180 ft

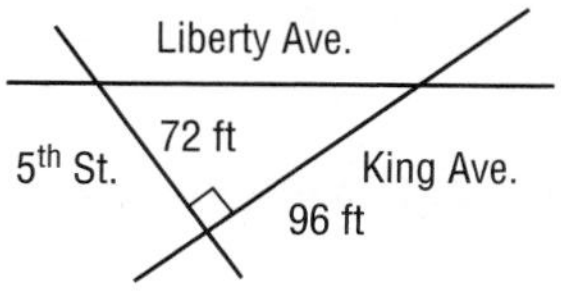

62. SHORT RESPONSE Rhiannon is paid \$52 for working 4 hours. At this rate, how many hours will it take her to earn \$845?

63. GEOMETRY Classify the triangle.

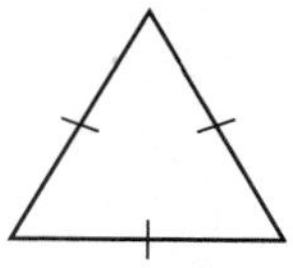

A right
B parallel
C obtuse
D equilateral

Spiral Review

Solve each inequality. Check your solution. (Lesson 5-2)

64. $\frac{y}{2} \leq -5$ **65.** $12b > -48$ **66.** $-\frac{2}{3}t \leq -30$

Solve each inequality. Check your solution, and graph it on a number line. (Lesson 5-1)

67. $6 - h > -8$ **68.** $p - 9 < 2$ **69.** $3 \geq 4 - m$

Solve each equation by graphing. Verify your answer algebraically. (Lesson 3-2)

70. $2x - 7 = 4x + 9$ **71.** $5 + 3x = 7x - 11$ **72.** $2(x - 3) = 5x + 12$

73. THEME PARKS In a recent year, 70.9 million people visited the top 5 theme parks in North America. That represents an increase of about 1.14% in the number of visitors from the prior year. About how many people visited the top 5 theme parks in North America in the prior year? (Lesson 2-7)

If $f(x) = 4x - 3$ and $g(x) = 2x^2 + 5$, find each value. (Lesson 1-7)

74. $f(-2)$ **75.** $g(2) - 5$ **76.** $f(c + 3)$

77. COSMETOLOGY On average, a barber received a tip of \$4 for each of 12 haircuts. Write and evaluate an expression to determine the total amount that she earned. (Lesson 1-4)

Skills Review

Graph each set of numbers on a number line.

78. $\{-4, -2, 2, 4\}$
79. $\{-3, 0, 1, 5\}$
80. {integers less than 3}
81. {integers greater than or equal to -2}
82. {integers between -3 and 4}
83. {integers less than -1}

CHAPTER 5

Mid-Chapter Quiz

Lessons 5-1 through 5-3

Solve each inequality. Then graph it on a number line. (Lesson 5-1)

1. $x - 8 > 4$

2. $m + 2 \geq 6$

3. $p - 4 < -7$

4. $12 \leq t - 9$

5. **CONCERTS** Lupe's allowance for the month is \$60. She wants to go to a concert for which a ticket costs \$45. (Lesson 5-1)

a. Write and solve an inequality that shows how much money she can spend that month after buying a concert ticket.

b. She spends \$9.99 on music downloads and \$2 on lunch in the cafeteria. Write and solve an inequality that shows how much she can spend after these purchases and the concert ticket.

Define a variable, write an inequality, and solve each problem. Check your solution. (Lesson 5-1)

6. The sum of a number and -2 is no more than 6.

7. A number decreased by 4 is more than -1.

8. Twice a number increased by 3 is less than the number decreased by 4.

9. **MULTIPLE CHOICE** Jane is saving money to buy a new cell phone that costs no more than \$90. So far, she has saved \$52. How much more money does Jane need to save? (Lesson 5-1)

A \$38

B more than \$38

C no more than \$38

D at least \$38

Solve each inequality. Check your solution. (Lesson 5-2)

10. $\frac{1}{3}y \geq 5$

11. $4 < \frac{c}{5}$

12. $-8x > 24$

13. $2m \leq -10$

14. $\frac{x}{2} < \frac{5}{8}$

15. $-9a \geq -45$

16. $\frac{w}{6} > -3$

17. $\frac{k}{7} < -2$

18. **ANIMALS** The world's heaviest flying bird is the great bustard. A male bustard can be up to 4 feet long and weigh up to 40 pounds. (Lesson 5-2)

a. Write inequalities to describe the ranges of lengths and weights of male bustards.

b. Male bustards are usually about four times as heavy as females. Write and solve an inequality that describes the range of weights of female bustards.

19. **GARDENING** Bill is building a fence around a square garden to keep deer out. He has 60 feet of fencing. Find the maximum length of a side of the garden. (Lesson 5-2)

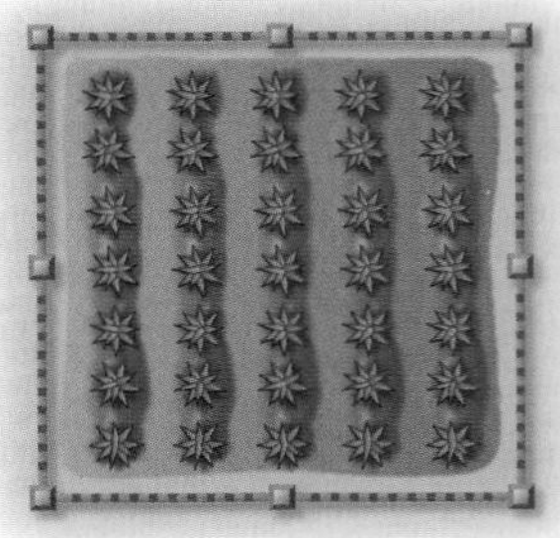

Solve each inequality. Check your solution. (Lesson 5-3)

20. $4a - 2 > 14$

21. $2x + 11 \leq 5x - 10$

22. $-p + 4 < -9$

23. $\frac{d}{4} + 1 \geq -3$

24. $-2(4b + 1) < -3b + 8$

Define a variable, write an inequality, and solve each problem. Check your solution. (Lesson 5-3)

25. Three times a number increased by 8 is no more than the number decreased by 4.

26. Two thirds of a number plus 5 is greater than 17.

27. **MULTIPLE CHOICE** Shoe rental costs \$2, and each game bowled costs \$3. How many games can Kyle bowl without spending more than \$15? (Lesson 5-3)

F 2

G 3

H 4

J 5

EXPLORE 5-4

Algebra Lab: Reading Compound Statements

A compound statement is made up of two simple statements connected by the word *and* or *or*. Before you can determine whether a compound statement is true or false, you must understand what the words *and* and *or* mean.

A spider has eight legs, *and* a dog has five legs.

For a compound statement connected by the word *and* to be true, both simple statements must be true.

A spider has eight legs. ⟶ true

A dog has five legs. ⟶ false

Since one of the statements is false, the compound statement is false.

A compound statement connected by the word *or* may be *exclusive* or *inclusive*. For example, the statement "With your lunch, you may have milk *or* juice," is exclusive. In everyday language, *or* means one or the other, but not both. However, in mathematics, *or* is inclusive. It means one or the other or both.

A spider has eight legs, *or* a dog has five legs.

For a compound statement connected by the word *or* to be true, at least one of the simple statements must be true. Since it is true that a spider has eight legs, the compound statement is true.

Exercises

Is each compound statement *true* or *false*? Explain.

1. Most top 20 movies in 2007 were rated PG-13, *or* most top 20 movies in 2005 were rated G.
2. In 2008 more top 20 movies were rated PG than were rated G, *and* more were rated PG than rated PG-13.
3. For the years shown most top 20 movies are rated PG-13, *and* the least top 20 movies are rated G.
4. No top 20 movies in 2008 were rated G, *or* most top 20 movies in 2008 were *not* rated PG.

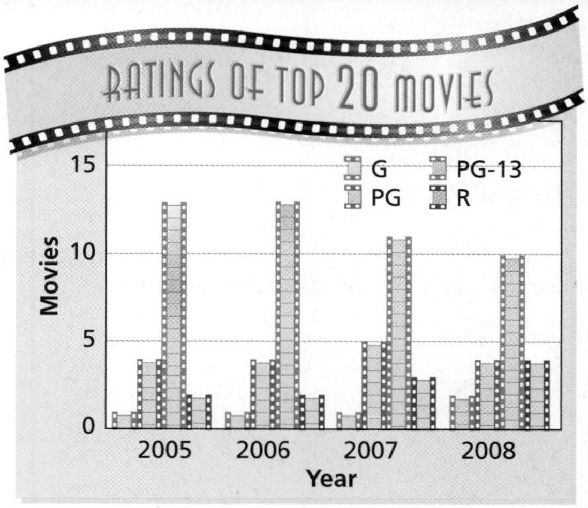

Source: National Association of Theater Owners

5. $11 < 5$ or $9 < 7$
6. $-2 > 0$ and $3 < 7$
7. $5 > 0$ and $-3 < 0$
8. $-2 > -3$ or $0 = 0$
9. $8 \neq 8$ or $-2 > -5$
10. $5 > 10$ and $4 > -2$

LESSON 5-4 Solving Compound Inequalities

Then

- You solved absolute value equations with two cases.

Now

1. Solve compound inequalities containing the word *and*, and graph their solution set.
2. Solve compound inequalities containing the word *or*, and graph their solution set.

Why?

- To ride certain roller coasters, you must be at least 52 inches tall, and your height cannot exceed 72 inches. If h represents the height of a rider, we can write two inequalities to represent this.

at least 52 inches	cannot exceed 72 inches
$h \geq 52$	$h \leq 72$

The inequalities $h \geq 52$ and $h \leq 72$ can be combined and written without using *and* as $52 \leq h \leq 72$.

NewVocabulary
compound inequality
intersection
union

Common Core State Standards

Content Standards

A.CED.1 Create equations and inequalities in one variable and use them to solve problems.

A.REI.3 Solve linear equations and inequalities in one variable, including equations with coefficients represented by letters.

Mathematical Practices

1 Make sense of problems and persevere in solving them.

8 Look for and express regularity in repeated reasoning.

1 Inequalities Containing *and*

When considered together, two inequalities such as $h \geq 52$ and $h \leq 72$ form a **compound inequality**. A compound inequality containing *and* is only true if both inequalities are true. Its graph is where the graphs of the two inequalities overlap. This is called the **intersection** of the two graphs.

The intersection can be found by graphing each inequality and then determining where the graphs intersect.

$x \geq 3$

$x < 7$

$x \geq 3$ and $x < 7$
$3 \leq x < 7$

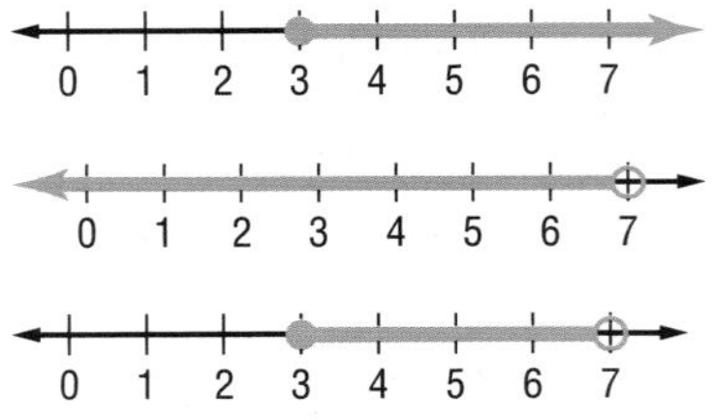

The statement $3 \leq x < 7$ can be read as *x is greater than or equal to 3 and less than 7* or *x is between 3 and 7 including 3.*

Example 1 Solve and Graph an Intersection

Solve $-2 \leq x - 3 < 4$. Then graph the solution set.

First, express $-2 \leq x - 3 < 4$ using *and*. Then solve each inequality.

$-2 \leq x - 3$	**and**	$x - 3 < 4$	Write the inequalities.
$-2 + 3 \leq x - 3 + 3$		$x - 3 + 3 < 4 + 3$	Add 3 to each side.
$1 \leq x$		$x < 7$	Simplify.

The solution set is $\{x \mid 1 \leq x < 7\}$. Now graph the solution set.

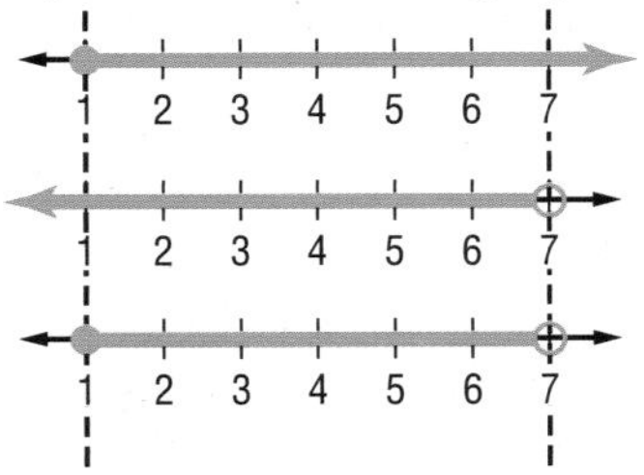

Graph $1 \leq x$ or $x \geq 1$.

Graph $x < 7$.

Find the intersection of the graphs.

Mark Segal/Index Stock/age fotostock

GuidedPractice

Solve each compound inequality. Then graph the solution set.

1A. $y - 3 \geq -11$ and $y - 3 \leq -8$ **1B.** $6 \leq r + 7 < 10$

2 Inequalities Containing *or*

Another type of compound inequality contains the word *or*. A compound inequality containing *or* is true if at least one of the inequalities is true. Its graph is the **union** of the graphs of two inequalities.

$x > 2$

$x \leq -1$

$x > 2$ or $x \leq -1$

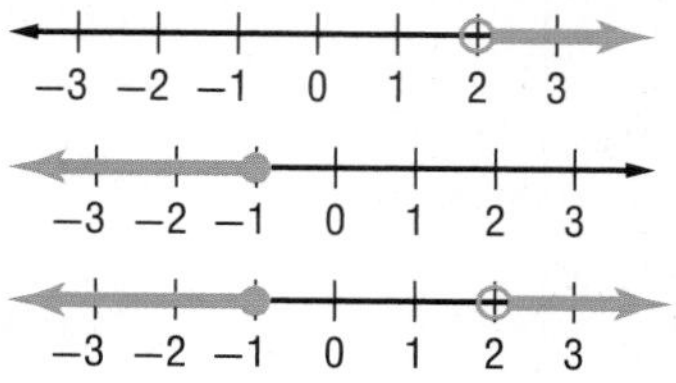

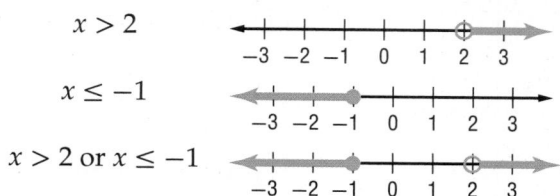

When solving problems involving inequalities, *within* is meant to be inclusive, so use ≥ or ≤. *Between* is meant to be exclusive, so use < or >.

Real-World Example 2 Write and Graph a Compound Inequality

SOUND The human ear can only detect sounds between the frequencies 20 Hertz and 20,000 Hertz. Write and graph a compound inequality that describes the frequency of sounds humans cannot hear.

The problem states that humans can hear the frequencies between 20 Hz and 20,000 Hz. We are asked to find the frequencies humans cannot hear.

Words	The frequency	is at most	20 Hertz	or	The frequency	is at least	20,000 Hertz.
Variable	Let f be the frequency.						
Inequality	f	$\leq$	20	or	f	$\geq$	20,000

ReadingMath

At Most The phrase *at most* in Example 2 indicates ≤. It could also have been phrased as *no more than* or *less than or equal to.*

Now, graph the solution set.

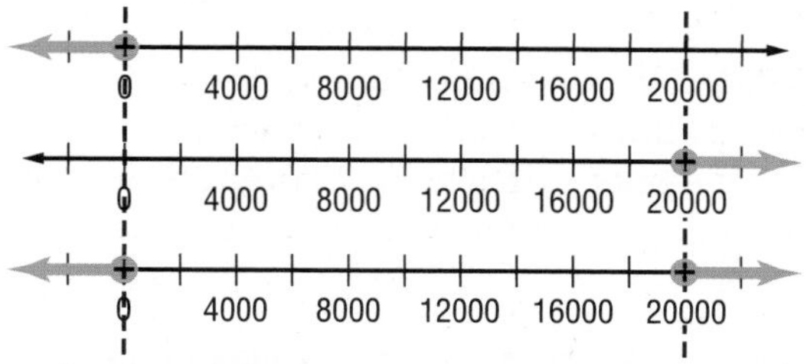

Graph $f < 20$.

Graph $f > 20{,}000$.

Find the union.

Notice that the graphs do not intersect. Humans cannot hear sounds at a frequency less than 20 Hertz or greater than 20,000 Hertz. The compound inequality is $\{f \mid f < 20 \text{ or } f > 20{,}000\}$.

GuidedPractice

2. MANUFACTURING A company is manufacturing an action figure that must be at least 11.2 centimeters and at most 11.4 centimeters tall. Write and graph a compound inequality that describes how tall the action figure can be.

Example 3 Solve and Graph a Union

Solve $-2m + 7 \leq 13$ or $5m + 12 > 37$. Then graph the solution set.

$-2m + 7 \leq 13$	or	$5m + 12 > 37$
$-2m + 7 - 7 \leq 13 - 7$	Subtract.	$5m + 12 - 12 > 37 - 12$
$-2m \leq 6$	Simplify.	$5m > 25$
$\frac{-2m}{-2} \geq \frac{6}{-2}$	Divide.	$\frac{5m}{5} > \frac{25}{5}$
$m \geq -3$	Simplify.	$m > 5$

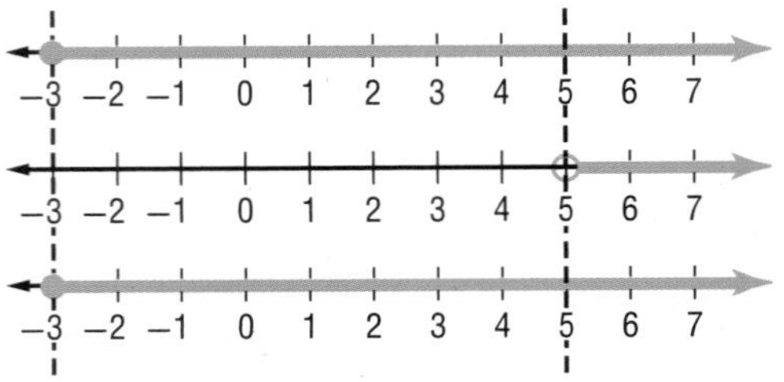

Graph $m \geq -3$.

Graph $m > 5$.

Find the union.

Notice that the graph of $m \geq -3$ contains every point in the graph of $m > 5$. So, the union is the graph of $m \geq -3$. The solution set is $\{m \mid m \geq -3\}$.

StudyTip

Intersections and Unions The graphs of compound inequalities containing *and* will be an intersection. The graphs of compound inequalities containing *or* will be a union.

Guided Practice

Solve each compound inequality. Then graph the solution set.

3A. $a + 1 < 4$ or $a - 1 \geq 3$

3B. $x \leq 9$ or $2 + 4x < 10$

Check Your Understanding

Examples 1, 3 **Solve each compound inequality. Then graph the solution set.**

1. $4 \leq p - 8$ and $p - 14 \leq 2$
2. $r + 6 < -8$ or $r - 3 > -10$
3. $4a + 7 \geq 31$ or $a > 5$
4. $2 \leq g + 4 < 7$

Example 2

5. CCSS **SENSE-MAKING** The recommended air pressure for the tires of a mountain bike is at least 35 pounds per square inch (psi), but no more than 80 pounds per square inch. If a bike's tires have 24 pounds per square inch, what is the recommended range of air that should be put into the tires?

Practice and Problem Solving

Examples 1, 3 **Solve each compound inequality. Then graph the solution set.**

6. $f - 6 < 5$ and $f - 4 \geq 2$
7. $n + 2 \leq -5$ and $n + 6 \geq -6$
8. $y - 1 \geq 7$ or $y + 3 < -1$
9. $t + 14 \geq 15$ or $t - 9 < -10$
10. $-5 < 3p + 7 \leq 22$
11. $-3 \leq 7c + 4 < 18$
12. $5h - 4 \geq 6$ and $7h + 11 < 32$
13. $22 \geq 4m - 2$ or $5 - 3m \leq -13$
14. $-4a + 13 \geq 29$ and $10 < 6a - 14$
15. $-y + 5 \geq 9$ or $3y + 4 < -5$

Practice and Problem Solving

Example 2

16. SPEED The posted speed limit on an interstate highway is shown. Write an inequality that represents the sign. Graph the inequality.

17. NUMBER THEORY Find all sets of two consecutive positive odd integers with a sum that is at least 8 and less than 24.

Write a compound inequality for each graph.

18. 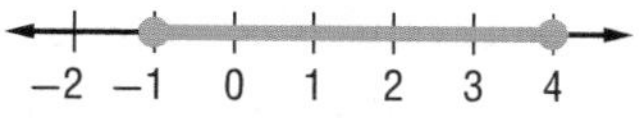

19. 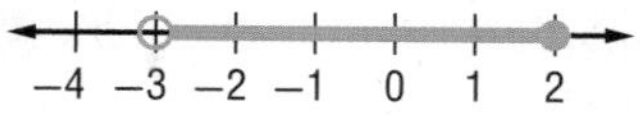

20. 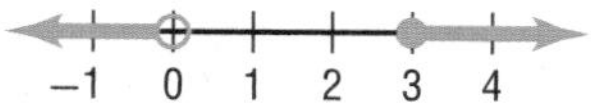

21. 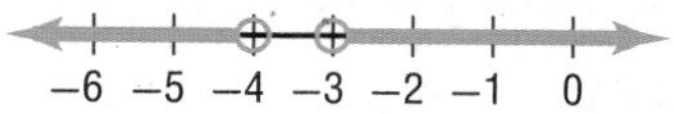

22. 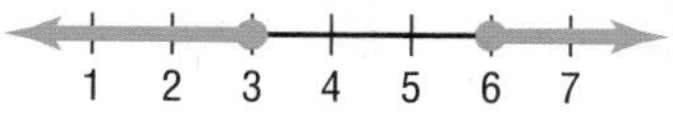

23. 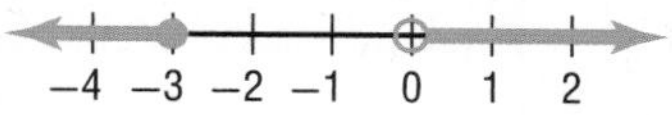

Solve each compound inequality. Then graph the solution set.

24. $3b + 2 < 5b - 6 \leq 2b + 9$

25. $-2a + 3 \geq 6a - 1 > 3a - 10$

26. $10m - 7 < 17m$ or $-6m > 36$

27. $5n - 1 < -16$ or $-3n - 1 < 8$

28. COUPON Juanita has a coupon for 10% off any digital camera at a local electronics store. She is looking at digital cameras that range in price from \$100 to \$250.

a. How much are the cameras after the coupon is used?

b. If the tax amount is 6.5%, how much should Juanita expect to spend?

Define a variable, write an inequality, and solve each problem. Then check your solution.

29. Eight less than a number is no more than 14 and no less than 5.

30. The sum of 3 times a number and 4 is between −8 and 10.

31. The product of −5 and a number is greater than 35 or less than 10.

32. One half a number is greater than 0 and less than or equal to 1.

33. SNAKES Most snakes live where the temperature ranges from 75°F to 90°F, inclusive. Write an inequality for temperatures where snakes will *not* thrive.

34. FUNDRAISING Yumas is selling gift cards to raise money for a class trip. He can earn prizes depending on how many cards he sells. So far, he has sold 34 cards. How many more does he need to sell to earn a prize in category 4?

Cards	Prize
1–15	1
16–30	2
31–45	3
46–60	4
+61	5

35. TURTLES Atlantic sea turtle eggs that incubate below 23°C or above 33°C rarely hatch. Write the temperature requirements in two ways: as a pair of simple inequalities, and as a compound inequality.

36. CCSS STRUCTURE The *Triangle Inequality Theorem* states that the sum of the measures of any two sides of a triangle is greater than the measure of the third side.

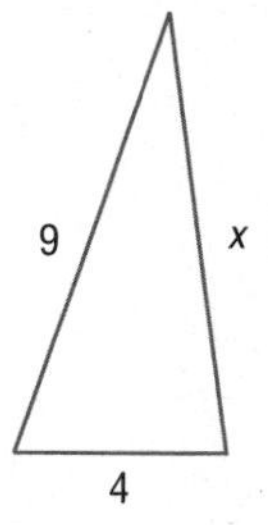

a. Write and solve three inequalities to express the relationships among the measures of the sides of the triangle shown at the right.

b. What are four possible lengths for the third side of the triangle?

c. Write a compound inequality for the possible values of x.

37. **HURRICANES** The Saffir-Simpson Hurricane Scale rates hurricanes on a scale from 1 to 5 based on their wind speed.

Category	Wind Speed (mph)	Example (year)
1	74–95	Gaston (2004)
2	96–110	Frances (2004)
3	111–130	Ivan (2004)
4	131–155	Charley (2004)
5	> 155	Andrew (1992)

a. Write a compound inequality for the wind speeds of a category 3 and a category 4 hurricane.

b. What is the intersection of the two graphs of the inequalities you found in part **a**?

38. **MULTIPLE REPRESENTATIONS** In this problem, you will investigate measurements. The **absolute error** of a measurement is equal to one half the unit of measure. The **relative error** of a measure is the ratio of the absolute error to the expected measure.

a. Tabular Copy and complete the table.

Measure	Absolute Error	Relative Error
14.3 cm	$\frac{1}{2}(0.1) = 0.05$ cm	$\frac{\text{absolute error}}{\text{expected measure}} = \frac{0.05 \text{ cm}}{14.3 \text{ cm}}$ ≈ 0.0035 or 0.4%
1.85 cm		
61.2 cm		
237 cm		

b. Analytical You measured a length of 12.8 centimeters. Compute the absolute error and then write the range of possible measures.

c. Logical To what precision would you have to measure a length in centimeters to have an absolute error of less than 0.05 centimeter?

d. Analytical To find the relative error of an area or volume calculation, add the relative errors of each linear measure. If the measures of the sides of a rectangular box are 6.5 centimeters, 7.2 centimeters, and 10.25 centimeters, what is the relative error of the volume of the box?

H.O.T. Problems Use Higher-Order Thinking Skills

39. **ERROR ANALYSIS** Chloe and Jonas are solving $3 < 2x - 5 < 7$. Is either of them correct? Explain your reasoning.

Chloe

$3 < 2x - 5 < 7$

$3 < 2x < 12$

$\frac{3}{2} < x < 6$

Jonas

$3 < 2x - 5 < 7$

$8 < 2x < 7$

$4 < x < \frac{7}{2}$

40. **CCSS PERSEVERANCE** Solve each inequality for x. Assume a is constant and $a > 0$.

a. $-3 < ax + 1 \leq 5$

b. $-\frac{1}{a}x + 6 < 1$ or $2 - ax > 8$

41. **OPEN ENDED** Create an example of a compound inequality containing *or* that has infinitely many solutions.

42. **CHALLENGE** Determine whether the following statement is *always, sometimes,* or *never* true. Explain. *The graph of a compound inequality that involves an* or *statement is bounded on the left and right by two values of x.*

43. **WRITING IN MATH** Give an example of a compound inequality you might encounter at an amusement park. Does the example represent an intersection or a union?

Standardized Test Practice

44. What is the solution set of the inequality $-7 < x + 2 < 4$?

A $\{x \mid -5 < x < 6\}$ C $\{x \mid -9 < x < 2\}$
B $\{x \mid -5 < x < 2\}$ D $\{x \mid -9 < x < 6\}$

45. GEOMETRY What is the surface area of the rectangular solid?

F 249.6 cm^2
G 278.4 cm^2
H 313.6 cm^2
J 371.2 cm^2

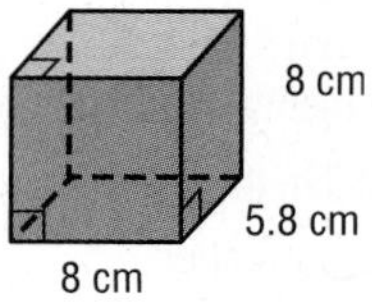

46. GRIDDED RESPONSE What is the next term in the sequence?

$$\frac{13}{2}, \frac{18}{5}, \frac{23}{8}, \frac{28}{11}, \frac{33}{14}, \ldots$$

47. After paying a \$15 membership fee, members of a video club can rent movies for \$2. Nonmembers can rent movies for \$4. What is the least number of movies which must be rented for it to be less expensive for members?

A 9 C 7
B 8 D 6

Spiral Review

48. BABYSITTING Marilyn earns \$150 per month delivering newspapers plus \$7 an hour babysitting. If she wants to earn at least \$300 this month, how many hours will she have to babysit? (Lesson 5-3)

49. MAGAZINES Carlos has earned more than \$260 selling magazine subscriptions. Each subscription was sold for \$12. How many did Carlos sell? (Lesson 5-2)

50. PUNCH Raquel is mixing lemon-lime soda and a fruit juice blend that is 45% juice. If she uses 3 quarts of soda, how many quarts of fruit juice must be added to produce punch that is 30% juice? (Lesson 2-9)

Solve each proportion. If necessary, round to the nearest hundredth. (Lesson 2-6)

51. $\frac{14}{x} = \frac{20}{8}$ **52.** $\frac{0.47}{6} = \frac{1.41}{m}$ **53.** $\frac{16}{7} = \frac{9}{b}$

54. $\frac{2+y}{5} = \frac{10}{3}$ **55.** $\frac{8}{9} = \frac{2r-3}{4}$ **56.** $\frac{6-2y}{8} = \frac{2}{18}$

Determine whether each relation is a function. Explain. (Lesson 1-7)

57.

Domain	2	6	10	7
Range	5	0	5	0

58.

Domain	−5	2	−3	2
Range	−10	−7	−5	−3

59. $\{(-4, 11), (-2, 7), (1, 3), (-4, -1)\}$ **60.** $\{(2, 7), (5, -3), (7, 6), (10, 7)\}$

Evaluate each expression. (Lesson 1-2)

61. $5 + (4 - 2^2)$ **62.** $\frac{3}{8}[8 \div (7 - 4)]$ **63.** $2(4 \cdot 9 - 3) + 5 \cdot \frac{1}{5}$

Skills Review

Solve each equation.

64. $4p - 2 = -6$ **65.** $18 = 5p + 3$ **66.** $9 = 1 + \frac{m}{7}$

67. $1.5a - 8 = 11$ **68.** $20 = -4c - 8$ **69.** $\frac{b+4}{-2} = -17$

70. $\frac{n-3}{8} = 20$ **71.** $6y - 16 = 44$ **72.** $130 = 11k + 9$

LESSON 5-5 Inequalities Involving Absolute Value

:: Then	:: Now	:: Why?
• You solved equations involving absolute value.	• **1** Solve and graph absolute value inequalities (<). **2** Solve and graph absolute value inequalities (>).	• Some companies use absolute value inequalities to control the quality of their product. To make baby carrots, long carrots are sliced into 3-inch sections and peeled. If the machine is accurate to within $\frac{1}{8}$ of an inch, the length ranges from $2\frac{7}{8}$ inches to $3\frac{1}{8}$ inches.

CCSS Common Core State Standards

Content Standards

A.CED.1 Create equations and inequalities in one variable and use them to solve problems.

A.REI.3 Solve linear equations and inequalities in one variable, including equations with coefficients represented by letters.

Mathematical Practices

3 Construct viable arguments and critique the reasoning of others.

7 Look for and make use of structure.

1 Absolute Value Inequalities (<)

The inequality $|x| < 3$ means that the distance between x and 0 is less than 3.

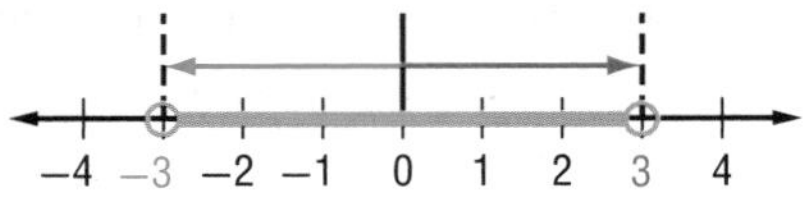

So, $x > -3$ and $x < 3$. The solution set is $\{x \mid -3 < x < 3\}$.

When solving absolute value inequalities, there are two cases to consider.

Case 1 The expression inside the absolute value symbols is nonnegative.

Case 2 The expression inside the absolute value symbols is negative.

The solution is the intersection of the solutions of these two cases.

Example 1 Solve Absolute Value Inequalities (<)

Solve each inequality. Then graph the solution set.

a. $|m + 2| < 11$

Rewrite $|m + 2| < 11$ for Case 1 *and* Case 2.

Case 1 $m + 2$ is nonnegative. **and** **Case 2** $m + 2$ is negative.

Case 1:

$$m + 2 < 11$$
$$m + 2 - 2 < 11 - 2$$
$$m < 9$$

Case 2:

$$-(m + 2) < 11$$
$$m + 2 > -11$$
$$m + 2 - 2 > -11 - 2$$
$$m > -13$$

So, $m < 9$ and $m > -13$. The solution set is $\{m \mid -13 < m < 9\}$.

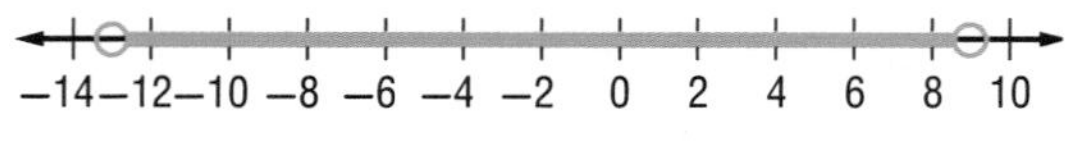

b. $|y - 1| < -2$

$|y - 1|$ cannot be negative. So it is not possible for $|y - 1|$ to be less than −2. Therefore, there is no solution, and the solution set is the empty set, Ø.

Guided Practice

1A. $|n - 8| \leq 2$

1B. $|2c - 5| < -3$

Real-World Example 2 Apply Absolute Value Inequalities

INTERNET A recent survey showed that 65% of young adults watched online video clips. The margin of error was within 3 percentage points. Find the range of young adults who use video sharing sites.

The difference between the actual number of viewers and the number from the survey is less than or equal to 3. Let x be the actual number of viewers. Then $|x - 65| \leq 3$.

Solve each case of the inequality.

Case 1 $x - 65$ is nonnegative. **and** **Case 2** $x - 65$ is negative.

Case 1	Case 2
$x - 65 \leq 3$	$-(x - 65) \leq 3$
$x - 65 + 65 \leq 3 + 65$	$x - 65 \geq -3$
$x \leq 68$	$x \geq 62$

The range of young adults who use video sharing sites is $\{x \mid 62 \leq x \leq 68\}$.

Real-WorldLink

One in five Americans use the Internet to view videos. Young adults tend to watch funny videos, while other age groups tend to watch the news.

Source: Pew Internet and American Life Project

GuidedPractice

2. CHEMISTRY The melting point of ice is 0°C. During a chemistry experiment, Jill observed ice melting within 2°C of this measurement. Write the range of temperatures that Jill observed.

2 Absolute Value Inequalities (>)

The inequality $|x| > 3$ means that the distance between x and 0 is greater than 3.

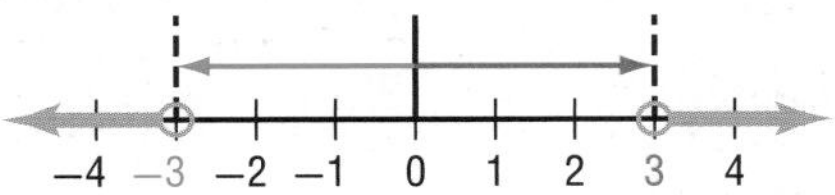

So, $x < -3$ or $x > 3$. The solution set is $\{x \mid x < -3 \text{ or } x > 3\}$.

As in the previous example, we must consider both cases.

Case 1 The expression inside the absolute value symbols is nonnegative.

Case 2 The expression inside the absolute value symbols is negative.

Example 3 Solve Absolute Value Inequalities (>)

Solve $|3n + 6| \geq 12$. Then graph the solution set.

Rewrite $|3n + 6| \geq 12$ for Case 1 *or* Case 2.

Case 1 $3n + 6$ is nonnegative. **or** **Case 2** $3n + 6$ is negative.

Case 1	Case 2
$3n + 6 \geq 12$	$-(3n + 6) \geq 12$
$3n + 6 - 6 \geq 12 - 6$	$3n + 6 \leq -12$
$3n \geq 6$	$3n \leq -18$
$n \geq 2$	$n \leq -6$

So, $n \geq 2$ or $n \leq -6$. The solution set is $\{n \mid n \geq 2 \text{ or } n \leq -6\}$.

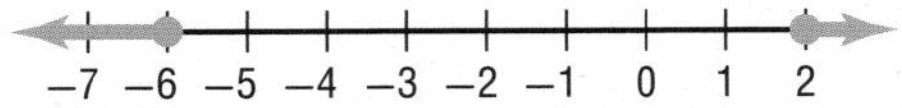

StudyTip

CCSS Structure For $|a| \geq b$, where a is any linear expression in one variable and b is a negative number, the solution set will always be the set of all real numbers. Since $|a|$ is always greater than or equal to zero, $|a|$ is always greater than or equal to b.

GuidedPractice

Solve each inequality. Then graph the solution set.

3A. $|2k + 1| > 7$

3B. $|r - 6| \geq -5$

Check Your Understanding

⬤ = Step-by-Step Solutions begin on page R13.

Examples 1–3 **Solve each inequality. Then graph the solution set.**

1. $|a - 5| < 3$
2. $|u + 3| < 7$
3. $|t + 4| \leq -2$
4. $|c + 2| > -2$
5. $|n + 5| \geq 3$
6. $|p - 2| \geq 8$

Example 2

7. **FINANCIAL LITERACY** Jerome bought stock in his favorite fast-food restaurant chain at \$70.85. However, it has fluctuated up to \$0.75 in a day. Find the range of prices for which the stock could trade in a day.

Practice and Problem Solving

Extra Practice is on page R5.

Examples 1–3 **Solve each inequality. Then graph the solution set.**

8. $|x + 8| < 16$
9. $|r + 1| \leq 2$
10. $|2c - 1| \leq 7$
11. $|3h - 3| < 12$
12. $|m + 4| < -2$
13. $|w + 5| < -8$
14. $|r + 2| > 6$
15. $|k - 4| > 3$
16. $|2h - 3| \geq 9$
17. $|4p + 2| \geq 10$
18. $|5v + 3| > -9$
19. $|-2c - 3| > -4$

Example 2

20. **SCUBA DIVING** The pressure of a scuba tank should be within 500 pounds per square inch (psi) of 2500 psi. Write the range of optimum pressures.

Solve each inequality. Then graph the solution set.

21. $|4n + 3| \geq 18$
22. $|5t - 2| \leq 6$
23. $\left|\frac{3h + 1}{2}\right| < 8$
24. $\left|\frac{2p - 8}{4}\right| \geq 9$
25. $\left|\frac{7c + 3}{2}\right| \leq -5$
26. $\left|\frac{2g + 3}{2}\right| > -7$
27. $|-6r - 4| < 8$
28. $|-3p - 7| > 5$
29. $|-h + 1.5| < 3$

30. **MUSIC DOWNLOADS** Kareem is allowed to download \$10 worth of music each month. This month he has spent within \$3 of his allowance.
    a. What is the range of money he has spent on music downloads this month?
    b. Graph the range of the money that he spent.

31. **CHEMISTRY** Water can be present in our atmosphere as a solid, liquid, or gas. Water freezes at 32°F and vaporizes at 212°F.
    a. Write the range of temperatures in which water is not a liquid.
    b. Graph this range.
    c. Write the absolute value inequality that describes this situation.

CCSS REGULARITY **Write an open sentence involving absolute value for each graph.**

32.
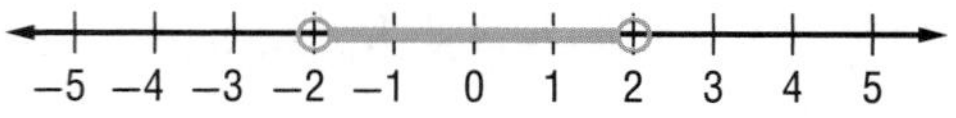

33.
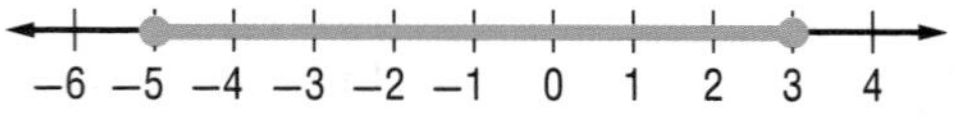

34.
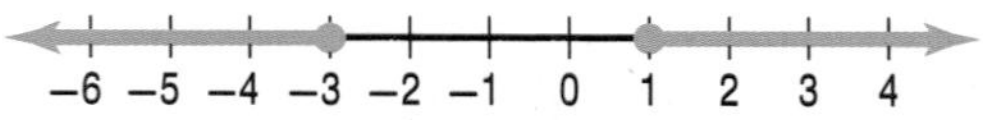

35.
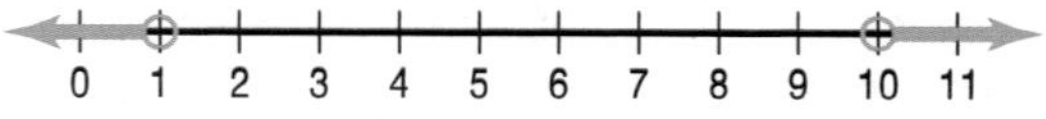

36. **ANIMALS** A sheep's normal body temperature is 39°C. However, a healthy sheep may have body temperatures 1°C above or below this temperature. What is the range of body temperatures for a sheep?

37. **MINIATURE GOLF** Ginger's score was within 5 strokes of her average score of 52. Determine the range of scores for Ginger's game.

Express each statement using an inequality involving absolute value. Do *not* solve.

38. The pH of a swimming pool must be within 0.3 of a pH of 7.5.

39. The temperature inside a refrigerator should be within 1.5 degrees of 38°F.

40. Ramona's bowling score was within 6 points of her average score of 98.

41. The cruise control of a car should keep the speed within 3 miles per hour of 55.

42. **MULTIPLE REPRESENTATIONS** In this problem, you will investigate the graphs of linear inequalities on a coordinate plane.

 a. **Tabular** Copy and complete the table. Substitute the x and $f(x)$ values for each point into each inequality. Mark whether the resulting statement is *true* or *false*.

Point	$f(x) \geq x - 1$	true/false	$f(x) \leq x - 1$	true/false
(−4, 2)				
(−2, 2)				
(0, 2)				
(2, 2)				
(4, 2)				

 b. **Graphical** Graph $f(x) = x - 1$.

 c. **Graphical** Plot each point from the table that made $f(x) \geq x - 1$ a true statement on the graph in red. Plot each point that made $f(x) \leq x - 1$ a true statement in blue.

 d. **Logical** Make a conjecture about what the graphs of $f(x) \geq x - 1$ and $f(x) \leq x - 1$ look like. Complete the table with other points to verify your conjecture.

 e. **Logical** Use what you discovered to describe the graph of a linear inequality.

H.O.T. Problems Use Higher-Order Thinking Skills

43. **ERROR ANALYSIS** Lucita sketched a graph of her solution to $|2a - 3| > 1$. Is she correct? Explain your reasoning.

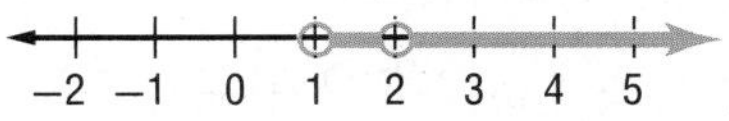

44. **REASONING** The graph of an absolute value inequality is *sometimes, always,* or *never* the union of two graphs. Explain.

45. **CCSS ARGUMENTS** Demonstrate why the solution of $|t| > 0$ is not all real numbers. Explain your reasoning.

46. **WRITING IN MATH** How are symbols used to represent mathematical ideas? Use an example to justify your reasoning.

47. **WRITING IN MATH** Explain how to determine whether an absolute value inequality uses a compound inequality with *and* or a compound inequality with *or*. Then summarize how to solve absolute value inequalities.

Standardized Test Practice

48. The formula for acceleration in a circle is $a = \frac{v^2}{r}$. Which of the following shows the equation solved for r?

A $r = v$

B $r = \frac{v^2}{a}$

C $r = av^2$

D $r = \frac{\sqrt{a}}{v}$

49. An engraver charges a \$3 set-up fee and \$0.25 per word. Which table shows the total price p for w words?

F

w	p
15	\$3
20	\$4.25
25	\$5.50
30	\$7.75

H

w	p
15	\$3.75
20	\$5
25	\$6.25
30	\$8.50

G

w	p
15	\$6.75
20	\$7
25	\$7.25
30	\$7.50

J

w	p
15	\$6.75
20	\$8
25	\$9.25
30	\$10.50

50. SHORT RESPONSE The table shows the items in stock at the school store the first day of class. What is the probability that an item chosen at random was a notebook?

Item	Number Purchased
pencil	57
pen	38
eraser	6
folder	25
notebook	18

51. Solve for n.

$$|2n - 3| = 5$$

A $\{-4, -1\}$

B $\{-1, 4\}$

C $\{1, 1\}$

D $\{4, 4\}$

Spiral Review

Solve each compound inequality. Then graph the solution set. (Lesson 5-4)

52. $b + 3 < 11$ and $b + 2 > -3$

53. $6 \leq 2t - 4 \leq 8$

54. $2c - 3 \geq 5$ or $3c + 7 \leq -5$

55. FINANCIAL LITERACY Jackson's bank charges him a monthly service fee of \$6 for his checking account and \$2 for each out-of-network ATM withdrawal. Jackson's account balance is \$87. Write and solve an inequality to find how many out-of-network ATM withdrawals of \$20 Jackson can make without overdrawing his account. (Lesson 5-3)

56. GEOMETRY One angle of a triangle measures 10° more than the second. The measure of the third angle is twice the sum of the measure of the first two angles. Find the measure of each angle. (Lesson 2-4)

Solve each equation. Then check your solution. (Lesson 2-2)

57. $c - 7 = 11$

58. $2w = 24$

59. $9 + p = -11$

60. $\frac{t}{5} = 20$

Skills Review

Graph each equation.

61. $y = 4x - 1$

62. $y - x = 3$

63. $2x - y = -4$

64. $3y + 2x = 6$

65. $4y = 4x - 16$

66. $2y - 2x = 8$

67. $-9 = -3x - y$

68. $-10 = 5y - 2x$

LESSON 5-6 Graphing Inequalities in Two Variables

Then	Now	Why?
You graphed linear equations.	1 Graph linear inequalities on the coordinate plane. 2 Solve inequalities by graphing.	Hannah has budgeted $35 every three months for car maintenance. From this she must buy oil costing $3 and filters that cost $7 each. How much oil and how many filters can Hannah buy and stay within her budget?

NewVocabulary
boundary
half-plane
closed half-plane
open half-plane

Common Core State Standards

Content Standards

A.CED.3 Represent constraints by equations or inequalities, and by systems of equations and/or inequalities, and interpret solutions as viable or nonviable options in a modeling context.

A.REI.12 Graph the solutions to a linear inequality in two variables as a halfplane (excluding the boundary in the case of a strict inequality), and graph the solution set to a system of linear inequalities in two variables as the intersection of the corresponding half-planes.

Mathematical Practices

5 Use appropriate tools strategically.

1 Graph Linear Inequalities

The graph of a linear inequality is the set of points that represent all of the possible solutions of that inequality. An equation defines a **boundary**, which divides the coordinate plane into two **half-planes**.

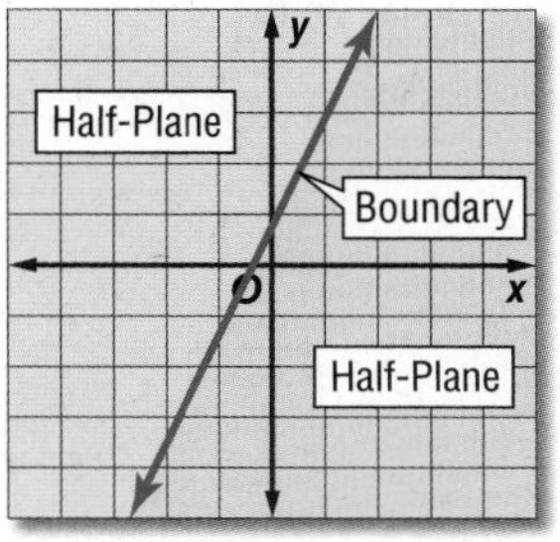

The boundary may or may not be included in the solution. When it is included, the solution is a **closed half-plane**. When not included, the solution is an **open half-plane**.

KeyConcept Graphing Linear Inequalities

Step 1 Graph the boundary. Use a solid line when the inequality contains $\leq$ or $\geq$. Use a dashed line when the inequality contains $<$ or $>$.

Step 2 Use a test point to determine which half-plane should be shaded.

Step 3 Shade the half-plane that contains the solution.

Example 1 Graph an Inequality ($<$ or $>$)

Graph $3x - y < 2$.

Step 1 First, solve for y in terms of x.

$$3x - y < 2$$
$$-y < -3x + 2$$
$$y > 3x - 2$$

Then, graph $y = 3x - 2$. Because the inequality involves $>$, graph the boundary with a dashed line.

Step 2 Select (0, 0) as a test point.

$3x - y < 2$ Original inequality

$3(0) - 0 < 2$ $x = 0$ and $y = 0$

$0 < 2$ true

Step 3 So, the half-plane containing the origin is the solution. Shade this half-plane.

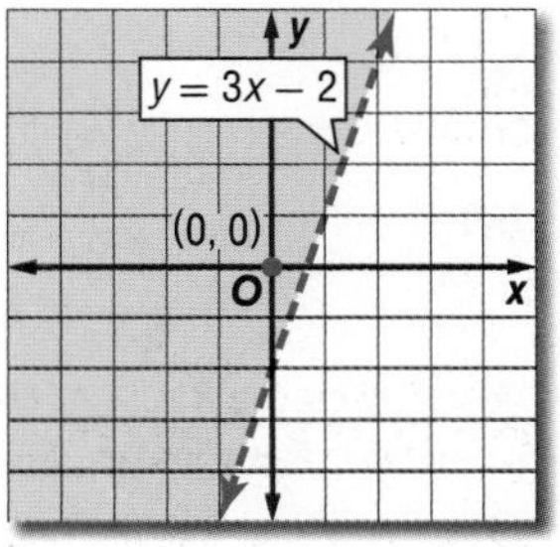

GuidedPractice Graph each inequality.

1A. $y > \frac{1}{2}x + 3$

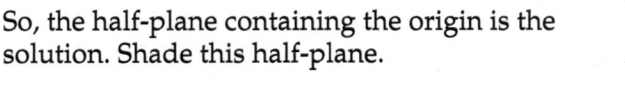

1B. $x - 1 > y$

Example 2 Graph an Inequality ($\leq$ or $\geq$)

Graph $x + 5y \leq 10$.

Step 1 Solve for y in terms of x.

$x + 5y \leq 10$ Original inequality

$5y \leq -x + 10$ Subtract x from each side and simplify.

$y \leq -\frac{1}{5}x + 2$ Divide each side by 5.

Graph $y = -\frac{1}{5}x + 2$. Because the inequality symbol is $\leq$, graph the boundary with a solid line.

StudyTip

Selecting a Test Point When selecting a test point, a standard choice is the origin because it offers easy calculations. However, if it lies on the border, you must choose another point that is not on the border.

Step 2 Select a test point. Let's use (3, 3). Substitute the values into the original inequality.

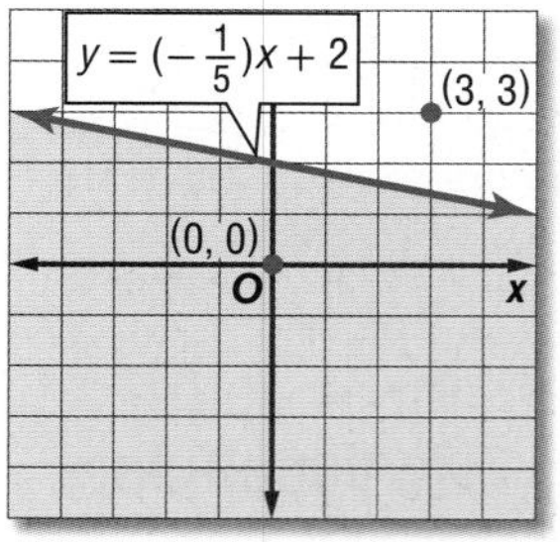

$x + 5y \leq 10$ Original inequality

$3 + 5(3) \leq 10$ $x = 3$ and $y = 3$

$18 \not\leq 10$ Simplify.

Step 3 Since this statement is false, shade the other half-plane.

Guided Practice

Graph each inequality.

2A. $x - y \leq 3$

2B. $2x + 3y \geq 18$

2 Solve Linear Inequalities

We can use a coordinate plane to solve inequalities with one variable.

Example 3 Solve Inequalities From Graphs

Use a graph to solve $3x + 5 < 14$.

Step 1 First graph the boundary, which is the related equation. Replace the inequality sign with an equals sign, and solve for x.

$3x + 5 < 14$ Original inequality

$3x + 5 = 14$ Change $<$ to $=$.

$3x = 9$ Subtract 5 from each side and simplify.

$x = 3$ Divide each side by 3.

Graph $x = 3$ with a dashed line.

Step 2 Choose (0, 0) as a test point. These values in the original inequality give us $5 < 14$.

Step 3 Since this statement is true, shade the half-plane that contains the point (0, 0).

Notice that the x-intercept of the graph is at 3. Since the half-plane to the left of the x-intercept is shaded, the solution is $x < 3$.

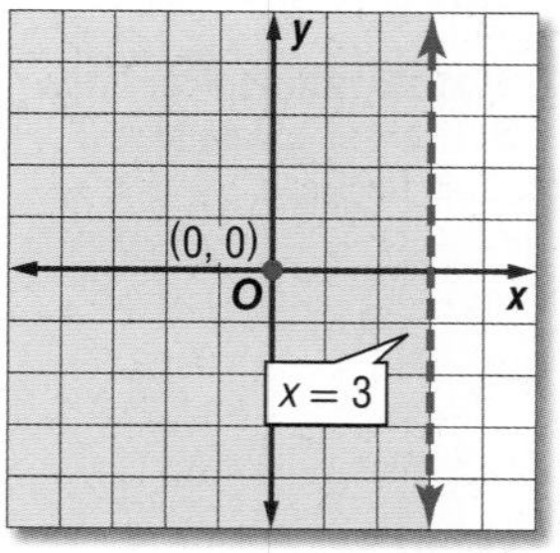

Guided Practice

Use a graph to solve each inequality.

3A. $4x - 3 \geq 17$

3B. $-2x + 6 > 12$

An inequality can be viewed as a constraint in a problem situation. Each solution of the inequality represents a combination that meets the constraint. In real-world problems, the domain and range are often restricted to nonnegative or whole numbers.

Real-WorldLink

As a supplement to traditional yearbooks, many schools are producing digital versions. They include features that allow you to click on a picture and see a short video clip.

Source: eSchool News

Real-World Example 4 Write and Solve an Inequality

CLASS PICNIC A yearbook company promises to give the junior class a picnic if they spend at least \$28,000 on yearbooks and class rings. Each yearbook costs \$35, and each class ring costs \$140. How many yearbooks and class rings must the junior class buy to get their picnic?

Understand You know the cost of each item and the minimum amount the class needs to spend.

Plan Let x = the number of yearbooks and y = the number of class rings the class must buy. Write an inequality.

\$35	times	the number of yearbooks	plus	\$140	times	the number of rings	is at least	\$28,000.
35	$\cdot$	x	$+$	140	$\cdot$	y	$\geq$	28,000

Solve Solve for y in terms of x.

$$35x + 140y - 35x \geq 28{,}000 - 35x \quad \text{Subtract } 35x \text{ from each side.}$$

$$140y \geq -35x + 28{,}000 \quad \text{Divide each side by 140.}$$

$$\frac{140y}{140} \geq \frac{-35x}{140} + \frac{28000}{140} \quad \text{Simplify.}$$

$$y \geq -0.25x + 200 \quad \text{Simplify.}$$

Because the yearbook company cannot sell a negative number of items, the domain and range must be nonnegative numbers. Graph the boundary with a solid line. If we test (0, 0), the result is $0 \geq 28{,}000$, which is false. Shade the closed half-plane that does not include (0, 0).

One solution is (500, 100), or 500 yearbooks and 100 class rings.

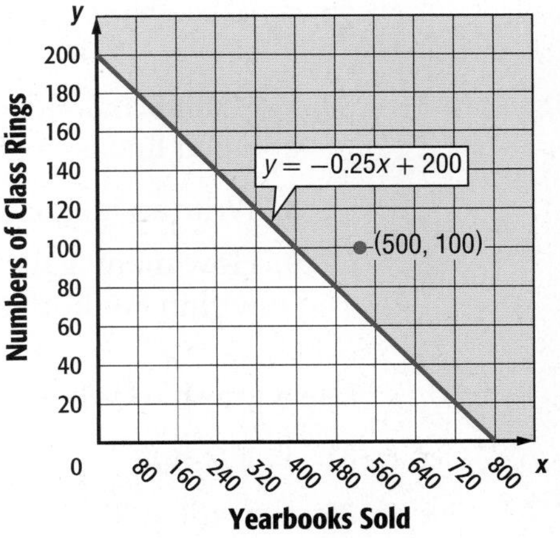

Check If we test (500, 100), the result is $100 \geq 75$, which is true. Because the company cannot sell a fraction of an item, only points with whole-number coordinates can be solutions.

Problem-SolvingTip

Use a Graph You can use a graph to visualize data, analyze trends, and make predictions.

GuidedPractice

4. MARATHONS Neil wants to run a marathon at a pace of at least 6 miles per hour. Write and graph an inequality for the miles y he will run in x hours.

Check Your Understanding

= Step-by-Step Solutions begin on page R13.

Examples 1–2 **Graph each inequality.**

1. $y > x + 3$
2. $y \geq -8$
3. $x + y > 1$
4. $y \leq x - 6$
5. $y < 2x - 4$
6. $x - y \leq 4$

Example 3 **Use a graph to solve each inequality.**

7. $7x + 1 < 15$
8. $-3x - 2 \geq 11$
9. $3y - 5 \leq 34$
10. $4y - 21 > 1$

Example 4

11. **FINANCIAL LITERACY** The surf shop has a weekly overhead of \$2300.
    - **a.** Write an inequality to represent the number of skimboards and longboards the shop sells each week to make a profit.
    - **b.** How many skimboards and longboards must the shop sell each week to make a profit?

Practice and Problem Solving

Extra Practice is on page R5.

Examples 1–2 **Graph each inequality.**

12. $y < x - 3$
13. $y > x + 12$
14. $y \geq 3x - 1$
15. $y \leq -4x + 12$
16. $6x + 3y > 12$
17. $2x + 2y < 18$
18. $5x + y > 10$
19. $2x + y < -3$
20. $-2x + y \geq -4$
21. $8x + y \leq 6$
22. $10x + 2y \leq 14$
23. $-24x + 8y \geq -48$

Example 3 **Use a graph to solve each inequality.**

24. $10x - 8 < 22$
25. $20x - 5 > 35$
26. $4y - 77 \geq 23$
27. $5y + 8 \leq 33$
28. $35x + 25 < 6$
29. $14x - 12 > -31$

Example 4

30. CCSS **MODELING** Sybrina is decorating her bedroom. She has \$300 to spend on paint and bed linens. A gallon of paint costs \$14, while a set of bed linens costs \$60.
    - **a.** Write an inequality for this situation.
    - **b.** How many gallons of paint and bed linen sets can Sybrina buy and stay within her budget?

Use a graph to solve each inequality.

31. $3x + 2 < 0$
32. $4x - 1 > 3$
33. $-6x - 8 \geq -4$
34. $-5x + 1 < 3$
35. $-7x + 13 < 10$
36. $-4x - 4 \leq -6$

37. **SOCCER** The girls' soccer team wants to raise \$2000 to buy new goals. How many of each item must they sell to buy the goals?
    - **a.** Write an inequality that represents this situation.
    - **b.** Graph this inequality.
    - **c.** Make a table of values that shows at least five possible solutions.
    - **d.** Plot the solutions from part **c**.

Graph each inequality. Determine which of the ordered pairs are part of the solution set for each inequality.

38. $y \geq 6$; $\{(0, 4), (-2, 7), (4, 8), (-4, -8), (1, 6)\}$

39. $x < -4$; $\{(2, 1), (-3, 0), (0, -3), (-5, -5), (-4, 2)\}$

40. $2x - 3y \leq 1$; $\{(2, 3), (3, 1), (0, 0), (0, -1), (5, 3)\}$

41. $5x + 7y \geq 10$; $\{(-2, -2), (1, -1), (1, 1), (2, 5), (6, 0)\}$

42. $-3x + 5y < 10$; $\{(3, -1), (1, 1), (0, 8), (-2, 0), (0, 2)\}$

43. $2x - 2y \geq 4$; $\{(0, 0), (0, 7), (7, 5), (5, 3), (2, -5)\}$

44. **RECYCLING** Mr. Jones would like to spend no more than \$37.50 per week on recycling. A curbside recycling service will remove up to 50 pounds of plastic bottles and paper products per week. They charge \$0.25 per pound of plastic and \$0.75 per pound of paper products.

a. Write an inequality that describes the number of pounds of each product that can be included in the curbside service.

b. Write an inequality that describes Mr. Jones' weekly cost for the service if he stays within his budget.

c. Graph an inequality for the weekly costs for the service.

45. **MULTIPLE REPRESENTATIONS** Use inequalities A and B to investigate graphing compound inequalities on a coordinate plane.

A. $7(y + 6) \leq 21x + 14$ **B.** $-3y \leq 3x - 12$

a. Numerical Solve each inequality for y.

b. Graphical Graph both inequalities on one graph. Shade the half-plane that makes A true in red. Shade the half-plane that makes B true in blue.

c. Verbal What does the overlapping region represent?

H.O.T. Problems Use Higher-Order Thinking Skills

46. **ERROR ANALYSIS** Reiko and Kristin are solving $4y \leq \frac{8}{3}x$ by graphing. Is either of them correct? Explain your reasoning.

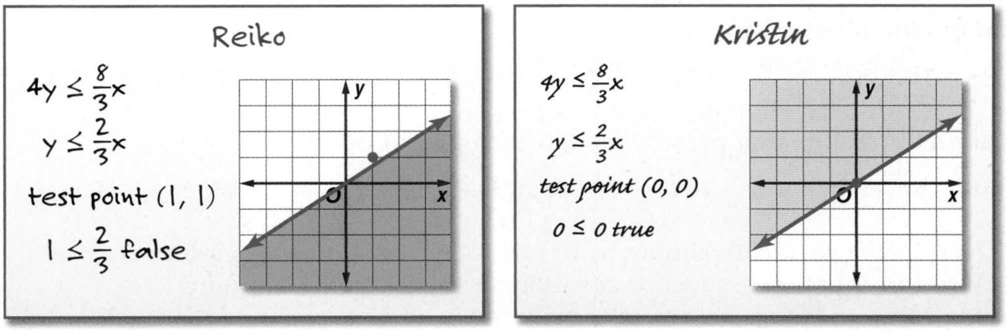

47. **CCSS TOOLS** Write a linear inequality for which $(-1, 2)$, $(0, 1)$, and $(3, -4)$ are solutions but $(1, 1)$ is not.

48. **REASONING** Explain why a point on the boundary should not be used as a test point.

49. **OPEN ENDED** Write a two-variable inequality with a restricted domain and range to represent a real-world situation. Give the domain and range, and explain why they are restricted.

50. **WRITING IN MATH** Summarize the steps to graph an inequality in two variables.

Standardized Test Practice

51. What is the domain of this function?

A $\{x \mid 0 \leq x \leq 3\}$
B $\{x \mid 0 \leq x \leq 9\}$
C $\{y \mid 0 \leq y \leq 9\}$
D $\{y \mid 0 \leq y \leq 3\}$

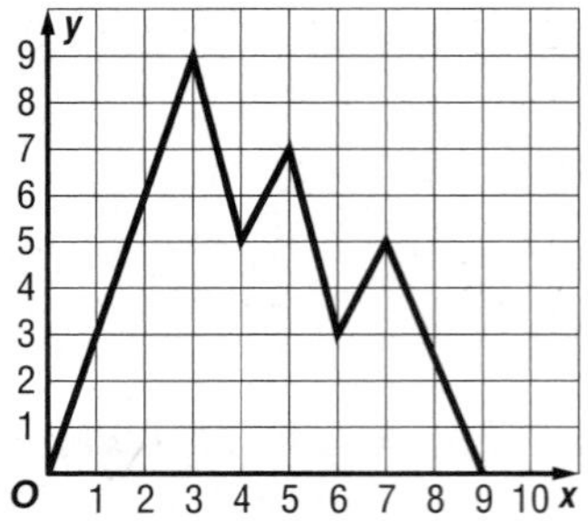

52. EXTENDED RESPONSE An arboretum will close for the winter when all of the trees have lost their leaves. The table shows the number of trees each day that still have leaves.

Day	5	10	15	20
Trees with Leaves	325	260	195	130

a. Write an equation that represents the number of trees with leaves y after d days.

b. Find the y-intercept. What does it mean in the context of this problem?

c. After how many days will the arboretum close? Explain how you got your answer.

53. Which inequality best represents the statement below?

A jar contains 832 gumballs. Ebony's guess was within 46 pieces.

F $|g - 832| \leq 46$
G $|g + 832| \leq 46$
H $|g - 832| \geq 46$
J $|g + 832| \geq 46$

54. GEOMETRY If the rectangular prism has a volume of 10,080 cm^3, what is the value of x?

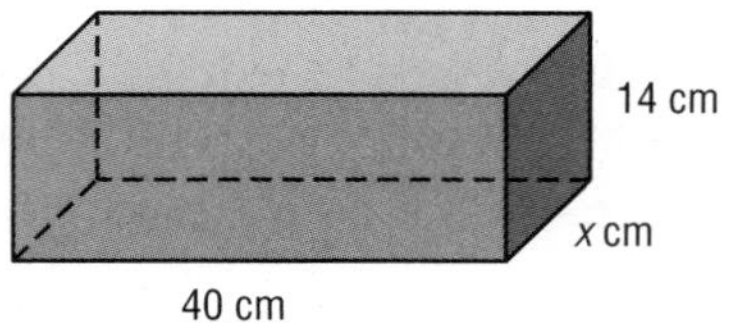

A 12
B 14
C 16
D 18

Spiral Review

Solve each open sentence. (Lesson 5-5)

55. $|y - 2| > 4$

56. $|t - 6| \leq 5$

57. $|3 + d| < -4$

Solve each compound inequality. (Lesson 5-4)

58. $4c - 4 < 8c - 16 < 6c - 6$

59. $5 < \frac{1}{2}p + 3 < 8$

60. $0.5n \geq -7$ or $2.5n + 2 \leq 9$

Write an equation of the line that passes through each pair of points. (Lesson 4-2)

61. $(1, -3)$ and $(2, 5)$

62. $(-2, -4)$ and $(-7, 3)$

63. $(-6, -8)$ and $(-8, -5)$

64. FITNESS The table shows the maximum heart rate to maintain during aerobic activities. Write an equation in function notation for the relation. Determine what would be the maximum heart rate to maintain in aerobic training for an 80-year-old. (Lesson 3-5)

Age (yr)	20	30	40	50	60	70
Pulse rate (beats/min)	175	166	157	148	139	130

Skills Review

65. WORK The formula $s = \frac{w - 10r}{m}$ is used to find keyboarding speeds. In the formula, s represents the speed in words per minute, w the number of words typed, r the number of errors, and m the number of minutes typed. Solve for r.

EXTEND 5-6

Graphing Technology Lab
Graphing Inequalities

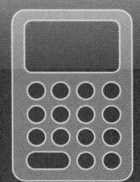

You can use a graphing calculator to investigate the graphs of inequalities.

CCSS **Common Core State Standards**
Content Standards
A.REI.12 Graph the solutions to a linear inequality in two variables as a halfplane (excluding the boundary in the case of a strict inequality), and graph the solution set to a system of linear inequalities in two variables as the intersection of the corresponding half-planes.
Mathematical Practices
5 Use appropriate tools strategically.

Activity 1 Less Than

Graph $y \leq 2x + 5$.

Clear all functions from the **Y=** list.

KEYSTROKES: Y= CLEAR

Graph $y \leq 2x + 5$ in a standard viewing window.

KEYSTROKES: 2 X,T,θ,*n* + 5 ◀ ◀ ◀ ◀ ◀ ◀ ENTER ENTER ENTER ZOOM 6

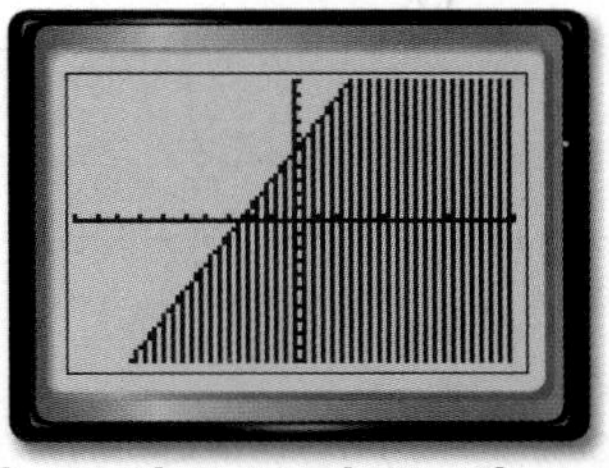

[−10, 10] scl: 1 by [−10, 10] scl: 1

All ordered pairs for which y is *less than or equal to* $2x + 5$ lie *below or on* the line and are solutions.

Activity 2 Greater Than

Graph $y - 2x \geq 5$.

Clear the graph that is currently displayed.

KEYSTROKES: Y= CLEAR

Rewrite $y - 2x \geq 5$ as $y \geq 2x + 5$ and graph it.

KEYSTROKES: 2 X,T,θ,*n* + 5 ◀ ◀ ◀ ◀ ◀ ◀ ENTER ENTER ZOOM 6

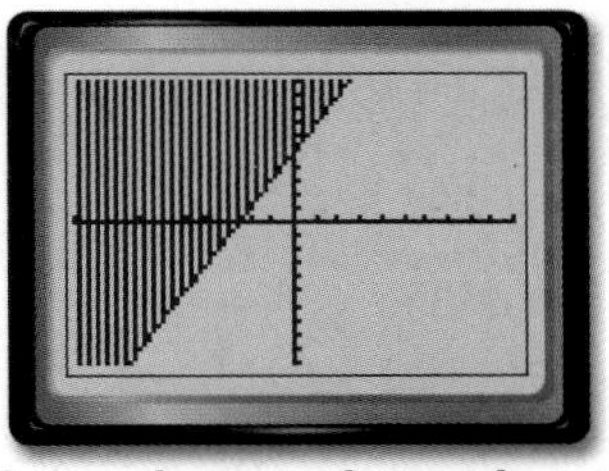

[−10, 10] scl: 1 by [−10, 10] scl: 1

All ordered pairs for which y is *greater than or equal to* $2x + 5$ lie *above or on* the line and are solutions.

Exercises

1. Compare and contrast the two graphs shown above.

2. Graph $y \geq -3x + 1$ in the standard viewing window. Using your graph, name four solutions of the inequality.

3. Suppose student water park tickets cost \$16, and adult water park tickets cost \$20. You would like to buy at least 10 tickets but spend no more than \$200.

a. Let x = number of student tickets and y = number of adult tickets. Write two inequalities, one representing the total number of tickets and the other representing the total cost of the tickets.

b. Graph the inequalities. Use the viewing window [0, 20] scl: 1 by [0, 20] scl: 1.

c. Name four possible combinations of student and adult tickets.

CHAPTER 5 Study Guide and Review

Study Guide

KeyConcepts

Solving One-Step Inequalities (Lessons 5-1 and 5-2)

For all numbers a, b, and c, the following are true.

- If $a > b$ and c is positive, $ac > bc$.
- If $a > b$ and c is negative, $ac < bc$.

Multi-Step and Compound Inequalities (Lessons 5-3 and 5-4)

- Multi-step inequalities can be solved by undoing the operations in the same way you would solve a multi-step equation.
- A compound inequality containing *and* is only true if both inequalities are true.
- A compound inequality containing *or* is true if at least one of the inequalities is true.

Absolute Value Inequalities (Lesson 5-5)

- The absolute value of any number x is its distance from zero on a number line and is written as $|x|$. If $x \geq 0$, then $|x| = x$. If $x < 0$, then $|x| = -x$.
- If $|x| < n$ and $n > 0$, then $-n < x < n$.
- If $|x| > n$ and $n > 0$, then $x > n$ or $x < -n$.

Inequalities in Two Variables (Lesson 5-6)

To graph an inequality:

Step 1 Graph the boundary. Use a solid line when the inequality contains $\leq$ or $\geq$. Use a dashed line when the inequality contains $<$ or $>$.

Step 2 Use a test point to determine which half-plane should be shaded.

Step 3 Shade the half-plane.

FOLDABLES StudyOrganizer

Be sure the Key Concepts are noted in your Foldable.

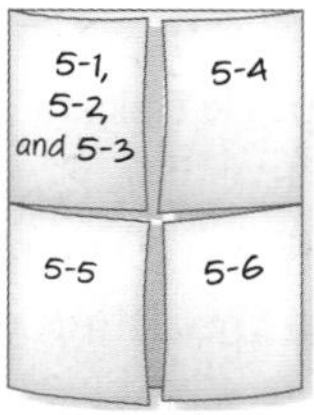

KeyVocabulary

boundary (p. 317)

closed half-plane (p. 317)

compound inequality (p. 306)

half-plane (p. 317)

inequality (p. 285)

intersection (p. 306)

open half-plane (p. 317)

set-builder notation (p. 286)

union (p. 307)

VocabularyCheck

State whether each sentence is *true* or *false*. If *false*, replace the underlined term to make a true sentence.

1. Set-builder notation is a less concise way of writing a solution set.
2. There are two types of compound inequalities.
3. The graph of a compound inequality containing *and* shows the union of the individual graphs.
4. A compound inequality containing *or* is true if one or both of the inequalities is true. Its graph is the union of the graphs of the two inequalities.
5. The graph of an inequality of the form $y < ax + b$ is a region on the coordinate plane called a half-plane.
6. A point defines the boundary of an open half-plane.
7. The boundary is the graph of the equation of the line that defines the edge of each half-plane.
8. The solution set to the inequality $y \geq x$ includes the boundary.
9. When solving an inequality, multiplying each side by a negative number reverses the inequality symbol.
10. The graph of a compound inequality that contains *and* is the intersection of the graphs of the two inequalities.

Lesson-by-Lesson Review

5-1 Solving Inequalities by Addition and Subtraction

Solve each inequality. Then graph it on a number line.

11. $w - 4 > 9$

12. $x + 8 \leq 3$

13. $6 + h < 1$

14. $-5 < a + 2$

15. $13 - p \geq 15$

16. $y + 1 \leq 8$

17. **FIELD TRIP** A bus can hold 44 people. If there are 35 students in Samantha's class, how many more people can ride on the bus?

Example 1

Solve $x - 9 < -4$. Then graph it on a number line.

$x - 9 < -4$ Original inequality

$x - 9 + 9 < -4 + 9$ Add 9 to each side.

$x < 5$ Simplify.

The solution set is $\{x \mid x < 5\}$.

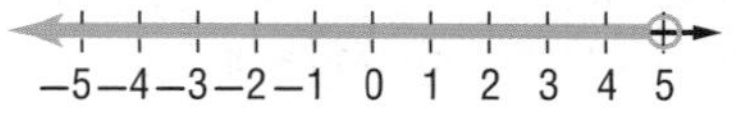

5-2 Solving Inequalities by Multiplication and Division

Solve each inequality. Graph the solution on a number line.

18. $\frac{1}{3}x > 6$

19. $\frac{1}{5}g \geq -4$

20. $4p < 32$

21. $-55 \leq -5w$

22. $-2m > 100$

23. $\frac{2}{3}t < -48$

24. **MOVIE RENTAL** Jack has no more than \$24 to spend on DVDs for a party. Each DVD rents for \$4. Find the maximum number of DVDs Jack can rent for his party.

Example 2

Solve $-14h < 56$. Check your solution.

$-14h < 56$ Original inequality

$\frac{-14h}{-14} > \frac{56}{-14}$ Divide each side by −14.

$h > -4$ Simplify.

$\{h \mid h > -4\}$

CHECK To check, substitute three different values into the original inequality: −4, a number less than −4, and a number greater than −4.

5-3 Solving Multi-Step Inequalities

Solve each inequality. Graph the solution on a number line.

25. $3h - 7 < 14$

26. $4 + 5b > 34$

27. $18 \leq -2x + 8$

28. $\frac{t}{3} - 6 > -4$

29. Four times a number decreased by 6 is less than −2. Define a variable, write an inequality, and solve for the number.

30. **TICKET SALES** The drama club collected \$160 from ticket sales for the spring play. They need to collect at least \$400 to pay for new lighting for the stage. If tickets sell for \$3 each, how many more tickets need to be sold?

Example 3

Solve $-6y - 13 > 29$. Check your solution.

$-6y - 13 > 29$ Original inequality

$-6y - 13 + 13 > 29 + 13$ Add 13 to each side.

$-6y > 42$ Simplify.

$\frac{-6y}{-6} < \frac{42}{-6}$ Divide each side by −6 and change > to <.

$y < -7$ Simplify.

The solution set is $\{y \mid y < -7\}$.

CHECK $-6y - 13 > 29$ Original inequality

$-6(-10) - 13 \overset{?}{>} 29$ Substitute −10 for y.

$47 > 29$ ✓ Simplify.

5-4 Solving Compound Inequalities

Solve each compound inequality. Then graph the solution set.

31. $m - 3 < 6$ and $m + 2 > 4$

32. $-4 < 2t - 6 < 8$

33. $3x + 2 \leq 11$ or $5x - 8 > 22$

34. KITES A kite can be flown in wind speeds no less than 7 miles per hour and no more than 16 miles per hour. Write an inequality for the wind speeds at which the kite can fly.

Example 4

Solve $-3w + 4 > -8$ and $2w - 11 > -19$. Then graph the solution set.

$-3w + 4 > -8$ and $2w - 11 > -19$

$w < 4$ $\qquad$ $w > -4$

To graph the solution set, graph $w < 4$ and graph $w > -4$. Then find the intersection.

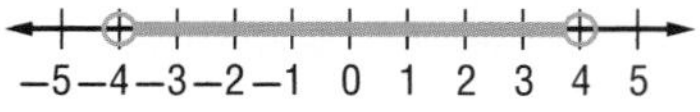

5-5 Inequalities Involving Absolute Value

Solve each inequality. Then graph the solution set.

35. $|x - 4| < 9$

36. $|p + 2| > 7$

37. $|2c + 3| \leq 11$

38. $|f - 9| \geq 2$

39. $|3d - 1| \leq 8$

40. $\left|\frac{4b - 2}{3}\right| < 12$

41. $\left|\frac{2t + 6}{2}\right| > 10$

42. $|-4y - 3| < 13$

43. $|m + 19| \leq 1$

44. $|-k - 7| \geq 4$

Example 5

Solve $|x - 6| < 9$. Then graph the solution set.

Case 1 $x - 6$ is nonnegative.

$x - 6 < 9$

$x < 15$

Case 2 $x - 6$ is negative.

$-(x - 6) < 9$

$x > -3$

The solution set is $\{x \mid -3 < x < 15\}$.

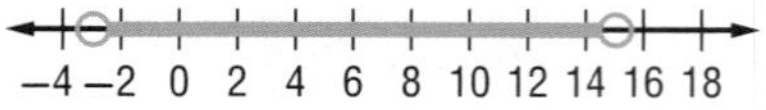

5-6 Graphing Inequalities in Two Variables

Graph each inequality.

45. $y > x - 3$

46. $y < 2x + 1$

47. $3x - y \leq 4$

48. $y \geq -2x + 6$

49. $5x - 2y < 10$

50. $3x + 4y > 12$

Graph each inequality. Determine which of the ordered pairs are part of the solution set for each inequality.

51. $y \leq 4$; $\{(3, 6), (1, 2), (-4, 8), (3, -2), (1, 7)\}$

52. $-2x + 3y \geq 12$; $\{(-2, 2), (-1, 1), (0, 4), (2, 2)\}$

53. BAKERY Ben has $24 to spend on cookies and cupcakes. Write and graph an inequality that represents what Ben can buy.

Example 6

Graph $2x - y > 3$.

Solve for y in terms of x.

$2x - y > 3$ $\quad$ Original inequality

$-y > -2x + 3$ $\quad$ Subtract $2x$ from each side.

$y < 2x - 3$ $\quad$ Multiply each side by -1.

Graph the boundary using a dashed line. Choose (0, 0) as a test point.

$2(0) - 0 \overset{?}{>} 3$

$0 \not> 3$

Since 0 is not greater than 3, shade the plane that does not contain (0, 0).

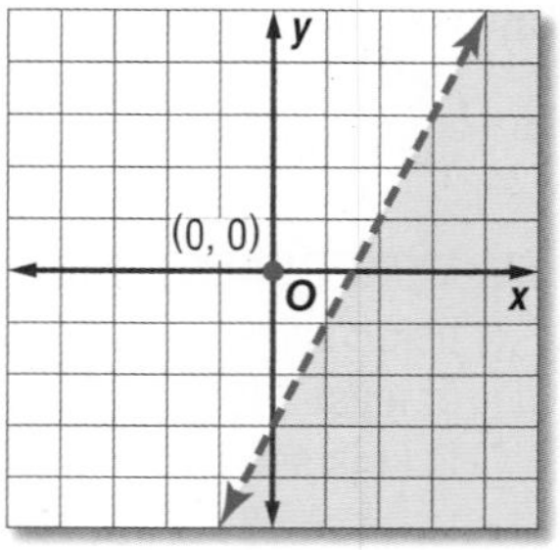

CHAPTER 5

Practice Test

Solve each inequality. Then graph it on a number line.

1. $x - 9 < -4$
2. $6p \geq 5p - 3$

3. **MULTIPLE CHOICE** Drew currently has 31 comic books in his collection. His friend Connor has 58 comic books. How many more comic books does Drew need to add to his collection in order to have a larger collection than Connor?

   A no more than 21

   B 27

   C at least 28

   D more than 30

Solve each inequality. Graph the solution on a number line.

4. $\frac{1}{5}h > 3$
5. $7w \leq -42$
6. $-\frac{2}{3}t \geq 24$
7. $-9m < -36$
8. $3c - 7 < 11$
9. $\frac{g}{4} + 3 \leq -9$
10. $-2(x - 4) > 5x - 13$

11. **ZOO** The 8th grade science class is going to the zoo. The class can spend up to $300 on admission.

Zoo Admission	
Visitor	**Cost**
student	$8
adult	$10

   a. Write an inequality for this situation.

   b. If there are 32 students in the class and 1 adult will attend for every 8 students, can the entire class go to the zoo?

Solve each compound inequality. Then graph the solution set.

12. $y - 8 < -3$ or $y + 5 > 19$
13. $-11 \leq 2h - 5 \leq 13$
14. $3z - 2 > -5$ and $7z + 4 < -17$

Define a variable, write an inequality, and solve the problem. Check your solution.

15. The difference of a number and 4 is no more than 8.

16. Nine times a number decreased by four is at least twenty-three.

17. **MULTIPLE CHOICE** Write a compound inequality for the graph shown below.

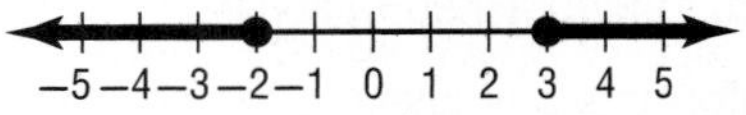

   F $-2 \leq x < 3$ **H** $x < -2$ or $x \geq 3$

   G $x \leq -2$ or $x \geq 3$ **J** $-2 < x \leq 3$

Solve each inequality. Then graph the solution set.

18. $|p - 5| < 3$
19. $|2f + 7| \geq 21$
20. $|-4m + 3| \leq 15$
21. $\left|\frac{x - 3}{4}\right| > 5$

22. **RETAIL** A sporting goods store is offering a $15 coupon on any pair of shoes.

   a. The most and least expensive pairs of shoes are $149.95 and $24.95. What is the range of costs for customers with coupons?

   b. When buying a pair of $109.95 shoes, you can use a coupon or a 15% discount. Which option is best?

Graph each inequality.

23. $y < 4x - 1$
24. $2x + 3y \geq 12$

25. Graph $y > -2x + 5$. Then determine which of the ordered pairs in $\{(-2, 0), (-1, 5), (2, 3), (7, 3)\}$ are in the solution set.

26. **PRESCHOOL** Mrs. Jones is buying new books and puzzles for her preschool classroom. Each book costs $6, and each puzzle costs $4. Write and graph an inequality to determine how many books and puzzles she can buy for $96.

CHAPTER 5

Preparing for Standardized Tests

Write and Solve an Inequality

Many multiple-choice items will require writing and solving inequalities. Follow the steps below to help you successfully solve these types of problems.

Strategies for Writing and Solving Inequalities

Step 1

Read the problem statement carefully.

Ask yourself:

- What am I being asked to solve?
- What information is given in the problem?
- What are the unknowns for which I need to solve?

Step 2

Translate the problem statement into an inequality.

- Assign variables to the unknown(s).
- Write the word sentence as a mathematical number sentence looking for words such as *greater than, less than, no more than, up to*, or *at least* to indicate the type of inequality as well as where to place the inequality sign.

Step 3

Solve the inequality.

- Solve for the unknowns in the inequality.
- Remember that multiplying or dividing each side by a negative number reverses the direction of the inequality.
- Check your answer to make sure it makes sense.

Standardized Test Example

Read the problem. Identify what you need to know. Then use the information in the problem to solve. Show your work.

Pedro has earned scores of 89, 74, 79, 85, and 88 on his tests this semester. He needs a test average of at least 85 in order to earn an A for the semester. There will be one more test given this semester.

A Write an inequality to model the situation.

B What score must he have on his final test to earn an A for the semester?

Read the problem carefully. You are given Pedro's first 5 test scores and told that he needs an average of *at least* 85 after his next test to earn an A for the semester.

a. Write the inequality.

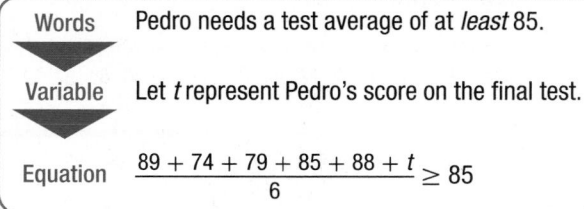

Words	Pedro needs a test average of at *least* 85.
Variable	Let t represent Pedro's score on the final test.
Equation	$\frac{89 + 74 + 79 + 85 + 88 + t}{6} \geq 85$

b. Solve the inequality for *t*.

$$\frac{89 + 74 + 79 + 85 + 88 + t}{6} \geq 85$$
$$89 + 74 + 79 + 85 + 88 + t \geq 85(6)$$
$$415 + t \geq 510$$
$$t \geq 95$$

So, Pedro's final test score must be greater than or equal to 95 in order for him to earn an A for the semester.

Exercises

Read each problem. Identify what you need to know. Then use the information in the problem to solve.

1. Craig has $20 to order a pizza. The pizza costs $12.50 plus $0.95 per topping. If there is also a $3 delivery fee, how many toppings can Craig order?

2. To join an archery club, Nina had to pay an initiation fee of $75, plus $40 per year in membership dues.

a. Write an equation to model the total cost, y, of belonging to the club for x years.

b. How many years will it take her to spend more than $400 to belong to the club?

3. The area of the triangle below is no more than 84 square millimeters. What is the height of the triangle?

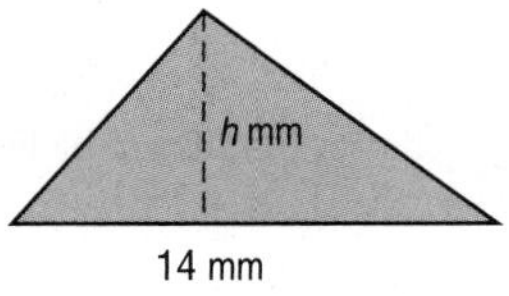

4. Rosa earns $200 a month delivering newspapers, plus an average of $11 per hour babysitting. If her goal is to earn at least $295 this month, how many hours will she have to babysit?

5. To earn money for a new bike, Ethan is selling some of his baseball cards. He has saved $245. If the bike costs $1400, and he can sell 154 cards, for how much money will he need to sell each card to reach his goal?

6. In a certain lacrosse league, there can be no more than 22 players on each team, and no more than 10 teams per age group. There are 6 age groups.

a. Write an inequality to represent this situation.

b. What is the greatest number of players that can play lacrosse in this league?

7. Sarah has $120 to shop for herself and to buy some gifts for 6 of her friends. She has purchased a shirt for herself for $32. Assuming that she spends an equal amount on each friend, what is the maximum that she can spend per person?

CHAPTER 5

Standardized Test Practice

Cumulative, Chapters 1 through 5

Multiple Choice

Read each question. Then fill in the correct answer on the answer document provided by your teacher or on a sheet of paper.

1. Miguel received a $100 gift certificate for a graduation gift. He wants to buy a CD player that costs $38 and CDs that cost $12 each. Which of the following inequalities represents how many CDs Miguel can buy?

A $n \leq 6$

B $n \geq 5$

C $n < 5$

D $n \leq 5$

2. Craig is paid time-and-a-half for any additional hours over 40 that he works.

Time	Pay Rate
Up to 40 hours	$12.80/hr
Additional hours worked over 40	$19.20/hr

If Craig's goal is to earn at least $600 next week, what is the minimum number of hours he needs to work?

F 43 hours

G 44 hours

H 45 hours

J 46 hours

3. Which equation has a slope of $-\frac{2}{3}$ and a y-intercept of 6?

A $y = 6x + \frac{2}{3}$

B $y = -\frac{2}{3}x - 6$

C $y = -\frac{2}{3}x + 6$

D $y = 6x - \frac{2}{3}$

4. The highest score that is on record on a video game is 10,219 points. The lowest score on record is 257 points. Which of the following inequalities best shows the range of scores recorded on the game?

F $x \leq 10{,}219$

G $x \geq 257$

H $257 < x < 10{,}219$

J $257 \leq x \leq 10{,}219$

5. The current temperature is 82°. If the temperature rises more than 4 degrees, there will be a new record high for the date. Which number line represents the temperatures that would set a new record high?

A

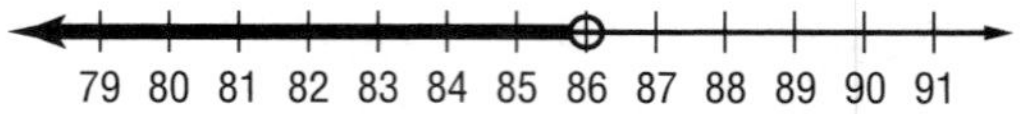

B

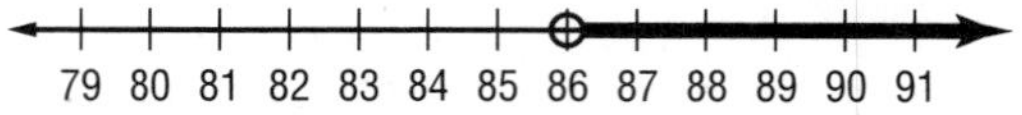

C

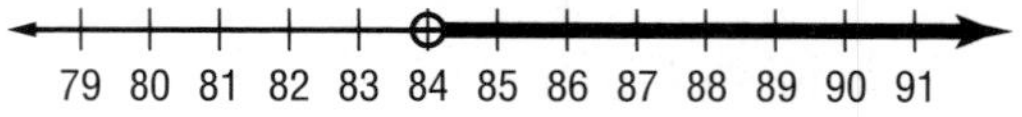

D

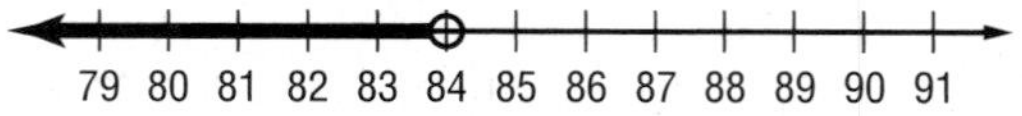

6. The girls' volleyball team is selling T-shirts and pennants to raise money for new uniforms. The team hopes to raise more than $250.

Item	Price
T-shirt	$10
Pennant	$4

Which of the following combinations of items sold would meet this goal?

F 16 T-shirts and 20 pennants

G 20 T-shirts and 12 pennants

H 18 T-shirts and 18 pennants

J 15 T-shirts and 20 pennants

7. What type of line does not have a defined slope?

A horizontal

B parallel

C perpendicular

D vertical

8. Which expression below illustrates the Associative Property?

F $abc = bac$

G $2(x - 3) = 2x - 6$

H $(p + 3) - t = p + (3 - t)$

J $5 + (-5) = 0$

Test-Taking Tip

Question 2 You can check your answer by finding Craig's earnings for the hours worked.

Short Response/Gridded Response

Record your answers on the answer sheet provided by your teacher or on a sheet of paper.

9. Solve $-4 < 3x + 8 \leq 23$.

10. GRIDDED RESPONSE Tien is saving money for a new television. She needs to save at least $720 to pay for her expenses. Each week Tien saves $50 toward her new television. How many weeks will it take so she can pay for the television?

11. Write an inequality that best represents the graph.

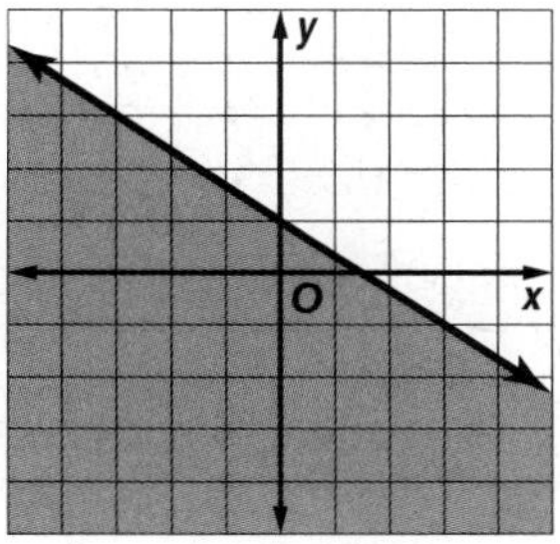

12. Solve $|x - 4| < 2$.

13. GRIDDED RESPONSE Daniel wants to ship a set of golf clubs and several boxes of golf balls in a box that can hold up to 20 pounds. If the set of clubs weighs 9 pounds and each box of golf balls weighs 12 ounces, how many boxes of golf balls can Daniel ship?

14. Graph the solution set for the inequality $3x - 6 \leq 4x - 4 \leq 3x + 1$.

15. Write an equation that represents the data in the table.

x	y
3	12.5
4	16
5	19.5
6	23
7	26.5

16. A sporting goods company near the beach rents bicycles for $10 plus $5 per hour. Write an equation in slope-intercept form that shows the total cost, y, of renting a bicycle for x hours. How much would it cost Emily to rent a bicycle for 6 hours?

Extended Response

Record your answers on a sheet of paper. Show your work.

17. Theresa is saving money for a vacation. She needs to save at least $640 to pay for her expenses. Each week, she puts $35 towards her vacation savings.

a. Let w represent the number of weeks Theresa saves money. Write an inequality to model the situation.

b. Solve the inequality from part a. What is the minimum number of weeks Theresa must save money in order to reach her goal?

c. If Theresa were to save $45 each week instead, by how many weeks would the minimum savings time be decreased?

Need ExtraHelp?

If you missed Question...	1	2	3	4	5	6	7	8	9	10	11	12	13	14	15	16	17
Go to Lesson...	5-3	5-3	4-1	5-4	5-1	5-6	3-3	1-3	5-4	5-2	5-6	5-5	5-3	5-4	3-5	4-2	5-2

CHAPTER 6

Systems of Linear Equations and Inequalities

Then

You solved linear equations in one variable.

Now

In this chapter, you will:

- Solve systems of linear equations by graphing, substitution, and elimination.
- Solve systems of linear inequalities by graphing.

Why? ▲

MUSIC $1500 worth of tickets were sold for a marching band competition. Adult tickets were $12 each, and student tickets were $8 each. If you knew how many total tickets were sold, you could use a system of equations to determine how many adult tickets and how many student tickets were sold.

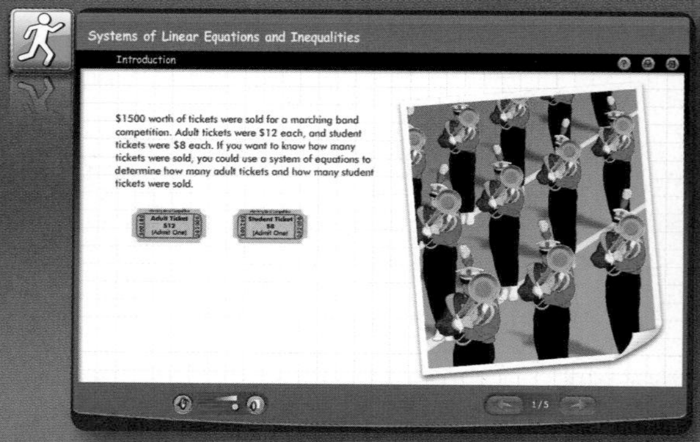

connectED.mcgraw-hill.com Your Digital Math Portal

Animation | Vocabulary | eGlossary | Personal Tutor | Virtual Manipulatives | Graphing Calculator | Audio | Foldables | Self-Check Practice | Worksheets

Get Ready for the Chapter

Diagnose Readiness | You have two options for checking prerequisite skills.

1 Textbook Option Take the Quick Check below. Refer to the Quick Review for help.

QuickCheck

Name the ordered pair for each point on the coordinate plane.

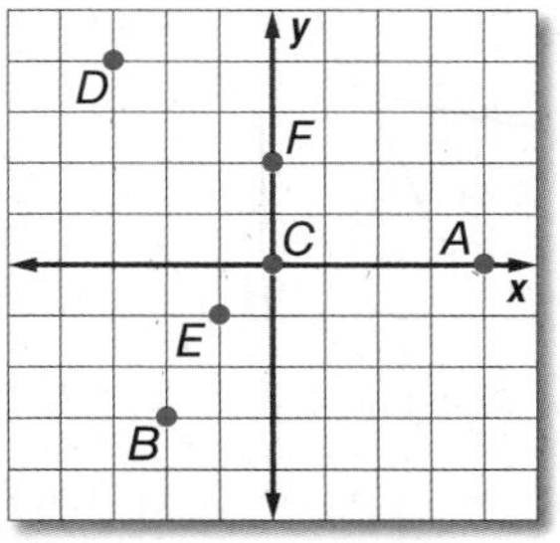

1. A
2. D
3. B
4. C
5. E
6. F

Solve each equation or formula for the variable specified.

7. $2x + 4y = 12$, for x
8. $x = 3y - 9$, for y
9. $m - 2n = 6$, for m
10. $y = mx + b$, for x
11. $P = 2\ell + 2w$, for ℓ
12. $5x - 10y = 40$, for y
13. **GEOMETRY** The formula for the area of a triangle is $A = \frac{1}{2}bh$, where A represents the area, b is the base, and h is the height of the triangle. Solve the equation for b.

QuickReview

Example 1

Name the ordered pair for Q on the coordinate plane.

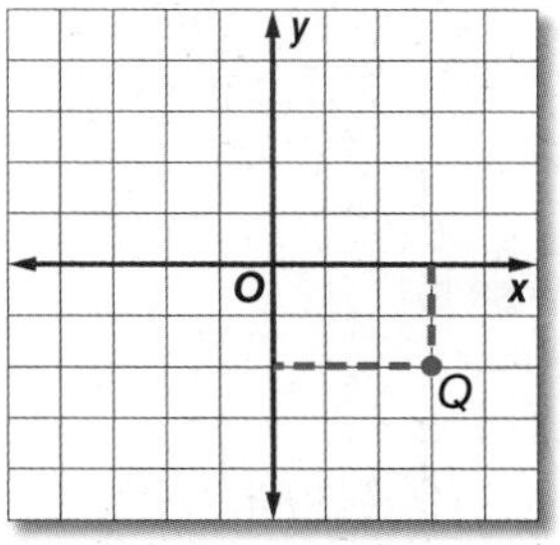

Follow a vertical line from the point to the x-axis. This gives the x-coordinate, 3.

Follow a horizontal line from the point to the y-axis. This gives the y-coordinate, -2.

The ordered pair is $(3, -2)$.

Example 2

Solve $12x + 3y = 36$ for y.

$12x + 3y = 36$	Original equation
$12x + 3y - 12x = 36 - 12x$	Subtract $12x$ from each side.
$3y = 36 - 12x$	Simplify.
$\frac{3y}{3} = \frac{36 - 12x}{3}$	Divide each side by 3.
$y = 12 - 4x$	Simplify.

2 Online Option Take an online self-check Chapter Readiness Quiz at connectED.mcgraw-hill.com.

Get Started on the Chapter

You will learn several new concepts, skills, and vocabulary terms as you study Chapter 6. To get ready, identify important terms and organize your resources. You may wish to refer to Chapter 0 to review prerequisite skills.

FOLDABLES® StudyOrganizer

Linear Functions Make this Foldable to help you organize your Chapter 6 notes about solving systems of equations and inequalities. Begin with a sheet of notebook paper.

1 **Fold** lengthwise to the holes.

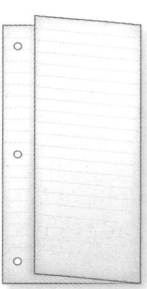

2 **Cut** 6 tabs.

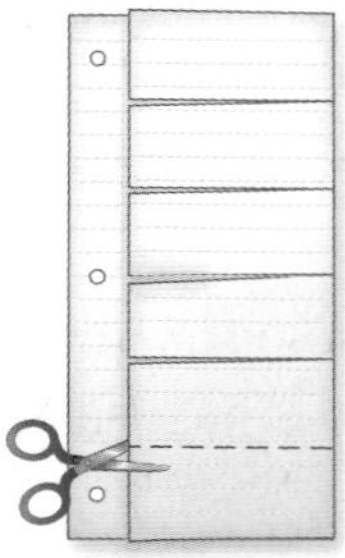

3 **Label** the tabs using the lesson titles.

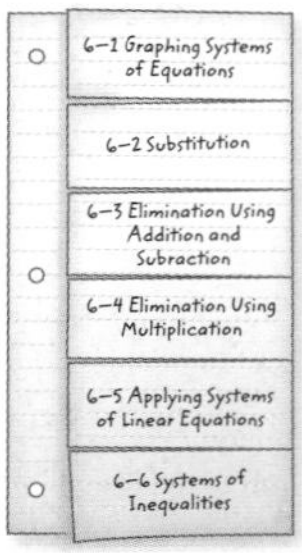

NewVocabulary

English		Español
system of equations	p. 335	sistema de ecuaciones
consistent	p. 335	consistente
independent	p. 335	independiente
dependent	p. 335	dependiente
inconsistent	p. 335	inconsistente
substitution	p. 344	sustitución
elimination	p. 350	eliminación
matrix	p. 370	matriz
element	p. 370	elemento
dimension	p. 370	dimensión
augmented matrix	p. 370	matriz ampliada
row reduction	p. 371	reducción de fila
identity matrix	p. 371	matriz
system of inequalities	p. 372	sistema de desigualdades

ReviewVocabulary

domain dominio the set of the first numbers of the ordered pairs in a relation

intersection intersección the graph of a compound inequality containing *and*; the solution is the set of elements common to both graphs

proportion proporción an equation stating that two ratios are equal

Proportion

÷ 6

$$\frac{24}{30} = \frac{4}{5}$$

÷ 6

LESSON 6-1 Graphing Systems of Equations

Then	Now	Why?
• You graphed linear equations.	**1** Determine the number of solutions a system of linear equations has. **2** Solve systems of linear equations by graphing.	• The cost to begin production on a band's CD is \$1500. Each CD costs \$4 to produce and will sell for \$10. The band wants to know how many CDs they will have to sell to earn a profit. Graphing a system can show when a company makes a profit. The cost of producing the CD can be modeled by the equation $y = 4x + 1500$, where y represents the cost of production and x is the number of CDs produced.

Band's CD Sales

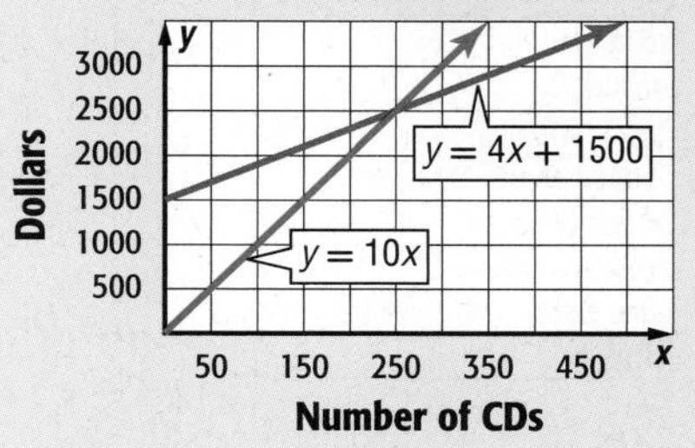

NewVocabulary
system of equations
consistent
independent
dependent
inconsistent

Common Core State Standards

Content Standards

A.CED.3 Represent constraints by equations or inequalities, and by systems of equations and/or inequalities, and interpret solutions as viable or nonviable options in a modeling context.

A.REI.6 Solve systems of linear equations exactly and approximately (e.g., with graphs), focusing on pairs of linear equations in two variables.

Mathematical Practices

3 Construct viable arguments and critique the reasoning of others.

8 Look for and express regularity in repeated reasoning.

1 Possible Number of Solutions

The income from the CDs sold can be modeled by the equation $y = 10x$, where y represents the total income of selling the CDs, and x is the number of CDs sold.

If we graph these equations, we can see at which point the band begins making a profit. The point where the two graphs intersect is where the band breaks even. This happens when the band sells 250 CDs. If the band sells more than 250 CDs, they will make a profit.

The two equations, $y = 4x + 1500$ and $y = 10x$, form a **system of equations**. The ordered pair that is a solution of both equations is the solution of the system. A system of two linear equations can have one solution, an infinite number of solutions, or no solution.

- If a system has at least one solution, it is said to be **consistent**. The graphs intersect at one point or are the same line.
- If a consistent system has exactly one solution, it is said to be **independent**. If it has an infinite number of solutions, it is **dependent**. This means that there are unlimited solutions that satisfy both equations.
- If a system has no solution, it is said to be **inconsistent**. The graphs are parallel.

ConceptSummary Possible Solutions

Number of Solutions	exactly one	infinite	no solution
Terminology	consistent and independent	consistent and dependent	inconsistent
Graph	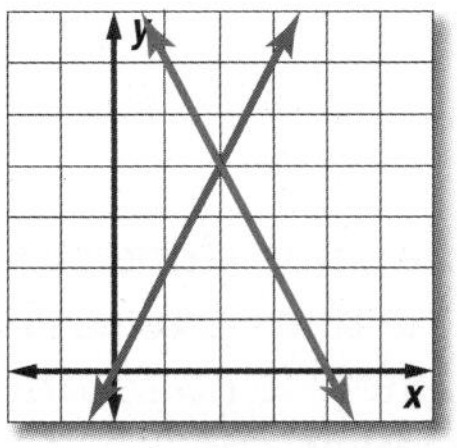	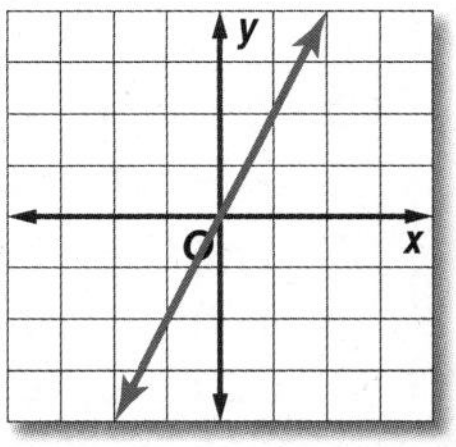	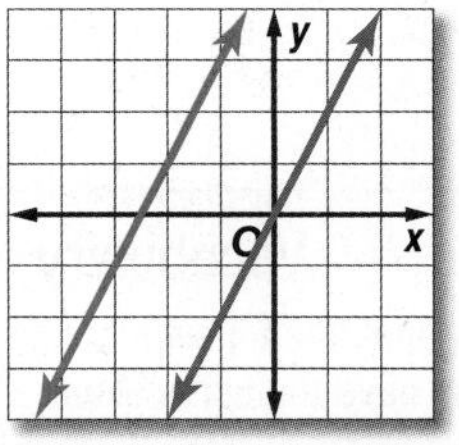

StudyTip

Number of Solutions When both equations are of the form $y = mx + b$, the values of m and b can determine the number of solutions.

Compare m and b	Number of Solutions
different m values	one
same m value, but different b values	none
same m value, and same b value	infinite

Example 1 Number of Solutions

Use the graph at the right to determine whether each system is *consistent* or *inconsistent* and if it is *independent* or *dependent*.

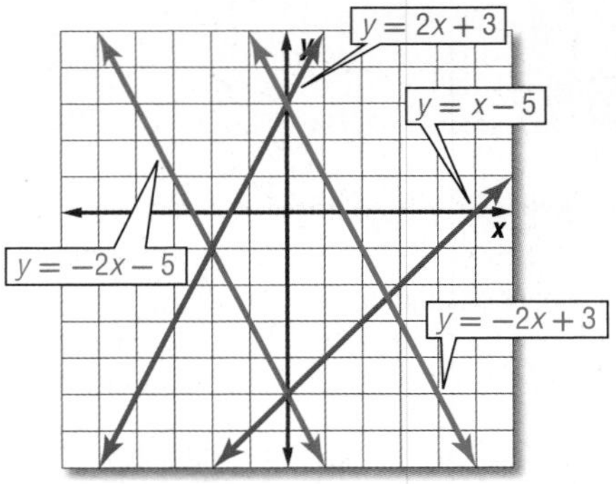

a. $y = -2x + 3$
$y = x - 5$

Since the graphs of these two lines intersect at one point, there is exactly one solution. Therefore, the system is consistent and independent.

b. $y = -2x - 5$
$y = -2x + 3$

Since the graphs of these two lines are parallel, there is no solution of the system. Therefore, the system is inconsistent.

GuidedPractice

1A. $y = 2x + 3$
$y = -2x - 5$

1B. $y = x - 5$
$y = -2x - 5$

2 Solve by Graphing

One method of solving a system of equations is to graph the equations carefully on the same coordinate grid and find their point of intersection. This point is the solution of the system.

Example 2 Solve by Graphing

Graph each system and determine the number of solutions that it has. If it has one solution, name it.

a. $y = -3x + 10$
$y = x - 2$

The graphs appear to intersect at the point (3, 1). You can check this by substituting 3 for x and 1 for y.

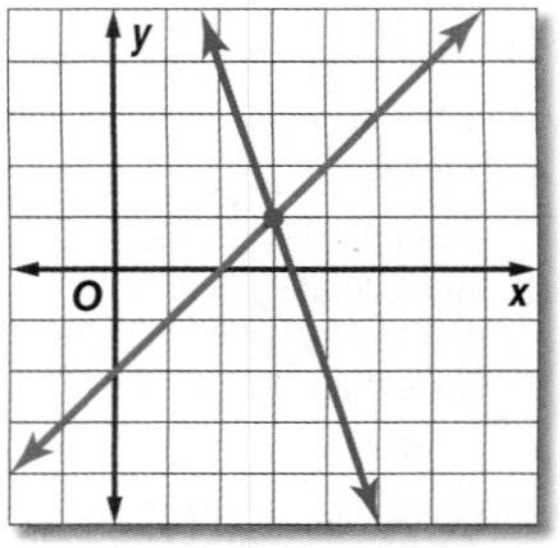

CHECK

$y = -3x + 10$	Original equation
$1 \stackrel{?}{=} -3(3) + 10$	Substitution
$1 \stackrel{?}{=} -9 + 10$	Multiply.
$1 = 1$ ✓	
$y = x - 2$	Original equation
$1 \stackrel{?}{=} 3 - 2$	Substitution
$1 = 1$ ✓	Multiply.

The solution is (3, 1).

b. $2x - y = -1$
$4x - 2y = 6$

The lines have the same slope but different y-intercepts, so the lines are parallel. Since they do not intersect, there is no solution of this system. The system is inconsistent.

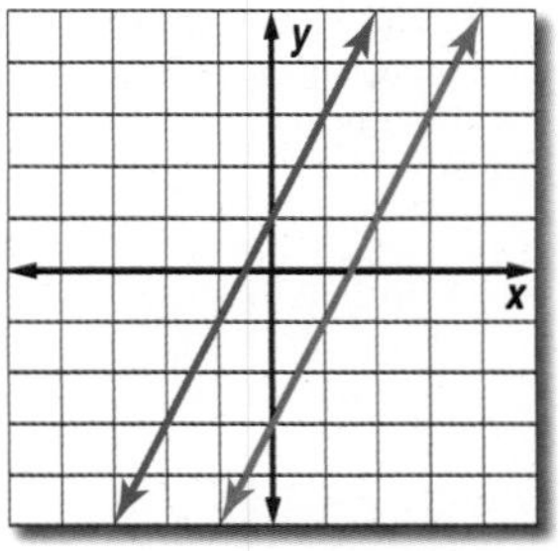

ReviewVocabulary

parallel lines never intersect and have the same slope

Guided Practice

Graph each system and determine the number of solutions that it has. If it has one solution, name it.

2A. $x - y = 2$
$3y + 2x = 9$

2B. $y = -2x - 3$
$6x + 3y = -9$

We can use what we know about systems of equations to solve many real-world problems involving constraints that are modeled by two or more different functions.

Real-WorldLink

In 2008, 3.1 million girls participated in high school sports. This was an all-time high for female participation.

Source: National Federation of State High School Associations

Real-World Example 3 Write and Solve a System of Equations

SPORTS The number of girls participating in high school soccer and track and field has steadily increased over the past few years. Use the information in the table to predict the approximate year when the number of girls participating in these two sports will be the same.

High School Sport	Number of Girls Participating in 2008 (thousands)	Average rate of increase (thousands per year)
soccer	345	8
track and field	458	3

Source: National Federation of State High School Associations

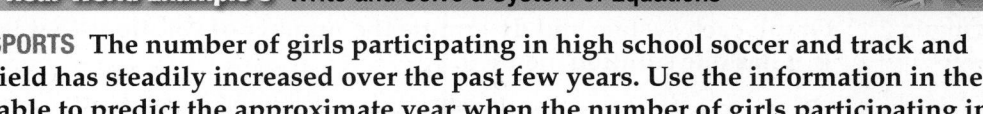

Words	Number of girls participating	equals	rate of increase	times	number of years after 2008	plus	number participating in 2008.

Variables: Let y = number of girls competing. Let x = number of years after 2008.

Equations:

Soccer:	y	=	8	•	x	+	345
Track and field:	y	=	3	•	x	+	458

Graph $y = 8x + 345$ and $y = 3x + 458$. The graphs appear to intersect at approximately (22.5, 525).

CHECK Use substitution to check this answer.

$y = 8x + 345$ $\qquad$ $y = 3x + 458$

$525 \stackrel{?}{=} 8(22.5) + 345$ $\qquad$ $525 \stackrel{?}{=} 3(22.5) + 458$

$525 = 525$ ✓ $\qquad$ $525 \approx 525.5$ ✓

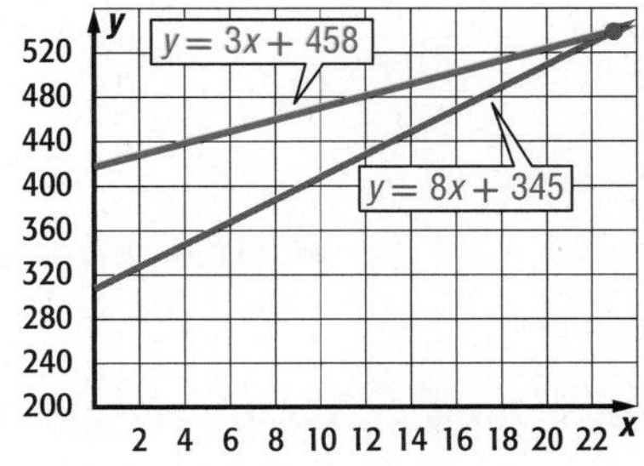

The solution means that approximately 22 years after 2008, or in 2030, the number of girls participating in high school soccer and track and field will be the same, about 525,000.

Guided Practice

3. VIDEO GAMES Joe and Josh each want to buy a video game. Joe has \$14 and saves \$10 a week. Josh has \$26 and saves \$7 a week. In how many weeks will they have the same amount?

Check Your Understanding

Example 1 **Use the graph at the right to determine whether each system is *consistent* or *inconsistent* and if it is *independent* or *dependent*.**

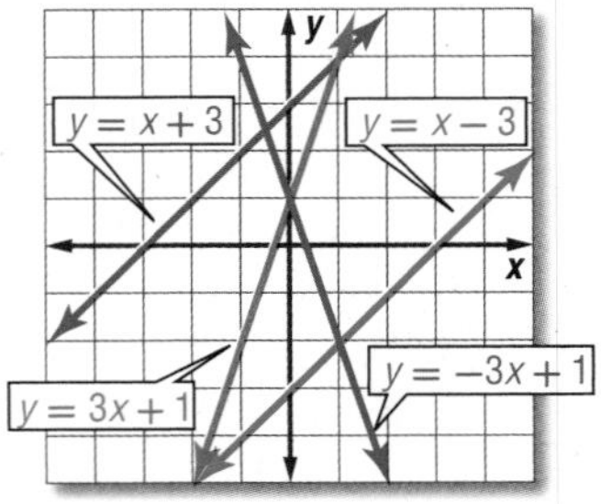

1. $y = -3x + 1$
$y = 3x + 1$

2. $y = 3x + 1$
$y = x - 3$

3. $y = x - 3$
$y = x + 3$

4. $y = x + 3$
$x - y = -3$

5. $x - y = -3$
$y = -3x + 1$

6. $y = -3x + 1$
$y = x - 3$

Example 2 **Graph each system and determine the number of solutions that it has. If it has one solution, name it.**

7. $y = x + 4$
$y = -x - 4$

8. $y = x + 3$
$y = 2x + 4$

Example 3

9. **CCSS MODELING** Alberto and Ashanti are reading a graphic novel.

a. Write an equation to represent the pages each boy has read.

b. Graph each equation.

c. How long will it be before Alberto has read more pages than Ashanti? Check and interpret your solution.

Practice and Problem Solving

Example 1 **Use the graph at the right to determine whether each system is *consistent* or *inconsistent* and if it is *independent* or *dependent*.**

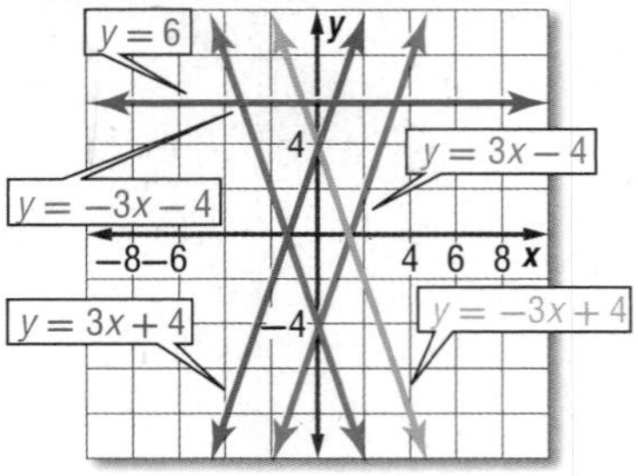

10. $y = 6$
$y = 3x + 4$

11. $y = 3x + 4$
$y = -3x + 4$

12. $y = -3x + 4$
$y = -3x - 4$

13. $y = -3x - 4$
$y = 3x - 4$

14. $3x - y = -4$
$y = 3x + 4$

15. $3x - y = 4$
$3x + y = 4$

Example 2 **Graph each system and determine the number of solutions that it has. If it has one solution, name it.**

16. $y = -3$
$y = x - 3$

17. $y = 4x + 2$
$y = -2x - 3$

18. $y = x - 6$
$y = x + 2$

19. $x + y = 4$
$3x + 3y = 12$

20. $x - y = -2$
$-x + y = 2$

21. $x + 2y = 3$
$x = 5$

22. $2x + 3y = 12$
$2x - y = 4$

23. $2x + y = -4$
$y + 2x = 3$

24. $2x + 2y = 6$
$5y + 5x = 15$

Example 3

25. **SCHOOL DANCE** Akira and Jen are competing to see who can sell the most tickets for the Winter Dance. On Monday, Akira sold 22 and then sold 30 per day after that. Jen sold 53 on Monday and then sold 20 per day after that.

a. Write equations for the number of tickets each person has sold.

b. Graph each equation.

c. Solve the system of equations. Check and interpret your solution.

26. **CCSS MODELING** If x is the number of years since 2000 and y is the percent of people using travel services, the following equations represent the percent of people using travel agents and the percent of people using the Internet to plan travel.

Travel agents: $y = -2x + 30$ Internet: $y = 6x + 41$

a. Graph the system of equations.

b. Estimate the year travel agents and the Internet were used equally.

Graph each system and determine the number of solutions that it has. If it has one solution, name it.

27. $y = \frac{1}{2}x$, $y = x + 2$

28. $y = 6x + 6$, $y = 3x + 6$

29. $y = 2x - 17$, $y = x - 10$

30. $8x - 4y = 16$, $-5x - 5y = 5$

31. $3x + 5y = 30$, $3x + y = 18$

32. $-3x + 4y = 24$, $4x - y = 7$

33. $2x - 8y = 6$, $x - 4y = 3$

34. $4x - 6y = 12$, $-2x + 3y = -6$

35. $2x + 3y = 10$, $4x + 6y = 12$

36. $3x + 2y = 10$, $2x + 3y = 10$

37. $3y - x = -2$, $y - \frac{1}{3}x = 2$

38. $\frac{8}{5}y = \frac{2}{5}x + 1$, $\frac{2}{5}y = \frac{1}{10}x + \frac{1}{4}$

39. $\frac{1}{3}x + \frac{1}{3}y = 1$, $x + y = 1$

40. $\frac{3}{4}x + \frac{1}{2}y = \frac{1}{4}$, $\frac{2}{3}x + \frac{1}{6}y = \frac{1}{2}$

41. $\frac{5}{6}x + \frac{2}{3}y = \frac{1}{2}$, $\frac{2}{5}x + \frac{1}{5}y = \frac{3}{5}$

42. **PHOTOGRAPHY** Suppose x represents the number of cameras sold and y represents the number of years since 2000. Then the number of digital cameras sold each year since 2000, in millions, can be modeled by the equation $y = 12.5x + 10.9$. The number of film cameras sold each year since 2000, in millions, can be modeled by the equation $y = -9.1x + 78.8$.

a. Graph each equation.

b. In which year did digital camera sales surpass film camera sales?

c. In what year did film cameras stop selling altogether?

d. What are the domain and range of each of the functions in this situation?

Graph each system and determine the number of solutions that it has. If it has one solution, name it.

43. $2y = 1.2x - 10$, $4y = 2.4x$

44. $x = 6 - \frac{3}{8}y$, $4 = \frac{2}{3}x + \frac{1}{4}y$

45 **WEB SITES** Personal publishing site *Lookatme* had 2.5 million visitors in 2009. Each year after that, the number of visitors rose by 13.1 million. Online auction site *Buyourstuff* had 59 million visitors in 2009, but each year after that the number of visitors fell by 2 million.

a. Write an equation for each of the companies.

b. Make a table of values for 5 years for each of the companies.

c. Graph each equation.

d. When will *Lookatme* and *Buyourstuff's* sites have the same number of visitors?

e. Name the domain and range of these functions in this situation.

46. **MULTIPLE REPRESENTATIONS** In this problem, you will explore different methods for finding the intersection of the graphs of two linear equations.

a. Algebraic Use algebra to solve the equation $\frac{1}{2}x + 3 = -x + 12$.

b. Graphical Use a graph to solve $y = \frac{1}{2}x + 3$ and $y = -x + 12$.

c. Analytical How is the equation in part **a** related to the system in part **b**?

d. Verbal Explain how to use the graph in part **b** to solve the equation in part **a**.

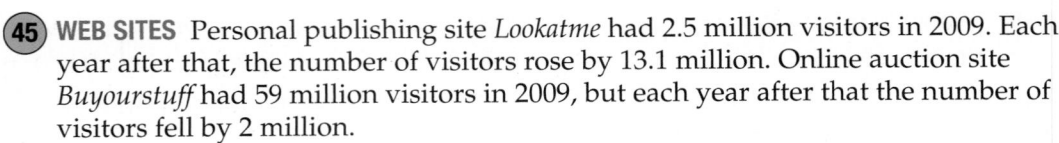

H.O.T. Problems Use Higher-Order Thinking Skills

47. **ERROR ANALYSIS** Store A is offering a 10% discount on the purchase of all electronics in their store. Store B is offering $10 off all the electronics in their store. Francisca and Alan are deciding which offer will save them more money. Is either of them correct? Explain your reasoning.

Francisca	Alan
You can't determine which store has the better offer unless you know the price of the items you want to buy.	*Store A has the better offer because 10% of the sale price is a greater discount than $10.*

48. **CHALLENGE** Use graphing to find the solution of the system of equations $2x + 3y = 5$, $3x + 4y = 6$, and $4x + 5y = 7$.

49. **CCSS ARGUMENTS** Determine whether a system of two linear equations with (0, 0) and (2, 2) as solutions *sometimes, always,* or *never* has other solutions. Explain.

50. **WHICH ONE DOESN'T BELONG?** Which one of the following systems of equations doesn't belong with the other three? Explain your reasoning.

$4x - y = 5$ $-2x + y = -1$	$-x + 4y = 8$ $3x - 6y = 6$	$4x + 2y = 14$ $12x + 6y = 18$	$3x - 2y = 1$ $2x + 3y = 18$

51. **OPEN ENDED** Write three equations such that they form three systems of equations with $y = 5x - 3$. The three systems should be inconsistent, consistent and independent, and consistent and dependent, respectively.

52. **WRITING IN MATH** Describe the advantages and disadvantages to solving systems of equations by graphing.

Standardized Test Practice

53. SHORT RESPONSE Certain bacteria can reproduce every 20 minutes, doubling the population. If there are 450,000 bacteria in a population at 9:00 A.M., how many bacteria will be in the population at 2:00 P.M.?

54. GEOMETRY An 84-centimeter piece of wire is cut into equal segments and then attached at the ends to form the edges of a cube. What is the volume of the cube?

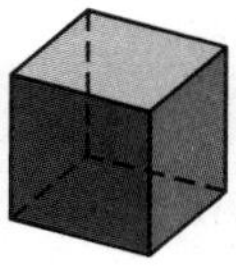

A 294 cm^3
B 343 cm^3
C 1158 cm^3
D 2744 cm^3

55. What is the solution of the inequality $-9 < 2x + 3 < 15$?

F $-x \geq 0$
G $x \leq 0$
H $-6 < x < 6$
J $-5 < x < 5$

56. What is the solution of the system of equations?

$$x + 2y = -1$$
$$2x + 4y = -2$$

A $(-1, -1)$
B $(2, 1)$
C no solution
D infinitely many solutions

Spiral Review

Graph each inequality. (Lesson 5-6)

57. $3x + 6y > 0$

58. $4x - 2y < 0$

59. $3y - x \leq 9$

60. $4y - 3x \geq 12$

61. $y < -4x - 8$

62. $3x - 1 > y$

63. LIBRARY To get a grant from the city's historical society, the number of history books must be within 25 of 1500. What is the range of the number of historical books that must be in the library? (Lesson 5-5)

64. SCHOOL Camilla's scores on three math tests are shown in the table. The fourth and final test of the grading period is tomorrow. She needs an average of at least 92 to receive an A for the grading period. (Lesson 5-3)

Test	Score
1	91
2	95
3	88

a. If m represents her score on the fourth math test, write an inequality to represent this situation.

b. If Camilla wants an A in math, what must she score on the test?

c. Is your solution reasonable? Explain.

Write the slope-intercept form of an equation for the line that passes through the given point and is perpendicular to the graph of the equation. (Lesson 4-4)

65. $(-3, 1), y = \frac{1}{3}x + 2$

66. $(6, -2), y = \frac{3}{5}x - 4$

67. $(2, -2), 2x + y = 5$

68. $(-3, -3), -3x + y = 6$

Skills Review

Find the solution of each equation using the given replacement set.

69. $f - 14 = 8$; $\{12, 15, 19, 22\}$

70. $15(n + 6) = 165$; $\{3, 4, 5, 6, 7\}$

71. $23 = \frac{d}{4}$; $\{91, 92, 93, 94, 95\}$

72. $36 = \frac{t - 9}{2}$; $\{78, 79, 80, 81\}$

Evaluate each expression if $a = 2$, $b = -3$, and $c = 11$.

73. $a + 6b$

74. $7 - ab$

75. $(2c + 3a) \div 4$

76. $b^2 + (a^3 - 8)5$

EXTEND

6-1 Graphing Technology Lab
Systems of Equations

You can use a graphing calculator to graph and solve a system of equations.

Activity 1 Solve a System of Equations

Solve the system of equations. State the decimal solution to the nearest hundredth.

$5.23x + y = 7.48$
$6.42x - y = 2.11$

Common Core State Standards
Content Standards
A.REI.6 Solve systems of linear equations exactly and approximately (e.g., with graphs), focusing on pairs of linear equations in two variables.
A.REI.11 Explain why the x-coordinates of the points where the graphs of the equations $y = f(x)$ and $y = g(x)$ intersect are the solutions of the equation $f(x) = g(x)$; find the solutions approximately, e.g., using technology to graph the functions, make tables of values, or find successive approximations. Include cases where $f(x)$ and/or $g(x)$ are linear, polynomial, rational, absolute value, exponential, and logarithmic functions.
Mathematical Practices
5 Use appropriate tools strategically.

Step 1 Solve each equation for y to enter them into the calculator.

$5.23x + y = 7.48$	First equation
$5.23x + y - 5.23x = 7.48 - 5.23x$	Subtract $5.23x$ from each side.
$y = 7.48 - 5.23x$	Simplify.
$6.42x - y = 2.11$	Second equation
$6.42x - y - 6.42x = 2.11 - 6.42x$	Subtract $6.42x$ from each side.
$-y = 2.11 - 6.42x$	Simplify.
$(-1)(-y) = (-1)(2.11 - 6.42x)$	Multiply each side by −1.
$y = -2.11 + 6.42x$	Simplify.

Step 2 Enter these equations in the **Y=** list and graph in the standard viewing window.

KEYSTROKES: [Y=] 7.48 [−] 5.23 [X,T,θ,n] [ENTER] [(−)] 2.11 [+] 6.42 [X,T,θ,n] [ZOOM] 6

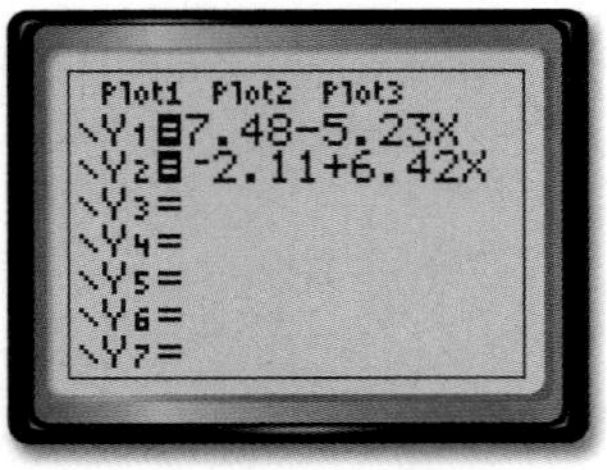

Step 3 Use the **CALC** menu to find the point of intersection.

KEYSTROKES: [2nd] **[CALC]** 5 [ENTER] [ENTER] [ENTER]

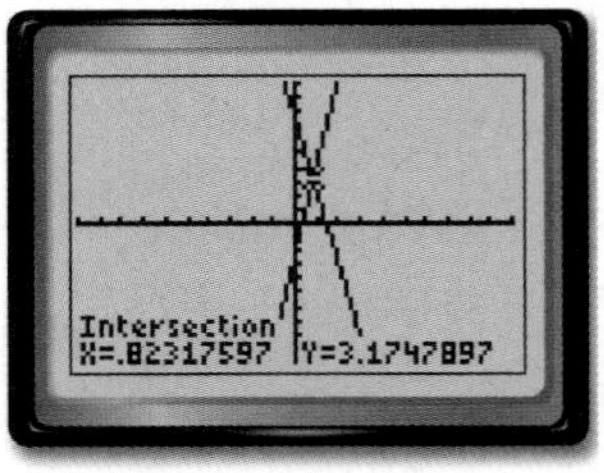

[−10, 10] scl: 1 by [−10, 10] scl: 1

The solution is approximately (0.82, 3.17).

When you solve a system of equations with $y = f(x)$ and $y = g(x)$, the solution is an ordered pair that satisfies both equations. The solution always occurs when $f(x) = g(x)$. Thus, the x-coordinate of the solution is the value of x where $f(x) = g(x)$.

One method you can use to solve an equation with one variable is by graphing and solving a system of equations based on the equation. To do this, write a system using both sides of the equation. Then use a graphing calculator to solve the system.

Activity 2 Use a System to Solve a Linear Equation

Use a system of equations to solve $5x + 6 = -4$.

Step 1 Write a system of equations. Set each side of the equation equal to y.

$y = 5x + 6$ First equation
$y = -4$ Second equation

Step 2 Enter these equations in the **Y=** list and graph.

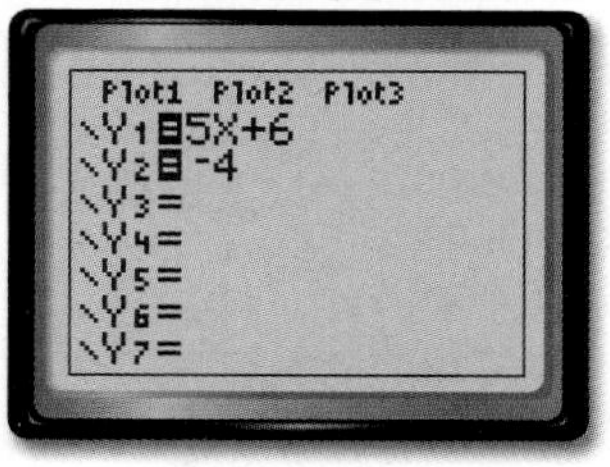

Step 3 Use the **CALC** menu to find the point of intersection.

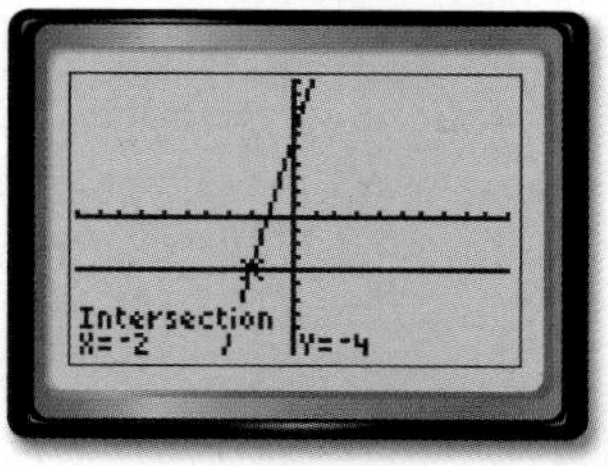

[−10, 10] scl: 1 by [−10, 10] scl: 1

The solution is −2.

Exercises

Use a graphing calculator to solve each system of equations. Write decimal solutions to the nearest hundredth.

1. $y = 2x - 3$
$y = -0.4x + 5$

2. $y = 6x + 1$
$y = -3.2x - 4$

3. $x + y = 9.35$
$5x - y = 8.75$

4. $2.32x - y = 6.12$
$4.5x + y = -6.05$

5. $5.2x - y = 4.1$
$1.5x + y = 6.7$

6. $1.8 = 5.4x - y$
$y = -3.8 - 6.2x$

7. $7x - 2y = 16$
$11x + 6y = 32.3$

8. $3x + 2y = 16$
$5x + y = 9$

9. $0.62x + 0.35y = 1.60$
$-1.38x + y = 8.24$

10. $75x - 100y = 400$
$33x - 10y = 70$

Use a graphing calculator to solve each equation. Write decimal solutions to the nearest hundredth.

11. $4x - 2 = -6$

12. $3 = 1 + \frac{x}{2}$

13. $\frac{x + 4}{-2} = -1$

14. $\frac{3}{2}x + \frac{1}{2} = 2x - 3$

15. $4x - 9 = 7 + 7x$

16. $-2 + 10x = 8x - 1$

17. **WRITING IN MATH** Explain why you can solve an equation like $r = ax + b$ by solving the system of equations $y = r$ and $y = ax + b$.

LESSON 6-2 Substitution

:: Then	:: Now	:: Why?
You solved systems of equations by graphing.	**1** Solve systems of equations by using substitution. **2** Solve real-world problems involving systems of equations by using substitution.	Two movies were released at the same time. Movie A earned \$31 million in its opening week, but fell to \$15 million the following week. Movie B opened earning \$21 million and fell to \$11 million the following week. If the earnings for each movie continue to decrease at the same rate, when will they earn the same amount?

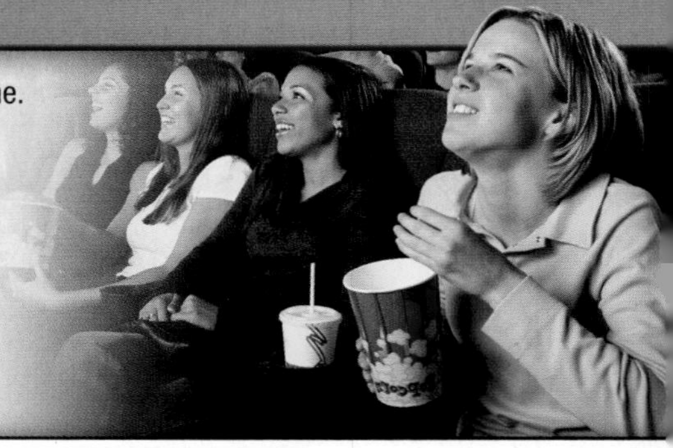

NewVocabulary
substitution

Common Core State Standards

Content Standards
A.CED.3 Represent constraints by equations or inequalities, and by systems of equations and/or inequalities, and interpret solutions as viable or nonviable options in a modeling context.
A.REI.6 Solve systems of linear equations exactly and approximately (e.g., with graphs), focusing on pairs of linear equations in two variables.

Mathematical Practices
2 Reason abstractly and quantitatively.

1 Solve by Substitution

You can use a system of equations to find when the movie earnings are the same. One method of finding an exact solution of a system of equations is called **substitution**.

KeyConcept Solving by Substitution

Step 1 When necessary, solve at least one equation for one variable.

Step 2 Substitute the resulting expression from Step 1 into the other equation to replace the variable. Then solve the equation.

Step 3 Substitute the value from Step 2 into either equation, and solve for the other variable. Write the solution as an ordered pair.

Example 1 Solve a System by Substitution

Use substitution to solve the system of equations.

$y = 2x + 1$ ← **Step 1** The first equation is already solved for y.
$3x + y = -9$

Step 2 Substitute $2x + 1$ for y in the second equation.

$3x + y = -9$	Second equation
$3x + 2x + 1 = -9$	Substitute $2x + 1$ for y.
$5x + 1 = -9$	Combine like terms.
$5x = -10$	Subtract 1 from each side.
$x = -2$	Divide each side by 5.

Step 3 Substitute -2 for x in either equation to find y.

$y = 2x + 1$	First equation
$= 2(-2) + 1$	Substitute -2 for x.
$= -3$	Simplify.

The solution is $(-2, -3)$.

CHECK You can check your solution by graphing.

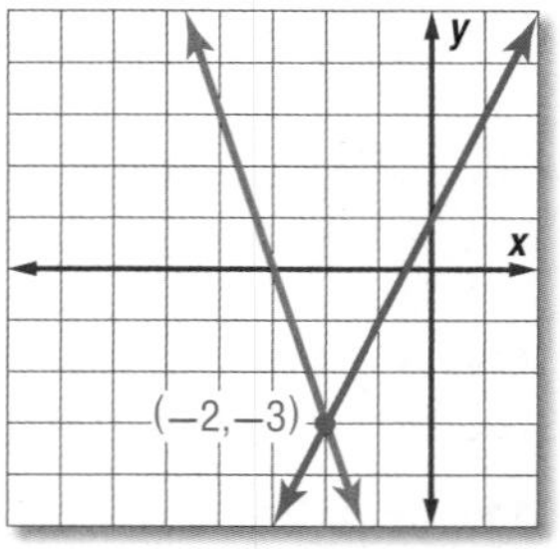

GuidedPractice

1A. $y = 4x - 6$
$5x + 3y = -1$

1B. $2x + 5y = -1$
$y = 3x + 10$

John Eder/Stone/Getty Images

StudyTip

Slope-Intercept Form If both equations are in the form $y = mx + b$, they can simply be set equal to each other and then solved for x. The solution for x can then be used to find the value of y.

If a variable is not isolated in one of the equations in a system, solve an equation for a variable first. Then you can use substitution to solve the system.

Example 2 Solve and then Substitute

PT

Use substitution to solve the system of equations.

$$x + 2y = 6$$
$$3x - 4y = 28$$

Step 1 Solve the first equation for x since the coefficient is 1.

$x + 2y = 6$	First equation
$x + 2y - 2y = 6 - 2y$	Subtract $2y$ from each side.
$x = 6 - 2y$	Simplify.

Step 2 Substitute $6 - 2y$ for x in the second equation to find the value of y.

$3x - 4y = 28$	Second equation
$3(6 - 2y) - 4y = 28$	Substitute $6 - 2y$ for x.
$18 - 6y - 4y = 28$	Distributive Property
$18 - 10y = 28$	Combine like terms.
$18 - 10y - 18 = 28 - 18$	Subtract 18 from each side.
$-10y = 10$	Simplify.
$y = -1$	Divide each side by -10.

Step 3 Find the value of x.

$x + 2y = 6$	First equation
$x + 2(-1) = 6$	Substitute -1 for y.
$x - 2 = 6$	Simplify.
$x = 8$	Add 2 to each side.

GuidedPractice

2A. $4x + 5y = 11$
$y - 3x = -13$

2B. $x - 3y = -9$
$5x - 2y = 7$

Generally, if you solve a system of equations and the result is a false statement such as $3 = -2$, there is no solution. If the result is an identity, such as $3 = 3$, then there are an infinite number of solutions.

Example 3 No Solution or Infinitely Many Solutions

PT

Use substitution to solve the system of equations.

$$y = 2x - 4$$
$$-6x + 3y = -12$$

Substitute $2x - 4$ for y in the second equation.

$-6x + 3y = -12$	Second equation
$-6x + 3(2x - 4) = -12$	Substitute $2x - 4$ for y.
$-6x + 6x - 12 = -12$	Distributive Property
$-12 = -12$	Combine like terms.

This statement is an identity. Thus, there are an infinite number of solutions.

StudyTip

Dependent Systems There are infinitely many solutions of the system in Example 3 because the equations in slope-intercept form are equivalent, and they have the same graph.

GuidedPractice **Use substitution to solve each system of equations.**

3A. $2x - y = 8$
$y = 2x - 3$

3B. $4x - 3y = 1$
$6y - 8x = -2$

2 Solve Real-World Problems

You can use substitution to find the solution of a real-world problem involving constraints modeled by a system of equations.

Real-WorldLink

Sound Engineering Technician Sound engineering technicians record, synchronize, mix, and reproduce music, voices, and sound effects in recording studios, sporting arenas, and theater, movie, or video productions. They need to have at least a 2-year associate's degree in electronics.

Real-World Example 4 Write and Solve a System of Equations

PT

MUSIC A store sold a total of 125 car stereo systems and speakers in one week. The stereo systems sold for $104.95, and the speakers sold for $18.95. The sales from these two items totaled $6926.75. How many of each item were sold?

Number of Units Sold	c	t	125
Sales ($)	$104.95c$	$18.95t$	6926.75

Let c = the number of car stereo systems sold, and let t = the number of speakers sold.

So, the two equations are $c + t = 125$ and $104.95c + 18.95t = 6926.75$.

Notice that $c + t = 125$ represents combinations of car stereo systems and speakers with a sum of 125. The equation $104.95c + 18.95t = 6926.75$ represents the combinations of car stereo systems and speakers with a sales of $6926.75. The solution of the system of equations represents the option that meets both of the constraints.

Step 1 Solve the first equation for c.

$c + t = 125$ First equation

$c + t - t = 125 - t$ Subtract t from each side.

$c = 125 - t$ Simplify.

Step 2 Substitute $125 - t$ for c in the second equation.

$104.95c + 18.95t = 6926.75$ Second equation

$104.95(125 - t) + 18.95t = 6926.75$ Substitute $125 - t$ for c.

$13{,}118.75 - 104.95t + 18.95t = 6926.75$ Distributive Property

$13{,}118.75 - 86t = 6926.75$ Combine like terms.

$-86t = -6192$ Subtract 13118.75 from each side.

$t = 72$ Divide each side by −86.

Step 3 Substitute 72 for t in either equation to find the value of c.

$c + t = 125$ First equation

$c + 72 = 125$ Substitute 72 for t.

$c = 53$ Subtract 72 from each side.

The store sold 53 car stereo systems and 72 speakers.

GuidedPractice

4. BASEBALL As of 2009, the New York Yankees and the Cincinnati Reds together had won a total of 32 World Series. The Yankees had won 5.4 times as many as the Reds. How many World Series had each team won?

Jessica Peterson/RubberBall/age fotostock

Check Your Understanding

Examples 1–3 **Use substitution to solve each system of equations.**

1. $y = x + 5$
   $3x + y = 25$

2. $x = y - 2$
   $4x + y = 2$

3. $3x + y = 6$
   $4x + 2y = 8$

4. $2x + 3y = 4$
   $4x + 6y = 9$

5. $x - y = 1$
   $3x = 3y + 3$

6. $2x - y = 6$
   $-3y = -6x + 18$

Example 4

7. **GEOMETRY** The sum of the measures of angles X and Y is 180°. The measure of angle X is 24° greater than the measure of angle Y.
   a. Define the variables, and write equations for this situation.
   b. Find the measure of each angle.

Practice and Problem Solving

Examples 1–3 **Use substitution to solve each system of equations.**

8. $y = 5x + 1$
   $4x + y = 10$

9. $y = 4x + 5$
   $2x + y = 17$

10. $y = 3x - 34$
    $y = 2x - 5$

11. $y = 3x - 2$
    $y = 2x - 5$

12. $2x + y = 3$
    $4x + 4y = 8$

13. $3x + 4y = -3$
    $x + 2y = -1$

14. $y = -3x + 4$
    $-6x - 2y = -8$

15. $-1 = 2x - y$
    $8x - 4y = -4$

16. $x = y - 1$
    $-x + y = -1$

17. $y = -4x + 11$
    $3x + y = 9$

18. $y = -3x + 1$
    $2x + y = 1$

19. $3x + y = -5$
    $6x + 2y = 10$

20. $5x - y = 5$
    $-x + 3y = 13$

21. $2x + y = 4$
    $-2x + y = -4$

22. $-5x + 4y = 20$
    $10x - 8y = -40$

Example 4

23. **ECONOMICS** In 2000, the demand for nurses was 2,000,000, while the supply was only 1,890,000. The projected demand for nurses in 2020 is 2,810,414, while the supply is only projected to be 2,001,998.
    a. Define the variables, and write equations to represent these situations.
    b. Use substitution to determine during which year the supply of nurses was equal to the demand.

24. **CCSS REASONING** The table shows the approximate number of tourists in two areas of the world during a recent year and the average rates of change in tourism.

Destination	Number of Tourists	Average Rates of Change in Tourists (millions per year)
South America and the Caribbean	40.3 million	increase of 0.8
Middle East	17.0 million	increase of 1.8

    a. Define the variables, and write an equation for each region's tourism rate.
    b. If the trends continue, in how many years would you expect the number of tourists in the regions to be equal?

25. **SPORTS** The table shows the winning times for the Triathlon World Championship.

Year	Men's	Women's
2000	1:51:39	1:54:43
2009	1:44:51	1:59:14

a. The times are in hours, minutes, and seconds. Rewrite the times rounded to the nearest minute.

b. Let the year 2000 be 0. Assume that the rate of change remains the same for years after 2000. Write an equation to represent each of the men's and women's winning times y in any year x.

c. If the trend continues, when would you expect the men's and women's winning times to be the same? Explain your reasoning.

26. **CONCERT TICKETS** Booker is buying tickets online for a concert. He finds tickets for himself and his friends for $65 each plus a one-time fee of $10. Paula is looking for tickets to the same concert. She finds them at another Web site for $69 and a one-time fee of $13.60.

a. Define the variables, and write equations to represent this situation.

b. Create a table of values for 1 to 5 tickets for each person's purchase.

c. Graph each of these equations.

d. Use the graph to determine who received the better deal. Explain why.

H.O.T. Problems Use Higher-Order Thinking Skills

27. **ERROR ANALYSIS** In the system $a + b = 7$ and $1.29a + 0.49b = 6.63$, a represents pounds of apples and b represents pounds of bananas. Guillermo and Cara are finding and interpreting the solution. Is either of them correct? Explain.

Guillermo

$$1.29a + 0.49b = 6.63$$
$$1.29a + 0.49(a + 7) = 6.63$$
$$1.29 + 0.49a + 3.43 = 6.63$$
$$0.49a = 3.2$$
$$a = 1.9$$

$a + b = 7$, so $b = 5$. The solution (2, 5) means that 2 pounds of apples and 5 pounds of bananas were bought.

Cara

$$1.29a + 0.49b = 6.63$$
$$1.29(7 - b) + 0.49b = 6.63$$
$$9.03 - 1.29b + 0.49b = 6.63$$
$$-0.8b = -2.4$$
$$b = 3$$

The solution $b = 3$ means that 3 pounds of apples and 3 pounds of bananas were bought.

28. CCSS **PERSEVERANCE** A local charity has 60 volunteers. The ratio of boys to girls is 7:5. Find the number of boy and the number of girl volunteers.

29. **REASONING** Compare and contrast the solution of a system found by graphing and the solution of the same system found by substitution.

30. **OPEN ENDED** Create a system of equations that has one solution. Illustrate how the system could represent a real-world situation and describe the significance of the solution in the context of the situation.

31. **WRITING IN MATH** Explain how to determine what to substitute when using the substitution method of solving systems of equations.

Standardized Test Practice

32. The debate team plans to make and sell trail mix. They can spend \$34.

Item	Cost Per Pound
sunflower seeds	\$4.00
raisins	\$1.50

The pounds of raisins in the mix are to be 3 times the pounds of sunflower seeds. Which system can be used to find r, the pounds of raisins, and p, pounds of sunflower seeds, they should buy?

A $3p = r$
$4p + 1.5r = 34$

B $3p = r$
$4r + 1.5p = 34$

C $3r = p$
$4p + 1.5r = 34$

D $3r = p$
$4r + 1.5p = 34$

33. GRIDDED RESPONSE The perimeters of two similar polygons are 250 centimeters and 300 centimeters, respectively. What is the scale factor between the first and second polygons?

34. Based on the graph, which statement is true?

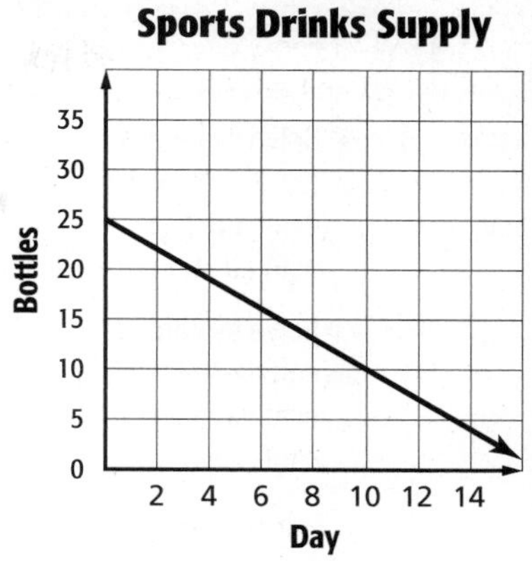

F Mary started with 30 bottles.
G On day 10, Mary will have 10 bottles left.
H Mary will be out of sports drinks on day 14.
J Mary drank 5 bottles the first two days.

35. If p is an integer, which of the following is the solution set for $2|p| = 16$?

A $\{0, 8\}$
B $\{-8, 0\}$
C $\{-8, 8\}$
D $\{-8, 0, 8\}$

Spiral Review

Graph each system and determine how many solutions it has. If it has one solution, name it. (Lesson 6-1)

36. $y = -5$
$3x + y = 1$

37. $x = 1$
$2x - y = 7$

38. $y = x + 5$
$y = x - 2$

39. $x + y = 1$
$3y + 3x = 3$

40. ENTERTAINMENT Coach Ross wants to take the soccer team out for pizza after their game. Her budget is at most \$70. (Lesson 5-6)

a. Using the sign, write an inequality that represents this situation.

b. Are there any restrictions on the variables? Explain.

Solve each inequality. Check your solution. (Lesson 5-3)

41. $6v + 1 \geq -11$

42. $24 > 18 + 2n$

43. $-11 \geq \frac{2}{5}q + 5$

44. $\frac{a}{8} - 10 > -3$

45. $-3t + 9 \leq 0$

46. $54 > -10 - 8n$

Skills Review

Rewrite each product using the Distributive Property. Then simplify.

47. $10b + 5(3 + 9b)$

48. $5(3t^2 + 4) - 8t$

49. $7h^2 + 4(3h + h^2)$

50. $-2(7a + 5b) + 5(2a - 7b)$

LESSON 6-3

Elimination Using Addition and Subtraction

Then

- You solved systems of equations by using substitution.

Now

1. Solve systems of equations by using elimination with addition.
2. Solve systems of equations by using elimination with subtraction.

Why?

- In Chicago, Illinois, there are two more months *a* when the mean high temperature is below 70°F than there are months *b* when it is above 70°F. The system of equations, $a + b = 12$ and $a - b = 2$, represents this situation.

NewVocabulary
elimination

Common Core State Standards

Content Standards
A.CED.2 Create equations in two or more variables to represent relationships between quantities; graph equations on coordinate axes with labels and scales.
A.REI.6 Solve systems of linear equations exactly and approximately (e.g., with graphs), focusing on pairs of linear equations in two variables.

Mathematical Practices
7 Look for and make use of structure.

1 Elimination Using Addition

If you add these equations, the variable b will be eliminated. Using addition or subtraction to solve a system is called **elimination**.

KeyConcept Solving by Elimination

Step 1 Write the system so like terms with the same or opposite coefficients are aligned.

Step 2 Add or subtract the equations, eliminating one variable. Then solve the equation.

Step 3 Substitute the value from Step 2 into one of the equations and solve for the other variable. Write the solution as an ordered pair.

PT

Example 1 Elimination Using Addition

Use elimination to solve the system of equations.

$4x + 6y = 32$
$3x - 6y = 3$ ← **Step 1** $6y$ and $-6y$ have opposite coefficients.

Step 2 Add the equations.

$$\begin{aligned} 4x + 6y &= 32 \\ (+)\ 3x - 6y &= 3 \\ \hline 7x \qquad &= 35 && \text{The variable } y \text{ is eliminated.} \\ \frac{7x}{7} &= \frac{35}{7} && \text{Divide each side by 7.} \\ x &= 5 && \text{Simplify.} \end{aligned}$$

Step 3 Substitute 5 for x in either equation to find the value of y.

$$\begin{aligned} 4x + 6y &= 32 && \text{First equation} \\ 4(5) + 6y &= 32 && \text{Replace } x \text{ with 5.} \\ 20 + 6y &= 32 && \text{Multiply.} \\ 20 + 6y - 20 &= 32 - 20 && \text{Subtract 20 from each side.} \\ 6y &= 12 && \text{Simplify.} \\ \frac{6y}{6} &= \frac{12}{6} && \text{Divide each side by 6.} \\ y &= 2 && \text{Simplify.} \end{aligned}$$

The solution is (5, 2).

Guided Practice

1A. $-4x + 3y = -3$
$4x - 5y = 5$

1B. $4y + 3x = 22$
$3x - 4y = 14$

We can use elimination to find specific numbers that are described as being related to each other.

Example 2 Write and Solve a System of Equations

Negative three times one number plus five times another number is −11. Three times the first number plus seven times the other number is −1. Find the numbers.

Negative three times one number	plus	five times another number	is	−11.
$-3x$	$+$	$5y$	$=$	-11
Three times the first number	plus	seven times the other number	is	−1.
$3x$	$+$	$7y$	$=$	-1

Steps 1 and 2 Write the equations vertically and add.

$$\begin{aligned} -3x + 5y &= -11 \\ (+)\ 3x + 7y &= -1 \\ \hline 12y &= -12 \end{aligned}$$ The variable x is eliminated.

$$\frac{12y}{12} = \frac{-12}{12}$$ Divide each side by 12.

$$y = -1$$ Simplify.

Step 3 Substitute -1 for y in either equation to find the value of x.

$3x + 7y = -1$ Second equation

$3x + 7(-1) = -1$ Replace y with -1.

$3x + (-7) = -1$ Simplify.

$3x + (-7) + 7 = -1 + 7$ Add 7 to each side.

$3x = 6$ Simplify.

$\frac{3x}{3} = \frac{6}{3}$ Divide each side by 3.

$x = 2$ Simplify.

The numbers are 2 and -1.

CHECK

$-3x + 5y = -11$ First equation

$-3(2) + 5(-1) \stackrel{?}{=} -11$ Substitute 2 for x and -1 for y.

$-11 = -11$ ✓ Simplify.

$3x + 7y = -1$ Second equation

$3(2) + 7(-1) \stackrel{?}{=} -1$ Substitute 2 for x and -1 for y.

$-1 = -1$ ✓ Simplify.

Guided Practice

2. The sum of two numbers is -10. Negative three times the first number minus the second number equals 2. Find the numbers.

Study Tip

Coefficients When the coefficients of a variable are the same, subtracting the equations will eliminate the variable. When the coefficients are opposites, adding the equations will eliminate the variable.

Problem-Solving Tip

CCSS Perseverance

Checking your answers in both equations of a system helps ensure there are no calculation errors.

2 Elimination Using Subtraction

Sometimes we can eliminate a variable by subtracting one equation from another.

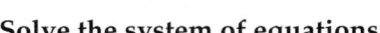

Standardized Test Example 3

Solve the system of equations. $2t + 5r = 6$
$9r + 2t = 22$

A $(-7, 15)$ **B** $\left(7, \frac{8}{9}\right)$ **C** $(4, -7)$ **D** $\left(4, -\frac{2}{5}\right)$

Read the Test Item

Since both equations contain $2t$, use elimination by subtraction.

Solve the Test Item

Step 1 Subtract the equations.

$$\begin{aligned} 5r + 2t &= 6 && \text{Write the system so like terms are aligned.} \\ (-)\ 9r + 2t &= 22 && \\ -4r &= -16 && \text{The variable } t \text{ is eliminated.} \\ r &= 4 && \text{Simplify.} \end{aligned}$$

Step 2 Substitute 4 for r in either equation to find the value of t.

$$\begin{aligned} 5r + 2t &= 6 && \text{First equation} \\ 5(4) + 2t &= 6 && r = 4 \\ 20 + 2t &= 6 && \text{Simplify.} \\ 20 + 2t - 20 &= 6 - 20 && \text{Subtract 20 from each side.} \\ 2t &= -14 && \text{Simplify.} \\ t &= -7 && \text{Simplify.} \end{aligned}$$

The solution is $(4, -7)$. The correct answer is C.

GuidedPractice

3. Solve the system of equations. $8b + 3c = 11$
$8b + 7c = 7$

F $(1.5, -1)$ **G** $(1.75, -1)$ **H** $(1.75, 1)$ **J** $(1.5, 1)$

Real-WorldLink

The five most dangerous jobs for teens are: delivery and other driving jobs, working alone in cash-based businesses, traveling youth crews, cooking, and construction.

Source: National Consumers League

Real-World Example 4 Write and Solve a System of Equations

JOBS Cheryl and Jackie work at an ice cream shop. Cheryl earns \$8.50 per hour and Jackie earns \$7.50 per hour. During a typical week, Cheryl and Jackie earn \$299.50 together. One week, Jackie doubles her work hours, and the girls earn \$412. How many hours does each girl work during a typical week?

Understand You know how much Cheryl and Jackie each earn per hour and how much they earned together.

Plan Let c = Cheryl's hours and j = Jackie's hours.

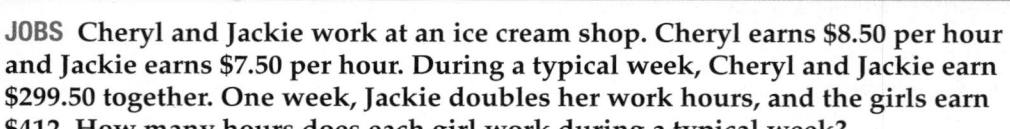

Cheryl's pay	plus	Jackie's pay	equals	\$299.50.
$8.50c$	$+$	$7.50j$	$=$	299.50
Cheryl's pay	plus	Jackie's pay	equals	\$412.
$8.50c$	$+$	$7.50(2)j$	$=$	412

Photodisc/Alamy

StudyTip

Another Method Instead of subtracting the equations, you could also multiply one equation by -1 and then add the equations.

Solve Subtract the equations to eliminate one of the variables. Then solve for the other variable.

$$\begin{array}{rl} 8.50c + 7.50j = 299.50 & \text{Write the equations vertically.} \\ (-)\ 8.50c + 7.50(2)j = 412 & \end{array}$$

$$\begin{array}{rl} 8.50c + 7.50j = 299.50 & \\ (-)\ 8.50c + 15j = 412 & \text{Simplify.} \\ -7.50j = -112.50 & \text{Subtract. The variable } c \text{ is eliminated.} \\ \frac{-7.50j}{-7.50} = \frac{-112.50}{-7.50} & \text{Divide each side by } -7.50. \\ j = 15 & \text{Simplify.} \end{array}$$

Now substitute 15 for j in either equation to find the value of c.

$$\begin{array}{rl} 8.50c + 7.50j = 299.50 & \text{First equation} \\ 8.50c + 7.50(\mathbf{15}) = 299.50 & \text{Substitute 15 for } j. \\ 8.50c + 112.50 = 299.50 & \text{Simplify.} \\ 8.50c = 187 & \text{Subtract 112.50 from each side.} \\ c = 22 & \text{Divide each side by 8.50.} \end{array}$$

Check Substitute both values into the other equation to see if the equation holds true. If $c = 22$ and $j = 15$, then $8.50(22) + 15(15)$ or 412.

Cheryl works 22 hours, while Jackie works 15 hours during a typical week.

GuidedPractice

4. **PARTIES** Tamera and Adelina are throwing a birthday party for their friend. Tamera invited 5 fewer friends than Adelina. Together they invited 47 guests. How many guests did each girl invite?

Check Your Understanding

Examples 1, 3 **Use elimination to solve each system of equations.**

1. $5m - p = 7$
   $7m - p = 11$

2. $8x + 5y = 38$
   $-8x + 2y = 4$

3. $7f + 3g = -6$
   $7f - 2g = -31$

4. $6a - 3b = 27$
   $2a - 3b = 11$

Example 2

5. **CCSS REASONING** The sum of two numbers is 24. Five times the first number minus the second number is 12. What are the two numbers?

Example 4

6. **RECYCLING** The recycling and reuse industry employs approximately 1,025,000 more workers than the waste management industry. Together they provide 1,275,000 jobs. How many jobs does each industry provide?

Practice and Problem Solving

Examples 1, 3 **Use elimination to solve each system of equations.**

7. $-v + w = 7$
$v + w = 1$

8. $y + z = 4$
$y - z = 8$

9. $-4x + 5y = 17$
$4x + 6y = -6$

10. $5m - 2p = 24$
$3m + 2p = 24$

11. $a + 4b = -4$
$a + 10b = -16$

12. $6r - 6t = 6$
$3r - 6t = 15$

13. $6c - 9d = 111$
$5c - 9d = 103$

14. $11f + 14g = 13$
$11f + 10g = 25$

15. $9x + 6y = 78$
$3x - 6y = -30$

16. $3j + 4k = 23.5$
$8j - 4k = 4$

17. $-3x - 8y = -24$
$3x - 5y = 4.5$

18. $6x - 2y = 1$
$10x - 2y = 5$

Example 2

19. The sum of two numbers is 22, and their difference is 12. What are the numbers?

20. Find the two numbers with a sum of 41 and a difference of 9.

21 Three times a number minus another number is −3. The sum of the numbers is 11. Find the numbers.

22. A number minus twice another number is 4. Three times the first number plus two times the second number is 12. What are the numbers?

Example 4

23. **TOURS** The Blackwells and Joneses are going to Hershey's Really Big 3D Show in Pennsylvania. Find the adult price and the children's price of the show.

Family	Number of Adults	Number of Children	Total Cost
Blackwell	2	5	\$31.65
Jones	2	3	\$23.75

Use elimination to solve each system of equations.

24. $4(x + 2y) = 8$
$4x + 4y = 12$

25. $3x - 5y = 11$
$5(x + y) = 5$

26. $4x + 3y = 6$
$3x + 3y = 7$

27. $6x - 7y = -26$
$6x + 5y = 10$

28. $\frac{1}{2}x + \frac{2}{3}y = 2\frac{3}{4}$
$\frac{1}{4}x - \frac{2}{3}y = 6\frac{1}{4}$

29. $\frac{3}{5}x + \frac{1}{2}y = 8\frac{1}{3}$
$-\frac{3}{5}x + \frac{3}{4}y = 8\frac{1}{3}$

30. CCSS **SENSE-MAKING** The total height of an office building b and the granite statue that stands on top of it g is 326.6 feet. The difference in heights between the building and the statue is 295.4 feet.

a. How tall is the statue?

b. How tall is the building?

31. **BIKE RACING** Professional Mountain Bike Racing currently has 66 teams. The number of non-U.S. teams is 30 more than the number of U.S. teams.

a. Let x represent the number of non-U.S. teams and y represent the number of U.S. teams. Write a system of equations that represents the number of U.S. teams and non-U.S. teams.

b. Use elimination to find the solution of the system of equations.

c. Interpret the solution in the context of the situation.

d. Graph the system of equations to check your solution.

32. **SHOPPING** Let x represent the number of years since 2004 and y represent the number of catalogs.

Catalogs	Number in 2004	Growth Rate (number per year)
online	7440	1293
print	3805	−1364

Source: MediaPost Publications

a. Write a system of equations to represent this situation.

b. Use elimination to find the solution to the system of equations.

c. Analyze the solution in terms of the situation. Determine the reasonableness of the solution.

33. **MULTIPLE REPRESENTATIONS** Collect 9 pennies and 9 paper clips. For this game, you use 9 objects to score points. Each paper clip is worth 1 point and each penny is worth 3 points. Let p represent the number of pennies and c represent the number of paper clips.

9 points = 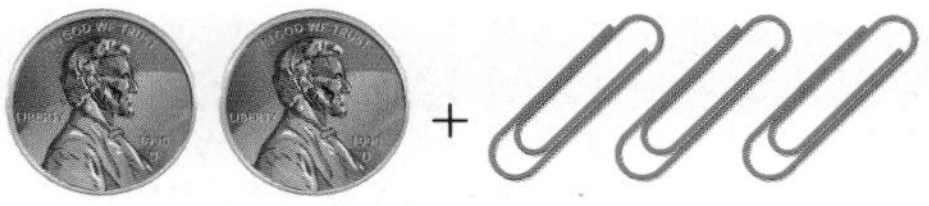 $= 3p + c$

$= 3(2) + 3$

a. Concrete Choose a combination of 9 objects and find your score.

b. Analytical Write and solve a system of equations to find the number of paper clips and pennies used for 15 points.

c. Tabular Make a table showing the number of paper clips used and the total number of points when the number of pennies is 0, 1, 2, 3, 4, or 5.

d. Verbal Does the result in the table match the results in part **b**? Explain.

H.O.T. Problems Use Higher-Order Thinking Skills

34. **REASONING** Describe the solution of a system of equations if after you added two equations the result was $0 = 0$.

35. **REASONING** What is the solution of a system of equations if the sum of the equations is $0 = 2$?

36. **OPEN ENDED** Create a system of equations that can be solved by using addition to eliminate one variable. Formulate a general rule for creating such systems.

37. CCSS **STRUCTURE** The solution of a system of equations is $(-3, 2)$. One equation in the system is $x + 4y = 5$. Find a second equation for the system. Explain how you derived this equation.

38. **CHALLENGE** The sum of the digits of a two-digit number is 8. The result of subtracting the units digit from the tens digit is -4. Define the variables and write the system of equations that you would use to find the number. Then solve the system and find the number.

39. **WRITING IN MATH** Describe when it would be most beneficial to use elimination to solve a system of equations.

Standardized Test Practice

40. SHORT RESPONSE Martina is on a train traveling at a speed of 188 mph between two cities 1128 miles apart. If the train has been traveling for an hour, how many more hours is her train ride?

41. GEOMETRY Ms. Miller wants to tile her rectangular kitchen floor. She knows the dimensions of the floor. Which formula should she use to find the area?

A $A = \ell w$ **C** $P = 2\ell + 2w$
B $V = Bh$ **D** $c^2 = a^2 + b^2$

42. If the pattern continues, what is the 8th number in the sequence?

$$2, 3, \frac{9}{2}, \frac{27}{4}, \frac{81}{8}, \ldots$$

F $\frac{2187}{64}$ **G** $\frac{2245}{64}$ **H** $\frac{2281}{64}$ **J** $\frac{2445}{64}$

43. What is the solution of this system of equations?

$$x + 4y = 1$$
$$2x - 3y = -9$$

A $(2, -8)$ **C** no solution
B $(-3, 1)$ **D** infinitely many solutions

Spiral Review

Use substitution to solve each system of equations. If the system does not have exactly one solution, state whether it has no solution or infinitely many solutions. (Lesson 6-2)

44. $y = 6x$; $2x + 3y = 40$

45. $x = 3y$; $2x + 3y = 45$

46. $x = 5y + 6$; $x = 3y - 2$

47. $y = 3x + 2$; $y = 4x - 1$

48. $3c = 4d + 2$; $c = d - 1$

49. $z = v + 4$; $2z - v = 6$

50. FINANCIAL LITERACY Gregorio and Javier each want to buy a bicycle. Gregorio has already saved \$35 and plans to save \$10 per week. Javier has \$26 and plans to save \$13 per week. (Lesson 6-1)

a. In how many weeks will Gregorio and Javier have saved the same amount of money?

b. How much will each person have saved at that time?

51. GEOMETRY A *parallelogram* is a quadrilateral in which opposite sides are parallel. Determine whether *ABCD* is parallelogram. Explain your reasoning. (Lesson 4-4)

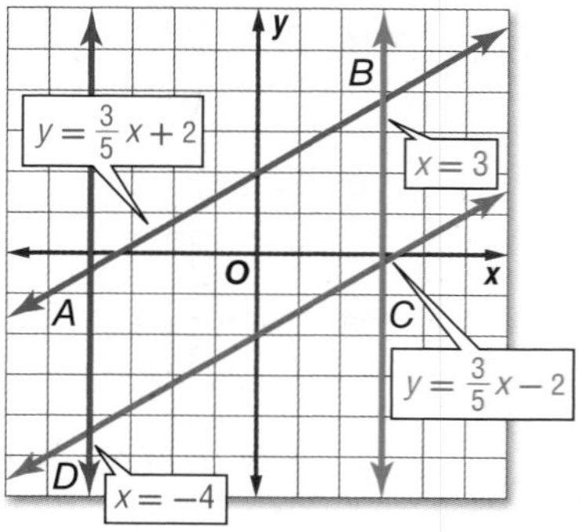

Solve each equation. Check your solution. (Lesson 2-2)

52. $6u = -48$

53. $75 = -15p$

54. $\frac{2}{3}a = 8$

55. $-\frac{3}{4}d = 15$

Skills Review

Simplify each expression. If not possible, write *simplified*.

56. $6q - 3 + 7q + 1$

57. $7w^2 - 9w + 4w^2$

58. $10(2 + r) + 3r$

59. $5y - 7(y + 5)$

LESSON

6-4 Elimination Using Multiplication

Then

- You used elimination with addition and subtraction to solve systems of equations.

Now

1. Solve systems of equations by using elimination with multiplication.
2. Solve real-world problems involving systems of equations.

Why?

- The table shows the number of cars at Scott's Auto Repair Shop for each type of service.

Item	Repairs	Maintenance
body	3	4
engine	2	2

The manager has allotted 1110 minutes for body work and 570 minutes for engine work. The system $3r + 4m = 1110$ and $2r + 2m = 570$ can be used to find the average time for each service.

CCSS Common Core State Standards

Content Standards

A.REI.5 Prove that, given a system of two equations in two variables, replacing one equation by the sum of that equation and a multiple of the other produces a system with the same solutions.

A.REI.6 Solve systems of linear equations exactly and approximately (e.g., with graphs), focusing on pairs of linear equations in two variables.

Mathematical Practices

1 Make sense of problems and persevere in solving them.

1 Elimination Using Multiplication

In the system above, neither variable can be eliminated by adding or subtracting. You can use multiplication to solve.

KeyConcept Solving by Elimination

Step 1 Multiply at least one equation by a constant to get two equations that contain opposite terms.

Step 2 Add the equations, eliminating one variable. Then solve the equation.

Step 3 Substitute the value from Step 2 into one of the equations and solve for the other variable. Write the solution as an ordered pair.

Example 1 Multiply One Equation to Eliminate a Variable

PT

Use elimination to solve the system of equations.

$5x + 6y = -8$

$2x + 3y = -5$

Steps 1 and 2

$5x + 6y = -8$

$2x + 3y = -5$ — Multiply each term by −2. →

$$\begin{array}{rrcr} & 5x + 6y & = & -8 \\ (+) & -4x - 6y & = & 10 \\ \hline & x & = & 2 \end{array}$$

Add. y is eliminated.

Step 3

$2x + 3y = -5$	Second equation
$2(2) + 3y = -5$	Substitution, $x = 2$
$4 + 3y = -5$	Simplify.
$3y = -9$	Subtract 4 from each side and simplify.
$y = -3$	Divide each side by 3 and simplify.

The solution is $(2, -3)$.

GuidedPractice

1A. $6x - 2y = 10$
$3x - 7y = -19$

1B. $9r + q = 13$
$3r + 2q = -4$

Sometimes you have to multiply each equation by a different number in order to solve the system.

Example 2 Multiply Both Equations to Eliminate a Variable

Use elimination to solve the system of equations.

$4x + 2y = 8$

$3x + 3y = 9$

Method 1 Eliminate x.

$4x + 2y = 8$ → Multiply by 3. → $12x + 6y = 24$

$3x + 3y = 9$ → Multiply by −4. → $(+)\ -12x - 12y = -36$

$-6y = -12$ Add equations. x is eliminated.

$\frac{-6y}{-6} = \frac{-12}{-6}$ Divide each side by −6.

$y = 2$ Simplify.

Now substitute 2 for y in either equation to find the value of x.

$3x + 3y = 9$	Second equation
$3x + 3(2) = 9$	Substitute 2 for y.
$3x + 6 = 9$	Simplify.
$3x = 3$	Subtract 6 from each side and simplify.
$\frac{3x}{3} = \frac{3}{3}$	Divide each side by 3.
$x = 1$	The solution is (1, 2).

Method 2 Eliminate y.

$4x + 2y = 8$ → Multiply by 3. → $12x + 6y = 24$

$3x + 3y = 9$ → Multiply by −2. → $(+)\ -6x - 6y = -18$

$6x = 6$ Add equations. y is eliminated.

$\frac{6x}{6} = \frac{6}{6}$ Divide each side by 6.

$x = 1$ Simplify.

Now substitute 1 for x in either equation to find the value of y.

$3x + 3y = 9$	Second equation
$3(1) + 3y = 9$	Substitute 1 for x.
$3 + 3y = 9$	Simplify.
$3y = 6$	Subtract 3 from each side and simplify.
$\frac{3y}{3} = \frac{6}{3}$	Divide each side by 3.
$y = 2$	Simplify.

The solution is (1, 2), which matches the result obtained with Method 1.

CHECK Substitute 1 for x and 2 for y in the first equation.

$4x + 2y = 8$	Original equation
$4(1) + 2(2) \stackrel{?}{=} 8$	Substitute (1, 2) for (x, y).
$4 + 4 \stackrel{?}{=} 8$	Multiply.
$8 = 8$ ✓	Add.

GuidedPractice

2A. $5x - 3y = 6$
$2x + 5y = -10$

2B. $6a + 2b = 2$
$4a + 3b = 8$

StudyTip

Choosing a Variable to Eliminate Unless the problem is asking for the value of a specific variable, you may use multiplication to eliminate either variable.

Math HistoryLink

Leonardo Pisano **(1170–1250)** Leonardo Pisano is better known by his nickname *Fibonacci*. His book introduced the Hindu-Arabic place-valued decimal system. Systems of linear equations are studied in this work.

Jean-Luc Tokyo/Photographers Direct

2 Solve Real-World Problems

Sometimes it is necessary to use multiplication before elimination in real-world problem solving too.

Real-World Example 3 Solve a System of Equations

FLIGHT A personal aircraft traveling with the wind flies 520 miles in 4 hours. On the return trip, the airplane takes 5 hours to travel the same distance. Find the speed of the airplane if the air is still.

You are asked to find the speed of the airplane in still air.

Let a = the rate of the airplane if the air is still.
Let w = the rate of the wind.

	r	t	d	$r \cdot t = d$
With the Wind	$a + w$	4	520	$(a + w)4 = 520$
Against the Wind	$a - w$	5	520	$(a - w)5 = 520$

So, our two equations are $4a + 4w = 520$ and $5a - 5w = 520$.

$4a + 4w = 520$ — Multiply by 5. → $20a + 20w = 2600$

$5a - 5w = 520$ — Multiply by 4. → $(+)\ 20a - 20w = 2080$

$40a = 4680$ — w is eliminated.

$\frac{40a}{40} = \frac{4680}{40}$ — Divide each side by 40.

$a = 117$ — Simplify.

The rate of the airplane in still air is 117 miles per hour.

Guided Practice

3. **CANOEING** A canoeist travels 4 miles downstream in 1 hour. The return trip takes the canoeist 1.5 hours. Find the rate of the boat in still water.

Check Your Understanding

Examples 1–2 **Use elimination to solve each system of equations.**

1. $2x - y = 4$
   $7x + 3y = 27$

2. $2x + 7y = 1$
   $x + 5y = 2$

3. $4x + 2y = -14$
   $5x + 3y = -17$

4. $9a - 2b = -8$
   $-7a + 3b = 12$

Example 3

5. **CCSS SENSE-MAKING** A kayaking group with a guide travels 16 miles downstream, stops for a meal, and then travels 16 miles upstream. The speed of the current remains constant throughout the trip. Find the speed of the kayak in still water.

Leave 10:00 A.M.
Stop for meal 12:00 noon
Return 1:00 P.M.
Finish 5:00 P.M.

6. **PODCASTS** Steve subscribed to 10 podcasts for a total of 340 minutes. He used his two favorite tags, Hobbies and Recreation and Soliloquies. Each of the Hobbies and Recreation episodes lasted about 32 minutes. Each Soliloquies episode lasted 42 minutes. To how many of each tag did Steve subscribe?

Practice and Problem Solving

Examples 1–2 **Use elimination to solve each system of equations.**

7. $x + y = 2$
$-3x + 4y = 15$

8. $x - y = -8$
$7x + 5y = 16$

9. $x + 5y = 17$
$-4x + 3y = 24$

10. $6x + y = -39$
$3x + 2y = -15$

11. $2x + 5y = 11$
$4x + 3y = 1$

12. $3x - 3y = -6$
$-5x + 6y = 12$

13. $3x + 4y = 29$
$6x + 5y = 43$

14. $8x + 3y = 4$
$-7x + 5y = -34$

15. $8x + 3y = -7$
$7x + 2y = -3$

16. $4x + 7y = -80$
$3x + 5y = -58$

17. $12x - 3y = -3$
$6x + y = 1$

18. $-4x + 2y = 0$
$10x + 3y = 8$

Example 3

19. **NUMBER THEORY** Seven times a number plus three times another number equals negative one. The sum of the two numbers is negative three. What are the numbers?

20. **FOOTBALL** A field goal is 3 points and the extra point after a touchdown is 1 point. In a recent post-season, Adam Vinatieri of the Indianapolis Colts made a total of 21 field goals and extra point kicks for 49 points. Find the number of field goals and extra points that he made.

Use elimination to solve each system of equations.

21. $2.2x + 3y = 15.25$
$4.6x + 2.1y = 18.325$

22. $-0.4x + 0.25y = -2.175$
$2x + y = 7.5$

23. $\frac{1}{4}x + 4y = 2\frac{3}{4}$
$3x + \frac{1}{2}y = 9\frac{1}{4}$

24. $\frac{2}{5}x + 6y = 24\frac{1}{5}$
$3x + \frac{1}{2}y = 3\frac{1}{2}$

25. **CCSS MODELING** A staffing agency for in-home nurses and support staff places necessary personnel at locations on a daily basis. Each placed nurse works 240 minutes per day at a daily rate of \$90. Each support staff employee works 360 minutes per day at a daily rate of \$120.

a. On a given day, 3000 total minutes are worked by the nurses and support staff that were placed. Write an equation that represents this relationship.

b. On the same day, earnings for placed nurses and support staff totaled \$1050. Write an equation that represents this relationship.

c. Solve the system of equations, and interpret the solution in the context of the situation.

26. **GEOMETRY** The graphs of $x + 2y = 6$ and $2x + y = 9$ contain two of the sides of a triangle. A vertex of the triangle is at the intersection of the graphs.

a. What are the coordinates of the vertex?

b. Draw the graph of the two lines. Identify the vertex of the triangle.

c. The line that forms the third side of the triangle is the line $x - y = -3$. Draw this line on the previous graph.

d. Name the other two vertices of the triangle.

27. **ENTERTAINMENT** At an entertainment center, two groups of people bought batting tokens and miniature golf games, as shown in the table.

Group	Number of Batting Tokens	Number of Miniature Golf Games	Total Cost
A	16	3	\$30
B	22	5	\$43

a. Define the variables, and write a system of linear equations from this situation.

b. Solve the system of equations, and explain what the solution represents.

28. **TESTS** Mrs. Henderson discovered that she had accidentally reversed the digits of a test score and did not give a student 36 points. Mrs. Henderson told the student that the sum of the digits was 14 and agreed to give the student his correct score plus extra credit if he could determine his actual score. What was his correct score?

H.O.T. Problems Use Higher-Order Thinking Skills

29. **REASONING** Explain how you could recognize a system of linear equations with infinitely many solutions.

30. **CCSS CRITIQUE** Jason and Daniela are solving a system of equations. Is either of them correct? Explain your reasoning.

Jason

$2r + 7t = 11$

$r - 9t = -7$

$2r + 7t = 11$

$(-)\ 2r - 18t = -14$

$25t = 25$

$t = 1$

$2r + 7t = 11$

$2r + 7(1) = 11$

$2r + 7 = 11$

$2r = 4$

$\frac{2r}{2} = \frac{4}{2}$

$r = 2$

The solution is (2, 1).

Daniela

$2r + 7t = 11$

$(-)\ r - 9t = -7$

$r = 18$

$2r + 7t = 11$

$2(18) + 7t = 11$

$36 + 7t = 11$

$7t = -25$

$\frac{7t}{7} = -\frac{25}{7}$

$t = -3.6$

The solution is (18, -3.6).

31. **OPEN ENDED** Write a system of equations that can be solved by multiplying one equation by -3 and then adding the two equations together.

32. **CHALLENGE** The solution of the system $4x + 5y = 2$ and $6x - 2y = b$ is $(3, a)$. Find the values of a and b. Discuss the steps that you used.

33. **WRITING IN MATH** Why is substitution sometimes more helpful than elimination, and vice versa?

Standardized Test Practice

34. What is the solution of this system of equations?

$2x - 3y = -9$
$-x + 3y = 6$

A (3, 3) **C** (−3, 1)
B (−3, 3) **D** (1, −3)

35. A buffet has one price for adults and another for children. The Taylor family has two adults and three children, and their bill was \$40.50. The Wong family has three adults and one child. Their bill was \$38. Which system of equations could be used to determine the price for an adult and for a child?

F $x + y = 40.50$, $x + y = 38$
H $2x + 3y = 40.50$, $x + 3y = 38$
G $2x + 3y = 40.50$, $3x + y = 38$
J $2x + 2y = 40.50$, $3x + y = 38$

36. SHORT RESPONSE A customer at the paint store has ordered 3 gallons of ivy green paint. Melissa mixes the paint in a ratio of 3 parts blue to one part yellow. How many quarts of blue paint does she use?

37. PROBABILITY The table shows the results of a number cube being rolled. What is the experimental probability of rolling a 3?

Outcome	Frequency
1	4
2	8
3	2
4	0
5	5
6	1

A $\frac{2}{3}$ **B** $\frac{1}{3}$ **C** 0.2 **D** 0.1

Spiral Review

Use elimination to solve each system of equations. (Lesson 6-3)

38. $f + g = -3$, $f - g = 1$

39. $6g + h = -7$, $6g + 3h = -9$

40. $5j + 3k = -9$, $3j + 3k = -3$

41. $2x - 4z = 6$, $x - 4z = -3$

42. $-5c - 3v = 9$, $5c + 2v = -6$

43. $4b - 6n = -36$, $3b - 6n = -36$

44. JOBS Brandy and Adriana work at an after-school child care center. Together they cared for 32 children this week. Brandy cared for 0.6 times as many children as Adriana. How many children did each girl care for? (Lesson 6-2)

Solve each inequality. Then graph the solution set. (Lesson 5-5)

45. $|m - 5| \leq 8$

46. $|q + 11| < 5$

47. $|2w + 9| > 11$

48. $|2r + 1| \geq 9$

Skills Review

Translate each sentence into a formula.

49. The area A of a triangle equals one half times the base b times the height h.

50. The circumference C of a circle equals the product of 2, π, and the radius r.

51. The volume V of a rectangular box is the length ℓ times the width w multiplied by the height h.

52. The volume of a cylinder V is the same as the product of π and the radius r to the second power multiplied by the height h.

53. The area of a circle A equals the product of π and the radius r squared.

54. Acceleration A equals the increase in speed s divided by time t in seconds.

CHAPTER 6

Mid-Chapter Quiz

Lessons 6-1 through 6-4

Use the graph to determine whether each system is *consistent* or *inconsistent* and if it is *independent* or dependent. (Lesson 6-1)

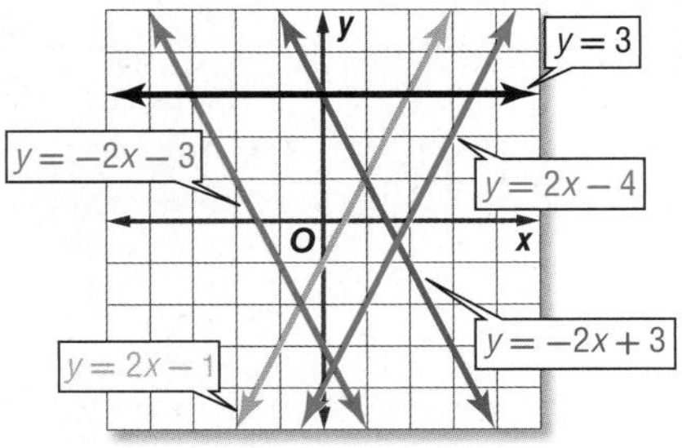

1. $y = 2x - 1$
$y = -2x + 3$

2. $y = -2x + 3$
$y = -2x - 3$

Graph each system and determine the number of solutions that it has. If it has one solution, name it. (Lesson 6-1)

3. $y = 2x - 3$
$y = x + 4$

4. $x + y = 6$
$x - y = 4$

5. $x + y = 8$
$3x + 3y = 24$

6. $x - 4y = -6$
$y = -1$

7. $3x + 2y = 12$
$3x + 2y = 6$

8. $2x + y = -4$
$5x + 3y = -6$

Use substitution to solve each system of equations. (Lesson 6-2)

9. $y = x + 4$
$2x + y = 16$

10. $y = -2x - 3$
$x + y = 9$

11. $x + y = 6$
$x - y = 8$

12. $y = -4x$
$6x - y = 30$

13. FOOD The cost of two meals at a restaurant is shown in the table below. (Lesson 6-2)

Meal	Total Cost
3 tacos, 2 burritos	\$7.40
4 tacos, 1 burrito	\$6.45

a. Define variables to represent the cost of a taco and the cost of a burrito.

b. Write a system of equations to find the cost of a single taco and a single burrito.

c. Solve the systems of equations, and explain what the solution means.

d. How much would a customer pay for 2 tacos and 2 burritos?

14. AMUSEMENT PARKS The cost of two groups going to an amusement park is shown in the table. (Lesson 6-3)

Group	Total Cost
4 adults, 2 children	\$184
4 adults, 3 children	\$200

a. Define variables to represent the cost of an adult ticket and the cost of a child ticket.

b. Write a system of equations to find the cost of an adult ticket and a child ticket.

c. Solve the system of equations, and explain what the solution means.

d. How much will a group of 3 adults and 5 children be charged for admission?

15. MULTIPLE CHOICE Angelina spent \$16 for 12 pieces of candy to take to a meeting. She has \$16. Each chocolate bar costs \$2, and each lollipop costs \$1. Determine how many of each she bought. (Lesson 6-3)

A 6 chocolate bars, 6 lollipops

B 4 chocolate bars, 8 lollipops

C 7 chocolate bars, 5 lollipops

D 3 chocolate bars, 9 lollipops

Use elimination to solve each system of equations. (Lessons 6-3 and 6-4)

16. $x + y = 9$
$x - y = -3$

17. $x + 3y = 11$
$x + 7y = 19$

18. $9x - 24y = -6$
$3x + 4y = 10$

19. $-5x + 2y = -11$
$5x - 7y = 1$

20. MULTIPLE CHOICE The Blue Mountain High School Drama Club is selling tickets to their spring musical. Adult tickets are \$4 and student tickets are \$1. A total of 285 tickets are sold for \$765. How many of each type of ticket are sold? (Lesson 6-4)

F 145 adult, 140 student

G 120 adult, 165 student

H 180 adult, 105 student

J 160 adult, 125 student

LESSON 6-5 Applying Systems of Linear Equations

Then

- You solved systems of equations by using substitution and elimination.

Now

1. Determine the best method for solving systems of equations.
2. Apply systems of equations.

Why?

- In speed skating, competitors race two at a time on a double track. Indoor speed skating rinks have two track sizes for race events: an official track and a short track.

Speed Skating Tracks	
official track	x
short track	y

The total length of the two tracks is 511 meters. The official track is 44 meters less than four times the short track. The total length is represented by $x + y = 511$. The length of the official track is represented by $x = 4y - 44$.

You can solve the system of equations to find the length of each track.

Mike Powell/Allsport Concepts/Getty Images

CCSS Common Core State Standards

Content Standards

A.REI.6 Solve systems of linear equations exactly and approximately (e.g., with graphs), focusing on pairs of linear equations in two variables.

Mathematical Practices

2 Reason abstractly and quantitatively.

4 Model with mathematics.

1 Determine the Best Method

You have learned five methods for solving systems of linear equations. The table summarizes the methods and the types of systems for which each method works best.

ConceptSummary Solving Systems of Equations

Method	The Best Time to Use
Graphing	To estimate solutions, since graphing usually does not give an exact solution.
Substitution	If one of the variables in either equation has a coefficient of 1 or −1.
Elimination Using Addition	If one of the variables has opposite coefficients in the two equations.
Elimination Using Subtraction	If one of the variables has the same coefficient in the two equations.
Elimination Using Multiplication	If none of the coefficients are 1 or −1 and neither of the variables can be eliminated by simply adding or subtracting the equations.

Substitution and elimination are algebraic methods for solving systems of equations. An algebraic method is best for an exact solution. Graphing, with or without technology, is a good way to estimate a solution.

A system of equations can be solved using each method. To determine the best approach, analyze the coefficients of each term in each equation.

Example 1 Choose the Best Method

Determine the best method to solve the system of equations. Then solve the system.

$4x - 4y = 8$
$-8x + y = 19$

Understand To determine the best method to solve the system of equations, look closely at the coefficients of each term.

Plan Neither the coefficients of x nor y are the same or additive inverses, so you cannot add or subtract to eliminate a variable. Since the coefficient of y in the second equation is 1, you can use substitution.

Solve First, solve the second equation for y.

$-8x + y = 19$	Second equation
$-8x + y + 8x = 19 + 8x$	Add $8x$ to each side.
$y = 19 + 8x$	Simplify.

Next, substitute $19 + 8x$ for y in the first equation.

$4x - 4y = 8$	First equation
$4x - 4(19 + 8x) = 8$	Substitution
$4x - 76 - 32x = 8$	Distributive Property
$-28x - 76 = 8$	Simplify.
$-28x - 76 + 76 = 8 + 76$	Add 76 to each side.
$-28x = 84$	Simplify.
$\frac{-28x}{-28} = \frac{84}{-28}$	Divide each side by -28.
$x = -3$	Simplify.

Last, substitute -3 for x in the second equation.

$-8x + y = 19$	Second equation
$-8(-3) + y = 19$	$x = -3$
$y = -5$	Simplify.

The solution of the system of equations is $(-3, -5)$.

Check Use a graphing calculator to check your solution. If your algebraic solution is correct, then the graphs will intersect at $(-3, -5)$.

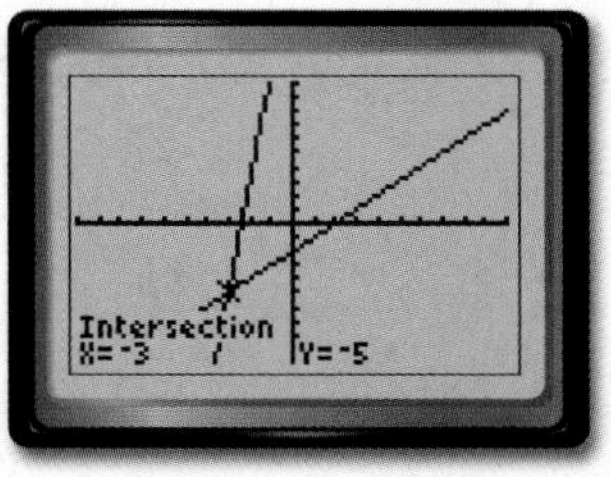

[−10, 10] scl: 1 [−10, 10] scl: 1

StudyTip

CCSS Reasoning The system of equations in Example 1 can also be solved by using elimination with multiplication. You can multiply the first equation by 2 and then add to eliminate the x-term.

Guided Practice

1A. $5x + 7y = 2$
$-2x + 7y = 9$

1B. $3x - 4y = -10$
$5x + 8y = -2$

1C. $x - y = 9$
$7x + y = 7$

1D. $5x - y = 17$
$3x + 2y = 5$

2 Apply Systems of Linear Equations

When applying systems of linear equations to problems, it is important to analyze each solution in the context of the situation.

Real-WorldLink

Four species of penguins are on the endangered species list. The colonies of penguins that live closest to human inhabitants are the most at risk for extinction.

Source: PBS

Real-World Example 2 Apply Systems of Linear Equations

PT

PENGUINS Of the 17 species of penguins in the world, the largest species is the emperor penguin. One of the smallest is the Galapagos penguin. The total height of the two penguins is 169 centimeters. The emperor penguin is 22 centimeters more than twice the height of the Galapagos penguin. Find the height of each penguin.

The total height of the two species can be represented by $p + g = 169$, where p represents the height of the emperor penguin and g the height of the Galapagos penguin. Next write an equation to represent the height of the emperor penguin.

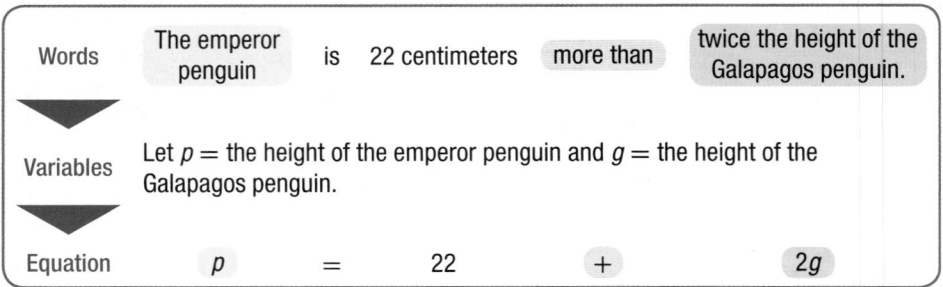

Words	The emperor penguin	is	22 centimeters	more than	twice the height of the Galapagos penguin.
Variables	Let p = the height of the emperor penguin and g = the height of the Galapagos penguin.				
Equation	p	$=$	22	$+$	$2g$

First rewrite the second equation.

$p = 22 + 2g$ Second equation

$p - 2g = 22$ Subtract $2g$ from each side.

You can use elimination by subtraction to solve this system of equations.

$p + g = 169$ First equation

$(-)\ p - 2g = 22$ Subtract the second equation.

$3g = 147$ Eliminate p.

$\frac{3g}{3} = \frac{147}{3}$ Divide each side by 3.

$g = 49$ Simplify.

Next substitute 49 for g in one of the equations.

$p = 22 + 2g$ Second equation

$= 22 + 2(49)$ $g = 49$

$= 120$ Simplify.

The height of the emperor penguin is 120 centimeters, and the height of the Galapagos penguin is 49 centimeters.

Does the solution make sense in the context of the problem?

Check by verifying the given information. The penguins' heights added together would be $120 + 49$ or 169 centimeters and $22 + 2(49)$ is 120 centimeters.

GuidedPractice

2. VOLUNTEERING Jared has volunteered 50 hours and plans to volunteer 3 hours in each coming week. Clementine is a new volunteer who plans to volunteer 5 hours each week. Write and solve a system of equations to find how long it will be before they will have volunteered the same number of hours.

Check Your Understanding

Example 1 **Determine the best method to solve each system of equations. Then solve the system.**

1. $2x + 3y = -11$
   $-8x - 5y = 9$

2. $3x + 4y = 11$
   $2x + y = -1$

3. $3x - 4y = -5$
   $-3x + 2y = 3$

4. $3x + 7y = 4$
   $5x - 7y = -12$

Example 2

5. **SHOPPING** At a sale, Salazar bought 4 T-shirts and 3 pairs of jeans for $181. At the same store, Jenna bought 1 T-shirt and 2 pairs of jeans for $94. The T-shirts were all the same price, and the jeans were all the same price.
   - **a.** Write a system of equations that can be used to represent this situation.
   - **b.** Determine the best method to solve the system of equations.
   - **c.** Solve the system.

Practice and Problem Solving

Example 1 **Determine the best method to solve each system of equations. Then solve the system.**

6. $-3x + y = -3$
   $4x + 2y = 14$

7. $2x + 6y = -8$
   $x - 3y = 8$

8. $3x - 4y = -5$
   $-3x - 6y = -5$

9. $5x + 8y = 1$
   $-2x + 8y = -6$

10. $y + 4x = 3$
    $y = -4x - 1$

11. $-5x + 4y = 7$
    $-5x - 3y = -14$

Example 2

12. **FINANCIAL LITERACY** For a Future Teachers of America fundraiser, Denzell sold food as shown in the table. He sold 11 more subs than pizzas and earned a total of $233. Write and solve a system of equations to represent this situation. Then describe what the solution means.

Item	Selling Price
pizza	$5.00
sub	$3.00

13. **DVDs** Manuela has a total of 40 DVDs of movies and television shows. The number of movies is 4 less than 3 times the number of television shows. Write and solve a system of equations to find the numbers of movies and television shows that she has on DVD.

14. **CAVES** The Caverns of Sonora have two different tours: the Crystal Palace tour and the Horseshoe Lake tour. The total length of both tours is 3.25 miles. The Crystal Palace tour is a half-mile less than twice the distance of the Horseshoe Lake tour. Determine the length of each tour.

15. **CCSS MODELING** The *break-even point* is the point at which income equals expenses. Ridgemont High School is paying $13,200 for the writing and research of their yearbook plus a printing fee of $25 per book. If they sell the books for $40 each, how many will they have to sell to break even? Explain.

16. **PAINTBALL** Clara and her friends are planning a trip to a paintball park. Find the cost of lunch and the cost of each paintball. What would be the cost for 400 paintballs and lunch?

17 **RECYCLING** Mara and Ling each recycled aluminum cans and newspaper, as shown in the table. Mara earned \$3.77, and Ling earned \$4.65.

Materials	Pounds Recycled	
	Mara	Ling
aluminum cans	9	9
newspaper	26	114

a. Define variables and write a system of linear equations from this situation.

b. What was the price per pound of aluminum? Determine the reasonableness of your solution.

18. **BOOKS** The library is having a book sale. Hardcover books sell for \$4 each, and paperback books are \$2 each. If Connie spends \$26 for 8 books, how many hardcover books did she buy?

19. **MUSIC** An online music club offers individual songs for one price or entire albums for another. Kendrick pays \$14.90 to download 5 individual songs and 1 album. Geoffrey pays \$21.75 to download 3 individual songs and 2 albums.

a. How much does the music club charge to download a song?

b. How much does the music club charge to download an entire album?

20. **CANOEING** Malik canoed against the current for 2 hours and then with the current for 1 hour before resting. Julio traveled against the current for 2.5 hours and then with the current for 1.5 hours before resting. If they traveled a total of 9.5 miles against the current, 20.5 miles with the current, and the current is 3 miles per hour, how fast do Malik and Julio travel in still water?

H.O.T. Problems Use Higher-Order Thinking Skills

21. **OPEN ENDED** Formulate a system of equations that represents a situation in your school. Describe the method that you would use to solve the system. Then solve the system and explain what the solution means.

22. CCSS **REASONING** In a system of equations, x represents the time spent riding a bike, and y represents the distance traveled. You determine the solution to be $(-1, 7)$. Use this problem to discuss the importance of analyzing solutions in the context of real-world problems.

23. **CHALLENGE** Solve the following system of equations by using three different methods. Show your work.

$$4x + y = 13$$
$$6x - y = 7$$

24. **WRITE A QUESTION** A classmate says that elimination is the best way to solve a system of equations. Write a question to challenge his conjecture.

25. **WHICH ONE DOESN'T BELONG?** Which system is different? Explain.

$x - y = 3$ $x + \frac{1}{2}y = 1$	$-x + y = 0$ $5x = 2y$	$y = x - 4$ $y = \frac{2}{x}$	$y = x + 1$ $y = 3x$

26. **WRITING IN MATH** How do you know what method to use when solving a system of equations?

Standardized Test Practice

27. If $5x + 3y = 12$ and $4x - 5y = 17$, what is y?

A -1 **B** 3 **C** $(-1, 3)$ **D** $(3, -1)$

28. STATISTICS The scatter plot shows the number of hay bales used on the Bostwick farm during the last year.

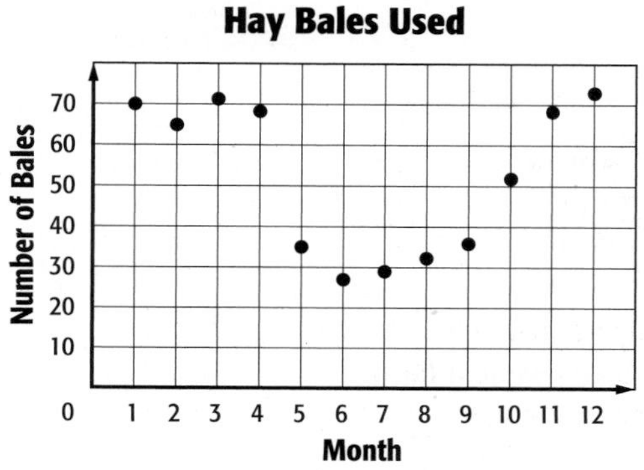

Which is an invalid conclusion?

F The Bostwicks used less hay in the summer than they did in the winter.

G The Bostwicks used about 629 bales of hay during the year.

H On average, the Bostwicks used about 52 bales each month.

J The Bostwicks used the most hay in February.

29. SHORT RESPONSE At noon, Cesar cast a shadow 0.15 foot long. Next to him a streetlight cast a shadow 0.25 foot long. If Cesar is 6 feet tall, how tall is the streetlight?

30. The graph shows the solution to which of the following systems of equations?

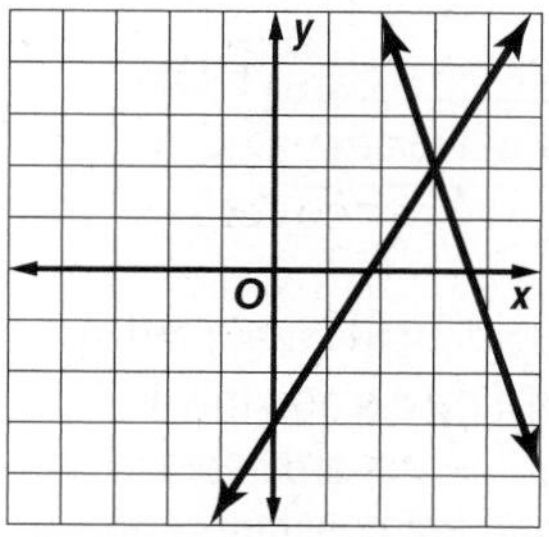

A $y = -3x + 11$
$3y = 5x - 9$

B $y = 5x - 15$
$2y = x + 7$

C $y = -3x + 11$
$2y = 4x - 5$

D $y = 5x - 15$
$3y = 2x + 18$

Spiral Review

Use elimination to solve each system of equations. (Lesson 6-4)

31. $x + y = 3$
$3x - 4y = -12$

32. $-4x + 2y = 0$
$2x - 3y = 16$

33. $4x + 2y = 10$
$5x - 3y = 7$

34. TRAVELING A youth group is traveling in two vans to visit an aquarium. The number of people in each van and the cost of admission for that van are shown. What are the adult and student prices? (Lesson 6-3)

Van	Number of Adults	Number of Students	Total Cost
A	2	5	\$77
B	2	7	\$95

Graph each inequality. (Lesson 5-6)

35. $y < 4$

36. $x \geq 3$

37. $7x + 12y > 0$

38. $y - 3x \leq 4$

Skills Review

Find each sum or difference.

39. $(-3.81) + (-8.5)$

40. $12.625 + (-5.23)$

41. $21.65 + (-15.05)$

42. $(-4.27) + 1.77$

43. $(-78.94) - 14.25$

44. $(-97.623) - (-25.14)$

EXTEND 6-5 Algebra Lab
Using Matrices to Solve Systems of Equations

A **matrix** is a rectangular arrangement of numbers, called **elements**, in rows and columns enclosed in brackets. Usually named using an uppercase letter, a matrix can be described by its **dimensions** or by the number of rows and columns in the matrix. A matrix with m rows and n columns is an $m \times n$ matrix (read "m by n").

CCSS Common Core State Standards
Content Standards
A.REI.6 Solve systems of linear equations exactly and approximately (e.g., with graphs), focusing on pairs of linear equations in two variables.

$$A = \begin{bmatrix} 7 & -9 & 5 & 3 \\ -1 & 3 & -3 & 6 \\ 0 & -4 & 8 & 2 \end{bmatrix}$$

A is a 3×4 matrix.

3 rows

4 columns

The element 2 is in Row 3, Column 4.

You can use an augmented matrix to solve a system of equations. An **augmented matrix** consists of the coefficients and the constant terms of a system of equations. Make sure that the coefficients of the x-terms are listed in one column, the coefficients of the y-terms are in another column, and the constant terms are in a third column. The coefficients and constant terms are usually separated by a dashed line.

Linear System	Augmented Matrix
$x - 3y = 8$ $-9x + 2y = -4$	$\left[\begin{array}{rr:r} 1 & -3 & 8 \\ -9 & 2 & -4 \end{array}\right]$

Activity 1 Write an Augmented Matrix

Write an augmented matrix for each system of equations.

a. $\mathbf{-2x + 7y = 11}$
$\mathbf{6x - 4y = 2}$

Place the coefficients of the equations and the constant terms into a matrix.

$$\begin{aligned} -2x + 7y &= 11 \\ 6x - 4y &= 2 \end{aligned} \quad \rightarrow \quad \left[\begin{array}{rr:r} -2 & 7 & 11 \\ 6 & -4 & 2 \end{array}\right]$$

b. $\mathbf{x - 2y = 5}$
$\mathbf{y = -4}$

$$\begin{aligned} x - 2y &= 5 \\ y &= -4 \end{aligned} \quad \rightarrow \quad \left[\begin{array}{rr:r} 1 & -2 & 5 \\ 0 & 1 & -4 \end{array}\right]$$

You can solve a system of equations by using an augmented matrix. By performing row operations, you can change the form of the matrix. The operations are the same as the ones used when working with equations.

KeyConcept Elementary Row Operations

The following operations can be performed on an augmented matrix.

- Interchange any two rows.
- Multiply all entries in a row by a nonzero constant.
- Replace one row with the sum of that row and a multiple of another row.

Row operations produce a matrix equivalent to the original system. **Row reduction** is the process of performing elementary row operations on an augmented matrix to solve a system.

The goal is to get the coefficients portion of the matrix to have the form $\begin{bmatrix} 1 & 0 \\ 0 & 1 \end{bmatrix}$, which is called the **identity matrix**. The first row will give you the solution for *x*, because the coefficient of *y* is 0. The second row will give you the solution for *y*, because the coefficient of *x* is 0.

Activity 2 Use Row Operations to Solve a System

Use an augmented matrix to solve the system of equations.

$-5x + 3y = 6$
$x - y = 4$

Step 1 Write the augmented matrix: $\left[\begin{array}{cc|c} -5 & 3 & 6 \\ 1 & -1 & 4 \end{array}\right]$.

Step 2 Notice that the first element in the second row is 1. Interchange the rows so 1 can be in the upper left-hand corner.

$$\left[\begin{array}{cc|c} -5 & 3 & 6 \\ 1 & -1 & 4 \end{array}\right] \xrightarrow{\text{Interchange } R_1 \text{ and } R_2.} \left[\begin{array}{cc|c} 1 & -1 & 4 \\ -5 & 3 & 6 \end{array}\right]$$

Step 3 To make the first element in the second row a 0, multiply the first row by 5 and add the result to row 2.

$$\left[\begin{array}{cc|c} 1 & -1 & 4 \\ -5 & 3 & 6 \end{array}\right] \xrightarrow{5R_1 + R_2} \left[\begin{array}{cc|c} 1 & -1 & 4 \\ 0 & -2 & 26 \end{array}\right]$$

$1(5) + (-5) = 0$; $-1(5) + 3 = -2$; $4(5) + 6 = 26$

Step 4 To make the second element in the second row a 1, multiply the second row by $-\frac{1}{2}$.

$$\left[\begin{array}{cc|c} 1 & -1 & 4 \\ 0 & -2 & 26 \end{array}\right] \xrightarrow{-\frac{1}{2}R_2} \left[\begin{array}{cc|c} 1 & -1 & 4 \\ 0 & 1 & -13 \end{array}\right]$$

$0\left(-\frac{1}{2}\right) = 0$; $-2\left(-\frac{1}{2}\right) = 1$; $26\left(-\frac{1}{2}\right) = -13$

Step 5 To make the second element in the second row a 0, add the rows together.

$$\left[\begin{array}{cc|c} 1 & -1 & 4 \\ 0 & 1 & -13 \end{array}\right] \xrightarrow{R_2 + R_1} \left[\begin{array}{cc|c} 1 & 0 & -9 \\ 0 & 1 & -13 \end{array}\right]$$

$1 + 0 = 1$; $-1 + 1 = 0$; $4 + (-13) = -9$

The solution is $(-9, -13)$.

Model and Analyze

Write an augmented matrix for each system of equations. Then solve the system.

1. $x + y = -3$
$x - y = 1$

2. $x - y = -2$
$2x + 2y = 12$

3. $3x - 4y = -27$
$x + 2y = 11$

4. $x + 4y = -6$
$2x - 5y = 1$

5. $x - 3y = -2$
$4x + y = 31$

6. $x + 2y = 3$
$-3x + 3y = 27$

LESSON 6-6 Systems of Inequalities

Then

- You graphed and solved linear inequalities.

Now

1. Solve systems of linear inequalities by graphing.
2. Apply systems of linear inequalities.

Why?

- Jacui is beginning an exercise program that involves an intense cardiovascular workout. Her trainer recommends that for a person her age, her heart rate should stay within the following range as she exercises.
  - It should be higher than 102 beats per minute.
  - It should not exceed 174 beats per minute.

  The graph shows the maximum and minimum target heart rate for people ages 0 to 30 as they exercise. If the preferred range is in light green, how old do you think Jacui is?

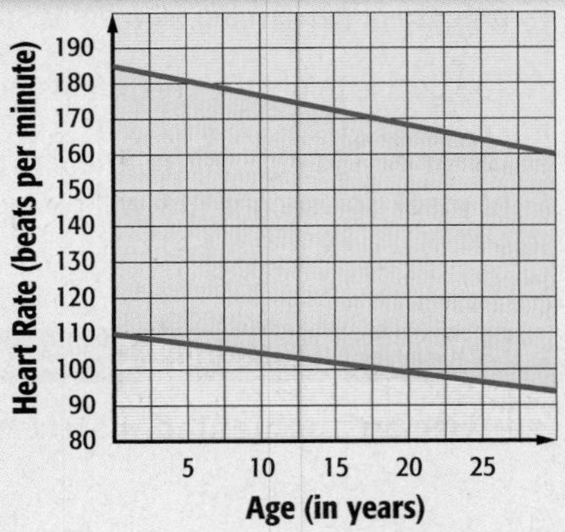

NewVocabulary
system of inequalities

Common Core State Standards

Content Standards
A.REI.12 Graph the solutions to a linear inequality in two variables as a halfplane (excluding the boundary in the case of a strict inequality), and graph the solution set to a system of linear inequalities in two variables as the intersection of the corresponding half-planes.

Mathematical Practices
1 Make sense of problems and persevere in solving them.
6 Attend to precision.

1 Systems of Inequalities

The graph above is a graph of two inequalities. A set of two or more inequalities with the same variables is called a **system of inequalities**.

The solution of a system of inequalities with two variables is the set of ordered pairs that satisfy all of the inequalities in the system. The solution set is represented by the overlap, or intersection, of the graphs of the inequalities.

Example 1 Solve by Graphing

Solve the system of inequalities by graphing.

$y > -2x + 1$
$y \leq x + 3$

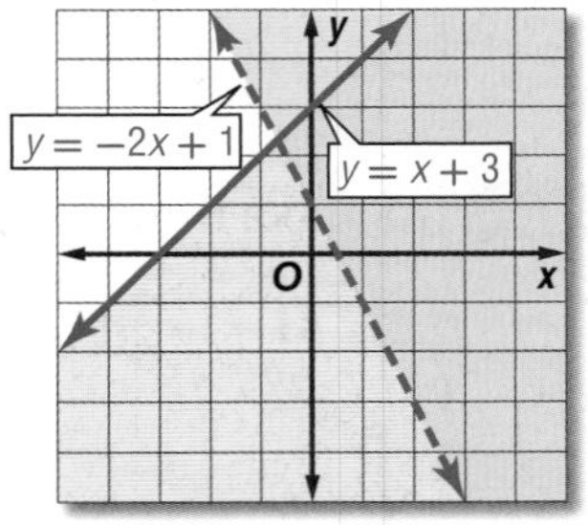

The graph of $y = -2x + 1$ is dashed and is not included in the graph of the solution. The graph of $y = x + 3$ is solid and is included in the graph of the solution.

The solution of the system is the set of ordered pairs in the intersection of the graphs of $y > -2x + 1$ and $y \leq x + 3$. This region is shaded in green.

When graphing more than one region, it is helpful to use two different colored pencils or two different patterns for each region. This will make it easier to see where the regions intersect and find possible solutions.

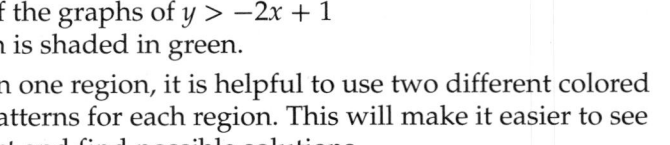

GuidedPractice

1A. $y \leq 3$
$x + y \geq 1$

1B. $2x + y \geq 2$
$2x + y < 4$

1C. $y \geq -4$
$3x + y \leq 2$

1D. $x + y > 2$
$-4x + 2y < 8$

Sometimes the regions never intersect. When this happens, there is no solution because there are no points in common.

StudyTip

Parallel Boundaries A system of equations represented by parallel lines does not have a solution. However, a system of inequalities with parallel boundaries can have a solution. For example:

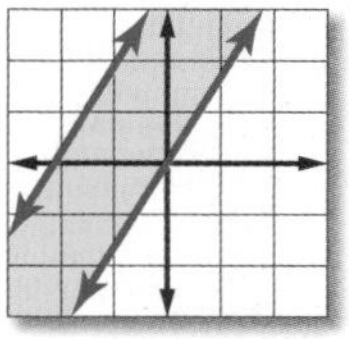

Example 2 No Solution

Solve the system of inequalities by graphing.

$3x - y \geq 2$
$3x - y < -5$

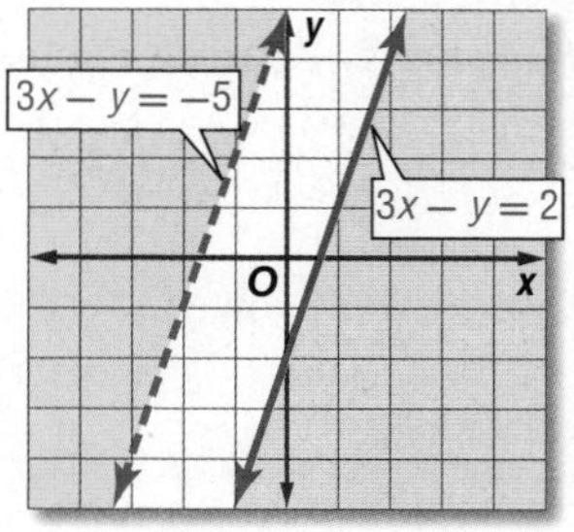

The graphs of $3x - y = 2$ and $3x - y = -5$ are parallel lines. The two regions do not intersect at any point, so the system has no solution.

GuidedPractice

2A. $y > 3$
$y < 1$

2B. $x + 6y \leq 2$
$y \geq -\frac{1}{6}x + 7$

2 Apply Systems of Inequalities

When using a system of inequalities to describe constraints on the possible combinations in a real-world problem, sometimes only whole-number solutions will make sense.

Real-WorldLink

Student government might be a good activity for you if you like to bring about change, plan events, and work with others.

Real-World Example 3 Whole-Number Solutions

ELECTIONS Monifa is running for student council. The election rules say that for the election to be valid, at least 80% of the 900 students must vote. Monifa knows that she needs more than 330 votes to win.

a. Define the variables, and write a system of inequalities to represent this situation. Then graph the system.

Let r = the number of votes required by the election rules; 80% of 900 students is 720 students. So $r \geq 720$.

Let v = the number of votes that Monifa needs to win. So $v > 330$.

The system of inequalities is $r \geq 720$ and $v > 330$.

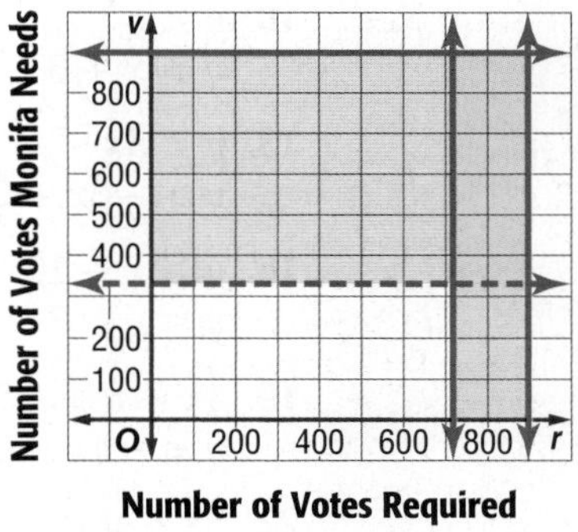

b. Name one viable option.

Only whole-number solutions make sense in this problem. One possible solution is (800, 400); 800 students voted and Monifa received 400 votes.

GuidedPractice

3. FUNDRAISING The Theater Club is selling shirts. They have only enough supplies to print 120 shirts. They will sell sweatshirts for \$22 and T-shirts for \$15, with a goal of at least \$2000 in sales.

A. Define the variables, and write a system of inequalities to represent this situation.

B. Then graph the system.

C. Name one possible solution.

D. Is (45, 30) a solution? Explain.

Check Your Understanding

Examples 1–2 **Solve each system of inequalities by graphing.**

1. $x \geq 4$
$y \leq x - 3$

2. $y > -2$
$y \leq x + 9$

3. $y < 3x + 8$
$y \geq 4x$

4. $3x - y \geq -1$
$2x + y \geq 5$

5. $y \leq 2x - 7$
$y \geq 2x + 7$

6. $y > -2x + 5$
$y \geq -2x + 10$

7. $2x + y \leq 5$
$2x + y \leq 7$

8. $5x - y < -2$
$5x - y > 6$

Example 3

9. **AUTO RACING** At a racecar driving school there are safety requirements.

a. Define the variables, and write a system of inequalities to represent the height and weight requirements in this situation. Then graph the system.

b. Name one possible solution.

c. Is (50, 180) a solution? Explain.

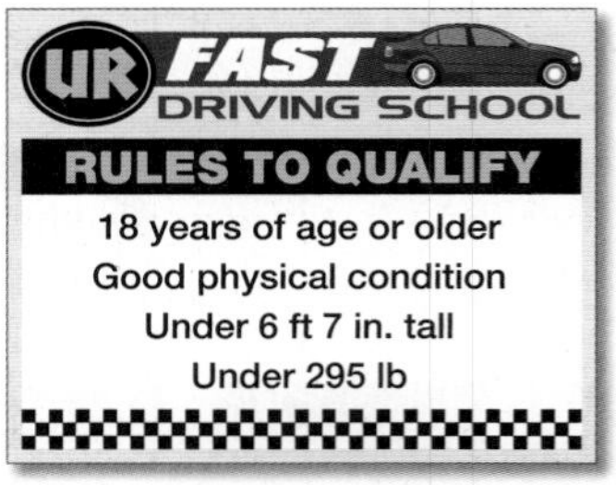

Practice and Problem Solving

Examples 1–2 **Solve each system of inequalities by graphing.**

10. $y < 6$
$y > x + 3$

11. $y \geq 0$
$y \leq x - 5$

12. $y \leq x + 10$
$y > 6x + 2$

13. $y < 5x - 2$
$y > -6x + 2$

14. $2x - y \leq 6$
$x - y \geq -1$

15. $3x - y > -5$
$5x - y < 9$

16. $y \geq x + 10$
$y \leq x - 3$

17. $y < 5x - 5$
$y > 5x + 9$

18. $y \geq 3x - 5$
$3x - y > -4$

19. $4x + y > -1$
$y < -4x + 1$

20. $3x - y \geq -2$
$y < 3x + 4$

21. $y > 2x - 3$
$2x - y \geq 1$

22. $5x - y < -6$
$3x - y \geq 4$

23. $x - y \leq 8$
$y < 3x$

24. $4x + y < -2$
$y > -4x$

Example 3

25. **ICE RINKS** Ice resurfacers are used for rinks of at least 1000 square feet and up to 17,000 square feet. The price ranges from as little as \$10,000 to as much as \$150,000.

a. Define the variables, and write a system of inequalities to represent this situation. Then graph the system.

b. Name one possible solution.

c. Is (15,000, 30,000) a solution? Explain.

26. CCSS **MODELING** Josefina works between 10 and 30 hours per week at a pizzeria. She earns \$6.50 an hour, but can earn tips when she delivers pizzas.

a. Write a system of inequalities to represent the dollars d she could earn for working h hours in a week.

b. Graph this system.

c. If Josefina received \$17.50 in tips and earned a total of \$180 for the week, how many hours did she work?

Solve each system of inequalities by graphing.

27. $x + y \geq 1$
$x + y \leq 2$

28. $3x - y < -2$
$3x - y < 1$

29. $2x - y \leq -11$
$3x - y \geq 12$

30. $y < 4x + 13$
$4x - y \geq 1$

31. $4x - y < -3$
$y \geq 4x - 6$

32. $y \leq 2x + 7$
$y < 2x - 3$

33. $y > -12x + 1$
$y \leq 9x + 2$

34. $2y \geq x$
$x - 3y > -6$

35. $x - 5y > -15$
$5y \geq x - 5$

36. CLASS PROJECT An economics class formed a company to sell school supplies. They would like to sell at least 20 notebooks and 50 pens per week, with a goal of earning at least $60 per week.

a. Define the variables, and write a system of inequalities to represent this situation.

b. Graph the system.

c. Name one possible solution.

37. FINANCIAL LITERACY Opal makes $15 per hour working for a photographer. She also coaches a competitive soccer team for $10 per hour. Opal needs to earn at least $90 per week, but she does not want to work more than 20 hours per week.

a. Define the variables, and write a system of inequalities to represent this situation.

b. Graph this system.

c. Give two possible solutions to describe how Opal can meet her goals.

d. Is (2, 2) a solution? Explain.

H.O.T. Problems Use Higher-Order Thinking Skills

38. CHALLENGE Create a system of inequalities equivalent to $|x| \leq 4$.

39. REASONING State whether the following statement is *sometimes*, *always*, or *never* true. Explain your answer with an example or counterexample.

Systems of inequalities with parallel boundaries have no solutions.

40. REASONING Describe the graph of the solution of this system without graphing.
$6x - 3y \leq -5$
$6x - 3y \geq -5$

41. OPEN ENDED One inequality in a system is $3x - y > 4$. Write a second inequality so that the system will have no solution.

42. CCSS PRECISION Graph the system of inequalities. Estimate the area of the solution.
$y \geq 1$
$y \leq x + 4$
$y \leq -x + 4$

43. WRITING IN MATH Refer to the beginning of the lesson. Explain what each colored region of the graph represents. Explain how shading in various colors can help to clearly show the solution set of a system of inequalities.

Standardized Test Practice

44. EXTENDED RESPONSE To apply for a scholarship, you must have a minimum of 20 hours of community service and a grade-point average of at least 3.75. Another scholarship requires at least 40 hours of community service and a minimum grade-point average of 3.0.

a. Write a system of inequalities to represent the credentials you must have to apply for both scholarships.

b. Graph the system of inequalities.

c. If you are eligible for both scholarships, give one possible solution.

45. GEOMETRY What is the measure of $\angle 1$?

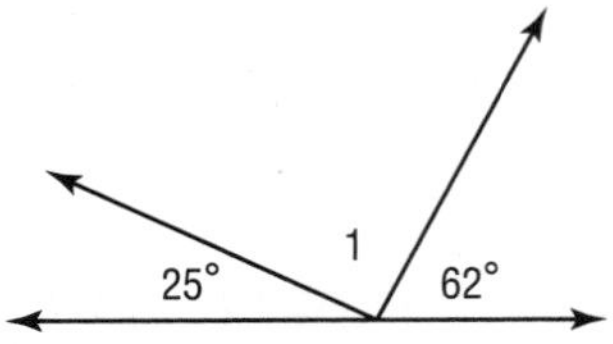

A 83°
B 87°
C 90°
D 93°

46. GEOMETRY What is the volume of the triangular prism?

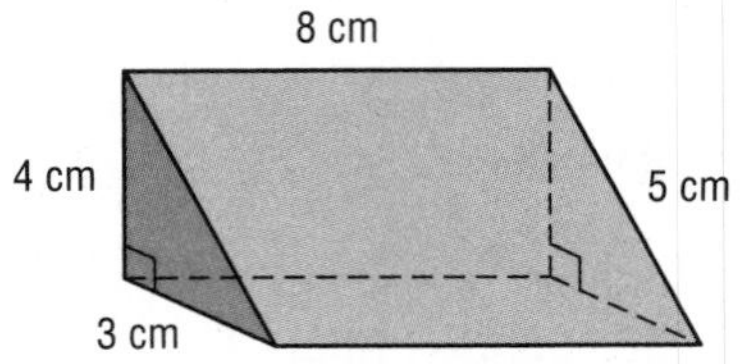

F 120 cm^3
G 96 cm^3
H 48 cm^3
J 30 cm^3

47. Ten pounds of fresh tomatoes make about 15 cups of cooked tomatoes. How many cups of cooked tomatoes does one pound of fresh tomatoes make?

A $1\frac{1}{2}$ cups
B 3 cups
C 4 cups
D 5 cups

Spiral Review

48. CHEMISTRY Orion Labs needs to make 500 gallons of 34% acid solution. The only solutions available are a 25% acid solution and a 50% acid solution. Write and solve a system of equations to find the number of gallons of each solution that should be mixed to make the 34% solution. (Lesson 6-5)

Use elimination to solve each system of equations. (Lesson 6-4)

49. $x + y = 7$
$2x + y = 11$

50. $a - b = 9$
$7a + b = 7$

51. $q + 4r = -8$
$3q + 2r = 6$

52. ENTERTAINMENT A group of 11 adults and children bought tickets for the baseball game. If the total cost was $156, how many of each type of ticket did they buy? (Lesson 6-4)

Graph each inequality. (Lesson 5-6)

53. $4x - 2 \geq 2y$

54. $9x - 3y < 0$

55. $2y \leq -4x - 6$

Skills Review

Evaluate each expression.

56. 3^3

57. 2^4

58. $(-4)^3$

EXTEND 6-6 Graphing Technology Lab
Systems of Inequalities

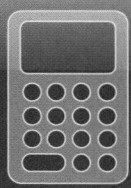

You can use TI-Nspire technology to explore systems of inequalities. To prepare your calculator, add a new **Graphs** page from the Home screen.

CCSS Common Core State Standards
Content Standards
A.REI.12 Graph the solutions to a linear inequality in two variables as a half-plane (excluding the boundary in the case of a strict inequality), and graph the solution set to a system of linear inequalities in two variables as the intersection of the corresponding half-planes.

Activity Graph Systems of Inequalities

Mr. Jackson owns a car washing and detailing business. It takes 20 minutes to wash a car and 60 minutes to detail a car. He works at most 8 hours per day and does at most 4 details per day. Write a system of linear inequalities to represent this situation.

First, write a linear inequality that represents the time it takes for car washing and car detailing. Let x represent the number of car washes, and let y represent the number of car details. Then $20x + 60y \leq 480$.

To graph this using a graphing calculator, solve for y.

$20x + 60y \leq 480$	Original inequality
$60y \leq -20x + 480$	Subtract $20x$ from each side and simplify.
$y \leq -\frac{1}{3}x + 8$	Divide each side by 60 and simplify.

Mr. Jackson does at most 4 details per day. This means that $y \leq 4$.

Step 1 Adjust the viewing window and then graph $y \leq 4$. Use the **Window Settings** option from the **Window/Zoom** menu to adjust the window to -4 to 30 for x and -2 to 10 for y. Keep the scales as **Auto**. Then enter **del** $\leq$ 4 **enter**.

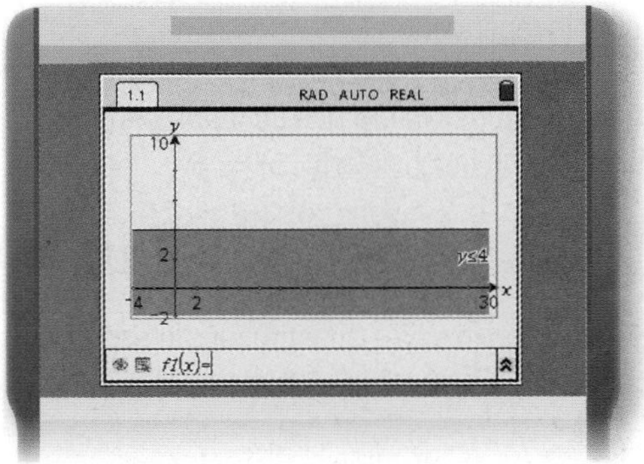

Step 2 Graph $y \leq -\frac{1}{3}x + 8$. Press **tab del** $\leq$ and then enter $-\frac{1}{3}x + 8$.

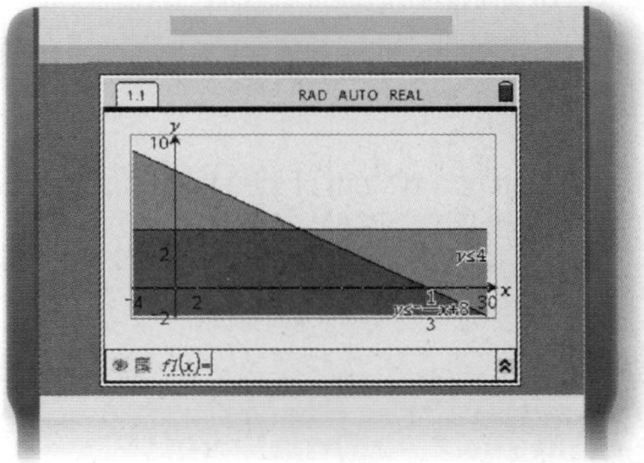

The darkest shaded region of the graph represents the solutions.

Analyze the Results

1. If Mr. Jackson charges \$75 for each car he details and \$25 for each car wash, what is the maximum amount of money he could earn in one day?
2. What is the greatest number of car washes that Mr. Jackson could do in a day? Explain your reasoning.

CHAPTER 6 Study Guide and Review

Study Guide

KeyConcepts

Systems of Equations (Lessons 6-1 through 6-5)

- A system with a graph of two intersecting lines has one solution and is *consistent and independent*.
- Graphing a system of equations can only provide approximate solutions. For exact solutions, you must use algebraic methods.
- In the substitution method, one equation is solved for a variable and the expression substituted into the second equation to find the value of another variable.
- In the elimination method, one variable is eliminated by adding or subtracting the equations.
- Sometimes multiplying one or both equations by a constant makes it easier to use the elimination method.
- The best method for solving a system of equations depends on the coefficients of the variables.

Systems of Inequalities (Lesson 6-6)

- A system of inequalities is a set of two or more inequalities with the same variables.
- The solution of a system of inequalities is the intersection of the graphs.

FOLDABLES® StudyOrganizer

Be sure the Key Concepts are noted in your Foldable.

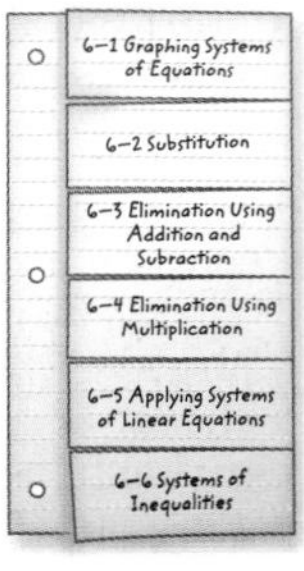

KeyVocabulary

augmented matrix (p. 370)
consistent (p. 335)
dependent (p. 335)
dimension (p. 370)
element (p. 370)
elimination (p. 350)
inconsistent (p. 335)
independent (p. 335)
matrix (p. 370)
substitution (p. 344)
system of equations (p. 335)
system of inequalities (p. 372)

VocabularyCheck

State whether each sentence is *true* or *false*. If *false*, replace the underlined term to make a true sentence.

1. If a system has at least one solution, it is said to be consistent.
2. If a consistent system has exactly two solution(s), it is said to be independent.
3. If a consistent system has an infinite number of solutions, it is said to be inconsistent.
4. If a system has no solution, it is said to be inconsistent.
5. Substitution involves substituting an expression from one equation for a variable in the other.
6. In some cases, dividing two equations in a system together will eliminate one of the variables. This process is called elimination.
7. A set of two or more inequalities with the same variables is called a system of equations.
8. When the graphs of the inequalities in a system of inequalities do not intersect, there are no solutions to the system.

Lesson-by-Lesson Review

6-1 Graphing Systems of Equations

Graph each system and determine the number of solutions that it has. If it has one solution, name it.

9. $x - y = 1$
$x + y = 5$

10. $y = 2x - 4$
$4x + y = 2$

11. $2x - 3y = -6$
$y = -3x + 2$

12. $-3x + y = -3$
$y = x - 3$

13. $x + 2y = 6$
$3x + 6y = 8$

14. $3x + y = 5$
$6x = 10 - 2y$

15. **MAGIC NUMBERS** Sean is trying to find two numbers with a sum of 14 and a difference of 4. Define two variables, write a system of equations, and solve by graphing.

Example 1

Graph the system and determine the number of solutions it has. If it has one solution, name it.

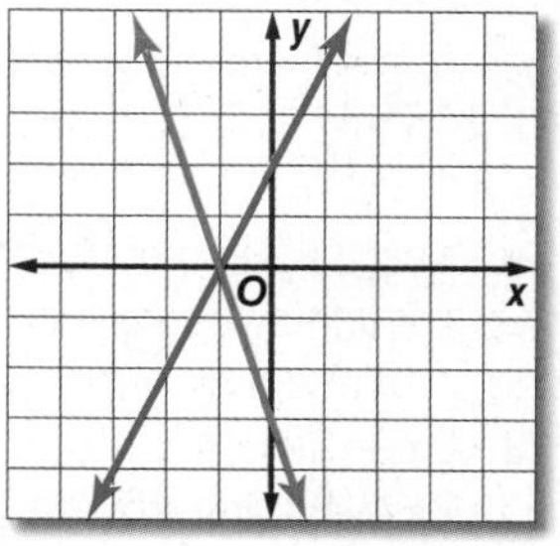

$y = 2x + 2$
$y = -3x - 3$

The lines appear to intersect at the point (−1, 0). You can check this by substituting −1 for x and 0 for y.

CHECK

$y = 2x + 2$	Original equation
$0 \stackrel{?}{=} 2(-1) + 2$	Substitution
$0 \stackrel{?}{=} -2 + 2$	Multiply.
$0 = 0$ ✓	
$y = -3x - 3$	Original equation
$0 \stackrel{?}{=} -3(-1) - 3$	Substitution
$0 \stackrel{?}{=} 3 - 3$	Multiply.
$0 = 0$ ✓	

The solution is (−1, 0).

6-2 Substitution

Use substitution to solve each system of equations.

16. $x + y = 3$
$x = 2y$

17. $x + 3y = -28$
$y = -5x$

18. $3x + 2y = 16$
$x = 3y - 2$

19. $x - y = 8$
$y = -3x$

20. $y = 5x - 3$
$x + 2y = 27$

21. $x + 3y = 9$
$x + y = 1$

22. **GEOMETRY** The perimeter of a rectangle is 48 inches. The length is 6 inches greater than the width. Define the variables, and write equations to represent this situation. Solve the system by using substitution.

Example 2

Use substitution to solve the system.

$3x - y = 18$
$y = x - 4$

$3x - y = 18$	First equation
$3x - (x - 4) = 18$	Substitute $x - 4$ for y.
$2x + 4 = 18$	Simplify.
$2x = 14$	Subtract 4 from each side.
$x = 7$	Divide each side by 2.

Use the value of x and either equation to find the value for y.

$y = x - 4$	Second equation
$= 7 - 4$ or 3	Substitute and simplify.

The solution is (7, 3).

Study Guide and Review *Continued*

6-3 Elimination Using Addition and Subtraction

Use elimination to solve each system of equations.

23. $x + y = 13$
$x - y = 5$

24. $-3x + 4y = 21$
$3x + 3y = 14$

25. $x + 4y = -4$
$x + 10y = -16$

26. $2x + y = -5$
$x - y = 2$

27. $6x + y = 9$
$-6x + 3y = 15$

28. $x - 4y = 2$
$3x + 4y = 38$

29. $2x + 2y = 4$
$2x - 8y = -46$

30. $3x + 2y = 8$
$x + 2y = 2$

31. BASEBALL CARDS Cristiano bought 24 baseball cards for $50. One type cost $1 per card, and the other cost $3 per card. Define the variables, and write equations to find the number of each type of card he bought. Solve by using elimination.

Example 3

Use elimination to solve the system of equations.

$\mathbf{3x - 5y = 11}$
$\mathbf{x + 5y = -3}$

$$\begin{array}{rl} 3x - 5y = 11 & \\ (+)\quad x + 5y = -3 & \\ \hline 4x \qquad = 8 & \text{The variable } y \text{ is eliminated.} \\ x = 2 & \text{Divide each side by 4.} \end{array}$$

Now, substitute 2 for x in either equation to find the value of y.

$$\begin{array}{rl} 3x - 5y = 11 & \text{First equation} \\ 3(2) - 5y = 11 & \text{Substitute.} \\ 6 - 5y = 11 & \text{Multiply.} \\ -5y = 5 & \text{Subtract 6 from each side.} \\ y = -1 & \text{Divide each side by } -5. \end{array}$$

The solution is (2, −1).

6-4 Elimination Using Multiplication

Use elimination to solve each system of equations.

32. $x + y = 4$
$-2x + 3y = 7$

33. $x - y = -2$
$2x + 4y = 38$

34. $3x + 4y = 1$
$5x + 2y = 11$

35. $-9x + 3y = -3$
$3x - 2y = -4$

36. $8x - 3y = -35$
$3x + 4y = 33$

37. $2x + 9y = 3$
$5x + 4y = 26$

38. $-7x + 3y = 12$
$2x - 8y = -32$

39. $8x - 5y = 18$
$6x + 6y = -6$

40. BAKE SALE On the first day, a total of 40 items were sold for $356. Define the variables, and write a system of equations to find the number of cakes and pies sold. Solve by using elimination.

Example 4

Use elimination to solve the system of equations.

$\mathbf{3x + 6y = 6}$
$\mathbf{2x + 3y = 5}$

Notice that if you multiply the second equation by −2, the coefficients of the y-terms are additive inverses.

$$\begin{array}{lcr} 3x + 6y = 6 & & 3x + 6y = 6 \\ 2x + 3y = 5 & \xrightarrow{\text{Multiply by } -2.} & (+)\ -4x - 6y = -10 \\ \hline & & -x \qquad = -4 \\ & & x = 4 \end{array}$$

Now, substitute 4 for x in either equation to find the value of y.

$$\begin{array}{rl} 2x + 3y = 5 & \text{Second equation} \\ 2(4) + 3y = 5 & \text{Substitution} \\ 8 + 3y = 5 & \text{Multiply.} \\ 3y = -3 & \text{Subtract 8 from both sides.} \\ y = -1 & \text{Divide each side by 3.} \end{array}$$

The solution is (4, −1).

6-5 Applying Systems of Linear Equations

Determine the best method to solve each system of equations. Then solve the system.

41. $y = x - 8$
$y = -3x$

42. $y = -x$
$y = 2x$

43. $x + 3y = 12$
$x = -6y$

44. $x + y = 10$
$x - y = 18$

45. $3x + 2y = -4$
$5x + 2y = -8$

46. $6x + 5y = 9$
$-2x + 4y = 14$

47. $3x + 4y = 26$
$2x + 3y = 19$

48. $11x - 6y = 3$
$5x - 8y = -25$

49. **COINS** Tionna has saved dimes and quarters in her piggy bank. Define the variables, and write a system of equations to determine the number of dimes and quarters. Then solve the system using the best method for the situation.

50. **FAIR** At a county fair, the cost for 4 slices of pizza and 2 orders of French fries is $21.00. The cost of 2 slices of pizza and 3 orders of French fries is $16.50. To find out how much a single slice of pizza and an order of French fries costs, define the variables and write a system of equations to represent the situation. Determine the best method to solve the system of equations. Then solve the system. (Lesson 6-5)

Example 5

Determine the best method to solve the system of equations. Then solve the system.

$3x + 5y = 4$
$4x + y = -6$

The coefficient of y is 1 in the second equation. So solving by substitution is a good method. Solve the second equation for y.

$4x + y = -6$	Second equation
$y = -6 - 4x$	Subtract $4x$ from each side.

Substitute $-6 - 4x$ for y in the first equation.

$3x + 5(-6 - 4x) = 4$	Substitute.
$3x - 30 - 20x = 4$	Distributive Property
$-17x - 30 = 4$	Simplify.
$-17x = 34$	Add 30 to each side.
$x = -2$	Divide by -17.

Last, substitute -2 for x in either equation to find y.

$4x + y = -6$	Second equation
$4(-2) + y = -6$	Substitute.
$-8 + y = -6$	Multiply.
$y = 2$	Add 8 to each side.

The solution is $(-2, 2)$.

Study Guide and Review *Continued*

6-6 Systems of Inequalities

Solve each system of inequalities by graphing.

51. $x > 3$
$y < x + 2$

52. $y \leq 5$
$y > x - 4$

53. $y < 3x - 1$
$y \geq -2x + 4$

54. $y \leq -x - 3$
$y \geq 3x - 2$

55. **JOBS** Kishi makes $7 an hour working at the grocery store and $10 an hour delivering newspapers. She cannot work more than 20 hours per week. Graph two inequalities that Kishi can use to determine how many hours she needs to work at each job if she wants to earn at least $90 per week.

Example 6

Solve the system of inequalities by graphing.

$y < 3x + 1$
$y \geq -2x + 3$

The solution set of the system is the set of ordered pairs in the intersection of the two graphs. This portion is shaded in the graph below.

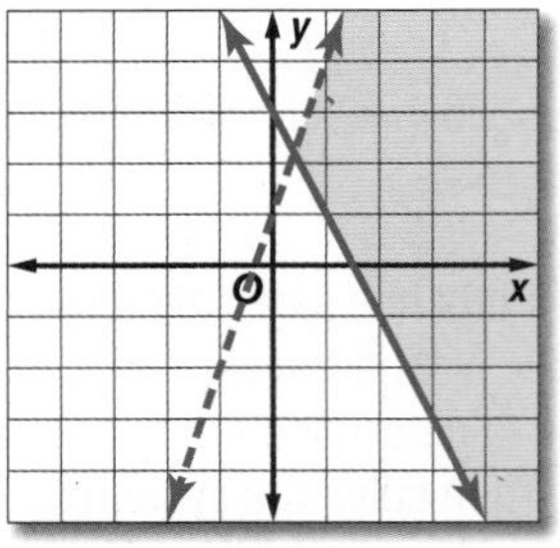

CHAPTER 6

Practice Test

Graph each system and determine the number of solutions that it has. If it has one solution, name it.

1. $y = 2x$
   $y = 6 - x$
2. $y = x - 3$
   $y = -2x + 9$
3. $x - y = 4$
   $x + y = 10$
4. $2x + 3y = 4$
   $2x + 3y = -1$

Use substitution to solve each system of equations.

5. $y = x + 8$
   $2x + y = -10$
6. $x = -4y - 3$
   $3x - 2y = 5$

7. **GARDENING** Corey has 42 feet of fencing around his garden. The garden is rectangular in shape, and its length is equal to twice the width minus 3 feet. Define the variables, and write a system of equations to find the length and width of the garden. Solve the system by using substitution.

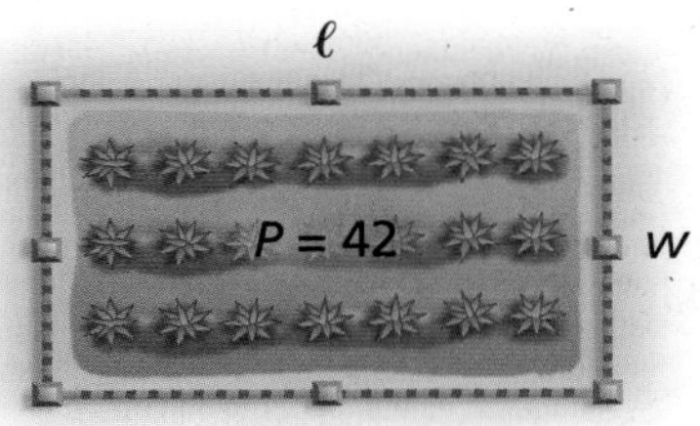

8. **MULTIPLE CHOICE** Use elimination to solve the system.

$$6x - 4y = 6$$
$$-6x + 3y = 0$$

   A (5, 6)

   B (−3, −6)

   C (1, 0)

   D (4, −8)

9. **SHOPPING** Shelly has $175 to shop for jeans and sweaters. Each pair of jeans costs $25, each sweater costs $20, and she buys 8 items. Determine the number of pairs of jeans and sweaters Shelly bought.

Use elimination to solve each system of equations.

10. $x + y = 13$
    $x - y = 5$
11. $3x + 7y = 2$
    $3x - 4y = 13$
12. $x + y = 8$
    $x - 3y = -4$
13. $2x + 6y = 18$
    $3x + 2y = 13$

14. **MAGAZINES** Julie subscribes to a sports magazine and a fashion magazine. She received 24 issues this year. The number of fashion issues is 6 less than twice the number of sports issues. Define the variables, and write a system of equations to find the number of issues of each magazine.

Determine the best method to solve each system of equations. Then solve the system.

15. $y = 3x$
    $x + 2y = 21$
16. $x + y = 12$
    $y = x - 4$
17. $x + y = 15$
    $x - y = 9$
18. $3x + 5y = 7$
    $2x - 3y = 11$

19. **OFFICE SUPPLIES** At a sale, Ricardo bought 24 reams of paper and 4 inkjet cartridges for $320. Britney bought 2 reams of paper and 1 inkjet cartridge for $50. The reams of paper were all the same price and the inkjet cartridges were all the same price. Write a system of equations to represent this situation. Determine the best method to solve the system of equations. Then solve the system.

Solve each system of inequalities by graphing.

20. $x > 2$
    $y < 4$
21. $x + y \leq 5$
    $y \geq x + 2$
22. $3x - y > 9$
    $y > -2x$
23. $y \geq 2x + 3$
    $-4x - 3y > 12$

Preparing for Standardized Tests

Guess and Check

It is very important to pace yourself and keep track of how much time you have when taking a standardized test. If time is running short, or if you are unsure how to solve a problem, the guess and check strategy may help you determine the correct answer quickly.

Strategies for Guessing and Checking

Step 1

Carefully look over each possible answer choice, and evaluate for reasonableness. Eliminate unreasonable answers.

Ask yourself:

- Are there any answer choices that are clearly incorrect?
- Are there any answer choices that are not in the proper format?
- Are there any answer choices that do not have the proper units for the correct answer?

Step 2

For the remaining answer choices, use the guess and check method.

- **Equations:** If you are solving an equation, substitute the answer choice for the variable and see if this results in a true number sentence.
- **Inequalities:** Likewise, you can substitute the answer choice for the variable and see if it satisfies the inequality.
- **System of Equations:** Find the answer choice that satisfies both equations of the system.

Step 3

Choose an answer choice and see if it satisfies the constraints of the problem statement. Identify the correct answer.

- If the answer choice you are testing does not satisfy the problem, move on to the next reasonable guess and check it.
- When you find the correct answer choice, stop. You do not have to check the other answer choices.

Sean Justice/Stone/Getty Images

Standardized Test Example

Read the problem. Identify what you need to know. Then use the information in the problem to solve.

Solve $\begin{cases} 4x - 8y = 20 \\ -3x + 5y = -14 \end{cases}$.

A (5, 0) **C** (3, −1)

B (4, −2) **D** (−6, −5)

The solution of a system of equations is an ordered pair, (x, y). Since all four answer choices are of this form, they are all possible correct answers and must be checked. Begin with the first answer choice and substitute it in each equation. Continue until you find the ordered pair that satisfies both equations of the system.

	First Equation	Second Equation
Guess: (5, 0)	$4x - 8y = 20$ $4(5) - 8(0) = 20$ ✓	$-3x + 5y = -14$ $-3(5) + 5(0) \neq -14$ ✗

	First Equation	Second Equation
Guess: (4, −2)	$4x - 8y = 20$ $4(4) - 8(-2) \neq 20$ ✗	$-3x + 5y = -14$ $-3(4) + 5(-2) \neq -14$ ✗

	First Equation	Second Equation
Guess: (3, −1)	$4x - 8y = 20$ $4(3) - 8(-1) = 20$ ✓	$-3x + 5y = -14$ $-3(3) + 5(-1) = -14$ ✓

The ordered pair (3, −1) satisfies both equations of the system. So, the correct answer is C.

Exercises

Read each problem. Eliminate any unreasonable answers. Then use the information in the problem to solve.

1. Gina bought 5 hot dogs and 3 soft drinks at the ball game for \$11.50. Renaldo bought 4 hot dogs and 2 soft drinks for \$8.50. How much does a single hot dog and a single drink cost?

A hot dogs: \$1.25
soft drinks: \$1.50

B hot dogs: \$1.25
soft drinks: \$1.75

C hot dogs: \$1.50
soft drinks: \$1.25

D hot dogs: \$1.50
soft drinks: \$1.75

2. The bookstore hopes to sell at least 30 binders and calculators each week. The store also hopes to have sales revenue of at least \$200 in binders and calculators. How many binders and calculators could be sold to meet both of these sales goals?

Store Prices

Item	Price
binders	\$3.65
calculators	\$14.80

F 25 binders,
5 calculators

G 12 binders,
15 calculators

H 22 binders,
9 calculators

J 28 binders,
6 calculators

CHAPTER 6

Standardized Test Practice

Cumulative, Chapters 1 through 6

Multiple Choice

Read each question. Then fill in the correct answer on the answer document provided by your teacher or on a sheet of paper.

1. Which of the following terms ***best*** describes the system of equations shown in the graph?

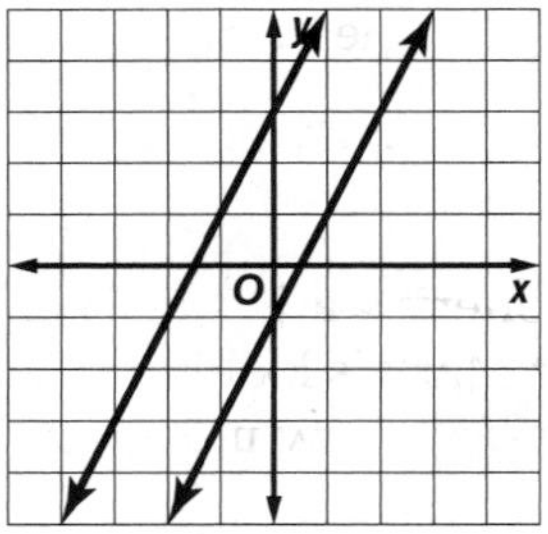

A consistent

B consistent and dependent

C consistent and independent

D inconsistent

2. Use substitution to solve the system of equations below.

$$\begin{cases} y = 4x - 7 \\ 3x - 2y = -1 \end{cases}$$

F (3, 5) **H** (5, −2)

G (4, −1) **J** (−6, 2)

3. Which ordered pair is the solution of the system of linear equations shown below?

$$\begin{cases} 3x - 8y = -50 \\ 3x - 5y = -38 \end{cases}$$

A $\left(\frac{5}{8}, \frac{3}{2}\right)$ **C** $\left(-\frac{2}{7}, \frac{4}{9}\right)$

B (4, −9) **D** (−6, 4)

4. A home goods store received $881 from the sale of 4 table saws and 9 electric drills. If the receipts from the saws exceeded the receipts from the drills by $71, what is the price of an electric drill?

F $45 **H** $108

G $59 **J** $119

5. A region is defined by this system.

$$y > -\frac{1}{2}x - 1$$
$$y > -x + 3$$

In which quadrant(s) of the coordinate plane is the region located?

A I and IV only **C** I, II, and IV only

B III only **D** II and III only

6. Which of the following terms ***best*** describes the system of equations shown in the graph?

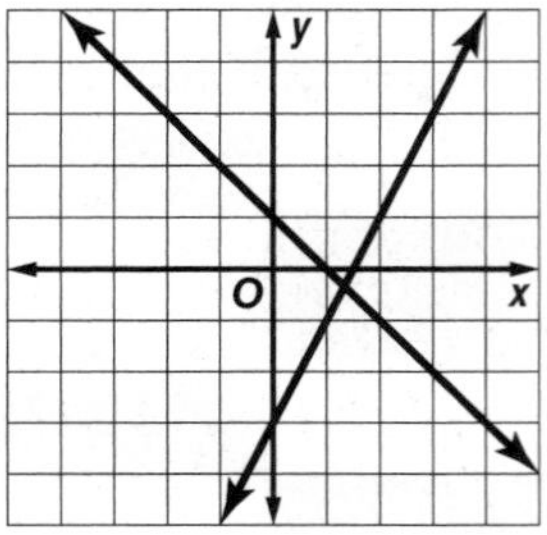

F consistent

G consistent and independent

H consistent and dependent

J inconsistent

7. Use elimination to solve the system of equations below.

$$3x + 2y = -2$$
$$2x - 2y = -18$$

A (1, 3) **C** (−2, −3)

B (7, −4) **D** (−4, 5)

8. What is the solution of the following system of equations?

$$\begin{cases} y = 6x - 1 \\ y = 6x + 1 \end{cases}$$

F (2, 11) **H** (7, 5)

G (−3, −14) **J** no solution

Test-Taking Tip

Question 8 You can subtract the second equation from the first equation to eliminate the *x*-variable. Then solve for *y*.

Short Response/Gridded Response

Record your answers on the answer sheet provided by your teacher or on a sheet of paper.

9. GRIDDED RESPONSE Angie and her sister have $15 to spend on pizza. A medium pizza costs $11.50 plus $0.75 per topping. What is the maximum number of toppings Angie and her sister can get on their pizza?

10. Write an inequality for the graph below.

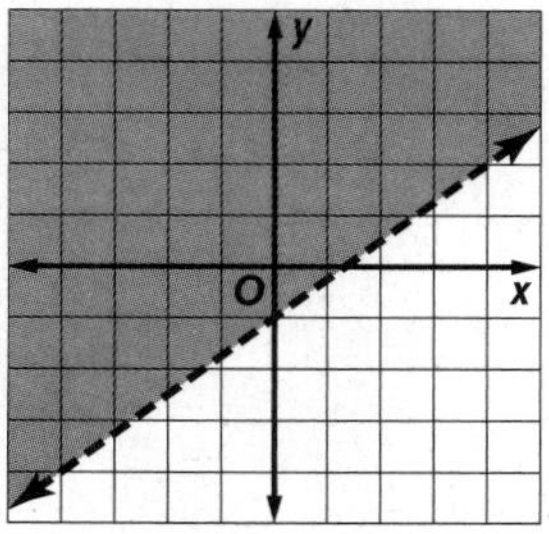

11. GRIDDED RESPONSE Christy is taking a road trip. After she drives 12 more miles, she will have driven at least half of the 108-mile trip. What is the least number of miles she has driven so far?

12 Write an equation in slope-intercept form with a slope of $-\frac{2}{3}$ and a y-intercept of 6.

13. A rental company charges $9.50 per hour for a scooter plus a $15 fee. Write an equation in slope-intercept form for the total rental cost C of renting a scooter for h hours.

14. GRIDDED RESPONSE A computer supplies store is having a storewide sale this weekend. An inkjet printer that normally sells for $179.00 is on sale for $143.20. What is the percent discount of the sale price?

15. In 1980, the population of Kentucky was about 3.66 million people. By 2000, this number had grown to about 4.04 million people. What was the annual rate of change in population from 1980 to 2000?

16. Joseph's cell phone service charges him $0.15 per text. Write an equation that represents the cost C of his cell phone service for t texts sent each month.

17. A store is offering a $15 mail-in-rebate on all printers. If Mark is looking at printers that range from $45 to $89, how much can he expect to pay?

Extended Response

Record your answers on a sheet of paper. Show your work.

18. The table shows how many canned goods were collected during the first day of a charity food drive.

Food Drive Day 1 Results	
Class	Number Collected
10th graders	78
11th graders	80
12th graders	92

a. Estimate how many canned goods will be collected during the 5-day food drive. Explain your answer.

b. Is this estimate a reasonable expectation? Explain.

Need ExtraHelp?

If you missed Question...	1	2	3	4	5	6	7	8	9	10	11	12	13	14	15	16	17	18
Go to Lesson...	6-1	6-2	6-3	6-3	6-6	6-1	6-3	6-3	5-3	5-6	5-3	4-2	4-2	2-7	2-7	3-4	5-4	1-4

CHAPTER 7

Exponents and Exponential Functions

Then

You evaluated expressions involving exponents.

Now

In this chapter, you will:

- Simplify and perform operations on expressions involving exponents.
- Extend the properties of integer exponents to rational exponents.
- Use scientific notation.
- Graph and use exponential functions.

Why? ▲

SPACE The Very Large Array is an arrangement of 27 radio antennas in a Y pattern. The data the antennas collect is used by astronomers around the world to study the planets and stars. Astrophysicists use and apply properties of exponents to model the distance and orbit of celestial bodies.

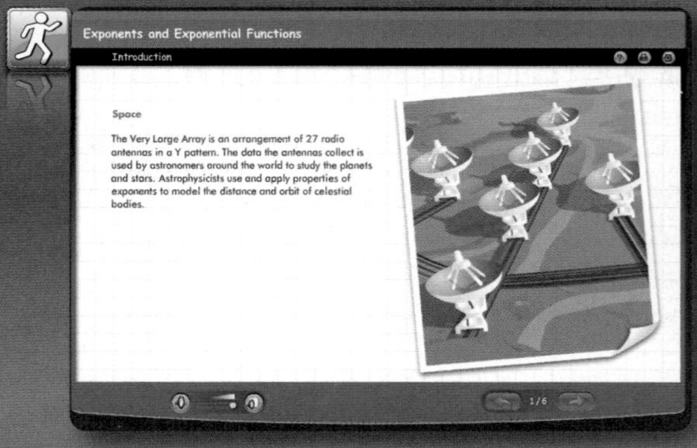

connectED.mcgraw-hill.com **Your Digital Math Portal**

Animation | Vocabulary | eGlossary | Personal Tutor | Virtual Manipulatives | Graphing Calculator | Audio | Foldables | Self-Check Practice | Worksheets

Get Ready for the Chapter

Diagnose Readiness | You have two options for checking prerequisite skills.

1 Textbook Option Take the Quick Check below. Refer to the Quick Review for help.

QuickCheck

Write each expression using exponents.

1. $4 \cdot 4 \cdot 4 \cdot 4 \cdot 4$
2. $y \cdot y \cdot y$
3. $6 \cdot 6$
4. $2 \cdot 2 \cdot 2 \cdot 2 \cdot 2 \cdot 2 \cdot 2 \cdot 2 \cdot 2$
5. $b \cdot b \cdot b \cdot b \cdot b \cdot b$
6. $m \cdot m \cdot m \cdot p \cdot p \cdot p \cdot p \cdot p \cdot p$
7. $\frac{1}{3} \cdot \frac{1}{3} \cdot \frac{1}{3} \cdot \frac{1}{3} \cdot \frac{1}{3} \cdot \frac{1}{3} \cdot \frac{1}{3} \cdot \frac{1}{3}$
8. $\frac{x}{y} \cdot \frac{x}{y} \cdot \frac{x}{y} \cdot \frac{x}{y} \cdot \frac{w}{z} \cdot \frac{w}{z}$

Find the area or volume of each figure.

9. 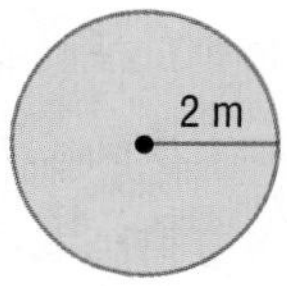

10. 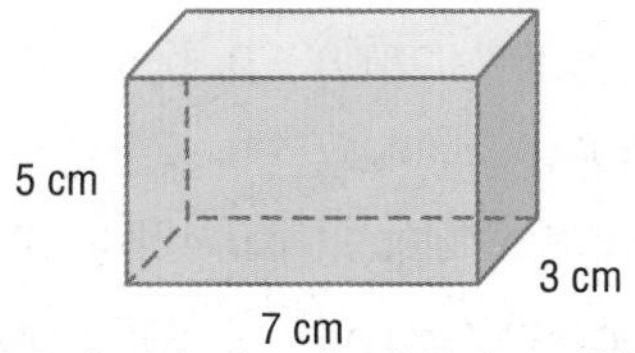

11. **PHOTOGRAPHY** A photo is 4 inches by 6 inches. What is the area of the photo?

Evaluate each expression.

12. 2^3
13. $(-5)^2$
14. 3^3
15. $(-4)^3$
16. $\left(\frac{2}{3}\right)^2$
17. $\left(\frac{1}{2}\right)^4$
18. **SCHOOL** The probability of guessing correctly on 5 true-false questions is $\left(\frac{1}{2}\right)^5$. Express this probability as a fraction without exponents.

QuickReview

Example 1

Write $5 \cdot 5 \cdot 5 \cdot 5 + x \cdot x \cdot x$ using exponents.

4 factors of 5 is 5^4.

3 factors of x is x^3.

So, $5 \cdot 5 \cdot 5 \cdot 5 + x \cdot x \cdot x = 5^4 + x^3$.

Example 2

Find the volume of the figure.

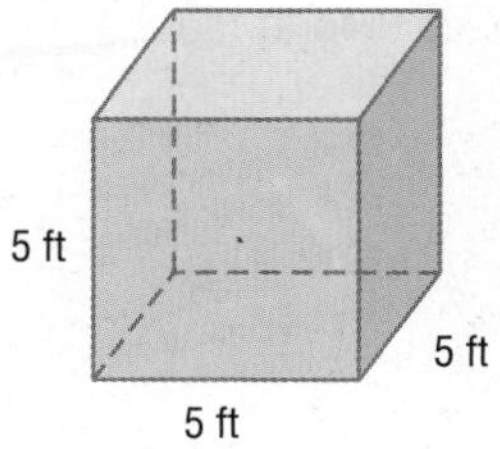

$V = \ell wh$ — Volume of a rectangular prism

$= 5 \cdot 5 \cdot 5$ or 125 — $\ell = 5$, $w = 5$, and $h = 5$

The volume is 125 cubic feet.

Example 3

Evaluate $\left(\frac{5}{7}\right)^2$.

$\left(\frac{5}{7}\right)^2 = \frac{5^2}{7^2}$ — Power of a Quotient

$= \frac{25}{49}$ — Simplify.

2 Online Option Take an online self-check Chapter Readiness Quiz at connectED.mcgraw-hill.com.

Get Started on the Chapter

You will learn several new concepts, skills, and vocabulary terms as you study Chapter 7. To get ready, identify important terms and organize your resources. You may wish to refer to Chapter 0 to review prerequisite skills.

FOLDABLES® StudyOrganizer

Exponents and Exponential Functions Make this Foldable to help you organize your Chapter 7 notes about exponents and exponential functions. Begin with nine sheets of notebook paper.

1 **Arrange** the paper into a stack.

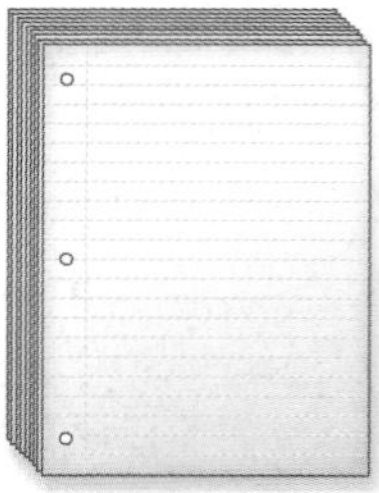

2 **Staple** along the left side. Starting with the second sheet of paper, cut along the right side to form tabs.

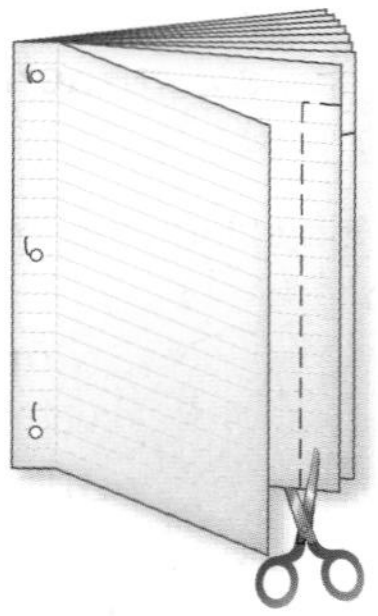

3 **Label** the cover sheet "Exponents and Exponential Functions" and label each tab with a lesson number.

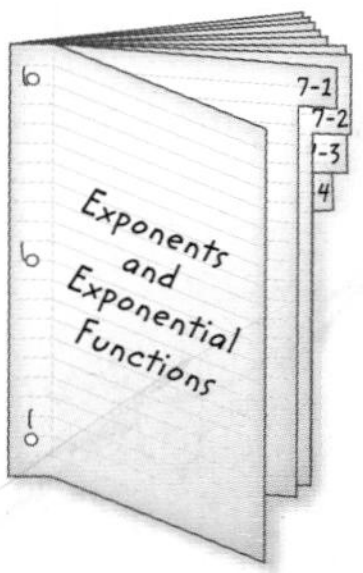

NewVocabulary

English		Español
monomial	p. 391	monomio
constant	p. 391	constante
zero exponent	p. 399	cero exponente
negative exponent	p. 400	exponente negativo
order of magnitude	p. 401	orden de magnitud
rational exponent	p. 406	exponent racional
cube root	p. 407	raíz cúbica
*n*th root	p. 407	raíz enésima
exponential equation	p. 409	ecuación exponencial
scientific notation	p. 414	notación científica
exponential function	p. 424	functión exponencial
exponential growth	p. 424	crecimiento exponencial
exponential decay	p. 424	desintegración exponencial
compound interest	p. 433	interés es compuesta
geometric sequence	p. 438	secuencia geométrica
common ratio	p. 438	proporción común
recursive formula	p. 445	fórmula recursiva

ReviewVocabulary

base base In an expression of the form x^n, the base is x.

Distributive Property Propiedad distributiva For any numbers a, b, and c, $a(b + c) = ab + ac$ and $a(b - c) = ab - ac$.

exponent exponente In an expression of the form x^n, the exponent is n. It indicates the number of times x is used as a factor.

exponent

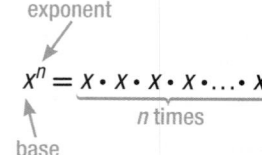

$$x^n = \underbrace{x \cdot x \cdot x \cdot x \cdot \ldots \cdot x}_{n \text{ times}}$$

base

LESSON 7-1

Multiplication Properties of Exponents

:: Then	:: Now	:: Why?
● You evaluated expressions with exponents.	● **1** Multiply monomials using the properties of exponents. **2** Simplify expressions using the multiplication properties of exponents.	● Many formulas contain *monomials*. For example, the formula for the horsepower of a car is $H = w\left(\frac{v}{234}\right)^3$. *H* represents the horsepower produced by the engine, *w* equals the weight of the car with passengers, and *v* is the velocity of the car at the end of a quarter of a mile. As the velocity increases, the horsepower increases.

NewVocabulary
monomial
constant

CCSS Common Core State Standards

Content Standards
A.SSE.2 Use the structure of an expression to identify ways to rewrite it.
F.IF.8b Use the properties of exponents to interpret expressions for exponential functions.

Mathematical Practices
8 Look for and express regularity in repeated reasoning.

1 Multiply Monomials

A **monomial** is a number, a variable, or the product of a number and one or more variables with nonnegative integer exponents. It has only one term. In the formula to calculate the horsepower of a car, the term $w\left(\frac{v}{234}\right)^3$ is a monomial.

An expression that involves division by a variable, like $\frac{ab}{c}$, is not a monomial.

A **constant** is a monomial that is a real number. The monomial $3x$ is an example of a *linear expression* since the exponent of x is 1. The monomial $2x^2$ is a *nonlinear expression* since the exponent is a positive number other than 1.

Example 1 Identify Monomials

Determine whether each expression is a monomial. Write *yes* or *no*. Explain your reasoning.

a. 10 — Yes; this is a constant, so it is a monomial.

b. $f + 24$ — No; this expression has addition, so it has more than one term.

c. h^2 — Yes; this expression is a product of variables.

d. j — Yes; single variables are monomials.

GuidedPractice

1A. $-x + 5$

1B. $23abcd^2$

1C. $\frac{xyz^2}{2}$

1D. $\frac{mp}{n}$

Recall that an expression of the form x^n is called a *power* and represents the result of multiplying x by itself n times. x is the *base*, and n is the *exponent*. The word *power* is also used sometimes to refer to the exponent.

exponent → 4; base → 3; 4 factors

$$3^4 = \underbrace{3 \cdot 3 \cdot 3 \cdot 3}_{\text{4 factors}} = 81$$

By applying the definition of a power, you can find the product of powers. Look for a pattern in the exponents.

$$2^2 \cdot 2^4 = \overbrace{2 \cdot 2}^{2 \text{ factors}} \cdot \overbrace{2 \cdot 2 \cdot 2 \cdot 2}^{4 \text{ factors}} \quad (2 + 4 = 6 \text{ factors})$$

$$4^3 \cdot 4^2 = \overbrace{4 \cdot 4 \cdot 4}^{3 \text{ factors}} \cdot \overbrace{4 \cdot 4}^{2 \text{ factors}} \quad (3 + 2 = 5 \text{ factors})$$

These examples demonstrate the property for the product of powers.

KeyConcept Product of Powers

Words	To multiply two powers that have the same base, add their exponents.
Symbols	For any real number a and any integers m and p, $a^m \cdot a^p = a^{m+p}$.
Examples	$b^3 \cdot b^5 = b^{3+5}$ or b^8 $\quad g^4 \cdot g^6 = g^{4+6}$ or g^{10}

PT

Example 2 Product of Powers

Simplify each expression.

a. $(6n^3)(2n^7)$

$(6n^3)(2n^7) = (6 \cdot 2)(n^3 \cdot n^7)$ — Group the coefficients and the variables.

$= (6 \cdot 2)(n^{3+7})$ — Product of Powers

$= 12n^{10}$ — Simplify.

b. $(3pt^3)(p^3t^4)$

$(3pt^3)(p^3t^4) = (3 \cdot 1)(p \cdot p^3)(t^3 \cdot t^4)$ — Group the coefficients and the variables.

$= (3 \cdot 1)(p^{1+3})(t^{3+4})$ — Product of Powers

$= 3p^4t^7$ — Simplify.

StudyTip

Coefficients and Powers of 1 A variable with no exponent or coefficient shown can be assumed to have an exponent and coefficient of 1. For example, $x = 1x^1$.

GuidedPractice

2A. $(3y^4)(7y^5)$

2B. $(-4rx^2t^3)(-6r^5x^2t)$

We can use the Product of Powers Property to find the power of a power. In the following examples, look for a pattern in the exponents.

$$(3^2)^4 = \overbrace{(3^2)(3^2)(3^2)(3^2)}^{4 \text{ factors}} = 3^{2+2+2+2} = 3^8$$

$$(r^4)^3 = \overbrace{(r^4)(r^4)(r^4)}^{3 \text{ factors}} = r^{4+4+4} = r^{12}$$

These examples demonstrate the property for the power of a power.

KeyConcept Power of a Power

Words	To find the power of a power, multiply the exponents.
Symbols	For any real number a and any integers m and p, $(a^m)^p = a^{m \cdot p}$.
Examples	$(b^3)^5 = b^{3 \cdot 5}$ or b^{15} $\quad (g^6)^7 = g^{6 \cdot 7}$ or g^{42}

StudyTip

CCSS Regularity The power rules are general methods. If you are unsure about when to multiply the exponents and when to add the exponents, write the expression in expanded form.

Standardized Test Example 3 Power of a Power

Simplify $[(2^3)^2]^4$.

A 2^{24} **B** 2^{12} **C** 2^{10} **D** 2^9

Read the Test Item

You need to apply the power of a power rule.

Solve the Test Item

$[(2^3)^2]^4 = (2^{3 \cdot 2})^4$	Power of a Power
$= (2^6)^4$	Simplify.
$= 2^{6 \cdot 4}$ or 2^{24}	Power of a Power

The correct choice is A.

GuidedPractice

3. Simplify $[(2^2)^2]^4$.

F 2^8 **G** 2^{10} **H** 2^{16} **J** 2^{24}

We can use the Product of Powers Property and the Power of a Power Property to find the power of a product. Look for a pattern in the exponents below.

$$(tw)^3 = \underbrace{(tw)(tw)(tw)}_{\text{3 factors}}$$
$$= (t \cdot t \cdot t)(w \cdot w \cdot w)$$
$$= t^3w^3$$

$$(2yz^2)^3 = \underbrace{(2yz^2)(2yz^2)(2yz^2)}_{\text{3 factors}}$$
$$= (2 \cdot 2 \cdot 2)(y \cdot y \cdot y)(z^2 \cdot z^2 \cdot z^2)$$
$$= 2^3y^3z^6 \text{ or } 8y^3z^6$$

These examples demonstrate the property for the power of a product.

KeyConcept Power of a Product

Words	To find the power of a product, find the power of each factor and multiply.
Symbols	For any real numbers a and b and any integer m, $(ab)^m = a^mb^m$.
Example	$(-2xy^3)^5 = (-2)^5x^5y^{15}$ or $-32x^5y^{15}$

Math-HistoryLink

Albert Einstein

(1879–1955) Albert Einstein is perhaps the most well-known scientist of the 20th century. His formula $E = mc^2$, where E represents the energy, m is the mass of the material, and c is the speed of light, shows that if mass is accelerated enough, it could be converted into usable energy.

Example 4 Power of a Product

GEOMETRY Express the area of the circle as a monomial.

Area $= \pi r^2$	Formula for the area of a circle
$= \pi(2xy^2)^2$	Replace r with $2xy^2$.
$= \pi(2^2x^2y^4)$	Power of a Product
$= 4x^2y^4\pi$	Simplify.

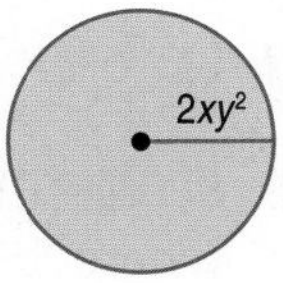

The area of the circle is $4x^2y^4\pi$ square units.

GuidedPractice

4A. Express the area of a square with sides of length $3xy^2$ as a monomial.

4B. Express the area of a triangle with height $4a$ and base $5ab^2$ as a monomial.

2 Simplify Expressions

We can combine and use these properties to simplify expressions involving monomials.

KeyConcept Simplify Expressions

To simplify a monomial expression, write an equivalent expression in which:

- each variable base appears exactly once,
- there are no powers of powers, and
- all fractions are in simplest form.

StudyTip

Simplify When simplifying expressions with multiple grouping symbols, begin at the innermost expression and work outward.

PT

Example 5 Simplify Expressions

Simplify $(3xy^4)^2[(-2y)^2]^3$.

$(3xy^4)^2[(-2y)^2]^3 = (3xy^4)^2(-2y)^6$	Power of a Power
$= (3)^2x^2(y^4)^2(-2)^6y^6$	Power of a Product
$= 9x^2y^8(64)y^6$	Power of a Power
$= 9(64)x^2 \cdot y^8 \cdot y^6$	Commutative
$= 576x^2y^{14}$	Product of Powers

GuidedPractice

5. Simplify $\left(\frac{1}{2}a^2b^2\right)^3[(-4b)^2]^2$.

Check Your Understanding

Example 1 **Determine whether each expression is a monomial. Write *yes* or *no*. Explain your reasoning.**

1. 15 **2.** $2 - 3a$ **3.** $\frac{5c}{d}$

4. $-15g^2$ **5.** $\frac{r}{2}$ **6.** $7b + 9$

Examples 2–3 **Simplify each expression.**

7. $k(k^3)$ **8.** $m^4(m^2)$ **9.** $2q^2(9q^4)$

10. $(5u^4v)(7u^4v^3)$ **11.** $[(3^2)^2]^2$ **12.** $(xy^4)^6$

13. $(4a^4b^9c)^2$ **14.** $(-2f^2g^3h^2)^3$ **15.** $(-3p^5t^6)^4$

Example 4 **16.** **GEOMETRY** The formula for the surface area of a cube is $SA = 6s^2$, where SA is the surface area and s is the length of any side.

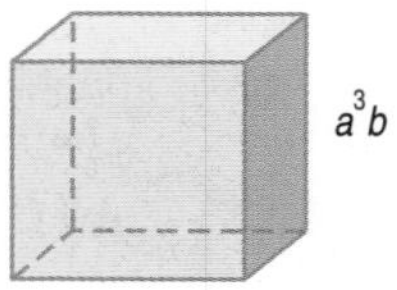

a. Express the surface area of the cube as a monomial.

b. What is the surface area of the cube if $a = 3$ and $b = 4$?

Example 5 **Simplify each expression.**

17. $(5x^2y)^2(2xy^3z)^3(4xyz)$ **18.** $(-3d^2f^3g)^2[(-3d^2f)^3]^2$

19. $(-2g^3h)(-3gj^4)^2(-ghj)^2$ **20.** $(-7ab^4c)^3[(2a^2c)^2]^3$

Practice and Problem Solving

Example 1 **Determine whether each expression is a monomial. Write *yes* or *no*. Explain your reasoning.**

21. 122
22. $3a^4$
23. $2c + 2$
24. $\frac{-2g}{4h}$
25. $\frac{5k}{10}$
26. $6m + 3n$

Examples 2–3 **Simplify each expression.**

27. $(q^2)(2q^4)$
28. $(-2u^2)(6u^6)$
29. $(9w^2x^8)(w^6x^4)$
30. $(y^6z^9)(6y^4z^2)$
31. $(b^8c^6d^5)(7b^6c^2d)$
32. $(14fg^2h^2)(-3f^4g^2h^2)$
33. $(j^5k^7)^4$
34. $(n^3p)^4$
35. $[(2^2)^2]^2$
36. $[(3^2)^2]^4$
37. $[(4r^2t)^3]^2$
38. $[(-2xy^2)^3]^2$

Example 4 **GEOMETRY** **Express the area of each triangle as a monomial.**

39.

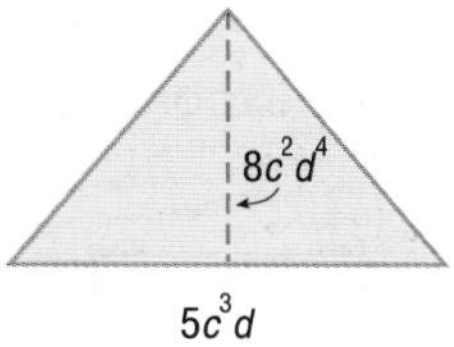

40.

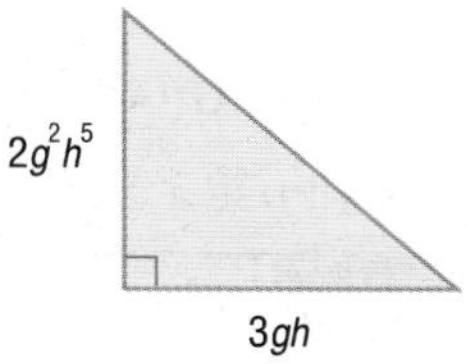

Example 5 **Simplify each expression.**

41. $(2a^3)^4(a^3)^3$
42. $(c^3)^2(-3c^5)^2$
43. $(2gh^4)^3[(-2g^4h)^3]^2$
44. $(5k^2m)^3[(4km^4)^2]^2$
45. $(p^5r^2)^4(-7p^3r^4)^2(6pr^3)$
46. $(5x^2y)^2(2xy^3z)^3(4xyz)$
47. $(5a^2b^3c^4)(6a^3b^4c^2)$
48. $(10xy^5z^3)(3x^4y^6z^3)$
49. $(0.5x^3)^2$
50. $(0.4h^5)^3$
51. $\left(-\frac{3}{4}c\right)^3$
52. $\left(\frac{4}{5}a^2\right)^2$
53. $(8y^3)(-3x^2y^2)\left(\frac{3}{8}xy^4\right)$
54. $\left(\frac{4}{7}m\right)^2(49m)(17p)\left(\frac{1}{34}p^5\right)$
55. $(-3r^3w^4)^3(2rw)^2(-3r^2)^3(4rw^2)^3(2r^2w^3)^4$
56. $(3ab^2c)^2(-2a^2b^4)^2(a^4c^2)^3(a^2b^4c^5)^2(2a^3b^2c^4)^3$

57. **FINANCIAL LITERACY** Cleavon has money in an account that earns 3% simple interest. The formula for computing simple interest is $I = Prt$, where I is the interest earned, P represents the principal that he put into the account, r is the interest rate (in decimal form), and t represents time in years.
    a. Cleavon makes a deposit of \$2c and leaves it for 2 years. Write a monomial that represents the interest earned.
    b. If c represents a birthday gift of \$250, how much will Cleavon have in this account after 2 years?

CCSS TOOLS **Express the volume of each solid as a monomial.**

58.

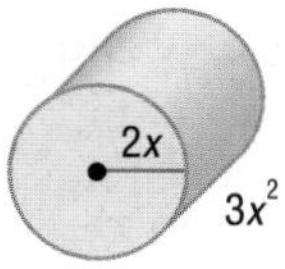

59.

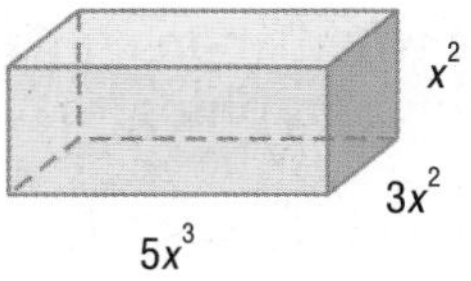

60.

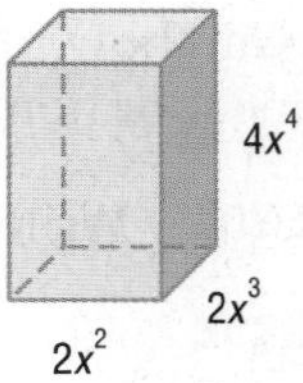

61. **PACKAGING** For a commercial art class, Aiko must design a new container for individually wrapped pieces of candy. The shape that she chose is a cylinder. The formula for the volume of a cylinder is $V = \pi r^2 h$.

a. The radius that Aiko would like to use is $2p^3$, and the height is $4p^3$. Write a monomial that represents the volume of her container.

b. Make a table for five possible measures for the radius and height of a cylinder having the same volume.

c. What is the volume of Aiko's container if the height is doubled?

62. **ENERGY** Albert Einstein's formula $E = mc^2$ shows that if mass is accelerated enough, it could be converted into usable energy. Energy E is measured in joules, mass m in kilograms, and the speed c of light is about 300 million meters per second.

a. Complete the calculations to convert 3 kilograms of gasoline completely into energy.

b. What happens to the energy if the amount of gasoline is doubled?

63. **MULTIPLE REPRESENTATIONS** In this problem, you will explore exponents.

a. Tabular Copy and use a calculator to complete the table.

Power	3^4	3^3	3^2	3^1	3^0	3^{-1}	3^{-2}	3^{-3}	3^{-4}
Value						$\frac{1}{3}$	$\frac{1}{9}$	$\frac{1}{27}$	$\frac{1}{81}$

b. Analytical What do you think the values of 5^0 and 5^{-1} are? Verify your conjecture using a calculator.

c. Analytical Complete: For any nonzero number a and any integer n, $a^{-n} =$ ______.

d. Verbal Describe the value of a nonzero number raised to the zero power.

H.O.T. Problems Use Higher-Order Thinking Skills

64. **CCSS PERSEVERANCE** For any nonzero real numbers a and b and any integers m and t, simplify the expression $\left(-\frac{a^m}{b^t}\right)^{2t}$ and describe each step.

65. **REASONING** Copy the table below.

Equation	Related Expression	Power of x	Linear or Nonlinear
$y = x$			
$y = x^2$			
$y = x^3$			

a. For each equation, write the related expression and record the power of x.

b. Graph each equation using a graphing calculator.

c. Classify each graph as *linear* or *nonlinear*.

d. Explain how to determine whether an equation, or its related expression, is linear or nonlinear without graphing.

66. **OPEN ENDED** Write three different expressions that can be simplified to x^6.

67. **WRITING IN MATH** Write two formulas that have monomial expressions in them. Explain how each is used in a real-world situation.

Standardized Test Practice

68. Which of the following is not a monomial?

A $-6xy$

B $\frac{1}{2}a^2$

C $-\frac{1}{2b^3}$

D $5gh^4$

69. GEOMETRY The accompanying diagram shows the transformation of $\triangle XYZ$ to $\triangle X'Y'Z'$.

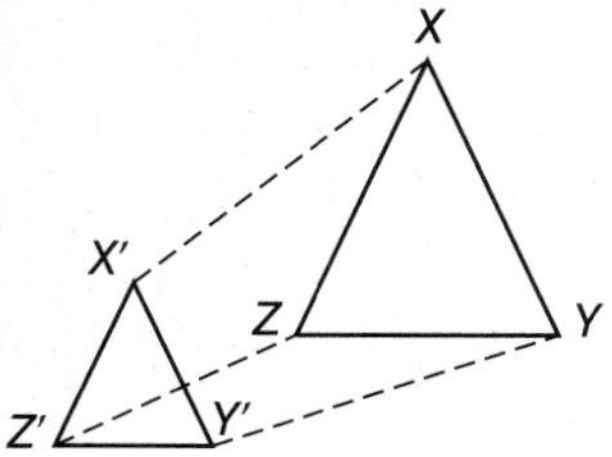

This transformation is an example of a

F dilation

G line reflection

H rotation

J translation

70. CARS In 2002, the average price of a new domestic car was \$19,126. In 2008, the average price was \$28,715. Based on a linear model, what is the predicted average price for 2014?

A \$45,495

B \$38,304

C \$35,906

D \$26,317

71. SHORT RESPONSE If a line has a positive slope and a negative y-intercept, what happens to the x-intercept if the slope and the y-intercept are both doubled?

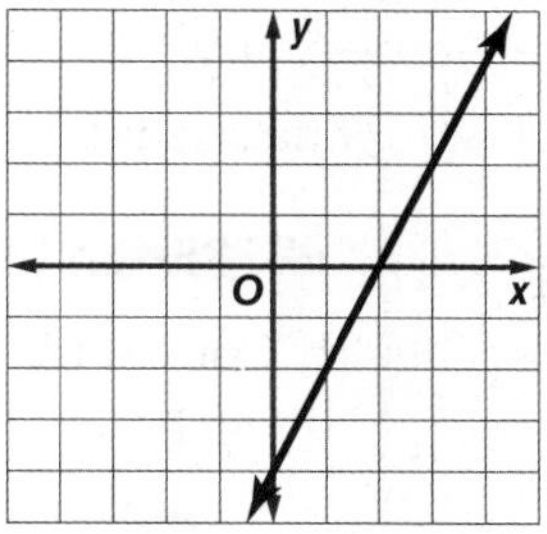

Spiral Review

Solve each system of inequalities by graphing. (Lesson 6-6)

72. $y < 4x$
$2x + 3y \geq -21$

73. $y \geq 2$
$2y + 2x \leq 4$

74. $y > -2x - 1$
$2y \leq 3x + 2$

75. $3x + 2y < 10$
$2x + 12y < -6$

76. SPORTS In the 2006 Winter Olympic Games, the total number of gold and silver medals won by the U.S. was 18. The total points scored for gold and silver medals was 45. Write and solve a system of equations to find how many gold and silver medals were won by the U.S. (Lesson 6-5)

77. DRIVING Tires should be kept within 2 pounds per square inch (psi) of the manufacturer's recommended tire pressure. If the recommendation for a tire is 30 psi, what is the range of acceptable pressures? (Lesson 5-5)

78. BABYSITTING Alexis charges \$10 plus \$4 per hour to babysit. Alexis needs at least \$40 more to buy a television for which she is saving. Write an inequality for this situation. Will she be able to get her television if she babysits for 5 hours? (Lesson 5-6)

Skills Review

Find each quotient.

79. $-64 \div (-8)$

80. $-78 \div 1.3$

81. $42.3 \div (-6)$

82. $-23.94 \div 10.5$

83. $-32.5 \div (-2.5)$

84. $-98.44 \div 4.6$

LESSON 7-2 Division Properties of Exponents

Then	Now	Why?
You multiplied monomials using the properties of exponents.	**1** Divide monomials using the properties of exponents. **2** Simplify expressions containing negative and zero exponents.	The tallest redwood tree is 112 meters or about 10^2 meters tall. The average height of a redwood tree is 15 meters. The closest power of ten to 15 is 10^1, so an average redwood is about 10^1 meters tall. The ratio of the tallest tree's height to the average tree's height is $\frac{10^2}{10^1}$ or 10^1. This means the tallest redwood tree is approximately 10 times as tall as the average redwood tree.

Getty Images/Taxi/Getty Images

NewVocabulary
zero exponent
negative exponent
order of magnitude

Common Core State Standards

Content Standards
A.SSE.2 Use the structure of an expression to identify ways to rewrite it.
F.IF.8b Use the properties of exponents to interpret expressions for exponential functions.

Mathematical Practices
2 Reason abstractly and quantitatively.

1 Divide Monomials

We can use the principles for reducing fractions to find quotients of monomials like $\frac{10^2}{10^1}$. In the following examples, look for a pattern in the exponents.

$$\frac{2^7}{2^4} = \frac{\overbrace{\cancel{2}\cdot\cancel{2}\cdot\cancel{2}\cdot\cancel{2}\cdot 2\cdot 2\cdot 2}^{7\text{ factors}}}{\underbrace{\cancel{2}\cdot\cancel{2}\cdot\cancel{2}\cdot\cancel{2}}_{4\text{ factors}}} = 2\cdot 2\cdot 2 \text{ or } 2^3$$

$$\frac{t^4}{t^3} = \frac{\overbrace{\cancel{t}\cdot\cancel{t}\cdot\cancel{t}\cdot t}^{4\text{ factors}}}{\underbrace{\cancel{t}\cdot\cancel{t}\cdot\cancel{t}}_{3\text{ factors}}} = t$$

These examples demonstrate the Quotient of Powers Rule.

KeyConcept Quotient of Powers

Words	To divide two powers with the same base, subtract the exponents.
Symbols	For any nonzero number a, and any integers m and p, $\frac{a^m}{a^p} = a^{m-p}$.
Examples	$\frac{c^{11}}{c^8} = c^{11-8}$ or c^3 $\frac{r^5}{r^2} = r^{5-2} = r^3$

Example 1 Quotient of Powers

Simplify $\frac{g^3h^5}{gh^2}$. Assume that no denominator equals zero.

$\frac{g^3h^5}{gh^2} = \left(\frac{g^3}{g}\right)\left(\frac{h^5}{h^2}\right)$ Group powers with the same base.

$= \left(g^{3-1}\right)\left(h^{5-2}\right)$ Quotient of Powers

$= g^2h^3$ Simplify.

GuidedPractice

Simplify each expression. Assume that no denominator equals zero.

1A. $\frac{x^3y^4}{x^2y}$

1B. $\frac{k^7m^{10}p}{k^5m^3p}$

We can use the Product of Powers Rule to find the powers of quotients for monomials. In the following example, look for a pattern in the exponents.

$$\left(\frac{3}{4}\right)^3 = \overbrace{\left(\frac{3}{4}\right)\left(\frac{3}{4}\right)\left(\frac{3}{4}\right)}^{\text{3 factors}} = \frac{\overbrace{3 \cdot 3 \cdot 3}^{\text{3 factors}}}{\underbrace{4 \cdot 4 \cdot 4}_{\text{3 factors}}} = \frac{3^3}{4^3}$$

$$\left(\frac{c}{d}\right)^2 = \overbrace{\left(\frac{c}{d}\right)\left(\frac{c}{d}\right)}^{\text{2 factors}} = \frac{\overbrace{c \cdot c}^{\text{2 factors}}}{\underbrace{d \cdot d}_{\text{2 factors}}} = \frac{c^2}{d^2}$$

StudyTip

Power Rules with Variables The power rules apply to variables as well as numbers. For example, $\left(\frac{3a}{4b}\right)^3 = \frac{(3a)^3}{(4b)^3}$ or $\frac{27a^3}{64b^3}$.

KeyConcept Power of a Quotient

Words To find the power of a quotient, find the power of the numerator and the power of the denominator.

Symbols For any real numbers a and $b \neq 0$, and any integer m, $\left(\frac{a}{b}\right)^m = \frac{a^m}{b^m}$.

Examples $\left(\frac{3}{5}\right)^4 = \frac{3^4}{5^4}$ $\left(\frac{r}{t}\right)^5 = \frac{r^5}{t^5}$

Real-WorldCareer

Astronomer An astronomer studies the universe and analyzes space travel and satellite communications. To be a technician or research assistant, a bachelor's degree is required.

Example 2 Power of a Quotient

PT

Simplify $\left(\frac{3p^3}{7}\right)^2$.

$$\left(\frac{3p^3}{7}\right)^2 = \frac{(3p^3)^2}{7^2} \quad \text{Power of a Quotient}$$

$$= \frac{3^2(p^3)^2}{7^2} \quad \text{Power of a Product}$$

$$= \frac{9p^6}{49} \quad \text{Power of a Power}$$

GuidedPractice

Simplify each expression.

2A. $\left(\frac{3x^4}{4}\right)^3$ **2B.** $\left(\frac{5x^5y}{6}\right)^2$ **2C.** $\left(\frac{2y^2}{3z^3}\right)^2$ **2D.** $\left(\frac{4x^3}{5y^4}\right)^3$

A calculator can be used to explore expressions with 0 as the exponent. There are two methods to explain why a calculator gives a value of 1 for 3^0.

Method 1

$\frac{3^5}{3^5} = 3^{5-5}$ Quotient of Powers

$= 3^0$ Simplify.

Method 2

$\frac{3^5}{3^5} = \frac{\cancel{3} \cdot \cancel{3} \cdot \cancel{3} \cdot \cancel{3} \cdot \cancel{3}}{\cancel{3} \cdot \cancel{3} \cdot \cancel{3} \cdot \cancel{3} \cdot \cancel{3}}$ Definition of powers

$= 1$ Simplify.

Since $\frac{3^5}{3^5}$ can only have one value, we can conclude that $3^0 = 1$. A **zero exponent** is any nonzero number raised to the zero power.

KeyConcept Zero Exponent Property

Words	Any nonzero number raised to the zero power is equal to 1.
Symbols	For any nonzero number a, $a^0 = 1$.
Examples	$15^0 = 1$ $\quad \left(\frac{b}{c}\right)^0 = 1$ $\quad \left(\frac{2}{7}\right)^0 = 1$

PT

Example 3 Zero Exponent

Simplify each expression. Assume that no denominator equals zero.

a. $\left(-\frac{4n^2q^5r^2}{9n^3q^2r}\right)^0$

$$\left(-\frac{4n^2q^5r^2}{9n^3q^2r}\right)^0 = 1 \qquad a^0 = 1$$

b. $\frac{x^5y^0}{x^3}$

$$\frac{x^5y^0}{x^3} = \frac{x^5(1)}{x^3} \qquad a^0 = 1$$

$$= x^2 \qquad \text{Quotient of Powers}$$

GuidedPractice

3A. $\frac{b^4c^2d^0}{b^2c}$

3B. $\left(\frac{2f^4g^7h^3}{15f^3g^9h^6}\right)^0$

> **StudyTip**
> Zero Exponent Be careful of parentheses. The expression $(5x)^0$ is 1 but $5x^0 = 5$.

2 Negative Exponents

Any nonzero real number raised to a negative power is a **negative exponent**. To investigate the meaning of a negative exponent, we can simplify expressions like $\frac{c^2}{c^5}$ using two methods.

Method 1

$\frac{c^2}{c^5} = c^{2-5}$ Quotient of Powers

$= c^{-3}$ Simplify.

Method 2

$\frac{c^2}{c^5} = \frac{\cancel{c} \cdot \cancel{c}}{\cancel{c} \cdot \cancel{c} \cdot c \cdot c \cdot c}$ Definition of powers

$= \frac{1}{c^3}$ Simplify.

Since $\frac{c^2}{c^5}$ can only have one value, we can conclude that $c^{-3} = \frac{1}{c^3}$.

KeyConcept Negative Exponent Property

Words	For any nonzero number a and any integer n, a^{-n} is the reciprocal of a^n. Also, the reciprocal of a^{-n} is a^n.
Symbols	For any nonzero number a and any integer n, $a^{-n} = \frac{1}{a^n}$.
Examples	$2^{-4} = \frac{1}{2^4} = \frac{1}{16}$ $\quad \frac{1}{j^{-4}} = j^4$

An expression is considered simplified when it contains only positive exponents, each base appears exactly once, there are no powers of powers, and all fractions are in simplest form.

PT

Example 4 Negative Exponents

Simplify each expression. Assume that no denominator equals zero.

a. $\dfrac{n^{-5}p^4}{r^{-2}}$

$$\frac{n^{-5}p^4}{r^{-2}} = \left(\frac{n^{-5}}{1}\right)\left(\frac{p^4}{1}\right)\left(\frac{1}{r^{-2}}\right) \quad \text{Write as a product of fractions.}$$

$$= \left(\frac{1}{n^5}\right)\left(\frac{p^4}{1}\right)\left(\frac{r^2}{1}\right) \quad a^{-n} = \frac{1}{a^n} \text{ and } \frac{1}{a^{-n}} = a^n$$

$$= \frac{p^4r^2}{n^5} \quad \text{Multiply.}$$

StudyTip

Negative Signs Be aware of where a negative sign is placed.

$5^{-1} = \frac{1}{5}$, while $-5^1 \neq \frac{1}{5}$.

b. $\dfrac{5r^{-3}t^4}{-20r^2t^7u^{-5}}$

$$\frac{5r^{-3}t^4}{-20r^2t^7u^{-5}} = \left(\frac{5}{-20}\right)\left(\frac{r^{-3}}{r^2}\right)\left(\frac{t^4}{t^7}\right)\left(\frac{1}{u^{-5}}\right) \quad \text{Group powers with the same base.}$$

$$= \left(-\frac{1}{4}\right)(r^{-3-2})(t^{4-7})(u^5) \quad \text{Quotient of Powers and Negative Exponents Property}$$

$$= -\frac{1}{4}r^{-5}t^{-3}u^5 \quad \text{Simplify.}$$

$$= -\frac{1}{4}\left(\frac{1}{r^5}\right)\left(\frac{1}{t^3}\right)(u^5) \quad \text{Negative Exponent Property}$$

$$= -\frac{u^5}{4r^5t^3} \quad \text{Multiply.}$$

c. $\dfrac{2a^2b^3c^{-5}}{10a^{-3}b^{-1}c^{-4}}$

$$\frac{2a^2b^3c^{-5}}{10a^{-3}b^{-1}c^{-4}} = \left(\frac{2}{10}\right)\left(\frac{a^2}{a^{-3}}\right)\left(\frac{b^3}{b^{-1}}\right)\left(\frac{c^{-5}}{c^{-4}}\right) \quad \text{Group powers with the same base.}$$

$$= \left(\frac{1}{5}\right)(a^{2-(-3)})(b^{3-(-1)})(c^{-5-(-4)}) \quad \text{Quotient of Powers and Negative Exponents Property}$$

$$= \frac{1}{5}a^5b^4c^{-1} \quad \text{Simplify.}$$

$$= \frac{1}{5}(a^5)(b^4)\left(\frac{1}{c}\right) \quad \text{Negative Exponent Property}$$

$$= \frac{a^5b^4}{5c} \quad \text{Multiply.}$$

Blend Images/SuperStock

GuidedPractice

Simplify each expression. Assume that no denominator equals zero.

4A. $\dfrac{v^{-3}wx^2}{wy^{-6}}$ **4B.** $\dfrac{32a^{-8}b^3c^{-4}}{4a^3b^5c^{-2}}$ **4C.** $\dfrac{5j^{-3}k^2m^{-6}}{25k^{-4}m^{-2}}$

Real-WorldLink

An adult human weighs about 70 kilograms and an adult dairy cow weighs about 700 kilograms. Their weights differ by 1 order of magnitude.

Order of magnitude is used to compare measures and to estimate and perform rough calculations. The **order of magnitude** of a quantity is the number rounded to the nearest power of 10. For example, the power of 10 closest to 95,000,000,000 is 10^{11}, or 100,000,000,000. So the order of magnitude of 95,000,000,000 is 10^{11}.

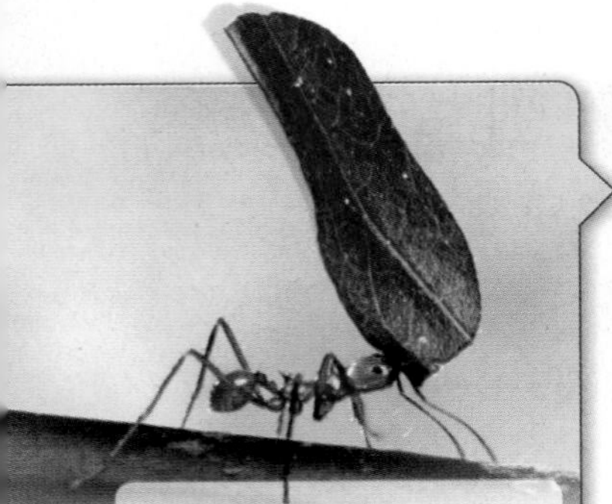

Real-WorldLink

There are over 14,000 species of ants living all over the world. Some ants can carry objects that are 50 times their own weight.

Source: Maine Animal Coalition

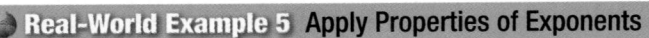

Real-World Example 5 Apply Properties of Exponents

HEIGHT Suppose the average height of a man is about 1.7 meters, and the average height of an ant is 0.0008 meter. How many orders of magnitude as tall as an ant is a man?

Understand We must find the order of magnitude of the heights of the man and ant. Then find the ratio of the orders of magnitude of the man's height to that of the ant's height.

Plan Round each height to the nearest power of ten. Then find the ratio of the height of the man to the height of the ant.

Solve The average height of a man is close to 1 meter. So, the order of magnitude is 10^0 meter. The average height of an ant is about 0.001 meter. So, the order of magnitude is 10^{-3} meters.

The ratio of the height of a man to the height of an ant is about $\frac{10^0}{10^{-3}}$.

$$\frac{10^0}{10^{-3}} = 10^{0-(-3)} \quad \text{Quotient of Powers}$$
$$= 10^3 \quad 0-(-3) = 0+3 \text{ or } 3$$
$$= 1000 \quad \text{Simplify.}$$

So, a man is approximately 1000 times as tall as an ant, or a man is 3 orders of magnitude as tall as an ant.

Check The ratio of the man's height to the ant's height is $\frac{1.7}{0.0008} = 2125$. The order of magnitude of 2125 is 10^3. ✓

GuidedPractice

5. **ASTRONOMY** The order of magnitude of the mass of Earth is about 10^{27}. The order of magnitude of the Milky Way galaxy is about 10^{44}. How many orders of magnitude as big is the Milky Way galaxy as Earth?

Check Your Understanding

Examples 1–4 **Simplify each expression. Assume that no denominator equals zero.**

1. $\frac{t^5u^4}{t^2u}$
2. $\frac{a^6b^4c^{10}}{a^3b^2c}$
3. $\frac{m^6r^5p^3}{m^5r^2p^3}$
4. $\frac{b^4c^6f^8}{b^4c^3f^5}$
5. $\frac{g^8h^2m}{hg^7}$
6. $\frac{r^4t^7v^2}{t^7v^2}$
7. $\frac{x^3y^2z^6}{z^5x^2y}$
8. $\frac{n^4q^4w^6}{q^2n^3w}$
9. $\left(\frac{2a^3b^5}{3}\right)^2$
10. $\frac{r^3v^{-2}}{t^{-7}}$
11. $\left(\frac{2c^3d^5}{5g^2}\right)^5$
12. $\left(-\frac{3xy^4z^2}{x^3yz^4}\right)^0$
13. $\left(\frac{3f^4gh^4}{32f^3g^4h}\right)^0$
14. $\frac{4r^2v^0t^5}{2rt^3}$
15. $\frac{f^{-3}g^2}{h^{-4}}$
16. $\frac{-8x^2y^8z^{-5}}{12x^4y^{-7}z^7}$
17. $\frac{2a^2b^{-7}c^{10}}{6a^{-3}b^2c^{-3}}$

Example 5

18. **FINANCIAL LITERACY** The gross domestic product (GDP) for the United States in 2008 was \$14.204 trillion, and the GDP per person was \$47,580. Use order of magnitude to approximate the population of the United States in 2008.

Scott Bauer/USDA

Practice and Problem Solving

Examples 1–4 **Simplify each expression. Assume that no denominator equals zero.**

19. $\frac{m^4p^2}{m^2p}$

20. $\frac{p^{12}t^3r}{p^2tr}$

21. $\frac{3m^{-3}r^4p^2}{12t^4}$

22. $\frac{c^4d^4f^3}{c^2d^4f^3}$

23. $\left(\frac{3xy^4}{5z^2}\right)^2$

24. $\left(\frac{3t^6u^2v^5}{9tuv^{21}}\right)^0$

25. $\left(\frac{p^2t^7}{10}\right)^3$

26. $\frac{x^{-4}y^9}{z^{-2}}$

27. $\frac{a^7b^8c^8}{a^5bc^7}$

28. $\left(\frac{3np^3}{7q^2}\right)^2$

29. $\left(\frac{2r^3t^6}{5u^9}\right)^4$

30. $\left(\frac{3m^5r^3}{4p^8}\right)^4$

31. $\left(-\frac{5f^9g^4h^2}{fg^2h^3}\right)^0$

32. $\frac{p^{12}t^7r^2}{p^2t^7r}$

33. $\frac{p^4t^{-3}}{r^{-2}}$

34. $-\frac{5c^2d^5}{8cd^5f^0}$

35. $\frac{-2f^3g^2h^0}{8f^2g^2}$

36. $\frac{12m^{-4}p^2}{-15m^3p^{-9}}$

37. $\frac{k^4m^3p^2}{k^2m^2}$

38. $\frac{14f^{-3}g^2h^{-7}}{21k^3}$

39. $\frac{39t^4uv^{-2}}{13t^{-3}u^7}$

40. $\left(\frac{a^{-2}b^4c^5}{a^{-4}b^{-4}c^3}\right)^2$

41. $\frac{r^3t^{-1}x^{-5}}{tx^5}$

42. $\frac{g^0h^7j^{-2}}{g^{-5}h^0j^{-2}}$

Example 5

43. **INTERNET** In a recent year, there were approximately 3.95 million Internet hosts. Suppose there were 208 million Internet users. Determine the order of magnitude for the Internet hosts and Internet users. Using the orders of magnitude, how many Internet users were there compared to Internet hosts?

44. **PROBABILITY** The probability of rolling a die and getting an even number is $\frac{1}{2}$. If you roll the die twice, the probability of getting an even number both times is $\left(\frac{1}{2}\right)\left(\frac{1}{2}\right)$ or $\left(\frac{1}{2}\right)^2$.
   a. What does $\left(\frac{1}{2}\right)^4$ represent?
   b. Write an expression to represent the probability of rolling a die d times and getting an even number every time. Write the expression as a power of 2.

Simplify each expression. Assume that no denominator equals zero.

45. $\frac{-4w^{12}}{12w^3}$

46. $\frac{13r^7}{39r^4}$

47. $\frac{(4k^3m^2)^3}{(5k^2m^{-3})^{-2}}$

48. $\frac{3wy^{-2}}{(w^{-1}y)^3}$

49. $\frac{20qr^{-2}t^{-5}}{4q^0r^4t^{-2}}$

50. $\frac{-12c^3d^0f^{-2}}{6c^5d^{-3}f^4}$

51. $\frac{(2g^3h^{-2})^2}{(g^2h^0)^{-3}}$

52. $\frac{(5pr^{-2})^{-2}}{(3p^{-1}r)^3}$

53. $\left(\frac{-3x^{-6}y^{-1}z^{-2}}{6x^{-2}yz^{-5}}\right)^{-2}$

54. $\left(\frac{2a^{-2}b^4c^2}{-4a^{-2}b^{-5}c^{-7}}\right)^{-1}$

55. $\frac{(16x^2y^{-1})^0}{(4x^0y^{-4}z)^{-2}}$

56. $\left(\frac{4^0c^2d^3f}{2c^{-4}d^{-5}}\right)^{-3}$

57. **CCSS SENSE-MAKING** The processing speed of an older desktop computer is about 10^8 instructions per second. A new computer can process about 10^{10} instructions per second. The newer computer is how many times as fast as the older one?

58. **ASTRONOMY** The brightness of a star is measured in magnitudes. The lower the magnitude, the brighter the star. A magnitude 9 star is 2.51 times as bright as a magnitude 10 star. A magnitude 8 star is $2.51 \cdot 2.51$ or 2.51^2 times as bright as a magnitude 10 star.

a. How many times as bright is a magnitude 3 star as a magnitude 10 star?

b. Write an expression to compare a magnitude m star to a magnitude 10 star.

c. A full moon is considered to be magnitude −13, approximately. Does your expression make sense for this magnitude? Explain.

59. **PROBABILITY** The probability of rolling a die and getting a 3 is $\frac{1}{6}$. If you roll the die twice, the probability of getting a 3 both times is $\frac{1}{6} \cdot \frac{1}{6}$ or $\left(\frac{1}{6}\right)^2$.

a. Write an expression to represent the probability of rolling a die d times and getting a 3 each time.

b. Write the expression as a power of 6.

60. **MULTIPLE REPRESENTATIONS** To find the area of a circle, use $A = \pi r^2$. The formula for the area of a square is $A = s^2$.

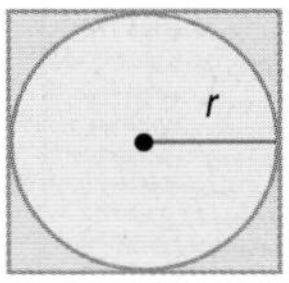

a. Algebraic Find the ratio of the area of the circle to the area of the square.

b. Algebraic If the radius of the circle and the length of each side of the square are doubled, find the ratio of the area of the circle to the square.

c. Tabular Copy and complete the table.

Radius	Area of Circle	Area of Square	Ratio
r			
$2r$			
$3r$			
$4r$			
$5r$			
$6r$			

d. Analytical What conclusion can be drawn from this?

H.O.T. Problems Use Higher-Order Thinking Skills

61. **REASONING** Is $x^y \cdot x^z = x^{yz}$ *sometimes*, *always*, or *never* true? Explain.

62. **OPEN ENDED** Name two monomials with a quotient of $24a^2b^3$.

63. **CHALLENGE** Use the Quotient of Powers Property to explain why $x^{-n} = \frac{1}{x^n}$.

64. **CCSS REGULARITY** Write a convincing argument to show why $3^0 = 1$.

65. **WRITING IN MATH** Explain how to use the Quotient of Powers property and the Power of a Quotient property.

Standardized Test Practice

66. What is the perimeter of the figure in meters?

A 40 meters
B 80 meters
C 160 meters
D 400 meters

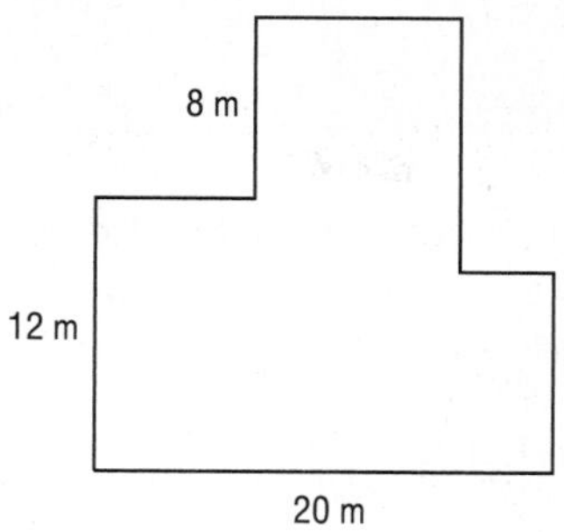

67. In researching her science project, Leigh learned that light travels at a constant rate and that it takes 500 seconds for light to travel the 93 million miles from the Sun to Earth. Mars is 142 million miles from the Sun. About how many seconds will it take for light to travel from the Sun to Mars?

F 235 seconds
G 327 seconds
H 642 seconds
J 763 seconds

68. EXTENDED RESPONSE Jessie and Jonas are playing a game using the spinners below. Each spinner is equally likely to stop on any of the four numbers. In the game, a player spins both spinners and calculates the product of the two numbers on which the spinners have stopped.

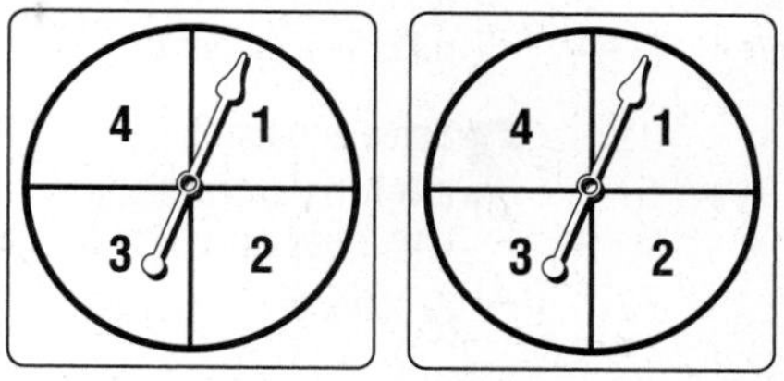

a. What product has the greatest probability of occurring?

b. What is the probability of that product occurring?

69. Simplify $(4^{-2} \cdot 5^0 \cdot 64)^3$.

A $\frac{1}{64}$ **C** 320
B 64 **D** 1024

Spiral Review

70. GEOMETRY A rectangular prism has a width of $7x^3$ units, a length of $4x^2$ units, and a height of $3x$ units. What is the volume of the prism? (Lesson 7-1)

Solve each system of inequalities by graphing. (Lesson 6-6)

71. $y \geq 1$, $x < -1$

72. $y \geq -3$, $y - x < 1$

73. $y < 3x + 2$, $y \geq -2x + 4$

74. $y - 2x < 2$, $y - 2x > 4$

Solve each inequality. Check your solution. (Lesson 5-3)

75. $5(2h - 6) > 4h$

76. $22 \geq 4(b - 8) + 10$

77. $5(u - 8) \leq 3(u + 10)$

78. $8 + t \leq 3(t + 4) + 2$

79. $9n + 3(1 - 6n) \leq 21$

80. $-6(b + 5) > 3(b - 5)$

81. GRADES In a high school science class, a test is worth three times as much as a quiz. What is the student's average grade? (Lesson 2-9)

Science Grades

Tests	Quizzes
85	82
92	75
	95

Skills Review

Evaluate each expression.

82. 9^2

83. 11^2

84. 10^6

85. 10^4

86. 3^5

87. 5^3

88. 12^3

89. 4^6

LESSON 7-3

Rational Exponents

Then

- You used the laws of exponents to find products and quotients of monomials.

Now

1. Evaluate and rewrite expressions involving rational exponents.
2. Solve equations involving expressions with rational exponents.

Why?

- It's important to protect your skin with sunscreen to prevent damage. The sun protection factor (SPF) of a sunscreen indicates how well it protects you. Sunscreen with an SPF of f absorbs about p percent of the UV-B rays, where $p = 50f^{0.2}$.

NewVocabulary

rational exponent
cube root
nth root
exponential equation

Common Core State Standards

Content Standards

N.RN.1 Explain how the definition of the meaning of rational exponents follows from extending the properties of integer exponents to those values, allowing for a notation for radicals in terms of rational exponents.

N.RN.2 Rewrite expressions involving radicals and rational exponents using the properties of exponents.

Mathematical Practices

5 Use appropriate tools strategically.

1 Rational Exponents

You know that an exponent represents the number of times that the base is used as a factor. But how do you evaluate an expression with an exponent that is not an integer like the one above? Let's investigate **rational exponents** by assuming that they behave like integer exponents.

$$\left(b^{\frac{1}{2}}\right)^2 = b^{\frac{1}{2}} \cdot b^{\frac{1}{2}}$$ Write as a multiplication expression.

$$= b^{\frac{1}{2}+\frac{1}{2}}$$ Product of Powers

$$= b^1 \text{ or } b$$ Simplify.

Thus, $b^{\frac{1}{2}}$ is a number with a square equal to b. So $b^{\frac{1}{2}} = \sqrt{b}$.

KeyConcept $b^{\frac{1}{2}}$

Words	For any nonnegative real number b, $b^{\frac{1}{2}} = \sqrt{b}$.
Examples	$16^{\frac{1}{2}} = \sqrt{16}$ or 4 $38^{\frac{1}{2}} = \sqrt{38}$

Example 1 Radical and Exponential Forms

Write each expression in radical form, or write each radical in exponential form.

a. $25^{\frac{1}{2}}$

$25^{\frac{1}{2}} = \sqrt{25}$ Definition of $b^{\frac{1}{2}}$

$= 5$ Simplify.

b. $\sqrt{18}$

$\sqrt{18} = 18^{\frac{1}{2}}$ Definition of $b^{\frac{1}{2}}$

c. $5x^{\frac{1}{2}}$

$5x^{\frac{1}{2}} = 5\sqrt{x}$ Definition of $b^{\frac{1}{2}}$

d. $\sqrt{8p}$

$\sqrt{8p} = (8p)^{\frac{1}{2}}$ Definition of $b^{\frac{1}{2}}$

GuidedPractice

1A. $a^{\frac{1}{2}}$ **1B.** $\sqrt{22}$ **1C.** $(7w)^{\frac{1}{2}}$ **1D.** $2\sqrt{x}$

Ariel Skelley/Blend Images/Getty Images

You know that to find the square root of a number a you find a number with a square of a. In the same way, you can find other roots of numbers. If $a^3 = b$, then a is the **cube root** of b, and if $a^n = b$ for a positive integer n, then a is an **nth root** of b.

KeyConcept *n*th Root

Words	For any real numbers a and b and any positive integer n, if $a^n = b$, then a is an nth root of b.
Symbols	If $a^n = b$, then $\sqrt[n]{b} = a$.
Example	Because $2^4 = 16$, 2 is a fourth root of 16; $\sqrt[4]{16} = 2$.

StudyTip

CCSS Tools You can use a graphing calculator to find nth roots. Enter n, then press MATH and choose $\sqrt[x]{\ }$.

Since $3^2 = 9$ and $(-3)^2 = 9$, both 3 and -3 are square roots of 9. Similarly, since $2^4 = 16$ and $(-2)^4 = 16$, both 2 and -2 are fourth roots of 16. The positive roots are called *principal roots.* Radical symbols indicate principal roots, so $\sqrt[4]{16} = 2$.

Example 2 *n*th roots

PT

Simplify.

a. $\sqrt[3]{27}$

$$\sqrt[3]{27} = \sqrt[3]{3 \cdot 3 \cdot 3}$$
$$= 3$$

b. $\sqrt[5]{32}$

$$\sqrt[5]{32} = \sqrt[5]{2 \cdot 2 \cdot 2 \cdot 2 \cdot 2}$$
$$= 2$$

GuidedPractice

2A. $\sqrt[3]{64}$

2B. $\sqrt[4]{10{,}000}$

Like square roots, nth roots can be represented by rational exponents.

$\left(b^{\frac{1}{n}}\right)^n = \underbrace{b^{\frac{1}{n}} \cdot b^{\frac{1}{n}} \cdot \ldots \cdot b^{\frac{1}{n}}}_{n \text{ factors}}$ Write as a multiplication expression.

$= b^{\frac{1}{n} + \frac{1}{n} + \ldots + \frac{1}{n}}$ Product of Powers

$= b^1$ or b Simplify.

Thus, $b^{\frac{1}{n}}$ is a number with an nth power equal to b. So $b^{\frac{1}{n}} = \sqrt[n]{b}$.

KeyConcept $b^{\frac{1}{n}}$

Words	For any positive real number b and any integer $n > 1$, $b^{\frac{1}{n}} = \sqrt[n]{b}$.
Example	$8^{\frac{1}{3}} = \sqrt[3]{8} = \sqrt[3]{2 \cdot 2 \cdot 2}$ or 2

StudyTip

Rational Exponents on a Calculator Use parentheses to evaluate expressions involving rational exponents on a graphing calculator. For example to find $125^{\frac{1}{3}}$, press 125 [^] [(] 1 [÷] 3 [)] [ENTER].

Example 3 Evaluate $b^{\frac{1}{n}}$ Expressions

Simplify.

a. $125^{\frac{1}{3}}$

$125^{\frac{1}{3}} = \sqrt[3]{125}$ — $b^{\frac{1}{n}} = \sqrt[n]{b}$

$= \sqrt[3]{5 \cdot 5 \cdot 5}$ — $125 = 5^3$

$= 5$ — Simplify.

b. $1296^{\frac{1}{4}}$

$1296^{\frac{1}{4}} = \sqrt[4]{1296}$ — $b^{\frac{1}{n}} = \sqrt[n]{b}$

$= \sqrt[4]{6 \cdot 6 \cdot 6 \cdot 6}$ — $1296 = 6^4$

$= 6$ — Simplify.

GuidedPractice

3A. $27^{\frac{1}{3}}$

3B. $256^{\frac{1}{4}}$

The Power of a Power property allows us to extend the definition of $b^{\frac{1}{n}}$ to $b^{\frac{m}{n}}$.

$b^{\frac{m}{n}} = \left(b^{\frac{1}{n}}\right)^m$ — Power of a Power

$= \left(\sqrt[n]{b}\right)^m$ or $\sqrt[n]{b^m}$ — $b^{\frac{1}{n}} = \sqrt[n]{b}$

KeyConcept $b^{\frac{m}{n}}$

Words For any positive real number b and any integers m and $n > 1$,

$$b^{\frac{m}{n}} = \left(\sqrt[n]{b}\right)^m \text{ or } \sqrt[n]{b^m}.$$

Example $8^{\frac{2}{3}} = \left(\sqrt[3]{8}\right)^2 = 2^2$ or 4

Example 4 Evaluate $b^{\frac{m}{n}}$ Expressions

Simplify.

a. $64^{\frac{2}{3}}$

$64^{\frac{2}{3}} = \left(\sqrt[3]{64}\right)^2$ — $b^{\frac{m}{n}} = \left(\sqrt[n]{b}\right)^m$

$= \left(\sqrt[3]{4 \cdot 4 \cdot 4}\right)^2$ — $64 = 4^3$

$= 4^2$ or 16 — Simplify.

b. $36^{\frac{3}{2}}$

$36^{\frac{3}{2}} = \left(\sqrt[2]{36}\right)^3$ — $b^{\frac{m}{n}} = \left(\sqrt[n]{b}\right)^m$

$= 6^3$ — $\sqrt{36} = 6$

$= 216$ — Simplify.

GuidedPractice

4A. $27^{\frac{2}{3}}$

4B. $256^{\frac{5}{4}}$

2 Solve Exponential Equations

In an **exponential equation**, variables occur as exponents. The Power Property of Equality and the other properties of exponents can be used to solve exponential equations.

KeyConcept Power Property of Equality

Words	For any real number $b > 0$ and $b \neq 1$, $b^x = b^y$ if and only if $x = y$.
Examples	If $5^x = 5^3$, then $x = 3$. If $n = \frac{1}{2}$, then $4^n = 4^{\frac{1}{2}}$.

Example 5 Solve Exponential Equations

PT

Solve each equation.

a. $6^x = 216$

$6^x = 216$	Original equation
$6^x = 6^3$	Rewrite 216 as 6^3.
$x = 3$	Property of Equality

CHECK $6^x = 216$

$6^3 \stackrel{?}{=} 216$

$216 = 216$ ✓

b. $25^{x-1} = 5$

$25^{x-1} = 5$	Original equation
$(5^2)^{x-1} = 5$	Rewrite 25 as 5^2.
$5^{2x-2} = 5^1$	Power of a Power, Distributive Property
$2x - 2 = 1$	Power Property of Equality
$2x = 3$	Add 2 to each side.
$x = \frac{3}{2}$	Divide each side by 2.

CHECK $25^{x-1} = 5$

$25^{\frac{3}{2}-1} \stackrel{?}{=} 5$

$25^{\frac{1}{2}} = 5$ ✓

GuidedPractice

5A. $5^x = 125$

5B. $12^{2x+3} = 144$

Real-World Example 6 Solve Exponential Equations

PT

SUNSCREEN Refer to the beginning of the lesson. Find the SPF that absorbs 100% of UV-B rays.

$p = 50f^{0.2}$	Original equation
$100 = 50f^{0.2}$	$p = 100$
$2 = f^{0.2}$	Divide each side by 50.
$2 = f^{\frac{1}{5}}$	$0.2 = \frac{1}{5}$
$(2^5)^{\frac{1}{5}} = f^{\frac{1}{5}}$	$2 = 2^1 = (2^5)^{\frac{1}{5}}$
$2^5 = f$	Power Property of Equality
$32 = f$	Simplify.

GuidedPractice

6. CHEMISTRY The radius r of the nucleus of an atom of mass number A is $r = 1.2A^{\frac{1}{3}}$ femtometers. Find A if $r = 3.6$ femtometers.

Real-WorldLink

Use extra caution near snow, water, and sand because they reflect the damaging rays of the Sun, which can increase your chance of sunburn.

Source: American Academy of Dermatology

Ryan McVay/Digital Vision/Getty Images

Check Your Understanding

Example 1 **Write each expression in radical form, or write each radical in exponential form.**

1. $12^{\frac{1}{2}}$ **2.** $3x^{\frac{1}{2}}$ **3.** $\sqrt{33}$ **4.** $\sqrt{8n}$

Examples 2–4 **Simplify.**

5. $\sqrt[3]{512}$ **6.** $\sqrt[5]{243}$ **7.** $343^{\frac{1}{3}}$ **8.** $\left(\frac{1}{16}\right)^{\frac{1}{4}}$

9. $343^{\frac{2}{3}}$ **10.** $81^{\frac{3}{4}}$ **11** $216^{\frac{4}{3}}$ **12.** $\left(\frac{1}{49}\right)^{\frac{3}{2}}$

Example 5 **Solve each equation.**

13. $8^x = 4096$ **14.** $3^{3x+1} = 81$ **15.** $4^{x-3} = 32$

Example 6 **16.** **CCSS TOOLS** A weir is used to measure water flow in a channel. For a rectangular broad crested weir, the flow Q in cubic feet per second is related to the weir length L in feet and height H of the water by $Q = 1.6LH^{\frac{3}{2}}$. Find the water height for a weir that is 3 feet long and has flow of 38.4 cubic feet per second.

Practice and Problem Solving

Example 1 **Write each expression in radical form, or write each radical in exponential form.**

17. $15^{\frac{1}{2}}$ **18.** $24^{\frac{1}{2}}$ **19.** $4k^{\frac{1}{2}}$ **20.** $(12y)^{\frac{1}{2}}$

21. $\sqrt{26}$ **22.** $\sqrt{44}$ **23.** $2\sqrt{ab}$ **24.** $\sqrt{3xyz}$

Examples 2–4 **Simplify.**

25. $\sqrt[3]{8}$ **26.** $\sqrt[5]{1024}$ **27.** $\sqrt[3]{216}$ **28.** $\sqrt[4]{10{,}000}$

29. $\sqrt[3]{0.001}$ **30.** $\sqrt[4]{\frac{16}{81}}$ **31.** $1331^{\frac{1}{3}}$ **32.** $64^{\frac{1}{6}}$

33. $3375^{\frac{1}{3}}$ **34.** $512^{\frac{1}{9}}$ **35.** $\left(\frac{1}{81}\right)^{\frac{1}{4}}$ **36.** $\left(\frac{3125}{32}\right)^{\frac{1}{5}}$

37. $8^{\frac{2}{3}}$ **38.** $625^{\frac{3}{4}}$ **39.** $729^{\frac{5}{6}}$ **40.** $256^{\frac{3}{8}}$

41. $125^{\frac{4}{3}}$ **42.** $49^{\frac{5}{2}}$ **43.** $\left(\frac{9}{100}\right)^{\frac{3}{2}}$ **44.** $\left(\frac{8}{125}\right)^{\frac{4}{3}}$

Example 5 **Solve each equation.**

45. $3^x = 243$
46. $12^x = 144$
47. $16^x = 4$
48. $27^x = 3$
49. $9^x = 27$
50. $32^x = 4$
51. $2^{x-1} = 128$
52. $4^{2x+1} = 1024$
53. $6^{x-4} = 1296$
54. $9^{2x+3} = 2187$
55. $4^{3x} = 512$
56. $128^{3x} = 8$

Example 6

57. CONSERVATION Water collected in a rain barrel can be used to water plants and reduce city water use. Water flowing from an open rain barrel has velocity $v = 8h^{\frac{1}{2}}$, where v is in feet per second and h is the height of the water in feet. Find the height of the water if it is flowing at 16 feet per second.

58. ELECTRICITY The radius r in millimeters of a platinum wire L centimeters long with resistance 0.1 ohm is $r = 0.059L^{\frac{1}{2}}$. How long is a wire with radius 0.236 millimeter?

Write each expression in radical form, or write each radical in exponential form.

59. $17^{\frac{1}{3}}$
60. $q^{\frac{1}{4}}$
61. $7b^{\frac{1}{3}}$
62. $m^{\frac{2}{3}}$
63. $\sqrt[3]{29}$
64. $\sqrt[5]{h}$
65. $2\sqrt[3]{a}$
66. $\sqrt[3]{xy^2}$

Simplify.

67. $\sqrt[3]{0.027}$
68. $\sqrt[4]{\frac{n^4}{16}}$
69. $a^{\frac{1}{3}} \cdot a^{\frac{2}{3}}$
70. $c^{\frac{1}{2}} \cdot c^{\frac{3}{2}}$
71. $(8^2)^{\frac{2}{3}}$
72. $\left(y^{\frac{3}{4}}\right)^{\frac{1}{2}}$
73. $9^{-\frac{1}{2}}$
74. $16^{-\frac{3}{2}}$
75. $(3^2)^{-\frac{3}{2}}$
76. $\left(81^{\frac{1}{4}}\right)^{-2}$
77. $k^{-\frac{1}{2}}$
78. $\left(d^{\frac{4}{3}}\right)^0$

Solve each equation.

79. $2^{5x} = 8^{2x-4}$
80. $81^{2x-3} = 9^{x+3}$
81. $2^{4x} = 32^{x+1}$
82. $16^x = \frac{1}{2}$
83. $25^x = \frac{1}{125}$
84. $6^{8-x} = \frac{1}{216}$

85. CCSS MODELING The frequency f in hertz of the nth key on a piano is $f = 440\left(2^{\frac{1}{12}}\right)^{n-49}$.

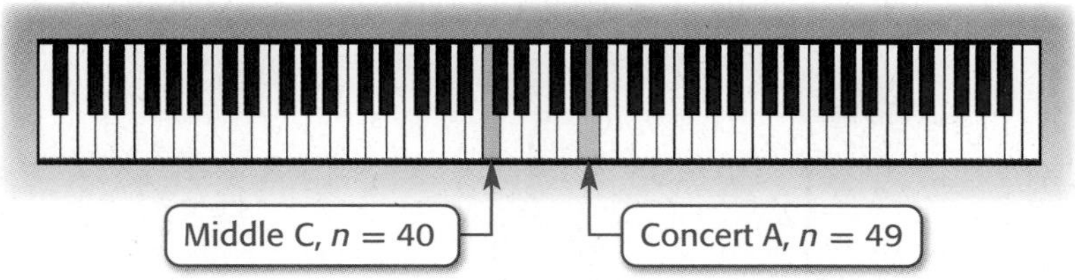

a. What is the frequency of Concert A?
b. Which note has a frequency of 220 Hz?

86. **RANDOM WALKS** Suppose you go on a walk where you choose the direction of each step at random. The path of a molecule in a liquid or a gas, the path of a foraging animal, and a fluctuating stock price are all modeled as random walks. The number of possible random walks w of n steps where you choose one of d directions at each step is $w = d^n$.

a. How many steps have been taken in a 2-direction random walk if there are 4096 possible walks?

b. How many steps have been taken in a 4-direction random walk if there are 65,536 possible walks?

c. If a walk of 7 steps has 2187 possible walks, how many directions could be taken at each step?

87. **SOCCER** The radius r of a ball that holds V cubic units of air is modeled by $r = 0.62V^{\frac{1}{3}}$. What are the possible volumes of each size soccer ball?

Soccer Ball Dimensions	
Size	**Diameter (in.)**
3	7.3–7.6
4	8.0–8.3
5	8.6–9.0

88. **MULTIPLE REPRESENTATIONS** In this problem, you will explore the graph of an exponential function.

a. TABULAR Copy and complete the table below.

x	-2	$-\frac{3}{2}$	-1	$-\frac{1}{2}$	0	$\frac{1}{2}$	1	$\frac{3}{2}$	2
$f(x) = 4^x$									

b. GRAPHICAL Graph $f(x)$ by plotting the points and connecting them with a smooth curve.

c. VERBAL Describe the shape of the graph of $f(x)$. What are its key features? Is it linear?

H.O.T. Problems Use Higher-Order Thinking Skills

89. **OPEN ENDED** Write two different expressions with rational exponents equal to $\sqrt{2}$.

90. **CCSS ARGUMENTS** Determine whether each statement is *always, sometimes,* or *never* true. Assume that x is a nonnegative real number. Explain your reasoning.

a. $x^2 = x^{\frac{1}{2}}$ **b.** $x^{-2} = x^{\frac{1}{2}}$ **c.** $x^{\frac{1}{3}} = x^{\frac{1}{2}}$

d. $\sqrt{x} = x^{\frac{1}{2}}$ **e.** $\left(x^{\frac{1}{2}}\right)^2 = x$ **f.** $x^{\frac{1}{2}} \cdot x^2 = x$

91. **CHALLENGE** For what values of x is $x = x^{\frac{1}{3}}$?

92. **ERROR ANALYSIS** Anna and Jamal are solving $128^x = 4$. Is either of them correct? Explain your reasoning.

Anna	Jamal
$128^x = 4$	$128^x = 4$
$(2^7)^x = 2^2$	$(2^7)^x = 4$
$2^{7x} = 2^2$	$2^{7x} = 4^1$
$7x = 2$	$7x = 1$
$x = \frac{2}{7}$	$x = \frac{1}{7}$

93. **WRITING IN MATH** Explain why 2 is the principal fourth root of 16.

Standardized Test Practice

94. What is the value of $16^{\frac{3}{4}} + 9^{\frac{3}{2}}$?

A 5
B 11
C 25
D 35

95. At a movie theater, the costs for various numbers of popcorn and hot dogs are shown.

Hot Dogs	Boxes of Popcorn	Total Cost
1	1	$8.50
2	4	$21.60

Which pair of equations can be used to find p, the cost of a box of popcorn, and h, the cost of a hot dog?

F $p + h = 8.5$
$p + 2h = 10.8$

G $p + h = 8.5$
$2h + 4p = 21.6$

H $p + h = 8.5$
$2p + 4h = 21.6$

J $p + h = 8.5$
$2p + 2h = 21.6$

96. SHORT RESPONSE Find the dimensions of the rectangle if its perimeter is 52 inches.

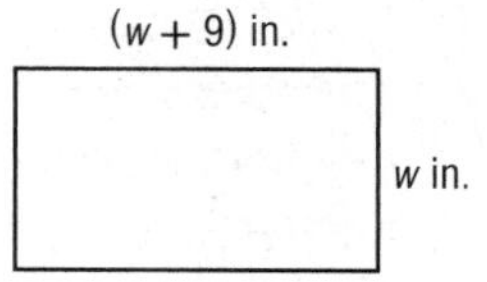

97. If $3^4 = 9^x$, then $x =$

A 1
B 2
C 4
D 5

Spiral Review

Simplify each expression. Assume that no denominator equals zero. (Lesson 7-2)

98. $\frac{a^3b^5}{ab^3}$

99. $\frac{c^8d^{11}}{c^4d^5}$

100. $\frac{4x^3y^3z^6}{xyz^5}$

101. $\frac{a^5b^3c}{a^5bc}$

102. $\left(\frac{3m^4}{4p^2}\right)^2$

103. $\left(\frac{3df^2}{9d^2f}\right)^0$

104. GARDENING Felipe is planting a flower garden that is shaped like a trapezoid as shown at the right. Use the formula $A = \frac{1}{2}h(b_1 + b_2)$ to find the area of the garden. (Lesson 7-1)

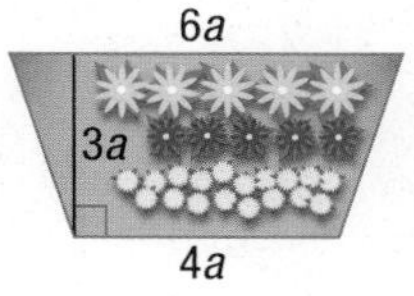

Write each equation in slope-intercept form. (Lesson 4-2)

105. $y - 2 = 3(x - 1)$

106. $y - 5 = 6(x + 1)$

107. $y + 2 = -2(x + 5)$

108. $y + 3 = \frac{1}{2}(x + 4)$

109. $y - 1 = \frac{2}{3}(x + 9)$

110. $y + 3 = -\frac{1}{4}(x + 2)$

Skills Check

Find each power.

111. 10^3

112. 10^5

113. 10^{-1}

114. 10^{-4}

LESSON 7-4 Scientific Notation

Then

- You used the laws of exponents to find products and quotients of monomials.

Now

1. Express numbers in scientific notation.
2. Find products and quotients of numbers expressed in scientific notation.

Why?

- Space tourism is a multibillion dollar industry. For a price of $20 million, a civilian can travel on a rocket or shuttle and visit the International Space Station (ISS) for a week.

NewVocabulary
scientific notation

Common Core State Standards

Content Standards
A.SSE.2 Use the structure of an expression to identify ways to rewrite it.

Mathematical Practices
3 Construct viable arguments and critique the reasoning of others.
6 Attend to precision.

1 Scientific Notation

Very large and very small numbers such as $20 million can be cumbersome to use in calculations. For this reason, numbers are often expressed in scientific notation. A number written in **scientific notation** is of the form $a \times 10^n$, where $1 \leq a < 10$ and n is an integer.

KeyConcept Standard Form to Scientific Notation

Step 1 Move the decimal point until it is to the right of the first nonzero digit. The result is a real number a.

Step 2 Note the number of places n and the direction that you moved the decimal point.

Step 3 If the decimal point is moved left, write the number as $a \times 10^n$. If the decimal point is moved right, write the number as $a \times 10^{-n}$.

Step 4 Remove the unnecessary zeros.

Example 1 Standard Form to Scientific Notation

Express each number in scientific notation.

a. 201,000,000

Step 1 201,000,000 ⟶ 2.01000000 $a = 2.01000000$

Step 2 The decimal point moved 8 places to the left, so $n = 8$.

Step 3 $201{,}000{,}000 = 2.01000000 \times 10^8$

Step 4 2.01×10^8

b. 0.000051

Step 1 0.000051 ⟶ 00005.1 $a = 00005.1$

Step 2 The decimal point moved 5 places to the right, so $n = 5$.

Step 3 $0.000051 = 00005.1 \times 10^{-5}$

Step 4 5.1×10^{-5}

GuidedPractice

1A. 68,700,000,000 **1B.** 0.0000725

Brand X Pictures/PunchStock

You can also rewrite numbers in scientific notation in standard form.

WatchOut!

Negative Signs Be careful about the placement of negative signs. A negative sign in the exponent means that the number is between 0 and 1. A negative sign before the number means that it is less than 0.

KeyConcept Scientific Notation to Standard Form

Step 1 In $a \times 10^n$, note whether $n > 0$ or $n < 0$.

Step 2 If $n > 0$, move the decimal point n places right.
If $n < 0$, move the decimal point $-n$ places left.

Step 3 Insert zeros, decimal point, and commas as needed for place value.

Example 2 Scientific Notation to Standard Form

Express each number in standard form.

a. 6.32×10^9

Step 1 The exponent is 9, so $n = 9$.

Step 2 Since $n > 0$, move the decimal point 9 places to the right.
$6.32 \times 10^9 \longrightarrow 6320000000$

Step 3 $6.32 \times 10^9 = 6{,}320{,}000{,}000$ Rewrite; insert commas.

b. 4×10^{-7}

Step 1 The exponent is -7, so $n = -7$.

Step 2 Since $n < 0$, move the decimal point 7 places to the left.
$4 \times 10^{-7} \longrightarrow 0000004$

Step 3 $4 \times 10^{-7} = 0.0000004$ Rewrite; insert a 0 before the decimal point.

GuidedPractice

2A. 3.201×10^6

2B. 9.03×10^{-5}

2 Product and Quotients in Scientific Notation

You can use scientific notation to simplify multiplying and dividing very large and very small numbers.

Problem-SolvingTip

CCSS Tools Estimating an answer before computing the solution can help you determine if your answer is reasonable.

Example 3 Multiply with Scientific Notation

Evaluate $(3.5 \times 10^{-3})(7 \times 10^5)$. Express the result in both scientific notation and standard form.

$(3.5 \times 10^{-3})(7 \times 10^5)$	Original expression
$= (3.5 \times 7)(10^{-3} \times 10^5)$	Commutative and Associative Properties
$= 24.5 \times 10^2$	Product of Powers
$= (2.45 \times 10^1) \times 10^2$	$24.5 = 2.45 \times 10$
$= 2.45 \times 10^3$ or 2450	Product of Powers

GuidedPractice

Evaluate each product. Express the results in both scientific notation and standard form.

3A. $(6.5 \times 10^{12})(8.7 \times 10^{-15})$

3B. $(7.8 \times 10^{-4})^2$

PT

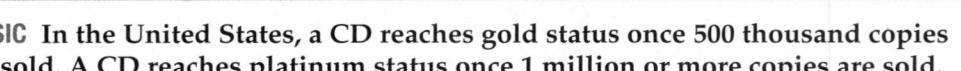

Example 4 Divide with Scientific Notation

Evaluate $\frac{3.066 \times 10^8}{7.3 \times 10^3}$. Express the result in both scientific notation and standard form.

$\frac{3.066 \times 10^8}{7.3 \times 10^3} = \left(\frac{3.066}{7.3}\right)\left(\frac{10^8}{10^3}\right)$	Product rule for fractions
$= 0.42 \times 10^5$	Quotient of Powers
$= 4.2 \times 10^{-1} \times 10^5$	$0.42 = 4.2 \times 10^{-1}$
$= 4.2 \times 10^4$	Product of Powers
$= 42{,}000$	Standard form

StudyTip

Quotient of Powers
Recall that the Quotient of Powers Property is only valid for powers that have the same base. Since 10^8 and 10^3 have the same base, the property applies.

GuidedPractice

Evaluate each quotient. Express the results in both scientific notation and standard form.

4A. $\frac{2.3958 \times 10^3}{1.98 \times 10^8}$

4B. $\frac{1.305 \times 10^3}{1.45 \times 10^{-4}}$

Real-WorldLink

The platinum award was created in 1976. In 2004, the criteria for the award was extended to digital sales. The top-selling artist of all time is the Beatles with 170 million units sold.

Source: Recording Industry Association of America

PT

Real-World Example 5 Use Scientific Notation

MUSIC In the United States, a CD reaches gold status once 500 thousand copies are sold. A CD reaches platinum status once 1 million or more copies are sold.

a. Express the number of copies of CDs that need to be sold to reach each status in standard notation.

gold status: 500 thousand = 500,000; platinum status: 1 million = 1,000,000

b. Write each number in scientific notation.

gold status: $500{,}000 = 5 \times 10^5$; platinum status: $1{,}000{,}000 = 1 \times 10^6$

c. How many copies of a CD have sold if it has gone platinum 13 times? Write your answer in scientific notation and standard form.

A CD reaches platinum status once it sells 1 million records. Since the CD has gone platinum 13 times, we need to multiply by 13.

$(13)(1 \times 10^6)$	Original expression
$= (13 \times 1)(10^6)$	Associative Property
$= 13 \times 10^6$	$13 \times 1 = 13$
$= (1.3 \times 10^1) \times 10^6$	$13 = 1.3 \times 10$
$= 1.3 \times 10^7$	Product of Powers
$= 13{,}000{,}000$	Standard form

GuidedPractice

5. SATELLITE RADIO Suppose a satellite radio company earned $125.4 million in one year.

A. Write this number in standard form.

B. Write this number in scientific notation.

C. If the following year the company earned 2.5 times the amount earned the previous year, determine the amount earned. Write your answer in scientific notation and standard form.

Check Your Understanding

Example 1 **Express each number in scientific notation.**

1. 185,000,000
2. 1,902,500,000
3. 0.000564
4. 0.00000804

MONEY **Express each number in scientific notation.**

5. Teens spend $13 billion annually on clothing.
6. Teens have an influence on their families' spending habit. They control about $1.5 billion of discretionary income.

Example 2 **Express each number in standard form.**

7. 1.98×10^7
8. 4.052×10^6
9. 3.405×10^{-8}
10. 6.8×10^{-5}

Example 3 **Evaluate each product. Express the results in both scientific notation and standard form.**

11. $(1.2 \times 10^3)(1.45 \times 10^{12})$
12. $(7.08 \times 10^{14})(5 \times 10^{-9})$
13. $(5.18 \times 10^2)(9.1 \times 10^{-5})$
14. $(2.18 \times 10^{-2})^2$

Example 4 **Evaluate each quotient. Express the results in both scientific notation and standard form.**

15. $\dfrac{1.035 \times 10^8}{2.3 \times 10^4}$
16. $\dfrac{2.542 \times 10^5}{4.1 \times 10^{-10}}$
17. $\dfrac{1.445 \times 10^{-7}}{1.7 \times 10^5}$
18. $\dfrac{2.05 \times 10^{-8}}{4 \times 10^{-2}}$

Example 5

19. **CCSS PRECISION** Salvador bought an air purifier to help him deal with his allergies. The filter in the purifier will stop particles as small as one hundredth of a micron. A micron is one millionth of a millimeter.
    a. Write one hundredth and one micron in standard form.
    b. Write one hundredth and one micron in scientific notation.
    c. What is the smallest size particle in meters that the filter will stop? Write the result in both standard form and scientific notation.

Practice and Problem Solving

Example 1 **Express each number in scientific notation.**

20. 1,220,000
21. 58,600,000
22. 1,405,000,000,000
23. 0.0000013
24. 0.000056
25. 0.000000000709

EMAIL **Express each number in scientific notation.**

26. Approximately 100 million emails sent to the President are put into the National Archives.
27. By 2015, the email security market will generate $6.5 billion.

Example 2 **Express each number in standard form.**

28. 1×10^{12}
29. 9.4×10^7
30. 8.1×10^{-3}
31. 5×10^{-4}
32. 8.73×10^{11}
33. 6.22×10^{-6}

Example 2 INTERNET **Express each number in standard form.**

34. About 2.1×10^7 people, aged 12 to 17, use the Internet.
35. Approximately 1.1×10^7 teens go online daily.

Examples 3–4 **Evaluate each product or quotient. Express the results in both scientific notation and standard form.**

36. $(3.807 \times 10^3)(5 \times 10^2)$
37. $\frac{9.6 \times 10^3}{1.2 \times 10^{-4}}$
38. $\frac{2.88 \times 10^3}{1.2 \times 10^{-5}}$
39. $(6.5 \times 10^7)(7.2 \times 10^{-2})$
40. $(9.5 \times 10^{-18})(9 \times 10^9)$
41. $\frac{8.8 \times 10^3}{4 \times 10^{-4}}$
42. $\frac{9.15 \times 10^{-3}}{6.1 \times 10}$
43. $(1.4 \times 10^6)^2$
44. $(2.58 \times 10^2)(3.6 \times 10^6)$
45. $\frac{5.6498 \times 10^{10}}{8.2 \times 10^4}$
46. $\frac{1.363 \times 10^{16}}{2.9 \times 10^6}$
47. $(5 \times 10^3)(1.8 \times 10^{-7})$
48. $(2.3 \times 10^{-3})^2$
49. $\frac{6.25 \times 10^{-4}}{1.25 \times 10^2}$
50. $\frac{3.75 \times 10^{-9}}{1.5 \times 10^{-4}}$
51. $(7.2 \times 10^7)^2$
52. $\frac{8.6 \times 10^4}{2 \times 10^{-6}}$
53. $(6.3 \times 10^{-5})^2$

Example 5

54. **ASTRONOMY** The distance between Earth and the Sun varies throughout the year. Earth is closest to the Sun in January when the distance is 91.4 million miles. In July, the distance is greatest at 94.4 million miles.
    a. Write 91.4 million in both standard form and in scientific notation.
    b. Write 94.4 million in both standard form and in scientific notation.
    c. What is the percent increase in distance from January to July? Round to the nearest tenth of a percent.

Evaluate each product or quotient. Express the results in both scientific notation and standard form.

55. $(4.65 \times 10^{-2})(5.91 \times 10^6)$
56. $\frac{2.548 \times 10^5}{2.8 \times 10^{-2}}$
57. $\frac{2.135 \times 10^5}{3.5 \times 10^{12}}$
58. $(3.16 \times 10^{-2})^2$
59. $(2.01 \times 10^{-4})(8.9 \times 10^{-3})$
60. $\frac{5.184 \times 10^{-5}}{7.2 \times 10^3}$
61. $(9.04 \times 10^6)(5.2 \times 10^{-4})$
62. $\frac{1.032 \times 10^{-4}}{8.6 \times 10^{-5}}$

LIGHT **The speed of light is approximately 3×10^8 meters per second.**

63. Write an expression to represent the speed of light in kilometers per second.
64. Write an expression to represent the speed of light in kilometers per hour.
65. Make a table to show how many kilometers light travels in a day, a week, a 30-day month, and a 365-day year. Express your results in scientific notation.
66. **CCSS MODELING** A recent cell phone study showed that company A's phone processes up to 7.95×10^5 bits of data every second. Company B's phone processes up to 1.41×10^6 bits of data every second. Evaluate and interpret $\frac{1.41 \times 10^6}{7.95 \times 10^5}$.

67 **EARTH** The population of Earth is about 6.623×10^9. The land surface of Earth is 1.483×10^8 square kilometers. What is the population density for the land surface area of Earth?

68. **RIVERS** A drainage basin separated from adjacent basins by a ridge, hill, or mountain is known as a watershed. The watershed of the Amazon River is 2,300,000 square miles. The watershed of the Mississippi River is 1,200,000 square miles.

a. Write each of these numbers in scientific notation.

b. How many times as large is the Amazon River watershed as the Mississippi River watershed?

69. **AGRICULTURE** In a recent year, farmers planted approximately 92.9 million acres of corn. They also planted 64.1 million acres of soybeans and 11.1 million acres of cotton.

a. Write each of these numbers in scientific notation and in standard form.

b. How many times as much corn was planted as soybeans? Write your results in standard form and in scientific notation. Round your answer to four decimal places.

c. How many times as much corn was planted as cotton? Write your results in standard form and in scientific notation. Round your answer to four decimal places.

H.O.T. Problems Use Higher-Order Thinking Skills

70. **REASONING** Which is greater, 100^{10} or 10^{100}? Explain your reasoning.

71. **ERROR ANALYSIS** Syreeta and Pete are solving a division problem with scientific notation. Is either of them correct? Explain your reasoning.

Syreeta

$$\frac{3.65 \times 10^{-12}}{5 \times 10^5} = 0.73 \times 10^{-17}$$
$$= 7.3 \times 10^{-16}$$

Pete

$$\frac{3.65 \times 10^{-12}}{5 \times 10^5} = 0.73 \times 10^{-17}$$
$$= 7.3 \times 10^{-18}$$

72. **CHALLENGE** Order these numbers from least to greatest without converting them to standard form.

$$5.46 \times 10^{-3}, 6.54 \times 10^3, 4.56 \times 10^{-4}, -5.64 \times 10^4, -4.65 \times 10^5$$

73. CCSS **ARGUMENTS** Determine whether the statement is *always, sometimes,* or *never* true. Give examples or a counterexample to verify your reasoning.

When multiplying two numbers written in scientific notation, the resulting number can have no more than two digits to the left of the decimal point.

74. **OPEN ENDED** Write two numbers in scientific notation with a product of 1.3×10^{-3}. Then name two numbers in scientific notation with a quotient of 1.3×10^{-3}.

75. **WRITING IN MATH** Write the steps that you would use to divide two numbers written in scientific notation. Then describe how you would write the results in standard form. Demonstrate by finding $\frac{a}{b}$ for $a = 2 \times 10^3$ and $b = 4 \times 10^5$.

Standardized Test Practice

76. Which number represents 0.05604×10^8 written in standard form?

A 0.0000000005604
B 560,400
C 5,604,000
D 50,604,000

77. Toni left school and rode her bike home. The graph below shows the relationship between her distance from the school and time.

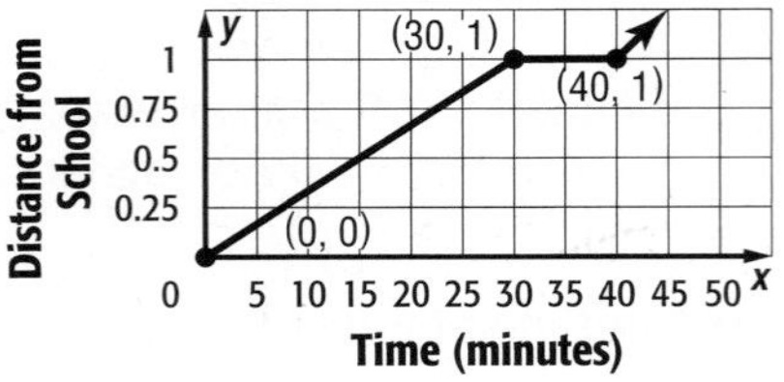

Which explanation could account for the section of the graph from $x = 30$ to $x = 40$?

F Toni rode her bike down a hill.
G Toni ran all the way home.
H Toni stopped at a friend's house on her way home.
J Toni returned to school to get her mathematics book.

78. SHORT RESPONSE In his first four years of coaching football, Coach Delgato's team won 5 games the first year, 10 games the second year, 8 games the third year, and 7 games the fourth year. How many games does the team need to win during the fifth year to have an average of 8 wins per year?

79. The table shows the relationship between Calories and grams of fat contained in an order of fried chicken from various restaurants.

Calories	305	410	320	500	510	440
Fat (g)	28	34	28	41	42	38

Assuming that the data can best be described by a linear model, about how many grams of fat would you expect to be in a 275-Calorie order of fried chicken?

A 22
B 25
C 28
D 30

Spiral Review

80. HEALTH A ponderal index p is a measure of a person's body based on height h in centimeters and mass m in kilograms. One such formula is $p = 100m^{\frac{1}{3}}h^{-1}$. If a person who is 182 centimeters tall has a ponderal index of about 2.2, how much does the person weigh in kilograms? (Lesson 7-3)

Simplify. Assume that no denominator is equal to zero. (Lesson 7-2)

81. $\frac{8^9}{8^6}$

82. $\frac{6^5}{6^3}$

83. $\frac{r^8t^{12}}{r^2t^7}$

84. $\left(\frac{3a^4b^4}{8c^2}\right)^4$

85. $\left(\frac{5d^3g^2}{3h^4}\right)^2$

86. $\left(\frac{4n^2p^4}{8p^3}\right)^3$

87. CHEMISTRY Lemon juice is 10^2 times as acidic as tomato juice. Tomato juice is 10^3 times as acidic as egg whites. How many times as acidic is lemon juice as egg whites? (Lesson 7-2)

Skills Review

Evaluate $a(b^x)$ for each of the given values.

88. $a = 1, b = 2, x = 4$

89. $a = 4, b = 1, x = 7$

90. $a = 5, b = 3, x = 0$

91. $a = 0, b = 6, x = 8$

92. $a = -2, b = 3, x = 1$

93. $a = -3, b = 5, x = 2$

CHAPTER 7

Mid-Chapter Quiz

Lessons 7-1 through 7-4

Simplify each expression. (Lesson 7-1)

1. $(x^3)(4x^5)$

2. $(m^2p^5)^3$

3. $[(2xy^3)^2]^3$

4. $(6ab^3c^4)(-3a^2b^3c)$

5. MULTIPLE CHOICE Express the volume of the solid as a monomial. (Lesson 7-1)

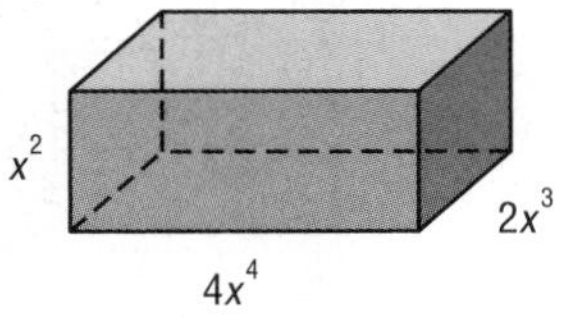

A $6x^9$

B $8x^9$

C $8x^{24}$

D $7x^{24}$

Simplify each expression. Assume that no denominator equals 0. (Lesson 7-2)

6. $\left(\frac{2a^4b^3}{c^6}\right)^3$

7. $\frac{2xy^0}{6x}$

8. $\frac{m^7n^4p}{m^3n^3p}$

9. $\frac{p^4t^{-2}}{r^{-5}}$

10. ASTRONOMY Physicists estimate that the number of stars in the universe has an order of magnitude of 10^{21}. The number of stars in the Milky Way galaxy is around 100 billion. Using orders of magnitude, how many times as many stars are there in the universe as the Milky Way? (Lesson 7-2)

Write each expression in radical form, or write each radical in exponential form. (Lesson 7-3)

11. $42^{\frac{1}{2}}$

12. $11x^{\frac{1}{2}}$

13. $(11g)^{\frac{1}{2}}$

14. $\sqrt{55}$

15. $\sqrt{5k}$

16. $4\sqrt{p}$

Simplify. (Lesson 7-3)

17. $\sqrt[3]{729}$

18. $\sqrt[4]{625}$

19. $1331^{\frac{1}{3}}$

20. $\left(\frac{16}{81}\right)^{\frac{1}{4}}$

21. $8^{\frac{2}{3}}$

22. $625^{\frac{3}{4}}$

23. $216^{\frac{5}{3}}$

24. $\left(\frac{1}{4}\right)^{\frac{3}{2}}$

Solve each equation. (Lesson 7-3)

25. $4^x = 4096$

26. $5^{2x+1} = 125$

27. $4^{x-3} = 128$

Express each number in scientific notation. (Lesson 7-4)

28. 0.00000054

29. 0.0042

30. 234,000

31. 418,000,000

Express each number in standard form. (Lesson 7-4)

32. 4.1×10^{-3}

33. 2.74×10^5

34. 3×10^9

35. 9.1×10^{-5}

Evaluate each product or quotient. Express the results in scientific notation. (Lesson 7-4)

36. $(2.13 \times 10^2)(3 \times 10^5)$

37. $(7.5 \times 10^6)(2.5 \times 10^{-2})$

38. $\frac{7.5 \times 10^8}{2.5 \times 10^4}$

39. $\frac{6.6 \times 10^5}{2 \times 10^{-3}}$

40. MAMMALS A blue whale has been caught that was 4.2×10^5 pounds. The smallest mammal is a bumblebee bat, which is about 0.0044 pound. (Lesson 7-4)

a. Write the whale's weight in standard form.

b. Write the bat's weight in scientific notation.

c. How many orders of magnitude as big as a blue whale is a bumblebee bat?

EXPLORE 7-5

Graphing Technology Lab
Family of Exponential Functions

An **exponential function** is a function of the form $y = ab^x$, where $a \neq 0$, $b > 0$, and $b \neq 1$. You have studied the effects of changing parameters in linear functions. You can use a graphing calculator to analyze how changing the parameters a and b affects the graphs in the family of exponential functions.

CCSS Common Core State Standards
Content Standards
F.IF.7e Graph exponential and logarithmic functions, showing intercepts and end behavior, and trigonometric functions, showing period, midline, and amplitude.
F.BF.3 Identify the effect on the graph of replacing $f(x)$ by $f(x) + k$, $kf(x)$, $f(kx)$, and $f(x + k)$ for specific values of k (both positive and negative); find the value of k given the graphs. Experiment with cases and illustrate an explanation of the effects on the graph using technology.

Activity 1 b in $y = b^x$, $b > 1$

Graph the set of equations on the same screen. Describe any similarities and differences among the graphs.

$y = 2^x, y = 3^x, y = 6^x$

Enter the equations in the [Y=] list and graph.

There are many similarities in the graphs. The domain for each function is all real numbers, and the range is all positive real numbers. The functions are increasing over the entire domain. The graphs do not display any line symmetry.

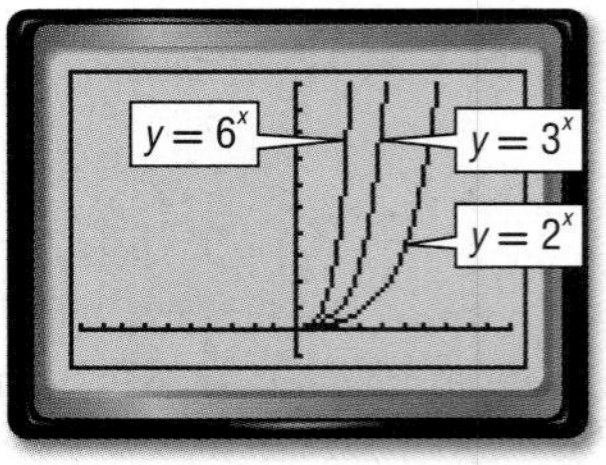

[−10, 10] scl: 1 by [−10, 100] scl: 10

Use the [ZOOM] feature to investigate the key features of the graphs.

Zooming in twice on a point near the origin allows closer inspection of the graphs. The y-intercept is 1 for all three graphs.

Tracing along the graphs reveals that there are no x-intercepts, no maxima and no minima.

The graphs are different in that the graphs for the equations in which b is greater are steeper.

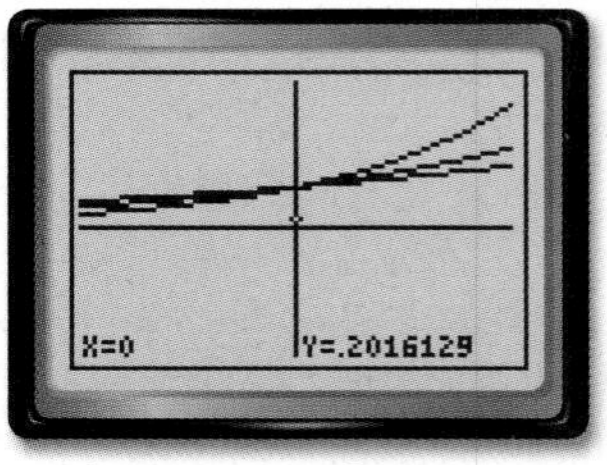

[−0.625, 0.625] scl: 1 by [−3.25..., 3.63...] scl: 10

The effect of b on the graph is different when $0 < b < 1$.

Activity 2 b in $y = b^x$, $0 < b < 1$

Graph the set of equations on the same screen. Describe any similarities and differences among the graphs.

$y = \left(\frac{1}{2}\right)^x, y = \left(\frac{1}{3}\right)^x, y = \left(\frac{1}{6}\right)^x$

The domain for each function is all real numbers, and the range is all positive real numbers. The function values are all positive and the functions are decreasing over the entire domain. The graphs display no line symmetry. There are no x-intercepts, and the y-intercept is 1 for all three graphs. There are no maxima or minima.

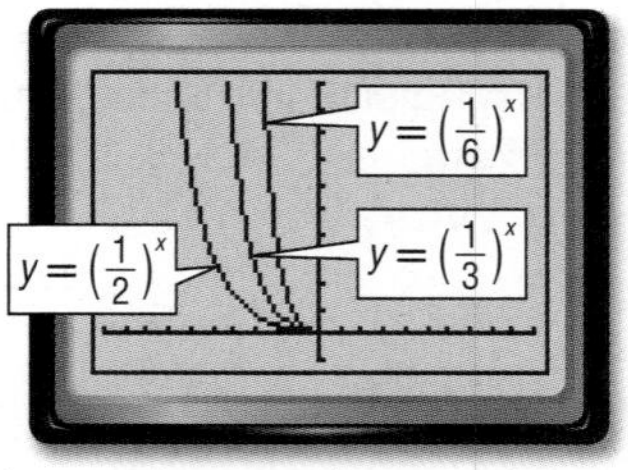

[−10, 10] scl: 1 by [−10, 100] scl: 10

However, the graphs in which b is lesser are steeper.

Activity 3 a in $y = ab^x$, $a > 0$

Graph each set of equations on the same screen. Describe any similarities and differences among the graphs.

$y = 2^x$, $y = 3(2^x)$, $y = \frac{1}{6}(2^x)$

The domain for each function is all real numbers, and the range is all positive real numbers. The functions are increasing over the entire domain. The graphs do not display any line symmetry.

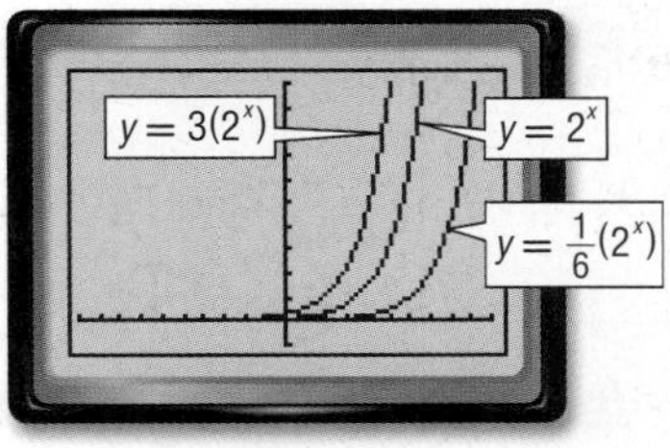

[−10, 10] scl: 1 by [−10, 100] scl: 10

Use the ZOOM feature to investigate the key features of the graphs.

Zooming in twice on a point near the origin allows closer inspection of the graphs.

Tracing along the graphs reveals that there are no x-intercepts, no maxima and no minima.

However, the graphs in which a is greater are steeper. The y-intercept is 1 in the graph of $y = 2^x$, 3 in $y = 3(2^x)$, and $\frac{1}{6}$ in $y = \frac{1}{6}(2^x)$.

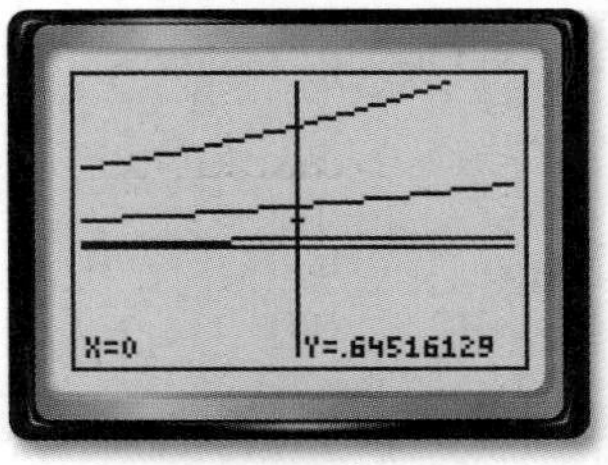

[−0.625, 0.625] scl: 1 by [−2.79..., 4.08...] scl: 10

Activity 4 a in $y = ab^x$, $a < 0$

Graph each set of equations on the same screen. Describe any similarities and differences among the graphs.

$y = -2^x$, $y = -3(2^x)$, $y = -\frac{1}{6}(2^x)$

The domain for each function is all real numbers, and the range is all negative real numbers. The functions are decreasing over the entire domain. The graphs do not display any line symmetry.

There are no x-intercepts, no maxima and no minima.

However, the graphs in which the absolute value of a is greater are steeper. The y-intercept is -1 in the graph of $y = -2^x$, -3 in $y = -3(2^x)$, and $-\frac{1}{6}$ in $y = -\frac{1}{6}(2^x)$.

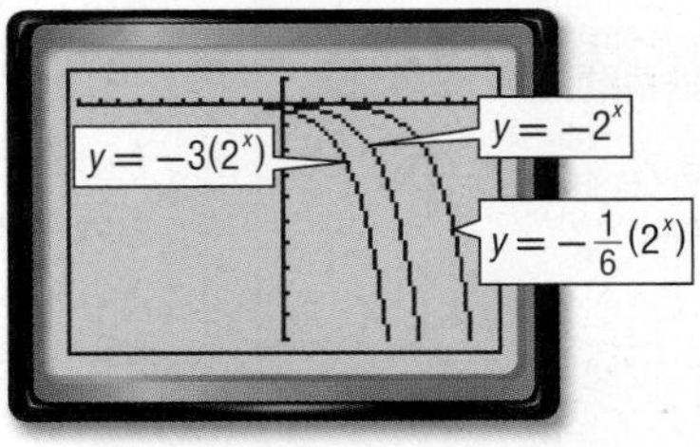

[−10, 10] scl: 1 by [−100, 10] scl: 10

Model and Analyze

1. How does b affect the graph of $y = ab^x$ when $b > 1$ and when $0 < b < 1$? Give examples.
2. How does a affect the graph of $y = ab^x$ when $a > 0$ and when $a < 0$? Give examples.
3. CCSS **REGULARITY** Make a conjecture about the relationship of the graphs of $y = 3^x$ and $y = \left(\frac{1}{3}\right)^x$. Verify your conjecture by graphing both functions.

LESSON
7-5 Exponential Functions

Then

- You evaluated numerical expressions involving exponents.

Now

1. Graph exponential functions.
2. Identify data that display exponential behavior.

Why?

- Tarantulas can appear scary with their large hairy bodies and legs, but they are harmless to humans. The graph shows a tarantula spider population that increases over time. Notice that the graph is not linear.

  The graph represents the function $y = 3(2)^x$. This is an example of an *exponential* function.

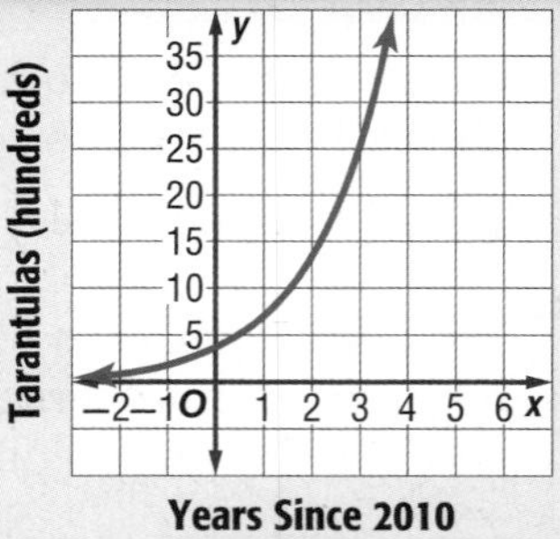

NewVocabulary
exponential function
exponential growth function
exponential decay function

Common Core State Standards

Content Standards

F.IF.7e Graph exponential and logarithmic functions, showing intercepts and end behavior, and trigonometric functions, showing period, midline, and amplitude.

F.LE.2 Construct linear and exponential functions, including arithmetic and geometric sequences, given a graph, a description of a relationship, or two input-output pairs (include reading these from a table).

Mathematical Practices

1 Make sense of problems and persevere in solving them.

1 Graph Exponential Functions

An **exponential function** is a function of the form $y = ab^x$, where $a \neq 0$, $b > 0$, and $b \neq 1$. Notice that the base is a constant and the exponent is a variable. Exponential functions are nonlinear.

KeyConcept Exponential Function

Words An exponential function is a function that can be described by an equation of the form $y = ab^x$, where $a \neq 0$, $b > 0$, and $b \neq 1$.

Examples $y = 2(3)^x$ $y = 4^x$ $y = \left(\frac{1}{2}\right)^x$

Example 1 Graph with $a > 0$ and $b > 1$

Graph $y = 3^x$. Find the y-intercept, and state the domain and range.

The graph crosses the y-axis at 1, so the y-intercept is 1. The domain is all real numbers, and the range is all positive real numbers.

Notice that the graph approaches the x-axis but there is no x-intercept. The graph is increasing on the entire domain.

x	3^x	y
−2	3^{-2}	$\frac{1}{9}$
−1	3^{-1}	$\frac{1}{3}$
0	3^0	1
$\frac{1}{2}$	$3^{\frac{1}{2}}$	≈1.73
1	3^1	3
2	3^2	9

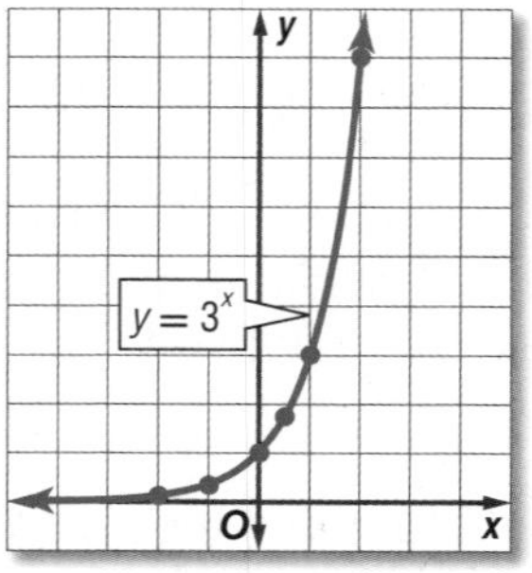

GuidedPractice

1. Graph $y = 7^x$. Find the y-intercept, and state the domain and range.

Functions of the form $y = ab^x$, where $a > 0$ and $b > 1$, are called **exponential growth functions** and all have the same shape as the graph in Example 1. Functions of the form $y = ab^x$, where $a > 0$ and $0 < b < 1$ are called **exponential decay functions** and also have the same general shape.

StudyTip

$a < 0$ If the value of a is less than 0, the graph will be reflected across the x-axis.

Example 2 Graph with $a > 0$ and $0 < b < 1$

Graph $y = \left(\frac{1}{3}\right)^x$. Find the y-intercept, and state the domain and range.

The y-intercept is 1. The domain is all real numbers, and the range is all positive real numbers. Notice that as x increases, the y-values decrease less rapidly.

x	$\left(\frac{1}{3}\right)^x$	y
−2	$\left(\frac{1}{3}\right)^{-2}$	9
0	$\left(\frac{1}{3}\right)^0$	1
2	$\left(\frac{1}{3}\right)^2$	$\frac{1}{9}$

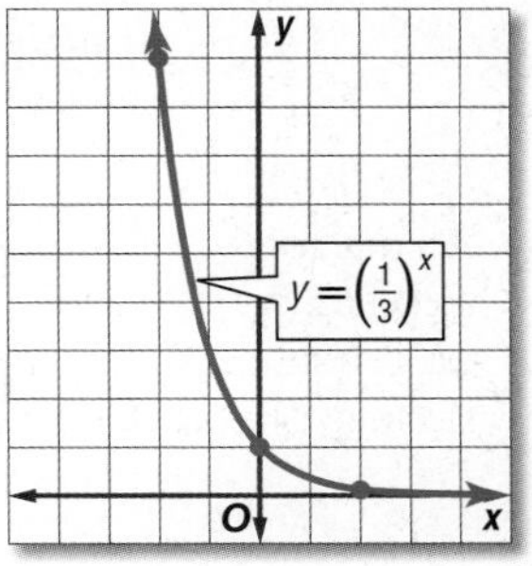

GuidedPractice

2. Graph $y = \left(\frac{1}{2}\right)^x - 1$. Find the y-intercept, and state the domain and range.

The key features of the graphs of exponential functions can be summarized as follows.

KeyConcept Graphs of Exponential Functions

Exponential Growth Functions

Equation: $f(x) = ab^x$, $a > 0$, $b > 1$

Domain, Range: all reals; all positive reals

Intercepts: one y-intercept, no x-intercepts

End behavior: as x increases, $f(x)$ increases; as x decreases, $f(x)$ approaches 0

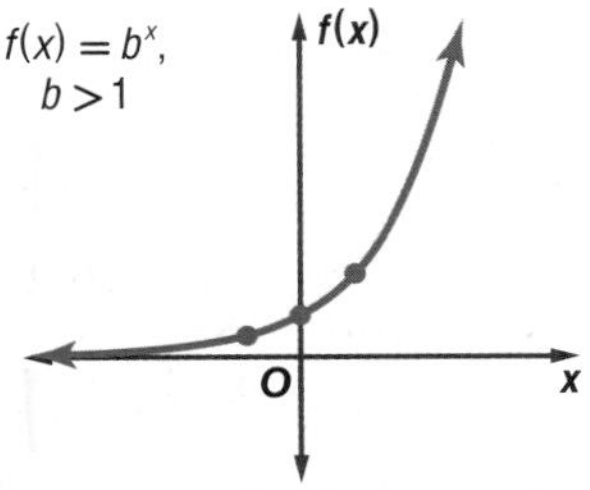

Exponential Decay Functions

Equation: $f(x) = ab^x$, $a > 0$, $0 < b < 1$

Domain, Range: all reals; all negative reals

Intercepts: one y-intercept, no x-intercepts

End behavior: as x increases, $f(x)$ approaches 0; as x decreases, $f(x)$ increases

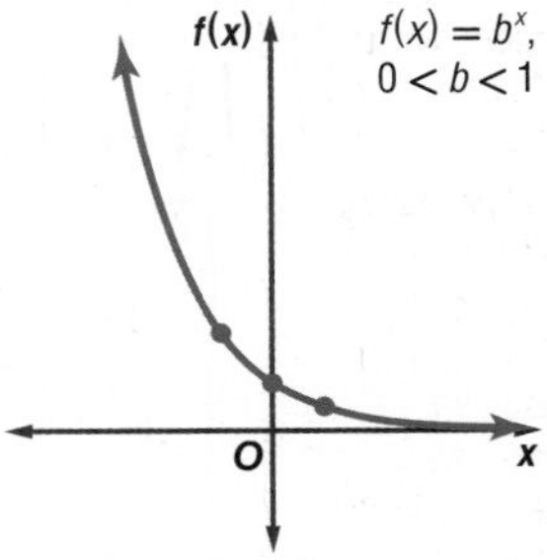

Exponential functions occur in many real world situations.

Real-WorldLink

The United States is the largest soda consumer in the world. In a recent year, the United States accounted for one third of the world's total soda consumption.

Source: Worldwatch Institute

BananaStock/PunchStock

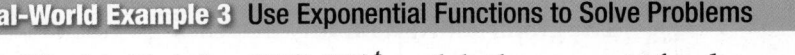

Real-World Example 3 Use Exponential Functions to Solve Problems

SODA The function $C = 179(1.029)^t$ models the amount of soda consumed in the world, where C is the amount consumed in billions of liters and t is the number of years since 2000.

a. Graph the function. What values of C and t are meaningful in the context of the problem?

Since t represents time, $t > 0$. At $t = 0$, the consumption is 179 billion liters. Therefore, in the context of this problem, $C > 179$ is meaningful.

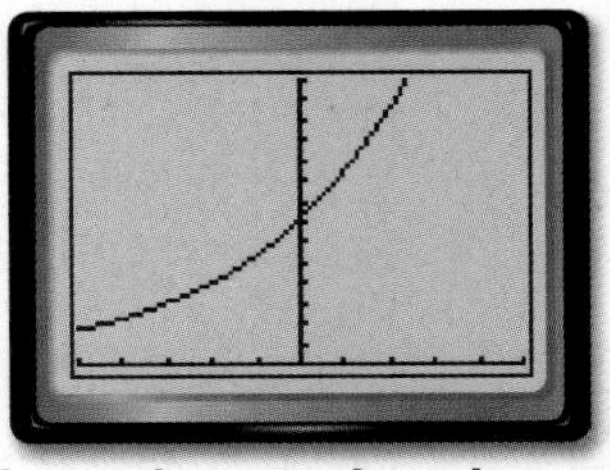

[−50, 50] scl: 10 by [0, 350] scl: 25

b. How much soda was consumed in 2005?

$C = 179(1.029)^t$ — Original equation

$= 179(1.029)^5$ — $t = 5$

≈ 206.5 — Use a calculator.

The world soda consumption in 2005 was approximately 206.5 billion liters.

GuidedPractice

3. **BIOLOGY** A certain bacteria population doubles every 20 minutes. Beginning with 10 cells in a culture, the population can be represented by the function $B = 10(2)^t$, where B is the number of bacteria cells and t is the time in 20 minute increments. How many will there be after 2 hours?

2 Identify Exponential Behavior

Recall from Lesson 3-3 that linear functions have a constant rate of change. Exponential functions do not have constant rates of change, but they do have constant ratios.

Problem-SolvingTip

Make an Organized List Making an organized list of x-values and corresponding y-values is helpful in graphing the function. It can also help you identify patterns in the data.

Example 4 Identify Exponential Behavior

Determine whether the set of data shown below displays exponential behavior. Write *yes* or *no*. Explain why or why not.

x	0	5	10	15	20	25
y	64	32	16	8	4	2

Method 1 Look for a pattern.

The domain values are at regular intervals of 5. Look for a common factor among the range values.

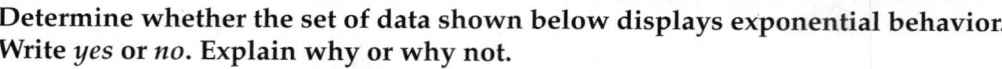

64 32 16 8 4 2

$\times\frac{1}{2}$ $\times\frac{1}{2}$ $\times\frac{1}{2}$ $\times\frac{1}{2}$ $\times\frac{1}{2}$

The range values differ by the common factor of $\frac{1}{2}$.

Since the domain values are at regular intervals and the range values differ by a positive common factor, the data are probably exponential. Its equation may involve $\left(\frac{1}{2}\right)^x$.

Method 2 Graph the data.

Plot the points and connect them with a smooth curve. The graph shows a rapidly decreasing value of y as x increases. This is a characteristic of exponential behavior in which the base is between 0 and 1.

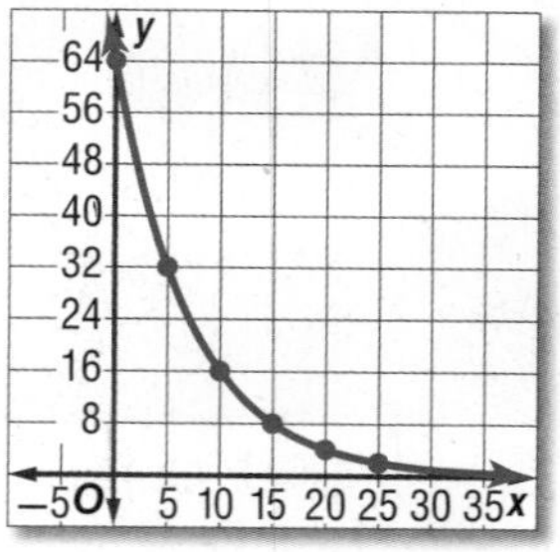

GuidedPractice

4. Determine whether the set of data shown below displays exponential behavior. Write *yes* or *no*. Explain why or why not.

x	0	3	6	9	12	15
y	12	16	20	24	28	32

Check Your Understanding

Examples 1–2 **Graph each function. Find the y-intercept and state the domain and range.**

1. $y = 2^x$ **2.** $y = -5^x$ **3.** $y = -\left(\frac{1}{5}\right)^x$

4. $y = 3\left(\frac{1}{4}\right)^x$ **5.** $f(x) = 6^x + 3$ **6.** $f(x) = 2 - 2^x$

Example 3

7. BIOLOGY The function $f(t) = 100(1.05)^t$ models the growth of a fruit fly population, where $f(t)$ is the number of flies and t is time in days.

a. What values for the domain and range are reasonable in the context of this situation? Explain.

b. After two weeks, approximately how many flies are in this population?

Example 4 **Determine whether the set of data shown below displays exponential behavior. Write *yes* or *no*. Explain why or why not.**

8.

x	1	2	3	4	5	6
y	−4	−2	0	2	4	6

9.

x	2	4	6	8	10	12
y	1	4	16	64	256	1024

Practice and Problem Solving

Examples 1–2 **Graph each function. Find the y-intercept and state the domain and range.**

10. $y = 2 \cdot 8^x$ **11.** $y = 2 \cdot \left(\frac{1}{6}\right)^x$ **12.** $y = \left(\frac{1}{12}\right)^x$

13. $y = -3 \cdot 9^x$ **14.** $y = -4 \cdot 10^x$ **15.** $y = 3 \cdot 11^x$

16. $y = 4^x + 3$ **17.** $y = \frac{1}{2}(2^x - 8)$

18. $y = 5(3^x) + 1$ **19.** $y = -2(3^x) + 5$

Example 3

20. CCSS **MODELING** A population of bacteria in a culture increases according to the model $p = 300(2.7)^{0.02t}$, where t is the number of hours and $t = 0$ corresponds to 9:00 A.M.

a. Use this model to estimate the number of bacteria at 11 A.M.

b. Graph the function and name the p-intercept. Describe what the p-intercept represents, and describe a reasonable domain and range for this situation.

Example 4 **Determine whether the set of data shown below displays exponential behavior. Write *yes* or *no*. Explain why or why not.**

21.

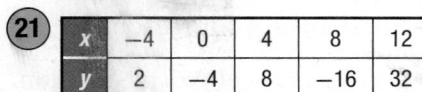

x	−4	0	4	8	12
y	2	−4	8	−16	32

22.

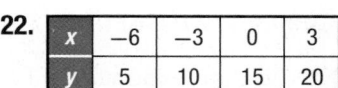

x	−6	−3	0	3
y	5	10	15	20

23.

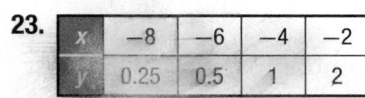

x	−8	−6	−4	−2
y	0.25	0.5	1	2

24.

x	20	30	40	50	60
y	1	0.4	0.16	0.064	0.0256

25. PHOTOGRAPHY Jameka is enlarging a photograph to make a poster for school. She will enlarge the picture repeatedly at 150%. The function $P = 1.5^x$ models the new size of the picture being enlarged, where x is the number of enlargements. How many times as big is the picture after 4 enlargements?

26. FINANCIAL LITERACY Daniel deposited \$500 into a savings account and after 8 years, his investment is worth \$807.07. The equation $A = d(1.005)^{12t}$ models the value of Daniel's investment A after t years with an initial deposit d.

a. What would the value of Daniel's investment be if he had deposited \$1000?

b. What would the value of Daniel's investment be if he had deposited \$250?

c. Interpret $d(1.005)^{12t}$ to explain how the amount of the original deposit affects the value of Daniel's investment.

Identify each function as *linear, exponential,* or *neither.*

27.

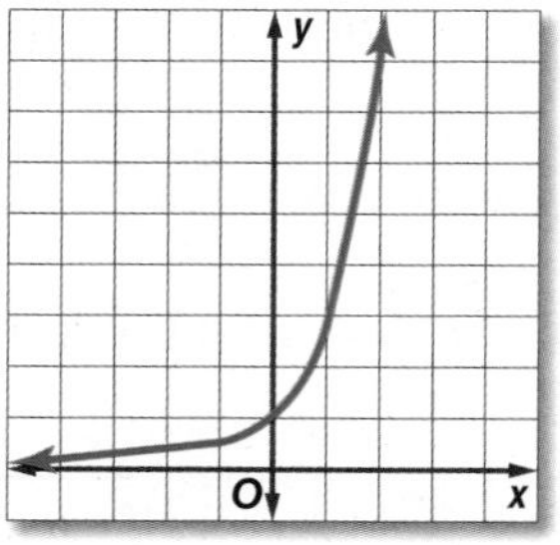

28.

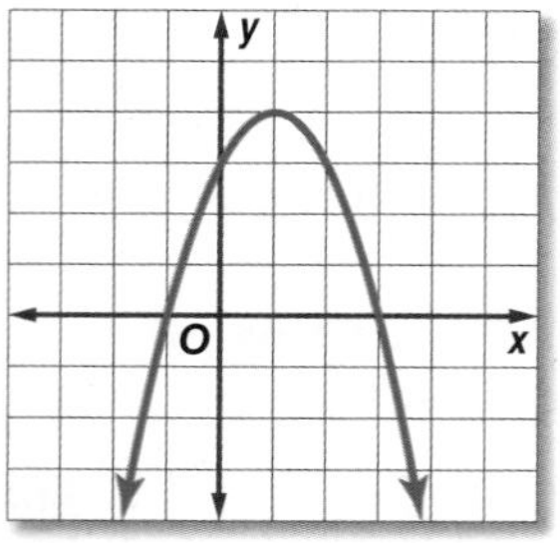

29.

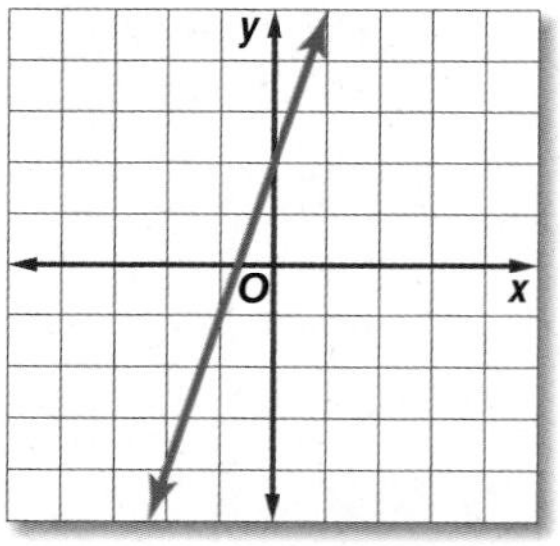

30. $y = 4^x$

31. $y = 2x(x - 1)$

32. $5x + y = 8$

33. GRADUATION The number of graduates at a high school has increased by a factor of 1.055 every year since 2001. In 2001, 110 students graduated. The function $N = 110(1.055)^t$ models the number of students N expected to graduate t years after 2001. How many students will graduate in 2012?

Describe the graph of each equation as a transformation of the graph of $y = 2^x$.

34. $y = 2^x + 6$

35. $y = 3(2)^x$

36. $y = -\frac{1}{4}(2)^x$

37. $y = -3 + 2^x$

38. $y = \left(\frac{1}{2}\right)^x$

39. $y = -5(2)^x$

40. DEER The deer population at a national park doubles every year. In 2000, there were 25 deer in the park. The function $N = 25(2)^t$ models the number of deer N in the park t years after 2000. What will the deer population be in 2015?

H.O.T. Problems Use Higher-Order Thinking Skills

41. CCSS PERSEVERANCE Write an exponential function for which the graph passes through the points at $(0, 3)$ and $(1, 6)$.

42. REASONING Determine whether the graph of $y = ab^x$, where $a \neq 0$, $b > 0$, and $b \neq 1$, *sometimes, always,* or *never* has an x-intercept. Explain your reasoning.

43. OPEN ENDED Find an exponential function that represents a real-world situation, and graph the function. Analyze the graph, and explain why the situation is modeled by an exponential function rather than a linear function.

44. REASONING Use tables and graphs to compare and contrast an exponential function $f(x) = ab^x + c$, where $a \neq 0$, $b > 0$, and $b \neq 1$, and a linear function $g(x) = ax + c$. Include intercepts, intervals where the functions are increasing, decreasing, positive, or negative, relative maxima and minima, symmetry, and end behavior.

45. WRITING IN MATH Explain how to determine whether a set of data displays exponential behavior.

Standardized Test Practice

46. SHORT RESPONSE What are the x-intercepts of the function graphed below?

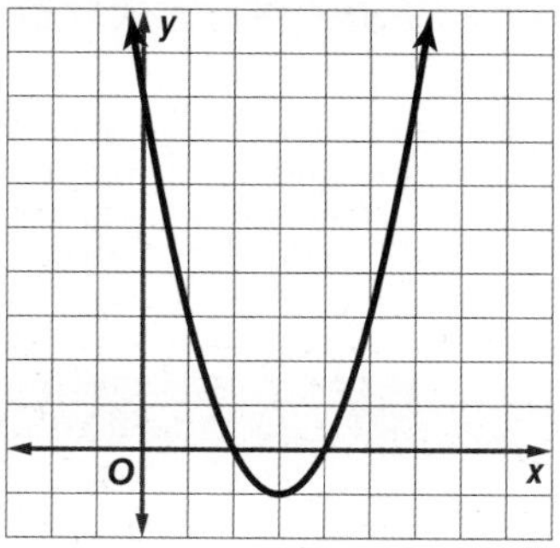

47. Hinto invested \$300 into a savings account. The equation $A = 300(1.005)^{12t}$ models the amount in Hinto's account A after t years. How much will be in Hinto's account after 7 years?

A \$25,326 **C** \$385.01
B \$456.11 **D** \$301.52

48. GEOMETRY Ayana placed a circular piece of paper on a square picture as shown below. If the picture extends 4 inches beyond the circle on each side, what is the perimeter of the square picture?

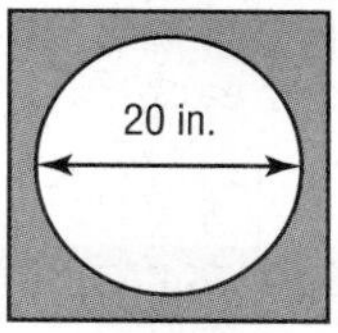

F 64 in. **H** 94 in.
G 80 in. **J** 112 in.

49. The points with coordinates (0, −3) and (2, 7) are on line l. Line p contains (3, −1) and is perpendicular to line l. What is the x-coordinate of the point where l and p intersect?

A $\frac{1}{2}$ **B** $-\frac{2}{5}$
C $-\frac{1}{2}$ **D** −3

Spiral Review

Evaluate each product. Express the results in both scientific notation and standard form. (Lesson 7-4)

50. $(1.9 \times 10^2)(4.7 \times 10^6)$ **51.** $(4.5 \times 10^{-3})(5.6 \times 10^4)$ **52.** $(3.8 \times 10^{-4})(6.4 \times 10^{-8})$

Simplify. (Lesson 7-3)

53. $\sqrt[3]{343}$ **54.** $\sqrt[6]{729}$ **55.** $\left(\frac{1}{32}\right)^{\frac{1}{5}}$

56. $729^{\frac{5}{6}}$ **57.** $216^{\frac{5}{3}}$ **58.** $\left(\frac{1}{81}\right)^{\frac{3}{2}}$

59. DEMOLITION DERBY When a car hits an object, the damage is measured by the collision impact. For a certain car the collision impact I is given by $I = 2v^2$, where v represents the speed in kilometers per minute. What is the collision impact if the speed of the car is 4 kilometers per minute? (Lesson 7-1)

Use elimination to solve each system of equations. (Lesson 6-3)

60. $x + y = -3$
$x - y = 1$

61. $3a + b = 5$
$2a + b = 10$

62. $3x - 5y = 16$
$-3x + 2y = -10$

Skills Review

Find the next three terms of each arithmetic sequence.

63. 1, 3, 5, 7, … **64.** −6, −4, −2, 0, … **65.** 6.5, 9, 11.5, 14, …

66. 10, 3, −4, −11, … **67.** $\frac{1}{2}, \frac{5}{4}, 2, \frac{11}{4}, \ldots$ **68.** $1, \frac{3}{4}, \frac{1}{2}, \frac{1}{4}, \ldots$

EXTEND 7-5

Graphing Technology Lab
Solving Exponential Equations and Inequalities

You can use TI-Nspire Technology to solve exponential equations and inequalities by graphing and by using tables.

CCSS Common Core State Standards
Content Standards
A.REI.11 Explain why the *x*-coordinates of the points where the graphs of the equations $y = f(x)$ and $y = g(x)$ intersect are the solutions of the equation $f(x) = g(x)$; find the solutions approximately, e.g., using technology to graph the functions, make tables of values, or find successive approximations. Include cases where $f(x)$ and/or $g(x)$ are linear, polynomial, rational, absolute value, exponential, and logarithmic functions.
Mathematical Practices
5 Use appropriate tools strategically.

Activity 1 Graph an Exponential Equation

Graph $y = 3^x + 4$ using a graphing calculator.

Step 1 Add a new **Graphs** page.

Step 2 Enter $3^x + 4$ as **f1(*x*)**.

Step 3 Use the **Window Settings** option from the **Window/Zoom** menu to adjust the window so that x is from -10 to 10 and y is from -100 to 100. Keep the scales as **Auto**.

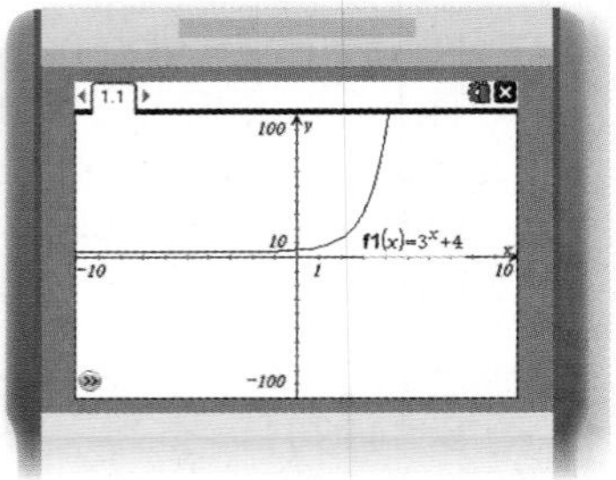

To solve an equation by graphing, graph both sides of the equation and locate the point(s) of intersection.

Activity 2 Solve an Exponential Equation by Graphing

Solve $2^{x-2} = \frac{3}{4}$.

Step 1 Add a new **Graphs** page.

Step 2 Enter 2^{x-2} as **f1(*x*)** and $\frac{3}{4}$ as **f2(*x*)**.

Step 3 Use the **Intersection Point(s)** tool from the **Points & Lines** menu to find the intersection of the two graphs. Select the graph of **f1(*x*) enter** and then the graph of **f2(*x*) enter**.

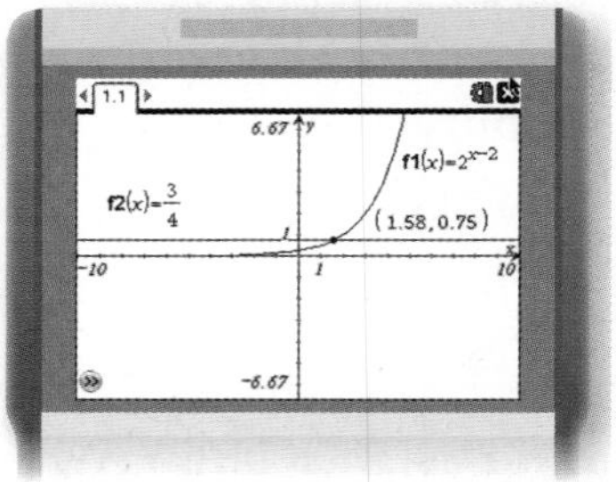

The graphs intersect at about (1.58, 0.75). Therefore, the solution of $2^{x-2} = \frac{3}{4}$ is 1.58.

Exercises

CCSS TOOLS Use a graphing calculator to solve each equation.

1. $\left(\frac{1}{3}\right)^{x-1} = \frac{3}{4}$
2. $2^{2x-1} = 2x$
3. $\left(\frac{1}{2}\right)^{2x} = 2^{2x}$
4. $5^{\frac{1}{3}x+2} = -x$
5. $\left(\frac{1}{8}\right)^{2x} = -2x + 1$
6. $2^{\frac{1}{4}x-1} = 3^{x+1}$
7. $2^{3x-1} = 4^x$
8. $4^{2x-3} = 5^{-x+1}$
9. $3^{2x-4} = 2^x + 1$

Activity 3 Solve an Exponential Equation by Using a Table

Solve $2\left(\frac{1}{2}\right)^{x+2} = \frac{1}{4}$ using a table.

Step 1 Add a new **Lists & Spreadsheet** page.

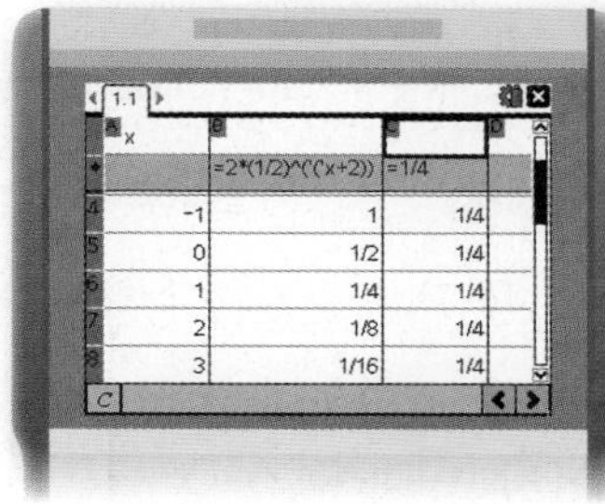

Step 2 Label column A as x. Enter values from -4 to 4 in cells A1 to A9.

Step 3 In column B in the formula row, enter the left side of the rational equation. In column C of the formula row, enter $= \frac{1}{4}$. Specify **Variable Reference** when prompted.

Scroll until you see where the values in Columns B and C are equal. This occurs at $x = 1$. Therefore, the solution of $2\left(\frac{1}{2}\right)^{x+2} = \frac{1}{4}$ is 1.

You can also use a graphing calculator to solve exponential inequalities.

Activity 4 Solve an Exponential Inequality

Solve $4^{x-3} \leq \left(\frac{1}{4}\right)^{2x}$.

Step 1 Add a new **Graphs** page.

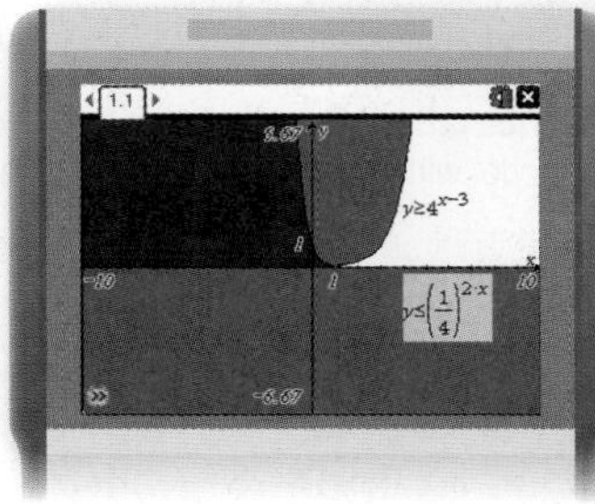

Step 2 Enter the left side of the inequality into **f1(*x*)**. Press **del**, select $\geq$, and enter 4^{x-3}. Enter the right side of the inequality into **f2(*x*)**. Press **tab del** $\leq$, and enter $\left(\frac{1}{4}\right)^{2x}$.

The x-values of the points in the region where the shading overlap is the solution set of the original inequality. Therefore the solution of $4^{x-3} \leq \left(\frac{1}{4}\right)^{2x}$ is $x \leq 1$.

Exercises

CCSS TOOLS Use a graphing calculator to solve each equation or inequality.

10. $\left(\frac{1}{3}\right)^{3x} = 3^x$

11. $\left(\frac{1}{6}\right)^{2x} = 4^x$

12. $3^{1-x} \leq 4^x$

13. $4^{3x} \leq 2x + 1$

14. $\left(\frac{1}{4}\right)^x > 2^{x+4}$

15. $\left(\frac{1}{3}\right)^{x-1} \geq 2^x$

LESSON 7-6 Growth and Decay

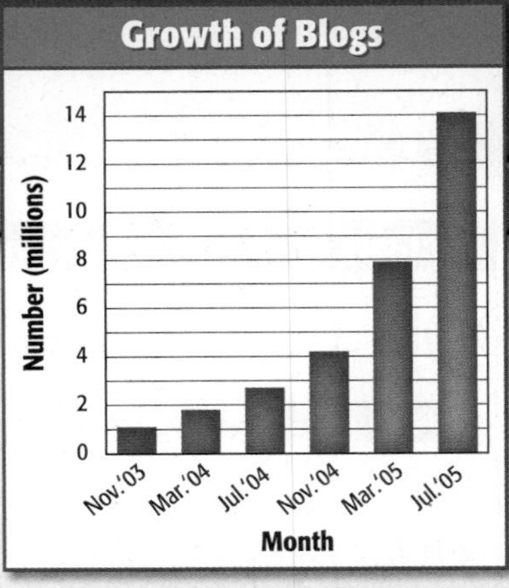

Then

- You analyzed exponential functions.

Now

1. Solve problems involving exponential growth.
2. Solve problems involving exponential decay.

Why?

- The number of Weblogs or blogs increased at a monthly rate of about 13.7% over 21 months. The average number of blogs per month can be modeled by $y = 1.1(1 + 0.137)^t$ or $y = 1.1(1.137)^t$, where y represents the total number of blogs in millions and t is the number of months since November 2003.

NewVocabulary

compound interest

Common Core State Standards

Content Standards

F.IF.8b Use the properties of exponents to interpret expressions for exponential functions.

F.LE.2 Construct linear and exponential functions, including arithmetic and geometric sequences, given a graph, a description of a relationship, or two input-output pairs (include reading these from a table).

Mathematical Practices

4 Model with mathematics.

1 Exponential Growth

The equation for the number of blogs is in the form $y = a(1 + r)^t$. This is the general equation for exponential growth.

KeyConcept Equation for Exponential Growth

$$y = a(1 + r)^t$$

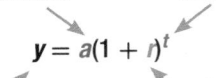

- a is the initial amount.
- t is time.
- y is the final amount.
- r is the rate of change expressed as a decimal, $r > 0$.

Real-World Example 1 Exponential Growth

PT

CONTEST The prize for a radio station contest begins with a \$100 gift card. Once a day, a name is announced. The person has 15 minutes to call or the prize increases by 2.5% for the next day.

a. Write an equation to represent the amount of the gift card in dollars after t days with no winners.

$y = a(1 + r)^t$	Equation for exponential growth
$y = 100(1 + 0.025)^t$	$a = 100$ and $r = 2.5\%$ or 0.025
$y = 100(1.025)^t$	Simplify.

In the equation $y = 100(1.025)^t$, y is the amount of the gift card and t is the number of days since the contest began.

b. How much will the gift card be worth if no one wins after 10 days?

$y = 100(1.025)^t$	Equation for amount of gift card
$= 100(1.025)^{10}$	$t = 10$
≈ 128.01	Use a calculator.

In 10 days, the gift card will be worth \$128.01.

GuidedPractice

1. TUITION A college's tuition has risen 5% each year since 2000. If the tuition in 2000 was \$10,850, write an equation for the amount of the tuition t years after 2000. Predict the cost of tuition for this college in 2015.

Compound interest is interest earned or paid on both the initial investment and previously earned interest. It is an application of exponential growth.

KeyConcept Equation for Compound Interest

$$A = P\left(1 + \frac{r}{n}\right)^{nt}$$

A is the current amount.

n is the number of times the interest is compounded each year, and *t* is time in years.

P is the principal or initial amount.

r is the annual interest rate expressed as a decimal, $r > 0$.

Real-WorldCareer

Financial Advisor Financial advisors help people plan their financial futures. A good financial advisor has mathematical, problem-solving, and communication skills. A bachelor's degree is strongly preferred but not required.

Real-World Example 2 Compound Interest

PT

FINANCE Maria's parents invested $14,000 at 6% per year compounded monthly. How much money will there be in the account after 10 years?

$A = P\left(1 + \frac{r}{n}\right)^{nt}$	Compound interest equation
$= 14{,}000\left(1 + \frac{0.06}{12}\right)^{12(10)}$	$P = 14{,}000$, $r = 6\%$ or 0.06, $n = 12$, and $t = 10$
$= 14{,}000(1.005)^{120}$	Simplify.
$\approx 25{,}471.55$	Use a calculator.

There will be about $25,471.55 in 10 years.

GuidedPractice

2. FINANCE Determine the amount of an investment if $300 is invested at an interest rate of 3.5% compounded monthly for 22 years.

2 Exponential Decay

Exponential Decay In exponential decay, the original amount decreases by the same percent over a period of time. A variation of the growth equation can be used as the general equation for exponential decay.

StudyTip

Growth and Decay Since *r* is added to 1, the value inside the parentheses will be greater than 1 for exponential growth functions. For exponential decay functions, this value will be less than 1 since *r* is subtracted from 1.

KeyConcept Equation for Exponential Decay

$$y = a(1 - r)^t$$

a is the initial amount.

t is time.

y is the final amount.

r is the rate of decay expressed as a decimal, $0 < r < 1$.

Real-World Example 3 Exponential Decay

PT

SWIMMING A fully inflated child's raft for a pool is losing 6.6% of its air every day. The raft originally contained 4500 cubic inches of air.

a. Write an equation to represent the loss of air.

$y = a(1 - r)^t$	Equation for exponential decay
$= 4500(1 - 0.066)^t$	$a = 4500$ and $r = 6.6\%$ or 0.066
$= 4500(0.934)^t$	Simplify.

$y = 4500(0.934)^t$, where y is the air in the raft in cubic inches after t days.

b. Estimate the amount of air in the raft after 7 days.

$y = 4500(0.934)^t$ — Equation for air loss

$= 4500(0.934)^7$ — $t = 7$

≈ 2790 — Use a calculator.

The amount of air in the raft after 7 days will be about 2790 cubic inches.

GuidedPractice

3. POPULATION The population of Campbell County, Kentucky, has been decreasing at an average rate of about 0.3% per year. In 2000, its population was 88,647. Write an equation to represent the population since 2000. If the trend continues, predict the population in 2010.

Check Your Understanding

Example 1

1. SALARY Ms. Acosta received a job as a teacher with a starting salary of $34,000. According to her contract, she will receive a 1.5% increase in her salary every year. How much will Ms. Acosta earn in 7 years?

Example 2

2. MONEY Paul invested $400 into an account with a 5.5% interest rate compounded monthly. How much will Paul's investment be worth in 8 years?

Example 3

3. ENROLLMENT In 2000, 2200 students attended Polaris High School. The enrollment has been declining 2% annually.

a. Write an equation for the enrollment of Polaris High School t years after 2000.

b. If this trend continues, how many students will be enrolled in 2015?

Practice and Problem Solving

Example 1

4. MEMBERSHIPS The Work-Out Gym sold 550 memberships in 2001. Since then the number of memberships sold has increased 3% annually.

a. Write an equation for the number of memberships sold at Work-Out Gym t years after 2001.

b. If this trend continues, predict how many memberships the gym will sell in 2020.

5. COMPUTERS The number of people who own computers has increased 23.2% annually since 1990. If half a million people owned a computer in 1990, predict how many people will own a computer in 2015.

6. COINS Camilo purchased a rare coin from a dealer for $300. The value of the coin increases 5% each year. Determine the value of the coin in 5 years.

Example 2

7. INVESTMENTS Theo invested $6600 at an interest rate of 4.5% compounded monthly. Determine the value of his investment in 4 years.

8. COMPOUND INTEREST Paige invested $1200 at an interest rate of 5.75% compounded quarterly. Determine the value of her investment in 7 years.

9. CCSS PRECISION Brooke is saving money for a trip to the Bahamas that costs $295.99. She puts $150 into a savings account that pays 7.25% interest compounded quarterly. Will she have enough money in the account after 4 years? Explain.

Example 3

10. INVESTMENTS Jin's investment of $4500 has been losing its value at a rate of 2.5% each year. What will his investment be worth in 5 years?

11. **POPULATION** In the years from 2010 to 2015, the population of the District of Columbia is expected decrease about 0.9% annually. In 2010, the population was about 530,000. What is the population of the District of Columbia expected to be in 2015?

12. **CARS** Leonardo purchases a car for \$18,995. The car depreciates at a rate of 18% annually. After 6 years, Manuel offers to buy the car for \$4500. Should Leonardo sell the car? Explain.

13. **HOUSING** The median house price in the United States increased an average of 1.4% each year between 2005 and 2007. Assume that this pattern continues.

    a. Write an equation for the median house price for t years after 2007.

    b. Predict the median house price in 2018.

Median House Price	
2005	\$240,900
2006	\$246,500
2007	\$247,900

Source: *Real Estate Journal*

14. **ELEMENTS** A radioactive element's half-life is the time it takes for one half of the element's quantity to decay. The half-life of Plutonium-241 is 14.4 years. The number of grams A of Plutonium-241 left after t years can be modeled by $A = p(0.5)^{\frac{t}{14.4}}$, where p is the original amount of the element.

    a. How much of a 0.2-gram sample remains after 72 years?

    b. How much of a 5.4-gram sample remains after 1095 days?

15. **COMBINING FUNCTIONS** A swimming pool holds a maximum of 20,500 gallons of water. It evaporates at a rate of 0.5% per hour. The pool currently contains 19,000 gallons of water.

    a. Write an exponential function $w(t)$ to express the amount of water remaining in the pool after time t where t is the number of hours after the pool has reached 19,000 gallons.

    b. At this same time, a hose is turned on to refill the pool at a rate of 300 gallons per hour. Write a function $p(t)$, where t is the time in hours the hose is running, to express the amount of water that is pumped into the pool.

    c. Find $C(t) = p(t) + w(t)$. What does this new function represent?

    d. Use the graph of $C(t)$ to determine how long the hose must run to fill the pool to its maximum capacity.

H.O.T. Problems Use Higher-Order Thinking Skills

16. **REASONING** Determine the growth rate (as a percent) of a population that quadruples every year. Explain.

17. **CCSS PRECISION** Santos invested \$1200 into an account with an interest rate of 8% compounded monthly. Use a calculator to approximate how long it will take for Santos's investment to reach \$2500.

18. **REASONING** The amount of water in a container doubles every minute. After 8 minutes, the container is full. After how many minutes was the container half full? Explain.

19. **E? WRITING IN MATH** What should you consider when using exponential models to make decisions?

20. **WRITING IN MATH** Compare and contrast the exponential growth formula and the exponential decay formula.

Standardized Test Practice

21. GEOMETRY The parallelogram has an area of 35 square inches. Find the height h of the parallelogram.

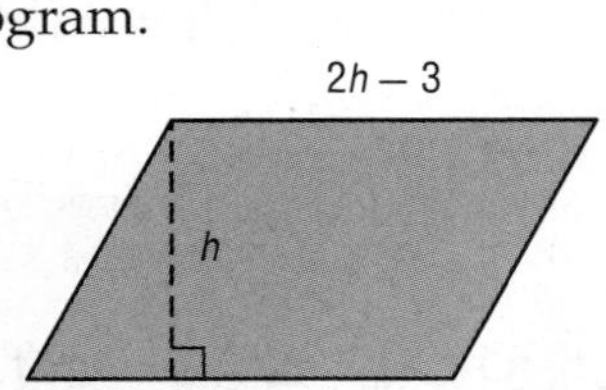

A 3.5 inches
B 4 inches
C 5 inches
D 7 inches

22. Which is greater than $64^{\frac{1}{3}}$?

F 2^2
G $64^{\frac{1}{6}}$
H $64^{\frac{1}{2}}$
J 64^{-3}

23. Thi purchased a car for $22,900. The car depreciated at an annual rate of 16%. Which of the following equations models the value of Thi's car after 5 years?

A $A = 22{,}900(1.16)^5$
B $A = 22{,}900(0.16)^5$
C $A = 16(22{,}900)^5$
D $A = 22{,}900(0.84)^5$

24. GRIDDED RESPONSE A deck measures 12 feet by 18 feet. If a painter charges $2.65 per square foot, including tax, how much will it cost in dollars to have the deck painted?

Spiral Review

Graph each function. Find the y-intercept and state the domain and range. (Lesson 7-5)

25. $y = 3^x$

26. $y = \left(\frac{1}{2}\right)^x$

27. $y = 6^x$

Evaluate each product. Express the results in both scientific notation and standard form. (Lesson 7-4)

28. $(4.2 \times 10^3)(3.1 \times 10^{10})$

29. $(6.02 \times 10^{23})(5 \times 10^{-14})$

30. $(7 \times 10^5)^2$

31. $(1.1 \times 10^{-2})^2$

32. $(9.1 \times 10^{-2})(4.2 \times 10^{-7})$

33. $(3.14 \times 10^2)(6.1 \times 10^{-3})$

34. EVENT PLANNING A hall does not charge a rental fee as long as at least $4000 is spent on food. For the prom, the hall charges $28.95 per person for a buffet. How many people must attend the prom to avoid a rental fee for the hall? (Lesson 5-2)

Determine whether the graphs of each pair of equations are *parallel, perpendicular,* or *neither.* (Lesson 4-4)

35. $y = -2x + 11$
$y + 2x = 23$

36. $3y = 2x + 14$
$-3x - 2y = 2$

37. $y = -5x$
$y = 5x - 18$

38. AGES The table shows equivalent ages for horses and humans. Write an equation that relates human age to horse age and find the equivalent horse age for a human who is 16 years old. (Lesson 3-4)

Horse age (x)	0	1	2	3	4	5
Human age (y)	0	3	6	9	12	15

Find the total price of each item. (Lesson 2-7)

39. umbrella: $14.00
tax: 5.5%

40. sandals: $29.99
tax: 5.75%

41. backpack: $35.00
tax: 7%

Skills Review

Graph each set of ordered pairs.

42. (3, 0), (0, 1), (−4, −6)

43. (0, −2), (−1, −6), (3, 4)

44. (2, 2), (−2, −3), (−3, −6)

EXTEND 7-6

Algebra Lab: Transforming Exponential Expressions

You can use the properties of rational exponents to transform exponential functions into other forms in order to solve real-world problems.

CCSS Common Core State Standards
Content Standards
A.SSE.3c Use the properties of exponents to transform expressions for exponential functions.
F.IF.8b Use the properties of exponents to interpret expressions for exponential functions.

Activity Write Equivalent Exponential Expressions

Monique is trying to decide between two savings account plans. Plan A offers a monthly compounding interest rate of 0.25%, while Plan B offers 2.5% interest compounded annually. Which is the better plan? Explain.

In order to compare the plans, we must compare rates with the same compounding frequency. One way to do this is to compare the approximate monthly interest rates of each plan, also called the *effective* monthly interest rate. While you can use the compound interest formula to find this rate, you can also use the properties of exponents.

Write a function to represent the amount A Monique would earn after t years with Plan B. For convenience, let the initial amount of Monique's investment be $1.

$y = a(1 + r)^t$	Equation for exponential growth
$A(t) = 1(1 + 0.025)^t$	$y = A(t)$, $a = 1$, $r = 2.5\%$ or 0.025
$= 1.025^t$	Simplify.

Now write a function equivalent to $A(t)$ that represents 12 compoundings per year, with a power of $12t$, instead of 1 per year, with a power of $1t$.

$A(t) = 1.025^{1t}$	Original function
$= 1.025^{\left(\frac{1}{12} \cdot 12\right)t}$	$1 = \frac{1}{12} \cdot 12$
$= \left(1.025^{\frac{1}{12}}\right)^{12t}$	Power of a Power
$\approx 1.0021^{12t}$	$(1.025)^{\frac{1}{12}} = \sqrt[12]{1.025}$ or about 1.0021

From this equivalent function, we can determine that the effective monthly interest by Plan B is about 0.0021 or about 0.21% per month. This rate is less than the monthly interest rate of 0.25% per month offered by Plan A, so Plan A is the better plan.

Model and Analyze

1. Use the compound interest formula $A = P\left(1 + \frac{r}{n}\right)^{nt}$ to determine the effective monthly interest rate for Plan B. How does this rate compare to the rate calculated using the method in the Activity above?

2. Write a function to represent the amount A Monique would earn after t months by Plan A. Then use the properties of exponents to write a function equivalent to $A(t)$ that represents the amount earned after t years.

3. From the expression you wrote in Exercise 2, identify the effective annual interest rate by Plan A. Use this rate to explain why Plan A is the better plan.

4. Suppose Plan A offered a quarterly compounded interest rate of 1.5%. Use the properties of exponents to explain which is the better plan.

LESSON 7-7

Geometric Sequences as Exponential Functions

Then

- You related arithmetic sequences to linear functions.

Now

1. Identify and generate geometric sequences.
2. Relate geometric sequences to exponential functions.

Why?

- You send a chain email to a friend who forwards the email to five more people. Each of these five people forwards the email to five more people. The number of new email generated forms a geometric sequence.

NewVocabulary
geometric sequence
common ratio

Common Core State Standards

Content Standards

F.BF.2 Write arithmetic and geometric sequences both recursively and with an explicit formula, use them to model situations, and translate between the two forms.

F.LE.1 Distinguish between situations that can be modeled with linear functions and with exponential functions.

a. Prove that linear functions grow by equal differences over equal intervals, and that exponential functions grow by equal factors over equal intervals.

b. Recognize situations in which one quantity changes at a constant rate per unit interval relative to another.

c. Recognize situations in which a quantity grows or decays by a constant percent rate per unit interval relative to another.

Mathematical Practices

7 Look for and make use of structure.

1 Recognize Geometric Sequences

The first person generates 5 emails. If each of these people sends the email to 5 more people, 25 emails are generated. If each of the 25 people sends 5 emails, 125 emails are generated. The sequence of emails generated, 1, 5, 25, 125, … is an example of a **geometric sequence**.

In a geometric sequence, the first term is nonzero and each term after the first is found by multiplying the previous term by a nonzero constant r called the **common ratio**. The common ratio can be found by dividing any term by its previous term.

Example 1 Identify Geometric Sequences

PT

Determine whether each sequence is *arithmetic*, *geometric*, or *neither*. Explain.

a. 256, 128, 64, 32, …

Find the ratios of consecutive terms.

256 128 64 32

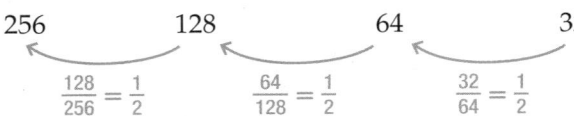

$\frac{128}{256} = \frac{1}{2}$ $\frac{64}{128} = \frac{1}{2}$ $\frac{32}{64} = \frac{1}{2}$

Since the ratios are constant, the sequence is geometric. The common ratio is $\frac{1}{2}$.

b. 4, 9, 12, 18, …

Find the ratios of consecutive terms.

4 9 12 18

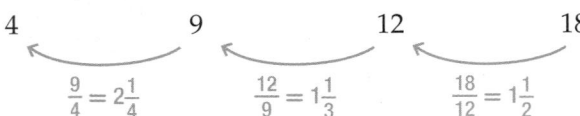

$\frac{9}{4} = 2\frac{1}{4}$ $\frac{12}{9} = 1\frac{1}{3}$ $\frac{18}{12} = 1\frac{1}{2}$

The ratios are not constant, so the sequence is not geometric.

Find the differences of consecutive terms.

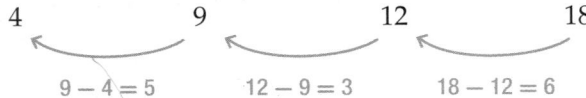

4 9 12 18

$9 - 4 = 5$ $12 - 9 = 3$ $18 - 12 = 6$

There is no common difference, so the sequence is not arithmetic.
Thus, the sequence is neither geometric nor arithmetic.

GuidedPractice

1A. 1, 3, 9, 27, … **1B.** −20, −15, −10, −5, … **1C.** 2, 8, 14, 22, …

Karen Moskowitz/Taxi/Getty Images

Once the common ratio is known, more terms of a sequence can be generated. The formula can be rewritten as $a_n = a_1 r^{n-1}$, where n is a counting number and r is the common ratio.

Example 2 Find Terms of Geometric Sequences

Find the next three terms in each geometric sequence.

a. 1, −4, 16, −64, ...

Step 1 Find the common ratio.

1   −4   16   −64

$\frac{-4}{1} = -4$ $\frac{16}{-4} = -4$ $\frac{-64}{16} = -4$

Step 2 Multiply each term by the common ratio to find the next three terms.

−64   256   −1024   4096

×(−4)   ×(−4)   ×(−4)

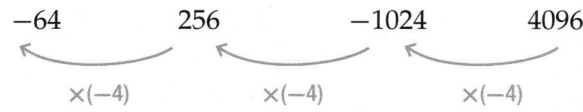

The next three terms are 256, −1024, and 4096.

b. 9, 3, 1, $\frac{1}{3}$ ...

Step 1 Find the common ratio.

9   3   1 $\frac{1}{3}$

$\frac{3}{9} = \frac{1}{3}$ $\frac{1}{3} = \frac{1}{3}$ $\frac{\frac{1}{3}}{3} = \frac{1}{3}$

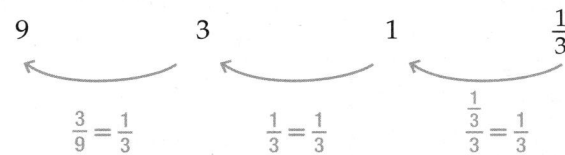

The value of r is $\frac{1}{3}$.

Step 2 Multiply each term by the common ratio to find the next three terms.

$\frac{1}{3}$ $\frac{1}{9}$ $\frac{1}{27}$ $\frac{1}{81}$

$\times\frac{1}{3}$ $\times\frac{1}{3}$ $\times\frac{1}{3}$

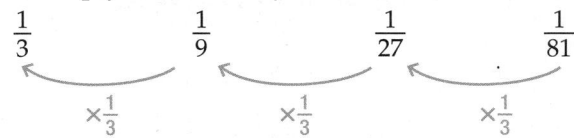

The next three terms are $\frac{1}{9}$, $\frac{1}{27}$, and $\frac{1}{81}$.

GuidedPractice

2A. −3, 15, −75, 375, ...

2B. 24, 36, 54, 81, ...

StudyTip

CCSS Structure If the terms of a geometric sequence alternate between positive and negative terms or vice versa, the common ratio is negative.

Math HistoryLink

Thomas Robert Malthus (1766–1834) Malthus studied populations and had pessimistic views about the future population of the world. In his work, he stated: "Population increases in a geometric ratio, while the means of subsistence increases in an arithmetic ratio."

2 Geometric Sequences and Functions

Finding the nth term of a geometric sequence would be tedious if we used the above method. The table below shows a rule for finding the nth term of a geometric sequence.

Position, n	1	2	3	4	...	n
Term, a_n	a_1	a_1r	a_1r^2	a_1r^3	...	a_1r^{n-1}

Notice that the common ratio between the terms is r. The table shows that to get the nth term, you multiply the first term by the common ratio r raised to the power $n - 1$. A geometric sequence can be defined by an exponential function in which n is the independent variable, a_n is the dependent variable, and r is the base. The domain is the counting numbers.

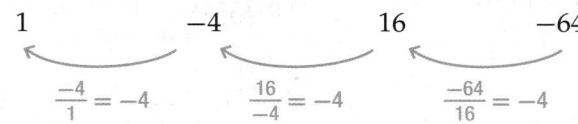

KeyConcept *n*th term of a Geometric Sequence

The *n*th term a_n of a geometric sequence with first term a_1 and common ratio r is given by the following formula, where n is any positive integer and $a_1, r \neq 0$.

$$a_n = a_1r^{n-1}$$

Example 3 Find the *n*th Term of a Geometric Sequence

a. Write an equation for the *n*th term of the sequence −6, 12, −24, 48, … .

The first term of the sequence is −6. So, $a_1 = -6$. Now find the common ratio.

$$-6 \quad 12 \quad -24 \quad 48$$

$$\frac{12}{-6} = -2 \qquad \frac{-24}{12} = -2 \qquad \frac{48}{-24} = -2$$

The common ratio is −2.

$a_n = a_1r^{n-1}$ Formula for *n*th term

$a_n = -6(-2)^{n-1}$ $a_1 = -6$ and $r = 2$

b. Find the ninth term of this sequence.

$a_n = a_1r^{n-1}$ Formula for *n*th term

$a_9 = -6(-2)^{9-1}$ For the *n*th term, $n = 9$.

$= -6(-2)^8$ Simplify.

$= -6(256)$ $(-2)^8 = 256$

$= -1536$

WatchOut!

Negative Common Ratio If the common ratio is negative, as in Example 3, make sure to enclose the common ratio in parentheses. $(-2)^8 \neq -2^8$

GuidedPractice

3. Write an equation for the *n*th term of the geometric sequence 96, 48, 24, 12, … . Then find the tenth term of the sequence.

Real-World Example 4 Graph a Geometric Sequence

BASKETBALL The NCAA women's basketball tournament begins with 64 teams. In each round, one half of the teams are left to compete, until only one team remains. Draw a graph to represent how many teams are left in each round.

Compared to the previous rounds, one half of the teams remain. So, $r = \frac{1}{2}$. Therefore, the geometric sequence that models this situation is 64, 32, 16, 8, 4, 2, 1. So in round two, 32 teams compete, in round three 16 teams compete and so forth. Use this information to draw a graph.

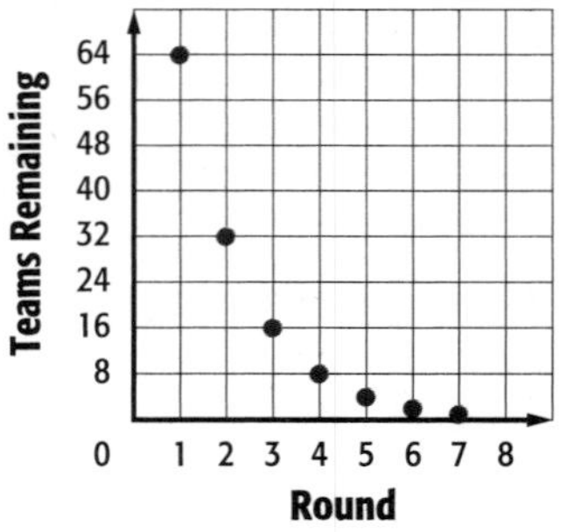

Real-WorldLink

The first NCAA Division I women's basketball tournament was held in 1982. The University of Tennessee has won the most national titles with 8 titles as of 2010.

Source: NCAA Sports

GuidedPractice

4. **TENNIS** A tennis ball is dropped from a height of 12 feet. Each time the ball bounces back to 80% of the height from which it fell. Draw a graph to represent the height of the ball after each bounce.

Check Your Understanding

Example 1 **Determine whether each sequence is *arithmetic*, *geometric*, or *neither*. Explain.**

1. 200, 40, 8, … **2.** 2, 4, 16, … **3.** −6, −3, 0, 3, … **4.** 1, −1, 1, −1, …

Example 2 **Find the next three terms in each geometric sequence.**

5. 10, 20, 40, 80, … **6.** 100, 50, 25, … **7.** $4, -1, \frac{1}{4}, \ldots$ **8.** −7, 21, −63, …

Example 3 **Write an equation for the *n*th term of each geometric sequence, and find the indicated term.**

9. the fifth term of −6, −24, −96, …

10. the seventh term of −1, 5, −25, …

11. the tenth term of 72, 48, 32, …

12. the ninth term of 112, 84, 63, …

Example 4 **13. EXPERIMENT** In a physics class experiment, Diana drops a ball from a height of 16 feet. Each bounce has 70% the height of the previous bounce. Draw a graph to represent the height of the ball after each bounce.

Practice and Problem Solving

Example 1 **Determine whether each sequence is *arithmetic*, *geometric*, or *neither*. Explain.**

14. 4, 1, 2, … **15.** 10, 20, 30, 40, … **16.** 4, 20, 100, …

17. 212, 106, 53, … **18.** −10, −8, −6, −4, … **19.** 5, −10, 20, 40, …

Example 2 **Find the next three terms in each geometric sequence.**

20. 2, −10, 50, … **21** 36, 12, 4, … **22.** 4, 12, 36, …

23. 400, 100, 25, … **24.** −6, −42, −294, … **25.** 1024, −128, 16, …

Example 3 **26.** The first term of a geometric series is 1 and the common ratio is 9. What is the 8th term of the sequence?

27. The first term of a geometric series is 2 and the common ratio is 4. What is the 14th term of the sequence?

28. What is the 15th term of the geometric sequence −9, 27, −81, …?

29. What is the 10th term of the geometric sequence 6, −24, 96, …?

Example 4 **30. PENDULUM** The first swing of a pendulum is shown. On each swing after that, the arc length is 60% of the length of the previous swing. Draw a graph that represents the arc length after each swing.

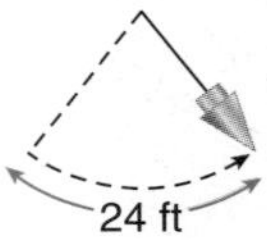

31. Find the eighth term of a geometric sequence for which $a_3 = 81$ and $r = 3$.

32. CCSS **REASONING** At an online mapping site, Mr. Mosley notices that when he clicks a spot on the map, the map zooms in on that spot. The magnification increases by 20% each time.

a. Write a formula for the *n*th term of the geometric sequence that represents the magnification of each zoom level. (*Hint:* The common ratio is not just 0.2.)

b. What is the fourth term of this sequence? What does it represent?

33. **ALLOWANCE** Danielle's parents have offered her two different options to earn her allowance for a 9-week period over the summer. She can either get paid \$30 each week or \$1 the first week, \$2 for the second week, \$4 for the third week, and so on.

a. Does the second option form a geometric sequence? Explain.

b. Which option should Danielle choose? Explain.

34. **SIERPINSKI'S TRIANGLE** Consider the inscribed equilateral triangles at the right. The perimeter of each triangle is one half of the perimeter of the next larger triangle. What is the perimeter of the smallest triangle?

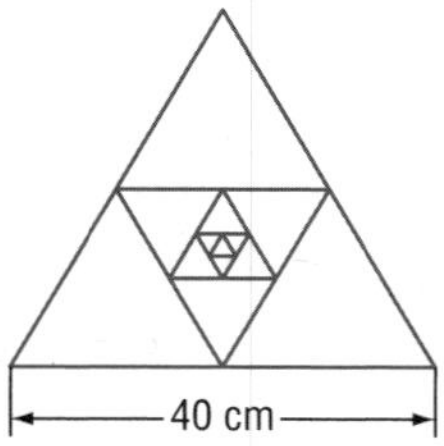

35. If the second term of a geometric sequence is 3 and the third term is 1, find the first and fourth terms of the sequence.

36. If the third term of a geometric sequence is −12 and the fourth term is 24, find the first and fifth terms of the sequence.

37. **EARTHQUAKES** The Richter scale is used to measure the force of an earthquake. The table shows the increase in magnitude for the values on the Richter scale.

Richter Number (x)	Increase in Magnitude (y)	Rate of Change (slope)
1	1	–
2	10	9
3	100	
4	1000	
5	10,000	

a. Copy and complete the table. Remember that the rate of change is the change in y divided by the change in x.

b. Plot the ordered pairs (Richter number, increase in magnitude).

c. Describe the graph that you made of the Richter scale data. Is the rate of change between any two points the same?

d. Write an exponential equation that represents the Richter scale.

H.O.T. Problems Use Higher-Order Thinking Skills

38. **CHALLENGE** Write a sequence that is both geometric and arithmetic. Explain your answer.

39. **CCSS CRITIQUE** Haro and Matthew are finding the ninth term of the geometric sequence −5, 10, −20, … . Is either of them correct? Explain your reasoning.

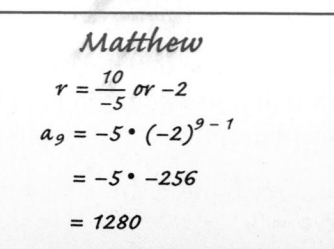

Haro	Matthew
$r = \frac{10}{-5}$ or -2	$r = \frac{10}{-5}$ or -2
$a_9 = -5(-2)^{9-1}$	$a_9 = -5 \cdot (-2)^{9-1}$
$= -5(512)$	$= -5 \cdot -256$
$= -2560$	$= 1280$

40. **REASONING** Write a sequence of numbers that form a pattern but are neither arithmetic nor geometric. Explain the pattern.

41. **WRITING IN MATH** How are graphs of geometric sequences and exponential functions similar? different?

42. **WRITING IN MATH** Summarize how to find a specific term of a geometric sequence.

Standardized Test Practice

43. Find the eleventh term of the sequence 3, −6, 12, −24, … .

A 6144 **C** 33

B 3072 **D** −6144

44. What is the total amount of the investment shown in the table below if interest is compounded monthly?

Principal	$500
Length of Investment	4 years
Annual Interest Rate	5.25%

F $613.56 **H** $616.56

G $616.00 **J** $718.75

45. SHORT RESPONSE Gloria has $6.50 in quarters and dimes. If she has 35 coins in total, how many of each coin does she have?

46. What are the domain and range of the function $y = 4(3^x) - 2$?

A D = {all real numbers}, R = $\{y \mid y > -2\}$

B D = {all real numbers}, R = $\{y \mid y > 0\}$

C D = {all integers}, R = $\{y \mid y > -2\}$

D D = {all integers}, R = $\{y \mid y > 0\}$

Spiral Review

Find the next three terms in each geometric sequence. (Lesson 7-6)

47. 2, 6, 18, 54, …

48. −5, −10, −20, −40, …

49. $1, -\frac{1}{2}, \frac{1}{4}, -\frac{1}{8}, \ldots$

50. −3, 1.5, −0.75, 0.375, …

51. 1, 0.6, 0.36, 0.216, …

52. 4, 6, 9, 13.5, …

Graph each function. Find the *y*-intercept and state the domain and range. (Lesson 7-5)

53. $y = \left(\frac{1}{4}\right)^x - 5$

54. $y = 2(4)^x$

55. $y = \frac{1}{2}(3^x)$

56. LANDSCAPING A blue spruce grows an average of 6 inches per year. A hemlock grows an average of 4 inches per year. If a blue spruce is 4 feet tall and a hemlock is 6 feet tall, write a system of equations to represent their growth. Find and interpret the solution in the context of the situation. (Lesson 6-2)

57. MONEY City Bank requires a minimum balance of $1500 to maintain free checking services. If Mr. Hayashi is going to write checks for the amounts listed in the table, how much money should he start with in order to have free checking? (Lesson 5-1)

Check	Amount
750	$1300
751	$947

Write an equation in slope-intercept form of the line with the given slope and *y*-intercept. (Lesson 4-2)

58. slope: 4, *y*-intercept: 2

59. slope: −3, *y*-intercept: $-\frac{2}{3}$

60. slope: $-\frac{1}{4}$, *y*-intercept: −5

61. slope: $\frac{1}{2}$, *y*-intercept: −9

62. slope: $-\frac{2}{5}$, *y*-intercept: $\frac{3}{4}$

63. slope: −6, *y*-intercept: −7

Skills Review

Simplify each expression. If not possible, write *simplified*.

64. $3u + 10u$

65. $5a - 2 + 6a$

66. $6m^2 - 8m$

67. $4w^2 + w + 15w^2$

68. $13(5 + 4a)$

69. $(4t - 6)16$

EXTEND 7-7

Algebra Lab
Average Rate of Change of Exponential Functions

You know that the rate of change of a linear function is the same for any two points on the graph. The rate of change of an exponential function is not constant.

CCSS Common Core State Standards
Content Standards
F.IF.6 Calculate and interpret the average rate of change of a function (presented symbolically or as a table) over a specified interval. Estimate the rate of change from a graph.

Activity Evaluating Investment Plans

John has \$2000 to invest in one of two plans. Plan 1 offers to increase his principal by \$75 each year, while Plan 2 offers to pay 3.6% interest compounded monthly. The dollar value of each investment after t years is given by $A_1 = 2000 + 75t$ and $A_2 = 2000(1.003)^{12t}$, respectively. Use the function values, the average rate of change, and the graphs of the equations to interpret and compare the plans.

Step 1 Copy and complete the table below by finding the missing values for A_1 and A_2.

t	0	1	2	3	4	5
A_1						
A_2						

Step 2 Find the average rate of change for each plan from $t = 0$ to 1, $t = 3$ to 4, and $t = 0$ to 5.

Plan 1: $\frac{2075 - 2000}{1 - 0}$ or 75 $\qquad \frac{2300 - 2225}{4 - 3}$ or 75 $\qquad \frac{2375 - 2000}{5 - 0}$ or 75

Plan 2: $\frac{2073.2 - 2000}{1 - 0}$ or 73.2 $\qquad \frac{2309.27 - 2227.74}{4 - 3}$ or about 82 $\qquad \frac{2393.79 - 2000}{5 - 0}$ or about 79

Step 3 Graph the ordered pairs for each function. Connect each set of points with a smooth curve.

Step 4 Use the graph and the rates of change to compare the plans. Both graphs have a rate of change for the first year of about \$75 per year. From year 3 to 4, Plan 1 continues to increase at \$75 per year, but Plan 2 grows at a rate of more than \$81 per year. The average rate of change over the first five years for Plan 1 is \$75 per year and for Plan 2 is over \$78 per year. This indicates that as the number of years increases, the investment in Plan 2 grows at an increasingly faster pace. This is supported by the widening gap between their graphs.

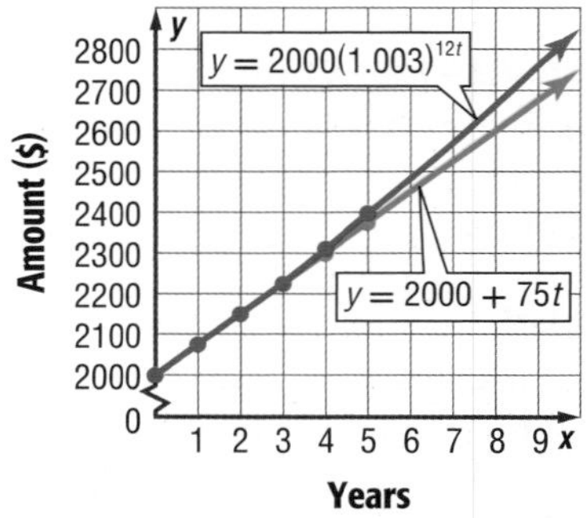

Exercises

The value of a company's piece of equipment decreases over time due to depreciation. The function $y = 16{,}000(0.985)^{2t}$ represents the value after t years.

1. What is the average rate of change over the first five years?
2. What is the average rate of change of the value from year 5 to year 10?
3. What conclusion about the value can we make based on these average rates of change?
4. **CCSS REGULARITY** Copy and complete the table for $y = x^4$.

x	−3	−2	−1	0	1	2	3
y							

Compare and interpret the average rate of change for $x = -3$ to 0 and for $x = 0$ to 3.

LESSON 7-8 Recursive Formulas

::Then	::Now	::Why?
You wrote explicit formulas to represent arithmetic and geometric sequences.	**1** Use a recursive formula to list terms in a sequence. **2** Write recursive formulas for arithmetic and geometric sequences.	Clients of a shuttle service get picked up from their homes and driven to premium outlet stores for shopping. The total cost of the service depends on the total number of customers. The costs for the first six customers are shown.

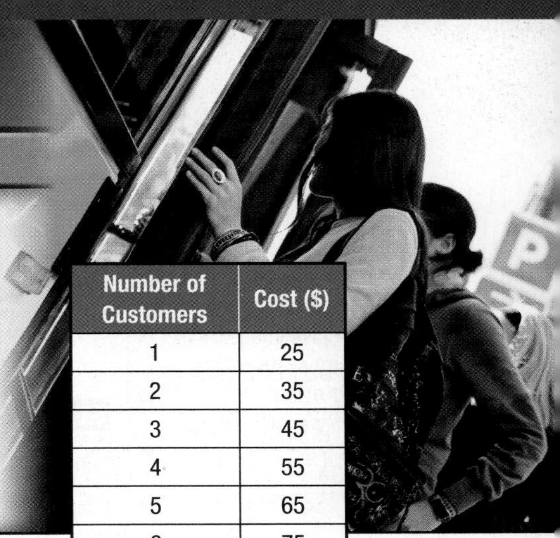

Number of Customers	Cost ($)
1	25
2	35
3	45
4	55
5	65
6	75

NewVocabulary
recursive formula

Common Core State Standards

Content Standards
F.IF.3 Recognize that sequences are functions, sometimes defined recursively, whose domain is a subset of the integers.
F.BF.2 Write arithmetic and geometric sequences both recursively and with an explicit formula, use them to model situations, and translate between the two forms.

Mathematical Practices
3 Construct viable arguments and critique the reasoning of others.

1 Using Recursive Formulas

An explicit formula allows you to find any term a_n of a sequence by using a formula written in terms of n. For example, $a_n = 2n$ can be used to find the fifth term of the sequence 2, 4, 6, 8, ...: $a_5 = 2(5)$ or 10.

A **recursive formula** allows you to find the nth term of a sequence by performing operations to one or more of the preceding terms. Since each term in the sequence above is 2 greater than the term that preceded it, we can add 2 to the fourth term to find that the fifth term is 8 + 2 or 10. We can then write a recursive formula for a_n.

$$\begin{aligned} a_1 &= & &= 2 \\ a_2 &= & a_1 + 2 \text{ or } 2 + 2 &= 4 \\ a_3 &= & a_2 + 2 \text{ or } 4 + 2 &= 6 \\ a_4 &= & a_3 + 2 \text{ or } 6 + 2 &= 8 \\ &\vdots & \vdots & \\ a_n &= & a_{n-1} + 2 & \end{aligned}$$

A recursive formula for the sequence above is $a_1 = 2$, $a_n = a_{n-1} + 2$, for $n \geq 2$ where n is an integer. The term denoted a_{n-1} represents the term immediately before a_n. Notice that the first term a_1 is given, along with the domain for n.

PT

Example 1 Use a Recursive Formula

Find the first five terms of the sequence in which $a_1 = 7$ and $a_n = 3a_{n-1} - 12$, if $n \geq 2$.

Use $a_1 = 7$ and the recursive formula to find the next four terms.

$$\begin{aligned} a_2 &= 3a_{2-1} - 12 && n = 2 \\ &= 3a_1 - 12 && \text{Simplify.} \\ &= 3(7) - 12 \text{ or } 9 && a_1 = 7 \end{aligned}$$

$$\begin{aligned} a_3 &= 3a_{3-1} - 12 && n = 3 \\ &= 3a_2 - 12 && \text{Simplify.} \\ &= 3(9) - 12 \text{ or } 15 && a_2 = 9 \end{aligned}$$

$$\begin{aligned} a_4 &= 3a_{4-1} - 12 && n = 4 \\ &= 3a_3 - 12 && \text{Simplify.} \\ &= 3(15) - 12 \text{ or } 33 && a_3 = 15 \end{aligned}$$

$$\begin{aligned} a_5 &= 3a_{5-1} - 12 && n = 5 \\ &= 3a_4 - 12 && \text{Simplify.} \\ &= 3(33) - 12 \text{ or } 87 && a_4 = 33 \end{aligned}$$

The first five terms are 7, 9, 15, 33, and 87.

GuidedPractice

1. Find the first five terms of the sequence in which $a_1 = -2$ and $a_n = (-3)\,a_{n-1} + 4$, if $n \geq 2$.

Patrick SheÀndell o' carroll/PhotoAlto

2 Writing Recursive Formulas

To write a recursive formula for an arithmetic or geometric sequence, complete the following steps.

KeyConcept Writing Recursive Formulas

Step 1 Determine if the sequence is arithmetic or geometric by finding a common difference or a common ratio.

Step 2 Write a recursive formula.

Arithmetic Sequences $a_n = a_{n-1} + d$, where d is the common difference

Geometric Sequences $a_n = r \cdot a_{n-1}$, where r is the common ratio

Step 3 State the first term and domain for n.

StudyTip

Defining n For the nth term of a sequence, the value of n must be a positive integer. Although we must still state the domain of n, from this point forward, we will assume that n is an integer.

Example 2 Write Recursive Formulas

PT

Write a recursive formula for each sequence.

a. 17, 13, 9, 5, ...

Step 1 First subtract each term from the term that follows it.

$13 - 17 = -4$ $9 - 13 = -4$ $5 - 9 = -4$

There is a common difference of -4. The sequence is arithmetic.

Step 2 Use the formula for an arithmetic sequence.

$a_n = a_{n-1} + d$ Recursive formula for arithmetic sequence

$a_n = a_{n-1} + (-4)$ $d = -4$

Step 3 The first term a_1 is 17, and $n \geq 2$.

A recursive formula for the sequence is $a_1 = 17$, $a_n = a_{n-1} - 4$, $n \geq 2$.

StudyTip

Domain The domain for n is decided by the given terms. Since the first term is already given, it makes sense that the first term to which the formula would apply is the 2nd term of the sequence, or when $n = 2$.

b. 6, 24, 96, 384, ...

Step 1 First subtract each term from the term that follows it.

$24 - 6 = 18$ $96 - 24 = 72$ $384 - 96 = 288$

There is no common difference. Check for a common ratio by dividing each term by the term that precedes it.

$\frac{24}{6} = 4$ $\frac{96}{24} = 4$ $\frac{384}{96} = 4$

There is a common ratio of 4. The sequence is geometric.

Step 2 Use the formula for a geometric sequence.

$a_n = r \cdot a_{n-1}$ Recursive formula for geometric sequence

$a_n = 4a_{n-1}$ $r = 4$

Step 3 The first term a_1 is 6, and $n \geq 2$.

A recursive formula for the sequence is $a_1 = 6$, $a_n = 4a_{n-1}$, $n \geq 2$.

GuidedPractice

2A. 4, 10, 25, 62.5, ...

2B. 9, 36, 63, 90, ...

A sequence can be represented by both an explicit formula and a recursive formula.

Real-WorldCareer

Transportation The number of jobs in the transportation industry is expected to grow by an estimated 1.1 million between 2004 and 2014. The specific fields dictate the educational requirements, which include a high school diploma and some form of specialized training.

Source: United States Department of Labor

PT

Example 3 Write Recursive and Explicit Formulas

COST Refer to the beginning of the lesson. Let n be the number of customers.

a. Write a recursive formula for the sequence.

Steps 1 and 2 First subtract each term from the term that follows it.

$35 - 25 = 10$ $\quad 45 - 35 = 10$ $\quad 55 - 45 = 10$

There is a common difference of 10. The sequence is arithmetic.

Step 3 Use the formula for an arithmetic sequence.

$a_n = a_{n-1} + d$ Recursive formula for arithmetic sequence

$a_n = a_{n-1} + 10$ $d = 10$

Step 4 The first term a_1 is 25, and $n \geq 2$.

A recursive formula for the sequence is $a_1 = 25$, $a_n = a_{n-1} + 10$, $n \geq 2$.

b. Write an explicit formula for the sequence.

Step 1 The common difference is 10.

Step 2 Use the formula for the nth term of an arithmetic sequence.

$a_n = a_1 + (n - 1)d$ Formula for the nth term

$= 25 + (n - 1)10$ $a_1 = 25$ and $d = 10$

$= 25 + 10n - 10$ Distributive Property

$= 10n + 15$ Simplify.

An explicit formula for the sequence is $a_n = 10n + 15$.

GuidedPractice

3. SAVINGS The money that Ronald has in his savings account earns interest each year. He does not make any withdrawals or additional deposits. The account balance at the beginning of each year is \$10,000, \$10,300, \$10,609, \$10,927.27, and so on. Write a recursive formula and an explicit formula for the sequence.

If several successive terms of a sequence are needed, a recursive formula may be useful, whereas if just the nth term of a sequence is needed, an explicit formula may be useful. Thus, it is sometimes beneficial to translate between the two forms.

PT

Example 4 Translate between Recursive and Explicit Formulas

a. Write a recursive formula for $a_n = 6n + 3$.

$a_n = 6n + 3$ is an explicit formula for an arithmetic sequence with $d = 6$ and $a_1 = 6(1) + 3$ or 9. Therefore, a recursive formula for a_n is $a_1 = 9$, $a_n = a_{n-1} + 6$, $n \geq 2$.

b. Write an explicit formula for $a_1 = 120$, $a_n = 0.8a_{n-1}$, $n \geq 2$.

$a_n = 0.8a_{n-1}$ is a recursive formula for a geometric sequence with $a_1 = 120$ and $r = 0.8$. Therefore, an explicit formula for a_n is $a_n = 120(0.8)^{n-1}$.

StudyTip

Geometric Sequence Recall that the formula for the nth term of a geometric sequence is $a_n = a_1r^{n-1}$.

GuidedPractice

4A. Write a recursive formula for $a_n = 4(3)^{n-1}$.

4B. Write an explicit formula for $a_1 = -16$, $a_n = a_{n-1} - 7$, $n \geq 2$.

John A. Rizzo/Photodisc/Getty Images

Check Your Understanding

Example 1 **Find the first five terms of each sequence.**

1. $a_1 = 16, a_n = a_{n-1} - 3, n \geq 2$

2. $a_1 = -5, a_n = 4a_{n-1} + 10, n \geq 2$

Example 2 **Write a recursive formula for each sequence.**

3. 1, 6, 11, 16, ...

4. 4, 12, 36, 108, ...

Example 3 **5. BALL** A ball is dropped from an initial height of 10 feet. The maximum heights the ball reaches on the first three bounces are shown.

a. Write a recursive formula for the sequence.

b. Write an explicit formula for the sequence.

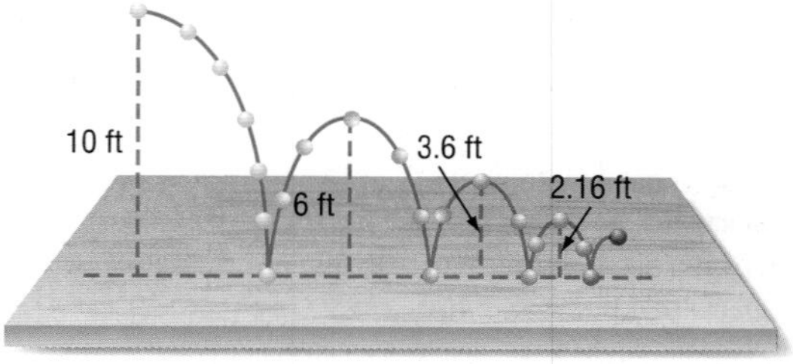

Example 4 **For each recursive formula, write an explicit formula. For each explicit formula, write a recursive formula.**

6. $a_1 = 4, a_n = a_{n-1} + 16, n \geq 2$

7 $a_n = 5n + 8$

8. $a_n = 15(2)^{n-1}$

9. $a_1 = 22, a_n = 4a_{n-1}, n \geq 2$

Practice and Problem Solving

Example 1 **Find the first five terms of each sequence.**

10. $a_1 = 23, a_n = a_{n-1} + 7, n \geq 2$

11. $a_1 = 48, a_n = -0.5a_{n-1} + 8, n \geq 2$

12. $a_1 = 8, a_n = 2.5a_{n-1}, n \geq 2$

13. $a_1 = 12, a_n = 3a_{n-1} - 21, n \geq 2$

14. $a_1 = 13, a_n = -2a_{n-1} - 3, n \geq 2$

15. $a_1 = \frac{1}{2}, a_n = a_{n-1} + \frac{3}{2}, n \geq 2$

Example 2 **Write a recursive formula for each sequence.**

16. 12, −1, −14, −27, ...

17. 27, 41, 55, 69, ...

18. 2, 11, 20, 29, ...

19. 100, 80, 64, 51.2, ...

20. 40, −60, 90, −135, ...

21. 81, 27, 9, 3, ...

Example 3 **22. CCSS MODELING** A landscaper is building a brick patio. Part of the patio includes a pattern constructed from triangles. The first four rows of the pattern are shown.

a. Write a recursive formula for the sequence.

b. Write an explicit formula for the sequence.

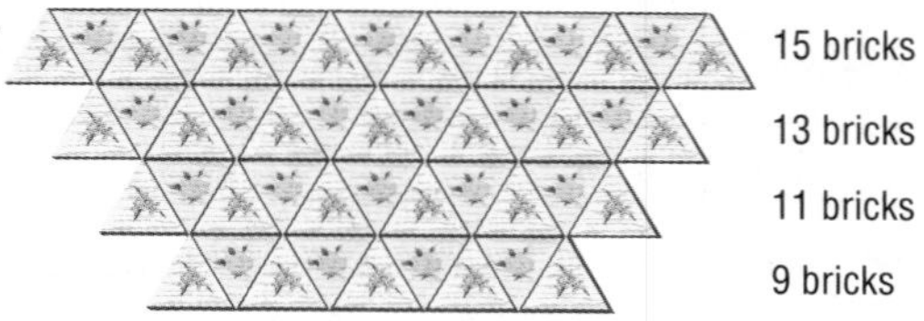

Example 4 **For each recursive formula, write an explicit formula. For each explicit formula, write a recursive formula.**

23. $a_n = 3(4)^{n-1}$

24. $a_1 = -2, a_n = a_{n-1} - 12, n \geq 2$

25. $a_1 = 38, a_n = \frac{1}{2}a_{n-1}, n \geq 2$

26. $a_n = -7n + 52$

27 **TEXTING** Barbara received a chain text that she forwarded to five of her friends. Each of her friends forwarded the text to five more friends, and so on.

a. Find the first five terms of the sequence representing the number of people who receive the text in the nth round.

b. Write a recursive formula for the sequence.

c. If Barbara represents a_1, find a_8.

28. **GEOMETRY** Consider the pattern below. The number of blue boxes increases according to a specific pattern.

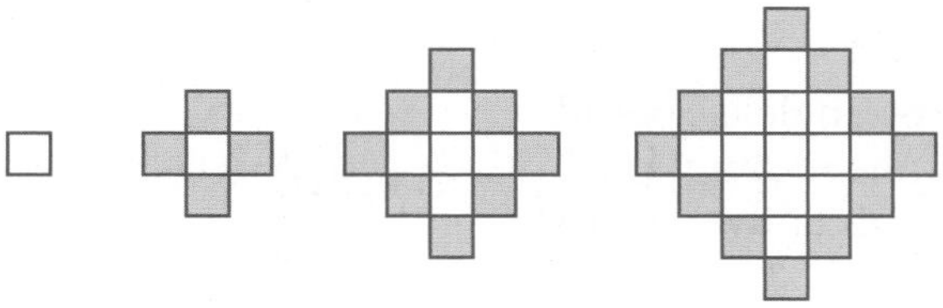

a. Write a recursive formula for the sequence of the number of blue boxes in each figure.

b. If the first box represents a_1, find the number of blue boxes in a_8.

29. **TREE** The growth of a certain type of tree slows as the tree continues to age. The heights of the tree over the past four years are shown.

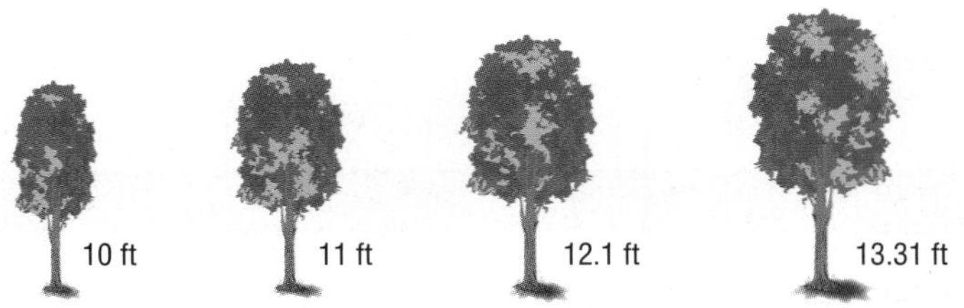

a. Write a recursive formula for the height of the tree.

b. If the pattern continues, how tall will the tree be in two more years? Round your answer to the nearest tenth of a foot.

30. **MULTIPLE REPRESENTATIONS** The Fibonacci sequence is neither arithmetic nor geometric and can be defined by a recursive formula. The first terms are 1, 1, 2, 3, 5, 8, …

a. **Logical** Determine the relationship between the terms of the sequence. What are the next five terms in the sequence?

b. **Algebraic** Write a formula for the nth term if $a_1 = 1$, $a_2 = 1$, and $n \geq 3$.

c. **Algebraic** Find the 15th term.

d. **Analytical** Explain why the Fibonacci sequence is not an arithmetic sequence.

H.O.T. Problems Use Higher-Order Thinking Skills

31. **ERROR ANALYSIS** Patrick and Lynda are working on a math problem that involves the sequence 2, −2, 2, −2, 2, … . Patrick thinks that the sequence can be written as a recursive formula. Lynda believes that the sequence can be written as an explicit formula. Is either of them correct? Explain.

32. **CHALLENGE** Find a_1 for the sequence in which $a_4 = 1104$ and $a_n = 4a_{n-1} + 16$.

33. CCSS **ARGUMENTS** Determine whether the following statement is *true* or *false.* Justify your reasoning.

There is only one recursive formula for every sequence.

34. **CHALLENGE** Find a recursive formula for 4, 9, 19, 39, 79, …

35. **WRITING IN MATH** Explain the difference between an explicit formula and a recursive formula.

Standardized Test Practice

36. Find a recursive formula for the sequence 12, 24, 36, 48, … .

A $a_1 = 12, a_n = 2a_{n-1}, n \geq 2.$

B $a_1 = 12, a_n = 4a_{n-1} - 24, n \geq 2.$

C $a_1 = 12, a_n = a_{n-1} + 12, n \geq 2.$

D $a_1 = 12, a_n = 12a_{n-1} + 12, n \geq 2.$

37. GEOMETRY The area of a rectangle is $36m^4n^6$ square feet. The length of the rectangle is $6m^3n^3$ feet. What is the width of the rectangle?

F $216m^7n^9$ ft

G $6mn^3$ ft

H $42m^7n^3$ ft

J $30mn^3$ ft

38. Find an inequality for the graph shown.

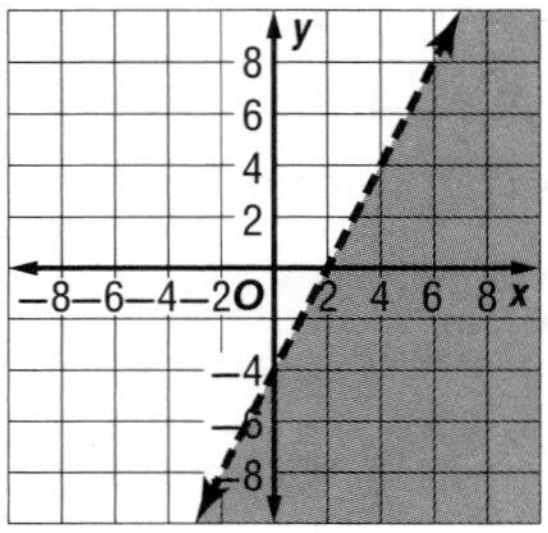

A $y > 2x - 4$ **C** $y < 2x - 4$

B $y \geq 2x - 4$ **D** $y \leq 2x - 4$

39. Write an equation of the line that passes through (−2, −20) and (4, 58).

F $y = 13x + 6$ **H** $y = 19x + 18$

G $y = 19x - 18$ **J** $y = 13x - 6$

Spiral Review

Find the next three terms in each geometric sequence. (Lesson 7-7)

40. 675, 225, 75, …

41. 16, −24, 36, …

42. 6, 18, 54, …

43. 512, −256, 128, …

44. 125, 25, 5, …

45. 12, 60, 300, …

46. INVESTMENT Nicholas invested \$2000 with a 5.75% interest rate compounded monthly. How much money will Nicholas have after 5 years? (Lesson 7-6)

47. TOURS The Snider family and the Rollins family are traveling together on a trip to visit a candy factory. The number of people in each family and the total cost are shown in the table below. Find the adult and children's admission prices. (Lesson 6-3)

Family	Number of Adults	Number of Children	Total Cost
Snider	2	3	\$58
Rollins	2	1	\$38

Write each equation in standard form. (Lesson 4-3)

48. $y + 6 = -3(x + 2)$

49. $y - 12 = 4(x - 7)$

50. $y + 9 = 5(x - 3)$

51. $y - 1 = \frac{1}{3}(x + 15)$

52. $y + 10 = \frac{2}{5}(x - 6)$

53. $y - 4 = -\frac{2}{7}(x + 1)$

Skills Review

Simplify each expression. If not possible, write *simplified*.

54. $8x + 3y^2 + 7x - 2y$

55. $4(x - 16) + 6x$

56. $4n - 3m + 9m - n$

57. $6r^2 + 7r$

58. $-2(4g - 5h) - 6g$

59. $9x^2 - 7x + 16y^2$

CHAPTER 7

Study Guide and Review

Study Guide

KeyConcepts

Multiplication and Division Properties of Exponents (Lessons 7-1 and 7-2)

For any nonzero real numbers a and b and any integers m, n, and p, the following are true.

- Product of Powers: $a^m \cdot a^n = a^{m+n}$
- Power of a Power: $(a^m)^n = a^{m \cdot n}$
- Power of a Product: $(ab)^m = a^m b^m$
- Quotient of Powers: $\frac{a^m}{a^p} = a^{m-p}$
- Power of a Quotient: $\left(\frac{a}{b}\right)^m = \frac{a^m}{b^m}$
- Zero Exponent: $a^0 = 1$
- Negative Exponent: $a^{-n} = \frac{1}{a^n}$ and $\frac{1}{a^{-n}} = a^n$

Rational Exponents (Lesson 7-3)

For any positive real number b and any integers m and $n > 1$, the following are true.

$b^{\frac{1}{2}} = \sqrt{b}$ $\quad b^{\frac{1}{n}} = \sqrt[n]{b}$ $\quad b^{\frac{m}{n}} = \left(\sqrt[n]{b}\right)^m$ or $\sqrt[n]{b^m}$

Scientific Notation (Lesson 7-4)

- A number is in scientific notation if it is in the form $a \times 10^n$, where $1 \leq a < 10$.
- To write in standard form:
  - If $n > 0$, move the decimal n places right.
  - If $n < 0$, move the decimal n places left.

Exponential Functions (Lessons 7-5 and 7-6)

- The equation for exponential growth is $y = a(1 + r)^t$, where $r > 0$. The equation for exponential decay is $y = a(1 - r)^t$, where $0 < r < 1$. y is the final amount, a is the initial amount, r is the rate of change, and t is the time in years.

FOLDABLES® StudyOrganizer

Be sure the Key Concepts are noted in your Foldable.

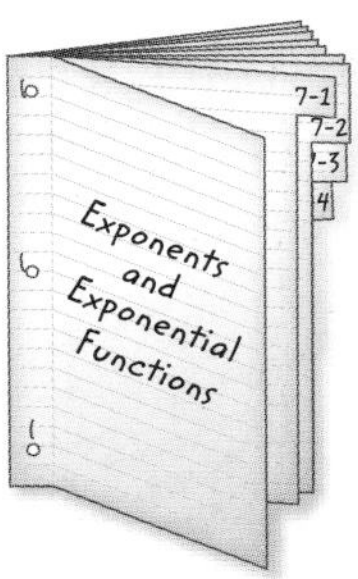

KeyVocabulary

common ratio (p. 438)
compound interest (p. 433)
constant (p. 391)
cube root (p. 407)
exponential decay (p. 424)
exponential equation (p. 409)
exponential function (p. 424)
exponential growth (p. 424)
geometric sequence (p. 438)
monomial (p. 391)
negative exponent (p. 400)
*n*th root (p. 407)
order of magnitude (p. 401)
rational exponent (p. 406)
recursive formula (p. 445)
scientific notation (p. 414)
zero exponent (p. 399)

VocabularyCheck

Choose the word or term that best completes each sentence.

1. $7xy^4$ is an example of a(n) ____________.
2. The ____________ of 95,234 is 10^5.
3. 2 is a(n) ____________ of 8.
4. The rules for operations with exponents can be extended to apply to expressions with a(n) ____________ such as $7^{\frac{2}{3}}$.
5. A number written in ____________ is of the form $a \times 10^n$, where $1 \leq a < 10$ and n is an integer.
6. $f(x) = 3^x$ is an example of a(n) ____________.
7. $a_1 = 4$ and $a_n = 3a_{n-1} - 12$, if $n \geq 2$, is a(n) ____________ for the sequence 4, −8, −20, −32, ….
8. $2^{3x-1} = 16$ is an example of a(n) ____________.
9. The equation for ____________ is $y = C(1 - r)^t$.
10. If $a^n = b$ for a positive integer n, then a is a(n) ____________ of b.

Lesson-by-Lesson Review

7-1 Multiplication Properties of Exponents

Simplify each expression.

11. $x \cdot x^3 \cdot x^5$

12. $(2xy)(-3x^2y^5)$

13. $(-4ab^4)(-5a^5b^2)$

14. $(6x^3y^2)^2$

15. $\left[(2r^3t)^3\right]^2$

16. $(-2u^3)(5u)$

17. $(2x^2)^3(x^3)^3$

18. $\frac{1}{2}(2x^3)^3$

19. GEOMETRY Use the formula $V = \pi r^2 h$ to find the volume of the cylinder.

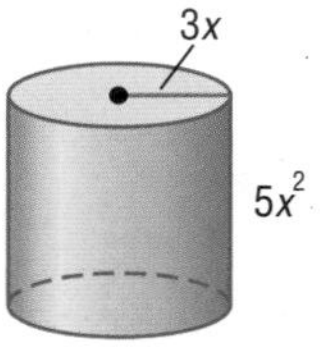

Example 1

Simplify $(5x^2y^3)(2x^4y)$.

$(5x^2y^3)(2x^4y)$

$= (5 \cdot 2)(x^2 \cdot x^4)(y^3 \cdot y)$ Commutative Property

$= 10x^6y^4$ Product of Powers

Example 2

Simplify $(3a^2b^4)^3$.

$(3a^2b^4)^3 = 3^3(a^2)^3(b^4)^3$ Power of a Product

$= 27a^6b^{12}$ Simplify.

7-2 Division Properties of Exponents

Simplify each expression. Assume that no denominator equals zero.

20. $\frac{(3x)^0}{2a}$

21. $\left(\frac{3xy^3}{2z}\right)^3$

22. $\frac{12y^{-4}}{3y^{-5}}$

23. $a^{-3}b^0c^6$

24. $\frac{-15x^7y^8z^4}{-45x^3y^5z^3}$

25. $\frac{(3x^{-1})^{-2}}{(3x^2)^{-2}}$

26. $\left(\frac{6xy^{11}z^9}{48x^6yz^{-7}}\right)^0$

27. $\left(\frac{12}{2}\right)\left(\frac{x}{y^5}\right)\left(\frac{y^4}{x^4}\right)$

28. GEOMETRY The area of a rectangle is $25x^2y^4$ square feet. The width of the rectangle is $5xy$ feet. What is the length of the rectangle?

Example 3

Simplify $\frac{2k^4m^3}{4k^2m}$. Assume that no denominator equals zero.

$\frac{2k^4m^3}{4k^2m} = \left(\frac{2}{4}\right)\left(\frac{k^4}{k^2}\right)\left(\frac{m^3}{m}\right)$ Group powers with the same base.

$= \left(\frac{1}{2}\right)k^{4-2}m^{3-1}$ Quotient of Powers

$= \frac{k^2m^2}{2}$ Simplify.

Example 4

Simplify $\frac{t^4uv^{-2}}{t^{-3}u^7}$. Assume that no denominator equals zero.

$\frac{t^4uv^{-2}}{t^{-3}u^7} = \left(\frac{t^4}{t^{-3}}\right)\left(\frac{u}{u^7}\right)(v^{-2})$ Group the powers with the same base.

$= (t^{4+3})(u^{1-7})(v^{-2})$ Quotient of Powers

$= t^7u^{-6}v^{-2}$ Simplify.

$= \frac{t^7}{u^6v^2}$ Simplify.

7-3 Rational Exponents

Simplify.

29. $\sqrt[3]{343}$

30. $\sqrt[6]{729}$

31. $625^{\frac{1}{4}}$

32. $\left(\frac{8}{27}\right)^{\frac{1}{3}}$

33. $256^{\frac{3}{4}}$

34. $32^{\frac{2}{5}}$

35. $343^{\frac{4}{3}}$

36. $\left(\frac{4}{49}\right)^{\frac{3}{2}}$

Solve each equation.

37. $6^x = 7776$

38. $4^{4x-1} = 32$

Example 5

Simplify $125^{\frac{2}{3}}$.

$125^{\frac{2}{3}} = \left(\sqrt[3]{125}\right)^2$ — $b^{\frac{m}{n}} = \left(\sqrt[n]{b}\right)^m$

$= \left(\sqrt[3]{5 \cdot 5 \cdot 5}\right)^2$ — $64 = 4^3$

$= 5^2$ or 25 — Simplify.

Example 6

Solve $9^{x-1} = 729$.

$9^{x-1} = 729$ — Original equation

$9^{x-1} = 9^3$ — Rewrite 729 as 9^3.

$x - 1 = 3$ — Power Property of Equality

$x = 4$ — Add 1 to each side.

7-4 Scientific Notation

Express each number in scientific notation.

39. 2,300,000

40. 0.0000543

41. **ASTRONOMY** Earth has a diameter of about 8000 miles. Jupiter has a diameter of about 88,000 miles. Write in scientific notation the ratio of Earth's diameter to Jupiter's diameter.

Example 7

Express 300,000,000 in scientific notation.

Step 1 300,000,000 ⟶ 3.00000000

Step 2 The decimal point moved 8 places to the left, so $n = 8$.

Step 3 $300{,}000{,}000 = 3 \times 10^8$

7-5 Exponential Functions

Graph each function. Find the *y*-intercept, and state the domain and range.

42. $y = 2^x$

43. $y = 3^x + 1$

44. $y = 4^x + 2$

45. $y = 2^x - 3$

46. **BIOLOGY** The population of bacteria in a petri dish increases according to the model $p = 550(2.7)^{0.008t}$, where t is the number of hours and $t = 0$ corresponds to 1:00 P.M. Use this model to estimate the number of bacteria in the dish at 5:00 P.M.

Example 8

Graph $y = 3^x + 6$. Find the *y*-intercept, and state the domain and range.

x	$3^x + 6$	y
−3	$3^{-3} + 6$	6.04
−2	$3^{-2} + 6$	6.11
−1	$3^{-1} + 6$	6.33
0	$3^0 + 6$	7
1	$3^1 + 6$	9

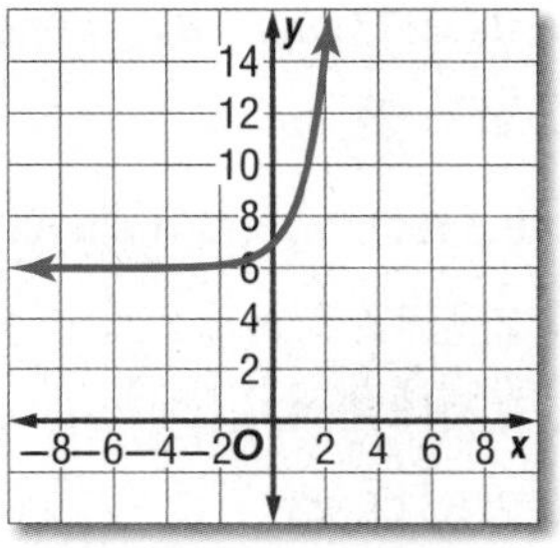

The *y*-intercept is (0, 7). The domain is all real numbers, and the range is all real numbers greater than 6.

Study Guide and Review *Continued*

7-6 Growth and Decay

47. Find the final value of \$2500 invested at an interest rate of 2% compounded monthly for 10 years.

48. **COMPUTERS** Zita's computer is depreciating at a rate of 3% per year. She bought the computer for \$1200.

a. Write an equation to represent this situation.

b. What will the computer's value be after 5 years?

Example 9

Find the final value of \$2000 invested at an interest rate of 3% compounded quarterly for 8 years.

$A = P\left(1 + \frac{r}{n}\right)^{nt}$ — Compound interest equation

$= 2000\left(1 + \frac{0.03}{4}\right)^{4(8)}$ — $P = 2000$, $r = 0.03$, $n = 4$, and $t = 8$

$\approx \$2540.22$ — Use a calculator.

7-7 Geometric Sequences as Exponential Functions

Find the next three terms in each geometric sequence.

49. $-1, 1, -1, 1, \ldots$

50. $3, 9, 27, \ldots$

51. $256, 128, 64, \ldots$

Write the equation for the *n*th term of each geometric sequence.

52. $-1, 1, -1, 1, \ldots$

53. $3, 9, 27, \ldots$

54. $256, 128, 64, \ldots$

55. **SPORTS** A basketball is dropped from a height of 20 feet. It bounces to $\frac{1}{2}$ its height after each bounce. Draw a graph to represent the situation.

Example 10

Find the next three terms in the geometric sequence 2, 6, 18, … .

Step 1 Find the common ratio. Each number is 3 times the previous number, so $r = 3$.

Step 2 Multiply each term by the common ratio to find the next three terms.

$18 \times 3 = 54$, $54 \times 3 = 162$, $162 \times 3 = 486$

The next three terms are 54, 162, and 486.

Example 11

Write the equation for the *n*th term of the geometric sequence −3, 12, −48, … .

The common ratio is -4. So $r = -4$.

$a_n = a_1 r^{n-1}$ — Formula for the *n*th term

$a_n = -3(-4)^{n-1}$ — $a_1 = -3$ and $r = -4$

7-8 Recursive Formulas

Find the first five terms of each sequence.

56. $a_1 = 11$, $a_n = a_{n-1} - 4$, $n \geq 2$

57. $a_1 = 3$, $a_n = 2a_{n-1} + 6$, $n \geq 2$

Write a recursive formula for each sequence.

58. $2, 7, 12, 17, \ldots$

59. $32, 16, 8, 4, \ldots$

60. $2, 5, 11, 23, \ldots$

Example 12

Write a recursive formula for 3, 1, −1, −3, ….

Step 1 First subtract each term from the term that follows it.

$1 - 3 = -2$, $-1 - 1 = -2$, $-3 - (-1) = -2$

There is a common difference of -2. The sequence is arithmetic.

Step 2 Use the formula for an arithmetic sequence.

$a_n = a_{n-1} + d$ — Recursive formula

$a_n = a_{n-1} + (-2)$ — $d = -2$

Step 3 The first term a_1 is 3, and $n \geq 2$.

A recursive formula is $a_1 = 3$, $a_n = a_{n-1} - 2$, $n \geq 2$.

CHAPTER 7

Practice Test

Simplify each expression.

1. $(x^2)(7x^8)$
2. $(5a^7bc^2)(-6a^2bc^5)$

3. **MULTIPLE CHOICE** Express the volume of the solid as a monomial.

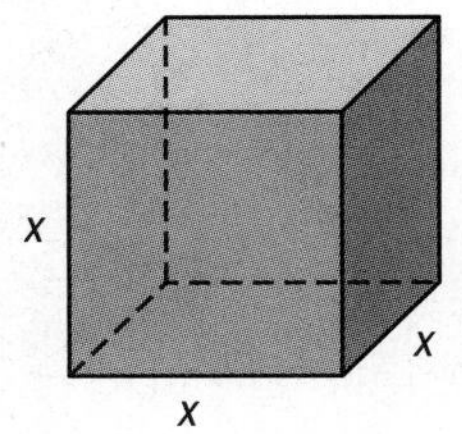

A x^3
B $6x$
C $6x^3$
D x^6

Simplify each expression. Assume that no denominator equals 0.

4. $\dfrac{x^6y^8}{x^2}$
5. $\left(\dfrac{2a^4b^3}{c^6}\right)^0$
6. $\dfrac{2xy^{-7}}{8x}$

Simplify.

7. $\sqrt[3]{1000}$
8. $\sqrt[5]{3125}$
9. $1728^{\frac{1}{3}}$
10. $\left(\dfrac{16}{81}\right)^{\frac{1}{2}}$
11. $27^{\frac{2}{3}}$
12. $10{,}000^{\frac{3}{4}}$
13. $27^{\frac{5}{3}}$
14. $\left(\dfrac{1}{121}\right)^{\frac{3}{2}}$

Solve each equation.

15. $12^x = 1728$
16. $7^{x-1} = 2401$
17. $9^{x-3} = 729$

Express each number in scientific notation.

18. 0.00021
19. 58,000

Express each number in standard form.

20. 2.9×10^{-5}
21. 9.1×10^6

Evaluate each product or quotient. Express the results in scientific notation.

22. $(2.5 \times 10^3)(3 \times 10^4)$
23. $\dfrac{8.8 \times 10^2}{4 \times 10^{-4}}$

24. **ASTRONOMY** The average distance from Mercury to the Sun is 35,980,000 miles. Express this distance in scientific notation.

Graph each function. Find the y-intercept, and state the domain and range.

25. $y = 2(5)^x$
26. $y = -3(11)^x$
27. $y = 3^x + 2$

Find the next three terms in each geometric sequence.

28. 2, −6, 18, …
29. 1000, 500, 250, …
30. 32, 8, 2, …

31. **MULTIPLE CHOICE** Lynne invested $500 into an account with a 6.5% interest rate compounded monthly. How much will Lynne's investment be worth in 10 years?

F $600.00
G $938.57
H $956.09
J $957.02

32. **INVESTMENTS** Shelly's investment of $3000 has been losing value at a rate of 3% each year. What will her investment be worth in 6 years?

Find the first five terms of each sequence.

33. $a_1 = 18, a_n = a_{n-1} - 4, n \geq 2$
34. $a_1 = -2, a_n = 4a_{n-1} + 5, n \geq 2$

CHAPTER 7

Preparing for Standardized Tests

Using a Scientific or Graphing Calculator

Scientific and graphing calculators are powerful problem-solving tools. There are times when a calculator can be used to make computations faster and easier, such as computations with very large numbers. However, there are times when using a calculator is necessary, like the estimation of irrational numbers.

Strategies for Using a Scientific or Graphing Calculator

Step 1

Familiarize yourself with the various functions of a scientific or graphing calculator as well as when they should be used:

- **Exponents** scientific notation, calculating with large or small numbers
- **Pi** solving circle problems, like circumference and area
- **Square roots** distance on a coordinate plane, Pythagorean theorem
- **Graphs** analyzing paired data in a scatter plot, graphing functions, finding roots of equations

Step 2

Use your scientific or graphing calculator to solve the problem.

- Remember to work as efficiently as possible. Some steps may be done mentally or by hand, while others should be completed using your calculator.
- If time permits, check your answer.

Standardized Test Example

Read the problem. Identify what you need to know. Then use the information in the problem to solve.

The distance from the Sun to Jupiter is approximately 7.786×10^{11} meters. If the speed of light is about 3×10^{8} meters per second, how long does it take for light from the Sun to reach Jupiter? Round to the nearest minute.

A about 43 minutes

B about 51 minutes

C about 1876 minutes

D about 2595 minutes

Read the problem carefully. You are given the approximate distance from the Sun to Jupiter as well as the speed of light. Both quantities are given in scientific notation. You are asked to find how many minutes it takes for light from the Sun to reach Jupiter. Use the relationship distance = rate × time to find the amount of time.

$d = r \times t$

$\frac{d}{r} = t$

To find the amount of time, divide the distance by the rate. Notice, however, that the units for time will be seconds.

$\frac{7.786 \times 10^{11} \text{ m}}{3 \times 10^{8} \text{ m/s}} = t$ seconds

Use a scientific calculator to quickly find the quotient. On most scientific calculators, the EE key is used to enter numbers in scientific notation.

KEYSTROKES: (7.786 2nd [EE] 11) ÷ (3 2nd [EE] 8) ENTER

The result is 2595.33333333 seconds. To convert this number to minutes, use your calculator to divide the result by 60. This gives an answer of about 43.2555 minutes. The answer is A.

Exercises

Read each problem. Identify what you need to know. Then use the information in the problem to solve.

1. Since its creation 5 years ago, approximately 2.504×10^7 items have been sold or traded on a popular online website. What is the average daily number of items sold or traded over the 5-year period?

   A about 9640 items per day

   B about 13,720 items per day

   C about 1,025,000 items per day

   D about 5,008,000 items per day

2. Evaluate $\sqrt{ab}$ if $a = 121$ and $b = 23$.

   F about 5.26

   G about 9.90

   H about 12

   J about 52.75

3. The population of the United States is about 3.034×10^8 people. The land area of the country is about 3.54×10^6 square miles. What is the average *population density* (number of people per square mile) of the United States?

   A about 136.3 people per square mile

   B about 112.5 people per square mile

   C about 94.3 people per square mile

   D about 85.7 people per square mile

4. Eleece is making a cover for the marching band's bass drum. The drum has a diameter of 20 inches. Estimate the area of the face of the bass drum.

   F 31.41 square inches

   G 62.83 square inches

   H 78.54 square inches

   J 314.16 square inches

Chapter 7 Standardized Test Practice

Cumulative, Chapters 1 through 7

Multiple Choice

Read each question. Then fill in the correct answer on the answer document provided by your teacher or on a sheet of paper.

1. Express the area of the triangle below as a monomial.

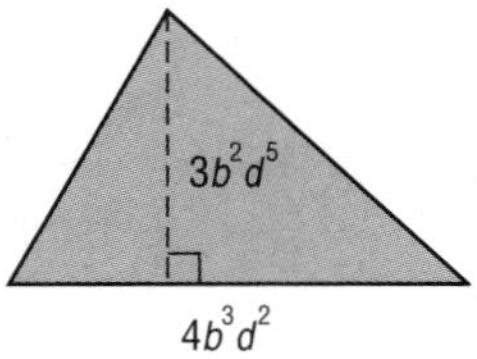

A $12b^5d^7$

B $12b^6d^{10}$

C $6b^6d^{10}$

D $6b^5d^7$

2. Simplify the following expression.

$$\left(\frac{2w^2z^5}{3y^4}\right)^3$$

F $\frac{2w^5z^8}{3y^7}$

G $\frac{8w^6z^{15}}{27y^{12}}$

H $\frac{8w^5z^8}{27y^7}$

J $\frac{2w^6z^{15}}{3y^{12}}$

3. Which equation of a line is perpendicular to $y = \frac{3}{5}x - 3$?

A $y = -\frac{5}{3}x + 2$

B $y = -\frac{3}{5}x + 2$

C $y = \frac{5}{3}x - 2$

D $y = \frac{3}{5}x - 2$

Test-Taking Tip

Question 2 Use the laws of exponents to simplify the expression. Remember, to find the power of a power, multiply the exponents.

4. Write a recursive formula for the sequence of the number of squares in each figure.

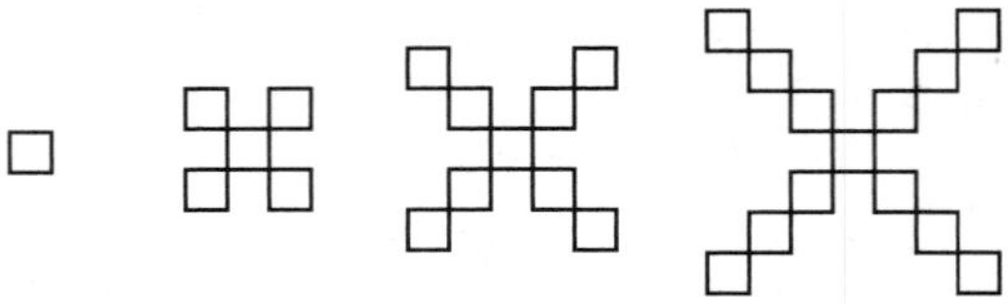

F $a_1 = 1, a_n = 4a_{n-1} - 3, n \geq 1$

G $a_1 = 1, a_n = 4a_{n-1}, n \geq 2$

H $a_1 = 1, a_n = a_{n-1} + 4, n \geq 2$

J $a_1 = 1, a_n = 4a_{n-1} + 4, n \geq 2$

5. Evaluate $(4.2 \times 10^6)(5.7 \times 10^8)$.

A 2.394×10^{15}

B 23.94×10^{14}

C 9.9×10^{14}

D 2.394×10^{48}

6. Which inequality is shown in the graph?

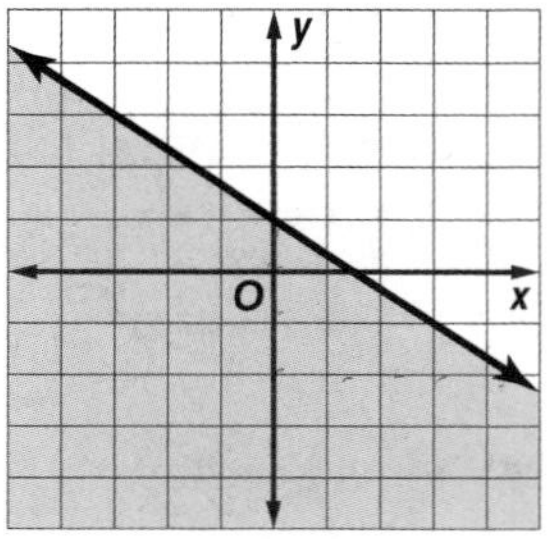

F $y \leq -\frac{2}{3}x - 1$

G $y \leq -\frac{3}{4}x - 1$

H $y \leq -\frac{2}{3}x + 1$

J $y \leq -\frac{3}{4}x + 1$

Short Response/Gridded Response

7. Jaden created a Web site for the Science Olympiad team. The total number of hits the site has received is shown.

Day	Total Hits	Day	Total Hits
3	5	17	27
6	7	21	33
10	12	26	40
13	17	34	55

a. Find an equation for the regression line.

b. Predict the number of total hits that the Web site will have received on day 46.

8. Find the value of x so that the figures have the same area.

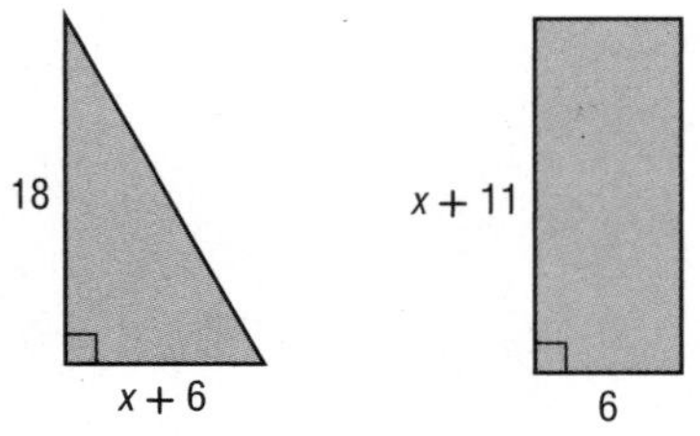

9. What is the solution to the following system of equations? Show your work.

$$\begin{cases} y = 6x - 1 \\ y = 6x + 4 \end{cases}$$

10. GRIDDED RESPONSE At a family fun center, the Wilson and Sanchez families each bought video game tokens and batting cage tokens as shown in the table.

Family	Wilson	Sanchez
Number of Video Game Tokens	25	30
Number of Batting Cage Tokens	8	6
Total Cost	$26.50	$25.50

What is the cost in dollars of a batting cage token at the family fun center?

Extended Response

Record your answers on a sheet of paper. Show your work.

11. The table below shows the distances from the Sun to Mercury, Earth, Mars, and Saturn. Use the data to answer each question.

Planet	Distance from Sun (km)
Mercury	5.79×10^7
Earth	1.50×10^8
Mars	2.28×10^8
Saturn	1.43×10^9

a. Of the planets listed, which one is the closest to the Sun?

b. About how many times as far from the Sun is Mars as Earth?

Need ExtraHelp?

If you missed Question...	1	2	3	4	5	6	7	8	9	10	11
Go to Lesson...	7-1	7-2	4-4	6-6	7-3	5-6	4-6	2-4	6-2	6-4	7-4

CHAPTER 8

Radical Functions, Rational Functions, and Geometry

Then

You solved quadratic and exponential equations.

Now

In this chapter, you will:

- Graph and transform radical functions.
- Simplify, add, subtract, and multiply radical expressions.
- Solve radical equations.
- Use the Pythagorean Theorem.
- Find trigonometric ratios.

Why? ▲

OCEANS Tsunamis, or large waves, are generated by undersea earthquakes. A radical equation can be used to find the speed of a tsunami in meters per second or the depth of the ocean in meters.

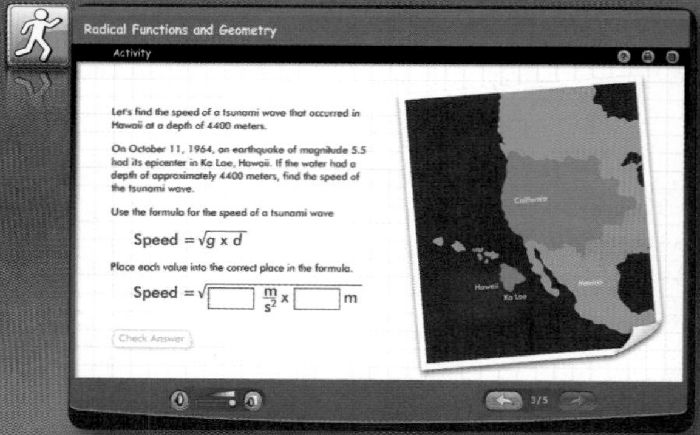

connectED.mcgraw-hill.com **Your Digital Math Portal**

Animation | Vocabulary | eGlossary | Personal Tutor | Virtual Manipulatives | Graphing Calculator | Audio | Foldables | Self-Check Practice | Worksheets

Get Ready for the Chapter

Diagnose Readiness | You have two options for checking prerequisite skills.

1 Textbook Option Take the Quick Check below. Refer to the Quick Review for help.

QuickCheck

Rewrite each expression using the Distributive Property. Then simplify.

1. $a(a + 5)$
2. $2(3 + x)$
3. $n(n - 3n^2 + 2)$
4. $-6(x^2 - 5x + 6)$
5. **FINANCIAL LITERACY** Five friends will pay \$9 per ticket, \$3 per drink, and \$6 per popcorn at the movies. Write an expression that could be used to determine the cost for them to go to the movies.

Simplify each expression. If not possible, write *simplified.*

6. $3u + 10u$
7. $5a - 2 + 6a$
8. $6m^2 - 8m$
9. $4w^2 + w + 15w^2$
10. $2x^2 + 5 - 11x^2$
11. $8v^3 - 27$
12. $4k^2 + 2k - 2k + 1$
13. $a^2 - 4a - 4a + 16$
14. $6y^2 + 2y - 3y - 1$
15. $9g^2 - 3g - 6g + 2$

Simplify.

16. $b(b^6)$
17. $4n^3(n^2)$
18. $8m(4m^2)$
19. $-5z^4(3z^5)$
20. $5xy(4x^3y)$
21. $(-2a^4c^5)(7ac^4)$
22. **GEOMETRY** A square is $6x^3$ inches on each side. What is the area of the square?

QuickReview

Example 1

Rewrite $6x(-3x - 5x - 5x^2 + x^3)$ using the Distributive Property. Then simplify.

$$\begin{aligned}
&6x(-3x - 5x - 5x^2 + x^3)\\
&= 6x(-3x) + 6x(-5x) + 6x(-5x^2) + 6x(x^3)\\
&= -18x^2 - 30x^2 - 30x^3 + 6x^4\\
&= -48x^2 - 30x^3 + 6x^4
\end{aligned}$$

Example 2

Simplify $8c + 6 - 4c + 2c^2$.

$$\begin{aligned}
8c + 6 - 4c + 2c^2 &= 2c^2 + 8c - 4c + 6\\
&= 2c^2 + (8 - 4)c + 6\\
&= 2c^2 + 4c + 6
\end{aligned}$$

Example 3

Simplify $(-2y^3)(9y^4)$.

$$\begin{aligned}
(9y^3)(-2y^4) &= (-2 \cdot 9)(y^3 \cdot y^4)\\
&= (-2 \cdot 9)(y^{3+4})\\
&= -18y^7
\end{aligned}$$

2 Online Option Take an online self-check Chapter Readiness Quiz at connectED.mcgraw-hill.com.

Get Started on the Chapter

You will learn several new concepts, skills, and vocabulary terms as you study Chapter 8. To get ready, identify important terms and organize your resources. You may wish to refer to Chapter 0 to review prerequisite skills.

FOLDABLES® StudyOrganizer

Radical Functions and Geometry Make this Foldable to help you organize your Chapter 8 notes about radical functions and geometry. Begin with four sheets of grid paper.

1 **Fold** in half along the width.

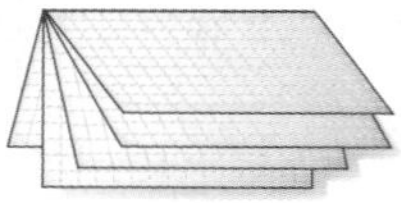

2 **Staple** along the fold.

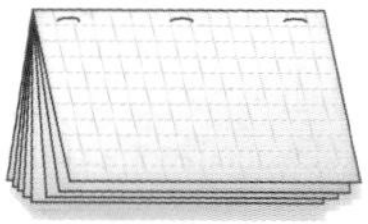

3 **Turn** the fold to the left and write the title of the chapter on the front. On each left-hand page of the booklet, write the title of a lesson from the chapter.

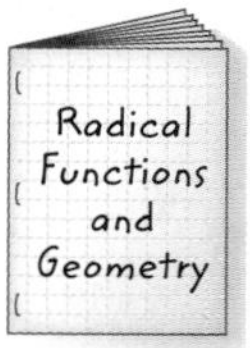

NewVocabulary

English		Español
square root function	p. 463	función radical
radicand	p. 463	radicando
radical function	p. 463	función radicales
radical expression	p. 470	expresión radical
conjugate	p. 472	conjugado
rationalize the denominator	p. 472	racionalizar el denominador
closed	p. 476	cerrado
radical equations	p. 482	ecuaciones radicales
extraneous solutions	p. 483	soluciones extrañas
inverse variation	p. 488	variación inversa
product rule	p. 489	regla del producto
rational function	p. 496	función racional
excluded values	p. 496	valores excluidos
asympote	p. 497	asíntota
rational equation	p. 502	ecuacion radical
work problem	p. 504	problema de trabajo
rate problem	p. 505	problema de tasa

ReviewVocabulary

FOIL method metodo FOIL to multiply two binomials, find the sum of the products of the First terms, Outer terms, Inner terms, and Last terms

perfect square cuadrado perfecto a number with a square root that is a rational number

proportion proporcion an equation of the form $\frac{a}{b} = \frac{c}{d}$, $b \neq 0$, $d \neq 0$ stating that two ratios are equivalent

$$\frac{a}{b} \times \frac{c}{d}$$
$$ad = bc$$

LESSON 8-1

Square Root Functions

Then

- You graphed and analyzed linear, exponential, and quadratic functions.

Now

1. Graph and analyze dilations of radical functions.
2. Graph and analyze reflections and translations of radical functions.

Why?

- Scientists use sounds of whales to track their movements. The distance to a whale can be found by relating time to the speed of sound in water.

The speed of sound in water can be described by the *square root function* $c = \sqrt{\frac{E}{d}}$, where E represents the bulk modulus elasticity of the water and d represents the density of the water.

NewVocabulary
square root function
radical function
radicand

Common Core State Standards

Content Standards

F.IF.4 For a function that models a relationship between two quantities, interpret key features of graphs and tables in terms of the quantities, and sketch graphs showing key features given a verbal description of the relationship.

F.IF.7b Graph square root, cube root, and piecewise-defined functions, including step functions and absolute value functions.

Mathematical Practices
6 Attend to precision.

1 Dilations of Radical Functions

A **square root function** contains the square root of a variable. Square root functions are a type of **radical function**. The expression under the radical sign is called the **radicand**. For a square root to be a real number, the radicand cannot be negative. Values that make the radicand negative are not included in the domain.

KeyConcept Square Root Function

Parent Function:	$f(x) = \sqrt{x}$
Type of Graph:	curve
Domain:	$\{x \mid x \geq 0\}$
Range:	$\{y \mid y \geq 0\}$

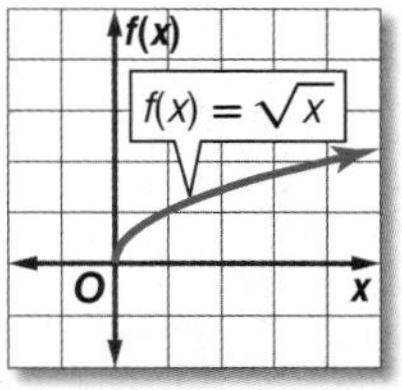

Example 1 Dilation of the Square Root Function

Graph $f(x) = 2\sqrt{x}$. State the domain and range.

Step 1 Make a table.

x	0	0.5	1	2	3	4
$f(x)$	0	≈ 1.4	2	≈ 2.8	≈ 3.5	4

Step 2 Plot points. Draw a smooth curve.

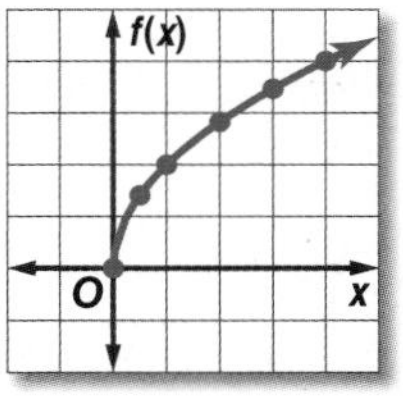

The domain is $\{x \mid x \geq 0\}$, and the range is $\{y \mid y \geq 0\}$. Notice that the graph is increasing on the entire domain, the minimum value is 0, and there is no symmetry.

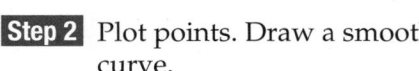

GuidedPractice

1A. $g(x) = 4\sqrt{x}$

1B. $h(x) = 6\sqrt{x}$

2 Reflections and Translations of Radical Functions

Recall that when the value of a is negative in the quadratic function $f(x) = ax^2$, the graph of the parent function is reflected across the x-axis.

StudyTip

Graphing Radical Functions Choose perfect squares for x-values that will result in coordinates that are easy to plot.

KeyConcept Graphing $y = a\sqrt{x + h} + k$

Step 1 Draw the graph of $y = a\sqrt{x}$. The graph starts at the origin and passes through $(1, a)$. If $a > 0$, the graph is in quadrant I. If $a < 0$, the graph is reflected across the x-axis and is in quadrant IV.

Step 2 Translate the graph k units up if $k > 0$ and $|k|$ units down if $k < 0$.

Step 3 Translate the graph h units left if $h > 0$ and $|h|$ units right if $h < 0$.

Example 2 Reflection of the Square Root Function

Graph $y = -3\sqrt{x}$. Compare to the parent graph. State the domain and range.

Make a table of values. Then plot the points on a coordinate system and draw a smooth curve that connects them.

x	0	0.5	1	4
y	0	≈ -2.1	-3	-6

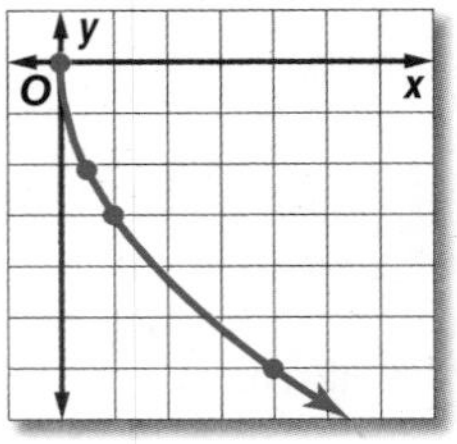

Notice that the graph is in the 4th quadrant. It is obtained by stretching the graph of $y = \sqrt{x}$ vertically and then reflecting across the x-axis. The domain is $\{x \mid x \geq 0\}$, and the range is $\{y \mid y \leq 0\}$.

GuidedPractice

2A. $y = -2\sqrt{x}$

2B. $y = -4\sqrt{x}$

StudyTip

Translating Radical Functions If $h > 0$, a radical function $f(x) = \sqrt{x - h}$ is a horizontal translation h units to the right. $f(x) = \sqrt{x + h}$ is a horizontal translation h units to the left.

Example 3 Translation of the Square Root Function

Graph each function. Compare to the parent graph. State the domain and range.

a. $g(x) = \sqrt{x} + 1$

x	0	0.5	1	4	9
y	0	≈ 1.7	2	3	4

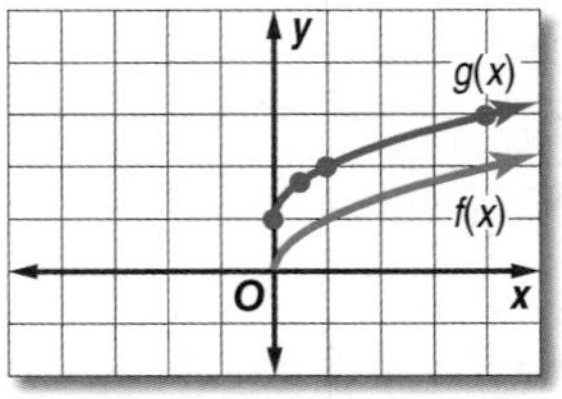

Notice that the values of $g(x)$ are 1 greater than those of $f(x) = \sqrt{x}$. This is a vertical translation 1 unit up from the parent function. The domain is $\{x \mid x \geq 0\}$, and the range is $\{y \mid y \geq 1\}$.

b. $h(x) = \sqrt{x - 2}$

x	2	3	4	6
y	0	1	≈ 1.4	2

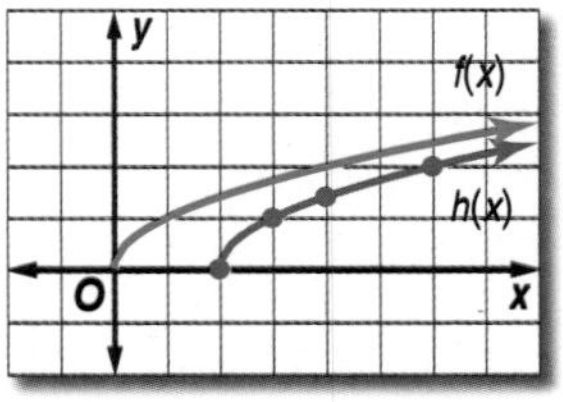

This is a horizontal translation 2 units to the right of the parent function. The domain is $\{x \mid x \geq 2\}$, and the range is $\{y \mid y \geq 0\}$.

Guided Practice

3A. $g(x) = \sqrt{x} - 4$

3B. $h(x) = \sqrt{x + 3}$

Physical phenomena such as motion can be modeled by radical functions. Often these functions are transformations of the parent square root function.

Real-WorldLink

Approximately 39 million cars cross the Golden Gate Bridge in San Francisco each year.

Source: San Francisco Convention and Visitors Bureau

Real-World Example 4 Analyze a Radical Function

BRIDGES The Golden Gate Bridge is about 67 meters above the water. The velocity v of a freely falling object that has fallen h meters is given by $v = \sqrt{2gh}$, where g is the constant 9.8 meters per second squared. Graph the function. If an object is dropped from the bridge, what is its velocity when it hits the water?

Use a graphing calculator to graph the function. To find the velocity of the object, substitute 67 meters for h.

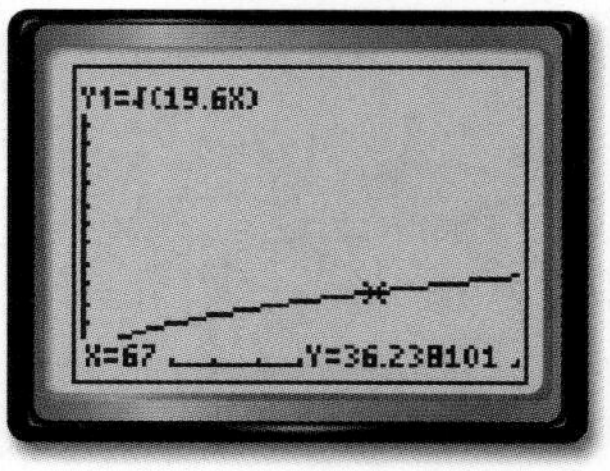

$v = \sqrt{2gh}$ Original function

$= \sqrt{2(9.8)(67)}$ $g = 9.8$ and $h = 67$

$= \sqrt{1313.2}$ Simplify.

$\approx 36.2 \text{ m/s}$ Use a calculator.

The velocity of the object is about 36.2 meters per second after dropping 67 meters.

Guided Practice

4. Use the graph above to estimate the initial height of an object if it is moving at 20 meters per second when it hits the water.

Transformations such as reflections, translations, and dilations can be combined in one equation.

Example 5 Transformations of the Square Root Function

Graph $y = -2\sqrt{x} + 1$, and compare to the parent graph. State the domain and range.

x	0	1	4	9
y	1	−1	−3	−5

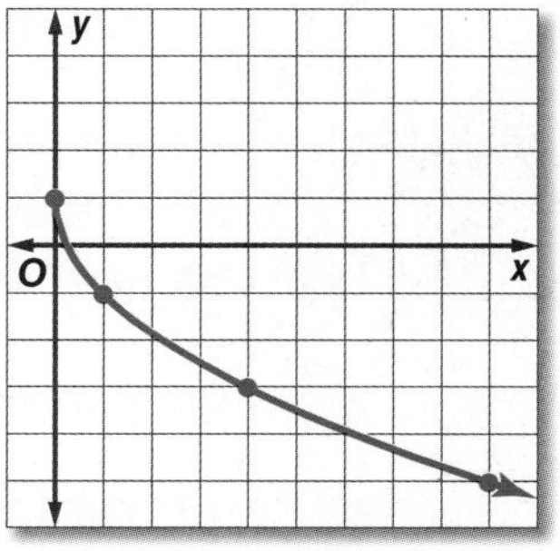

This graph is the result of a vertical stretch of the graph of $y = \sqrt{x}$ followed by a reflection across the x-axis, and then a translation 1 unit up. The domain is $\{x \mid x \geq 0\}$, and the range is $\{y \mid y \leq 1\}$.

Guided Practice

5A. $y = \frac{1}{2}\sqrt{x} - 1$

5B. $y = -2\sqrt{x - 1}$

Didier Dorval/Masterfile

Check Your Understanding

Examples 1–3 **Graph each function. Compare to the parent graph. State the domain and range.**

1. $y = 3\sqrt{x}$
2. $y = -5\sqrt{x}$
3. $y = \frac{1}{3}\sqrt{x}$
4. $y = -\frac{1}{2}\sqrt{x}$
5. $y = \sqrt{x} + 3$
6. $y = \sqrt{x} - 2$
7. $y = \sqrt{x + 2}$
8. $y = \sqrt{x - 3}$

Example 4

9. **FREE FALL** The time t, in seconds, that it takes an object to fall a distance d, in feet, is given by the function $t = \frac{1}{4}\sqrt{d}$ (assuming zero air resistance). Graph the function, and state the domain and range.

Example 5 **Graph each function, and compare to the parent graph. State the domain and range.**

10. $y = \frac{1}{2}\sqrt{x} + 2$
11. $y = -\frac{1}{4}\sqrt{x} - 1$
12. $y = -2\sqrt{x + 1}$
13. $y = 3\sqrt{x - 2}$

Practice and Problem Solving

Examples 1–3 **Graph each function. Compare to the parent graph. State the domain and range.**

14. $y = 5\sqrt{x}$
15. $y = \frac{1}{2}\sqrt{x}$
16. $y = -\frac{1}{3}\sqrt{x}$
17. $y = 7\sqrt{x}$
18. $y = -\frac{1}{4}\sqrt{x}$
19. $y = -\sqrt{x}$
20. $y = -\frac{1}{5}\sqrt{x}$
21. $y = -7\sqrt{x}$
22. $y = \sqrt{x} + 2$
23. $y = \sqrt{x} + 4$
24. $y = \sqrt{x} - 1$
25. $y = \sqrt{x} - 3$
26. $y = \sqrt{x} + 1.5$
27. $y = \sqrt{x} - 2.5$
28. $y = \sqrt{x + 4}$
29. $y = \sqrt{x - 4}$
30. $y = \sqrt{x + 1}$
31. $y = \sqrt{x - 0.5}$
32. $y = \sqrt{x + 5}$
33. $y = \sqrt{x - 1.5}$

Example 4

34. **GEOMETRY** The perimeter of a square is given by the function $P = 4\sqrt{A}$, where A is the area of the square.
    a. Graph the function.
    b. Determine the perimeter of a square with an area of 225 m^2.
    c. When will the perimeter and the area be the same value?

Example 5 **Graph each function, and compare to the parent graph. State the domain and range.**

35. $y = -2\sqrt{x} + 2$
36. $y = -3\sqrt{x} - 3$
37. $y = \frac{1}{2}\sqrt{x + 2}$
38. $y = -\sqrt{x - 1}$
39. $y = \frac{1}{4}\sqrt{x - 1} + 2$
40. $y = \frac{1}{2}\sqrt{x - 2} + 1$

41. **ENERGY** An object has kinetic energy when it is in motion. The velocity in meters per second of an object of mass m kilograms with an energy of E joules is given by the function $v = \sqrt{\frac{2E}{m}}$. Use a graphing calculator to graph the function that represents the velocity of a basketball with a mass of 0.6 kilogram.

42. GEOMETRY The radius of a circle is given by $r = \sqrt{\frac{A}{\pi}}$, where A is the area of the circle.

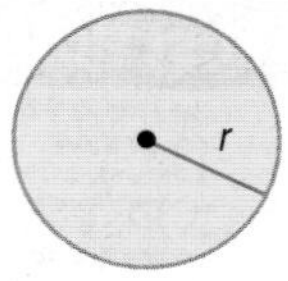

a. Graph the function.

b. Use a graphing calculator to determine the radius of a circle that has an area of 27 in^2.

43. SPEED OF SOUND The speed of sound in air is determined by the temperature of the air. The speed c in meters per second is given by $c = 331.5\sqrt{1 + \frac{t}{273.15}}$, where t is the temperature of the air in degrees Celsius.

a. Use a graphing calculator to graph the function.

b. How fast does sound travel when the temperature is 55°C?

c. How is the speed of sound affected when the temperature increases to 65°C?

44. MULTIPLE REPRESENTATIONS In this problem, you will explore the relationship between the graphs of square root functions and parabolas.

a. Graphical Graph $y = x^2$ on a coordinate system.

b. Algebraic Write a piecewise-defined function to describe the graph of $y^2 = x$ in each quadrant.

c. Graphical On the same coordinate system, graph $y = \sqrt{x}$ and $y = -\sqrt{x}$.

d. Graphical On the same coordinate system, graph $y = x$. Plot the points (2, 4), (4, 2), and (1, 1).

e. Analytical Compare the graph of the parabola to the graphs of the square root functions.

H.O.T. Problems Use Higher-Order Thinking Skills

CHALLENGE Determine whether each statement is *true* or *false*. Provide an example or counterexample to support your answer.

45. Numbers in the domain of a radical function will always be nonnegative.

46. Numbers in the range of a radical function will always be nonnegative.

47. WRITING IN MATH Why are there limitations on the domain and range of square root functions?

48. CCSS TOOLS Write a radical function with a domain of all real numbers greater than or equal to 2 and a range of all real numbers less than or equal to 5.

49. WHICH DOES NOT BELONG? Identify the equation that does not belong. Explain.

$y = 3\sqrt{x}$	$y = 0.7\sqrt{x}$	$y = \sqrt{x} + 3$	$y = \frac{\sqrt{x}}{6}$

50. OPEN ENDED Write a function that is a reflection, translation, and a dilation of the parent graph $y = \sqrt{x}$.

51. REASONING If the range of the function $y = a\sqrt{x}$ is $\{y \mid y \leq 0\}$, what can you conclude about the value of a? Explain your reasoning.

52. WRITING IN MATH Compare and contrast the graphs of $f(x) = \sqrt{x} + 2$ and $g(x) = \sqrt{x + 2}$.

Standardized Test Practice

53.

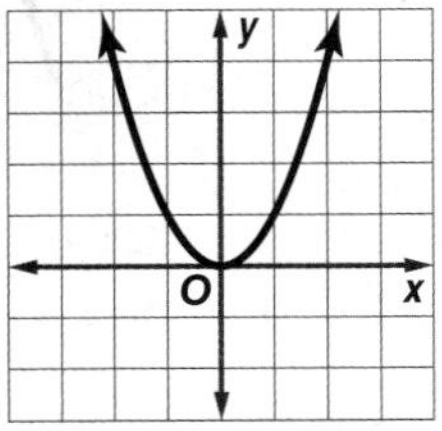

Which function *best* represents the graph?

A $y = x^2$ C $y = \sqrt{x}$
B $y = 2^x$ D $y = x$

54. The statement "$x < 10$ and $3x - 2 \geq 7$" is true when x is equal to what?

F 0 H 8
G 2 J 12

55. Which of the following is the equation of a line parallel to $y = -\frac{1}{2}x + 3$ and passing through $(-2, -1)$?

A $y = \frac{1}{2}x$ C $y = -\frac{1}{2}x + 2$

B $y = 2x + 3$ D $y = -\frac{1}{2}x - 2$

56. SHORT RESPONSE A landscaper needs to mulch 6 rectangular flower beds that are 8 feet by 4 feet and 4 circular flower beds each with a radius of 3 feet. One bag of mulch covers 25 square feet. How many bags of mulch are needed to cover the flower beds?

Spiral Review

57. HEALTH Aida exercises every day by walking and jogging at least 3 miles. Aida walks at a rate of 4 miles per hour and jogs at a rate of 8 miles per hour. Suppose she has at most one half-hour to exercise today. (Lesson 6-6)

a. Draw a graph showing the possible amounts of time she can spend walking and jogging today.

b. List three possible solutions.

58. NUTRITION Determine whether the graph shows a *positive*, *negative*, or *no* correlation. If there is a positive or negative correlation, describe its meaning in the situation. (Lesson 4-5)

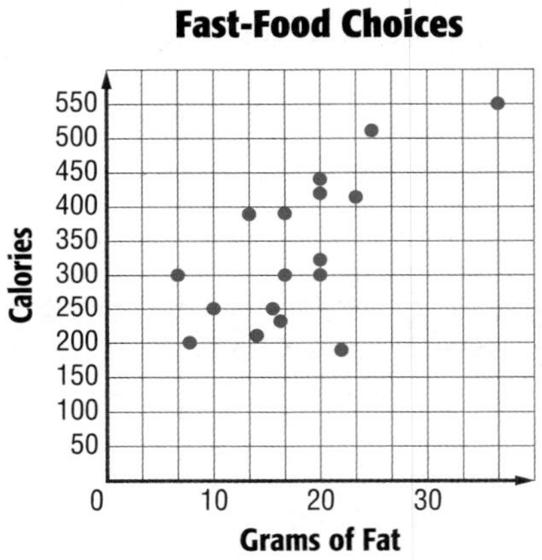

Skills Review

Factor each monomial completely.

59. $28n^3$

60. $-33a^2b$

61. $150rt$

62. $-378nq^2r^2$

63. $225a^3b^2c$

64. $-160x^2y^4$

EXTEND 8-1

Graphing Technology Lab
Graphing Square Root Functions

For a square root to be a real number, the radicand cannot be negative. When graphing a radical function, determine when the radicand would be negative and exclude those values from the domain.

CCSS Common Core State Standards
Content Standards
F.IF.7b Graph square root, cube root, and piecewise-defined functions, including step functions and absolute value functions.
Mathematical Practices
5 Use appropriate tools strategically.

Activity 1 Parent Function

Graph $y = \sqrt{x}$.

Enter the equation in the **Y=** list, and graph in the standard viewing window.

KEYSTROKES: [Y=] [2nd] [√] [X,T,θ,n] [ )] [ZOOM] 6

1A. Examine the graph. What is the domain of the function?

1B. What is the range of the function?

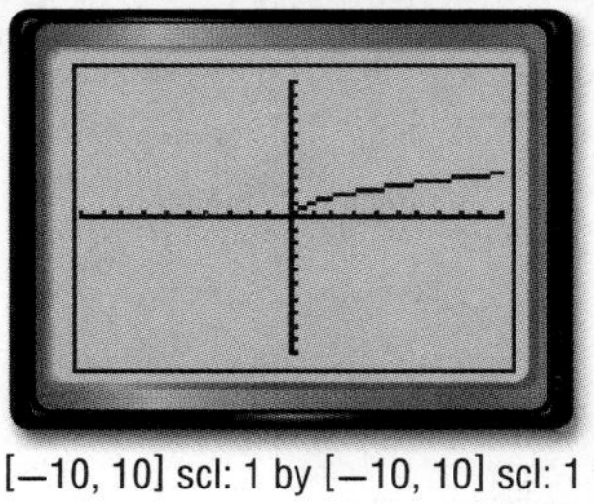

[−10, 10] scl: 1 by [−10, 10] scl: 1

Activity 2 Translation of Parent Function

Graph $y = \sqrt{x - 2}$.

Enter the equation in the **Y=** list, and graph in the standard viewing window.

KEYSTROKES: [Y=] [2nd] [√] [X,T,θ,n] [−] 2 [ )] [ZOOM] 6

2A. What are the domain and range of the function?

2B. How does the graph of $y = \sqrt{x - 2}$ compare to the graph of the parent function $y = \sqrt{x}$?

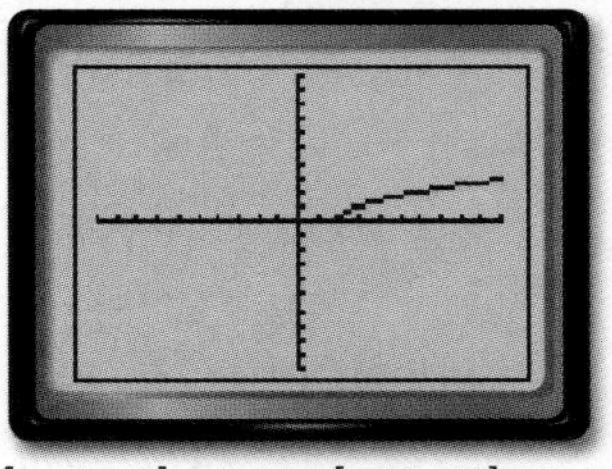

[−10, 10] scl: 1 by [−10, 10] scl: 1

Exercises

Graph each equation, and sketch the graph on your paper. State the domain and range. Describe how the graph differs from that of the parent function $y = \sqrt{x}$.

1. $y = \sqrt{x - 1}$

2. $y = \sqrt{x + 3}$

3. $y = \sqrt{x} - 2$

4. $y = \sqrt{-x}$

5. $y = -\sqrt{x}$

6. $y = \sqrt{2x}$

7. $y = \sqrt{2 - x}$

8. $y = \sqrt{x - 3} + 2$

Solve each equation for y. Does the equation represent a function? Explain your reasoning.

9. $x = y^2$

10. $x^2 + y^2 = 4$

11. $x^2 + y^2 = 2$

Write a function with a graph that translates $y = \sqrt{x}$ in each way.

12. shifted 4 units to the left

13. shifted up 7 units

14. shifted down 6 units

15. shifted 5 units to the right and up 3 units

LESSON 8-2 Simplifying Radical Expressions

Then

- You simplified radicals.

Now

1. Simplify radical expressions by using the Product Property of Square Roots.
2. Simplify radical expressions by using the Quotient Property of Square Roots.

Why?

- The Sunshine Skyway Bridge across Florida's Tampa Bay is supported by 21 steel cables, each 9 inches in diameter.

To find the diameter a steel cable should have to support a given weight, you can use the equation $d = \sqrt{\frac{w}{8}}$, where d is the diameter of the cable in inches and w is the weight in tons.

NewVocabulary
radical expression
rationalizing the denominator
conjugate

Common Core State Standards

Content Standards
A.REI.4a Use the method of completing the square to transform any quadratic equation in x into an equation of the form $(x - p)^2 = q$ that has the same solutions. Derive the quadratic formula from this form.

Mathematical Practices
7 Look for and make use of structure.
8 Look for and express regularity in repeated reasoning.

1 Product Property of Square Roots

A **radical expression** contains a radical, such as a square root. Recall the expression under the radical sign is called the radicand. A radicand is in simplest form if the following three conditions are true.

- No radicands have perfect square factors other than 1.
- No radicands contain fractions.
- No radicals appear in the denominator of a fraction.

The following property can be used to simplify square roots.

KeyConcept Product Property of Square Roots

Words	For any nonnegative real numbers a and b, the square root of ab is equal to the square root of a times the square root of b.
Symbols	$\sqrt{ab} = \sqrt{a} \cdot \sqrt{b}$, if $a \geq 0$ and $b \geq 0$
Examples	$\sqrt{4 \cdot 9} = \sqrt{36}$ or 6 $\sqrt{4 \cdot 9} = \sqrt{4} \cdot \sqrt{9} = 2 \cdot 3$ or 6

Example 1 Simplify Square Roots

PT

Simplify $\sqrt{80}$.

$\sqrt{80} = \sqrt{2 \cdot 2 \cdot 2 \cdot 2 \cdot 5}$ Prime factorization of 80

$= \sqrt{2^2} \cdot \sqrt{2^2} \cdot \sqrt{5}$ Product Property of Square Roots

$= 2 \cdot 2 \cdot \sqrt{5}$ or $4\sqrt{5}$ Simplify.

GuidedPractice

1A. $\sqrt{54}$

1B. $\sqrt{180}$

SuperStock/age fotostock

Example 2 Multiply Square Roots

Simplify $\sqrt{2} \cdot \sqrt{14}$.

$\sqrt{2} \cdot \sqrt{14} = \sqrt{2} \cdot \sqrt{2} \cdot \sqrt{7}$ — Product Property of Square Roots

$= \sqrt{2^2} \cdot \sqrt{7}$ or $2\sqrt{7}$ — Product Property of Square Roots

GuidedPractice

2A. $\sqrt{5} \cdot \sqrt{10}$

2B. $\sqrt{6} \cdot \sqrt{8}$

Consider the expression $\sqrt{x^2}$. It may seem that $x = \sqrt{x^2}$, but when finding the principal square root of an expression containing variables, you have to be sure that the result is not negative. Consider $x = -3$.

$\sqrt{x^2} \stackrel{?}{=} x$

$\sqrt{(-3)^2} \stackrel{?}{=} -3$ — Replace x with -3.

$\sqrt{9} \stackrel{?}{=} -3$ — $(-3)^2 = 9$

$3 \neq -3$ — $\sqrt{9} = 3$

Notice in this case, if the right hand side of the equation were $|x|$, the equation would be true. For expressions where the exponent of the variable inside a radical is even and the simplified exponent is odd, you must use absolute value.

$$\sqrt{x^2} = |x| \qquad \sqrt{x^3} = x\sqrt{x} \qquad \sqrt{x^4} = x^2 \qquad \sqrt{x^6} = |x^3|$$

Example 3 Simplify a Square Root with Variables

Simplify $\sqrt{90x^3y^4z^5}$.

$\sqrt{90x^3y^4z^5} = \sqrt{2 \cdot 3^2 \cdot 5 \cdot x^3 \cdot y^4 \cdot z^5}$ — Prime factorization

$= \sqrt{2} \cdot \sqrt{3^2} \cdot \sqrt{5} \cdot \sqrt{x^2} \cdot \sqrt{x} \cdot \sqrt{y^4} \cdot \sqrt{z^4} \cdot \sqrt{z}$ — Product Property

$= \sqrt{2} \cdot 3 \cdot \sqrt{5} \cdot x \cdot \sqrt{x} \cdot y^2 \cdot z^2 \cdot \sqrt{z}$ — Simplify.

$= 3y^2z^2x\sqrt{10xz}$ — Simplify.

GuidedPractice

3A. $\sqrt{32r^2k^4t^5}$

3B. $\sqrt{56xy^{10}z^5}$

2 Quotient Property of Square Roots

To divide square roots and simplify radical expressions, you can use the Quotient Property of Square Roots.

ReadingMath

Fractions in the Radicand The expression $\sqrt{\frac{a}{b}}$ is read *the square root of a over b,* or *the square root of the quantity of a over b.*

KeyConcept Quotient Property of Square Roots

Words For any real numbers a and b, where $a \geq 0$ and $b > 0$, the square root of $\frac{a}{b}$ is equal to the square root of a divided by the square root of b.

Symbols $\sqrt{\frac{a}{b}} = \frac{\sqrt{a}}{\sqrt{b}}$

You can use the properties of square roots to **rationalize the denominator** of a fraction with a radical. This involves multiplying the numerator and denominator by a factor that eliminates radicals in the denominator.

Standardized Test Example 4 Rationalize a Denominator

Which expression is equivalent to $\sqrt{\frac{35}{15}}$?

A $\frac{5\sqrt{21}}{15}$ B $\frac{\sqrt{21}}{3}$ C $\frac{\sqrt{525}}{15}$ D $\frac{\sqrt{35}}{15}$

Test-Taking Tip

CCSS Structure Look at the radicand to see if it can be simplified first. This may make your computations simpler.

Read the Test Item The radical expression needs to be simplified.

Solve the Test Item

$\sqrt{\frac{35}{15}} = \sqrt{\frac{7}{3}}$ Reduce $\frac{35}{15}$ to $\frac{7}{3}$.

$= \frac{\sqrt{7}}{\sqrt{3}}$ Quotient Property of Square Roots

$= \frac{\sqrt{7}}{\sqrt{3}} \cdot \frac{\sqrt{3}}{\sqrt{3}}$ Multiply by $\frac{\sqrt{3}}{\sqrt{3}}$.

$= \frac{\sqrt{21}}{3}$ Product Property of Square Roots

The correct choice is B.

Guided Practice

4. Simplify $\frac{\sqrt{6y}}{\sqrt{12}}$.

F $\frac{\sqrt{y}}{2}$ G $\frac{\sqrt{y}}{4}$ H $\frac{\sqrt{2y}}{2}$ J $\frac{\sqrt{2y}}{4}$

Binomials of the form $a\sqrt{b} + c\sqrt{d}$ and $a\sqrt{b} - c\sqrt{d}$, where a, b, c, and d are rational numbers, are called **conjugates**. For example, $2 + \sqrt{7}$ and $2 - \sqrt{7}$ are conjugates. The product of two conjugates is a rational number and can be found using the pattern for the difference of squares.

Example 5 Use Conjugates to Rationalize a Denominator

Simplify $\frac{3}{5 + \sqrt{2}}$.

$\frac{3}{5 + \sqrt{2}} = \frac{3}{5 + \sqrt{2}} \cdot \frac{5 - \sqrt{2}}{5 - \sqrt{2}}$ The conjugate of $5 + \sqrt{2}$ is $5 - \sqrt{2}$.

$= \frac{3(5 - \sqrt{2})}{5^2 - (\sqrt{2})^2}$ $(a - b)(a + b) = a^2 - b^2$

$= \frac{15 - 3\sqrt{2}}{25 - 2}$ or $\frac{15 - 3\sqrt{2}}{23}$ $(\sqrt{2})^2 = 2$

Guided Practice **Simplify each expression.**

5A. $\frac{3}{2 + \sqrt{2}}$ **5B.** $\frac{7}{3 - \sqrt{7}}$

Check Your Understanding

Examples 1–3 **Simplify each expression.**

1. $\sqrt{24}$
2. $3\sqrt{16}$
3. $2\sqrt{25}$
4. $\sqrt{10} \cdot \sqrt{14}$
5. $\sqrt{3} \cdot \sqrt{18}$
6. $3\sqrt{10} \cdot 4\sqrt{10}$
7. $\sqrt{60x^4y^7}$
8. $\sqrt{88m^3p^2r^5}$
9. $\sqrt{99ab^5c^2}$

Example 4

10. **MULTIPLE CHOICE** Which expression is equivalent to $\sqrt{\frac{45}{10}}$?

A $\frac{5\sqrt{2}}{10}$ B $\frac{\sqrt{45}}{10}$ C $\frac{\sqrt{50}}{10}$ D $\frac{3\sqrt{2}}{2}$

Example 5 **Simplify each expression.**

11. $\frac{3}{3 + \sqrt{5}}$
12. $\frac{5}{2 - \sqrt{6}}$
13. $\frac{2}{1 - \sqrt{10}}$
14. $\frac{1}{4 + \sqrt{12}}$
15. $\frac{4}{6 - \sqrt{7}}$
16. $\frac{6}{5 + \sqrt{11}}$

Practice and Problem Solving

Examples 1–3 **Simplify each expression.**

17. $\sqrt{52}$
18. $\sqrt{56}$
19. $\sqrt{72}$
20. $3\sqrt{18}$
21. $\sqrt{243}$
22. $\sqrt{245}$
23. $\sqrt{5} \cdot \sqrt{10}$
24. $\sqrt{10} \cdot \sqrt{20}$
25. $3\sqrt{8} \cdot 2\sqrt{7}$
26. $4\sqrt{2} \cdot 5\sqrt{8}$
27. $3\sqrt{25t^2}$
28. $5\sqrt{81q^5}$
29. $\sqrt{28a^2b^3}$
30. $\sqrt{75qr^3}$
31. $7\sqrt{63m^3p}$
32. $4\sqrt{66g^2h^4}$
33. $\sqrt{2ab^2} \cdot \sqrt{10a^5b}$
34. $\sqrt{4c^3d^3} \cdot \sqrt{8c^3d}$

35. **ROLLER COASTER** Starting from a stationary position, the velocity v of a roller coaster in feet per second at the bottom of a hill can be approximated by $v = \sqrt{64h}$, where h is the height of the hill in feet.

    a. Simplify the equation.

    b. Determine the velocity of a roller coaster at the bottom of a 134-foot hill.

36. **CCSS PRECISION** When fighting a fire, the velocity v of water being pumped into the air is modeled by the function $v = \sqrt{2hg}$, where h represents the maximum height of the water and g represents the acceleration due to gravity (32 ft/s^2).

    a. Solve the function for h.

    b. The Hollowville Fire Department needs a pump that will propel water 80 feet into the air. Will a pump advertised to project water with a velocity of 70 feet per second meet their needs? Explain.

    c. The Jackson Fire Department must purchase a pump that will propel water 90 feet into the air. Will a pump that is advertised to project water with a velocity of 77 feet per second meet the fire department's need? Explain.

Examples 4–5 **Simplify each expression.**

37. $\sqrt{\frac{32}{t^4}}$

38. $\sqrt{\frac{27}{m^5}}$

39. $\frac{\sqrt{68ac^3}}{\sqrt{27a^2}}$

40. $\frac{\sqrt{h^3}}{\sqrt{8}}$

41. $\sqrt{\frac{3}{16}} \cdot \sqrt{\frac{9}{5}}$

42. $\sqrt{\frac{7}{2}} \cdot \sqrt{\frac{5}{3}}$

43. $\frac{7}{5+\sqrt{3}}$

44. $\frac{9}{6-\sqrt{8}}$

45. $\frac{3\sqrt{3}}{-2+\sqrt{6}}$

46. $\frac{3}{\sqrt{7}-\sqrt{2}}$

47. $\frac{5}{\sqrt{6}+\sqrt{3}}$

48. $\frac{2\sqrt{5}}{2\sqrt{7}+3\sqrt{3}}$

49. **ELECTRICITY** The amount of current in amperes I that an appliance uses can be calculated using the formula $I = \sqrt{\frac{P}{R}}$, where P is the power in watts and R is the resistance in ohms.
    a. Simplify the formula.
    b. How much current does an appliance use if the power used is 75 watts and the resistance is 5 ohms?

50. **KINETIC ENERGY** The speed v of a ball can be determined by the equation $v = \sqrt{\frac{2k}{m}}$, where k is the kinetic energy and m is the mass of the ball.
    a. Simplify the formula if the mass of the ball is 3 kilograms.
    b. If the ball is traveling 7 meters per second, what is the kinetic energy of the ball in Joules?

51. **SUBMARINES** The greatest distance d in miles that a lookout can see on a clear day is modeled by the formula shown. Determine how high the submarine would have to raise its periscope to see a ship, if the submarine is the given distances away from the ship.

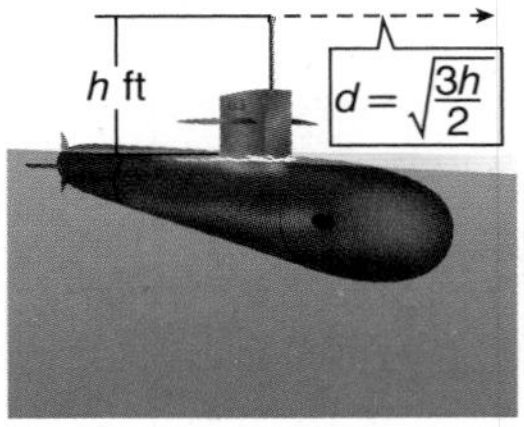

Distance	3	6	9	12	15
Height					

H.O.T. Problems Use Higher-Order Thinking Skills

52. **CCSS STRUCTURE** Explain how to solve $\frac{\sqrt{3}+2}{x} = \frac{\sqrt{3}-1}{\sqrt{3}}$.

53. **CHALLENGE** Simplify each expression.
    a. $\sqrt[3]{27}$ **b.** $\sqrt[3]{40}$ **c.** $\sqrt[3]{750}$

54. **REASONING** Marge takes a number, subtracts 4, multiplies by 4, takes the square root, and takes the reciprocal to get $\frac{1}{2}$. What number did she start with? Write a formula to describe the process.

55. **OPEN ENDED** Write two binomials of the form $a\sqrt{b} + c\sqrt{f}$ and $a\sqrt{b} - c\sqrt{f}$. Then find their product.

56. **CHALLENGE** Use the Quotient Property of Square Roots to derive the Quadratic Formula by solving the quadratic equation $ax^2 + bx + c = 0$. (*Hint*: Begin by completing the square.)

57. **WRITING IN MATH** Summarize how to write a radical expression in simplest form.

Standardized Test Practice

58. Jerry's electric bill is \$23 less than his natural gas bill. The two bills are a total of \$109. Which of the following equations can be used to find the amount of his natural gas bill?

A $g + g = 109$ **C** $g - 23 = 109$

B $23 + 2g = 109$ **D** $2g - 23 = 109$

59. Solve $a^2 - 2a + 1 = 25$.

F $-4, -6$ **H** $-4, 6$

G $4, -6$ **J** $4, 6$

60. The expression $\sqrt{160x^2y^5}$ is equivalent to which of the following?

A $16|x|y^2\sqrt{10y}$ **C** $4|x|y^2\sqrt{10y}$

B $|x|y^2\sqrt{160y}$ **D** $10|x|y^2\sqrt{4y}$

61. GRIDDED RESPONSE Miki earns \$10 an hour and 10% commission on sales. If Miki worked 38 hours and had a total sales of \$1275 last week, how much did she make?

Spiral Review

Graph each function. Compare to the parent graph. State the domain and range. (Lesson 8-1)

62. $y = 2\sqrt{x} - 1$

63. $y = \frac{1}{2}\sqrt{x}$

64. $y = 2\sqrt{x + 2}$

65. $y = -\sqrt{x + 1}$

66. $y = -3\sqrt{x - 3}$

67. $y = -2\sqrt{x} + 1$

68. POPULATION The country of Latvia has been experiencing a 1.1% annual decrease in population. In 2009, its population was 2,261,294. If the trend continues, predict Latvia's population in 2019. (Lesson 7-6)

69. TOMATOES There are more than 10,000 varieties of tomatoes. One seed company produces seed packages for 200 varieties of tomatoes. For how many varieties do they not provide seeds? (Lesson 5-1)

Skills Review

Write the prime factorization of each number.

70. 24

71. 88

72. 180

73. 31

74. 60

75. 90

EXTEND 8-2

Algebra Lab Rational and Irrational Numbers

A set is **closed** under an operation if for any numbers in the set, the result of the operation is also in the set. A set may be closed under one operation and not closed under another.

CCSS Common Core State Standards
Content Standards
N.RN.3 Explain why the sum or product of two rational numbers is rational; that the sum of a rational number and an irrational number is irrational; and that the product of a nonzero rational number and an irrational number is irrational.
Mathematical Practices
7 Look for and make use of structure.

Activity 1 Closure of Rational Numbers and Irrational Numbers

Are the sets of rational and irrational numbers closed under multiplication? under addition?

Step 1 To determine if each set is closed under multiplication, examine several products of two rational factors and then two irrational factors.

Rational: $5 \times 2 = 10$; $-3 \times 4 = -12$; $3.7 \times 0.5 = 1.85$; $\frac{3}{4} \times \frac{2}{3} = \frac{1}{2}$

Irrational: $\pi \times \sqrt{2} = \sqrt{2}\pi$; $\sqrt{3} \times \sqrt{7} = \sqrt{21}$; $\sqrt{5} \times \sqrt{5} = 5$

The product of each pair of rational numbers is rational. However, the products of pairs of irrational numbers are both irrational and rational. Thus, it appears that the set of rational numbers is closed under multiplication, but the set of irrational numbers is not.

Step 2 Repeat this process for addition.

Rational: $3 + 8 = 11$; $-4 + 7 = 3$; $3.7 + 5.82 = 9.52$; $\frac{2}{5} + \frac{1}{4} = \frac{13}{20}$

Irrational: $\sqrt{3} + \pi = \sqrt{3} + \pi$; $3\sqrt{5} + 6\sqrt{5} = 9\sqrt{5}$; $\sqrt{12} + \sqrt{50} = 2\sqrt{3} + 5\sqrt{2}$

The sum of each pair of rational numbers is rational, and the sum of each pair of irrational numbers is irrational. Both sets are closed under addition.

Activity 2 Rational and Irrational Numbers

What kind of numbers are the product and sum of a rational and irrational number?

Step 1 Examine the products of several pairs of rational and irrational numbers.

$3 \times \sqrt{8} = 6\sqrt{2}$; $\frac{3}{4} \times \sqrt{2} = \frac{3\sqrt{2}}{4}$; $1 \times \sqrt{7} = \sqrt{7}$; $0 \times \sqrt{5} = 0$

The product is rational only when the rational factor is 0. The product of each nonzero rational number and irrational number is irrational.

Step 2 Find the sums of several pairs of a rational and irrational number.

$5 + \sqrt{3} = 5 + \sqrt{3}$; $\frac{2}{3} + \sqrt{5} = \frac{2 + 3\sqrt{5}}{3}$; $-4 + \sqrt{6} = -1(4 - \sqrt{6})$

The sum of each rational and irrational number is irrational.

Analyze the Results

1. What kinds of numbers are the difference of two unique rational numbers, two unique irrational numbers, and a rational and an irrational number?
2. Is the quotient of every rational and irrational number always another rational or irrational number? If not, provide a counterexample.
3. **CHALLENGE** Recall that rational numbers are numbers that can be written in the form $\frac{a}{b}$, where a and b are integers and $b \neq 0$. Using $\frac{a}{b}$ and $\frac{c}{d}$ show that the sum and product of two rational numbers must always be a rational number.

LESSON 8-3

Operations with Radical Expressions

Then

- You simplified radical expressions.

Now

1. Add and subtract radical expressions.
2. Multiply radical expressions.

Why?

- Conchita is going to run in her neighborhood to get ready for the soccer season. She plans to run the course that she has laid out three times each day.

How far does Conchita have to run to complete the course that she laid out?

How far does she run every day?

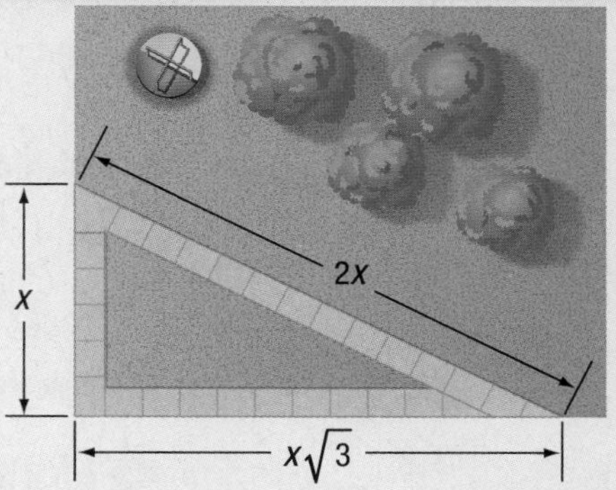

Common Core State Standards

Content Standards

N.RN.2 Rewrite expressions involving radicals and rational exponents using the properties of exponents.

Mathematical Practices

2 Reason abstractly and quantitatively.

1 Add or Subtract Radical Expressions

To add or subtract radical expressions, the radicands must be alike in the same way that monomial terms must be alike to add or subtract.

Monomials	Radical Expressions
$4a + 2a = (4 + 2)a$	$4\sqrt{5} + 2\sqrt{5} = (4 + 2)\sqrt{5}$
$= 6a$	$= 6\sqrt{5}$
$9b - 2b = (9 - 2)b$	$9\sqrt{3} - 2\sqrt{3} = (9 - 2)\sqrt{3}$
$= 7b$	$= 7\sqrt{3}$

Notice that when adding and subtracting radical expressions, the radicand does not change. This is the same as when adding or subtracting monomials.

Example 1 Add and Subtract Expressions with Like Radicands

Simplify each expression.

a. $\mathbf{5\sqrt{2} + 7\sqrt{2} - 6\sqrt{2}}$

$5\sqrt{2} + 7\sqrt{2} - 6\sqrt{2} = (5 + 7 - 6)\sqrt{2}$ Distributive Property

$= 6\sqrt{2}$ Simplify.

b. $\mathbf{10\sqrt{7} + 5\sqrt{11} + 4\sqrt{7} - 6\sqrt{11}}$

$10\sqrt{7} + 5\sqrt{11} + 4\sqrt{7} - 6\sqrt{11} = (10 + 4)\sqrt{7} + (5 - 6)\sqrt{11}$ Distributive Property

$= 14\sqrt{7} - \sqrt{11}$ Simplify.

Guided Practice

1A. $3\sqrt{2} - 5\sqrt{2} + 4\sqrt{2}$

1B. $6\sqrt{11} + 2\sqrt{11} - 9\sqrt{11}$

1C. $15\sqrt{3} - 14\sqrt{5} + 6\sqrt{5} - 11\sqrt{3}$

1D. $4\sqrt{3} + 3\sqrt{7} - 6\sqrt{3} + 3\sqrt{7}$

Not all radical expressions have like radicands. Simplifying the expressions may make it possible to have like radicands so that they can be added or subtracted.

StudyTip

Simplify First Simplify each radical term first. Then perform the operations needed.

Example 2 Add and Subtract Expressions with Unlike Radicands

Simplify $2\sqrt{18} + 2\sqrt{32} + \sqrt{72}$.

$$2\sqrt{18} + 2\sqrt{32} + \sqrt{72} = 2(\sqrt{3^2} \cdot \sqrt{2}) + 2(\sqrt{4^2} \cdot \sqrt{2}) + (\sqrt{6^2} \cdot \sqrt{2})$$ Product Property

$$= 2(3\sqrt{2}) + 2(4\sqrt{2}) + (6\sqrt{2})$$ Simplify.

$$= 6\sqrt{2} + 8\sqrt{2} + 6\sqrt{2}$$ Multiply.

$$= 20\sqrt{2}$$ Simplify.

GuidedPractice

2A. $4\sqrt{54} + 2\sqrt{24}$

2B. $4\sqrt{12} - 6\sqrt{48}$

2C. $3\sqrt{45} + \sqrt{20} - \sqrt{245}$

2D. $\sqrt{24} - \sqrt{54} + \sqrt{96}$

2 Multiply Radical Expressions

Multiplying radical expressions is similar to multiplying monomial algebraic expressions. Let $x \geq 0$.

Monomials	Radical Expressions
$(2x)(3x) = 2 \cdot 3 \cdot x \cdot x$	$(2\sqrt{x})(3\sqrt{x}) = 2 \cdot 3 \cdot \sqrt{x} \cdot \sqrt{x}$
$= 6x^2$	$= 6x$

You can also apply the Distributive Property to radical expressions.

Example 3 Multiply Radical Expressions

WatchOut!

Multiplying Radicands Make sure that you multiply the radicands when multiplying radical expressions. A common mistake is to add the radicands rather than multiply.

Simplify each expression.

a. $3\sqrt{2} \cdot 2\sqrt{6}$

$$3\sqrt{2} \cdot 2\sqrt{6} = (3 \cdot 2)(\sqrt{2} \cdot \sqrt{6})$$ Associative Property

$$= 6(\sqrt{12})$$ Multiply.

$$= 6(2\sqrt{3})$$ Simplify.

$$= 12\sqrt{3}$$ Multiply.

b. $3\sqrt{5}(2\sqrt{5} + 5\sqrt{3})$

$$3\sqrt{5}(2\sqrt{5} + 5\sqrt{3}) = (3\sqrt{5} \cdot 2\sqrt{5}) + (3\sqrt{5} \cdot 5\sqrt{3})$$ Distributive Property

$$= [(3 \cdot 2)(\sqrt{5} \cdot \sqrt{5})] + [(3 \cdot 5)(\sqrt{5} \cdot \sqrt{3})]$$ Associative Property

$$= [6(\sqrt{25})] + [15(\sqrt{15})]$$ Multiply.

$$= [6(5)] + [15(\sqrt{15})]$$ Simplify.

$$= 30 + 15\sqrt{15}$$ Multiply.

GuidedPractice

3A. $2\sqrt{6} \cdot 7\sqrt{3}$

3B. $9\sqrt{5} \cdot 11\sqrt{15}$

3C. $3\sqrt{2}(4\sqrt{3} + 6\sqrt{2})$

3D. $5\sqrt{3}(3\sqrt{2} - \sqrt{3})$

You can also multiply radical expressions with more than one term in each factor. This is similar to multiplying two algebraic binomials with variables.

Real-World Example 4 Multiply Radical Expressions

GEOMETRY Find the area of the rectangle in simplest form.

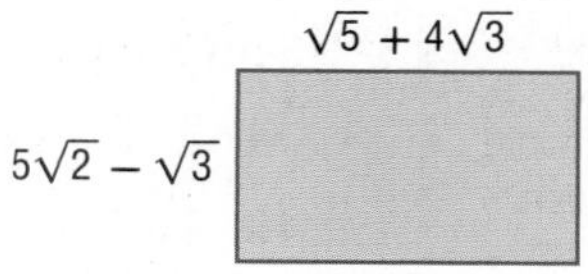

$A = (5\sqrt{2} - \sqrt{3})(\sqrt{5} + 4\sqrt{3})$ $\qquad A = \ell \cdot w$

$= \underbrace{(5\sqrt{2})(\sqrt{5})}_{\text{First Terms}} + \underbrace{(5\sqrt{2})(4\sqrt{3})}_{\text{Outer Terms}} + \underbrace{(-\sqrt{3})(\sqrt{5})}_{\text{Inner Terms}} + \underbrace{(\sqrt{3})(4\sqrt{3})}_{\text{Last Terms}}$

$= 5\sqrt{10} + 20\sqrt{6} - \sqrt{15} - 4\sqrt{9}$ Multiply.

$= 5\sqrt{10} + 20\sqrt{6} - \sqrt{15} - 12$ Simplify.

Review Vocabulary

FOIL Method Multiply two binomials by finding the sum of the products of the First terms, the Outer terms, the Inner terms, and the Last terms.

Guided Practice

4. GEOMETRY The area A of a rhombus can be found using the equation $A = \frac{1}{2}d_1d_2$, where d_1 and d_2 are the lengths of the diagonals. What is the area of the rhombus at the right?

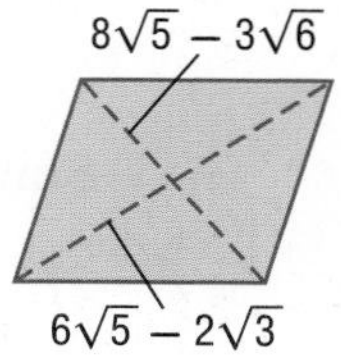

Concept Summary Operations with Radical Expressions

Operation	Symbols	Example
addition, $b \geq 0$	$a\sqrt{b} + c\sqrt{b} = (a + c)\sqrt{b}$ like radicands	$4\sqrt{3} + 6\sqrt{3} = (4 + 6)\sqrt{3}$ $= 10\sqrt{3}$
subtraction, $b \geq 0$	$a\sqrt{b} - c\sqrt{b} = (a - c)\sqrt{b}$ like radicands	$12\sqrt{5} - 8\sqrt{5} = (12 - 8)\sqrt{5}$ $= 4\sqrt{5}$
multiplication, $b \geq 0, g \geq 0$	$a\sqrt{b}(f\sqrt{g}) = af\sqrt{bg}$ Radicands do not have to be like radicands.	$3\sqrt{2}(5\sqrt{7}) = (3 \cdot 5)(\sqrt{2 \cdot 7})$ $= 15\sqrt{14}$

Check Your Understanding

Examples 1–3 **Simplify each expression.**

1. $3\sqrt{5} + 6\sqrt{5}$
2. $8\sqrt{3} + 5\sqrt{3}$
3. $\sqrt{7} - 6\sqrt{7}$
4. $10\sqrt{2} - 6\sqrt{2}$
5. $4\sqrt{5} + 2\sqrt{20}$
6. $\sqrt{12} - \sqrt{3}$
7. $\sqrt{8} + \sqrt{12} + \sqrt{18}$
8. $\sqrt{27} + 2\sqrt{3} - \sqrt{12}$
9. $9\sqrt{2}(4\sqrt{6})$
10. $4\sqrt{3}(8\sqrt{3})$
11. $\sqrt{3}(\sqrt{7} + 3\sqrt{2})$
12. $\sqrt{5}(\sqrt{2} + 4\sqrt{2})$

Example 4

13. **GEOMETRY** The area A of a triangle can be found by using the formula $A = \frac{1}{2}bh$, where b represents the base and h is the height. What is the area of the triangle at the right?

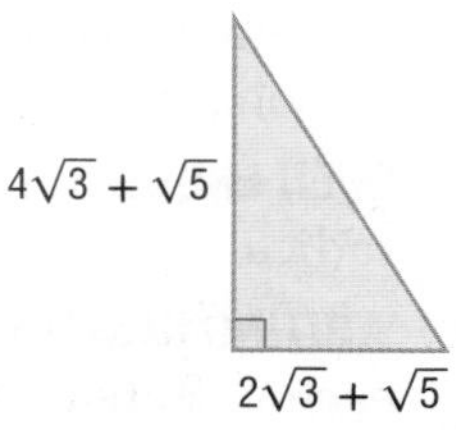

Practice and Problem Solving

Examples 1–3 **Simplify each expression.**

14. $7\sqrt{5} + 4\sqrt{5}$
15. $2\sqrt{6} + 9\sqrt{6}$
16. $3\sqrt{5} - 2\sqrt{20}$
17. $3\sqrt{50} - 3\sqrt{32}$
18. $7\sqrt{3} - 2\sqrt{2} + 3\sqrt{2} + 5\sqrt{3}$
19. $\sqrt{5}(\sqrt{2} + 4\sqrt{2})$
20. $\sqrt{6}(2\sqrt{10} + 3\sqrt{2})$
21. $4\sqrt{5}(3\sqrt{5} + 8\sqrt{2})$
22. $5\sqrt{3}(6\sqrt{10} - 6\sqrt{3})$
23. $(\sqrt{3} - \sqrt{2})(\sqrt{15} + \sqrt{12})$
24. $(3\sqrt{11} + 3\sqrt{15})(3\sqrt{3} - 2\sqrt{2})$
25. $(5\sqrt{2} + 3\sqrt{5})(2\sqrt{10} - 5)$

Example 4

26. **GEOMETRY** Find the perimeter and area of a rectangle with a width of $2\sqrt{7} - 2\sqrt{5}$ and a length of $3\sqrt{7} + 3\sqrt{5}$.

Simplify each expression.

27. $\sqrt{\frac{1}{5}} - \sqrt{5}$
28. $\sqrt{\frac{2}{3}} + \sqrt{6}$
29. $2\sqrt{\frac{1}{2}} + 2\sqrt{2} - \sqrt{8}$
30. $8\sqrt{\frac{5}{4}} + 3\sqrt{20} - 10\sqrt{\frac{1}{5}}$
31. $(3 - \sqrt{5})^2$
32. $(\sqrt{2} + \sqrt{3})^2$

33. **ROLLER COASTERS** The velocity v in feet per second of a roller coaster at the bottom of a hill is related to the vertical drop h in feet and the velocity v_0 of the coaster at the top of the hill by the formula $v_0 = \sqrt{v^2 - 64h}$.
   a. What velocity must a coaster have at the top of a 225-foot hill to achieve a velocity of 120 feet per second at the bottom?
   b. Explain why $v_0 = v - 8\sqrt{h}$ is not equivalent to the formula given.

34. **FINANCIAL LITERACY** Tadi invests \$225 in a savings account. In two years, Tadi has \$232 in his account. You can use the formula $r = \sqrt{\frac{v_2}{v_0}} - 1$ to find the average annual interest rate r that the account has earned. The initial investment is v_0, and v_2 is the amount in two years. What was the average annual interest rate that Tadi's account earned?

35. **ELECTRICITY** Electricians can calculate the electrical current in amps A by using the formula $A = \frac{\sqrt{w}}{\sqrt{r}}$, where w is the power in watts and r the resistance in ohms. How much electrical current is running through a microwave oven that has 850 watts of power and 5 ohms of resistance? Write the number of amps in simplest radical form, and then estimate the amount of current to the nearest tenth.

H.O.T. Problems Use Higher-Order Thinking Skills

36. **CHALLENGE** Determine whether the following statement is *true* or *false*. Provide a proof or counterexample to support your answer.

$$x + y > \sqrt{x^2 + y^2} \text{ when } x > 0 \text{ and } y > 0$$

37. **CCSS ARGUMENTS** Make a conjecture about the sum of a rational number and an irrational number. Is the sum *rational* or *irrational*? Is the product of a nonzero rational number and an irrational number *rational* or *irrational*? Explain your reasoning.

38. **OPEN ENDED** Write an equation that shows a sum of two radicals with different radicands. Explain how you could combine these terms.

39. **WRITING IN MATH** Describe step by step how to multiply two radical expressions, each with two terms. Write an example to demonstrate your description.

Standardized Test Practice

40. SHORT RESPONSE The population of a town is 13,000 and is increasing by about 250 people per year. This can be represented by the equation $p = 13{,}000 + 250y$, where y is the number of years from now and p represents the population. In how many years will the population of the town be 14,500?

41. GEOMETRY Which expression represents the sum of the lengths of the 12 edges on this rectangular solid?

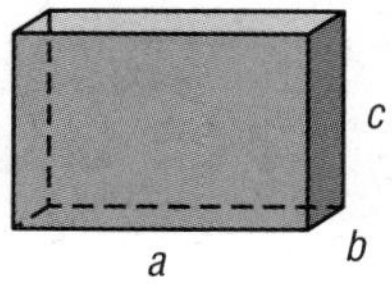

A $2(a + b + c)$
B $3(a + b + c)$
C $4(a + b + c)$
D $12(a + b + c)$

42. Evaluate $\sqrt{n-9}$ and $\sqrt{n} - \sqrt{9}$ for $n = 25$.

F 4; 4
G 4; 2
H 2; 4
J 2; 2

43. The current I in a simple electrical circuit is given by the formula $I = \frac{V}{R}$, where V is the voltage and R is the resistance of the circuit. If the voltage remains unchanged, what effect will doubling the resistance of the circuit have on the current?

A The current will remain the same.
B The current will double its previous value.
C The current will be half its previous value.
D The current will be two units more than its previous value.

Spiral Review

Simplify. (Lesson 8-2)

44. $\sqrt{18}$
45. $\sqrt{24}$
46. $\sqrt{60}$
47. $\sqrt{50a^3b^5}$
48. $\sqrt{169x^4y^7}$
49. $\sqrt{63c^3d^4f^5}$

Graph each function. Compare to the parent graph. State the domain and range. (Lesson 8-1)

50. $y = 2\sqrt{x}$
51. $y = -3\sqrt{x}$
52. $y = \sqrt{x+1}$
53. $y = \sqrt{x-4}$
54. $y = \sqrt{x} + 3$
55. $y = \sqrt{x} - 2$

56. FINANCIAL LITERACY Determine the value of an investment if \$400 is invested at an interest rate of 7.25% compounded quarterly for 7 years. (Lesson 7-6)

Skills Review

Solve each equation. Round each solution to the nearest tenth, if necessary.

57. $-4c - 1.2 = 0.8$
58. $-2.6q - 33.7 = 84.1$
59. $0.3m + 4 = 9.6$
60. $-10 - \frac{n}{5} = 6$
61. $\frac{-4h - (-5)}{-7} = 13$
62. $3.6t + 6 - 2.5t = 8$

LESSON 8-4

Radical Equations

Then

- You added, subtracted, and multiplied radical expressions.

Now

1. Solve radical equations.
2. Solve radical equations with extraneous solutions.

Why?

- The waterline length of a sailboat is the length of the line made by the water's edge when the boat is full. A sailboat's hull speed is the fastest speed that it can travel.

  You can estimate hull speed h by using the formula $h = 1.34\sqrt{\ell}$, where ℓ is the length of the sailboat's waterline.

NewVocabulary
radical equations
extraneous solutions

Common Core State Standards

Content Standards
N.RN.2 Rewrite expressions involving radicals and rational exponents using the properties of exponents.
A.CED.2 Create equations in two or more variables to represent relationships between quantities; graph equations on coordinate axes with labels and scales.

Mathematical Practices
3 Construct viable arguments and critique the reasoning of others.
4 Model with mathematics.

1 Radical Equations

Equations that contain variables in the radicand, like $h = 1.34\sqrt{\ell}$, are called **radical equations**. To solve, isolate the desired variable on one side of the equation first. Then square each side of the equation to eliminate the radical.

KeyConcept Power Property of Equality

Words	If you square both sides of a true equation, the resulting equation is still true.
Symbols	If $a = b$, then $a^2 = b^2$.
Examples	If $\sqrt{x} = 4$, then $(\sqrt{x})^2 = 4^2$.

PT

Real-World Example 1 Variable as a Radicand

SAILING Idris and Sebastian are sailing in a friend's sailboat. They measure the hull speed at 9 nautical miles per hour. Find the length of the sailboat's waterline. Round to the nearest foot.

Understand You know how fast the boat will travel and that it relates to the length.

Plan The boat travels at 9 nautical miles per hour. The formula for hull speed is $h = 1.34\sqrt{\ell}$.

Solve

$h = 1.34\sqrt{\ell}$	Formula for hull speed
$9 = 1.34\sqrt{\ell}$	Substitute 9 for h.
$\frac{9}{1.34} = \frac{1.34\sqrt{\ell}}{1.34}$	Divide each side by 1.34.
$6.72 \approx \sqrt{\ell}$	Simplify.
$(6.72)^2 \approx (\sqrt{\ell})^2$	Square each side of the equation.
$45.16 \approx \ell$	Simplify.

The sailboat's waterline length is about 45 feet.

Check Check by substituting the estimate into the original formula.

$h = 1.34\sqrt{\ell}$	Formula for hull speed
$9 \stackrel{?}{=} 1.34\sqrt{45}$	$h = 9$ and $\ell = 45$
$9 \approx 8.98899327$ ✓	Multiply.

Guided Practice

1. **DRIVING** The equation $v = \sqrt{2.5r}$ represents the maximum velocity that a car can travel safely on an unbanked curve when v is the maximum velocity in miles per hour and r is the radius of the turn in feet. If a road is designed for a maximum speed of 65 miles per hour, what is the radius of the turn?

To solve a radical equation, isolate the radical first. Then square both sides of the equation.

Example 2 Expression as a Radicand

PT

Solve $\sqrt{a+5}+7=12$.

$\sqrt{a+5}+7=12$	Original equation
$\sqrt{a+5}=5$	Subtract 7 from each side.
$\left(\sqrt{a+5}\right)^2=5^2$	Square each side.
$a+5=25$	Simplify.
$a=20$	Subtract 5 from each side.

Watch Out!

Squaring Each Side Remember that when you square each side of the equation, you must square the entire side of the equation, even if there is more than one term on the side.

Guided Practice

Solve each equation.

2A. $\sqrt{c-3}-2=4$

2B. $4+\sqrt{h+1}=14$

2 Extraneous Solutions

Squaring each side of an equation sometimes produces a solution that is not a solution of the original equation. These are called **extraneous solutions**. Therefore, you must check all solutions in the original equation.

Example 3 Variable on Each Side

PT

Solve $\sqrt{k+1}=k-1$. Check your solution.

$\sqrt{k+1}=k-1$	Original equation
$\left(\sqrt{k+1}\right)^2=(k-1)^2$	Square each side.
$k+1=k^2-2k+1$	Simplify.
$0=k^2-3k$	Subtract k and 1 from each side.
$0=k(k-3)$	Factor.
$k=0$ or $k-3=0$	Zero Product Property
$k=3$	Solve.

CHECK

$\sqrt{k+1}=k-1$	Original equation	$\sqrt{k+1}=k-1$	Original equation
$\sqrt{0+1}\stackrel{?}{=}0-1$	$k=0$	$\sqrt{3+1}\stackrel{?}{=}3-1$	$k=3$
$\sqrt{1}\stackrel{?}{=}-1$	Simplify.	$\sqrt{4}\stackrel{?}{=}2$	Simplify.
$1\neq-1$ ✗	False	$2=2$ ✓	True

Since 0 does not satisfy the original equation, 3 is the only solution.

Study Tip

Extraneous Solutions When checking solutions for extraneous solutions, we are only interested in principal roots.

Guided Practice

Solve each equation. Check your solution.

3A. $\sqrt{t+5}=t+3$

3B. $x-3=\sqrt{x-1}$

Check Your Understanding

Example 1

1. **GEOMETRY** The surface area of a basketball is x square inches. What is the radius of the basketball if the formula for the surface area of a sphere is $SA = 4\pi r^2$?

Examples 2–3 **Solve each equation. Check your solution.**

2. $\sqrt{10h} + 1 = 21$
3. $\sqrt{7r + 2} + 3 = 7$
4. $5 + \sqrt{g - 3} = 6$
5. $\sqrt{3x - 5} = x - 5$
6. $\sqrt{2n + 3} = n$
7. $\sqrt{a - 2} + 4 = a$

Practice and Problem Solving

Example 1

8. **EXERCISE** Suppose the function $S = \pi\sqrt{\frac{9.8\ell}{1.6}}$, where S represents speed in meters per second and ℓ is the leg length of a person in meters, can approximate the maximum speed that a person can run.
   a. What is the maximum running speed of a person with a leg length of 1.1 meters to the nearest tenth of a meter?
   b. What is the leg length of a person with a running speed of 6.7 meters per second to the nearest tenth of a meter?
   c. As leg length increases, does maximum speed increase or decrease? Explain.

Examples 2–3 **Solve each equation. Check your solution.**

9. $\sqrt{a} + 11 = 21$
10. $\sqrt{t} - 4 = 7$
11. $\sqrt{n - 3} = 6$
12. $\sqrt{c + 10} = 4$
13. $\sqrt{h - 5} = 2\sqrt{3}$
14. $\sqrt{k + 7} = 3\sqrt{2}$
15. $y = \sqrt{12 - y}$
16. $\sqrt{u + 6} = u$
17. $\sqrt{r + 3} = r - 3$
18. $\sqrt{1 - 2t} = 1 + t$
19. $5\sqrt{a - 3} + 4 = 14$
20. $2\sqrt{x - 11} - 8 = 4$
21. **RIDES** The amount of time t, in seconds, that it takes a simple pendulum to complete a full swing is called the *period*. It is given by $t = 2\pi\sqrt{\frac{\ell}{32}}$, where ℓ is the length of the pendulum, in feet.
   a. The Giant Swing completes a period in about 8 seconds. About how long is the pendulum's arm? Round to the nearest foot.
   b. Does increasing the length of the pendulum increase or decrease the period? Explain.

Solve each equation. Check your solution.

22. $\sqrt{6a - 6} = a + 1$
23. $\sqrt{x^2 + 9x + 15} = x + 5$
24. $6\sqrt{\frac{5k}{4}} - 3 = 0$
25. $\sqrt{\frac{5y}{6}} - 10 = 4$
26. $\sqrt{2a^2 - 121} = a$
27. $\sqrt{5x^2 - 9} = 2x$
28. CCSS **REASONING** The formula for the slant height c of a cone is $c = \sqrt{h^2 + r^2}$, where h is the height of the cone and r is the radius of its base. Find the height of the cone if the slant height is 4 units and the radius is 2 units. Round to the nearest tenth.

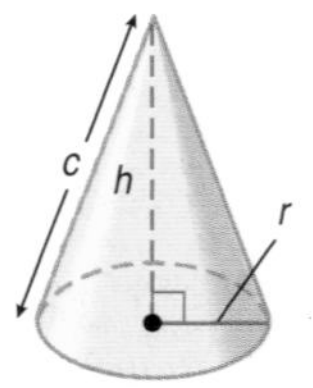

29. **MULTIPLE REPRESENTATIONS** Consider $\sqrt{2x - 7} = x - 7$.

a. **Graphical** Clear the **Y=** list. Enter the left side of the equation as $Y1 = \sqrt{2x - 7}$. Enter the right side of the equation as $Y2 = x - 7$. Press GRAPH.

b. **Graphical** Sketch what is shown on the screen.

c. **Analytical** Use the **intersect** feature on the **CALC** menu to find the point of intersection.

d. **Analytical** Solve the radical equation algebraically. How does your solution compare to the solution from the graph?

30. **PACKAGING** A cylindrical container of chocolate drink mix has a volume of 162 cubic inches. The radius r of the container can be found by using the formula $r = \sqrt{\frac{V}{\pi h}}$, where V is the volume of the container and h is the height.

a. If the radius is 2.5 inches, find the height of the container. Round to the nearest hundredth.

b. If the height of the container is 10 inches, find the radius. Round to the nearest hundredth.

H.O.T. Problems Use Higher-Order Thinking Skills

31. **CCSS CRITIQUE** Jada and Fina solved $\sqrt{6 - b} = \sqrt{b + 10}$. Is either of them correct? Explain.

Jada

$\sqrt{6 - b} = \sqrt{b + 10}$

$(\sqrt{6 - b})^2 = (\sqrt{b + 10})^2$

$6 - b = b + 10$

$-2b = 4$

$b = -2$

Check $\sqrt{6 - (-2)} \stackrel{?}{=} \sqrt{(-2) + 10}$

$\sqrt{8} = \sqrt{8}$ ✓

Fina

$\sqrt{6 - b} = \sqrt{b + 10}$

$(\sqrt{6 - b})^2 = (\sqrt{b + 10})^2$

$6 - b = b + 10$

$2b = 4$

$b = 2$

Check $\sqrt{6 - (2)} \stackrel{?}{=} \sqrt{(2) + 10}$

$\sqrt{4} \neq \sqrt{12}$ ✗

no solution

32. **REASONING** Which equation has the same solution set as $\sqrt{4} = \sqrt{x + 2}$? Explain.

A. $\sqrt{4} = \sqrt{x} + \sqrt{2}$ **B.** $4 = x + 2$ **C.** $2 - \sqrt{2} = \sqrt{x}$

33. **REASONING** Explain how solving $5 = \sqrt{x} + 1$ is different from solving $5 = \sqrt{x + 1}$.

34. **OPEN ENDED** Write a radical equation with a variable on each side. Then solve the equation.

35. **REASONING** Is the following equation *sometimes*, *always* or *never* true? Explain.

$$\sqrt{(x - 2)^2} = x - 2$$

36. **CHALLENGE** Solve $\sqrt{x + 9} = \sqrt{3} + \sqrt{x}$.

37. **WRITING IN MATH** Write some general rules about how to solve radical equations. Demonstrate your rules by solving a radical equation.

Standardized Test Practice

38. SHORT RESPONSE Zack needs to drill a hole at A, B, C, D, and E on circle P.

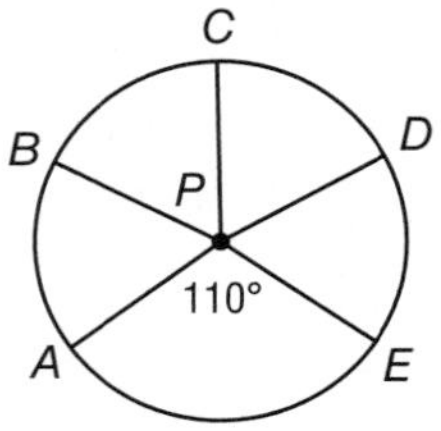

If Zack drills holes so that $m\angle APE = 110°$ and the other four angles are congruent, what is $m\angle CPD$?

39. Which expression is undefined when $w = 3$?

A $\frac{w-3}{w+1}$

B $\frac{w^2-3w}{3w}$

C $\frac{w+1}{w^2-3w}$

D $\frac{3w}{3w^2}$

40. What is the slope of a line that is parallel to the line?

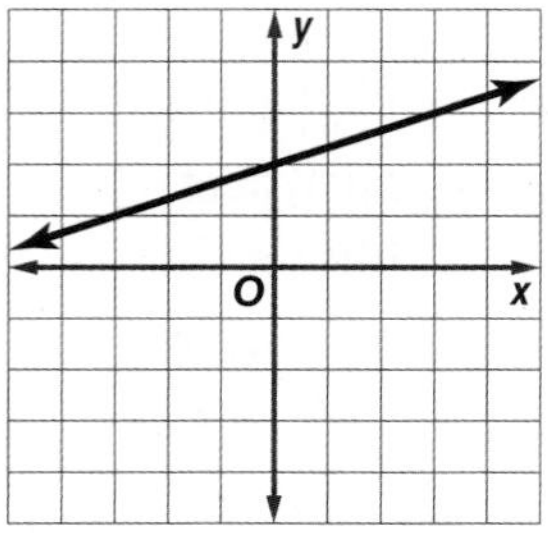

F -3

G $-\frac{1}{3}$

H $\frac{1}{3}$

J 3

41. What are the solutions of $\sqrt{x+3} - 1 = x - 4$?

A 1, 6

B $-1, -6$

C 1

D 6

Spiral Review

42. ELECTRICITY The voltage V required for a circuit is given by $V = \sqrt{PR}$, where P is the power in watts and R is the resistance in ohms. How many more volts are needed to light a 100-watt light bulb than a 75-watt light bulb if the resistance of both is 110 ohms? (Lesson 8-3)

Simplify each expression. (Lesson 8-2)

43. $\sqrt{6} \cdot \sqrt{8}$

44. $\sqrt{3} \cdot \sqrt{6}$

45. $7\sqrt{3} \cdot 2\sqrt{6}$

46. $\sqrt{\frac{27}{a^2}}$

47. $\sqrt{\frac{5c^5}{4d^5}}$

48. $\frac{\sqrt{9x^3y}}{\sqrt{16x^2y^2}}$

Determine whether each expression is a monomial. Write *yes* or *no*. Explain. (Lesson 7-1)

49. 12

50. $4x^3$

51. $a - 2b$

52. $4n + 5p$

53. $\frac{x}{y^2}$

54. $\frac{1}{5}$

Skills Review

Simplify.

55. 9^2

56. 10^6

57. 4^5

58. $(8v)^2$

59. $\left(\frac{w^3}{9}\right)^2$

60. $(10y^2)^3$

CHAPTER 8 Mid-Chapter Quiz

Lessons 8-1 through 8-4

Graph each function. Compare to the parent graph. State the domain and range. (Lesson 8-1)

1. $y = 2\sqrt{x}$
2. $y = -4\sqrt{x}$
3. $y = \frac{1}{2}\sqrt{x}$
4. $y = \sqrt{x} - 3$
5. $y = \sqrt{x - 1}$
6. $y = 2\sqrt{x - 2}$

7. **MULTIPLE CHOICE** The length of the side of a square is given by the function $s = \sqrt{A}$, where A is the area of the square. What is the length of the side of a square that has an area of 121 square inches? (Lesson 8-1)

   A 121 inches **C** 44 inches
   B 11 inches **D** 10 inches

Simplify each expression. (Lesson 8-2)

8. $2\sqrt{25}$
9. $\sqrt{12} \cdot \sqrt{8}$
10. $\sqrt{72xy^5z^6}$
11. $\dfrac{3}{1 + \sqrt{5}}$
12. $\dfrac{1}{5 - \sqrt{7}}$

13. **SATELLITES** A satellite is launched into orbit 200 kilometers above Earth. The orbital velocity of a satellite is given by the formula $v = \sqrt{\frac{Gm_E}{r}}$. v is velocity in meters per second, G is a given constant, m_E is the mass of Earth, and r is the radius of the satellite's orbit in meters. (Lesson 8-2)

    a. The radius of Earth is 6,380,000 meters. What is the radius of the satellite's orbit in meters?

    b. The mass of Earth is 5.97×10^{24} kilograms, and the constant G is $6.67 \times 10^{-11}\ \mathrm{N} \cdot \frac{\mathrm{m}^2}{\mathrm{kg}^2}$ where N is in Newtons. Use the formula to find the orbital velocity of the satellite in meters per second.

14. **MULTIPLE CHOICE** Which expression is equivalent to $\sqrt{\frac{16}{32}}$? (Lesson 8-2)

    F $\frac{1}{2}$

    G $\frac{\sqrt{2}}{2}$

    H 2

    J 4

Simplify each expression. (Lesson 8-3)

15. $3\sqrt{2} + 5\sqrt{2}$
16. $\sqrt{11} - 3\sqrt{11}$
17. $6\sqrt{2} + 4\sqrt{50}$
18. $\sqrt{27} - \sqrt{48}$
19. $4\sqrt{3}(2\sqrt{6})$
20. $3\sqrt{20}(2\sqrt{5})$
21. $(\sqrt{5} + \sqrt{7})(\sqrt{20} + \sqrt{3})$

22. **GEOMETRY** Find the area of the rectangle. (Lesson 8-3)

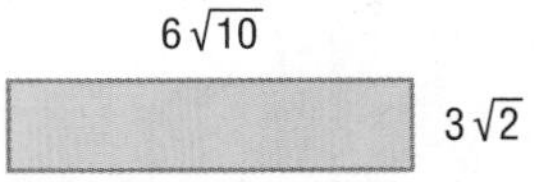

Solve each equation. Check your solution. (Lesson 8-4)

23. $\sqrt{5x} - 1 = 4$
24. $\sqrt{a - 2} = 6$
25. $\sqrt{15 - x} = 4$
26. $\sqrt{3x^2 - 32} = x$
27. $\sqrt{2x - 1} = 2x - 7$
28. $\sqrt{x + 1} + 2 = 4$

29. **GEOMETRY** The lateral surface area S of a cone can be found by using the formula $S = \pi r\sqrt{r^2 + h^2}$, where r is the radius of the base and h is the height of the cone. Find the height of the cone. (Lesson 8-4)

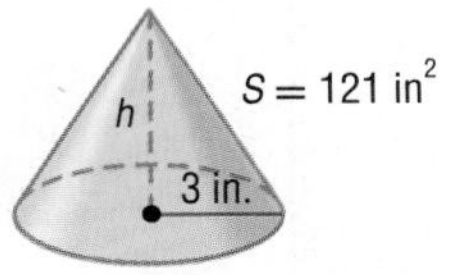

LESSON 8-5

Inverse Variation

Then

- You solved problems involving direct variation.

Now

1. Identify and use inverse variations.
2. Graph inverse variations.

Why?

- The time it takes a runner to finish a race is inversely proportional to the average pace of the runner. The runner's time decreases as the pace of the runner increases. So, these quantities are *inversely proportional.*

NewVocabulary

inverse variation
product rule

Common Core State Standards

Mathematical Practices

1 Make sense of problems and persevere in solving them.

1 Identify and Use Inverse Variations

An **inverse variation** can be represented by the equation $y = \frac{k}{x}$ or $xy = k$.

KeyConcept Inverse Variation

y varies inversely as x if there is some nonzero constant k such that $y = \frac{k}{x}$ or $xy = k$, where $x, y \neq 0$.

In an inverse variation, the product of two values remains constant. Recall that a relationship of the form $y = kx$ is a *direct variation*. The constant k is called the *constant of variation* or the *constant of proportionality*.

Example 1 Identify Inverse and Direct Variations

PT

Determine whether each table or equation represents an *inverse* or a *direct variation*. Explain.

a.

x	y
1	16
2	8
4	4

In an inverse variation, xy equals a constant k. Find xy for each ordered pair in the table.

$1 \cdot 16 = 16 \quad 2 \cdot 8 = 16 \quad 4 \cdot 4 = 16$

The product is constant, so the table represents an inverse variation.

b.

x	y
1	3
2	6
3	9

Notice that xy is not constant. So, the table does not represent an indirect variation.

$3 = k(1) \qquad 6 = k(2) \qquad 9 = k(3)$

$3 = k \qquad 3 = k \qquad 3 = k$

The table of values represents the direct variation $y = 3x$.

c. $x = 2y$

The equation can be written as $y = \frac{1}{2}x$. Therefore, it represents a direct variation.

d. $2xy = 10$

$2xy = 10$ Write the equation.

$xy = 5$ Divide each side by 2.

The equation represents an inverse variation.

GuidedPractice

1A.

x	1	2	5
y	10	5	2

1B. $-2x = y$

Stockbyte/Getty Images

You can use $xy = k$ to write an inverse variation equation that relates x and y.

ReadingMath

Variation Equations For direct variation equations, you say that *y varies directly* as *x*. For inverse variation equations, you say that *y varies inversely* as *x*.

PT

Example 2 Write an Inverse Variation

Assume that y varies inversely as x. If $y = 18$ when $x = 2$, write an inverse variation equation that relates x and y.

$xy = k$	Inverse variation equation
$2(18) = k$	$x = 2$ and $y = 18$
$36 = k$	Simplify.

The constant of variation is 36. So, an equation that relates x and y is $xy = 36$ or $y = \frac{36}{x}$.

GuidedPractice

2. Assume that y varies inversely as x. If $y = 5$ when $x = -4$, write an inverse variation equation that relates x and y.

If (x_1, y_1) and (x_2, y_2) are solutions of an inverse variation, then $x_1y_1 = k$ and $x_2y_2 = k$.

$x_1y_1 = k$ and $x_2y_2 = k$	
$x_1y_1 = x_2y_2$	Substitute x_2y_2 for k.

The equation $x_1y_1 = x_2y_2$ is called the **product rule** for inverse variations.

KeyConcept Product Rule for Inverse Variations

Words	If (x_1, y_1) and (x_2, y_2) are solutions of an inverse variation, then the products x_1y_1 and x_2y_2 are equal.
Symbols	$x_1y_1 = x_2y_2$ or $\frac{x_1}{x_2} = \frac{y_2}{y_1}$

PT

Example 3 Solve for *x* or *y*

Assume that y varies inversely as x. If $y = 3$ when $x = 12$, find x when $y = 4$.

$x_1y_1 = x_2y_2$	Product rule for inverse variations
$12 \cdot 3 = x_2 \cdot 4$	$x_1 = 12$, $y_1 = 3$, and $y_2 = 4$
$36 = x_2 \cdot 4$	Simplify.
$\frac{36}{4} = x_2$	Divide each side by 4.
$9 = x_2$	Simplify.

So, when $y = 4$, $x = 9$.

GuidedPractice

3. If y varies inversely as x and $y = 4$ when $x = -8$, find y when $x = -4$.

The product rule for inverse variations can be used to write an equation to solve real-world problems.

Real-WorldLink

A standard hockey puck is 1 inch thick and 3 inches in diameter. Its mass is between approximately 156 and 170 grams.

Source: *NHL Rulebook*

Real-World Example 4 Use Inverse Variations

PHYSICS The acceleration a of a hockey puck is inversely proportional to its mass m. Suppose a hockey puck with a mass of 164 grams is hit so that it accelerates 122 m/s^2. Find the acceleration of a 158-gram hockey puck if the same amount of force is applied.

Make a table to organize the information.

Let $m_1 = 164$, $a_1 = 122$, and $m_2 = 164$. Solve for a_2.

Puck	Mass	Acceleration
1	164 g	122 m/s^2
2	158 g	a_2

$m_1a_1 = m_2a_2$ — Use the product rule to write an equation.

$164 \cdot 122 = 158a_2$ — $m_1 = 164$, $a_1 = 122$, and $m_2 = 158$

$20{,}008 = 158a_2$ — Simplify.

$126.6 \approx a_2$ — Divide each side by 158 and simplify.

The 158-gram puck has an acceleration of approximately 126.6 m/s^2.

GuidedPractice

4. RACING Manuel runs an average of 8 miles per hour and finishes a race in 0.39 hour. Dyani finished the race in 0.35 hour. What was her average pace?

2 Graph Inverse Variations

The graph of an inverse variation is not a straight line like the graph of a direct variation.

Problem-SolvingTip

CCSS Sense-Making

Sometimes it is necessary to break a problem into parts, solve each part, and then combine them to find the solution to the problem.

Example 5 Graph an Inverse Variation

Graph an inverse variation equation in which $y = 8$ when $x = 3$.

Step 1 Write an inverse variation equation.

$xy = k$ — Inverse variation equation

$3(8) = k$ — $x = 3, y = 8$

$24 = k$ — Simplify.

The inverse variation equation is $xy = 24$ or $y = \frac{24}{x}$.

Step 2 Choose values for x and y that have a product of 24.

Step 3 Plot each point and draw a smooth curve that connects the points.

x	y
−12	−2
−8	−3
−4	−6
−2	−12
0	undefined
2	12
3	8
6	4
12	2

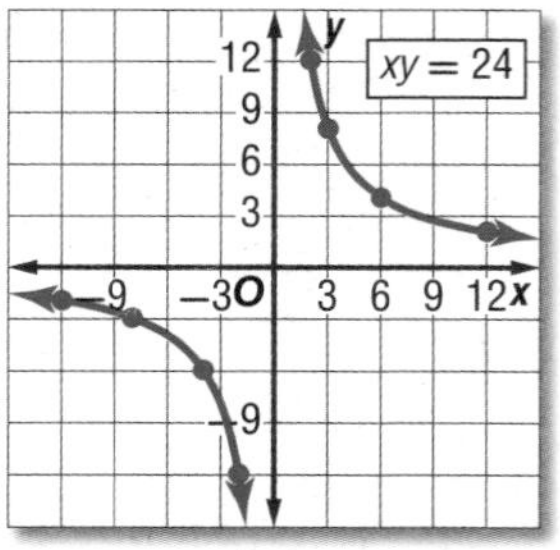

Notice that since y is undefined when $x = 0$, there is no point on the graph when $x = 0$. This graph is called a hyperbola.

GuidedPractice

5. Graph an inverse variation equation in which $y = 16$ when $x = 4$.

ConceptSummary Direct and Inverse Variations

Direct Variation	Inverse Variation
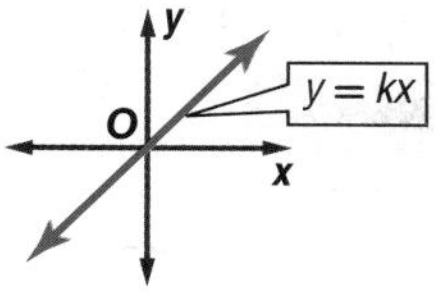 $k>0$ 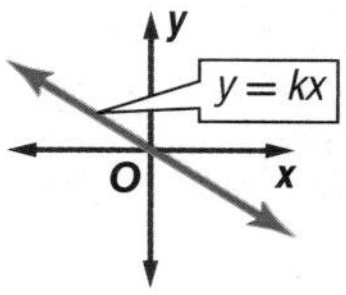 $k<0$	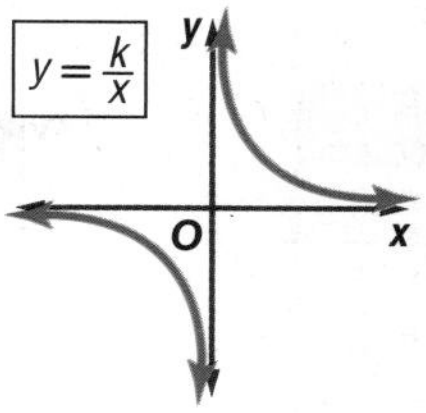 $k>0$ 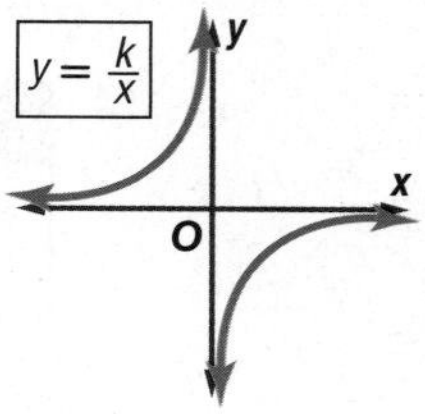 $k<0$
• $y = kx$ • y varies directly as x. • The ratio $\frac{y}{x}$ is a constant.	• $y = \frac{k}{x}$ • y varies inversely as x. • The product xy is a constant.

Check Your Understanding

Example 1 **Determine whether each table or equation represents an *inverse* or a *direct* variation. Explain.**

1.

x	1	4	8	12
y	2	8	16	24

2.

x	1	2	3	4
y	24	12	8	6

3. $xy = 4$

4. $y = \frac{x}{10}$

Examples 2, 5 **Assume that y varies inversely as x. Write an inverse variation equation that relates x and y. Then graph the equation.**

5. $y = 8$ when $x = 6$

6. $y = 2$ when $x = 5$

7. $y = 3$ when $x = -10$

8. $y = -1$ when $x = -12$

Example 3 **Solve. Assume that y varies inversely as x.**

9. If $y = 8$ when $x = 4$, find x when $y = 2$.

10. If $y = 7$ when $x = 6$, find y when $x = -21$.

11. If $y = -5$ when $x = 9$, find y when $x = 6$.

Example 4

12. **RACING** The time it takes to complete a go-cart race course is inversely proportional to the average speed of the go-cart. One rider has an average speed of 73.3 feet per second and completes the course in 30 seconds. Another rider completes the course in 25 seconds. What was the average speed of the second rider?

13. **OPTOMETRY** When a person does not have clear vision, an optometrist can prescribe lenses to correct the condition. The power P of a lens, in a unit called diopters, is equal to 1 divided by the focal length f, in meters, of the lens.
    a. Graph the inverse variation $P = \frac{1}{f}$.
    b. Find the powers of lenses with focal lengths $+0.2$ to -0.4 meters.

Practice and Problem Solving

Example 1 **Determine whether each table or equation represents an *inverse* or a *direct* variation. Explain.**

14.

x	y
1	30
2	15
5	6
6	5

15.

x	y
2	−6
3	−9
4	−12
5	−15

16.

x	y
−4	−2
−2	−1
2	1
4	2

17.

x	y
−5	8
−2	20
4	−10
8	−5

18. $5x - y = 0$ **19.** $xy = \frac{1}{4}$ **20.** $x = 14y$ **21.** $\frac{y}{x} = 9$

Examples 2, 5 **Assume that y varies inversely as x. Write an inverse variation equation that relates x and y. Then graph the equation.**

22. $y = 2$ when $x = 20$ **23.** $y = 18$ when $x = 4$ **24.** $y = -6$ when $x = -3$

25. $y = -4$ when $x = -3$ **26.** $y = -4$ when $x = 16$ **27.** $y = 12$ when $x = -9$

Example 3 **Solve. Assume that y varies inversely as x.**

28. If $y = 12$ when $x = 3$, find x when $y = 6$.

29. If $y = 5$ when $x = 6$, find x when $y = 2$.

30. If $y = 4$ when $x = 14$, find x when $y = -5$.

31. If $y = 9$ when $x = 9$, find y when $x = -27$.

32. If $y = 15$ when $x = -2$, find y when $x = 3$.

33. If $y = -8$ when $x = -12$, find y when $x = 10$.

Example 4 **34.** **EARTH SCIENCE** The water level in a river varies inversely with air temperature. When the air temperature was 90° Fahrenheit, the water level was 11 feet. If the air temperature was 110° Fahrenheit, what was the level of water in the river?

35 **MUSIC** When under equal tension, the frequency of a vibrating string in a piano varies inversely with the string length. If a string that is 420 millimeters in length vibrates at a frequency of 523 cycles a second, at what frequency will a 707-millimeter string vibrate?

Determine whether each situation is an example of an *inverse* or a *direct* variation. Justify your reasoning.

36. The drama club can afford to purchase 10 wigs at \$2 each or 5 wigs at \$4 each.

37. The Spring family buys several lemonades for \$1.50 each.

38. Nicole earns \$14 for babysitting 2 hours, and \$21 for babysitting 3 hours.

39. Thirty video game tokens are divided evenly among a group of friends.

Determine whether each table or graph represents an *inverse* or a *direct* variation. Explain.

40.

x	y
5	1
8	1.6
11	2.2

41.

x	y
−3	−7
−2	−10.5
4	5.25

42.

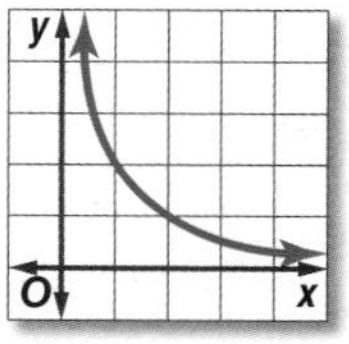

43.

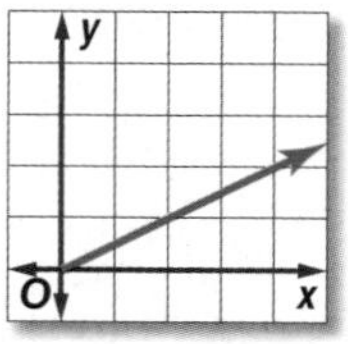

44. **PHYSICAL SCIENCE** When two people are balanced on a seesaw, their distances from the center of the seesaw are inversely proportional to their weights. If a 118-pound person sits 1.8 meters from the center of the seesaw, how far should a 125-pound person sit from the center to balance the seesaw?

Solve. Assume that y varies inversely as x.

45. If $y = 9.2$ when $x = 6$, find x when $y = 3$.

46. If $y = 3.8$ when $x = 1.5$, find x when $y = 0.3$.

47. If $y = \frac{1}{5}$ when $x = -20$, find y when $x = -\frac{8}{5}$.

48. If $y = -6.3$ when $x = \frac{2}{3}$, find y when $x = 8$.

49. **SWIMMING** Logan and Brianna each bought a pool membership. Their average cost per day is inversely proportional to the number of days that they go to the pool. Logan went to the pool 25 days for an average cost per day of \$5.60. Brianna went to the pool 35 days. What was her average cost per day?

50. **PHYSICAL SCIENCE** The amount of force required to do a certain amount of work in moving an object is inversely proportional to the distance that the object is moved. Suppose 90 N of force is required to move an object 10 feet. Find the force needed to move another object 15 feet if the same amount of work is done.

51. **DRIVING** Lina must practice driving 40 hours with a parent or guardian before she is allowed to take the test to get her driver's license. She plans to practice the same number of hours each week.

a. Let h represent the number of hours per week that she practices driving. Make a table showing the number of weeks w that she will need to practice for the following values of h: 1, 2, 4, 5, 8, and 10.

b. Describe how the number of weeks changes as the number of hours per week increases.

c. Write and graph an equation that shows the relationship between h and w.

H.O.T. Problems Use Higher-Order Thinking Skills

52. **CCSS CRITIQUE** Christian and Trevor found an equation such that x and y vary inversely, and $y = 10$ when $x = 5$. Is either of them correct? Explain.

Christian

$k = \frac{y}{x}$

$= \frac{10}{2}$ or 5

$y = 5x$

Trevor

$k = xy$

$= (5)(10)$ or 50

$y = \frac{50}{x}$

53. **CHALLENGE** Suppose f varies inversely with g, and g varies inversely with h. What is the relationship between f and h?

54. **REASONING** Does $xy = -k$ represent an inverse variation when $k \neq 0$? Explain.

55. **OPEN ENDED** Give a real-world situation or phenomena that can be modeled by an inverse variation equation. Use the correct terminology to describe your example and explain why this situation is an inverse variation.

56. **WRITING IN MATH** Compare and contrast direct and inverse variation. Include a description of the relationship between slope and the graphs of a direct and inverse variation.

Standardized Test Practice

57. Given a constant force, the acceleration of an object varies inversely with its mass. Assume that a constant force is acting on an object with a mass of 6 pounds resulting in an acceleration of 10 ft/s^2. The same force acts on another object with a mass of 12 pounds. What would be the resulting acceleration?

A 4 ft/s^2 **C** 6 ft/s^2

B 5 ft/s^2 **D** 7 ft/s^2

58. Fiona had an average of 56% on her first seven tests. What would she have to make on her eighth test to average 60% on 8 tests?

F 82% **H** 98%

G 88% **J** 100%

59. Anthony takes a picture of a 1-meter snake beside a brick wall. When he develops the pictures, the 1-meter snake is 2 centimeters long and the wall is 4.5 centimeters high. What was the actual height of the brick wall?

A 2.25 cm

B 22.5 cm

C 225 cm

D 2250 cm

60. SHORT RESPONSE Find the area of the rectangle.

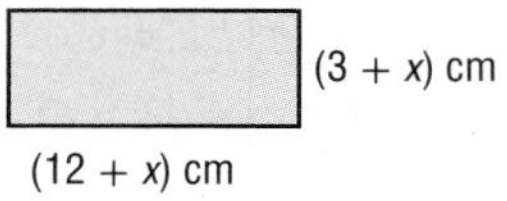

Spiral Review

61. TESTS Determine whether the graph at the right shows a *positive*, *negative*, or *no* correlation. If there is a correlation, describe its meaning. (Lesson 4-5)

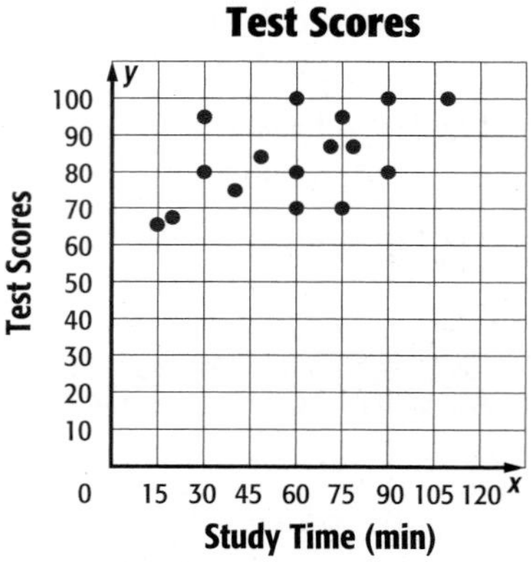

Suppose y varies directly as x. (Lesson 3-4)

62. If $y = 2.5$ when $x = 0.5$, find y when $x = 20$.

63. If $y = -6.6$ when $x = 9.9$, find y when $x = 6.6$.

64. If $y = 2.6$ when $x = 0.25$, find y when $x = 1.125$.

65. If $y = 6$ when $x = 0.6$, find x when $y = 12$.

66. FINANCIAL LITERACY A salesperson is paid $32,000 a year plus 5% of the amount in sales made. What is the amount of sales needed to have an annual income greater than $45,000? (Lesson 5-3)

Skills Review

Simplify. Assume that no denominator is equal to zero.

67. $\dfrac{7^8}{7^6}$

68. $\dfrac{x^8y^{12}}{x^2y^7}$

69. $\dfrac{5pq^7}{10p^6q^3}$

70. $\left(\dfrac{2c^3d}{7z^2}\right)^3$

71. $\left(\dfrac{4a^2b}{2c^3}\right)^2$

72. $y^0(y^5)(y^{-9})$

73. $\dfrac{(4m^{-3}n^5)^0}{mn}$

74. $\dfrac{(3x^2y^5)^0}{(21x^5y^2)^0}$

EXPLORE

8-6 Graphing Technology Lab
Family of Rational Functions

You can use a graphing calculator to analyze how changing the parameters a and b in $y = \frac{a}{x - b} + c$ affects the graphs in the family of rational functions.

Activity Change Parameters

Graph each set of equations on the same screen in the standard viewing window. Describe any similarities and differences among the graphs.

a. $y = \frac{1}{x}, y = \frac{1}{x} + 2, y = \frac{1}{x} - 4$

Enter the equations in the Y= list and graph in the standard viewing window.

The graphs have the same shape. Each graph approaches the y-axis on both sides. However, the graphs have different vertical positions.

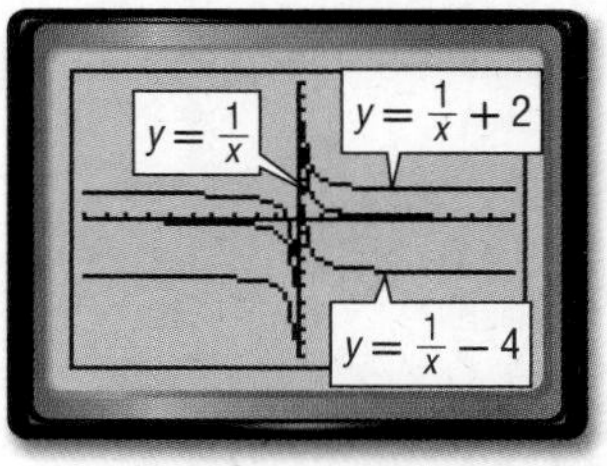

[−10, 10] scl: 1 by [−10, 10] scl: 1

b. $y = \frac{1}{x}, y = \frac{1}{x + 2}, y = \frac{1}{x - 4}$

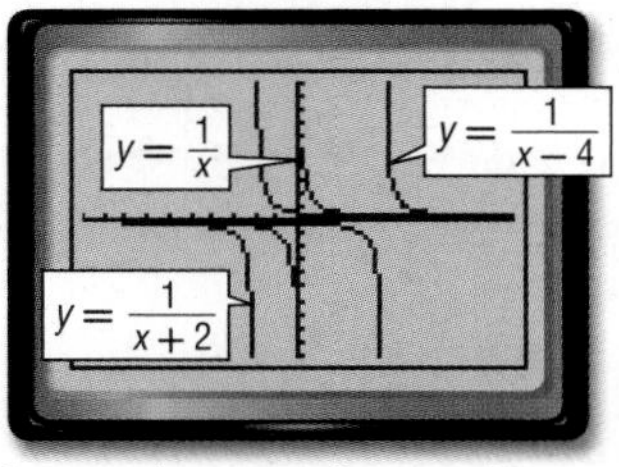

[−10, 10] scl: 1 by [−10, 10] scl: 1

The graphs have the same shape, and all approach the x-axis from both sides. However, the graphs have different horizontal positions.

c. $y = \frac{1}{x}, y = \frac{2}{x}, y = \frac{4}{x}$

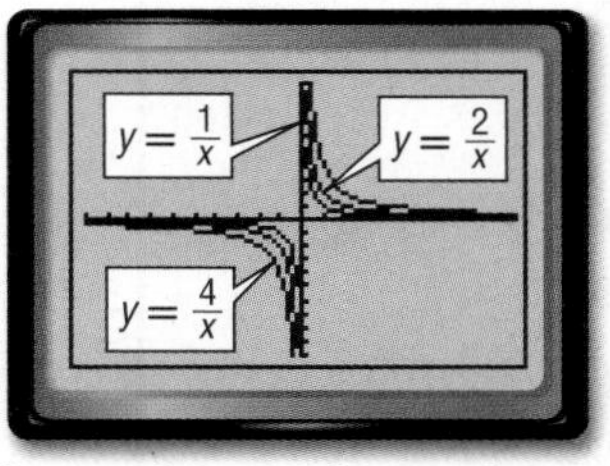

[−10, 10] scl: 1 by [−10, 10] scl: 1

The graphs all approach the x-axis and the y-axis from both sides. However, the graphs have different shapes.

Model and Analyze

1. How do a, b, and c affect the graph of $y = \frac{a}{x - b} + c$? Give examples.

Examine each pair of equations and predict the similarities and differences in their graphs. Use a graphing calculator to confirm your predictions. Write a sentence or two comparing the two graphs.

2. $y = \frac{1}{x}, y = \frac{1}{x} + 2$

3. $y = \frac{1}{x}, y = \frac{1}{x + 5}$

4. $y = \frac{1}{x}, y = \frac{3}{x}$

LESSON 8-6

Rational Functions

:: Then	:: Now	:: Why?
• You wrote inverse variation equations.	• **1** Identify excluded values. **2** Identify and use asymptotes to graph rational functions.	• Trina is reading a 300-page book. The average number of pages she reads each day y is given by $y = \frac{300}{x}$, where x is the number of days that she reads.

NewVocabulary
rational function
excluded value
asymptote

Common Core State Standards

Content Standards
A.CED.2 Create equations in two or more variables to represent relationships between quantities; graph equations on coordinate axes with labels and scales.

Mathematical Practices
3 Construct viable arguments and critique the reasoning of others.
7 Look for and make use of structure.

1 Identify Excluded Values

The function $y = \frac{300}{x}$ is an example of a **rational function**. This function is nonlinear.

KeyConcept Rational Functions

Words A rational function can be described by an equation of the form $y = \frac{p}{q}$, where p and q are polynomials and $q \neq 0$.

Parent function:	$f(x) = \frac{1}{x}$
Type of graph:	hyperbola
Domain:	$\{x \mid x \neq 0\}$
Range:	$\{y \mid y \neq 0\}$

Graph

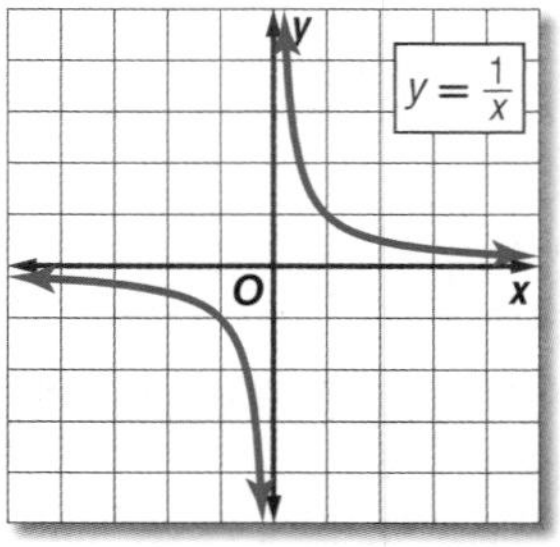

Since division by zero is undefined, any value of a variable that results in a denominator of zero in a rational function is excluded from the domain of the function. These are called **excluded values** for the rational function.

Example 1 Find Excluded Values

PT

State the excluded value for each function.

a. $y = -\frac{2}{x}$

The denominator cannot equal 0. So, the excluded value is $x = 0$.

b. $y = \frac{2}{x+1}$

$x + 1 = 0$ Set the denominator equal to 0.

$x = -1$

The excluded value is $x = -1$.

c. $y = \frac{5}{4x - 8}$

$4x - 8 = 0$

$4x = 8$

$x = 2$

The excluded value is $x = 2$.

GuidedPractice

1A. $y = \frac{5}{2x}$ **1B.** $y = \frac{x}{x-7}$ **1C.** $y = \frac{4}{3x+9}$

Depending on the real-world situation, in addition to excluding x-values that make a denominator zero from the domain of a rational function, additional values might have to be excluded from the domain as well.

Real-World Example 2 Graph Real-Life Rational Functions

BALLOONS If there are x people in the basket of a hot air balloon, the function $y = \frac{20}{x}$ represents the number of square feet y per person. Graph this function.

Since the number of people cannot be zero or less, it is reasonable to exclude negative values and only use positive values for x.

Number of People x	2	4	5	10
Square Feet per Person y	10	5	4	2

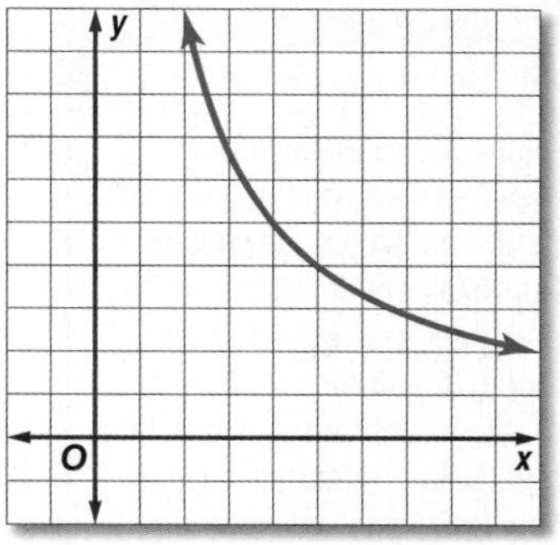

Notice that as x increases y approaches 0. This is reasonable since as the number of people increases, the space per person gets closer to 0.

Real-WorldLink

As the temperature of the gas inside a hot air balloon increases, the density of the gas decreases. A hot air balloon rises because the density of the air inside it is less than the density of the air outside.

Source: Goddard Space Flight Center

GuidedPractice

2. GEOMETRY A rectangle has an area of 18 square inches. The function $\ell = \frac{18}{w}$ shows the relationship between the length and width. Graph the function.

2 Identify and Use Asymptotes

In Example 2, an excluded value is $x = 0$. Notice that the graph approaches the vertical line $x = 0$, but never touches it.

The graph also approaches but never touches the horizontal line $y = 0$. The lines $x = 0$ and $y = 0$ are called *asymptotes*. An **asymptote** is a line that the graph of a function approaches.

StudyTip

Use Asymptotes Asymptotes are helpful for graphing rational functions. However, they are not part of the graph.

KeyConcept Asymptotes

Words A rational function in the form $y = \frac{a}{x - b} + c$, $a \neq 0$, has a vertical asymptote at the x-value that makes the denominator equal zero, $x = b$. It has a horizontal asymptote at $y = c$.

Model

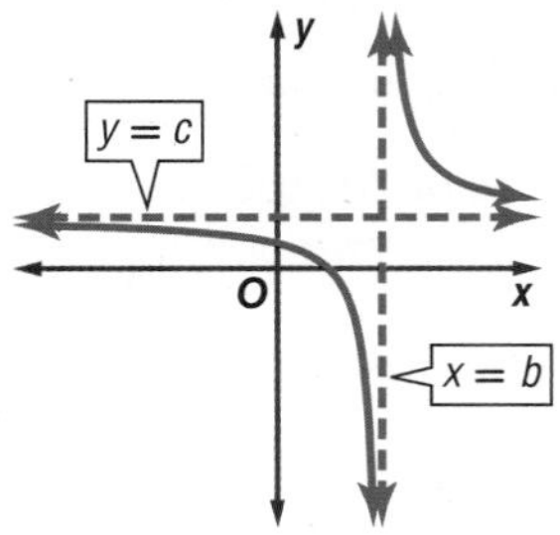

Example

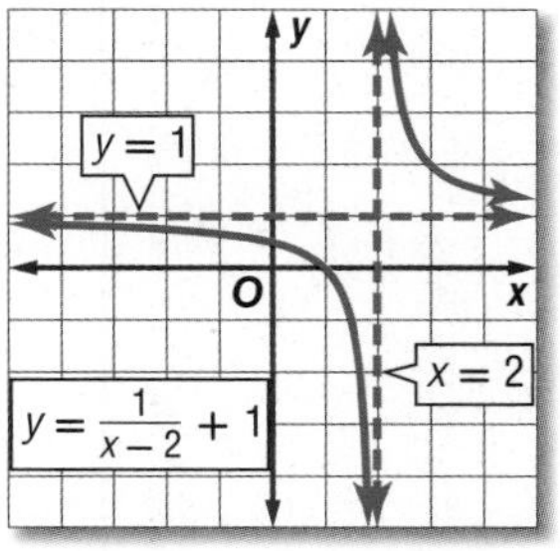

The domain of $y = \frac{a}{x - b} + c$ is all real numbers except $x = b$. The range is all real numbers except $y = c$. Rational functions cannot be traced with a pencil that never leaves the paper, so choose x-values on both sides of the vertical asymptote to graph both portions of the function.

Math HistoryLink

Evelyn Boyd Granville

(1924–) Granville majored in mathematics and physics at Smith College in 1945, where she graduated summa cum laude. She earned an M.A. in mathematics and physics and a Ph.D. in mathematics from Yale University. Granville's doctoral work focused on functional analysis.

Example 3 Identify and Use Asymptotes to Graph Functions

Identify the asymptotes of each function. Then graph the function.

a. $y = \frac{2}{x} - 4$

Step 1 Identify and graph the asymptotes using dashed lines.

vertical asymptote: $x = 0$

horizontal asymptote: $y = -4$

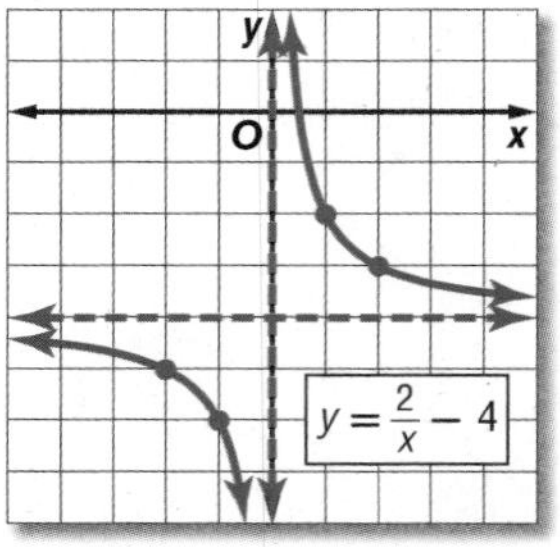

Step 2 Make a table of values and plot the points. Then connect them.

x	−2	−1	1	2
y	−5	−6	−2	−3

b. $y = \frac{1}{x+1}$

Step 1 To find the vertical asymptote, find the excluded value.

$x + 1 = 0$ Set the denominator equal to 0.

$x = -1$ Subtract 1 from each side.

vertical asymptote: $x = -1$

horizontal asymptote: $y = 0$

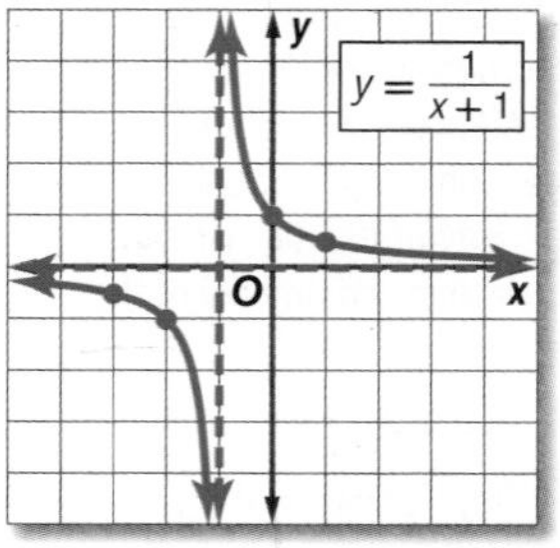

Step 2

x	−3	−2	0	1
y	−0.5	−1	1	0.5

GuidedPractice

3A. $y = -\frac{6}{x}$ **3B.** $y = \frac{1}{x-3}$ **3C.** $y = \frac{2}{x+2} + 1$

Four types of nonlinear functions are shown below.

ConceptSummary Families of Functions

Quadratic	Exponential	Radical	Rational
Parent function: $y = x^2$	Parent function: varies	Parent function: $y = \sqrt{x}$	Parent function: $y = \frac{1}{x}$
General form: $y = ax^2 + bx + c$	General form: $y = ab^x$	General form: $y = \sqrt{x - b} + c$	General form: $y = \frac{a}{x - b} + c$
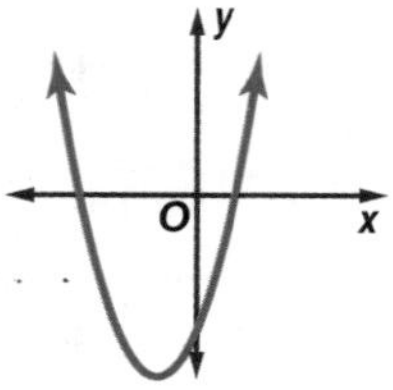	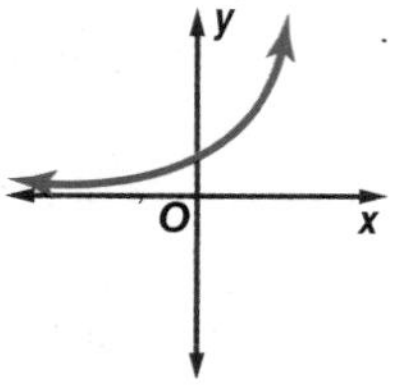	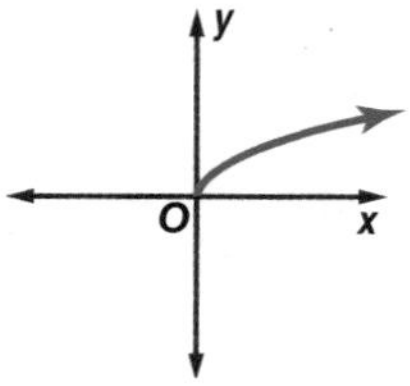	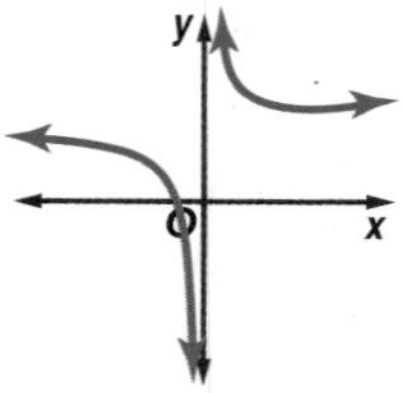

Check Your Understanding

Example 1 **State the excluded value for each function.**

1. $y = \frac{5}{x}$ **2.** $y = \frac{1}{x+3}$ **3.** $y = \frac{x+2}{x-1}$ **4.** $y = \frac{x}{2x-8}$

Example 2 **5. PARTY PLANNING** The cost of decorations for a party is $32. This is split among a group of friends. The amount each person pays y is given by $y = \frac{32}{x}$, where x is the number of people. Graph the function.

Example 3 **Identify the asymptotes of each function. Then graph the function.**

6. $y = \frac{2}{x}$ **7.** $y = \frac{3}{x} - 1$ **8.** $y = \frac{1}{x-2}$

9. $y = \frac{-4}{x+2}$ **10.** $y = \frac{3}{x-1} + 2$ **11.** $y = \frac{2}{x+1} - 5$

Practice and Problem Solving

Example 1 **State the excluded value for each function.**

12. $y = \frac{-1}{x}$ **13.** $y = \frac{8}{x-8}$ **14.** $y = \frac{x}{x+2}$ **15.** $y = \frac{4}{x+6}$

16. $y = \frac{x+1}{x-3}$ **17.** $y = \frac{2x+5}{x+5}$ **18.** $y = \frac{7}{5x-10}$ **19.** $y = \frac{x}{2x+14}$

Example 2 **20. ANTELOPES** A pronghorn antelope can run 40 miles without stopping. The average speed is given by $y = \frac{40}{x}$, where x is the time it takes to run the distance.

a. Graph $y = \frac{40}{x}$.

b. Describe the asymptotes.

21. CYCLING A cyclist rides 10 miles each morning. Her average speed y is given by $y = \frac{10}{x}$, where x is the time it takes her to ride 10 miles. Graph the function.

Example 3 **Identify the asymptotes of each function. Then graph the function.**

22. $y = \frac{5}{x}$ **23.** $y = \frac{-3}{x}$ **24.** $y = \frac{2}{x} + 3$

25. $y = \frac{1}{x} - 2$ **26.** $y = \frac{1}{x+3}$ **27.** $y = \frac{1}{x-2}$

28. $y = \frac{-2}{x+1}$ **29.** $y = \frac{4}{x-1}$ **30.** $y = \frac{1}{x-2} + 1$

31. $y = \frac{3}{x-1} - 2$ **32.** $y = \frac{2}{x+1} - 4$ **33.** $y = \frac{-1}{x+4} + 3$

34. READING Refer to the application at the beginning of the lesson.

a. Graph the function. Interpret key features of the graph in terms of the situation.

b. Choose a point on the graph, and describe what it means in the context of the situation.

35. CCSS STRUCTURE The graph shows a translation of the graph of $y = \frac{1}{x}$.

a. Describe the asymptotes.

b. Write a possible function for the graph.

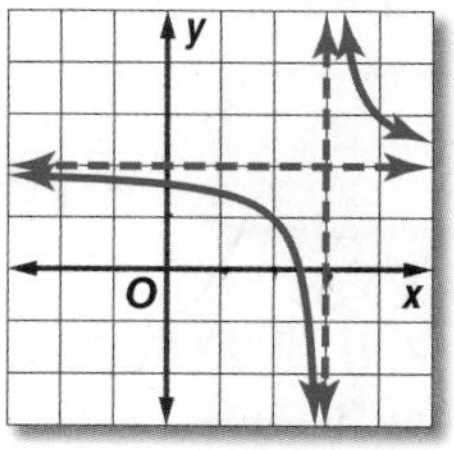

36. BIRDS A long-tailed jaeger is a sea bird that can migrate 5000 miles or more each year. The average rate in miles per hour r can be given by the function $r = \frac{5000}{t}$, where t is the time in hours. Use the function to determine the average rate of the bird if it spends 250 hours flying.

37. CLASS TRIP The freshmen class is going to a science museum. As part of the trip, each person is also contributing an equal amount of money to name a star.

Welcome to
The Museum
Admission $8.50
As a special memory
of your visit,
name a star for $95

a. Write a verbal description for the cost per person.

b. Write an equation to represent the total cost y per person if p people go to the museum.

c. Use a graphing calculator to graph the equation. Interpret key features of the graph in terms of the situation.

d. Estimate the number of people needed for the total cost of the trip to be about $15.

Graph each function. Identify the asymptotes.

38. $y = \frac{4x + 3}{2x - 4}$

39. $y = \frac{x^2}{x^2 - 1}$

40. $y = \frac{x}{x^2 - 9}$

41. GEOMETRY The equation $h = \frac{2(64)}{b_1 + 8}$ represents the height h of a trapezoid with an area of 64 square units. The trapezoid has two opposite sides that are parallel and h units apart; one is b_1 units long and another is 8 units long.

a. Describe a reasonable domain and range for the function.

b. Graph the function in the first quadrant.

c. Use the graph to estimate the value of h when $b_1 = 10$.

H.O.T. Problems Use Higher-Order Thinking Skills

42. CHALLENGE Graph $y = \frac{1}{x^2 - 4}$. State the domain and the range of the function.

43. REASONING Without graphing, describe the transformation that takes place between the graph of $y = \frac{1}{x}$ and the graph of $y = \frac{1}{x + 5} - 2$.

44. OPEN ENDED Write a rational function if the asymptotes of the graph are at $x = 3$ and $y = 1$. Explain how you found the function.

45. CCSS ARGUMENTS Is the following statement *true* or *false*? If false, give a counterexample.

The graph of a rational function will have at least one intercept.

46. WHICH ONE DOESN'T BELONG Identify the function that does not belong with the other three. Explain your reasoning.

$y = \frac{4}{x}$	$y = \frac{6}{x + 1}$	$y = \frac{8}{x} + 1$	$y = \frac{10}{2x}$

47. WRITING IN MATH How are the properties of a rational function reflected in its graph?

Standardized Test Practice

48. Simplify $\frac{2a^2d}{3bc} \cdot \frac{9b^2c}{16ad^2}$.

A $\frac{abd}{c}$

B $\frac{ab}{d}$

C $\frac{6a}{4bd}$

D $\frac{3ab}{8d}$

49. SHORT RESPONSE One day Lola ran 100 meters in 15 seconds, 200 meters in 45 seconds, and 300 meters over low hurdles in one and a half minutes. How many more seconds did it take her to run 300 meters over low hurdles than the 200-meter dash?

50. Scott and Ian started a T-shirt printing business. The total start-up costs were $450. It costs $5.50 to print one T-shirt. Write a rational function $A(x)$ for the average cost of producing x T-shirts.

F $A(x) = \frac{450 + 5.5x}{x}$

G $A(x) = \frac{450}{x} + 5.5$

H $A(x) = 450x + 5.5$

J $A(x) = 450 + 5.5x$

51. GEOMETRY Which of the following is a quadrilateral with exactly one pair of parallel sides?

A parallelogram

B rectangle

C square

D trapezoid

Spiral Review

52. TRAVEL The Brooks family can drive to the beach, which is 220 miles away, in 4 hours if they drive 55 miles per hour. Kendra says that they would save at least a half an hour if they were to drive 65 miles per hour. Is Kendra correct? Explain. (Lesson 8-5)

53. SIGHT The formula $d = \sqrt{\frac{3h}{2}}$ represents the distance d in miles that a person h feet high can see. Irene is standing on a cliff that is 310 feet above sea level. How far can Irene see from the cliff? Write a simplified radical expression and a decimal approximation. (Lesson 8-2)

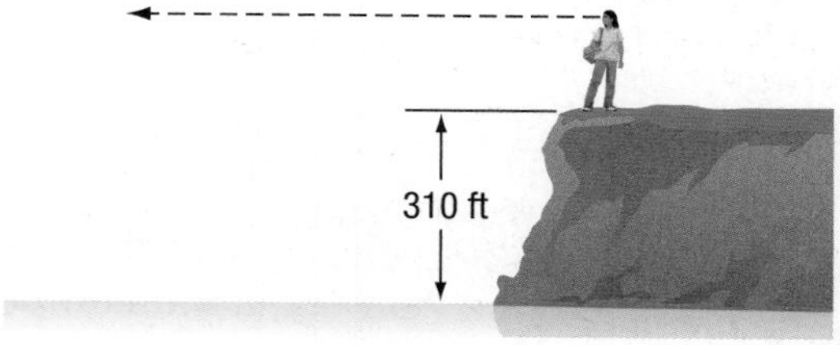

Skills Review

Factor each trinomial.

54. $x^2 + 11x + 24$

55. $w^2 + 13w - 48$

56. $p^2 - 2p - 35$

57. $72 + 27a + a^2$

58. $c^2 + 12c + 35$

59. $d^2 - 7d + 10$

60. $g^2 - 19g + 60$

61. $n^2 + 3n - 54$

62. $5x^2 + 27x + 10$

63. $24b^2 - 14b - 3$

64. $12a^2 - 13a - 35$

65. $6x^2 - 14x - 12$

LESSON 8-7

Rational Equations

Then

- You solved proportions.

Now

1. Solve rational equations.
2. Use rational equations to solve problems.

Why?

- Oceanic species of dolphins can swim 5 miles per hour faster than coastal species of dolphins. An oceanic dolphin can swim 3 miles in the same time that it takes a coastal dolphin to swim 2 miles.

Dolphins

Species	Distance	Rate	Time
coastal	2 miles	x mph	t hours
oceanic	3 miles	$x + 5$ mph	t hours

Since time $= \frac{\text{distance}}{\text{rate}}$, the equation below represents this situation.

Time an oceanic dolphin swims 3 miles equals time a coastal dolphin swims 2 miles.

$$\frac{3}{x+5} = \frac{2}{x}$$

(distance → 3, rate → $x + 5$; 2 ← distance, x ← rate)

NewVocabulary

rational equation
extraneous solution
work problem
rate problem

Common Core State Standards

Content Standards

A.CED.2 Create equations in two or more variables to represent relationships between quantities; graph equations on coordinate axes with labels and scales.

Mathematical Practices

2 Reason abstractly and quantitatively.
4 Model with mathematics.

1 Solve Rational Equations

A **rational equation** contains one or more rational expressions. When a rational equation is a proportion, you can use cross products to solve it.

PT

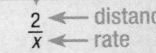

Real-World Example 1 Use Cross Products to Solve Equations

DOLPHINS Refer to the information above. Solve $\frac{3}{x+5} = \frac{2}{x}$ to find the speed of a coastal dolphin. Check the solution.

$\frac{3}{x+5} = \frac{2}{x}$	Original equation
$3x = 2(x + 5)$	Find the cross products.
$3x = 2x + 10$	Distributive Property
$x = 10$	Subtract $2x$ from each side.

So, a coastal dolphin can swim 10 miles per hour.

CHECK

$\frac{3}{x+5} = \frac{2}{x}$	Original equation
$\frac{3}{10+5} \stackrel{?}{=} \frac{2}{10}$	Replace x with 10.
$\frac{3}{15} \stackrel{?}{=} \frac{1}{5}$	Simplify.
$\frac{1}{5} = \frac{1}{5}$ ✓	Simplify.

GuidedPractice

Solve each equation. Check the solution.

1A. $\frac{7}{y-3} = \frac{3}{y+1}$

1B. $\frac{13}{10} = \frac{2f + 0.2}{7}$

Another method that can be used to solve any rational equation is to find the LCD of all the fractions in the equation. Then multiply each side of the equation by the LCD to eliminate the fractions.

PT

Example 2 Use the LCD to Solve Rational Equations

Solve $\frac{4}{y} + \frac{5y}{y+1} = 5$. Check the solution.

Step 1 Find the LCD.

The LCD of $\frac{4}{y}$ and $\frac{5y}{y+1}$ is $y(y + 1)$.

Step 2 Multiply each side of the equation by the LCD.

$$\frac{4}{y} + \frac{5y}{y+1} = 5$$ Original equation

$$y(y+1)\left(\frac{4}{y} + \frac{5y}{y+1}\right) = y(y+1)(5)$$ Multiply each side by the LCD, $y(y + 1)$.

$$\left(\frac{\overset{1}{\cancel{y}}(y+1)}{1} \cdot \frac{4}{\underset{1}{\cancel{y}}}\right) + \left(\frac{y\overset{1}{\cancel{(y+1)}}}{1} \cdot \frac{5y}{\underset{1}{\cancel{y+1}}}\right) = y(y+1)(5)$$ Distributive Property

$$(y+1)4 + y(5y) = y(y+1)(5)$$ Simplify.

$$4y + 4 + 5y^2 = 5y^2 + 5y$$ Multiply.

$$4y + 4 + 5y^2 - 5y^2 = 5y^2 - 5y^2 + 5y$$ Subtract $5y^2$ from each side.

$$4y + 4 = 5y$$ Simplify.

$$4y - 4y + 4 = 5y - 4y$$ Subtract $4y$ from each side.

$$4 = y$$ Simplify.

CHECK $\frac{4}{y} + \frac{5y}{y+1} = 5$ Original equation

$$\frac{4}{4} + \frac{5(4)}{4+1} \stackrel{?}{=} 5$$ Replace y with 4.

$$1 + 4 \stackrel{?}{=} 5$$ Simplify.

$$5 = 5 \checkmark$$ Simplify.

> **StudyTip**
> **Solutions** It is important to check the solutions of rational equations to be sure that they satisfy the original equation.

GuidedPractice

Solve each equation. Check your solutions.

2A. $\frac{2b-5}{b-2} - 2 = \frac{3}{b+2}$

2B. $1 + \frac{1}{c+2} = \frac{28}{c^2+2c}$

2C. $\frac{y+2}{y-2} - \frac{2}{y+2} = -\frac{7}{3}$

2D. $\frac{n}{3n+6} - \frac{n}{5n+10} = \frac{2}{5}$

> **VocabularyLink**
> **extraneous**
> Everyday Use
> irrelevant or unimportant
> **extraneous solution**
> Math Use a result that is not a solution of the original equation

Recall that any value of a variable that makes the denominator of a rational expression zero must be excluded from the domain.

In the same way, when a solution of a rational equation results in a zero in the denominator, that solution must be excluded. Such solutions are also called **extraneous solutions**.

$$\frac{4+x}{x-5} + \frac{1}{x} = \frac{2}{x+1}$$ 5, 0, and −1 cannot be solutions.

Example 3 Extraneous Solutions

Solve $\frac{2n}{n-5} + \frac{4n-30}{n-5} = 5$. State any extraneous solutions.

$$\frac{2n}{n-5} + \frac{4n-30}{n-5} = 5$$ Original equation

$$(n-5)\left(\frac{2n}{n-5} + \frac{4n-30}{n-5}\right) = (n-5)5$$ Multiply each side by the LCD, $n-5$.

$$\left(\frac{\cancel{n-5}^{1}}{1} \cdot \frac{2n}{\cancel{n-5}_{1}}\right) + \left(\frac{\cancel{n-5}^{1}}{1} \cdot \frac{4n-30}{\cancel{n-5}_{1}}\right) = (n-5)5$$ Distributive Property

$$2n + 4n - 30 = 5n - 25$$ Simplify.

$$6n - 30 = 5n - 25$$ Add like terms.

$$6n - 5n - 30 = 5n - 5n - 25$$ Subtract $5n$ from each side.

$$n - 30 = -25$$ Simplify.

$$n - 30 + 30 = -25 + 30$$ Add 30 to each side.

$$n = 5$$ Simplify.

Since $n = 5$ results in a zero in the denominator of the original equation, it is an extraneous solution. So, the equation has no solution.

Study Tip

Solutions It is possible to get both a valid solution and an extraneous solution when solving a rational equation.

Guided Practice

3. Solve $\frac{n^2 - 3n}{n^2 - 4} - \frac{10}{n^2 - 4} = 2$. State any extraneous solutions.

2 Use Rational Equations to Solve Problems

You can use rational equations to solve **work problems**, or problems involving work rates.

Real-World Example 4 Work Problem

JOBS At his part-time job at the zoo, Ping can clean the bird area in 2 hours. Natalie can clean the same area in 1 hour and 15 minutes. How long would it take them if they worked together?

Understand It takes Ping 2 hours to complete the job and Natalie $1\frac{1}{4}$ hours.

You need to find the rate that each person works and the total time t that it will take them if they work together.

Plan Find the fraction of the job that each person can do in an hour.

Ping's rate ⟶ $\frac{1 \text{ job}}{2 \text{ hours}} = \frac{1}{2}$ job per hour

Natalie's rate ⟶ $\frac{1 \text{ job}}{1\frac{1}{4} \text{ hours}}$ or $\frac{1 \text{ job}}{\frac{5}{4} \text{ hours}} = \frac{4}{5}$ job per hour

Since rate • time = fraction of job done, multiply each rate by the time t to represent the amount of the job done by each person.

StudyTip

CCSS Reasoning When solving work problems, remember that each term should represent the portion of a job completed in one unit of time.

Solve

Fraction of job Ping completes	plus	fraction of job Natalie completes	equals	1 job.
$\frac{1}{2}t$	$+$	$\frac{4}{5}t$	$=$	1

$10\left(\frac{1}{2}t + \frac{4}{5}t\right) = 10(1)$ Multiply each side by the LCD, 10.

$10\left(\frac{1}{2}t\right) + 10\left(\frac{4}{5}t\right) = 10$ Distributive Property

$5t + 8t = 10$ Simplify.

$t = \frac{10}{13}$ Add like terms and divide each side by 13.

So, it would take them $\frac{10}{13}$ hour or about 46 minutes to complete the job if they work together.

Check In $\frac{10}{13}$ hour, Ping would complete $\frac{1}{2} \cdot \frac{10}{13}$ or $\frac{5}{13}$ of the job and Natalie would complete $\frac{4}{5} \cdot \frac{10}{13}$ or $\frac{8}{13}$ of the job. Together, they complete $\frac{5}{13} + \frac{8}{13}$ or 1 whole job. So, the answer is reasonable. ✓

GuidedPractice

4. RAKING Jenna can rake the leaves in 2 hours. It takes her brother Benjamin 3 hours. How long would it take them if they worked together?

Rational equations can also be used to solve **rate problems**.

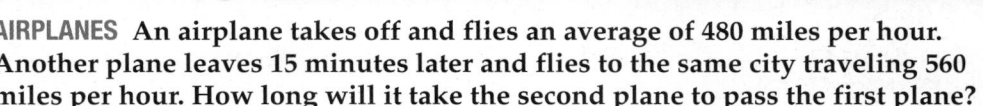

Real-World Example 5 Rate Problem

AIRPLANES An airplane takes off and flies an average of 480 miles per hour. Another plane leaves 15 minutes later and flies to the same city traveling 560 miles per hour. How long will it take the second plane to pass the first plane?

Record the information that you know in a table.

Plane	Distance	Rate	Time
1	d miles	480 mi/h	t hours
2	d miles	560 mi/h	$t - \frac{1}{4}$ hours

← Plane 2 took off 15 minutes, or $\frac{1}{4}$ hour, after Plane 1

Since both planes will have traveled the same distance when Plane 2 passes Plane 1, you can write the following equation.

Distance for Plane 1 = Distance for Plane 2

$480 \cdot t = 560 \cdot \left(t - \frac{1}{4}\right)$ distance = rate • time

$480t = (560 \cdot t) - \left(560 \cdot \frac{1}{4}\right)$ Distributive Property

$480t = 560t - 140$ Simplify.

$-80t = -140$ Subtract $560t$ from each side.

$t = 1.75$ Divide each side by −80.

So, the second plane passes the first plane after 1.75 hours.

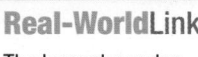

Real-WorldLink

The longest nonstop commercial flight was 13,422 miles from Hong Kong Airport in China to London Heathrow in the United Kingdom. It took 22 hours and 42 minutes.

Source: *Guinness Book of World Records*

Jeff Hunter/The Image Bank/Getty Images

GuidedPractice

5. Lenora leaves the house walking at 3 miles per hour. After 10 minutes, her mother leaves the house riding a bicycle at 10 miles per hour. In how many minutes will Lenora's mother catch her?

Check Your Understanding

Examples 1–3 **Solve each equation. State any extraneous solutions.**

1. $\frac{2}{x+1} = \frac{4}{x}$
2. $\frac{t+3}{5} = \frac{2t+3}{9}$
3. $\frac{a+3}{a} - \frac{6}{5a} = \frac{1}{a}$
4. $4 - \frac{p}{p-1} = \frac{2}{p-1}$
5. $\frac{2t}{t+1} + \frac{4}{t-1} = 2$
6. $\frac{x+3}{x^2-1} - \frac{2x}{x-1} = 1$

Example 4

7. **WEEDING** Maurice can weed the garden in 45 minutes. Olinda can weed the garden in 50 minutes. How long would it take them to weed the garden if they work together?

Example 5

8. **LANDSCAPING** Hunter is filling a 3.5-gallon bucket to water plants at a faucet that flows at a rate of 1.75 gallons a minute. If he were to add a hose that flows at a rate of 1.45 gallons per minute, how many minutes would it take him to fill the bucket? Round to the nearest tenth.

Practice and Problem Solving

Examples 1–3 **Solve each equation. State any extraneous solutions.**

9. $\frac{8}{n} = \frac{3}{n-5}$
10. $\frac{6}{t+2} = \frac{4}{t}$
11. $\frac{3g+2}{12} = \frac{g}{2}$
12. $\frac{5h}{4} + \frac{1}{2} = \frac{3h}{8}$
13. $\frac{2}{3w} = \frac{2}{15} + \frac{12}{5w}$
14. $\frac{c-4}{c+1} = \frac{c}{c-1}$
15. $\frac{x-1}{x+1} - \frac{2x}{x-1} = -1$
16. $\frac{y+4}{y-2} + \frac{6}{y-2} = \frac{1}{y+3}$
17. $\frac{a}{a+3} + \frac{a^2}{a+3} = 2$
18. $\frac{12}{a+3} + \frac{6}{a^2-9} = \frac{8}{a+3}$
19. $\frac{3n}{n-1} + \frac{6n-9}{n-1} = 6$
20. $\frac{n^2-n-6}{n^2-n} - \frac{n-5}{n-1} = \frac{n-3}{n^2-n}$

Example 4

21. **PAINTING** It takes Noah 3 hours to paint one side of a fence. It takes Gilberto 5 hours. How long would it take them if they worked together?

22. **DISHWASHING** Ron works as a dishwasher and can wash 500 plates in two hours and 15 minutes. Chris can finish the 500 plates in 3 hours. About how long would it take them to finish all of the plates if they work together?

Example 5

23. **ICE** A hotel has two ice machines in its kitchen. How many hours would it take both machines to make 60 pounds of ice? Round to the nearest tenth.

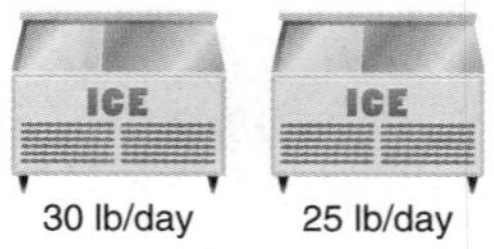

24. **CYCLING** Two cyclists travel in opposite directions around a 5.6-mile circular trail. They start at the same time. The first cyclist completes the trail in 22 minutes and the second in 28 minutes. At what time do they pass each other?

GRAPHING CALCULATOR **For each function, a) describe the shape of the graph, b) use factoring to simplify the function, and c) find the zeros of the function.**

25. $f(x) = \frac{x^2 - x - 30}{x - 6}$
26. $f(x) = \frac{x^3 + x^2 - 2x}{x + 2}$
27. $f(x) = \frac{x^3 + 6x^2 + 12x}{x}$

28. CCSS **REASONING** Morgan can paint a standard-sized house in about 5 days. For his latest job, Morgan hires two assistants. At what rate must these assistants work for Morgan to meet a deadline of two days?

29. **AIRPLANES** Headwinds push against a plane and reduce its total speed, while tailwinds push on a plane and increase its total speed. Let w equal the speed of the wind, r equal the speed set by the pilot, and s equal the total speed.

a. Write an equation for the total speed with a headwind and an equation for the total speed with a tailwind.

b. Use the rate formula to write an equation for the distance traveled by a plane with a headwind and another equation for the distance traveled by a plane with a tailwind. Then solve each equation for time instead of distance.

30. **MIXTURES** A pitcher of fruit juice has 3 pints of pineapple juice and 2 pints of orange juice. Erin wants to add more orange juice so that the fruit juice mixture is 60% orange juice. Let x equal the pints of orange juice that she needs to add.

a. Copy and complete the table below.

Juice	Pints of Orange Juice	Total Pints of Juice	Percent of Orange Juice
original mixture		5	
final mixture	$2 + x$		0.6

b. Write and solve an equation to find the pints of orange juice to add.

31. **DORMITORIES** The number of hours h it takes to clean a dormitory varies inversely with the number of people cleaning it c and directly with the number of people living there p.

a. Write an equation showing how h, c, and p are related. (*Hint*: Include the constant k.)

b. It takes 8 hours for 5 people to clean the dormitory when there are 100 people there. How long will it take to clean the dormitory if there are 10 people cleaning and the number of people living in the dorm stays the same?

Solve each equation. State any extraneous solutions.

32. $\frac{4b+2}{b^2-3b} + \frac{b+2}{b} = \frac{b-1}{b}$

33. $\frac{x^2-x-6}{x+2} + \frac{x^3+x^2}{x} = 3$

34. $\frac{y^2+5y-6}{y^3-2y^2} = \frac{5}{y} - \frac{6}{y^3-2y^2}$

35. $\frac{x-\frac{6}{5}}{x} - \frac{x-10\frac{1}{2}}{x-5} = \frac{x+21}{x^2-5x}$

H.O.T. Problems Use Higher-Order Thinking Skills

36. **CHALLENGE** Solve $\frac{2x}{x-2} + \frac{x^2+3x}{(x+1)(x-2)} = \frac{2}{(x+1)(x-2)}$.

37. **REASONING** How is an excluded value of a rational expression related to an extraneous solution of a corresponding rational equation? Explain.

38. **WRITING IN MATH** Why should you check solutions of rational equations?

39. **CCSS ARGUMENTS** Find a counterexample for the following statement.

The solution of a rational equation can never be zero.

40. **WRITING IN MATH** Describe the steps for solving a rational equation that is not a proportion.

Standardized Test Practice

41. It takes Cheng 4 hours to build a fence. If he hires Odell to help him, they can do the job in 3 hours. If Odell built the same fence alone, how long would it take him?

A $1\frac{5}{7}$ hours **C** 8 hours

B $3\frac{2}{3}$ hours **D** 12 hours

42. In the 1000-meter race, Zoe finished 35 meters ahead of Taryn and 53 meters ahead of Evan. How far was Taryn ahead of Evan?

F 18 m **G** 35 m **H** 53 m **J** 88 m

43. Twenty gallons of lemonade were poured into two containers of different sizes. Express the amount of lemonade poured into the smaller container in terms of g, the amount poured into the larger container.

A $g + 20$ **C** $g - 20$

B $20 + g$ **D** $20 - g$

44. GRIDDED RESPONSE The gym has 2-kilogram and 5-kilogram disks for weight lifting. They have fourteen disks in all. The total weight of the 2-kilogram disks is the same as the total weight of the 5-kilogram disks. How many 2-kilogram disks are there?

Spiral Review

45. POPULATION The country of Latvia has been experiencing a 1.1% annual decrease in population. In 2009, its population was 2,261,294. If the trend continues, predict Latvia's population in 2019. (Lesson 7-6)

46. TOMATOES There are more than 10,000 varieties of tomatoes. One seed company produces seed packages for 200 varieties of tomatoes. For how many varieties do they not provide seeds? (Lesson 5-1)

47. DRIVING Tires should be kept within 2 pounds per square inch (psi) of the manufacturer's recommended tire pressure. If the recommendation for a tire is 30 psi, what is the range of acceptable pressures? (Lesson 5-5)

Express each number in scientific notation. (Lesson 7-4)

48. 12,300 **49.** 0.0000375 **50.** 1,255,000

51. FINANCIAL LITERACY Ruben has $13 to order pizza. The pizza costs $7.50 plus $1.25 per topping. He plans to tip 15% of the total cost. Write and solve an inequality to find out how many toppings he can order. (Lesson 5-3)

Solve each inequality. Check your solution. (Lesson 5-2)

52. $\frac{b}{10} \leq 5$ **53.** $-7 > -\frac{r}{7}$ **54.** $\frac{5}{8}y \geq -15$

Skills Review

Determine the probability of each event if you randomly select a marble from a bag containing 9 red marbles, 6 blue marbles, and 5 yellow marbles.

55. P(blue) **56.** P(red) **57.** P(not yellow)

EXTEND 8-7

Graphing Technology Lab
Solving Rational Equations

You can use TI-Nspire Technology to solve rational equations by graphing, by using tables, and by using a computer algebra system (CAS).

To solve by graphing, graph both sides of the equation and locate the point(s) of intersection.

CCSS Common Core State Standards
Content Standards
A.REI.11 Explain why the *x*-coordinates of the points where the graphs of the equations $y = f(x)$ and $y = g(x)$ intersect are the solutions of the equation $f(x) = g(x)$; find the solutions approximately, e.g., using technology to graph the functions, make tables of values, or find successive approximations. Include cases where $f(x)$ and/or $g(x)$ are linear, polynomial, rational, absolute value, exponential, and logarithmic functions.

Mathematical Practices
5 Use appropriate tools strategically.

Activity 1 Solve a Rational Equation by Graphing

Solve $\frac{5}{x+2} = \frac{3}{x}$ by graphing.

Step 1 Add a new **Graphs** page.

Step 2 Use the **Window Settings** option from the **Window/Zoom** menu to adjust the window to −20 to 20 for both x and y. Set both scales to 2.

Step 3 Enter $\frac{5}{x+2}$ into **f1(*x*)** and $\frac{3}{x}$ into **f2(*x*)**.

Step 4 Change the thickness of the graph of **f1(*x*)** by selecting the graph of **f1(*x*)** and the **ctrl menu Attributes** option.

Step 5 Use the **Intersection Point(s)** tool from the **Points & Lines** menu to find the intersection of the two graphs. Select the graph of **f1(*x*) enter** and then the graph of **f2(*x*) enter**.

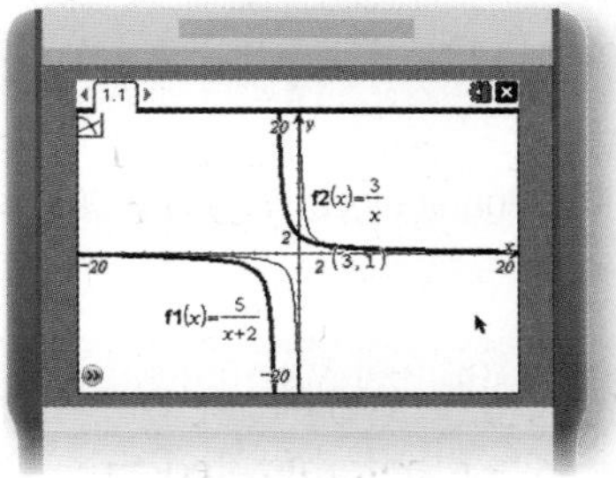

[−20, 20] scl: 2 by [−20, 20] scl: 2

The graphs intersect at (3, 1). This means that $\frac{5}{x+2}$ and $\frac{3}{x}$ both equal 1 when $x = 3$. Thus, the solution of $\frac{5}{x+2} = \frac{3}{x}$ is $x = 3$.

Exercises

Use a graphing calculator to solve each equation.

1. $\frac{5}{x} + \frac{4}{x} = 10$

2. $\frac{12}{x} + \frac{3}{4} = \frac{3}{2}$

3. $\frac{6}{x} + \frac{3}{2x} = 12$

4. $\frac{4}{x} + \frac{3}{4x} = \frac{1}{8}$

5. $\frac{4}{x} + \frac{x-2}{2x} = x$

6. $\frac{3}{3x-2} + \frac{5}{x} = 0$

7. $\frac{2x+1}{2} + \frac{3}{2x} = \frac{2}{x}$

8. $\frac{x}{x+2} + x = \frac{5x+8}{x+2}$

9. $\frac{1}{2x} + \frac{5}{x} = \frac{3}{x-1}$

10. $\frac{4x-3}{x-2} + \frac{2x+5}{x-2} = 6$

Graphing Technology Lab
Solving Rational Equations *Continued*

Activity 2 Solve a Rational Equation by Using a Table

Solve $\frac{2x+1}{3} = \frac{x+2}{2}$ using a table.

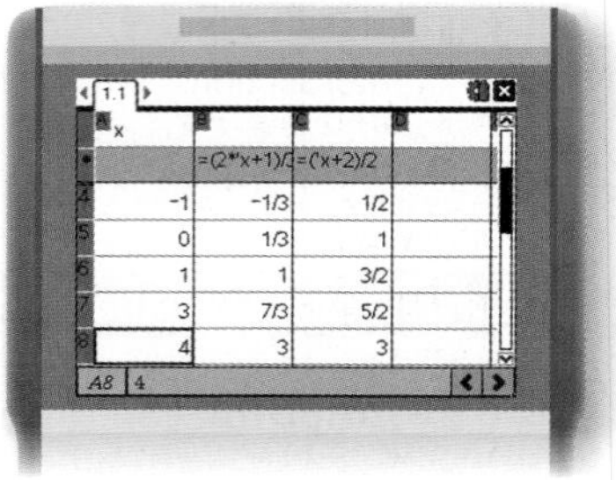

Step 1 Add a new **Lists & Spreadsheet** page.

Step 2 Label column A as x. Enter values from -4 to 4 in cells A1 to A9.

Step 3 In column B in the formula row, enter the left side of the rational equation, with parenthesis around the binomials. In column C in the formula row, enter the right side of the rational equation, with parenthesis around the binomials. Specify **Variable Reference** when prompted.

Scroll until you see where the values in Columns B and C are equal. This occurs at $x = 4$. Therefore the solution of $\frac{2x+1}{3} = \frac{x+2}{2}$ is 4.

You can also use a computer algebra system (CAS) to solve rational equations.

Activity 3 Solve a Rational Equation by Using a CAS

Solve $\frac{x-3}{x} - \frac{x-4}{x-2} = \frac{1}{x}$ using a CAS.

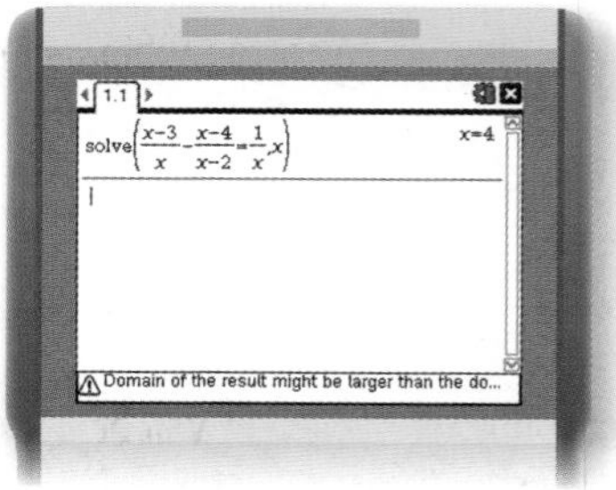

Step1 Add a new **Calculator** page.

Step 2 To solve, select the **Solve** tool from the **Algebra** menu. Enter the left side of the equation with parenthesis around the binomials. Enter = and the right side of the equation. Then type a comma, followed by x, and then **enter**.

The solution of 4 is displayed.

Exercises

Use a table or CAS to solve each equation.

11. $\frac{2}{x} + \frac{2+x}{2} = \frac{x+3}{2}$

12. $\frac{4}{x-2} = -\frac{1}{x+3}$

13. $\frac{3}{x+2} + \frac{4}{x-1} = 0$

14. $\frac{1}{x+1} + \frac{2}{x-1} = 0$

15. $\frac{2}{x+4} + \frac{4}{x-1} = 0$

16. $\frac{1}{x-2} + \frac{x+2}{4} = 2x$

17. $\frac{2x}{x+3} + \frac{x+1}{2} = x$

18. $\frac{2}{x-3} + \frac{3}{x-2} = \frac{4}{x}$

19. $\frac{x^2}{x+1} + \frac{x}{x-1} = x$

Study Guide and Review

Study Guide

KeyConcepts

Square Root Functions (Lesson 8-1)

- A square root function contains the square root of a variable.
- The parent function of the family of square root functions is $f(x) = \sqrt{x}$.

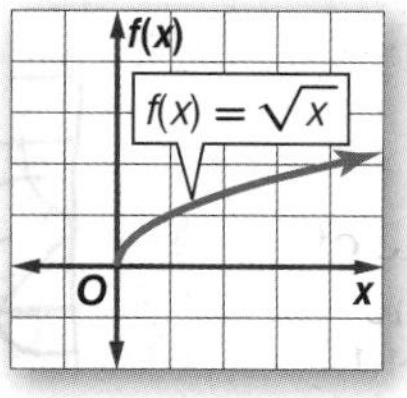

Simplifying Radical Expressions (Lesson 8-2)

- A radical expression is in simplest form when
  - no radicands have perfect square factors other than 1,
  - no radicals contain fractions,
  - and no radicals appear in the denominator of a fraction.

Operations with Radical Expressions and Equations
(Lessons 8-3 and 8-4)

- Radical expressions with like radicals can be added or subtracted.
- Use the FOIL method to multiply radical expressions.

Inverse Variation (Lesson 8-5)

- You can use $\frac{x_1}{x_2} = \frac{y_2}{y_1}$ to solve problems involving inverse variation.

Rational Functions (Lesson 8-6)

- Excluded values are values of a variable that result in a denominator of zero.
- If vertical asymptotes occur, it will be at excluded values.

Solving Rational Equations (Lesson 8-7)

- Use cross products to solve rational equations with a single fraction on each side of the equals sign.

FOLDABLES® StudyOrganizer

Be sure the Key Concepts are noted in your Foldable.

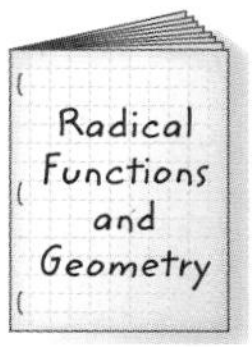

KeyVocabulary

asymptote (p. 497)
closed (p. 476)
conjugate (p. 472)
excluded value (p. 496)
extraneous solution (p. 483)
inverse variation (p. 488)
product rule (p. 489)
radical equations (p. 482)
radical expression (p. 470)
radical function (p. 463)
radicand (p. 463)
rate problem (p. 505)
rationalize the denominator (p. 472)
rational function (p. 496)
rational equation (p. 502)
square root function (p. 463)
work problem (p. 504)

VocabularyCheck

State whether each sentence is *true* or *false*. If *false*, replace the underlined word, phrase, expression, or number to make a true sentence.

1. The expressions $12\sqrt{4}$ and $\sqrt{288}$ are equivalent.
2. The expressions $2 + \sqrt{5}$ and $2 - \sqrt{5}$ are conjugates.
3. In the expression $-5\sqrt{2}$, the radicand is 2.
4. If the product of two variables is a nonzero constant, the relationship is an inverse variation.
5. If the line $x = a$ is a vertical asymptote of a rational function, then a is an excluded value.
6. The excluded values for $\frac{x}{x^2 + 5x + 6}$ are -2 and -3.
7. The equation $\frac{3x}{x - 2} = \frac{6}{x - 2}$ has an extraneous solution, 2.

Study Guide and Review *Continued*

Lesson-by-Lesson Review

8-1 Square Root Functions

Graph each function. Compare to the parent graph. State the domain and range.

8. $y = \sqrt{x} - 3$
9. $y = \sqrt{x} + 2$
10. $y = -5\sqrt{x}$
11. $y = \sqrt{x} - 6$
12. $y = \sqrt{x - 1}$
13. $y = \sqrt{x} + 5$

14. **GEOMETRY** The function $s = \sqrt{A}$ can be used to find the length of a side of a square given its area. Use this function to determine the length of a side of a square with an area of 90 square inches. Round to the nearest tenth if necessary.

Example 1

Graph $y = -3\sqrt{x}$. Compare to the parent graph. State the domain and range.

Make a table. Choose nonnegative values for x.

x	0	1	2	3	4
y	0	−3	≈−4.2	≈−5.2	−6

Plot points and draw a smooth curve.

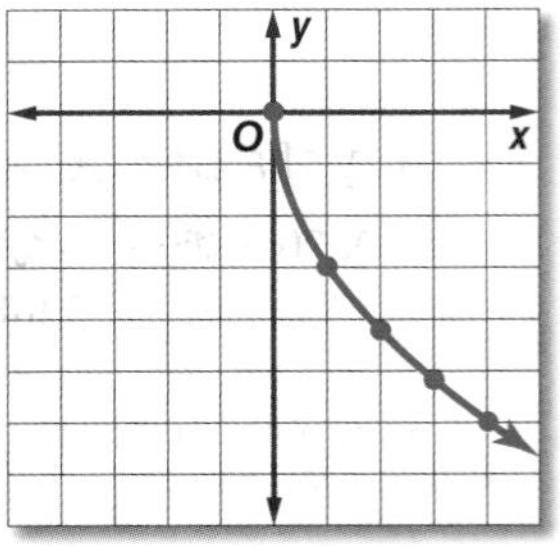

The graph of $y = \sqrt{x}$ is stretched vertically and is reflected across the x-axis.

The domain is $\{x | x \geq 0\}$.

The range is $\{y | y \leq 0\}$.

8-2 Simplifying Radical Expressions

Simplify.

15. $\sqrt{36x^2y^7}$
16. $\sqrt{20ab^3}$
17. $\sqrt{3} \cdot \sqrt{6}$
18. $2\sqrt{3} \cdot 3\sqrt{12}$
19. $(4 - \sqrt{5})^2$
20. $(1 + \sqrt{2})^2$
21. $\sqrt{\frac{50}{a^2}}$
22. $\sqrt{\frac{2}{5}} \cdot \sqrt{\frac{3}{4}}$
23. $\frac{3}{2 - \sqrt{5}}$
24. $\frac{5}{\sqrt{7} + 6}$
25. **WEATHER** To estimate how long a thunderstorm will last, use $t = \sqrt{\frac{d^3}{216}}$, where t is the time in hours and d is the diameter of the storm in miles. A storm is 10 miles in diameter. How long will it last?

Example 2

Simplify $\frac{2}{4 + \sqrt{3}}$.

$\frac{2}{4 + \sqrt{3}}$ — Original expression

$= \frac{2}{4 + \sqrt{3}} \cdot \frac{4 - \sqrt{3}}{4 - \sqrt{3}}$ — Rationalize the denominator.

$= \frac{2(4) - 2\sqrt{3}}{4^2 - (\sqrt{3})^2}$ — $(a - b)(a + b) = a^2 - b^2$

$= \frac{8 - 2\sqrt{3}}{16 - 3}$ — $(\sqrt{3})^2 = 3$

$= \frac{8 - 2\sqrt{3}}{13}$ — Simplify.

8-3 Operations with Radical Expressions

Simplify each expression.

26. $\sqrt{6} - \sqrt{54} + 3\sqrt{12} + 5\sqrt{3}$

27. $2\sqrt{6} - \sqrt{48}$

28. $4\sqrt{3x} - 3\sqrt{3x} + 3\sqrt{3x}$

29. $\sqrt{50} + \sqrt{75}$

30. $\sqrt{2}(5 + 3\sqrt{3})$

31. $(2\sqrt{3} - \sqrt{5})(\sqrt{10} + 4\sqrt{6})$

32. $(6\sqrt{5} + 2)(4\sqrt{2} + \sqrt{3})$

33. **MOTION** The velocity of a dropped object when it hits the ground can be found using $v = \sqrt{2gd}$, where v is the velocity in feet per second, g is the acceleration due to gravity, and d is the distance in feet the object drops. Find the speed of a penny when it hits the ground, after being dropped from 984 feet. Use 32 feet per second squared for g.

Example 3

Simplify $2\sqrt{6} - \sqrt{24}$.

$2\sqrt{6} - \sqrt{24} = 2\sqrt{6} - \sqrt{4 \cdot 6}$		Product Property
$= 2\sqrt{6} - 2\sqrt{6}$		Simplify.
$= 0$		Simplify.

Example 4

Simplify $(\sqrt{3} - \sqrt{2})(\sqrt{3} + 2\sqrt{2})$.

$$(\sqrt{3} - \sqrt{2})(\sqrt{3} + 2\sqrt{2})$$
$$= (\sqrt{3})(\sqrt{3}) + (\sqrt{3})(2\sqrt{2}) + (-\sqrt{2})(\sqrt{3}) + (\sqrt{2})(2\sqrt{2})$$
$$= 3 + 2\sqrt{6} - \sqrt{6} + 4$$
$$= 7 + \sqrt{6}$$

8-4 Radical Equations

Solve each equation. Check your solution.

34. $10 + 2\sqrt{x} = 0$

35. $\sqrt{5 - 4x} - 6 = 7$

36. $\sqrt{a + 4} = 6$

37. $\sqrt{3x} = 2$

38. $\sqrt{x + 4} = x - 8$

39. $\sqrt{3x - 14} + x = 6$

40. **FREE FALL** Assuming no air resistance, the time t in seconds that it takes an object to fall h feet can be determined by $t = \frac{\sqrt{h}}{4}$. If a skydiver jumps from an airplane and free falls for 10 seconds before opening the parachute, how many feet does she free fall?

Example 5

Solve $\sqrt{7x + 4} - 18 = 5$.

$\sqrt{7x + 4} - 18 = 5$	Original equation
$\sqrt{7x + 4} = 23$	Add 18 to each side.
$(\sqrt{7x + 4})^2 = 23^2$	Square each side.
$7x + 4 = 529$	Simplify.
$7x = 525$	Subtract 4 from each side.
$x = 75$	Divide each side by 7.

CHECK

$\sqrt{7x + 4} - 18 = 5$	Original equation
$\sqrt{7(75) + 4} - 18 \stackrel{?}{=} 5$	$x = 75$
$\sqrt{525 + 4} - 18 \stackrel{?}{=} 5$	Multiply.
$\sqrt{529} - 18 \stackrel{?}{=} 5$	Add.
$23 - 18 \stackrel{?}{=} 5$	Simplify.
$5 = 5$ ✓	True.

Lesson-by-Lesson Review

8-5 Inverse Variation

Solve. Assume that y varies inversely as x.

41. If $y = 4$ when $x = 1$, find x when $y = 12$

42. If $y = -1$ when $x = -3$, find y when $x = -9$

43. If $y = 1.5$ when $x = 6$, find y when $x = -16$

44. PHYSICS A 135-pound person sits 5 feet from the center of a seesaw. How far from the center should a 108-pound person sit to balance the seesaw?

Example 6

If y varies inversely as x and $y = 28$ when $x = 42$, find y when $x = 56$.

Let $x_1 = 42$, $x_2 = 56$, and $y_1 = 28$. Solve for y_2.

$$\frac{x_1}{x_2} = \frac{y_2}{y_1} \quad \text{Proportion for inverse variation}$$

$$\frac{42}{56} = \frac{y_2}{28} \quad \text{Substitution}$$

$$1176 = 56y_2 \quad \text{Cross multiply.}$$

$$21 = y_2$$

Thus, $y = 21$ when $x = 56$.

8-6 Rational Functions

State the excluded value for each function.

45. $y = \frac{1}{x-3}$

46. $y = \frac{2}{2x-5}$

47. $y = \frac{3}{3x-6}$

48. $y = \frac{-1}{2x+8}$

49. PIZZA PARTY Katelyn ordered pizza and soda for her study group for \$38. The cost per person y is given by $y = \frac{38}{x}$, where x is the number of people in the study group. Graph the function and describe the asymptotes.

Example 7

State the excluded value for the function $y = \frac{1}{4x+16}$.

Set the denominator equal to zero.

$$4x + 16 = 0$$

$$4x + 16 - 16 = 0 - 16 \quad \text{Subtract 16 from each side.}$$

$$4x = -16 \quad \text{Simplify.}$$

$$x = -4 \quad \text{Divide each side by 4.}$$

8-7 Rational Equations

Solve each equation. State any extraneous solutions.

50. $\frac{5n}{6} + \frac{1}{n-2} = \frac{n+1}{3(n-2)}$

51. $\frac{4x}{3} + \frac{7}{2} = \frac{7x}{12} - 14$

52. $\frac{11}{2x} + \frac{2}{4x} = \frac{1}{4}$

53. $\frac{1}{x+4} - \frac{1}{x-1} = \frac{2}{x^2+3x-4}$

54. $\frac{1}{n-2} = \frac{n}{8}$

55. PAINTING Anne can paint a room in 6 hours. Oljay can paint a room in 4 hours. How long will it take them to paint the room working together?

Example 8

Solve $\frac{3}{x^2+3x} + \frac{x+2}{x+3} = \frac{1}{x}$.

$$\frac{3}{x^2+3x} + \frac{x+2}{x+3} = \frac{1}{x}$$

$$x(x+3)\left(\frac{3}{x(x+3)}\right) + x(x+3)\left(\frac{x+2}{x+3}\right) = x(x+3)\left(\frac{1}{x}\right)$$

$$3 + x(x+2) = 1(x+3)$$

$$3 + x^2 + 2x = x + 3$$

$$x^2 + x = 0$$

$$x(x+1) = 0$$

$$x = 0 \text{ or } x = -1$$

The solution is -1, and there is an extraneous solution of 0.

CHAPTER 8 Practice Test

Graph each function, and compare to the parent graph. State the domain and range.

1. $y = -\sqrt{x}$

2. $y = \frac{1}{4}\sqrt{x}$

3. $y = \sqrt{x} + 5$

4. $y = \sqrt{x + 4}$

5. MULTIPLE CHOICE The length of the side of a square is given by the function $s = \sqrt{A}$, where A is the area of the square. What is the perimeter of a square that has an area of 64 square inches?

A 64 inches

B 8 inches

C 32 inches

D 16 inches

Simplify each expression.

6. $5\sqrt{36}$

7. $\frac{3}{1 - \sqrt{2}}$

8. $2\sqrt{3} + 7\sqrt{3}$

9. $3\sqrt{6}(5\sqrt{2})$

10. MULTIPLE CHOICE Find the area of the rectangle.

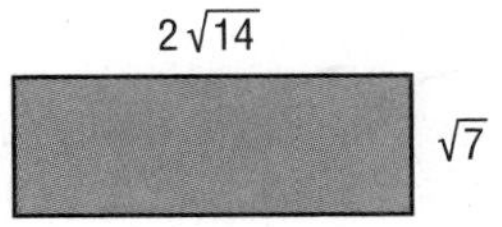

F $7\sqrt{2}$

G 14

H $14\sqrt{2}$

J $98\sqrt{2}$

Solve each equation. Check your solution.

11. $\sqrt{10x} = 20$

12. $\sqrt{4x - 3} = 6 - x$

13. PACKAGING A cylindrical container of chocolate drink mix has a volume of about 162 in^3. The radius of the container can be found by using the formula $r = \sqrt{\frac{V}{\pi h}}$, where r is the radius and h is the height. If the height is 8.25 inches, find the radius of the container.

Determine whether each table represents an inverse variation. Explain.

14.

x	y
2	10
4	12
8	14

Solve. Assume that y varies inversely as x.

15. If $y = 3$ when $x = 9$, find x when $y = 1$.

16. If $y = 2$ when $x = 0.5$, find y when $x = 3$.

Assume that y varies inversely as x. Write an inverse variation equation that relates x and y.

17. $y = 2$ when $x = 8$

18. $y = -3$ when $x = 1$

19. MULTIPLE CHOICE Lee can shovel the driveway in 3 hours, and Susan can shovel the driveway in 2 hours. How long will it take them working together?

F 6 hours

G 5 hours

H $\frac{3}{2}$ hours

J $\frac{6}{5}$ hours

20. PAINTING Sydney can paint a 60-square foot wall in 40 minutes. Working with her friend Cleveland, the two of them can paint the wall in 25 minutes. How long would it take Cleveland to do the job himself?

CHAPTER 8

Preparing for Standardized Tests

Draw a Picture

Sometimes it is easier to visualize how to solve a problem if you draw a picture first. You can sketch your picture on scrap paper or in your test booklet (if allowed). Be careful not make any marks on your answer sheet other than your answers.

Strategies for Drawing a Picture

Step 1

Read the problem statement carefully.

Ask yourself:

- What am I being asked to solve?
- What information is given in the problem?
- What is the unknown quantity for which I need to solve?

Step 2

Sketch and label your picture.

- Draw your picture as clearly and accurately as possible.
- Label the picture carefully. Be sure to include all of the information given in the problem statement.

Step 3

Solve the problem.

- Use your picture to help you model the problem situation with an equation. Then solve the equation.
- Check your answer to make sure it is reasonable.

Standardized Test Example

Read the problem. Identify what you need to know. Then use the information in the problem to solve. Show your work.

An 18-foot ladder is leaning against a building. For stability, the base of the ladder must be 36 inches away from the wall. How far up the wall does the ladder reach?

Jupiterimages/Polka Dot/Alamy

Scoring Rubric	
Criteria	Score
Full Credit: The answer is correct and a full explanation is provided that shows each step.	2
Partial Credit: • The answer is correct, but the explanation is incomplete. • The answer is incorrect, but the explanation is correct.	1
No Credit: Either an answer is not provided or the answer does not make sense.	0

Read the problem statement carefully. You know the height of the ladder leaning against the building and you know that the base of the ladder must be 36 inches away from the wall. You need to find how far up the wall the ladder reaches.

Example of a 2-point response:

First convert all measurements to feet.

36 inches = 3 feet

Use a right triangle to find how high the ladder reaches. Draw and label a triangle to represent the situation.

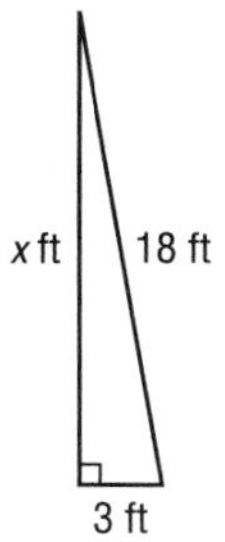

You know the measures of a leg and the hypotenuse, and need to know the length of the other leg. So you can use the Pythagorean Theorem.

$$c^2 = a^2 + b^2$$
$$18^2 = 3^2 + b^2$$
$$324 = 9 + b^2$$
$$315 = b^2$$
$$\pm 315 = b$$
$$17.7 \approx b$$

The ladder reaches about 17.7 feet or about 17 feet 9 inches.

Exercises

Read each problem. Identify what you need to know. Then use the information in the problem to solve. Show your work.

1. A building casts a 15-foot shadow, while a billboard casts a 4.5-foot shadow. If the billboard is 26 feet high, what is the height of the building? Round to the nearest tenth if necessary.

2. A space shuttle is directed toward the Moon, but drifts 1.2° from its intended course. The distance from Earth to the Moon is about 240,000 miles. If the pilot doesn't get the shuttle back on course, how far will the shuttle have drifted from its intended landing position?

CHAPTER 8 Standardized Test Practice

Cumulative, Chapters 1 through 8

Multiple Choice

Read each question. Then fill in the correct answer on the answer document provided by your teacher or on a sheet of paper.

1. Each year a local country club sponsors a tennis tournament. Play starts with 256 participants. During each round, half of the players are eliminated. How many players remain after 6 rounds?

A 128

B 64

C 16

D 4

2. Evaluate $\frac{5^6 - 5^5}{4}$.

F 5^6

G 5^5

H $\frac{5}{4}$

J $\frac{25}{4}$

3. Which of the following numbers is less than zero?

A 1.03×10^{-21}

B 7.5×10^2

C 8.21543×10^{10}

D none of the above

4. Write an equation in slope-intercept form with a slope of $\frac{9}{10}$ and y-intercept of 3.

F $y = 3x + \frac{9}{10}$

G $y = \frac{9}{10}x + 3$

H $y = \frac{9}{10}x - 3$

J $y = 3x - \frac{9}{10}$

5. Jason is playing games at a family fun center. So far he has won 38 prize tickets. How many more tickets would he need to win to place him in the gold prize category?

Number of Tickets	Prize Category
1–20	bronze
21–40	silver
41–60	gold
61–80	platinum

F $2 \leq t \leq 22$

G $3 \leq t \leq 22$

H $1 \leq t \leq 20$

J $3 \leq t \leq 20$

6. Which of the following is an equation of the line perpendicular to $4x - 2y = 6$ and passing through $(4, -4)$?

F $y = -\frac{3}{4}x + 3$

G $y = -\frac{3}{4}x - 1$

H $y = -\frac{1}{2}x - 4$

J $y = -\frac{1}{2}x - 2$

Short Response/Gridded Response

Record your answers on the answer sheet provided by your teacher or on a sheet of paper.

7. GRIDDED RESPONSE Mr. Branson bought a total of 9 tickets to the zoo. He bought children tickets at the rate of $6.50 and adult tickets for $9.25 each. If he spent $69.50 altogether, how many adult tickets did Mr. Branson purchase?

8. What is the domain of the following relation? $\{(2, -1), (4, 3), (7, 6)\}$

9. Ken just added 15 more songs to his digital media player, making the total number of songs more than 84. Draw a number line that represents the original number of songs he had on his digital media player.

10. Carlos bought a rare painting in 1995 for $14,200. By 2003, the painting was worth $17,120. Assuming that a linear relationship exists, write an equation in slope-intercept form that represents the value V of the painting after t years.

11. Marcel spent $24.50 on peanuts and walnuts for a party. He bought 1.5 pounds more peanuts than walnuts. How many pounds of peanuts and walnuts did he buy?

Product	Price per pound
Peanuts p	$3.80
Cashews c	$6.90
Walnuts w	$5.60

12. GRIDDED RESPONSE Misty purchased a car several years ago for $21,459. The value of the car depreciated at a rate of 15% annually. What was the value of the car after 5 years? Round your answer to the nearest whole dollar.

13. GRIDDED RESPONSE The amount of money that Humberto earns varies directly as the number of hours that he works as shown in the graph. How much money will he earn for working 40 hours next week? Express your answer in dollars.

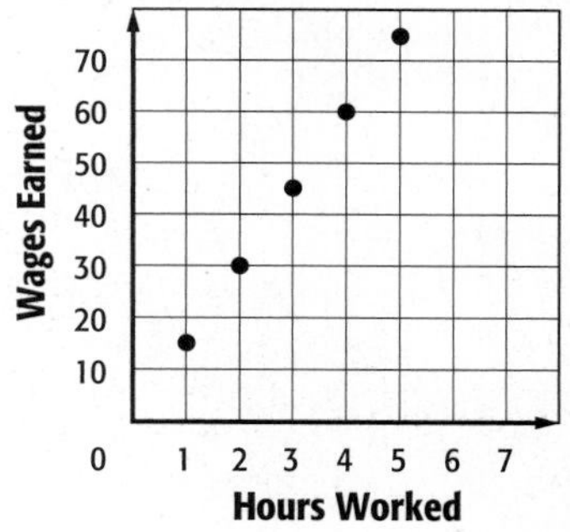

Extended Response

Record your answers on a sheet of paper. Show your work.

14. The Fare charged by a taxi drive is a $3 fixed charge plus $0.35 per mile. Beth pays $10 for a ride of m miles.

Part A Write an equation that can be used to find m. Show your work.

Part B Use the equation in Part A to find how many miles Beth rode. Show your work.

Need ExtraHelp?

If you missed Question...	1	2	3	4	5	6	7	8	9	10	11	12	13	14
Go to Lesson...	7-7	7-2	7-4	4-2	5-1	4-4	6-5	1-6	5-1	4-2	2-9	7-6	3-4	4-1

CHAPTER 9

Statistics and Probability

Then

You calculated simple probability.

Now

In this chapter, you will:

- Design surveys and evaluate results.
- Use permutations and combinations.
- Find probabilities of compound events.
- Design and use simulations.

Why? ▲

RESTAURANTS A restaurant may ask their customers to complete a survey about their visit. The survey data can be analyzed using statistical methods. The restaurant staff can learn more about their customers and how to improve their experiences in the restaurant.

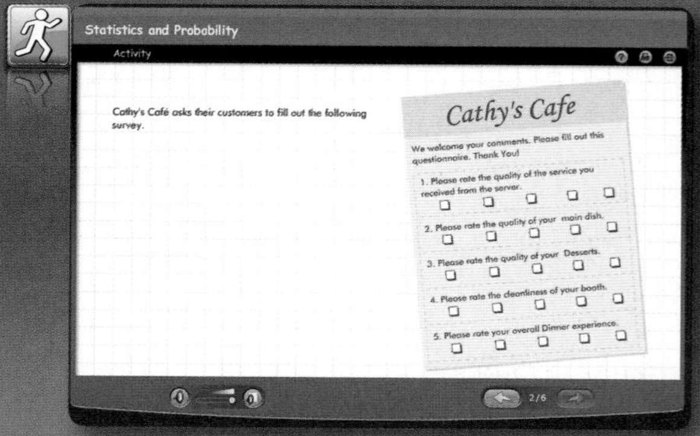
Statistics and Probability
Activity
Cathy's Café asks their customers to fill out the following survey.
Cathy's Cafe
We welcome your comments. Please fill out this questionnaire. Thank You!
1. Please rate the quality of the service you received from the server.
2. Please rate the quality of your main dish.
3. Please rate the quality of your Desserts.
4. Please rate the cleanliness of your booth.
5. Please rate your overall Dinner experience.
2/6

connectED.mcgraw-hill.com Your Digital Math Portal

Animation | Vocabulary | eGlossary | Personal Tutor | Virtual Manipulatives | Graphing Calculator | Audio | Foldables | Self-Check Practice | Worksheets

Richard Nowitz/National Geographic/Getty Ima

Get Ready for the Chapter

Diagnose Readiness | **You have two options for checking prerequisite skills.**

1 Textbook Option

Take the Quick Check below. Refer to the Quick Review for help.

QuickCheck

Determine the probability of each event if you randomly select a cube from a bag containing 6 red cubes, 4 yellow cubes, 3 blue cubes, and 1 green cube.

1. *P*(red)

2. *P*(blue)

3. *P*(not red)

4. *P*(white)

5. Jim rolls a die with 6 sides. What is the probability of rolling a 5?

6. Malika spins a spinner that is divided into 8 equal sections. Each section is a different color, including blue. What is the probability the spinner lands on the blue section?

Find each product.

7. $\frac{5}{4} \cdot \frac{2}{3}$

8. $\frac{4}{19} \cdot \frac{7}{20}$

9. $\frac{4}{32} \cdot \frac{7}{32}$

10. $\frac{5}{12} \cdot \frac{6}{11}$

11. $\frac{56}{100} \cdot \frac{24}{100}$

12. $\frac{9}{34} \cdot \frac{17}{27}$

Write each fraction as a percent. Round to the nearest tenth.

13. $\frac{14}{17}$

14. $\frac{7}{8}$

15. $\frac{107}{125}$

16. $\frac{625}{1024}$

17. SHOPPERS At the mall, 700 of the 2000 people shopping were under the age of 21. What percent of the shoppers were under 21?

QuickReview

Example 1

Determine the probability of selecting a green cube if you randomly select a cube from a bag containing 6 red cubes, 4 yellow cubes, and 1 green cube.

There is 1 green cube and a total of 11 cubes in the bag.

$$\frac{1}{11} = \frac{\text{number of green cubes}}{\text{total number of cubes}}$$

The probability of selecting a green cube is $\frac{1}{11}$.

Example 2

Find $\frac{4}{5} \cdot \frac{3}{4}$.

$\frac{4}{5} \cdot \frac{3}{4} = \frac{4 \cdot 3}{5 \cdot 4}$ Multiply the numerators and the denominators.

$= \frac{12}{20}$ Simplify.

$= \frac{3}{5}$ Rename in simplest form.

Example 3

Write the fraction $\frac{33}{80}$ as a percent. Round to the nearest tenth.

$\frac{33}{80} \approx 0.413$ Simplify and round.

$0.413 \cdot 100 = 41.3$ Multiply the decimal by 100.

$\frac{33}{80}$ written as a percent is about 41.3%.

2 Online Option

Take an online self-check Chapter Readiness Quiz at connectED.mcgraw-hill.com.

Get Started on the Chapter

You will learn several new concepts, skills, and vocabulary terms as you study Chapter 9. To get ready, identify important terms and organize your resources. You may wish to refer to Chapter 0 to review prerequisite skills.

FOLDABLES® Study Organizer

Statistics and Probability Make this Foldable to help you organize your Chapter 9 notes about Statistics and Probability. Begin with 8 sheets of $8\frac{1}{2}$" by 11" paper.

1 **Fold** each sheet of paper in half. Cut 1 inch from the end to the fold. Then cut 1 inch along the fold.

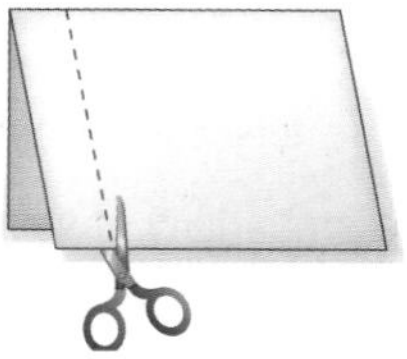

2 **Label** 7 of the 8 sheets with the lesson number and title.

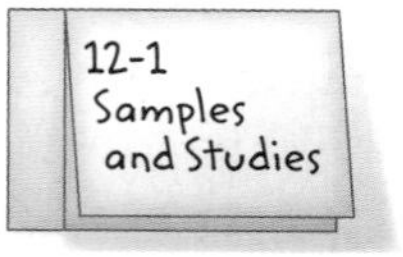

3 **Label** the inside of each sheet with Definitions and Examples.

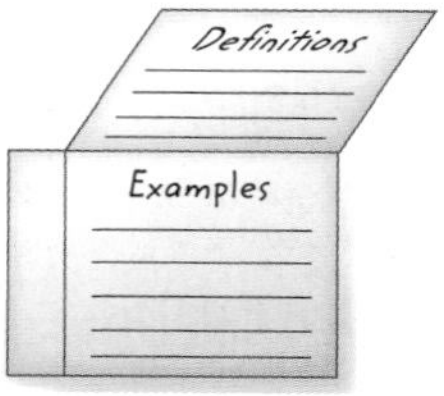

4 **Stack** the sheets. Staple along the left side. Write the title of the chapter on the first page.

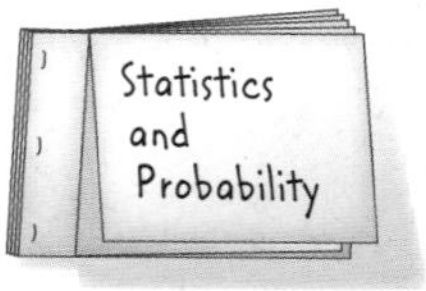

New Vocabulary

English		Español
statistical inference	p. 523	inferencía estadistica
statistic	p. 523	estadística
parameter	p. 523	parámetro
mean absolute deviation(MAD)	p. 524	desviación absolute media
standard deviation	p. 525	desviación estándar
variance	p. 525	varianza
distribution	p. 530	distribución
symmetric distribution	p. 530	distribución simétrica
linear transformation	p. 537	interpolación lineal
relative frequency	p. 547	frecuancia relativa

Review Vocabulary

probability probilidad the ratio of favorable outcomes to the total possible outcomes

sample space espacio muestral the list of all possible outcomes

LESSON 9-1

Statistics and Parameters

Then

- You analyzed data collection techniques.

Now

- **1** Identify sample statistics and population parameters.
- **2** Analyze data sets using statistics.

Why?

- At the start of every class period for one week, each of Mr. Day's algebra students randomly draws 9 pennies from a jar of 1000 pennies. Each student calculates the mean age of the random sample of pennies drawn and then returns the pennies to the jar.

How does the mean age for 9 pennies compare to the mean age of all 1000 pennies?

NewVocabulary

statistical inference
statistic
parameter
mean absolute deviation (MAD)
standard deviation
variance

Common Core State Standards

Content Standards
S.ID.2 Use statistics appropriate to the shape of the data distribution to compare center (median, mean) and spread (interquartile range, standard deviation) of two or more different data sets.

Mathematical Practices
2 Reason abstractly and quantitatively.
6 Attend to precision.

1 Statistics and Parameters

The statistics of a sample are used to draw conclusions about the entire population. This is called **statistical inference**. In the scenario above, each student takes a random sample of pennies from the jar. The jar of 1000 pennies represents the population.

A **statistic** is a measure that describes a characteristic of a sample. A **parameter** is a measure that describes a characteristic of a population. Parameters are fixed values that can be determined by the entire population, but are typically estimated based on the statistics of a carefully chosen random sample. A statistic can and usually will vary from sample to sample. A parameter will not change, for it represents the entire population.

Example 1 Statistics and Parameters

PT

Identify the sample and the population for each situation. Then describe the sample statistic and the population parameter.

a. At a local university, a random sample of 40 scholarship applicants is selected. The mean grade-point average of the 40 applicants is calculated.

Sample:	the group of 40 scholarship applicants
Population:	all applicants
Sample statistic:	mean grade-point average of the sample
Population parameter:	mean grade-point average of all applicants

b. A stratified random sample of registered nurses is selected from all hospitals in a three-county area, and the median salary is calculated.

Sample:	randomly selected registered nurses from hospitals in the three-county area
Population:	all nurses at the hospitals in the same region
Sample statistic:	median salary of nurses in the sample
Population parameter:	median salary of all nurses in all hospitals in a three-county area.

GuidedPractice

1. **CEREAL** Starting with a randomly selected box of cereal from the manufacturing line, every 50th box of cereal is removed and weighed. The mode weight of a day's sample is calculated.

Halfdark/Getty Images

2 Statistical Analysis

2 Statistical Analysis Univariate data can be represented by measures of central tendency, such as the mean, median, and mode. Univariate data can also be represented by measures of variation that assess the variability of the data. Some examples are the range, quartiles, interquartile range, mean absolute deviation, and standard deviation.

The **mean absolute deviation (MAD)** is the average of the absolute values of the differences between the mean and each value in the data set. The mean absolute deviation is used to predict errors and judge how well the mean represents the data.

KeyConcept Mean Absolute Deviation

Step 1 Find the mean, $\bar{x}$.

Step 2 Find the absolute value of the difference between each data value x_n and the mean, $|\bar{x} - x_n|$.

Step 3 Find the sum of all of the values in Step 2.

Step 4 Divide the sum by the number of values in the set of data n.

Formula $\text{MAD} = \dfrac{|\bar{x} - x_1| + |\bar{x} - x_2| + \ldots + |\bar{x} - x_n|}{n}$

StudyTip

Ellipsis The ellipsis in the formula for MAD denotes "and so on". All of the terms between the second term and the last term are implied to save space.

PT

Example 2 Mean Absolute Deviation

MARKETING Each person who visited the Comic Book Shoppe's Web site was asked to enter the number of comic books they buy each month. They received the following responses in one day: {2, 2, 3, 4, 14}. Find and interpret the mean absolute deviation.

Step 1 Find the mean.

$$\bar{x} = \frac{2 + 2 + 3 + 4 + 14}{5} \text{ or } 5$$

Step 2 Find the absolute values of the differences.

$x_1 = 2: |\bar{x} - x_1| = |5 - 2| \text{ or } 3$ $\quad$ $x_2 = 2: |\bar{x} - x_2| = |5 - 2| \text{ or } 3$

$x_3 = 3: |\bar{x} - x_3| = |5 - 3| \text{ or } 2$ $\quad$ $x_4 = 4: |\bar{x} - x_4| = |5 - 4| \text{ or } 1$

$x_5 = 14: |\bar{x} - x_5| = |5 - 14| \text{ or } 9$

Step 3 Find the sum.

$$3 + 3 + 2 + 1 + 9 = 18$$

Step 4 Find the mean absolute deviation.

$$\text{MAD} = \frac{|\bar{x} - x_1| + |\bar{x} - x_2| + \ldots + |\bar{x} - x_n|}{n}$$ Formula for Mean Absolute Deviation

$$= \frac{18}{5} \text{ or } 3.6$$ The sum is 18 and $n = 5$.

A mean absolute deviation of 3.6 indicates that the data, on average, are 3.6 units away from the mean. This value is significantly influenced by the outlier 14. Without the outlier, the data set would have a mean of 2.75 and a mean absolute deviation of 0.75.

GuidedPractice

2. DANCES The prom committee kept a count of how many tickets it sold each day during lunch: {12, 32, 36, 41, 22, 47, 51, 33, 37, 49}. Find and interpret the mean absolute deviation of these data.

In a set of data, the **standard deviation** shows how the data deviate from the mean. A low standard deviation indicates that the data tend to be very close to the mean, while a high standard deviation indicates that the data are spread out over a larger range of values.

The standard deviation is represented by the lowercase Greek letter sigma, σ. The **variance** σ^2 of the data is the square of the standard deviation.

StudyTip

Symbols The mean of a sample and the mean of a population are calculated the same way. $\bar{x}$ refers to the mean of a sample and μ refers to the mean of a population. In this text, $\bar{x}$ will refer to both.

KeyConcept Standard Deviation

Step 1 Find the mean, $\bar{x}$.

Step 2 Find the square of the difference between each data value x_n and the mean, $(\bar{x} - x_n)^2$.

Step 3 Find the sum of all of the values in Step 2.

Step 4 Divide the sum by the number of values in the set of data n. This value is the variance.

Step 5 Take the square root of the variance.

Formula $\sigma = \sqrt{\dfrac{(\bar{x} - x_1)^2 + (\bar{x} - x_2)^2 + \ldots + (\bar{x} - x_n)^2}{n}}$

PT

Example 3 Variance and Standard Deviation

ELECTRONICS Ed surveys his classmates to find out how many electronic gadgets each person has in their home. Find and interpret the standard deviation of the data set.

{9, 10, 11, 6, 9, 11, 9, 8, 11, 8, 7, 9, 11, 11, 5}

Step 1 Find the mean.

$$\bar{x} = \frac{9 + 10 + 11 + 6 + 9 + 11 + 9 + 8 + 11 + 8 + 7 + 9 + 11 + 11 + 5}{15} \text{ or } 9$$

Step 2 Find the square of the differences, $(\bar{x} - x_n)^2$.

$(9 - 9)^2 = 0$	$(9 - 10)^2 = 1$	$(9 - 11)^2 = 4$	$(9 - 6)^2 = 9$	$(9 - 9)^2 = 0$
$(9 - 11)^2 = 4$	$(9 - 9)^2 = 0$	$(9 - 8)^2 = 1$	$(9 - 11)^2 = 4$	$(9 - 8)^2 = 1$
$(9 - 7)^2 = 4$	$(9 - 9)^2 = 0$	$(9 - 11)^2 = 4$	$(9 - 11)^2 = 4$	$(9 - 5)^2 = 16$

Step 3 Find the sum.

$$0 + 1 + 4 + 9 + 0 + 4 + 0 + 1 + 4 + 1 + 4 + 0 + 4 + 4 + 16 = 52$$

Step 4 Find the variance.

$$\sigma^2 = \frac{(\bar{x} - x_1)^2 + (\bar{x} - x_2)^2 + \ldots + (\bar{x} - x_n)^2}{n}$$ Formula for Variance

$$= \frac{52}{15} \text{ or about } 3.47$$ The sum is 52 and $n = 15$.

Step 5 Find the standard deviation.

$$\sigma = \sqrt{\sigma^2}$$ Square Root of the Variance

$$\approx \sqrt{3.47} \text{ or about } 1.86$$

A standard deviation of 1.86 is small compared to the mean of 9. This suggests that most of the data values are relatively close to the mean.

GuidedPractice

3. DIET Caleb tracked his Calorie intake for a week. Find and interpret the standard deviation of his Calorie intake.

1950, 2000, 2100, 2000, 1900, 2100, 2000

The mean and standard deviation can be used to compare two different sets of data.

StudyTip

Symbols The standard deviation of a sample s and the standard deviation of a population σ are calculated in different ways. In this text, you will calculate the standard deviation of a population.

Example 4 Compare Two Sets of Data

Miguel plays golf at Table Rock and Blackhawk golf courses. Compare the means and standard deviations of each set of Miguel's scores.

Table Rock				
81	78	79	82	80
80	79	83	81	80

Blackhawk				
84	79	86	78	77
88	85	79	87	86

Use a graphing calculator to find the mean and standard deviation. Clear all lists. Then press STAT ENTER, and enter each data value into **L1**. To view the statistics, press STAT ▶ 1 ENTER.

Table Rock

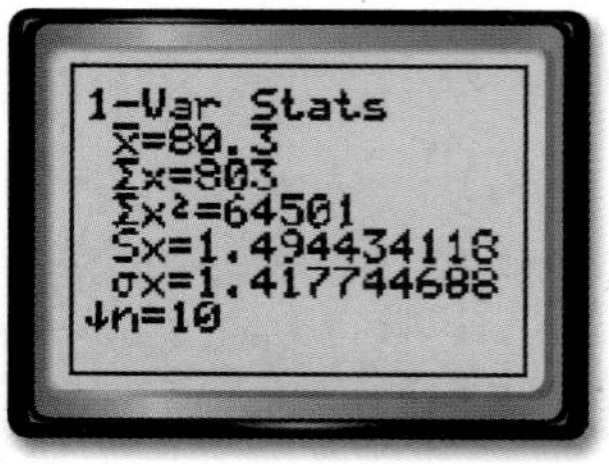

Blackhawk

Miguel's mean score at Table Rock is 80.3 with a standard deviation of about 1.4. His mean score at Blackhawk is 82.9 with a standard deviation of about 4.0. Therefore, he tends to score lower at Table Rock. The greater standard deviation at Blackhawk indicates that there is greater variability to his scores at that course, but he is more consistent at Table Rock.

GuidedPractice

4. SWIMMING Anna is considering two different lineups for her 4×100 relay team. Below are the times in minutes recorded for each lineup. Compare the means and standard deviations of each set of data.

Lineup A				
4.25	4.31	4.19	4.40	4.23
4.18	4.71	4.56	4.32	4.39

Lineup B				
4.47	4.68	4.25	4.41	4.49
4.18	4.27	4.69	4.32	4.44

Check Your Understanding

Example 1

1. BOOKS A stratified random sample of 1000 college students in the United States is surveyed about how much money they spend on books per year. Identify the sample and the population. Then describe the sample statistic and the population parameter.

Example 2

2. AMUSEMENT PARKS An amusement park manager kept track of how many bags of cotton candy they sold each hour on a Saturday: {16, 24, 15, 17, 22, 16, 18, 24, 17, 13, 25, 21}. Find and interpret the mean absolute deviation.

Example 3

3. PART-TIME JOBS Ms. Johnson asks all of the members of the girls' tennis team to find the number of hours each week they work at part-time jobs: {10, 12, 0, 6, 9, 15, 12, 10, 11, 20}. Find and interpret the standard deviation of the data set.

Example 4

4. CCSS MODELING Mr. Jones recorded the number of pull-ups done by his students. Compare the means and standard deviations of each group.
Boys: {5, 16, 3, 8, 4, 12, 2, 15, 0, 1, 9, 3} Girls: {2, 4, 0, 3, 5, 4, 6, 1, 3, 8, 3, 4}

Practice and Problem Solving

Example 1 **Identify the sample and the population for each situation. Then describe the sample statistic and the population parameter.**

5. POLITICS A random sample of 1003 Mercy County voters is asked if they would vote for the incumbent for governor. The percent responding *yes* is calculated.

6. ACTIVITIES A stratified random sample of high school students from each school in the county was polled about the time spent each week on extracurricular activities.

7. MONEY A stratified random sample of 2500 high school students across the country was asked how much money they spent each month.

Example 2 **8. DVDS** A math teacher asked all of his students to count the number of DVDs they owned. Find and interpret the mean absolute deviation.

Number of DVDs					
26	39	5	82	12	14
0	3	15	19	41	6
2	0	11	1	19	29

9. SWIMMING The owner of a public swimming pool tracked the daily attendance. Find and interpret the mean absolute deviation.

Daily Attendance					
86	45	91	104	95	88
111	85	79	102	166	103
89	94	79	103	88	84

Example 3 **10. CCSS REASONING** Samantha wants to see if she is getting a fair wage for babysitting at \$8.50 per hour. She takes a survey of her friends to see what they charge per hour. The results are {\$8.00, \$8.50, \$9.00, \$7.50, \$15.00, \$8.25, \$8.75}. Find and interpret the standard deviation of the data.

11. ARCHERY Carla participates in competitive archery. Each competition allows a maximum of 90 points. Carla's results for the last 8 competitions are {76, 78, 81, 75, 80, 80, 76, 77}. Find and interpret the standard deviation of the data.

Example 4 **12. BASKETBALL** The coach of the Wildcats basketball team is comparing the number of fouls called against his team with the number called against their rivals, the Trojans. He records the number of fouls called against each team for each game of the season. Compare the means and standard deviations of each set of data.

Wildcats			
15	12	13	9
11	12	14	12
8	16	9	9
11	13	12	14

Trojans			
9	10	14	13
7	8	10	10
9	7	11	9
12	11	13	8

13. MOVIE RATINGS Two movies were rated by the same group of students. Ratings were from 1 to 10, with 10 being the best.

a. Compare the means and standard deviations of each set of data.

b. Provide an argument for why Movie A would be preferred. Movie B?

Movie A			
7	8	7	6
8	6	7	8
6	8	8	6
7	7	8	8

Movie B			
9	5	10	6
3	10	9	4
8	3	9	9
2	8	10	3

14. PENNIES Mr. Day has another jar of pennies on his desk. There are 30 pennies in this jar. Theo chooses 5 pennies from the jar. Lola chooses 10 pennies, and Peter chooses 20 pennies. Pennies are chosen and replaced.

Years of Pennies in Jar					
2001	1990	2000	1982	1991	1975
2007	1981	2005	2007	2003	2005
1997	1974	1992	1994	1991	1992
2000	1995	1999	2005	2006	2005
2004	2004	1998	2001	2002	2006

a. Theo's pennies are {1974, 1975, 1981, 1999, 1992}. Find the mean absolute deviation.

b. Lola's pennies are {2004, 1999, 2004, 2005, 1991, 2003, 2005, 2000, 2001, 1998}. Find the mean absolute deviation.

c. Peter's pennies are {2007, 2005, 1975, 2003, 2005, 1997, 1992, 1994, 1991, 1992, 2000, 1999, 2005, 1982, 2005, 2004, 1998, 2001, 2002, 2006}. Find the mean absolute deviation.

d. Find the mean absolute deviation for all of the pennies in the jar. Which sample most accurately reflected the population mean? Explain.

15. RUNNING The results of a 5K race are published in a local paper. Over a thousand people participated, but only the times of the top 15 finishers are listed.

15th Annual 5K Road Race					
Place	**Time (min:s)**	**Place**	**Time (min:s)**	**Place**	**Time (min:s)**
1	15:56	6	16:34	11	17:14
2	16:06	7	16:41	12	17:46
3	16:11	8	16:54	13	17:56
4	16:21	9	17:00	14	17:57
5	16:26	10	17:03	15	18:03

a. Find the mean and standard deviation of the top 15 running times. (*Hint*: Convert each time to seconds.)

b. Identify the sample and population.

c. Analyze the sample. Classify the data as *quantitative* or *qualitative*. Can a statistical analysis of the sample be applied to the population? Explain your reasoning.

H.O.T. Problems Use Higher-Order Thinking Skills

16. CCSS CRITIQUE Jennifer and Megan are determining one way to decrease the size of the standard deviation of a set of data. Is either of them correct? Explain.

Jennifer	Megan
Remove the outliers from the data set.	*Add data values to the data set that are equal to the mean.*

17. REASONING Determine whether the statement *Two random samples taken from the same population will have the same mean and standard deviation* is *sometimes*, *always*, or *never* true. Explain.

18. OPEN ENDED Describe a situation in which it would be useful to use a sample mean to help estimate a population mean. How could you collect a random sample?

19. CHALLENGE Write a set of data with a standard deviation that is equal to the mean absolute deviation.

WRITING IN MATH **Compare and contrast each of the following.**

20. statistics and parameters

21. standard deviation and mean absolute deviation

Standardized Test Practice

22. Melina bought a shirt that was marked 20% off. She paid \$15.75. What was the original price?

A \$16.69 **C** \$18.69
B \$17.69 **D** \$19.69

23. SHORT RESPONSE A group of student ambassadors visited the Capitol building. Twenty students met with the local representative. This was 16% of the students. How many student ambassadors were there altogether?

24. The tallest 7 trees in a park have heights in meters of 19, 24, 17, 26, 24, 20, and 18. Find the mean absolute deviation of their heights.

F 3.0 **H** 3.4
G 3.2 **J** 21

25. It takes 3 hours for a boat to travel 27 miles upstream. The same boat can travel 30 miles downstream in 2 hours. Find the speed of the boat.

A 3 mph **C** 12 mph
B 5 mph **D** 14 mph

Spiral Review

26. GEOMETRY If the side length of a cube is s, the volume is represented by s^3, and the surface area is represented by $6s^2$. (Lessons 7-1 and 7-2)

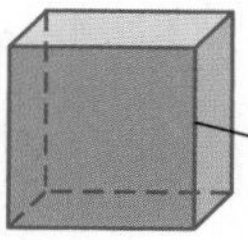

a. Are the expressions for volume and surface area monomials? Explain.

b. If the side of a cube measures 3 feet, find the volume and surface area.

c. Find a side length s such that the volume and surface area have the same measure.

d. The volume of a cylinder can be found by multiplying the radius squared times the height times π, or $V = \pi r^2 h$. Suppose you have two cylinders. Each measure of the second is twice the measure of the first, so $V = \pi(2r)^2(2h)$. What is the ratio of the volume of the first cylinder to the second cylinder?

Skills Review

Find the range, median, lower quartile, and upper quartile for each set of data.

27. {15, 23, 46, 36, 15, 19}

28. {55, 57, 39, 72, 46, 53, 81}

29. {21, 25, 19, 18, 22, 16, 27}

30. {52, 29, 72, 64, 33, 49, 51, 68}

31. {8, 12, 9, 11, 11, 10, 14, 18}

32. {133, 119, 147, 94, 141, 106, 118, 149}

LESSON 9-2

Distributions of Data

Then

- You calculated measures of central tendency and variation.

Now

1. Describe the shape of a distribution.
2. Use the shapes of distributions to select appropriate statistics.

Why?

- While training for the 100-meter dash, Sarah pulled a muscle in her lower back. After being cleared for practice, she continued to train. Sarah's median time was about 12.34 seconds, but her average time dropped to about 12.53 seconds.

NewVocabulary

distribution
negatively skewed distribution
symmetric distribution
positively skewed distribution

Common Core State Standards

Content Standards

S.ID.2 Use statistics appropriate to the shape of the data distribution to compare center (median, mean) and spread (interquartile range, standard deviation) of two or more different data sets.

S.ID.3 Interpret differences in shape, center, and spread in the context of the data sets, accounting for possible effects of extreme data points (outliers).

Mathematical Practices

5 Use appropriate tools strategically.

1 Describing Distributions

A **distribution** of data shows the observed or theoretical frequency of each possible data value. Recall that a histogram is a type of bar graph used to display data that have been organized into equal intervals. A histogram is useful when viewing the overall distribution of the data within a set over its range. You can see the shape of the distribution by drawing a curve over the histogram.

KeyConcept Symmetric and Skewed Distributions

Negatively Skewed Distribution

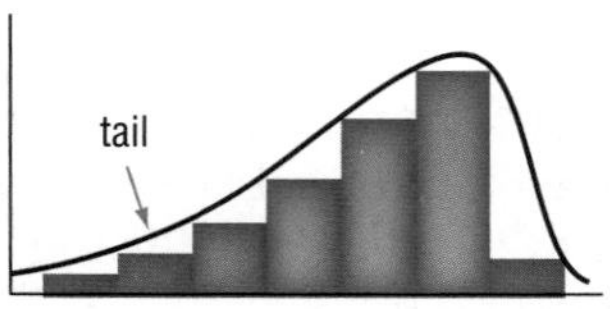

The majority of the data are on the right.

Symmetric Distribution

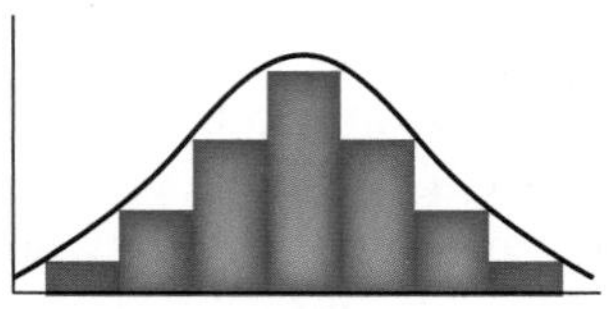

The data are evenly distributed.

Positively Skewed Distribution

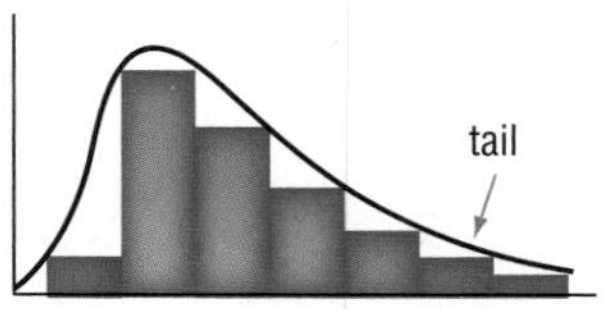

The majority of the data are on the left.

Example 1 Distribution Using a Histogram

Use a graphing calculator to construct a histogram for the data, and use it to describe the shape of the distribution.

25, 22, 31, 25, 26, 35, 18, 39, 22, 32, 34, 26, 42, 23, 40, 36, 18, 30
26, 30, 37, 23, 19, 33, 24, 29, 39, 21, 43, 25, 34, 24, 26, 30, 21, 22

First, press [STAT] [ENTER] and enter each data value. Then, press [2nd] [STAT PLOT] [ENTER] [ENTER] and choose ⊪. Press [ZOOM] [ZoomStat] to adjust the window.

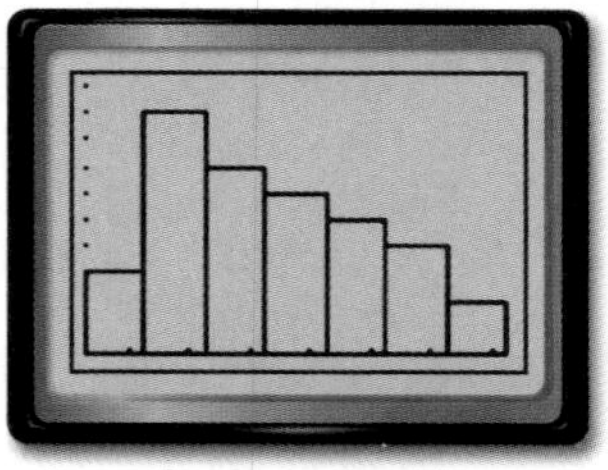

[17, 45] scl: 4 by [0, 10] scl: 1

The graph is high on the left and has a tail on the right. Therefore, the distribution is positively skewed.

GuidedPractice

1. Use a graphing calculator to construct a histogram for the data, and use it to describe the shape of the distribution.

8, 11, 15, 25, 21, 26, 20, 12, 32, 20, 31, 14, 19, 27, 22, 21, 14, 8
6, 23, 18, 16, 28, 25, 16, 20, 29, 24, 17, 35, 20, 27, 10, 16, 22, 12

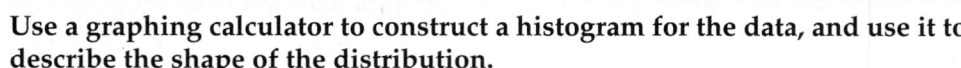

A box-and-whisker plot can also be used to identify the shape of a distribution. Recall from Lesson 0-13 that a box-and-whisker plot displays the spread of a data set by dividing it into four quartiles. The data from Example 1 are displayed below.

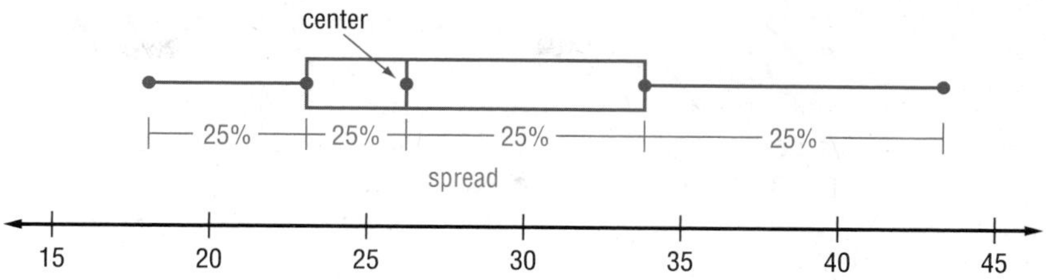

Notice that the left whisker is shorter than the right whisker, and that the line representing the median is closer to the left whisker. This represents a peak on the left and a tail to the right.

KeyConcept Symmetric and Skewed Box-and-Whisker Plots

Negatively Skewed	Symmetric	Positively Skewed
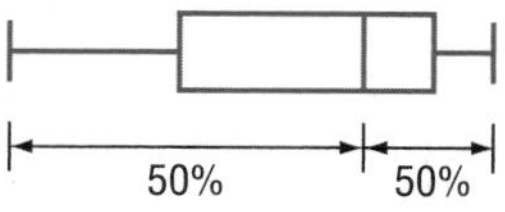	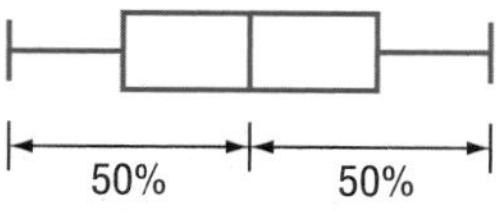	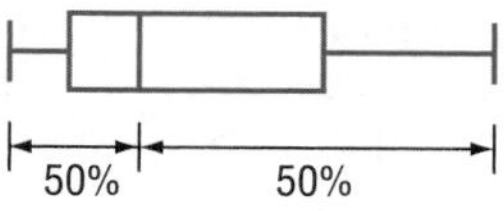
The left whisker is longer than the right. The median is closer to the shorter whisker.	The whiskers are the same length. The median is in the center of the data.	The right whisker is longer than the left. The median is closer to the shorter whisker.

Example 2 Distribution Using a Box-and-Whisker Plot

Use a graphing calculator to construct a box-and-whisker plot for the data, and use it to determine the shape of the distribution.

9, 17, 15, 10, 16, 2, 17, 19, 10, 18, 14, 8, 20, 20, 3, 21, 12, 11
5, 26, 15, 28, 12, 5, 27, 26, 15, 53, 12, 7, 22, 11, 8, 16, 22, 15

StudyTip

Outliers In Example 2, notice that the outlier does not affect the shape of the distribution.

Enter the data as **L1**. Press [2nd] **[STAT PLOT]** [ENTER] [ENTER] and choose ·□··. Adjust the window to the dimensions shown.

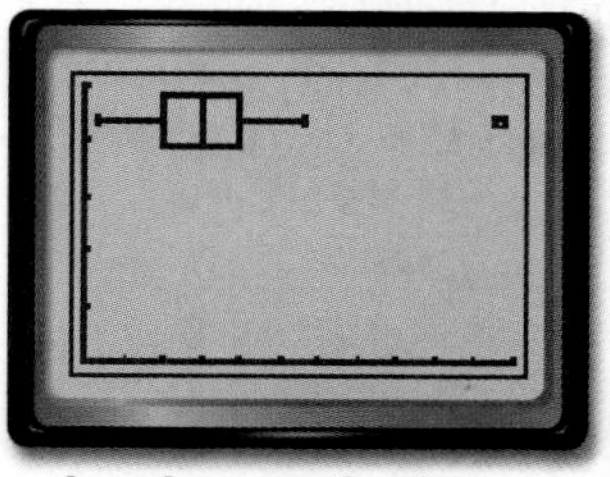

[0, 55] scl: 5 by [0, 5] scl: 1

The lengths of the whiskers are approximately equal, and the median is in the middle of the data. This indicates that the data are equally distributed to the left and right of the median. Thus, the distribution is symmetric.

GuidedPractice

2. Use a graphing calculator to construct a box-and-whisker plot for the data, and use it to describe the shape of the distribution.

40, 50, 35, 48, 43, 31, 52, 42, 54, 38, 50, 46, 49, 43, 40, 50, 32, 53
51, 43, 47, 41, 49, 50, 34, 54, 51, 44, 54, 39, 47, 35, 51, 44, 48, 37

2 Analyzing Distributions

You have learned that data can be described using statistics. The mean and median describe the center. The standard deviation and quartiles describe the spread. You can use the shape of the distribution to choose the most appropriate statistics that describe the center and spread of a set of data.

When a distribution is symmetric, the mean accurately reflects the center of the data. However, when a distribution is skewed, this statistic is not as reliable.

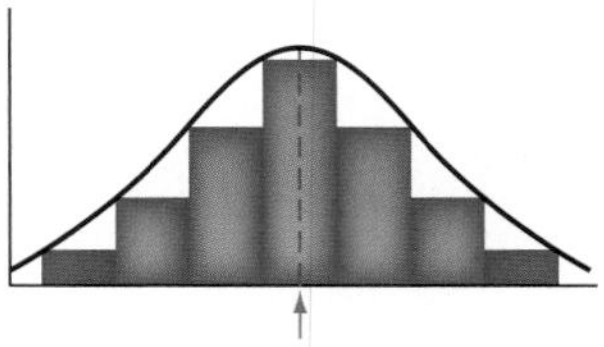

Outliers can have a strong effect on the mean of a data set, while the median is less affected. So, when a distribution is skewed, the mean lies away from the majority of the data toward the tail. The median is less affected and stays near the majority of the data.

Negatively Skewed Distribution

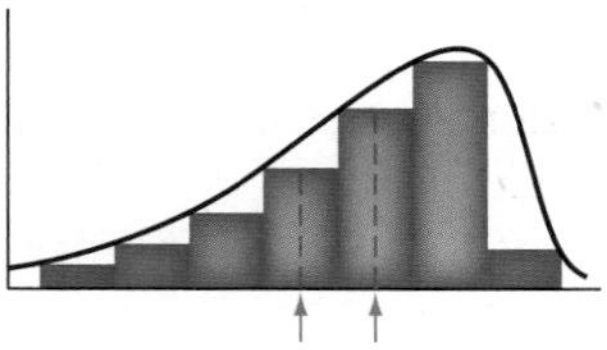

Positively Skewed Distribution

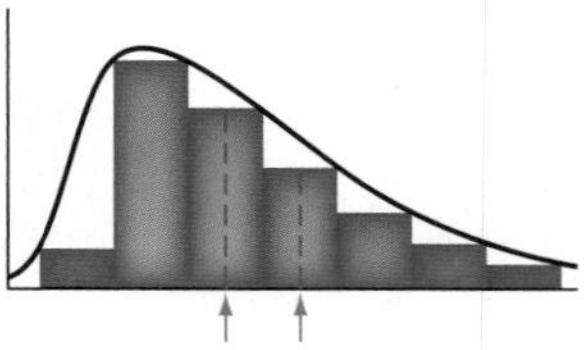

When choosing appropriate statistics to represent a set of data, first determine the shape of the distribution.

- If the distribution is relatively symmetric, the mean and standard deviation can be used.
- If the distribution is skewed or has outliers, use the five-number summary.

Example 3 Choose Appropriate Statistics

Describe the center and spread of the data using either the mean and standard deviation or the five-number summary. Justify your choice by constructing a histogram for the data.

21, 28, 16, 30, 25, 34, 21, 47, 18, 36, 24, 28, 30, 15, 33, 24, 32, 22
27, 38, 23, 29, 15, 27, 33, 19, 34, 29, 23, 26, 19, 30, 25, 13, 20, 25

Use a graphing calculator to create a histogram. The graph is high in the middle and low on the left and right. Therefore, the distribution is symmetric.

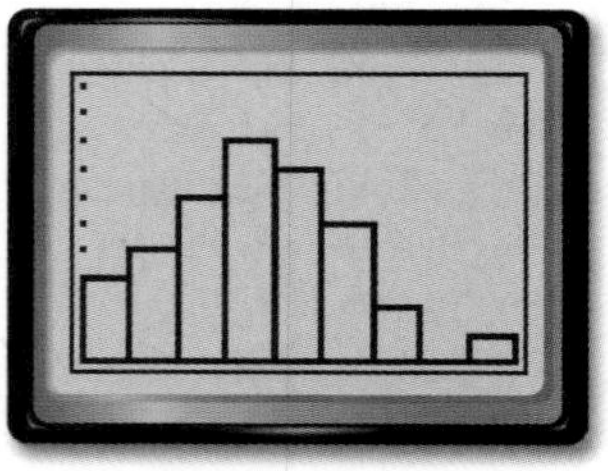

[12, 48] scl: 4 by [0, 10] scl: 1

The distribution is symmetric, so use the mean and standard deviation to describe the center and spread. Press STAT ▶ ENTER ENTER.

The mean $\bar{x}$ is about 26.1 with standard deviation σ of about 7.1.

> **Technology Tip**
>
> **CCSS Tools** On a graphing calculator, each bar is called a *bin*. The width of each bin can be adjusted by pressing WINDOW and changing Xscl. View the histogram using different bin widths and compare the results to determine the appropriate bin width.

Guided Practice

3. Describe the center and spread of the data using either the mean and standard deviation or the five-number summary. Justify your choice by creating a histogram for the data.

19, 2, 25, 14, 24, 20, 27, 30, 14, 25, 19, 32, 21, 31, 25, 16, 24, 22
29, 6, 26, 32, 17, 26, 24, 26, 32, 10, 28, 19, 26, 24, 11, 23, 19, 8

A box-and-whisker plot is helpful when viewing a skewed distribution since it is constructed using the five-number summary.

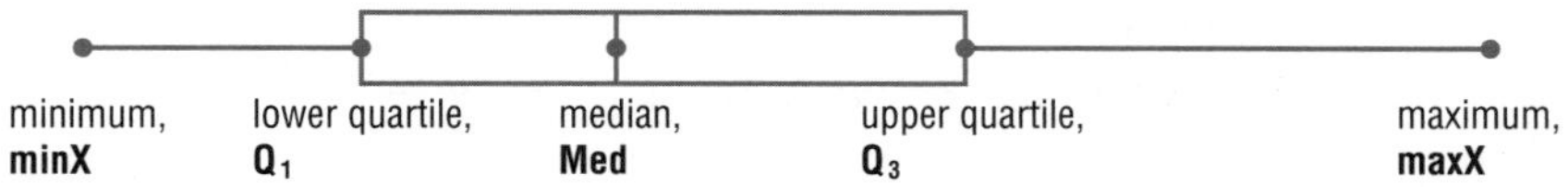

Real-World Link

Volunteers in the Peace Corps must be at least 18 years old, and more than 90% of volunteers have college degrees. Volunteers work in another country for 27 months and are placed in host countries that have the greatest needs for skilled volunteers.

Source: Peace Corps

Real-World Example 4 Choose Appropriate Statistics

COMMUNITY SERVICE The number of community service hours each of Ms. Tucci's students completed is shown. Describe the center and spread of the data using either the mean and standard deviation or the five-number summary. Justify your choice by constructing a box-and-whisker plot for the data.

Community Service Hours												
6	13	8	7	19	12	2	19	11	22	7	33	13
3	8	10	5	25	16	6	14	7	20	10	30	

Use a graphing calculator to create a box-and-whisker plot. The right whisker is longer than the left and the median is closer to the left whisker. Therefore, the distribution is positively skewed.

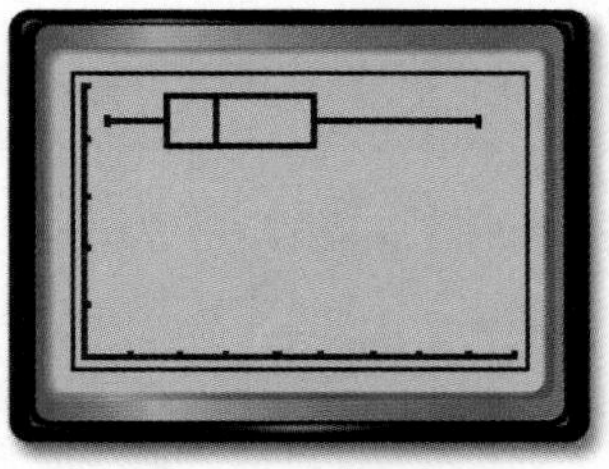

[0, 36] scl: 4 by [0, 5] scl: 1

The distribution is positively skewed, so use the five-number summary. The range is 33 – 2 or 31. The median number of hours completed is 11, and half of the students completed between 7 and 19 hours.

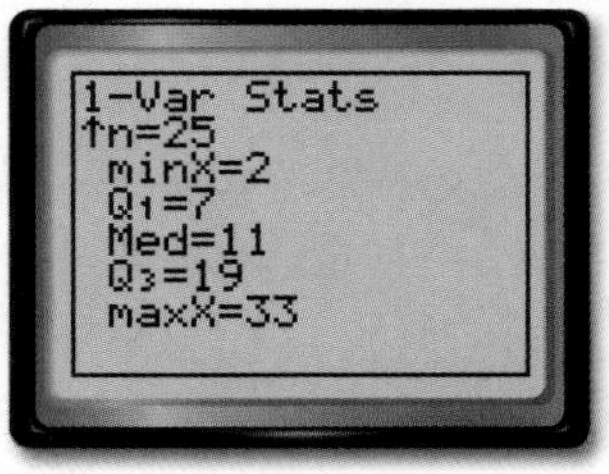

Guided Practice

4. FUNDRAISER The money raised per student in Mr. Bulanda's 5th period class is shown. Describe the center and spread of the data using either the mean and standard deviation or the five-number summary. Justify your choice by creating a box-and-whisker plot for the data.

Money Raised per Student (dollars)									
41	27	52	18	42	32	16	95	27	65
36	45	5	34	50	15	62	38	57	20
38	21	33	58	25	42	31	8	40	28

Check Your Understanding

Examples 1–2 **Use a graphing calculator to construct a histogram and a box-and-whisker plot for the data. Then describe the shape of the distribution.**

1. 80, 84, 68, 64, 57, 88, 61, 72, 76, 80, 83, 77, 78, 82, 65, 70, 83, 78
73, 79, 70, 62, 69, 66, 79, 80, 86, 82, 73, 75, 71, 81, 74, 83, 77, 73

2. 30, 24, 35, 84, 60, 42, 29, 16, 68, 47, 22, 74, 34, 21, 48, 91, 66, 51
33, 29, 18, 31, 54, 75, 23, 45, 25, 32, 57, 40, 23, 32, 47, 67, 62, 23

Example 3 **Describe the center and spread of the data using either the mean and standard deviation or the five-number summary. Justify your choice by constructing a histogram for the data.**

3. 58, 66, 52, 75, 60, 56, 78, 63, 59, 54, 60, 67, 72, 80, 68, 88, 55, 60
59, 61, 82, 70, 67, 60, 58, 86, 74, 61, 92, 76, 58, 62, 66, 74, 69, 64

Example 4

4. **PRESENTATIONS** The length of the students' presentations in Ms. Monroe's 2nd period class are shown. Describe the center and spread of the data using either the mean and standard deviation or the five-number summary. Justify your choice by constructing a box-and-whisker plot for the data.

Presentations

20, 18, 15, 17, 18, 10, 15
10, 18, 19, 17, 19, 12, 6
19, 15, 21, 10, 9, 18

Practice and Problem Solving

Examples 1–2 **Use a graphing calculator to construct a histogram and a box-and-whisker plot for the data. Then describe the shape of the distribution.**

5. 55, 65, 70, 73, 25, 36, 33, 47, 52, 54, 55, 60, 45, 39, 48, 55, 46, 38
50, 54, 63, 31, 49, 54, 68, 35, 27, 45, 53, 62, 47, 41, 50, 76, 67, 49

6. 42, 48, 51, 39, 47, 50, 48, 51, 54, 46, 49, 36, 50, 55, 51, 43, 46, 37
50, 52, 43, 40, 33, 51, 45, 53, 44, 40, 52, 54, 48, 51, 47, 43, 50, 46

Example 3 **Describe the center and spread of the data using either the mean and standard deviation or the five-number summary. Justify your choice by constructing a histogram for the data.**

7. 32, 44, 50, 49, 21, 12, 27, 41, 48, 30, 50, 23, 37, 16, 49, 53, 33, 25
35, 40, 48, 39, 50, 24, 15, 29, 37, 50, 36, 43, 49, 44, 46, 27, 42, 47

8. 82, 86, 74, 90, 70, 81, 89, 88, 75, 72, 69, 91, 96, 82, 80, 78, 74, 94
85, 77, 80, 67, 76, 84, 80, 83, 88, 92, 87, 79, 84, 96, 85, 73, 82, 83

Example 4

9. **WEATHER** The daily low temperatures for New Carlisle over a 30-day period are shown. Describe the center and spread of the data using either the mean and standard deviation or the five-number summary. Justify your choice by constructing a box-and-whisker plot for the data.

Temperature (°F)														
48	50	55	53	57	53	44	61	57	49	51	58	46	54	57
50	55	47	57	48	58	53	49	56	59	52	48	55	53	51

10. TRACK Refer to the beginning of the lesson. Sarah's 100-meter dash times are shown.

a. Use a graphing calculator to create a box-and-whisker plot. Describe the center and spread of the data.

b. Sarah's slowest time prior to pulling a muscle was 12.50 seconds. Use a graphing calculator to create a box-and-whisker plot that *does not* include the times that she ran after pulling the muscle. Then describe the center and spread of the new data set.

100-meter dash (seconds)				
12.20	12.35	13.60	12.24	12.72
12.18	12.06	12.41	12.28	13.06
12.87	12.04	12.38	12.20	13.12
12.30	13.27	12.93	12.16	12.02
12.50	12.14	11.97	12.24	13.09
12.46	12.33	13.57	11.96	13.34

c. What effect does removing the times recorded after Sarah pulled a muscle have on the shape of the distribution and on how you should describe the center and spread?

11. MENU The prices for entrees at a restaurant are shown.

a. Use a graphing calculator to create a box-and-whisker plot. Describe the center and spread of the data.

b. The owner of the restaurant decides to eliminate all entrees that cost more than $15. Use a graphing calculator to create a box-and-whisker plot that reflects this change. Then describe the center and spread of the new data set.

Entree Prices ($)				
9.00	11.25	16.50	9.50	13.00
18.50	7.75	11.50	13.75	9.75
8.00	16.50	12.50	10.25	17.75
13.00	10.75	16.75	8.50	11.50

H.O.T. Problems Use Higher-Order Thinking Skills

CHALLENGE Identify the box-and-whisker plot that corresponds to each of the following histograms.

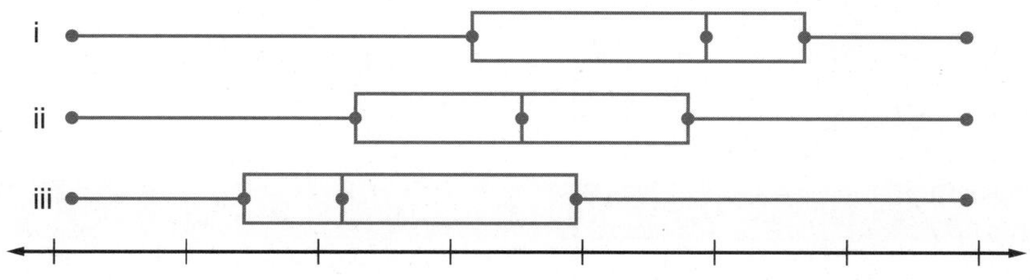

12.

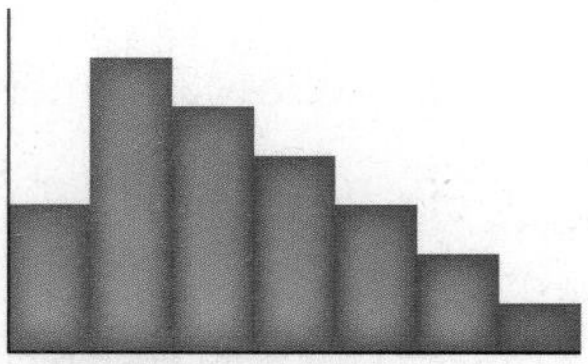

13.

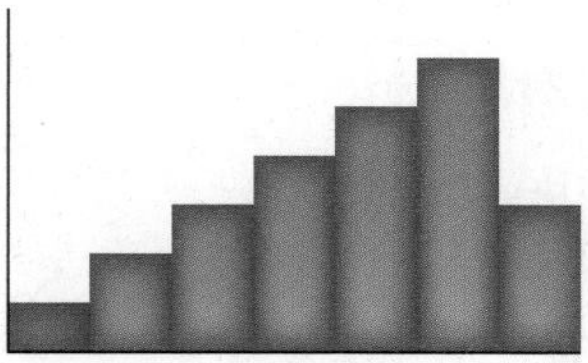

14.

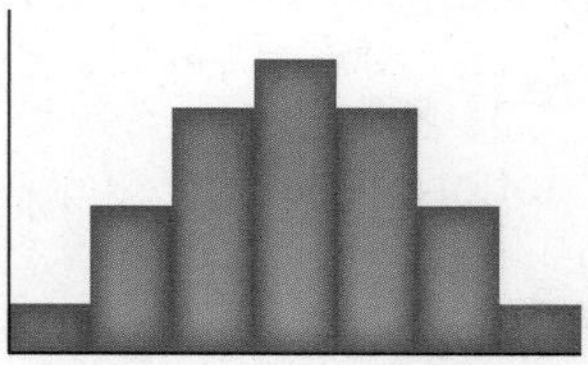

15. CCSS ARGUMENTS Research and write a definition for a *bimodal distribution*. How can the measures of center and spread of a bimodal distribution be described?

16. OPEN ENDED Give an example of a set of real-world data with a distribution that is symmetric and one with a distribution that is not symmetric.

17. WRITING IN MATH Explain why the mean and standard deviation are used to describe the center and spread of a symmetrical distribution and the five-number summary is used to describe the center and spread of a skewed distribution.

Standardized Test Practice

18. At the county fair, 1000 tickets were sold. Adult tickets cost \$8.50, children's tickets cost \$4.50, and a total of \$7300 was collected. How many children's tickets were sold?

A 700   C 400

B 600   D 300

19. Edward has 20 dimes and nickels, which together total \$1.40. How many nickels does he have?

F 12   H 8

G 10   J 6

20. If 4.5 kilometers is about 2.8 miles, about how many miles is 6.1 kilometers?

A 3.2 miles   C 3.8 miles

B 3.6 miles   D 4.0 miles

21. EXTENDED RESPONSE Three times the width of a certain rectangle exceeds twice its length by three inches, and four times its length is twelve more than its perimeter.

a. Translate the sentences into equations.

b. Find the dimensions of the rectangle.

c. What is the area of the rectangle?

Spiral Review

Identify the sample and the population for each situation. Then describe the sample statistic and the population parameter. (Lesson 9-1)

22. AMUSEMENT PARK A systematic sample of 250 guests is asked how much money they spent on concessions inside the park. The median amount of money is calculated.

23. PROM A random sample of 100 high school seniors at North Boyton High School is surveyed, and the mean amount of money spent on prom by a senior is calculated.

Find the inverse of each function. (Lesson 4-7)

24. $f(x) = 2x - 14$

25. $f(x) = 17 - 5x$

26. $f(x) = \frac{1}{4}x + 3$

27. $f(x) = -\frac{1}{7}x - 1$

28. $f(x) = \frac{2}{3}x + 6$

29. $f(x) = 12 - \frac{3}{5}x$

Skills Review

A bowl contains 3 red chips, 6 green chips, 5 yellow chips, and 8 orange chips. A chip is drawn randomly. Find each probability.

30. red

31. orange

32. yellow or green

33. not orange

34. not green

35. red or orange

LESSON

9-3 Comparing Sets of Data

Then

- You calculated measures of central tendency and variation.

Now

1. Determine the effect that transformations of data have on measures of central tendency and variation.
2. Compare data using measures of central tendency and variation.

Why?

- Tom gets paid hourly to do landscaping work. Because he is such a good employee, Tom is planning to ask his boss for a bonus. Tom's initial pay for a month is shown. He is trying to decide whether he should ask for an extra $5 per day or a 10% increase in his daily wages.

Tom's Pay ($)		
44	52	50
40	48	46
44	52	54
58	42	52
54	50	52
42	52	46
56	48	44
50	42	

NewVocabulary
linear transformation

Common Core State Standards

Content Standards

S.ID.2 Use statistics appropriate to the shape of the data distribution to compare center (median, mean) and spread (interquartile range, standard deviation) of two or more different data sets.

S.ID.3 Interpret differences in shape, center, and spread in the context of the data sets, accounting for possible effects of extreme data points (outliers).

Mathematical Practices

1 Make sense of problems and persevere in solving them.

1 Transformations of Data

To see the effect that an extra $5 per day would have on Tom's daily pay, we can find the new daily pay values and compare the measures of center and variation for the two sets of data. The new data can be found by performing a *linear transformation*. A **linear transformation** is an operation performed on a data set that can be written as a linear function. Tom's daily pay after the $5 bonus can be found using $y = 5 + x$, where x represents his original daily pay and y represents his daily pay after the bonus.

Tom's Earnings Before Extra $5

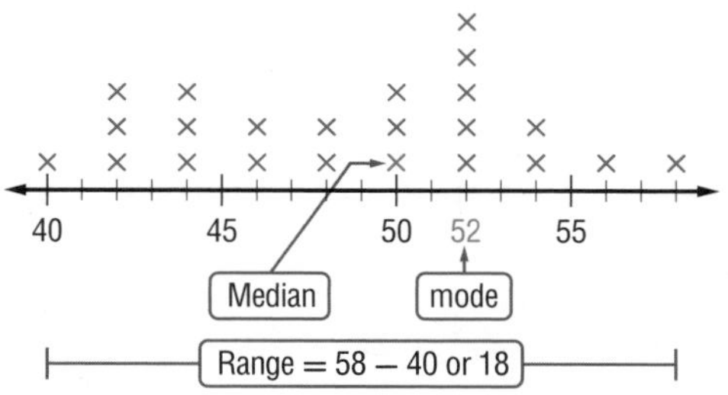

Mean 48.6 **Standard Deviation** 4.9

Tom's Earnings With Extra $5

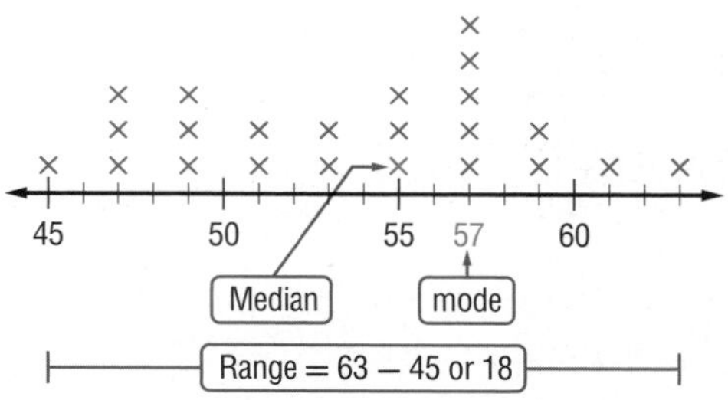

Mean 53.6 **Standard Deviation** 4.9

Notice that each value was translated 5 units to the right. Thus, the mean, median, and mode increased by 5. Since the new minimum and maximum values also increased by 5, the range remained the same. The standard deviation is unchanged because the amount by which each value deviates from the mean stayed the same.

These results occur when any positive or negative number is added to every value in a set of data.

KeyConcept Transformations Using Addition

If a real number k is added to every value in a set of data, then:

- the mean, median, and mode of the new data set can be found by adding k to the mean, median, and mode of the original data set, and
- the range and standard deviation will not change.

Peter Cade/Iconica/Getty Images

Example 1 Transformation Using Addition

Find the mean, median, mode, range, and standard deviation of the data set obtained after adding 7 to each value.

13, 5, 8, 12, 7, 4, 5, 8, 14, 11, 13, 8

> **TechnologyTip**
> 1-Var Stats To quickly calculate the mean $\bar{x}$, median **Med**, standard deviation σ, and range of a data set, enter the data as **L1** in a graphing calculator, and then press [STAT] [▶] [ENTER] [ENTER]. Subtract **minX** from **maxX** to find the range.

Method 1 Find the mean, median, mode, range, and standard deviation of the original data set.

Mean	9	Mode	8	Standard Deviation	3.3
Median	8	Range	10		

Add 7 to the mean, median, and mode. The range and standard deviation are unchanged.

Mean	16	Mode	15	Standard Deviation	3.3
Median	15	Range	10		

Method 2 Add 7 to each data value.

20, 12, 15, 19, 14, 11, 12, 15, 21, 18, 20, 15

Find the mean, median, mode, range, and standard deviation of the new data set.

Mean	16	Mode	15	Standard Deviation	3.3
Median	15	Range	10		

GuidedPractice

1. Find the mean, median, mode, range, and standard deviation of the data set obtained after adding −4 to each value.

27, 41, 15, 36, 26, 40, 53, 38, 37, 24, 45, 26

To see the effect that a daily increase of 10% has on the data set, we can multiply each value by 1.10 and recalculate the measures of center and variation.

Tom's Earnings Before Extra 10%

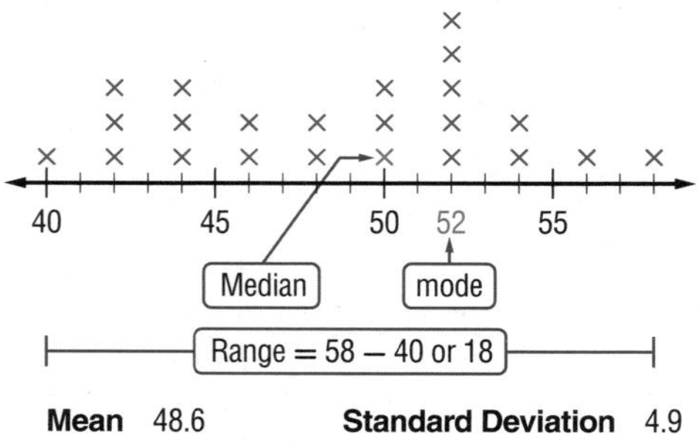

Tom's Earnings With Extra 10%

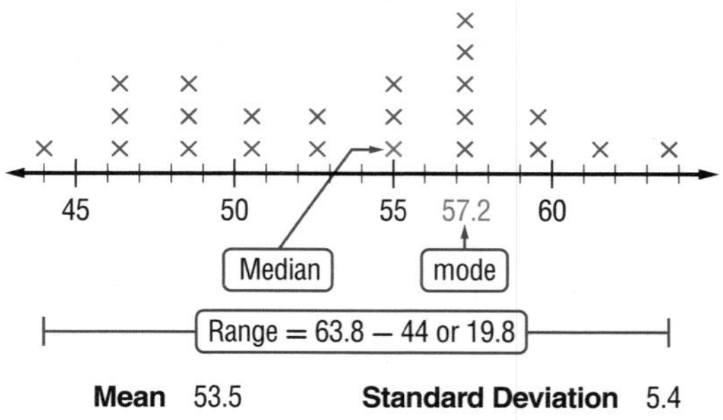

Notice that each value did not increase by the same amount, but did increase by a factor of 1.10. Thus, the mean, median, and mode increased by a factor of 1.10. Since each value was increased by a constant percent and not by a constant amount, the range and standard deviation both changed, also increasing by a factor of 1.10.

> **KeyConcept Transformations Using Multiplication**
> If every value in a set of data is multiplied by a constant k, $k > 0$, then the mean, median, mode, range, and standard deviation of the new data set can be found by multiplying each original statistic by k.

Since the medians for both bonuses are equal and the means are approximately equal, Tom should ask for the bonus that he thinks he has the best chance of receiving.

PT

Example 2 Transformation Using Multiplication

Find the mean, median, mode, range, and standard deviation of the data set obtained after multiplying each value by 3.

21, 12, 15, 18, 16, 10, 12, 19, 17, 18, 12, 22

Find the mean, median, mode, range, and standard deviation of the original data set.

Mean	16	Mode	12	Standard Deviation	3.7
Median	16.5	Range	12		

Multiply the mean, median, mode, range, and standard deviation by 3.

Mean	48	Mode	36	Standard Deviation	11.1
Median	49.5	Range	36		

GuidedPractice

2. Find the mean, median, mode, range, and standard deviation of the data set obtained after multiplying each value by 0.8.

63, 47, 54, 60, 55, 46, 51, 60, 58, 50, 56, 60

2 Comparing Distributions

Comparing Distributions Recall that when choosing appropriate statistics to represent data, you should first analyze the shape of the distribution. The same is true when comparing distributions.

- Use the mean and standard deviation to compare two symmetric distributions.
- Use the five-number summaries to compare two skewed distributions or a symmetric distribution and a skewed distribution.

PT

Example 3 Compare Data Using Histograms

QUIZ SCORES **Robert and Elaine's quiz scores for the first semester of Algebra 1 are shown below.**

Robert's Quiz Scores
85, 95, 70, 87, 78, 82, 84, 84, 85, 99, 88, 74, 75, 89, 79, 80, 92, 91, 96, 81

Elaine's Quiz Scores
89, 76, 87, 86, 92, 77, 78, 83, 83, 82, 81, 82, 84, 85, 85, 86, 89, 93, 77, 85

TechnologyTip

Histograms To create a histogram for a set of data in **L2**, press 2nd [STAT PLOT] ENTER ENTER, choose the histogram icon, and enter **L2** for **Xlist**.

a. Use a graphing calculator to construct a histogram for each set of data. Then describe the shape of each distribution.

Enter Robert's quiz scores as **L1** and Elaine's quiz scores as **L2**.

Robert's Quiz Scores

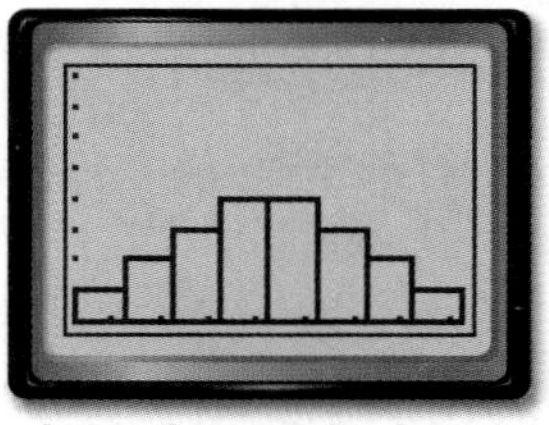

[69, 101] scl: 4 by [0, 8] scl: 1

Elaine's Quiz Scores

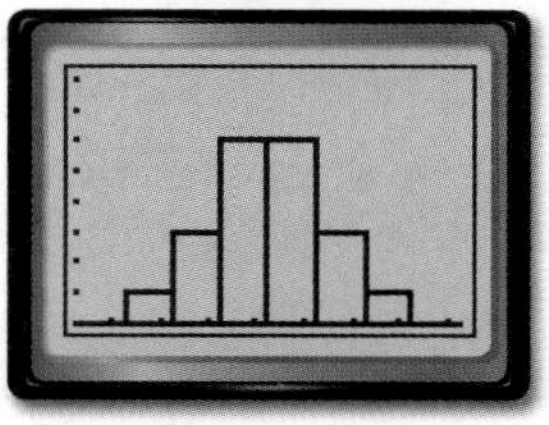

[69, 101] scl: 4 by [0, 8] scl: 1

Both distributions are high in the middle and low on the left and right. Therefore, both distributions are symmetric.

TechnologyTip

Multiple Data Sets In order to calculate statistics for a set of data in **L2**, press STAT ▶ ENTER 2nd [L2] ENTER.

b. Compare the data sets using either the means and standard deviations or the five-number summaries. Justify your choice.

Both distributions are symmetric, so use the means and standard deviations to describe the centers and spreads.

Robert's Quiz Scores

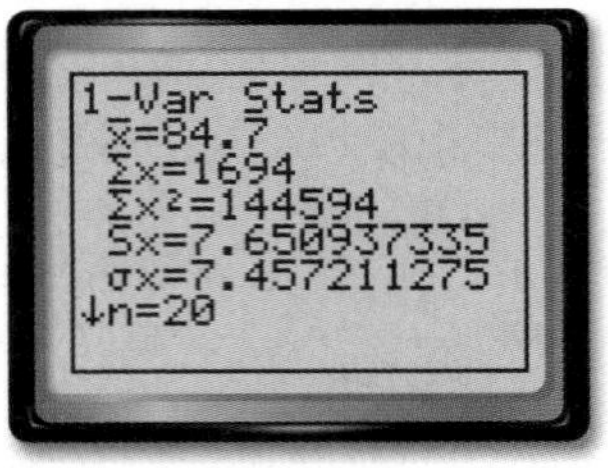

Elaine's Quiz Scores

The means for the students' quiz scores are approximately equal, but Robert's quiz scores have a much higher standard deviation than Elaine's quiz scores. This means that Elaine's quiz scores are generally closer to her mean than Robert's quiz scores are to his mean.

GuidedPractice

COMMUTE The students in two of Mr. Martin's classes found the average number of minutes that they each spent traveling to school each day.

3A. Use a graphing calculator to construct a histogram for each set of data. Then describe the shape of each distribution.

3B. Compare the data sets using either the means and standard deviations or the five-number summaries. Justify your choice.

2nd Period (minutes)
8, 4, 18, 7, 13, 26, 12, 6, 20, 5, 9, 24, 8, 16, 31, 13, 17, 10, 8, 22, 12, 25, 13, 11, 18, 12, 16, 22, 25, 33

7th Period (minutes)
21, 4, 20, 13, 22, 6, 10, 23, 13, 25, 14, 16, 19, 21, 19, 8, 20, 18, 9, 14, 21, 17, 19, 22, 4, 19, 21, 26

Box-and-whisker plots are useful for comparisons of data because they can be displayed on the same screen.

Real-WorldLink

Only about 2% of all high school student athletes attend a college with the aid of athletic scholarships.

Source: NCAA

Real-World Example 4 Compare Data Using Box-and-Whisker Plots

FOOTBALL Kurt's total rushing yards per game for his junior and senior seasons are shown.

Junior Season (yards)					
16	20	72	4	25	18
34	10	42	17	56	12

Senior Season (yards)					
77	54	109	60	156	72
39	83	73	101	46	80

a. Use a graphing calculator to construct a box-and-whisker plot for each set of data. Then describe the shape of each distribution.

Enter Kurt's rushing yards from his junior season as **L1** and his rushing yards from his senior season as **L2**. Graph both box-and-whisker plots on the same screen by graphing **L1** as **Plot1** and **L2** as **Plot2**.

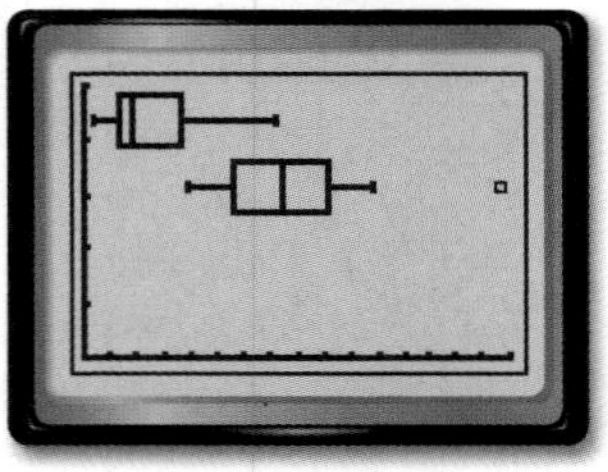

[0, 160] scl: 10 by [0, 5] scl: 1

For Kurt's junior season, the right whisker is longer than the left, and the median is closer to the left whisker. The distribution is positively skewed.

For Kurt's senior season, the lengths of the whiskers are approximately equal, and the median is in the middle of the data. The distribution is symmetric.

b. Compare the data sets using either the means and standard deviations or the five-number summaries. Justify your choice.

One distribution is symmetric and the other is skewed, so use the five-number summaries to compare the data.

The upper quartile for Kurt's junior season was 38, while the minimum for his senior season was 39. This means that Kurt rushed for more yards in every game during his senior season than 75% of the games during his junior season.

The maximum for Kurt's junior season was 72, while his median for his senior season was 75. This means that in half of his games during his senior year, he rushed for more yards than in any game during his junior season. Overall, we can conclude that Kurt rushed for many more yards during his senior season than during his junior season.

StudyTip

Box-and-Whisker Plots Recall that a box-and-whisker plot displays the spread of a data set by dividing it into four quartiles. Each quartile accounts for 25% of the data.

GuidedPractice

BASKETBALL The points Vanessa scored per game during her junior and senior seasons are shown.

4A. Use a graphing calculator to construct a histogram for each set of data. Then describe the shape of each distribution.

4B. Compare the data sets using either the means and standard deviations or the five-number summaries. Justify your choice.

Junior Season (points)
10, 12, 6, 10, 13, 8, 12, 3, 21, 14, 7, 0, 15, 6, 16, 8, 17, 3, 17, 2

Senior Season (points)
10, 32, 3, 22, 20, 30, 26, 24, 5, 22, 28, 32, 26, 21, 6, 20, 24, 18, 12, 25

Check Your Understanding

Example 1 **Find the mean, median, mode, range, and standard deviation of each data set that is obtained after adding the given constant to each value.**

1. 10, 13, 9, 8, 15, 8, 13, 12, 7, 8, 11, 12; + (−7)

2. 38, 36, 37, 42, 31, 44, 37, 45, 29, 42, 30, 42; + 23

Example 2 **Find the mean, median, mode, range, and standard deviation of each data set that is obtained after multiplying each value by the given constant.**

3. 6, 10, 3, 7, 4, 9, 3, 8, 5, 11, 2, 1; × 3

4. 42, 39, 45, 44, 37, 42, 38, 37, 41, 49, 42, 36; × 0.5

Example 3

5. TRACK Mark and Kyle's long jump distances are shown.

Kyle's Distances (ft)
17.2, 18.28, 18.56, 17.28, 17.36, 18.08, 17.43, 17.71, 17.46, 18.26, 17.51, 17.58, 17.41, 18.21, 17.34, 17.63, 17.55, 17.26, 17.18, 17.78, 17.51, 17.83, 17.92, 18.04, 17.91

Mark's Distances (ft)
18.88, 19.24, 17.63, 18.69, 17.74, 19.18, 17.92, 18.96, 18.19, 18.21, 18.46, 17.47, 18.49, 17.86, 18.93, 18.73, 18.34, 18.67, 18.56, 18.79, 18.47, 18.84, 18.87, 17.94, 18.7

a. Use a graphing calculator to construct a histogram for each set of data. Then describe the shape of each distribution.

b. Compare the data sets using either the means and standard deviations or the five-number summaries. Justify your choice.

Example 4

6. **TIPS** Miguel and Stephanie are servers at a restaurant. The tips that they earned to the nearest dollar over the past 15 workdays are shown.

Miguel's Tips ($)
14, 68, 52, 21, 63, 32, 43, 35, 70, 37, 42, 16, 47, 38, 48

Stephanie's Tips ($)
34, 52, 43, 39, 41, 50, 46, 36, 37, 47, 39, 49, 44, 36, 50

a. Use a graphing calculator to construct a box-and-whisker plot for each set of data. Then describe the shape of each distribution.

b. Compare the data sets using either the means and standard deviations or the five-number summaries. Justify your choice.

Practice and Problem Solving

Example 1

Find the mean, median, mode, range, and standard deviation of each data set that is obtained after adding the given constant to each value.

7. 52, 53, 49, 61, 57, 52, 48, 60, 50, 47; $+ 8$

8. 101, 99, 97, 88, 92, 100, 97, 89, 94, 90; $+ (-13)$

9. 27, 21, 34, 42, 20, 19, 18, 26, 25, 33; $+ (-4)$

10. 72, 56, 71, 63, 68, 59, 77, 74, 76, 66; $+ 16$

Example 2

Find the mean, median, mode, range, and standard deviation of each data set that is obtained after multiplying each value by the given constant.

11. 11, 7, 3, 13, 16, 8, 3, 11, 17, 3; $\times 4$

12. 64, 42, 58, 40, 61, 67, 58, 52, 51, 49; $\times 0.2$

13. 33, 37, 38, 29, 35, 37, 27, 40, 28, 31; $\times 0.8$

14. 1, 5, 4, 2, 1, 3, 6, 2, 5, 1; $\times 6.5$

Example 3

15. **BOOKS** The page counts for the books that the students chose are shown.

1st Period
388, 439, 206, 438, 413, 253, 311, 427, 258, 511, 283, 578, 291, 358, 297, 303, 325, 506, 331, 482, 343, 372, 456, 267, 484, 227

6th Period
357, 294, 506, 392, 296, 467, 308, 319, 485, 333, 352, 405, 359, 451, 378, 490, 379, 401, 409, 421, 341, 438, 297, 440, 500, 312, 502

a. Use a graphing calculator to construct a histogram for each set of data. Then describe the shape of each distribution.

b. Compare the data sets using either the means and standard deviations or the five-number summaries. Justify your choice.

16. **TELEVISIONS** The prices for a sample of televisions are shown.

The Electronics Superstore
46, 25, 62, 45, 30, 43, 40, 46, 33, 53, 35, 38, 39, 40, 52, 42, 44, 48, 50, 35, 32, 55, 28, 58

Game Central
53, 49, 26, 61, 40, 50, 42, 35, 45, 48, 31, 48, 33, 50, 35, 55, 38, 50, 42, 53, 44, 54, 48, 58

a. Use a graphing calculator to construct a histogram for each set of data. Then describe the shape of each distribution.

b. Compare the data sets using either the means and standard deviations or the five-number summaries. Justify your choice.

Example 4

17. **BRAINTEASERS** The time that it took Leon and Cassie to complete puzzles is shown.

Leon's Times (minutes)
4.5, 1.8, 3.2, 5.1, 2.0, 2.6, 4.8, 2.4, 2.2, 2.8, 1.8, 2.2, 3.9, 2.3, 3.3, 2.4

Cassie's Times (minutes)
2.3, 5.8, 4.8, 3.3, 5.2, 4.6, 3.6, 5.7, 3.8, 4.2, 5.0, 4.3, 5.5, 4.9, 2.4, 5.2

a. Use a graphing calculator to construct a box-and-whisker plot for each set of data. Then describe the shape of each distribution.

b. Compare the data sets using either the means and standard deviations or the five-number summaries. Justify your choice.

18. DANCE The total amount of money that a sample of students spent to attend the homecoming dance is shown.

Boys (dollars)
114, 98, 131, 83, 91, 64, 94, 77, 96, 105, 72, 108, 87, 112, 58, 126

Girls (dollars)
124, 74, 105, 133, 85, 162, 90, 109, 94, 102, 98, 171, 138, 89, 154, 76

a. Use a graphing calculator to construct a box-and-whisker plot for each set of data. Then describe the shape of each distribution.

b. Compare the data sets using either the means and standard deviations or the five-number summaries. Justify your choice.

19. LANDSCAPING Refer to the beginning of the lesson. Rhonda, another employee that works with Tom, earned the following over the past month.

Rhonda's Pay ($)		
45	55	53
47	53	54
44	56	59
63	47	53
60	57	62
44	50	45
60	53	49
62	47	

a. Find the mean, median, mode, range, and standard deviation of Rhonda's earnings.

b. A $5 bonus had been added to each of Rhonda's daily earnings. Find the mean, median, mode, range, and standard deviation of Rhonda's earnings before the $5 bonus.

20. SHOPPING The items Lorenzo purchased are shown.

Baseball hat	$14.98
Jeans	$24.61
T-shirt	$12.84
T-shirt	$16.05
Backpack	$42.80
Folders	$2.14
Sweatshirt	$19.26

a. Find the mean, median, mode, range, and standard deviation of the prices.

b. A 7% sales tax was added to the price of each item. Find the mean, median, mode, range, and standard deviation of the items without the sales tax.

H.O.T. Problems Use Higher-Order Thinking Skills

21. CHALLENGE A salesperson has 15 SUVs priced between $33,000 and $37,000 and 5 luxury cars priced between $44,000 and $48,000. The average price for all of the vehicles is $39,250. The salesperson decides to reduce the prices of the SUVs by $2000 per vehicle. What is the new average price for all of the vehicles?

22. REASONING If every value in a set of data is multiplied by a constant k, $k < 0$, then how can the mean, median, mode, range, and standard deviation of the new data set be found?

23. WRITING IN MATH Compare and contrast the benefits of displaying data using histograms and box-and-whisker plots.

24. CCSS REGULARITY If k is added to every value in a set of data, and then each resulting value is multiplied by a constant m, $m > 0$, how can the mean, median, mode, range, and standard deviation of the new data set be found? Explain your reasoning.

25. WRITING IN MATH Explain why the mean and standard deviation are used to compare the center and spread of two symmetrical distributions and the five-number summary is used to compare the center and spread of two skewed distributions or a symmetric distribution and a skewed distribution.

Standardized Test Practice

26. A store manager recorded the number of customers each day for a week: {46, 57, 63, 78, 91, 110, 101}. Find the mean absolute deviation.

A 16.8 **C** 19.4
B 18.1 **D** 22.7

27. SHORT RESPONSE Solve the right triangle. Round each side length to the nearest tenth.

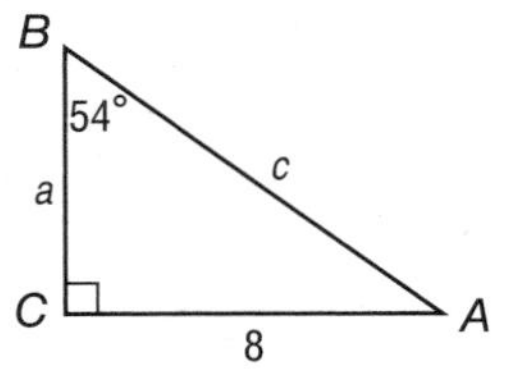

28. A research company divides a group of volunteers by age, and then randomly selects volunteers from each group to complete a survey. What type of sample is this?

F simple **H** self-selected
G systematic **J** stratified

29. Which set of measures can be the measures of the sides of a right triangle?

A 6, 7, 9
B 9, 12, 19
C 12, 15, 17
D 14, 48, 50

Spiral Review

30. Use a graphing calculator to construct a histogram for the data, and use it to describe the shape of the distribution. (Lesson 9-2)

23, 45, 50, 22, 37, 24, 36, 46, 24, 52, 25, 42, 25, 26, 54, 47, 27, 55
63, 28, 29, 30, 45, 31, 55, 43, 32, 34, 30, 23, 30, 35, 27, 35, 38, 40

31. SUBSCRIPTIONS Ms. Wilson's students are selling magazine subscriptions. Her students recorded the total number of subscriptions they each sold: {8, 12, 10, 7, 4, 3, 0, 4, 9, 0, 5, 3, 23, 6, 2}. Find and interpret the standard deviation of the data set. (Lesson 9-1)

Skills Review

Find the degree of each polynomial.

32. $2x^2 + 5y - 21$

33. $16xy^3 - 17x^2y - 16z^3$

34. $3ac^3d + 14a^2$

35. 18

36. $3a^2b^3 + 11ab^2c$

37. $7x + 11$

CHAPTER 9

Mid-Chapter Quiz

Lessons 9-1 through 9-3

Identify each sample, and suggest a population from which it was selected. Then classify the sample as *simple, systematic, self-selected, convenience,* or *stratified.* Explain your reasoning. (Lesson 9-1)

1. **NUTRITION** The table shows the number of Calories in twelve different snacks. Find the mean absolute deviation (Lesson 9-1).

Number of Calories in Snacks			
122	91	149	121
64	138	342	72
179	105	99	114

F 46 **H** 1.5

G 43 **J** 0.8

2. **GRIDDED RESPONSE** Find the standard deviation of the set of data below to the nearest tenth. (Lesson 9-1)

14	11	9	6
10	16	15	13
9	12	19	10

Identify the sample and the population for each situation. Then describe the sample statistic and the population parameter. (Lesson 9-2)

3. **DINING** At a restaurant, a random sample of 15 diners is selected. The amount of money spent on each meal is recorded.

4. **POOLS** A random sample of 25 children at a community pool is asked if they visit the pool at least once each week. The percent responding *yes* is calculated.

5. **PLAY AREA** Ian listed the ages of the children playing at the play area at the mall. Find and interpret the standard deviation of the data set. (Lesson 9-2)

{2, 3, 2, 2, 4, 2, 3, 2, 8, 3, 4, 2}

6. **MULTIPLE CHOICE** Several friends are chipping in to buy a gift for their teacher. Indigo is keeping track of how much each friend spends. Find the mean absolute deviation. (Lesson 9-2)

{$10, $5, $3, $6, $7, $8}

A $1.83 **C** $2.40

B $2.22 **D** $6.50

7. Use a graphing calculator to construct a histogram for the data, and use it to describe the shape of the distribution. (Lesson 9-3)

19, 36, 26, 36, 40, 31, 30, 33, 23, 38, 23, 46

8. Describe the center and spread of the data using either the mean and standard deviation or the five-number summary. Justify your choice by constructing a box-and-whisker plot for the data. (Lesson 9-3)

9, 11, 2, 6, 8, 10, 6, 3, 10, 11, 9, 8, 3,
8, 5, 11, 14, 6, 8, 6, 11, 5, 9, 10, 8

9. **MULTIPLE CHOICE** Which pair of box-and-whisker plots depicts two positively skewed sets of data in which 75% of one set of data is larger than 75% of the other set of data? (Lesson 9-3)

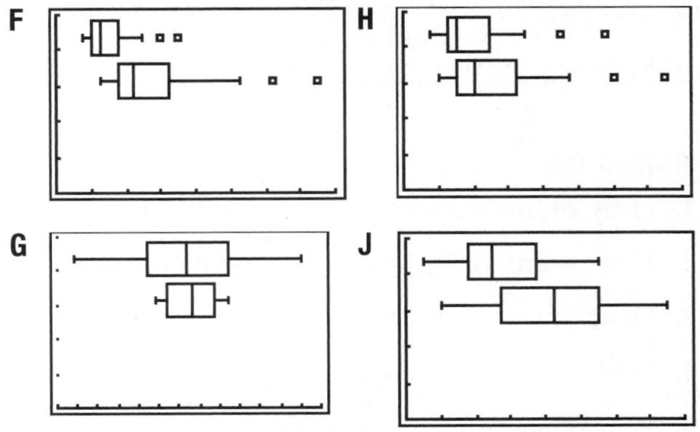

Find the mean, median, mode, range, and standard deviation of each data set that is obtained after adding the given constant to each value. (Lesson 9-3)

10. 6, 9, 0, 15, 9, 14, 11, 13, 9, 5, 8, 6; $+(-3)$

11. 19, 22, 10, 17, 26, 24, 12, 22, 18, 17; $+8$

EXTEND

9-3A Algebra Lab: Two-Way Frequency Tables

Joana sent out a survey to the freshmen and sophomores, asking if they were planning on attending the dance. One way of organizing her responses is to use a two-way frequency table. A **two-way frequency table** or *contingency table* is used to show the frequencies of data from a survey or experiment classified according to two categories, with the rows indicating one category and the columns indicating the other.

For Joana's survey, the two categories are *class* and *attendance*. These categories can be split into subcategories: *freshman* and *sophomore* for *class*, and *attending* and *not attending* for *attendance*.

Class	Attending	Not Attending	Totals
Freshman			
Sophomore			
Totals			

subcategories

Activity 1 Two-Way Frequency Table

DANCE Sixty-six freshmen responded to the survey, with 32 saying that they would be attending. Of the 84 sophomores that responded, 46 said they would attend. Organize the data in a two-way table.

Step 1 Find the values for every combination of subcategories. One combination is freshmen/not attending. Since 32 of 66 freshmen are attending, 66 − 32 or 34 freshmen are *not* attending. These combinations are called **joint frequencies.**

Step 2 Place every combination in the corresponding cell.

Step 3 Find the totals of each subcategory and place them in their corresponding cell. These values are called **marginal frequencies.**

Step 4 Find the sum of each set of marginal frequencies. These two sums should be equal. Place the value in the bottom right corner.

Class	Attending	Not Attending	Totals
Freshman	32	34	66
Sophomore	46	38	84
Totals	78	72	150

joint frequencies

marginal frequencies

marginal frequencies

Analyze the Results

1. How many students responded to the survey?
2. How many of the students that were surveyed are attending the dance?
3. How many of the surveyed sophomores are not attending the dance?
4. What does each of the joint frequencies represent?
5. What does each of the marginal frequencies represent?
6. **WORK** Heather sent out a survey asking who was working during the holiday. Of the 50 boys who responded, 34 said *yes*. Of the 45 girls who responded, 21 said *no*. Create a two-way frequency table of the results.
7. **SOCCER** Pamela asked if anyone would be interested in a co-ed soccer team. Of the 28 boys who responded, 18 said that they would play and 4 were undecided. Of the 22 girls who responded, 6 said they did not want to play and 3 were undecided. Create a two-way frequency table of the results.

Two-Way Frequency Tables *Continued*

A **relative frequency** is the ratio of the number of observations in a category to the total number of observations. Relative frequencies are also probabilities. To create a relative frequency two-way table, divide each of the values by the total number of observations and replace them with their corresponding decimals or percents.

Class	Attending	Not Attending	Totals
Freshman	$\frac{32}{150} \approx 21.3\%$	22.7%	44%
Sophomore	30.7%	25.3%	56%
Totals	52%	48%	100%

A **conditional relative frequency** is the ratio of the joint frequency to the marginal frequency. For example, given that a student is a freshman, what is the conditional relative frequency that he or she is going to the dance? In other words, what is the probability that a freshman is going to the dance?

Activity 2 Two-Way Conditional Relative Frequency Table

DANCE Joana wants to determine the conditional relative frequencies (or probabilities) given the fact that she knows the class of the respondents.

Step 1 Refer to the table in Activity 1. A total of 66 freshmen responded, and 32 said *yes*. Therefore, the conditional relative frequency that a respondent said *yes* given that the respondent is a freshman is $\frac{32}{66}$.

Step 2 Place every conditional relative frequency in the corresponding cell.

Step 3 The conditional relative frequencies for each row should sum to 100%.

Conditional Relative Frequencies by Class			
Class	Attending	Not Attending	Totals
Freshman	$\frac{32}{66} \approx 48\%$	$\frac{34}{66} \approx 52\%$	100%
Sophomore	$\frac{46}{84} \approx 55\%$	$\frac{38}{84} \approx 45\%$	100%

Analyze the Results

8. Given that a respondent was a sophomore, what is the probability that he or she said *no*?

9. What does each of the conditional relative frequencies represent?

10. Why do you think that the columns do not sum to 100%?

11. Create a two-way conditional relative frequency table for the category *attendance*.

12. Given that a respondent was not attending, what is the probability that he or she is a freshman?

13. **ACTIVITIES** The managers, staff, and assistants were given three options for the holiday activity: a potluck, a dinner at a restaurant, and a gift exchange. Five of the 11 managers want a dinner, while 3 want a potluck. Eleven of the 45 staff members want a gift exchange, while 18 want a dinner. Ten of the 32 assistants want a dinner, while 8 of them want a gift exchange.

a. Create a two-way frequency table.

b. Convert the two-way frequency table into a relative frequency table.

c. Create two conditional relative frequency tables: one for the activities and one for the employees.

EXTEND 9-3B

Graphing Technology Lab
The Normal Curve

When there are a large number of values in a data set, the frequency distribution tends to cluster around the mean of the set in a distribution (or shape) called a **normal distribution**. The graph of a normal distribution is called a **normal curve**. Since the shape of the graph resembles a bell, the graph is also called a *bell curve*.

Data sets that have a normal distribution include reaction times of drivers that are the same age, achievement test scores, and the heights of people that are the same age.

CCSS Common Core State Standards
Content Standards
S.ID.2 Use statistics appropriate to the shape of the data distribution to compare center (median, mean) and spread (interquartile range, standard deviation) of two or more different data sets.

Mathematical Practices
2 Reason abstractly and quantitatively.

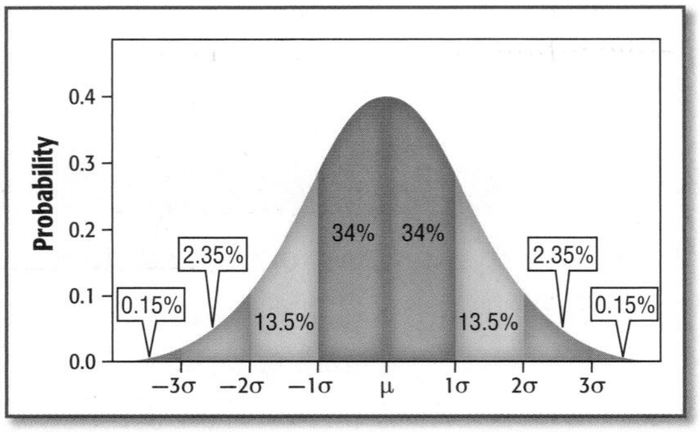

You can use a graphing calculator to graph and analyze a normal distribution if the mean and standard deviation of the data are known.

Activity 1 Graph a Normal Distribution

HEIGHT **The mean height of 15-year-old boys in the city where Isaac lives is 67 inches, with a standard deviation of 2.8 inches. Use a normal distribution to represent these data.**

Step 1 Set the viewing window. [WINDOW]

- Xmin = 67 [−] 3 [×] 2.8 [ENTER] **58.6**
- Xmax = 67 [+] 3 [×] 2.8 [ENTER] **75.4**
- Xscl = 2.8 [ENTER]
- Ymin = 0 [ENTER]
- Ymax = 1 [÷] [(] 2 [×] 2.8 [)] [ENTER] **.17857142...**
- Yscale = 1 [ENTER]

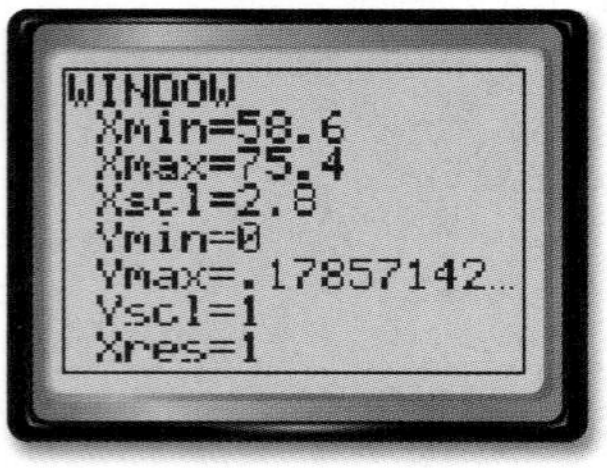

Step 2 By entering the mean and standard deviation into the calculator, we can graph the corresponding normal curve. Enter the values using the following keystrokes.

KEYSTROKES: [Y=] [2nd] **[DISTR]** [ENTER] [X,T,θ,*n*] [,] 67 [,] 2.8 [)] [GRAPH]

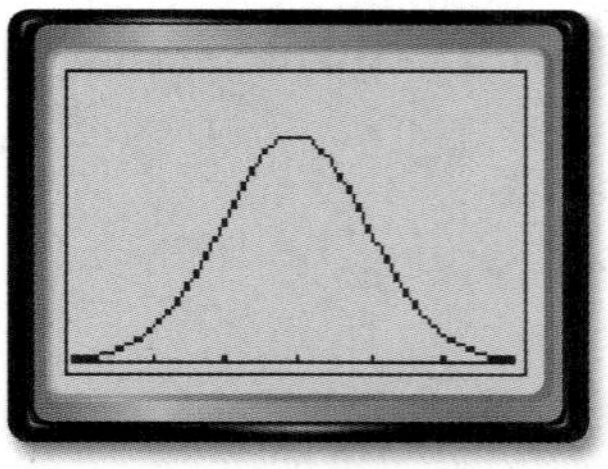

[58.6, 75.4] scl: 2.8 by [0, 0.17857142] scl: 1

Graphing Technology Lab
The Normal Curve *Continued*

The probability of a range of values is the area under the curve.

Activity 2 Analyze a Normal Distribution

Use the graph to answer questions about the data. What is the probability that Isaac will be at most 67 inches tall when he is 15?

The sum of all the y-values up to $x = 67$ would give us the probability that Isaac's height will be less than or equal to 67 inches. This is also the area under the curve. We will shade the area under the curve from negative infinity to 67 inches and find the area of the shaded portion of the graph.

Step 1 Use the **ShadeNorm** function.

KEYSTROKES: [2nd] [DISTR] [▶] [ENTER]

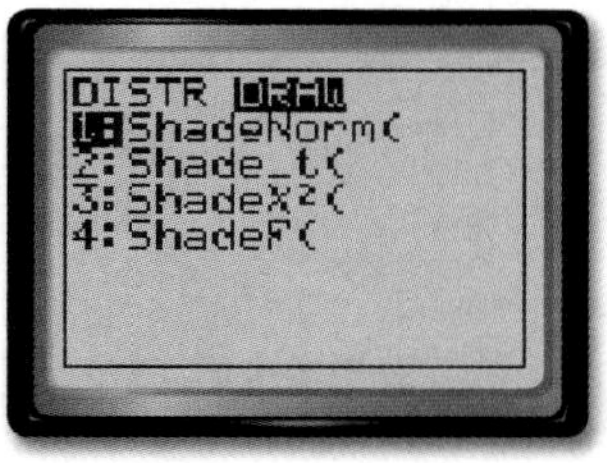

Step 2 Shade the graph.

Next enter the lowest value, highest value, mean, and standard deviation.

On the TI-84 Plus, -1×10^{99} represents negative infinity.

KEYSTROKES: [(−)] 1 [2nd] [EE] 99 [,] 67 [,] 67 [,] 2.8 [)] [ENTER]

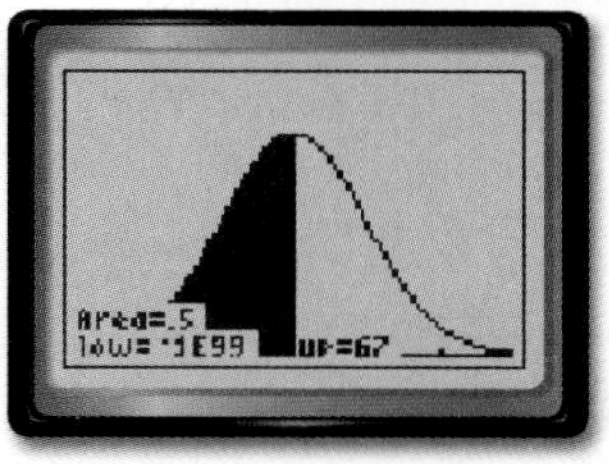

[58.6, 75.4] scl: 2.8 by [0, 0.17857142] scl: 1

The area is given as 0.5. The probability that Isaac will be 67 inches tall is 0.5 or 50%. Since the mean value is 67, we expect the probability to be 50%.

Exercises

1. What is the probability that Isaac will be at least 6 feet tall when he is 15?
2. What is the probability that Isaac will be between 65 and 68 inches?
3. The **z-score** represents the number of standard deviations that a given data value is from the mean. The z-score for a data value X is given by $z = \frac{X - \mu}{\sigma}$, where μ is the mean and σ is the standard deviation. Find and interpret the z-score of a height of 73 inches.
4. Find and interpret the z-score of a height of 61 inches.

Extension

Refer to the curve at the right.

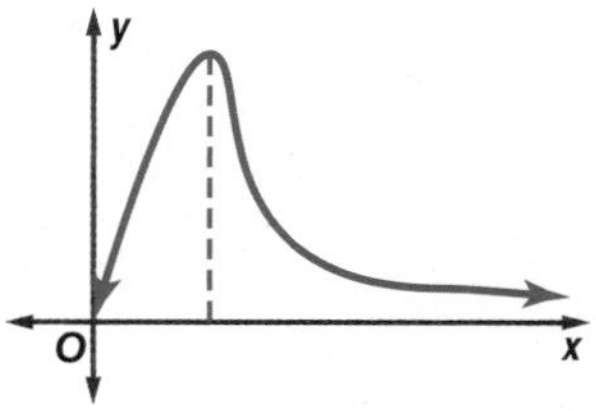

5. Compare this curve to the normal curve in Activity 1.
6. Describe where an outlier of the data set would be graphed on this curve.

CHAPTER 9 Study Guide and Review

Study Guide

KeyConcepts

Statistics and Parameters (Lesson 9-1)

- The mean absolute deviation is used to predict errors and judge how well the mean represents the data.
- A low standard deviation indicates that the data tend to be very close to the mean, while a high standard deviation indicates that the data are spread out over a larger range.

Distributions of Data and Comparing Sets of Data (Lessons 9-2 and 9-3)

- In a negatively skewed distribution, the majority of the data are on the right. In a positively skewed distribution, the majority of the data are on the left. In a symmetric distribution, the data are evenly distributed.

Probability Distributions (Lesson 9-3B)

- For each value of X, $0 \leq P(X) \leq 1$. The sum of the probabilities of each value of X is 1.
- The expected value $E(X)$ of a discrete random variable of a probability distribution is its weighted average.

FOLDABLES® StudyOrganizer

Be sure the Key Concepts are noted in your Foldable.

KeyVocabulary

conditional relative frequency (p. 547)
distribution (p. 530)
joint frequency table (p. 546)
linear transformation (p. 537)
marginal frequency (p. 546)
mean absolute deviation (MAD) (p. 524)
normal curve (p. 548)
normal distribution (p. 548)
parameter (p. 523)
statistic (p. 523)
statistical inference (p. 523)
standard deviation (p. 525)
symmetric distribution (p. 530)
two-way frequency table (p. 546)
variance (p. 525)
z-score (p. 549)

VocabularyCheck

Choose the term that best completes each sentence.

1. An arrangement in which order is important is called a (combination, permutation).
2. A (sample, population) consists of all of the members of a group.
3. (Experimental probability, Theoretical probability) is the ratio of the number of favorable outcomes to the total number of outcomes.
4. A variable with a value that is the numerical outcome of a random event is called a (discrete random variable, random variable).

Lesson-by-Lesson Review

9-1 Statistics and Parameters

1. **SHOVELING** Ben shovels sidewalks to raise money. The number of sidewalks he shovels each day is {2, 4, 3, 5, 3}. Find and interpret the mean absolute deviation.

2. **CANDY BARS** Luci is keeping track of the number of candy bars each member of the drill team sold. The results are {20, 25, 30, 50, 40, 60, 20, 10, 42}. Find and interpret the mean absolute deviation.

3. **FOOD** A fast food company polls a random sample of its day and night customers to find how many times a month they eat out. Compare the means and standard deviations of each data set.

Day Customers	Night Customers
10, 3, 12, 15, 7, 8, 4, 12, 9, 14, 12, 9	15, 12, 13, 9, 11, 12, 14, 12, 8, 16, 9, 9

Example 1

GIFTS Joshua is collecting money from his family for a Mother's Day gift. He keeps track of how much each person has donated: {10, 5, 20, 15, 10}. Find and interpret the mean absolute deviation.

Step 1 Find the mean: $\bar{x} = \frac{10 + 5 + 20 + 15 + 10}{5}$ or 12.

Step 2 Find the absolute values of the differences.

$x_1 = 10$: $|12 - 10|$ or 2 $x_2 = 5$: $|12 - 5|$ or 7

$x_3 = 20$: $|12 - 20|$ or 8 $x_4 = 15$: $|12 - 15|$ or 3

$x_5 = 10$: $|12 - 10|$ or 2

Step 3 Find the sum: $2 + 7 + 8 + 3 + 2 = 22$.

Step 4 Find the mean absolute deviation.

$$\text{MAD} = \frac{|\bar{x} - x_1| + |\bar{x} - x_2| + \ldots + |\bar{x} - x_n|}{n}$$

$$= \frac{22}{5} \text{ or } 4.4$$

A mean absolute deviation of 4.4 indicates that the data, on average, are 4.4 units away from the mean.

9-2 Distributions of Data

Use a graphing calculator to construct a histogram for the data. Then describe the shape of the distribution.

4. 55, 62, 32, 56, 31, 59, 19, 61, 8, 48, 41, 69, 32, 63, 48, 60, 43, 66, 71, 70, 49, 56, 21, 67

5. 4, 19, 62, 28, 26, 59, 33, 39, 36, 72, 46, 48, 49, 44, 72, 76, 55, 53, 55, 62, 66, 69, 71, 74

6. **MILK** A grocery store manager tracked the amount of milk in gallons sold each day. Describe the center and spread of the data using either the mean and standard deviation or the five-number summary. Justify your choice by constructing a box-and-whisker plot for the data.

Gallons of Milk Sold Per Day					
383	296	354	288	195	372
421	367	411	355	296	321
403	357	432	229	180	266

Example 2

DRIVING TESTS Several driving test results are shown. Describe the center and spread of the data using either the mean and standard deviation or the five-number summary. Justify your choice by constructing a box-and-whisker plot for the data.

Driving Test Scores					
80	95	100	95	95	100
100	90	75	60	90	80

Use a graphing calculator to create a box-and-whisker plot.

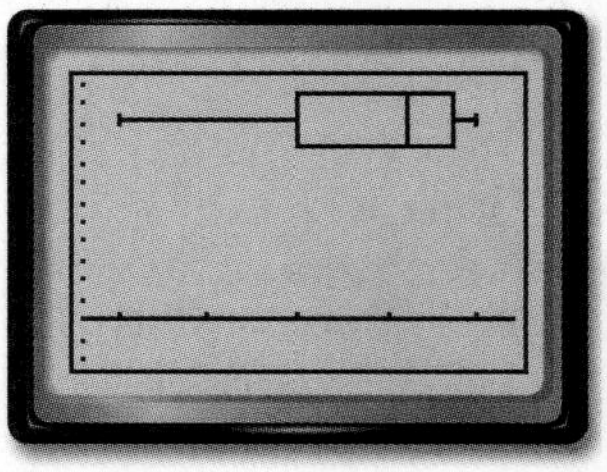

[56, 104] scl: 10 by [−2, 12] scl: 1

The left whisker is longer than the right and the median is closer to the right whisker. Therefore, the distribution is negatively skewed.

Use the five-number summary. The range is 40. The median score is 92.5, and half of the drivers scored between 80 and 97.5.

Study Guide and Review *Continued*

9-3 Comparing Sets of Data

Find the mean, median, mode, range, and standard deviation of each data set that is obtained after adding the given constant to each value.

7. 27, 21, 34, 42, 20, 19, 18, 26, 25, 33; $+(-4)$
8. 72, 56, 71, 63, 68, 59, 77, 74, 76, 66; $+16$
9. **SCHOOL** Principal Andrews tracked the number of disciplinary actions given by Ms. Miller and Ms. Anderson to their students each week.

Ms. Miller
9, 16, 12, 11, 12, 9, 10, 14, 13, 10, 9, 10, 11, 9, 12, 10, 11, 12

Ms. Anderson
7, 1, 0, 4, 2, 1, 6, 2, 2, 1, 4, 3, 0, 7, 0, 2, 5, 0

a. Use a graphing calculator to construct a histogram for each set of data. Then describe the shape of each distribution.
b. Compare the data sets using either the means and standard deviations or the five-number summaries. Justify your choice.

Example 3

Find the mean, median, mode, range, and standard deviation of the data set obtained after adding 6 to each value.

12, 15, 11, 12, 14, 16, 15, 12, 10, 13

Find the mean, median, mode, range, and standard deviation of the original data set.

Mean 13	Mode 12	Standard Deviation 1.8
Median 12.5	Range 6	

Add 6 to the mean, median, and mode. The range and standard deviation are unchanged.

Mean 19	Mode 18	Standard Deviation 1.8
Median 18.5	Range 6	

CHAPTER 9 Practice Test

1. **SPORTS** A quarterback threw 18 completed passes out of 30 attempts. Find the experimental probability of making a completed pass. Express the probability as a percent.

2. A die is rolled 200 times. What is the experimental probability of rolling less than 3?

Outcome	Frequency
1	30
2	26
3	44
4	38
5	22
6	40

3. **MULTIPLE CHOICE** Use a graphing calculator to construct a histogram for the data, and use it to describe the shape of the distribution.

16, 18, 14, 31, 19, 18, 10, 29,
12, 12, 28, 19, 17, 26, 15, 20

A positively skewed **C** symmetric

B negatively skewed **D** none of the above

4. **SALES** Nate is keeping track of how much people spent at the school bookstore in one day. Find and interpret the mean absolute deviation for the data: 1, 1, 2, 3, 4, 5, 12.

5. **EDUCATION** Kristin surveys 200 people in her school to determine how many nights per week students do homework. The results are shown in the table.

Number of Nights	Number of Students
0	10
1	30
2	50
3	90
4	10
5 or more	10

a. Find the probability that a randomly chosen student will have studied more than 4 nights.

b. Find the probability that a randomly chosen student will have studied no more than 3 nights.

One letter from the word MISSISSIPPI is chosen at random. Find each probability.

6. P(M)

7. P(I)

8. P(Constant)

9. P(Vowe)

CHAPTER 9

Preparing for Standardized Tests

Organize Data

Sometimes you may be given a set of data that you need to analyze in order to solve problems on a standardized test. Use this lesson to practice organizing data to help you solve problems.

image100/CORBIS

Strategies for Organizing Data

Step 1

When you are given a problem statement containing data, consider:

- **making a list** of the data.
- **using a table** to organize the data.
- **using a data display** (such as a bar graph, Venn diagram, circle graph, line graph, or box-and-whisker plot) to organize the data.

Step 2

Organize the data.

- Create your table, list, or data display.
- If possible, fill in any missing values that can be found by intermediate computations.

Step 3

Analyze the data to solve the problem.

- Reread the problem statement to determine what you are being asked to solve.
- Use the properties of algebra to work with the organized data and solve the problem.
- If time permits, go back and check your answer.

Standardized Test Example

Read the problem. Identify what you need to know. Then use the information in the problem to solve. Show your work.

Of the 24 students in a music class, 10 play the flute, 14 play the piano, and 13 play the guitar. Two students play the flute only, 5 the piano only, and 7 the guitar only. One student plays the flute and the guitar but not the piano. Two students play the piano and guitar but not the flute. Three students play all the instruments. If a student is selected at random, what is the probability that he or she plays the piano and flute, but not the guitar?

Scoring Rubric	
Criteria	**Score**
Full Credit: The answer is correct and a full explanation is provided that shows each step.	2
Partial Credit: • The answer is correct, but the explanation is incomplete. • The answer is incorrect, but the explanation is correct.	1
No Credit: Either an answer is not provided or the answer does not make sense.	0

Read the problem carefully. The data is difficult to analyze as it is presented. Use a Venn diagram to organize the data and solve the problem.

Example of a 2-point response:

Use a Venn diagram to organize the data. Fill in all of the information given in the problem statement. There are 14 students who play the piano, so $14 - 5 - 2 - 3$ or 4 students play the piano and the flute, but not the guitar. Find the probability.

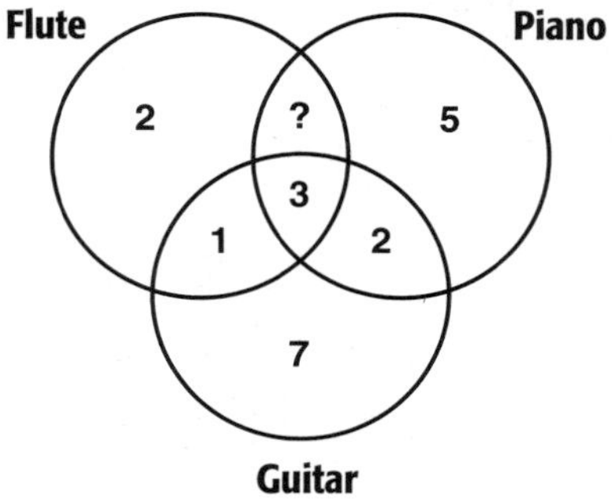

$P(\text{piano and flute}) = \frac{4}{24}$ or $\frac{1}{6}$

So, the probability that a randomly selected student plays the piano and flute but not the guitar is $\frac{1}{6}$.

Exercises

Read the problem. Identify what you need to know. Then use the information in the problem to solve. Show your work.

1. There are 40 students, 9 camp counselors, and 5 teachers at Camp Kern. Each person is assigned to one activity this afternoon. There are 9 students going hiking and 17 students going horseback riding. Of the camp counselors, 2 will supervise the hike and 3 will help with the canoe trip. There are 2 teachers helping with the canoe trip and 2 going horseback riding. Suppose a person is selected at random during the afternoon activities. What is the probability that the one selected is a student on the canoe trip or a camp counselor on a horse? Express your answer as a fraction.

2. The table shows the number of coins in a piggy bank.

Coin	Number
Penny	16
Nickel	18
Dime	20
Quarter	10

a. Find the probability that a randomly selected coin will be a dime.

b. Find the probability that a randomly selected coin will be either a nickel or a quarter.

3. It takes Craig 40 minutes to mow his family's lawn. His brother Jacob can do the same job in 50 minutes. How long would it take them to mow the lawn together? Round your answer to the nearest tenth of a minute.

CHAPTER 9

Standardized Test Practice

Cumulative, Chapters 1 through 9

Multiple Choice

Read each question. Then fill in the correct answer on the answer document provided by your teacher or a sheet of paper.

1. What is the equation of the square root function graphed below?

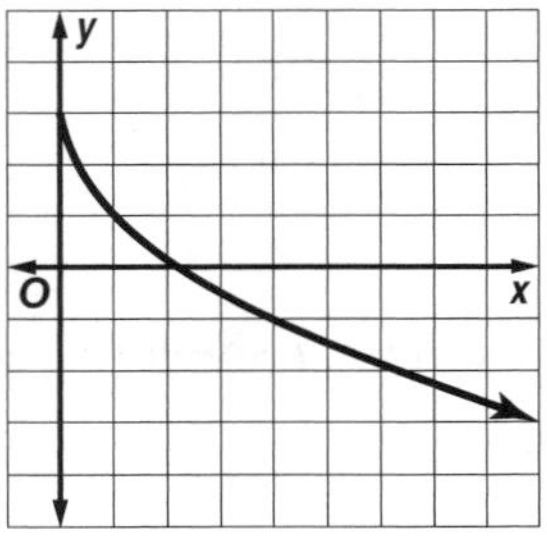

A $y = -2\sqrt{x} + 1$

B $y = -2\sqrt{x} + 3$

C $y = 2\sqrt{x} + 3$

D $y = 2\sqrt{x} + 1$

2. Simplify $\frac{1}{4 + \sqrt{2}}$.

F $\frac{4 + \sqrt{2}}{14}$

G $\frac{2 - \sqrt{2}}{7}$

H $\frac{4 - \sqrt{2}}{14}$

J $\frac{2 + \sqrt{2}}{7}$

3. What is the area of the triangle below?

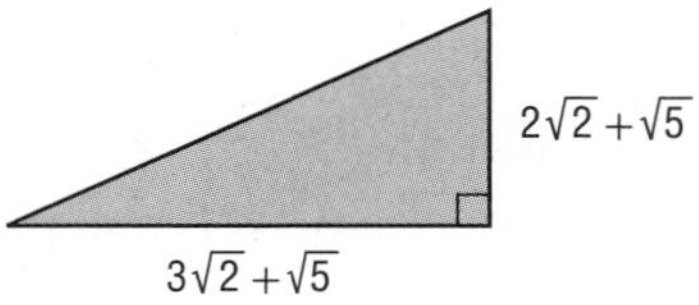

A $3\sqrt{2} + 10\sqrt{5}$

B $17 + 5\sqrt{10}$

C $12\sqrt{2} + 8\sqrt{5}$

D $8.5 + 2.5\sqrt{10}$

4. The formula for the slant height c of a cone is $c = \sqrt{h^2 + r^2}$, where h is the height of the cone and r is the radius of its base. What is the radius of the cone below? Round to the nearest tenth.

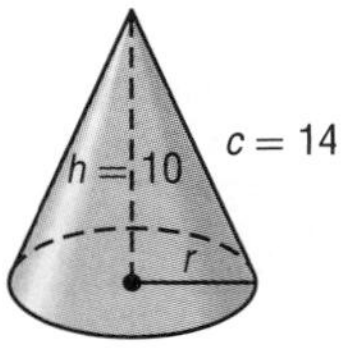

F 4.9

G 6.3

H 9.8

J 10.2

5. In 1990, the population of a country was about 3.66 million people. By 2010, this number had grown to about 4.04 million people. What was the annual rate of change in population from 1990 to 2010?

F about 15,000 people per year

G about 19,000 people per year

H about 24,000 people per year

J about 38,000 people per year

6. The scale on a map shows that 1.5 centimeters is equivalent to 40 miles. If the distance on the map between two cities is 8 centimeters, about how many miles apart are the cities?

A 178 miles

B 213 miles

C 224 miles

D 275 miles

Test-Taking Tip

Question 4 Substitute for c and h in the formula. Then solve for r.

Short Response/Gridded Response

Record your answers on the answer sheet provided by your teacher or on a sheet of paper.

7. **GRIDDED RESPONSE** Misty purchased a car several years ago for $21,459. The value of the car depreciated at a rate of 15% annually. What was the value of the car after 5 years? Round your answer to the nearest whole dollar.

8. The cost of 5 notebooks and 3 pens is $9.75. The cost of 4 notebooks and 6 pens is $10.50.
   - **a.** Write a system of equations to model the situation.
   - **b.** Solve the system of equations. How much does each item cost?

9. The table shows the total cost of renting a canoe for n hours.

Number of Hours (n)	Rental Cost (C)
1	$15
2	$20
3	$25
4	$30

   - **a.** Write a function to represent the situation.
   - **b.** How much would it cost to rent the canoe for 7 hours?

10. **GRIDDED RESPONSE** In football, a field goal is worth 3 points, and the extra point after a touchdown is worth 1 point. During the 2006 season, John Kasay of the Carolina Panthers scored a total of 100 points for his team by making a total of 52 field goals and extra points. How many field goals did he make?

Extended Response

Record your answers on a sheet of paper. Show your work.

11. Consider $2x + 3y = 15$

Part A Make a table of at least six pairs of values that satisfy the equation above.

Part B Use your table from Part A to graph the equation.

Need ExtraHelp?

If you missed Question...	1	2	3	4	5	6	7	8	9	10	11
Go to Lesson...	8-1	8-2	8-3	8-4	3-3	2-6	7-6	6-4	3-5	6-4	3-1

CHAPTER 10

Tools of Geometry

Then

You graphed points on the coordinate plane and evaluated mathematical expressions.

Now

In this chapter, you will:

- Find distances between points and midpoints of line segments.
- Identify angle relationships.
- Find perimeters, areas, surface areas, and volumes.

Why? ▲

MAPS Geometric figures and terms can be used to represent and describe real-world situations. On a map, locations of cities can be represented by points, highways or streets by lines, and national parks by polygons that have both perimeter and area. The map itself is representative of a plane.

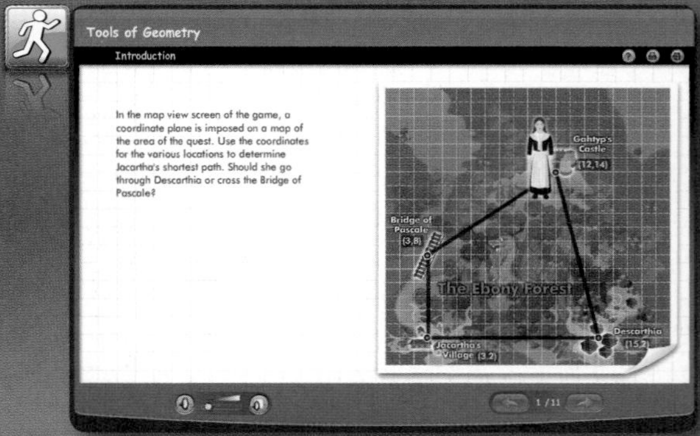

Tools of Geometry
Introduction
In the map view screen of the game, a coordinate plane is imposed on a map of the area of the quest. Use the coordinates for the various locations to determine Jocartha's shortest path. Should she go through Descarthia or cross the Bridge of Pascale?
Gohtyp's Castle (12,14)
Bridge of Pascale (3,8)
The Ebony Forest
Descarthia (15,2)
Jocartha's Village (3,2)
1/11

connectED.mcgraw-hill.com **Your Digital Math Portal**

Animation | Vocabulary | eGlossary | Personal Tutor | Virtual Manipulatives | Graphing Calculator | Audio | Foldables | Self-Check Practice | Worksheets

Get Ready for the Chapter

Diagnose Readiness | **You have two options for checking Prerequisite Skills.**

1 Textbook Option Take the Quick Check below. Refer to the Quick Review for help.

QuickCheck

Graph and label each point in the coordinate plane.

1. $W(5, 2)$

2. $X(0, 6)$

3. $Y(-3, -1)$

4. $Z(4, -2)$

5. GAMES Carolina is using the diagram to record her chess moves. She moves her knight 2 spaces up and 1 space to the left from f3. What is the location of the knight after Carolina completes her turn?

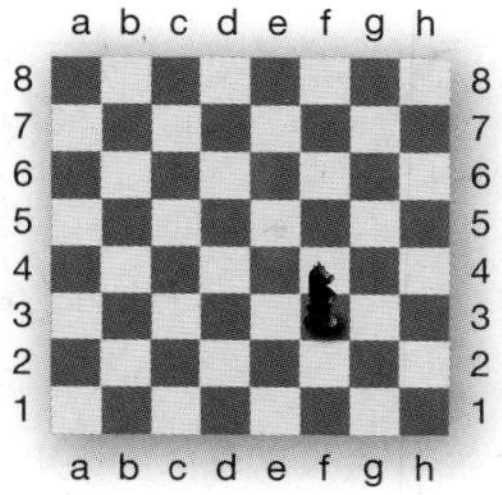

Find each sum or difference.

6. $\frac{2}{3} + \frac{5}{6}$

7. $2\frac{1}{18} + 4\frac{3}{4}$

8. $\frac{13}{18} - \frac{5}{9}$

9. $14\frac{3}{5} - 9\frac{7}{15}$

10. FOOD Alvin ate $\frac{1}{3}$ of a pizza for dinner and took $\frac{1}{6}$ of it for lunch the next day. How much of the pizza does he have left?

Evaluate each expression.

11. $(-4 - 5)^2$

12. $(6 - 10)^2$

13. $(8 - 5)^2 + [9 - (-3)]^2$

Solve each equation.

14. $6x + 5 + 2x - 11 = 90$

15. $8x - 7 = 53 - 2x$

QuickReview

Example 1

Graph and label the point $Q(-3, 4)$ in the coordinate plane.

Start at the origin. Since the x-coordinate is negative, move 3 units to the left. Then move 4 units up since the y-coordinate is positive. Draw a dot and label it Q.

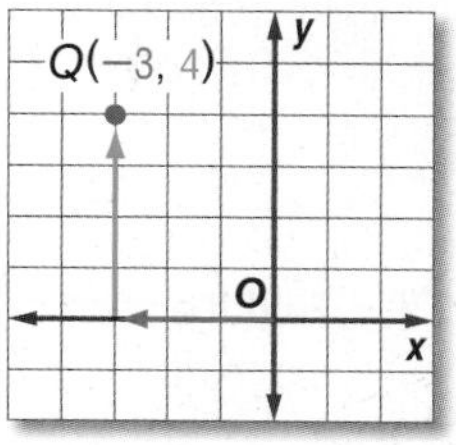

Example 2

Find $3\frac{1}{6} + 2\frac{3}{4}$.

$3\frac{1}{6} + 2\frac{3}{4} = \frac{19}{6} + \frac{11}{4}$	Write as improper fractions.
$= \frac{19}{6}\left(\frac{2}{2}\right) + \frac{11}{4}\left(\frac{3}{3}\right)$	The LCD is 12.
$= \frac{38}{12} + \frac{33}{12}$	Multiply.
$= \frac{71}{12}$ or $5\frac{11}{12}$	Simplify.

Example 3

Evaluate the expression $[-2 - (-7)]^2 + (1 - 8)^2$.

Follow the order of operations.

$[-2 - (-7)]^2 + (1 - 8)^2$

$= 5^2 + (-7)^2$	Subtract.
$= 25 + 49$	$5^2 = 25, (-7)^2 = 49$
$= 74$	Add.

2 Online Option Take an online self-check Chapter Readiness Quiz at connectED.mcgraw-hill.com.

Get Started on the Chapter

You will learn several new concepts, skills, and vocabulary terms as you study Chapter 10. To get ready, identify important terms and organize your resources.

FOLDABLES® Study Organizer

Tools of Geometry Make this Foldable to help you organize your Chapter 10 notes about points, lines, and planes; angles and angle relationships; and formulas and notes for distance, midpoint, perimeter, area, and volume. Begin with a sheet of 11″ × 17″ paper.

1 **Fold** the short sides to meet in the middle.

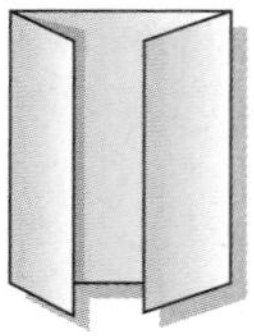

2 **Fold** the booklet in thirds lengthwise.

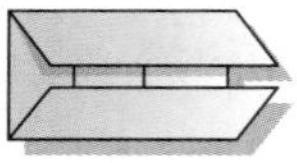

3 **Open and cut** the booklet in thirds lengthwise.

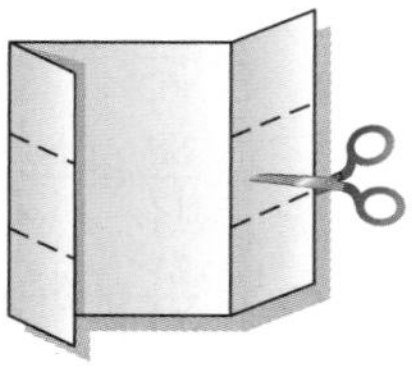

4 **Label** the tabs as shown.

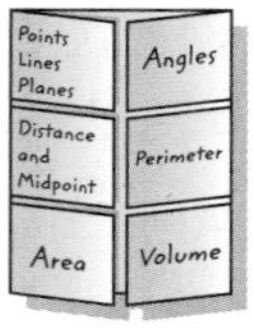

New Vocabulary

English		Español
collinear	p. 561	colineal
coplanar	p. 561	coplanar
congruent	p. 572	congruente
midpoint	p. 583	punto medio
segment bisector	p. 585	bisectriz de segmento
angle	p. 592	angulo
vertex	p. 592	vertice
angle bisector	p. 595	bisectriz de un angulo
polygon	p. 603	poligono
perimeter	p. 605	perimetro

Review Vocabulary

ordered pair par ordenado a set of numbers or coordinates used to locate any point on a coordinate plane, written in the form (x, y)

origin origen the point where the two axes intersect at their zero points

quadrants cuadrantes the four regions into which the x-axis and y-axis separate the coordinate plane

x-coordinate coordenada x the first number in an ordered pair

y-coordinate coordenada y the second number in an ordered pair

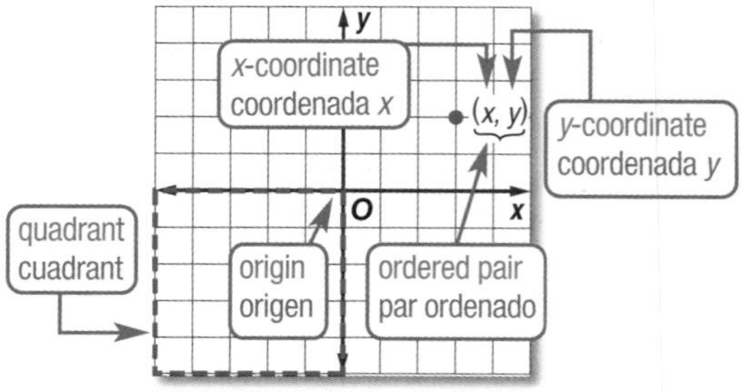

LESSON 10-1 Points, Lines, and Planes

∴ Then	∴ Now	∴ Why?
You used basic geometric concepts and properties to solve problems.	**1** Identify and model points, lines, and planes. **2** Identify intersecting lines and planes.	On a subway map, the locations of stops are represented by *points*. The route the train can take is modeled by a series of connected paths that look like *lines*. The flat surface of the map on which these points and lines lie is representative of a *plane*.

NewVocabulary

undefined term
point
line
plane
collinear
coplanar
intersection
definition
defined term
space

Common Core State Standards

Content Standards

G.CO.1 Know precise definitions of angle, circle, perpendicular line, parallel line, and line segment, based on the undefined notions of point, line, distance along a line, and distance around a circular arc.

Mathematical Practices

4 Model with mathematics.
6 Attend to precision.

Brand X Pictures/Jupiterimages

1 Points, Lines, and Planes Unlike the real-world objects that they model, shapes, points, lines, and planes do not have any actual size. In geometry, *point*, *line*, and *plane* are considered **undefined terms** because they are only explained using examples and descriptions.

You are already familiar with the terms point, line, and plane from algebra. You graphed on a coordinate *plane* and found ordered pairs that represented *points* on *lines*. In geometry, these terms have a similar meaning.

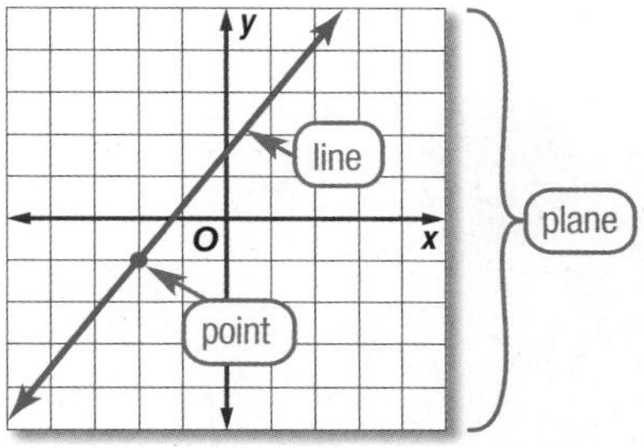

The phrase *exactly one* in a statement such as, "There is exactly one line through any two points," means that there is *one and only one*.

KeyConcept Undefined Terms

A **point** is a location. It has neither shape nor size.

Named by a capital letter

Example point *A*

A **line** is made up of points and has no thickness or width. There is exactly one line through any two points.

Named by the letters representing two points on the line or a lowercase script letter

Example line *m*, line *PQ* or $\overleftrightarrow{PQ}$, line *QP* or $\overleftrightarrow{QP}$

A **plane** is a flat surface made up of points that extends infinitely in all directions. There is exactly one plane through any three points not on the same line.

Named by a capital script letter or by the letters naming three points that are not all on the same line

Example plane $\mathcal{K}$, plane *BCD*, plane *CDB*, plane *DCB*, plane *DBC*, plane *CBD*, plane *BDC*

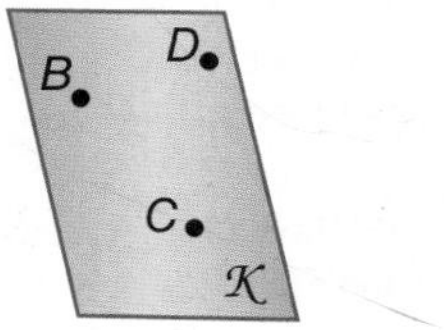

Collinear points are points that lie on the same line. *Noncollinear* points do not lie on the same line. **Coplanar** points are points that lie in the same plane. *Noncoplanar* points do not lie in the same plane.

Example 1 Name Lines and Planes

Use the figure to name each of the following.

a. a line containing point *W*

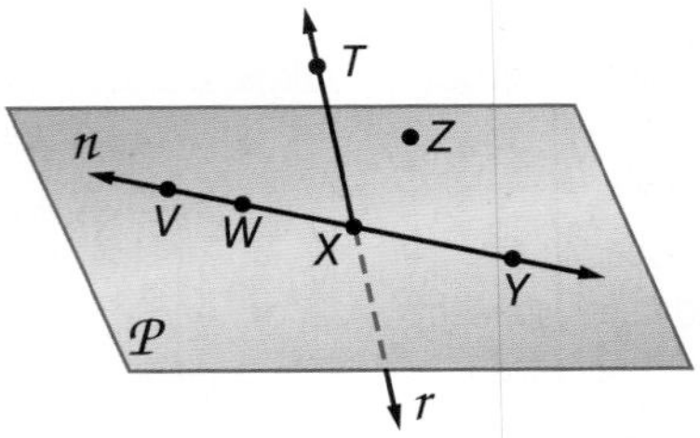

The line can be named as line *n*, or any two of the four points on the line can be used to name the line.

$\overleftrightarrow{VW}$ $\overleftrightarrow{WV}$ $\overleftrightarrow{VX}$ $\overleftrightarrow{XV}$ $\overleftrightarrow{VY}$ $\overleftrightarrow{YV}$
$\overleftrightarrow{WX}$ $\overleftrightarrow{XW}$ $\overleftrightarrow{WY}$ $\overleftrightarrow{YW}$ $\overleftrightarrow{XY}$ $\overleftrightarrow{YX}$

b. a plane containing point *X*

One plane that can be named is plane $\mathcal{P}$. You can also use the letters of any three *noncollinear* points to name this plane.

plane *XZY*	plane *VZW*	plane *VZX*
plane *VZY*	plane *WZX*	plane *WZY*

The letters of each of these names can be reordered to create other acceptable names for this plane. For example, *XZY* can also be written as *XYZ*, *ZXY*, *ZYX*, *YXZ*, and *YZX*. In all, there are 36 different three-letter names for this plane.

StudyTip

Additional Planes Although not drawn in Example 1b, there is another plane that contains point *X*. Since points *W*, *T*, and *X* are noncollinear, point *X* is also in plane *WTX*.

GuidedPractice

1A. a plane containing points *T* and *Z*

1B. a line containing point *T*

Real-World Example 2 Model Points, Lines, and Planes

MESSAGE BOARD **Name the geometric terms modeled by the objects in the picture.**

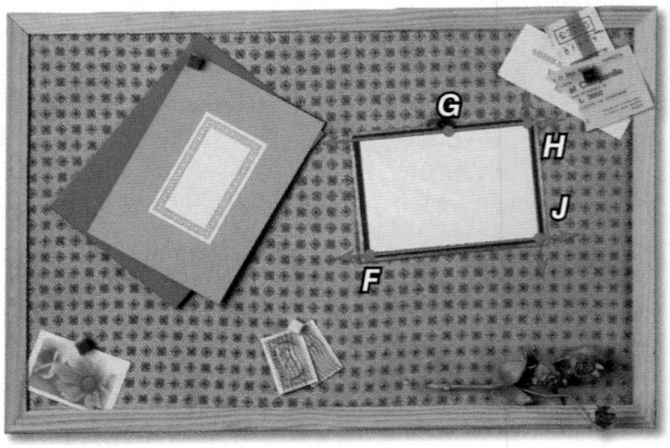

The push pin models point *G*.

The maroon border on the card models line *GH*.

The edge of the card models line *HJ*.

The card itself models plane *FGJ*.

GuidedPractice

Name the geometric term modeled by each object.

2A. stripes on a sweater

2B. the corner of a box

Real-WorldCareer

Drafter Drafters use perspective to create drawings to build everything from toys to school buildings. Drafters need skills in math and computers. They get their education at trade schools, community colleges, and some 4-year colleges. Refer to Exercises 50 and 51.

2 Intersections of Lines and Planes

The **intersection** of two or more geometric figures is the set of points they have in common. Two lines intersect in a point. Lines can intersect planes, and planes can intersect each other.

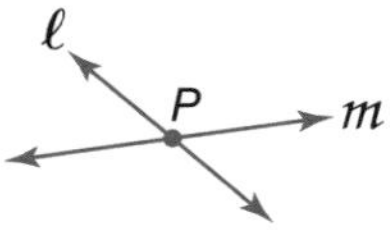

P represents the intersection of lines ℓ and *m*.

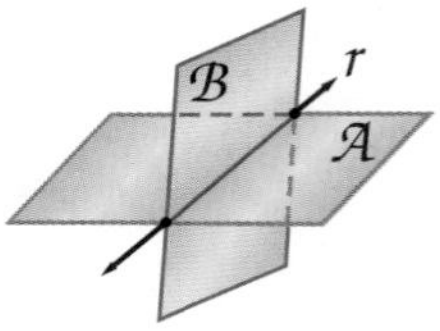

Line *r* represents the intersection of planes $\mathcal{A}$ and $\mathcal{B}$.

Example 3 Draw Geometric Figures

Draw and label a figure for each relationship.

a. ALGEBRA Lines AB and CD intersect at E for $A(-2, 4)$, $B(0, -2)$, $C(-3, 0)$, and $D(3, 3)$ on a coordinate plane. Point F is coplanar with these points, but not collinear with $\overleftrightarrow{AB}$ or $\overleftrightarrow{CD}$.

Graph each point and draw $\overleftrightarrow{AB}$ and $\overleftrightarrow{CD}$.

Label the intersection point as E.

An infinite number of points are coplanar with A, B, C, D and E but not collinear with $\overleftrightarrow{AB}$ and $\overleftrightarrow{CD}$. In the graph, one such point is $F(2, -3)$.

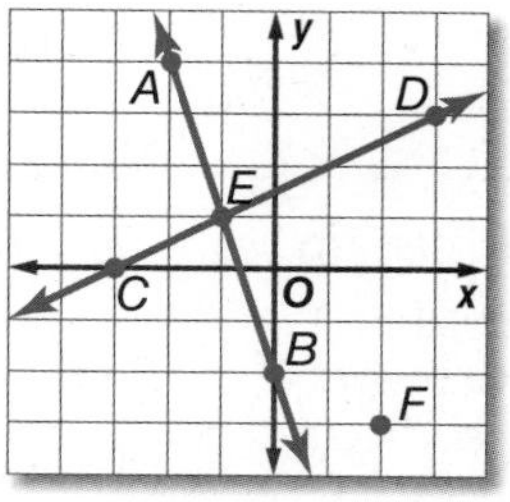

b. QR intersects plane $\mathcal{T}$ at point S.

Draw a surface to represent plane $\mathcal{T}$ and label it.

Draw a dot for point S anywhere on the plane and a dot that is not on plane $\mathcal{T}$ for point Q.

Draw a line through points Q and S. Dash the line to indicate the portion hidden by the plane. Then draw another dot on the line and label it R.

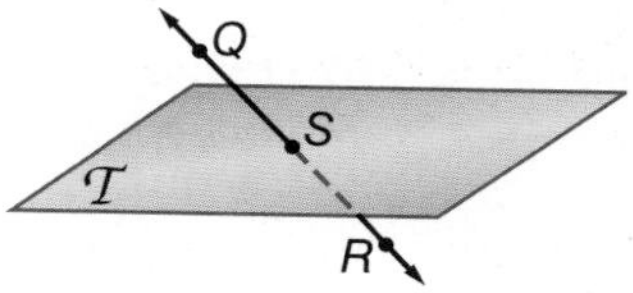

StudyTip

Three-Dimensional Drawings Because it is impossible to show an entire plane in a figure, edged shapes with different shades of color are used to represent planes.

GuidedPractice

3A. Points $J(-4, 2)$, $K(3, 2)$, and L are collinear.

3B. Line p lies in plane $\mathcal{N}$ and contains point L.

Definitions or **defined terms** are explained using undefined terms and/or other defined terms. **Space** is defined as a boundless, three-dimensional set of all points. Space can contain lines and planes.

Example 4 Interpret Drawings

a. How many planes appear in this figure?

Six: plane $\mathcal{X}$, plane JDH, plane JDE, plane EDF, plane FDG, and plane HDG.

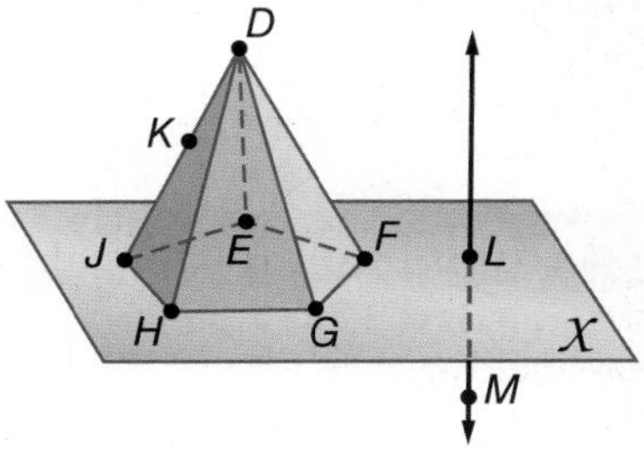

b. Name three points that are collinear.

Points J, K, and D are collinear.

c. Name the intersection of plane HDG with plane $\mathcal{X}$.

Plane HDG intersects plane $\mathcal{X}$ in $\overleftrightarrow{HG}$.

d. At what point do $\overleftrightarrow{LM}$ and $\overleftrightarrow{EF}$ intersect? Explain.

It does not appear that these lines intersect. $\overleftrightarrow{EF}$ lies in plane $\mathcal{X}$, but only point L of $\overleftrightarrow{LM}$ lies in $\mathcal{X}$.

StudyTip

CCSS Precision A point has no dimension. A line exists in one dimension. However, a circle is two-dimensional, and a pyramid is three-dimensional.

GuidedPractice

Explain your reasoning.

4A. Are points E, D, F, and G coplanar?

4B. At what point or in what line do planes JDH, JDE, and EDF intersect?

Check Your Understanding

Example 1 **Use the figure to name each of the following.**

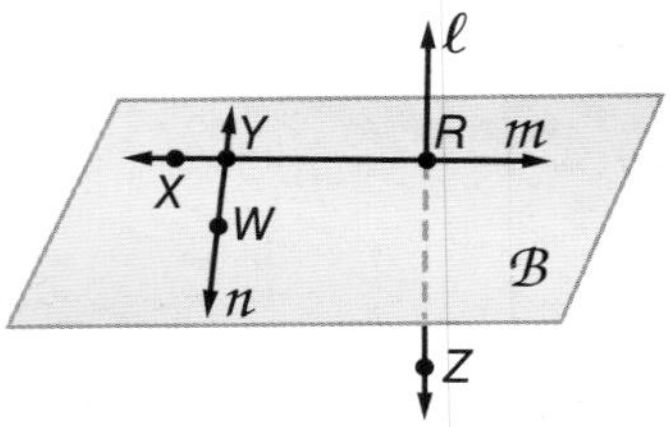

1. a line containing point X
2. a line containing point Z
3. a plane containing points W and R

Example 2 **Name the geometric term modeled by each object.**

4. a beam from a laser
5. a floor

Example 3 **Draw and label a figure for each relationship.**

6. A line in a coordinate plane contains $A(0, -5)$ and $B(3, 1)$ and a point C that is not collinear with $\overleftrightarrow{AB}$.
7. Plane Z contains lines x, y, and w. Lines x and y intersect at point V and lines x and w intersect at point P.

Example 4 **Refer to the figure.**

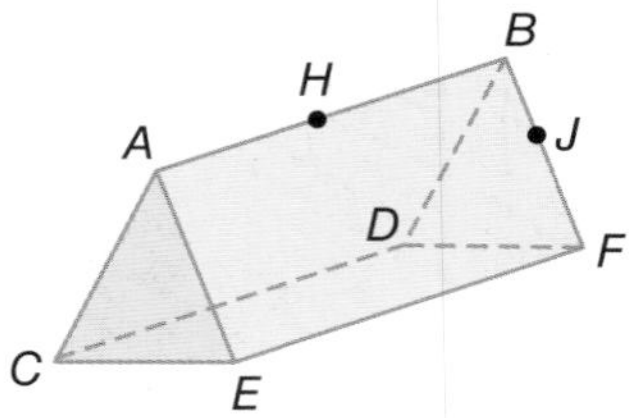

8. How many planes are shown in the figure?
9. Name three points that are collinear.
10. Are points A, H, J, and D coplanar? Explain.
11. Are points B, D, and F coplanar? Explain.
12. **ASTRONOMY** Ursa Minor, or the Little Dipper, is a constellation made up of seven stars in the northern sky including the star Polaris.
    a. What geometric figures are modeled by the stars?
    b. Are Star 1, Star 2, and Star 3 collinear on the constellation map? Explain.
    c. Are Polaris, Star 2, and Star 6 coplanar on the map?

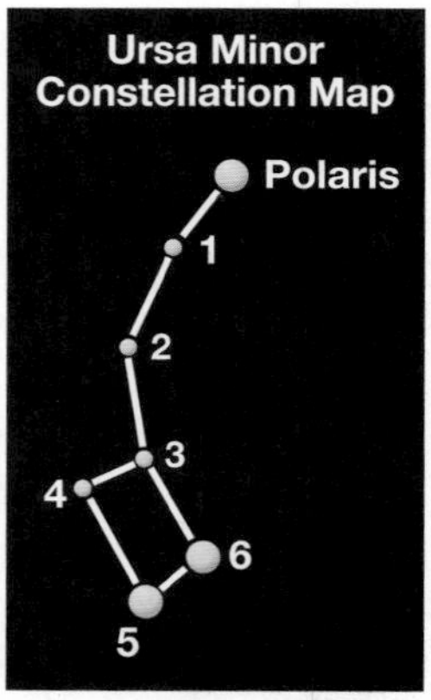

Practice and Problem Solving

Example 1 **Refer to the figure.**

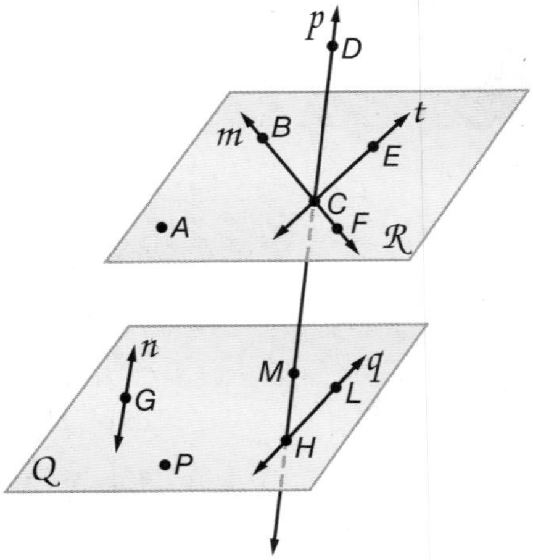

13. Name the lines that are only in plane Q.
14. How many planes are labeled in the figure?
15. Name the plane containing the lines m and t.
16. Name the intersection of lines m and t.
17. Name a point that is not coplanar with points A, B, and C.
18. Are points F, M, G, and P coplanar? Explain.
19. Name the points not contained in a line shown.
20. What is another name for line t?
21. Does line n intersect line q? Explain.

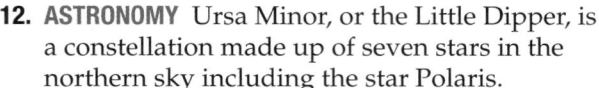

Example 2 **Name the geometric term(s) modeled by each object.**

22.

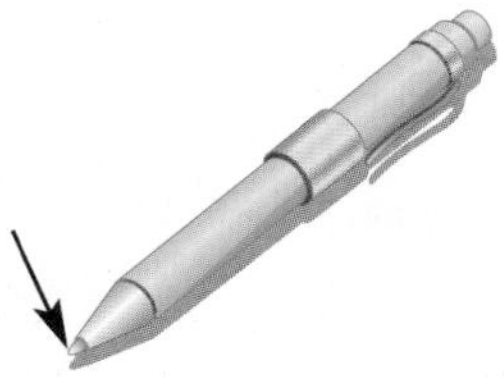

23.

24.

25.

26. a blanket

27. a knot in a rope

28. a telephone pole

29. the edge of a desk

30. two connected walls

31. a partially opened folder

Example 3 **Draw and label a figure for each relationship.**

32. Line m intersects plane $\mathcal{R}$ at a single point.

33. Two planes do not intersect.

34. Points X and Y lie on $\overleftrightarrow{CD}$.

35. Three lines intersect at point J but do not all lie in the same plane.

36. Points $A(2, 3)$, $B(2, -3)$, C and D are collinear, but A, B, C, D, and F are not.

37. Lines $\overleftrightarrow{LM}$ and $\overleftrightarrow{NP}$ are coplanar but do not intersect.

38. $\overleftrightarrow{FG}$ and $\overleftrightarrow{JK}$ intersect at $P(4, 3)$, where point F is at $(-2, 5)$ and point J is at $(7, 9)$.

39. Lines s and t intersect, and line v does not intersect either one.

Example 4 CCSS **MODELING** **When packing breakable objects such as glasses, movers frequently use boxes with inserted dividers like the one shown.**

40. How many planes are modeled in the picture?

41. What parts of the box model lines?

42. What parts of the box model points?

Refer to the figure at the right.

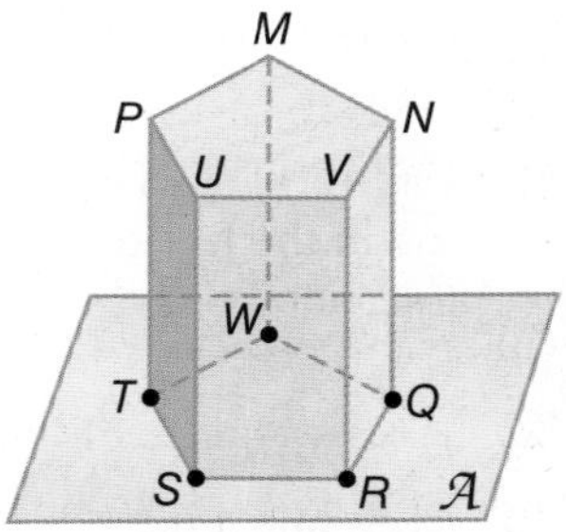

43. Name two collinear points.

44. How many planes appear in the figure?

45. Do plane $\mathcal{A}$ and plane MNP intersect? Explain.

46. In what line do planes $\mathcal{A}$ and QRV intersect?

47. Are points T, S, R, Q, and V coplanar? Explain.

48. Are points T, S, R, Q, and W coplanar? Explain.

49. **FINITE PLANES** A *finite plane* is a plane that has boundaries, or does not extend indefinitely. The street signs shown are finite planes.

a. If the pole models a line, name the geometric term that describes the intersection between the signs and the pole.

b. What geometric term(s) describes the intersection between the two finite planes? Explain your answer with a diagram if necessary.

50. **ONE-POINT PERSPECTIVE** One-point perspective drawings use lines to convey depth. Lines representing horizontal lines in the real object can be extended to meet at a single point called the *vanishing point*. Suppose you want to draw a tiled ceiling in the room below with nine tiles across.

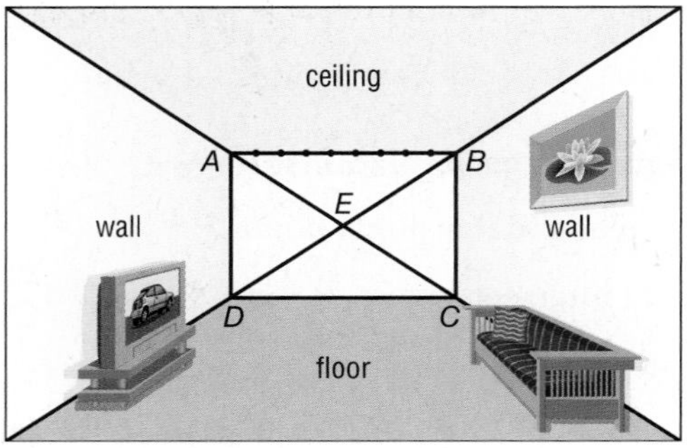

a. What point represents the vanishing point in the drawing?

b. Trace the figure. Then draw lines from the vanishing point through each of the eight points between *A* and *B*. Extend these lines to the top edge of the drawing.

c. How could you change the drawing to make the back wall of the room appear farther away?

51. **TWO-POINT PERSPECTIVE** Two-point perspective drawings use two vanishing points to convey depth.

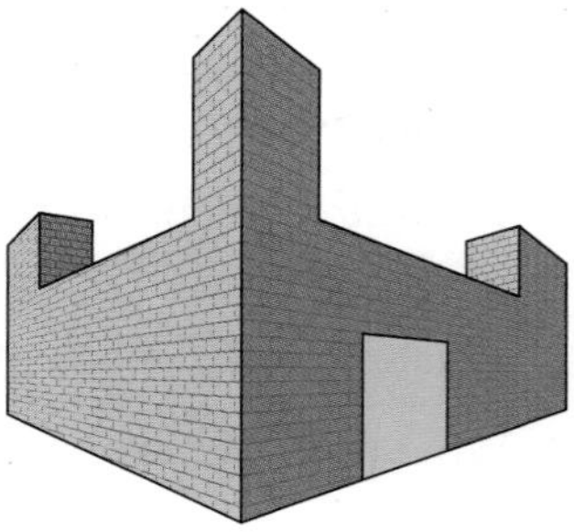

a. Trace the drawing of the castle shown. Draw five of the vertical lines used to create the drawing.

b. Draw and extend the horizontal lines to locate the vanishing points and label them.

c. What do you notice about the vertical lines as they get closer to the vanishing point?

d. Draw a two-point perspective of a home or a room in a home.

52. CCSS **ARGUMENTS** Name two points on the same line in the figure. How can you support your assertion?

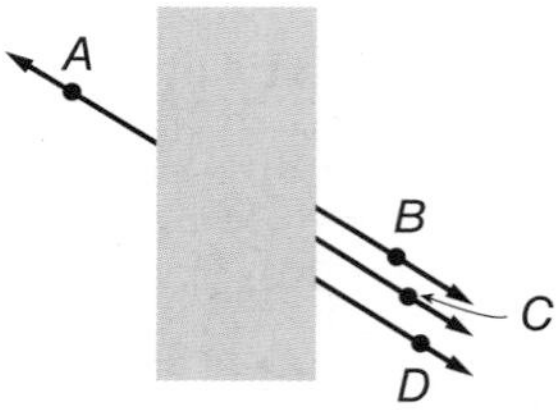

ThinkStock/Jupiterimages

53. **TRANSPORTATION** When two cars enter an intersection at the same time on opposing paths, one of the cars must adjust its speed or direction to avoid a collision. Two airplanes, however, can cross paths while traveling in different directions without colliding. Explain how this is possible.

54. **MULTIPLE REPRESENTATIONS** Another way to describe a group of points is called a locus. A **locus** is a set of points that satisfy a particular condition. In this problem, you will explore the locus of points that satisfy an equation.

   a. **Tabular** Represent the locus of points satisfying the equation $2 + x = y$ using a table of at least five values.

   b. **Graphical** Represent this same locus of points using a graph.

   c. **Verbal** Describe the geometric figure that the points suggest.

55. **PROBABILITY** Three of the labeled points are chosen at random.

   a. What is the probability that the points chosen are collinear?

   b. What is the probability that the points chosen are coplanar?

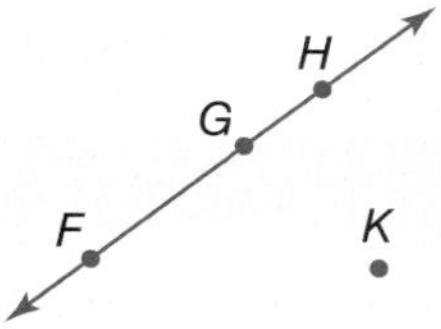

56. **MULTIPLE REPRESENTATIONS** In this problem, you will explore the locus of points that satisfy an inequality.

   a. **Tabular** Represent the locus of points satisfying the inequality $y < -3x - 1$ using a table of at least ten values.

   b. **Graphical** Represent this same locus of points using a graph.

   c. **Verbal** Describe the geometric figure that the points suggest.

H.O.T. Problems Use Higher-Order Thinking Skills

57. **OPEN ENDED** Sketch three planes that intersect in a line.

58. **ERROR ANALYSIS** Camille and Hiroshi are trying to determine the most number of lines that can be drawn using any two of four random points. Is either correct? Explain.

Camille	Hiroshi
Since there are four points, $4 \cdot 3$ or 12 lines can be drawn between the points.	You can draw $3 \cdot 2 \cdot 1$ or 6 lines between the points.

59. **CCSS ARGUMENTS** What is the greatest number of planes determined using any three of the points A, B, C, and D if no three points are collinear?

60. **REASONING** Is it possible for two points on the surface of a prism to be neither collinear nor coplanar? Justify your answer.

61. **WRITING IN MATH** Refer to Exercise 49. Give a real-life example of a finite plane. Is it possible to have a real-life object that is an infinite plane? Explain your reasoning.

Standardized Test Practice

62. Which statement about the figure below is *not* true?

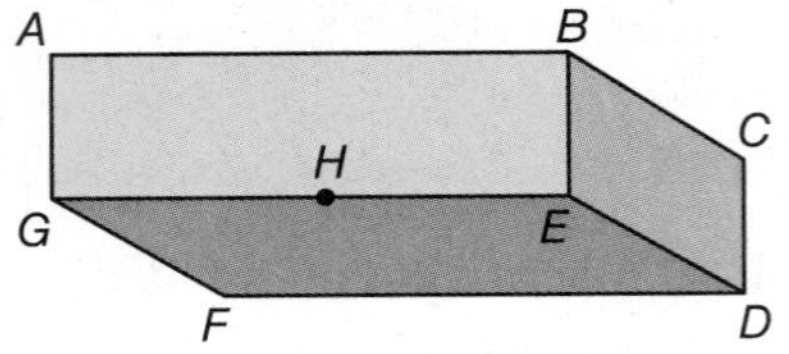

A Point H lies in planes AGE and GED.

B Planes GAB, GFD and BED intersect at point E.

C Points F, E, and B are coplanar.

D Points A, H, and D are collinear.

63. ALGEBRA What is the value of x if $3x + 2 = 8$?

F -2 **G** 0 **H** 2 **J** 6

64. GRIDDED RESPONSE An ice chest contains 3 types of drinks: 10 apple juices, 15 grape juices, and 15 bottles of water. What is the probability that a drink selected randomly from the ice chest does *not* contain fruit juice?

65. SAT/ACT A certain school's enrollment increased 6% this year over last year's enrollment. If the school now has 1378 students enrolled, how many students were enrolled last year?

A 1295 **C** 1350 **E** 1500

B 1300 **D** 1460

Spiral Review

Use elimination to solve each system of equations. (Lesson 6-4)

66. $2x + y = 5$
$3x - 2y = 4$

67. $4x - 3y = 12$
$x + 2y = 14$

68. $2x - 3y = 2$
$5x + 4y = 28$

69. HEALTH About 20% of the time you sleep is spent in rapid eye movement (REM), which is associated with dreaming. If an adult sleeps 7 to 8 hours, how much time is spent in REM sleep? (Lesson 5-4)

Simplify. Assume that no denominator is equal to zero. (Lesson 7-2)

70. $\frac{a^6}{a^3}$

71. $\frac{4^7}{4^5}$

72. $\frac{c^3d^4}{cd^7}$

73. $\left(\frac{4h^{-2}g}{2g^5}\right)^0$

74. $\frac{5q^{-2}t^6}{10q^2t^{-4}}$

75. $b^3(m^{-3})(b^{-6})$

Solve each open sentence. (Lesson 5-5)

76. $|y - 2| > 7$

77. $|z + 5| < 3$

78. $|2b + 7| \leq -6$

79. $|3 - 2y| \geq 8$

80. $|9 - 4m| < -1$

81. $|5c - 2| \leq 13$

Skills Review

Replace each ● with >, <, or = to make a true statement.

82. $\frac{1}{4}$ in. ● $\frac{1}{2}$ in.

83. $\frac{3}{4}$ in. ● $\frac{5}{8}$ in.

84. $\frac{3}{8}$ in. ● $\frac{6}{16}$ in.

85. 18 mm ● 2 cm

86. 32 mm ● 3.2 cm

87. 0.8 m ● 8 cm

EXTEND 10-1

Geometry Lab
Describing What You See

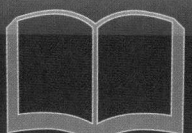

When you are learning geometric concepts, it is critical to have accurate drawings to represent the information. It is helpful to know what words and phrases can be used to describe figures. Likewise, it is important to know how to read a geometric description and be able to draw the figure it describes.

CCSS Common Core State Standards
Content Standards
G.MG.1 Use geometric shapes, their measures, and their properties to describe objects (e.g., modeling a tree trunk or a human torso as a cylinder). ★
Mathematical Practices 6

The figures and descriptions below help you visualize and write about points, lines, and planes.

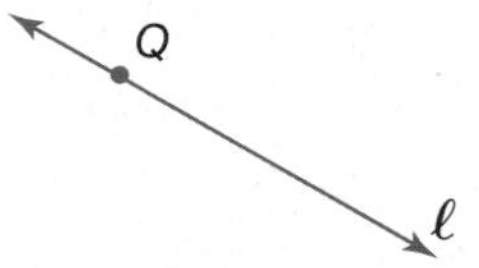

Point Q is on ℓ.

Line ℓ contains Q.

Line ℓ passes through Q.

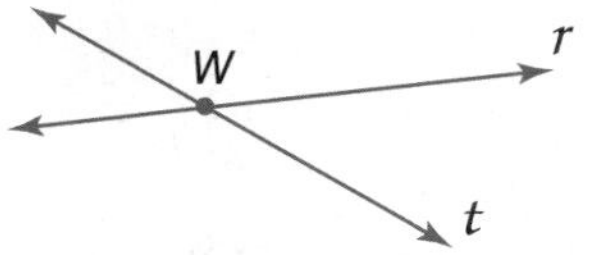

Lines r and t intersect at W.

Point W is the intersection of r and t.

Point W is on r. Point W is on t.

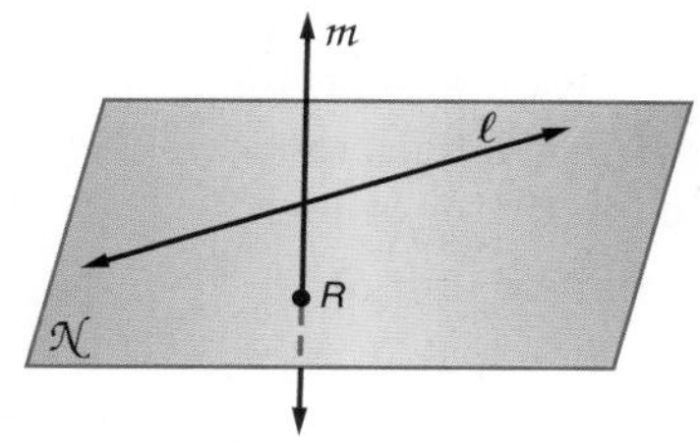

Line ℓ and point R are in $\mathcal{N}$.

Point R lies in $\mathcal{N}$.

Plane $\mathcal{N}$ contains R and ℓ.

Line m intersects $\mathcal{N}$ at R.

Point R is the intersection of m with $\mathcal{N}$.

Lines ℓ and m do not intersect.

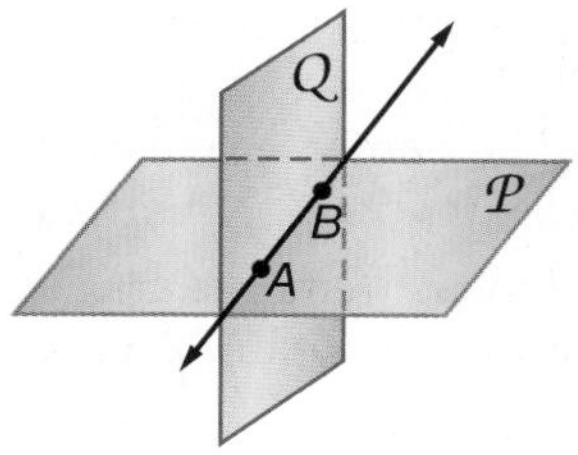

$\overleftrightarrow{AB}$ is in $\mathcal{P}$ and $\mathcal{Q}$.

Points A and B lie in both $\mathcal{P}$ and $\mathcal{Q}$.

Planes $\mathcal{P}$ and $\mathcal{Q}$ both contain $\overleftrightarrow{AB}$.

Planes $\mathcal{P}$ and $\mathcal{Q}$ intersect in $\overleftrightarrow{AB}$.

$\overleftrightarrow{AB}$ is the intersection of $\mathcal{P}$ and $\mathcal{Q}$.

Exercises

Write a description for each figure.

1.

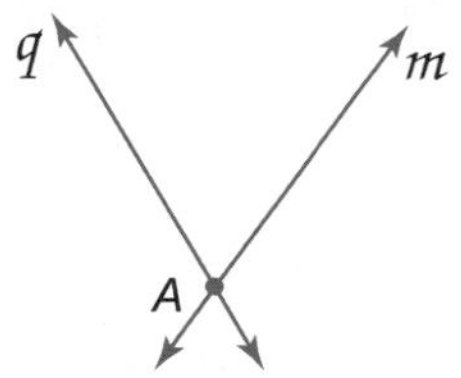

2.

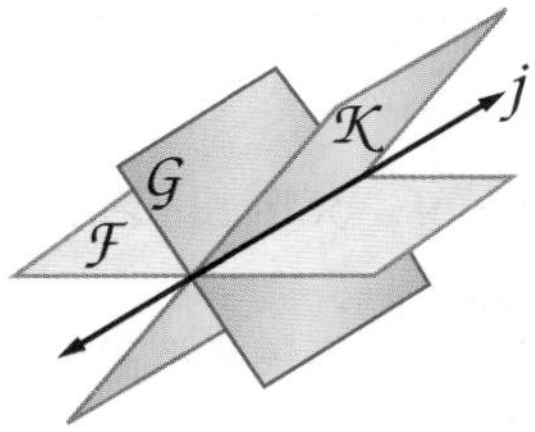

3.

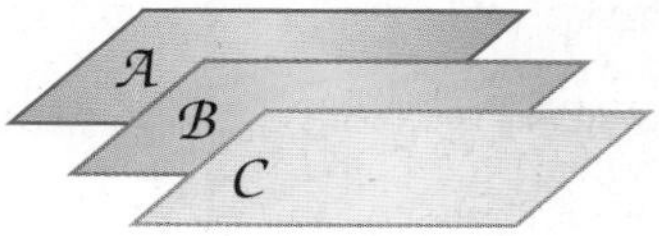

4. Draw and label a figure for the statement Planes $\mathcal{N}$ and $\mathcal{P}$ contain line a.

LESSON 10-2 Linear Measure

Then

- You identified and modeled points, lines, and planes.

Now

1. Measure segments.
2. Calculate with measures.

Why?

- When the ancient Egyptians found a need for a measurement system, they used the human body as a guide. The cubit was the length of an arm from the elbow to the fingertips. Eventually the Egyptians standardized the length of a cubit, with ten *royal cubits* equivalent to one *rod*.

NewVocabulary

line segment
betweenness of points
between
congruent segments
construction

Common Core State Standards

Content Standards

G.CO.1 Know precise definitions of angle, circle, perpendicular line, parallel line, and line segment, based on the undefined notions of point, line, distance along a line, and distance around a circular arc.

G.CO.12 Make formal geometric constructions with a variety of tools and methods (compass and straightedge, string, reflective devices, paper folding, dynamic geometric software, etc.).

Mathematical Practices

6 Attend to precision.

1 Measure Line Segments

Unlike a line, a **line segment**, or *segment*, can be measured because it has two endpoints. A segment with endpoints A and B can be named as $\overline{AB}$ or $\overline{BA}$. The *measure* of $\overline{AB}$ is written as AB. The length or measure of a segment always includes a unit of measure, such as meter or inch.

All measurements are approximations dependent upon the smallest unit of measure available on the measuring instrument.

Example 1 Length in Metric Units

Find the length of $\overline{AB}$ using each ruler.

a.

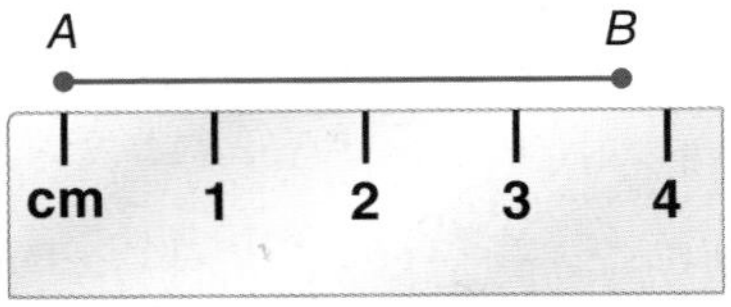

The ruler is marked in centimeters. Point B is closer to the 4-centimeter mark than to 3 centimeters.

Thus, $\overline{AB}$ is about 4 centimeters long.

b.

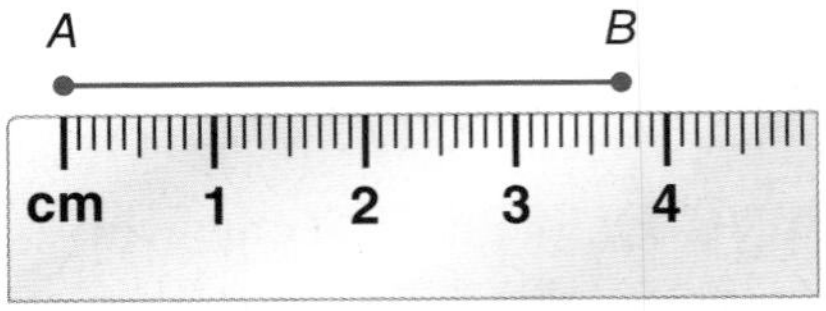

The long marks are centimeters, and the shorter marks are millimeters. There are 10 millimeters for each centimeter.

Thus, $\overline{AB}$ is about 3.7 centimeters long.

GuidedPractice

1A. Measure the length of a dollar bill in centimeters.

1B. Measure the length of a pencil in millimeters.

1C. Find the length of $\overline{CD}$.

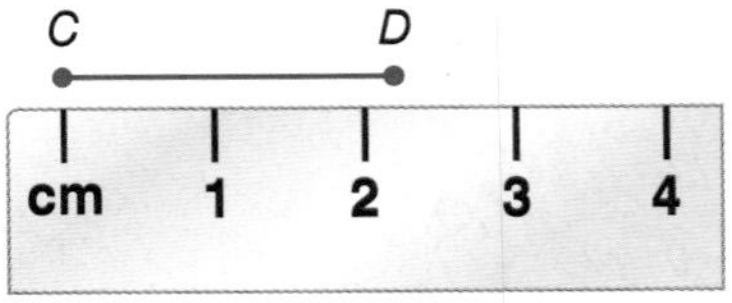

StudyTip

Using a Ruler The zero point on a ruler may not be clearly marked. For some rulers, zero is the left edge of the ruler. On others, it may be a fine line farther in on the scale. If it is not clear where the zero is, align one endpoint on 1 and subtract 1 from the measurement at the other endpoint.

Example 2 Length in Standard Units

Find the length of $\overline{CD}$ using each ruler.

a.

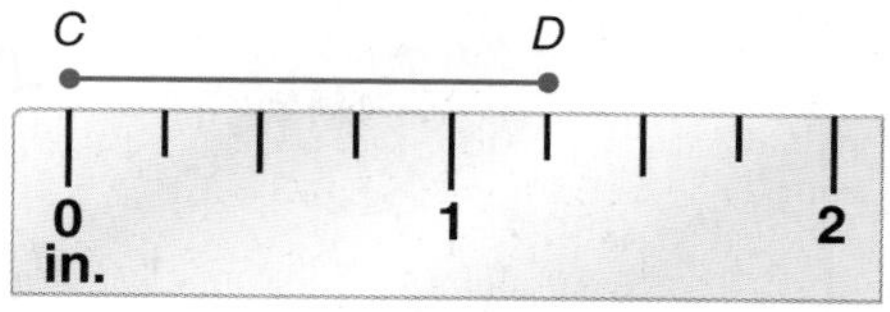

Each inch is divided into fourths.

Point D is closer to the $1\frac{1}{4}$-inch mark.

$\overline{CD}$ is about $1\frac{1}{4}$ inches long.

b.

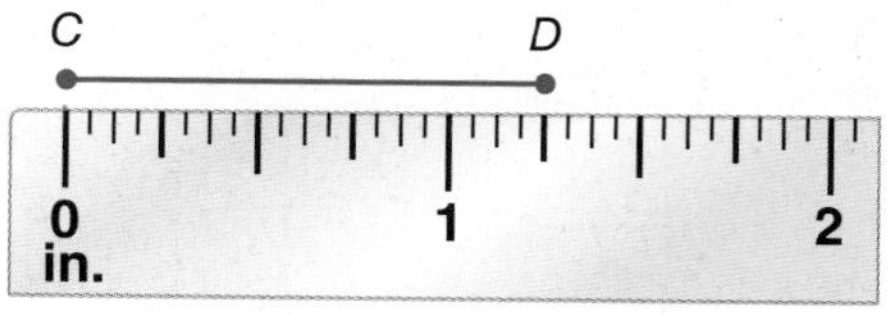

Each inch is divided into sixteenths.

Point D is closer to the $1\frac{4}{16}$-inch mark.

$\overline{CD}$ is about $1\frac{4}{16}$ or $1\frac{1}{4}$ inches long.

GuidedPractice

2A. Measure the length of a dollar bill in inches.

2B. Measure the length of a pencil in inches.

2 Calculate Measures

Recall that for any two real numbers a and b, there is a real number n that is *between* a and b such that $a < n < b$. This relationship also applies to points on a line and is called **betweenness of points**. In the figure, point N is between points A and B, but points R and P are not.

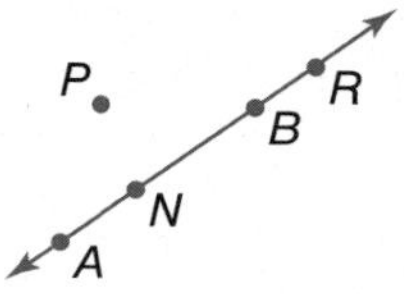

Measures are real numbers, so all arithmetic operations can be used with them. You know that the whole usually equals the sum of its parts. That is also true of line segments in geometry.

KeyConcept Betweenness of Points

Words

Point M is **between** points P and Q if and only if P, Q, and M are collinear and $PM + MQ = PQ$.

Model

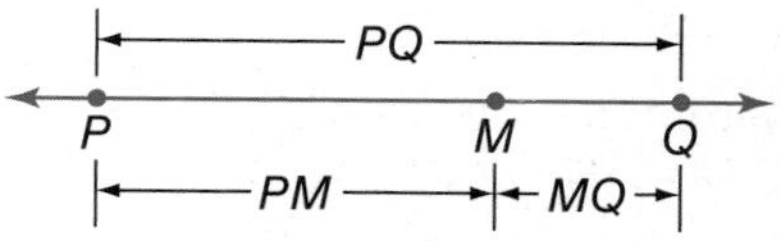

StudyTip

Comparing Measures Because measures are real numbers, you can compare them. If points X, Y, and Z are collinear in that order, then one of these statements is true: $XY = YZ$, $XY > YZ$, or $XY < YZ$.

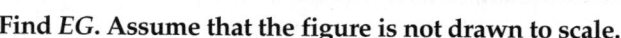

Example 3 Find Measurements by Adding

Find EG. Assume that the figure is not drawn to scale.

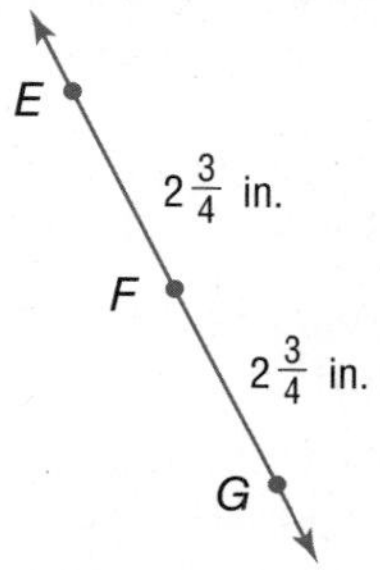

EG is the measure of $\overline{EG}$. Point F is between E and G. Find EG by adding EF and FG.

$EF + FG = EG$	Betweenness of points
$2\frac{3}{4} + 2\frac{3}{4} = EG$	Substitution
$5\frac{1}{2}$ in. $= EG$	Add.

GuidedPractice

3. Find JL. Assume that the figure is not drawn to scale.

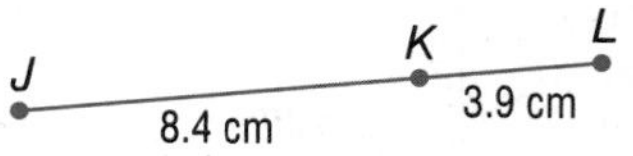

Example 4 Find Measurements by Subtracting

Find AB. Assume that the figure is not drawn to scale.

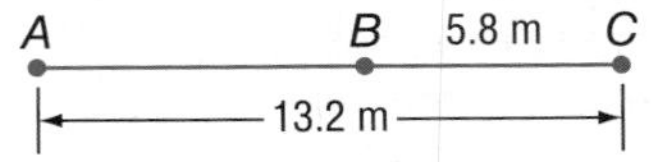

Point B is between A and C.

$AB + BC = AC$	Betweenness of points
$AB + 5.8 = 13.2$	Substitution
$AB + 5.8 - 5.8 = 13.2 - 5.8$	Subtract 5.8 from each side.
$AB = 7.4$ m	Simplify.

GuidedPractice

4. Find QR. Assume that the figure is not drawn to scale.

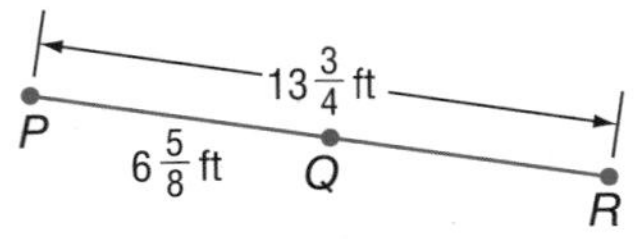

Example 5 Write and Solve Equations to Find Measurements

ALGEBRA Find the value of a and XY if Y is between X and Z, $XY = 3a$, $XZ = 5a - 4$, and $YZ = 14$.

Draw a figure to represent this information.

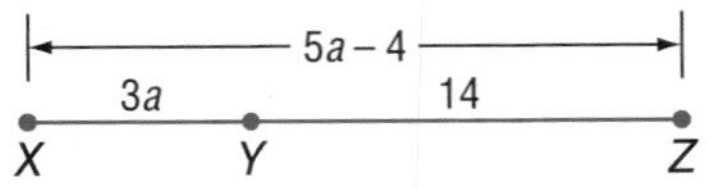

$XZ = XY + YZ$	Betweenness of points
$5a - 4 = 3a + 14$	Substitution
$5a - 4 - 3a = 3a + 14 - 3a$	Subtract 3a from each side.
$2a - 4 = 14$	Simplify.
$2a - 4 + 4 = 14 + 4$	Add 4 to each side.
$2a = 18$	Simplify.
$\frac{2a}{2} = \frac{18}{2}$	Divide each side by 2.
$a = 9$	Simplify.

Now find XY.

$XY = 3a$	Given
$= 3(9)$ or 27	$a = 9$

GuidedPractice

5. Find x and BC if B is between A and C, $AC = 4x - 12$, $AB = x$, and $BC = 2x + 3$.

Segments that have the same measure are called **congruent segments**.

WatchOut!

Equal vs. Congruent Lengths are equal and segments are congruent. It is correct to say that $AB = CD$ and $\overline{AB} \cong \overline{CD}$. However, it is *not* correct to say that $\overline{AB} = \overline{CD}$ or that $AB \cong CD$.

KeyConcept Congruent Segments

Words Congruent segments have the same measure.

Symbols $\cong$ is read *is congruent to*. Red slashes on the figure also indicate congruence.

Example $\overline{AB} \cong \overline{CD}$

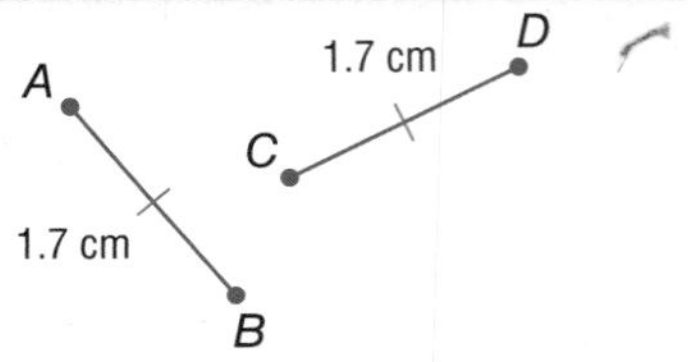

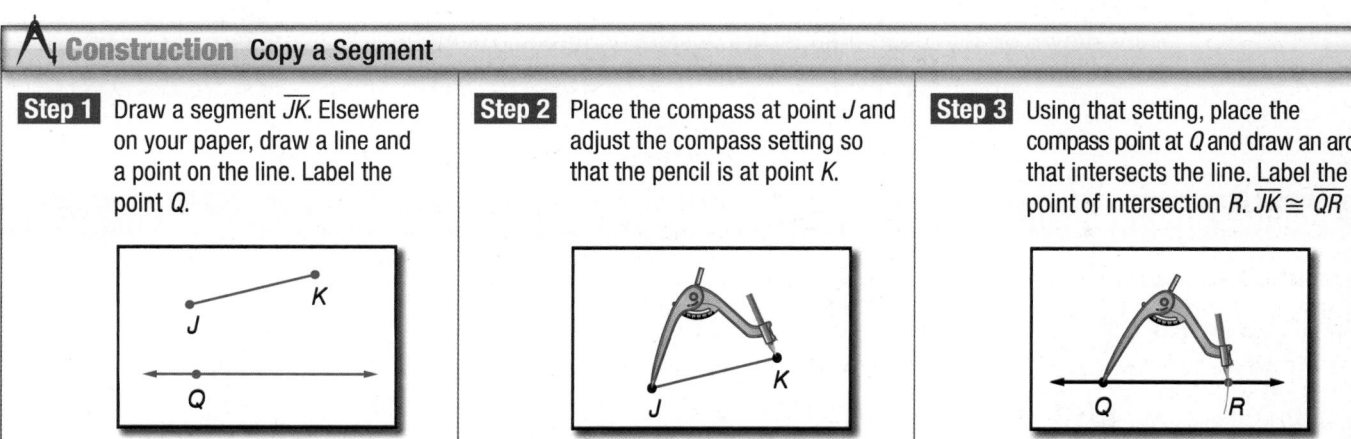

Real-WorldLink

The first commercial skateboard was introduced in 1959. Now there are more than 500 skate parks in the United States.

Source: *Encyclopaedia Britannica*

Real-World Example 6 Congruent Segments

PT

SKATE PARKS In the graph, suppose a segment was drawn along the top of each bar. Which states would have segments that are congruent? Explain.

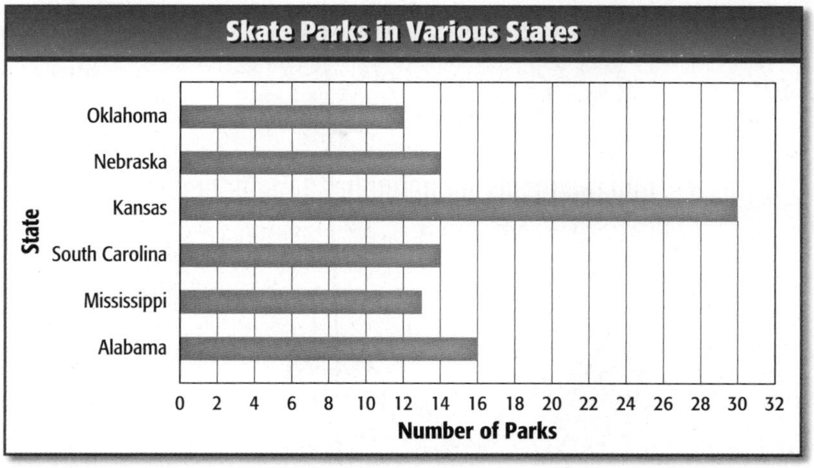

Source: SITE Design Group, Inc.

The segments on the bars for Nebraska and South Carolina would be congruent because they both represent the same number of skate parks.

GuidedPractice

6A. Suppose Oklahoma added another skate park. The segment drawn along the bar representing Oklahoma would be congruent to which other segment?

6B. Name the congruent segments in the sign shown.

Drawings of geometric figures are created using measuring tools such as a ruler and protractor. **Constructions** are methods of creating these figures without the benefit of measuring tools. Generally, only a pencil, straightedge, and compass are used in constructions. *Sketches* are created without the use of any of these tools.

You can construct a segment that is congruent to a given segment.

Construction Copy a Segment

Step 1 Draw a segment $\overline{JK}$. Elsewhere on your paper, draw a line and a point on the line. Label the point Q.

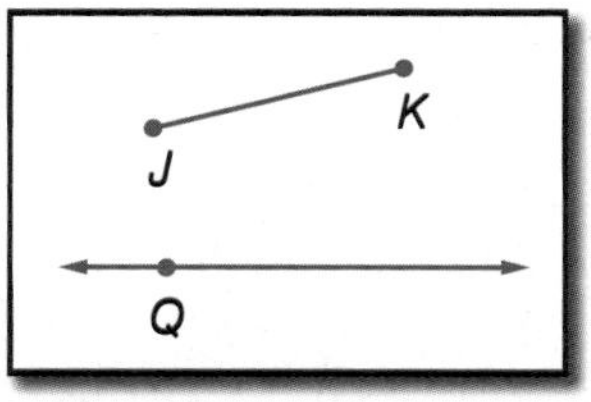

Step 2 Place the compass at point J and adjust the compass setting so that the pencil is at point K.

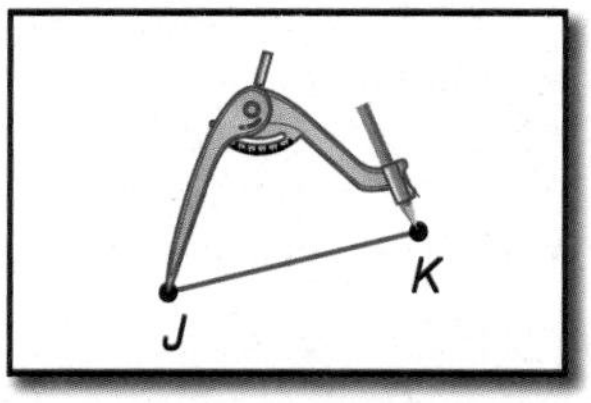

Step 3 Using that setting, place the compass point at Q and draw an arc that intersects the line. Label the point of intersection R. $\overline{JK} \cong \overline{QR}$

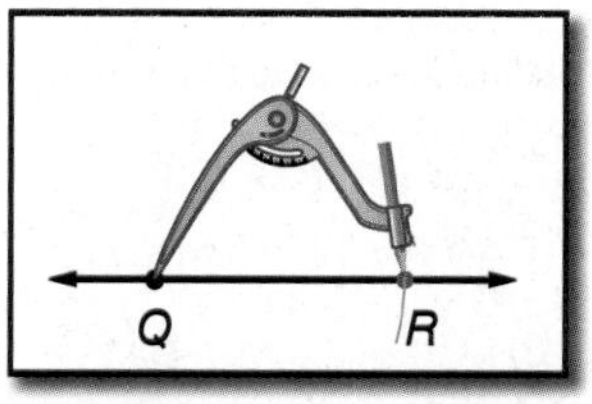

Matt Stroshane/Getty Images

Check Your Understanding

Example 1 **Find the length of each line segment or object.**

1.

2.

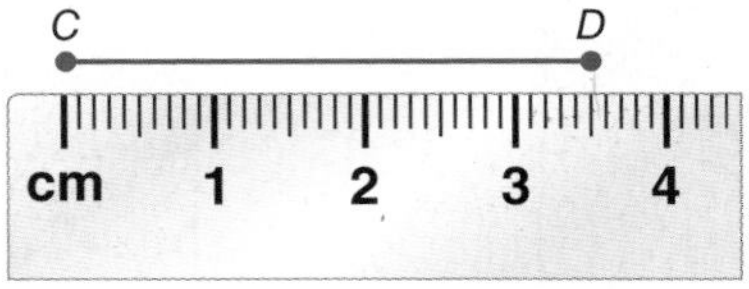

Example 2

3.

4.

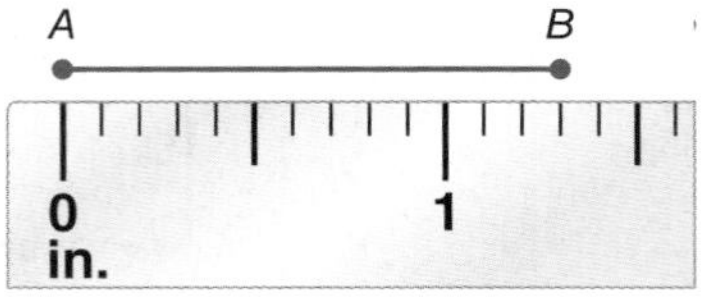

Examples 3–4 **Find the measurement of each segment. Assume that each figure is not drawn to scale.**

5. $\overline{CD}$

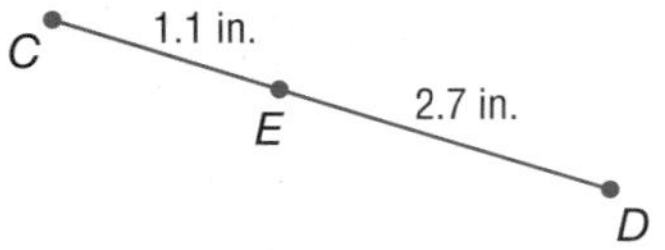

6. $\overline{RS}$

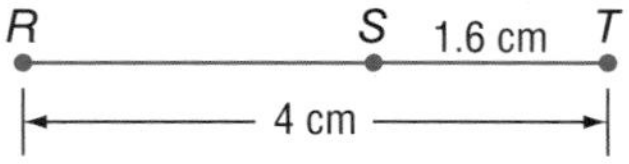

Example 5 **ALGEBRA Find the value of x and BC if B is between C and D.**

7. $CB = 2x$, $BD = 4x$, and $BD = 12$

8. $CB = 4x - 9$, $BD = 3x + 5$, and $CD = 17$

Example 6

9. **CCSS STRUCTURE** The Indiana State Flag was adopted in 1917. The measures of the segments between the stars and the flame are shown on the diagram in inches. List all of the congruent segments in the figure.

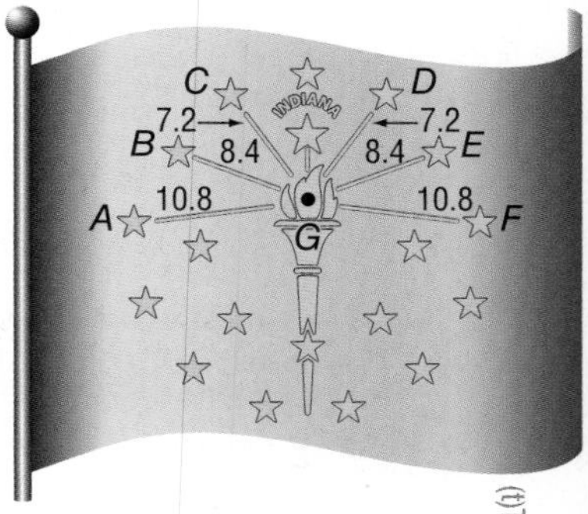

Practice and Problem Solving

Examples 1–2 **Find the length of each line segment.**

10.

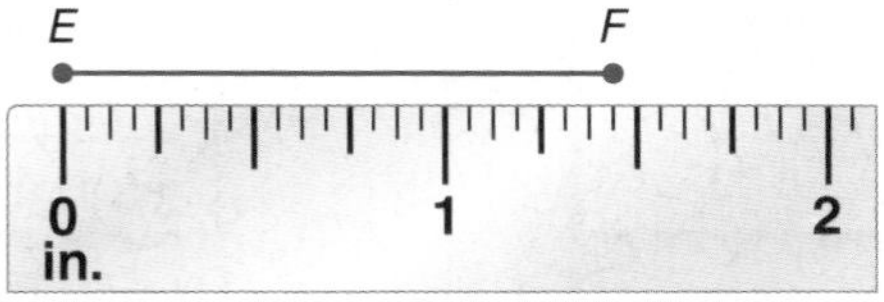

11.

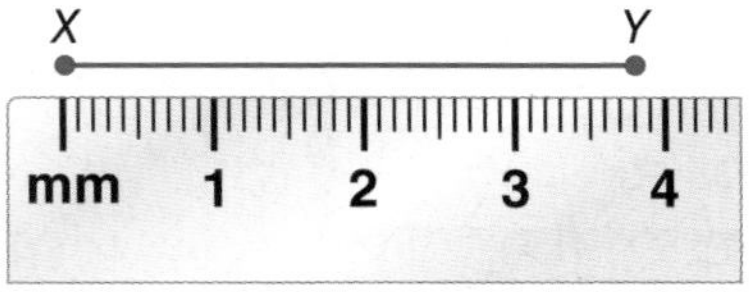

12.

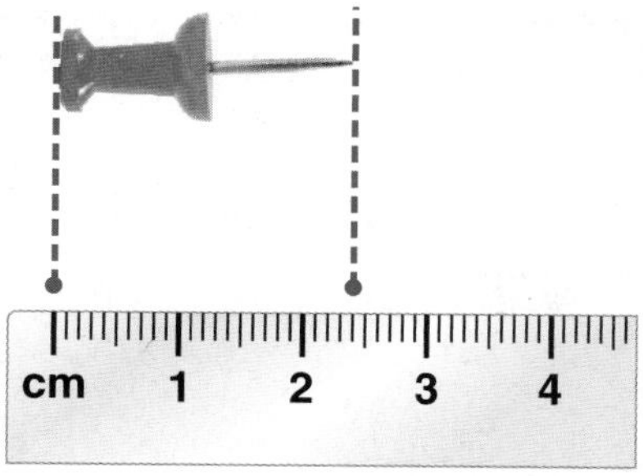

13.

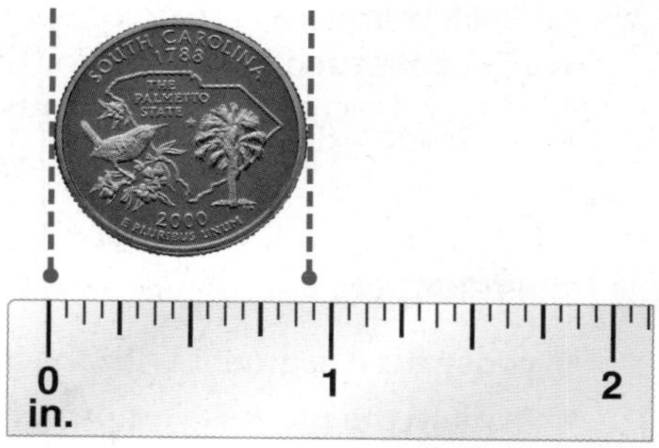

Examples 3–4 **Find the measurement of each segment. Assume that each figure is not drawn to scale.**

14. $\overline{EF}$

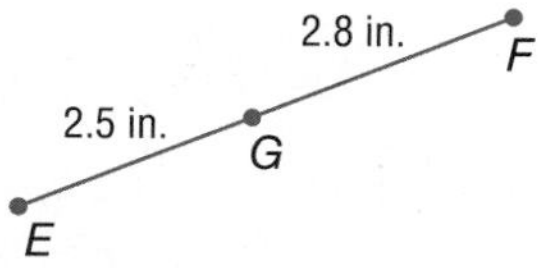

15. $\overline{JL}$

16. $\overline{PR}$

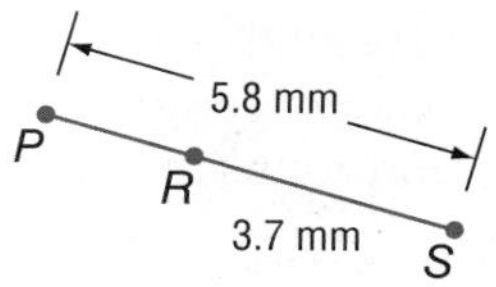

17. $\overline{SV}$

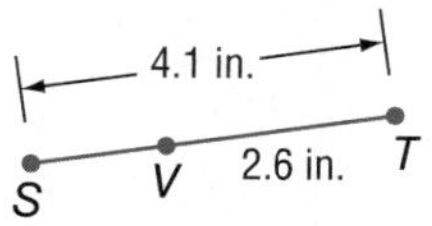

18. $\overline{WY}$

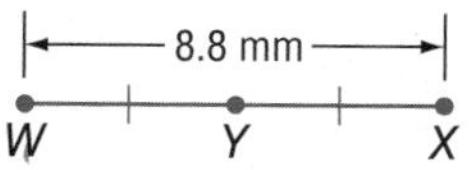

19. $\overline{FG}$

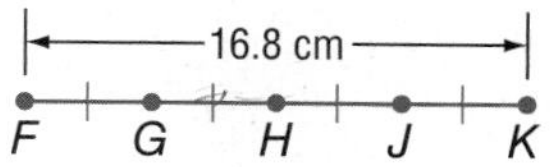

20. **CCSS SENSE-MAKING** The stacked bar graph shows the number of canned food items donated by the girls and the boys in a homeroom class over three years. Use the concept of betweenness of points to find the number of cans donated by the boys for each year. Explain your method.

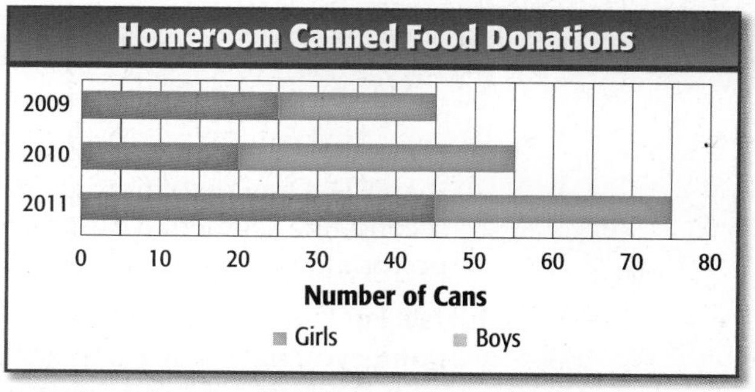

Example 5 **ALGEBRA Find the value of the variable and YZ if Y is between X and Z.**

21. $XY = 11$, $YZ = 4c$, $XZ = 83$

22. $XY = 6b$, $YZ = 8b$, $XZ = 175$

23. $XY = 7a$, $YZ = 5a$, $XZ = 6a + 24$

24. $XY = 11d$, $YZ = 9d - 2$, $XZ = 5d + 28$

25. $XY = 4n + 3$, $YZ = 2n - 7$, $XZ = 22$

26. $XY = 3a - 4$, $YZ = 6a + 2$, $XZ = 5a + 22$

Example 6 **Determine whether each pair of segments is congruent.**

27. $\overline{KJ}$, $\overline{HL}$

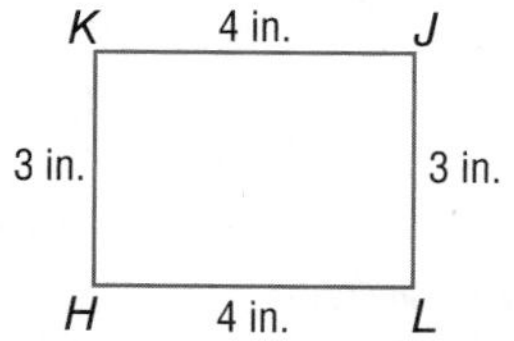

28. $\overline{AC}$, $\overline{BD}$

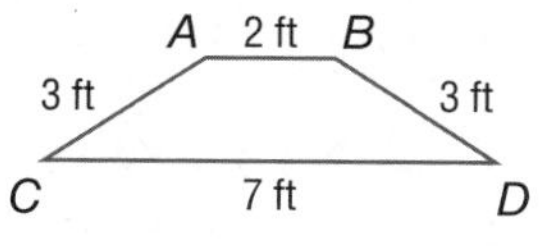

29. $\overline{EH}$, $\overline{FG}$

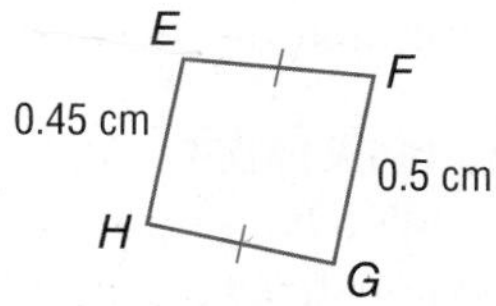

30. $\overline{VW}$, $\overline{UZ}$

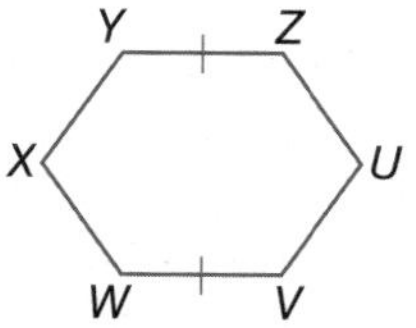

31. $\overline{MN}$, $\overline{RQ}$

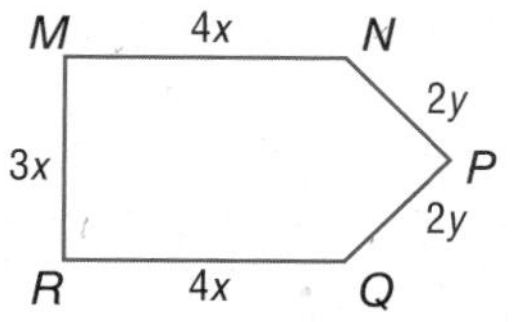

32. $\overline{SU}$, $\overline{VT}$

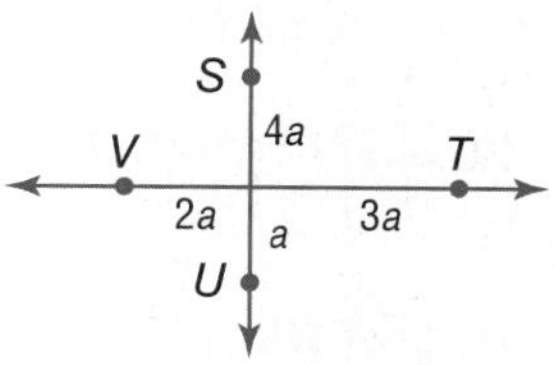

33. TRUSSES A truss is a structure used to support a load over a span, such as a bridge or the roof of a house. List all of the congruent segments in the figure.

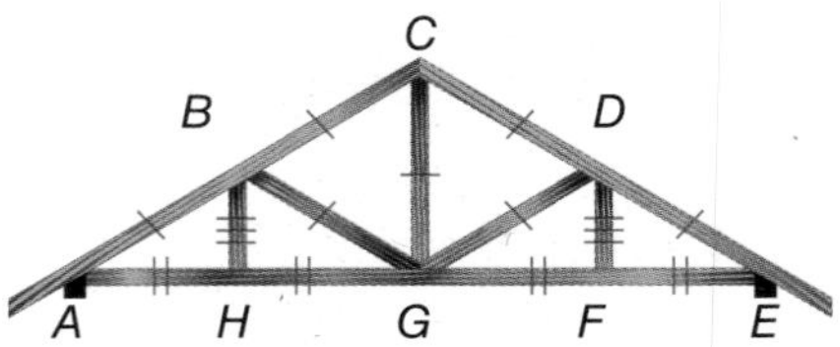

34. CONSTRUCTION For each expression:

- construct a segment with the given measure,
- explain the process you used to construct the segment, and
- verify that the segment you constructed has the given measure.

a. $2(XY)$ **b.** $6(WZ) - XY$

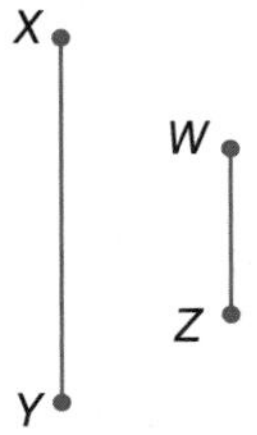

35. BLUEPRINTS Use a ruler to determine at least five pairs of congruent segments with labeled endpoints in the blueprint at the right.

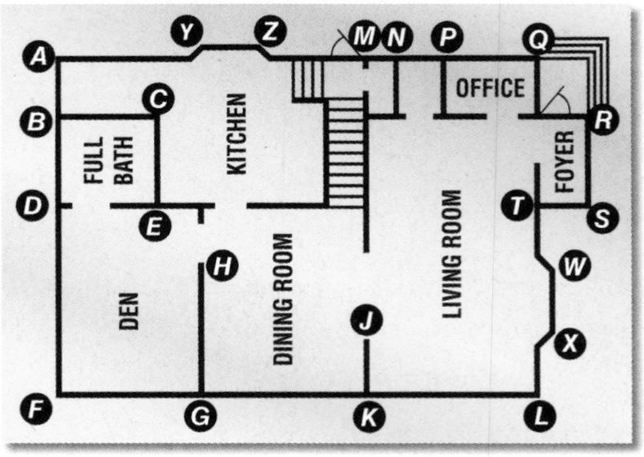

36. MULTIPLE REPRESENTATIONS Betweenness of points ensures that a line segment may be divided into an infinite number of line segments.

a. Geometric Use a ruler to draw a line segment 3 centimeters long. Label the endpoints *A* and *D*. Draw two more points along the segment and label them *B* and *C*. Draw a second line segment 6 centimeters long. Label the endpoints *K* and *P*. Add four more points along the line and label them *L*, *M*, *N*, and *O*.

b. Tabular Use a ruler to measure the length of the line segment between each of the points you have drawn. Organize the lengths of the segments in $\overline{AD}$ and $\overline{KP}$ into a table. Include a column in your table to record the sum of these measures.

c. Algebraic Give an equation that could be used to find the lengths of $\overline{AD}$ and $\overline{KP}$. Compare the lengths determined by your equation to the actual lengths.

H.O.T. Problems Use Higher-Order Thinking Skills

37. WRITING IN MATH If point *B* is between points *A* and *C*, explain how you can find *AC* if you know *AB* and *BC*. Explain how you can find *BC* if you know *AB* and *AC*.

38. OPEN ENDED Draw a segment $\overline{AB}$ that measures between 2 and 3 inches long. Then sketch a segment $\overline{CD}$ congruent to $\overline{AB}$, draw a segment $\overline{EF}$ congruent to $\overline{AB}$, and construct a segment $\overline{GH}$ congruent to $\overline{AB}$. Compare your methods.

39. CHALLENGE Point *K* is between points *J* and *L*. If $JK = x^2 - 4x$, $KL = 3x - 2$, and $JL = 28$, write and solve an equation to find the lengths of *JK* and *KL*.

40. CCSS REASONING Determine whether the statement *If point M is between points C and D, then CD is greater than either CM or MD* is *sometimes*, *never*, or *always* true. Explain.

41. WRITING IN MATH Why is it important to have a standard of measure?

Standardized Test Practice

42. SHORT RESPONSE A 36-foot-long ribbon is cut into three pieces. The first piece of ribbon is half as long as the second piece of ribbon. The third piece is 1 foot longer than twice the length of the second piece of ribbon. How long is the longest piece of ribbon?

43. In the figure, points A, B, C, D, and E are collinear. If $AE = 38$, $BD = 15$, and $\overline{BC} \cong \overline{CD} \cong \overline{DE}$, what is the length of $\overline{AD}$?

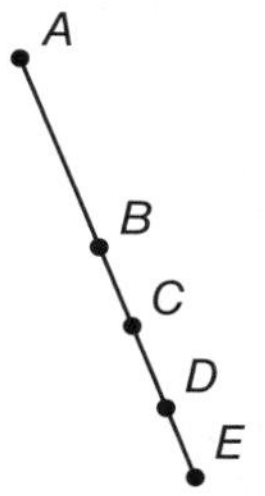

A 7.5

B 15

C 22.5

D 30.5

44. SAT/ACT If $f(x) = 7x^2 - 4x$, what is the value of $f(2)$?

F −8

G 2

H 6

J 17

K 20

45. ALGEBRA
Simplify $(3x^2 - 2)(2x + 4) - 2x^2 + 6x + 7$.

A $4x^2 + 14x - 1$

B $4x^2 - 14x + 15$

C $6x^3 + 12x^2 + 2x - 1$

D $6x^3 + 10x^2 + 2x - 1$

Spiral Review

Refer to the figure. (Lesson 10-1)

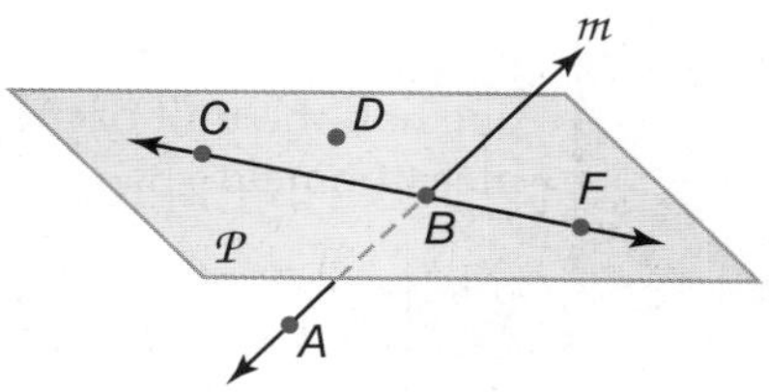

46. What are two other names for $\overleftrightarrow{AB}$?

47. Give another name for plane $\mathcal{P}$.

48. Name the intersection of plane $\mathcal{P}$ and $\overleftrightarrow{AB}$.

49. Name three collinear points.

50. Name two points that are not coplanar.

51. GEOMETRY Supplementary angles are two angles with measures that have a sum of 180°. For the supplementary angles in the figure, the measure of the larger angle is 24° greater than the measure of the smaller angle. Write and solve a system of equations to find these measures. (Lesson 6-5)

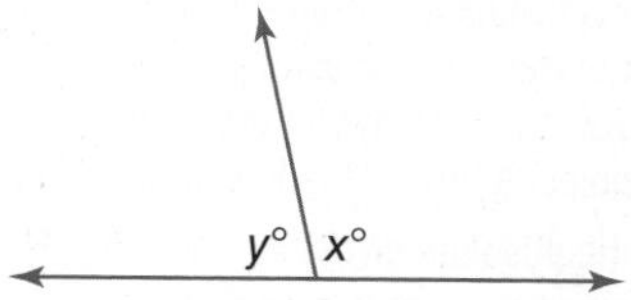

Write an equation in point-slope form for the line that passes through each point with the given slope. (Lesson 4-3)

52. $(2, 5)$, $m = 3$

53. $(-3, 6)$, $m = -7$

54. $(-1, -2)$, $m = -\frac{1}{2}$

Skills Review

Evaluate each expression if $a = -7$, $b = 4$, $c = -3$, and $d = 5$.

55. $b - c$

56. $|a - d|$

57. $|d - c|$

58. $\frac{b - a}{2}$

59. $(a - c)^2$

60. $\sqrt{(a - b)^2 + (c - d)^2}$

EXTEND

10-2 Extension Lesson
Precision and Accuracy

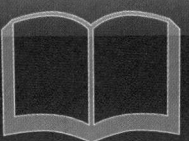

Objective

1 Determine precision of measurements.

2 Determine accuracy of measurements.

As stated in Lesson 10-2, all measurements are approximations. Two main factors are considered when determining the quality of such an approximation.

- How *precise* is the measure?
- How *accurate* is the measure?

1 Precision

Precision refers to the clustering of a group of measurements. It depends only on the smallest unit of measure available on a measuring tool. Suppose you are told that a segment measures 8 centimeters. The length, to the nearest centimeter, of each segment shown below is 8 centimeters.

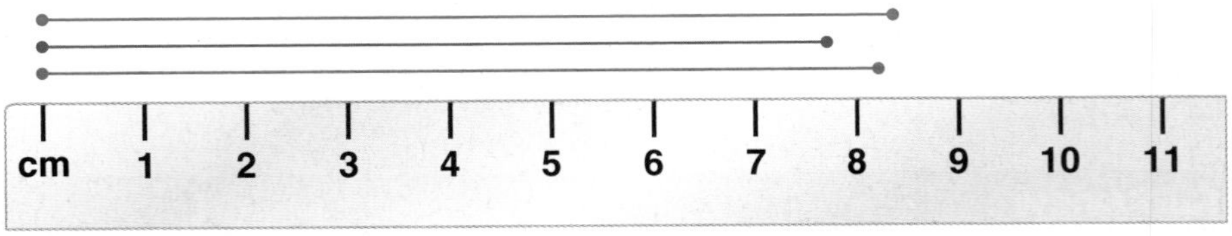

Notice that the exact length of each segment above is between 7.5 and 8.5 centimeters, or within 0.5 centimeter of 8 centimeters. The **absolute error** of a measurement is equal to one half the unit of measure. The smaller the unit of measure, the more precise the measurement.

StudyTip

Precision The absolute error of a measurement in customary units is determined before reducing the fraction. For example, if you measure the length of an object to be $1\frac{4}{16}$ inches, then the absolute error measurement is precise to within $\frac{1}{32}$ inch.

Example 1 Find Absolute Error

Find the absolute error of each measurement. Then explain its meaning.

a. 6.4 centimeters

The measure is given to the nearest 0.1 centimeter, so the absolute error of this measurement is $\frac{1}{2}(0.1)$ or 0.05 centimeter. Therefore, the exact measurement could be between 6.35 and 6.45 centimeters. The two segments below measure 6.4 ± 0.05 centimeters.

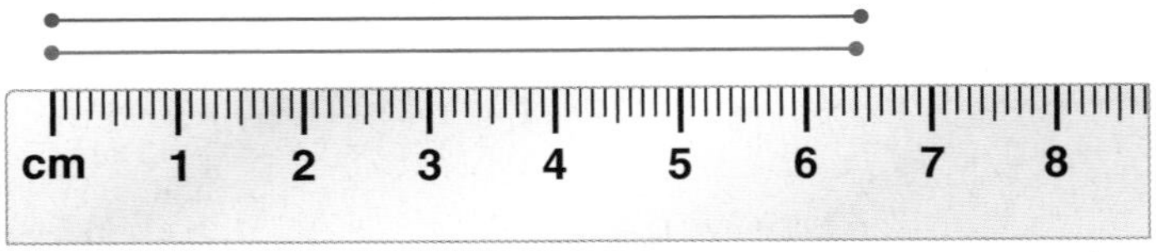

b. $2\frac{1}{4}$ inches

The measure is given to the nearest $\frac{1}{4}$ inch, so the absolute error of this measurement is $\frac{1}{2}\left(\frac{1}{4}\right)$ or $\frac{1}{8}$ inch. Therefore, the exact measurement could be between $2\frac{1}{8}$ and $2\frac{3}{8}$ inches. The two segments below measure $2\frac{1}{4} \pm \frac{1}{8}$ inches.

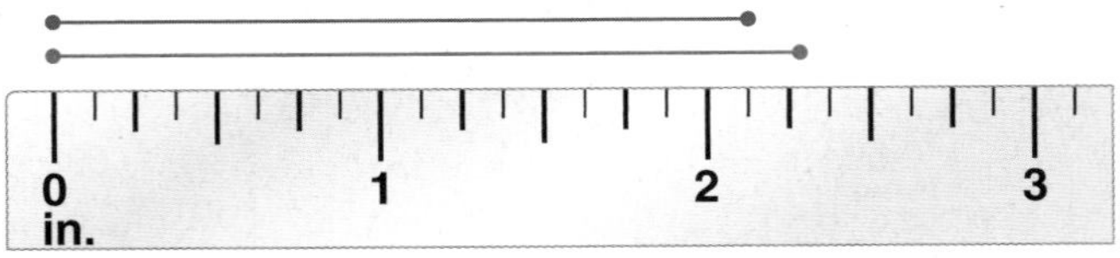

GuidedPractice

1A. $1\frac{1}{2}$ inches

1B. 4 centimeters

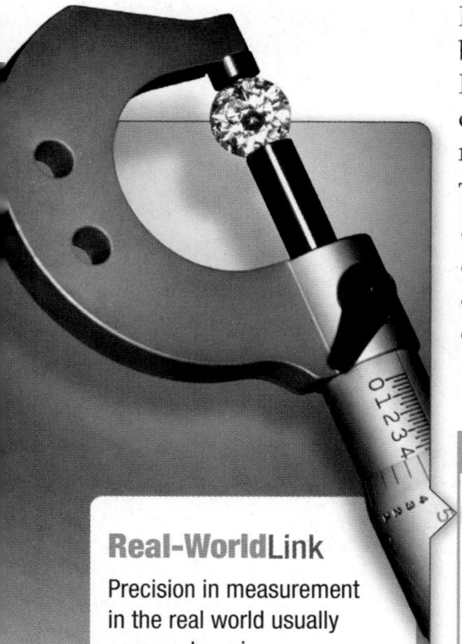

Precision in a measurement is usually expressed by the number of **significant digits** reported. Reporting that the measure of $\overline{AB}$ is 4 centimeters is *less precise* than reporting that the measure of $\overline{AB}$ is 4.1 centimeters.

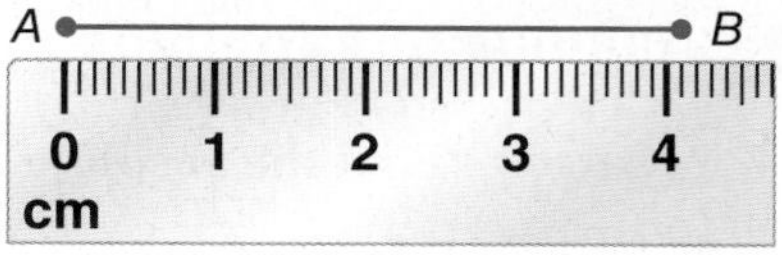

To determine whether digits are considered significant, use the following rules.

- Nonzero digits are always significant.
- In whole numbers, zeros are significant if they fall between nonzero digits.
- In decimal numbers greater than or equal to 1, every digit is significant.
- In decimal numbers less than 1, the first nonzero digit and every digit to the right are significant.

Example 2 Significant Digits

Determine the number of significant digits in each measurement.

a. 430.008 meters

Since this is a decimal number greater than 1, every digit is significant. So, this measurement has 6 significant digits.

b. 0.00750 centimeter

This is a decimal number less than 1. The first nonzero digit is 7, and there are two digits to the right of 7, 5 and 0. So, this measurement has 3 significant digits.

Guided Practice

2A. 779,000 mi **2B.** 50,008 ft **2C.** 230.004500 m

Real-World Link

Precision in measurement in the real world usually comes at a price.

- Precision in a process to 3 significant digits, commercial quality, can cost $100.
- Precision in a process to 4 significant digits, industrial quality, can cost $500.
- Precision in a process to 5 significant digits, scientific quality, can cost $2500.

Source: Southwest Texas Junior College

2 Accuracy

Accuracy refers to how close a measured value comes to the actual or desired value. Consider the target practice results shown below.

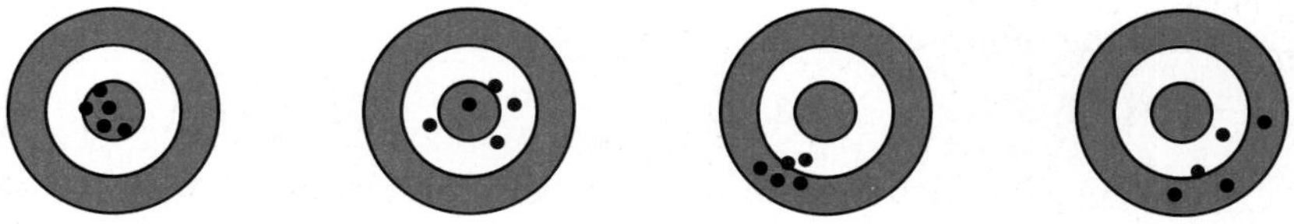

accurate and precise | accurate but not precise | precise but not accurate | not accurate and not precise

The relative error of a measure is the ratio of the absolute error to the expected measure. A measurement with a smaller relative error is said to be more accurate.

Example 3 Find Relative Error

MANUFACTURING **A manufacturer measures each part for a piece of equipment to be 23 centimeters in length. Find the relative error of this measurement.**

$$\text{relative error} = \frac{\text{absolute error}}{\text{expected measure}} = \frac{0.5 \text{ cm}}{23 \text{ cm}} \approx 0.022 \text{ or } 2.2\%$$

Guided Practice

Find the relative error of each measurement.

3A. 3.2 mi **3B.** 1 ft **3C.** 26 ft

Study Tip

Accuracy The accuracy or relative error of a measurement depends on both the absolute error and the size of the object being measured.

Practice and Problem Solving

Find the absolute error of each measurement. Then explain its meaning.

1. 12 yd **2.** $50\frac{4}{16}$ in. **3.** 3.28 ft **4.** 2.759 cm

5. **ERROR ANALYSIS** In biology class, Manuel and Jocelyn measure a beetle as shown. Manuel says that the beetle measures between $1\frac{5}{8}$ and $1\frac{3}{4}$ inches. Jocelyn says that it measures between $1\frac{9}{16}$ and $1\frac{5}{8}$ inches. Is either of their statements about the beetle's measure correct? Explain your reasoning.

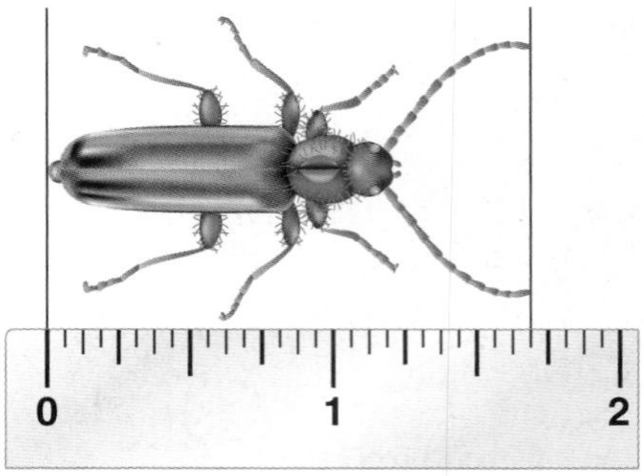

6. **PYRAMIDS** Research suggests that the design dimensions of the Great Pyramid of Giza in Egypt were 440 by 440 royal cubits. The sides of the pyramid are precise within 0.05%. What are the greatest and least possible lengths of the sides?

Determine the number of significant digits in each measurement.

7. 4.05 in. **8.** 53,000 mi **9.** 0.0005 mm **10.** 750,001 ft

11. **VOLUME** When multiplying or dividing measures, the product or quotient should have only as many significant digits as the multiplied or divided measurement showing the least number of significant digits. To how many significant digits should the volume of the rectangle prism shown be reported? Report the volume to this number of significant digits.

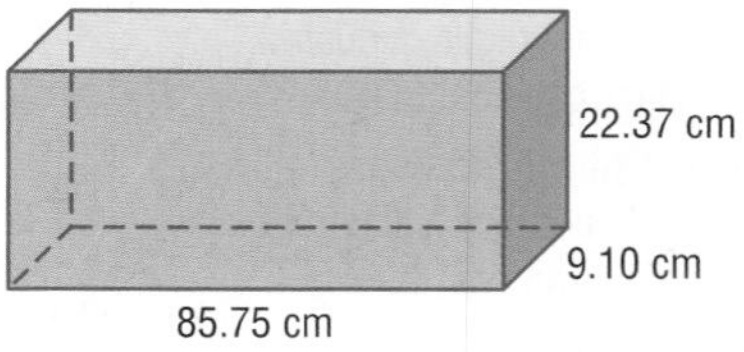

Find the relative error of each measurement.

12. 48 in. **13.** 2.0 mi **14.** 11.14 cm **15.** 0.6 m

Determine which measurement is more precise and which is more accurate. Explain your reasoning.

16. 22.4 ft; 5.82 ft **17.** 25 mi; 8 mi **18.** 9.2 cm; 42 mm **19.** $18\frac{1}{4}$ in.; 125 yd

For each situation, determine the level of accuracy needed. Explain.

20. You are estimating the height of a person. Which unit of measure should you use: 1 foot, 1 inch, or $\frac{1}{16}$ inch?

21. You are estimating the height of a mountain. Which unit of measure should you use: 1 foot, 1 inch, or $\frac{1}{16}$ inch?

22. **PERIMETER** The *perimeter* of a geometric figure is the sum of the lengths of its sides. Jermaine uses a ruler divided into inches and measures the sides of a rectangle to be $2\frac{1}{4}$ inches and $4\frac{3}{4}$ inches. What are the least and greatest possible perimeters of the rectangle? Explain.

23. **WRITING IN MATH** How precise is precise enough?

LESSON 10-3

Distance and Midpoints

Then

- You graphed points on the coordinate plane.

Now

1. Find the distance between two points.
2. Find the midpoint of a segment.

Why?

- The location of a city on a map is given in degrees of latitude and longitude. For short distances, the Pythagorean Theorem can be used to approximate the distance between two locations.

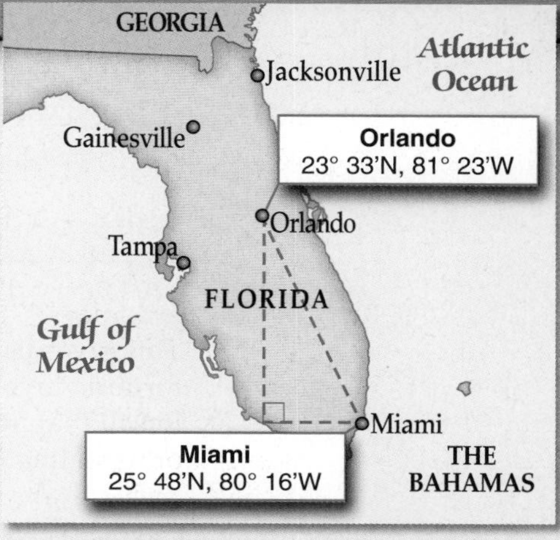

NewVocabulary

distance
irrational number
midpoint
segment bisector

Common Core State Standards

Content Standards

G.CO.1 Know precise definitions of angle, circle, perpendicular line, parallel line, and line segment, based on the undefined notions of point, line, distance along a line, and distance around a circular arc.

G.CO.12 Make formal geometric constructions with a variety of tools and methods (compass and straightedge, string, reflective devices, paper folding, dynamic geometric software, etc.).

Mathematical Practices

2 Reason abstractly and quantitatively.

7 Look for and make use of structure.

1 Distance Between Two Points

The **distance** between two points is the length of the segment with those points as its endpoints. The coordinates of the points can be used to find this length. Because $\overline{PQ}$ is the same as $\overline{QP}$, the order in which you name the endpoints is not important when calculating distance.

KeyConcept Distance Formula (on Number Line)

Words The distance between two points is the absolute value of the difference between their coordinates.

Symbols If P has coordinate x_1 and Q has coordinate x_2, $PQ = |x_2 - x_1|$ or $|x_1 - x_2|$.

Example 1 Find Distance on a Number Line

Use the number line to find *BE*.

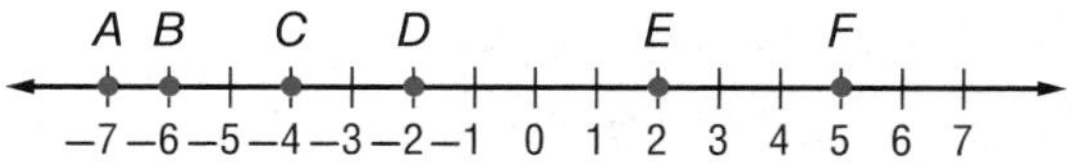

The coordinates of B and E are -6 and 2.

$BE = |x_2 - x_1|$ Distance Formula

$= |2 - (-6)|$ $x_1 = -6$ and $x_2 = 2$

$= 8$ Simplify.

GuidedPractice

Use the number line above to find each measure.

1A. AC **1B.** CF **1C.** FB

To find the distance between two points A and B in the coordinate plane, you can form a right triangle with $\overline{AB}$ as its hypotenuse and point C as its vertex as shown. Then use the Pythagorean Theorem to find AB.

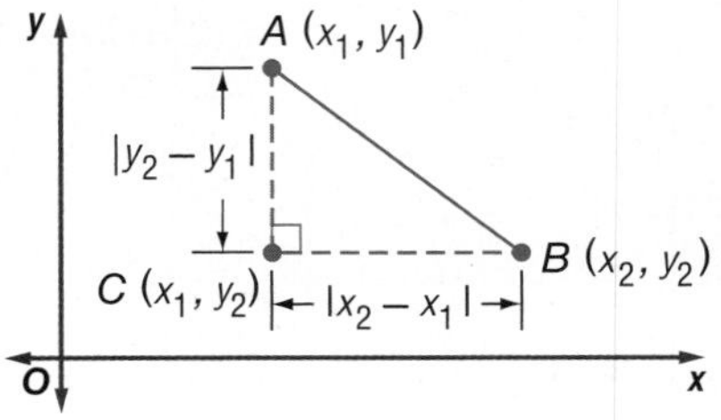

$$(CB)^2 + (AC)^2 = (AB)^2$$ Pythagorean Theorem

$$(|x_2 - x_1|)^2 + (|y_2 - y_1|)^2 = (AB)^2$$ $CB = |x_2 - x_1|$, $AC = |y_2 - y_1|$

$$(x_2 - x_1)^2 + (y_2 - y_1)^2 = (AB)^2$$ The square of a number is always positive.

$$\sqrt{(x_2 - x_1)^2 + (y_2 - y_1)^2} = AB$$ Take the positive square root of each side.

This gives us a Distance Formula for points in the coordinate plane. Because this formula involves taking the square root of a real number, distances can be irrational. Recall that an **irrational number** is a number that cannot be expressed as a terminating or repeating decimal.

KeyConcept Distance Formula (in Coordinate Plane)

If P has coordinates (x_1, y_1) and Q has coordinates (x_2, y_2), then

$$PQ = \sqrt{(x_2 - x_1)^2 + (y_2 - y_1)^2}.$$

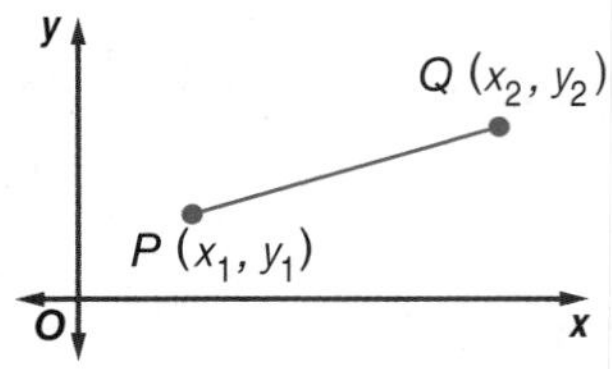

The order of the x- and y-coordinates in each set of parentheses is not important.

Example 2 Find Distance on a Coordinate Plane

Find the distance between $C(-4, -6)$ and $D(5, -1)$.

$CD = \sqrt{(x_2 - x_1)^2 + (y_2 - y_1)^2}$ Distance Formula

$= \sqrt{[5 - (-4)]^2 + [-1 - (-6)]^2}$ $(x_1, y_1) = (-4, -6)$ and $(x_2, y_2) = (5, -1)$

$= \sqrt{9^2 + 5^2}$ or $\sqrt{106}$ Subtract.

The distance between C and D is $\sqrt{106}$ units. Use a calculator to find that $\sqrt{106}$ units is approximately 10.3 units.

CHECK Graph the ordered pairs and check by using the Pythagorean Theorem.

$(CD)^2 \stackrel{?}{=} (EC)^2 + (ED)^2$

$(CD)^2 \stackrel{?}{=} 5^2 + 9^2$

$(CD)^2 \stackrel{?}{=} 106$

$CD = \sqrt{106}$ ✓

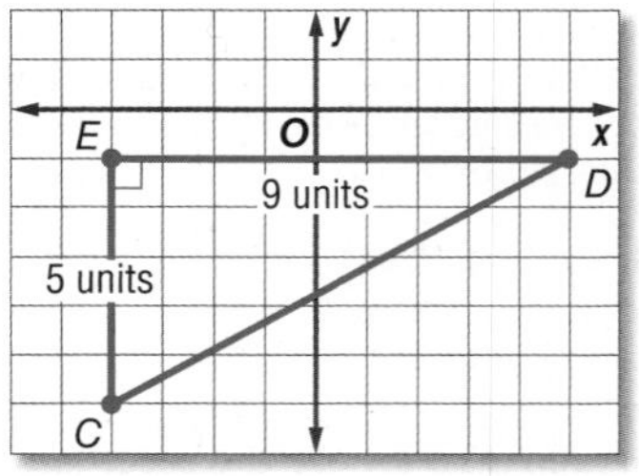

GuidedPractice

Find the distance between each pair of points.

2A. $E(-5, 6)$ and $F(8, -4)$

2B. $J(4, 3)$ and $K(-3, -7)$

2 Midpoint of a Segment

The **midpoint** of a segment is the point halfway between the endpoints of the segment. If X is the midpoint of $\overline{AB}$, then $AX = XB$ and $\overline{AX} \cong \overline{XB}$. You can find the midpoint of a segment on a number line by finding the *mean*, or the average, of the coordinates of its endpoints.

KeyConcept Midpoint Formula (on Number Line)

If $\overline{AB}$ has endpoints at x_1 and x_2 on a number line, then the midpoint M of $\overline{AB}$ has coordinate

$$\frac{x_1 + x_2}{2}.$$

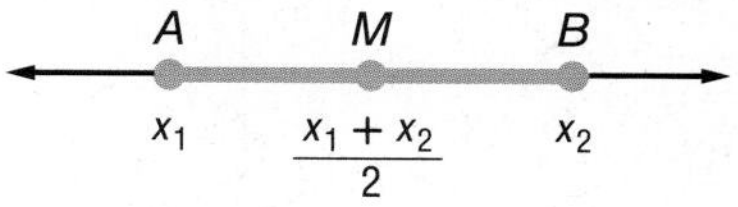

StudyTip

Alternative Method In Example 3, the coordinate of the midpoint could also have been located by first finding the length of AB, which is $37.5 - 15$ or 22.5 inches. Half of this measure is the distance from one endpoint to the point midway between A and B, $\frac{22.5}{2}$ or 11.25. Add this distance to point A's distance from the left wall. So the midpoint between A and B is $15 + 11.25$ or 26.25 inches from the left wall.

Real-World Example 3 Find Midpoint on a Number Line

PT

DECORATING Jacinta hangs a picture 15 inches from the left side of a wall. How far from the edge of the wall should she mark the location for the nail the picture will hang on if the right edge is 37.5 inches from the wall's left side?

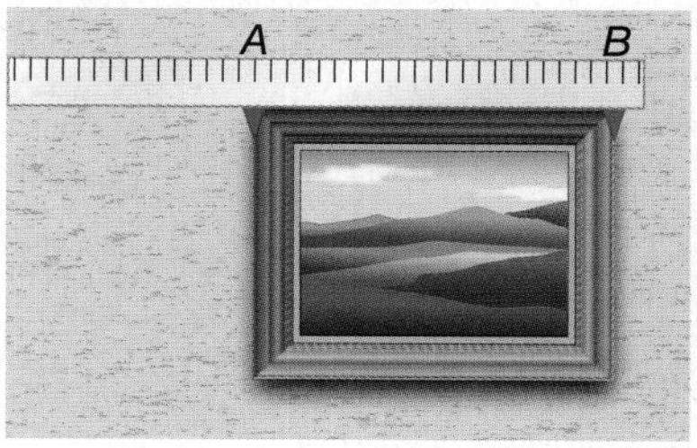

The coordinates of the endpoints of the top of the picture frame are 15 inches and 37.5 inches. Let M be the midpoint of $\overline{AB}$.

$M = \frac{x_1 + x_2}{2}$ — Midpoint Formula

$= \frac{15 + 37.5}{2}$ — $x_1 = 15$, $x_2 = 37.5$

$= \frac{52.5}{2}$ or 26.25 — Simplify.

The midpoint is located at 26.25 or $26\frac{1}{4}$ inches from the left edge of the wall.

GuidedPractice

3. TEMPERATURE The temperature on a thermometer dropped from a reading of 25° to −8°. Find the midpoint of these temperatures.

You can find the midpoint of a segment on the coordinate plane by finding the average of the x-coordinates and of the y-coordinates of the endpoints.

KeyConcept Midpoint Formula (in Coordinate Plane)

If $\overline{PQ}$ has endpoints at $P(x_1, y_1)$ and $Q(x_2, y_2)$ in the coordinate plane, then the midpoint M of $\overline{PQ}$ has coordinates

$$M\left(\frac{x_1 + x_2}{2}, \frac{y_1 + y_2}{2}\right).$$

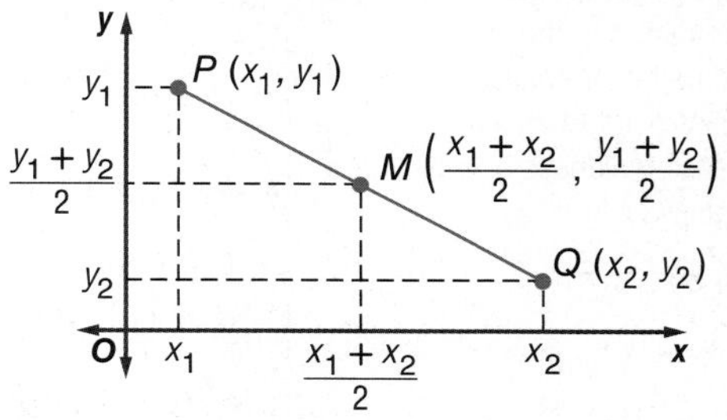

When finding the midpoint of a segment, the order of the coordinates of the endpoints is not important.

Example 4 Find Midpoint in Coordinate Plane

Find the coordinates of M, the midpoint of $\overline{ST}$, for $S(-6, 3)$ and $T(1, 0)$.

$$M = \left(\frac{x_1 + x_2}{2}, \frac{y_1 + y_2}{2}\right) \quad \text{Midpoint Formula}$$

$$= \left(\frac{-6 + 1}{2}, \frac{3 + 0}{2}\right) \quad (x_1, y_1) = S(-6, 3), (x_2, y_2) = T(1, 0)$$

$$= \left(\frac{-5}{2}, \frac{3}{2}\right) \text{ or } M\left(-2\tfrac{1}{2}, 1\tfrac{1}{2}\right) \quad \text{Simplify.}$$

CHECK Graph S, T, and M. The distance from S to M does appear to be the same as the distance from M to T, so our answer is reasonable.

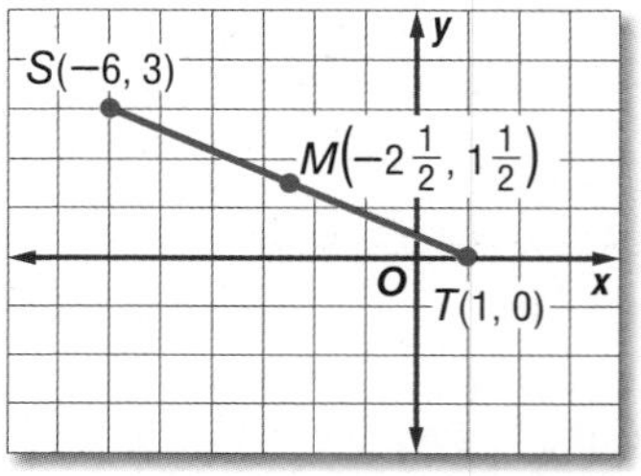

GuidedPractice

Find the coordinates of the midpoint of a segment with the given coordinates.

4A. $A(5, 12)$, $B(-4, 8)$

4B. $C(-8, -2)$, $D(5, 1)$

You can also find the coordinates of the endpoint of a segment if you know the coordinates of its other endpoint and its midpoint.

Example 5 Find the Coordinates of an Endpoint

Find the coordinates of J if $K(-1, 2)$ is the midpoint of $\overline{JL}$ and L has coordinates $(3, -5)$.

Step 1 Let J be (x_1, y_1) and L be (x_2, y_2) in the Midpoint Formula.

$$K\left(\frac{x_1 + 3}{2}, \frac{y_1 + (-5)}{2}\right) = K(-1, 2) \quad (x_2, y_2) = (3, -5)$$

Step 2 Write two equations to find the coordinates of J.

$\frac{x_1 + 3}{2} = -1$	Midpoint Formula	$\frac{y_1 + (-5)}{2} = 2$	Midpoint Formula
$x_1 + 3 = -2$	Multiply each side by 2.	$y_1 - 5 = 4$	Multiply each side by 2.
$x_1 = -5$	Subtract 3 from each side.	$y_1 = 9$	Add 5 to each side.

The coordinates of J are $(-5, 9)$.

StudyTip

Check for Reasonableness Always graph the given information and the calculated coordinates of the third point to check the reasonableness of your answer.

CHECK Graph J, K, and L. The distance from J to K does appear to be the same as the distance from K to L, so our answer is reasonable.

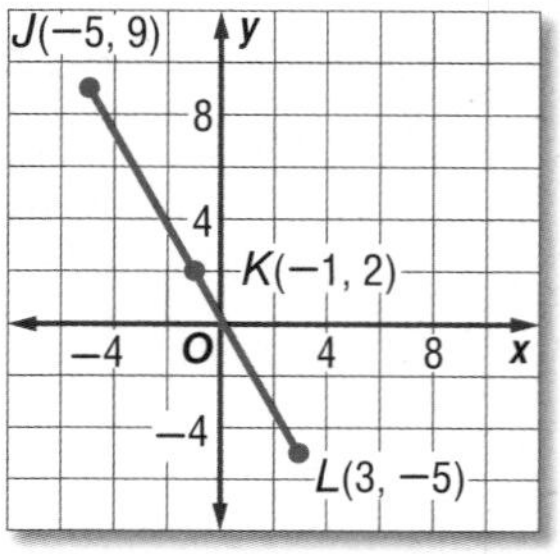

GuidedPractice

Find the coordinates of the missing endpoint if P is the midpoint of $\overline{EG}$.

5A. $E(-8, 6)$, $P(-5, 10)$

5B. $P(-1, 3)$, $G(5, 6)$

You can use algebra to find a missing measure or value in a figure that involves the midpoint of a segment.

StudyTip

CCSS Sense-Making and Perseverance The four-step problem solving plan is a tool for making sense of any problem. When making and executing your plan, continually ask yourself, "Does this make sense?" Monitor and evaluate your progress and change course if necessary.

Example 6 Use Algebra to Find Measures

ALGEBRA Find the measure of $\overline{PQ}$ if Q is the midpoint of $\overline{PR}$.

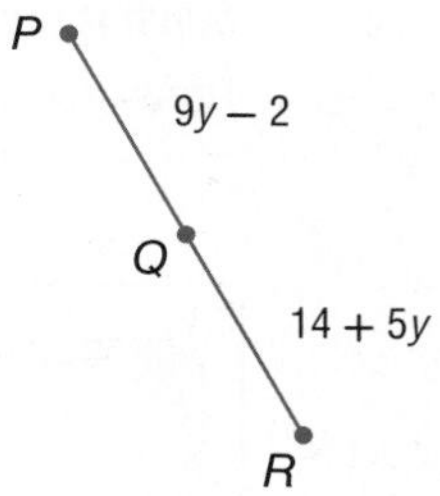

Understand You know that Q is the midpoint of $\overline{PR}$.
You are asked to find the measure of $\overline{PQ}$.

Plan Because Q is the midpoint, you know that $PQ = QR$. Use this equation to find a value for y.

Solve

$PQ = QR$	Definition of midpoint
$9y - 2 = 14 + 5y$	$PQ = 9y - 2$, $QR = 14 + 5y$
$4y - 2 = 14$	Subtract $5y$ from each side.
$4y = 16$	Add 2 to each side.
$y = 4$	Divide each side by 4.

Now substitute 4 for y in the expression for PQ.

$PQ = 9y - 2$	Original measure
$= 9(4) - 2$	$y = 4$
$= 36 - 2$ or 34	Simplify.

The measure of $\overline{PQ}$ is 34.

Check Since $PQ = QR$, when the expression for QR is evaluated for 4, it should also be 34.

$QR = 14 + 5y$	Original measure
$\stackrel{?}{=} 14 + 5(4)$	$y = 4$
$= 34$ ✓	Simplify.

GuidedPractice

6A. Find the measure of $\overline{YZ}$ if Y is the midpoint of $\overline{XZ}$ and $XY = 2x - 3$ and $YZ = 27 - 4x$.

6B. Find the value of x if C is the midpoint of $\overline{AB}$, $AC = 4x + 5$, and $AB = 78$.

StudyTip

Segment Bisectors There can be an infinite number of bisectors and each must contain the midpoint of the segment.

Any segment, line, or plane that intersects a segment at its midpoint is called a **segment bisector**. In the figure at the right, M is the midpoint of $\overline{PQ}$. Plane $\mathcal{A}$, $\overline{MJ}$, $\overleftrightarrow{KM}$, and point M are all bisectors of $\overline{PQ}$. We say that they *bisect* $\overline{PQ}$.

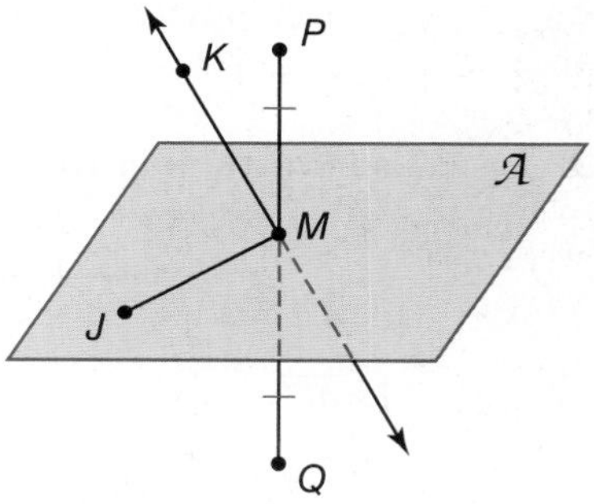

The construction on the following page shows how to construct a line that bisects a segment to find the midpoint of a given segment.

Construction Bisect a Segment

Step 1 Draw a segment and name it $\overline{AB}$. Place the compass at point A. Adjust the compass so that its width is greater than $\frac{1}{2}AB$. Draw arcs above and below $\overline{AB}$.

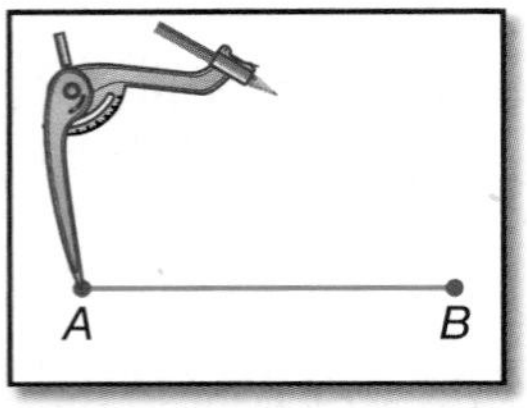

Step 2 Using the same compass setting, place the compass at point B and draw arcs above and below $\overline{AB}$ so that they intersect the two arcs previously drawn. Label the points of the intersection of the arcs as C and D.

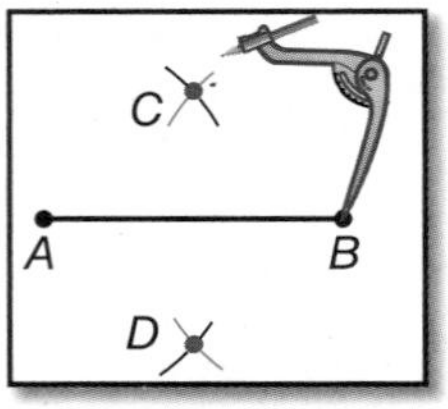

Step 3 Use a straightedge to draw $\overline{CD}$. Label the point where it intersects $\overline{AB}$ as M. Point M is the midpoint of $\overline{AB}$, and $\overline{CD}$ is a bisector of $\overline{AB}$.

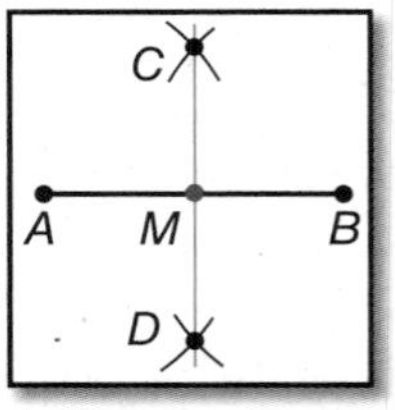

Check Your Understanding

Example 1 **Use the number line to find each measure.**

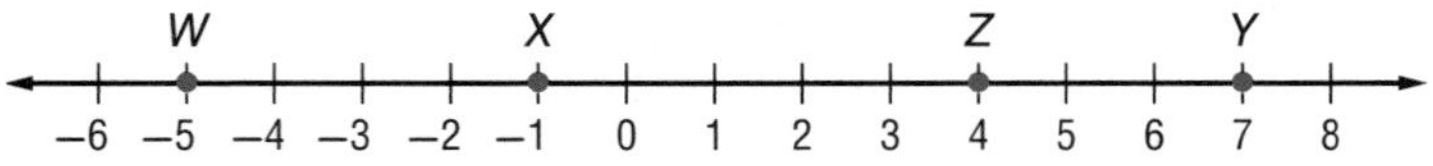

1. XY **2.** WZ

Example 2 **TIME CAPSULE** Graduating classes have buried time capsules on the campus of East Side High School for over twenty years. The points on the diagram show the position of three time capsules. Find the distance between each pair of time capsules.

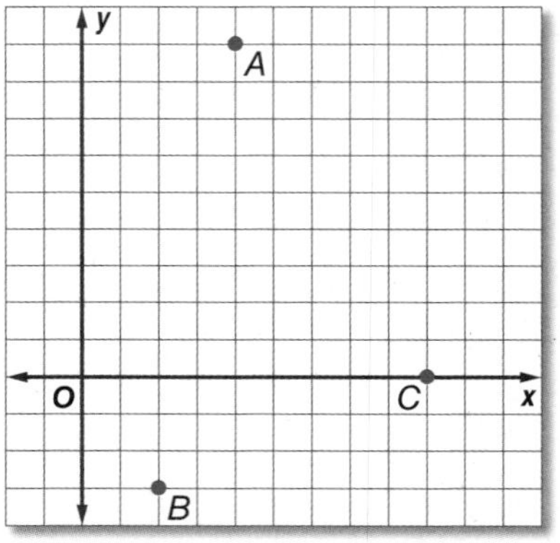

3. $A(4, 9)$, $B(2, -3)$

4. $A(4, 9)$, $C(9, 0)$

5. $B(2, -3)$, $C(9, 0)$

6. CCSS **REASONING** Which two time capsules are the closest to each other? Which are farthest apart?

Example 3 **Use the number line to find the coordinate of the midpoint of each segment.**

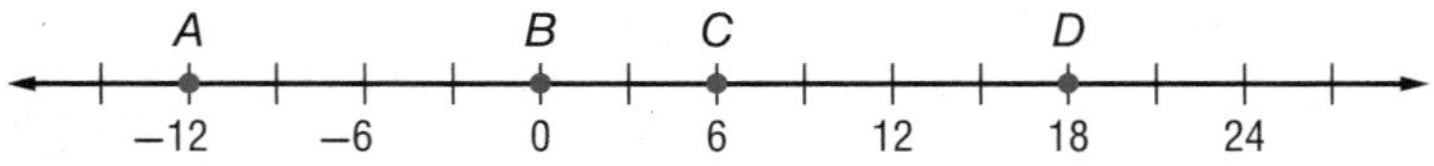

7. $\overline{AC}$ **8.** $\overline{BD}$

Example 4 **Find the coordinates of the midpoint of a segment with the given endpoints.**

9. $J(5, -3)$, $K(3, -8)$ **10.** $M(7, 1)$, $N(4, -1)$

Example 5 **11** Find the coordinates of G if $F(1, 3.5)$ is the midpoint of $\overline{GJ}$ and J has coordinates $(6, -2)$.

Example 6 **12.** **ALGEBRA** Point M is the midpoint of $\overline{CD}$. What is the value of a in the figure?

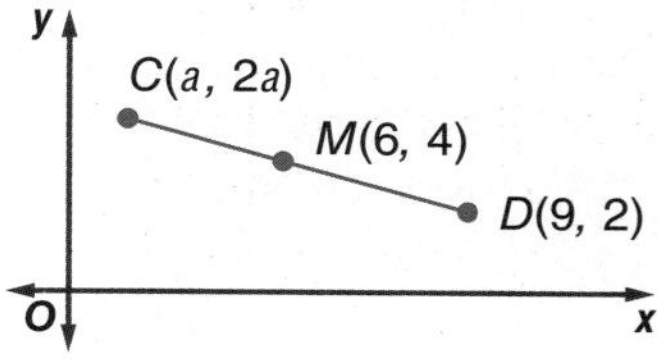

Practice and Problem Solving

Example 1 **Use the number line to find each measure.**

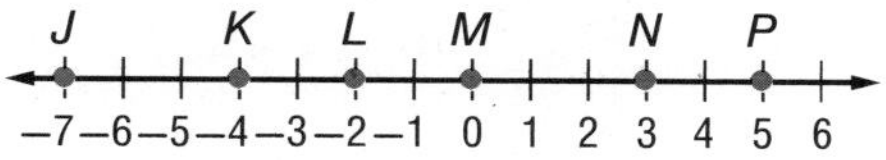

13. JL **14.** JK **15.** KP

16. NP **17.** JP **18.** LN

Example 2 **Find the distance between each pair of points.**

19.

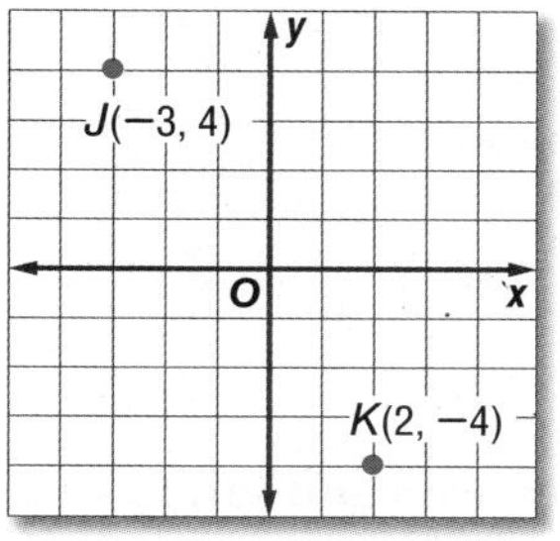

20.

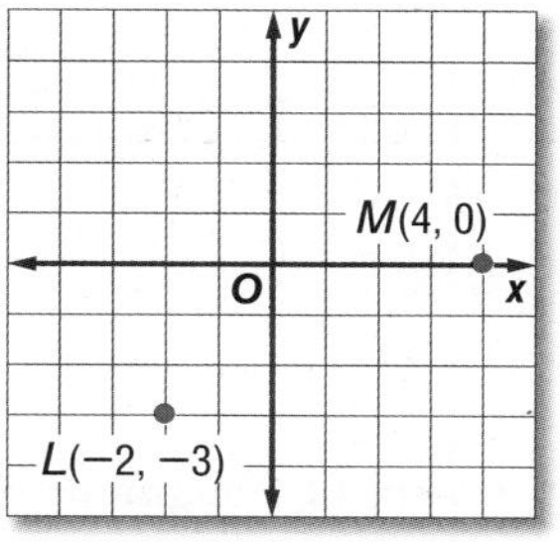

21.

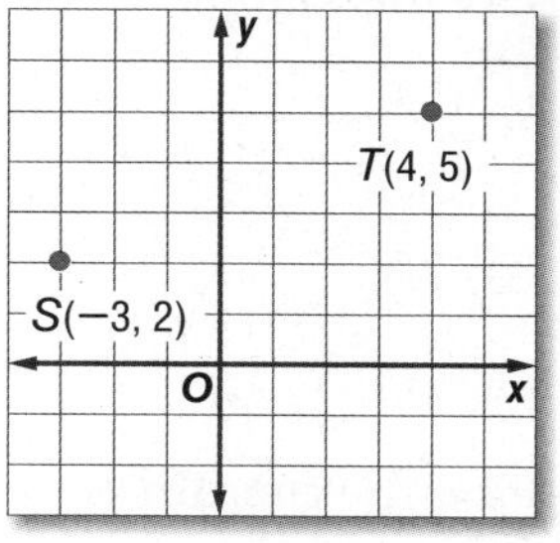

22.

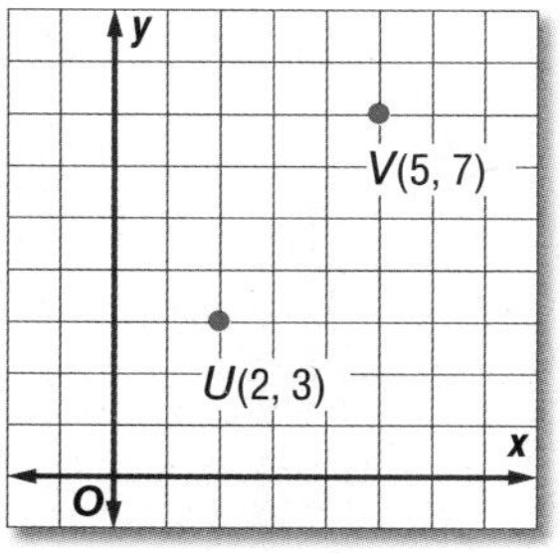

23.

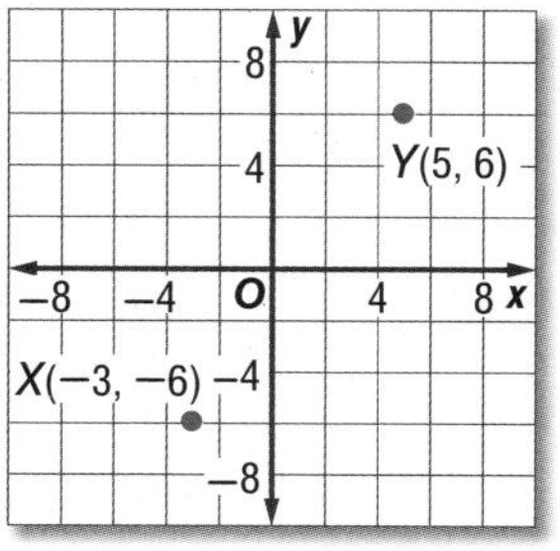

24.

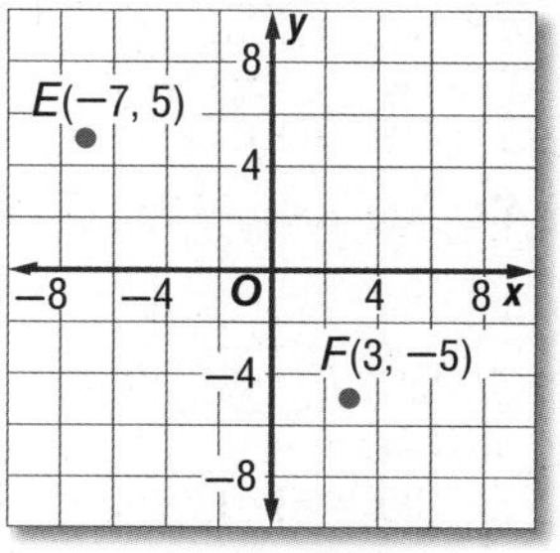

25. $X(1, 2)$, $Y(5, 9)$ **26.** $P(3, 4)$, $Q(7, 2)$ **27.** $M(-3, 8)$, $N(-5, 1)$

28. $Y(-4, 9)$, $Z(-5, 3)$ **29.** $A(2, 4)$, $B(5, 7)$ **30.** $C(5, 1)$, $D(3, 6)$

31. **CCSS REASONING** Vivian is planning to hike to the top of Humphreys Peak on her family vacation. The coordinates of the peak of the mountain and of the base of the trail are shown. If the trail can be approximated by a straight line, estimate the length of the trail. (*Hint:* 1 mi = 5280 ft)

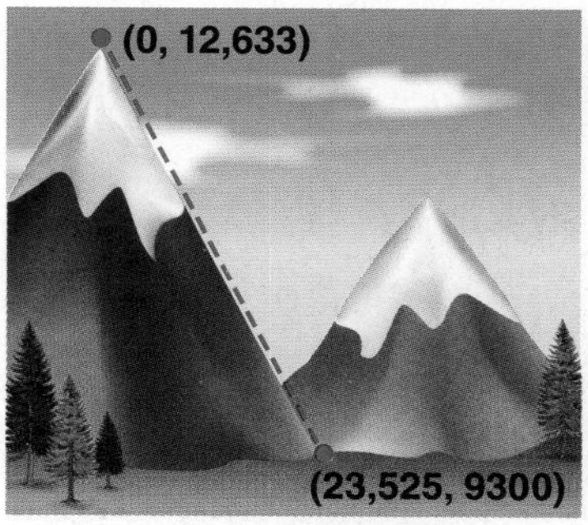

32. **CCSS MODELING** Penny and Akiko live in the locations shown on the map below.

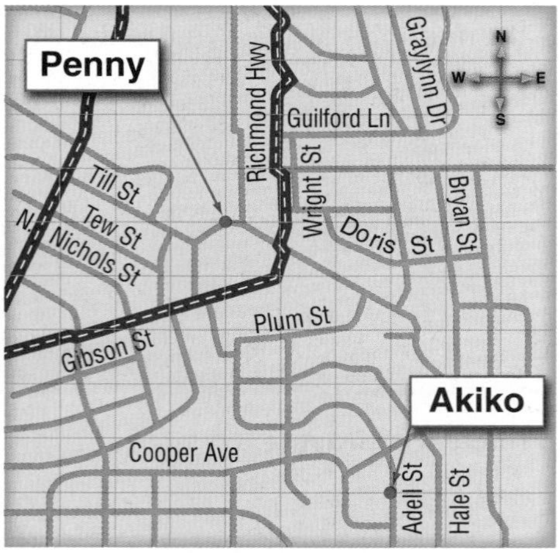

a. If each square on the grid represents one block and the bottom left corner of the grid is the location of the origin, what is the straight-line distance from Penny's house to Akiko's?

b. If Penny moves three blocks to the north and Akiko moves 5 blocks to the west, how far apart will they be?

Example 3 **Use the number line to find the coordinate of the midpoint of each segment.**

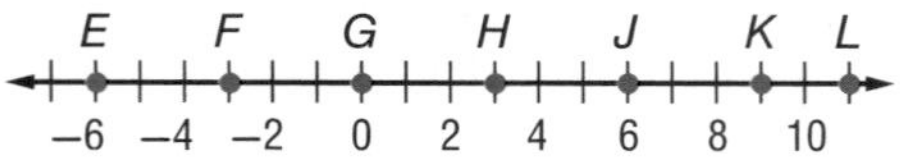

33. $\overline{HK}$

34. $\overline{JL}$

35. $\overline{EF}$

36. $\overline{FG}$

37. $\overline{FK}$

38. $\overline{EL}$

Example 4 **Find the coordinates of the midpoint of a segment with the given endpoints.**

39. $C(22, 4)$, $B(15, 7)$

40. $W(12, 2)$, $X(7, 9)$

41. $D(-15, 4)$, $E(2, -10)$

42. $V(-2, 5)$, $Z(3, -17)$

43. $X(-2.4, -14)$, $Y(-6, -6.8)$

44. $J(-11.2, -3.4)$, $K(-5.6, -7.8)$

45.

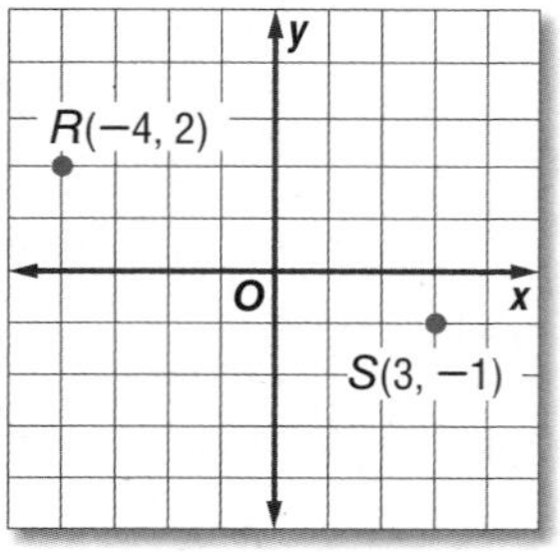

46.

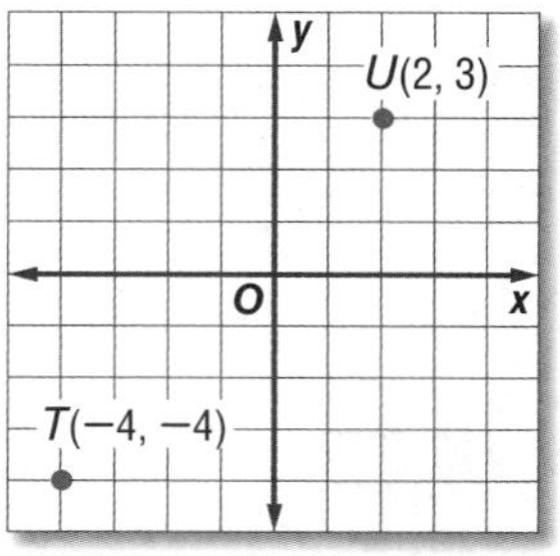

Example 5 **Find the coordinates of the missing endpoint if B is the midpoint of $\overline{AC}$.**

47. $C(-5, 4)$, $B(-2, 5)$

48. $A(1, 7)$, $B(-3, 1)$

49. $A(-4, 2)$, $B(6, -1)$

50. $C(-6, -2)$, $B(-3, -5)$

51. $A(4, -0.25)$, $B(-4, 6.5)$

52. $C\left(\frac{5}{3}, -6\right)$, $B\left(\frac{8}{3}, 4\right)$

Example 6 **ALGEBRA Suppose M is the midpoint of $\overline{FG}$. Use the given information to find the missing measure or value.**

53. $FM = 3x - 4$, $MG = 5x - 26$, $FG = ?$

54. $FM = 5y + 13$, $MG = 5 - 3y$, $FG = ?$

55. $MG = 7x - 15$, $FG = 33$, $x = ?$

56. $FM = 8a + 1$, $FG = 42$, $a = ?$

57 **BASKETBALL** The dimensions of a basketball court are shown below. Suppose a player throws the ball from a corner to a teammate standing at the center of the court.

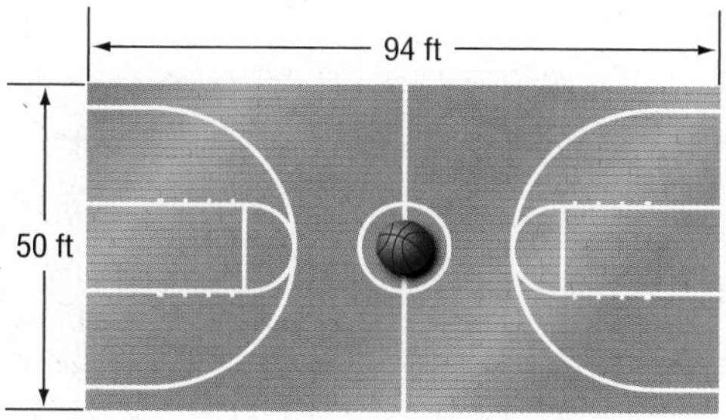

a. If center court is located at the origin, find the ordered pair that represents the location of the player in the bottom right corner.

b. Find the distance that the ball travels.

CCSS TOOLS **Spreadsheets can be used to perform calculations quickly. The spreadsheet below can be used to calculate the distance between two points. Values are used in formulas by using a specific cell name. The value of x_1 is used in a formula using its cell name, A2.**

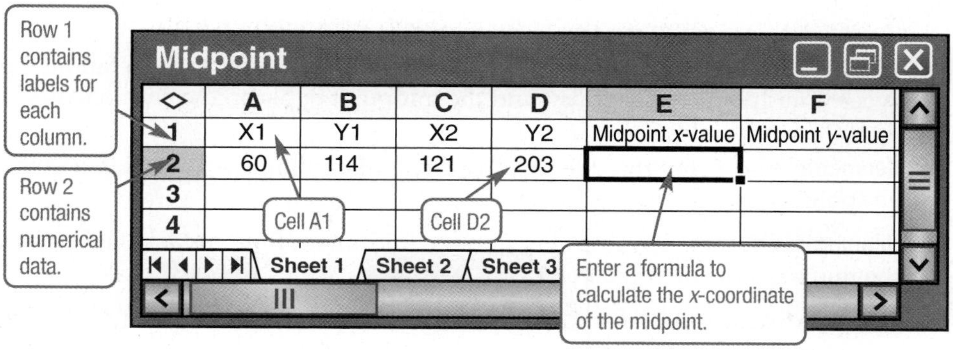

Write a formula for the indicated cell that could be used to calculate the indicated value using the coordinates (x_1, y_1) and (x_2, y_2) as the endpoint of a segment.

58. E2; the x-value of the midpoint of the segment

59. F2; the y-value of the midpoint of the segment

60. G2; the length of the segment

Name the point(s) that satisfy the given condition.

61. two points on the x-axis that are 10 units from (1, 8)

62. two points on the y-axis that are 25 units from (−24, 3)

63. **COORDINATE GEOMETRY** Find the coordinates of B if B is the midpoint of $\overline{AC}$ and C is the midpoint of $\overline{AD}$.

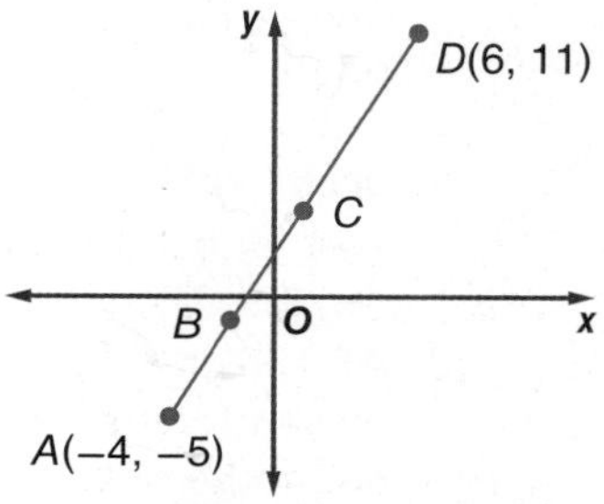

ALGEBRA **Determine the value(s) of n.**

64. $J(n, n + 2), K(3n, n - 1), JK = 5$

65. $P(3n, n - 7), Q(4n, n + 5), PQ = 13$

66. CCSS **PERSEVERANCE** Wilmington, North Carolina, is located at (34.3°, 77.9°), which represents north latitude and west longitude. Winston-Salem is in the northern part of the state at (36.1°, 80.2°).

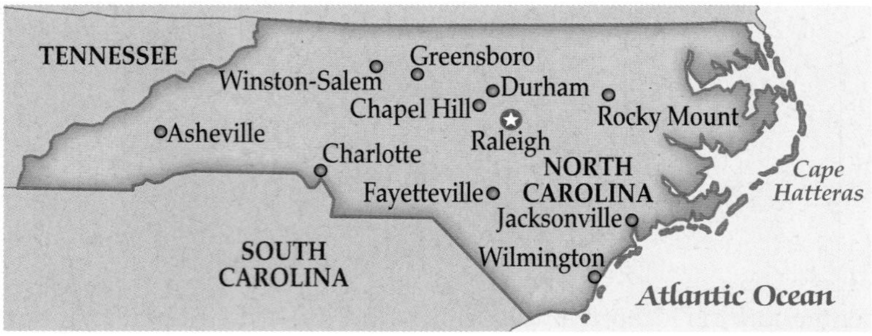

a. Find the latitude and longitude of the midpoint of the segment between Wilmington and Winston-Salem.

b. Use an atlas or the Internet to find a city near the location of the midpoint.

c. If Winston-Salem is the midpoint of the segment with one endpoint at Wilmington, find the latitude and longitude of the other endpoint.

d. Use an atlas or the Internet to find a city near the location of the other endpoint.

67 **MULTIPLE REPRESENTATIONS** In this problem, you will explore the relationship between a midpoint of a segment and the midpoint between the endpoint and the midpoint.

a. Geometric Use a straightedge to draw three different line segments. Label the endpoints A and B.

b. Geometric On each line segment, find the midpoint of $\overline{AB}$ and label it C. Then find the midpoint of $\overline{AC}$ and label it D.

c. Tabular Measure and record AB, AC, and AD for each line segment. Organize your results into a table.

d. Algebraic If $AB = x$, write an expression for the measures AC and AD.

e. Verbal Make a conjecture about the relationship between AB and each segment if you were to continue to find the midpoints of a segment and a midpoint you previously found.

H.O.T. Problems Use Higher-Order Thinking Skills

68. WRITING IN MATH Explain how the Pythagorean Theorem and the Distance Formula are related.

69. REASONING Is the point one third of the way from (x_1, y_1) to (x_2, y_2) *sometimes, always,* or *never* the point $\left(\frac{x_1 + x_2}{3}, \frac{y_1 + y_2}{3}\right)$? Explain.

70. CHALLENGE Point P is located on the segment between point $A(1, 4)$ and point $D(7, 13)$. The distance from A to P is twice the distance from P to D. What are the coordinates of point P?

71. OPEN ENDED Draw a segment and name it $\overline{AB}$. Using only a compass and a straightedge, construct a segment $\overline{CD}$ such that $CD = 5\frac{1}{4}AB$. Explain and then justify your construction.

72. WRITING IN MATH Describe a method of finding the midpoint of a segment that has one endpoint at (0, 0). Give an example using your method, and explain why your method works.

Standardized Test Practice

73. Which of the following best describes the first step in bisecting $\overline{AB}$?

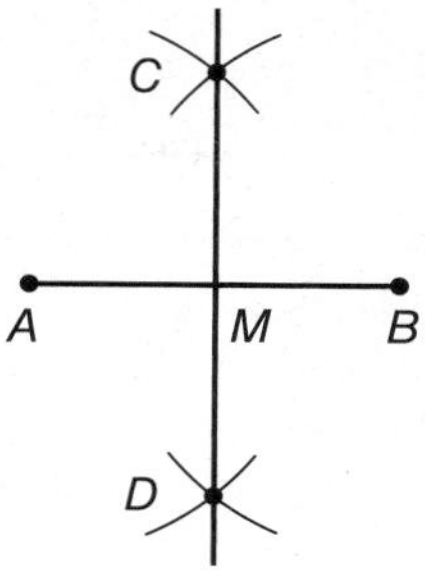

A From point A, draw equal arcs on $\overline{CD}$ using the same compass width.

B From point A, draw equal arcs above and below $\overline{AB}$ using a compass width of $\frac{1}{3}\overline{AB}$.

C From point A, draw equal arcs above and below $\overline{AB}$ using a compass width greater than $\frac{1}{2}\overline{AB}$.

D From point A, draw equal arcs above and below $\overline{AB}$ using a compass width less than $\frac{1}{2}\overline{AB}$.

74. ALGEBRA Beth paid \$74.88 for 3 pairs of jeans. All 3 pairs of jeans were the same price. How much did each pair of jeans cost?

F \$24.96 **H** \$74.88

G \$37.44 **J** \$224.64

75. SAT/ACT If $5^{2x-3} = 1$, then $x =$

A 0.4 **D** 1.6

B 0.6 **E** 2

C 1.5

76. GRIDDED RESPONSE One endpoint of $\overline{AB}$ has coordinates $(-3, 5)$. If the coordinates of the midpoint of $\overline{AB}$ are $(2, -6)$, what is the approximate length of $\overline{AB}$?

Spiral Review

Find the length of each object. (Lesson 10-2)

77.

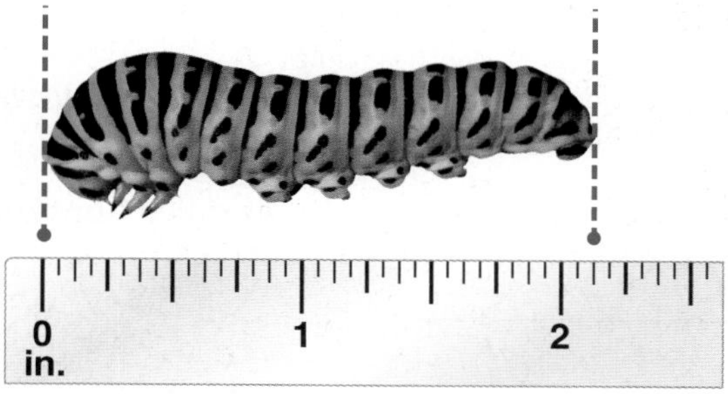

78.

Draw and label a figure for each relationship. (Lesson 10-1)

79. $\overleftrightarrow{FG}$ lies in plane M and contains point H.

80. Lines r and s intersect at point W.

Skills Review

Solve each equation.

81. $8x - 15 = 5x$

82. $5y - 3 + y = 90$

83. $16a + 21 = 20a - 9$

84. $9k - 7 = 21 - 3k$

85. $11z - 13 = 3z + 17$

86. $15 + 6n = 4n + 23$

LESSON 10-4 Angle Measure

:: Then	:: Now	:: Why?
• You measured line segments.	• **1** Measure and classify angles. **2** Identify and use congruent angles and the bisector of an angle.	• One of the skills Dale must learn in carpentry class is how to cut a *miter* joint. This joint is created when two boards are cut at an angle to each other. He has learned that one miscalculation in angle measure can result in mitered edges that do not fit together.

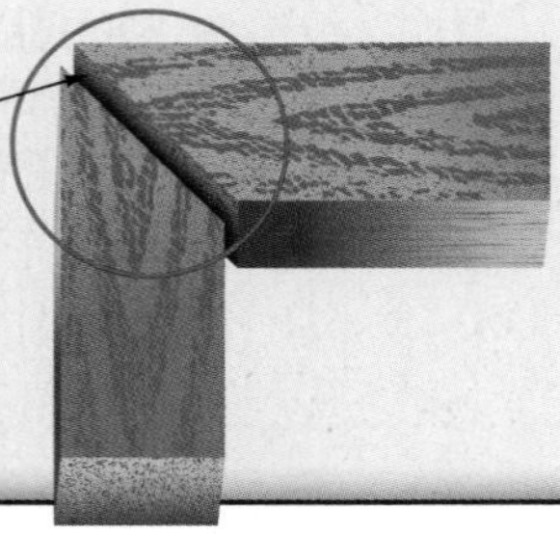

NewVocabulary

ray
opposite rays
angle
side
vertex
interior
exterior
degree
right angle
acute angle
obtuse angle
angle bisector

Common Core State Standards

Content Standards

G.CO.1 Know precise definitions of angle, circle, perpendicular line, parallel line, and line segment, based on the undefined notions of point, line, distance along a line, and distance around a circular arc.

G.CO.12 Make formal geometric constructions with a variety of tools and methods (compass and straightedge, string, reflective devices, paper folding, dynamic geometric software, etc.).

Mathematical Practices

5 Use appropriate tools strategically.
6 Attend to precision.

1 Measure and Classify Angles

A **ray** is a part of a line. It has one endpoint and extends indefinitely in one direction. Rays are named by stating the endpoint first and then any other point on the ray. The ray shown cannot be named as $\overrightarrow{OM}$ because O is not the endpoint of the ray.

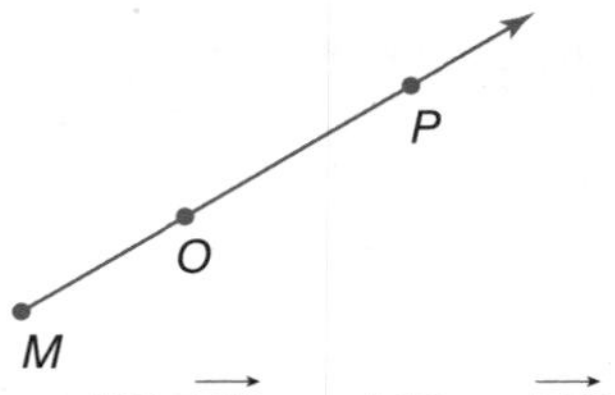

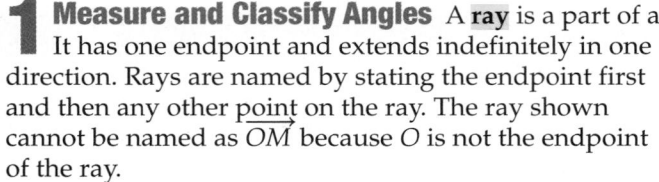

ray MP, $\overrightarrow{MP}$, ray MO, or $\overrightarrow{MO}$

If you choose a point on a line, that point determines exactly two rays called **opposite rays**. Since both rays share a common endpoint, opposite rays are collinear

$\overrightarrow{JH}$ and $\overrightarrow{JK}$ are opposite rays.

An **angle** is formed by two *noncollinear* rays that have a common endpoint. The rays are called **sides** of the angle. The common endpoint is the **vertex**.

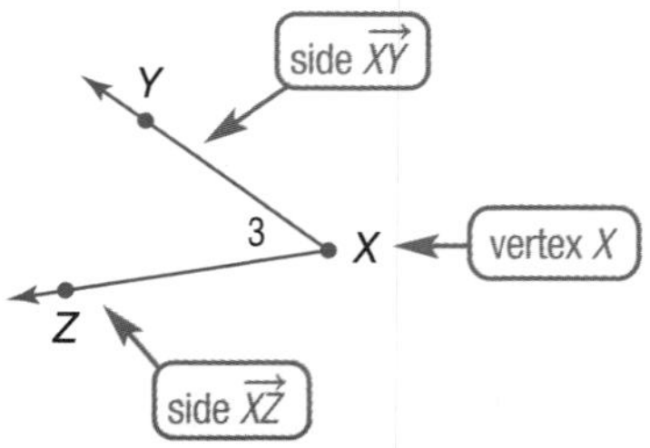

When naming angles using three letters, the vertex must be the second of the three letters. You can name an angle using a single letter only when there is exactly one angle located at that vertex. The angle shown can be named as $\angle X$, $\angle YXZ$, $\angle ZXY$, or $\angle 3$.

An angle divides a plane into three distinct parts.

- Points Q, M, and N lie on the angle.
- Points S and R lie in the **interior** of the angle.
- Points P and O lie in the **exterior** of the angle.

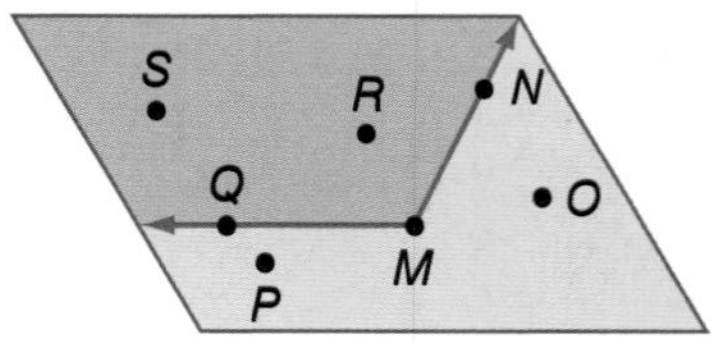

Real-World Example 1 Angles and Their Parts

MAPS **Use the map of a high school shown.**

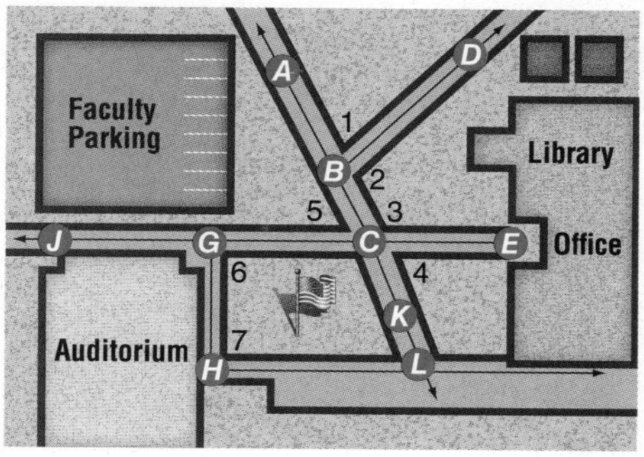
Faculty Parking
Library
Office
Auditorium
A
B
C
D
E
G
H
J
K
L
1
2
3
4
5
6
7

a. Name all angles that have *B* as a vertex.

∠1 or ∠*ABD*, and ∠2 or ∠*DBC*

b. Name the sides of ∠3.

$\overrightarrow{CA}$ and $\overline{CE}$ or $\overrightarrow{CB}$ and $\overline{CE}$

c. What is another name for ∠*GHL*?

∠7, ∠*H*, or ∠*LHG*

d. Name a point in the interior of ∠*DBK*.

Point *E*

> **StudyTip**
> **Segments as Sides** Because a ray can contain a line segment, the side of an angle can be a segment.

Guided Practice

1A. What is the vertex of ∠5?

1B. Name the sides of ∠5.

1C. Write another name for ∠*ECL*.

1D. Name a point in the exterior of ∠*CLH*.

Angles are measured in units called degrees. The **degree** results from dividing the distance around a circle into 360 parts.

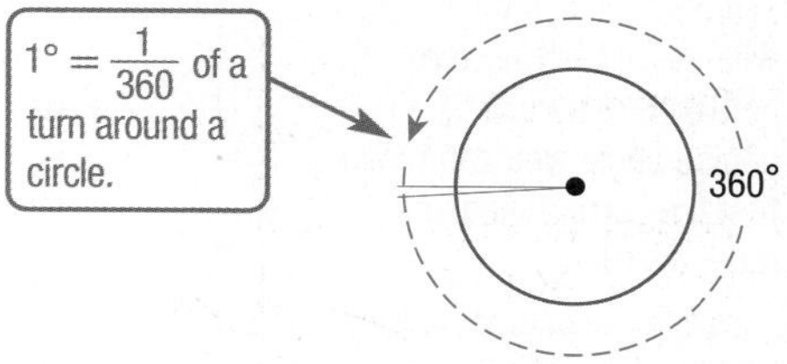
$1° = \frac{1}{360}$ of a turn around a circle.
360°

To measure an angle, you can use a *protractor*. Angle *DEF* below is a 50 degree (50°) angle. We say that the *degree measure* of ∠*DEF* is 50, or *m*∠*DEF* = 50.

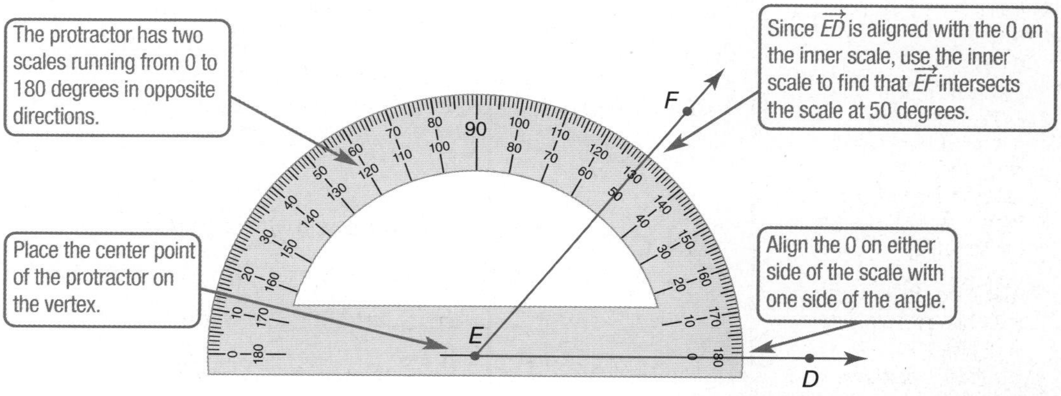
The protractor has two scales running from 0 to 180 degrees in opposite directions.
Since $\overrightarrow{ED}$ is aligned with the 0 on the inner scale, use the inner scale to find that $\overrightarrow{EF}$ intersects the scale at 50 degrees.
Place the center point of the protractor on the vertex.
Align the 0 on either side of the scale with one side of the angle.
F
E
D
90

ReadingMath

Straight Angle Opposite rays with the same vertex form a *straight angle*. Its measure is 180. Unless otherwise specified in this book, however, the term *angle* means a nonstraight angle.

Angles can be classified by their measures as shown below.

KeyConcept Classify Angles

right angle	acute angle	obtuse angle
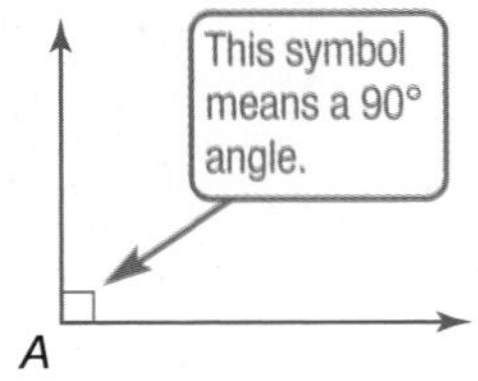	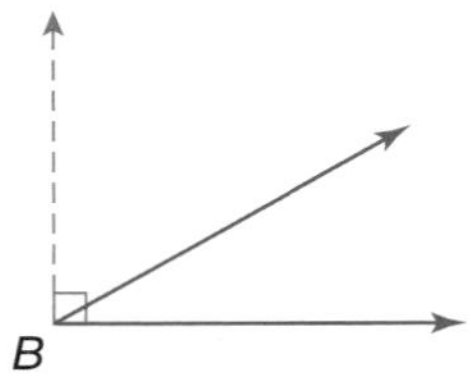	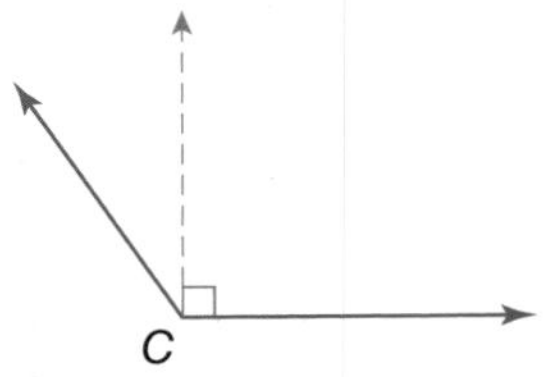
$m\angle A = 90$	$m\angle B < 90$	$180 > m\angle C > 90$

Example 2 Measure and Classify Angles

Copy the diagram below, and extend each ray. Classify each angle as *right*, *acute*, or *obtuse*. Then use a protractor to measure the angle to the nearest degree.

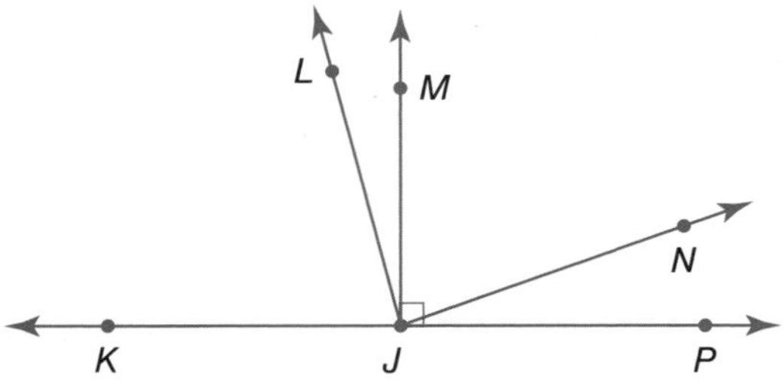

a. $\angle MJP$

$\angle MJP$ is marked as a right angle, so $m\angle MJP = 90$.

b. $\angle LJP$

Point L on angle $\angle LJP$ lies on the exterior of right angle $\angle MJP$, so $\angle LJP$ is an obtuse angle. Use a protractor to find that $m\angle LJP = 105$

CHECK Since $105 > 90$, $\angle LJP$ is an obtuse angle. ✓

c. $\angle NJP$

Point N on angle $\angle NJP$ lies on the interior of right angle $\angle MJP$, so $\angle NJP$ is an acute angle. Use a protractor to find that $m\angle NJP = 20$.

CHECK Since $20 < 90$, $\angle NJP$ is an acute angle. ✓

WatchOut!

Classify Before Measuring Classifying an angle before measuring it can prevent you from choosing the wrong scale on your protractor. In Example 2b, you must decide whether $\angle LJP$ measures 75 or 105. Since $\angle LJP$ is an obtuse angle, you can reason that the correct measure must be 105.

GuidedPractice

2A. $\angle AFB$

2B. $\angle CFA$

2C. $\angle AFD$

2D. $\angle CFD$

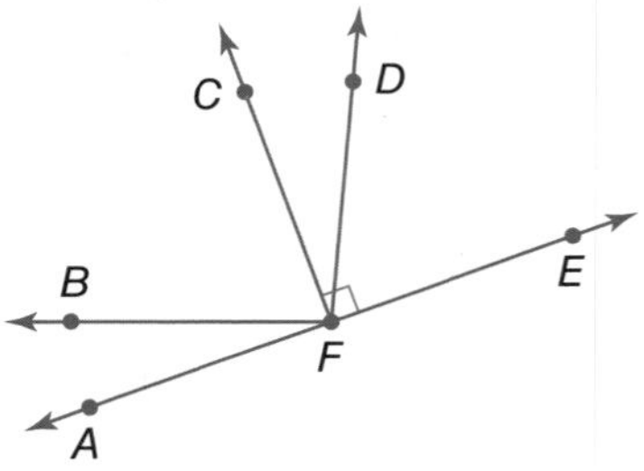

2 Congruent Angles

Just as segments that have the same measure are congruent segments, angles that have the same measure are *congruent angles*.

In the figure, since $m\angle ABC = m\angle FED$, then $\angle ABC \cong \angle FED$. Matching numbers of arcs on a figure also indicate congruent angles, so $\angle CBE \cong \angle DEB$.

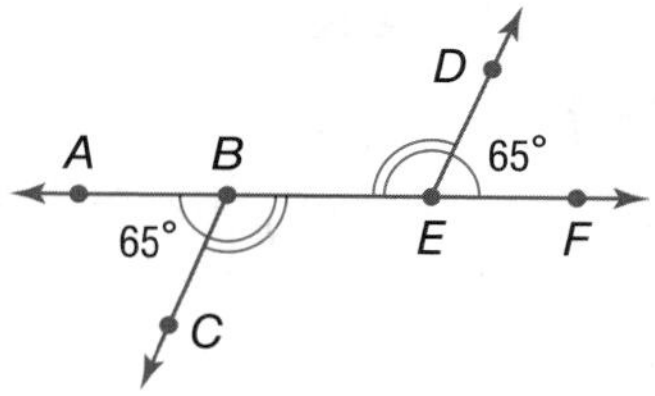

You can produce an angle congruent to a given angle using a construction.

Construction Copy an Angle

Step 1 Draw an angle like $\angle B$ on your paper. Use a straightedge to draw a ray on your paper. Label its endpoint G.

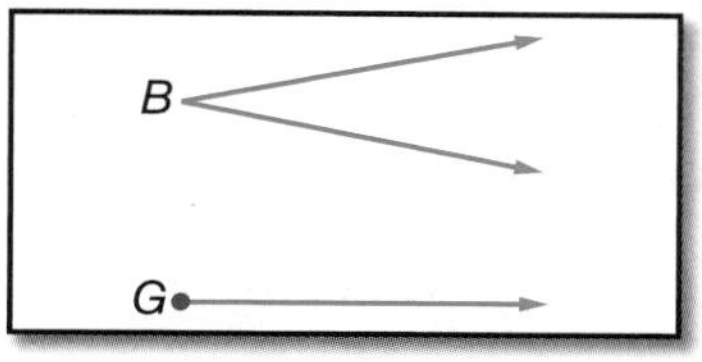

Step 2 Place the tip of the compass at point B and draw a large arc that intersects both sides of $\angle B$. Label the points of intersection A and C.

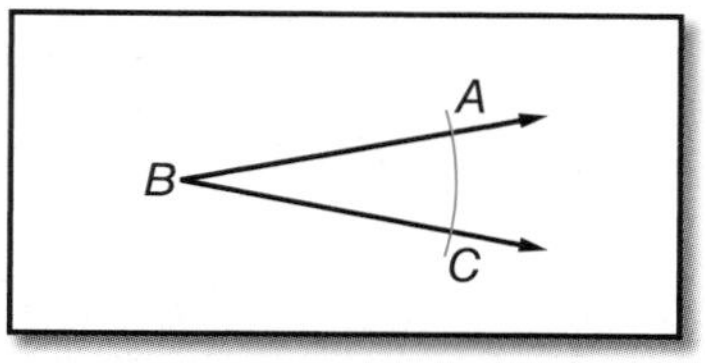

Step 3 Using the same compass setting, put the compass at point G and draw a large arc that starts above the ray and intersects the ray. Label the point of intersection H.

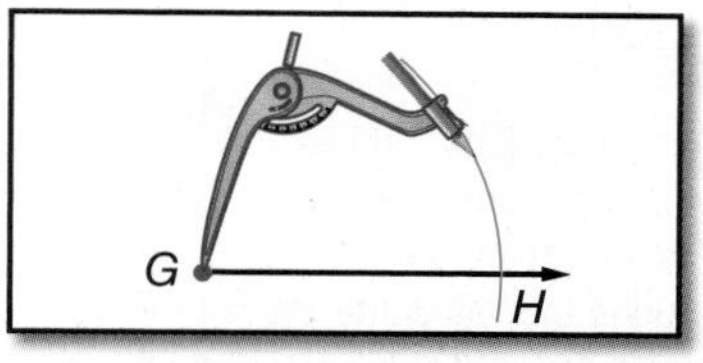

Step 4 Place the point of your compass on C and adjust so that the pencil tip is on A.

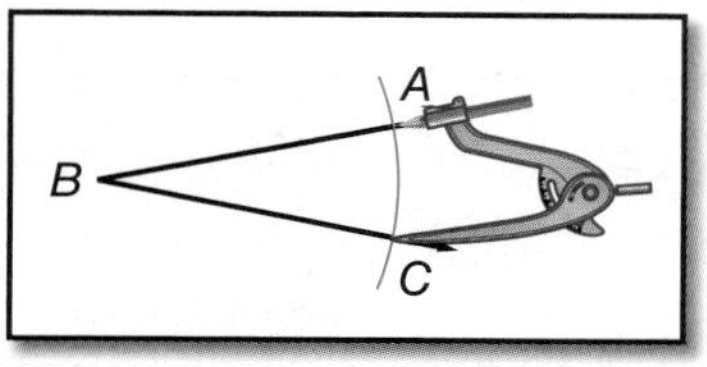

Step 5 Without changing the setting, place the compass at point H and draw an arc to intersect the larger arc you drew in Step 4. Label the point of intersection F.

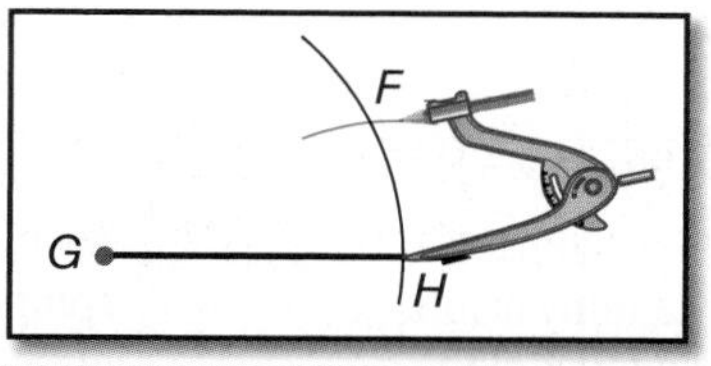

Step 6 Use a straightedge to draw $\overrightarrow{GF}$. $\angle ABC \cong \angle FGH$

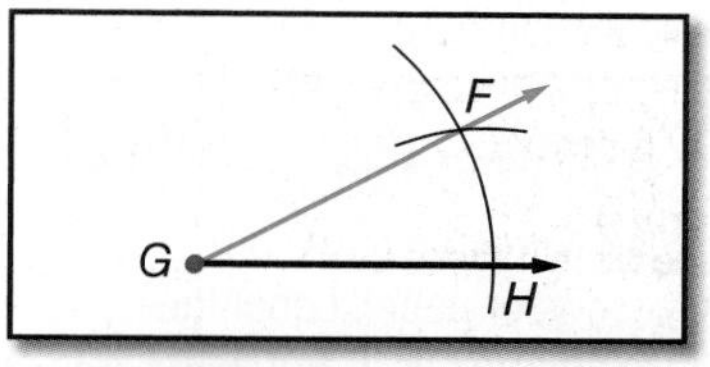

StudyTip

Segments A line segment can also bisect an angle.

A ray that divides an angle into two congruent angles is called an **angle bisector**. If $\overrightarrow{YW}$ is the angle bisector of $\angle XYZ$, then point W lies in the interior of $\angle XYZ$ and $\angle XYW \cong \angle WYZ$.

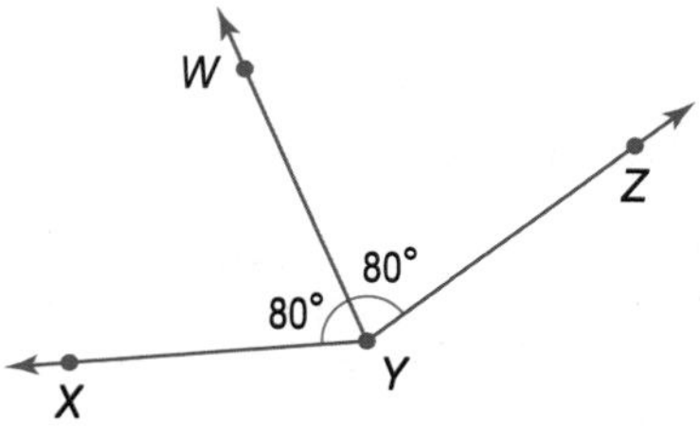

Just as with segments, when a line, segment, or ray divides an angle into smaller angles, the sum of the measures of the smaller angles equals the measure of the largest angle. So in the figure, $m\angle XYW + m\angle WYZ = m\angle XYZ$.

Example 3 Measure and Classify Angles

ALGEBRA In the figure, $\overrightarrow{KJ}$ and $\overrightarrow{KM}$ are opposite rays, and $\overrightarrow{KN}$ bisects $\angle JKL$. If $m\angle JKN = 8x - 13$ and $m\angle NKL = 6x + 11$, find $m\angle JKN$.

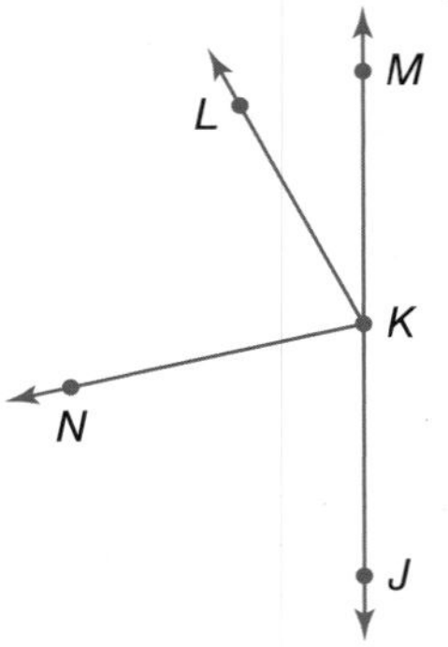

Step 1 Solve for x.

Since $\overrightarrow{KN}$ bisects $\angle JKL$, $\angle JKN \cong \angle NKL$.

$m\angle JKN = m\angle NKL$	Definition of congruent angles
$8x - 13 = 6x + 11$	Substitution
$8x = 6x + 24$	Add 13 to each side.
$2x = 24$	Subtract $6x$ from each side.
$x = 12$	Divide each side by 2.

Step 2 Use the value of x to find $m\angle JKN$.

$m\angle JKN = 8x - 13$	Given
$= 8(12) - 13$	$x = 12$
$= 96 - 13$ or 83	Simplify.

StudyTip

Checking Solutions Check that you have computed the value of x correctly by substituting the value into the expression for $\angle NKL$. If you don't get the same measure as $\angle JKN$, you have made an error.

GuidedPractice

3. Suppose $m\angle JKL = 9y + 15$ and $m\angle JKN = 5y + 2$. Find $m\angle JKL$.

You can produce the angle bisector of any angle without knowing the measure of the angle.

Construction Bisect an Angle

Step 1 Draw an angle on your paper. Label the vertex as P. Put your compass at point P and draw a large arc that intersects both sides of $\angle P$. Label the points of intersection Q and R.

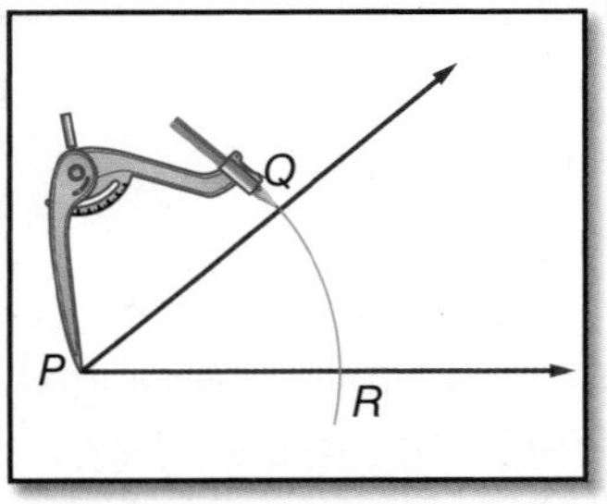

Step 2 With the compass at point Q, draw an arc in the interior of the angle.

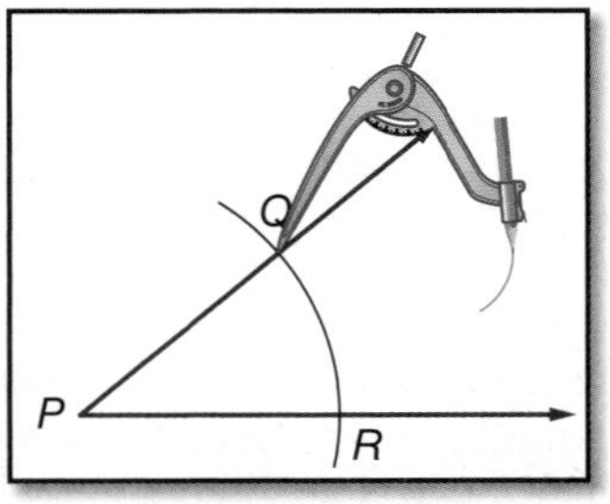

Step 3 Keeping the same compass setting, place the compass at point R and draw an arc that intersects the arc drawn in Step 2. Label the point of intersection T.

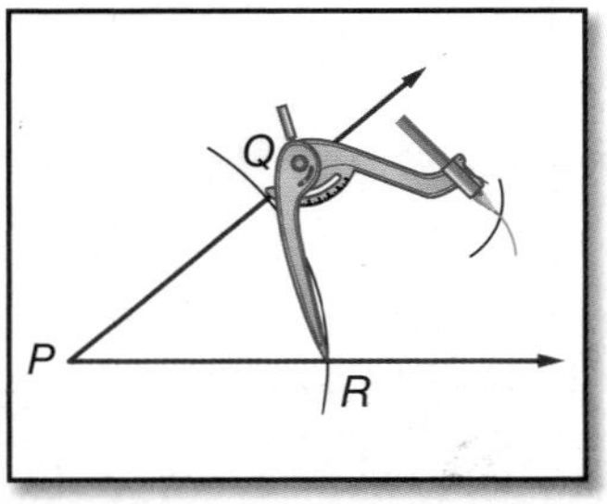

Step 4 Draw $\overrightarrow{PT}$. $\overrightarrow{PT}$ is the bisector of $\angle P$.

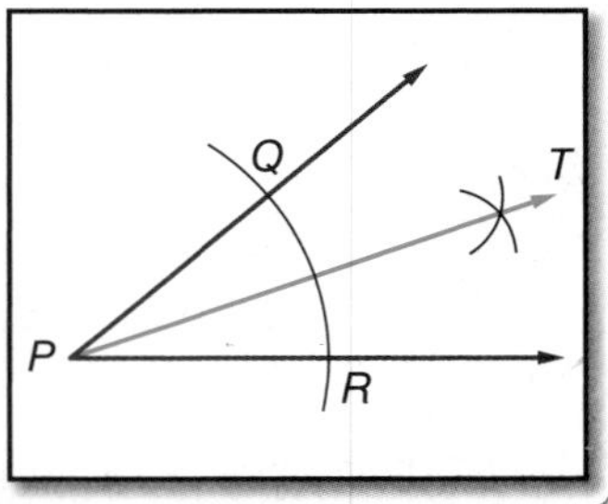

Check Your Understanding

Example 1 **Use the figure at the right.**

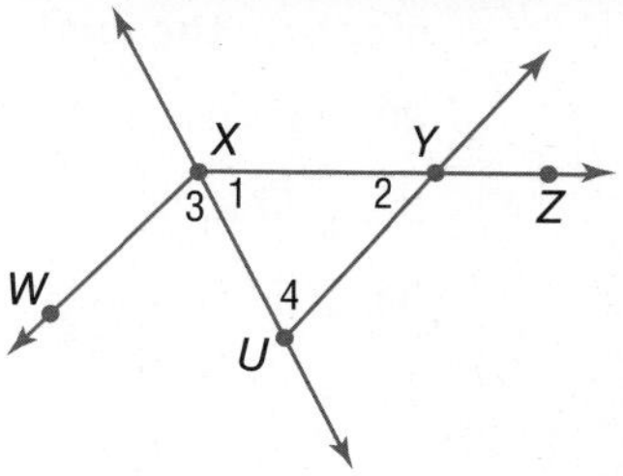

1. Name the vertex of ∠4.
2. Name the sides of ∠3.
3. What is another name for ∠2?
4. What is another name for ∠*UXY*?

Example 2 **Copy the diagram shown, and extend each ray. Classify each angle as *right*, *acute*, or *obtuse*. Then use a protractor to measure the angle to the nearest degree.**

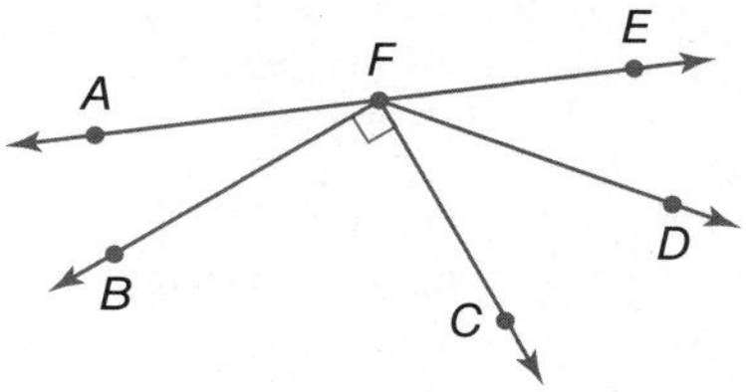

5. ∠*CFD*
6. ∠*AFD*
7. ∠*BFC*
8. ∠*AFB*

Example 3 **ALGEBRA In the figure, $\overrightarrow{KJ}$ and $\overrightarrow{KL}$ are opposite rays. $\overrightarrow{KN}$ bisects ∠*LKM*.**

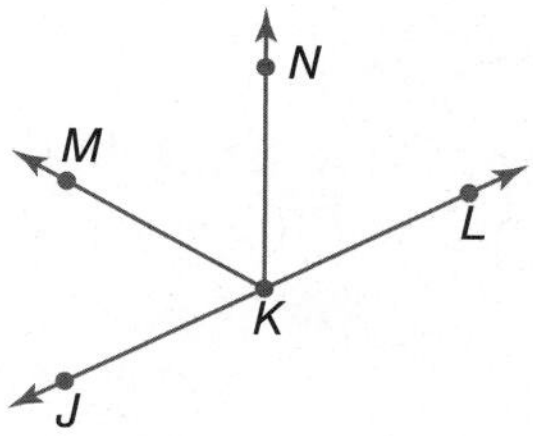

9. If $m\angle LKM = 7x - 5$ and $m\angle NKM = 3x + 9$, find $m\angle LKM$.
10. If $m\angle NKL = 7x - 9$ and $m\angle JKM = x + 3$, find $m\angle JKN$.
11. **CCSS PRECISION** A miter cut is used to build picture frames with corners that meet at right angles.

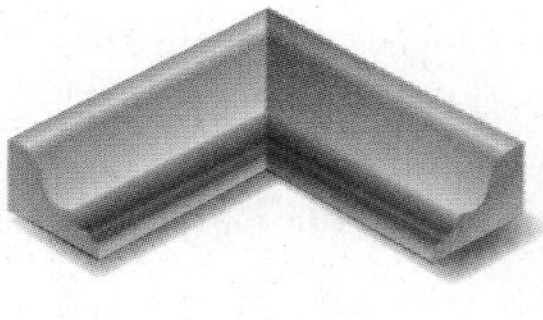

   a. José miters the ends of some wood for a picture frame at congruent angles. What is the degree measure of his cut? Explain and classify the angle.

   b. What does the joint represent in relation to the angle formed by the two pieces?

Practice and Problem Solving

Example 1 **For Exercises 12–29, use the figure at the right.**

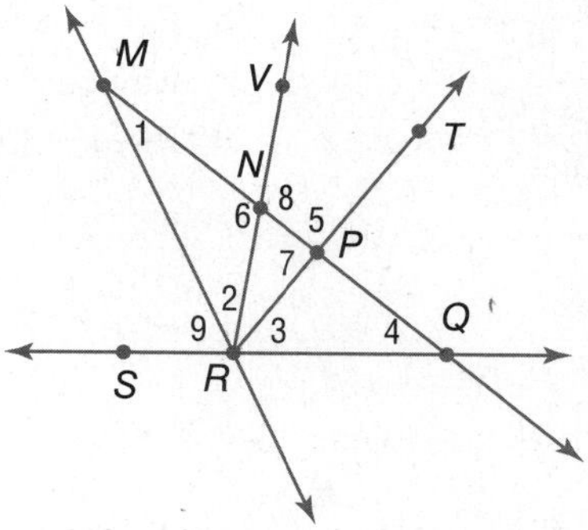

Name the vertex of each angle.

12. ∠4
13. ∠7
14. ∠2
15. ∠1

Name the sides of each angle.

16. ∠*TPQ*
17. ∠*VNM*
18. ∠6
19. ∠3

Write another name for each angle.

20. ∠9
21. ∠*QPT*
22. ∠*MQS*
23. ∠5
24. Name an angle with vertex *N* that appears obtuse.
25. Name an angle with vertex *Q* that appears acute.
26. Name a point in the interior of ∠*VRQ*.
27. Name a point in the exterior of ∠*MRT*.
28. Name a pair of angles that share exactly one point.
29. Name a pair of angles that share more than one point.

Example 2 **Copy the diagram shown, and extend each ray. Classify each angle as *right, acute,* or *obtuse*. Then use a protractor to measure the angle to the nearest degree.**

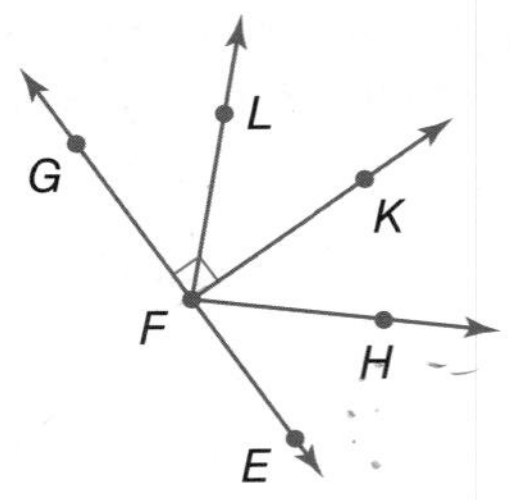

30. $\angle GFK$

31. $\angle EFK$

32. $\angle LFK$

33. $\angle EFH$

34. $\angle GFH$

35. $\angle EFL$

36. **CLOCKS** Determine at least three different times during the day when the hands on a clock form each of the following angles. Explain.

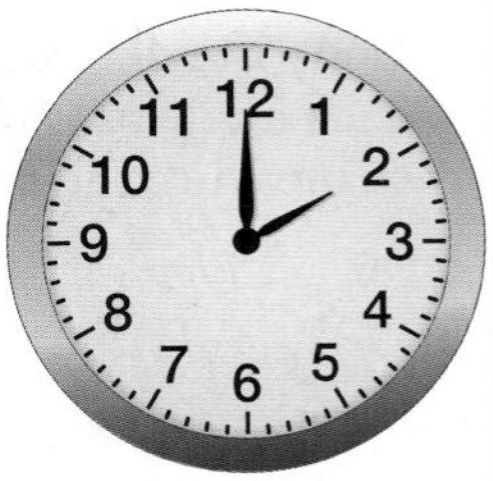

a. right angle

b. obtuse angle

c. congruent acute angles

Example 3 **ALGEBRA** **In the figure, $\overrightarrow{BA}$ and $\overrightarrow{BC}$ are opposite rays. $\overrightarrow{BH}$ bisects $\angle EBC$.**

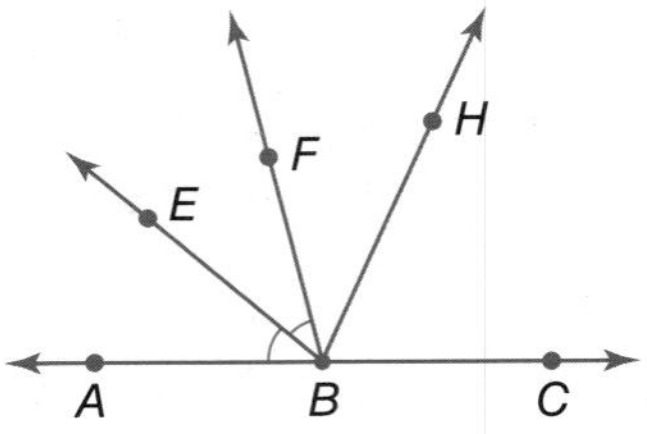

37. If $m\angle ABE = 2n + 7$ and $m\angle EBF = 4n - 13$, find $m\angle ABE$.

38. If $m\angle EBH = 6x + 12$ and $m\angle HBC = 8x - 10$, find $m\angle EBH$.

39. If $m\angle ABF = 7b - 24$ and $m\angle ABE = 2b$, find $m\angle EBF$.

40. If $m\angle EBC = 31a - 2$ and $m\angle EBH = 4a + 45$, find $m\angle HBC$.

41. If $m\angle ABF = 8s - 6$ and $m\angle ABE = 2(s + 11)$, find $m\angle EBF$.

42. If $m\angle EBC = 3r + 10$ and $m\angle ABE = 2r - 20$, find $m\angle EBF$.

43. **MAPS** **Estimate the measure of the angle formed by each city or location listed, the North Pole, and the Prime Meridian.**

a. Nuuk, Greenland

b. Fairbanks, Alaska

c. Reykjavik, Iceland

d. Prime Meridian

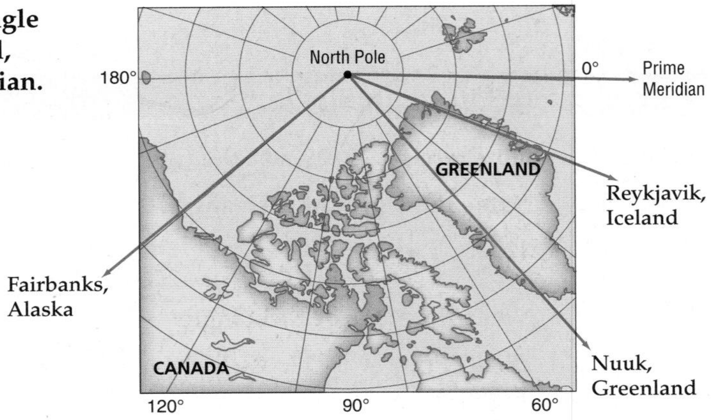

44. **CCSS TOOLS** A compass rose is a design on a map that shows directions. In addition to the directions of north, south, east, and west, a compass rose can have as many as 32 markings.

a. With the center of the compass as its vertex, what is the measure of the angle between due west and due north?

b. What is the measure of the angle between due north and north-west?

c. How does the north-west ray relate to the angle in part **a**?

Plot the points in a coordinate plane and sketch ∠*XYZ*. Then classify it as *right, acute,* or *obtuse*.

45. $X(5, -3)$, $Y(4, -1)$, $Z(6, -2)$

46. $X(6, 7)$, $Y(2, 3)$, $Z(4, 1)$

47 **PHYSICS** When you look at a pencil in water, it looks bent. This illusion is due to *refraction*, or the bending of light when it moves from one substance to the next.

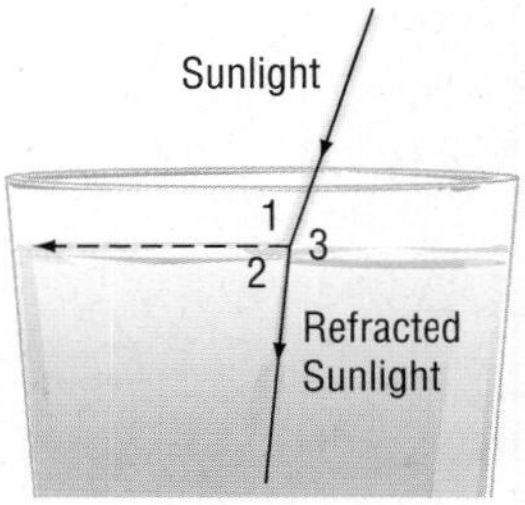

a. What is $m\angle 1$? Classify this angle as *acute, right,* or *obtuse*.

b. What is $m\angle 2$? Classify this angle as *acute, right,* or *obtuse*.

c. Without measuring, determine how many degrees the path of the light changes after it enters the water. Explain your reasoning.

48. **MULTIPLE REPRESENTATIONS** In this problem, you will explore the relationship of angles that compose opposite rays.

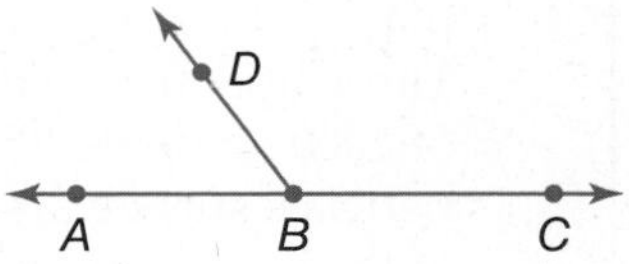

a. Geometric Draw four lines, each with points *A*, *B*, and *C*. Draw $\overrightarrow{BD}$ for each line, varying the placement of point *D*. Use a protractor to measure ∠*ABD* and ∠*DBC* for each figure.

b. Tabular Organize the measures for each figure into a table. Include a row in your table to record the sum of these measures.

c. Verbal Make a conjecture about the sum of the measures of the two angles. Explain your reasoning.

d. Algebraic If x is the measure of ∠*ABD* and y is the measure of ∠*DBC*, write an equation that relates the two angle measures.

H.O.T. Problems Use Higher-Order Thinking Skills

49. **OPEN ENDED** Draw an obtuse angle named *ABC*. Measure ∠*ABC*. Construct an angle bisector $\overrightarrow{BD}$ of ∠*ABC*. Explain the steps in your construction and justify each step. Classify the two angles formed by the angle bisector.

50. **CHALLENGE** Describe how you would use a protractor to measure the angle shown.

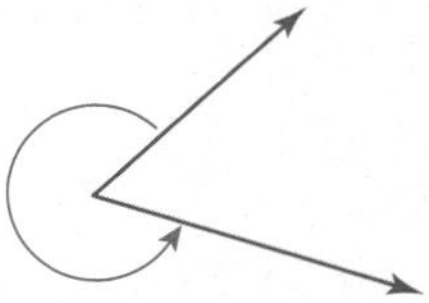

51. **CCSS ARGUMENTS** The sum of two acute angles is *sometimes*, *always*, or *never* an obtuse angle. Explain.

52. **CHALLENGE** $\overrightarrow{MP}$ bisects ∠*LMN*, $\overrightarrow{MQ}$ bisects ∠*LMP*, and $\overrightarrow{MR}$ bisects ∠*QMP*. If $m\angle RMP = 21$, find $m\angle LMN$. Explain your reasoning.

53. **WRITING IN MATH** Rashid says that he can estimate the measure of an acute angle using a piece of paper to within six degrees of accuracy. Explain how this would be possible. Then use this method to estimate the measure of the angle shown.

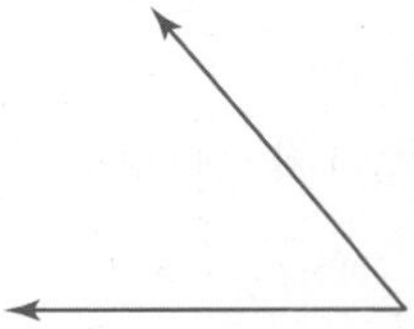

Standardized Test Practice

54. Which of the following angles measures closest to 60°?

A

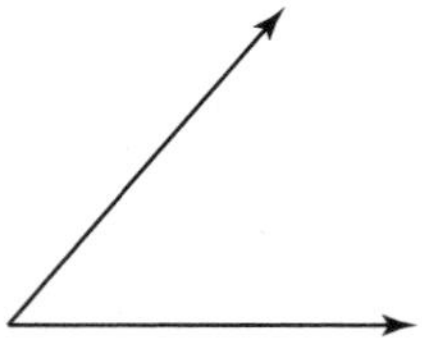

C

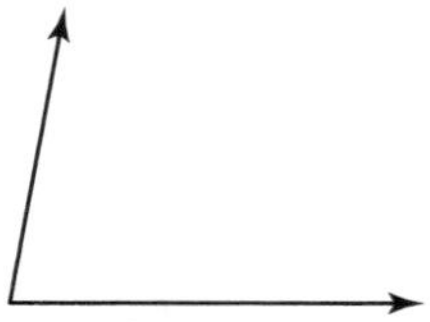

B

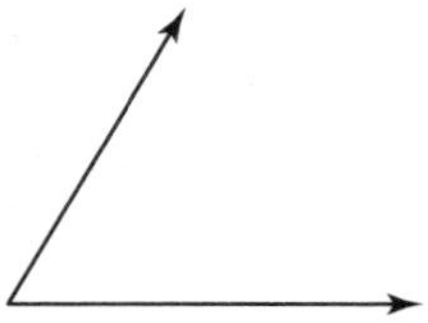

D

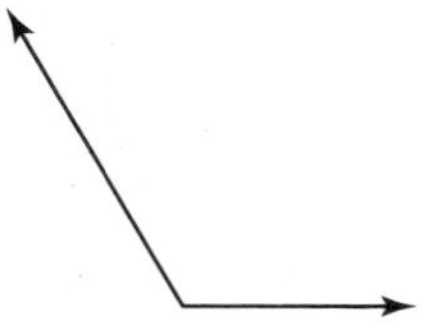

55. SHORT RESPONSE Leticia surveyed 50 English majors at a university to see if the school should play jazz music in the cafeteria during lunch. The school has 75 different majors and a total of 2000 students. Explain why the results of Leticia's survey are or are not representative of the entire student body.

56. In the figure below, if $m\angle BAC = 38$, what must be the measure of $\angle BAD$ in order for $\overrightarrow{AC}$ to be an angle bisector?

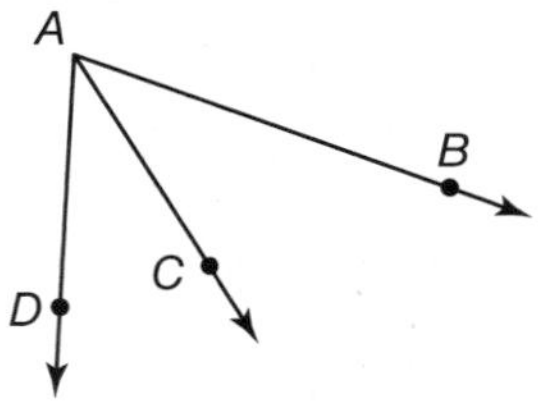

F 142 **H** 52

G 76 **J** 38

57. SAT/ACT If n is divisible by 2, 5, and 14, which of the following is also divisible by these numbers?

A $n + 7$ **D** $n + 20$

B $n + 10$ **E** $n + 70$

C $n + 14$

Spiral Review

Find the distance between each pair of points. Round to the nearest hundredth. (Lesson 10-3)

58. $A(-1, -8), B(3, 4)$

59. $C(0, 1), D(-2, 9)$

60. $E(-3, -12), F(5, 4)$

61. $G(4, -10), H(9, -25)$

62. $J\left(1, \frac{1}{4}\right), K\left(-3, \frac{7}{4}\right)$

63. $L\left(-5, \frac{8}{5}\right), M\left(5, \frac{2}{5}\right)$

Find the value of the variable and ST if S is between R and T. (Lesson 10-2)

64. $RS = 7a, ST = 12a, RT = 76$

65. $RS = 12, ST = 2x, RT = 34$

66. PHOTOGRAPHY Photographers often place their cameras on tripods. In the diagram, the tripod is placed on an inclined surface, and the length of each leg is adjusted so that the camera remains level with the horizon. Are the feet of the tripod coplanar? Explain your reasoning. (Lesson 10-1)

Skills Review

Solve each equation.

67. $(90 - x) - x = 18$

68. $(5x + 3) + 7x = 180$

69. $(13x + 10) + 2x = 90$

70. $(180 - x) - 4x = 56$

71. $(4n + 17) + (n - 2) = 180$

72. $(8a - 23) + (9 - 2a) = 90$

CHAPTER 10

Mid-Chapter Quiz

Lessons 10-1 through 10-4

Use the figure to complete each of the following. (Lesson 10-1)

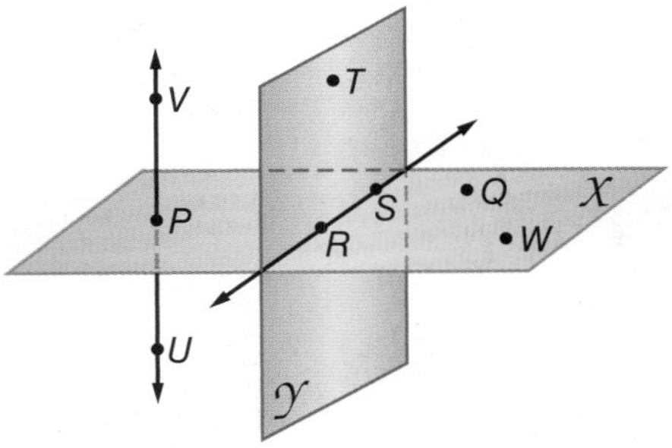

1. Name another point that is collinear with points U and V.
2. What is another name for plane Y?
3. Name a line that is coplanar with points P, Q, and W.

Find the value of x and AC if B is between points A and C. (Lesson 10-2)

4. $AB = 12$, $BC = 8x - 2$, $AC = 10x$
5. $AB = 5x$, $BC = 9x - 2$, $AC = 11x + 7.6$
6. Find CD and the coordinate of the midpoint of $\overline{CD}$.

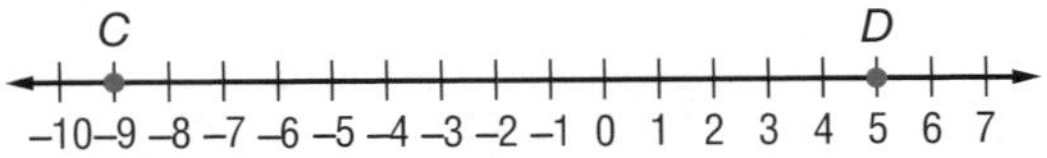

Find the coordinates of the midpoint of each segment. Then find the length of each segment. (Lesson 10-3)

7.

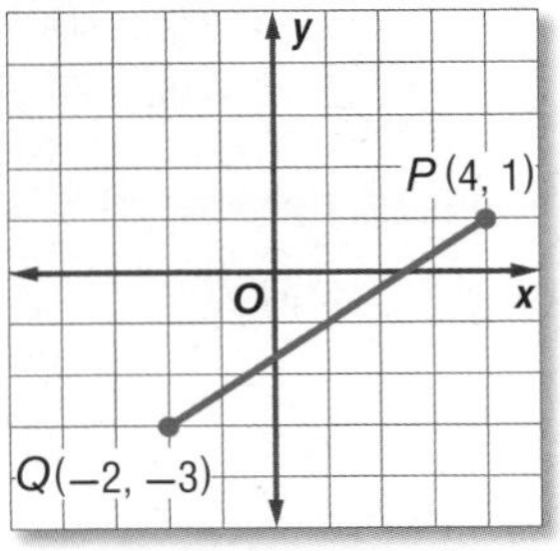

8.

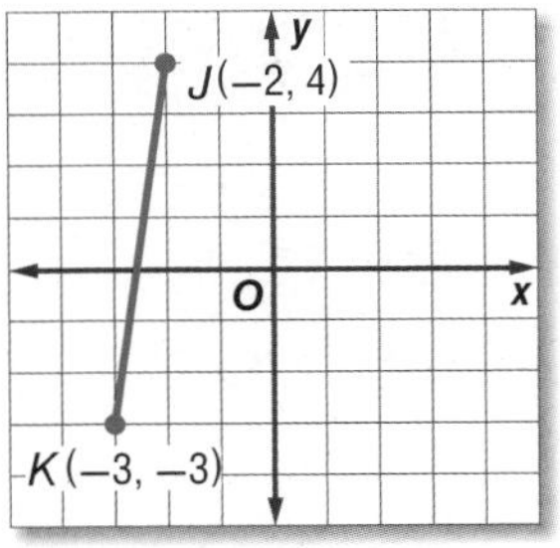

Find the coordinates of the midpoint of a segment with the given endpoints. Then find the distance between each pair of points. (Lesson 10-3)

9. $P(26, 12)$ and $Q(8, 42)$
10. $M(6, -41)$ and $N(-18, -27)$

11. **MAPS** A map of a town is drawn on a coordinate grid. The high school is found at point (3, 1) and town hall is found at (–5, 7). (Lesson 10-3)
    a. If the high school is at the midpoint between the town hall and the town library, at which ordered pair should you find the library?
    b. If one unit on the grid is equivalent to 50 meters, how far is the high school from town hall?

12. **MULTIPLE CHOICE** The vertex of $\angle ABC$ is located at the origin. Point A is located at (5, 0) and Point C is located at (0, 2). How can $\angle ABC$ be classified?

A acute	**C** right
B obtuse	**D** scalene

In the figure, $\overrightarrow{XA}$ and $\overrightarrow{XE}$ are opposite rays, and $\angle AXC$ is bisected by $\overrightarrow{XB}$. (Lesson 10-4)

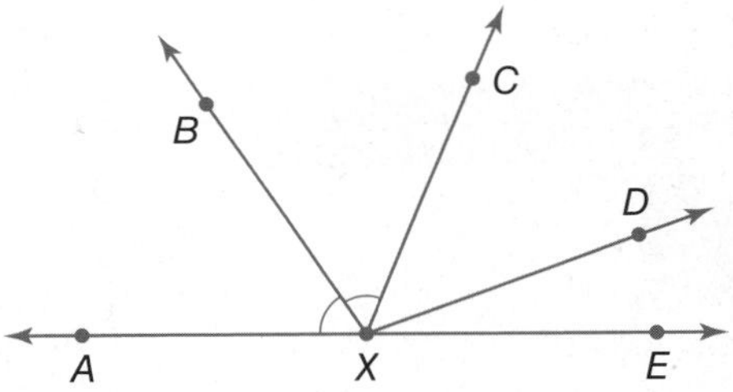

13. If $m\angle AXC = 8x - 7$ and $m\angle AXB = 3x + 10$, find $m\angle AXC$.
14. If $m\angle CXD = 4x + 6$, $m\angle DXE = 3x + 1$, and $m\angle CXE = 8x - 2$, find $m\angle DXE$.

Classify each angle as *acute*, *right*, or *obtuse*. (Lesson 10-4)

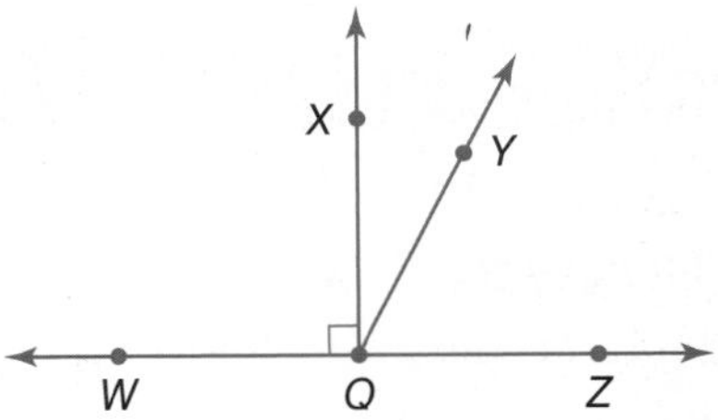

15. $\angle WQY$
16. $\angle YQZ$

EXTEND 10-5

Geometry Lab
Constructing Perpendiculars

You can use a compass and a straightedge to construct a line perpendicular to a given line through a point on the line, or through a point *not* on the line.

CCSS Common Core State Standards
Content Standards
G.CO.12 Make formal geometric constructions with a variety of tools and methods (compass and straightedge, string, reflective devices, paper folding, dynamic geometric software, etc.).
Mathematical Practices 5

Activity Construct a Perpendicular

a. Construct a line perpendicular to line ℓ and passing through point P on ℓ.

Step 1

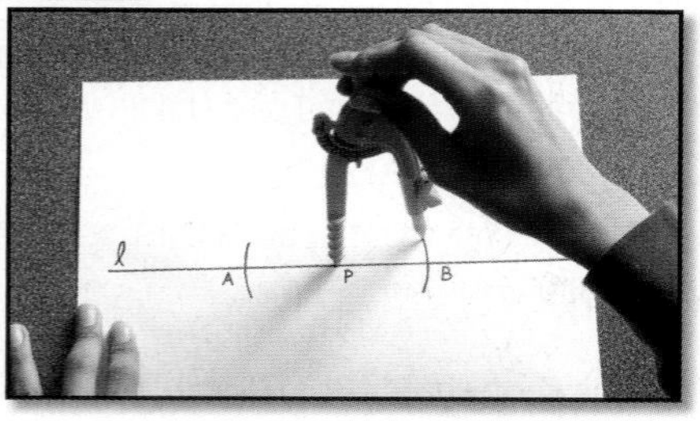

Place the compass at P. Draw arcs to the right and left of P that intersect line ℓ using the same compass setting. Label the points of intersection A and B.

Step 2

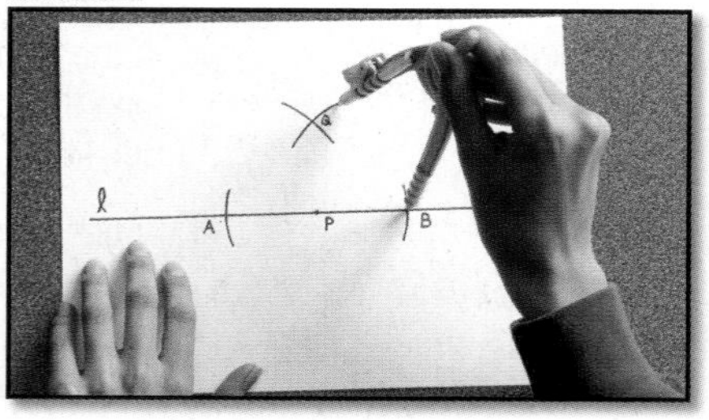

With the compass at A, draw an arc above line ℓ using a setting greater than AP. Using the same compass setting, draw an arc from B that intersects the previous arc. Label the intersection Q.

Step 3

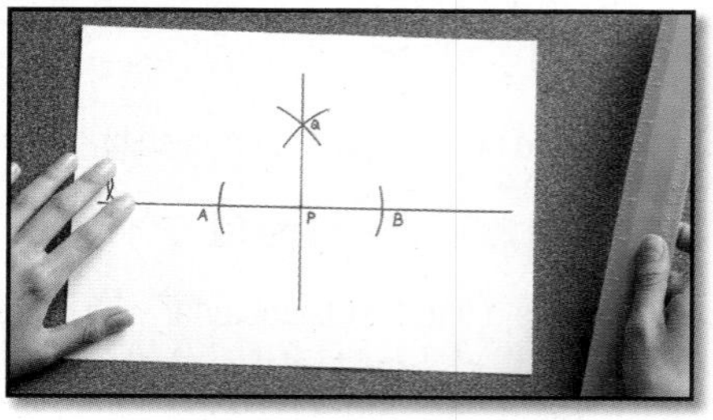

Use a straightedge to draw $\overleftrightarrow{QP}$.

b. Construct a line perpendicular to line k and passing through point P *not* on k.

Step 1

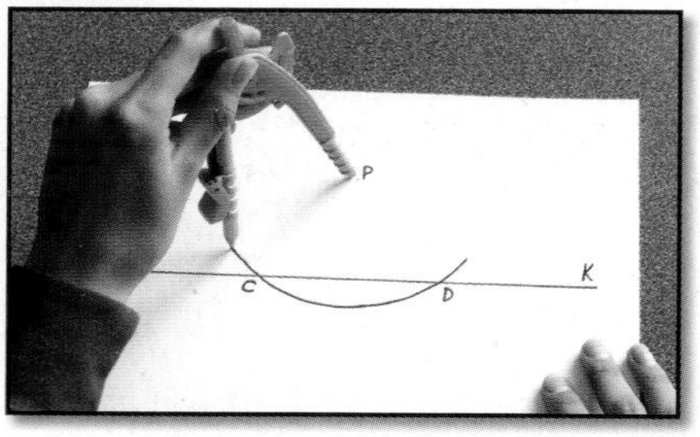

Place the compass at P. Draw an arc that intersects line k in two different places. Label the points of intersection C and D.

Step 2

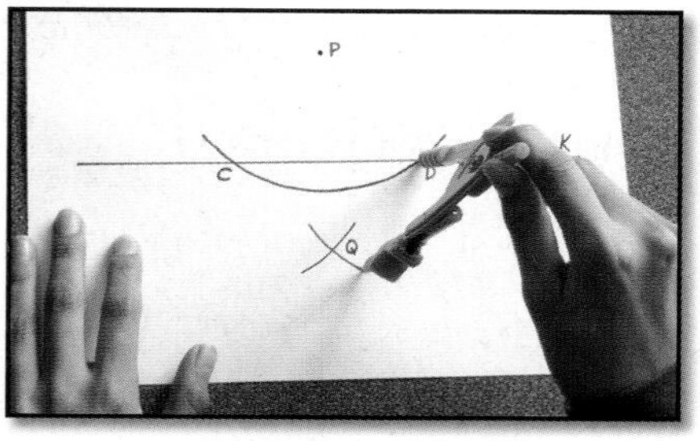

With the compass at C, draw an arc below line k using a setting greater than $\frac{1}{2}CD$. Using the same compass setting, draw an arc from D that intersects the previous arc. Label the intersection Q.

Step 3

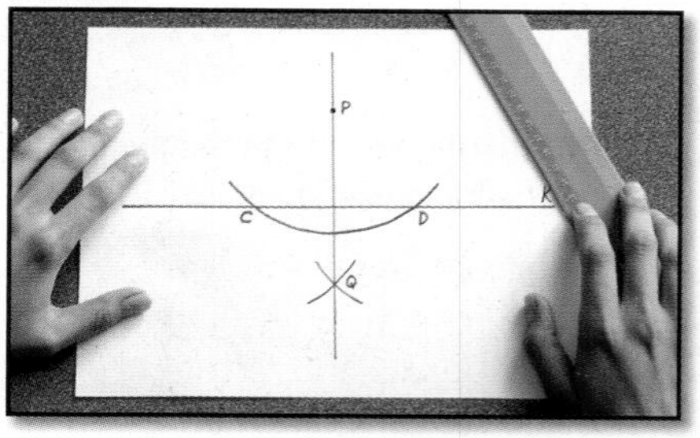

Use a straightedge to draw $\overleftrightarrow{PQ}$.

Model and Analyze the Results

1. Draw a line and construct a line perpendicular to it through a point on the line.
2. Draw a line and construct a line perpendicular to it through a point not on the line.
3. How is the second construction similar to the first one?

LESSON 10-6 Two-Dimensional Figures

Then	Now	Why?
You measured one-dimensional figures.	**1** Identify and name polygons. **2** Find perimeter, circumference, and area of two-dimensional figures.	Mosaics are patterns or pictures created using small bits of colored glass or stone. They are usually set into a wall or floor and often make use of polygons.

NewVocabulary
- polygon
- vertex of a polygon
- concave
- convex
- *n*-gon
- equilateral polygon
- equiangular polygon
- regular polygon
- perimeter
- circumference
- area

Common Core State Standards

Content Standards
G.GPE.7 Use coordinates to compute perimeters of polygons and areas of triangles and rectangles, e.g., using the distance formula.

Mathematical Practices
2 Reason abstractly and quantitatively.
6 Attend to precision.

1 Identify Polygons

Most of the closed figures shown in the mosaic are polygons. The term *polygon* is derived from a Greek word meaning *many angles.*

KeyConcept Polygons

A **polygon** is a closed figure formed by a finite number of coplanar segments called *sides* such that

- the sides that have a common endpoint are noncollinear, and
- each side intersects exactly two other sides, but only at their endpoints.

The vertex of each angle is a **vertex of the polygon**. A polygon is named by the letters of its vertices, written in order of consecutive vertices.

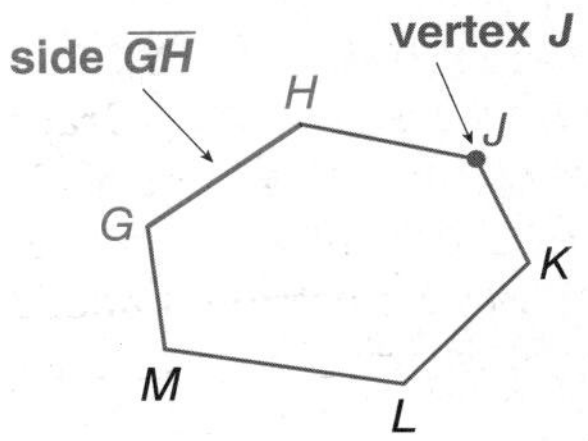

polygon *GHJKLM*

The table below shows some additional examples of polygons and some examples of figures that are not polygons.

Polygons	Not Polygons
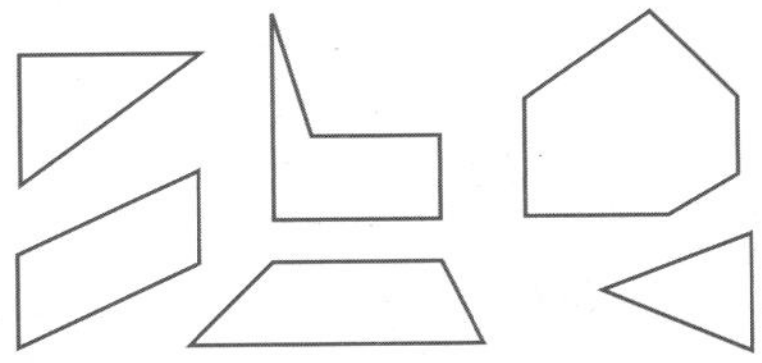	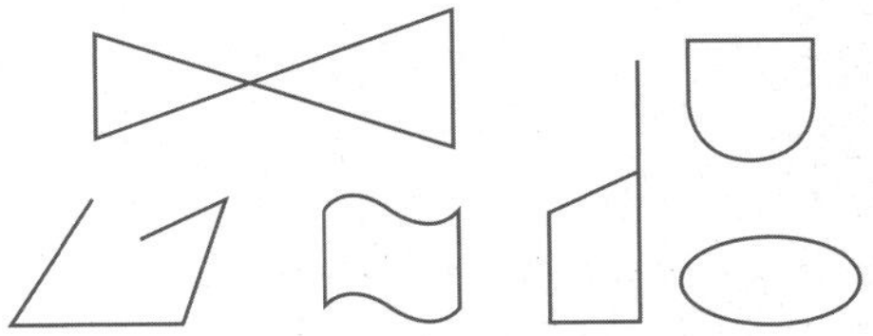

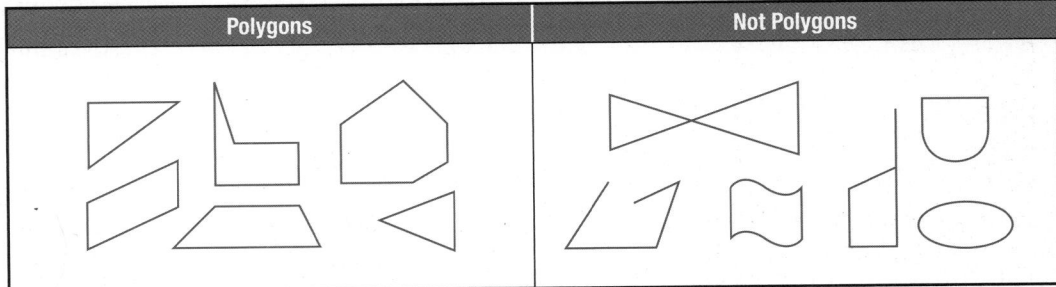

Polygons can be **concave** or **convex**. Suppose the line containing each side is drawn. If any of the lines contain any point in the interior of the polygon, then it is concave. Otherwise it is convex.

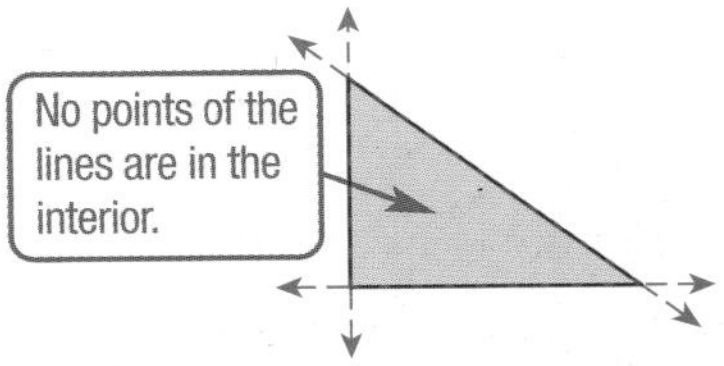

convex polygon

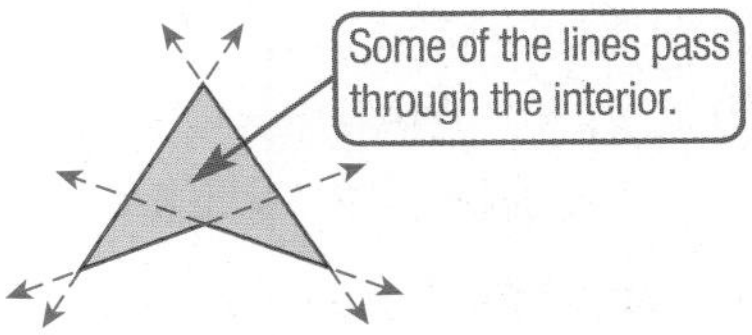

concave polygon

> **StudyTip**
> **Naming Polygons** The Greek prefixes used to name polygons are also used to denote number. For example a *bi*cycle has two wheels, and a *tri*pod has three legs.

In general, a polygon is classified by its number of sides. The table lists some common names for various categories of polygon. A polygon with n sides is an ***n*-gon**. For example, a polygon with 15 sides is a 15-gon.

Number of Sides	Polygon
3	triangle
4	quadrilateral
5	pentagon
6	hexagon
7	heptagon
8	octagon
9	nonagon
10	decagon
11	hendecagon
12	dodecagon
n	n-gon

An **equilateral polygon** is a polygon in which all sides are congruent. An **equiangular polygon** is a polygon in which all angles are congruent.

A convex polygon that is both equilateral and equiangular is called a **regular polygon**. An *irregular polygon* is a polygon that is *not* regular.

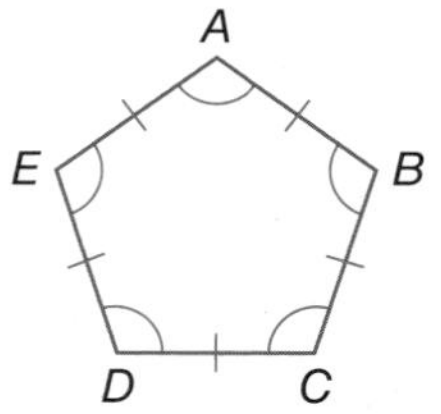

regular pentagon *ABCDE*

Example 1 Name and Classify Polygons

Name each polygon by its number of sides. Then classify it as *convex* or *concave* and *regular* or *irregular*.

a.

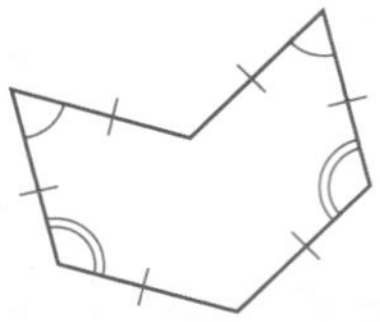

The polygon has 6 sides, so it is a hexagon.

Two of the lines containing the sides of the polygon will pass through the interior of the hexagon, so it is concave.

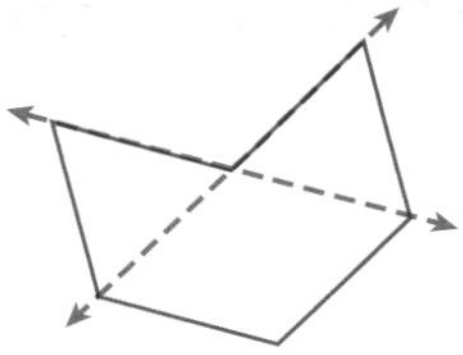

Only convex polygons can be regular, so this is an irregular hexagon.

b.

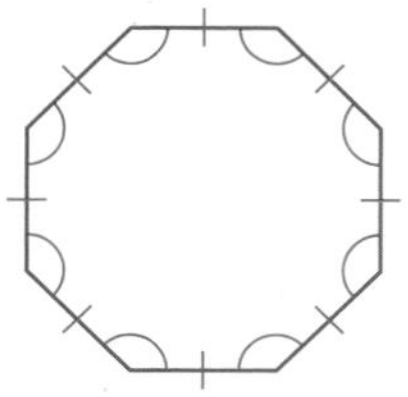

There are 8 sides, so this is an octagon.

No line containing any of the sides will pass through the interior of the octagon, so it is convex.

All of the sides are congruent, so it is equilateral. All of the angles are congruent, so it is equiangular.

Since the polygon is convex, equilateral, and equiangular, it is regular. So this is a regular octagon.

> **ReadingMath**
> **Simple Closed Curves** Polygons and circles are examples of *simple closed curves*. Such a curve begins and ends at the same point without crossing itself. The figures below are *not* simple closed curves.
>
> 

GuidedPractice

1A.

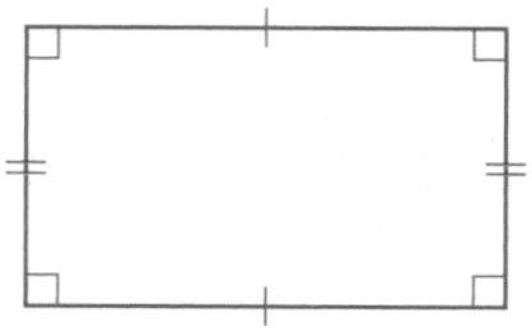

1B.

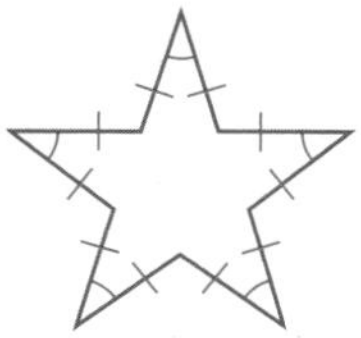

1C.

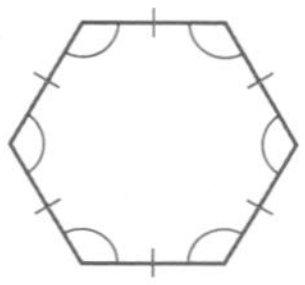

2 Perimeter, Circumference, and Area

The **perimeter** of a polygon is the sum of the lengths of the sides of the polygon. Some shapes have special formulas for perimeter, but all are derived from the basic definition of perimeter. The **circumference** of a circle is the distance around the circle.

The **area** of a figure is the number of square units needed to cover a surface. Review the formulas for the perimeter and area of three common polygons and circle given below.

KeyConcept Perimeter, Circumference, and Area

Triangle	Square	Rectangle	Circle
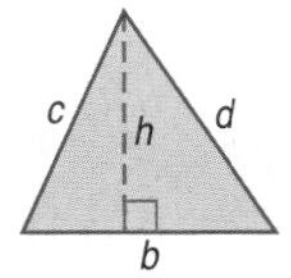	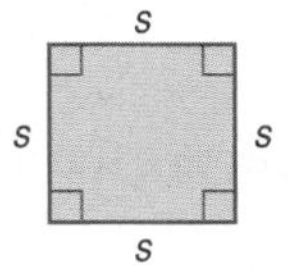	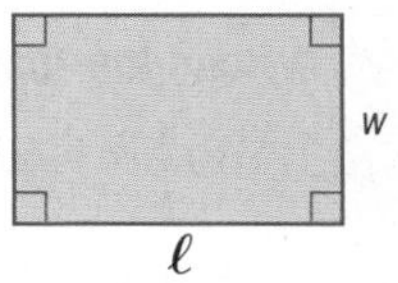	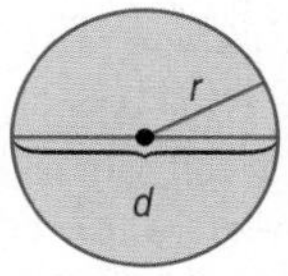
$P = b + c + d$	$P = s + s + s + s$ $= 4s$	$P = \ell + w + \ell + w$ $= 2\ell + 2w$	$C = 2\pi r$ or $C = \pi d$
$A = \frac{1}{2}bh$	$A = s^2$	$A = \ell w$	$A = \pi r^2$

P = perimeter of polygon; A = area of figure; C = circumference

b = base, h = height; ℓ = length, w = width; r = radius, d = diameter

ReadingMath

Pi The symbol π is read *pi*. This is not a variable but an irrational number. The most accurate way to perform a calculation with π is to use a calculator. If no calculator is available, 3.14 is a good estimate for π.

PT

Example 2 Find Perimeter and Area

Find the perimeter or circumference and area of each figure.

a.

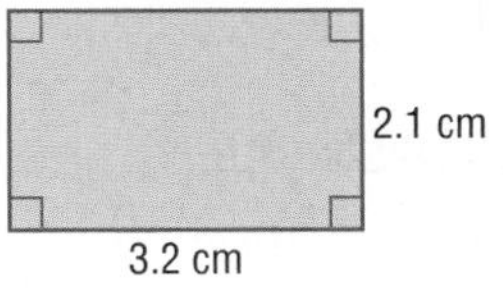

$P = 2\ell + 2w$ Perimeter of rectangle

$= 2(3.2) + 2(2.1)$ $\ell = 3.2$, $w = 2.1$

$= 10.6$ Simplify.

The perimeter is 10.6 centimeters.

$A = \ell w$ Area of rectangle

$= (3.2)(2.1)$ $\ell = 3.2$, $w = 2.1$

$= 6.72$ Simplify.

The area is about 6.7 square centimeters.

b.

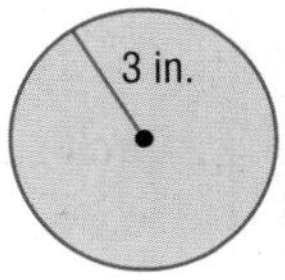

$C = 2\pi r$ Circumference

$= 2\pi(3)$ $r = 3$

≈ 18.85 Use a calculator.

The circumference is about 18.9 inches.

$A = \pi r^2$ Area of circle

$= \pi(3)^2$ $r = 3$

≈ 28.3 Use a calculator.

The area is about 28.3 square inches.

StudyTip

Perimeter vs. Area Since calculating the area of a figure involves multiplying two dimensions (unit × unit), *square units* are used. There is only one dimension used when finding the perimeter (the distance around), thus, it is given simply in *units*.

GuidedPractice

2A.

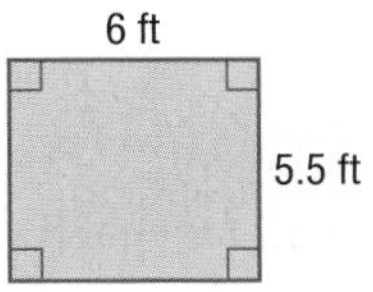

2B.

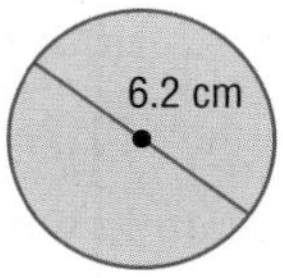

2C.

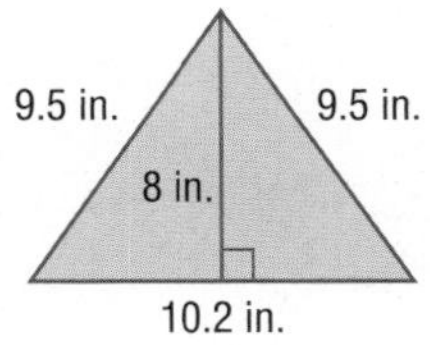

Standardized Test Example 3 Largest Area

Yolanda has 26 centimeters of cording to frame a photograph in her scrapbook. Which of these shapes would use *most* or all of the cording and enclose the *largest* area?

A right triangle with each leg about 7 centimeters long

B circle with a radius of about 4 centimeters

C rectangle with a length of 8 centimeters and a width of 4.5 centimeters

D square with a side length of 6 centimeters

Test-Taking Tip

Mental Math When you are asked to compare measures for varying figures, it can be helpful to use mental math. Estimate the perimeter or area of each figure, and then check your calculations.

Read the Test Item

You are asked to compare the area and perimeter of four different shapes.

Solve the Test Item

Find the perimeter and area of each shape.

Right Triangle

Use the Pythagorean Theorem to find the length of the hypotenuse.

$c^2 = a^2 + b^2$ — Pythagorean Theorem

$c^2 = 7^2 + 7^2$ or 98 — $a = 7, b = 7$

$c = \sqrt{98}$ or about 9.9 — Simplify.

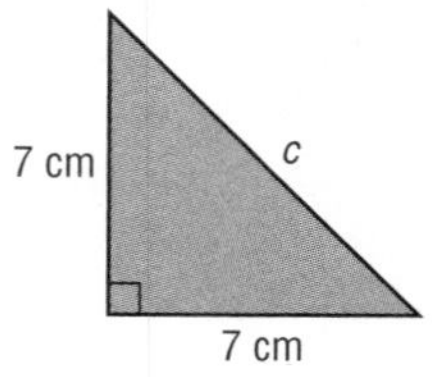

$P = a + b + c$ — Perimeter of a triangle

$\approx 7 + 7 + 9.9$ or about 23.9 cm — Substitution

$A = \frac{1}{2}bh$ — Area of a triangle

$= \frac{1}{2}(7)(7)$ or 24.5 cm^2 — Substitution

Study Tip

Irrational Measures Notice that the triangle perimeter given in Example 3 is only an approximation. Because the length of the hypotenuse is an irrational number, the actual perimeter of the triangle is the irrational measure $(14 + \sqrt{98})$ centimeters.

Circle

$C = 2\pi r$

$= 2\pi(4)$

≈ 25.1 cm

$A = \pi r^2$

$= \pi(4)^2$

≈ 50.3 cm^2

Rectangle

$P = 2\ell + 2w$

$= 2(8) + 2(4.5)$

$= 25$ cm

$A = \ell w$

$= (8)(4.5)$

$= 36$ cm^2

Square

$P = 4s$

$= 4(6)$

$= 24$ cm

$A = s^2$

$= 6^2$

$= 36$ cm^2

The shape that uses the most cording and encloses the largest area is the circle. The answer is B.

Guided Practice

3. Dasan has 32 feet of fencing to fence in a play area for his dog. Which shape of play area uses *most* or all of the fencing and encloses the *largest* area?

F circle with radius of about 5 feet

G rectangle with length 5 feet and width 10 feet

H right triangle with legs of length 10 feet each

J square with side length 8 feet

You can use the Distance Formula to find the perimeter of a polygon graphed on a coordinate plane.

Example 4 Perimeter and Area on the Coordinate Plane

COORDINATE GEOMETRY Find the perimeter and area of $\triangle PQR$ with vertices $P(-1, 3)$, $Q(-3, -1)$, and $R(4, -1)$.

Step 1 Find the perimeter of $\triangle PQR$.

Graph $\triangle PQR$.

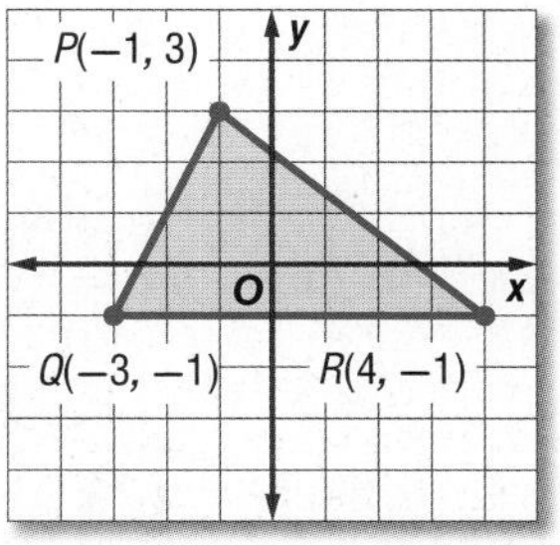

To find the perimeter of $\triangle PQR$, first find the lengths of each side. Counting the squares on the grid, we find that $QR = 7$ units. Use the Distance Formula to find the lengths of $\overline{PQ}$ and $\overline{PR}$.

$\overline{PQ}$ has endpoints at $P(-1, 3)$ and $Q(-3, -1)$.

$PQ = \sqrt{(x_2 - x_1)^2 + (y_2 - y_1)^2}$ Distance Formula

$= \sqrt{[-1 - (-3)]^2 + [3 - (-1)]^2}$ Substitute.

$= \sqrt{2^2 + 4^2}$ Subtract.

$= \sqrt{20}$ or about 4.5 Simplify.

$\overline{PR}$ has endpoints at $P(-1, 3)$ and $R(4, -1)$.

$PR = \sqrt{(x_2 - x_1)^2 + (y_2 - y_1)^2}$ Distance Formula

$= \sqrt{(-1 - 4)^2 + [3 - (-1)]^2}$ Substitute.

$= \sqrt{(-5)^2 + 4^2}$ Subtract.

$= \sqrt{41}$ or about 6.4 Simplify.

The perimeter of $\triangle PQR$ is $7 + \sqrt{20} + \sqrt{41}$ or about 17.9 units.

Step 2 Find the area of $\triangle PQR$.

To find the area of the triangle, find the lengths of the height and base. The height is the perpendicular distance from P to $\overline{QR}$. Counting squares on the graph, the height is 4 units. The length of $\overline{QR}$ is 7 units.

$A = \frac{1}{2}bh$ Area of a triangle

$= \frac{1}{2}(7)(4)$ or 14 Substitute and simplify.

The area of $\triangle PQR$ is 14 square units.

GuidedPractice

4. Find the perimeter and area of $\triangle ABC$ with vertices $A(-1, 4)$, $B(-1, -1)$, and $C(6, -1)$.

StudyTip

Linear and Square Units Remember to use linear units with perimeter and square units with area.

Check Your Understanding

Example 1 **Name each polygon by its number of sides. Then classify it as *convex* or *concave* and *regular* or *irregular*.**

1.

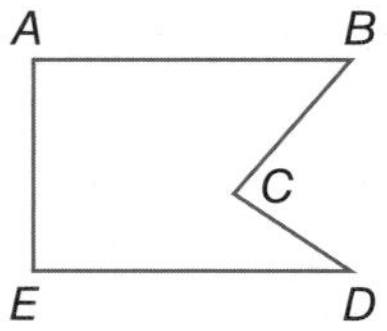

2.

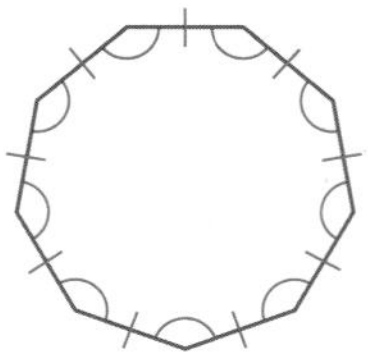

SIGNS **Identify the shape of each traffic sign and classify it as *regular* or *irregular*.**

3. stop

4. caution or warning

5. slow moving vehicle

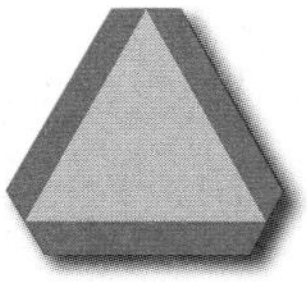

Example 2 **Find the perimeter or circumference and area of each figure. Round to the nearest tenth.**

6.

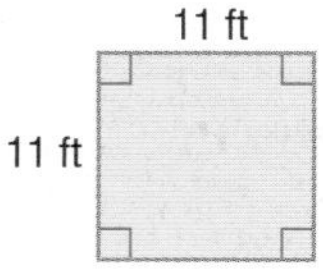

7.

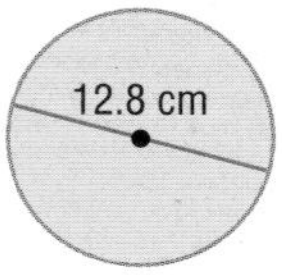

8.

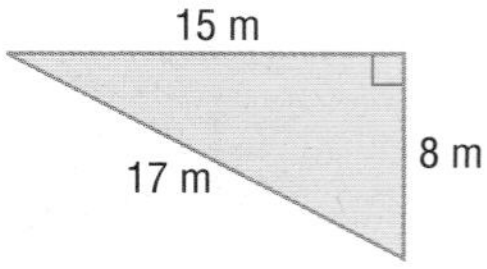

Example 3 **9. MULTIPLE CHOICE** Vanesa is making a banner for the game. She has 20 square feet of fabric. What shape will use *most* or all of the fabric?

A a square with a side length of 4 feet

B a rectangle with a length of 4 feet and a width of 3.5 feet

C a circle with a radius of about 2.5 feet

D a right triangle with legs of about 5 feet each

Example 4 **10.** CCSS **REASONING** Find the perimeter and area of $\triangle ABC$ with vertices $A(-1, 2)$, $B(3, 6)$, and $C(3, -2)$.

Practice and Problem Solving

Example 1 **Name each polygon by its number of sides. Then classify it as *convex* or *concave* and *regular* or *irregular*.**

11.

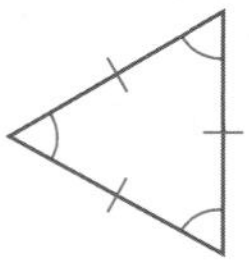

12.

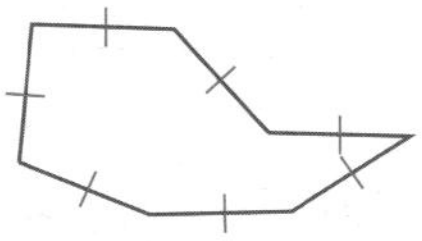

13.

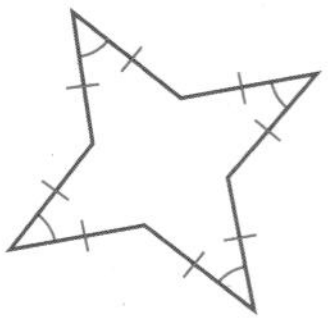

14.

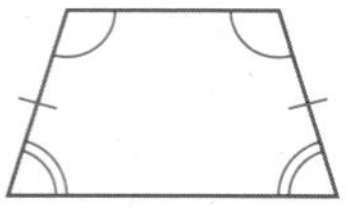

15.

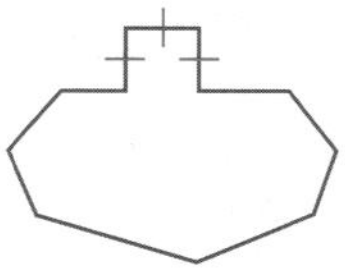

16.

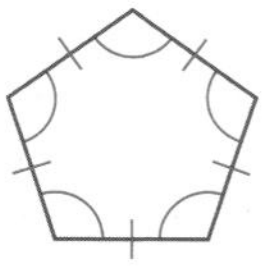

Examples 2–3 **Find the perimeter or circumference and area of each figure. Round to the nearest tenth.**

17.

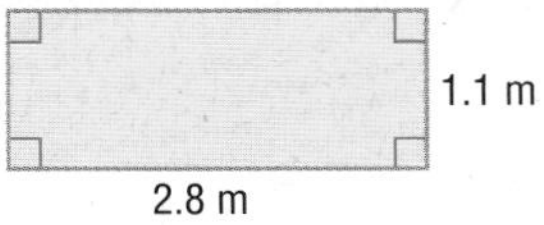

18.

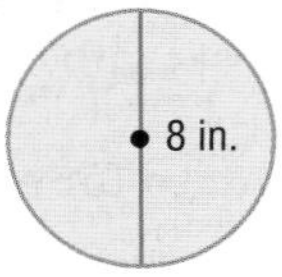

19.

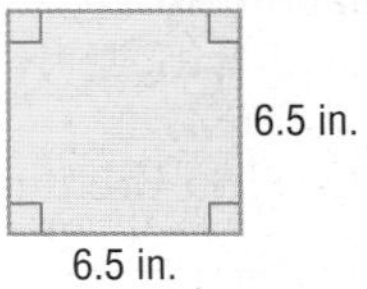

20.

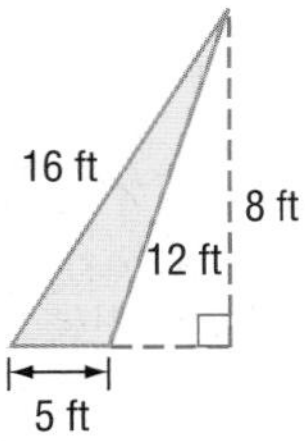

21

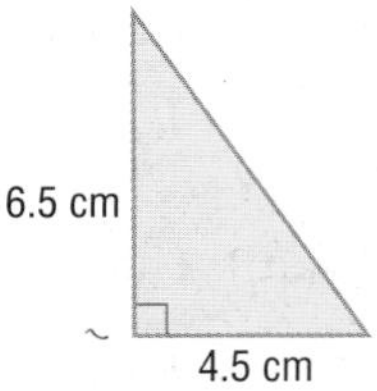

22.

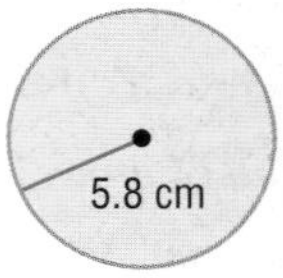

23. **CRAFTS** Joy has a square picture that is 4 inches on each side. The picture is framed with a length of ribbon. She wants to use the same piece of ribbon to frame a circular picture. What is the maximum radius of the circular frame?

24. **LANDSCAPING** Mr. Jackson has a circular garden with a diameter of 10 feet surrounded by edging. Using the same length of edging, he is going to create a square garden. What is the maximum side length of the square?

Example 4

CCSS REASONING **Graph each figure with the given vertices and identify the figure. Then find the perimeter and area of the figure.**

25. $D(-2, -2)$, $E(-2, 3)$, $F(2, -1)$

26. $J(-3, -3)$, $K(3, 2)$, $L(3, -3)$

27. $P(-1, 1)$, $Q(3, 4)$, $R(6, 0)$, $S(2, -3)$

28. $T(-2, 3)$, $U(1, 6)$, $V(5, 2)$, $W(2, -1)$

29. **CHANGING DIMENSIONS** Use the rectangle at the right.

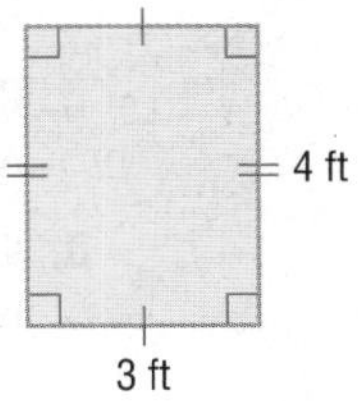

a. Find the perimeter of the rectangle.

b. Find the area of the rectangle.

c. Suppose the length and width of the rectangle are doubled. What effect would this have on the perimeter? the area? Justify your answer.

d. Suppose the length and width of the rectangle are halved. What effect does this have on the perimeter? the area? Justify your answer.

30. **CHANGING DIMENSIONS** Use the triangle at the right.

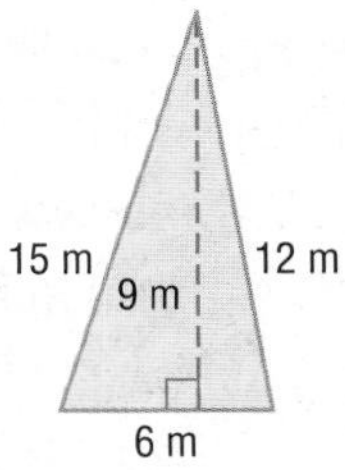

a. Find the perimeter of the triangle.

b. Find the area of the triangle.

c. Suppose the side lengths and height of the triangle were doubled. What effect would this have on the perimeter? the area? Justify your answer.

d. Suppose the side lengths and height of the triangle were divided by three. What effect would this have on the perimeter? the area? Justify your answer.

31. **ALGEBRA** A rectangle of area 360 square yards is 10 times as long as it is wide. Find its length and width.

32. **ALGEBRA** A rectangle of area 350 square feet is 14 times as wide as it is long. Find its length and width.

33 **DISC GOLF** The diameter of the most popular brand of flying disc used in disc golf measures between 8 and 10 inches. Find the range of possible circumferences and areas for these flying discs to the nearest tenth.

ALGEBRA **Find the perimeter or circumference for each figure described.**

34. The area of a square is 36 square units.

35. The length of a rectangle is half the width. The area is 25 square meters.

36. The area of a circle is 25π square units.

37. The area of a circle is 32π square units.

38. A rectangle's length is 3 times its width. The area is 27 square inches.

39. A rectangle's length is twice its width. The area is 48 square inches.

CCSS PRECISION **Find the perimeter in inches and area in square inches of each figure. Round to the nearest hundredth, if necessary.**

40.

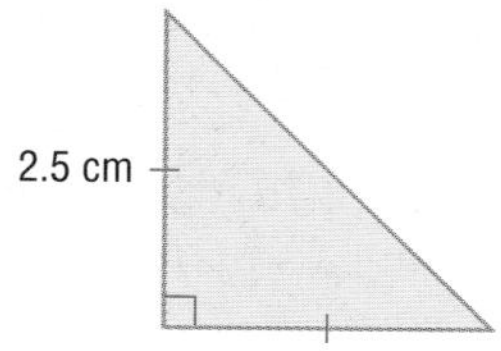

41.

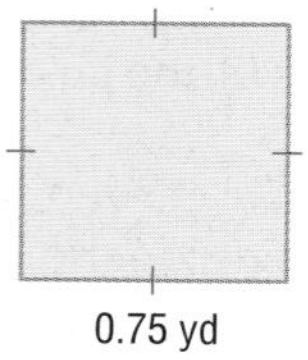

42.

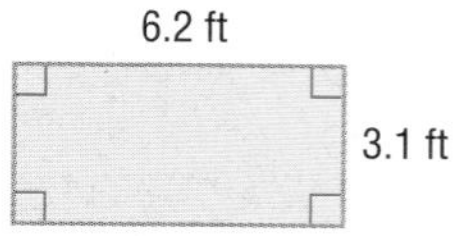

43. **MULTIPLE REPRESENTATIONS** Collect and measure the diameter and circumference of ten round objects using a millimeter measuring tape.

Object	d	C	$\frac{C}{d}$
1			
2			
3			
⋮			
10			

a. Tabular Record the measures in a table as shown.

b. Algebraic Compute the value of $\frac{C}{d}$ to the nearest hundredth for each object and record the result.

c. Graphical Make a scatter plot of the data with d-values on the horizontal axis and C-values on the vertical axis.

d. Verbal Find an equation for a line of best fit for the data. What does this equation represent? What does the slope of the line represent?

H.O.T. Problems Use Higher-Order Thinking Skills

44. **WHICH ONE DOESN'T BELONG?** Identify the term that does not belong with the other three. Explain your reasoning.

square | circle | triangle | pentagon

45. **CHALLENGE** The vertices of a rectangle with side lengths of 10 and 24 units are on a circle of radius 13 units. Find the area between the figures.

46. **REASONING** Name a polygon that is always regular and a polygon that is sometimes regular. Explain your reasoning.

47. **OPEN ENDED** Draw a pentagon. Is your pentagon *convex* or *concave*? Is your pentagon *regular* or *irregular*? Justify your answers.

48. **CHALLENGE** A rectangular room measures 20 feet by 12.5 feet. How many 5-inch square tiles will it take to cover the floor of this room? Explain.

49. **WRITING IN MATH** Describe two possible ways that a polygon can be equiangular but not a regular polygon.

Standardized Test Practice

50. Find the perimeter of the figure.

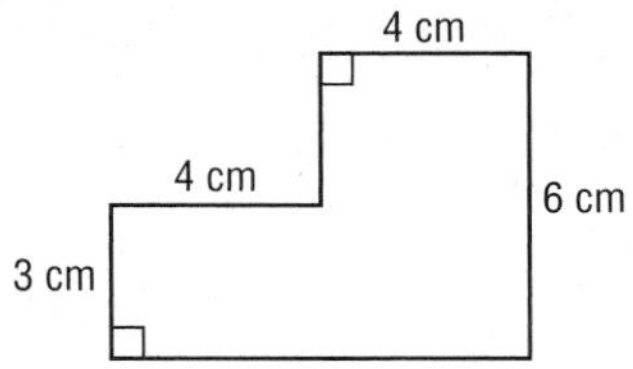

A 17 cm

B 25 cm

C 28 cm

D 31 cm

51. PROBABILITY In three successive rolls of a fair number cube, Matt rolls a 6. What is the probability of Matt rolling a 6 if the number cube is rolled a fourth time?

F $\frac{1}{6}$

G $\frac{1}{4}$

H $\frac{1}{3}$

J 1

52. SHORT RESPONSE Miguel is planning a party for 80 guests. According to the pattern in the table, how many gallons of ice cream should Miguel buy?

Number of Guests	Gallons of Ice Cream
8	2
16	4
24	6
32	8

53. SAT/ACT A frame 2 inches wide surrounds a painting that is 18 inches wide and 14 inches tall. What is the area of the frame?

A 68 in^2

B 84 in^2

C 144 in^2

D 252 in^2

E 396 in^2

Spiral Review

Determine whether each statement can be assumed from the figure. Explain. (Lesson 10-5)

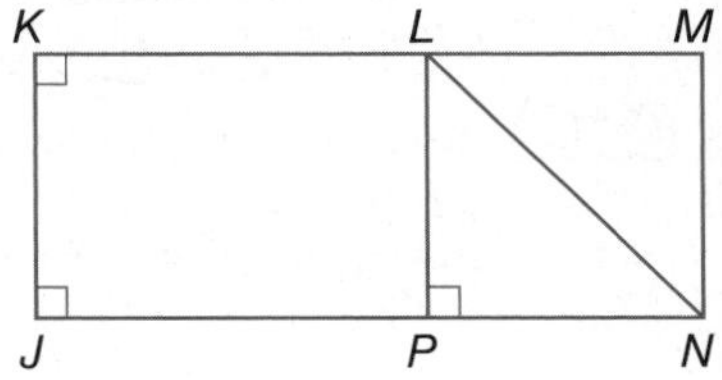

54. $\angle KJN$ is a right angle.

55. $\angle PLN \cong \angle NLM$

56. $\angle PNL$ and $\angle MNL$ are complementary.

57. $\angle KLN$ and $\angle MLN$ are supplementary.

58. TABLE TENNIS The diagram shows the angle of play for a table tennis player. If a right-handed player has a strong forehand, he should stand to the left of the center line of his opponent's angle of play. (Lesson 10-4)

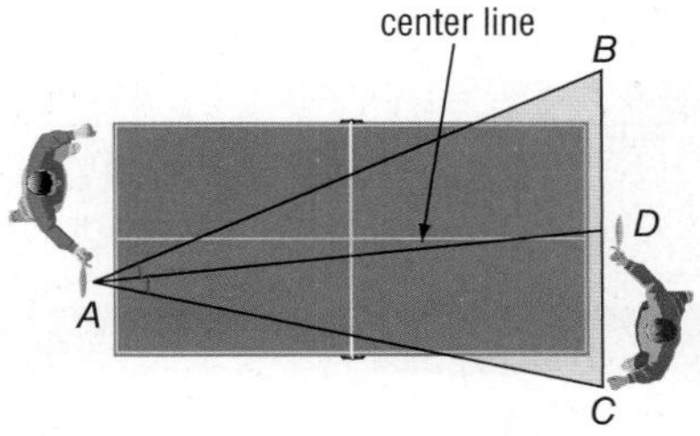

a. What geometric term describes the center line?

b. If the angle of play shown in the diagram measures 43°, what is $m\angle BAD$?

Skills Review

Evaluate each expression if $P = 10$, $B = 12$, $h = 6$, $r = 3$, and $\ell = 5$. Round to the nearest tenth, if necessary.

59. $\frac{1}{2}P\ell + B$

60. $\frac{1}{3}Bh$

61. $\frac{1}{3}\pi r^2 h$

62. $2\pi rh + 2\pi r^2$

EXTEND

10-6 Geometry Software Lab: Two-Dimensional Figures

You can use The Geometer's Sketchpad® to draw and investigate polygons.

CCSS Common Core State Standards
Content Standards
G.CO.12 Make formal geometric constructions with a variety of tools and methods (compass and straightedge, string, reflective devices, paper folding, dynamic geometric software, etc.).
Mathematical Practices 5

Activity 1 Draw a Polygon

Draw △*XYZ*.

Step 1 Select the segment tool from the toolbar, and click to set the first endpoint X of side $\overline{XY}$. Then drag the cursor, and click again to set the other endpoint Y.

Step 2 Click on point Y to set the endpoint of $\overline{YZ}$. Drag the cursor and click to set point Z.

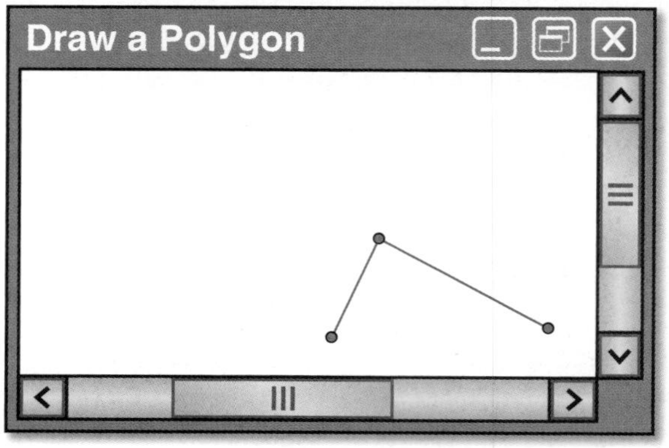

Step 3 Click on point Z to set the endpoint of $\overline{ZX}$. Then move the cursor to highlight point X. Click on X to draw $\overline{ZX}$.

Step 4 Use the pointer tool to click on points X, Y, and Z. Under the **Display** menu, select **Show Labels** to label the vertices of your triangle.

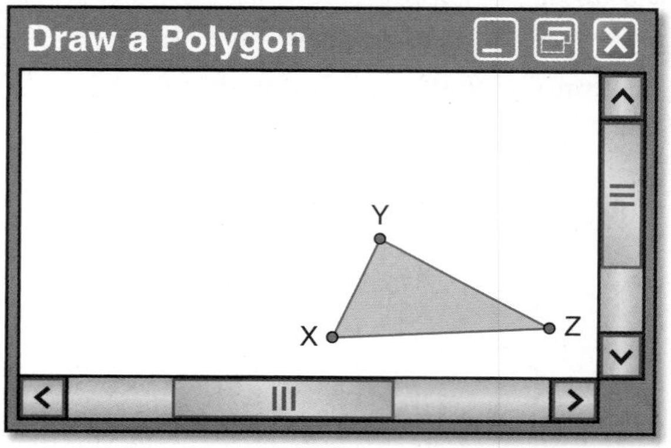

Activity 2 Measure Sides

Find *XY*, *YZ*, and *ZX*.

Step 1 Use the pointer tool to select $\overline{XY}$, $\overline{YZ}$, and $\overline{ZX}$.

Step 2 Select the **Length** command under the **Measure** menu to display the lengths of $\overline{XY}$, $\overline{YZ}$, and $\overline{ZX}$.

$XY = 1.79$ cm

$YZ = 3.11$ cm

$ZX = 3.48$ cm

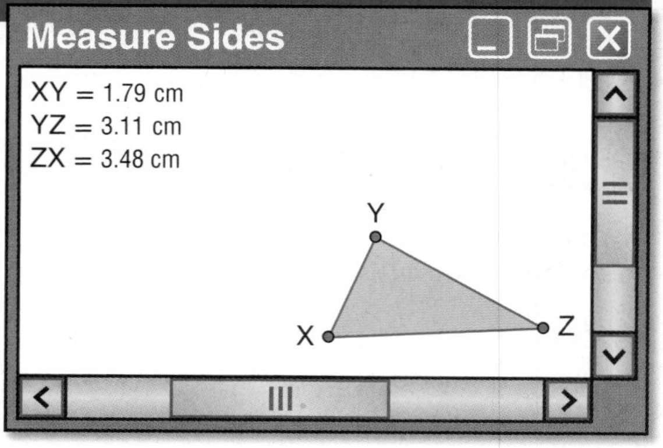

(continued on the next page)

Geometry Software Lab
Two-Dimensional Figures *Continued*

Activity 3 Find Perimeter

Find the perimeter of $\triangle XYZ$.

Step 1 Use the pointer tool to select points X, Y, and Z.

Step 2 Under the **Construct** menu, select **Triangle Interior**. The triangle will now be shaded.

Step 3 Select the triangle interior using the pointer.

Step 4 Choose the **Perimeter** command under the **Measure** menu to find the perimeter of $\triangle XYZ$.

The perimeter of $\triangle XYZ$ is 8.38 centimeters.

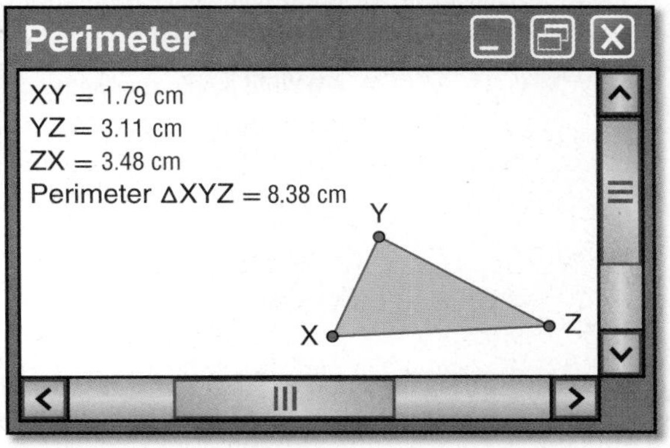

Activity 4 Measure Angles

Find $m\angle X$, $m\angle Y$, and $m\angle Z$.

Step 1 Recall that $\angle X$ can also be named $\angle YXZ$ or $\angle ZXY$. Use the pointer to select points Y, X, and Z in order.

Step 2 Select the **Angle** command from the **Measure** menu to find $m\angle X$.

Step 3 Select points X, Y, and Z. Find $m\angle Y$.

Step 4 Select points X, Z, and Y. Find $m\angle Z$.

$m\angle X = 63.16$, $m\angle Y = 86.05$, and $m\angle Z = 30.8$.

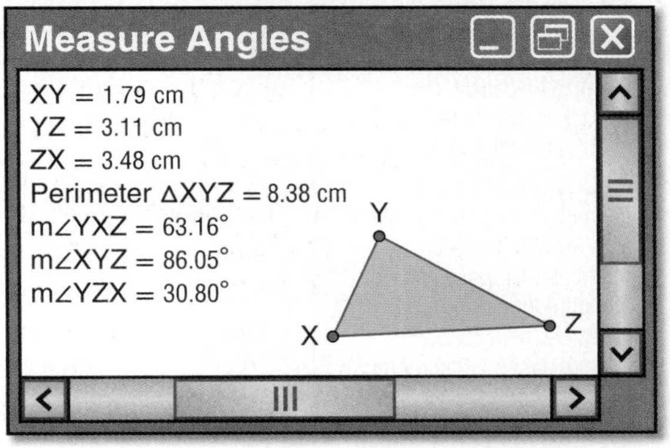

Analyze the Results

1. Add the side measures from Activity 2. How does this compare to the result in Activity 3?
2. What is the sum of the angle measures of $\triangle XYZ$?
3. Repeat the activities for each figure.

   a. irregular quadrilateral **b.** square **c.** pentagon **d.** hexagon

4. Draw another quadrilateral and find its perimeter. Then enlarge your figure using the **Dilate** command. How does changing the sides affect the perimeter?
5. Compare your results with those of your classmates.
6. Make a conjecture about the sum of the measures of the angles in any triangle.
7. What is the sum of the measures of the angles of a quadrilateral? pentagon? hexagon?
8. How are the sums of the angles of polygons related to the number of sides?
9. Test your conjecture on other polygons. Does your conjecture hold? Explain.
10. When the sides of a polygon are changed by a common factor, does the perimeter of the polygon change by the same factor as the sides? Explain.

LESSON 10-7 Proving Segment Relationships

Then

- You wrote algebraic and two-column proofs.

Now

1. Write proofs involving segment addition.
2. Write proofs involving segment congruence.

Why?

- Emma works at a fabric store after school. She measures a length of fabric by holding the straight edge of the fabric against a yardstick. To measure lengths such as 39 inches, which is longer than the yardstick, she marks a length of 36 inches. From the end of that mark, she measures an additional length of 3 inches. This ensures that the total length of fabric is 36 + 3 inches or 39 inches.

CCSS Common Core State Standards

Content Standards

G.CO.9 Prove theorems about lines and angles.

G.CO.12 Make formal geometric constructions with a variety of tools and methods (compass and straightedge, string, reflective devices, paper folding, dynamic geometric software, etc.).

Mathematical Practices

2 Reason abstractly and quantitatively.

3 Construct viable arguments and critique the reasoning of others.

1 Ruler Postulate In Lesson 10-2, you measured segments with a ruler by matching the mark for zero with one endpoint and then finding the number on the ruler that corresponded to the other endpoint. This illustrates the Ruler Postulate.

Postulate 10.1 Ruler Postulate

Words The points on any line or line segment can be put into one-to-one correspondence with real numbers.

Symbols Given any two points *A* and *B* on a line, if *A* corresponds to zero, then *B* corresponds to a positive real number.

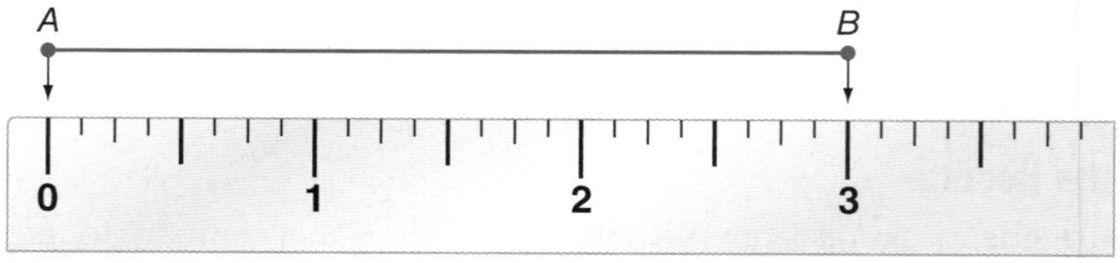

In Lesson 10-2, you also learned about what it means for a point to be *between* two other points. This relationship can be expressed as the Segment Addition Postulate.

Postulate 10.2 Segment Addition Postulate

Words If *A*, *B*, and *C* are collinear, then point *B* is between *A* and *C* if and only if $AB + BC = AC$.

Symbols

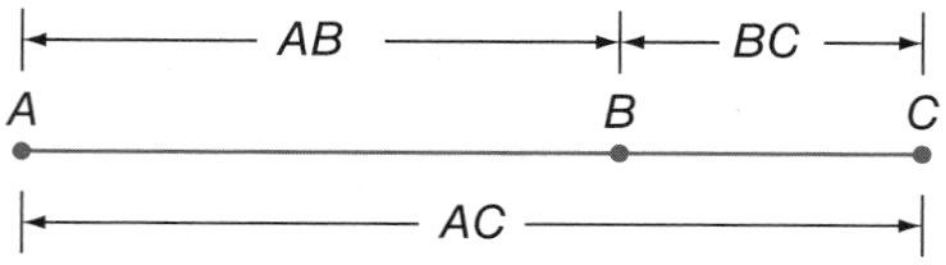

The Segment Addition Postulate is used as a justification in many geometric proofs.

Richard Nowitz/National Geographic/Getty Images

Example 1 Use the Segment Addition Postulate

Prove that if $\overline{CE} \cong \overline{FE}$ and $\overline{ED} \cong \overline{EG}$ then $\overline{CD} \cong \overline{FG}$.

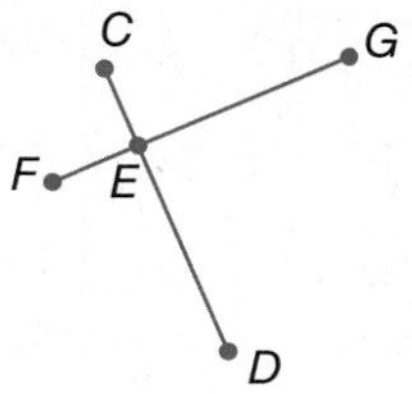

Given: $\overline{CE} \cong \overline{FE}$; $\overline{ED} \cong \overline{EG}$

Prove: $\overline{CD} \cong \overline{FG}$

Proof:

Statements	Reasons
1. $\overline{CE} \cong \overline{FE}$; $\overline{ED} \cong \overline{EG}$	**1.** Given
2. $CE = FE$; $ED = EG$	**2.** Definition of congruence
3. $CE + ED = CD$	**3.** Segment Addition Postulate
4. $FE + EG = CD$	**4.** Substitution (Steps 2 & 3)
5. $FE + EG = FG$	**5.** Segment Addition Postulate
6. $CD = FG$	**6.** Substitution (Steps 4 & 5)
7. $\overline{CD} \cong \overline{FG}$	**7.** Definition of congruence

ReadingMath

Substitution Property The Substitution Property of Equality is often just written as *Substitution*.

GuidedPractice

Copy and complete the proof.

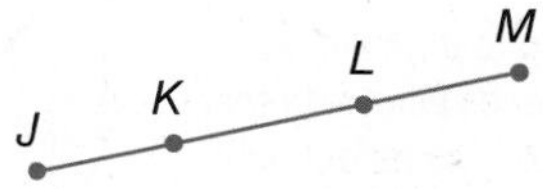

1. Given: $\overline{JL} \cong \overline{KM}$

Prove: $\overline{JK} \cong \overline{LM}$

Proof:

Statements	Reasons
a. $\overline{JL} \cong \overline{KM}$	**a.** Given
b. $JL = KM$	**b.** ___?___
c. $JK + KL =$ ___?___; $KL + LM =$ ___?___	**c.** Segment Addition Postulate
d. $JK + KL = KL + LM$	**d.** ___?___
e. $JK + KL - KL = KL + LM - KL$	**e.** Subtraction Property of Equality
f. ___?___	**f.** Substitution
g. $\overline{JK} \cong \overline{LM}$	**g.** Definition of congruence

2 Segment Congruence

Segment Congruence Segments with the same measure are congruent. Congruence of segments is also reflexive, symmetric, and transitive.

Theorem 10.1 Properties of Segment Congruence

Reflexive Property of Congruence	$\overline{AB} \cong \overline{AB}$
Symmetric Property of Congruence	If $\overline{AB} \cong \overline{CD}$, then $\overline{CD} \cong \overline{AB}$.
Transitive Property of Congruence	If $\overline{AB} \cong \overline{CD}$ and $\overline{CD} \cong \overline{EF}$, then $\overline{AB} \cong \overline{EF}$.

You will prove the Symmetric and Reflexive Properties in Exercises 6 and 7, respectively.

VocabularyLink

Symmetric

Everyday Use balanced or proportional

Math Use If $a = b$, then $b = a$.

Proof Transitive Property of Congruence

Given: $\overline{AB} \cong \overline{CD}$; $\overline{CD} \cong \overline{EF}$

Prove: $\overline{AB} \cong \overline{EF}$

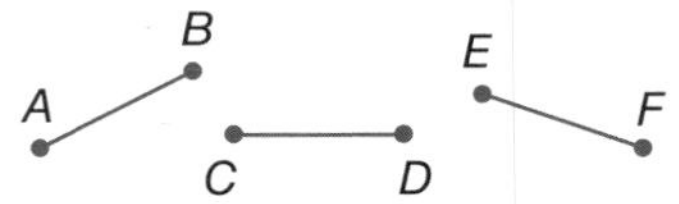

Paragraph Proof:

Since $\overline{AB} \cong \overline{CD}$ and $\overline{CD} \cong \overline{EF}$, $\overline{AB} = \overline{CD}$ and $\overline{CD} = \overline{EF}$ by the definition of congruent segments. By the Transitive Property of Equality, $\overline{AB} = \overline{EF}$. Thus, $\overline{AB} \cong \overline{EF}$ by the definition of congruence.

Real-WorldLink

According to a recent poll, 70% of teens who volunteer began doing so before age 12. Others said they would volunteer if given more opportunities to do so.

Source: Youth Service America

Real-World Example 2 Proof Using Segment Congruence

PT

VOLUNTEERING The route for a charity fitness run is shown. Checkpoints *X* and *Z* are the midpoints between the starting line and Checkpoint *Y* and Checkpoint *Y* and the finish line *F*, respectively. If Checkpoint *Y* is the same distance from Checkpoints *X* and *Z*, prove that the route from Checkpoint *Z* to the finish line is congruent to the route from the starting line to Checkpoint *X*.

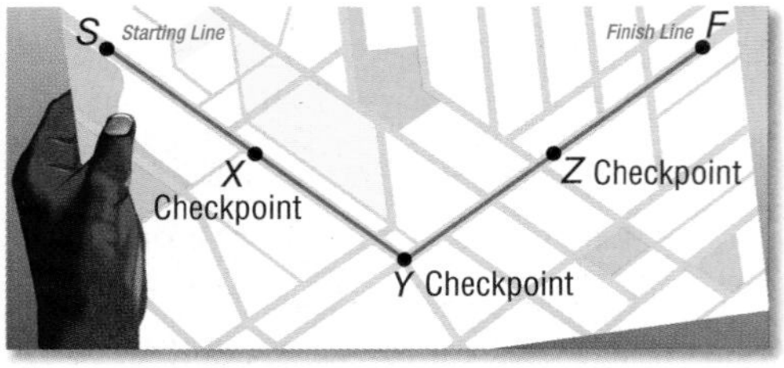

Given: X is the midpoint of $\overline{SY}$. Z is the midpoint of $\overline{YF}$. $XY = YZ$

Prove: $\overline{ZF} \cong \overline{SX}$

Two-Column Proof:

Statements	**Reasons**
1. X is the midpoint of $\overline{SY}$. Z is the midpoint of $\overline{YF}$. $XY = YZ$	**1.** Given
2. $\overline{SX} \cong \overline{XY}$; $\overline{YZ} \cong \overline{ZF}$	**2.** Definition of midpoint
3. $\overline{XY} \cong \overline{YZ}$	**3.** Definition of congruence
4. $\overline{SX} \cong \overline{YZ}$	**4.** Transitive Property of Congruence
5. $\overline{SX} \cong \overline{ZF}$	**5.** Transitive Property of Congruence
6. $\overline{ZF} \cong \overline{SX}$	**6.** Symmetric Property of Congruence

GuidedPractice

2. CARPENTRY A carpenter cuts a 2″ × 4″ board to a desired length. He then uses this board as a pattern to cut a second board congruent to the first. Similarly, he uses the second board to cut a third board and the third board to cut a fourth board. Prove that the last board cut has the same measure as the first.

Jim West/age fotostock

Check Your Understanding

Example 1

1. Copy and complete the proof.

Given: $\overline{AB} \cong \overline{FE}$, $\overline{BC} \cong \overline{ED}$

Prove: $\overline{AC} \cong \overline{FD}$

Proof:

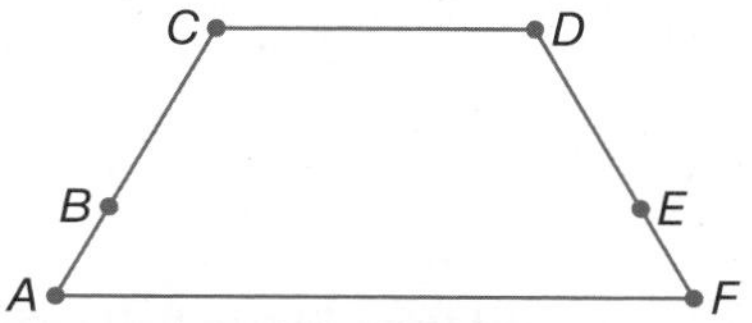

Statements	Reasons
a. $\overline{AB} \cong \overline{FE}$, $\overline{BC} \cong \overline{ED}$	**a.**
b.	**b.** Definition of congruent segments
c. $AB + FE = BC + ED$	**c.**
d.	**d.** Segment Addition Postulate
e. $AC = FD$	**e.**
f. $\overline{AC} \cong \overline{FD}$	**f.**

Example 2

2. PROOF Prove the following.

Given: $\overline{JK} \cong \overline{LM}$

Prove: $\overline{JL} \cong \overline{KM}$

3. PLIERS Refer to the diagram shown. $\overline{WP}$ is congruent to $\overline{YP}$, $\overline{ZP}$ is congruent to $\overline{XP}$. Prove that $WP + ZP = YP + XP$

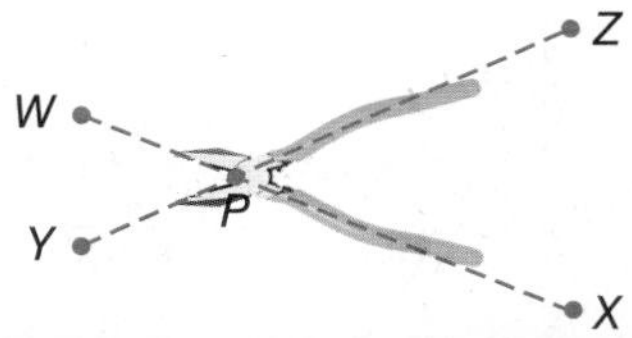

Practice and Problem Solving

Example 1

4. Copy and complete the proof.

Given: K is the midpoint of $\overline{HW}$

A is the midpoint of $\overline{ME}$

$\overline{HW} \cong \overline{ME}$

Prove: $\overline{HK} \cong \overline{MA}$

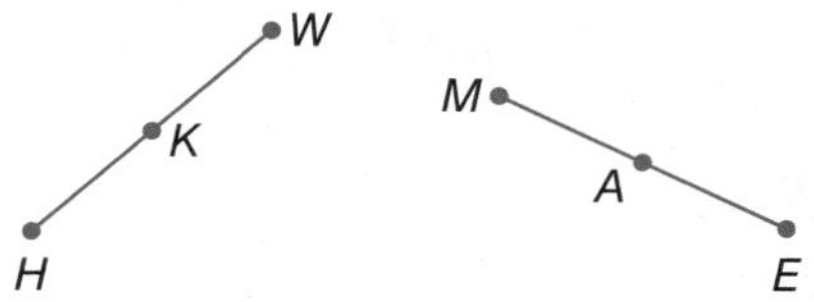

Statements	Reasons
a.	**a.** Given
b. $HK = KW$, $MA = AE$	**b.**
c. $HW = ME$	**c.**
d.	**d.** Segment Addition Postulate
e. $HK + KW = MA + AE$	**e.**
f. $HK + HK = MA + MA$	**f.**
g.	**g.** Simplify.
h.	**h.** Division Property of Equality
i. $\overline{HK} \cong \overline{MA}$	**i.**

Prove each theorem.

5. Symmetric Property of Congruence Theorem 10.1

6. Reflexive Property of Congruence Theorem 10.1

7. **TRAVEL** Kadoka, Rapid City, Sioux Falls, Alexandria, South Dakota are all connected by Interstate 90.

- Sioux Falls is 256 miles from Kadoka and 352 miles from Rapid City
- Rapid City is 96 miles from Kadoka and 292 miles from Alexandria

a. Draw a diagram to represent the locations of the cities in relation to each other and the distances between each city. Assume that Interstate 90 is straight.

b. Write a paragraph proof to support your conclusion.

PROOF **Prove the following.**

8. If $\overline{XW} \cong \overline{YZ}$ and $\overline{YZ} \cong \overline{ZX}$, then $\overline{XW} \cong \overline{ZX}$.

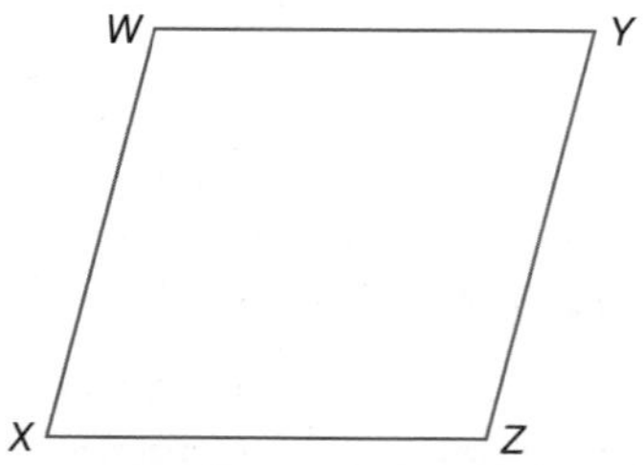

9. If $\overline{AC} \cong \overline{AD}$ and $\overline{ED} \cong \overline{BC}$, then $\overline{AE} \cong \overline{AB}$.

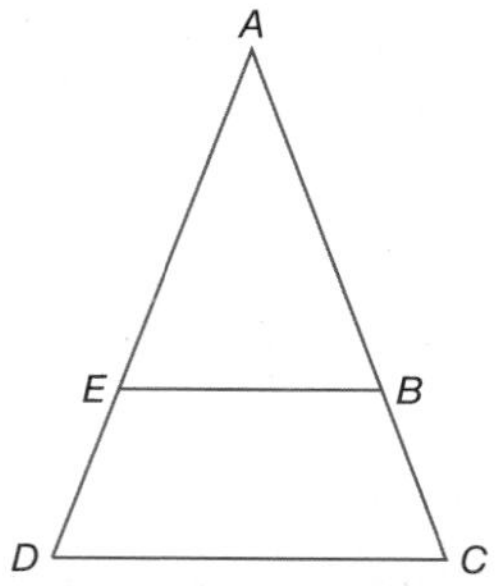

10. If R is the midpoint of $\overline{QS}$ and $\overline{PQ} \cong \overline{ST}$, then $\overline{PA} \cong \overline{RT}$.

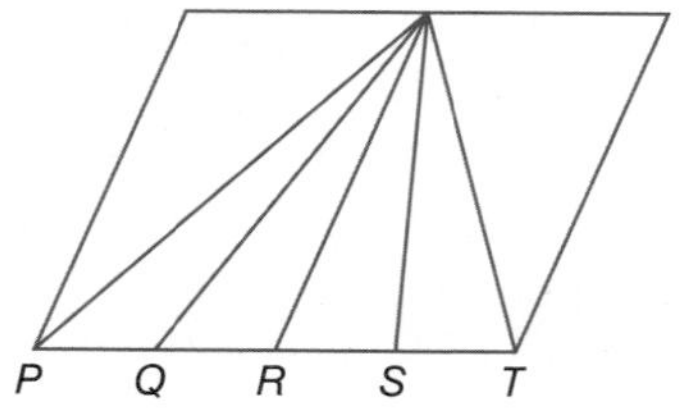

11. If Q is the midpoint of $\overline{PR}$, S is the midpoint of $\overline{RT}$, and $\overline{QR} \cong \overline{RS}$, then $PT = 4QR$.

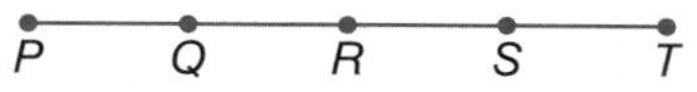

Example 1

12.

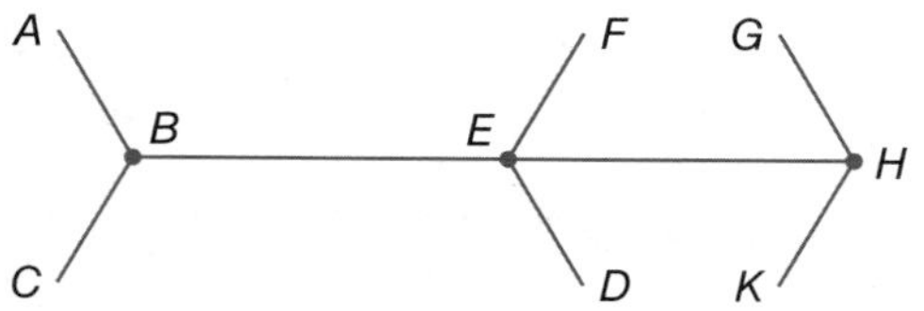

Given: $\overline{AB} \cong \overline{FE}$, $\overline{ED} \cong \overline{HK}$ and $AB + BE + ED = EF + EH + HK$

Prove: $\overline{BE} \cong \overline{EH}$

13. **CONSTRUCTION** Construct a segment that is twice as long as $\overline{PQ}$. Explain how the Segment Addition Postulate can be used to justify your construction.

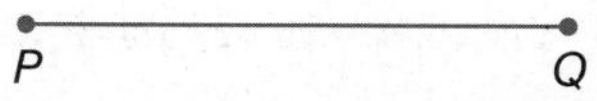

14. **MULTIPLE REPRESENTATIONS** A is the midpoint of $\overline{PQ}$, B is the midpoint of $\overline{PA}$, and C is the midpoint of $\overline{PB}$.

   a. **Geometric** Make a sketch to represent this situation.

   b. **Algebraic** Make a conjecture as to the algebraic relationship between PC and PQ.

   c. **Geometric** Copy segment $\overline{PQ}$ from your sketch. Then construct points B and C on $\overline{PQ}$. Explain how you can use your construction to support your conjecture.

   d. **Concrete** Use a ruler to draw a segment congruent to $\overline{PQ}$ from your sketch and to draw points B and C on $\overline{PQ}$. Use your drawing to support your conjecture.

   e. **Logical** Prove your conjecture.

H.O.T. Problems Use Higher-Order Thinking Skills

15. **CCSS ERROR ANALYSIS** In the diagram, $\overline{AB} \cong \overline{BC}$ and $\overline{BC} \cong \overline{DG}$. Examine the conclusions made by Mary and Susan. Is either of them correct? Explain your reasoning.

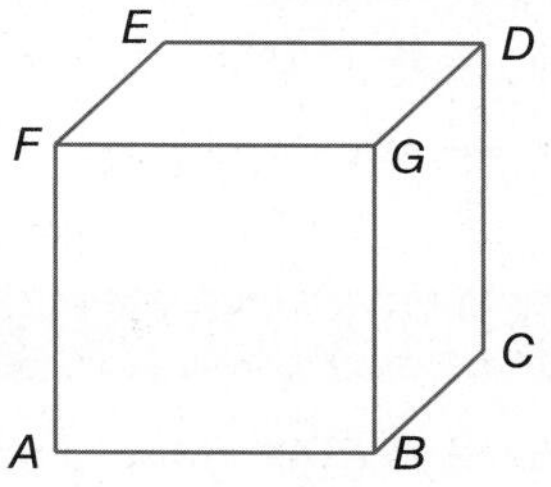

Mary

Since $\overline{AB} \cong \overline{BC}$ and $\overline{BC} \cong \overline{DG}$, then $\overline{AB} \cong \overline{DE}$ by the Transitive Property of Congruence.

Susan

Since $\overline{AB} \cong \overline{BC}$ and $\overline{BC} \cong \overline{DG}$, then $\overline{AB} \cong \overline{DG}$ by the Reflexive Property of Congruence.

16. **CHALLENGE** $ABCD$ is a rectangle. Prove that $\overline{AC} \cong \overline{BD}$.

17. **WRITING IN MATH** Does there exist a Subtraction Property of Congruence? Explain.

18. **REASONING** Classify the following statement as true or false. If false, provide a counterexample.

If, A, B, C, D, and E are collinear with B being the midpoint between A and C, C being the midpoint between B and D, and D being the midpoint between C and E, then AB = BC = DE.

19. **OPEN ENDED** Draw a representation of the Segment Addition Postulate in which the segment is $1\frac{1}{2}$ inches long, contains four collinear points, and contains no congruent segments.

20. **WRITING IN MATH** Compare and contrast paragraph proofs and two-column proofs.

Standardized Test Practice

21. ALGEBRA The chart below shows annual recycling by material in the United States. About how many pounds of aluminum are recycled each year?

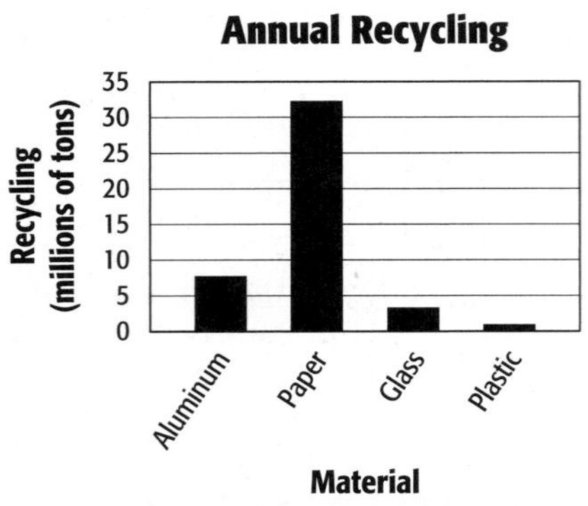

A 7.5

B 15,000

C 7,500,000

D 15,000,000,000

22. ALGEBRA Which expression is equivalent to $\frac{12x^{-4}}{4x^{-8}}$?

F $\frac{1}{3x^4}$

G $3x^4$

H $8x^2$

J $\frac{x^4}{3}$

23. SHORT RESPONSE The measures of two complementary angles are in the ratio 4 : 1. What is the measure of the smaller angle?

24. SAT/ACT Julie can word process 40 words per minute. How many minutes will it take Julie to word process 200 words?

A 0.5

B 2

C 5

D 10

E 12

Spiral Review

25. GEOMETRY If the side length of a cube is s, the volume is represented by s^3, and the surface area is represented by $6s^2$. (Lessons 7-1 and 7-2)

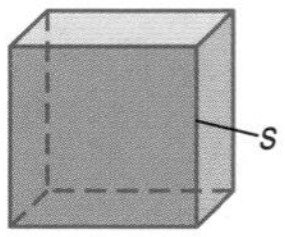

a. Are the expressions for volume and surface area monomials? Explain.

b. If the side of a cube measures 3 feet, find the volume and surface area.

c. Find a side length s such that the volume and surface area have the same measure.

d. The volume of a cylinder can be found by multiplying the radius squared times the height times π, or $V = \pi r^2 h$. Suppose you have two cylinders. Each measure of the second is twice the measure of the first, so $V = \pi(2r)^2(2h)$. What is the ratio of the volume of the first cylinder to the second cylinder?

26. PATTERN BLOCKS Pattern blocks can be arranged to fit in a circular pattern without leaving spaces. Remember that the measurement around a full circle is 360°. Determine the degree measure of the numbered angles shown below. (Lesson 1-4)

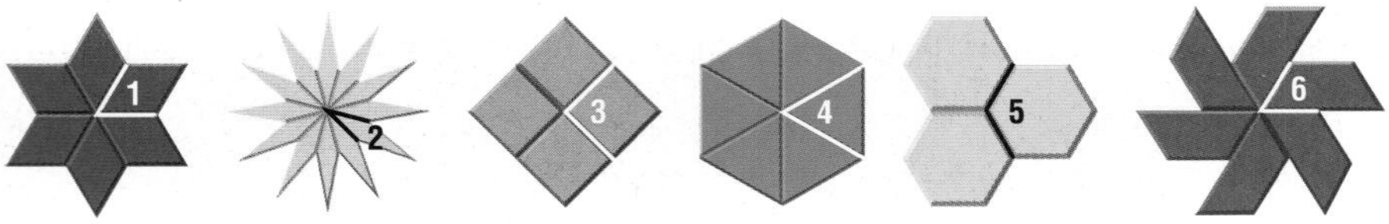

Skills Review

ALGEBRA Find x.

27.

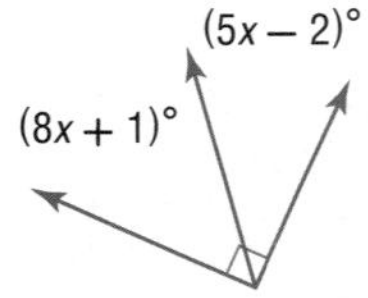

28.

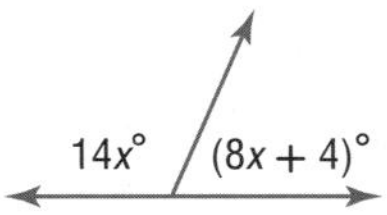

29.

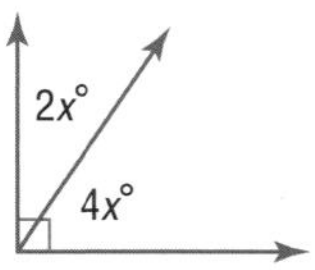

LESSON

10-8 Proving Angle Relationships

Then

- You identified and used special pairs of angles.

Now

1. Write proofs involving supplementary and complementary angles.
2. Write proofs involving congruent and right angles.

Why?

- Jamal's school is building a walkway that will include bricks with the names of graduates from each class. All of the bricks are rectangular, so when the bricks are laid, all of the angles form linear pairs.

CCSS Common Core State Standards

Content Standards

G.CO.9 Prove theorems about lines and angles.

Mathematical Practices

3 Construct viable arguments and critique the reasoning of others.

6 Attend to precision.

1 Supplementary and Complementary Angles

The Protractor Postulate illustrates the relationship between angle measures and real numbers.

Postulate 10.3 Protractor Postulate

Words Given any angle, the measure can be put into one-to-one correspondence with real numbers between 0 and 180.

Example If $\overrightarrow{BA}$ is placed along the protractor at 0°, then the measure of $\angle ABC$ corresponds to a positive real number.

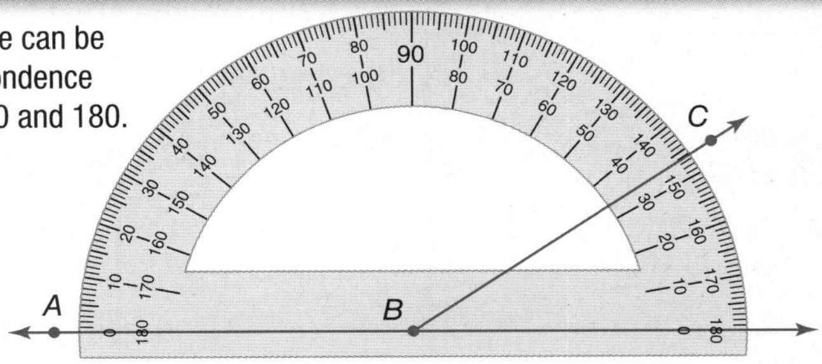

In Lesson 10-7, you learned about the Segment Addition Postulate. A similar relationship exists between the measures of angles.

Postulate 10.4 Angle Addition Postulate

D is in the interior of $\angle ABC$ if and only if $m\angle ABD + m\angle DBC = m\angle ABC$.

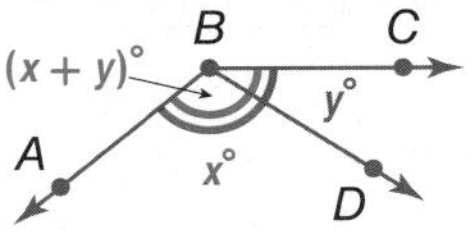

PT

Example 1 Use the Angle Addition Postulate

Find $m\angle 1$ if $m\angle 2 = 56$ and $m\angle JKL = 145$.

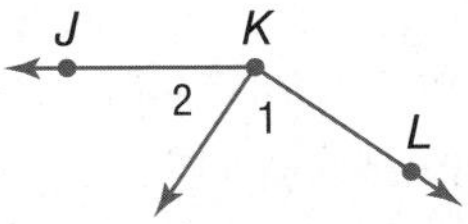

$m\angle 1 + m\angle 2 = m\angle JKL$	Angle Addition Postulate
$m\angle 1 + 56 = 145$	$m\angle 2 = 56$ $m\angle JKL = 145$
$m\angle 1 + 56 - 56 = 145 - 56$	Subtraction Property of Equality
$m\angle 1 = 89$	Substitution

GuidedPractice

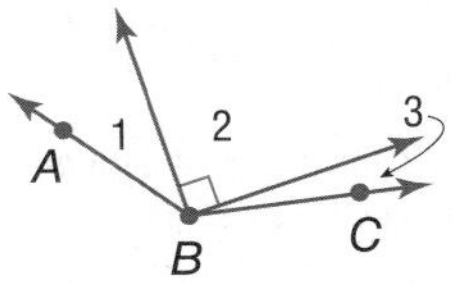

1. If $m\angle 1 = 23$ and $m\angle ABC = 131$, find the measure of $\angle 3$. Justify each step.

The Angle Addition Postulate can be used with other angle relationships to provide additional theorems relating to angles.

StudyTip

Linear Pair Theorem The Supplement Theorem may also be known as the *Linear Pair Theorem.*

Theorems

10.2 Supplement Theorem If two angles form a linear pair, then they are supplementary angles.

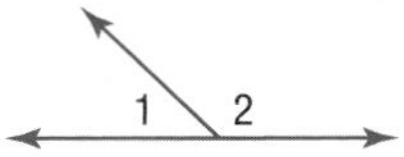

Example $m\angle 1 + m\angle 2 = 180$

10.3 Complement Theorem If the noncommon sides of two adjacent angles form a right angle, then the angles are complementary angles.

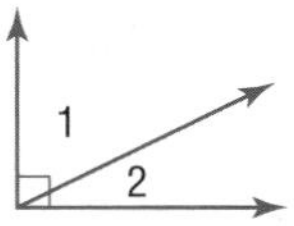

Example $m\angle 1 + m\angle 2 = 90$

You will prove Theorems 10.2 and 10.3 in Exercises 16 and 17, respectively.

PT

Real-World Example 2 Use Supplement or Complement

SURVEYING Using a transit, a surveyor sights the top of a hill and records an angle measure of about 73°. What is the measure of the angle the top of the hill makes with the horizon? Justify each step.

Understand Make a sketch of the situation. The surveyor is measuring the angle of his line of sight below the vertical. Draw a vertical ray and a horizontal ray from the point where the surveyor is sighting the hill, and label the angles formed. We know that the vertical and horizontal rays form a right angle.

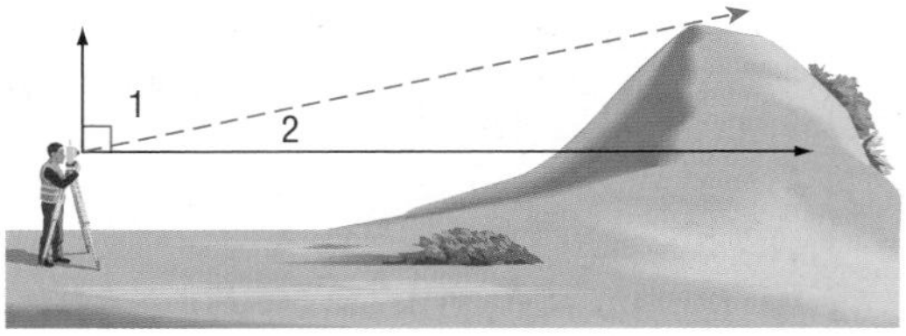

Plan Since $\angle 1$, and $\angle 2$ form a right angle, you can use the Complement Theorem.

Solve

$m\angle 1 + m\angle 2 = 90$	Complement Theorem
$73 + m\angle 2 = 90$	$m\angle 1 = 73$
$73 + m\angle 2 - 73 = 90 - 73$	Subtraction Property of Equality
$m\angle 2 = 17$	Substitution

The top of the hill makes a 17° angle with the horizon.

Check Since we know that the sum of the angles should be 90, check your math. The sum of 17 and 73 is 90. ✓

ReviewVocabulary

supplementary angles two angles with measures that add to 180

complementary angles two angles with measures that add to 90

linear pair a pair of adjacent angles with noncommon sides that are opposite rays

GuidedPractice

2. $\angle 6$ and $\angle 7$ form linear pair. If $m\angle 6 = 3x + 32$ and $m\angle 7 = 5x + 12$, find x, $m\angle 6$, and $m\angle 7$. Justify each step.

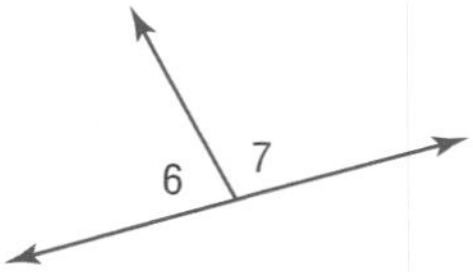

2 Congruent Angles The properties of algebra that applied to the congruence of segments and the equality of their measures also hold true for the congruence of angles and the equality of their measures.

Theorem 10.4 Properties of Angle Congruence

Reflexive Property of Congruence

$\angle 1 \cong \angle 1$

Symmetric Property of Congruence

If $\angle 1 \cong \angle 2$, then $\angle 2 \cong \angle 1$.

Transitive Property of Congruence

If $\angle 1 \cong \angle 2$ and $\angle 2 \cong \angle 3$, then $\angle 1 \cong \angle 3$.

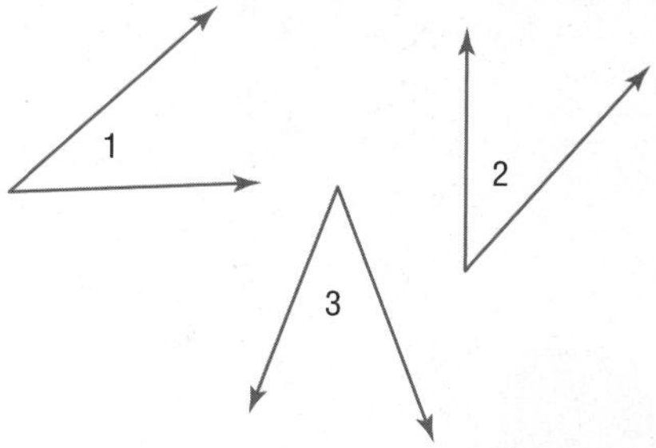

You will prove the Reflexive and Transitive Properties of Congruence in Exercises 18 and 19, respectively.

Proof Symmetric Property of Congruence

Given: $\angle A \cong \angle B$

Prove: $\angle B \cong \angle A$

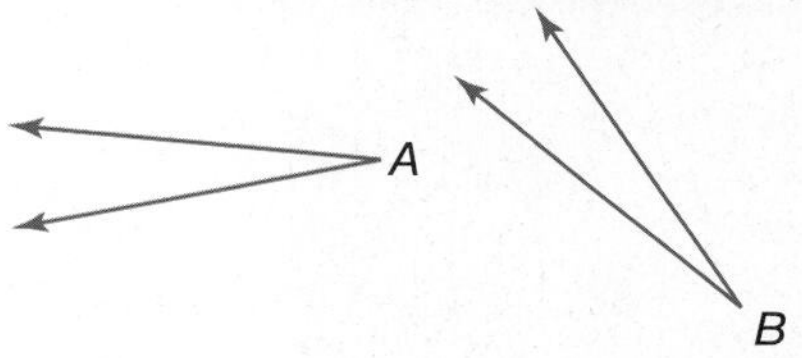

Paragraph Proof:

We are given $\angle A \cong \angle B$. By the definition of congruent angles, $m\angle A = m\angle B$. Using the Symmetric Property of Equality, $m\angle B = m\angle A$. Thus, $\angle B \cong \angle A$ by the definition of congruent angles.

Algebraic properties can be applied to prove theorems for congruence relationships involving supplementary and complementary angles.

Theorems

10.5 Congruent Supplements Theorem
Angles supplementary to the same angle or to congruent angles are congruent.

Abbreviation *$\angle$s suppl. to same $\angle$ or $\cong$ $\angle$s are $\cong$.*

Example If $m\angle 1 + m\angle 2 = 180$ and $m\angle 2 + m\angle 3 = 180$, then $\angle 1 \cong \angle 3$.

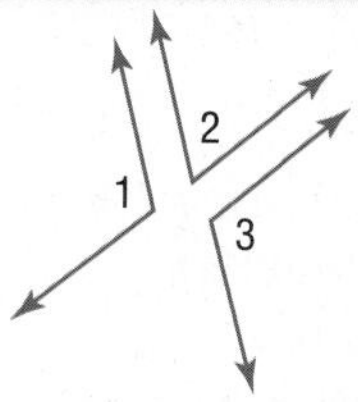

10.6 Congruent Complements Theorem
Angles complementary to the same angle or to congruent angles are congruent.

Abbreviation *$\angle$s compl. to same $\angle$ or $\cong$ $\angle$s are $\cong$.*

Example If $m\angle 4 + m\angle 5 = 90$ and $m\angle 5 + m\angle 6 = 90$, then $\angle 4 \cong \angle 6$.

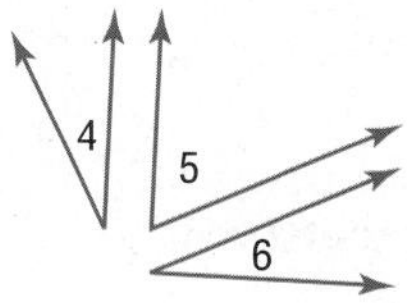

ReadingMath

Abbreviations and Symbols
The notation $\angle$s means *angles.*

You will prove one case of Theorem 10.5 in Exercise 6.

Proof One Case of the Congruent Supplements Theorem

Given: ∠1 and ∠2 are supplementary.
∠2 and ∠3 are supplementary.

Prove: ∠1 ≅ ∠3

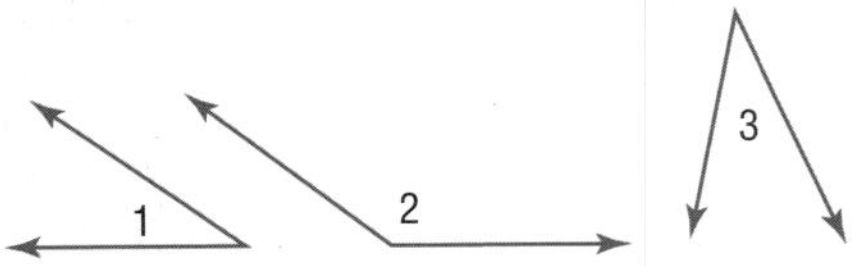

Proof:

Statements	Reasons
1. ∠1 and ∠2 are supplementary. ∠2 and ∠3 are supplementary.	**1.** Given
2. $m\angle 1 + m\angle 2 = 180$; $m\angle 2 + m\angle 3 = 180$	**2.** Definition of supplementary angles
3. $m\angle 1 + m\angle 2 = m\angle 2 + m\angle 3$	**3.** Substitution
4. $m\angle 2 = m\angle 2$	**4.** Reflexive Property
5. $m\angle 1 = m\angle 3$	**5.** Subtraction Property
6. ∠1 ≅ ∠3	**6.** Definition of congruent angles

Real-WorldLink

The 100-story John Hancock Building uses huge X-braces in its design. These diagonals are connected to the exterior columns, making it possible for strong wind forces to be carried from the braces to the exterior columns and back.

Source: PBS

Example 3 Proofs Using Congruent Comp. or Suppl. Theorems

PT

Prove that vertical angles 2 and 4 in the photo at the left are congruent.

Given: ∠2 and ∠4 are vertical angles.

Prove: ∠2 ≅ ∠4

Proof:

Statements	Reasons
1. ∠2 and ∠4 are vertical angles.	**1.** Given
2. ∠2 and ∠4 are nonadjacent angles formed by intersecting lines.	**2.** Definition of vertical angles
3. ∠2 and ∠3 from a linear pair. ∠3 and ∠4 form a linear pair.	**3.** Definition of a linear pair
4. ∠2 and ∠3 are supplementary. ∠3 and ∠4 are supplementary.	**4.** Supplement Theorem
5. ∠2 ≅ ∠4	**5.** ∠s suppl. to same ∠ or ≅ ∠s are ≅.

GuidedPractice

3. In the figure, ∠*ABE* and ∠*DBC* are right angles. Prove that ∠*ABD* ≅ ∠*EBC*.

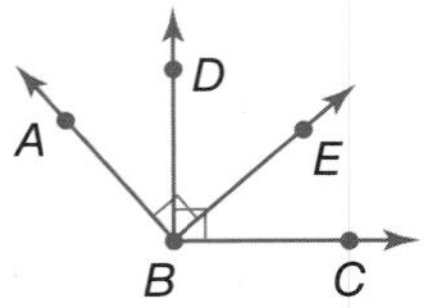

ReviewVocabulary

Vertical Angles two nonadjacent angles formed by intersecting lines

Note that in Example 3, ∠1 and ∠3 are vertical angles. The conclusion in the example supports the following Vertical Angles Theorem.

Theorem 10.7 Vertical Angles Theorem

If two angles are vertical angles, then they are congruent.

Abbreviation *Vert. ∠s are ≅.*

Example ∠1 ≅ ∠3 and ∠2 ≅ ∠4

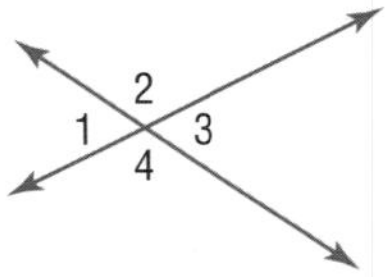

You will prove Theorem 10.7 in Exercise 28.

Yannis Emmanuel Mavromatakis/Alamy

Example 4 Use Vertical Angles

Prove that if $\overrightarrow{DB}$ bisects $\angle ADC$, then $\angle 2 \cong \angle 3$.

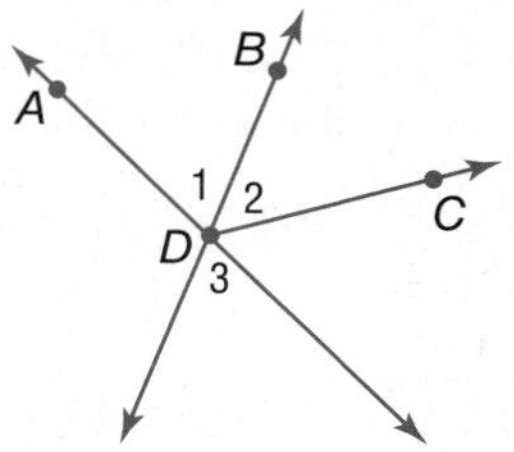

Given: $\overrightarrow{DB}$ bisects $\angle ADC$.

Prove: $\angle 2 \cong \angle 3$

Proof:

Statements	Reasons
1. $\overrightarrow{DB}$ bisects $\angle ADC$.	**1.** Given
2. $\angle 1 \cong \angle 2$	**2.** Definition of angle bisector
3. $\angle 1$ and $\angle 3$ are vertical angles.	**3.** Definition of vertical angles
4. $\angle 3 \cong \angle 1$	**4.** Vert. ∠s are ≅.
5. $\angle 3 \cong \angle 2$	**5.** Transitive Property of Congruence
6. $\angle 2 \cong \angle 3$	**6.** Symmetric Property of Congruence

Guided Practice

4. If $\angle 3$ and $\angle 4$ are vertical angles, $m\angle 3 = 6x + 2$, and $m\angle 4 = 8x - 14$, find $m\angle 3$ and $m\angle 4$. Justify each step.

The theorems in this lesson can be used to prove the following right angle theorems.

Theorems Right Angle Theorems

Theorem	Example
10.8 Perpendicular lines intersect to form four right angles. **Example** If $\overrightarrow{AC} \perp \overrightarrow{DB}$, then $\angle 1$, $\angle 2$, $\angle 3$, and $\angle 4$ are rt. ∠s.	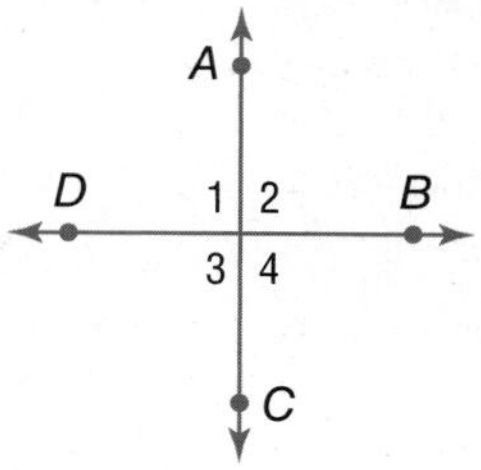
10.9 All right angles are congruent. **Example** If $\angle 1$, $\angle 2$, $\angle 3$, and $\angle 4$ are rt. ∠s, then $\angle 1 \cong \angle 2 \cong \angle 3 \cong \angle 4$.	
10.10 Perpendicular lines form congruent adjacent angles. **Example** If $\overrightarrow{AC} \perp \overrightarrow{DB}$, then $\angle 1 \cong \angle 2$, $\angle 2 \cong \angle 4$, $\angle 3 \cong \angle 4$, and $\angle 1 \cong \angle 3$.	
10.11 If two angles are congruent and supplementary, then each angle is a right angle. **Example** If $\angle 5 \cong \angle 6$ and $\angle 5$ is suppl. to $\angle 6$, then $\angle 5$ and $\angle 6$ are rt. ∠s.	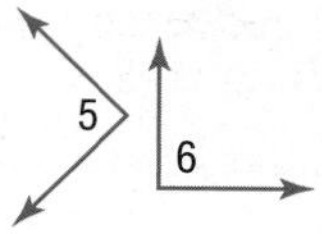
10.12 If two congruent angles form a linear pair, then they are right angles. **Example** If $\angle 7$ and $\angle 8$ form a linear pair, then $\angle 7$ and $\angle 8$ are rt. ∠s.	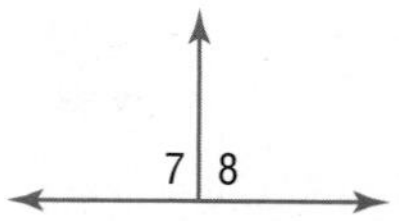

You will prove Theorems 10.8–10.12 in Exercises 22–26.

Check Your Understanding

Example 1 **Find the measure of each numbered angle, and name the theorems that justify your work.**

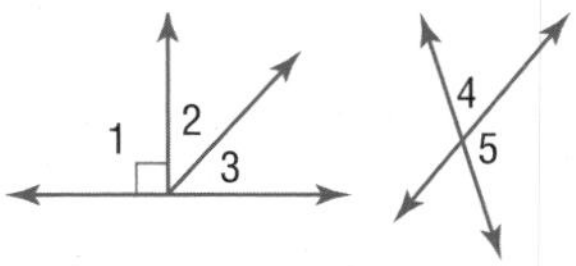

1. $m\angle 2 = 36$

2. $m\angle 2 = x$, $m\angle 3 = x + 6$

3. $m\angle 4 = 2x$, $m\angle 5 = x - 9$

4. $m\angle 4 = 3(x - 7)$, $m\angle 5 = x + 25$

Example 2 **5. PARKING** Refer to the diagram of the parking lot at the right. Given that $\angle 1 \cong \angle 5$, prove that $\angle 3 \cong \angle 7$.

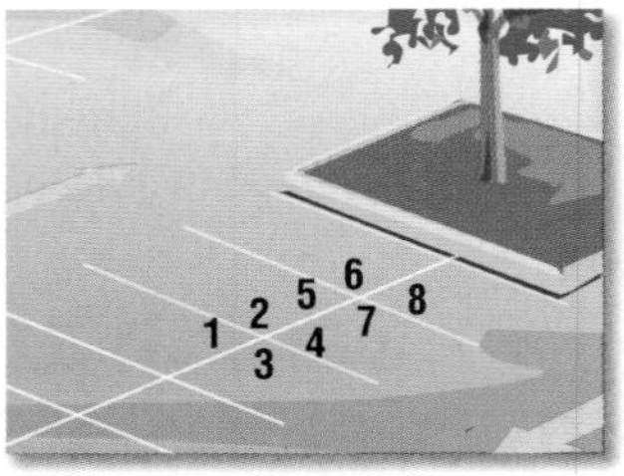

Example 3 **6. PROOF** **Copy and complete the proof of one case of Theorem 10.6.**

Given: $\angle 1$ and $\angle 3$ are complementary.
$\angle 2$ and $\angle 4$ are complementary.
$\angle 3 \cong \angle 4$

Prove: $\angle 1 \cong \angle 2$

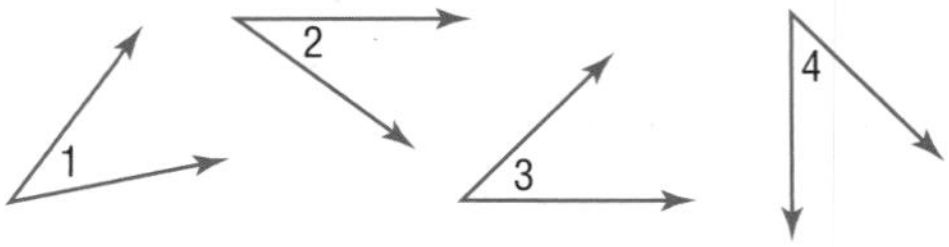

Proof:

Statements	Reasons
a. $\angle 1$ and $\angle 3$ are complementary $\angle 2$ and $\angle 4$ are complementary $\angle 3 \cong \angle 4$	**a.**
b. $m\angle 1 + m\angle 3 = 90$ $m\angle 2 + m\angle 4 = 90$	**b.**
c. $m\angle 1 + m\angle 3 = m\angle 2 + m\angle 4$	**c.**
d.	**d.** Definition of Congruence
e. $m\angle 1 = m\angle 2$	**e.**
f. $\angle 1 \cong \angle 2$	**f.**

Example 4 **7. PROOF** **Write a two-column proof.**

Given: $\angle 4 \cong \angle 6$

Prove: $\angle 5 \cong \angle 7$

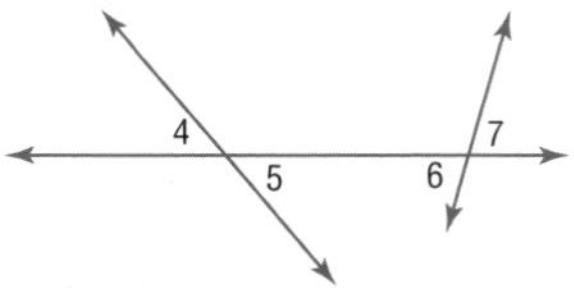

Practice and Problem Solving

Examples 1–3 **Find the measure of each numbered angle, and name the theorems used that justify your work.**

8. $m\angle 1 = m\angle 2$

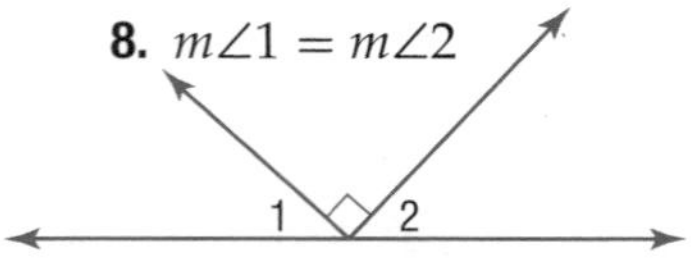

9. $\angle 2$ and $\angle 3$ are complementary $\angle 1 \cong \angle 4$ and $m\angle 3 = 18$

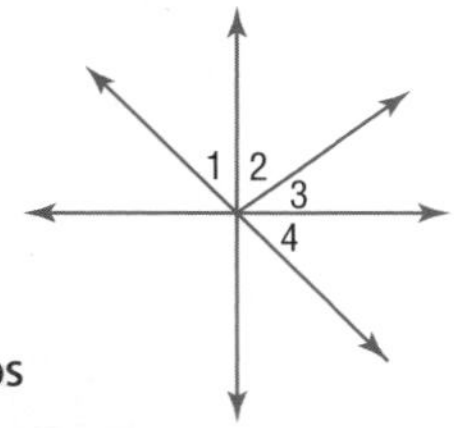

10. $\angle 2$ and $\angle 4$ and $\angle 4$ and $\angle 5$ are supplementary $m\angle 4 = 110$

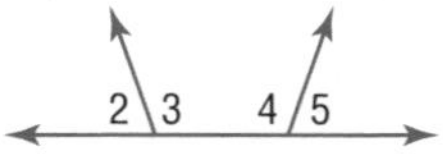

Find the measure of each numbered angle and name the theorems used that justify your work.

11. $m\angle 9 = 3x - 24$

$m\angle 10 = x + 12$

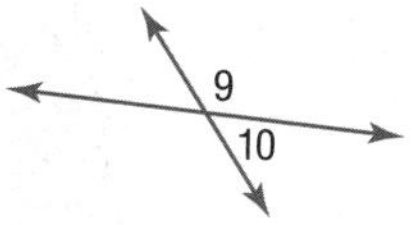

12. $m\angle 3 = 2x + 23$

$m\angle 4 = 5x - 70$

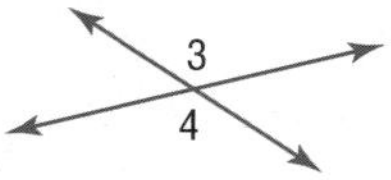

13. $m\angle 6 = 2x - 29$

$m\angle 7 = 3x - 21$

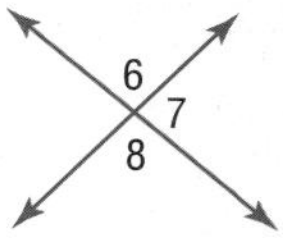

Example 4 **PROOF** **Write a two-column proof.**

14. Given: $\angle ABC$ is a straight angle. D is in the interior of $\angle ABC$

Prove: $\angle ABD$ and $\angle CBD$ are supplementary.

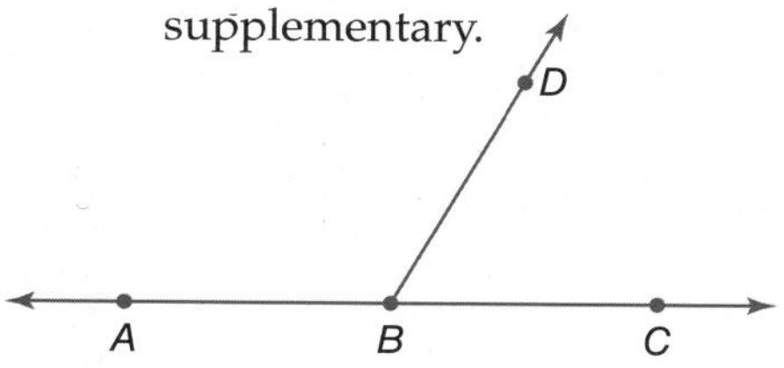

15. Given: $\angle 4 \cong \angle 7$

Prove: $\angle 4$ and $\angle 6$ are supplementary.

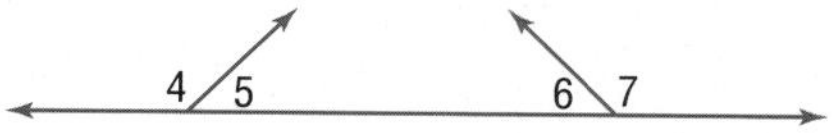

Write a proof for each theorem.

16. Supplement Theorem 10.2

17. Complement Theorem 10.3

18. Reflexive Property of Angle Congruence 10.4

19. Transitive Property of Angle Congruence 10.4

20. **FLAGS** Refer to the Jamaican flag at the right. Prove that the sum of the four angle measures is 360.

21. CCSS **NATURE** The diamondback rattlesnake has a diamond pattern on its back. An enlargement of the snake is shown below. If $\angle 1 \cong \angle 4$, prove that $\angle 2 \cong \angle 3$.

PROOF **Use the figure to write a proof of each theorem.**

22. Theorem 10.8

23. Theorem 10.9

24. Theorem 10.10

25. Theorem 10.11

26. Theorem 10.12

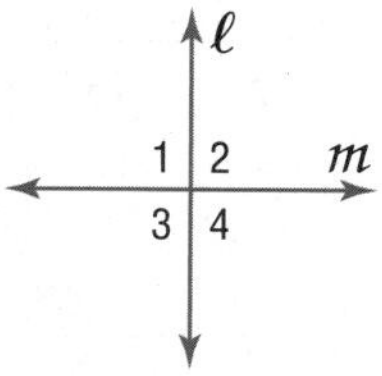

27. CCSS **TIME** To mark time, the arm on a metronome is adjusted so that it swings at a specific rate. Suppose $\angle ABC$ in the photo is at a right angle. If $m\angle 1 = 45$, write a paragraph proof to show that $\overrightarrow{BR}$ bisects $\angle ABC$.

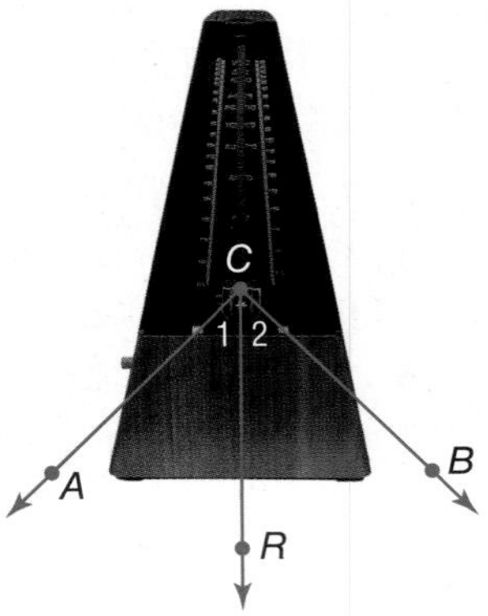

28. **PROOF** Write a proof of Theorem 10.7.

29. **SPORTS** The infield of a baseball diamond is a square. Drawing the diagonals from first base to third base and second base to home plate, four angles are formed. If $\angle 2$ is a right angle, prove that lines l and m are perpendicular.

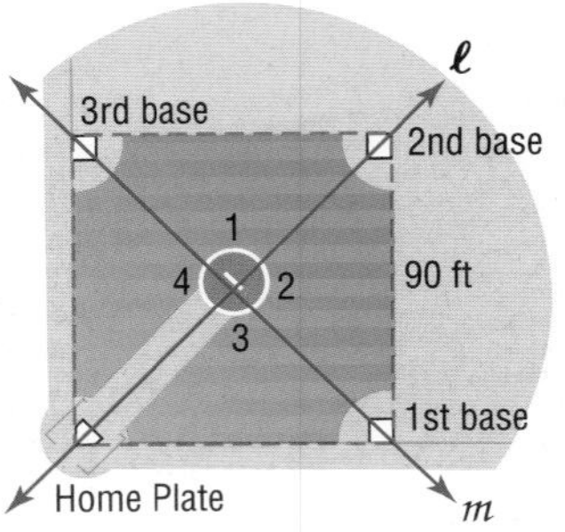

30. **MULTIPLE REPRESENTATIONS** In this problem, you will explore angle relationships.

   a. **Geometric** Draw a right angle XYZ. Place point W in the interior of this angle and draw $\overrightarrow{YW}$. Draw $\overleftrightarrow{AB}$ and construct $\angle CAB$ congruent to $\angle XYW$.

   b. **Verbal** Make a conjecture as to the relationship between $\angle CAB$ and $\angle WYZ$.

   c. **Logical** Prove your conjecture.

H.O.T. Problems Use Higher-Order Thinking Skills

31. **OPEN ENDED** Draw an $\angle ABC$ such that $m\angle ABC = 90$. Construct $\angle DBC$ congruent to $\angle ABC$. Make a conjecture as to the measure of $\angle ABD$, and then prove your conjecture.

32. **WRITING IN MATH** Write the steps that you would use to complete the proof below.

Given: $\overline{XY} \cong \overline{YZ}$, $WX = \frac{1}{2}XZ$

Prove: $\overline{WX} \cong \overline{YZ}$

33. **CHALLENGE** In this lesson, one case of the Congruent Supplements Theorem was proven. In Exercise 6, you proved one case for the Congruent Complements Theorem. Explain why there is another case for each of these theorems. Then write a proof of the second case for each theorem.

34. **REASONING** Determine whether the following statement is *sometimes*, *always*, or *never* true. Explain your reasoning.

If one of the angles formed by two intersecting lines is obtuse, then the other three angles formed are acute angles.

35. **WRITING IN MATH** Explain how you can use your protractor to quickly find the measure of the supplement of an angle.

Standardized Test Practice

36. GRIDDED RESPONSE What is the mode of this set of data?

4, 3, −2, 1, 4, 0, 1, 4

37. Find the measure of $\angle CFD$.

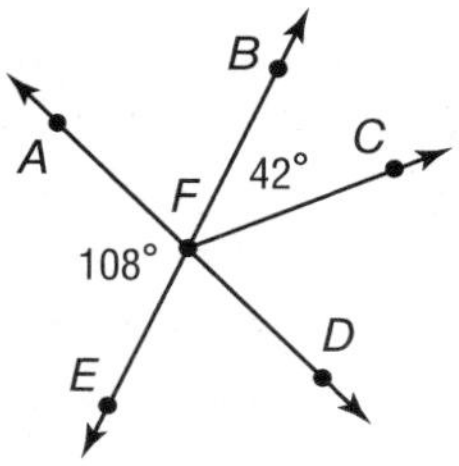

A 66°

B 72°

C 108°

D 138°

38. ALGEBRA Simplify.

$$4(3x - 2)(2x + 4) + 3x^2 + 5x - 6$$

F $9x^2 + 3x - 14$

G $9x^2 + 13x - 14$

H $27x^2 + 37x - 38$

J $27x^2 + 27x - 26$

39. SAT/ACT On a coordinate grid where each unit represents 1 mile, Isabel's house is located at (3, 0) and a mall is located at (0, 4). What is the distance between Isabel's house and the mall?

A 3 miles

B 5 miles

C 12 miles

D 13 miles

E 25 miles

Spiral Review

40. MAPS On a U.S. map, there is a scale that lists kilometers on the top and miles on the bottom.

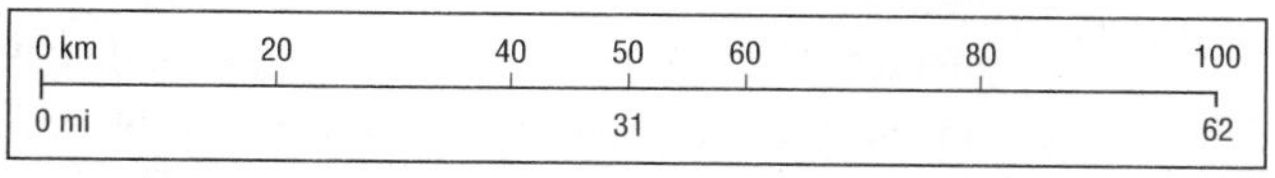

Suppose $\overline{AB}$ and $\overline{CD}$ are segments on this map. If $AB = 100$ kilometers and $CD = 62$ miles, is $\overline{AB} \cong \overline{CD}$? Explain. (Lesson 10-7)

State the property that justifies each statement. (Lesson 10-6)

41. If $y + 7 = 5$, then $y = -2$.

42. If $MN = PQ$, then $PQ = MN$.

43. If $a - b = x$ and $b = 3$, then $a - 3 = x$.

44. If $x(y + z) = 4$, then $xy + xz = 4$.

Determine the truth value of the following statement for each set of conditions.
If you have a fever, then you are sick. (Lesson 10-3)

45. You do not have a fever, and you are sick.

46. You have a fever, and you are not sick.

47. You do not have a fever, and you are not sick.

48. You have a fever, and you are sick.

Skills Review

Refer to the figure.

49. Name a line that contains point P.

50. Name the intersection of lines n and m.

51. Name a point not contained in lines ℓ, m, or n.

52. What is another name for line n?

53. Does line ℓ intersect line m or line n? Explain.

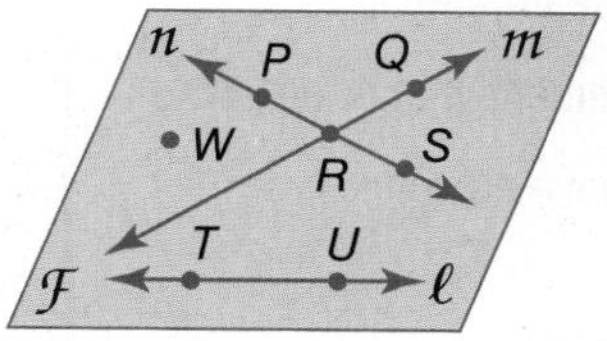

CHAPTER 10 Study Guide and Review

Study Guide

KeyConcepts

Points, Lines, and Planes (Lesson 10-1)

- There is exactly one line through any two points.
- There is exactly one plane through any three noncollinear points.

Distance and Midpoints (Lesson 10-3)

- On a number line, the measure of a segment with endpoint coordinates a and b is $|a - b|$.
- In the coordinate plane, the distance between two points (x_1, y_1) and (x_2, y_2) is given by $d = \sqrt{(x_2 - x_1)^2 + (y_2 - y_1)^2}$.
- On a number line, the coordinate of the midpoint of a segment with endpoints a and b is $\frac{a + b}{2}$.
- In the coordinate plane, the coordinates of the midpoint of a segment with endpoints that are (x_1, y_1) and (x_2, y_2) are $\left(\frac{x_1 + x_2}{2}, \frac{y_1 + y_2}{2}\right)$.

Angles (Lessons 10-3, 10-4, and 10-5)

- An angle is formed by two noncollinear rays that have a common endpoint, called its vertex. Angles can be classified by their measures.
- Adjacent angles are two coplanar angles that lie in the same plane and have a common vertex and a common side but no common interior points.
- Vertical angles are two nonadjacent angles formed by two intersecting lines.
- A linear pair is a pair of adjacent angles with noncommon sides that are opposite rays.
- Complementary angles are two angles with measures that have a sum of 90.
- Supplementary angles are two angles with measures that have a sum of 180.

Proof (Lessons 10-7 and 10-8)

Step 1 List the given information and draw a diagram, if possible.

Step 2 State what is to be proved.

Step 3 Create a deductive argument.

Step 4 Justify each statement with a reason.

Step 5 State what you have proved.

KeyVocabulary

acute angle (p. 594)
angle (p. 592)
angle bisector (p. 595)
area (p. 605)
between (p. 571)
circumference (p. 605)
collinear (p. 561)
concave (p. 603)
congruent (p. 572)
construction (p. 573)
convex (p. 603)
coplanar (p. 561)
degree (p. 593)
distance (p. 581)
equiangular polygon (p. 604)
equilateral polygon (p. 604)
exterior (p. 592)
interior (p. 592)
intersection (p. 562)
line (p. 561)
line segment (p. 570)
midpoint (p. 583)
***n*-gon** (p. 604)
obtuse angle (p. 594)
opposite rays (p. 592)
perimeter (p. 605)
plane (p. 561)
point (p. 561)
polygon (p. 603)
ray (p. 592)
regular polygon (p. 604)
right angle (p. 594)
segment bisector (p. 585)
side (p. 592)
space (p. 563)
undefined term (p. 561)
vertex (p. 592)
vertex of a polygon (p. 603)

FOLDABLES® StudyOrganizer

Be sure the Key Concepts are noted in your Foldable.

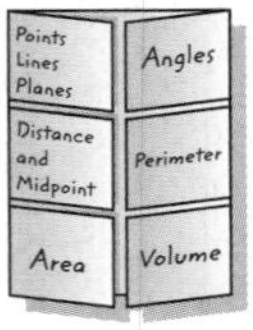

Lesson-by-Lesson Review

10-1 Points, Lines, and Planes

Use the figure to complete each of the following.

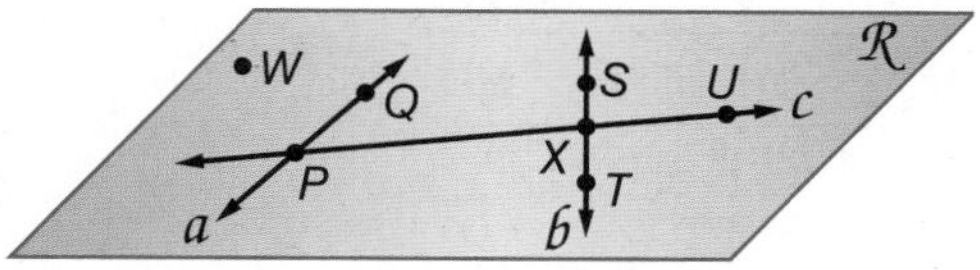

1. Name the intersection of lines a and c.
2. Give another name for line b.
3. Name a point that is not contained in any of the three lines a, b, or c.
4. Give another name for plane WPX.

Name the geometric term that is best modeled by each item.

5. 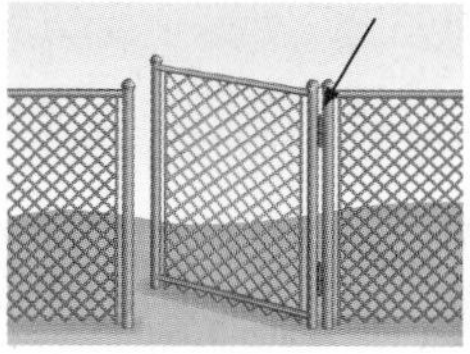

6. 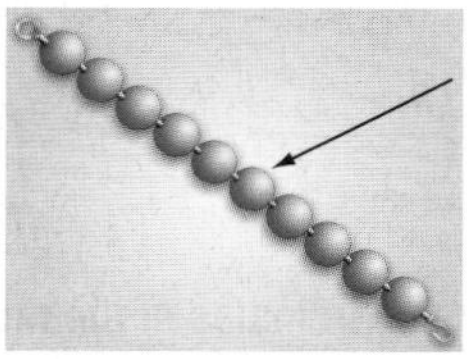

Example 1

Draw and label a figure for the relationship below.

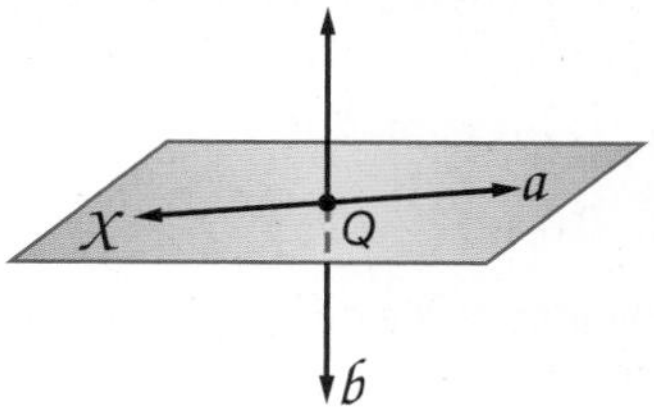

Plane $\mathcal{X}$ contains line a, line b intersects line a at point Q, but line b is not in plane $\mathcal{X}$.

Draw a surface to represent plane $\mathcal{X}$ and label it.

Draw a line in plane $\mathcal{X}$ and label it line a.

Draw a line b intersecting both the plane and line a and label the point of intersection Q.

10-2 Linear Measure

Find the value of the variable and XP, if X is between P and Q.

7. $XQ = 13$, $XP = 5x - 3$, $PQ = 40$
8. $XQ = 3k$, $XP = 7k - 2$, $PQ = 6k + 16$

Determine whether each pair of segments is congruent.

9. $\overline{AB}$, $\overline{CD}$

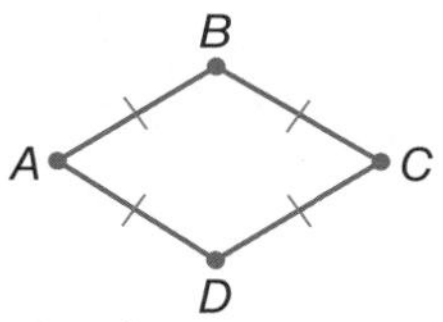

10. $\overline{XY}$, $\overline{YZ}$

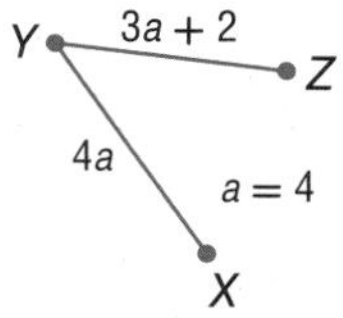

11. **DISTANCE** The distance from Salvador's job to his house is 3 times greater than the distance from his house to school. If his house is between his job and school and the distance from his job to school is 6 miles, how far is it from Salvador's house to school?

Example 2

Use the figure to find the value of the variable and the length of $\overline{YZ}$.

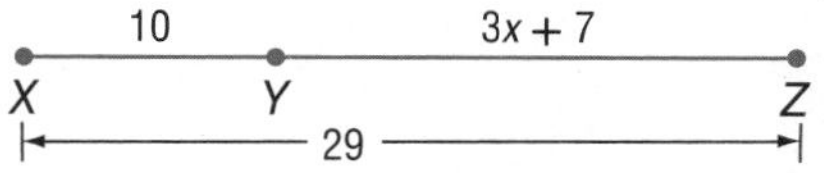

$XZ = XY + YZ$	Betweenness of points
$29 = 10 + 3x + 7$	Substitution
$29 = 3x + 17$	Simplify.
$12 = 3x$	Subtract 17 from each side.
$4 = x$	Divide each side by 3.
$YZ = 3x + 7$	Given
$= 3(4) + 7$ or 19	Substitution

So, $x = 4$ and $YZ = 19$.

10-3 Distance and Midpoints

Find the distance between each pair of points.

12. $A(-3, 1)$, $B(7, 13)$

13. $P(2, -1)$, $Q(10, -7)$

Find the coordinates of the midpoint of a segment with the given endpoints.

14. $L(-3, 16)$, $M(17, 4)$

15. $C(32, -1)$, $D(0, -12)$

Find the coordinates of the missing endpoint if M is the midpoint of $\overline{XY}$.

16. $X(-11, -6)$, $M(15, 4)$

17. $M(-4, 8)$, $Y(19, 0)$

18. **HIKING** Carol and Marita are hiking in a state park and decide to take separate trails. The map of the park is set up on a coordinate grid. Carol's location is at the point (7, 13) and Marita is at (3, 5).

a. Find the distance between them.

b. Find the coordinates of the point midway between their locations.

Example 3

Find the distance between $X(5, 7)$ and $Y(-7, 2)$.

Let $(x_1, y_1) = (5, 7)$ and $(x_2, y_2) = (-7, 2)$.

$$d = \sqrt{(x_2 - x_1)^2 + (y_2 - y_1)^2}$$
$$= \sqrt{(-7 - 5)^2 + (2 - 7)^2}$$
$$= \sqrt{(-12)^2 + (-5)^2}$$
$$= \sqrt{169} \text{ or } 13$$

The distance from X to Y is 13 units.

Example 4

Find the coordinates of the midpoint between $P(-4, 13)$ and $Q(6, 5)$.

Let $(x_1, y_1) = (-4, 13)$ and $(x_2, y_2) = (6, 5)$.

$$M\left(\frac{x_1 + x_2}{2}, \frac{y_1 + y_2}{2}\right) = M\left(\frac{-4 + 6}{2}, \frac{13 + 5}{2}\right)$$
$$= M(1, 9)$$

The coordinates of the midpoint are (1, 9).

10-4 Angle Measure

For Exercises 19–22, refer to the figure below.

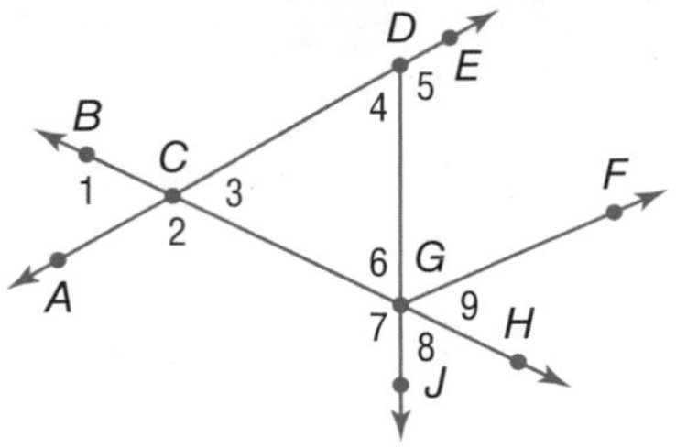

19. Name the vertex of $\angle 7$.

20. Write another name for $\angle 4$.

21. Name the sides of $\angle 2$.

22. Name a pair of opposite rays.

23. **SIGNS** A sign at West High School has the shape shown. Measure each of the angles and classify them as *right*, *acute*, or *obtuse*.

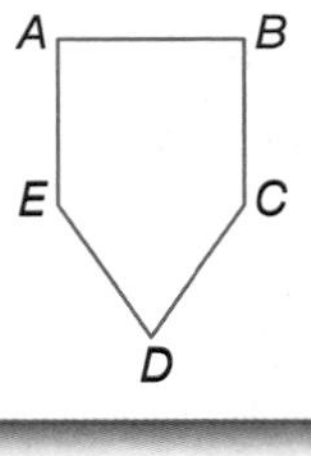

Example 5

Refer to the figure below. Name all angles that have Q as a vertex.

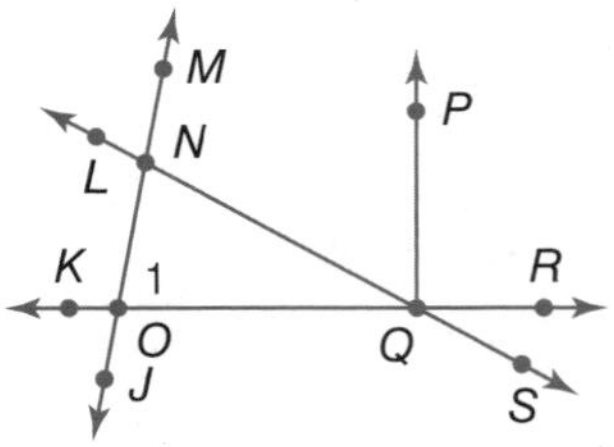

$\angle OQN$, $\angle NQP$, $\angle PQR$, $\angle RQS$, $\angle SQO$, $\angle OQP$, $\angle NQR$, $\angle PQS$, $\angle OQR$

Example 6

In the figure above, list all other names for $\angle 1$.

$\angle NOQ$, $\angle QON$, $\angle MOQ$, $\angle QOM$, $\angle MOR$, $\angle ROM$, $\angle NOR$, $\angle RON$

10-6 Two-Dimensional Figures

Name each polygon by its number of sides. Then classify it as *convex* or *concave* and *regular* or *irregular*.

24. 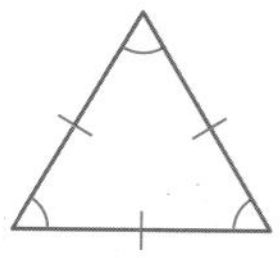

25. 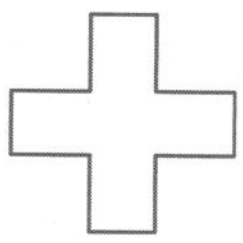

26. Find the perimeter of quadrilateral *ABCD* with vertices $A(-3, 5)$, $B(0, 5)$, $C(2, 0)$, and $D(-5, 0)$.

27. PARKS Westside Park received 440 feet of chain-link fencing as a donation to build an enclosed play area for dogs. The park administrators need to decide what shape the area should have. They have three options: (1) a rectangle with length of 100 feet and width of 120 feet, (2) a square with sides of length 110 feet, or (3) a circle with radius of approximately 70 feet. Find the areas of all three enclosures and determine which would provide the largest area for the dogs.

Example 7

Name the polygon by its number of sides. Then classify it as *convex* or *concave* and *regular* or *irregular*.

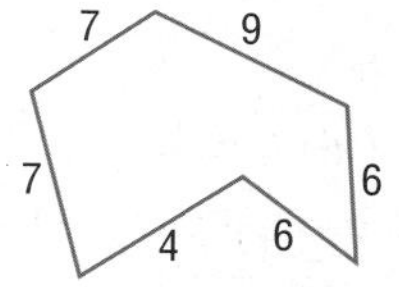

There are 6 sides, so this is a hexagon. If two of the sides are extended to make lines, they will pass through the interior of the hexagon, so it is concave. Since it is concave, it cannot be regular.

Example 8

Find the perimeter of the polygon in the figure above.

$P = s_1 + s_2 + s_3 + s_4 + s_5 + s_6$	Definition of perimeter
$= 7 + 7 + 9 + 6 + 6 + 4$	Substitution
$= 39$	Simplify.

The perimeter of the polygon is 39 units.

10-7 Proving Segment Relationships

Write a two-column proof.

28. Given: *X* is the midpoint of $\overline{WY}$ and $\overline{VZ}$.

Prove: $VW = ZY$

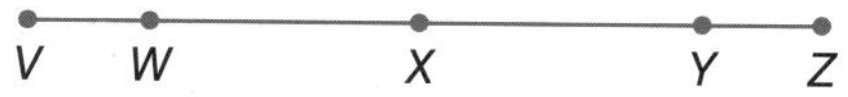

29. Given: $AB = DC$

Prove: $AC = DB$

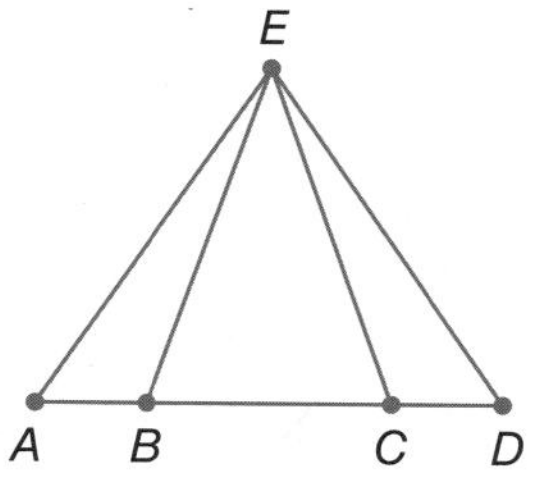

30. GEOGRAPHY Leandro is planning to drive from Kansas City to Minneapolis along Interstate 35. The map he is using gives the distance from Kansas City to Des Moines as 194 miles and from Des Moines to Minneapolis as 243 miles. What allows him to conclude that the distance he will be driving is 437 miles from Kansas City to Minneapolis? Assume that Interstate 35 forms a straight line.

Example 9

Write a two-column proof.

Given: *B* is the midpoint of $\overline{AC}$.

C is the midpoint of $\overline{BD}$.

Prove: $\overline{AB} \cong \overline{CD}$

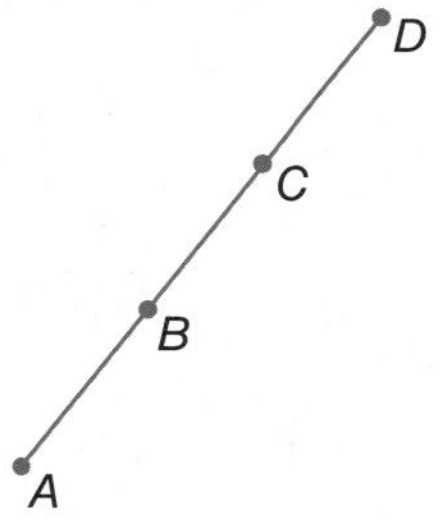

Proof:

Statements	**Reasons**
1. *B* is the midpoint of $\overline{AC}$.	**1.** Given
2. $\overline{AB} \cong \overline{BC}$	**2.** Definition of midpoint
3. *C* is the midpoint of $\overline{BD}$.	**3.** Given
4. $\overline{BC} \cong \overline{CD}$	**4.** Definition of midpoint
5. $\overline{AB} \cong \overline{CD}$	**5.** Transitive Property of Equality

Study Guide and Review *Continued*

10-8 Proving Angle Relationships

Find the measure of each angle.

31. ∠5

32. ∠6

33. ∠7

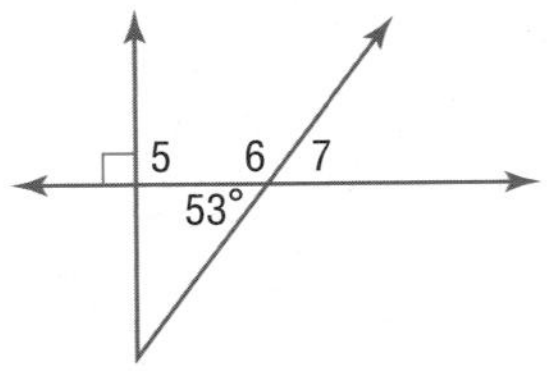

34. PROOF Write a two-column proof.
Given: ∠1 ≅ ∠4, ∠2 ≅ ∠3
Prove: ∠*AFC* ≅ ∠*EFC*

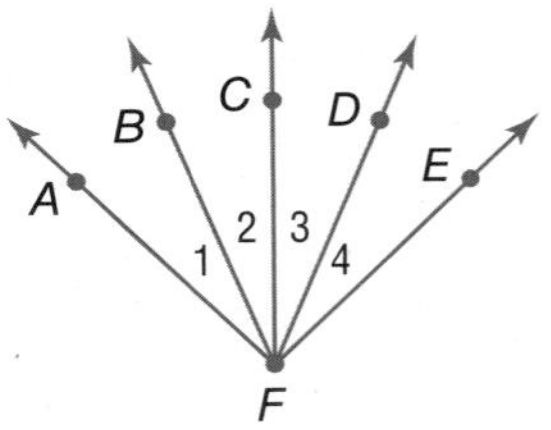

Example 10

Find the measure of each numbered angle if $m\angle 1 = 72$ and $m\angle 3 = 26$.

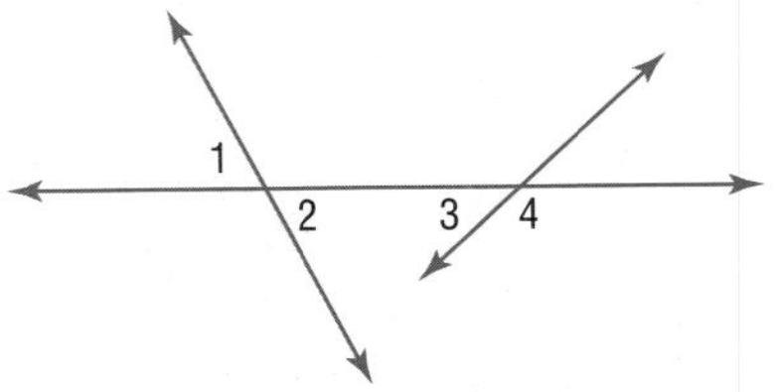

$m\angle 2 = 72$, since ∠1 and ∠2 are vertical angles.

∠3 and ∠4 form a linear pair and must be supplementary angles.

$26 + m\angle 4 = 180$	Definition of supplementary angles
$m\angle 4 = 154$	Subtract 26 from each side.

CHAPTER 10

Practice Test

Use the figure to name each of the following.

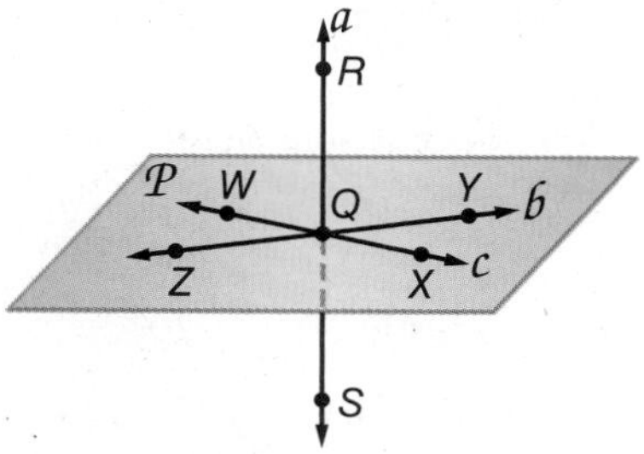

1. the line that contains points Q and Z
2. two points that are coplanar with points W, X, and Y
3. the intersection of lines a and b

Find the value of the variable if P is between J and K.

4. $JP = 2x$, $PK = 7x$, $JK = 27$
5. $JP = 3y + 1$, $PK = 12y - 4$, $JK = 75$
6. $JP = 8z - 17$, $PK = 5z + 37$, $JK = 17z - 4$

Find the coordinates of the midpoint of a segment with the given endpoints.

7. (16, 5) and (28, −13)
8. (−11, 34) and (47, 0)
9. (−4, −14) and (−22, 9)

Find the distance between each pair of points.

10. (43, −15) and (29, −3)
11. (21, 5) and (28, −1)
12. (0, −5) and (18, −10)

13. **ALGEBRA** The measure of $\angle X$ is 18 more than three times the measure of its complement. Find the measure of $\angle X$.

14. Find the value of x that will make lines a and b perpendicular in the figure below.

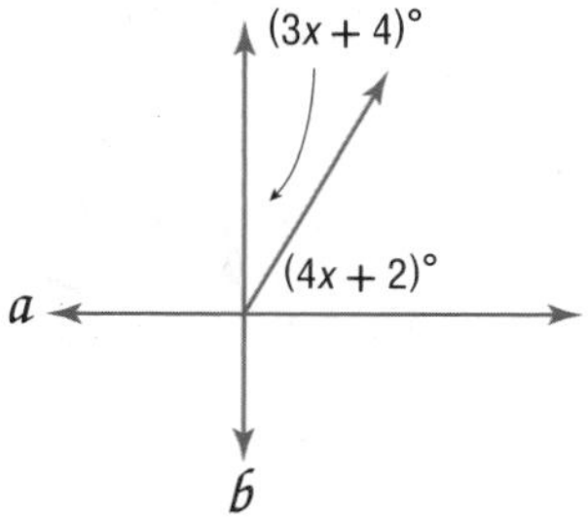

For Exercises 15–18, use the figure below.

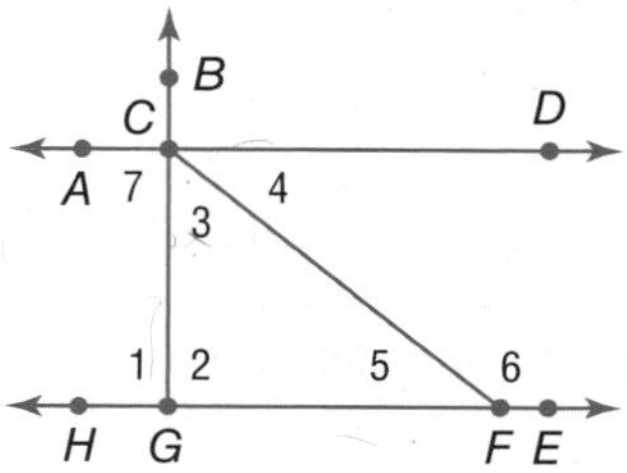

15. Name the vertex of $\angle 3$.
16. Name the sides of $\angle 1$.
17. Write another name for $\angle 6$.
18. Name a pair of angles that share exactly one point.

19. **MULTIPLE CHOICE** If $m\angle 1 = m\angle 2$, which of the following statements is true?

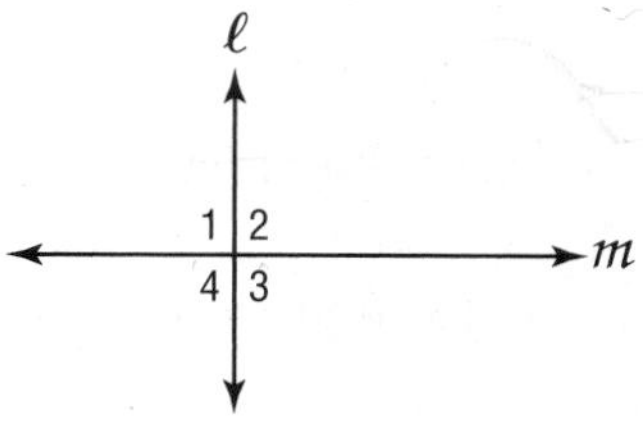

A $\angle 2 \cong \angle 4$

B $\angle 2$ is a right angle.

C $\ell \perp m$

D All of the above

Find the perimeter of each polygon.

20. triangle XYZ with vertices $X(3, 7)$, $Y(-1, -5)$, and $Z(6, -4)$
21. rectangle $PQRS$ with vertices $P(0, 0)$, $Q(0, 7)$, $R(12, 7)$, and $S(12, 0)$

22. **SAFETY** A severe weather siren in a local city can be heard within a radius of 1.3 miles. If the mayor of the city wants a new siren that will cover double the area of the old siren, what should the radius of the new siren be? Round to the nearest tenth of a mile.

23. **PROOF** Write a paragraph proof.

Given: $\overline{JK} \cong \overline{CB}$, $\overline{KL} \cong \overline{AB}$

Prove: $\overline{JL} \cong \overline{AC}$

Preparing for Standardized Tests

Solving Math Problems

Strategies for Solving Math Problems

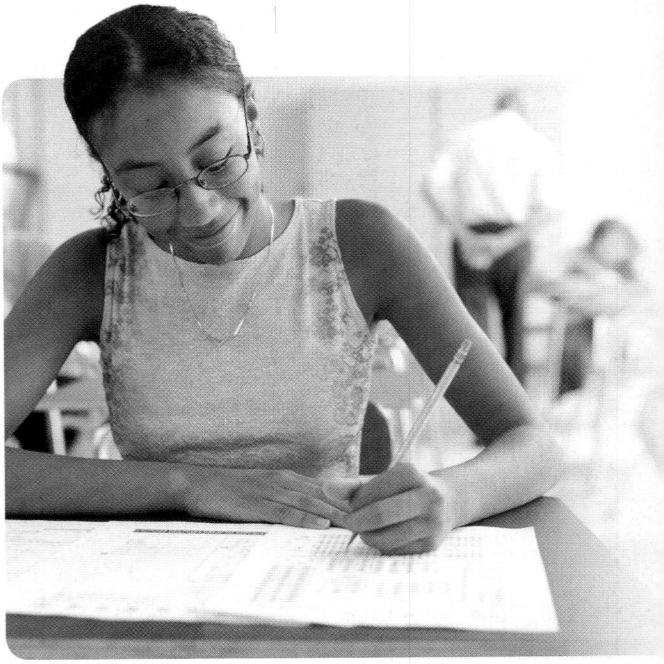

Sean Justice/Stone/Getty Images

The first step to solving any math problem is to read the problem. When reading a math problem to get the information you need to solve, it is helpful to use special reading strategies.

Step 1

Read the problem to determine what information is given.

- **Analyze:** Determine what exactly the problem is asking you to solve.
- **Underline:** If you are able to write in your test book, underline any important information.

Step 2

Reread the problem to determine what information is needed to solve the problem.

- **Think:** How does the information fit together?
- **Key Words:** Are there any key words, variables or mathematical terms in the problem?
- **Diagrams:** Do you need to use a diagram, list or table?
- **Formulas:** Do you need a formula or an equation to solve the problem?

Step 3

Devise a plan and solve the problem. Use the information you found in Steps 1 and 2.

- **Question:** What problem are you solving?
- **Estimate:** Estimate an answer.
- **Eliminate:** Eliminate all answers that do not make sense and/or vary greatly from your estimate.

Step 4

Check your answer.

- **Reread:** Quickly reread the problem to make sure you solved the whole problem.
- **Reasonableness:** Is your answer reasonable?
- **Units:** Make sure your answer has the correct units of measurement.

Standardized Test Example

Read the problem. Identify what you need to know. Then use the information in the problem to solve.

Carmen is using a coordinate grid to make a map of her backyard. She plots the swing set at point $S(2, 5)$ and the big oak tree at point $O(-3, -6)$. If each unit on the grid represents 5 feet, what is the distance between the swing set and the oak tree? Round your answer to the nearest whole foot.

A 12 ft **B** 25 ft **C** 60 ft **D** 74 ft

Determine what exactly the problem is asking you to solve. Underline any important information.

Carmen is using a coordinate grid to make a map of her backyard. She plots the swing set at point $S(2, 5)$ and the big oak tree at point $O(-3, -6)$. If each unit on the grid represents 5 feet, what is the distance between the swing set and the oak tree? Round your answer to the nearest whole foot.

The problem is asking for the distance between the swing set and the oak tree. The key word is distance, so you know you will need to use the Distance Formula.

$d = \sqrt{(x_2 - x_1)^2 + (y_2 - y_1)^2}$ Distance Formula

$= \sqrt{(-3 - 2)^2 + (-6 - 5)^2}$ $(x_1, y_1) = (2, 5)$ and $(x_2, y_2) = (-3, -6)$

$= \sqrt{(-5)^2 + (-11)^2}$ Subtract.

$= \sqrt{25 + 121}$ or $\sqrt{146}$ Simplify.

The distance between swing set and the oak tree is $\sqrt{146}$ units. Use a calculator to find that $\sqrt{146}$ units is approximately 12.08 units.

Since each unit on the grid represents 5 feet, the distance is $(12.08) \cdot (5)$ or 60.4 ft. Therefore, the correct answer is C.

Check your answer to make sure it is reasonable, and that you have used the correct units.

Exercises

Read each question. Then fill in the correct answer on the answer document provided by your teacher or on a sheet of paper.

1. A regular pentagon has a perimeter of 24 inches. What is the measure of each side?

A 3 inches **C** 4 inches

B 3.8 inches **D** 4.8 inches

2. What is the value of x in the figure at the right?

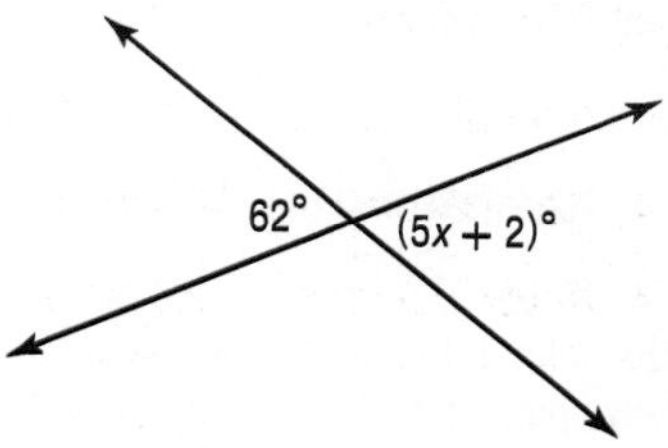

F 10

G 12

H 14

J 15

CHAPTER 10

Standardized Test Practice

Cumulative, Chapters 1 through 10

Multiple Choice

Read each question. Then fill in the correct answer on the answer document provided by your teacher or on a sheet of paper.

1. Ricky's Rentals rented 12 more bicycles than scooters last weekend for a total revenue of $2,125. How many scooters were rented?

Item	Rental Fee
Bicycle	$20
Scooter	$45

A 26 **C** 37

B 29 **D** 41

2. Find the distance between $M(-3, 1)$ and $N(2, 8)$ on a coordinate plane.

F 6.1 units

G 6.9 units

H 7.3 units

J 8.6 units

3. Which of the following terms best describes points *F*, *G*, and *H*?

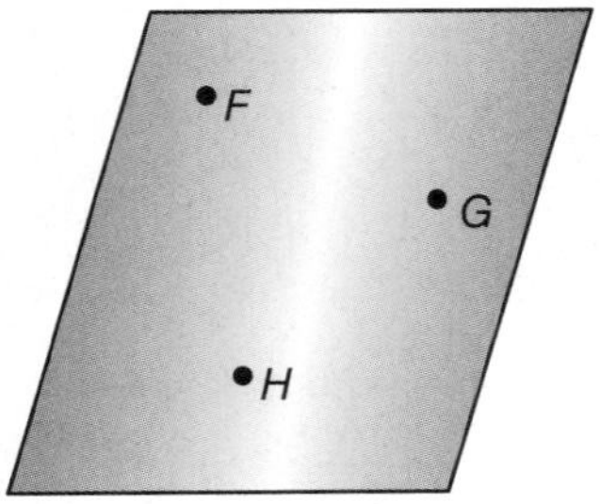

A collinear **C** coplanar

B congruent **D** skew

Test-Taking Tip

Question 3 Understanding the terms of geometry can help you solve problems. The term *congruent* refers to geometric figures, and *skew* refers to lines, therefore both answers can be eliminated.

4. What is the length of segment *BD*?

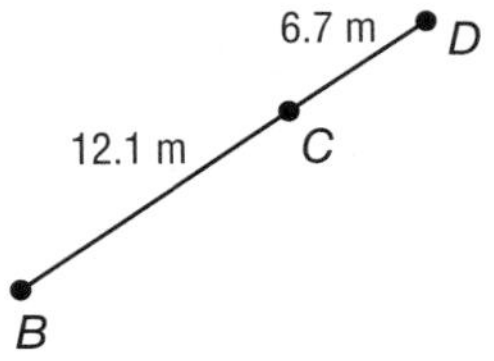

F 17.4 m **H** 18.8 m

G 18.3 m **J** 19.1 m

5. In the figure below, what is the measure of angle *CDN*?

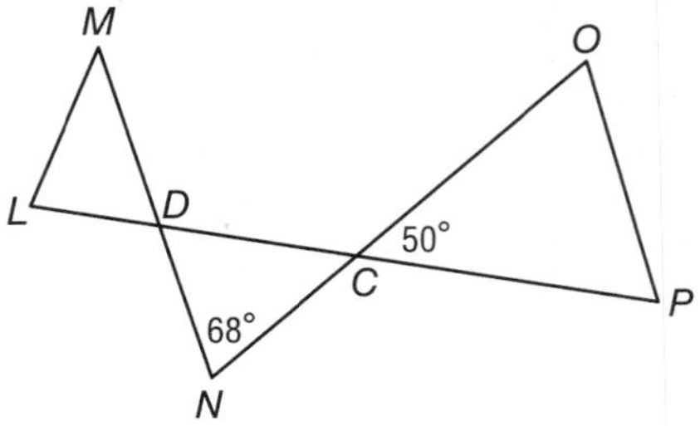

A 58° **C** 68°

B 62° **D** 70°

6. Find the perimeter of the figure below.

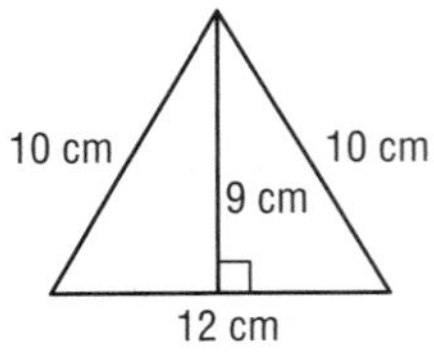

F 20 cm **H** 32 cm

G 29 cm **J** 41 cm

7. What is the relationship of ∠1 and ∠2?

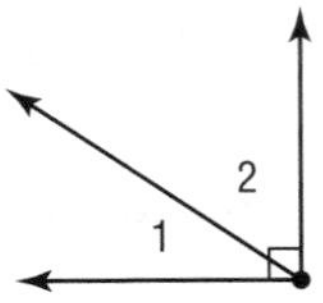

A complementary angles

B linear pair

C supplementary angles

D vertical angles

Short Response/Gridded Response

Record your answers on the answer sheet provided by your teacher or on a sheet of paper.

8. Find the distance between points *R* and *S* on the coordinate grid below. Round to the nearest tenth.

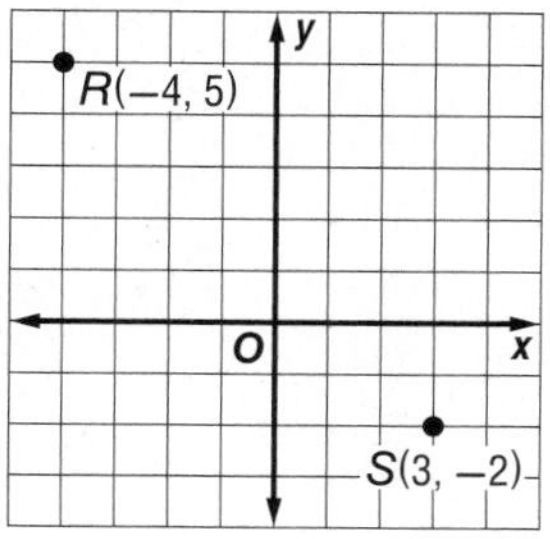

9. SHORT RESPONSE Find the value of *x* and *AB* if *B* is between *A* and *C*, $AB = 2x$, $AC = 6x - 5$, and $BC = 7$.

10. Suppose two lines intersect in a plane.

a. What do you know about the two pairs of vertical angles formed?

b. What do you know about the pairs of adjacent angles formed?

11. GRIDDED RESPONSE How many planes are shown in the figure below?

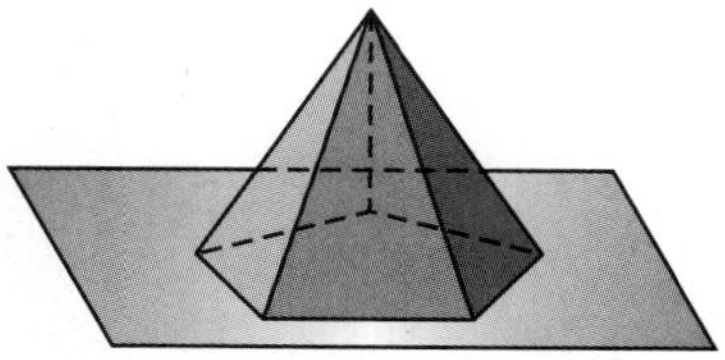

12. Jason received a \$50 gift certificate for his birthday. He wants to buy a DVD and a poster from a media store. (Assume that sales tax is included in the prices.) Write and solve a linear inequality to show how much he would have left to spend after making these purchases.

13. GRIDDED RESPONSE What is the value of *x* in the figure?

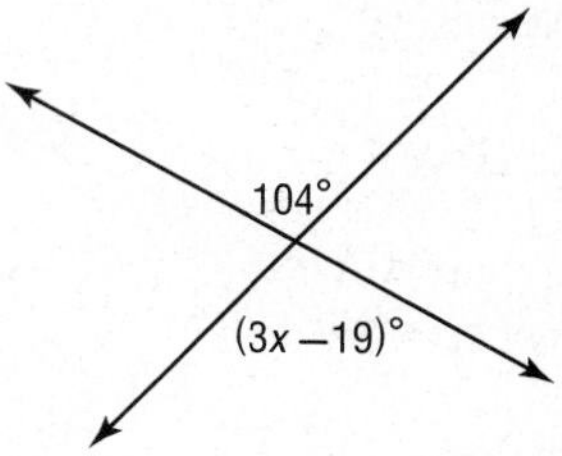

Extended Response

Record your answers on a sheet of paper. Show your work.

14. Julie's room has the dimensions shown in the figure.

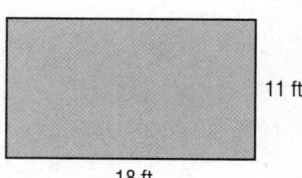

a. Find the perimeter of her room.

b. Find the area of her room.

c. If the length and width doubled, what effect would it have on the perimeter?

d. What effect would it have on the area?

Need ExtraHelp?

If you missed Question...	1	2	3	4	5	6	7	8	9	10	11	12	13	14
Go to Lesson...	6-2	10-3	10-1	10-2	10-4	10-6	10-5	10-3	10-2	10-5	10-1	5-1	10-4	10-6

CHAPTER 11

Parallel and Perpendicular Lines

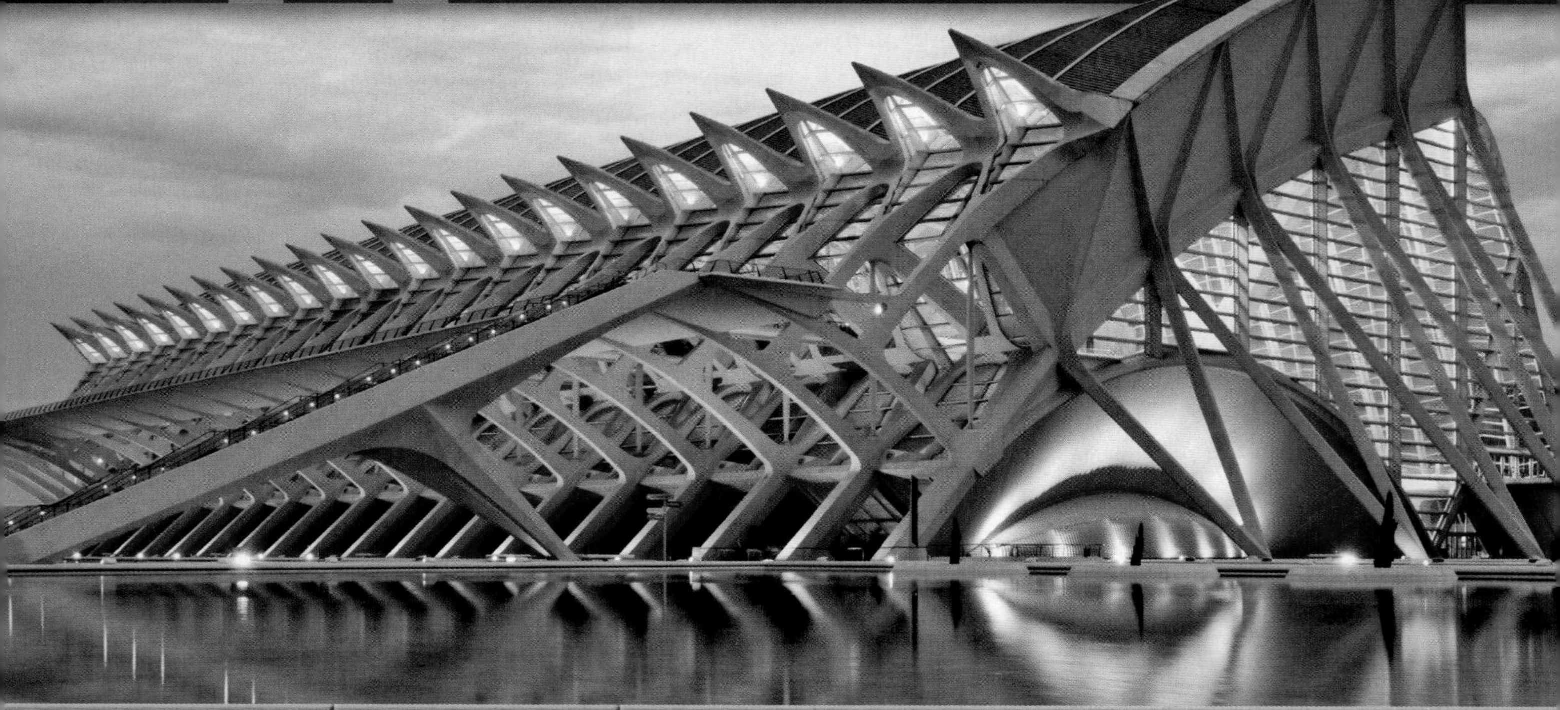

Then

You learned about lines and angles and writing geometric proofs.

Now

In this chapter, you will:

- Identify and prove angle relationships that occur with parallel lines and a transversal.
- Use slope to analyze a line and to write its equation.
- Find the distance between a point and a line and between two parallel lines.

Why? ▲

CONSTRUCTION and ENGINEERING Architects, carpenters, and engineers use parallel and perpendicular lines to design buildings, furniture, and machines.

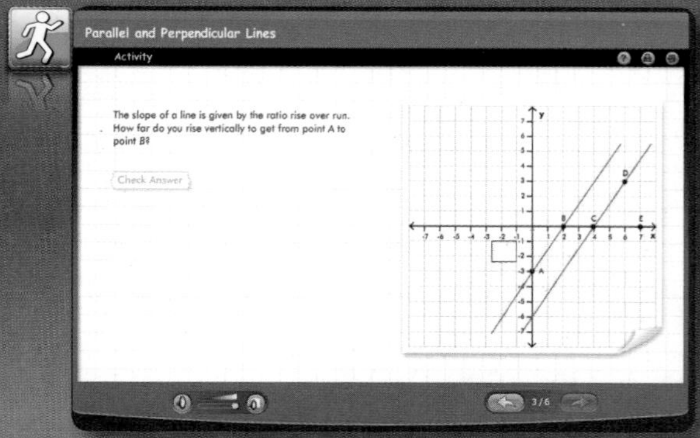

connectED.mcgraw-hill.com Your Digital Math Portal

Animation | Vocabulary | eGlossary | Personal Tutor | Virtual Manipulatives | Graphing Calculator | Audio | Foldables | Self-Check Practice | Worksheets

Get Ready for the Chapter

Diagnose Readiness | You have two options for checking prerequisite skills.

1 Textbook Option
Take the Quick Check below. Refer to the Quick Review for help.

QuickCheck

Refer to the figure to identify each of the following.

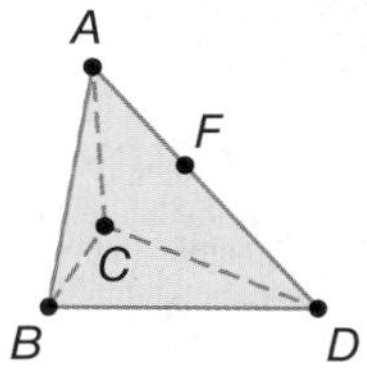

1. How many planes are shown in this figure?

2. Name three points that are collinear.

3. Are points C and D coplanar? Explain.

4. **PHOTOGRAPHY** Tina is taking a picture of her friends. If she sets a tripod level on the ground, will the bottom of each of the three legs of the tripod be coplanar?

QuickReview

Example 1

Refer to the figure.

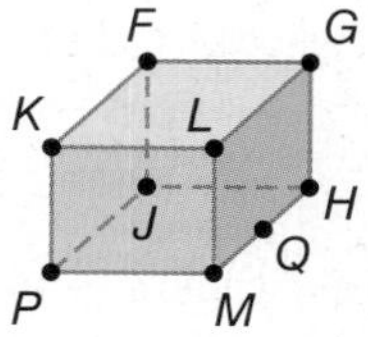

a. How many planes are shown in this figure?
Six: plane *FGLK*, plane *JHMP*, plane *FKPJ*, plane *GLMH*, plane *FGHJ*, and plane *KLMP*

b. Name three points that are collinear.
Points *M*, *Q*, and *H* are collinear.

c. Are points *F*, *K*, and *J* coplanar? Explain.
Yes. Points *F*, *K*, and *J* all lie in plane *FKPJ*.

Find each angle measure.

5. $\angle 1$

6. $\angle 2$

7. $\angle 3$

8. $\angle 4$

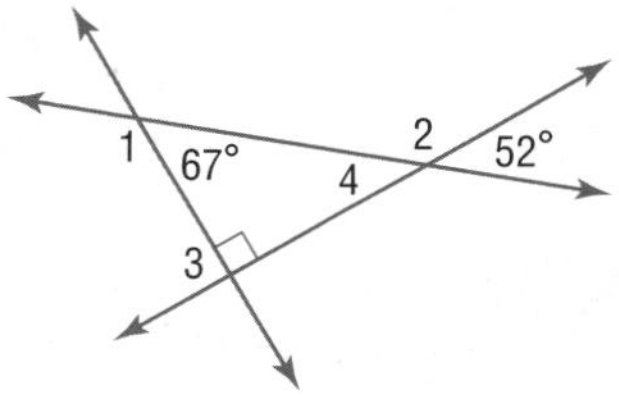

Example 2

Find $m\angle 1$.

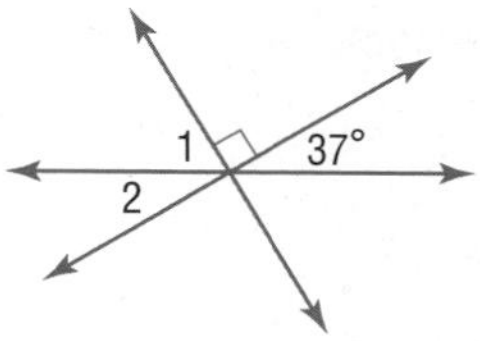

$m\angle 1 + 37 + 90 = 180$ Add.

$m\angle 1 = 53$ Simplify.

Find the value of x for the given values of a and b.

9. $a + 8 = -4(x - b)$, for $a = 8$ and $b = 3$

10. $b = 3x + 4a$, for $a = -9$ and $b = 12$

11. $\frac{a + 2}{b + 13} = 5x$, for $a = 18$ and $b = -1$

12. **MINIATURE GOLF** A miniature golf course offers a $1 ice cream cone with each round of golf purchased. If five friends each had a cone after golfing and spend a total of $30, how much does one round of golf cost?

Example 3

Find x in $a + 8 = b(x - 7)$ if $a = 12$ and $b = 10$.

$a + 8 = b(x - 7)$	Write the equation.
$12 + 8 = 10(x - 7)$	$a = 12$ and $b = 10$
$20 = 10x - 70$	Simplify.
$90 = 10x$	Add.
$x = 9$	Divide.

2 Online Option
Take an online self-check Chapter Readiness Quiz at connectED.mcgraw-hill.com.

Get Started on the Chapter

You will learn several new concepts, skills, and vocabulary terms as you study Chapter 11. To get ready, identify important terms and organize your resources. You may wish to refer to Chapter 0 to review prerequisite skills.

FOLDABLES StudyOrganizer

Parallel and Perpendicular Lines Make this Foldable to help you organize your Chapter 11 notes about relationships between lines. Begin with a sheet of 11″ × 17″ paper and six index cards.

1 **Fold** lengthwise about 3″ from the bottom.

2 **Fold** the paper in thirds.

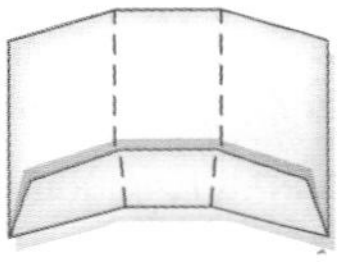

3 **Open** and staple the edges on either side to form three pockets.

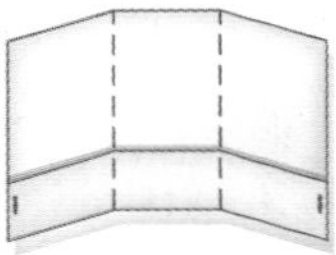

4 **Label** the pockets as shown. Place two index cards in each pocket.

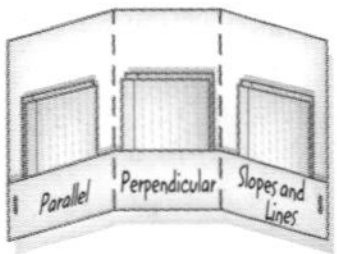

NewVocabulary

English		Español
parallel lines	p. 643	rectas paralelas
skew lines	p. 643	rectas alabeadas
parallel planes	p. 643	planos paralelos
transversal	p. 644	transversal
interior angles	p. 644	ángulos interiores
exterior angles	p. 644	ángulos externos
corresponding angles	p. 644	ángulos correspondientes
slope	p. 658	pendiente
rate of change	p. 659	tasa de cambio
slope-intercept form	p. 668	forma pendiente-intersección
point-slope form	p. 668	forma punto-pendiente
equidistant	p. 688	equidistante

ReviewVocabulary

congruent angles ángulos congruentes two angles that have the same degree measure

perpendicular perpendicular two lines, segments, or rays that intersect to form right angles

vertical angles ángulos opusetos por el vértice two nonadjacent angles formed by intersecting lines

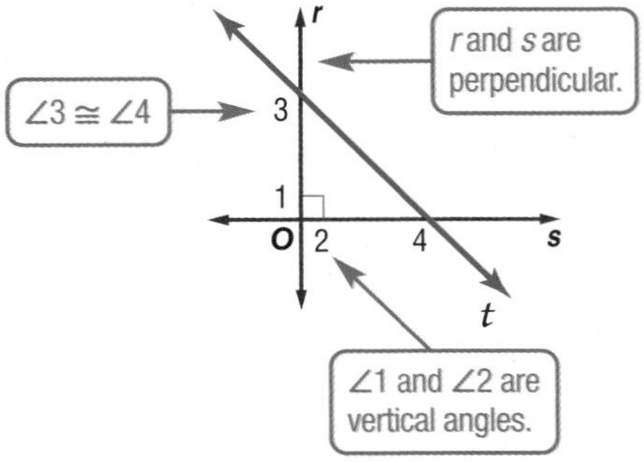

LESSON 11-1

Parallel Lines and Transversals

∴ Then	∴ Now	∴ Why?
● You used angle and line segment relationships to prove theorems.	● **1** Identify the relationships between two lines or two planes. **2** Name angle pairs formed by parallel lines and transversals.	● An Ames room creates the illusion that a person standing in the right corner is much larger than a person standing in the left corner. From a front viewing hole the front and back walls appear parallel, when in fact they are slanted. The ceiling and floor appear horizontal, but are actually tilted.

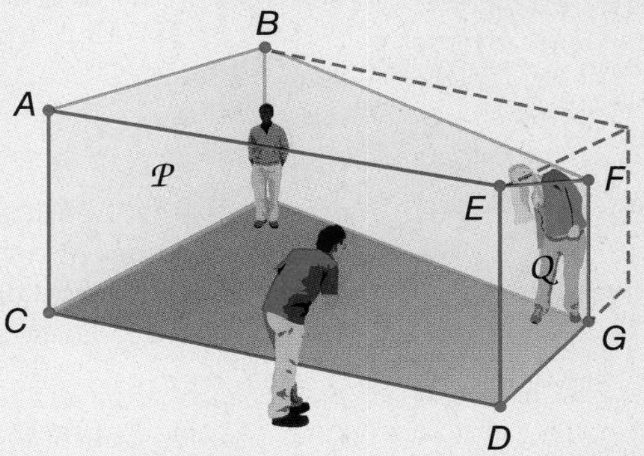

NewVocabulary

parallel lines
skew lines
parallel planes
transversal
interior angles
exterior angles
consecutive interior angles
alternate interior angles
alternate exterior angles
corresponding angles

Common Core State Standards

Content Standards

G.CO.1 Know precise definitions of angle, circle, perpendicular line, parallel line, and line segment, based on the undefined notions of point, line, distance along a line, and distance around a circular arc.

Mathematical Practices

1 Make sense of problems and persevere in solving them.

3 Construct viable arguments and critique the reasoning of others.

1 Relationships Between Lines and Planes

The construction of the Ames room above makes use of intersecting, parallel, and skew lines, as well as intersecting and parallel planes, to create an optical illusion.

KeyConcepts Parallel and Skew

Parallel lines are coplanar lines that do not intersect.

Example $\overleftrightarrow{JK} \parallel \overleftrightarrow{LM}$

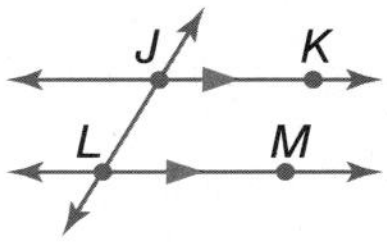

Arrows are used to indicate that lines are parallel.

Skew lines are lines that do not intersect and are not coplanar.

Example Lines ℓ and m are skew.

Parallel planes are planes that do not intersect.

Example Planes $\mathcal{A}$ and $\mathcal{B}$ are parallel.

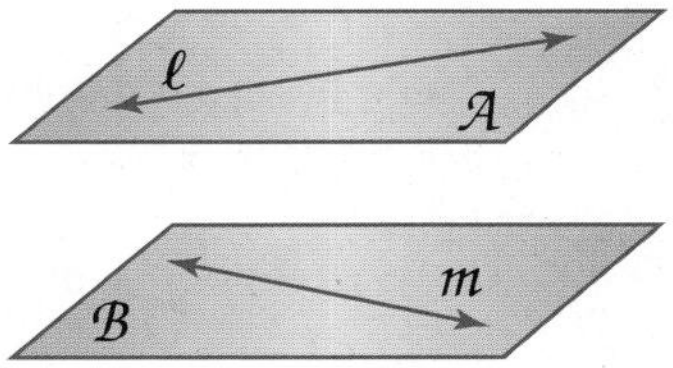

$\overleftrightarrow{JK} \parallel \overleftrightarrow{LM}$ is read as *line JK is parallel to line LM.*

If segments or rays are contained within lines that are parallel or skew, then the segments or rays are parallel or skew.

Real-World Example 1 Identify Parallel and Skew Relationships

Identify each of the following using the wedge of cheese below.

a. all segments parallel to $\overline{JP}$

$\overline{KQ}$ and $\overline{LR}$

b. a segment skew to $\overline{KL}$

$\overline{JP}$, $\overline{PQ}$, or $\overline{PR}$

c. a plane parallel to plane *PQR*

Plane *JKL* is the only plane parallel to plane *PQR*.

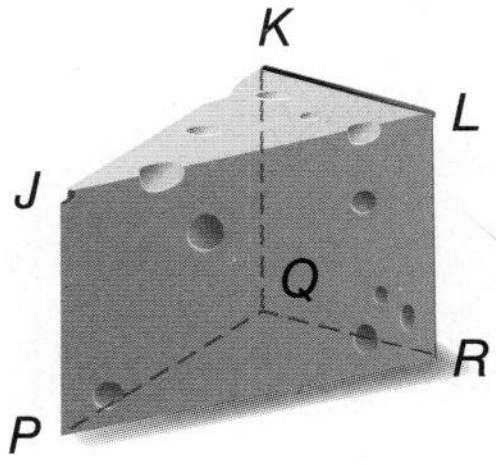

WatchOut!

Parallel vs. Skew In Check Your Progress 1A, $\overleftrightarrow{FE}$ is *not* skew to $\overleftrightarrow{BC}$. Instead, these lines are parallel in plane *BCF*.

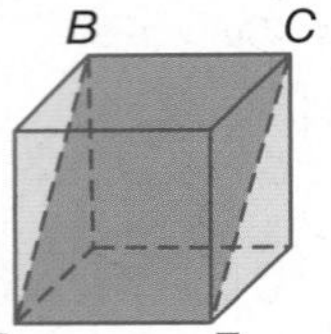

GuidedPractice

Identify each of the following using the cube shown.

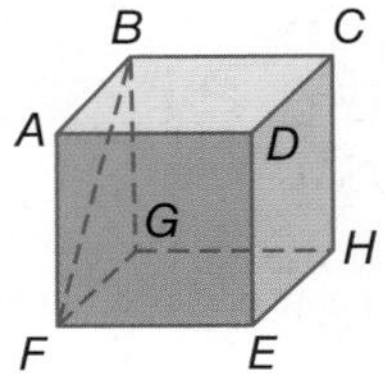

1A. all segments skew to $\overleftrightarrow{BC}$

1B. a segment parallel to $\overleftrightarrow{EH}$

1C. all planes parallel to plane *DCH*

2 Transversal Angle Pair Relationships

A line that intersects two or more coplanar lines at two different points is called a **transversal**. In the diagram below, line *t* is a transversal of lines *q* and *r*. Notice that line *t* forms a total of eight angles with lines *q* and *r*. These angles, and specific pairings of these angles, are given special names.

ReadingMath

Same-Side Interior Angles Consecutive interior angles are also called *same-side interior angles*.

KeyConcept Transversal Angle Pair Relationships

Four **interior angles** lie in the region between lines *q* and *r*.	∠3, ∠4, ∠5, ∠6
Four **exterior angles** lie in the two regions that are not between lines *q* and *r*.	∠1, ∠2, ∠7, ∠8
Consecutive interior angles are interior angles that lie on the same side of transversal *t*.	∠4 and ∠5, ∠3 and ∠6
Alternate interior angles are nonadjacent interior angles that lie on opposite sides of transversal *t*.	∠3 and ∠5, ∠4 and ∠6
Alternate exterior angles are nonadjacent exterior angles that lie on opposite sides of transversal *t*.	∠1 and ∠7, ∠2 and ∠8
Corresponding angles lie on the same side of transversal *t* and on the same side of lines *q* and *r*.	∠1 and ∠5, ∠2 and ∠6 ∠3 and ∠7, ∠4 and ∠8

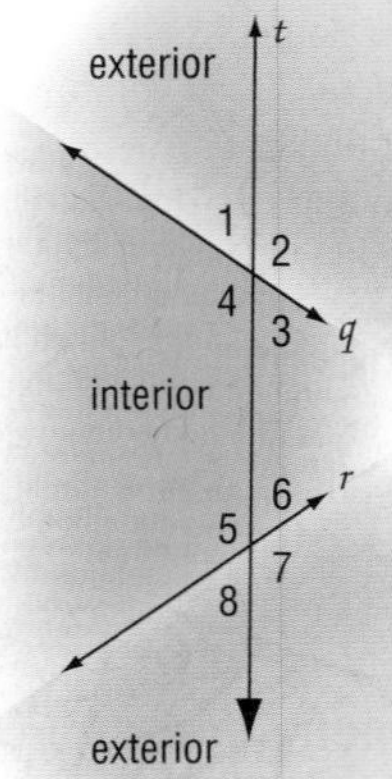

Example 2 Classify Angle Pair Relationships

Refer to the figure below. Classify the relationship between each pair of angles as *alternate interior, alternate exterior, corresponding,* or *consecutive interior* angles.

a. ∠1 and ∠5
alternate exterior

b. ∠6 and ∠7
consecutive interior

c. ∠2 and ∠4
corresponding

d. ∠2 and ∠6
alternate interior

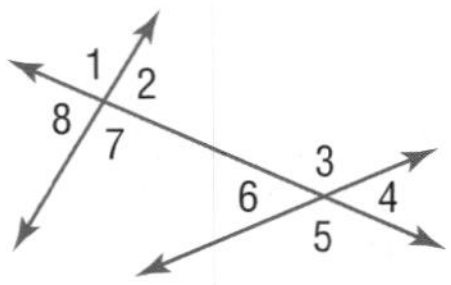

GuidedPractice

2A. ∠3 and ∠7 **2B.** ∠5 and ∠7 **2C.** ∠4 and ∠8 **2D.** ∠2 and ∠3

When more than one line can be considered a transversal, first identify the transversal for a given angle pair by locating the line that connects the vertices of the angles.

StudyTip

Nonexample In the figure below, line c is *not* a transversal of lines a and b, since line c intersects lines a and b in only one point.

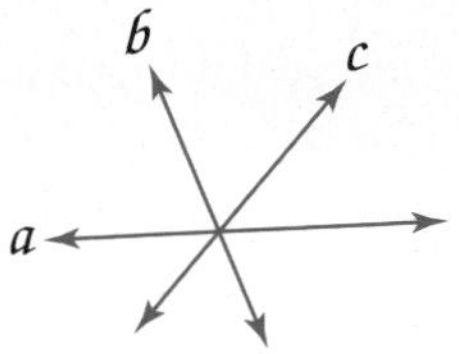

Example 3 Identify Transversals and Classify Angle Pairs

PT

Identify the transversal connecting each pair of angles in the photo. Then classify the relationship between each pair of angles.

a. **∠1 and ∠3**

The transversal connecting ∠1 and ∠3 is line h. These are alternate exterior angles.

b. **∠5 and ∠6**

The transversal connecting ∠5 and ∠6 is line k. These are consecutive interior angles.

c. **∠2 and ∠6**

The transversal connecting ∠2 and ∠6 is line ℓ. These are corresponding angles.

GuidedPractice

3A. ∠3 and ∠5

3B. ∠2 and ∠8

3C. ∠5 and ∠7

3D. ∠2 and ∠9

Check Your Understanding

Example 1 **Refer to the figure at the right to identify each of the following.**

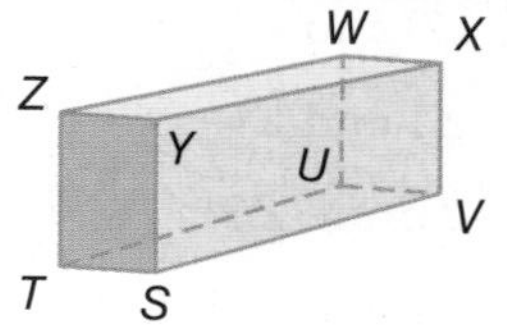

1. a plane parallel to plane ZWX

2. a segment skew to $\overline{TS}$ that contains point W

3. all segments parallel to $\overline{SV}$

4. **CONSTRUCTION** Use the diagram of the partially framed storage shed shown to identify each of the following.

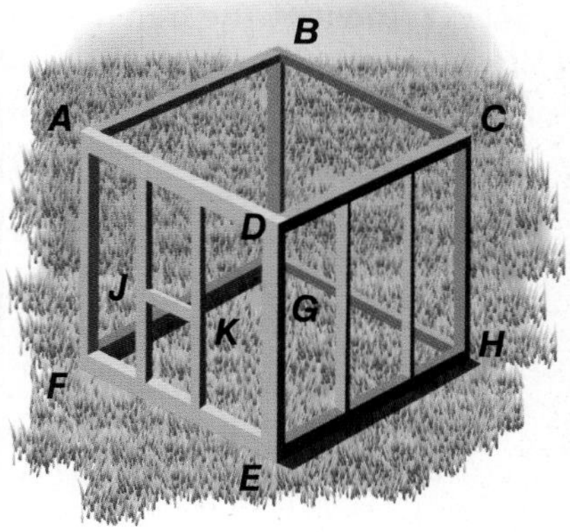

a. Name three pairs of parallel planes.

b. Name three segments parallel to $\overline{DE}$.

c. Name two segments parallel to $\overline{FE}$.

d. Name two pairs of skew segments.

Example 2 **Classify the relationship between each pair of angles as *alternate interior, alternate exterior, corresponding,* or *consecutive interior* angles.**

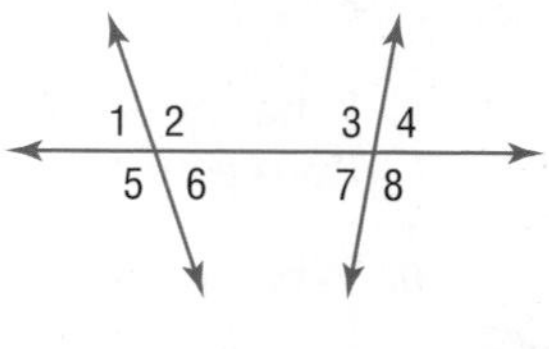

5. ∠1 and ∠8

6. ∠2 and ∠4

7. ∠3 and ∠6

8. ∠6 and ∠7

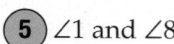

Example 3 **Identify the transversal connecting each pair of angles. Then classify the relationship between each pair of angles.**

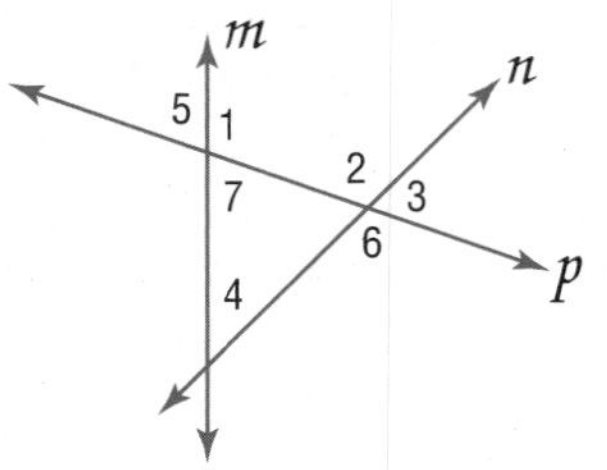

9. $\angle 2$ and $\angle 4$ **10.** $\angle 5$ and $\angle 6$

11. $\angle 4$ and $\angle 7$ **12.** $\angle 2$ and $\angle 7$

Practice and Problem Solving

Example 1 **Refer to the figure to identify each of the following.**

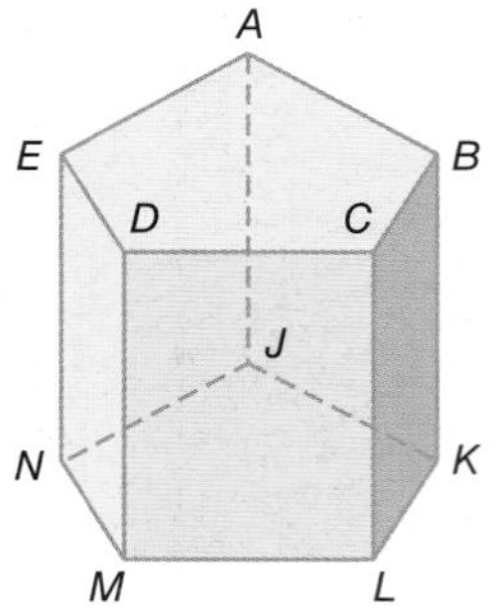

13. all segments parallel to $\overline{DM}$

14. a plane parallel to plane ACD

15 a segment skew to $\overline{BC}$

16. all planes intersecting plane EDM

17. all segments skew to $\overline{AE}$

18. a segment parallel to $\overline{EN}$

19. a segment parallel to $\overline{AB}$ through point J

20. a segment skew to $\overline{CL}$ through point E

Examples 2–3 CCSS **PRECISION** **Identify the transversal connecting each pair of angles. Then classify the relationship between each pair of angles as *alternate interior, alternate exterior, corresponding,* or *consecutive interior* angles.**

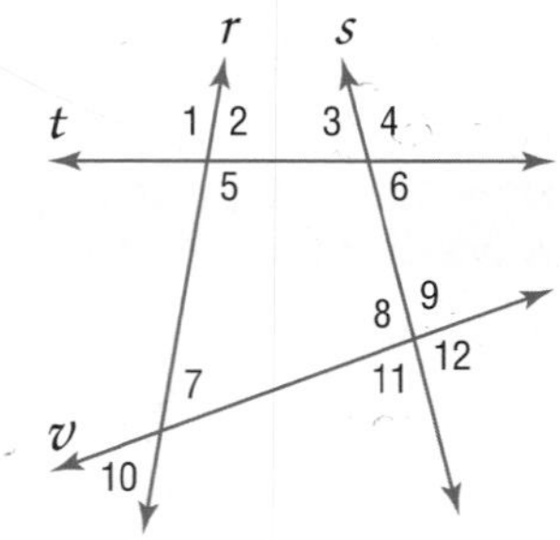

21. $\angle 4$ and $\angle 9$ **22.** $\angle 5$ and $\angle 7$

23. $\angle 3$ and $\angle 5$ **24.** $\angle 10$ and $\angle 11$

25. $\angle 1$ and $\angle 6$ **26.** $\angle 6$ and $\angle 8$

27. $\angle 2$ and $\angle 3$ **28.** $\angle 9$ and $\angle 10$

29. $\angle 4$ and $\angle 11$ **30.** $\angle 7$ and $\angle 11$

Example 3 **SAFETY** **Identify the transversal connecting each pair of angles in the photo of a fire escape shown. Then classify the relationship between each pair of angles.**

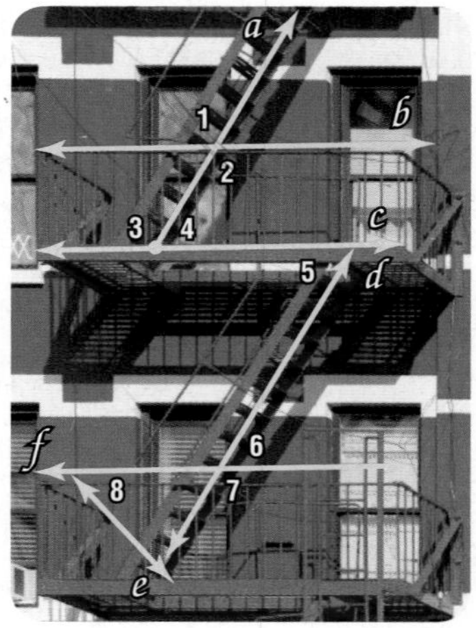

31. $\angle 1$ and $\angle 2$ **32.** $\angle 2$ and $\angle 4$

33. $\angle 4$ and $\angle 5$ **34.** $\angle 6$ and $\angle 7$

35. $\angle 7$ and $\angle 8$ **36.** $\angle 2$ and $\angle 3$

37. **POWER** Power lines are not allowed to intersect.

a. What must be the relationship between power lines p and m? Explain your reasoning.

b. What is the relationship between line q and lines p and m?

Describe the relationship between each pair of segments as *parallel, skew,* or *intersecting*.

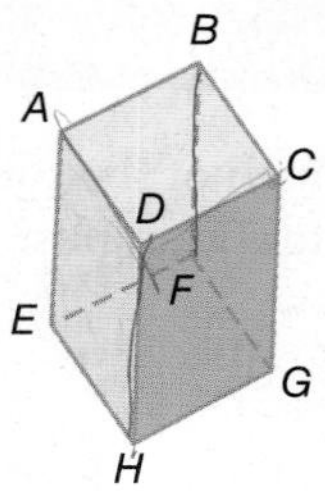

38. $\overline{FG}$ and $\overline{BC}$

39. $\overline{AB}$ and $\overline{CG}$

40. $\overline{DH}$ and $\overline{HG}$

41. $\overline{DH}$ and $\overline{BF}$

42. $\overline{EF}$ and $\overline{BC}$

43. $\overline{CD}$ and $\overline{AD}$

44. **CCSS SENSE-MAKING** The illusion at the right is created using squares and straight lines.

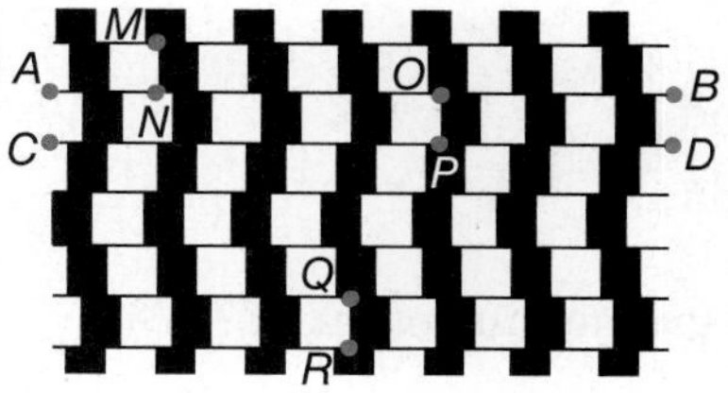

a. How are $\overline{AB}$ and $\overline{CD}$ related? Justify your reasoning.

b. How are $\overline{MN}$ and $\overline{QR}$ related? $\overline{AB}$, $\overline{CD}$, and $\overline{OP}$?

45 **ESCALATORS** Escalators consist of steps on a continuous loop that is driven by a motor. At the top and bottom of the platform, the steps collapse to provide a level surface for entrance and exit.

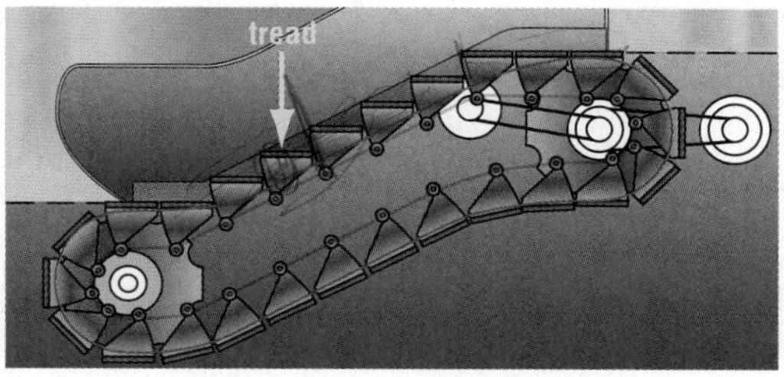

a. What is the relationship between the treads of the ascending stairs?

b. What is the relationship between the treads of the two steps at the top of the incline?

c. How do the treads of the steps on the incline of the escalator relate to the treads of the steps on the bottom of the escalator?

H.O.T. Problems Use Higher-Order Thinking Skills

46. **OPEN ENDED** Plane $\mathcal{P}$ contains lines a and b. Line c intersects plane $\mathcal{P}$ at point J. Lines a and b are parallel, lines a and c are skew, and lines b and c are not skew. Draw a figure based upon this description.

47. **CHALLENGE** Suppose points A, B, and C lie in plane $\mathcal{P}$, and points D, E, and F lie in plane Q. Line m contains points D and F and does not intersect plane $\mathcal{P}$. Line n contains points A and E.

a. Draw a diagram to represent the situation.

b. What is the relationship between planes $\mathcal{P}$ and Q?

c. What is the relationship between lines m and n?

REASONING **Plane $\mathcal{X}$ and plane $\mathcal{Y}$ are parallel and plane $\mathcal{Z}$ intersects plane $\mathcal{X}$. Line $\overleftrightarrow{AB}$ is in plane $\mathcal{X}$, line $\overleftrightarrow{CD}$ is in plane $\mathcal{Y}$, and line $\overleftrightarrow{EF}$ is in plane $\mathcal{Z}$. Determine whether each statement is *always, sometimes,* or *never* true. Explain.**

48. $\overleftrightarrow{AB}$ is skew to $\overleftrightarrow{CD}$.

49. $\overleftrightarrow{AB}$ intersects $\overleftrightarrow{EF}$.

50. **WRITING IN MATH** Can a pair of planes be described as skew? Explain.

Standardized Test Practice

51. Which of the following angle pairs are alternate exterior angles?

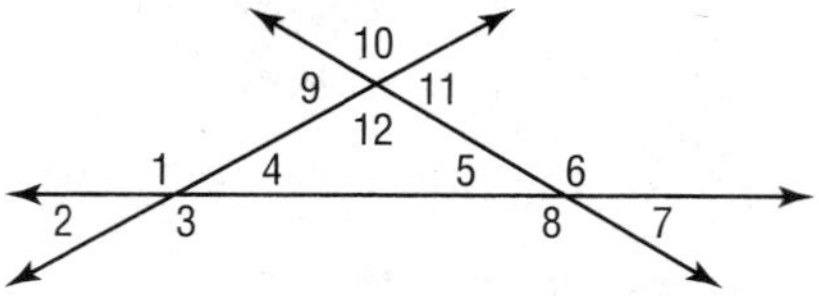

A $\angle 1$ and $\angle 5$ **C** $\angle 2$ and $\angle 10$

B $\angle 2$ and $\angle 6$ **D** $\angle 5$ and $\angle 9$

52. What is the measure of $\angle XYZ$?

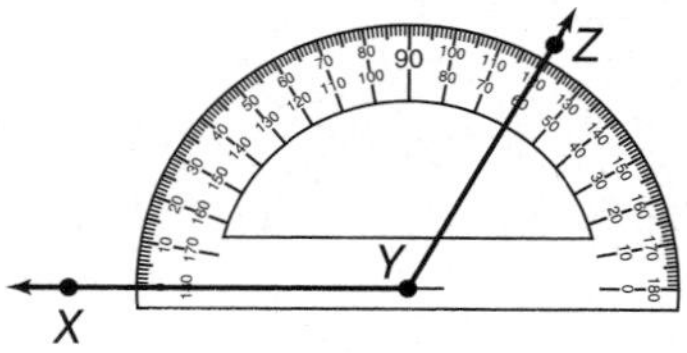

F 30° **H** 120°

G 60° **J** 150°

53. SHORT RESPONSE Name the coordinates of the points representing the x- and y-intercepts of the graph shown below.

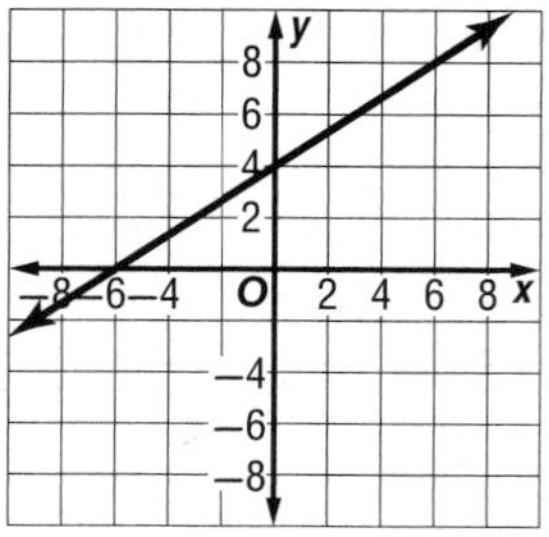

54. SAT/ACT Of the following, the one that is *not* equivalent to 485 is:

A $(3 \times 100) + (4 \times 10) + 145$
B $(3 \times 100) + (18 \times 10) + 5$
C $(4 \times 100) + (8 \times 10) + 15$
D $(4 \times 100) + (6 \times 10) + 25$
E $(4 \times 100) + (5 \times 10) + 35$

Spiral Review

Find the measure of each numbered angle. (Lesson 10-8)

55. $m\angle 9 = 2x - 4$, $m\angle 10 = 2x + 4$

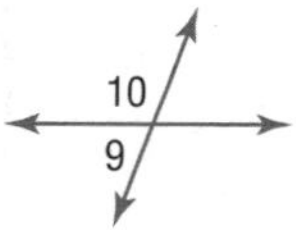

56. $m\angle 11 = 4x$, $m\angle 12 = 2x - 6$

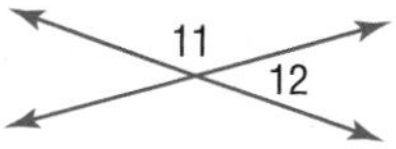

57. $m\angle 19 = 100 + 20x$, $m\angle 20 = 20x$

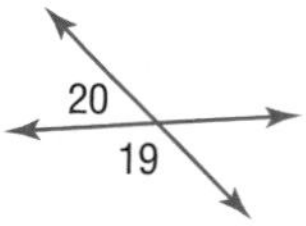

58. PROOF Prove the following. (Lesson 10-7)

Given: $\overline{WY} \cong \overline{ZX}$
A is the midpoint of $\overline{WY}$.
A is the midpoint of $\overline{ZX}$.

Prove: $\overline{WA} \cong \overline{ZA}$

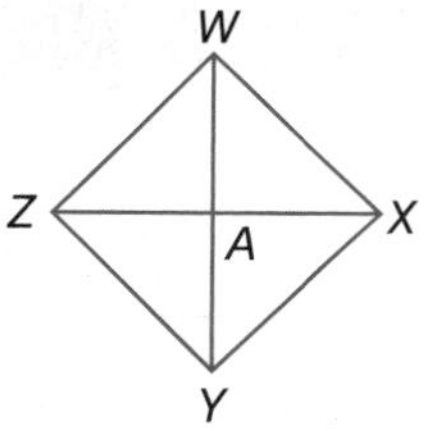

Skills Review

Find x.

59.

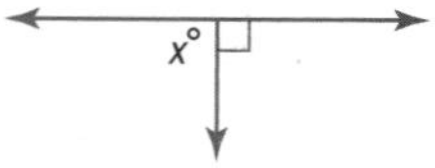

60.

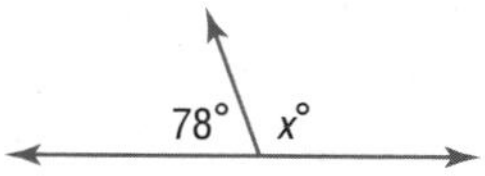

61.

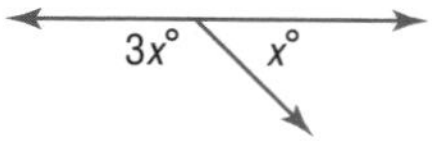

EXPLORE

11-2 Geometry Software Lab
Angles and Parallel Lines

You can use The Geometer's Sketchpad® to explore the angles formed by two parallel lines and a transversal.

CCSS Common Core State Standards
Content Standards
G.CO.12 Make formal geometric constructions with a variety of tools and methods (compass and straightedge, string, reflective devices, paper folding, dynamic geometric software, etc.).
Mathematical Practices 5

Activity Parallel Lines and a Transversal

Step 1 **Draw a line.**

Draw and label points F and G. Then use the line tool to draw $\overleftrightarrow{FG}$.

Step 2 **Draw a parallel line.**

Draw a point that is not on $\overleftrightarrow{FG}$ and label it J. Select $\overleftrightarrow{FG}$ and point J, and then choose **Parallel Line** from the **Construct** menu. Draw and label a point K on this parallel line.

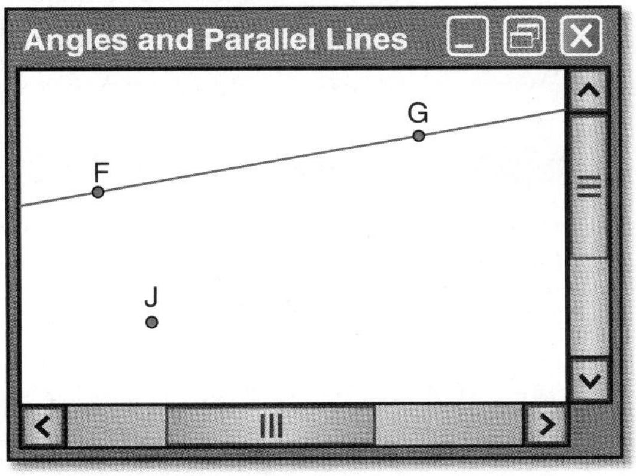

Step 3 **Draw a transversal.**

Draw and label point A on $\overleftrightarrow{FG}$ and point B on $\overleftrightarrow{JK}$. Select A and B and then choose **Line** from the **Construct** menu to draw transversal $\overleftrightarrow{AB}$. Then draw and label points C and D on $\overleftrightarrow{AB}$ as shown.

Step 4 **Measure each angle.**

Measure all eight angles formed by these lines. For example, select points F, A, then C, and choose **Angle** from the **Measure** menu to find $m\angle FAC$.

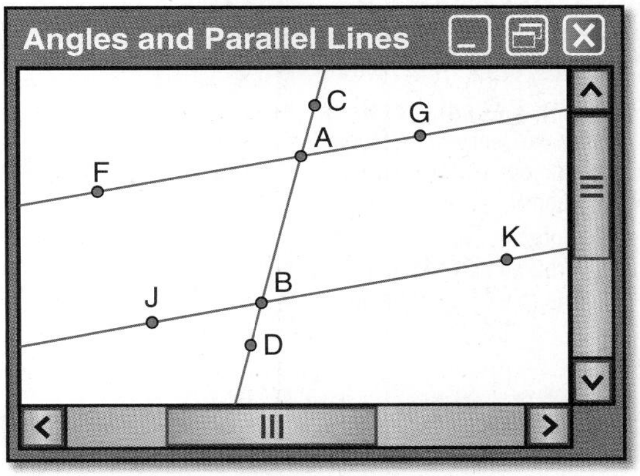

Analyze the Results

1. Record the measures from Step 4 in a table like this one. Which angles have the same measure?

Angle	$\angle FAC$	$\angle CAG$	$\angle GAB$	$\angle FAB$	$\angle JBA$	$\angle ABK$	$\angle KBD$	$\angle JBD$
1st Measure								

2. Drag point C or D to move transversal $\overleftrightarrow{AB}$ so that it intersects the two parallel lines at a different angle. Add a row **2nd Measure** to your table and record the new measures. Repeat these steps until your table has 3rd, 4th, and 5th Measure rows of data.

3. Using the angles listed in the table, identify and describe the relationship between all angle pairs that have the following special names. Then write a conjecture in if-then form about each angle pair when formed by any two parallel lines cut by a transversal.

   a. corresponding **b.** alternate interior **c.** alternate exterior **d.** consecutive interior

4. Drag point C or D so that the measure of any of the angles is 90.

   a. What do you notice about the measures of the other angles?

   b. Make a conjecture about a transversal that is perpendicular to one of two parallel lines.

LESSON 11-2 Angles and Parallel Lines

Then

- You named angle pairs formed by parallel lines and transversals.

Now

1. Use theorems to determine the relationships between specific pairs of angles.
2. Use algebra to find angle measurements.

Why?

- Construction and maintenance workers often use an access scaffold. This structure provides support and access to elevated areas. The transversal *t* shown provides structural support to the two parallel working areas.

CCSS Common Core State Standards

Content Standards

G.CO.1 Know precise definitions of angle, circle, perpendicular line, parallel line, and line segment, based on the undefined notions of point, line, distance along a line, and distance around a circular arc.

G.CO.9 Prove theorems about lines and angles.

Mathematical Practices

1 Make sense of problems and persevere in solving them.

3 Construct viable arguments and critique the reasoning of others.

1 Parallel Lines and Angle Pairs

In the photo, line *t* is a transversal of lines *a* and *b*, and ∠1 and ∠2 are corresponding angles. Since lines *a* and *b* are parallel, there is a special relationship between corresponding angle pairs.

Postulate 11.1 Corresponding Angles Postulate

If two parallel lines are cut by a transversal, then each pair of corresponding angles is congruent.

Examples ∠1 ≅ ∠3, ∠2 ≅ ∠4, ∠5 ≅ ∠7, ∠6 ≅ ∠8

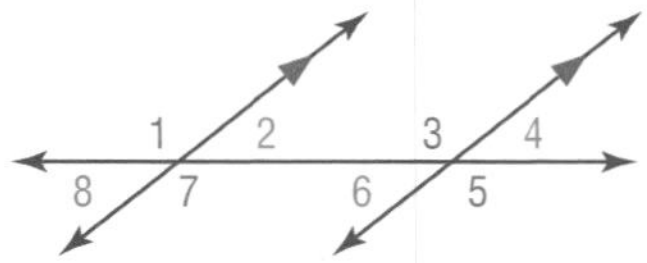

PT

Example 1 Use Corresponding Angles Postulate

In the figure, $m\angle 5 = 72$. Find the measure of each angle. Tell which postulate(s) or theorem(s) you used.

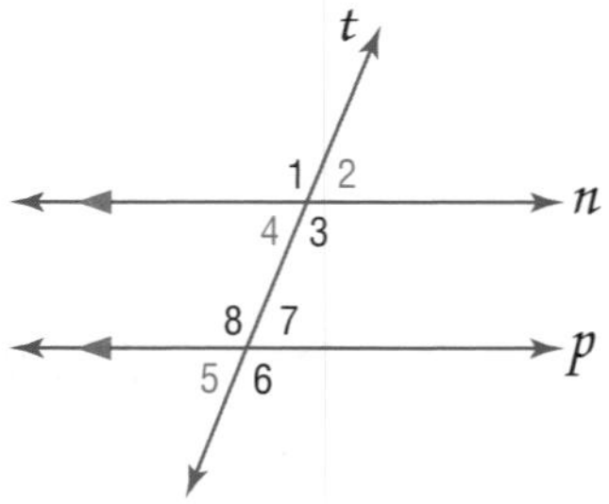

a. ∠4

∠4 ≅ ∠5	Corresponding Angles Postulate
$m\angle 4 = m\angle 5$	Definition of congruent angles
$m\angle 4 = 72$	Substitution

b. ∠2

∠2 ≅ ∠4	Vertical Angles Theorem
∠4 ≅ ∠5	Corresponding Angles Postulate
∠2 ≅ ∠5	Transitive Property of Congruence
$m\angle 2 = m\angle 5$	Definition of congruent angles
$m\angle 2 = 72$	Substitution

Guided Practice

In the figure, suppose that $m\angle 8 = 105$. Find the measure of each angle. Tell which postulate(s) or theorem(s) you used.

1A. ∠1 **1B.** ∠2 **1C.** ∠3

In Example 1, ∠2 and ∠5 are congruent alternate exterior angles. This and other examples suggest the following theorems about the other angle pairs formed by two parallel lines cut by a transversal.

StudyTip

Angle Relationships These theorems generalize the relationships between specific pairs of angles. If you get confused about the relationships, you can verify them with the methods you used in Example 1, using only corresponding, vertical, and supplementary angles.

Theorems Parallel Lines and Angle Pairs

11.1 Alternate Interior Angles Theorem If two parallel lines are cut by a transversal, then each pair of alternate interior angles is congruent.

Examples $\angle 1 \cong \angle 3$ and $\angle 2 \cong \angle 4$

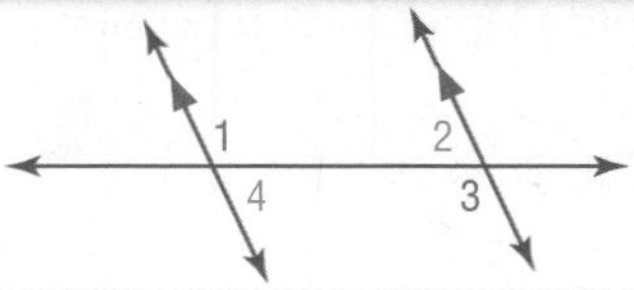

11.2 Consecutive Interior Angles Theorem If two parallel lines are cut by a transversal, then each pair of consecutive interior angles is supplementary.

Examples $\angle 1$ and $\angle 2$ are supplementary.
$\angle 3$ and $\angle 4$ are supplementary.

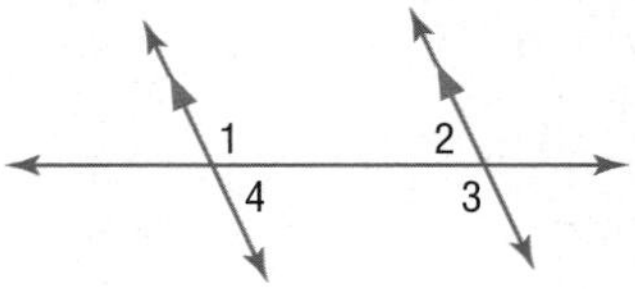

11.3 Alternate Exterior Angles Theorem If two parallel lines are cut by a transversal, then each pair of alternate exterior angles is congruent.

Examples $\angle 5 \cong \angle 7$ and $\angle 6 \cong \angle 8$

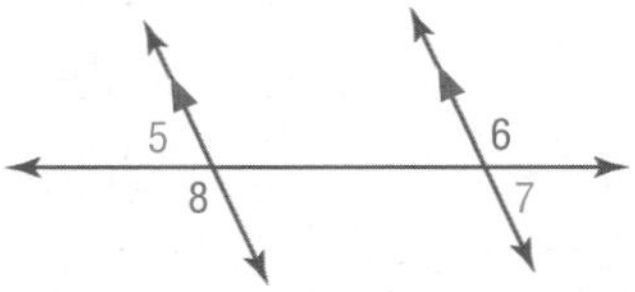

You will prove Theorems 11.2 and 11.3 in Exercises 30 and 35, respectively.

Since postulates are accepted without proof, you can use the Corresponding Angles Postulate to prove each of the theorems above.

Proof Alternate Interior Angles Theorem

Given: $a \parallel b$
t is a transversal of a and b.

Prove: $\angle 4 \cong \angle 5$, $\angle 3 \cong \angle 6$

Paragraph Proof: We are given that $a \parallel b$ with a transversal t. By the Corresponding Angles Postulate, corresponding angles are congruent. So, $\angle 2 \cong \angle 4$ and $\angle 6 \cong \angle 8$. Also, $\angle 5 \cong \angle 2$ and $\angle 8 \cong \angle 3$ because vertical angles are congruent. Therefore, $\angle 5 \cong \angle 4$ and $\angle 3 \cong \angle 6$ since congruence of angles is transitive.

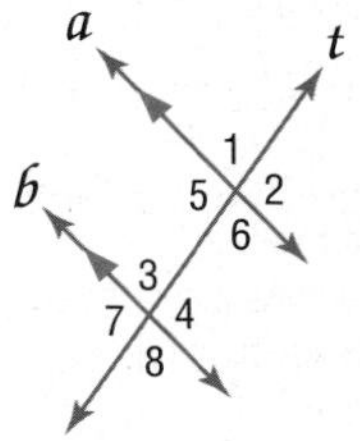

PT

Real-World Example 2 Use Theorems about Parallel Lines

COMMUNITY PLANNING Redding Lane and Creek Road are parallel streets that intersect Park Road along the west side of Wendell Park. If $m\angle 1 = 118$, find $m\angle 2$.

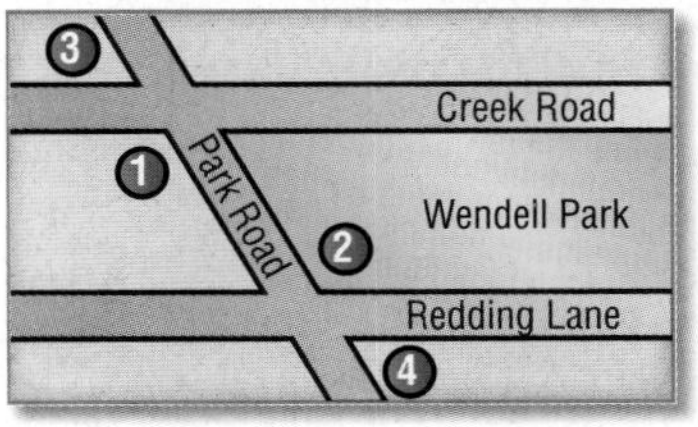

$\angle 2 \cong \angle 1$	Alternate Interior Angles Postulate
$m\angle 2 = m\angle 1$	Definition of congruent angles
$m\angle 2 = 118$	Substitution

Glowimages/Getty Images

Real-WorldLink

Some cities require that streets in newly planned subdivisions intersect at no less than a 60° angle.

GuidedPractice

COMMUNITY PLANNING Refer to the diagram above to find each angle measure. Tell which postulate(s) or theorem(s) you used.

2A. If $m\angle 1 = 100$, find $m\angle 4$.

2B. If $m\angle 3 = 70$, find $m\angle 4$.

2 Algebra and Angle Measures

The special relationships between the angles formed by two parallel lines and a transversal can be used to find unknown values.

Example 3 Find Values of Variables

PT

ALGEBRA Use the figure at the right to find the indicated variable. Explain your reasoning.

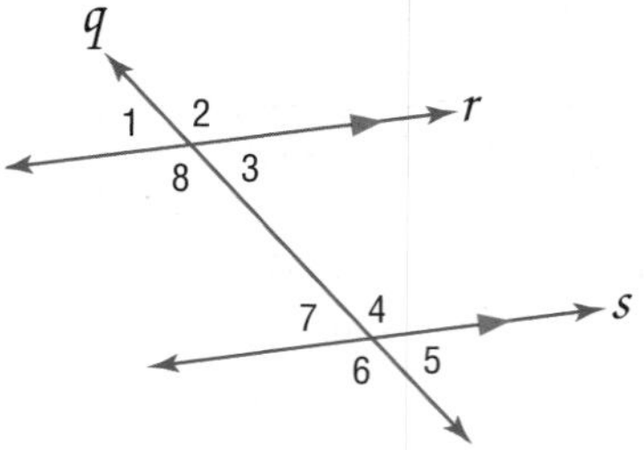

a. If $m\angle 4 = 2x - 17$ and $m\angle 1 = 85$, find x.

$\angle 3 \cong \angle 1$	Vertical Angles Theorem
$m\angle 3 = m\angle 1$	Definition of congruent angles
$m\angle 3 = 85$	Substitution

Since lines r and s are parallel, $\angle 4$ and $\angle 3$ are supplementary by the Consecutive Interior Angles Theorem.

$m\angle 3 + m\angle 4 = 180$	Definition of supplementary angles
$85 + 2x - 17 = 180$	Substitution
$2x + 68 = 180$	Simplify.
$2x = 112$	Subtract 68 from each side.
$x = 56$	Divide each side by 2.

b. Find y if $m\angle 3 = 4y + 30$ and $m\angle 7 = 7y + 6$.

$\angle 3 \cong \angle 7$	Alternate Interior Angles Theorem
$m\angle 3 = m\angle 7$	Definition of congruent angles
$4y + 30 = 7y + 6$	Substitution
$30 = 3y + 6$	Subtract 4y from each side.
$24 = 3y$	Subtract 6 from each side.
$8 = y$	Divide each side by 3.

StudyTip

CCSS Precision The postulates and theorems you will be studying in this lesson only apply to *parallel* lines cut by a transversal. You should assume that lines are parallel only if the information is given or the lines are marked with parallel arrows.

GuidedPractice

3A. If $m\angle 2 = 4x + 7$ and $m\angle 7 = 5x - 13$, find x.

3B. Find y if $m\angle 5 = 68$ and $m\angle 3 = 3y - 2$.

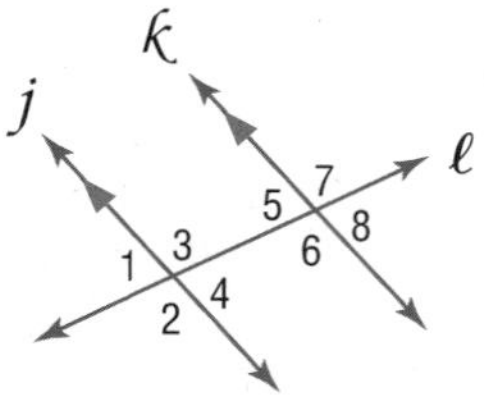

A special relationship exists when the transversal of two parallel lines is a perpendicular line.

Theorem 11.4 Perpendicular Transversal Theorem

In a plane, if a line is perpendicular to one of two parallel lines, then it is perpendicular to the other.

Examples If line $a \parallel$ line b and line $a \perp$ line t, then line $b \perp$ line t.

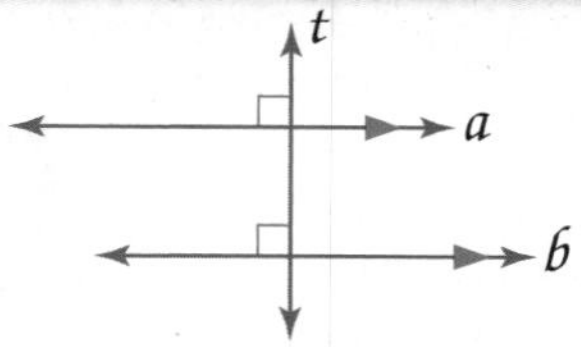

You will prove Theorem 11.4 in Exercise 37.

Check Your Understanding

Example 1 **In the figure, $m\angle 2 = 85$. Find the measure of each angle. Tell which postulate(s) or theorem(s) you used.**

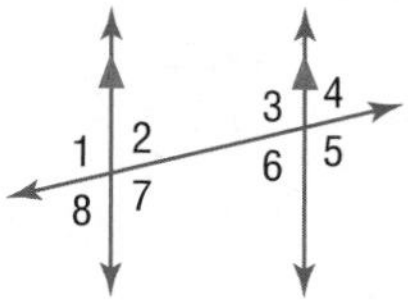

1. $\angle 4$ **2.** $\angle 6$ **3.** $\angle 7$

Example 2 **In the figure, $m\angle 6 = 110$. Find the measure of each angle. Tell which postulate(s) or theorem(s) you used.**

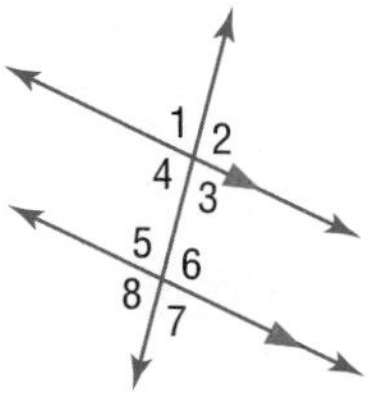

4. $\angle 4$ **5.** $\angle 3$ **6.** $\angle 1$

Example 3 **Find the value of the variable(s) in each figure. Explain your reasoning.**

7.

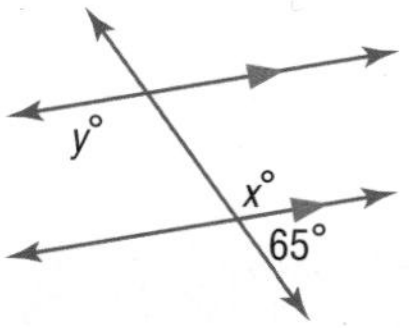

8.

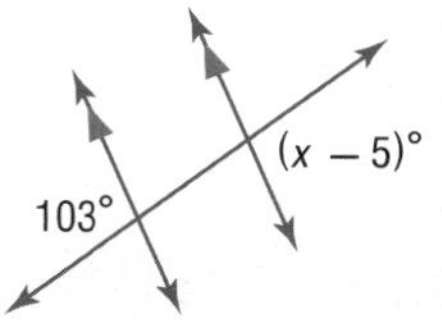

9.

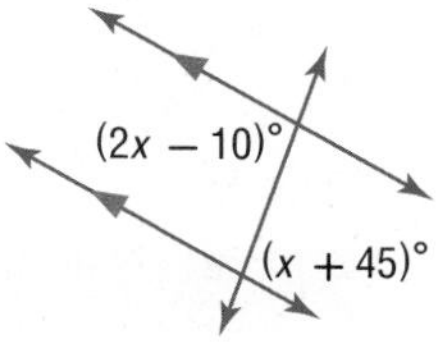

Practice and Problem Solving

Examples 1–2 **In the figure, $m\angle 11 = 23$ and $m\angle 14 = 17$. Find the measure of each angle. Tell which postulate(s) or theorem(s) you used.**

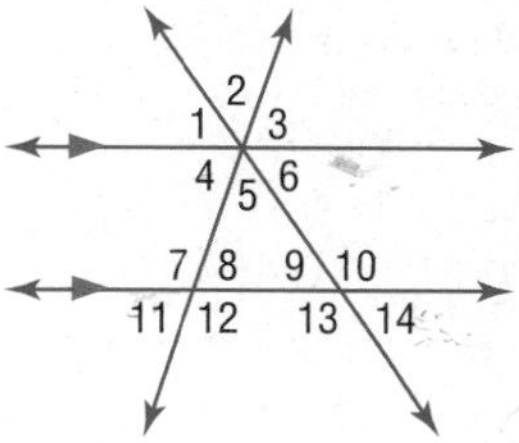

10. $\angle 4$ **11.** $\angle 3$ **12.** $\angle 12$

13. $\angle 8$ **14.** $\angle 6$ **15.** $\angle 2$

16. $\angle 10$ **17.** $\angle 5$ **18.** $\angle 1$

Example 3 **CCSS SATELLITE RECEIVER** A television dish collects the signal by directing the radiation from the satellite to a receiver located at the focal point of the dish. Assume that the radiation rays are parallel. Determine the relationship between each pair of angles, and explain your reasoning.

19. $\angle 1$ and $\angle 2$ **20.** $\angle 1$ and $\angle 3$ **21.** $\angle 2$ and $\angle 4$ **22.** $\angle 1$ and $\angle 4$

Find the value of the variable(s) in each figure. Explain your reasoning.

23.

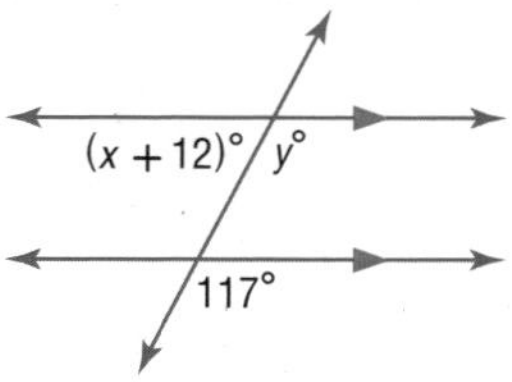

24.

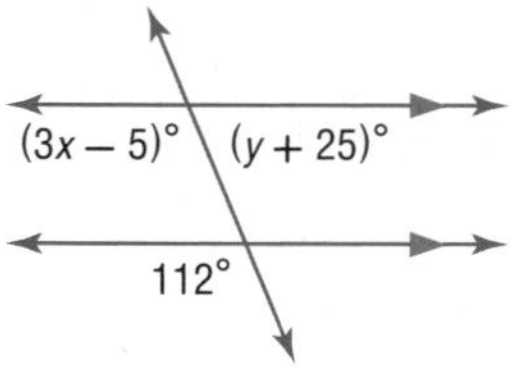

25.

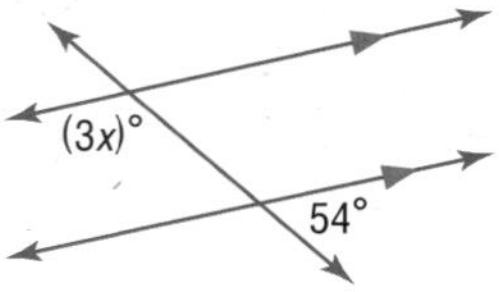

26.

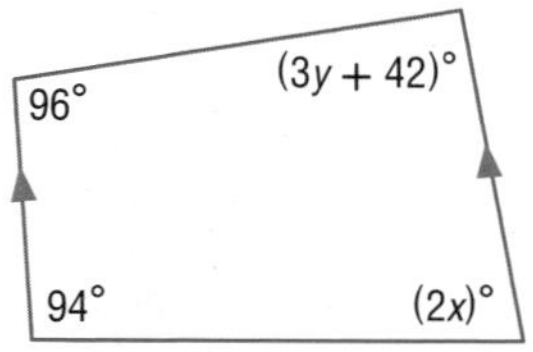

27.

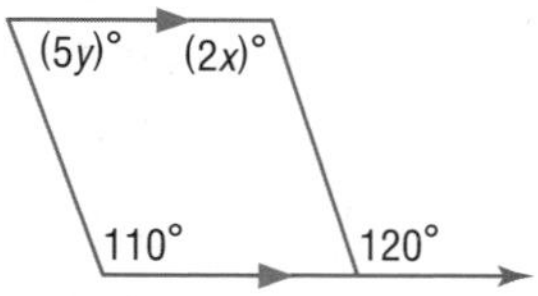

28.

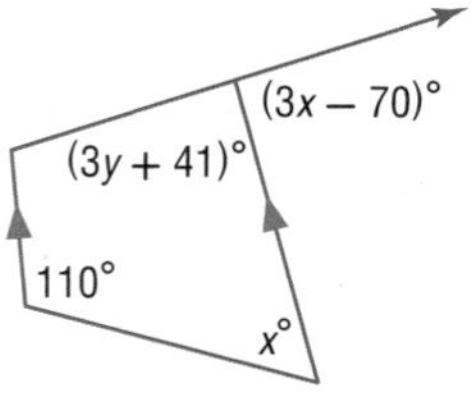

29. PROOF Copy and complete the proof of Theorem 11.2.

Given: $m \parallel n$; ℓ is a transversal.

Prove: $\angle 1$ and $\angle 2$ are supplementary; $\angle 3$ and $\angle 4$ are supplementary.

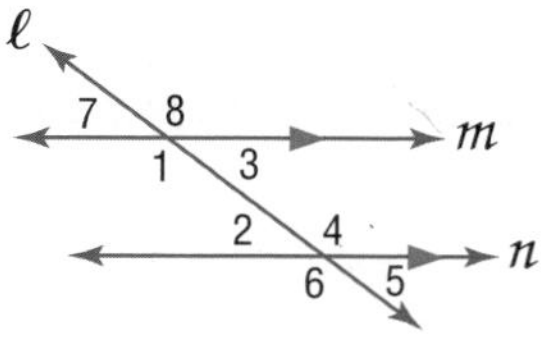

Proof:

Statements	Reasons
a. ?	**a.** Given
b. $\angle 1$ and $\angle 3$ form a linear pair; $\angle 2$ and $\angle 4$ form a linear pair.	**b.** ?
c. ?	**c.** If two angles form a linear pair, then they are supplementary.
d. $\angle 1 \cong \angle 4$, $\angle 2 \cong \angle 3$	**d.** ?
e. $m\angle 1 = m\angle 4$, $m\angle 2 = m\angle 3$	**e.** Definition of Congruence
f. ?	**f.** ?

STORAGE When industrial shelving needs to be accessible from either side, additional support is provided on the side by transverse members. Determine the relationship between each pair of angles and explain your reasoning.

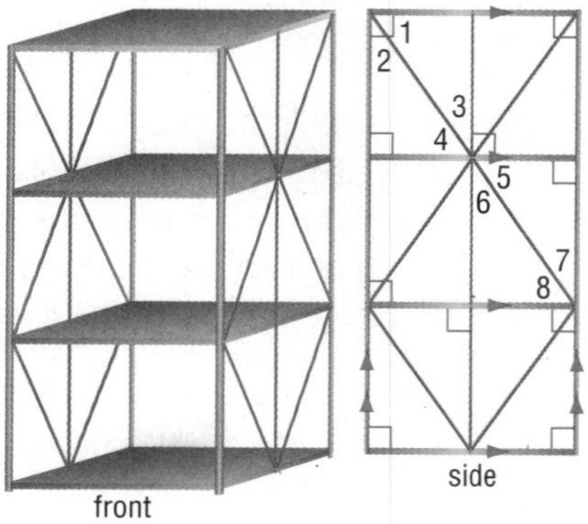

30. $\angle 2$ and $\angle 7$

31. $\angle 3$ and $\angle 7$

32. $\angle 4$ and $\angle 5$

33. $\angle 5$ and $\angle 6$

34. PROOF Write a two-column proof of the Alternate Exterior Angles Theorem. (Theorem 11.3)

35. BRIDGES Refer to the diagram of the Truss Bridge at the right. The two horizontal supports of the bridge are parallel.

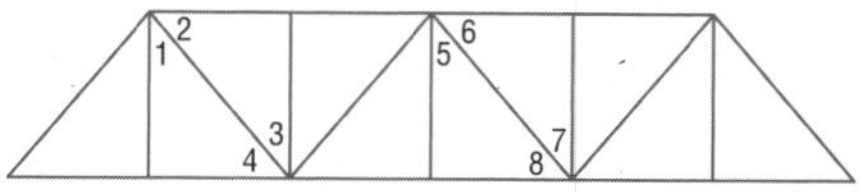

a. Write a conjecture about the even-numbered angles. Explain your reasoning.

b. Write a conjecture about the odd-numbered angles. Explain your reasoning.

c. Write a conjecture about any pair of angles in which one of the angles is an odd-numbered and the other is even-numbered. Explain your reasoning.

36. PROOF In a plane, prove that if a line is perpendicular to one of two parallel lines, then it is perpendicular to the other. (Theorem 11.4)

CCSS TOOLS Find x. (*Hint:* Draw an auxiliary line.)

37.

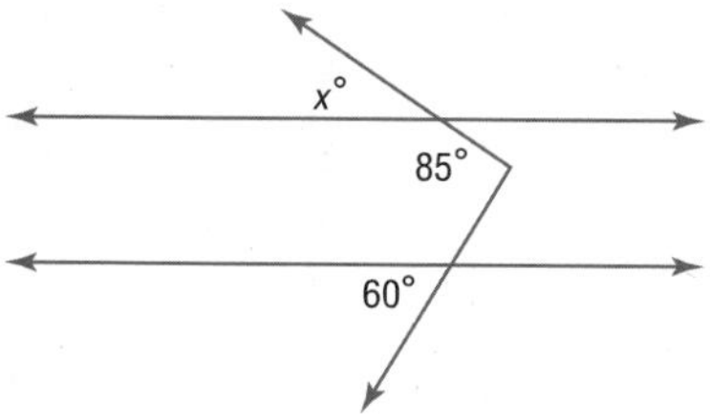

38.

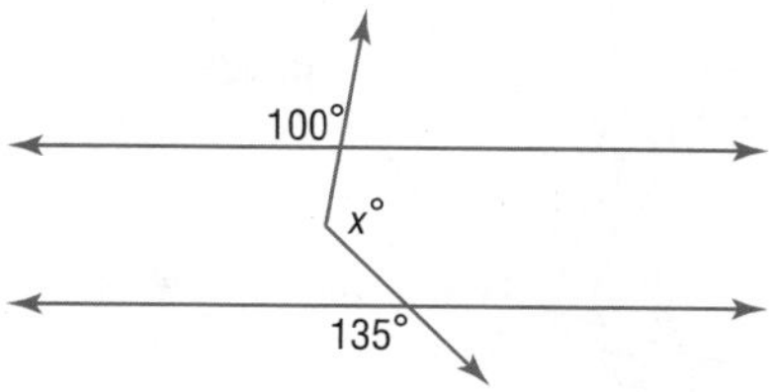

39. Draw a pair of parallel lines, x and y. Draw a line, w, that is a transversal that cuts through lines x and y. Number the angles so that the odd numbers are on one side of the transversal and the even numbers are on the other side of the transversal.

a. List all possible pairs of even-numbered angles. State the relationship between each pair.

b. List all possible pairs of odd-numbered angles. State the relationship between each pair.

c. If you were to select two angles at random, how many possible angle pairings are there?

d. What are the possible relationship(s) between the angle pairings?

e. What is the probability of selecting a pair of congruent angles?

f. What is the probability of selecting a pair of supplementary angles?

H.O.T. Problems Use Higher-Order Thinking Skills

40. WRITING IN MATH If line a is parallel to line b and $\angle 5 \cong \angle 6$, describe the relationship between lines a and c. Explain your reasoning.

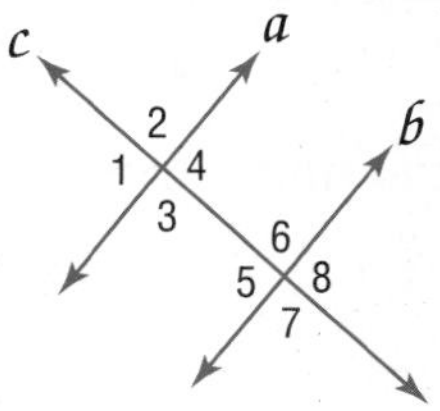

41. WRITING IN MATH Compare and contrast the Alternate Exterior Angles Theorem and the Consecutive Exterior Angles Theorem.

42. OPEN ENDED Draw a pair of parallel lines cut by a transversal and measure the two exterior angles on the same side of the transversal. Include the measures on your drawing. Based on the pattern you have seen for naming other pairs of angles, what do you think the name of the pair you measured would be?

43. CHALLENGE Find x and y.

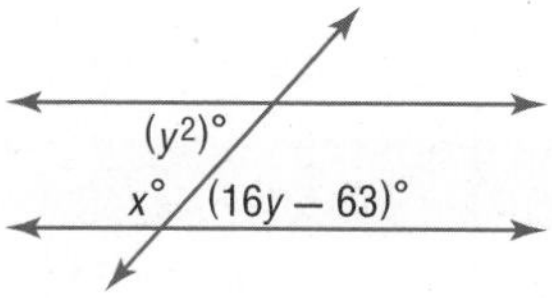

REASONING Determine whether the following statement is *sometimes, always, or never* true. Explain your reasoning.

44. If two parallel lines are cut by a transversal and the measure of one of the angles is known, then the measure of all of the other angles are also known.

Standardized Test Practice

45. Suppose $\angle 4$ and $\angle 5$ form a linear pair. If $m\angle 1 = 2x$, $m\angle 2 = 3x - 20$, and $m\angle 3 = x - 4$, what is $m\angle 3$?

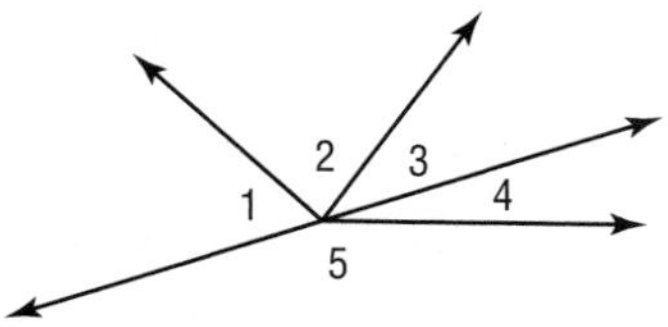

A 26°
B 28°
C 30°
D 32°

46. SAT/ACT A farmer raises chickens and pigs. If his animals have a total of 120 heads and a total of 300 feet, how many chickens does the farmer have?

F 60
G 70
H 80
J 90

47. SHORT RESPONSE If $m \parallel n$, then which of the following statements must be true?

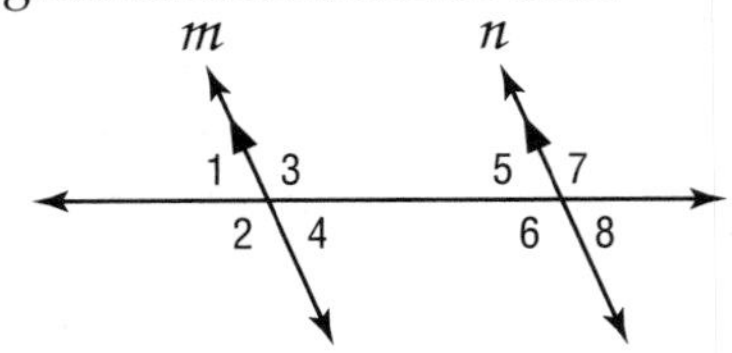

I. $\angle 3$ and $\angle 6$ are Alternate Interior Angles.
II. $\angle 4$ and $\angle 6$ are Consecutive Interior Angles.
III. $\angle 1$ and $\angle 7$ are Alternate Exterior Angles.

48. ALGEBRA If $-2 + x = -6$, then $-17 - x = ?$

A -13
B -4
C 9
D 13
E 21

Spiral Review

49. AVIATION Airplanes are assigned an altitude level based on the direction they are flying. If one airplane is flying northwest at 34,000 feet and another airplane is flying east at 25,000 feet, describe the type of lines formed by the paths of the airplanes. Explain your reasoning. (Lesson 11-1)

Use the given statement to find the measure of each numbered angle. (Lesson 10-8)

50. $\angle 1$ and $\angle 2$ form a linear pair and $m\angle 2 = 67$.

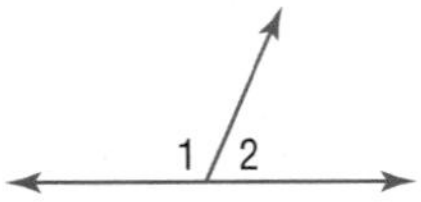

51. $\angle 6$ and $\angle 8$ are complementary and $m\angle 8 = 47$.

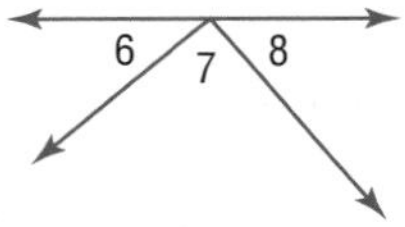

52. $m\angle 4 = 32$

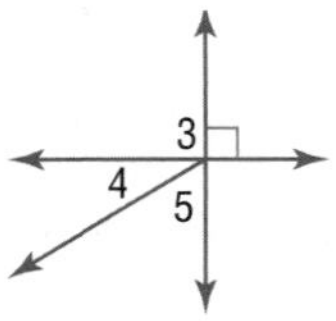

Skills Review

Simplify each expression.

53. $\dfrac{6-5}{4-2}$

54. $\dfrac{-5-2}{4-7}$

55. $\dfrac{-11-4}{12-(-9)}$

56. $\dfrac{16-12}{15-11}$

57. $\dfrac{10-22}{8-17}$

58. $\dfrac{8-17}{12-(-3)}$

EXPLORE 11-3

Graphing Technology Lab
Investigating Slope

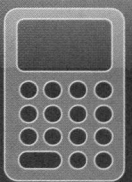

The rate of change of the steepness of a line is called the *slope*. Slope can be used to investigate the relationship between real-world quantities.

CCSS Common Core State Standards
Content Standards
G.GPE.5 Prove the slope criteria for parallel and perpendicular lines and use them to solve geometric problems (e.g., find the equation of a line parallel or perpendicular to a given line that passes through a given point).
Mathematical Practices 5

Set Up the Lab

- Connect a data collection device to a graphing calculator. Place the device on a desk or table so that it can read the motion of a walker.
- Mark the floor at distances of 1 meter and 6 meters from the device.

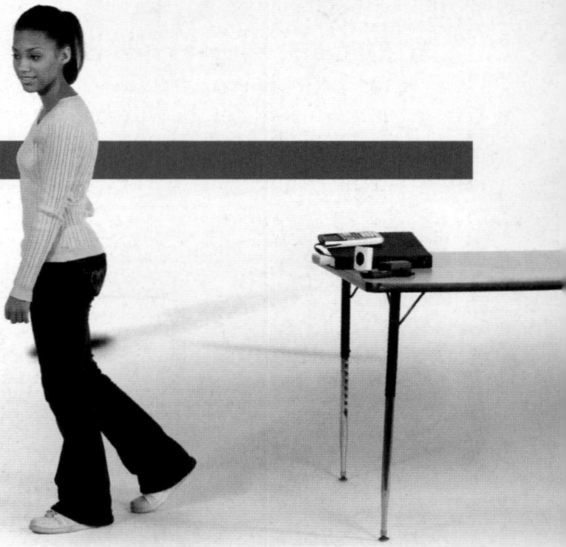

Activity

Step 1 Have one group member stand at the 1-meter mark. When another group member presses the button to begin collecting data, the walker should walk away from the device at a slow, steady pace.

Step 2 Stop collecting data when the walker passes the 6-meter mark. Save the data as Trial 1.

Step 3 Repeat the experiment, walking more quickly. Save the data as Trial 2.

Step 4 For Trial 3, repeat the experiment by slowly walking toward the data collection device.

Step 5 Repeat the experiment, walking quickly toward the device. Save the data as Trial 4.

Analyze the Results

1. Compare and contrast the graphs for Trials 1 and 2. How do the graphs for Trials 1 and 3 compare?

2. Use the **TRACE** feature of the calculator to find the coordinates of two points on each graph. Record the coordinates in a table like the one shown. Then use the points to find the slope of the line.

Trial	Point A (x_1, y_1)	Point B (x_2, y_2)	Slope $= \frac{y_2 - y_1}{x_2 - x_1}$
1			
2			
3			
4			

3. Compare and contrast the slopes for Trials 1 and 2. How do the slopes for Trials 1 and 2 compare to the slopes for Trials 3 and 4?

4. The slope of a line describes the rate of change of the quantities represented by the x- and y-values. What is represented by the rate of change in this experiment?

5. **MAKE A CONJECTURE** What would the graph look like if you were to collect data while the walker was standing still? Use the data collection device to test your conjecture.

LESSON 11-3

Slopes of Lines

Then

- You used the properties of parallel lines to determine congruent angles.

Now

1. Find slopes of lines.
2. Use slope to identify parallel and perpendicular lines.

Why?

- Ski resorts assign ratings to their ski trails according to their difficulty. A primary factor in determining this rating is a trail's steepness or *slope gradient*. A trail with a 6% or $\frac{6}{100}$ grade falls 6 feet vertically for every 100 feet traveled horizontally.

The easiest trails, labeled ●, have slopes ranging from 6% to 25%, while more difficult trails, labeled ♦ or ♦♦, have slopes of 40% or greater.

NewVocabulary
slope
rate of change

Common Core State Standards

Content Standards
G.GPE.5 Prove the slope criteria for parallel and perpendicular lines and use them to solve geometric problems (e.g., find the equation of a line parallel or perpendicular to a given line that passes through a given point).

Mathematical Practices
4 Model with mathematics.
7 Look for and make use of structure.
8 Look for and express regularity in repeated reasoning.

1 Slope of a Line

The steepness or slope of a hill is described by the ratio of the hill's vertical rise to its horizontal run. In algebra, you learned that the slope of a line in the coordinate plane can be calculated using any two points on the line.

KeyConcept Slope of a Line

In a coordinate plane, the **slope** of a line is the ratio of the change along the y-axis to the change along the x-axis between any two points on the line.

The slope m of a line containing two points with coordinates (x_1, y_1) and (x_2, y_2) is given by the formula

$$m = \frac{y_2 - y_1}{x_2 - x_1}, \text{ where } x_1 \neq x_2.$$

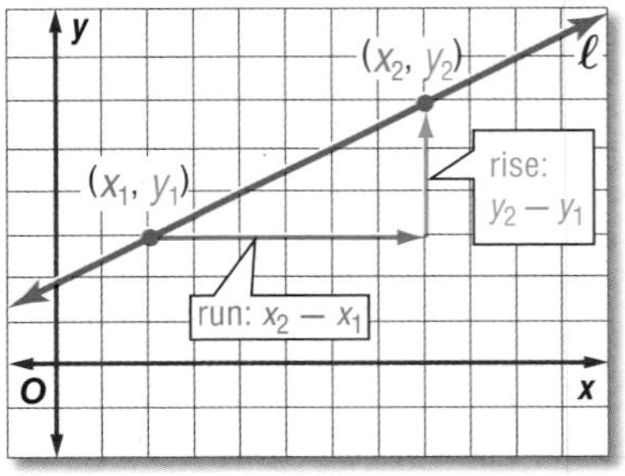

$$m = \frac{\text{rise}}{\text{run}} = \frac{y_2 - y_1}{x_2 - x_1}$$

Example 1 Find the Slope of a Line

Find the slope of each line.

a.

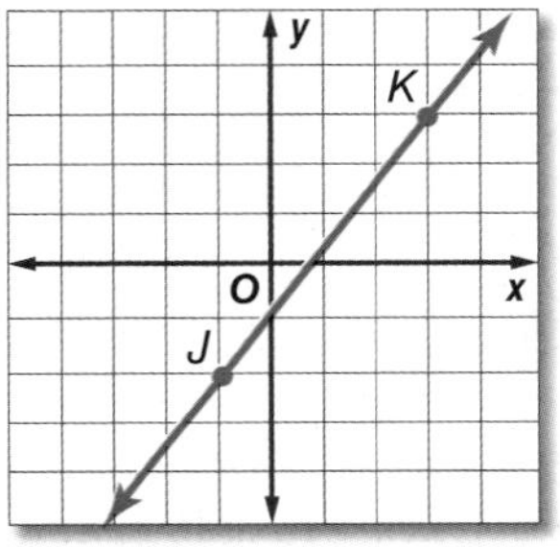

Substitute $(-1, -2)$ for (x_1, y_1) and $(3, 3)$ for (x_2, y_2).

$m = \frac{y_2 - y_1}{x_2 - x_1}$ Slope Formula

$= \frac{3 - (-2)}{3 - (-1)}$ Substitution

$= \frac{5}{4}$ Simplify.

b.

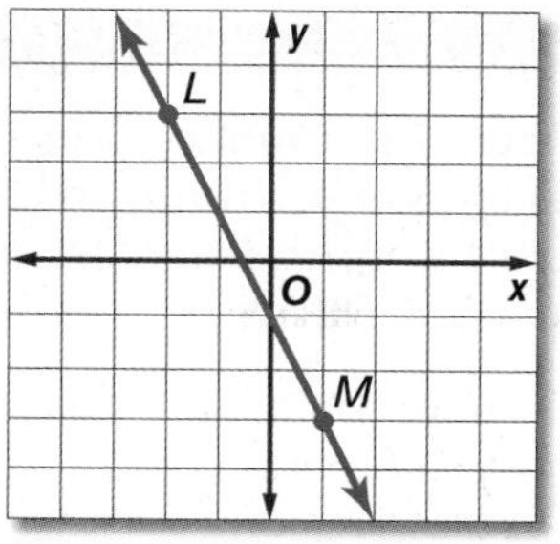

Substitute (−2, 3) for (x_1, y_1) and (1, −3) for (x_2, y_2).

$$m = \frac{y_2 - y_1}{x_2 - x_1} \quad \text{Slope Formula}$$
$$= \frac{-3 - 3}{1 - (-2)} \quad \text{Substitution}$$
$$= -2 \quad \text{Simplify.}$$

c.

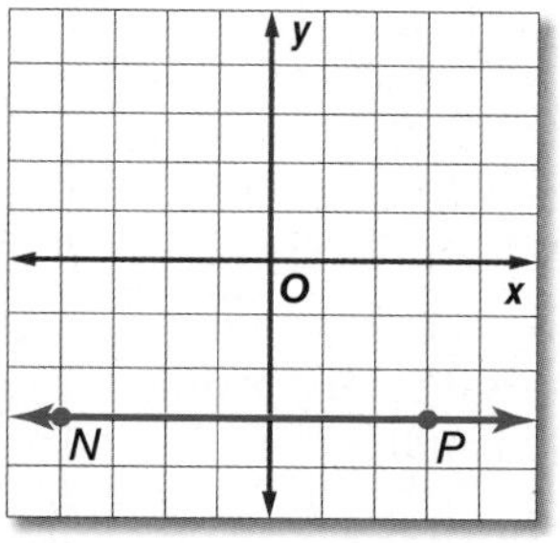

Substitute (−4, −3) for (x_1, y_1) and (3, −3) for (x_2, y_2).

$$m = \frac{y_2 - y_1}{x_2 - x_1} \quad \text{Slope Formula}$$
$$= \frac{-3 - (-3)}{3 - (-4)} \quad \text{Substitution}$$
$$= \frac{0}{7} \text{ or } 0 \quad \text{Simplify.}$$

d.

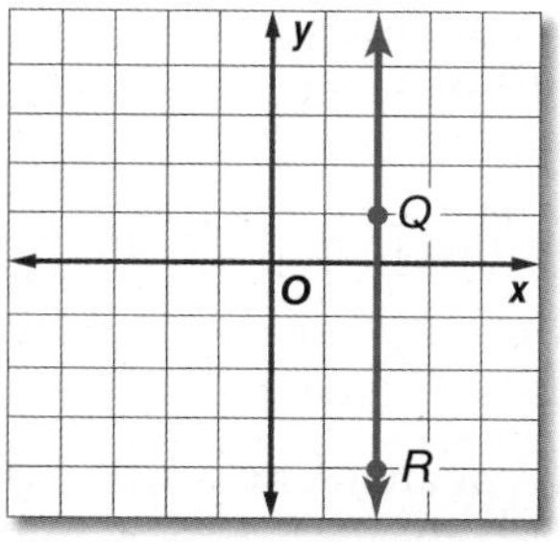

Substitute (2, 1) for (x_1, y_1) and (2, −4) for (x_2, y_2).

$$m = \frac{y_2 - y_1}{x_2 - x_1} \quad \text{Slope Formula}$$
$$= \frac{-4 - 1}{2 - 2} \quad \text{Substitution}$$
$$= \frac{-5}{0} \quad \text{Simplify.}$$

This slope is **undefined**.

StudyTip

Dividing by 0 The slope $\frac{-5}{0}$ is undefined because there is no number that you can multiply by 0 and get −5. Since this is true for any number, all numbers divided by 0 will have an undefined slope. All vertical lines have undefined slopes.

GuidedPractice

1A. the line containing (6, −2) and (−3, −5)

1B. the line containing (8, −3) and (−6, −2)

1C. the line containing (4, 2) and (4, −3)

1D. the line containing (−3, 3) and (4, 3)

Example 1 illustrates the four different types of slopes.

ConceptSummary Classifying Slopes

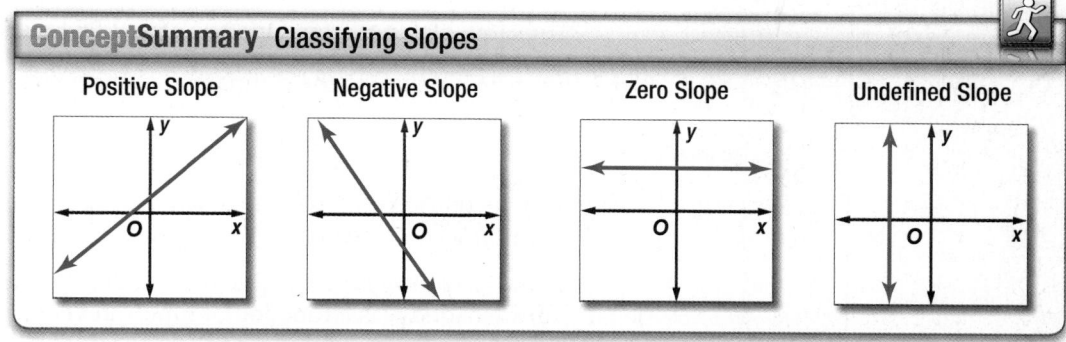

Positive Slope

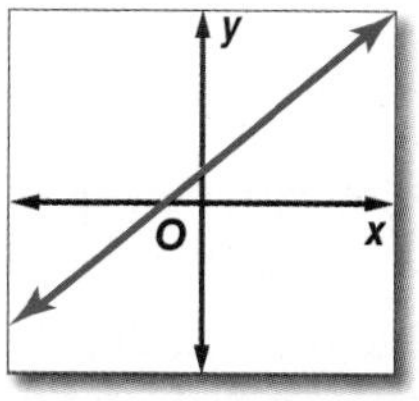

Negative Slope

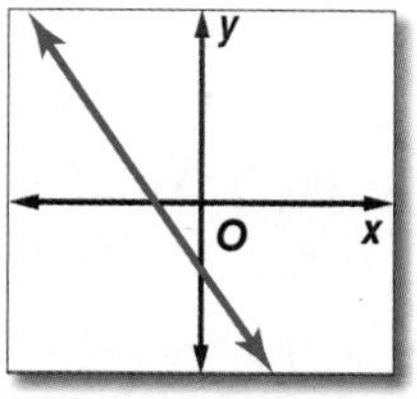

Zero Slope

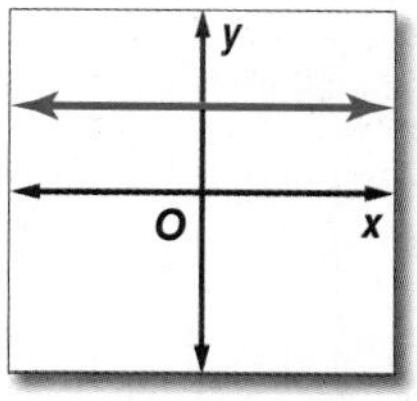

Undefined Slope

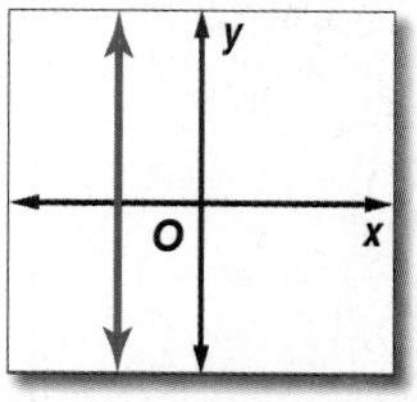

Slope can be interpreted as a **rate of change**, describing how a quantity *y* changes in relation to quantity *x*. The slope of a line can also be used to identify the coordinates of any point on the line.

Real-WorldCareer

Flight Attendants Flight attendants check tickets, assist passengers with boarding and carry-ons, and provide an overview of emergency equipment and procedures. A high school diploma is required, but airlines increasingly favor bi- or multi-lingual candidates with college degrees.

Real-World Example 2 Use Slope as Rate of Change

PT

TRAVEL A pilot flies a plane from Columbus, Ohio, to Orlando, Florida. After 0.5 hour, the plane reaches its cruising altitude and is 620 miles from Orlando. Half an hour later, the plane is 450 miles from Orlando. How far was the plane from Orlando 1.25 hours after takeoff?

Understand Use the data given to graph the line that models the distance from Orlando y in miles as a function of time x in hours.

Assume that speed is constant. Plot the points (0.5, 620) and (1.0, 450), and draw a line through them.

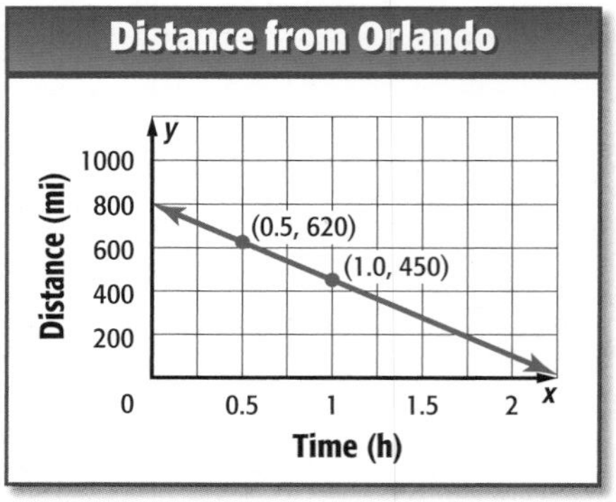

You want to find the distance from Orlando after 1.25 hours.

From the graph we can estimate that after 1.25 hours, the distance was a little less than 400 miles.

Plan Find the slope of the line graphed. Use this rate of change in the plane's distance from Orlando per hour to find the distance from Orlando after 1.25 hours.

Solve Use the Slope Formula to find the slope of the line.

$$m = \frac{y_2 - y_1}{x_2 - x_1} = \frac{(450 - 620)\text{ miles}}{(1.0 - 0.5)\text{ hours}} = \frac{-170\text{ miles}}{0.5\text{ hour}} \text{ or } -\frac{340\text{ miles}}{1\text{ hour}}$$

The plane traveled at an average speed of 340 miles per hour. The negative sign indicates a *decrease* in distance over time.

Use the slope of the line and one known point on the line to calculate the distance y when the time x is 1.25.

$m = \frac{y_2 - y_1}{x_2 - x_1}$	Slope Formula
$-340 = \frac{y_2 - 620}{1.25 - 0.5}$	$m = -340$, $x_1 = 0.5$, $y_1 = 620$, and $x_2 = 1.25$
$-340 = \frac{y_2 - 620}{0.75}$	Simplify.
$-255 = y_2 - 620$	Multiply each side by 0.75.
$365 = y_2$	Add 620 to each side.

Thus, the distance from Orlando after 1.25 hours is 365 miles.

Check Since 365 is close to the estimate, our answer is reasonable. ✓

GuidedPractice

2. DOWNLOADS In 2006, 500 million songs were legally downloaded from the Internet. In 2004, 200 million songs were legally downloaded.

A. Use the data given to graph the line that models the number of songs legally downloaded y as a function of time x in years.

B. Find the slope of the line, and interpret its meaning.

C. If this trend continues at the same rate, how many songs will be legally downloaded in 2020?

Jupiterimages/Creatas/Alamy

2 Parallel and Perpendicular Lines

You can use the slopes of two lines to determine whether the lines are parallel or perpendicular. Lines with the same slope are parallel.

Postulates Parallel and Perpendicular Lines

11.2 Slopes of Parallel Lines Two nonvertical lines have the same slope if and only if they are parallel. All vertical lines are parallel.

Example Parallel lines ℓ and m have the same slope, 4.

11.3 Slopes of Perpendicular Lines Two nonvertical lines are perpendicular if and only if the product of their slopes is -1. Vertical and horizontal lines are perpendicular.

Example line $m \perp$ line p

product of slopes $= 4 \cdot -\frac{1}{4}$ or -1

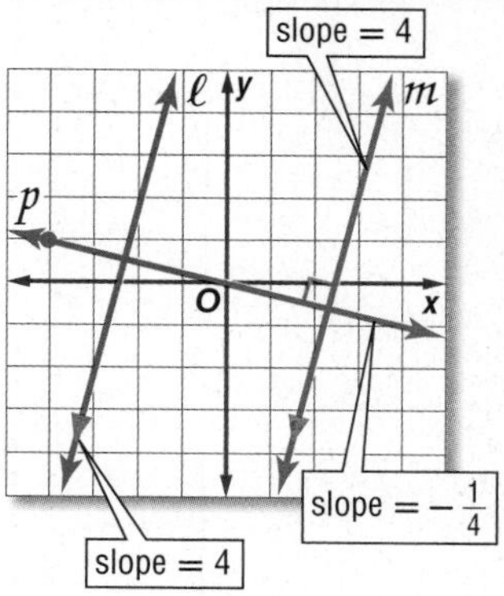

Example 3 Determine Line Relationships

Determine whether $\overleftrightarrow{AB}$ and $\overleftrightarrow{CD}$ are *parallel, perpendicular,* or *neither* for $A(1, 1)$, $B(-1, -5)$, $C(3, 2)$, and $D(6, 1)$. Graph each line to verify your answer.

Step 1 Find the slope of each line.

slope of $\overleftrightarrow{AB} = \frac{-5-1}{-1-1} = \frac{-6}{-2}$ or 3

slope of $\overleftrightarrow{CD} = \frac{1-2}{6-3}$ or $\frac{-1}{3}$

Step 2 Determine the relationship, if any, between the lines.

The two lines do not have the same slope, so they are *not* parallel. To determine if the lines are perpendicular, find the product of their slopes.

$3\left(-\frac{1}{3}\right) = -1$ Product of slopes for $\overleftrightarrow{AB}$ and $\overleftrightarrow{CD}$

Since the product of their slopes is -1, $\overleftrightarrow{AB}$ is perpendicular to $\overleftrightarrow{CD}$.

CHECK When graphed, the two lines appear to intersect and form four right angles. ✔

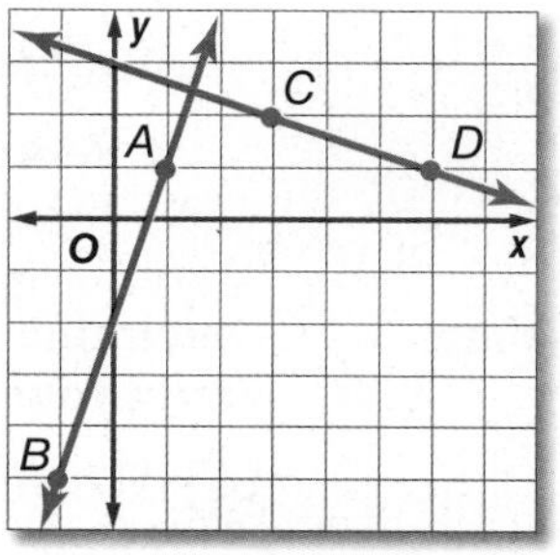

Study Tip

Slopes of Perpendiculars If a line ℓ has a slope of $\frac{a}{b}$, then the slope of a line perpendicular to line ℓ is the opposite reciprocal, $-\frac{b}{a}$, since $\frac{a}{b}\left(-\frac{b}{a}\right) = -1$.

Guided Practice

Determine whether $\overleftrightarrow{AB}$ and $\overleftrightarrow{CD}$ are *parallel, perpendicular,* or *neither.* Graph each line to verify your answer.

3A. $A(14, 13)$, $B(-11, 0)$, $C(-3, 7)$, $D(-4, -5)$

3B. $A(3, 6)$, $B(-9, 2)$, $C(5, 4)$, $D(2, 3)$

Example 4 Use Slope to Graph a Line

Graph the line that contains $A(-3, 0)$ and is perpendicular to $\overleftrightarrow{CD}$ with $C(-2, -3)$ and $D(2, 0)$.

The slope of $\overleftrightarrow{CD}$ is $\frac{0-(-3)}{2-(-2)}$ or $\frac{3}{4}$.

Since $\frac{3}{4}\left(\frac{4}{-3}\right) = -1$, the slope of the line perpendicular to $\overleftrightarrow{CD}$ through A is $-\frac{4}{3}$ or $\frac{-4}{3}$.

To graph the line, start at point A. Move down 4 units and then right 3 units. Label the point B and draw $\overleftrightarrow{AB}$.

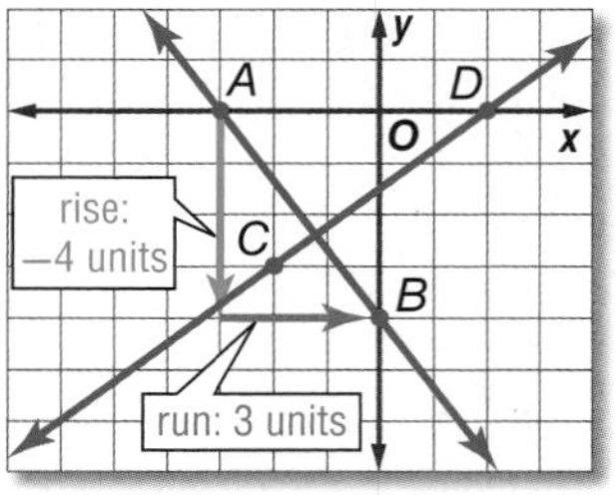

GuidedPractice

4. Graph the line that contains $P(0, 1)$ and is perpendicular to $\overleftrightarrow{QR}$ with $Q(-6, -2)$ and $R(0, -6)$.

Check Your Understanding

Example 1 **Find the slope of each line.**

1.

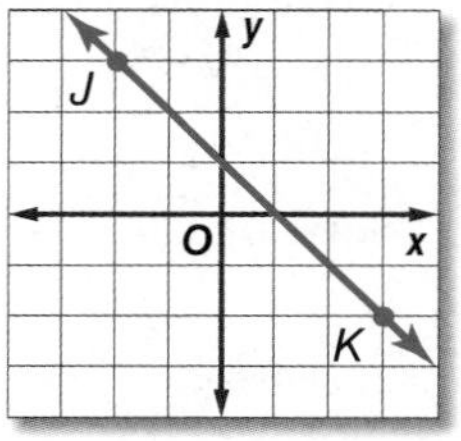

2.

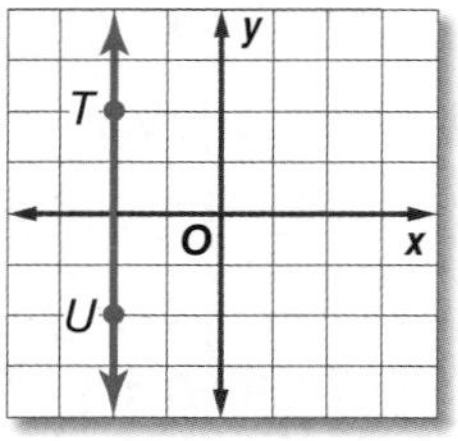

3.

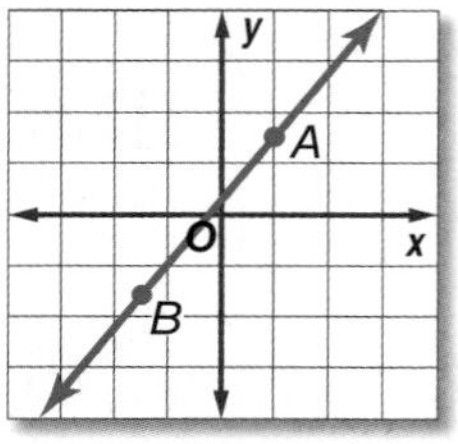

Example 2

4. BOTANY Kudzu is a fast-growing vine found in the southeastern United States. An initial measurement of the length of a kudzu vine was 0.5 meter. Seven days later the plant was 4 meters long.

a. Graph the line that models the length of the plant over time.

b. What is the slope of your graph? What does it represent?

c. Assuming that the growth rate of the plant continues, how long will the plant be after 15 days?

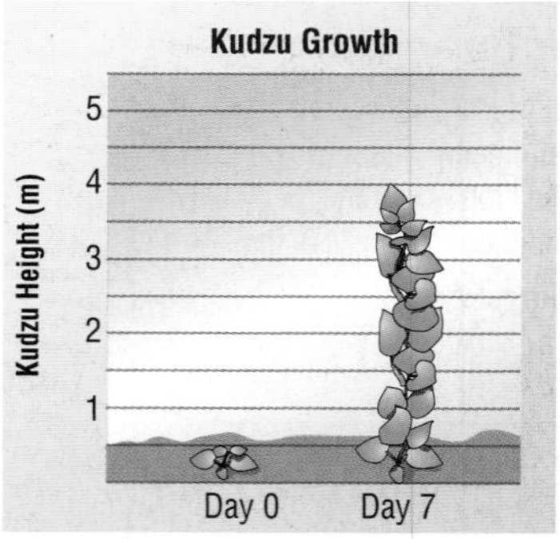

Example 3 **Determine whether $\overleftrightarrow{WX}$ and $\overleftrightarrow{YZ}$ are *parallel, perpendicular,* or *neither*. Graph each line to verify your answer.**

5. $W(2, 4)$, $X(4, 5)$, $Y(4, 1)$, $Z(8, -7)$

6. $W(1, 3)$, $X(-2, -5)$, $Y(-6, -2)$, $Z(8, 3)$

7. $W(-7, 6)$, $X(-6, 9)$, $Y(6, 3)$, $Z(3, -6)$

8. $W(1, -3)$, $X(0, 2)$, $Y(-2, 0)$, $Z(8, 2)$

Example 4 **Graph the line that satisfies each condition.**

9. passes through $A(3, -4)$, parallel to $\overleftrightarrow{BC}$ with $B(2, 4)$ and $C(5, 6)$

10. slope = 3, passes through $A(-1, 4)$

11. passes through $P(7, 3)$, perpendicular to $\overleftrightarrow{LM}$ with $L(-2, -3)$ and $M(-1, 5)$

Practice and Problem Solving

Example 1 **Find the slope of each line.**

12.

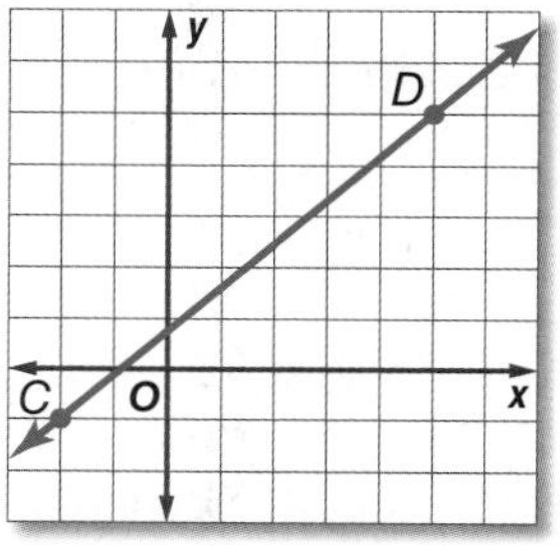

13.

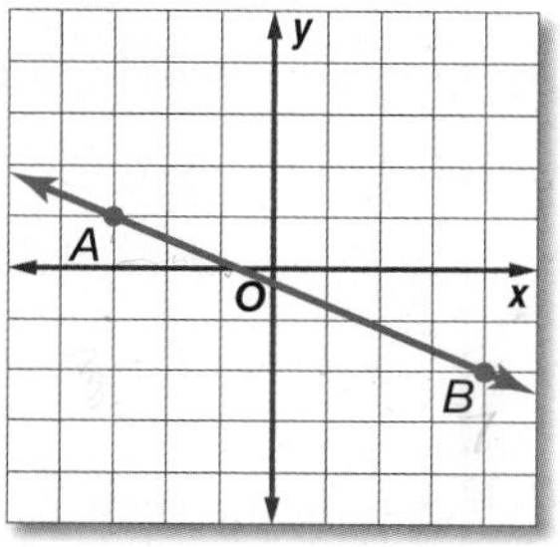

14.

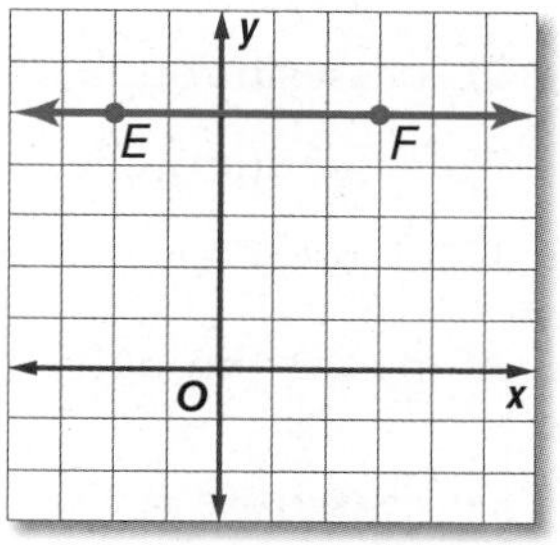

15.

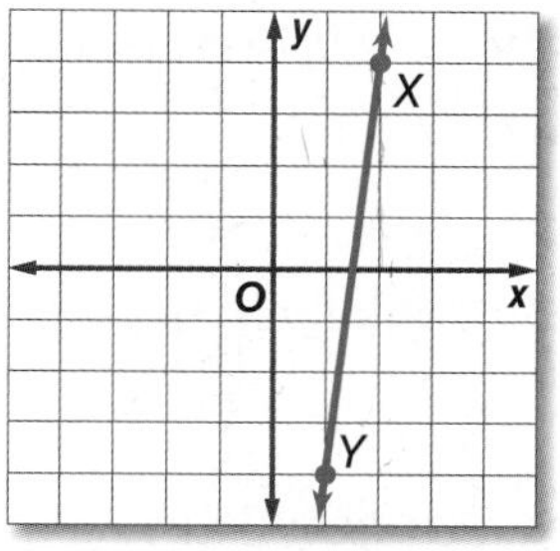

16.

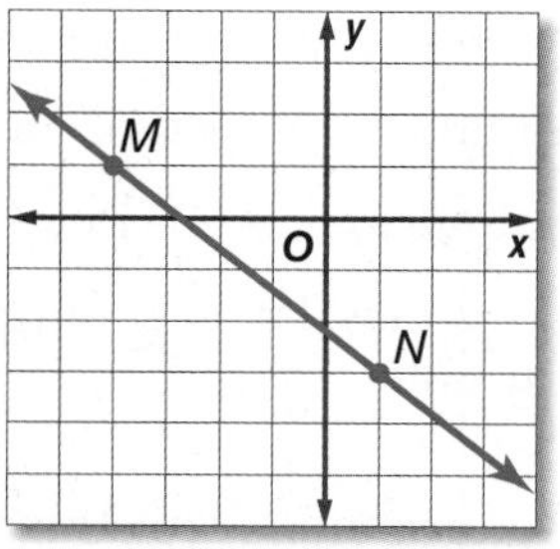

17.

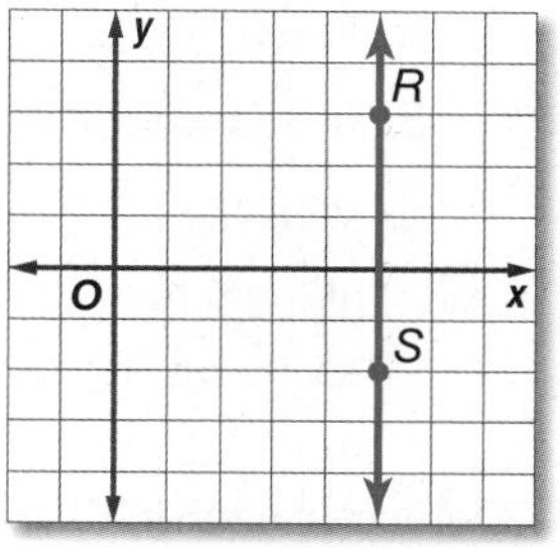

Determine the slope of the line that contains the given points.

18. $C(3, 1), D(-2, 1)$

19. $E(5, -1), F(2, -4)$

20. $G(-4, 3), H(-4, 7)$

21. $J(7, -3), K(-8, -3)$

22. $L(8, -3), M(-4, -12)$

23. $P(-3, -5), Q(-3, -1)$

24. $R(2, -6), S(-6, 5)$

25. $T(-6, -11), V(-12, -10)$

Example 2

26. **CCSS MODELING** In 2004, 8 million Americans over the age of 7 participated in mountain biking, and in 2006, 8.5 million participated.

a. Create a graph to show the number of participants in mountain biking based on the change in participation from 2004 to 2006.

b. Based on the data, what is the growth per year of the sport?

c. If participation continues at the same rate, what will be the participation in 2013 to the nearest 10,000?

27. **FINANCIAL LITERACY** Suppose an MP3 player cost \$499 in 2003 and \$249.99 in 2009.

a. Graph a trend line to predict the price of the MP3 player for 2003 through 2009.

b. Based on the data, how much does the price drop per year?

c. If the trend continues, what will be the cost of an MP3 player in 2013?

Example 3 **Determine whether $\overleftrightarrow{AB}$ and $\overleftrightarrow{CD}$ are *parallel, perpendicular,* or *neither.* Graph each line to verify your answer.**

28. $A(1, 5), B(4, 4), C(9, -10), D(-6, -5)$

29. $A(-6, -9), B(8, 19), C(0, -4), D(2, 0)$

30. $A(4, 2), B(-3, 1), C(6, 0), D(-10, 8)$

31. $A(8, -2), B(4, -1), C(3, 11), D(-2, -9)$

32. $A(8, 4), B(4, 3), C(4, -9), D(2, -1)$

33. $A(4, -2), B(-2, -8), C(4, 6), D(8, 5)$

Example 4 **Graph the line that satisfies each condition.**

34. passes through $A(2, -5)$, parallel to $\overleftrightarrow{BC}$ with $B(1, 3)$ and $C(4, 5)$

35. slope $= -2$, passes through $H(-2, -4)$

36. passes through $K(3, 7)$, perpendicular to $\overleftrightarrow{LM}$ with $L(-1, -2)$ and $M(-4, 8)$

37. passes through $X(1, -4)$, parallel to $\overleftrightarrow{YZ}$ with $Y(5, 2)$ and $Z(-3, -5)$

38. slope $= \frac{2}{3}$, passes through $J(-5, 4)$

39. passes through $D(-5, -6)$, perpendicular to $\overleftrightarrow{FG}$ with $F(-2, -9)$ and $G(1, -5)$

40. **STADIUMS** Before it was demolished, the RCA Dome was home to the Indianapolis Colts. The attendance in 2001 was 450,746, and the attendance in 2005 was 457,373.

a. What is the approximate rate of change in attendance from 2001 to 2005?

b. If this rate of change continues, predict the attendance for 2012.

c. Will the attendance continue to increase indefinitely? Explain.

d. The Colts have now built a new, larger stadium. Do you think their decision was reasonable? Why or why not?

Determine which line passing through the given points has a steeper slope.

41. Line 1: (0, 5) and (6, 1)
Line 2: (−4, 10) and (8, −5)

42. Line 1: (0, −4) and (2, 2)
Line 2: (0, −4) and (4, 5)

43. Line 1: (−9, −4) and (7, 0)
Line 2: (0, 1) and (7, 4)

44. Line 1: (−6, 7) and (9, −3)
Line 2: (−9, 9) and (3, 5)

45. **CCSS MODELING** Michigan provides habitat for two endangered species, the bald eagle and the gray wolf. The graph shows the Michigan population of each species in 1992 and 2006.

a. Which species experienced a greater rate of change in population?

b. Make a line graph showing the growth of both populations.

c. If both species continue to grow at their respective rates, what will the population of each species be in 2012?

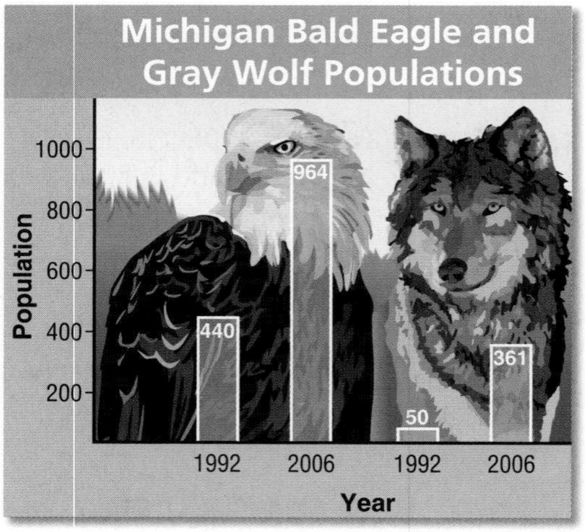

Find the value of x or y that satisfies the given conditions. Then graph the line.

46. The line containing $(4, -1)$ and $(x, -6)$ has a slope of $-\frac{5}{2}$.

47. The line containing $(-4, 9)$ and $(4, 3)$ is parallel to the line containing $(-8, 1)$ and $(4, y)$.

48. The line containing $(8, 7)$ and $(7, -6)$ is perpendicular to the line containing $(2, 4)$ and $(x, 3)$.

49. The line containing $(1, -3)$ and $(3, y)$ is parallel to the line containing $(5, -6)$ and $(9, y)$.

50. **SCHOOLS** In 2000, Jefferson High School had 1125 students. By 2006, the student body had increased to 1425 students. When Fairview High School was built in 2001, it had 1275 students. How many students did Fairview High School have in 2006 if the student body grew at the same rate as Jefferson High School?

51 **MUSIC** Maggie and Mikayla want to go to the music store near Maggie's house after school. They can walk 3.5 miles per hour and ride their bikes 10 miles per hour.

a. Create a table to show how far Maggie and Mikayla can travel walking and riding their bikes. Include distances for 0, 1, 2, 3, and 4 hours.

b. Create a graph to show how far Maggie and Mikayla can travel based on time for both walking and riding their bikes. Be sure to label the axes of your graph.

c. What does the slope represent in your graph?

d. Maggie's mom says they can only go if they can make it to the music store and back in less than two hours. If they want to spend at least 30 minutes in the music store and it is four miles away, can they make it? Should they walk or ride their bikes? Explain your reasoning.

H.O.T. Problems Use Higher-Order Thinking Skills

52. **WRITE A QUESTION** A classmate says that all lines have positive or negative slope. Write a question that would challenge his conjecture.

53. **ERROR ANALYSIS** Terrell and Hale calculated the slope of the line passing through the points $Q(3, 5)$ and $R(-2, 2)$. Is either of them correct? Explain your reasoning.

Terrell

$$m = \frac{5 - 2}{3 - (-2)} = \frac{3}{5}$$

Hale

$$m = \frac{5 - 2}{-2 - 3} = -\frac{3}{5}$$

54. **CCSS REASONING** Draw a square $ABCD$ with opposite vertices at $A(2, -4)$ and $C(10, 4)$.

a. Find the other two vertices of the square and label them B and D.

b. Show that $\overline{AD} \parallel \overline{BC}$ and $\overline{AB} \parallel \overline{DC}$.

c. Show that the measure of each angle inside the square is equal to 90.

55. **WRITING IN MATH** Describe the slopes of the Sears Tower and the Leaning Tower of Pisa.

Sears Tower

Leaning Tower of Pisa

56. **CHALLENGE** In this lesson you learned that $m = \frac{y_2 - y_1}{x_2 - x_1}$. Use an algebraic proof to show that the slope can also be calculated using the equation $m = \frac{y_1 - y_2}{x_1 - x_2}$.

57. **WRITING IN MATH** Find two additional points that lie along the same line as $X(3, -1)$ and $Y(-1, 7)$. Generalize a method you can use to find additional points on the line from any given point.

Standardized Test Practice

58. What is the slope of a line perpendicular to the line through the points $(-1, 6)$ and $(3, -4)$?

A $m = -\frac{5}{2}$

B $m = -1$

C $m = -\frac{2}{5}$

D $m = \frac{2}{5}$

59. SHORT RESPONSE A set of 25 cards is randomly placed face down on a table. 15 cards have only the letter A written on the face, and 10 cards have only the letter B. Patrick turned over 1 card. What is the probability of this card having the letter B written on its face?

60. ALGEBRA Jamie is collecting money to buy an \$81 gift for her teacher. She has already contributed \$24. She will collect \$3 from each contributing student. How many other students must contribute?

F 3 students

G 9 students

H 12 students

J 19 students

61. SAT/ACT The area of a circle is 20π square centimeters. What is its circumference?

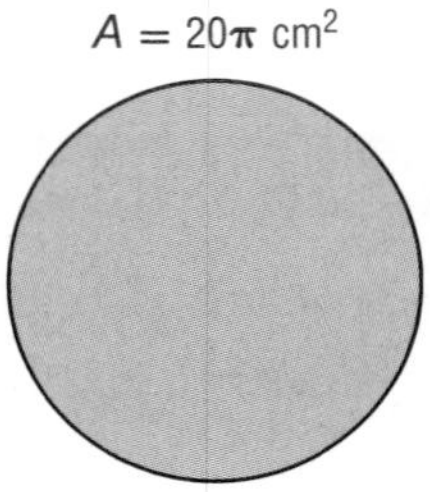

A $\sqrt{5}\pi$ cm

B $2\sqrt{5}\pi$ cm

C $4\sqrt{5}\pi$ cm

D 20π cm

E 40π cm

Spiral Review

In the figure, $a \parallel b$, $c \parallel d$, and $m\angle 4 = 57$. Find the measure of each angle. (Lesson 11-2)

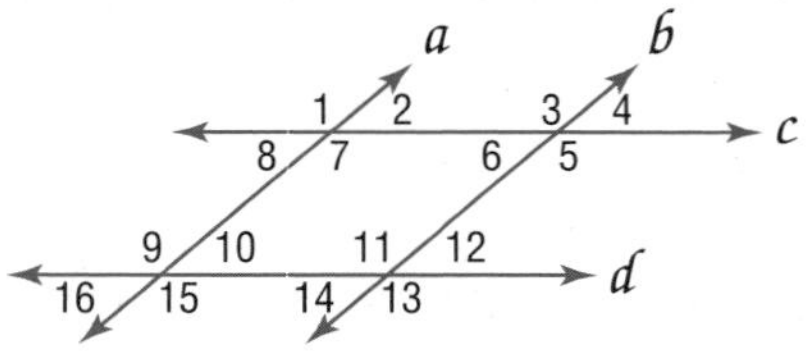

62. $\angle 5$

63. $\angle 1$

64. $\angle 8$

65. $\angle 10$

Refer to the diagram at the right. (Lesson 11-1)

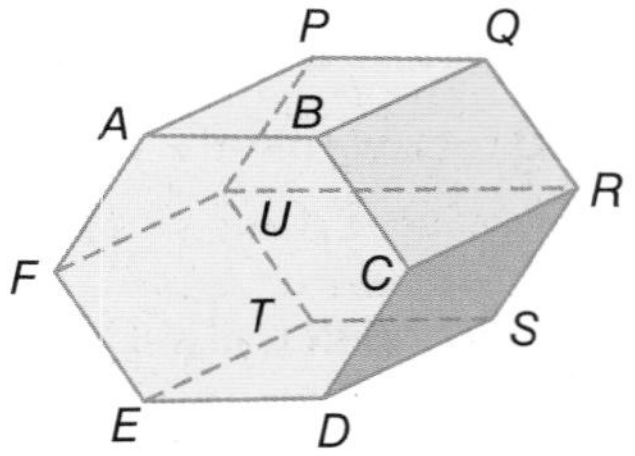

66. Name all segments parallel to $\overline{TU}$.

67. Name all planes intersecting plane BCR.

68. Name all segments skew to $\overline{DE}$.

69. CONSTRUCTION There are four buildings on the Mansfield High School Campus, no three of which stand in a straight line. How many sidewalks need to be built so that each building is directly connected to every other building? (Lesson 10-1)

Skills Review

Solve for *y*.

70. $3x + y = 5$

71. $4x + 2y = 6$

72. $4y - 3x = 5$

CHAPTER 11 Mid-Chapter Quiz

Lessons 11-1 through 11-3

Identify the transversal connecting each pair of angles. Then classify the relationship between each pair of angles as *alternate interior, alternate exterior, corresponding,* or *consecutive interior* angles. (Lesson 11-1)

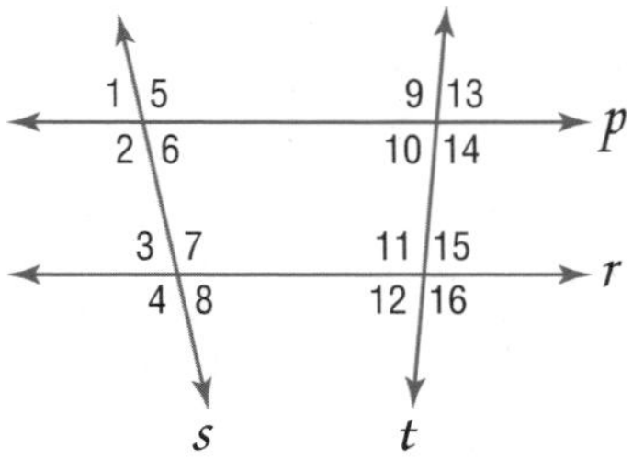

1. ∠6 and ∠3
2. ∠1 and ∠14
3. ∠10 and ∠11
4. ∠5 and ∠7

Refer to the figure to identify each of the following. (Lesson 11-1)

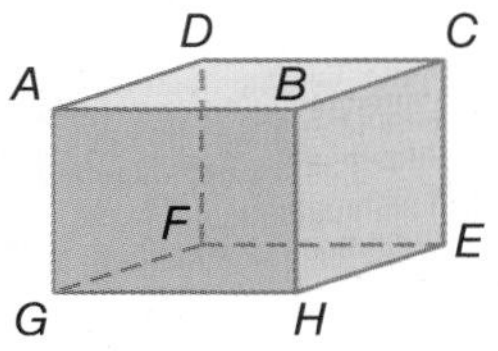

5. a plane parallel to plane *ABCD*
6. a segment skew to $\overline{GH}$ that contains point *D*
7. all segments parallel to $\overline{HE}$

8. **MULTIPLE CHOICE** Which term best describes ∠4 and ∠8? (Lesson 11-1)

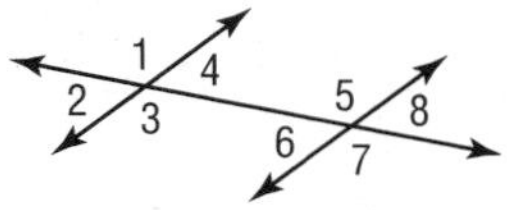

A corresponding
B alternate exterior
C alternate interior
D consecutive interior

In the figure, $m\angle 4 = 104$, $m\angle 14 = 118$. Find the measure of each angle. Tell which postulate(s) or theorem(s) you used. (Lesson 11-2)

9. ∠2
10. ∠9
11. ∠10
12. ∠7

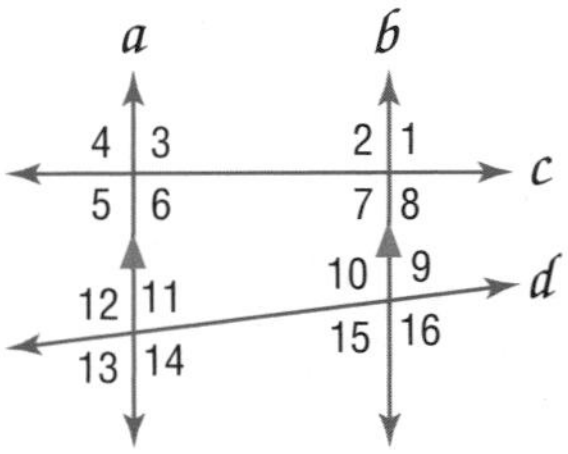

13. Find *x*. (Lesson 11-2)

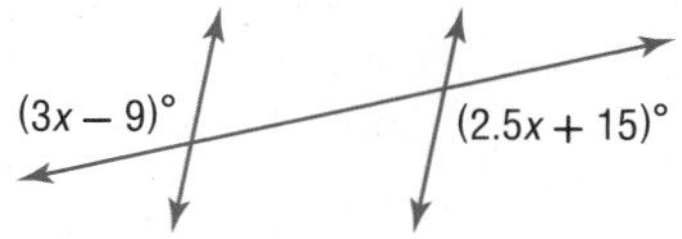

14. **MODEL TRAINS** Amy is setting up two parallel train tracks so that a third track runs diagonally across the first two. To properly place a switch, she needs the angle between the diagonal and the top right portion of the second track to be twice as large as the angle between the diagonal and bottom right portion of the first track. What is the measure of the angle between the diagonal and the top right portion of the second track? (Lesson 11-2)

Determine whether $\overleftrightarrow{AB}$ and $\overleftrightarrow{XY}$ are *parallel, perpendicular,* or *neither.* Graph each line to verify your answer. (Lesson 11-3)

15. $A(2, 0)$, $B(4, -5)$, $X(-3, 3)$, $Y(-5, 8)$
16. $A(1, 1)$, $B(6, -9)$, $X(4, -10)$, $Y(7, -4)$

Find the slope of each line. (Lesson 11-3)

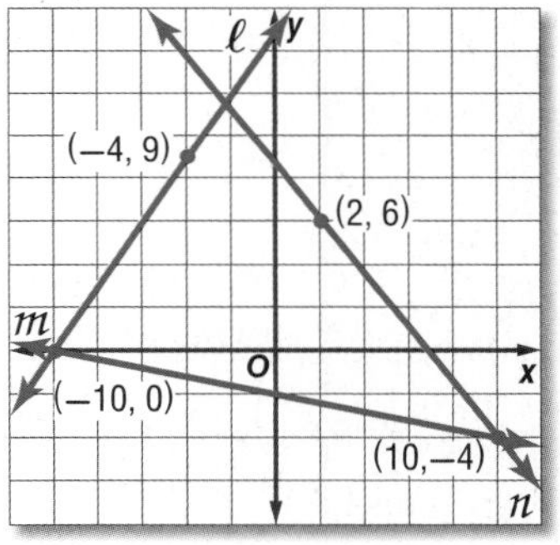

17. line ℓ
18. a line parallel to *m*
19. a line perpendicular to *n*

20. **SALES** The 2008 and 2011 sales figures for Vaughn Electronics are in the table below. (Lesson 11-3)

Year	Approximate Sales ($)
2008	240,000
2011	330,000

a. What is the rate of change in approximate sales from 2008 to 2011?

b. If this rate of change continues, predict the approximate sales for the year 2015.

LESSON 11-4 Equations of Lines

Then

- You found the slopes of lines.

Now

1. Write an equation of a line given information about the graph.
2. Solve problems by writing equations.

Why?

- On an interstate near Lauren's hometown, the minimum fine for speeding ten or fewer miles per hour over the speed limit of 65 miles per hour is $42.50. There is an additional charge of $2 for each mile per hour over this initial ten miles per hour. The total charge, not including court costs, can be represented by the equation $C = 42.5 + 2m$.

Cost of Speeding

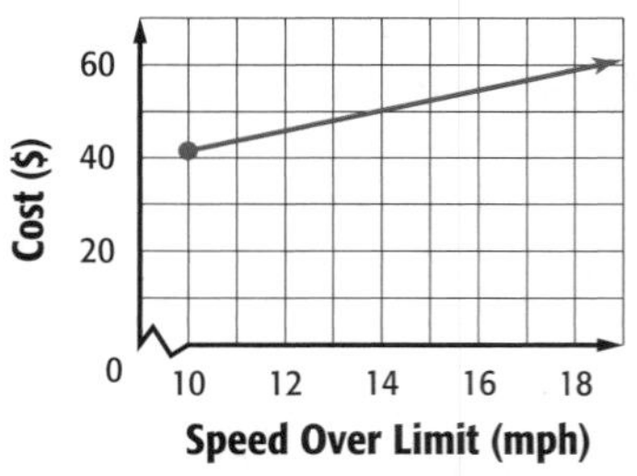

NewVocabulary
slope-intercept form
point-slope form

Common Core State Standards

Content Standards
G.GPE.5 Prove the slope criteria for parallel and perpendicular lines and use them to solve geometric problems (e.g., find the equation of a line parallel or perpendicular to a given line that passes through a given point).

Mathematical Practices
4 Model with mathematics.
8 Look for and express regularity in repeated reasoning.

1 Write Equations of Lines

You may remember from algebra that an equation of a nonvertical line can be written in different but equivalent forms.

KeyConcept Nonvertical Line Equations

The **slope-intercept form** of a linear equation is $y = mx + b$, where m is the slope of the line and b is the y-intercept.

$y = mx + b$ — $y = 3x + 8$ (slope: 3; y-intercept: 8)

The **point-slope form** of a linear equation is $y - y_1 = m(x - x_1)$, where (x_1, y_1) is any point on the line and m is the slope of the line.

$y - 5 = -2(x - 3)$ (point on line (3, 5); slope: −2)

When given the slope and either the y-intercept or a point on a line, you can use these forms to write the equation of the line.

Example 1 Slope and y-intercept

Write an equation in slope-intercept form of the line with slope 3 and y-intercept of −2. Then graph the line.

$y = mx + b$	Slope-intercept form
$y = 3x + (-2)$	$m = 3, b = -2$
$y = 3x - 2$	Simplify.

Plot a point at the y-intercept, −2. Use the slope of 3 or $\frac{3}{1}$ to find another point 3 units up and 1 unit to the right of the y-intercept. Then draw the line through these two points.

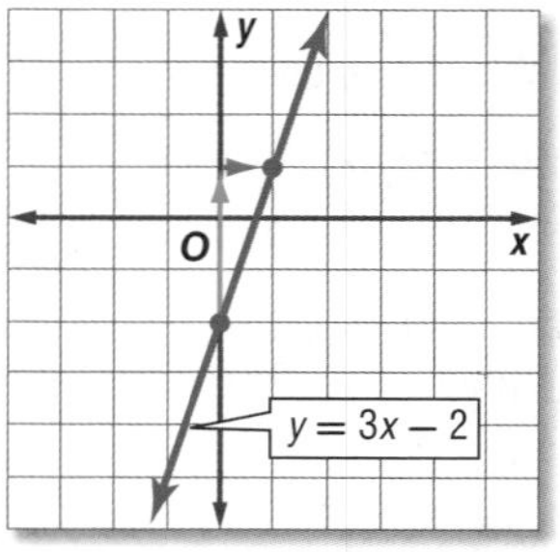

GuidedPractice

1. Write an equation in slope-intercept form of the line with slope $-\frac{1}{2}$ and y-intercept of 8. Then graph the line.

Example 2 Slope and a Point on the Line

Write an equation in point-slope form of the line with slope $-\frac{3}{4}$ that contains (−2, 5). Then graph the line.

$y - y_1 = m(x - x_1)$ — Point-Slope form

$y - 5 = -\frac{3}{4}[x - (-2)]$ — $m = -\frac{3}{4}$, $(x_1, y_1) = (-2, 5)$

$y - 5 = -\frac{3}{4}(x + 2)$ — Simplify.

Graph the given point (−2, 5). Use the slope $-\frac{3}{4}$ or $\frac{-3}{4}$ to find another point 3 units down and 4 units to the right. Then draw the line through these two points.

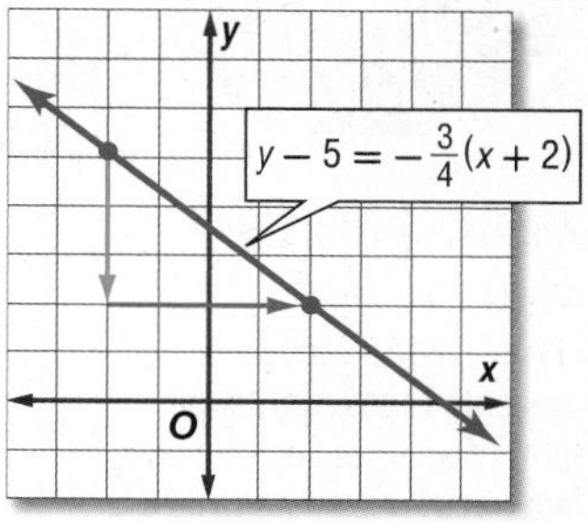

WatchOut!

Substituting Negative Coordinates When substituting negative coordinates, use parentheses to avoid making errors with the signs.

GuidedPractice

2. Write an equation in point-slope form of the line with slope 4 that contains (−3, −6). Then graph the line.

When the slope of a line is not given, use two points on the line to calculate the slope. Then use the point-slope or slope-intercept form to write an equation of the line.

Example 3 Two Points

Write an equation of the line through each pair of points in slope-intercept form.

a. (0, 3) and (−2, −1)

Step 1 Find the slope of the line through the points.

$m = \frac{y_2 - y_1}{x_2 - x_1} = \frac{-1 - 3}{-2 - 0} = \frac{-4}{-2}$ or 2 — Use the Slope Formula.

Step 2 Write an equation of the line.

$y = mx + b$ — Slope-Intercept form

$y = 2x + 3$ — $m = 2$; (0, 3) is the y-intercept.

b. (−7, 4) and (9, −4)

Step 1 $m = \frac{y_2 - y_1}{x_2 - x_1} = \frac{-4 - 4}{9 - (-7)} = \frac{-8}{16}$ or $-\frac{1}{2}$ — Use the Slope Formula.

Step 2 $y - y_1 = m(x - x_1)$ — Point-Slope form

$y - 4 = -\frac{1}{2}[x - (-7)]$ — $m = -\frac{1}{2}$, $(x_1, y_1) = (-7, 4)$

$y - 4 = -\frac{1}{2}(x + 7)$ — Simplify.

$y - 4 = -\frac{1}{2}x - \frac{7}{2}$ — Distribute.

$y = -\frac{1}{2}x + \frac{1}{2}$ — Add 4 to each side: $-\frac{7}{2} + 4 = -\frac{7}{2} + \frac{8}{2} = \frac{1}{2}$

StudyTip

CCSS Perseverance In Example 3b, you could also use the slope-intercept form and one point to find the y-intercept and write the equation.

$y = mx + b$

$4 = -\frac{1}{2}(-7) + b$

$4 = \frac{7}{2} + b$

$4 - \frac{7}{2} = b$

$b = \frac{1}{2}$

So, $y = -\frac{1}{2}x + \frac{1}{2}$.

GuidedPractice

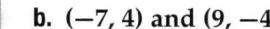

3A. (−2, 4) and (8, 10)

3B. (−1, 3) and (7, 3)

Math HistoryLink

Gaspard Monge **(1746–1818)** Monge presented the point-slope form of an equation of a line in a paper published in 1784.

Example 4 Horizontal Line

Write an equation of the line through (−2, 6) and (5, 6) in slope-intercept form.

Step 1 $m = \frac{y_2 - y_1}{x_2 - x_1} = \frac{6 - 6}{5 - (-2)} = \frac{0}{7}$ or 0 This is a horizontal line.

Step 2

$y - y_1 = m(x - x_1)$ Point-Slope form

$y - 6 = 0[x - (-2)]$ $m = -\frac{1}{2}, (x_1, y_1) = (-2, 6)$

$y - 6 = 0$ Simplify.

$y = 6$ Add 6 to each side.

GuidedPractice

4. Write an equation of the line through (5, 0) and (−1, 0) in slope-intercept form.

The equations of horizontal and vertical lines involve only one variable.

KeyConcepts Horizontal and Vertical Line Equations

The equation of a horizontal line is $y = b$, where b is the y-intercept of the line.

Example $y = -3$

The equation of a vertical line is $x = a$, where a is the x-intercept of the line.

Example $x = -2$

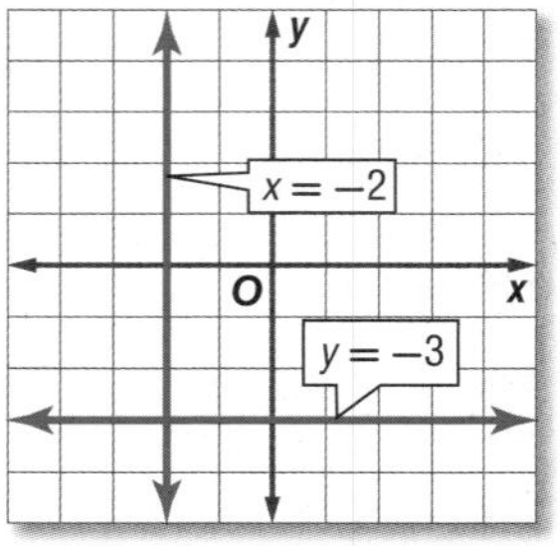

Parallel lines that are not vertical have equal slopes. Two nonvertical lines are perpendicular if the product of their slope is −1. Vertical and horizontal lines are always perpendicular to one another.

Example 5 Write Equations of Parallel or Perpendicular Lines

Write an equation in slope-intercept form for a line perpendicular to $y = -3x + 2$ containing (4, 0).

The slope of $y = -3x + 2$ is −3, so the slope of a line perpendicular to it is $\frac{1}{3}$.

$y = mx + b$ Slope-Intercept form

$0 = \frac{1}{3}(4) + b$ $m = \frac{1}{3}$ and $(x, y) = (4, 0)$

$0 = \frac{4}{3} + b$ Simplify.

$-\frac{4}{3} = b$ Subtract $\frac{4}{3}$ from each side.

So, the equation is $y = \frac{1}{3}x + \left(-\frac{4}{3}\right)$ or $y = \frac{1}{3}x - 1\frac{1}{3}$.

GuidedPractice

5. Write an equation in slope-intercept form for a line parallel to $y = -\frac{3}{4}x + 3$ containing (−3, 6).

ReadingMath

Linear The word *linear* indicates a line. The graph of a linear equation is a line.

2 Write Equations to Solve Problems

Many real-world situations can be modeled using a linear equation.

Real-World Example 6 Write Linear Equations

FINANCIAL LITERACY Benito's current wireless phone plan, Plan X, costs \$39.95 per month for unlimited calls and \$0.05 per text message. He is considering switching to Plan Y, which costs \$35 per month for unlimited calls plus \$0.10 for each text message. Which plan offers him the better rate?

Understand Plan X costs \$39.95 per month plus \$0.05 per text message. Plan Y costs \$35 per month plus \$0.10 per text message. You want to compare the two plans to determine when the cost of one plan is less than the other.

Plan Write an equation to model the total monthly cost C of each plan for t text messages sent or received. Then graph the equations in order to compare the two plans.

Solve The rates of increase, or slopes m, in the total costs are 0.05 for Plan X and 0.10 for Plan Y. When the number of text messages is 0, the total charge is just the monthly fee. So, the y-intercept b is 39.95 for Plan X and 35 for Plan Y.

Plan X		Plan Y
$C = mt + b$	Slope-intercept form	$C = mt + b$
$C = 0.05t + 39.95$	Substitute for m and b.	$C = 0.10t + 35$

Graph the two equations on the same coordinate plane.

Cost of Text Messages

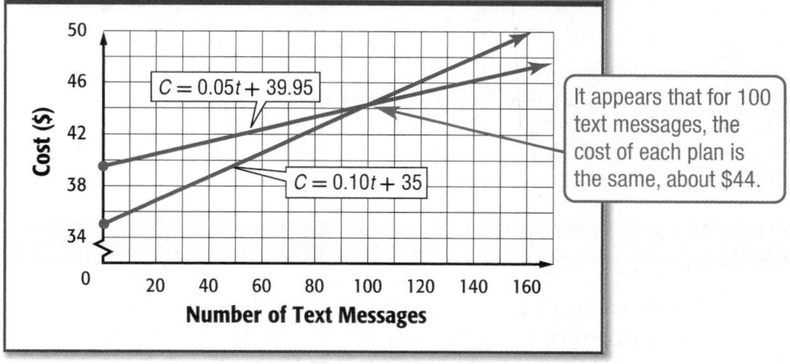

From the graph, it appears that if Benito sends or receives less than about 100 text messages, Plan Y offers the lower rate. For more than 100 messages, Plan X is lower.

Check Check your estimate. For 100 text messages, Plan X costs $0.05(100) + 39.95$ or 44.95, and Plan Y costs $0.1(100) + 35$ or 45. Adjusting our estimate, we find that when the number of messages is 99, both plans cost \$44.90. ✓

Problem-SolvingTip

Draw a Graph In Example 6, although Plan Y has a lower monthly fee, the charge per text message is higher. This makes the plans more difficult to compare. A graph can often give you a better comparison of two linear situations.

GuidedPractice

6. Suppose the rate for Plan Y was \$44 a month and \$0.02 per text message. Which plan would offer Benito the better rate? Justify your answer.

Check Your Understanding

Example 1 **Write an equation in slope-intercept form of the line having the given slope and y-intercept. Then graph the line.**

1. m: 4, y-intercept: -3 **2.** m: $\frac{1}{2}$, y-intercept: -1 **3.** m: $-\frac{2}{3}$, y-intercept: 5

Example 2 **Write an equation in point-slope form of the line having the given slope that contains the given point. Then graph the line.**

4. $m = 5, (3, -2)$ **5.** $m = \frac{1}{4}, (-2, -3)$ **6.** $m = -4.25, (-4, 6)$

Examples 3–4 **Write an equation of the line through each pair of points in slope-intercept form.**

7.

x	y
0	−1
4	4

8.

x	y
4	3
1	−6

9.

x	y
6	5
−1	−4

Example 5

10. Write an equation in slope-intercept form for a line perpendicular to $y = -2x + 6$ containing (3, 2).

11. Write an equation in slope-intercept form for a line parallel to $y = 4x - 5$ containing $(-1, 5)$.

Example 6

12. CCSS **MODELING** Kameko currently subscribes to Ace Music, an online music service, but she is considering switching to another online service, Orange Tunes. The plan for each online music service is shown.

a. Write an equation to represent the total monthly cost for each plan.

b. Graph the equations.

c. If Kameko downloads 15 songs per month, should she keep her current plan, or change to the other plan? Explain.

Practice and Problem Solving

Example 1 **Write an equation in slope-intercept form of the line having the given slope and y-intercept or points. Then graph the line.**

13. m: -5, y-intercept: -2 **14.** m: -7, b: -4 **15.** m: 9, b: 2

16. m: 12, y-intercept: $\frac{4}{5}$ **17.** m: $-\frac{3}{4}$, (0, 4) **18.** m: $\frac{5}{11}$, (0, −3)

Example 2 **Write an equation in point-slope form of the line having the given slope that contains the given point. Then graph the line.**

19. $m = 2, (3, 11)$ **20.** $m = 4, (-4, 8)$ **21.** $m = -7, (1, 9)$

22. $m = \frac{5}{7}, (-2, -5)$ **23.** $m = -\frac{4}{5}, (-3, -6)$ **24.** $m = -2.4, (14, -12)$

Examples 3–4 **Write an equation of the line through each pair of points in slope-intercept form.**

25. $(-1, -4)$ and $(3, -4)$ **26.** $(2, -1)$ and $(2, 6)$

27. $(-3, -2)$ and $(-3, 4)$ **28.** $(0, 5)$ and $(3, 3)$

29. $(-12, -6)$ and $(8, 9)$ **30.** $(2, 4)$ and $(-4, -11)$

Write an equation in slope-intercept form for each line shown or described.

31.

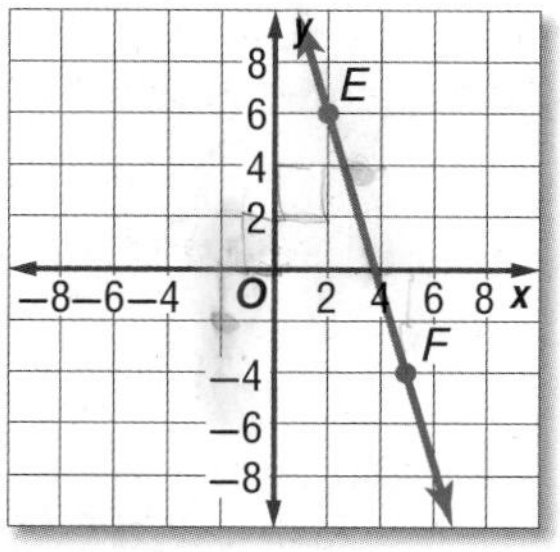

32.

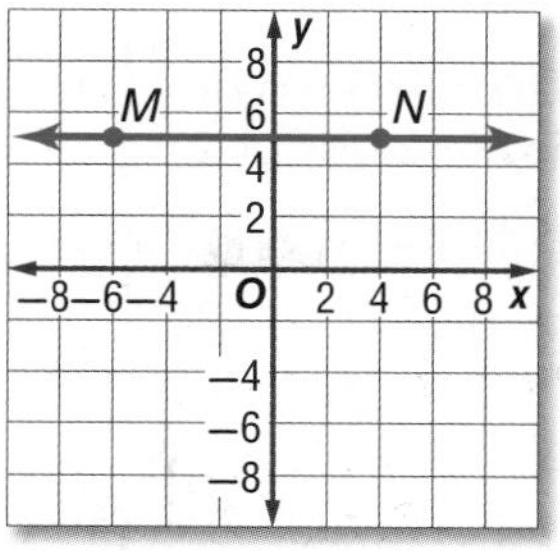

33.

x	−1	3
y	−2	4

34.

x	−4	−8
y	−5	−13

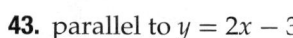

35. x-intercept $= 3$, y-intercept $= -2$

36. x-intercept $= -\frac{1}{2}$, y-intercept $= 4$

Example 5 **Write an equation in slope-intercept form for each line described.**

37. passes through $(-7, -4)$, perpendicular to $y = \frac{1}{2}x + 9$

38. passes through $(-1, -10)$, parallel to $y = 7$

39. passes through $(6, 2)$, parallel to $y = -\frac{2}{3}x + 1$

40. passes through $(-2, 2)$, perpendicular to $y = -5x - 8$

Example 6 **41** **PLANNING** Karen is planning a graduation party for the senior class. She plans to rent a meeting room at the convention center that costs \$400. There is an additional fee of \$5.50 for each person who attends the party.

a. Write an equation to represent the cost y of the party if x people attend.

b. Graph the equation.

c. There are 285 people in Karen's class. If $\frac{2}{3}$ of these people attend, how much will the party cost?

d. If the senior class has raised \$2000 for the party, how many people can attend?

42. **CCSS MODELING** Victor is saving his money to buy a new satellite radio for his car. He wants to save enough money for the radio and one year of satellite radio service before he makes the purchase. He started saving for the radio with \$50 that he got for his birthday. Since then, he has been adding \$15 every week after he cashes his paycheck.

a. Write an equation to represent Victor's savings y after x weeks.

b. Graph the equation.

c. How long will it take Victor to save \$150?

d. A satellite radio costs \$180. Satellite radio service costs \$10 per month. If Victor started saving two weeks ago, how much longer will it take him to save enough money? Explain.

Name the line(s) on the graph shown that match each description.

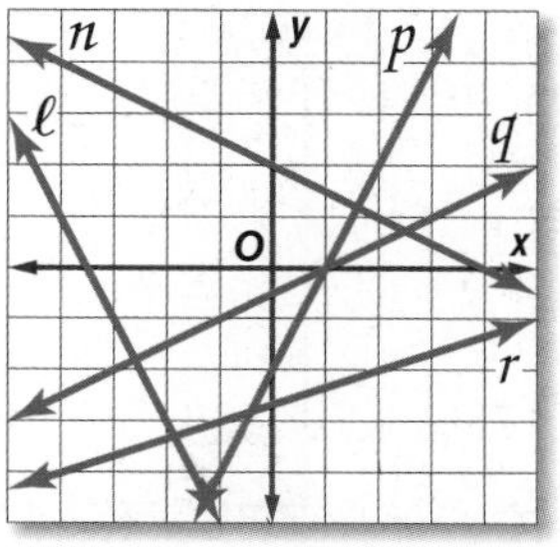

43. parallel to $y = 2x - 3$

44. perpendicular to $y = \frac{1}{2}x + 7$

45. intersecting, but not perpendicular to $y = \frac{1}{2}x - 5$

Determine whether the lines are *parallel*, *perpendicular*, or *neither*.

46. $y = 2x + 4$, $y = 2x - 10$

47. $y = -\frac{1}{2}x - 12$, $y = 2x + 7$

48. $y - 4 = 3(x + 5)$, $y + 3 = -\frac{1}{3}(x + 1)$

49. $y - 3 = 6(x + 2)$, $y + 3 = -\frac{1}{3}(x - 4)$

50. Write an equation in slope-intercept form for a line containing (4, 2) that is parallel to the line $y - 2 = 3(x + 7)$.

51. Write an equation for a line containing (−8, 12) that is perpendicular to the line containing the points (3, 2) and (−7, 2).

52. Write an equation in slope-intercept form for a line containing (5, 3) that is parallel to the line $y + 11 = \frac{1}{2}(4x + 6)$.

53. POTTERY A community center offers pottery classes. A $40 enrollment fee covers supplies and materials, including one bag of clay. Extra bags of clay cost $15 each. Write an equation to represent the cost of the class and *x* bags of clay.

54. MULTIPLE REPRESENTATIONS In Algebra 1, you learned that the solution of a system of two linear equations is an ordered pair that is a solution of both equations. Consider lines *q*, *r*, *s*, and *t* with the equations given.

line *q*: $y = 3x + 2$ line *r* : $y = 0.5x - 3$ line *s*: $2y = x - 6$ line *t*: $y = 3x - 3$

a. Tabular Make a table of values for each equation for $x = -3, -2, -1, 0, 1, 2,$ and 3. Which pairs of lines appear to represent a system of equations with one solution? no solution? infinitely many solutions? Use your tables to explain your reasoning.

b. Graphical Graph the equations on the same coordinate plane. Describe the geometric relationship between each pair of lines, including points of intersection.

c. Analytical How could you have determined your answers to part **a** using only the equations of the lines?

d. Verbal Explain how to determine whether a given system of two linear equations has one solution, no solution, or infinitely many solutions using a table, a graph, or the equations of the lines.

H.O.T. Problems Use Higher-Order Thinking Skills

55. CHALLENGE Find the value of *n* so that the line perpendicular to the line with the equation $-2y + 4 = 6x + 8$ passes through the points at (*n*, −4) and (2, −8).

56. REASONING Determine whether the points at (−2, 2), (2, 5), and (6, 8) are collinear. Justify your answer.

57. OPEN ENDED Write equations for two different pairs of perpendicular lines that intersect at the point at (−3, −7).

58. CCSS CRITIQUE Mark and Josefina wrote an equation of a line with slope −5 that passes through the point (−2, 4). Is either of them correct? Explain your reasoning.

Mark	Josefina
$y - 4 = -5(x - (-2))$	$y - 4 = -5(x - (-2))$
$y - 4 = -5(x + 2)$	$y - 4 = -5(x + 2)$
$y - 4 = -5x - 10$	
$y = -5x - 6$	

59. WRITING IN MATH When is it easier to use the point-slope form to write an equation of a line and when is it easier to use the slope-intercept form?

Standardized Test Practice

60. Which graph best represents a line passing through the point $(-2, -3)$?

A
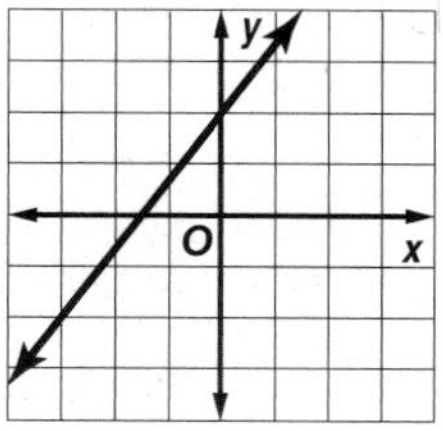

C
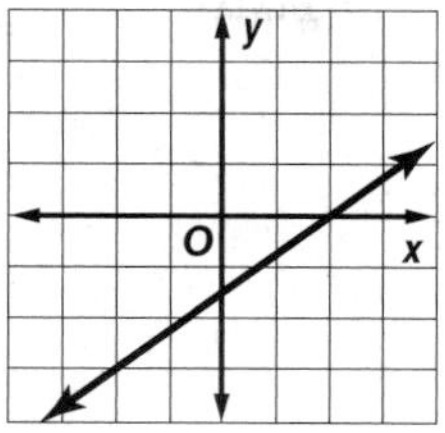

B
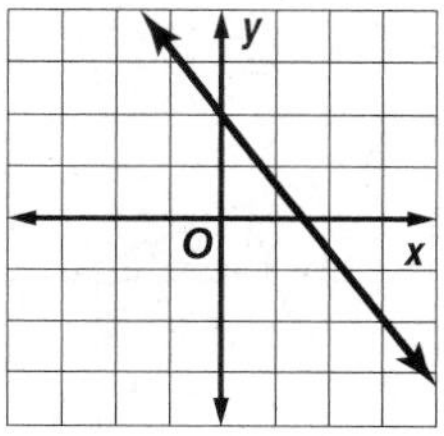

D
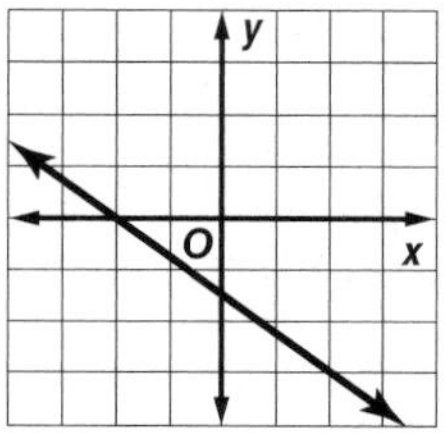

61. Which equation describes the line that passes through the point at $(-2, 1)$ and is perpendicular to the line $y = \frac{1}{3}x + 5$?

F $y = 3x + 7$

G $y = \frac{1}{3}x + 7$

H $y = -3x - 5$

J $y = -\frac{1}{3}x - 5$

62. GRIDDED RESPONSE At Jefferson College, 80% of students have cell phones. Of the students who have cell phones, 70% have computers. What percent of the students at Jefferson College have both a cell phone and a computer?

63. SAT/ACT Which expression is equivalent to $4(x - 6) - \frac{1}{2}(x^2 + 8)$?

A $4x^2 + 4x - 28$

B $-\frac{1}{2}x^2 + 4x - 20$

C $-\frac{1}{2}x^2 + 6x - 24$

D $3x - 20$

E $-\frac{1}{2}x^2 + 4x - 28$

Spiral Review

Determine the slope of the line that contains the given points. (Lesson 11-3)

64. $J(4, 3), K(5, -2)$

65. $X(0, 2), Y(-3, -4)$

66. $A(2, 5), B(5, 1)$

Find x and y in each figure. (Lesson 11-2)

67.

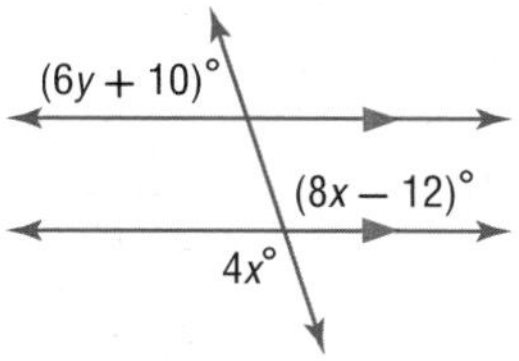

68.

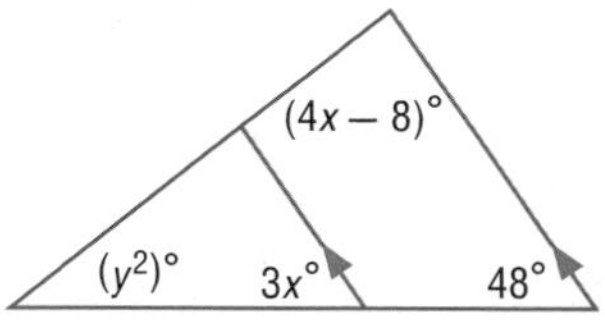

69. DRIVING Lacy's home is located at the midpoint between Newman's Gas Station and Gas-O-Rama. Newman's Gas Station is a quarter mile away from Lacy's home. How far away is Gas-O-Rama from Lacy's home? How far apart are the two gas stations? (Lesson 10-3)

Skills Review

Determine the relationship between each pair of angles.

70. $\angle 1$ and $\angle 12$

71. $\angle 7$ and $\angle 10$

72. $\angle 4$ and $\angle 8$

73. $\angle 2$ and $\angle 11$

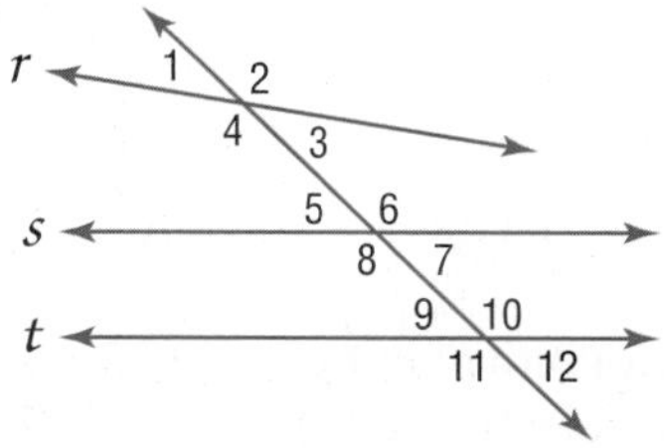

EXTEND 11-4

Geometry Lab
Equations of Perpendicular Bisectors

You can apply what you have learned about slope and equations of lines to geometric figures on a plane.

CCSS Common Core State Standards
Content Standards
G.GPE.5 Prove the slope criteria for parallel and perpendicular lines and use them to solve geometric problems (e.g., find the equation of a line parallel or perpendicular to a given line that passes through a given point).
Mathematical Practices 8

Activity

Find the equation of a line that is a perpendicular bisector of a segment AB with endpoints $A(-3, 3)$ and $B(4, 0)$.

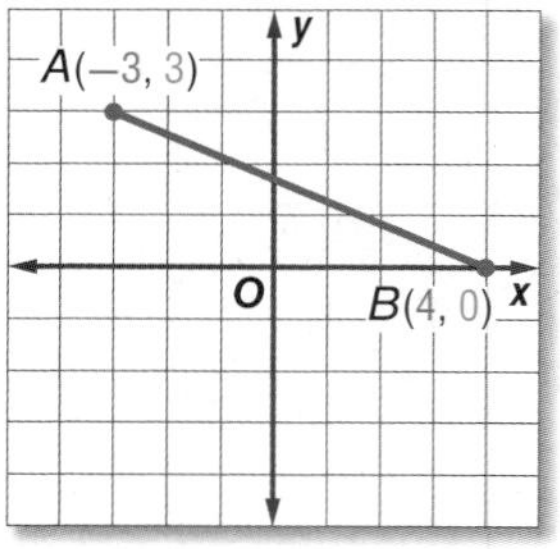

Step 1 A segment bisector contains the midpoint of the segment. Use the Midpoint Formula to find the midpoint M of $\overline{AB}$.

$$M\left(\frac{x_1 + x_2}{2}, \frac{y_1 + y_2}{2}\right) = M\left(\frac{-3 + 4}{2}, \frac{3 + 0}{2}\right)$$
$$= M\left(\frac{1}{2}, \frac{3}{2}\right)$$

Step 2 A perpendicular bisector is perpendicular to the segment through the midpoint. In order to find the slope of the bisector, first find the slope of $\overline{AB}$.

$m = \frac{y_2 - y_1}{x_2 - x_1}$ Slope Formula

$= \frac{0 - 3}{4 - (-3)}$ $x_1 = -3, x_2 = 4, y_1 = 3, y_2 = 0$

$= -\frac{3}{7}$ Simplify.

Step 3 Now use the point-slope form to write the equation of the line. The slope of the bisector is $\frac{7}{3}$ since $-\frac{3}{7}\left(\frac{7}{3}\right) = -1$

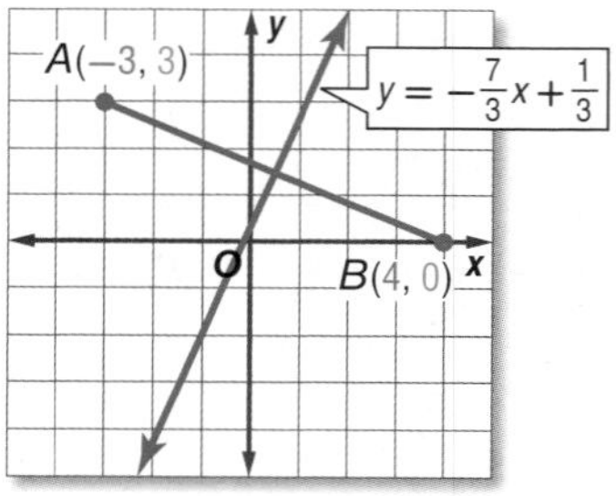

$y - y_1 = m(x - x_1)$ Point-slope form

$y - \frac{3}{2} = \frac{7}{3}\left(x - \frac{1}{2}\right)$ $m = \frac{7}{3}, (x_1, y_1) = \left(\frac{1}{2}, \frac{3}{2}\right)$

$y - \frac{3}{2} = \frac{7}{3}x - \frac{7}{6}$ Distributive Property

$y = \frac{7}{3}x + \frac{1}{3}$ Add $\frac{3}{2}$ to each side.

Exercises

Find the equation of a line that is the perpendicular bisector $\overline{PQ}$ for the given endpoints.

1. $P(5, 2), Q(7, 4)$
2. $P(-3, 9), Q(-1, 5)$
3. $P(-6, -1), Q(8, 7)$
4. $P(-2, 1), Q(0, -3)$
5. $P(0, 1.6), Q(0.5, 2.1)$
6. $P(-7, 3), Q(5, 3)$
7. **CHALLENGE** Find the equations of the lines that contain the sides of $\triangle XYZ$ with vertices $X(-2, 0)$, $Y(1, 3)$, and $Z(3, -1)$.

LESSON 11-5

Proving Lines Parallel

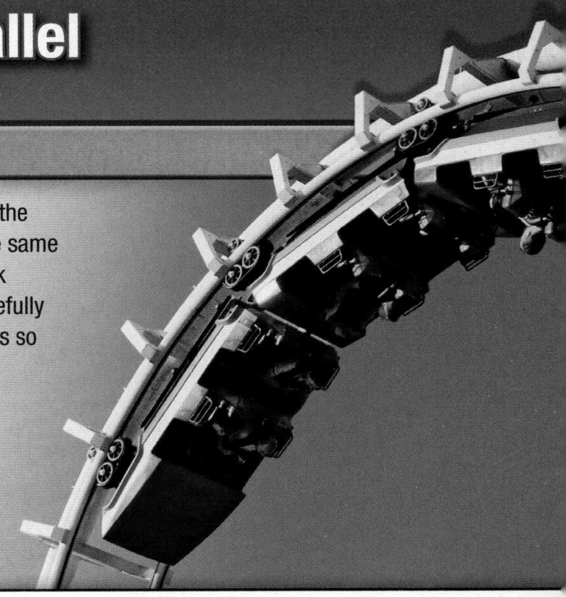

Then

- You used slopes to identify parallel and perpendicular lines.

Now

1. Recognize angle pairs that occur with parallel lines.
2. Prove that two lines are parallel.

Why?

- When you see a roller coaster track, the two sides of the track are always the same distance apart, even though the track curves and turns. The tracks are carefully constructed to be parallel at all points so that the car is secure on the track.

CCSS Common Core State Standards

Content Standards

G.CO.9 Prove theorems about lines and angles.

G.CO.12 Make formal geometric constructions with a variety of tools and methods (compass and straightedge, string, reflective devices, paper folding, dynamic geometric software, etc.).

Mathematical Practices

1 Make sense of problems and persevere in solving them.

3 Construct viable arguments and critique the reasoning of others.

1 Identify Parallel Lines

The two sides of the track of a roller coaster are parallel, and all of the supports along the track are also parallel. Each of the angles formed between the track and the supports are corresponding angles. We have learned that corresponding angles are congruent when lines are parallel. The converse of this relationship is also true.

Postulate 11.4 Converse of Corresponding Angles Postulate

If two lines are cut by a transversal so that corresponding angles are congruent, then the lines are parallel.

Examples If $\angle 1 \cong \angle 3$, $\angle 2 \cong \angle 4$, $\angle 5 \cong \angle 7$, $\angle 6 \cong \angle 8$, then $a \parallel b$.

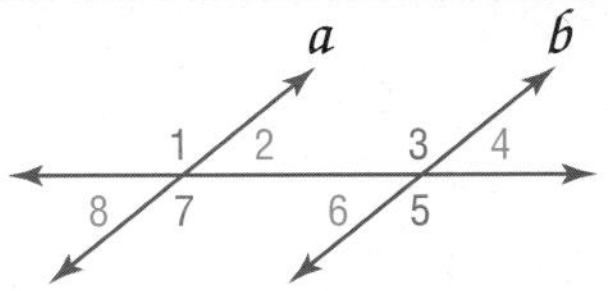

The Converse of the Corresponding Angles Postulate can be used to construct parallel lines.

Construction Parallel Line Through a Point Not on the Line

Step 1 Use a straightedge to draw $\overleftrightarrow{AB}$. Draw a point C that is not on $\overleftrightarrow{AB}$. Draw $\overleftrightarrow{CA}$.

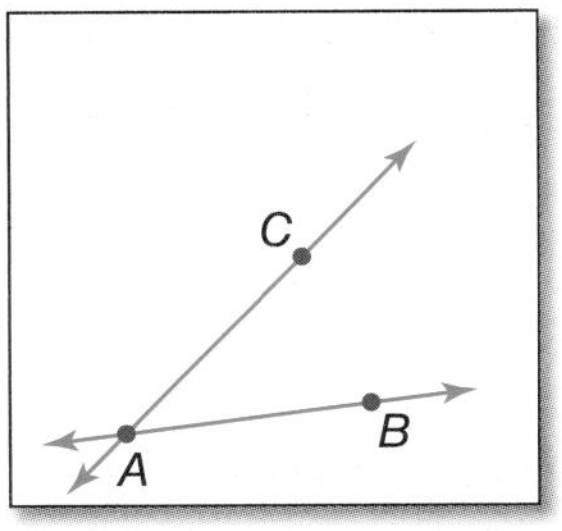

Step 2 Copy $\angle CAB$ so that C is the vertex of the new angle. Label the intersection points D and E.

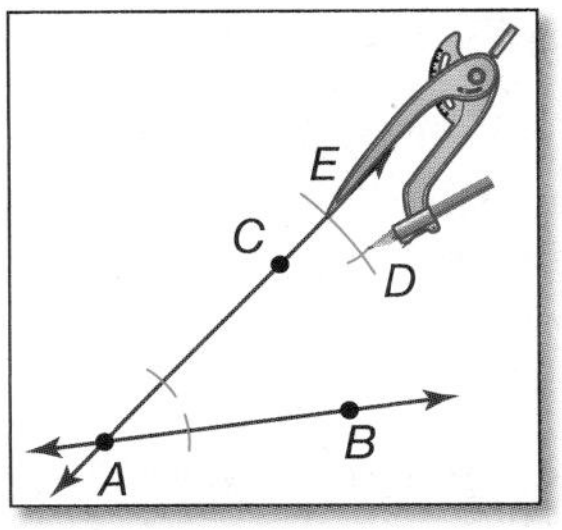

Step 3 Draw CD. Because $\angle ECD \cong \angle CAB$ by construction and they are corresponding angles, $\overleftrightarrow{AB} \parallel \overleftrightarrow{CD}$.

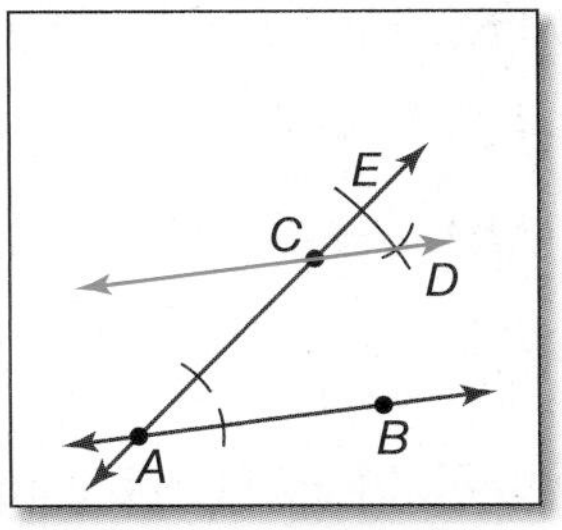

The construction establishes that there is *at least* one line through C that is parallel to $\overleftrightarrow{AB}$. The following postulate guarantees that this line is the *only* one.

> **StudyTip**
> **Euclid's Postulates** The father of modern geometry, Euclid (c. 300 B.C.) realized that only a few postulates were needed to prove the theorems in his day. Postulate 11.5 is one of Euclid's five original postulates.

Postulate 11.5 Parallel Postulate

If given a line and a point not on the line, then there exists exactly one line through the point that is parallel to the given line.

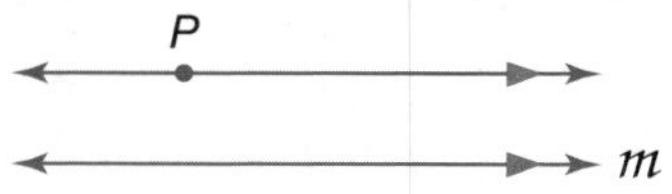

Parallel lines that are cut by a transversal create several pairs of congruent angles. These special angle pairs can also be used to prove that a pair of lines are parallel.

Theorems Proving Lines Parallel

11.5 Alternate Exterior Angles Converse
If two lines in a plane are cut by a transversal so that a pair of alternate exterior angles is congruent, then the two lines are parallel.

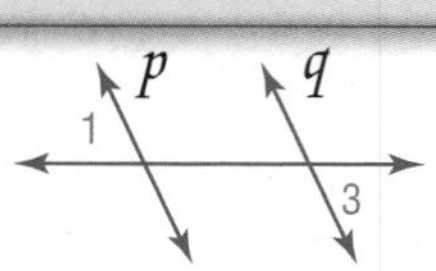

If $\angle 1 \cong \angle 3$, then $p \parallel q$.

11.6 Consecutive Interior Angles Converse
If two lines in a plane are cut by a transversal so that a pair of consecutive interior angles is supplementary, then the lines are parallel.

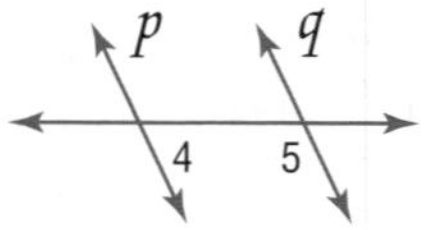

If $m\angle 4 + m\angle 5 = 180$, then $p \parallel q$.

11.7 Alternate Interior Angles Converse
If two lines in a plane are cut by a transversal so that a pair of alternate interior angles is congruent, then the lines are parallel.

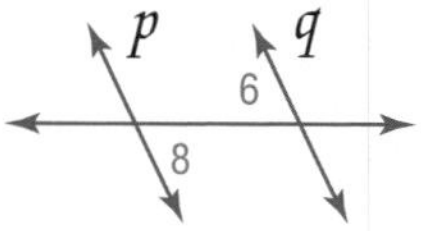

If $\angle 6 \cong \angle 8$, then $p \parallel q$.

11.8 Perpendicular Transversal Converse
In a plane, if two lines are perpendicular to the same line, then they are parallel.

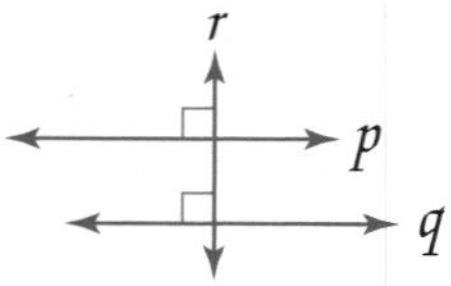

If $p \perp r$ and $q \perp r$, then $p \parallel q$.

You will prove Theorems 11.5, 11.6, 11.7, and 11.8 in Exercises 6, 23, 31, and 30, respectively.

Example 1 Identify Parallel Lines

PT

Given the following information, determine which lines, if any, are parallel. State the postulate or theorem that justifies your answer.

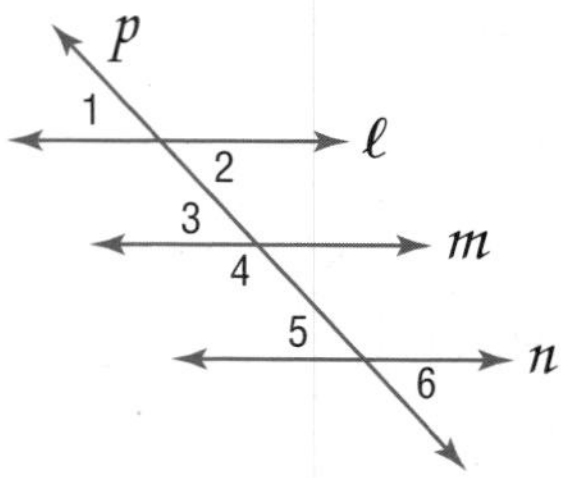

a. $\angle 1 \cong \angle 6$

$\angle 1$ and $\angle 6$ are alternate exterior angles of lines ℓ and n.

Since $\angle 1 \cong \angle 6$, $\ell \parallel n$ by the Converse of the Alternate Exterior Angles Theorem.

b. $\angle 2 \cong \angle 3$

$\angle 2$ and $\angle 3$ are alternate interior angles of lines ℓ and m.

Since $\angle 2 \cong \angle 3$, $\ell \parallel m$ by the Converse of the Alternate Interior Angles Theorem.

GuidedPractice

1A. $\angle 2 \cong \angle 8$

1B. $\angle 3 \cong \angle 11$

1C. $\angle 12 \cong \angle 14$

1D. $\angle 1 \cong \angle 15$

1E. $m\angle 8 + m\angle 13 = 180$

1F. $\angle 8 \cong \angle 6$

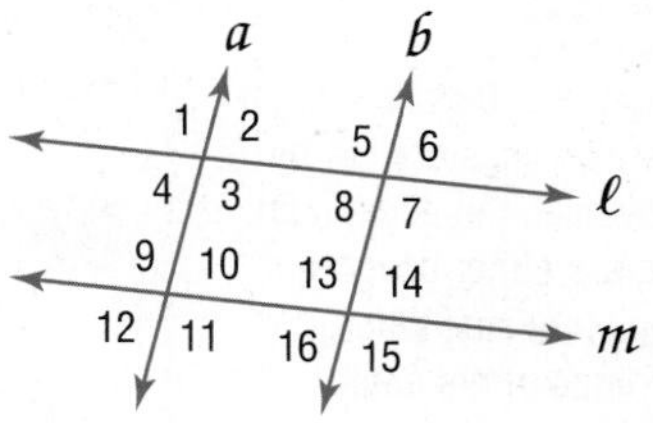

Angle relationships can be used to solve problems involving unknown values.

Standardized Test Example 2 Use Angle Relationships

PT

OPEN ENDED Find $m\angle MRQ$ so that $a \parallel b$. Show your work.

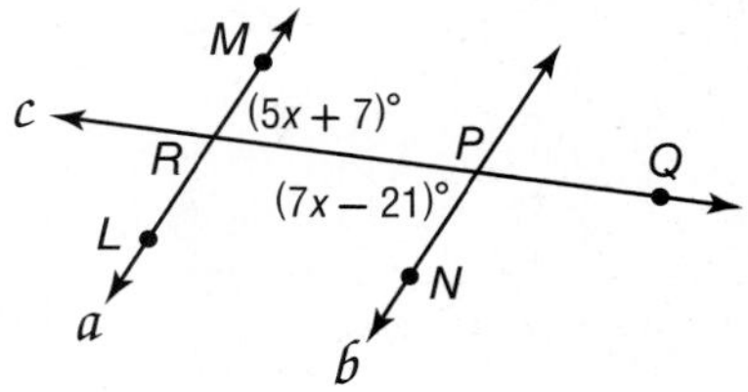

Read the Test Item

From the figure, you know that $m\angle MRQ = 5x + 7$ and $m\angle RPN = 7x - 21$. You are asked to find the measure of $\angle MRQ$.

Solve the Test Item

$\angle MRQ$ and $\angle RPN$ are alternate interior angles. For lines a and b to be parallel, alternate interior angles must be congruent, so $\angle MRQ \cong \angle RPN$. By the definition of congruence, $m\angle MRQ = m\angle RPN$. Substitute the given angle measures into this equation and solve for x.

$m\angle MRQ = m\angle RPN$	Alternate interior angles
$5x + 7 = 7x - 21$	Substitution
$7 = 2x - 21$	Subtract $5x$ from each side.
$28 = 2x$	Add 21 to each side.
$14 = x$	Divide each side by 2.

Now, use the value of x to find $\angle MRQ$.

$m\angle MRQ = 5x + 7$	Substitution
$= 5(14) + 7$	$x = 14$
$= 77$	Simplify.

CHECK Check your answer by using the value of x to find $m\angle RPN$.

$m\angle RPN = 7x - 21$

$= 7(14) - 21$ or 77 ✓

Since $m\angle MRQ = m\angle RPN$, $\angle MRQ \cong \angle RPN$ and $a \parallel b$. ✓

StudyTip

Finding What Is Asked For Be sure to reread test questions carefully to be sure you are answering the question that was asked. In Example 2, a common error would be to stop after you have found the value of x and say that the solution of the problem is 14.

GuidedPractice

2. Find y so that $e \parallel f$. Show your work.

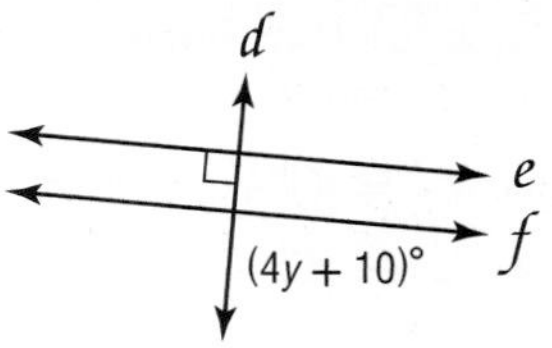

StudyTip

Proving Lines Parallel When two parallel lines are cut by a transversal, the angle pairs formed are either congruent or supplementary. When a pair of lines forms angles that do not meet this criterion, the lines cannot possibly be parallel.

2 Prove Lines Parallel

The angle pair relationships formed by a transversal can be used to prove that two lines are parallel.

PT

Real-World Example 3 Prove Lines Parallel

HOME FURNISHINGS In the ladder shown, each rung is perpendicular to the two rails. Is it possible to prove that the two rails are parallel and that all of the rungs are parallel? If so, explain how. If not, explain why not.

Since both rails are perpendicular to each rung, the rails are parallel by the Perpendicular Transversal Converse. Since any pair of rungs is perpendicular to the rails, they are also parallel.

GuidedPractice

3. **ROWING** In order to move in a straight line with maximum efficiency, rower's oars should be parallel. Refer to the photo at the right. Is it possible to prove that any of the oars are parallel? If so, explain how. If not, explain why not.

Check Your Understanding

Example 1 **Given the following information, determine which lines, if any, are parallel. State the postulate or theorem that justifies your answer.**

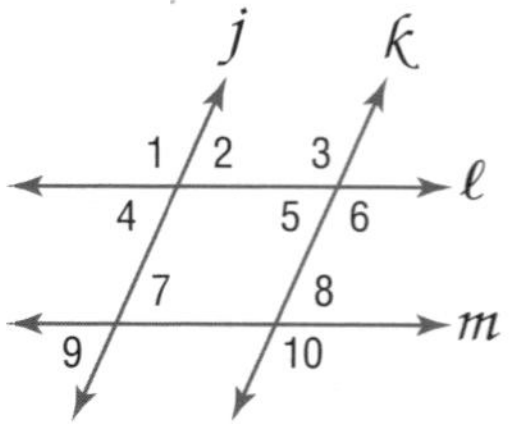

1. $\angle 6 \cong \angle 10$
2. $\angle 4 \cong \angle 7$
3. $\angle 1 \cong \angle 6$
4. $m\angle 2 + m\angle 3 = 180$

Example 2

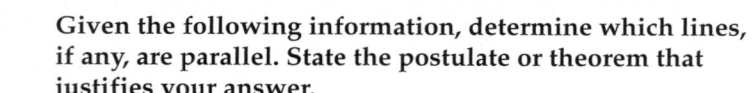

5. **SHORT RESPONSE** Find x so that $m \parallel n$. Show your work.

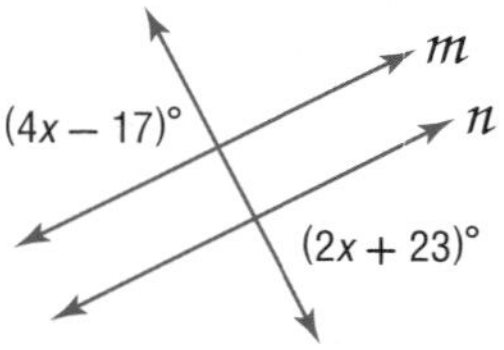

Example 3

6. **PROOF** Copy and complete the proof of Theorem 11.5.

Given: $\angle 1 \cong \angle 2$

Prove: $\ell \parallel m$

Proof:

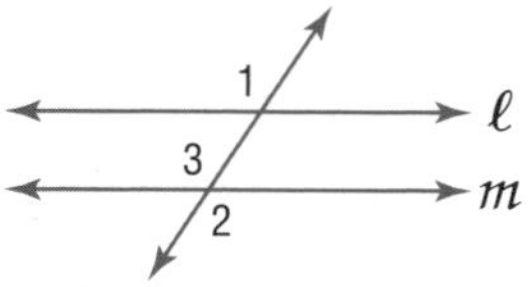

Statements	Reasons
a. $\angle 1 \cong \angle 2$	a. Given
b. $\angle 2 \cong \angle 3$	b. ___?___
c. $\angle 1 \cong \angle 3$	c. Transitive Property
d. ___?___	d. ___?___

7. **CONSTRUCTION** Is it possible to prove that the benches on this picnic table are parallel to each other? If so, explain. If not, explain why not.

Practice and Problem Solving

Example 1 **Given the following information, determine which lines, if any, are parallel. State the postulate or theorem that justifies your answer.**

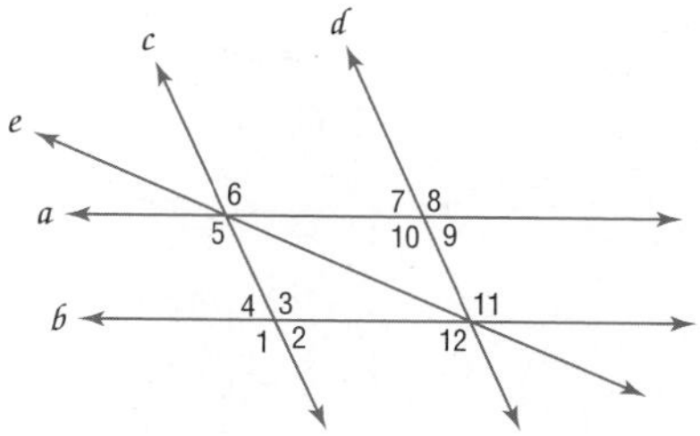

8. $\angle 8 \cong \angle 11$
9. $\angle 8 \cong \angle 12$
10. $\angle 3 \cong \angle 5$
11. $m\angle 2 + m\angle 12 = 180$
12. $m\angle 4 + m\angle 5 = 180$
13. $\angle 6 \cong \angle 10$
14. $\angle 1 \cong \angle 9$
15. $\angle 6 \cong \angle 8$

Example 2 **Find x so that $m \parallel n$. Identify the postulate or theorem you used.**

16.
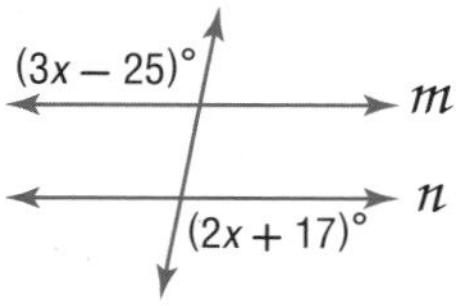

17.
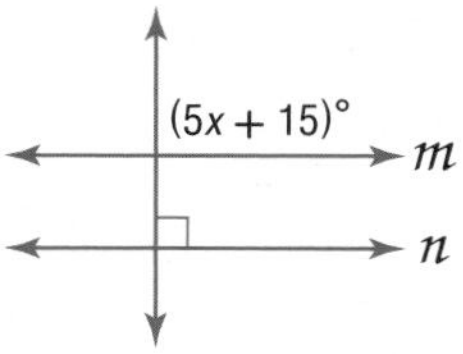

18.
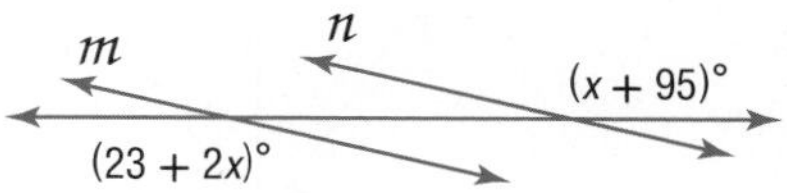

19.
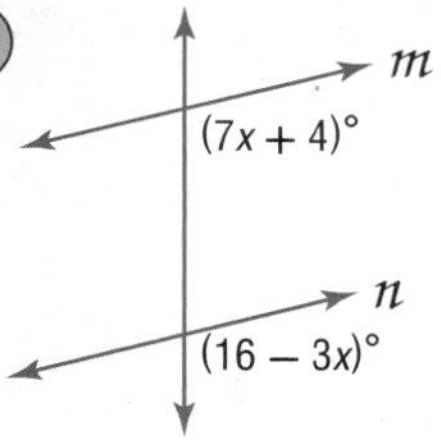

20.
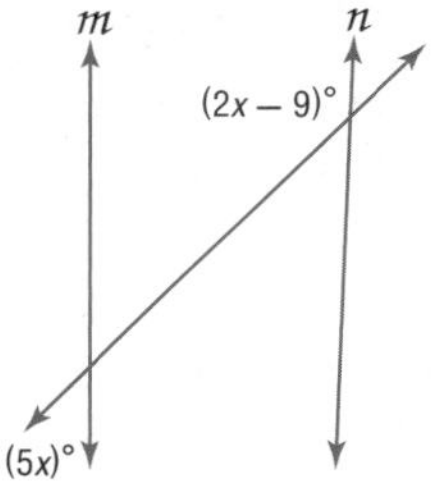

21.
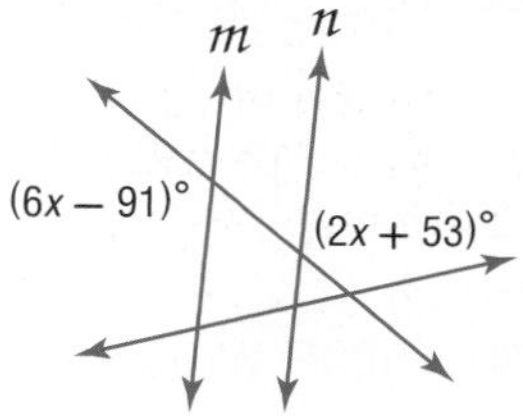

22. **CCSS FRAMING** Wooden door frames are often constructed using a miter box or miter saw. These tools allow you to cut at an angle of a given size. If each of the three pieces of framing material is cut at a 45° angle, will the sides of the door frame be parallel? Explain your reasoning.

Example 3 23. **PROOF** Copy and complete the proof of Theorem 11.6.

Given: $\angle 1$ and $\angle 2$ are supplementary.

Prove: $\ell \parallel m$

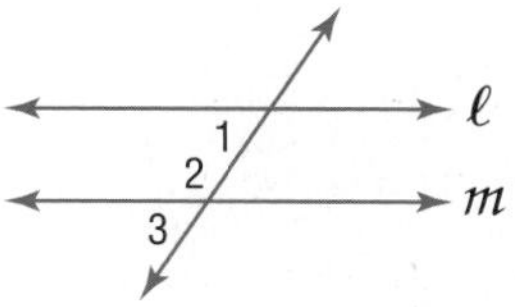

Proof:

Statements	Reasons
a. ___?___	a. Given
b. $\angle 2$ and $\angle 3$ form a linear pair.	b. ___?___
c. ___?___	c. ___?___
d. $\angle 1 \cong \angle 3$	d. ___?___
e. $\ell \parallel m$	e. ___?___

24. Jacqui is making a frame for her favorite poster. She bought a sectional frame kit. As she is putting the frame together, she notices that the corners are cut at 45° angles. How does she know that the corners are right angles and that each pair of opposite sides are parallel?

PROOF **Write a two-column proof for each of the following.**

25. **Given:** $\angle 1 \cong \angle 3$

$\overline{AB} \parallel \overline{CD}$

Prove: $\overline{AC} \parallel \overline{BD}$

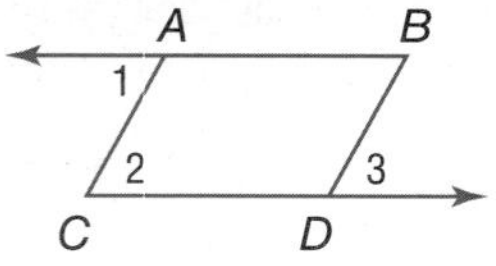

26. **Given:** $\overline{WY} \parallel \overline{XZ}$

$\angle 2 \cong \angle 4$

Prove: $\overline{WX} \parallel \overline{YZ}$

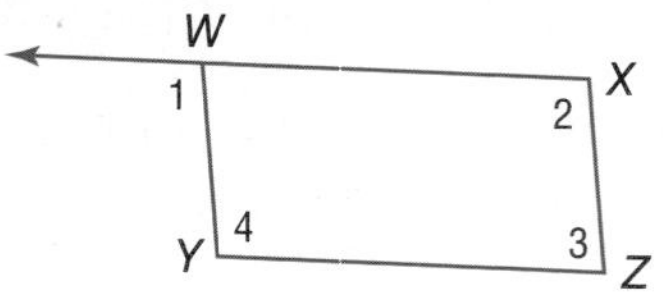

27. **Given:** $\angle TQR \cong \angle TSR$

$m\angle R + m\angle TSR = 180$

Prove: $\overline{QT} \parallel \overline{RS}$

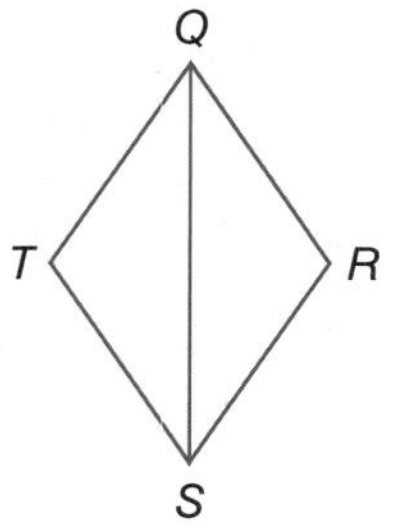

28. **Given:** $\angle DAB \cong \angle DCB$

$\overline{AD} \perp \overline{AB}$

Prove: $\overline{DC} \perp \overline{BC}$

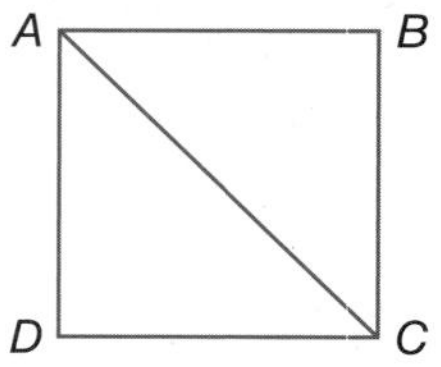

29. **STORAGE** Small parts are often kept in drawers to make finding the correct size easier. In the storage box shown, the frame for each drawer is perpendicular to each of the sides. What can you conclude about the drawers? Explain your reasoning.

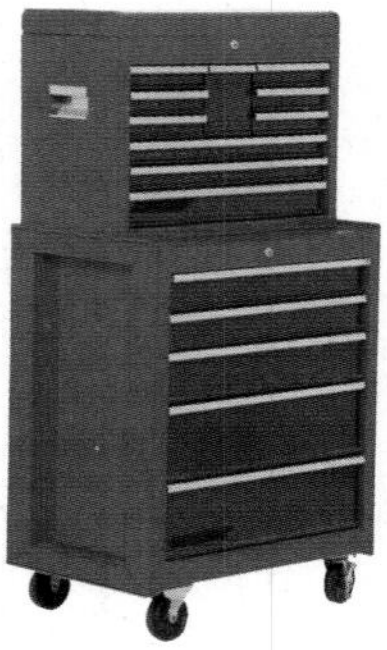

30. **PROOF** Write a paragraph proof of Theorem 11.8.

31. **PROOF** Write a two-column proof of Theorem 11.7.

32. **CCSS** **LADDER RUNGS** Based upon the information given in the photo of the ladder at the right, what is the relationship between each rung? Explain your answer.

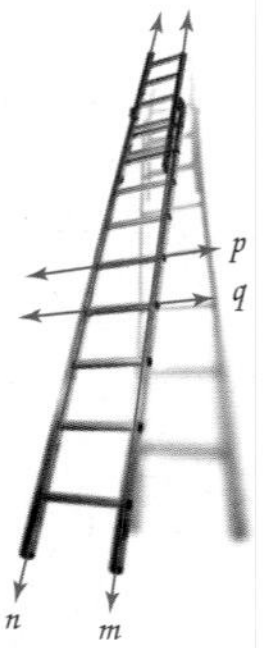

Determine whether lines *r* and *s* are parallel. Justify your answer.

33.

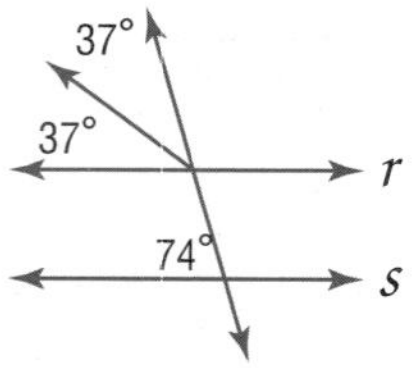

34.

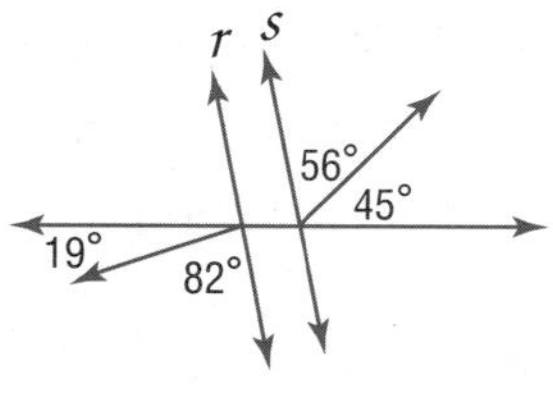

35.

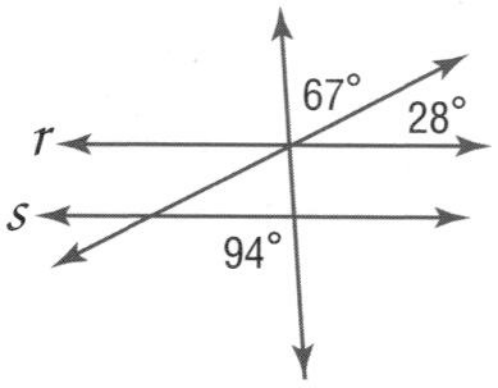

36. **ERROR ANALYSIS** Sumi and Daniela were told that in the figure on the right $\overline{AD} \parallel \overline{BC}$. Sumi says that this is only true if $\angle 1 \cong \angle 4$. Daniela disagrees and says that this is only true if $\angle 2 \cong \angle 3$. Is either of them correct? Explain.

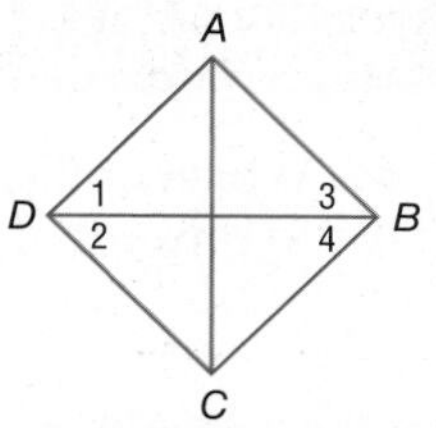

H.O.T. Problems Use Higher-Order Thinking Skills

37. **CHALLENGE** The distance from a line to a point is the length of the line segment perpendicular to the line from the point. The distance between two parallel lines is the distance between any point on one of the lines and the other line. Find the distance between the lines $y = 2x + 5$ and $y = 2x - 1$. Hint: Use the distance formula.

38. **CCSS REASONING** Is Theorem 11.8 still true if the two lines are not coplanar? Draw a figure to justify your answer.

39. **CHALLENGE** Use the figure at the right to prove that two lines parallel to a third line are parallel to each other.

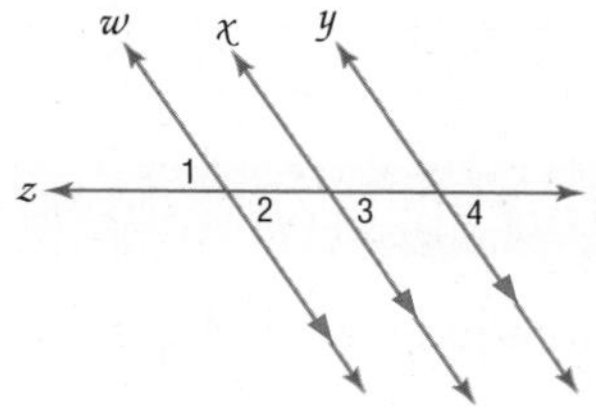

40. Copy the figure on the right on to your paper.

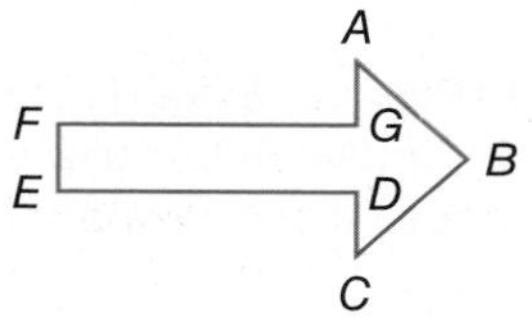

a. Construct a line that is parallel to $\overline{FG}$ through point A.

b. Use measurement to justify that the line you constructed is parallel to $\overline{FG}$.

c. Construct a line that is parallel to $\overline{FG}$ through point C.

d. Make a conjecture about the relationship between the two lines that you constructed. Explain.

41. **CHALLENGE** Refer to the figure at the right.

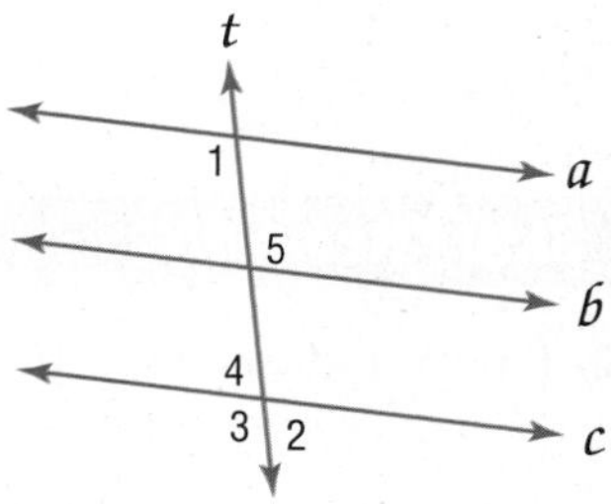

a. If $m\angle 5 + m\angle 2 = 180$, prove that $b \parallel c$.

b. Given $a \parallel b$, if $m\angle 1 + m\angle 5 = 180$, prove that $t \perp b$.

42. **WRITING IN MATH** Summarize the five methods used in this lesson to prove that two lines are parallel.

REASONING Determine whether the statement is *sometimes*, *always*, or *never* true. Explain your reasoning.

43. A linear pair of angles is both supplementary and congruent.

Standardized Test Practice

44. Which of the following facts would be sufficient to prove that line d is parallel to $\overline{XZ}$?

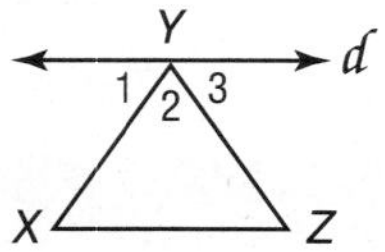

A $\angle 1 \cong \angle 3$
B $\angle 3 \cong \angle Z$
C $\angle 1 \cong \angle Z$
D $\angle 2 \cong \angle X$

45. ALGEBRA The expression $\sqrt{52} + \sqrt{117}$ is equivalent to

F 13
G $5\sqrt{13}$
H $6\sqrt{13}$
J $13\sqrt{13}$

46. What is the approximate surface area of the figure?

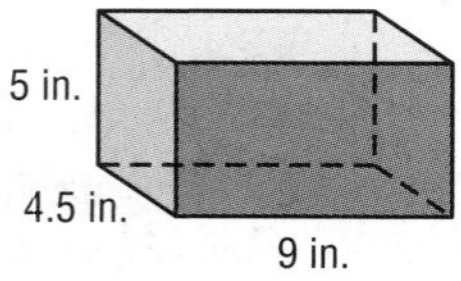

A 101.3 in^2
B 108 in^2
C 202.5 in^2
D 216 in^2

47. SAT/ACT If $x^2 = 25$ and $y^2 = 9$, what is the greatest possible value of $(x - y)^2$?

F 4
G 16
H 58
J 64
K 70

Spiral Review

Write an equation in slope-intercept form of the line having the given slope and y-intercept. (Lesson 11-4)

48. m: 2.5, (0, 0.5)

49. m: $\frac{4}{5}$, (0, −9)

50. m: $-\frac{7}{8}$, $\left(0, -\frac{5}{6}\right)$

51. ROAD TRIP Anne is driving 400 miles to visit Niagara Falls. She manages to travel the first 100 miles of her trip in two hours. If she continues at this rate, how long will it take her to drive the remaining distance? (Lesson 11-3)

Find the perimeter or circumference and area of each figure. Round to the nearest tenth. (Lesson 10-6)

52.

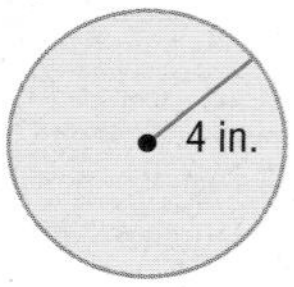

53.

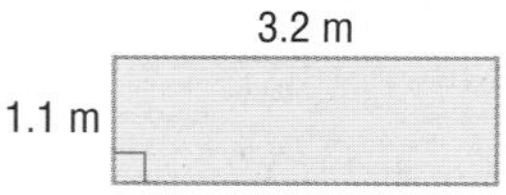

54.

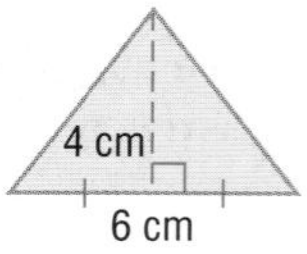

Skills Review

55. Find x and y so that $\overline{BE}$ and $\overline{AD}$ are perpendicular.

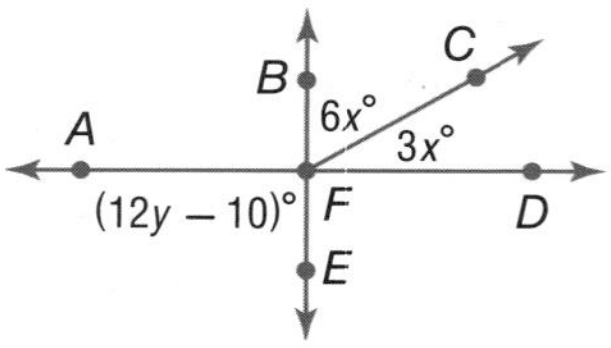

LESSON 11-6

Perpendiculars and Distance

Then

- You proved that two lines are parallel using angle relationships.

Now

1. Find the distance between a point and a line.
2. Find the distance between parallel lines.

Why?

- A *plumb bob* is made of string with a specially designed weight. When the weight is suspended and allowed to swing freely, the point of the bob is precisely below the point to which the string is fixed.

  The plumb bob is useful in establishing what is the true vertical or *plumb* when constructing a wall or when hanging wallpaper.

NewVocabulary
equidistant

Common Core State Standards

Content Standards

G.CO.12 Make formal geometric constructions with a variety of tools and methods (compass and straightedge, string, reflective devices, paper folding, dynamic geometric software, etc.).

G.MG.3 Apply geometric methods to solve problems (e.g., designing an object or structure to satisfy physical constraints or minimize cost; working with typographic grid systems based on ratios). ★

Mathematical Practices

2 Reason abstractly and quantitatively.

4 Model with mathematics.

1 Distance From a Point to a Line

The plumb bob also indicates the shortest distance between the point at which it is attached on the ceiling and a level floor below. This perpendicular distance between a point and a line is the shortest in all cases.

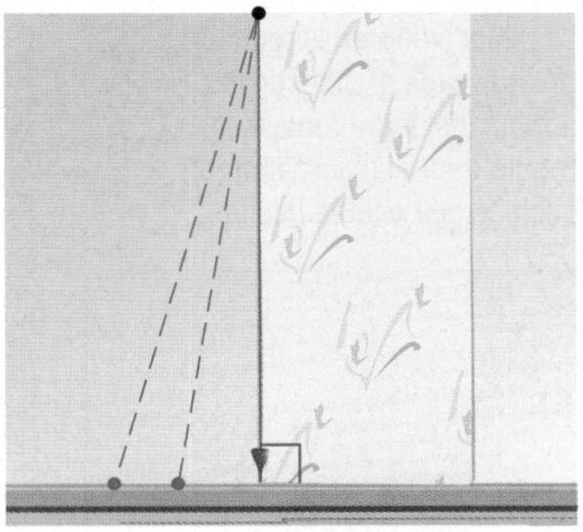

KeyConcept Distance Between a Point and a Line

Words The distance between a line and a point not on the line is the length of the segment perpendicular to the line from the point.

Model

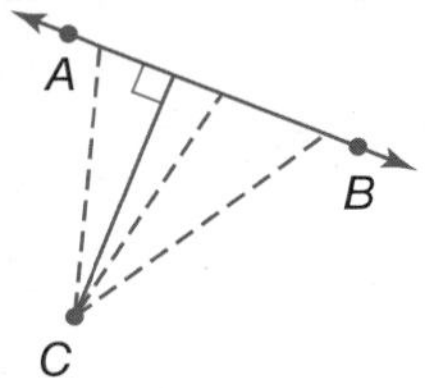

The construction of a line perpendicular to an existing line through a point not on the existing line in Extend Lesson 10-5 establishes that there is *at least one* line through a point P that is perpendicular to a line AB. The following postulate states that this line is the *only* line through P perpendicular to $\overleftrightarrow{AB}$.

Postulate 11.6 Perpendicular Postulate

Words If given a line and a point not on the line, then there exists exactly one line through the point that is perpendicular to the given line.

Model

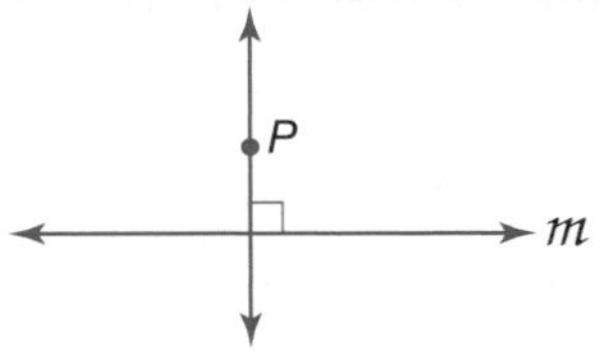

Real-World Career

Landscape Architect Landscape architects enjoy working with their hands and possess strong analytical skills. Creative vision and artistic talent are also desirable qualities. Typically, a bachelor's degree is required of landscape architects, but a master's degree may be required for specializations such as golf course design.

Real-World Example 1 Construct Distance From a Point to a Line

LANDSCAPING A landscape architect notices that one part of a yard does not drain well. She wants to tap into an existing underground drain represented by line m. Construct and name the segment with the length that represents the shortest amount of pipe she will need to lay to connect this drain to point A.

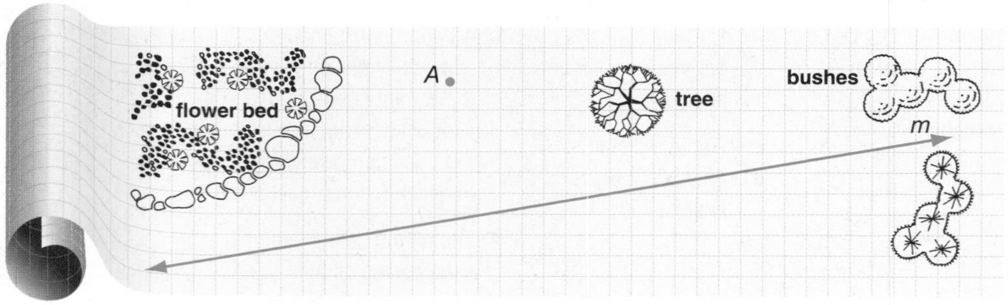

The distance from a line to a point not on the line is the length of the segment perpendicular to the line from the point. Locate points B and C on line m equidistant from point A.

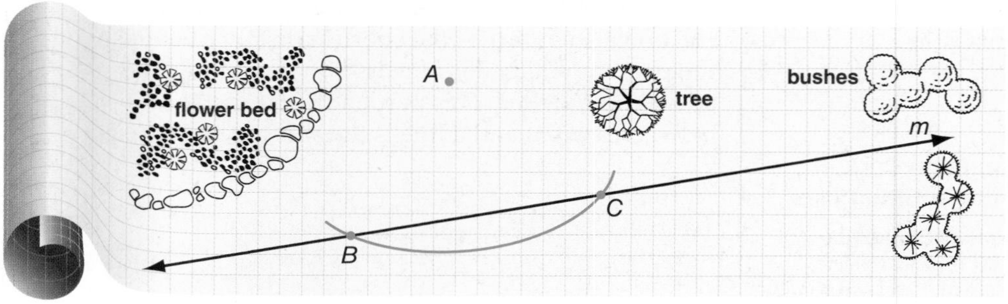

Locate a third point on line m equidistant from B and C. Label this point D. Then draw $\overleftrightarrow{AD}$ so that $\overleftrightarrow{AD} \perp \overleftrightarrow{BC}$.

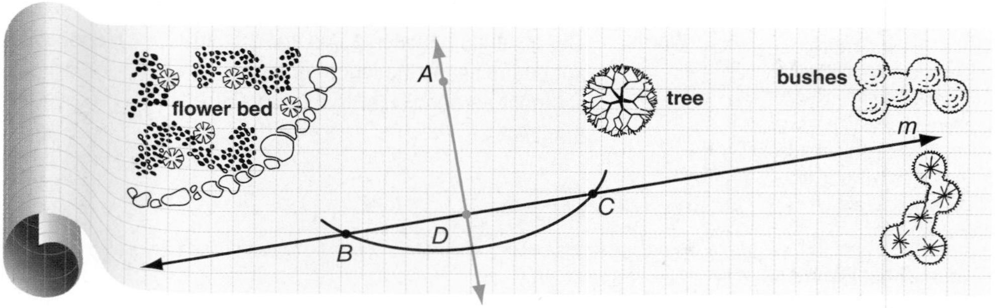

The measure of $\overline{AD}$ represents the shortest amount of pipe the architect will need to lay to connect the drain to point A.

Study Tip

Drawing the Shortest Distance You can use tools like the corner of a piece of paper to help you draw a perpendicular segment from a point to a line, but only a compass and a straightedge can be used to construct this segment.

Guided Practice

1. Copy the figure. Then construct and name the segment that represents the distance from Q to $\overleftrightarrow{PR}$.

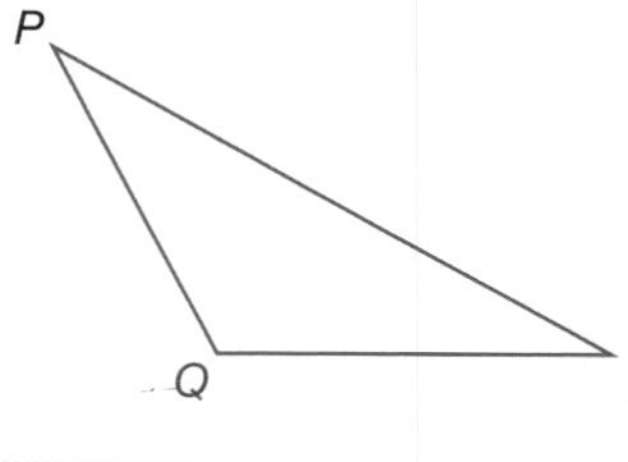

StudyTip

Distance to Axes Note that the distance from a point to the x-axis can be determined by looking at the y-coordinate, and the distance from a point to the y-axis can be determined by looking at the x-coordinate.

Example 2 Distance from a Point to a Line on Coordinate Plane

COORDINATE GEOMETRY Line ℓ contains points at (−5, 3) and (4, −6). Find the distance between line ℓ and point $P(2, 4)$.

Step 1 Find the equation of the line ℓ.

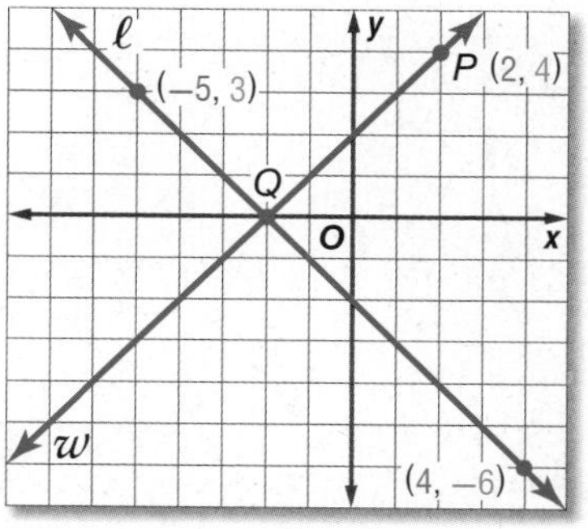

Begin by finding the slope of the line through points (−5, 3) and (4, −6).

$$m = \frac{y_2 - y_1}{x_2 - x_1} = \frac{-6 - 3}{4 - (-5)} = \frac{-9}{9} \text{ or } -1$$

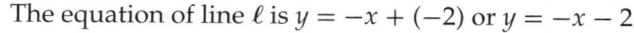

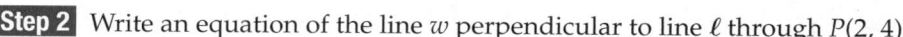

Then write the equation of this line using the point (4, −6) on the line.

$y = mx + b$	Slope-intercept form
$-6 = -1(4) + b$	$m = -1$, $(x, y) = (4, -6)$
$-6 = -4 + b$	Simplify.
$-2 = b$	Add 4 to each side.

The equation of line ℓ is $y = -x + (-2)$ or $y = -x - 2$.

Step 2 Write an equation of the line w perpendicular to line ℓ through $P(2, 4)$.

Since the slope of line ℓ is −1, the slope of a line p is 1. Write the equation of line w through $P(2, 4)$ with slope 1.

$y = mx + b$	Slope-intercept form
$4 = 1(2) + b$	$m = -1$, $(x, y) = (2, 4)$
$4 = 2 + b$	Simplify.
$2 = b$	Subtract 2 from each side.

The equation of line w is $y = x + 2$.

Step 3 Solve the system of equations to determine the point of intersection.

line ℓ: $y = -x - 2$

line w: $(+)\ y = x + 2$

$2y = 0$	Add the two equations.
$y = 0$	Divide each side by 2.

Solve for x.

$0 = x + 2$	Substitute 0 for y in the second equation.
$-2 = x$	Subtract 2 from each side.

The point of intersection is (−2, 0). Let this be point Q.

Step 4 Use the Distance Formula to determine the distance between $P(2, 4)$ and $Q(-2, 0)$.

$d = \sqrt{(x_2 - x_1)^2 + (y_2 - y_1)^2}$	Distance formula
$= \sqrt{(-2 - 2)^2 + (0 - 4)^2}$	$x_2 = -2$, $x_1 = 2$, $y_2 = 0$, $y_1 = 4$
$= \sqrt{32}$	Simplify.

The distance between the point and the line is $\sqrt{32}$ or about 5.66 units.

GuidedPractice

2. Line ℓ contains points at (1, 2) and (5, 4). Construct a line perpendicular to ℓ through $P(1, 7)$. Then find the distance from P to ℓ.

2 Distance Between Parallel Lines

By definition, parallel lines do not intersect. An alternate definition states that two lines in a plane are parallel if they are everywhere **equidistant**. Equidistant means that the distance between two lines measured along a perpendicular line to the lines is always the same.

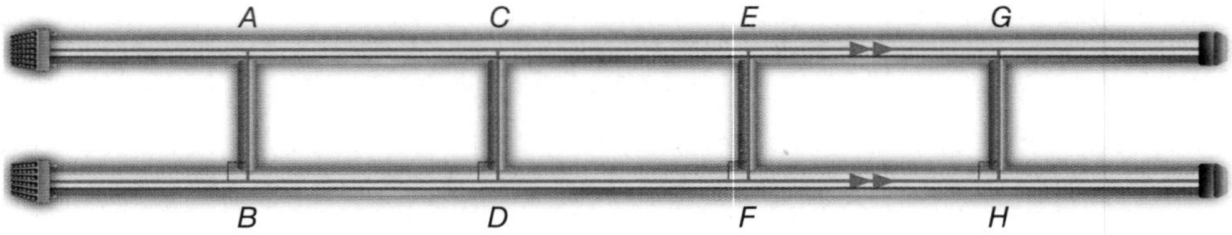

$AB = CD = EF = GH$

This leads to the definition of the distance between two parallel lines.

KeyConcept Distance Between Parallel Lines

The distance between two parallel lines is the perpendicular distance between one of the lines and any point on the other line.

Recall from Lesson 10-1 that a *locus* is the set of all points that satisfy a given condition. Parallel lines can be described as the locus of points in a plane equidistant from a given line.

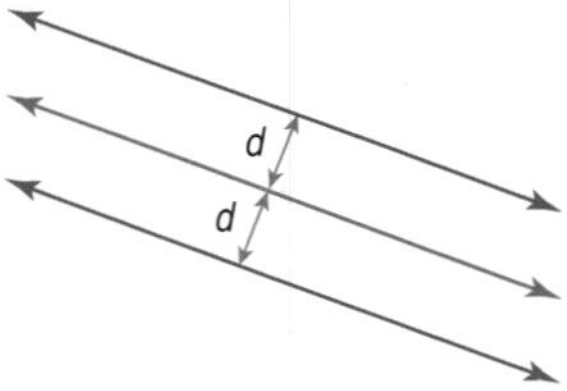

> **StudyTip**
>
> **Locus of Points Equidistant from Two Parallel Lines** Conversely, the locus of points in a plane that are equidistant from two parallel lines is a third line that is parallel to and centered between the two parallel lines.
>
> 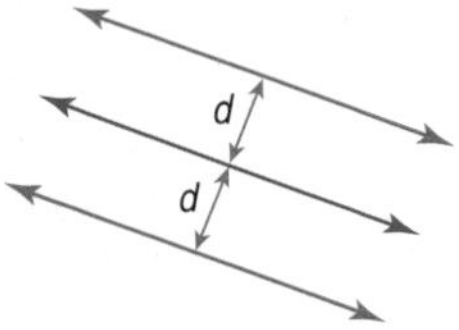
>

Theorem 11.9 Two Lines Equidistant from a Third

In a plane, if two lines are each equidistant from a third line, then the two lines are parallel to each other.

You will prove Theorem 11.9 in Exercise 30.

Example 3 Distance Between Parallel Lines

Find the distance between the parallel lines ℓ and m with equations $y = 2x + 1$ and $y = 2x - 3$, respectively.

You will need to solve a system of equations to find the endpoints of a segment that is perpendicular to both ℓ and m. From their equations, we know that the slope of line ℓ and line m is 2.

Sketch line p through the y-intercept of line m, (0, −3), perpendicular to lines m and ℓ.

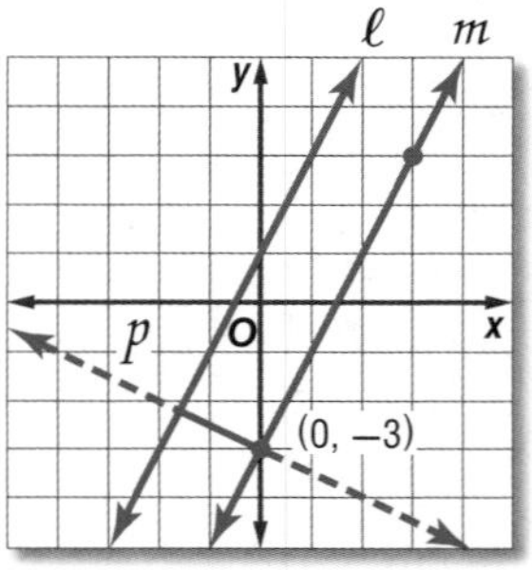

Step 1 Write an equation of line p. The slope of p is the opposite reciprocal of 2, or $-\frac{1}{2}$. Use the y-intercept of line m, (0, −3), as one of the endpoints of the perpendicular segment.

$$(y - y_1) = m(x - x_1)$$ Point-slope form

$$[y - (-3)] = -\frac{1}{2}(x - 0)$$ $x_1 = 0$, $y_1 = 3$, and $m = -\frac{1}{2}$

$$y + 3 = -\frac{1}{2}x$$ Simplify.

$$y = -\frac{1}{2}x - 3$$ Subtract 3 from each side.

Step 2 Use a system of equations to determine the point of intersection of lines ℓ and p.

ℓ: $y = 2x + 1$

p: $y = -\frac{1}{2}x - 3$

$$2x + 1 = -\frac{1}{2}x - 3$$ Substitute $2x + 1$ for y in the second equation.

$$2x + \frac{1}{2}x = -3 - 1$$ Group like terms on each side.

$$\frac{5}{2}x = -4$$ Simplify on each side.

$$x = -\frac{8}{5}$$ Multiply each side by $\frac{2}{5}$.

$$y = -\frac{1}{2}\left(-\frac{8}{5}\right) - 3$$ Substitute $-\frac{8}{5}$ for x in the equation for p.

$$= -\frac{11}{5}$$ Simplify.

The point of intersection is $\left(-\frac{8}{5}, -\frac{11}{5}\right)$ or (−1.6, −2.2).

Step 3 Use the Distance Formula to determine the distance between (0, −3) and (−1.6, −2.2).

$$d = \sqrt{(x_2 - x_1)^2 + (y_2 - y_1)^2}$$ Distance Formula

$$= \sqrt{(-1.6 - 0)^2 + [-2.2 - (-3)]^2}$$ $x_2 = -1.6$, $x_1 = 0$, $y_2 = -2.2$, and $y_1 = -3$

$$\approx 1.8$$ Simplify using a calculator.

The distance between the lines is about 1.8 units.

GuidedPractice

3A. Find the distance between the parallel lines r and s whose equations are $y = -3x - 5$ and $y = -3x + 6$, respectively.

3B. Find the distance between parallel lines a and b with equations $x + 3y = 6$ and $x + 3y = -14$, respectively.

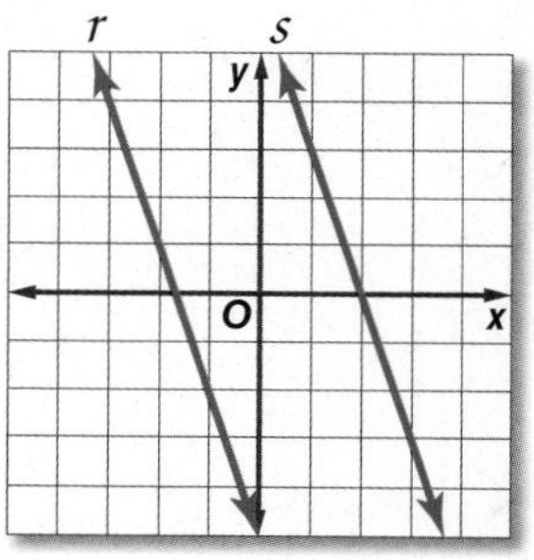

Check Your Understanding

Example 1 **Copy each figure. Construct the segment that represents the distance indicated.**

1. Y to $\overleftrightarrow{TS}$

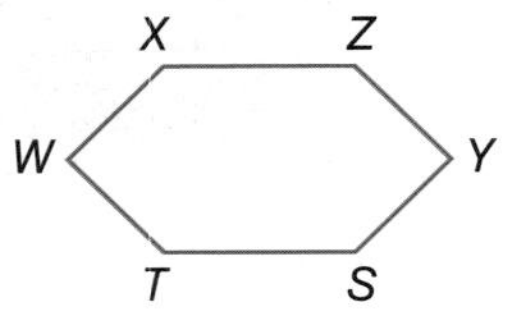

2. C to $\overleftrightarrow{AB}$

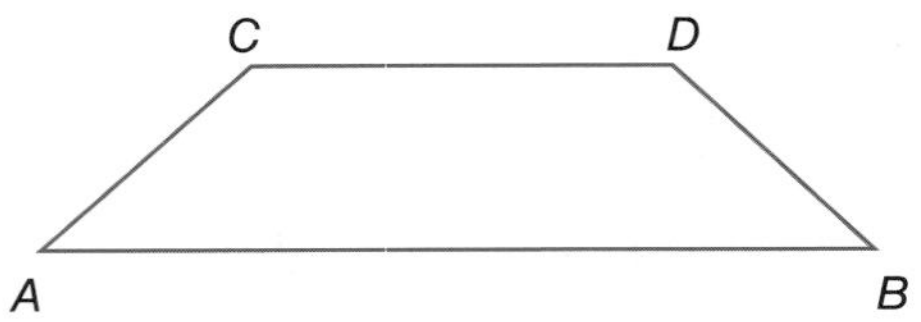

3. **CCSS STRUCTURE** After forming a line, every even member of a marching band turns to face the home team's end zone and marches 5 paces straight forward. At the same time, every odd member turns in the opposite direction and marches 5 paces straight forward. Assuming that each band member covers the same distance, what formation should result? Justify your answer.

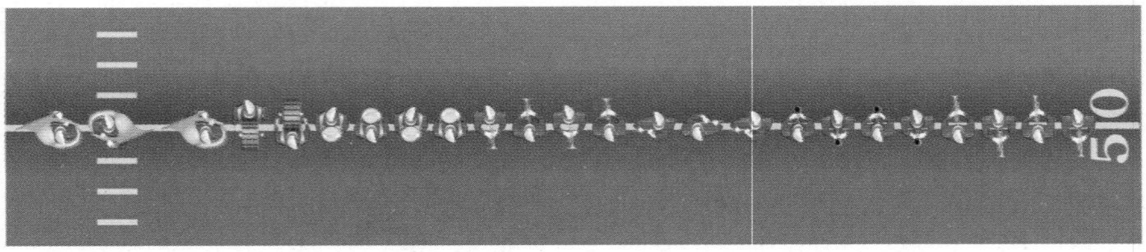

Example 2 **COORDINATE GEOMETRY** **Find the distance from P to ℓ.**

4. Line ℓ contains points (4, 3) and (−2, 0). Point P has coordinates (3, 10).

5. Line ℓ contains points (−6, 1) and (9, −4). Point P has coordinates (4, 1).

6. Line ℓ contains points (4, 18) and (−2, 9). Point P has coordinates (−9, 5).

Example 3 **Find the distance between each pair of parallel lines with the given equations.**

7. $y = -2x + 4$
$y = -2x + 14$

8. $y = 7$
$y = -3$

Practice and Problem Solving

Example 1 **Copy each figure. Construct the segment that represents the distance indicated.**

9. Q to $\overline{RS}$

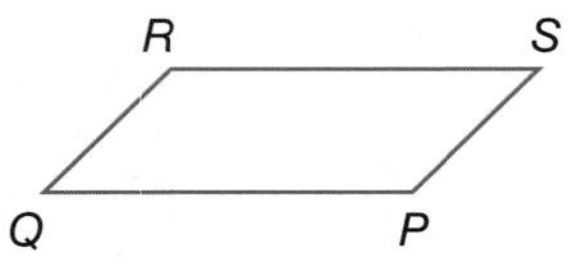

10. A to $\overline{BC}$

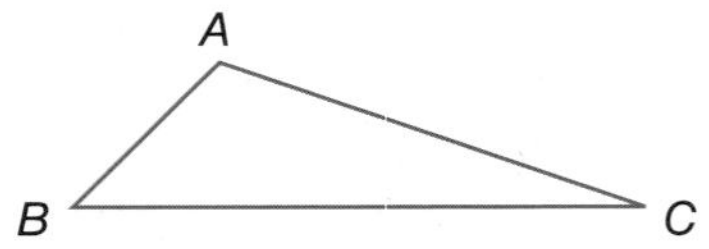

11. H to $\overline{FG}$

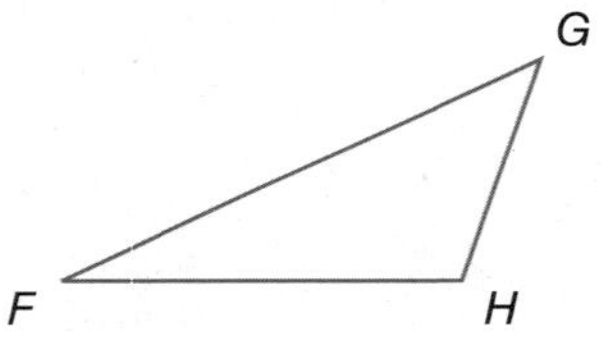

12. K to $\overline{LM}$

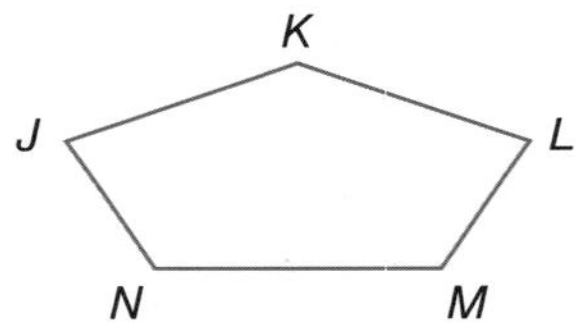

13. **DRIVEWAYS** In the diagram at the right, is the driveway shown the shortest possible one from the house to the road? Explain why or why not.

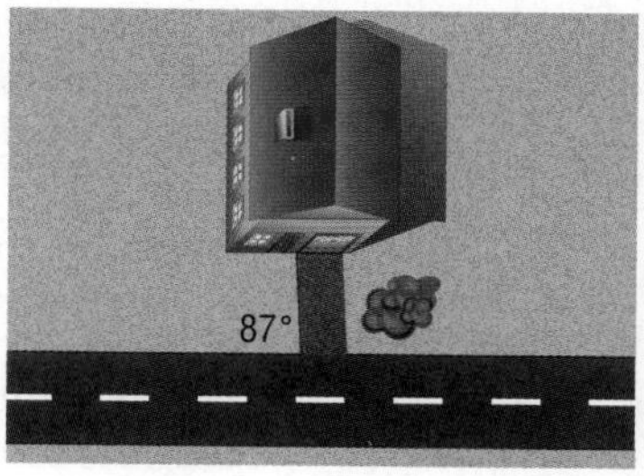

14. **CCSS MODELING** Rondell is crossing the courtyard in front of his school. Three possible paths are shown in the diagram at the right. Which of the three paths shown is the shortest? Explain your reasoning.

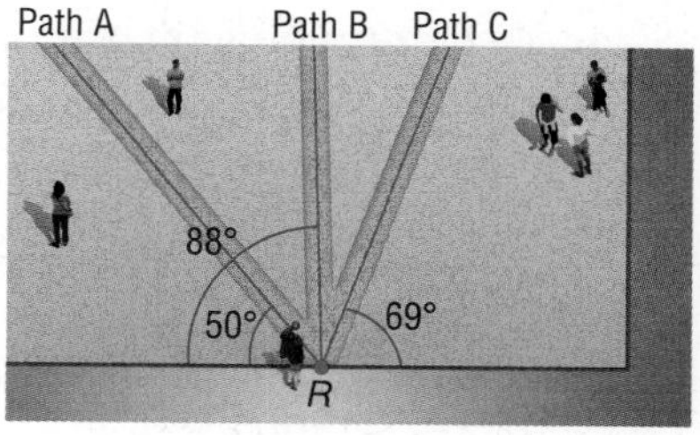

Example 2 **COORDINATE GEOMETRY Find the distance from P to ℓ.**

15. Line ℓ contains points (0, −3) and (7, 4). Point P has coordinates (4, 3).
16. Line ℓ contains points (11, −1) and (−3, −11). Point P has coordinates (−1, 1).
17. Line ℓ contains points (−2, 1) and (4, 1). Point P has coordinates (5, 7).
18. Line ℓ contains points (4, −1) and (4, 9). Point P has coordinates (1, 6).
19. Line ℓ contains points (1, 5) and (4, −4). Point P has coordinates (−1, 1).
20. Line ℓ contains points (−8, 1) and (3, 1). Point P has coordinates (−2, 4).

Example 3 **Find the distance between each pair of parallel lines with the given equations.**

21. $y = -2$, $y = 4$

22. $x = 3$, $x = 7$

23. $y = 5x - 22$, $y = 5x + 4$

24. $y = \frac{1}{3}x - 3$, $y = \frac{1}{3}x + 2$

25. $x = 8.5$, $x = -12.5$

26. $y = 15$, $y = -4$

27. $y = \frac{1}{4}x + 2$, $4y - x = -60$

28. $3x + y = 3$, $y + 17 = -3x$

29. $y = -\frac{5}{4}x + 3.5$, $4y + 10.6 = -5x$

30. **PROOF** Write a two-column proof of Theorem 11.9.

Find the distance from the line to the given point.

31. $y = -3$, (5, 2)

32. $y = \frac{1}{6}x + 6$, (−6, 5)

33. $x = 4$, (−2, 5)

34. **POSTERS** Alma is hanging two posters on the wall in her room as shown. How can Alma use perpendicular distances to confirm that the posters are parallel?

35 **SCHOOL SPIRIT** Brock is decorating a hallway bulletin board to display pictures of students demonstrating school spirit. He cuts off one length of border to match the width of the top of the board, and then uses that strip as a template to cut a second strip that is exactly the same length for the bottom.

When stapling the bottom border in place, he notices that the strip he cut is about a quarter of an inch too short. Describe what he can conclude about the bulletin board. Explain your reasoning.

CONSTRUCTION **Line ℓ contains points at (−4, 3) and (2, −3). Point P at (−2, 1) is on line ℓ. Complete the following construction.**

Step 1

Graph line ℓ and point P, and put the compass at point P. Using the same compass setting, draw arcs to the left and right of P. Label these points A and B.

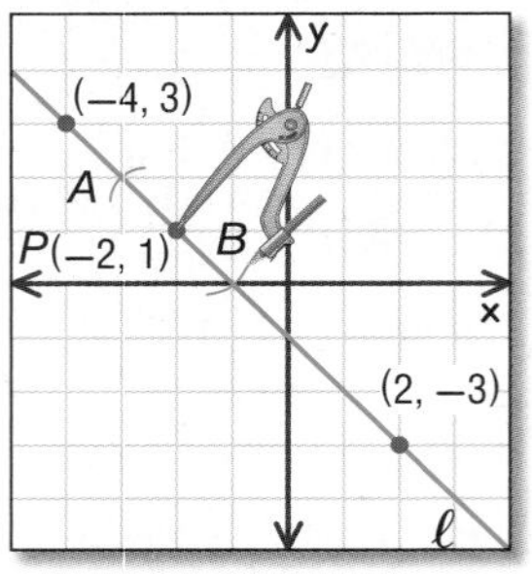

Step 2

Open the compass to a setting greater than AP. Put the compass at point A and draw an arc above line ℓ.

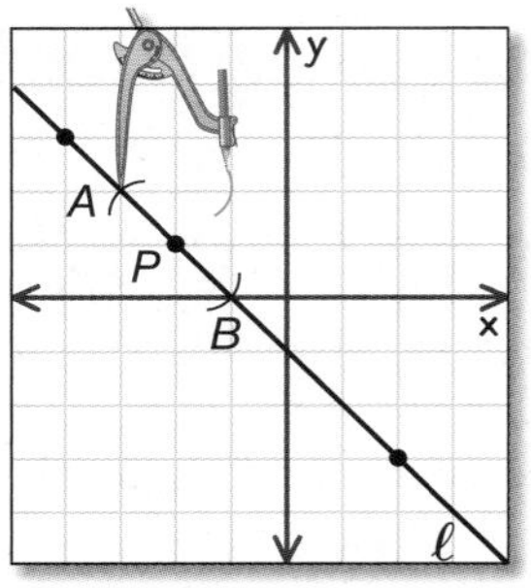

Step 3

Using the same compass setting, put the compass at point B and draw an arc above line ℓ. Label the point of intersection Q. Then draw $\overleftrightarrow{PQ}$.

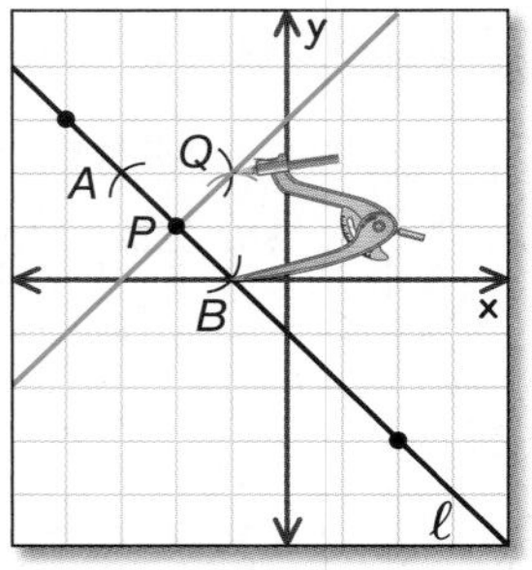

36. What is the relationship between line ℓ and $\overleftrightarrow{PQ}$? Verify your conjecture using the slopes of the two lines.

37. Repeat the construction above using a different line and point on that line.

38. CCSS **SENSE-MAKING** $\overline{AB}$ has a slope of 2 and midpoint $M(3, 2)$. A segment perpendicular to $\overline{AB}$ has midpoint $P(4, -1)$ and shares endpoint B with $\overline{AB}$.

a. Graph the segments.

b. Find the coordinates of A and B.

39. **MULTIPLE REPRESENTATIONS** In this problem, you will explore the areas of triangles formed by points on parallel lines.

a. Geometric Draw two parallel lines and label them as shown.

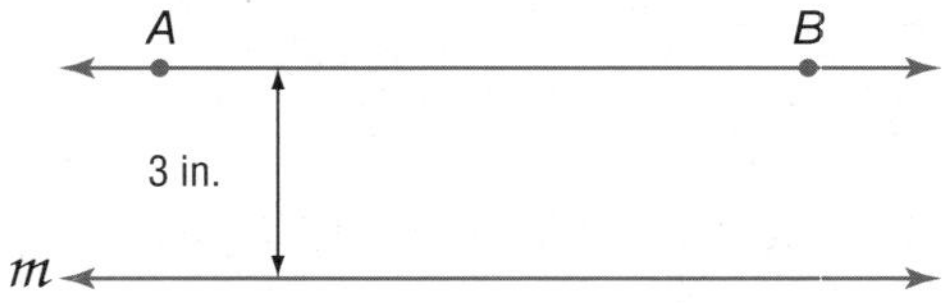

b. Verbal Where would you place point C on line m to ensure that triangle ABC would have the largest area? Explain your reasoning.

c. Analytical If $AB = 11$ inches, what is the maximum area of $\triangle ABC$?

40. PERPENDICULARITY AND PLANES Make a copy of the diagram below to answer each question, marking the diagram with the given information.

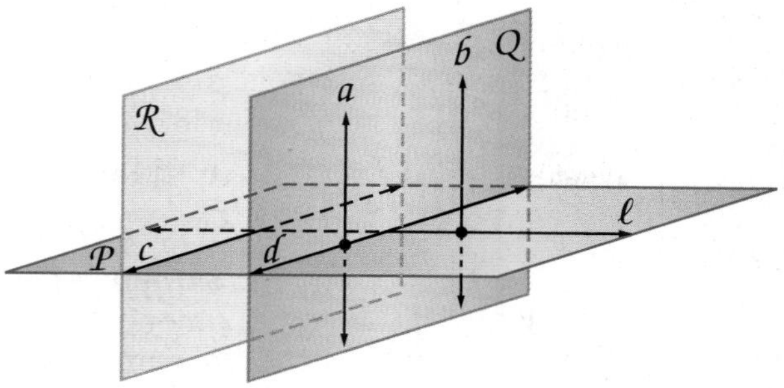

a. If two lines are perpendicular to the same plane, then they are coplanar. If both line a and line b are perpendicular to plane $\mathcal{P}$, what must also be true?

b. If a plane intersects two parallel planes, then the intersections form two parallel lines. If planes $\mathcal{R}$ and $\mathcal{Q}$ are parallel and they intersect plane $\mathcal{P}$, what must also be true?

c. If two planes are perpendicular to the same line, then they are parallel. If both plane $\mathcal{Q}$ and plane $\mathcal{R}$ are perpendicular to line ℓ, what must also be true?

H.O.T. Problems Use Higher-Order Thinking Skills

41. ERROR ANALYSIS Han draws the segments $\overline{AB}$ and $\overline{CD}$ shown below using a straightedge. He claims that these two lines, if extended, will never intersect. Shenequa claims that they will. Is either of them correct? Justify your answer.

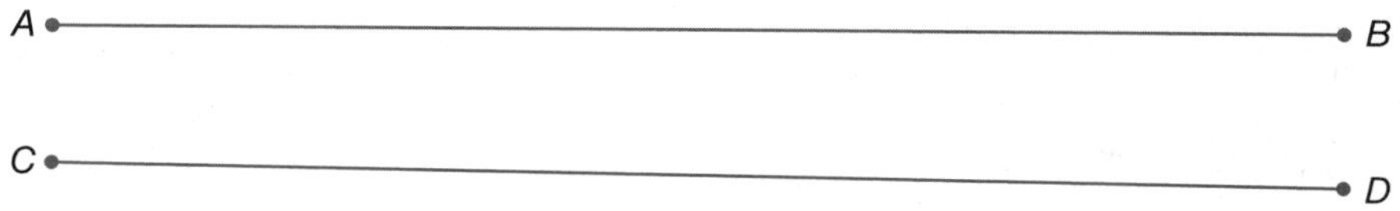

42. CHALLENGE Describe the locus of points that are equidistant from two intersecting lines, and sketch an example.

43. CHALLENGE Suppose a line perpendicular to a pair of parallel lines intersects the lines at the points $(a, 4)$ and $(0, 6)$. If the distance between the parallel lines is $\sqrt{5}$, find the value of a and the equations of the parallel lines.

44. REASONING Determine whether the following statement is *sometimes*, *always*, or *never* true. Explain.

The distance between a line and a plane can be found.

45. OPEN ENDED Draw an irregular convex pentagon using a straightedge.

a. Use a compass and straightedge to construct a line between one vertex and a side opposite the vertex.

b. Use measurement to justify that the line constructed is perpendicular to the side chosen.

c. Use mathematics to justify this conclusion.

46. CCSS SENSE-MAKING Rewrite Theorem 11.9 in terms of two planes that are equidistant from a third plane. Sketch an example.

47. WRITING IN MATH Summarize the steps necessary to find the distance between a pair of parallel lines given the equations of the two lines.

Standardized Test Practice

48. EXTENDED RESPONSE Segment *AB* is perpendicular to segment *CD*. Segment *AB* and segment *CD* bisect each other at point *X*.

a. Draw a figure to represent the problem.

b. Find $\overline{BD}$ if $AB = 12$ and $CD = 16$.

c. Find $\overline{BD}$ if $AB = 24$ and $CD = 18$.

49. A city park is square and has an area of 81,000 square feet. Which of the following is the closest to the length of one side of the park?

A 100 ft **C** 300 ft
B 200 ft **D** 400 ft

50. ALGEBRA Pablo bought a sweater on sale for 25% off the original price and another 40% off the discounted price. If the sweater originally cost $48, what was the final price of the sweater?

F $14.40 **H** $31.20
G $21.60 **J** $36.00

51. SAT/ACT After *N* cookies are divided equally among 8 children, 3 remain. How many would remain if $(N + 6)$ cookies were divided equally among the 8 children?

A 0 **C** 2 **E** 6
B 1 **D** 4

Spiral Review

52. Refer to the figure at the right. Determine whether $a \parallel b$. Justify your answer. (Lesson 11-5)

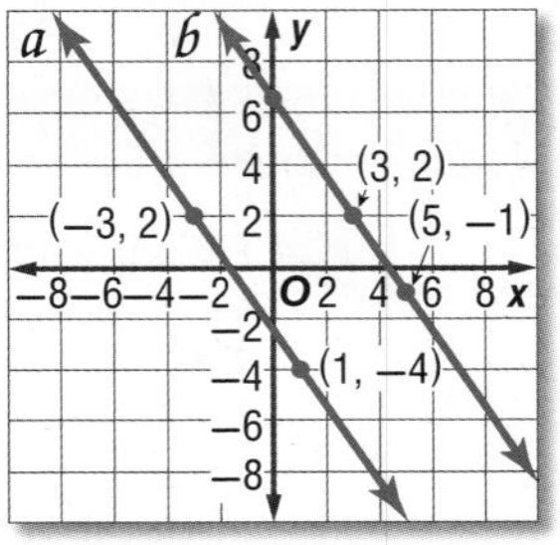

Write an equation in point-slope form of the line having the given slope that contains the given point. (Lesson 11-4)

53. m: $\frac{1}{4}$, $(3, -1)$

54. m: 0, $(-2, 6)$

55. m: -1, $(-2, 3)$

56. m: -2, $(-6, -7)$

Prove the following. (Lesson 10-7)

57. If $AB = BC$, then $AC = 2BC$.

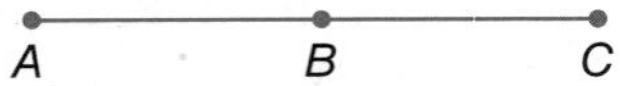

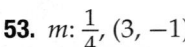

58. Given: $\overline{JK} \cong \overline{KL}$, $\overline{HJ} \cong \overline{GH}$, $\overline{KL} \cong \overline{HJ}$

Prove: $\overline{GH} \cong \overline{JK}$

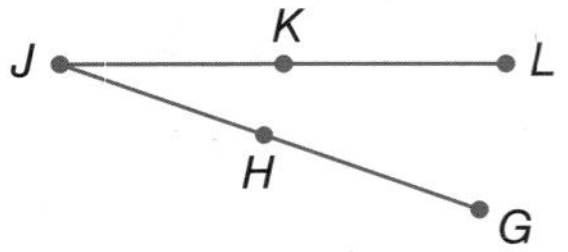

Skills Review

Use the Distance Formula to find the distance between each pair of points.

59. $A(0, 0)$, $B(15, 20)$

60. $O(-12, 0)$, $P(-8, 3)$

61. $C(11, -12)$, $D(6, 2)$

62. $R(-2, 3)$, $S(3, 15)$

63. $M(1, -2)$, $N(9, 13)$

64. $Q(-12, 2)$, $T(-9, 6)$

CHAPTER 11

Study Guide and Review

Study Guide

KeyConcepts

Transversals (Lessons 11-1 and 11-2)

- When a transversal intersects two lines, the following types of angles are formed: exterior, interior, consecutive interior, alternate interior, alternate exterior, and corresponding.
- If two parallel lines are cut by a transversal, then:
  - each pair of corresponding angles is congruent,
  - each pair of alternate interior angles is congruent,
  - each pair of consecutive interior angles is supplementary, and
  - each pair of alternate exterior angles is congruent.

Slope (Lessons 11-3 and 11-4)

- The slope m of a line containing two points with coordinates (x_1, y_1) and (x_2, y_2) is $m = \frac{y_2 - y_1}{x_2 - x_1}$, where $x_1 \neq x_2$.

Proving Lines Parallel (Lesson 11-5)

- If two lines in a plane are cut by a transversal so that any one of the following is true, then the two lines are parallel:
  - a pair of corresponding angles is congruent,
  - a pair of alternate exterior angles is congruent,
  - a pair of alternate interior angles is congruent, or
  - a pair of consecutive interior angles is supplementary.
- In a plane, if two lines are perpendicular to the same line, then they are parallel.

Distance (Lesson 11-6)

- The distance from a line to a point not on the line is the length of the segment perpendicular to the line from the point.
- The distance between two parallel lines is the perpendicular distance between one of the lines and any point on the other line.

FOLDABLES StudyOrganizer

Be sure the Key Concepts are noted in your Foldable.

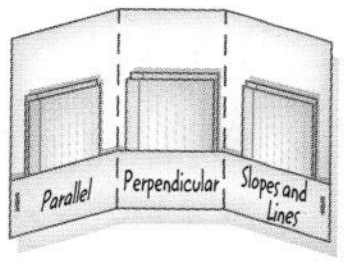

KeyVocabulary

alternate exterior angles (p. 644)
alternate interior angles (p. 644)
consecutive interior angles (p. 644)
corresponding angles (p. 644)
equidistant (p. 688)
parallel lines (p. 643)
parallel planes (p. 643)
point-slope form (p. 668)
rate of change (p. 659)
skew lines (p. 643)
slope (p. 658)
slope-intercept form (p. 668)
transversal (p. 644)

VocabularyCheck

State whether each sentence is *true* or *false*. If *false*, replace the underlined word or number to make a true sentence.

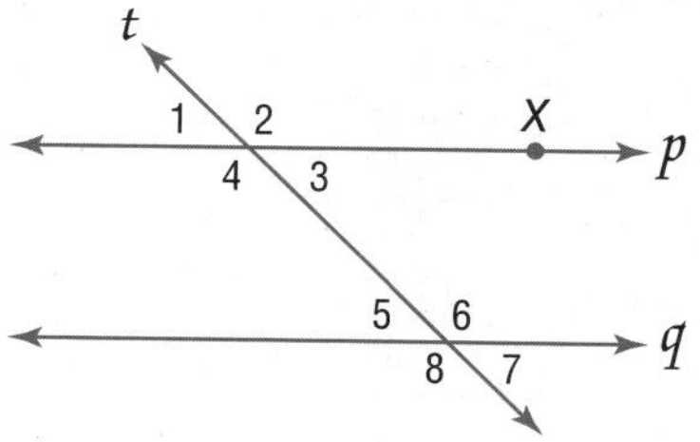

1. If $\angle 1 \cong \angle 5$, then lines p and q are skew lines.
2. Angles 4 and 6 are alternate interior angles.
3. Angles 1 and 7 are alternate exterior angles.
4. If lines p and q are parallel, then angles 3 and 6 are congruent.
5. The distance from point X to line q is the length of the segment perpendicular to line q from X.
6. Line t is called the transversal for lines p and q.
7. If $p \parallel q$, then $\angle 2$ and $\angle 8$ are supplementary.
8. Angles 4 and 8 are corresponding angles.

Lesson-by-Lesson Review

11-1 Parallel Lines and Transversals

Classify the relationship between each pair of angles as *alternate interior, alternate exterior, corresponding,* or *consecutive interior* angles.

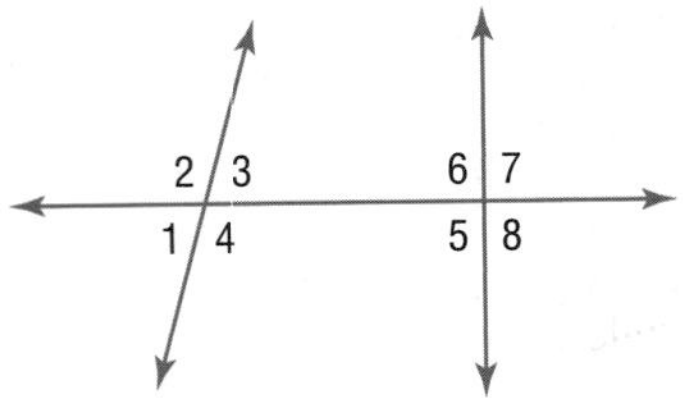

9. $\angle 1$ and $\angle 5$

10. $\angle 4$ and $\angle 6$

11. $\angle 2$ and $\angle 8$

12. $\angle 4$ and $\angle 5$

13. **BRIDGES** The Roebling Suspension Bridge extends over the Ohio River connecting Cincinnati, Ohio, to Covington, Kentucky. Describe the type of lines formed by the bridge and the river.

Example 1

Refer to the figure below. Classify the relationship between each pair of angles as *alternate interior, alternate exterior, corresponding,* or *consecutive interior* angles.

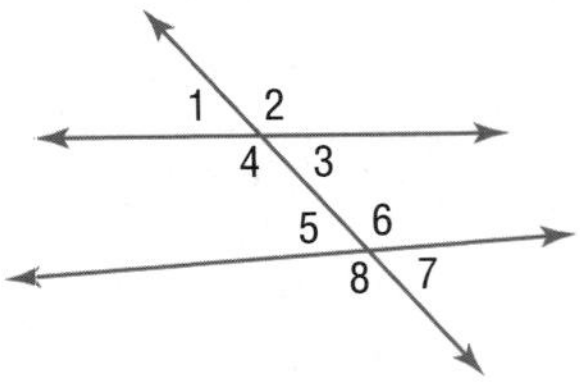

a. $\angle 3$ and $\angle 6$
consecutive interior

b. $\angle 2$ and $\angle 6$
corresponding

c. $\angle 1$ and $\angle 7$
alternate exterior

d. $\angle 3$ and $\angle 5$
alternate interior

11-2 Angles and Parallel Lines

In the figure, $m\angle 1 = 123$. Find the measure of each angle. Tell which postulate(s) or theorem(s) you used.

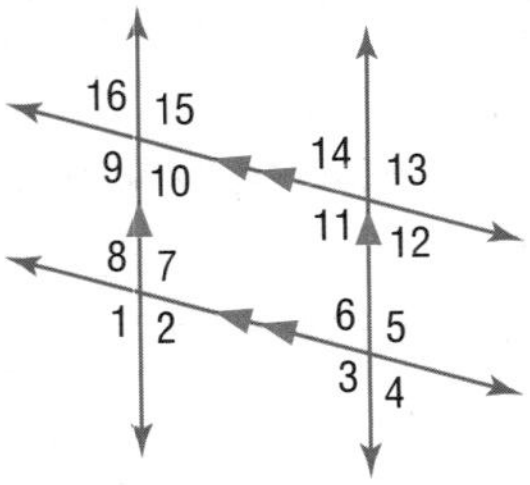

14. $\angle 5$

15. $\angle 14$

16. $\angle 16$

17. $\angle 11$

18. $\angle 4$

19. $\angle 6$

20. **MAPS** The diagram shows the layout of Elm, Plum, and Oak streets. Find the value of *x*.

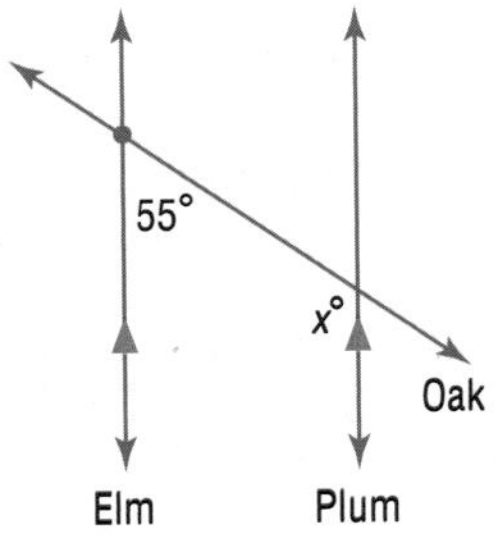

Example 2

ALGEBRA If $m\angle 5 = 7x - 5$ and $m\angle 4 = 2x + 23$, find *x*. Explain your reasoning.

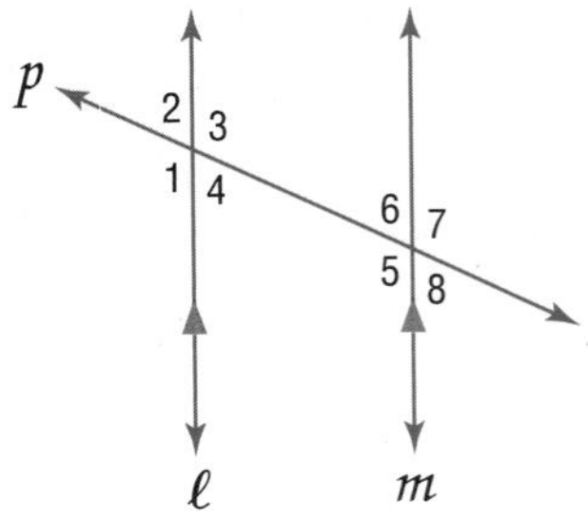

$m\angle 4 + m\angle 5 = 180$	Def. of Supp. $\angle$s
$(2x + 23) + (7x - 5) = 180$	Substitution
$9x + 18 = 180$	Simplify.
$9x = 162$	Subtract.
$x = 18$	Divide.

Since lines ℓ and m are parallel, $\angle 4$ and $\angle 5$ are supplementary by the Consecutive Interior Angles Theorem.

11-3 Slopes of Lines

Determine whether $\overleftrightarrow{AB}$ and $\overleftrightarrow{XY}$ are *parallel, perpendicular,* or *neither.* Graph each line to verify your answer.

21. $A(5, 3)$, $B(8, 0)$, $X(-7, 2)$, $Y(1, 10)$

22. $A(-3, 9)$, $B(0, 7)$, $X(4, 13)$, $Y(-5, 7)$

23. $A(8, 1)$, $B(-2, 7)$, $X(-6, 2)$, $Y(-1, -1)$

Graph the line that satisfies each condition.

24. contains $(-3, 4)$ and is parallel to $\overleftrightarrow{AB}$ with $A(2, 5)$ and $B(9, 2)$

25. contains $(1, 3)$ and is perpendicular to $\overleftrightarrow{PQ}$ with $P(4, -6)$ and $Q(6, -1)$

26. **AIRPLANES** Two Oceanic Airlines planes are flying at the same altitude. Using satellite imagery, each plane's position can be mapped onto a coordinate plane. Flight 815 was mapped at (23, 17) and (5, 11) while Flight 44 was mapped at (3, 15) and (9, 17). Determine whether their paths are *parallel, perpendicular,* or *neither.*

Example 3

Graph the line that contains $C(0, -4)$ and is perpendicular to $\overleftrightarrow{AB}$ with $A(5, -4)$ and $B(0, -2)$.

The slope of $\overleftrightarrow{AB}$ is $\frac{-2-(-4)}{0-5}$ or $-\frac{2}{5}$.

Since $-\frac{2}{5}\left(\frac{5}{2}\right) = -1$, the slope of the line perpendicular to $\overleftrightarrow{AB}$ through C is $\frac{5}{2}$.

To graph the line, start at C. Move up 5 units and then right 2 units. Label the point D and draw $\overleftrightarrow{CD}$.

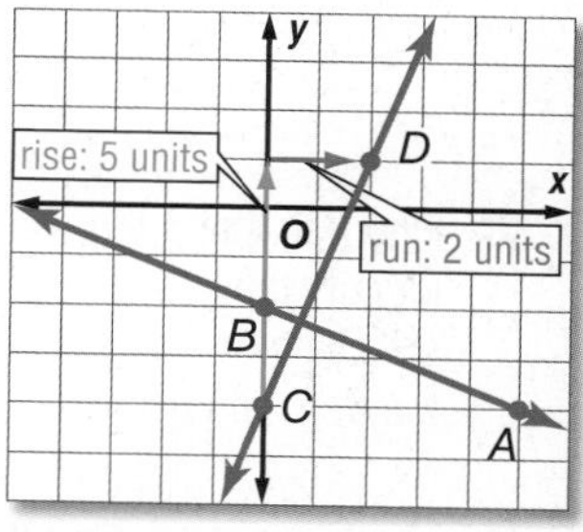

11-4 Equations of Lines

Write an equation in point-slope form of the line having the given slope that contains the given point.

27. $m = 2, (4, -9)$

28. $m = -\frac{3}{4}, (8, -1)$

Write an equation in slope-intercept form of the line having the given slope and y-intercept.

29. m: 5, y-intercept: -3

30. m: $\frac{1}{2}$, y-intercept: 4

Write an equation in slope-intercept form for each line.

31. $(-3, 12)$ and $(15, 0)$

32. $(-7, 2)$ and $(5, 8)$

33. **WINDOW CLEANING** Ace Window Cleaning Service charges \$50 for the service call and \$20 for each hour spent on the job. Write an equation in slope-intercept form that represents the total cost C in terms of the number of hours h.

Example 4

Write an equation of the line through (2, 5) and (6, 3) in slope-intercept form.

Step 1 Find the slope of the line through the points.

$m = \frac{y_2 - y_1}{x_2 - x_1}$ — Slope Formula

$= \frac{3-5}{6-2}$ — $x_1 = 2$, $y_1 = 5$, $x_2 = 6$, and $y_2 = 3$

$= \frac{-2}{4}$ or $-\frac{1}{2}$ — Simplify.

Step 2 Write an equation of the line.

$y - y_1 = m(x - x_1)$ — Point-slope form

$y - 5 = -\frac{1}{2}[x - (2)]$ — $m = -\frac{1}{2}$, $(x_1, y_1) = (2, 5)$

$y - 5 = -\frac{1}{2}x + 1$ — Simplify.

$y = -\frac{1}{2}x + 6$ — Add 5 to each side.

Study Guide and Review *Continued*

11-5 Proving Lines Parallel

Given the following information, determine which lines, if any, are parallel. State the postulate or theorem that justifies your answer.

34. $\angle 7 \cong \angle 10$

35. $\angle 2 \cong \angle 10$

36. $\angle 1 \cong \angle 3$

37. $\angle 3 \cong \angle 11$

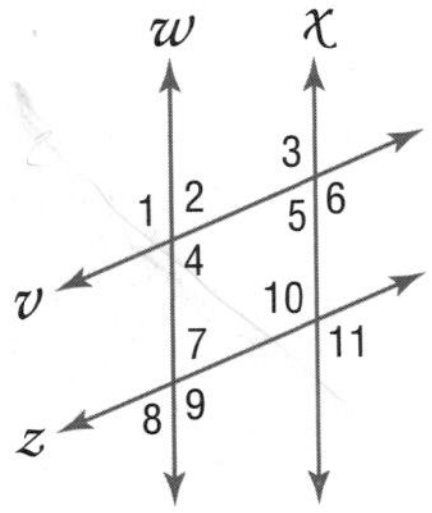

38. Find x so that $p \parallel q$. Identify the postulate or theorem you used.

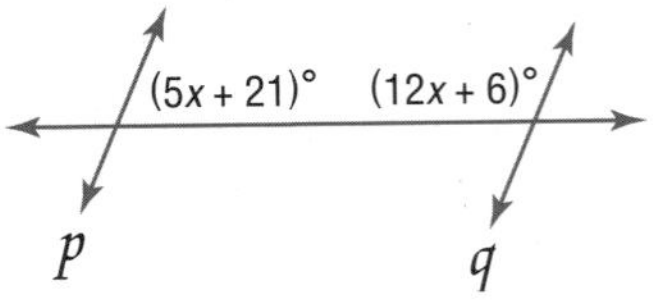

39. LANDSCAPING Find the measure needed for $m\angle ADC$ that will make $\overline{AB} \parallel \overline{CD}$ if $m\angle BAD = 45$.

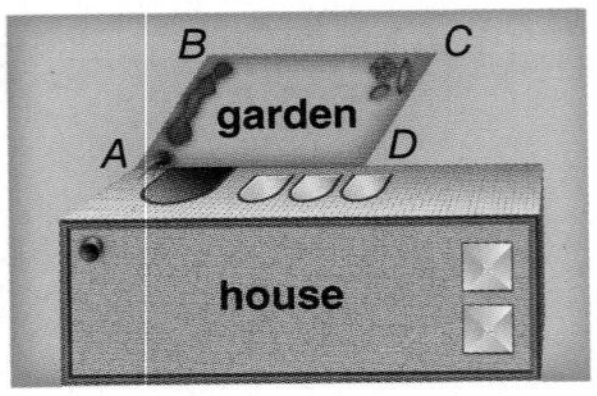

Example 5

Given the following information, determine which lines, if any, are parallel. State the postulate or theorem that justifies your answer.

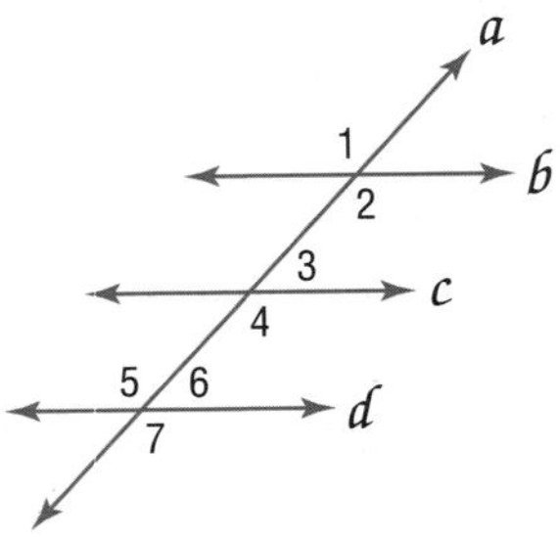

a. $\angle 1 \cong \angle 7$

$\angle 1$ and $\angle 7$ are alternate exterior angles of lines b and d.

Since $\angle 1 \cong \angle 7$, $b \parallel d$ by the Converse of the Alternate Exterior Angles Theorem.

b. $\angle 4 \cong \angle 5$

$\angle 4$ and $\angle 5$ are alternate interior angles of lines c and d.

Since $\angle 4 \cong \angle 5$, $c \parallel d$ by the Converse of the Alternate Interior Angles Theorem.

11-6 Perpendiculars and Distance

Copy each figure. Draw the segment that represents the distance indicated.

40. X to $\overline{VW}$

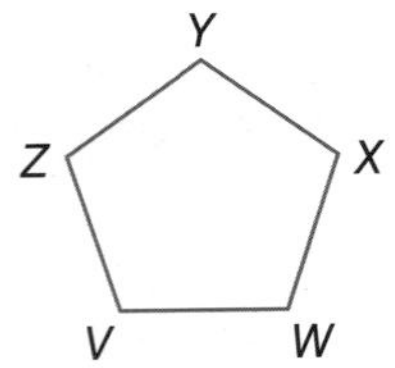

41. L to $\overline{JK}$

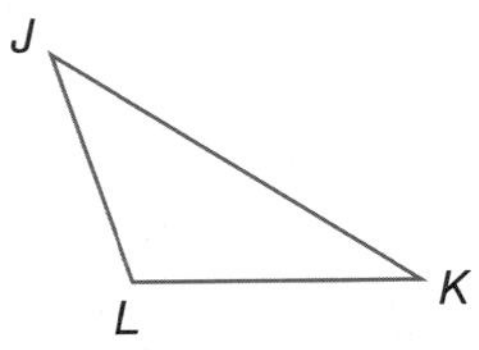

42. HOME DÉCOR Scott wants to hang two rows of framed pictures in parallel lines on his living room wall. He first spaces the nails on the wall in a line for the top row. Next, he hangs a weighted plumb line from each nail and measures an equal distance below each nail for the second row. Why does this ensure that the two rows of pictures will be parallel?

Example 6

Copy the figure. Draw the segment that represents the distance from point A to $\overline{CD}$.

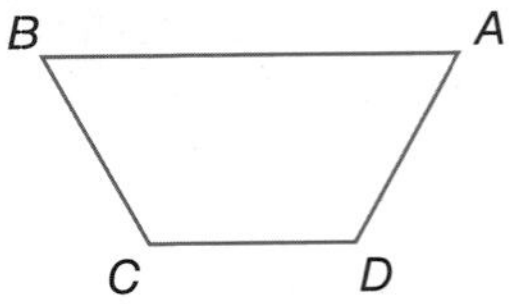

The distance from a line to a point not on the line is the length of the segment perpendicular to the line that passes through the point.

Extend $\overline{CD}$ and draw the segment perpendicular to $\overline{CD}$ from A.

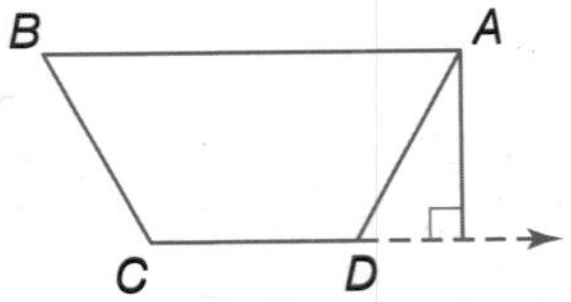

CHAPTER 11 Practice Test

Classify the relationship between each pair of angles as *alternate interior, alternate exterior, corresponding,* or *consecutive interior* angles.

1. $\angle 6$ and $\angle 3$
2. $\angle 4$ and $\angle 7$
3. $\angle 5$ and $\angle 4$

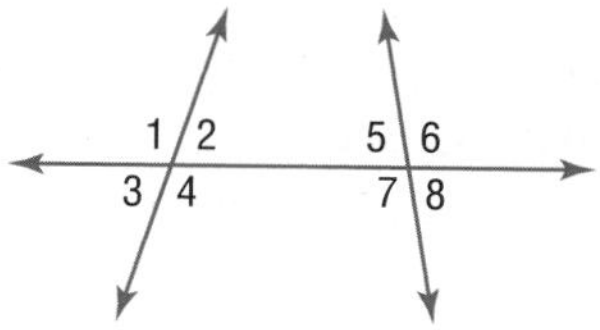

Determine the slope of the line that contains the given points.

4. $G(8, 1), H(8, -6)$
5. $A(0, 6), B(4, 0)$
6. $E(6, 3), F(-6, 3)$
7. $E(5, 4), F(8, 1)$

In the figure, $m\angle 8 = 96$ and $m\angle 12 = 42$. Find the measure of each angle. Tell which postulate(s) or theorem(s) you used.

8. $\angle 9$
9. $\angle 11$
10. $\angle 6$

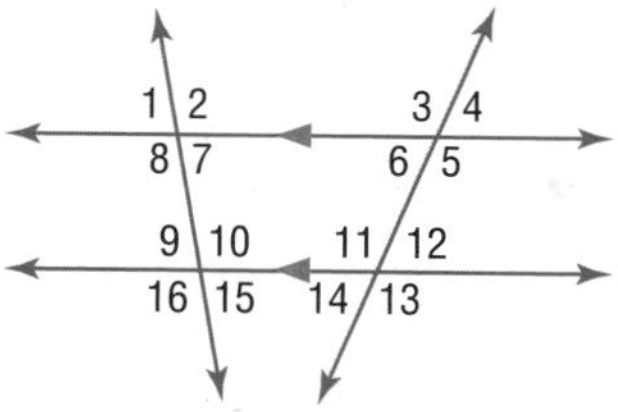

11. Find the value of x in the figure below.

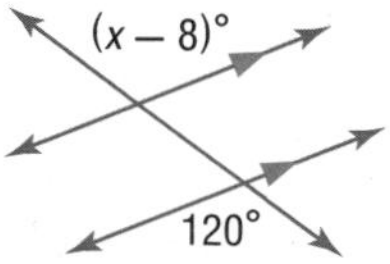

12. **FITNESS** You would like to join a fitness center. Fit-N-Trim charges \$80 per month. Fit-For-Life charges a one-time membership fee of \$75 and \$55 per month.
    a. Write and graph two equations in slope-intercept form to represent the cost y to attend each fitness center for x months.
    b. Are the lines you graphed in part **a** parallel? Explain why or why not.
    c. Which fitness center offers the better rate? Explain.

Write an equation in slope-intercept form for each line described.

13. passes through $(-8, 1)$, perpendicular to $y = 2x - 17$
14. passes through $(0, 7)$, parallel to $y = 4x - 19$
15. passes through $(-12, 3)$, perpendicular to $y = -\frac{2}{3}x - 11$

Find the distance between each pair of parallel lines with the given equations.

16. $y = x - 11$
    $y = x - 7$
17. $y = -2x + 1$
    $y = -2x + 16$

18. **MULTIPLE CHOICE** Which segment is skew to $\overline{CD}$?

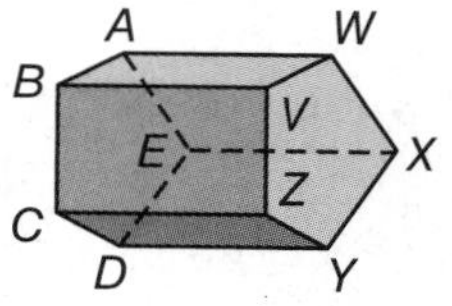

A $\overline{ZY}$

B $\overline{AB}$

C $\overline{DE}$

D $\overline{VZ}$

19. Find x so that $a \parallel b$. Identify the postulate or theorem you used.

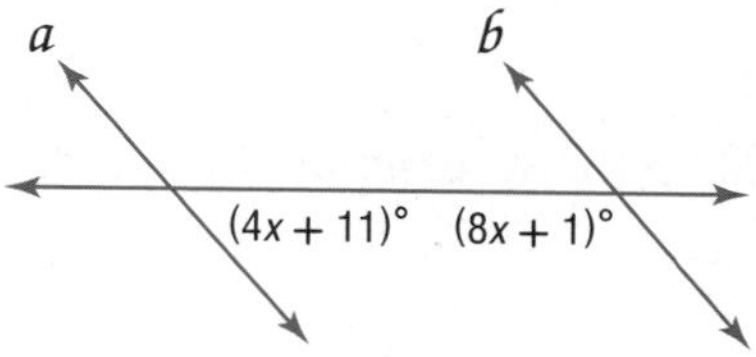

COORDINATE GEOMETRY Find the distance from P to ℓ.

20. Line ℓ contains points $(-4, 2)$ and $(3, -5)$. Point P has coordinates $(1, 2)$.
21. Line ℓ contains points $(6, 5)$ and $(2, 3)$. Point P has coordinates $(2, 6)$.

Given the following information, determine which lines, if any, are parallel. State the postulate or theorem that justifies your answer.

22. $\angle 4 \cong \angle 10$
23. $\angle 9 \cong \angle 6$
24. $\angle 7 \cong \angle 11$

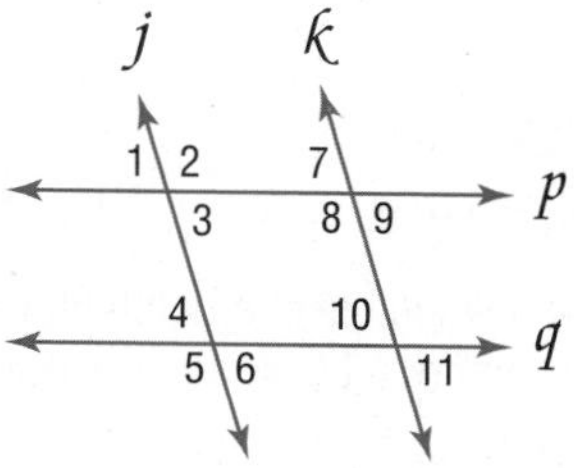

25. **JOBS** Hailey works at a gift shop. She is paid \$10 per hour plus a 15% commission on merchandise she sells. Write an equation in slope-intercept form that represents her earnings in a week if she sold \$550 worth of merchandise.

CHAPTER 11 Preparing for Standardized Tests

Gridded Response Questions

In addition to multiple-choice, short-answer, and extended-response questions, you will likely encounter gridded-response questions on standardized tests. After solving a gridded-response question, you must print your answer on an answer sheet and mark in the correct circles on the grid to match your answer. Answers to gridded-response questions may be whole numbers, decimals, or fractions.

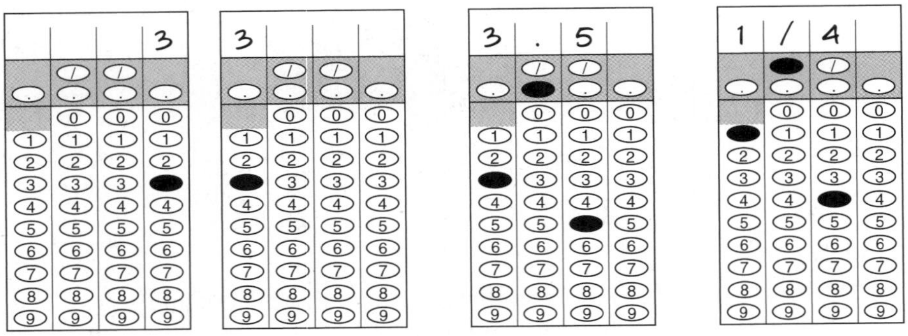

Whole Numbers Decimals Fractions

Strategies for Solving Gridded-Response Questions

Step 1

Read the problem carefully and solve.

- Be sure your answer makes sense.
- If time permits, check your answer.

Step 2

Print your answer in the answer boxes.

- Print only one digit or symbol in each answer box.
- Do not write any digits or symbols outside the answer boxes.
- Write answer as a whole number, decimal, or fraction.

Step 3

Fill in the grid.

- Fill in only one bubble for every answer box that you have written in. Be sure not to fill in a bubble under a blank answer box.
- Fill in each bubble completely and clearly.

Standardized Test Example

Read the problem. Identify what you need to know. Then use the information in the problem to solve.

GRIDDED RESPONSE In the figure below, $\angle ABC$ is intersected by parallel lines ℓ and m. What is the measure of $\angle ABC$? Express your answer in degrees.

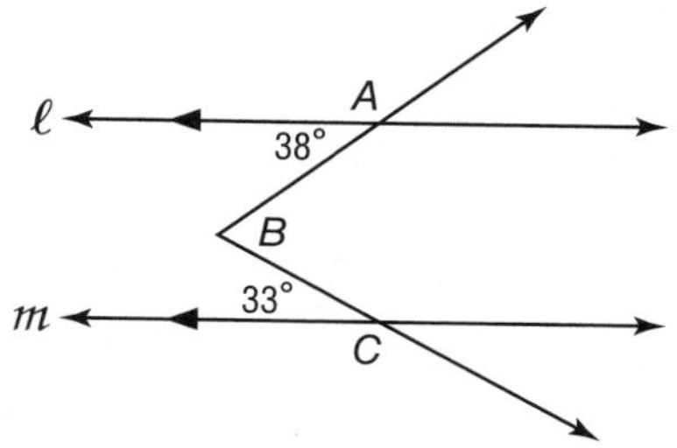

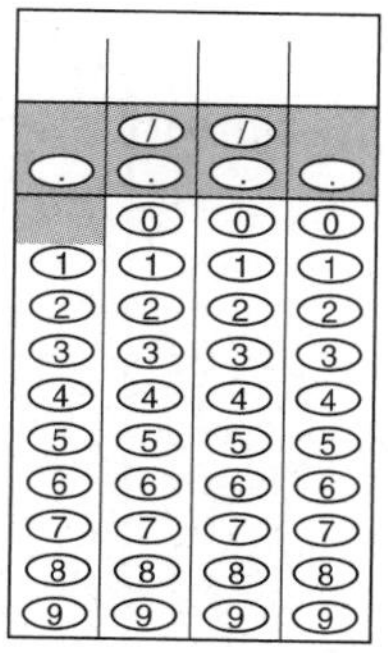

Redraw the figure and add a third line parallel to lines ℓ and m through point B. Find the angle measures using alternate interior angles.

Solve the Problem

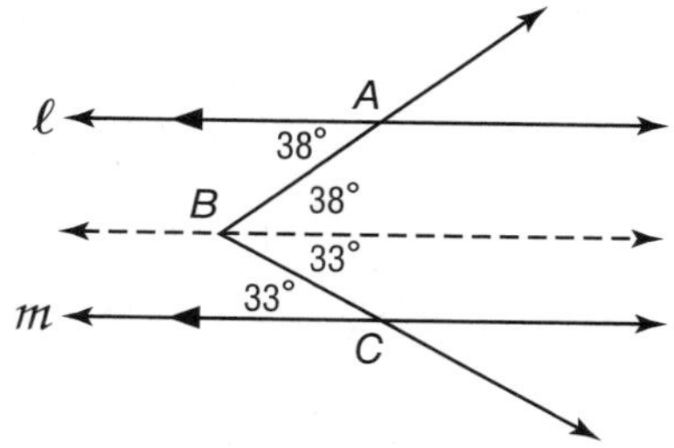

$m\angle ABC = 38 + 33 = 71$

Fill in the Grid

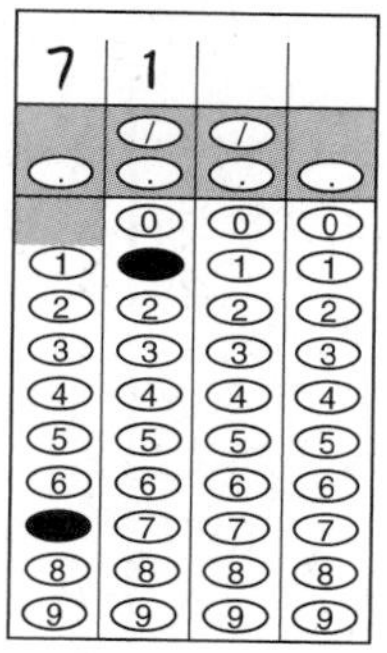

Print your answer in the answer box and fill in the grid.

Exercises

Read each question. Then fill in the correct answer on the answer document provided by your teacher or on a sheet of paper.

1. **GRIDDED RESPONSE** What is the slope of the line that contains the points $R(-2, 1)$ and $S(10, 6)$? Express your answer as a fraction.

2. **GRIDDED RESPONSE** Solve for x in the figure below.

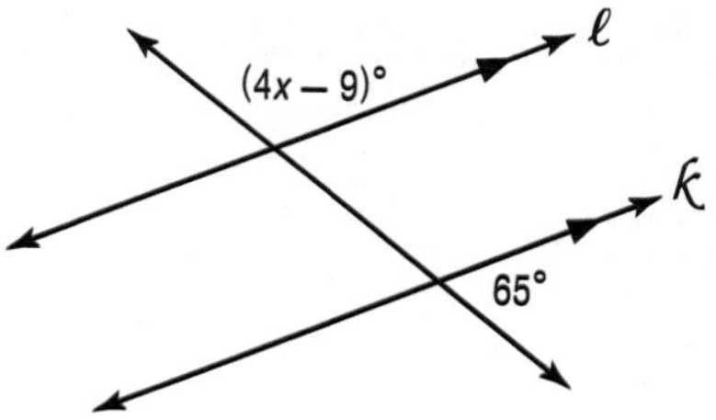

Standardized Test Practice

Cumulative, Chapters 1 through 11

Multiple Choice

Read each question. Then fill in the correct answer on the answer document provided by your teacher or on a sheet of paper.

1. If $a \parallel b$ in the diagram below, which of the following may *not* be true?

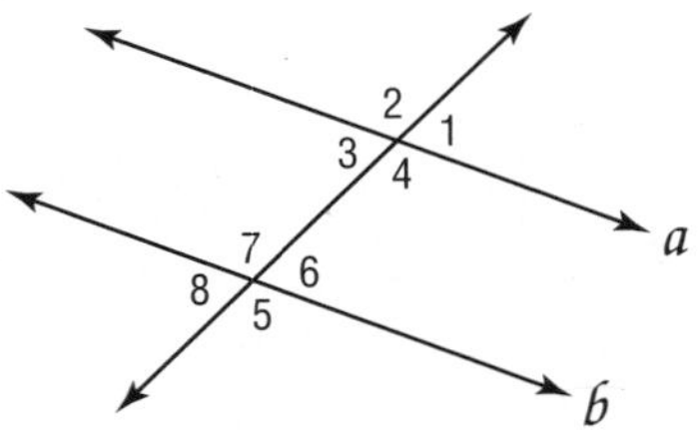

A $\angle 1 \cong \angle 3$

B $\angle 4 \cong \angle 7$

C $\angle 2 \cong \angle 5$

D $\angle 8 \cong \angle 2$

2. At a museum, each child admission costs \$5.75 and each adult costs \$8.25. How much does it cost a family that consists of 2 adults and 4 children?

A \$34.50

B \$39.50

C \$44.50

D \$49.50

3. What is the slope of the line?

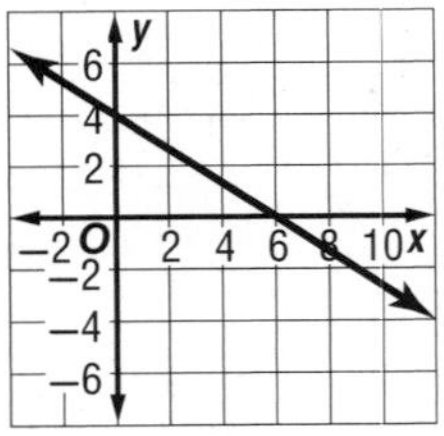

A $-\frac{2}{3}$

B $-\frac{1}{2}$

C $-\frac{2}{5}$

D $-\frac{1}{6}$

4. Line k contains points at (4, 1) and (−5, −5). Find the distance between line k and point $F(-4, 0)$.

F 3.3 units

G 3.6 units

H 4.0 units

J 4.2 units

5. The graph of which equation passes through the points (−1, −3) and (−2, 3)?

F $y = -6x - 9$

G $y = -\frac{1}{4}x + 3$

H $y = 4x - 5$

J $y = \frac{2}{3}x + 1$

6. What is $m\angle 1$ in the figure below?

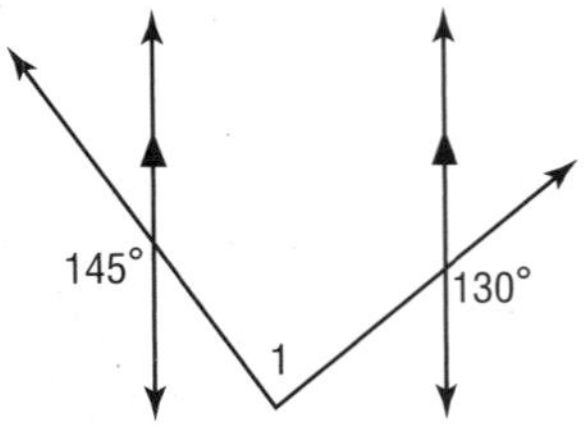

F 85

G 90

H 95

J 100

7. Jason is saving money to buy a car stereo. He has \$45 saved, and he can save \$15 per week. If the stereo that he wants is \$210, how many weeks will it take Jason to buy the stereo?

A 10

B 11

C 12

D 13

Test-Taking Tip

Question 6 *Drawing a diagram* can help you solve problems. Draw a third parallel line through the vertex of angle 1. Then use the properties of parallel lines and transversals to solve the problem.

Short Response/Gridded Response

Record your answers on the answer sheet provided by your teacher or on a sheet of paper.

8. GRIDDED RESPONSE For a given line and a point not on the line, how many lines exist that pass through the point and are parallel to the given line?

9. GRIDDED RESPONSE Find the slope of the line that contains the points (4, 3) and (−2, −5).

10. Complete the proof.

Given: $\angle 1 \cong \angle 2$

Prove: $a \parallel b$

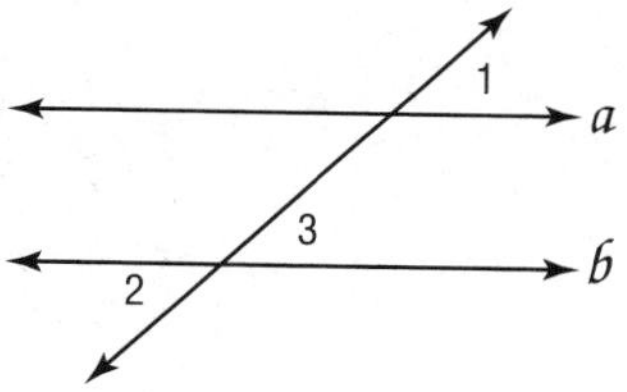

Proof:

Statements	Reasons
1. $\angle 1 \cong \angle 2$	1. Given
2. $\angle 2 \cong \angle 3$	2. ___?___
3. $\angle 1 \cong \angle 3$	3. Transitive Prop.
4. $a \parallel b$	4. If corresponding angles are congruent, then the lines are parallel.

11. Write an expression that describes the area in square units of a triangle with a height of $4c^3d^2$ and a base of $3cd^4$.

Extended Response

Record your answers on a sheet of paper. Show your work.

12. Refer to the figure to identify each of the following.

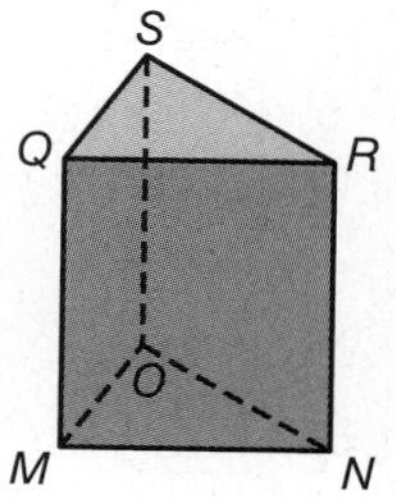

a. all segments parallel to $\overline{MQ}$

b. all planes intersecting plane *SRN*

c. a segment skew to $\overline{ON}$

13. Use this graph to answer each question.

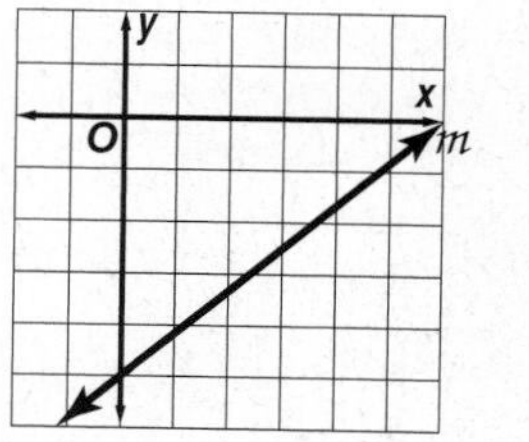

a. What is the equation of line *m*?

b. What is the slope of a line that is parallel to line *m*?

c. What is the slope of a line that is perpendicular to *line m*?

Need ExtraHelp?

If you missed Question...	1	2	3	4	5	6	7	8	9	10	11	12	13
Go to Lesson...	11-2	10-3	11-3	11-6	4-2	11-2	11-4	11-6	11-3	11-1	7-1	11-1	11-4

CHAPTER 12

Congruent Triangles

Then

You learned about segments, angles, and discovered relationships between their measures.

Now

In this chapter, you will:

- Apply special relationships about the interior and exterior angles of triangles.
- Identify corresponding parts of congruent triangles and prove triangles congruent.
- Learn about the special properties of isosceles and equilateral triangles

Why? ▲

FITNESS Triangles are used to add strength to many structures, including fitness equipment such as bike frames.

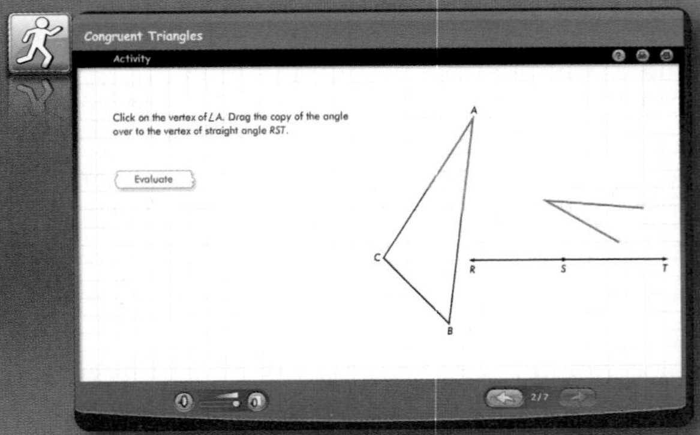

connectED.mcgraw-hill.com **Your Digital Math Portal**

Animation | Vocabulary | eGlossary | Personal Tutor | Virtual Manipulatives | Graphing Calculator | Audio | Foldables | Self-Check Practice | Worksheets

Get Ready for the Chapter

Diagnose Readiness | **You have two options for checking prerequisite skills.**

1 Textbook Option
Take the Quick Check below. Refer to the Quick Review for help.

QuickCheck

Classify each angle as *right*, *acute*, or *obtuse*.

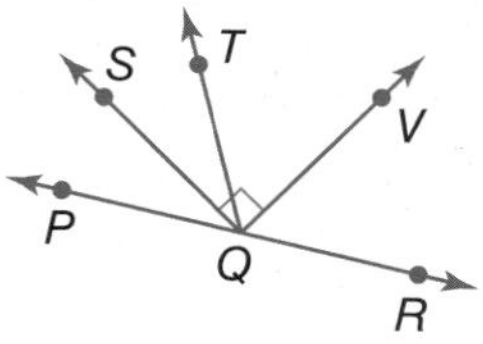

1. $m\angle VQS$ **2.** $m\angle TQV$ **3.** $m\angle PQV$

4. **ORIGAMI** The origami fold involves folding a strip of paper so that the lower edge of the strip forms a right angle with itself. Identify each angle as *right*, *acute*, or *obtuse*.

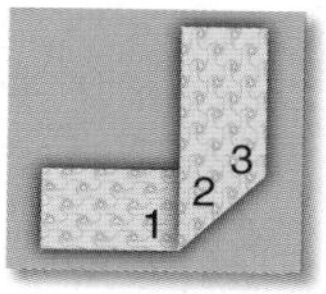

QuickReview

Example 1

Classify each angle as *right*, *acute*, or *obtuse*.

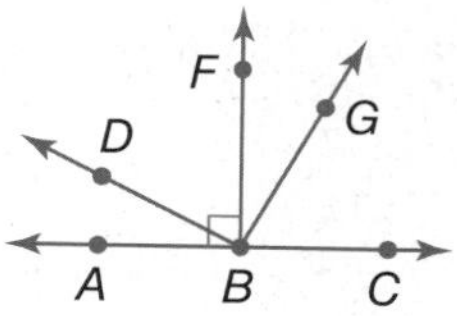

a. $m\angle ABG$

Point G on angle $\angle ABG$ lies on the exterior of right angle $\angle ABF$, so $\angle ABG$ is an obtuse angle.

b. $m\angle DBA$

Point D on angle $\angle DBA$ lies on the interior of right angle $\angle FBA$, so $\angle DBA$ is an acute angle.

ALGEBRA Use the figure to find the indicated variable(s). Explain your reasoning.

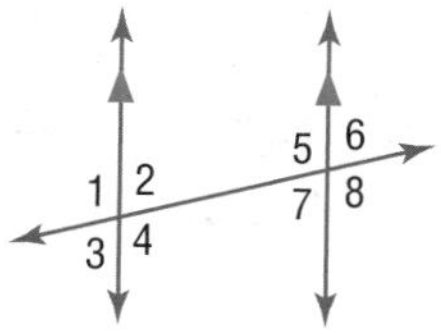

5. Find x if $m\angle 3 = x - 12$ and $m\angle 6 = 72$.

6. If $m\angle 4 = 2y + 32$ and $m\angle 5 = 3y - 3$, find y.

Example 2

In the figure, $m\angle 4 = 42$. Find $m\angle 7$.

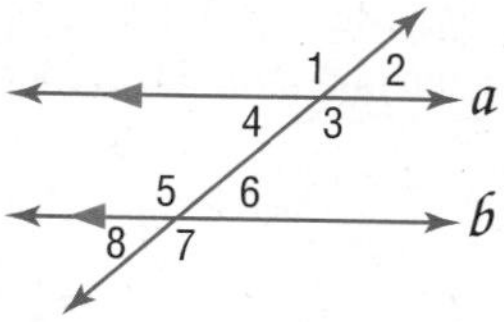

$\angle 7$ and $\angle 1$ are alternate interior angles, so they are congruent. $\angle 1$ and $\angle 4$ are a linear pair, so they are supplementary. Therefore, $\angle 7$ is supplementary to $\angle 1$. The measure of $\angle 7$ is $180 - 42$ or 138.

Find the distance between each pair of points.

7. $F(3, 6)$, $G(7, -4)$ **8.** $X(-2, 5)$, $Y(1, 11)$

9. $R(8, 0)$, $S(-9, 6)$ **10.** $A(14, -3)$, $B(9, -9)$

11. **MAPS** Miranda laid a coordinate grid on a map of a state where each 1 unit is equal to 10 miles. If her city is located at $(-8, -12)$ and the state capital is at $(0, 0)$, find the distance from her city to the capital to the nearest tenth of a mile.

Example 3

Find the distance between $J(5, 2)$ and $K(11, -7)$.

$JK = \sqrt{(x_2 - x_1)^2 + (y_2 - y_1)^2}$	Distance Formula
$= \sqrt{(11 - 5)^2 + [(-7) - 2]^2}$	Substitute.
$= \sqrt{6^2 + (-9)^2}$	Subtract.
$= \sqrt{36 + 81}$ or $\sqrt{117}$	Simplify.

2 Online Option
Take an online self-check Chapter Readiness Quiz at connectED.mcgraw-hill.com.

Get Started on the Chapter

You will learn several new concepts, skills, and vocabulary terms as you study Chapter 12. To get ready, identify important terms and organize your resources. You may wish to refer to Chapter 0 to review prerequisite skills.

FOLDABLES® StudyOrganizer

Congruent Triangles Make this Foldable to help you organize your Chapter 12 notes about congruent triangles. Begin with a sheet of $8\frac{1}{2}$" × 11" paper.

1 **Fold** into a taco forming a square. Cut off the excess paper strip formed by the square.

2 **Open** the fold and refold it the opposite way forming another taco and an X fold pattern.

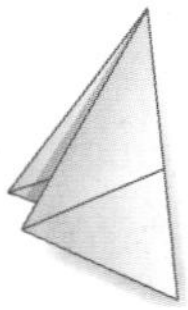

3 **Open** and fold the corners toward the center point of the X forming a small square.

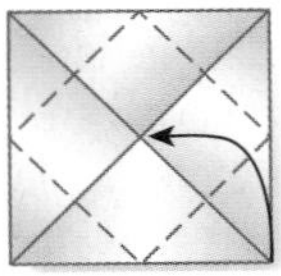

4 **Label** the flaps as shown.

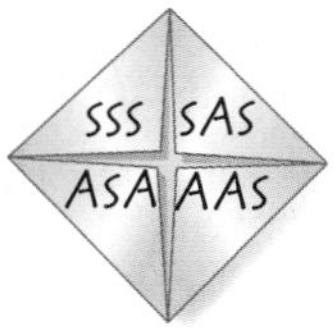
SSS
SAS
ASA
AAS

NewVocabulary

English		Español
equiangular triangle	p. 707	triángulo equiangular
equilateral triangle	p. 708	triángulo equilátero
isosceles triangle	p. 708	triángulo isósceles
scalene triangle	p. 708	triángulo escaleno
auxiliary line	p. 716	linea auxiliar
congruent	p. 725	congruente
congruent polygons	p. 725	polígonos congruentes
corresponding parts	p. 725	partes correspondientes
included angle	p. 736	ángulo incluido
included side	p. 745	lado incluido
base angle	p. 755	ángulo de la base
transformation	p. 766	transformación
preimage	p. 766	preimagen
image	p. 766	imagen
reflection	p. 766	reflexión
translation	p. 766	traslación
rotation	p. 766	rotación

ReviewVocabulary

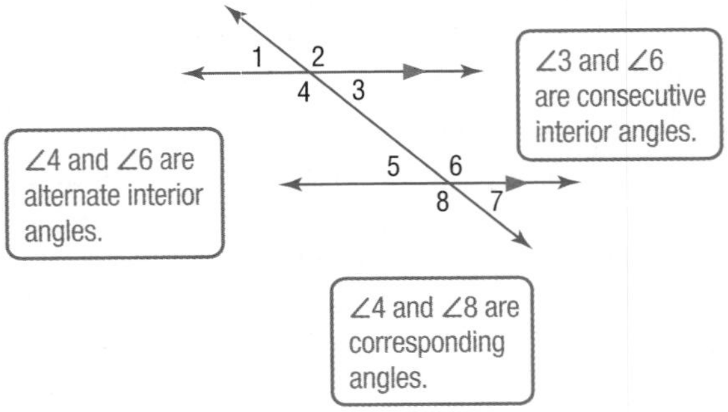
1
2
4
3
5
6
8
7
∠3 and ∠6 are consecutive interior angles.
∠4 and ∠6 are alternate interior angles.
∠4 and ∠8 are corresponding angles.

LESSON 12-1

Classifying Triangles

Then	Now	Why?
● You measured and classified angles.	**1** Identify and classify triangles by angle measures. **2** Identify and classify triangles by side measures.	● Radio transmission towers are designed to support antennas for broadcasting radio or television signals. The structure of the tower shown reveals a pattern of triangular braces.

NewVocabulary

- acute triangle
- equiangular triangle
- obtuse triangle
- right triangle
- equilateral triangle
- isosceles triangle
- scalene triangle

CCSS Common Core State Standards

Content Standards

G.CO.12 Make formal geometric constructions with a variety of tools and methods (compass and straightedge, string, reflective devices, paper folding, dynamic geometric software, etc.).

Mathematical Practices

2 Reason abstractly and quantitatively.

6 Attend to precision.

1 Classify Triangles by Angles

Recall that a triangle is a three-sided polygon. Triangle ABC, written $\triangle ABC$, has parts that are named using A, B, and C.

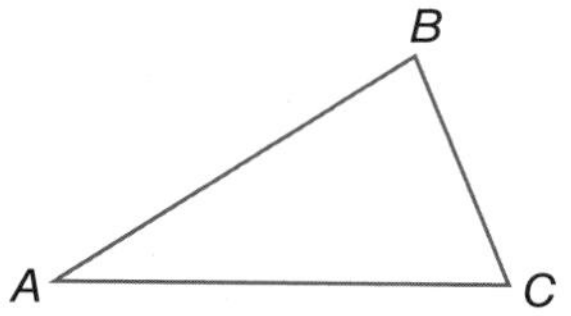

The sides of $\triangle ABC$ are $\overline{AB}$, $\overline{BC}$, and $\overline{CA}$.

The vertices are points A, B, and C.

The angles are $\angle BAC$ or $\angle A$, $\angle ABC$ or $\angle B$, and $\angle BCA$ or $\angle C$.

Triangles can be classified in two ways—by their angles or by their sides. All triangles have at least two acute angles, but the third angle is used to classify the triangle.

KeyConcept Classifications of Triangles by Angles

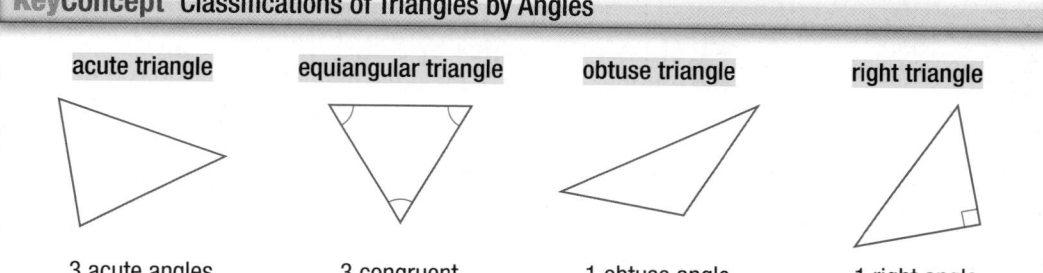

acute triangle

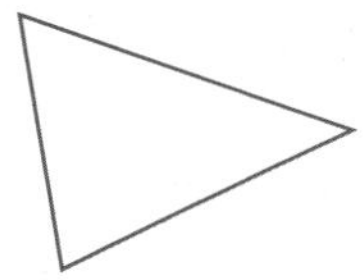

3 acute angles

equiangular triangle

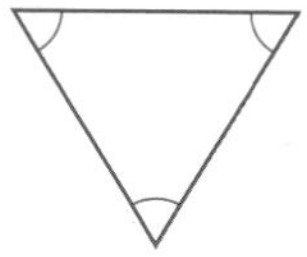

3 congruent acute angles

obtuse triangle

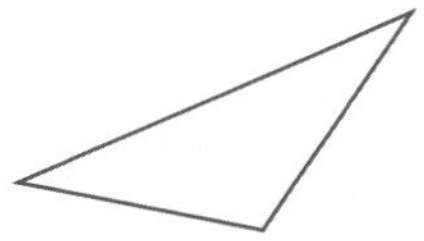

1 obtuse angle

right triangle

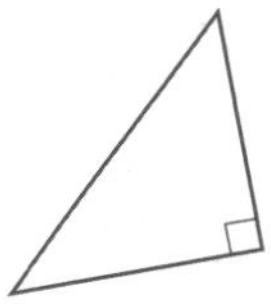

1 right angle

An equiangular triangle is a special kind of acute triangle.

When classifying triangles, be as specific as possible. While a triangle with three congruent acute angles is an acute triangle, it is more specific to classify it as an equiangular triangle.

Example 1 Classify Triangles by Angles

PT

Classify each triangle as *acute, equiangular, obtuse,* or *right*.

a.

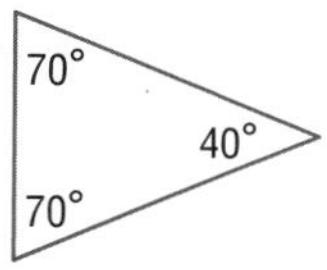

The triangle has three acute angles that are not all equal. It is an acute triangle.

b.

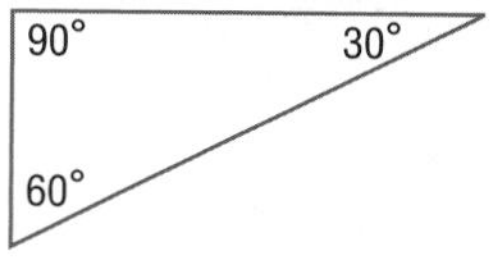

One angle of the triangle measures 90, so it is a right angle. Since the triangle has a right angle, it is a right triangle.

ReviewVocabulary

acute angle an angle with a degree measure less than 90

right angle an angle with a degree measure of 90

obtuse angle an angle with a degree measure greater than 90

GuidedPractice

Classify each triangle as *acute, equiangular, obtuse,* or *right.*

1A.

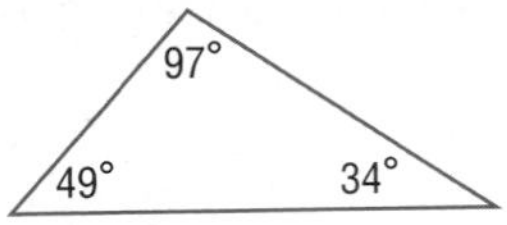

1B.

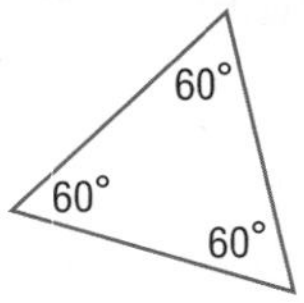

PT

Example 2 Classify Triangles by Angles Within Figures

Classify $\triangle PQR$ as *acute, equiangular, obtuse,* or *right*. Explain your reasoning.

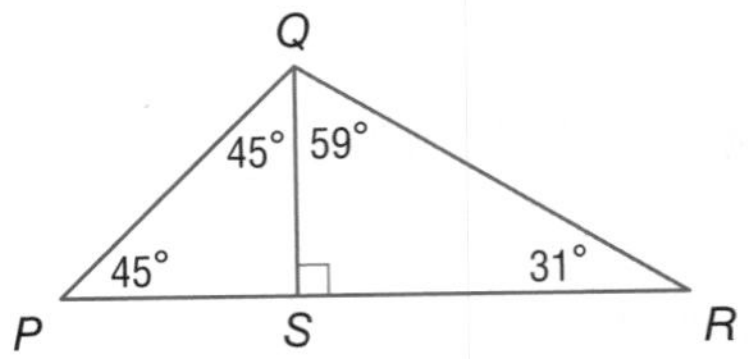

Point S is in the interior of $\angle PQR$, so by the Angle Addition Postulate, $m\angle PQR = m\angle PQS + m\angle SQR$. By substitution, $m\angle PQR = 45 + 59$ or 104.

Since $\triangle PQR$ has one obtuse angle, it is an obtuse triangle.

GuidedPractice

2. Use the diagram to classify $\triangle PQS$ as *acute, equiangular, obtuse* or *right*. Explain your reasoning.

2 Classify Triangles by Sides

Triangles can also be classified according to the number of congruent sides they have. To indicate that sides of a triangle are congruent, an equal number of hash marks is drawn on the corresponding sides.

KeyConcept Classifications of Triangles by Sides

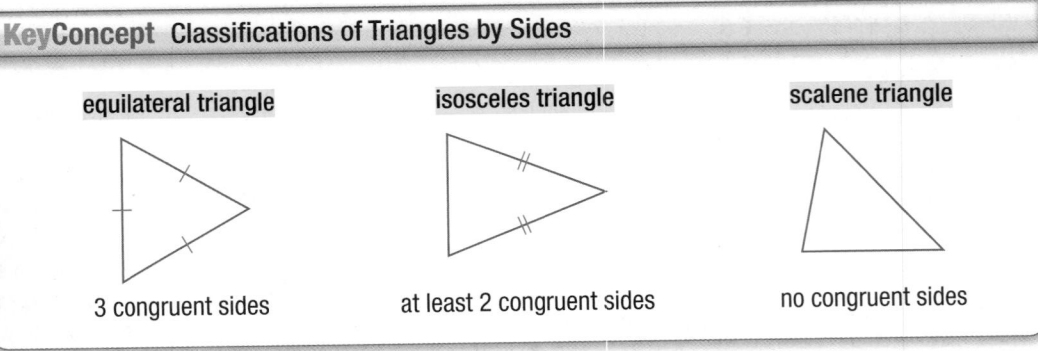

equilateral triangle

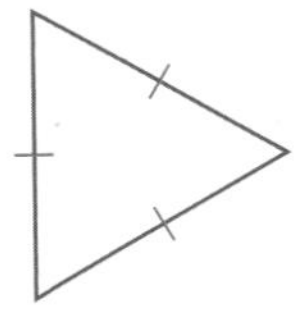

3 congruent sides

isosceles triangle

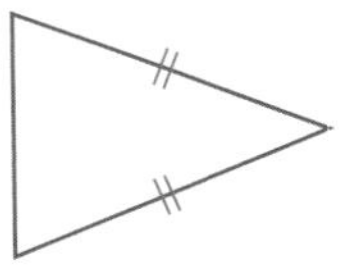

at least 2 congruent sides

scalene triangle

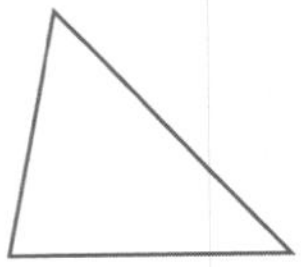

no congruent sides

An equilateral triangle is a special kind of isosceles triangle.

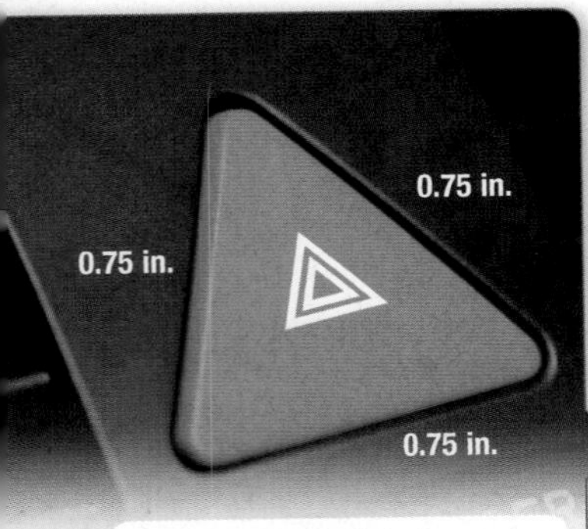

Real-WorldLink

In many cars, hazard lights are activated by pushing a small button located near the steering column. The switch is usually an icon shaped like an equilateral triangle.

Source: General Motors

PT

Example 3 Classify Triangles by Sides

A triangle has the following side lengths: 16 in., 18.5 in., and 16 in. Classify it as *equilateral, isosceles,* or *scalene.*

Two sides have the same measure, 16 inches, so the triangle has two congruent sides. The triangle is isosceles.

GuidedPractice

3. DRIVING SAFETY Classify the button in the picture at the left by its sides.

Example 4 Classify Triangles by Sides Within Figures

PT

If point M is the midpoint of $\overline{JL}$, classify $\triangle JKM$ as *equilateral, isosceles,* or *scalene*. Explain your reasoning.

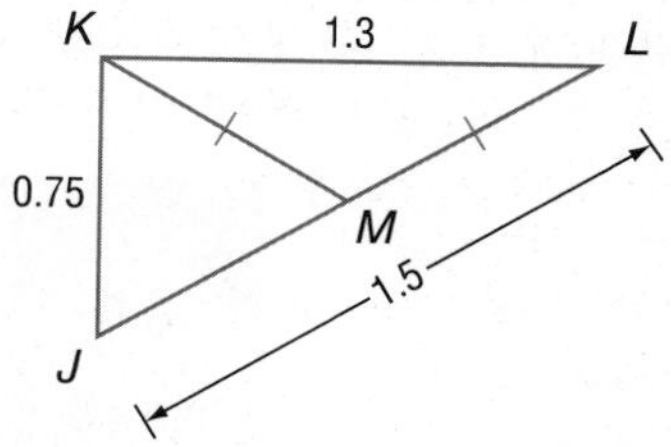

By the definition of midpoint, $JM = ML$.

$JM + ML = JL$	Segment Addition Postulate
$ML + ML = 1.5$	Substitution
$2ML = 1.5$	Simplify.
$ML = 0.75$	Divide each side by 2.

$JM = ML$ or 0.75. Since $\overline{KM} \cong \overline{ML}$, $KM = ML$ or 0.75.

Since $KJ = JM = KM = 0.75$, the triangle has three sides with the same measure. Therefore, the triangle has three congruent sides, so it is equilateral.

GuidedPractice

4. Classify $\triangle KML$ as *equilateral, isosceles,* or *scalene*. Explain your reasoning.

You can also use the properties of isosceles and equilateral triangles to find missing values.

Example 5 Finding Missing Values

PT

ALGEBRA Find the measures of the sides of isosceles triangle ABC.

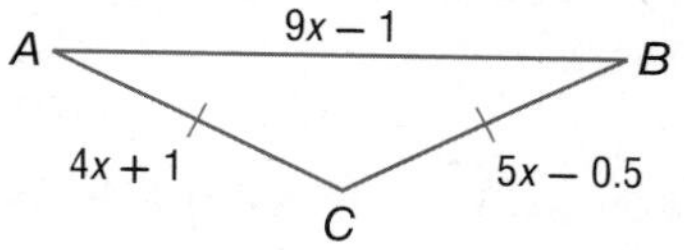

Step 1 Find x.

$AC = CB$	Given
$4x + 1 = 5x - 0.5$	Substitution
$1 = x - 0.5$	Subtract $4x$ from each side.
$1.5 = x$	Add 0.5 to each side.

Step 2 Substitute to find the length of each side.

$AC = 4x + 1$	Given
$= 4(1.5) + 1$ or 7	$x = 1.5$
$CB = AC$	Given
$= 7$	$AC = 7$
$AB = 9x - 1$	Given
$= 9(1.5) - 1$	$x = 1.5$
$= 12.5$	Simplify.

StudyTip

CCSS Perseverance In Example 5, to check your answer, test to see if $CB = AC$ when 1.5 is substituted for x in the expression for CB, $5x - 0.5$.

$CB = 5x - 0.5$
$= 5(1.5) - 0.5$ or 7 ✔

GuidedPractice

5. Find the measures of the sides of equilateral triangle FGH.

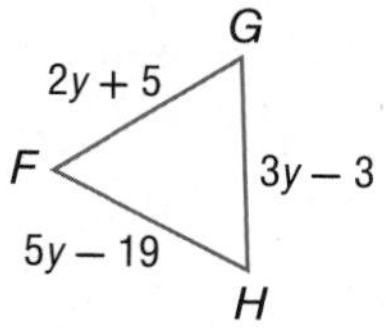

Check Your Understanding

Example 1 **ARCHITECTURE** **Use the best description to classify each triangle: *acute, equiangular, obtuse,* or *right*.**

1.

2.

3.

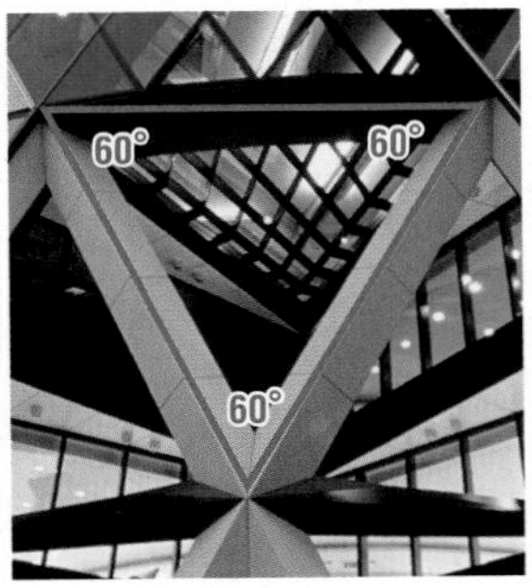

Example 2 **Classify each triangle as *acute, equiangular, obtuse,* or *right*. Explain your reasoning.**

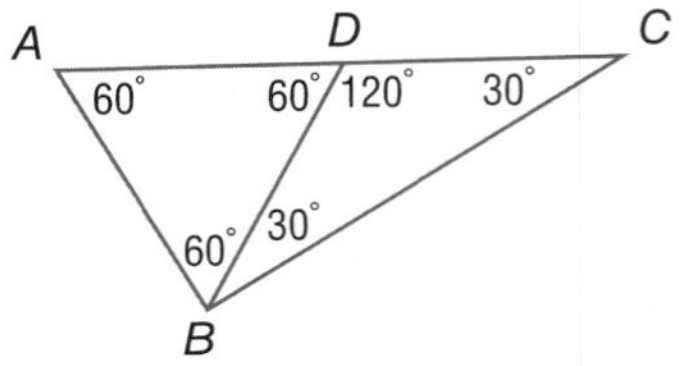

4. $\triangle ABD$

5. $\triangle BDC$

6. $\triangle ABC$

Example 3 **CCSS PRECISION** **Classify each triangle as *equilateral, isosceles,* or *scalene*.**

7.

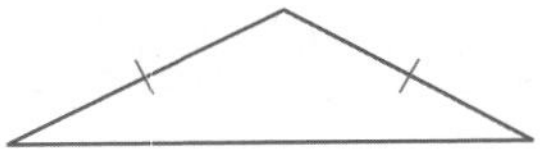

8.

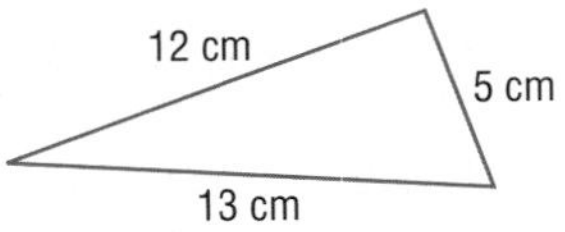

Example 4 **If point K is the midpoint of $\overline{FH}$, classify each triangle in the figure at the right as *equilateral, isosceles,* or *scalene*.**

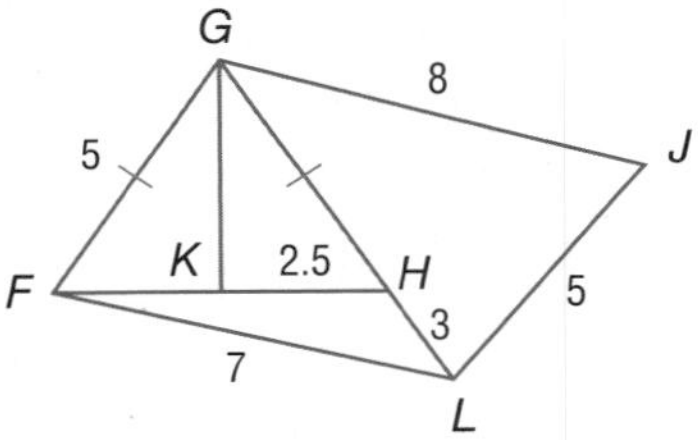

9. $\triangle FGH$

10. $\triangle GJL$

11. $\triangle FHL$

Example 5 **ALGEBRA** **Find x and the measures of the unknown sides of each triangle.**

12.

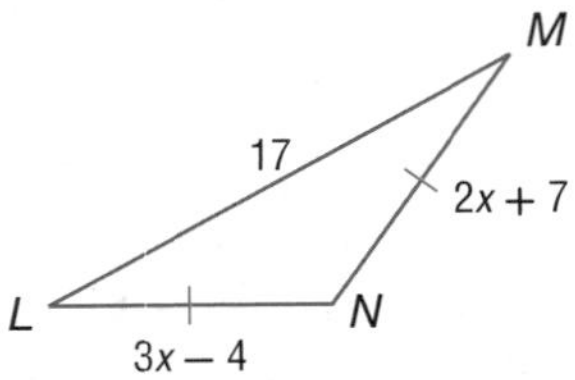

13.

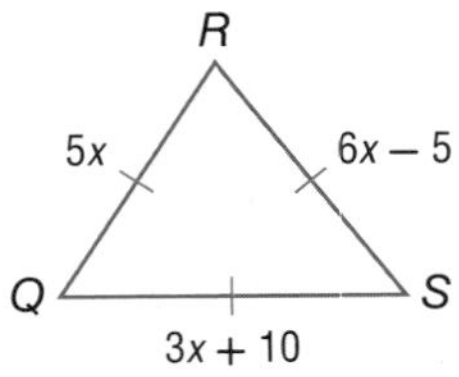

14. **JEWELRY** Suppose you are bending stainless steel wire to make the earring shown. The triangular portion of the earring is an isosceles triangle. If 1.5 centimeters are needed to make the hook portion of the earring, how many earrings can be made from 45 centimeters of wire? Explain your reasoning.

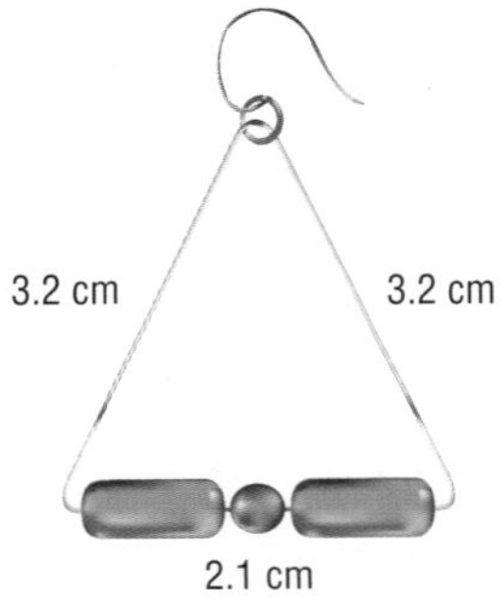

Practice and Problem Solving

Example 1 **Use the best description to classify each triangle:** ***acute, equiangular, obtuse,*** **or** ***right.***

15.

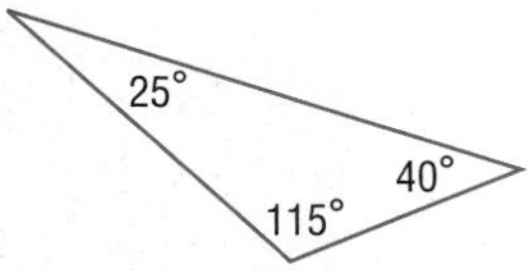

16.

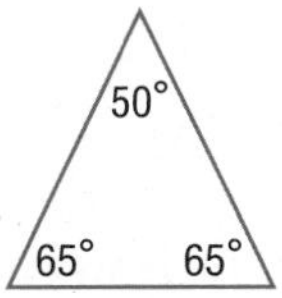

17.

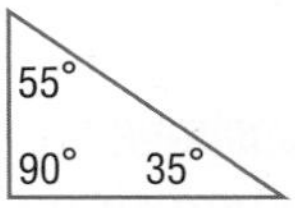

18.

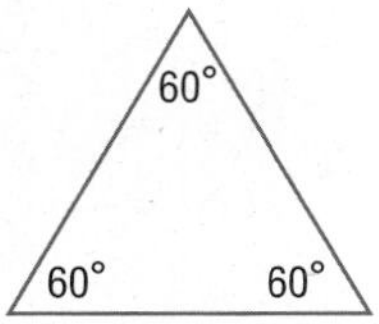

19.

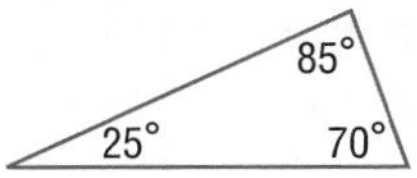

20.

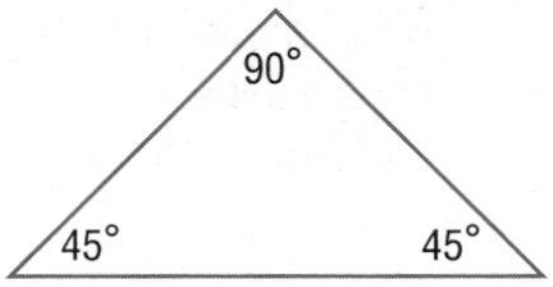

Example 2 **CCSS PRECISION** **Classify each triangle as** ***acute, equiangular, obtuse,*** **or** ***right.***

21. $\triangle UYZ$

22. $\triangle BCD$

23. $\triangle ADB$

24. $\triangle UXZ$

25. $\triangle UWZ$

26. $\triangle UXY$

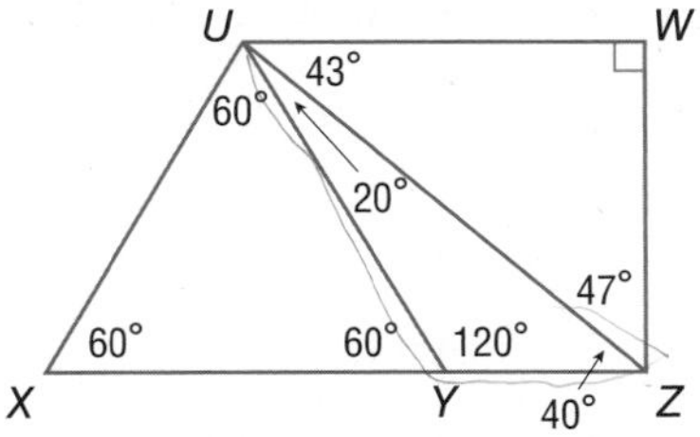

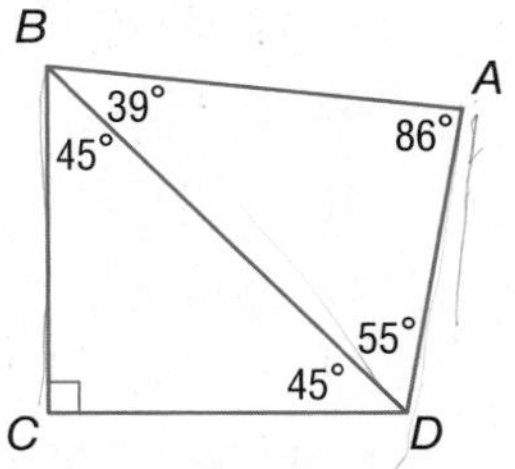

Example 3 **Classify each triangle as** ***equilateral, isosceles,*** **or** ***scalene.***

27.

28.

29.

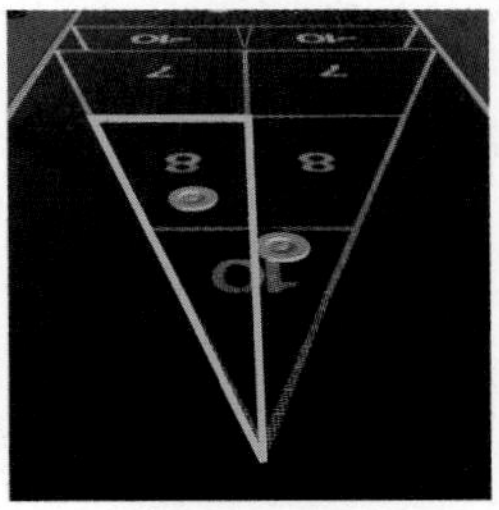

Example 4 **If point *C* is the midpoint of $\overline{BD}$ and point *E* is the midpoint of $\overline{DF}$, classify each triangle as** ***equilateral, isosceles,*** **or** ***scalene.***

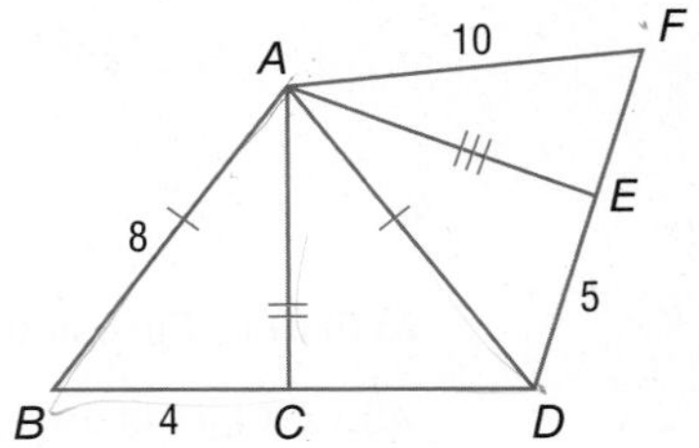

30. $\triangle ABC$

31. $\triangle AEF$

32. $\triangle ADF$

33. $\triangle ACD$

34. $\triangle AED$

35. $\triangle ABD$

Example 5

36. ALGEBRA Find x and the length of each side if $\triangle ABC$ is an isosceles triangle with $\overline{AB} \cong \overline{BC}$.

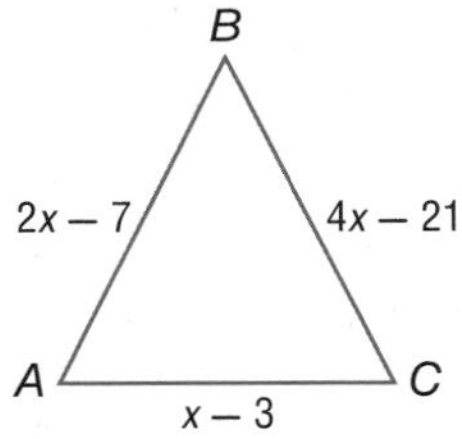

37. ALGEBRA Find x and the length of each side if $\triangle FGH$ is an equilateral triangle.

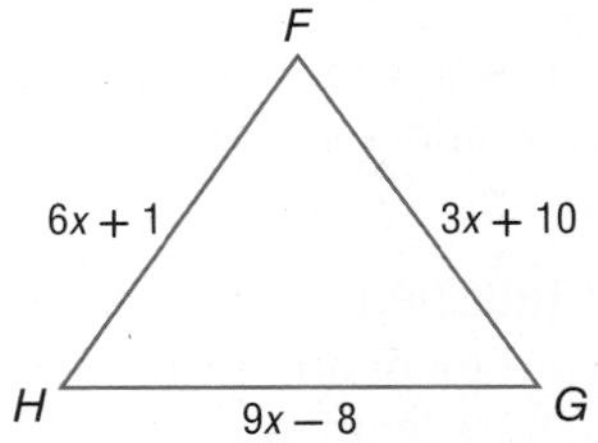

38. GRAPHIC ART Refer to the illustration shown. Classify each numbered triangle in *Kat* by its angles and by its sides. Use the corner of a sheet of notebook paper to classify angle measures and a ruler to measure sides.

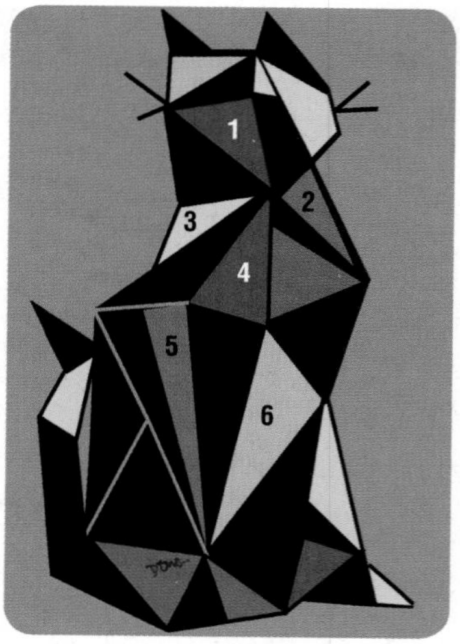

Kat, 2002, by Diana Ong, computer graphic

39. KALEIDOSCOPE Josh is building a kaleidoscope using PVC pipe, cardboard, bits of colored paper, and a 12-inch square mirror tile. The mirror tile is to be cut into strips and arranged to form an open prism with a base like that of an equilateral triangle. Make a sketch of the prism, giving its dimensions. Explain your reasoning.

CCSS PRECISION **Classify each triangle in the figure by its angles and sides.**

40. $\triangle ABE$

41. $\triangle EBC$

42. $\triangle BDC$

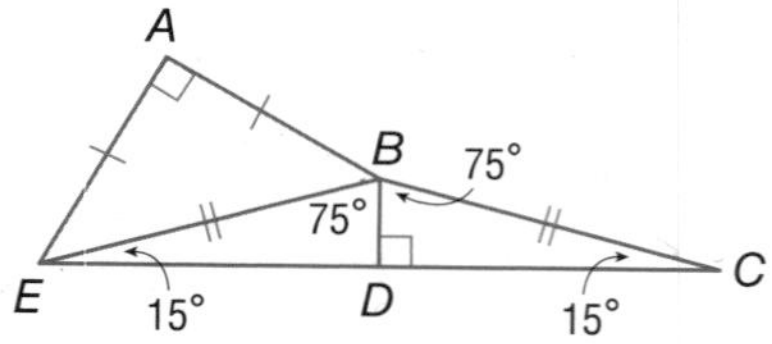

COORDINATE GEOMETRY **Find the measures of the sides of $\triangle XYZ$ and classify each triangle by its sides.**

43. $X(-5, 9), Y(2, 1), Z(-8, 3)$

44. $X(7, 6), Y(5, 1), Z(9, 1)$

45. $X(3, -2), Y(1, -4), Z(3, -4)$

46. $X(-4, -2), Y(-3, 7), Z(4, -2)$

47. PROOF Write a paragraph proof to prove that $\triangle DBC$ is an acute triangle if $m\angle ADC = 120$ and $\triangle ABC$ is acute.

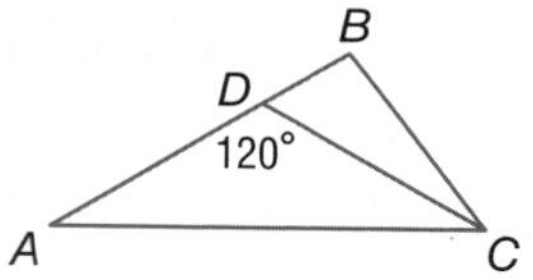

48. PROOF Write a two-column proof to prove that $\triangle BCD$ is equiangular if $\triangle ACE$ is equiangular and $\overline{BD} \parallel \overline{AE}$.

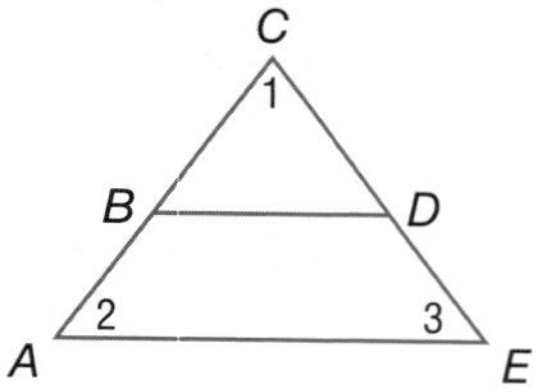

ALGEBRA **For each triangle, find x and the measure of each side.**

49. $\triangle FGH$ is an equilateral triangle with $FG = 3x - 10$, $GH = 2x + 5$, and $HF = x + 20$.

50. $\triangle JKL$ is isosceles with $\overline{JK} \cong \overline{KL}$, $JK = 4x - 1$, $KL = 2x + 5$, and $LJ = 2x - 1$.

51. $\triangle MNP$ is isosceles with $\overline{MN} \cong \overline{NP}$. MN is two less than five times x, NP is seven more than two times x, and PM is two more than three times x.

52. $\triangle RST$ is equilateral. RS is three more than four times x, ST is seven more than two times x, and TR is one more than five times x.

53. CONSTRUCTION Construct an equilateral triangle. Verify your construction using measurement and justify it using mathematics. (*Hint:* Use the construction for copying a segment.)

54. STOCKS Technical analysts use charts to identify patterns that can suggest future activity in stock prices. Symmetrical triangle charts are most useful when the fluctuation in the price of a stock is decreasing over time.

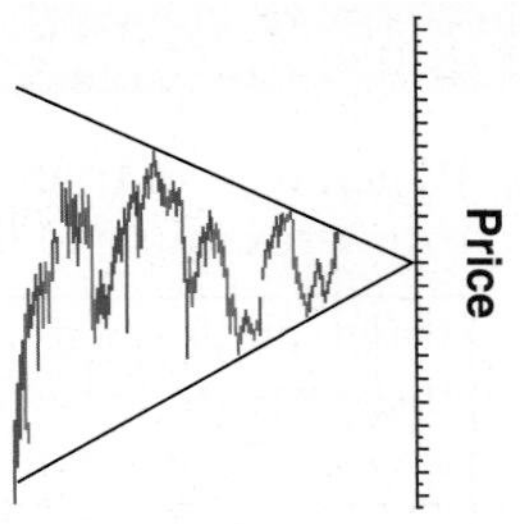

a. Classify by its sides and angles the triangle formed if a vertical line is drawn at any point on the graph.

b. How would the price have to fluctuate in order for the data to form an obtuse triangle? Draw an example to support your reasoning.

55 **MULTIPLE REPRESENTATIONS** In the diagram, the vertex *opposite* side $\overline{BC}$ is $\angle A$.

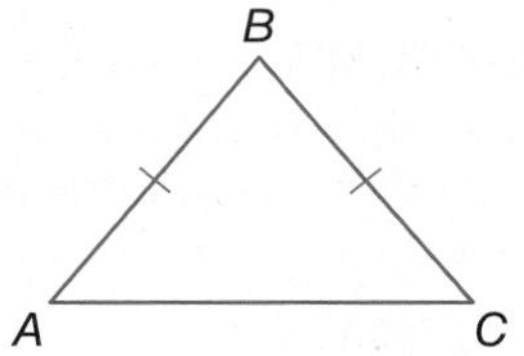

a. Geometric Draw four isosceles triangles, including one acute, one right, and one obtuse isosceles triangle. Label the vertices opposite the congruent sides as A and C. Label the remaining vertex B. Then measure the angles of each triangle and label each angle with its measure.

b. Tabular Measure all the angles of each triangle. Organize the measures for each triangle into a table. Include a column in your table to record the sum of these measures.

c. Verbal Make a conjecture about the measures of the angles that are opposite the congruent sides of an isosceles triangle. Then make a conjecture about the sum of the measures of the angles of an isosceles triangle.

d. Algebraic If x is the measure of one of the angles opposite one of the congruent sides in an isosceles triangle, write expressions for the measures of each of the other two angles in the triangle. Explain.

H.O.T. Problems Use Higher-Order Thinking Skills

56. ERROR ANALYSIS Elaina says that $\triangle DFG$ is obtuse. Ines disagrees, explaining that the triangle has more acute angles than obtuse angles so it must be acute. Is either of them correct? Explain your reasoning.

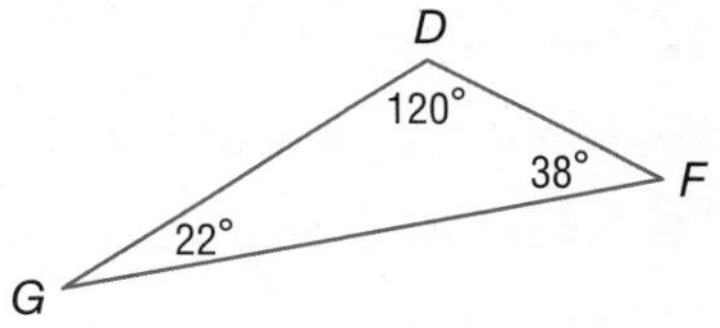

CCSS PRECISION **Determine whether the statements below are *sometimes*, *always*, or *never* true. Explain your reasoning.**

57. Equiangular triangles are also right triangles.

58. Equilateral triangles are isosceles.

59. Right triangles are equilateral.

60. CHALLENGE An equilateral triangle has sides that measure $5x + 3$ units and $7x - 5$ units. What is the perimeter of the triangle? Explain.

OPEN ENDED **Draw an example of each type of triangle below using a protractor and a ruler. Label the sides and angles of each triangle with their measures. If not possible, explain why not.**

61. scalene right **62.** isosceles obtuse **63.** equilateral obtuse

64. WRITING IN MATH Explain why classifying an equiangular triangle as an *acute* equiangular triangle is unnecessary.

Standardized Test Practice

65. Which type of triangle can serve as a counterexample to the conjecture below?

If two angles of a triangle are acute, then the measure of the third angle must be greater than or equal to 90.

A equilateral
B obtuse
C right
D scalene

66. ALGEBRA A baseball glove originally cost \$84.50. Kenji bought it at 40% off. How much was deducted from the original price?

F \$50.70
G \$44.50
H \$33.80
J \$32.62

67. GRIDDED RESPONSE Jorge is training for a 20-mile race. Jorge runs 7 miles on Monday, Tuesday, and Friday, and 12 miles on Wednesday and Saturday. After 6 weeks of training, Jorge will have run the equivalent of how many races?

68. SAT/ACT What is the slope of the line determined by the equation $2x + y = 5$?

A $-\frac{5}{2}$
B -2
C -1
D 2
E $\frac{5}{2}$

Spiral Review

Find the distance between each pair of parallel lines with the given equations. (Lesson 11-6)

69. $x = -2$
$x = 5$

70. $y = -6$
$y = 1$

71. $y = 2x + 3$
$y = 2x - 7$

72. $y = x + 2$
$y = x - 4$

73. FOOTBALL When striping the practice football field, Mr. Hawkins first painted the sidelines. Next he marked off 10-yard increments on one sideline. He then constructed lines perpendicular to the sidelines at each 10-yard mark. Why does this guarantee that the 10-yard lines will be parallel? (Lesson 11-5)

Refer to the figure at the right. (Lesson 10-1)

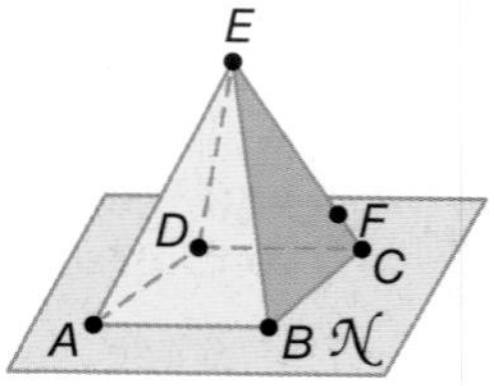

74. How many planes appear in this figure?

75. Name the intersection of plane AEB with plane $\mathcal{N}$.

76. Name three points that are collinear.

77. Are points D, E, C, and B coplanar?

Skills Review

Identify each pair of angles as *alternate interior, alternate exterior, corresponding,* or *consecutive interior angles.*

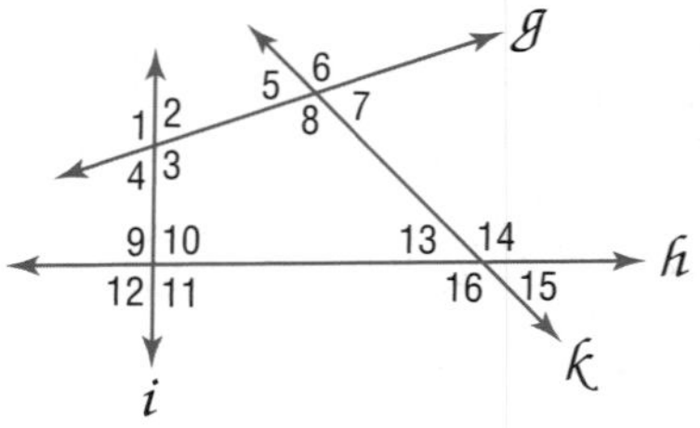

78. ∠5 and ∠3

79. ∠9 and ∠4

80. ∠11 and ∠13

81. ∠1 and ∠11

EXPLORE

12-2 Geometry Lab
Angles of Triangles

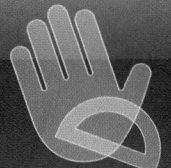

In this lab, you will find special relationships among the angles of a triangle.

CCSS Common Core State Standards
Content Standards
G.CO.12 Make formal geometric constructions with a variety of tools and methods (compass and straightedge, string, reflective devices, paper folding, dynamic geometric software, etc.).
Mathematical Practices 5

Activity 1 Interior Angles of a Triangle

Step 1

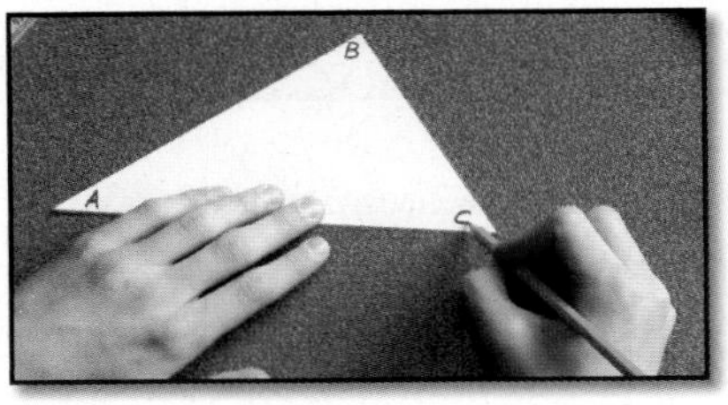

Draw and cut out several different triangles. Label the vertices *A*, *B*, and *C*.

Step 2

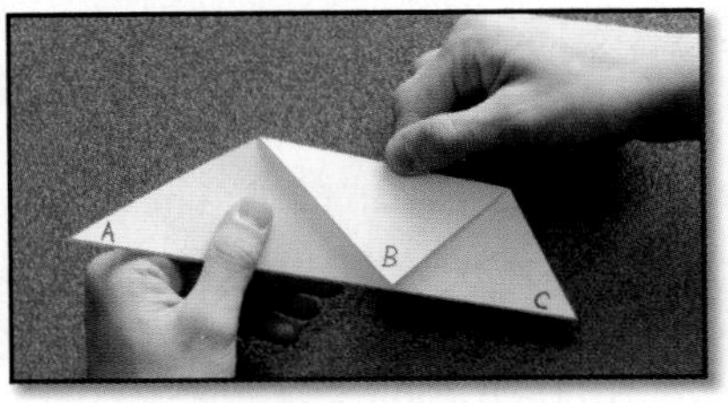

For each triangle, fold vertex *B* down so that the fold line is parallel to $\overline{AC}$. Relabel as vertex *B*.

Step 3

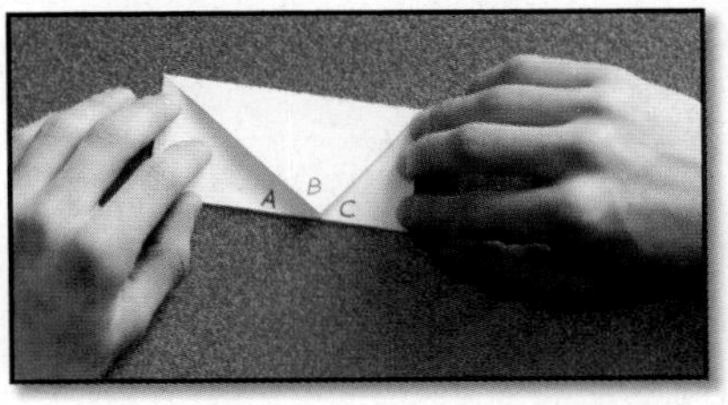

Then fold vertices *A* and *C* so that they meet vertex *B*. Relabel as vertices *A* and *C*.

Analyze the Results

1. Angles *A*, *B*, and *C* are called *interior angles* of triangle *ABC*. What type of figure do these three angles form when joined together in Step 3?
2. **Make a conjecture** about the sum of the measures of the interior angles of a triangle.

Activity 2 Exterior Angles of a Triangle

Step 1

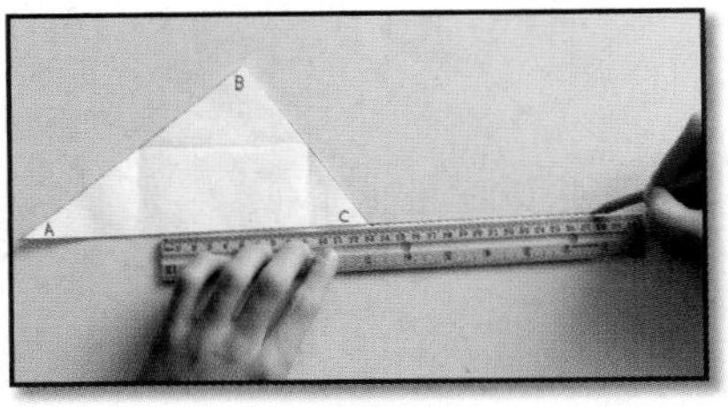

Unfold each triangle from Activity 1 and place each on a separate piece of paper. Extend $\overline{AC}$ as shown.

Step 2

For each triangle, tear off $\angle A$ and $\angle B$.

Step 3

Arrange $\angle A$ and $\angle B$ so that they fill the angle adjacent to $\angle C$ as shown.

Model and Analyze the Results

3. The angle adjacent to $\angle C$ is called an *exterior angle* of triangle *ABC*. **Make a conjecture** about the relationship among $\angle A$, $\angle B$, and the exterior angle at *C*.
4. Repeat the steps in Activity 2 for the exterior angles of $\angle A$ and $\angle B$ in each triangle.
5. **Make a conjecture** about the measure of an exterior angle and the sum of the measures of its nonadjacent interior angles.

LESSON 12-2

Angles of Triangles

Then

- You classified triangles by their side or angle measures.

Now

1. Apply the Triangle Angle-Sum Theorem.
2. Apply Exterior Angle Theorem.

Why?

- Massachusetts Institute of Technology (MIT) sponsors the annual *Design 2.007* contest in which students design and build a robot.

One test of a robot's movements is to program it to move in a triangular path. The sum of the measures of the pivot angles through which the robot must turn will always be the same.

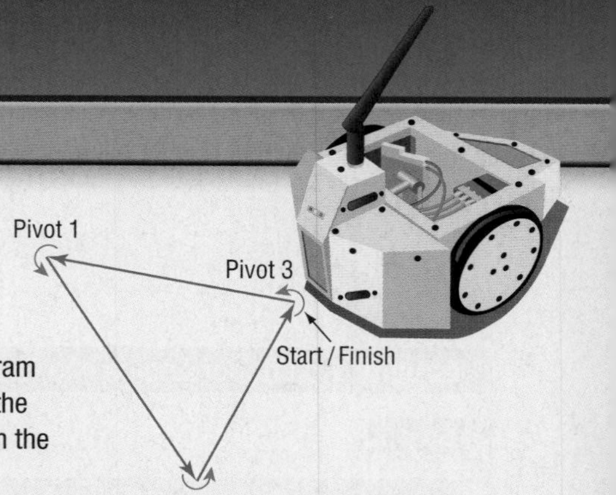

NewVocabulary
auxiliary line
exterior angle
remote interior angles
flow proof
corollary

Common Core State Standards

Content Standards
G.CO.10 Prove theorems about triangles.

Mathematical Practices
1 Make sense of problems and persevere in solving them.
3 Construct viable arguments and critique the reasoning of others.

1 Triangle Angle-Sum Theorem

The Triangle Angle-Sum Theorem gives the relationship among the interior angle measures of any triangle.

Theorem 12.1 Triangle Angle-Sum Theorem

Words The sum of the measures of the angles of a triangle is 180.

Example $m\angle A + m\angle B + m\angle C = 180$

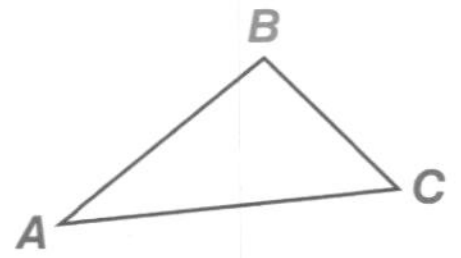

The proof of the Triangle Angle-Sum Theorem requires the use of an auxiliary line. An **auxiliary line** is an extra line or segment drawn in a figure to help analyze geometric relationships. As with any statement in a proof, you must justify any properties of an auxiliary line that you have drawn.

Proof Triangle Angle-Sum Theorem

Given: $\triangle ABC$

Prove: $m\angle 1 + m\angle 2 + m\angle 3 = 180$

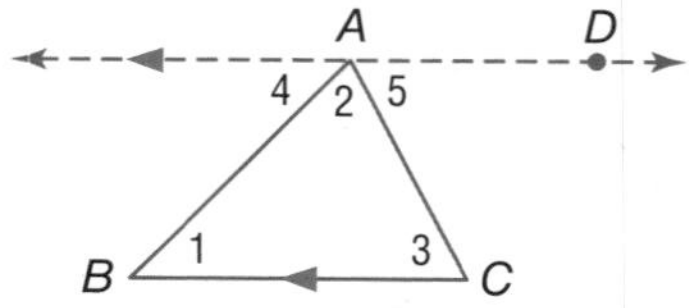

Proof:

Statements	Reasons
1. $\triangle ABC$	1. Given
2. Draw $\overleftrightarrow{AD}$ through A parallel to $\overline{BC}$.	2. Parallel Postulate
3. $\angle 4$ and $\angle BAD$ form a linear pair.	3. Def. of a linear pair
4. $\angle 4$ and $\angle BAD$ are supplementary.	4. If 2 ∡ form a linear pair, they are supplementary.
5. $m\angle 4 + m\angle BAD = 180$	5. Def. of suppl. ∡
6. $m\angle BAD = m\angle 2 + m\angle 5$	6. Angle Addition Postulate
7. $m\angle 4 + m\angle 2 + m\angle 5 = 180$	7. Substitution
8. $\angle 4 \cong \angle 1$, $\angle 5 \cong \angle 3$	8. Alt. Int. ∡ Theorem
9. $m\angle 4 = m\angle 1$, $m\angle 5 = m\angle 3$	9. Def. of $\cong$ ∡
10. $m\angle 1 + m\angle 2 + m\angle 3 = 180$	10. Substitution

The Triangle Angle-Sum Theorem can be used to determine the measure of the third angle of a triangle when the other two angle measures are known.

Real-World Example 1 Use the Triangle Angle-Sum Theorem

PT

SOCCER The diagram shows the path of the ball in a passing drill created by four friends. Find the measure of each numbered angle.

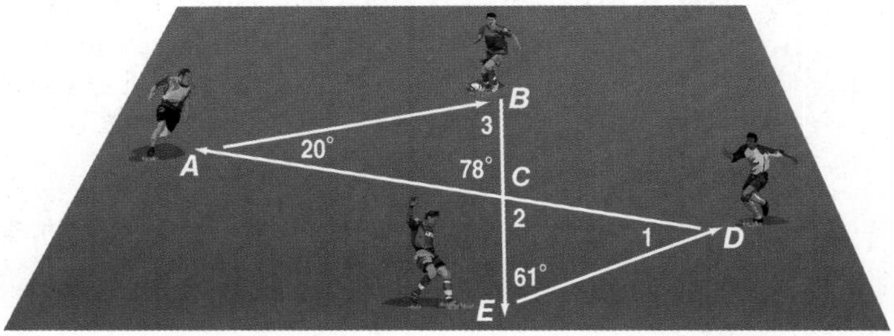

Understand Examine the information given in the diagram. You know the measures of two angles of one triangle and only one measure of another. You also know that $\angle ACB$ and $\angle 2$ are vertical angles.

Plan Find $m\angle 3$ using the Triangle Angle-Sum Theorem, because the measures of two angles of $\angle ABC$ are known. Use the Vertical Angles Theorem to find $m\angle 2$. Then you will have enough information to find the measure of $\angle 1$ in $\triangle CDE$.

Solve

$m\angle 3 + m\angle BAC + m\angle ACB = 180$	Triangle Angle-Sum Theorem
$m\angle 3 + 20 + 78 = 180$	Substitution
$m\angle 3 + 98 = 180$	Simplify.
$m\angle 3 = 82$	Subtract 98 from each side.

$\angle ACB$ and $\angle 2$ are congruent vertical angles. So, $m\angle 2 = 78$.

Use $m\angle 2$ and $\angle CED$ of $\triangle CDE$ to find $m\angle 1$.

$m\angle 1 + m\angle 2 + m\angle CED = 180$	Triangle Angle-Sum Theorem
$m\angle 1 + 78 + 61 = 180$	Substitution
$m\angle 1 + 139 = 180$	Simplify.
$m\angle 1 = 41$	Subtract 139 from each side.

Check The sums of the measures of the angles of $\triangle ABC$ and $\triangle CDE$ should be 180.

$\triangle ABC$: $m\angle 3 + m\angle BAC + m\angle ACB = 82 + 20 + 78$ or 180 ✓

$\triangle CDE$: $m\angle 1 + m\angle 2 + m\angle CED = 41 + 78 + 61$ or 180 ✓

GuidedPractice

Find the measures of each numbered angle.

1A.

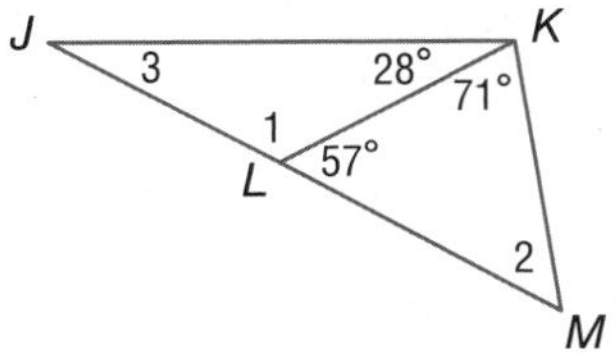

1B.

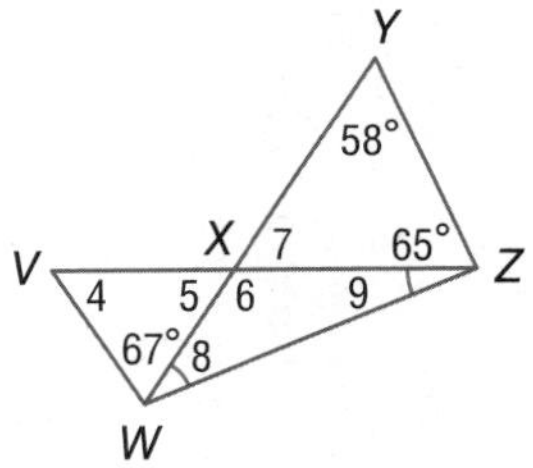

Real-WorldLink

The pass-and-move soccer drill incorporates several fundamental aspects of passing. All passes in this drill are made in a triangle, which is the basis of all ball movement. Additionally, the players are forced to move immediately after passing the ball.

Problem-SolvingTip

CCSS Sense-Making Often a complex problem can be more easily solved if you first break it into more manageable parts. In Example 1, before you can find $m\angle 1$, you must first find $m\angle 2$.

2 Exterior Angle Theorem

In addition to its three interior angles, a triangle can have **exterior angles** formed by one side of the triangle and the extension of an adjacent side. Each exterior angle of a triangle has two **remote interior angles** that are not adjacent to the exterior angle.

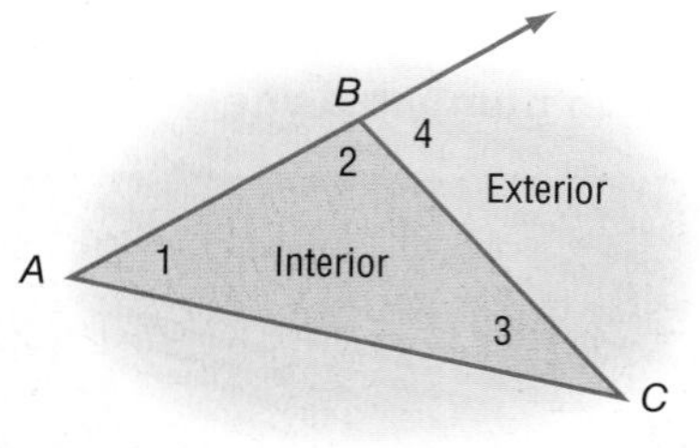

$\angle 4$ is an exterior angle of $\triangle ABC$. Its two remote interior angles are $\angle 1$ and $\angle 3$.

Theorem 12.2 Exterior Angle Theorem

The measure of an exterior angle of a triangle is equal to the sum of the measures of the two remote interior angles.

Example $m\angle A + m\angle B = m\angle 1$

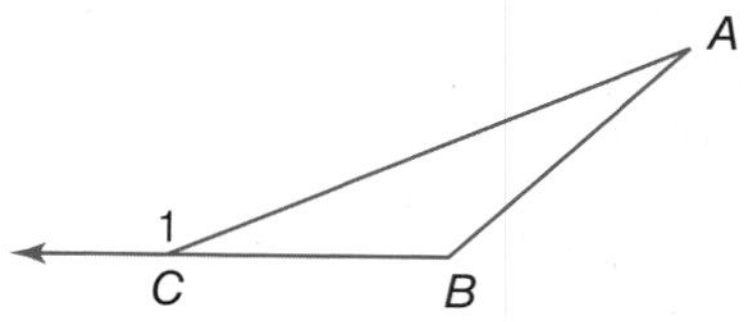

ReadingMath

Flowchart Proof A flow proof is sometimes called a *flowchart* proof.

A **flow proof** uses statements written in boxes and arrows to show the logical progression of an argument. The reason justifying each statement is written below the box. You can use a flow proof to prove the Exterior Angle Theorem.

StudyTip

Flow Proofs Flow proofs can be written vertically or horizontally.

Proof Exterior Angle Theorem

Given: $\triangle ABC$

Prove: $m\angle A + m\angle B = m\angle 1$

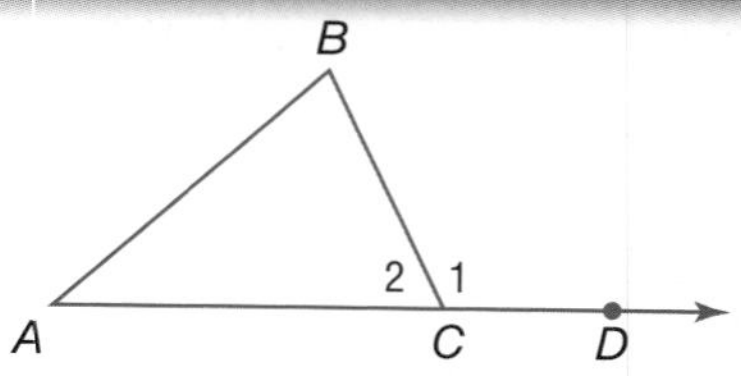

Flow Proof:

$\triangle ABC$
Given
↓
$m\angle A + m\angle B + m\angle 2 = 180$
Triangle Angle-Sum Theorem

$\angle 2$ and $\angle 1$ form a linear pair.
Definition of a linear pair
↓
$\angle 2$ and $\angle 1$ are supplementary.
If 2 ∠s form a linear pair, they are supplementary.
↓
$m\angle 2 + m\angle 1 = 180$
Definition of supplementary

↘ ↙

$m\angle A + m\angle B + m\angle 2 = m\angle 2 + m\angle 1$
Substitution
↓
$m\angle A + m\angle B = m\angle 1$
Subtraction Property of Equality

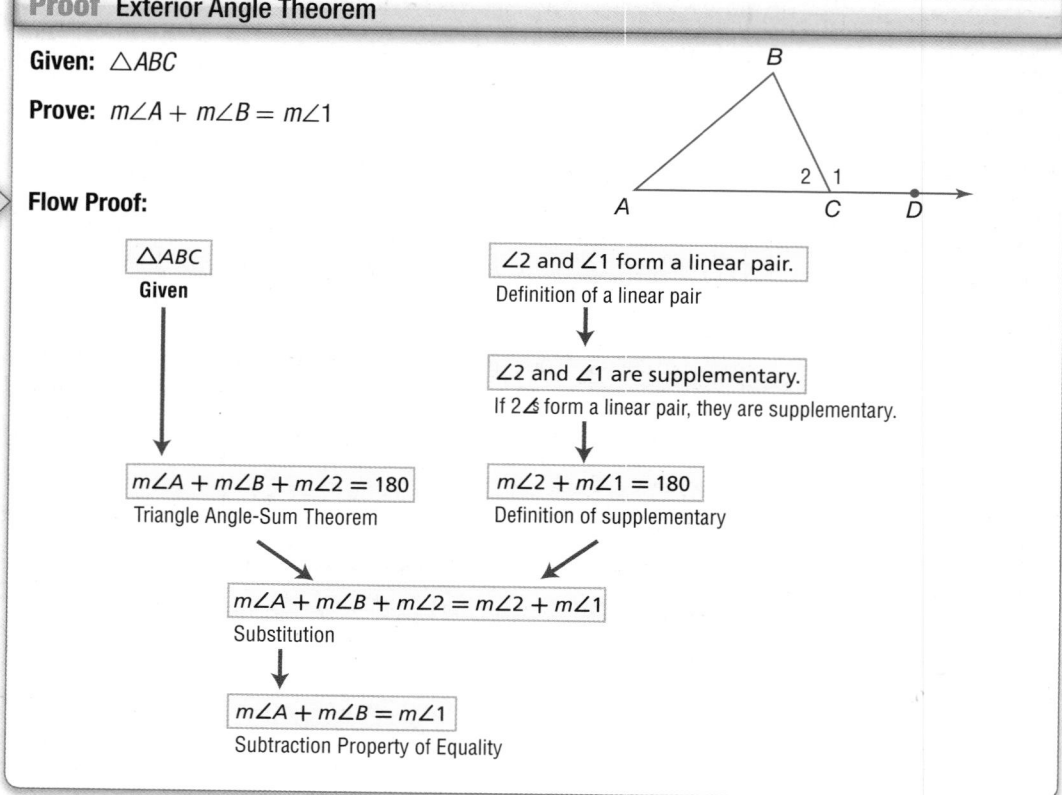

The Exterior Angle Theorem can also be used to find missing measures.

Real-World Example 2 Use the Exterior Angle Theorem

FITNESS **Find the measure of $\angle JKL$ in the Triangle Pose shown.**

$m\angle KLM + m\angle LMK = m\angle JKL$	Exterior Angle Theorem
$x + 50 = 2x - 15$	Substitution
$50 = x - 15$	Subtract x from each side.
$65 = x$	Add 15 to each side.

So, $m\angle JKL = 2(65) - 15$ or 115.

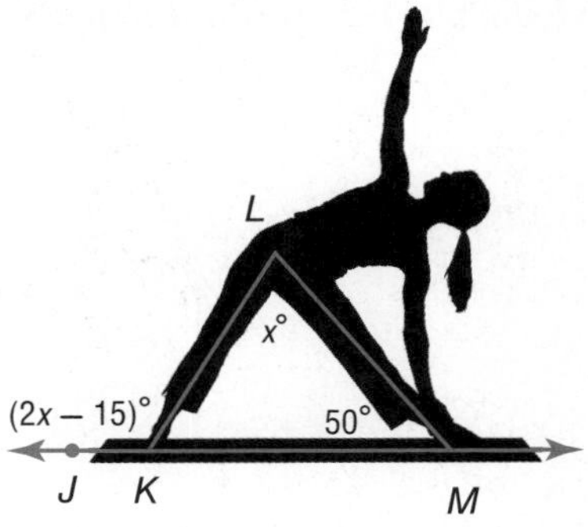

Real-World Career

Personal Trainer Personal trainers instruct and motivate individuals in exercise activities. They demonstrate various exercises and help clients improve their exercise techniques. Personal trainers must obtain certification in the fitness field.

Guided Practice

2. CLOSET ORGANIZING Tanya mounts the shelving bracket shown to the wall of her closet. What is the measure of $\angle 1$, the angle that the bracket makes with the wall?

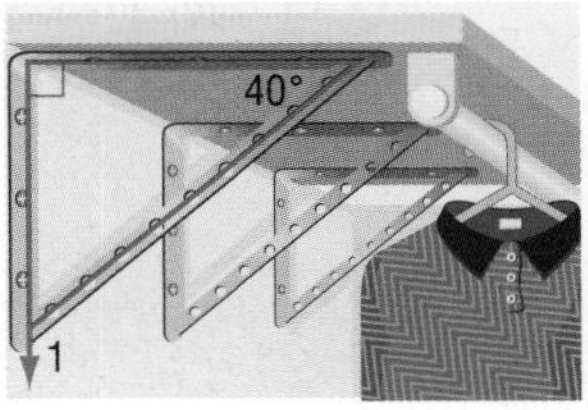

A **corollary** is a theorem with a proof that follows as a direct result of another theorem. As with a theorem, a corollary can be used as a reason in a proof. The corollaries below follow directly from the Triangle Angle-Sum Theorem.

Corollaries Triangle Angle-Sum Corollaries

12.1 The acute angles of a right triangle are complementary.

Abbreviation: *Acute ∡ of a rt. △ are comp.*

Example: If $\angle C$ is a right angle, then $\angle A$ and $\angle B$ are complementary.

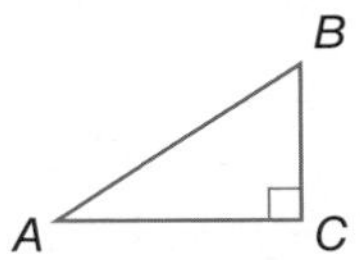

12.2 There can be at most one right or obtuse angle in a triangle.

Example: If $\angle L$ is a right or an obtuse angle, then $\angle J$ and $\angle K$ must be acute angles.

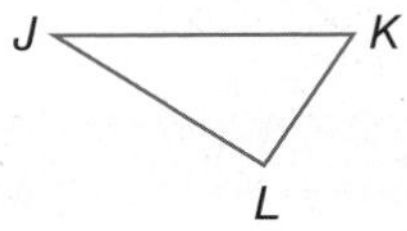

You will prove Corollaries 12.1 and 12.2 in Exercises 34 and 35.

Example 3 Find Angle Measures in Right Triangles

Find the measures of each numbered angle.

$m\angle 1 + m\angle TYZ = 90$	Acute ∡ of a rt. △ are comp.
$m\angle 1 + 52 = 90$	Substitution
$m\angle 1 = 38$	Subtract 52 from each side.

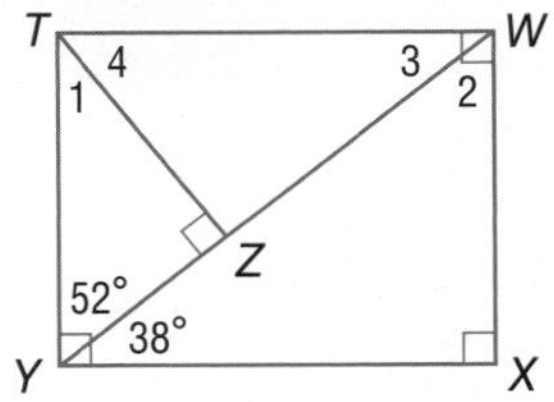

Study Tip

Check for Reasonableness When you are solving for the measure of one or more angles of a triangle, always check to make sure that the sum of the angle measures is 180.

Guided Practice

3A. $\angle 2$ **3B.** $\angle 3$ **3C.** $\angle 4$

Check Your Understanding

Example 1 Find the measures of each numbered angle.

1.

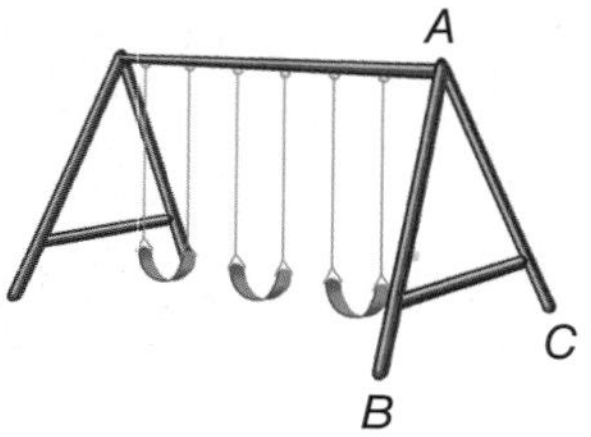

2.

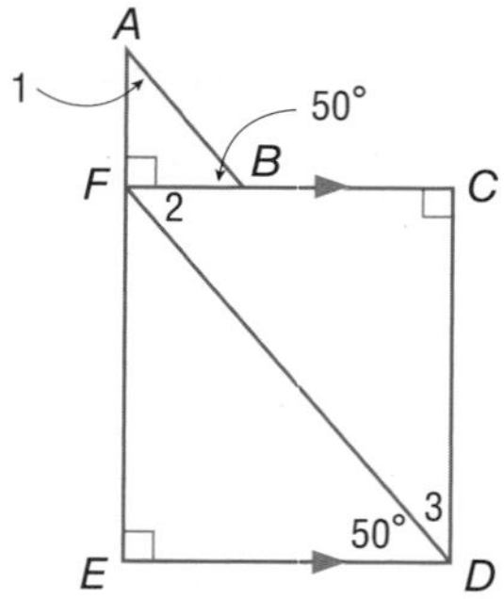

Example 2 Find each measure.

3. $m\angle 2$

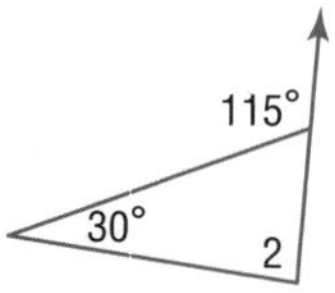

4. $m\angle MPQ$

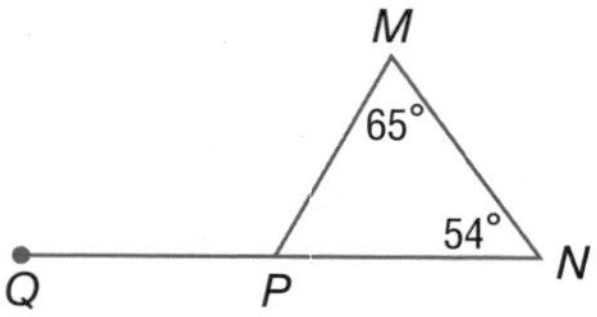

DECK CHAIRS The brace of this deck chair forms a triangle with the rest of the chair's frame as shown. If $m\angle 1 = 105$ and $m\angle 3 = 48$, find each measure.

5. $m\angle 4$

6. $m\angle 6$

7. $m\angle 2$

8. $m\angle 5$

Example 3 CCSS **REGULARITY** Find each measure.

9. $m\angle 1$

10. $m\angle 3$

11. $m\angle 2$

Practice and Problem Solving

Example 1 Find the measure of each numbered angle.

12.

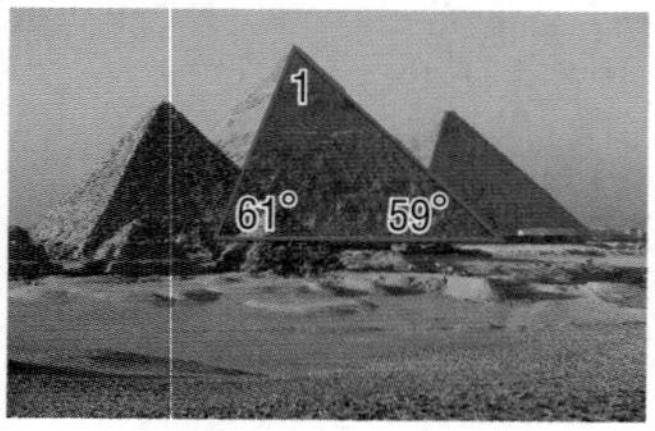

13.

14.

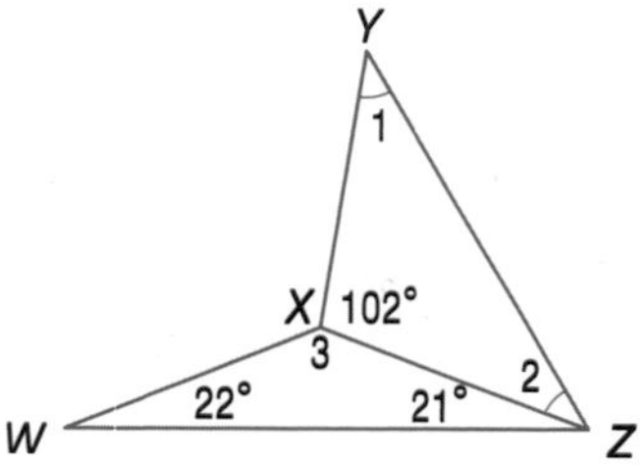

15.

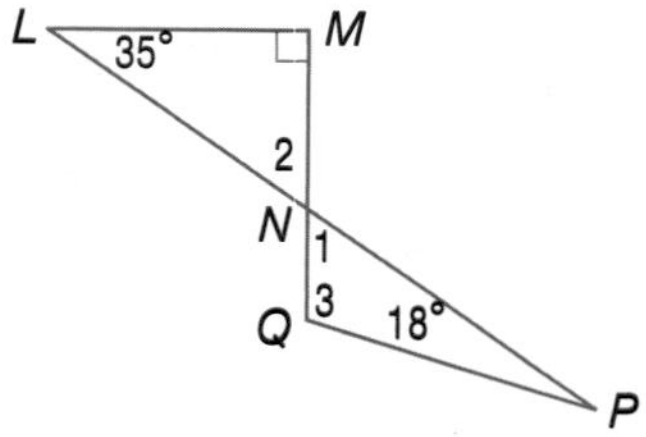

(t)Peter Adams/Digital Vision/Getty Images, (bl)Goodshoot/Masterfile, (br)Jonelle Weaver/Taxi/Getty Images

16. AIRPLANES The path of an airplane can be modeled using two sides of a triangle as shown. The distance covered during the plane's ascent is equal to the distance covered during its descent.

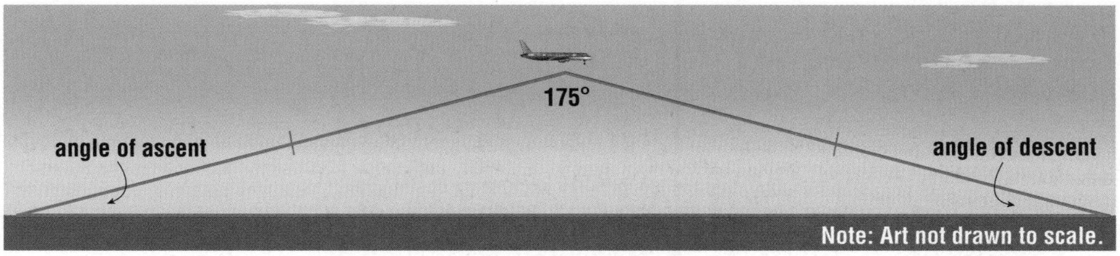

a. Classify the model using its sides and angles.

b. The angles of ascent and descent are congruent. Find their measures.

Example 2 **Find each measure.**

17. $m\angle 1$

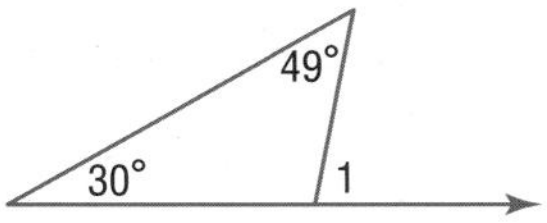

18. $m\angle 3$

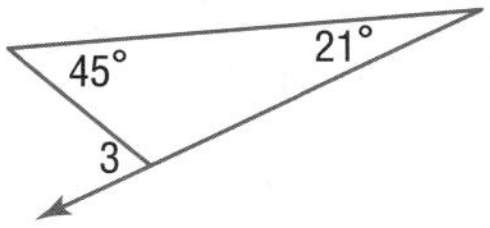

19. $m\angle 2$

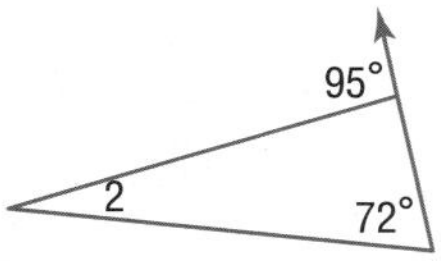

20. $m\angle 4$

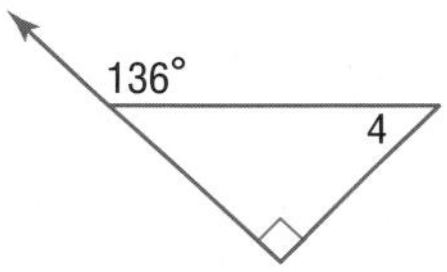

21. $m\angle ABC$

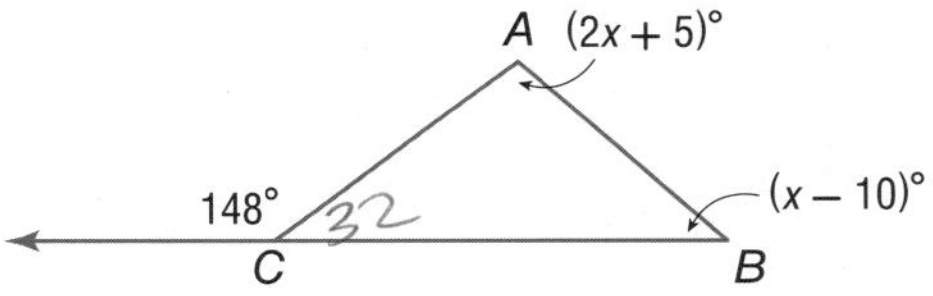

22. $m\angle JKL$

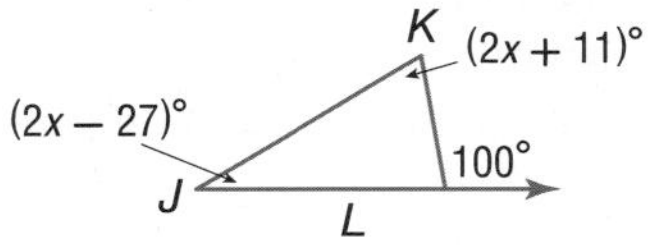

Example 3 **23. WHEELCHAIR RAMP** Suppose the wheelchair ramp shown makes a 12° angle with the ground. What is the measure of the angle the ramp makes with the van door?

CCSS REGULARITY **Find each measure.**

24. $m\angle 1$ **25.** $m\angle 2$

26. $m\angle 3$ **27.** $m\angle 4$

28. $m\angle 5$ **29.** $m\angle 6$

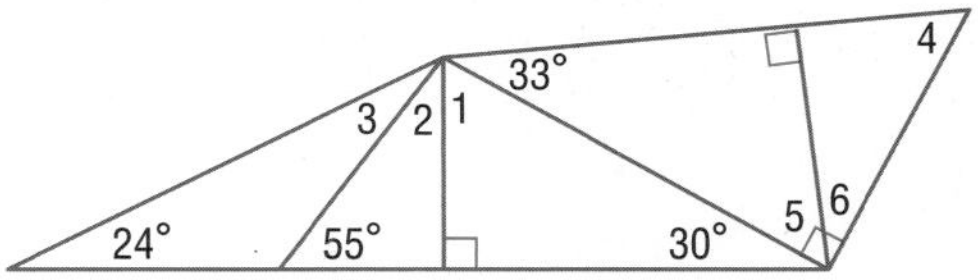

ALGEBRA **Find the value of x. Then find the measure of each angle.**

30.

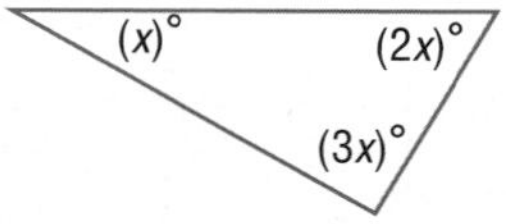

31.

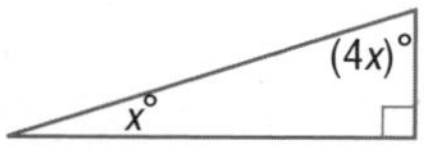

32.

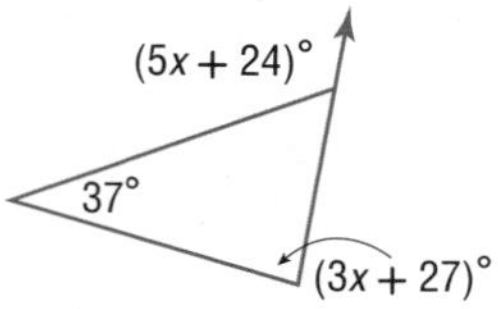

33. **GARDENING** A landscaper decides to include a triangular shaped greenhouse within a garden. The shape is an isosceles triangle in which the apex angle is one-fourth the size of a base angle. What should be the measure of each angle?

PROOF **Write the specified type of proof.**

34. flow proof of Corollary 12.1

35. paragraph proof of Corollary 12.2

CCSS REGULARITY **Find the measure of each numbered angle.**

36.

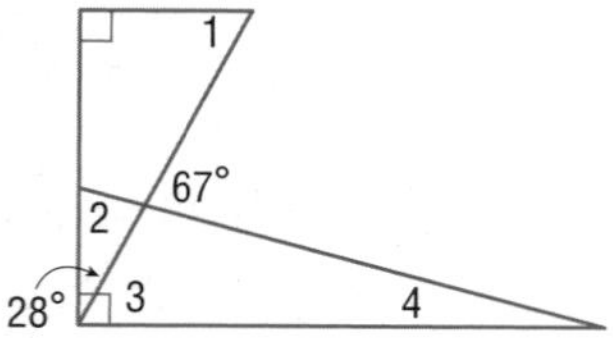

37.

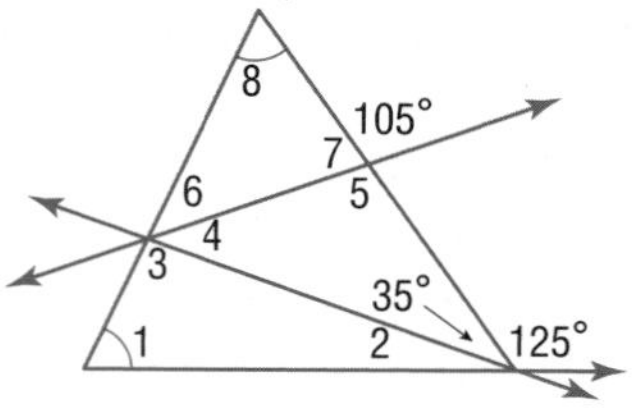

38. **ALGEBRA** Classify the triangle shown by its angles. Explain your reasoning.

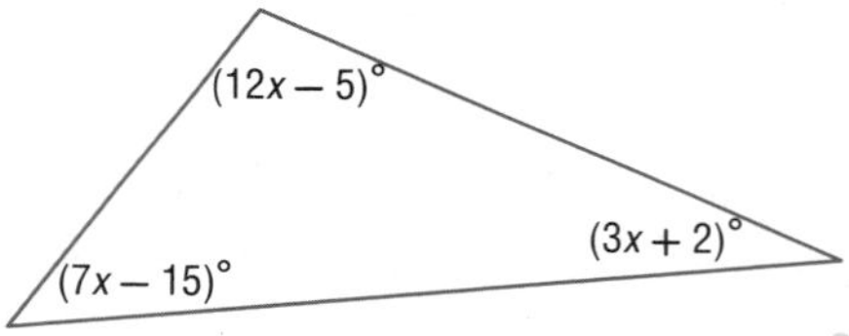

39. **ALGEBRA** The measure of the larger acute angle in a right triangle is twelve degrees less than four times the measure of the smaller acute angle. Find the measure of each angle.

40. Determine whether the following statement is *true* or *false*. If false, give a counterexample. If true, give an argument to support your conclusion.

If the sum of two acute angles of a triangle is less than 90, then the triangle is acute.

41. **ALGEBRA** In $\triangle XYZ$, $m\angle X = 152$, $m\angle Y = y$, and $m\angle Z = z$. Write an inequality to describe the possible measures of $\angle Z$. Explain your reasoning.

42. **CARS** Refer to the photo at the right.

a. Find $m\angle 1$ and $m\angle 2$.

b. If the support for the hood were longer than the one shown, how would $m\angle 1$ change? Explain.

c. If the support for the hood were longer than the one shown, how would $m\angle 2$ change? Explain.

Jan Stromme/Photonica/Getty Images

PROOF **Write the specified type of proof.**

43. two-column proof

Given: *ABCDEF* is a hexagon.
Prove: $m\angle B + m\angle BCD + m\angle CDE + m\angle DEF + m\angle F + m\angle FAB = 720$

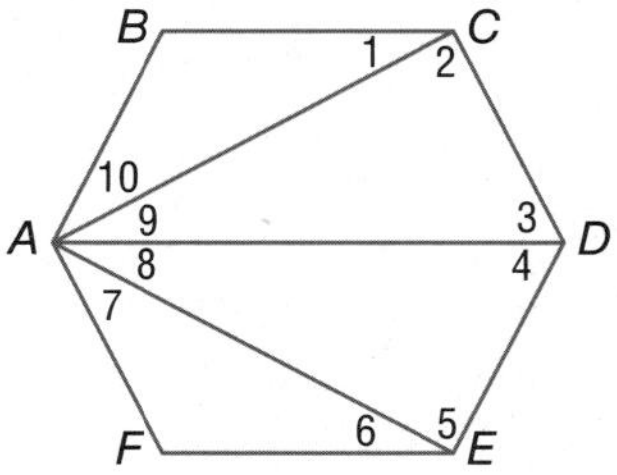

44. paragraph proof

Given: Picture on the Right
Prove: $m\angle 1 + m\angle 2 = m\angle 5 + m\angle 6$

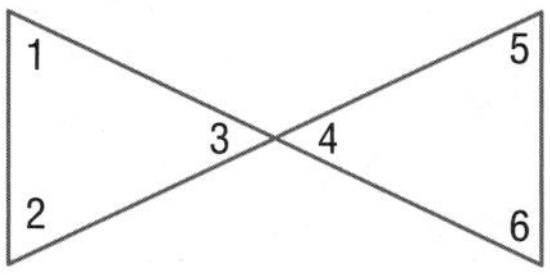

45. **MULTIPLE REPRESENTATIONS** In this problem, you will explore the sum of the measures of the exterior angles of a triangle.

a. Geometric Draw five different triangles, extending the sides and labeling the angles as shown. Be sure to include at least one obtuse, one right, and one acute triangle.

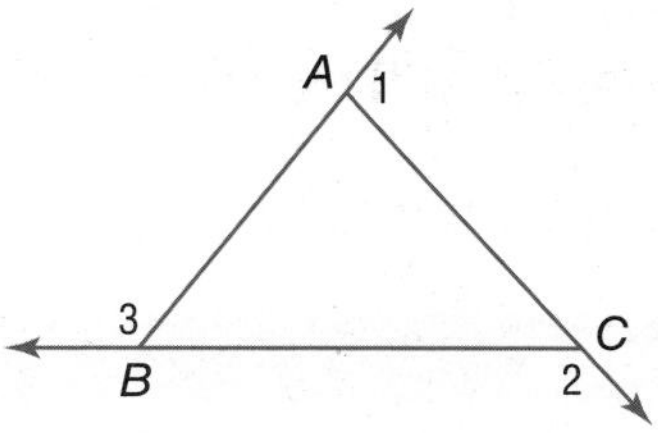

b. Tabular Measure the exterior angles of each triangle. Record the measures for each triangle and the sum of these measures in a table.

c. Verbal Make a conjecture about the sum of the exterior angles of a triangle. State your conjecture using words.

d. Algebraic State the conjecture you wrote in part **c** algebraically.

e. Analytical Write a paragraph proof of your conjecture.

H.O.T. Problems Use Higher-Order Thinking Skills

46. **CCSS ERROR ANALYSIS** Curtis measured and labeled the angles of the triangle as shown. Arnoldo says that at least one of his measures is incorrect. Explain in at least two different ways how Arnoldo knows that this is true.

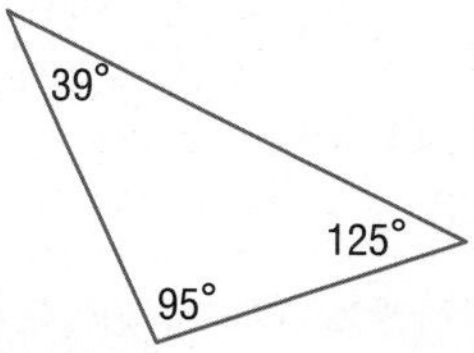

47. **WRITING IN MATH** Explain how you would find the missing measures in the figure shown.

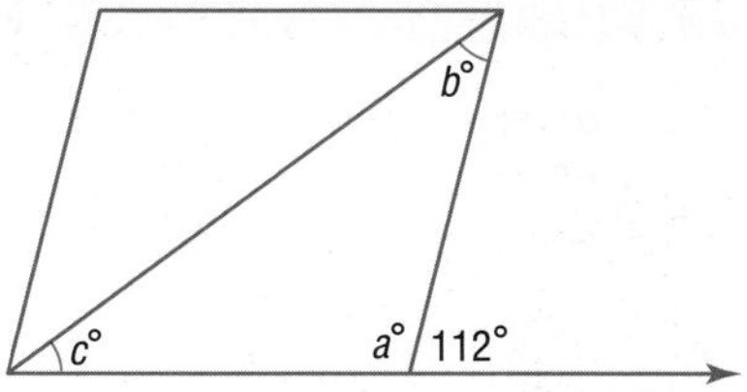

48. **CHALLENGE** Find the values of x and y in the figure below.

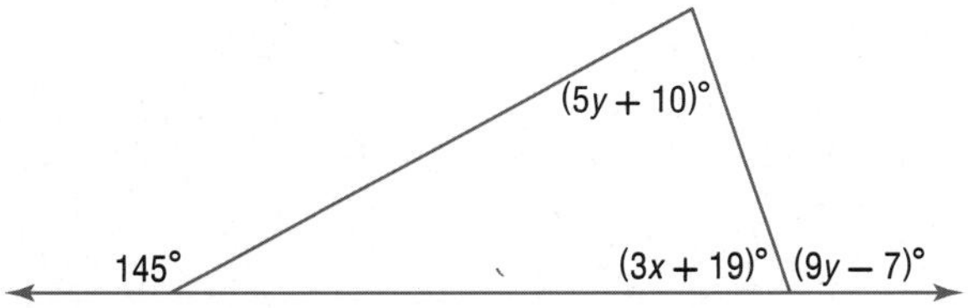

49. **REASONING** If an exterior angle adjacent to $\angle A$ is obtuse, is $\triangle ABC$ acute, right, or obtuse, or can its classification not be determined? Explain your reasoning.

50. **WRITING IN MATH** Explain why a triangle cannot have an obtuse, acute, and a right interior angle.

Standardized Test Practice

51. PROBABILITY Mr. Glover owns a video store and wants to survey his customers to find what type of movies he should buy. Which of the following options would be the best way for Mr. Glover to get accurate survey results?

A surveying customers who come in from 9 P.M. until 10 P.M.

B surveying customers who come in on the weekend

C surveying the male customers

D surveying at different times of the week and day

52. SHORT RESPONSE Two angles of a triangle have measures of 35° and 80°. Find the values of the exterior angle measures of the triangle.

53. ALGEBRA Which equation is equivalent to $7x - 3(2 - 5x) = 8x$?

F $2x - 6 = 8$

G $22x - 6 = 8x$

H $-8x - 6 = 8x$

J $22x + 6 = 8x$

54. SAT/ACT Joey has 4 more video games than Solana and half as many as Melissa. If together they have 24 video games, how many does Melissa have?

A 7

B 9

C 12

D 13

E 14

Spiral Review

Classify each triangle as *acute, equiangular, obtuse,* or *right.* (Lesson 12-1)

55.

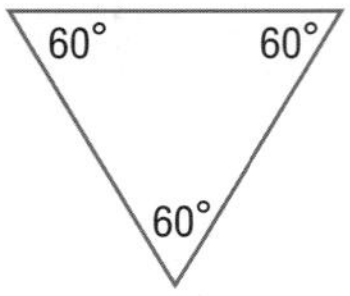

56.

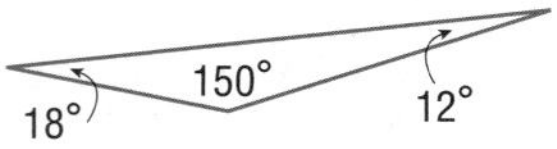

57.

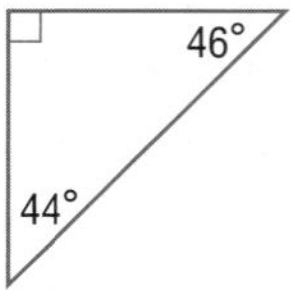

COORDINATE GEOMETRY Find the distance from *P* to ℓ. (Lesson 11-6)

58. Line ℓ contains points $(0, -2)$ and $(1, 3)$. Point P has coordinates $(-4, 4)$.

59. Line ℓ contains points $(-3, 0)$ and $(3, 0)$. Point P has coordinates $(4, 3)$.

Skills Review

State the property that justifies each statement.

60. If $\frac{x}{2} = 7$, then $x = 14$.

61. If $x = 5$ and $b = 5$, then $x = b$.

62. If $XY - AB = WZ - AB$, then $XY = WZ$.

63. If $m\angle A = m\angle B$ and $m\angle B = m\angle C$, $m\angle A = m\angle C$.

64. If $m\angle 1 + m\angle 2 = 90$ and $m\angle 2 = m\angle 3$, then $m\angle 1 + m\angle 3 = 90$.

LESSON 12-3
Congruent Triangles

Then	Now	Why?
You identified and used congruent angles.	**1** Name and use corresponding parts of congruent polygons. **2** Prove triangles congruent using the definition of congruence.	As an antitheft device, many manufacturers make car stereos with removable faceplates. The shape and size of the faceplate and of the space where it fits must be exactly the same for the faceplate to properly attach to the car's dashboard.

NewVocabulary
congruent
congruent polygons
corresponding parts

Common Core State Standards
Content Standards
G.CO.7 Use the definition of congruence in terms of rigid motions to show that two triangles are congruent if and only if corresponding pairs of sides and corresponding pairs of angles are congruent.
G.SRT.5 Use congruence and similarity criteria for triangles to solve problems and to prove relationships in geometric figures.

Mathematical Practices
6 Attend to precision.
3 Construct viable arguments and critique the reasoning of others.

1 Congruence and Corresponding Parts If two geometric figures have exactly the same shape and size, they are **congruent**.

Congruent	Not Congruent
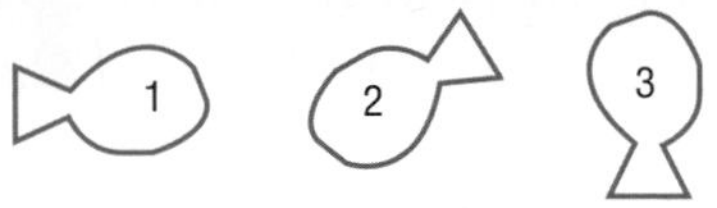	
While positioned differently, Figures 1, 2, and 3 are exactly the same shape and size.	Figures 4 and 5 are exactly the same shape but not the same size. Figures 5 and 6 are the same size but not exactly the same shape.

In two **congruent polygons**, all of the parts of one polygon are congruent to the **corresponding parts** or matching parts of the other polygon. These corresponding parts include *corresponding angles* and *corresponding sides*.

KeyConcept Definition of Congruent Polygons

Words Two polygons are congruent if and only if their corresponding parts are congruent.

Example Corresponding Angles
$\angle A \cong \angle H$ $\quad \angle B \cong \angle J$ $\quad \angle C \cong \angle K$

Corresponding Sides
$\overline{AB} \cong \overline{HJ}$ $\quad \overline{BC} \cong \overline{JK}$ $\quad \overline{AC} \cong \overline{HK}$

Congruence Statement
$\triangle ABC \cong \triangle HJK$

Model

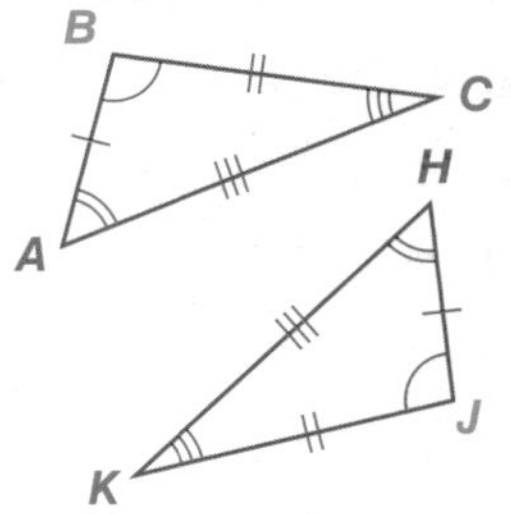

Other congruence statements for the triangles above exist. Valid congruence statements for congruent polygons list corresponding vertices in the same order.

Valid Statement	Not a Valid Statement
$\triangle BCA \cong \triangle JKH$	$\triangle ABC \cong \triangle HKJ$

Math HistoryLink

Johann Carl Friedrich Gauss (1777–1855) Gauss developed the congruence symbol to show that two sides of an equation were the same even if they weren't equal. He made many advances in math and physics, including a proof of the fundamental theorem of algebra.

Source: The Granger Collection, New York

Example 1 Identify Corresponding Congruent Parts

PT

Show that the polygons are congruent by identifying all the congruent corresponding parts. Then write a congruence statement.

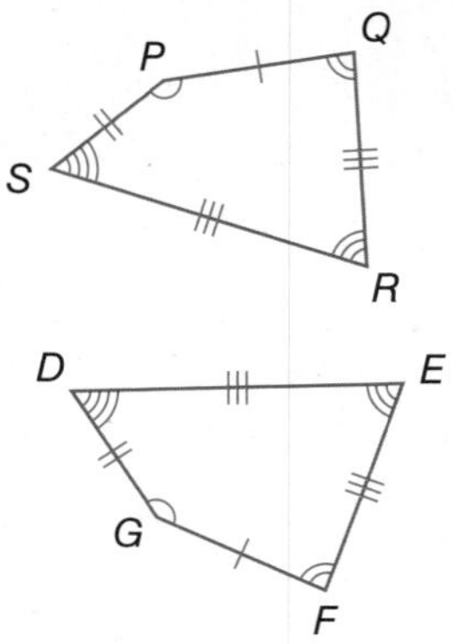

Angles: $\angle P \cong \angle G$, $\angle Q \cong \angle F$, $\angle R \cong \angle E$, $\angle S \cong \angle D$

Sides: $\overline{PQ} \cong \overline{GF}$, $\overline{QR} \cong \overline{FE}$, $\overline{RS} \cong \overline{ED}$, $\overline{SP} \cong \overline{DG}$

All corresponding parts of the two polygons are congruent. Therefore, polygon *PQRS* ≅ polygon *GFED*.

GuidedPractice

1A.

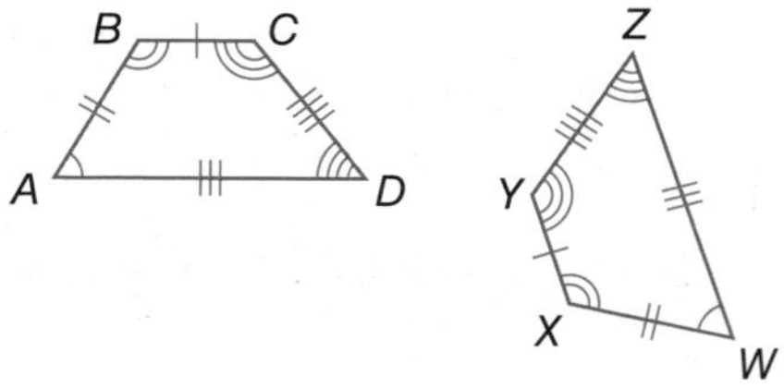

1B.

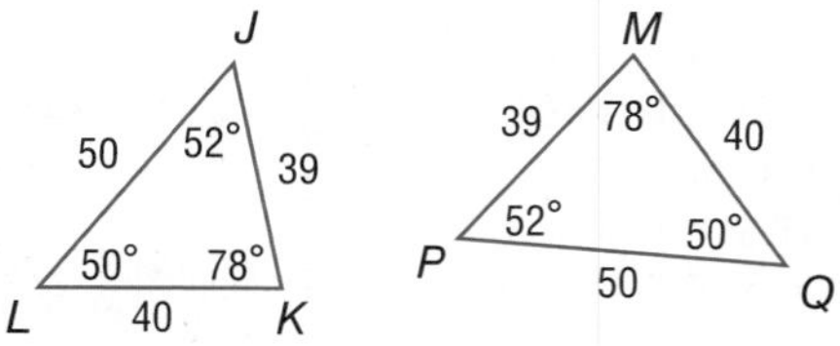

The phrase "if and only if" in the congruent polygon definition means that both the conditional and its converse are true. So, if two polygons are congruent, then their corresponding parts are congruent. For triangles, we say *Corresponding parts of congruent triangles are congruent,* or CPCTC.

Example 2 Use Corresponding Parts of Congruent Triangles

PT

StudyTip

Using a Congruence Statement You can use a congruence statement to help you correctly identify corresponding sides.

$\triangle \mathbf{ABC} \cong \triangle \mathbf{DFE}$

$\overline{\mathbf{BC}} \cong \overline{\mathbf{FE}}$

In the diagram, $\triangle ABC \cong \triangle DFE$. Find the values of x and y.

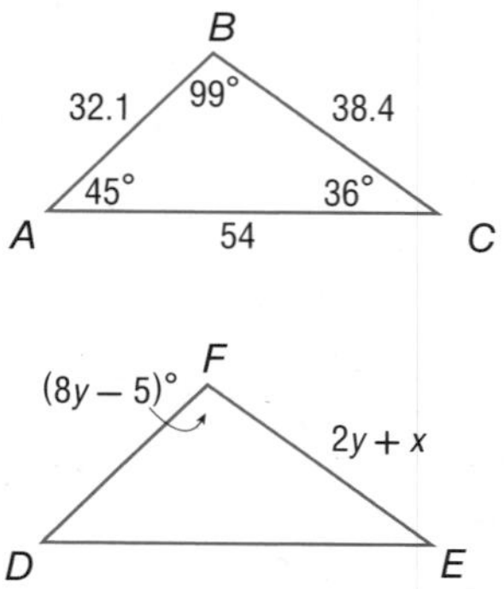

$\angle F \cong \angle B$	CPCTC
$m\angle F = m\angle B$	Definition of congruence
$8y - 5 = 99$	Substitution
$8y = 104$	Add 5 to each side.
$y = 13$	Divide each side by 8.
$\overline{FE} \cong \overline{BC}$	CPCTC
$FE = BC$	Definition of congruence
$2y + x = 38.4$	Substitution
$2(13) + x = 38.4$	Substitution
$26 + x = 38.4$	Simplify.
$x = 12.4$	Subtract 26 from each side.

GuidedPractice

2. In the diagram, $\triangle RSV \cong \triangle TVS$. Find the values of x and y.

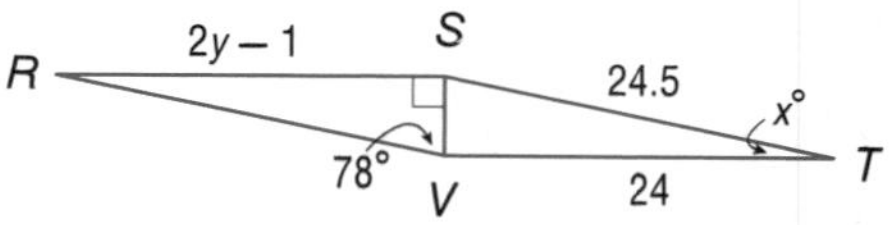

2 Prove Triangles Congruent

The Triangle Angle-Sum Theorem you learned in Lesson 4-2 leads to another theorem about the angles in two triangles.

Theorem 12.3 Third Angles Theorem

Words: If two angles of one triangle are congruent to two angles of a second triangle, then the third angles of the triangles are congruent.

Example: If $\angle C \cong \angle K$ and $\angle B \cong \angle J$, then $\angle A \cong \angle L$.

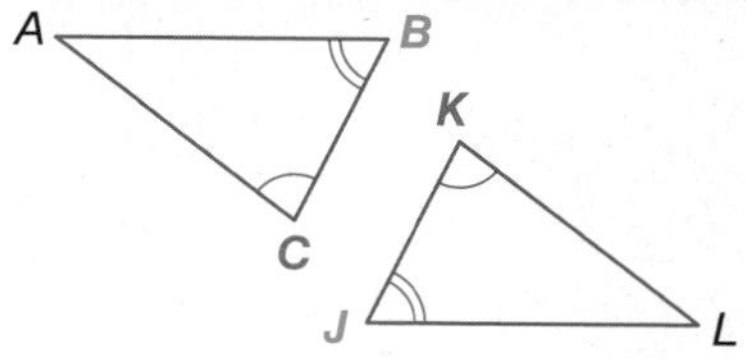

You will prove this theorem in Exercise 21.

Real-WorldLink

Using some basic skills with napkin folding can add an elegant touch to any party. Many of the folds use triangles.

Real-World Example 3 Use the Third Angles Theorem

PT

PARTY PLANNING **The planners of the Senior Banquet decide to fold the dinner napkins using the Triangle Pocket Fold so that they can place a small gift in the pocket. If $\angle NPQ \cong \angle RST$, and $m\angle NPQ = 40$, find $m\angle SRT$.**

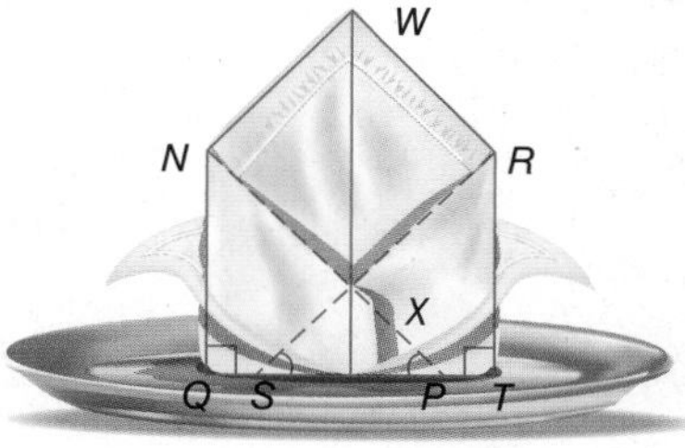

$\angle NPQ \cong \angle RST$, and since all right angles are congruent, $\angle NQP \cong \angle RTS$. So by the Third Angles Theorem, $\angle QNP \cong \angle SRT$. By the definition of congruence, $m\angle QNP = m\angle TRS$.

$m\angle QNP + m\angle NPQ = 90$	The acute angles of a right triangle are complementary.
$m\angle QNP + 40 = 90$	Substitution
$m\angle QNP = 50$	Subtract 40 from each side.

By substitution, $m\angle SRT = m\angle QNP$ or 50.

GuidedPractice

3. In the diagram above, if $\angle WNX \cong \angle WRX$, $\overline{WX}$ bisects $\angle NXR$, $m\angle WNX = 88$, and $m\angle NXW = 49$, find $m\angle NWR$. Explain your reasoning.

Example 4 Prove That Two Triangles are Congruent

PT

Write a two-column proof.

Given: $\overline{DE} \cong \overline{GE}$, $\overline{DF} \cong \overline{GF}$, $\angle D \cong \angle G$, $\angle DFE \cong \angle GFE$

Prove: $\triangle DEF \cong \triangle GEF$

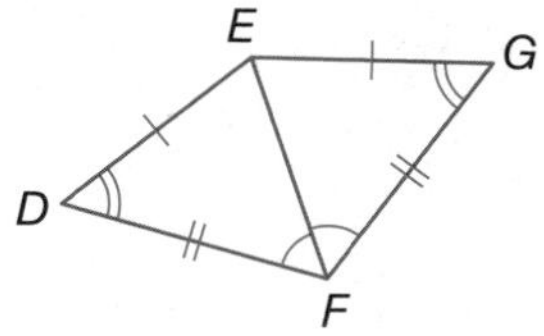

Proof:

Statements	Reasons
1. $\overline{DE} \cong \overline{GE}$, $\overline{DF} \cong \overline{GF}$	**1.** Given
2. $\overline{EF} \cong \overline{EF}$	**2.** Reflexive Property of Congruence
3. $\angle D \cong \angle G$, $\angle DFE \cong \angle GFE$	**3.** Given
4. $\angle DEF \cong \angle GEF$	**4.** Third Angles Theorem
5. $\triangle DEF \cong \triangle GEF$	**5.** Definition of Congruent Polygons

StudyTip

Reflexive Property When two triangles share a common side, use the Reflexive Property of Congruence to establish that the common side is congruent to itself.

A Kompatscher/age fotostock

Guided Practice

4. Write a two column proof.

Given: $\angle J \cong \angle P$, $\overline{JK} \cong \overline{PM}$, $\overline{JL} \cong \overline{PL}$, and L bisects $\overline{KM}$.

Prove: $\triangle JLK \cong \triangle PLM$

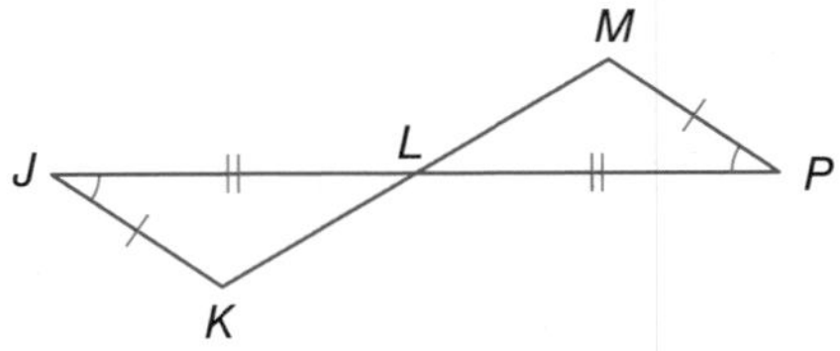

Like congruence of segments and angles, congruence of triangles is reflexive, symmetric, and transitive.

Theorem 12.4 Properties of Triangle Congruence

Reflexive Property of Triangle Congruence

$\triangle ABC \cong \triangle ABC$

Symmetric Property of Triangle Congruence

If $\triangle ABC \cong \triangle EFG$, then $\triangle EFG \cong \triangle ABC$.

Transitive Property of Triangle Congruence

If $\triangle ABC \cong \triangle EFG$ and $\triangle EFG \cong \triangle JKL$, then $\triangle ABC \cong \triangle JKL$.

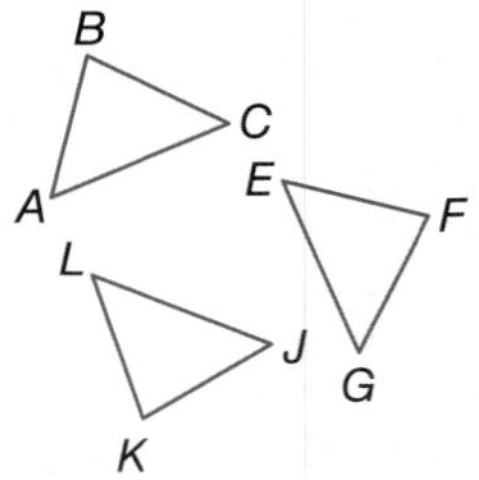

You will prove the reflexive, symmetric, and transitive parts of Theorem 12.4 in Exercises 27, 22, and 26, respectively.

Check Your Understanding

Example 1 **Show that polygons are congruent by identifying all congruent corresponding parts. Then write a congruence statement.**

1.

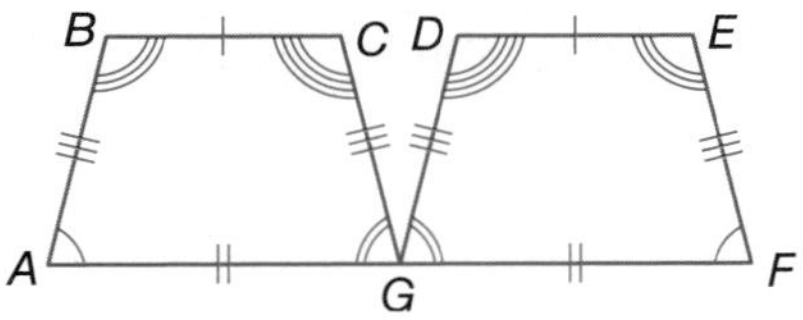

2.

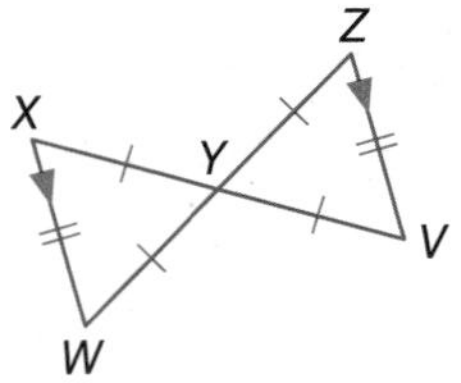

Example 2 **In the figure, $\triangle LMN \cong \triangle QRS$.**

3. Find x.

4. Find y.

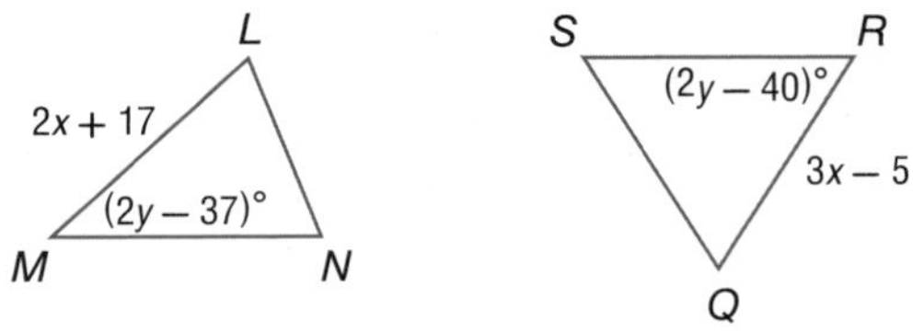

Example 3 **CCSS REGULARITY** **Find x. Explain your reasoning.**

5.

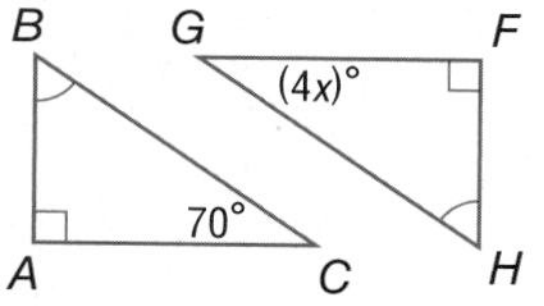

6.

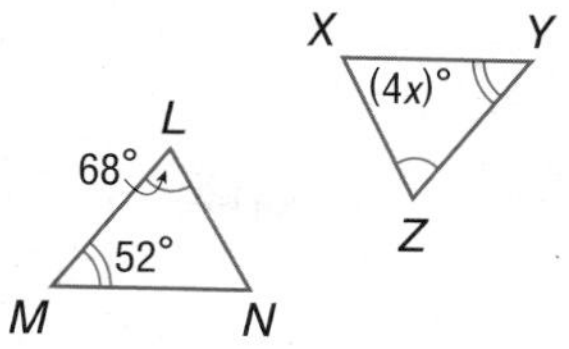

Example 4 **7. PROOF** Write a paragraph proof.

Given: Y is the midpoint of $\overline{XV}$ and $\overline{WZ}$, $\overline{WX} \parallel \overline{ZV}$; $\overline{WX} \cong \overline{ZV}$

Prove: $\triangle WYX \cong \triangle ZYV$

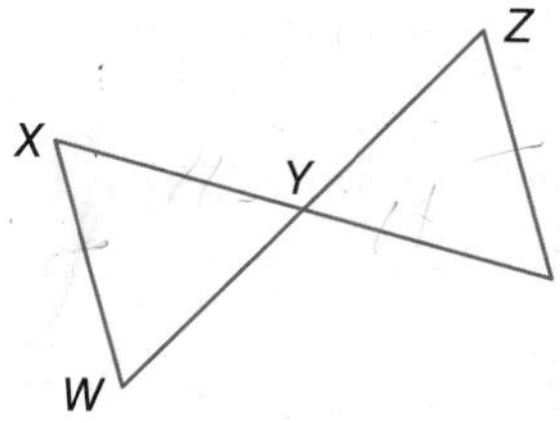

Practice and Problem Solving

Example 1 **Show that polygons are congruent by identifying all congruent corresponding parts. Then write a congruence statement.**

8.

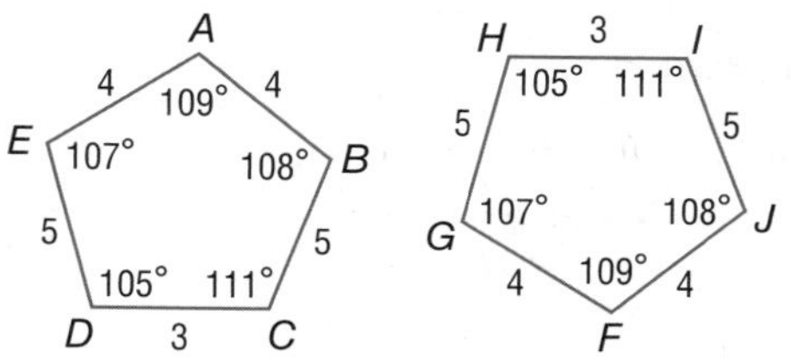

9.

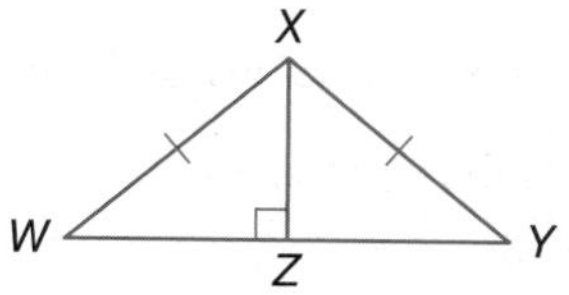

10.

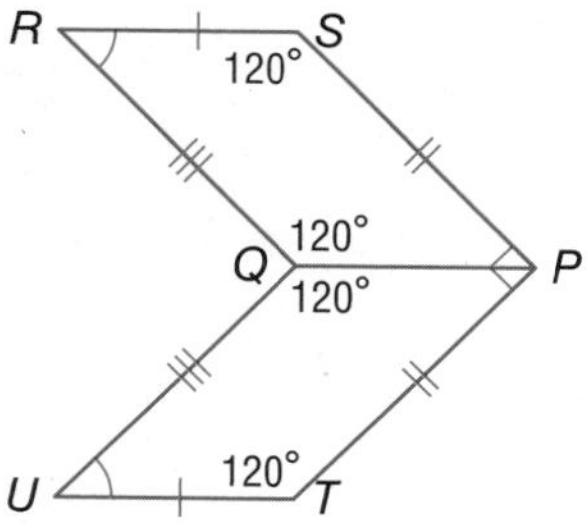

11.

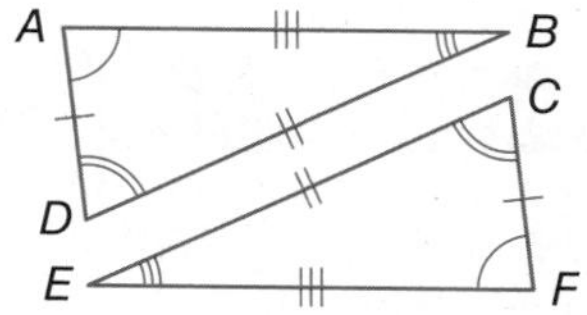

Example 2 **Polygon $BCDE \cong$ polygon $RSTU$. Find each value.**

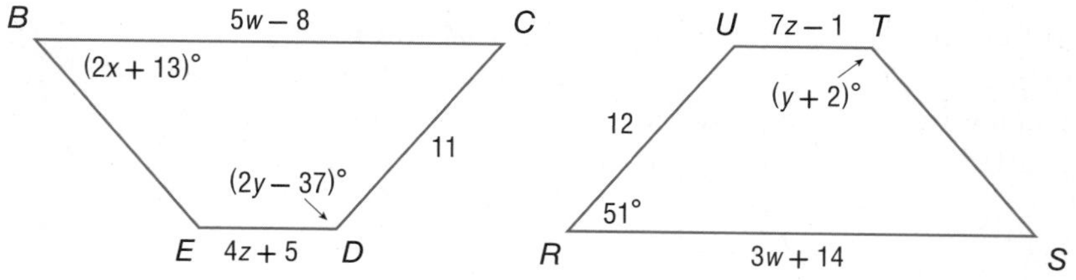

12. x

13. y

14. z

15. w

Example 3 **Find x and y.**

16.

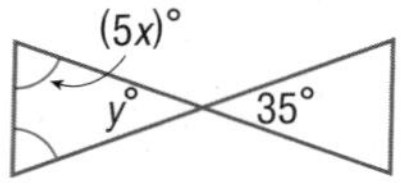

17.

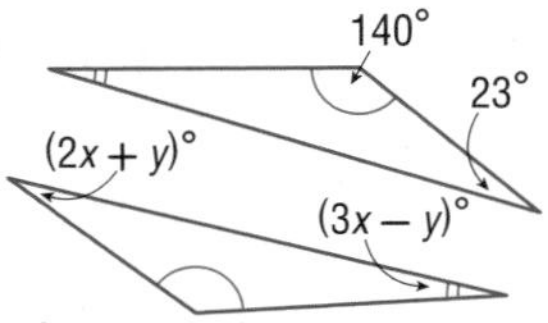

18.

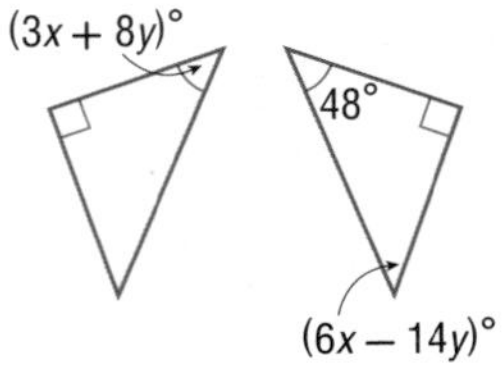

Example 4 **19. PROOF** Write a two-column proof of Theorem 12.3.

20. PROOF Put the statements used to prove the statement below in the correct order. Provide the reasons for each statement.

Congruence of triangles is symmetric. (Theorem 12.4)

Given: $\triangle RST \cong \triangle XYZ$

Prove: $\triangle XYZ \cong \triangle RST$

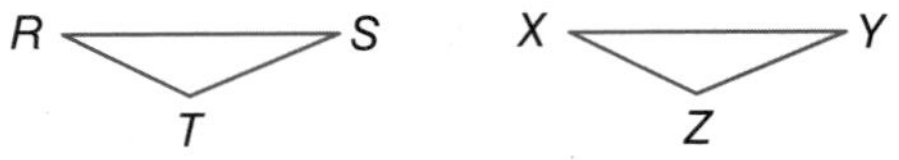

Proof:

$\angle X \cong \angle R$, $\angle Y \cong \angle S$, $\angle Z \cong \angle T$, $\overline{XY} \cong \overline{RS}$, $\overline{YZ} \cong \overline{ST}$, $\overline{XZ} \cong \overline{RT}$	$\angle R \cong \angle X$, $\angle S \cong \angle Y$, $\angle T \cong \angle Z$, $\overline{RS} \cong \overline{XY}$, $\overline{ST} \cong \overline{YZ}$, $\overline{RT} \cong \overline{XZ}$	$\triangle RST \cong \triangle XYZ$	$\triangle XYZ \cong \triangle RST$
?	?	?	?

CCSS ARGUMENTS **Write a two-column proof.**

21. Given: Parallelogram $PQRS$
Prove: $\triangle PQS \cong \triangle RSQ$

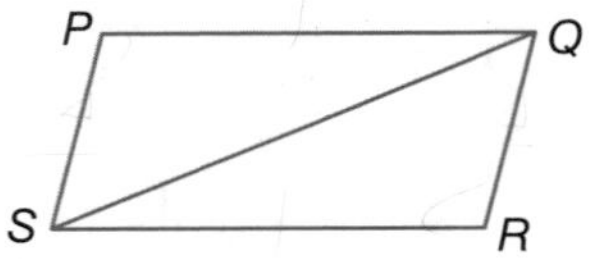

22. Given: $\angle A \cong \angle C$; $\angle ABD \cong \angle CBD$;
$\angle ADB \cong \angle CDB$
$\overline{AB} \cong \overline{CB}$; $\overline{CD} \cong \overline{AD}$

Prove: $\triangle ABD \cong \triangle CBD$

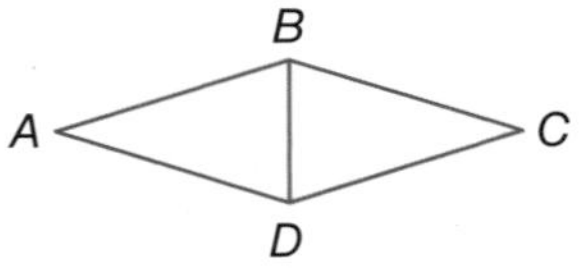

23. T-SHIRT PRINTING Susan loves math. She wanted to print T-shirts for her friends. She went to a company that prints custom shirts. Her design is on the right. What property guarantees that the printed designs are congruent?

Enrique Algarra/age fotostock

PROOF **Write the specified type of proof of the indicated part of Theorem 12.4.**

24. Congruence of triangles is transitive. (paragraph proof)

25. Congruence of triangles is reflexive. (flow proof)

ALGEBRA **Draw and label a figure to represent the congruent triangles. Then find x and y.**

26. $\triangle ABC \cong \triangle DEF$, $AB = 11$, $AC = 17 + x$, $DF = 2x + 13$, and $DE = 3y + 2$

27. $\triangle LMN \cong \triangle RST$, $m\angle L = 51$, $m\angle M = 9y$, $m\angle S = 72$, and $m\angle T = 4x + 15$

28. $\triangle JKL \cong \triangle MNP$, $JK = 12$, $LJ = 7$, $PM = 3x - 2$, $m\angle L = 67$, $m\angle K = y + 9$ and $m\angle N = 2y - 4$

29. PENNANTS Scott is in charge of roping off an area of 100 square feet for the band to use during a pep rally. He is using a string of pennants that are congruent isosceles triangles.

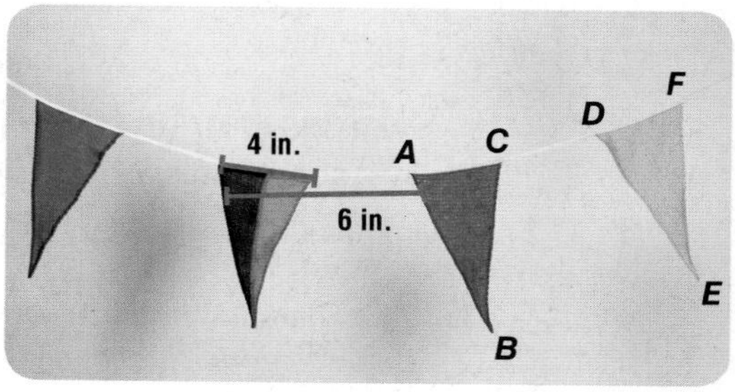

a. List seven pairs of congruent segments in the photo.

b. If the area he ropes off is a square, how long will the pennant string need to be?

c. How many pennants will be on the string?

30. MULTIPLE REPRESENTATIONS In this problem, you will explore the statement *The perimeters of congruent triangles are equal.*

a. Verbal Write a conditional statement to represent the relationship between the perimeters of a pair of congruent triangles.

b. Verbal Write the converse of your conditional statement. Is the converse true or false? Explain your reasoning.

c. Geometric Geometric If possible, draw two triangles that have the same perimeter but are not congruent. If not possible, explain why not.

d. Geometric If possible, draw two rectangles that have the same perimeter but are not congruent. If not, explain why not.

31. PATTERNS The Flying Geese is a block that is often used in quilting.

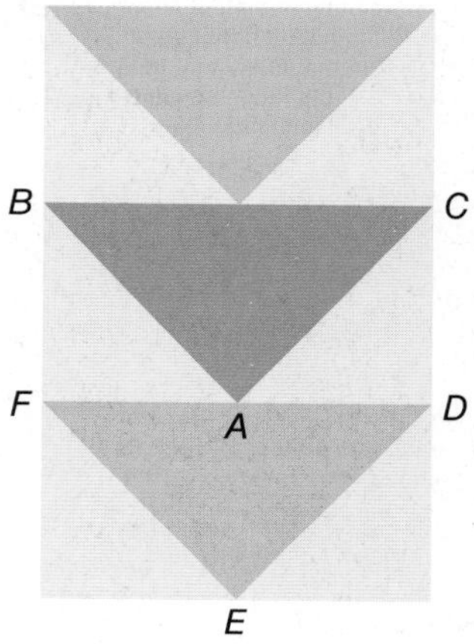

a. What two polygons are used to create the pattern?

b. Name a pair of congruent triangles.

c. Name a pair of corresponding angles.

d. If $BC = 4$, what is FD? Explain.

e. What is the measure of $\angle E$? Explain.

32. MUSIC Bass drum hoops can be used to repair bass drums. They must all be the same size. What measure(s) would you use to prove that all of the hoops are congruent? Explain your reasoning.

H.O.T. Problems Use Higher-Order Thinking Skills

33. **WRITING IN MATH** Explain why the order of the vertices is important when naming congruent triangles. Give an example to support your answer.

34. **ERROR ANALYSIS** Jasmine and Will are evaluating the congruent figures below. Jasmine says that $\triangle ABC \cong \triangle XYZ$ and Will says that $\triangle CAB \cong \triangle XYZ$. Is either of them correct? Explain.

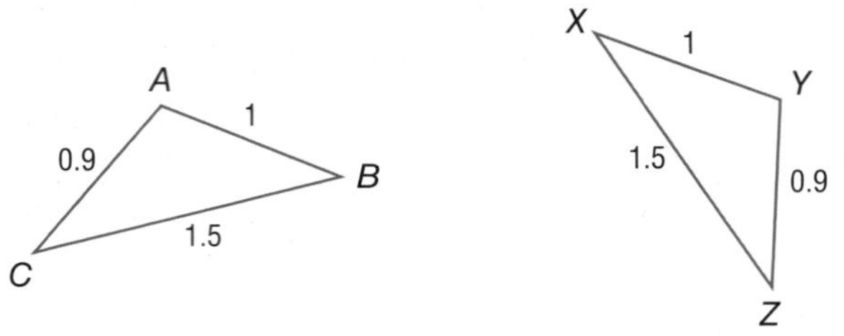

CCSS WRITING IN MATH Determine whether the following statement is always, sometimes, or never true. Explain your reasoning.

35. *Equiangular triangles are congruent.*

36. *Two triangles with two pairs of congruent corresponding sides and one pair of congruent angles are congruent.*

37. *Two triangles with three pairs of corresponding congruent sides are congruent.*

38. *Two right triangles with the two pairs of legs congruent are congruent.*

39. **CHALLENGE** Find x and y if $\triangle PQS \cong \triangle RQS$

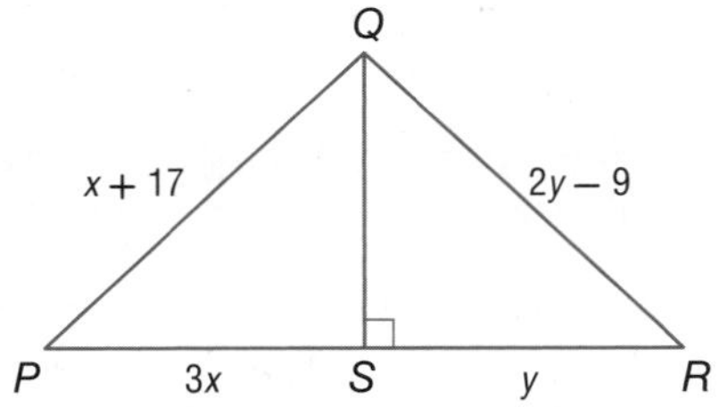

40. **CHALLENGE** Write a paragraph proof to prove that the four triangles that are formed by the diagonals of a square are congruent to each other.

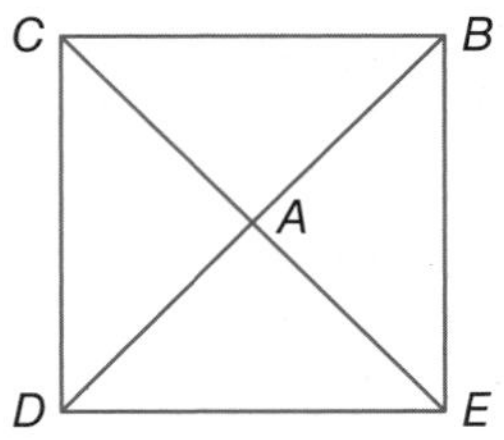

Standardized Test Practice

41. Barrington cut four congruent triangles off the corners of a rectangle to make an octagon as shown below. What is the area of the octagon?

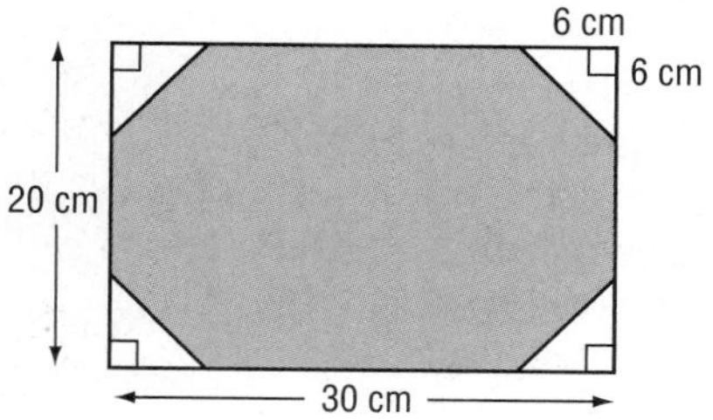

A 456 cm^2 C 552 cm^2
B 528 cm^2 D 564 cm^2

42. GRIDDED RESPONSE Triangle ABC is congruent to $\triangle HIJ$. The vertices of $\triangle ABC$ are $A(-1, 2)$, $B(0, 3)$ and $C(2, -2)$. What is the measure of side $\overline{HJ}$?

43. ALGEBRA Which is a factor of $x^2 + 19x - 42$?

F $x + 14$ H $x - 2$
G $x + 2$ J $x - 14$

44. SAT/ACT Mitsu travels a certain distance at 30 miles per hour and returns the same route at 65 miles per hour. What is his average speed in miles per hour for the round trip?

A 32.5 D 47.5
B 35.0 E 55.3
C 41.0

Spiral Review

Find each measure in the triangle at the right. (Lesson 12-2)

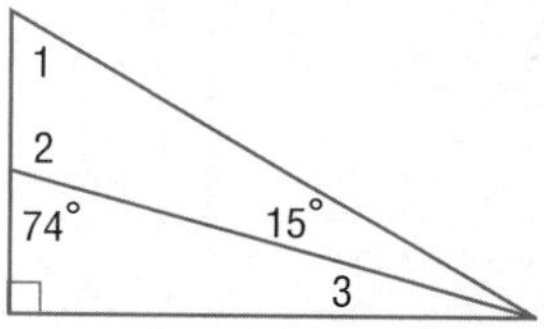

45. $m\angle 2$ **46.** $m\angle 1$ **47.** $m\angle 3$

COORDINATE GEOMETRY **Find the measures of the sides of $\triangle JKL$ and classify each triangle by the measures of its sides.** (Lesson 12-1)

48. $J(-7, 10)$, $K(15, 0)$, $L(-2, -1)$

49. $J(9, 9)$, $K(12, 14)$, $L(14, 6)$

50. $J(4, 6)$, $K(4, 11)$, $L(9, 6)$

51. $J(16, 14)$, $K(7, 6)$, $L(-5, -14)$

Skills Review

52. Copy and complete the proof.

Given: $\overline{MN} \cong \overline{PQ}$, $\overline{PQ} \cong \overline{RS}$
Prove: $\overline{MN} \cong \overline{RS}$
Proof:

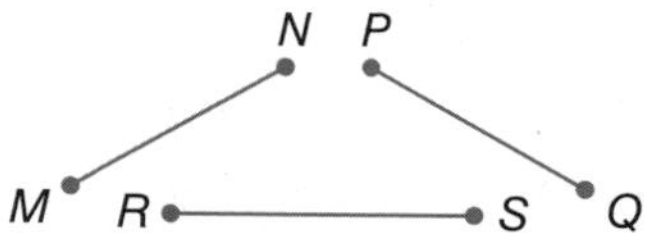

Statements	Reasons
a. ___?___	**a.** Given
b. $MN = PQ$, $PQ = RS$	**b.** ___?___
c. ___?___	**c.** ___?___
d. $\overline{MN} \cong \overline{RS}$	**d.** Definition of congruent segments

LESSON 12-4

Proving Triangles Congruent–SSS, SAS

Then

- You proved triangles congruent using the definition of congruence.

Now

1. Use the SSS Postulate to test for triangle congruence.
2. Use the SAS Postulate to test for triangle congruence.

Why?

- An A-frame sandwich board is a convenient way to display information. Not only does it fold flat for easy storage, but with each sidearm locked into place, the frame is extremely sturdy. With the sidearms the same length and positioned the same distance from the top on either side, the open frame forms two congruent triangles.

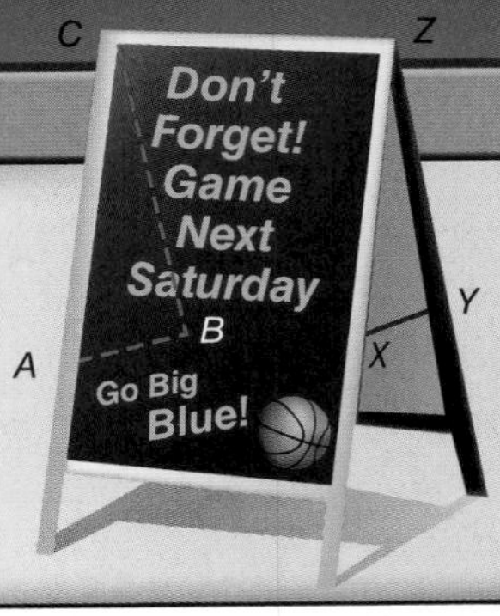

NewVocabulary
included angle

Common Core State Standards

Content Standards
G.CO.10 Prove theorems about triangles.
G.SRT.5 Use congruence and similarity criteria for triangles to solve problems and to prove relationships in geometric figures.

Mathematical Practices
3 Construct viable arguments and critique the reasoning of others.
1 Make sense of problems and persevere in solving them.

1 SSS Postulate

In Lesson 12-3, you proved that two triangles were congruent by showing that all six pairs of corresponding parts were congruent. It is possible to prove two triangles congruent using fewer pairs.

The sandwich board demonstrates that if two triangles have the same three side lengths, then they are congruent. This is expressed in the postulate below.

Postulate 12.1 Side-Side-Side (SSS) Congruence

If three sides of one triangle are congruent to three sides of a second triangle, then the triangles are congruent.

Example If Side $\overline{AB} \cong \overline{DE}$,
Side $\overline{BC} \cong \overline{EF}$, and
Side $\overline{AC} \cong \overline{DF}$,
then $\triangle ABC \cong \triangle DEF$.

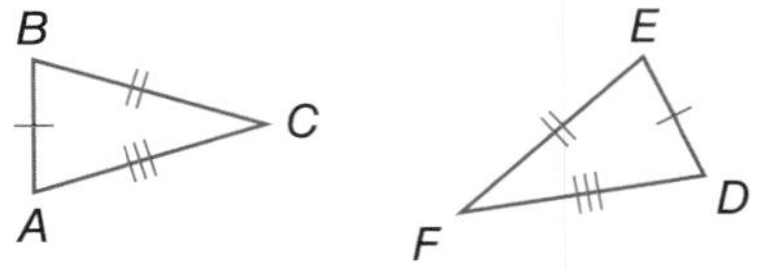

PT

Example 1 Use SSS to Prove Triangles Congruent

Write a flow proof.

Given: $\overline{GH} \cong \overline{KJ}$, $\overline{HL} \cong \overline{JL}$, and L is the midpoint of $\overline{GK}$.

Prove: $\triangle GHL \cong \triangle KJL$

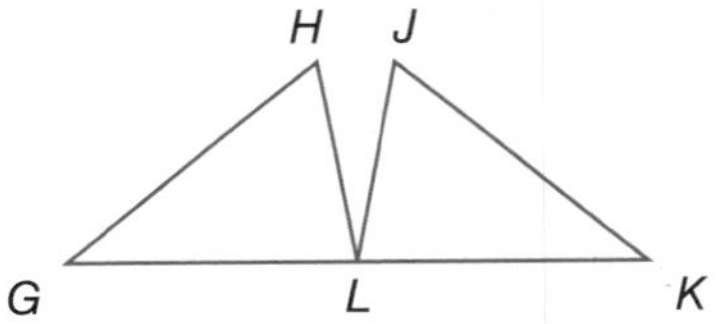

Flow Proof:

- $\overline{GH} \cong \overline{KJ}$ — Given
- $\overline{HL} \cong \overline{JL}$ — Given
- L is the midpoint of $\overline{GK}$. — Given → $\overline{GL} \cong \overline{KL}$ — Midpoint Theorem
- → $\triangle GHL \cong \triangle KJL$ — SSS

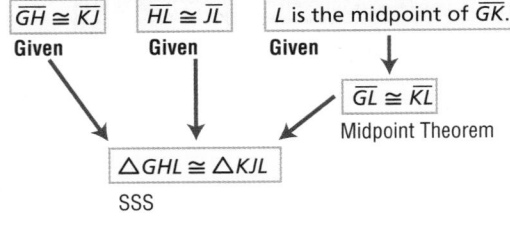

GuidedPractice

1. Write a flow proof.

Given: $\triangle QRS$ is isosceles with $\overline{QR} \cong \overline{SR}$. $\overline{RT}$ bisects $\overline{QS}$ at point T.

Prove: $\triangle QRT \cong \triangle SRT$

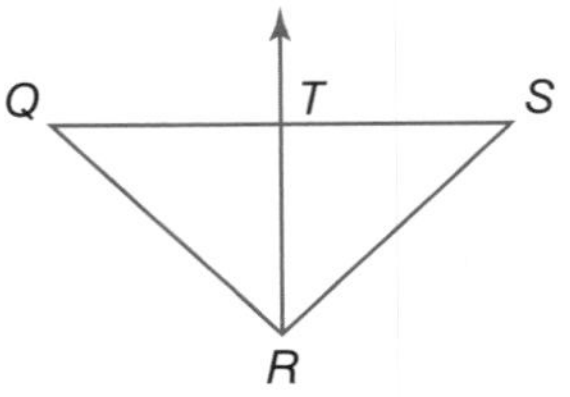

Standardized Test Example 2 SSS on the Coordinate Plane

PT

EXTENDED RESPONSE **Triangle *ABC* has vertices *A*(1, 1), *B*(0, 3), and *C*(2, 5). Triangle *EFG* has vertices *E*(1, −1), *F*(2, −5), and *G*(4, −4).**

a. **Graph both triangles on the same coordinate plane.**

b. **Use your graph to make a conjecture as to whether the triangles are congruent. Explain your reasoning.**

c. **Write a logical argument using coordinate geometry to support the conjecture you made in part b.**

Read the Test Item

You are asked to do three things in this problem. In part **a**, you are to graph $\triangle ABC$ and $\triangle EFG$ on the same coordinate plane. In part **b**, you should make a conjecture that $\triangle ABC \cong \triangle EFG$ or $\triangle ABC \ncong \triangle EFG$ based on your graph. Finally, in part **c**, you are asked to prove your conjecture.

Test-Taking Tip

CCSS Tools When you are solving problems using the coordinate plane, remember to use tools like the Distance, Midpoint, and Slope Formulas to solve problems and to check your solutions.

Solve the Test Item

a.

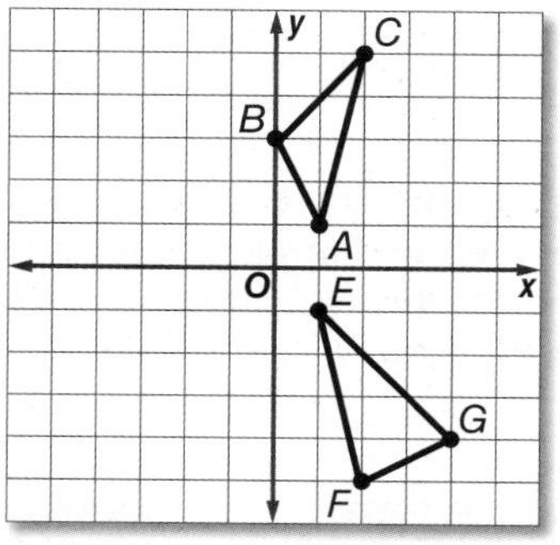

b. From the graph, it appears that the triangles do not have the same shape, so we can conjecture that they are not congruent.

c. Use the Distance Formula to show that not all corresponding sides have the same measure.

$AB = \sqrt{(0-1)^2 + (3-1)^2}$

$= \sqrt{1+4}$ or $\sqrt{5}$

$EF = \sqrt{(2-1)^2 + [-5-(-1)]^2}$

$= \sqrt{1+16}$ or $\sqrt{17}$

$BC = \sqrt{(2-0)^2 + (5-3)^2}$

$= \sqrt{4+4}$ or $\sqrt{8}$

$FG = \sqrt{(4-2)^2 + [-4-(-5)]^2}$

$= \sqrt{4+1}$ or $\sqrt{5}$

$AC = \sqrt{(2-1)^2 + (5-1)^2}$

$= \sqrt{1+16}$ or $\sqrt{17}$

$EG = \sqrt{(4-1)^2 + [-4-(-1)]^2}$

$= \sqrt{9+9}$ or $\sqrt{18}$

While $AB = FG$ and $AC = EF$, $BC \neq EG$. Since SSS congruence is not met, $\triangle ABC \ncong \triangle EFG$.

Reading Math

Symbols $\triangle ABC \ncong \triangle EFG$ is read as *triangle ABC is not congruent to triangle EFG.*

Guided Practice

2. Triangle *JKL* has vertices *J*(2, 5), *K*(1, 1), and *L*(5, 2). Triangle *NPQ* has vertices *N*(−3, 0), *P*(−7, 1), and *Q*(−4, 4).

a. Graph both triangles on the same coordinate plane.

b. Use your graph to make a conjecture as to whether the triangles are congruent. Explain your reasoning.

c. Write a logical argument using coordinate geometry to support the conjecture you made in part **b**.

Construction Congruent Triangles Using Sides

Draw a triangle and label it $\triangle ABC$. Then use the SSS Postulate to construct $\triangle XYZ \cong \triangle ABC$.

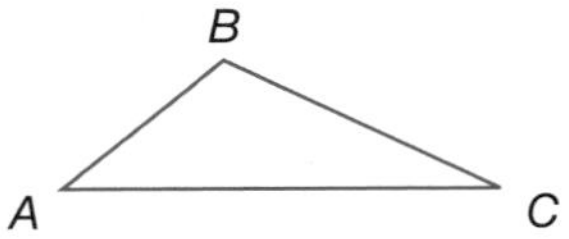

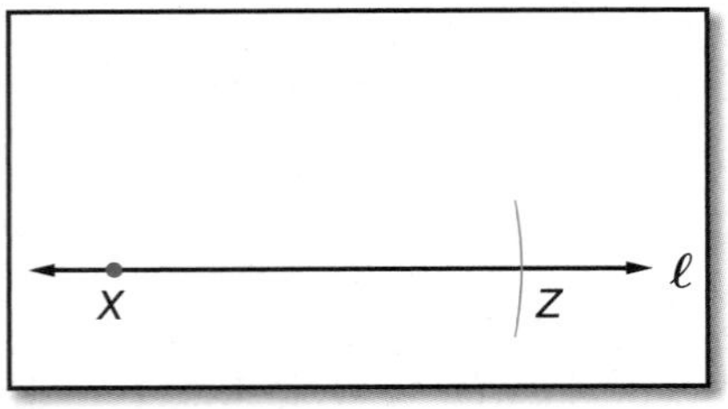

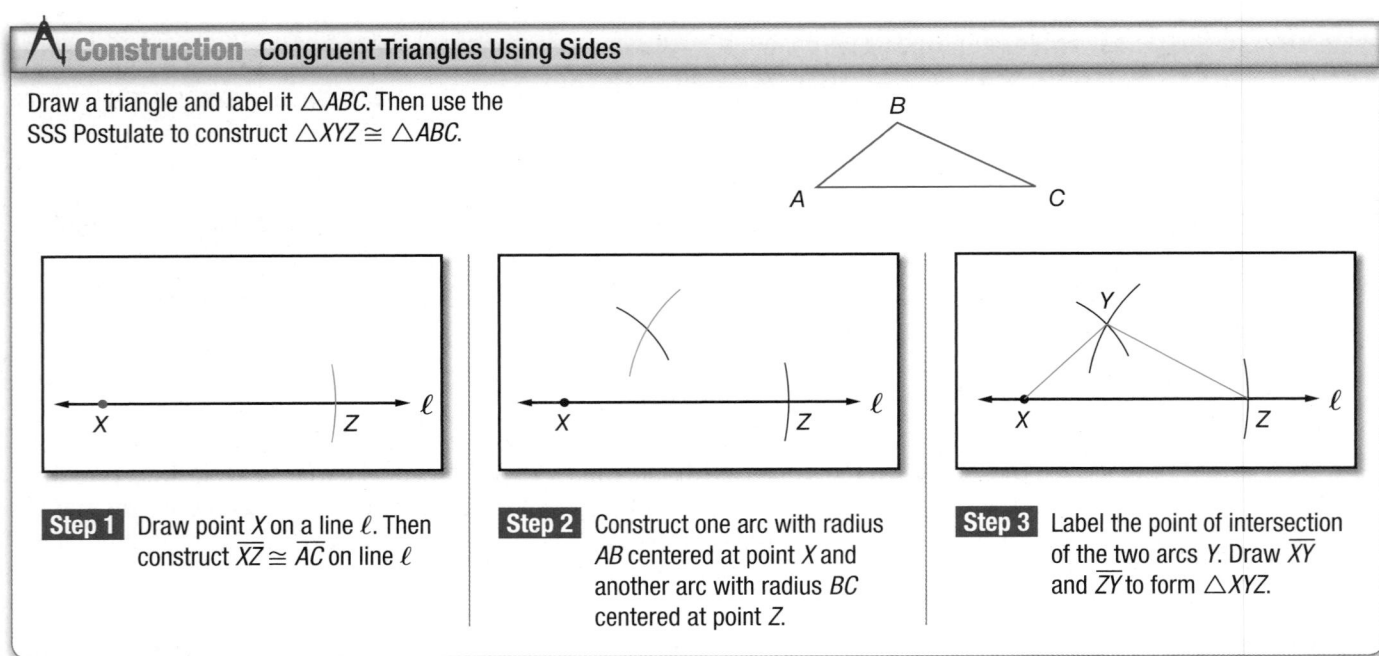

Step 1 Draw point X on a line ℓ. Then construct $\overline{XZ} \cong \overline{AC}$ on line ℓ

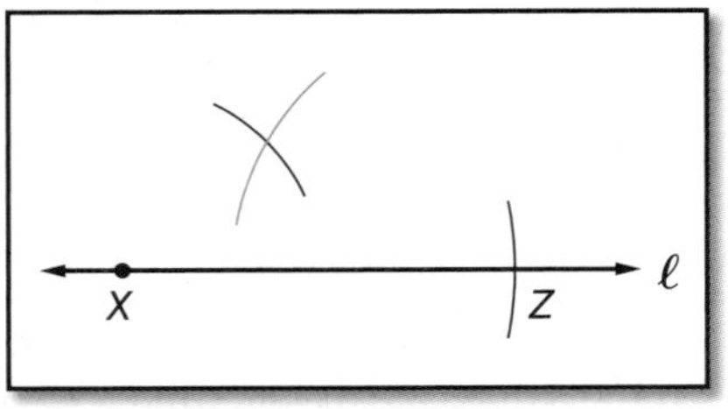

Step 2 Construct one arc with radius AB centered at point X and another arc with radius BC centered at point Z.

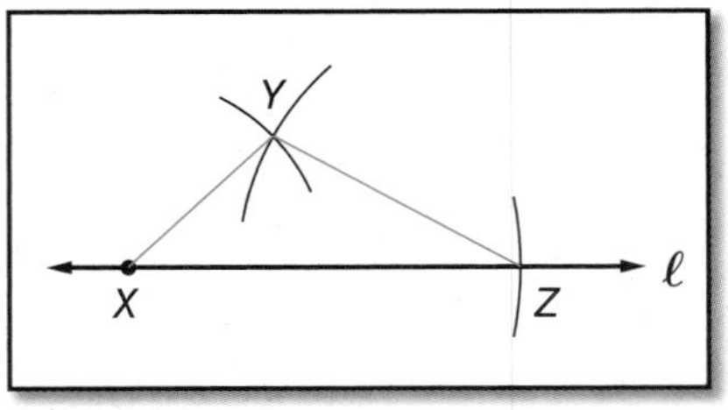

Step 3 Label the point of intersection of the two arcs Y. Draw $\overline{XY}$ and $\overline{ZY}$ to form $\triangle XYZ$.

2 SAS Postulate

The angle formed by two adjacent sides of a polygon is called an **included angle**. Consider included angle JKL formed by the hands on the first clock shown below. Any time the hands form an angle with the same measure, the distance between the ends of the hands $\overline{JL}$ and $\overline{PR}$ will be the same.

$\triangle PKR \cong \triangle JKL$

Any two triangles formed using the same side lengths and included angle measure will be congruent. This illustrates the following postulate.

StudyTip

Side-Side-Angle The measures of two sides and a nonincluded angle are not sufficient to prove two triangles congruent.

Postulate 12.2 Side-Angle-Side (SAS) Congruence

Words If two sides and the included angle of one triangle are congruent to two sides and the included angle of a second triangle, then the triangles are congruent.

Example If Side $\overline{AB} \cong \overline{DE}$,
Angle $\angle B \cong \angle E$, and
Side $\overline{BC} \cong \overline{EF}$,
then $\triangle ABC \cong \triangle DEF$.

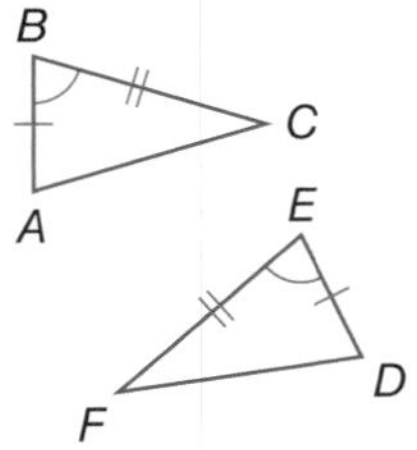

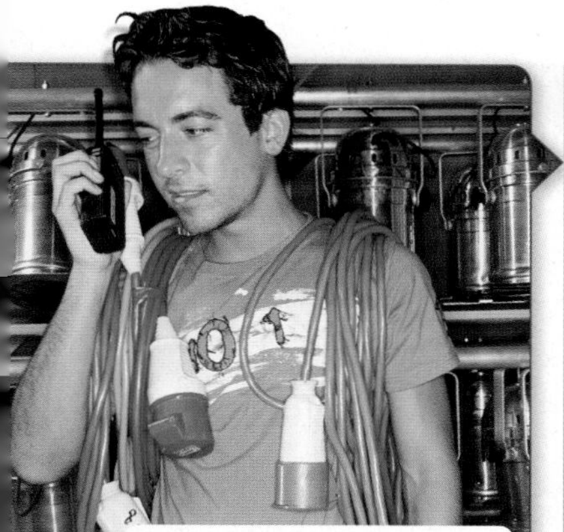

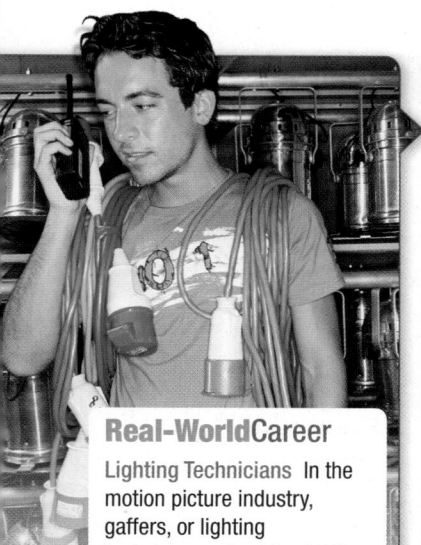

Real-World Example 3 Use SAS to Prove Triangles are Congruent

LIGHTING The scaffolding for stage lighting shown appears to be made up of congruent triangles. If $\overline{WX} \cong \overline{YZ}$ and $\overline{WX} \parallel \overline{ZY}$, write a two-column proof to prove that $\triangle WXZ \cong \triangle YZX$.

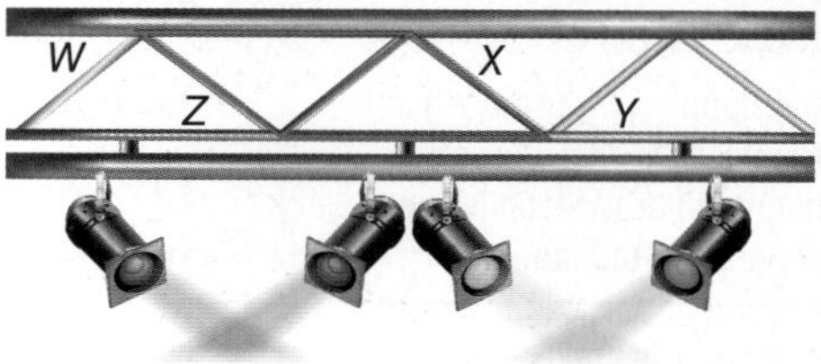

Proof:

Statements	Reasons
1. $\overline{WX} \cong \overline{YZ}$	**1.** Given
2. $\overline{WX} \parallel \overline{ZY}$	**2.** Given
3. $\angle WXZ \cong \angle XZY$	**3.** Alternate Interior Angle Theorem
4. $\overline{XZ} \cong \overline{ZX}$	**4.** Reflexive Property of Congruence
5. $\triangle WXZ \cong \triangle YZX$	**5.** SAS

GuidedPractice

3. EXTREME SPORTS The wings of the hang glider shown appear to be congruent triangles. If $\overline{FG} \cong \overline{GH}$ and $\overline{JG}$ bisects $\angle FGH$, prove that $\triangle FGJ \cong \triangle HGJ$.

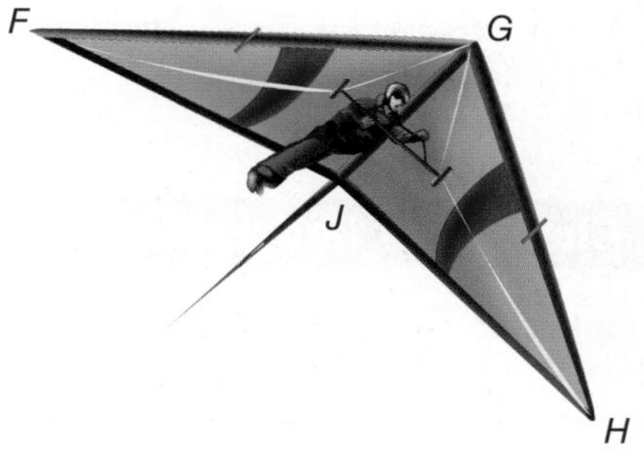

Real-WorldCareer

Lighting Technicians In the motion picture industry, gaffers, or lighting technicians, place the lighting required for a film. Gaffers make sure the angles the lights form are in the correct positions. They may have college or technical school degrees, or they may have completed a formal training program.

You can also construct congruent triangles given two sides and the included angle.

Construction Congruent Triangles Using Two Sides and the Included Angle

Draw a triangle and label it $\triangle ABC$. Then use the SAS Postulate to construct $\triangle RST \cong \triangle ABC$.

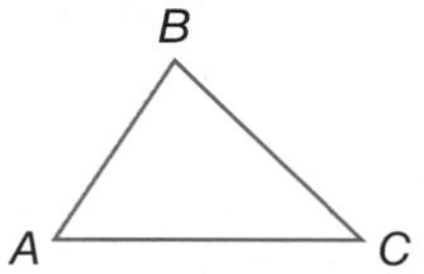

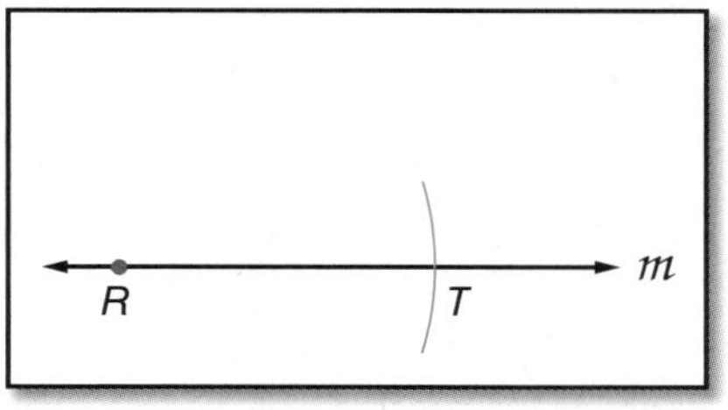

Step 1 Draw point R on a line m. Then construct $\overline{RT} \cong \overline{AC}$ on line m.

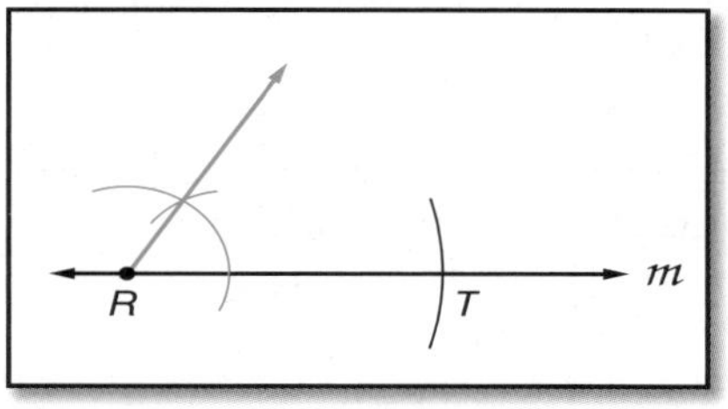

Step 2 Construct $\angle R \cong \angle A$ using $\overline{RT}$ as a side of the angle and point R.

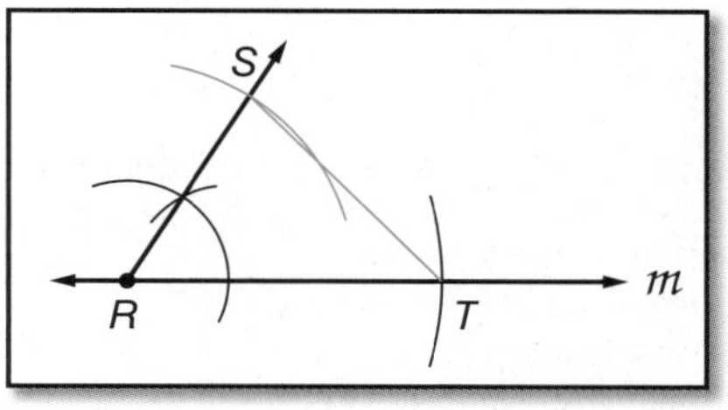

Step 3 Construct $\overline{RS} \cong \overline{AB}$. Then draw $\overline{ST}$ to form $\triangle RST$.

Example 4 Use SAS or SSS in Proofs

Write a paragraph proof.

Given: $\overline{BC} \cong \overline{DC}$, $\angle BCF \cong \angle DCE$, $\overline{FC} \cong \overline{EC}$

Prove: $\angle CFD \cong \angle CEB$

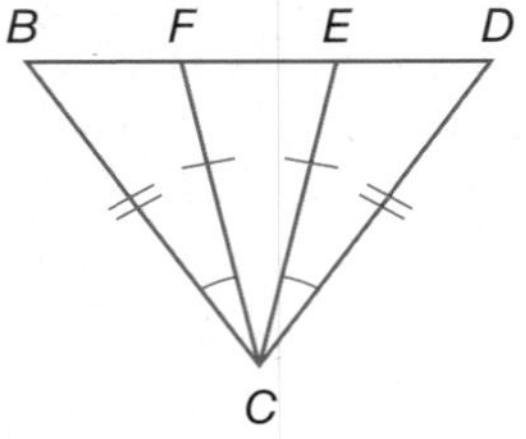

Proof:

Since $\overline{BC} \cong \overline{DC}$, $\angle BCF \cong \angle DCE$, and $\overline{FC} \cong \overline{EC}$, then $\triangle BCF \cong \triangle DCE$ by SAS. By CPCTC, $\angle CFB \cong \angle CED$. $\angle CFD$ forms a linear pair with $\angle CFB$, and $\angle CEB$ forms a linear pair with $\angle CED$. By the Congruent Supplements Theorem, $\angle CFD$ is supplementary to $\angle CFB$ and $\angle CEB$ is supplementary to $\angle CED$. Since angles supplementary to the same angle or congruent angles are congruent, $\angle CFD \cong \angle CEB$.

StudyTip

Overlapping Figures When triangles overlap, it can be helpful to draw each triangle separately and label the congruent parts. In Example 4, the figure could have been separated as shown.

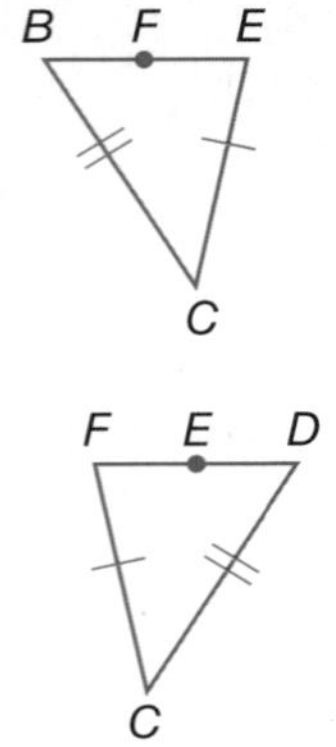

GuidedPractice

4. Write a two-column proof.

Given: $\overline{MN} \cong \overline{PN}$, $\overline{LM} \cong \overline{LP}$

Prove: $\angle LNM \cong \angle LNP$

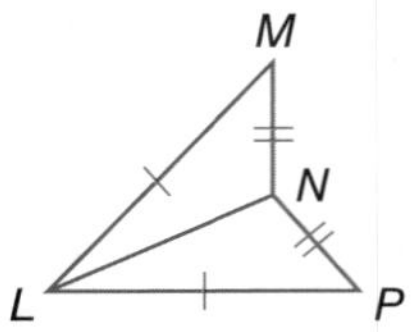

Check Your Understanding

Example 1

1. ARCHITECTURE Triangles are widely used in architecture because they are "rigid" figures. How does the property of triangle congruence explain this property? Other than roofs, give at least one example of triangle congruence in your home.

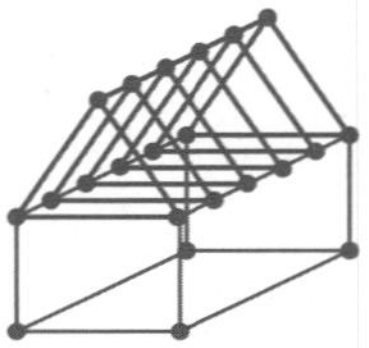

Example 2

2. EXTENDED RESPONSE Triangle ABC has vertices $A(-4, 1)$, $B(-1, 1)$, and $C(-1, 5)$. Triangle XYZ has vertices $X(4, -1)$, $Y(1, -1)$, and $Z(1, -5)$.

a. Graph both triangles on the same coordinate plane.

b. Use your graph to make a conjecture as to whether the triangles are congruent. Explain your reasoning.

c. Write a logical argument using coordinate geometry to support your conjecture.

Example 3

3. In the diagram, $\triangle TQR$ is equilateral, $\angle RSQ \cong \angle UTQ$; $\overline{SR} \cong \overline{TU}$. Write a paragraph proof to show that $\triangle RSQ \cong \triangle UTQ$.

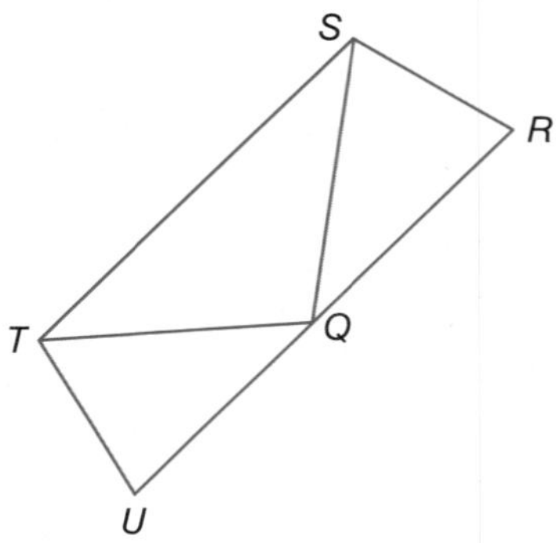

Example 4

4. Write a two-column proof.

Given: $\overline{JK} \cong \overline{LM}$; $\angle KJL \cong \angle MLJ$

Prove: $\overline{JM} \cong \overline{LK}$

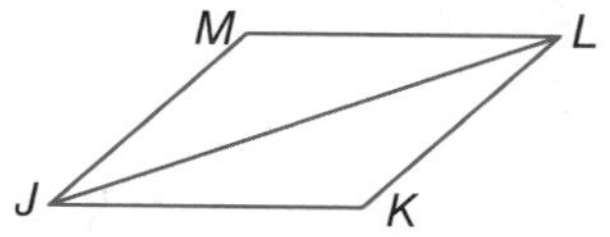

Practice and Problem Solving

Example 1

PROOF Write the specified type of proof.

5. paragraph proof

Given: $\overline{XY} \cong \overline{ZW}$; $\overline{XW} \cong \overline{ZY}$

Prove: $\triangle XYZ \cong \triangle ZWX$

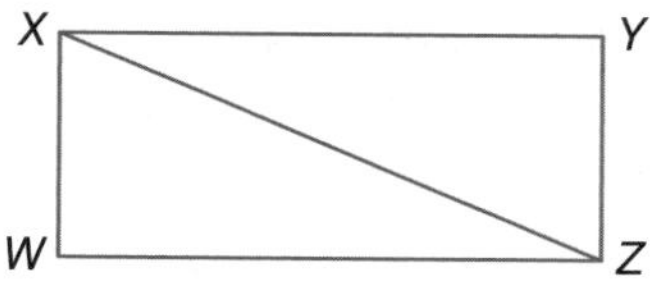

6. two-column proof

Given: C is the midpoint of both $\overline{BE}$ and $\overline{AD}$.

Prove: $\triangle ABC \cong \triangle DCE$

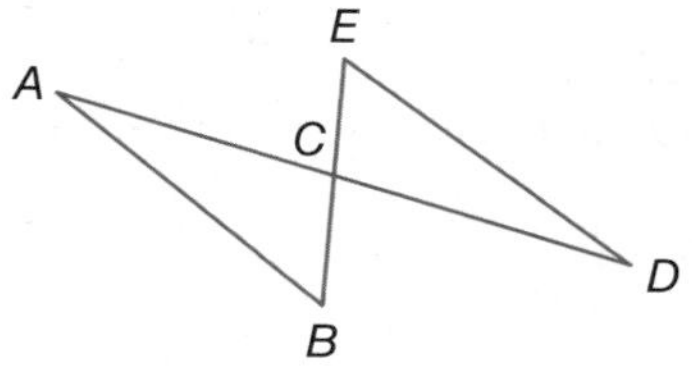

7. BRIDGES The suspension bridge below is in Yicang, China, Hubei province. It is supported using steel cables suspended from concrete supports. If the supports are the same height above the roadway and perpendicular to the roadway, and the topmost cables meet at a point midway between the supports, prove that the two triangles shown in the photo are congruent.

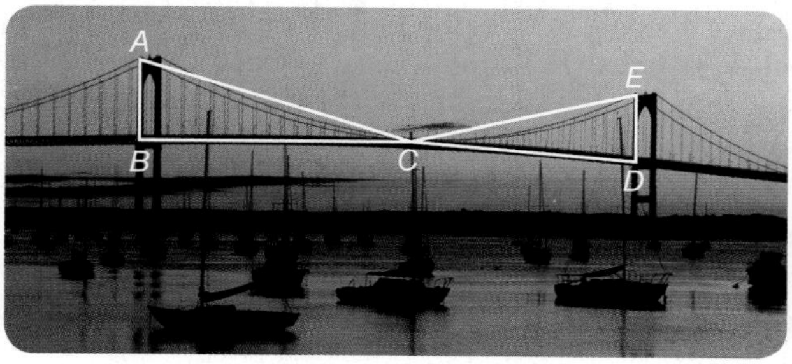

Example 2

CCSS SENSE-MAKING Determine whether $\triangle MNO \cong \triangle QRS$. Explain.

8. $M(2, 5)$, $N(5, 2)$, $O(1, 1)$, $Q(-4, -4)$, $R(-7, -1)$, $S(-3, 0)$

9. $M(0, -1)$, $N(-1, -4)$, $O(-4, -3)$, $Q(-3, 3)$, $R(-4, 4)$, $S(-3, 7)$

10. $M(0, -3)$, $N(0, 2)$, $O(-3, 1)$, $Q(4, -1)$, $R(6, 1)$, $S(9, -1)$

11. $M(4, 7)$, $N(5, 4)$, $O(2, 3)$, $Q(2, 3)$, $R(3, 0)$, $S(0, -1)$

Example 3

PROOF Write the specified type of proof.

12. two-column proof

Given: $\overline{KG}$ is the perpendicular bisector of $\overline{FH}$

Prove: $\triangle KGH \cong \triangle KGF$

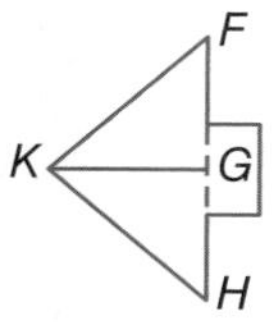

13. paragraph proof

Given: Rectangle $ABDE$; C is the midpoint of $\overline{BD}$

Prove: $\triangle ABC \cong \triangle EDC$

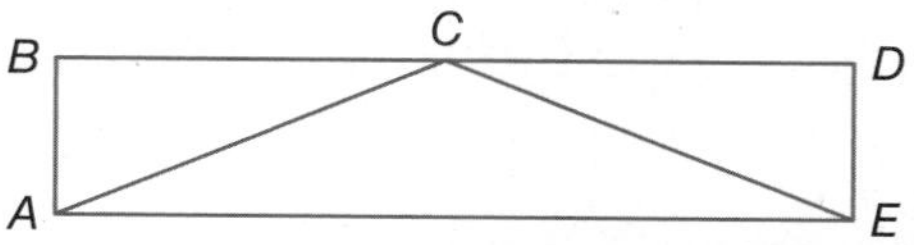

Example 4 **PROOF** **Write the specified type of proof.**

14. two-column proof

Given: K is the midpoint of $\overline{JL}$; P is the midpoint of $\overline{JN}$; M is the midpoint of $\overline{NL}$; $\triangle JLN$ is equilateral

Prove: $\triangle NPM \cong \triangle LKM$

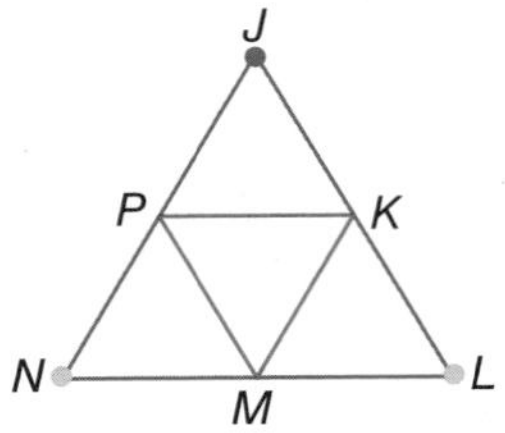

15. paragraph proof

Given: $\overline{AB}$ and $\overline{WP}$ bisect each other

Prove: $\angle A \cong \angle B$

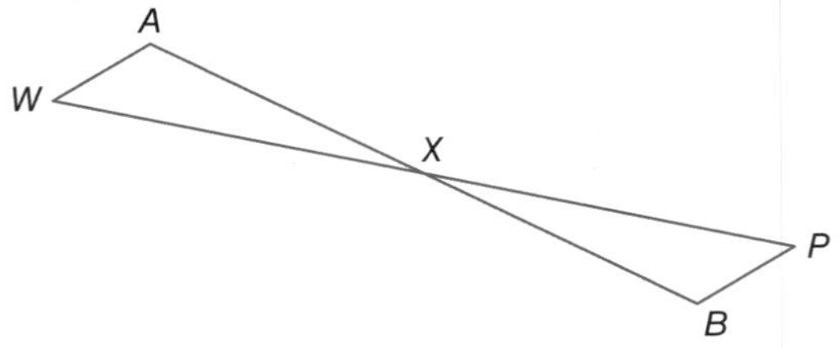

CCSS ARGUMENTS **Determine which postulate can be used to prove that the triangles are congruent. If it is not possible to prove congruence, write *not possible*.**

16.

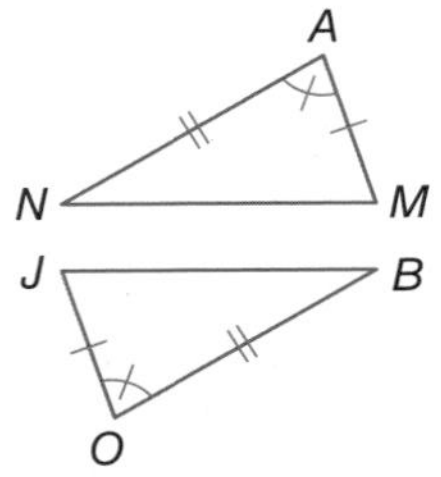

17.

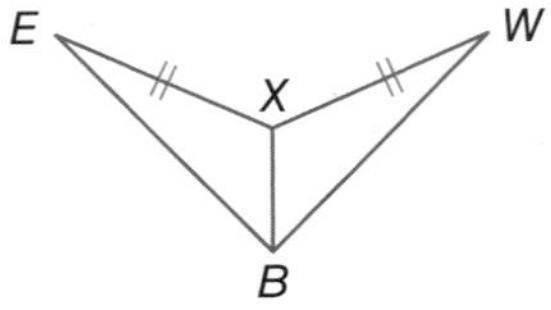

18.

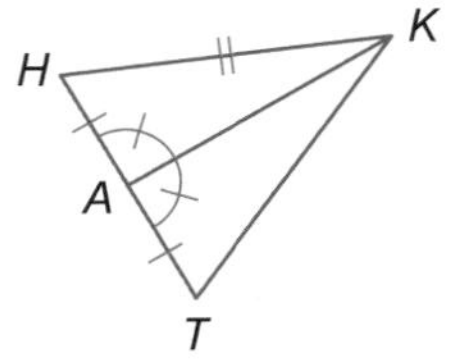

19.

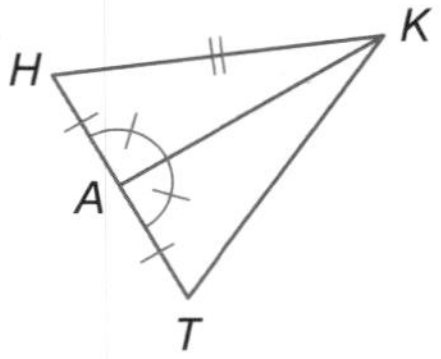

20. **MUSIC** To make a specific tempo, the weight on the pendulum of a metronome is adjusted so that it swings at a specific rate. Prove that the triangles formed by the swinging of the pendulum are congruent; i.e. prove $\triangle ABR \cong \triangle CBR$.

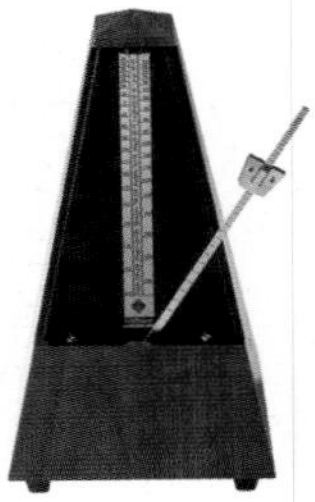

PROOF **Write a flow proof.**

21. **Given:** $\overline{XB}$ bisects $\angle EBW$; $\overline{EB} \cong \overline{WB}$

Prove: $\angle E \cong \angle W$

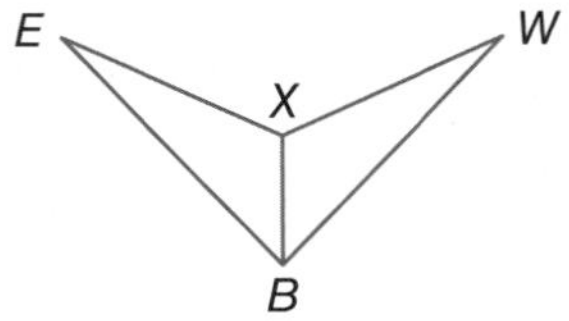

22. **Given:** Isosceles trapezoid $PQRS$

Prove: $\triangle PQR \cong \triangle SRQ$

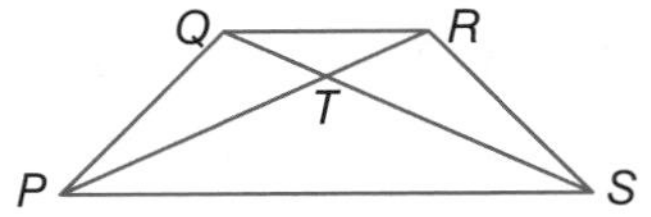

23. **BASEBALL** Use the diagram of a baseball field shown.

a. Write a two column proof to prove that the distance from first base to third base is the same as the distance from home plate to second base.

b. Write a two-column proof to prove that the angle formed between second base, home plate, and third base is the same as the angle formed between second base, home plate, and first base.

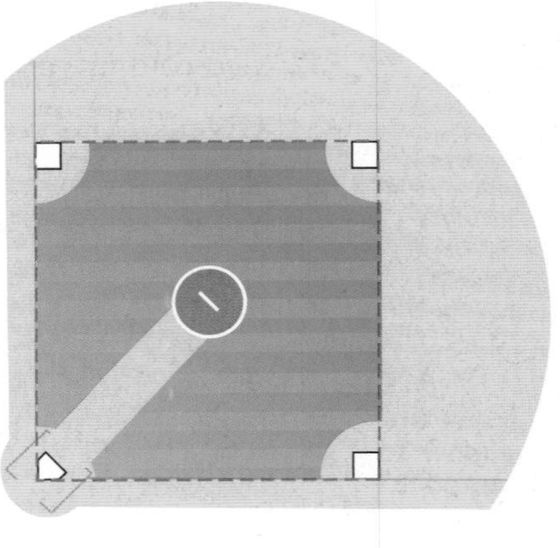

24. **Given:** $\overline{XW} \cong \overline{ZW}$, $\overline{XY} \cong \overline{ZY}$

Prove: $\angle X \cong \angle Z$

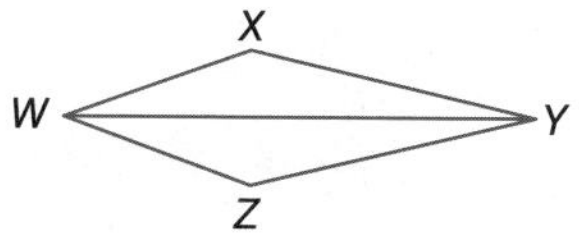

25. **Given:** $\triangle EAB \cong \triangle DCB$

Prove: $\triangle ADE \cong \triangle CED$

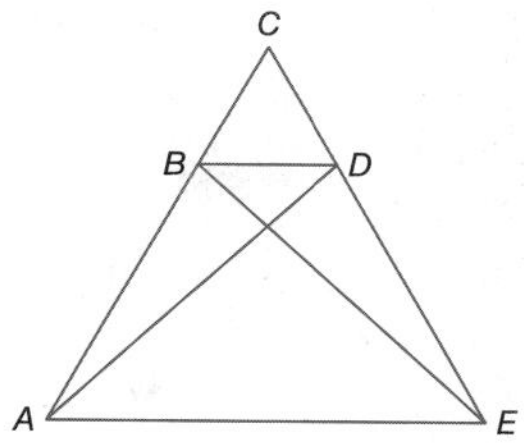

26. CCSS **ARGUMENTS** Write a paragraph proof.

Given: $\overline{BF} \cong \overline{DF}$; $\overline{FE} \cong \overline{FA}$;
$\overline{AB} \cong \overline{ED}$

Prove: $\triangle ABE \cong \triangle EDA$

ALGEBRA **Using CPCTC, find the value of the variables that yields congruent triangles.**

27. $\triangle WXY \cong \triangle WXZ$

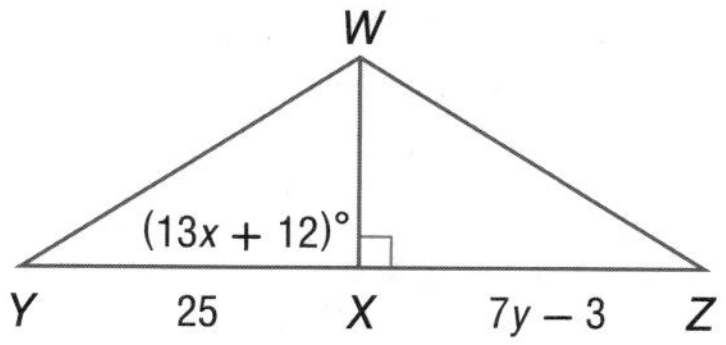

28. $\triangle ABC \cong \triangle FGH$

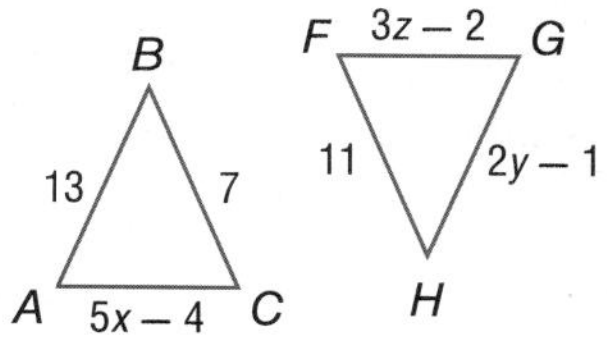

H.O.T. Problems Use Higher-Order Thinking Skills

29. **CHALLENGE** **Refer to the graph shown.**

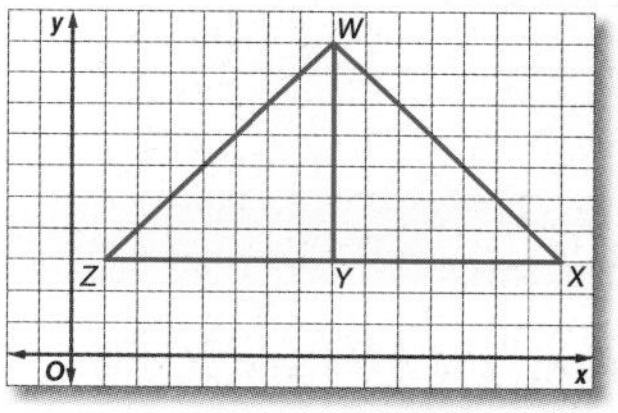

a. Describe two methods you could use to prove that $\triangle WYZ$ is congruent to $\triangle WYX$. You may not use a ruler or a protractor. Which method do you think is more efficient? Explain.

b. Are $\triangle WYZ$ and $\triangle WYX$ congruent? Explain your reasoning.

30. **REASONING** Determine whether the following statement is *true* or *false*. If true, explain your reasoning. If *false*, provide a counterexample.

If the base angles in one isosceles triangle have the same measure as the base angles in another isosceles triangle, then the triangles are congruent.

31. **ERROR ANALYSIS** Bonnie says that $\triangle ABC \cong \triangle CAD$ by SSS. Shada disagrees. She says that they are congruent by SAS. Is either of them correct? Explain.

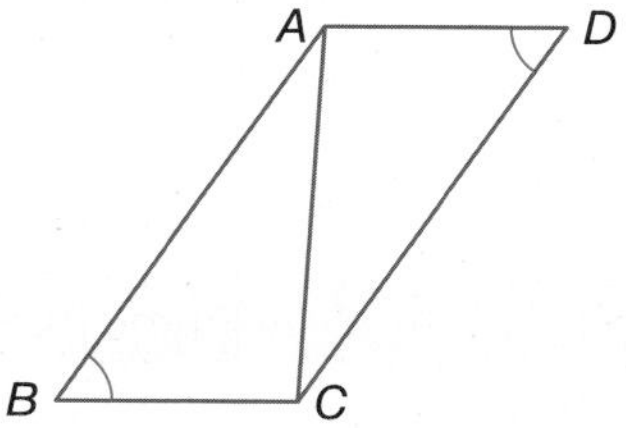

32. **OPEN ENDED** Use a straightedge to draw obtuse triangle ABC. Then construct $\triangle XYZ$ so that it is congruent to $\triangle ABC$ using either SSS or SAS. Justify your construction mathematically and verify it using measurement.

33. **WRITING IN MATH** Determine whether the following statement is always, sometimes, or never true. Explain your reasoning.
If two pairs of corresponding sides of two right triangles are congruent, then the triangles are congruent.

Standardized Test Practice

34. ALGEBRA The Ross Family drove 300 miles to visit their grandparents. Mrs. Ross drove 70 miles per hour for 65% of the trip and 35 miles per hour or less for 20% of the trip that was left. Assuming that Mrs. Ross never went over 70 miles per hour, how many miles did she travel at a speed between 35 and 70 miles per hour?

A 195 **C** 21

B 84 **D** 18

35. In the figure, $\angle C \cong \angle Z$ and $\overline{AC} \cong \overline{XZ}$.

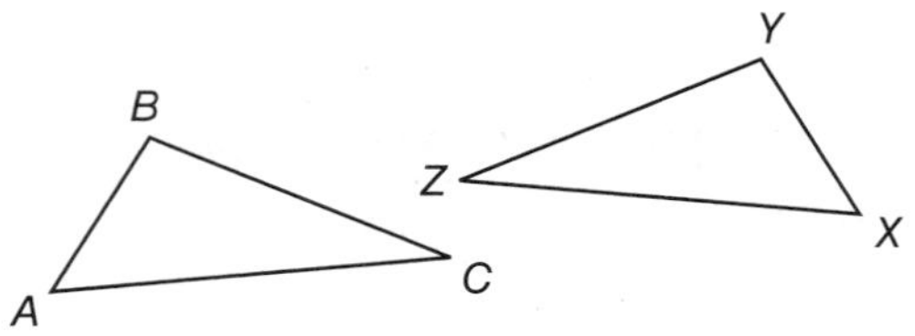

What additional information could be used to prove that $\triangle ABC \cong \triangle XYZ$?

F $\overline{BC} \cong \overline{YZ}$

G $\overline{AB} \cong \overline{XY}$

H $\overline{BC} \cong \overline{XZ}$

J $\overline{XZ} \cong \overline{XY}$

36. EXTENDED RESPONSE The graph below shows the eye colors of all of the students in a class. What is the probability that a student chosen at random from this class will have blue eyes? Explain your reasoning.

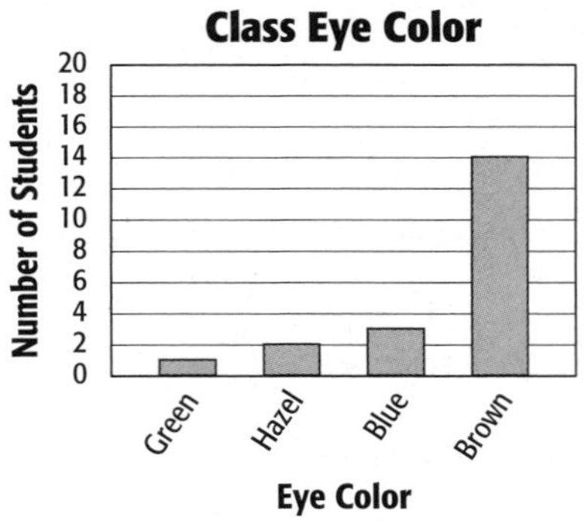

37. SAT/ACT If $4a + 6b = 6$ and $-2a + b = -7$, what is the value of a?

A -2

B -1

C 2

D 3

E 4

Spiral Review

In the diagram, $\triangle LMN \cong \triangle QRS$. (Lesson 12-3)

38. Find x.

39. Find y.

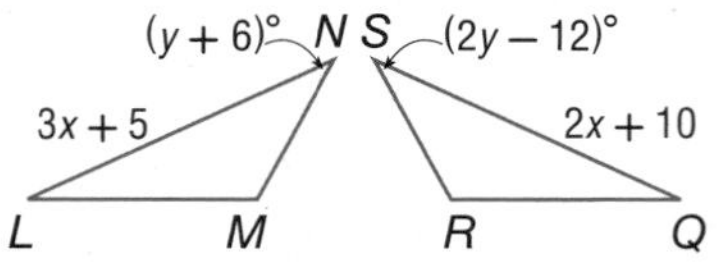

40. ASTRONOMY The Big Dipper is a part of the larger constellation Ursa Major. Three of the brighter stars in the constellation form $\triangle RSA$. If $m\angle R = 41$ and $m\angle S = 109$, find $m\angle A$. (Lesson 12-2)

Write an equation in slope-intercept form for each line. (Lesson 11-4)

41. $(-5, -3)$ and $(10, -6)$

42. $(4, -1)$ and $(-2, -1)$

43. $(-4, -1)$ and $(-8, -5)$

Skills Review

State the property that justifies each statement.

44. $AB = AB$

45. If $EF = GH$ and $GH = JK$, then $EF = JK$.

46. If $a^2 = b^2 - c^2$, then $b^2 - c^2 = a^2$.

47. If $XY + 20 = YW$ and $XY + 20 = DT$, then $YW = DT$.

EXTEND 12-4

Geometry Lab
Proving Constructions

When you perform a construction using a straightedge and compass, you assume that segments constructed using the same compass setting are congruent. You can use this information, along with definitions, postulates, and theorems to prove constructions.

CCSS Common Core State Standards
Content Standards
G.CO.12 Make formal geometric constructions with a variety of tools and methods (compass and straightedge, string, reflective devices, paper folding, dynamic geometric software, etc.).
G.SRT.5 Use congruence and similarity criteria for triangles to solve problems and to prove relationships in geometric figures.
Mathematical Practices 3, 5

Activity

Follow the steps below to bisect an angle. Then prove the construction.

Step 1

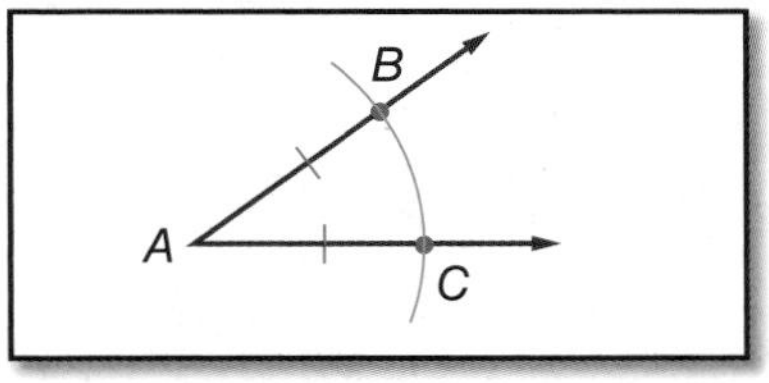

Draw any angle with vertex A. Place the compass point at A and draw an arc that intersects both sides of $\angle A$. Label the points B and C. Mark the congruent segments.

Step 2

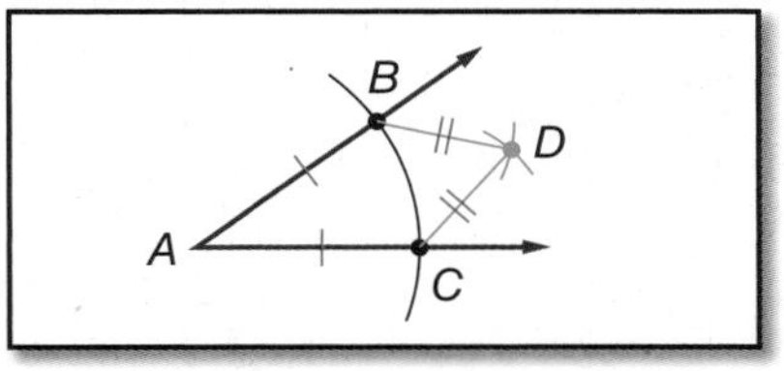

With the compass point at B, draw an arc in the interior of $\angle A$. With the same radius, draw an arc from C intersecting the first arc at D. Draw the segments $\overline{BD}$ and $\overline{CD}$. Mark the congruent segments.

Step 3

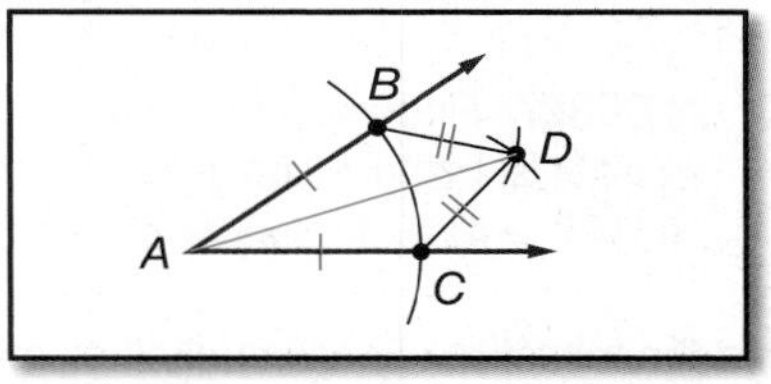

Draw $\overline{AD}$.

Given: Description of steps and diagram of construction

Prove: $\overline{AD}$ bisects $\angle BAC$.

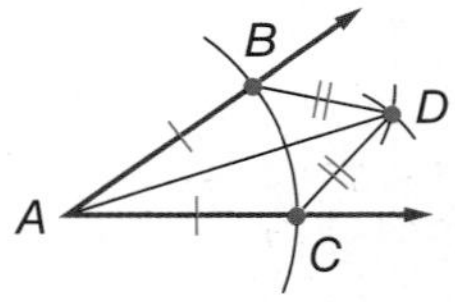

Proof:

Statements	Reasons
1. $\overline{AB} \cong \overline{AC}$	1. The same compass setting was used from point A to construct points B and C.
2. $\overline{BD} \cong \overline{CD}$	2. The same compass setting was used from points B and C to construct point D.
3. $\overline{AD} \cong \overline{AD}$	3. Reflexive Property
4. $\triangle ABD \cong \triangle ACD$	4. SSS Postulate
5. $\angle BAD \cong \angle CAD$	5. CPCTC
6. $\overline{AD}$ bisects $\angle BAC$.	6. Definition of angle bisector

Exercises

1. Construct a line parallel to a given line through a given point. Write a two-column proof of your construction.

2. Construct an equilateral triangle. Write a paragraph proof of your construction.

3. **CHALLENGE** Construct the bisector of a segment that is also perpendicular to the segment and write a two-column proof of your construction. (*Hint:* You will need to use more than one pair of congruent triangles.).

Mid-Chapter Quiz

Lessons 12-1 through 12-4

1. **COORDINATE GEOMETRY** Classify $\triangle ABC$ with vertices $A(-2, -1)$, $B(-1, 3)$, and $C(2, 0)$ as *scalene*, *equilateral*, or *isosceles*. (Lesson 12-1)

2. **MULTIPLE CHOICE** Which of the following are the measures of the sides of isosceles triangle QRS? (Lesson 12-1)

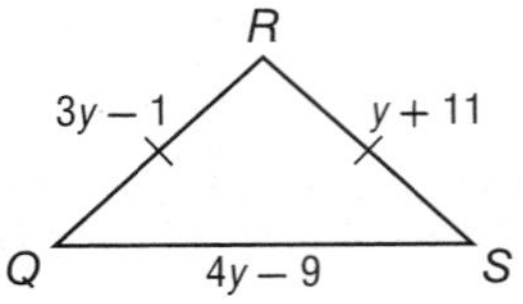

A 17, 17, 15 **C** 14, 15, 14

B 15, 15, 16 **D** 14, 14, 16

3. **ALGEBRA** Find x and the length of each side if $\triangle WXY$ is an equilateral triangle with sides $\overline{WX} = 6x - 12$, $\overline{XY} = 2x + 10$, and $\overline{WY} = 4x - 1$. (Lesson 12-1)

Find the measure of each angle indicated. (Lesson 12-2)

4. $m\angle 1$
5. $m\angle 2$
6. $m\angle 3$

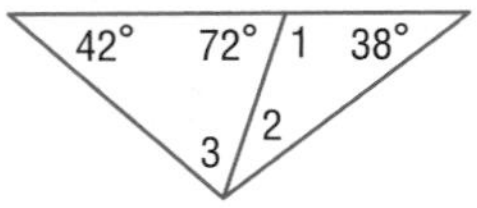

7. **ASTRONOMY** Leo is a constellation that represents a lion. Three of the brighter stars in the constellation form $\triangle LEO$. If the angles have measures as shown in the figure, find $m\angle OLE$. (Lesson 12-2)

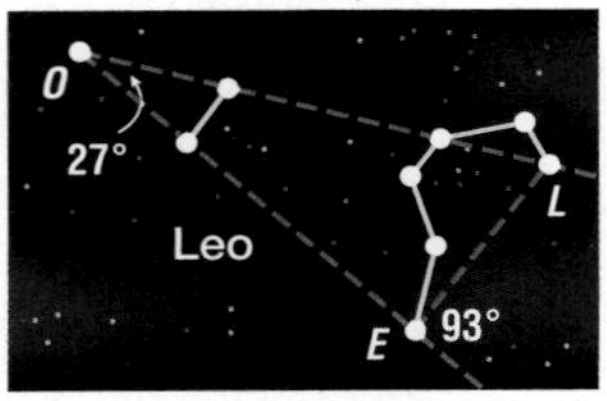

Find the measure of each numbered angle. (Lesson 12-2)

8. $m\angle 4$
9. $m\angle 5$
10. $m\angle 6$
11. $m\angle 7$

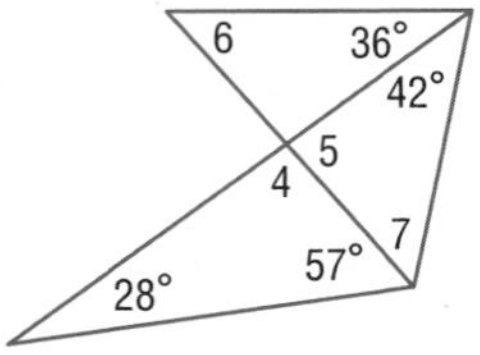

In the diagram, $\triangle RST \cong \triangle ABC$. (Lesson 12-3)

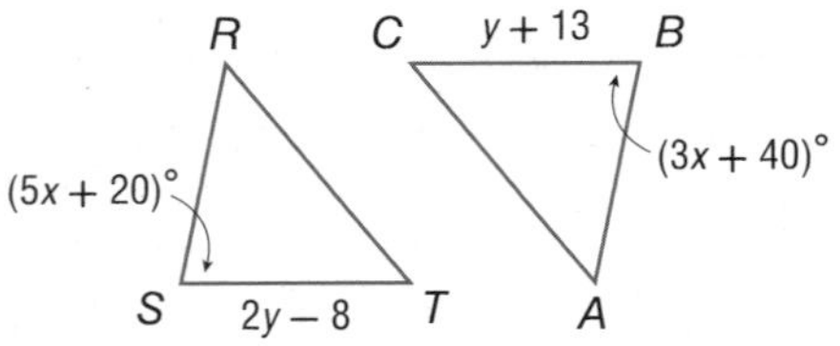

12. Find x.

13. Find y.

14. **ARCHITECTURE** The diagram shows an A-frame house with various points labeled. Assume that segments and angles that appear to be congruent in the diagram are congruent. Indicate which triangles are congruent. (Lesson 12-3)

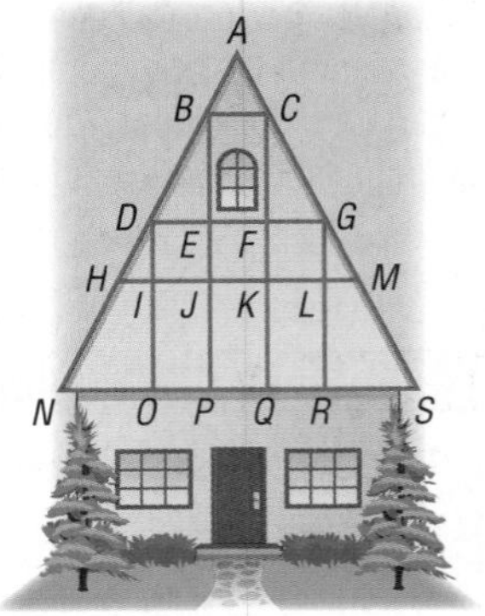

15. **MULTIPLE CHOICE** Determine which statement is true given that $\triangle CBX \cong \triangle SML$. (Lesson 12-3)

F $\overline{MO} \cong \overline{SL}$ **H** $\angle X \cong \angle S$

G $\overline{XC} \cong \overline{ML}$ **J** $\angle XCB \cong \angle LSM$

16. **BRIDGES** A bridge truss is shown in the diagram below, where $\overline{AC} \perp \overline{BD}$ and B is the midpoint of $\overline{AC}$. What method can be used to prove that $\triangle ABD \cong \triangle CBD$? (Lesson 12-4)

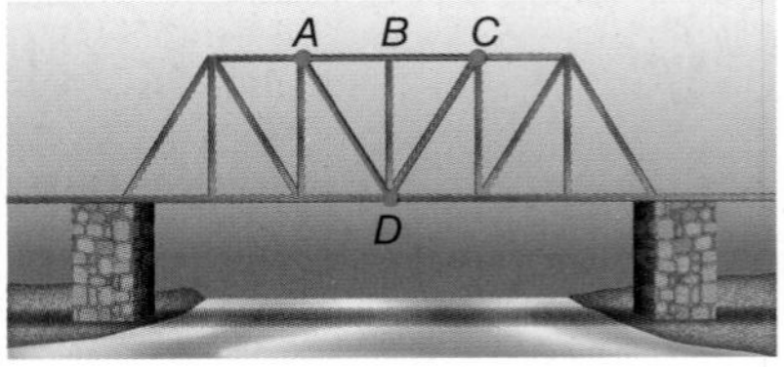

Determine whether $\triangle PQR \cong \triangle XYZ$. (Lesson 12-4)

17. $P(3, -5)$, $Q(11, 0)$, $R(1, 6)$, $X(5, 1)$, $Y(13, 6)$, $Z(3, 12)$

18. $P(-3, -3)$, $Q(-5, 1)$, $R(-2, 6)$, $X(2, -6)$, $Y(3, 3)$, $Z(5, -1)$

19. $P(8, 1)$, $Q(-7, -15)$, $R(9, -6)$, $X(5, 11)$, $Y(-10, -5)$, $Z(6, 4)$

20. **Write a two-column proof.** (Lesson 12-4)

Given: $\triangle LMN$ is isos. with $\overline{LM} \cong \overline{NM}$, and $\overline{MO}$ bisects $\angle LMN$.

Prove: $\triangle MLO \cong \triangle MNO$

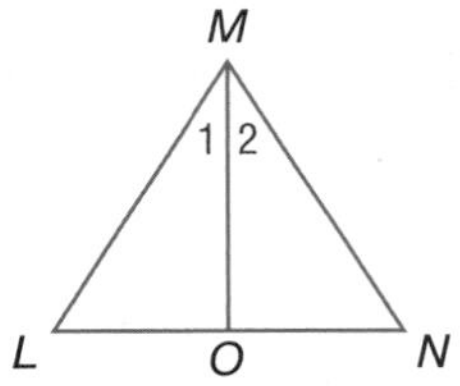

LESSON 12-5 Proving Triangles Congruent—ASA, AAS

Then

- You proved triangles congruent using SSS and SAS.

Now

- **1** Use the ASA Postulate to test for congruence.
- **2** Use the AAS Theorem to test for congruence.

Why?

- Competitive sweep rowing, also called *crew*, involves two or more people who sit facing the stern of the boat, with each rower pulling one oar. In high school competitions, a race, called a *regatta*, usually requires a body of water that is more than 1500 meters long. Congruent triangles can be used to measure distances that are not easily measured directly, like the length of a regatta course.

NewVocabulary
included side

Common Core State Standards

Content Standards
G.CO.10 Prove theorems about triangles.
G.SRT.5 Use congruence and similarity criteria for triangles to solve problems and to prove relationships in geometric figures.

Mathematical Practices
3 Construct viable arguments and critique the reasoning of others.
5 Use appropriate tools strategically.

1 ASA Postulate

An **included side** is the side located between two consecutive angles of a polygon. In $\triangle ABC$ at the right, $\overline{AC}$ is the included side between $\angle A$ and $\angle C$.

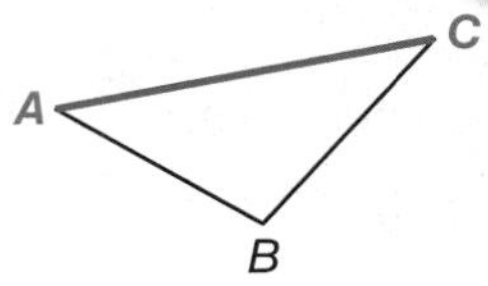

Postulate 12.3 Angle-Side-Angle (ASA) Congruence

If two angles and the included side of one triangle are congruent to two angles and the included side of another triangle, then the triangles are congruent.

Example If **A**ngle $\angle A \cong \angle D$,
Side $\overline{AB} \cong \overline{DE}$, and
Angle $\angle B \cong \angle E$,
then $\triangle ABC \cong \triangle DEF$.

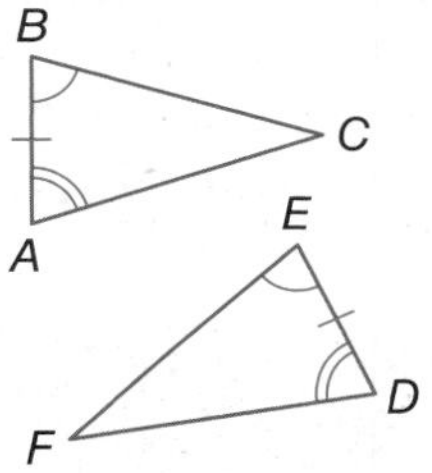

Construction Congruent Triangles Using Two Angles and Included Side

Draw a triangle and label it $\triangle ABC$. Then use the ASA Postulate to construct $\triangle XYZ \cong \triangle ABC$.

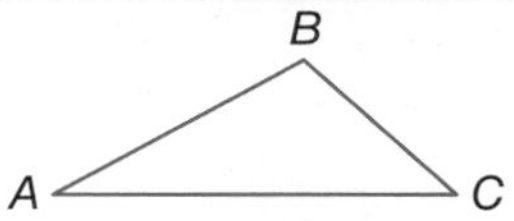

Step 1

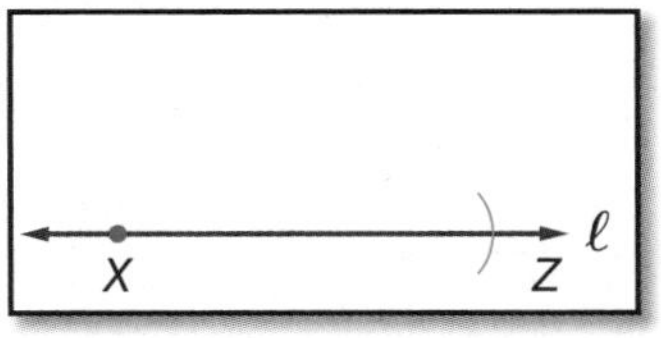

Draw a line ℓ and select a point X. Construct $\overline{XZ}$ such that $\overline{XZ} \cong \overline{AC}$.

Step 2

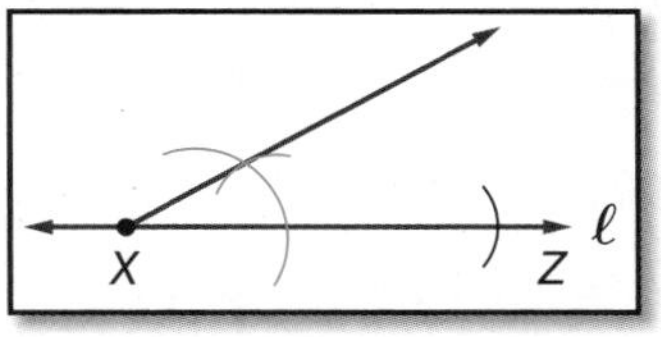

Construct an angle congruent to $\angle A$ at X using $\overleftrightarrow{XZ}$ as a side of the angle.

Step 3

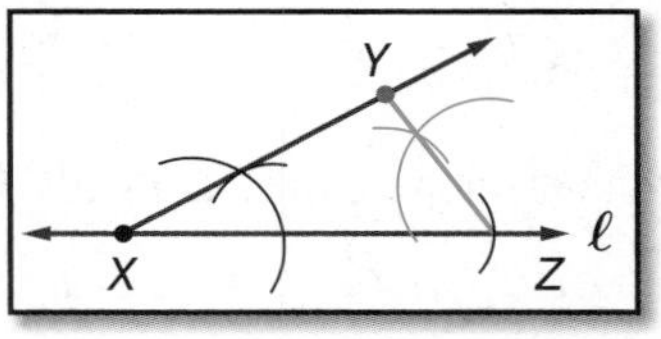

Construct an angle congruent to $\angle C$ at Z using $\overleftrightarrow{XZ}$ as a side of the angle. Label the point where the new sides of the angles meet as Y.

Example 1 Use ASA to Prove Triangles Congruent

Write a two-column proof.

Given: $\overline{QS}$ bisects $\angle PQR$;
$\angle PSQ \cong \angle RSQ$.

Prove: $\triangle PQS \cong \triangle RQS$

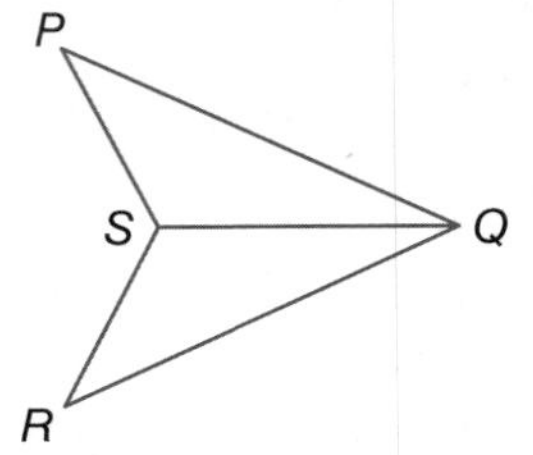

Proof:

Statements	Reasons
1. $\overline{QS}$ bisects $\angle PQR$; $\angle PSQ \cong \angle RSQ$.	**1.** Given
2. $\angle PQS \cong \angle RQS$	**2.** Definition of Angle Bisector
3. $\overline{QS} \cong \overline{QS}$	**3.** Reflexive Property of Congruence
4. $\triangle PQS \cong \triangle RQS$	**4.** ASA

Guided Practice

1. Write a flow proof.
Given: $\overline{ZX}$ bisects $\angle WZY$; $\overline{XZ}$ bisects $\angle YXW$.
Prove: $\triangle WXZ \cong \triangle XZY$

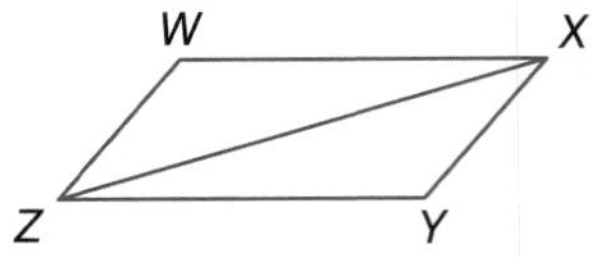

2 AAS Theorem

The congruence of two angles and a nonincluded side are also sufficient to prove two triangles congruent. This congruence relationship is a theorem because it can be proved using the Third Angles Theorem.

Theorem 12.5 Angle-Angle-Side (AAS) Congruence

If two angles and the nonincluded side of one triangle are congruent to the corresponding two angles and side of a second triangle, then the two triangles are congruent.

Example If Angle $\angle A \cong \angle D$,
Angle $\angle B \cong \angle E$, and
Side $\overline{BC} \cong \overline{EF}$,
then $\triangle ABC \cong \triangle DEF$.

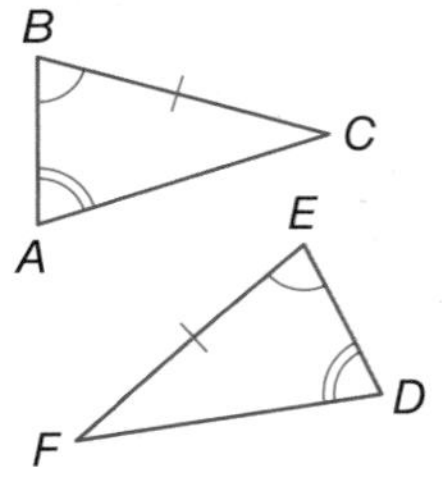

Proof Angle-Angle-Side Theorem

Given: $\angle L \cong \angle Q$, $\angle M \cong \angle R$, $\overline{MN} \cong \overline{RS}$

Prove: $\triangle LMN \cong \triangle QRS$

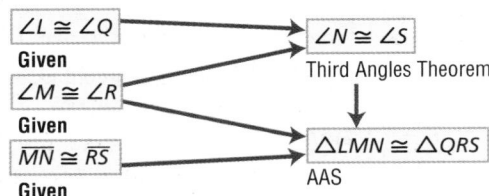

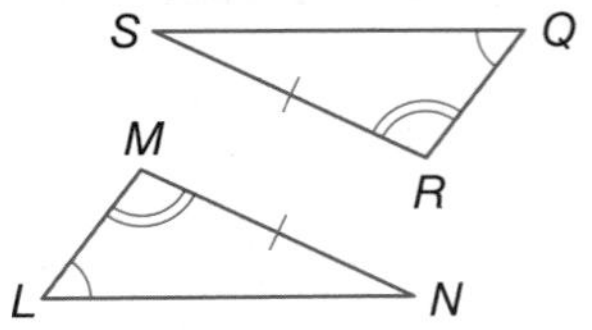

Proof:

$\angle L \cong \angle Q$ — **Given**
$\angle M \cong \angle R$ — **Given**
$\overline{MN} \cong \overline{RS}$ — **Given**

$\angle L \cong \angle Q$ and $\angle M \cong \angle R$ → $\angle N \cong \angle S$ — Third Angles Theorem

$\angle N \cong \angle S$, $\angle M \cong \angle R$, and $\overline{MN} \cong \overline{RS}$ → $\triangle LMN \cong \triangle QRS$ — AAS

Example 2 Use AAS to Prove Triangles Congruent

PT

Write a two-column proof.

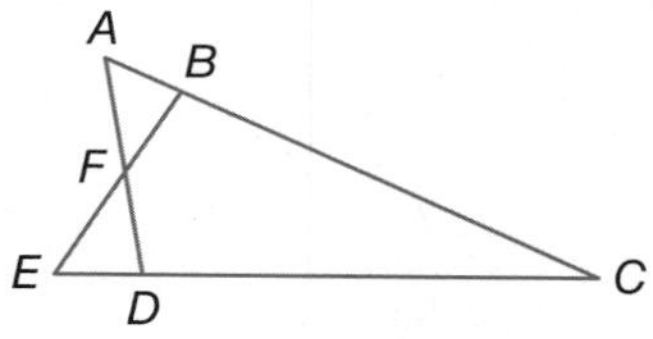

Given: $\angle DAC \cong \angle BEC$
$\overline{DC} \cong \overline{BC}$

Prove: $\triangle ACD \cong \triangle ECB$

Proof: We are given that $\angle DAC \cong \angle BEC$ and $\overline{DC} \cong \overline{BC}$. $\angle C \cong \angle C$ by the Reflexive Property. By AAS, $\triangle ACD \cong \triangle ECB$.

GuidedPractice

2. Write a flow proof.
Given: $\overline{RQ} \cong \overline{ST}$ and $\overline{RQ} \parallel \overline{ST}$
Prove: $\triangle RUQ \cong \triangle TUS$

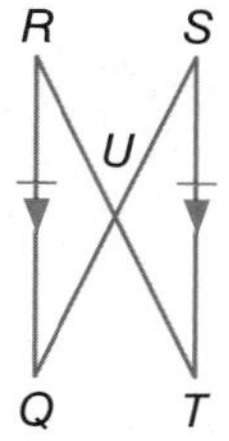

You can use congruent triangles to measure distances that are difficult to measure directly.

Real-World Example 3 Apply Triangle Congruence

PT

COMMUNITY SERVICE Jeremias is working with a community service group to build a bridge across a creek at a local park. The bridge will span the creek between points *C* and *B*. Jeremias located a fixed point *D* to use as a reference point so that the segments have the relationships shown. *A* is the midpoint of $\overline{CD}$ and *DE* is 15 feet. How long does the bridge need to be?

In order to determine the length of $\overline{CB}$, we must first prove that the two triangles Jeremias has created are congruent.

- Since $\overline{CD}$ is perpendicular to both $\overline{CB}$ and $\overline{DE}$, the segments form right angles as shown on the diagram.
- All right angles are congruent, so $\angle BCA \cong \angle EDA$.
- Point *A* is the midpoint of $\overline{CD}$, so $\overline{CA} \cong \overline{AD}$.
- $\angle BAC$ and $\angle EAD$ are vertical angles, so they are congruent.

Therefore, by ASA, $\triangle BAC \cong \triangle EAD$.

Since $\triangle BAC \cong \triangle EAD$, $\overline{DE} \cong \overline{CB}$ by CPCTC. Since the measure of $\overline{DE}$ is 15 feet, the measure of $\overline{CB}$ is also 15 feet. Therefore, the bridge needs to be 15 feet long.

> **StudyTip**
> **Angle-Angle-Angle** In Example 3, $\angle B$ and $\angle E$ are congruent by the Third Angles Theorem. Congruence of all three corresponding angles is not sufficient, however, to prove two triangles congruent.

GuidedPractice

3. In the sign scaffold shown at the right, $\overline{BC} \perp \overline{AC}$ and $\overline{DE} \perp \overline{CE}$. $\angle BAC \cong \angle DCE$, and $\overline{AB} \cong \overline{CD}$. Write a paragraph proof to show that $\overline{BC} \cong \overline{DE}$.

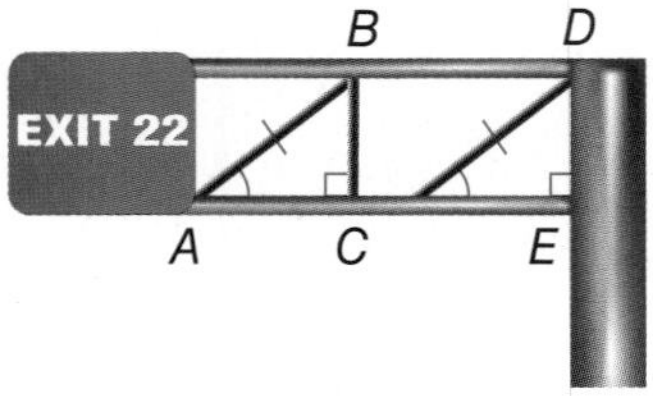

You have learned several methods for proving triangle congruence.

ConceptSummary Proving Triangles Congruent

SSS	SAS	ASA	AAS
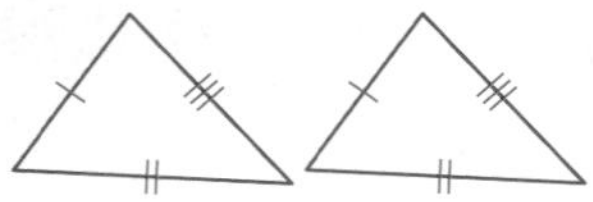	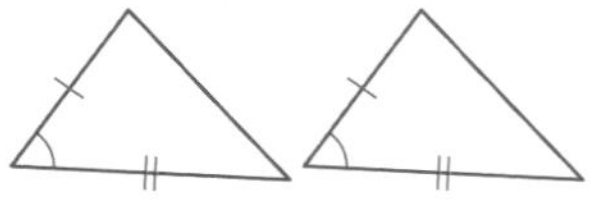	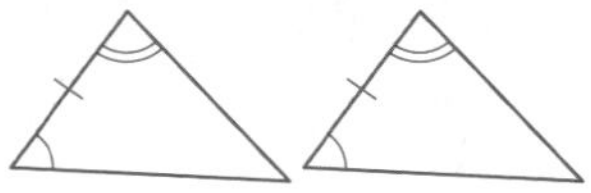	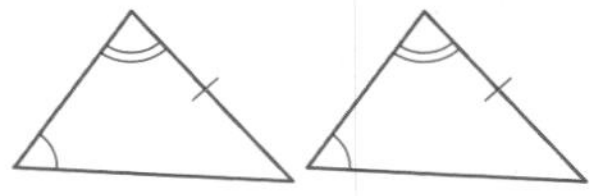
Three pairs of corresponding sides are congruent.	Two pairs of corresponding sides and their included angles are congruent.	Two pairs of corresponding angles and their included sides are congruent.	Two pairs of corresponding angles and the corresponding nonincluded sides are congruent.

Check Your Understanding

Example 1 **PROOF** **Write the specified type of proof.**

1. flow proof

Given: regular pentagon *ABCDE*

Prove: $\overline{AD} \cong \overline{DB}$

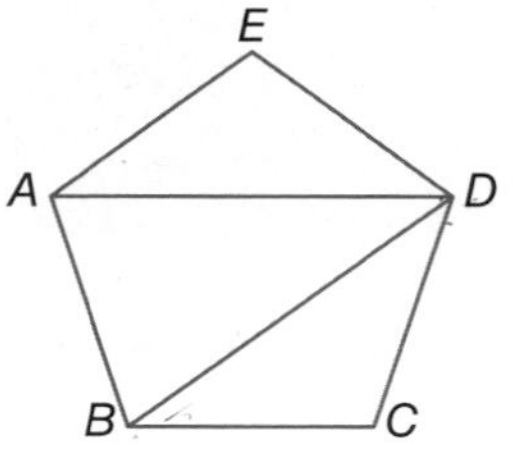

2. two-column proof

Given: $\overline{WT} \parallel \overline{NE}$; $\overline{TO} \cong \overline{EO}$

Prove: $\triangle WOT \cong \triangle NOE$

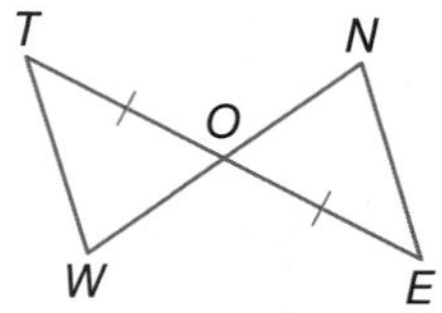

Example 2

3. paragraph proof

Given: $\overline{RV} \parallel \overline{TW}$; $\overline{RT} \parallel \overline{VW}$

Prove: $\triangle RWV \cong \triangle WRT$

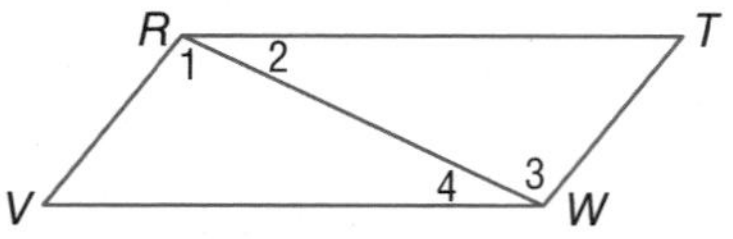

4. two-column proof

Given: $\overline{EX} \cong \overline{WX}$; $\overline{XB}$ bisects $\angle EBW$ and $\angle EXW$

Prove: $\triangle EXB \cong \triangle WXB$

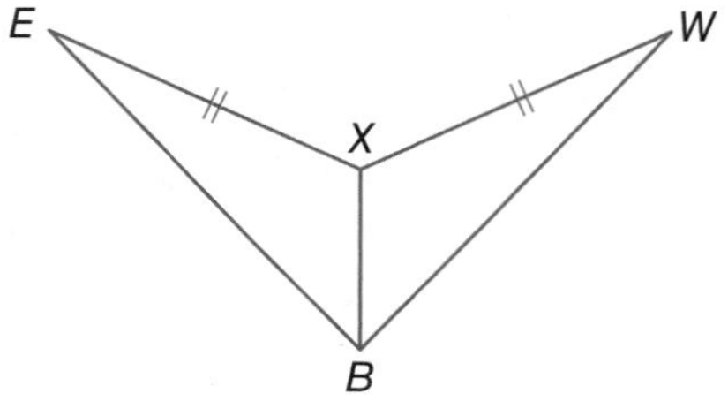

Example 3

5. **BRIDGE BUILDING** A surveyor needs to find the distance from point A to point B across a canyon. She places a stake at A, and a coworker places a stake at B on the other side of the canyon. The surveyor then locates C on the same side of the canyon as A such that $\overline{CA} \perp \overline{AB}$. A fourth stake is placed at E, the midpoint of $\overline{CA}$. Finally, a stake is placed at D such that $\overline{CD} \perp \overline{CA}$ and D, E, and B are sited as lying along the same line.

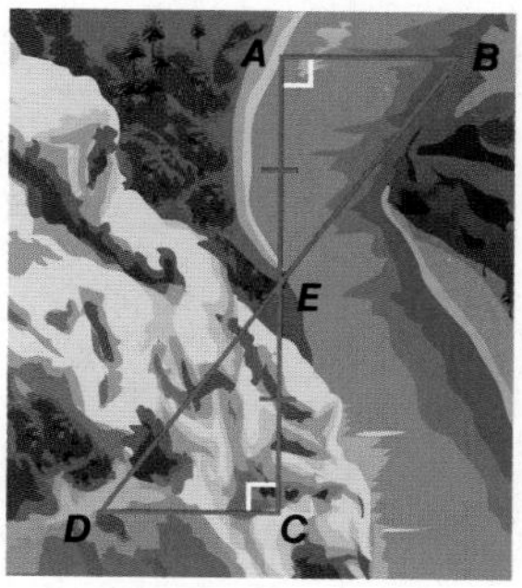

   a. Explain how the surveyor can use the triangles formed to find AB.

   b. If $AC = 1500$ meters, $DC = 690$ meters, and $DE = 973.5$ meters, what is AB? Explain your reasoning.

Practice and Problem Solving

Example 1

PROOF **Write a paragraph proof.**

6. **Given:** $\overline{WY}$ bisects both $\angle XWZ$ and $\angle XYZ$

   Prove: $\triangle WYX \cong \triangle YWZ$

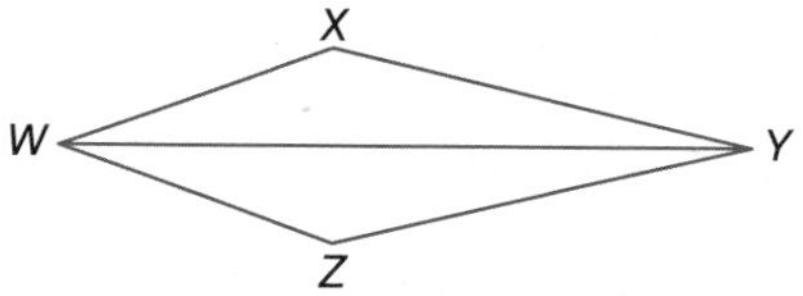

7. **Given:** $\overline{AB} \perp \overline{BC}$; $\overline{AB} \perp \overline{AD}$

   Prove: $\triangle ACD \cong \triangle CAB$

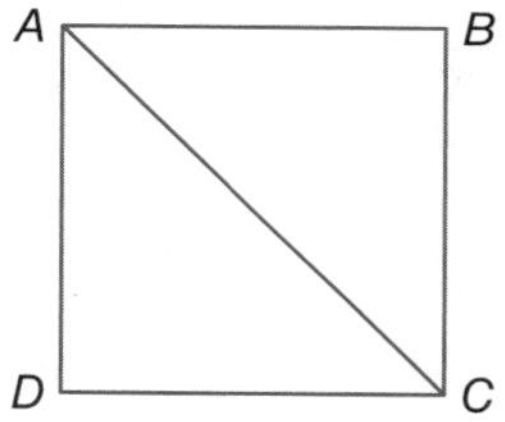

8. **GAMES** The picture on the right shows a house cards. A house of cards is a structure created by stacking playing card on top of each other. Explain how parallel lines and congruent triangles would help someone who was trying to build a house of cards.

Example 2

PROOF **Write a two-column proof.**

9. **Given:** $\overline{HZ} \parallel \overline{ET}$; $\overline{AG} \cong \overline{BD}$; $\angle A \cong \angle B$

   Prove: $\triangle ADE \cong \triangle BGZ$

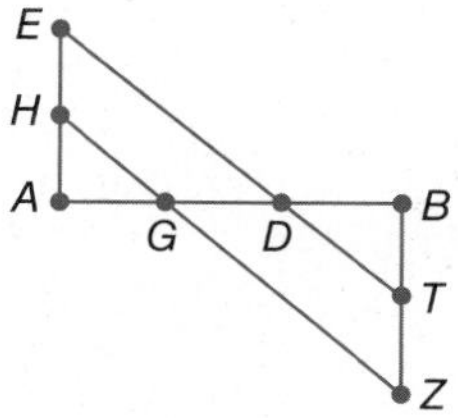

10. **Given:** $\triangle CDB \cong \triangle CDA$

    Prove: $\triangle ADE \cong \triangle BDF$

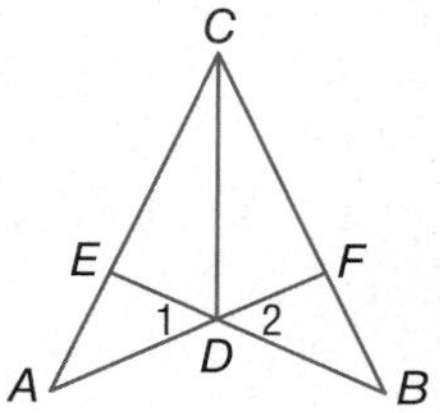

11. **CCSS ARGUMENTS** Write a flow proof.

    Given: $\overline{AY} \cong \overline{BA}$; $\overline{ZX} \parallel \overline{BC}$

    Prove: $\overline{YZ} \cong \overline{BC}$

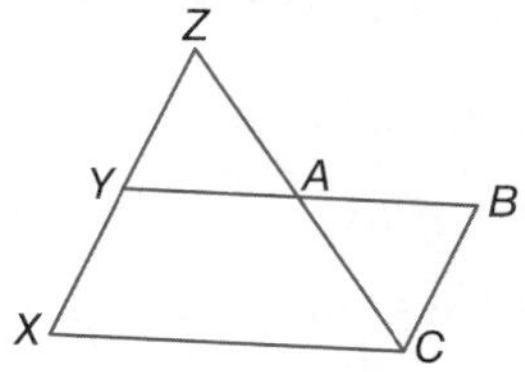

12. PROOF Write a flow proof.

Given: $\overline{XZ}$ is the perpendicular bisector of $\overline{WY}$

Prove: $\angle W \cong \angle Y$

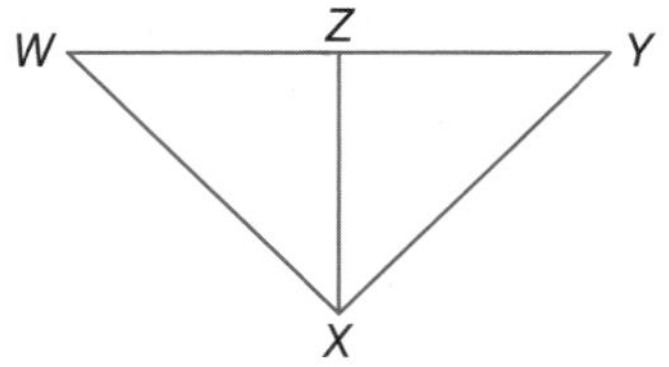

Example 3

13. CCSS MODELING A high school wants to hold a 1500-meter regatta on Lake Powell but is unsure if the lake is long enough. To measure the distance across the lake, the crew members locate the vertices of the triangles below and find the measures of the lengths of $\triangle HJK$ as shown below.

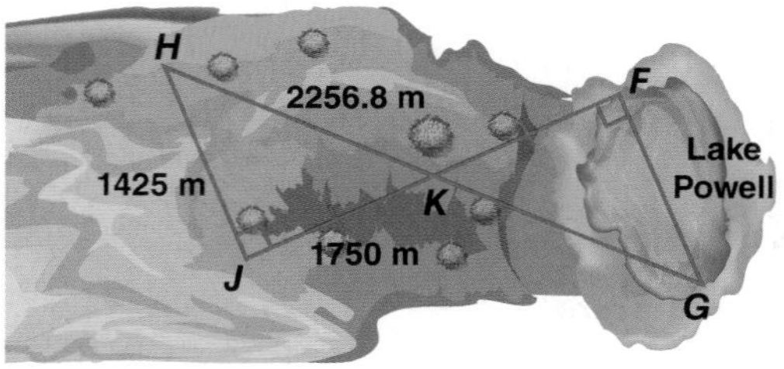

a. Explain how the crew team can use the triangles formed to estimate the distance FG across the lake.

b. Using the measures given, is the lake long enough for the team to use as the location for their regatta? Explain your reasoning.

ALGEBRA Find the value of the variable that yields congruent triangles.

14. $\triangle BCD \cong \triangle WXY$

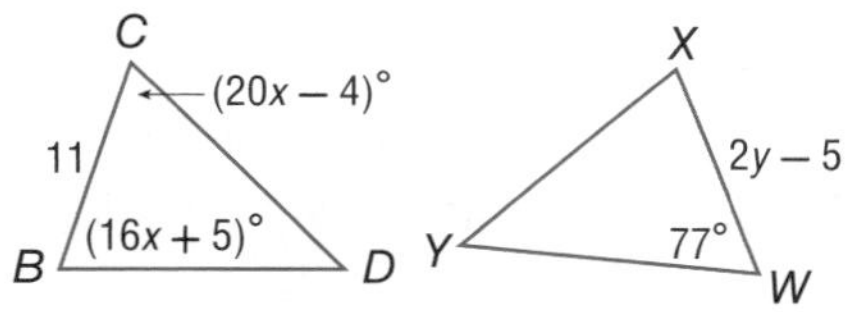

15. $\triangle MHJ \cong \triangle PQJ$

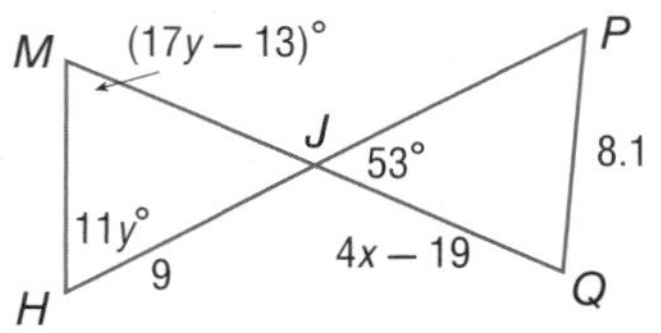

16. THEATER DESIGN The trusses of the roof of the outdoor theater shown below appear to be several different pairs of congruent triangles. Assume that trusses that appear to lie on the same line actually lie on the same line.

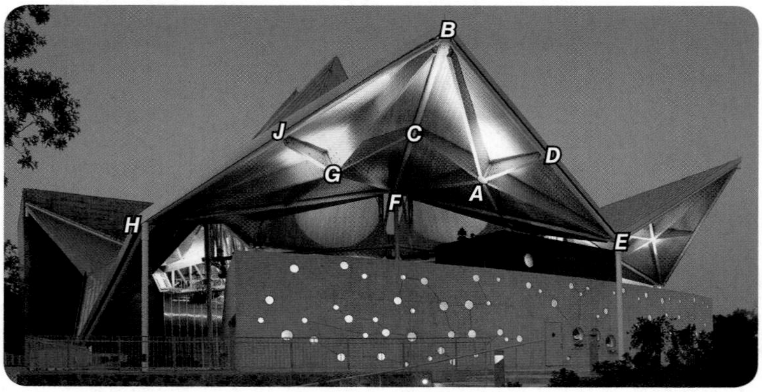

a. If $\overline{AB}$ bisects $\angle CBD$ and $\angle CAD$, prove that $\triangle ABC \cong \triangle ABD$.

b. If $\triangle ABC \cong \triangle ABD$ and $\angle FCA \cong \angle EDA$, prove that $\triangle CAF \cong \triangle DAE$.

c. If $\overline{HB} \cong \overline{EB}$, $\angle BHG \cong \angle BEA$, $\angle HGJ \cong \angle EAD$, and $\angle JGB \cong \angle DAB$, prove that $\triangle BHG \cong \triangle BEA$.

Greg Murphey

PROOF **Write a paragraph proof.**

17. **Given:** $\overleftrightarrow{RS}$ bisects $\angle CSA$ and $\angle CHA$

Prove: $\triangle CHS \cong \triangle AHS$

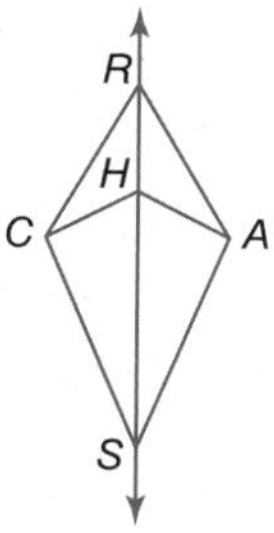

18. **Given:** $\triangle BDF$ is eqilateral; $\angle DEB \cong \angle BAD$

Prove: $\triangle BAD \cong \triangle DEB$

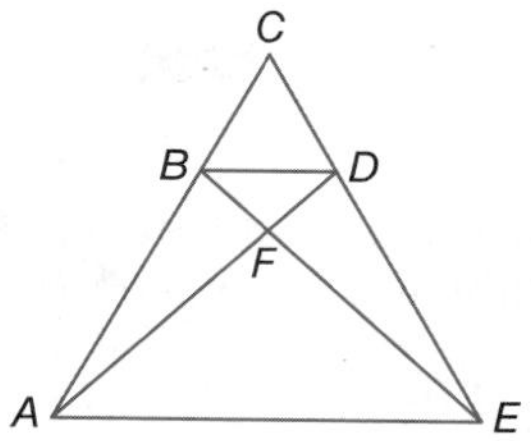

PROOF **Write a two-column proof.**

19. **Given:** $\angle CED \cong \angle CFD$; $\overline{CD}$ bisects $\angle ECF$

Prove: $\triangle CED \cong \triangle CFD$

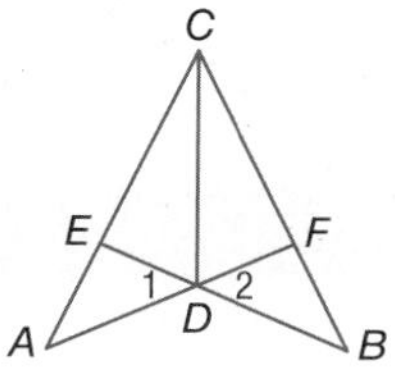

20. **Given:** $\overline{VK} \perp \overline{KX}$; $\overline{EM} \perp \overline{MX}$; $\overline{KX} \cong \overline{MX}$

Prove: $\angle V \cong \angle E$

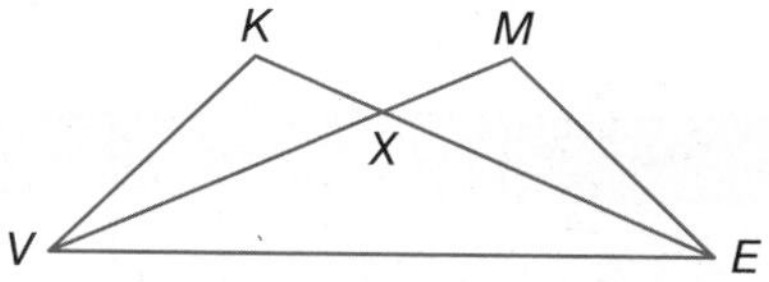

21. **TRICYCLE** The sketch below is an aerial perspective drawing of a tricycle frame.

a. Make a conjecture about the two types of triangles that are used to make the basic frame.

b. What information would you need to prove that the triangles are congruent?

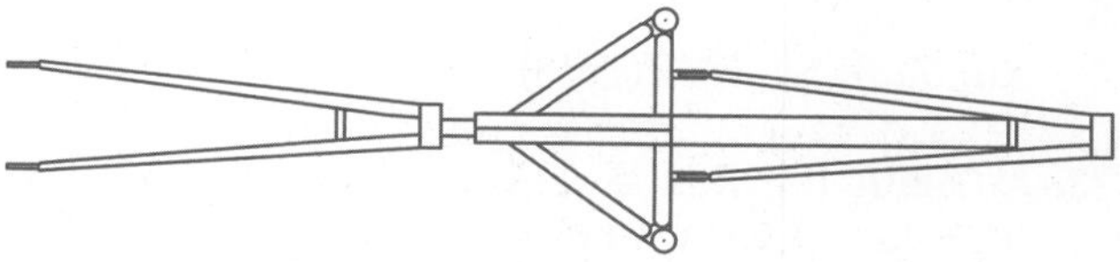

H.O.T. Problems Use Higher-Order Thinking Skills

22. **WRITING IN MATH** Given a rectangle, explain at least 2 different ways to prove that the diagonal divides the rectangle into two congruent triangles.

23. **CCSS ERROR ANALYSIS** Tyrone says it is possible to show that $\triangle ACB \cong \triangle ADE$ but Lorenzo disagrees. Is either of them correct? Explain your reasoning.

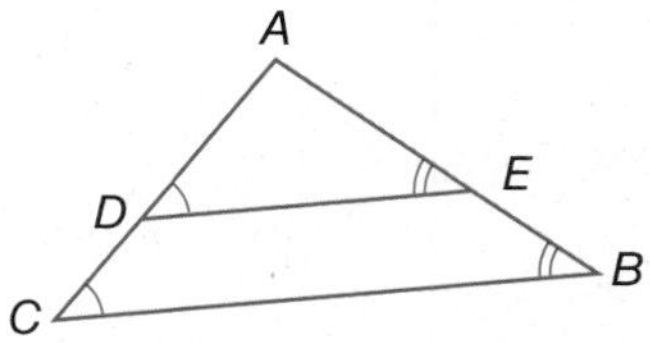

24. **REASONING** Determine whether SSA (Side-Side-Angle) can be used to prove the congruence of two triangles. Explain your reasoning.

25. **CHALLENGE** Using the information given in the diagram, write a flow proof to show that $\triangle PVT \cong \triangle SVQ$.

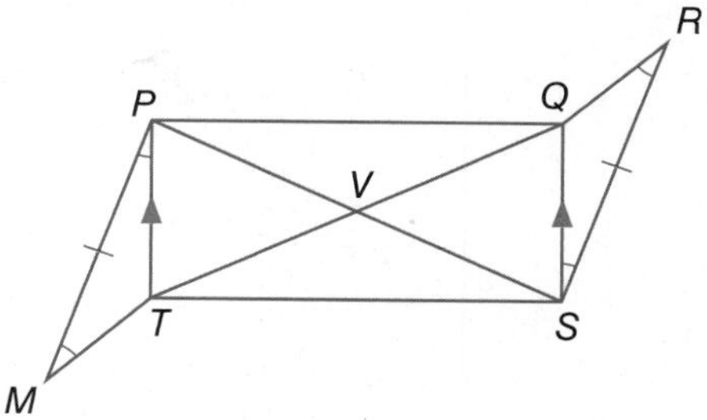

26. **WRITING IN MATH** How do you know what method (SSS, SAS, etc.) to use when proving triangle congruence? Use a chart to explain your reasoning.

Standardized Test Practice

27. Given: $\overline{BC}$ is perpendicular to $\overline{AD}$; $\angle 1 \cong \angle 2$.

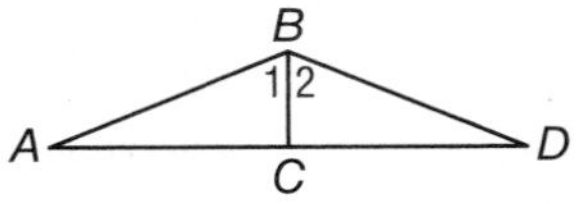

Which theorem or postulate could be used to prove $\triangle ABC \cong \triangle DBC$?

A AAS

B ASA

C SAS

D SSS

28. SHORT RESPONSE Write an expression that can be used to find the values of $s(n)$ in the table.

n	−8	−4	−1	0	1
$s(n)$	1.00	2.00	2.75	3.00	3.25

29. ALGEBRA If −7 is multiplied by a number greater than 1, which of the following describes the result?

F a number greater than 7

G a number between −7 and 7

H a number greater than −7

J a number less than −7

30. SAT/ACT $\sqrt{121 + 104} = ?$

A 15

B 21

C 25

D 125

E 225

Spiral Review

Determine whether $\triangle ABC \cong \triangle XYZ$. Explain. (Lesson 12-4)

31. $A(6, 4), B(1, -6), C(-9, 5),$
$X(0, 7), Y(5, -3), Z(15, 8)$

32. $A(0, 5), B(0, 0), C(-2, 0),$
$X(4, 8), Y(4, 3), Z(6, 3)$

33. ALGEBRA If $\triangle RST \cong \triangle JKL$, $RS = 7$, $ST = 5$, $RT = 9 + x$, $JL = 2x - 10$, and $JK = 4y - 5$, draw and label a figure to represent the congruent triangles. Then find x and y. (Lesson 12-3)

34. FINANCIAL LITERACY Maxine charges \$5 to paint a mailbox and \$4 per hour to mow a lawn. Write an equation to represent the amount of money Maxine can earn from a homeowner who has his or her mailbox painted and lawn mowed. (Lesson 11-4)

Skills Review

PROOF Write a two-column proof for each of the following.

35. Given: $\angle 2 \cong \angle 1$
$\angle 1 \cong \angle 3$

Prove: $\overline{AB} \parallel \overline{DE}$

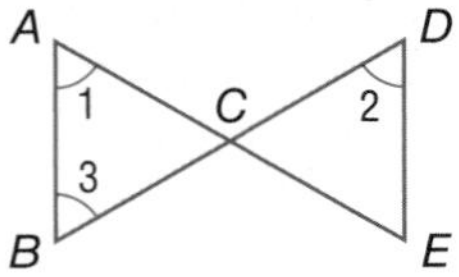

36. Given: $\angle MJK \cong \angle KLM$
$\angle LMJ$ and $\angle KLM$ are supplementary.

Prove: $\overline{KJ} \parallel \overline{LM}$

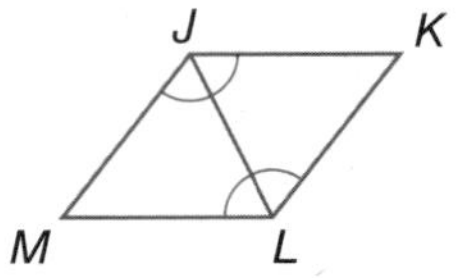

Geometry Lab: Congruence in Right Triangles

In Lessons 12-4 and 12-5, you learned theorems and postulates to prove triangles congruent. How do these theorems and postulates apply to right triangles?

CCSS Common Core State Standards
Content Standards
G.SRT.5 Use congruence and similarity criteria for triangles to solve problems and to prove relationships in geometric figures.
Mathematical Practices 5

Study each pair of right triangles.

a.

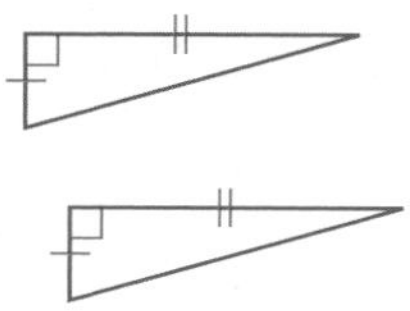

b.

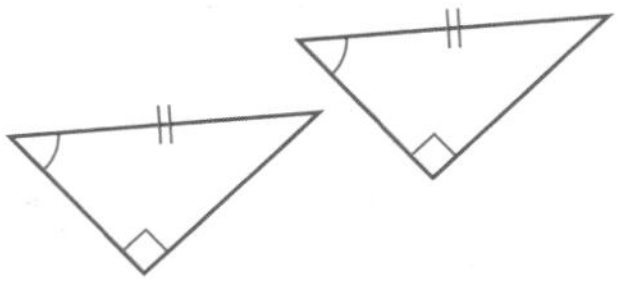

c.

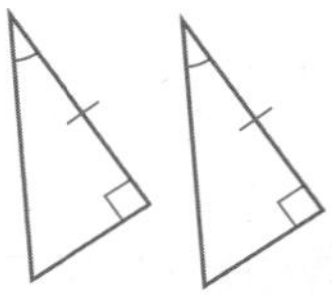

Analyze

1. Is each pair of triangles congruent? If so, which congruence theorem or postulate applies?
2. Rewrite the congruence rules from Exercise 1 using *leg*, (L), or *hypotenuse*, (H), to replace *side*. Omit the *A* for any right angle since we know that all right triangles contain a right angle and all right angles are congruent.
3. **MAKE A CONJECTURE** If you know that the corresponding legs of two right triangles are congruent, what other information do you need to declare the triangles congruent? Explain.

In Lesson 12-5, you learned that SSA is not a valid test for determining triangle congruence. Can SSA be used to prove right triangles congruent?

Activity SSA and Right Triangles

Step 1

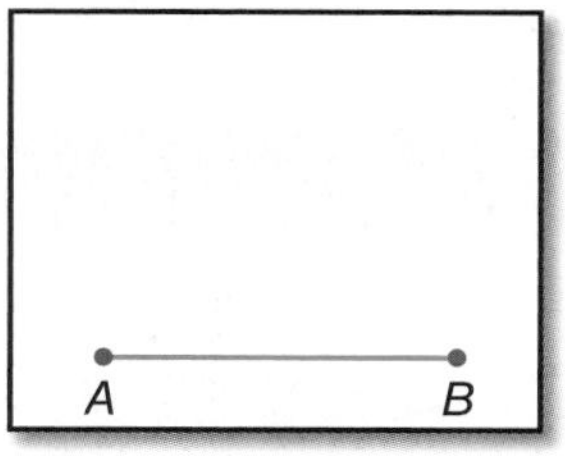

Draw $\overline{AB}$ so that $AB = 6$ centimeters.

Step 2

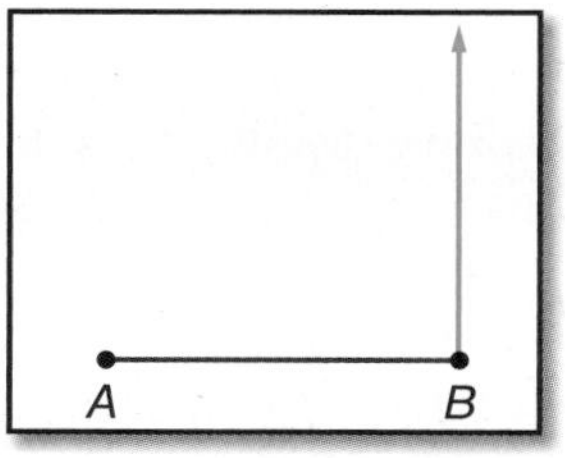

Use a protractor to draw a ray from *B* that is perpendicular to $\overline{AB}$.

Step 3

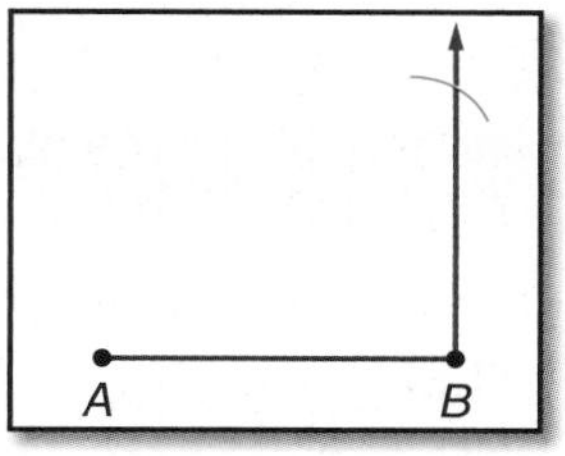

Open your compass to a width of 8 centimeters. Place the point at *A* and draw an arc to intersect the ray.

Step 4

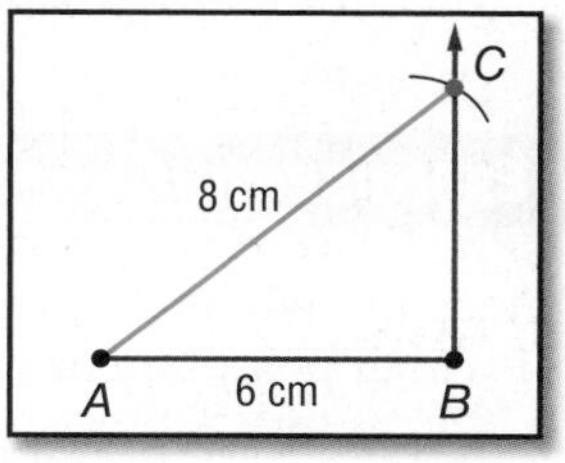

Label the intersection *C* and draw $\overline{AC}$ to complete $\triangle ABC$.

Analyze

4. Does the model yield a unique triangle?
5. Can you use the lengths of the hypotenuse and a leg to show right triangles are congruent?
6. **Make a conjecture** about the case of SSA that exists for right triangles.

(continued on the next page)

Geometry Lab

Congruence in Right Triangles *Continued*

Your work on the previous page provides evidence for four ways to prove right triangles congruent.

Theorem Right Triangle Congruence

Theorem 12.6 Leg-Leg Congruence
If the legs of one right triangle are congruent to the corresponding legs of another right triangle, then the triangles are congruent.

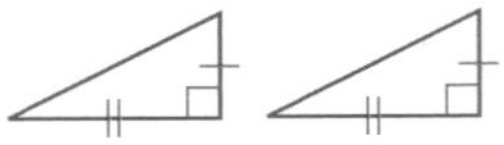

Abbreviation *LL*

Theorem 12.7 Hypotenuse-Angle Congruence
If the hypotenuse and acute angle of one right triangle are congruent to the hypotenuse and corresponding acute angle of another right triangle, then the two triangles are congruent.

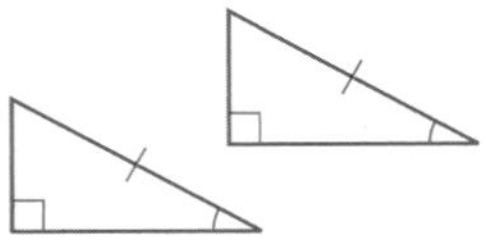

Abbreviation *HA*

Theorem 12.8 Leg-Angle Congruence
If one leg and an acute angle of one right triangle are congruent to the corresponding leg and acute angle of another right triangle, then the triangles are congruent.

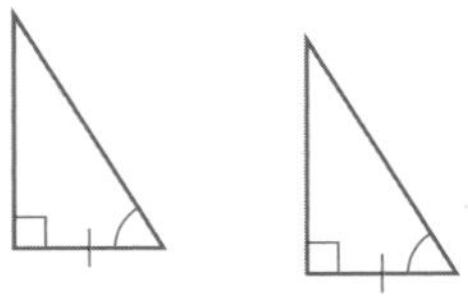

Abbreviation *LA*

Theorem 12.9 Hypotenuse-Leg Congruence
If the hypotenuse and a leg of one right triangle are congruent to the hypotenuse and corresponding leg of another right triangle, then the triangles are congruent.

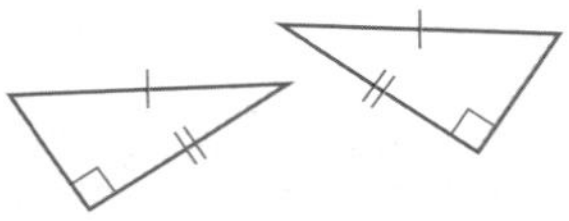

Abbreviation *HL*

Exercises

Determine whether each pair of triangles is congruent. If yes, tell which postulate or theorem applies.

7.

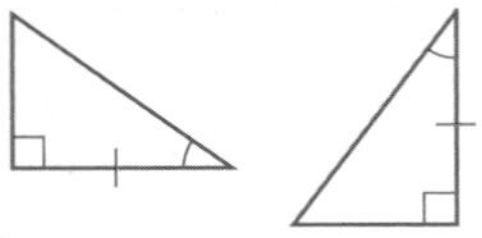

8.

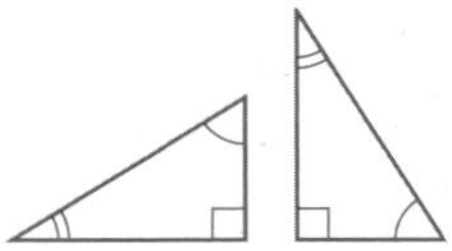

9.

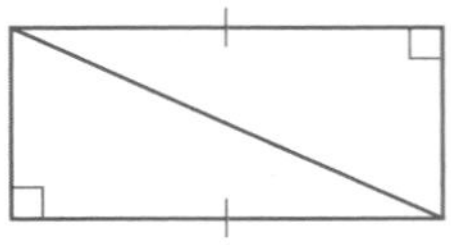

PROOF Write a proof for each of the following.

10. Theorem 12.6

11. Theorem 12.7

12. Theorem 12.8 (*Hint*: There are two possible cases.)

13. Theorem 12.9 (*Hint*: Use the Pythagorean Theorem.)

Use the figure at the right.

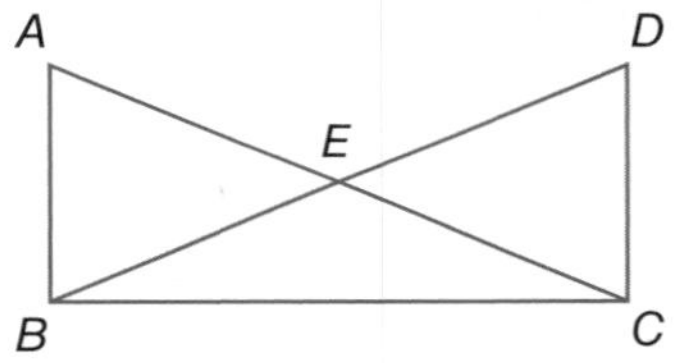

14. Given: $\overline{AB} \perp \overline{BC}$, $\overline{DC} \perp \overline{BC}$
$\overline{AC} \cong \overline{BD}$

Prove: $\overline{AB} \cong \overline{DC}$

15. Given: $\overline{AB} \parallel \overline{DC}$, $\overline{AB} \perp \overline{BC}$
E is the midpoint of $\overline{AC}$ and $\overline{BD}$.

Prove: $\overline{AC} \cong \overline{DB}$

LESSON 12-6

Isosceles and Equilateral Triangles

Then	Now	Why?
• You identified isosceles and equilateral triangles.	**1** Use properties of isosceles triangles. **2** Use properties of equilateral triangles.	• The tracks on the roller coaster have triangular reinforcements between the tracks for support and stability. The triangle supports in the photo are isosceles triangles.

NewVocabulary
legs of an isosceles triangle
vertex angle
base angles

Common Core State Standards

Content Standards
G.CO.10 Prove theorems about triangles.
G.CO.12 Make formal geometric constructions with a variety of tools and methods (compass and straightedge, string, reflective devices, paper folding, dynamic geometric software, etc.).

Mathematical Practices
2 Reason abstractly and quantitatively.
3 Construct viable arguments and critique the reasoning of others.

Rodolfo Arpia/age fotostock

1 Properties of Isosceles Triangles

Recall that isosceles triangles have at least two congruent sides. The parts of an isosceles triangle have special names.

The two congruent sides are called the **legs of an isosceles triangle**, and the angle with sides that are the legs is called the **vertex angle**. The side of the triangle opposite the vertex angle is called the *base*. The two angles formed by the base and the congruent sides are called the **base angles**.

$\angle 1$ is the vertex angle.
$\angle 2$ and $\angle 3$ are the base angles.

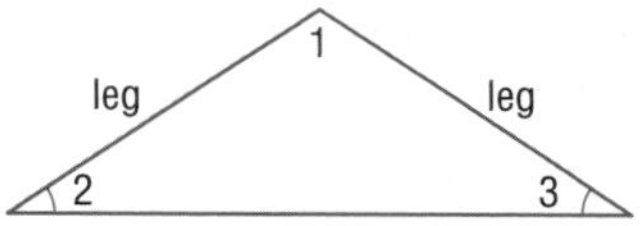

Theorems Isosceles Triangle

12.10 Isosceles Triangle Theorem
If two sides of a triangle are congruent, then the angles opposite those sides are congruent.

Example If $\overline{AC} \cong \overline{BC}$, then $\angle 2 \cong \angle 1$.

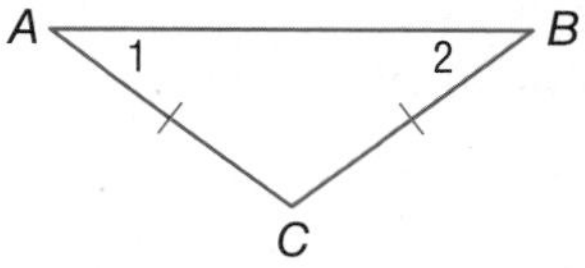

12.11 Converse of Isosceles Triangle Theorem
If two angles of a triangle are congruent, then the sides opposite those angles are congruent.

Example If $\angle 1 \cong \angle 2$, then $\overline{FE} \cong \overline{DE}$.

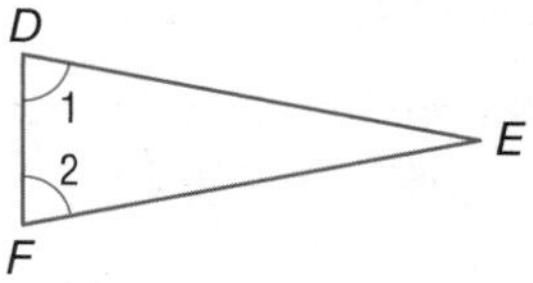

You will prove Theorem 12.11 in Exercise 37.

Example 1 Congruent Segments and Angles

PT

a. Name two unmarked congruent angles.
$\angle ACB$ is opposite $\overline{AB}$ and $\angle B$ is opposite $\overline{AC}$, so $\angle ACB \cong \angle B$.

b. Name two unmarked congruent segments.
$\overline{AD}$ is opposite $\angle ACD$ and $\overline{AC}$ is opposite $\angle D$, so $\overline{AD} \cong \overline{AC}$.

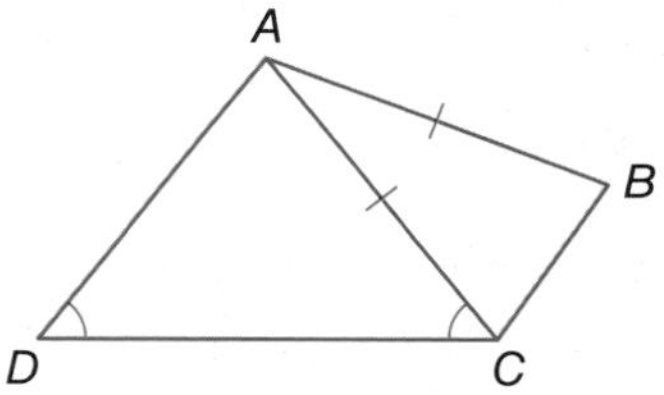

GuidedPractice

1A. Name two unmarked congruent angles.

1B. Name two unmarked congruent segments.

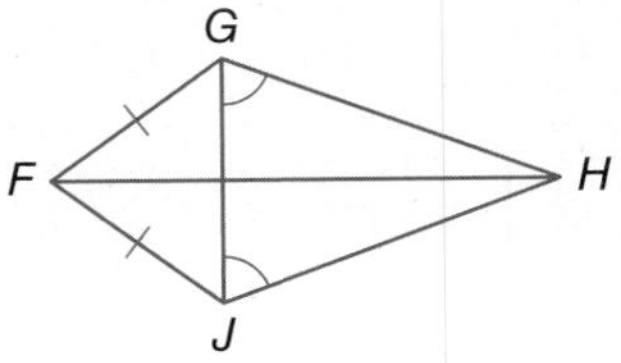

To prove the Isosceles Triangle Theorem, draw an auxiliary line and use the two triangles formed.

Proof Isosceles Triangle Theorem

Given: $\triangle LMP$; $\overline{LM} \cong \overline{LP}$

Prove: $\angle M \cong \angle P$

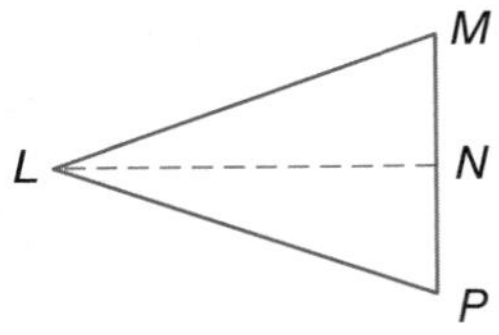

Proof:

Statements	Reasons
1. Let N be the midpoint of $\overline{MP}$.	**1.** Every segment has exactly one midpoint.
2. Draw an auxiliary segment $\overline{LN}$.	**2.** Two points determine a line.
3. $\overline{MN} \cong \overline{PN}$	**3.** Midpoint Theorem
4. $\overline{LN} \cong \overline{LN}$	**4.** Reflexive Property of Congruence
5. $\overline{LM} \cong \overline{LP}$	**5.** Given
6. $\triangle LMN \cong \triangle LPN$	**6.** SSS
7. $\angle M \cong \angle P$	**7.** CPCTC

2 Properties of Equilateral Triangles

Properties of Equilateral Triangles The Isosceles Triangle Theorem leads to two corollaries about the angles of an equilateral triangle.

ReviewVocabulary

equilateral triangle a triangle with three congruent sides

Corollaries Equilateral Triangle

12.3 A triangle is equilateral if and only if it is equiangular.

Example If $\angle A \cong \angle B \cong \angle C$, then $\overline{AB} \cong \overline{BC} \cong \overline{CA}$.

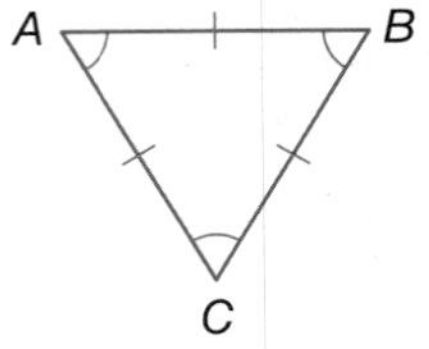

12.4 Each angle of an equilateral triangle measures 60.

Example If $\overline{DE} \cong \overline{EF} \cong \overline{FE}$, then $m\angle A = m\angle B = m\angle C = 60$.

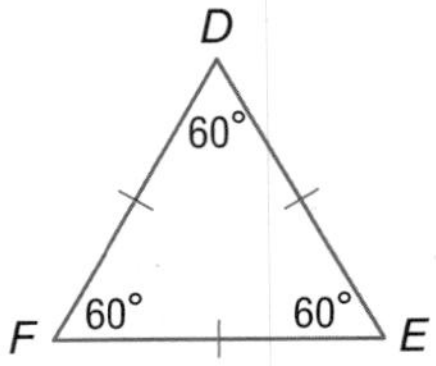

You will prove Corollaries 12.3 and 12.4 in Exercises 35 and 36.

Example 2 Find Missing Measures

Find each measure.

a. $m\angle Y$

Since $XY = XZ$, $\overline{XY} \cong \overline{XZ}$. By the Isosceles Triangle Theorem, base angles Z and Y are congruent, so $m\angle Z = m\angle Y$. Use the Triangle Sum Theorem to write and solve an equation to find $m\angle Y$.

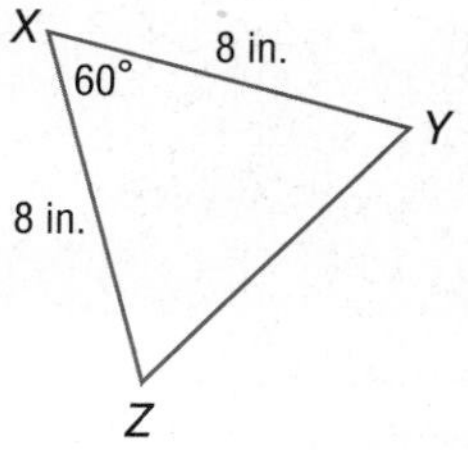

$m\angle X + m\angle Y + m\angle Z = 180$	Triangle Sum Theorem
$60 + m\angle Y + m\angle Y = 180$	$m\angle X = 60$, $m\angle Z = m\angle Y$
$60 + 2(m\angle Y) = 180$	Simplify.
$2(m\angle Y) = 120$	Subtract 60 from each side.
$m\angle Y = 60$	Divide each side by 2.

b. YZ

$m\angle Z = m\angle Y$, so $m\angle Z = 60$ by substitution. Since $m\angle X = 60$, all three angles measure 60, so the triangle is equiangular. Because an equiangular triangle is also equilateral, $XY = XZ = ZY$. Since $XY = 8$ inches, $YZ = 8$ inches by substitution.

StudyTip

Isosceles Triangles As you discovered in Example 2, any isosceles triangle that has one 60° angle must be an equilateral triangle.

GuidedPractice

2A. $m\angle M$

2B. PN

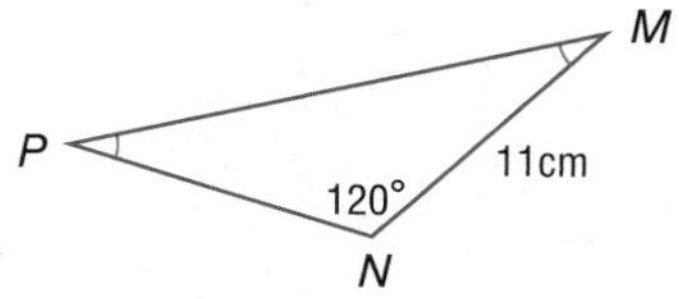

You can use the properties of equilateral triangles and algebra to find missing values.

Example 3 Find Missing Values

ALGEBRA Find the value of each variable.

Since $\angle B = \angle A$, $\overline{AC} \cong \overline{BC}$ by the Converse of the Isosceles Triangle Theorem. All of the sides of the triangle are congruent, so the triangle is equilateral. Each angle of an equilateral triangle measures 60°, so $2x = 60$ and $x = 30$.

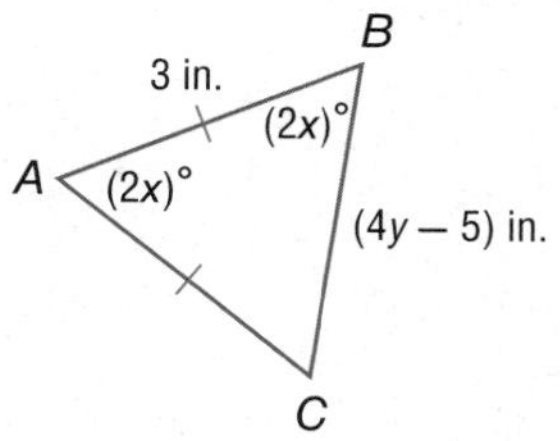

The triangle is equilateral, so all of the sides are congruent, and the lengths of all of the sides are equal.

$AB = BC$	Definition of equilateral triangle
$3 = 4y - 5$	Substitution
$8 = 4y$	Add 5 to each side.
$2 = y$	Divide each side by 4.

GuidedPractice

3. Find the value of each variable.

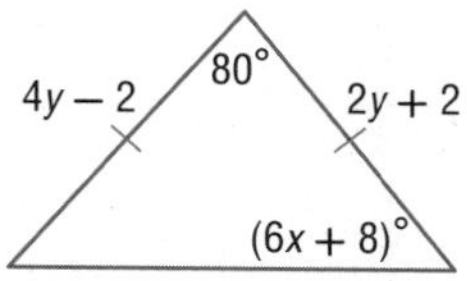

Real-WorldLink

Biosphere II is the largest totally enclosed ecosystem ever built, covering 3.14 acres in Oracle, Arizona. The controlled-environment facility is 91 feet at its highest point, and it has 6500 windows that enclose a volume of 7.2 million cubic feet.

Source: University of Arizona

PT

Real-World Example 4 Apply Triangle Congruence

ENVIRONMENT Refer to the photo of Biosphere II at the right. $\triangle ACE$ is an equilateral triangle. F is the midpoint of $\overline{AE}$, D is the midpoint of $\overline{EC}$, and B is the midpoint of $\overline{CA}$. Prove that $\triangle FBD$ is also equilateral.

Given: $\triangle ACE$ is equilateral. F is the midpoint of $\overline{AE}$, D is the midpoint of $\overline{EC}$, and B is the midpoint of $\overline{CA}$.

Prove: $\triangle FBD$ is equilateral.

Proof:

Statements	Reasons
1. $\triangle ACE$ is equilateral.	**1.** Given
2. F is the midpoint of AE, D is the midpoint of EC, and B is the midpoint of CA.	**2.** Given
3. $m\angle A = 60, m\angle C = 60, m\angle E = 60$	**3.** Each angle of an equilateral triangle measures 60.
4. $\angle A \cong \angle C \cong \angle E$	**4.** Definition of congruence and substitution
5. $\overline{AE} \cong \overline{EC} \cong \overline{CA}$	**5.** Definition of equilateral triangle
6. $AE = EC = CA$	**6.** Definition of congruence
7. $\overline{AF} \cong \overline{FE}, \overline{ED} \cong \overline{DC}, \overline{CB} \cong \overline{BA}$	**7.** Midpoint Theorem
8. $AF = FE, ED = DC, CB = BA$	**8.** Definition of congruence
9. $AF + FE = AE, ED + DC = EC, CB + BA = CA$	**9.** Segment Addition Postulate
10. $AF + AF = AE, FE + FE = AE, ED + ED = EC, DC + DC = EC, CB + CB = CA, BA + BA = CA$	**10.** Substitution
11. $2AF = AE, 2FE = AE, 2ED = EC, 2DC = EC, 2CB = CA, 2BA = CA$	**11.** Addition Property
12. $2AF = AE, 2FE = AE, 2ED = AE, 2DC = AE, 2CB = AE, 2BA = AE$	**12.** Substitution Property
13. $2AF = 2ED = 2CB, 2FE = 2DC = 2BA$	**13.** Transitive Property
14. $AF = ED = CB, FE = DC = BA$	**14.** Division Property
15. $\overline{AF} \cong \overline{ED} \cong \overline{CB}, \overline{FE} \cong \overline{DC} \cong \overline{BA}$	**15.** Definition of congruence
16. $\triangle AFB \cong \triangle EDF \cong \triangle CBD$	**16.** SAS
17. $\overline{DF} \cong \overline{FB} \cong \overline{BD}$	**17.** CPCTC
18. $\triangle FBD$ is equilateral.	**18.** Definition of equilateral triangle

GuidedPractice

4. Given that $\triangle ACE$ is equilateral, $\overline{FB} \parallel \overline{EC}$, $\overline{FD} \parallel \overline{BC}$, $\overline{BD} \parallel \overline{EF}$, and D is the midpoint of $\overline{EC}$, prove that $\triangle FED \cong \triangle BDC$.

Check Your Understanding

Example 1 **Refer to the figure at the right.**

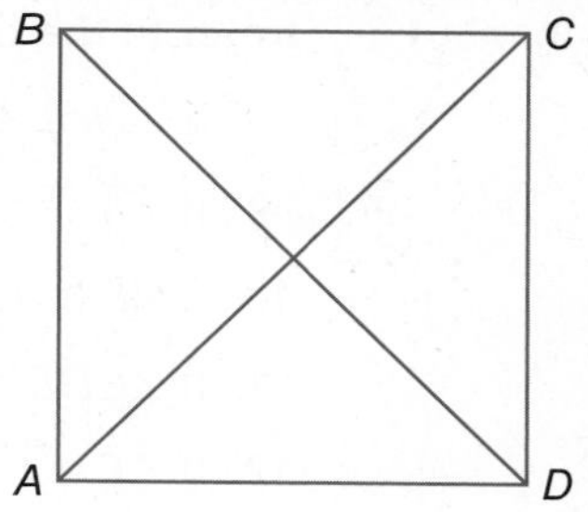

1. If $\overline{AB} \cong \overline{AD}$, name two congruent angles.

2. If $\angle CAD \cong \angle ACD$, name two congruent segments.

Example 2 **Find each measure.**

3. DE

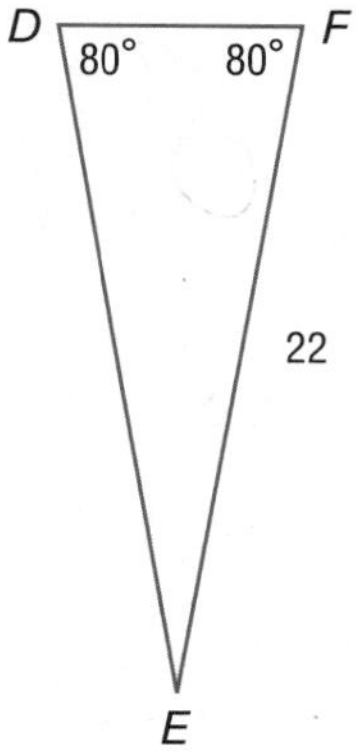

4. $m\angle MPN$

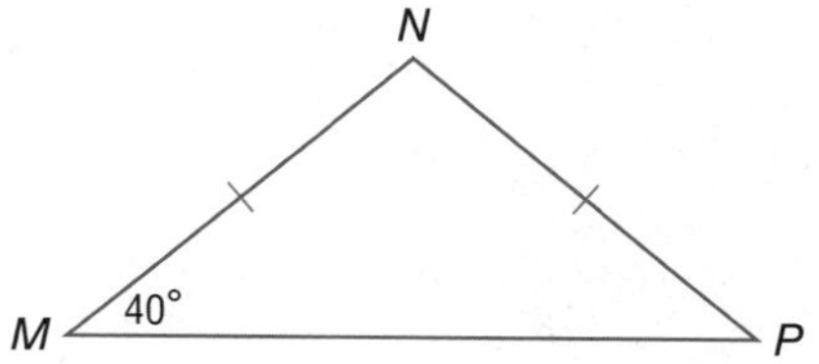

Example 3 **CCSS ALGEBRA** **Find the value of each variable.**

5.

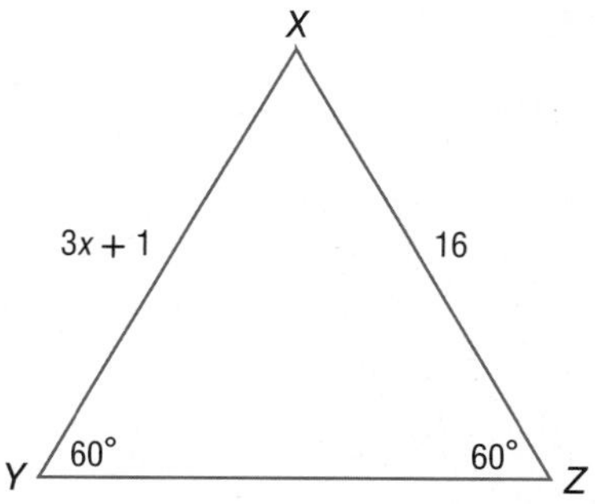

6.

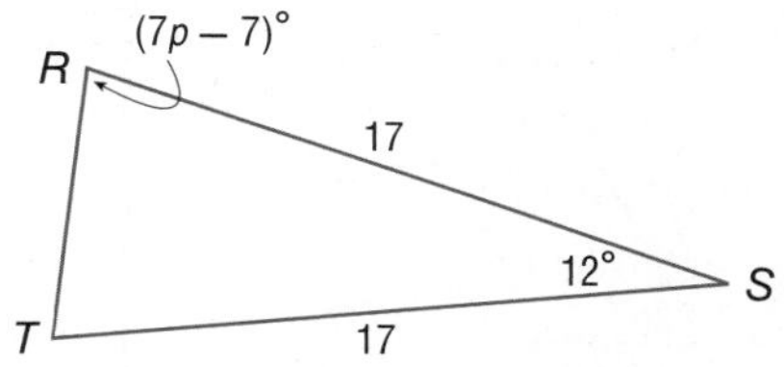

Example 4 **7. PROOF** **Write a two-column proof.**

Given: $m\angle ABC = 60$, $\overline{DA} \cong \overline{DC}$, $\angle BAD \cong \angle BCD$.

Prove: $\triangle ABC$ is equilateral.

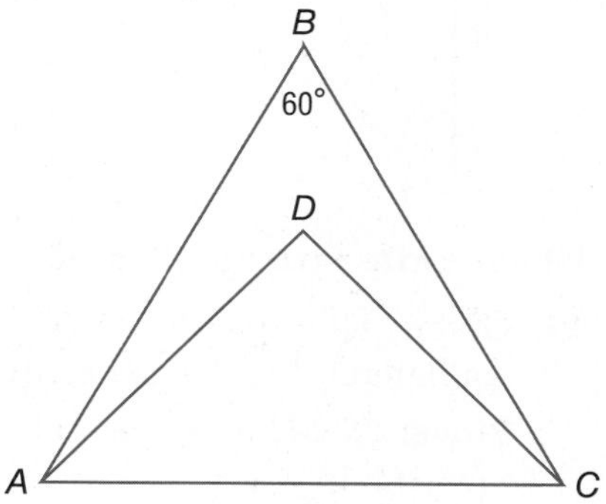

Practice and Problem Solving

Example 1 **Refer to the figure at the right.**

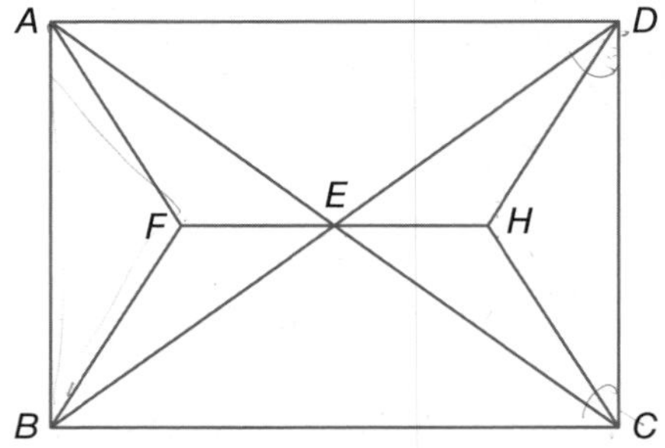

8. If $\angle DAE \cong \angle ADE$, name two congruent segments.

9. If $\angle BAF \cong \angle ABF$, name two congruent segments.

10. If $\overline{CE} \cong \overline{BE}$, name two congruent angles.

11. If $\angle CDE \cong \angle DCE$, name two congruent segments.

12. If $\overline{AE} \cong \overline{DE}$, name two congruent angles.

13. If $\overline{DH} \cong \overline{CH}$, name two congruent angles.

Example 2 **Find each measure.**

14. AB

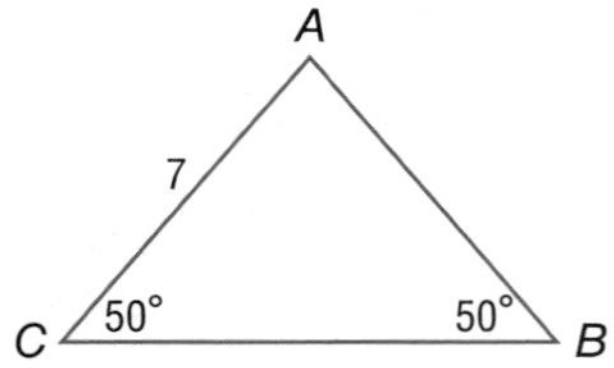

15. HG

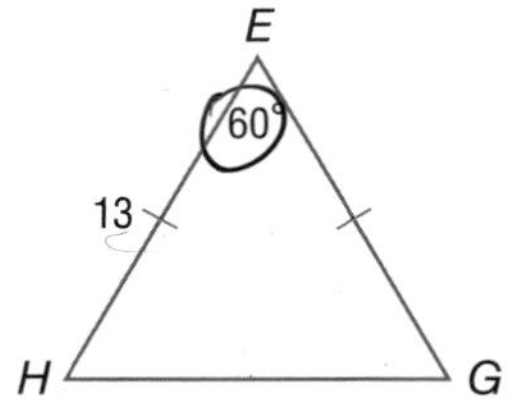

16. $m\angle NMP$

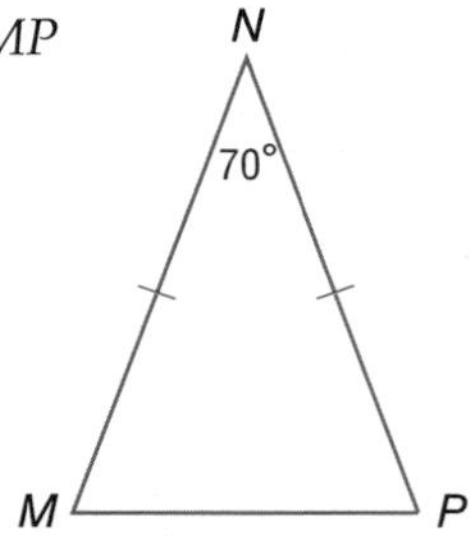

17. $m\angle RST$

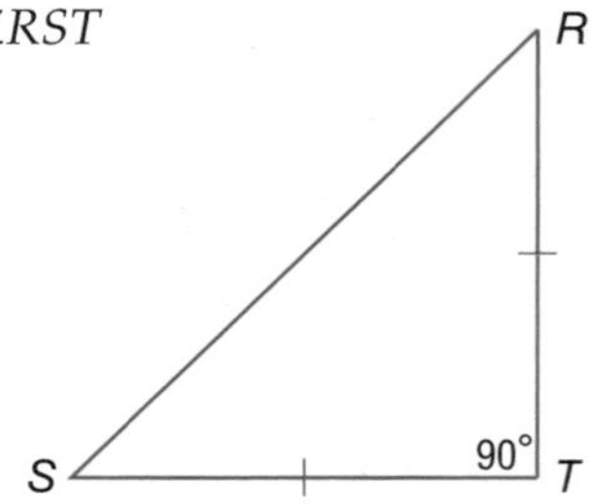

Example 3 **CCSS ALGEBRA** **Find the value of each variable.**

18.

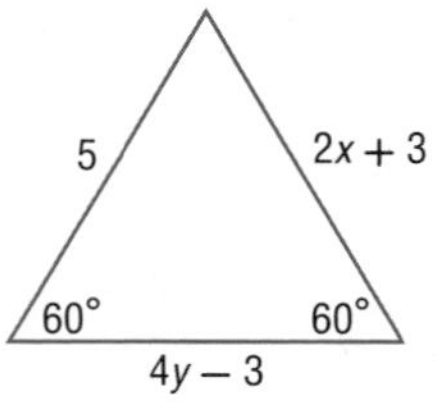

19.

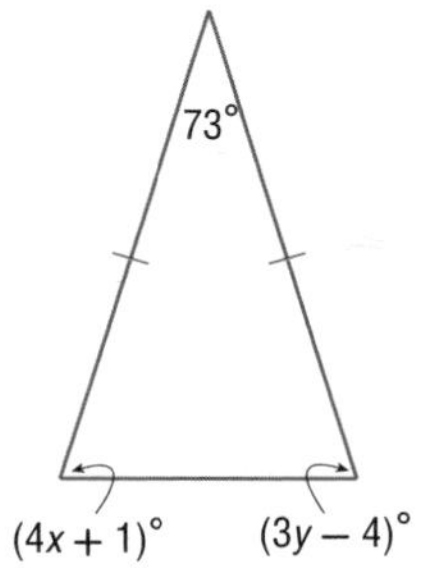

20.

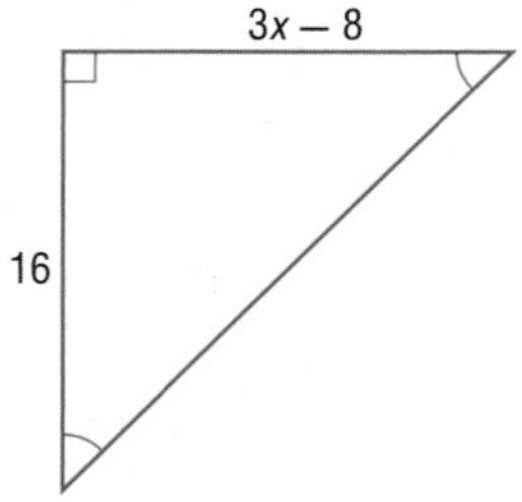

21.

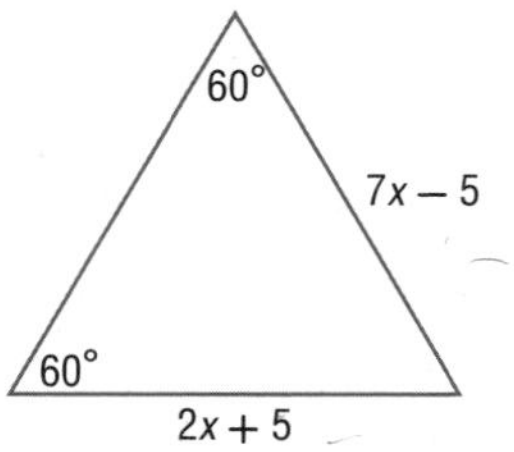

Example 4 **PROOF** **Write a paragraph proof.**

22. Given: $\overline{DE}$ is parallel to $\overline{BC}$, $\triangle ABC$ is equilateral, and $\triangle DEH$ is equilateral.

Prove: $\triangle DBH$ is equilateral.

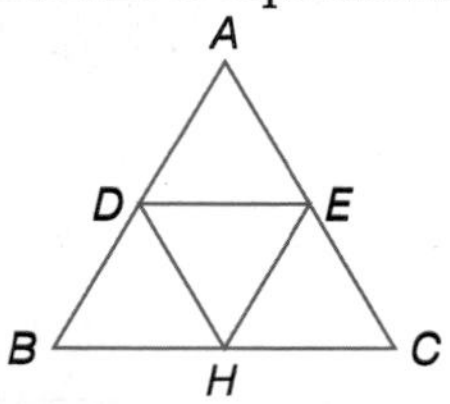

23. Given: $\triangle HNJ \cong \triangle HMP$, $\triangle JNK \cong \triangle MPL$

Prove: $m\angle HKL = m\angle HLK$

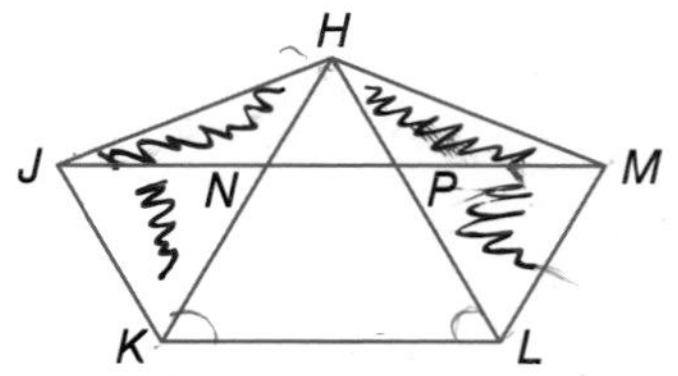

24. PYRAMIDS The pyramid shown consists of 4 triangles. If :△*JKL*, :△*JLM*, and :△*JMN* are all isosceles triangles, prove that :△*JKN* is also an isosceles triangle.

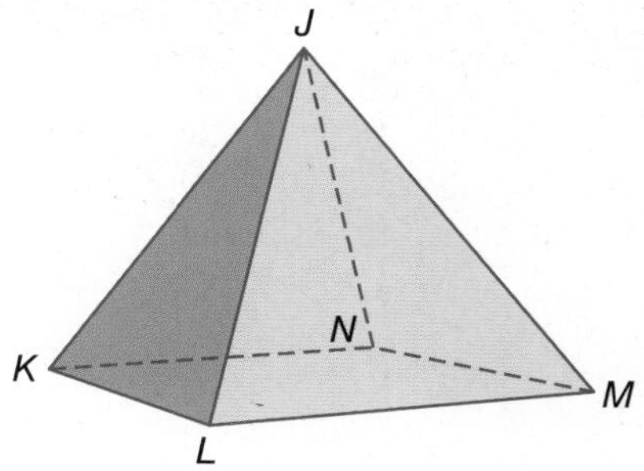

25. CONSTRUCTION Construct three different equilateral triangles. Explain your method. Then verify your constructions using measurement and mathematics. Then construct the angle bisectors of one of the angles in each triangle.

26. PROOF Based on your construction in Exercise 27, make and prove a conjecture about the relationship between the angle bisector and the side of the triangle it intersects.

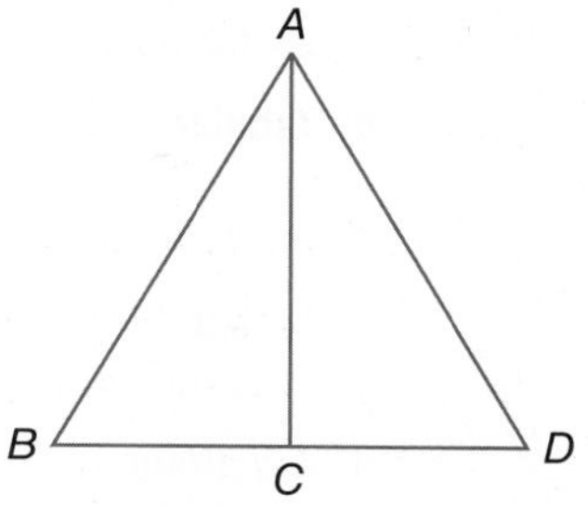

CCSS Find each measure.

27. $m\angle JLM$

28. $m\angle HJM$

29. $m\angle JKL$

30. $m\angle JLK$

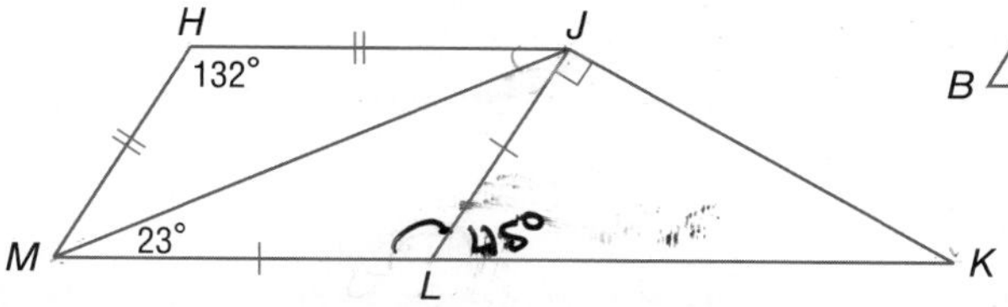

31. BIRD WATCHING Kay and Charlie are both watching a bird building a nest in a tree. If Kay and Charlie have to use the same angle of elevation to see the bird, show that the tree is half way between Kay and Charlie.

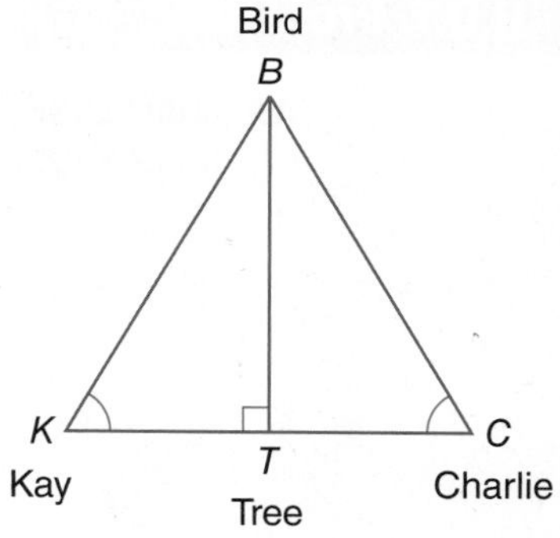

32. Given: △*ACD* and △*ABD* are isosceles and $\overline{AB}$ is parallel to $\overline{CD}$.

Prove: ∠*BAC* and ∠*ABD* are supplementary.

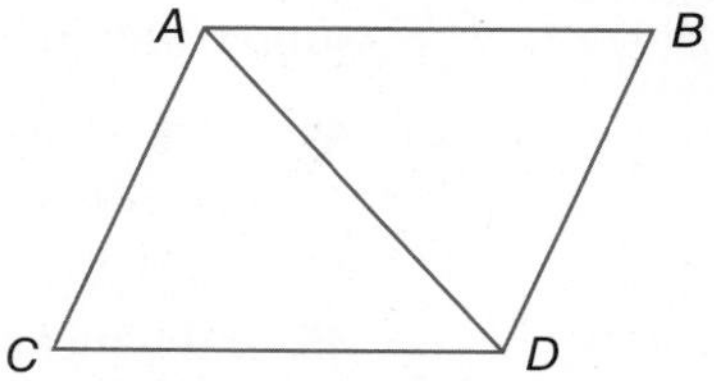

PROOF Write a two-column proof of each corollary or theorem.

33. Corollary 12.3

34. Corollary 12.4

35. Theorem 12.11

Find the value of each variable.

36.

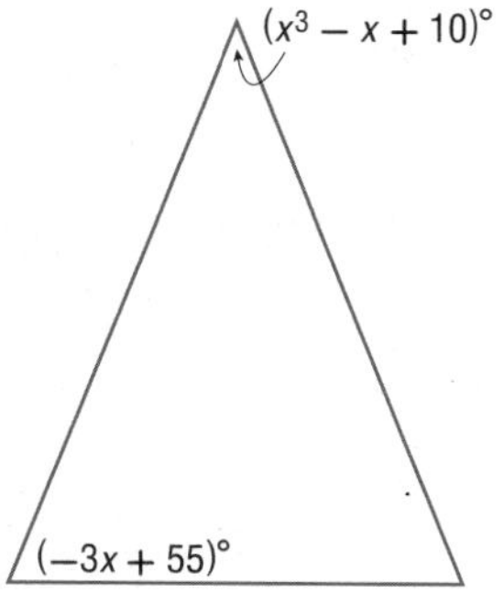

37.

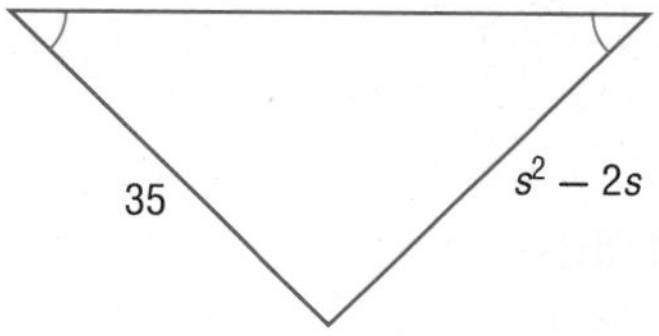

RECREATION **Use the diagram of a kite shown to find each measure**

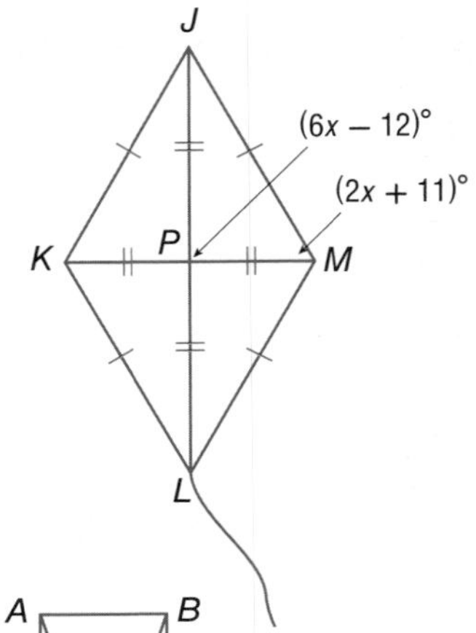

38. $m\angle JMP$

39. $m\angle MJK$

40. $m\angle MKL$

41. $m\angle KLM$

42. **MULTIPLE REPRESENTATIONS** In this problem, you will explore the triangles formed by the diagonals of a rectangle.

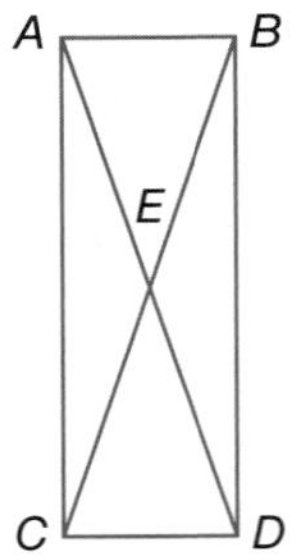

a. **Geometric** Use a ruler and protractor to draw three different rectangles and their diagonals. Label as shown.

b. **Tabular** Use a protractor to measure and record $m\angle CAE$ and $m\angle ACE$. Use these measurements to find $m\angle AEC$, $m\angle AEB$, $m\angle BAE$, and $m\angle ABE$. Organize your results in a table.

c. **Verbal** Explain how you used $m\angle CAE$ and $m\angle ACE$. to find $m\angle AEC$, $m\angle AEB$, $m\angle BAE$, and $m\angle ABE$.

d. **Algebraic** If $m\angle CAE = x$, write an expression for the measures of $m\angle AEC$, $m\angle AEB$, $m\angle BAE$, and $m\angle ABE$.

H.O.T. Problems Use Higher-Order Thinking Skills

43. **CHALLENGE** ΔXYZ is inscribed in a circle with center C as show. If $m\angle YCZ = 120$ and $\overline{CZ}$ bisects $\angle XZY$, prove ΔXYZ is equilateral.

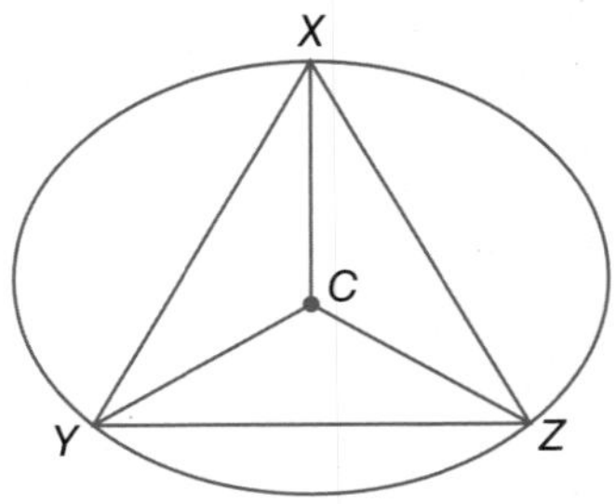

CCSS REASONING **Determine whether the following statements are** ***sometimes, always,*** **or** ***never*** **true. Explain.**

44. If the measure of the vertex angle of an isosceles triangle is an integer, then the measure of each base angle is even.

45. If the measure of the base angles of an isosceles triangle is even, then the measure of the vertex angle is odd.

46. **ERROR ANALYSIS** Zoe and Theodore are finding the value of x in the figure show. Zoe says that $x = 5$, while Theodore says that $x = 8$. Is either of them correct? Explain your reasoning.

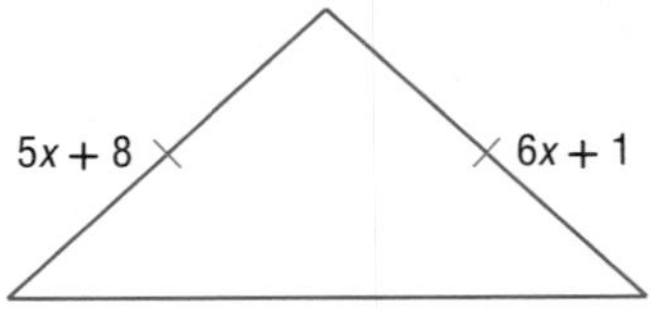

47. **REASONING** If you are given a diagram of an isosceles triangle, how many angles do you need to be given the measure of to find the measure of each of the angles? Explain your reasoning.

48. **WRITING IN MATH** Where do you see symmetry in isosceles and equilateral triangles?

Standardized Test Practice

49. ALGEBRA What quantity should be added to both sides of this equation to complete the square?

$$x^2 - 10x = 3$$

A -25 **C** 5
B -5 **D** 25

50. SHORT RESPONSE In a school of 375 students, 150 students play sports and 70 students are involved in the community service club. 30 students play sports and are involved in the community service club. How many students are *not* involved in either sports or the community service club?

51. In the figure $\overline{AE}$ and $\overline{BD}$ bisect each other at point C.

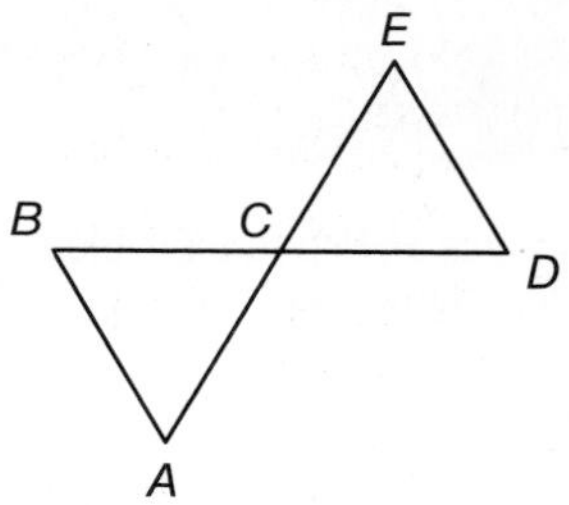

Which additional piece of information would be enough to prove that $\overline{DE} \cong \overline{DC}$?

F $\angle A \cong \angle BCA$ **H** $\angle ACB \cong \angle EDC$
G $\angle B \cong \angle D$ **J** $\angle A \cong \angle B$

52. SAT/ACT If $x = -3$, then $4x^2 - 7x + 5 =$

A 2 **C** 20 **E** 62
B 14 **D** 42

Spiral Review

53. If $m\angle ADC = 35$, $m\angle ABC = 35$, $m\angle DAC = 26$, and $m\angle BAC = 26$, determine whether $\triangle ADC \cong \triangle ABC$. (Lesson 12-5)

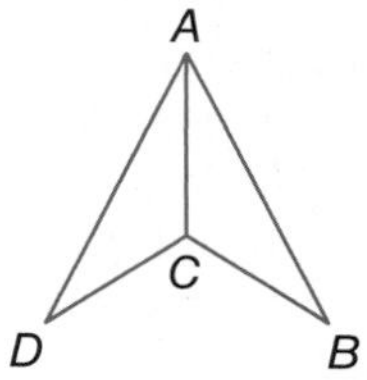

Determine whether $\triangle STU \cong \triangle XYZ$. Explain. (Lesson 12-4)

54. $S(0, 5)$, $T(0, 0)$, $U(1, 1)$, $X(4, 8)$, $Y(4, 3)$, $Z(6, 3)$

55. $S(2, 2)$, $T(4, 6)$, $U(3, 1)$, $X(-2, -2)$, $Y(-4, 6)$, $Z(-3, 1)$

56. PHOTOGRAPHY Film is fed through a traditional camera by gears that catch the perforation in the film. The distance from A to C is the same as the distance from B to D. Show that the two perforated strips are the same width. (Lesson 10-7)

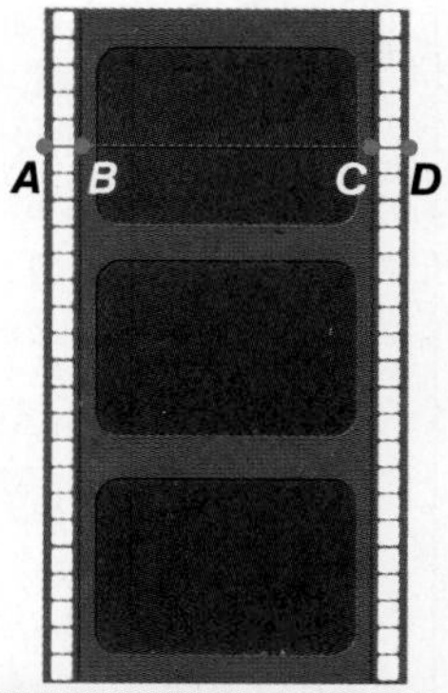

Refer to the figure at the right. (Lesson 10-1)

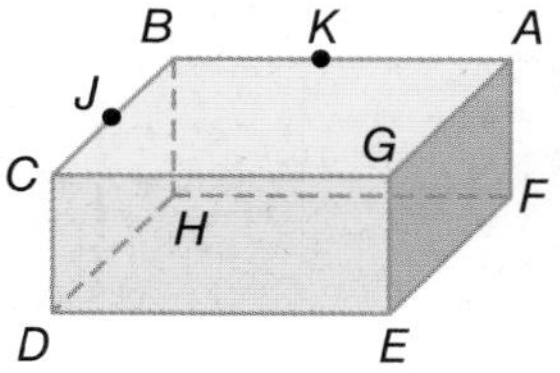

57. How many planes appear in this figure?

58. Name three points that are collinear.

59. Are points A, C, D, and J coplanar?

Skills Review

60. PROOF If $\angle ACB \cong \angle ABC$, then $\angle XCA \cong \angle YBA$.

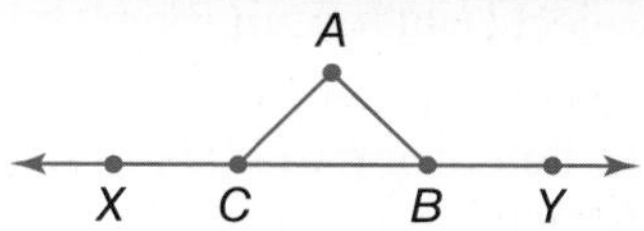

EXPLORE 12-7 Graphing Technology Lab
Congruence Transformations

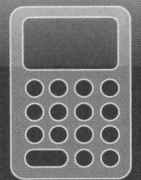

You can use TI-Nspire technology to perform *transformations* on triangles in the coordinate plane and test for congruence.

CCSS Common Core State Standards
Content Standards
G.CO.5 Given a geometric figure and a rotation, reflection, or translation, draw the transformed figure using, e.g., graph paper, tracing paper, or geometry software. Specify a sequence of transformations that will carry a given figure onto another.
G.CO.6 Use geometric descriptions of rigid motions to transform figures and to predict the effect of a given rigid motion on a given figure; given two figures, use the definition of congruence in terms of rigid motions to decide if they are congruent.
Mathematical Practices 5

PT

Activity 1 Translate a Triangle and Test for Congruence

Step 1 Open a new **Graphs** page. Select **Show Grid** from the **View** menu. Use the **Window/Zoom** menu to adjust the window size.

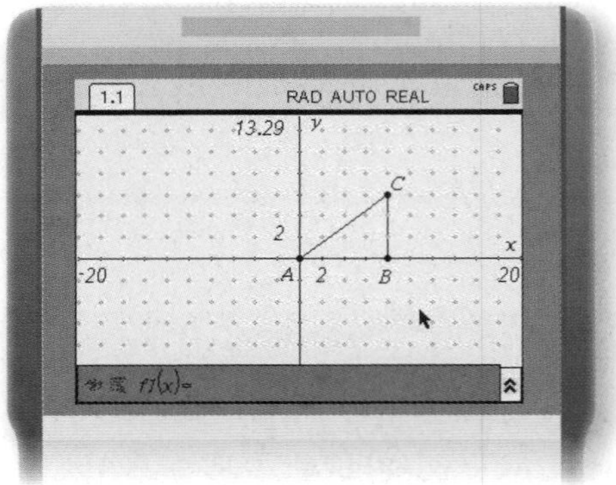

Step 2 Select **Triangle** from the **Shapes** menu and draw a right triangle with legs measuring 6 units and 8 units as shown by placing the first point at (0, 0), the second point at (8, 0), and the third point at (8, 6). Use the **Text** tool under the **Actions** menu to label the vertices of the triangle as A, B, and C.

Step 3 Select **Translation** from the **Transformation** menu. Then select $\triangle ABC$ and point A. Translate or *slide* the right triangle 8 units down and 14 units left. Label the corresponding vertices of the image as A', B', and C'.

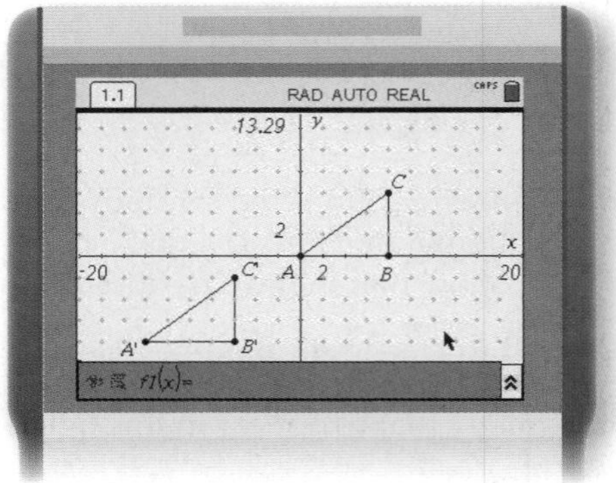

Step 4 To verify that $\triangle A'B'C'$ is congruent to $\triangle ABC$, select **Length** from the **Measurement** menu. Then select any two endpoints and press the **ENTER** key to determine the length of the segment. Repeat this for each segment of each triangle.

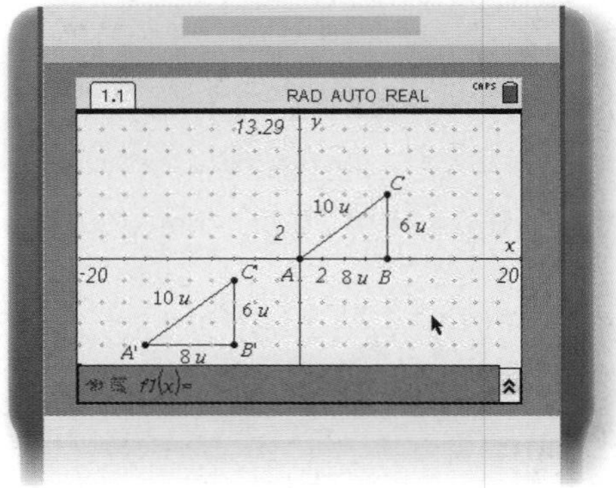

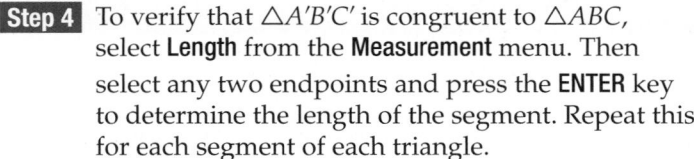

In addition to measuring lengths, the TI-Nspire can also be used to measure angles. This will allow you to use other tests for triangle congruence that involve angle measure.

Activity 2 Reflect a Triangle and Test for Congruence

Step 1 Open a new **Graphs** page, show the grid, and redraw $\triangle ABC$ from Activity 1.

Step 2 Select **Reflection** from the **Transformation** menu. Then select $\triangle ABC$ and then the y-axis to reflect or *flip* $\triangle ABC$ in the y-axis. Label the corresponding vertices of the image as A', B', and C'.

Step 3 Use the **Angle** tool from the **Measurement** menu to find $m\angle A$ and $m\angle A'$. Use the **Length** tool from the **Measurement** menu to find AB, $A'B'$, AC, and $A'C'$.

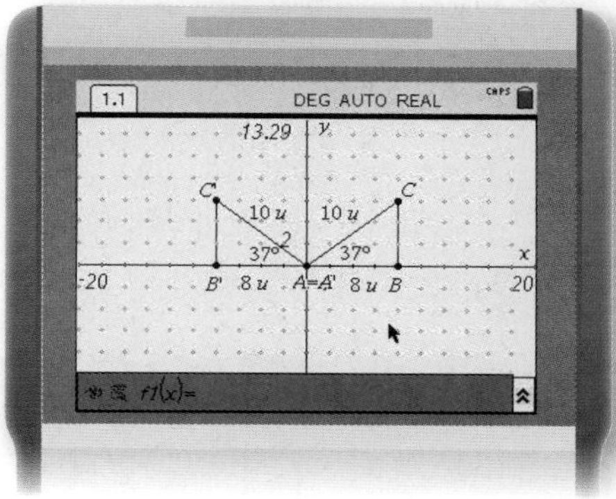

To rotate a figure about the origin using TI-Nspire technology, use the Rotation tool to select the figure, then the point (0, 0), then draw an angle of rotation.

Activity 3 Rotate a Triangle and Test for Congruence

Step 1 Open a new **Graphs** page, show the grid, and redraw $\triangle ABC$ from Activity 1.

Step 2 Select **Rotation** from the **Transformation** menu. Then select $\triangle ABC$, select the origin, and type in a number for the angle of rotation.

Step 3 Use the **Angle** tool from the **Measurement** menu to find $m\angle A$, $m\angle A'$, $m\angle C$, and $m\angle C'$. Use the **Length** tool from the **Measurement** menu to find AC and $A'C'$.

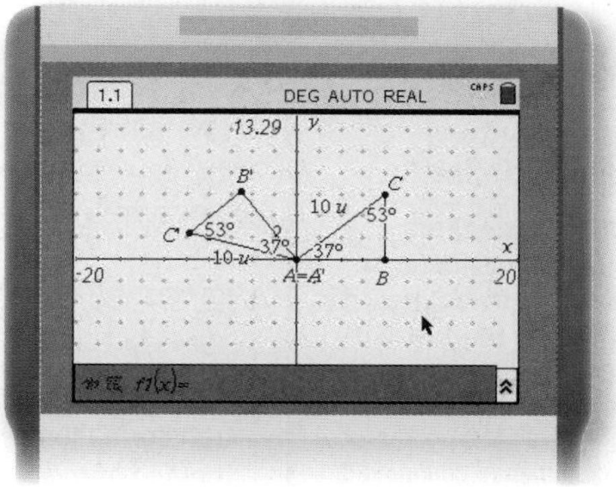

Analyze the Results

Determine whether $\triangle ABC$ and $\triangle A'B'C'$ are congruent. Explain your reasoning.

1. Activity 1
2. Activity 2
3. Activity 3
4. Explain why $\triangle A'B'C'$ in Activity 3 does not appear to be congruent to $\triangle ABC$.
5. **MAKE A CONJECTURE** Repeat Activities 1–3 using a different triangle XYZ. Analyze your results and compare them to those found in Exercises 1–3. Make a conjecture as to the relationship between a triangle and its transformed image under a translation, reflection, or a rotation.
6. Do the measurements and observations you made in Activities 1–3 constitute a proof of the conjecture you made in Exercise 5? Explain.

LESSON

12-7 Congruence Transformations

Then

- You proved whether two triangles were congruent.

Now

1. Identify reflections, translations, and rotations.
2. Verify congruence after a congruence transformation.

Why?

- The fashion industry often uses prints that display patterns. Many of these patterns are created by taking one figure and sliding it to create another figure in a different location, flipping the figure to create a mirror image of the original, or turning the original figure to create a new one.

NewVocabulary

transformation
preimage
image
congruence transformation
isometry
reflection
translation
rotation

Common Core State Standards

Content Standards

G.CO.6 Use geometric descriptions of rigid motions to transform figures and to predict the effect of a given rigid motion on a given figure; given two figures, use the definition of congruence in terms of rigid motions to decide if they are congruent.

G.CO.7 Use the definition of congruence in terms of rigid motions to show that two triangles are congruent if and only if corresponding pairs of sides and corresponding pairs of angles are congruent.

Mathematical Practices

1 Make sense of problems and persevere in solving them.

7 Look for and make use of structure.

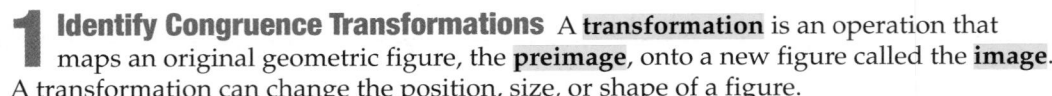

1 Identify Congruence Transformations

A **transformation** is an operation that maps an original geometric figure, the **preimage**, onto a new figure called the **image**. A transformation can change the position, size, or shape of a figure.

A transformation can be noted using an arrow. The transformation statement $\triangle ABC \rightarrow \triangle XYZ$ tells you that A is mapped to X, B is mapped to Y, and C is mapped to Z.

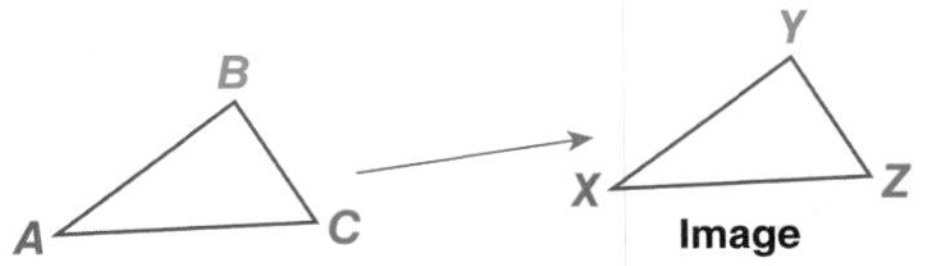

A **congruence transformation**, also called a *rigid transformation* or an **isometry**, is one in which the position of the image may differ from that of the preimage, but the two figures remain congruent. The three main types of congruence transformations are shown below.

KeyConcept Reflections, Translations, and Rotations

A **reflection** or *flip* is a transformation over a line called the *line of reflection.* Each point of the preimage and its image are the same distance from the line of reflection.

Example

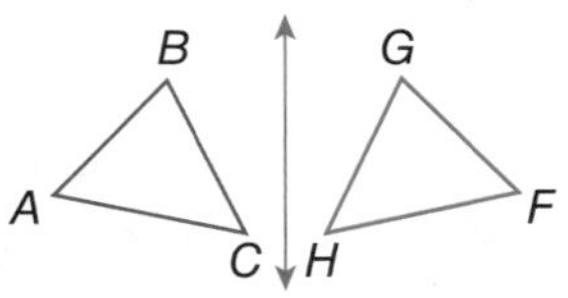

$\triangle ABC \longrightarrow \triangle FGH$

A **translation** or *slide* is a transformation that moves all points of the original figure the same distance in the same direction.

Example

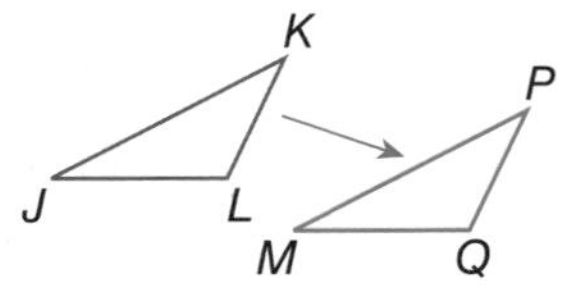

$\triangle JKL \longrightarrow \triangle MPQ$

A **rotation** or *turn* is a transformation around a fixed point called the *center of rotation*, through a specific angle, and in a specific direction. Each point of the original figure and its image are the same distance from the center.

Example

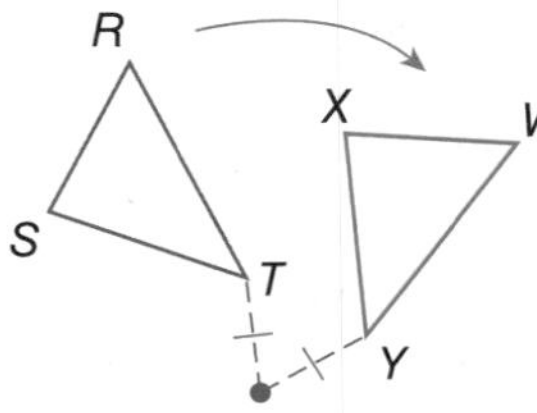

$\triangle RST \longrightarrow \triangle WXY$

StudyTip

Transformations Not all transformations preserve congruence. Only transformations that do not change the size or shape of the figure are congruence transformations.

Example 1 Identify Congruence Transformations

Identify the type of congruence transformation shown as a *reflection, translation,* or *rotation.*

a.

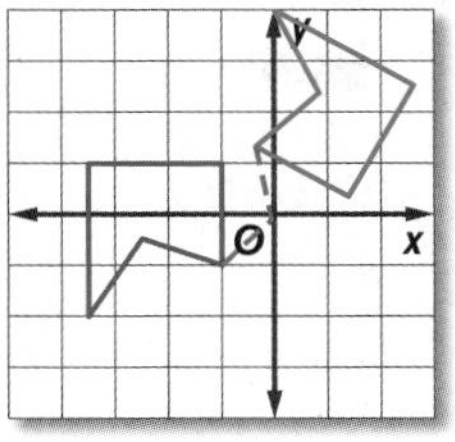

Each vertex and its image are the same distance from the origin. The angles formed by each pair of corresponding points and the origin are congruent. This is a rotation.

b.

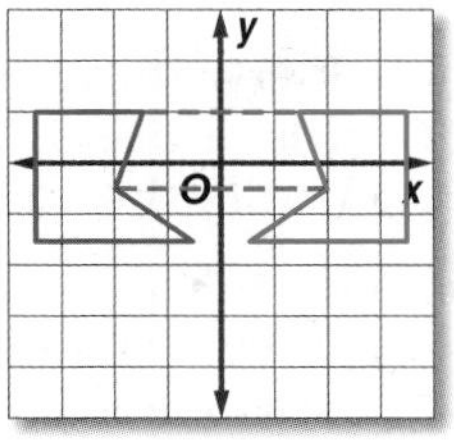

Each vertex and its image are the same distance from the y-axis. This is a reflection.

c.

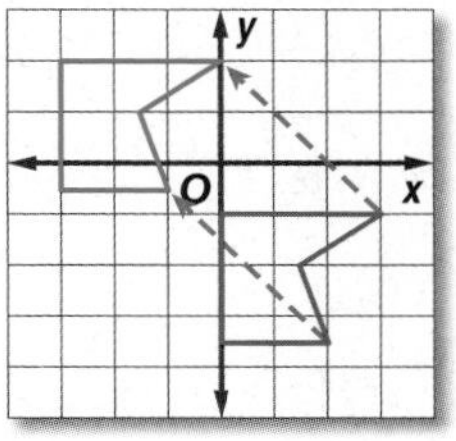

Each vertex and its image are in the same position, just 3 units left and 3 units up. This is a translation.

GuidedPractice

1A.

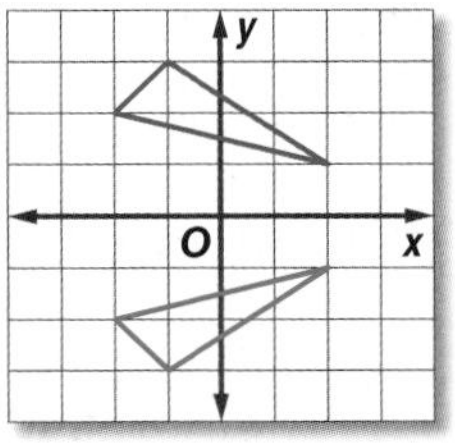

1B.

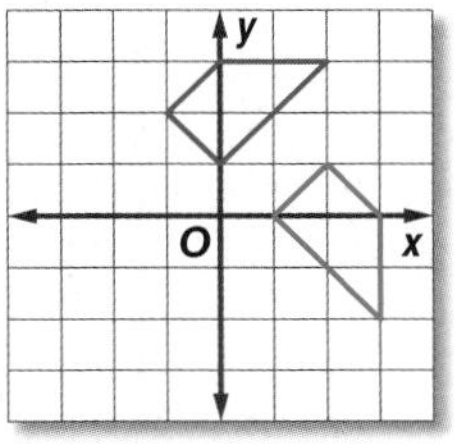

1C.

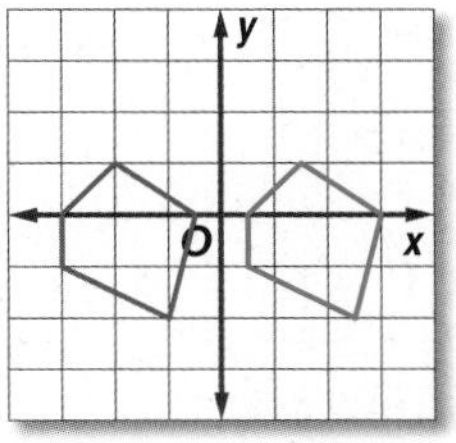

Some real-world motions or objects can be represented by transformations.

Real-WorldLink

The game shown above involves a weight attached to a ring that you can place around your ankle. As the rope passes in front of your other foot, you skip over it.

Corbis/SuperStock

Real-World Example 2 Identify a Real-World Transformation

GAMES Refer to the information at the left. Identify the type of congruence transformation shown in the diagram as a *reflection, translation,* or *rotation.*

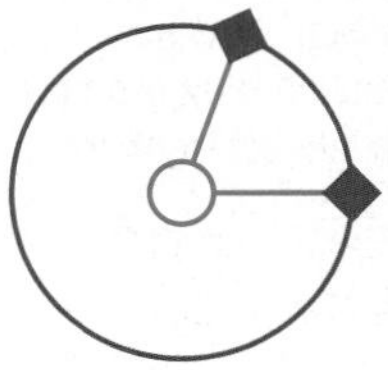

The position of the weight at different times is an example of a rotation. The center of rotation is the person's ankle.

GuidedPractice

Identify the type of congruence transformation shown as a *reflection, translation,* or *rotation.*

2A.

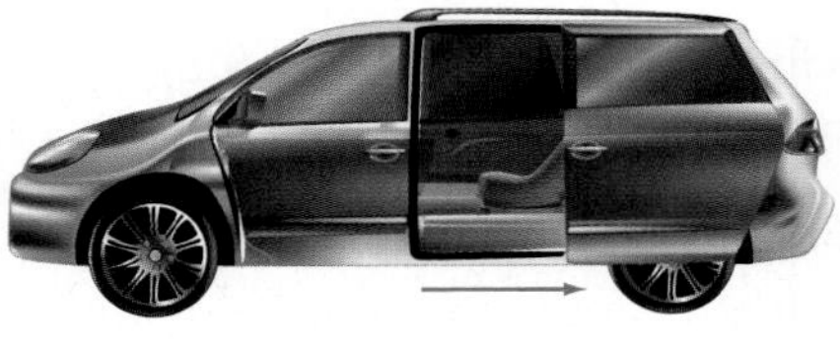

2B.

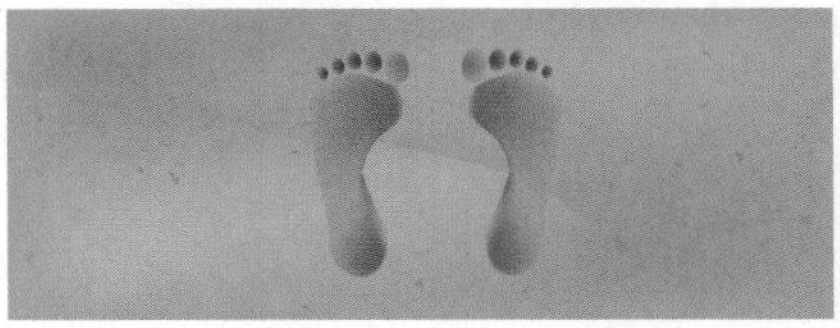

2 Verify Congruence

Verify Congruence You can verify that reflections, translations, and rotations of triangles produce congruent triangles using SSS.

Example 3 Verify Congruence after a Transformation

Triangle XZY with vertices $X(2, -8)$, $Z(6, -7)$, and $Y(4, -2)$ is a transformation of $\triangle ABC$ with vertices $A(2, 8)$, $B(6, 7)$, and $C(4, 2)$. Graph the original figure and its image. Identify the transformation and verify that it is a congruence transformation.

Understand You are asked to identify the type of transformation—reflection, translation, or rotation. Then, you need to show that the two figures are congruent.

Plan Use the Distance Formula to find the measure of each side. Then show that the two triangles are congruent by SSS.

Solve Graph each figure. The transformation appears to be a reflection over the x-axis. Find the measures of the sides of each triangle.

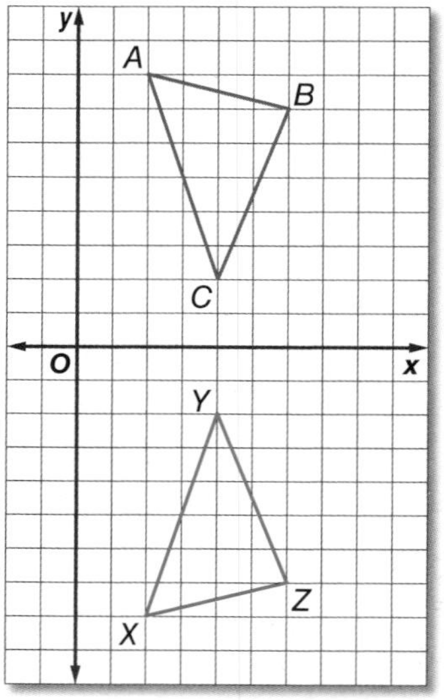

$AB = \sqrt{(6-2)^2 + (7-8)^2}$ or $\sqrt{17}$

$BC = \sqrt{(6-4)^2 + (7-2)^2}$ or $\sqrt{29}$

$AC = \sqrt{(4-2)^2 + (2-8)^2}$ or $\sqrt{40}$

$XZ = \sqrt{(6-2)^2 + [-7-(-8)]^2}$ or $\sqrt{17}$

$ZY = \sqrt{(6-4)^2 + [-7-(-2)]^2}$ or $\sqrt{29}$

$XY = \sqrt{(2-4)^2 + [-8-(-2)]^2}$ or $\sqrt{40}$

Since $AB = XZ$, $BC = ZY$, and $AC = XY$, $\overline{AB} \cong \overline{XZ}$, $\overline{BC} \cong \overline{ZY}$, and $\overline{AC} \cong \overline{XY}$. By SSS, $\triangle ABC \cong \triangle XZY$.

Check Use the definition of a reflection. Use a ruler to measure and compare the segments connecting each vertex and its image to the line of symmetry. These segments are congruent, so the triangles are congruent. ✓

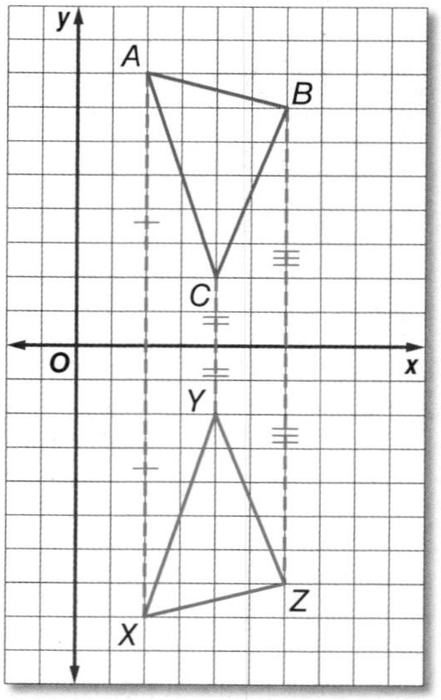

StudyTip

Isometry While an isometry preserves congruence, a *direct isometry* also preserves orientation or order of lettering. An *indirect* or *opposite isometry* changes this order, such as from clockwise to counterclockwise. The reflection shown in Example 3 is an example of an indirect isometry.

GuidedPractice

3. Triangle JKL with vertices $J(-2, 2)$, $K(-8, 5)$, and $L(-4, 6)$ is a transformation of $\triangle PQR$ with vertices $P(2, -2)$, $Q(8, -5)$, and $R(4, -6)$. Graph the original figure and its image. Identify the transformation and verify that it is a congruence transformation.

Check Your Understanding

Example 1 Identify the type of congruence transformation shown as a *reflection, translation,* or *rotation.*

1.

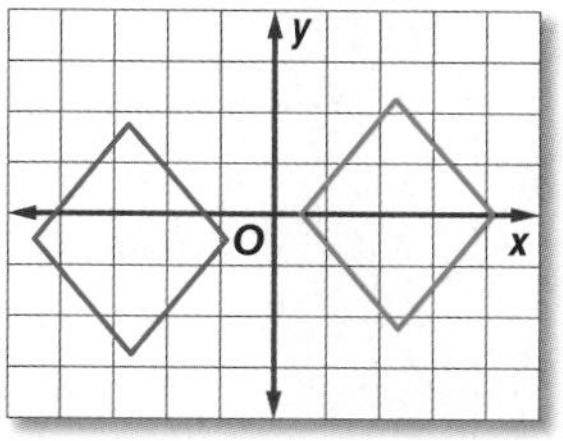

2.

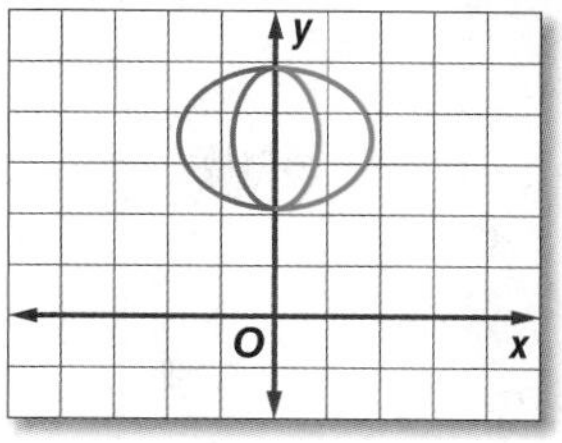

Example 2

3.

4.

Example 3 COORDINATE GEOMETRY Identify each transformation and verify that it is a congruence transformation.

5.

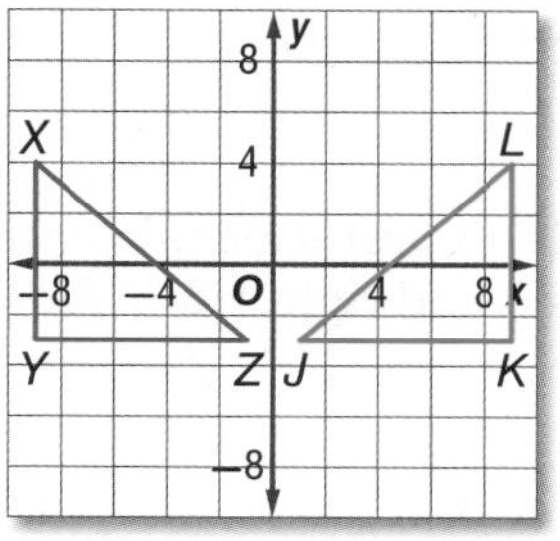

6.

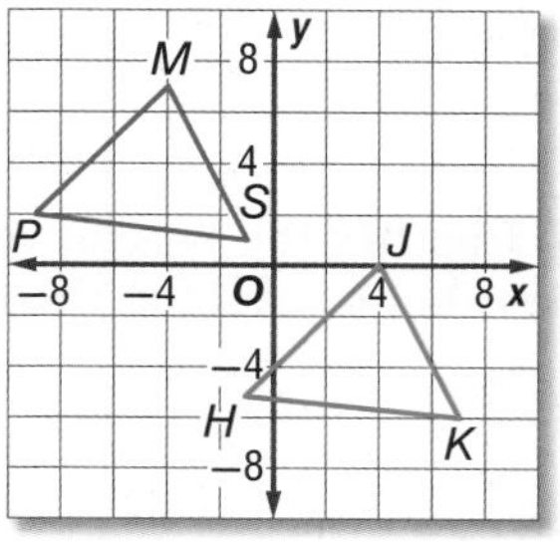

Practice and Problem Solving

Example 1 CCSS STRUCTURE Identify the type of congruence transformation shown as a *reflection, translation,* or *rotation.*

7.

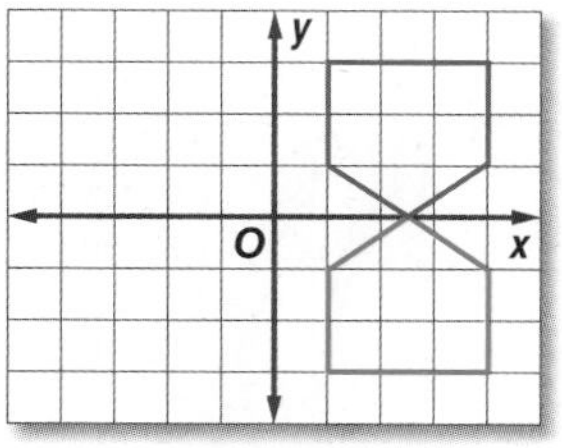

8.

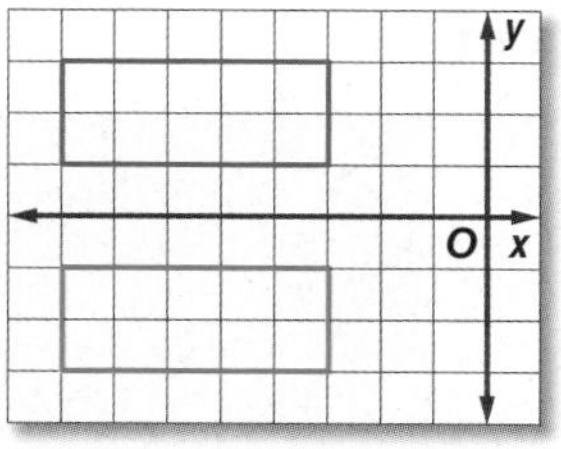

9

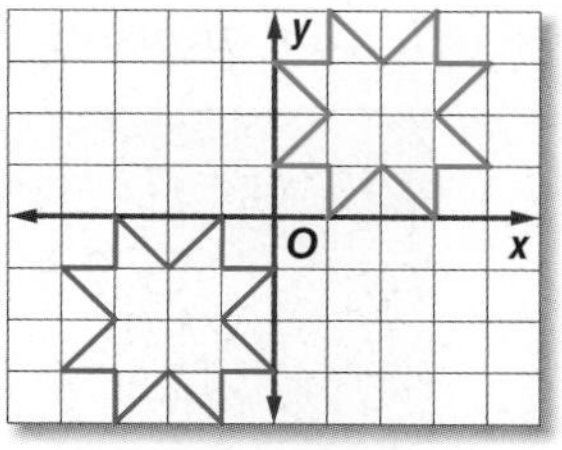

10.

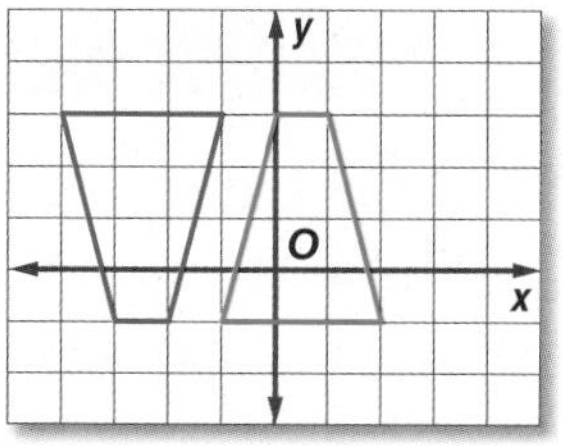

11.

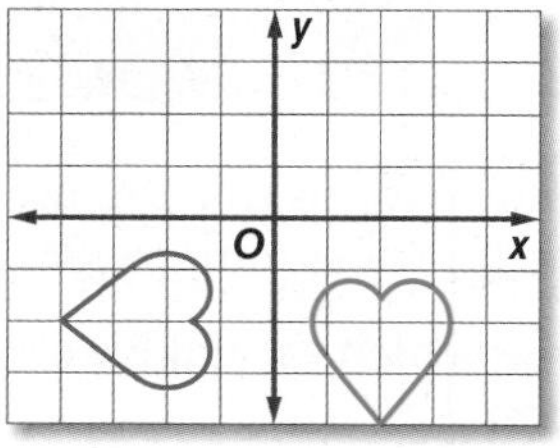

12.

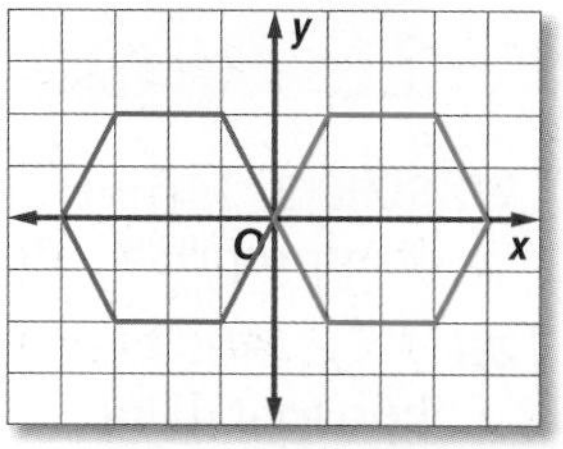

Example 2 **Identify the type of congruence transformation shown in each picture as a *reflection, translation,* or *rotation.***

13.

14.

15.

16.

Example 3 **COORDINATE GEOMETRY** **Graph each pair of triangles with the given vertices. Then, identify the transformation, and verify that it is a congruence transformation.**

17. $M(-7, -1)$, $P(-7, -7)$, $R(-1, -4)$; $T(7, -1)$, $V(7, -7)$, $S(1, -4)$

18. $A(3, 9)$, $B(3, 7)$, $C(7, 7)$; $S(3, 5)$, $T(3, 3)$, $R(7, 3)$

19. $A(-4, 5)$, $B(0, 2)$, $C(-4, 2)$; $X(-5, -4)$, $Y(-2, 0)$, $Z(-2, -4)$

20. $A(2, 2)$, $B(4, 7)$, $C(6, 2)$; $D(2, -2)$, $F(4, -7)$, $G(6, -2)$

CONSTRUCTION **Identify the type of congruence transformation performed on each given triangle to generate the other triangle in the truss with matching left and right sides shown below.**

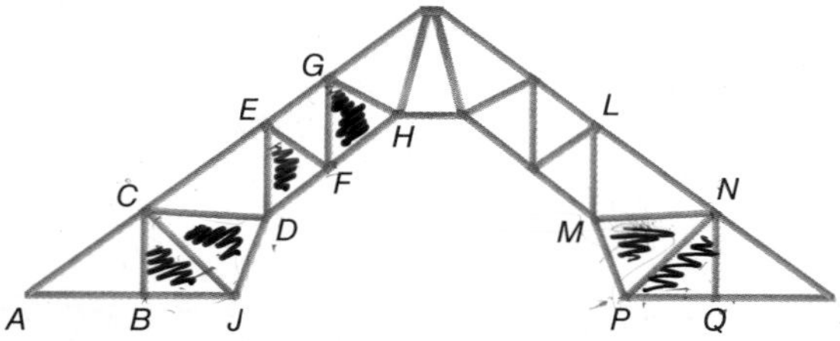

21. $\triangle NMP$ to $\triangle CJD$

22. $\triangle EFD$ to $\triangle GHF$

23. $\triangle CBJ$ to $\triangle NQP$

AMUSEMENT RIDES **Identify the type of congruence transformation shown in each picture as a *reflection, translation,* or *rotation.***

24.

25.

26.

27. **SCHOOL** Identify the transformations that are used to open a combination lock on a locker. If appropriate, identify the line of symmetry or center of rotation.

28. **CCSS STRUCTURE** Determine which capital letters of the alphabet have vertical and/or horizontal lines of reflection.

(tl)moodboard/CORBIS, (tr)Robert Harding Picture Library/SuperStock, (cl)Car Culture/Collection Mix: Subjects/Getty Images, (cr)Steve Allen/Brand X Picture/Getty Images, (bl)Kristy-Anne Glubish/Design Pics/CORBIS, (bc)NielsDK/Photolibrary, (br)KFS/age footstock

29 **DECORATING** Tionne is redecorating her bedroom. She can use stencils or a stamp to create the design shown.

a. If Tionne used the stencil, what type of transformation was used to produce each flower in the design?

b. What type of transformation was used if she used the stamp to produce each flower in the design?

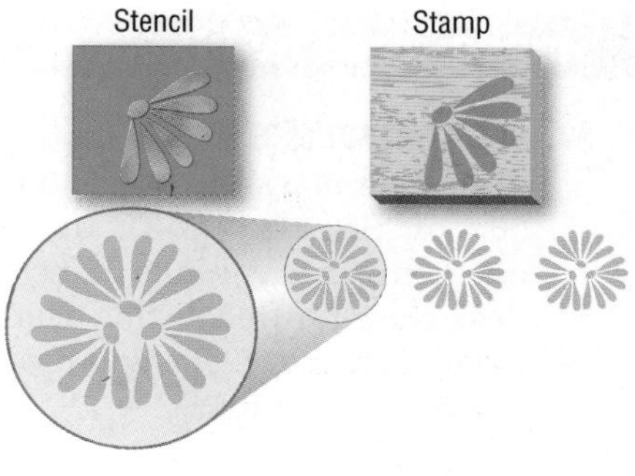

30. **MULTIPLE REPRESENTATIONS** In this problem, you will investigate the relationship between the ordered pairs of a figure and its translated image.

a. Geometric Draw congruent rectangles *ABCD* and *WXYZ* on a coordinate plane.

b. Verbal How do you get from a vertex on *ABCD* to the corresponding vertex on *WXYZ* using only horizontal and vertical movement?

c. Tabular Copy the table shown. Use your rectangles to fill in the x-coordinates, the y-coordinates, and the unknown value in the transformation column.

Rectangle *ABCD*	Transformation	Rectangle *WXYZ*
A(?, ?)	$(x_1 + ?, y_1 + ?)$	*W*(?, ?)
B(?, ?)	$(x_1 + ?, y_1 + ?)$	*X*(?, ?)
C(?, ?)	$(x_1 + ?, y_1 + ?)$	*Y*(?, ?)
D(?, ?)	$(x_1 + ?, y_1 + ?)$	*Z*(?, ?)

d. Algebraic Function notation $(x, y) \rightarrow (x + a, y + b)$, where a and b are real numbers, represents a mapping from one set of coordinates onto another. Complete the following notation that represents the rule for the translation $ABCD \rightarrow WXYZ$: $(x, y) \rightarrow (x + a, y + b)$.

H.O.T. Problems Use Higher-Order Thinking Skills

31. **CHALLENGE** Use the diagram at the right.

a. Identify two transformations of Triangle 1 that can result in Triangle 2.

b. What must be true of the triangles in order for more than one transformation on a preimage to result in the same image? Explain your reasoning.

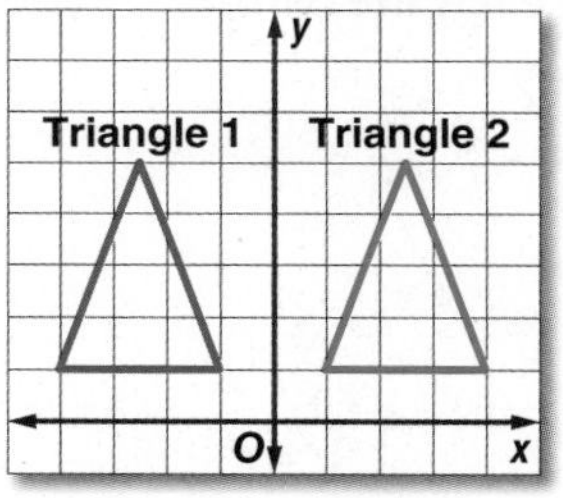

32. **CCSS REASONING** A *dilation* is another type of transformation. In the diagram, a small paper clip has been dilated to produce a larger paper clip. Explain why dilations are not a congruence transformation.

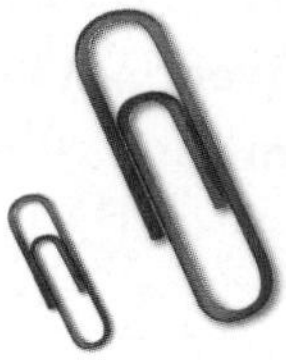

OPEN ENDED **Describe a real-world example of each of the following, other than those given in this lesson.**

33. reflection

34. translation

35. rotation

36. **WRITING IN MATH** In the diagram at the right $\triangle DEF$ is called a *glide reflection* of $\triangle ABC$. Based on the diagram, define a glide reflection. Is a glide reflection a congruence transformation? Include a definition of congruence transformation in your response. Explain your reasoning.

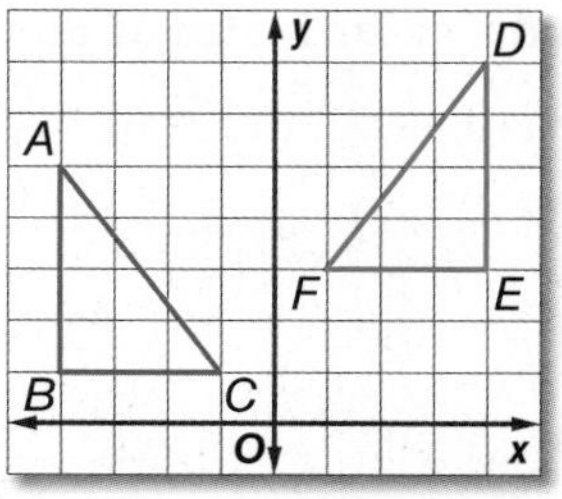

Standardized Test Practice

37. SHORT RESPONSE Cindy is shopping for a new desk chair at a store where the desk chairs are 50% off. She also has a coupon for 50% off any one item. Cindy thinks that she can now get the desk chair for free. Is this true? If not, what will be the percent off she will receive with both the sale and the coupon?

38. Identify the congruence transformation shown.

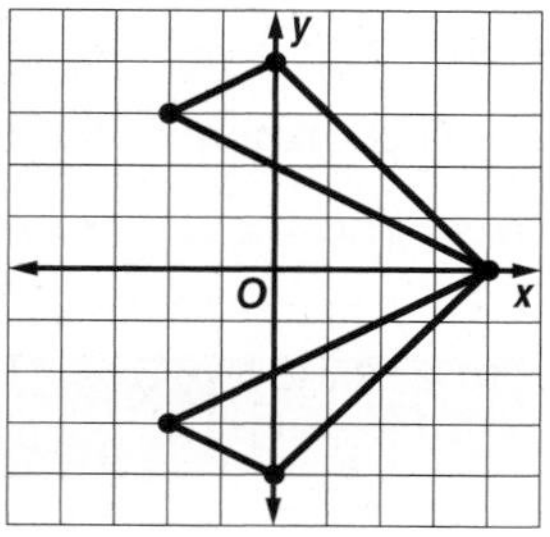

A dilation
B reflection
C rotation
D translation

39. Look at the graph below. What is the slope of the line shown?

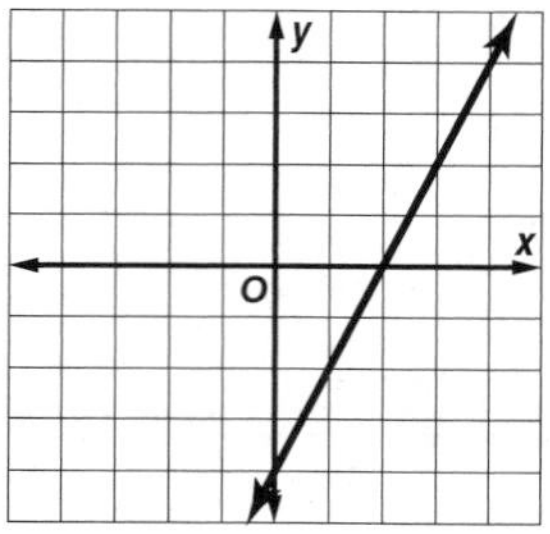

F -2
G -1
H 1
J 2

40. SAT/ACT What is the y-intercept of the line determined by the equation $3x - 4 = 12y - 3$?

A -12
B $-\frac{1}{12}$
C $\frac{1}{12}$
D $\frac{1}{4}$
E 12

Spiral Review

Find each measure. (Lesson 12-6)

41. YZ

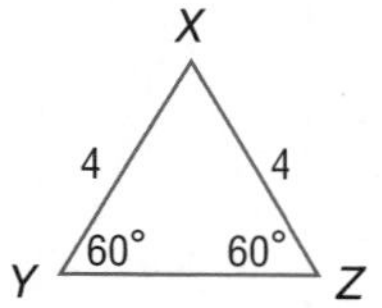

42. $m\angle JLK$

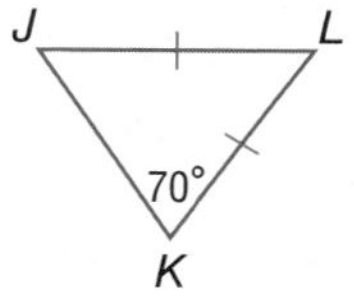

43. AB

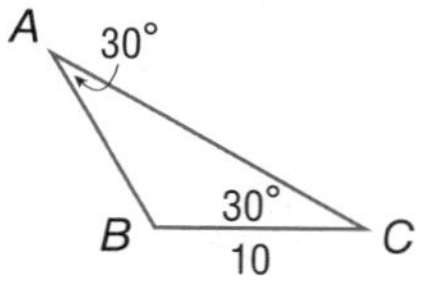

44. PROOF Write a paragraph proof. (Lesson 12-5)

Given: $\angle YWZ \cong \angle XZW$ and $\angle YZW \cong \angle XWZ$

Prove: $\triangle WXZ \cong \triangle ZYW$

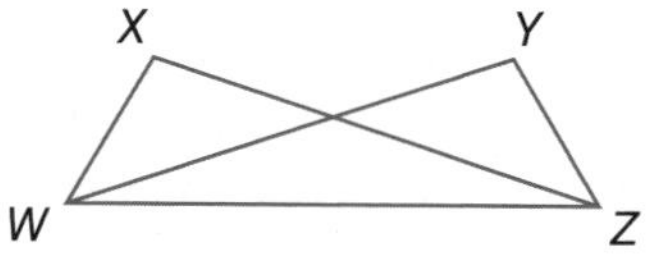

Skills Review

Find the coordinates of the midpoint of a segment with the given endpoints.

45. $A(10, -12), C(5, -6)$

46. $A(13, 14), C(3, 5)$

47. $A(-28, 8), C(-10, 2)$

48. $A(-12, 2), C(-3, 5)$

49. $A(0, 0), C(3, -4)$

50. $A(2, 14), C(0, 5)$

LESSON

12-8 Triangles and Coordinate Proof

Then

- You used coordinate geometry to prove triangle congruence.

Now

1. Position and label triangles for use in coordinate proofs.
2. Write coordinate proofs.

Why?

- A global positioning system (GPS) receives transmissions from satellites that allow the exact location of a car to be determined. The information can be used with navigation software to provide driving directions.

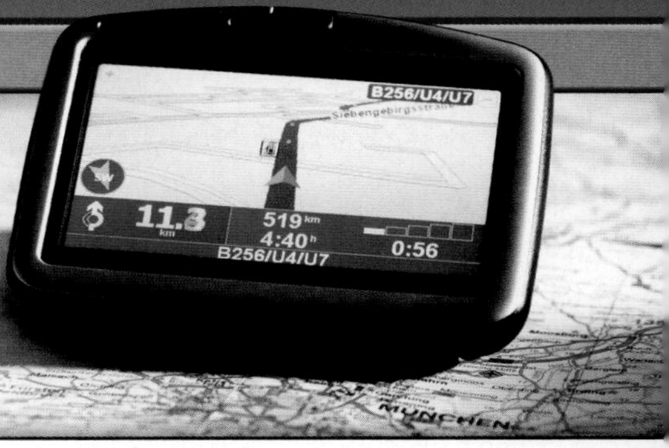

NewVocabulary
coordinate proof

Common Core State Standards

Content Standards
G.CO.10 Prove theorems about triangles.
G.GPE.4 Use coordinates to prove simple geometric theorems algebraically.

Mathematical Practices
3 Construct viable arguments and critique the reasoning of others.
2 Reason abstractly and quantitatively.

1 Position and Label Triangles

As with global positioning systems, knowing the coordinates of a figure in a coordinate plane allows you to explore its properties and draw conclusions about it. **Coordinate proofs** use figures in the coordinate plane and algebra to prove geometric concepts. The first step in a coordinate proof is placing the figure on the coordinate plane.

PT

Example 1 Position and Label a Triangle

Position and label right triangle MNP on the coordinate plane so that leg $\overline{MN}$ is a units long and leg $\overline{NP}$ is b units long.

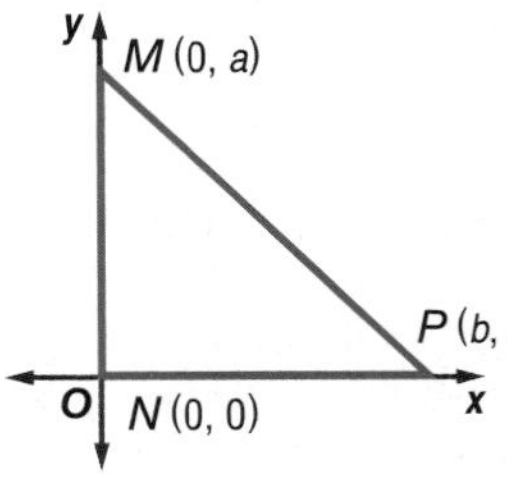

- The length(s) of the side(s) that are along the axes will be easier to determine than the length(s) of side(s) that are not along an axis. Since this is a right triangle, two sides can be located on an axis.
- Placing the right angle of the triangle, $\angle N$, at the origin will allow the two legs to be along the x- and y-axes.
- Position the triangle in the first quadrant.
- Since M is on the y-axis, its x-coordinate is 0. Its y-coordinate is a because the leg is a units long.
- Since P is on the x-axis, its y-coordinate is 0. Its x-coordinate is b because the leg is b units long.

GuidedPractice

1. Position and label isosceles triangle JKL on the coordinate plane so that its base $\overline{JL}$ is a units long, vertex K is on the y-axis, and the height of the triangle is b units.

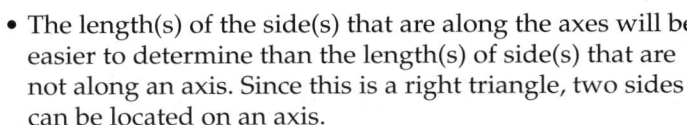

KeyConcept Placing Triangles on Coordinate Plane

Step 1 Use the origin as a vertex or center of the triangle.

Step 2 Place at least one side of a triangle on an axis.

Step 3 Keep the triangle within the first quadrant if possible.

Step 4 Use coordinates that make computations as simple as possible.

Example 2 Identify Missing Coordinates

Name the missing coordinates of isosceles triangle *XYZ*.

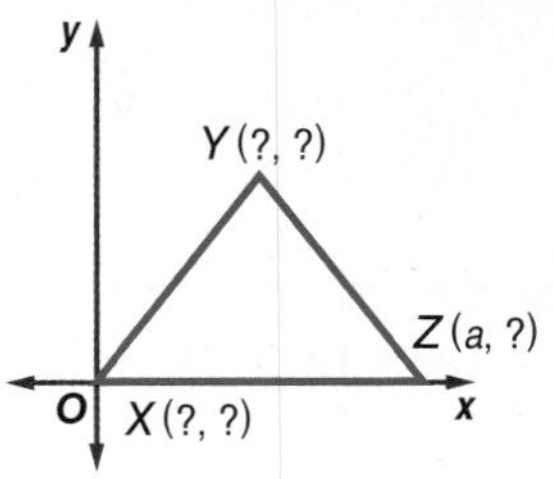

Vertex X is positioned at the origin; its coordinates are (0, 0).

Vertex Z is on the x-axis, so its y-coordinate is 0. The coordinates of vertex Z are $(a, 0)$.

$\triangle XYZ$ is isosceles, so using a vertical segment from Y to the x-axis and the Hypotenuse-Leg Theorem shows that the x-coordinate of Y is halfway between 0 and a or $\frac{a}{2}$. We cannot write the y-coordinate in terms of a, so call it b. The coordinates of point Y are $\left(\frac{a}{2}, b\right)$.

StudyTip

Right Angle The intersection of the x- and y-axis forms a right angle, so it is a convenient place to locate the right angle of a figure such as a right triangle.

GuidedPractice

2. Name the missing coordinates of isosceles right triangle ABC.

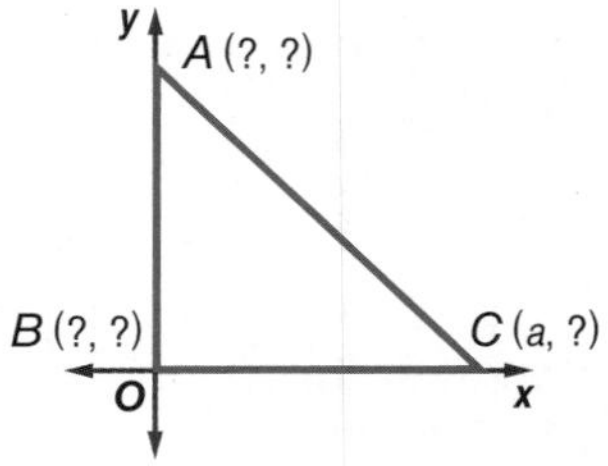

2 Write Coordinate Proofs

Write Coordinate Proofs After a triangle is placed on the coordinate plane and labeled, we can use coordinate proofs to verify properties and to prove theorems.

StudyTip

Coordinate Proof The guidelines and methods used in this lesson apply to all polygons, not just triangles.

Example 3 Write a Coordinate Proof

Write a coordinate proof to show that a line segment joining the midpoints of two sides of a triangle is parallel to the third side.

Place a vertex at the origin and label it A. Use coordinates that are multiples of 2 because the Midpoint Formula involves dividing the sum of the coordinates by 2.

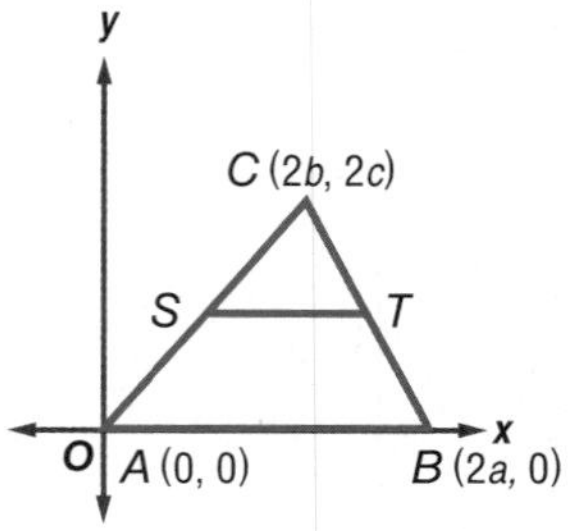

Given: $\triangle ABC$
S is the midpoint of $\overline{AC}$.
T is the midpoint of $\overline{BC}$.

Prove: $\overline{ST} \parallel \overline{AB}$

Proof:

By the Midpoint Formula, the coordinates of S are $\left(\frac{2b+0}{2}, \frac{2c+0}{2}\right)$ or (b, c) and the coordinates of T are $\left(\frac{2a+2b}{2}, \frac{0+2c}{2}\right)$ or $(a + b, c)$.

By the Slope Formula, the slope of $\overline{ST}$ is $\frac{c-c}{a+b-b}$ or 0 and the slope of $\overline{AB}$ is $\frac{0-0}{2a-0}$ or 0.

Since $\overline{ST}$ and $\overline{AB}$ have the same slope, $\overline{ST} \parallel \overline{AB}$.

GuidedPractice

3. Write a coordinate proof to show that $\triangle ABX \cong \triangle CDX$.

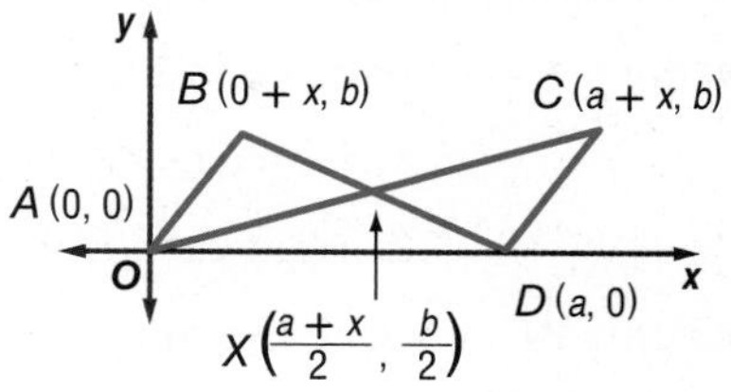

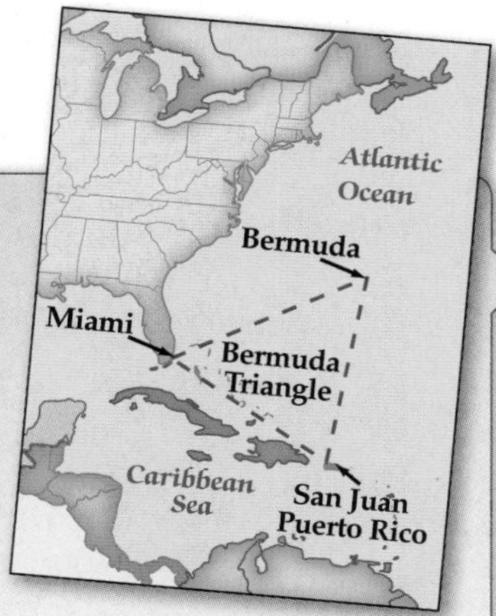

Real-WorldLink

More than 50 ships and 20 airplanes have mysteriously disappeared from a section of the North Atlantic Ocean off of North America commonly referred to as the Bermuda Triangle.

Source: *Encyclopaedia Britannica*

The techniques used for coordinate proofs can be used to solve real-world problems.

Real-World Example 4 Classify Triangles

PT

GEOGRAPHY The Bermuda Triangle is a region formed by Miami, Florida, San Jose, Puerto Rico, and Bermuda. The approximate coordinates of each location, respectively, are 25.8°N 80.27°W, 18.48°N 66.12°W, and 33.37°N 64.68°W. Write a coordinate proof to prove that the Bermuda Triangle is scalene.

The first step is to label the coordinates of each location. Let M represent Miami, B represent Bermuda, and P represent Puerto Rico.

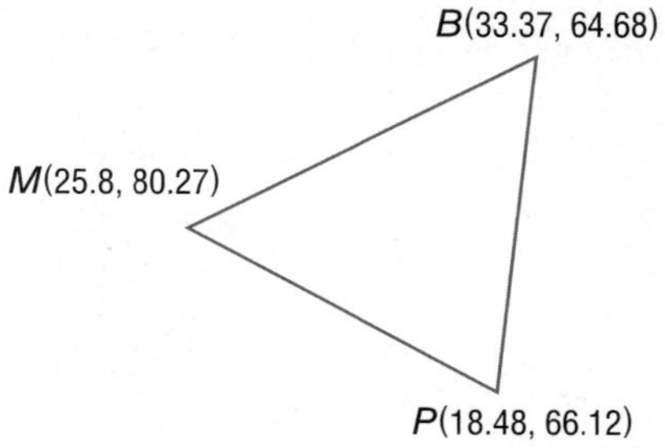

If no two sides of $\triangle MPB$ are congruent, then the Bermuda Triangle is scalene. Use the Distance Formula and a calculator to find the distance between each location.

$$MB = \sqrt{(33.37 - 25.8)^2 + (64.68 - 80.27)^2}$$

$$\approx 17.33$$

$$MP = \sqrt{(25.8 - 18.48)^2 + (80.27 - 66.12)^2}$$

$$\approx 15.93$$

$$PB = \sqrt{(33.37 - 18.48)^2 + (64.68 - 66.12)^2}$$

$$\approx 14.96$$

Since each side is a different length, $\triangle MPB$ is scalene. Therefore, the Bermuda Triangle is scalene.

GuidedPractice

4. **GEOGRAPHY** In 2006, a group of art museums collaborated to form the West Texas Triangle to promote their collections. This region is formed by the cities of Odessa, Albany, and San Angelo. The approximate coordinates of each location, respectively, are 31.9°N 102.3°W, 32.7°N 99.3°W, and 31.4°N 100.5°W. Write a coordinate proof to prove that the West Texas Triangle is approximately isosceles.

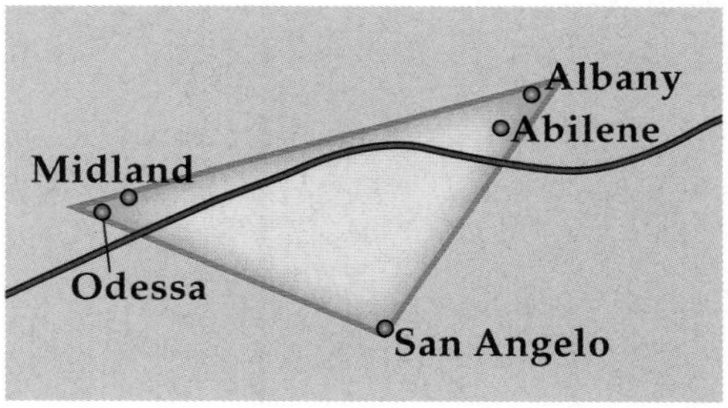

Check Your Understanding

Example 1 **Position and label each triangle on the coordinate plane.**

1. isosceles $\triangle ABC$ with base $\overline{BC}$ that is $4a$ units long.
2. right $\triangle FGH$ with legs $\overline{FG}$ and $\overline{GH}$ so that $\overline{FG}$ is $3a$ units long and leg $\overline{GH}$ is $5b$ units long

Example 2 **Name the missing coordinate(s) of each triangle.**

3.

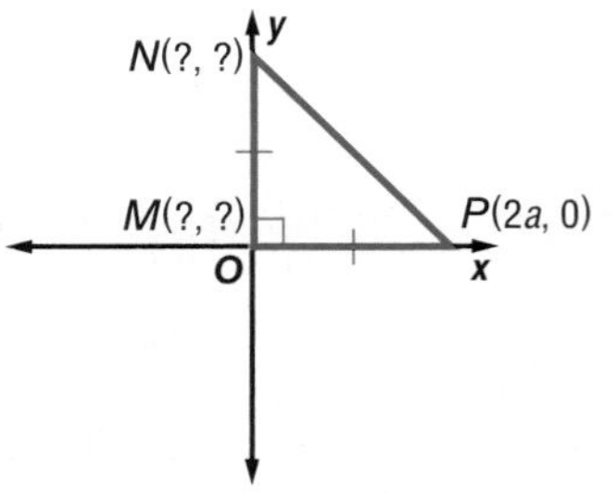

4.

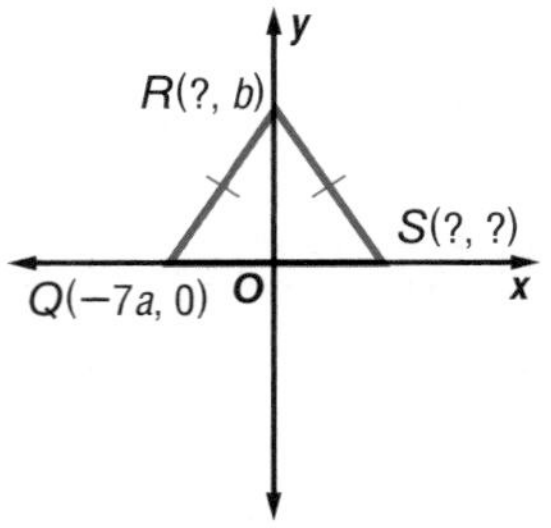

Example 3

5. Write a coordinate proof to show that $\triangle TXZ$ is similar to $\triangle WXY$.

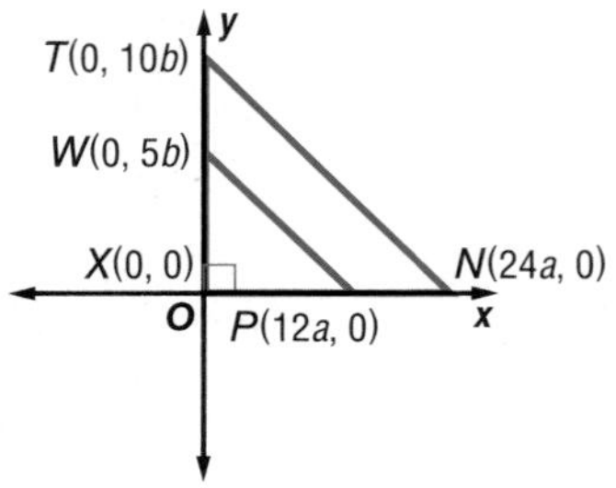

Example 4

6. **OLYMPICS** During it's trip from Olympia, Greece to the 2010 Winter Games the Olympic torch passed through London, England, Niagara Falls, Ontario and ended up in Vancouver, British Columbia. The approximate coordinates of each location respectively, are 42.9°N, 81.2°W, 43.1°N 79.1°W, and 49.3°N 123.1°W. Write a coordinate proof to prove that these three points on the torches journey form a scalene triangle.

Practice and Problem Solving

Example 1 **Position and label each triangle on the coordinate plane.**

7. equilateral $\triangle ABC$ with sides $5a$ units long.
8. isosceles right $\triangle RST$ with hypotenuse $\overline{RS}$ $4d$ units long.

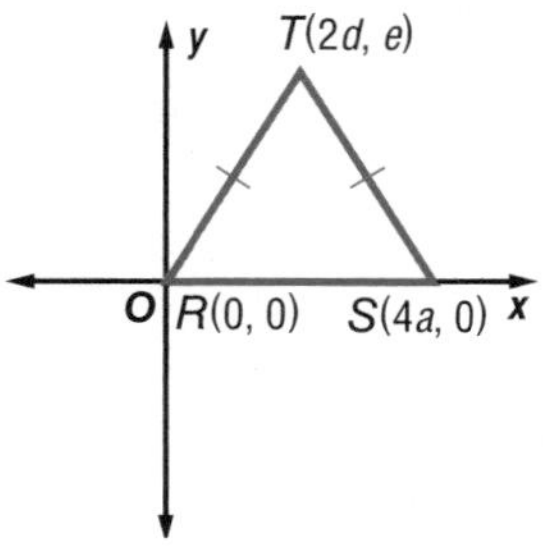

9. right $\triangle JKL$ with legs $\overline{JK}$ and $\overline{JL}$, so that $\overline{JK}$ is a units and $\overline{JL}$ is 4 times the length of $\overline{JK}$.

10. equilateral $\triangle XYZ$ with sides $\frac{1}{4}c$ units long.

11. isosceles ΔDEF with legs $\overline{DE}$ and $\overline{DF}$, with a base that is $6d$ units long.

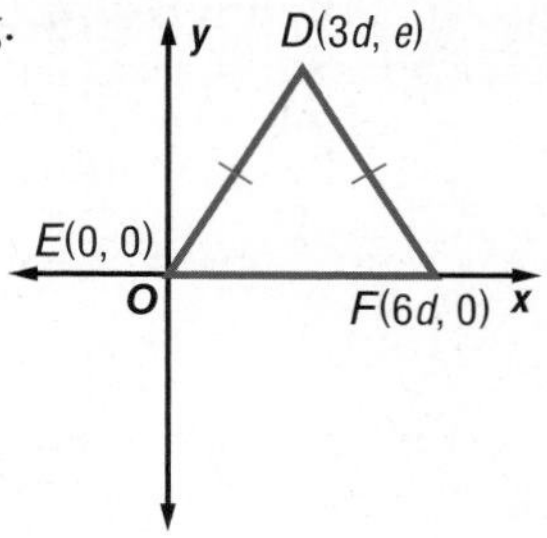

12. right ΔMNP with hypotenuse $\overline{MN}$. The length of $\overline{MP}$ is $2a$ units long and the length of $\overline{NP}$ is $4b$ units long.

Example 2 **Name the missing coordinate(s) of each triangle.**

13.

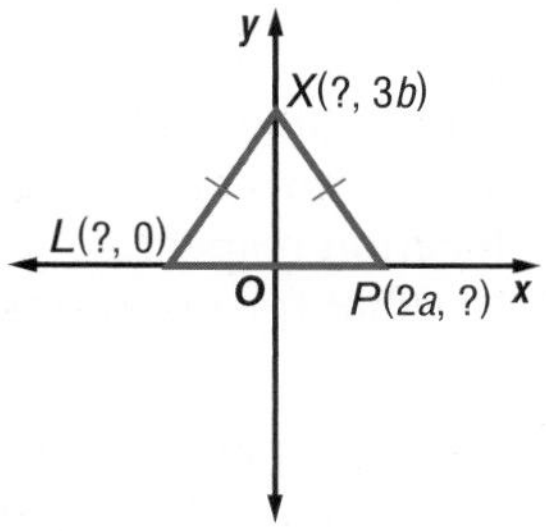

14.

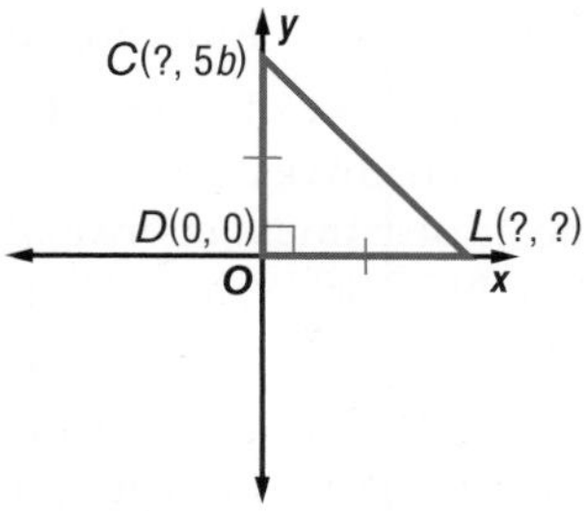

15.

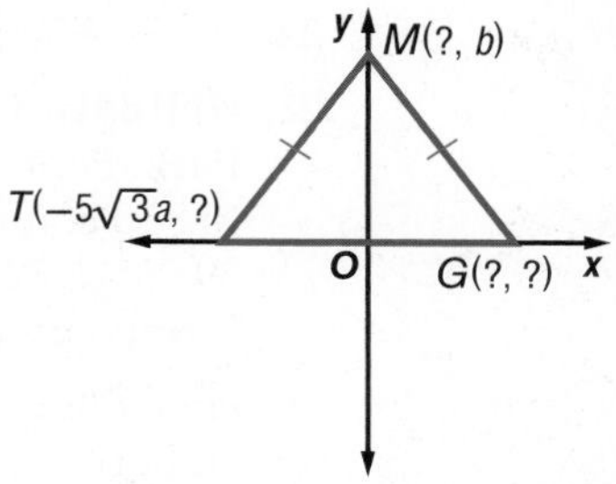

16.

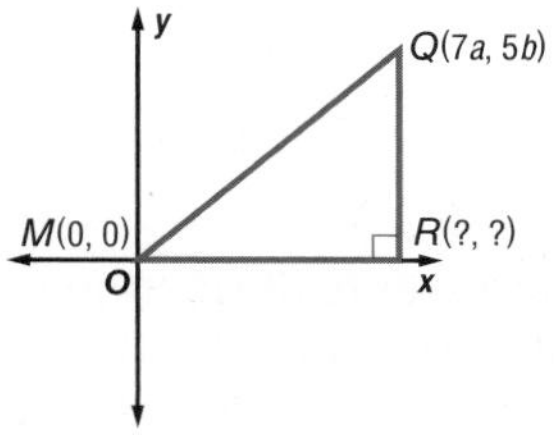

17.

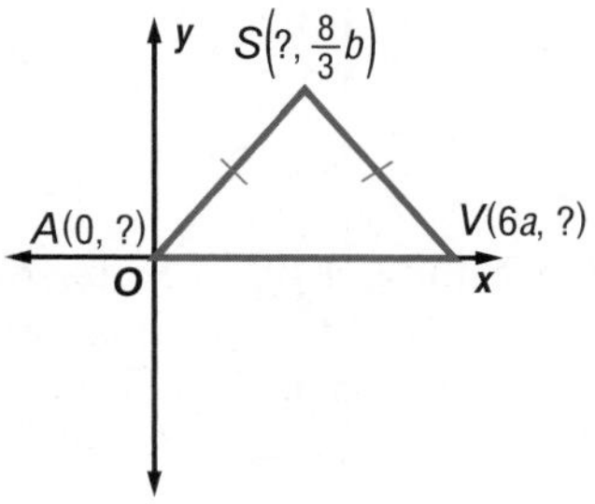

18.

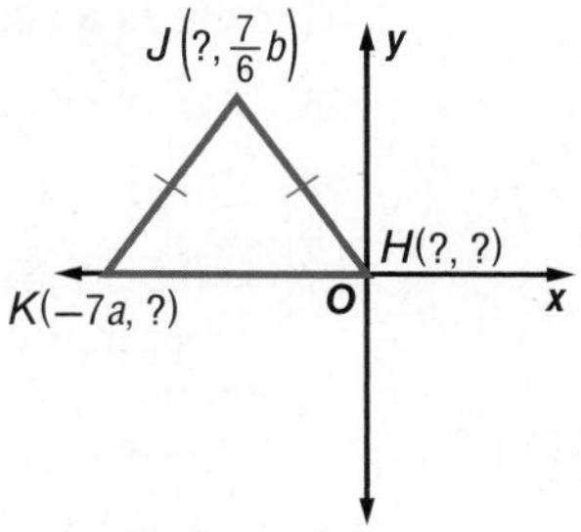

Example 3 **PROOF** **Write a coordinate proof for each statement.**

19. When the altitude is drawn in an isosceles triangle, two congruent triangles are formed.

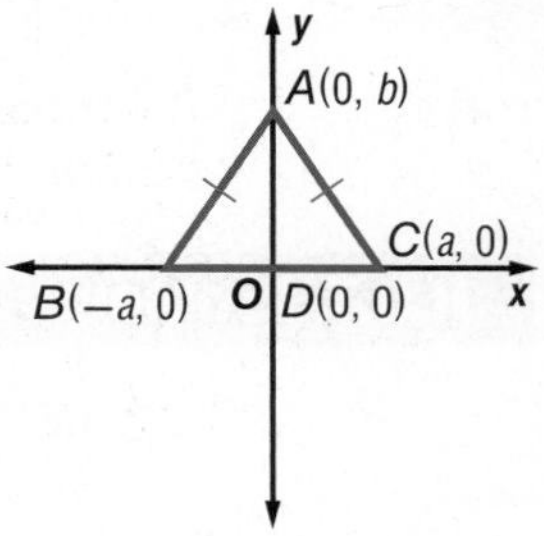

20. The segment joining the midpoint of the two legs of a right triangle is parallel to the hypotenuse.

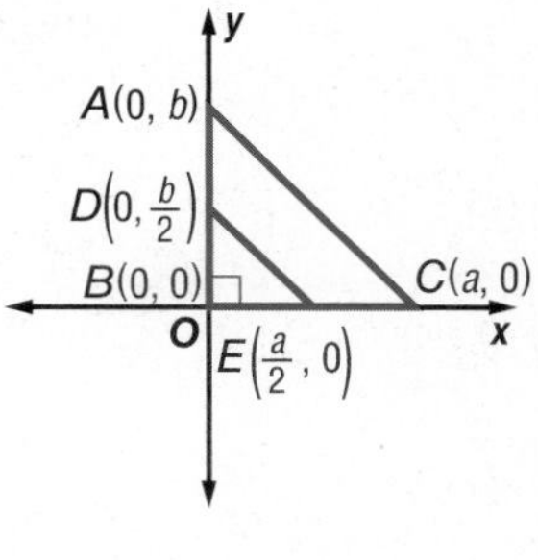

Example 4 **PROOF Write a coordinate proof for each statement.**

21. ΔXYZ is similar to ΔRSZ.

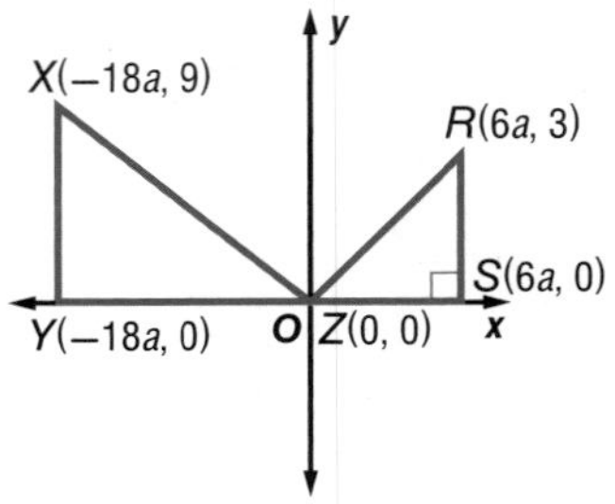

22. R(−3, −3), S(3, −3), T(0, $3\sqrt{3} - 3$) form an equilateral triangle.

23. **FOOTBALL** Ohio State located in Columbus, Ohio, Penn State located in University Park, Pennsylvania, and Northwestern located in Evanston, Illinois are all part of the Big 10 conference. The approximate coordinates of each location respectively are 39.98°N, 82.98°W, 40.79°N 77.86°W, and 41.88°N, 87.62°W. What type of triangle is formed by these three cities?

24. **PAINTBALL** Nick, Joey and Arif are on the same team in a game of paintball. Joey is stationed at the origin, Nick is at (4, 3), an Arif is at (0, 5). Write a coordinate proof to prove that the triangle formed by their paintball team is isosceles.

Draw ΔXYZ and find the slope of each side of the triangle. Determine whether the triangle is a right triangle. Explain.

25. $X(0, 0)$, $Y(2a, 3b)$, $Z(3a, 2b)$

26. $X(0, 0)$, $Y(7c, 3)$, $Z(-3c, 7c^2)$

27. **AMUSEMENT PARKS** Matthew is at an amusement park and wants to ride the roller coaster, merry go round, and the bumper cars. If the roller coaster is located at (2, −1), the merry go round is located at (3, 3), and the bumper cars are located at (−2, 0), write a coordinate proof to prove that the figure formed by the three rides is a right triangle.

28. **PROOF** Write a coordinate proof to prove ΔABC is a scalene triangle if the vertices are A(0, 0), B($3a$, $5a$), C($-2a$, $8a$).

29. **TRIATHLON** Sandy is participating in a triathlon. The starting point is at the origin. During the first leg of the triathlon Sandy runs 10 kilometers to the east, then she bikes 40 kilometers due north. The final leg is a swim 1.5 kilometers north. Write a coordinate proof to prove that the triangle formed by the starting point, beginning of the cycling, and end of the swim form a scalene triangle.

H.O.T. Problems Use Higher-Order Thinking Skills

30. **REASONING** If the origin is the midpoint of the hypotenuse of a right triangle and two vertices are at (−4, 2) and (4, 2), find the third vertex.

31. **CHALLENGE** Write a coordinate proof to show that if you multiply each x-coordinate and each y-coordinate by 2, the resulting figure is similar to the original triangle.

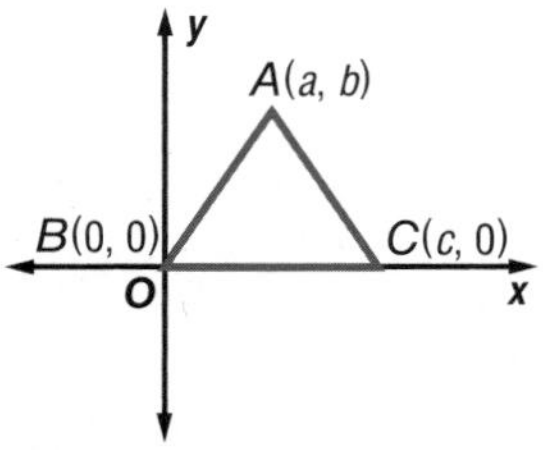

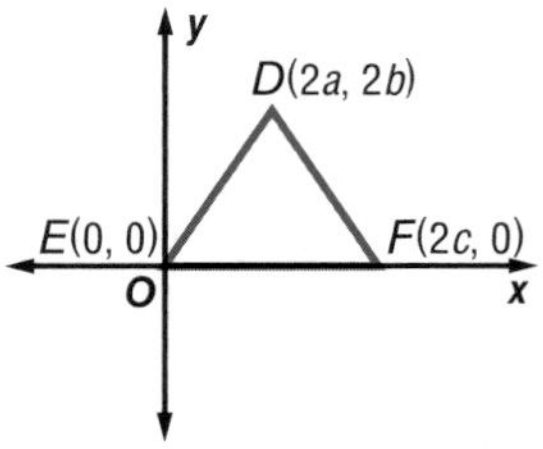

32. **REASONING** If ΔABC is an isosceles right triangle and the coordinates are $A(0, 0)$, $B(4, 0)$, how many different points can C be located at on the coordinate plane?

Standardized Test Practice

33. GRIDDED RESPONSE In the figure below, $m\angle B = 76$. The measure of $\angle A$ is half the measure of $\angle B$. What is $m\angle C$?

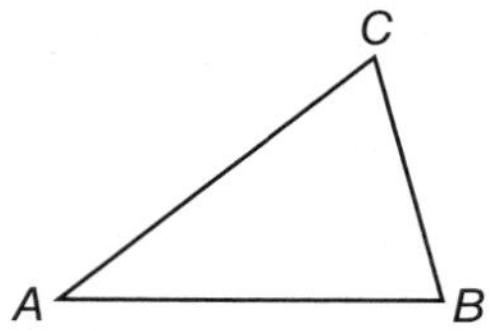

34. ALGEBRA What is the x-coordinate of the solution to the system of equations shown below?

$$\begin{cases} 2x - 3y = 3 \\ -4x + 2y = -18 \end{cases}$$

A -6 **C** 3

B -3 **D** 6

35. What are the coordinates of point R in the triangle?

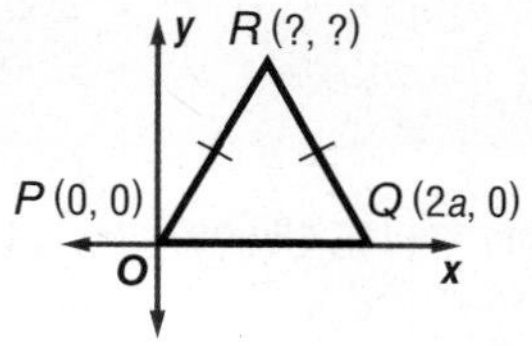

F $\left(\frac{a}{2}, a\right)$ **H** $\left(\frac{b}{2}, a\right)$

G (a, b) **J** $\left(\frac{b}{2}, \frac{a}{2}\right)$

36. SAT/ACT For all x,
$17x^5 + 3x^2 + 2 - (-4x^5 + 3x^3 - 2) =$

A $13x^5 + 3x^3 + 3x^2$

B $13x^5 + 6x^2 + 4$

C $21x^5 - 3x^3 + 3x^2 + 4$

D $21x^5 + 3x^2 + 3x^3$

E $21x^5 + 3x^3 + 3x^2 + 4$

Spiral Review

Refer to the figure at the right. (Lesson 12-6)

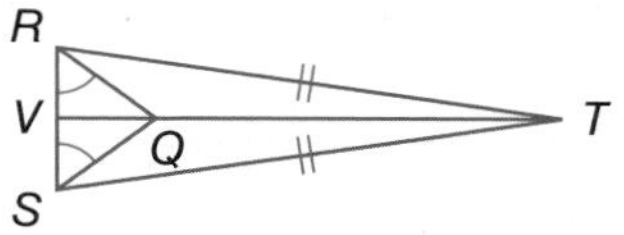

37. Name two congruent angles.

38. Name two congruent segments.

39. Name a pair of congruent triangles.

40. RAMPS The Americans with Disabilities Act requires that wheelchair ramps have at least a 12-inch run for each rise of 1 inch. (Lesson 12-3)

a. Determine the slope represented by this requirement.

b. The maximum length that the law allows for a ramp is 30 feet. How many inches tall is the highest point of this ramp?

Skills Review

Find the distance between each pair of points. Round to the nearest tenth.

41. $X(5, 4)$ and $Y(2, 1)$

42. $A(1, 5)$ and $B(-2, -3)$

43. $J(-2, 6)$ and $K(1, 4)$

EXPLORE 12-9A

Geometry Lab
Constructing Bisectors

Paper folding can be used to construct special segments in triangles.

CCSS Common Core State Standards
Content Standards
G.CO.12 Make formal geometric constructions with a variety of tools and methods (compass and straightedge, string, reflective devices, paper folding, dynamic geometric software, etc.).
Mathematical Practices 5

Construction Perpendicular Bisector

Construct a perpendicular bisector of the side of a triangle.

Step 1

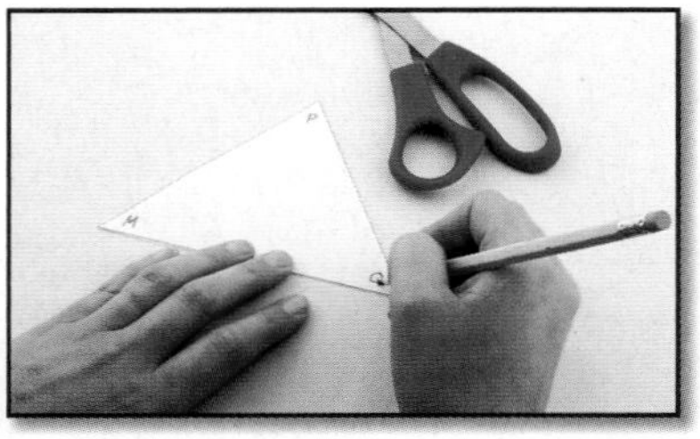

Draw, label, and cut out $\triangle MPQ$.

Step 2

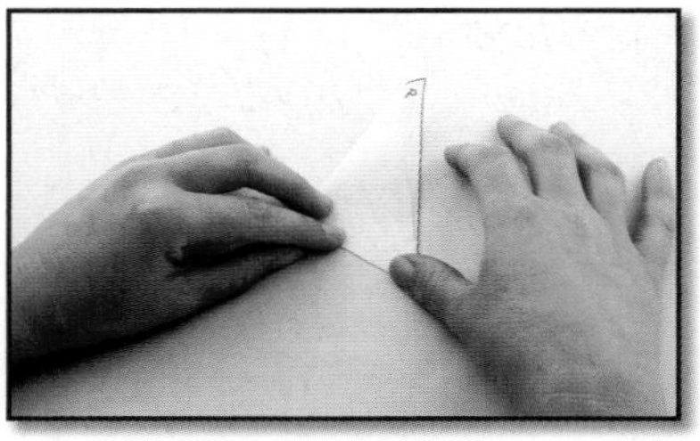

Fold the triangle in half along $\overline{MQ}$ so that vertex M touches vertex Q.

Step 3

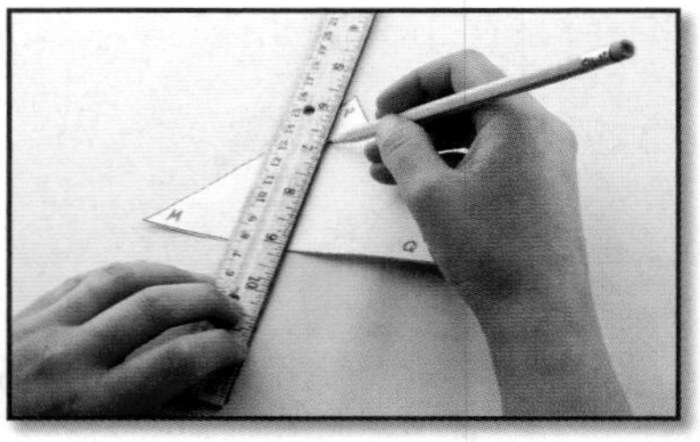

Use a straightedge to draw $\overline{AB}$ along the fold. $\overline{AB}$ is the perpendicular bisector of $\overline{MQ}$.

An angle bisector in a triangle is a line containing a vertex of the triangle and bisecting that angle.

Construction Angle Bisector

Construct an angle bisector of a triangle.

Step 1

Draw, label, and cut out $\triangle ABC$.

Step 2

Fold the triangle in half through vertex A, such that sides $\overline{AC}$ and $\overline{AB}$ are aligned.

Step 3

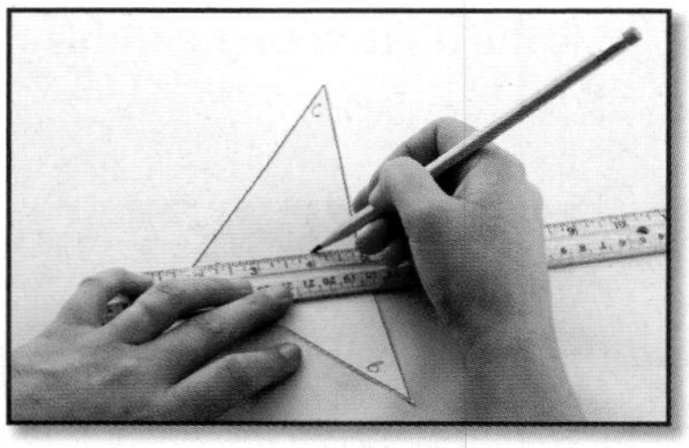

Label point L at the crease along edge $\overline{BC}$. Use a straightedge to draw $\overline{AL}$ along the fold. $\overline{AL}$ is an angle bisector of $\triangle ABC$.

Model and Analyze

1. Construct the perpendicular bisectors and angle bisectors of the other two sides and angles of $\triangle MPQ$. What do you notice about their intersections?

Repeat the two constructions for each type of triangle.

2. acute
3. obtuse
4. right

EXPLORE 12-9B

Geometry Lab
Constructing Medians and Altitudes

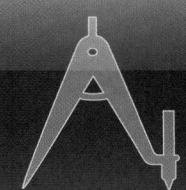

A *median* of a triangle is a segment with endpoints that are a vertex and the midpoint of the side opposite that vertex. You can use the construction for the midpoint of a segment to construct a median.

Wrap the end of string around a pencil. Use a thumbtack to fix the string to a vertex.

CCSS Common Core State Standards
Content Standards
G.CO.12 Make formal geometric constructions with a variety of tools and methods (compass and straightedge, string, reflective devices, paper folding, dynamic geometric software, etc.).
Mathematical Practices 5

Construction 1 Median of a Triangle

Step 1

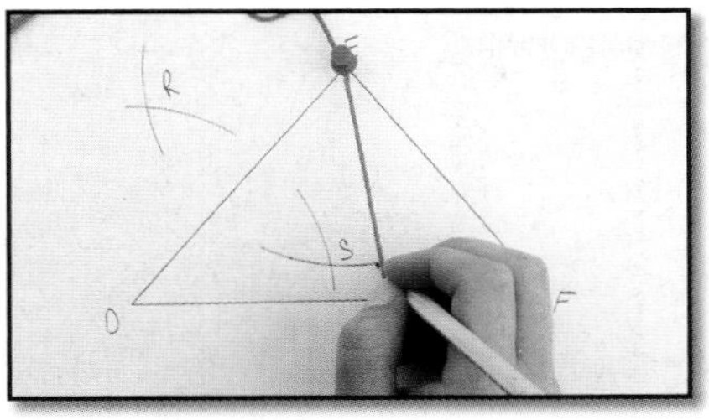

Place the thumbtack on vertex *D* and then on vertex *E* to draw intersecting arcs above and below $\overline{DE}$. Label the points of intersection *R* and *S*.

Step 2

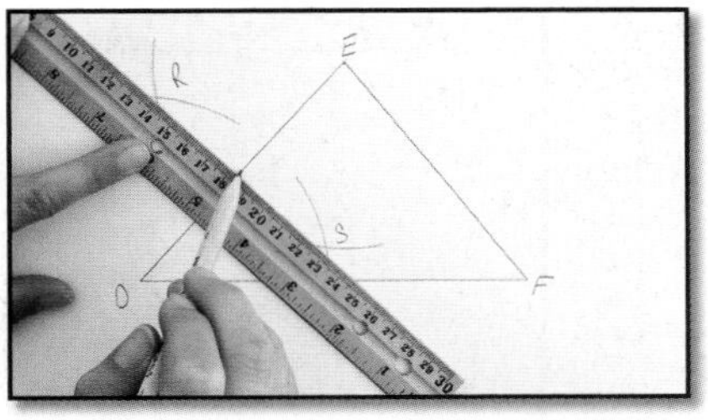

Use a straightedge to find the point where $\overline{RS}$ intersects $\overline{DE}$. Label the point *M*. This is the midpoint of $\overline{DE}$.

Step 3

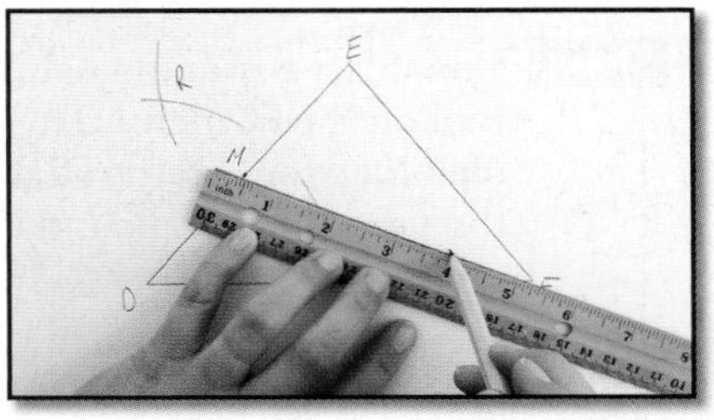

Draw a line through *F* and *M*. $\overline{FM}$ is a median of $\triangle DEF$.

An *altitude* of a triangle is a segment from a vertex of the triangle to the opposite side and is perpendicular to the opposite side.

Construction 2 Altitude of a Triangle

Step 1

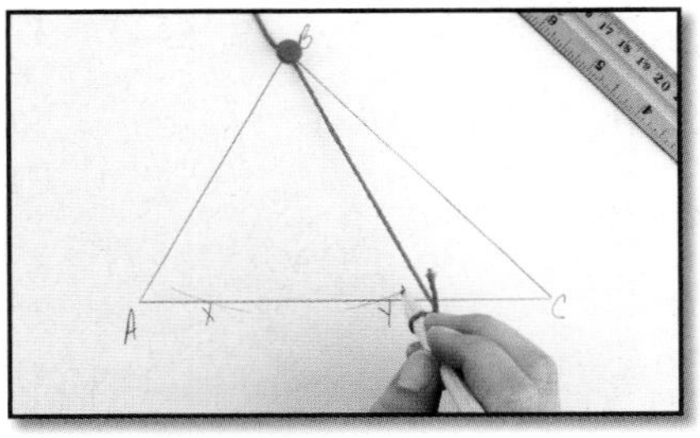

Place the thumbtack on vertex *B* and draw two arcs intersecting $\overline{AC}$. Label the points where the arcs intersect the sides as *X* and *Y*.

Step 2

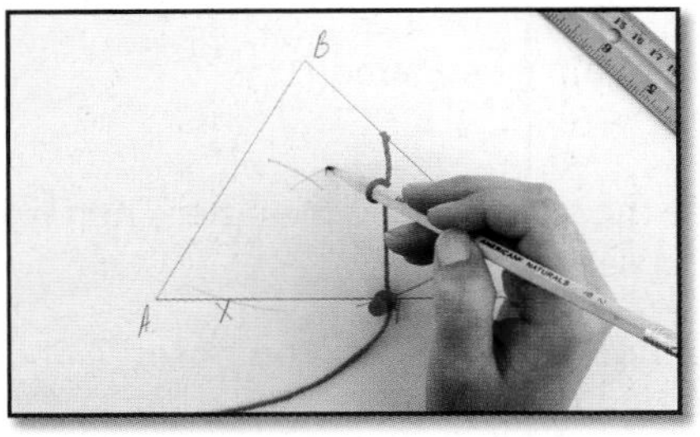

Adjust the length of the string so that it is greater than $\frac{1}{2}XY$. Place the tack on *X* and draw an arc above $\overline{AC}$. Use the same length of string to draw an arc from *Y*. Label the points of intersection of the arcs *H*.

Step 3

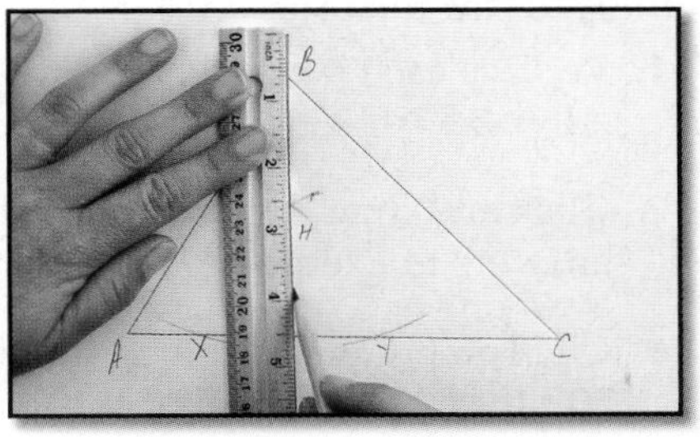

Use a straightedge to draw $\overleftrightarrow{BH}$. Label the point where $\overleftrightarrow{BH}$ intersects $\overline{AC}$ as *D*. $\overline{BD}$ is an altitude of $\triangle ABC$ and is perpendicular to $\overline{AC}$.

Model and Analyze

1. Construct the medians of the other two sides of $\triangle DEF$. What do you notice about the medians of a triangle?
2. Construct the altitudes to the other two sides of $\triangle ABC$. What do you observe?

EXPLORE

12-9C Graphing Technology Lab: The Triangle Inequality

You can use the Cabri™ Jr. application on a TI-83/84 Plus graphing calculator to discover properties of triangles.

CCSS Common Core State Standards
Content Standards
G.CO.12 Make formal geometric constructions with a variety of tools and methods (compass and straightedge, string, reflective devices, paper folding, dynamic geometric software, etc.).
Mathematical Practices 5

Activity 1

Construct a triangle. Observe the relationship between the sum of the lengths of two sides and the length of the other side.

Step 1 Construct a triangle using the triangle tool on the **F2** menu. Then use the **Alph-Num** tool on the **F5** menu to label the vertices as *A*, *B*, and *C*.

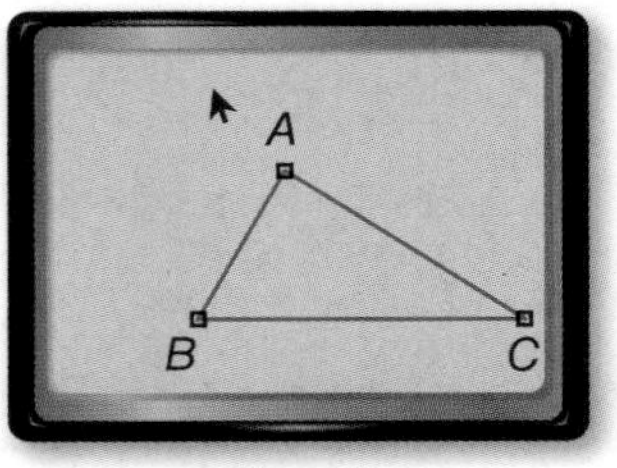

Step 1

Step 2 Access the **distance & length** tool, shown as **D. & Length**, under **Measure** on the **F5** menu. Use the tool to measure each side of the triangle.

Step 3 Display $AB + BC$, $AB + CA$, and $BC + CA$ by using the **Calculate** tool on the **F5** menu. Label the measures.

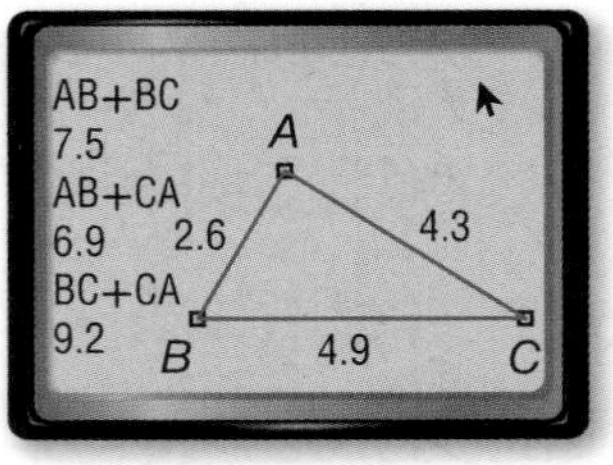

Steps 2 and 3

Step 4 Click and drag the vertices to change the shape of the triangle.

Analyze the Results

1. Replace each ● with <, >, or = to make a true statement.
   $AB + BC$ ● CA $\quad$ $AB + CA$ ● BC $\quad$ $BC + CA$ ● AB

2. Click and drag the vertices to change the shape of the triangle. Then review your answers to Exercise 1. What do you observe?

3. Click on point *A* and drag it to lie on line *BC*. What do you observe about *AB*, *BC*, and *CA*? Are *A*, *B*, and *C* the vertices of a triangle? Explain.

4. **Make a conjecture** about the sum of the lengths of two sides of a triangle and the length of the third side.

5. Do the measurements and observations you made in the Activity and in Exercises 1–3 constitute a proof of the conjecture you made in Exercise 4? Explain.

6. Replace each ● with <, >, or = to make a true statement.
   $|AB - BC|$ ● CA $\quad$ $|AB - CA|$ ● BC $\quad$ $|BC - CA|$ ● AB

   Then click and drag the vertices to change the shape of the triangle and review your answers. What do you observe?

7. How could you use your observations to determine the possible lengths of the third side of a triangle if you are given the lengths of the other two sides?

LESSON 12-9 Areas of Parallelograms and Triangles

Then

- You found areas of rectangles and squares.

Now

1. Find perimeters and areas of parallelograms.
2. Find perimeters and areas of triangles.

Why?

- A tangram is an ancient Chinese puzzle that can be rearranged to form different images, such as the animals shown. The area of the puzzle, before and after being rearranged, remains the same. It is the sum of all the areas of its pieces.

NewVocabulary
base of a parallelogram
height of a parallelogram
base of a triangle
height of a triangle

Common Core State Standards

Content Standards
G.GPE.7 Use coordinates to compute perimeters of polygons and areas of triangles and rectangles, e.g., using the distance formula.

Mathematical Practices
1 Make sense of problems and persevere in solving them.
7 Look for and make use of structure.

1 Areas of Parallelograms

A *parallelogram* is a quadrilateral with both pairs of opposite sides parallel. Any side of a parallelogram can be called the **base of a parallelogram**. The **height of a parallelogram** is the perpendicular distance between any two parallel bases.

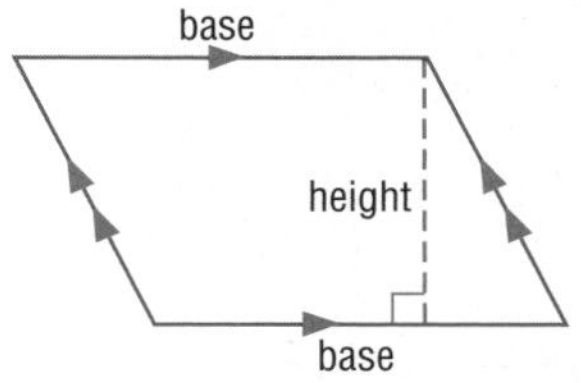

You can use the following postulate to develop the formula for the area of a parallelogram.

Postulate 12.4 Area Addition Postulate

The area of a region is the sum of the areas of its nonoverlapping parts.

In the figures below, a right triangle is cut off from one side of a parallelogram and translated to the other side as shown to form a rectangle with the same base and height.

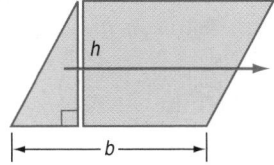

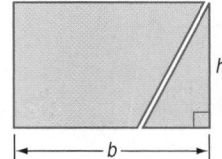

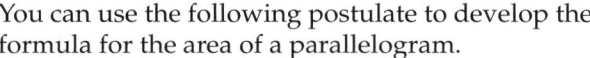

Recall from Lesson 10-6 that the area of a rectangle is the product of its base and height. By the Area Addition Postulate, a parallelogram with base b and height h has the same area as a rectangle with base b and height h.

KeyConcept Area of a Parallelogram

Words The area A of a parallelogram is the product of a base b and its corresponding height h.

Symbols $A = bh$

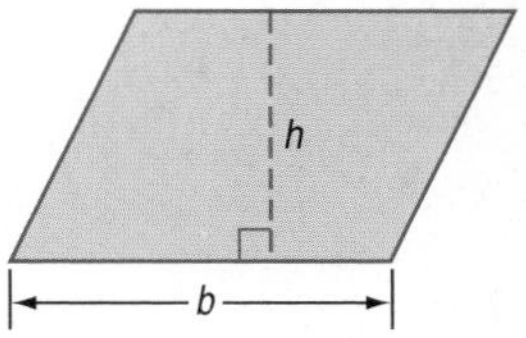

Example 1 Perimeter and Area of a Parallelogram

Find the perimeter and area of $\square ABCD$.

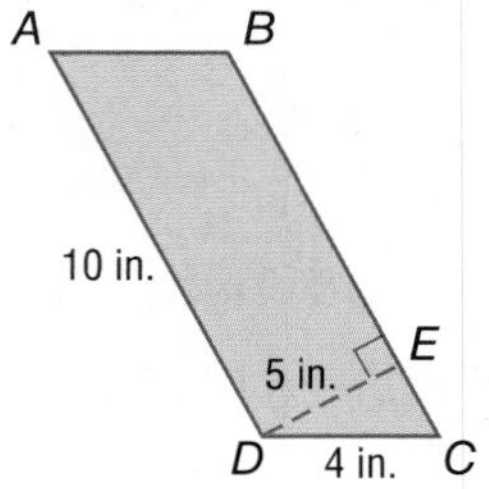

Perimeter

Since opposite sides of a parallelogram are congruent, $\overline{AB} \cong \overline{DC}$ and $\overline{BC} \cong \overline{AD}$. So $AB = 4$ inches and $BC = 10$ inches.

Perimeter of $\square ABCD = AB + BC + DC + AD$
$= 4 + 10 + 4 + 10$ or 28 in.

Area

The height given, DE, is 5 inches. $\overline{BC}$ is the base, which measures 10 inches.

$A = bh$ — Area of a parallelogram

$= (10)(5)$ or 50 in^2 — $b = 10$ and $h = 5$

> **StudyTip**
>
> **Heights of Figures** The height of a figure can be measured by extending a base. In Example 1, the height of $\square ABCD$ that corresponds to base $\overline{DC}$ can be measured by extending $\overline{DC}$.
>
> 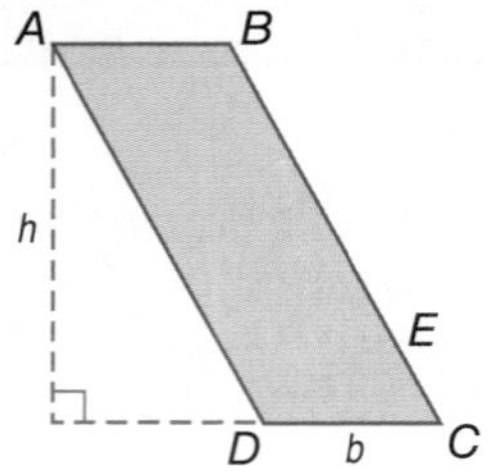
>

GuidedPractice

Find the perimeter and area of each parallelogram.

1A.

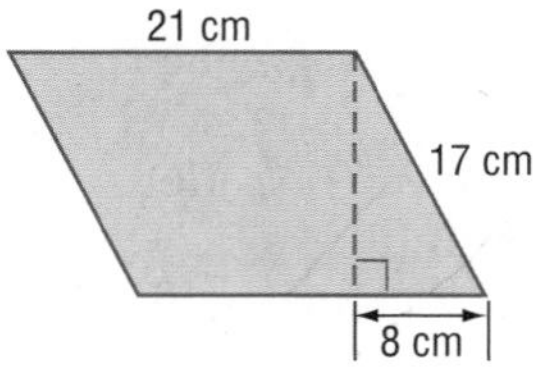

1B.

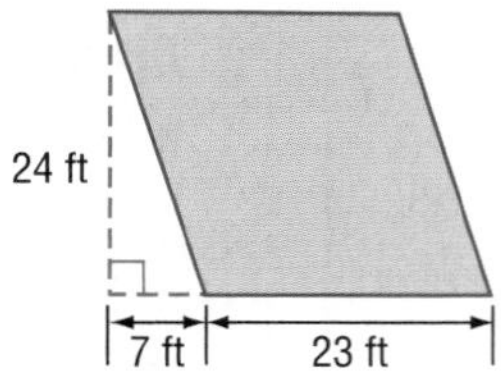

You may need to use trigonometry to find the area of a parallelogram.

Example 2 Area of a Parallelogram

Find the area of $\square EFGH$.

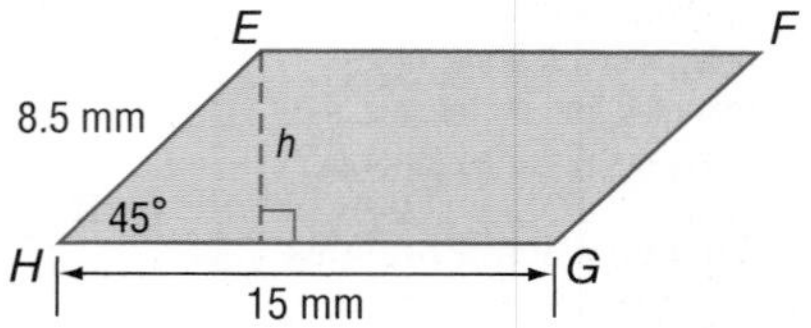

Step 1 Use a 45°-45°-90° triangle to find the height h of the parallelogram.

Recall that if the measure of the leg opposite the 45° angle is h, then the measure of the hypotenuse is $h\sqrt{2}$.

$h\sqrt{2} = 8.5$ — Substitute 8.5 for the measure of the hypotenuse.

$h = \frac{8.5}{\sqrt{2}}$ or about 6 mm — Divide each side by $\sqrt{2}$.

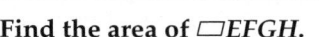

Step 2 Find the area.

$A = bh$ — Area of a parallelogram

$\approx (15)(6)$ or 90 mm^2 — $b = 15$ and $h \approx 6$

> **WatchOut!**
>
> **CCSS Precision** Remember that perimeter is measured in linear units such as inches and centimeters. Area is measured in square units such as square feet and square millimeters.

GuidedPractice

Find the area of each parallelogram. Round to the nearest tenth if necessary.

2A.

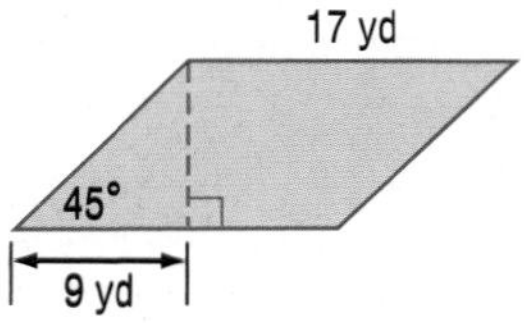

2B.

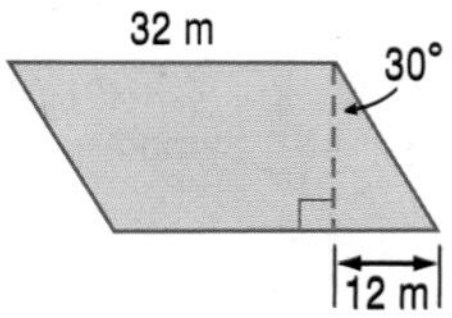

ReviewVocabulary

altitude of a triangle
a segment from a vertex of a triangle to the line containing the opposite side and perpendicular to the line containing that side

2 Areas of Triangles

Like the base of a parallelogram, the **base of a triangle** can be any side. The **height of a triangle** is the length of an altitude drawn to a given base.

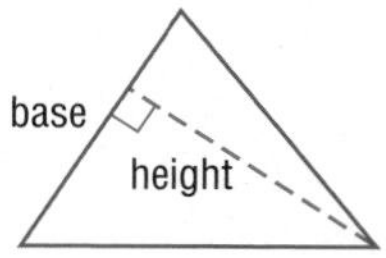

You can use the following postulate to develop the formula for the area of a triangle.

Postulate 12.5 Area Congruence Postulate

If two figures are congruent, then they have the same area.

In the figures below, a parallelogram is cut in half along a diagonal to form two congruent triangles with the same base and height.

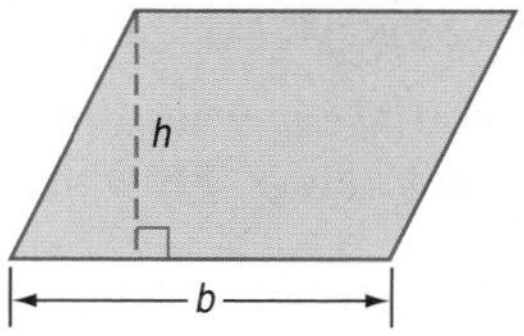

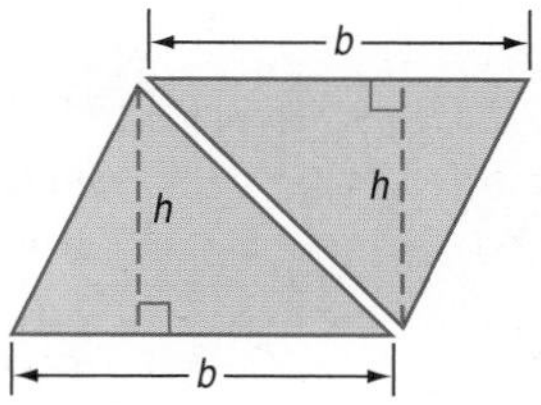

By the Area Congruence Postulate, the two congruent triangles have the same area. So, one triangle with base b and height h has half the area of a parallelogram with base b and height h.

KeyConcept Area of a Triangle

Words The area A of a triangle is one half the product of a base b and its corresponding height h.

Symbols $A = \frac{1}{2}bh$ or $A = \frac{bh}{2}$

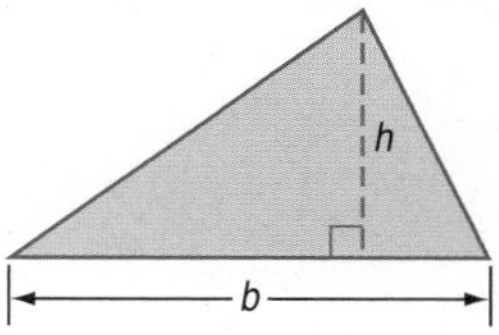

Real-WorldLink

Triangular gardens can serve as focal points in landscaping or simply result from intersecting walkways.

Janet Johnson/Gap Photos/age fotostock

Real-World Example 3 Perimeter and Area of a Triangle

GARDENING D'Andre needs enough mulch to cover the triangular garden shown and enough paving stones to border it. If one bag of mulch covers 12 square feet and one paving stone provides a 4-inch border, how many bags of mulch and how many stones does he need to buy?

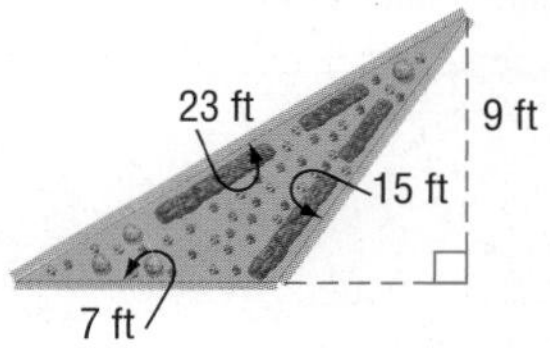

Step 1 Find the perimeter of the garden.

Perimeter of garden = 23 + 15 + 7 or 45 ft

Step 2 Find the area of the garden.

$A = \frac{1}{2}bh$ Area of a triangle

$= \frac{1}{2}(7)(9)$ or $31.5\ \text{ft}^2$ $b = 7$ and $h = 9$

Step 3 Use unit analysis to determine how many of each item are needed.

Bags of Mulch

$31.5\ \cancel{\text{ft}^2} \cdot \frac{1\ \text{bag}}{12\ \cancel{\text{ft}^2}} = 2.625\ \text{bags}$

Paving Stones

$45\ \cancel{\text{ft}} \cdot \frac{12\ \cancel{\text{in.}}}{1\ \cancel{\text{ft}}} \cdot \frac{1\ \text{stone}}{4\ \cancel{\text{in.}}} = 135\ \text{stones}$

Round the number of bags up so there is enough mulch. He will need 3 bags of mulch and 135 paving stones.

GuidedPractice

Find the perimeter and area of each triangle.

3A.

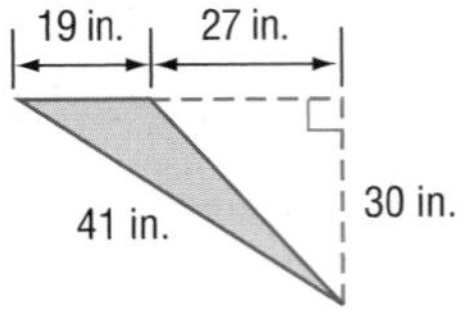

3B.

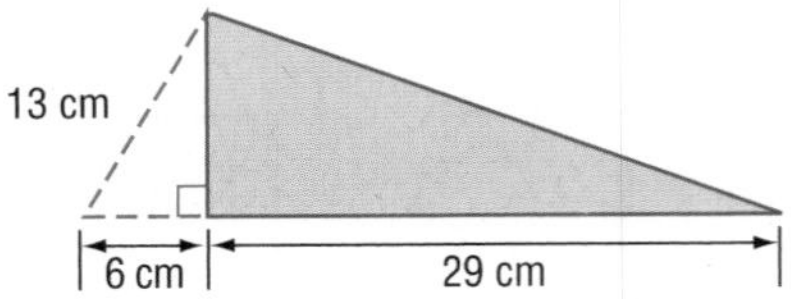

You can use algebra to solve for unknown measures in parallelograms and triangles.

Example 4 Use Area to Find Missing Measures

ALGEBRA The height of a triangle is 5 centimeters more than its base. The area of the triangle is 52 square centimeters. Find the base and height.

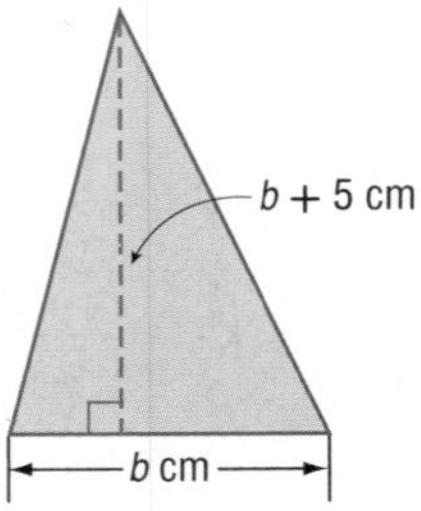

Step 1 Write expressions to represent each measure.

Let b represent the base of the triangle. Then the height is $b + 5$.

Step 2 Use the formula for the area of a triangle to find b.

$A = \frac{1}{2}bh$	Area of a triangle
$52 = \frac{1}{2}b(b + 5)$	Replace A with 52 and h with $b + 5$.
$104 = b(b + 5)$	Multiply each side by 2.
$104 = b^2 + 5b$	Distributive Property
$0 = b^2 + 5b - 104$	Subtract 104 from each side.
$0 = (b + 13)(b - 8)$	Factor.
$b + 13 = 0$ and $b - 8 = 0$	Zero Product Property
$b = -13$ $\quad$ $b = 8$	Solve for b.

StudyTip

Zero Product Property If the product of two factors is 0, then at least one of the factors must be 0.

Step 3 Use the expressions from Step 1 to find each measure.

Since a length cannot be negative, the base measures 8 centimeters and the height measures 8 + 5 or 13 centimeters.

GuidedPractice

ALGEBRA Find x.

4A. $A = 148\ m^2$

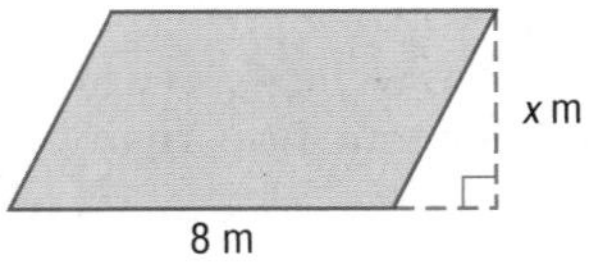

4B. $A = 357\ in^2$

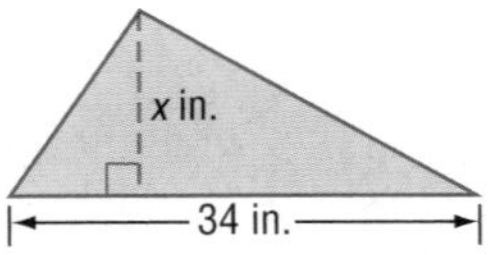

4C. ALGEBRA The base of a parallelogram is twice its height. If the area of the parallelogram is 72 square feet, find its base and height.

Check Your Understanding

Examples 1–3 **Find the perimeter and area of each parallelogram or triangle. Round to the nearest tenth if necessary.**

1.

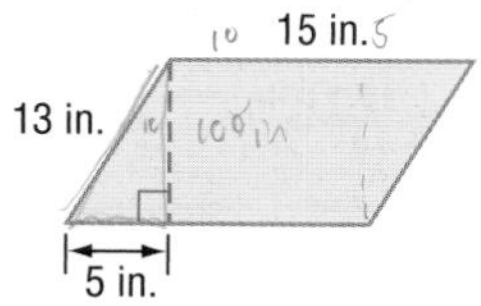

2.

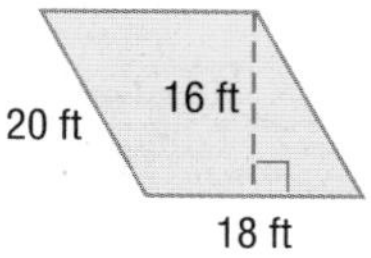

3.

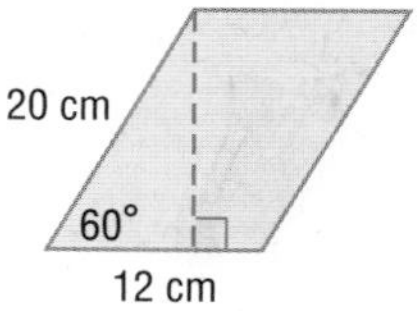

4.

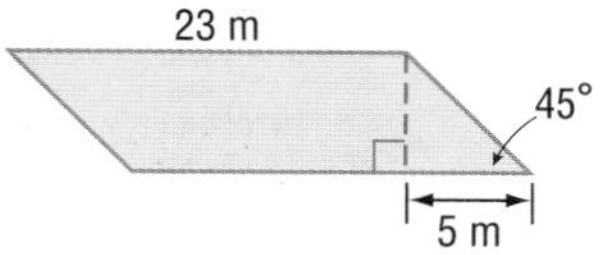

5.

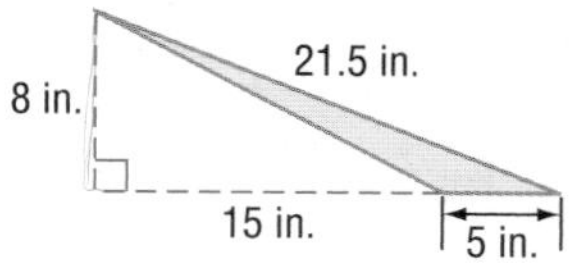

6.

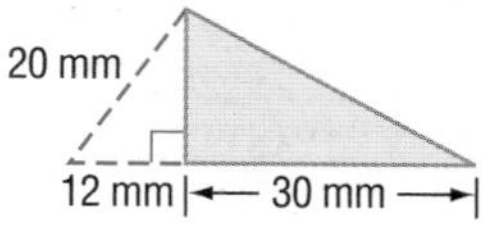

7. CRAFTS Marquez and Victoria are making pinwheels. Each pinwheel is composed of 4 triangles with the dimensions shown. Find the perimeter and area of one triangle.

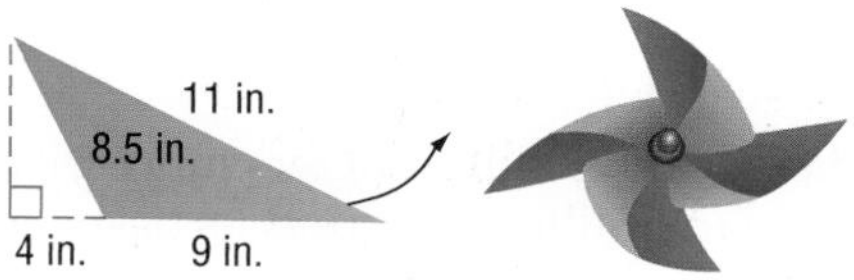

Example 4 **Find *x*.**

8. $A = 153 \text{ in}^2$

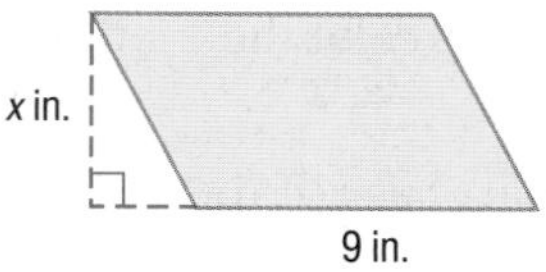

9. $A = 165 \text{ cm}^2$

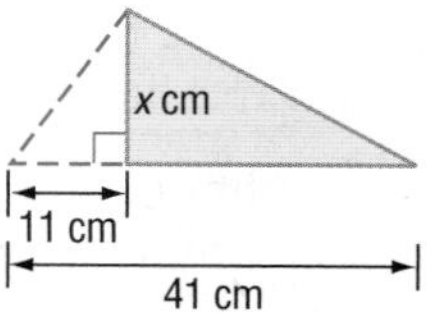

Practice and Problem Solving

Examples 1–3 **CCSS STRUCTURE** **Find the perimeter and area of each parallelogram or triangle. Round to the nearest tenth if necessary.**

10.

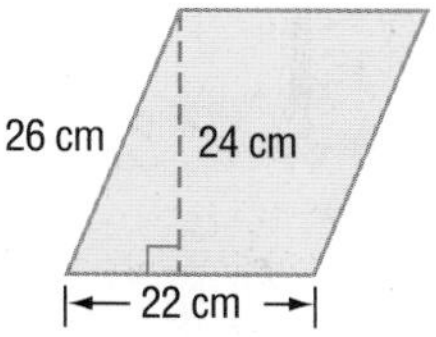

11

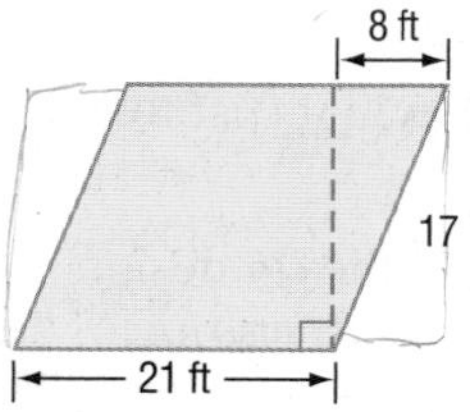

12.

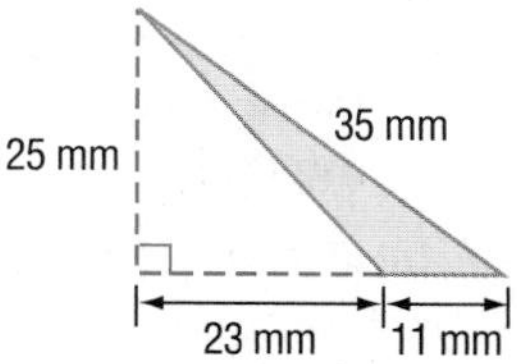

13.

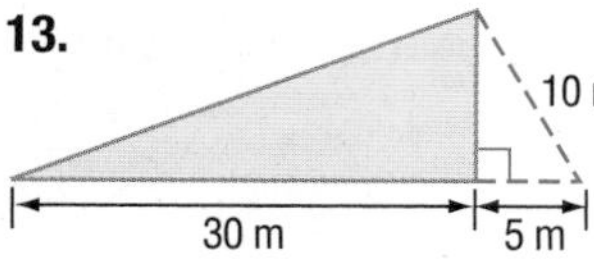

14.

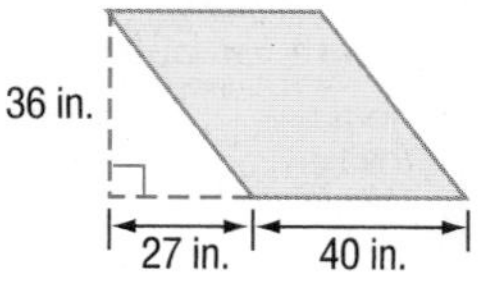

15.

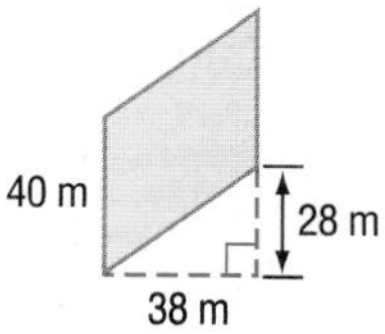

16. TANGRAMS The tangram shown is a 4-inch square.

a. Find the perimeter and area of the purple triangle. Round to the nearest tenth.

b. Find the perimeter and area of the blue parallelogram. Round to the nearest tenth.

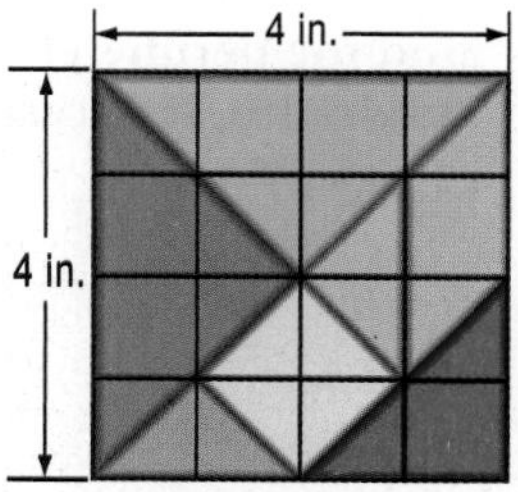

Example 2 **CCSS STRUCTURE** **Find the area of each parallelogram. Round to the nearest tenth if necessary.**

17.

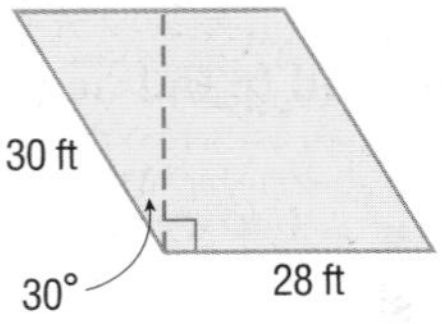

18.

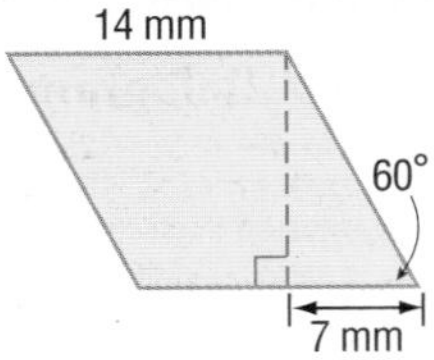

19.

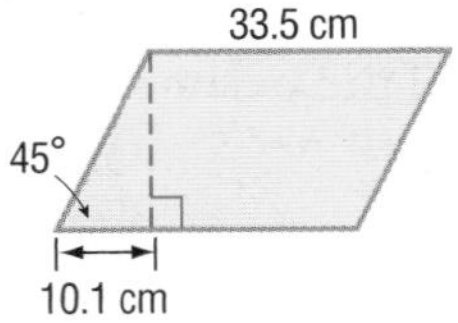

20.

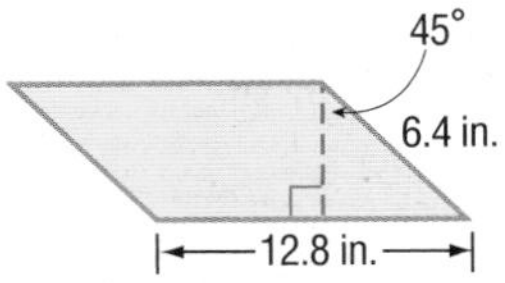

21.

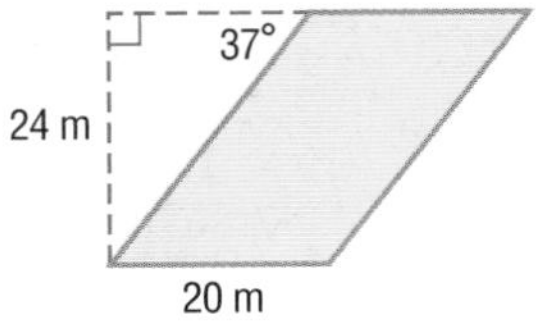

22.

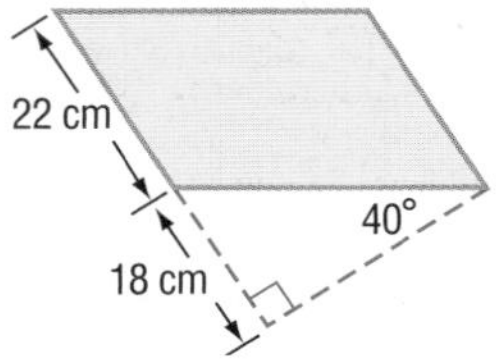

23 **WEATHER** Tornado watch areas are often shown on weather maps using parallelograms. What is the area of the region affected by the tornado watch shown? Round to the nearest square mile.

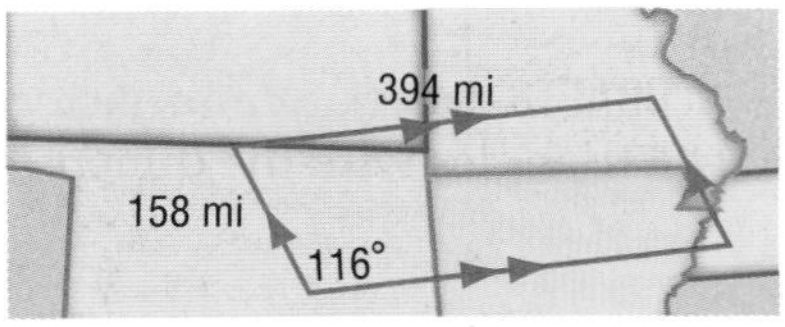

Example 4

24. The height of a parallelogram is 4 millimeters more than its base. If the area of the parallelogram is 221 square millimeters, find its base and height.

25. The height of a parallelogram is one fourth of its base. If the area of the parallelogram is 36 square centimeters, find its base and height.

26. The base of a triangle is twice its height. If the area of the triangle is 49 square feet, find its base and height.

27. The height of a triangle is 3 meters less than its base. If the area of the triangle is 44 square meters, find its base and height.

28. **FLAGS** Omar wants to make a replica of Guyana's national flag.

a. What is the area of the piece of fabric he will need for the red region? for the yellow region?

b. If the fabric costs $3.99 per square yard for each color and he buys exactly the amount of fabric he needs, how much will it cost to make the flag?

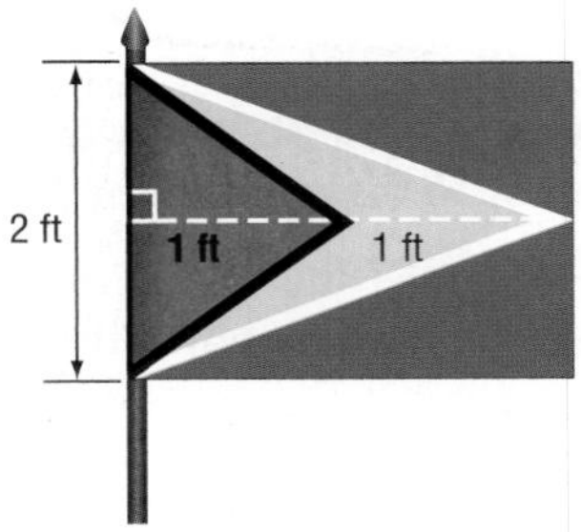

29. **DRAMA** Madison is in charge of the set design for her high school's rendition of *Romeo and Juliet*. One pint of paint covers 80 square feet. How many pints will she need of each color if the roof and tower each need 3 coats of paint?

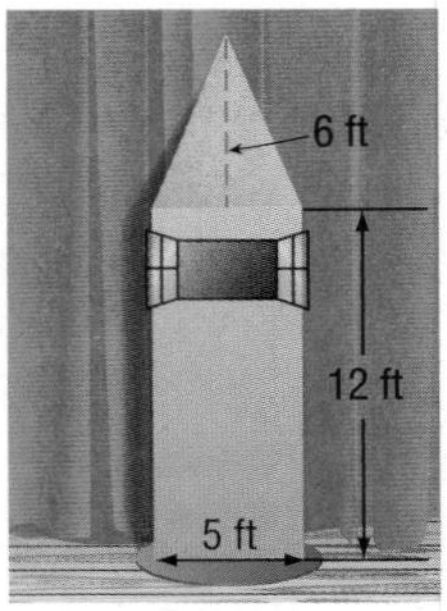

Find the perimeter and area of each figure. Round to the nearest hundredth, if necessary.

30.

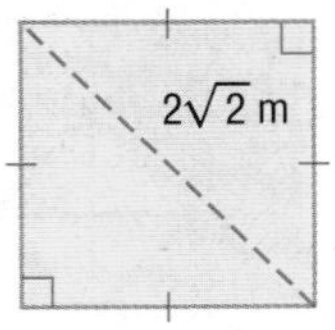

31.

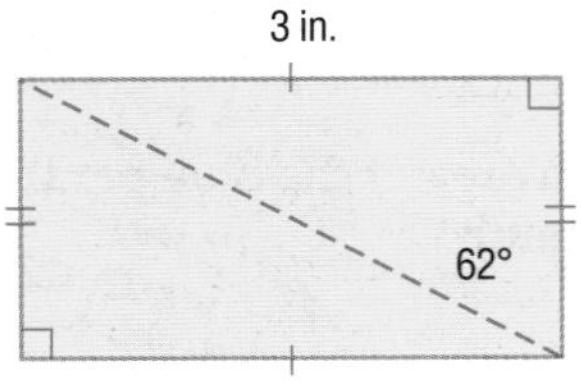

32.

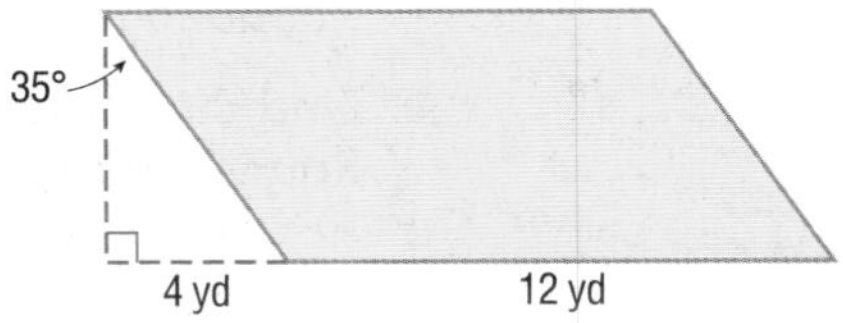

COORDINATE GEOMETRY **Find the area of each figure. Explain the method that you used.**

33. ▱$ABCD$ with $A(4, 7)$, $B(2, 1)$, $C(8, 1)$, and $D(10, 7)$

34. $\triangle RST$ with $R(-8, -2)$, $S(-2, -2)$, and $T(-3, -7)$

35. **HERON'S FORMULA** Heron's Formula relates the lengths of the sides of a triangle to the area of the triangle. The formula is $A = \sqrt{s(s-a)(s-b)(s-c)}$, where s is the *semiperimeter*, or one half the perimeter, of the triangle and a, b, and c are the side lengths.

a. Use Heron's Formula to find the area of a triangle with side lengths 7, 10, and 4.

b. Show that the areas found for a 5-12-13 right triangle are the same using Heron's Formula and using the triangle area formula you learned earlier in this lesson.

36. **MULTIPLE REPRESENTATIONS** In this problem, you will investigate the relationship between the area and perimeter of a rectangle.

a. Algebraic A rectangle has a perimeter of 12 units. If the length of the rectangle is x and the width of the rectangle is y, write equations for the perimeter and area of the rectangle.

b. Tabular Tabulate all possible whole-number values for the length and width of the rectangle, and find the area for each pair.

c. Graphical Graph the area of the rectangle with respect to its length.

d. Verbal Describe how the area of the rectangle changes as its length changes.

e. Analytical For what whole-number values of length and width will the area be greatest? least? Explain your reasoning.

H.O.T. Problems Use Higher-Order Thinking Skills

37. **CHALLENGE** Find the area of $\triangle ABC$ graphed at the right. Explain your method.

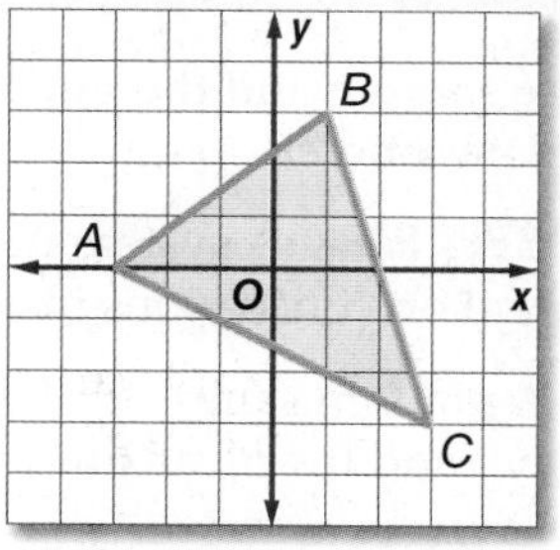

38. **CCSS ARGUMENTS** Will the perimeter of a nonrectangular parallelogram *always*, *sometimes*, or *never* be greater than the perimeter of a rectangle with the same area and the same height? Explain.

39. **WRITING IN MATH** Points J and L lie on line m. Point K lies on line p. If lines m and p are parallel, describe how the area of $\triangle JKL$ will change as K moves along line p.

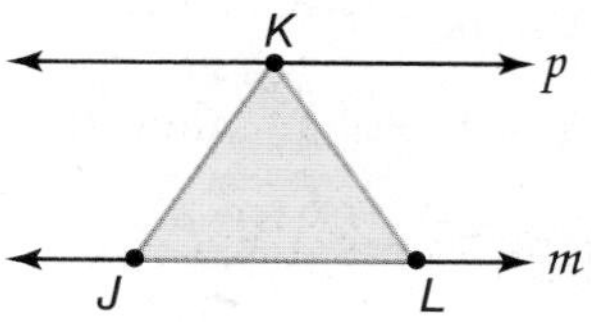

40. **OPEN ENDED** The area of a polygon is 35 square units. The height is 7 units. Draw three different triangles and three different parallelograms that meet these requirements. Label the base and height on each.

41. **WRITING IN MATH** Describe two different ways you could use measurement to find the area of parallelogram $PQRS$.

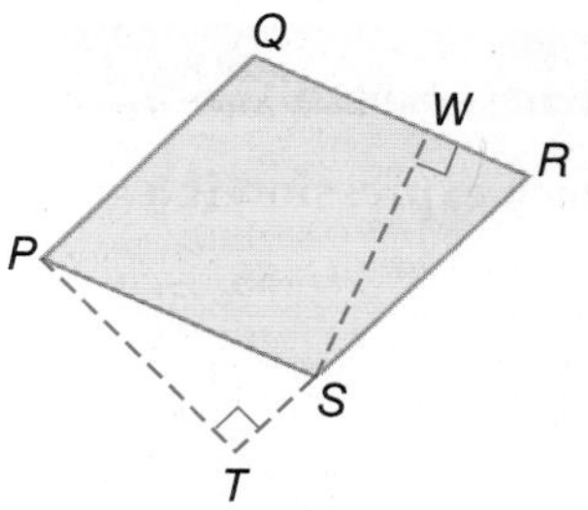

Standardized Test Practice

42. What is the area, in square units, of the parallelogram shown?

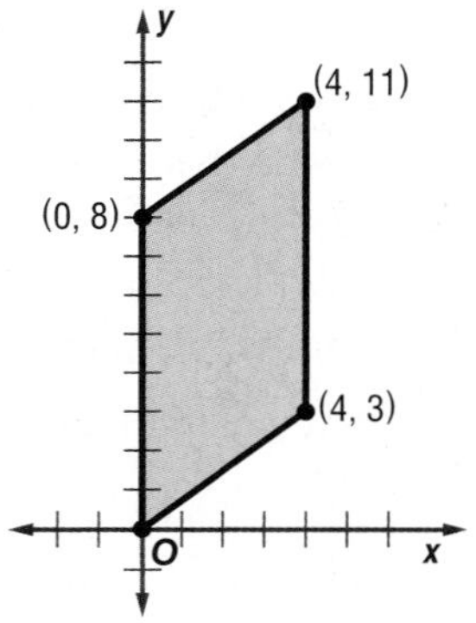

A 12 **B** 20 **C** 32 **D** 40

43. GRIDDED RESPONSE In parallelogram $ABCD$, $\overline{BD}$ and $\overline{AC}$ intersect at E. If $AE = 9$, $BE = 3x - 7$, and $DE = x + 5$, find x.

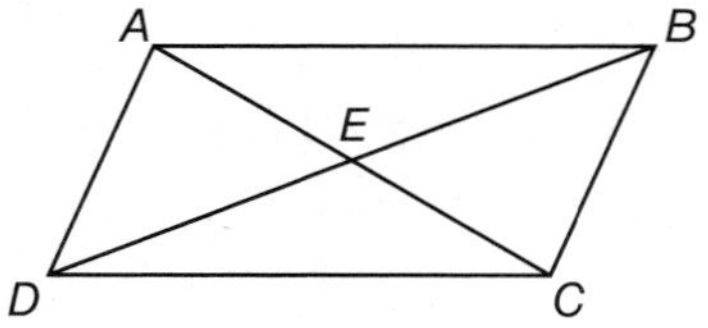

44. A wheelchair ramp is built that is 20 inches high and has a length of 12 feet as shown. What is the measure of the angle x that the ramp makes with the ground, to the *nearest* degree?

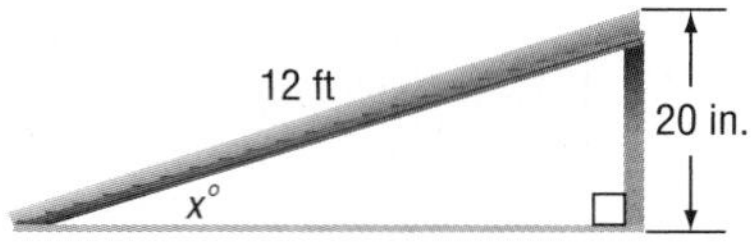

Note: Not drawn to scale.

F 8 **G** 16 **H** 37 **J** 53

45. SAT/ACT The formula for converting a Celsius temperature to a Fahrenheit temperature is $F = \frac{9}{5}C + 32$, where F is the temperature in degrees Fahrenheit and C is the temperature in degrees Celsius. Which of the following is the Celsius equivalent to a temperature of 86° Fahrenheit?

A 15.7° C
B 30° C
C 65.5° C
D 122.8° C
E 186.8° C

Spiral Review

Identify the sample and the population for each situation. Then describe the sample statistic and the population parameter. (Lesson 12-2)

46. AMUSEMENT PARK A systematic sample of 250 guests is asked how much money they spent on concessions inside the park. The mean amount of money is calculated.

47. PROM A random sample of 100 high school seniors at North Boyton High School is surveyed, and the mean amount of money spent on prom by a senior is calculated.

Find the inverse of each function. (Lesson 4-7)

48. $f(x) = 2x - 14$

49. $f(x) = 17 - 5x$

50. $f(x) = \frac{1}{4}x + 3$

51. $f(x) = -\frac{1}{7}x - 1$

52. $f(x) = \frac{2}{3}x + 6$

53. $f(x) = 12 - \frac{3}{5}x$

Skills Review

Evaluate each expression if $a = 2$, $b = 6$, and $c = 3$.

54. $\frac{1}{2}ac$

55. $\frac{1}{2}cb$

56. $\frac{1}{2}b(2a + c)$

57. $\frac{1}{2}c(b + a)$

58. $\frac{1}{2}a(2c + b)$

Study Guide and Review

Study Guide

KeyConcepts

Classifying Triangles (Lesson 12-1)

- Triangles can be classified by their angles as acute, obtuse, or right, and by their sides as scalene, isosceles, or equilateral.

Angles of Triangles (Lesson 12-2)

- The measure of an exterior angle is equal to the sum of its two remote interior angles.

Congruent Triangles (Lesson 12-3 through 12-5)

- SSS: If all of the corresponding sides of two triangles are congruent, then the triangles are congruent.
- SAS: If two pairs of corresponding sides of two triangles and the included angles are congruent, then the triangles are congruent.
- ASA: If two pairs of corresponding angles of two triangles and the included sides are congruent, then the triangles are congruent.
- AAS: If two pairs of corresponding angles of two triangles are congruent, and a corresponding pair of nonincluded sides is congruent, then the triangles are congruent.

Isosceles and Equilateral Triangles (Lesson 12-6)

- The base angles of an isosceles triangle are congruent and a triangle is equilateral if it is equiangular.

Transformations and Coordinate Proofs
(Lessons 12-7 and 12-8)

- In a congruence transformation, the position of the image may differ from the preimage, but the two figures remain congruent.
- Coordinate proofs use algebra to prove geometric concepts.

FOLDABLES® StudyOrganizer

Be sure the Key Concepts are noted in your Foldable.

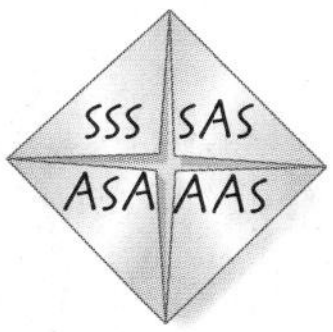

KeyVocabulary

VocabularyCheck

State whether each sentence is *true* or *false*. If *false*, replace the underlined word or phrase to make a true sentence.

1. An equiangular triangle is also an example of an <u>acute</u> triangle.
2. A triangle with an angle that measures greater than 90° is a <u>right</u> triangle.
3. An <u>equilateral</u> triangle is always equiangular.
4. A <u>scalene</u> triangle has at least two congruent sides.
5. The <u>vertex</u> angles of an isosceles triangle are congruent.
6. An <u>included</u> side is the side located between two consecutive angles of a polygon.
7. The three types of <u>congruence transformations</u> are rotation, reflection, and translation.
8. A <u>rotation</u> moves all points of a figure the same distance and in the same direction.
9. A <u>flow proof</u> uses figures in the coordinate plane and algebra to prove geometric concepts.
10. The measure of an <u>exterior angle</u> of a triangle is equal to the sum of the measures of its two remote interior angles.

CHAPTER 12 Study Guide and Review *Continued*

Lesson-by-Lesson Review

12-1 Classifying Triangles

Use the best description to classify each triangle: *acute, equiangular, obtuse,* or *right.*

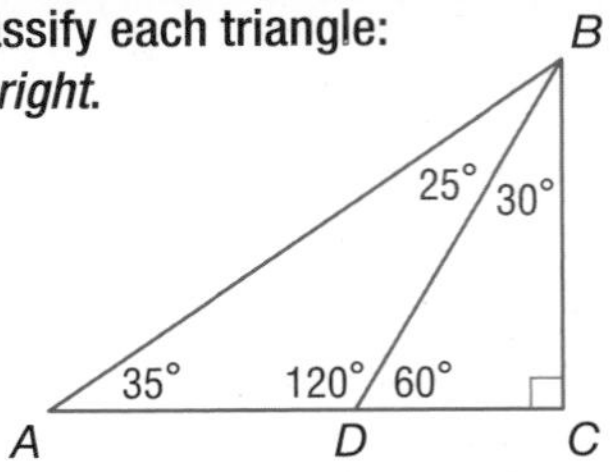

11. △*ADB*

12. △*BCD*

13. △*ABC*

ALGEBRA Find *x* and the measures of the unknown sides of each triangle.

14.

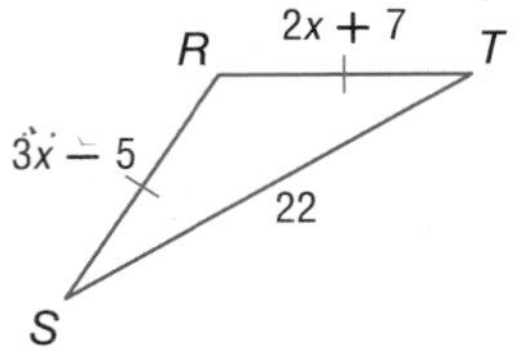

15.

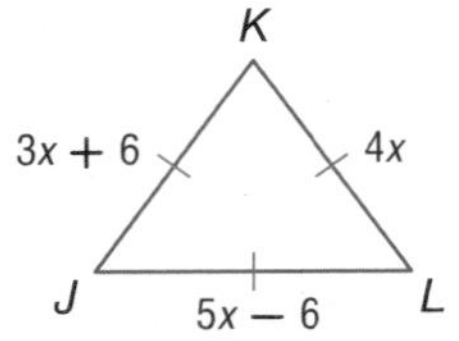

16. MAPS The distance from Chicago to Cleveland to Cincinnati and back to Chicago is 900 miles. The distance from Chicago to Cleveland is 50 miles more than the distance from Cincinnati to Chicago, and the distance from Cleveland to Cincinnati is 50 miles less than the distance from Cincinnati to Chicago. Find each distance and classify the triangle formed by the three cities.

Example 1

Classify each triangle as *acute*, *equiangular*, *obtuse*, or *right.*

a.

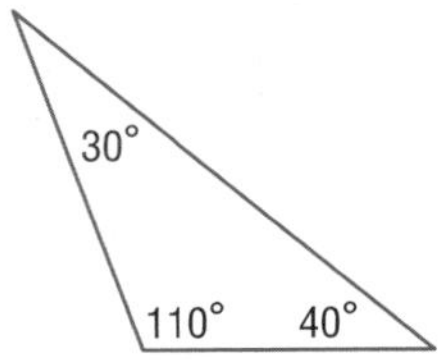

Since the triangle has one obtuse angle, it is an obtuse triangle.

b.

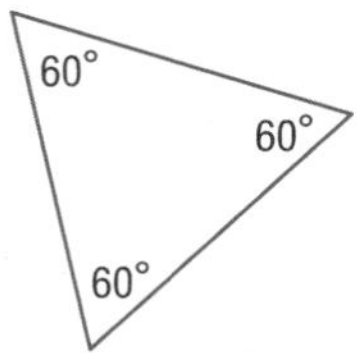

The triangle has three acute angles that are all equal. It is an equiangular triangle.

12-2 Angles of Triangles

Find the measure of each numbered angle.

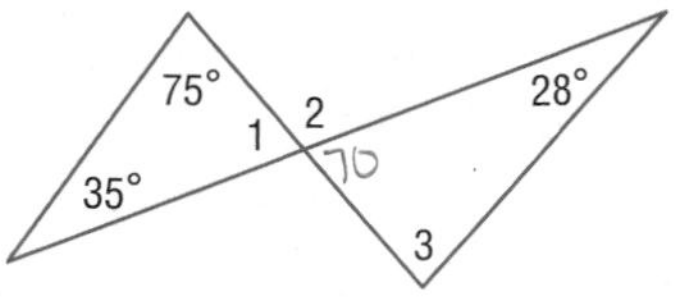

17. $\angle 1$

18. $\angle 2$

19. $\angle 3$

20. HOUSES The roof support on Lamar's house is in the shape of an isosceles triangle with base angles of 38°. Find *x*.

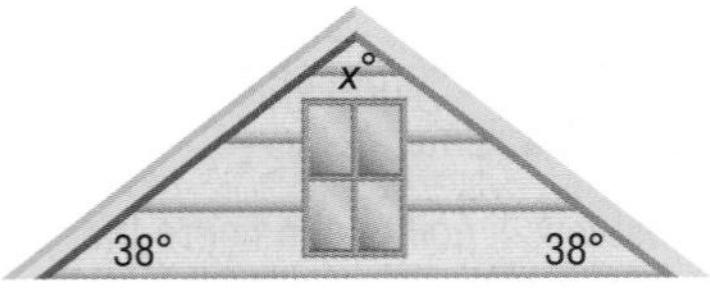

Example 2

Find the measure of each numbered angle.

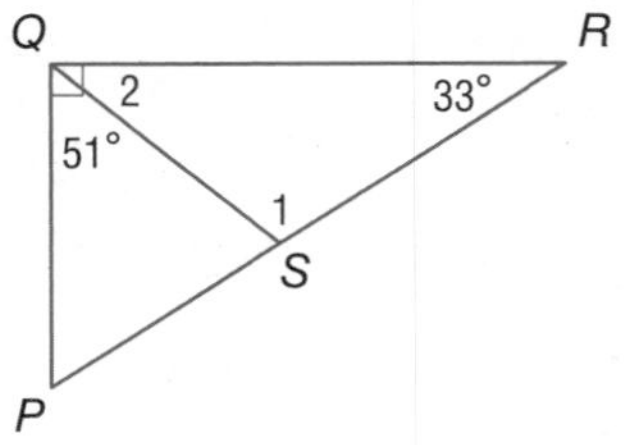

$m\angle 2 + m\angle PQS = 90$	
$m\angle 2 + 51 = 90$	Substitution
$m\angle 2 = 39$	Subtract 51 from each side.
$m\angle 1 + m\angle 2 + 33 = 180$	Triangle Sum Theorem
$m\angle 1 + 39 + 33 = 180$	Substitution
$m\angle 1 + 72 = 180$	Simplify.
$m\angle 1 = 108$	Subtract.

12-3 Congruent Triangles

Show that the polygons are congruent by identifying all congruent corresponding parts. Then write a congruence statement.

21.

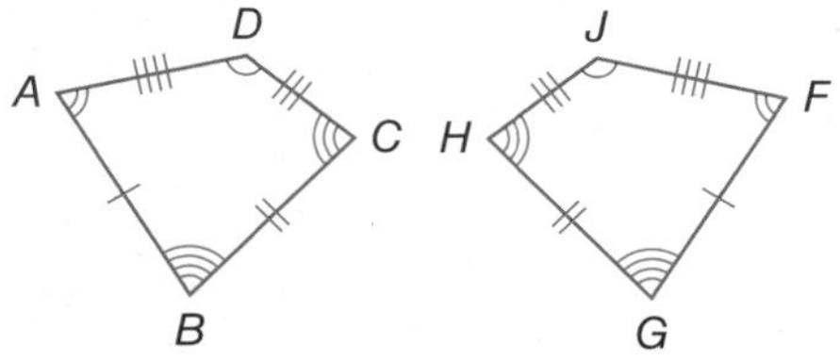

22.

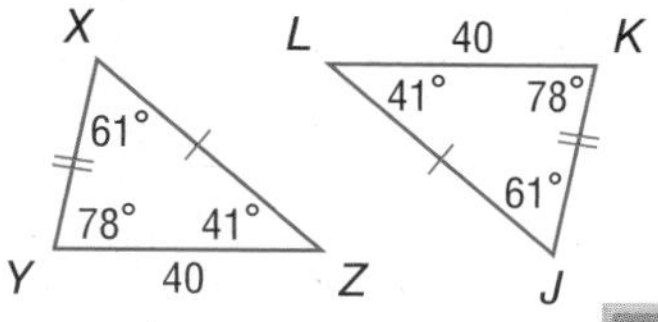

23. **MOSAIC TILING** A section of a mosaic tiling is shown. Name the triangles that appear to be congruent.

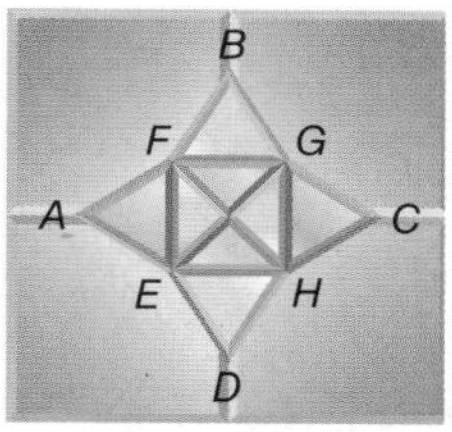

Example 3

Show that the polygons are congruent by identifying all the congruent corresponding parts. Then write a congruence statement.

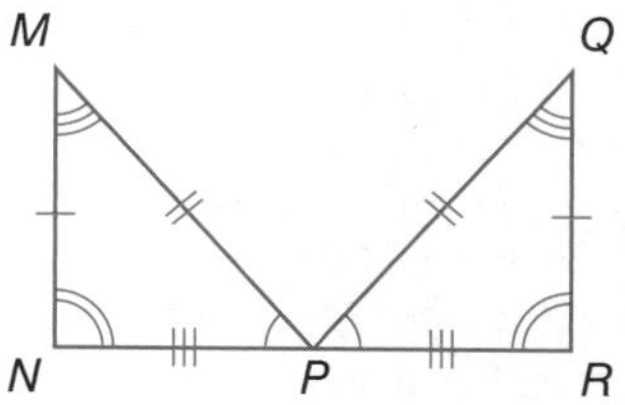

Angles: $\angle N \cong \angle R$, $\angle M \cong \angle Q$, $\angle MPN \cong \angle QPR$

Sides: $\overline{MN} \cong \overline{QR}$, $\overline{MP} \cong \overline{QP}$, $\overline{NP} \cong \overline{RP}$

All corresponding parts of the two triangles are congruent. Therefore, $\triangle MNP \cong \triangle QRP$.

12-4 Proving Triangles Congruent—SSS, SAS

Write a two-column proof.

24. Given: $\overline{AB} \parallel \overline{DC}$, $\overline{AB} \cong \overline{DC}$

Prove: $\triangle ABE \cong \triangle CDE$

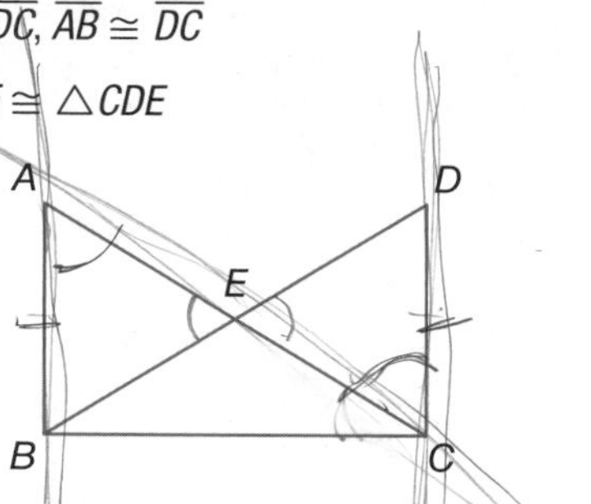

25. **KITES** Denise's kite is shown in the figure at the right. Given that $\overline{WY}$ bisects both $\angle XWZ$ and $\angle XYZ$, prove that $\triangle WXY \cong \triangle WZY$.

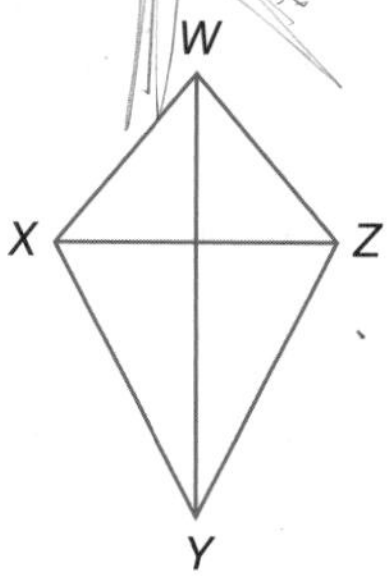

Example 4

Write a flow proof.

Given: $\overline{PQ}$ *bisects* $\angle RPS$.
$\angle R \cong \angle S$

Prove: $\triangle RPQ \cong \triangle SPQ$

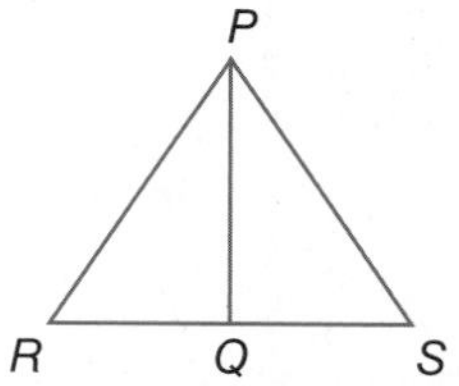

Flow Proof:

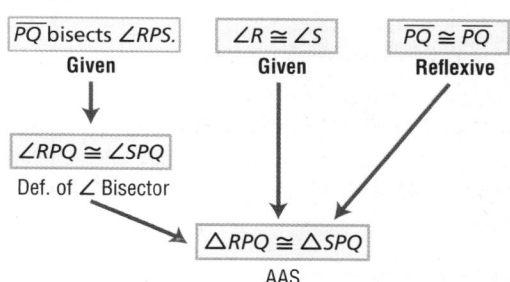

$\overline{PQ}$ bisects $\angle RPS$. — **Given**

$\angle R \cong \angle S$ — **Given**

$\overline{PQ} \cong \overline{PQ}$ — **Reflexive**

$\angle RPQ \cong \angle SPQ$ — Def. of $\angle$ Bisector

$\triangle RPQ \cong \triangle SPQ$ — AAS

CHAPTER 12

Study Guide and Review *Continued*

12-6 Isosceles and Equilateral Triangles

Find the value of each variable.

26.

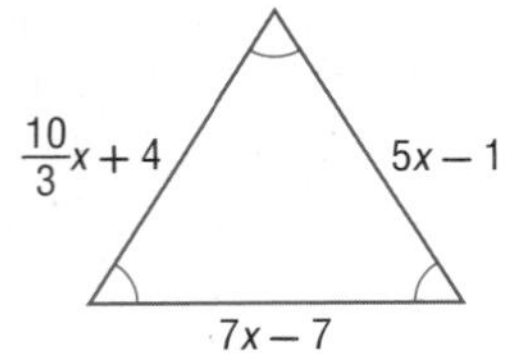

27.

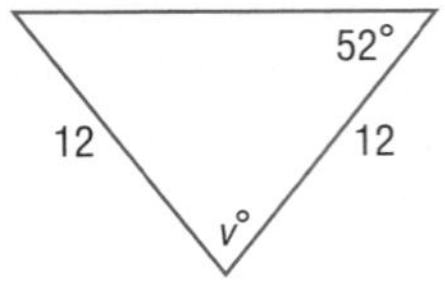

28. **PAINTING** Pam is painting using a wooden easel. The support bar on the easel forms an isosceles triangle with the two front supports. According to the figure below, what are the measures of the base angles of the triangle?

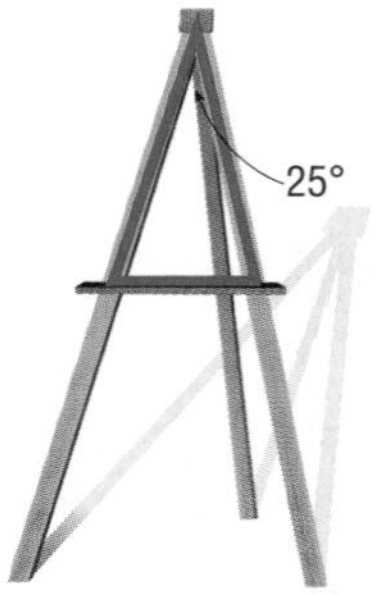

Example 5

Find each measure.

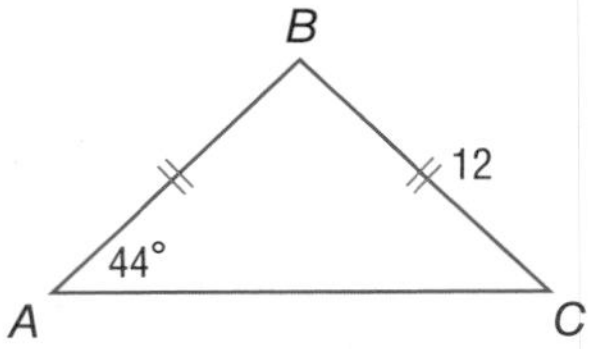

a. $m\angle B$

Since $AB = BC$, $\overline{AB} \cong \overline{BC}$. By the Isosceles Triangle Theorem, base angles A and C are congruent, so $m\angle A = m\angle C$. Use the Triangle Sum Theorem to write and solve an equation to find $m\angle B$.

$m\angle A + m\angle B + m\angle C = 180$	△ Sum Theorem
$44 + m\angle B + 44 = 180$	$m\angle A = m\angle C = 44$
$88 + m\angle B = 180$	Simplify.
$m\angle B = 92$	Subtract.

b. AB

$AB = BC$, so $\triangle ABC$ is isosceles. Since $BC = 12$, $AB = 12$ by substitution.

12-7 Congruence Transformations

Identify the type of congruence transformation shown as a *reflection*, *translation*, or *rotation*.

29.

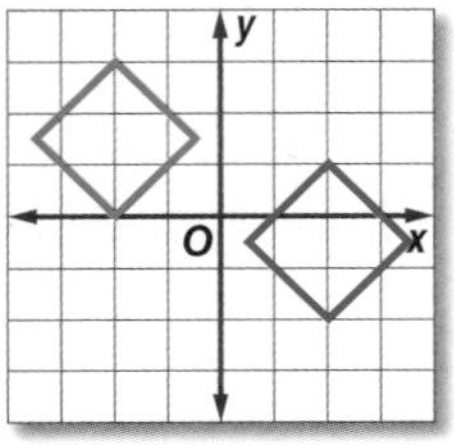

30.

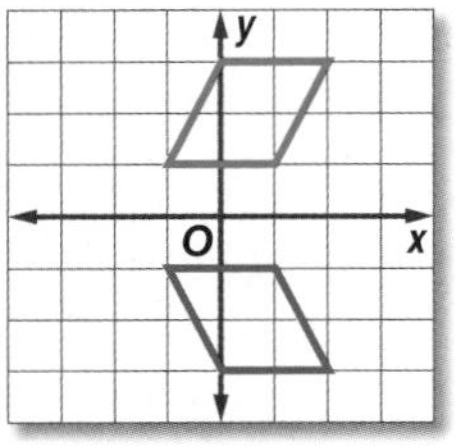

31.

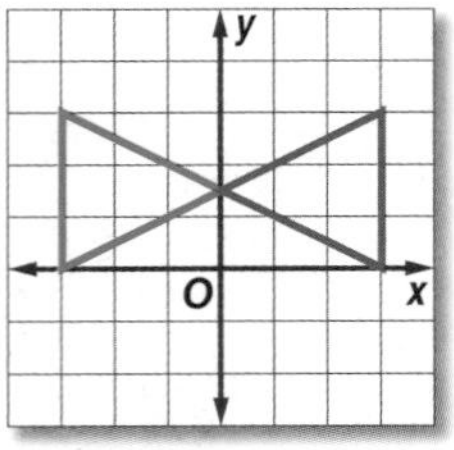

32.

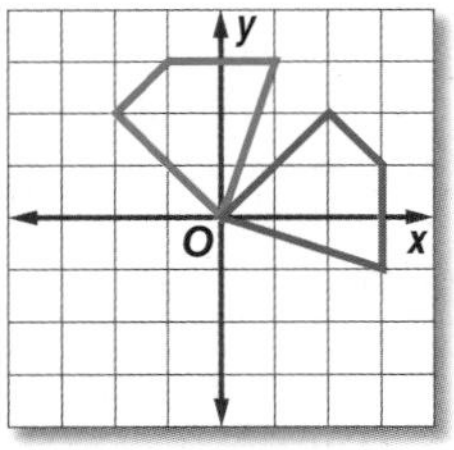

33. Triangle ABC with vertices $A(1, 1)$, $B(2, 3)$, and $C(3, -1)$ is a transformation of $\triangle MNO$ with vertices $M(-1, 1)$, $N(-2, 3)$, and $O(-3, -1)$. Graph the original figure and its image. Identify the transformation and verify that it is a congruence transformation.

Example 6

Triangle RST with vertices $R(4, 1)$, $S(2, 5)$, and $T(-1, 0)$ is a transformation of $\triangle CDF$ with vertices $C(1, -3)$, $D(-1, 1)$, and $F(-4, -4)$. Identify the transformation and verify that it is a congruence transformation.

Graph each figure. The transformation appears to be a translation. Find the lengths of the sides of each triangle.

$RS = \sqrt{(4-2)^2 + (1-5)^2}$ or $\sqrt{20}$

$TS = \sqrt{(-1-2)^2 + (0-5)^2}$ or $\sqrt{34}$

$RT = \sqrt{(-1-4)^2 + (0-1)^2}$ or $\sqrt{26}$

$CD = \sqrt{(-1-1)^2 + [1-(-3)]^2}$ or $\sqrt{20}$

$DF = \sqrt{[-4-(-1)]^2 + (-4-1)^2}$ or $\sqrt{34}$

$CF = \sqrt{(-4-1)^2 + [-4-(-3)]^2}$ or $\sqrt{26}$

Since each vertex of $\triangle CDF$ has undergone a transformation 3 units to the right and 4 units up, this is a translation.

Since $RS = CD$, $TS = DF$, and $RT = CF$, $\overline{RS} \cong \overline{CD}$, $\overline{TS} \cong \overline{DF}$, and $\overline{RT} \cong \overline{CF}$. By SSS, $\triangle RST \cong \triangle CDF$.

CHAPTER 12 Practice Test

Use the best description to classify each triangle: *acute, equiangular, obtuse,* or *right*.

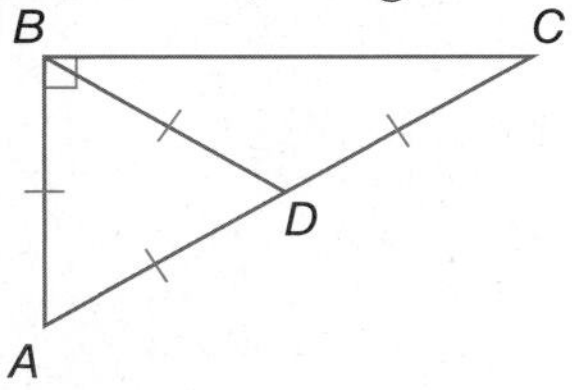

1. $\triangle ABD$
2. $\triangle ABC$
3. $\triangle BDC$

Find the measure of each numbered angle.

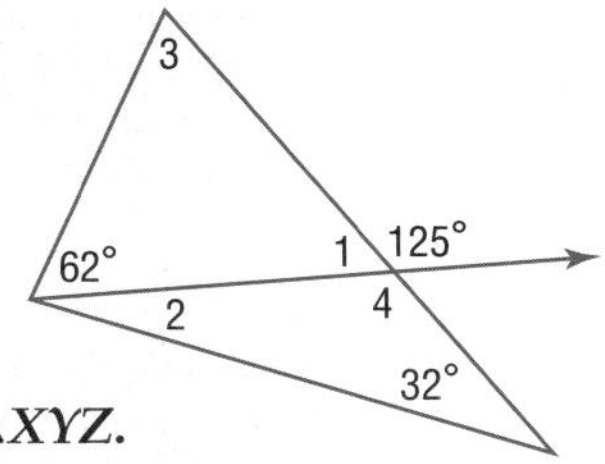

4. $\angle 1$
5. $\angle 2$
6. $\angle 3$
7. $\angle 4$

In the diagram, $\triangle RST \cong \triangle XYZ$.

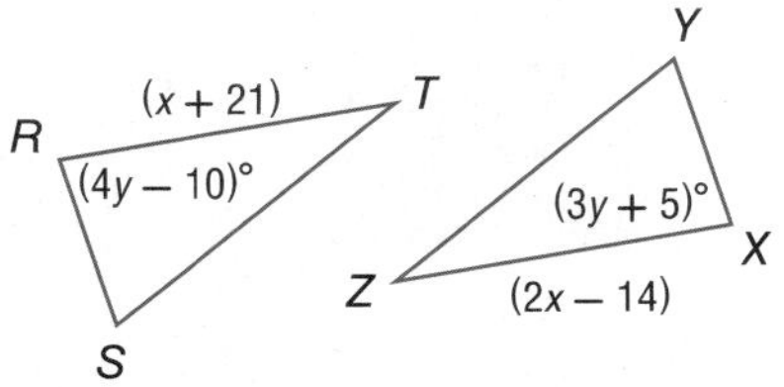

8. Find x.
9. Find y.

10. **PROOF** Write a flow proof.

Given: $\overline{XY} \parallel \overline{WZ}$ and $\overline{XW} \parallel \overline{YZ}$

Prove: $\triangle XWZ \cong \triangle ZYX$

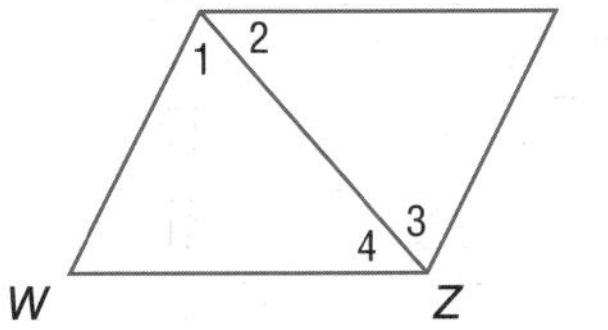

11. **MULTIPLE CHOICE** Find x.

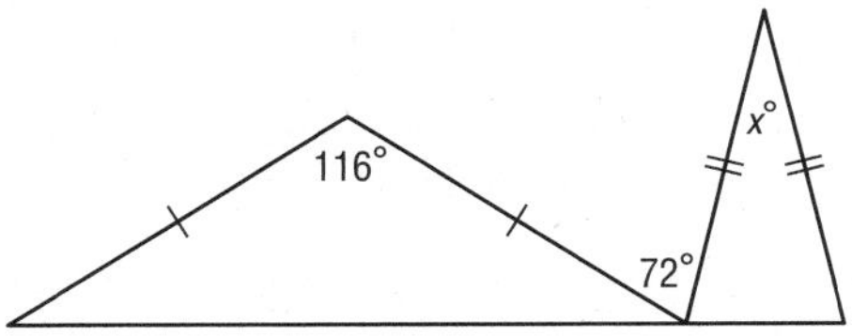

A 36
B 32
C 28
D 22

12. Determine whether $\triangle TJD \cong \triangle SEK$ given $T(-4, -2)$, $J(0, 5)$, $D(1, -1)$, $S(-1, 3)$, $E(3, 10)$, and $K(4, 4)$. Explain.

Determine which postulate or theorem can be used to prove each pair of triangles congruent. If it is not possible to prove them congruent, write *not possible*.

13.

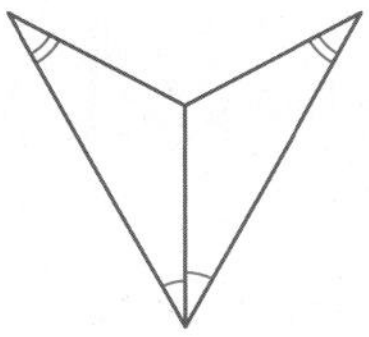

14.

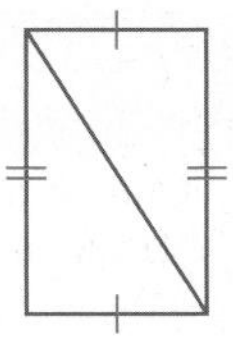

15.

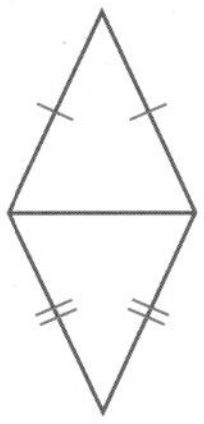

16.

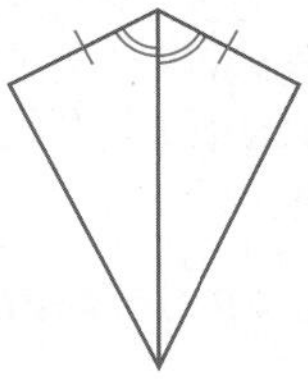

17. **LANDSCAPING** Angie has laid out a design for a garden consisting of two triangular areas as shown below. The points are $A(0, 0)$, $B(0, 5)$, $C(3, 5)$, $D(6, 5)$, and $E(6, 0)$. Name the type of congruence transformation for the preimage $\triangle ABC$ to $\triangle EDC$.

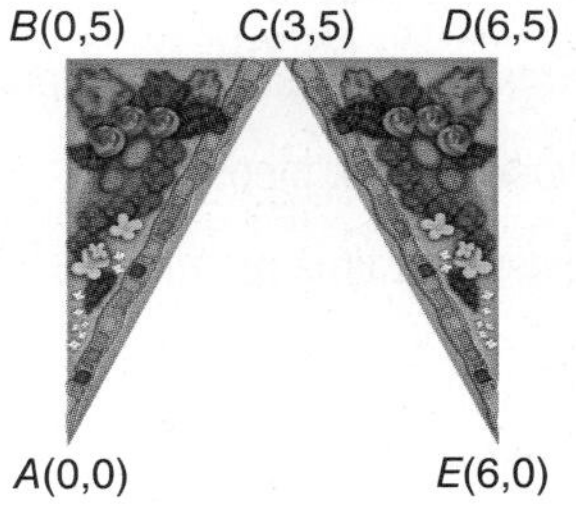

Find the measure of each numbered angle.

18. $\angle 1$
19. $\angle 2$

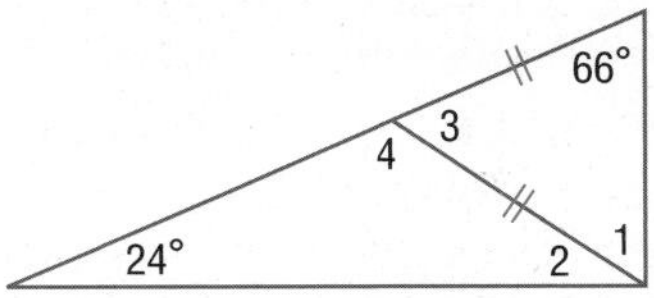

20. **PROOF** $\triangle ABC$ is a right isosceles triangle with hypotenuse $\overline{AB}$. M is the midpoint of $\overline{AB}$. Write a coordinate proof to show that $\overline{CM}$ is perpendicular to $\overline{AB}$.

CHAPTER 12

Preparing for Standardized Tests

Short-Answer Questions

Short-answer questions require you to provide a solution to the problem, along with a method, explanation, and/or justification used to arrive at the solution.

Short-answer questions are typically graded using a **rubric**, or a scoring guide.

The following is an example of a short-answer question scoring rubric.

Scoring Rubric		
	Criteria	Score
Full Credit	The answer is correct and a full explanation is provided that shows each step.	2
Partial Credit	• The answer is correct, but the explanation is incomplete. • The answer is incorrect, but the explanation is correct.	1 1
No Credit	Either an answer is not provided or the answer does not make sense.	0

Strategies for Solving Short-Answer Questions

Step 1

Read the problem to gain an understanding of what you are trying to solve.

- Identify relevant facts.
- Look for key words and mathematical terms.

Step 2

Make a plan and solve the problem.

- Explain your reasoning or state your approach to solving the problem.
- Show all of your work or steps.
- Check your answer if time permits.

Standardized Test Example

Read the problem. Identify what you need to know. Then use the information in the problem to solve. Show your work.

Triangle ABC is an isosceles triangle with base $\overline{BC}$. What is the perimeter of the triangle?

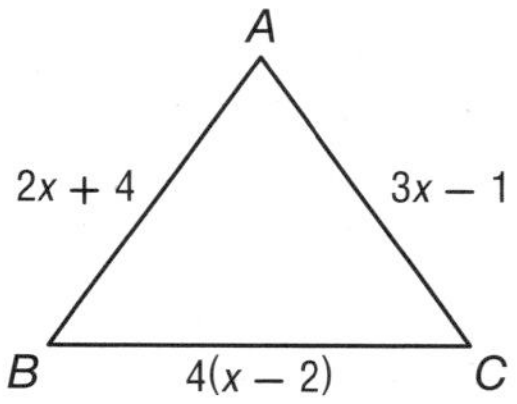

Rob Gage/Taxi/Getty Images

Read the problem carefully. You are told that $\triangle ABC$ is isosceles with base $\overline{BC}$. You are asked to find the perimeter of the triangle.

Make a plan and solve the problem.

The legs of an isosceles triangle are congruent. So, $\overline{AB} \cong \overline{AC}$ or $AB = AC$. Solve for x.

$$\begin{aligned} AB &= AC \\ 2x + 4 &= 3x - 1 \\ 2x - 3x &= -1 - 4 \\ -x &= -5 \\ x &= 5 \end{aligned}$$

Next, find the length of each side.

$AB = 2(5) + 4 = 14$ units

$AC = 3(5) - 1 = 14$ units

$BC = 4(5 - 2) = 12$ units

The perimeter of $\triangle ABC$ is $14 + 14 + 12 = 40$ units.

The steps, calculations, and reasoning are clearly stated. The student also arrives at the correct answer. So, this response is worth the full 2 points.

Exercises

Read each problem. Identify what you need to know. Then use the information in the problem to solve. Show your work.

1. Classify $\triangle DEF$ according to its angle measures.

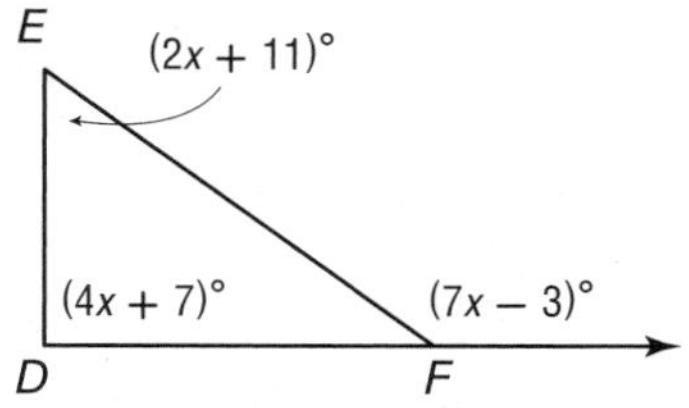

2. In the figure below, $\triangle RST \cong \triangle VUT$. What is the area of $\triangle RST$?

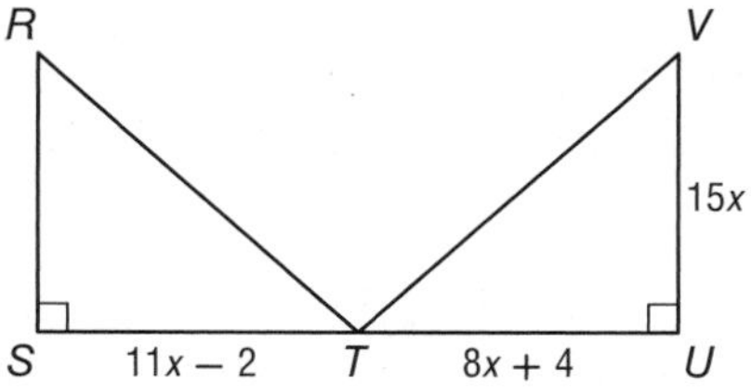

3. A farmer needs to make a 48-square-foot rectangular enclosure for chickens. He wants to save money by purchasing the least amount of fencing possible to enclose the area. What whole-number dimensions will require the least amount of fencing?

4. What is $m\angle 1$ in degrees?

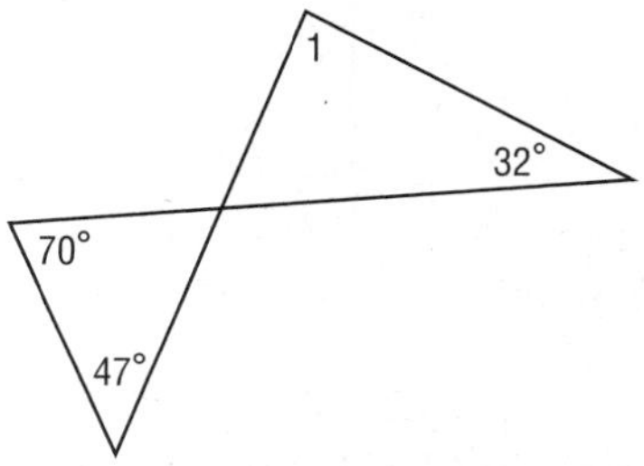

5. Write an equation of the line containing the points (2, 4) and (0, −2).

CHAPTER 12

Standardized Test Practice

Cumulative, Chapters 1 through 12

Multiple Choice

Read each question. Then fill in the correct answer on the answer document provided by your teacher or on a sheet of paper.

1. If $m\angle 1 = 110°$, what must $m\angle 2$ equal for lines x and z to be parallel?

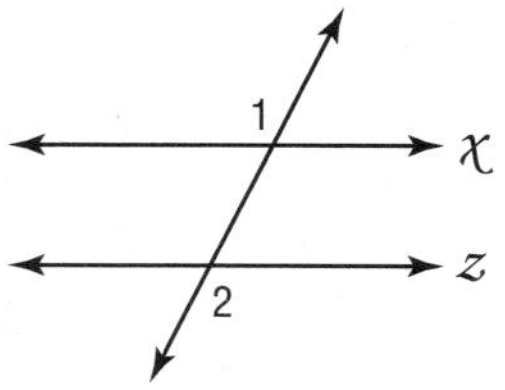

A 30° **B** 60° **C** 70° **D** 110°

2. Which of the following terms best describes the transformation below?

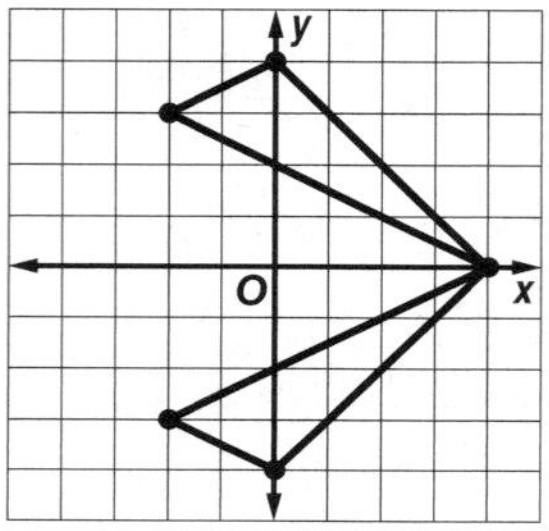

F dilation

G reflection

H rotation

J translation

3. Classify the triangle below according to its side lengths.

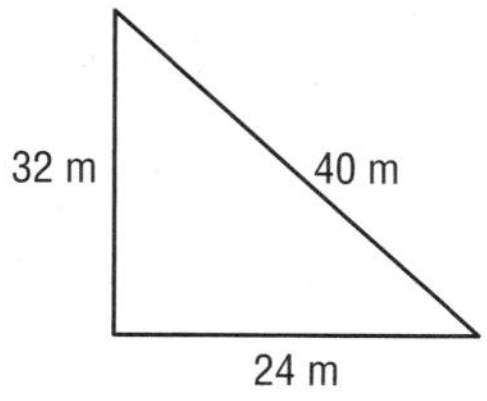

A equilateral

B isosceles

C right

D scalene

Test-Taking Tip

Question 3 Read the problem statement carefully to make sure you select the correct answer.

4. Given: $\overline{WX} \cong \overline{JK}$, $\overline{YX} \cong \overline{IK}$, $\angle X \cong \angle K$

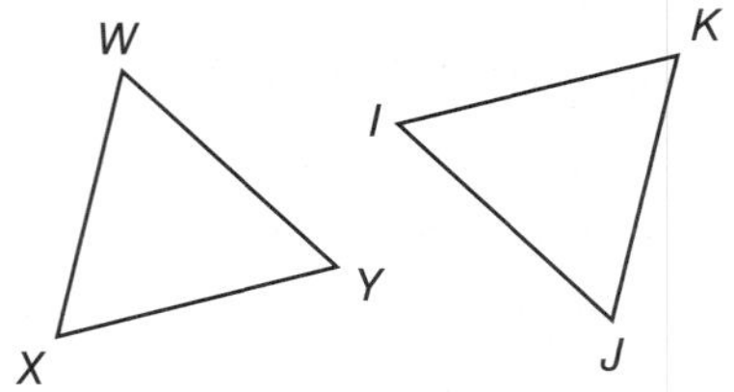

Which of the following lists the correct triangle congruence?

F $\triangle WXY \cong \triangle KIJ$

G $\triangle WXY \cong \triangle IKJ$

H $\triangle WXY \cong \triangle JKI$

J $\triangle WXY \cong \triangle IJK$

5. What is the area of the triangle below? Round your answer to the nearest tenth if necessary.

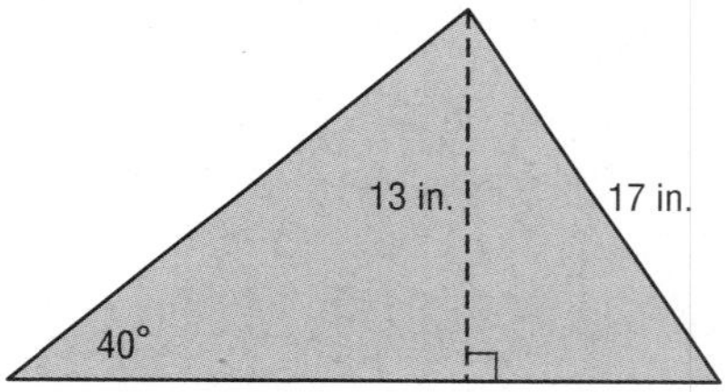

A 110.5 in^2

B 144.2 in^2

C 164.5 in^2

D 171.9 in^2

6. What is the measure of angle R below?

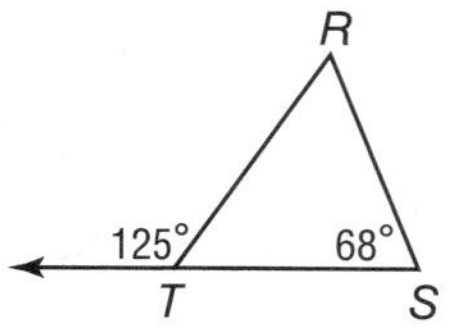

F 57° **G** 59° **H** 65° **J** 68°

7. Suppose one base angle of an isosceles triangle has a measure of 44°. What is the measure of the vertex angle?

A 108°

B 92°

C 56°

D 44°

Short Response/Gridded Response

Record your answers on the answer sheet provided by your teacher or on a sheet of paper.

8. **GRIDDED RESPONSE** In the figure below, $\triangle NDG \cong \triangle LGD$. What is the value of x?

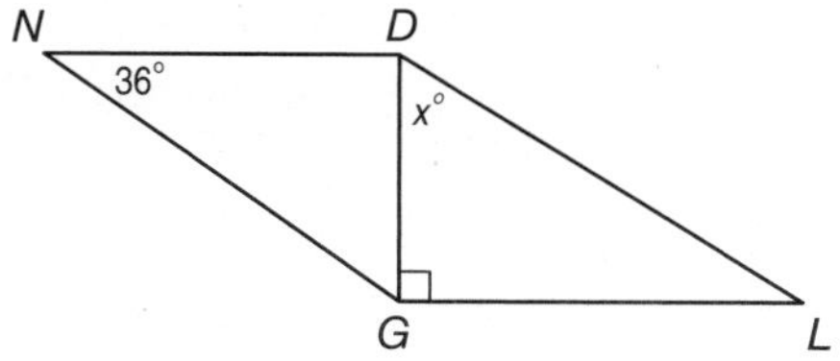

9. **GRIDDED RESPONSE** Suppose line ℓ contains points A, B, and C. If $AB = 7$ inches, $AC = 32$ inches, and point B is between points A and C, what is the length of $\overline{BC}$? Express your answer in inches.

10. Use the figure and the given information below.

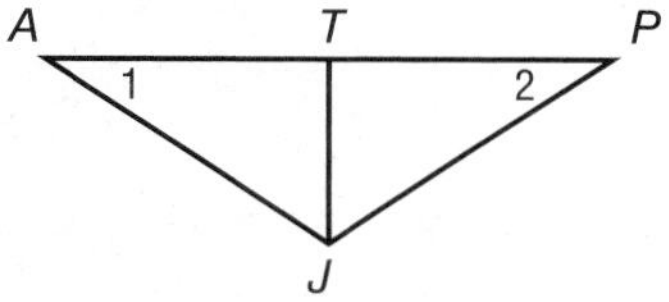

Given: $\overline{JT} \perp \overline{AP}$

$\angle 1 \cong \angle 2$

Which congruence theorem could you use to prove $\triangle PTJ \cong \triangle ATJ$ with only the information given? Explain.

11. Write an equation in slope intercept form for the line which goes through the points (0, 3) and (4, −5).

12. **GRIDDED RESPONSE** Find $m\angle TUV$ in the figure.

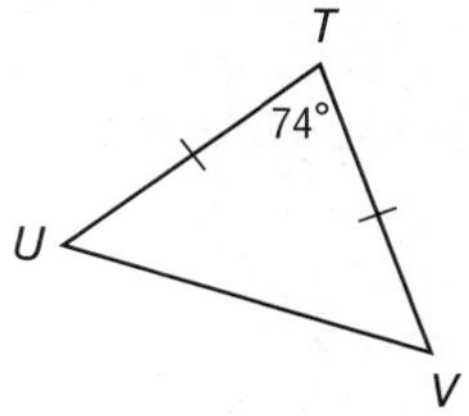

13. Suppose two sides of triangle ABC are congruent to two sides of triangle MNO. Also, suppose one of the nonincluded angles of $\triangle ABC$ is congruent to one of the nonincluded angles of $\triangle MNO$. Are the triangles congruent? If so, write a paragraph proof showing the congruence. If not, sketch a counterexample.

Extended Response

Record your answers on a sheet of paper. Show your work.

14. Use a coordinate grid to write a coordinate proof of the following statement.

If the vertices of a triangle are A(0, 0), B(2a, b), and C(4a, 0), then the triangle is isosceles.

a. Plot the vertices on a coordinate grid to model the problem.

b. Use the Distance Formula to write an expression for AB.

c. Use the Distance Formula to write an expression for BC.

d. Use your results from parts **b** and **c** to draw a conclusion about $\triangle ABC$.

Need ExtraHelp?

If you missed Question...	1	2	3	4	5	6	7	8	9	10	11	12	13	14
Go to Lesson...	11-2	12-7	12-1	12-3	12-9	12-2	12-6	12-3	10-2	12-5	11-4	12-6	12-4	12-8

CHAPTER 13

Quadrilaterals

Then

You classified polygons. You recognized and applied properties of polygons.

Now

In this chapter, you will:

- Find and use the sum of the measures of the interior and exterior angles of a polygon.
- Recognize and apply properties of quadrilaterals.
- Compare quadrilaterals.

Why? ▲

FUN AND GAMES The properties of quadrilaterals can be used to find various angle measures and side lengths such as the measures of angles in game equipment, playing fields, and game boards.

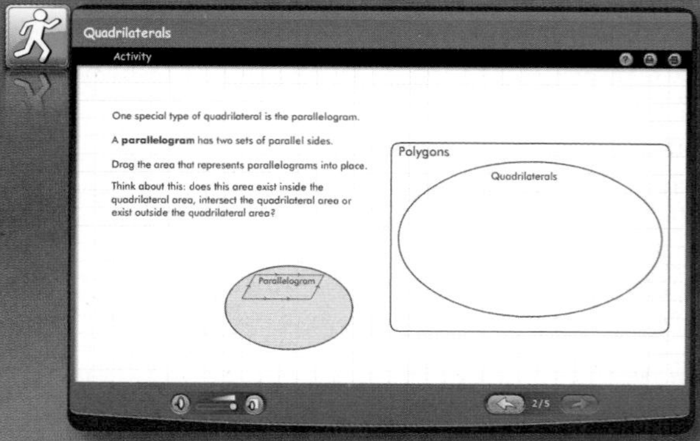

connectED.mcgraw-hill.com Your Digital Math Portal

Animation | Vocabulary | eGlossary | Personal Tutor | Virtual Manipulatives | Graphing Calculator | Audio | Foldables | Self-Check Practice | Worksheets

Rubberball/Erik Isakson/Getty Ima

Get Ready for the Chapter

Diagnose Readiness | You have two options for checking Prerequisite Skills.

1 **Textbook Option** Take the Quick Check below. Refer to the Quick Review for help.

QuickCheck

Find x to the nearest tenth.

1.

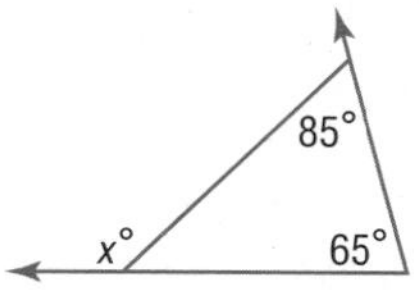

2.

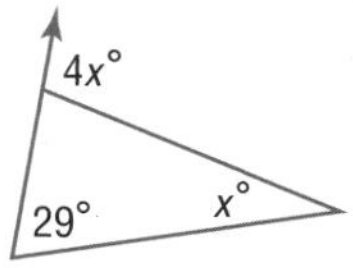

SPEED SKATING A speed skater forms at least two sets of triangles and exterior angles as she skates. Find each measure.

3. $m\angle 1$

4. $m\angle 2$

5. $m\angle 3$

6. $m\angle 4$

ALGEBRA Find x and the measures of the unknown sides of each triangle.

7.

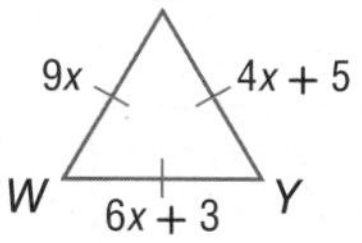

8.

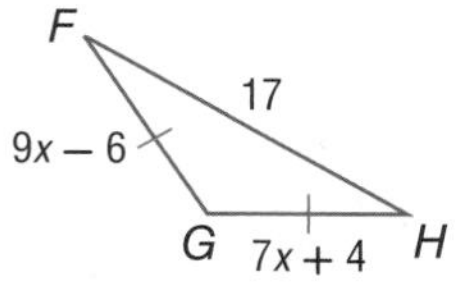

9. TRAVEL A plane travels from Des Moines to Phoenix, on to Atlanta, and back to Des Moines, as shown below. Find the distance in miles from Des Moines to Phoenix if the total trip was 3482 miles.

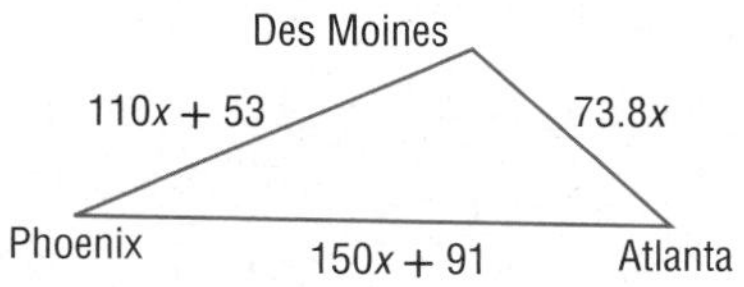

QuickReview

Example 1

Find the measure of each numbered angle.

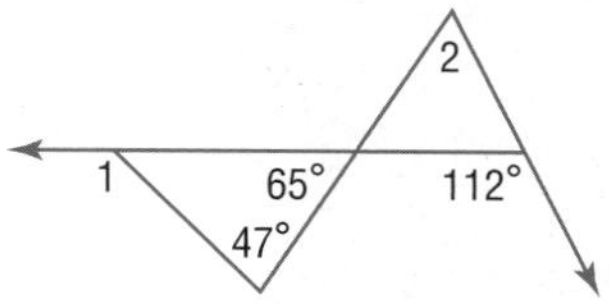

a. $m\angle 1$

$m\angle 1 = 65 + 47$ — Exterior ∠ Theorem

$m\angle 1 = 112$ — Add.

b. $m\angle 2$

$180 = m\angle 2 + 68 + 65$ — Triangle Sum Theorem

$180 = m\angle 2 + 133$ — Simplify.

$m\angle 2 = 47$ — Subtract.

Example 2

ALGEBRA Find the measures of the sides of isosceles $\triangle XYZ$.

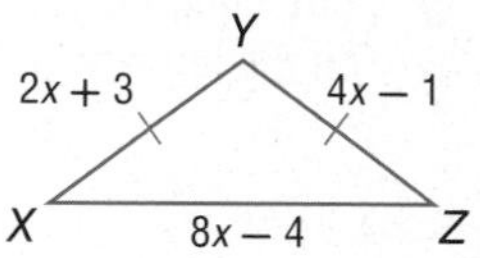

$XY = YZ$ — Given

$2x + 3 = 4x - 1$ — Substitution

$-2x = -4$ — Subtract.

$x = 2$ — Simplify.

$XY = 2x + 3$ — Given

$= 2(2) + 3$ or 7 — $x = 2$

$YZ = XY$ — Given

$= 7$ — $XY = 7$

$XZ = 8x - 4$ — Given

$= 8(2) - 4$ or 12 — $x = 2$

2 **Online Option** Take an online self-check Chapter Readiness Quiz at connectED.mcgraw-hill.com.

Get Started on the Chapter

You will learn several new concepts, skills, and vocabulary terms as you study Chapter 13. To get ready, identify important terms and organize your resources. You may wish to refer to Chapter 0 to review prerequisite skills.

FOLDABLES® Study Organizer

Quadrilaterals Make this Foldable to help you organize your Chapter 6 notes about quadrilaterals. Begin with one sheet of notebook paper.

1 **Fold** lengthwise to the holes.

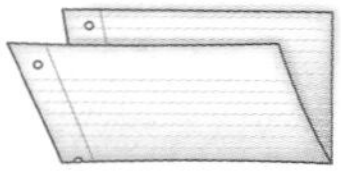

2 **Fold** along the width of the paper twice and unfold the paper.

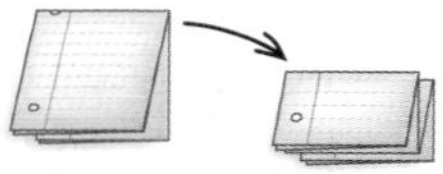

3 **Cut** along the fold marks on the left side of the paper.

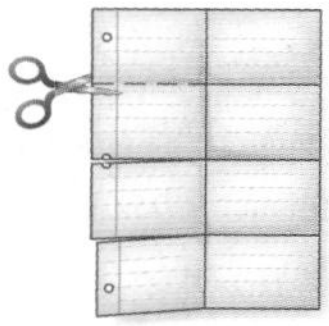

4 **Label** as shown.

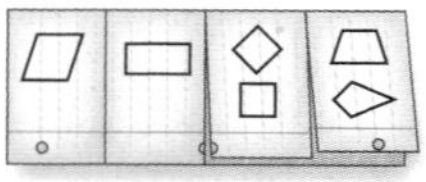

New Vocabulary

English		Español
parallelogram	p. 803	paralelogramo
rectangle	p. 823	rectángulo
rhombus	p. 830	rombo
square	p. 831	cuadrado
trapezoid	p. 839	trapecio
base	p. 839	base
legs	p. 839	catetos
isosceles trapezoid	p. 839	trapecio isósceles
midsegment of a trapezoid	p. 841	segmento medio de un trapecio

Review Vocabulary

exterior angle ángulo externo an angle formed by one side of a triangle and the extension of another side

remote interior angle ángulos internos no adyacentes the angles of a triangle that are not adjacent to a given exterior angle

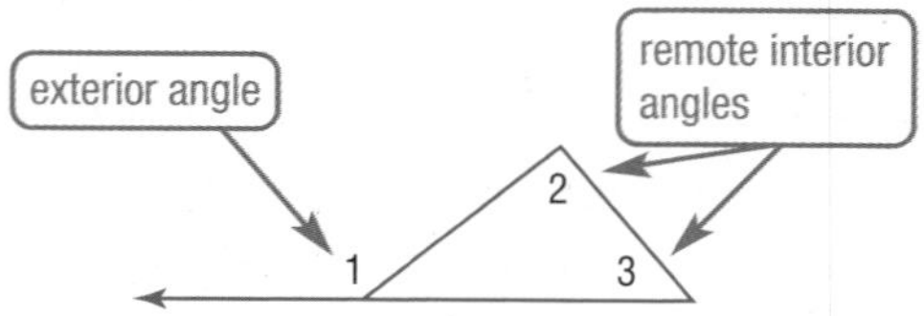

slope pendiente for a (nonvertical) line containing two points (x_1, y_1) and (x_2, y_2), the number m given by the formula $m = \frac{(y_2 - y_1)}{(x_2 - x_1)}$ where $x_2 \neq x_1$

LESSON 13-1

Parallelograms

:: Then	:: Now	:: Why?
You classified polygons with four sides as quadrilaterals.	**1** Recognize and apply properties of the sides and angles of parallelograms. **2** Recognize and apply properties of the diagonals of parallelograms.	The arm of the basketball goal shown can be adjusted to a height of 10 feet or 5 feet. Notice that as the height is adjusted, each pair of opposite sides of the quadrilateral formed by the arms remains parallel.

NewVocabulary
parallelogram

Common Core State Standards

Content Standards
G.CO.11 Prove theorems about parallelograms.
G.GPE.4 Use coordinates to prove simple geometric theorems algebraically.

Mathematical Practices
4 Model with mathematics.
3 Construct viable arguments and critique the reasoning of others.

1 Sides and Angles of Parallelograms

A **parallelogram** is a quadrilateral with both pairs of opposite sides parallel. To name a parallelogram, use the symbol ▱. In ▱$ABCD$, $\overline{BC} \parallel \overline{AD}$ and $\overline{AB} \parallel \overline{DC}$ by definition.

Other properties of parallelograms are given in the theorems below.

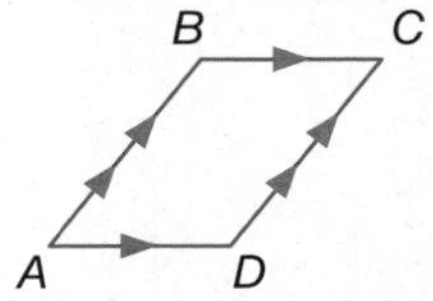

▱$ABCD$

Theorem Properties of Parallelograms

13.1 If a quadrilateral is a parallelogram, then its opposite sides are congruent.

Abbreviation *Opp. sides of a ▱ are ≅.*

Example If $JKLM$ is a parallelogram, then $\overline{JK} \cong \overline{ML}$ and $\overline{JM} \cong \overline{KL}$.

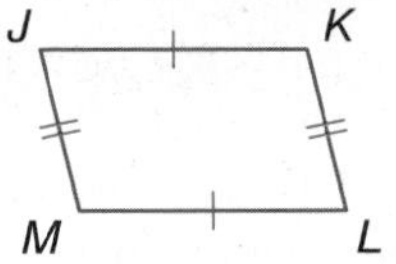

13.2 If a quadrilateral is a parallelogram, then its opposite angles are congruent.

Abbreviation *Opp. ∠s of a ▱ are ≅.*

Example If $JKLM$ is a parallelogram, then $\angle J \cong \angle L$ and $\angle K \cong \angle M$.

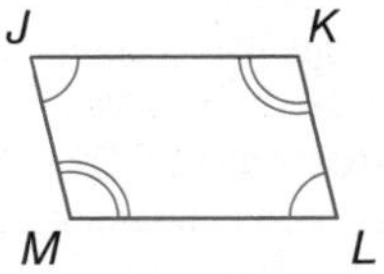

13.3 If a quadrilateral is a parallelogram, then its consecutive angles are supplementary.

Abbreviation *Cons. ∠s in a ▱ are supplementary.*

Example If $JKLM$ is a parallelogram, then $x + y = 180$.

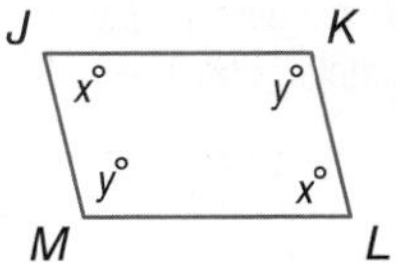

13.4 If a parallelogram has one right angle, then it has four right angles.

Abbreviation *If a ▱ has 1 rt. ∠, it has 4 rt. ∠s.*

Example In ▱$JKLM$, if $\angle J$ is a right angle, then $\angle K$, $\angle L$, and $\angle M$ are also right angles.

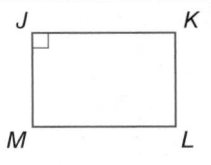

You will prove Theorems 13.1, 13.3, and 13.4 in Exercises 28, 26, and 7, respectively.

StudyTip

Including a Figure Theorems are presented in general terms. In a proof, you must include a drawing so that you can refer to segments and angles specifically.

Proof Theorem 13.2

Write a two-column proof of Theorem 13.4.

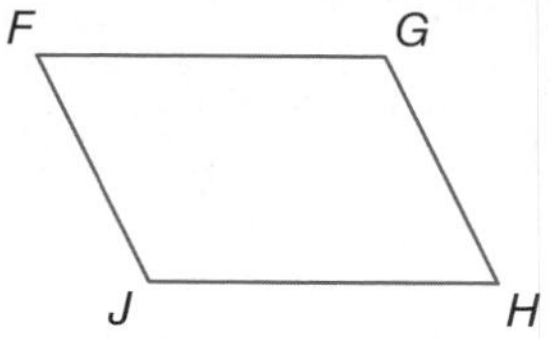

Given: ▱$FGHJ$

Prove: $\angle F \cong \angle H$, $\angle J \cong \angle G$

Proof:

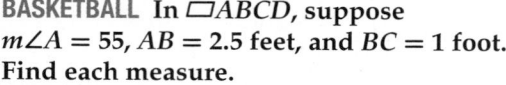

Statements	Reasons
1. ▱$FGHJ$	**1.** Given
2. $\overline{FG} \parallel \overline{JH}$; $\overline{FJ} \parallel \overline{GH}$	**2.** Definition of parallelogram
3. $\angle F$ and $\angle J$ are supplementary. $\angle J$ and $\angle H$ are supplementary. $\angle H$ and $\angle G$ are supplementary.	**3.** If parallel lines are cut by a transversal, consecutive interior angles are supplementary.
4. $\angle F \cong \angle H$, $\angle J \cong \angle G$	**4.** Supplements of the same angles are congruent.

Real-WorldCareer

Coach Coaches organize amateur and professional atheletes, teaching them the fundamentals of a sport. They manage teams during both practice sessions and competitions. Additional tasks may include selecting and issuing sports equiment, materials, and supplies. Head coaches at public secondary schools usually have a bachelor's degree.

Real-World Example 1 Use Properties of Parallelograms

PT

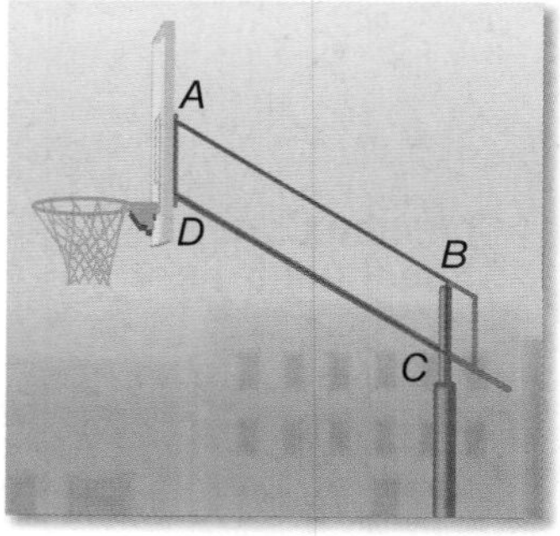

BASKETBALL **In ▱$ABCD$, suppose $m\angle A = 55$, $AB = 2.5$ feet, and $BC = 1$ foot. Find each measure.**

a. DC

$DC = AB$ — Opp. sides of a ▱ are ≅.

$= 2.5$ ft — Substitution

b. $m\angle B$

$m\angle B + m\angle A = 180$ — Cons. ⦣ in a ▱ are supplementary.

$m\angle B + 55 = 180$ — Substitution

$m\angle B = 125$ — Subtract 55 from each side.

c. $m\angle C$

$m\angle C = m\angle A$ — Opp. ⦣ of a ▱ are ≅.

$= 55$ — Substitution

GuidedPractice

1. MIRRORS The wall-mounted mirror shown uses parallelograms that change shape as the arm is extended. In ▱$JKLM$, suppose $m\angle J = 47$. Find each measure.

A. $m\angle L$ **B.** $m\angle M$

C. Suppose the arm was extended further so that $m\angle J = 90$. What would be the measure of each of the other angles? Justify your answer.

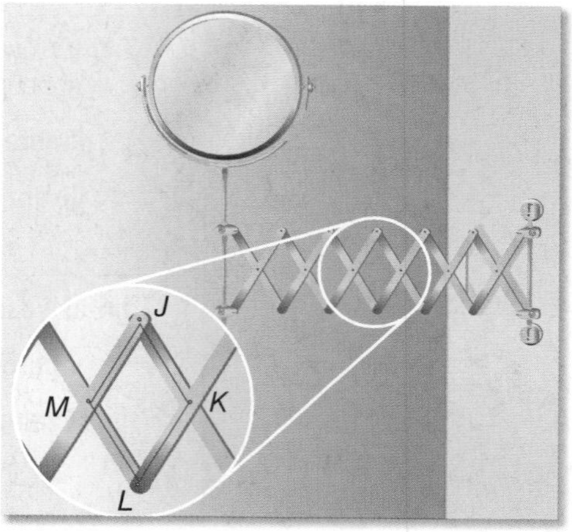

2 Diagonals of Parallelograms

The diagonals of a parallelogram have special properties as well.

Theorem Diagonals of Parallelograms

13.5 If a quadrilateral is a parallelogram, then its diagonals bisect each other.

Abbreviation *Diag. of a ▱ bisect each other.*

Example If *ABCD* is a parallelogram, then $\overline{AP} \cong \overline{PC}$ and $\overline{DP} \cong \overline{PB}$.

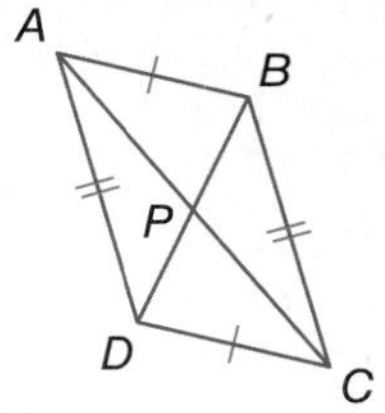

13.6 If a quadrilateral is a parallelogram, then each diagonal separates the parallelogram into two congruent triangles.

Abbreviation *Diag. separates a ▱ into 2 ≅ △s.*

Example If *ABCD* is a parallelogram, then $\triangle ABD \cong \triangle CDB$.

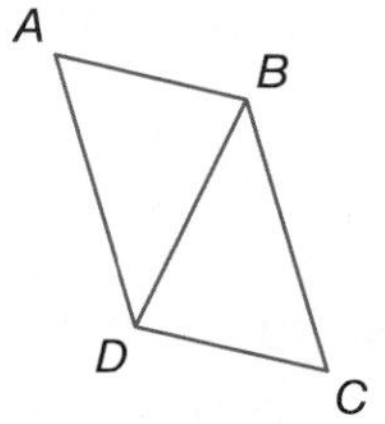

You will prove Theorems 13.5 and 13.6 in Exercises 29 and 27, respectively.

Example 2 Use Properties of Parallelograms and Algebra

ALGEBRA If *QRST* is a parallelogram, find the value of the indicated variable.

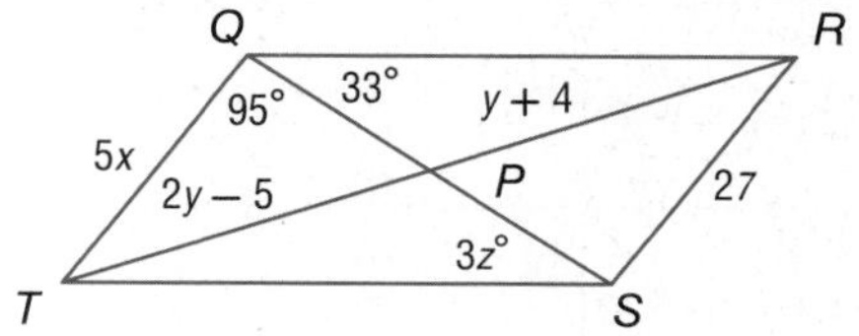

a. x

$\overline{QT} \cong \overline{RS}$	Opp. sides of a ▱ are ≅.
$QT = RS$	Definition of congruence
$5x = 27$	Substitution
$x = 5.4$	Divide each side by 5.

b. y

$\overline{TP} \cong \overline{PR}$	Diag. of a ▱ bisect each other.
$TP = PR$	Definition of congruence
$2y - 5 = y + 4$	Substitution
$y = 9$	Subtract y and add 5 to each side.

c. z

$\triangle TQS \cong \triangle RSQ$	Diag. separates a ▱ into 2 ≅ △s.
$\angle QST \cong \angle SQR$	CPCTC
$m\angle QST = m\angle SQR$	Definition of congruence
$3z = 33$	Substitution
$z = 11$	Divide each side by 3.

StudyTip

Congruent Triangles
A parallelogram with two diagonals divides the figure into two pairs of congruent triangles.

GuidedPractice

Find the value of each variable in the given parallelogram.

2A.

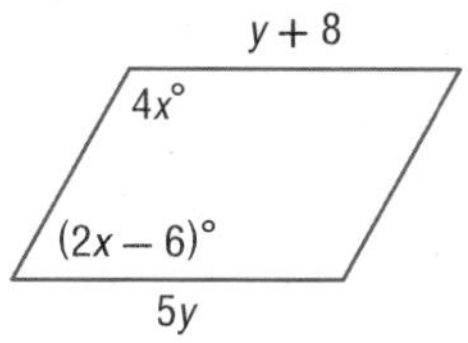

2B.

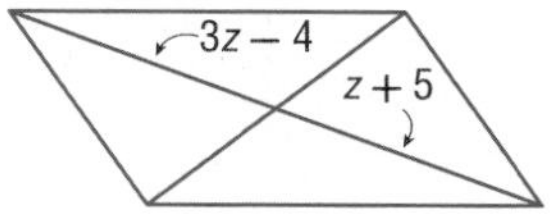

You can use Theorem 13.5 to determine the coordinates of the intersection of the diagonals of a parallelogram on a coordinate plane given the coordinates of the vertices.

Example 3 Parallelograms and Coordinate Geometry

COORDINATE GEOMETRY Determine the coordinates of the intersection of the diagonals of ▱*FGHJ* with vertices *F*(−2, 4), *G*(3, 5), *H*(2, −3), and *J*(−3, −4).

Since the diagonals of a parallelogram bisect each other, their intersection point is the midpoint of $\overline{FH}$ and $\overline{GJ}$. Find the midpoint of $\overline{FH}$ with endpoints (−2, 4) and (2, −3).

$$\left(\frac{x_1 + x_2}{2}, \frac{y_1 + y_2}{2}\right) = \left(\frac{-2 + 2}{2}, \frac{4 + (-3)}{2}\right) \quad \text{Midpoint Formula}$$

$$= (0, 0.5) \quad \text{Simplify.}$$

The coordinates of the intersection of the diagonals of ▱*FGHJ* are (0, 0.5).

CHECK Find the midpoint of $\overline{GJ}$ with endpoints (3, 5) and (−3, −4).

$$\left(\frac{3 + (-3)}{2}, \frac{5 + (-4)}{2}\right) = (0, 0.5) \checkmark$$

StudyTip

CCSS Regularity Graph the parallelogram in Example 3 and the point of intersection of the diagonals you found. Draw the diagonals. The point of intersection appears to be correct.

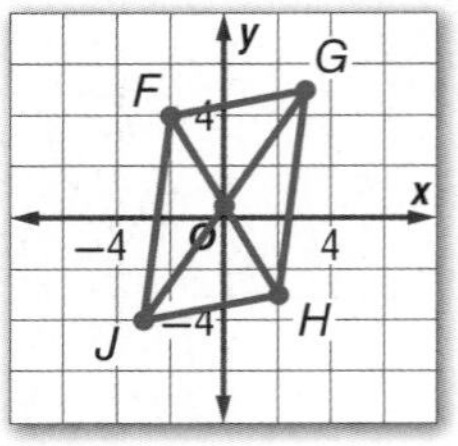

GuidedPractice

3. COORDINATE GEOMETRY Determine the coordinates of the intersection of the diagonals of *RSTU* with vertices *R*(−8, −2), *S*(−6, 7), *T*(6, 7), and *U*(4, −2).

You can use the properties of parallelograms and their diagonals to write proofs.

Example 4 Proofs Using the Properties of Parallelograms

Write a paragraph proof.

Given: ▱*ABDG*, $\overline{AF} \cong \overline{CF}$

Prove: ∠*BDG* ≅ ∠*C*

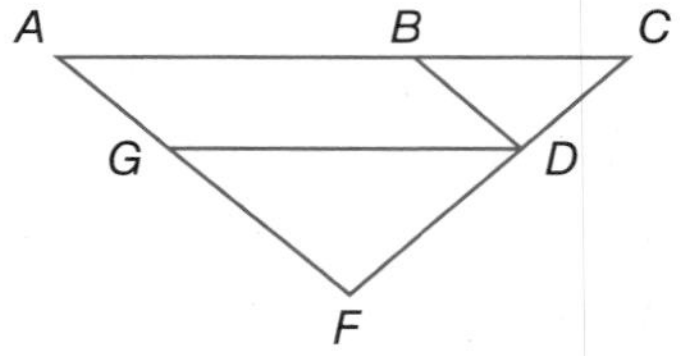

Proof:

We are given *ABDG* is a parallelogram. Since opposite angles in a parallelogram are congruent, ∠*BDG* ≅ ∠*A*. We are also given that $\overline{AF} \cong \overline{CF}$. By the Isosceles Triangle Theorem, ∠*A* ≅ ∠*C*. So, by the Transitive Property of Congruence, ∠*BDG* ≅ ∠*C*.

GuidedPractice

4. Write a two-column proof.

Given: ▱*HJKP* and ▱*PKLM*

Prove: $\overline{HJ} \cong \overline{ML}$

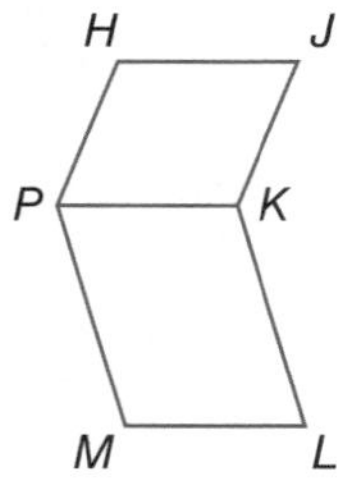

Check Your Understanding

Example 1

1. **BICYCLES** The frame of some bicycles is a parallelogram. In ▱$ABCD$, suppose $m\angle ABC = 45$, $AB = 61$ cm, and $AD = 58$ cm. Find each measure.

   a. $m\angle BAD$

   b. BC

   c. $m\angle ADC$

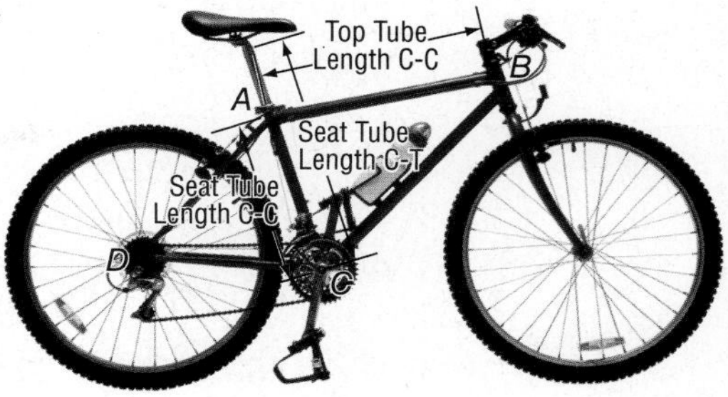

Example 2

ALGEBRA Find the value of each variable in each parallelogram.

2.

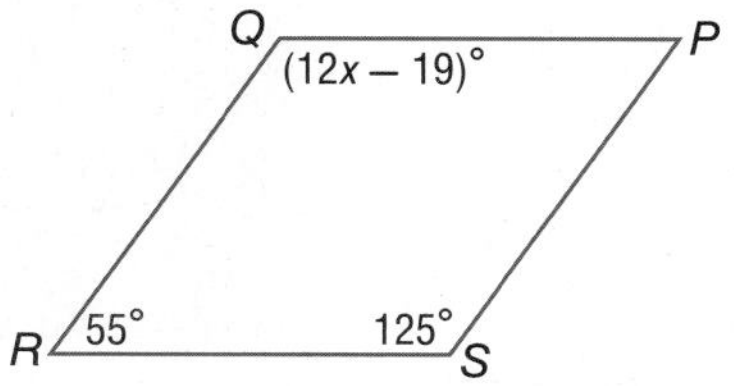

3.

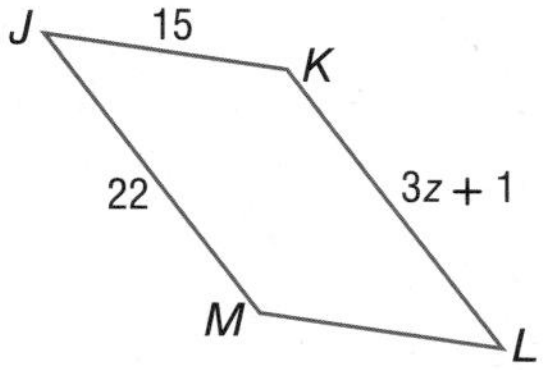

4.

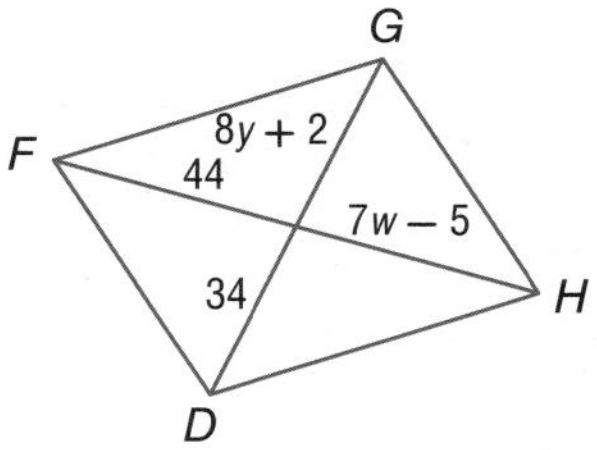

5.

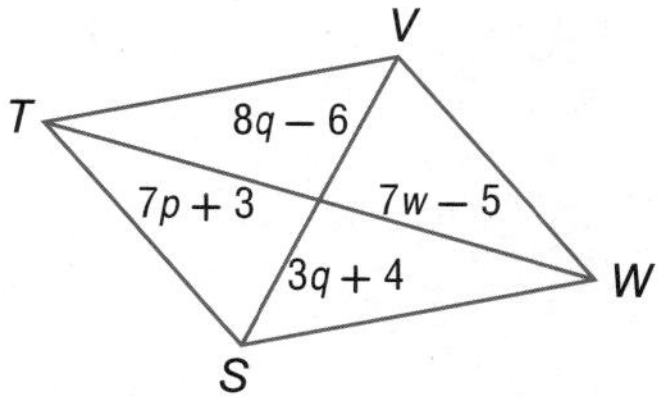

Example 3

6. **COORDINATE GEOMETRY** Determine the coordinates of the intersection of the diagonals of ▱$FGHJ$ with vertices $F(-7, 6)$, $G(-1, 10)$, $H(3, 0)$, and $J(-3, -4)$.

Example 4

CCSS PROOF Write the indicated type of proof.

7. two-column

   Given: ▱$ABCD$, $\angle A$ is a right angle.

   Prove: $\angle B$, $\angle C$, and $\angle D$ are right angles. (Theorem 13.6)

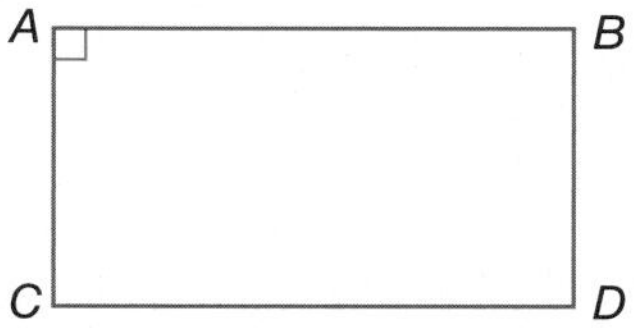

8. paragraph

   Given: $XYZW$ and $YRSW$ are parallelograms.

   Prove: $\overline{XW} \cong \overline{RS}$

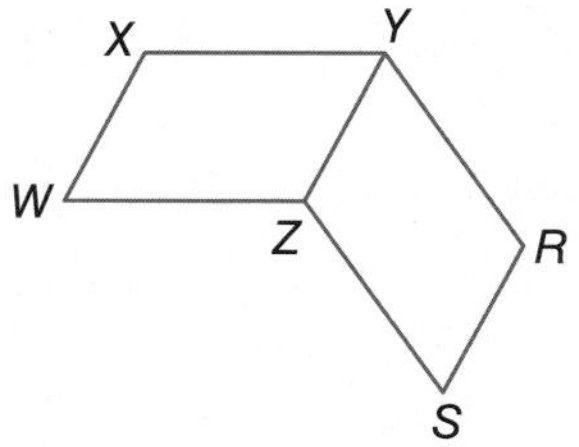

Practice and Problem Solving

Example 1

Use ▱$LMNP$ to find each measure.

9. $m\angle L$

10. MP

11. $m\angle M$

12. LM

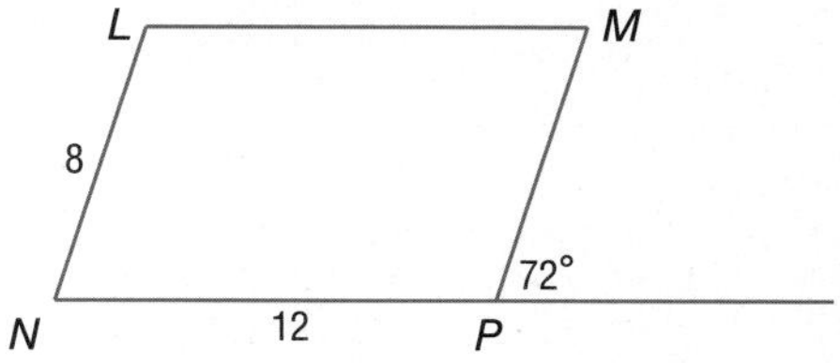

Masterfile

13. **DRAWING** 3 parallelograms are used to draw a cube that appears to be 3-D. In ▱$FGHD$, $FD = \frac{1}{2}$ inch, $FG = 1$ inch, and $m\angle GFD = 132$. Find each measure.

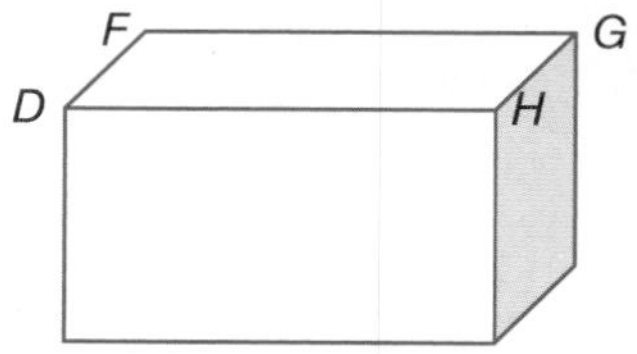

a. DH

b. GH

c. $m\angle GHD$

d. $m\angle FDH$

14. **CCSS ARCHITECTURE** The awning of this building is constructed using a parallelogram..

a. Identify two pairs of congruent segments.

b. Identify two pairs of supplementary angles.

Example 2 **ALGEBRA** **Find the value of each variable in each parallelogram.**

15.

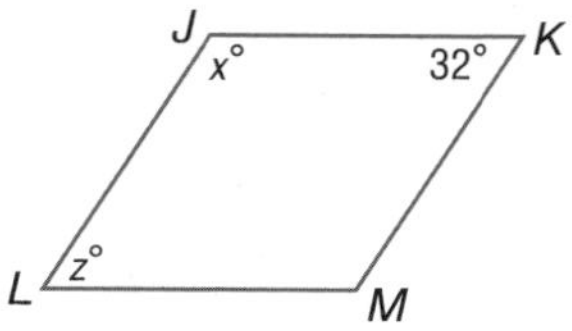

16.

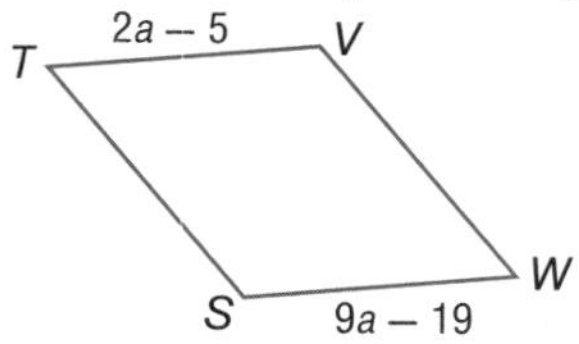

17.

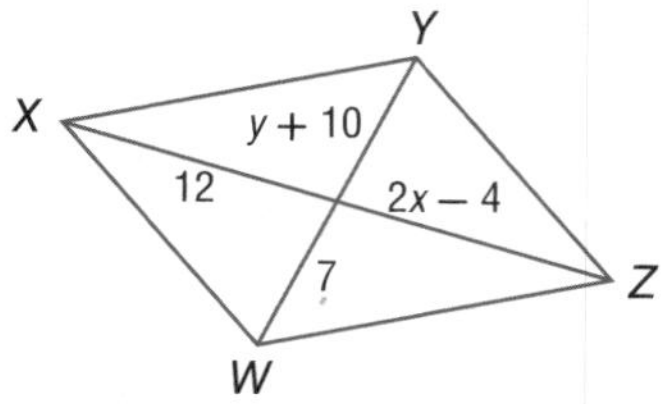

18.

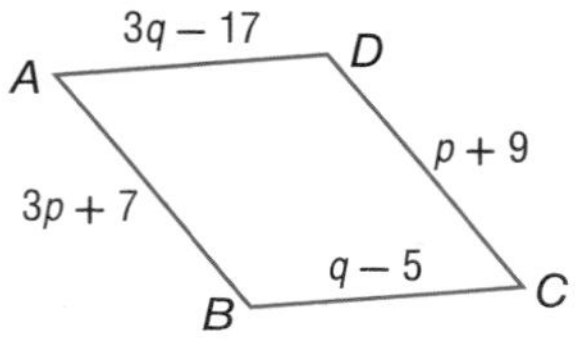

19.

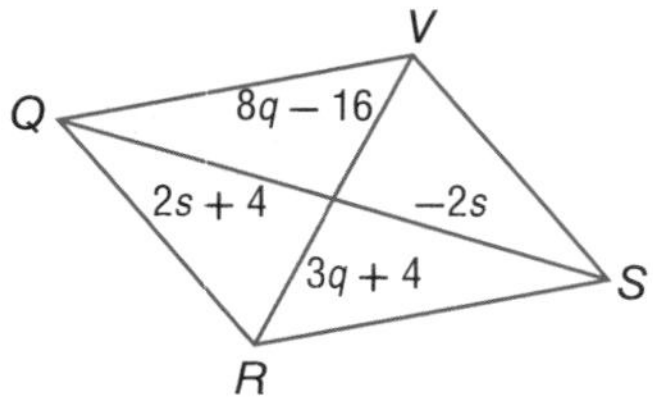

20.

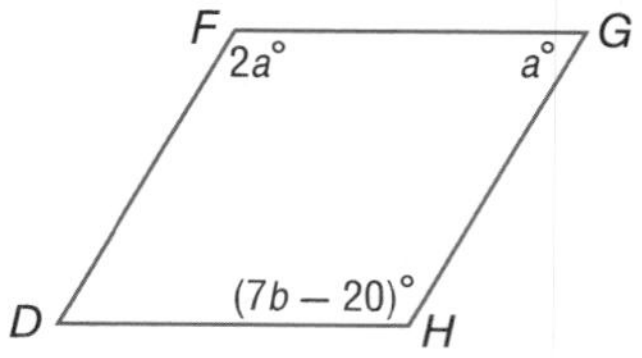

Example 3 **COORDINATE GEOMETRY** **Find the coordinates of the intersection of the diagonals of ▱$WXYZ$ with the given vertices.**

21. $W(-3, 5)$, $X(1, 7)$, $Y(3, 1)$, $Z(-1, -1)$

22. $W(1, 2)$, $X(4, 7)$, $Y(6, 5)$, $Z(3, 0)$

Example 4 **PROOF** **Write a two-column proof.**

23. **Given:** $ABDE$ and $ABCD$ are parallelograms.
Prove: $\triangle ADE \cong \triangle BCD$

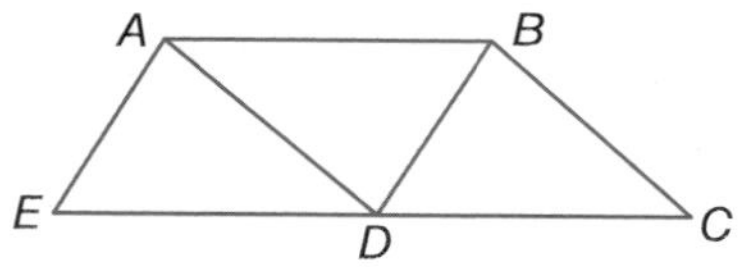

24. **Given:** $\triangle LMN$ is isosceles. $KLNP$ is a parallelogram.
Prove: $\angle KPN$ is supplementary to $\angle LMN$.

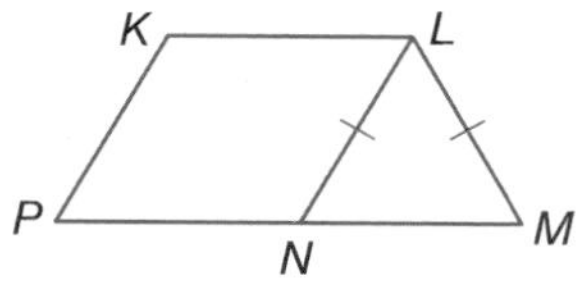

PROOF **Write the indicated type of proof.**

25. two-column
Given: ▱$GKLM$
Prove: $\angle G$ and $\angle K$, $\angle K$ and $\angle L$, $\angle L$ and $\angle M$, and $\angle M$ and $\angle G$ are supplementary. (Theorem 13.3)

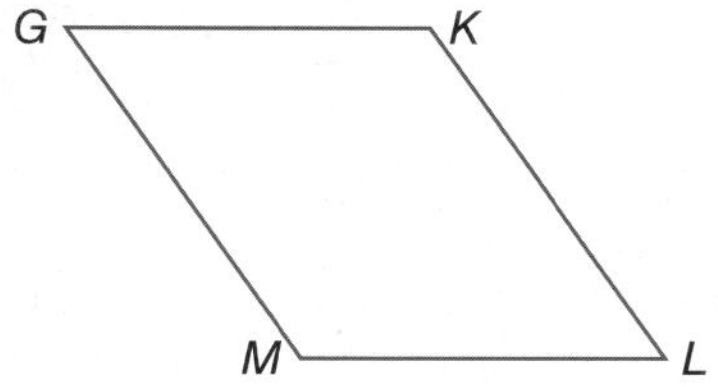

26. two-column
Given: ▱$WXYZ$
Prove: $\triangle WXZ \cong \triangle YZX$ (Theorem 13.6)

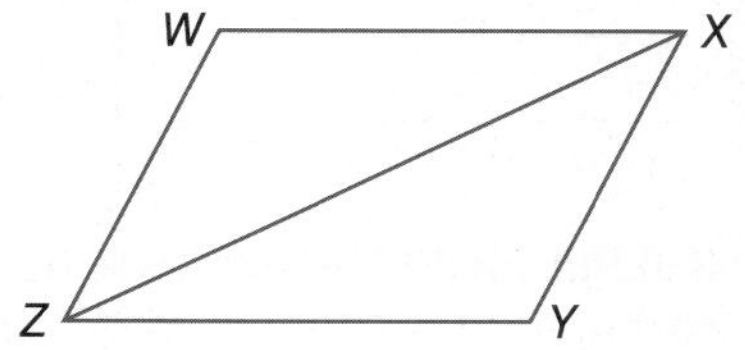

27. two-column
Given: ▱$PQRS$
Prove: $\overline{PQ} \cong \overline{RS}$, $\overline{QR} \cong \overline{SP}$ (Theorem 13.1)

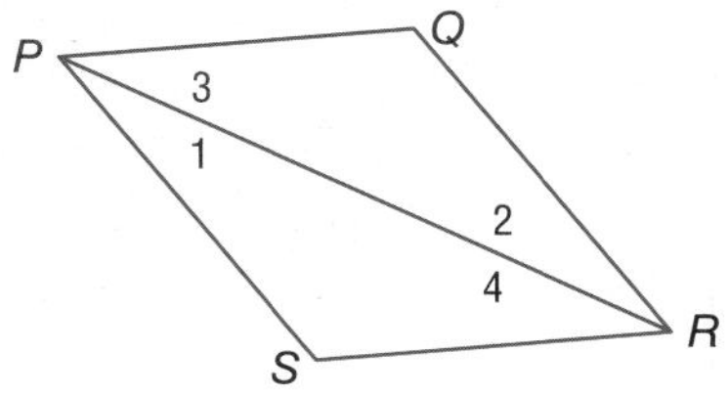

28. paragraph
Given: ▱$ACDE$ is a parallelogram.
Prove: $\overline{EC}$ bisects $\overline{AD}$. (Theorem 13.5)

29. COORDINATE GEOMETRY Use graph shown.

a. Use the distance formula to show $\overline{QP} \cong \overline{WY}$ and $\overline{QW} \cong \overline{PY}$.

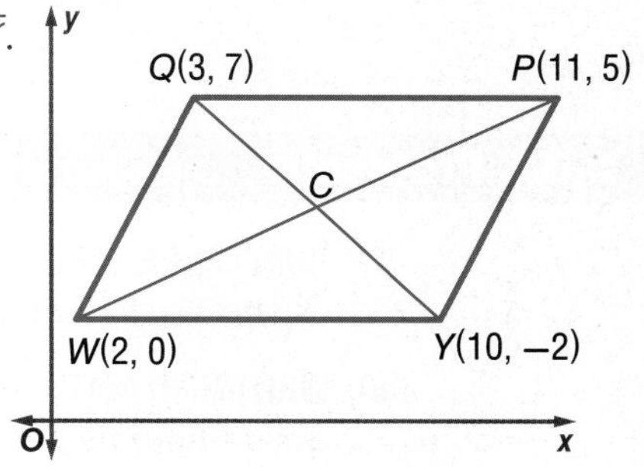

b. Find the coordinates of C if $\overline{QY}$ bisects $\overline{WP}$

c. Use slopes to determine if $QPWY$ is a parallelogram.

ALGEBRA **Use ▱$FGHD$ to find each measure or value.**

30. z

31. $m\angle FHJ$

32. $m\angle FHJ$

33. p

34. $m\angle GHK$

35. $m\angle FJH$

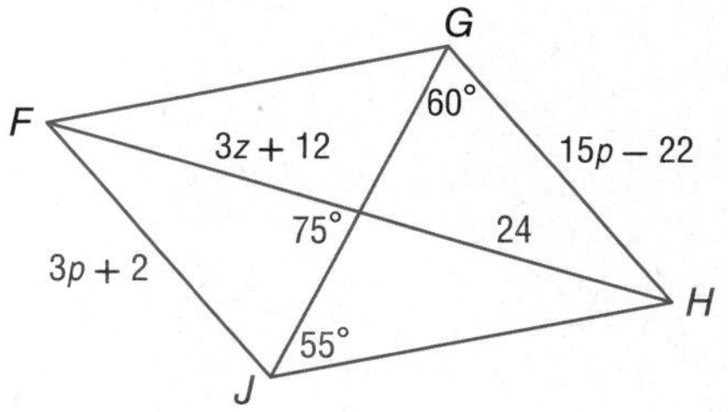

36. **COORDINATE GEOMETRY** If ▱$WXYZ$ has vertices $W(3, -2)$, $X(-2, -4)$, $Y(1, 1)$, determine the coordinates of vertex Z if it is located in Quadrant I.

PROOF **Write a two-column proof.**

37. **Given:** ▱$EFGH$, $\overline{HJ}$ bisects $\overline{EF}$, $\overline{EK}$ bisects $\overline{HG}$.
Prove: $\triangle EJH \cong \triangle GKF$

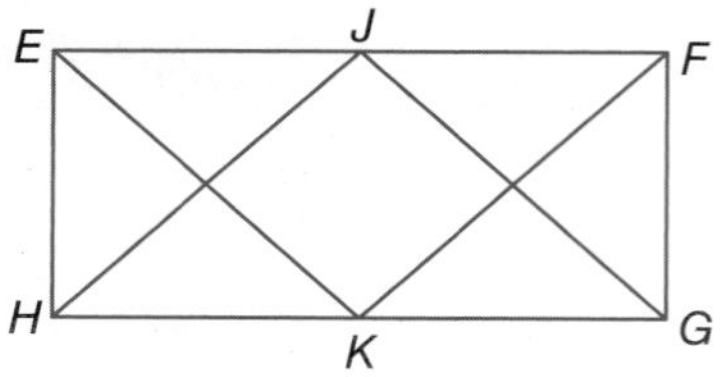

38. **MULTIPLE REPRESENTATIONS** In this problem you will explore the angle formed by extending one side of a parallelogram.

a. **Geometric** Construct 3 parallelograms and extend one side. Label each parallelogram $ABCD$ as shown. Measure and label the sides and angles of the parallelogram.

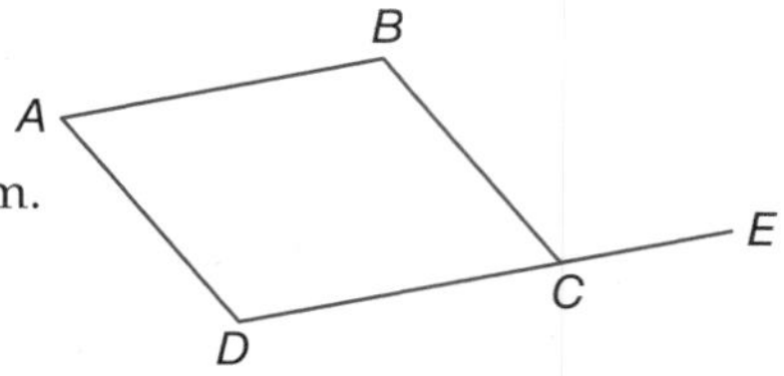

b. **Tabular** Copy and complete the table below.

Parallelogram	$m\angle A$	$m\angle D$	$m\angle BCE$
Parallelogram 1			
Parallelogram 2			
Parallelogram 3			

c. **Verbal** Make a conjecture about the angle formed by extending one side of a parallelogram and the other angles.

H.O.T. Problems Use Higher-Order Thinking Skills

39. **CHALLENGE** $ABCD$ is a parallelogram with diagonals as shown. List all pairs of congruent triangles.

40. **WRITING IN MATH** Explain what makes parallelograms special types of quadrilaterals.

41. **OPEN ENDED** Provide a counter example to show that parallelograms are not always congruent if their corresponding angles are congruent.

42. **CCSS REASONING** If $A(-1, 2)$, $B(2, 1)$, and $C(1, -1)$ are three vertices of a parallelogram, what points could be used as the fourth vertex?

43. **WRITING IN MATH** Explain why rectangles are *always* parallelograms, but parallelograms are *sometimes* rectangles.

Standardized Test Practice

44. Two consecutive angles of a parallelogram measure $3x + 42$ and $9x - 18$. What are the measures of the angles?

A 13, 167 **C** 39, 141

B 58.5, 31.5 **D** 81, 99

45. GRIDDED RESPONSE Parallelogram *MNPQ* is shown. What is the value of *x*?

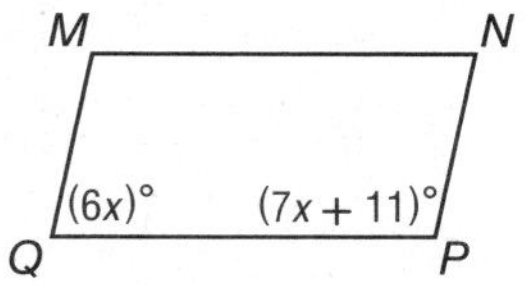

46. ALGEBRA In a history class with 32 students, the ratio of girls to boys is 5 to 3. How many more girls are there than boys?

F 2 **G** 8 **H** 12 **J** 15

47. SAT/ACT The table shows the heights of the tallest buildings in Kansas City, Missouri. To the nearest tenth, what is the positive difference between the median and the mean of the data?

Name	Height (m)
One Kansas City Place	193
Town Pavillion	180
Hyatt Regency	154
Power and Light Building	147
City Hall	135
1201 Walnut	130

A 5

B 6

C 7

D 8

E 10

Spiral Review

In the figure, $a \parallel b$, $c \parallel d$, and $m\angle 4 = 57$. Find the measure of each angle. (Lesson 11-2)

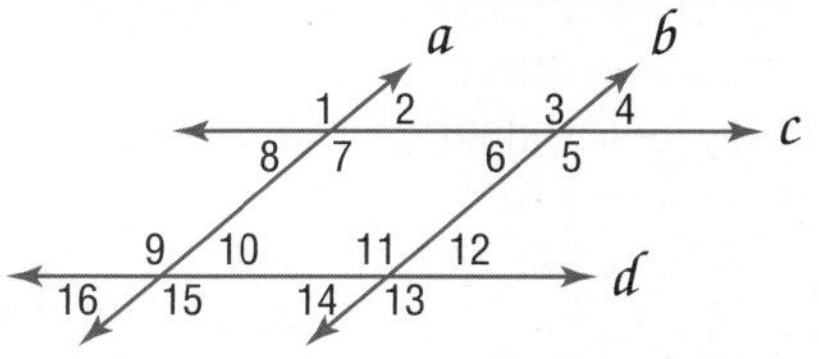

48. $\angle 5$ **49.** $\angle 1$

50. $\angle 8$ **51.** $\angle 10$

Refer to the diagram at the right. (Lesson 11-1)

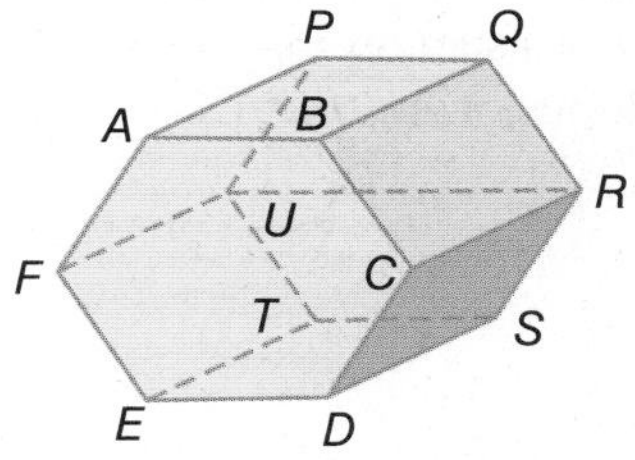

52. Name all segments parallel to $\overline{TU}$.

53. Name all planes intersecting plane *BCR*.

54. Name all segments skew to $\overline{DE}$.

55. CONSTRUCTION There are four buildings on the Mansfield High School Campus, no three of which stand in a straight line. How many sidewalks need to be built so that each building is directly connected to every other building? (Lesson 10-1)

Skills Review

The vertices of a quadrilateral are $W(3, -1)$, $X(4, 2)$, $Y(-2, 3)$ and $Z(-3, 0)$. Determine whether each segment is a side or diagonal of the quadrilateral, and find the slope of each segment.

56. $\overline{YZ}$ **57.** $\overline{YW}$ **58.** $\overline{ZW}$

EXPLORE 13-2

Graphing Technology Lab
Parallelograms

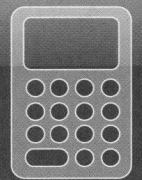

You can use the Cabri™ Jr. application on a TI-83/84 Plus graphing calculator to discover properties of parallelograms.

CCSS Common Core State Standards
Content Standards
G.CO.12 Make formal geometric constructions with a variety of tools and methods (compass and straightedge, string, reflective devices, paper folding, dynamic geometric software, etc.).
Mathematical Practices 5

Activity

Construct a quadrilateral with one pair of sides that are both parallel and congruent.

Step 1 Construct a segment using the **Segment** tool on the **F2** menu. Label the segment $\overline{AB}$. This is one side of the quadrilateral.

Step 2 Use the **Parallel** tool on the **F3** menu to construct a line parallel to the segment. Pressing ENTER will draw the line and a point on the line. Label the point *C*.

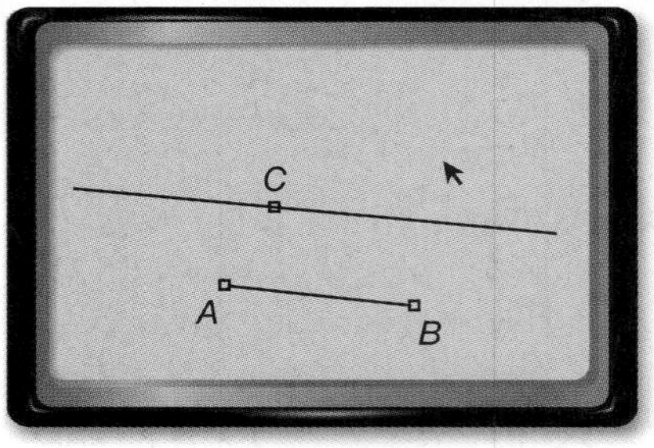

Steps 1 and 2

Step 3 Access the **Compass** tool on the **F3** menu. Set the compass to the length of $\overline{AB}$ by selecting one endpoint of the segment and then the other. Construct a circle centered at *C*.

Step 4 Use the **Point Intersection** tool on the **F2** menu to draw a point at the intersection of the line and the circle. Label the point *D*. Then use the **Segment** tool on the **F2** menu to draw $\overline{AC}$ and $\overline{BD}$.

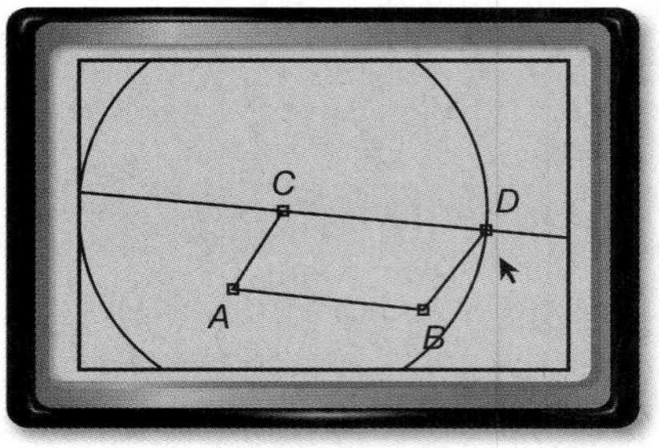

Steps 3 and 4

Step 5 Use the **Hide/Show** tool on the **F5** menu to hide the circle. Then access **Slope** tool under **Measure** on the **F5** menu. Display the slopes of $\overline{AB}$, $\overline{BD}$, $\overline{CD}$, and $\overline{AC}$.

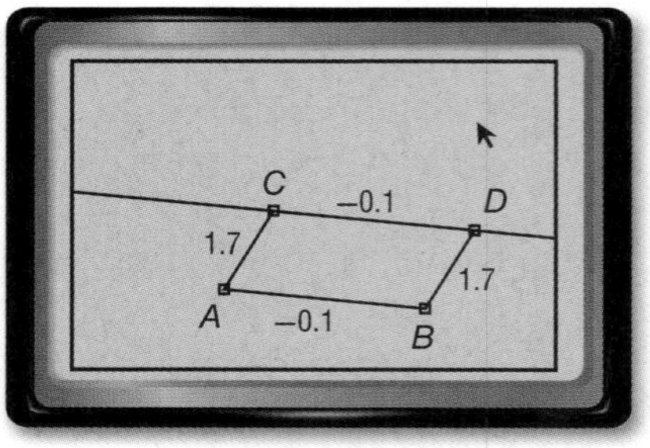

Step 5

Analyze the Results

1. What is the relationship between sides $\overline{AB}$ and $\overline{CD}$? Explain how you know.
2. What do you observe about the slopes of opposite sides of the quadrilateral? What type of quadrilateral is *ABDC*? Explain.
3. Click on point *A* and drag it to change the shape of *ABDC*. What do you observe?
4. Make a conjecture about a quadrilateral with a pair of opposite sides that are both congruent and parallel.
5. Use a graphing calculator to construct a quadrilateral with both pairs of opposite sides congruent. Then analyze the slopes of the sides of the quadrilateral. Make a conjecture based on your observations.

LESSON 13-2

Tests for Parallelograms

Then

- You recognized and applied properties of parallelograms.

Now

1. Recognize the conditions that ensure a quadrilateral is a parallelogram.
2. Prove that a set of points forms a parallelogram in the coordinate plane.

Why?

- Lexi and Rosalinda cut strips of bulletin board paper at an angle to form the hallway display shown. Their friends asked them how they cut the strips so that their sides were parallel without using a protractor.

Rosalinda explained that since the left and right sides of the paper were parallel, she only needed to make sure that the sides were cut to the same length to guarantee that a strip would form a parallelogram.

CCSS Common Core State Standards

Content Standards

G.CO.11 Prove theorems about parallelograms.

G.GPE.4 Use coordinates to prove simple geometric theorems algebraically.

Mathematical Practices

3 Construct viable arguments and critique the reasoning of others.

2 Reason abstractly and quantitatively.

1 Conditions for Parallelograms

If a quadrilateral has each pair of opposite sides parallel, it is a parallelogram by definition.

This is not the only test, however, that can be used to determine if a quadrilateral is a parallelogram.

Theorems Conditions for Parallelograms

13.7 If both pairs of opposite sides of a quadrilateral are congruent, then the quadrilateral is a parallelogram.

Abbreviation *If both pairs of opp. sides are ≅, then quad. is a ▱.*

Example If $\overline{AB} \cong \overline{DC}$ and $\overline{AD} \cong \overline{BC}$, then $ABCD$ is a parallelogram.

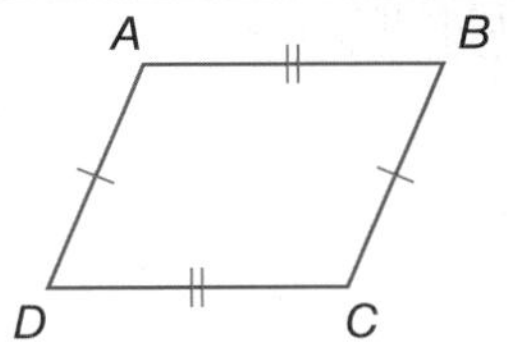

13.8 If both pairs of opposite angles of a quadrilateral are congruent, then the quadrilateral is a parallelogram.

Abbreviation *If both pairs of opp. ∠s are ≅, then quad. is a ▱.*

Example If $\angle A \cong \angle C$ and $\angle B \cong \angle D$, then $ABCD$ is a parallelogram.

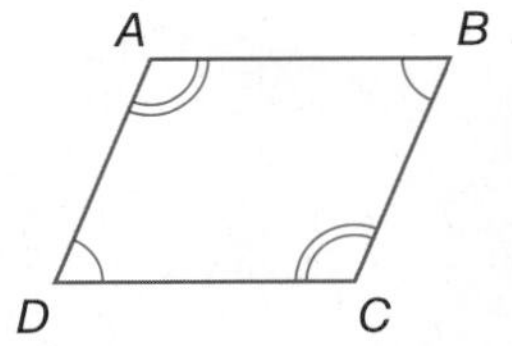

13.9 If the diagonals of a quadrilateral bisect each other, then the quadrilateral is a parallelogram.

Abbreviation *If diag. bisect each other, then quad. is a ▱.*

Example If $\overline{AC}$ and $\overline{DB}$ bisect each other, then $ABCD$ is a parallelogram.

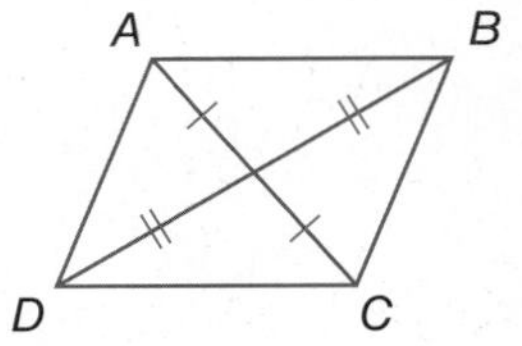

13.10 If one pair of opposite sides of a quadrilateral is both parallel and congruent, then the quadrilateral is a parallelogram.

Abbreviation *If one pair of opp. sides is ≅ and ||, then the quad. is a ▱.*

Example If $\overline{AB} \parallel \overline{DC}$ and $\overline{AB} \cong \overline{DC}$, then $ABCD$ is a parallelogram.

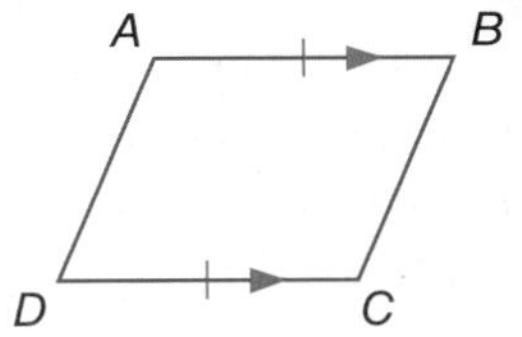

You will prove Theorems 13.8, 13.10, and 13.11 in Exercises 30, 32, and 33, respectively.

Proof Theorem 13.7

Write a paragraph proof of Theorem 13.7.

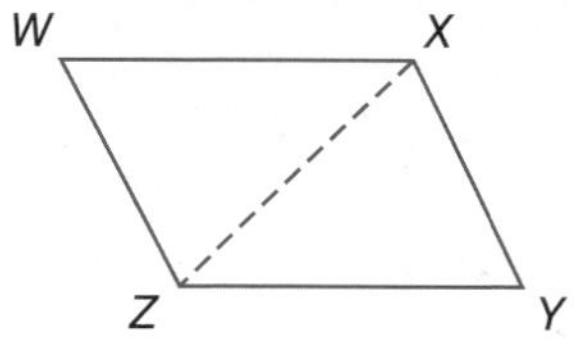

Given: $\overline{WX} \cong \overline{ZY}$, $\overline{WZ} \cong \overline{XY}$

Prove: $WXYZ$ is a parallelogram.

Paragraph Proof:

Two points determine a line, so we can draw auxiliary line $\overline{ZX}$ to form $\triangle ZWX$ and $\triangle XYZ$. We are given that $\overline{WX} \cong \overline{ZY}$ and $\overline{WZ} \cong \overline{XY}$. Also, $\overline{ZX} \cong \overline{XZ}$ by the Reflexive Property of Congruence. So $\triangle ZWX \cong \triangle XYZ$ by SSS. By CPCTC, $\angle WXZ \cong \angle YZX$ and $\angle WZX \cong \angle YXZ$. This means that $\overline{WX} \parallel \overline{ZY}$ and $\overline{WZ} \parallel \overline{XY}$ by the Alternate Interior Angles Converse. Opposite sides of $WXYZ$ are parallel, so by definition $WXYZ$ is a parallelogram.

Example 1 Identify Parallelograms

Determine whether the quadrilateral is a parallelogram. Justify your answer.

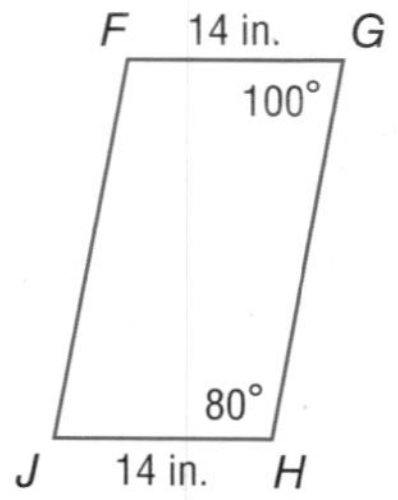

Opposite sides $\overline{FG}$ and $\overline{JH}$ are congruent because they have the same measure. Also, since $\angle FGH$ and $\angle GHJ$ are supplementary consecutive interior angles, $\overline{FG} \parallel \overline{JH}$. Therefore, by Theorem 13.7, $FGHJ$ is a parallelogram.

GuidedPractice

1A.

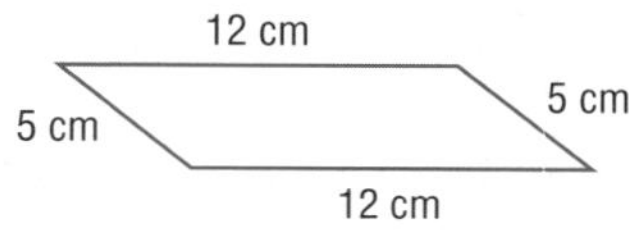

1B.

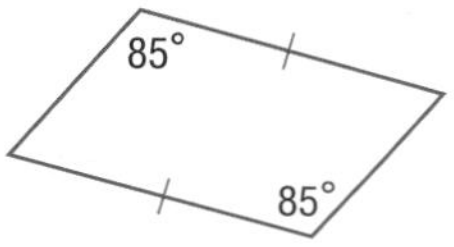

You can use the conditions of parallelograms to prove relationships in real-world situations.

Real-WorldLink

A 2- or 3-cantilever tackle box is often used to organize lures and other fishing supplies. The trays lift up and away so that all items in the box are easily accessible.

Real-World Example 2 Use Parallelograms to Prove Relationships

FISHING The diagram shows a side view of the tackle box at the left. In the diagram, $PQ = RS$ and $PR = QS$. Explain why the upper and middle trays remain parallel no matter to what height the trays are raised or lowered.

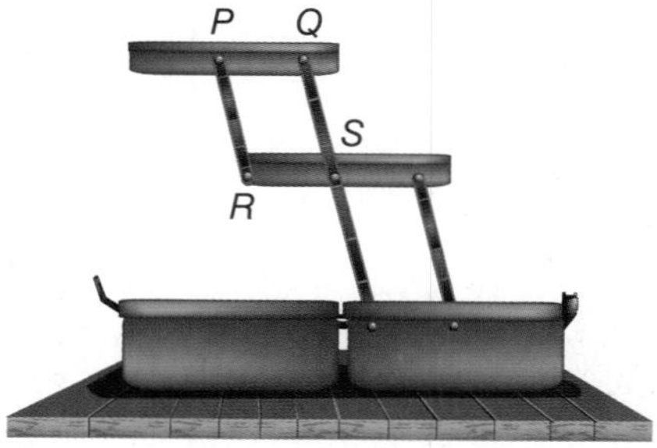

Since both pairs of opposite sides of quadrilateral $PQSR$ are congruent, $PQRS$ is a parallelogram by Theorem 13.7. By the definition of a parallelogram, opposite sides are parallel, so $\overline{PQ} \parallel \overline{RS}$. Therefore, no matter the vertical position of the trays, they will always remain parallel.

GuidedPractice

2. BANNERS In the example at the beginning of the lesson, explain why the cuts made by Lexi and Rosalinda are parallel.

You can also use the conditions of parallelograms along with algebra to find missing values that make a quadrilateral a parallelogram.

WatchOut!

Parallelograms In Example 3, if x is 4, then y must be 2.5 in order for $FGHJ$ to be a parallelogram. In other words, if x is 4 and y is 1, then $FGHJ$ is not a parallelogram.

Example 3 Use Parallelograms and Algebra to Find Values

PT

If $FK = 3x - 1$, $KG = 4y + 3$, $JK = 6y - 2$, and $KH = 2x + 3$, find x and y so that the quadrilateral is a parallelogram.

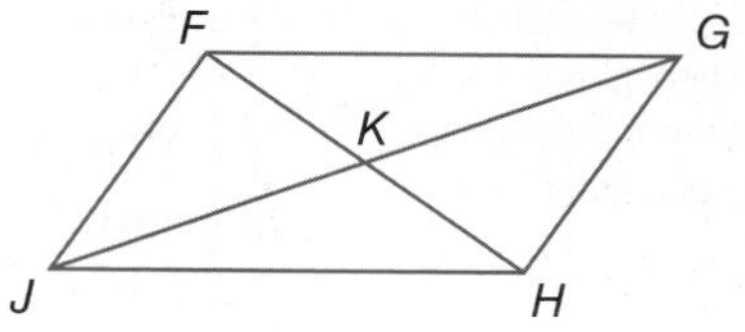

By Theorem 13.9, if the diagonals of a quadrilateral bisect each other, then it is a parallelogram. So find x such that $\overline{FK} \cong \overline{KH}$ and y such that $\overline{JK} \cong \overline{KG}$.

$FK = KH$	Definition of $\cong$
$3x - 1 = 2x + 3$	Substitution
$x - 1 = 3$	Subtract $2x$ from each side.
$x = 4$	Add 1 to each side.
$JK = KG$	Definition of $\cong$
$6y - 2 = 4y + 3$	Substitution
$2y - 2 = 3$	Subtract $4y$ from each side.
$2y = 5$	Add 2 to each side.
$y = 2.5$	Divide each side by 2.

So, when x is 4 and y is 2.5, quadrilateral $FGHJ$ is a parallelogram.

GuidedPractice

Find x and y so that each quadrilateral is a parallelogram.

3A.

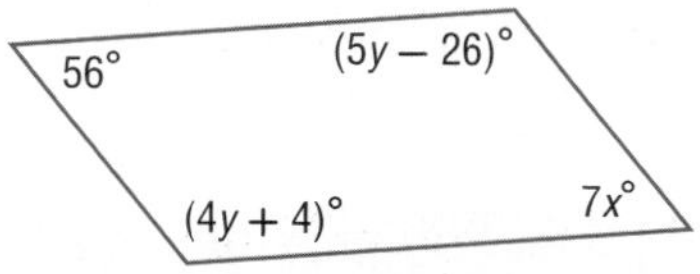

3B.

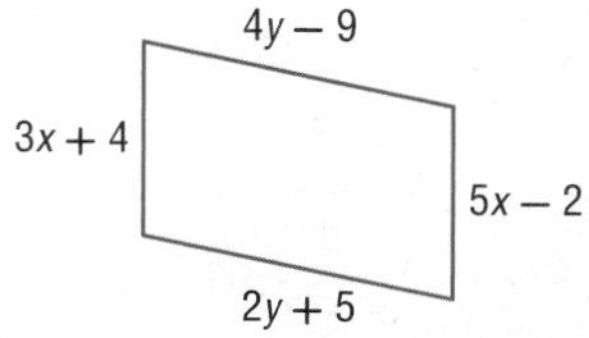

You have learned the conditions of parallelograms. The following list summarizes how to use the conditions to prove a quadrilateral is a parallelogram.

Concept Summary

Prove that a Quadrilateral Is a Parallelogram

- Show that both pairs of opposite sides are parallel. (Definition)
- Show that both pairs of opposite sides are congruent. (Theorem 13.7)
- Show that both pairs of opposite angles are congruent. (Theorem 13.8)
- Show that the diagonals bisect each other. (Theorem 13.9)
- Show that a pair of opposite sides is both parallel and congruent. (Theorem 13.10)

StudyTip

Midpoint Formula
To show that a quadrilateral is a parallelogram, you can also use the Midpoint Formula. If the midpoint of each diagonal is the same point, then the diagonals bisect each other.

2 Parallelograms on the Coordinate Plane

We can use the Distance, Slope, and Midpoint Formulas to determine whether a quadrilateral in the coordinate plane is a parallelogram.

Example 4 Parallelograms and Coordinate Geometry

COORDINATE GEOMETRY Graph quadrilateral *KLMN* with vertices *K*(2, 3), *L*(8, 4), *M*(7, −2), and *N*(1, −3). Determine whether the quadrilateral is a parallelogram. Justify your answer using the Slope Formula.

If the opposite sides of a quadrilateral are parallel, then it is a parallelogram.

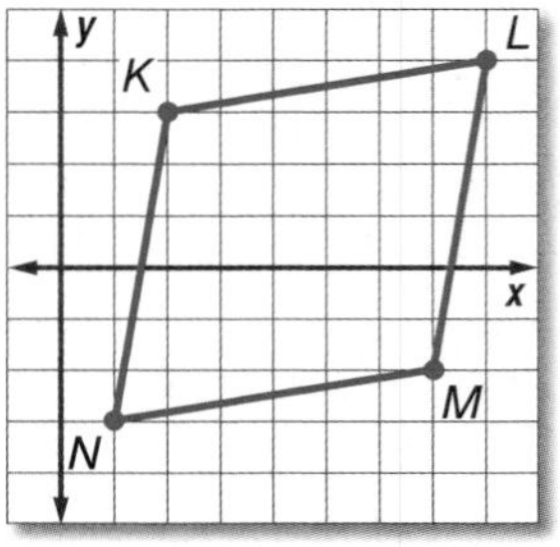

slope of $\overline{KL} = \frac{4-3}{8-2}$ or $\frac{1}{6}$

slope of $\overline{NM} = \frac{-2-(-3)}{7-1}$ or $\frac{1}{6}$

slope of $\overline{KN} = \frac{-3-3}{1-2} = \frac{-6}{-1}$ or 6

slope of $\overline{LM} = \frac{-2-4}{7-8} = \frac{-6}{-1}$ or 6

Since opposite sides have the same slope, $\overline{KL} \parallel \overline{NM}$ and $\overline{KN} \parallel \overline{LM}$. Therefore, *KLMN* is a parallelogram by definition.

GuidedPractice

Determine whether the quadrilateral is a parallelogram. Justify your answer using the given formula.

4A. *A*(3, 3), *B*(8, 2), *C*(6, −1), *D*(1, 0); Distance Formula

4B. *F*(−2, 4), *G*(4, 2), *H*(4, −2), *J*(−2, −1); Midpoint Formula

In Chapter 12, you learned that variable coordinates can be assigned to the vertices of triangles. Then the Distance, Slope, and Midpoint Formulas were used to write coordinate proofs of theorems. The same can be done with quadrilaterals.

ReviewVocabulary

coordinate proof a proof that uses figures in the coordinate plane and algebra to prove geometric concepts

Example 5 Parallelograms and Coordinate Proofs

Write a coordinate proof for the following statement.

If one pair of opposite sides of a quadrilateral is both parallel and congruent, then the quadrilateral is a parallelogram.

Step 1 Position quadrilateral *ABCD* on the coordinate plane such that $\overline{AB} \parallel \overline{DC}$ and $\overline{AB} \cong \overline{DC}$.

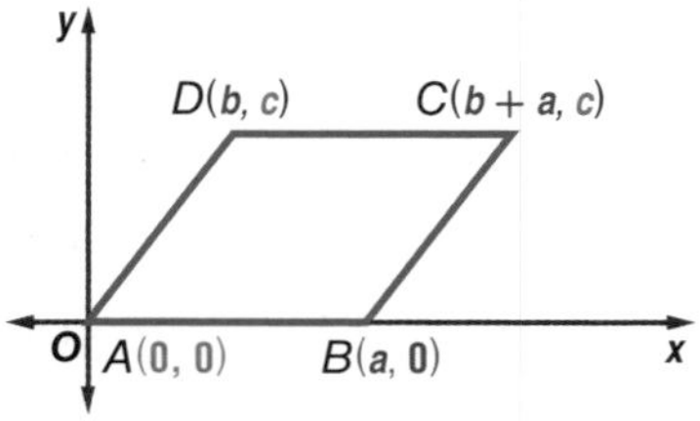

- Begin by placing the vertex *A* at the origin.
- Let $\overline{AB}$ have a length of *a* units. Then *B* has coordinates (*a*, 0).
- Since horizontal segments are parallel, position the endpoints of $\overline{DC}$ so that they have the same *y*-coordinate, *c*.
- So that the distance from *D* to *C* is also *a* units, let the *x*-coordinate of *D* be *b* and of *C* be *b* + *a*.

Math HistoryLink

René Descartes

(1596–1650)

René Descartes was a French mathematician who was the first to use a coordinate grid. It has been said that he first thought of locating a point on a plane with a pair of numbers when he was watching a fly on the ceiling, but this is a myth.

Step 2 Use your figure to write a proof.

Given: quadrilateral $ABCD$, $\overline{AB} \parallel \overline{DC}$, $\overline{AB} \cong \overline{DC}$

Prove: $ABCD$ is a parallelogram.

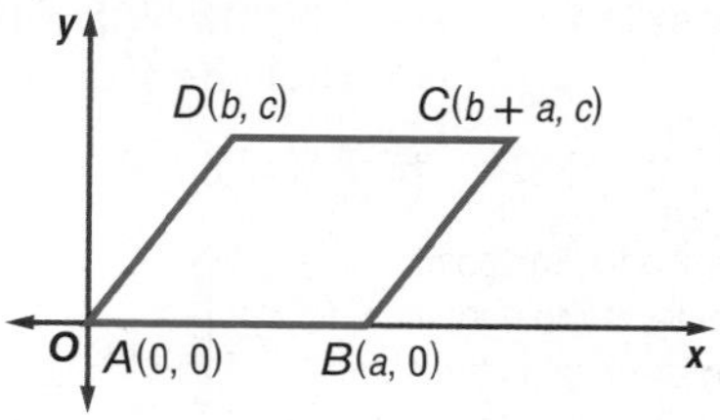

Coordinate Proof:

By definition, a quadrilateral is a parallelogram if opposite sides are parallel. We are given that $\overline{AB} \parallel \overline{DC}$, so we need only show that $\overline{AD} \parallel \overline{BC}$.

Use the Slope Formula.

slope of $\overline{AD} = \frac{c-0}{b-0} = \frac{c}{b}$ slope of $\overline{BC} = \frac{c-0}{b+a-a} = \frac{c}{b}$

Since $\overline{AD}$ and $\overline{BC}$ have the same slope, $\overline{AD} \parallel \overline{BC}$. So quadrilateral $ABCD$ is a parallelogram because opposite sides are parallel.

GuidedPractice

5. Write a coordinate proof of this statement: *If a quadrilateral is a parallelogram, then opposite sides are congruent.*

Check Your Understanding

Example 1 **Determine whether each quadrilateral is a parallelogram. Justify your answer.**

1.

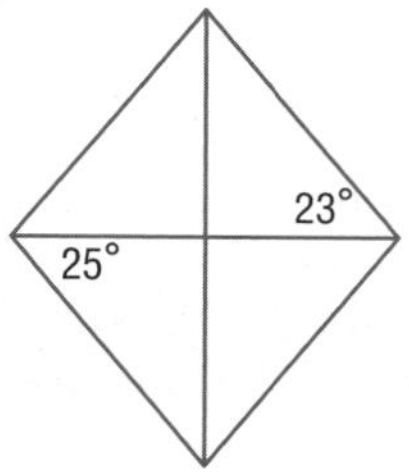

2.

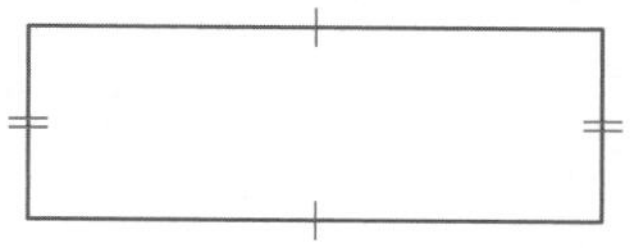

Example 2

3. CARPENTRY Luke is building a table and needs to make sure the points where the legs meet the floor form a parallelogram with right angles. How can Luke use the table top to prove the legs form a parallelogram?

Example 3 **ALGEBRA Find x and y so that the quadrilateral is a parallelogram.**

4.

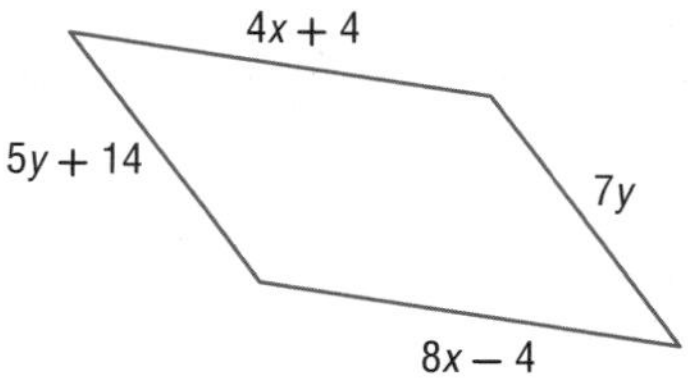

5.

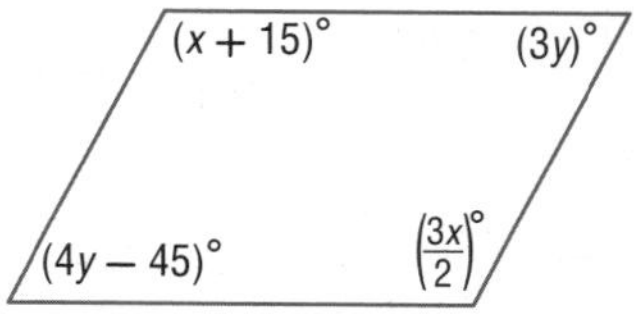

Example 4 **COORDINATE GEOMETRY** **Graph each quadrilateral with the given vertices. Determine whether the figure is a parallelogram. Justify your answer with the method indicated.**

6. $K(1, 1)$, $L(4, -2)$, $M(-7, -3)$, $N(-2, 4)$; midpoint formula

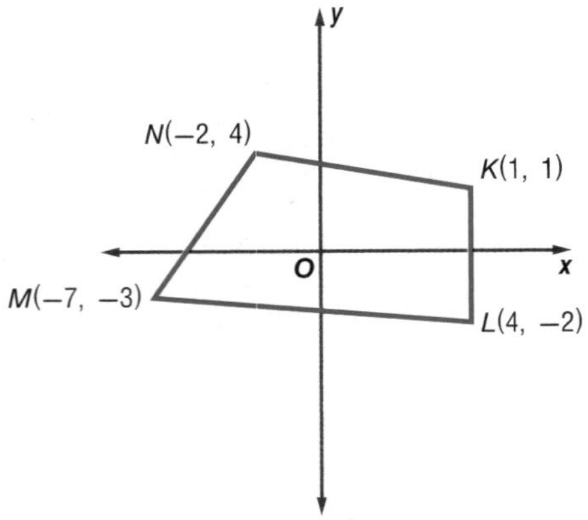

7. $F(-5, 3)$, $G(-1, 2)$, $H(-3, 0)$, $J(-7, 1)$; slope formula

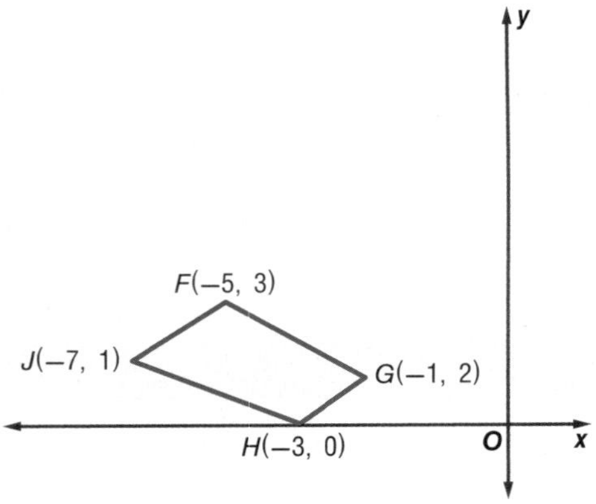

Example 5 **8.** Write a coordinate proof for the statement: *If a quadrilateral is a parallelogram, then its diagonals bisect each other.*

Practice and Problem Solving

Example 1 **Determine whether each quadrilateral is a parallelogram. Justify your answer.**

9.

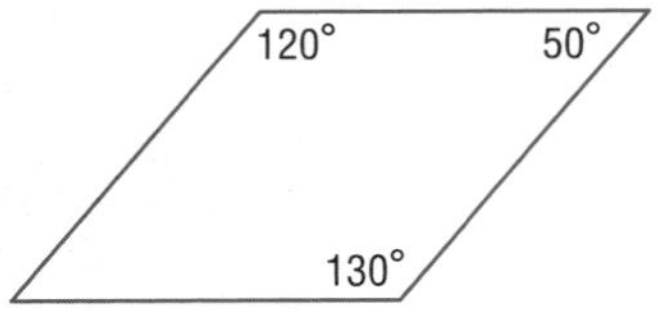

10.

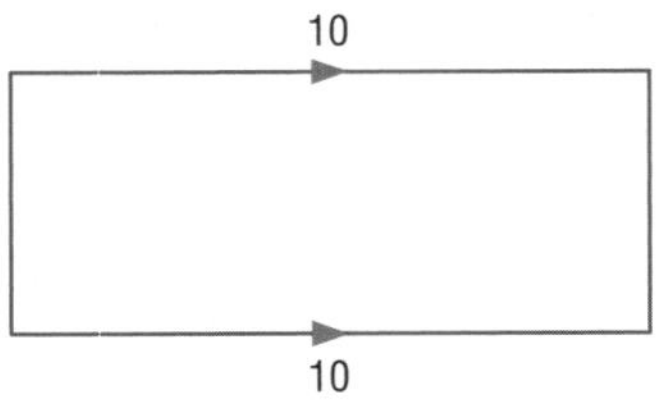

11.

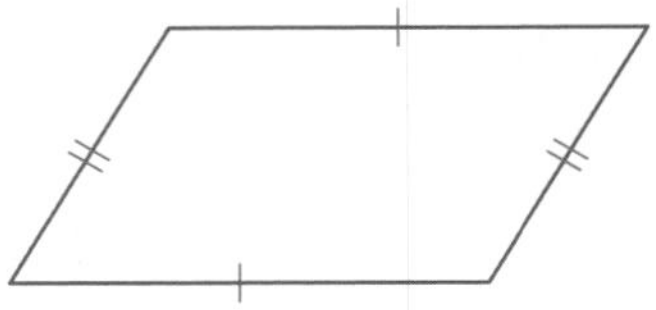

12.

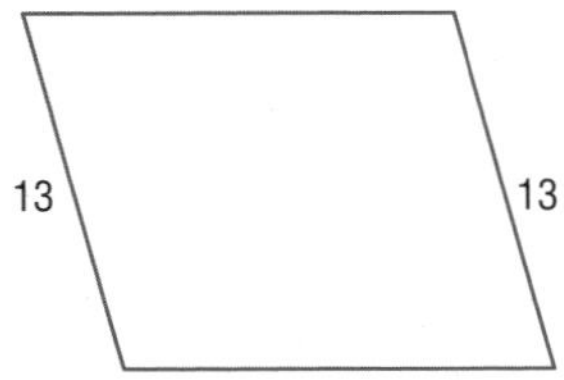

13.

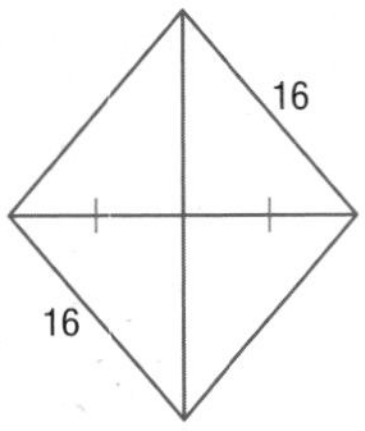

14.

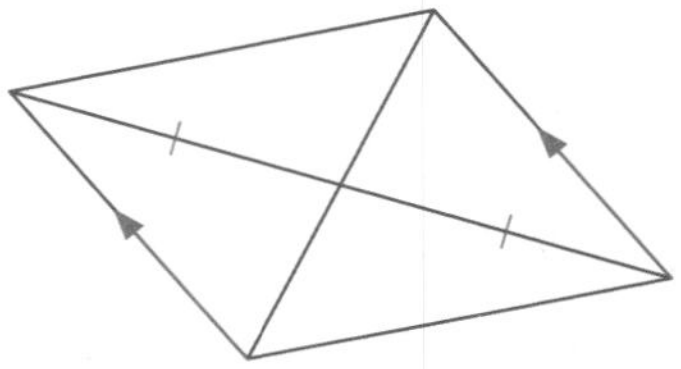

Example 2 **COORDINATE GEOMETRY** **Graph each quadrilateral with the given vertices. Determine whether the figure is a parallelogram. Justify your answer with the method indicated.**

15. $Q(-1, 2)$, $R(4, 3)$, $S(2, -1)$, $T(-2, -1)$; slope formula

16. $J(1, 4)$, $K(4, 0)$, $L(-4, -6)$, $M(-7, -2)$; slope formula

17. $A(-5, 8)$, $B(-3, 7)$, $C(-2, 1)$, $D(-4, 0)$; distance formula

18. $V(10, 4)$, $W(15, 3)$, $X(13, 0)$, $Y(8, 1)$; distance and slope formulas

Example 3 **19.** Write a coordinate proof for the statement: *If both pairs of opposite sides of a quadrilateral are congruent, then the quadrilateral is a parallelogram.*

20. Write a coordinate proof for the statement: *If a parallelogram has one right angle, it has four right angles.*

21. PROOF Write a paragraph proof of Theorem 13.8

22. TOURISM While taking a trip to New York City, Carol wants to visit four tourist destinations: The Empire State Building, the Statue of Liberty, Central Park, and Time's Square. If the gps coordinates of the Empire State building are 40.74°N 73.99°W, the GPS coordinates of the Statue of Liberty are 30.69°N 74.05°W, the GPS coordinates for Central Park ar 40.78°N 73.97°W, and the GPS coordinates for Time's Square are 40.75°N 73.99°W, determine if these four destinations form a parallelogram.

PROOF Write a two-column proof.

23. Theorem 13.9

24. Theorem 13.10

25. Explain how you can use Theorem 13.9 to construct a parallelogram. Then construct a parallelogram using your method.

Name the missing coordinates for each parallelogram.

26.

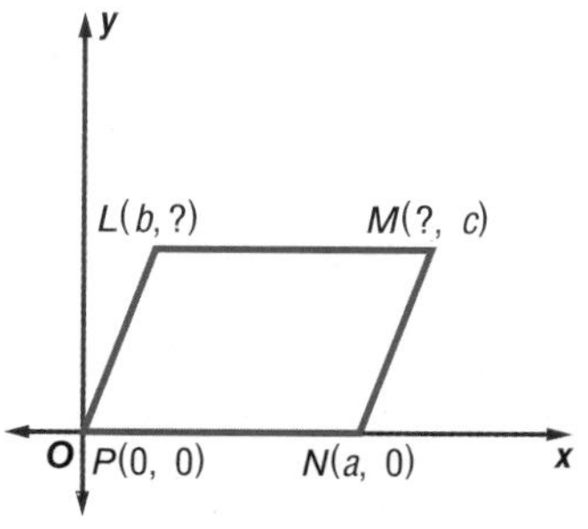

27.

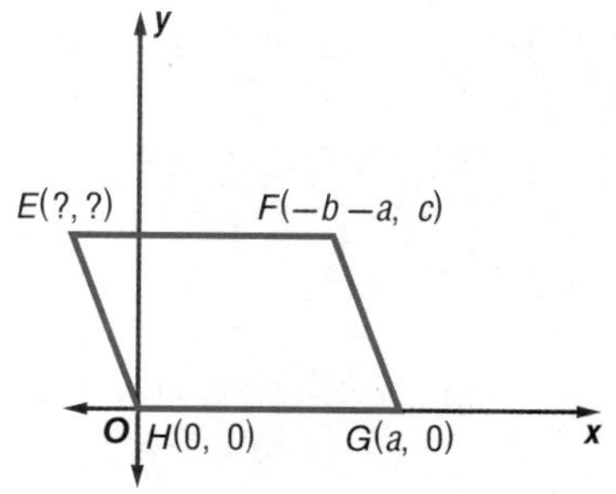

28. DRIVING Bruce is painting lines for a new parking lot. What is the fewest number of measurements Bruce needs to take with a protractor and measuring tape to make sure the lines form a parallelogram?

29. **PROOF** Write a coordinate proof to prove the diagonals of a parallelogram create two sets of congruent triangles.

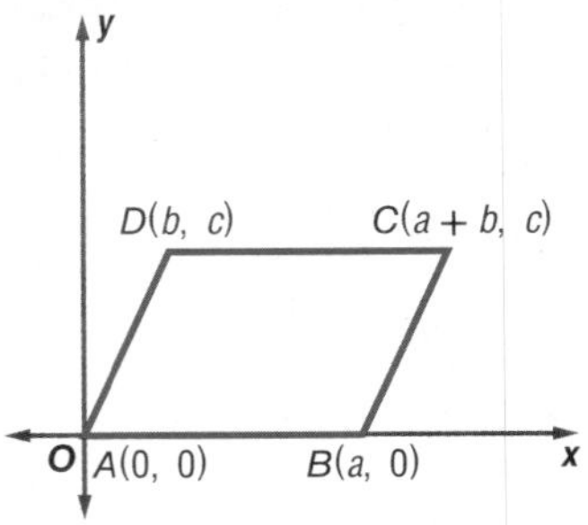

30. **MULTIPLE REPRESENTATIONS** In this problem, you will explore the properties of rectangles.

a. Geometric Plot the following points on three separate coordinate planes. The four points on each graph form a rectangle.

Rectangle 1: $A(0, 0)$, $B(0, 4)$, $C(5, 4)$, $D(5, 0)$

Rectangle 2: $A(-2, 4)$, $B(5, 5)$, $C(6, -2)$, $D(-1, 3)$

Rectangle 3: $A(-2, 2)$, $B(0, -2)$, $C(6, 1)$, $D(4, 5)$

b. Tabular Copy the table below. Use the slopes of $\overline{AB}$, $\overline{BC}$, $\overline{CD}$, and $\overline{AD}$ to complete the table.

Rectangle	$m\angle A$	$m\angle B$	$m\angle C$	$m\angle D$	*ABCD* a parallelogram?
Rectangle 1					
Rectangle 2					
Rectangle 3					

c. Verbal Make a conjecture about the definition of a rectangle.

H.O.T. Problems Use Higher-Order Thinking Skills

31. **ERROR ANALYSIS** Sarah says quadrilateral *ABCD* is a parallelogram, but Sandy says it is not a parallelogram. Which one is correct? Explain your reasoning.

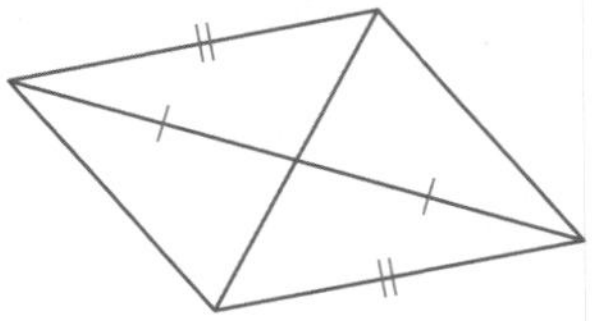

32. **WRITING IN MATH** Explain the different ways of using parallel sides to prove a quadrilateral is a parallelogram.

33. **CCSS REASONING** If two parallelograms have four congruent corresponding sides, are the parallelograms *sometimes*, *always*, or *never* congruent?

34. **OPEN ENDED** Position and label a parallelogram on the coordinate plane without any of the vertices being located at the origin.

35. **CHALLENGE** Find the values of a, b, and c if *ABCD* is a parallelogram.

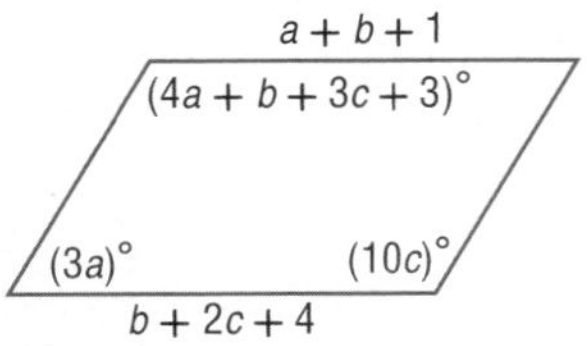

36. **WRITING IN MATH** Compare and contrast theorems 13.5 and 13.9.

Standardized Test Practice

37. If sides $\overline{AB}$ and $\overline{DC}$ of quadrilateral $ABCD$ are parallel, which additional information would be sufficient to prove that quadrilateral $ABCD$ is a parallelogram?

A $\overline{AB} \cong \overline{AC}$ **C** $\overline{AC} \cong \overline{BD}$

B $\overline{AB} \cong \overline{DC}$ **D** $\overline{AD} \cong \overline{BC}$

38. SHORT RESPONSE Quadrilateral $ABCD$ is shown. AC is 40 and BD is $\frac{3}{5}AC$. $\overline{BD}$ bisects $\overline{AC}$. For what value of x is $ABCD$ a parallelogram?

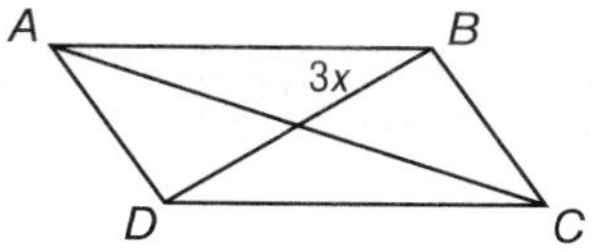

39. ALGEBRA Jarod's average driving speed for a 5-hour trip was 58 miles per hour. During the first 3 hours, he drove 50 miles per hour. What was his average speed in miles per hour for the last 2 hours of his trip?

F 70 **H** 60

G 66 **J** 54

40. SAT/ACT A parallelogram has vertices at (0, 0), (3, 5), and (0, 5). What are the coordinates of the fourth vertex?

A (0, 3) **D** (0, −3)

B (5, 3) **E** (3, 0)

C (5, 0)

Spiral Review

COORDINATE GEOMETRY Find the coordinates of the intersection of the diagonals of ▱$ABCD$ with the given vertices. (Lesson 13-1)

41. $A(-3, 5)$, $B(6, 5)$, $C(5, -4)$, $D(-4, -4)$

42. $A(2, 5)$, $B(10, 7)$, $C(7, -2)$, $D(-1, -4)$

Determine the slope of the line that contains the given points. (Lesson 11-3)

43. $J(4, 3)$, $K(5, -2)$

44. $X(0, 2)$, $Y(-3, -4)$

45. $A(2, 5)$, $B(5, 1)$

Find x and y in each figure. (Lesson 11-2)

46.

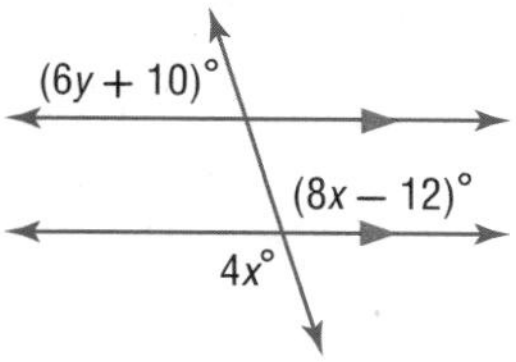

47.

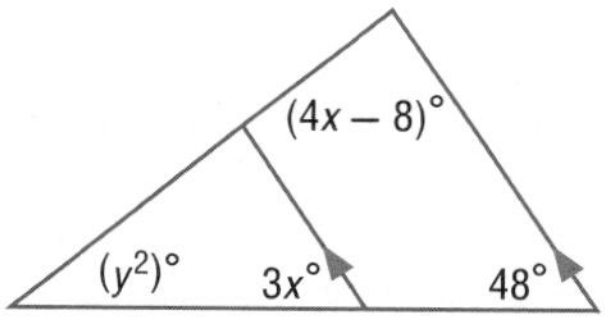

PROOF Write a flow proof. (Lesson 12-5)

48. Given: $\overline{EJ} \parallel \overline{FK}$, $\overline{JG} \parallel \overline{KH}$, $\overline{EF} \cong \overline{GH}$
Prove: $\triangle EJG \cong \triangle FKH$

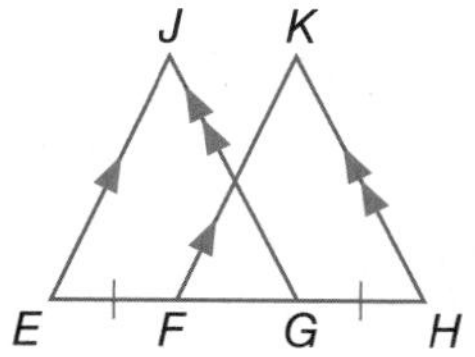

49. Given: $\overline{MN} \cong \overline{PQ}$, $\angle M \cong \angle Q$, $\angle 2 \cong \angle 3$
Prove: $\triangle MLP \cong \triangle QLN$

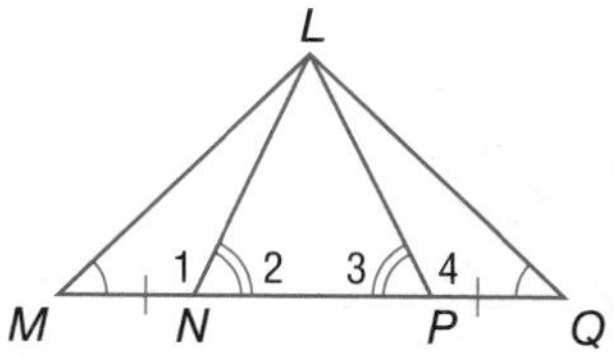

Skills Review

Use slope to determine whether XY and YZ are *perpendicular* or *not perpendicular*.

50. $X(-2, 2)$, $Y(0, 1)$, $Z(4, 1)$

51. $X(4, 1)$, $Y(5, 3)$, $Z(6, 2)$

CHAPTER 13

Mid-Chapter Quiz

Lessons 13-1 through 13-2

Use ▱$WXYZ$ to find each measure. (Lesson 13-1)

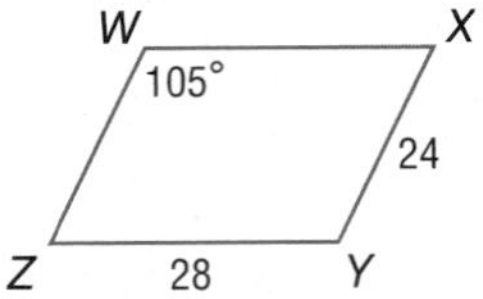

1. $m\angle WZY$
2. WZ
3. $m\angle XYZ$

4. **DESIGN** Describe two ways to ensure that the pieces of the design at the right would fit properly together. (Lesson 13-1)

ALGEBRA **Find the value of each variable in each parallelogram.** (Lesson 13-1)

5.

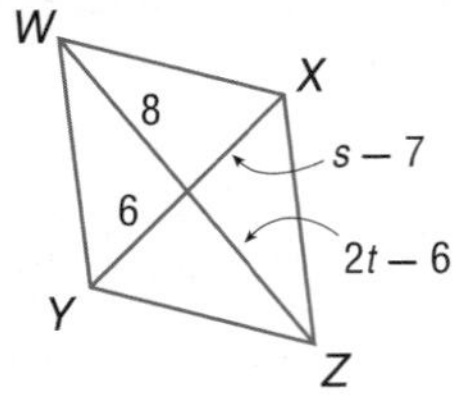

6.

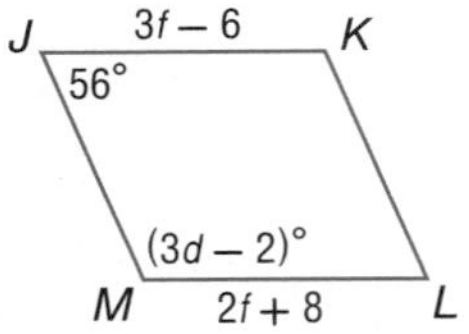

7. **PROOF** Write a two-column proof. (Lesson 13-1)
   Given: ▱$GFBA$ and ▱$HACD$
   Prove: $\angle F \cong \angle D$

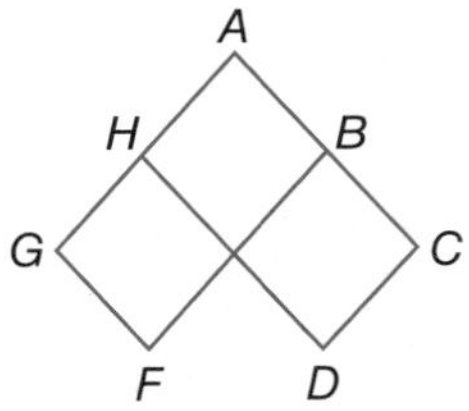

Find x and y so that each quadrilateral is a parallelogram. (Lesson 13-2)

8.

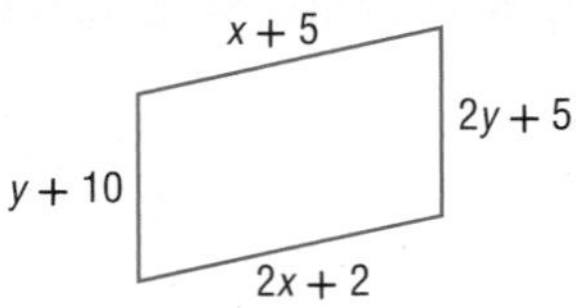

9.

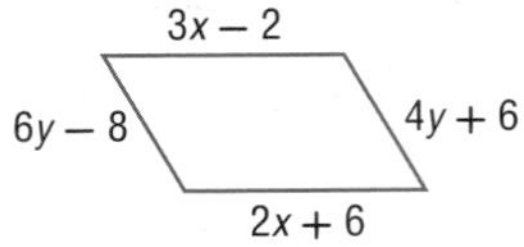

10. **MUSIC** Why will the keyboard stand with legs joined at the midpoints always remain parallel to the floor? (Lesson 13-2)

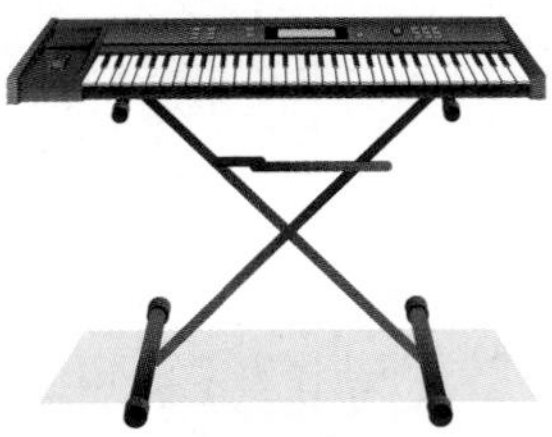

11. **MULTIPLE CHOICE** Which of the following quadrilaterals is not a parallelogram? (Lesson 13-2)

A

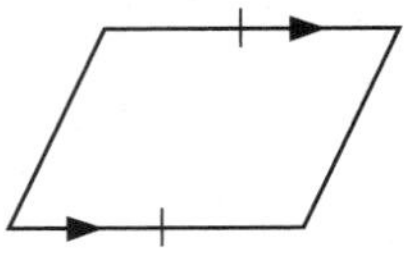

C

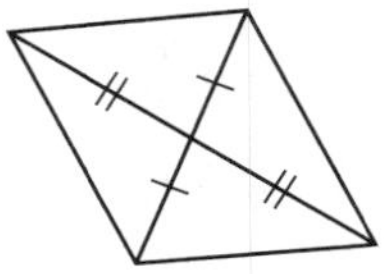

B

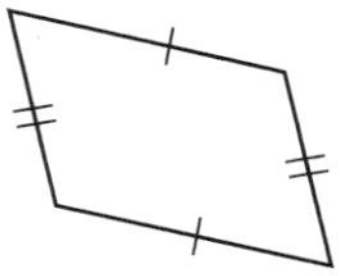

D

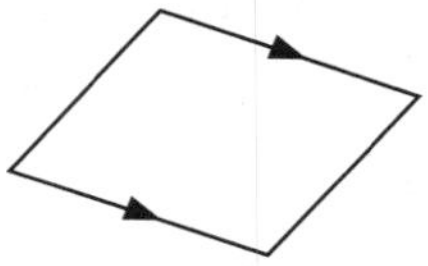

COORDINATE GEOMETRY **Determine whether the figure is a parallelogram. Justify your answer with the method indicated.** (Lesson 13-2)

12. $A(-6, -5)$, $B(-1, -4)$, $C(0, -1)$, $D(-5, -2)$; Distance Formula

13. $Q(-5, 2)$, $R(-3, -6)$, $S(2, 2)$, $T(-1, 6)$; Slope Formula

14. **COORDINATE GEOMETRY** Find the coordinates of the intersection of the diagonals of ▱$ABCD$ with vertices $A(1, 3)$, $B(6, 2)$, $C(4, -2)$, and $D(-1, -1)$. (Lesson 13-2)

LESSON 13-3 Rectangles

Then

- You used properties of parallelograms and determined whether quadrilaterals were parallelograms.

Now

1. Recognize and apply properties of rectangles.
2. Determine whether parallelograms are rectangles.

Why?

- Leonardo is in charge of set design for a school play. He needs to use paint to create the appearance of a doorway on a lightweight solid wall. The doorway is to be a rectangle 36 inches wide and 80 inches tall. How can Leonardo be sure that he paints a rectangle?

NewVocabulary
rectangle

Common Core State Standards

Content Standards
G.CO.11 Prove theorems about parallelograms.
G.GPE.4 Use coordinates to prove simple geometric theorems algebraically.

Mathematical Practices
3 Construct viable arguments and critique the reasoning of others.
5 Use appropriate tools strategically.

1 Properties of Rectangles

A **rectangle** is a parallelogram with four right angles. By definition, a rectangle has the following properties.

- All four angles are right angles.
- Opposite sides are parallel and congruent.
- Opposite angles are congruent.
- Consecutive angles are supplementary.
- Diagonals bisect each other.

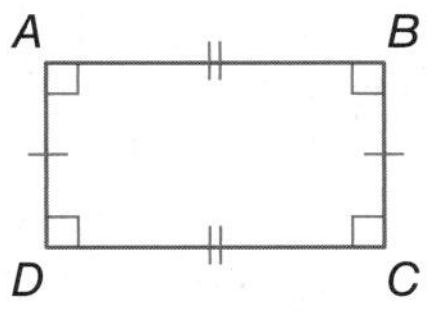

Rectangle *ABCD*

In addition, the diagonals of a rectangle are congruent.

Theorem 13.11 Diagonals of a Rectangle

If a parallelogram is a rectangle, then its diagonals are congruent.

Abbreviation *If a ▱ is a rectangle, diag. are ≅.*

Example If ▱*JKLM* is a rectangle, then $\overline{JL} \cong \overline{MK}$.

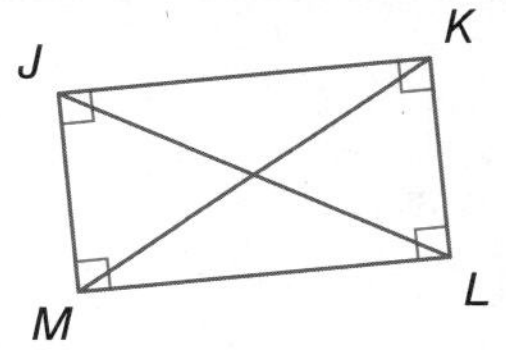

You will prove Theorem 13.11 in Exercise 33.

Real-World Example 1 Use Properties of Rectangles

EXERCISE A rectangular park has two walking paths as shown. If $PS = 180$ meters and $PR = 200$ meters, find QT.

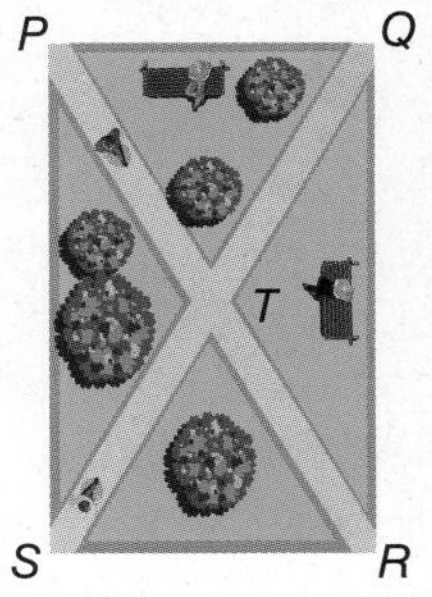

$\overline{QS} \cong \overline{PR}$	If a ▱ is a rectangle, diag. are ≅.
$QS = PR$	Definition of congruence
$QS = 200$	Substitution

Since *PQRS* is a rectangle, it is a parallelogram. The diagonals of a parallelogram bisect each other, so $QT = ST$.

$QT + ST = QS$	Segment Addition
$QT + QT = QS$	Substitution
$2QT = QS$	Simplify.
$QT = \frac{1}{2}QS$	Divide each side by 2.
$QT = \frac{1}{2}(200)$ or 100	Substitution

GuidedPractice **Refer to the figure in Example 1.**

1A. If $TS = 120$ meters, find *PR*.

1B. If $m\angle PRS = 64$, find $m\angle SQR$.

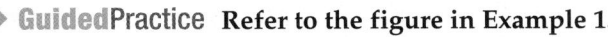

Siri Stafford/Stone/Getty Images

You can use the properties of rectangles along with algebra to find missing values.

Example 2 Use Properties of Rectangles and Algebra

ALGEBRA Quadrilateral $JKLM$ is a rectangle. If $m\angle KJL = 2x + 4$ and $m\angle JLK = 7x + 5$, find x.

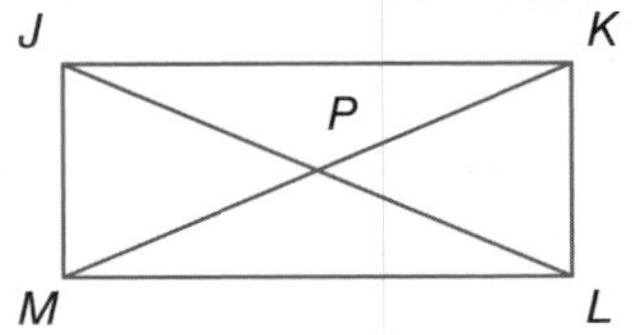

Since $JKLM$ is a rectangle, it has four right angles. So, $m\angle MLK = 90$. Since a rectangle is a parallelogram, opposite sides are parallel. Alternate interior angles of parallel lines are congruent, so $\angle JLM \cong \angle KJL$ and $m\angle JLM = m\angle KJL$.

$m\angle JLM + m\angle JLK = 90$	Angle Addition
$m\angle KJL + m\angle JLK = 90$	Substitution
$2x + 4 + 7x + 5 = 90$	Substitution
$9x + 9 = 90$	Add like terms.
$9x = 81$	Subtract 9 from each side.
$x = 9$	Divide each side by 9.

Guided Practice

2. Refer to the figure in Example 2. If $JP = 3y - 5$ and $MK = 5y + 1$, find y.

Study Tip

Right Angles Recall from Theorem 13.4 that if a parallelogram has one right angle, then it has four right angles.

2 Prove that Parallelograms are Rectangles

The converse of Theorem 13.11 is also true.

Theorem 13.12 Diagonals of a Rectangle

If the diagonals of a parallelogram are congruent, then the parallelogram is a rectangle.

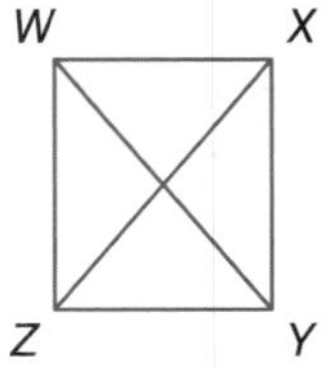

Abbreviation *If diag. of a ▱ are ≅, then ▱ is a rectangle.*

Example If $\overline{WY} \cong \overline{XZ}$ in ▱$WXYZ$, then ▱$WXYZ$ is a rectangle.

You will prove Theorem 13.12 in Exercise 34.

Real-World Example 3 Providing Rectangle Relationships

DODGEBALL A community recreation center has created an outdoor dodgeball playing field. To be sure that it meets the ideal playing field requirements, they measure the sides of the field and its diagonals. If $AB = 60$ feet, $BC = 30$ feet, $CD = 60$ feet, $AD = 30$ feet, $AC = 67$ feet, and $BD = 67$ feet, explain how the recreation center can be sure that the playing field is rectangular.

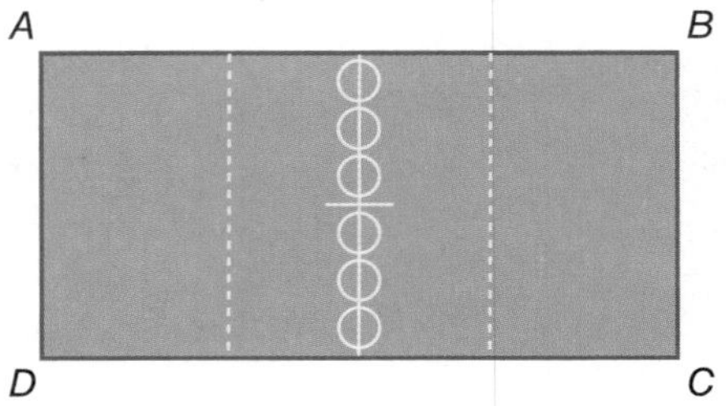

Since $AB = CD$, $BC = AD$, and $AC = BD$, $\overline{AB} \cong \overline{CD}$, $\overline{BC} \cong \overline{AD}$, and $\overline{AC} \cong \overline{BD}$. Because $\overline{AB} \cong \overline{CD}$ and $\overline{BC} \cong \overline{AD}$, $ABCD$ is a parallelogram. Since $\overline{AC}$ and $\overline{BD}$ are congruent diagonals in ▱$ABCD$, ▱$ABCD$ is a rectangle.

Real-World Link

The game of dodgeball is played on a rectangular playing field ideally 60 feet long and 30 feet wide. The field is divided into two equal sections by a center-line and attack-lines that are 3 meters (9.8 feet) from, and parallel to, the centerline.

Source: National Amateur Dodgeball Assoc.

Micah Walter/Getty Images

Real-WorldLink

The Mosaic Youth Theater in Detroit, Michigan, is a professional performing arts training program for young people ages 12 to 18. Students are involved in all aspects of performances, including set and lighting design, set construction, stage management, sound, and costumes.

GuidedPractice

3. SET DESIGN Refer to the beginning of the lesson. Leonardo measures the sides of his figure and confirms that they have the desired measures as shown. Using a carpenter's square, he also confirms that the measure of the bottom left corner of the figure is a right angle. Can he conclude that the figure is a rectangle? Explain.

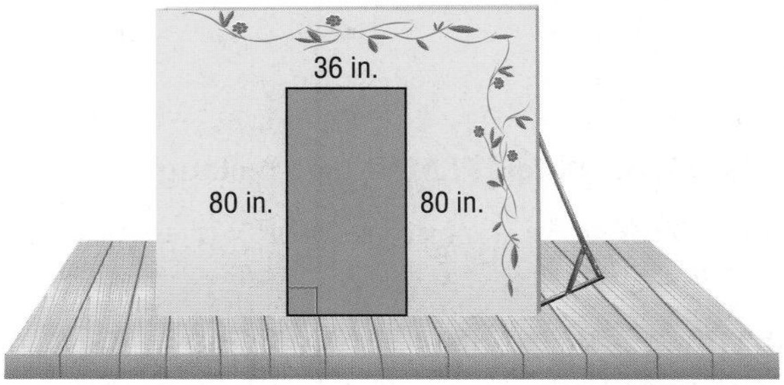

You can also use the properties of rectangles to prove that a quadrilateral positioned on a coordinate plane is a rectangle given the coordinates of the vertices.

Example 4 Rectangles and Coordinate Geometry

COORDINATE GEOMETRY Quadrilateral *PQRS* has vertices *P*(−5, 3), *Q*(1, −1), *R*(−1, −4), and *S*(−7, 0). Determine whether *PQRS* is a rectangle by using the Distance Formula.

Step 1 Use the Distance Formula to determine whether *PQRS* is a parallelogram by determining if opposite sides are congruent.

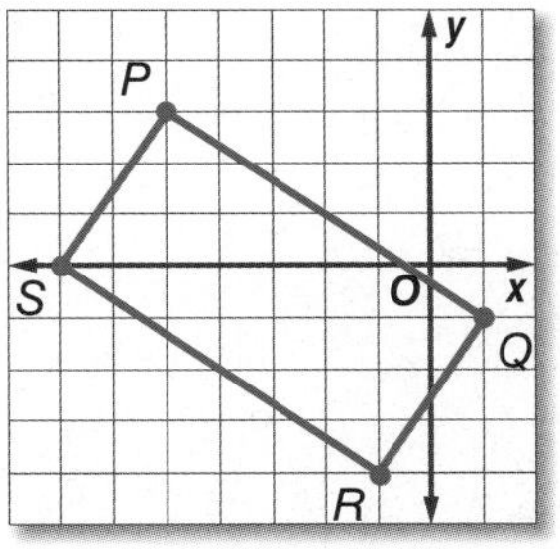

$PQ = \sqrt{(-5-1)^2 + [3-(-1)]^2}$ or $\sqrt{52}$

$RS = \sqrt{[-1-(-7)]^2 + (-4-0)^2}$ or $\sqrt{52}$

$PS = \sqrt{[-5-(-7)]^2 + (3-0)^2}$ or $\sqrt{13}$

$QR = \sqrt{[1-(-1)]^2 + [-1-(-4)]^2}$ or $\sqrt{13}$

Since opposite sides of the quadrilateral have the same measure, they are congruent. So, quadrilateral *PQRS* is a parallelogram.

StudyTip

Rectangles and Parallelograms A rectangle is a parallelogram, but a parallelogram is not necessarily a rectangle.

Step 2 Determine whether the diagonals of ▱*PQRS* are congruent.

$PR = \sqrt{[-5-(-1)]^2 + [3-(-4)]^2}$ or $\sqrt{65}$

$QS = \sqrt{[1-(-7)]^2 + (-1-0)^2}$ or $\sqrt{65}$

Since the diagonals have the same measure, they are congruent. So, ▱*PQRS* is a rectangle.

GuidedPractice

4. Quadrilateral *JKLM* has vertices *J*(−10, 2), *K*(−8, −6), *L*(5, −3), and *M*(2, 5). Determine whether *JKLM* is a rectangle using the Slope Formula.

Check Your Understanding

Example 1 **FLAGS** **The Jamaican flag is show to the right. If AE is 5.8 feet, AD is 3 feet, and $m\angle EDC = 33$, find each measure.**

1. BC

2. BD

3. $m\angle ADE$

4. $m\angle ABE$

Example 2 **ALGEBRA** **Quadrilateral $LMNP$ is a rectangle.**

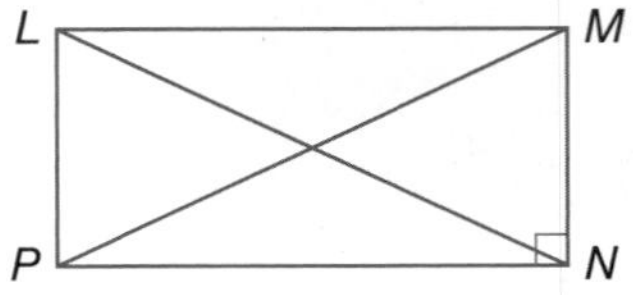

5. If $m\angle MLN = 5x + y$ and $m\angle NLP = x + 10y - 1$, find $m\angle MLN$.

6. If $MN = 5x + 2$ and $LP = 4x - 3$, find MN.

Example 3 **7.** **PROOF** If $DEFG$ is a rectangle and $HJ \parallel GF$, prove $DEJH$ is a rectangle.

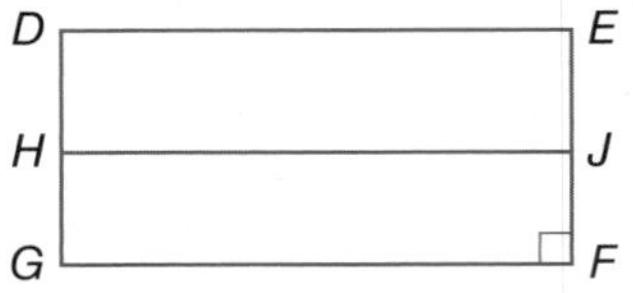

Example 4 **COORDINATE GEOMETRY** **Graph each quadrilateral with the given vertices. Determine whether the figure is a rectangle. Justify your answer using the indicated formula.**

8. $R(7, 9)$, $S(8, 0)$, $T(-2, -4)$, $Q(-3, 5)$, slope formula

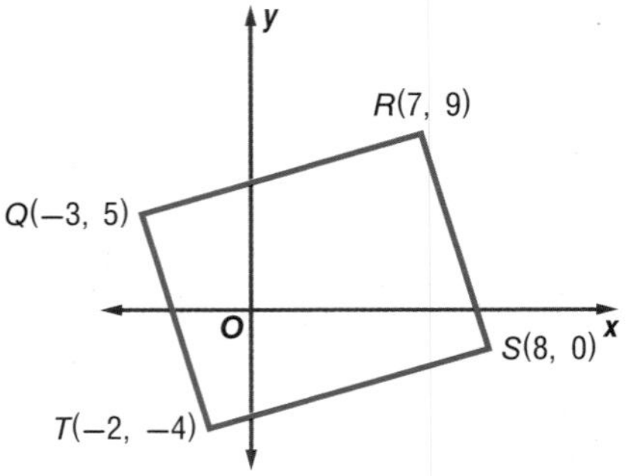

9. $C(2, 9)$, $D(5, 10)$, $E(6, 7)$, $F(3, 6)$, distance formula

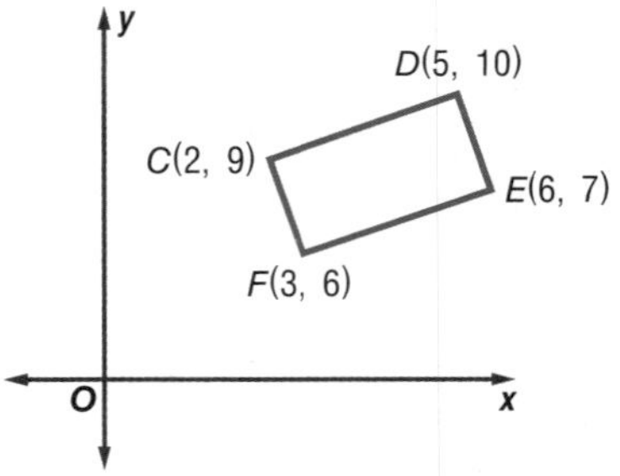

Practice and Problem Solving

Example 1 **MUSIC** **The stand show holds a keyboard. $KLMN$ forms a rectangle. IF $NM = 30$ inches, $NP = 13$ inches and $m\angle LPK = 25$, find each measure.**

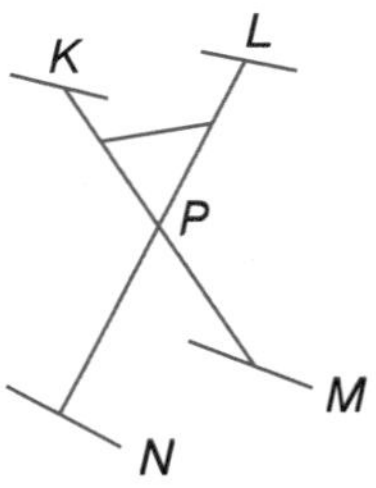

10. KL

11. KP

12. LN

13. $m\angle LPM$

Example 2 **ALGEBRA** **Quadrilateral *ABCD* is a rectangle.**

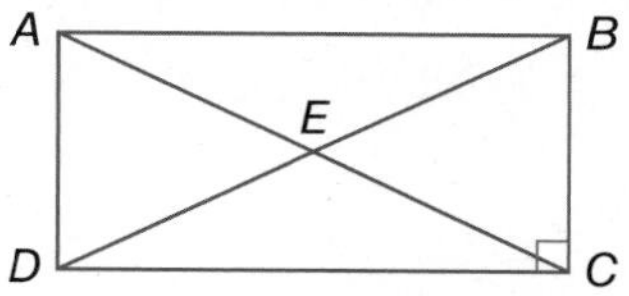

14. If $m\angle BAC = 7x - 7$ and $m\angle CAD = 8x - 8$, find $m\angle BAC$.

15. If $m\angle ADB = 8x$ and $m\angle BDC = 3x - 9$, find $m\angle DBC$.

16. If $AD = 3x + 6$ and $BC = 9x - 7$, find AD.

17. If $DE = 4x + 3$ and $EC = 5x - 1$, find AE.

18. If $m\angle BDC = 6x - 1$ and $m\angle CBD = 11x - 11$, find $m\angle ADB$.

19. If $BE = 2x - 3$ and $AC = 3x + 1$, find AC.

Example 3 **PROOF** **Write a two-column proof.**

20. **Given:** *VWXY* is a parallelogram. $\triangle XZY$ is isosceles and $\triangle VZY \cong \triangle WZX$.

Prove: *VWXY* is a rectangle.

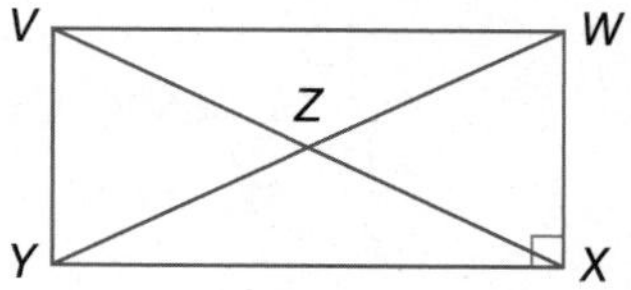

21. **Given:** *ABCD* is a rectangle. *M* is the midpoint of $\overline{AB}$, *N* is the midpoint of $\overline{BC}$, *O* is the midpoint of $\overline{DC}$, and *P* is the midpoint of $\overline{AD}$.

Prove: *MNOP* is a parallelogram.

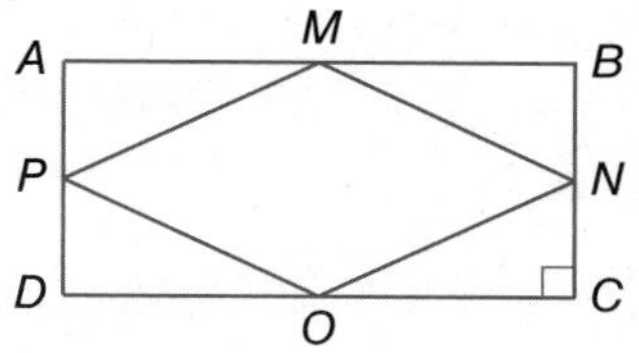

Example 4 **COORDINATE GEOMETRY** **Graph each quadrilateral with the given vertices. Determine whether the figure is a rectangle. Justify your answer using the indicated formula.**

22. $G(-3, 4)$, $H(1, 7)$, $J(4, 3)$, $K(0, 0)$; slope formula

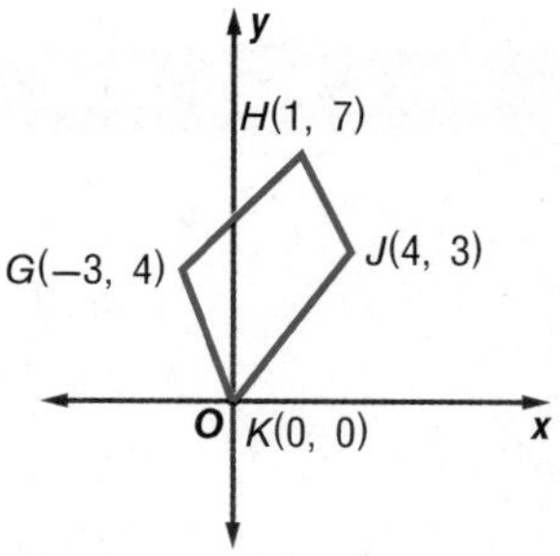

23. $J(4, -1)$, $K(8, -2)$, $L(6, -4)$, $M(0, -7)$; slope formula

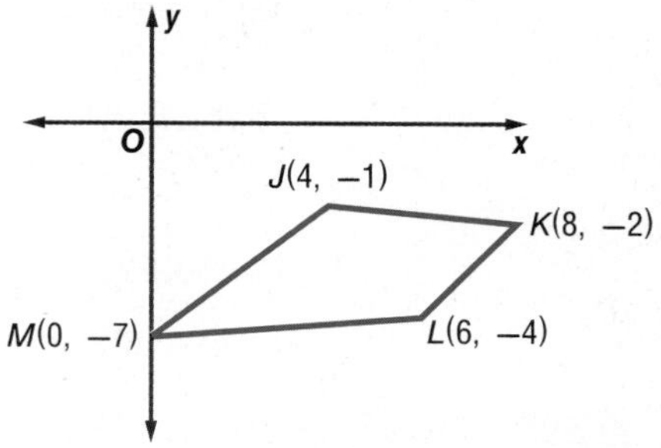

24. $W(-10, 2)$, $X(4, 2)$, $Y(5, -1)$, $Z(-10, -4)$; distance formula

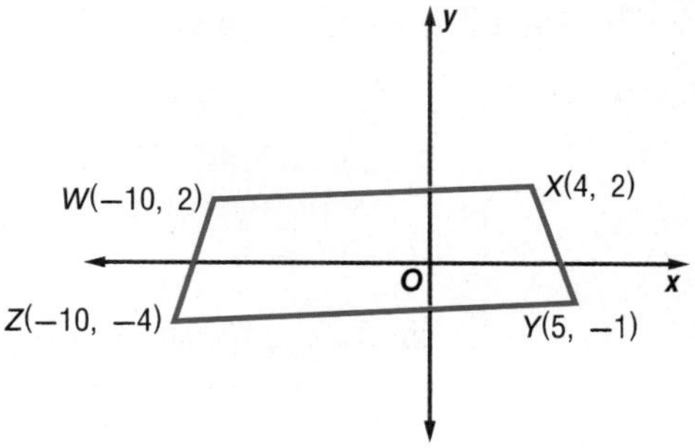

25. $Q(-6, -2)$, $R(-1, -4)$, $S(-3, -9)$, $T(-8, -7)$; distance formula

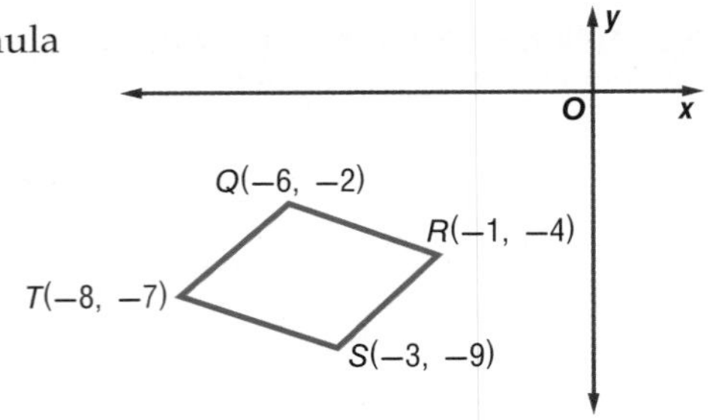

Quadrilateral *WXYZ* is a rectangle. Find each measure if $m\angle 6 = 110$.

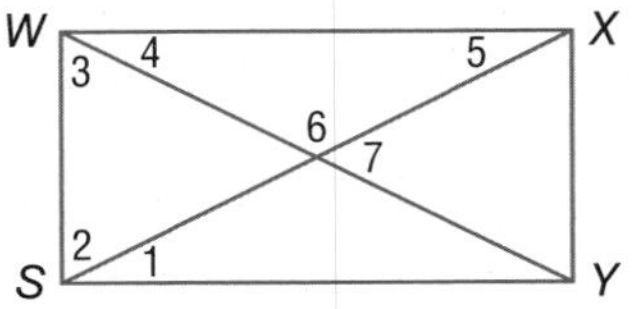

26. $m\angle 1$ **27.** $m\angle 2$ **28.** $m\angle 3$

29. $m\angle 4$ **30.** $m\angle 5$ **31.** $m\angle 7$

ALGEBRA Quadrilateral *CDEF* is a rectangle.

32. If $CF = 5$, $FE = 12$, find DF.

33. If $DE = 8$ and $DF = 10$, find CD.

34. CONSTRUCTION Explain how to use congruent sides and perpendicular lines to construct a rectangle.

35. GARDENING Ariel is constructing a rectangular flower box to use in her garden. Explain how Ariel can confirm the bottom of the box is rectangular using only a tape measure.

H.O.T. Problems Use Higher-Order Thinking Skills

36. CHALLENGE In rectangle *CDEF*, $m\angle EBF = 11x + 4y$, $m\angle DCB = \frac{3x}{2} + 5y - 1$, and $m\angle DBE = 65$. Find the values of x and y.

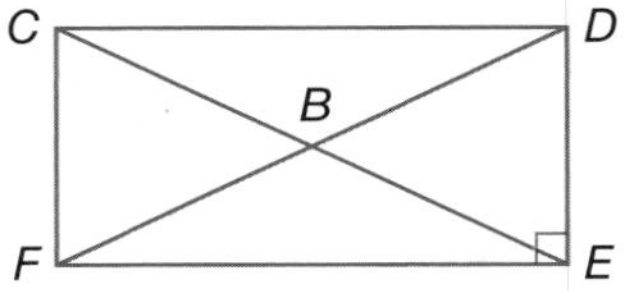

37. CCSS ERROR ANALYSIS In the figure of rectangle *KLNM*, Scott claims that $\angle KLM \cong \angle LMN$, but Patrick claims that $\angle KLM \cong \angle KML$. Is either of them correct? Explain your reasoning.

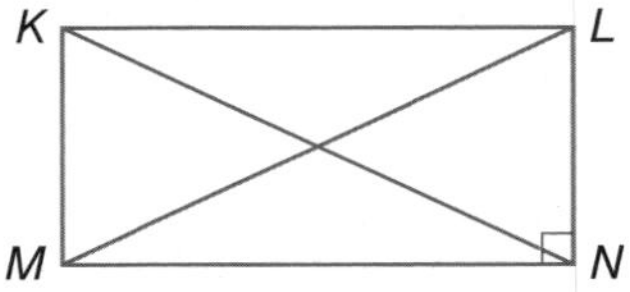

38. REASONING

a. List all right triangles in the figure of rectangle *ABCD*.

b. List all isosceles triangles in the figure of rectangle ABCD.

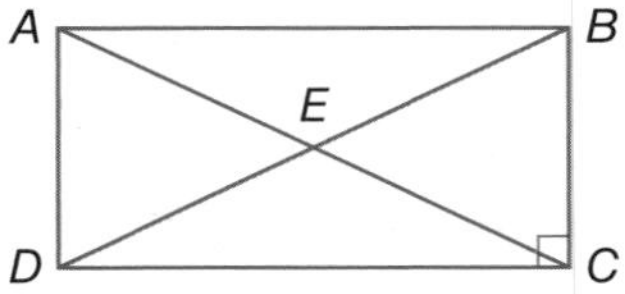

39. OPEN ENDED Find coordinates of vertices of a rectangle whose diagonal has length 5.

40. WRITING IN MATH Explain how you can use the lengths of the sides of a rectangle to find the length of the diagonals of a rectangle.

Standardized Test Practice

41. If $FJ = -3x + 5y$, $FM = 3x + y$, $GH = 11$, and $GM = 13$, what values of x and y make parallelogram $FGHJ$ a rectangle?

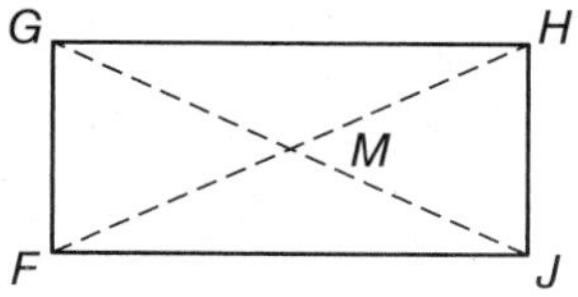

A $x = 3, y = 4$

B $x = 4, y = 3$

C $x = 7, y = 8$

D $x = 8, y = 7$

42. ALGEBRA A rectangular playground is surrounded by an 80-foot fence. One side of the playground is 10 feet longer than the other. Which of the following equations could be used to find r, the shorter side of the playground?

F $10r + r = 80$

G $4r + 10 = 80$

H $r(r + 10) = 80$

J $2(r + 10) + 2r = 80$

43. SHORT RESPONSE What is the measure of $\angle APB$?

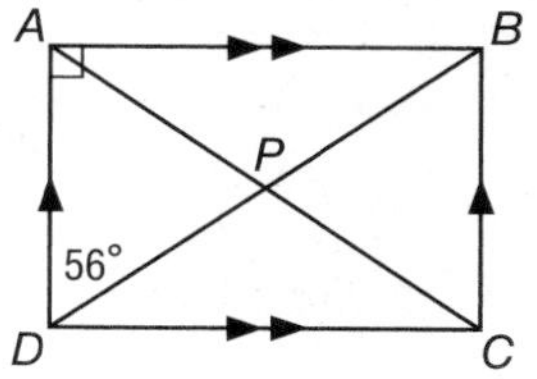

44. SAT/ACT If p is odd, which of the following must also be odd?

A $2p$

B $2p + 2$

C $\frac{p}{2}$

D $2p - 2$

E $p + 2$

Spiral Review

ALGEBRA **Find x and y so that the quadrilateral is a parallelogram.** (Lesson 13-3)

45.

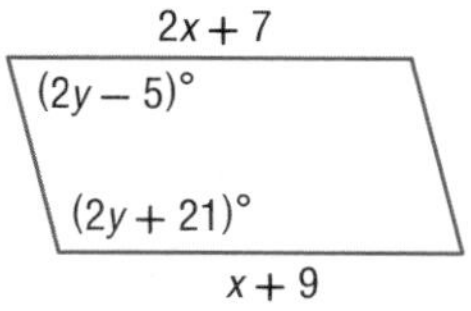

46.

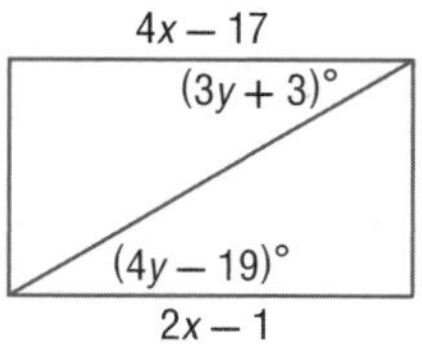

47.

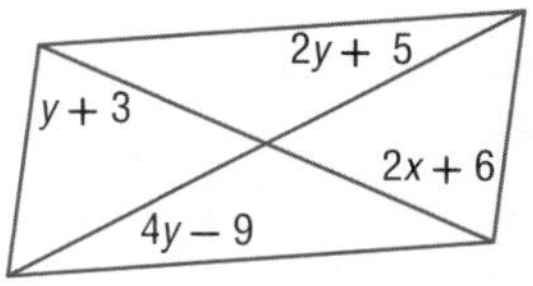

48. COORDINATE GEOMETRY Find the coordinates of the intersection of the diagonals of ▱$ABCD$ with vertices $A(1, 3)$, $B(6, 2)$, $C(4, -2)$, and $D(-1, -1)$. (Lesson 13-2)

Refer to the figure at the right. (Lesson 12-6)

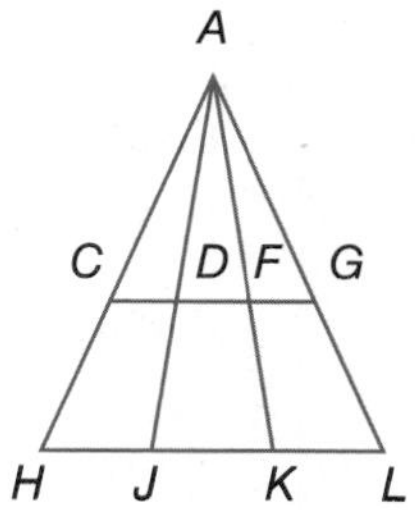

49. If $\overline{AC} \cong \overline{AF}$, name two congruent angles.

50. If $\angle AHJ \cong \angle AJH$, name two congruent segments.

51. If $\angle AJL \cong \angle ALJ$, name two congruent segments.

52. If $\overline{JA} \cong \overline{KA}$, name two congruent angles.

Skills Review

Find the distance between each pair of points.

53. $(4, 2), (2, -5)$

54. $(0, 6), (-1, -4)$

55. $(-4, 3), (3, -4)$

LESSON 13-4 Rhombi and Squares

Then

- You determined whether quadrilaterals were parallelograms and/or rectangles.

Now

1. Recognize and apply the properties of rhombi and squares.
2. Determine whether quadrilaterals are rectangles, rhombi, or squares.

Why?

- Some fruits, nuts, and vegetables are packaged using bags made out of rhombus-shaped tubular netting. Similar shaped nylon netting is used for goals in such sports as soccer, hockey, and football. A rhombus and a square are both types of equilateral parallelograms.

NewVocabulary
rhombus
square

Common Core State Standards

Content Standards
G.CO.11 Prove theorems about parallelograms.
G.GPE.4 Use coordinates to prove simple geometric theorems algebraically.

Mathematical Practices
3 Construct viable arguments and critique the reasoning of others.
2 Reason abstractly and quantitatively.

1 Properties of Rhombi and Squares

A **rhombus** is a parallelogram with all four sides congruent. A rhombus has all the properties of a parallelogram and the two additional characteristics described in the theorems below.

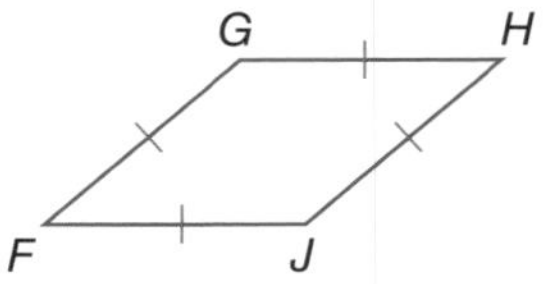

Theorems Diagonals of a Rhombus

13.13 If a parallelogram is a rhombus, then its diagonals are perpendicular.

Example If ▱*ABCD* is a rhombus, then $\overline{AC} \perp \overline{BD}$.

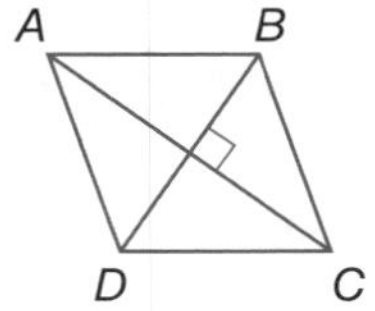

13.14 If a parallelogram is a rhombus, then each diagonal bisects a pair of opposite angles.

Example If ▱*NPQR* is a rhombus, then $\angle 1 \cong \angle 2$, $\angle 3 \cong \angle 4$, $\angle 5 \cong \angle 6$, and $\angle 7 \cong \angle 8$.

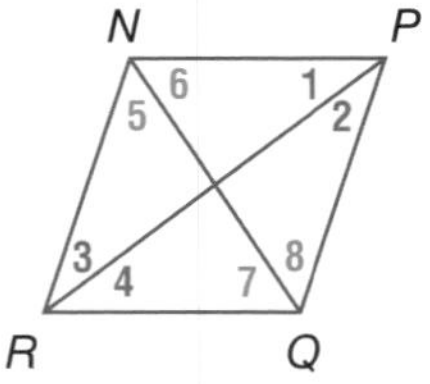

You will prove Theorem 13.14 in Exercise 34.

Proof Theorem 13.13

Given: *ABCD* is a rhombus.

Prove: $\overline{AC} \perp \overline{BD}$

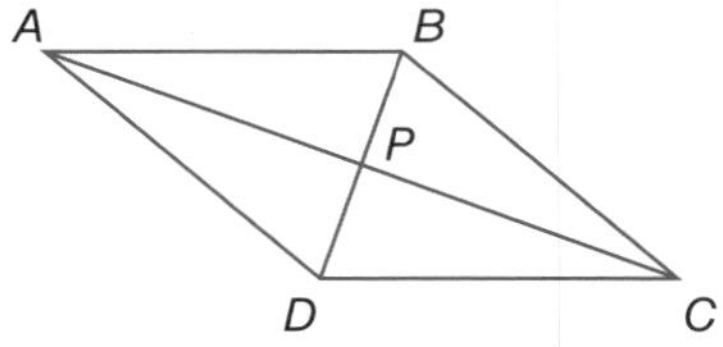

Paragraph Proof:

Since *ABCD* is a rhombus, by definition $\overline{AB} \cong \overline{BC}$. A rhombus is a parallelogram and the diagonals of a parallelogram bisect each other, so $\overline{BD}$ bisects $\overline{AC}$ at *P*. Thus, $\overline{AP} \cong \overline{PC}$. $\overline{BP} \cong \overline{BP}$ by the Reflexive Property. So, $\triangle APB \cong \triangle CPB$ by SSS. $\angle APB \cong \angle CPB$ by CPCTC. $\angle APB$ and $\angle CPB$ also form a linear pair. Two congruent angles that form a linear pair are right angles. $\angle APB$ is a right angle, so $\overline{AC} \perp \overline{BD}$ by the definition of perpendicular lines.

ReadingMath

Rhombi The plural form of rhombus is *rhombi*, pronounced ROM-bye.

Example 1 Use Properties of a Rhombus

The diagonals of rhombus $FGHJ$ intersect at K. Use the given information to find each measure or value.

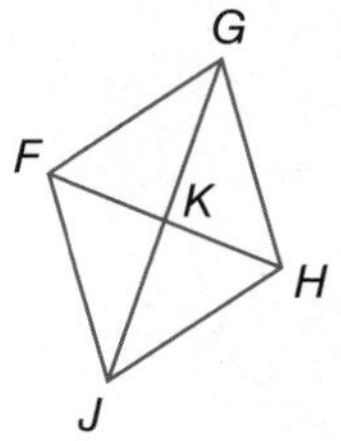

a. If $m\angle FJH = 82$, find $m\angle KHJ$.

Since $FGHJ$ is a rhombus, diagonal $\overline{JG}$ bisects $\angle FJH$. Therefore, $m\angle KJH = \frac{1}{2}m\angle FJH$. So $m\angle KJH = \frac{1}{2}(82)$ or 41. Since the diagonals of a rhombus are perpendicular, $m\angle JKH = 90$ by the definition of perpendicular lines.

$m\angle KJH + m\angle JKH + m\angle KHJ = 180$	Triangle Sum Theorem
$41 + 90 + m\angle KHJ = 180$	Substitution
$131 + m\angle KHJ = 180$	Simplify.
$m\angle KHJ = 49$	Subtract 131 from each side.

b. ALGEBRA If $GH = x + 9$ and $JH = 5x - 2$, find x.

$\overline{GH} \cong \overline{JH}$	By definition, all sides of a rhombus are congruent.
$GH = JH$	Definition of congruence
$x + 9 = 5x - 2$	Substitution
$9 = 4x - 2$	Subtract x from each side.
$11 = 4x$	Add 2 to each side.
$2.75 = x$	Divide each side by 4.

GuidedPractice

Refer to rhombus $FGHJ$ above.

1A. If $FK = 5$ and $FG = 13$, find KJ.

1B. ALGEBRA If $m\angle JFK = 6y + 7$ and $m\angle KFG = 9y - 5$, find y.

A **square** is a parallelogram with four congruent sides and four right angles. Recall that a parallelogram with four right angles is a rectangle, and a parallelogram with four congruent sides is a rhombus. Therefore, a parallelogram that is both a rectangle and a rhombus is also a square.

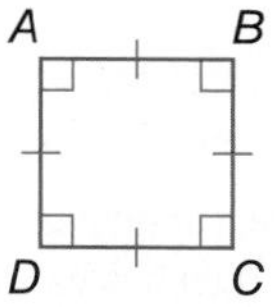

Square $ABCD$

The Venn diagram summarizes the relationships among parallelograms, rhombi, rectangles, and squares.

ConceptSummary Parallelograms

Parallelograms (Opp. sides are ∥.)

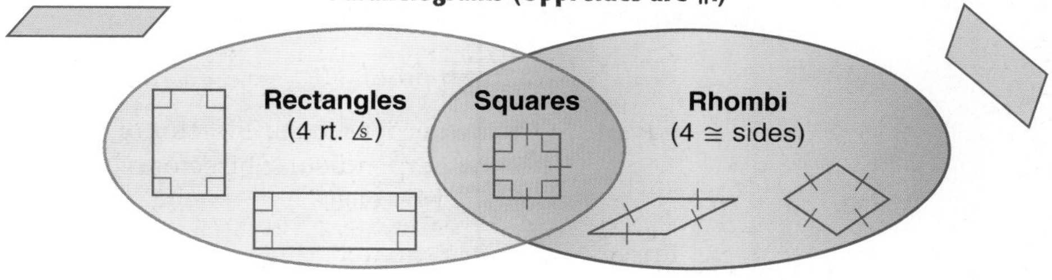

All of the properties of parallelograms, rectangles, and rhombi apply to squares. For example, the diagonals of a square bisect each other (parallelogram), are congruent (rectangle), and are perpendicular (rhombus).

2 Prove that Quadrilaterals are Rhombi or Squares

The theorems below provide conditions for rhombi and squares.

> **StudyTip**
> **Common Misconception** Theorems 13.15, 13.16, and 13.17 apply only if you already know that a quadrilateral is a parallelogram.

Theorems Conditions for Rhombi and Squares

13.15 If the diagonals of a parallelogram are perpendicular, then the parallelogram is a rhombus. (Converse of Theorem. 13.13)

Example If $\overline{JL} \perp \overline{KM}$, then $\square JKLM$ is a rhombus.

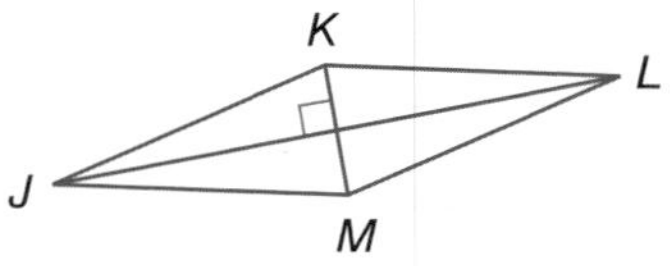

13.16 If one diagonal of a parallelogram bisects a pair of opposite angles, then the parallelogram is a rhombus. (Converse of Theorem. 13.14)

Example If $\angle 1 \cong \angle 2$ and $\angle 3 \cong \angle 4$, or $\angle 5 \cong \angle 6$ and $\angle 7 \cong \angle 8$, then $\square WXYZ$ is a rhombus.

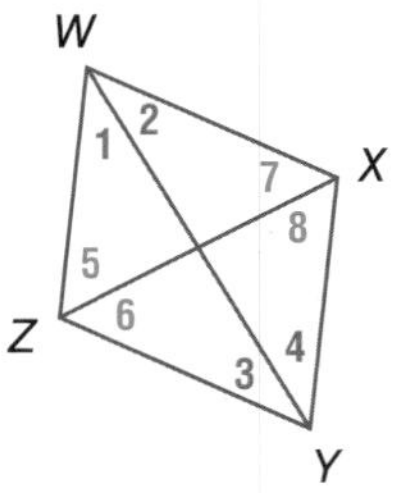

13.17 If one pair of consecutive sides of a parallelogram are congruent, the parallelogram is a rhombus.

Example If $\overline{AB} \cong \overline{BC}$, then $\square ABCD$ is a rhombus.

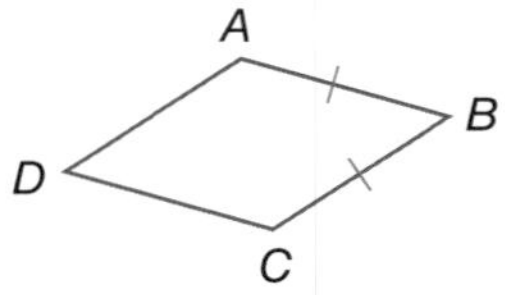

13.18 If a quadrilateral is both a rectangle and a rhombus, then it is a square.

You will prove Theorems 13.15–13.18 in Exercises 35–38, respectively.

You can use the properties of rhombi and squares to write proofs.

Example 2 Proofs Using Properties of Rhombi and Squares

> **StudyTip**
> **Congruent Triangles** Since a rhombus has four congruent sides, one diagonal separates the rhombus into two congruent isosceles triangles. Drawing two diagonals separates the rhombus into four congruent right triangles.

Write a paragraph proof.

Given: *JKLM* is a parallelogram.

$\triangle JKL$ is isosceles.

Prove: *JKLM* is a rhombus.

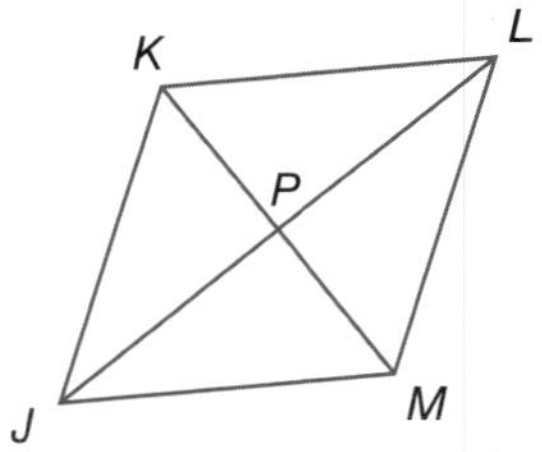

Paragraph Proof:

Since it is given that $\triangle JKL$ is isosceles, $\overline{KL} \cong \overline{JK}$ by definition. These are consecutive sides of the given parallelogram *JKLM*. So, by Theorem 13.17, *JKLM* is a rhombus.

GuidedPractice

2. Write a paragraph proof.

Given: $\overline{SQ}$ is the perpendicular bisector of $\overline{PR}$.
$\overline{PR}$ is the perpendicular bisector of $\overline{SQ}$.
$\triangle RMS$ is isosceles.

Prove: *PQRS* is a square.

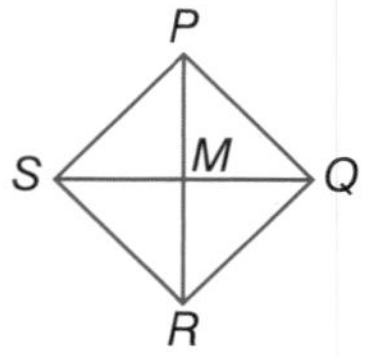

Real-WorldLink

Archaeology is the study of artifacts that provide information about human life and activities in the past. Since humans only began writing about 5000 years ago, information from periods before that time must be gathered from the objects that archeologists locate.

Source: Encyclopeadia Britannica

Real-World Example 3 Use Conditions for Rhombi and Squares

PT

ARCHAEOLOGY The key to the successful excavation of an archaeological site is accurate mapping. How can archaeologists be sure that the region they have marked off is a 1-meter by 1-meter square?

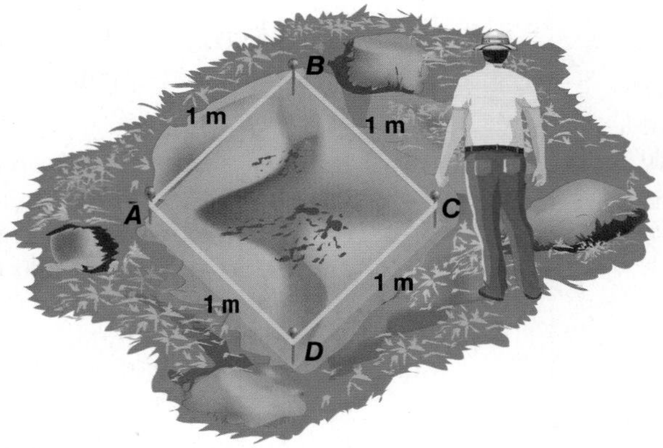

Each side of quadrilateral *ABCD* measures 1 meter. Since opposite sides are congruent, *ABCD* is a parallelogram. Since consecutive sides of ▱*ABCD* are congruent, it is a rhombus. If the archaeologists can show that ▱*ABCD* is also a rectangle, then by Theorem 13.20, ▱*ABCD* is a square.

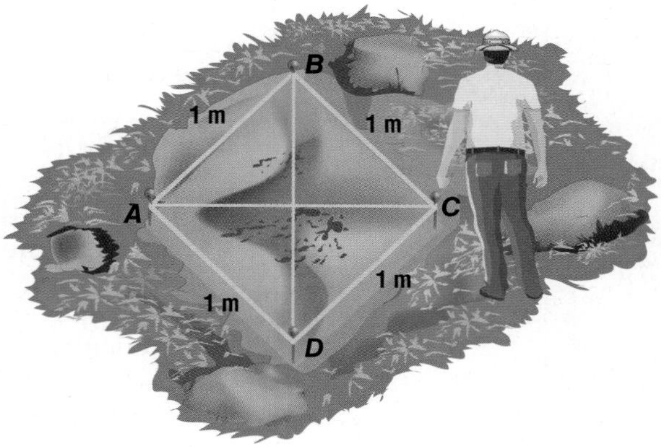

If the diagonals of a parallelogram are congruent, then the parallelogram is a rectangle. So if the archeologists measure the length of string needed to form each diagonal and find that these lengths are equal, then *ABCD* is a square.

GuidedPractice

3. QUILTING Kathy is designing a quilt with blocks like the one shown.

A. If she marks the diagonals of each yellow piece and determines that each pair of diagonals is perpendicular, can she conclude that each yellow piece is a rhombus? Explain.

B. If all four angles of the green piece have the same measure and the bottom and left sides have the same measure, can she conclude that the green piece is a square? Explain.

In Chapter 12, you used coordinate geometry to classify triangles. Coordinate geometry can also be used to classify quadrilaterals.

Example 4 Classify Quadrilaterals Using Coordinate Geometry

COORDINATE GEOMETRY Determine whether ▱*JKLM* with vertices *J*(−7, −2), *K*(0, 4), *L*(9, 2), and *M*(2, −4) is a *rhombus*, a *rectangle*, or a *square*. List all that apply. Explain.

> **Problem-Solving Tip**
> Make a Graph When analyzing a figure using coordinate geometry, graph the figure to help formulate a conjecture and also to help check the reasonableness of the answer you obtain algebraically.

Understand Plot and connect the vertices on a coordinate plane.

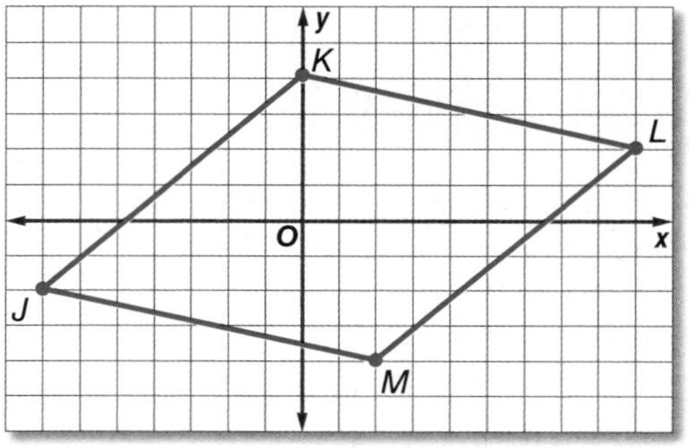

It appears from the graph that the parallelogram has four congruent sides, but no right angles. So, it appears that the figure is a rhombus, but not a square or a rectangle.

Plan If the diagonals of the parallelogram are congruent, then it is a rectangle. If they are perpendicular, then it is a rhombus. If they are both congruent and perpendicular, the parallelogram is a rectangle, a rhombus, and a square.

Solve **Step 1** Use the Distance Formula to compare the diagonal lengths.

$KM = \sqrt{(2-0)^2 + (-4-4)^2} = \sqrt{68}$ or $2\sqrt{17}$

$JL = \sqrt{[9-(-7)]^2 + [2-(-2)]^2} = \sqrt{272}$ or $4\sqrt{17}$

Since $2\sqrt{17} \neq 4\sqrt{17}$, the diagonals are not congruent. So, ▱*JKLM* is *not* a rectangle. Since the figure is not a rectangle, it also *cannot* be a square.

> **Study Tip**
> Square and Rhombus
> A square is a rhombus, but a rhombus is not necessarily a square.

Step 2 Use the Slope Formula to determine whether the diagonals are perpendicular.

slope of $\overline{KM} = \frac{-4-4}{2-0} = \frac{-8}{2}$ or -4

slope of $\overline{JL} = \frac{2-(-2)}{9-(-7)} = \frac{4}{16}$ or $\frac{1}{4}$

Since the product of the slopes of the diagonals is −1, the diagonals are perpendicular, so ▱*JKLM* is a rhombus.

Check $JK = \sqrt{[4-(-2)]^2 + [0-(-7)]^2}$ or $\sqrt{85}$

$KL = \sqrt{(9-0)^2 + (2-4)^2}$ or $\sqrt{85}$

So, ▱*JKLM* is a rhombus by Theorem 13.20.

Since the slope of $\overline{JK} = \frac{4-(-2)}{0-(-7)}$ or $\frac{6}{7}$, the slope of $\overline{KL} = \frac{2-4}{9-0}$ or $-\frac{2}{9}$, and the product of these slopes is not −1, consecutive sides $\overline{JK}$ and $\overline{KL}$ are not perpendicular. Therefore, ∠*JKL* is not a right angle. So ▱*JKLM* is not a rectangle or a square. ✔

Guided Practice

4. Given *J*(5, 0), *K*(8, −11), *L*(−3, −14), *M*(−6, −3), determine whether parallelogram *JKLM* is a *rhombus*, a *rectangle*, or a *square*. List all that apply. Explain.

Check Your Understanding

Example 1 **CCSS ALGEBRA Quadrilateral *ABCD* is a rhombus. Find each value or measure.**

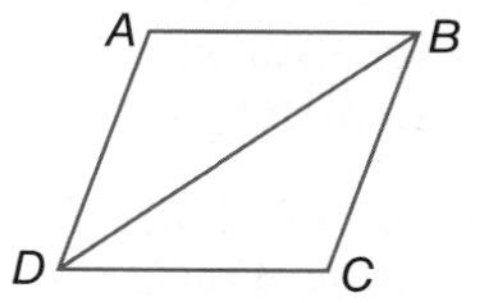

1. If $BC = 4x + 3$ and $AB = 8x - 5$, find AD.

2. If $m\angle ADC = 70$, find $m\angle ABD$.

Examples 2

3. PROOF If *LMNP* is a rhombus, write a two-column proof to prove $\triangle LQM \cong \triangle NQM$.

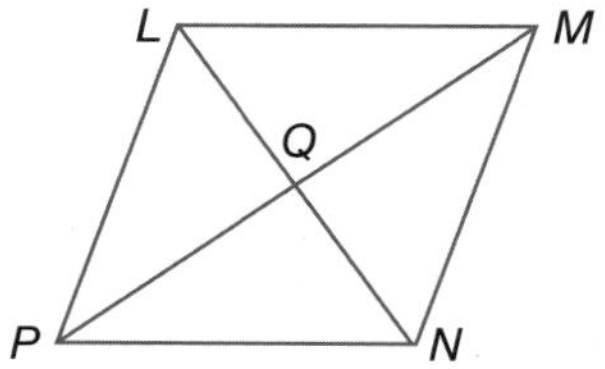

4. PHOTOGRAPHY Russell is creating a photo collage using 36 congruent squares. Use this information to prove the collage itself is a square.

Example 4 **COORDINATE GEOMETRY Given each set of vertices, determine whether ▱*XYWZ* is a rhombus, a rectangle, or a square. List all that apply. Explain your reasoning.**

5. $X(-2, 1)$, $Y(0, -3)$, $W(4, -1)$, $Z(2, 3)$

6. $X(4, -1)$, $Y(-1, 0)$, $W(0, 3)$, $Z(5, 2)$

Practice and Problem Solving

Example 1 **CCSS ALGEBRA Quadrilateral *ABCD* is a rhombus. Find each value or measure.**

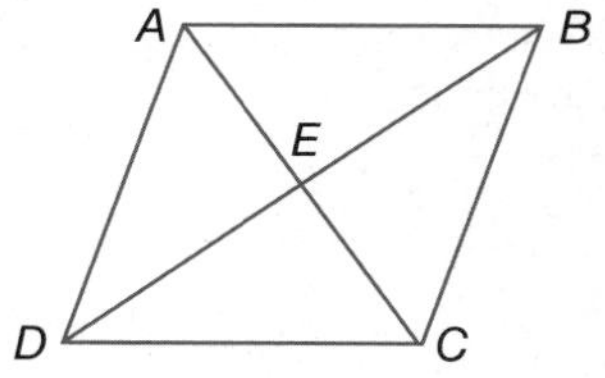

7. If $m\angle DAE = 25$, find $m\angle DAB$.

8. If $DC = 12$, find AD

9. If $m\angle EDC = 6x - 6$ and $m\angle DBC = 5x + 6$, find $m\angle DCB$.

10. If $m\angle EAD = 5x + 5$ and $m\angle BCE = 7x - 9$, find $m\angle EAD$.

11. If $AE = 5x - 3$ and $AC = 7x$, find BD

12. If $BC = 6x - 3$ and $AD = 7x - 6$, find DC.

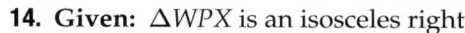

Example 2 **PROOF Write a two-column proof.**

13. Given: $m\angle LMQ = m\angle QPN$, $m\angle NMQ = m\angle LPQ$, $\overline{LM} \cong \overline{MN}$
Prove: *LMNP* is a rhombus.

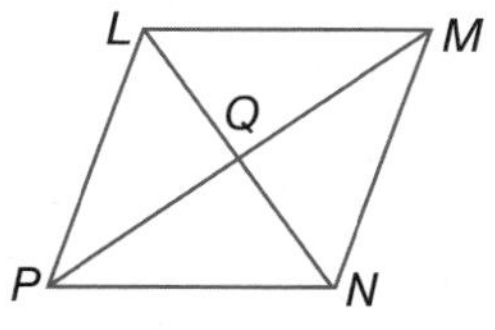

14. Given: $\triangle WPX$ is an isosceles right triangle, $\triangle WPX \cong \triangle ZPY$
Prove: *WXYZ* is a square.

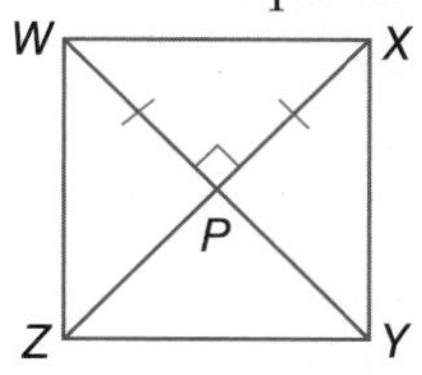

15. Given: *LMPQ* is a parallelogram. *K* bisects *LM*, *N* bisects *MO*, *P* bisects *OQ* and *R* bisects *LQ*. $\angle L \cong \angle M$
Prove: *KNPR* is a rhombus.

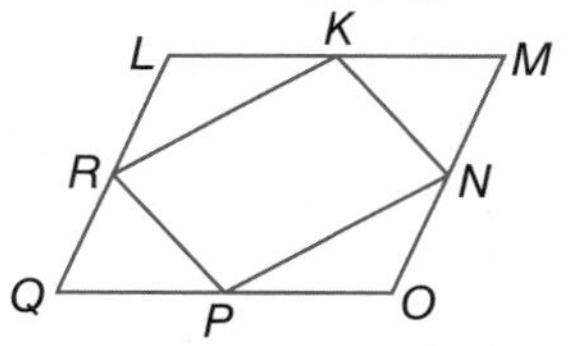

16. Given: *ABDE* is a square, $\triangle ABE \cong \triangle BCD$
Prove: *BCDE* is a parallelogram

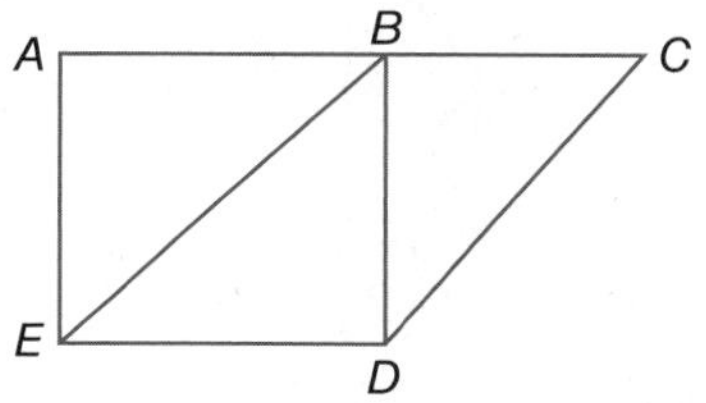

17. **ORIGAMI** Lisa is cutting a piece of paper to use for origami. If Lisa has used a protractor to confirm the corners have measures of 90, and the diagonals are the same length, can Lisa be sure it is a square? Explain your reasoning.

Example 4 **COORDINATE GEOMETRY Given each set of vertices, determine whether *ABCD* is a rhombus, a rectangle, or a square. List all that apply. Explain your reasoning.**

18. $A(-2, 1), B(3, 1), C(6, 5), D(1, 5)$

19. $A(-6, -5), B(-1, -5), C(2, -1), D(-3, -1)$

20. $A(2, 3), B(0, 7), C(5, 9), D(7, 5)$

21. $A(-5, -4), B(0, -3), C(0, 2), D(-5, 2)$

***JKLM* is a rhombus. If $JK = 8$, $CM = 4$, and $m\angle CJM = 30$, find each measure.**

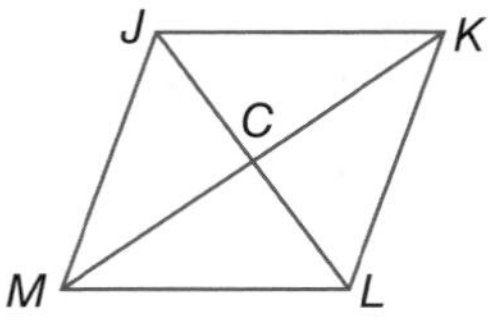

22. JC

23. $m\angle JKL$

24. MK

25. $m\angle CJK$

***RSTV* is a square. If $RP = 7$, find each measure.**

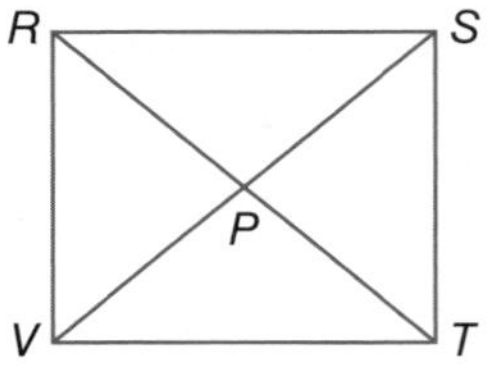

26. VT

27. SV

28. $m\angle SPR$

29. $m\angle PST$

Classify each quadrilateral.

30. 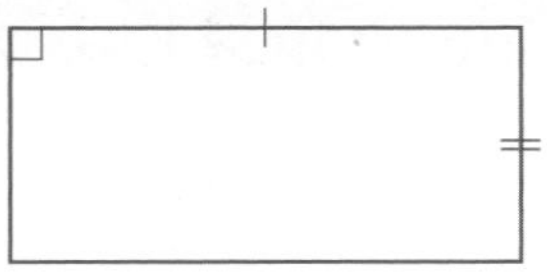

31. 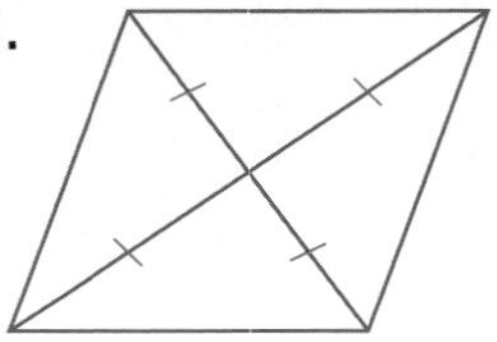

32. 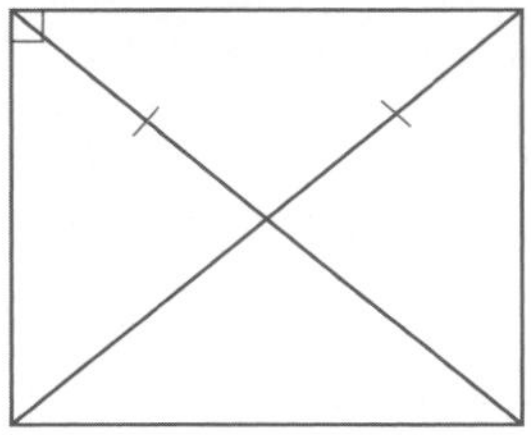

PROOF Write a paragraph proof.

33. Theorem 13.14

34. Theorem 13.15

35. Theorem 13.16

36. Theorem 13.17

37. Theorem 13.18

CONSTRUCTION Use diagonals to construct each figure. Justify each construction.

38. rhombus

39. square

PROOF Write a coordinate proof of each statement.

40. The diagonals of a rhombus are perpendicular.

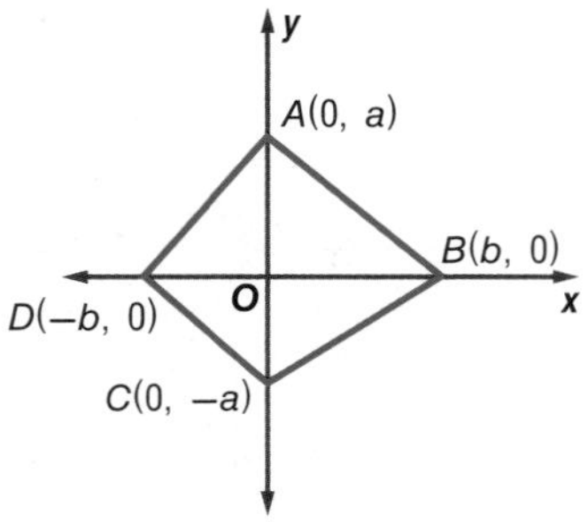

41. The diagonals of a square create 4 congruent triangles.

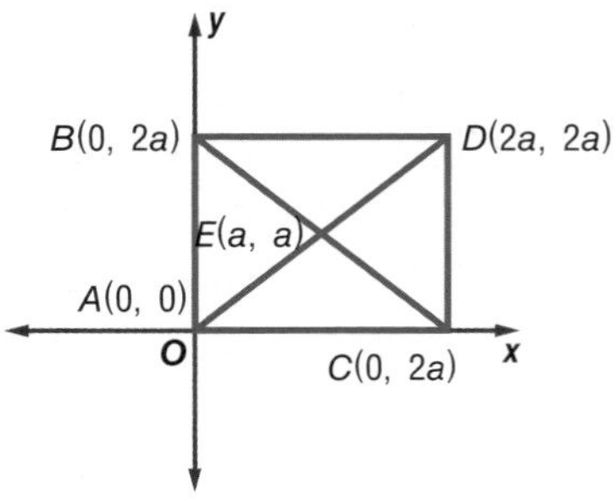

42. **SPORTS** The diagram below is of a tennis court. If the court is symmetric across the net, classify the quadrilateral *ABCD*. Explain your reasoning.

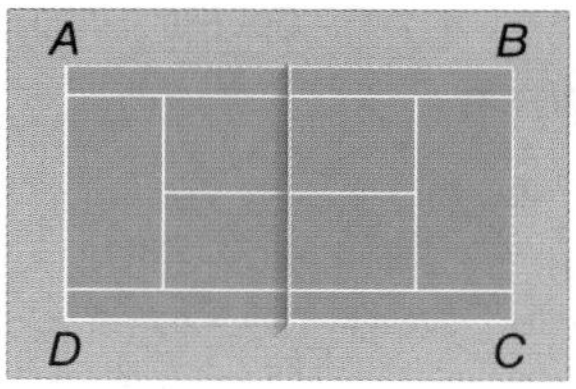

43. **BAKING** The diagram of a baking pan shown below. If this pan is being used to bake a batch of brownies. If the pan will be cut into 9 brownies, what will the dimensions of each brownie be?

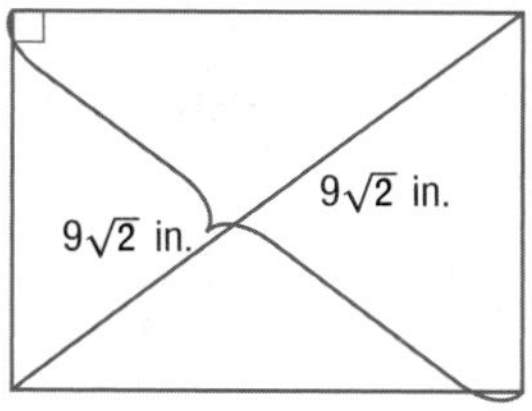

44. **MULTIPLE REPRESENTATIONS** In this problem you will explore the properties of Kites, which are quadrilaterals with exactly two distinct pairs of adjacent sides.

   a. **Geometric** Draw 3 kites with varying side lengths. Label one of them *ABCD*, one of them *QRSP*, and one of them *WXYZ*.

   b. **Tabular** Use a protractor to measure the angles of each kite. Organize these results in a table.

   c. **Verbal** Make a conjecture about the diagonals of a kite.

H.O.T. Problems Use Higher-Order Thinking Skills

45. **REASONING** Determine whether the statement is true or false. The write the converse, inverse, and contrapositive of the statement and determine the truth value of each. Explain your reasoning.

   If a quadrilateral is a rhombus, then it is a square.

46. **CHALLENGE** The figure to the right is a cube. If $\overline{AD} = 5$, find $\overline{AH}$.

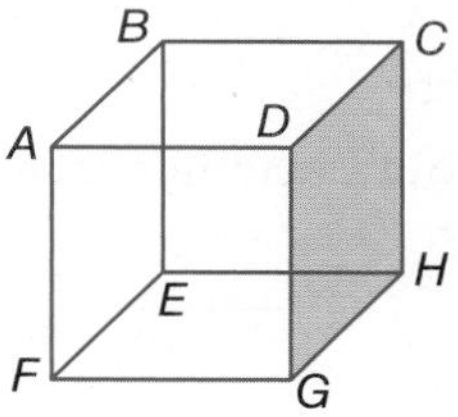

47. **ERROR ANALYSIS** In parallelogram *ABCD*, $m\angle CAB = 45$ and $\overline{AE} = \overline{ED}$. Melissa thinks the parallelogram is a square, but Kojo thinks it is only a rhombus. Is either of them correct? Explain your reasoning.

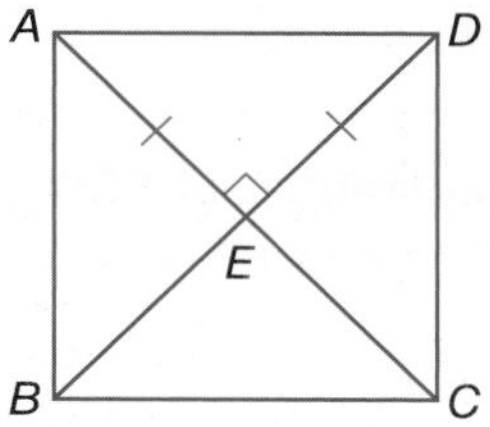

48. **OPEN ENDED** Write down the equations of two perpendicular lines. Find the vertices of a square with diagonals contained in the lines you wrote down.

49. **WRITING IN MATH** Explain the ways to prove a parallelogram is a square.

Standardized Test Practice

50. *JKLM* is a rhombus. If $CK = 8$ and $JK = 10$, find *JC*.

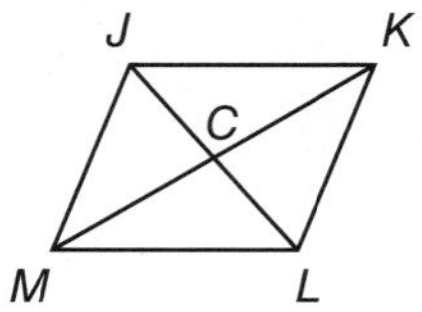

A 4   C 8
B 6   D 10

51. EXTENDED RESPONSE The sides of square *ABCD* are extended by sides of equal length to form square *WXYZ*.

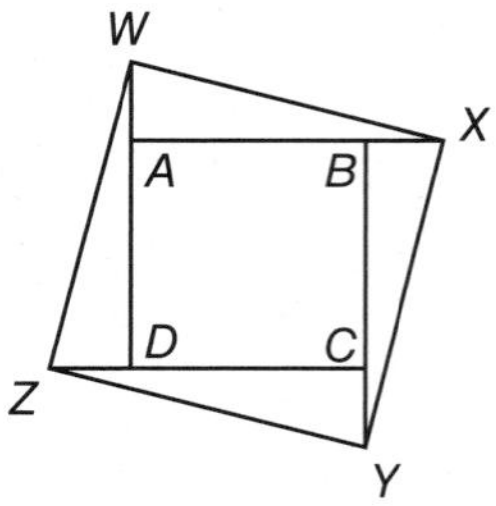

a. If $CY = 3$ cm and the area of *ABCD* is 81 cm^2, find the area of *WXYZ*.

b. If the areas of *ABCD* and *WXYZ* are 49 cm^2 and 169 cm^2 respectively, find *DZ*.

c. If $AB = 2CY$ and the area of $ABCD = g$ square meters, find the area of *WXYZ* in square meters.

52. ALGEBRA What values of x and y make quadrilateral *ABCD* a parallelogram?

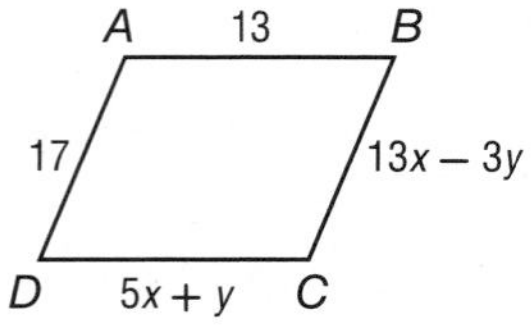

F $x = 3, y = 2$
G $x = \frac{3}{2}, y = -1$
H $x = 2, y = 3$
J $x = 3, y = -1$

53. SAT/ACT What is 6 more than the product of -3 and a certain number x?

A $-3x - 6$ D $-3x + 6$
B $-3x$ E $6 + 3x$
C $-x$

Spiral Review

Quadrilateral *ABDC* is a rectangle. Find each measure if $m\angle 1 = 38$. (Lesson 13-4)

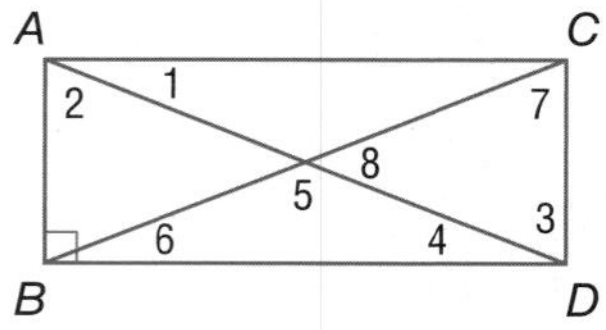

54. $m\angle 2$ **55.** $m\angle 5$ **56.** $m\angle 6$

Determine whether each quadrilateral is a parallelogram. Justify your answer. (Lesson 6-3)

57.

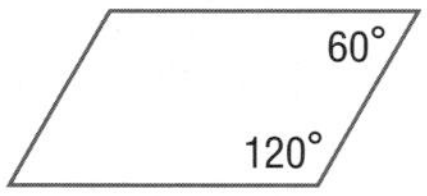

58.

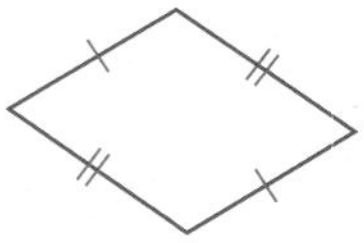

59.

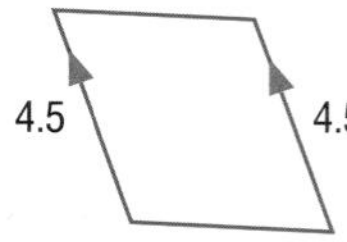

60. COORDINATE GEOMETRY Identify the transformation and verify that it is a congruence transformation. (Lesson 12-7)

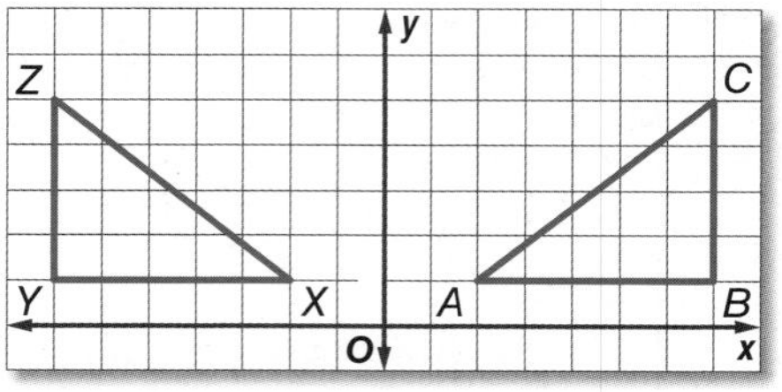

Skills Review

Solve each equation.

61. $\frac{1}{2}(5x + 7x - 1) = 11.5$ **62.** $\frac{1}{2}(10x + 6x + 2) = 7$ **63.** $\frac{1}{2}(12x + 6 - 8x + 7) = 9$

LESSON

13-5 Trapezoids and Kites

∴ Then	∴ Now	∴ Why?
• You used properties of special parallelograms.	• **1** Apply properties of trapezoids. **2** Apply properties of kites.	• In gymnastics, vaulting boxes made out of high compression foam are used as spotting platforms, vaulting horses, and steps. The left and right side of each section is a *trapezoid*.

NewVocabulary

trapezoid
bases
legs of a trapezoid
base angles
isosceles trapezoid
midsegment of a trapezoid
kite

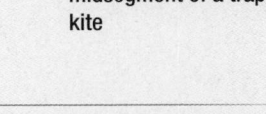

Common Core State Standards

Content Standards

G.GPE.4 Use coordinates to prove simple geometric theorems algebraically.

G.MG.3 Apply geometric methods to solve problems (e.g., designing an object or structure to satisfy physical constraints or minimize cost; working with typographic grid systems based on ratios). ★

Mathematical Practices

1 Make sense of problems and persevere in solving them.
2 Reason abstractly and quantitatively.

1 Properties of Trapezoids

A **trapezoid** is a quadrilateral with exactly one pair of parallel sides. The parallel sides are called **bases**. The nonparallel sides are called **legs**. The **base angles** are formed by the base and one of the legs. In trapezoid *ABCD*, $\angle A$ and $\angle B$ are one pair of base angles and $\angle C$ and $\angle D$ are the other pair. If the legs of a trapezoid are congruent, then it is an **isosceles trapezoid**.

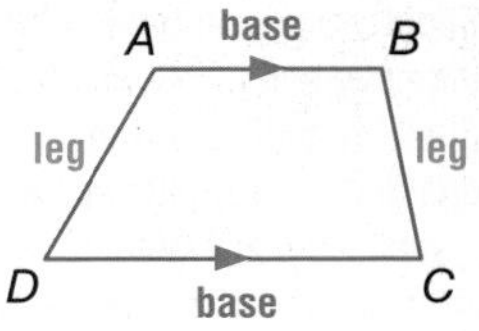

Theorems Isosceles Trapezoids

13.19 If a trapezoid is isosceles, then each pair of base angles is congruent.

Example If trapezoid *FGHJ* is isosceles, then $\angle G \cong \angle H$ and $\angle F \cong \angle J$.

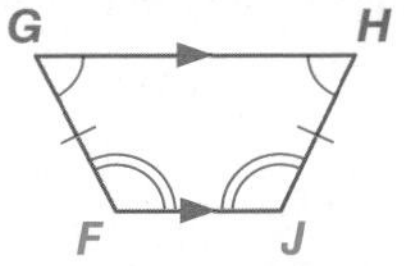

13.20 If a trapezoid has one pair of congruent base angles, then it is an isosceles trapezoid.

Example If $\angle L \cong \angle M$, then trapezoid *KLMP* is isosceles.

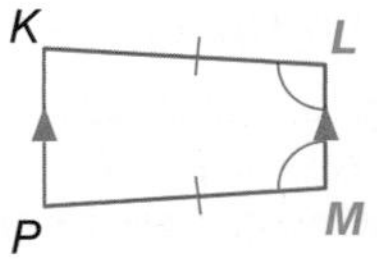

13.21 A trapezoid is isosceles if and only if its diagonals are congruent.

Example If trapezoid *QRST* is isosceles, then $\overline{QS} \cong \overline{RT}$. Likewise, if $\overline{QS} \cong \overline{RT}$, then trapezoid *QRST* is isosceles.

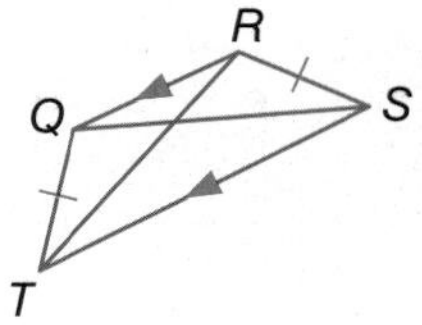

You will prove Theorem 13.19, Theorem 13.20, and the other part of Theorem 13.21 in Exercises 28, 29, and 30.

Proof Part of Theorem 13.21

Given: *ABCD* is an isosceles trapezoid.

Prove: $\overline{AC} \cong \overline{BD}$

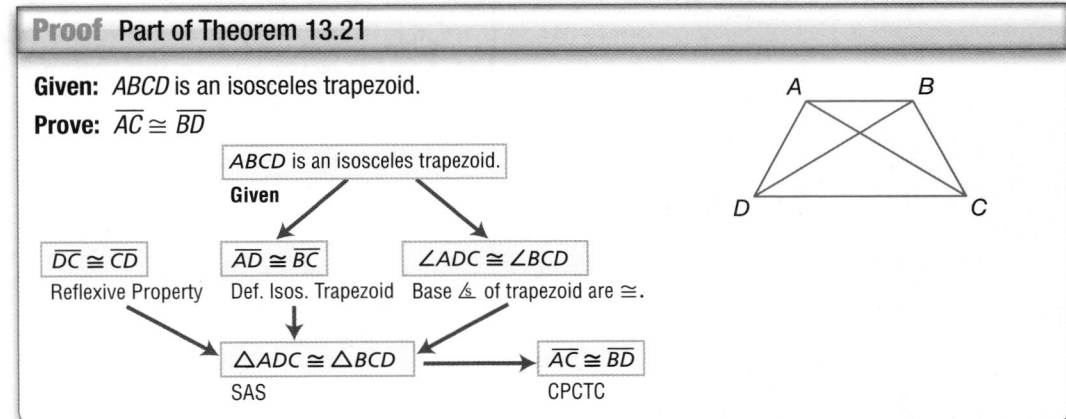

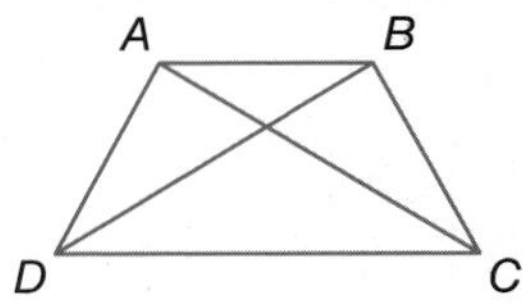

ABCD is an isosceles trapezoid. — **Given**

$\overline{DC} \cong \overline{CD}$ — Reflexive Property

$\overline{AD} \cong \overline{BC}$ — Def. Isos. Trapezoid

$\angle ADC \cong \angle BCD$ — Base ∠s of trapezoid are ≅.

$\triangle ADC \cong \triangle BCD$ — SAS

$\overline{AC} \cong \overline{BD}$ — CPCTC

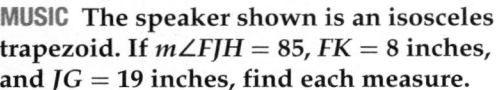

Real-World Example 1 Use Properties of Isosceles Trapezoids

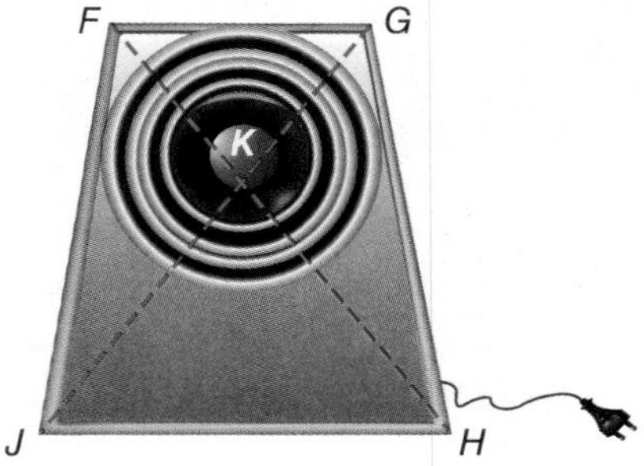

MUSIC The speaker shown is an isosceles trapezoid. If $m\angle FJH = 85$, $FK = 8$ inches, and $JG = 19$ inches, find each measure.

a. $m\angle FGH$

Since $FGHJ$ is an isosceles trapezoid, $\angle FJH$ and $\angle GHJ$ are congruent base angles. So, $m\angle GHJ = m\angle FJH = 85$.

Since $FGHJ$ is a trapezoid, $\overline{FG} \parallel \overline{JH}$.

$m\angle FGH + m\angle GHJ = 180$	Consecutive Interior Angles Theorem
$m\angle FGH + 85 = 180$	Substitution
$m\angle FGH = 95$	Subtract 85 from each side.

b. KH

Since $FGHJ$ is an isosceles trapezoid, diagonals $\overline{FH}$ and $\overline{JG}$ are congruent.

$FH = JG$	Definition of congruent
$FK + KH = JG$	Segment Addition
$8 + KH = 19$	Substitution
$KH = 11$ cm	Subtract 8 from each side.

Real-WorldLink

Speakers are amplifiers that intensify sound waves so that they are audible to the unaided ear. Amplifiers exist in devices such as televisions, stereos, and computers.

Source: How Stuff Works

GuidedPractice

1. CAFETERIA TRAYS To save space at a square table, cafeteria trays often incorporate trapezoids into their design. If $WXYZ$ is an isosceles trapezoid and $m\angle YZW = 45$, $WV = 15$ centimeters, and $VY = 10$ centimeters, find each measure.

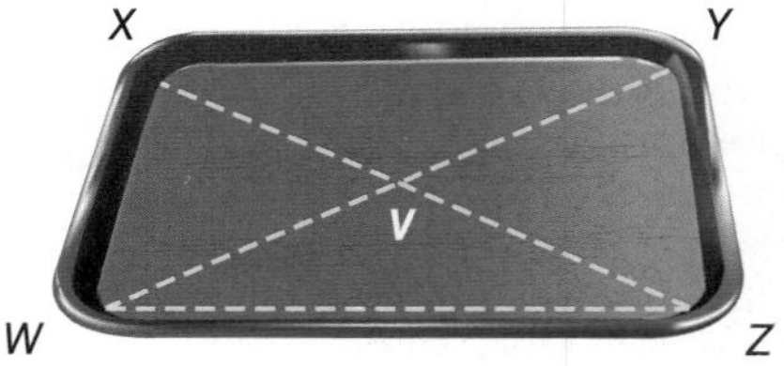

A. $m\angle XWZ$ **B.** $m\angle WXY$

C. XZ **D.** XV

StudyTip

Isosceles Trapezoids The base angles of a trapezoid are only congruent if the trapezoid is isosceles.

You can use coordinate geometry to determine whether a trapezoid is an isosceles trapezoid.

Example 2 Isosceles Trapezoids and Coordinate Geomerty

COORDINATE GEOMETRY Quadrilateral $ABCD$ has vertices $A(-3, 4)$, $B(2, 5)$, $C(3, 3)$, and $D(-1, 0)$. Show that $ABCD$ is a trapezoid and determine whether it is an isosceles trapezoid.

Graph and connect the vertices of $ABCD$.

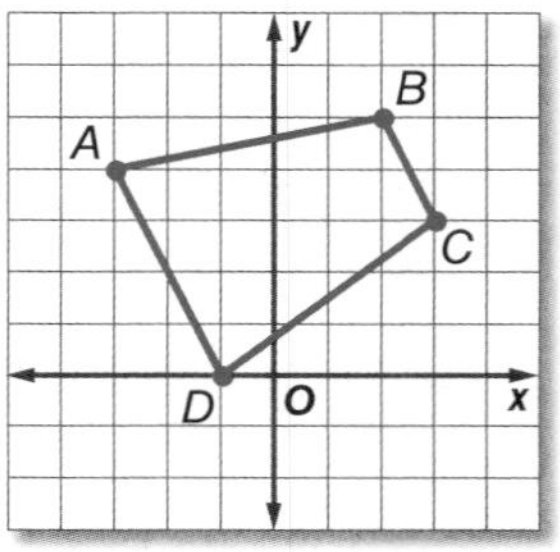

Step 1 Use the Slope Formula to compare the slopes of opposite sides $\overline{BC}$ and $\overline{AD}$ and of opposite sides $\overline{AB}$ and $\overline{DC}$. A quadrilateral is a trapezoid if exactly one pair of opposite sides are parallel.

Opposite sides $\overline{BC}$ and $\overline{AD}$:

slope of $\overline{BC} = 3 - \frac{5}{3} - 2 = -\frac{2}{1}$ or -2

slope of $\overline{AD} = \frac{0 - 4}{-1 - (-3)} = \frac{-4}{2}$ or -2

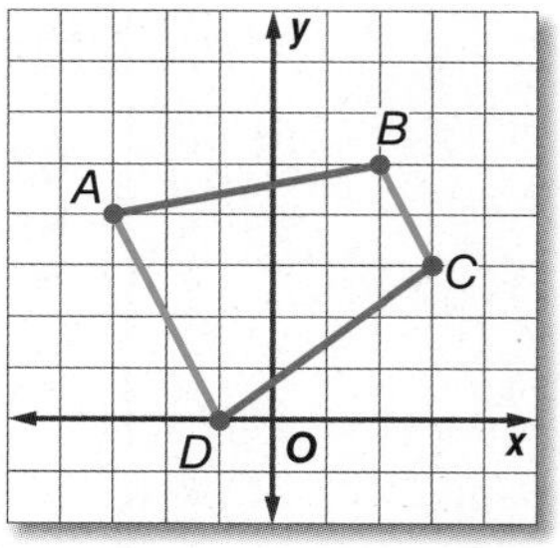

Since the slopes of $\overline{BC}$ and $\overline{AD}$ are equal, $\overline{BC} \parallel \overline{AD}$.

Opposite sides $\overline{AB}$ and $\overline{DC}$:

slope of $\overline{AB} = \frac{5 - 4}{2 - (-3)} = \frac{1}{5}$ slope of $\overline{DC} = \frac{0 - 3}{-1 - 3} = \frac{-3}{-4}$ or $\frac{3}{4}$

Since the slopes of $\overline{AB}$ and $\overline{DC}$ are *not* equal, $\overline{BC} \nparallel \overline{AD}$. Since quadrilateral *ABCD* has only one pair of opposite sides that are parallel, quadrilateral *ABCD* is a trapezoid.

ReadingMath

Symbols Recall that the symbol $\nparallel$ means *is not parallel to.*

Step 2 Use the Distance Formula to compare the lengths of legs $\overline{AB}$ and $\overline{DC}$. A trapezoid is isosceles if its legs are congruent.

$AB = \sqrt{(-3 - 2)^2 + (4 - 5)^2}$ or $\sqrt{26}$

$DC = \sqrt{(-1 - 3)^2 + (0 - 3)^2} = \sqrt{25}$ or 5

Since $AB \neq DC$, legs $\overline{AB}$ and $\overline{DC}$ are *not* congruent. Therefore, trapezoid *ABCD* is not isosceles.

GuidedPractice

2. Quadrilateral *QRST* has vertices $Q(-8, -4)$, $R(0, 8)$, $S(6, 8)$, and $T(-6, -10)$. Show that *QRST* is a trapezoid and determine whether *QRST* is an isosceles trapezoid.

ReadingMath

Midsegment A midsegment of a trapezoid can also be called a *median.*

The **midsegment of a trapezoid** is the segment that connects the midpoints of the legs of the trapezoid.

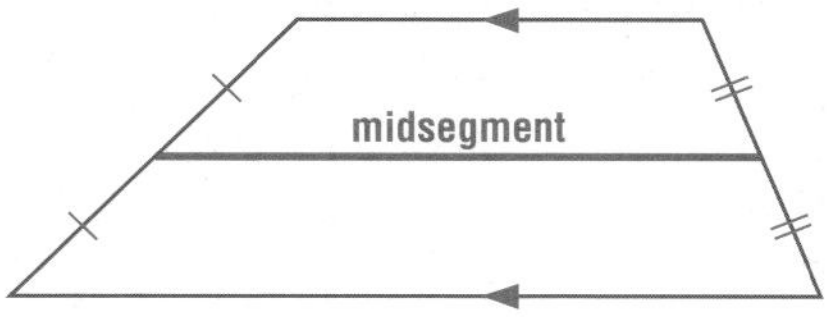

The theorem below relates the midsegment and the bases of a trapezoid.

Theorem 13.22 Trapezoid Midsegment Theorem

The midsegment of a trapezoid is parallel to each base and its measure is one half the sum of the lengths of the bases.

Example If $\overline{BE}$ is the midsegment of trapezoid *ACDF*, then $\overline{AF} \parallel \overline{BE}$, $\overline{CD} \parallel \overline{BE}$, and $BE = \frac{1}{2}(AF + CD)$.

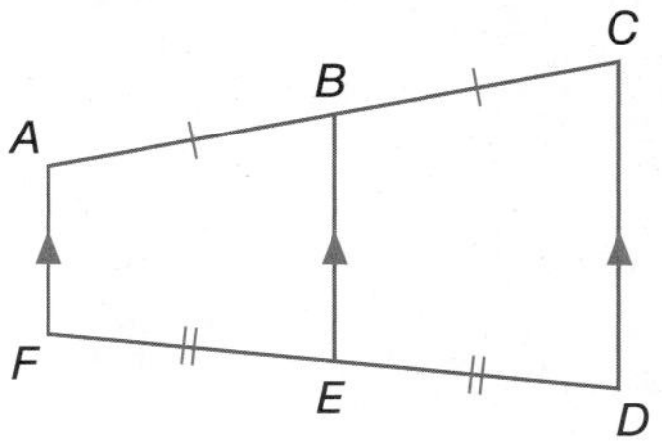

You will prove Theorem 13.22 in Exercise 33.

Standardized Test Example 3 Midsegment of a Trapezoid

GRIDDED RESPONSE **In the figure, $\overline{LH}$ is the midsegment of trapezoid *FGJK*. What is the value of *x*?**

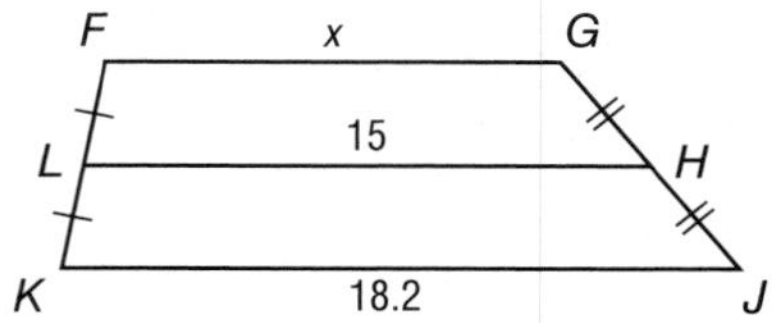

Note: The figure is not drawn to scale.

Read the Test Item

You are given the measure of the midsegment of a trapezoid and the measure of one of its bases. You are asked to find the measure of the other base.

Solve the Test Item

$LH = \frac{1}{2}(FG + KJ)$	Trapezoid Midsegment Theorem
$5 = \frac{1}{2}(x + 18.2)$	Substitution
$30 = x + 18.2$	Multiply each side by 2.
$11.8 = x$	Subtract 18.2 from each side.

Grid In Your Answer

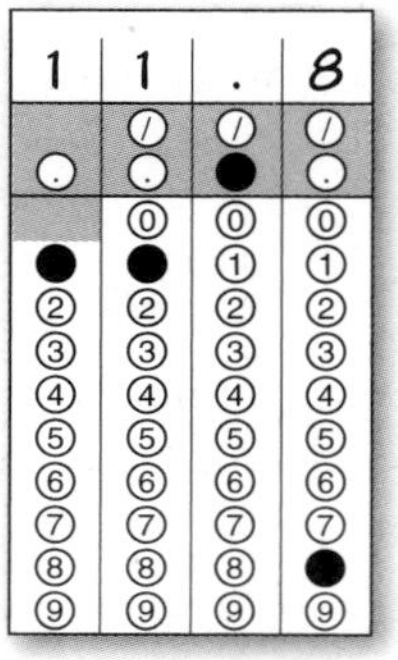

- You can align the numerical answer by placing the first digit in the left answer box or by putting the last digit in the right answer box.
- Do not leave blank boxes in the middle of an answer.
- Fill in **one** bubble for each filled answer box. Do not fill more than one bubble for an answer box. Do not fill in a bubble for blank answer boxes.

Test-Taking Tip

Gridded Responses Rational answers can often be gridded in more than one way. An answer such as $\frac{8}{5}$ could be gridded as 8/5 or 1.6, but not as 1 3/5.

Guided Practice

3. GRIDDED RESPONSE Trapezoid *ABCD* is shown below. If $\overline{FG}$ is parallel to $\overline{AD}$, what is the *x*-coordinate of point *G*?

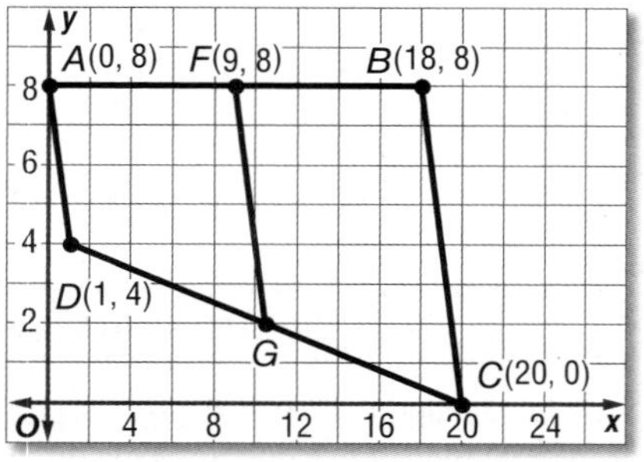

2 Properties of Kites

A **kite** is a quadrilateral with exactly two pairs of consecutive congruent sides. Unlike a parallelogram, the opposite sides of a kite are not congruent or parallel.

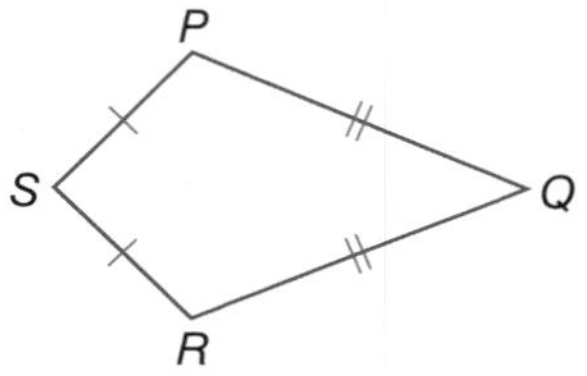

StudyTip

Kites The congruent angles of a kite are included by the non-congruent adjacent sides.

Theorems Kites

13.23 If a quadrilateral is a kite, then its diagonals are perpendicular.

Example If quadrilateral $ABCD$ is a kite, then $\overline{AC} \perp \overline{BD}$.

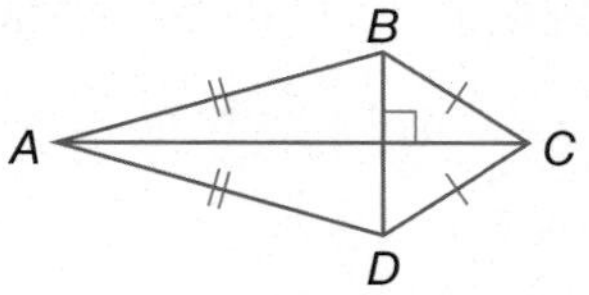

13.24 If a quadrilateral is a kite, then exactly one pair of opposite angles is congruent.

Example If quadrilateral $JKLM$ is a kite, $\overline{JK} \cong \overline{KL}$, and $\overline{JM} \cong \overline{LM}$, then $\angle J \cong \angle L$ and $\angle K \not\cong \angle M$.

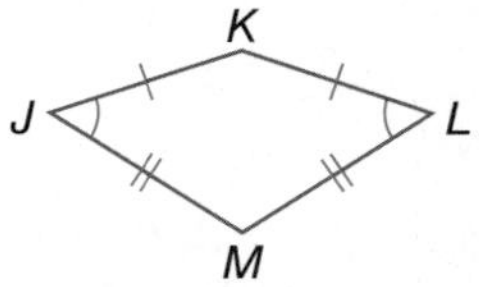

You will prove Theorems 13.23 and 13.24 in Exercises 31 and 32, respectively.

You can use the theorems above, the Pythagorean Theorem, and the Polygon Interior Angles Sum Theorem to find missing measures in kites.

Real-WorldLink

The fastest recorded speed of a kite is over 120 miles per hour. The record for the highest single kite flown is 12,471 feet.

Source: Borealis Kites

Example 4 Use Properties of Kites

a. If $FGHJ$ is a kite, find $m\angle GFJ$.

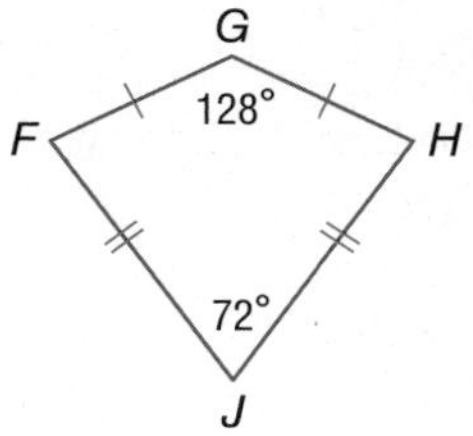

Since a kite can only have one pair of opposite congruent angles and $\angle G \not\cong \angle J$, then $\angle F \cong \angle H$. So, $m\angle F = m\angle H$. Write and solve an equation to find $m\angle F$.

$m\angle F + m\angle G + m\angle H + m\angle J = 360$	Polygon Interior Angles Sum Theorem
$m\angle F + 128 + m\angle F + 72 = 360$	Substitution
$2m\angle F + 200 = 360$	Simplify.
$2m\angle F = 160$	Subtract 200 from each side.
$m\angle F = 80$	Divide each side by 2.

b. If $WXYZ$ is a kite, find ZY.

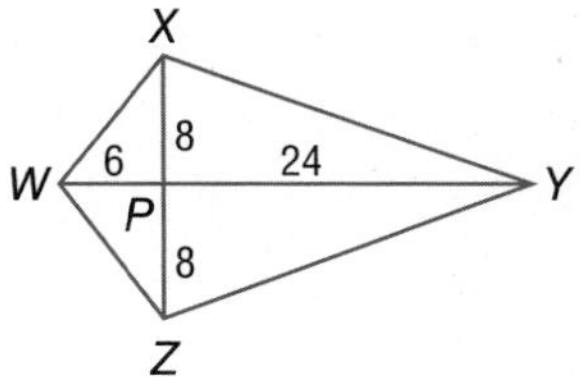

Since the diagonals of a kite are perpendicular, they divide $WXYZ$ into four right triangles. Use the Pythagorean Theorem to find ZY, the length of the hypotenuse of right $\triangle YPZ$.

$PZ^2 + PY^2 = ZY^2$	Pythagorean Theorem
$8^2 + 24^2 = ZY^2$	Substitution
$640 = ZY^2$	Simplify.
$\sqrt{640} = ZY$	Take the square root of each side.
$8\sqrt{10} = ZY$	Simplify.

GuidedPractice

4A. If $m\angle BAD = 38$ and $m\angle BCD = 50$, find $m\angle ADC$.

4B. If $BT = 5$ and $TC = 8$, find CD.

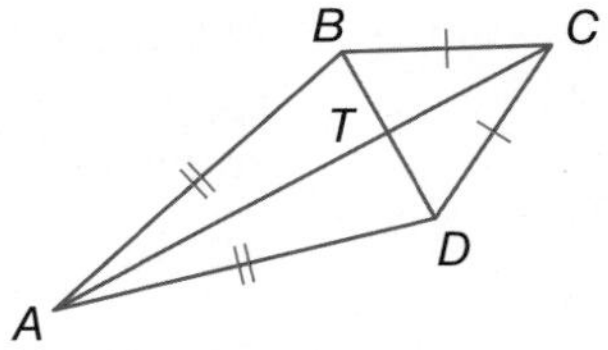

Guy Grenier/Masterfile

Check Your Understanding

Example 1 **Find each measure.**

1. $m\angle C$

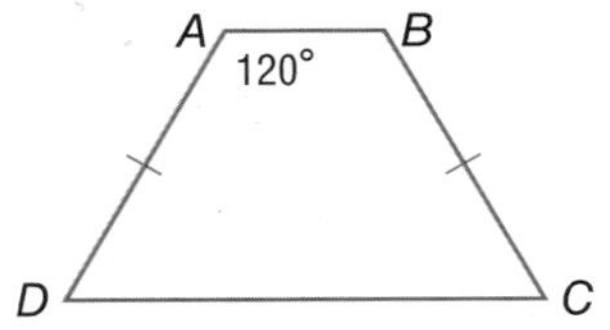

2. EJ, if $HF = 50$ and $JG = 10$

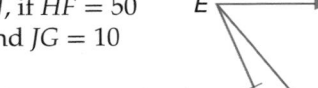

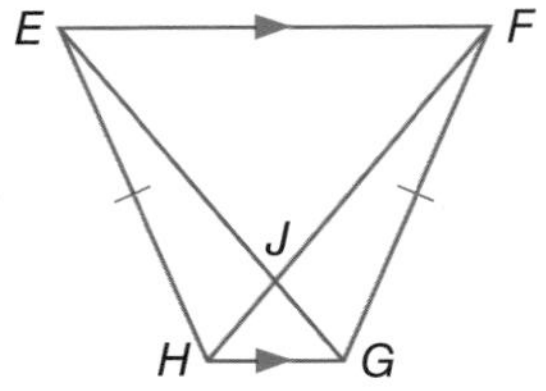

Example 2 **COORDINATE GEOMETRY** **Quadrilateral *JKLM* has vertices *J*(3, 10), *K*(2, 6), *L*(11, 6), and *M*(8, 10).**

3. Verify that $JKLM$ is a trapezoid.

4. Determine whether $JKLM$ is an isosceles trapezoid. Explain.

Example 3

5. **GRIDDED RESPONSE** In the figure at the right, $\overline{ST}$ is the midsegment of trapezoid $NPQR$. Determine the value of x.

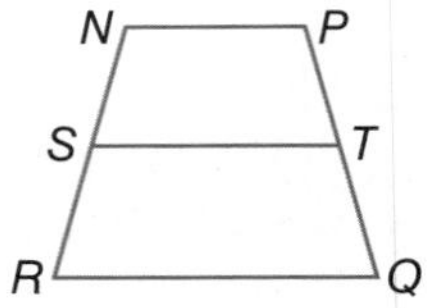

Example 4 CCSS **If *VWXY* is a kite, find each measure.**

6. $m\angle W$

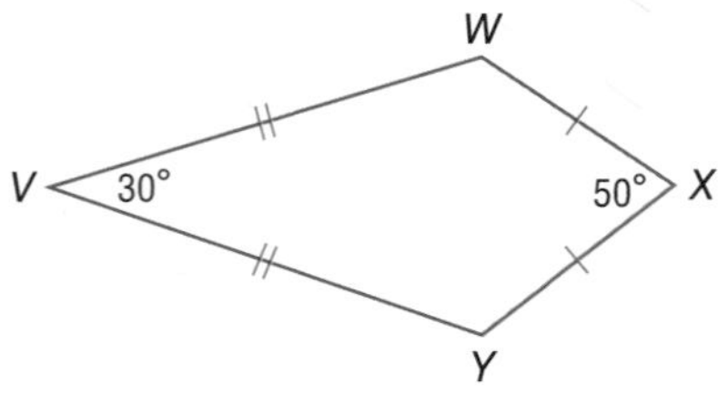

7. YX

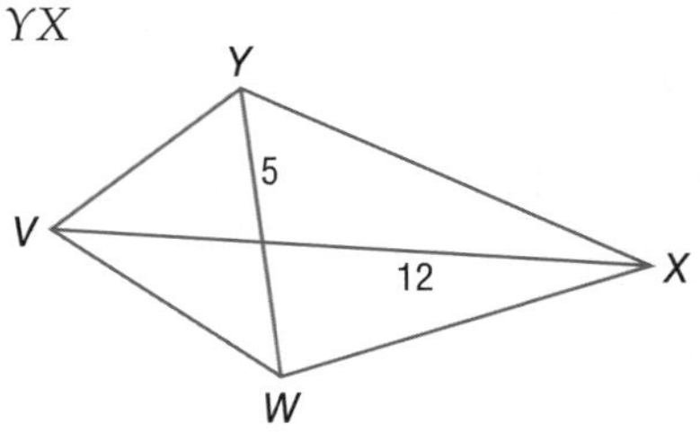

Practice and Problem Solving

Example 1 **Find each measure.**

8. AC, if $BE = 12$ and $ED = 24$

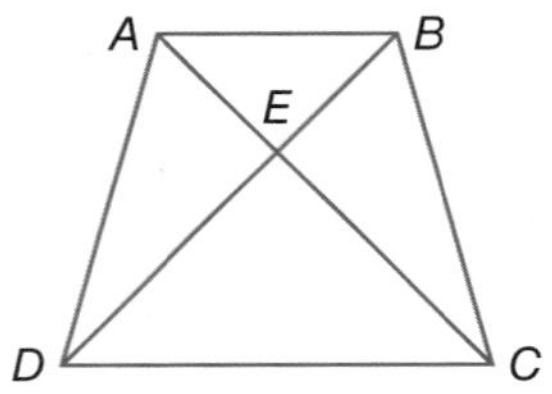

9. $m\angle F$

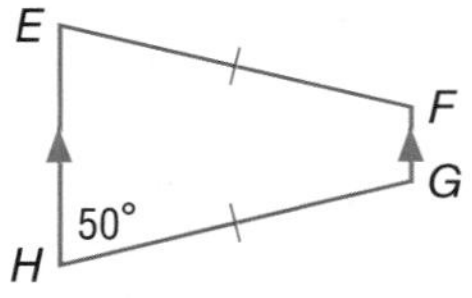

10. JC, if $MK = 18$ and $CL = 12$

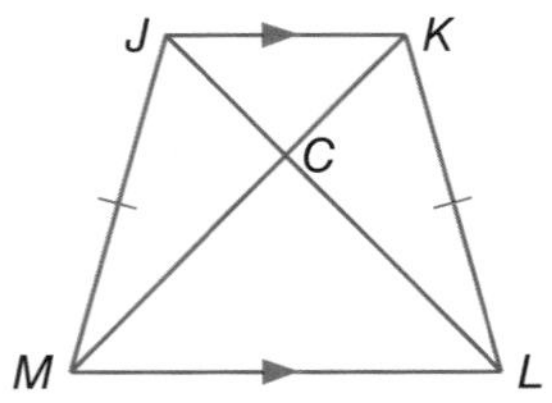

11. $m\angle P$

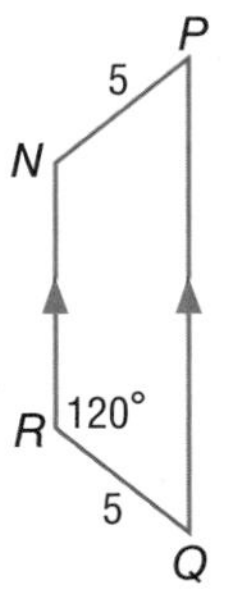

Example 2 **COORDINATE GEOMETRY** **For each quadrilateral with the given vertices, verify that the quadrilateral is a trapezoid and determine whether the figure is an isosceles trapezoid.**

12. $A(-6, -3), B(-4, 1), C(1, 1), D(3, -3)$

13. $E(0, 3), F(-4, -1), G(-3, -8), H(7, 2)$

14. $J(0, 4), K(3, 7), L(8, 6), M(10, 2)$

15. $N(2, 0), P(12, 8), Q(7, 9), R(2, 5)$

Example 3 **For trapezoid *ABCD*, *E* and *F* are midpoints of the legs.**

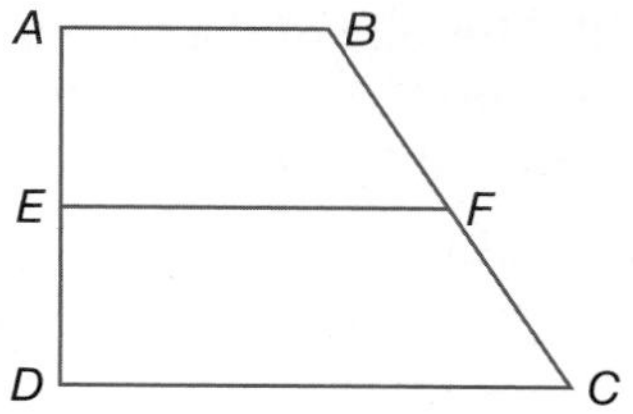

16. If $AB = 10$ and $CD = 14$, find EF.

17. If $EF = 7$ and $CD = 10$, find AB.

18. If $AB = 5$ and $EF = 10$, find DC.

19. If $EF = 13$ and $DC = 14$, find AB.

20. If $AB = 12$ and $EF = 14$, find DC.

21. If $AB = 7$ and $DC = 33$, find EF.

CCSS **If *MNPQ* is a kite, find each measure.**

22. QN

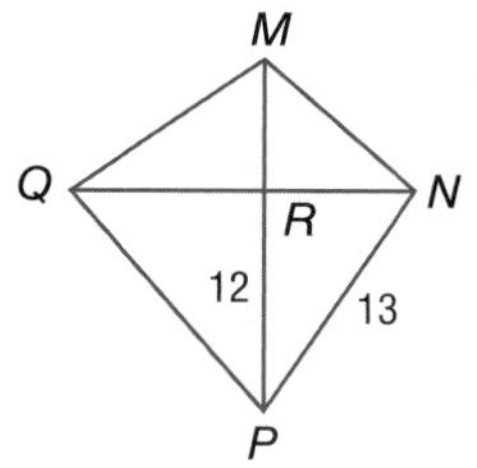

23. $m\angle P$

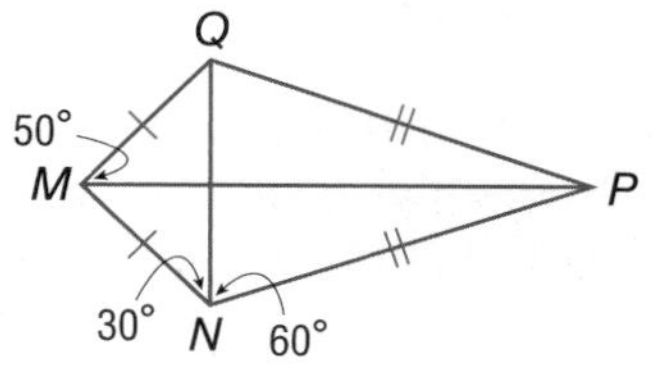

24. NR

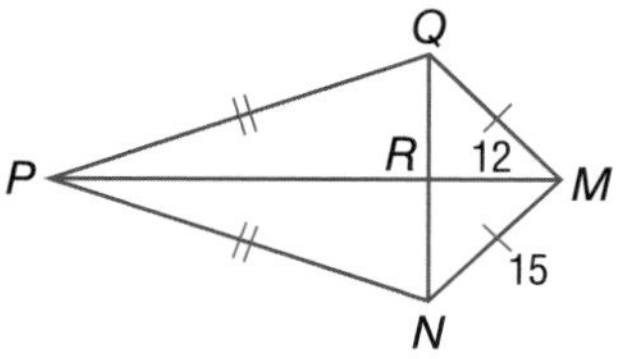

25. $m\angle Q$

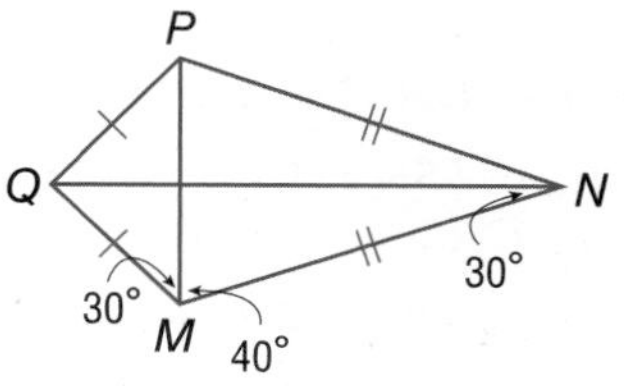

PROOF **Write a paragraph proof for each theorem.**

26. Theorem 13.19

27. Theorem 13.20

28. Theorem 13.21

29. Theorem 13.23

30. Theorem 13.24

31. **PROOF** Write a coordinate proof for Theorem 13.22.

32. **COORDINATE GEOMETRY** Refer to quadrilateral *RSTV*.

a. Determine whether the figure is a trapezoid. If so, is it isosceles? Explain.

b. Is the origin contained in the midsegment? Justify your answer.

c. Find the length of the midsegment.

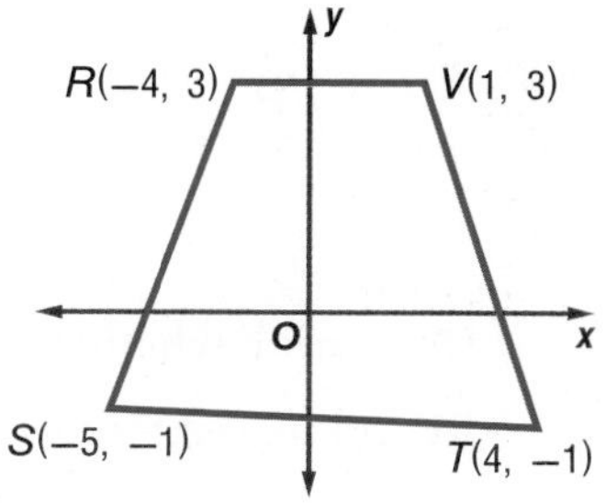

ALGEBRA ***WXYZ* is a trapezoid.**

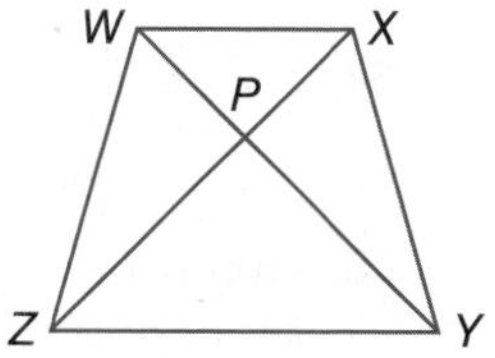

33. If $m\angle WZY = 4x + 10$, and $m\angle XYZ = 5x - 5$, find the value of x so that $WXYZ$ is isosceles.

34. If $WY = 4x + 1$ and $XZ = 5x - 3$, find the value of x so that $WXYZ$ is isosceles.

FOOD **The side of the take-out container shown is an isosceles trapezoid. If $AE = 3$ inches, $ED = 2$ inches, and $m\angle ABD = 75$, find each measure.**

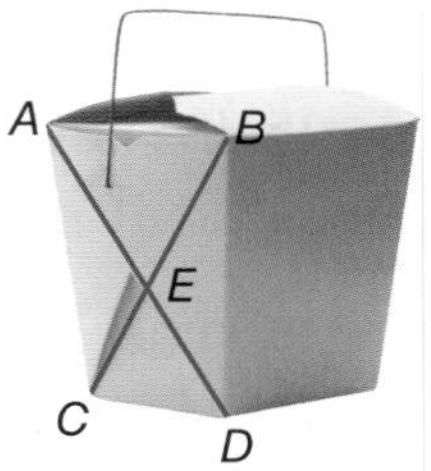

35. $m\angle BAC$

36. AD

37. $m\angle BDC$

38. BC

ALGEBRA **For trapezoid $ABCD$, E and F are the midpoints of the legs.**

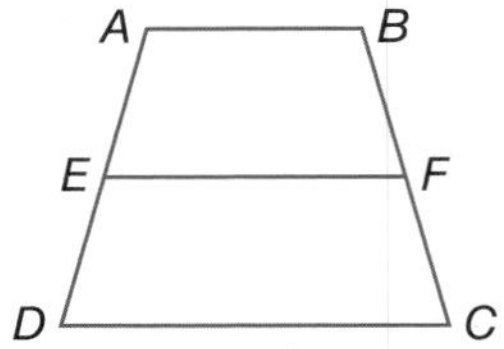

39. If $AB = x + 2$, $DC = 2x + 1$, and $EF = 9$, find x.

40. If $AB = 6$, $EF = 3x$, and $DC = 5x - 3$, find x.

41. If $AB = 3x - 6$, $EF = 4x - 8$, and $DC = 20$, find EF.

42. If $AB = x + 4$, $EF = 2x - 3$, and $DC = 2x - 1$, find the value of x.

MUSIC **The zither is a musical instrument that is often in the shape of an isosceles trapezoid. In the diagram shown, $LN = 24$ inches and $QP = 10$ inches, $m\angle LPQ = 65$, find each measure.**

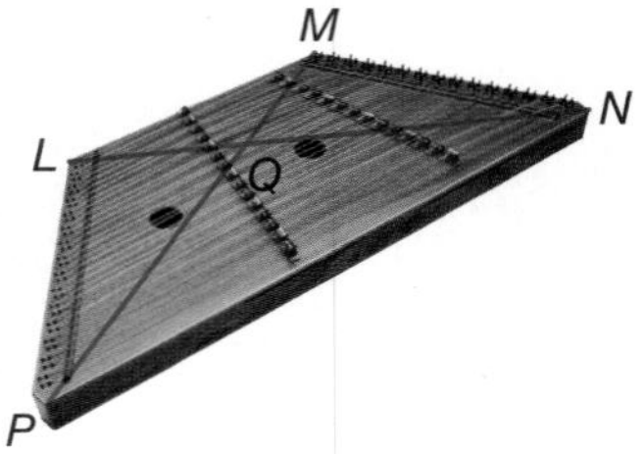

43. $m\angle MLP$

44. LQ

45. $m\angle MNP$

46. MP

ALGEBRA **$QRST$ is a kite.**

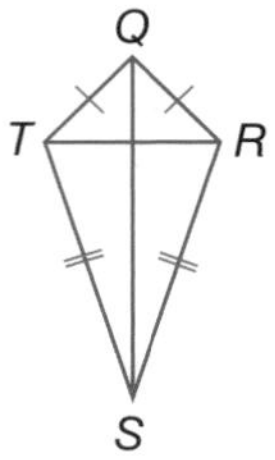

47. If $m\angle TSR = 40$, $m\angle TQR = 6x$, and $m\angle QRS = 7x + 10$, find $m\angle QTS$.

48. If $m\angle TQR = 60$, $m\angle RST = x - 3$, and $m\angle QTS = 7x$, find $m\angle QRS$.

CCSS **PROOF** **Write a two-column proof.**

49. Given: $\angle BAD \cong \angle EDA$, $\triangle AED \cong \triangle BCD$

Prove: $ABCE$ is an isosceles trapezoid.

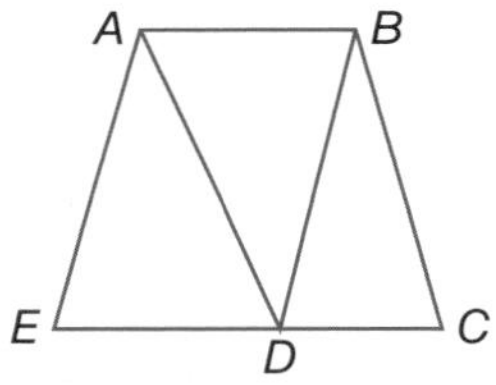

50. Given: $PMNO$ is a trapezoid

Prove: All angles of $\triangle LPM$ are congruent to the angles of $\triangle LON$

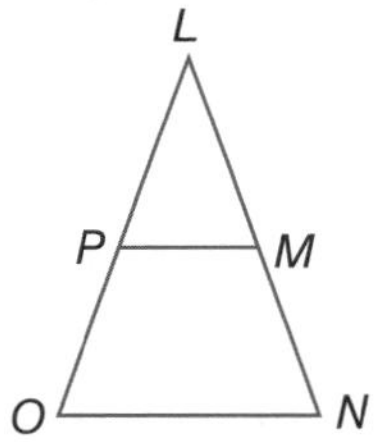

Determine whether each statement is *always*, *sometimes*, or *never* true.

51. A kite is a trapezoid.

52. Adjacent angles of a trapezoid are supplementary.

53. A quadrilateral is a parallelogram.

54. A square is a rectangle.

55. The diagonals of a kite are perpendicular.

56. **PROOF** Given $ABCD$ is a kite, write a paragraph proof to prove $\triangle ADC \cong \triangle ABC$.

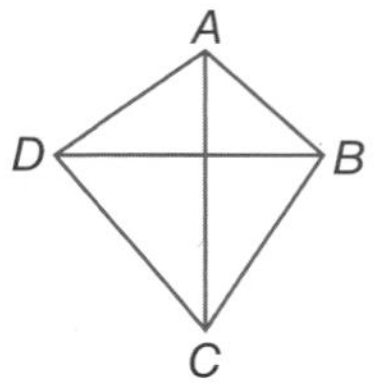

57. **TABLE** Fill in the following table. Listed in the left column are different types of quadrilaterals. In the right column fill in the other groups that the quadrilateral in the same row also fits into. Write none if it fits in no other categories.

Quadrilateral	Also is...
Example: Rectangle	Parallelogram
Rhombus	
Square	
Isosceles Trapezoid	
Trapezoid	
Kite	

COORDINATE GEOMETRY **Determine whether each figure is a trapezoid, a parallelogram, a square, a rhombus, or a quadrilateral. Choose the most specific term. Explain.**

58. $L(1, 1), M(0, -5), N(7, 0), P(6, -6)$

59. $A(2, 7), B(5, 9), C(6, 6), D(3, 4)$

60. **MULTIPLE REPRESENTATIONS** In this problem you will explore proportions in isosceles trapezoids.

a. **Geometric** Construct three different isosceles trapezoids. Label each one $ABCD$. Draw the diagonals and label the intersection R.

b. **Tabular** Copy the following table. Use a ruler to complete the table.

Trapezoid	AR	RC	$\frac{RC}{AR}$	DC	AB	$\frac{DC}{AB}$
Trapezoid 1						
Trapezoid 2						
Trapezoid 3						

c. **Verbal** Make a conjecture about the proportions of the diagonals and the bases of a perpendicular.

PROOF **Write a coordinate proof of each statement.**

61. The midsegments of a trapezoid is parallel to the bases.

62. The diagonals of a kite are perpendicular.

Given: $ABCD$ is a kite

Prove: AC is perpendicular to BD.

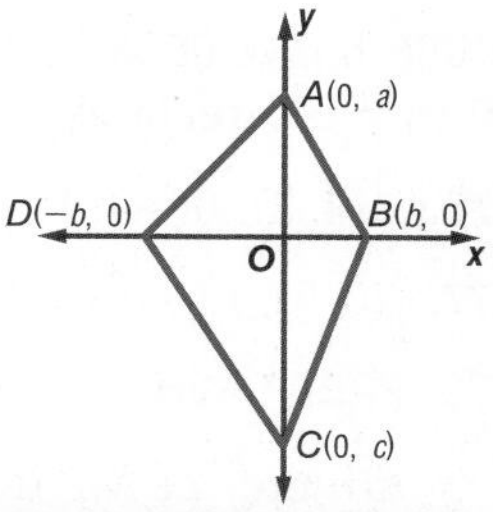

H.O.T. Problems Use Higher-Order Thinking Skills

63. **REASONING** Is it sometimes, always, or never true that a kite is also a rectangle?

64. **OPEN ENDED** Sketch two non-congruent kites $ABCD$ and $LMNP$ in which $\overline{AB} \cong \overline{LM}$.

65. **CCSS ERROR ANALYSIS** Robi and Matt are trying to determine $m\angle F$ in the trapezoid shown. Is either of them correct? Explain.

Robi	Matt
$m\angle F = 45$	$m\angle F = 135$

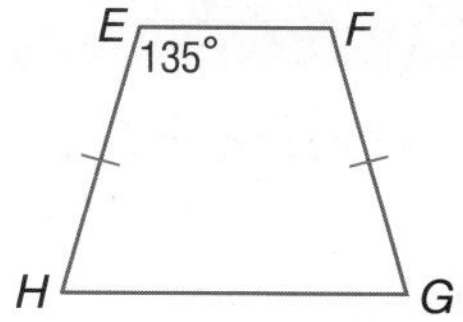

66. **CHALLENGE** $\triangle AED$, $\triangle ADB$, $\triangle DBC$ are all equilateral triangles. Prove $ABCE$ is an isosceles triangle.

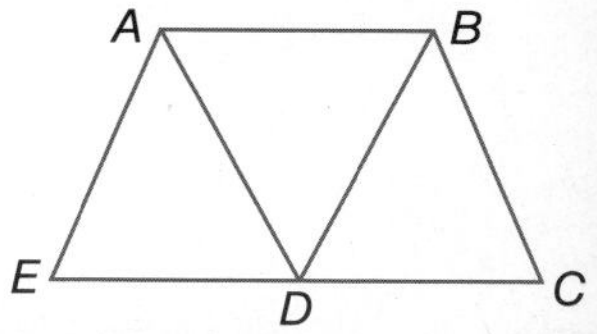

67. **WRITING IN MATH** Compare and contrast the properties of a parallelogram and the properties of a trapezoid.

Standardized Test Practice

68. ALGEBRA All of the items on a breakfast menu cost the same whether ordered with something else or alone. Two pancakes and one order of bacon costs \$4.92. If two orders of bacon cost \$3.96, what does one pancake cost?

A \$0.96 **C** \$1.98
B \$1.47 **D** \$2.94

69. GRIDDED RESPONSE If quadrilateral *ABCD* is a kite, what is $m\angle C$?

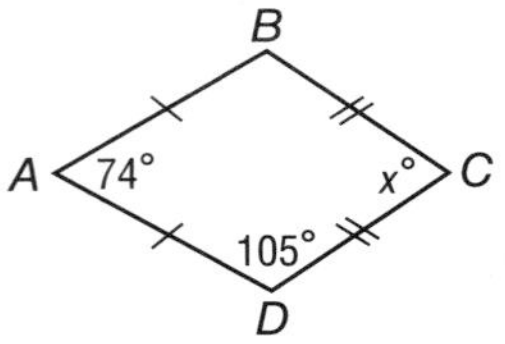

70. Which figure can serve as a counterexample to the conjecture below?

If the diagonals of a quadrilateral are congruent, then the quadrilateral is a rectangle.

F square **H** parallelogram
G rhombus **J** isosceles trapezoid

71. SAT/ACT In the figure below, what is the value of *x*?

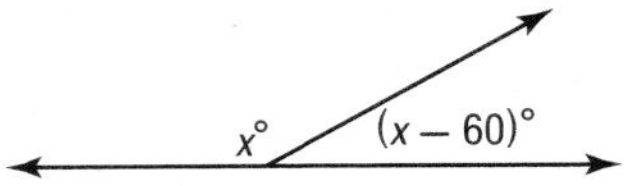

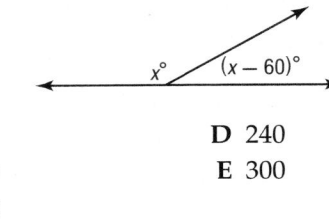

A 60 **D** 240
B 120 **E** 300
C 180

Spiral Review

ALGEBRA Quadrilateral *DFGH* is a rhombus. Find each value or measure. (Lesson 13-5)

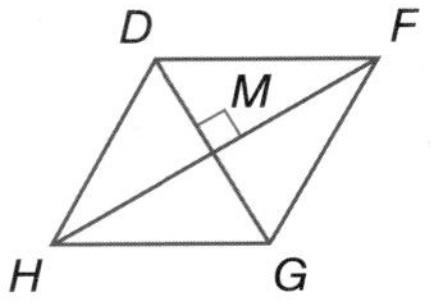

72. If $m\angle FGH = 118$, find $m\angle MHG$.

73. If $DM = 4x - 3$ and $MG = x + 6$, find DG.

74. If $DF = 10$, find FG.

75. If $HM = 12$ and $HD = 15$, find MG.

COORDINATE GEOMETRY Graph each quadrilateral with the given vertices. Determine whether the figure is a rectangle. Justify your answer using the indicated formula. (Lesson 13-4)

76. $A(4, 2)$, $B(-4, 1)$, $C(-3, -5)$, $D(5, -4)$; Distance Formula

77. $J(0, 7)$, $K(-8, 6)$, $L(-7, 0)$, $M(1, 1)$; Slope Formula

78. PROOF Write a two-column proof. (Lesson 12-5)

Given: $\angle CMF \cong \angle EMF$,
$\angle CFM \cong \angle EFM$
Prove: $\triangle DMC \cong \triangle DME$

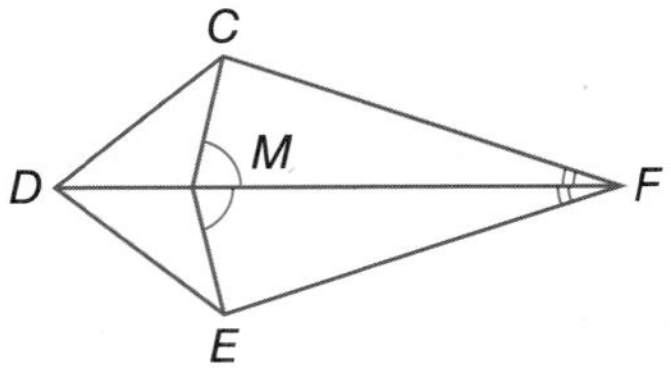

Skills Review

Write an expression for the slope of each segment given the coordinates and endpoints.

79. $(x, 4y)$, $(-x, 4y)$

80. $(-x, 5x)$, $(0, 6x)$

81. (y, x), (y, y)

CHAPTER 13

Study Guide and Review

Study Guide

KeyConcepts

Properties of Parallelograms (Lessons 13-1 and 13-2)

- Opposite sides are congruent and parallel.
- Opposite angles are congruent.
- Consecutive angles are supplementary.
- If a parallelogram has one right angle, it has four right angles.
- Diagonals bisect each other.

Properties of Rectangles, Rhombi, Squares, and Trapezoids (Lesson 13-3 through 13-5)

- A rectangle has all the properties of a parallelogram. Diagonals are congruent and bisect each other. All four angles are right angles.
- A rhombus has all the properties of a parallelogram. All sides are congruent. Diagonals are perpendicular. Each diagonal bisects a pair of opposite angles.
- A square has all the properties of a parallelogram, a rectangle, and a rhombus.
- In an isosceles trapezoid, both pairs of base angles are congruent and the diagonals are congruent.

FOLDABLES® StudyOrganizer

Be sure the Key Concepts are noted in your Foldable.

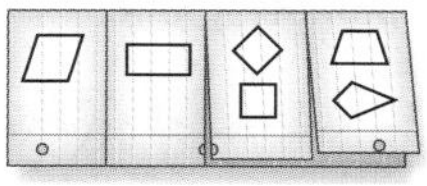

KeyVocabulary

base (p. 839)
base angle (p. 839)
isosceles trapezoid (p. 839)
kite (p. 842)
legs (p. 839)
midsegment of a trapezoid (p. 841)
parallelogram (p. 803)
rectangle (p. 823)
rhombus (p. 830)
square (p. 831)
trapezoid (p. 839)

VocabularyCheck

State whether each sentence is *true* or *false*. If *false*, replace the underlined word or phrase to make a true sentence.

1. <u>No</u> angles in an isosceles trapezoid are congruent.
2. If a parallelogram is a <u>rectangle</u>, then the diagonals are congruent.
3. A <u>midsegment of a trapezoid</u> is a segment that connects any two nonconsecutive vertices.
4. The base of a trapezoid is one of the <u>parallel</u> sides.
5. The diagonals of a <u>rhombus</u> are perpendicular.
6. A rectangle <u>is not always</u> a parallelogram.
7. A quadrilateral with only one set of parallel sides is a <u>parallelogram</u>.
8. A rectangle that is also a rhombus is a <u>square</u>.
9. The leg of a trapezoid is one of the <u>parallel</u> sides.

Lesson-by-Lesson Review

13-1 Parallelograms

Use ▱*ABCD* to find each measure.

10. $m\angle ADC$

11. AD

12. AB

13. $m\angle BCD$

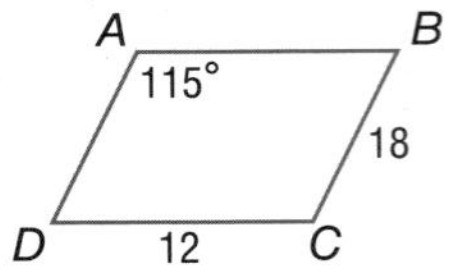

ALGEBRA Find the value of each variable in each parallelogram.

14.

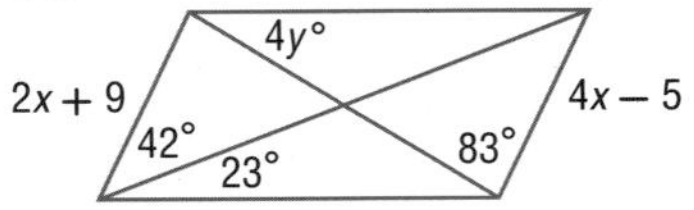

15.

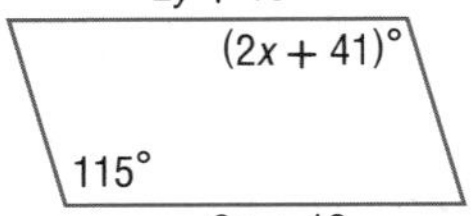

16. DESIGN What type of information is needed to determine whether the shapes that make up the stained glass window below are parallelograms?

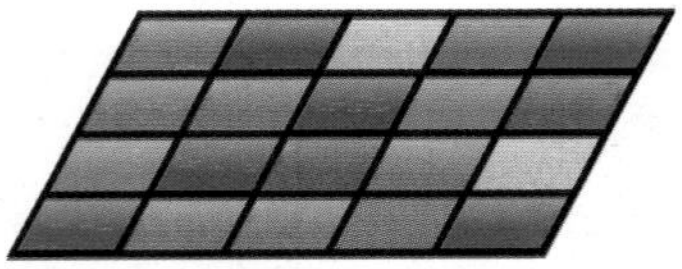

Example 1

ALGEBRA If *KLMN* is a parallelogram, find the value of the indicated variable.

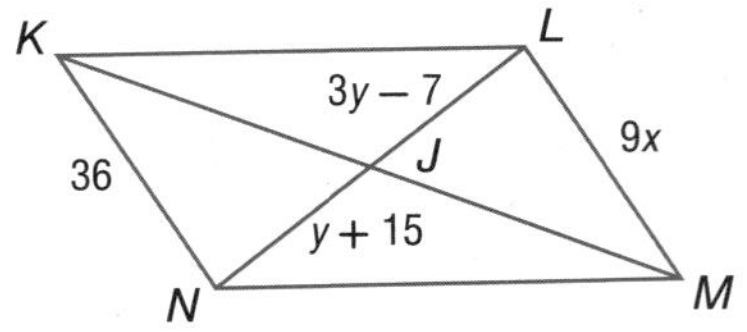

a. x

$\overline{KN} \cong \overline{LM}$	Opp. sides of a ▱ are ≅.
$KN = LM$	Definition of congruence
$36 = 9x$	Substitution
$4 = x$	Divide.

b. y

$\overline{NJ} \cong \overline{JL}$	Diag. of a ▱ bisect each other.
$NJ = JL$	Definition of congruence
$y + 15 = 3y - 7$	Substitution
$-2y = -22$	Subtract.
$y = 11$	Divide.

13-2 Tests for Parallelograms

Determine whether each quadrilateral is a parallelogram. Justify your answer.

17.

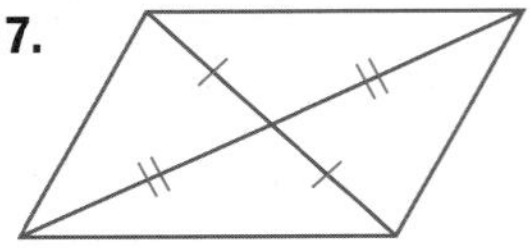

18.

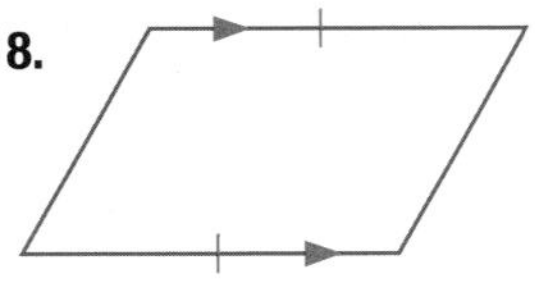

19. PROOF Write a two-column proof.

Given: ▱*ABCD*, $\overline{AE} \cong \overline{CF}$

Prove: Quadrilateral *EBFD* is a parallelogram.

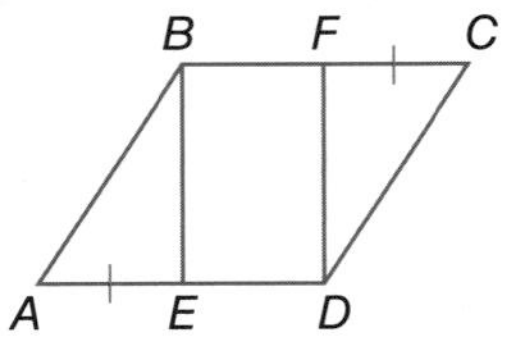

ALGEBRA Find x and y so that the quadrilateral is a parallelogram.

20.

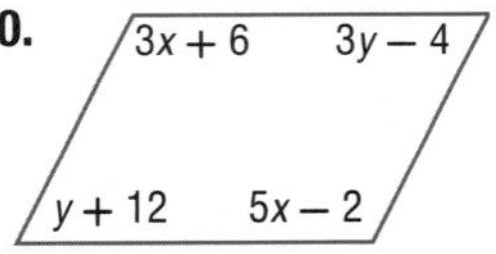

21.

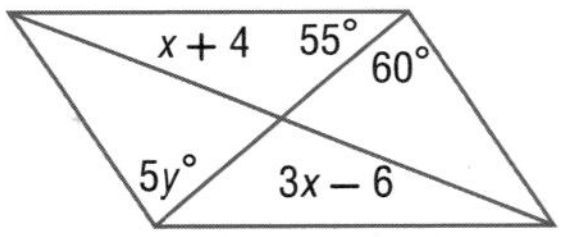

Example 2

If $TP = 4x + 2$, $QP = 2y - 6$, $PS = 5y - 12$, and $PR = 6x - 4$, find x and y so that the quadrilateral is a parallelogram.

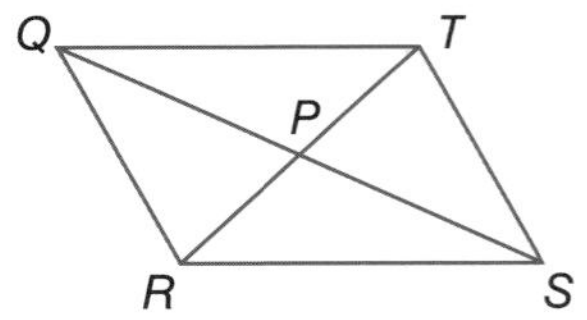

Find x such that $\overline{TP} \cong \overline{PR}$ and y such that $\overline{QP} \cong \overline{PS}$.

$TP = PR$	Definition of ≅
$4x + 2 = 6x - 4$	Substitution
$-2x = -6$	Subtract.
$x = 3$	Divide.

$QP = PS$	Definition of ≅
$2y - 6 = 5y - 12$	Substitution
$-3y = -6$	Subtract.
$y = 2$	Divide.

13-3 Rectangles

22. PARKING The lines of the parking space shown below are parallel. How wide is the space (in inches)?

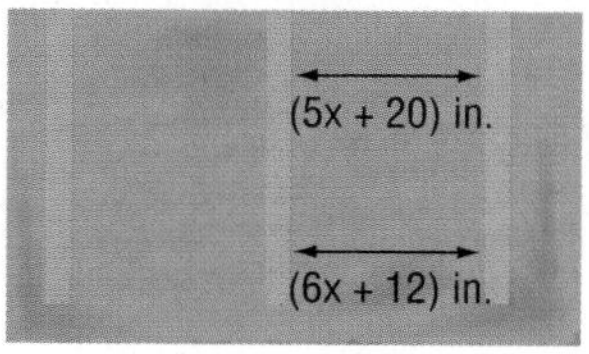

ALGEBRA Quadrilateral *EFGH* is a rectangle.

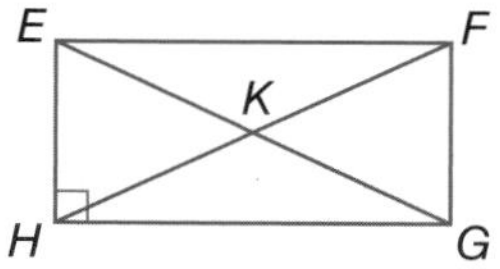

23. If $m\angle FEG = 57$, find $m\angle GEH$.

24. If $m\angle HGE = 13$, find $m\angle FGE$.

25. If $FK = 32$ feet, find EG.

26. Find $m\angle HEF + m\angle EFG$.

27. If $EF = 4x - 6$ and $HG = x + 3$, find EF.

Example 3

ALGEBRA Quadrilateral *ABCD* is a rectangle. If $m\angle ADB = 4x + 8$ and $m\angle DBA = 6x + 12$, find x.

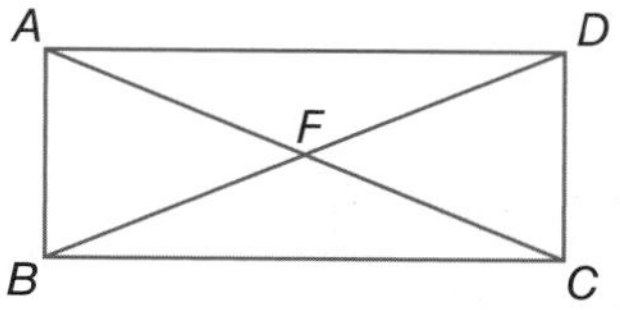

ABCD is a rectangle, so $m\angle ABC = 90$. Since the opposite sides of a rectangle are parallel, and the alternate interior angles of parallel lines are congruent, $\angle DBC \cong \angle ADB$ and $m\angle DBC = m\angle ADB$.

$m\angle DBC + m\angle DBA = 90$	Angle Addition
$m\angle ADB + m\angle DBA = 90$	Substitution
$4x + 8 + 6x + 12 = 90$	Substitution
$10x + 20 = 90$	Add.
$10x = 70$	Subtract.
$x = 7$	Divide.

13-4 Rhombi and Squares

ALGEBRA *ABCD* is a rhombus. If $EB = 9$, $AB = 12$ and $m\angle ABD = 55$, find each measure.

28. AE

29. $m\angle BDA$

30. CE

31. $m\angle ACB$

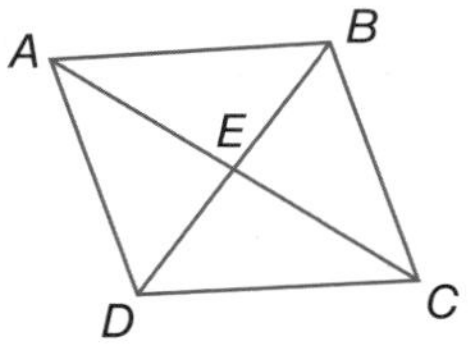

32. LOGOS A car company uses the symbol shown at the right for their logo. If the inside space of the logo is a rhombus, what is the length of FJ?

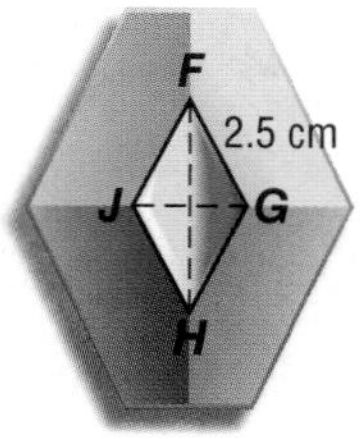

COORDINATE GEOMETRY Given each set of vertices, determine whether ▱*QRST* is a *rhombus*, a *rectangle*, or a *square*. List all that apply. Explain.

33. $Q(12, 0)$, $R(6, -6)$, $S(0, 0)$, $T(6, 6)$

34. $Q(-2, 4)$, $R(5, 6)$, $S(12, 4)$, $T(5, 2)$

Example 4

The diagonals of rhombus *QRST* intersect at *P*. Use the information to find each measure or value.

a. ALGEBRA If $QT = x + 7$ and $TS = 2x - 9$, find x.

$\overline{QT} \cong \overline{TS}$	Def. of rhombus
$QT = TS$	Def. of congruence
$x + 7 = 2x - 9$	Substitution
$-x = -16$	Subtract.
$x = 16$	Divide.

b. If $m\angle QTS = 76$, find $m\angle TSP$.

$\overline{TR}$ bisects $\angle QTS$. Therefore, $m\angle PTS = \frac{1}{2}m\angle QTS$. So $m\angle PTS = \frac{1}{2}(76)$ or 38. Since the diagonals of a rhombus are perpendicular, $m\angle TPS = 90$.

$m\angle PTS + m\angle TPS + m\angle TSP = 180$	△ Sum Thm.
$38 + 90 + m\angle TSP = 180$	Substitution
$128 + m\angle TSP = 180$	Add.
$m\angle TSP = 52$	Subtract.

CHAPTER 13 Study Guide and Review *Continued*

13-5 Trapezoids and Kites

Find each measure.

35. GH

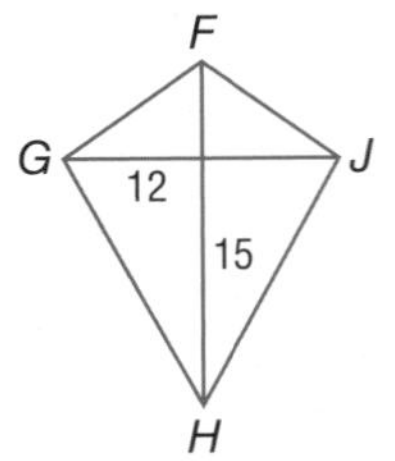

36. $m\angle Z$

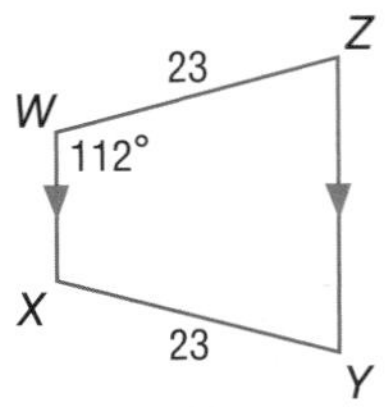

37. DESIGN Renee designed the square tile as an art project.

a. Describe a way to determine if the trapezoids in the design are isosceles.

b. If the perimeter of the tile is 48 inches and the perimeter of the red square is 16 inches, what is the perimeter of one of the trapezoids?

Example 5

If $QRST$ is a kite, find $m\angle RST$.

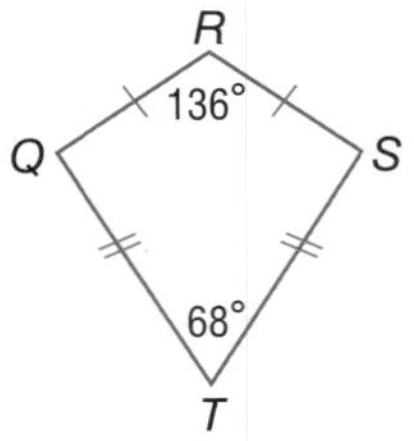

Since $\angle Q \cong \angle S$, $m\angle Q = m\angle S$. Write and solve an equation to find $m\angle S$.

$m\angle Q + m\angle R + m\angle S + m\angle T = 360$	Polygon Int. ∠s Sum Thm
$m\angle Q + 136 + m\angle S + 68 = 360$	Substitution
$2m\angle S + 204 = 360$	Simplify.
$2m\angle S = 156$	Subtract.
$m\angle S = 78$	Divide.

CHAPTER 13 Practice Test

1. **ART** Jen is making a frame to stretch a canvas over for a painting. She nailed four pieces of wood together at what she believes will be the four vertices of a square.

   a. How can she be sure that the canvas will be a square?

   b. If the canvas has the dimensions shown below, what are the missing measures?

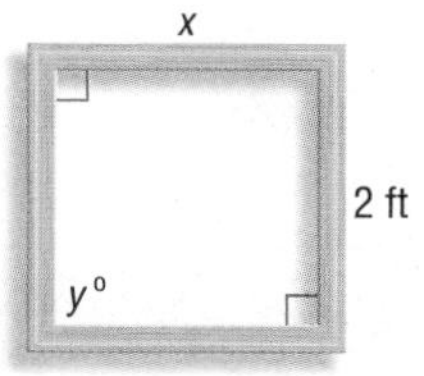

Quadrilateral *ABCD* is an isosceles trapezoid.

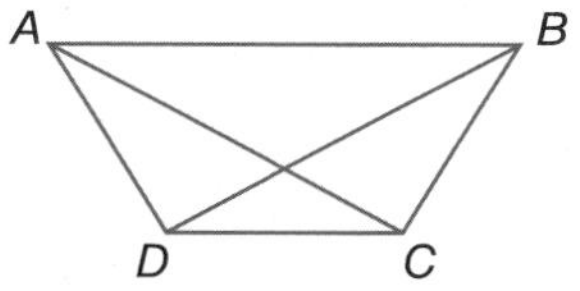

2. Which angle is congruent to $\angle C$?
3. Which side is parallel to $\overline{AB}$?
4. Which segment is congruent to $\overline{AC}$?

5. **MULTIPLE CHOICE** If *QRST* is a parallelogram, what is the value of x?

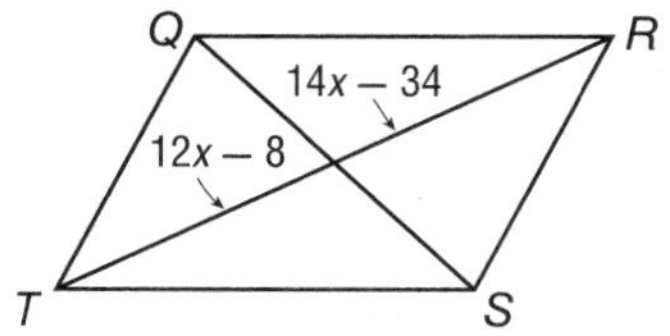

   A 11 **C** 13
   B 12 **D** 14

If *CDFG* is a kite, find each measure.

6. *GF*

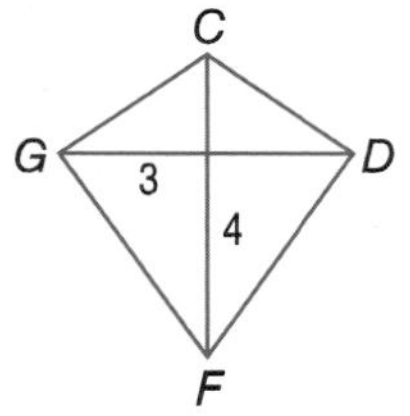

7. $m\angle D$

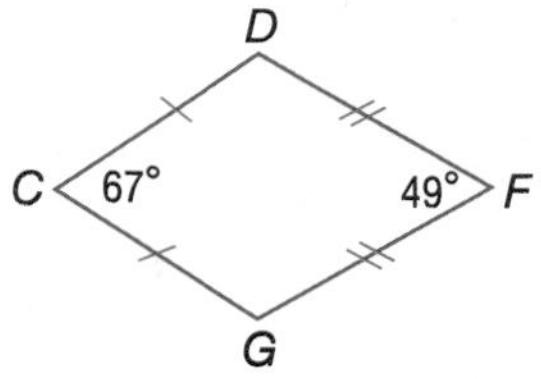

ALGEBRA Quadrilateral *MNOP* is a rhombus. Find each value or measure.

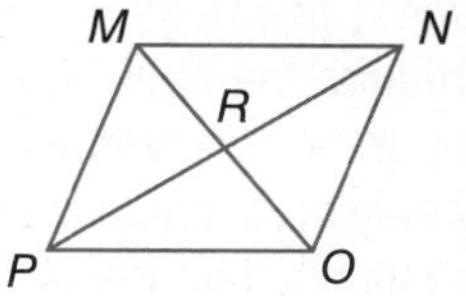

8. $m\angle MRN$
9. If $PR = 12$, find RN.
10. If $m\angle PON = 124$, find $m\angle POM$.

11. **CONSTRUCTION** The Smiths are building an addition to their house. Mrs. Smith is cutting an opening for a new window. If she measures to see that the opposite sides are congruent and that the diagonal measures are congruent, can Mrs. Smith be sure that the window opening is rectangular? Explain.

Use ▱*JKLM* to find each measure.

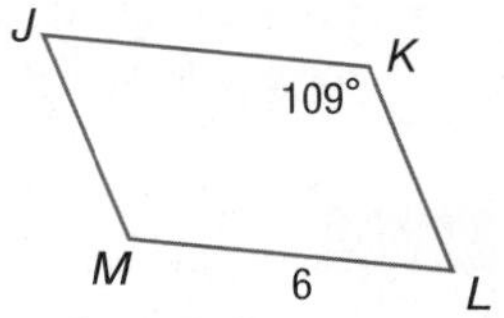

12. $m\angle JML$
13. JK
14. $m\angle KLM$

ALGEBRA Quadrilateral *DEFG* is a rectangle.

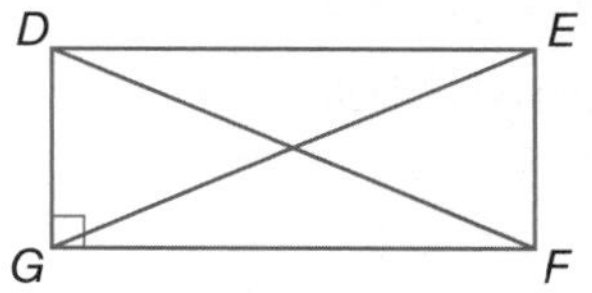

15. If $DF = 2(x + 5) - 7$ and $EG = 3(x - 2)$, find EG.
16. If $m\angle EDF = 5x - 3$ and $m\angle DFG = 3x + 7$, find $m\angle EDF$.
17. If $DE = 14 + 2x$ and $GF = 4(x - 3) + 6$, find GF.

Determine whether each quadrilateral is a parallelogram. Justify your answer.

18.

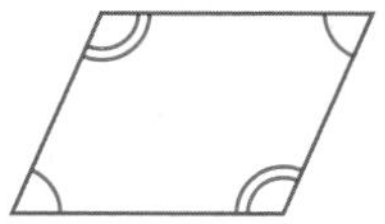

19.

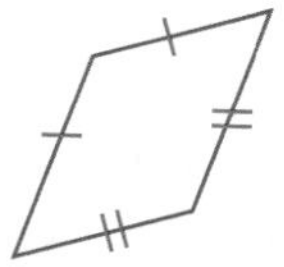

CHAPTER 13

Preparing for Standardized Tests

Apply Definitions and Properties

Many geometry problems on standardized tests require the application of definitions and properties in order to solve them. Use this section to practice applying definitions to help you solve extended-response test items.

Strategies for Applying Definitions and Properties

Step 1

Read the problem statement carefully.

- Determine what you are being asked to solve.
- Study any figures given in the problem.
- **Ask yourself:** What principles or properties of this figure can I apply to solve the problem?

Step 2

Solve the problem.

- Identify any definitions or geometric concepts you can use to help you find the unknowns in the problem.
- Use definitions and properties of figures to set up and solve an equation.

Step 3

- Check your answer.

Standardized Test Example

Read the problem. Identify what you need to know. Then use the information in the problem to solve. Show your work.

A performing arts group is building a theater in the round for upcoming productions. The stage will be a regular octagon with a perimeter of 76 feet.

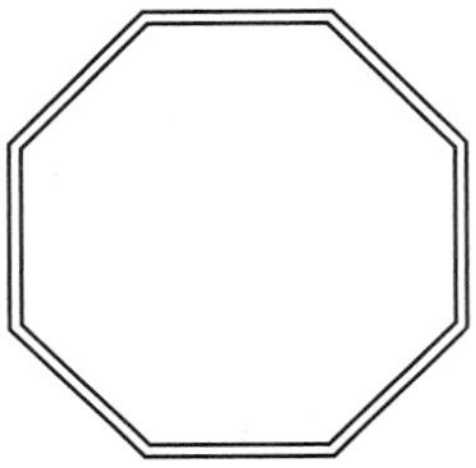

a. What length should each board be to form the sides of the stage?

b. What angle should the end of each board be cut so that they will fit together properly to form the stage? Explain.

Comstock/Jupiterimages

Read the problem carefully. You are told that the boards form a regular octagon with a perimeter of 76 feet. You need to find the length of each board and the angle that they should be cut to fit together properly.

To find the length of each board, divide the perimeter by the number of boards.

$76 \div 8 = 9.5$

So, each board should be 9.5 feet, or 9 feet 6 inches, long.

Use the property of the interior angle sum of convex polygons to find the measure of an interior angle of a regular octagon. First find the sum S of the interior angles.

$$S = (n - 2) \cdot 180$$
$$= (8 - 2) \cdot 180$$
$$= 1080$$

So, the measure of an interior angle of a regular octagon is $1080 \div 8$, or 135°. Since two boards are used to form each vertex of the stage, the end of each board should be cut at an angle of $135 \div 2$, or 67.5°.

Exercises

Read each problem. Identify what you need to know. Then use the information in the problem to solve. Show your work.

1. $\overline{RS}$ is the midsegment of trapezoid $MNOP$. What is the length of $\overline{RS}$?

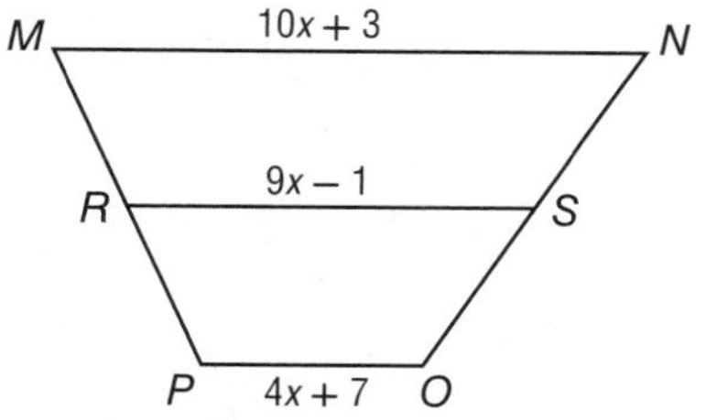

A 14 units

B 19 units

C 23 units

D 26 units

2. If $\overline{AB} \parallel \overline{DC}$, find x.

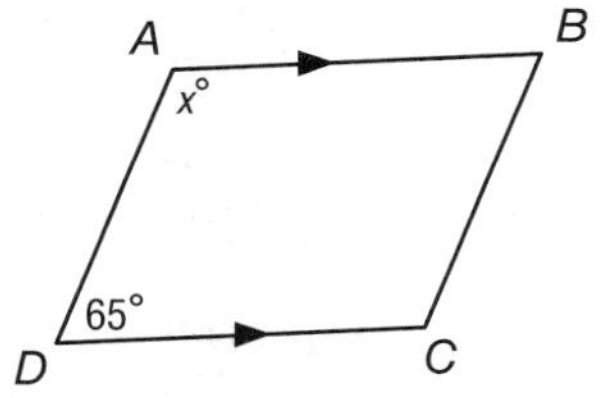

F 32.5

G 65

H 105

J 115

3. Use the graph shown below to answer each question.

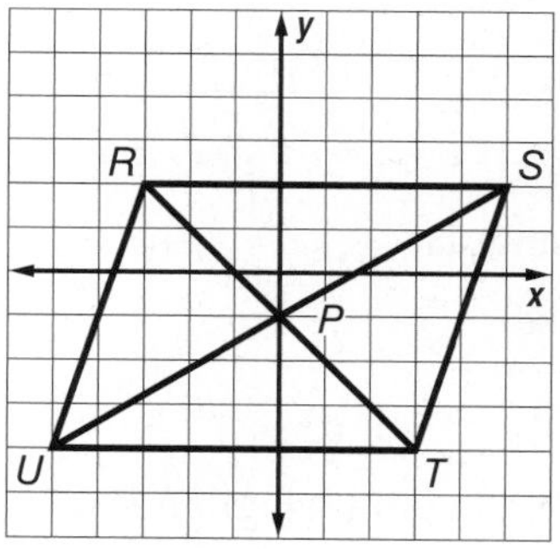

a. Do the diagonals of quadrilateral $RSTU$ bisect each other? Use the Distance Formula to verify your answer.

b. What type of quadrilateral is $RSTU$? Explain using the properties and/or definitions of this type of quadrilateral.

4. What is the sum of the measures of the exterior angles of a regular octagon?

A 45

B 135

C 360

D 1080

CHAPTER 13

Standardized Test Practice

Cumulative, Chapters 1 through 13

Multiple Choice

Read each question. Then fill in the correct answer on the answer document provided by your teacher or on a sheet of paper.

1. If $a \parallel b$, which of the following is *not* true?

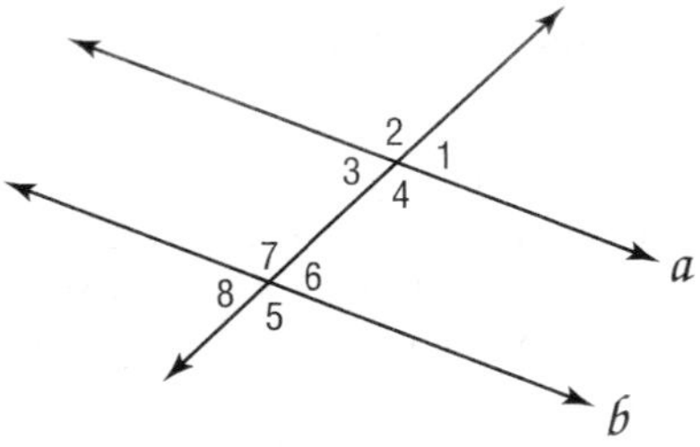

A $\angle 1 \cong \angle 3$

B $\angle 4 \cong \angle 7$

C $\angle 2 \cong \angle 5$

D $\angle 8 \cong \angle 2$

2. Classify the triangle below according to its angle measures. Choose the most appropriate term.

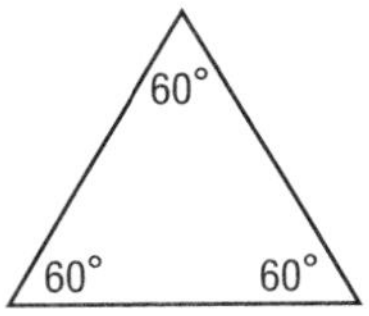

F acute

G equiangular

H obtuse

J right

3. Solve for x in parallelogram *RSTU*.

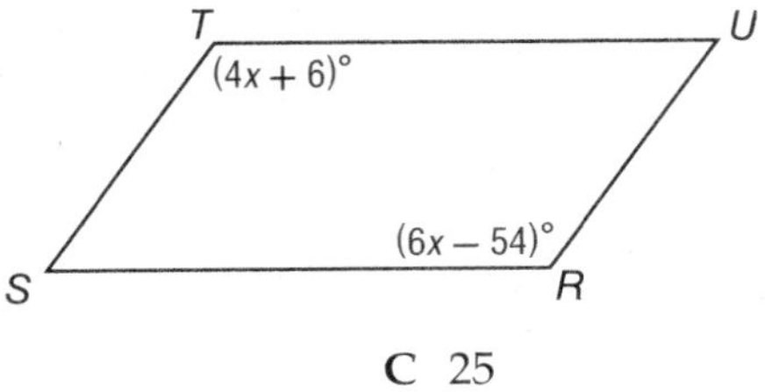

A 12

B 18

C 25

D 30

> **Test-Taking Tip**
>
> Question 3 Use the properties of parallelograms to solve the problem. Opposite angles are congruent.

4. The highest point in North Carolina is Mt. Mitchell at an elevation of 2,037 meters above sea level. Suppose the position of a hiker is given by the function $p(t) = -2.5t + 2{,}037$, where t is the number of minutes. Which of the following is the best interpretation of the slope of the function?

F The hiker's initial position was 2,037 feet below sea level.

G The hiker's initial position was 2,037 feet above sea level.

H The hiker is descending at a rate of 2.5 meters per minute.

J The hiker is ascending at a rate of 2.5 meters per minute

5. Quadrilateral *ABCD* is a rhombus. If $m\angle BCD = 120$, find $m\angle DAC$.

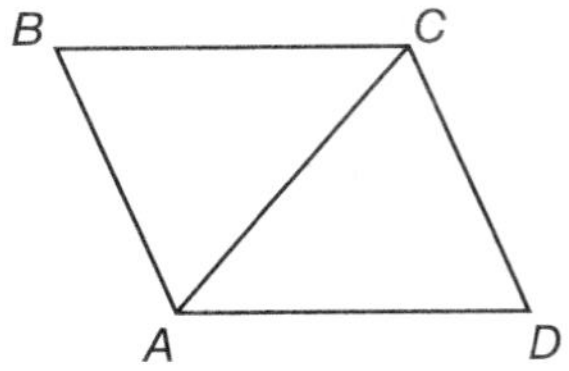

A 30

B 60

C 90

D 120

6. What is the value of x in the figure below?

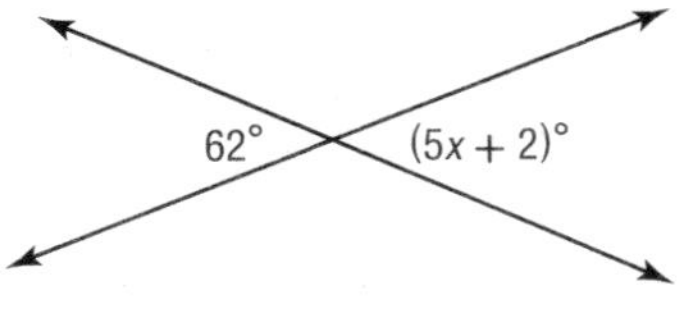

F 10

G 12

H 14

J 15

7. Which of the following statements is true?

A All rectangles are squares.

B All rhombi are squares.

C All rectangles are parallelograms.

D All parallelograms are rectangles.

Short Response/Gridded Response

Record your answers on the answer sheet provided by your teacher or a sheet of paper.

8. **GRIDDED RESPONSE** The distance required for a car to stop is directly proportional to the square of its velocity. If a car can stop in 242 meters at 22 kilometers per hour, how many meters are needed to stop at 30 kilometers per hour?

9. What are the coordinates of point O, the fourth vertex of an isosceles trapezoid? Show your work.

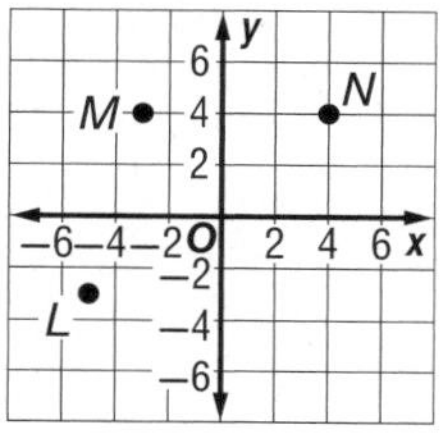

10. What do you know about a parallelogram if its diagonals are perpendicular? Explain.

11. Casey made 84 field goals during the basketball season for a total of 183 points. Each field goal was worth either 2 or 3 points. How many 2-point and 3-point field goals did Casey make during the season?

12. **GRIDDED RESPONSE** Solve for x in the figure below. Round to the nearest tenth if necessary.

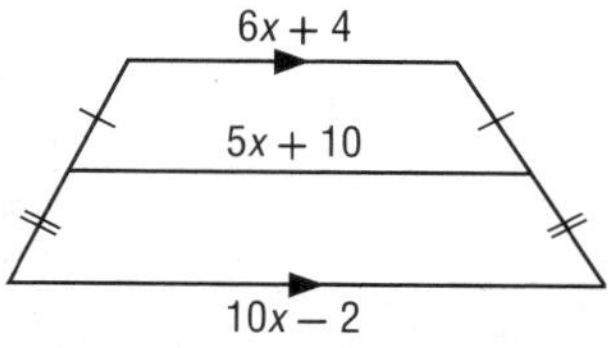

13. **GRIDDED RESPONSE** The booster club pays $180 to rent a concession stand at a football game. They purchase cans of soda for $0.25 and sell them at the game for $1.15. How many cans of soda must they sell to break even?

Extended Response

Record your answers on a sheet of paper. Show your work.

14. Determine whether you can prove each figure is a parallelogram. If not, tell what additional information would be needed to prove that it is a parallelogram. Explain your reasoning.

a.

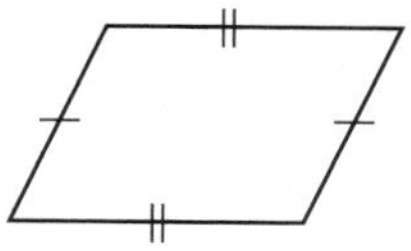

b.

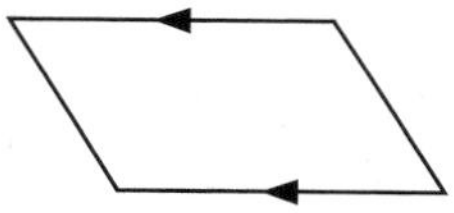

c.

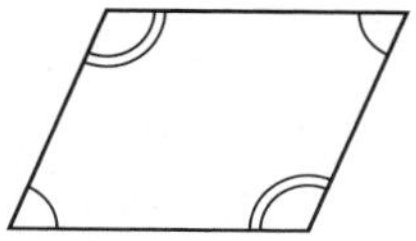

Need ExtraHelp?

If you missed Question...	1	2	3	4	5	6	7	8	9	10	11	12	13	14
Go to Lesson...	11-2	12-1	13-1	11-3	13-4	10-8	13-4	3-6	13-5	13-4	6-5	13-5	2-4	13-2

CHAPTER 14

Similarity, Transformations, and Symmetry

Then

You learned about ratios and proportions and applied them to real-world applications.

Now

In this chapter, you will:

- Identify similar polygons and use ratios and proportions to solve problems.
- Identify and apply similarity transformations.
- Use scale models and drawings to solve problems.

Why?

SPORTS Similar triangles can be used in sports to describe the path of a ball, such as a bounce pass from one person to another.

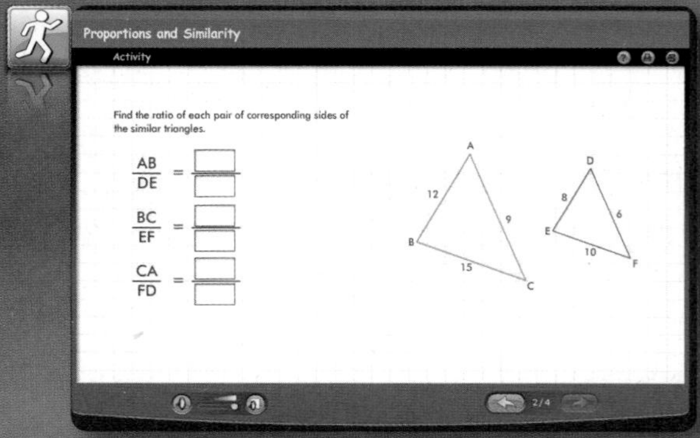

connectED.mcgraw-hill.com Your Digital Math Portal

Animation | Vocabulary | eGlossary | Personal Tutor | Virtual Manipulatives | Graphing Calculator | Audio | Foldables | Self-Check Practice | Worksheets

Barry Rosenthal/The Image Bank/Getty Ima

Get Ready for the Chapter

Diagnose Readiness | You have two options for checking prerequisite skills.

1 Textbook Option Take the Quick Check below. Refer to the Quick Review for help.

QuickCheck

Solve each equation.

1. $\frac{3x}{8} = \frac{6}{x}$
2. $\frac{7}{3} = \frac{x-4}{6}$
3. $\frac{x+9}{2} = \frac{3x-1}{8}$
4. $\frac{3}{2x} = \frac{3x}{8}$
5. **EDUCATION** The student to teacher ratio at Elder High School is 17 to 1. If there are 1088 students in the school, how many teachers are there?

ALGEBRA In the figure, $\overrightarrow{BA}$ and $\overrightarrow{BC}$ are opposite rays and $\overrightarrow{BD}$ bisects $\angle ABF$.

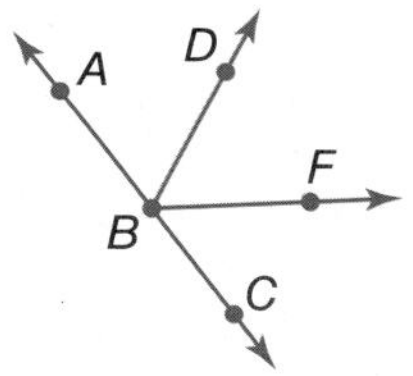

6. If $m\angle ABF = 3x - 8$ and $m\angle ABD = x + 14$, find $m\angle ABD$.
7. If $m\angle FBC = 2x + 25$ and $m\angle ABF = 10x - 1$, find $m\angle DBF$.
8. **LANDSCAPING** A landscape architect is planning to add sidewalks around a fountain as shown below. If $\overrightarrow{BA}$ and $\overrightarrow{BC}$ are opposite rays and $\overrightarrow{BD}$ bisects $\angle ABF$, find $m\angle FBC$.

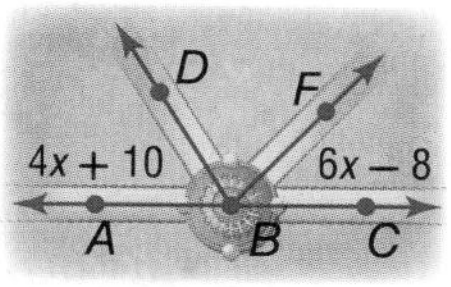

QuickReview

Example 1

Solve $\frac{4x-3}{5} = \frac{2x+11}{3}$.

$\frac{4x-3}{5} = \frac{2x+11}{3}$	Original equation
$3(4x-3) = 5(2x+11)$	Cross multiplication
$12x - 9 = 10x + 55$	Distributive Property
$2x = 64$	Add.
$x = 32$	Simplify.

Example 2

In the figure, $\overrightarrow{QP}$ and $\overrightarrow{QR}$ are opposite rays, and $\overrightarrow{QT}$ bisects $\angle SQR$. If $m\angle SQR = 6x + 8$ and $m\angle TQR = 4x - 14$, find $m\angle SQT$.

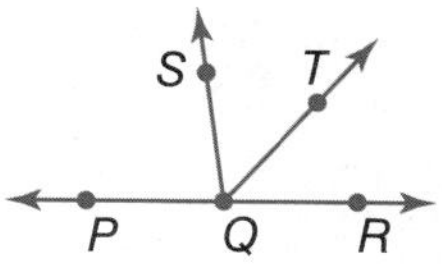

Since $\overrightarrow{TQ}$ bisects $\angle SQR$, $m\angle SQR = 2(m\angle TQR)$.

$m\angle SQR = 2(m\angle TQR)$	Def. of $\angle$ bisector
$6x + 8 = 2(4x - 14)$	Substitution
$6x + 8 = 8x - 28$	Distributive Property
$-2x = -36$	Subtract.
$x = 18$	Simplify.

Since $\overrightarrow{TQ}$ bisects $\angle SQR$, $m\angle SQT = m\angle TQR$.

$m\angle SQT = m\angle TQR$	Def. of $\angle$ bisector
$m\angle SQT = 4x - 14$	Substitution
$m\angle SQT = 58$	$x = 18$

2 Online Option Take an online self-check Chapter Readiness Quiz at connectED.mcgraw-hill.com.

Get Started on the Chapter

You will learn several new concepts, skills, and vocabulary terms as you study Chapter 14. To get ready, identify important terms and organize your resources. You may wish to refer to Chapter 0 to review prerequisite skills.

FOLDABLES® StudyOrganizer

Proportions and Similarity Make this Foldable to help you organize your Chapter 14 notes about proportions, similar polygons, and similarity transformations. Begin with four sheets of notebook paper.

1 **Fold** the four sheets of paper in half.

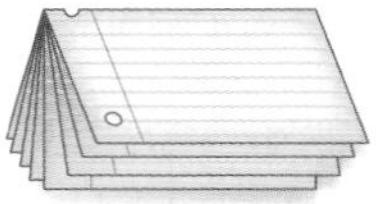

2 **Cut** along the top fold of the papers. Staple along the side to form a book.

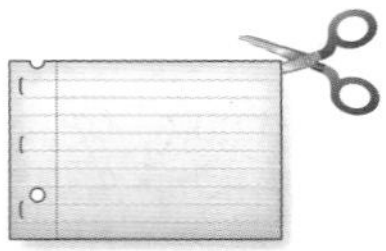

3 **Cut** the right sides of each paper to create a tab for each lesson.

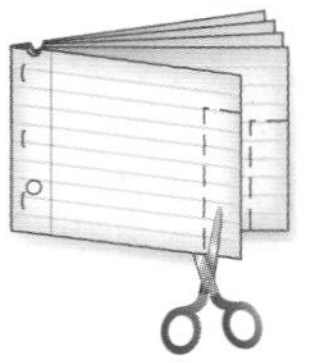

4 **Label** each tab with a lesson number, as shown.

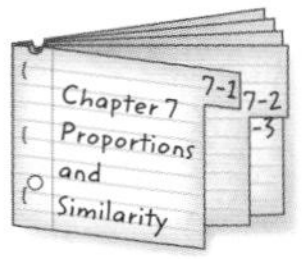

NewVocabulary

English		Español
dilation	p. 883	dilatación
similarity transformation	p. 883	transformación de semejanza
enlargement	p. 883	ampliación
reduction	p. 883	reducción
line of reflection	p. 890	línea de reflexión
center of rotation	p. 908	centro de rotación
angle of rotation	p. 908	ángulo de rotación
composition of transformations	p. 918	compasición de transformaciones
symmetry	p. 930	símetria
line symmetry	p. 930	símetria lineal
line of symmetry	p. 930	eje de símetria

ReviewVocabulary

altitude altura a segment drawn from a vertex of a triangle perpendicular to the line containing the other side

angle bisector bisectriz de un ángulo a ray that divides an angle into two congruent angles

median mediana a segment drawn from a vertex of a triangle to the midpoint of the opposite side

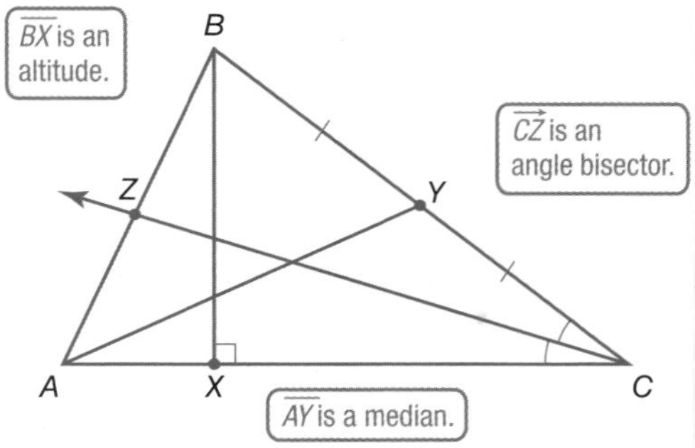

LESSON

14-1 Similar Triangles

Then

- You used the AAS, SSS, and SAS Congruence Theorems to prove triangles congruent.

Now

1. Identify similar triangles using the AA Similarity Postulate and the SSS and SAS Similarity Theorems.
2. Use similar triangles to solve problems.

Why?

- Julian wants to draw a similar version of his skate club's logo on a poster. He first draws a line at the bottom of the poster. Next, he uses a cutout of the original triangle to copy the two bottom angles. Finally, he extends the noncommon sides of the two angles.

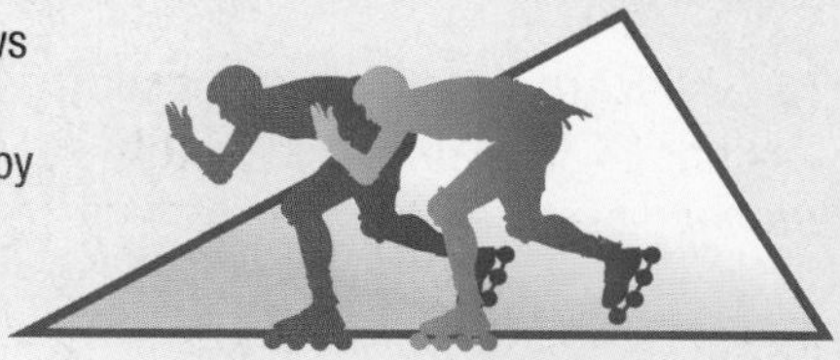

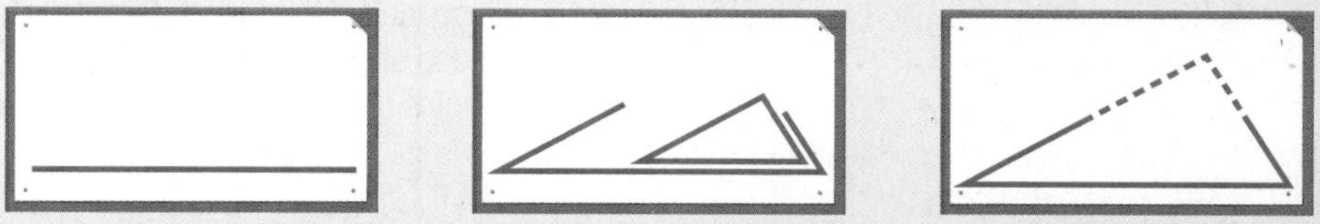

Common Core State Standards

Content Standards

G.SRT.4 Prove theorems about triangles.

G.SRT.5 Use congruence and similarity criteria for triangles to solve problems and to prove relationships in geometric figures.

Mathematical Practices

4 Model with mathematics.

7 Look for and make use of structure.

1 Identify Similar Triangles

The example suggests that two triangles are similar if two pairs of corresponding angles are congruent.

Postulate 14.1 Angle-Angle (AA) Similarity

If two angles of one triangle are congruent to two angles of another triangle, then the triangles are similar.

Example If $\angle A \cong \angle F$ and $\angle B \cong \angle G$, then $\triangle ABC \sim \triangle FGH$.

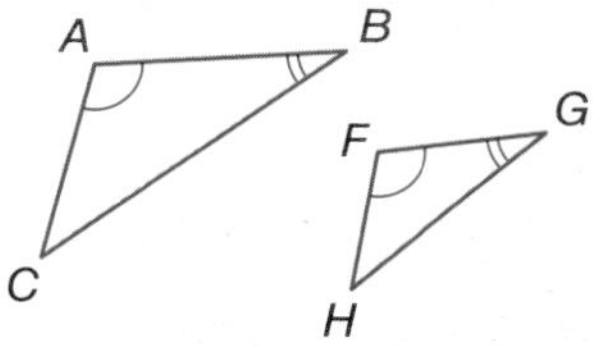

PT

Example 1 Use the AA Similarity Postulate

Determine whether the triangles are similar. If so, write a similarity statement. Explain your reasoning.

a.

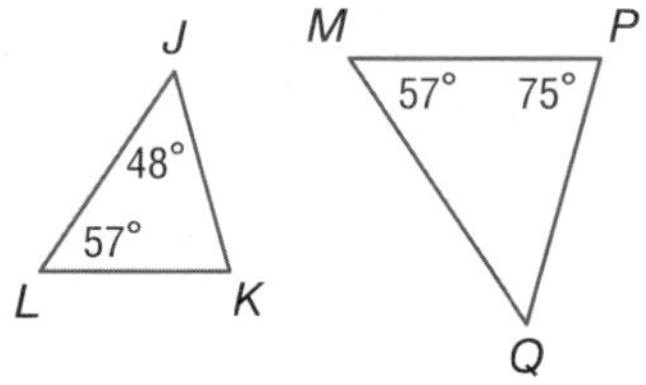

b.

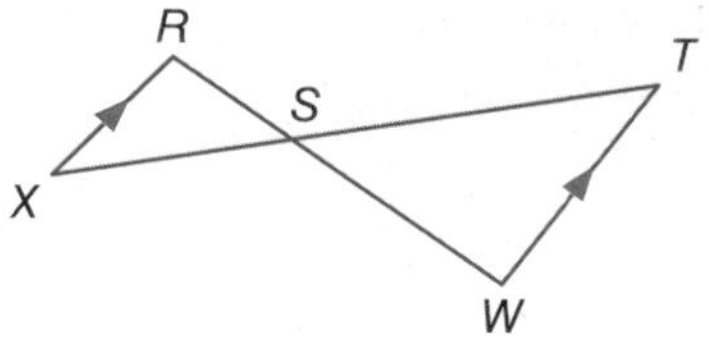

a. Since $m\angle L = m\angle M$, $\angle L \cong \angle M$. By the Triangle Sum Theorem, $57 + 48 + m\angle K = 180$, so $m\angle K = 75$. Since $m\angle P = 75$, $\angle K \cong \angle P$. So, $\triangle LJK \sim \triangle MQP$ by AA Similarity.

b. $\angle RSX \cong \angle WST$ by the Vertical Angles Theorem. Since $\overline{RX} \parallel \overline{TW}$, $\angle R \cong \angle W$. So, $\triangle RSX \sim \triangle WST$ by AA Similarity.

GuidedPractice

1A.

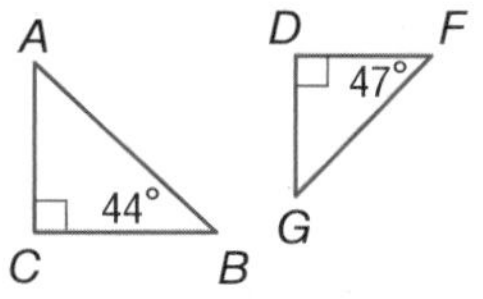

1B.

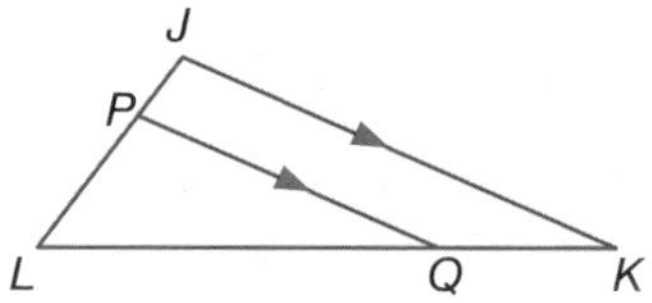

You can use the AA Similarity Postulate to prove the following two theorems.

Theorems Triangle Similarity

14.1 Side-Side-Side (SSS) Similarity
If the corresponding side lengths of two triangles are proportional, then the triangles are similar.

Example If $\frac{JK}{MP} = \frac{KL}{PQ} = \frac{LJ}{QM}$, then $\triangle JKL \sim \triangle MPQ$.

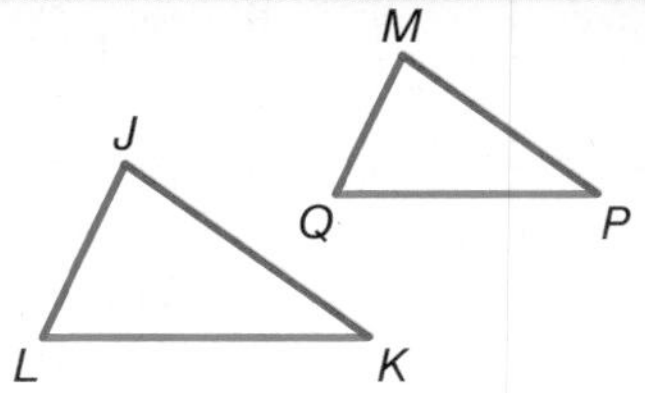

14.2 Side-Angle-Side (SAS) Similarity
If the lengths of two sides of one triangle are proportional to the lengths of two corresponding sides of another triangle and the included angles are congruent, then the triangles are similar.

Example If $\frac{RS}{XY} = \frac{ST}{YZ}$ and $\angle S \cong \angle Y$, then $\triangle RST \sim \triangle XYZ$.

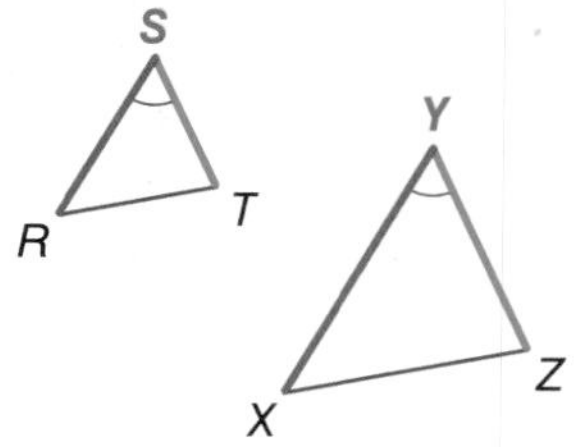

You will prove Theorem 14.2 in Exercise 25.

Proof Theorem 14.1

Given: $\frac{AB}{FG} = \frac{BC}{GH} = \frac{AC}{FH}$

Prove: $\triangle ABC \sim \triangle FGH$

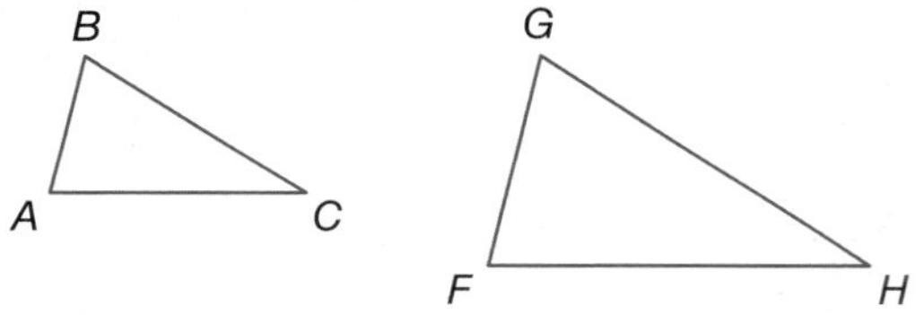

StudyTip

Corresponding Sides To determine which sides of two triangles correspond, begin by comparing the longest sides, then the next longest sides, and finish by comparing the shortest sides.

Paragraph Proof:

Locate J on $\overline{FG}$ so that $JG = AB$. Draw $\overline{JK}$ so that $\overline{JK} \parallel \overline{FH}$. Label $\angle GJK$ as $\angle 1$.

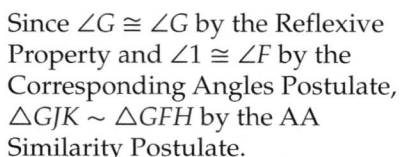

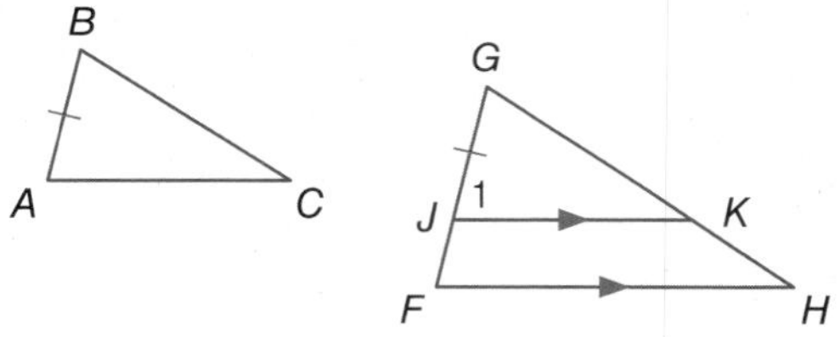

Since $\angle G \cong \angle G$ by the Reflexive Property and $\angle 1 \cong \angle F$ by the Corresponding Angles Postulate, $\triangle GJK \sim \triangle GFH$ by the AA Similarity Postulate.

By the definition of similar polygons, $\frac{JG}{FG} = \frac{GK}{GH} = \frac{JK}{FH}$. By substitution, $\frac{AB}{FG} = \frac{GK}{GH} = \frac{JK}{FH}$.

Since we are also given that $\frac{AB}{FG} = \frac{BC}{GH} = \frac{AC}{FH}$, we can say that $\frac{GK}{GH} = \frac{BC}{GH}$ and $\frac{JK}{FH} = \frac{AC}{FH}$. This means that $GK = BC$ and $JK = AC$, so $\overline{GK} \cong \overline{BC}$ and $\overline{JK} \cong \overline{AC}$.

By SSS, $\triangle ABC \cong \triangle JGK$.

By CPCTC, $\angle B \cong \angle G$ and $\angle A \cong \angle 1$. Since $\angle 1 \cong \angle F$, $\angle A \cong \angle F$ by the Transitive Property. By AA Similarity, $\triangle ABC \sim \triangle FGH$.

Example 2 Use the SSS and SAS Similarity Theorems

Determine whether the triangles are similar. If so, write a similarity statement. Explain your reasoning.

a.

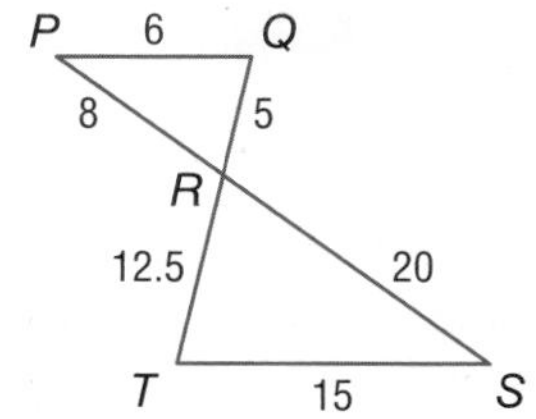

$\frac{PR}{SR} = \frac{8}{20}$ or $\frac{2}{5}$, $\frac{PQ}{ST} = \frac{6}{15}$ or $\frac{2}{5}$, and $\frac{QR}{TR} = \frac{5}{12.5} = \frac{50}{125}$ or $\frac{2}{5}$. So, $\triangle PQR \sim \triangle STR$ by the SSS Similarity Theorem.

b.

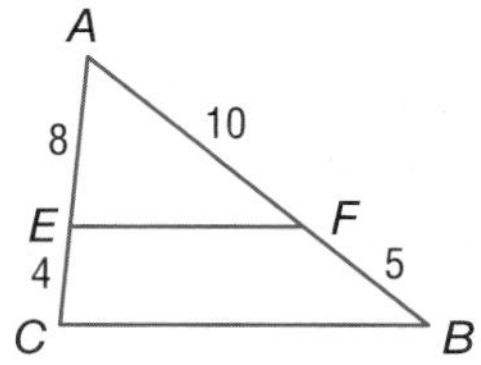

By the Reflexive Property, $\angle A \cong \angle A$.

$\frac{AF}{AB} = \frac{10}{10+5} = \frac{10}{15}$ or $\frac{2}{3}$ and $\frac{AE}{AC} = \frac{8}{8+4} = \frac{8}{12}$ or $\frac{2}{3}$.

Since the lengths of the sides that include $\angle A$ are proportional, $\triangle AEF \sim \triangle ACB$ by the SAS Similarity Theorem.

Study Tip

Draw Diagrams It is helpful to redraw similar triangles so that the corresponding side lengths have the same orientation.

Guided Practice

2A.

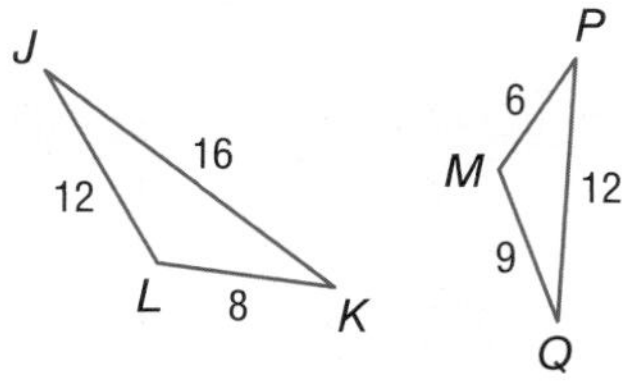

2B.

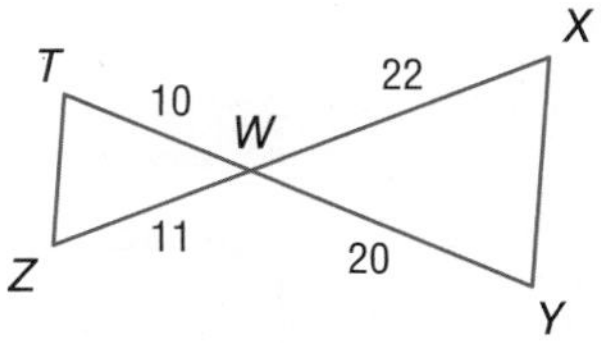

You can decide what is sufficient to prove that two triangles are similar.

Standardized Test Example 3 Sufficient Conditions

In the figure, $\angle ADB$ is a right angle. Which of the following would *not* be sufficient to prove that $\triangle ADB \sim \triangle CDB$?

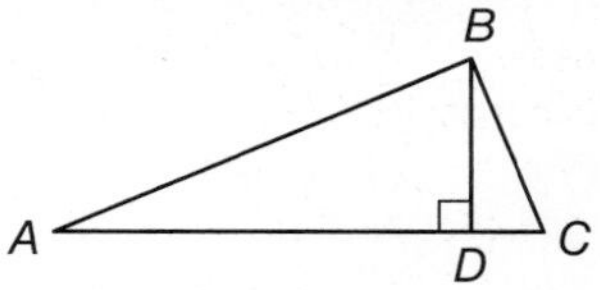

A $\frac{AD}{BD} = \frac{BD}{CD}$

B $\frac{AB}{BC} = \frac{BD}{CD}$

C $\angle ABD \cong \angle C$

D $\frac{AD}{BD} = \frac{BD}{CD} = \frac{AB}{BC}$

Read the Test Item

You are given that $\angle ADB$ is a right angle and asked to identify which additional information would not be enough to prove that $\triangle ADB \sim \triangle CDB$.

Solve the Test Item

Since $\angle ADB$ is a right angle, $\angle CDB$ is also a right angle. Since all right angles are congruent, $\angle ADB \cong \angle CDB$. Check each answer choice until you find one that does not supply a sufficient additional condition to prove that $\triangle ADB \sim \triangle CDB$.

Test-Taking Tip

Identifying Nonexamples Sometimes test questions require you to find a nonexample, as in this case. You must check each option until you find a valid nonexample. If you would like to check your answer, confirm that each additional option is correct.

Choice A: If $\frac{AD}{BD} = \frac{BD}{CD}$ and $\angle ADB \cong \angle CDB$, then $\triangle ADB \sim \triangle CDB$ by SAS Similarity.

Choice B: If $\frac{AB}{BC} = \frac{BD}{CD}$ and $\angle ADB \cong \angle CDB$, then we cannot conclude that $\triangle ADB \sim \triangle CDB$ because the included angle of side $\overline{AB}$ and $\overline{BD}$ is not $\angle ADB$. So the answer is B.

GuidedPractice

3. If $\triangle JKL$ and $\triangle FGH$ are two triangles such that $\angle J \cong \angle F$, which of the following would be sufficient to prove that the triangles are similar?

F $\frac{KL}{GH} = \frac{JL}{FH}$ **G** $\frac{JL}{JK} = \frac{FH}{FG}$ **H** $\frac{JK}{FG} = \frac{KL}{GH}$ **J** $\frac{JL}{JK} = \frac{GH}{FG}$

2 Use Similar Triangles

Like the congruence of triangles, similarity of triangles is reflexive, symmetric, and transitive.

Theorem 14.3 Properties of Similarity

Reflexive Property of Similarity	$\triangle ABC \sim \triangle ABC$
Symmetric Property of Similarity	If $\triangle ABC \sim \triangle DEF$, then $\triangle DEF \sim \triangle ABC$.
Transitive Property of Similarity	If $\triangle ABC \sim \triangle DEF$, and $\triangle DEF \sim \triangle XYZ$, then $\triangle ABC \sim \triangle XYZ$.

You will prove Theorem 14.3 in Exercise 26.

Example 4 Parts of Similar Triangles

Find *BE* and *AD*.

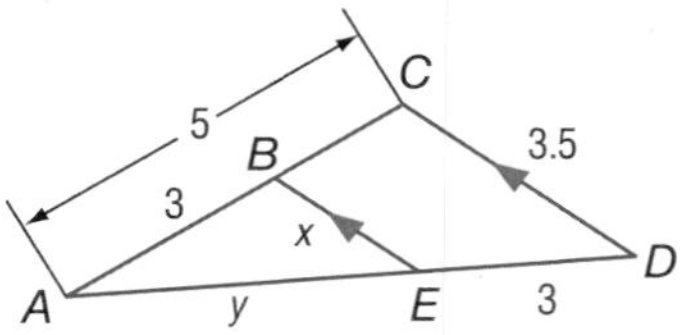

Since $\overline{BE} \parallel \overline{CD}$, $\angle ABE \cong \angle BCD$, and $\angle AEB \cong \angle EDC$ because they are corresponding angles. By AA Similarity, $\triangle ABE \sim \triangle ACD$.

StudyTip

Proportions An additional proportion that is true for Example 4 is $\frac{AC}{CD} = \frac{AB}{BE}$.

$\frac{AB}{AC} = \frac{BE}{CD}$	Definition of Similar Polygons
$\frac{3}{5} = \frac{x}{3.5}$	$AC = 5$, $CD = 3.5$, $AB = 3$, $BE = x$
$3.5 \cdot 3 = 5 \cdot x$	Cross Products Property
$2.1 = x$	BE is 2.1.
$\frac{AC}{AB} = \frac{AD}{AE}$	Definition of Similar Polygons
$\frac{5}{3} = \frac{y+3}{y}$	$AC = 5$, $AB = 3$, $AD = y + 3$, $AE = y$
$5 \cdot y = 3(y + 3)$	Cross Products Property
$5y = 3y + 9$	Distributive Property
$2y = 9$	Subtract $3y$ from each side.
$y = 4.5$	AD is $y + 3$ or 7.5.

GuidedPractice

Find each measure.

4A. QP and MP

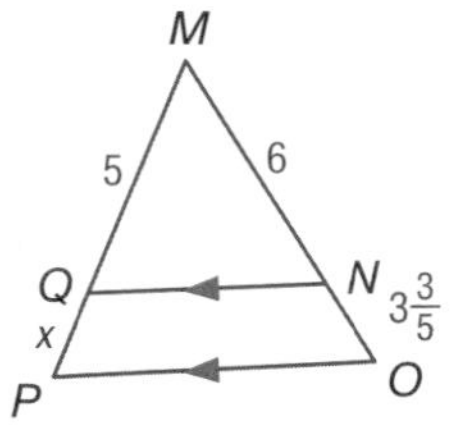

4B. WR and RT

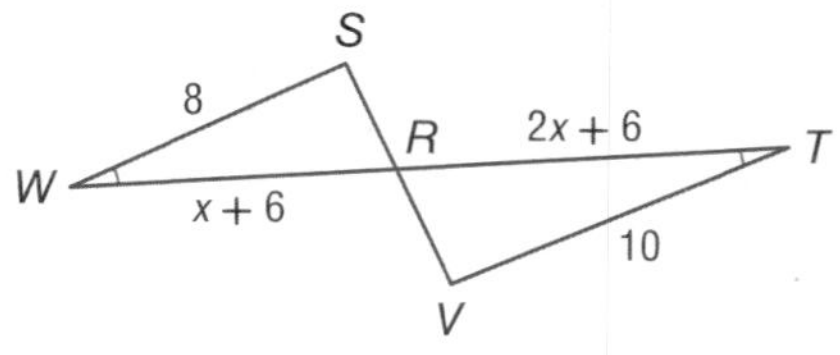

Real-World Example 5 Indirect Measurement

ROLLER COASTERS Hallie is estimating the height of the Superman roller coaster in Mitchellville, Maryland. She is 5 feet 3 inches tall and her shadow is 3 feet long. If the length of the shadow of the roller coaster is 40 feet, how tall is the roller coaster?

Understand Make a sketch of the situation. 5 feet 3 inches is equivalent to 5.25 feet.

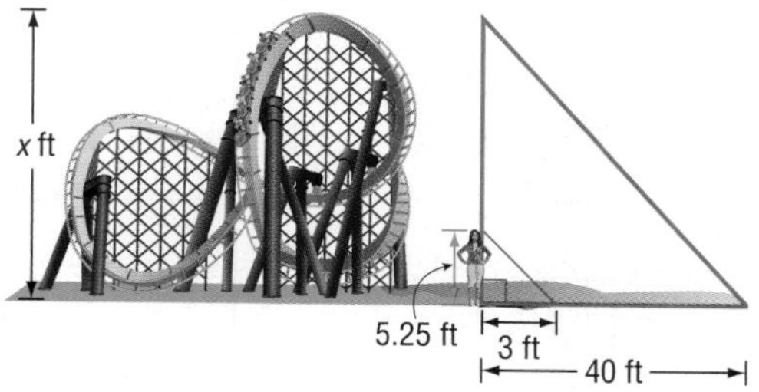

Plan In shadow problems, you can assume that the angles formed by the Sun's rays with any two objects are congruent and that the two objects form the sides of two right triangles.

Since two pairs of angles are congruent, the right triangles are similar by the AA Similarity Postulate. So, the following proportion can be written.

$$\frac{\text{Hallie's height}}{\text{coaster's height}} = \frac{\text{Hallie's shadow length}}{\text{coaster's shadow length}}$$

Solve Substitute the known values and let x = roller coaster's height.

$\frac{5.25}{x} = \frac{3}{40}$	Substitution
$3 \cdot x = 40(5.25)$	Cross Products Property
$3x = 210$	Simplify.
$x = 70$	Divide each side by 3.

The roller coaster is 70 feet tall.

Check The roller coaster's shadow length is $\frac{40 \text{ ft}}{3 \text{ ft}}$ or about 13.3 times Hallie's shadow length. Check to see that the roller coaster's height is about 13.3 times Hallie's height. $\frac{70 \text{ ft}}{5.25 \text{ ft}} \approx 13.3$ ✓

Problem-Solving Tip

Reasonable Answers When you have solved a problem, check your answer for reasonableness. In this example, Hallie's shadow is a little more than half her height. The coaster's shadow is also a little more than half of the height you calculated. Therefore, the answer is reasonable.

Guided Practice

5. **BUILDINGS** Adam is standing next to the Palmetto Building in Columbia, South Carolina. He is 6 feet tall and the length of his shadow is 9 feet. If the length of the shadow of the building is 322.5 feet, how tall is the building?

Concept Summary Triangle Similarity

AA Similarity Postulate

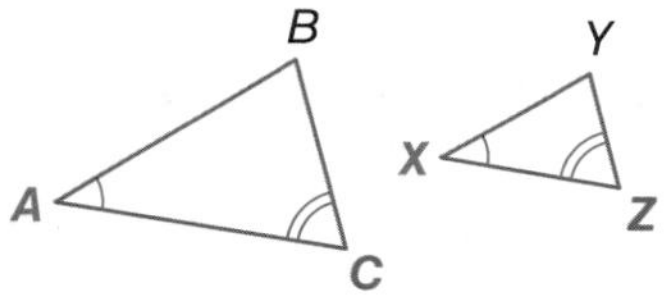

If $\angle A \cong \angle X$ and $\angle C \cong \angle Z$, then $\triangle ABC \sim \triangle XYZ$.

SSS Similarity Theorem

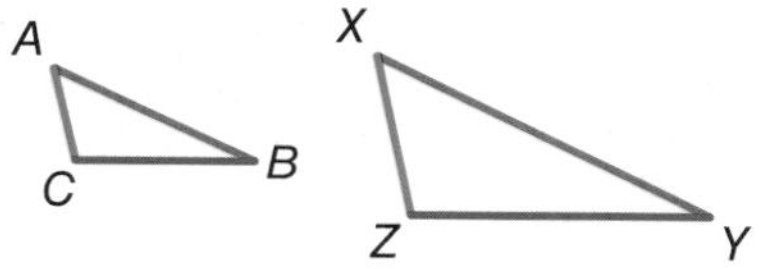

If $\frac{AB}{XY} = \frac{BC}{YZ} = \frac{CA}{ZX}$, then $\triangle ABC \sim \triangle XYZ$.

SAS Similarity Theorem

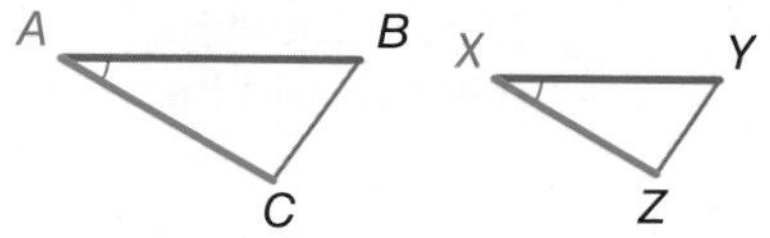

If $\angle A \cong \angle X$ and $\frac{AB}{XY} = \frac{CA}{ZX}$, then $\triangle ABC \sim \triangle XYZ$.

Check Your Understanding

Examples 1–2 **Determine whether the triangles are similar. If so, write a similarity statement. Explain your reasoning.**

1.

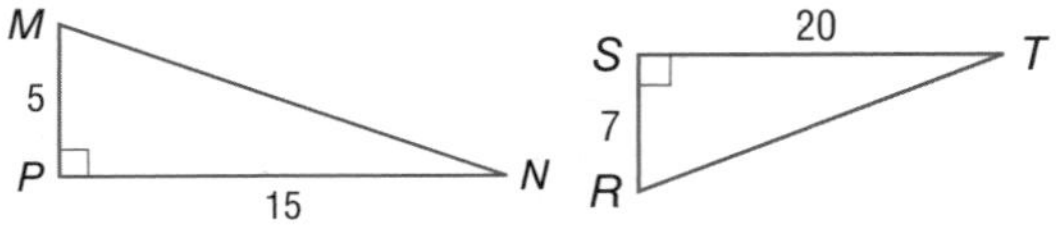

2.

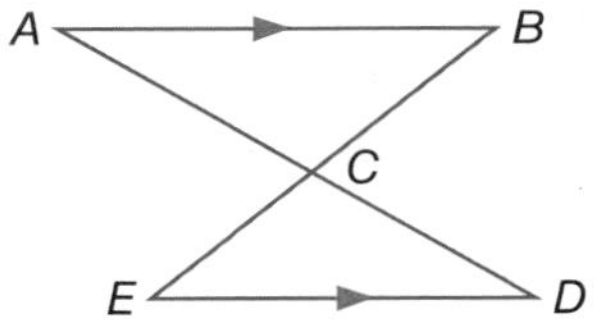

3.

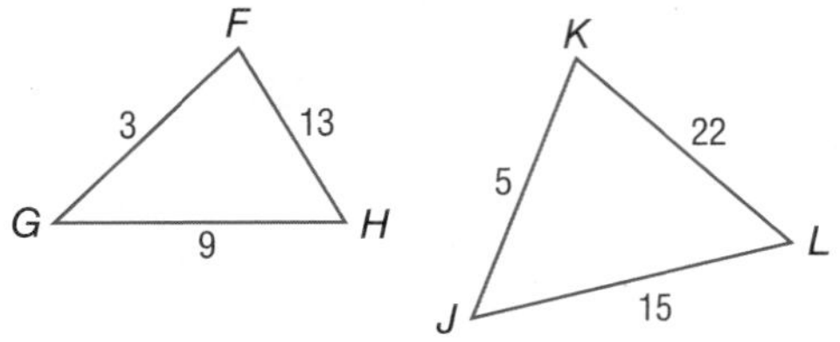

4.

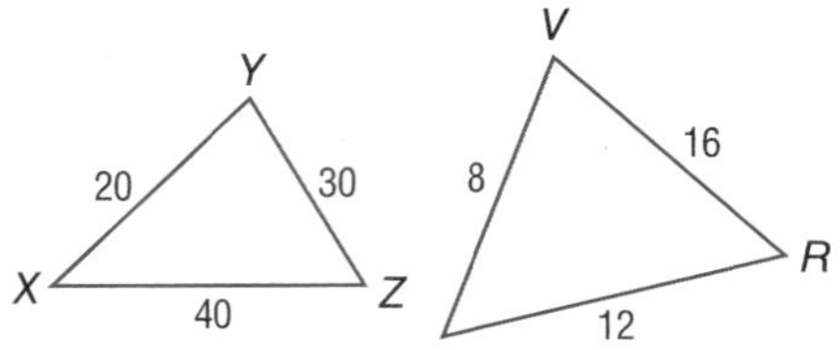

Example 3

5. MULTIPLE CHOICE In the figure, *AB* is perpendicular to *BD*. Which additional information would be enough to prove. $\triangle ABC \sim \triangle DEC$?

A $m\angle A = 60$

B $m\angle ABD = m\angle BDC$

C $\overline{AB} \cong \overline{BC}$

D *BD* bisects *AC*

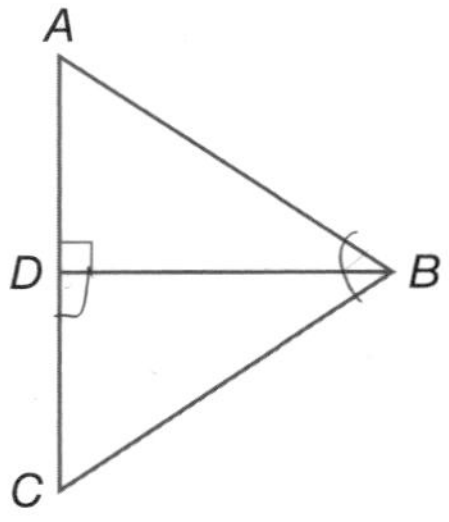

Example 4

CCSS ALGEBRA **Identify the similar triangles. Find each measure.**

6.

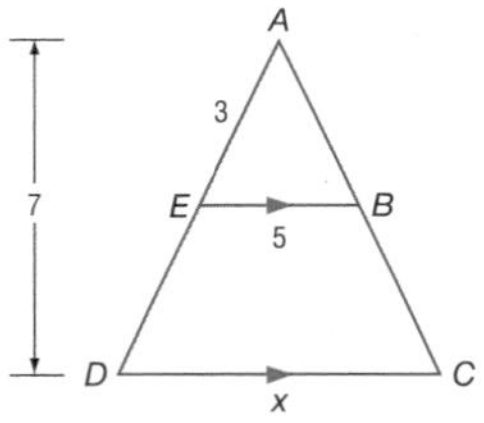

7.

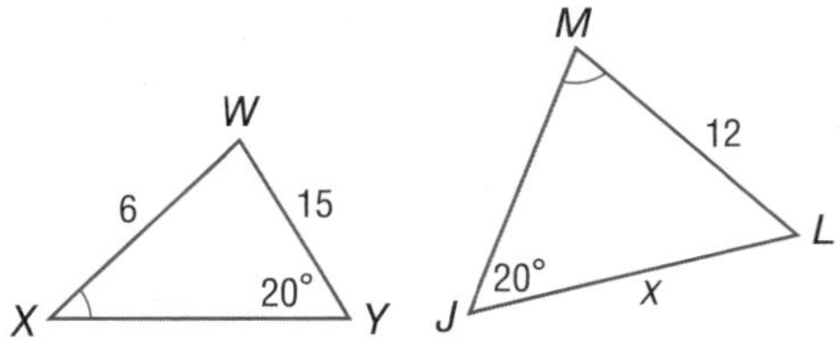

Example 5

8. PETS Emily is walking her dog Max. If Emily is 5 feet 4 inches tall and her shadow is 3 feet 2 inches long. If Max's shadow is 1 foot 6 inches long, how tall is Max?

Practice and Problem Solving

Examples 1–3 **Determine whether the triangles are similar. If so, write a similarity statement. If not, what would be sufficient to prove the triangles similar? Explain your reasoning.**

9.

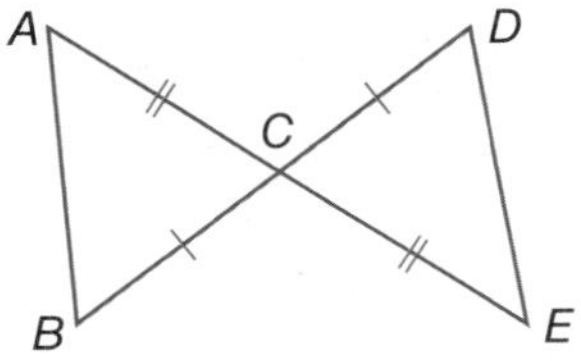

10.

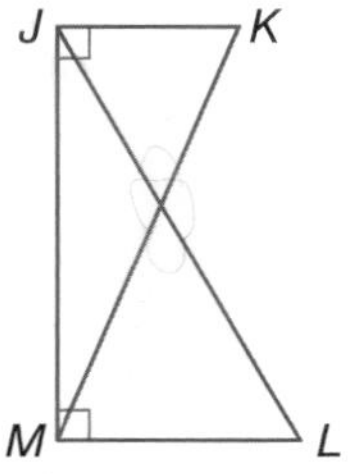

11.

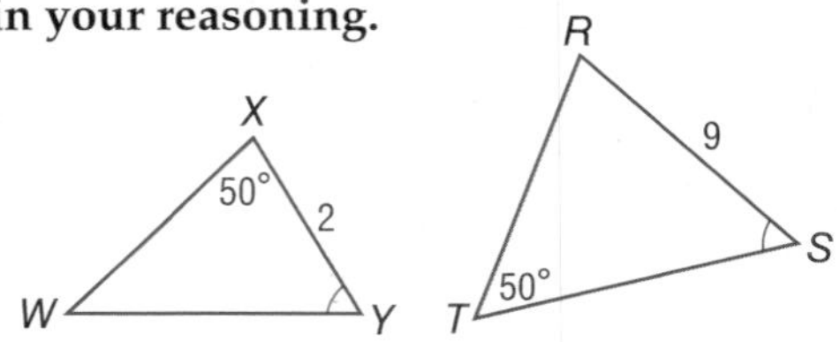

12.

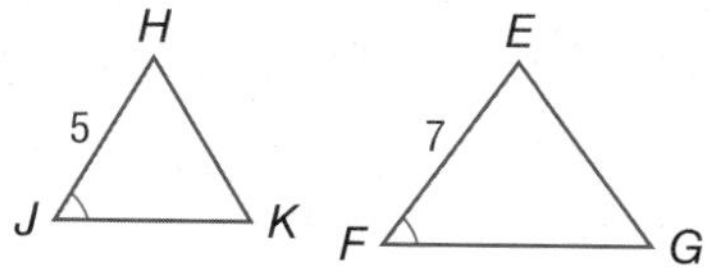

13.

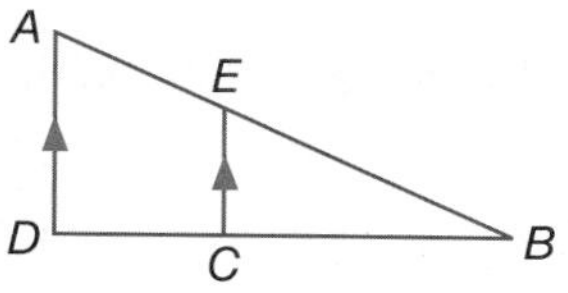

14.

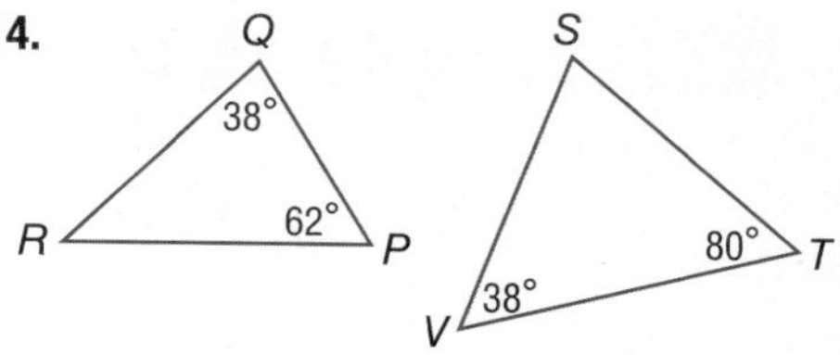

Example 4 **ALGEBRA** **Identify similar triangles. Then find each measure.**

15. DF

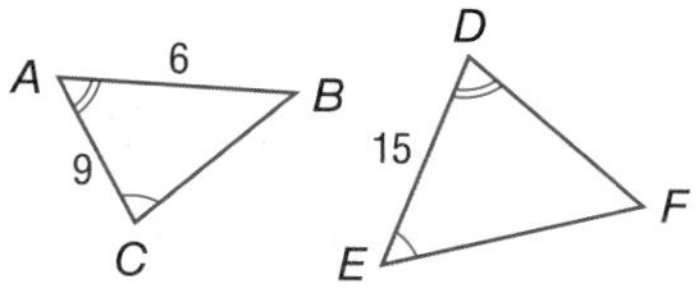

16. KJ

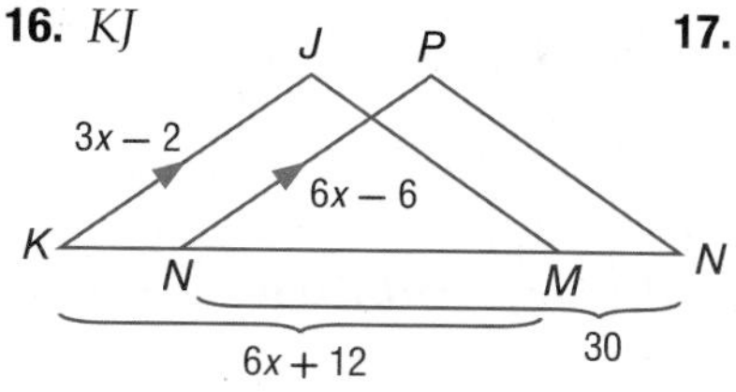

17. WX, XZ

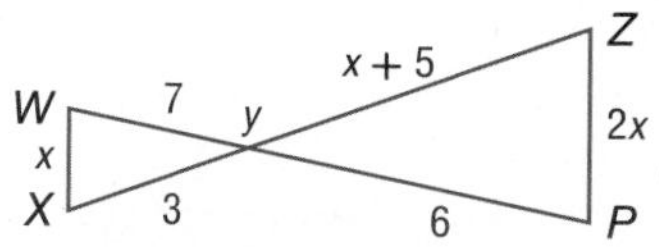

18. SR, TR

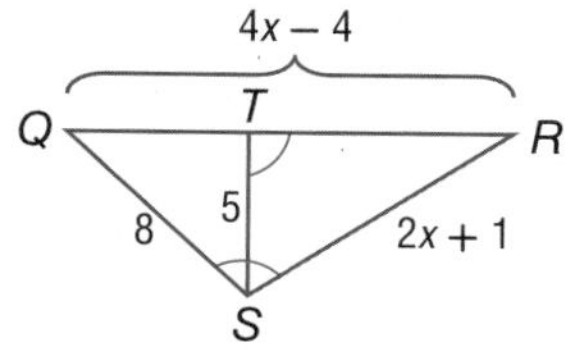

19. HJ, DK

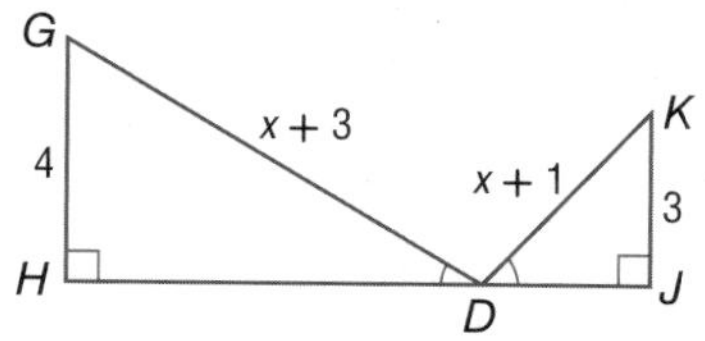

20. SR, QR

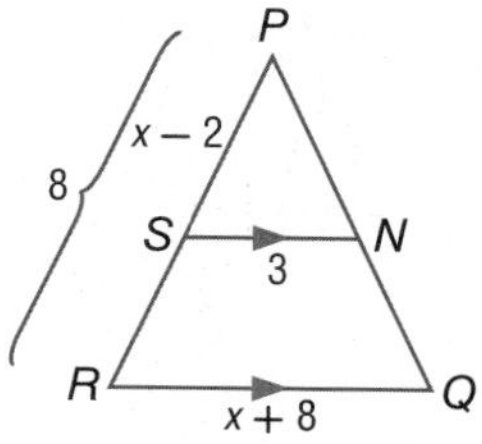

Example 5

21. **TOWERS** Gene is standing next to a cell phone tower. If Gene is 6 feet tall, his shadow is 18 inches long, and the tower's shadow is 55 feet long, how tall is the tower?

22. **FLAGS** When Anna, who is 5'3" stands next to a flag pole her shadow is 1"11" long, and the flag pole's shadow is 5'9", how tall is the flag pole?

23. **CONSTRUCTION** Dale uses ladders in his house painting business. With each ladder Dale wants the angle the ladder makes with the ground to be 65°. When the ladders are leaned against a house with this angle the 15 foot ladder reaches a height of 13.6 feet. How high can a 20 foot ladder reach?

PROOF **Write a two-column proof**

24. Theorem 14. 2

25. Theorem 14.3

PROOF **Write a two-column proof.**

26. **Given:** BD is perpendicular to AC, $\frac{AB}{BC} = \frac{AD}{BD}$

Prove: $\triangle ABD \sim \triangle BCD$

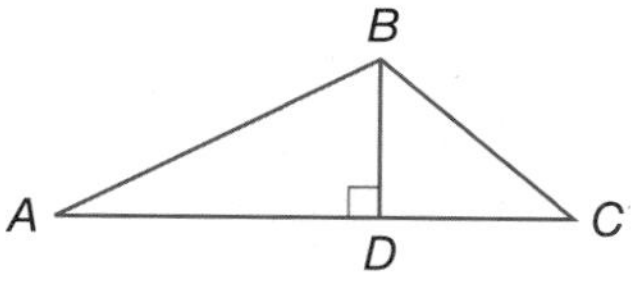

27. **Given:** $LMNP$ is a kite,

Prove: $\frac{AP}{AM} = \frac{PQ}{QM}$

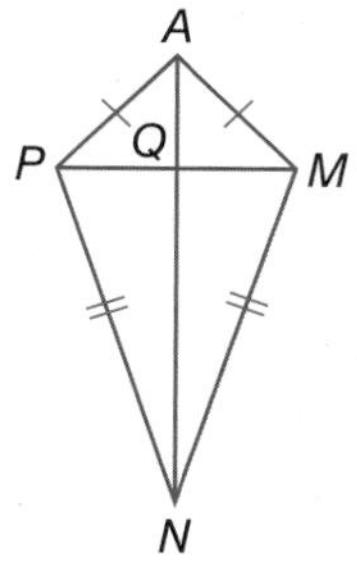

28. CCSS **CHORES** The height of the ironing board shown to the right is adjustable. If the ironing board is parallel to the floor, prove $\frac{AE}{EC} = \frac{AB}{DC}$

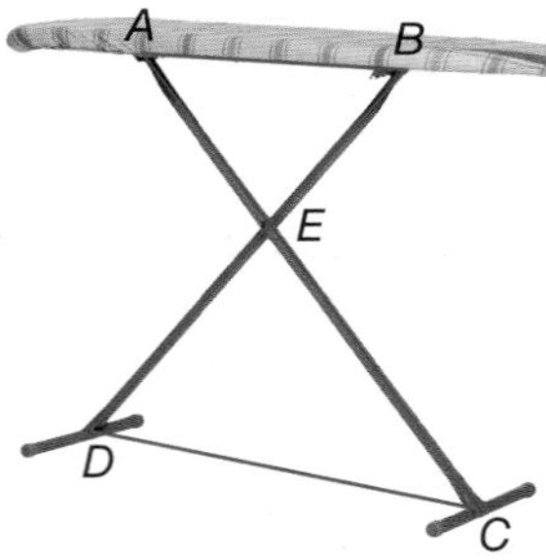

COORDINATE GEOMETRY **$\triangle ABC$ and $\triangle EBF$ have vertices $A(1, -7)$, $B(7, 5)$, $C(1, 8)$, $E(3, -3)$, $F(3, 7)$.**

29. Graph the triangles, and determine that $\triangle ABC \sim \triangle EBF$.

30. Find the scale factor and the ratios of the perimeters of the two triangles shown.

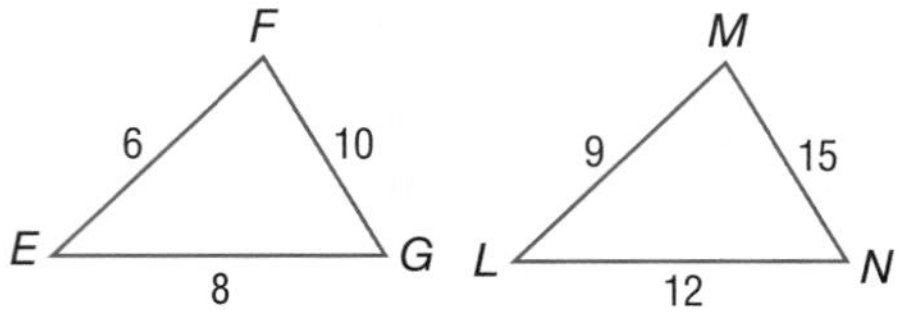

31. **SKIING** Scott is riding up a ski lift. In 20 feet on the lift he has risen 5 feet higher off the ground. Use similar triangles to figure out how high he is off of the ground after 50 feet on the lift.

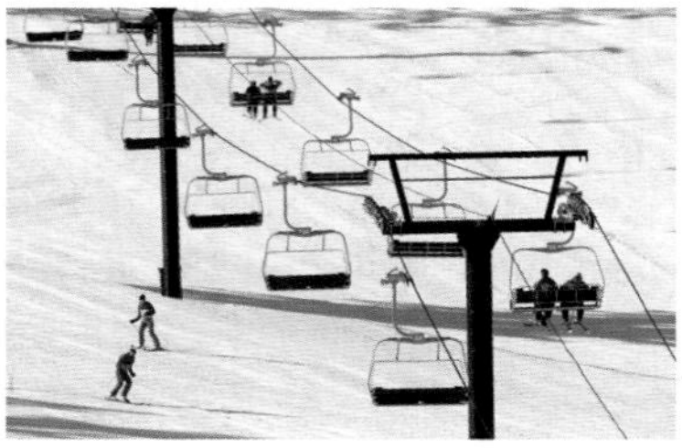

PROOF OR COUNTEREXAMPLE **Give a proof or counterexample to the following statements.**

32. All right triangles are similar.

33. All isosceles right triangles are similar.

34. All equilateral triangles are similar.

35. **MULTIPLE REPRESENTATIONS** In this problem you will explore the perimeters of similar triangles.

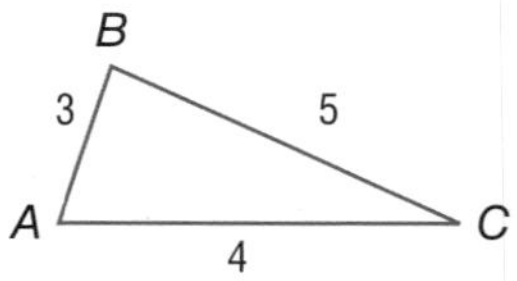

a. **Geometric** Draw three triangles similar to $\triangle ABC$. Label the triangles EFG, LMN, and XYZ. Label the lengths of each side.

b. Tabular Copy and complete the table below.

Perimeter of $\triangle EFG$	Perimeter of $\triangle EFG$	Perimeter of $\triangle EFG$	$\frac{\text{Perimeter of } \triangle EFG}{\text{Perimeter of } \triangle ABC}$
Perimeter of $\triangle LMN$	Perimeter of $\triangle LMN$	Perimeter of $\triangle LMN$	$\frac{\text{Perimeter of } \triangle LMN}{\text{Perimeter of } \triangle ABC}$
Perimeter of $\triangle XYZ$	Perimeter of $\triangle XYZ$	Perimeter of $\triangle XYZ$	$\frac{\text{Perimeter of } \triangle XYZ}{\text{Perimeter of } \triangle ABC}$

c. Verbal Make a conjecture about the relationship between perimeters of similar triangles.

H.O.T. Problems Use Higher-Order Thinking Skills

36. WRITING IN MATH Compare and contrast similar triangles and congruent triangles.

37. OPEN ENDED Draw two triangles that are similar to each other. Explain how you are sure they are similar to each other.

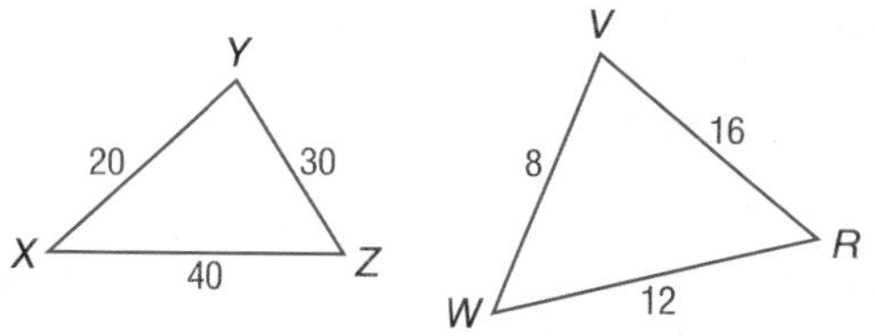

38. CCSS REASONING Decide whether the following statement is *sometimes, always,* or *never* true. Explain your reasoning.

Two congruent triangles are similar.

39. CHALLENGE If $\triangle ABC \sim \triangle XYZ$ use the diagram below to find the values of x and y.

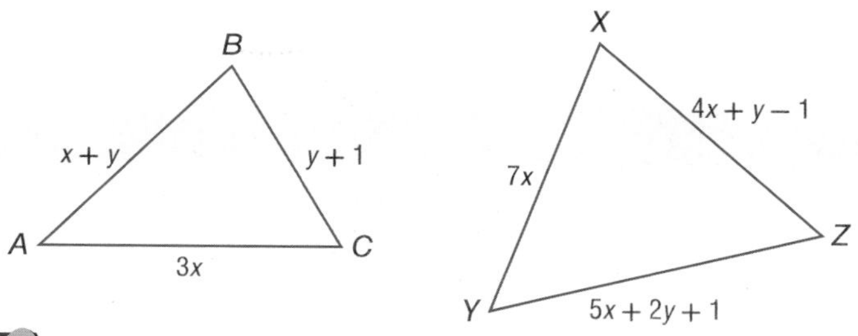

40. WRITING IN MATH Explain what information you need to prove two triangles are similar.

Standardized Test Practice

41. PROBABILITY $\frac{x!}{(x-3)!} =$

A 3.0

B 0.33

C $x^2 - 3x + 2$

D $x^3 - 3x^2 + 2x$

42. EXTENDED RESPONSE In the figure below, $\overline{EB} \parallel \overline{DC}$.

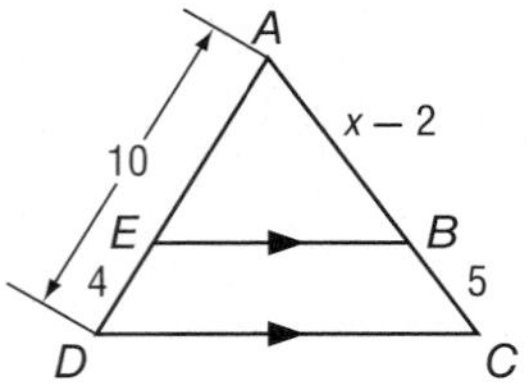

a. Write a proportion that could be used to find x.

b. Find the value of x and the measure of $\overline{AB}$.

43. ALGEBRA Which polynomial represents the area of the shaded region?

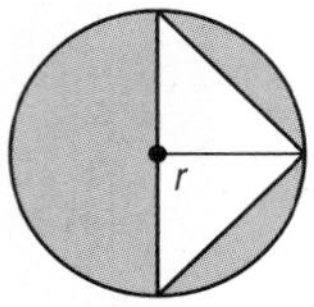

F πr^2

G $\pi r^2 + r^2$

H $\pi r^2 + r$

J $\pi r^2 - r^2$

44. SAT/ACT The volume of a certain rectangular solid is $16x$ cubic units. If the dimensions of the solid are integers x, y, and z units, what is the greatest possible value of z?

A 32

B 16

C 8

D 4

E 2

Spiral Review

Solve each compound inequality. Then graph the solution set. (Lesson 5-4)

45. $k + 2 > 12$ and $k + 2 \leq 18$

46. $d - 4 > 3$ or $d - 4 \leq 1$

47. $3 < 2x - 3 < 15$

48. $3t - 7 \geq 5$ and $2t + 6 \leq 12$

49. $h - 10 < -21$ or $h + 3 < 2$

50. $4 < 2y - 2 < 10$

51. FINANCIAL LITERACY A home security company provides security systems for \$5 per week, plus an installation fee. The total cost for installation and 12 weeks of service is \$210. Write the point-slope form of an equation to find the total fee y for any number of weeks x. What is the installation fee? (Lesson 4-3)

52. TANGRAMS A tangram set consists of seven pieces: a small square, two small congruent right triangles, two large congruent right triangles, a medium-sized right triangle, and a quadrilateral. How can you determine the shape of the quadrilateral? Explain. (Lesson 13-2)

Determine which postulate can be used to prove that the triangles are congruent. If it is not possible to prove congruence, write *not possible*. (Lesson 12-4)

53.

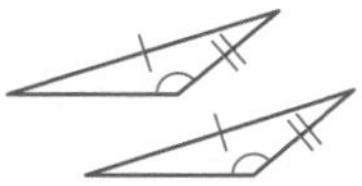

54.

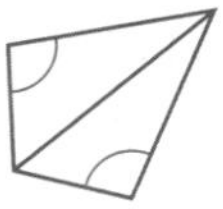

55.

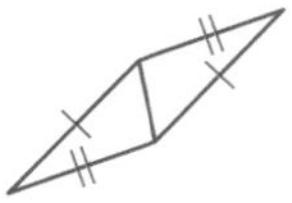

Skills Review

Write a two-column proof.

56. Given: $r \parallel t$; $\angle 5 \cong \angle 6$

Prove: $\ell \parallel m$

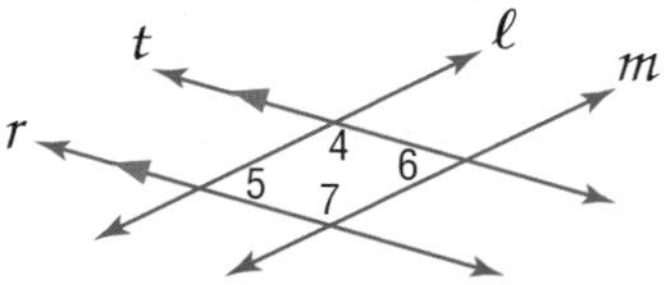

EXTEND 14-1

Geometry Lab
Proofs of Perpendicular and Parallel Lines

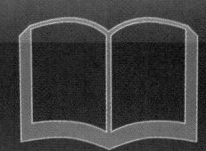

You have learned that two straight lines that are neither horizontal nor vertical are perpendicular if and only if the product of their slopes is –1. In this activity, you will use similar triangles to prove the first half of this theorem: if two straight lines are perpendicular, then the product of their slopes is –1.

CCSS Common Core State Standards
Content Standards
G.GPE.5 Prove the slope criteria for parallel and perpendicular lines and use them to solve geometric problems (e.g., find the equation of a line parallel or perpendicular to a given line that passes through a given point).
Mathematical Practices 3

Activity 1 Perpendicular Lines

Given: Slope of $\overleftrightarrow{AC} = m_1$, slope of $\overleftrightarrow{CE} = m_2$, and $\overleftrightarrow{AC} \perp \overleftrightarrow{CE}$.
Prove: $m_1 m_2 = -1$

Step 1 On a coordinate plane, construct $\overleftrightarrow{AC} \perp \overleftrightarrow{CE}$ and transversal $\overleftrightarrow{BD}$ parallel to the x-axis through C. Then construct right $\triangle ABC$ such that $\overline{AC}$ is the hypotenuse and right $\triangle EDC$ such that $\overline{CE}$ is the hypotenuse. The legs of both triangles should be parallel to the x-and y-axes, as shown.

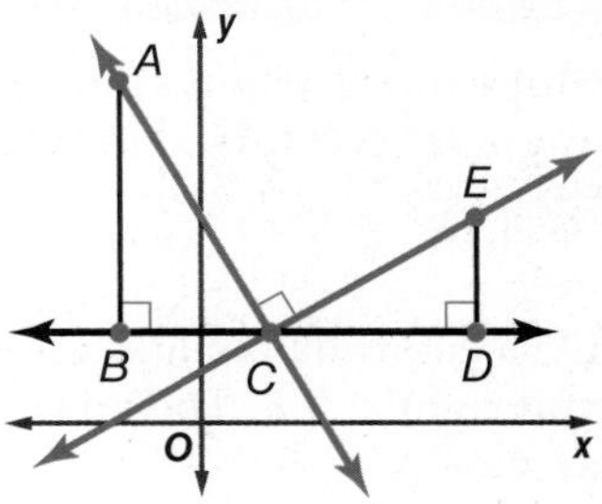

Step 2 Find the slopes of $\overleftrightarrow{AC}$ and $\overleftrightarrow{CE}$.

Slope of $\overleftrightarrow{AC}$		Slope of $\overleftrightarrow{CE}$	
$m_1 = \frac{\text{rise}}{\text{run}}$	Slope Formula	$m_2 = \frac{\text{rise}}{\text{run}}$	Slope Formula
$= \frac{-AB}{BC}$ or $-\frac{AB}{BC}$	rise $= -AB$, run $= BC$	$= \frac{DE}{CD}$	rise $= DE$, run $= CD$

Step 3 Show that $\triangle ABC \sim \triangle CDE$.

Since $\triangle ACB$ is a right triangle with right angle B, $\angle BAC$ is complementary to $\angle ACB$. It is given that $\overleftrightarrow{AC} \perp \overleftrightarrow{CE}$, so we know that $\triangle ACE$ is a right angle. By construction, $\angle BCD$ is a straight angle. So, $\angle ECD$ is complementary to $\angle ACB$. Since angles complementary to the same angle are congruent, $\angle BAC \cong \angle ECD$. Since right angles are congruent, $\angle B \cong \angle D$. Therefore, by AA Similarity, $\triangle ABC \sim \triangle CDE$.

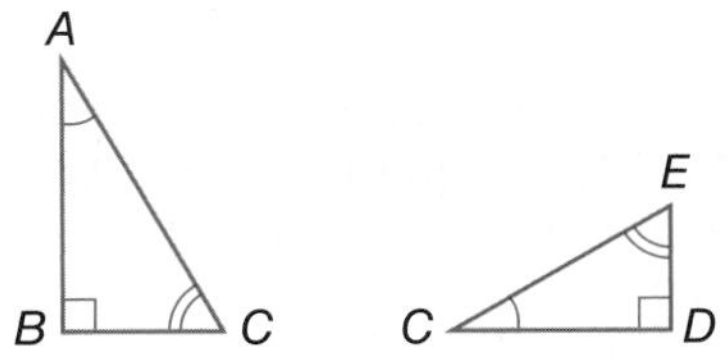

Step 4 Use the fact that $\triangle ABC \sim \triangle CDE$ to show that $m_1 m_2 = -1$.

Since $m_1 = -\frac{AB}{BC}$ and $m_2 = \frac{DE}{CD}$, $m_1 m_2 = \left(-\frac{AB}{BC}\right)\left(\frac{DE}{CD}\right)$. Since two similar polygons have proportional sides, $\frac{AB}{BC} = \frac{CD}{DE}$. Therefore, by substitution, $m_1 m_2 = \left(-\frac{CD}{DE}\right)\left(\frac{DE}{CD}\right)$ or -1.

Model

1. **PROOF** Use the diagram from Activity 1 to prove the second half of the theorem.

   Given: Slope of $\overleftrightarrow{CE} = m_1$, slope of $\overleftrightarrow{AC} = m_2$, and $m_1 m_2 = -1$. $\triangle ABC$ is a right triangle with right angle B. $\triangle CDE$ is a right triangle with right angle D.

   Prove: $\overleftrightarrow{CE} \perp \overleftrightarrow{AC}$

You can also use similar triangles to prove statements about parallel lines.

Activity 2 Parallel Lines

Given: Slope of $\overleftrightarrow{FG} = m_1$, slope of $\overleftrightarrow{JK} = m_2$, and $m_1 = m_2$. $\triangle FHG$ is a right triangle with right angle H. $\triangle JLK$ is a right triangle with right angle L.

Prove: $\overleftrightarrow{FG} \parallel \overleftrightarrow{JK}$

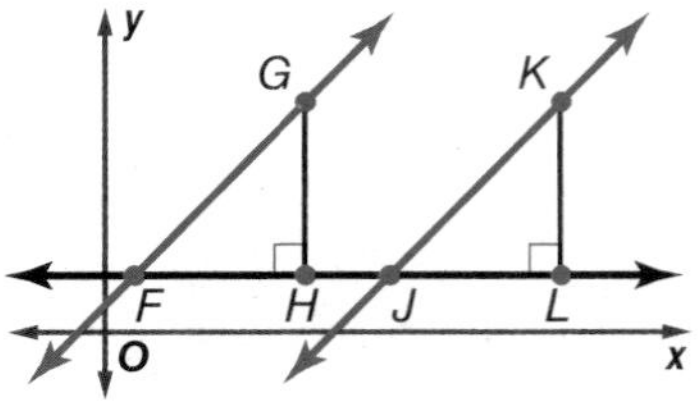

Step 1 On a coordinate plane, construct $\overleftrightarrow{FG}$ and $\overleftrightarrow{JK}$, right $\triangle FHG$, and right $\triangle JLK$. Then draw horizontal transversal $\overleftrightarrow{FL}$, as shown.

Step 2 Find the slopes of $\overleftrightarrow{FG}$ and $\overleftrightarrow{JK}$.

Slope of $\overleftrightarrow{FG}$		Slope of $\overleftrightarrow{JK}$	
$m_1 = \frac{\text{rise}}{\text{run}}$	Slope Formula	$m_2 = \frac{\text{rise}}{\text{run}}$	Slope Formula
$= \frac{GH}{HF}$	rise = GH, run = HF	$= \frac{KL}{LJ}$	rise = KL, run = LJ

Step 3 Show that $\triangle FHG \sim \triangle JLK$.

It is given that $m_1 = m_2$. By substitution, $\frac{GH}{HF} = \frac{KL}{LJ}$. This ratio can be rewritten as $\frac{GH}{KL} = \frac{HF}{LJ}$. Since $\angle H$ and $\angle L$ are right angles, $\angle H \cong \angle L$. Therefore, by SAS similarity, $\triangle FHG \sim \triangle JLK$.

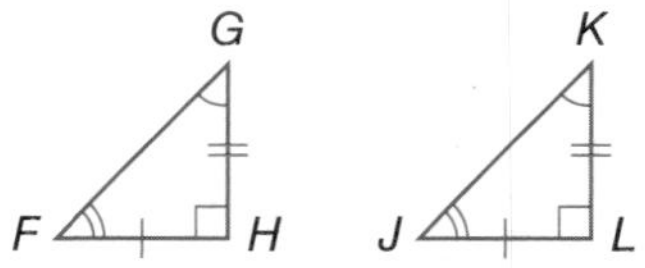

Step 4 Use the fact that $\triangle FHG \sim \triangle JLK$ to prove that $\overleftrightarrow{FG} \parallel \overleftrightarrow{JK}$.

Corresponding angles in similar triangles are congruent, so $\angle GFH \cong \angle KJL$. From the definition of congruent angles, $m\angle GFH = m\angle KJL$ (or $\angle GFH \cong \angle KJL$). By definition, $\angle KJH$ and $\angle KJL$ form a linear pair. Since linear pairs are supplementary, $m\angle KJH + m\angle KJL = 180$. So, by substitution, $m\angle KJH + m\angle GFH = 180$. By definition, $\angle KJH$ and $\angle GFH$ are supplementary. Since $\angle KJH$ and $\angle GFH$ are supplementary and are consecutive interior angles, $\overleftrightarrow{FG} \parallel \overleftrightarrow{JK}$.

Model

2. **PROOF** Use the diagram from Activity 2 to prove the following statement.

   Given: Slope of $\overleftrightarrow{FG} = m_1$, slope of $\overleftrightarrow{JK} = m_2$, and $\overleftrightarrow{FG} \parallel \overleftrightarrow{JK}$.

   Prove: $m_1 = m_2$

LESSON 14-2

Parallel Lines and Proportional Parts

Then

- You used proportions to solve problems between similar triangles.

Now

1. Use proportional parts within triangles.
2. Use proportional parts with parallel lines.

Why?

- Photographers have many techniques at their disposal that can be used to add interest to a photograph. One such technique is the use of a vanishing point perspective, in which an image with parallel lines, such as train tracks, is photographed so that the lines appear to converge at a point on the horizon.

NewVocabulary
midsegment of a triangle

Common Core State Standards

Content Standards
G.SRT.4 Prove theorems about triangles.
G.SRT.5 Use congruence and similarity criteria for triangles to solve problems and to prove relationships in geometric figures.

Mathematical Practices
1 Make sense of problems and persevere in solving them.
3 Construct viable arguments and critique the reasoning of others.

1 Proportional Parts Within Triangles

When a triangle contains a line that is parallel to one of its sides, the two triangles formed can be proved similar using the Angle-Angle Similarity Postulate. Since the triangles are similar, their sides are proportional.

Theorem 14.4 Triangle Proportionality Theorem

If a line is parallel to one side of a triangle and intersects the other two sides, then it divides the sides into segments of proportional lengths.

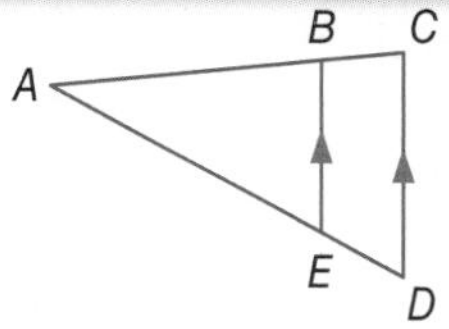

Example If $\overline{BE} \parallel \overline{CD}$, then $\frac{AB}{BC} = \frac{AE}{ED}$.

You will prove Theorem 14.4 in Exercise 30.

Example 1 Find the Length of a Side

PT

In $\triangle PQR$, $\overline{ST} \parallel \overline{RQ}$. If $PT = 7.5$, $TQ = 3$, and $SR = 2.5$, find PS.

Use the Triangle Proportionality Theorem.

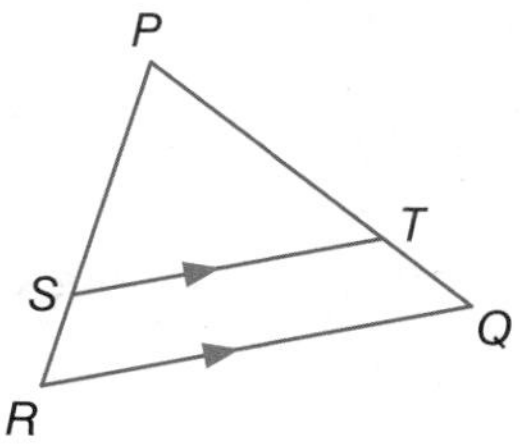

$\frac{PS}{SR} = \frac{PT}{TQ}$	Triangle Proportionality Theorem
$\frac{PS}{2.5} = \frac{7.5}{3}$	Substitute.
$PS \cdot 3 = (2.5)(7.5)$	Cross Products Property
$3PS = 18.75$	Multiply.
$PS = 6.25$	Divide each side by 3.

GuidedPractice

1. If $PS = 12.5$, $SR = 5$, and $PT = 15$, find TQ.

Gregory Warran/Flickr/Getty Images

Math HistoryLink

Galileo Galilei (1564–1642) Galileo was born in Pisa, Italy. He studied philosophy, astronomy, and mathematics. Galileo made essential contributions to all three disciplines. Refer to Exercise 39.

Source: *Encyclopaedia Britannica*

The converse of Theorem 14.4 is also true and can be proved using the proportional parts of a triangle.

Theorem 14.5 Converse of Triangle Proportionality Theorem

If a line intersects two sides of a triangle and separates the sides into proportional corresponding segments, then the line is parallel to the third side of the triangle.

Example If $\frac{AE}{EB} = \frac{CD}{DB}$, then $\overline{AC} \parallel \overline{ED}$.

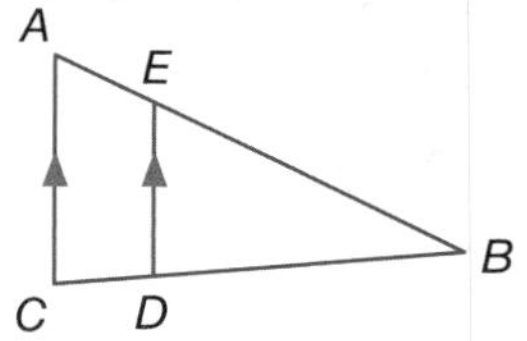

You will prove Theorem 14.5 in Exercise 31.

Example 2 Determine if Lines are Parallel

In $\triangle DEF$, $EH = 3$, $HF = 9$, and DG is one-third the length of $\overline{GF}$. Is $\overline{DE} \parallel \overline{GH}$?

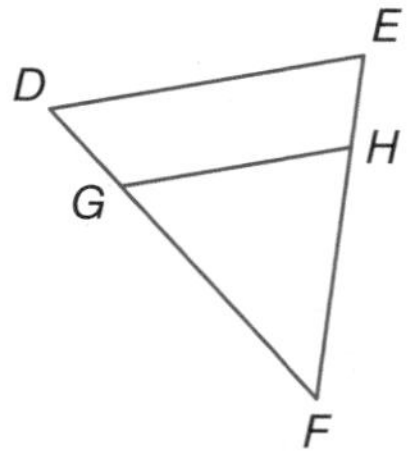

Using the converse of the Triangle Proportionality Theorem, in order to show that $\overline{DE} \parallel \overline{GH}$, we must show that $\frac{DG}{GF} = \frac{EH}{HF}$.

Find and simplify each ratio. Let $DG = x$.
Since DG is one-third of GF, $GF = 3x$.

$\frac{DG}{GF} = \frac{x}{3x}$ or $\frac{1}{3}$ $\qquad$ $\frac{EH}{HF} = \frac{3}{9}$ or $\frac{1}{3}$

Since $\frac{1}{3} = \frac{1}{3}$, the sides are proportional, so $\overline{DE} \parallel \overline{GH}$.

GuidedPractice

2. DG is half the length of $\overline{GF}$, $EH = 6$, and $HF = 10$. Is $\overline{DE} \parallel \overline{GH}$?

StudyTip

Midsegment Triangle The three midsegments of a triangle form the *midsegment triangle.*

A **midsegment of a triangle** is a segment with endpoints that are the midpoints of two sides of the triangle. Every triangle has three midsegments. The midsegments of $\triangle ABC$ are $\overline{RP}$, $\overline{PQ}$, $\overline{RQ}$.

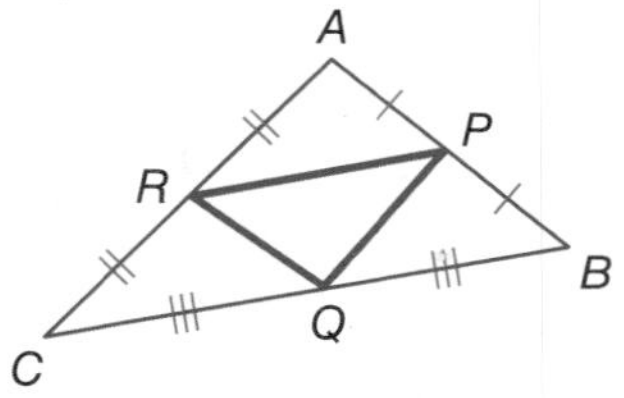

A special case of the Triangle Proportionality Theorem is the Triangle Midsegment Theorem.

Theorem 14.6 Triangle Midsegment Theorem

A midsegment of a triangle is parallel to one side of the triangle, and its length is one half the length of that side.

Example If J and K are midpoints of $\overline{FH}$ and $\overline{HG}$, respectively, then $\overline{JK} \parallel \overline{FG}$ and $JK = \frac{1}{2}FG$.

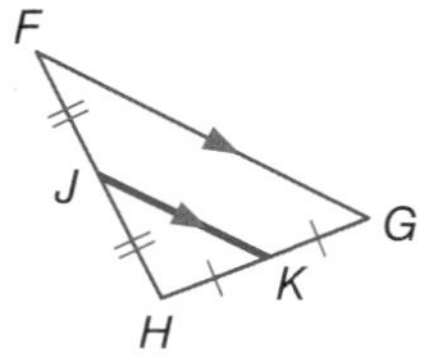

You will prove Theorem 14.6 in Exercise 32.

StudyTip

Midsegment The Triangle Midsegment Theorem is similar to the Trapezoid Midsegment Theorem, which states that the midsegment of a trapezoid is parallel to the bases and its length is one half the sum of the measures of the bases. (Lesson 6-6)

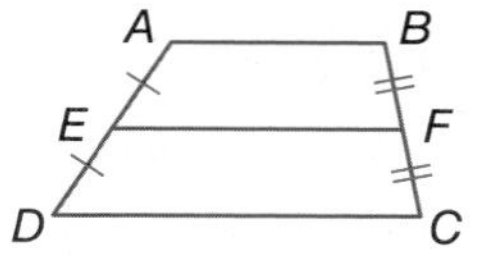

$\overline{EF} \parallel \overline{AB} \parallel \overline{DC}$

$EF = \frac{1}{2}(AB + DC)$

Example 3 Use the Triangle Midsegment Theorem

In the figure, $\overline{XY}$ and $\overline{XZ}$ are midsegments of $\triangle RST$. Find each measure.

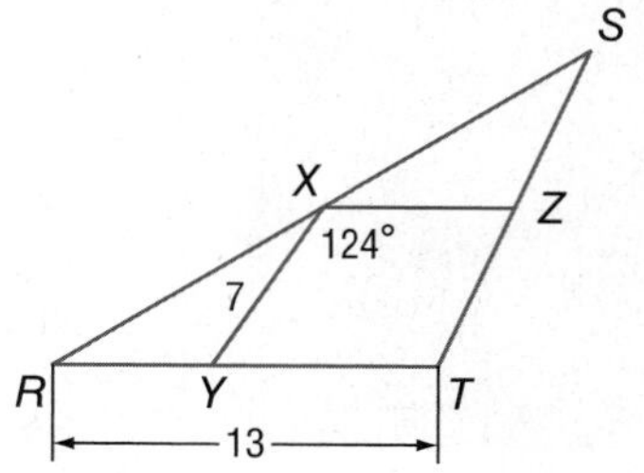

a. XZ

$XZ = \frac{1}{2}RT$ — Triangle Midsegment Theorem

$XZ = \frac{1}{2}(13)$ — Substitution

$XZ = 6.5$ — Simplify.

b. ST

$XY = \frac{1}{2}ST$ — Triangle Midsegment Theorem

$7 = \frac{1}{2}ST$ — Substitution

$14 = ST$ — Multiply each side by 2.

c. $m\angle RYX$

By the Triangle Midsegment Theorem, $\overline{XZ} \parallel \overline{RT}$.

$\angle RYX \cong \angle YXZ$ — Alternate Interior Angles Theorem

$m\angle RYX = m\angle YXZ$ — Definition of congruence

$m\angle RYX = 124$ — Substitution

GuidedPractice

Find each measure.

3A. DE

3B. DB

3C. $m\angle FED$

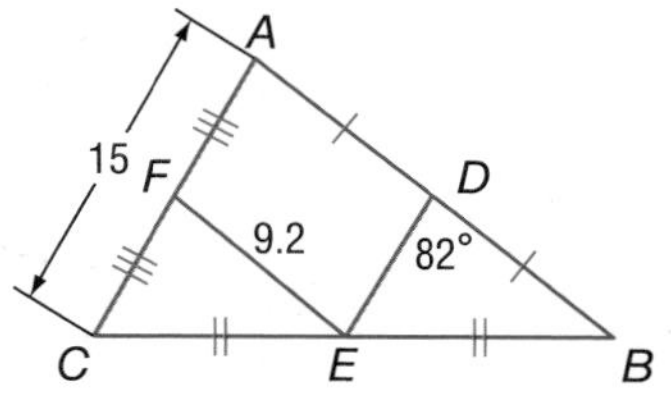

2 Proportional Parts with Parallel Lines

Another special case of the Triangle Proportionality Theorem involves three or more parallel lines cut by two transversals. Notice that if transversals a and b are extended, they form triangles with the parallel lines.

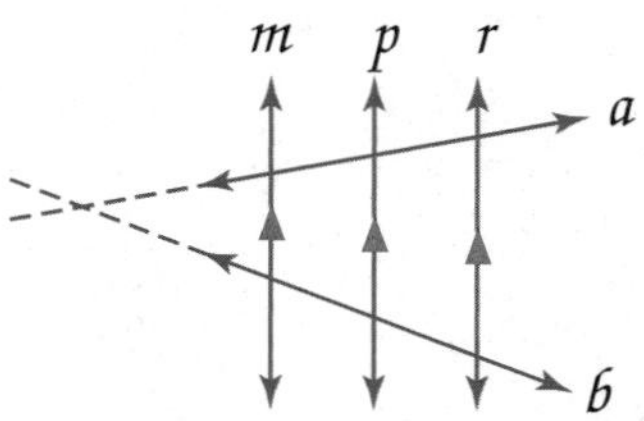

StudyTip

Other Proportions Two other proportions can be written for the example in Corollary 7.1.

$\frac{AB}{EF} = \frac{BC}{FG}$ and $\frac{AC}{BC} = \frac{EG}{FG}$

Corollary 14.1 Proportional Parts of Parallel Lines

If three or more parallel lines intersect two transversals, then they cut off the transversals proportionally.

Example If $\overline{AE} \parallel \overline{BF} \parallel \overline{CG}$, then $\frac{AB}{BC} = \frac{EF}{FG}$.

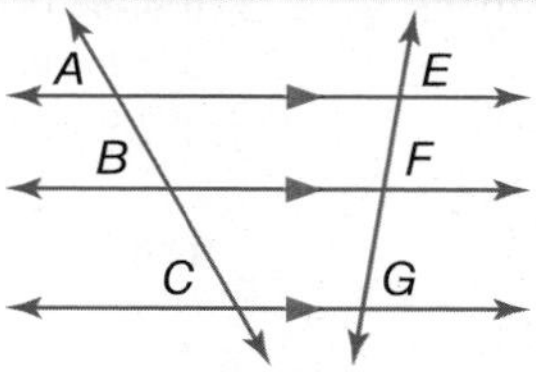

You will prove Corollary 14.1 in Exercise 28.

Real-WorldLink

To make a two-dimensional drawing appear three-dimensional, an artist provides several perceptual cues.

- *size* - faraway items look smaller
- *clarity* - closer objects appear more in focus
- *detail* - nearby objects have texture, while distant ones are roughly outlined

Source: Center for Media Literacy

Real-World Example 4 Use Proportional Segments of Transversals

PT

ART Megan is drawing a hallway in one-point perspective. She uses the guidelines shown to draw two windows on the left wall. If segments $\overline{AD}$, $\overline{BC}$, $\overline{WZ}$, and $\overline{XY}$ are all parallel, $AB = 8$ centimeters, $DC = 9$ centimeters, and $ZY = 5$ centimeters, find WX.

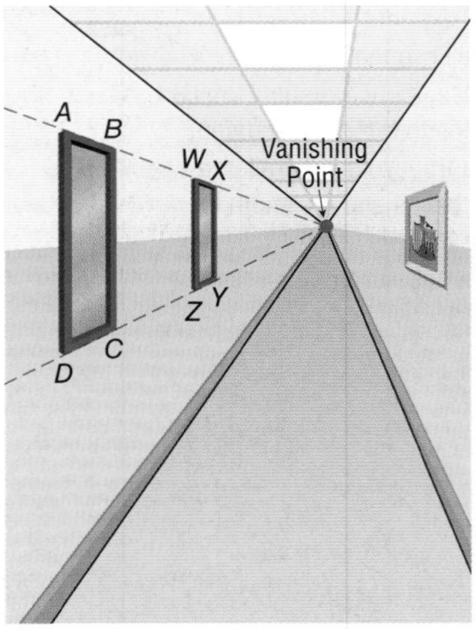

By Corollary 7.1, if $\overline{AD} \parallel \overline{BC} \parallel \overline{WZ} \parallel \overline{XY}$,

then $\frac{AB}{WX} = \frac{DC}{ZY}$.

$\frac{AB}{WX} = \frac{DC}{ZY}$	Corollary 14.1
$\frac{8}{WX} = \frac{9}{5}$	Substitute.
$WX \cdot 9 = 8 \cdot 5$	Cross Products Property
$9WX = 40$	Simplify.
$WX = \frac{40}{9}$	Divide each side by 4.

The distance between W and X should be $\frac{40}{9}$ or about 4.4 centimeters.

CHECK The ratio of DC to ZY is 9 to 5, which is about 10 to 5 or 2 to 1. The ratio of AB to WX is 8 to 4.4 or about 8 to 4 or 2 to 1 as well, so the answer is reasonable. ✓

GuidedPractice

4. REAL ESTATE *Frontage* is the measurement of a property's boundary that runs along the side of a particular feature such as a street, lake, ocean, or river. Find the ocean frontage for Lot A to the nearest tenth of a yard.

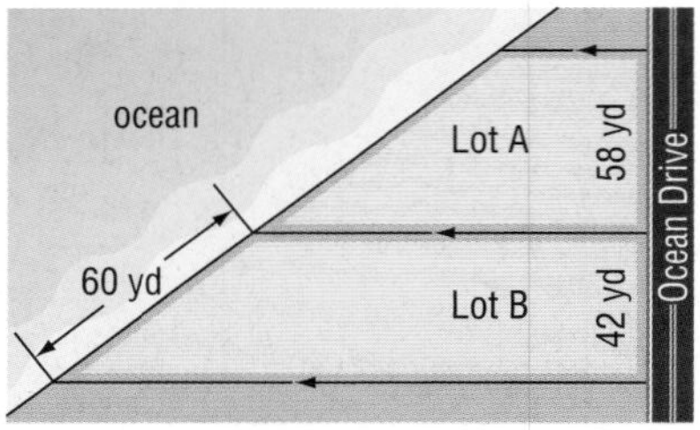

If the scale factor of the proportional segments is 1, they separate the transversals into congruent parts.

Corollary 14.2 Congruent Parts of Parallel Lines

If three or more parallel lines cut off congruent segments on one transversal, then they cut off congruent segments on every transversal.

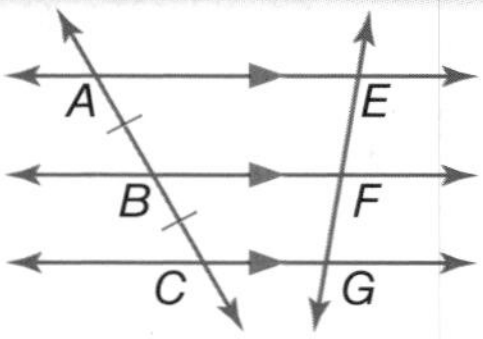

Example If $\overline{AE} \parallel \overline{BF} \parallel \overline{CG}$, and $\overline{AB} \cong \overline{BC}$, then $\overline{EF} \cong \overline{FG}$.

You will prove Corollary 14.2 in Exercise 29.

Real-World Example 5 Use Congruent Segments of Transversals

ALGEBRA Find x and y.

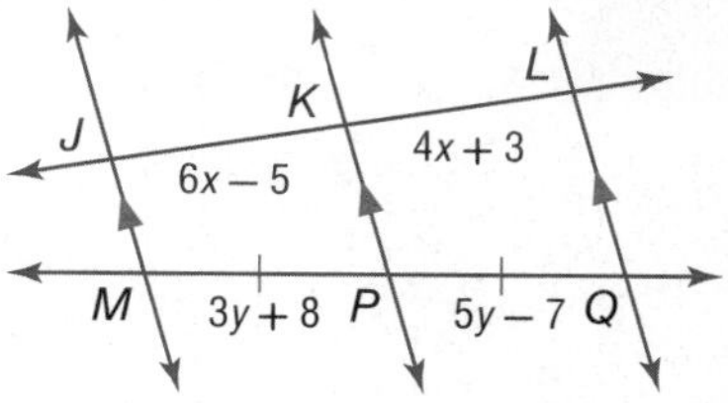

Since $\overleftrightarrow{JM} \parallel \overleftrightarrow{KP} \parallel \overleftrightarrow{LQ}$ and $\overline{MP} \cong \overline{PQ}$, then $\overline{JK} \cong \overline{KL}$ by Corollary 7.2.

$JK = KL$	Definition of congruence
$6x - 5 = 4x + 3$	Substitution
$2x - 5 = 3$	Subtract $4x$ from each side.
$2x = 8$	Add 5 to each side.
$x = 4$	Divide each side by 2.
$MP = PQ$	Definition of congruence
$3y + 8 = 5y - 7$	Substitution
$8 = 2y - 7$	Subtract $3y$ from each side.
$15 = 2y$	Add 7 to each side.
$7.5 = y$	Divide each side by 2.

GuidedPractice

5A.

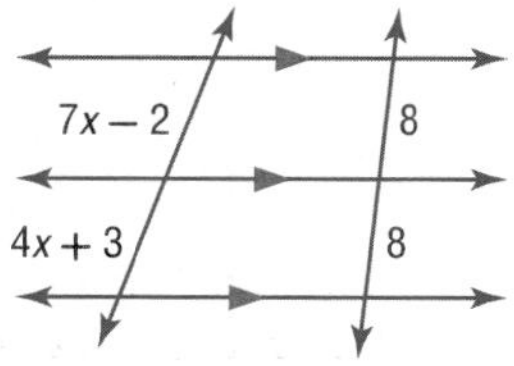

5B.

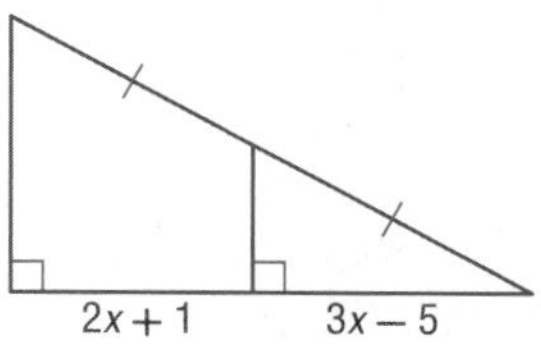

It is possible to separate a segment into two congruent parts by constructing the perpendicular bisector of a segment. However, a segment cannot be separated into three congruent parts by constructing perpendicular bisectors. To do this, you must use parallel lines and Corollary 14.2.

Construction Trisect a Segment

Draw a segment $\overline{AB}$. Then use Corollary 14.2 to trisect $\overline{AB}$.

Step 1 Draw $\overline{AC}$. Then with the compass at A, mark off an arc that intersects $\overline{AC}$ at X.

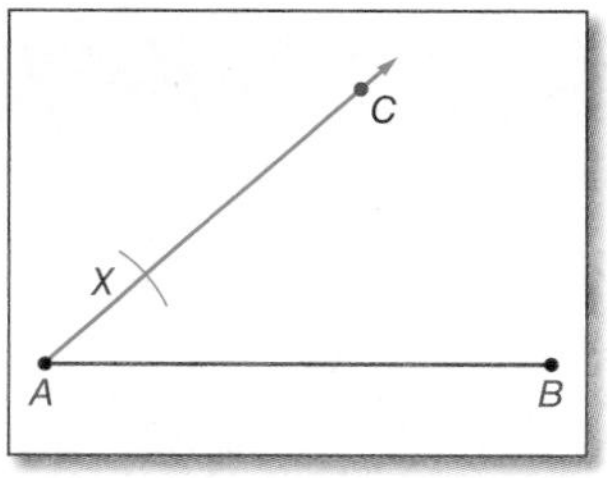

Step 2 Use the same compass setting to mark off Y and Z such that $\overline{AX} \cong \overline{XY} \cong \overline{YZ}$. Then draw ZB.

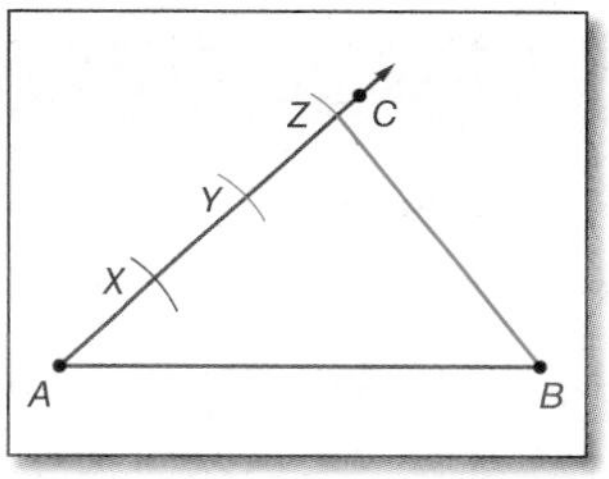

Step 3 Construct lines through Y and X that are parallel to $\overline{ZB}$. Label the intersection points on $\overline{AB}$ as J and K.

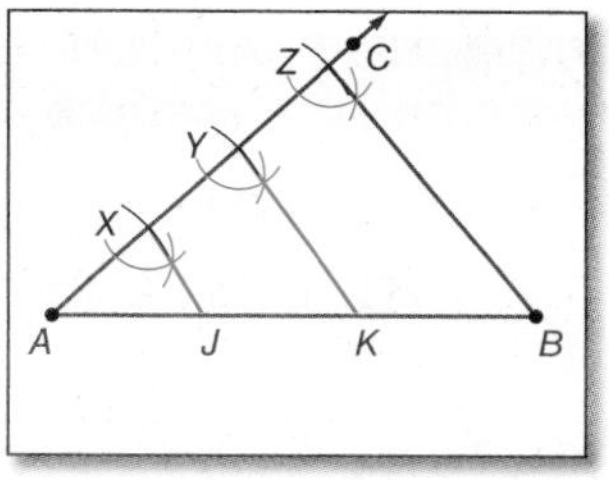

Conclusion: Since parallel lines cut off congruent segments on transversals, $\overline{AJ} \cong \overline{JK} \cong \overline{KB}$.

Check Your Understanding

Example 1

1. If $AB = 3$, $AE = 1$, and $ED = 12$, find AD.
2. If $AB = 7$, $AE = 2$, and $AC = 28$, find ED.

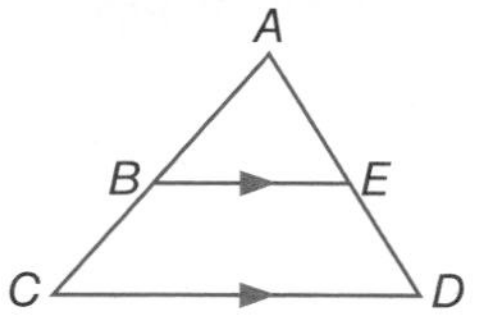

Example 2

3. In $\triangle XYZ$, if $XY = 76$, $XW = 19$, $XV = 18$, and $XZ = 72$, determine whether $\overline{VW} \parallel \overline{ZY}$. Justify your answer.

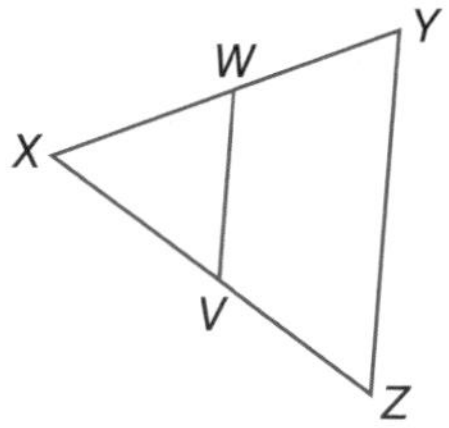

4. In $\triangle LMN$, if $LP = 6$, $LM = 22$, $LQ = 14$, and $LN = 43$, determine whether $\overline{QP} \parallel \overline{NP}$. Justify your answer.

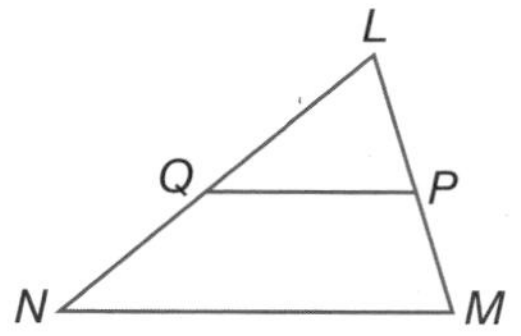

Example 3

$\overline{RS}$ is a midsegment of $\triangle MPQ$. Find the value of x.

5.

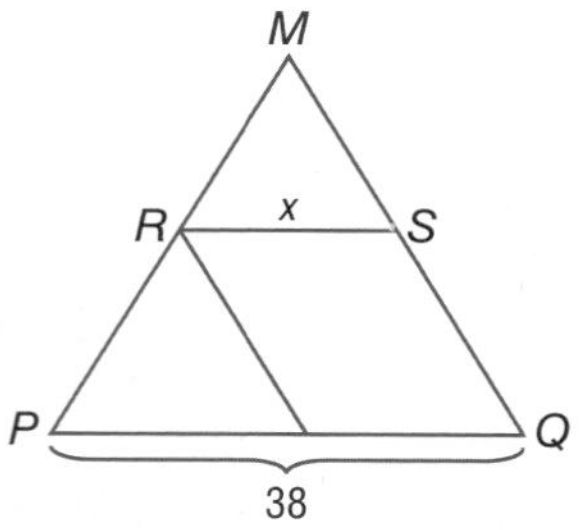

6.

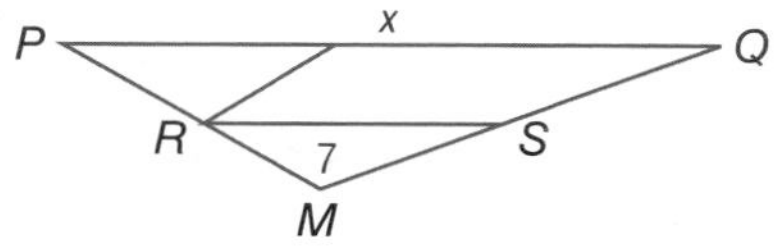

Example 4

7. **SPORTS** Just is building a bike ramp with dimensions as shown. If the support is parallel to the back of the ramp, what is the distance from the front of the ramp to the support?

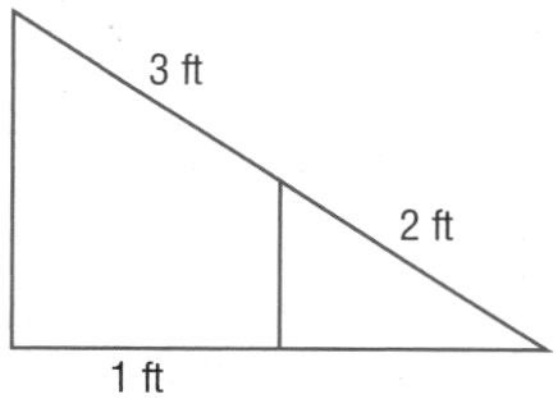

Example 5

ALGEBRA Find x and y.

8.

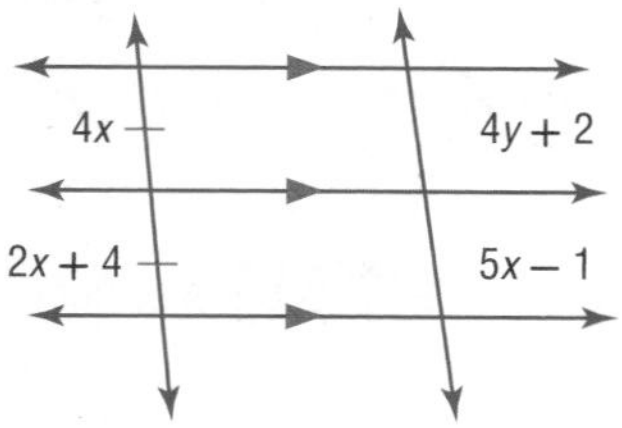

9.

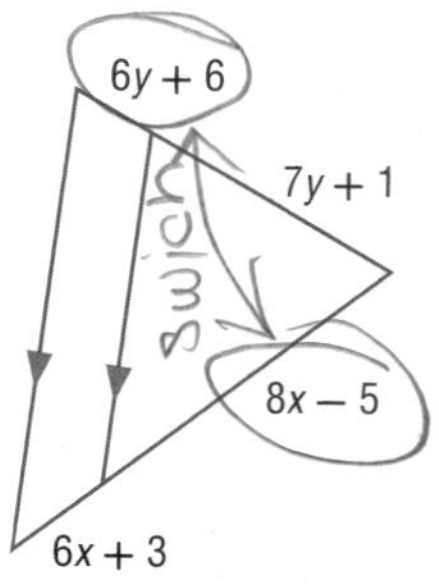

Practice and Problem Solving

Example 1

10. If $WQ = 37$, $MQ = 30$, and $RN = 74$, find PR.
11. If $MQ = 44$, $PR = 22$, and $RN = 34$, find MQ.
12. If $NQ = 18$, $MQ = 47$, and $PR = 94$, find RN.
13. If $MQ = 60$, $PR = 20$, and $RN = 30$, find QN.

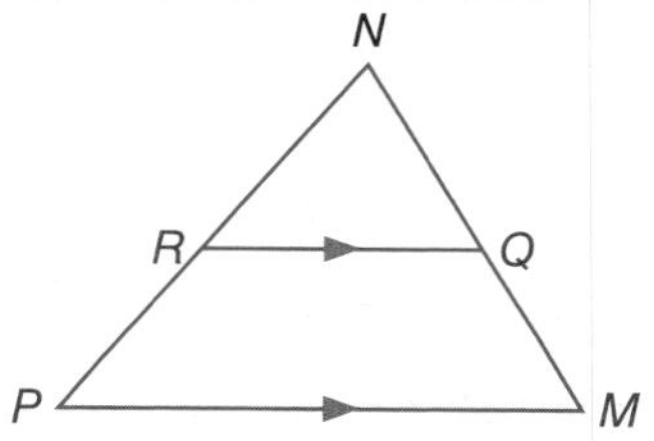

Example 2 **Determine whether $\overline{DE} \parallel \overline{BC}$. Justify your answer.**

14. $\overline{AD} = 19$, $\overline{DB} = 20$, $\overline{AE} = 76$, and $\overline{EC} = 80$

15. $\overline{AD} = 52$, $\overline{DB} = 25$, $\overline{AE} = 13$, and $\overline{EC} = 6$

16. $\overline{AD} = 26$, $\overline{DB} = 19$, $\overline{AE} = 52$, and $\overline{EC} = 18$

17. $\overline{AD} = 69$, $\overline{DB} = 42$, $\overline{AE} = 23$, and $\overline{EC} = 14$

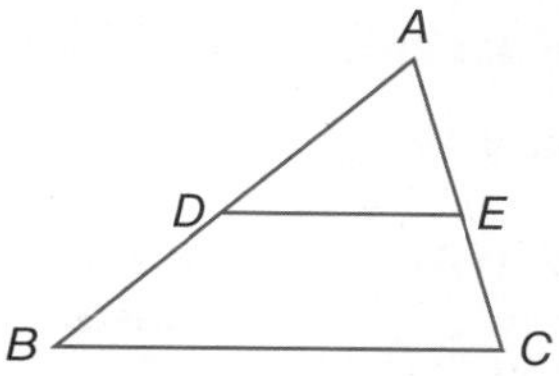

Example 3 **$\overline{TV}$, $\overline{VW}$, and $\overline{TW}$ are midsegments of $\triangle RQS$. Find the value of x.**

18.

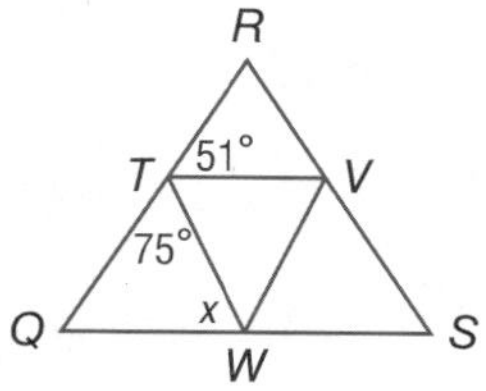

19.

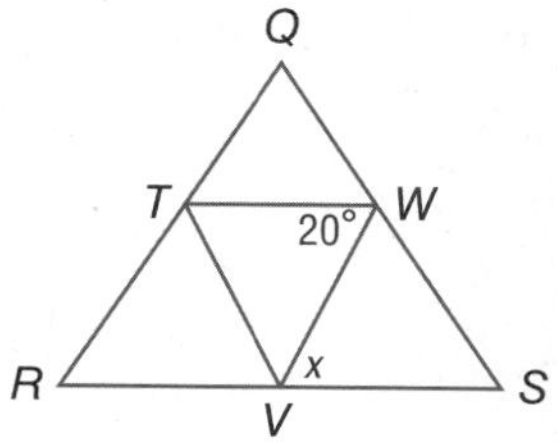

20.

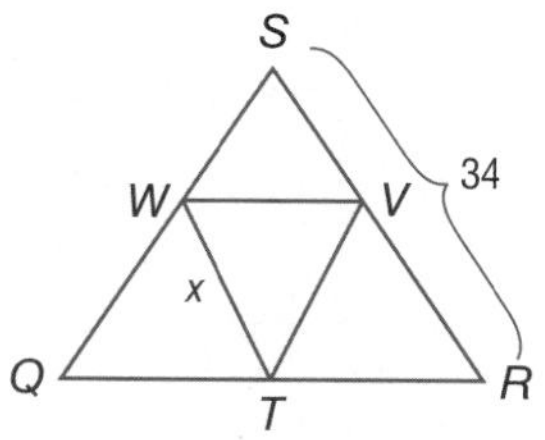

21.

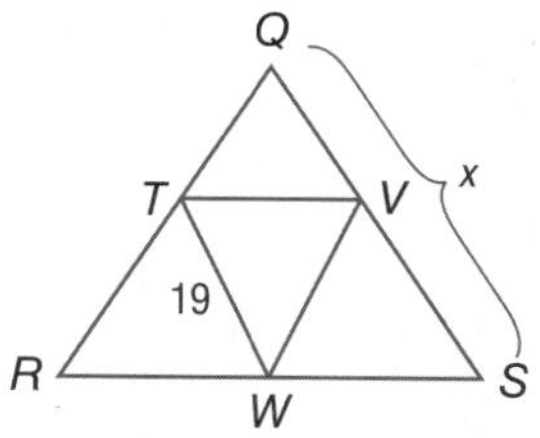

Example 4 **22.** **CCSS SCHOOL SPIRIT** Jessica is creating a banner for a school pep rally. If $PN \parallel LM$, $LP = 26''$, $PO = 50''$, and $MN = 13''$, find MO.

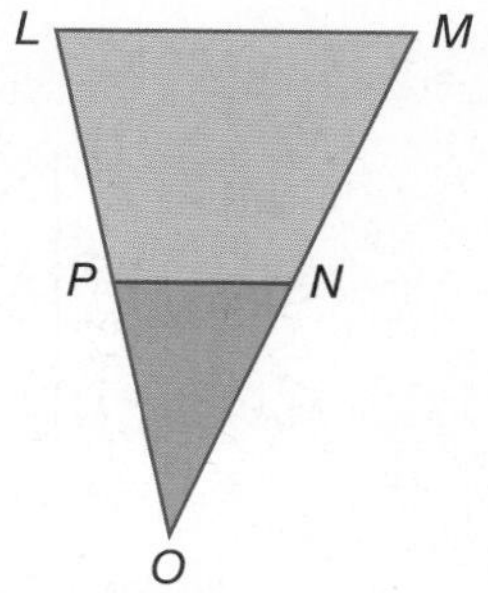

23. **CAMPING** While camping, Joe wants to set up his tent half-way between a tree and the fire pit. If the distance between the top of the tree and the top of his tent is 36 feet, How far is the top of his tent from the fire pit?

Example 5 **ALGEBRA** **Find x and y.**

24.

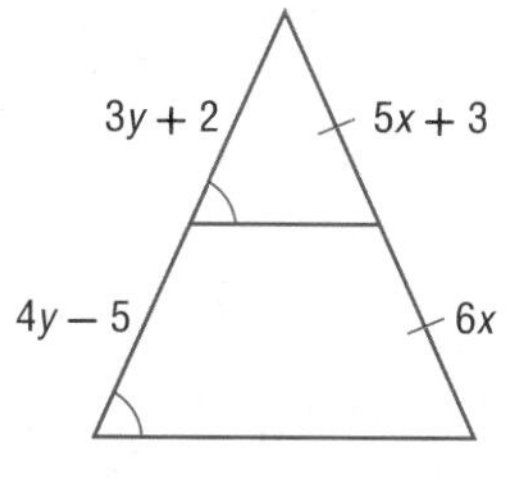

25.

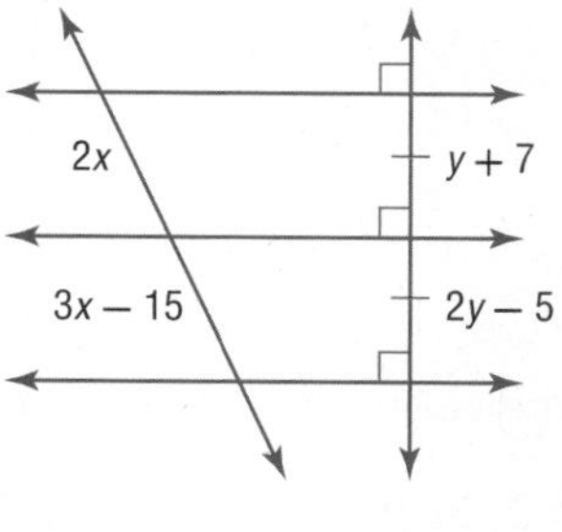

26.

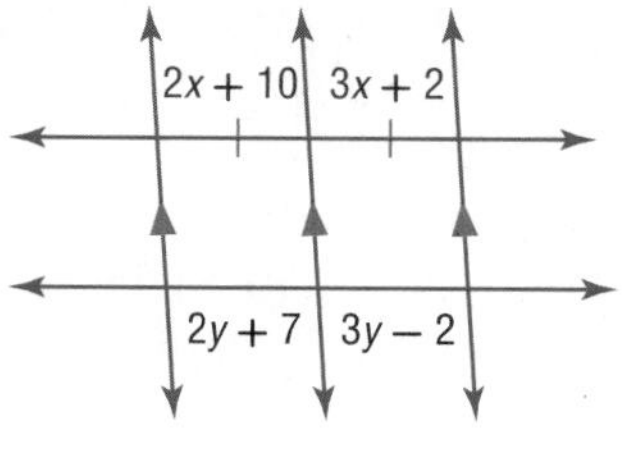

27.

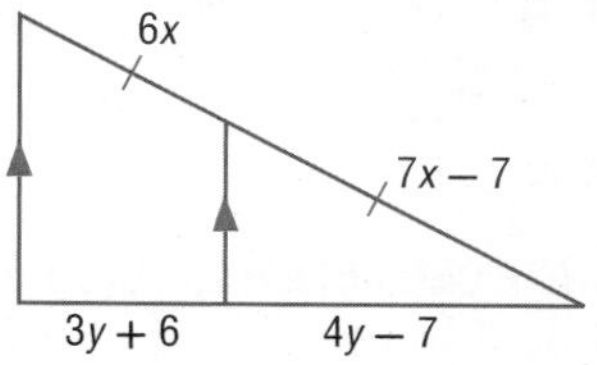

CCSS PROOF **Write a paragraph proof.**

28. Corollary 14.1

29. Corollary 14.2

30. Theorem 14.4

CCSS PROOF **Write a two-column proof.**

31. Theorem 14.5

32. Theorem 14.6

Refer to $\triangle LMP$.

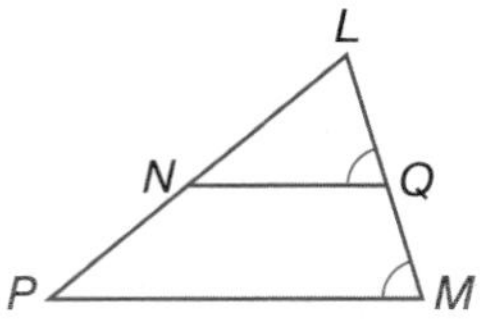

33. If $LQ = 42$, $QM = 42$, and $NQ = 50$, find PM.

34. If $LN = 12$, $NQ = 18$, and $PM = 36$, find LP.

35. If $WZ = 2x + 4$, $ZY = 2x + 1$, $WV = 68$, and $WX = 130$, find ZY and x.

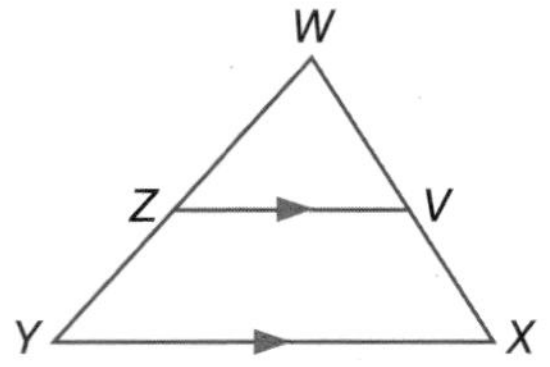

36. If $AB = t + 2$, $AC = 31$, $AE = 4t + 8$, and $ED = 2t - 4$, find AB.

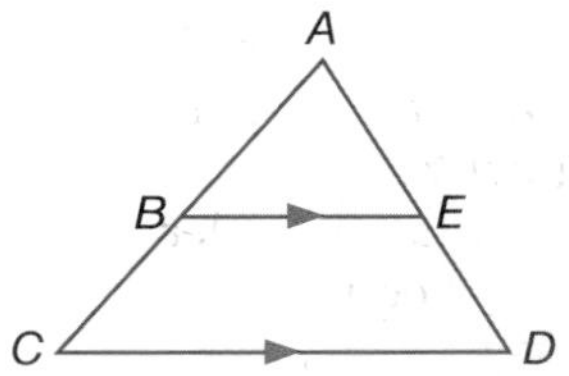

37. If $JK = 12$, $MN = 7$, $JQ = 24$, and $KM = 49$, find PO and QP.

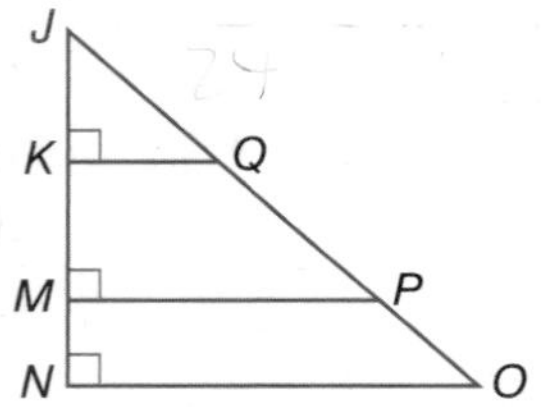

38. If $QR = 4$, $QW = 2$, $RS = 80$, $XY = 6$, and $TV = 28$, find WX, ST, and YZ.

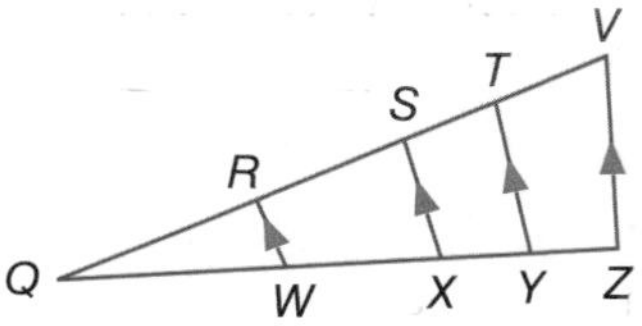

39. BABYSITTING While babysitting, Jacqueline notices the support bar on the swing set is a midsegment of $\triangle ABC$. If Jacqueline estimates the support bar to be 4 feet long, how far is point B from point C?

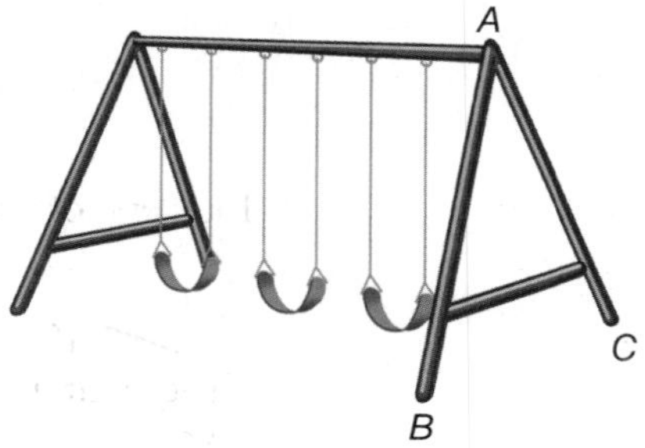

Determine the value of x so that $JM \parallel KL$.

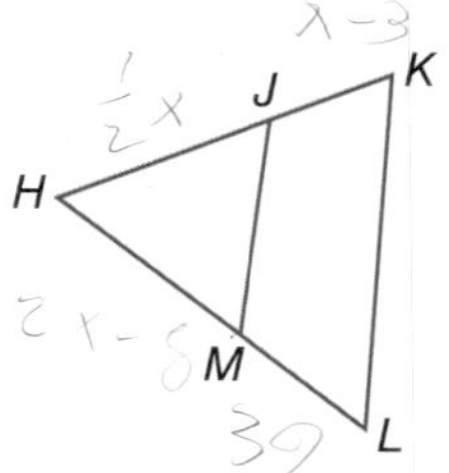

40. $HJ = 19x + 2$, $JK = 93$, $HM = 6x + 2$, and $ML = 7x + 3$

41. $HJ = \frac{1}{2}x$, $JK = x - 3$, $HM = 2x - 8$, and $ML = 39$

42. COORDINATE GEOMETRY $\triangle QRS$ has vertices $Q(-5, 10)$, $R(3, 4)$ and $S(10, 8)$. Draw $\triangle QRS$. Determine the coordinates of the midsegment that is parallel to RS. Justify your answer.

43 **ART** As part of an art project, Ahmed creates an isosceles triangle out of different strips of colored paper. If each strip of paper has the same width, find the lengths of BD, DE, EF, and FA.

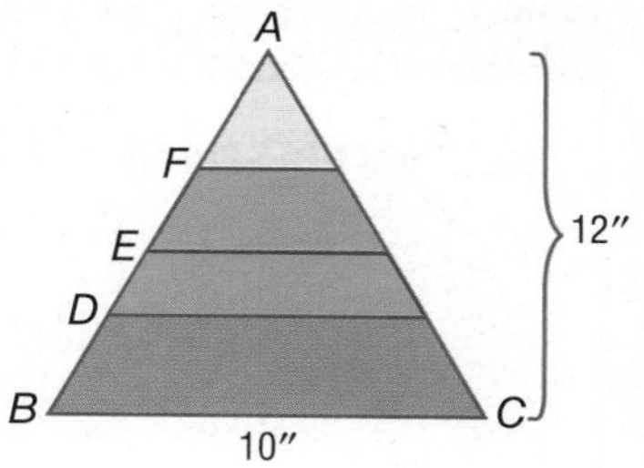

CONSTRUCTION **Construct each segment as directed.**

44. a segment separated into five congruent segments

45. a segment separated into two segments in which their lengths have a ratio of 1 to 3

46. a segment 3 inches long, separated into four congruent segments

47. **MULTIPLE REPRESENTATIONS** In this problem you will explore the midsegments of right triangles.

a. Geometric Draw three right triangles and their midsegments. Label the triangles ABC and the midsegments MLP. Measure and label the lengths of each midsegment.

b. Tabular Copy and complete the following table.

Triangle	ML	LP	MP	$\triangle MLP$ right triangle?
1				
2				
3				

c. Verbal Make a conjecture about the midsegments of a right triangle.

H.O.T. Problems Use Higher-Order Thinking Skills

48. **CCSS ERROR ANALYSIS** Justin and Adam are trying to figure out if $BE \parallel CD$. Justin thinks that BE is not parallel to CD, but Adam thinks they are parallel. Which one of them is correct? Explain your answer.

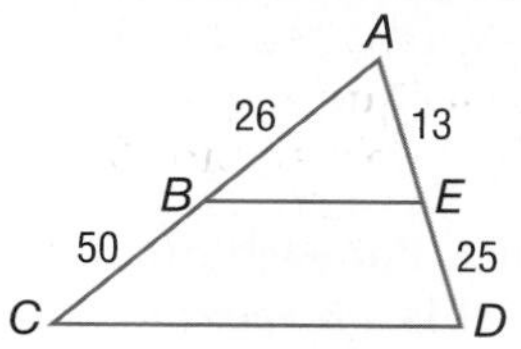

49. **REASONING** If $\triangle LMN$ is a right triangle with PQ as a midsegment, is $\angle LPQ$ a right angle? Explain your answer.

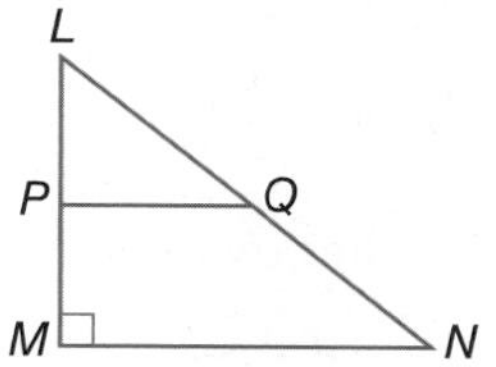

50. **CHALLENGE** Using the diagram shown, find x, y, and z.

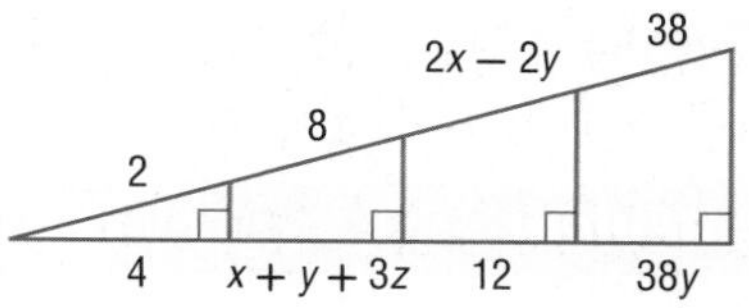

51. **OPEN ENDED** Draw and label triangle ABC with midsegment PQ which is parallel to BC, such that $AP = 3$ and $QC = 4$.

52. **WRITING IN MATH** Compare and contrast Corollary 14.1 and Corollary 14.2.

Standardized Test Practice

53. **SHORT RESPONSE** What is the value of x?

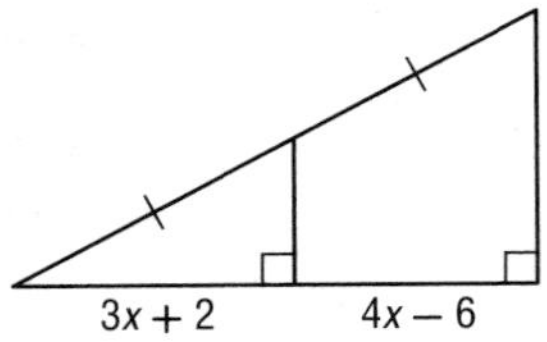

54. If the vertices of triangle JKL are (0, 0), (0, 10) and (10, 10), then the area of triangle JKL is

A 20 units^2
B 30 units^2
C 40 units^2
D 50 units^2

55. **ALGEBRA** A breakfast cereal contains wheat, rice, and oats in the ratio 2:4:1. If the manufacturer makes a mixture using 110 pounds of wheat, how many pounds of rice will be used?

F 120 lb
G 220 lb
H 240 lb
J 440 lb

56. **SAT/ACT** If the area of a circle is 16 square meters, what is its radius in meters?

A $\frac{4\sqrt{\pi}}{\pi}$
B $\frac{8}{\pi}$
C $\frac{16}{\pi}$
D 12π
E 16π

Spiral Review

ALGEBRA Identify the similar triangles. Then find the measure(s) of the indicated segment(s). (Lesson 14-1)

57. $\overline{AB}$

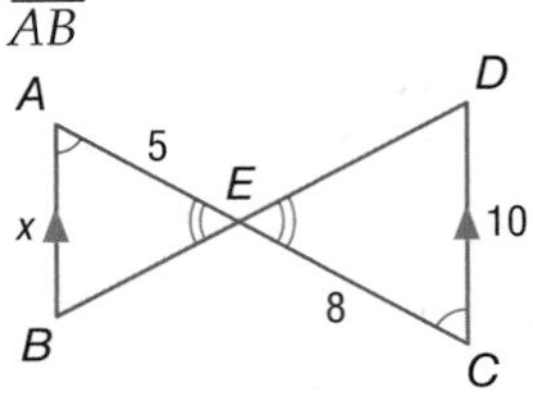

58. $\overline{RT}$, $\overline{RS}$

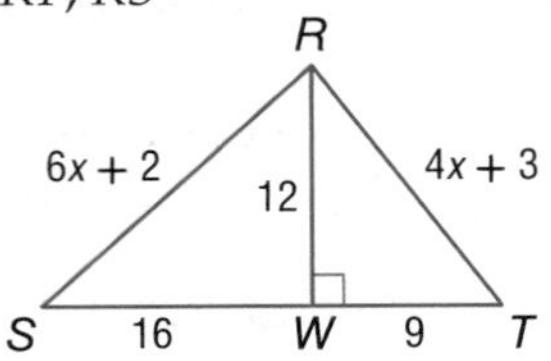

59. $\overline{TY}$

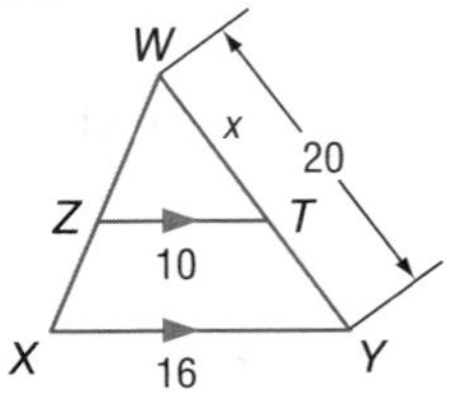

60. **BASKETBALL** In basketball, a free throw is 1 point and a field goal is either 2 or 3 points. In a season, Tim Duncan of the San Antonio Spurs scored a total of 1342 points. The total number of 2-point field goals and 3-point field goals was 517, and he made 305 of the 455 free throws that he attempted. Find the number of 2-point field goals and 3-point field goals Duncan made that season. (Lesson 6-4)

COORDINATE GEOMETRY For each quadrilateral with the given vertices, verify that the quadrilateral is a trapezoid and determine whether the figure is an isosceles trapezoid. (Lesson 13-5)

61. $Q(-12, 1)$, $R(-9, 4)$, $S(-4, 3)$, $T(-11, -4)$

62. $A(-3, 3)$, $B(-4, -1)$, $C(5, -1)$, $D(2, 3)$

Solve each inequality. Check your solution. (Lesson 5-3)

63. $3y - 4 > -37$

64. $-5q + 9 > 24$

65. $-2k + 12 < 30$

66. $5q + 7 \leq 3(q + 1)$

67. $\frac{z}{4} + 7 \geq -5$

68. $8c - (c - 5) > c + 17$

Skills Review

Solve each proportion.

69. $\frac{1}{3} = \frac{x}{2}$

70. $\frac{3}{4} = \frac{5}{x}$

71. $\frac{2.3}{4} = \frac{x}{3.7}$

72. $\frac{x - 2}{2} = \frac{4}{5}$

73. $\frac{x}{12 - x} = \frac{8}{3}$

LESSON 14-3 Similarity Transformations

Then

- You identified congruence transformations.

Now

1. Identify similarity transformations.
2. Verify similarity after a similarity transformation.

Why?

- Adriana uses a copier to enlarge a movie ticket to use as the background for a page in her movie ticket scrapbook. She places the ticket on the glass of the copier. Then she must decide what percentage to input in order to create an image that is three times as big as her original ticket.

NewVocabulary
dilation
similarity transformation
center of dilation
scale factor of a dilation
enlargement
reduction

Common Core State Standards

Content Standards
G.SRT.2 Given two figures, use the definition of similarity in terms of similarity transformations to decide if they are similar; explain using similarity transformations the meaning of similarity for triangles as the equality of all corresponding pairs of angles and the proportionality of all corresponding pairs of sides.
G.SRT.5 Use congruence and similarity criteria for triangles to solve problems and to prove relationships in geometric figures.

Mathematical Practices
6 Attend to precision.
4 Model with mathematics.

1 Identify Similarity Transformations

Recall from Lesson 12-7 that a *transformation* is an operation that maps an original figure, the *preimage*, onto a new figure called the *image*.

A **dilation** is a transformation that enlarges or reduces the original figure proportionally. Since a dilation produces a similar figure, a dilation is a type of **similarity transformation**.

Dilations are performed with respect to a fixed point called the **center of dilation**.

The **scale factor of a dilation** describes the extent of the dilation. The scale factor is the ratio of a length on the image to a corresponding length on the preimage.

The letter k usually represents the scale factor of a dilation. The value of k determines whether the dilation is an enlargement or a reduction.

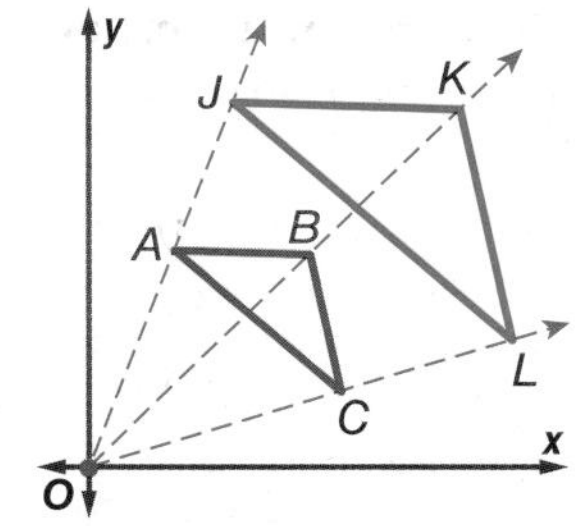

$\triangle JKL$ is a dilation of $\triangle ABC$.
Center of dilation: (0, 0)
Scale factor: $\frac{JK}{AB}$

ConceptSummary Types of Dilations

A dilation with a scale factor greater than 1 produces an **enlargement**, or an image that is larger than the original figure.

Symbols If $k > 1$, the dilation is an enlargement.

Example $\triangle FGH$ is dilated by a scale factor of 3 to produce $\triangle RST$. Since $3 > 1$, $\triangle RST$ is an enlargement of $\triangle FGH$.

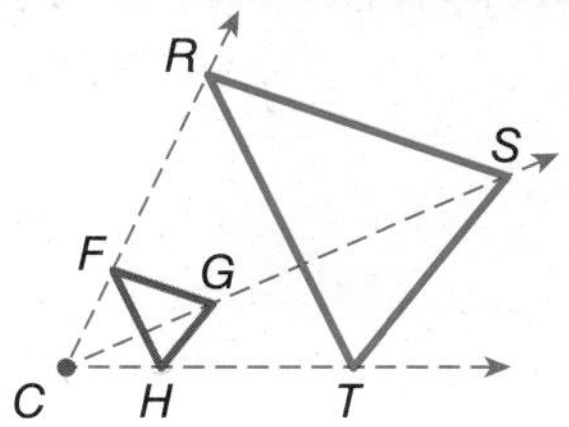

$k < 1$

A dilation with a scale factor between 0 and 1 produces a **reduction**, an image that is smaller than the original figure.

Symbols If $0 < k < 1$, the dilation is a reduction.

Example $ABCD$ is dilated by a scale factor of $\frac{1}{4}$ to produce $WXYZ$. Since $0 < \frac{1}{4} < 1$, $WXYZ$ is a reduction of $ABCD$.

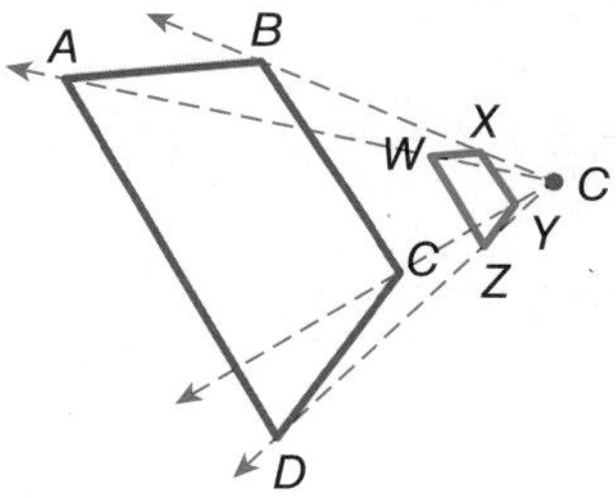

$0 < k < 1$

StudyTip

Multiple Representations The scale factor of a dilation can be represented as a fraction, a decimal, or as a percent. For example, a scale factor of $\frac{2}{5}$ can also be written as 0.4 or as 40%.

Example 1 Identify a Dilation and Find Its Scale Factor

Determine whether the dilation from A to B is an *enlargement* or a *reduction*. Then find the scale factor of the dilation.

a.

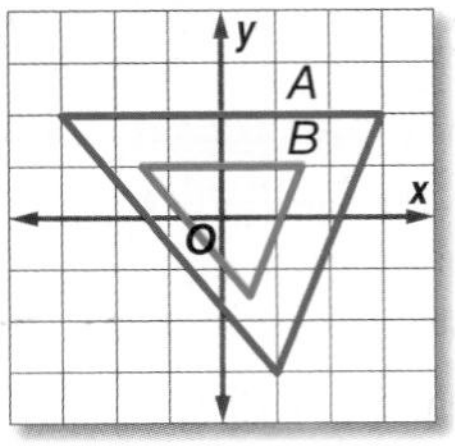

B is smaller than A, so the dilation is a reduction.

The distance between the vertices at $(-3, 2)$ and $(3, 2)$ for A is 6 and from the vertices at $(-1.5, 1)$ and $(1.5, 1)$ for B is 3. So the scale factor is $\frac{3}{6}$ or $\frac{1}{2}$.

b.

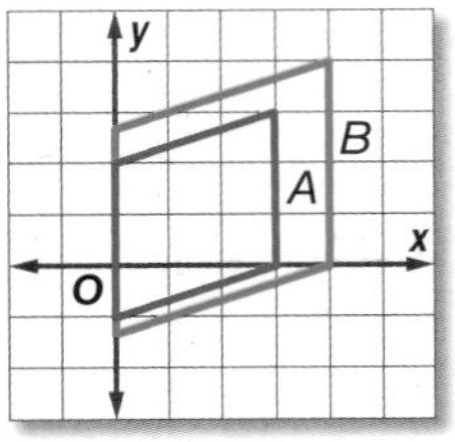

B is larger than A, so the dilation is an enlargement.

The distance between the vertices at $(3, 3)$ and $(3, 0)$ for A is 3 and between the vertices at $(4, 4)$ and $(4, 0)$ for B is 4. So the scale factor is $\frac{4}{3}$.

GuidedPractice

1A.

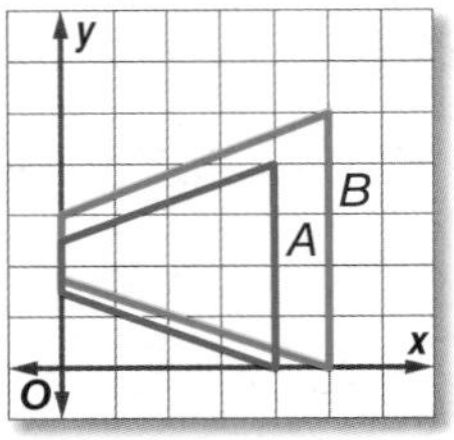

1B.

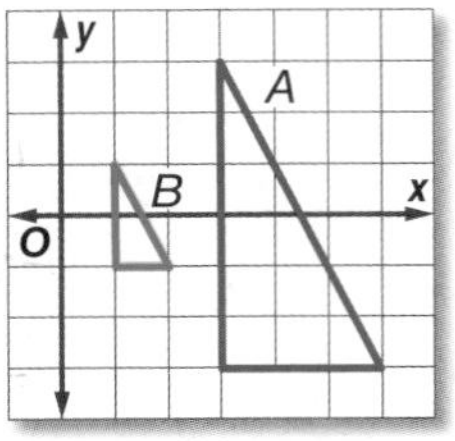

Dilations and their scale factors are used in many real-world situations.

Real-WorldLink

Hew Weng Fatt accepted a contest challenge to collect the most movie stubs from a certain popular fantasy movie. He collected 6561 movie stubs in 38 days!

Source: *Youth2, Star Publications*

Real-World Example 2 Find and Use a Scale Factor

COLLECTING Refer to the beginning of the lesson. By what percent should Adriana enlarge the ticket stub so that the dimensions of its image are 3 times that of her original? What will be the dimensions of the enlarged image?

Adriana wants to create a dilated image of her ticket stub using the copier. The scale factor of her enlargement is 3. Written as a percent, the scale factor is $(3 \cdot 100)\%$ or 300%. Now find the dimension of the enlarged image using the scale factor.

width: $5 \text{ cm} \cdot 300\% = 15 \text{ cm}$ length: $6.4 \text{ cm} \cdot 300\% = 19.2 \text{ cm}$

The enlarged ticket stub image will be 15 centimeters by 19.2 centimeters.

GuidedPractice

2. If the resulting ticket stub image was 1.5 centimeters wide by about 1.9 centimeters long instead, what percent did Adriana mistakenly use to dilate the original image? Explain your reasoning.

CORBIS/SuperStock

2 Verify Similarity

Verify Similarity You can verify that a dilation produces a similar figure by comparing corresponding sides and angles. For triangles, you can also use SAS Similarity.

Example 3 Verify Similarity after a Dilation

StudyTip

Center of Dilation Unless otherwise stated, all dilations on the coordinate plane use the origin as their center of dilation.

Graph the original figure and its dilated image. Then verify that the dilation is a similarity transformation.

a. original: $A(-6, -3)$, $B(3, 3)$, $C(3, -3)$; image: $X(-4, -2)$, $Y(2, 2)$, $Z(2, -2)$

Graph each figure. Since $\angle C$ and $\angle Z$ are both right angles, $\angle C \cong \angle Z$. Show that the lengths of the sides that include $\angle C$ and $\angle Z$ are proportional.

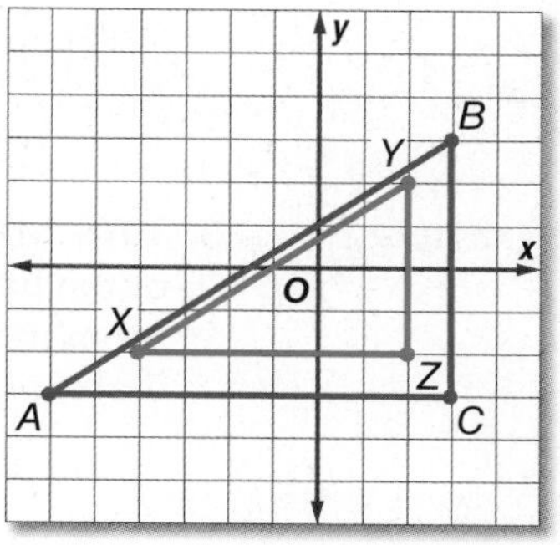

Use the coordinate grid to find the side lengths.

$\frac{XZ}{AC} = \frac{6}{9}$ or $\frac{2}{3}$, and $\frac{YZ}{BC} = \frac{4}{6}$ or $\frac{2}{3}$, so $\frac{XZ}{AC} = \frac{YZ}{BC}$.

Since the lengths of the sides that include $\angle C$ and $\angle Z$ are proportional, $\triangle XYZ \sim \triangle ABC$ by SAS Similarity.

b. original: $J(-6, 4)$, $K(6, 8)$, $L(8, 2)$, $M(-4, -2)$; image: $P(-3, 2)$, $Q(3, 4)$, $R(4, 1)$, $S(-2, -1)$

Use the Distance Formula to find the length of each side.

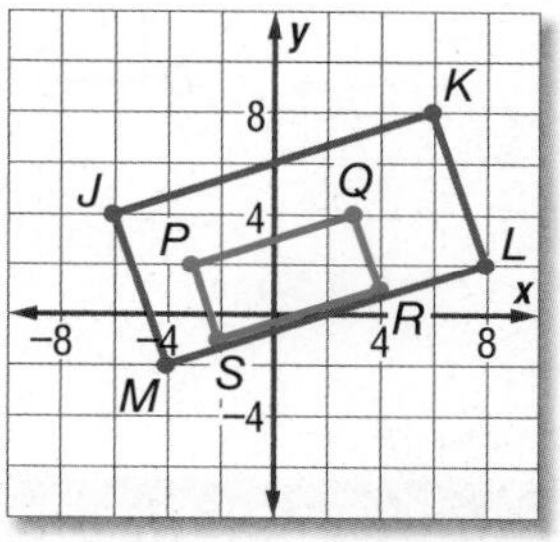

$JK = \sqrt{[6-(-6)]^2 + (8-4)^2} = \sqrt{160}$ or $4\sqrt{10}$

$PQ = \sqrt{[3-(-3)]^2 + (4-2)^2} = \sqrt{40}$ or $2\sqrt{10}$

$KL = \sqrt{(8-6)^2 + (2-8)^2} = \sqrt{40}$ or $2\sqrt{10}$

$QR = \sqrt{(4-3)^2 + (1-4)^2} = \sqrt{10}$

$LM = \sqrt{(-4-8)^2 + (-2-2)^2} = \sqrt{160}$ or $4\sqrt{10}$

$RS = \sqrt{(-2-4)^2 + (-1-1)^2} = \sqrt{40}$ or $2\sqrt{10}$

$MJ = \sqrt{[-6-(-4)]^2 + [4-(-2)]^2} = \sqrt{40}$ or $2\sqrt{10}$

$SP = \sqrt{[-3-(-2)]^2 + [2-(-1)]^2} = \sqrt{10}$

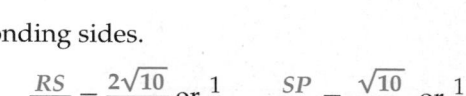

Find and compare the ratios of corresponding sides.

$\frac{PQ}{JK} = \frac{2\sqrt{10}}{4\sqrt{10}}$ or $\frac{1}{2}$ $\quad \frac{QR}{KL} = \frac{\sqrt{10}}{2\sqrt{10}}$ or $\frac{1}{2}$ $\quad \frac{RS}{LM} = \frac{2\sqrt{10}}{4\sqrt{10}}$ or $\frac{1}{2}$ $\quad \frac{SP}{MJ} = \frac{\sqrt{10}}{2\sqrt{10}}$ or $\frac{1}{2}$

PQRS and *JKLM* are both rectangles. This can be proved by showing that diagonals $\overline{PR} \cong \overline{SQ}$ and $\overline{JL} \cong \overline{KM}$ are congruent using the Distance Formula. Since they are both rectangles, their corresponding angles are congruent.

Since $\frac{PQ}{JK} = \frac{QR}{KL} = \frac{RS}{LM} = \frac{SP}{MJ}$ and corresponding angles are congruent, $PQRS \sim JKLM$.

GuidedPractice

3A. original: $A(2, 3)$, $B(0, 1)$, $C(3, 0)$
image: $D(4, 6)$, $F(0, 2)$, $G(6, 0)$

3B. original: $H(0, 0)$, $J(6, 0)$, $K(6, 4)$, $L(0, 4)$
image: $W(0, 0)$, $X(3, 0)$, $Y(3, 2)$, $Z(0, 2)$

Check Your Understanding

Example 1 **Determine whether the dilation from *A* to *B* is an *enlargement* or a *reduction*. Then find the scale factor of the dilation.**

1.

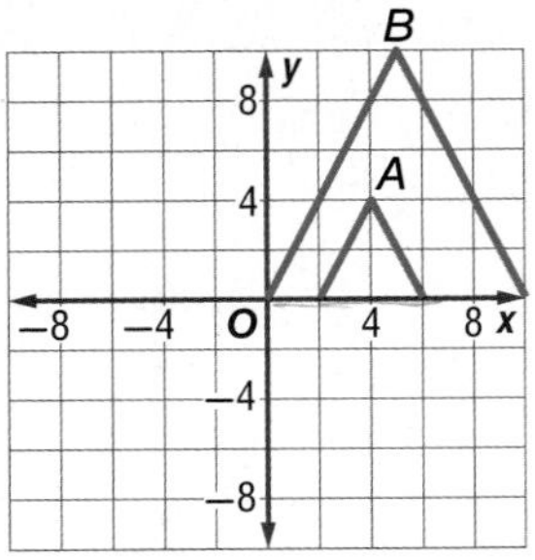

2.

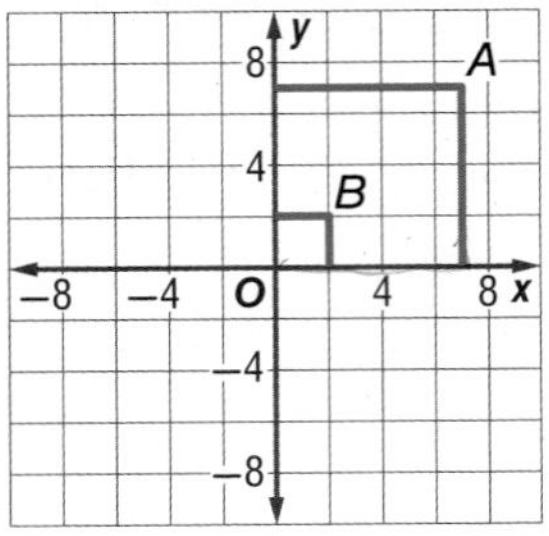

Example 2 **3. TOYS** Michaelea has a doll house that looks like the house she lives in. The dimensions of the living room in the house are 15 feet by 10 feet and the dimensions of the living room in the doll house are 9 inches by 6 inches. Is the living room in her doll house a dialation of the living room in her house? If so, what is the scale factor? Explain.

Example 3 **CCSS ARGUMENTS** **Verify that the dilation is a similarity transformation.**

4.

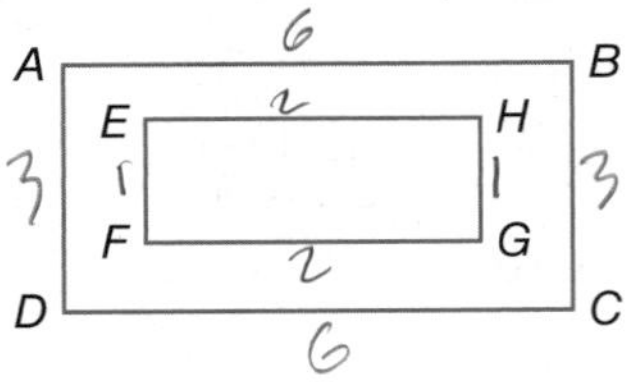

5.

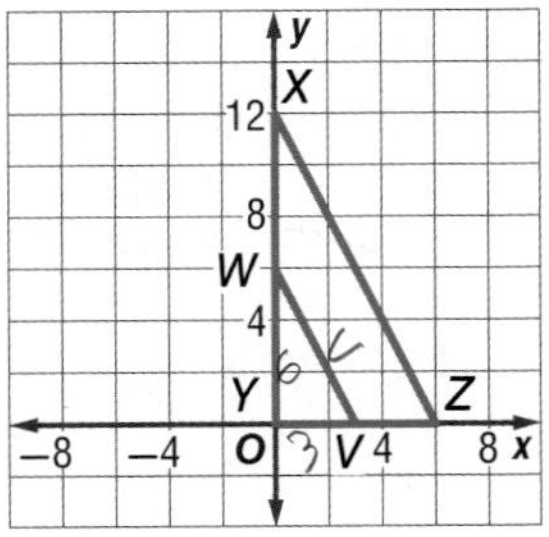

Practice and Problem Solving

Example 1 **Determine whether the dilation from *A* to *B* is an *enlargement* or a *reduction*. Then find the scale factor of the dilation.**

6.

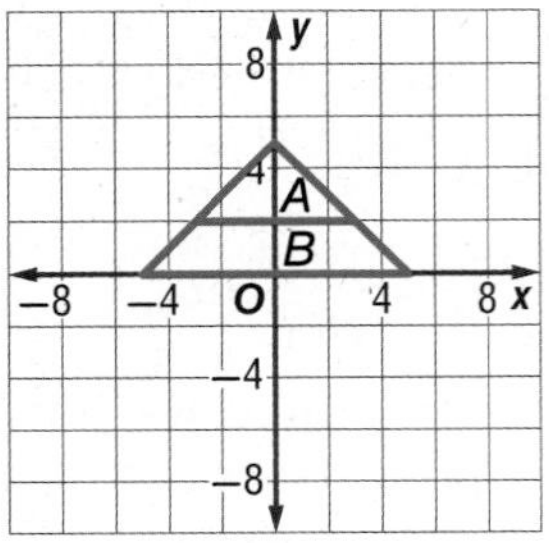

7.

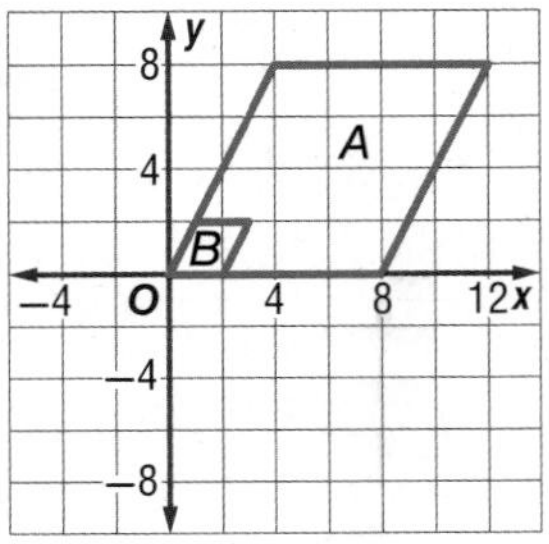

8.

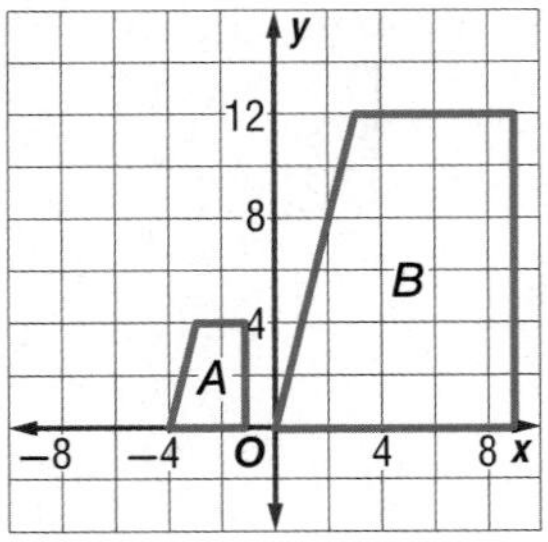

9.

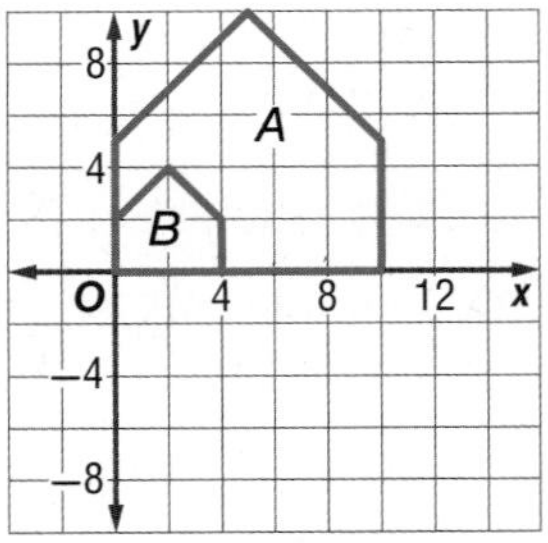

Determine whether each dilation is an *enlargement* or *reduction*.

10. Before After

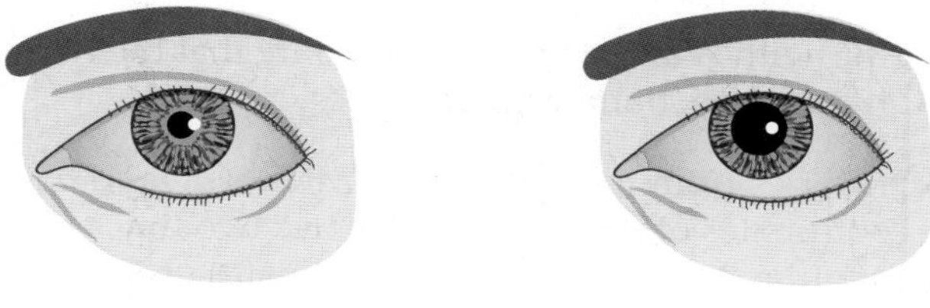

11.

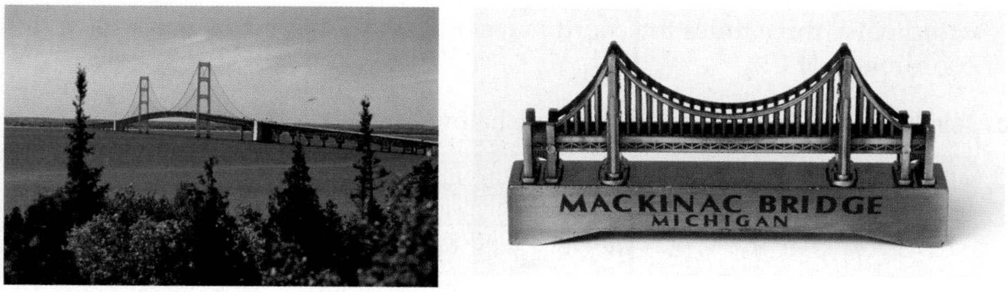

Example 2

12. MAPS John is looking at a large map of Central Park. If Central Park is 2.5 miles long and 0.5 miles wide and the map is 30 feet by 6 feet. Is the map a dilation of the park? If so, what is the scale factor?

13. CCSS REPLICAS The dimensions of the original Declaration of Independence is 29.75 inches by 24.5 inches. If a replica of the document is 20.5 inches by 14.5 inches is the replica a dilation of the original? If so, what is the scale factor?

Example 3

Graph the original figure and its dilated image. Then verify that the dilation is a similarity transformation.

14. $A(4, 7), B(5, 5), C(8, 8); E(0, 9), F(3, 3), G(12, 12)$

15. $X(-1, 1), Y(-3, 0), Z(-1, -1), W(-9, -3), V(-1, -7)$

16. $L(1, 8), M(-1, 4), N(5, 6); P(-2, 2), Q(7, 5)$

17. $J(-8, 6), K(4, 4), M(-4, 2), R(-14, 14), S(10, 10), T(-6, 6)$

If $\triangle LMN \sim \triangle LYZ$, find the missing coordinate.

18.

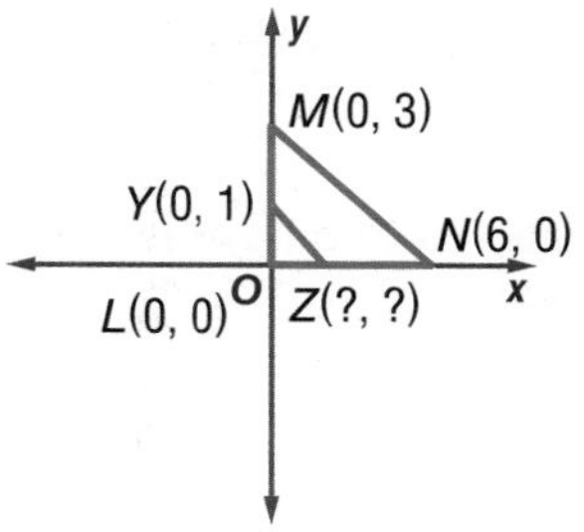

19.

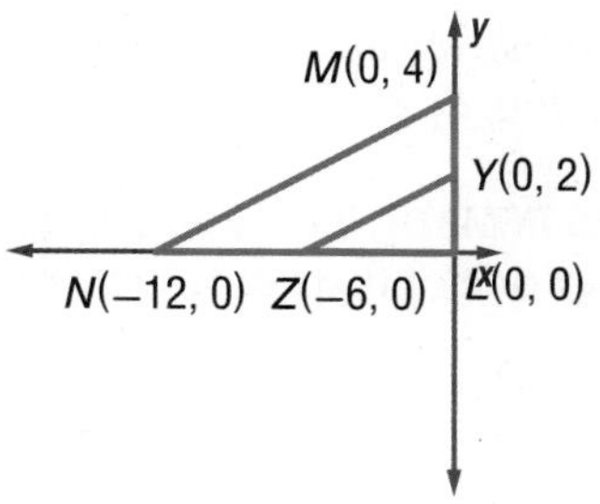

20. BAKING Matt used 2 cups of flour to bake a cake with dimensions 4" by 6" as a sample of a larger cake. The larger cake will be 16" by 24".

a. Explain why the larger cake is a dilation of the sample cake.

b. How many cups of flour will Matt need for the larger cake?

21. **MULTIPLE REPRESENTATIONS** In this problem you will investigate similarity of triangles in the coordinate plane.

a. **Geometric** Draw a triangle *ABC* with *A* at the origin. Make sure that the two additional vertices *B* and *C* have whole number coordinates.

b. **Geometric** Create two new triangles from *ABC*. Label one *MNP* where *M* has coordinates two times the coordinates of *A*, *N* has coordinates two times the coordinates of *B*, and *P* has coordinates two times the coordinates of *C*. Label the other triangle *XYZ* where *X* has coordinates three times the coordinates of *A*, *Y* has coordinates three times the coordinates of *B*, and *Z* has coordinates three times the coordinates of *C*.

c. **Tabular** Copy and complete the table below.

Triangle	Coordinates of point 1	Coordinates of point 2	Coordinates of point 3	Similar with $\triangle ABC$?	Scale factor with $\triangle ABC$.
$\triangle ABC$	$A(\ , \ )$	$B(\ , \ )$	$C(\ , \ )$		
$\triangle MNP$	$M(\ , \ )$	$N(\ , \ )$	$P(\ , \ )$		
$\triangle XYZ$	$X(\ , \ )$	$Y(\ , \ )$	$A(\ , \ )$		

d. **Verbal** Make a conjecture about how you could predict the coordinates of a dilated triangle with a scale factor of *n* if the two similar triangles share a corresponding vertex at the origin.

H.O.T. Problems Use Higher-Order Thinking Skills

22. **CHALLENGE** If rectangle *ABCD* has coordinates *A*(0, 0), *B*(0, 4), *C*(4, 3), *D*(0, 3), find the coordinates of rectangle *WXYZ* where *WXYZ* is a dilation of *ABCD* such that the area of *WXYZ* is one half the area of *ABCD*.

REASONING Determine if the following statements are *always*, *sometimes*, or *never* true.

23. Given two rectangles, one is a dilation of the other.

24. Given two squares, one is a dilation of the other.

25. Given a square and a rectangle one is a dilation of the other.

26. **OPEN ENDED** Write the coordinates of a triangle that is a dilation of the triangle *A*(0, 0), *B*(0, 4), *C*(2, 5).

27. **WRITING IN MATH** Explain how you can use the coordinates of two figures to determine whether one of the figures is a dilation of the other figure.

Standardized Test Practice

28. ALGEBRA Which equation describes the line that passes through $(-3, 4)$ and is perpendicular to $3x - y = 6$?

A $y = -\frac{1}{3}x + 4$　　**C** $y = 3x + 4$

B $y = -\frac{1}{3}x + 3$　　**D** $y = 3x + 3$

29. SHORT RESPONSE What is the scale factor of the dilation shown below?

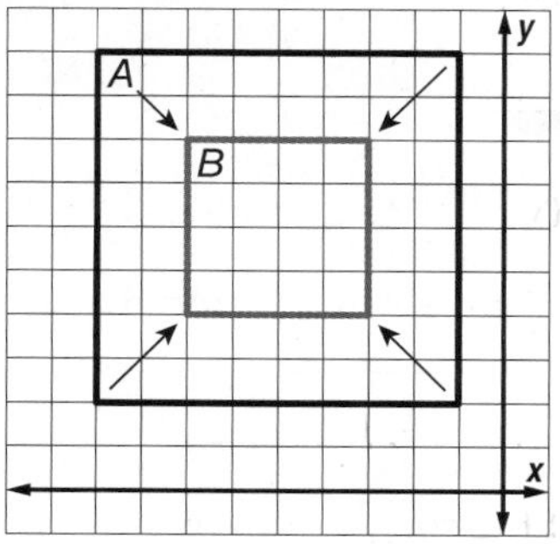

30. In the figure below, $\angle A \cong \angle C$.

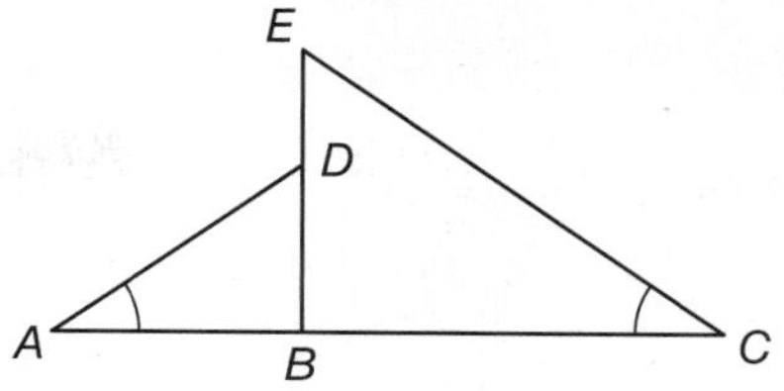

Which additional information would *not* be enough to prove that $\triangle ADB \sim \triangle CEB$?

F $\frac{AB}{DB} = \frac{CB}{EB}$　　**H** $\overline{ED} \cong \overline{DB}$

G $\angle ADB \cong \angle CEB$　　**J** $\overline{EB} \perp \overline{AC}$

31. SAT/ACT $x = \frac{6}{4p + 3}$ and $xy = \frac{3}{4p + 3}$. $y =$

A 4　　**C** 1　　**E** $\frac{1}{2}$

B 2　　**D** $\frac{3}{4}$

Spiral Review

Determine whether $\overline{AB} \parallel \overline{CD}$. Justify your answer. (Lesson 14-2)

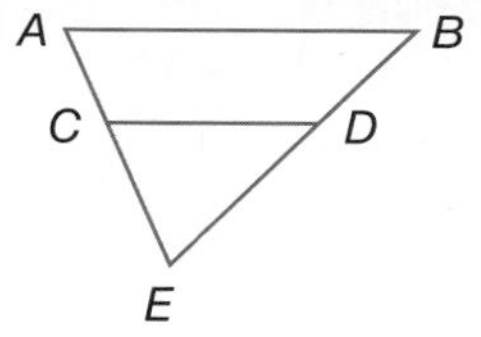

32. $AC = 8.4$, $BD = 6.3$, $DE = 4.5$, and $CE = 6$

33. $AC = 7$, $BD = 10.5$, $BE = 22.5$, and $AE = 15$

34. $AB = 8$, $AE = 9$, $CD = 4$, and $CE = 4$

If each figure is a kite, find each measure. (Lesson 13-5)

35. QR

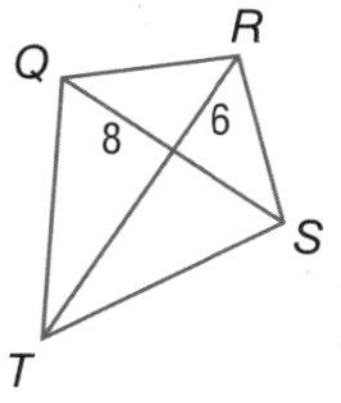

36. $m\angle K$

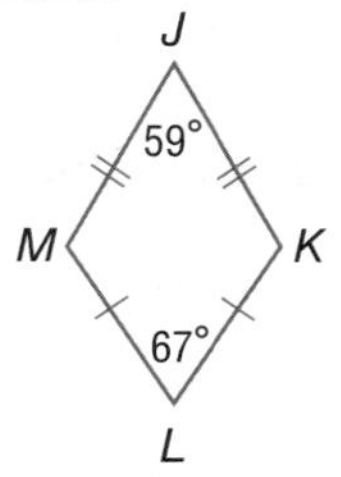

37. BC

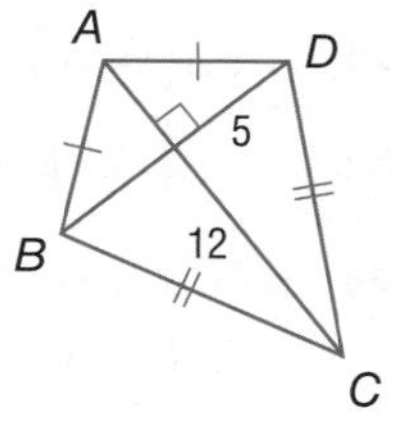

38. PROOF Write a coordinate proof for the following statement. (Lesson 12-8)
If a line segment joins the midpoints of two sides of a triangle, then it is parallel to the third side.

Skills Review

Solve each equation.

39. $145 = 29 \cdot t$

40. $216 = d \cdot 27$

41. $2r = 67 \cdot 5$

42. $100t = \frac{70}{240}$

43. $\frac{80}{4} = 14d$

44. $\frac{2t + 15}{t} = 92$

LESSON 14-4 Reflections

Then	Now	Why?
You identified reflections and verified them as congruence transformations.	**1** Draw reflections. **2** Draw reflections in the coordinate plane.	Notice in this water reflection that the distance a point lies above the water line appears the same as the distance its image lies below the water.

NewVocabulary
line of reflection

Common Core State Standards

Content Standards

G.CO.4 Develop definitions of rotations, reflections, and translations in terms of angles, circles, perpendicular lines, parallel lines, and line segments.

G.CO.5 Given a geometric figure and a rotation, reflection, or translation, draw the transformed figure using, e.g., graph paper, tracing paper, or geometry software. Specify a sequence of transformations that will carry a given figure onto another.

Mathematical Practices

5 Use appropriate tools strategically.

7 Look for and make use of structure.

1 Draw Reflections

In Lesson 12-7, you learned that a reflection or *flip* is a transformation in a line called the **line of reflection**. Each point of the preimage and its corresponding point on the image are the same distance from this line.

KeyConcept Reflection in a Line

A reflection in a line is a function that maps a point to its image such that

- if the point is on the line, then the image and preimage are the same point, or
- if the point is not on the line, the line is the perpendicular bisector of the segment joining the two points.

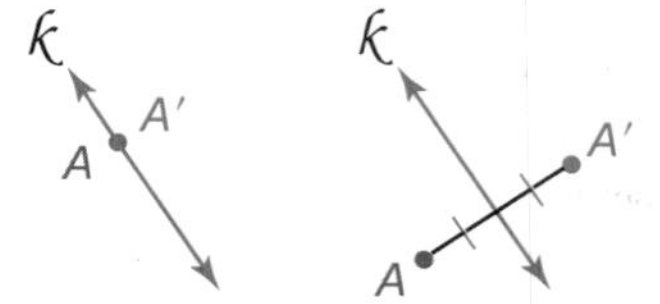

A is on line *k*. *A* is not on line *k*.

A′, *A″*, *A‴*, and so on, name corresponding points for one or more transformations.

To reflect a polygon in a line, reflect each of the polygon's vertices. Then connect these vertices to form the reflected image.

Example 1 Reflect a Figure in a Line

Copy the figure and the given line of reflection. Then draw the reflected image in this line using a ruler.

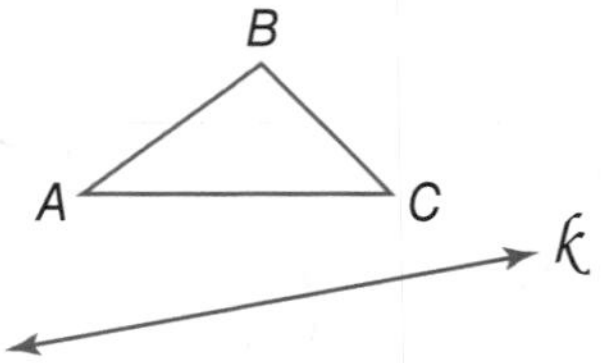

Step 1 Draw a line through each vertex that is perpendicular to line *k*.

Step 2 Measure the distance from point *A* to line *k*. Then locate *A′* the same distance from line *k* on the opposite side

Step 3 Repeat Step 2 to locate points *B′* and *C′*. Then connect vertices *A′*, *B′*, and *C′* to form the reflected image.

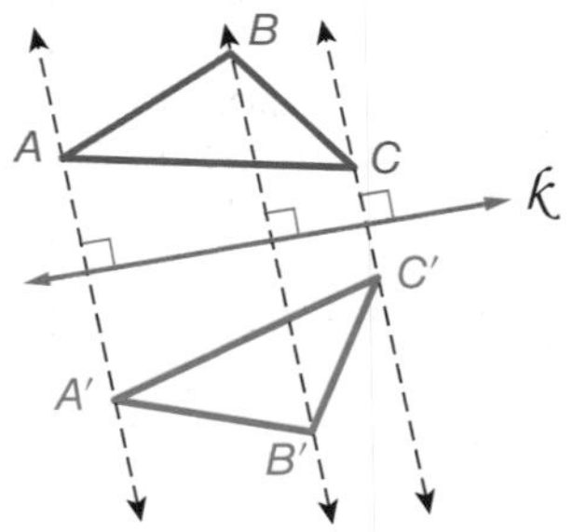

GuidedPractice

1A.

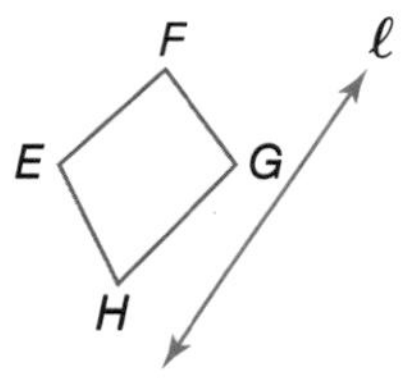

1B.

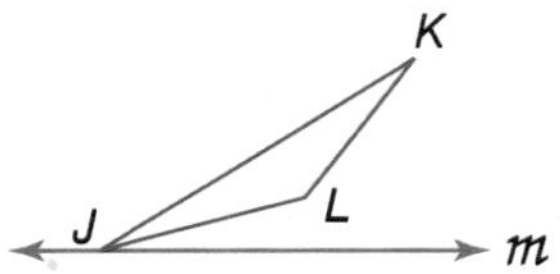

1C.

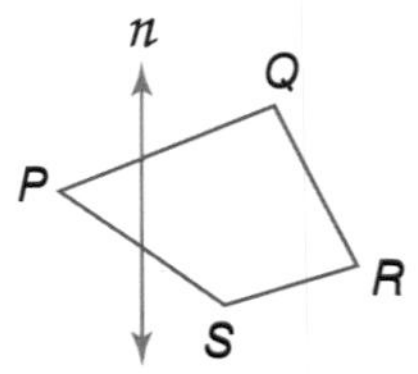

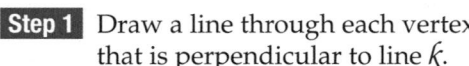

Recall that a reflection is a *congruence transformation* or *isometry*. In the figure in Example 1, $\triangle ABC \cong \triangle A'B'C'$.

Real-World Example 2 Minimize Distance by Using a Reflection

SHOPPING Suppose you are going to buy clothes in Store B, return to your car, and then buy shoes at Store G. Where along line s of parking spaces should you park to minimize the distance you will walk?

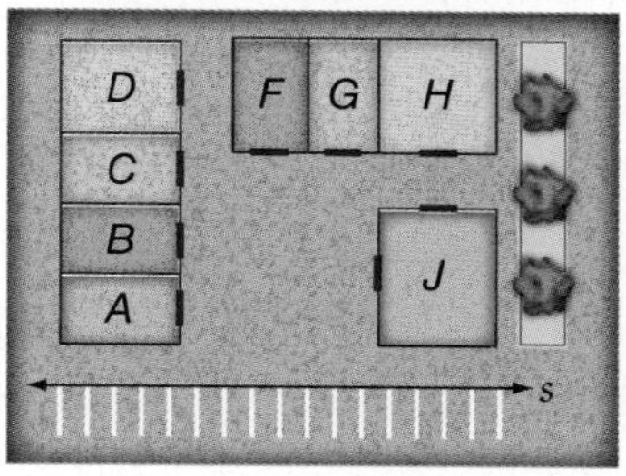

Understand You are asked to locate a point P on line s such that $BP + PG$ has the least possible value.

Plan The total distance from B to P and then from P to G is least when these three points are collinear. Use the reflection of point B in line s to find the location for point P.

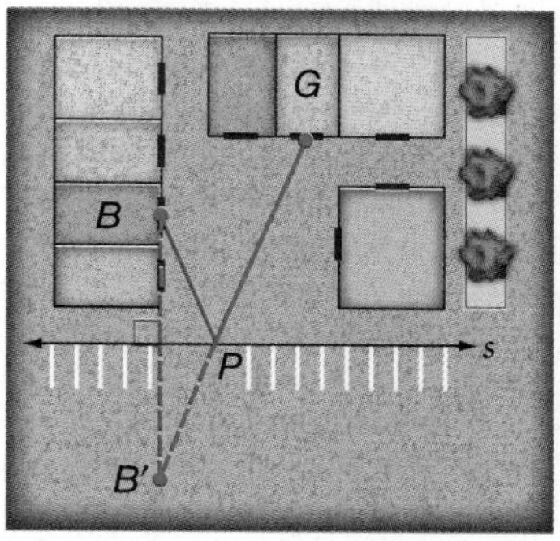

Solve Draw $\overline{B'G}$. Locate P at the intersection of line s and $\overline{B'G}$.

Check Compare the sum $BP + PG$ for each case to verify that the location found for P minimizes this sum.

Guided Practice

2. TICKET SALES Joy wants to select a good location to sell tickets for a dance. Locate point P such that the distance someone would have to walk from Hallway A, to point P on the wall, and then to their next class in Hallway B is minimized.

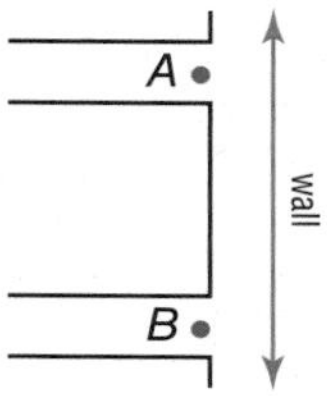

Real-World Career

Photographer Photographers take photos for a variety of reasons such as journalism, art, to record an event, or for scientific purposes. In some photography fields such as photojournalism and scientific photography, a bachelor's degree is required. For others, such as portrait photography, technical proficiency is the only requirement.

2 Draw Reflections in the Coordinate Plane

Reflections can also be performed in the coordinate plane by using the techniques presented in Example 3.

Study Tip

Characteristics of a Reflection Reflections, like all isometries, preserve distance, angle measure, betweenness of points, and collinearity. The orientation of a preimage and its image, however, are reversed.

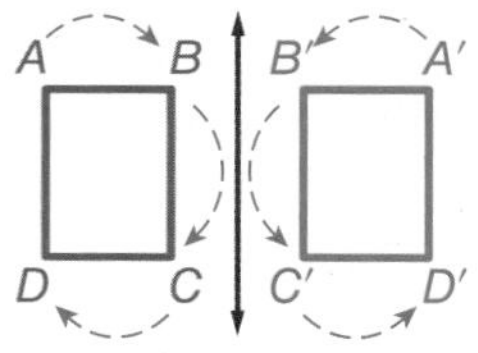

Example 3 Reflect a Figure in a Horizontal or Vertical Line

Triangle JKL has vertices $J(0, 3)$, $K(-2, -1)$, and $L(-6, 1)$. Graph $\triangle JKL$ and its image in the given line.

a. $x = -4$

Find a corresponding point for each vertex so that a vertex and its image are equidistant from the line $x = -4$.

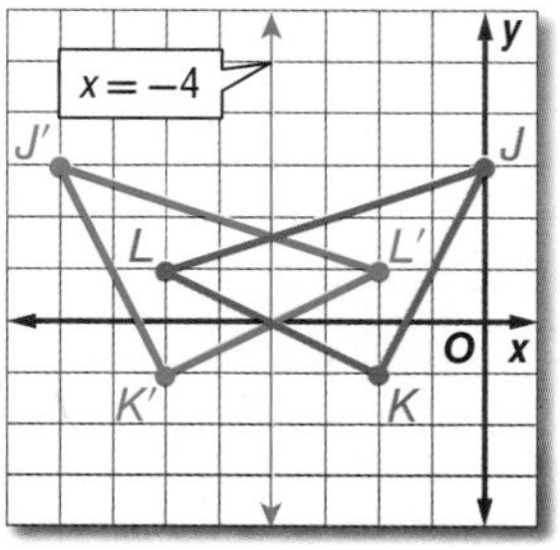

b. $y = 2$

Find a corresponding point for each vertex so that a vertex and its image are equidistant from the line $y = 2$.

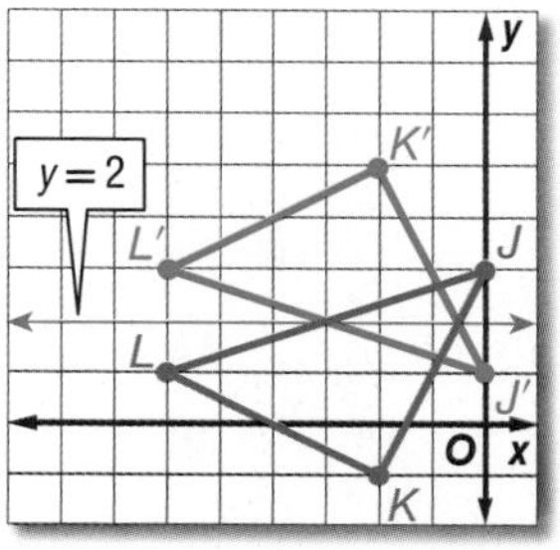

Masterfile

GuidedPractice

Trapezoid *RSTV* has vertices *R*(−1, 1), *S*(4, 1), *T*(4, −1), and *V*(−1, −3). Graph trapezoid *RSTV* and its image in the given line.

3A. $y = -3$ **3B.** $x = 2$

When the line of reflection is the *x*- or *y*-axis, you can use the following rule.

KeyConcept Reflection in the *x*- or *y*-axis

Reflection in the *x*-axis		Reflection in the *y*-axis	
Words	To reflect a point in the *x*-axis, multiply its *y*-coordinate by −1.	Words	To reflect a point in the *y*-axis, multiply its *x*-coordinate by −1.
Symbols	$(x, y) \rightarrow (x, -y)$	Symbols	$(x, y) \rightarrow (-x, y)$
Example	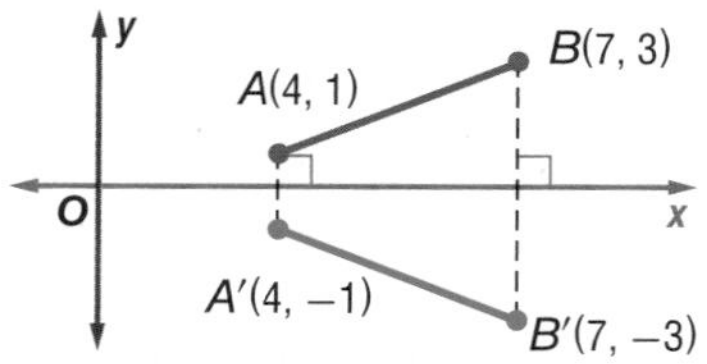	Example	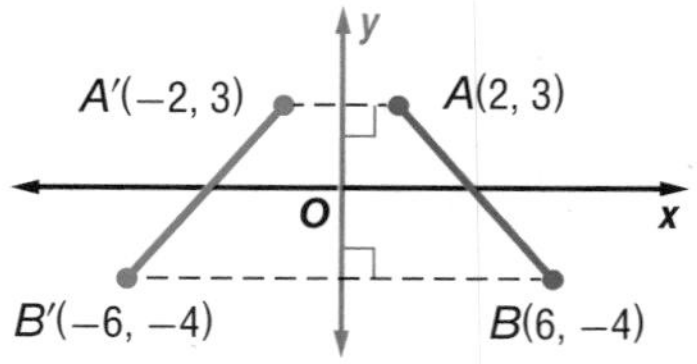

ReadingMath

Coordinate Function Notation The expression $P(a, b) \rightarrow P'(a, -b)$ can be read as *point P with coordinates a and b is mapped to new location P prime with coordinates a and negative b.*

Example 4 Reflect a Figure in the *x*- or *y*-axis

Graph each figure and its image under the given reflection.

a. △*ABC* with vertices *A*(−5, 3), *B*(2, 0), and *C*(1, 2) in the *x*-axis

Multiply the *y*-coordinate of each vertex by −1.

(x, y)	$\rightarrow$	$(x, -y)$
$A(-5, 3)$	$\rightarrow$	$A'(-5, -3)$
$B(2, 0)$	$\rightarrow$	$B'(2, 0)$
$C(1, 2)$	$\rightarrow$	$C'(1, -2)$

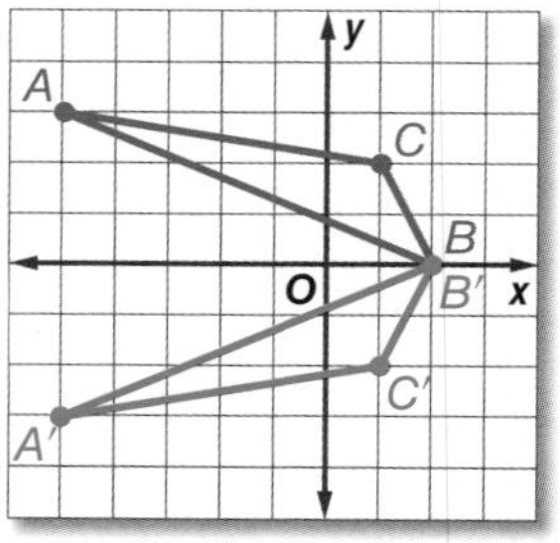

StudyTip

Invariant Points In Example 4a, point *B* is called an *invariant point* because it maps onto itself. Only points that lie on the line of reflection are invariant under a reflection.

b. parallelogram *PQRS* with vertices *P*(−4, 1), *Q*(2, 3), *R*(2, −1), and *S*(−4, −3) in the *y*-axis

Multiply the *x*-coordinate of each vertex by −1.

(x, y)	$\rightarrow$	$(-x, y)$
$P(-4, 1)$	$\rightarrow$	$P'(4, 1)$
$Q(2, 3)$	$\rightarrow$	$Q'(-2, 3)$
$R(2, -1)$	$\rightarrow$	$R'(-2, -1)$
$S(-4, -3)$	$\rightarrow$	$S'(4, -3)$

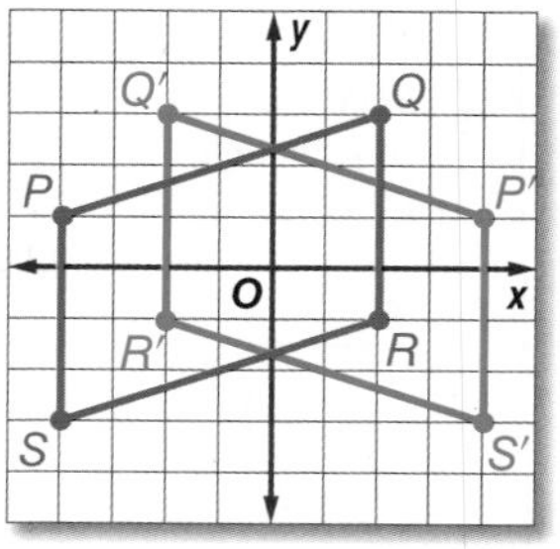

GuidedPractice

4A. rectangle with vertices *E*(−4, −1), *F*(2, 2), *G*(3, 0), and *H*(−3, −3) in the *x*-axis

4B. △*JKL* with vertices *J*(3, 2), *K*(2, −2), and *L*(4, −5) in the *y*-axis

ReviewVocabulary

Perpendicular Lines Two nonvertical lines are perpendicular if and only if the product of their slopes is -1.

You can also reflect an image in the line $y = x$.

The slope of $y = x$ is 1. In the graph shown, $\overline{CC'}$ is perpendicular to $y = x$, so its slope is -1. From $C(-3, 2)$, move right 2.5 units and down 2.5 units to reach $y = x$. From this point on $y = x$, move right 2.5 units and down 2.5 units to locate $C'(2, -3)$. Using a similar method, the image of $D(-3, -1)$ is found to be $D'(-1, -3)$.

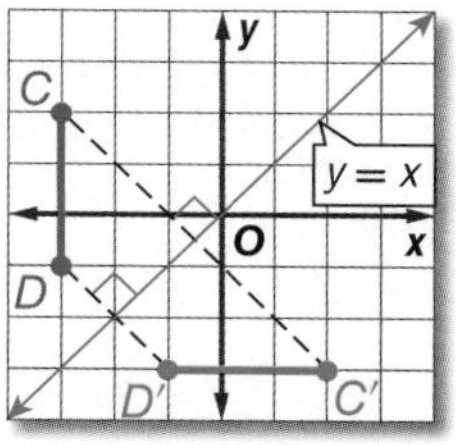

Comparing the coordinates of these and other examples leads to the following rule for reflections in the line $y = x$.

KeyConcept Reflection in Line $y = x$

Words To reflect a point in the line $y = x$, interchange the x- and y-coordinates.

Symbols $(x, y) \rightarrow (y, x)$

Example

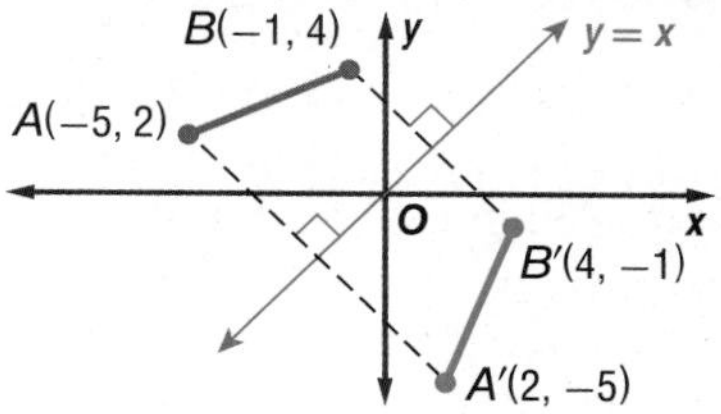

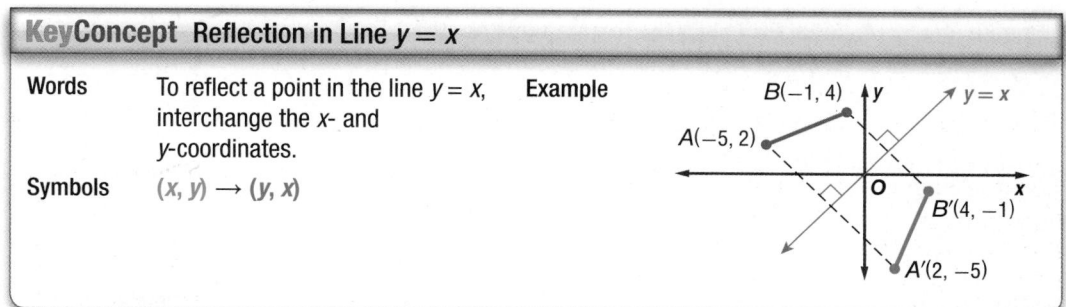

Example 5 Reflect a Figure in the Line $y = x$

Quadrilateral *JKLM* has vertices $J(2, 2)$, $K(4, 1)$, $L(3, -3)$, and $M(0, -4)$. Graph *JKLM* and its image $J'K'L'M'$ in the line $y = x$.

Interchange the x- and y-coordinates of each vertex.

(x, y)	$\rightarrow$	(y, x)
$J(2, 2)$	$\rightarrow$	$J'(2, 2)$
$K(4, 1)$	$\rightarrow$	$K'(1, 4)$
$L(3, -3)$	$\rightarrow$	$L'(-3, 3)$
$M(0, -4)$	$\rightarrow$	$M'(-4, 0)$

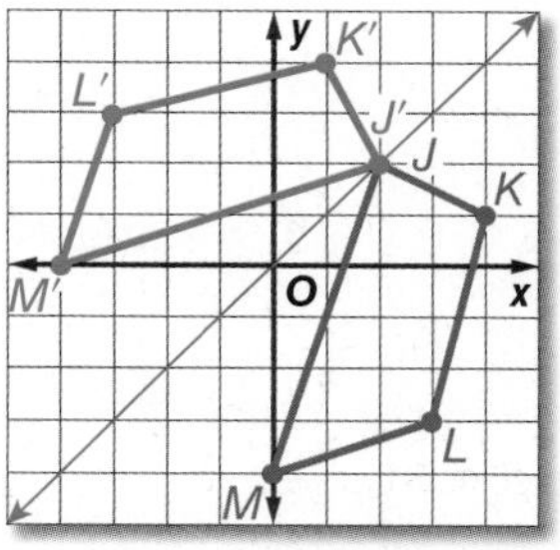

StudyTip

Preimage and Image In this book, the preimage will always be blue and the image will always be green.

GuidedPractice

5. $\triangle BCD$ has vertices $B(-3, 3)$, $C(1, 4)$, and $D(-2, -4)$. Graph $\triangle BCD$ and its image in the line $y = x$.

ConceptSummary Reflection in the Coordinate Plane

Reflection in the x-axis	Reflection in the y-axis	Reflection in the line $y = x$
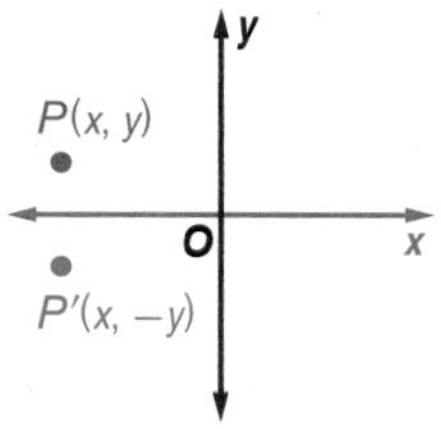	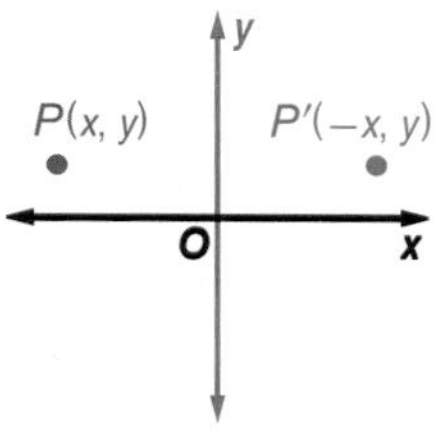	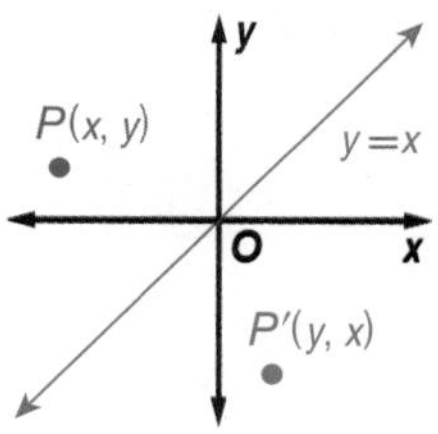
$(x, y) \rightarrow (x, -y)$	$(x, y) \rightarrow (-x, y)$	$(x, y) \rightarrow (y, x)$

Check Your Understanding

Example 1 **Copy the figure and the given line of reflection. Then draw the reflected image in this line using a ruler.**

1. 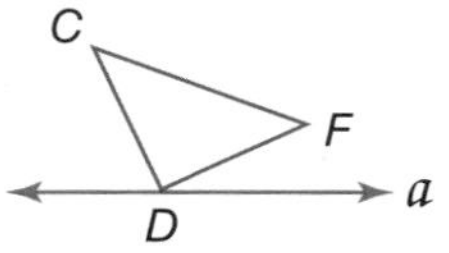

2. 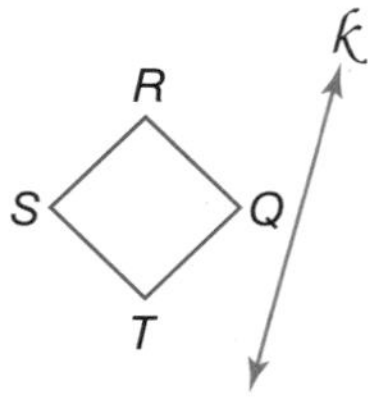

3. 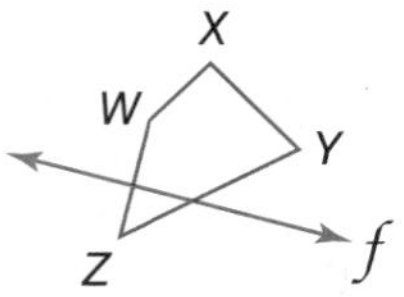

Example 2

4. **SPORTING EVENTS** Toru is waiting at a café for a friend to bring him a ticket to a sold-out sporting event. At what point *P* along the street should the friend try to stop his car to minimize the distance Toru will have to walk from the café, to the car, and then to the arena entrance? Draw a diagram.

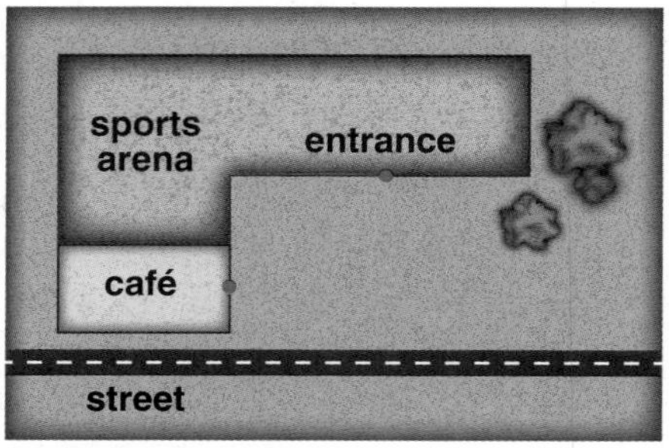

Example 3 **Graph $\triangle ABC$ and its image in the given line.**

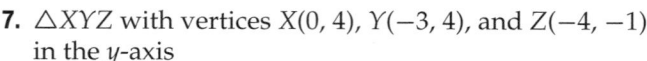

5. $y = -2$

6. $x = 3$

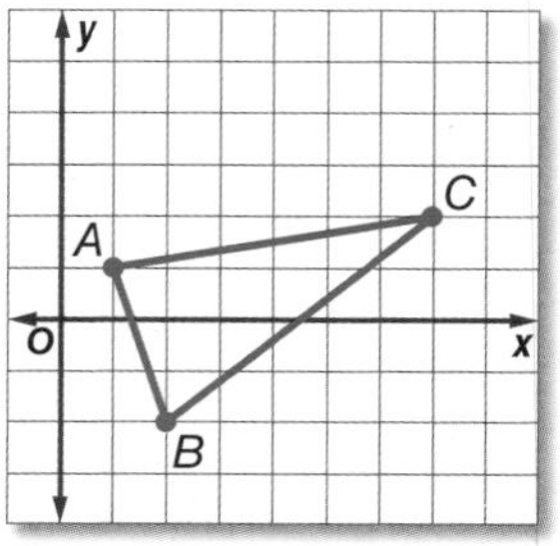

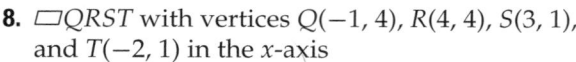

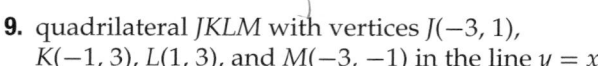

Examples 4–5 **Graph each figure and its image under the given reflection.**

7. $\triangle XYZ$ with vertices $X(0, 4)$, $Y(-3, 4)$, and $Z(-4, -1)$ in the y-axis

8. $\square QRST$ with vertices $Q(-1, 4)$, $R(4, 4)$, $S(3, 1)$, and $T(-2, 1)$ in the x-axis

9. quadrilateral $JKLM$ with vertices $J(-3, 1)$, $K(-1, 3)$, $L(1, 3)$, and $M(-3, -1)$ in the line $y = x$

Practice and Problem Solving

Example 1 **CCSS TOOLS** **Copy the figure and the given line of reflection. Then draw the reflected image in this line using a ruler.**

10. 

11. 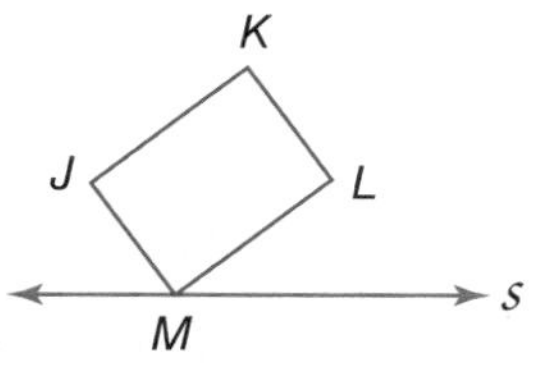

12. 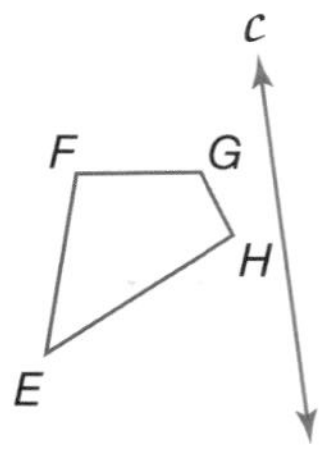

13. 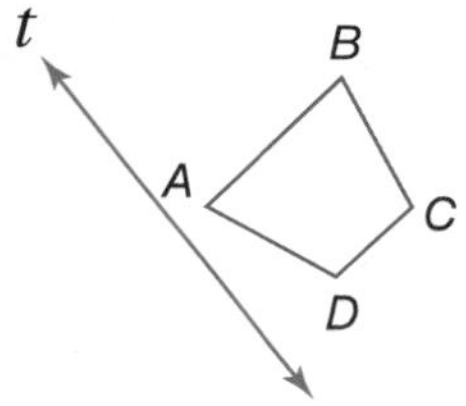

14. 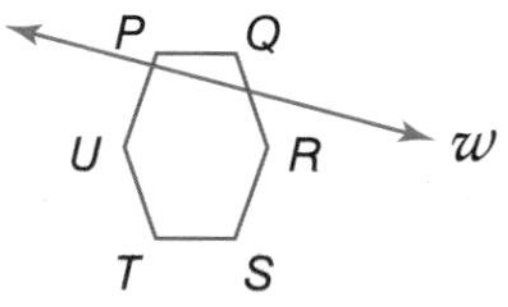

15. 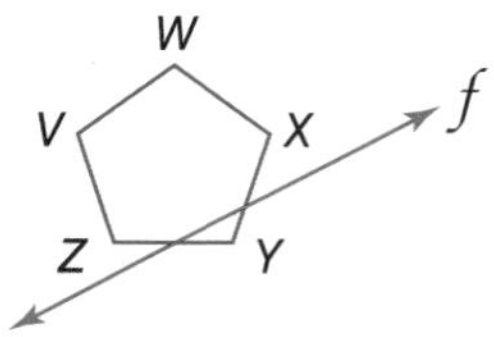

Example 2 **SPORTS When a ball is rolled or struck without spin against a wall, it bounces off the wall and travels in a ray that is the reflected image of the path of the ball if it had gone straight through the wall. Use this information in Exercises 16 and 17.**

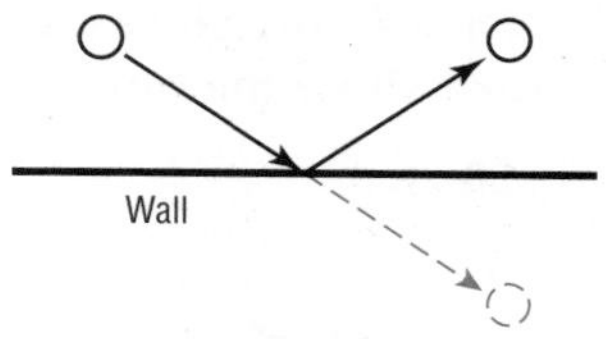

16. BILLIARDS Tadeo is playing billiards. He wants to strike the eight ball with the cue ball so that the eight ball bounces off the rail and rolls into the indicated pocket. If the eight ball moves with no spin, draw a diagram showing the exact point P along the right rail where the eight ball should hit after being struck by the cue ball.

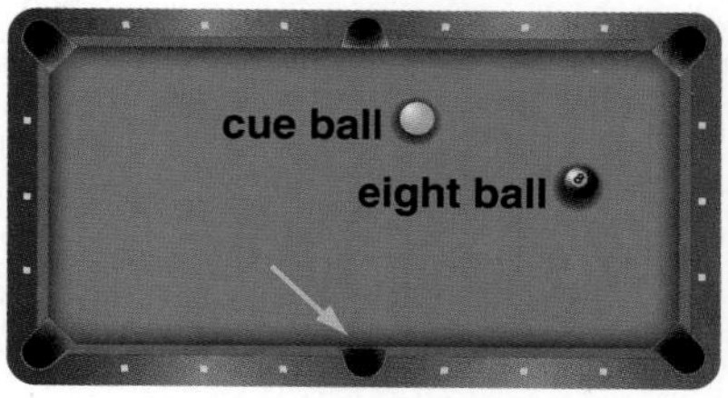

17. INDOOR SOCCER Abby is playing indoor soccer, and she wants to hit the ball to point C, but must avoid an opposing player at point B. She decides to hit the ball at point A so that it bounces off the side wall. Draw a diagram that shows the exact point along the top wall for which Abby should aim.

Example 3 **Graph each figure and its image in the given line.**

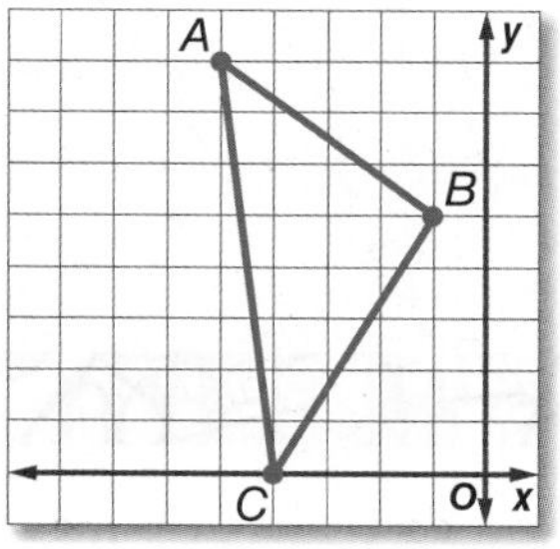

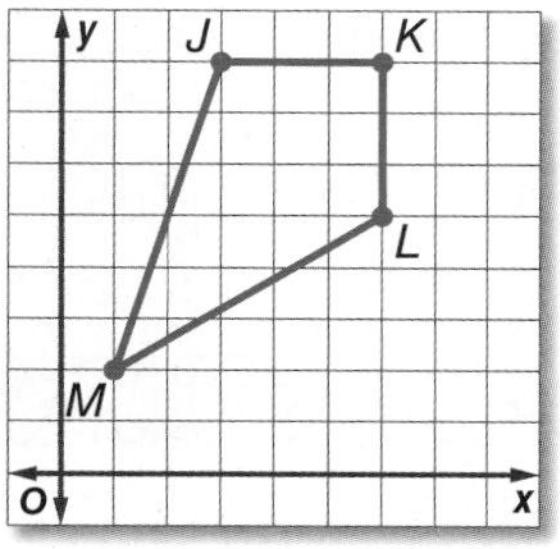

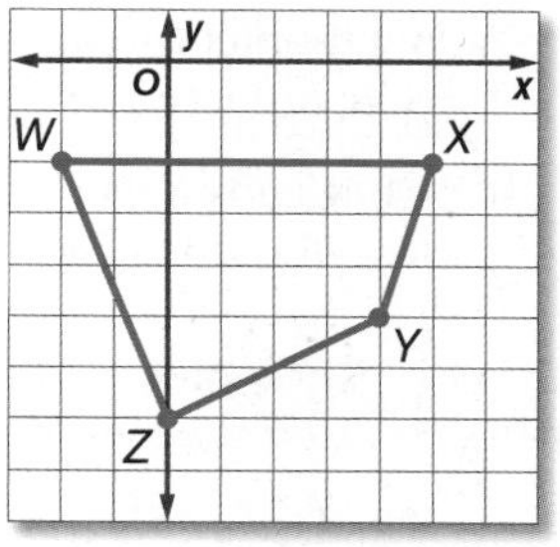

18. $\triangle ABC$; $y = 3$

19. $\triangle ABC$; $x = -1$

20. $JKLM$; $x = 1$

21. $JKLM$; $y = 4$

22. $WXYZ$; $y = -4$

23. $WXYZ$; $x = -2$

Examples 4–5 CCSS **STRUCTURE Graph each figure and its image under the given reflection.**

24. rectangle $ABCD$ with vertices $A(-5, 2)$, $B(1, 2)$, $C(1, -1)$, and $D(-5, -1)$ in the line $y = -2$

25. square $JKLM$ with vertices $J(-4, 6)$, $K(0, 6)$, $L(0, 2)$, and $M(-4, 2)$ in the y-axis

26. $\triangle FGH$ with vertices $F(-3, 2)$, $G(-4, -1)$, and $H(-6, -1)$ in the line $y = x$

27. $\square WXYZ$ with vertices $W(2, 3)$, $X(7, 3)$, $Y(6, -1)$, and $Z(1, -1)$ in the x-axis

28. trapezoid $PQRS$ with vertices $P(-1, 4)$, $Q(2, 4)$, $R(1, -1)$, and $S(-1, -1)$ in the y-axis

29. $\triangle STU$ with vertices $S(-3, -2)$, $T(-2, 3)$, and $U(2, 2)$ in the line $y = x$

Each figure shows a preimage and its reflected image in some line. Copy each figure and draw the line of reflection.

30.

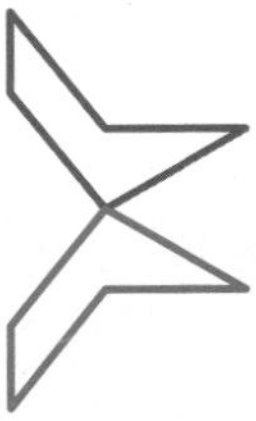

31.

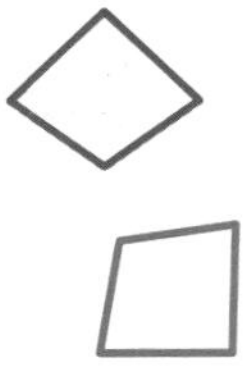

32.

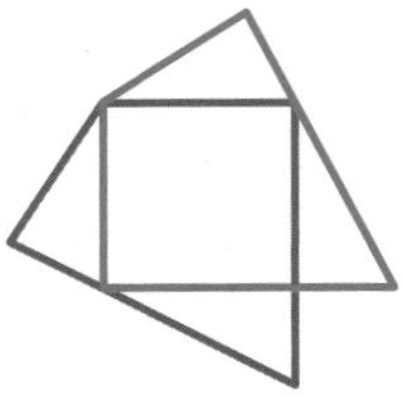

CONSTRUCTION To construct the reflection of a figure in a line using only a compass and a straightedge, you can use:

- the construction of a line perpendicular to a given line through a point not on the line, and
- the construction of a segment congruent to a given segment.

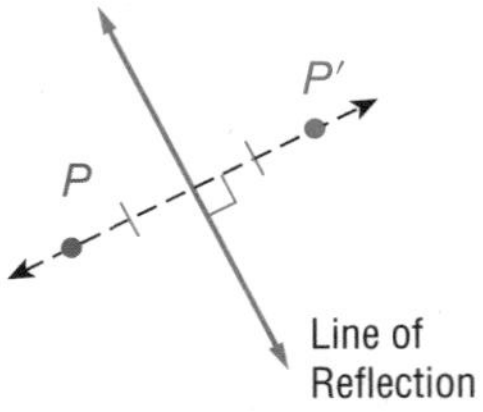

CCSS TOOLS Copy each figure and the given line of reflection. Then construct the reflected image.

33.

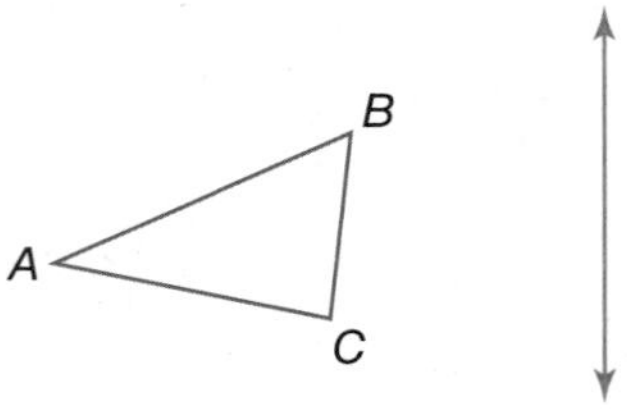

34.

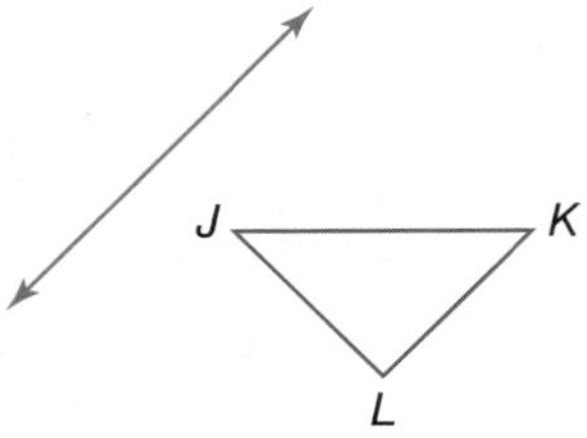

35 **PHOTOGRAPHY** Refer to the photo at the right.

a. What object separates the zebras and their reflections?

b. What geometric term can be used to describe this object?

ALGEBRA Graph the line $y = 2x - 3$ and its reflected image in the given line. What is the equation of the reflected image?

36. x-axis

37. y-axis

38. $y = x$

39. Reflect $\triangle CDE$ shown below in the line $y = 3x$.

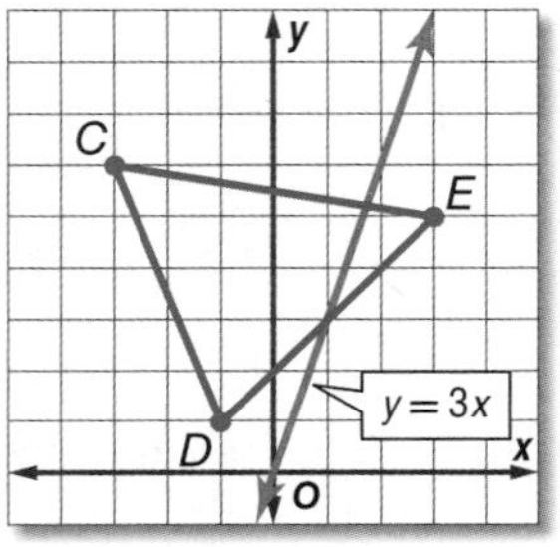

40. Relocate vertex C so that $ABCDE$ is convex, and all sides remain the same length.

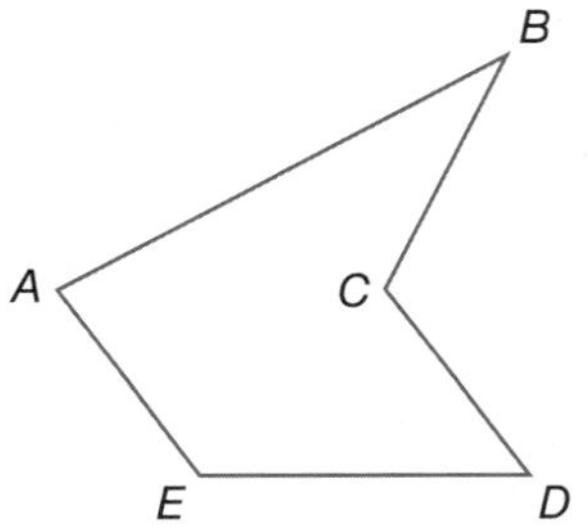

ALGEBRA Graph the reflection of each function in the given line. Then write the equation of the reflected image.

41. x-axis

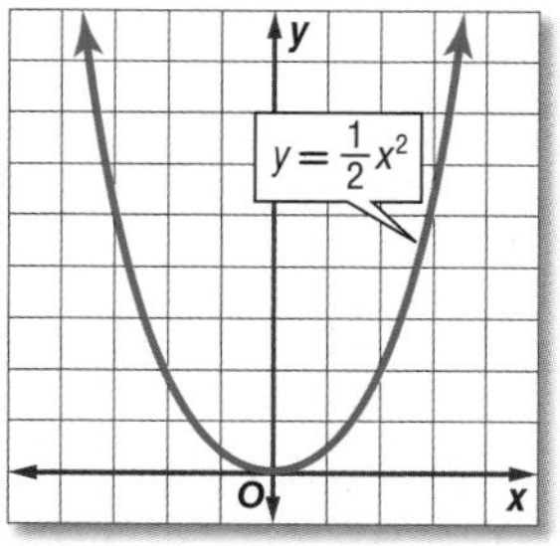

42. y-axis

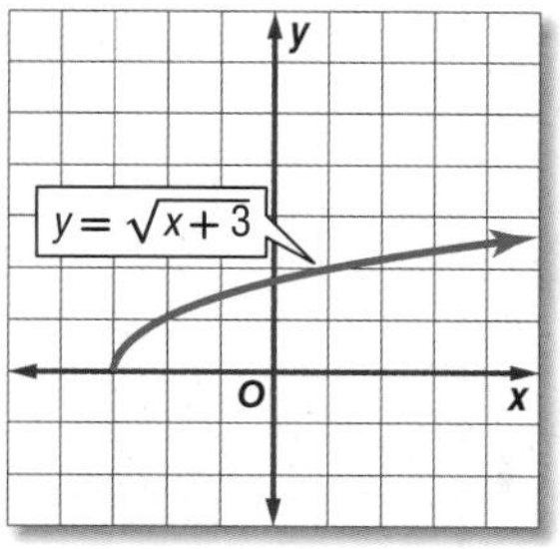

43. x-axis

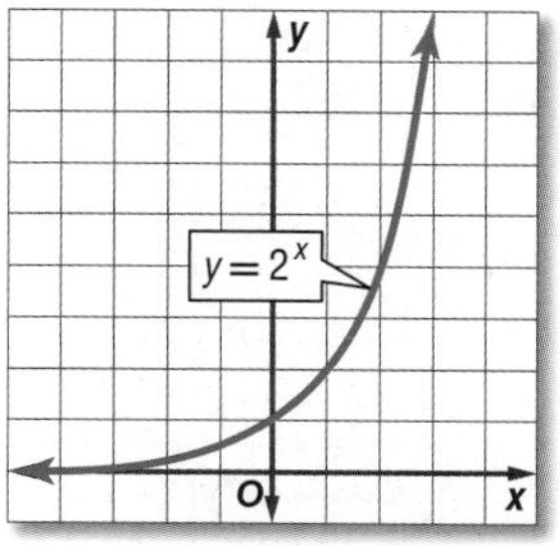

44. **MULTIPLE REPRESENTATIONS** In this problem, you will investigate a reflection in the origin.

a. Geometric Draw $\triangle ABC$ in the coordinate plane so that each vertex is a whole-number ordered pair.

b. Graphical Locate each reflected point A', B', and C' so that the reflected point, the original point, and the origin are collinear, and both the original point and the reflected point are equidistant from the origin.

c. Tabular Copy and complete the table below.

	$\triangle ABC$		$\triangle A'B'C'$	
Coordinates	A		A'	
	B		B'	
	C		C'	

d. Verbal Make a conjecture about the relationship between corresponding vertices of a figure reflected in the origin.

H.O.T. Problems Use Higher-Order Thinking Skills

45. **ERROR ANALYSIS** Jamil and Ashley are finding the coordinates of the image of (2, 3) after a reflection in the x-axis. Is either of them correct? Explain.

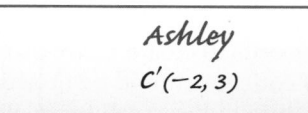

Jamil	Ashley
$C'(2, -3)$	$C'(-2, 3)$

46. **WRITING IN MATH** Describe how to reflect a figure not on the coordinate plane across a line.

47. **CHALLENGE** A point in the second quadrant with coordinates $(-a, b)$ is reflected in the x-axis. If the reflected point is then reflected in the line $y = -x$, what are the final coordinates of the image?

48. **OPEN ENDED** Draw a polygon on the coordinate plane that when reflected in the x-axis looks exactly like the original figure.

49. **CHALLENGE** When $A(4, 3)$ is reflected in a line, its image is $A'(-1, 0)$. Find the equation of the line of reflection. Explain your reasoning.

50. **CCSS PRECISION** The image of a point reflected in a line is *always, sometimes,* or *never* located on the other side of the line of reflection.

51. **WRITING IN MATH** Suppose points P, Q, and R are collinear, with point Q between points P and R. Describe a plan for a proof that the reflection of points P, Q, and R in a line preserves collinearity and betweenness of points.

Standardized Test Practice

52. SHORT RESPONSE If quadrilateral $WXYZ$ is reflected across the y-axis to become quadrilateral $W'X'Y'Z'$, what are the coordinates of X'?

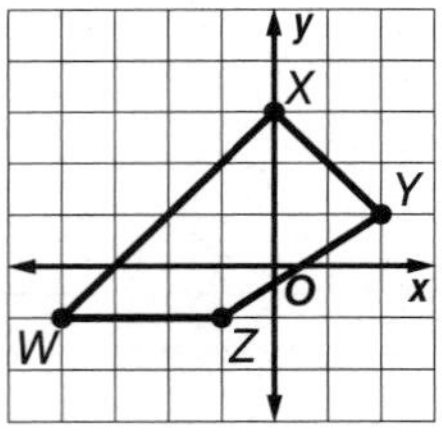

53. ALGEBRA If the arithmetic mean of $6x$, $3x$, and 27 is 18, then what is the value of x?

A 2 **C** 5
B 3 **D** 6

54. In $\triangle DEF$, $m\angle E = 108$, $m\angle F = 26$, and $f = 20$. Find d to the nearest whole number.

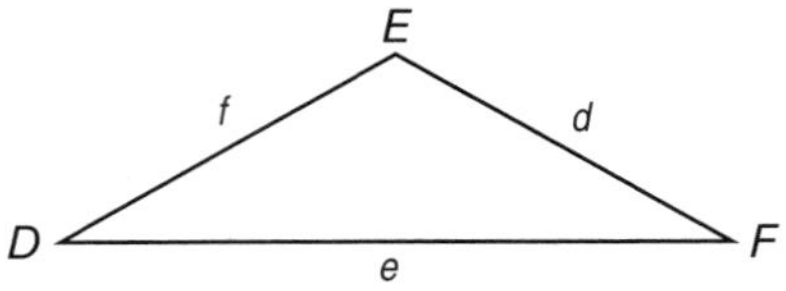

F 26 **G** 33 **H** 60 **J** 65

55. SAT/ACT In a coordinate plane, points A and B have coordinates $(-2, 4)$ and $(3, 3)$, respectively. What is the value of AB?

A $\sqrt{50}$ **D** $(1, -1)$
B $(1, 7)$ **E** $\sqrt{26}$
C $(5, -1)$

Spiral Review

56. COORDINATE GEOMETRY In $\triangle LMN$, $\overline{PR}$ divides $\overline{NL}$ and $\overline{MN}$ proportionally. If the vertices are $N(8, 20)$, $P(11, 16)$, and $R(3, 8)$ and $\frac{LP}{PN} = \frac{2}{1}$, find the coordinates of L and M. (Lesson 14-2)

57. BIOLOGY Each type of fish thrives in a specific range of temperatures. The best temperatures for sharks range from 18°C to 22°C, inclusive. Write a compound inequality to represent temperatures where sharks will not thrive. (Lesson 5-4)

Write an equation of the line that passes through each pair of points. (Lesson 4-2)

58. $(1, 1), (7, 4)$ **59.** $(5, 7), (0, 6)$ **60.** $(5, 1), (8, -2)$

61. COFFEE A coffee store wants to create a mix using two coffees. How many pounds of coffee A should be mixed with 9 pounds of coffee B to get a mixture that can sell for \$6.95 per pound? (Lesson 2-9)

Skills Review

Find the magnitude and direction of each vector.

62. $\overrightarrow{RS}$: $R(-3, 3)$ and $S(-9, 9)$
63. $\overrightarrow{JK}$: $J(8, 1)$ and $K(2, 5)$
64. $\overrightarrow{FG}$: $F(-4, 0)$ and $G(-6, -4)$
65. $\overrightarrow{AB}$: $A(-1, 10)$ and $B(1, -12)$

LESSON 14-5 Translations

:: Then	:: Now	:: Why?
You found the magnitude and direction of vectors.	**1** Draw translations. **2** Draw translations in the coordinate plane.	Stop-motion animation is a technique in which an object is moved by very small amounts between individually photographed frames. When the series of frames is played as a continuous sequence, the result is the illusion of movement.

NewVocabulary
translation vector

Common Core State Standards

Content Standards

G.CO.4 Develop definitions of rotations, reflections, and translations in terms of angles, circles, perpendicular lines, parallel lines, and line segments.

G.CO.5 Given a geometric figure and a rotation, reflection, or translation, draw the transformed figure using, e.g., graph paper, tracing paper, or geometry software. Specify a sequence of transformations that will carry a given figure onto another.

Mathematical Practices

5 Use appropriate tools strategically.

4 Model with mathematics.

1 Draw Translations

In Lesson 12-7, you learned that a translation or *slide* is a transformation that moves all points of a figure the same distance in the same direction. Since vectors can be used to describe both distance and direction, vectors can be used to define translations.

KeyConcept Translation

A translation is a function that maps each point to its image along a vector, called the **translation vector**, such that

- each segment joining a point and its image has the same length as the vector, and
- this segment is also parallel to the vector.

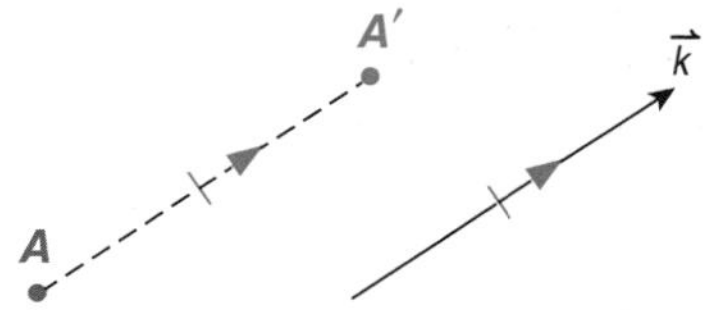

Point A' is a translation of point A along translation vector $\vec{k}$.

Example 1 Draw a Translation

Copy the figure and the given translation vector. Then draw the translation of the figure along the translation vector.

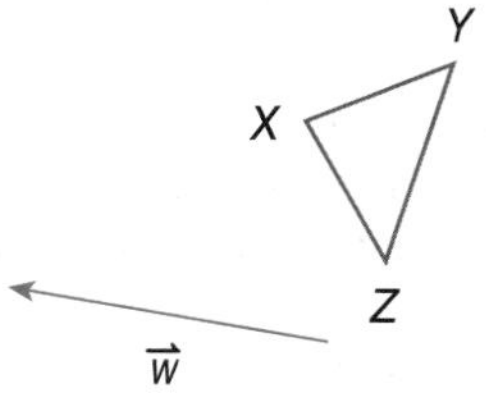

Step 1 Draw a line through each vertex parallel to vector $\vec{w}$

Step 2 Measure the length of vector $\vec{w}$. Locate point X' by marking off this distance along the line through vertex X, starting at X and in the same direction as the vector.

Step 3 Repeat Step 2 to locate points Y' and Z'. Then connect vertices X', Y', and Z' to form the translated image.

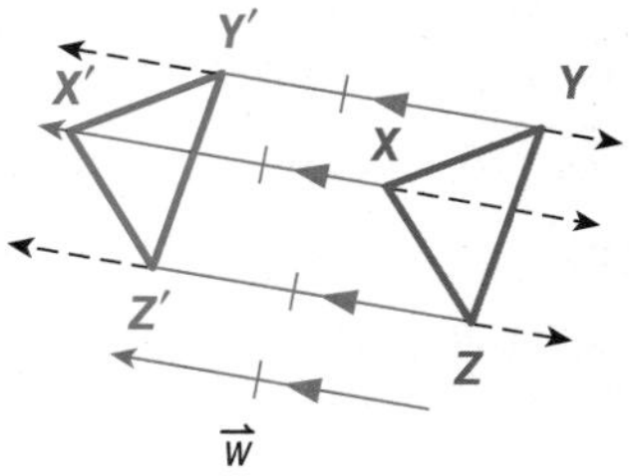

GuidedPractice

1A.

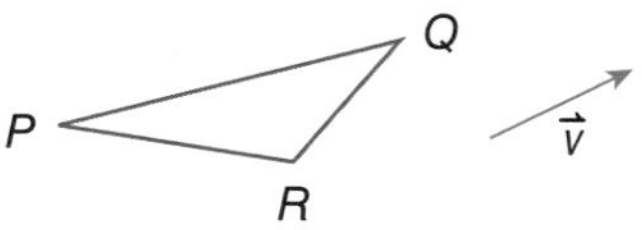

1B.

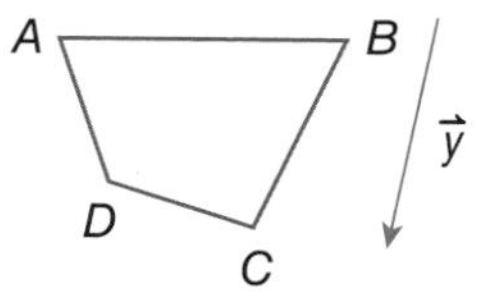

2 Draw Translations in the Coordinate Plane

Recall that a vector in the coordinate plane can be written as $\langle a, b\rangle$, where a represents the horizontal change and b is the vertical change from the vector's tip to its tail. $\overline{CD}$ is represented by the ordered pair $\langle 2, -4\rangle$.

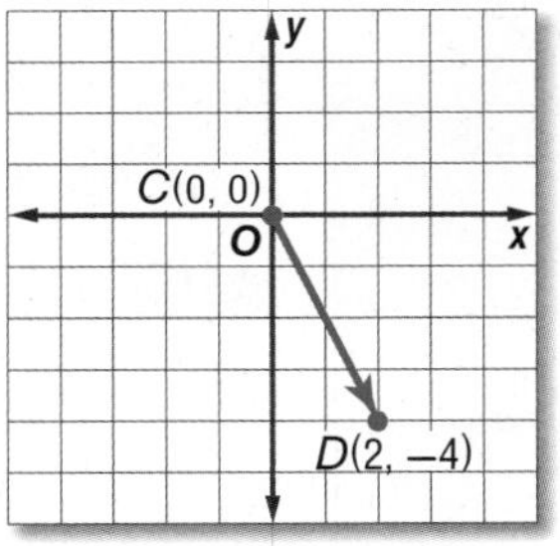

Written in this form, called the component form, a vector can be used to translate a figure in the coordinate plane.

ReadingMath

Horizontal and Vertical Translations When the translation vector is of the form $\langle a, 0\rangle$, the translation is horizontal only. When the translation vector is of the form $\langle 0, b\rangle$, the translation is vertical only.

KeyConcept Translation in the Coordinate Plane

Words To translate a point along vector $\langle a, b\rangle$, add a to the x-coordinate and b to the y-coordinate.

Symbols $(x, y) \rightarrow (x + a, y + b)$

Example The image of $P(-2, 3)$ translated along vector $\langle 7, 4\rangle$ is $P'(5, 7)$.

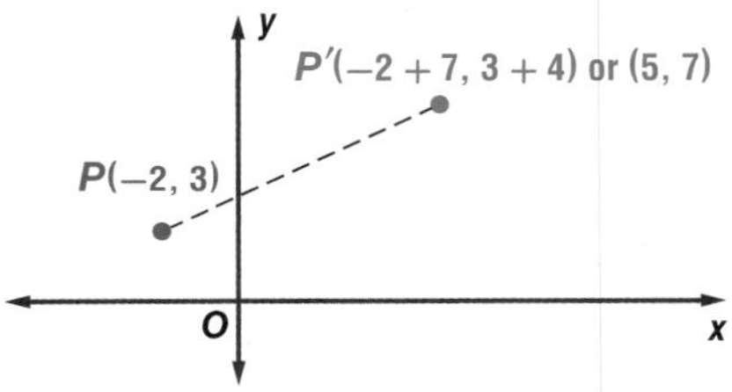

A translation is another type of congruence transformation or isometry.

Example 2 Translations in the Coordinate Plane

Graph each figure and its image along the given vector.

a. **$\triangle EFG$ with vertices $E(-7, -1)$, $F(-4, -4)$, and $G(-3, -1)$; $\langle 2, 5\rangle$**

The vector indicates a translation 2 units right and 5 units up.

$(x, y) \rightarrow (x + 2, y + 5)$

$E(-7, -1) \rightarrow E'(-5, 4)$

$F(-4, -4) \rightarrow F'(-2, 1)$

$G(-3, -1) \rightarrow G'(-1, 4)$

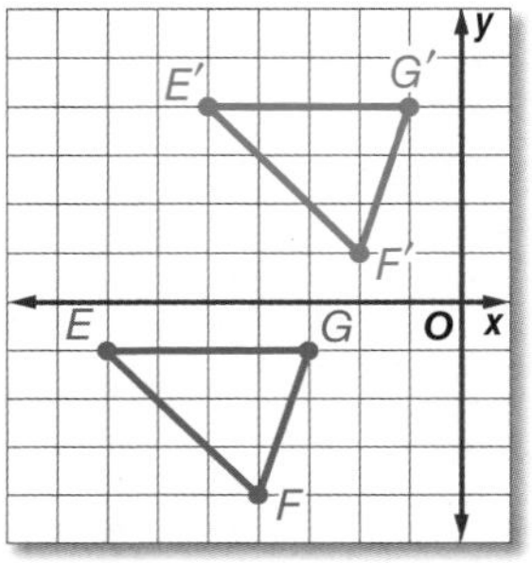

b. **square $JKLM$ with vertices $J(3, 4)$, $K(5, 2)$, $L(7, 4)$, and $M(5, 6)$; $\langle -3, -4\rangle$**

The vector indicates a translation 3 units left and 4 units down.

$(x, y) \rightarrow (x + (-3), y + (-4))$

$J(3, 4) \rightarrow J'(0, 0)$

$K(5, 2) \rightarrow K'(2, -2)$

$L(7, 4) \rightarrow L'(4, 0)$

$M(5, 6) \rightarrow M'(2, 2)$

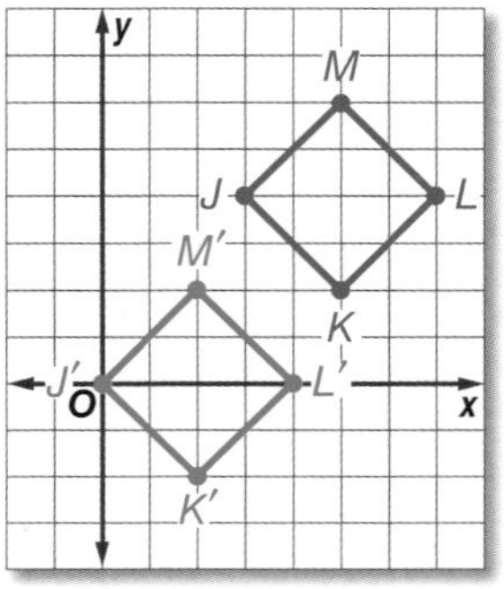

Math HistoryLink

Felix Klein (1849–1925) Klein's definition of geometry as the study of the properties of a space that remain invariant under a group of transformations allowed for the inclusion of both Euclidean and non-Euclidean geometry.

GuidedPractice

2A. $\triangle ABC$ with vertices $A(2, 6)$, $B(1, 1)$, and $C(7, 5)$; $\langle -4, -1\rangle$

2B. quadrilateral $QRST$ with vertices $Q(-8, -2)$, $R(-9, -5)$, $S(-4, -7)$, and $T(-4, -2)$; $\langle 7, 1\rangle$

Real-World Example 3 Describing Translations

MARCHING BAND In one part of a marching band's performance, a line of trumpet players starts at position 1, marches to position 2, and then to position 3. Each unit on the graph represents one step.

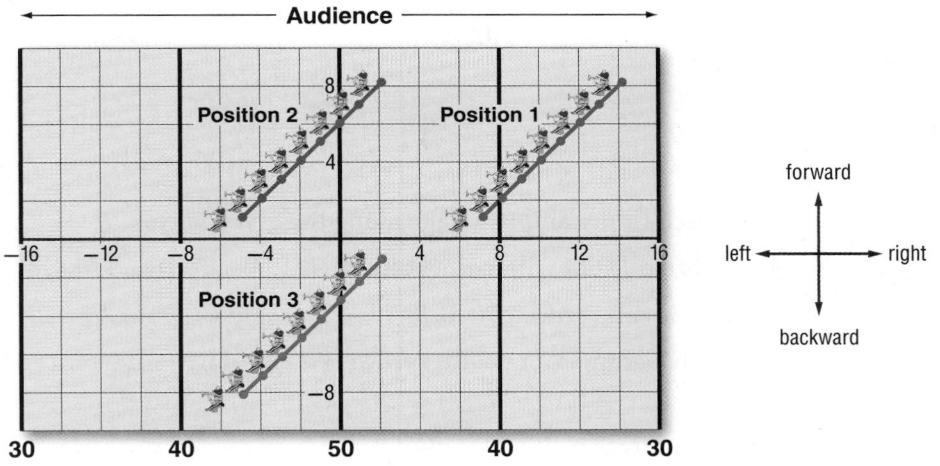

Real-WorldLink

Marching bands often make use of a series of formations that can include geometric shapes. Usually, each band member has an assigned position in each formation. *Floating* is the movement of a group of members together without changing the shape or size of their formation.

a. Describe the translation of the trumpet line from position 1 to position 2 in function notation and in words.

One point on the line in position 1 is (14, 8). In position 2, this point moves to (2, 8). Use the translations function $(x, y) \rightarrow (x + a, y + b)$ to write and solve equations to find a and b.

$(14 + a, 8 + b)$ or $(2, 8)$

$14 + a = 2$ $\qquad$ $8 + b = 8$

$a = -12$ $\qquad$ $b = 0$

function notation: $(x, y) \rightarrow (x + (-12), y + 0)$

So, the trumpet line is translated 12 steps *left* but no steps forward or backward from position 1 to position 2.

b. Describe the translation of the line from position 1 to position 3 using a translation vector.

$(14 + a, 8 + b)$ or $(2, -1)$

$14 + a = 2$ $\qquad$ $8 + b = -1$

$a = -12$ $\qquad$ $b = -9$

translation vector: $\langle -12, -9 \rangle$

GuidedPractice

3. ANIMATION A coin is filmed using stop-motion animation so that it appears to move.

A. Describe the translation from A to B in function notation and in words.

B. Describe the translation from A to C using a translation vector.

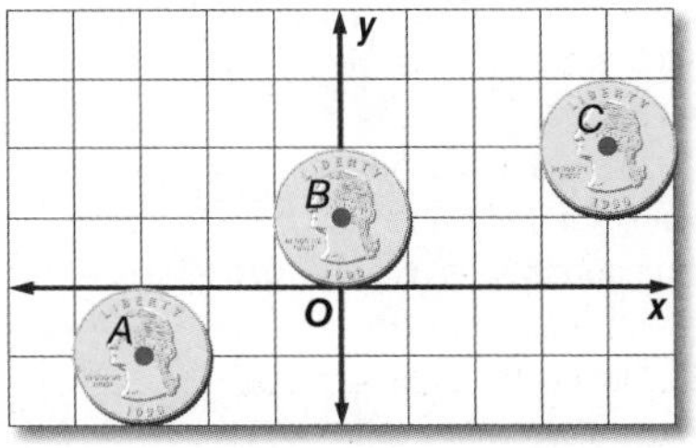

Check Your Understanding

Example 1 **Copy the figure and the given translation vector. Then draw the translation of the figure along the translation vector.**

1. 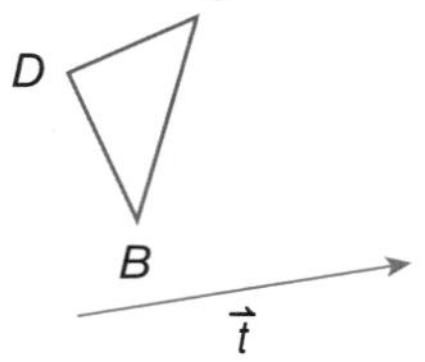

2. 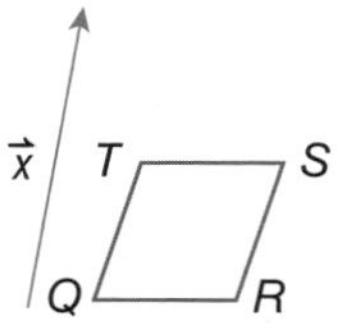

3. 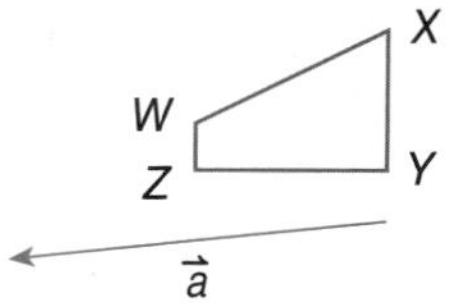

Example 2 **Graph each figure and its image along the given vector.**

4. trapezoid *JKLM* with vertices *J*(2, 4), *K*(1, 1), *L*(5, 1) and *M*(4, 4); ⟨7, 1⟩

5. △*DFG* with vertices *D*(−8, 8), *F*(−10, 4), and *G*(−7, 6); ⟨5, −2⟩

6. parallelogram *WXYZ* with vertices *W*(−6, −5), *X*(−2, −5), *Y*(−1, −8), and *Z*(−5, −8); ⟨−1, 4⟩

Example 3 **7. VIDEO GAMES** The object of the video game shown is to manipulate the colored tiles left or right as they fall from the top of the screen to completely fill each row without leaving empty spaces. If the starting position of the tile piece at the top of the screen is (*x*, *y*), use function notation to describe the translation that will fill the indicated row.

Practice and Problem Solving

Example 1 **CCSS TOOLS** **Copy the figure and the given translation vector. Then draw the translation of the figure along the translation vector.**

8. 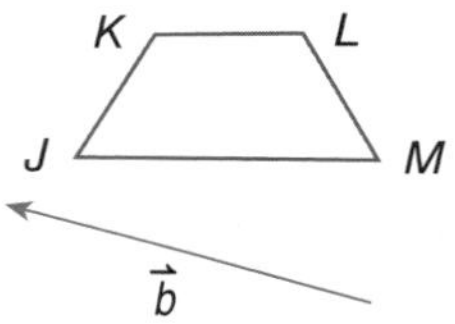

9. 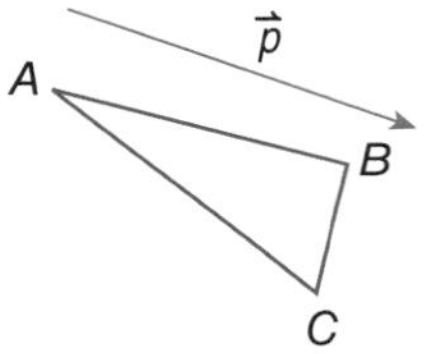

10. 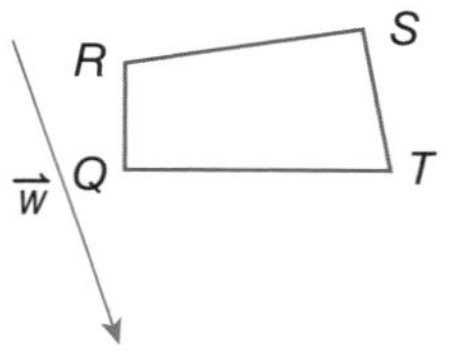

11. 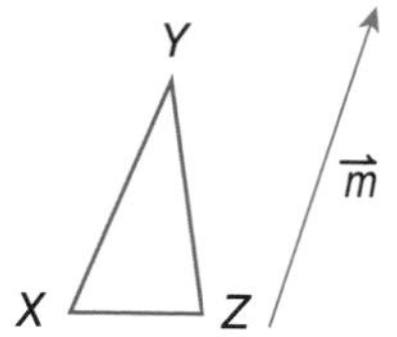

12. 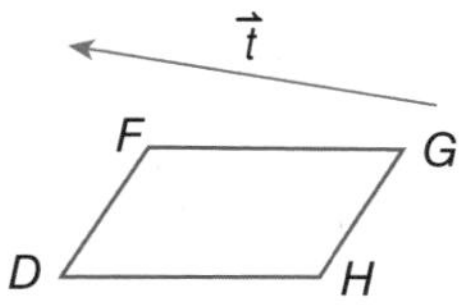

13. 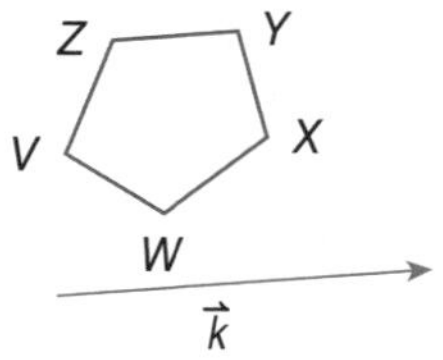

Example 2 **Graph each figure and its image along the given vector.**

14. △*ABC* with vertices *A*(1, 6), *B*(3, 2), and *C*(4, 7); ⟨4, −1⟩

15. △*MNP* with vertices *M*(4, −5), *N*(5, −8), and *P*(8, −6); ⟨−2, 5⟩

16. rectangle *QRST* with vertices *Q*(−8, 4), *R*(−8, 2), *S*(−3, 2), and *T*(−3, 4); ⟨2, 3⟩

17. quadrilateral *FGHJ* with vertices *F*(−4, −2), *G*(−1, −1), *H*(0, −4), and *J*(−3, −6); ⟨−5, −2⟩

18. ▱*WXYZ* with vertices *W*(−3, −1), *X*(1, −1), *Y*(2, −4), and *Z*(−2, −4); ⟨−3, 4⟩

19. trapezoid *JKLM* with vertices *J*(−4, −2), *K*(−1, −2), *L*(0, −5), and *M*(−5, −5); ⟨6, 5⟩

Example 3 **20.** CCSS **MODELING** Brittany's neighborhood is shown on the grid at the right.

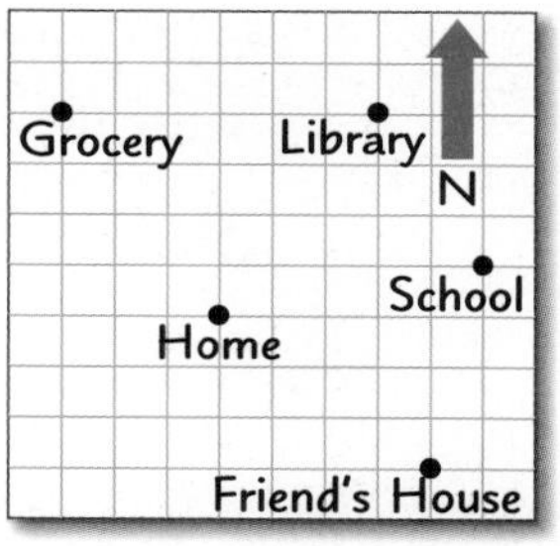

a. If she leaves home and travels 4 blocks north and 3 blocks east, what is her new location?

b. Use words to describe two possible translations that will take Brittany home from school.

21 **FOOTBALL** A wide receiver starts from his 15-yard line on the right hash mark and runs a route that takes him 12 yards to the left and down field for a gain of 17 yards. Write a translation vector to describe the receiver's route.

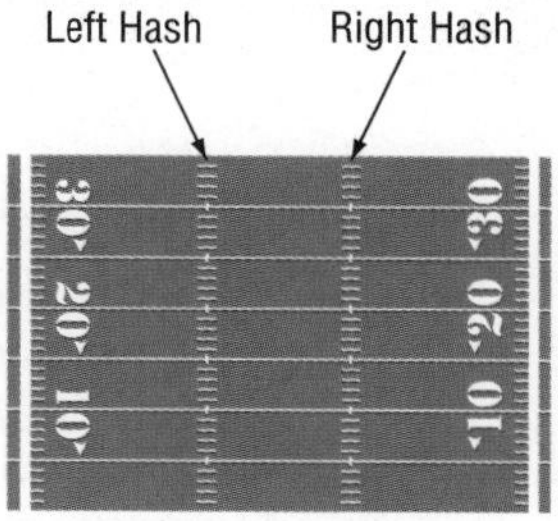

22. **CHESS** Each chess piece has a path that it can follow to move. The rook, which begins in square a8, can only move vertically or horizontally. The knight, which begins in square b8, can move two squares horizontally and then one square vertically, or two squares vertically and one square horizontally. The bishop, which begins in square f8, can only move diagonally.

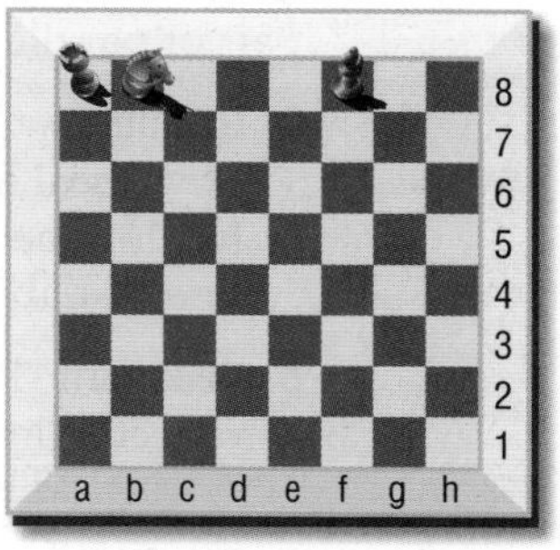

a. The knight moves 2 squares vertically and 1 square horizontally on its first move, then two squares horizontally and 1 square vertically on its second move. What are the possible locations for the knight after two moves?

b. After two moves, the rook is in square d3. Describe a possible translation to describe the two moves.

c. Describe a translation that can take the bishop to square a1. What is the minimum number of moves that can be used to accomplish this translation?

Write each translation vector.

23.

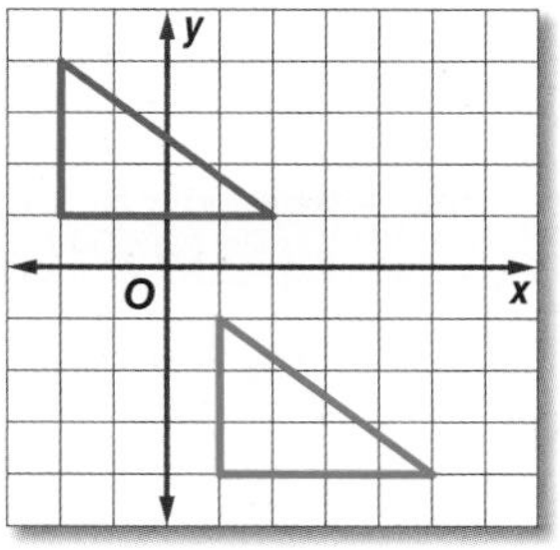

24.

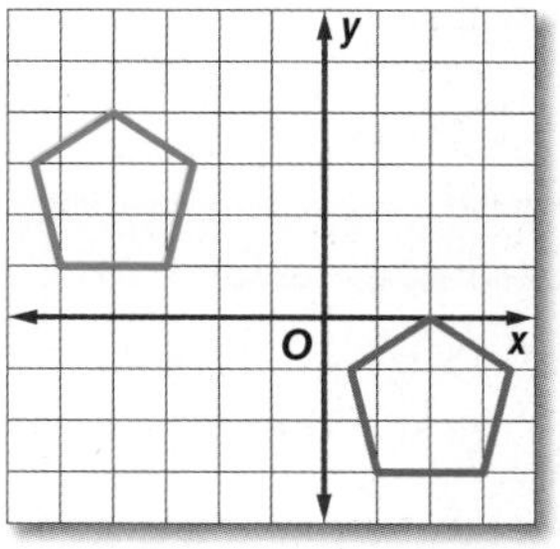

25. **CONCERTS** Dexter's family buys tickets every year for a concert. Last year they were in seats C3, C4, C5, and C6. This year, they will be in seats D16, D17, D18, and D19. Write a translation in words and using vector notation that can be used to describe the change in their seating.

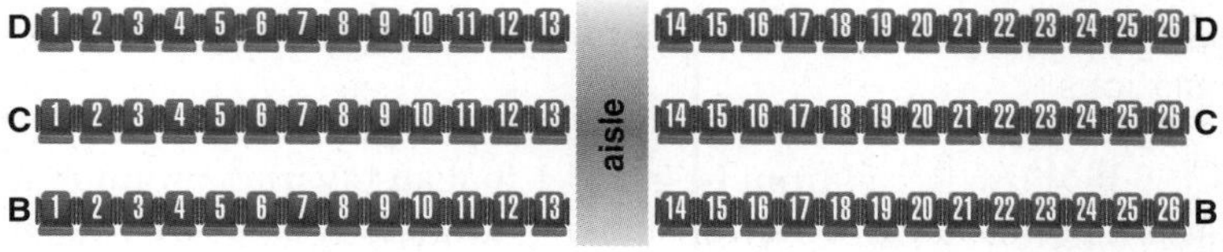

CCSS SENSE-MAKING **Graph the translation of each function along the given vector. Then write the equation of the translated image.**

26. $\langle 4, 1\rangle$

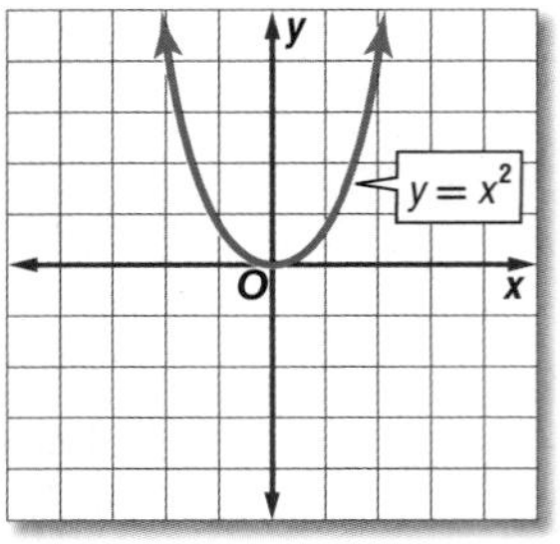

27. $\langle -2, 0\rangle$

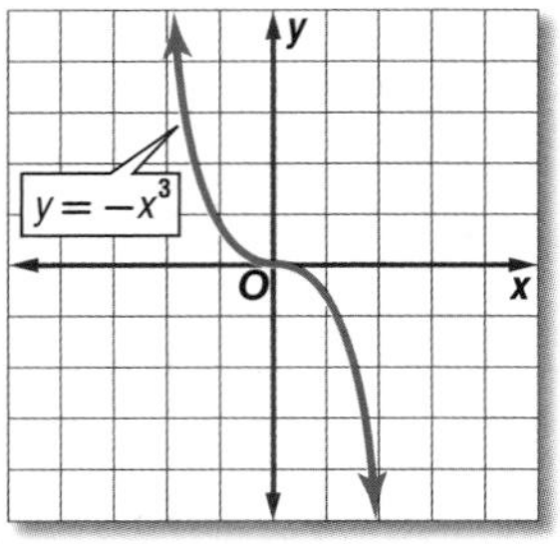

28. ROLLER COASTERS The length of the roller coaster track from the top of a hill to the bottom of the hill is 125 feet at a 53° angle with the vertical. If the position at the top of the hill is (x, y), use function notation to describe the translation to the bottom of the hill. Round to the nearest foot.

29. MULTIPLE REPRESENTATIONS In this problem, you will investigate reflections over a pair of parallel lines.

a. Geometric On patty paper, draw $\triangle ABC$ and a pair of vertical lines ℓ and m. Reflect $\triangle ABC$ in line ℓ by folding the patty paper. Then reflect $\triangle A'B'C'$, in line m. Label the final image $\triangle A''B''C''$.

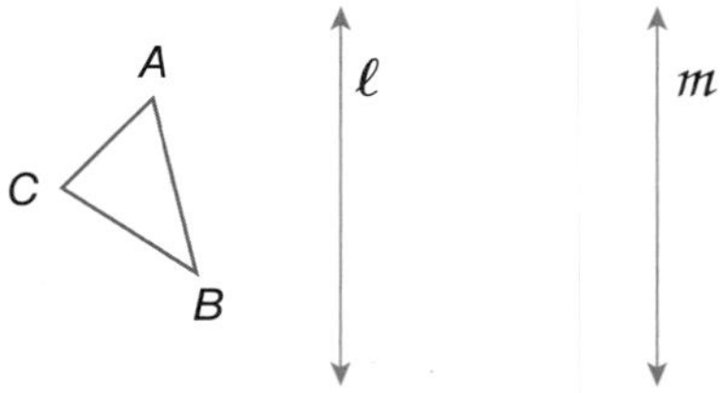

b. Geometric Repeat the process in part **a** for $\triangle DEF$ reflected in vertical lines n and p and $\triangle JKL$ reflected in vertical lines q and r.

c. Tabular Copy and complete the table below.

Distance Between Corresponding Points (cm)		Distance Between Vertical Lines (cm)	
A and A'', B and B'', C and C''		ℓ and m	
D and D'', E and E'', F and F''		n and p	
J and J'', K and K'', L and L''		q and r	

d. Verbal Describe the result of two reflections in two vertical lines using one transformation.

H.O.T. Problems Use Higher-Order Thinking Skills

30. REASONING Determine a rule to find the final image of a point that is translated along $\langle x + a, y + b\rangle$ and then $\langle x + c, y + d\rangle$.

31. CHALLENGE A line $y = mx + b$ is translated using the vector $\langle a, b\rangle$. Write the equation of the translated line. What is the value of the y-intercept?

32. OPEN ENDED Draw a figure on the coordinate plane so that the figure has the same orientation after it is reflected in the line $y = 1$. Explain what must be true in order for this to occur.

33. WRITING IN MATH Compare and contrast function notation and vector notation for translations.

34. WRITING IN MATH Recall from Lesson 9-1 that an invariant point maps onto itself. Can invariant points occur with translations? Explain why or why not.

Standardized Test Practice

35. Identify the location of point P under translation $(x + 3, y + 1)$.

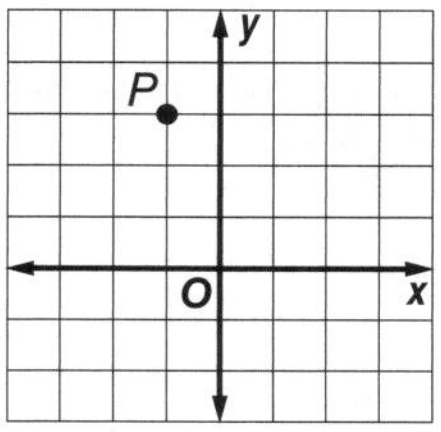

A (0, 6)
B (0, 3)
C (2, −4)
D (2, 4)

36. SHORT RESPONSE Which vector best describes the translation of $A(3, -5)$ to $A'(-2, -8)$?

37. ALGEBRA Over the next four days, Amanda plans to drive 160 miles, 235 miles, 185 miles, and 220 miles. If her car gets an average of 32 miles per gallon of gas, how many gallons of gas should she expect to use in all?

F 25 **G** 30 **H** 35 **J** 40

38. SAT/ACT A bag contains 5 red marbles, 2 blue marbles, 4 white marbles, and 1 yellow marble. If two marbles are chosen in a row, without replacement, what is the probability of getting 2 white marbles?

A $\frac{1}{66}$
B $\frac{1}{11}$
C $\frac{1}{9}$
D $\frac{5}{33}$
E $\frac{2}{5}$

Spiral Review

Graph each figure and its image under the given reflection. (Lesson 14-4)

39. $\overline{DJ}$ with endpoints $D(4, 4)$, $J(-3, 2)$ in the y-axis

40. $\triangle XYZ$ with vertices $X(0, 0)$, $Y(3, 0)$, and $Z(0, 3)$ in the x-axis

41. $\triangle ABC$ with vertices $A(-3, -1)$, $B(0, 2)$, and $C(3, -2)$, in the line $y = x$

42. quadrilateral $JKLM$ with vertices $J(-2, 2)$, $K(3, 1)$, $L(4, -1)$, and $M(-2, -2)$ in the origin

Write an equation in function notation for each relation. (Lesson 3-6)

43.

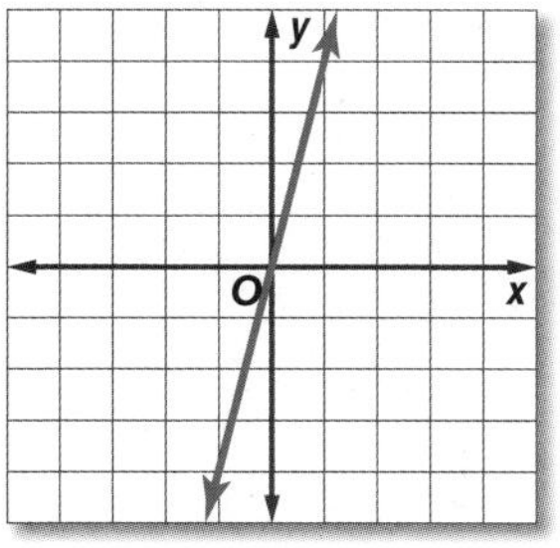

44.

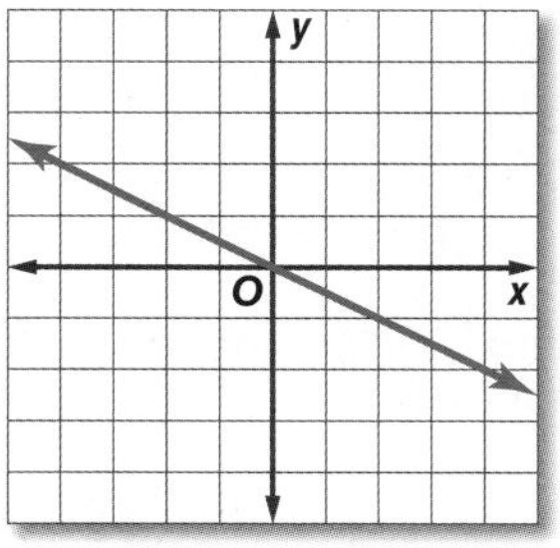

Use $\square JKLM$ to find each measure. (Lesson 13-1)

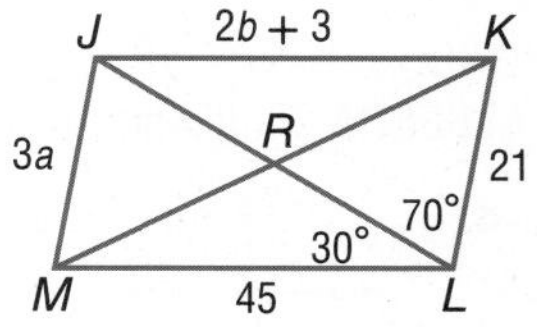

45. $m\angle MJK$

46. $m\angle JML$

47. $m\angle JKL$

48. $m\angle KJL$

Skills Review

Copy the diagram shown, and extend each ray. Classify each angle as *right*, *acute*, or *obtuse*. Then use a protractor to measure the angle to the nearest degree.

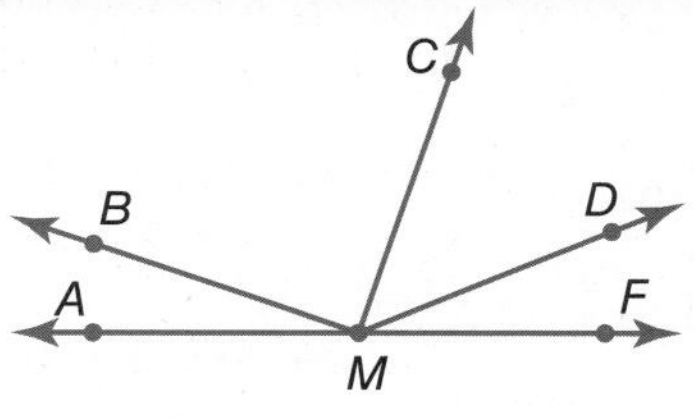

49. $\angle AMC$

50. $\angle BMD$

51. $\angle FMD$

52. $\angle CMB$

Mid-Chapter Quiz

Lessons 14-1 through 14-5

ALGEBRA Identify the similar triangles. Find each measure. (Lesson 14-1)

1. *SR*

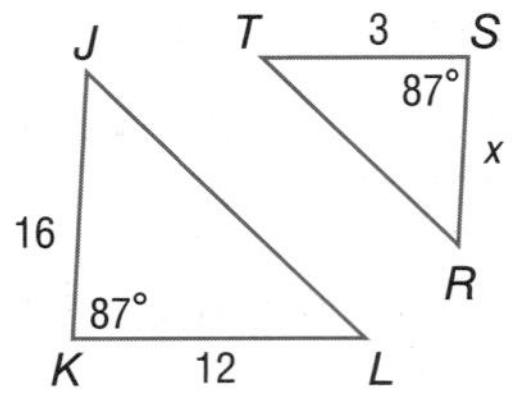

2. *AF*

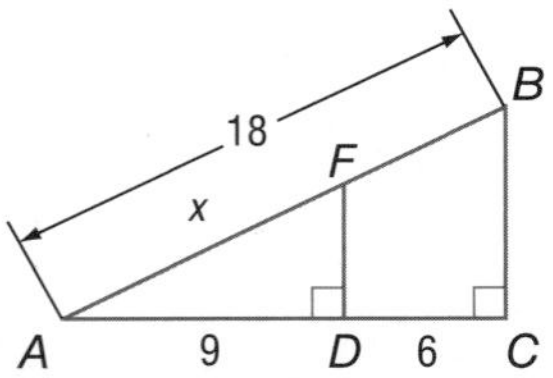

3. HISTORY In the fifteenth century, mathematicians and artists tried to construct the perfect letter. A square was used as a frame to design the letter "A," as shown below. The thickness of the major stroke of the letter was $\frac{1}{12}$ the height of the letter. (Lesson 14-1)

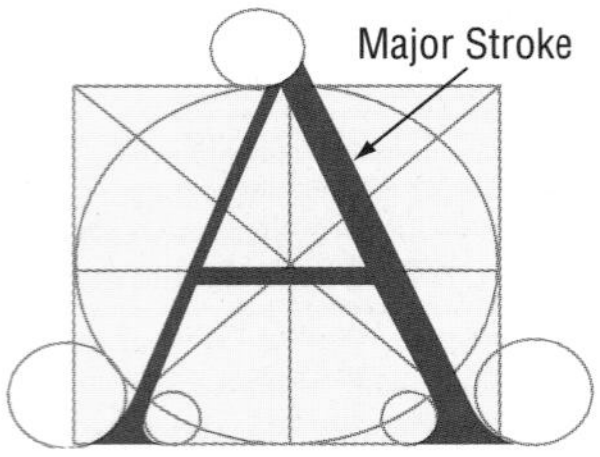

a. Explain why the bar through the middle of the A is half the length of the space between the outside bottom corners of the sides of the letter.

b. If the letter were 3 centimeters tall, how wide would the major stroke be?

ALGEBRA Find *x* and *y*. (Lesson 14-1)

4.

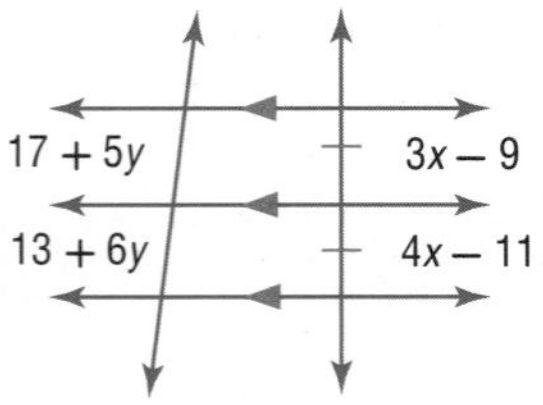

5.

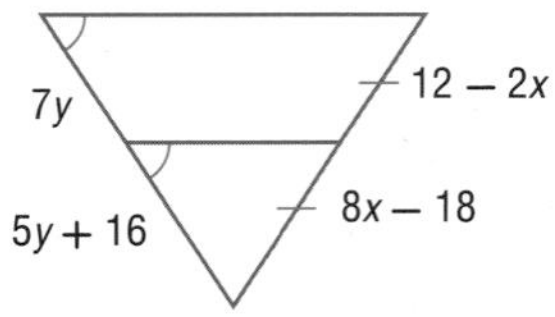

Copy the figure and the given line of reflection. Then draw the reflected image in this line using a ruler. (Lesson 14-4)

6.

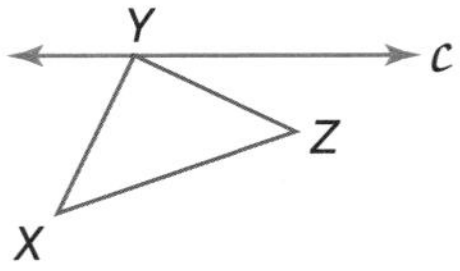

7.

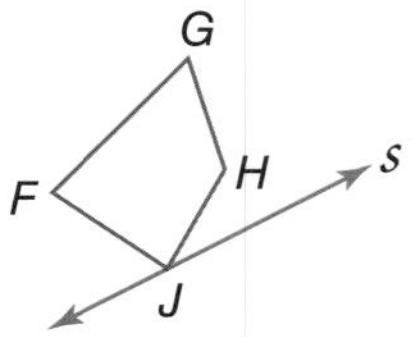

Graph each figure and its image after the specified reflection. (Lesson 14-4)

8. $\triangle FGH$ has vertices $F(-4, 3)$, $G(-2, 0)$, and $H(-1, 4)$; in the *y*-axis

9. rhombus *QRST* has vertices $Q(2, 1)$, $R(4, 3)$, $S(6, 1)$, and $T(4, -1)$; in the *x*-axis

10. CLUBS The drama club is selling candy during the intermission of a school play. Locate point *P* along the wall to represent the candy table so that people coming from either door *A* or door *B* would walk the same distance to the table. (Lesson 14-4)

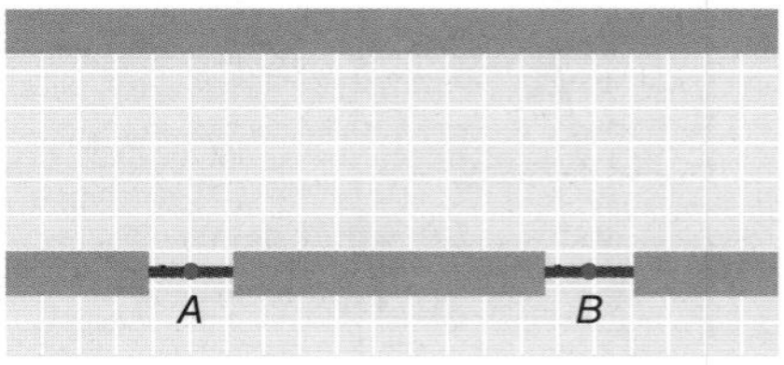

Graph each figure and its image after the specified translation. (Lesson 14-5)

11. $\triangle ABC$ with vertices $A(0, 0)$, $B(2, 1)$, $C(1, -3)$; $\langle 3, -1 \rangle$

12. rectangle *JKLM* has vertices $J(-4, 2)$, $K(-4, -2)$, $L(-1, -2)$, and $M(-1, 2)$; $\langle 5, -3 \rangle$

Copy the figure and the given translation vector. Then draw the translation of the figure along the translation vector. (Lesson 14-5)

13.

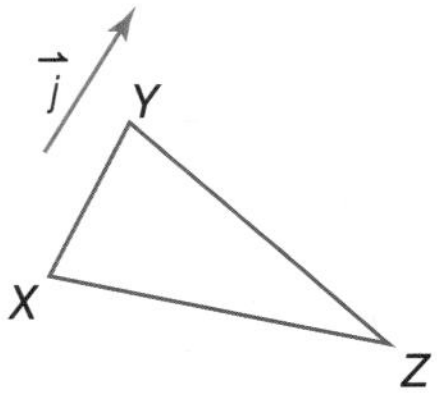

14.

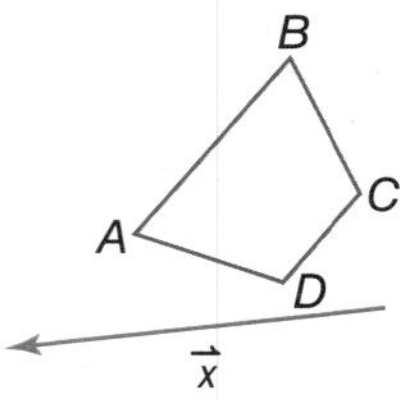

EXPLORE 14-6

Geometry Lab
Rotations

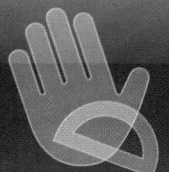

You learned that a rotation is a type of transformation that moves a figure about a fixed point, or center of rotation, through a specific angle and in a specific direction. In this activity you will use tracing paper to explore the properties of rotations.

CCSS Common Core State Standards
Content Standards
G.CO.2 Represent transformations in the plane using, e.g., transparencies and geometry software; describe transformations as functions that take points in the plane as inputs and give other points as outputs. Compare transformations that preserve distance and angle to those that do not (e.g., translation versus horizontal stretch).
G.CO.5 Given a geometric figure and a rotation, reflection, or translation, draw the transformed figure using, e.g., graph paper, tracing paper, or geometry software. Specify a sequence of transformations that will carry a given figure onto another.
Mathematical Practices 5

Activity Explore Rotations by Using Patty Paper

Step 1 On a piece of tracing paper, draw quadrilateral *ABCD* and a point *P*.

Step 2 On another piece of tracing paper, trace quadrilateral *ABCD* and point *P*. Label the new quadrilateral *A′B′C′D′* and the new point *P*.

Step 3 Position the tracing paper so that both points *P* coincide. Rotate the paper so that *ABCD* and *A′B′C′D′* do not overlap. Tape the two pieces of tracing paper together.

Step 4 Measure the distance between *A*, *B*, *C*, and *D* to point *P*. Repeat for quadrilateral *A′B′C′D′*. Then copy and complete the table below.

Quadrilateral	Length			
ABCD	*AP*	*BP*	*CP*	*DP*
A′B′C′D′	*A′P*	*B′P*	*C′P*	*D′P*

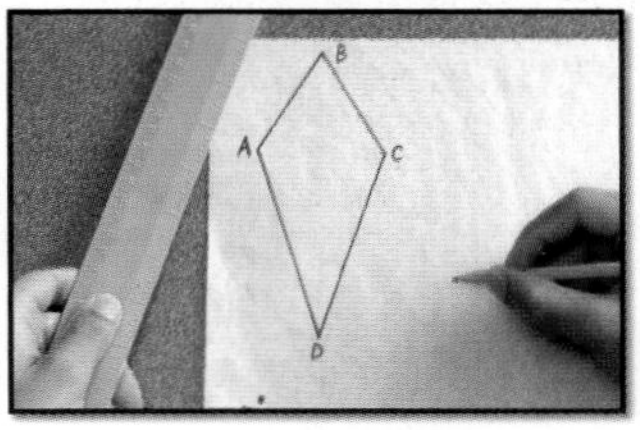

Step 1

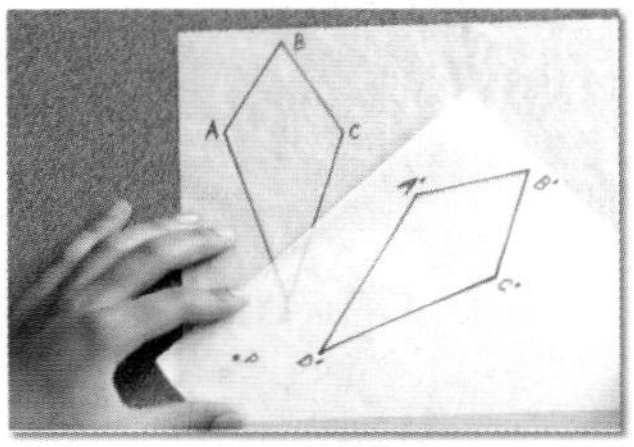

Steps 2 and 3

Exercises

1. Graph $\triangle JKL$ with vertices $J(1, 3)$, $K(2, 1)$, and $L(3, 4)$ on a coordinate plane, and then trace on tracing paper.
   a. Use a protractor to rotate each vertex 90° clockwise about the origin as shown in the figure at the right. What are the vertices of the rotated image?
   b. Rotate $\triangle JKL$ 180° about the origin. What are the vertices of the rotated image?
   c. Use the Distance Formula to find the distance from points *J*, *K*, and *L* to the origin. Repeat for *J′K′L′* and *J″K″L″*.

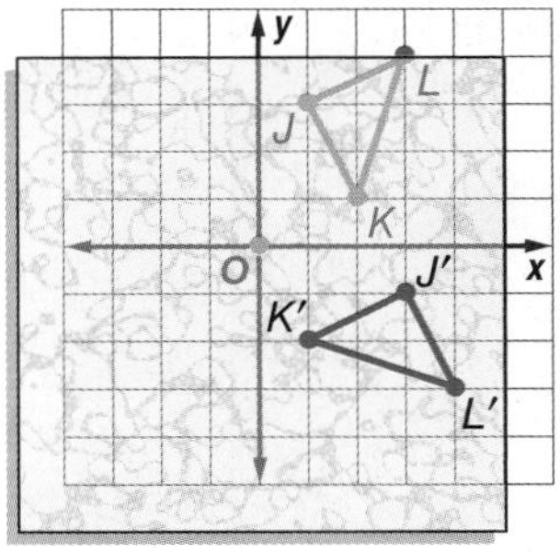

2. **WRITING IN MATH** If you rotate point (4, 2) 90° and 180° about the origin, how do the *x*- and *y*-coordinates change?
3. **MAKE A PREDICTION** What are the new coordinates of a point (x, y) that is rotated 270°?
4. **MAKE A CONJECTURE** Make a conjecture about the distances from the center of rotation *P* to each corresponding vertex of *ABCD* and *A′B′C′D′*.

LESSON

14-6 Rotations

∴ Then	∴ Now	∴ Why?
• You identified rotations and verified them as congruence transformations.	**1** Draw rotations. **2** Draw rotations in the coordinate plane.	• Modern windmill technology may be an important alternative to fossil fuels. Windmills convert the wind's energy into electricity through the rotation of turbine blades.

NewVocabulary
center of rotation
angle of rotation

Common Core State Standards

Content Standards

G.CO.4 Develop definitions of rotations, reflections, and translations in terms of angles, circles, perpendicular lines, parallel lines, and line segments.

G.CO.5 Given a geometric figure and a rotation, reflection, or translation, draw the transformed figure using, e.g., graph paper, tracing paper, or geometry software. Specify a sequence of transformations that will carry a given figure onto another.

Mathematical Practices

2 Reason abstractly and quantitatively.

5 Use appropriate tools strategically.

1 Draw Rotations

In Lesson 12-7, you learned that a rotation or *turn* moves every point of a preimage through a specified angle and direction about a fixed point.

KeyConcept Rotation

A rotation about a fixed point, called the **center of rotation**, through an angle of $x°$ is a function that maps a point to its image such that

- if the point is the center of rotation, then the image and preimage are the same point, or
- if the point is not the center of rotation, then the image and preimage are the same distance from the center of rotation and the measure of the **angle of rotation** formed by the preimage, center of rotation, and image points is x.

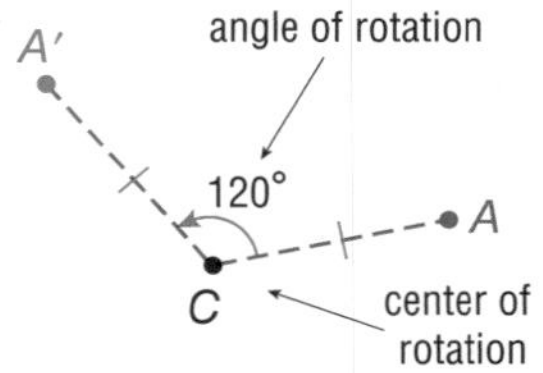

A' is the image of A after a 120° rotation about point C.

The direction of a rotation can be either clockwise or counterclockwise. Assume that all rotations are counterclockwise unless stated otherwise.

PT

Example 1 Draw a Rotation

Copy $\triangle ABC$ and point K. Then use a protractor and ruler to draw a 140° rotation of $\triangle ABC$ about point K.

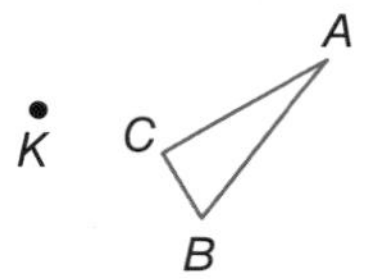

Step 1 Draw a segment from A to K.

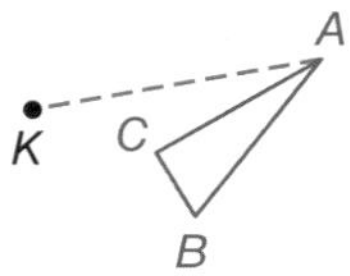

Step 2 Draw a 140° angle using $\overline{KA}$.

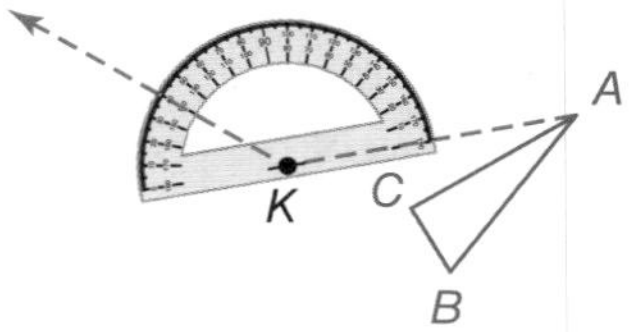

Step 3 Use a ruler to draw A' such that $KA' = KA$.

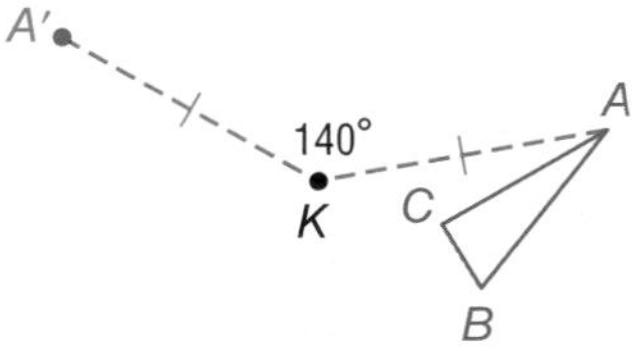

Step 4 Repeat Steps 1–3 for vertices B and C and draw $\triangle A'B'C'$.

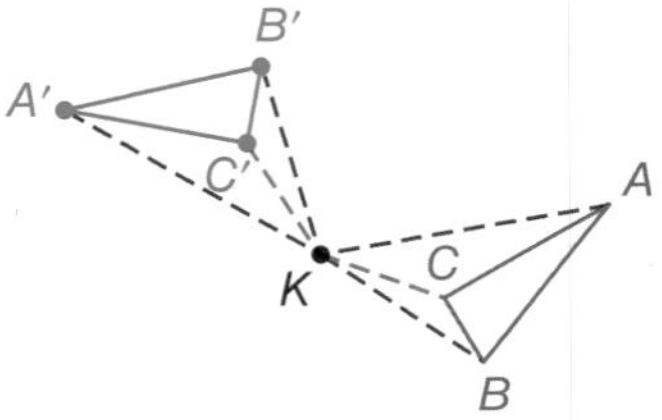

GuidedPractice

Copy each figure and point *K*. Then use a protractor and ruler to draw a rotation of the figure the given number of degrees about *K*.

1A. 65°

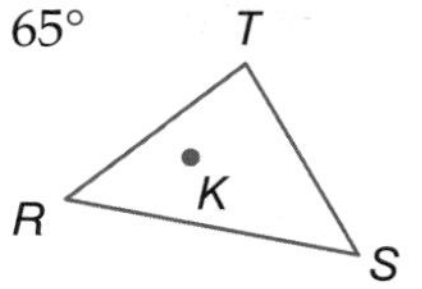

1B. 170°

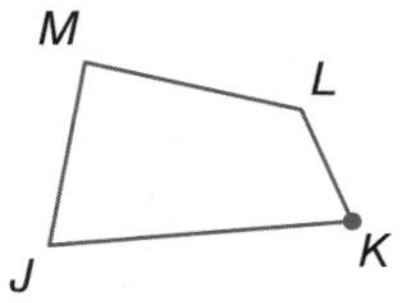

> **StudyTip**
> **Clockwise Rotation** Clockwise rotation can be designated by a negative angle measure. For example a rotation of −90° about the origin is a rotation 90° clockwise about the origin.

2 Draw Rotations in the Coordinate Plane

When a point is rotated 90°, 180°, or 270° counterclockwise about the origin, you can use the following rules.

KeyConcept Rotations in the Coordinate Plane

90° Rotation

To rotate a point 90° counterclockwise about the origin, multiply the *y*-coordinate by −1 and then interchange the *x*- and *y*-coordinates.

Symbols $(x, y) \rightarrow (-y, x)$

Example

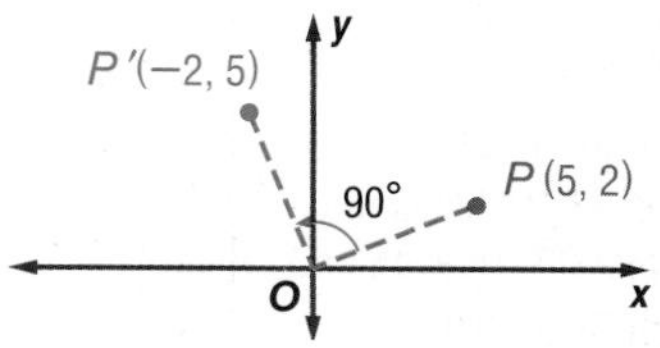

180° Rotation

To rotate a point 180° counterclockwise about the origin, multiply the *x*- and *y*-coordinates by −1.

Symbols $(x, y) \rightarrow (-x, -y)$

Example

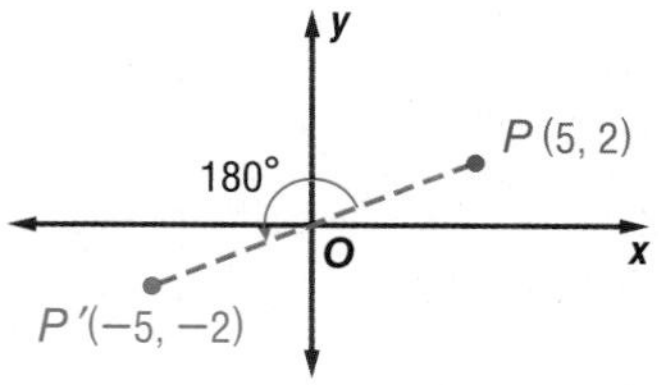

270° Rotation

To rotate a point 270° counterclockwise about the origin, multiply the *x*-coordinate by −1 and then interchange the *x*- and *y*-coordinates.

Symbols $(x, y) \rightarrow (y, -x)$

Example

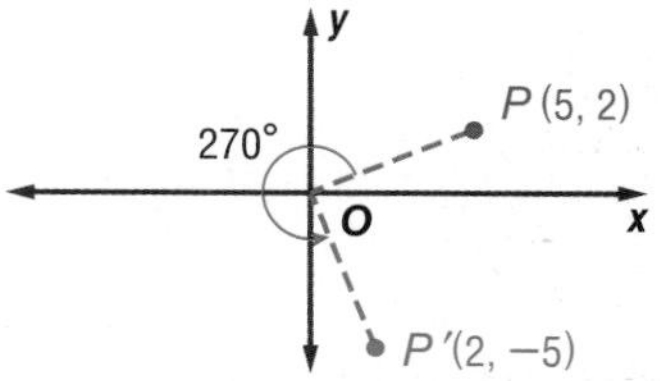

> **StudyTip**
> **360° Rotation** A rotation of 360° about a point returns a figure to its original position. That is, the image under a 360° rotation is equal to the preimage.

Example 2 Rotations in the Coordinate Plane

Triangle *PQR* has vertices *P*(1, 1), *Q*(4, 5), and *R*(5, 1). Graph △*PQR* and its image after a rotation 90° about the origin.

Multiply the *y*-coordinate of each vertex by −1 and interchange.

(x, y)	→	$(-y, x)$
$P(1, 1)$	→	$P'(-1, 1)$
$Q(4, 5)$	→	$Q'(-5, 4)$
$R(5, 1)$	→	$R'(-1, 5)$

Graph △*PQR* and its image △*P′Q′R′*.

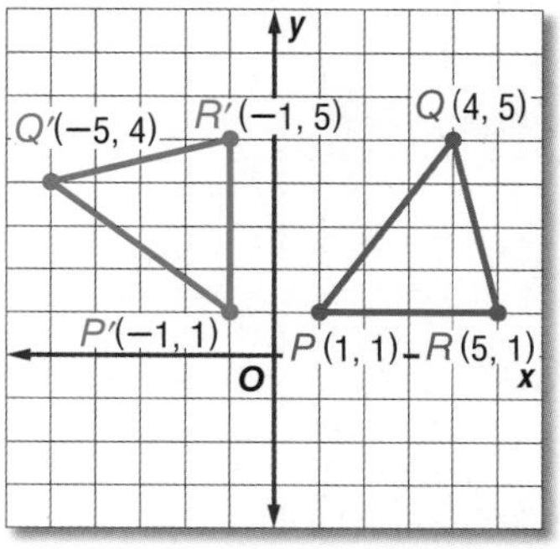

GuidedPractice

2. Parallelogram *FGHJ* has vertices *F*(2, 1), *G*(7, 1), *H*(6, −3), and *J*(1, −3). Graph *FGHJ* and its image after a rotation 180° about the origin.

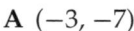

Standardized Test Example 3 Rotations in the Coordinate Plane

Triangle *JKL* is shown at the right. What is the image of point *J* after a rotation 270° counterclockwise about the origin?

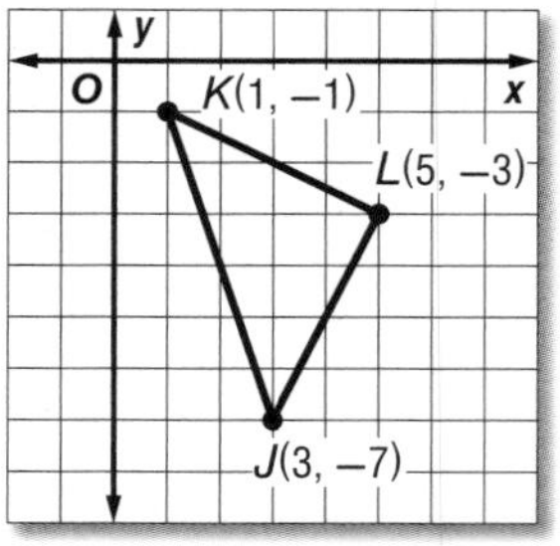

A (–3, –7)

B (–7, 3)

C (–7, –3)

D (7, –3)

Read the Test Item

You are given that △*JKL* has coordinates *J*(3, –7), *K*(1, –1), and *L*(5, –3) and are then asked to identify the coordinates of the image of point *J* after a 270° counterclockwise rotation about the origin.

StudyTip

270° Rotation You can complete a 270° rotation by performing a 90° rotation and a 180° rotation in sequence.

Solve the Test Item

To find the coordinates of point *J* after a 270° counterclockwise rotation about the origin, multiply the *x*-coordinate by –1 and then interchange the *x*- and *y*-coordinates.

$(x, y) \rightarrow (y, -x)$ $\qquad$ $(3, -7) \rightarrow (-7, -3)$

The answer is choice C.

GuidedPractice

3. Parallelogram *WXYZ* is rotated 180° counterclockwise about the origin. Which of these graphs represents the resulting image?

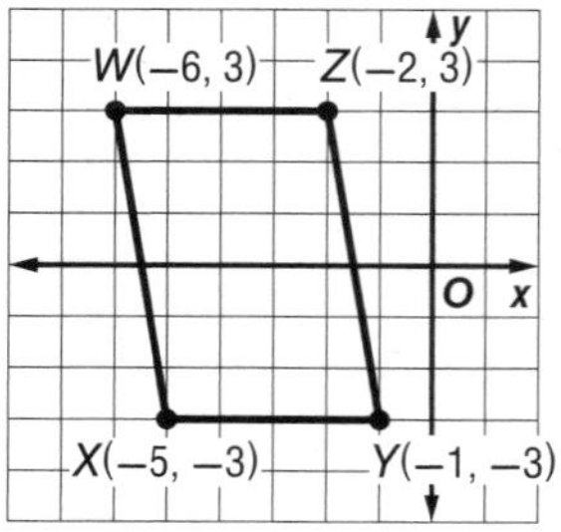

Test-TakingTip

CCSS Sense-Making

Instead of checking all four vertices of parallelogram *WXYZ* in each graph, check just one vertex, such as *X*.

F

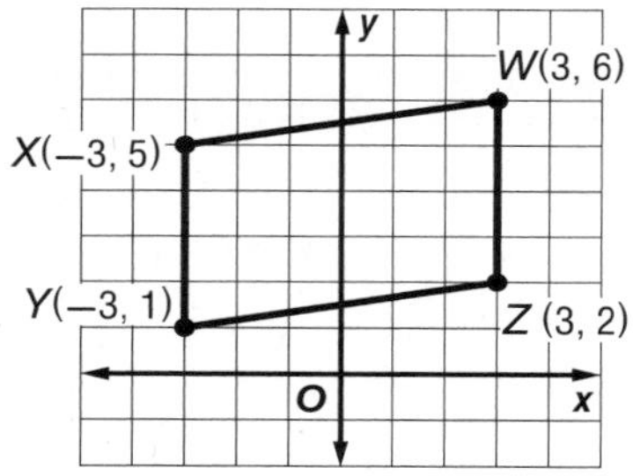

H

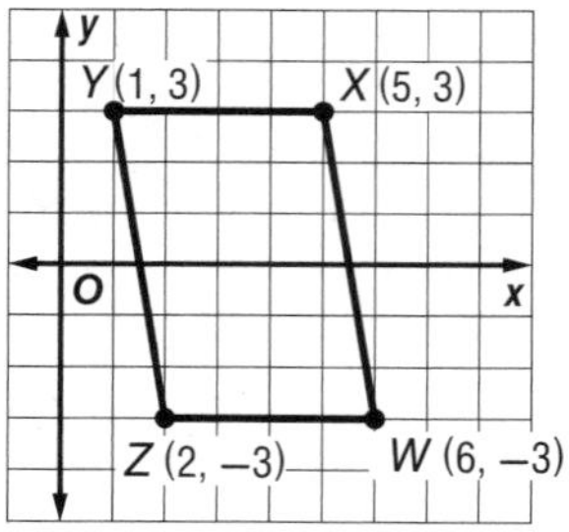

G

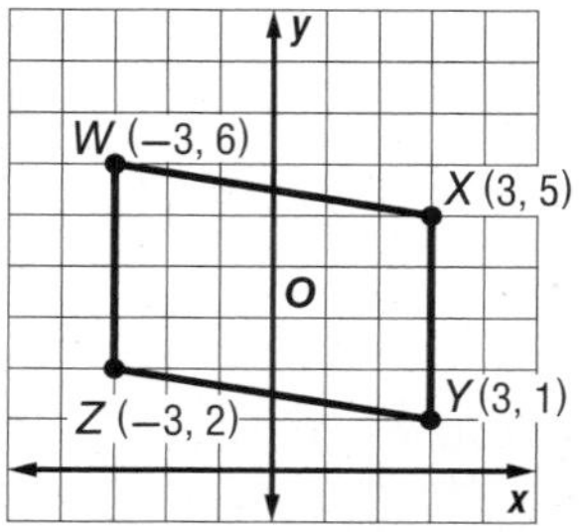

J

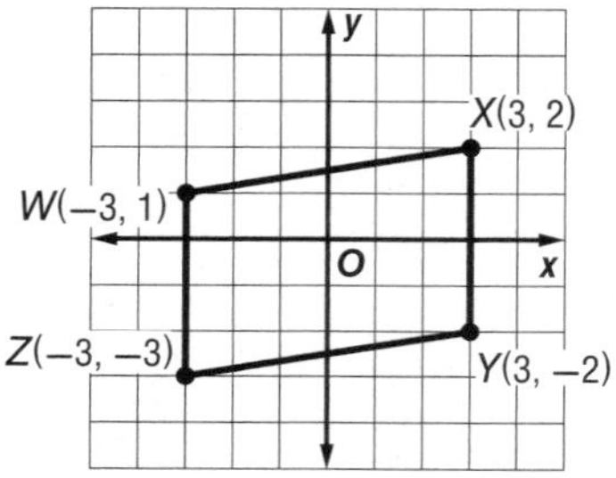

Check Your Understanding

Example 1 **Copy each polygon and point *K*. Then use a protractor and ruler to draw the specified rotation of each figure about point *K*.**

1. 45°

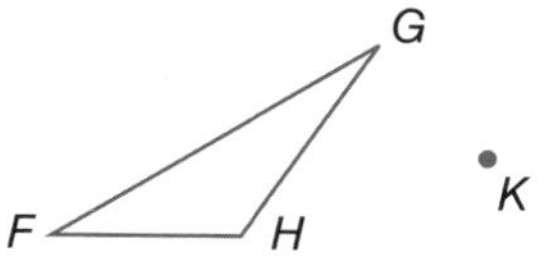

2. 120°

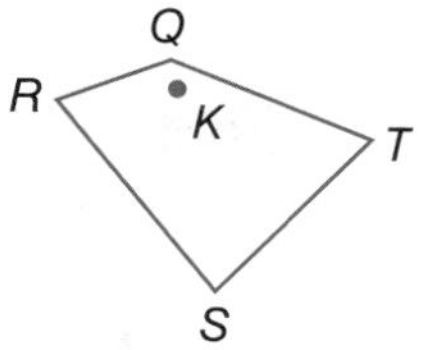

Example 2 **3** Triangle *DFG* has vertices *D*(−2, 6), *F*(2, 8), and *G*(2, 3). Graph △*DFG* and its image after a rotation 180° about the origin.

Example 3 **4. MULTIPLE CHOICE** For the transformation shown, what is the measure of the angle of rotation of *ABCD* about the origin?

A 90°

B 180°

C 270°

D 360°

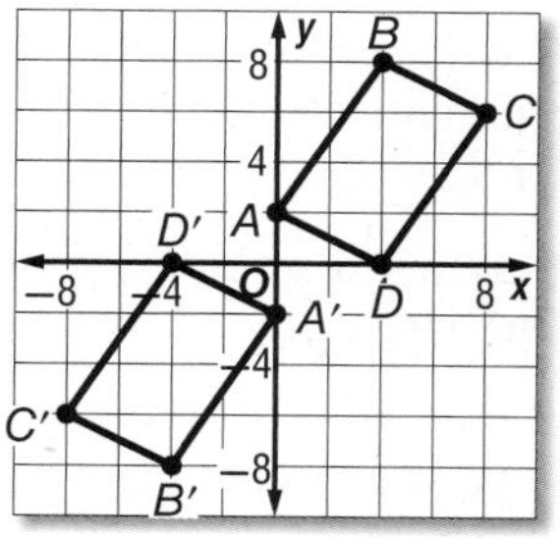

Practice and Problem Solving

Example 1 **CCSS TOOLS** **Copy each polygon and point *K*. Then use a protractor and ruler to draw the specified rotation of each figure about point *K*.**

5. 90°

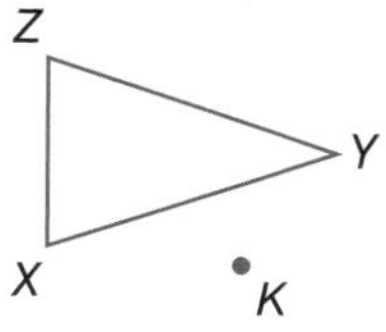

6. 15°

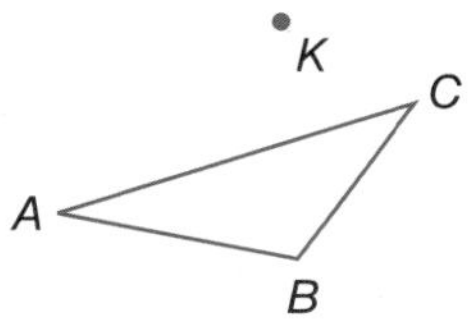

7. 145°

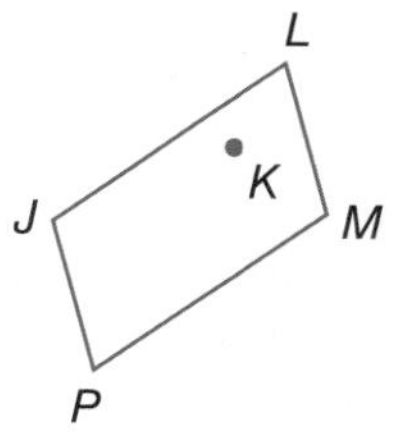

8. 30°

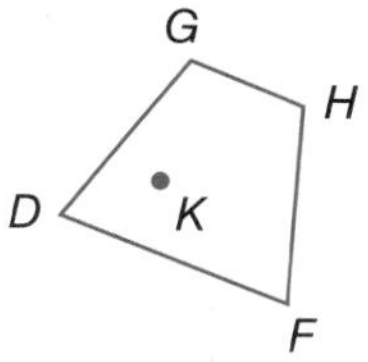

9. 260°

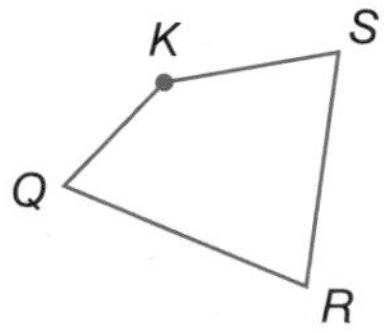

10. 50°

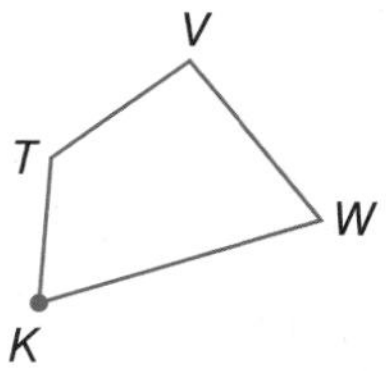

PINWHEELS **Find the angle of rotation to the nearest tenth of a degree that maps *P* onto *P′*. Explain your reasoning.**

11.

12.

13.

Examples 2–3 **Graph each figure and its image after the specified rotation about the origin.**

14. $\triangle JKL$ has vertices $J(2, 6)$, $K(5, 2)$, and $L(7, 5)$; 90°

15. rhombus $WXYZ$ has vertices $W(-3, 4)$, $X(0, 7)$, $Y(3, 4)$, and $Z(0, 1)$; 90°

16. $\triangle FGH$ has vertices $F(2, 4)$, $G(5, 6)$, and $H(7, 2)$; 180°

17. trapezoid $ABCD$ has vertices $A(-7, -2)$, $B(-6, -6)$, $C(-1, -1)$, and $D(-5, 0)$; 180°

18. $\triangle RST$ has vertices $R(-6, -1)$, $S(-4, -5)$, and $T(-2, -1)$; 270°

19. parallelogram $MPQV$ has vertices $M(-6, 3)$, $P(-2, 3)$, $Q(-3, -2)$, and $V(-7, -2)$; 270°

20. WEATHER A weathervane is used to indicate the direction of the wind. If the vane is pointing northeast and rotates 270°, what is the new wind direction?

21. CCSS MODELING The photograph of the Grande Roue, or Big Wheel, at the right appears blurred because of the camera's shutter speed—the length of time the camera's shutter was open. The diameter of the wheel is 60 meters.

a. Estimate the angle of rotation in the photo. (*Hint:* Use points A and A'.)

b. If the Ferris wheel makes one revolution per minute, use your estimate from part **a** to estimate the camera's shutter speed.

Each figure shows a preimage and its image after a rotation about point *P*. Copy each figure, locate point *P*, and find the angle of rotation.

22.

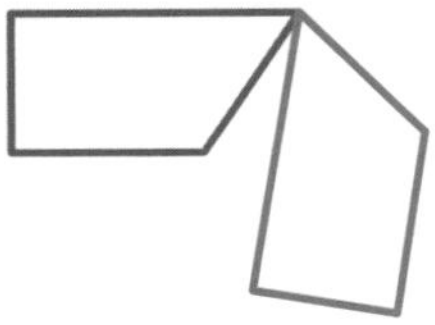

23.

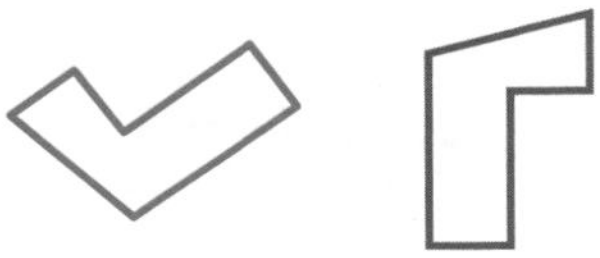

ALGEBRA **Give the equation of the line $y = -x - 2$ after a rotation about the origin through the given angle. Then describe the relationship between the equations of the image and preimage.**

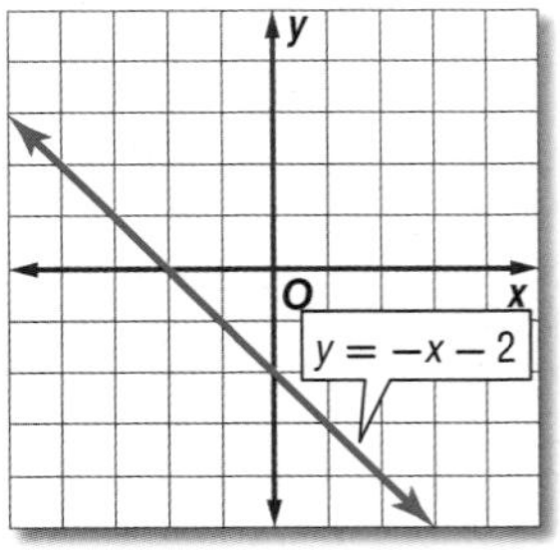

24. 90°

25. 180°

26. 270°

27. 360°

ALGEBRA **Rotate the line the specified number of degrees about the *x*- and *y*-intercepts and find the equation of the resulting image.**

28. $y = x - 5$; 90°

29. $y = 2x + 4$; 180°

30. $y = 3x - 2$; 270°

31 **RIDES** An amusement park ride consists of four circular cars. The ride rotates at a rate of 0.25 revolution per second. In addition, each car rotates 0.5 revolution per second. If Jane is positioned at point P when the ride begins, what coordinates describe her position after 31 seconds?

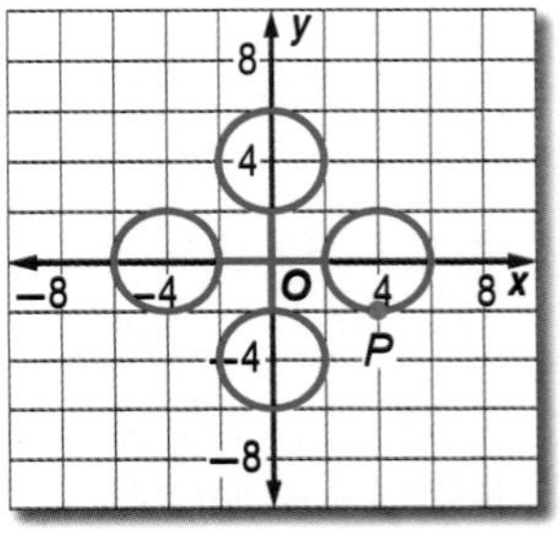

32. **BICYCLE RACING** Brandon and Nestor are participating in a bicycle race on a circular track with a radius of 200 feet.

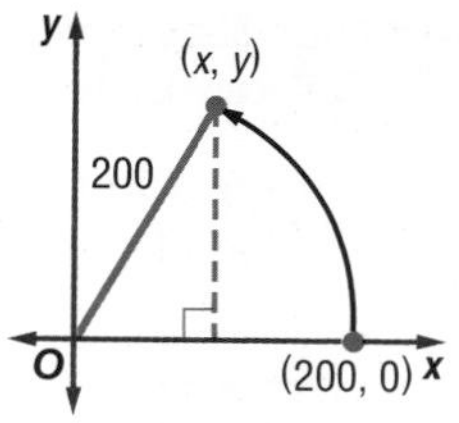

a. If the race starts at (200, 0) and both complete one rotation in 30 seconds, what are their coordinates after 5 seconds?

b. Suppose the length of race is 50 laps and Brandon continues the race at the same rate. If Nestor finishes in 26.2 minutes, who is the winner?

33. **MULTIPLE REPRESENTATIONS** In this problem, you will investigate reflections over a pair of intersecting lines.

a. **Geometric** On a coordinate plane, draw a triangle and a pair of intersecting lines. Label the triangle ABC and the lines ℓ and m. Reflect $\triangle ABC$ in the line ℓ. Then reflect $\triangle A'B'C'$ in the line m. Label the final image $A''B''C''$.

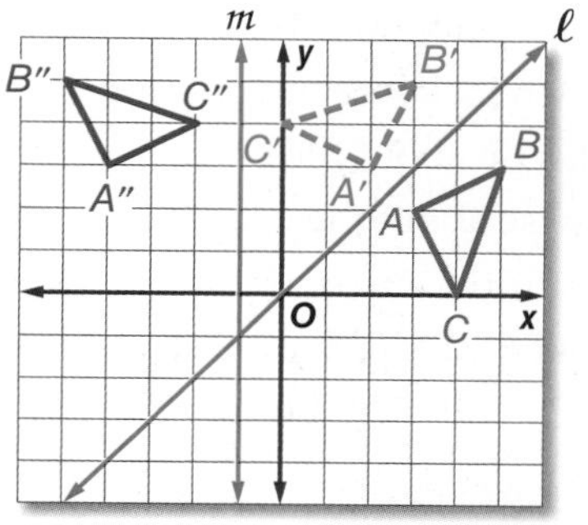

b. **Geometric** Repeat the process in part a two more times in two different quadrants. Label the second triangle DEF and reflect it in intersecting lines n and p. Label the third triangle MNP and reflect it in intersecting lines q and r.

c. **Tabular** Measure the angle of rotation of each triangle about the point of intersection of the two lines. Copy and complete the table below.

Angle of Rotation Between Figures		Angle Between Intersecting Lines	
$\triangle ABC$ and $\triangle A''B''C''$		ℓ and m	
$\triangle DEF$ and $\triangle D''E''F''$		n and p	
$\triangle MNP$ and $\triangle M''N''P''$		q and r	

d. **Verbal** Make a conjecture about the angle of rotation of a figure about the intersection of two lines after the figure is reflected in both lines.

H.O.T. Problems Use Higher-Order Thinking Skills

34. **WRITING IN MATH** Are collinearity and betweenness of points maintained under rotation? Explain.

35. **CHALLENGE** Point C has coordinates $C(5, 5)$. The image of this point after a rotation of 100° about a certain point is $C'(-5, 7.5)$. Use construction to estimate the coordinates of the center of this rotation. Explain.

36. **OPEN ENDED** Draw a figure on the coordinate plane. Describe a nonzero rotation that maps the image onto the preimage with no change in orientation.

37. **CCSS ARGUMENTS** Is the reflection of a figure in the x-axis equivalent to the rotation of that same figure 180° about the origin? Explain.

38. **WRITING IN MATH** Do invariant points *sometimes*, *always*, or *never* occur in a rotation? Explain your reasoning.

Standardized Test Practice

39. What rotation of trapezoid $QRST$ creates an image with point R' at (4, 3)?

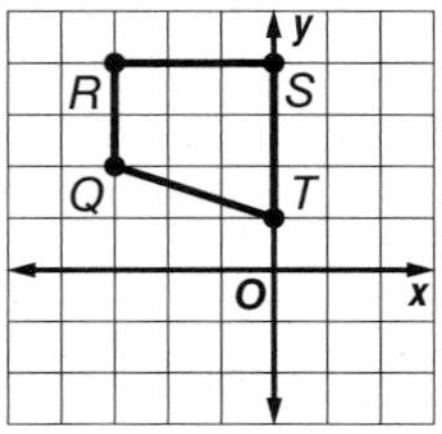

A 270° counterclockwise about point T
B 185° counterclockwise about point T
C 180° clockwise about the origin
D 90° clockwise about the origin

40. SHORT RESPONSE $\triangle XYZ$ has vertices $X(1, 7)$, $Y(0, 2)$, and $Z(-5, -2)$. What are the coordinates of X' after a rotation 270° counterclockwise about the origin?

41. ALGEBRA The population of the United States in July of 2007 was estimated to have surpassed 301,000,000. At the same time the world population was estimated to be over 6,602,000,000. What percent of the world population, to the nearest tenth, lived in the United States at this time?

F 3.1%
G 3.5%
H 4.2%
J 4.6%

42. SAT/ACT An 18-foot ladder is placed against the side of a house. The base of the ladder is positioned 8 feet from the house. How high up on the side of the house, to the nearest tenth of a foot, does the ladder reach?

A 10.0 ft
B 16.1 ft
C 19.7 ft
D 22.5 ft
E 26.0 ft

Spiral Review

43. VOLCANOES A cloud of dense gas and dust from a volcano blows 40 miles west and then 30 miles north. Make a sketch to show the translation of the dust particles. Then find the distance of the shortest path that would take the particles to the same position. (Lesson 14-5)

Copy the figure and the given line of reflection. Then draw the reflected image in this line using a ruler. (Lesson 14-4)

44.

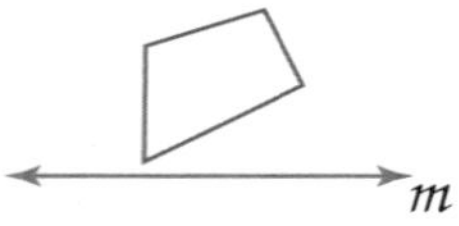

45.

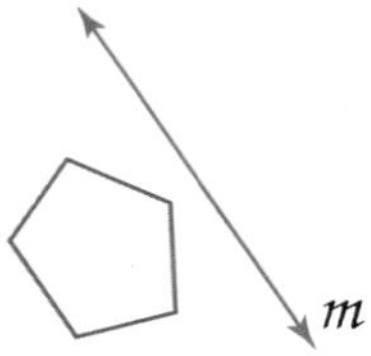

46.

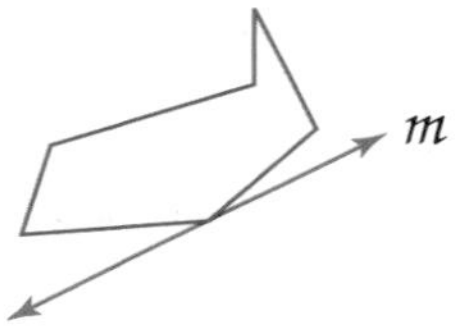

Skills Review

Identify the type of congruence transformation shown as a *reflection*, *translation*, or *rotation*.

47.

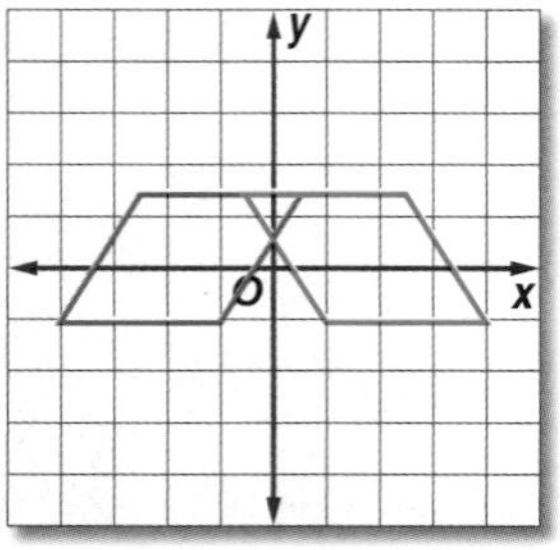

48.

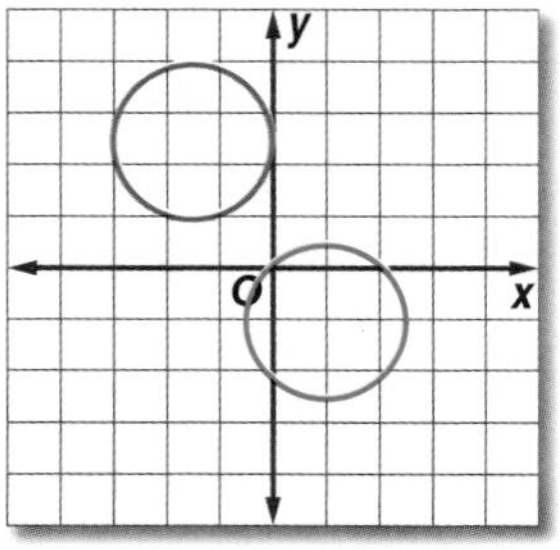

49.

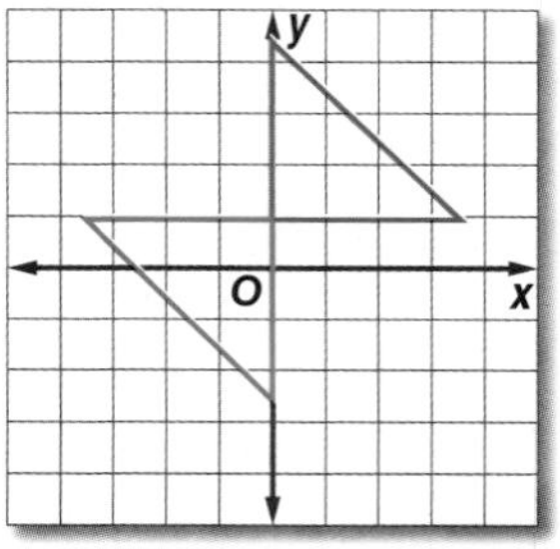

Stella Snead/Bruce Coleman, Inc./Photoshot

EXTEND

14-6 Geometry Lab
Solids of Revolution

A **solid of revolution** is a three-dimensional figure obtained by rotating a plane figure or curve about a line.

CCSS Common Core State Standards
Content Standards
G.GMD.4 Identify the shapes of two-dimensional cross-sections of three-dimensional objects, and identify three-dimensional objects generated by rotations of two-dimensional objects.
Mathematical Practices 5

Activity 1

Identify and sketch the solid formed by rotating the right triangle shown about line ℓ.

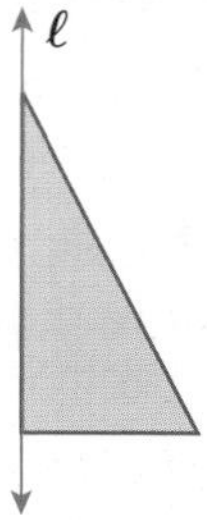

Step 1 Copy the triangle onto card stock or heavy construction paper and cut it out.

Step 2 Use tape to attach the triangle to a dowel rod or straw.

Step 3 Rotate the end of the straw quickly between your hands and observe the result.

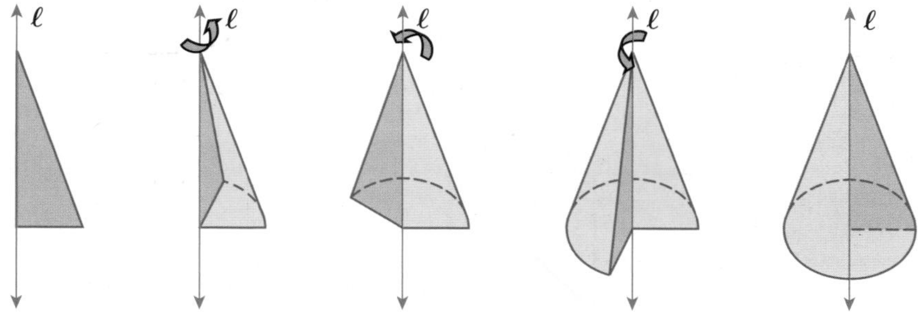

The blurred image you observe is that of a cone.

Model and Analyze

Identify and sketch the solid formed by rotating the two-dimensional shape about line ℓ.

1.

2.

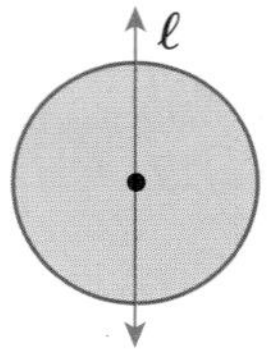

3.

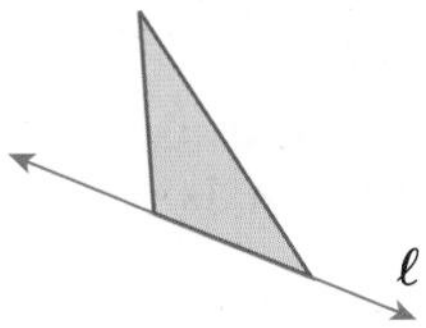

4. Sketch and identify the solid formed by rotating the rectangle shown about the line containing

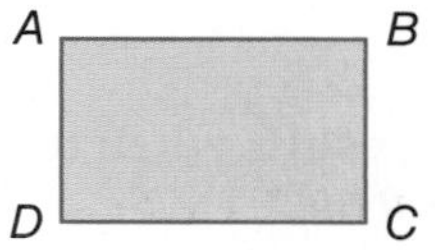

a. side $\overline{AB}$.

b. side $\overline{AD}$.

c. the midpoints of sides $\overline{AB}$ and $\overline{AD}$.

5. DESIGN Draw a two-dimensional figure that could be rotated to form the vase shown, including the line in which it should be rotated.

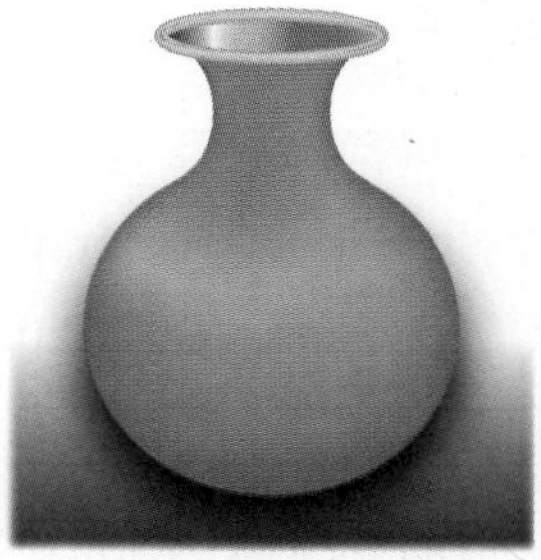

6. REASONING *True* or *false:* All solids can be formed by rotating a two-dimensional figure. Explain your reasoning.

Solids of Revolution *Continued*

In calculus, you will be asked to find the volumes of solids generated by revolving a region on the coordinate plane about the *x*- or *y*-axis. An important first step in solving these problems is visualizing the solids formed.

Activity 2

Sketch the solid that results when the region enclosed by $y = x$, $x = 4$, and $y = 0$ is revolved about the y-axis.

Step 1 Graph each equation to find the region to be rotated.

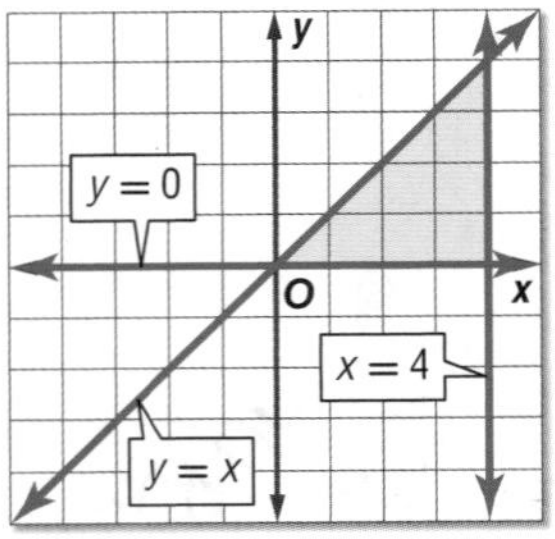

Step 2 Reflect the region about the y-axis.

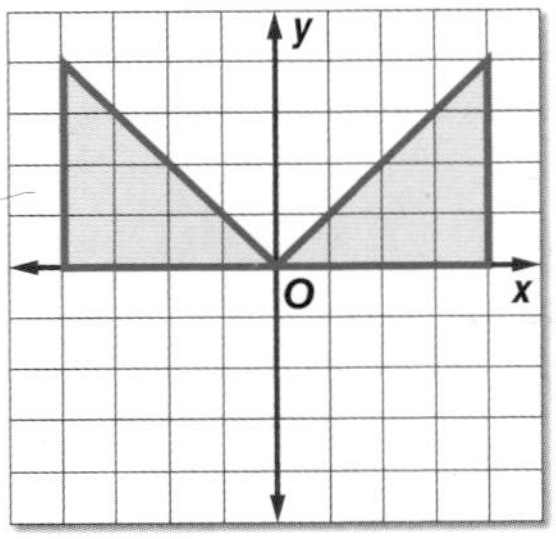

Step 3 Connect the vertices of the right triangles using curved lines.

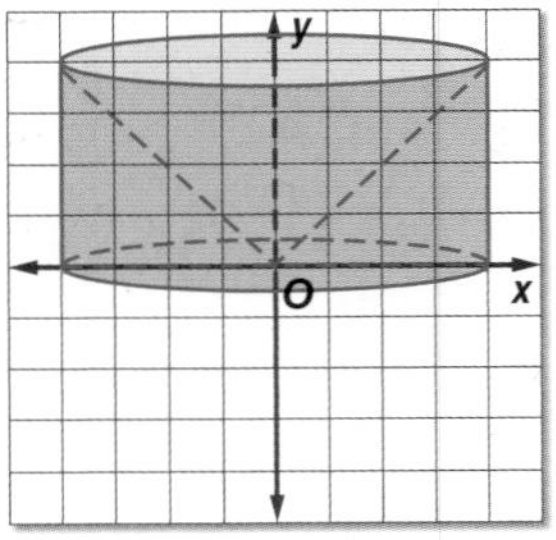

The solid is a cylinder with a cone cut out of its center.

Model and Analyze

Sketch the solid that results when the region enclosed by the given equations is revolved about the y-axis.

7. $y = -x + 4$
$x = 0$
$y = 0$

8. $y = x^2$
$y = 4$

9. $y = x^2$
$y = 2x$

Sketch the solid that results when the region enclosed by the given equations is revolved about the x-axis.

10. $y = -x + 4$
$x = 0$
$y = 0$

11. $y = x^2$
$y = 0$
$x = 2$

12. $y = x^2$
$y = 2x$

13. OPEN ENDED Graph a region in the first quadrant of the coordinate plane.

a. Sketch the graph of the region when revolved about the y-axis.

b. Sketch the graph of the region when revolved about the x-axis.

14. CHALLENGE Find equations that enclose a region such that when rotated about the x-axis, a solid is produced with a volume of 18π cubic units.

EXPLORE **14-7**

Geometry Software Lab
Compositions of Transformations

In this lab, you will use Geometer's Sketchpad to explore the effects of performing multiple transformations on a figure.

CCSS Common Core State Standards
Content Standards
G.CO.2 Represent transformations in the plane using, e.g., transparencies and geometry software; describe transformations as functions that take points in the plane as inputs and give other points as outputs. Compare transformations that preserve distance and angle to those that do not (e.g., translation versus horizontal stretch).
G.CO.5 Given a geometric figure and a rotation, reflection, or translation, draw the transformed figure using, e.g., graph paper, tracing paper, or geometry software. Specify a sequence of transformations that will carry a given figure onto another.
Mathematical Practices 5

Activity

Reflect a figure in two vertical lines.

Step 1 Use the line segment tool to construct a triangle with one vertex pointing to the left so that you can easily see changes as you perform transformations. Label the triangle *ABC*.

Step 2 Insert and label a line *m* to the right of $\triangle ABC$. Insert a point so that the distance from the point to line *m* is greater than the width of $\triangle ABC$. Draw the line parallel to line *m* through the point and label the new line *r*.

Step 3 Select line *m* and choose **Mark Mirror** from the **Transform** menu. Select all sides and vertices of $\triangle ABC$ and choose **Reflect** from the **Transform** menu.

Step 4 Repeat the process you used in Step 3 to reflect the new image in line *r*.

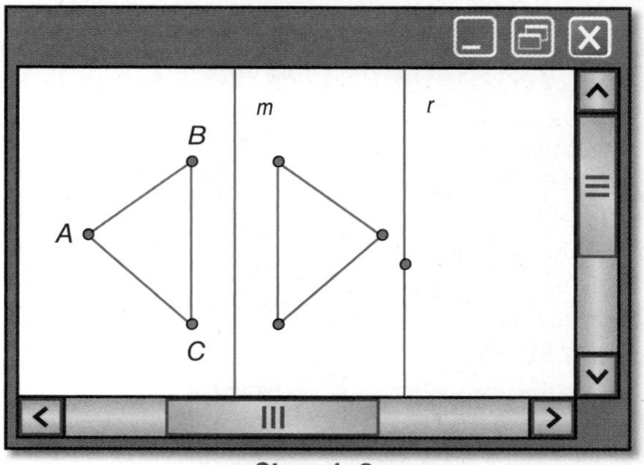

Steps 1–3

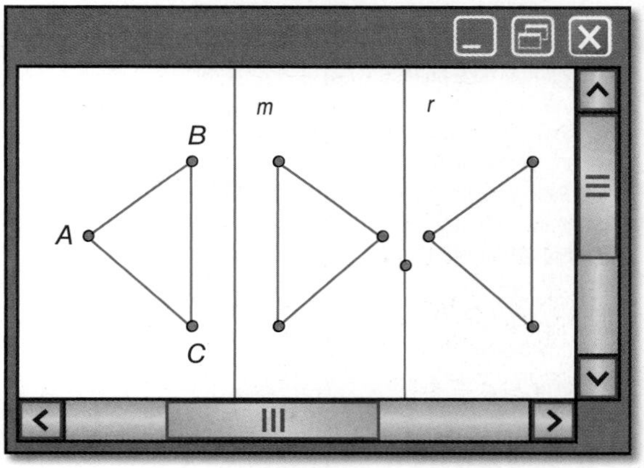

Step 4

Analyze the Results

1. How are the original figure and the final figure related?
2. What single transformation could be used to produce the final figure?
3. If you move line *m*, what happens? if you move line *r*?
4. **MAKE A CONJECTURE** If you reflected the figure in a third line, what single transformation do you think could be used to produce the final figure? Explain your reasoning.
5. Repeat the activity for a pair of perpendicular lines. What single transformation could be used to produce the same final figure?
6. **MAKE A CONJECTURE** If you reflected the figure from Exercise 5 in a third line perpendicular to the second line, what single transformation do you think could be used to produce the final figure? Explain your reasoning.

LESSON 14-7 Compositions of Transformations

Then

- You drew reflections, translations, and rotations.

Now

1. Draw glide reflections and other compositions of isometries in the coordinate plane.
2. Draw compositions of reflections in parallel and intersecting lines.

Why?

- The pattern of footprints left in the sand after a person walks along the edge of a beach illustrates the composition of two different transformations—translations and reflections.

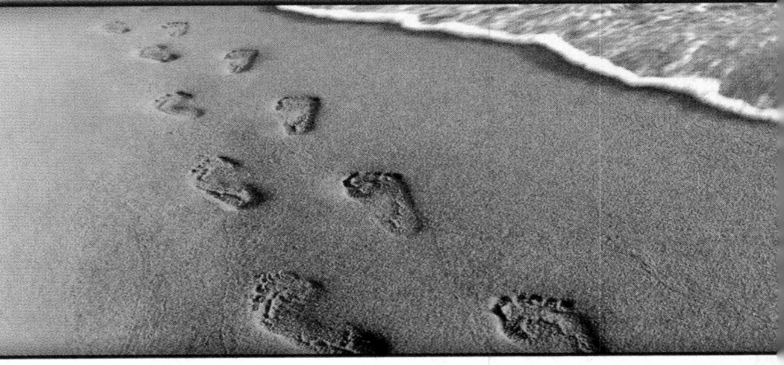

NewVocabulary
composition of transformations
glide reflection

Common Core State Standards

Content Standards

G.CO.2 Represent transformations in the plane using, e.g., transparencies and geometry software; describe transformations as functions that take points in the plane as inputs and give other points as outputs. Compare transformations that preserve distance and angle to those that do not (e.g., translation versus horizontal stretch).

G.CO.5 Given a geometric figure and a rotation, reflection, or translation, draw the transformed figure using, e.g., graph paper, tracing paper, or geometry software. Specify a sequence of transformations that will carry a given figure onto another.

Mathematical Practices

1 Make sense of problems and persevere in solving them.
4 Model with mathematics.

1 Glide Reflections

When a transformation is applied to a figure and then another transformation is applied to its image, the result is called a **composition of transformations**. A glide reflection is one type of composition of transformations.

KeyConcept Glide Reflection

A **glide reflection** is the composition of a translation followed by a reflection in a line parallel to the translation vector.

Example

The glide reflection shown is the composition of a translation along $\vec{w}$ followed by a reflection in line ℓ.

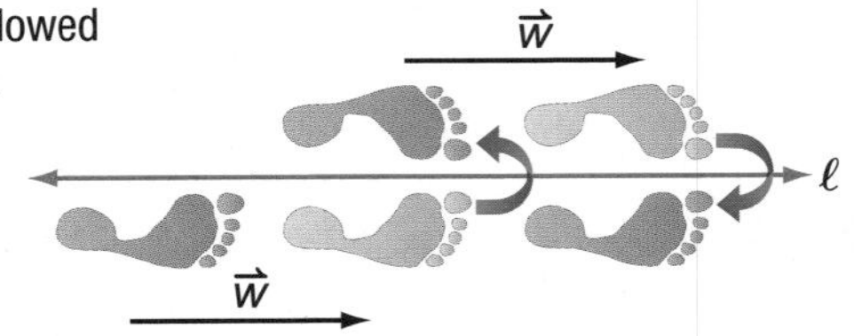

PT

Example 1 Graph a Glide Reflection

Triangle JKL has vertices $J(6, -1)$, $K(10, -2)$, and $L(5, -3)$. Graph $\triangle JKL$ and its image after a translation along $\langle 0, 4\rangle$ and a reflection in the y-axis.

Step 1 translation along $\langle 0, 4\rangle$

(x, y)	$\rightarrow$	$(x, y + 4)$
$J(6, -1)$	$\rightarrow$	$J'(6, 3)$
$K(10, -2)$	$\rightarrow$	$K'(10, 2)$
$L(5, -3)$	$\rightarrow$	$L'(5, 1)$

Step 2 reflection in the y-axis

(x, y)	$\rightarrow$	$(-x, y)$
$J'(6, 3)$	$\rightarrow$	$J''(-6, 3)$
$K'(10, 2)$	$\rightarrow$	$K''(-10, 2)$
$L'(5, 1)$	$\rightarrow$	$L''(-5, 1)$

Step 3 Graph $\triangle JKL$ and its image $\triangle J''K''L''$.

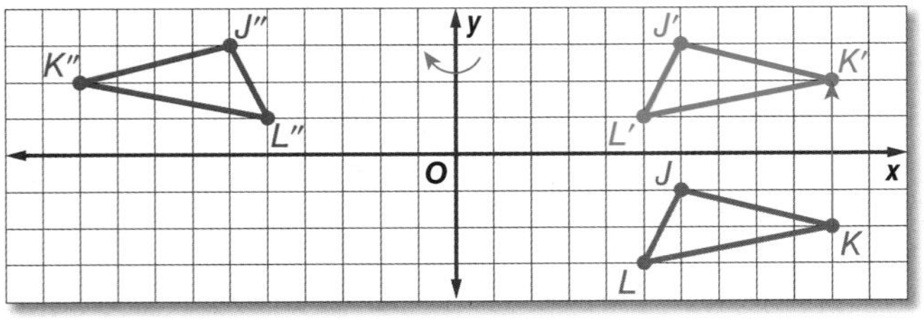

GuidedPractice

Triangle PQR has vertices $P(1, 1)$, $Q(2, 5)$, and $R(4, 2)$. Graph $\triangle PQR$ and its image after the indicated glide reflection.

1A. Translation: along $\langle -2, 0\rangle$
Reflection: in x-axis

1B. Translation: along $\langle -3, -3\rangle$
Reflection: in $y = x$

In Example 1, $\triangle JKL \cong \triangle J'K'L'$ and $\triangle J'K'L' \cong \triangle J''K''L''$. By the Transitive Property of Congruence, $\triangle JKL \cong \triangle J''K''L''$. This suggests the following theorem.

Theorem 14.7 Composition of Isometries

The composition of two (or more) isometries is an isometry.

You will prove one case of Theorem 14.7 in Exercise 30.

StudyTip

Rigid Motions Glide reflections, reflections, translations, and rotations are the only four *rigid motions* or isometries in a plane.

So, the composition of two or more isometries—reflections, translations, or rotations—results in an image that is congruent to its preimage.

Example 2 Graph Other Compositions of Isometries

PT

The endpoints of $\overline{CD}$ are $C(-7, 1)$ and $D(-3, 2)$. Graph $\overline{CD}$ and its image after a reflection in the x-axis and a rotation 90° about the origin.

Step 1 reflection in the x-axis

$(x, y) \rightarrow (x, -y)$

$C(-7, 1) \rightarrow C'(-7, -1)$

$D(-3, 2) \rightarrow D'(-3, -2)$

Step 2 rotation 90° about origin

$(x, y) \rightarrow (-y, x)$

$C'(-7, -1) \rightarrow C''(1, -7)$

$D'(-3, -2) \rightarrow D''(2, -3)$

Step 3 Graph $\overline{CD}$ and its image $\overline{C''D''}$.

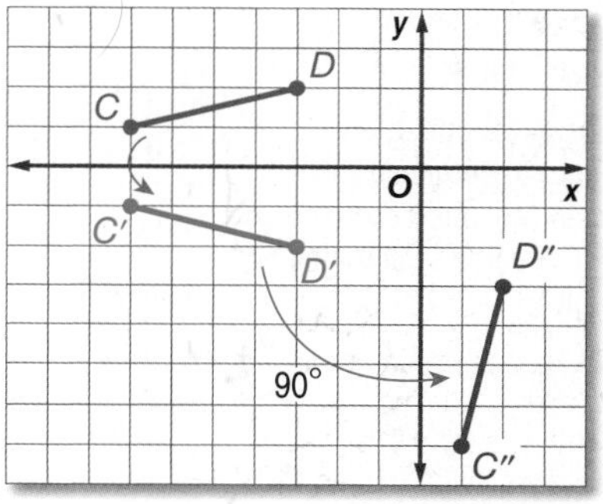

ReadingMath

Double Primes Double primes are used to indicate that a vertex is the image of a second transformation.

GuidedPractice

Triangle ABC has vertices $A(-6, -2)$, $B(-5, -5)$, and $C(-2, -1)$. Graph $\triangle ABC$ and its image after the composition of transformations in the order listed.

2A. Translation: along $\langle 3, -1 \rangle$
Reflection: in y-axis

2B. Rotation: 180° about origin
Translation: along $\langle -2, 4 \rangle$

2 Compositions of Two Reflections

The composition of two reflections in parallel lines is the same as a translation.

Theorem 14.8 Reflections in Parallel Lines

The composition of two reflections in parallel lines can be described by a translation vector that is

- perpendicular to the two lines, and
- twice the distance between the two lines.

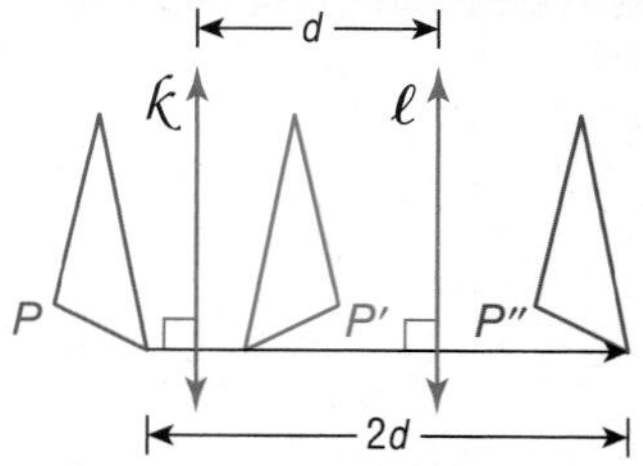

You will prove Theorem 14.8 in Exercise 36.

The composition of two reflections in intersecting lines is the same as a rotation.

Theorem 14.9 Reflections in Intersecting Lines

The composition of two reflections in intersecting lines can be described by a rotation

- about the point where the lines intersect and
- through an angle that is twice the measure of the acute or right angle formed by the lines.

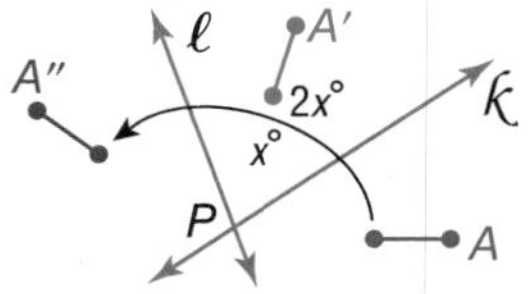

You will prove Theorem 14.9 in Exercise 37.

Example 3 Reflect a Figure in Two Lines

Copy and reflect figure A in line m and then line p. Then describe a single transformation that maps A onto A''.

a.

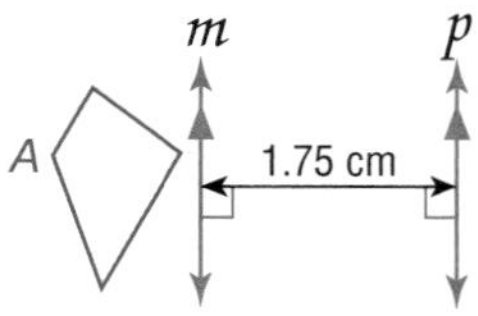

b.

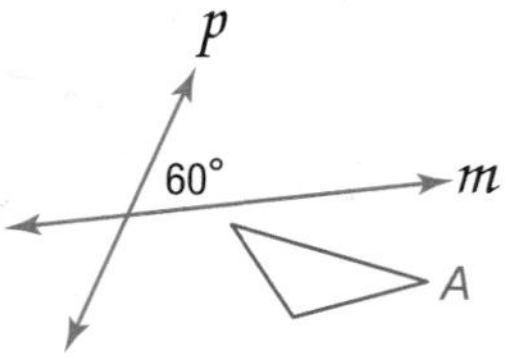

> **WatchOut!**
> Order of Composition
> Be sure to compose two transformations according to the order in which they are given.

a. Step 1 Reflect A in line m.

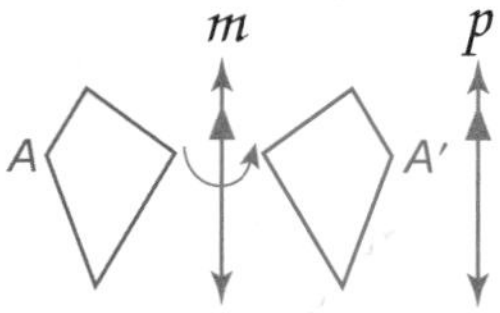

Step 2 Reflect A' in line p.

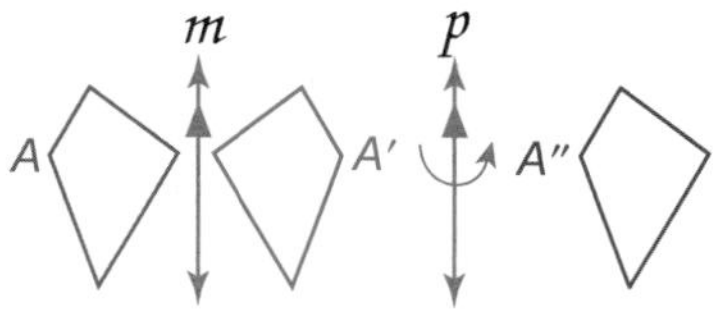

By Theorem 14.2, the composition of two reflections in parallel vertical lines m and p is equivalent to a horizontal translation right $2 \cdot 1.75$ or 3.5 centimeters.

b. Step 1

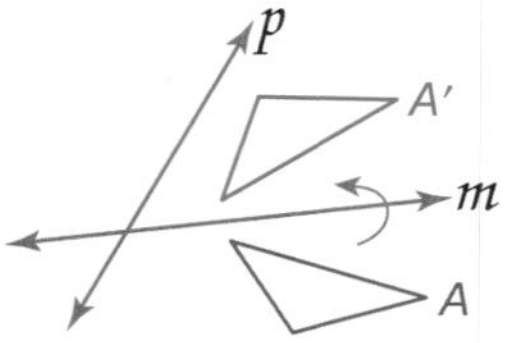

Step 2

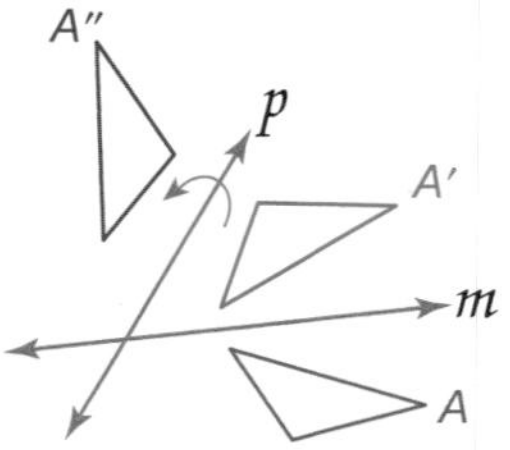

By Theorem 14.3, the composition of two reflections in intersecting lines m and p is equivalent to a $2 \cdot 60°$ or 120° counterclockwise rotation about the point where lines m and p intersect.

GuidedPractice

Copy and reflect figure B in line n and then line q. Then describe a single transformation that maps B onto B''.

3A.

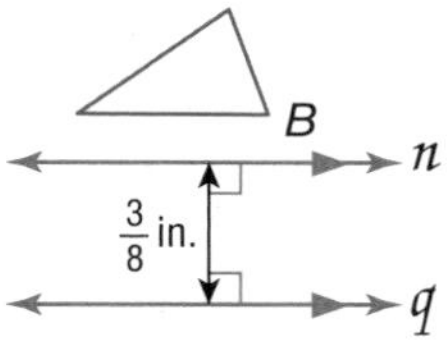

3B.

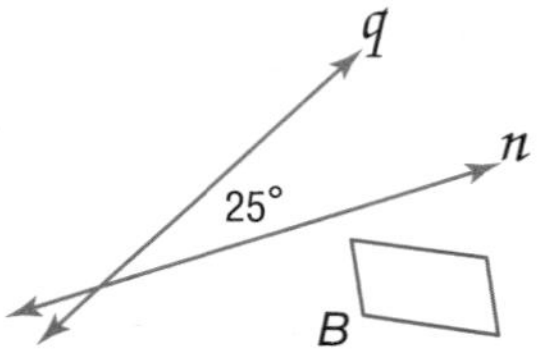

Many patterns in the real world are created using compositions of transformations.

Real-WorldLink

In carpets, border patterns result when any of several basic transformations are repeated in one direction. There are seven possible combinations: translations, horizontal reflections, vertical reflections, vertical followed by horizontal reflections, glide reflections, rotations, and reflections followed by glide reflections.

Source: The Textile Museum

Real-World Example 4 Describe Transformations

STATIONERY BORDERS **Describe the transformations that are combined to create each stationery border shown.**

a.

The pattern is created by successive translations of the first four potted plants. So this pattern can be created by combining two reflections in lines m and p as shown. Notice that line m goes through the center of the preimage.

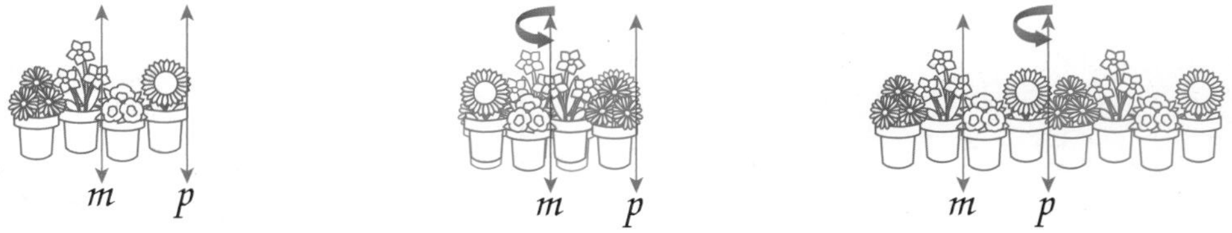

b.

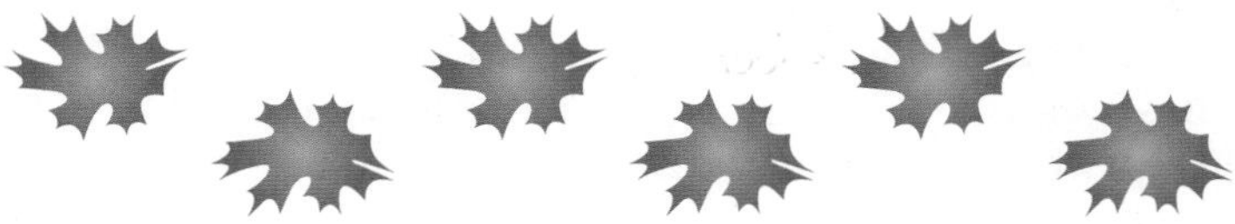

The pattern is created by glide reflection. So this pattern can be created by combining a translation along translation vector $\vec{v}$ followed by a reflection over horizontal line n as shown.

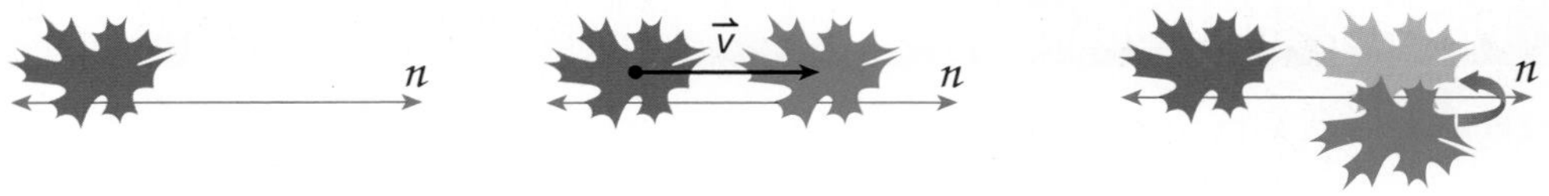

GuidedPractice

4. CARPET PATTERNS Describe the transformations that are combined to create each carpet pattern shown.

A.

B.

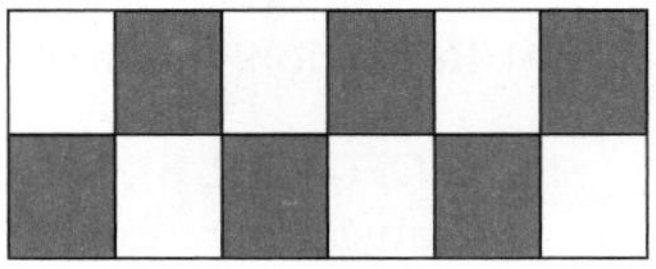

ConceptSummary Compositions of Translations

Glide Reflection	Translation	Rotation
the composition of a reflection and a translation	the composition of two reflections in parallel lines	the composition of two reflections in intersecting lines

Check Your Understanding

Example 1 **Triangle *CDE* has vertices $C(-5, -1)$, $D(-2, -5)$, and $E(-1, -1)$. Graph $\triangle CDE$ and its image after the indicated glide reflection.**

1. Translation: along $\langle 4, 0\rangle$
Reflection: in x-axis

2. Translation: along $\langle 0, 6\rangle$
Reflection: in y-axis

Example 2

3. The endpoints of $\overline{JK}$ are $J(2, 5)$ and $K(6, 5)$. Graph $\overline{JK}$ and its image after a reflection in the x-axis and a rotation 90° about the origin.

Example 3 **Copy and reflect figure *S* in line *m* and then line *p*. Then describe a single transformation that maps *S* onto *S*″.**

4.

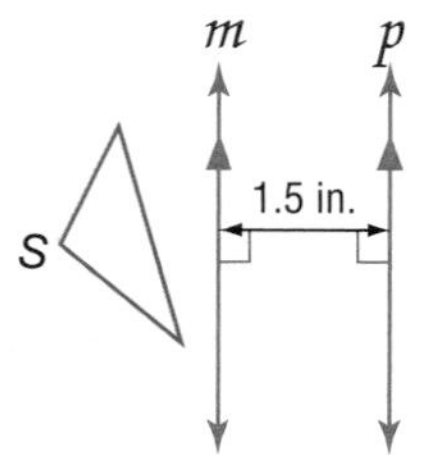

5.

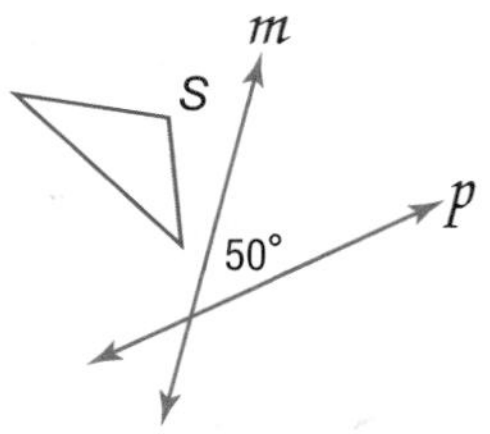

Example 4

6. TILE PATTERNS Viviana is creating a pattern for the top of a table with tiles in the shape of isosceles triangles. Describe the transformation combination that was used to transform the white triangle to the blue triangle.

Practice and Problem Solving

Example 1 **Graph each figure with the given vertices and its image after the indicated glide reflection.**

7. $\triangle RST$: $R(1, -4)$, $S(6, -4)$, $T(5, -1)$
Translation: along $\langle 2, 0\rangle$
Reflection: in x-axis

8. $\triangle JKL$: $J(1, 3)$, $K(5, 0)$, $L(7, 4)$
Translation: along $\langle -3, 0\rangle$
Reflection: in x-axis

9. $\triangle XYZ$: $X(-7, 2)$, $Y(-5, 6)$, $Z(-2, 4)$
Translation: along $\langle 0, -1\rangle$
Reflection: in y-axis

10. $\triangle ABC$: $A(2, 3)$, $B(4, 7)$, $C(7, 2)$
Translation: along $\langle 0, 4\rangle$
Reflection: in y-axis

11. $\triangle DFG$: $D(2, 8)$, $F(1, 2)$, $G(4, 6)$
Translation: along $\langle 3, 3\rangle$
Reflection: in $y = x$

12. $\triangle MPQ$: $M(-4, 3)$, $P(-5, 8)$, $Q(-1, 6)$
Translation: along $\langle -4, -4\rangle$
Reflection: in $y = x$

Example 2 CCSS **SENSE-MAKING** **Graph each figure with the given vertices and its image after the indicated composition of transformations.**

13. $\overline{WX}$: $W(-4, 6)$ and $X(-4, 1)$
Reflection: in x-axis
Rotation: 90° about origin

14. $\overline{AB}$: $A(-3, 2)$ and $B(3, 8)$
Rotation: 90° about origin
Translation: along $\langle 4, 4\rangle$

15. $\overline{FG}$: $F(1, 1)$ and $G(6, 7)$
Reflection: in x-axis
Rotation: 180° about origin

16. $\overline{RS}$: $R(2, -1)$ and $S(6, -5)$
Translation: along $\langle -2, -2\rangle$
Reflection: in y-axis

Example 3 **Copy and reflect figure D in line m and then line p. Then describe a single transformation that maps D onto D''.**

17.

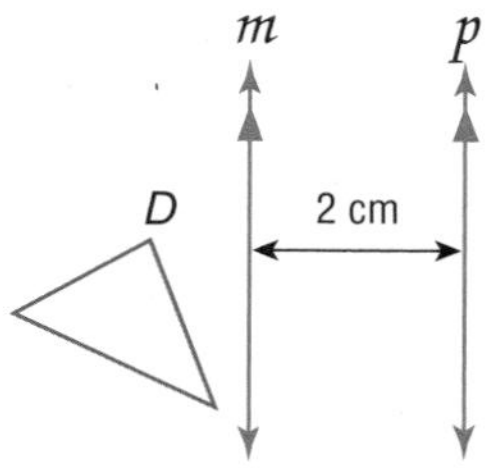

18.

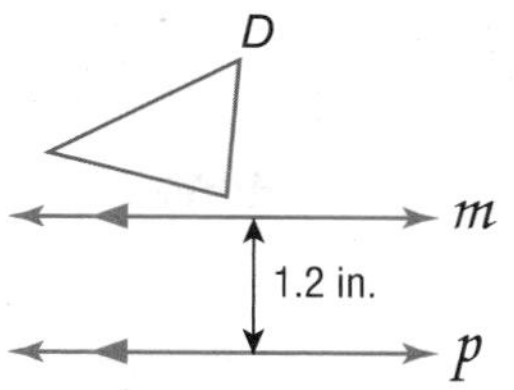

19.

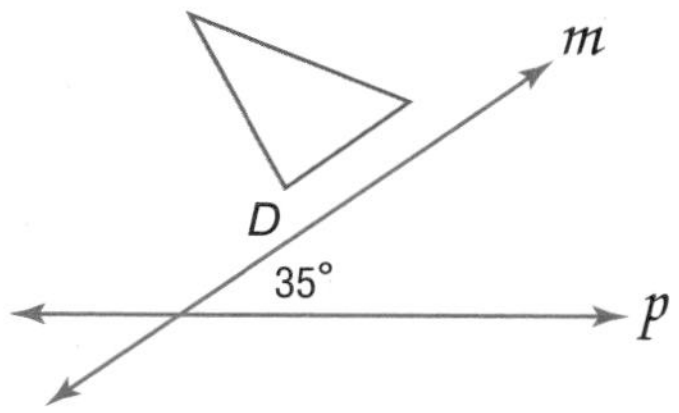

20.

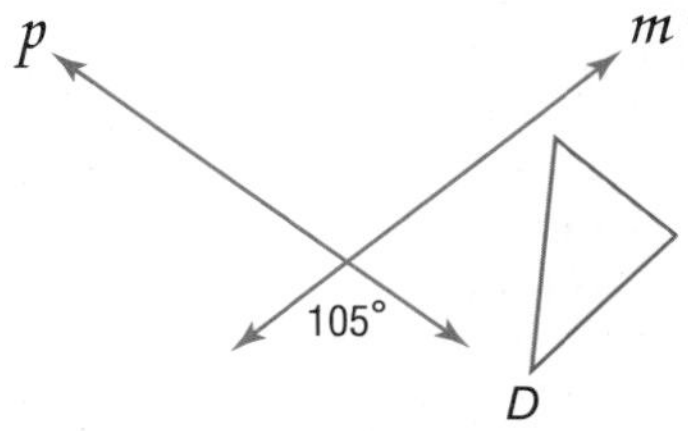

Example 4 **CCSS MODELING** **Describe the transformations combined to create the outlined kimono fabric pattern.**

21.

22.

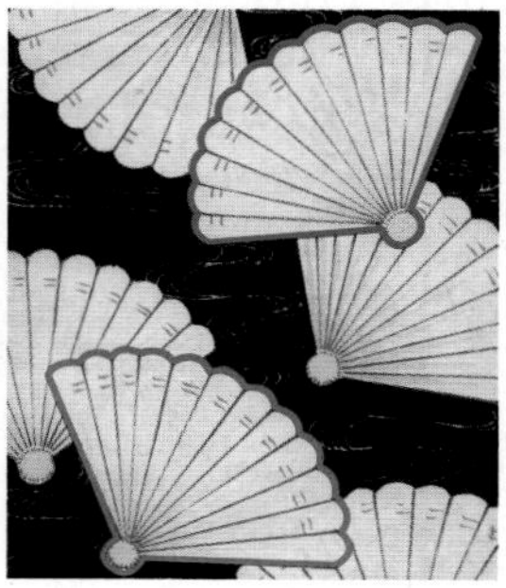

23.

24. SKATEBOARDS Elizabeth has airbrushed the pattern shown onto her skateboard. What combination of transformations did she use to create the pattern?

ALGEBRA **Graph each figure and its image after the indicated transformations.**

25. Rotation: 90° about the origin
Reflection: in x-axis

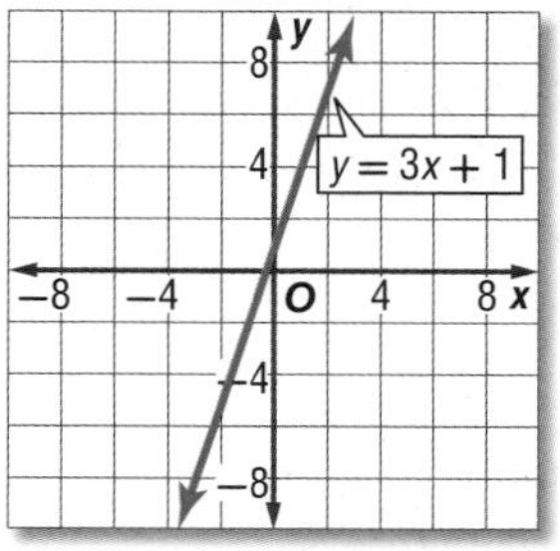

26. Reflection: in x-axis
Reflection: in y-axis

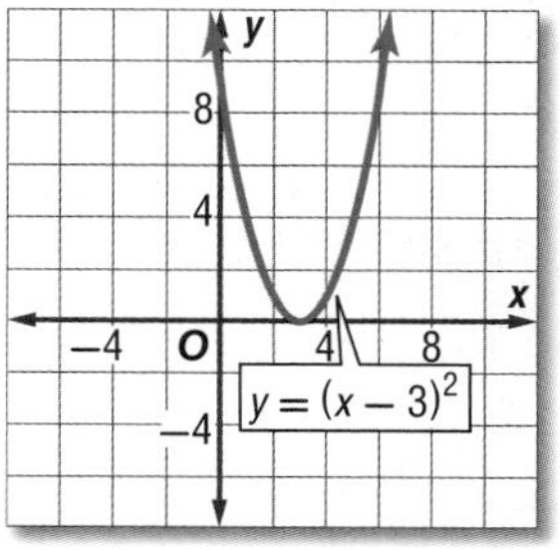

27. Find the coordinates of $\triangle A''B''C''$ after a reflection in the x-axis and a rotation of 180° about the origin if $\triangle ABC$ has vertices $A(-3, 1)$, $B(-2, 3)$, and $C(-1, 0)$.

28. **FIGURE SKATING** Kayla is practicing her figure skating routine. What combination of transformations is needed for Kayla to start at A, skate to A', and end up at A''?

29. **DANCING** Describe the transformations combined to go from Step 1 to Step 3.

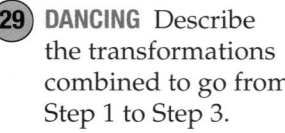

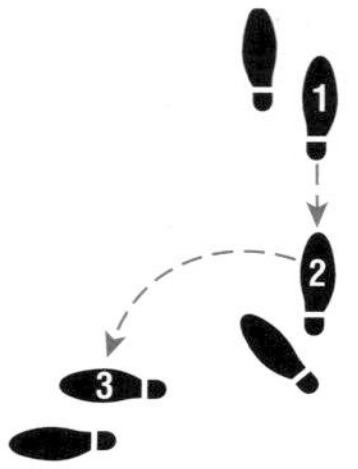

30. **PROOF** Write a paragraph proof for one case of the Composition of Isometries Theorem.

Given: A translation along $\langle a, b\rangle$ maps X to X' and Y to Y'. A reflection in z maps X' to X'' and Y' to Y''.

Prove: $\overline{XY} \cong \overline{X''Y''}$

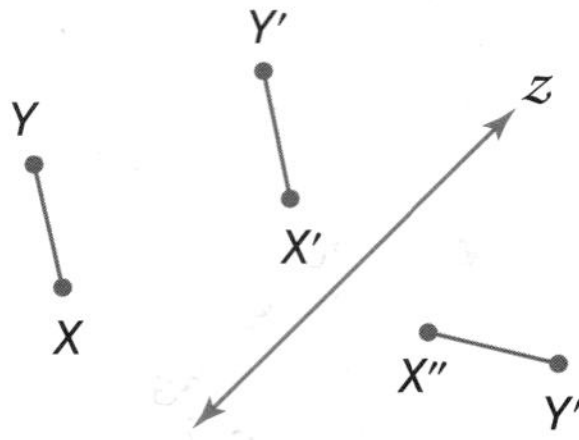

CCSS MODELING The length of an animal's stride is the distance between two consecutive tracks. The average stride length of a turkey is about 11 inches, and the average stride length of a duck is about 5 inches. Write a glide reflection that can be used to predict the location of the next track for each set of animal tracks.

31. turkey

32. duck

33. **KNITTING** Tonisha is knitting a scarf using the tumbling blocks pattern shown at the right. Describe the transformations combined to transform the red figure to the blue figure.

Describe the transformations that combined to map each figure.

34.

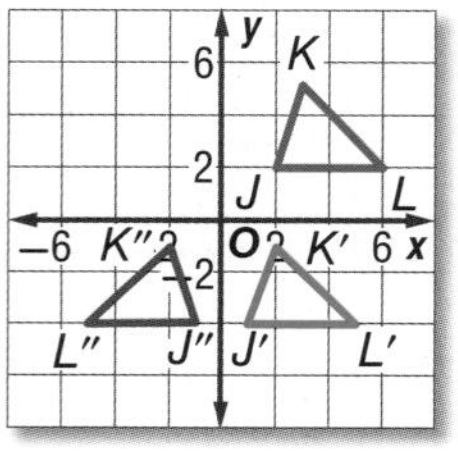

35.

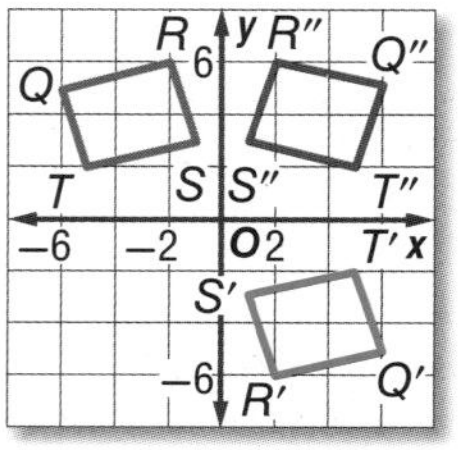

36. PROOF Write a two-column proof of Theorem 14.2.

Given: A reflection in line p maps $\overline{BC}$ to $\overline{B'C'}$.
A reflection in line q maps $\overline{B'C'}$ to $\overline{B''C''}$.
$p \parallel q$, $AD = x$

Prove: **a.** $\overline{BB''} \perp p$, $\overline{BB''} \perp q$

b. $BB'' = 2x$

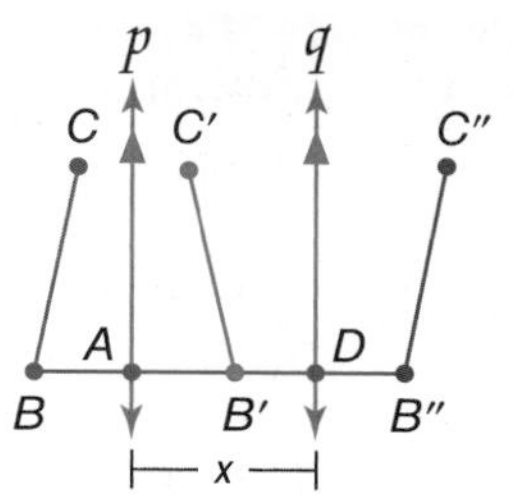

37. PROOF Write a paragraph proof of Theorem 14.3.

Given: Lines ℓ and m intersect at point P.
A is any point not on ℓ or m.

Prove: **a.** If you reflect point A in m, and then reflect its image A' in ℓ, A'' is the image of A after a rotation about point P.

b. $m\angle APA'' = 2(m\angle SPR)$

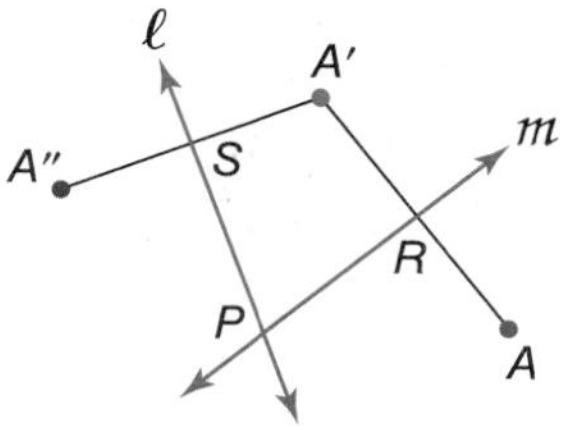

H.O.T. Problems Use Higher-Order Thinking Skills

38. ERROR ANALYSIS Daniel and Lolita are translating $\triangle XYZ$ along $\langle 2, 2 \rangle$ and reflecting it in the line $y = 2$. Daniel says that the transformation is a glide reflection. Lolita disagrees and says that the transformation is a composition of transformations. Is either of them correct? Explain your reasoning.

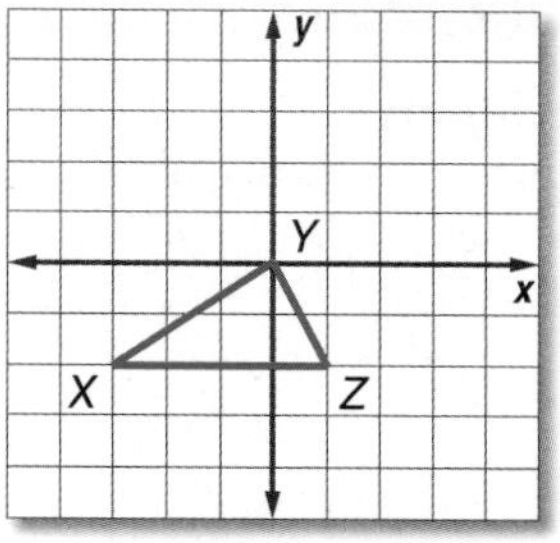

39. WRITING IN MATH Do any points remain invariant under glide reflections? under compositions of transformations? Explain.

40. CHALLENGE If $PQRS$ is translated along $\langle 3, -2 \rangle$, reflected in $y = -1$, and rotated 90° about the origin, what are the coordinates of $P'''Q'''R'''S'''$?

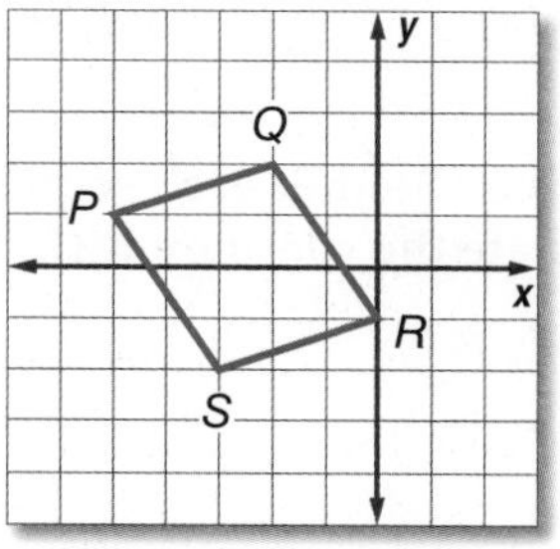

41. CCSS ARGUMENTS If an image is to be reflected in the line $y = x$ and the x-axis, does the order of the reflections affect the final image? Explain.

42. OPEN ENDED Write a glide reflection or composition of transformations that can be used to transform $\triangle ABC$ to $\triangle DEF$.

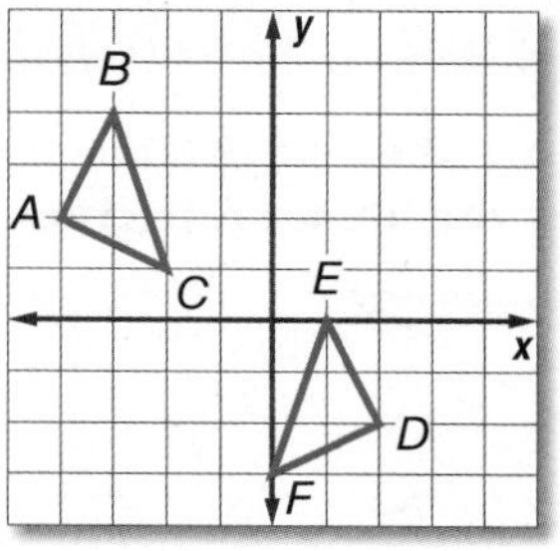

43 REASONING When two rotations are performed on a single image, does the order of the rotations *sometimes*, *always*, or *never* affect the location of the final image? Explain.

44. WRITING IN MATH Compare and contrast glide reflections and compositions of transformations.

Standardized Test Practice

45. $\triangle ABC$ is translated along the vector $\langle -2, 3\rangle$ and then reflected in the x-axis. What are the coordinates of A' after the transformation?

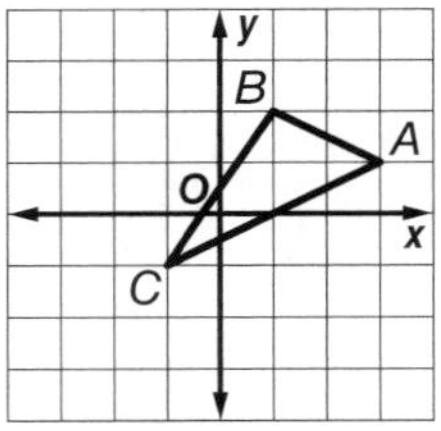

A $(1, -4)$
B $(1, 4)$
C $(-1, 4)$
D $(-1, -4)$

46. SHORT RESPONSE What are the coordinates of D'' if $\overline{CD}$ with vertices $C(2, 4)$ and $D(8, 7)$ is translated along $\langle -6, 2\rangle$ and then reflected over the y-axis?

47. ALGEBRA Write $\dfrac{18x^2 - 2}{3x^2 - 5x - 2}$ in simplest terms.

F $\dfrac{18}{3x + 1}$ **H** $\dfrac{2(3x - 1)}{x - 2}$

G $\dfrac{2(3x + 1)}{x - 2}$ **J** $2(3x - 1)$

48. SAT/ACT If $f(x) = x^3 - x^2 - x$, what is the value of $f(-3)$?

A -39 **D** -15
B -33 **E** -12
C -21

Spiral Review

Copy each polygon and point X. Then use a protractor and ruler to draw the specified rotation of each figure about point X. (Lesson 14-6)

49. 60°

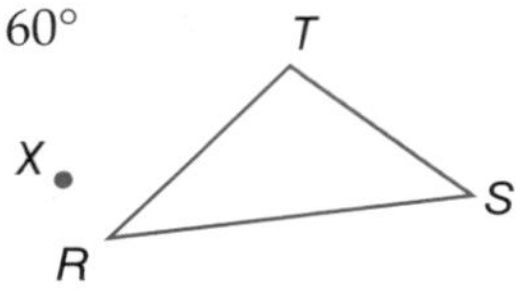

50. 120°

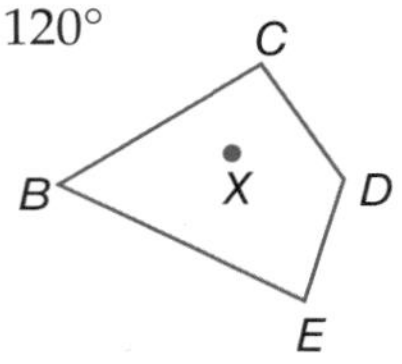

51. 180°

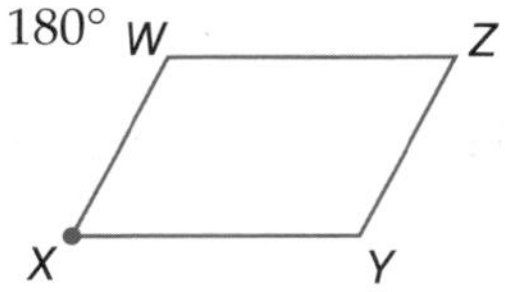

52. JOBS Kimi received an offer for a new job. She wants to compare the offer with her current job. What is total amount of sales that Kimi must get each month to make the same income at either job? (Lesson 6-2)

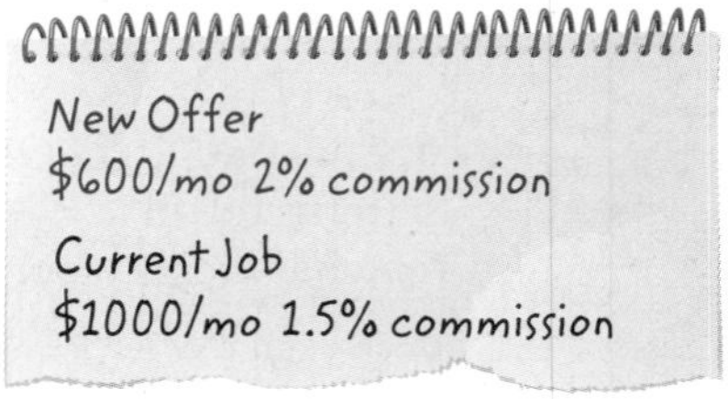

Determine whether each sequence is an arithmetic sequence. If it is, state the common difference. (Lesson 3-5)

53. 24, 16, 8, 0, …

54. $3\frac{1}{4}, 6\frac{1}{2}, 13, 26, \ldots$

55. 7, 6, 5, 4, …

56. 10, 12, 15, 18, …

57. $-15, -11, -7, -3, \ldots$

58. $-0.3, 0.2, 0.7, 1.2, \ldots$

Skills Review

Each figure shows a preimage and its reflected image in some line. Copy each figure and draw the line of reflection.

59.

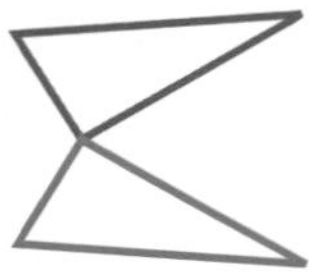

60.

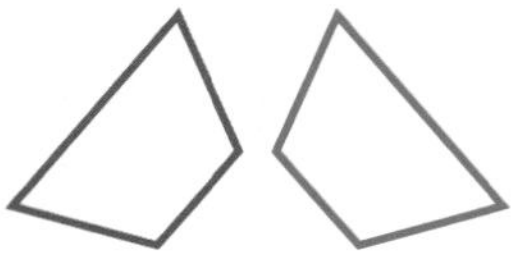

61.

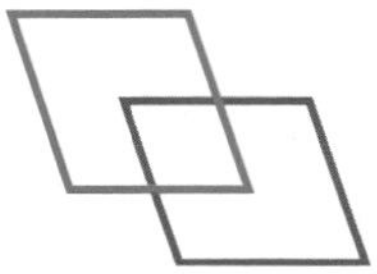

EXPLORE

14-7 Geometry Lab
Tessellations

A **tessellation** is a pattern of one or more figures that covers a plane so that there are no overlapping or empty spaces. The sum of the angles around the vertex of a tessellation is 360°.

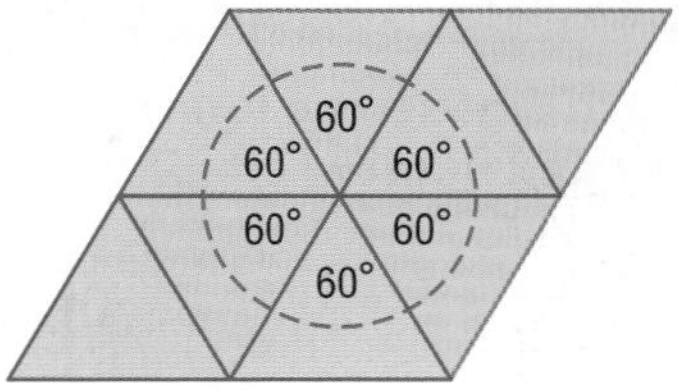

60°
60°
60°
60°
60°
60°

A **regular tessellation** is formed by only one type of regular polygon. A regular polygon will tessellate if it has an interior angle measure that is a factor of 360. A **semi-regular tessellation** is formed by two or more regular polygons.

Activity 1 Regular Tessellation

Determine whether each regular polygon will tessellate in the plane. Explain.

a. hexagon

Let x represent the measure of an interior angle of a regular hexagon.

$x = \frac{180(n-2)}{n}$ Interior Angle Formula

$= \frac{180(6-2)}{6}$ $n = 6$

$= 120$ Simplify.

Since 120 is a factor of 360, a regular hexagon will tessellate in the plane.

b. decagon

Let x represent the measure of an interior angle of a regular decagon.

$x = \frac{180(n-2)}{n}$ Interior Angle Formula

$= \frac{180(10-2)}{10}$ $n = 10$

$= 144$ Simplify.

Since 144 is not a factor of 360, a regular decagon will not tessellate in the plane.

A tessellation is **uniform** if it contains the same arrangement of shapes and angles at each vertex.

Uniform

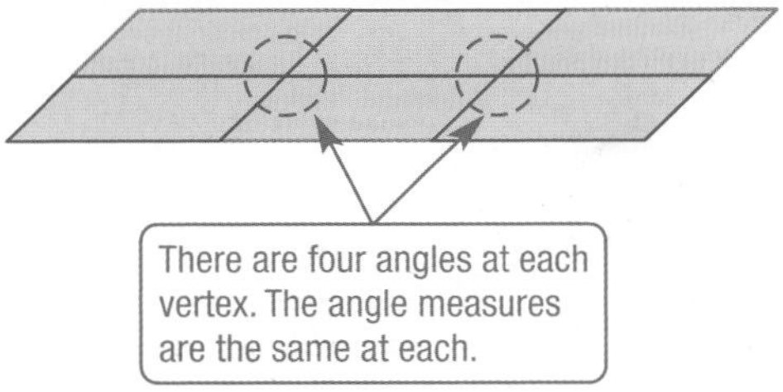

There are four angles at each vertex. The angle measures are the same at each.

Not Uniform

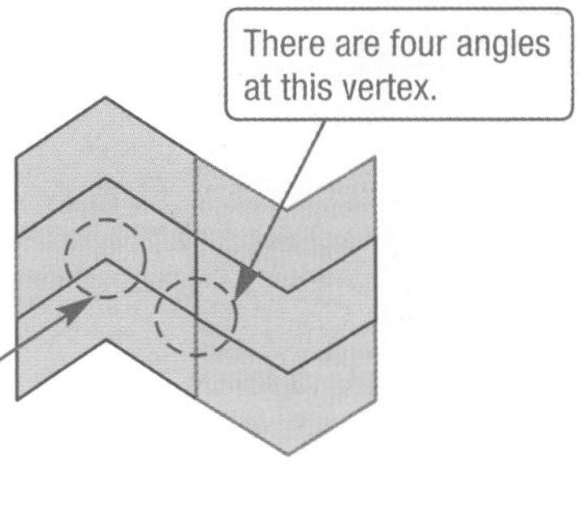

There are four angles at this vertex.
There are two angles at this vertex.

Tessellations *Continued*

Activity 2 Classify Tessellations

Determine whether each pattern is a tessellation. If so, describe it as *regular*, *semi-regular*, or *neither* and *uniform* or *not uniform*.

a.

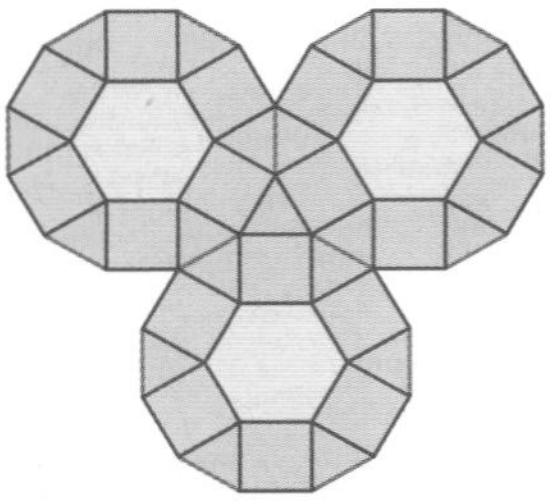

There is no unfilled space, and none of the figures overlap, so the pattern is a **tessellation**.

The tessellation consists of regular hexagons, squares and equilateral triangles, so it is **semi-regular**.

There are four angles around some of the vertices and five around others, so it is **not uniform**.

b.

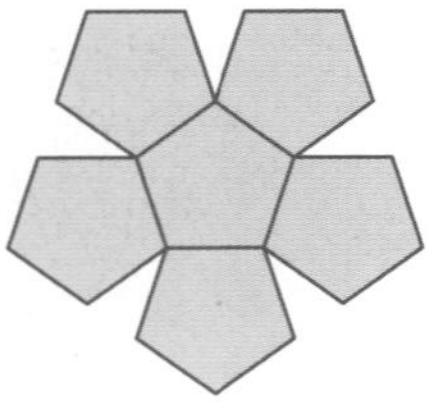

There is unfilled space, so the pattern is a **not a tessellation**.

c.

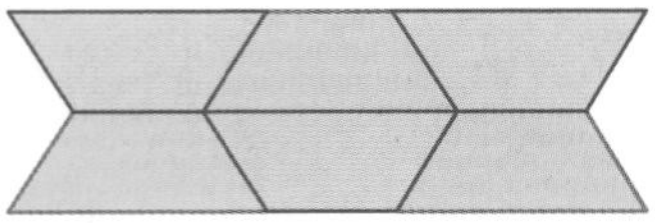

There is no unfilled space, and none of the figures overlap, so the pattern is a **tessellation**.

The tessellation consists of trapezoids, which are not regular polygons, so it is **neither** regular nor semi-regular.

There are four angles around each of the vertices and the angle measures are the same at each vertex, so it is **uniform**.

You can use the properties of tessellations to design and create tessellations.

Activity 3 Draw a Tessellation

Draw a triangle and use it to create a tessellation.

Step 1 Draw a triangle and find the midpoint of one side.

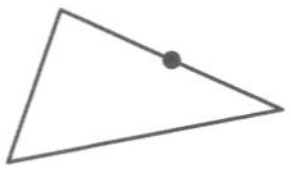

Step 2 Rotate the triangle 180° about the point.

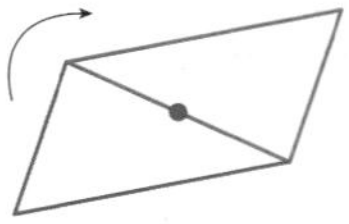

Step 3 Translate the pair of triangles to make a row.

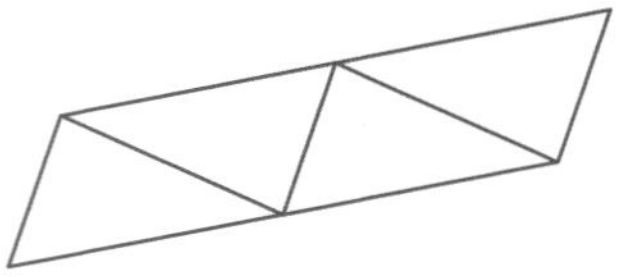

Step 4 Translate the row to make a tessellation.

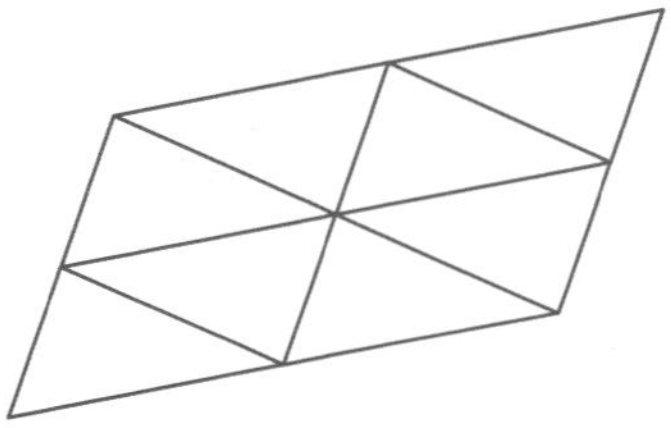

Activity 4 Tessellations using Technology

Use Geometer's Sketchpad to create a tessellation.

Step 1 Insert three points and construct a line through two of the points. Then construct the line parallel to the first line through the third point using the **Parallel Line** option from the **Construct** menu. Complete the parallelogram and label the points A, B, C, and D. Hide the lines.

Step 2 Insert another point E on the exterior of the parallelogram. Draw the segments between A and B, B and E, E and C, and C and D.

Step 3 Highlight B and then A. From the **Transform** menu, choose **Mark Vector**. Select the $\overline{BE}$, $\overline{EC}$, and point E. From the **Transform** menu, choose **Translate**.

Step 4 Starting with A, select all of the vertices around the perimeter of the polygon. Choose **Hexagon Interior** from the **Construct** menu.

Step 5 Choose point A and then point B and mark the vector as you did in Step 3. Select the interior of the polygon and choose **Translate** from the **Transform** menu. Continue the tessellation by marking vectors and translating the polygon. You can choose **Color** from the **Display** menu to create a color pattern.

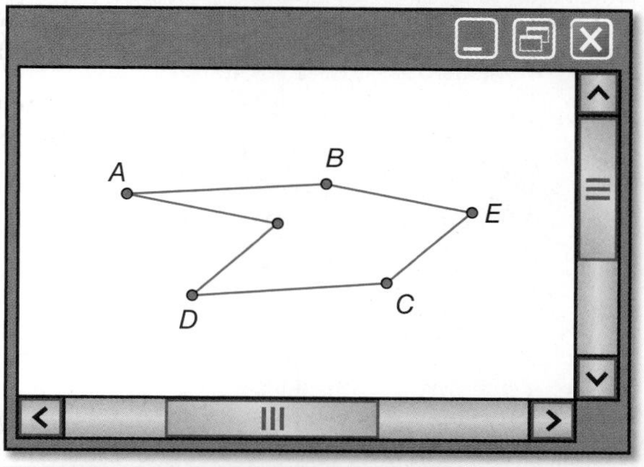

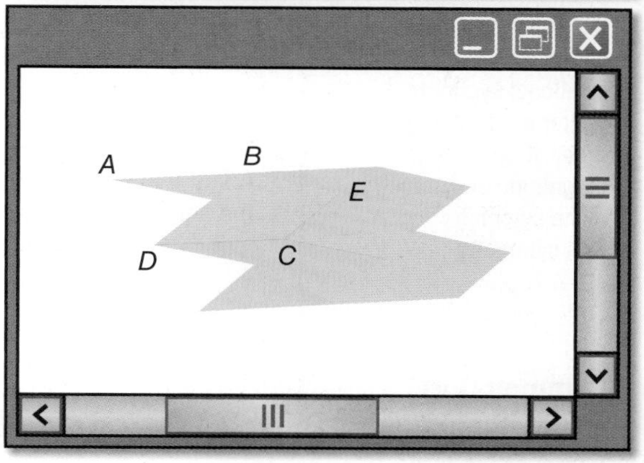

Exercises

Determine whether each regular polygon will tessellate in the plane. Write *yes* or *no*. Explain.

1. triangle
2. pentagon
3. 16-gon

Determine whether each pattern is a tessellation. Write *yes* or *no*. If so, describe it as *regular*, *semi-regular*, or *neither* and *uniform* or *not uniform*.

4. 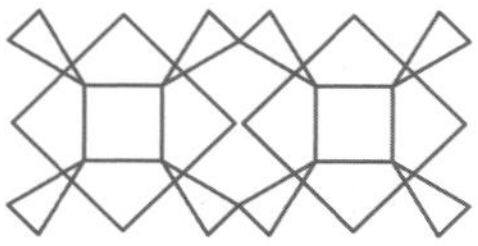

5. 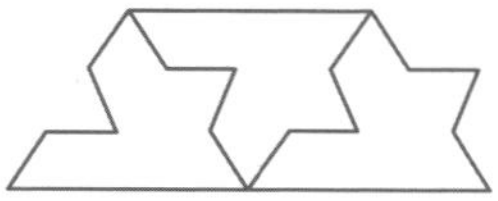

6. 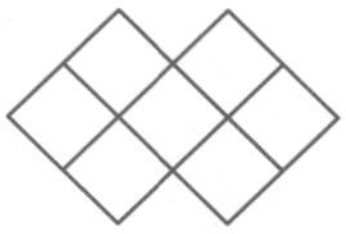

Draw a tessellation using the following shape(s).

7. octagon and square
8. hexagon and triangle
9. right triangle
10. trapezoid and a parallelogram
11. **WRITING IN MATH** Find examples of the use of tessellations in architecture, mosaics, and artwork. For each example, explain how tessellations were used.
12. **MAKE A CONJECTURE** Describe a figure that you think will tessellate in three-dimensional space. Explain your reasoning.

LESSON 14-8 Symmetry

Then

- You drew reflections and rotations of figures.

Now

1. Identify line and rotational symmetries in two-dimensional figures.
2. Identify plane and axis symmetries in three-dimensional figures.

Why?

- In the animal kingdom, the symmetry of an animal's body is often an indication of the animal's complexity. Animals displaying line symmetry, such as insects, are usually more complex life forms than those displaying rotational symmetry, like a jellyfish.

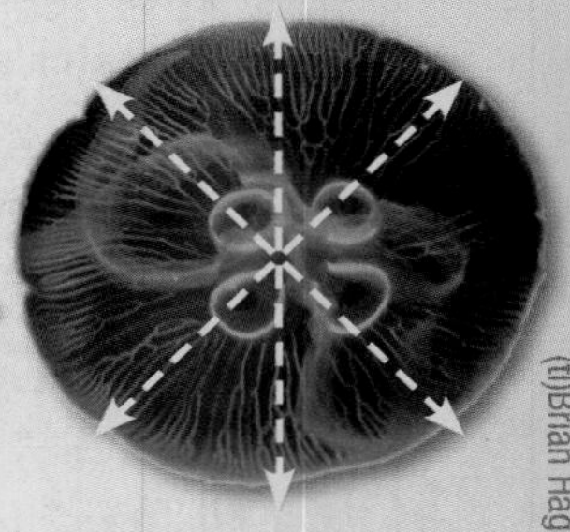

NewVocabulary
symmetry
line symmetry
line of symmetry
rotational symmetry
center of symmetry
order of symmetry
magnitude of symmetry
plane symmetry
axis symmetry

Common Core State Standards

Content Standards
G.CO.3 Given a rectangle, parallelogram, trapezoid, or regular polygon, describe the rotations and reflections that carry it onto itself.

Mathematical Practices
4 Model with mathematics.
8 Look for and express regularity in repeated reasoning.

1 Symmetry in Two-Dimensional Figures

A figure has **symmetry** if there exists a rigid motion—reflection, translation, rotation, or glide reflection—that maps the figure onto itself. One type of symmetry is line symmetry.

KeyConcept Line Symmetry

A figure in the plane has **line symmetry** (or *reflection symmetry*) if the figure can be mapped onto itself by a reflection in a line, called a **line of symmetry** (or *axis of symmetry*).

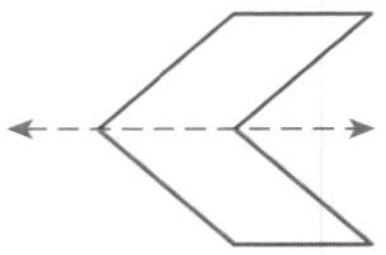

Real-World Example 1 Identify Line Symmetry

PT

BEACHES State whether the object appears to have line symmetry. Write *yes* or *no*. If so, copy the figure, draw all lines of symmetry, and state their number.

a.

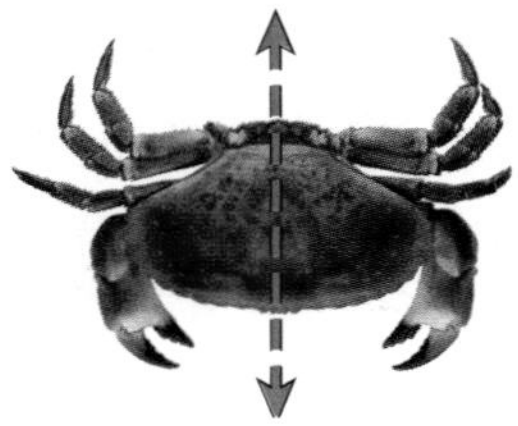

Yes; the crab has one line of symmetry.

b.

Yes; the starfish has five lines of symmetry.

c.

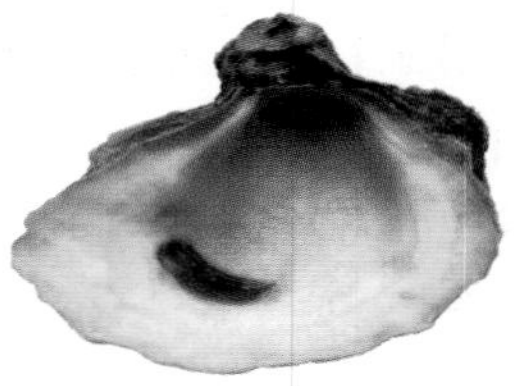

No; there is no line in which the oyster shell can be reflected so that it maps onto itself.

GuidedPractice

State whether the figure has line symmetry. Write *yes* or *no*. If so, copy the figure, draw all lines of symmetry, and state their number.

1A.

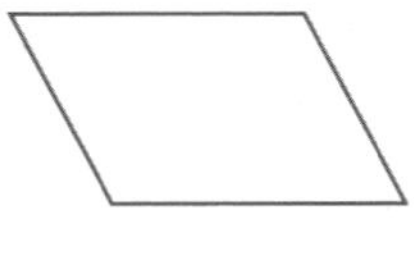

1B.

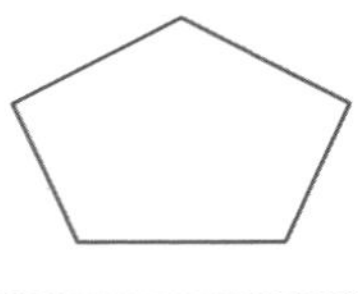

1C.

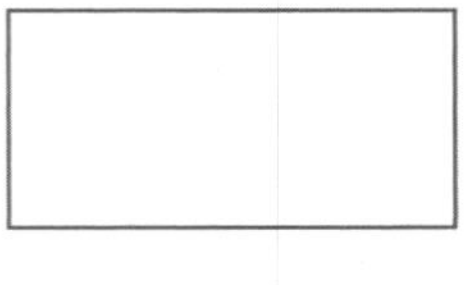

Another type of symmetry is rotational symmetry.

KeyConcept Rotational Symmetry

A figure in the plane has **rotational symmetry** (or *radial symmetry*) if the figure can be mapped onto itself by a rotation between 0° and 360° about the center of the figure, called the **center of symmetry** (or *point of symmetry*).

Examples The figure below has rotational symmetry because a rotation of 90°, 180°, or 270° maps the figure onto itself.

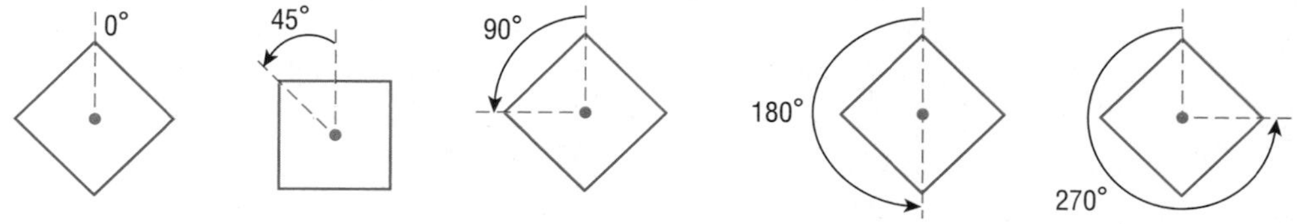

The number of times a figure maps onto itself as it rotates from 0° to 360° is called the **order of symmetry**. The **magnitude of symmetry** (or angle of rotation) is the smallest angle through which a figure can be rotated so that it maps onto itself. The order and magnitude of a rotation are related by the following equation.

$$\text{magnitude} = 360° \div \text{order}$$

The figure above has rotational symmetry of order 4 and magnitude 90°.

Example 2 Identify Rotational Symmetry

State whether the figure has rotational symmetry. Write *yes* or *no*. If so, copy the figure, locate the center of symmetry, and state the order and magnitude of symmetry.

a.

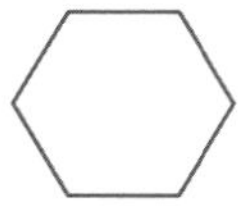

Yes; the regular hexagon has order 6 rotational symmetry and magnitude 360° ÷ 6 or 60°. The center is the intersection of the diagonals.

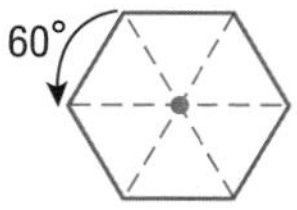

b.

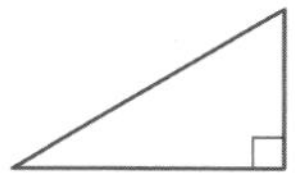

No; no rotation between 0° and 360° maps the right triangle onto itself.

c.

Yes; the figure has order 2 rotational symmetry and magnitude 360° ÷ 2 or 180°. The center is the intersection of the diagonals.

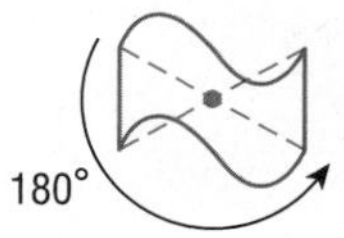

StudyTip

Point Symmetry A figure has *point symmetry* if the figure can be mapped onto itself by a rotation of 180°. A playing card exhibits point symmetry. It looks the same right-side up as upside down.

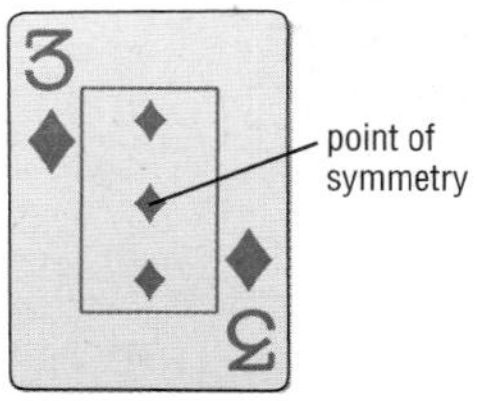

GuidedPractice

FLOWERS **State whether the flower appears to have rotational symmetry. Write *yes* or *no*. If so, copy the flower, locate the center of symmetry, and state the order and magnitude of symmetry.**

2A.

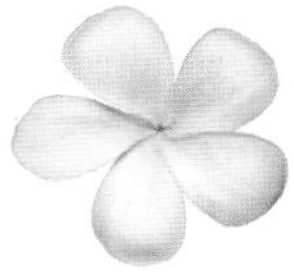

2B.

2C.

2 Symmetry in Three-Dimensional Figures

Three-dimensional figures can also have symmetry.

KeyConcept Three-Dimensional Symmetries

Plane Symmetry

A three-dimensional figure has **plane symmetry** if the figure can be mapped onto itself by a reflection in a plane.

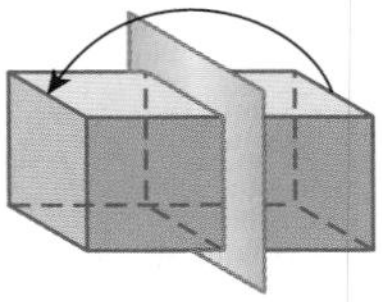

Axis Symmetry

A three-dimensional figure has **axis symmetry** if the figure can be mapped onto itself by a rotation between 0° and 360° in a line.

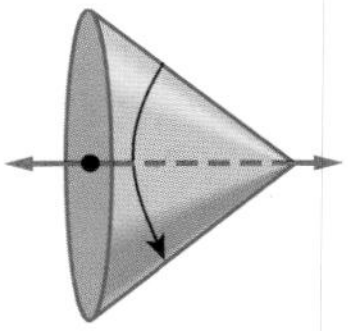

Example 3 Three-Dimensional Symmetry

State whether the figure has *plane* symmetry, *axis* symmetry, *both*, or *neither*.

a. L-shaped prism

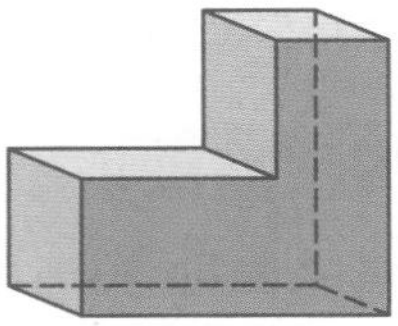

plane symmetry

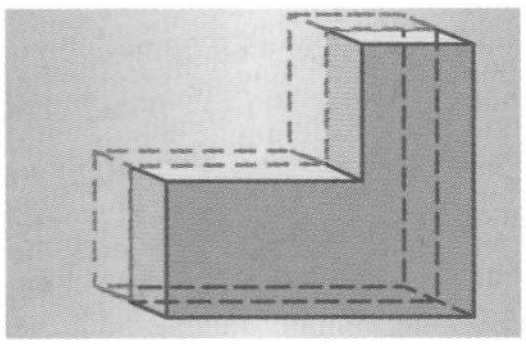

b. regular pentagonal prism

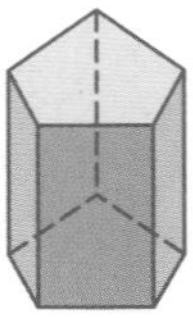

both plane symmetry and axis symmetry

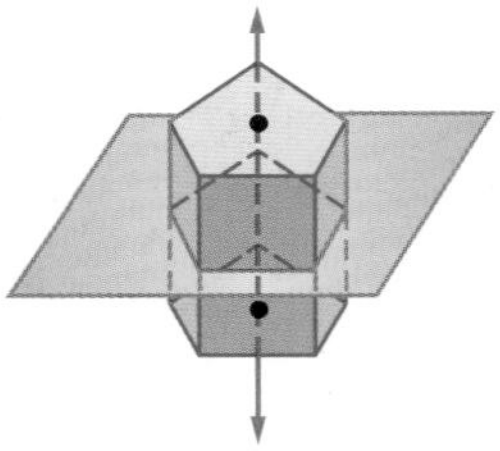

ReviewVocabulary

prism a polyhedron with two parallel congruent bases connected by parallelogram faces

Real-WorldLink

Aerodynamically designed to spin after it is thrown, a football's shape is that of a prolate spheroid. This means that one axis of symmetry is longer than its other axes.

Source: *Complete Idiot's Guide to Football*

GuidedPractice

SPORTS State whether each piece of sports equipment appears to have *plane* symmetry, *axis* symmetry, *both*, or *neither* (ignoring the equipment's stitching or markings).

3A.

3B.

3C.

3D.

Check Your Understanding

Example 1 **State whether the figure appears to have line symmetry. Write *yes* or *no*. If so, copy the figure, draw all lines of symmetry, and state their number.**

1.

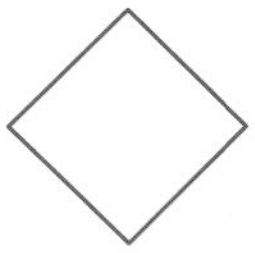

2.

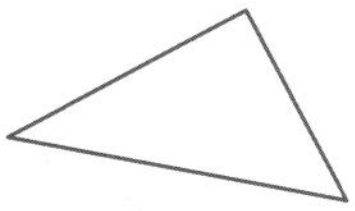

3.

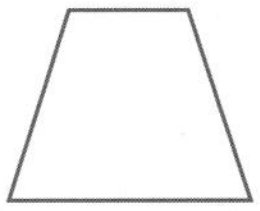

Example 2 **State whether the figure has rotational symmetry. Write *yes* or *no*. If so, copy the figure, locate the center of symmetry, and state the order and magnitude of symmetry.**

4.

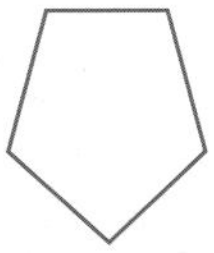

5.

6.

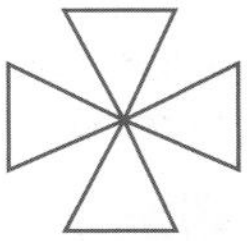

Examples 2–3 **7.** **U.S. CAPITOL** Completed in 1863, the dome is one of the most recent additions to the United States Capitol. It is supported by 36 iron ribs and has 108 windows, divided equally among three levels.

a. Excluding the spire of the dome, how many horizontal and vertical planes of symmetry does the dome appear to have?

b. Does the dome have axis symmetry? If so, state the order and magnitude of symmetry.

Example 3 **8.** State whether the figure has *plane* symmetry, *axis* symmetry, *both*, or *neither*.

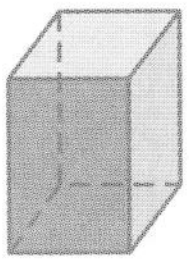

Practice and Problem Solving

Example 1 **CCSS REGULARITY** **State whether the figure appears to have line symmetry. Write *yes* or *no*. If so, copy the figure, draw all lines of symmetry, and state their number.**

9.

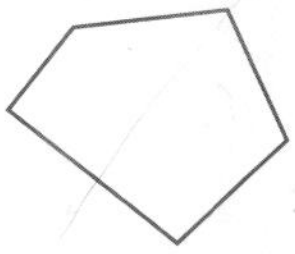

10.

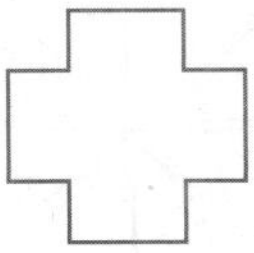

11.

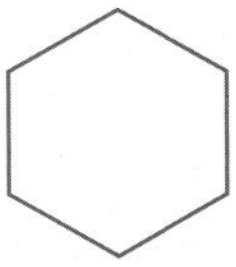

12.

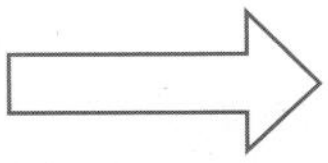

13.

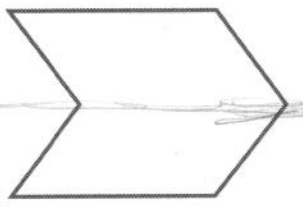

14.

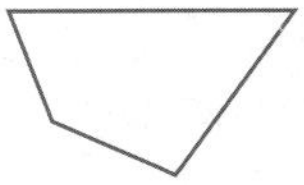

FLAGS **State whether each flag design appears to have line symmetry. Write *yes* or *no*. If so, copy the flag, draw all lines of symmetry, and state their number.**

15.

16.

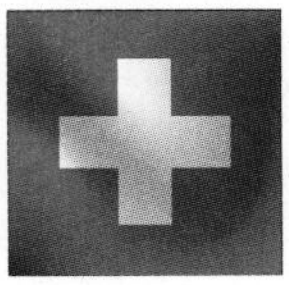

17.

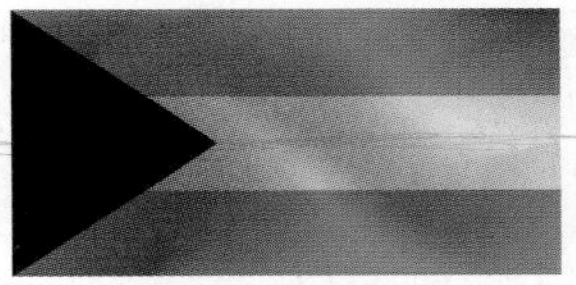

Hisham F. Ibrahim/Photodisc/Getty Images

Example 2 **State whether the figure has rotational symmetry. Write *yes* or *no*. If so, copy the figure, locate the center of symmetry, and state the order and magnitude of symmetry.**

18. 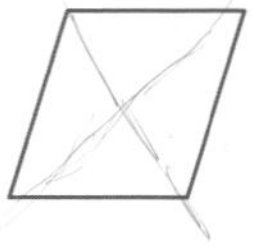

19. 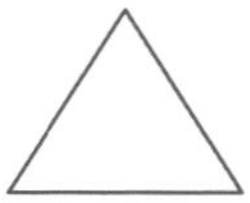

20. 

21. 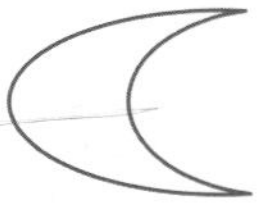

22. 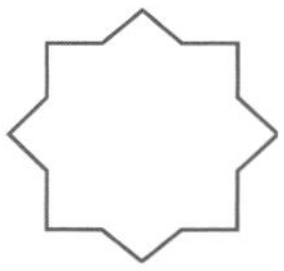

23. 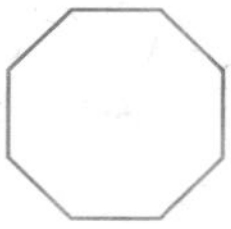

Example 2 WHEELS **State whether each wheel cover appears to have rotational symmetry. Write *yes* or *no*. If so, state the order and magnitude of symmetry.**

24. 

25. 

26. 

Example 3 **State whether the figure has *plane* symmetry, *axis* symmetry, *both*, or *neither*.**

27. 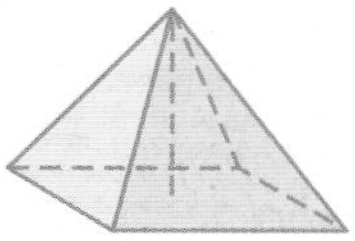

28. 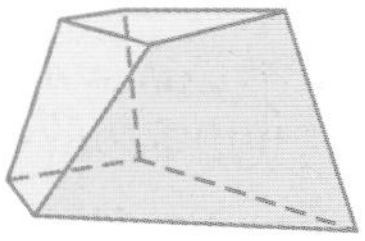

29. 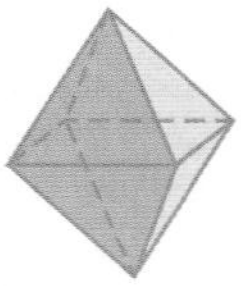

30. 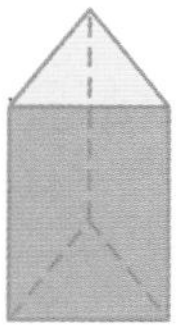

CONTAINERS **Determine the number of horizontal and vertical planes of symmetry for each container shown below.**

31. 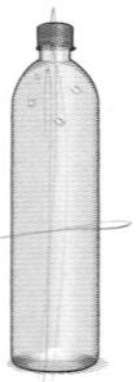

32. 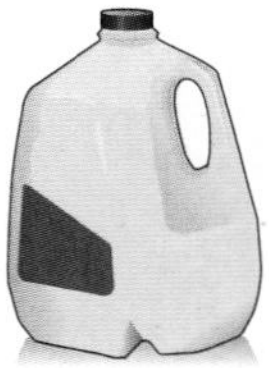

33. 

34. CCSS MODELING Symmetry is an important component of photography. Photographers often use reflection in water to create symmetry in photos. The photo at the right is a long exposure shot of the Eiffel tower reflected in a pool.

a. Describe the two-dimensional symmetry created by the photo.

b. Is three-dimensional symmetry applicable? Explain your reasoning.

COORDINATE GEOMETRY **Determine whether the figure with the given vertices has *line* symmetry and/or *rotational* symmetry.**

35. $A(-4, 0)$, $B(0, 4)$, $C(4, 0)$, $D(0, -4)$

36. $R(-3, 3)$, $S(-3, -3)$, $T(3, 3)$

37. $F(0, -4)$, $G(-3, -2)$, $H(-3, 2)$, $J(0, 4)$, $K(3, 2)$, $L(3, -2)$

38. $W(-2, 3)$, $X(-3, -3)$, $Y(3, -3)$, $Z(2, 3)$

ALGEBRA **Graph the function and determine whether the graph has *line* and/or *rotational* symmetry. If so, state the order and magnitude of symmetry, and write the equations of any lines of symmetry.**

39. $y = x$

40. $y = x^2 + 1$

41. $y = -x^3$

CRYSTALLOGRAPHY **Determine whether the crystals below have *plane* symmetry and/or *axis* symmetry. If so, state the magnitude of symmetry.**

42.

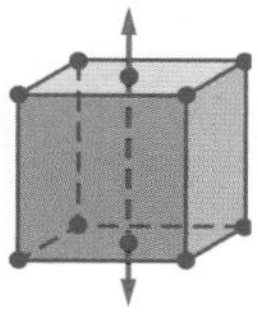

43.

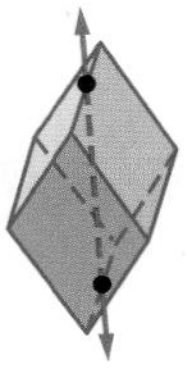

44.

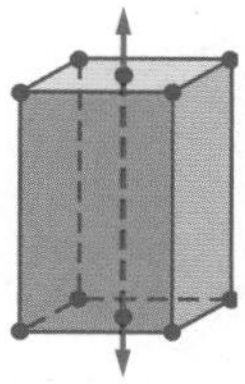

45. **MULTIPLE REPRESENTATIONS** In this problem, you will use dynamic geometric software to investigate line and rotational symmetry in regular polygons.

a. Geometric Use The Geometer's Sketchpad to draw an equilateral triangle. Use the reflection tool under the transformation menu to investigate and determine all possible lines of symmetry. Then record their number.

b. Geometric Use the rotation tool under the transformation menu to investigate the rotational symmetry of the figure in part **a**. Then record its order of symmetry.

c. Tabular Repeat the process in parts **a** and **b** for a square, regular pentagon, and regular hexagon. Record the number of lines of symmetry and the order of symmetry for each polygon.

d. Verbal Make a conjecture about the number of lines of symmetry and the order of symmetry for a regular polygon with n sides.

H.O.T. Problems Use Higher-Order Thinking Skills

46. CCSS **CRITIQUE** Jaime says that Figure A has only line symmetry, and Jewel says that Figure A has only rotational symmetry. Is either of them correct? Explain your reasoning.

Figure A

47. **CHALLENGE** A quadrilateral in the coordinate plane has exactly two lines of symmetry, $y = x - 1$ and $y = -x + 2$. Find possible vertices for the figure. Graph the figure and the lines of symmetry.

48. **REASONING** A regular polyhedron has axis symmetry of order 3, but does not have plane symmetry. What is the figure? Explain.

49. **OPEN ENDED** Draw a figure with line symmetry but not rotational symmetry. Explain.

50. **WRITING IN MATH** How are line symmetry and rotational symmetry related?

Standardized Test Practice

51. How many lines of symmetry can be drawn on the picture of the Canadian flag below?

A 0 **C** 2

B 1 **D** 4

52. GRIDDED RESPONSE What is the order of symmetry for the figure below?

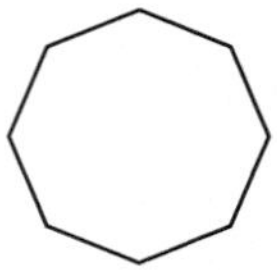

53. ALGEBRA A computer company ships computers in wooden crates that each weigh 45 pounds when empty. If each computer weighs no more than 13 pounds, which inequality *best* describes the total weight in pounds w of a crate of computers that contains c computers?

F $c \leq 13 + 45w$ **H** $w \leq 13c + 45$

G $c \geq 13 + 45w$ **J** $w \geq 13c + 45$

54. SAT/ACT What is the slope of the line determined by the linear equation $5x - 2y = 10$?

A -5 **D** $\frac{2}{5}$

B $-\frac{5}{2}$ **E** $\frac{5}{2}$

C $-\frac{2}{5}$

Spiral Review

Triangle JKL has vertices $J(1, 5)$, $K(3, 1)$, and $L(5, 7)$. Graph $\triangle JKL$ and its image after the indicated transformation. (Lesson 14-7)

55. Translation: along $\langle -7, -1 \rangle$
Reflection: in x-axis

56. Translation: along $\langle 1, 2 \rangle$
Reflection: in y-axis

57. Quadrilateral $QRST$ is shown at the right. What is the image of point R after a rotation 180° counterclockwise about the origin? (Lesson 14-6)

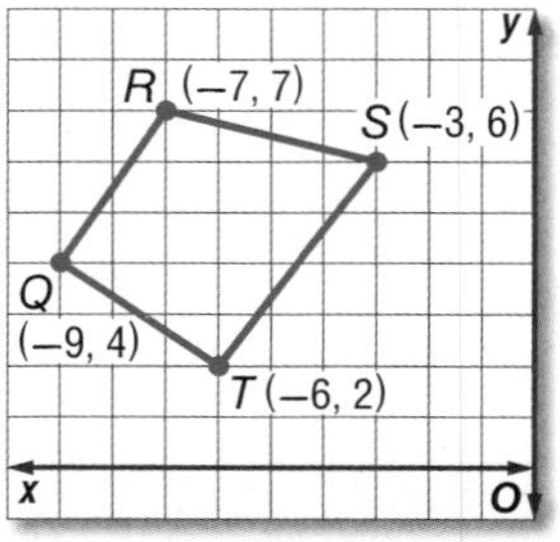

Skills Review

Determine whether the dilation from Figure A to Figure B is an *enlargement* or a *reduction*. Then find the scale factor of the dilation.

58.

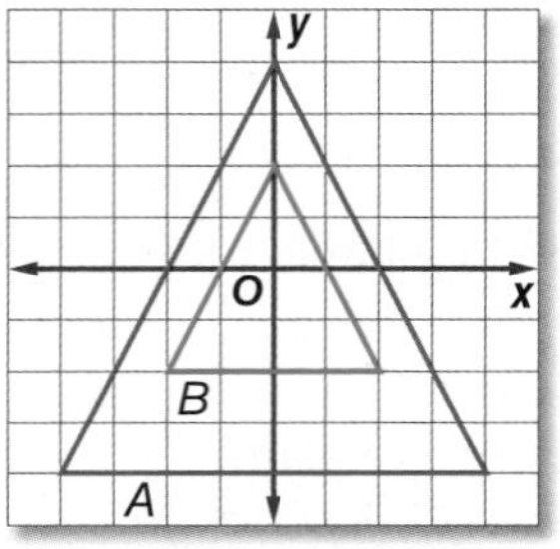

59.

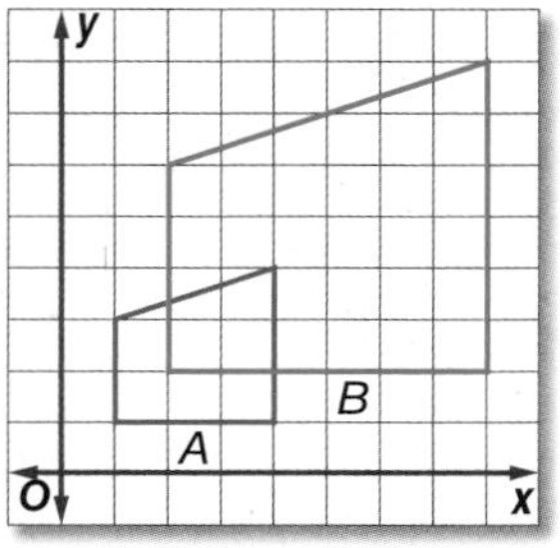

60.

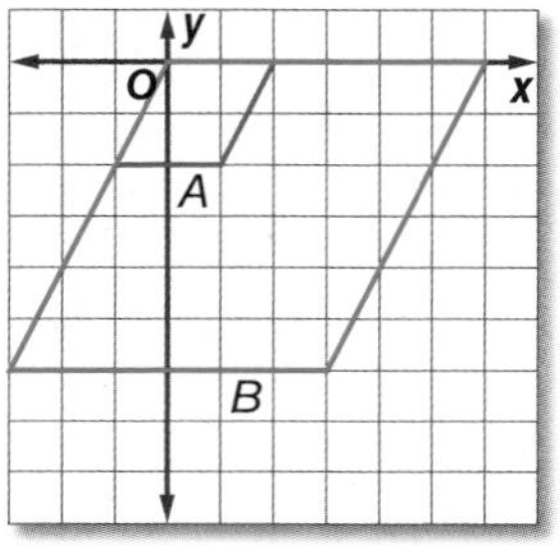

EXTEND 14-8

Geometry Lab
Exploring Constructions with a Reflective Device

A reflective device is a tool made of semitransparent plastic that reflects objects. It works best if you lay it on a flat service in a well-lit room. You can use a reflective device to transform geometric objects.

CCSS Common Core State Standards
Content Standards
G.CO.12 Make formal geometric constructions with a variety of tools and methods (compass and straightedge, string, reflective devices, paper folding, dynamic geometric software, etc.).
Mathematical Practices 5

Activity 1 Reflect a Triangle

Use a reflective device to reflect $\triangle ABC$ in w. Label the reflection $\triangle A'B'C'$.

Step 1 Draw $\triangle ABC$ and the line of reflection w.

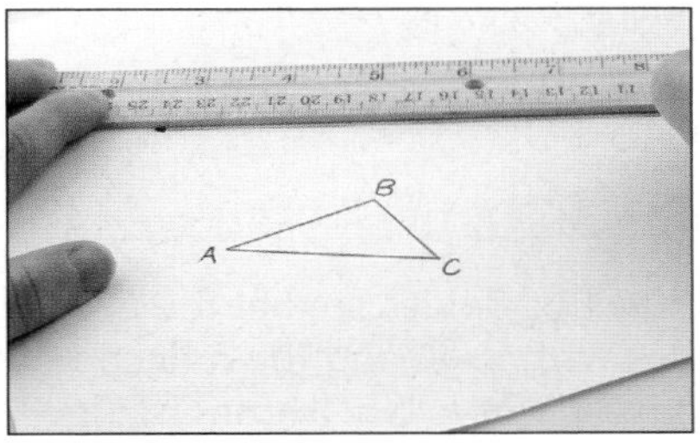

Step 2 With the reflective device on line w, draw points for the vertices of the reflection.

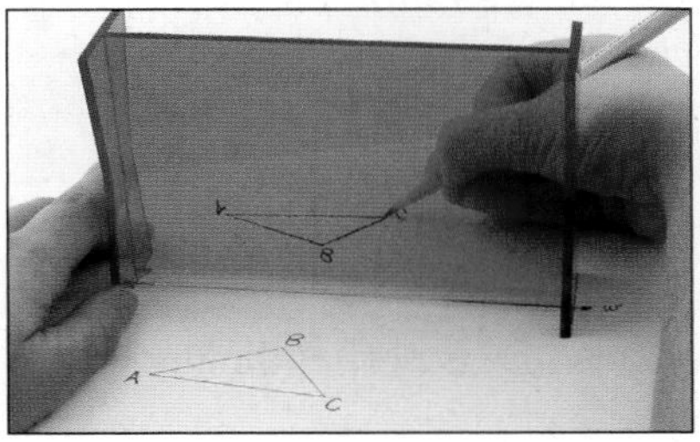

Step 3 Use a straightedge to connect the points to form $\triangle A'B'C'$.

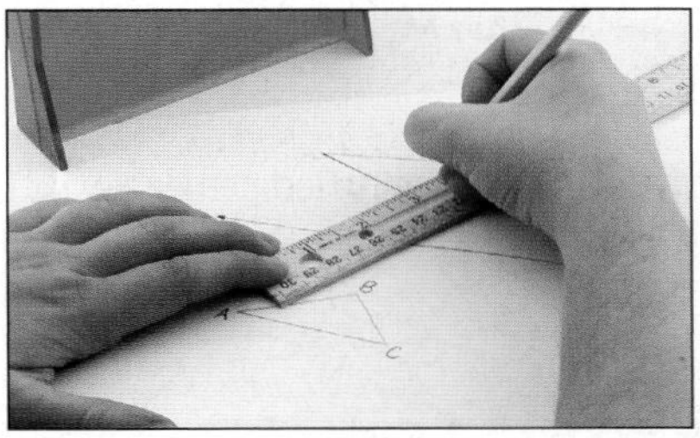

We have used a compass, straightedge, string, and paper folding to make geometric constructions. You can also use a reflective device for constructions.

Activity 2 Construct Lines of Symmetry

Use a reflective device to construct the lines of symmetry for a regular hexagon.

Step 1 Draw a regular hexagon. Place the reflective device on the shape and move it until one half of the shape matches the reflection of the other half. Draw the line of symmetry.

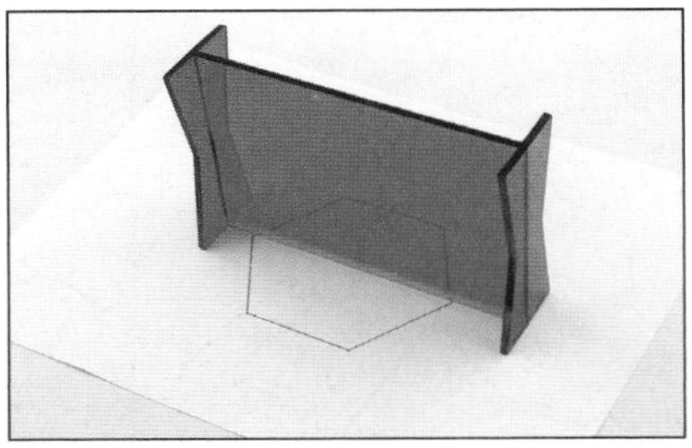

Step 2 Repeat Step 1 until you have found all the lines of symmetry.

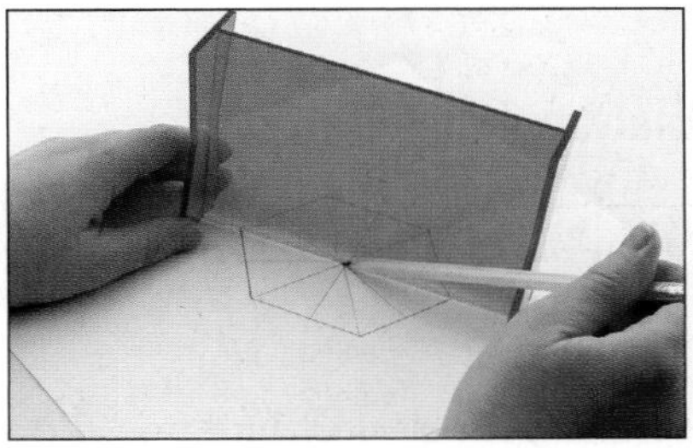

Exploring Constructions with a Reflective Device *Continued*

Activity 3 Construct a Parallel line

Use a reflective device to reflect line ℓ to line m that is parallel and passes through point P.

Step 1

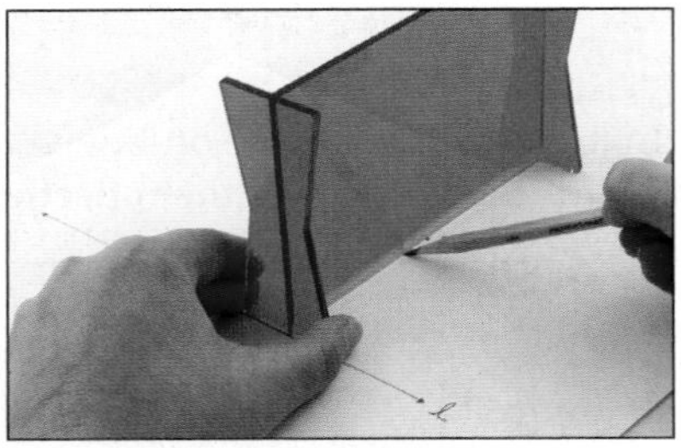

Draw line ℓ and point P. Place a short side of the reflective device on line ℓ and the long side on point P. Draw a line. This line is perpendicular to ℓ through P.

Step 2

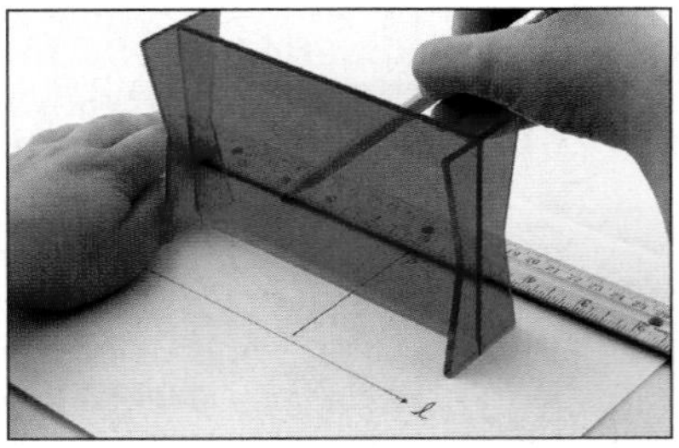

Place the reflective device so that the perpendicular line coincides with itself and the reflection of line ℓ passes through point P. Use a straightedge to draw the parallel line m through P.

In Explore Lesson 12-9A, we constructed perpendicular bisectors with paper folding. You can also use a reflective device to construct perpendicular bisectors of a triangle.

Activity 4 Construct Perpendicular Bisectors

Use a reflective device to find the circumcenter of $\triangle ABC$.

Step 1 Draw $\triangle ABC$. Place the reflective device between A and B and adjust it until A and B coincide. Draw the line of symmetry.

Step 2 Repeat Step 1 for sides $\overline{AC}$ and $\overline{BC}$. Then place a point at the intersection of the three perpendicular bisectors. This is the circumcenter of the triangle.

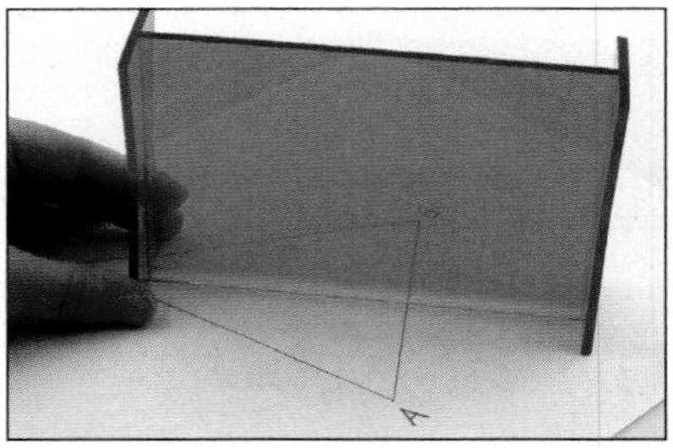

Model and Analyze

1. How do you know that the steps in Activity 4 give the actual perpendicular bisector and the circumcenter of $\triangle ABC$?

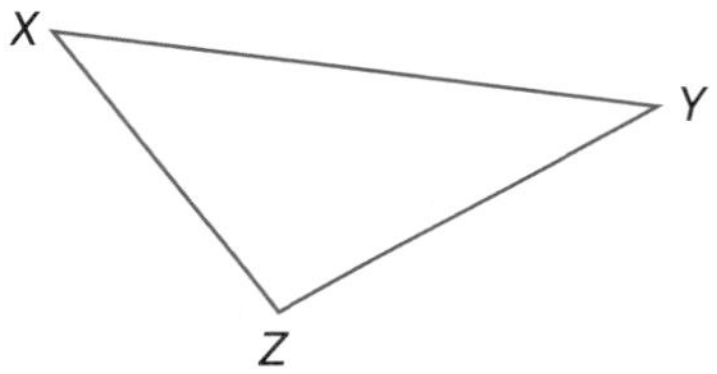

2. Construct the angle bisectors and find the incenter of $\triangle XYZ$. Describe how you used the reflective device for the construction.

EXPLORE 14-9

Graphing Technology Lab
Dilations

You can use TI-Nspire Technology to explore properties of dilations.

CCSS Common Core State Standards
Content Standards
G.SRT.1 Understand similarity in terms of similarity transformations. Verify experimentally the properties of dilations given by a center and a scale factor:
a. A dilation takes a line not passing through the center of the dilation to a parallel line, and leaves a line passing through the center unchanged.
b. The dilation of a line segment is longer or shorter in the ratio given by the scale factor.
Mathematical Practices 5

Activity 1 Dilation of a Triangle

Dilate a triangle by a scale factor of 1.5.

Step 1 Add a new **Geometry** page. Then, from the **Points & Lines** menu, use the **Point** tool to add a point and label it X.

Step 2 From the **Shapes** menu, select **Triangle** and specify three points. Label the points A, B, and C.

Step 3 From the **Actions** menu, use the **Text** tool to separately add the text *Scale Factor* and *1.5* to the page.

Step 4 From the **Transformation** menu, select **Dilation**. Then select point X, $\triangle ABC$, and the text *1.5*.

Step 5 Label the points on the image A', B', and C'.

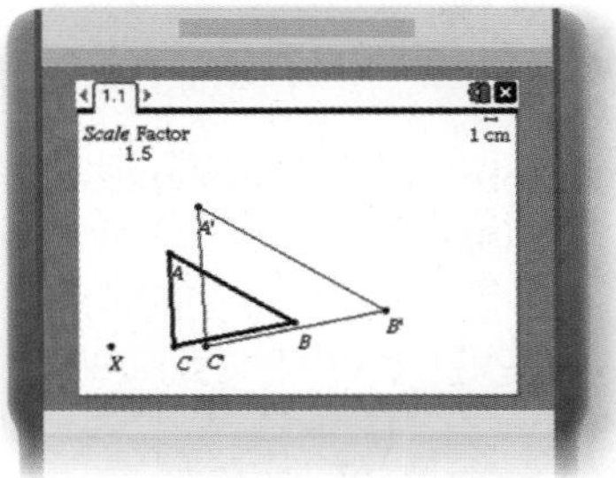

Analyze the Results

1. Using the **Slope** tool on the **Measurement** menu, describe the effect of the dilation on $\overline{AB}$. That is, how are the lines through $\overline{AB}$ and $\overline{A'B'}$ related?
2. What is the effect of the dilation on the line passing through side $\overline{CA}$?
3. What is the effect of the dilation on the line passing through side $\overline{CB}$?

Activity 2 Dilation of a Polygon

Dilate a polygon by a scale factor of −0.5.

Step 1 Add a new **Geometry** page and draw polygon $ABCDX$ as shown. Add the text *Scale Factor* and *−0.5* to the page.

Step 2 From the **Transformation** menu, select **Dilation**. Then select point X, polygon $ABCDX$, and the text *−0.5*.

Step 3 Label the points on the image A', B', C', and D'.

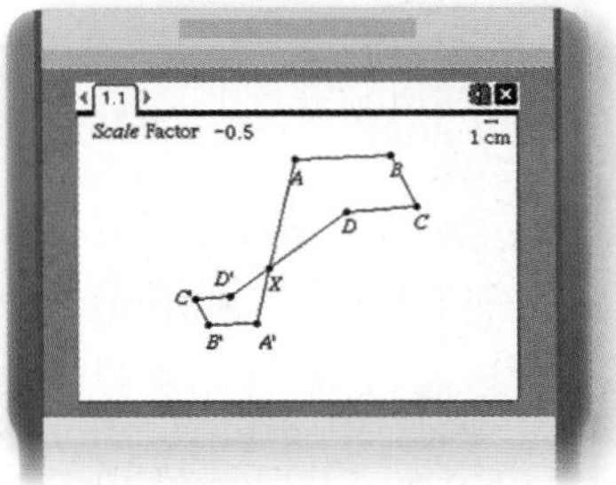

Model and Analyze

4. Analyze the effect of the dilation in Activity 2 on sides that contain the center of the dilation.
5. Analyze the effect of a dilation of trapezoid $ABCD$ shown with a scale factor of 0.75 and the center of the dilation at A.
6. **MAKE A CONJECTURE** Describe the effect of a dilation on segments that pass through the center of a dilation and segments that do not pass through the center of a dilation.

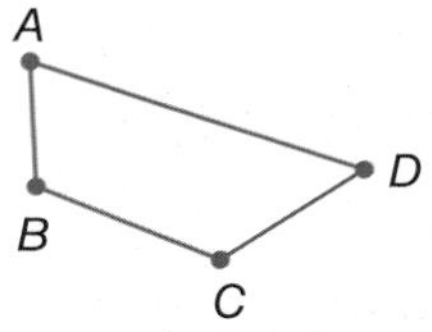

Graphing Technology Lab
Dilations *Continued*

Activity 3 Dilation of a Segment

Dilate a segment $\overline{AB}$ by the indicated scale factor.

a. scale factor: 0.75

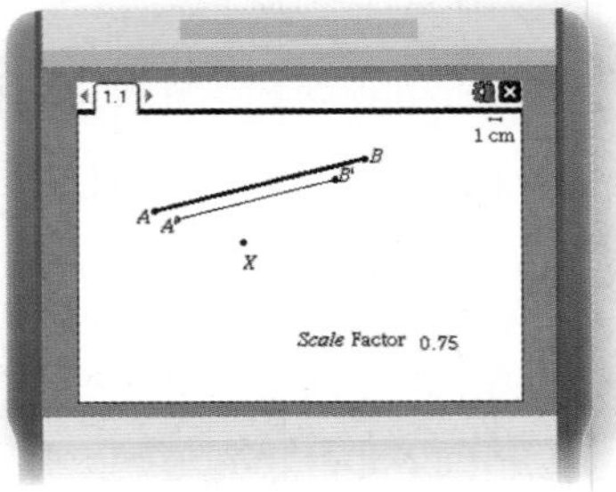

Step 1 On a new **Geometry** page, draw a line segment using the **Points & Lines** menu. Label the endpoints *A* and *B*. Then add and label a point *X*.

Step 2 Add the text *Scale Factor* and *0.75* to the page.

Step 3 From the **Transformation** menu, select **Dilation**. Then select point *X*, $\overline{AB}$, and the text *0.75*.

Step 4 Label the dilated segment $\overline{A'B'}$.

b. scale factor: 1.25

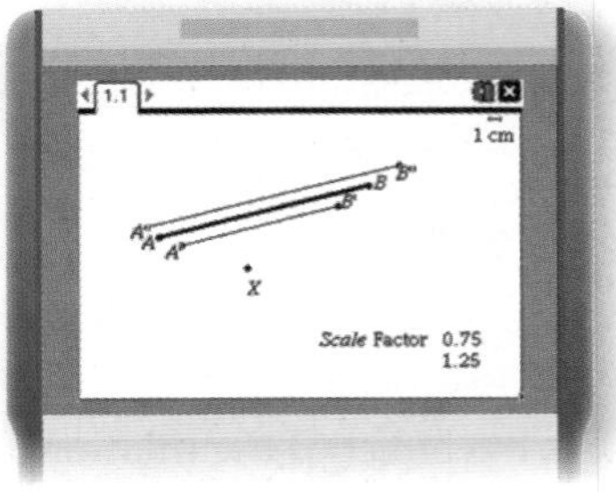

Step 1 Add the text *1.25* to the page.

Step 2 From the **Transformation** menu, select **Dilation**. Then select point *X*, $\overline{AB}$, and the text 1.25.

Step 3 Label the dilated segment $\overline{A''B''}$.

Model and Analyze

7. Using the **Length** tool on the **Measurement** menu, find the measures of $\overline{AB}$, $\overline{A'B'}$, and $\overline{A''B''}$.

8. What is the ratio of *A'B'* to *AB*? What is the ratio of *A''B''* to *AB*?

9. What is the effect of the dilation with scale factor 0.75 on segment $\overline{AB}$? What is the effect of the dilation with scale factor 1.25 on segment $\overline{AB}$?

10. Dilate segment $\overline{AB}$ in Activity 3 by scale factors of −0.75 and −1.25. Describe the effect on the length of each dilated segment.

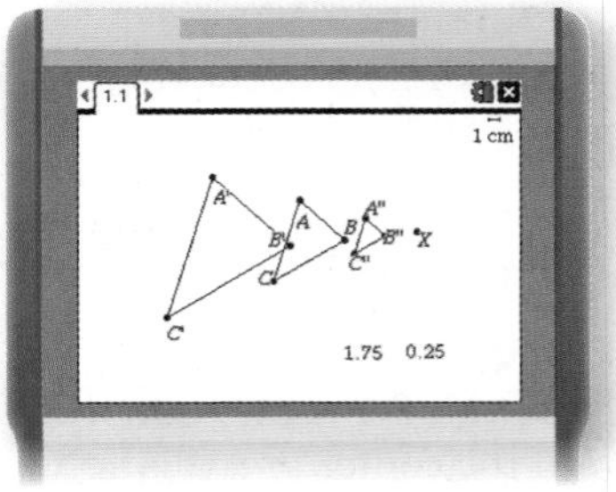

11. **MAKE A CONJECTURE** Describe the effect of a dilation on the length of a line segment.

12. Describe the dilation from $\overline{AB}$ to $\overline{A'B'}$ and $\overline{A'B'}$ to $\overline{A''B''}$ in the triangles shown.

LESSON

14-9 Dilations

Then	Now	Why?
You identified dilations and verified them as similarity transformations.	1 Draw dilations. 2 Draw dilations in the coordinate plane.	Some photographers still prefer traditional cameras and film to produce negatives. From these negatives, photographers can create scaled reproductions.

CCSS Common Core State Standards

Content Standards

G.CO.2 Represent transformations in the plane using, e.g., transparencies and geometry software; describe transformations as functions that take points in the plane as inputs and give other points as outputs. Compare transformations that preserve distance and angle to those that do not (e.g., translation versus horizontal stretch).

G.SRT.1 Understand similarity in terms of similarity transformations. Verify experimentally the properties of dilations given by a center and a scale factor:

a. A dilation takes a line not passing through the center of the dilation to a parallel line, and leaves a line passing through the center unchanged.

b. The dilation of a line segment is longer or shorter in the ratio given by the scale factor.

Mathematical Practices

1 Make sense of problems and persevere in solving them.

5 Use appropriate tools strategically.

1 Draw Dilations

A dilation or *scaling* is a similarity transformation that enlarges or reduces a figure proportionally with respect to a *center* point and a *scale* factor.

KeyConcept Dilation

A dilation with center C and positive scale factor k, $k \neq 1$, is a function that maps a point P in a figure to its image such that

- if point P and C coincide, then the image and preimage are the same point, or
- if point P is not the center of dilation, then P' lies on $\overrightarrow{CP}$ and $CP' = k(CP)$.

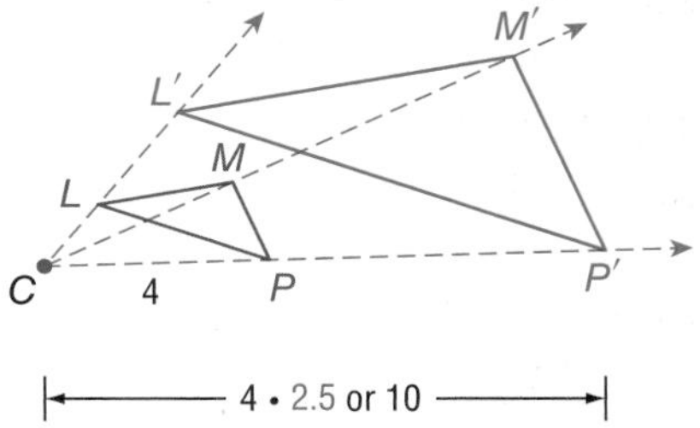

$\triangle L'M'P'$ is the image of $\triangle LMP$ under a dilation with center C and scale factor 2.5.

Example 1 Draw a Dilation

Copy $\triangle ABC$ and point D. Then use a ruler to draw the image of $\triangle ABC$ under a dilation with center D and scale factor $\frac{1}{2}$.

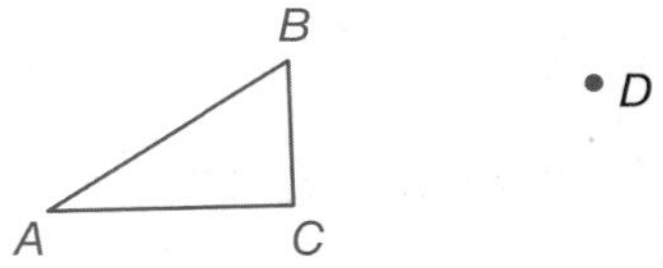

Step 1 Draw rays from D though each vertex.

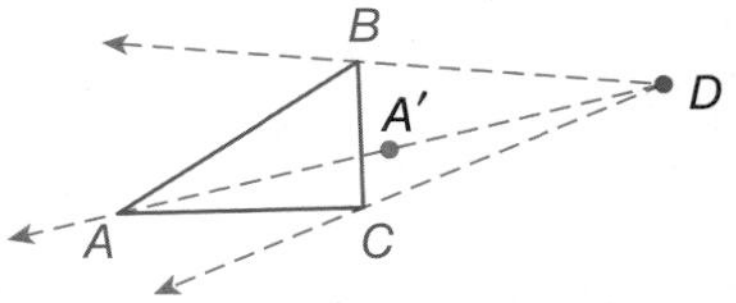

Step 2 Locate A' on $\overrightarrow{DA}$ such that $DA' = \frac{1}{2}DA$.

Step 3 Locate B' on $\overrightarrow{DB}$ and C' on $\overrightarrow{DC}$ in the same way. Then draw $\triangle A'B'C'$.

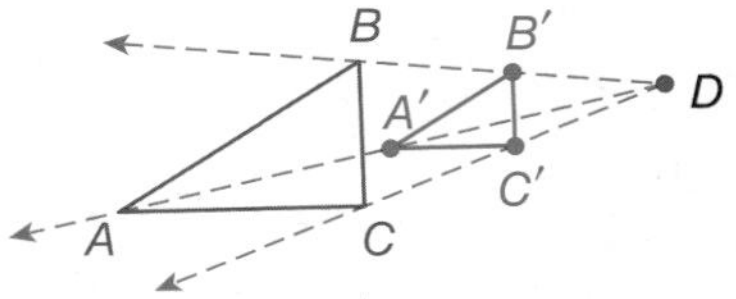

GuidedPractice

Copy the figure and point J. Then use a ruler to draw the image of the figure under a dilation with center J and the scale factor k indicated.

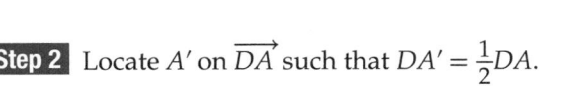

1A. $k = \frac{3}{2}$

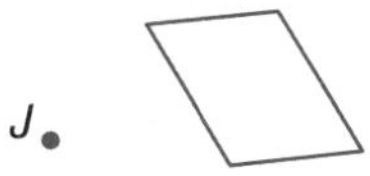

1B. $k = 0.75$

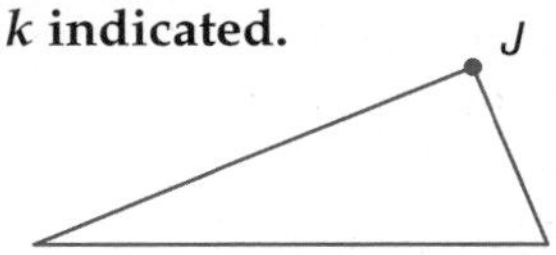

John Lund/Stone/Getty Images

In Lesson 14-3, you also learned that if $k > 1$, then the dilation is an *enlargement*. If $0 < k < 1$, then the dilation is a *reduction*. Since $\frac{1}{2}$ is between 0 and 1, the dilation in Example 1 is a reduction.

A dilation with a scale factor of 1 is called an *isometry dilation*. It produces an image that coincides with the preimage. The two figures are congruent.

PT

Real-World Example 2 Find the Scale Factor of a Dilation

PHOTOGRAPHY To create different-sized prints, you can adjust the distance between a film negative and the enlarged print by using a photographic enlarger. Suppose the distance between the light source *C* and the negative is 45 millimeters (*CP*). To what distance *PP′* should you adjust the enlarger to create a 22.75-centimeter wide print (*X′Y′*) from a 35-millimeter wide negative (*XY*)?

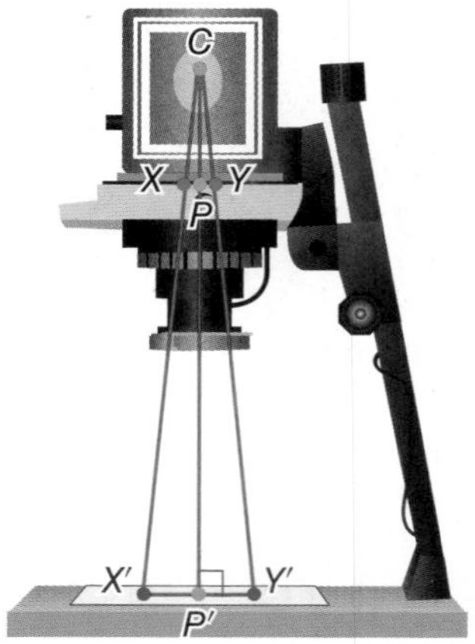

Understand This problem involves a dilation. The center of dilation is C, $XY = 35$ mm, $X'Y' = 22.75$ cm or 227.5 mm, and $CP = 45$ mm. You are asked to find PP'.

Plan Find the scale factor of the dilation from the preimage XY to the image $X'Y'$. Use the scale factor to find CP' and then use CP and CP' to find PP'.

Solve The scale factor k of the enlargement is the ratio of a length on the image to a corresponding length on the preimage.

$$k = \frac{\text{image length}}{\text{preimage length}} \quad \text{Scale factor of image}$$

$$= \frac{X'Y'}{XY} \quad \text{image} = X'Y', \text{ preimage} = XY$$

$$= \frac{227.5}{35} \text{ or } 6.5 \quad \text{Divide.}$$

Use this scale factor of 6.5 to find CP'.

$$CP' = k(CP) \quad \text{Definition of dilation}$$

$$= 6.5(45) \quad k = 6.5 \text{ and } CP = 45$$

$$= 292.5 \quad \text{Multiply.}$$

Use CP' and CP to find PP'.

$$CP + PP' = CP' \quad \text{Segment Addition}$$

$$45 + PP' = 292.5 \quad CP = 45 \text{ and } CP' = 292.5$$

$$PP' = 247.5 \quad \text{Subtract 45 from each side.}$$

So the enlarger should be adjusted so that the distance from the negative to the enlarged print (PP') is 247.5 millimeters or 24.75 centimeters.

Check Since the dilation is an enlargement, the scale factor should be greater than 1. Since $6.5 > 1$, the scale factor found is reasonable. ✓

> **Problem-Solving Tip**
>
> CCSS Perseverance
>
> To prevent careless errors in your calculations, estimate the answer to a problem before solving. In Example 2, you can estimate the scale factor of the dilation to be about $\frac{240}{40}$ or 6. Then CP' would be about $6 \cdot 50$ or 300 and PP' about $300 - 50$ or 250 millimeters, which is 25 centimeters. A measure of 24.75 centimeters is close to this estimate, so the answer is reasonable.

GuidedPractice

2. Determine whether the dilation from Figure Q to Q' is an *enlargement* or a *reduction*. Then find the scale factor of the dilation and x.

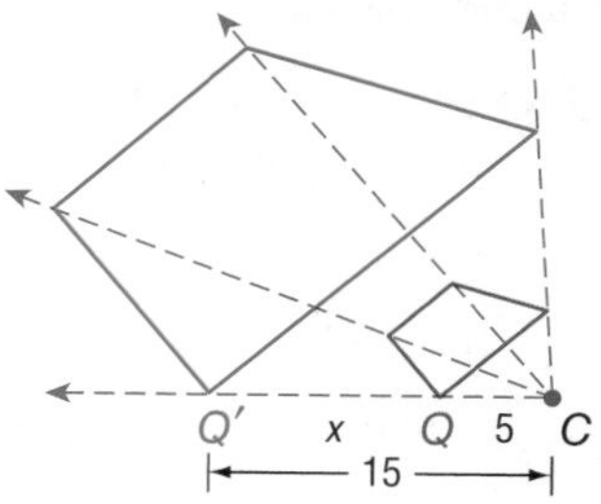

2 Dilations in the Coordinate Plane

You can use the following rules to find the image of a figure after a dilation centered at the origin.

StudyTip

Negative Scale Factors Dilations can also have negative scale factors. You will investigate this type of dilation in Exercise 36.

KeyConcept Dilations in the Coordinate Plane

Words To find the coordinates of an image after a dilation centered at the origin, multiply the x- and y-coordinates of each point on the preimage by the scale factor of the dilation, k.

Symbols $(x, y) \rightarrow (kx, ky)$

Example

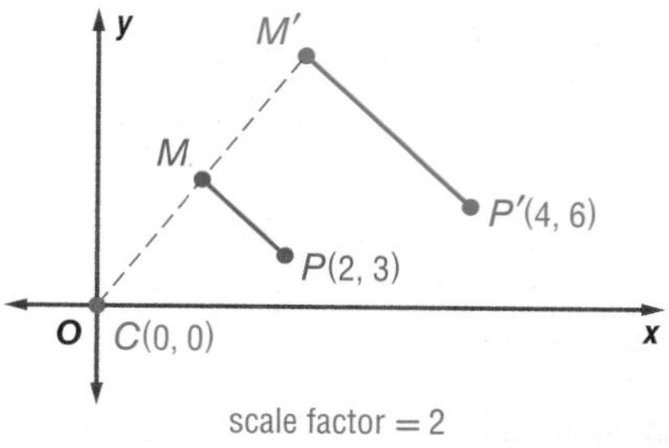

Example 3 Dilations in the Coordinate Plane

Quadrilateral $JKLM$ has vertices $J(-2, 4)$, $K(-2, -2)$, $L(-4, -2)$, and $M(-4, 2)$. Graph the image of $JKLM$ after a dilation centered at the origin with a scale factor of 2.5.

Multiply the x- and y-coordinates of each vertex by the scale factor, 2.5.

$(x, y) \rightarrow (2.5x, 2.5y)$

$J(-2, 4) \rightarrow J'(-5, 10)$

$K(-2, -2) \rightarrow K'(-5, -5)$

$L(-4, -2) \rightarrow L'(-10, -5)$

$M(-4, 2) \rightarrow M'(-10, 5)$

Graph $JKLM$ and its image $J'K'L'M'$.

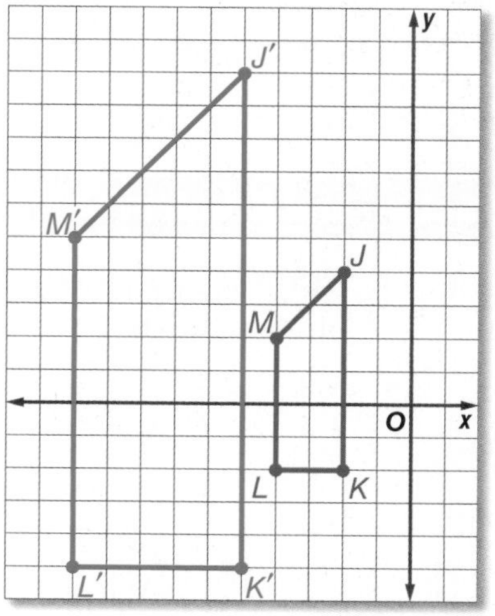

GuidedPractice

Find the image of each polygon with the given vertices after a dilation centered at the origin with the given scale factor.

3A. $Q(0, 6)$, $R(-6, -3)$, $S(6, -3)$; $k = \frac{1}{3}$

3B. $A(2, 1)$, $B(0, 3)$, $C(-1, 2)$, $D(0, 1)$; $k = 2$

Check Your Understanding

Example 1 **Copy the figure and point *M*. Then use a ruler to draw the image of the figure under a dilation with center *M* and the scale factor *k* indicated.**

1. $k = \frac{1}{4}$

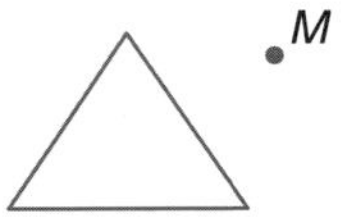

2. $k = 2$

Example 2

3 Determine whether the dilation from Figure *B* to *B′* is an *enlargement* or a *reduction*. Then find the scale factor of the dilation and *x*.

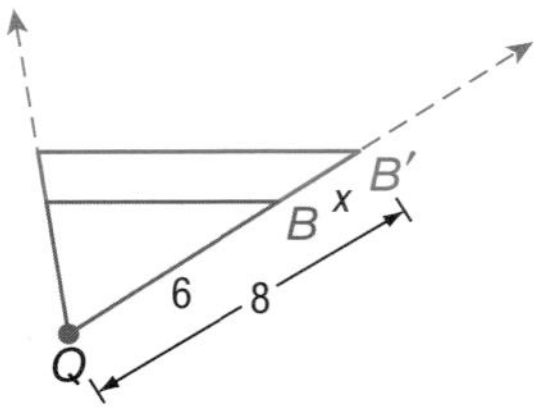

4. BIOLOGY Under a microscope, a single-celled organism 200 microns in length appears to be 50 millimeters long. If 1 millimeter = 1000 microns, what magnification setting (scale factor) was used? Explain your reasoning.

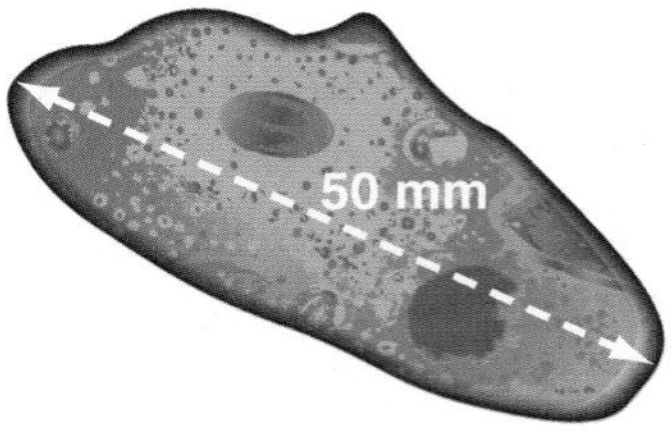

Example 3 **Graph the image of each polygon with the given vertices after a dilation centered at the origin with the given scale factor.**

5. $W(0, 0), X(6, 6), Y(6, 0); k = 1.5$

6. $Q(-4, 4), R(-4, -4), S(4, -4), T(4, 4); k = \frac{1}{2}$

7. $A(-1, 4), B(2, 4), C(3, 2), D(-2, 2); k = 2$

8. $J(-2, 0), K(2, 4), L(8, 0), M(2, -4); k = \frac{3}{4}$

Practice and Problem Solving

Example 1 **CCSS TOOLS** **Copy the figure and point *S*. Then use a ruler to draw the image of the figure under a dilation with center *S* and the scale factor *k* indicated.**

9. $k = \frac{5}{2}$

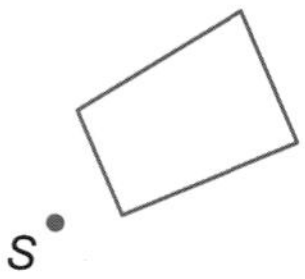

10. $k = 3$

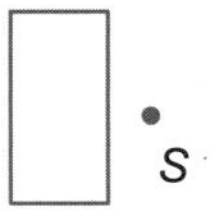

11. $k = 0.8$

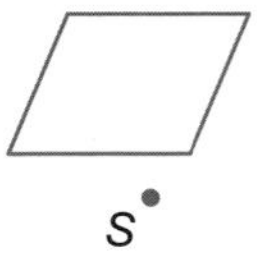

12. $k = \frac{1}{3}$

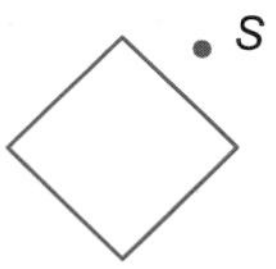

13. $k = 2.25$

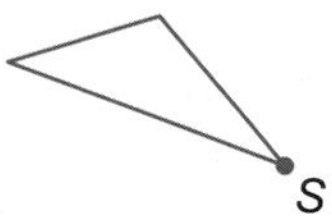

14. $k = \frac{7}{4}$

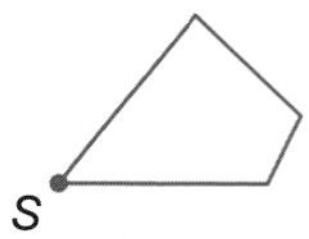

Example 2 **Determine whether the dilation from figure W to W' is an *enlargement* or a *reduction*. Then find the scale factor of the dilation and x.**

15.

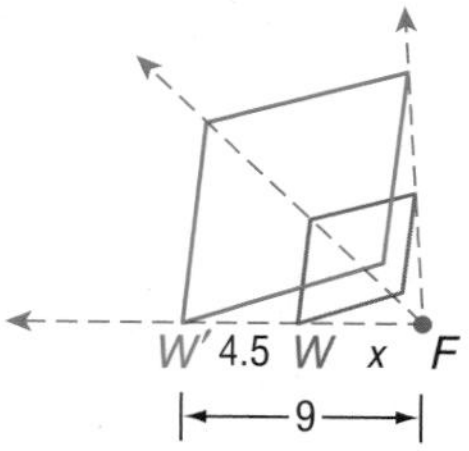

16.

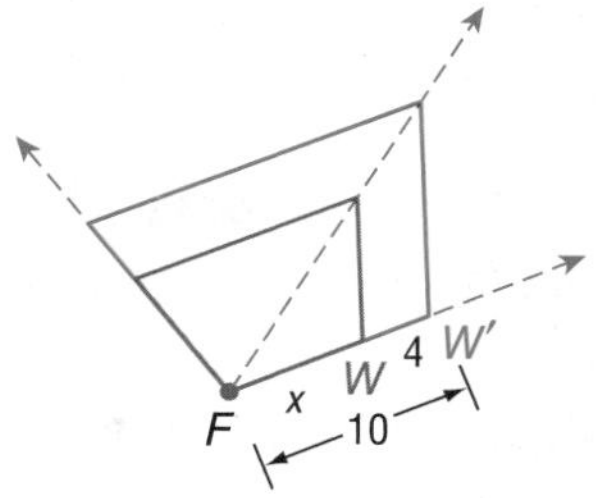

17.

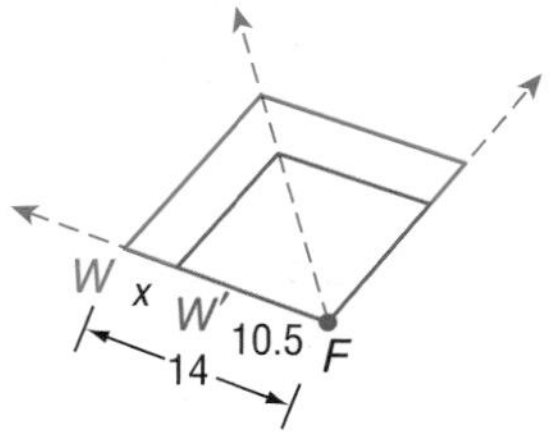

18.

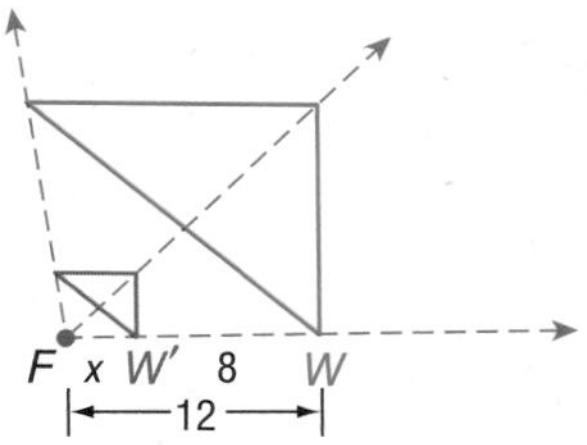

INSECTS **When viewed under a microscope, each insect has the measurement given on the picture. Given the actual measure of each insect, what magnification was used? Explain your reasoning.**

19.

Cat Flea
Actual Length: 2.5 mm

20.

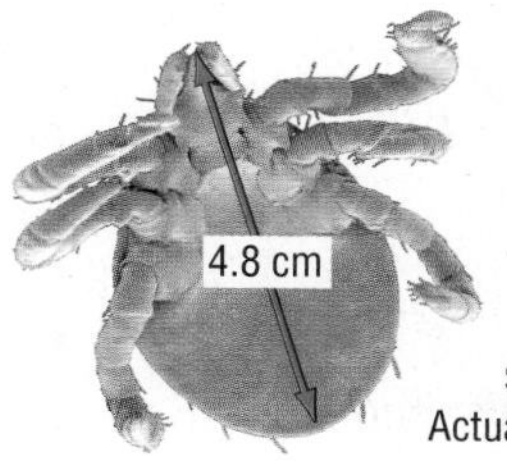

Spider Mite
Actual Length: 0.5 mm

Example 3 **CCSS SENSE-MAKING** **Find the image of each polygon with the given vertices after a dilation centered at the origin with the given scale factor.**

21 $J(-8, 0), K(-4, 4), L(-2, 0); k = 0.5$

22. $S(0, 0), T(-4, 0), V(-8, -8); k = 1.25$

23. $A(9, 9), B(3, 3), C(6, 0); k = \frac{1}{3}$

24. $D(4, 4), F(0, 0), G(8, 0); k = 0.75$

25. $M(-2, 0), P(0, 2), Q(2, 0), R(0, -2); k = 2.5$

26. $W(2, 2), X(2, 0), Y(0, 1), Z(1, 2); k = 3$

27. **COORDINATE GEOMETRY** Refer to the graph of $FGHJ$.

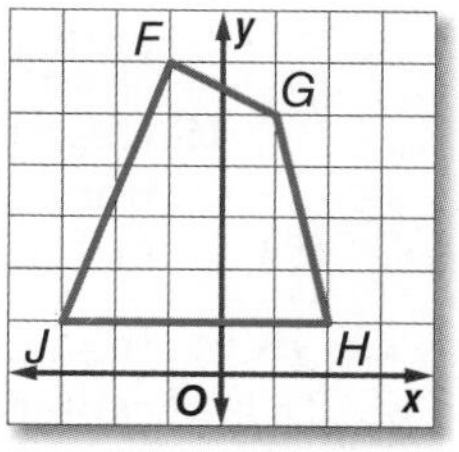

a. Dilate $FGHJ$ by a scale factor of $\frac{1}{2}$ centered at the origin, and then reflect the dilated image in the y-axis.

b. Complete the composition of transformations in part **a** in reverse order.

c. Does the order of the transformations affect the final image?

d. Will the order of a composition of a dilation and a reflection *always, sometimes,* or *never* affect the dilated image? Explain your reasoning.

28. **PHOTOGRAPHY AND ART** To make a scale drawing of a photograph, students overlay a $\frac{1}{4}$-inch grid on a 5-inch by 7-inch high contrast photo, overlay a $\frac{1}{2}$-inch grid on a 10-inch by 14-inch piece of drawing paper, and then sketch the image in each square of the photo to the corresponding square on the drawing paper.

a. What is the scale factor of the dilation?

b. To create an image that is 10 times as large as the original, what size grids are needed?

c. What would be the area of a grid drawing of a 5-inch by 7-inch photo that used 2-inch grids?

29. **MEASUREMENT** Determine whether the image shown is a dilation of *ABCD*. Explain your reasoning.

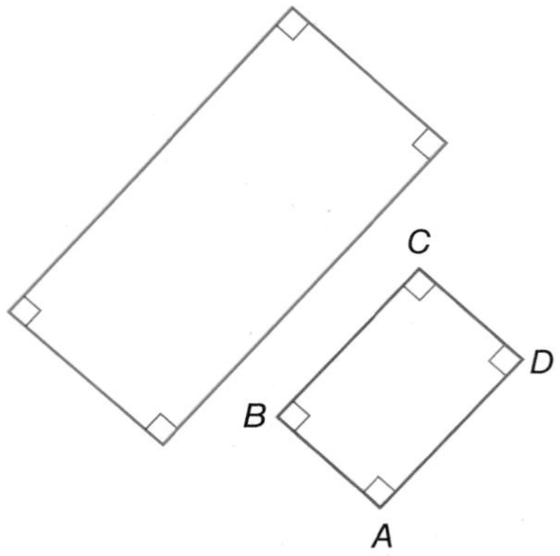

30. **COORDINATE GEOMETRY** *WXYZ* has vertices $W(6, 2)$, $X(3, 7)$, $Y(-1, 4)$, and $Z(4, -2)$.

a. Graph *WXYZ* and find the perimeter of the figure. Round to the nearest tenth.

b. Graph the image of *WXYZ* after a dilation of $\frac{1}{2}$ centered at the origin.

c. Find the perimeter of the dilated image. Round to the nearest tenth. How is the perimeter of the dilated image related to the perimeter of *WXYZ*?

31. **CHANGING DIMENSIONS** A three-dimensional figure can also undergo a dilation. Consider the rectangular prism shown.

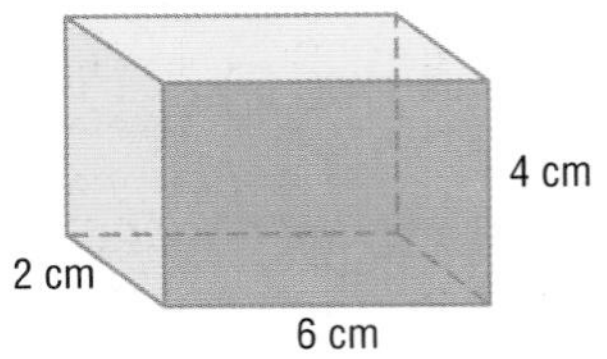

a. Find the surface area and volume of the prism.

b. Find the surface area and volume of the prism after a dilation with a scale factor of 2.

c. Find the surface area and volume of the prism after a dilation with a scale factor of $\frac{1}{2}$.

d. How many times as great is the surface area and volume of the image as the preimage after each dilation?

e. Make a conjecture as to the effect a dilation with a positive scale factor *r* would have on the surface area and volume of a prism.

32. **CCSS PERSEVERANCE** Refer to the graph of $\triangle DEF$.

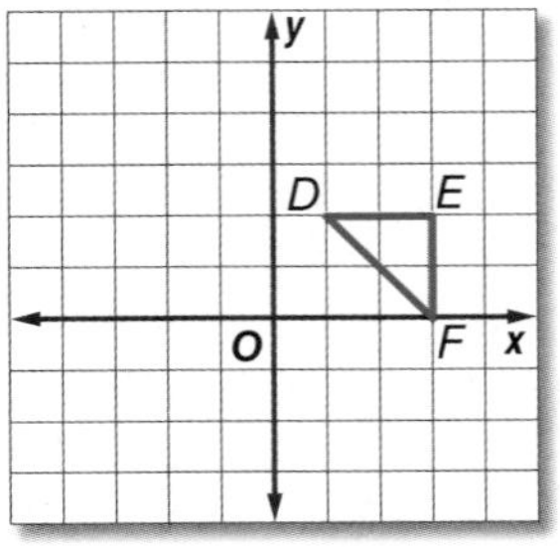

a. Graph the dilation of $\triangle DEF$ centered at point *D* with a scale factor of 3.

b. Describe the dilation as a composition of transformations including a dilation with a scale factor of 3 centered at the origin.

c. If a figure is dilated by a scale factor of 3 with a center of dilation (x, y), what composition of transformations, including a dilation with a scale factor of 3 centered at the origin, will produce the same final image?

33 **HEALTH** A coronary artery may be dilated with a balloon catheter as shown. The cross section of the middle of the balloon is a circle.

a. A surgeon inflates a balloon catheter in a patient's coronary artery, dilating the balloon from a diameter of 1.5 millimeters to 2 millimeters. Find the scale factor of this dilation.

b. Find the cross-sectional area of the balloon before and after the dilation.

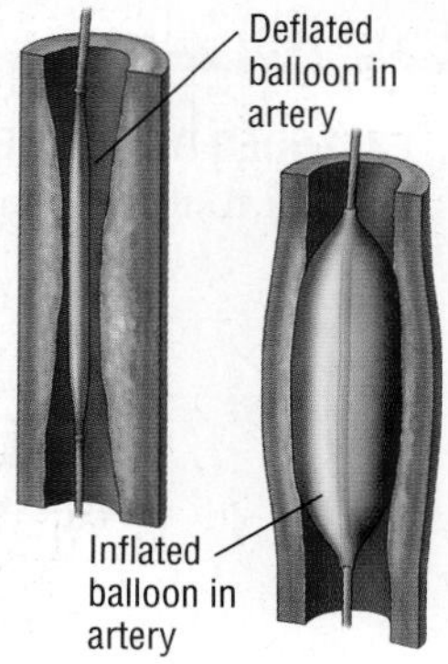

Each figure shows a preimage and its image after a dilation centered at point *P*. Copy each figure, locate point *P*, and estimate the scale factor.

34.

35.

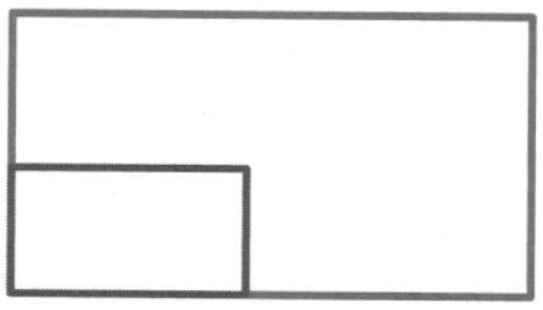

36. **MULTIPLE REPRESENTATIONS** In this problem, you will investigate dilations centered at the origin with negative scale factors.

a. Geometric Draw $\triangle ABC$ with points $A(-2, 0)$, $B(2, -4)$, and $C(4, 2)$. Then draw the image of $\triangle ABC$ after a dilation centered at the origin with a scale factor of -2. Repeat the dilation with scale factors of $-\frac{1}{2}$ and -3. Record the coordinates for each dilation.

b. Verbal Make a conjecture about the function relationship for a dilation centered at the origin with a negative scale factor.

c. Analytical Write the function rule for a dilation centered at the origin with a scale factor of $-k$.

d. Verbal Describe a dilation centered at the origin with a negative scale factor as a composition of transformations.

H.O.T. Problems Use Higher-Order Thinking Skills

37. **CHALLENGE** Find the equation for the dilated image of the line $y = 4x - 2$ if the dilation is centered at the origin with a scale factor of 1.5.

38. **WRITING IN MATH** Are parallel lines (parallelism) and collinear points (collinearity) preserved under all transformations? Explain.

39. **CCSS ARGUMENTS** Determine whether invariant points are *sometimes*, *always*, or *never* maintained for the transformations described below. If so, describe the invariant point(s). If not, explain why invariant points are not possible.

a. dilation of *ABCD* with scale factor 1

b. rotation of $\overline{AB}$ 74° about *B*

c. reflection of $\triangle MNP$ in the *x*-axis

d. translation of *PQRS* along $\langle 7, 3 \rangle$

e. dilation of $\triangle XYZ$ centered at the origin with scale factor 2

40. **OPEN ENDED** Graph a triangle. Dilate the triangle so that its area is four times the area of the original triangle. State the scale factor and center of your dilation.

41. **WRITING IN MATH** Can you use transformations to create congruent figures, similar figures, and equal figures? Explain.

Standardized Test Practice

42. EXTENDED RESPONSE Quadrilateral *PQRS* was dilated to form quadrilateral *WXYZ*.

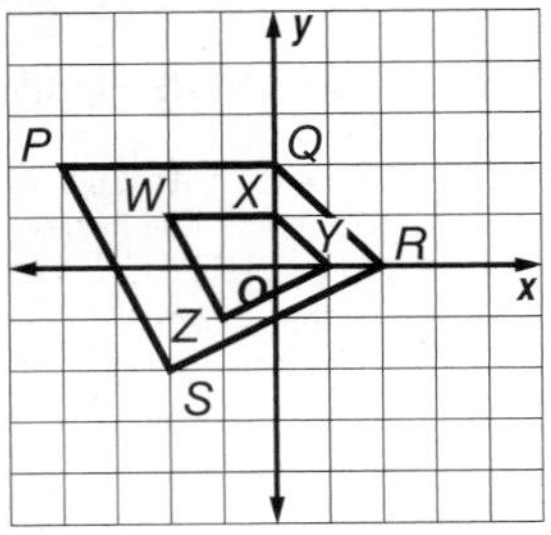

a. Is the dilation from *PQRS* to *WXYZ* an enlargement or reduction?

b. Which number *best* represents the scale factor for this dilation?

43. ALGEBRA How many ounces of pure water must a pharmacist add to 50 ounces of a 15% saline solution to make a solution that is 10% saline?

A 25 **C** 15

B 20 **D** 5

44. Tionna wants to replicate a painting in an art museum. The painting is 3 feet wide and 6 feet long. She decides on a dilation reduction factor of 0.25. What size paper should she use?

F 4 in. × 8 in. **H** 8 in. × 16 in.

G 6 in. × 12 in. **J** 10 in. × 20 in.

45. SAT/ACT For all x, $(x - 7)^2 = ?$

A $x^2 - 49$ **D** $x^2 - 14x + 49$

B $x^2 + 49$ **E** $x^2 + 14x - 49$

C $x^2 - 14x - 49$

Spiral Review

State whether the figure appears to have line symmetry. Write *yes* or *no*. If so, copy the figure, draw all lines of symmetry, and state their number. (Lesson 14-8)

46.

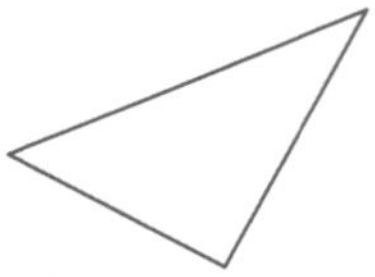

47.

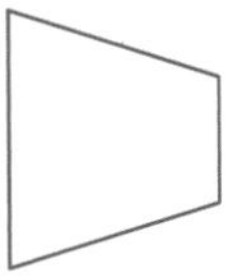

48.

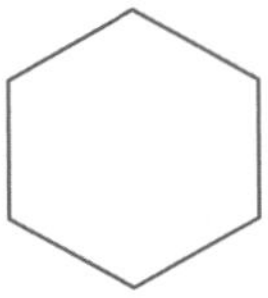

Describe the transformations that combined to map each figure. (Lesson 14-7)

49.

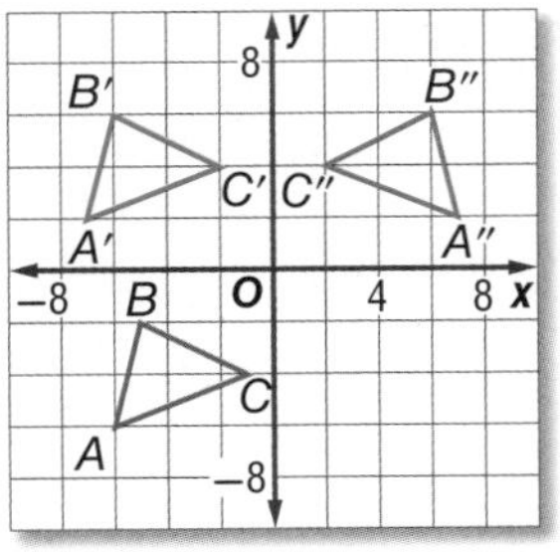

50.

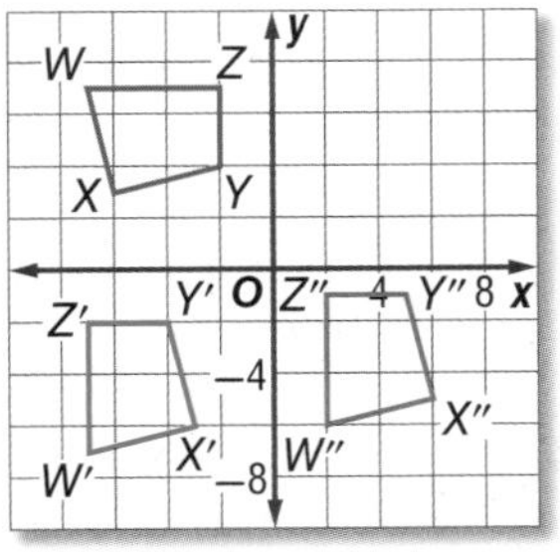

Skills Review

Find the value of x to the nearest tenth.

51. $58.9 = 2x$

52. $\frac{108.6}{\pi} = x$

53. $228.4 = \pi x$

54. $\frac{336.4}{x} = \pi$

EXTEND 14-9

Geometry Lab
Establishing Triangle Congruence and Similarity

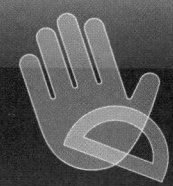

Two triangles are defined to be congruent if all of their corresponding parts are congruent and the criteria for proving triangle congruence (SAS, SSS, and ASA) are presented as postulates. Triangle congruence can also be defined in terms of rigid motions (reflections, translations, rotations).

The **principle of superposition** states that two figures are congruent if and only if there is a rigid motion or a series of rigid motions that maps one figure exactly onto the other. We can use the following assumed properties of rigid motions to establish the SAS, SSS, and ASA criteria for triangle congruence.

- The distance between points is preserved. Sides are mapped to sides of the same length.
- Angle measures are preserved. Angles are mapped to angles of the same measure.

CCSS Common Core State Standards
Content Standards
G.CO.8 Explain how the criteria for triangle congruence (ASA, SAS, and SSS) follow from the definition of congruence in terms of rigid motions.
G.SRT.3 Use the properties of similarity transformations to establish the AA criterion for two triangles to be similar.
Mathematical Practices 5

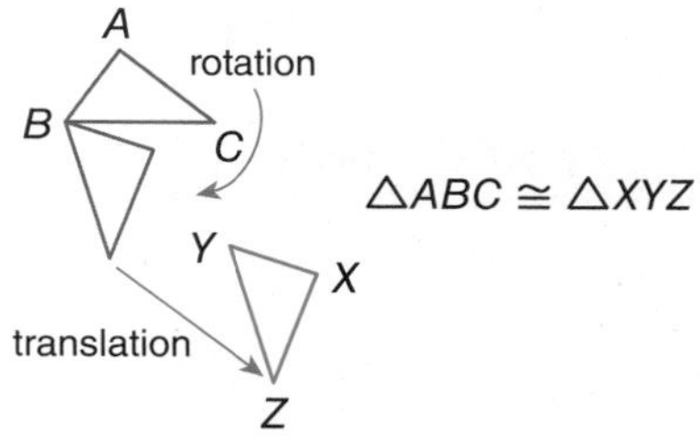

Activity 1 Establish Congruence

Use a rigid motion to map side $\overline{AB}$ of $\triangle ABC$ onto side $\overline{XY}$ of $\triangle XYZ$, $\angle A$ onto $\angle X$, and side $\overline{AC}$ onto side $\overline{XZ}$.

Step 1 Copy the triangles below onto a sheet of paper.

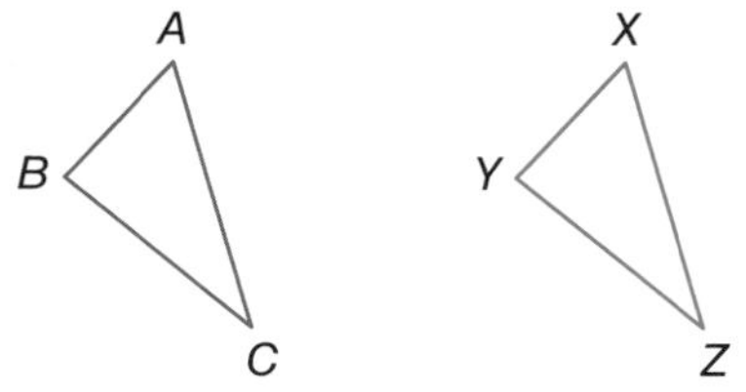

Step 2 Copy $\triangle ABC$ onto a sheet of tracing paper and label. Translate the paper until $\overline{AB}$, $\angle A$, and $\overline{AC}$ lie exactly on top of $\overline{XY}$, $\angle X$, and $\overline{XZ}$.

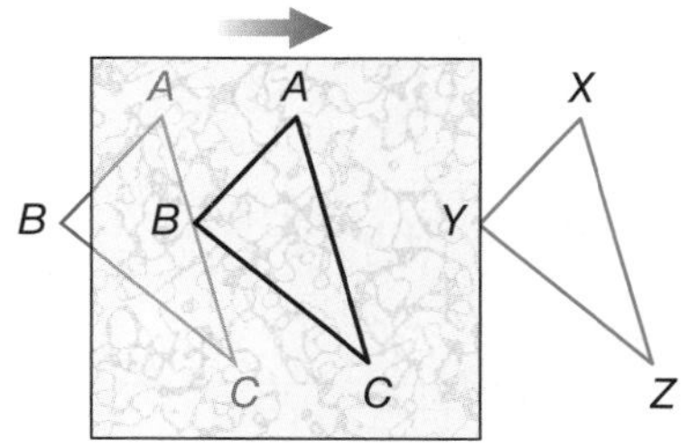

Analyze the Results

1. Use this activity to explain how the SAS criterion for triangle congruence follows from the definition of congruence in terms of rigid motions. (*Hint*: Extend lines on the tracing paper.)

2. Use the principle of superposition to explain why two triangles are congruent if and only if corresponding pairs of sides and corresponding pairs of angles are congruent.

Using the same triangles shown above, describe the steps in an activity to illustrate the indicated criterion for triangle congruence. Then explain how this criterion follows from the principle of superposition.

3. SSS

4. ASA

Geometry Lab

Establishing Triangle Congruence and Similarity *Continued*

Two figures are similar if there is a rigid motion, or a series of rigid motions, followed by a dilation, or vice versa, that map one figure exactly onto the other. We can use the following assumed properties of dilations to establish the AA criteria for triangle similarity.

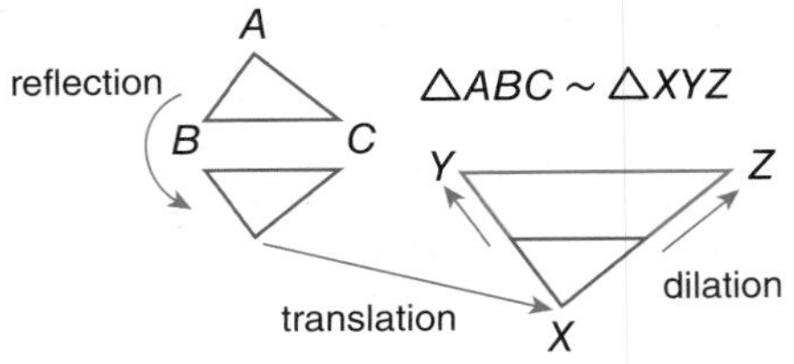

- Angle measures are preserved. Angles are mapped to angles of the same measure.
- Lines are mapped to parallel lines and sides are mapped to parallel sides that are longer or shorter in the ratio given by the scale factor.

Activity 2 Establish Similarity

Use a rigid motion followed by a dilation to map $\angle B$ onto $\angle Y$ and $\angle A$ onto $\angle X$.

Step 1 Copy the triangles below onto a sheet of paper.

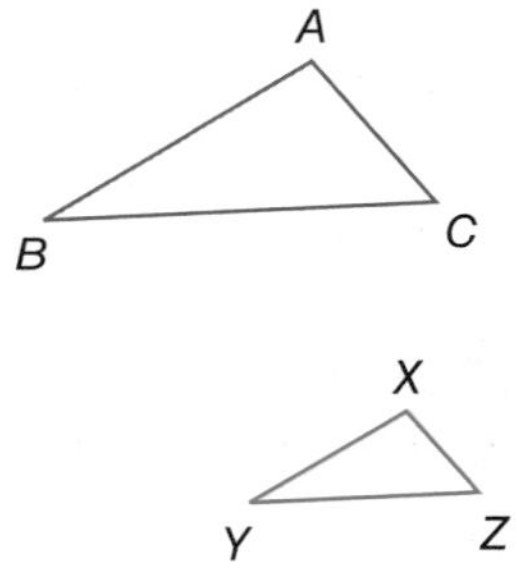

Step 2 Copy $\triangle ABC$ onto tracing paper and label.

Step 3 Translate the paper until $\angle B$ lies exactly on top of $\angle Y$. Tape this paper down so that it will not move.

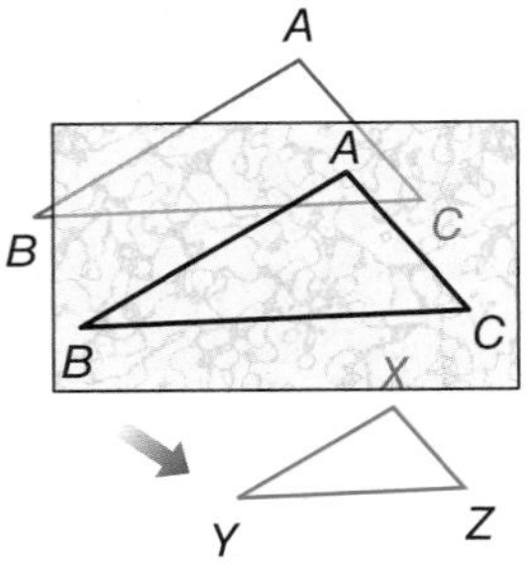

Step 4 On another sheet of tracing paper, copy and label $\angle A$.

Step 5 Translate this second sheet of tracing paper along the line from A to Y on the first sheet, until this second $\angle A$ lies exactly on top of $\angle X$.

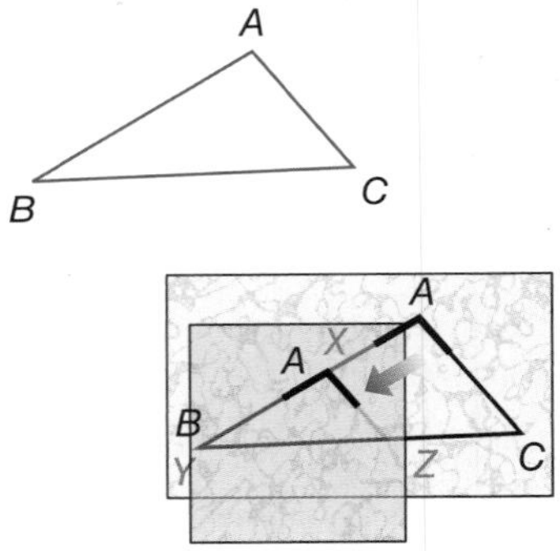

Analyze the Results

5. Use this activity to explain how the AA criterion for triangle similarity follows from the definition of similarity in terms of dilations. (*Hint*: Use parallel lines.)

6. Use the definition of similarity in terms of transformations to explain why two triangles are similar if all corresponding pairs of angles are congruent and all corresponding pairs of sides are proportional.

Use a series of rigid motions and/or dilations to determine whether $\triangle ABC$ and $\triangle XYZ$ are *congruent, similar,* or *neither*.

7.

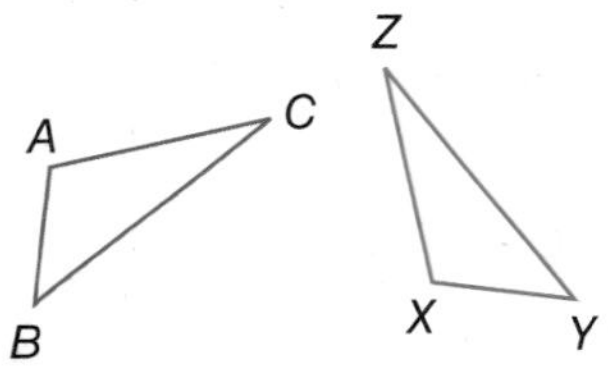

8.

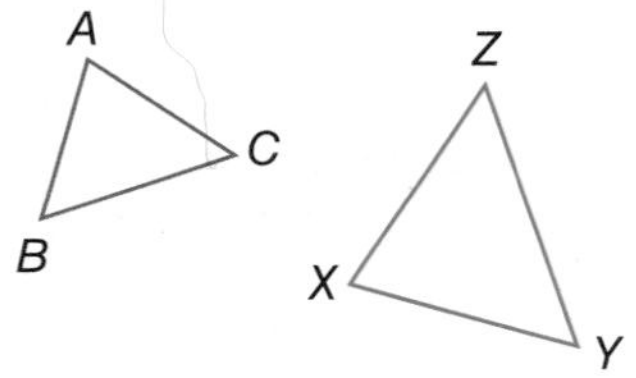

9.

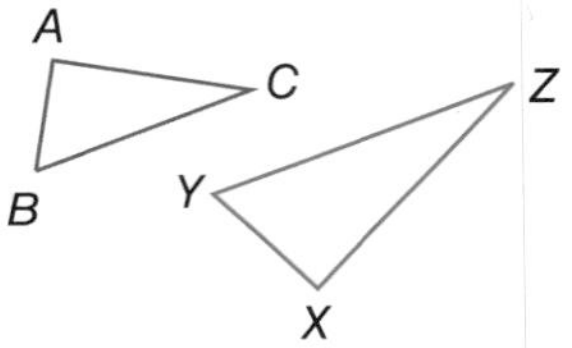

Study Guide and Review

Study Guide

KeyConcepts

Similar Triangles (Lesson 14-1)

- Two Triangles are similar if:
  AA: Two angles of one triangle are congruent to two angles of other triangle.
  SSS: The measures of the corresponding sides of the two triangles are proportional.
  SAS: The measures of two sides of one triangle are proportional to the measures of two corresponding sides of another triangle and their included angles are congruent.

Proportional Parts (Lesson 14-2)

- If a line is parallel to one side of a triangle and intersects the other two sides in two distinct points, then it separates these sides into segments of proportional length.
- A midsegment of a triangle is parallel to one side of the triangle and its length is one-half the length of that side.

Reflections (Lesson 14-4)

- A reflection is a transformation representing a flip of a figure over a point, line, or plane.

Translations (Lesson 14-5)

- A translation is a transformation that moves all points of a figure the same distance in the same direction.
- A translation maps each point to its image along a translation vector.

Rotations (Lesson 14-6)

- A rotation turns each point in a figure through the same angle about a fixed point.

Compositions of Transformations (Lesson 14-7)

- A translation can be represented as a composition of reflections in parallel lines and a rotation can be represented as a composition of reflections in intersecting lines.

Symmetry (Lesson 14-8)

- The line of symmetry in a figure is a line where the figure could be folded in half so that the two halves match exactly.
- The number of times a figure maps onto itself as it rotates from 0° to 360° is called the order of symmetry.
- The magnitude of symmetry is the smallest angle through which a figure can be rotated so that it maps onto itself.

Dilations (Lesson 14-9)

- Dilations enlarge or reduce figures proportionally.

KeyVocabulary

angle of rotation (p. 908)
axis symmetry (p. 932)
center of rotation (p. 908)
composition of transformation (p. 918)
dilation (p. 883)
enlargement (p. 883)
glide reflection (p. 918)
line of reflection (p. 890)
line of symmetry (p. 930)
line symmetry (p. 930)
magnitude of symmetry (p. 931)
midsegment of a triangle (p. 874)
order of symmetry (p. 931)
plane symmetry (p. 932)
reduction (p. 883)
rotational symmetry (p. 931)
similarity transformation (p. 883)
translation vector (p. 899)

VocabularyCheck

Choose the term that best completes each sentence.

1. When a transformation is applied to a figure, and then another transformation is applied to its image, this is a(n) (composition of transformations, order of symmetries).
2. If a figure is folded across a straight line and the halves match exactly, the fold line is called the (line of reflection, line of symmetry).
3. A (dilation, glide reflection) enlarges or reduces a figure proportionally.
4. The number of times a figure maps onto itself as it rotates from 0° to 360° is called the (magnitude of symmetry, order of symmetry).
5. A (line of reflection, translation vector) is the same distance from each point of a figure and its image.
6. A figure has (a center of rotation, symmetry) if it can be mapped onto itself by a rigid motion.
7. A glide reflection includes both a reflection and a (rotation, translation).
8. To rotate a point (90°, 180°) counterclockwise about the origin, multiply the *y*-coordinate by −1 and then interchange the *x*- and *y*-coordinates.
9. A (vector, reflection) is a congruence transformation.

Lesson-by-Lesson Review

14-1 Similar Triangles

Determine whether the triangles are similar. If so, write a similarity statement. Explain your reasoning.

10.

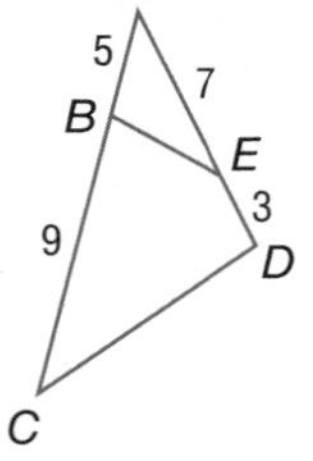

11.

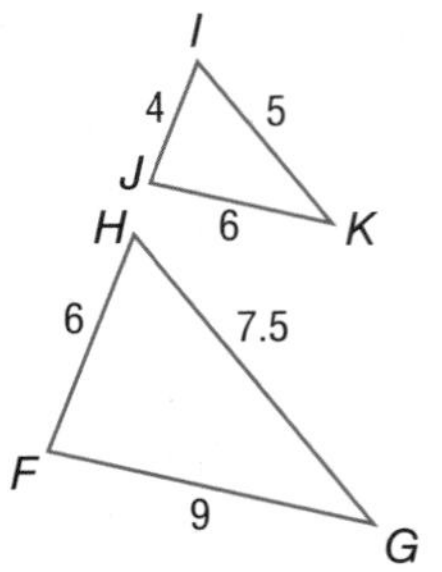

12.

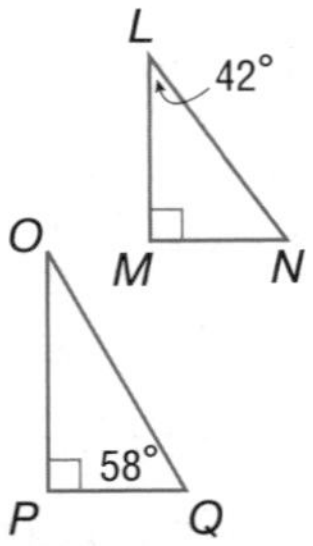

13.

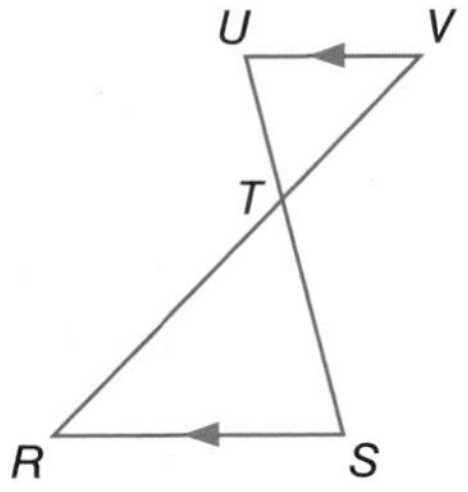

14. **TREES** To estimate the height of a tree, Dave stands in the shadow of the tree so that his shadow and the tree's shadow end at the same point. Dave is 6 feet 4 inches tall and his shadow is 15 feet long. If he is standing 66 feet away from the tree, what is the height of the tree?

Example 1

Determine whether the triangles are similar. If so, write a similarity statement. Explain your reasoning.

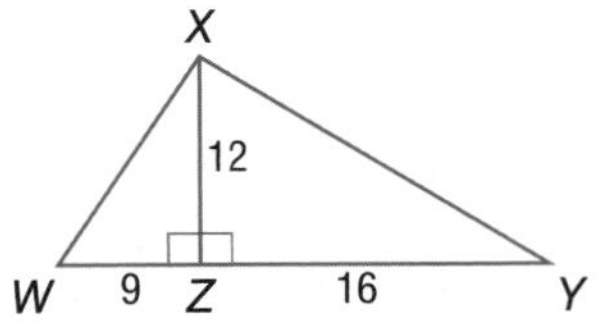

$\angle WZX \cong \angle XZY$ because they are both right angles. Now compare the ratios of the legs of the right triangles.

$$\frac{WZ}{XZ} = \frac{9}{12} = \frac{3}{4} \qquad \frac{XZ}{YZ} = \frac{12}{16} = \frac{3}{4}$$

Since two pairs of sides are proportional with the included angles congruent, $\triangle WZX \sim \triangle XZY$ by SAS Similarity.

14-2 Parallel Lines and Proportional Parts

Find *x*.

15.

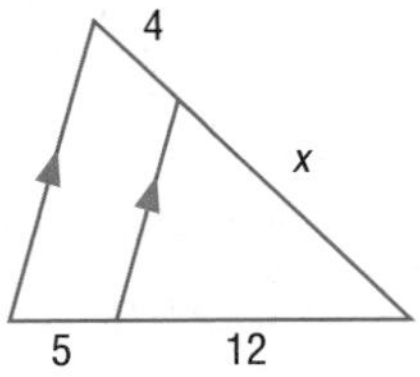

16.

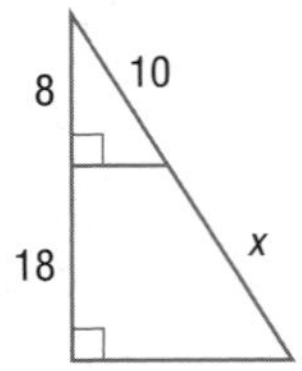

17. **STREETS** Find the distance along Broadway between 37th Street and 36th Street.

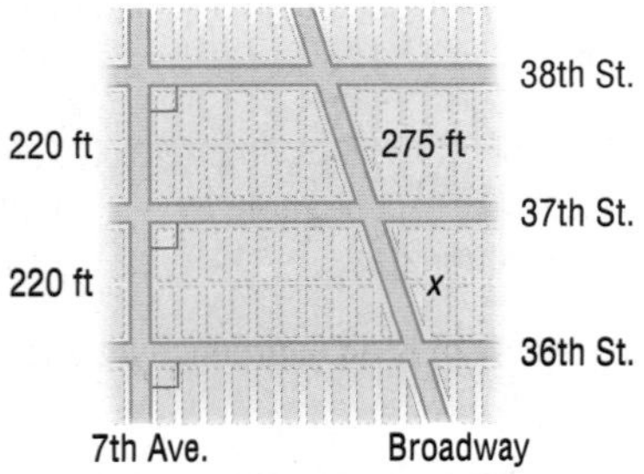

Example 2

ALGEBRA **Find *x* and *y*.**

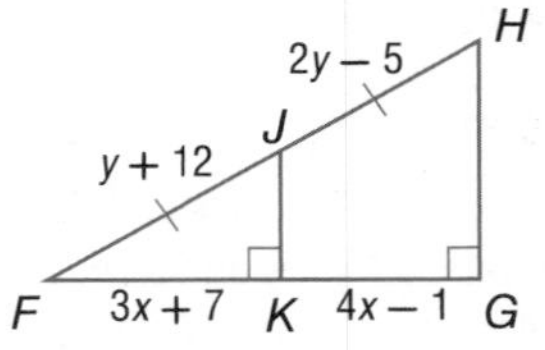

$FK = KG$

$3x + 7 = 4x - 1$

$-x = -8$

$x = 8$

$FJ = JH$	Definition of congruence
$y + 12 = 2y - 5$	Substitution
$-y = -17$	Subtract.
$y = 17$	Simplify.

14-3 Similarity Transformations

Determine whether the dilation from *A* to *B* is an *enlargement* or a *reduction*. Then find the scale factor of the dilation.

18.

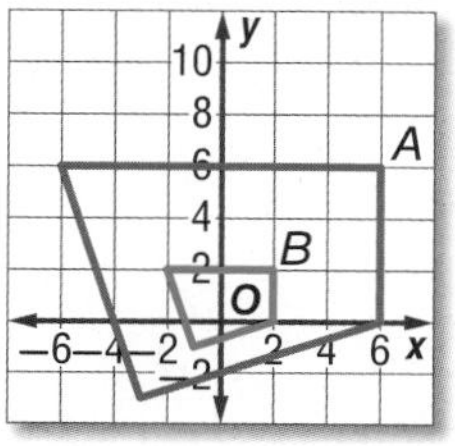

19.

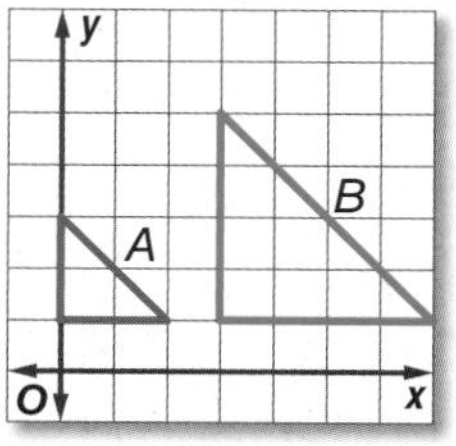

20. GRAPHIC DESIGN Jamie wants to use a photocopier to enlarge her design for the Honors Program at her school. She sets the copier to 250%. If the original drawing was 6 inches by 9 inches, find the dimensions of the enlargement.

Example 3

Determine whether the dilation from *A* to *B* is an *enlargement* or a *reduction*. Then find the scale factor of the dilation.

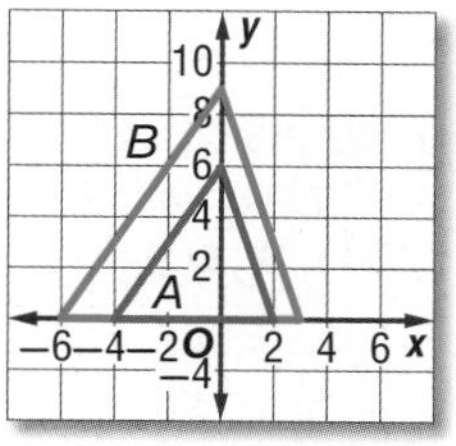

B is larger than *A*, so the dilation is an enlargement. The distance between the vertices at (−4, 0) and (2, 0) for *A* is 6 and the distance between the vertices at (−6, 0) and (3, 0) for *B* is 9. So the scale factor is $\frac{9}{6}$ or $\frac{3}{2}$.

14-4 Reflections

Graph each figure and its image under the given reflection.

21. rectangle *ABCD* with *A*(2, −4), *B*(4, −6), *C*(7, −3), and *D*(5, −1) in the *x*-axis

22. triangle *XYZ* with *X*(−1, 1), *Y*(−1, −2), and *Z*(3, −3) in the *y*-axis

23. quadrilateral *QRST* with *Q*(−4, −1), *R*(−1, 2), *S*(2, 2), and *T*(0, −4) in the line $y = x$

24. ART Anita is making the two-piece sculpture shown for a memorial garden. In her design, one piece of the sculpture is a reflection of the other, to be placed beside a sidewalk that would be located along the line of reflection. Copy the figures and draw the line of reflection.

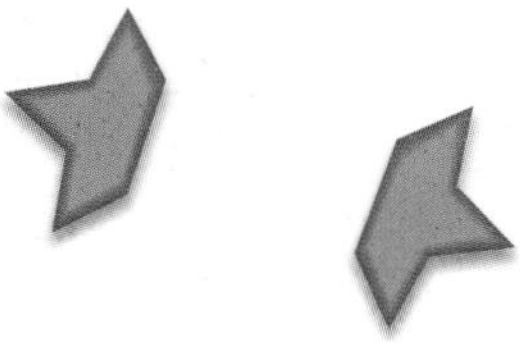

Example 4

Graph △*JKL* with vertices *J*(1, 4), *K*(2, 1), and *L*(6, 2) and its reflected image in the *x*-axis.

Multiply the *y*-coordinate of each vertex by −1.

(*x*, *y*)	→	(*x*, −*y*)
J(1, 4)	→	*J*′(1, −4)
K(2, 1)	→	*K*′(2, −1)
L(6, 2)	→	*L*′(6, −2)

Graph △*JKL* and its image △*J*′*K*′*L*′.

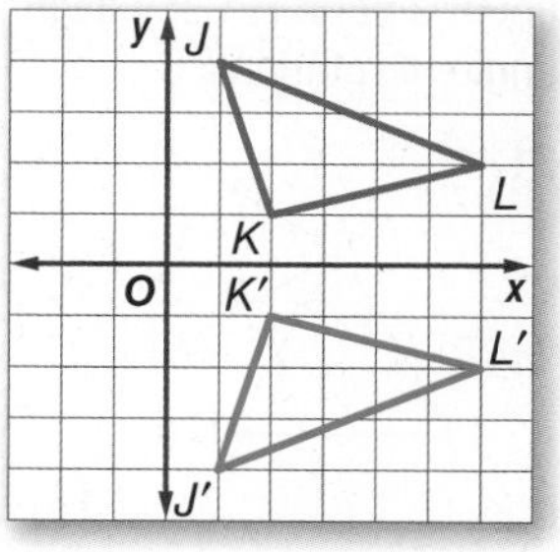

Study Guide and Review *Continued*

14-5 Translations

25. Graph $\triangle ABC$ with vertices $A(0, -1)$, $B(2, 0)$, $C(3, -3)$ and its image along $\langle -5, 4 \rangle$.

26. Copy the figure and the given translation vector. Then draw the translation of the figure along the translation vector.

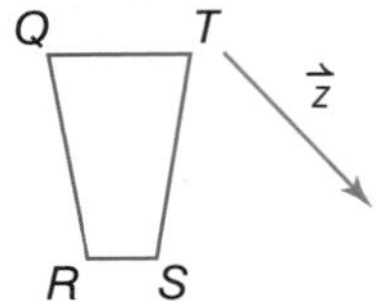

27. **DANCE** Five dancers are positioned onstage as shown. Dancers B, F, and C move along $\langle 0, -2 \rangle$, while dancer A moves along $\langle 5, -1 \rangle$. Draw the dancers' final positions.

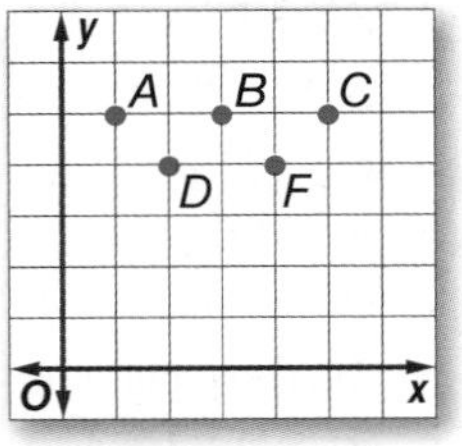

Example 5

Graph $\triangle XYZ$ with vertices $X(2, 2)$, $Y(5, 5)$, $Z(5, 3)$ and its image along $\langle -3, -5 \rangle$.

The vector indicates a translation 3 units left and 5 units down.

$(x, y) \rightarrow (x - 3, y - 5)$

$X(2, 2) \rightarrow X'(-1, -3)$

$Y(5, 5) \rightarrow Y'(2, 0)$

$Z(5, 3) \rightarrow Z'(2, -2)$

Graph $\triangle XYZ$ and its image $\triangle X'Y'Z'$.

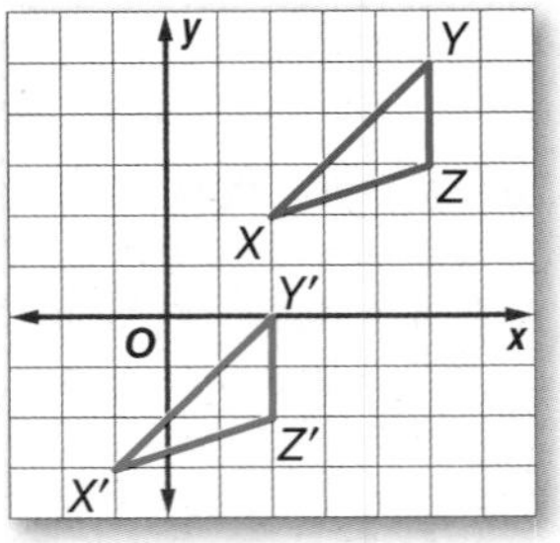

14-6 Rotations

28. Copy trapezoid $CDEF$ and point P. Then use a protractor and ruler to draw a 50° rotation of $CDEF$ about point P.

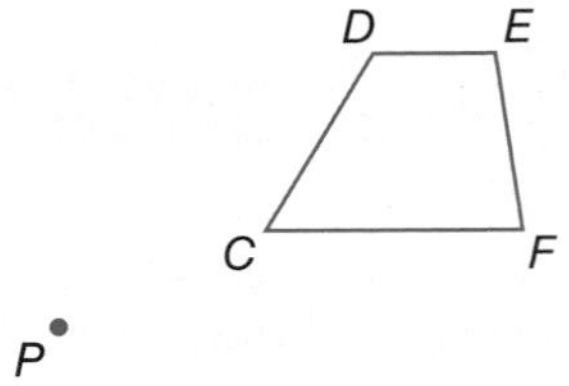

Graph each figure and its image after the specified rotation about the origin.

29. $\triangle MNO$ with vertices $M(-2, 2)$, $N(0, -2)$, $O(1, 0)$; 180°

30. $\triangle DGF$ with vertices $D(1, 2)$, $G(2, 3)$, $F(1, 3)$; 90°

Each figure shows a preimage and its image after a rotation about a point P. Copy each figure, locate point P, and find the angle of rotation.

31.

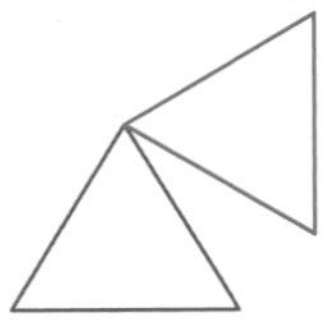

32.

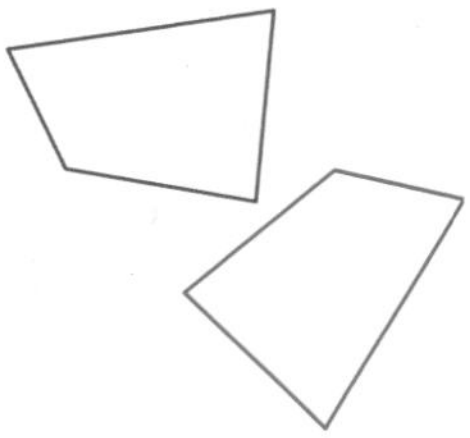

Example 6

Triangle ABC has vertices $A(-4, 0)$, $B(-3, 4)$, and $C(-1, 1)$. Graph $\triangle ABC$ and its image after a rotation 270° about the origin.

One method to solve this is to combine a 180° rotation with a 90° rotation. Multiply the x- and y-coordinates of each vertex by -1.

$(x, y) \rightarrow (-x, -y)$

$A(-4, 0) \rightarrow A'(4, 0)$

$B(-3, 4) \rightarrow B'(3, -4)$

$C(-1, 1) \rightarrow C'(1, -1)$

Multiply the y-coordinate of each vertex by -1 and interchange.

$(-x, -y) \rightarrow (y, -x)$

$A'(4, 0) \rightarrow A''(0, 4)$

$B'(3, -4) \rightarrow B''(4, 3)$

$C'(1, -1) \rightarrow C''(1, 1)$

Graph $\triangle ABC$ and its image $\triangle A''B''C''$.

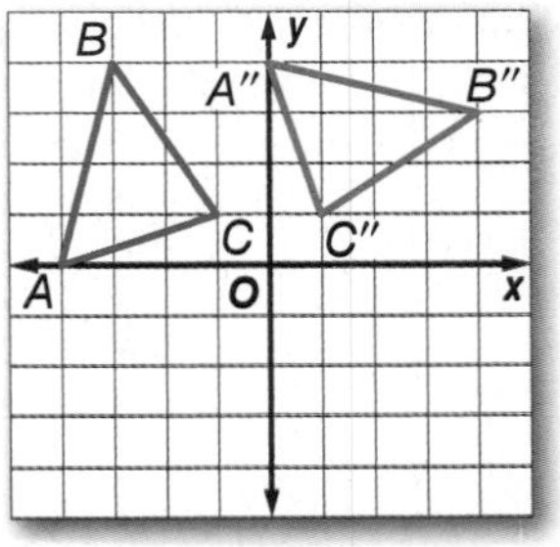

14-7 Compositions of Transformations

Graph each figure with the given vertices and its image after the indicated transformation.

33. $\overline{CD}$: $C(3, 2)$ and $D(1, 4)$
Reflection: in $y = x$
Rotation: 270° about the origin.

34. $\overline{GH}$: $G(-2, -3)$ and $H(1, 1)$
Translation: along $\langle 4, 2 \rangle$
Reflection: in the x-axis

35. **PATTERNS** Jeremy is creating a pattern for the border of a poster using a stencil. Describe the transformation combination that he used to create the pattern below.

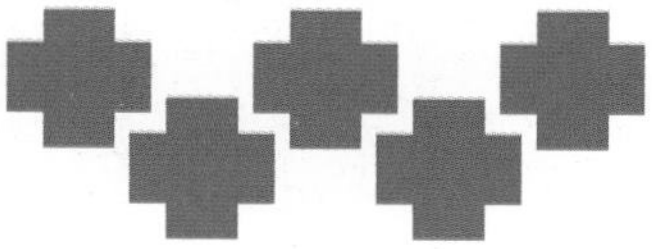

36. Copy and reflect figure T in line ℓ and then line m. Then describe a single transformation that maps T onto T''.

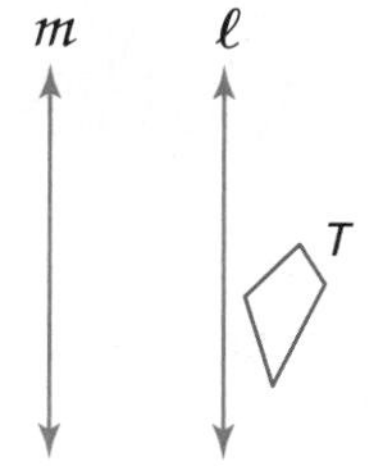

Example 7

The endpoints of $\overline{RS}$ are $R(4, 3)$ and $S(1, 1)$. Graph $\overline{RS}$ and its image after a translation along $\langle -5, -1 \rangle$ and a rotation 180° about the origin.

Step 1 translation along $\langle -5, -1 \rangle$

(x, y)	$\rightarrow$	$(x - 5, y - 1)$
$R(4, 3)$	$\rightarrow$	$R'(-1, 2)$
$S(1, 1)$	$\rightarrow$	$S'(-4, 0)$

Step 2 rotation 180° about origin

(x, y)	$\rightarrow$	$(-x, -y)$
$R'(-1, 2)$	$\rightarrow$	$R''(1, -2)$
$S'(-4, 0)$	$\rightarrow$	$S''(4, 0)$

Step 3 Graph $\overline{RS}$ and its image $\overline{R''S''}$.

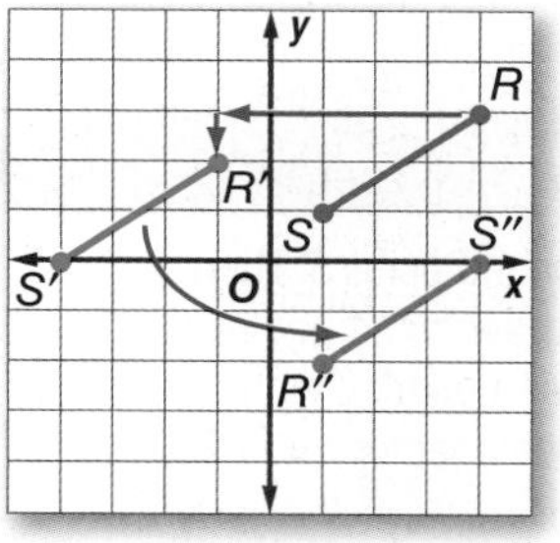

14-8 Symmetry

State whether each figure appears to have line symmetry. Write *yes* or *no*. If so, copy the figure, draw all lines of symmetry, and state their number.

37.

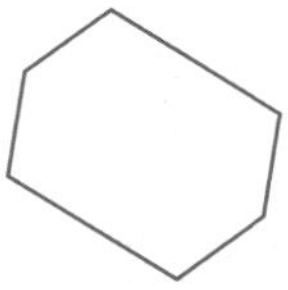

38.

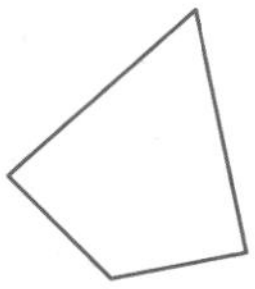

State whether each figure has rotational symmetry. Write *yes* or *no*. If so, copy the figure, locate the center of symmetry, and state the order and magnitude of symmetry.

39.

40.

Example 8

State whether each figure has *plane* symmetry, *axis* symmetry, *both*, or *neither*.

a.

The light bulb has both plane and axis symmetry.

41. KNITTING Amy is creating a pattern for a scarf she is knitting for her friend. How many lines of symmetry are there in the pattern?

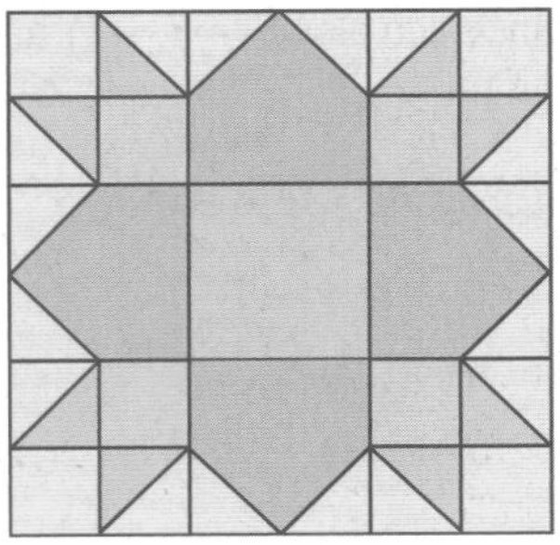

b.

The prism has plane symmetry.

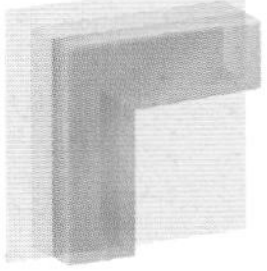

14-9 Dilations

42. Copy the figure and point S. Then use a ruler to draw the image of the figure under a dilation with center S and scale factor $r = 1.25$.

43. Determine whether the dilation from figure W to W' is an *enlargement* or a *reduction*. Then find the scale factor of the dilation and x.

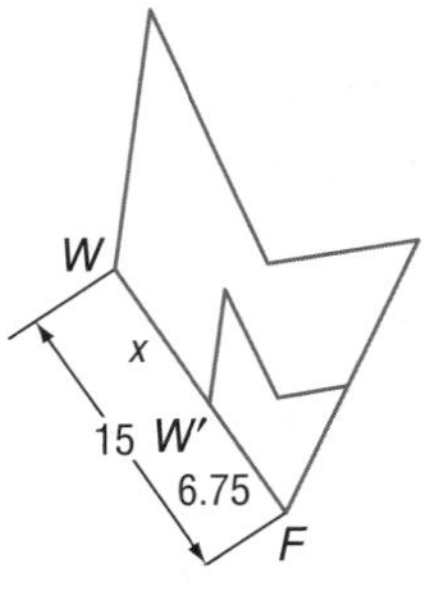

44. CLUBS The members of the Math Club use an overhead projector to make a poster. If the original image was 6 inches wide, and the image on the poster is 4 feet wide, what is the scale factor of the enlargement?

Example 9

Square $ABCD$ has vertices $A(0, 0)$, $B(0, 8)$, $C(8, 8)$, and $D(8, 0)$. Find the image of $ABCD$ after a dilation centered at the origin with a scale factor of 0.5.

Multiply the x- and y-coordinates of each vertex by the scale factor, 0.5.

$(x, y) \rightarrow (0.5x, 0.5y)$

$A(0, 0) \rightarrow A'(0, 0)$

$B(0, 8) \rightarrow B'(0, 4)$

$C(8, 8) \rightarrow C'(4, 4)$

$D(8, 0) \rightarrow D'(4, 0)$

Graph $ABCD$ and its image $A'B'C'D'$.

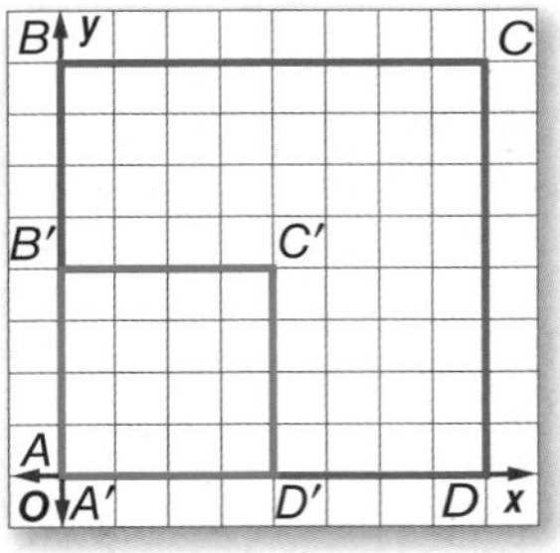

CHAPTER 14 Practice Test

Copy the figure and the given line of reflection. Then draw the reflected image in this line using a ruler.

1.

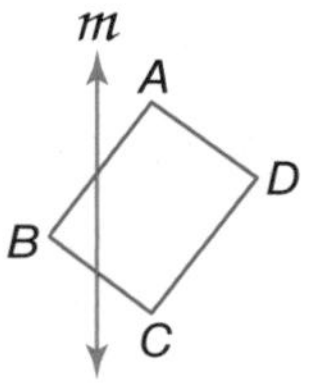

2.

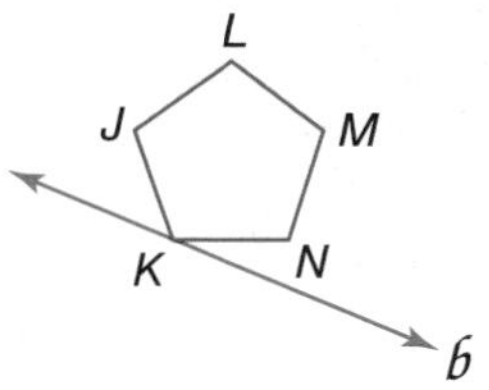

Find x.

3.

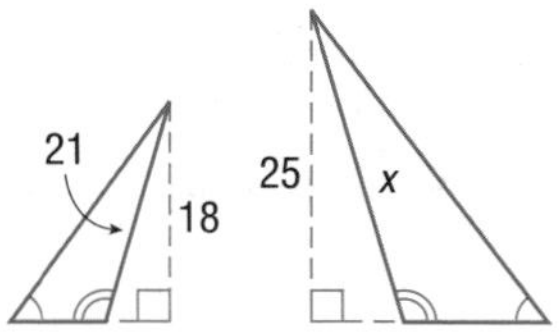

4.

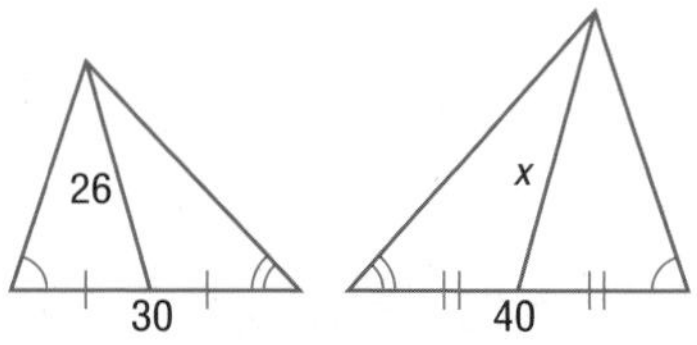

Copy the figure and point M. Then use a ruler to draw the image of the figure under a dilation with center M and the scale factor r indicated.

5. $r = 1.5$

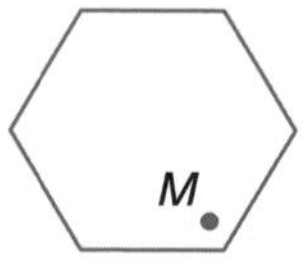

6. $r = \frac{1}{3}$

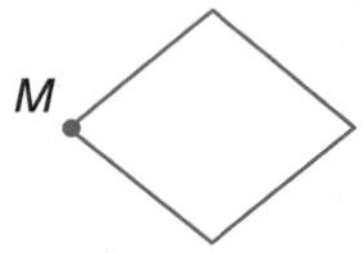

7. PARKS Isabel is on a ride at an amusement park that slides the rider to the right, and then rotates counterclockwise about its own center 60° every 2 seconds. How many seconds pass before Isabel returns to her starting position?

State whether each figure has *plane* symmetry, *axis* symmetry, *both*, or *neither*.

8.

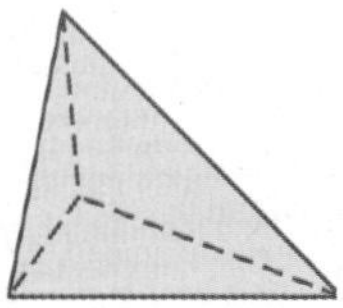

9.

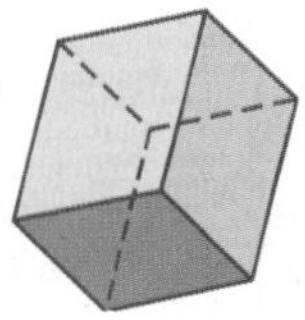

Graph each figure and its image under the given transformation.

10. ▱$FGHJ$ with vertices $F(-1, -1)$, $G(-2, -4)$, $H(1, -4)$, and $J(2, -1)$ in the x-axis

11. $\triangle ABC$ with vertices $A(0, -1)$, $B(2, 0)$, $C(3, -3)$; $\langle -5, 4 \rangle$

12. quadrilateral $WXYZ$ with vertices $W(2, 3)$, $X(1, 1)$, $Y(3, 0)$, $Z(5, 2)$; 180° about the origin

Copy the figure and the given translation vector. Then draw the translation of the figure along the translation vector.

13.

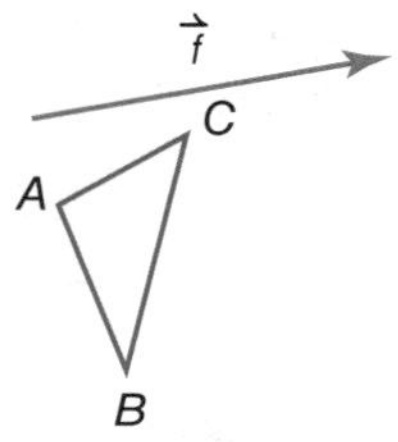

14.

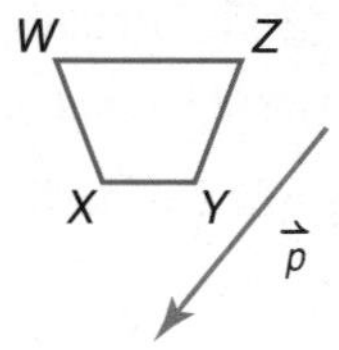

Determine whether the dilation from A to B is an *enlargement* or a *reduction*. Then find the scale factor of the dilation.

15.

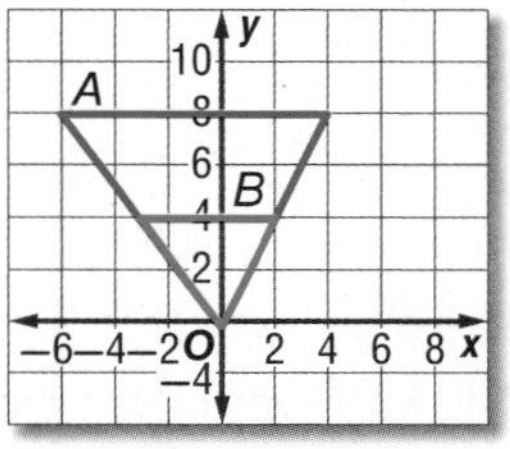

16.

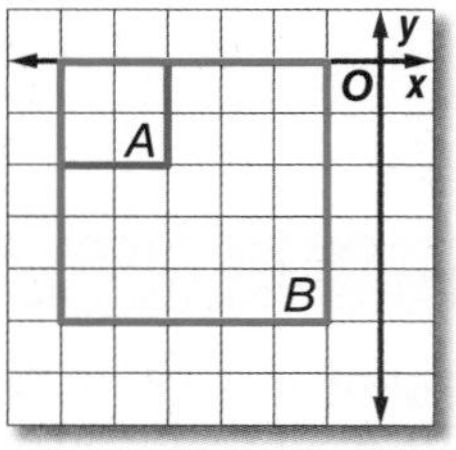

17. ALGEBRA Identify the similar triangles. Find WZ and UZ.

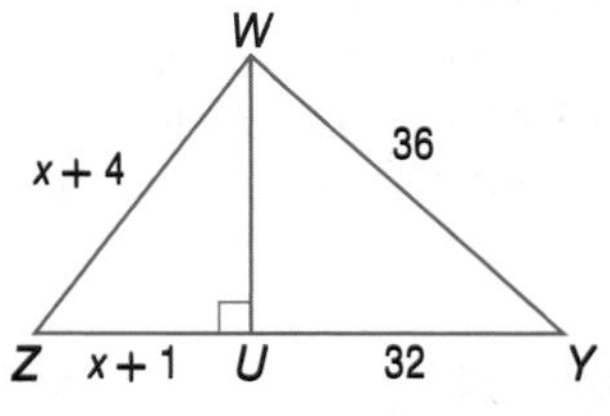

Preparing for Standardized Tests

Work Backward

In most problems, a set of conditions or facts is given and you must find the end result. However, some problems give you the end result and ask you to find something that happened earlier in the process. To solve problems like this, you must work backward.

Strategies for Working Backward

Step 1

Look for keywords that indicate you will need to work backward to solve the problem.

Sample Keywords:

- What was the **original**...?
- What was the value **before**...?
- Where was the **starting** or **beginning**...?

Step 2

Undo the steps given in the problem statement to solve.

- List the sequence of steps from the beginning to the end result.
- Begin with the end result. Retrace the steps in reverse order.
- "Undo" each step using inverses to get back to the original value.

Step 3

Check your solution if time permits.

- Make sure your answer makes sense.
- Begin with your answer and follow the steps in the problem statement forward to see if you get the same end result.

Standardized Test Example

Solve the problem below. Responses will be graded using the short-response scoring rubric shown.

Kelly is using a geometry software program to experiment with transformations on the coordinate grid. She began with a point and translated it 4 units up and 8 units left. Then she reflected the image in the x-axis. Finally, she dilated this new image by a scale factor of 0.5 with respect to the origin to arrive at $(-1, -4)$. What were the original coordinates of the point?

Scoring Rubric	
Criteria	**Score**
Full Credit: The answer is correct and a full explanation is provided that shows each step.	2
Partial Credit: • The answer is correct, but the explanation is incomplete. • The answer is incorrect, but the explanation is correct.	1
No Credit: Either an answer is not provided or the answer does not make sense.	0

Read the problem statement carefully. You are given a sequence of transformations of a point on a coordinate grid. You know the coordinates of the final image and are asked to find the original coordinates. Undo each transformation in reverse order to work backward and solve the problem.

Example of a 2-point response:

original point → translation → reflection → dilation → end result

Begin with the coordinates of the end result and work backward.

Dilate by 2 to undo the dilation by 0.5:

$(-1, -4) \rightarrow (-1 \times 2, -4 \times 2) = (-2, -8)$

Reflect back across the x-axis to undo the reflection:

$(-2, -8) \rightarrow (-2, 8)$

Translate 4 units down and 8 units right to undo the translation:

$(-2, 8) \rightarrow (-2 + 8, 8 - 4) = (6, 4)$

The original coordinates of the point were (6, 4).

The steps, calculations, and reasoning are clearly stated. The student also arrives at the correct answer. So, this response is worth the full 2 points.

Exercises

Solve each problem. Show your work. Responses will be graded using the short-response scoring rubric given at the beginning of the lesson.

1. A flea landed on a coordinate grid. The flea hopped across the x-axis and then across the y-axis in the form of two consecutive reflections. Then it walked 9 units to the right and 4 units down. If the flea's final position was at (4, –1), what point did it originally land on?

2. The coordinate grid below shows the final image when a point was rotated 90° clockwise about the origin, dilated by a scale factor of 2, and shifted 7 units right. What were the original coordinates?

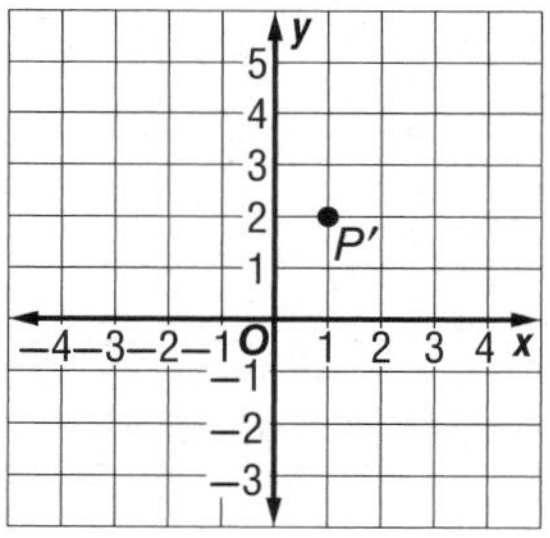

3. Figure $ABCD$ is an isosceles trapezoid.

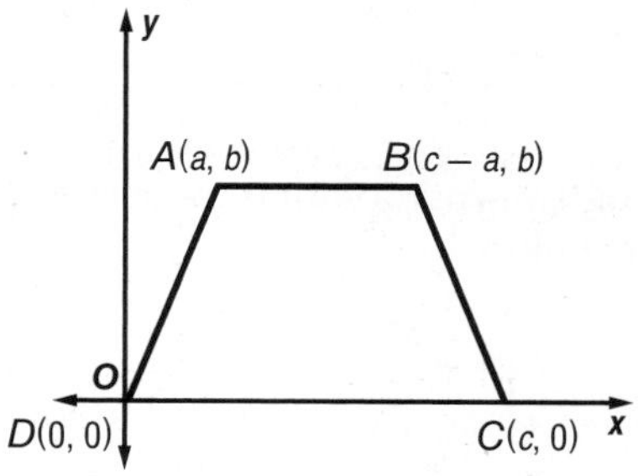

Which of the following are the coordinates of an endpoint of the median of $ABCD$?

A $\left(\frac{a+b}{2}, \frac{a+b}{2}\right)$ **C** $\left(\frac{c}{2}, 0\right)$

B $\left(\frac{2c-a}{2}, \frac{b}{2}\right)$ **D** $\left(\frac{c}{2}, b\right)$

4. If the measure of an interior angle of a regular polygon is 108, what type of polygon is it?

F octagon **H** pentagon

G hexagon **J** triangle

Standardized Test Practice
Cumulative, Chapters 1 through 14

Multiple Choice

Read each question. Then fill in the correct answer on the answer document provided by your teacher or on a sheet of paper.

1. Point N has coordinates (4, −3). What will the coordinates of its image be after a reflection across the y-axis?

A N′(−3, 4)

B $N'(-4, 3)$

C N′(4, 3)

D $N'(-4, -3)$

2. Which pair of figures shows a reflection across the line followed by a translation up?

F

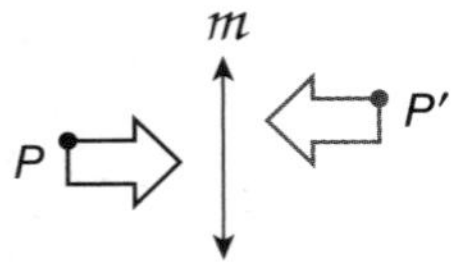

H

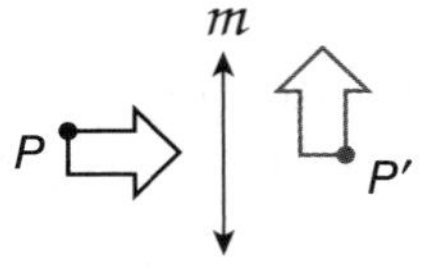

G

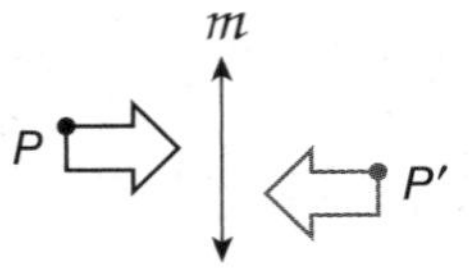

J

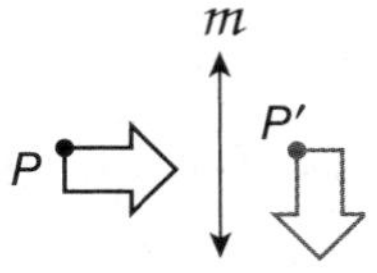

3. What is the angle of rotation that maps point T onto T' in the figure below?

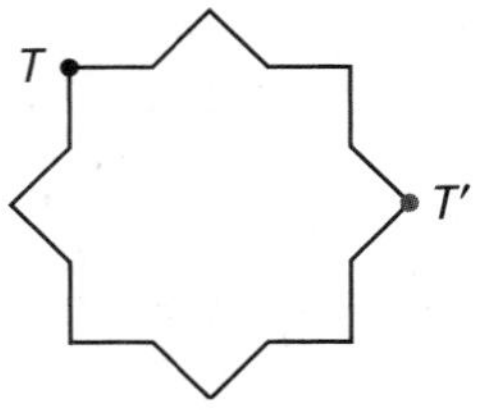

A 90°

B 120°

C 135°

D 145°

> **Test-Taking Tip**
>
> Question 3 How many points are there on the star? Divide 360° by this number to find the angle of rotation from one point to the next.

4. **Given:** $a \parallel b$

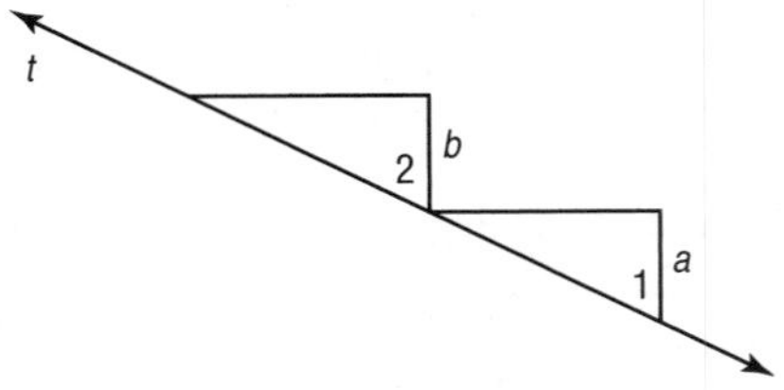

Which statement below justifies the conclusion that $\angle 1 \cong \angle 2$?

F If $a \parallel b$ and are cut by transversal t, then alternate exterior angles are congruent.

G If $a \parallel b$ and are cut by transversal t, then alternate interior angles are congruent.

H If $a \parallel b$ and are cut by transversal t, then corresponding angles are congruent.

J If $a \parallel b$ and are cut by transversal t, then vertical angles are congruent.

5. Which of the following is a side length in isosceles triangle DEF?

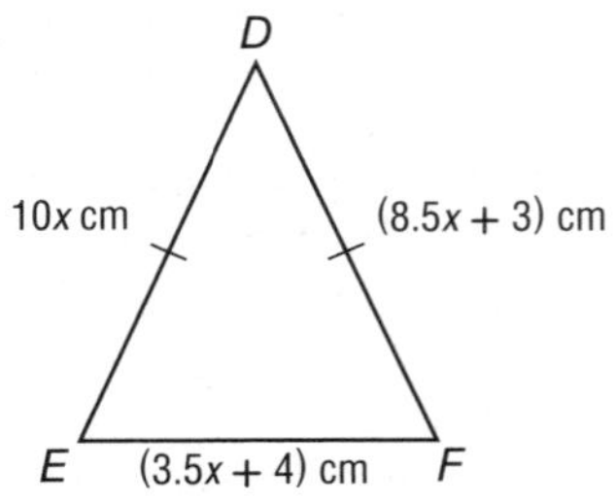

F 2 cm

G 8 cm

H 9 cm

J 11 cm

6. Which of the following has exactly two pairs of consecutive congruent sides?

A kite

B parallelogram

C rhombus

D trapezoid

Short Response/Gridded Response

Record your answers on the answer sheet provided by your teacher or on a sheet of paper.

7. State whether the figure has rotational symmetry. If so, copy the figure, locate the center, and state the order and magnitude of symmetry.

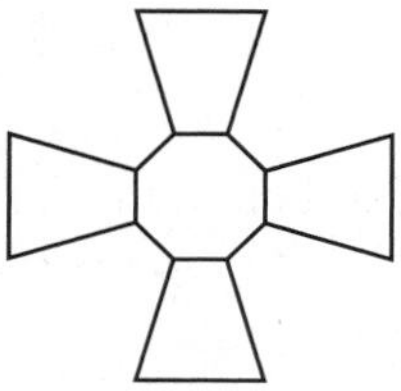

8. Dilate the figure shown on the coordinate grid by a scale factor of 1.5 centered at the origin.

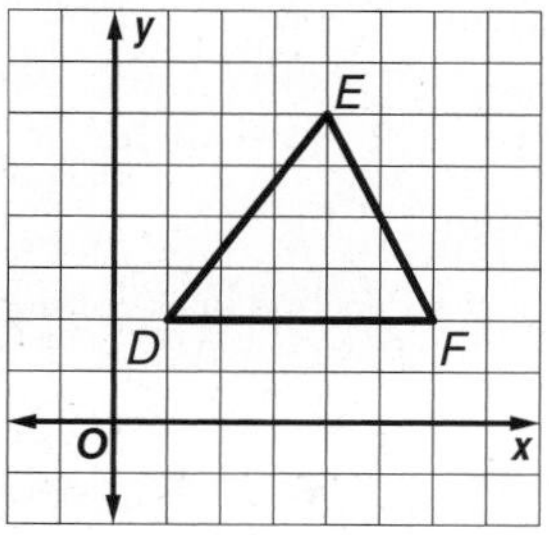

9. Quadrilateral *WXYZ* is a rhombus. If $m\angle XYZ = 110°$, find $m\angle ZWY$.

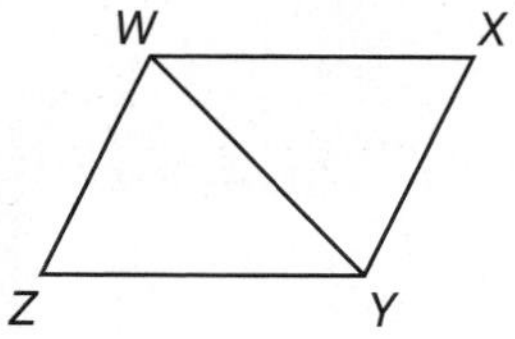

10. Regina left her office downtown and traveled 3 blocks west and 5 blocks north. Write a translation vector to describe her route.

11. GRIDDED RESPONSE In the triangle below, $\overline{MN} \parallel \overline{BC}$. Solve for x.

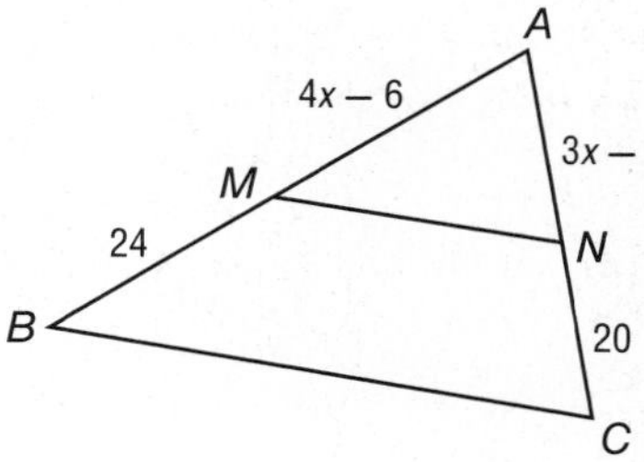

Extended Response

Record your answers on a sheet of paper. Show your work.

12. Refer to triangle *XYZ* to answer each question.

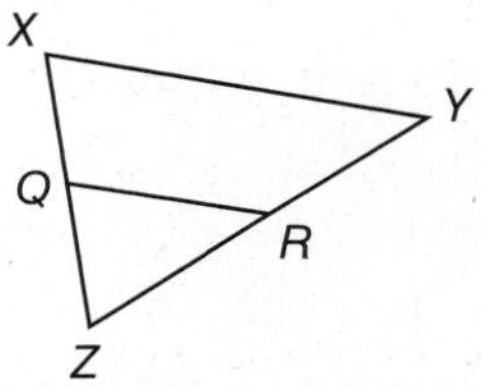

a. Suppose $\overline{QR} \parallel \overline{XY}$. What do you know about the relationship between segments *XQ*, *QZ*, *YR*, and *RZ*?

b. If $\overline{QR} \parallel \overline{XY}$, $XQ = 15$, $QZ = 12$, and $YR = 20$, what is the length of $\overline{RZ}$?

c. Suppose $\overline{QR} \parallel \overline{XY}$, $\overline{XQ} \cong \overline{QZ}$, and $QR = 9.5$ units. What is the length of $\overline{XY}$?

Need ExtraHelp?

If you missed Question...	1	2	3	4	5	6	7	8	9	10	11	12
Go to Lesson...	14-4	14-7	14-6	11-2	12-6	13-5	14-8	14-9	13-4	14-5	14-3	14-2

CHAPTER 15

Circles

Then

You learned about special segments and angle relationships in triangles.

Now

In this chapter, you will:

- Learn the relationships between central angles, arcs, and inscribed angles in a circle.
- Define and use secants and tangents.
- Use an equation to identify or describe a circle.

Why?

SCIENCE The actual shape of a rainbow is a complete circle. The portion of the circle that can be seen above the horizon is a special segment of a circle called an arc.

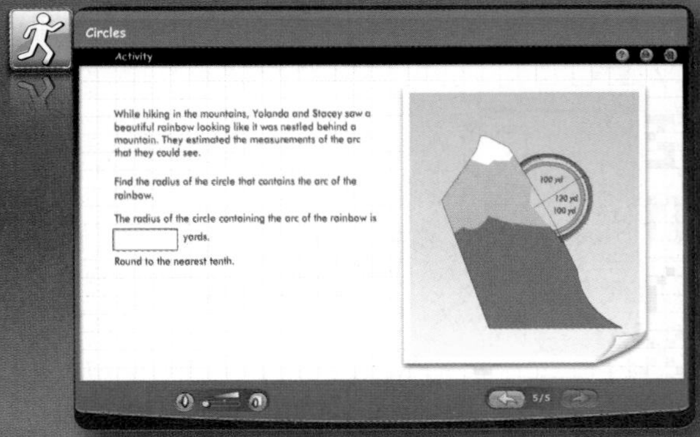

connectED.mcgraw-hill.com **Your Digital Math Portal**

Animation | Vocabulary | eGlossary | Personal Tutor | Virtual Manipulatives | Graphing Calculator | Audio | Foldables | Self-Check Practice | Worksheets

Get Ready for the Chapter

Diagnose Readiness | You have two options for checking prerequisite skills.

1 Textbook Option
Take the Quick Check below. Refer to the Quick Review for help.

QuickCheck

Find the percent of the given number.

1. 26% of 500 **2.** 79% of 623

3. 19% of 82 **4.** 10% of 180

5. 92% of 90 **6.** 65% of 360

7. **TIPPING** A couple ate dinner at an Italian restaurant where their bill was $32.50. If they want to leave an 18% tip, how much tip money should they leave?

8. Find *x*. Round to the nearest tenth.

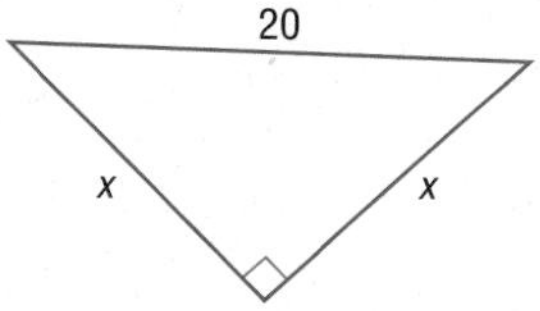

9. **CONSTRUCTION** Jennifer is putting a brace in a board, as shown at the right. Find the length of the board used for a brace.

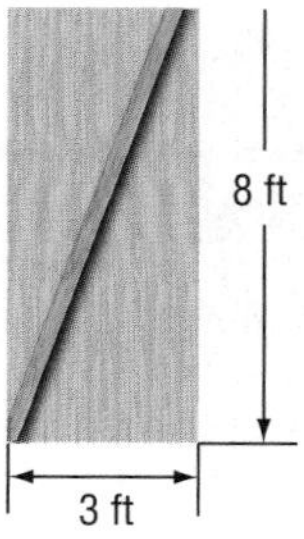

Solve each equation by using the Quadratic Formula. Round to the nearest tenth if necessary.

10. $5x^2 + 4x - 20 = 0$ **11.** $x^2 = x + 12$

12. **FIREWORKS** The Patriot Squad, a professional fireworks company, performed a show during a July 4th celebration. One of the rockets in the show followed the path modeled by $d = 80t - 16t^2$ where t is the time in seconds, but it failed to explode.

QuickReview

Example 1

Find the percent of the given number.

15% of $35 = (0.15)(35)$ Change the percent to a decimal.

$= 5.25$ Multiply.

So, 15% of 35 is 5.25.

Example 2

Find *x*. Round to the nearest tenth.

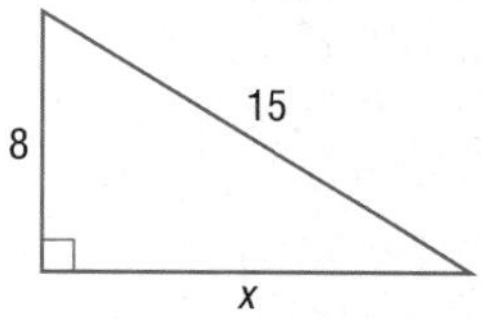

$a^2 + b^2 = c^2$ Pythagorean Theorem

$x^2 + 8^2 = 15^2$ Substitution

$x^2 + 64 = 225$ Simplify.

$x^2 = 161$ Subtract.

$x = \sqrt{161}$ or about 12.7

Example 3

Solve $x^2 + 3x - 40 = 0$ by using the Quadratic Formula. Round to the nearest tenth.

$x = \frac{-b \pm \sqrt{b^2 - 4ac}}{2a}$ Quadratic Formula

$= \frac{-3 \pm \sqrt{3^2 - 4(1)(-40)}}{2(1)}$ Substitution

$= \frac{-3 \pm \sqrt{169}}{2}$ Simplify.

$= 5$ or -8 Simplify.

2 Online Option
Take an online self-check Chapter Readiness Quiz at connectED.mcgraw-hill.com.

Get Started on the Chapter

You will learn several new concepts, skills, and vocabulary terms as you study Chapter 15. To get ready, identify important terms and organize your resources. You may wish to refer to Chapter 0 to review prerequisite skills.

FOLDABLES® StudyOrganizer

Circles Make this Foldable to help you organize your Chapter 10 notes on circles. Begin with nine sheets of paper.

1 **Trace** an 8-inch circle on each paper using a compass.

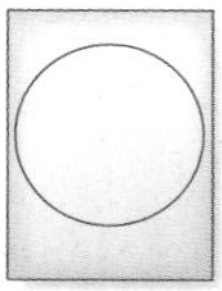

2 **Cut** out each of the circles.

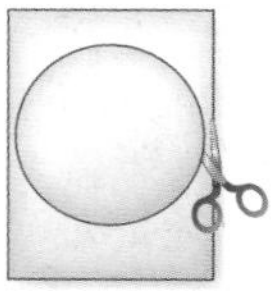

3 **Staple** an inch from the left side of the papers.

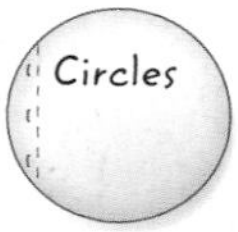
Circles

4 **Label** as shown.

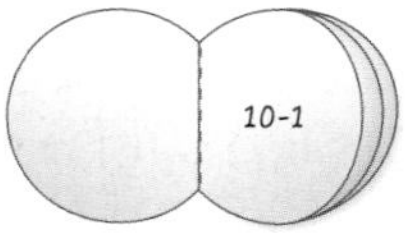
10-1

NewVocabulary

English		Español
circle	p. 965	círculo
center	p. 965	centro
radius	p. 965	radio
chord	p. 965	cuerda
diameter	p. 965	diámetro
circumference	p. 967	circunferencia
pi (π)	p. 967	pi (π)
inscribed	p. 968	inscrito
circumscribed	p. 968	circunscrito
tangent	p. 982	tangente

ReviewVocabulary

coplanar coplanar points that lie in the same plane

degree grado $\frac{1}{360}$ of the circular rotation about a point

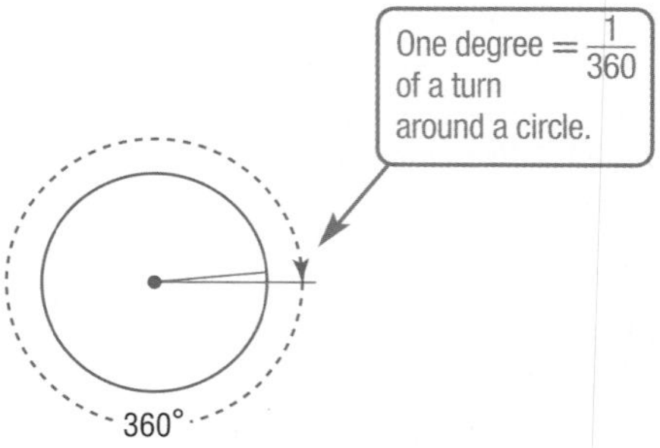
One degree = $\frac{1}{360}$ of a turn around a circle.
360°

LESSON

15-1 Circles and Circumference

∵ Then	∵ Now	∵ Why?
You identified and used parts of parallelograms.	**1** Identify and use parts of circles. **2** Solve problems involving the circumference of a circle.	The maxAir ride shown speeds back and forth and rotates counterclockwise. At times, the riders are upside down 140 feet above the ground experiencing "airtime"—a feeling of weightlessness. The ride's width, or *diameter,* is 44 feet. You can find the distance that a rider travels in one rotation by using this measure.

NewVocabulary

circle
center
radius
chord
diameter
concentric circles
circumference
pi (π)
inscribed
circumscribed

Common Core State Standards

Content Standards

G.CO.1 Know precise definitions of angle, circle, perpendicular line, parallel line, and line segment, based on the undefined notions of point, line, distance along a line, and distance around a circular arc.

G.C.1 Prove that all circles are similar.

Mathematical Practices

4 Model with mathematics.

1 Make sense of problems and persevere in solving them.

1 Segments in Circles

A **circle** is the locus or set of all points in a plane equidistant from a given point called the **center** of the circle.

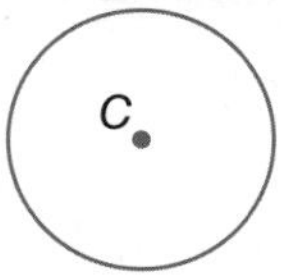

Circle C or $\odot C$

Segments that intersect a circle have special names.

KeyConcept Special Segments in a Circle

A **radius** (plural radii) is a segment with endpoints at the center and on the circle.

Examples $\overline{CD}$, $\overline{CE}$, and $\overline{CF}$ are radii of $\odot C$.

A **chord** is a segment with endpoints on the circle.

Examples $\overline{AB}$ and $\overline{DE}$ are chords of $\odot C$.

A **diameter** of a circle is a chord that passes through the center and is made up of collinear radii.

Example $\overline{DE}$ is a diameter of $\odot C$. Diameter $\overline{DE}$ is made up of collinear radii $\overline{CD}$ and $\overline{CE}$.

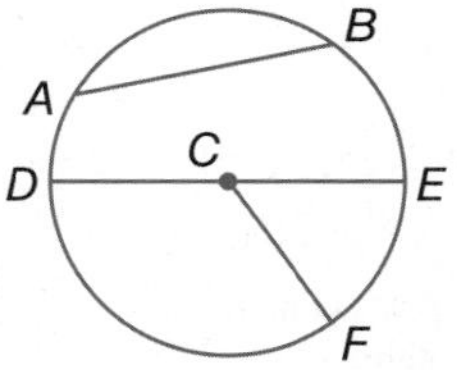

PT

Example 1 Identify Segments in a Circle

a. Name the circle and identify a radius.

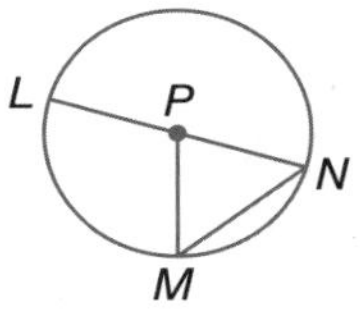

The circle has a center at P, so it is named circle P, or $\odot P$. Three radii are shown: $\overline{PL}$, $\overline{PN}$, and $\overline{PM}$.

b. Identify a chord and a diameter of the circle.

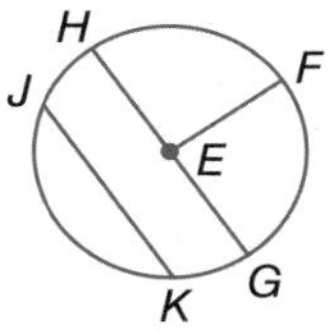

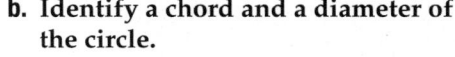

Two chords are shown: $\overline{JK}$ and $\overline{HG}$. $\overline{HG}$ goes through the center, so $\overline{HG}$ is a diameter.

GuidedPractice

1. Name the circle, a radius, a chord, and a diameter of the circle.

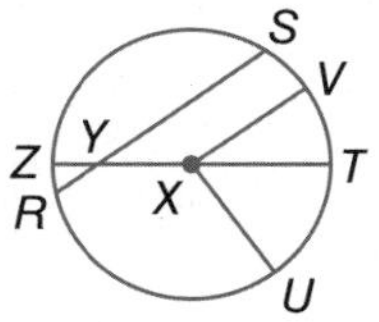

ReadingMath

CCSS Precision The words *radius* and *diameter* are used to describe lengths as well as segments. Since a circle has many different radii and diameters, the phrases *the radius* and *the diameter* refer to lengths rather than segments.

By definition, the distance from the center of a circle to any point on the circle is always the same. Therefore, all radii r of a circle are congruent. Since a diameter d is composed of two radii, all diameters of a circle are also congruent.

KeyConcept Radius and Diameter Relationships

If a circle has radius r and diameter d, the following relationships are true.

Radius Formula $r = \frac{d}{2}$ or $r = \frac{1}{2}d$ **Diameter Formula** $d = 2r$

Example 2 Find Radius and Diameter

If $QV = 8$ inches, what is the diameter of $\odot Q$?

$d = 2r$ Diameter Formula

$= 2(8)$ or 16 Substitute and simplify.

The diameter of $\odot Q$ is 16 inches.

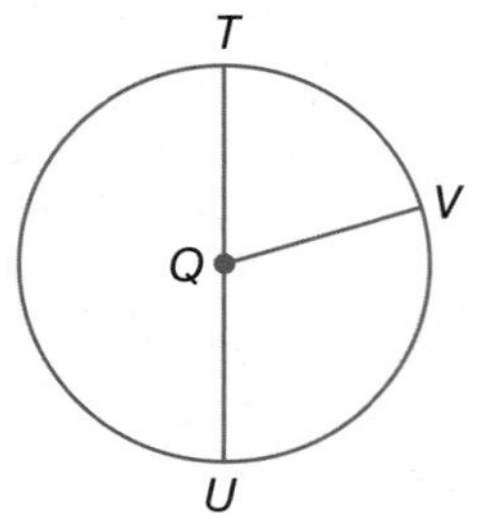

GuidedPractice

2A. If $TU = 14$ feet, what is the radius of $\odot Q$?

2B. If $QT = 11$ meters, what is QU?

As with other figures, pairs of circles can be congruent, similar, or share other special relationships.

ReviewVocabulary

coplanar points that lie in the same plane

KeyConcept Circle Pairs

Two circles are congruent if and only if they have congruent radii.

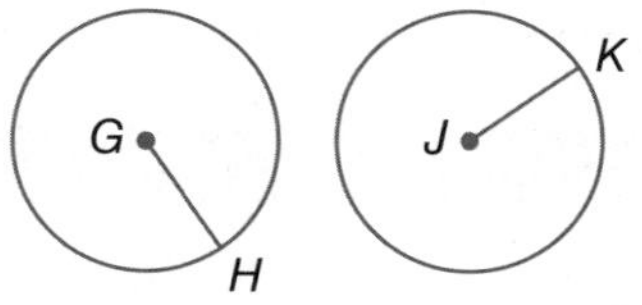

Example $\overline{GH} \cong \overline{JK}$, so $\odot G \cong \odot J$.

All circles are similar.

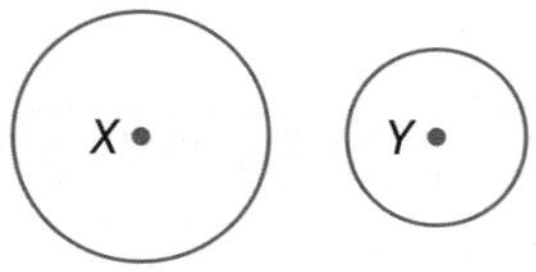

Example $\odot X \sim \odot Y$

Concentric circles are coplanar circles that have the same center.

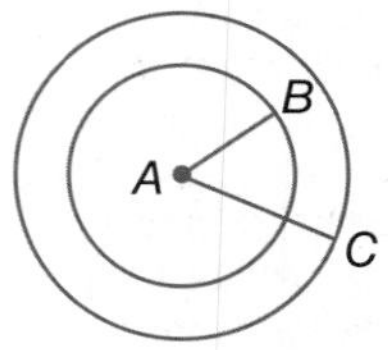

Example $\odot A$ with radius $\overline{AB}$ and $\odot A$ with radius $\overline{AC}$ are concentric.

You will prove that all circles are similar in Exercise 52.

Two circles can intersect in two different ways.

2 Points of Intersection	1 Point of Intersection	No Points of Intersection
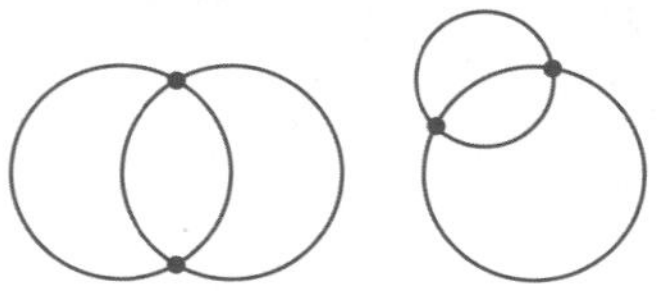		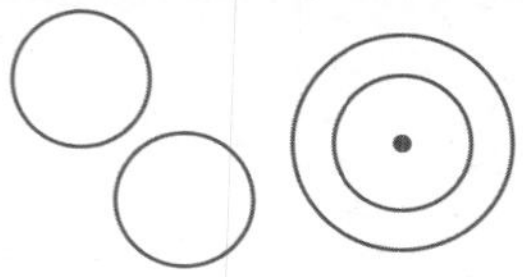

The segment connecting the centers of the two intersecting circles contains the radii of the two circles.

Example 3 Find Measures in Intersecting Circles

PT

The diameter of $\odot S$ is 30 units, the diameter of $\odot R$ is 20 units, and $DS = 9$ units. Find CD.

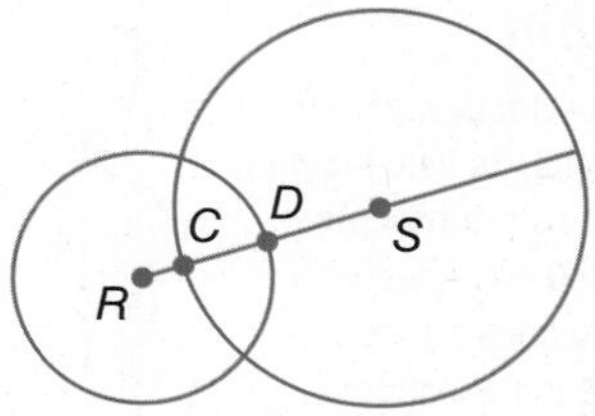

Since the diameter of $\odot S$ is 30, $CS = 15$.
$\overline{CD}$ is part of radius $\overline{CS}$.

$CD + DS = CS$	Segment Addition Postulate
$CD + 9 = 15$	Substitution
$CD = 6$	Subtract 9 from each side.

GuidedPractice

3. Use the diagram above to find RC.

2 Circumference

Circumference The **circumference** of a circle is the distance around the circle. By definition, the ratio $\frac{C}{d}$ is an irrational number called **pi (π)**. Two formulas for circumference can be derived by using this definition.

$\frac{C}{d} = \pi$	Definition of pi
$C = \pi d$	Multiply each side by d.
$C = \pi(2r)$	$d = 2r$
$C = 2\pi r$	Simplify.

KeyConcept Circumference

Words	If a circle has diameter d or radius r, the circumference C equals the diameter times pi or twice the radius times pi.
Symbols	$C = \pi d$ or $C = 2\pi r$

David Cannon/Getty Images Sport/Getty Images

Real-WorldLink

In 2005, Roger Federer and Andre Agassi played tennis on the helipad of the Burj Al Arab hotel in the United Arab Emirates. The helipad has a diameter of 79 feet and is nearly 700 feet high.

Source: Burj Al Arab, Emporis Buildings

Real-World Example 4 Find Circumference

PT

TENNIS Find the circumference of the helipad described at the left.

$C = \pi d$	Circumference formula
$= \pi(79)$	Substitution
$= 79\pi$	Simplify.
≈ 248.19	Use a calculator.

The circumference of the helipad is 79π feet or about 248.19 feet.

GuidedPractice

Find the circumference of each circle described. Round to the nearest hundredth.

4A. radius = 2.5 centimeters

4B. diameter = 16 feet

These circumference formulas can also be used to determine the diameter and radius of a circle when the circumference is given.

PT

Example 5 Find Diameter and Radius

Find the diameter and radius of a circle to the nearest hundredth if the circumference of the circle is 106.4 millimeters.

$C = \pi d$	Circumference Formula	$r = \frac{1}{2}d$	Radius Formula
$106.4 = \pi d$	Substitution	$\approx \frac{1}{2}(33.87)$	$d \approx 33.87$
$\frac{106.4}{\pi} = d$	Divide each side by π.	≈ 16.94 mm	Use a calculator.
$33.87 \text{ mm} \approx d$	Use a calculator.		

Guided Practice

5. Find the diameter and radius of a circle to the nearest hundredth if the circumference of the circle is 77.8 centimeters.

StudyTip

Levels of Accuracy Since π is irrational, its value cannot be given as a terminating decimal. Using a value of 3 for π provides a quick estimate in calculations. Using a value of 3.14 or $\frac{22}{7}$ provides a closer approximation. For the most accurate approximation, use the π key on a calculator. Unless stated otherwise, assume that in this text, a calculator with a π key was used to generate answers.

A polygon is **inscribed** in a circle if all of its vertices lie on the circle. A circle is **circumscribed** about a polygon if it contains all the vertices of the polygon.

- Quadrilateral *LMNP* is *inscribed in* $\odot K$.
- Circle *K* is *circumscribed about* quadrilateral *LMNP*.

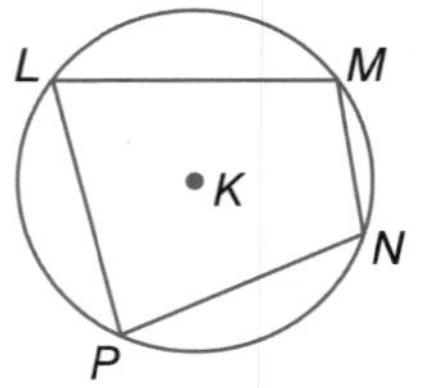

PT

Standardized Test Example 6 Circumference of Circumscribed Polygon

SHORT RESPONSE A square with side length of 9 inches is inscribed in $\odot J$. Find the exact circumference of $\odot J$.

Read the Test Item

You need to find the diameter of the circle and use it to calculate the circumference.

Solve the Test Item

First, draw a diagram. The diagonal of the square is the diameter of the circle and the hypotenuse of a right triangle.

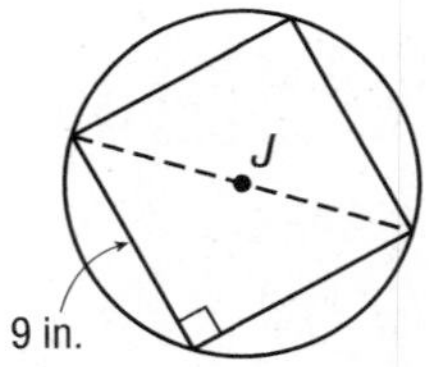

$a^2 + b^2 = c^2$	Pythagorean Theorem
$9^2 + 9^2 = c^2$	Substitution
$162 = c^2$	Simplify.
$9\sqrt{2} = c$	Take the positive square root of each side.

The diameter of the circle is $9\sqrt{2}$ inches.

Find the circumference in terms of π by substituting $9\sqrt{2}$ for d in $C = \pi d$. The exact circumference is $9\pi\sqrt{2}$ inches.

StudyTip

Circumcircle A *circumcircle* is a circle that passes through all of the vertices of a polygon.

Guided Practice

Find the exact circumference of each circle by using the given polygon.

6A. inscribed right triangle with legs 7 meters and 3 meters long

6B. circumscribed square with side 10 feet long

Check Your Understanding

Examples 1–2 **For Exercises 1–4, refer to ⊙*A*.**

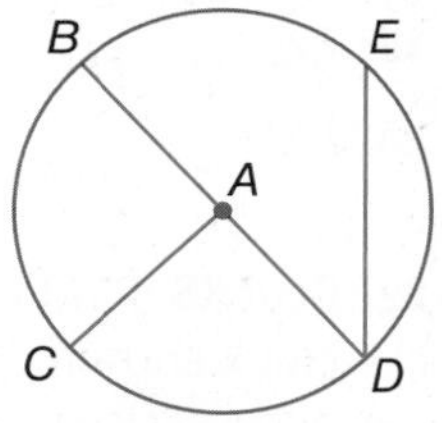

1. Name the circle.

2. Identify each.

a. a chord **b.** a diameter **c.** a radius

3. If $BA = 5$ inches, find CA.

4. If $CA = 7$ feet, what is the diameter of the circle?

Example 3 **The diameters of ⊙*E*, ⊙*F*, and ⊙*G*, are 14 feet, 5 feet, and 9 feet respectively. Find each measure.**

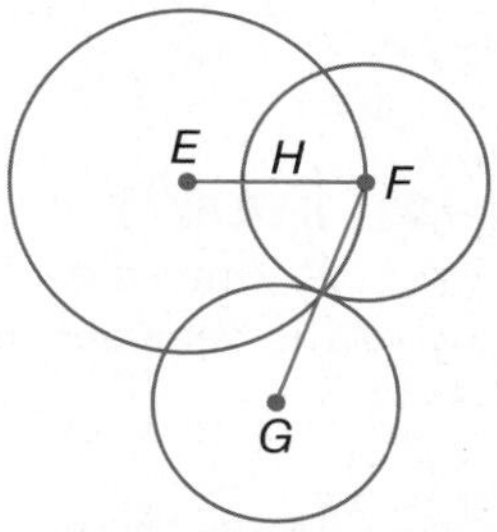

5. *FG*

6. *EH*

Example 4 **7. CAKES** The cake shown has a diameter of 8". What are the radius and circumference of the cake? Round to the nearest hundredth, if necessary.

8. CCSS **HISTORY** The circumference of the Roman Colosseum is 545 meters, what are the diameter and radius of the Colosseum? Round to the nearest hundredth.

Example 5 **9. SHORT RESPONSE** The triangle shown is inscribed in ⊙*Z*. Find the exact circumference of ⊙*Z*.

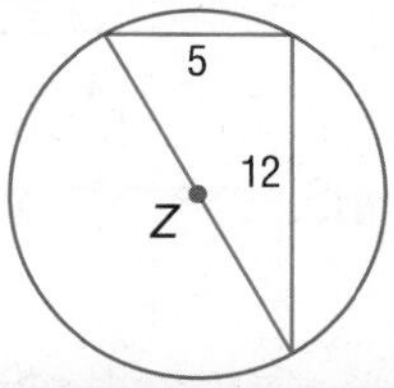

Practice and Problem Solving

Examples 1–2 **For Exercises 10–13, refer to ⊙*J*.**

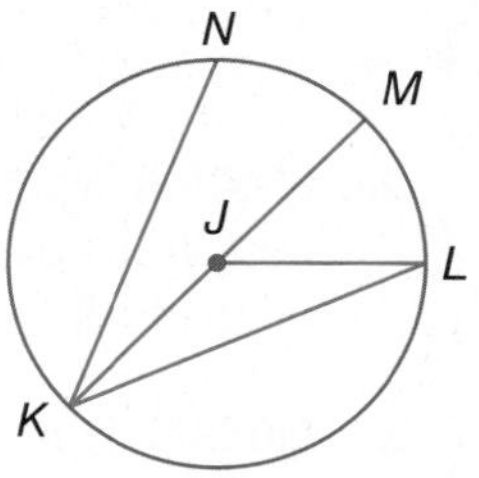

10. Name the center of the circle.

11. Identify a chord that is also a diameter.

12. Is LK a radius?

13. If $KM = 32$ cm, what is JL?

For Exercises 14–17, refer to ⊙*P*.

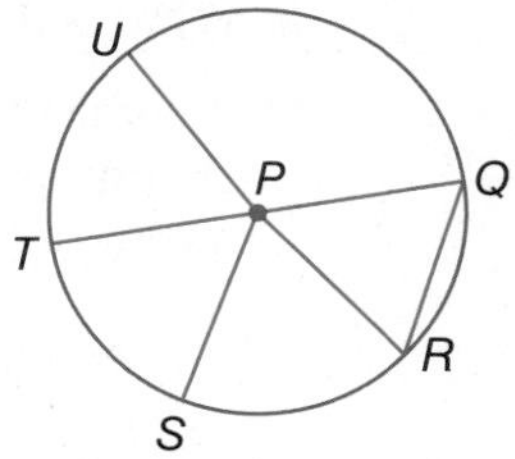

14. Identify a chord that is not a diameter.

15. If TP is 38 inches, what is the diameter of the circle?

16. Is $UP \cong TQ$? Explain.

17. If $TQ = 56$ cm, what is PR?

Example 3 **$\odot B$ has a radius of 3 units, $\odot D$ has a radius of 7 units, and $\odot G$ has a radius of 5 units. Find each measure.**

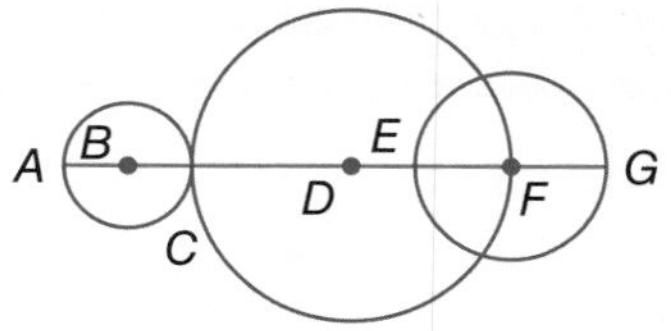

18. EF

19. BG

20. BD

21. AH

Example 4 **22. CLOCKS** Find the radius and circumference of the clock shown. Round to the nearest hundredth if necessary.

23. JEWLRY The bracelet shown has a circumference of 8". Find the radius and the diameter of the bracelet. Round to the nearest hundredth, if necessary.

Example 5 **Find the diameter and radius of a circle with the given circumference. Round to the nearest hundredth.**

24. 13 ft

25. 176 inches

26. 43.98 cm

27. 201.06 m

Example 6 **CCSS SENSE-MAKING** **Find the exact circumference of each circle by using the inscribed or circumscribed polygon.**

28.

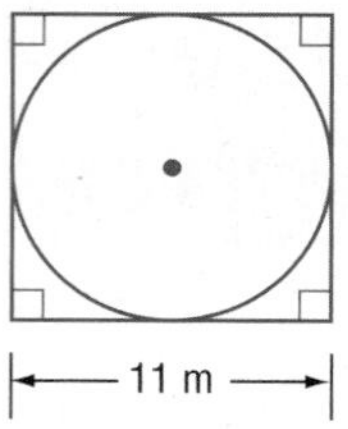

29.

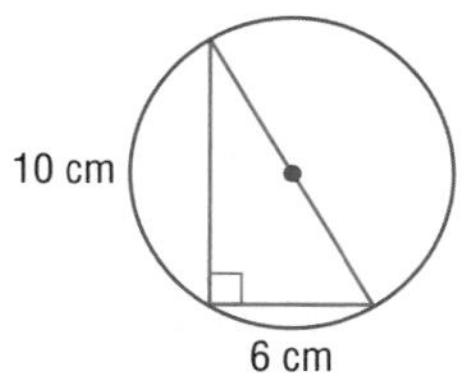

30.

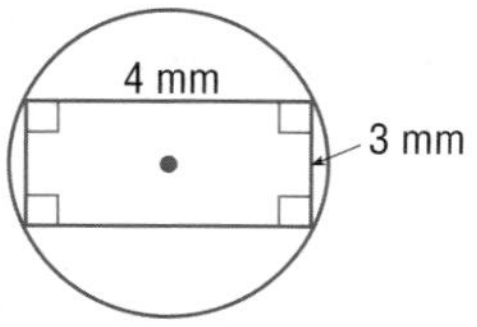

31.

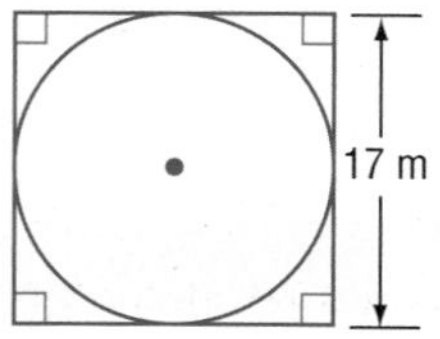

32.

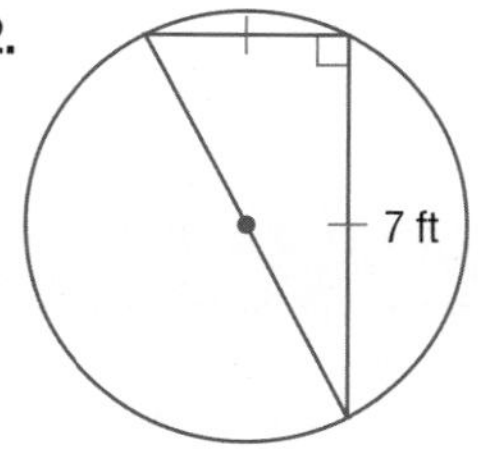

33.

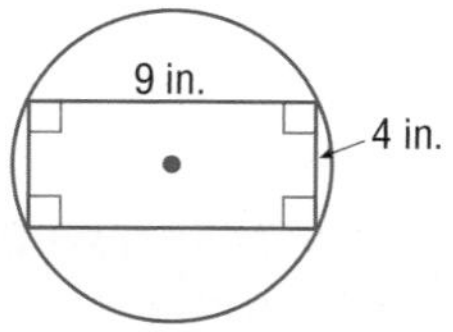

34. FENCING Max is fencing in his circular garden to keep out deer. The fencing costs $4 per foot. If his garden has a radius of 15 feet, find the total cost of the fence. Round to the nearest cent.

35. MOSAIC Sonia is designing a circular tile mosaic to decorate her bathroom. A diagram of the mosaic is shown.

a. What is the approximate circumference of the mosaic?

b. If Sonia changes the plans so that the inner circle has a circumference of 10 feet, what should the radius of the mosaic be to the nearest tenth of a foot?

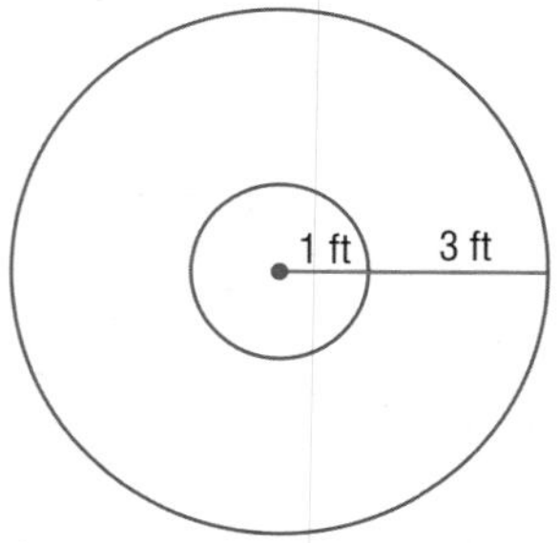

The radius, diameter, or circumference of a circle is given. Find each missing measure to the nearest hundredth.

36. $d = 16.5$ m $r = ?, C = ?$

37. $C = 72x$ yd, $d = ?, r = ?$

38. $r = 14.5$ ft, $d = ?, C = ?$

39. $d = 14x$ units , $r = ?, C = ?$

Determine whether the circles in the figures below appear to be congruent, concentric, or neither.

40.

41.

42.

43. RIDE The Star of Nanchang is a Ferris wheel in China with a diameter of 525 feet. If the gondolas that people ride in are approximately 27.49 feet apart, how many gondolas are there total on the ride?

44. CCSS **MIRRORS** If the radius of a mirror is 12 inches and the frame is 3 inches wide, what is the total circumference of the mirror?

45. MULTIPLE REPRESENTATIONS In this problem you will explore the ratio of circumferences of concentric circles.

a. Geometric Use a compass to draw three concentric circles in which the scale factor from each circle to the next is 1:2. Label the circles *A*, *B*, and *C*. Label the length of the radius of each circle.

b. Tabular Copy and Complete the following table.

Circle	Radius	Ratio of Radius and Radius of Circle *A*	Circumference	Ratio of Circumference and Circumference of Circle *A*
A				
B				
C				

c. Verbal Make a conjecture about the ratio between the circumference of two circles with different radii.

46. **BUFFON'S NEEDLE** Measure the length ℓ of a needle (or toothpick) in centimeters. Next, draw a set of horizontal lines that are ℓ centimeters apart on a sheet of plain white paper.

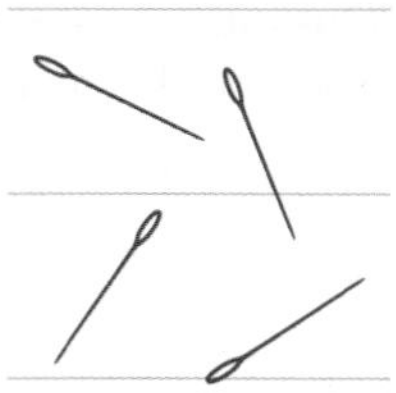

a. Drop the needle onto the paper. When the needle lands, record whether it touches one of the lines as a hit. Record the number of hits after 25, 50, and 100 drops.

b. Calculate the ratio of two times the total number of drops to the number of hits after 25, 50, and 100 drops.

c. How are the values you found in part **b** related to π?

47. **SPORTS** A target used in archery is shown. The labels on the diagram indicate the radius (in inches) of the rings.

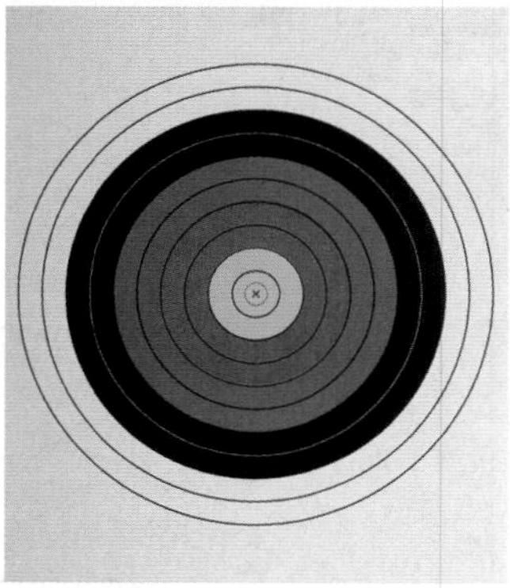

a. How much greater is the circumference of the white ring than the yellow ring?

b. If the radius of each circle increased by 1 inch, how much would the circumference of the target change?

H.O.T. Problems Use Higher-Order Thinking Skills

48. **WRITING IN MATH** Explain the differences between congruent and concentric circles.

49. **REASONING** Is the following statement *sometimes, always,* or *never* true: If a circle circumscribes a square the area of the circle is larger than the area of the square.

50. **CHALLENGE** In the figure, $\odot A$ is inscribed in equilateral triangle BCD. What is the circumference of $\odot A$?

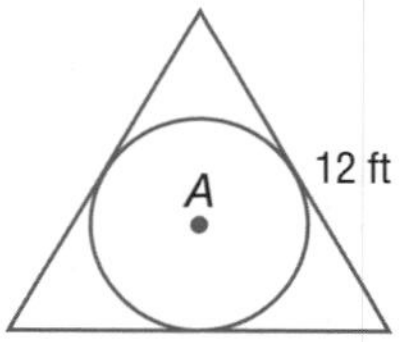

51. **REASONING** Is the distance from the any two points on a circle sometimes, always, or never smaller than the diameter of a circle?

52. **CCSS ERROR ANALYSIS** Zoe thinks the radius of a circle with circumference of 25 yards is approximately 8 yards, but Thomas thinks it is approximately 16 yards. Is either of them correct?

53. **CHALLENGE** If the sum of the circumferences of circles L, M, and N is 30π, find x.

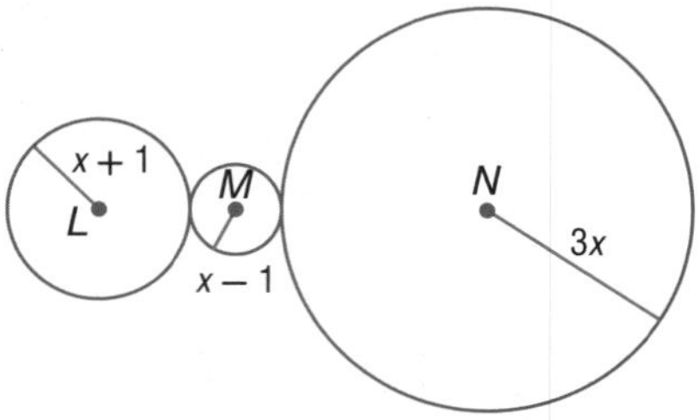

54. **WRITING IN MATH** Research and write about Archimedes' method of approximating Pi.

Standardized Test Practice

55. GRIDDED RESPONSE What is the circumference of $\odot T$? Round to the nearest tenth.

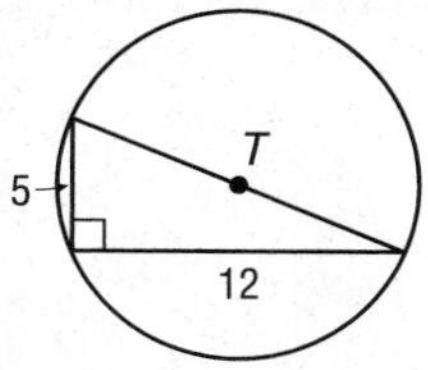

56. What is the radius of a table with a circumference of 10 feet?

A 1.6 ft
B 2.5 ft
C 3.2 ft
D 5 ft

57. ALGEBRA Bill is planning a circular vegetable garden with a fence around the border. If he can use up to 50 feet of fence, what radius can he use for the garden?

F 10 **G** 9 **H** 8 **J** 7

58. SAT/ACT What is the radius of a circle with an area of $\frac{\pi}{4}$ square units?

A 0.4 units
B 0.5 units
C 2 units
D 4 units
E 16 units

Spiral Review

Copy each figure and point B. Then use a ruler to draw the image of the figure under a dilation with center B and the scale factor r indicated. (Lesson 14-9)

59. $r = \frac{1}{5}$

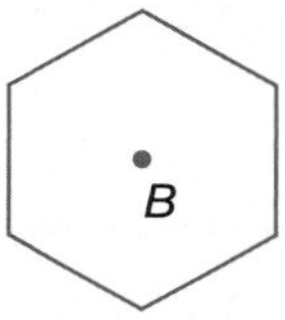

60. $r = \frac{2}{5}$

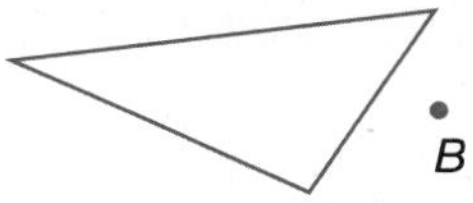

61. $r = 2$

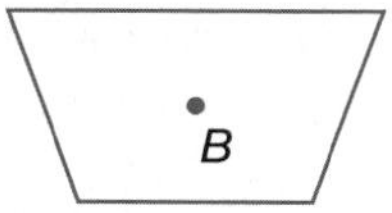

62. $r = 3$

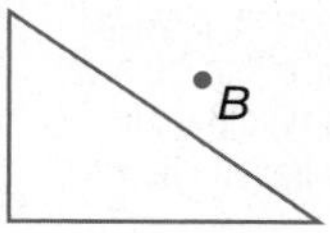

State whether each figure has rotational symmetry. If so, copy the figure, locate the center of symmetry, and state the order and magnitude of symmetry. (Lesson 14-8)

63.

64.

65.

66.

Skills Review

Find x.

67.

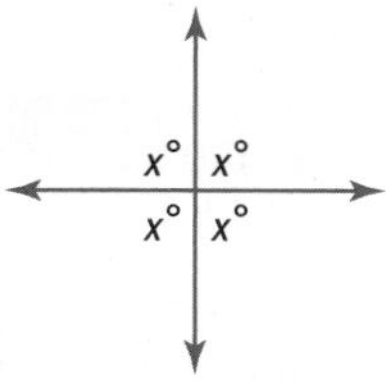

68.

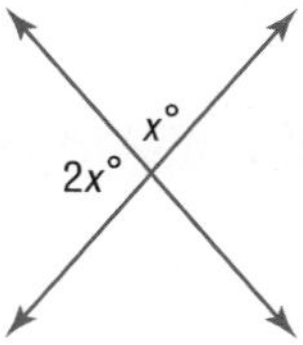

69.

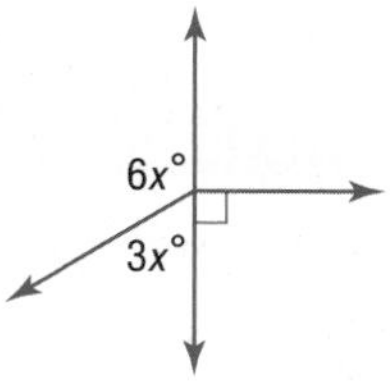

70.

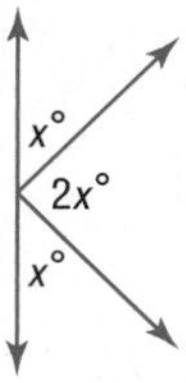

LESSON

15-2 Arcs and Chords

:: Then	:: Now	:: Why?
• You used the relationships between arcs and angles to find measures.	• **1** Recognize and use relationships between arcs and chords. **2** Recognize and use relationships between arcs, chords, and diameters.	• Embroidery hoops are used in sewing, quilting, and cross-stitching, as well as for embroidering. The endpoints of the snowflake shown are both the endpoints of a chord and the endpoints of an arc.

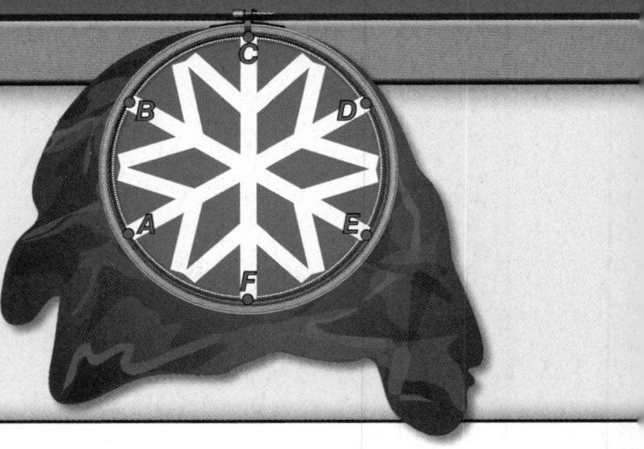

CCSS Common Core State Standards

Content Standards

G.C.2 Identify and describe relationships among inscribed angles, radii, and chords.

G.MG.3 Apply geometric methods to solve problems (e.g., designing an object or structure to satisfy physical constraints or minimize cost; working with typographic grid systems based on ratios). ★

Mathematical Practices

4 Model with mathematics.

3 Construct viable arguments and critique the reasoning of others.

1 Arcs and Chords

A *chord* is a segment with endpoints on a circle. If a chord is not a diameter, then its endpoints divide the circle into a major and a minor arc.

Theorem 15.1

Words In the same circle or in congruent circles, two minor arcs are congruent if and only if their corresponding chords are congruent.

Example $\widehat{FG} \cong \widehat{HJ}$ if and only if $\overline{FG} \cong \overline{HJ}$.

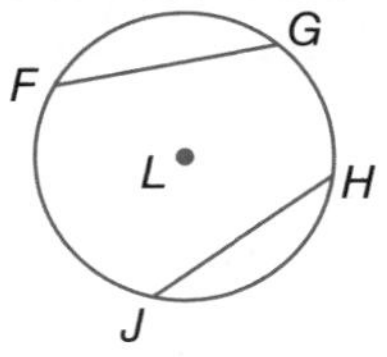

Proof Theorem 15.1 (part 1)

Given: $\odot P$; $\widehat{QR} \cong \widehat{ST}$

Prove: $\overline{QR} \cong \overline{ST}$

Proof:

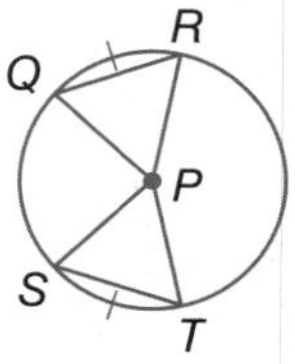

Statements	Reasons
1. $\odot P$, $\widehat{QR} \cong \widehat{ST}$	1. Given
2. $\angle QPR \cong \angle SPT$	2. If arcs are $\cong$, their corresponding central $\angle$s are $\cong$.
3. $\overline{QP} \cong \overline{PR} \cong \overline{SP} \cong \overline{PT}$	3. All radii of a circle are $\cong$.
4. $\triangle PQR \cong \triangle PST$	4. SAS
5. $\overline{QR} \cong \overline{ST}$	5. CPCTC

You will prove part 2 of Theorem 15.1 in Exercise 25.

PT

Real-World Example 1 Use Congruent Chords to Find Arc Measure

CRAFTS In the embroidery hoop, $\overline{AB} \cong \overline{CD}$ and $m\widehat{AB} = 60$. Find $m\widehat{CD}$.

$\overline{AB}$ and $\overline{CD}$ are congruent chords, so the corresponding arcs $\widehat{AB}$ and $\widehat{CD}$ are congruent. $m\widehat{AB} = m\widehat{CD} = 60$

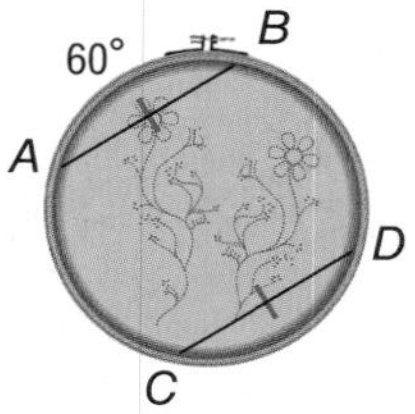

Guided Practice

1. If $m\widehat{AB} = 78$ in the embroidery hoop, find $m\widehat{CD}$.

Example 2 Use Congruent Arcs to Find Chord Lengths

ALGEBRA In the figures, $\odot J \cong \odot K$ and $\widehat{MN} \cong \widehat{PQ}$. Find PQ.

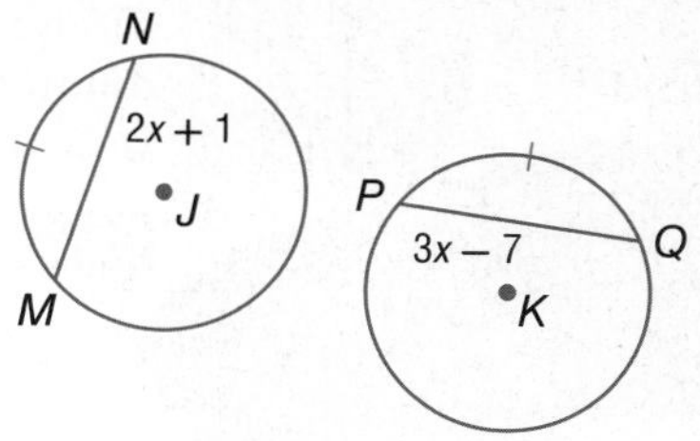

$\widehat{MN}$ and $\widehat{PQ}$ are congruent arcs in congruent circles, so the corresponding chords $\overline{MN}$ and $\overline{PQ}$ are congruent.

$MN = PQ$	Definition of congruent segments
$2x + 1 = 3x - 7$	Substitution
$8 = x$	Simplify.

So, $PQ = 3(8) - 7$ or 17.

GuidedPractice

2. In $\odot W$, $\widehat{RS} \cong \widehat{TV}$. Find RS.

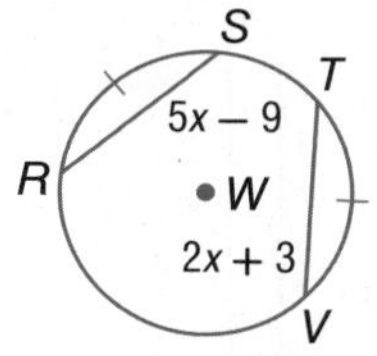

StudyTip

Arc Bisectors In the figure below, $\overline{FH}$ is an arc bisector of $\widehat{JG}$.

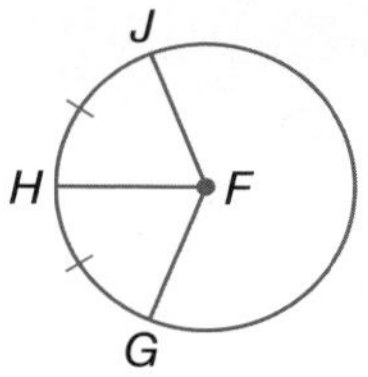

2 Bisecting Arcs and Chords

If a line, segment, or ray divides an arc into two congruent arcs, then it *bisects* the arc.

Theorems

15.2 If a diameter (or radius) of a circle is perpendicular to a chord, then it bisects the chord and its arc.

Example If diameter $\overline{AB}$ is perpendicular to chord $\overline{XY}$, then $\overline{XZ} \cong \overline{ZY}$ and $\widehat{XB} \cong \widehat{BY}$.

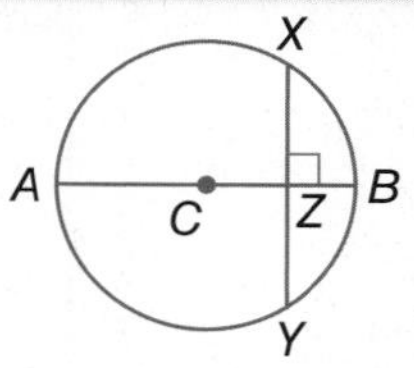

15.3 The perpendicular bisector of a chord is a diameter (or radius) of the circle.

Example If $\overline{AB}$ is a perpendicular bisector of chord $\overline{XY}$, then $\overline{AB}$ is a diameter of $\odot C$.

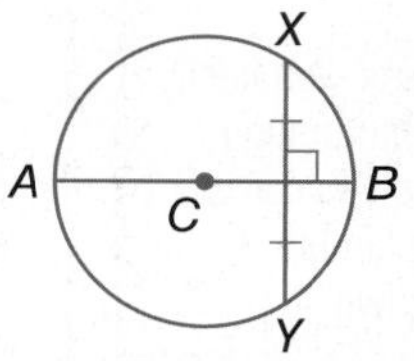

You will prove Theorems 15.2 and 15.3 in Exercises 26 and 28, respectively.

Example 3 Use a Radius Perpendicular to a Chord

In $\odot S$, $m\widehat{PQR} = 98$. Find $m\widehat{PQ}$.

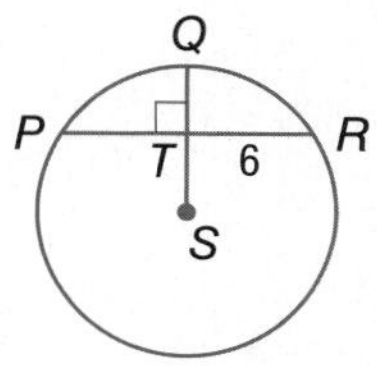

Radius $\overline{SQ}$ is perpendicular to chord $\overline{PR}$. So by Theorem 10.3, $\overline{SQ}$ bisects $\widehat{PQR}$. Therefore, $m\widehat{PQ} = m\widehat{QR}$. By substitution, $m\widehat{PQ} = \frac{98}{2}$ or 49.

GuidedPractice

3. In $\odot S$, find PR.

Real-World Example 4 Use a Diameter Perpendicular to a Chord

STAINED GLASS In the stained glass window, diameter $\overline{GH}$ is 30 inches long and chord $\overline{KM}$ is 22 inches long. Find JL.

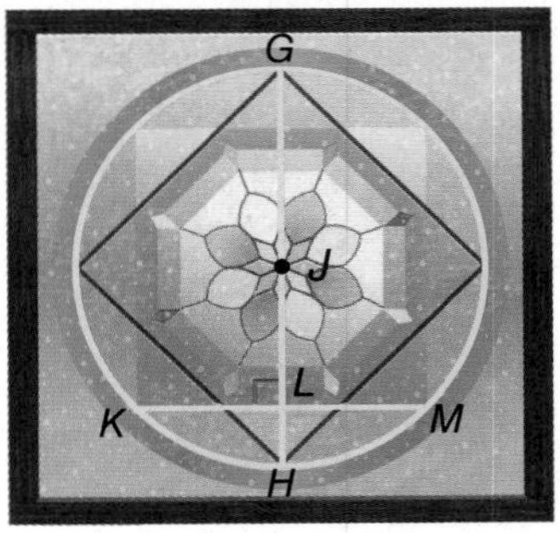

Step 1 Draw radius $\overline{JK}$.

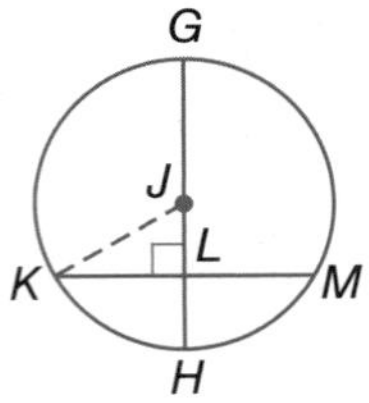

This forms right $\triangle JKL$.

Step 2 Find JK and KL.

Since $GH = 30$ inches, $JH = 15$ inches. All radii of a circle are congruent, so $JK = 15$ inches.

Since diameter $\overline{GH}$ is perpendicular to $\overline{KM}$, $\overline{GH}$ bisects chord $\overline{KM}$ by Theorem 10.3. So, $KL = \frac{1}{2}(22)$ or 11 inches.

Step 3 Use the Pythagorean Theorem to find JL.

$KL^2 + JL^2 = JK^2$	Pythagorean Theorem
$11^2 + JL^2 = 15^2$	$KL = 11$ and $JK = 15$
$121 + JL^2 = 225$	Simplify.
$JL^2 = 104$	Subtract 121 from each side.
$JL = \sqrt{104}$	Take the positive square root of each side.

So, JL is $\sqrt{104}$ or about 10.20 inches long.

Real-WorldLink

To make stained glass windows, glass is heated to a temperature of 2000 degrees, until it is the consistency of taffy. The colors are caused by the addition of metallic oxides.

Source: *Artistic Stained Glass by Regg*

GuidedPractice

4. In $\odot R$, find TV. Round to the nearest hundredth.

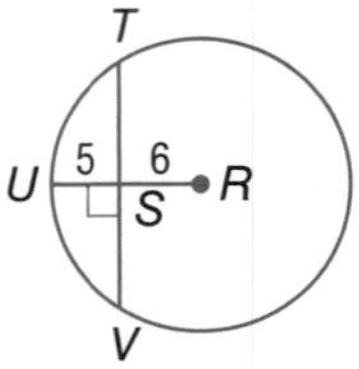

StudyTip

Drawing Segments You can add any known information to a figure to help you solve the problem. In Example 4, radius $\overline{JK}$ was drawn.

In addition to Theorem 15.1, you can use the following theorem to determine whether two chords in a circle are congruent.

Theorem 15.4

Words In the same circle or in congruent circles, two chords are congruent if and only if they are equidistant from the center.

Example $\overline{FG} \cong \overline{JH}$ if and only if $LX = LY$.

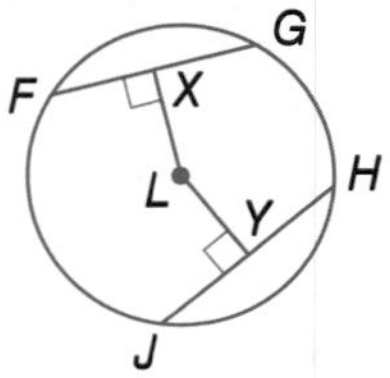

You will prove Theorem 15.4 in Exercises 29 and 30.

Example 5 Chords Equidistant from Center

ALGEBRA In $\odot A$, $WX = XY = 22$. Find AB.

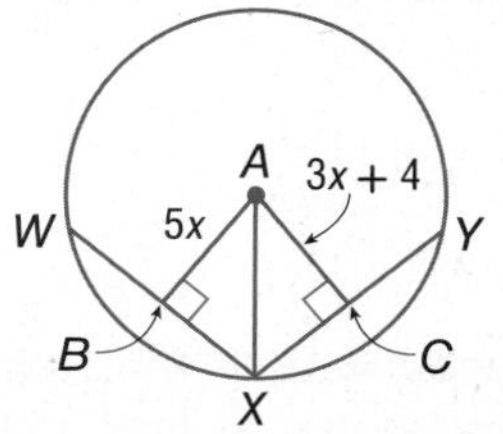

Since chords $\overline{WX}$ and $\overline{XY}$ are congruent, they are equidistant from A. So, $AB = AC$.

$AB = AC$

$5x = 3x + 4$ Substitution

$x = 2$ Simplify.

So, $AB = 5(2)$ or 10.

GuidedPractice

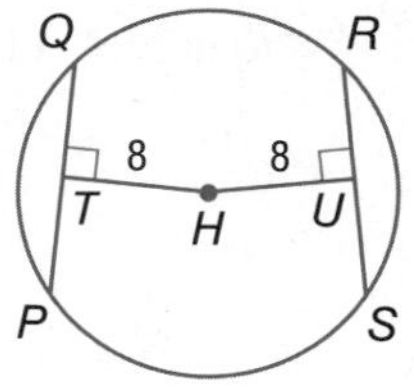

5. In $\odot H$, $PQ = 3x - 4$ and $RS = 14$. Find x.

You can use Theorem 15.4 to find the point equidistant from three noncollinear points.

Construction Circle Through Three Noncollinear Points

Step 1

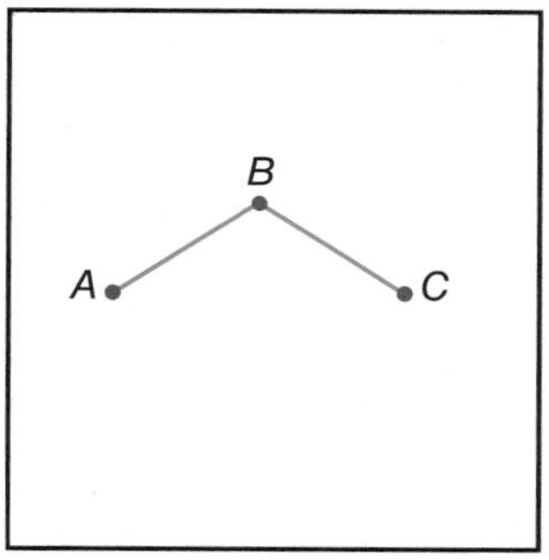

Draw three noncollinear points A, B, and C. Then draw segments $\overline{AB}$ and $\overline{BC}$.

Step 2

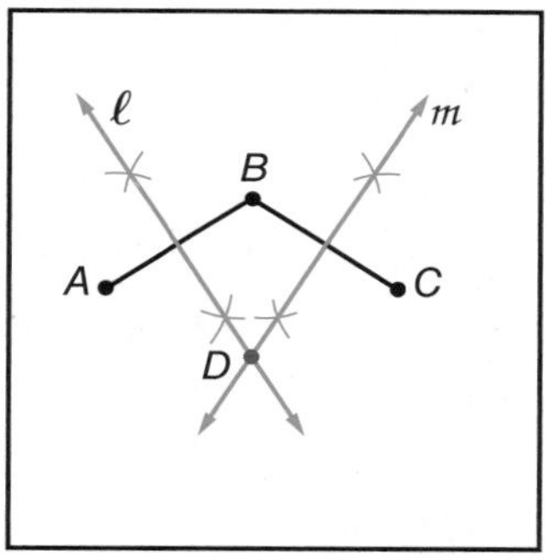

Construct the perpendicular bisectors ℓ and m of $\overline{AB}$ and $\overline{BC}$. Label the point of intersection D.

Step 3

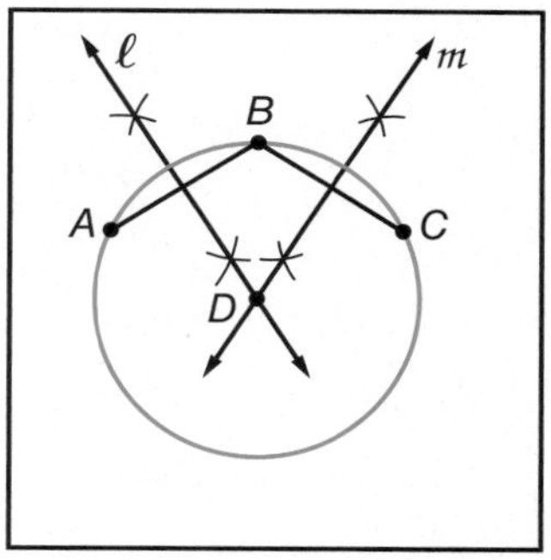

By Theorem 15.3, lines ℓ and m contain diameters of $\odot D$. With the compass at point D, draw a circle through points A, B, and C.

Check Your Understanding

Examples 1–2 ALGEBRA Find the value of x.

1.

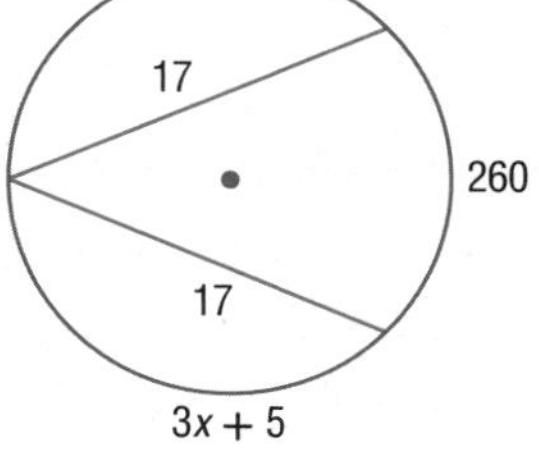

2.

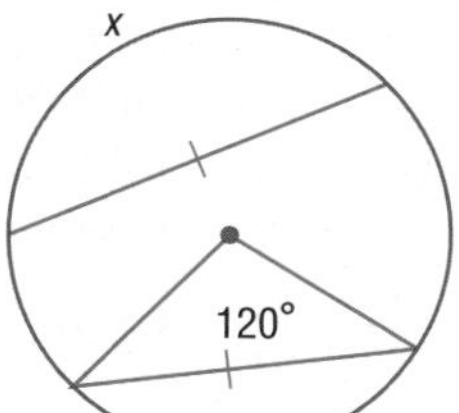

3.

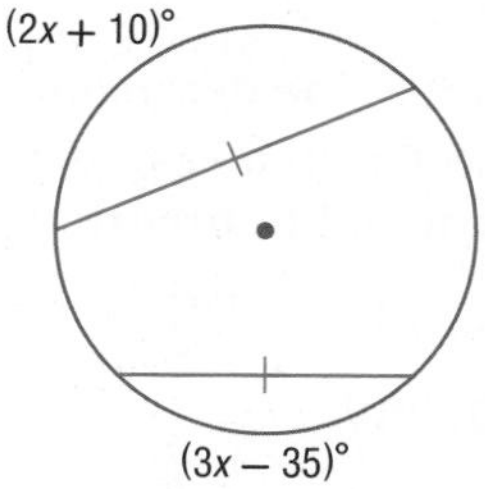

Examples 3–4 In $\odot A$, $CE = 12$ and $m\widehat{CBE} = 150$. Find each measure.

4. DE

5. $m\widehat{BE}$

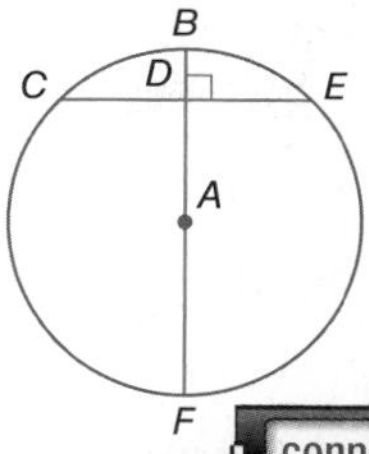

Example 5 **6.** In $\odot P$, $LM = 2x + 1$, $QN = 3x - 6$, Find x.

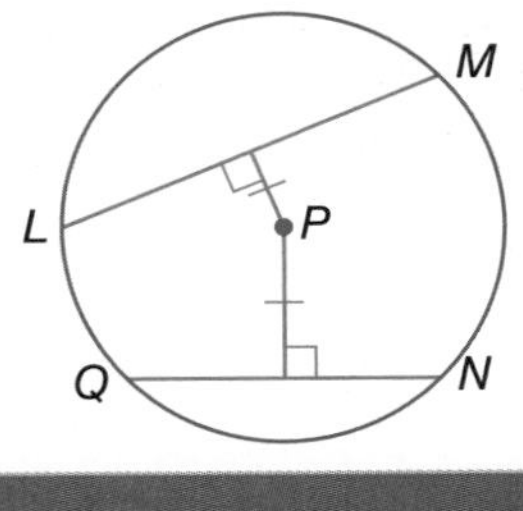

Practice and Problem Solving

Examples 1–2 **ALGEBRA** **Find the value of x.**

7.

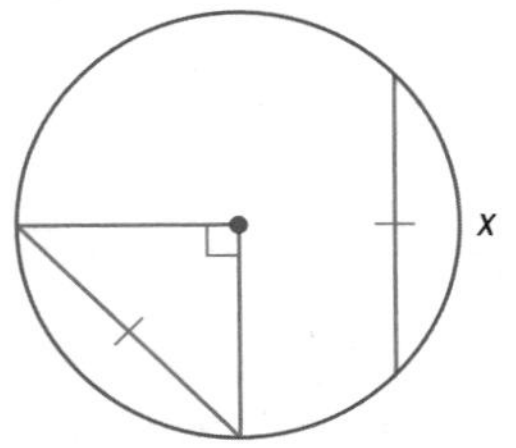

8.

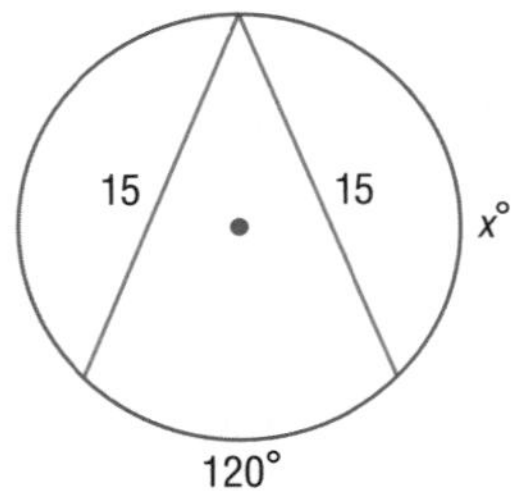

9.

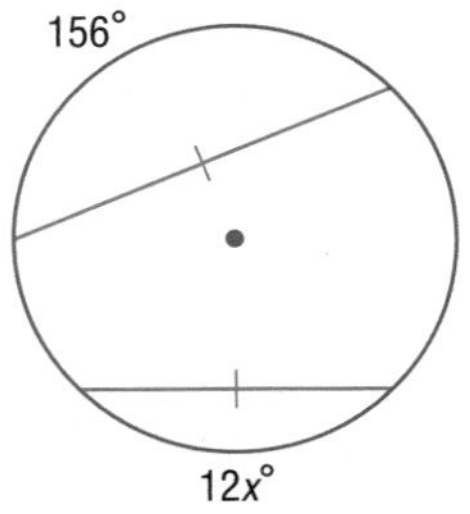

10.

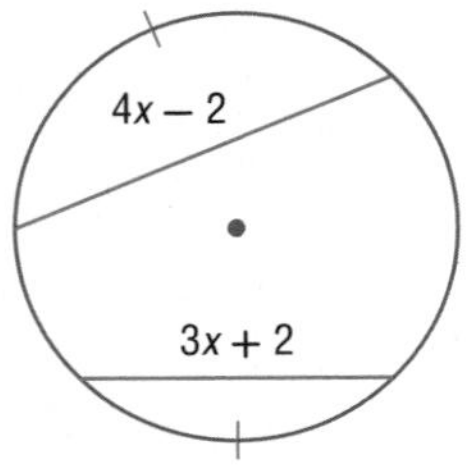

11.

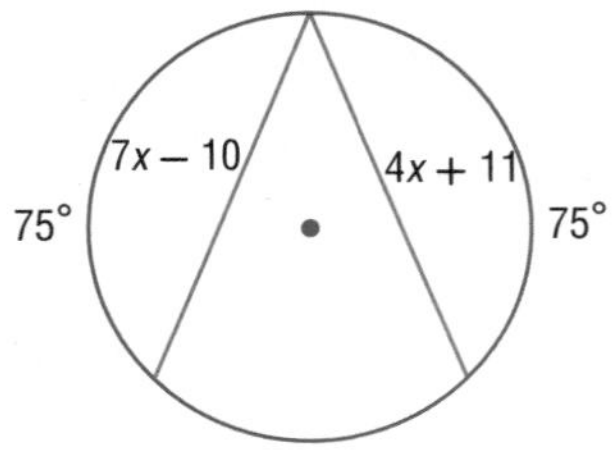

12.

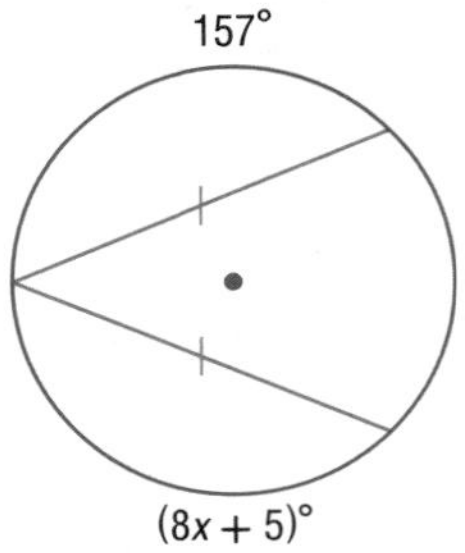

13. $\odot F \cong \odot G$

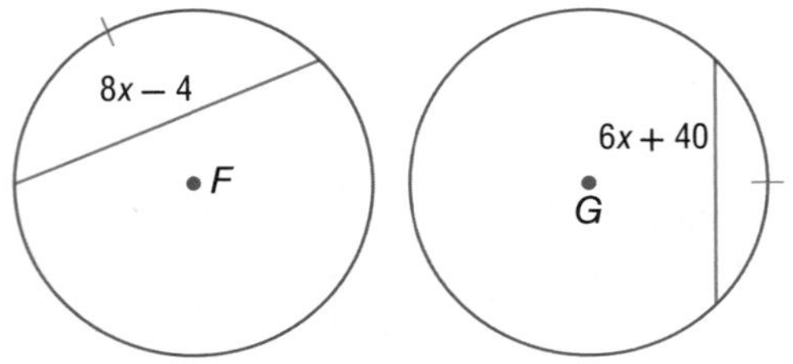

14. $\odot E \cong \odot M$

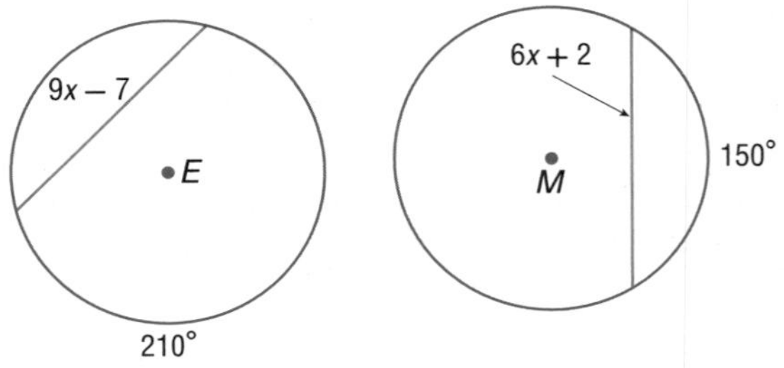

15. **CCSS MUSIC** Tiffany is learning how to play guitar. Her guitar has 6 strings that are stretched over a sound hole. If the outer strings are both E strings, and the strings are evenly spaced over the sound hole, is the length of both E strings over the sound hole the same? Explain.

Examples 3–4 **In $\odot S$, the diameter is 38, and $RV = 20$. Find each measure. Round to the nearest hundredth, if necessary.**

16. PV

17. PW

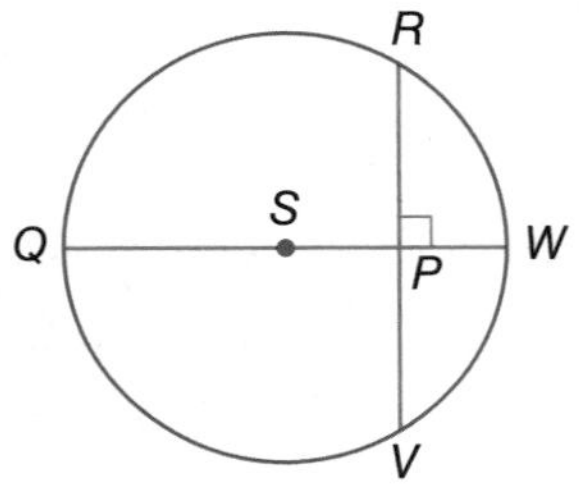

In $\odot M$, the diameter is 22, $FP = 18$, and $m\widehat{FP} = 76$. Find each measure. Round to the nearest hundredth, if necessary.

18. $m\widehat{CP}$

19. GM

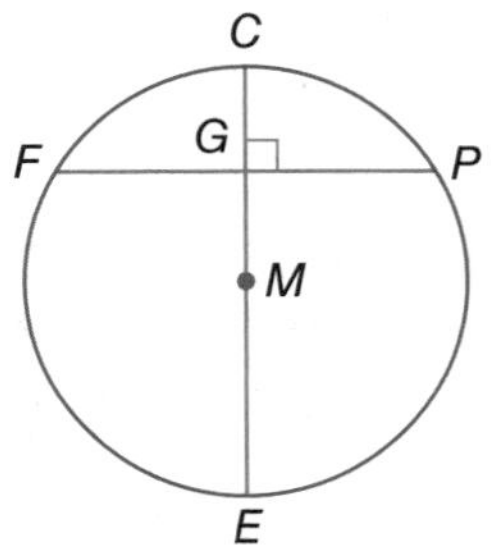

20. **GEOMETRY** Shown to the right is a triangle inscribed in $\odot M$. If $PM = MQ = MR$, write a paragraph proof proving that $\triangle JKL$ is equilateral.

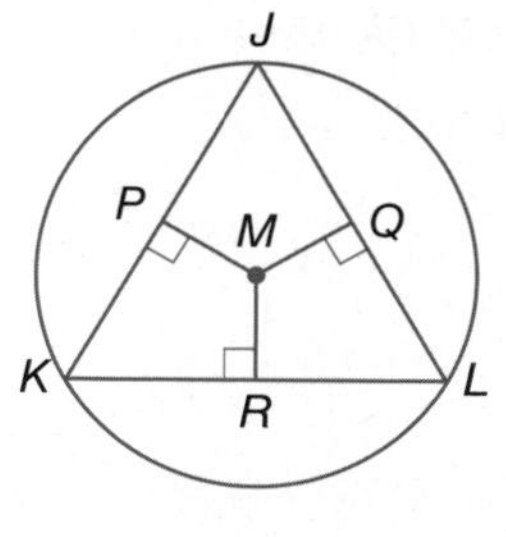

Example 5

21. **ALGEBRA** In $\odot G$, $EF \cong FG$, $BC = 12x - 26$, $AD = 9x + 28$. What is x?

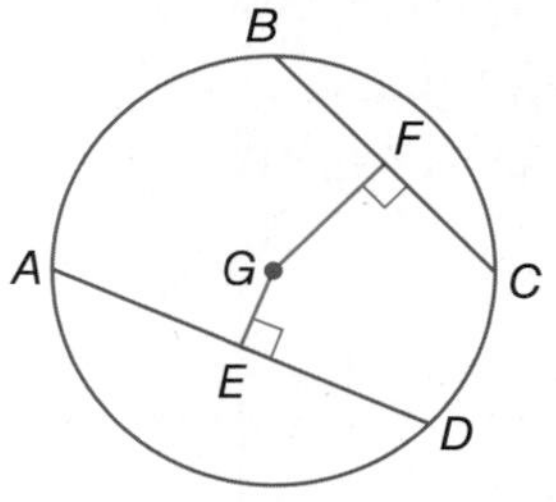

22. **ALGEBRA** In $\odot J$, $XY = 13x - 30$ and $YZ = 7x + 6$. What is x?

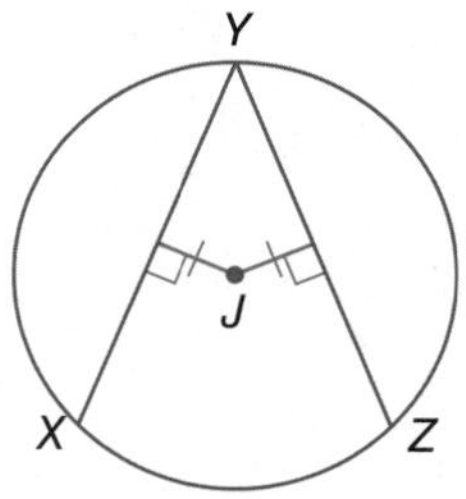

PROOF **Write a two-column proof.**

23. **Given:** $\odot P$, $\overline{KM} \perp \overline{JP}$

Prove: $\overline{JP}$ bisects $\overline{KM}$ and $\widehat{KM}$.

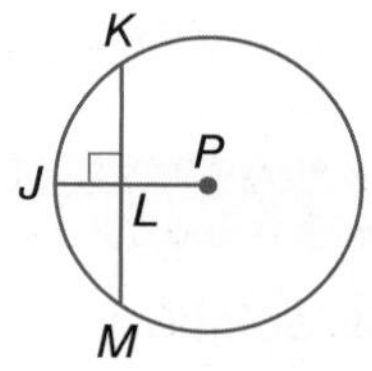

PROOF **Write the specified type of proof.**

24. paragraph proof of Theorem 15.1, part 2

Given: $\odot P$, $\overline{QR} \cong \overline{ST}$

Prove: $\widehat{QR} \cong \widehat{ST}$

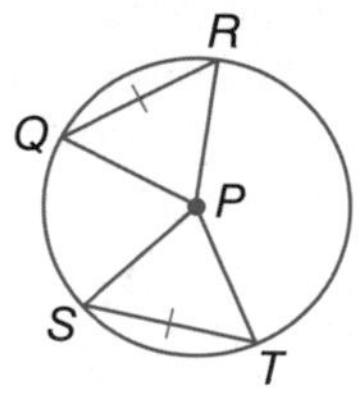

25. two-column proof of Theorem 15.2

Given: $\odot C$, $\overline{AB} \perp \overline{XY}$

Prove: $\overline{XZ} \cong \overline{YZ}$, $\widehat{XB} \cong \widehat{YB}$

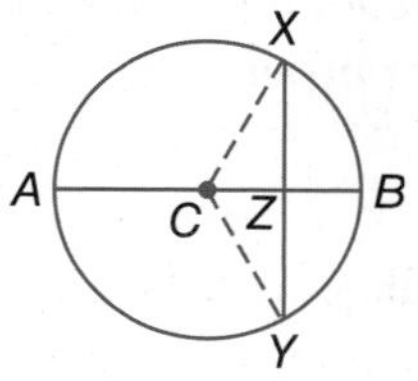

26. **CCSS** **PROOF** Write a two-column proof of Theorem 15.3.

CCSS **PROOF** **Write a two-column proof of the indicated part of Theorem 15.4**

27. In a circle, if two chords are equidistant from the center, then they are congruent.

28. In a circle, if two chords are congruent, then they are equidistant from the center.

ALGEBRA **Find the value of x.**

29. $\widehat{JML} \cong \widehat{JLM}$

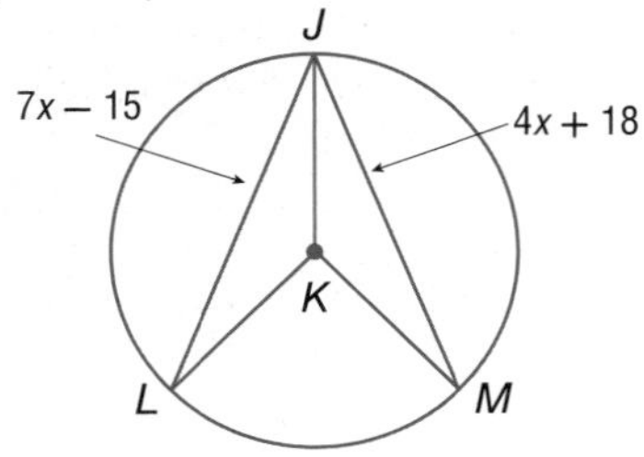

30. $\overline{XZ} \cong \overline{YW}$

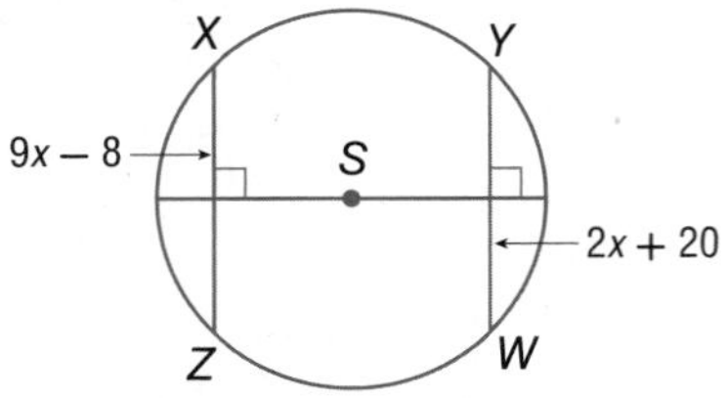

31. $\overline{AC} \cong \overline{BD}$

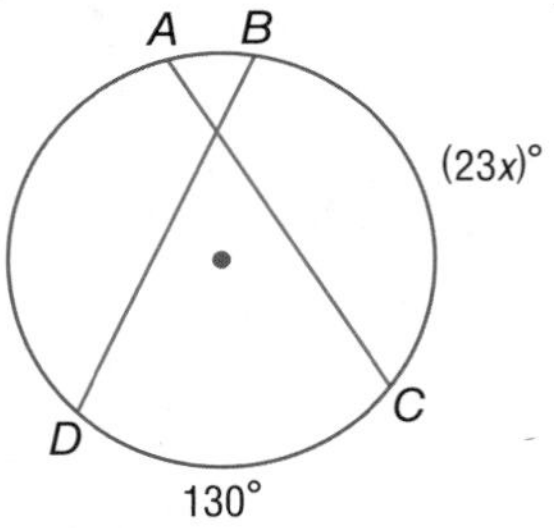

H.O.T. Problems Use Higher-Order Thinking Skills

32. REASONING If AB and CD are chords of $\odot E$ and the length of $AB = 2CD$, is CD *sometimes, always, or never* a radius of the circle?

33. ERROR ANALYSIS Eric and Alex are looking at the diagram of the circle to the right. The radius of the circle is 10 inches and the length of ED is 6 inches. Alex thinks the length of AC is 16 inches, while Eric thinks the length of AC is 8 inches. Is either of them correct? Explain your reasoning.

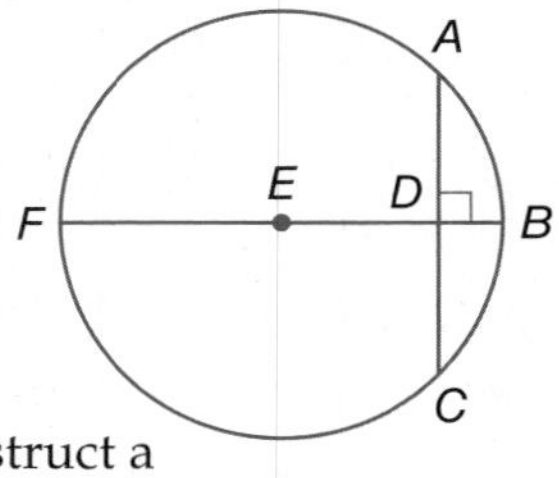

34. OPEN ENDED Draw a circle and label it $\odot A$. Draw a chord of the circle. Construct a diameter perpendicular to the chord. Measure the radius of the circle and the distance from the center of the circle to the chord. Find the length of the chord.

35. CHALLENGE $\odot M \cong \odot N$ and $MP = NQ$, find x and y.

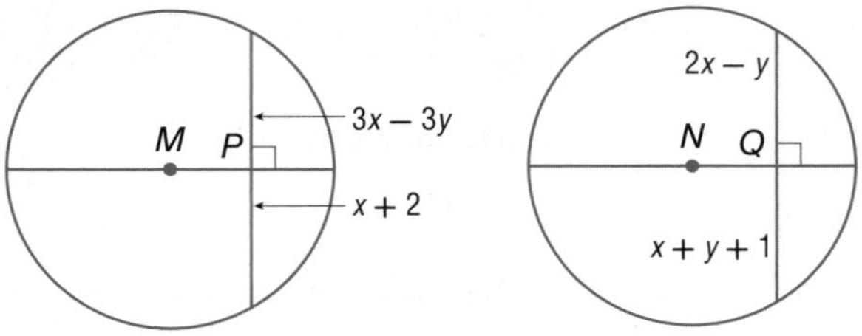

36. WRITING IN MATH Explain the different ways you know how to prove two chords of a circle are congruent.

Standardized Test Practice

37. If $CW = WF$ and $ED = 30$, what is DF?

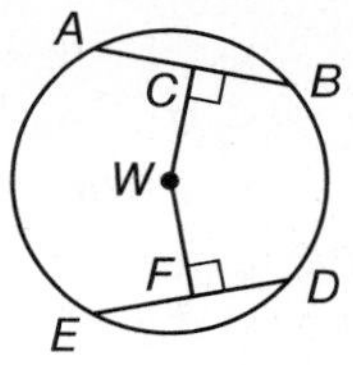

A 60

B 45

C 30

D 15

38. ALGEBRA Write the ratio of the area of the circle to the area of the square in simplest form.

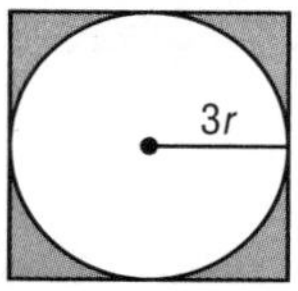

F $\frac{\pi}{4}$ **H** $\frac{3\pi}{4}$

G $\frac{\pi}{2}$ **J** π

39. SHORT RESPONSE The pipe shown is divided into five equal sections. How long is the pipe in feet (ft) and inches (in.)?

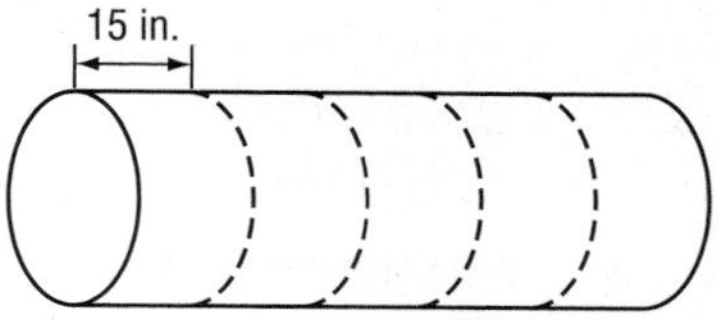

40. SAT/ACT Point *B* is the center of a circle, tangent to the *y*-axis, and the coordinates of Point *B* are (3, 1). What is the area of the circle?

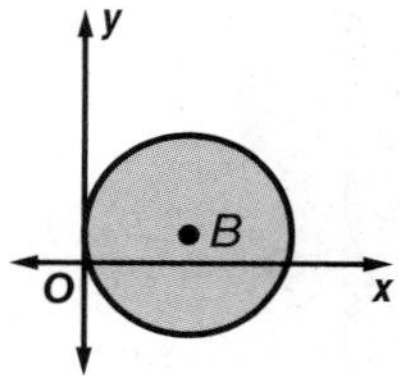

A π units^2 **D** 6π units^2

B 3π units^2 **E** 9π units^2

C 4π units^2

Spiral Review

41. CRAFTS Ruby created a pattern to sew flowers onto a quilt by first drawing a regular pentagon that was 3.5 inches long on each side. Then she added a semicircle onto each side of the pentagon to create the appearance of five petals. How many inches of gold trim does she need to edge 10 flowers? Round to the nearest inch. (Lesson 15-1)

42. INVESTMENTS Joey's investment of $2500 has been decreasing in value at a rate of 1.5% each year. What will his investment be worth in 5 years? (Lesson 7-7)

Write an equation for the *n*th term of each geometric sequence, and find the seventh term of each sequence. (Lesson 7-6)

43. 1, 2, 4, 8, … **44.** −20, −10, −5, … **45.** 4, −12, 36, …

46. 99, −33, 11, … **47.** 22, 44, 88, … **48.** $\frac{2}{3}, \frac{1}{3}, \frac{1}{6}, \ldots$

Skills Review

ALGEBRA Quadrilateral *WXZY* is a rhombus. Find each value or measure.

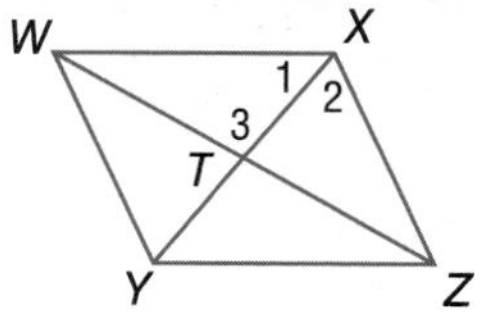

49. If $m\angle 3 = y^2 - 31$, find y.

50. If $m\angle XZY = 56$, find $m\angle YWZ$.

LESSON 15-3 Tangents

Then

- You used the Pythagorean Theorem to find side lengths of right triangles.

Now

1. Use properties of tangents.
2. Solve problems involving circumscribed polygons.

Why?

- The first bicycles were moved by pushing your feet on the ground. Modern bicycles use pedals, a chain, and gears. The chain loops around circular gears. The length of the chain between these gears is measured from the points of tangency.

NewVocabulary

tangent
point of tangency
common tangent

Common Core State Standards

Content Standards

G.CO.12 Make formal geometric constructions with a variety of tools and methods (compass and straightedge, string, reflective devices, paper folding, dynamic geometric software, etc.).

G.C.4 Construct a tangent line from a point outside a given circle to the circle.

Mathematical Practices

1. Make sense of problems and persevere in solving them.
2. Reason abstractly and quantitatively.

1 Tangents

A **tangent** is a line in the same plane as a circle that intersects the circle in exactly one point, called the **point of tangency**. $\overleftrightarrow{AB}$ is tangent to $\odot C$ at point *A*. $\overline{AB}$ and $\overrightarrow{AB}$ are also called tangents.

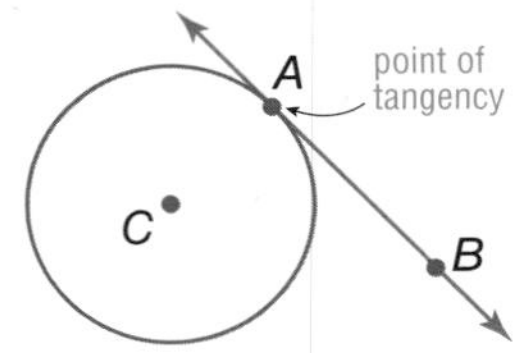

A **common tangent** is a line, ray, or segment that is tangent to two circles in the same plane. In each figure below, line ℓ is a common tangent of circles *F* and *G*.

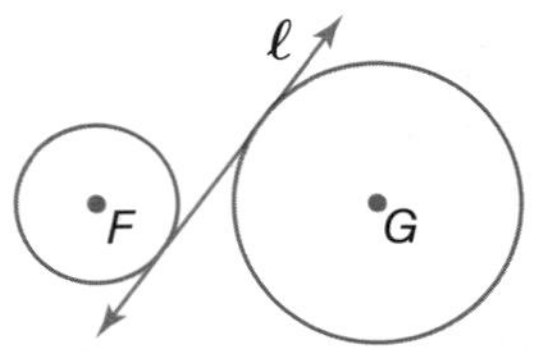

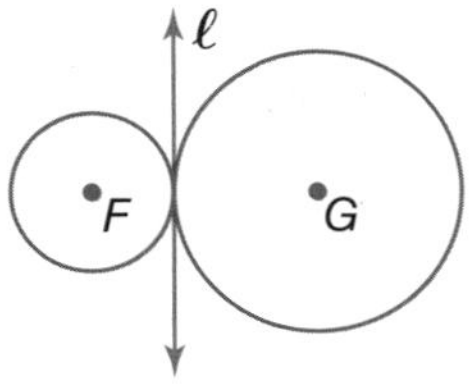

Example 1 Identify Common Tangents

Copy each figure and draw the common tangents. If no common tangent exists, state *no common tangent*.

a.

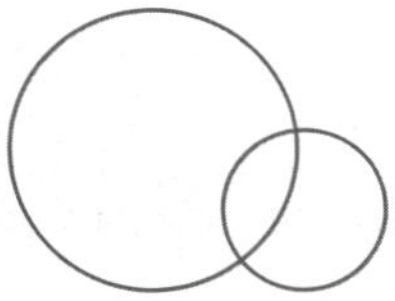

These circles have two common tangents.

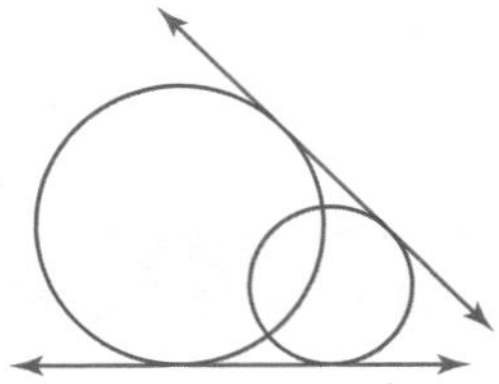

b.

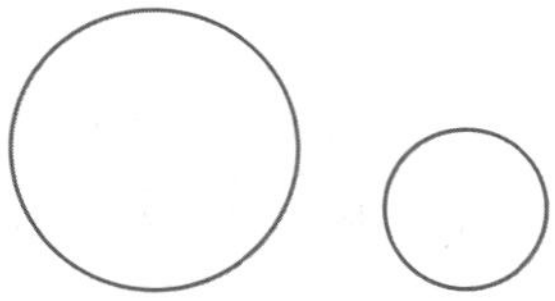

These circles have 4 common tangents.

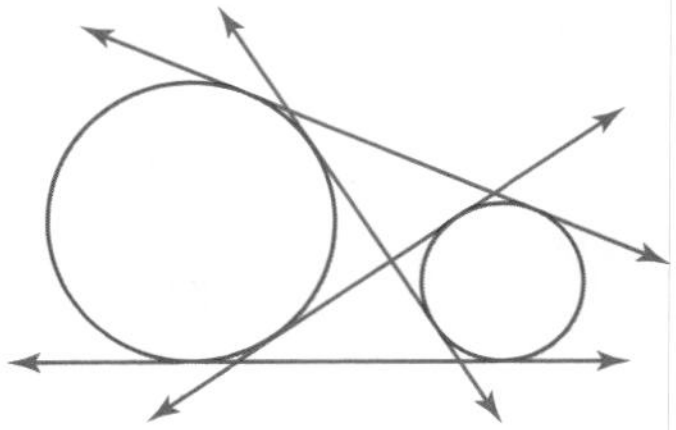

GuidedPractice

1A.

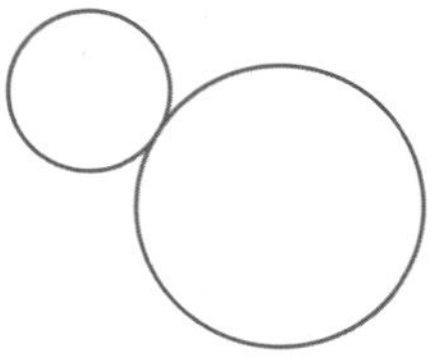

1B.

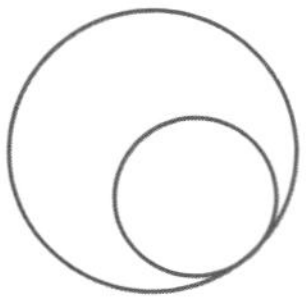

The shortest distance from a tangent to the center of a circle is the radius drawn to the point of tangency.

Theorem 15.5

Words In a plane, a line is tangent to a circle if and only if it is perpendicular to a radius drawn to the point of tangency.

Example Line ℓ is tangent to $\odot S$ if and only if $\ell \perp \overline{ST}$.

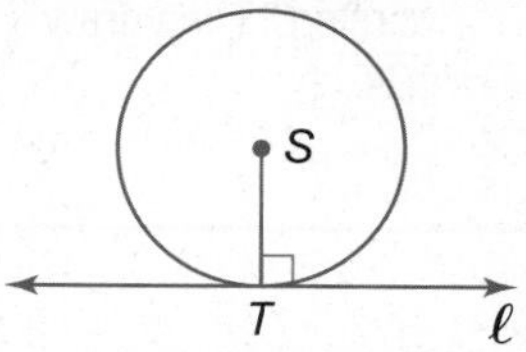

You will prove both parts of Theorem 15.5 in Exercises 32 and 33.

Example 2 Identify a Tangent

PT

$\overline{JL}$ is a radius of $\odot J$. Determine whether $\overline{KL}$ is tangent to $\odot J$. Justify your answer.

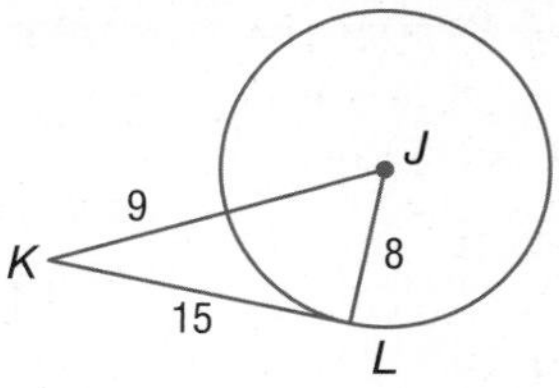

Test to see if $\triangle JKL$ is a right triangle.

$8^2 + 15^2 \stackrel{?}{=} (8 + 9)^2$ Pythagorean Theorem

$289 = 289$ ✓ Simplify.

$\triangle JKL$ is a right triangle with right angle JLK. So $\overline{KL}$ is perpendicular to radius $\overline{JL}$ at point L. Therefore, by Theorem 10.10, $\overline{KL}$ is tangent to $\odot J$.

Guided Practice

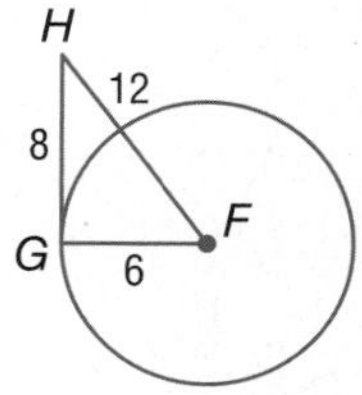

2. Determine whether $\overline{GH}$ is tangent to $\odot F$. Justify your answer.

You can also use Theorem 15.5 to identify missing values.

Problem-Solving Tip

CCSS Sense-Making You can use the *solve a simpler problem* strategy by sketching and labeling the right triangles without the circles. A drawing of the triangle in Example 3 is shown below.

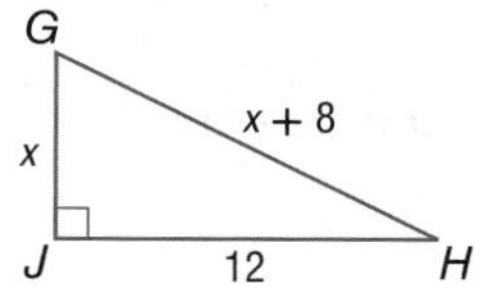

Example 3 Use a Tangent to Find Missing Values

PT

$\overline{JH}$ is tangent to $\odot G$ at J. Find the value of x.

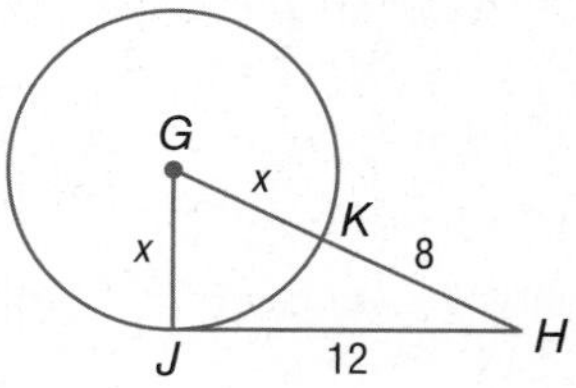

By Theorem 10.10, $\overline{JH} \perp \overline{GJ}$. So, $\triangle GHJ$ is a right triangle.

$GJ^2 + JH^2 = GH^2$ Pythagorean Theorem

$x^2 + 12^2 = (x + 8)^2$ $GJ = x$, $JH = 12$, and $GH = x + 8$

$x^2 + 144 = x^2 + 16x + 64$ Multiply.

$80 = 16x$ Simplify.

$5 = x$ Divide each side by 16.

Guided Practice

Find the value of x. Assume that segments that appear to be tangent are tangent.

3A.

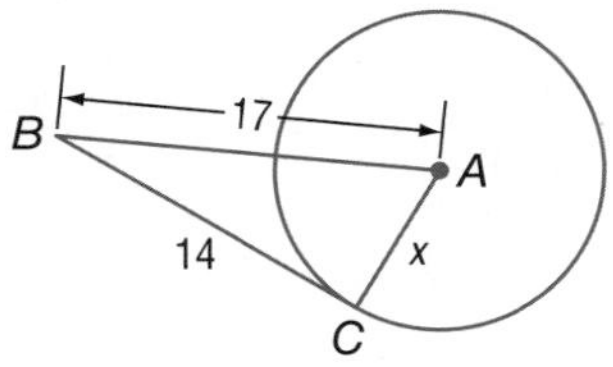

3B.

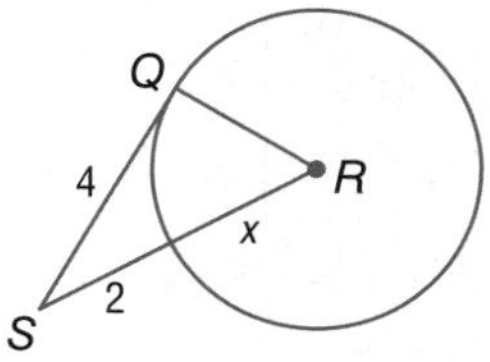

Construction Line Tangent to a Circle Through an External Point

Step 1 Use a compass to draw circle C and a point A outside of circle C. Then draw $\overline{CA}$.

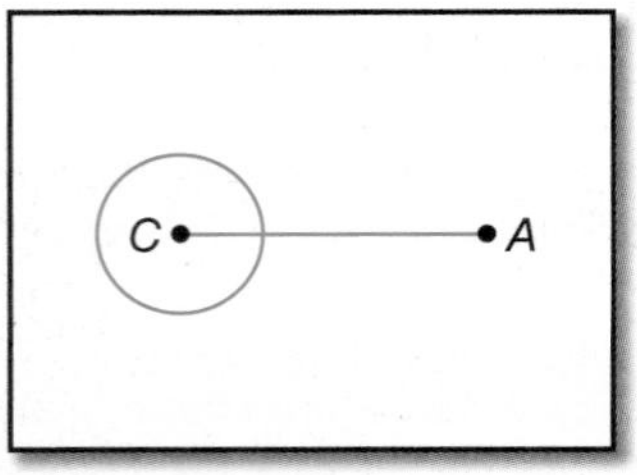

Step 2 Construct line ℓ, the perpendicular bisector of $\overline{CA}$. Label the point of intersection X.

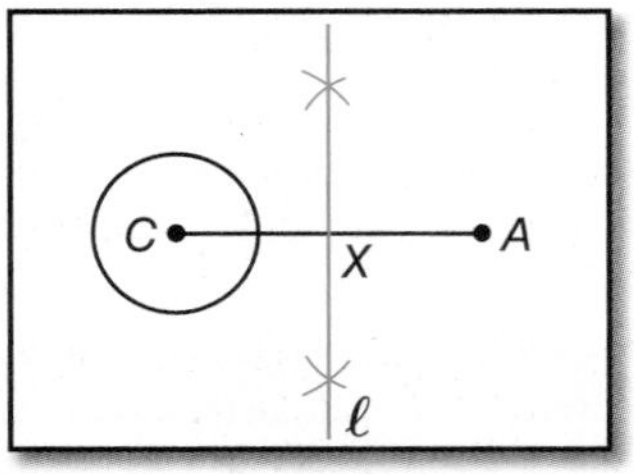

Step 3 Construct circle X with radius $\overline{XC}$. Label the points of intersection of the two circles D and E.

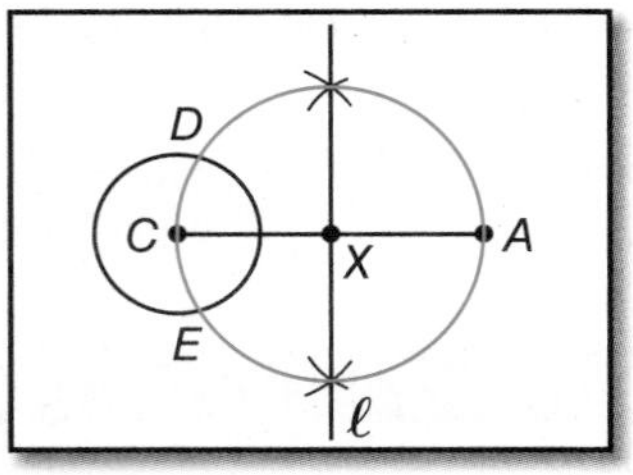

Step 4 Draw $\overleftrightarrow{AD}$ and $\overline{DC}$. $\triangle ADC$ is inscribed in a semicircle. So, $\angle ADC$ is a right angle and $\overleftrightarrow{AD}$ is tangent to $\odot C$.

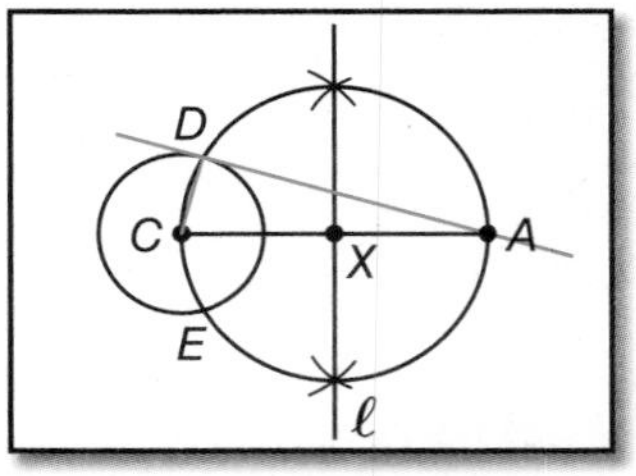

You will justify this construction in Exercise 36 and construct a line tangent to a circle through a point on the circle in Exercise 34.

More than one line can be tangent to the same circle.

Theorem 15.6

Words If two segments from the same exterior point are tangent to a circle, then they are congruent.

Example If $\overline{AB}$ and $\overline{CB}$ are tangent to $\odot D$, then $\overline{AB} \cong \overline{CB}$.

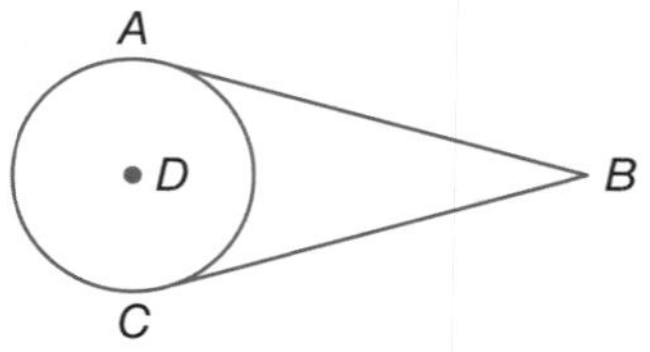

You will prove Theorem 15.6 in Exercise 28.

Example 4 Use Congruent Tangents to Find Measures

PT

ALGEBRA $\overline{AB}$ and $\overline{CB}$ are tangent to $\odot D$. Find the value of x.

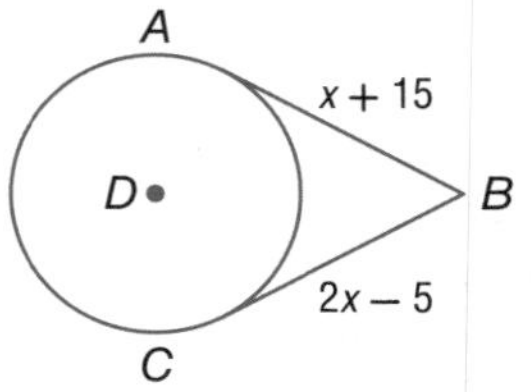

$AB = CB$	Tangents from the same exterior point are congruent.
$x + 15 = 2x - 5$	Substitution
$15 = x - 5$	Subtract x from each side.
$20 = x$	Add 5 to each side.

GuidedPractice

ALGEBRA Find the value of x. Assume that segments that appear to be tangent are tangent.

4A.

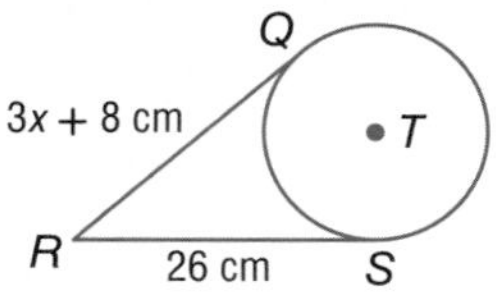

4B.

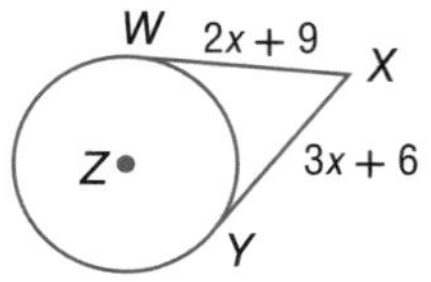

2 Circumscribed Polygons

A polygon is circumscribed about a circle if every side of the polygon is tangent to the circle.

WatchOut!

Identifying Circumscribed Polygons Just because the circle is tangent to one or more of the sides of a polygon does not mean that the polygon is circumscribed about the circle, as shown in the second set of figures.

Circumscribed Polygons	Polygons Not Circumscribed
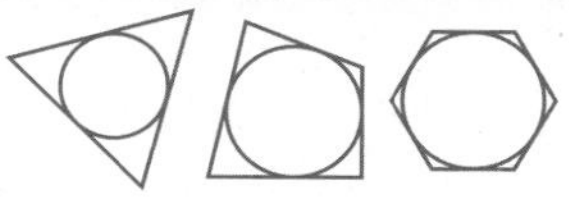	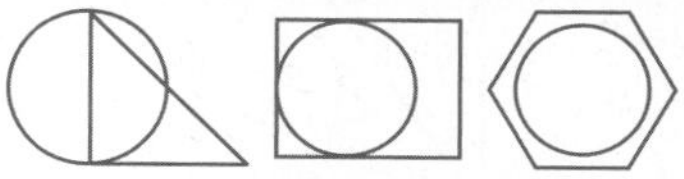

You can use Theorem 15.6 to find missing measures in circumscribed polygons.

Real-World Example 5 Find Measures in Circumscribed Polygons

GRAPHIC DESIGN A graphic designer is giving directions to create a larger version of the triangular logo shown. If $\triangle ABC$ is circumscribed about $\odot G$, find the perimeter of $\triangle ABC$.

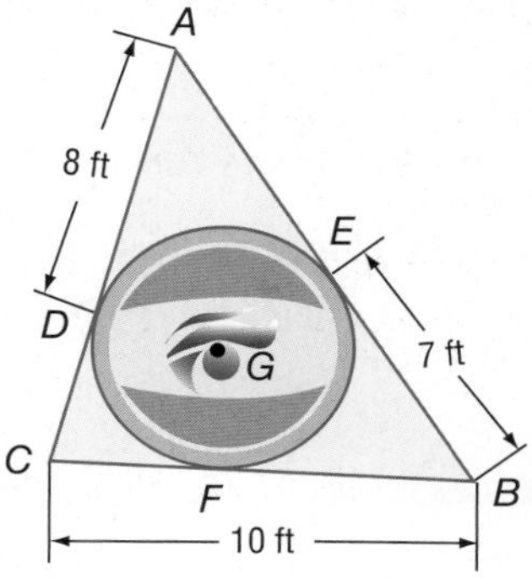

Step 1 Find the missing measures.

Since $\triangle ABC$ is circumscribed about $\odot G$, $\overline{AE}$ and $\overline{AD}$ are tangent to $\odot G$, as are $\overline{BE}$, $\overline{BF}$, $\overline{CF}$, and $\overline{CD}$. Therefore, $\overline{AE} \cong \overline{AD}$, $\overline{BF} \cong \overline{BE}$, and $\overline{CF} \cong \overline{CD}$.

So, $AE = AD = 8$ feet, $BF = BE = 7$ feet.

By Segment Addition, $CF + FB = CB$, so $CF = CB - FB = 10 - 7$ or 3 feet. So, $CD = CF = 3$ feet.

Step 2 Find the perimeter of $\triangle ABC$.

$$\text{perimeter} = AE + EB + BC + CD + DA$$
$$= 8 + 7 + 10 + 3 + 8 \text{ or } 36$$

So, the perimeter of $\triangle ABC$ is 36 feet.

GuidedPractice

5. Quadrilateral $RSTU$ is circumscribed about $\odot J$. If the perimeter is 18 units, find x.

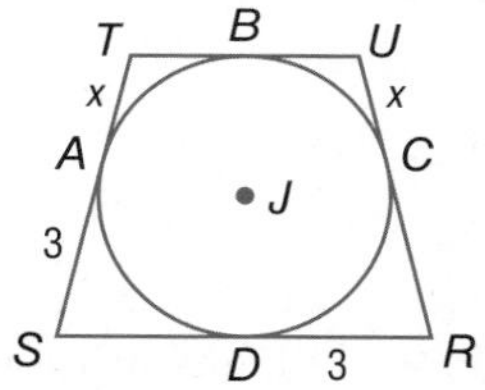

Check Your Understanding

Example 1

1. Copy the figure shown, and draw the common tangents. If no common tangent exists, *state no common tangent*.

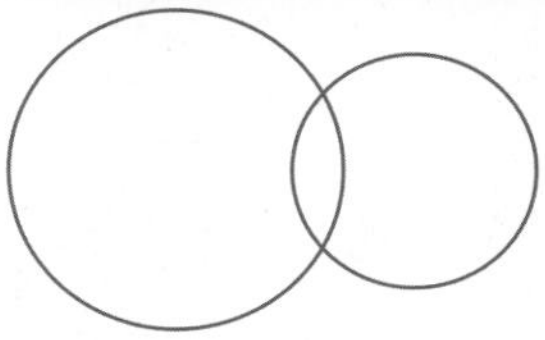

Example 2

Determine whether $\overline{AB}$ is tangent to $\odot C$. Justify your answer.

2.

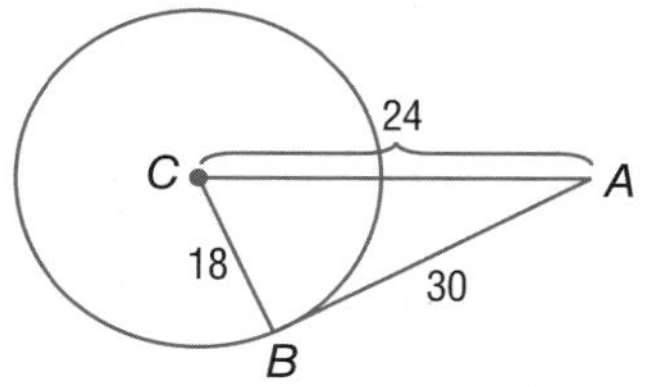

3

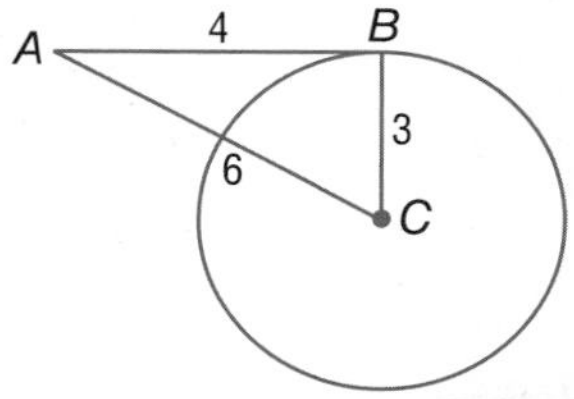

Examples 3–4 **Find x. Assume that segments that appear to be tangent are tangent.**

4.

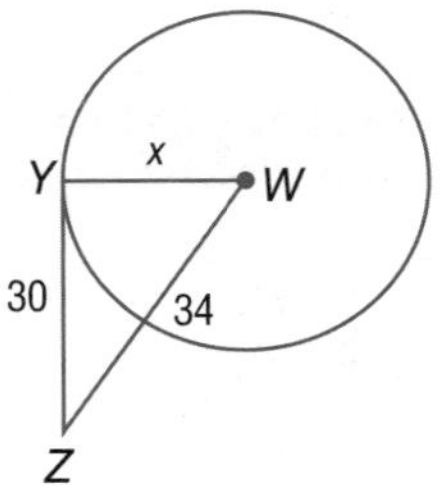

5.

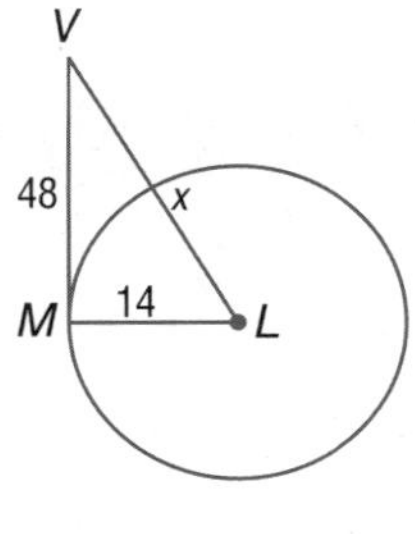

6.

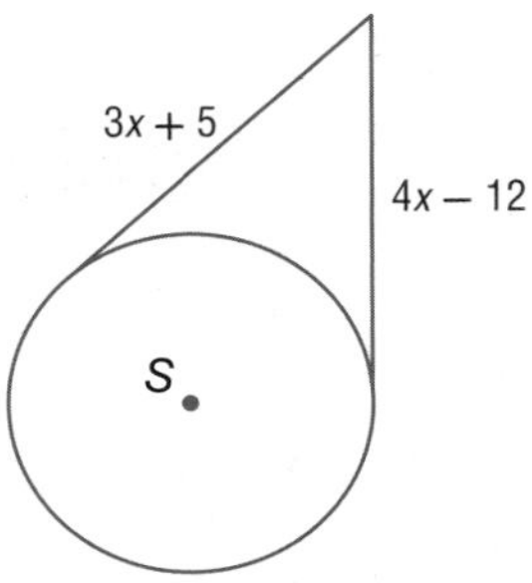

7. OLYMPICS Below is a picture of the symbol for the Olympic games, the Olympic rings. Copy the diagram of the rings and draw any common tangents of the blue ring and the green ring.

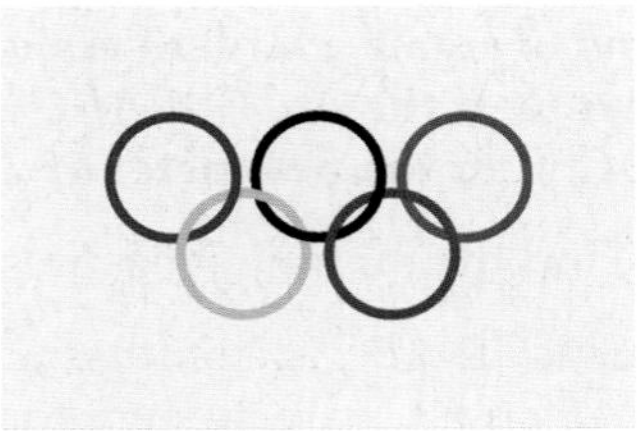

Example 5

8. CCSS **ALGEBRA** Quadrilateral $JKLM$ is circumscribed about $\odot S$.

a. Find x.

b. Find the perimeter of quadrilateral $\triangle JKLM$.

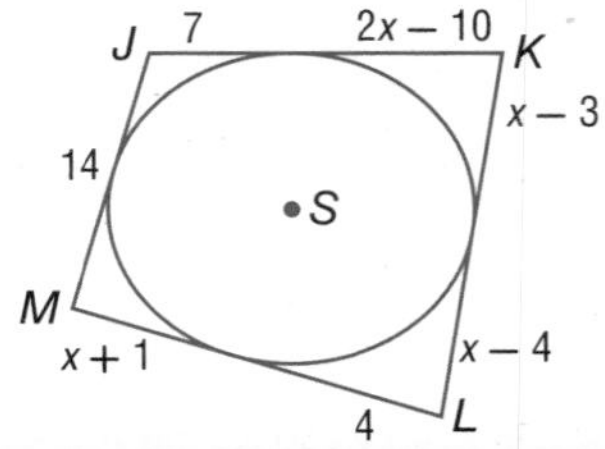

Practice and Problem Solving

Example 1 **Copy each figure shown, and draw the common tangents. If no common tangent exists, *state no common tangent*.**

9.

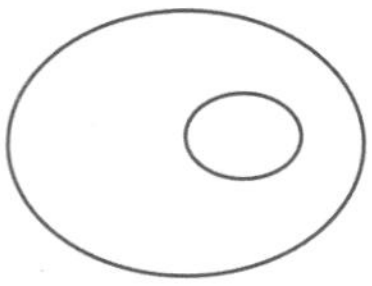

10.

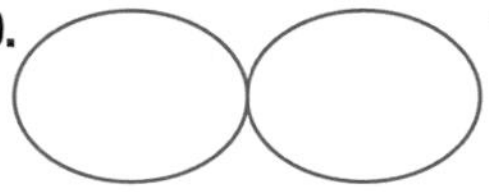

11.

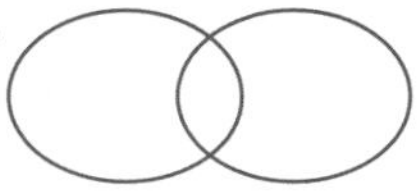

12.

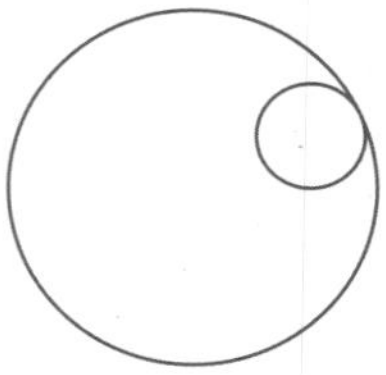

Example 2 **Determine whether each $\overline{JK}$ is tangent to the given circle. Justify your answer.**

13.

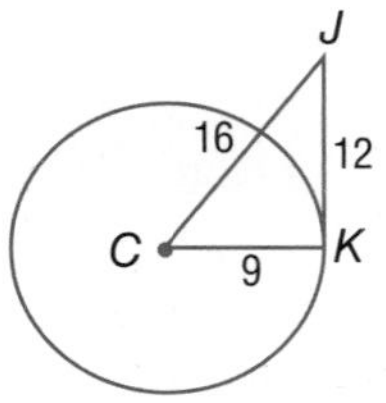

14.

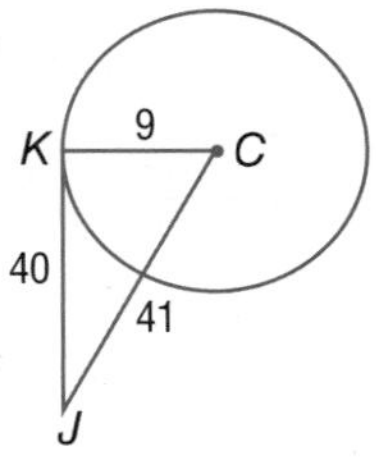

15.

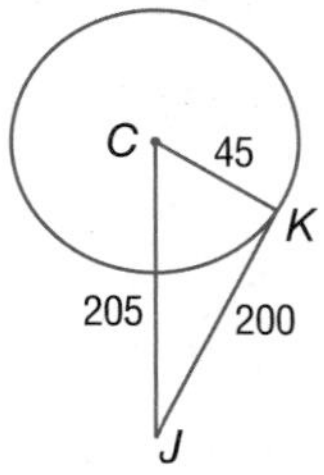

16.

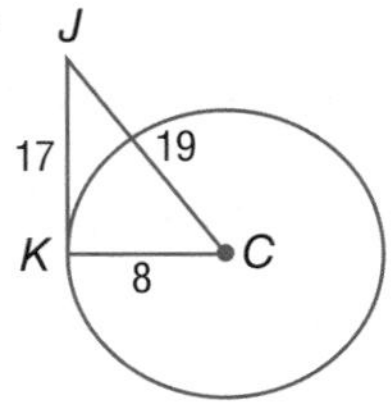

Examples 3–4 **Find x. Assume that segments that appear to be tangent are tangent.**

17.

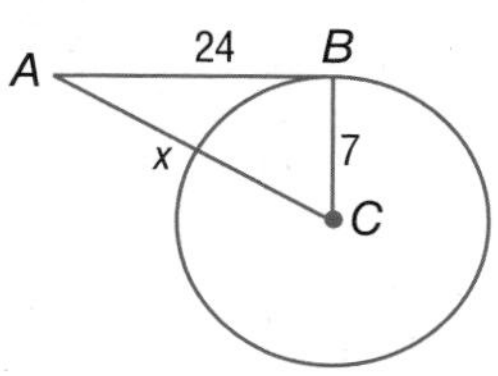

18.

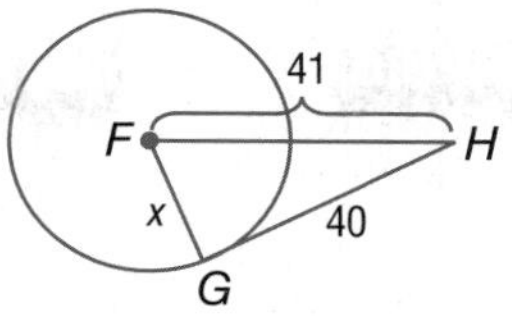

19.

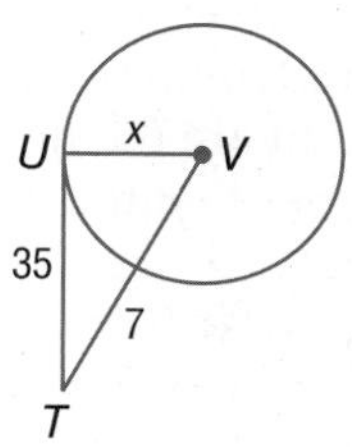

20.

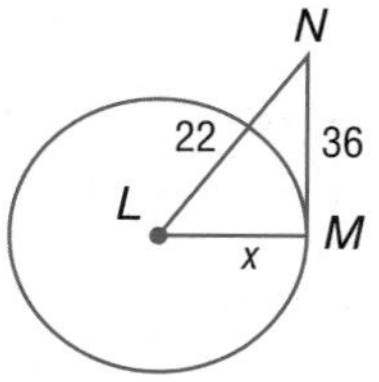

21.

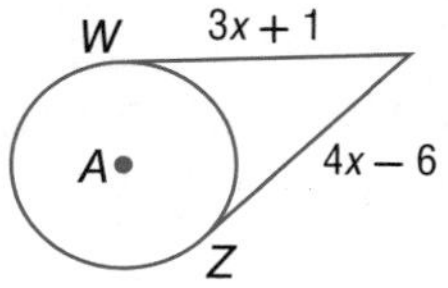

22.

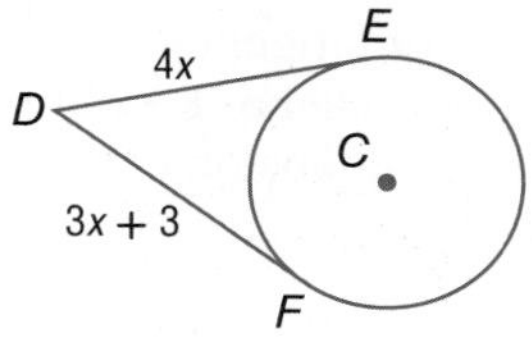

23. PEP RALLY Kristen is sketching a megaphone on a sign advertising a school pep rally. If AB and AC are tangent to the circle that forms the opening of the megaphone and AB is 10 inches long, how long is AC?

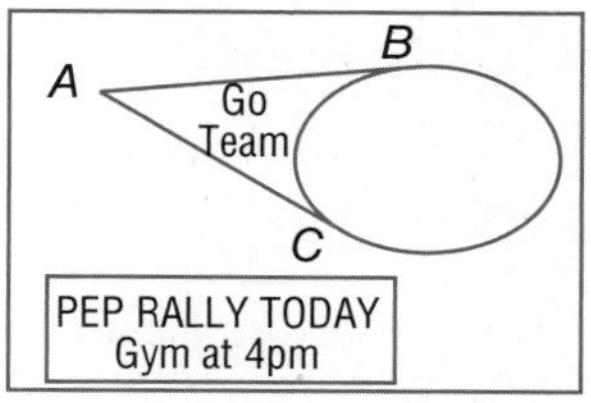

Example 5 CCSS **Find the value of x. Then find the perimeter.**

24.

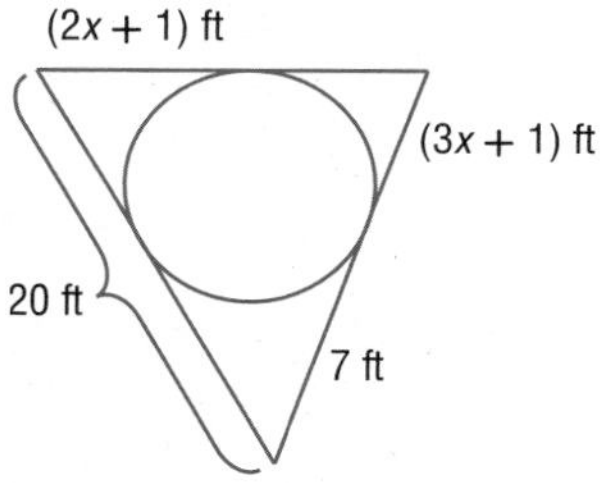

25.

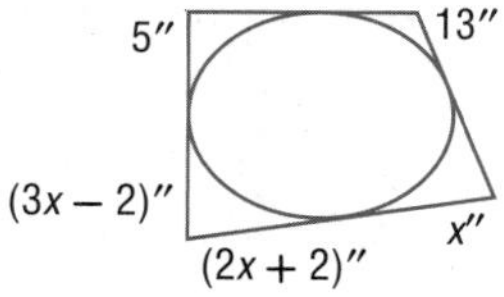

Find x to the nearest hundredth. Assume that segments that appear to be tangent are tangent.

26.

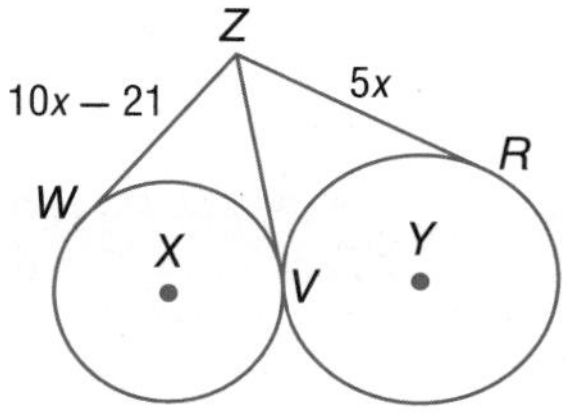

27.

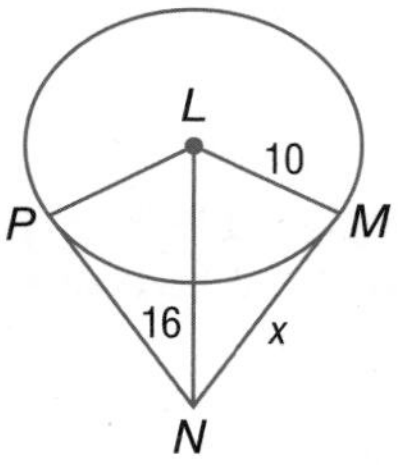

Write the specified type of proof.

28. two-column proof of Theorem 15.6

Given: $\overline{AC}$ is tangent to $\odot H$ at C.
$\overline{AB}$ is tangent to $\odot H$ at B.

Prove: $\overline{AC} \cong \overline{AB}$

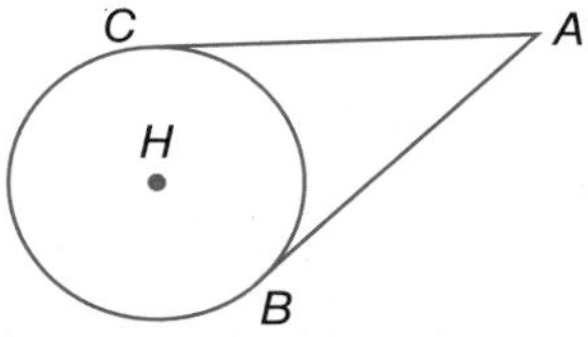

29. paragraph proof.

Given: Circle G is circumscribed by equilateral triangle ABC, D is the midpoint of AB.

Prove: ΔDEF is equilateral.

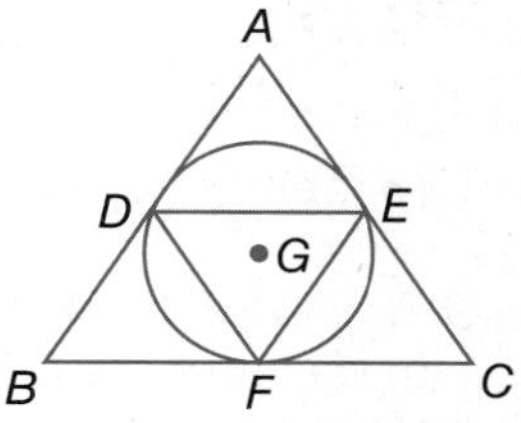

30. **CLUBS** The math club is designing a new logo that consists of a circle circumscribed in an equilateral triangle that is 3 inches long. What are the minimum dimensions of paper that the logo could fit on?

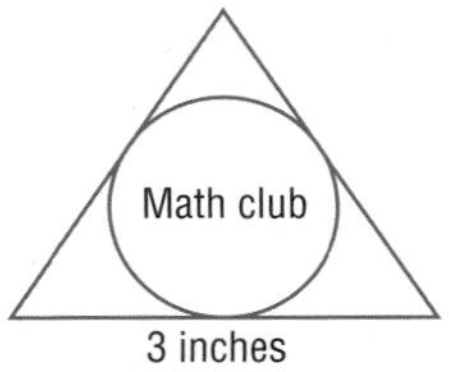

31. **QUILTING** Carol is creating a quilt that contains 9 circles all circumscribed in a square border. Create a diagram of the quilt and mark each point of tangency with an x. How many points of tangency are there in the diagram? If each circle has a radius of 6 inches, what are the dimensions of the quilt?

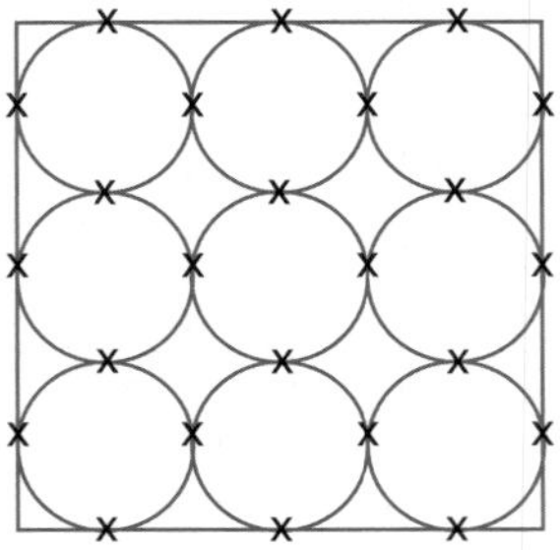

32. **PROOF** Write an indirect proof to show that if a line is tangent to a circle, then it is perpendicular to a radius of the circle. (Part 1 of Theorem 15.5)

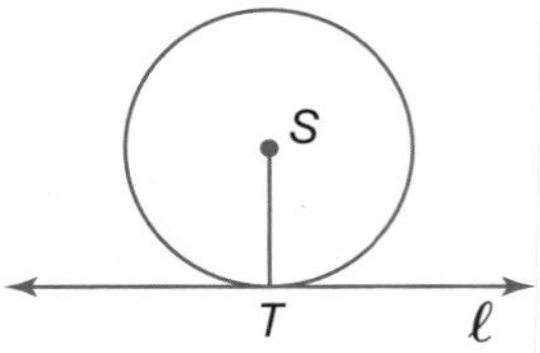

Given: ℓ is tangent to $\odot S$ at T; $\overline{ST}$ is a radius of $\odot S$.

Prove: $\ell \perp \overline{ST}$

(*Hint:* Assume ℓ is *not* $\perp$ to $\overline{ST}$.)

33. **PROOF** Write an indirect proof to show that if a line is perpendicular to the radius of a circle at its endpoint, then the line is a tangent of the circle. (Part 2 of Theorem 15.5)

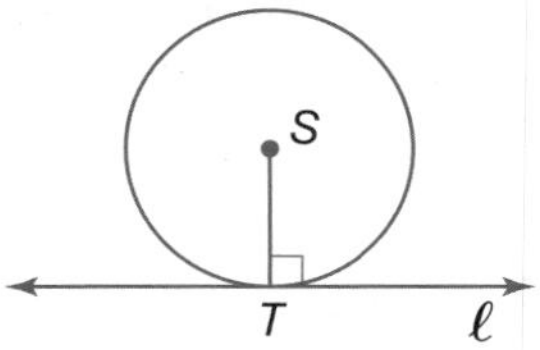

Given: $\ell \perp \overline{ST}$; $\overline{ST}$ is a radius of $\odot S$.

Prove: ℓ is tangent to $\odot S$.

(*Hint:* Assume ℓ is *not* tangent to $\odot S$.)

34. **CCSS TOOLS** Construct a line tangent to a circle through a point on the circle.

Use a compass to draw $\odot A$. Choose a point P on the circle and draw $\overleftrightarrow{AP}$. Then construct a segment through point P perpendicular to $\overleftrightarrow{AP}$. Label the tangent line t. Explain and justify each step.

H.O.T. Problems Use Higher-Order Thinking Skills

35. **OPEN ENDED** Draw a circle circumscribed in a pentagon.

36. **WRITING IN MATH** Explain and justify each step of the construction of a line tangent to a circle through an external point.

37. **CHALLENGE** If $\overline{AB}$ and $\overline{CB}$ are tangent to circles Q and R, find x and y.

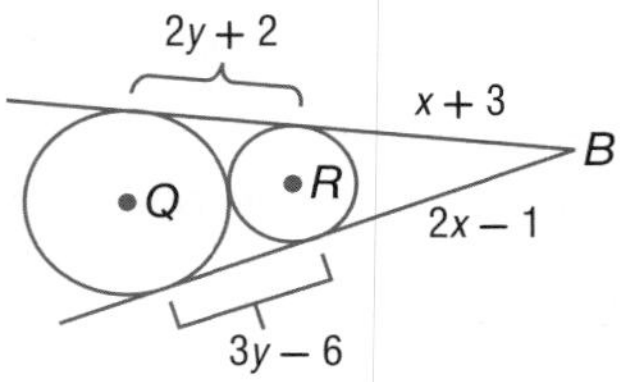

38. **REASONING** Determine if the following statements are *sometimes*, *always*, or *never* true.

a. Two concentric circles have a common tangent.

b. Two non-intersecting circles have a common tangent.

c. Two intersecting circles have a common tangent.

39. **WRITING IN MATH** Explain how the Pythagorean theorem can be used to determine if a line is tangent to a circle.

Standardized Test Practice

40. $\odot P$ has a radius of 10 centimeters, and $\overline{ED}$ is tangent to the circle at point D. F lies both on $\odot P$ and on segment $\overline{EP}$. If $ED = 24$ centimeters, what is the length of $\overline{EF}$?

A 10 cm **C** 21.8 cm
B 16 cm **D** 26 cm

41. SHORT RESPONSE A square is inscribed in a circle having a radius of 6 inches. Find the length of each side of the square.

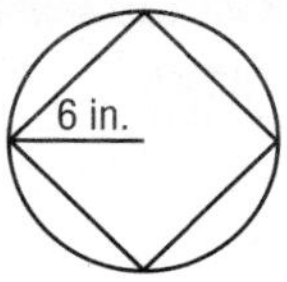

42. ALGEBRA Which of the following shows $25x^2 - 5x$ factored completely?

F $5x(x)$ **H** $x(x - 5)$
G $5x(5x - 1)$ **J** $x(5x - 1)$

43. SAT/ACT What is the perimeter of the triangle shown below?

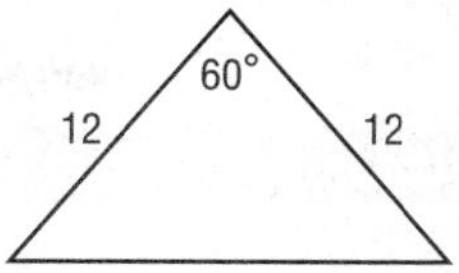

A 12 units **D** 36 units
B 24 units **E** 104 units
C 34.4 units

Spiral Review

In $\odot F$, $GK = 14$ and $m\widehat{GHK} = 142$. Find each measure. Round to the nearest hundredth. (Lesson 15-2)

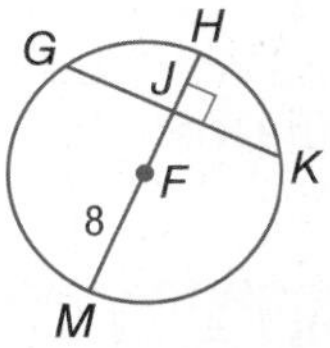

44. $m\widehat{GH}$ **45.** JK **46.** $m\widehat{KM}$

Determine whether the triangles are similar. If so, write a similarity statement. Explain your reasoning. (Lesson 14-1)

47.

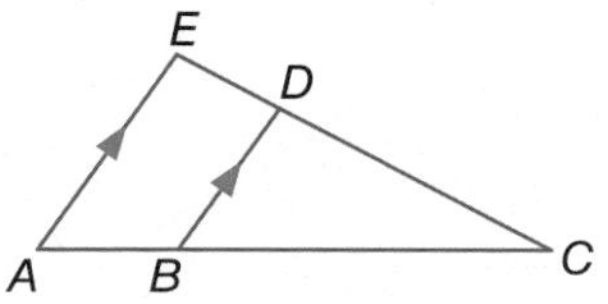

48.

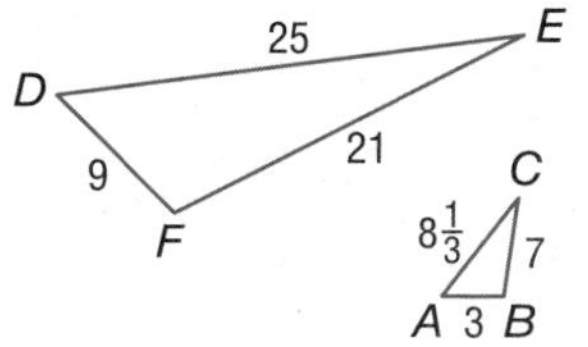

49. HEALTH Aida exercises every day by walking and jogging at least 3 miles. Aida walks at a rate of 4 miles per hour and jogs at a rate of 8 miles per hour. Suppose she has exactly one half-hour to exercise today. (Lesson 6-6)

a. Draw a graph showing the possible amounts of time she can spend walking and jogging.

b. List three possible solutions.

50. NUTRITION Determine whether the graph shows a *positive*, *negative*, or *no* correlation. If there is a positive or negative correlation, describe its meaning in the situation. (Lesson 4-5)

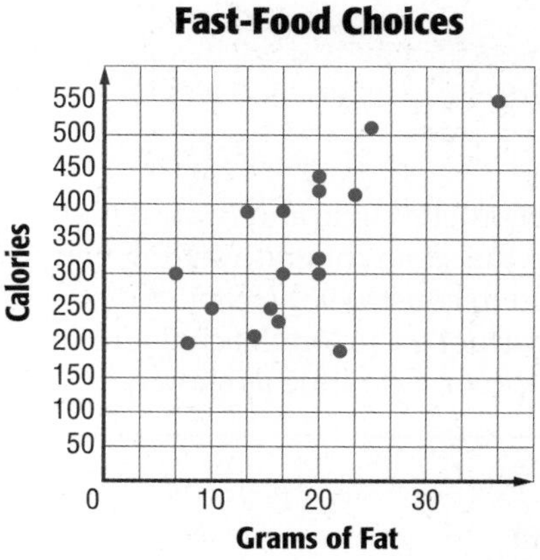

Skills Review

Solve each equation.

51. $15 = \frac{1}{2}[(360 - x) - 2x]$ **52.** $x + 12 = \frac{1}{2}[(180 - 120)]$ **53.** $x = \frac{1}{2}[(180 - 64)]$

Geometry Lab
Inscribed and Circumscribed Circles

In this lab, you will perform constructions that involve inscribing or circumscribing a circle.

CCSS Common Core State Standards
Content Standards
G.CO.13 Construct an equilateral triangle, a square, and a regular hexagon inscribed in a circle.
G.C.3 Construct the inscribed and circumscribed circles of a triangle, and prove properties of angles for a quadrilateral inscribed in a circle.
Mathematical Practices 5

Activity 1 Construct a Circle Inscribed in a Triangle

Step 1

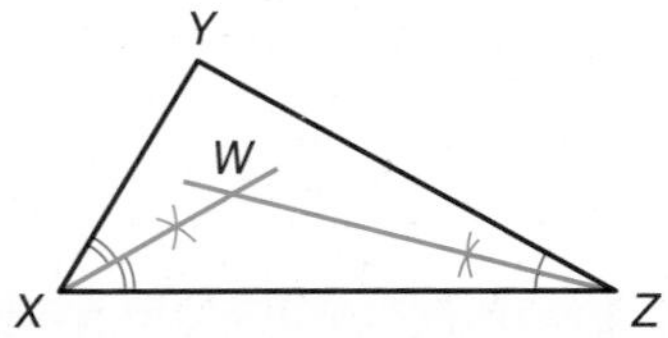

Draw a triangle XYZ and construct two angle bisectors of the triangle to locate the incenter W.

Step 2

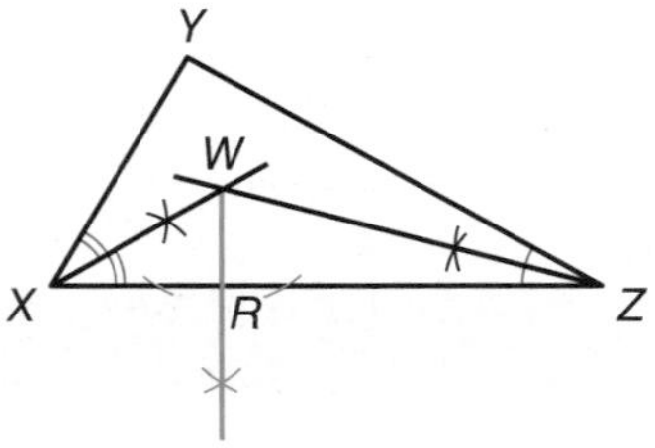

Construct a segment perpendicular to a side through the incenter. Label the intersection R.

Step 3

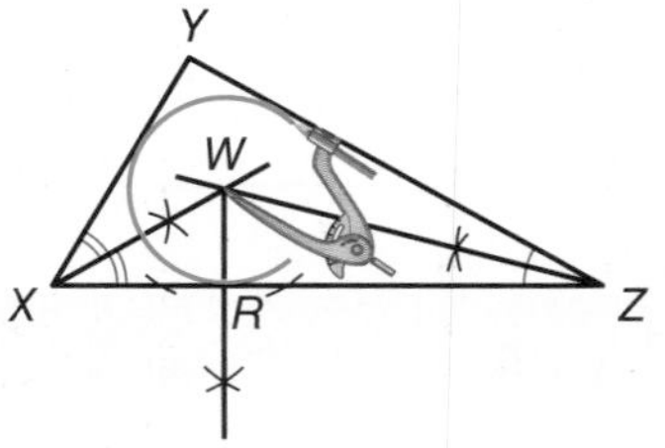

Set a compass of the length of $\overline{WR}$. Put the point of the compass on W and draw a circle with that radius.

Activity 2 Construct a Triangle Circumscribed About a Circle

Step 1

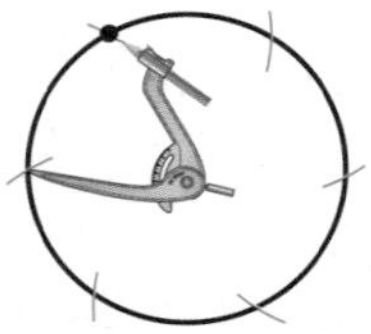

Construct a circle and draw a point. Use the same compass setting you used to construct the circle to construct an arc on the circle from the point. Continue as shown.

Step 2

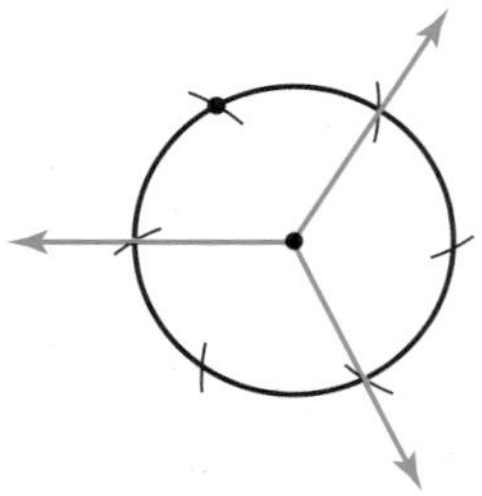

Draw rays from the center through every other arc.

Step 3

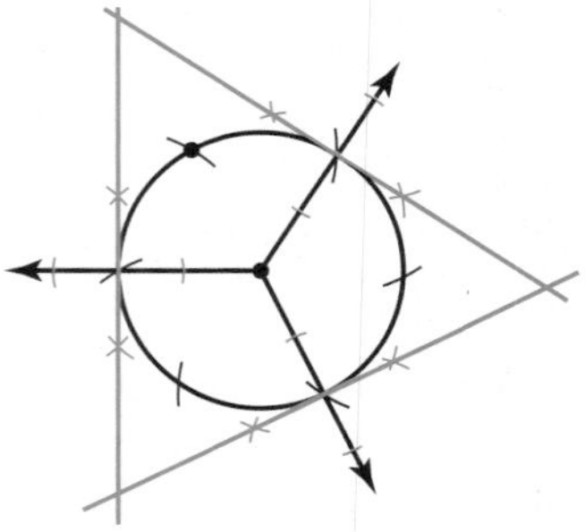

Construct a line perpendicular to each of the rays.

Model

1. Draw a right triangle and inscribe a circle in it.
2. Inscribe a regular hexagon in a circle. Then inscribe an equilateral triangle in a circle. (*Hint:* The first step of each construction is identical to Step 1 in Activity 2.)
3. Inscribe a square in a circle. Then circumscribe a square about a circle.
4. **CHALLENGE** Circumscribe a regular hexagon about a circle.

LESSON 15-4 Equations of Circles

Then	Now	Why?
You wrote equations of lines using information about their graphs.	**1** Write the equation of a circle. **2** Graph a circle on the coordinate plane.	Telecommunications towers emit radio signals that are used to transmit cellular calls. Each tower covers a circular area, and towers are arranged so that a signal is available at any location in the coverage area.

NewVocabulary
compound locus

Common Core State Standards

Content Standards
G.GPE.1 Derive the equation of a circle of given center and radius using the Pythagorean Theorem; complete the square to find the center and radius of a circle given by an equation.
G.GPE.6 Find the point on a directed line segment between two given points that partitions the segment in a given ratio.

Mathematical Practices
2 Reason abstractly and quantitatively.
7 Look for and make use of structure.

1 Equation of a Circle

Since all points on a circle are equidistant from the center, you can find an equation of a circle by using the Distance Formula.

Let (x, y) represent a point on a circle centered at the origin. Using the Pythagorean Theorem, $x^2 + y^2 = r^2$.

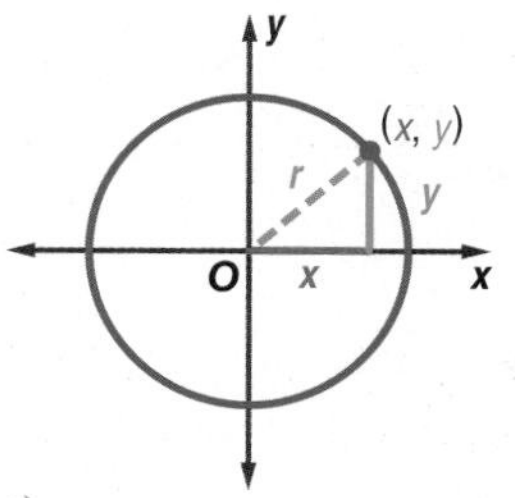

Now suppose that the center is not at the origin, but at the point (h, k). You can use the Distance Formula to develop an equation for the circle.

$d = \sqrt{(x_2 - x_1)^2 + (y_2 - y_1)^2}$ Distance Formula

$r = \sqrt{(x - h)^2 + (y - k)^2}$ $d = r, (x_1, y_1) = (h, k), (x_2, y_2) = (x, y)$

$r^2 = (x - h)^2 + (y - k)^2$ Square each side.

KeyConcept Equation of a Circle in Standard Form

The standard form of the equation of a circle with center at (h, k) and radius r is $(x - h)^2 + (y - k)^2 = r^2$.

The standard form of the equation of a circle is also called the *center-radius* form.

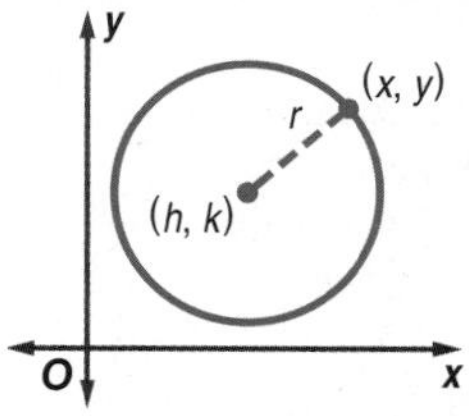

Example 1 Write an Equation Using the Center and Radius

Write the equation of each circle.

a. center at (1, −8), radius 7

$(x - h)^2 + (y - k)^2 = r^2$ Equation of a circle

$(x - 1)^2 + [y - (-8)]^2 = 7^2$ $(h, k) = (1, -8), r = 7$

$(x - 1)^2 + (y + 8)^2 = 49$ Simplify.

b. the circle graphed at the right

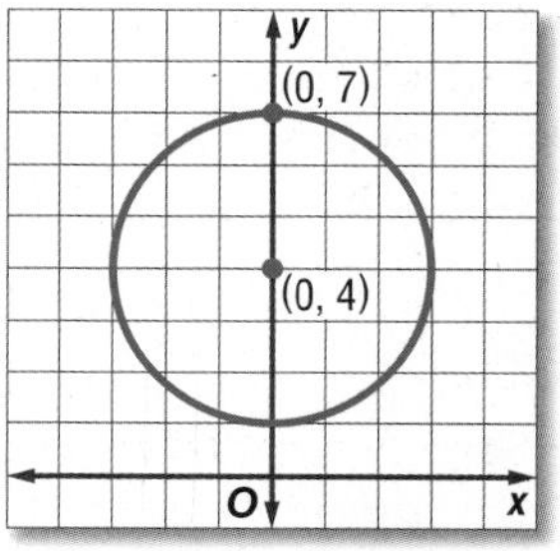

The center is at (0, 4) and the radius is 3.

$(x - h)^2 + (y - k)^2 = r^2$ Equation of a circle

$(x - 0)^2 + (y - 4)^2 = 3^2$ $(h, k) = (0, 4), r = 3$

$x^2 + (y - 4)^2 = 9$ Simplify.

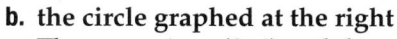

GuidedPractice

1A. center at origin, radius $\sqrt{10}$

1B. center at (4, −1), diameter 8

Greg Pease/Photographer's Choice/Getty Images

Example 2 Write an Equation Using the Center and a Point

Write the equation of the circle with center at (−2, 4), that passes through (−6, 7).

Step 1 Find the distance between the points to determine the radius.

$$r = \sqrt{(x_2 - x_1)^2 + (y_2 - y_1)^2}$$ Distance Formula

$$= \sqrt{[-6 - (-2)]^2 + (7 - 4)^2}$$ $(x_1, y_1) = (-2, 4)$ and $(x_2, y_2) = (-6, 7)$

$$= \sqrt{25} \text{ or } 5$$ Simplify.

Step 2 Write the equation using $h = -2$, $k = 4$, and $r = 5$.

$$(x - h)^2 + (y - k)^2 = r^2$$ Equation of a circle

$$[x - (-2)]^2 + (y - 4)^2 = 5^2$$ $h = -2$, $k = 4$, and $r = 5$

$$(x + 2)^2 + (y - 4)^2 = 25$$ Simplify.

GuidedPractice

2. Write the equation of the circle with center at (−3, −5) that passes through (0, 0).

2 Graph Circles

Graph Circles You can use the equation of a circle to graph it on a coordinate plane. To do so, you may need to write the equation in standard form first.

Example 3 Graph a Circle

The equation of a circle is $x^2 + y^2 - 8x + 2y = -8$. State the coordinates of the center and the measure of the radius. Then graph the equation.

Write the equation in standard form by completing the square.

$$x^2 + y^2 - 8x + 2y = -8$$ Original equation

$$x^2 - 8x + y^2 + 2y = -8$$ Isolate and group like terms.

$$x^2 - 8x + 16 + y^2 + 2y + 1 = -8 + 16 + 1$$ Complete the squares.

$$(x - 4)^2 + (y + 1)^2 = 9$$ Factor and simplify.

$$(x - 4)^2 + [y - (-1)]^2 = 3^2$$ Write +1 as − (−1) and 9 as 3^2.

> **StudyTip**
> **Completing the Square** To complete the square for any quadratic expression of the form $x^2 + bx$, follow these steps.
> **Step 1** Find one half of b.
> **Step 2** Square the result in Step 1.
> **Step 3** Add the result of Step 2 to $x^2 + bx$.

With the equation now in standard form, you can identify h, k, and r.

$$(x - 4)^2 + [y - (-1)]^2 = 3^2$$
$$\uparrow \quad \uparrow \quad \uparrow$$
$$(x - h)^2 + (y - k)^2 = r^2$$

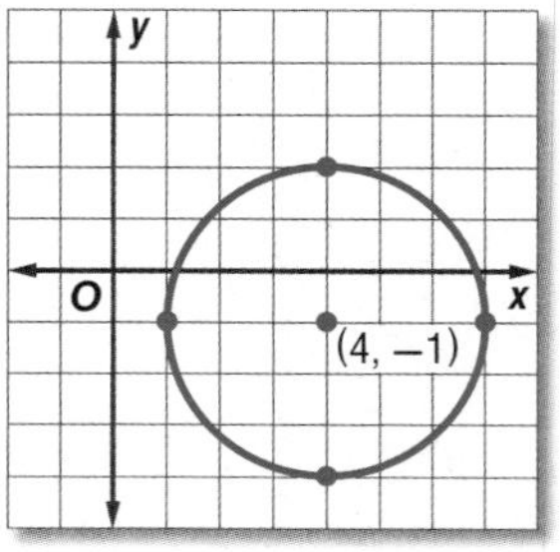

So, $h = 4$, $k = -1$, and $r = 3$. The center is at (4, −1), and the radius is 3. Plot the center and four points that are 3 units from this point. Sketch the circle through these four points.

GuidedPractice

For each circle with the given equation, state the coordinates of the center and the measure of the radius. Then graph the equation.

3A. $x^2 + y^2 - 4 = 0$

3B. $x^2 + y^2 + 8x - 14y + 40 = 0$

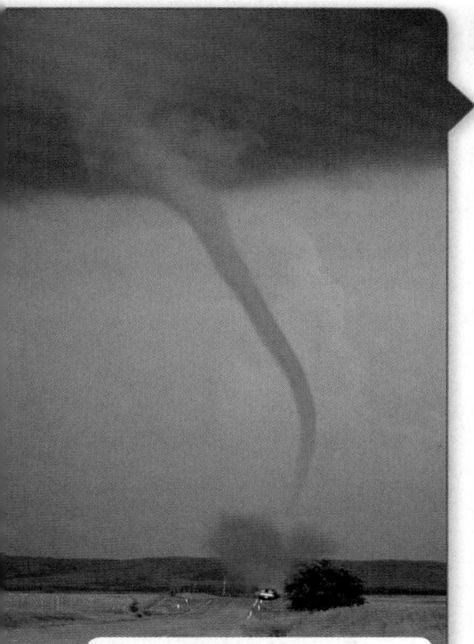

Real-World Example 4 Use Three Points to Write an Equation

TORNADOES Three tornado sirens are placed strategically on a circle around a town so they can be heard by all. Write the equation of the circle on which they are placed if the coordinates of the sirens are $A(-8, 3)$, $B(-4, 7)$, and $C(-4, -1)$.

Understand You are given three points that lie on a circle.

Plan Graph $\triangle ABC$. Construct the perpendicular bisectors of two sides to locate the center of the circle. Then find the radius.

Use the center and radius to write an equation.

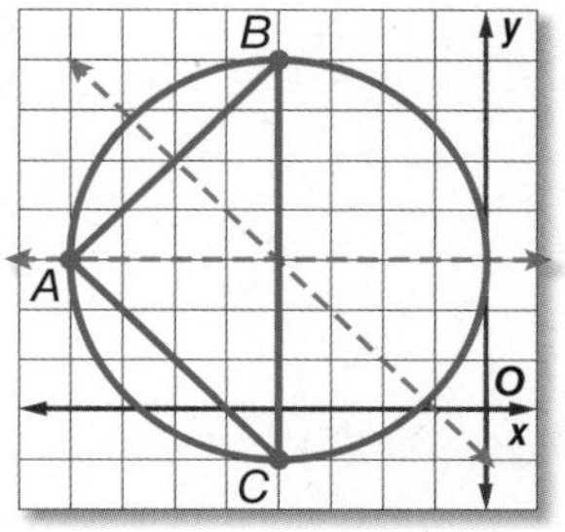

Solve The center appears to be at $(-4, 3)$. The radius is 4. Write an equation.

$$(x - h)^2 + (y - k)^2 = r^2$$

$$[x - (-4)]^2 + (y - 3)^2 = 4^2$$

$$(x + 4)^2 + (y - 3)^2 = 16$$

Check Verify the center by finding the equations of the two bisectors and solving the system of equations. Verify the radius by finding the distance between the center and another point on the circle. ✓

GuidedPractice

4. Write an equation of a circle that contains $R(1, 2)$, $S(-3, 4)$, and $T(-5, 0)$.

Real-WorldLink

About 1000 tornadoes are reported across the United States each year. The most violent tornadoes have wind speeds of 250 mph or more. Damage paths can be a mile wide and 50 miles long.

Source: National Oceanic & Atmospheric Administration

A line can intersect a circle in at most two points. You can find the point(s) of intersection between a circle and a line by applying techniques used to find the intersection between two lines and techniques used to solve quadratic equations.

Example 5 Intersections with Circles

Find the point(s) of intersection between $x^2 + y^2 = 4$ and $y = x$.

Graph these equations on the same coordinate plane. The points of intersection are solutions of both equations. You can estimate these points on the graph to be at about $(-1.4, -1.4)$ and $(1.4, 1.4)$. Use substitution to find the coordinates of these points algebraically.

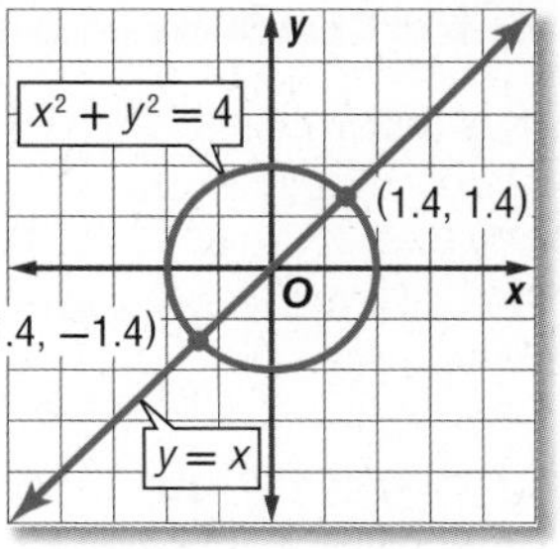

$x^2 + y^2 = 4$	Equation of circle
$x^2 + x^2 = 4$	Since $y = x$, substitute x for y.
$2x^2 = 4$	Simplify.
$x^2 = 2$	Divide each side by 2.
$x = \pm\sqrt{2}$	Take the square root of each side.

So $x = \sqrt{2}$ or $x = -\sqrt{2}$. Use the equation $y = x$ to find the corresponding y-values.

$y = x$	Equation of line	$y = x$
$y = \sqrt{2}$	$x = \sqrt{2}$ or $x = -\sqrt{2}$	$y = -\sqrt{2}$

The points of intersection are located at $(\sqrt{2}, \sqrt{2})$ and $(-\sqrt{2}, -\sqrt{2})$ or at about $(-1.4, -1.4)$ and $(1.4, 1.4)$. Check these solutions in both of the original equations.

StudyTip

Quadratic Techniques In addition to taking square roots, other quadratic techniques that you may need to apply in order to solve equations of the form $ax^2 + bx + c = 0$ include completing the square, factoring, and the Quadratic Formula, $x = \frac{-b \pm \sqrt{b^2 - 4ac}}{2a}$.

GuidedPractice

5. Find the point(s) of intersection between $x^2 + y^2 = 8$ and $y = -x$.

Check Your Understanding

Examples 1–2 **Write the equation of each circle.**

1. center at (4, 0), radius 3
2. center at (1, 3), diameter 18
3. center at origin, passes through (4, 4)
4. center at (3, −5), passes through (−4, 1)
5. 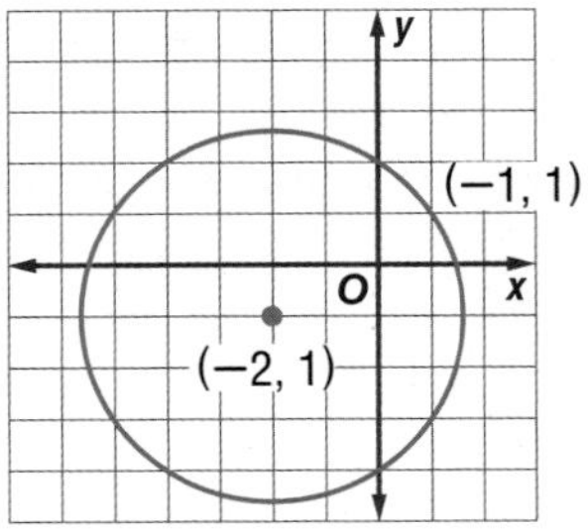

6. 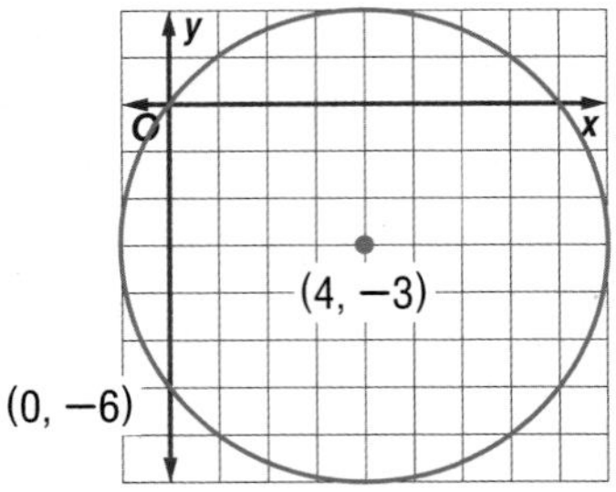

Example 3 **For each circle with the given equation, state the coordinates of the center and the measure of the radius. Then graph the equation.**

7. $x^2 + 2x + y^2 - 4y = 11$
8. $(x + 1)^2 + y^2 = 9$

Example 4

9. **RADIOS** Three radio towers are modeled by the points $R(2, -3)$, $S(-2, 5)$, and $T(2, 7)$. Determine the location of another tower equidistant from all three towers, and write an equation for the circle.

10. **COMMUNICATION** Three cell phone towers can be modeled by the points $X(3, 5)$, $Y(-3, 11)$, and $Z(-9, 5)$. Determine the location of another cell phone tower equidistant from the other three, and write an equation for the circle.

Example 5 **Find the point(s) of intersection, if any, between each circle and line with the equations given.**

11. $(x - 3)^2 + y^2 = 4$
    $y = x - 1$
12. $(x - 3)^2 + (y + 2)^2 = 18$
    $y = -3x + 1$

Practice and Problem Solving

Examples 1–2 **CCSS STRUCTURE** **Write the equation of each circle.**

13. center at origin, radius 7
14. center at (1, 6), radius 4
15. center at (0, −2), diameter 20
16. center at (−9, 8), radius $\sqrt{13}$
17. center at (6, −3), passes through (0, −3)
18. center at (−3, 1), passes through (−4, 3)
19. 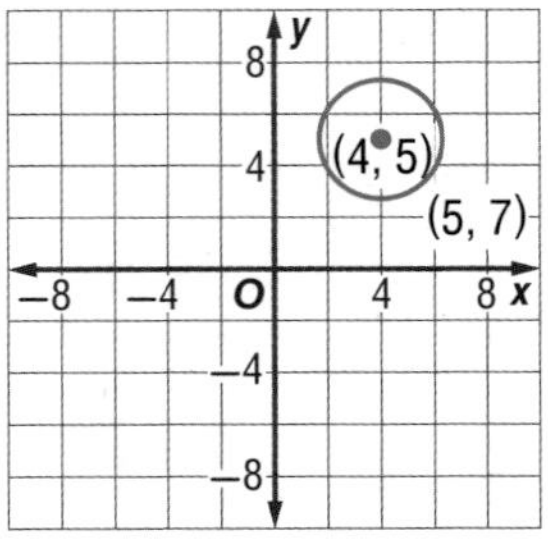

20. 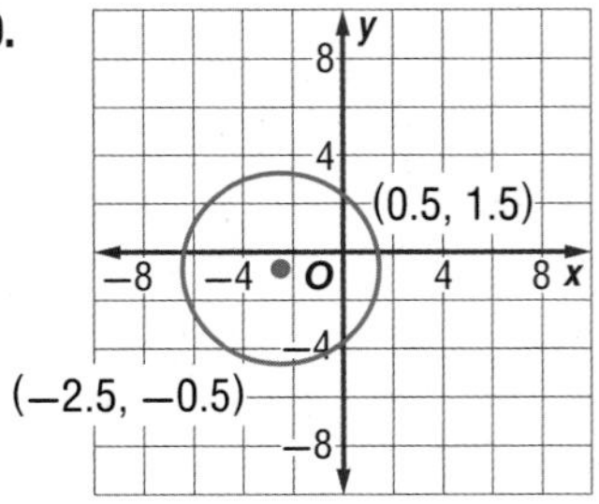

21. **WEATHER** A Doppler radar screen shows concentric rings around a storm. If the center of the radar screen is the origin and each ring is 15 miles farther from the center, what is the equation of the fourth ring?

22. **GARDENING** A sprinkler waters a circular area that has a diameter of 12 feet. The sprinkler is located 24 feet north of the house. If the house is located at the origin, what is the equation for the circle of area that is watered?

Example 3 **For each circle with the given equation, state the coordinates of the center and the measure of the radius. Then graph the equation.**

23. $x^2 + y^2 = 49$

24. $x^2 + y^2 - 10x + 4y = 31$

25. $x^2 + y^2 + 6x + 8y = 75$

26. $x^2 + y^2 - 10x = -15$

Example 4 **Write an equation of a circle that contains each set of points. Then graph the circle.**

27. $A(-2, -5), B(6, -5), C(2, -9)$

28. $F(-6, -4), G(0, -10), H(2, -8)$

Example 5 **Find the point(s) of intersection, if any, between each circle and line with the equations given.**

29. $x^2 + y^2 = 25$
$y = \frac{1}{2}x$

30. $x^2 + y^2 = 4$
$y = x - 2$

31. $x^2 + (y + 3)^2 = 8$
$y = -x - 3$

32. $(x + 2)^2 + y^2 = 16$
$y = 2x$

33. $x^2 + y^2 = 10$
$y = -2x$

34. $(x - 2)^2 + (y - 5)^2 = 7$
$y = -x$

Write the equation of each circle.

35. a circle with a diameter having endpoints at $(-2, 5)$ and $(4, -3)$

36. a circle with $d = 26$ and a center translated 5 units left and 7 units up from the origin

37. CCSS **MODELING** Different-sized engines will launch model rockets to different altitudes. The higher a rocket goes, the larger the circle of possible landing sites becomes. Under normal wind conditions, the landing radius is three times the altitude of the rocket.

a. Write the equation of the landing circle for a rocket that travels 200 feet in the air.

b. What would be the radius of the landing circle for a rocket that travels 1500 feet in the air? Assume the center of the circle is at the origin.

38. SKYDIVING Three of the skydivers in the circular formation shown have approximate coordinates of $G(-5, -4)$, $H(9, -4)$, and $J(2, 10)$.

a. What are the approximate coordinates of the center skydiver?

b. If each unit represents 1 foot, what is the diameter of the skydiving formation?

39. DELIVERY Pizza and Subs offers free delivery within 6 miles of the restaurant. The restaurant is located 5 miles west and 4 miles south of Consuela's house.

a. Write and graph an equation to represent this situation if Consuela's house is at the origin of the coordinate system.

b. Can Consuela get free delivery if she orders pizza from Pizza and Subs? Explain.

40. INTERSECTIONS OF CIRCLES Graph $x^2 + y^2 = 9$ and $(x + 3)^2 + y^2 = 9$ on the same coordinate plane.

a. Estimate the point(s) of intersection between the two circles.

b. Solve $x^2 + y^2 = 9$ for y.

c. Substitute the value you found in part **b** into $(x + 3)^2 + y^2 = 9$ and solve for x.

d. Substitute the value you found in part **c** into $x^2 + y^2 = 9$ and solve for y.

e. Use your answers to parts **c** and **d** to write the coordinates of the points of intersection. Compare these coordinates to your estimate from part **a**.

f. Verify that the point(s) you found in part **d** lie on both circles.

41. Prove or disprove that the point $(2, 2\sqrt{3})$ lies on a circle centered at the origin and containing the point $(0, -4)$.

42. **MULTIPLE REPRESENTATIONS** In this problem, you will investigate a compound locus for a pair of points. A **compound locus** satisfies more than one distinct set of conditions.

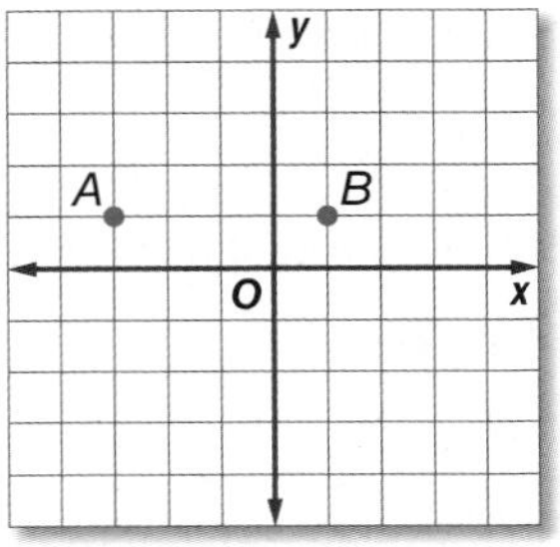

a. **Tabular** Choose two points A and B in the coordinate plane. Locate 5 coordinates from the locus of points equidistant from A and B.

b. **Graphical** Represent this same locus of points by using a graph.

c. **Verbal** Describe the locus of all points equidistant from a pair of points.

d. **Graphical** Using your graph from part **b**, determine and graph the locus of all points in a plane that are a distance of AB from B.

e. **Verbal** Describe the locus of all points in a plane equidistant from a single point. Then describe the locus of all points that are both equidistant from A and B and are a distance of AB from B. Describe the graph of the compound locus.

43. A circle with a diameter of 14 has its center in the second quadrant. The lines $y = -6$ and $x = 2$ are tangent to the circle. Write an equation of the circle.

H.O.T. Problems Use Higher-Order Thinking Skills

44. **CHALLENGE** Write a coordinate proof to show that if an inscribed angle intercepts the diameter of a circle, as shown, the angle is a right angle.

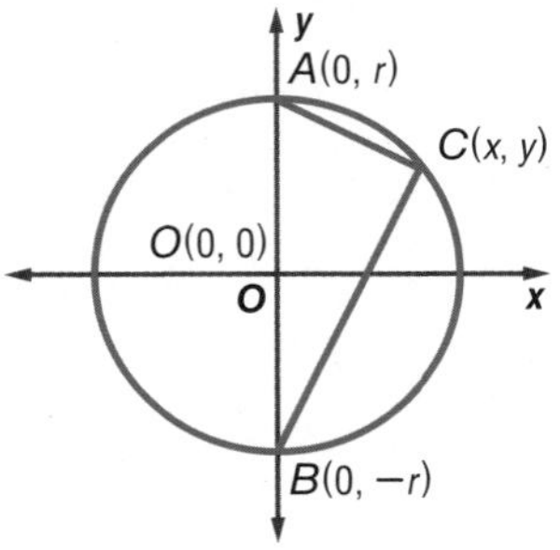

45. **CCSS REASONING** A circle has the equation $(x - 5)^2 + (y + 7)^2 = 16$. If the center of the circle is shifted 3 units left and 5 units up, what would be the equation of the new circle? Explain your reasoning.

46. **OPEN ENDED** Graph three noncollinear points and connect them to form a triangle. Then construct the circle that circumscribes it.

47. **WRITING IN MATH** Seven new radio stations must be assigned broadcast frequencies. The stations are located at $A(10, 3)$, $B(8, 7)$, $C(6, 2)$, $D(9, 0)$, $E(5, 5)$, $F(9, 6)$, and $G(4, 3)$, where 1 unit = 50 miles.

a. If stations that are more than 200 miles apart can share the same frequency, what is the least number of frequencies that can be assigned to these stations?

b. Describe two different beginning approaches to solving this problem.

c. Choose an approach, solve the problem, and explain your reasoning.

CHALLENGE Find the coordinates of point P on $\overrightarrow{AB}$ that partitions the segment into the given ratio AP to PB.

48. $A(0, 0)$, $B(-3, -4)$, 3 to 2

49. $A(0, 0)$, $B(8, -6)$, 1 to 4

50. **WRITING IN MATH** Describe how the equation for a circle changes if the circle is translated a units to the right and b units down.

Standardized Test Practice

51. Which of the following is the equation of a circle with center (6, 5) that passes through (2, 8)?

A $(x - 6)^2 + (y - 5)^2 = 5^2$
B $(x - 5)^2 + (y - 6)^2 = 7^2$
C $(x + 6)^2 + (y + 5)^2 = 5^2$
D $(x - 2)^2 + (y - 8)^2 = 7^2$

52. ALGEBRA What are the solutions of $n^2 - 4n = 21$?

F 3, 7
G 3, −7
H −3, 7
J −3, −7

53. SHORT RESPONSE Solve: $5(x - 4) = 16$.

Step 1: $5x - 4 = 16$
Step 2: $5x = 20$
Step 3: $x = 4$

Which is the first incorrect step in the solution shown above?

54. SAT/ACT The center of $\odot F$ is at (−4, 0) and has a radius of 4. Which point lies on $\odot F$?

A (4, 0)
B (0, 4)
C (4, 3)
D (−4, 4)
E (0, 8)

Spiral Review

Determine whether each graph shows a *positive*, a *negative*, or *no* correlation. If there is a positive or negative correlation, describe its meaning in the situation. (Lesson 4-5)

55. **Electronic Tax Returns**

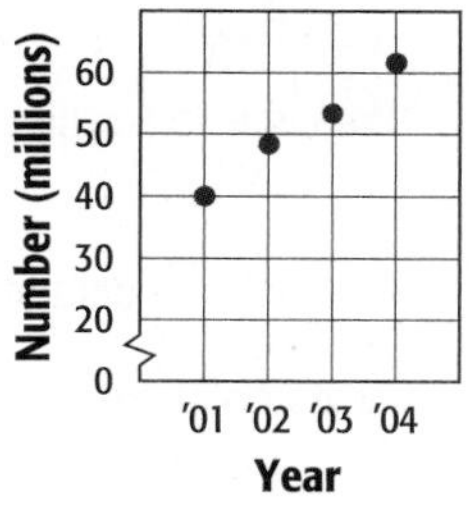

56. **Atlantic Hurricanes**

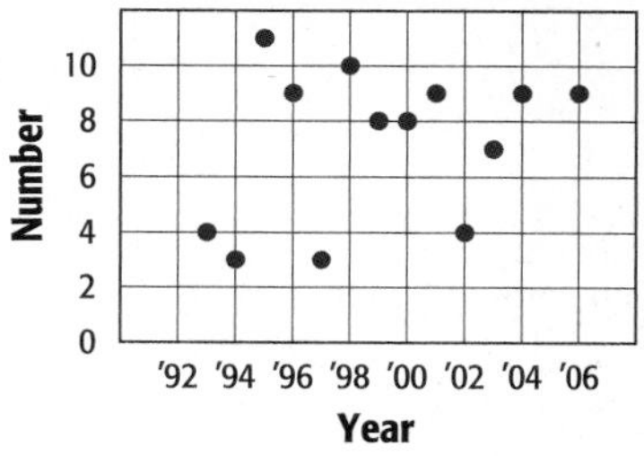

57. STREETS The neighborhood where Vincent lives has roundabouts where certain streets meet. If Vincent rides his bike once around the very edge of the grassy circle, how many feet will he have ridden? (Lesson 15-1)

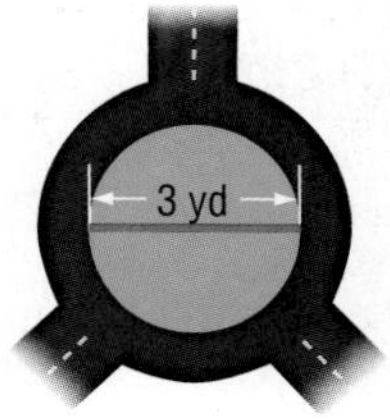

Skills Review

Find the perimeter and area of each figure.

58.

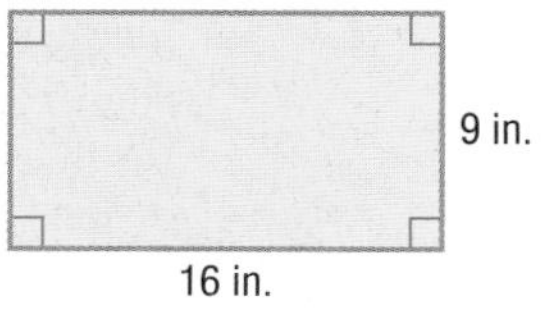

59.

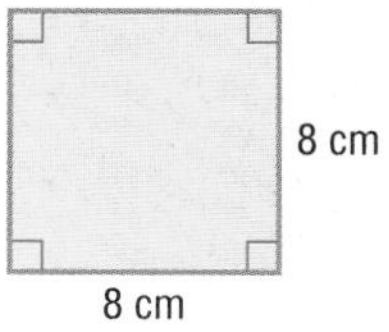

60.

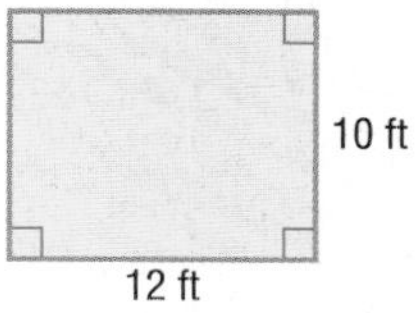

Study Guide and Review

Study Guide

KeyConcepts

Circles and Circumference (Lesson 15-1)

- The circumference of a circle is equal to πd or $2\pi r$.

Arcs and Chords (Lesson 15-2)

- The length of an arc is proportional to the length of the circumference.
- Diameters perpendicular to chords bisect chords and intercepted arcs.

Tangents (Lesson 15-3)

- A line that is tangent to a circle intersects the circle in exactly one point and is perpendicular to a radius.
- Two segments tangent to a circle from the same exterior point are congruent.

Equations of Circles (Lesson 15-4)

- The equation of a circle with center (h, k) and radius r is $(x - h)^2 - (y - k)^2 = r^2$.

FOLDABLES® StudyOrganizer

Be sure the Key Concepts are noted in your Foldable.

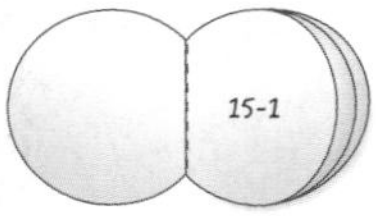

KeyVocabulary

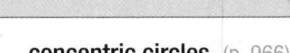

- **center** (p. 965)
- **chord** (p. 965)
- **circle** (p. 965)
- **circumference** (p. 967)
- **circumscribed** (p. 968)
- **common tangent** (p. 982)
- **compound locus** (p. 996)
- **concentric circles** (p. 966)
- **diameter** (p. 965)
- **inscribed** (p. 968)
- **pi (π)** (p. 967)
- **point of tangency** (p. 982)
- **radius** (p. 945)
- **tangent** (p. 982)

VocabularyCheck

State whether each sentence is *true* or *false.* If *false,* replace the underlined word or phrase to make a true sentence.

1. Any segment with both endpoints on the circle is a radius of the circle.
2. A chord passing through the center of a circle is a diameter.
3. A common tangent is the point at which a line in the same plane as a circle intersects the circle.
4. A secant segment is a segment of a diameter that has exactly one endpoint on the circle.
5. Two circles are concentric circles if and only if they have congruent radii.

Lesson-by-Lesson Review

15-1 Circles and Circumference

For Exercises 6–8, refer to ⊙*D*.

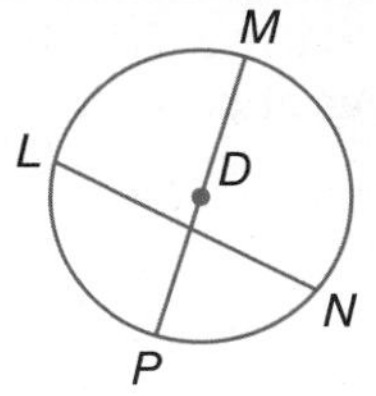

6. Name the circle.

7. Name a radius.

8. Name a chord that is not a diameter.

Find the diameter and radius of a circle with the given circumference. Round to the nearest hundredth.

9. $C = 43$ cm

10. $C = 108.5$ ft

11. $C = 26.7$ yd

12. $C = 225.9$ mm

Example 1

Find the circumference of ⊙*A*.

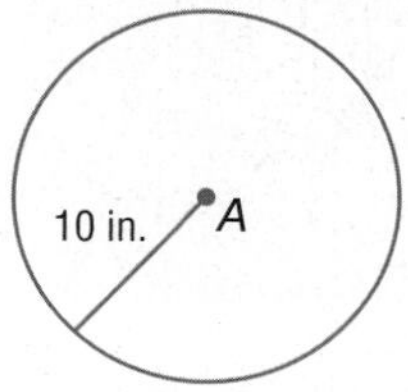

$C = 2\pi r$	Circumference formula
$= 2\pi(10)$	Substitution
≈ 62.83	Use a calculator.

The circumference of ⊙*A* is about 62.83 inches.

15-2 Arcs and Chords

13. Find the value of *x*.

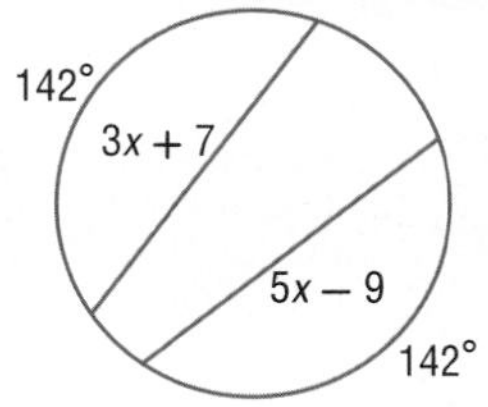

In ⊙*K*, $MN = 16$ and $m\overset{\frown}{MN} = 98$. Find each measure. Round to the nearest hundredth.

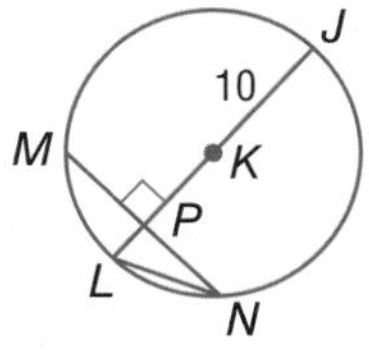

14. $m\overset{\frown}{NJ}$

15. *LN*

16. **GARDENING** The top of the trellis shown is an arc of a circle in which $\overline{CD}$ is part of the diameter and $\overline{CD} \perp \overline{AB}$. If $\overset{\frown}{ACB}$ is about 28% of a complete circle, what is $m\overset{\frown}{CB}$?

Example 2

ALGEBRA In ⊙*E*, $EG = EF$. Find *AB*.

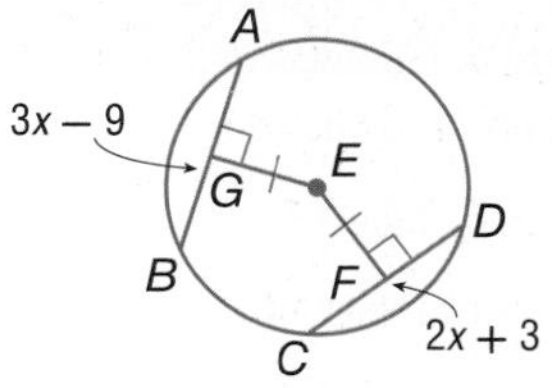

Since chords $\overline{EG}$ and $\overline{EF}$ are congruent, they are equidistant from *E*. So, $AB = CD$.

$AB = CD$	Theorem 10.5
$3x - 9 = 2x + 3$	Substitution
$3x = 2x + 12$	Add.
$x = 12$	Simplify.

So, $AB = 3(12) - 9$ or 27.

Study Guide and Review *Continued*

15-3 Tangents

17. SCIENCE FICTION In a story Todd is writing, instantaneous travel between a two-dimensional planet and its moon is possible when the time-traveler follows a tangent. Copy the figures below and draw all possible travel paths.

18. Find x and y. Assume that segments that appear to be tangent are tangent. Round to the nearest tenth if necessary.

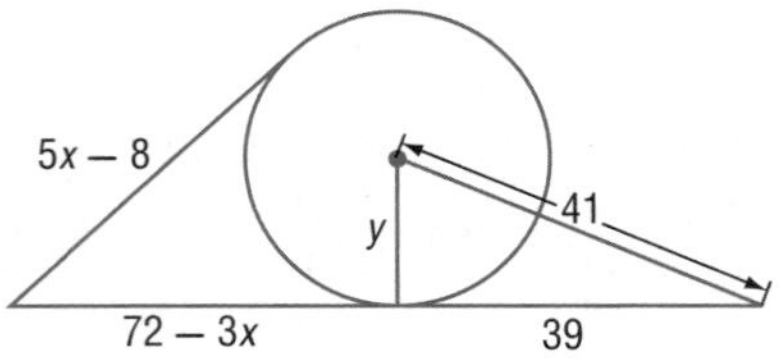

Example 3

In the figure, $\overline{KL}$ is tangent to $\odot M$ at K. Find the value of x.

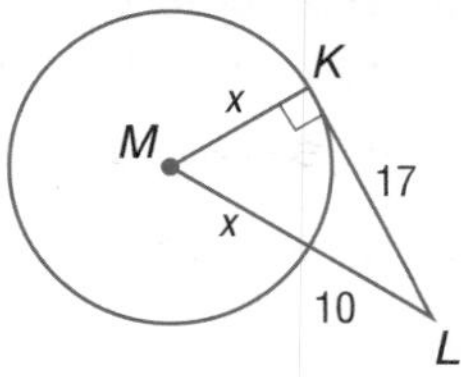

By Theorem 10.9, $\overline{MK} \perp \overline{KL}$. So, $\triangle MKL$ is a right triangle.

$KM^2 + KL^2 = ML^2$	Pythagorean Theorem
$x^2 + 17^2 = (x + 10)^2$	Substitution
$x^2 + 289 = x^2 + 20x + 100$	Multiply.
$289 = 20x + 100$	Simplify.
$189 = 20x$	Subtract.
$9.45 = x$	Divide.

15-4 Equations of Circles

Write the equation of each circle.

19. center at $(-2, 4)$, radius 5

20. center at $(1, 2)$, diameter 14

21. FIREWOOD In an outdoor training course, Kat learns a wood-chopping safety check that involves making a circle with her arm extended, to ensure she will not hit anything overhead as she chops. If her reach is 19 inches, the hatchet handle is 15 inches, and her shoulder is located at the origin, what is the equation of Kat's safety circle?

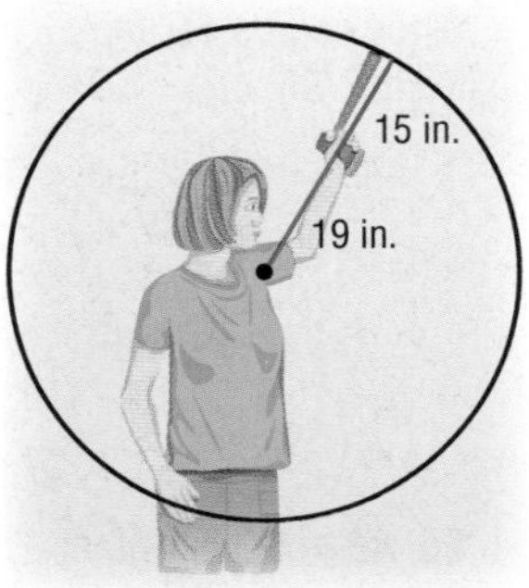

Example 4

Write the equation of the circle graphed below.

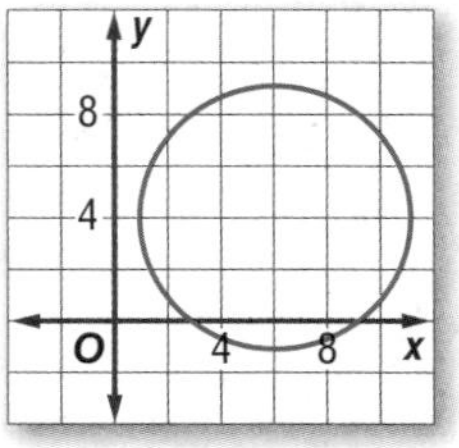

The center is at $(6, 4)$ and the radius is 5.

$(x - h)^2 + (y - k)^2 = r^2$	Equation of a circle
$(x - 6)^2 + (y - 4)^2 = 5^2$	$(h, k) = (6, 4)$ and $r = 5$
$(x - 6)^2 + (y - 4)^2 = 25$	Simplify.

CHAPTER 15 Practice Test

1. **POOLS** Amanda's family has a swimming pool that is 4 feet deep in their backyard. If the diameter of the pool is 25 feet, what is the circumference of the pool to the nearest foot?

2. Find the exact circumference of the circle below.

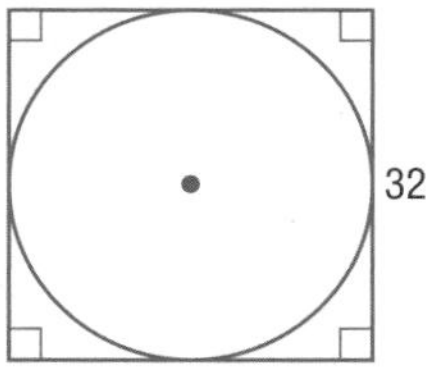

For Exercises 3–5, refer to $\odot A$.

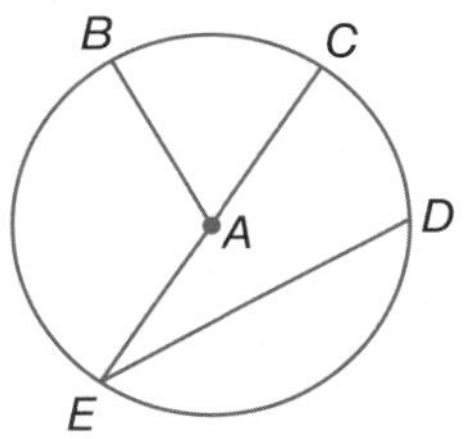

3. Name the circle.

4. Name a diameter.

5. Name a chord that is not a diameter.

6. **MULTIPLE CHOICE** What is ED?

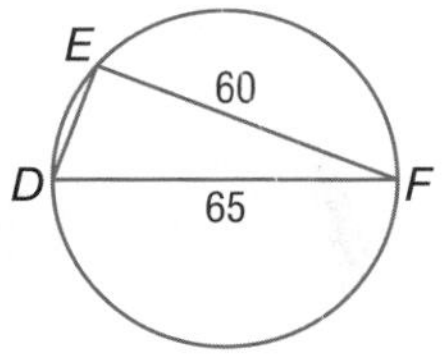

A 15 **C** 88.5

B 25 **D** not enough information

7. Find x if $\odot M \cong \odot N$.

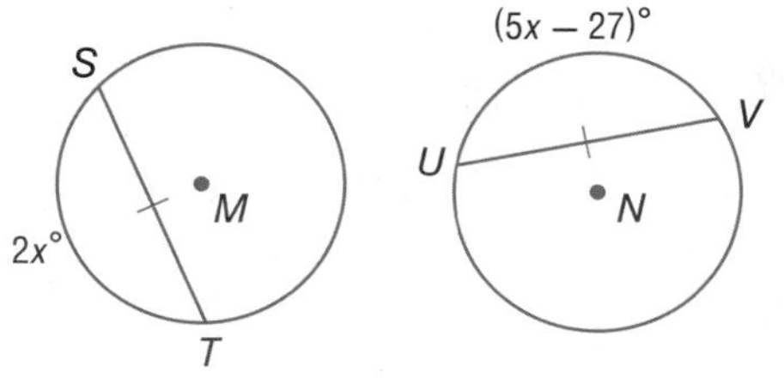

8. **MULTIPLE CHOICE** How many points are shared by concentric circles?

F 0 **H** 2

G 1 **J** infinite points

9. Determine whether $\overline{FG}$ is tangent to $\odot E$. Justify your answer.

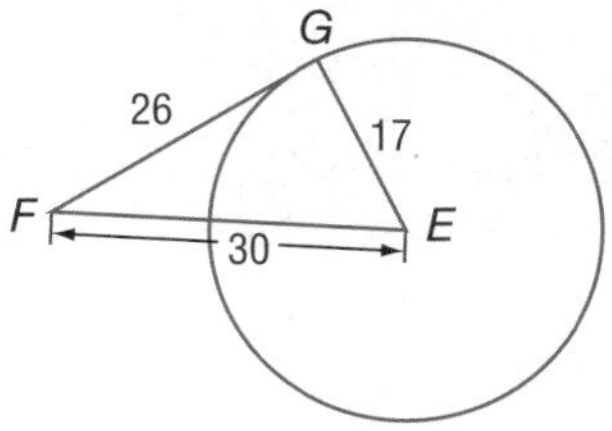

10. **BICYCLES** A bicycle has tires that are 24 inches in diameter.

 a. Find the circumference of one tire.

 b. How many inches does the tire travel after 100 rotations?

11. Find the perimeter of the triangle at the right. Assume that segments that appear to be tangent are tangent.

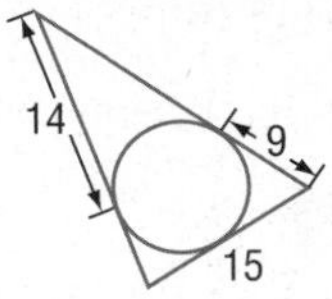

12. **FLOWERS** Hannah wants to encircle a tree trunk with a flower bed. If the center of the tree trunk is the origin and Hannah wants the flower bed to extend to 3 feet from the center of the tree, what is the equation that would represent the flower bed?

CHAPTER 15 Preparing for Standardized Tests

Properties of Circles

A circle is a unique shape in which the angles, arcs, and segments intersecting the circle have special properties and relationships. You should be able to identify the parts of a circle, write the equation of a circle, and solve for arc, angle, and segment measures in a circle.

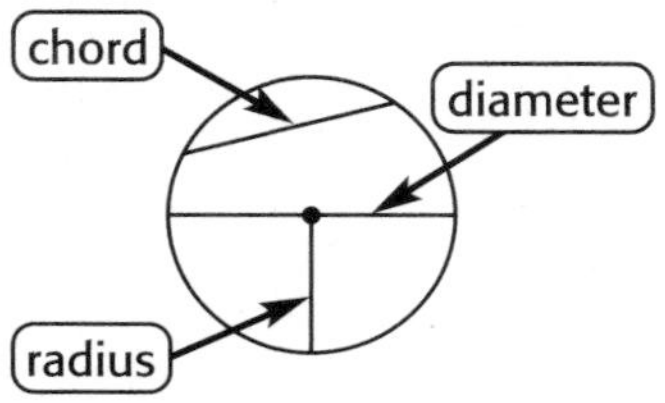

$r = \frac{1}{2}d$

$d = 2r$

$C = 2\pi r$ or πd

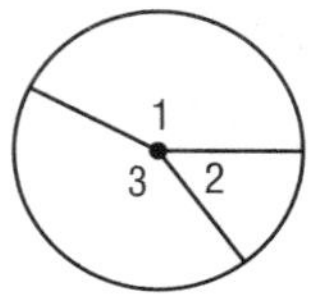

$m\angle 1 + m\angle 2 + m\angle 3 = 360$

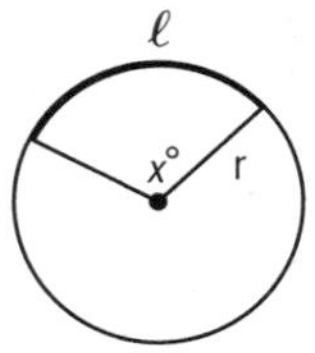

$\ell = \frac{x}{360} \cdot 2\pi r$

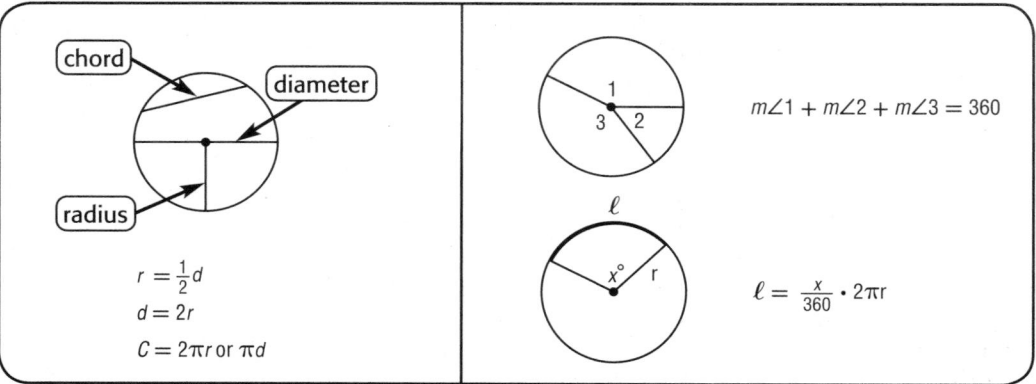

Strategies for Applying the Properties of Circles

Step 1

Review the parts of a circle and their relationships.

- Some key parts include: **radius, diameter, arc, chord, tangent, secant**
- Study the key theorems and the properties of circles as well as the relationships between the parts of a circle.

Step 2

Read the problem statement and study any figure you are given carefully.

- Determine what you are being asked to find.
- Fill in any information in the figure that you can.
- Determine which theorems or properties apply to the problem situation.

Step 3

Solve the problem and check your answer.

- Apply the theorems or properties to solve the problem.
- Check your answer to be sure it makes sense.

Standardized Test Example

Read the problem. Identify what you need to know. Then use the information in the problem to solve.

Solve for x in the figure.

A 2 **C** 4

B 3 **D** 6

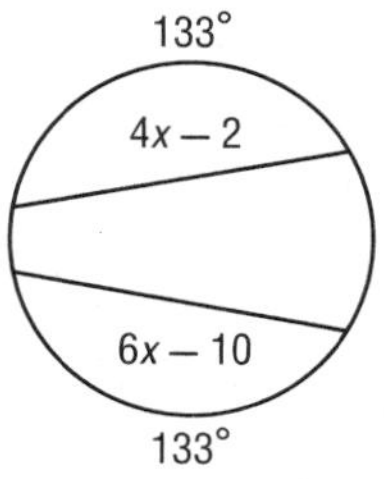

Read the problem statement and study the figure carefully. You are given a circle with two chords that correspond to congruent minor arcs. One important property of circles is that two chords are congruent if and only if their corresponding minor arcs are congruent. You can use this property to set up and solve an equation for x.

$4x - 2 = 6x - 10$	Definition of Congruent Segments
$4x - 6x = -10 + 2$	Subtract.
$-2x = -8$	Simplify.
$\frac{-2x}{-2} = \frac{-8}{-2}$	Divide each side by −2.
$x = 4$	Simplify.

So, the value of x is 4. The answer is C. You can check your answer by substituting 4 into each expression and making sure both chords have the same length.

Exercises

Read each problem. Identify what you need to know. Then use the information in the problem to solve.

1. Solve for x in the figure below.

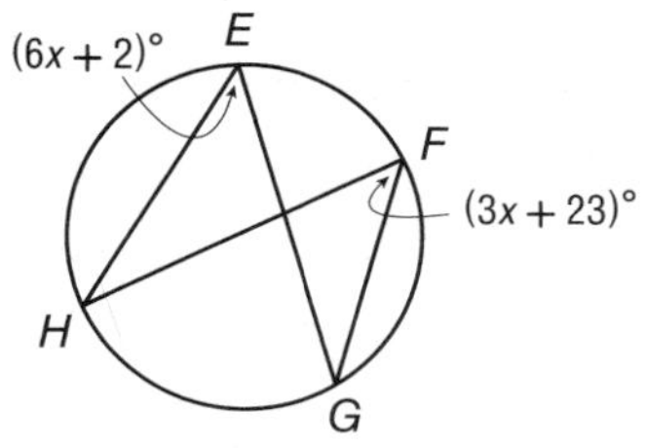

A 4 **C** 6

B 5 **D** 7

2. Triangle RST is circumscribed about the circle below. What is the perimeter of the triangle?

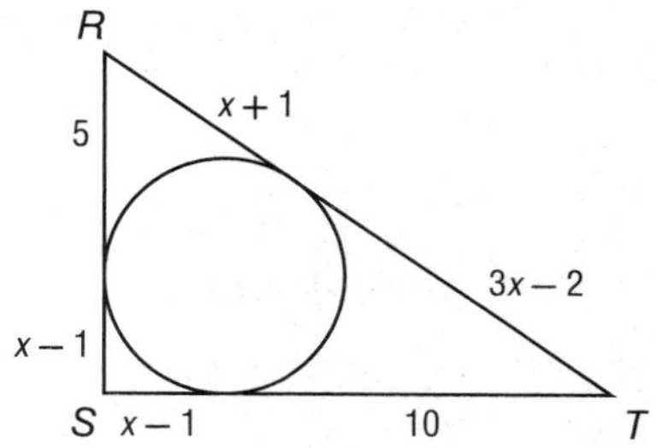

F 33 units **H** 37 units

G 36 units **J** 40 units

CHAPTER 15

Standardized Test Practice

Cumulative, Chapters 1 through 15

Multiple Choice

Read each question. Then fill in the correct answer on the answer document provided by your teacher or on a sheet of paper.

1. If $ABCD$ is a rhombus, and $m\angle ABC = 70°$, what is $m\angle 1$?

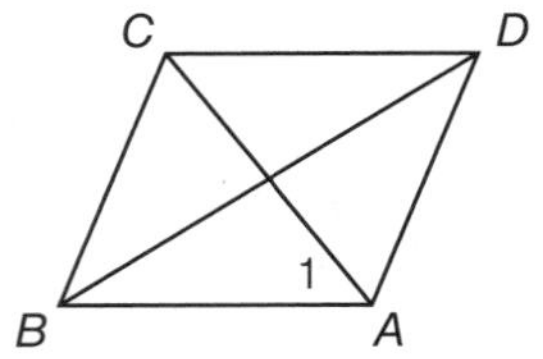

A 45°

B 55°

C 70°

D 125°

2. A bicycle has 2 tires that are 24 inches in diameter. Find the circumference of one tire.

A 25.2 in

B 48 in

C 75.4 in

D 12 in

3. Given $a \parallel b$, find $m\angle 1$.

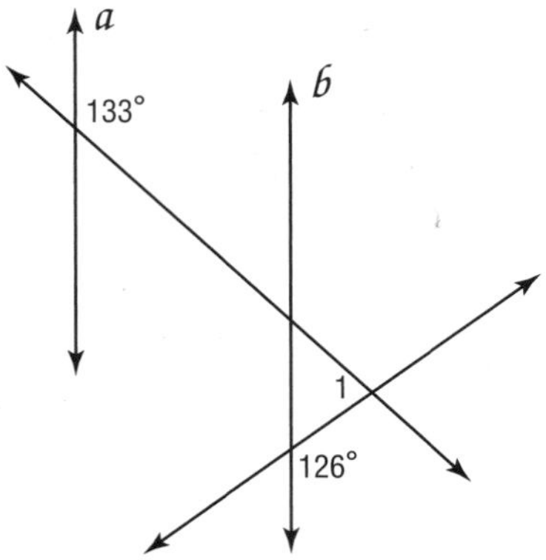

F 47°

G 54°

H 79°

J 101°

4. Which of the following conditions would *not* guarantee that a quadrilateral is a parallelogram?

A both pairs of opposite sides congruent

B both pairs of opposite angles congruent

C diagonals bisect each other

D one pair of opposite sides parallel

5. Find the value of x.

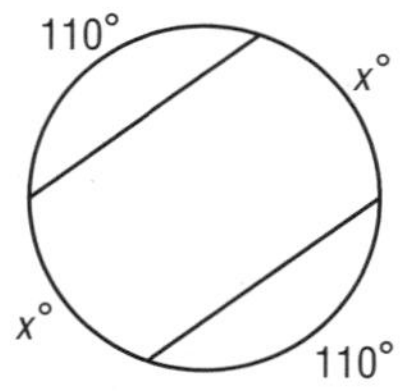

A 70

B 110

C 220

D 50

Short Response/Gridded Response

Record your answers on the answer sheet provided by your teacher or on a sheet of paper.

6. Does the figure shown have rotational symmetry? If so, give the order of symmetry.

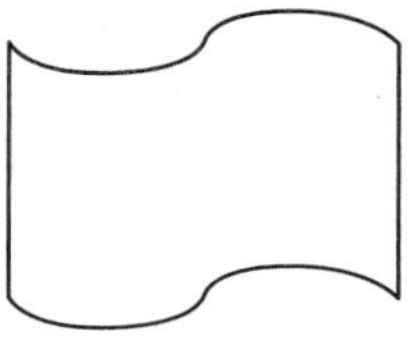

7. GRIDDED RESPONSE A square with 5-centimeter sides is inscribed in a circle. What is the circumference of the circle? Round your answer to the nearest tenth of a centimeter.

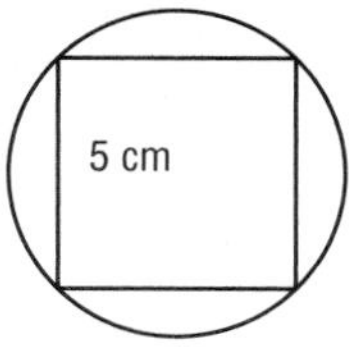

8. In ⊙B, $CE = 13.5$. Find BD. Round to the nearest hundredth.

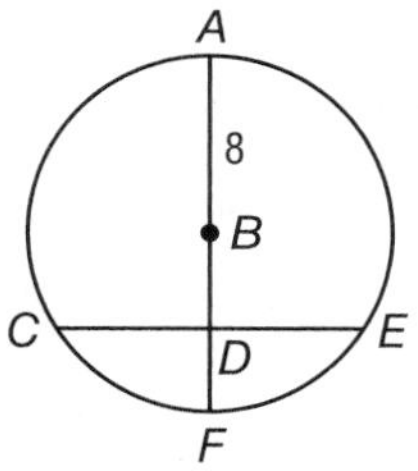

9. The two circles shown are congruent. Find x and the length of the chord.

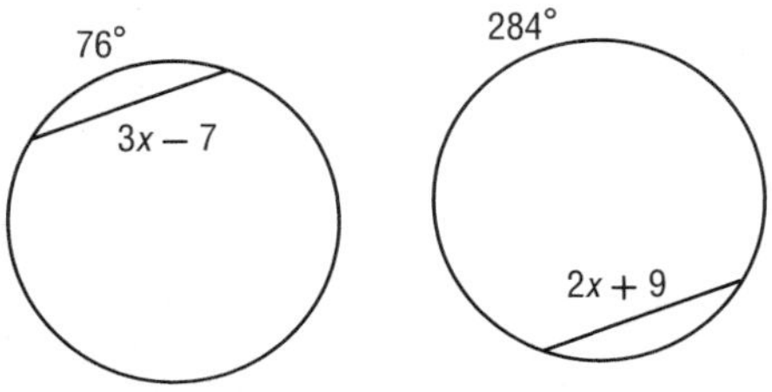

10. GRIDDED RESPONSE State the magnitude of rotational symmetry of the figure. Express your answer in degrees.

11. What is the length of $\overline{EF}$?

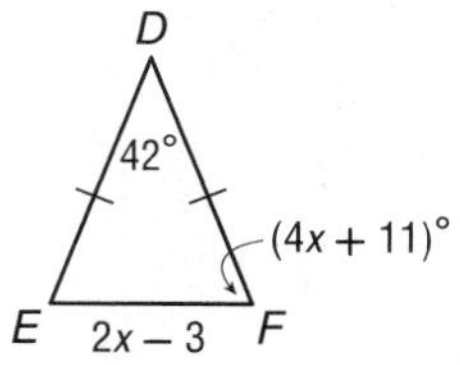

Extended Response

Record your answers on a sheet of paper. Show your work.

12. Use the circle shown to answer each question.

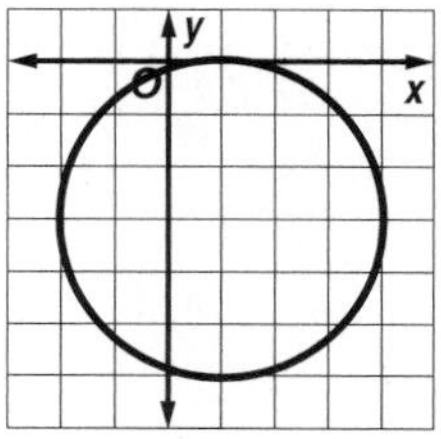

a. What is the center of the circle?

b. What is the radius of the circle?

c. Write an equation for the circle.

Need ExtraHelp?

If you missed Question...	1	2	3	4	5	6	7	8	9	10	11	12
Go to Lesson...	13-4	15-1	11-2	13-2	15-2	14-8	15-1	15-2	15-2	14-8	12-6	15-4

Student Handbook

This **Student Handbook** can help you answer these questions.

What if I Forget a Vocabulary Word?

What if I Need to Find Something Quickly?

What if I Forget a Formula?

Rubberball/Getty Images

Glossary/Glosario

MultilingualeGlossary

Go to connectED.mcgraw-hill.com for a glossary of terms in these additional languages:

Arabic	Chinese	Hmong	Spanish	Vietnamese
Bengali	English	Korean	Tagalog	
Brazilian Portugese	Haitian Creole	Russian	Urdu	

English

A

absolute value The distance a number is from zero on the number line.

acute angle An angle with a degree measure less than 90.

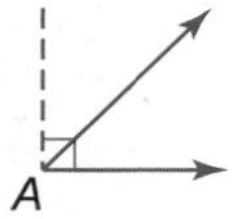

$0 < m\angle A < 90$

acute triangle A triangle in which all of the angles are acute angles.

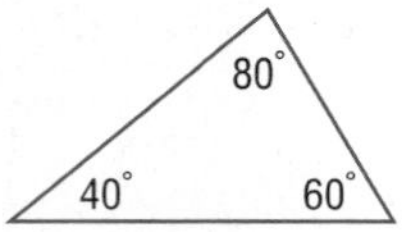

three acute angles

additive identity For any number a, $a + 0 = 0 + a = a$.

additive inverse Two integers, x and $-x$, are called additive inverses. The sum of any number and its additive inverse is zero.

adjacent angles Two angles that lie in the same plane, have a common vertex and a common side, but no common interior points.

algebraic expression An expression consisting of one or more numbers and variables along with one or more arithmetic operations.

alternate exterior angles In the figure, transversal t intersects lines ℓ and m. $\angle 5$ and $\angle 3$, and $\angle 6$ and $\angle 4$ are alternate exterior angles.

Español

valor aboluto Es la distancia que dista de cero en una recta numerica.

ángulo agudo Ángulo cuya medida en grados es menos de 90.

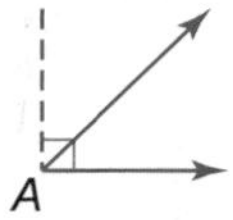

$0 < m\angle A < 90$

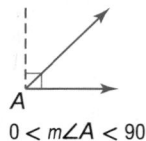

triángulo acutángulo Triángulo cuyos ángulos son todos agudos.

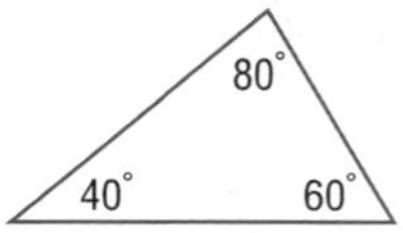

tres ángulos agudos

identidad de la adición Para cualquier número a, $a + 0 = 0 + a = a$.

inverso aditivo Dos enteros x y $-x$ reciben el nobre de inversos aditivos. La suma de cualquier número y su inverso aditivo es cero.

ángulos adyacentes Dos ángulos que yacen sobre el mismo plano, tienen el mismo vértice y un lado en común, pero ningún punto interior en común.

expresión algebraica Una expresión que consiste en uno o más números y variables, junto con una o más operaciones aritméticas.

ángulos alternos externos En la figura, la transversal t interseca las rectas ℓ y m. $\angle 5$ y $\angle 3$, y $\angle 6$ y $\angle 4$ son ángulos alternos externos.

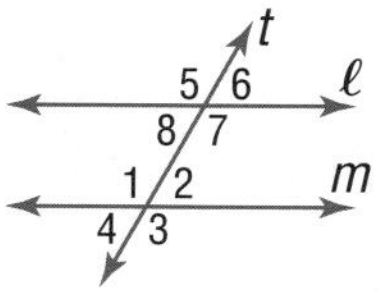

alternate interior angles In the figure at the bottom of page R115, transversal t intersects lines ℓ and m. $\angle 1$ and $\angle 7$, and $\angle 2$ and $\angle 8$ are alternate interior angles.

altitude **1.** In a triangle, a segment from a vertex of the triangle to the line containing the opposite side and perpendicular to that side. **2.** In a prism or cylinder, a segment perpendicular to the bases with an endpoint in each plane. **3.** In a pyramid or cone, the segment that has the vertex as one endpoint and is perpendicular to the base.

angle The intersection of two noncollinear rays at a common endpoint. The rays are called *sides* and the common endpoint is called the *vertex*.

angle bisector A ray that divides an angle into two congruent angles.

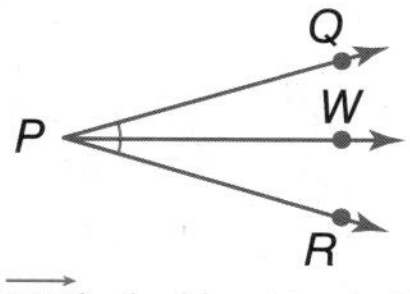

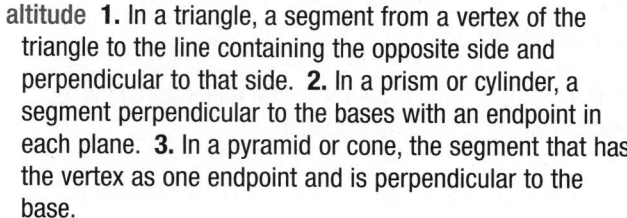

$\overrightarrow{PW}$ is the bisector of $\angle P$.

angle of rotation The angle through which a preimage is rotated to form the image.

area **1.** The measure of the surface enclosed by a geometric figure. **2.** The number of square units needed to cover a surface.

arithmetic sequence A numerical pattern that increases or decreases at a constant rate or value. The difference between successive terms of the sequence is constant.

asymptote A line that a graph approaches.

augmented matrix A coefficient matrix with an extra column containing the constant terms.

auxiliary line An extra line or segment drawn in a figure to help complete a proof.

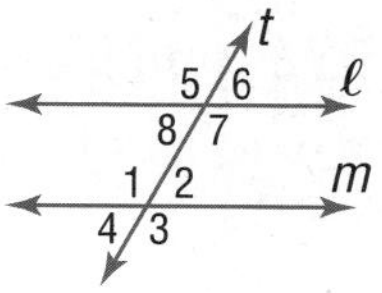

ángulos alternos internos En la figura anterior, la transversal t interseca las rectas ℓ y m. $\angle 1$ y $\angle 7$, y $\angle 2$ y $\angle 8$ son ángulos alternos internos.

altura **1.** En un triángulo, segmento trazado desde uno de los vértices del triángulo hasta el lado opuesto y que es perpendicular a dicho lado. **2.** En un prisma o un cilindro, segmento perpendicular a las bases con un extremo en cada plano. **3.** En una pirámide o un cono, segmento que tiene un extremo en el vértice y que es perpendicular a la base.

ángulo La intersección de dos rayos no colineales en un extremo común. Las rayos se llaman *lados* y el punto común se llama *vértice*.

bisectriz de un ángulo Rayo que divide un ángulo en dos ángulos congruentes.

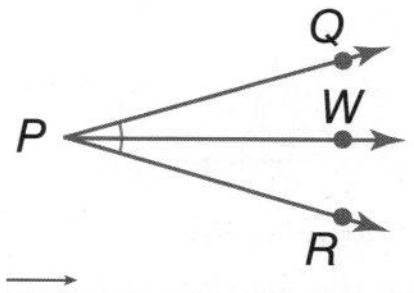

$\overrightarrow{PW}$ es la bisectriz del $\angle P$.

ángulo de rotación Ángulo a través del cual se rota una preimagen para formar la imagen.

área **1.** La medida de la superficie incluida por una figura geométrica. **2.** Número de unidades cuadradas para cubrir una superficie.

sucesión aritmética Un patrón numérico que aumenta o disminuye a una tasa o valor constante. La diferencia entre términos consecutivos de la sucesión es siempre la misma.

asíntota Una línea a que un gráfico acerca.

matriz aumentada una matriz del coeficiente con una columna adicional que contiene los términos de la constante.

línea auxiliar Recta o segmento de recta adicional que es traza en una figura para ayudar a completar una demostración.

B

bar graph A graphic form using bars to make comparisons of statistics.

gráfico de barra Forma gráfica usando barras para comparar estadísticas.

base In an expression of the form x^n, the base is x.

base En una expresión de la forma x^n, la base es x.

base angle of an isosceles triangle See *isosceles triangle* and isosceles trapezoid.

ángulo de la base de un triángulo isósceles Ver *triángulo isósceles* y trapecio isósceles.

base of parallelogram Any side of a parallelogram.

base de un paralelogramo Cualquier lado de un paralelogramo.

best-fit line The line that most closely approximates the data in a scatter plot.

recta de ajuste óptimo La recta que mejor aproxima los datos de una gráfica de dispersión.

between For any two points A and B on a line, there is another point C between A and B if and only if A, B, and C are collinear and $AC + CB = AB$.

entre Para cualquier par de puntos A y B de una recta, existe un punto C ubicado entre A y B si y sólo si A, B y C son colineales y $AC + CB = AB$.

betweenness of points See *between*.

intermediación de puntos Ver *entre*.

bivariate data Data with two variables.

datos bivariate Datos con dos variables.

boundary A line or curve that separates the coordinate plane into regions.

frontera Recta o curva que divide el plano de coordenadas en regiones.

box-and-whisker plot A diagram that divides a set of data into four parts using the median and quartiles. A box is drawn around the quartile values and whiskers extend from each quartile to the extreme data points.

diagrama de caja y patillas Diagram que divide un conjunto de datos en cuatro partes usando la mediana y los cuartiles. Se dibuja una caja alrededor de los cuartiles y se extienden patillas de cada uno de ellos a los valores extremos.

C

center The given point from which all points on the circle are the same distance.

centro Punto dado del cual equidistan todos los puntos de un circulo.

center of circle The central point where radii form a locus of points called a circle.

centro de un círculo Punto central desde el cual los radios forman un lugar geométrico de puntos llamado círculo.

center of dilation The center point from which dilations are performed.

centro de la homotecia Punto fijo en torno al cual se realizan las homotecias.

center of rotation A fixed point around which shapes move in a circular motion to a new position.

centro de rotación Punto fijo alrededor del cual gira una figura hasta alcanzar una posición dada.

center of symmetry See *point of symmetry*.

centro de la simetría Vea *el punto de simetría*.

chord **1.** For a given circle, a segment with endpoints that are on the circle. **2.** For a given sphere, a segment with endpoints that are on the sphere.

cuerda **1.** Para cualquier círculo, segmento cuyos extremos están en el círculo. **2.** Para cualquier esfera, segmento cuyos extremos están en la esfera.

circle The locus of all points in a plane equidistant from a given point called the *center* of the circle.

círculo Lugar geométrico formado por todos los puntos en un plano, equidistantes de un punto dado llamado *centro* del círculo.

P is the center of the circle.

circle The set of all points in a plane that are the same distance from a given point called the center.

circle graph A type of statistical graph used to compare parts of a whole.

circumference The distance around a circle.

circumscribed A circle is circumscribed about a polygon if the circle contains all the vertices of the polygon.

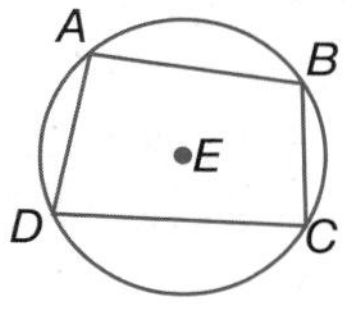

$\odot E$ is circumscribed about quadrilateral *ABCD*.

closed A set is closed under an operation if for any numbers in the set, the result of the operation is also in the set.

closed half-plane The solution of a linear inequality that includes the boundary line.

coefficient The numerical factor of a term.

collinear Points that lie on the same line.

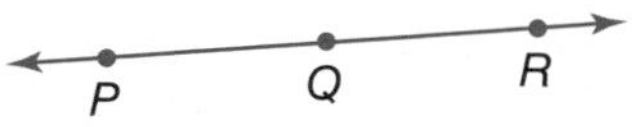

P, *Q*, and *R* are collinear.

common difference The difference between the terms in an arithmetic sequence.

common ratio The ratio of successive terms of a geometric sequence.

common tangent A line or segment that is tangent to two circles in the same plane.

complementary angles Two angles with measures that have a sum of 90.

complements One of two parts of a probability making a whole.

composition of transformations The resulting transformation when a transformation is applied to a figure and then another transformation is applied to its image.

P es el centro del círculo.

círculo Conjunto de todos los puntos del plano que están a la misma distancia de un punto dado del plano llamado centro.

gráfico del círculo Tipo de gráfica estadística que se usa para comparar las partes de un todo.

circunferencia Longitud del contorno de un círculo.

circunscrito Un polígono está circunscrito a un círculo si todos sus vértices están contenidos en el círculo.

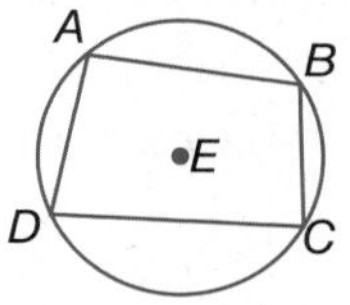

$\odot E$ está circunscrito al cuadrilátero *ABCD*.

cerrado Un conjunto es cerrado bajo una operación si para cualquier número en el conjunto, el resultado de la operación es también en el conjunto.

mitad-plano cerrado La solución de una desigualdad linear que incluye la línea de límite.

coeficiente Factor numérico de un término.

colineal Puntos que yacen sobre la misma recta.

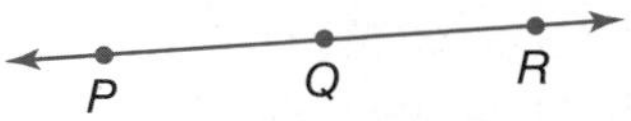

P, *Q* y *R* son colineales.

diferencia común Diferencia entre términos consecutivos de una sucesión aritmética.

razón común El razón de términos sucesivos de una secuencia geométrica.

tangente común Recta o segmento de recta tangente a dos círculos en el mismo plano.

ángulos complementarios Dos ángulos cuyas medidas suman 90.

complementos Una de dos partes de una probabilidad que forma un todo.

composición de transformaciones Transformación que resulta cuando se aplica una transformación a una figura y luego se le aplica otra transformación a su imagen.

compound inequality Two or more inequalities that are connected by the words *and* or *or*.

compound interest A special application of exponential growth.

concave polygon A polygon for which there is a line containing a side of the polygon that also contains a point in the interior of the polygon.

concentric circles Coplanar circles with the same center.

congruence transformations A mapping for which a geometric figure and its image are congruent.

congruent Having the same measure.

congruent polygons Polygons in which all matching parts are congruent.

conjugates Binomials of the form $a\sqrt{b} + c\sqrt{d}$ and $a\sqrt{b} - c\sqrt{d}$.

consecutive integers Integers in counting order.

consecutive interior angles In the figure, transversal t intersects lines ℓ and m. There are two pairs of consecutive interior angles: $\angle 8$ and $\angle 1$, and $\angle 7$ and $\angle 2$.

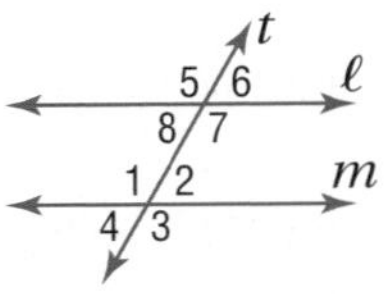

consistent A system of equations that has at least one ordered pair that satisfies both equations.

constant A monomial that is a real number.

constant function A linear function of the form $y = b$.

constant of variation The number k in equations of the form $y = kx$.

construction A method of creating geometric figures without the benefit of measuring tools. Generally, only a pencil, straightedge, and compass are used.

continuous function A function that can be graphed with a line or a smooth curve.

convex polygon A polygon for which there is no line that contains both a side of the polygon and a point in the interior of the polygon.

desigualdad compuesta Dos o más desigualdades que están unidas por las palabras *y* u *o*.

interés compuesto Aplicación especial de crecimiento exponencial.

polígono cóncavo Polígono para el cual existe una recta que contiene un lado del polígono y un punto en el interior del polígono.

círculos concéntricos Círculos coplanarios con el mismo centro.

transformaciones de congruencia Aplicación en la cual una figura geométrica y su imagen son congruentes.

congruente Que tienen la misma medida.

polígonos congruentes Polígonos cuyas partes correspondientes son todas congruentes.

conjugados Binomios de la forma $a\sqrt{b} + c\sqrt{d}$ and $a\sqrt{b} - c\sqrt{d}$.

enteros consecutivos Enteros en el orden de contar.

ángulos internos consecutivos En la figura, la transversal t interseca las rectas ℓ y m. La figura presenta dos pares de ángulos internos consecutivos; $\angle 8$ y $\angle 1$; y $\angle 7$ y $\angle 2$.

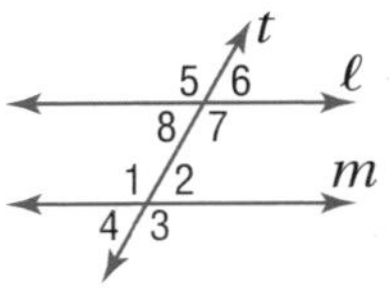

consistente Sistema de ecuaciones para el cual existe al menos un par ordenado que satisface ambas ecuaciones.

constante Monomio que es un número real.

función constante Función lineal de la forma $f(x) = b$.

constante de variación El número k en ecuaciones de la forma $y = kx$.

construcción Método para dibujar figuras geométricas sin el uso de instrumentos de medición. En general, sólo requiere de un lápiz, una regla y un compás.

función continua Función cuya gráfica puedes ser una recta o una curva suave.

polígono convexo Polígono para el cual no existe recta alguna que contenga un lado del polígono y un punto en el interior del polígono.

coordinate The number that corresponds to a point on a number line.

coordinate plane The plane containing the *x*- and *y*-axes.

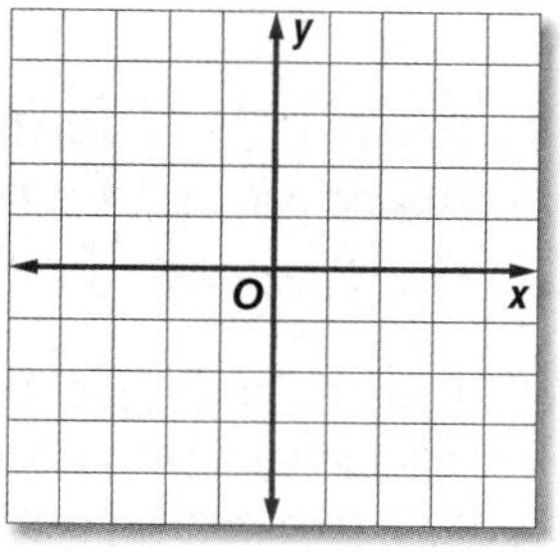

coordinate proofs Proofs that use figures in the coordinate plane and algebra to prove geometric concepts.

coordinate system The grid formed by the intersection of two number lines, the horizontal axis and the vertical axis.

coplanar Points that lie in the same plane.

corollary A statement that can be easily proved using a theorem is called a corollary of that theorem.

correlation coefficient A value that shows how close data points are to a line.

corresponding angles In the figure, transversal *t* intersects lines ℓ and *m*. There are four pairs of corresponding angles: $\angle 5$ and $\angle 1$, $\angle 8$ and $\angle 4$, $\angle 6$ and $\angle 2$, and $\angle 7$ and $\angle 3$.

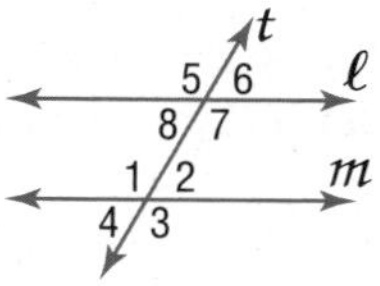

corresponding parts Matching parts of congruent polygons.

counterexample A specific case in which a statement is false.

cube root If $a^3 = b$, then *a* is the cube root of *b*.

coordenada Número que corresponde a un punto en una recta numérica.

plano de coordenadas Plano que contiene los ejes *x* y *y*.

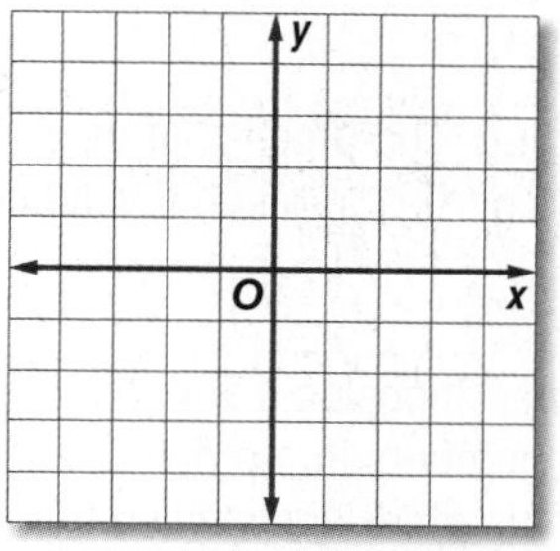

demostraciones en coordinadas Demostraciones que usan figuras en el plano de coordinados y álgebra para demostrar conceptos geométricos.

sistema de coordenadas Cuadriculado formado por la intersección de dos rectas numéricas: los ejes *x* y *y*.

coplanar Puntos que yacen en el mismo plano.

corolario Un enunciado que se puede demostrar fácilmente usando un teorema se conoce como corolario de dicho teorema.

coeficiente de correlación Un valor que demostraciones cómo los puntos de referencias cercanos están a una línea.

ángulos correspondientes En la figura, la transversal *t* interseca las rectas ℓ y *m*. La figura muestra cuatro pares de ángulos correspondientes: $\angle 5$ y $\angle 1$, $\angle 8$ y $\angle 4$, $\angle 6$ y $\angle 2$; y $\angle 7$ y $\angle 3$.

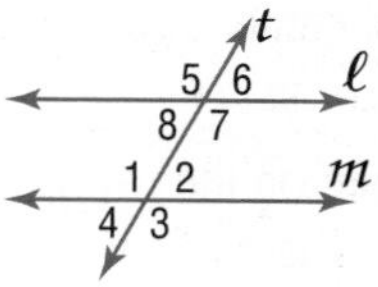

partes correspondientes Partes que coinciden de polígonos congruentes.

contraejemplo Ejemplo específico de la falsedad de un enunciado.

raíz cúbica Si $a^3 = b$, entonces *a* es la raíz cúbica de *b*.

D

decreasing The graph of a function goes down on a portion of its domain when viewed from left to right.

decreciente El gráfico de una función va abajo en una porción de su dominio cuando está visto de izquierda a derecha.

deductive reasoning The process of using facts, rules, definitions, or properties to reach a valid conclusion.

razonamiento deductivo Proceso de usar hechos, reglas, definiciones o propiedades para sacar conclusiones válidas.

defining a variable Choosing a variable to represent one of the unspecified numbers in a problem and using it to write expressions for the other unspecified numbers in the problem.

definir una variable Consiste en escoger una variable para representar uno de los números desconocidos en un problema y luego usarla para escribir expresiones para otros números desconocidos en el problema.

degree A unit of measure used in measuring angles and arcs. An arc of a circle with a measure of 1° is $\frac{1}{360}$ of the entire circle.

grado Unidad de medida que se usa para medir ángulos y arcos. El arco de un círculo que mide 1° equivale a $\frac{1}{360}$ del círculo completo.

dependent A system of equations that has an infinite number of solutions.

dependiente Sistema de ecuaciones que posee un número infinito de soluciones.

dependent variable The variable in a relation with a value that depends on the value of the independent variable.

variable dependiente La variable de una relación cuyo valor depende del valor de la variable independiente.

diameter **1.** In a circle, a chord that passes through the center of the circle. **2.** In a sphere, a segment that contains the center of the sphere, and has endpoints that are on the sphere. **3.** The distance across a circle through its center.

diámetro **1.** En un círculo cuerda que pasa por el centro. **2.** En una estera segmento que incluye el centro de la esfera y cuyos extremos están ubicados en la esfera. **3.** La distancia a través de un círculo a través de su centro.

dilation A transformation that enlarges or reduces the original figure proportionally. A dilation with center C and positive scale factor k, $k \neq 1$, is a function that maps a point P in a figure to its image such that

- if point P and C coincide, then the image and preimage are the same point, or
- if point P is not the center of dilation, then P' lies on $\overrightarrow{CP}$ and $CP' = k(CP)$.

If $k < 0$, P' is the point on the ray opposite $\overrightarrow{CP}$ such that $CP' = |k|(CP)$.

homotecia Transformación que amplía o disminuye proporcionalmente el tamaño de una figura. Una homotecia con centro C y factor de escala positivo k, $k \neq 1$, es una función que aplica un punto P a su imagen, de modo que si el punto P coincide con el punto C, entonces la imagen y la preimagen son el mismo punto, o si el punto P no es el centro de la homotecia, entonces P' yace sobre $\overrightarrow{CP}$ y $CP' = k(CP)$. Si $k < 0$, P' es el punto sobre el rayo opuesto a $\overrightarrow{CP}$, tal que $CP' = |k|(CP)$.

dimension The number of rows, m, and the number of columns, n, of a matrix written as $m \times n$.

dimension El número de filas, de m, y del número de la columna, n, de una matriz escrita como $m \times n$.

dimensional analysis The process of carrying units throughout a computation.

análisis dimensional Proceso de tomar en cuenta las unidades de medida al hacer cálculos.

direct isometry An isometry in which the image of a figure is found by moving the figure intact within the plane.

isometría directa Isometría en la cual se obtiene la imagen de una figura, al mover la figura intacta dentro del plano.

direct variation An equation of the form $y = kx$, where $k \neq 0$.

variación directa Una ecuación de la forma $y = kx$, donde $k \neq 0$.

discrete function A function of points that are not connected.

función discreta Función de puntos desconectados.

distance between two points The length of the segment between two points.

distancia entre dos puntos Longitud del segmento entre dos puntos.

distribution A graph or table that shows the theoretical frequency of each possible data value.

distrubución Un gráfico o una tabla que muestra la frecuencia teórica de cada valor de datos posible.

domain The set of the first numbers of the ordered pairs in a relation.

dominio Conjunto de los primeros números de los pares ordenados de una relación.

E

element Each entry in a matrix.

elemento Cada entrada de una matriz.

elimination The use of addition or subtraction to eliminate one variable and solve a system of equations.

eliminación El uso de la adición o la sustracción para eliminar una variable y resolver así un sistema de ecuaciones.

end behavior Describes how the values of a function behave at each end of the graph.

comportamiento extremo Describe como los valores de una función se comportan en el cada fin del gráfico.

enlargement An image that is larger that the original figure.

ampliación Imagen que es más grande que la figura original.

equally likely The outcomes of an experiment are equally likely if there are n outcomes and the probability of each is $\frac{1}{n}$.

igualmente probablemente Los resultados de un experimento son igualmente probables si hay resultados de n y la probabilidad de cada uno es $\frac{1}{n}$.

equation A mathematical sentence that contains an equals sign, $=$.

ecuación Enunciado matemático que contiene el signo de igualdad, $=$.

equiangular polygon A polygon with all congruent angles.

polígono equiangular Polígono cuyos ángulos son todos congruentes.

equiangular triangle A triangle with all angles congruent.

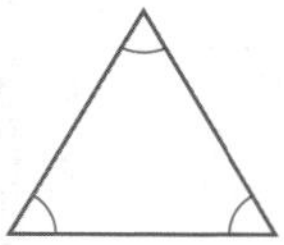

triángulo equiangular Triángulo cuyos ángulos son todos congruentes.

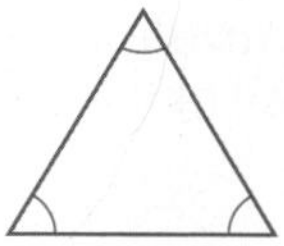

equidistant The distance between two lines measured along a perpendicular line is always the same.

equidistante La distancia entre dos rectas que siempre permanece constante cuando se mide a lo largo de una perpendicular.

equilateral polygon A polygon with all congruent sides.

polígono equilátero Polígono cuyos lados son todos congruentes.

equilateral triangle A triangle with all sides congruent.

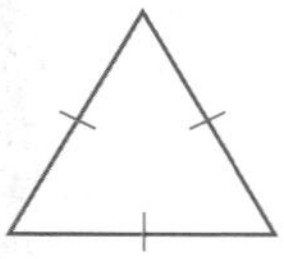

triángulo equilátero Triángulo cuyos lados son todos congruentes.

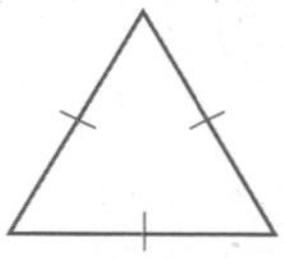

equivalent equations Equations that have the same solution.

equivalent expressions Expressions that denote the same value for all values of the variable(s).

evaluate To find the value of an expression.

excluded values Any values of a variable that result in a denominator of 0 must be excluded from the domain of that variable.

exponent In an expression of the form x^n, the exponent is n. It indicates the number of times x is used as a factor.

exponential decay When an initial amount decreases by the same percent over a given period of time.

exponential equation An equation in which the variables occur as exponents.

exponential function A function that can be described by an equation of the form $y = a^x$, where $a > 0$ and $a \neq 1$.

exponential growth When an initial amount increases by the same percent over a given period of time.

exterior A point is in the exterior of an angle if it is neither on the angle nor in the interior of the angle.

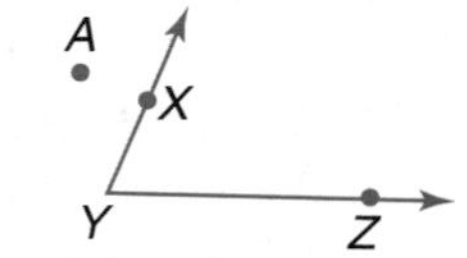

A is in the exterior of $\angle XYZ$.

exterior angles **1.** An angle that lies in the region that is not between two transversals that intersect the same line. **2.** An angle formed by one side of a triangle and the extension of another side.

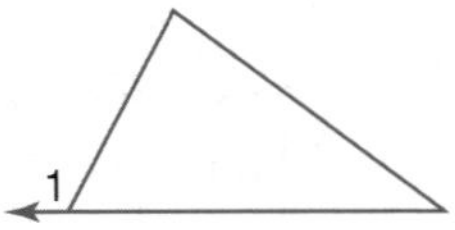

$\angle 1$ is an exterior angle.

extraneous solutions Results that are not solutions to the original equation.

extremes In the ratio $\frac{a}{b} = \frac{c}{d}$, a and d are the extremes.

ecuaciones equivalentes Ecuaciones que poseen la misma solución.

expresiones equivalentes Expresiones que denotan el mismo valor para todos los valores de la(s) variable(s).

evaluar Calcular el valor de una expresión.

valores excluidos Cualquier valor de una variable cuyo resultado sea un denominador igual a cero, debe excluirse del dominio de dicha variable.

exponente En una expresión de la forma x^n, el exponente es n. Éste indica cuántas veces se usa x como factor.

desintegración exponencial La cantidad inicial disminuye según el mismo porcentaje a lo largo de un período de tiempo dado.

ecuación exponencial Ecuación en que las variables aparecen en los exponentes.

función exponencial Función que puede describirse mediante una ecuación de la forma $y = a^x$, donde $a > 0$ y $a \neq 1$.

crecimiento exponencial La cantidad inicial aumenta según el mismo porcentaje a lo largo de un período de tiempo dado.

exterior Un punto yace en el exterior de un ángulo si no se ubica ni en el ángulo ni en el interior del ángulo.

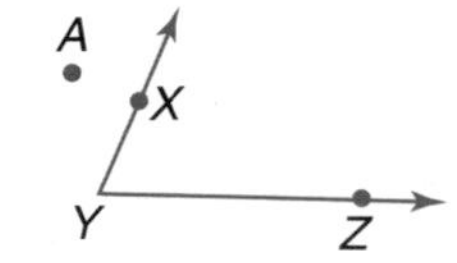

A está en el exterior del $\angle XYZ$.

ángulos externos **1.** Un ángulo que está en la región que no está entre dos transversals que cruzan la misma línea. **2.** Ángulo formado por un lado de un triángulo y la prolongación de otro de sus lados.

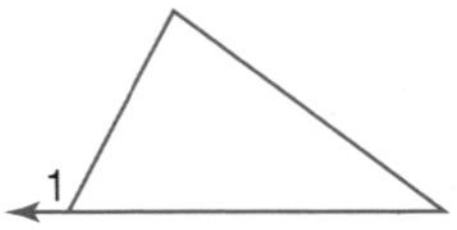

$\angle 1$ es un ángulo externo.

soluciones extrañas Resultados que no son soluciones de la ecuación original.

extremos En la razón $\frac{a}{b} = \frac{c}{d}$, a y d son los extremos.

F

factors In an algebraic expression, the quantities being multiplied are called factors.

factores En una expresión algebraica, los factores son las cantidades que se multiplican.

family of graphs Graphs and equations of graphs that have at least one characteristic in common.

familia de gráficas Gráficas y ecuaciones de gráficas que tienen al menos una característica común.

finite plane A plane that has boundaries or does not extend indefinitely.

plano finito Plano que tiene límites o que no se extiende indefinidamente.

flow proof A proof that organizes statements in logical order, starting with the given statements. Each statement is written in a box with the reason verifying the statement written below the box. Arrows are used to indicate the order of the statements.

demostración de flujo Demostración que organiza los enunciados en orden lógico, comenzando con los enunciados dados. Cada enunciado se escribe en una casilla y debajo de cada casilla se escribe el argumento que verifica dicho enunciado. El orden de los enunciados se indica con flechas.

formula An equation that states a rule for the relationship between certain quantities.

fórmula Ecuación que establece una relación entre ciertas cantidades.

four-step problem-solving plan
Step 1 Explore the problem.
Step 2 Plan the solution.
Step 3 Solve the problem.
Step 4 Check the solution.

plan de cuatro pasos para resolver problemas
Paso 1 Explora el problema.
Paso 2 Planifica la solución.
Paso 3 Resuelve el problema.
Paso 4 Examina la solución.

frequency table A chart that indicates the number of values in each interval.

Tabla de frecuencias Tabla que indica el número de valores en cada intervalo.

function A relation in which each element of the domain is paired with exactly one element of the range.

función Una relación en que a cada elemento del dominio le corresponde un único elemento del rango.

function notation A way to name a function that is defined by an equation. In function notation, the equation $y = 3x - 8$ is written as $f(x) = 3x - 8$.

notación funcional Una manera de nombrar una función definida por una ecuación. En notación funcional, la ecuación $y = 3x - 8$ se escribe $f(x) = 3x - 8$.

Fundamental Counting Principle If an event M can occur in m ways and is followed by an event N that can occur in n ways, then the event M followed by the event N can occur in $m \times n$ ways.

Principio fundamental de contar Si un evento M puede ocurrir de m maneras y lo sigue un evento N que puede ocurrir de n maneras, entonces el evento M seguido del evento N puede ocurrir de $m \times n$ maneras.

G

general equation for exponential decay $y = C(1 - r)^t$, where y is the final amount, C is the initial amount, r is the rate of decay expressed as a decimal, and t is time.

ecuación general de desintegración exponencial $y = C(1 - r)^t$, donde y es la cantidad final, C es la cantidad inicial, r es la tasa de desintegración escrita como decimal y t es el tiempo.

general equation for exponential growth $y = C(1 + r)^t$, where y is the final amount, C is the initial amount, r is the rate of change expressed as a decimal, and t is time.

ecuación general de crecimiento exponencial $y = C(1 + r)^t$, donde y es la cantidad final, C es la cantidad inicial, r es la tasa de cambio del crecimiento escrita como decimal y t es el tiempo.

geometric sequence A sequence in which each term after the first is found by multiplying the previous term by a constant r, called the common ratio.

secuencia geométrica Una secuencia en la cual cada término después de que la primera sea encontrada multiplicando el término anterior por un r constante, llamado el razón común.

glide reflection The composition of a translation followed by a reflection in a line parallel to the translation vector.

graph To draw, or plot, the points named by certain numbers or ordered pairs on a number line or coordinate plane.

H

half-plane The region of the graph of an inequality on one side of a boundary.

height of a parallelogram The length of an altitude of a parallelogram.

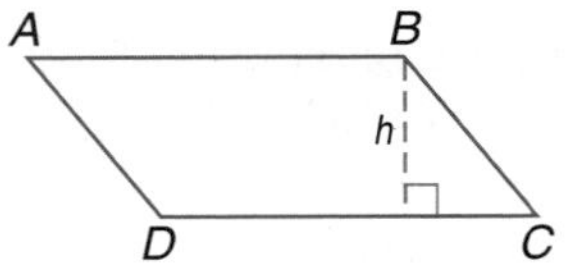

h is the height of parallelogram *ABCD*.

histogram A graphical display that uses bars to display numerical data that have been organized into equal intervals.

I

identity An equation that is true for every value of the variable.

identity function The function $y = x$.

identity matrix A square matrix that, when multiplied by another matrix, equals that same matrix. If A is any $n \times n$ matrix and I is the $n \times n$ identity matrix, then $A \cdot I = A$ and $I \cdot A = A$.

image A figure that results from the transformation of a geometric figure.

included angle In a triangle, the angle formed by two sides is the included angle for those two sides.

included side The side of a polygon that is a side of each of two angles.

inconsistent A system of equations with no ordered pair that satisfy both equations.

increasing The graph of a function goes up on a portion of its domain when viewed from left to right.

reflexión del deslizamiento Composición de una traslación seguida por una reflexión en una recta paralela al vector de la traslación.

graficar Marcar los puntos que denotan ciertos números en una recta numérica o ciertos pares ordenados en un plano de coordenadas.

semiplano Región de la gráfica de una desigualdad en un lado de la frontera.

altura de un paralelogramo Longitud del segmento perpendicular que va desde la base hasta el vértice opuesto a ella.

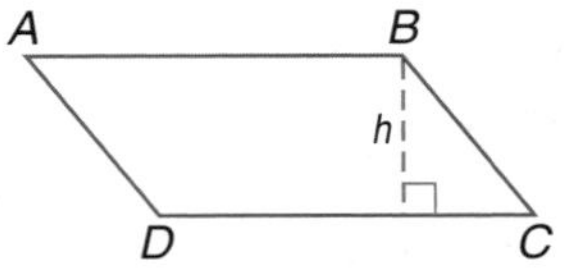

h es la altura del paralelogramo *ABCD*.

histograma Una exhibición gráfica que utiliza barras para exhibir los datos numéricos que se han organizado en intervalos iguales.

identidad Ecuación que es verdad para cada valor de la variable.

unción identidad La función $y = x$.

matriz de la identidad Una matriz cuadrada que, cuando es multiplicada por otra matriz, iguala que la misma matriz. Si A es alguna de la matriz $n \times n$ e I es la matriz de la identidad de $n \times n$, entonces $A \cdot I = A$ e $I \cdot A = A$.

imagen Figura que resulta de la transformación de una figura geométrica.

ángulo incluido En un triángulo, el ángulo formado por dos lados es el ángulo incluido de esos dos lados.

lado incluido Lado de un polígono común a dos de sus ángulos.

inconsistente Un sistema de ecuaciones para el cual no existe par ordenado alguno que satisfaga ambas ecuaciones.

crecciente El gráfico de una función va arriba en una porción de su dominio cuando está visto de izquierda a derecha.

independent A system of equations with exactly one solution.

independent variable The variable in a function with a value that is subject to choice.

indirect isometry An isometry that cannot be performed by maintaining the orientation of the points, as in a direct isometry.

inductive reasoning A conclusion based on a pattern of examples.

inequality An open sentence that contains the symbol $<$, $\leq$, $>$, or $\geq$.

inscribed A polygon is inscribed in a circle if each of its vertices lie on the circle.

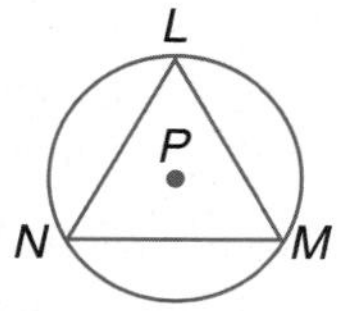

$\triangle LMN$ is inscribed in $\odot P$.

integers The set $\{\ldots, -2, -1, 0, 1, 2, \ldots\}$.

interior A point is in the interior of an angle if it does not lie on the angle itself and it lies on a segment with endpoints that are on the sides of the angle.

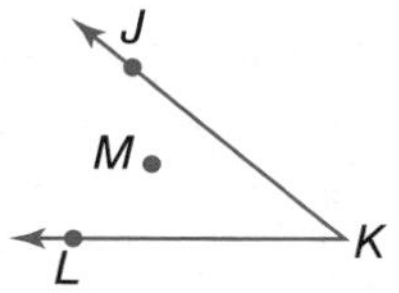

M is in the interior of $\angle JKL$.

interior angles Angles that lie between two transversals that intersect the same line.

interquartile range The range of the middle half of a set of data. It is the difference between the upper quartile and the lower quartile.

intersection **1.** The graph of a compound inequality containing and; the solution is the set of elements common to both inequalities. **2.** A set of points common to two or more geometric figures.

inverse variation An equation of the form $xy = k$, where $k \neq 0$.

irrational number A number that cannot be expressed as a terminating or repeating decimal.

independiente Un sistema de ecuaciones que posee una única solución.

variable independiente La variable de una función sujeta a elección.

isometría indirecta Tipo de isometría que no se puede obtener manteniendo la orientación de los puntos, como ocurre con la isometría directa.

razonamiento inductivo Conclusión basada en un patrón de ejemplos.

desigualdad Enunciado abierto que contiene uno o más de los símbolos $<$, $\leq$, $>$, o $\geq$.

inscrito Un polígono está inscrito en un círculo si todos sus vértices yacen en el círculo.

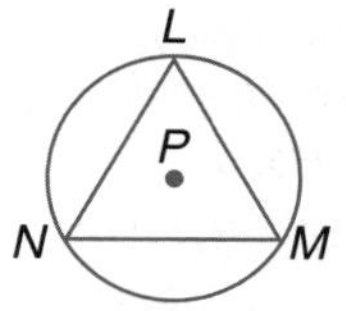

$\triangle LMN$ está inscrito en $\odot P$.

enteros El conjunto $\{\ldots, -2, -1, 0, 1, 2, \ldots\}$.

interior Un punto se encuenta en el interior de un ángulo si no yace en el ángulo como tal y si está en un segmento cuyos extremos están en los lados del ángulo.

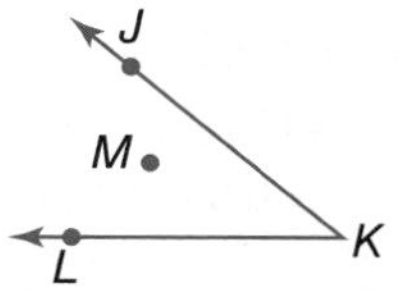

M está en el interior del $\angle JKL$.

ángulos interiores Ángulos que yacen entre dos transversales que intersecan la misma recta.

amplitud intercuartílica Amplitude de la mitad central de un conjunto de datos. Es la diferenccia entre el cuartil superior y el inferior.

intersección **1.** Gráfica de una desigualdad compuesta que contiene la palabra y; la solución es el conjunto de soluciones de ambas desigualdades. **2.** Conjunto de puntos comunes a dos o más figuras geométricas.

variación inversa Ecuación de la forma $xy = k$, donde $k \neq 0$.

número irracional Número que no se puede expresar como un decimal terminal o periódico.

irregular figure A polygon with sides and angles that are not all congruent.

isometry A mapping for which the original figure and its image are congruent.

isosceles trapezoid A trapezoid in which the legs are congruent, both pairs of base angles are congruent, and the diagonals are congruent.

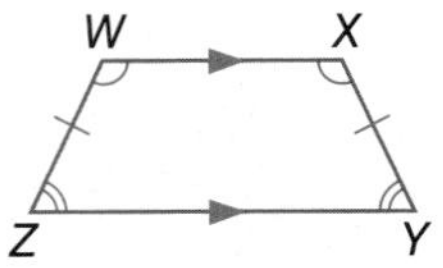

isosceles triangle A triangle with at least two sides congruent. The congruent sides are called *legs*. The angles opposite the legs are *base angles*. The angle formed by the two legs is the *vertex angle*. The side opposite the vertex angle is the *base*.

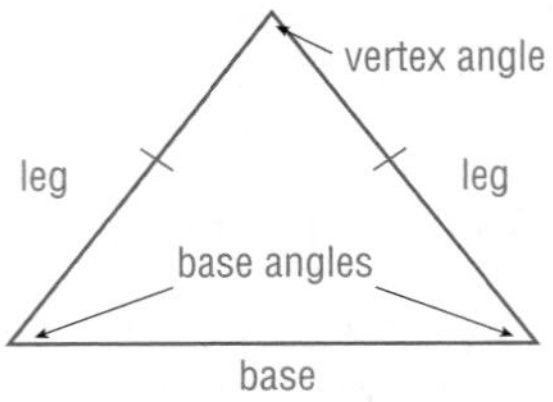

figura irregular Polígono cuyos lados y ángulos no son todo congruentes.

isometría Aplicación en la cual la figura original y su imagen son congruentes.

trapecio isósceles Trapecio cuyos catetos son congruentes, ambos pares de ángulos de las bases son congruentes y las diagonales son congruentes.

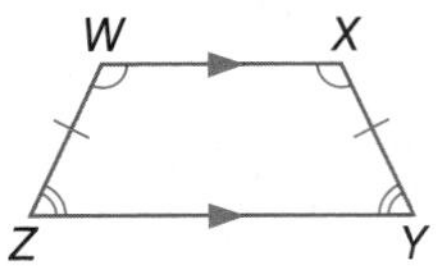

triángulo isósceles Triángulo que tiene por lo menos dos lados congruentes. Los lados congruentes se llaman *catetos*. Los ángulos opuestos a los catetos son los *ángulos de la base*. El ángulo formado por los dos catetos es el *ángulo del vértice*. El lado opuesto al ángulo del vértice es la *base*.

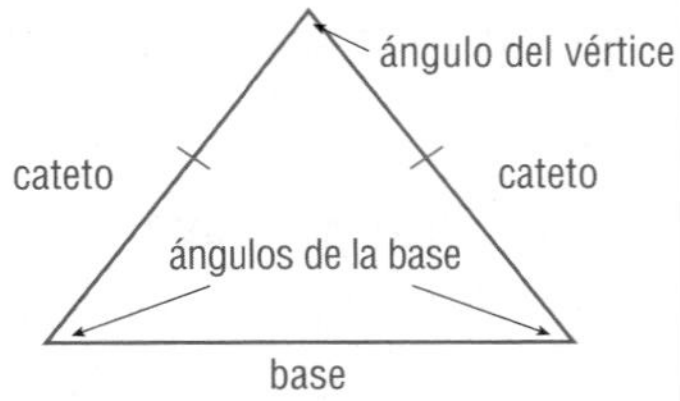

K

kite A quadrilateral with exactly two distinct pairs of adjacent congruent sides.

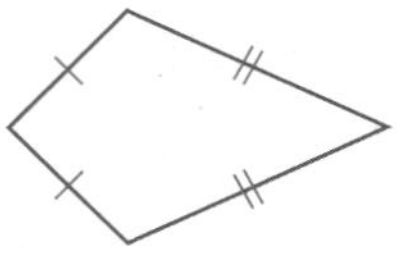

cometa Cuadrilátero que tiene exactamente dos pares differentes de lados congruentes y adyacentes.

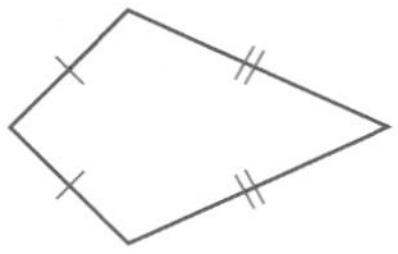

L

legs of a right triangle The shorter sides of a right triangle.

legs of a trapezoid The nonparallel sides of a trapezoid.

legs of an isosceles triangle The two congruent sides of an isosceles triangle.

like terms Terms that contain the same variables, with corresponding variables having the same exponent.

line A basic undefined term of geometry. A line is made up

catetos de un triángulo rectángulo Lados más cortos de un triángulo rectángulo.

catetos de un trapecio Los lados no paralelos de un trapecio.

catetos de un triángulo isósceles Las dos lados congruentes de un triángulo isósceles.

términos semejantes Expresiones que tienen las mismas variables, con las variables correspondientes elevadas a los mismos exponentes.

recta Término geometrico basico no definido. Una recta

of points and has no thickness or width. In a figure, a line is shown with an arrowhead at each end. Lines are usually named by lowercase script letters or by writing capital letters for two points on the line, with a double arrow over the pair of letters.

line of fit A line that describes the trend of the data in a scatter plot.

line of reflection A line in which each point on the preimage and its corresponding point on the image are the same distance from this line.

line of symmetry A line that can be drawn through a plane figure so that the figure on one side is the reflection image of the figure on the opposite side.

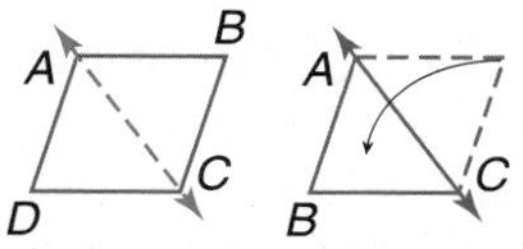

$\overleftrightarrow{AC}$ is a line of symmetry.

line segment A measurable part of a line that consists of two points, called endpoints, and all of the points between them.

linear equation An equation in the form $Ax + By = C$, with a graph that is a straight line.

linear extrapolation The use of a linear equation to predict values that are outside the range of data.

linear function A function with ordered pairs that satisfy a linear equation.

linear interpolation The use of a linear equation to predict values that are inside of the data range.

linear pair A pair of adjacent angles whose non-common sides are opposite rays.

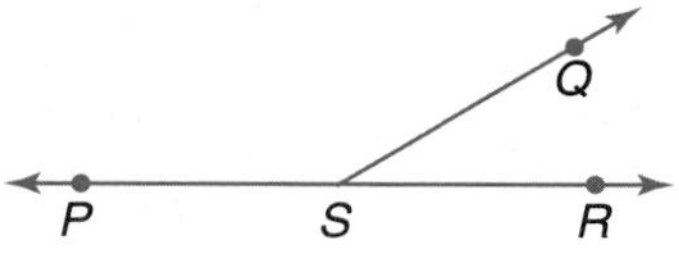

$\angle PSQ$ and $\angle QSR$ are a linear pair.

linear regression An algorithm to find a precise line of fit for a set of data.

linear transformation One or more operations performed on a set of data that can be written as a linear function.

literal equation A formula or equation with several variables.

está formada por puntos y carece de grosor o ancho. En una figura, una recta se representa con una flecha en cada extremo. Generalmente se designan con letras minúsculas o con las dos letras mayúsculas de dos puntos sobre la recta y una flecha doble sobre el par de letras.

recta de ajuste Recta que describe la tendencia de los datos en una gráfica de dispersión.

línea de reflexión Una línea en la cual cada punto en el preimage y el su corresponder senalañ en la imagen es la misma distancia de esta línea.

eje de simetría Recta que se traza a través de una figura plana, de modo que un lado de la figura es la imagen reflejada del lado opuesto.

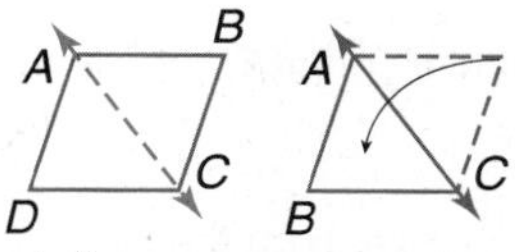

$\overleftrightarrow{AC}$ es un eje de simetría.

segmento de recta Sección medible de una recta que consta de dos puntos, llamados extremos, y todos los puntos entre ellos.

ecuación lineal Ecuación de la forma $Ax + By = C$, cuya gráfica es una recta.

extrapolación lineal Uso de una ecuación lineal para predecir valores fuera de la amplitud de los datos.

función lineal Función cuyos pares ordenados satisfacen una ecuación lineal.

interpolación lineal Uso de una ecuación lineal para predecir valores dentro de la amplitud de los datos.

par lineal Par de ángulos adyacentes cuyos lados no comunes forman rayos opuestos.

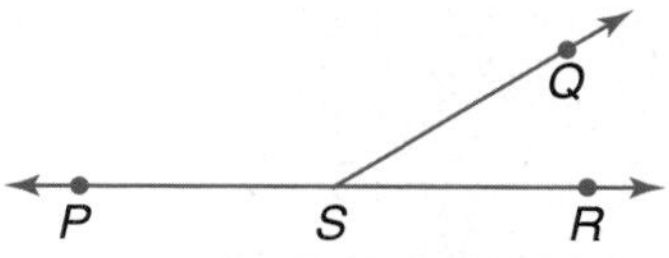

$\angle PSQ$ y $\angle QSR$ forman un par lineal.

regresión linear Un algoritmo para encontrar una línea exacta del ajuste para un sistema de datos.

transformación lineal Una o más operaciones que se hacen en un conjunto de datos y que se pueden escribir como una función lineal.

ecuación literal Un fórmula o ecuación con varias variables.

locus The set of points that satisfy a given condition.

lower quartile Divides the lower half of the data into two equal parts.

M

mapping Illustrates how each element of the domain is paired with an element in the range.

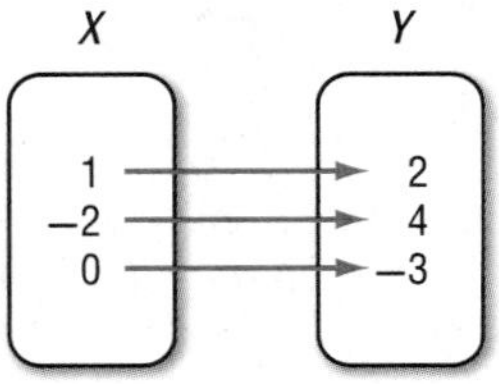

matrix Any rectangular arrangement of numbers in rows and columns.

maximum The highest point on the graph of a curve.

mean The sum of numbers in a set of data divided by the number of items in the data set.

mean absolute deviation The average of the absolute values of differences between the mean and each value in a data set. It is used to predict errors and to judge equality.

means The middle terms of the proportion.

measures of central tendency Numbers or pieces of data that can represent the whole set of data.

measures of position Measures that compare the position of a value relative to other values in a set.

measures of variation Used to describe the distribution of statistical data.

median The middle number in a set of data when the data are arranged in numerical order. If the data set has an even number, the median is the mean of the two middle numbers.

median fit line A type of best-fit line that is calculated using the medians of the coordinates of the data points.

metric A rule for assigning a number to some characteristic or attribute.

midpoint The point on a segment exactly halfway between the endpoints of the segment.

midsegment A segment with endpoints that are the midpoints of two sides of a triangle.

lugar geométrico Conjunto de puntos que satisfacen una condición dada.

cuartil inferior Éste divide en dos partes iguales la mitad inferior de un conjunto de datos.

aplicaciones Ilustra la correspondencia entre cada elemento del dominio con un elemento del rango.

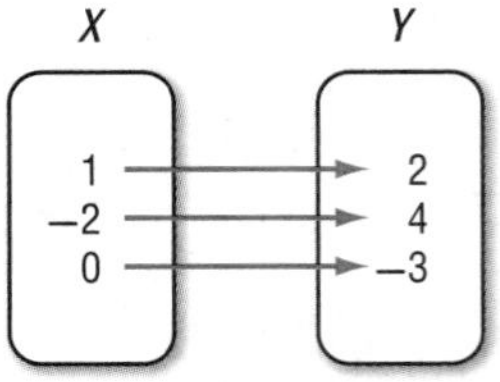

matriz Disposción rectangular de numeros colocados en filas y columnas.

máximo El punto más alto en la gráfica de una curva.

media La suma de los números de un conjunto de datos dividida entre el numero total de artículos.

desviación absoluta media El promedio de los valores absolutos de diferencias entre el medio y cada valor de un conjunto de datos. Ha usado para predecir errores y para juzgar igualdad.

medios Los términos centrales de una proporción.

medidas de tendencia central Números o fragmentos que pueden representar el conjunto de datos total de datos.

medidas de la posición Las medidas que comparar la posición de un valor relativo a otros valores de un conjunto.

medidas de variación Números que se usan para describir la distribución o separación de un conjunto de datos.

mediana El número central de conjunto de datos, una vezque los datos han sido ordenados numéricamente. Si hay un número par de datos, la mediana es el promedio de los datos centrales.

línea apta del punto medio Tipo de mejor-cupo la línea se calcula que usando los puntos medios de los coordenadas de los puntos de referencias.

métrico Una regla para asignar un número a alguna característica o atribuye.

punto medio Punto en un segmento que yace exactamente en la mitad, entre los extremos del segmento.

segmento medio Segmento cuyos extremos son los puntos medios de dos lados de un triángulo.

midsegment of trapezoid A segment that connects the midpoints of the legs of a trapezoid.

segmento medio de un trapecio Segmento que conecta los puntos medios de los catetos de un trapecio.

midsegment of triangle A segment with endpoints that are the midpoints of two sides of a triangle.

segmento medio de un triángulo Segmento cuyas extremos son los puntos medianos de dos lados de un triángulo.

minimum The lowest point on the graph of a curve.

mínimo El punto más bajo en la gráfica de una curva.

mixture problems Problems in which two or more parts are combined into a whole.

problemas de mezclas Problemas en que dos o más partes se combinan en un todo.

mode The number(s) that appear most often in a set of data.

moda El número(s) que aparece más frecuencia en un conjunto de datos.

monomial A number, a variable, or a product of a number and one or more variables.

monomio Número, variable o producto de un número por una o más variables.

multiplicative identity For any number a, $a \cdot 1 = 1 \cdot a = a$.

identidad de la multiplicación Para cualquier número $a \cdot 1 = 1 \cdot a = a$.

multiplicative inverses Two numbers with a product of 1.

inversos multiplicativos Dos números cuyo producto es igual a 1.

multi-step equation Equations with more than one operation.

ecuaciones de varios pasos Ecuaciones con más de una operación.

N

nth root If $a^n = b$ for a positive integer n, then a is an nth root of b.

raíz enésima Si $a^n = b$ para cualquier entero positive n, entonces a se llama una raíz enésima de b.

n-gon A polygon with n sides.

*en*ágono Polígono con n lados.

natural numbers The set $\{1, 2, 3, \ldots\}$.

números naturales El conjunto $\{1, 2, 3, \ldots\}$.

negative A function is negative on a portion of its domain where its graph lies below the x-axis.

negativo Una función es negativa en una porción de su dominio donde su gráfico está debajo del eje-x.

negative correlation In a scatter plot, as x increases, y decreases.

correlación negativa En una gráfica de dispersión, a medida que x aumenta, y disminuye.

negative exponent For any real number $a \neq 0$ and any integer n, $a^{-n} = \frac{1}{a^n}$ and $\frac{1}{a^{-n}} = a^n$.

exponiente negativo Para números reales, si $a \neq 0$, y cualquier número entero n, entonces $a^{-n} = \frac{1}{a^n}$ and $\frac{1}{a^{-n}} = a^n$.

negative number Any value less than zero.

número negativo Cualquier valor menor que cero.

nonlinear function A function with a graph that is not a straight line.

función no lineal Una función con un gráfica que no es una línea recta.

number theory The study of numbers and the relationships between them.

teoría del número El estudio de números y de las relaciones entre ellas.

obtuse angle An angle with degree measure greater than 90 and less than 180.

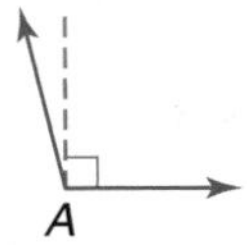

$90 < m\angle A < 180$

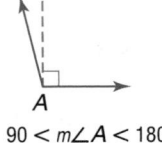

obtuse triangle A triangle with an obtuse angle.

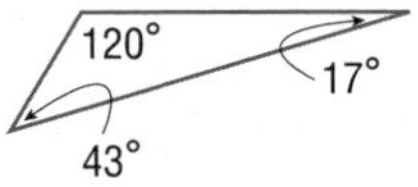

one obtuse angle

odds The ratio of the probability of the success of an event to the probability of its complement.

open half-plane The solution of a linear inequality that does not include the boundary line.

open sentence A mathematical statement with one or more variables.

opposite rays Two rays $\overrightarrow{BA}$ and $\overrightarrow{BC}$ such that B is between A and C.

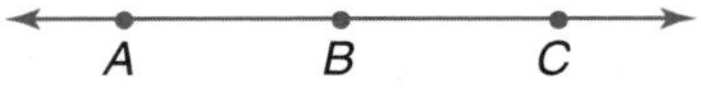

opposites Two numbers with the same absolute value but different signs.

ordered pair A set of numbers or coordinates used to locate any point on a coordinate plane, written in the form (x, y).

order of magnitude The order of magnitude of a quantity is the number rounded to the nearest power of 10.

order of operations
1. Evaluate expressions inside grouping symbols.
2. Evaluate all powers.
3. Do all multiplications and/or divisions from left to right.
4. Do all additions and/or subtractions from left to right.

order of symmetry The number of times a figure can map onto itself as it rotates from 0° to 360°.

ángulo obtuso Ángulo que mide más de 90 y menos de 180.

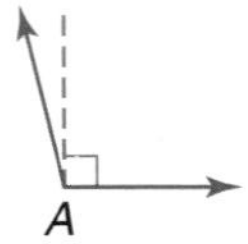

$90 < m\angle A < 180$

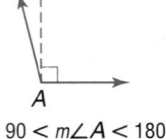

triángulo obtusángulo Triángulo con un ángulo obtuso.

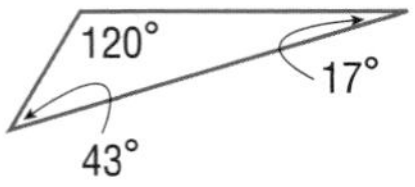

un ángulo obtuso

probabilidades El cociente de la probabilidad del éxito de un acontecimiento a la probabilidad de su complemento.

abra el mitad-plano La solución de una desigualdad linear que no incluya la línea de límite.

enunciado abierto Un enunciado matemático que contiene una o más variables.

rayos opuestos Dos rayos $\overrightarrow{BA}$ y $\overrightarrow{BC}$ donde B esta entre A y C.

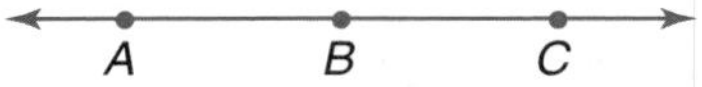

opuestos Dos números que tienen el mismo valor absolute, pero que tienen distintos signos.

par ordenado Un par de números que se usa para ubicar cualquier punto de un plano de coordenadas y que se escribe en la forma (x, y).

orden de magnitud de una cantidad Un número redondeado a la potencia más cercana de 10.

orden de las operaciones
1. Evalúa las expresiones dentro de los símbolos de agrupamiento.
2. Evalúa todas las potencias.
3. Multiplica o divide de izquierda a derecha.
4. Suma o resta de izquierda a derecha.

orden de la simetría Número de veces que una figura se puede aplicar sobre sí misma mientras gira de 0° a 360°.

origin The point where the two axes intersect at their zero points.

outliers Data that are more than 1.5 times the interquartile range beyond the quartiles.

P

parallel lines **1.** Lines in the same plane that do not intersect and either have the same slope or are vertical lines.

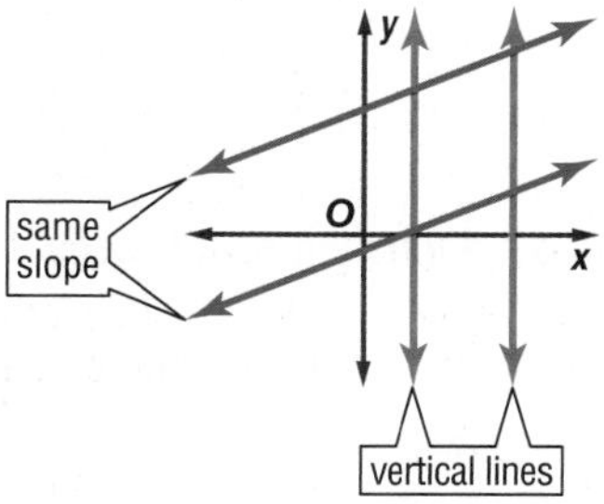

2. Coplanar lines that do not intersect.

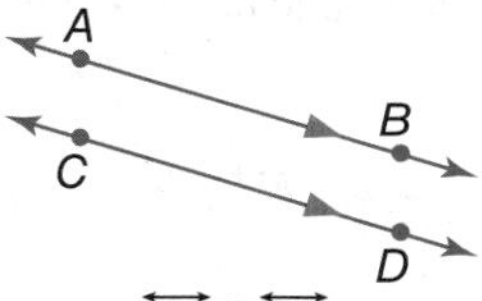

$\overleftrightarrow{AB} \parallel \overleftrightarrow{CD}$

parallel planes Planes that do not intersect.

parallelogram A quadrilateral with parallel opposite sides. Any side of a parallelogram may be called a *base*.

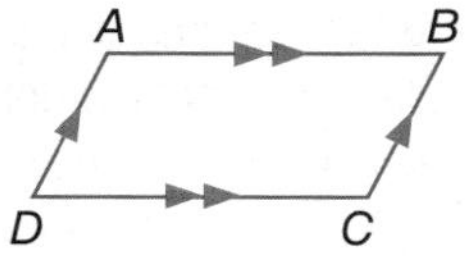

$\overline{AB} \parallel \overline{DC}$; $\overline{AD} \parallel \overline{BC}$

parameter A measure that describes a characteristic of the population as a whole.

parent function The simplest of functions in a family.

parent graph The simplest of the graphs in a family of graphs.

percent A ratio that compares a number to 100.

percent of change When an increase or decrease is expressed as a percent.

origen Punto donde se intersecan los dos ejes en sus puntos cero.

valores atípicos Datos que distan de los cuartiles más de 1.5 veces la amplitude intercuartílica.

rectas paralelas Rectas en el mismo plano que no se intersecan y que tienen pendientes iguales, o las mismas rectas verticales.

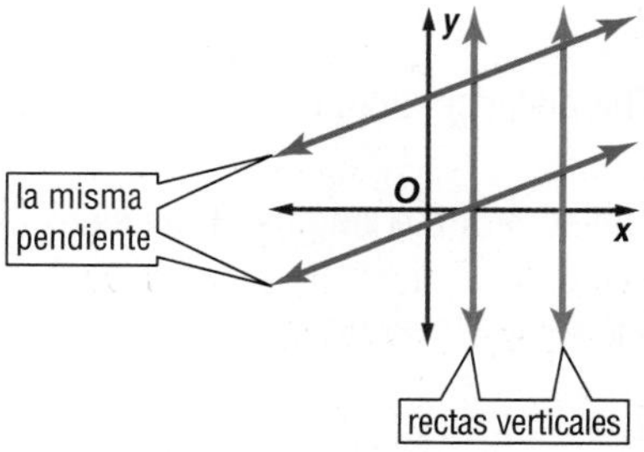

2. Rectas coplanares que no se intersecan.

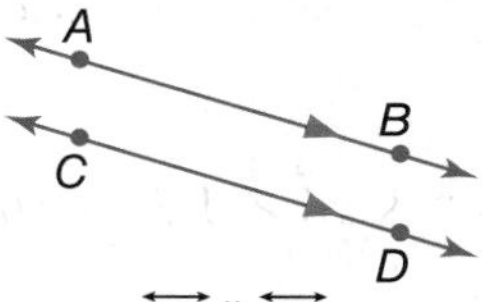

$\overleftrightarrow{AB} \parallel \overleftrightarrow{CD}$

planos paralelos Planos que no se intersecan.

paralelogramo Cuadrilátero cuyos lados opuestos son paralelos y cuya *base* puede ser cualquier de sus lados.

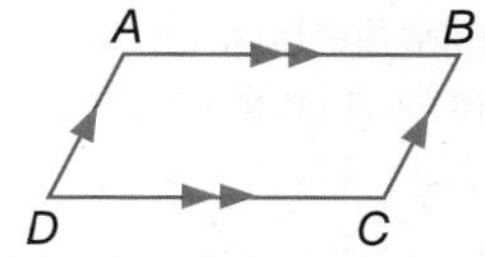

$\overline{AB} \parallel \overline{DC}$; $\overline{AD} \parallel \overline{BC}$

parámetro Una medida que se describe característica de la población en su totalidad.

función básica La función más fundamental de un familia de funciones.

gráfica madre La gráfica más sencilla en una familia de gráficas.

porcentaje Razón que compara un numero con 100.

porcentaje de cambio Cuando un aumento o disminución se escribe como un tanto por ciento.

percent of decrease The ratio of an amount of decrease to the previous amount, expressed as a percent.

percent of increase The ratio of an amount of increase to the previous amount, expressed as a percent.

percent proportion
$\frac{\text{part}}{\text{whole}} = \frac{\text{percent}}{100}$ or $\frac{a}{b} = \frac{P}{100}$

perfect square A number with a square root that is a rational number.

perimeter The distance around a geometric figure.

perimeter The sum of the lengths of the sides of a polygon.

perpendicular lines Lines that intersect to form a right angle.

perpendicular lines Lines that form right angles.

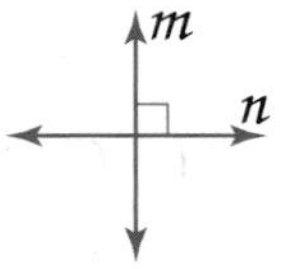

line $m \perp$ line n

pi (π) An irrational number represented by the ratio of the circumference of a circle to the diameter of the circle.

plane A basic undefined term of geometry. A plane is a flat surface made up of points that has no depth and extends indefinitely in all directions. In a figure, a plane is often represented by a shaded, slanted four-sided figure. Planes are usually named by a capital script letter or by three noncollinear points on the plane.

plane Euclidean geometry Geometry based on Euclid's axioms dealing with a system of points, lines, and planes.

plane symmetry Symmetry in a three- dimensional figure that occurs if the figure can be mapped onto itself by a reflection in a plane.

point A basic undefined term of geometry. A point is a location. In a figure, points are represented by a dot. Points are named by capital letters.

point of symmetry A figure that can be mapped onto itself by a rotation of 180°.

porcentaje de disminución Razón de la cantidad de disminución a la cantidad original, escrita como un tanto por ciento.

porcentaje de aumento Razón de la cantidad de aumento a la cantidad original, escrita como un tanto por ciento.

proporción porcentual
$\frac{\text{parte}}{\text{todo}} = \frac{\text{por ciento}}{100}$ or $\frac{a}{b} = \frac{P}{100}$

cuadrado perfecto Número cuya raíz cuadrada es un número racional.

perímetro Longitud alrededor una figura geométrica.

perímetro Suma de la longitud de los lados de un polígono.

recta perpendicular Recta que se intersecta formando un ángulo recto

rectas perpendiculares Rectas que forman ángulos rectos.

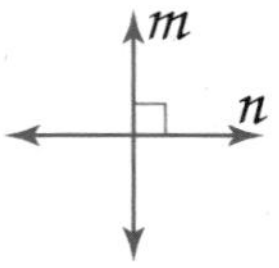

recta $m \perp$ recta n

pi (π) Número irracional representado por la razón de la circunferencia de un círculo al diámetro del mismo.

plano Término geométrico básico no definido. Superficie plana sin espesor formada por puntos y que se extiende hasta el infinito en todas direcciones. En una figura, los planos a menudo se representan con una figura inclinada y sombreada y se designan con una letra mayúscula o con tres puntos no colineales del plano.

geometría del plano euclidiano Geometría basada en los axiomas de Euclides, los cuales abarcan un sistema de puntos, rectas y planos.

simetría plana Simetría en una figura tridimensional que ocurre si la figura se puede aplicar sobre sí misma mediante una reflexión en el plano.

punto Término geométrico básico no definido. Un punto representa un lugar o ubicación. En una figura, se representa con una marca puntual y se designan con letras mayúsculas.

punto de simetría Una figura que se puede traz sobre sí mismo por una rotación de 180°.

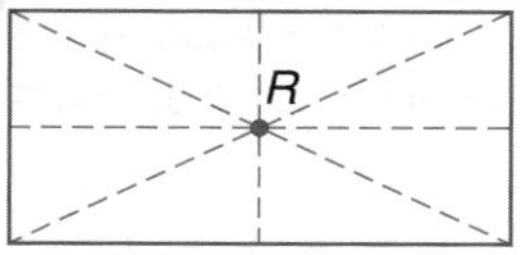

R is a point of symmetry.

point of tangency For a line that intersects a circle in only one point, the point at which they intersect.

point-slope form An equation of the form $y - y_1 = m(x - x_1)$, where (x_1, y_1) are the coordinates of any point on the line and m is the slope of the line.

polygon A closed figure formed by a finite number of coplanar segments called *sides* such that the following conditions are met:

1. The sides that have a common endpoint are noncollinear.
2. Each side intersects exactly two other sides, but only at their endpoints, called the *vertices*.

positive A function is positive on a portion of its domain where its graph lies above the x-axis.

positive correlation In a scatter plot, as x increases, y increases.

positive number Any value that is greater than zero.

power An expression of the form x^n, read *x to the nth power*.

preimage The graph of an object before a transformation.

principal square root The nonnegative square root of a number.

principle of superposition Two figures are congruent if and only if there is a rigid motion or a series of rigid motions that maps one figure exactly onto the other.

probability The ratio of the number of favorable equally likely outcomes to the number of possible equally likely outcomes.

probability graph A way to give the probability distribution for a random variable and obtain other data.

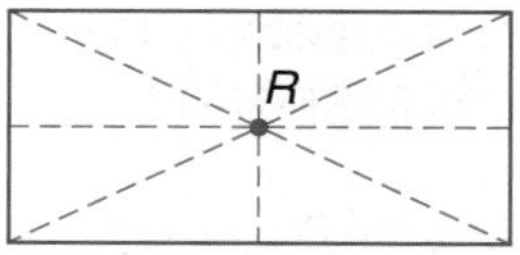

R es un punto de simetría.

punto de tangencia Punto de intersección de una recta en un círculo en un solo punto.

forma punto-pendiente Ecuación de la forma $y - y_1 = m(x - x_1)$, donde (x_1, y_1) representan las coordenadas de un punto cualquiera sobre la recta y m representa la pendiente de la recta.

polígono Figura cerrada formada por un número finito de segmentos coplanares llamados *lados*, tal que satisface las siguientes condiciones:

1. Los lados que tienen un extremo común son no colineales.
2. Cada lado interseca exactamente dos lados mas, pero sólo en sus extremos, llamados *vértices*.

positiva Una función es positiva en una porción de su dominio donde su gráfico está encima del eje-x.

correlación positiva En una gráfica de dispersión, a medida que x aumenta, y aumenta.

número positivos Cualquier valor mayor que cero.

potencia Una expresión de la forma x^n, se lee *x a la enésima potencia*.

preimagen Gráfica de una figura antes de una transformación.

raíz cuadrada principal La raíz cuadrada no negativa de un número.

principio de superposición Dos figuras son congruentes si y sólo si existe un movimiento rígido o una serie de movimientos rígidos que aplican una de las figuras exactamente sobre la otra.

probabilidad La razón del número de maneras en que puede ocurrir el evento al numero de resultados posibles.

gráfico probabilístico Una manera de exhibir la distribución de probabilidad de una variable aleatoria y obtener otros datos.

product In an algebraic expression, the result of quantities being multiplied is called the product.

producto En una expresión algebraica, se llama producto al resultado de las cantidades que se multiplican.

product rule If (x_1, y_1) and (x_2, y_2) are solutions to an inverse variation, then $y_1x_1 = y_2x_2$.

regla del producto Si (x_1, y_1) y (x_2, y_2) son soluciones de una variación inversa, entonces $y_1x_1 = y_2x_2$.

proportion An equation of the form $\frac{a}{b} = \frac{c}{d}$, where $b, d \neq 0$, stating that two ratios are equivalent.

proporción Ecuación de la forma $\frac{a}{b} = \frac{c}{d}$, donde $b, d \neq 0$, que afirma la equivalencia de dos razones.

Q

quartile The values that divide a set of data into four equal parts.

cuartile Valores que dividen en conjunto de datos en cuarto partes iguales.

R

radical equations Equations that contain radicals with variables in the radicand.

ecuaciones radicales Ecuaciones que contienen radicales con variables en el radicando.

radical expression An expression that contains a square root.

expresión radical Expresión que contiene una raíz cuadrada.

radical function A function that contains radicals with variables in the radicand.

ecuaciones radicales Ecuaciones que contienen radicales con variables en el radicando.

radicand The expression that is under the radical sign.

radicando La expresión debajo del signo radical.

radius **1.** In a circle, any segment with endpoints that are the center of the circle and a point on the circle. **2.** In a sphere, any segment with endpoints that are the center and a point on the sphere.

radio **1.** En un círculo, cualquier segmento cuyos extremos son en el centro y un punto del círculo. **2.** En una esfera, cualquier segmento cuyos extremos son el centro y un punto de la esfera.

range **1.** The set of second numbers of the ordered pairs in a relation. **2.** The difference between the greatest and least data values.

rango **1.** Conjunto de los segundos números de los pares ordenados de una relación. **2.** La diferencia entre los valores de datos más grande o menos.

rate The ratio of two measurements having different units of measure.

tasa Razón de dos medidas que tienen distintas unidades de medida.

rate of change **1.** How a quantity is changing with respect to a change in another quantity. **2.** Describes how a quantity is changing over time.

tasa de cambio **1.** Cómo cambia una cantidad con respecto a un cambio en otra cantidad. **2.** Describe cómo cambia una cantidad a través del tiempo.

rate problems Problems in which an object moves at a certain speed, or rate.

problemas de tasas Problemas en el que un objeto se mueue a una velocidad o tasa determinada.

ratio A comparison of two numbers by division.

razón Comparación de dos números mediante división.

rational exponent For any positive real number b and any integers m and $n > 1$, $b^{\frac{m}{n}} = \left(\sqrt[n]{b}\right)^m$ or $\sqrt[n]{b^m}$. $\frac{m}{n}$ is a rational exponent.

exponent racional Para cualquier número real no nulo b y cualquier entero m y $n > 1$, $b^{\frac{m}{n}} = \left(\sqrt[n]{b}\right)^m$ or $\sqrt[n]{b^m}$. $\frac{m}{n}$ es un exponent racional.

rational expression An algebraic fraction with a numerator and denominator that are polynomials.

expresión racional Fracción algebraica cuyo numerador y denominador son polinomios.

rational function An equation of the form $f(x) = \frac{p(x)}{q(x)}$, where $p(x)$ and $q(x)$ are polynomial functions, and $q(x) \neq 0$.

rationalizing the denominator A method used to eliminate radicals from the denominator of a fraction.

rational numbers The set of numbers expressed in the form of a fraction $\frac{a}{b}$, where a and b are integers and $b \neq 0$.

ray $\overrightarrow{PQ}$ is a ray if it is the set of points consisting of $\overline{PQ}$ and all points S for which Q is between P and S.

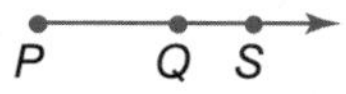

real numbers The set of rational numbers and the set of irrational numbers together.

reciprocal The multiplicative inverse of a number.

rectangle A quadrilateral with four right angles.

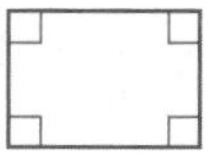

recursive formula Each term is formulated from one or more previous terms.

reduction An image that is smaller than the original figure.

reflection A transformation representing the flip of a figure over a point, line or plane. A reflection in a line is a function that maps a point to its image such that

- if the point is on the line, then the image and preimage are the same point, or
- if the point is not on the line, the line is the perpendicular bisector of the segment joining the two points.

regular polygon A convex polygon in which all of the sides are congruent and all of the angles are congruent.

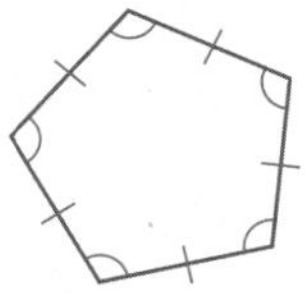

regular tessellation A tessellation formed by only one type of regular polygon.

relation A set of ordered pairs.

remote interior angles The angles of a triangle that are not adjacent to a given exterior angle.

función racional Ecuación de la forma $f(x) = \frac{p(x)}{q(x)}$, donde $p(x)$ y $q(x)$ son funciones polinomiales y $q(x) \neq 0$.

racionalizar el denominador Método que se usa para eliminar radicales del denominador de una fracción.

números racionales Conjunto de los números que pueden escribirse en forma de fracción $\frac{a}{b}$, donde a y b son enteros y $b \neq 0$.

rayo $\overrightarrow{PQ}$ es un rayo si se el conjunto de puntos formado por $\overline{PQ}$ y todos los puntos S para los cuales Q se ubica entre P y S.

números reales El conjunto de los números racionales junto con el conjunto de los números irracionales.

recíproco Inverso multiplicativo de un número.

rectángulo Cuadrilátero con cuatro ángulos rectos.

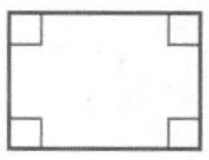

fórmula recursiva Cada tórmino proviene de uno o más terminos anteriores.

reducción Imagen más pequeña que la figura original.

reflexión Transformación en la cual una figura se "voltea" a través de un punto, una recta o un plano. Una reflexión en una recta es una función que aplica un punto a su imagen, de modo que si el punto yace sobre la recta, entonces la imagen y la preimagen son el mismo punto, o si el punto no yace sobre la recta, la recta es la mediatriz del segmento que une los dos puntos.

polígono regular Polígono convexo cuyos los lados y ángulos son congruentes.

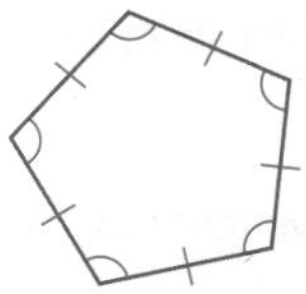

teselado regular Teselado formado por un solo tipo de polígono regular.

relación Conjunto de pares ordenados.

ángulos internos no adyacentes Ángulos de un triángulo que no son adyacentes a un ángulo exterior dado.

replacement set A set of numbers from which replacements for a variable may be chosen.

conjunto de sustitución Conjunto de números del cual se pueden escoger sustituciones para una variable.

residual The difference between an observed y-value and its predicted y-value on a regression line.

residual Diferencia entre el valor observado de y y el valor redicho de y en la recta de regression.

rhombus A quadrilateral with all four sides congruent.

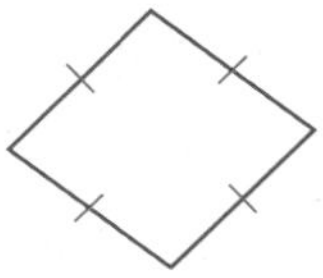

rombo Cuadrilátero con cuatro lados congruentes.

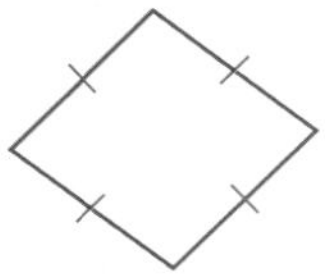

right angle An angle with a degree measure of 90.

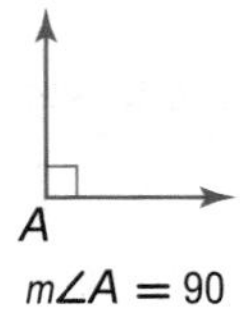

ángulo recto Ángulo que mide 90.

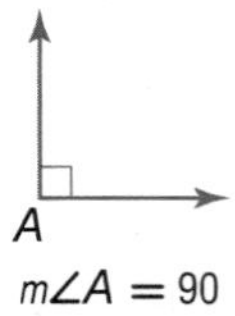

right triangle A triangle with a right angle. The side opposite the right angle is called the *hypotenuse*. The other two sides are called *legs*.

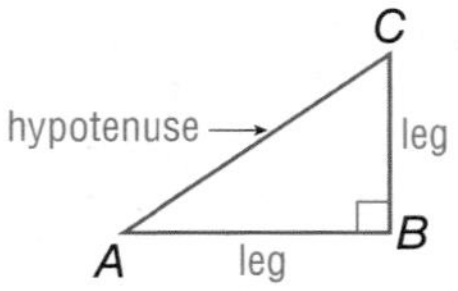

triángulo rectángulo Triángulo con un ángulo recto. El lado opuesto al ángulo recto se conoce como *hipotenusa*. Los otros dos lados se llaman *catetos*.

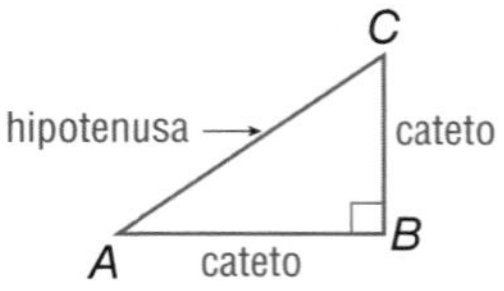

root The solutions of a quadratic equation.

raíces Las soluciones de una ecuación cuadrática

rotation A transformation that turns every point of a preimage through a specified angle and direction about a fixed point, called the *center of rotation*. A rotation about a fixed point through an angle of $x°$ is a function that maps a point to its image such that

- if the point is the center of rotation, then the image and preimage are the same point, or
- if the point is not the center of rotation, then the image and preimage are the same distance from the center of rotation and the measure of the angle of rotation formed by the preimage, center of rotation, and image points is x.

rotación Transformación en la cual se hace girar cada punto de la preimagen a través de un ángulo y una dirección determinadas alrededor de un punto llamado *centro de rotación*. La rotación de $x°$ es una función que aplica un punto a su imagen, de modo que si el punto es el centro de rotación, entonces la imagen y la preimagen están a la misma distancia del centro de rotación y la medida del ángulo formado por los puntos de la preimagen, centro de rotación e imagen es x.

rotational symmetry If a figure can be rotated less than 360° about a point so that the image and the preimage are indistinguishable, the figure has rotational symmetry.

simetría rotacional Si una imagen se puede girar menos de 360° alrededor de un punto, de modo que la imagen y la preimagen sean idénticas, entonces la figura tiene simetría rotacional.

row reduction The process of performing elementary row operations on an augmented matrix to solve a system.

reducción de la fila El proceso de realizar operaciones elementales de la fila en una matriz aumentada para solucionar un sistema.

S

sample space The list of all possible outcomes.

espacio muestral Lista de todos los resultados posibles.

scale The relationship between the measurements on a drawing or model and the measurements of the real object.

escala Relación entre las medidas de un dibujo o modelo y las medidas de la figura verdadera.

scale factor of dilation The ratio of a length on an image to a corresponding length on the preimage.

factor de escala de homotecia Razon de una longitud en la imagen a una longitud correspondiente en la preimagen.

scale model A model used to represent an object that is too large or too small to be built at actual size.

modelo a escala Modelo que se usa para representar un figura que es demasiado grande o pequeña como para ser construida de tamaño natural.

scalene triangle A triangle with no two sides congruent.

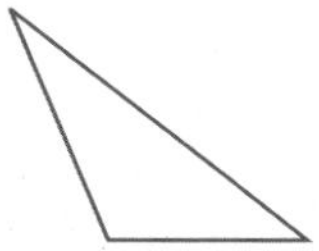

triángulo escaleno Triángulo que no tiene dos lados congruentes.

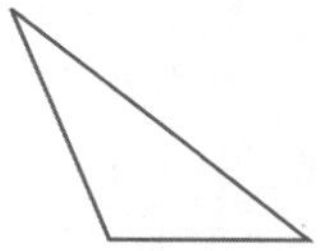

scatter plot A scatter plot shows the relationship between a set of data with two variables, graphed as ordered pairs on a coordinate plane.

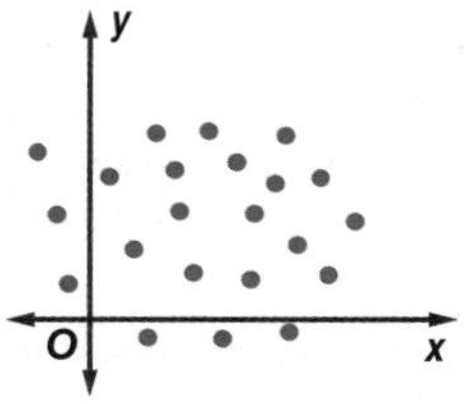

gráfica de dispersión Es un diagrama que muestra la relación entre un conjunto de datos con dos variables, graficados como pares ordenados en un plano coordenadas.

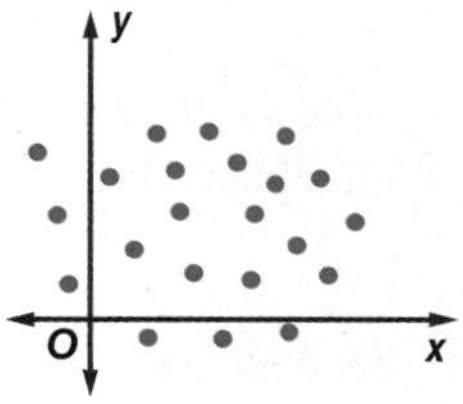

scientific notation A number in scientific notation is expressed as $a \times 10^n$, where $1 \leq a < 10$ and n is an integer.

notación científica Un numero en notación científica se escribe con $a \times 10^n$, donde $1 \leq a < 10$ y n es un número entero.

segment See *line segment.*

segmento Ver *segmento de recta.*

segment bisector A segment, line, or plane that intersects a segment at its midpoint.

bisector del segmento Segmento, recta o plano que interseca un segmento en su punto medio.

semi-regular tessellation A uniform tessellation formed using two or more regular polygons.

teselado semiregular Teselado uniforme compuesto por dos o más polígonos regulares.

sequence A set of numbers in a specific order.

sucesión Conjunto de números en un orden específico.

set-builder notation A concise way of writing a solution set. For example, $\{t \mid t < 17\}$ represents the set of all numbers t such that t is less than 17.

notación de construcción de conjuntos Manera concisa de escribir un conjunto solución. Por ejemplo, $\{t \mid t < 17\}$ representa el conjunto de todos los números t que son menores o iguales que 17.

sides of an angle The rays of an angle.

lados de un ángulo Los rayos de un ángulo.

Glossary/Glosario

similarity transformation When a figure and its transformation image are similar.

simplest form An expression is in simplest form when it is replaced by an equivalent expression having no like terms or parentheses.

skew lines Lines that do not intersect and are not coplanar.

slope For a (nonvertical) line containing two points (x_1, y_1) and (x_2, y_2), the number m given by the formula

$$m = \frac{y_2 - y_1}{x_2 - x_1} \text{ where } x_2 \neq x_1.$$

slope The ratio of the change in the y-coordinates (rise) to the corresponding change in the x-coordinates (run) as you move from one point to another along a line.

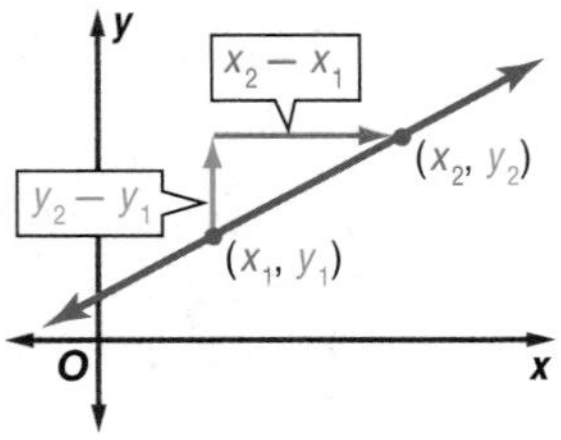

slope-intercept form A linear equation of the form $y = mx + b$. The graph of such an equation has slope m and y-intercept b.

slope-intercept form An equation of the form $y = mx + b$, where m is the slope and b is the y-intercept.

solid of revolution A three-dimensional figure obtained by rotating a plane figure about a line.

solution A replacement value for the variable in an open sentence.

solution set The set of elements from the replacement set that make an open sentence true.

solve an equation The process of finding all values of the variable that make the equation a true statement.

solving an open sentence Finding a replacement value for the variable that results in a true sentence or an ordered pair that results in a true statement when substituted into the equation.

space A boundless three-dimensional set of all points.

transformación de semejanza cuando una figura y su imagen transformada son semejantes.

forma reducida Una expresión está reducida cuando se puede sustituir por una expresión equivalente que no tiene ni términos semejantes ni paréntesis.

rectas alabeadas Rectas que no se intersecan y que no son coplanares.

pendiente Para una recta (no vertical) que contiene dos puntos (x_1, y_1) y (x_2, y_2), tel número m viene dado por la fórmula

$$m = \frac{y_2 - y_1}{x_2 - x_1} \text{ donde } x_2 \neq x_1.$$

pendiente Razón del cambio en la coordenada y (elevación) al cambio correspondiente en la coordenada x (desplazamiento) a medida que uno se mueve de un punto a otro en una recta.

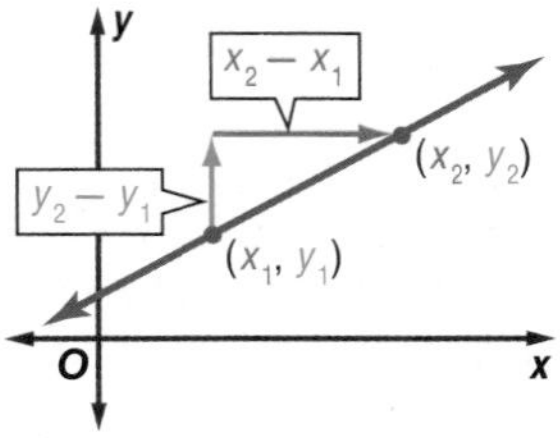

forma pendiente-intersección Ecuación lineal de la forma $y = mx + b$ donde, la pendiente es m y la intersección y es b.

forma pendiente-intersección Ecuación de la forma $y = mx + b$, donde m es la pendiente y b es la intersección y.

sólido de revolución Figura tridimensional que se obtiene al rotar una figura plana alrededor de una recta.

solución Valor de sustitución de la variable en un enunciado abierto.

conjunto solución Conjunto de elementos del conjunto de sustitución que hacen verdadero un enunciado abierto.

resolver una ecuación Proceso en que se hallan todos los valores de la variable que hacen verdadera la ecuación.

resolver un enunciado abierto Hallar un valor de sustitución de la variable que resulte en un enunciado verdadero o un par ordenado que resulte en una proposición verdadera cuando se lo sustituye en la ecuación.

espacio Conjunto tridimensional no acotado de todos los puntos.

square A quadrilateral with four right angles and four congruent sides.

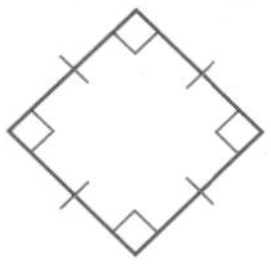

square root One of two equal factors of a number.

square root function Function that contains the square root of a variable.

standard deviation The square root of the variance.

standard form The standard form of a linear equation is $Ax + By = C$, where $A \geq 0$, A and B are not both zero, and A, B, and C are integers with a greatest common factor of 1.

statistic A quantity calculated from a sample.

statistical inference The statistics of a sample are used to draw conclusions about the population.

stem-and-leaf plot A system used to condense a set of data where the greatest place value of the data forms the stem and the next greatest place value forms the leaves.

substitution Use algebraic methods to find an exact solution of a system of equations.

supplementary angles Two angles with measures that have a sum of 180.

surface area The sum of the areas of all the surfaces of a three-dimensional figure.

symmetry **1.** A geometric property of figures that can be folded and each half matches the other exactly. **2.** A figure has symmetry if there exists a rigid motion—reflection, translation, rotation, or glide reflection—that maps the figure onto itself.

system of equations A set of equations with the same variables.

system of inequalities A set of two or more inequalities with the same variables.

cuadrado Cuadrilátero con cuatro ángulos rectos y cuatro lados congruentes.

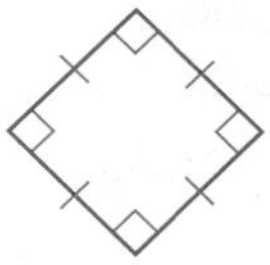

raíz cuadrada Uno de dos factores iguales de un número.

función radical Función que contiene la raíz cuadrada de una variable.

desviación tipica Calculada como la raíz cuadrada de la varianza.

forma estándar La forma estándar de una ecuación lineal es $Ax + By = C$, donde $A \geq 0$, ni A ni B son ambos cero, y A, B, y C son enteros cuyo máximo común divisor es 1.

estadística Una cantidad calculaba de una muestra.

inferencia estadística La estadística de una muestra se utiliza para dibujar conclusiones sobre la población.

diagrama de tallo y hojas Sistema que se usa para condensar un conjunto de datos, en que el valor de posición máximo de los datos forma el tallo y el segundo valor de posiciós máximo forma las hojas. El valor de posición máximo de los datos forma eld tallo y el segundo valor de posición máximo forma las hojas.

sustitución Usa métodos algebraicos para hallar una solución exacta a un sistema de ecuaciones.

ángulos suplementarios Dos ángulos cuya suma es igual a 180.

área de superficie Suma de las áreas de todas las superficies (caras) de una figura tridimensional.

simetría **1.** Propiedad geométrica de figuras que pueden plegarse de modo que cada mitad corresponde exactamente a la otra. **2.** Una figura tiene simetría si existe un movimiento rígido (reflexión, translación, rotación, o reflexión con deslizamiento) que aplica la figura sobre sí misma.

sistema de ecuaciones Conjunto de ecuaciones con las mismas variables.

sistema de desigualdades Conjunto de dos o más desigualdades con las mismas variables.

T

term A number, a variable, or a product or quotient of numbers and variables.

terms of a sequence The numbers in a sequence.

tessellation A pattern that covers a plane by transforming the same figure or set of figures so that there are no overlapping or empty spaces.

transformation In a plane, a mapping for which each point has exactly one image point and each image point has exactly one preimage point.

translation A transformation that moves a figure the same distance in the same direction. A translation is a function that maps each point to its image along a vector such that each segment joining a point and its image has the same length as the vector, and this segment is also parallel to the vector.

translation vector The vector in which a translation maps each point to its image.

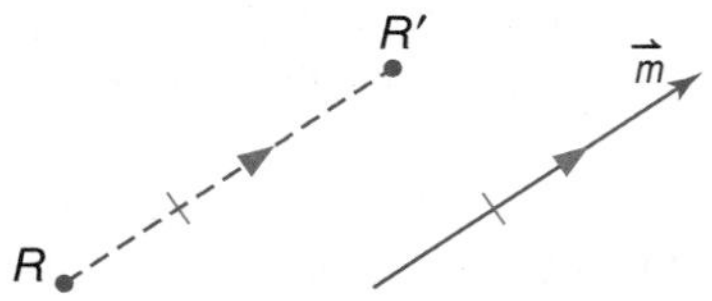

Point R', is a translation of point R along translation vector m.

transversal A line that intersects two or more lines in a plane at different points.

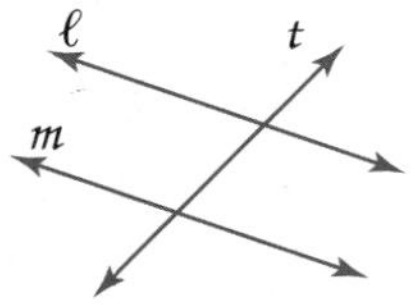

Line t is a transversal.

trapezoid A quadrilateral with exactly one pair of parallel sides. The parallel sides of a trapezoid are called *bases*. The nonparallel sides are called *legs*. The pairs of angles with their vertices at the endpoints of the same base are called *base angles*.

término Número, variable o producto, o cociente de números y variables.

términos Los números de una sucesión.

teselado Patrón con que se cubre un plano aplicando la misma figura o conjunto de figuras, sin que haya traslapes ni espacios vacíos.

transformación En un plano, aplicación para la cual cada punto del plano tiene un único punto de la imagen y cada punto de la imagen tiene un único punto de la preimagen.

traslación Transformación que mueve una figura la misma distancia en la misma dirección. Una traslación es una función que aplica cada punto a su imagen a lo largo de un vector, de modo que cada segmento que une un punto a su imagen tiene la misma longitud que el vector y este segmento es también paralelo al vector.

vector de traslación Vector en el cual una traslación aplica cada punto a su imagen.

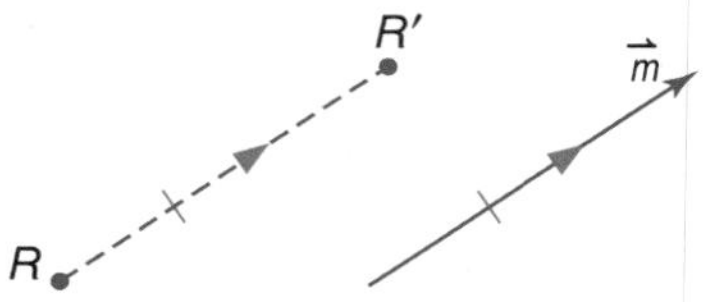

El punta R', es la traslación del punto R a lo largo del vector m de traslación.

transversal Recta que interseca dos o más rectas en el diferentes puntos del mismo plano.

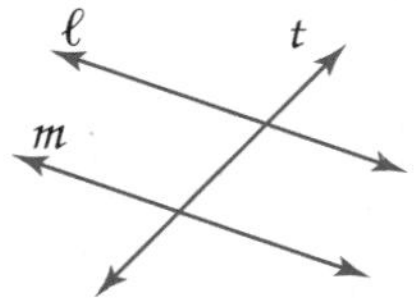

La recta t es una transversal.

trapecio Cuadrilátero con sólo un par de lados paralelos. Los lados paralelos del trapecio se llaman *bases*. Los lados no paralelos se llaman *catetos*. Los pares de ángulos cuyos vértices coinciden en los extremos de la misma base son los *ángulos de la base*.

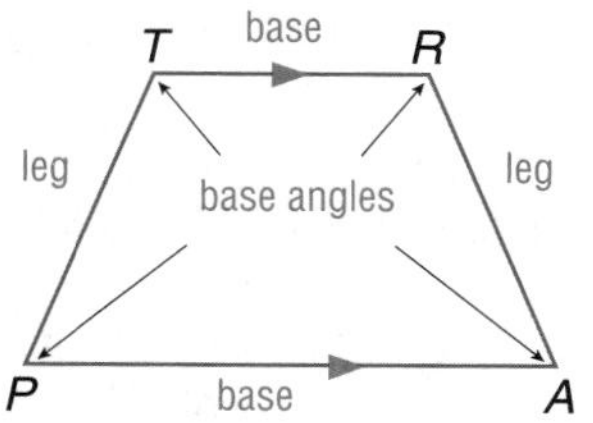

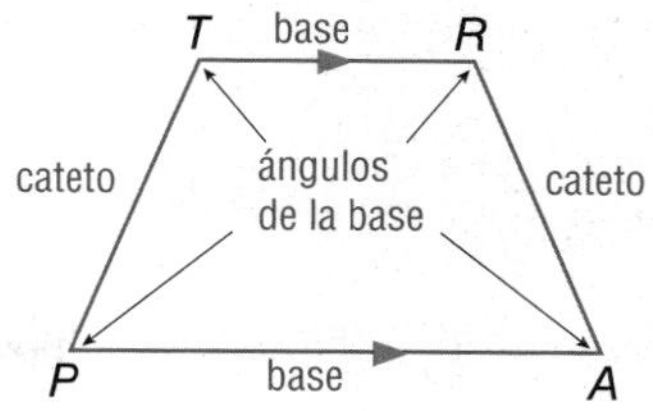

tree diagram A diagram used to show the total number of possible outcomes.

diagrama de árbol Diagrama que se usa para mostrar el número total de resultados posibles.

two-column proof A formal proof that contains statements and reasons organized in two columns. Each step is called a *statement*, and the properties that justify each step are called *reasons*.

demostración de dos columnas Demostración formal que contiene enunciados y razones organizadas en dos columnas. Cada paso se llama *enunciado* y las propiedades que lo justifican son las *razones*.

U

undefined term Words, usually readily understood, that are not formally explained by means of more basic words and concepts. The basic undefined terms of geometry are point, line, and plane.

término geométrico básico no definido Palabras que por lo general se entienden fácilmente y que no se explican formalmente mediante palabras o conceptos más básicos. Los términos geométricos básicos no definidos son el punto, la recta y el plano.

uniform motion problems Problems in which an object moves at a certain speed, or rate.

problemas de movimiento uniforme Problemas en que el cuerpo se mueve a cierta velocidad o tasa.

uniform tessellations Tessellations containing the same arrangement of shapes and angles at each vertex.

teselado uniforme Teselados con el mismo patrón de formas y ángulos en cada vértice.

union The graph of a compound inequality containing or; the solution is a solution of either inequality, not necessarily both.

unión Gráfica de una desigualdad compuesta que contiene la palabra o; la solución es el conjunto de soluciones de por lo menos una de las desigualdades, no necesariamente ambas.

unit analysis The process of including units of measurement when computing.

análisis de la unidad Proceso de incluir unidades de medida al computar.

unit rate A ratio of two quantities, the second of which is one unit.

tasa unitaria Tasa reducida que tiene denominador igual a 1.

univariate data Data with one variable.

datos univariate Datos con una variable.

upper quartile The median of the upper half of a set of data.

cuartil superior Mediana de la mitad superior de un conjunto de datos.

V

variable **1.** Symbols used to represent unspecified numbers or values. **2.** a characteristic of a group of people or objects that can assume different values

variable **1.** Símbolos que se usan para representar números o valores no especificados. **2.** una característica de un grupo de personas u objetos que pueden asumir valores diferentes

variance The mean of the squares of the deviations from the arithmetic mean.

varianza Media de los cuadrados de las desviaciones de la media aritmética.

vertex angle of an isosceles triangle See *isosceles triangle.*

vertex of an angle The common endpoint of an angle.

vertex of a polygon The vertex of each angle of a polygon.

vertical angles Two nonadjacent angles formed by two intersecting lines.

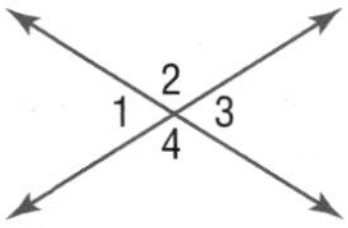

∠1 and ∠3 are vertical angles.
∠2 and ∠4 are vertical angles.

vertical line test If any vertical line passes through no more than one point of the graph of a relation, then the relation is a function.

volume The measure of space occupied by a solid region.

ángulo del vértice un triángulo isosceles Ver *triángulo isósceles.*

vértice de un ángulo Extremo común de un ángulo.

vertice de un polígono Vértice de cada ángulo de un polígono.

ángulos opuestos por el vértice Dos ángulos no adyacentes formados por dos rectas que se intersecan.

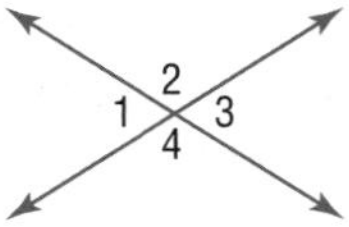

∠1 y ∠3 son ángulos opuestos por el vértice.
∠2 y ∠4 son ángulos opuestos por el vértice.

prueba de la recta vertical Si cualquier recta vertical pasa por un sólo punto de la gráfica de una relación, entonces la relación es una función.

volumen Medida del espacio que ocupa un solido.

W

weighted average The sum of the product of the number of units and the value per unit divided by the sum of the number of units, represented by *M*.

whole numbers The set {0, 1, 2, 3, …}.

promedio ponderado Suma del producto del número de unidades por el valor unitario dividida entre la suma del número de unidades y la cual se denota por *M*.

números enteros El conjunto {0, 1, 2, 3, …}.

X

***x*-axis** The horizontal number line on a coordinate plane.

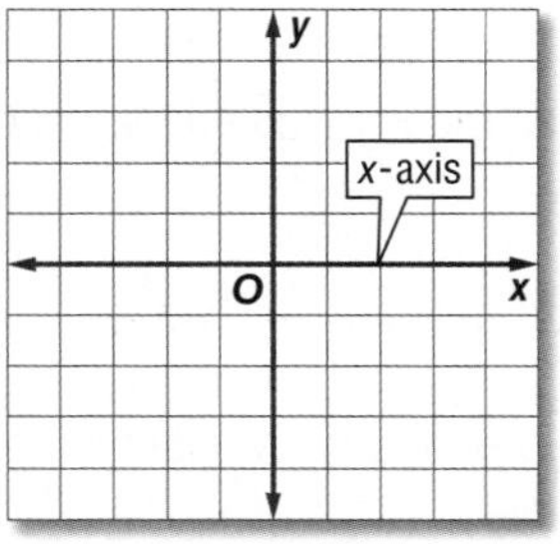

***x*-coordinate** The first number in an ordered pair.

***x*-intercept** The *x*-coordinate of a point where a graph crosses the *x*-axis.

eje *x* Recta numérica horizontal que forma parte de un plano de coordenadas.

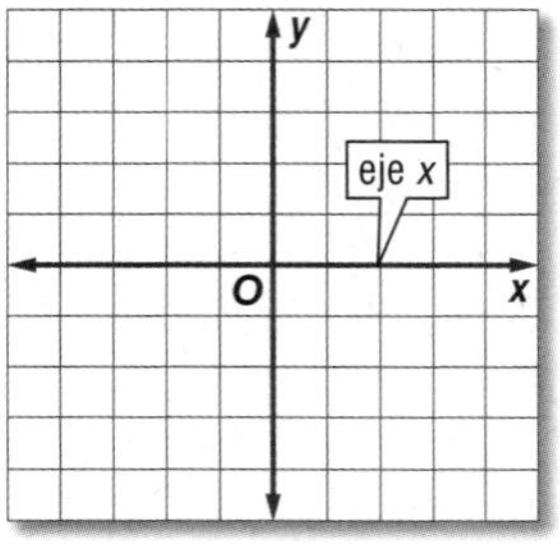

coordenada *x* El primer número de un par ordenado.

intersección *x* La coordenada *x* de un punto donde la gráfica corte al eje de *x*.

Y

***y*-axis** The vertical number line on a coordinate plane.

eje *y* Recta numérica vertical que forma parte de un plano de coordenadas.

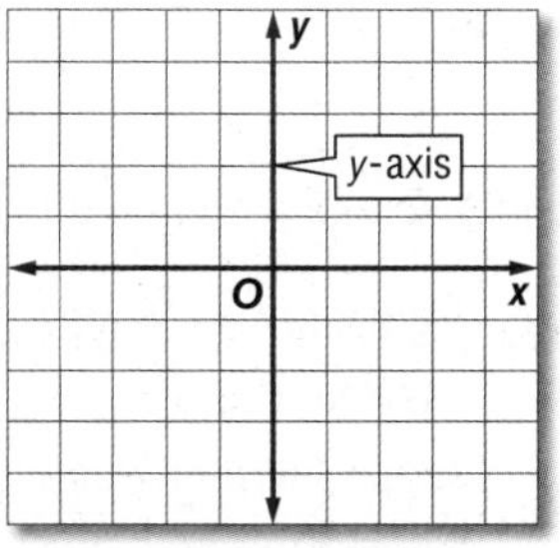

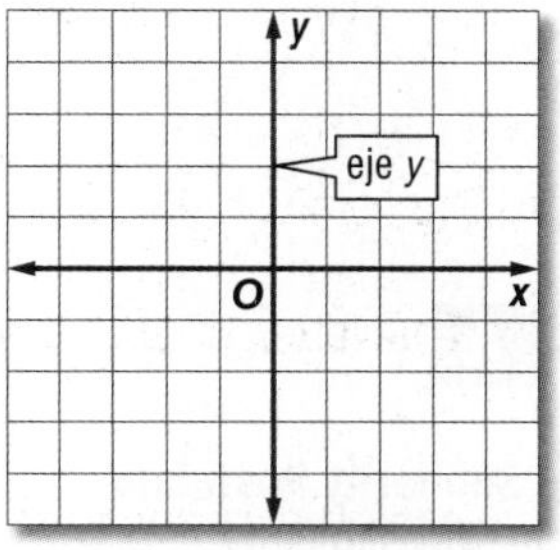

y-coordinate The second number in an ordered pair.

coordenada y El segundo número de un par ordenado.

y-intercept The y-coordinate of a point where a graph crosses the y-axis.

intersección y La coordenada y de un punto donde la grafica corta al eje de y.

Z

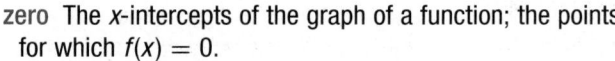

zero The x-intercepts of the graph of a function; the points for which $f(x) = 0$.

cero Las intersecciones x de la grafica de una función; los puntos x para los que $f(x) = 0$.

zero exponent For any nonzero number a, $a^0 = 1$. Any nonzero number raised to the zero power is equal to 1.

exponente cero Para cualquier número distinto a cero a, $a^0 = 1$. Cualquier número distinto a cero levantado al potente cero es igual a 1.

Index

A

B

C

F

J

K

L

N

O

Symbols

$\neq$	is not equal to	AB	measure of $\overline{AB}$
$\approx$	is approximately equal to	$\angle$	angle
$\sim$	is similar to	$\triangle$	triangle
$>, \geq$	is greater than, is greater than or equal to	$^\circ$	degree
$<, \leq$	is less than, is less than or equal to	π	pi
$-a$	opposite or additive inverse of a	$\sin x$	sine of x
$\|a\|$	absolute value of a	$\cos x$	cosine of x
$\sqrt{a}$	principal square root of a	$\tan x$	tangent of x
$a : b$	ratio of a to b	!	factorial
(x, y)	ordered pair	$P(a)$	probability of a
$f(x)$	f of x, the value of f at x	$P(n, r)$	permutation of n objects taken r at a time
$\overline{AB}$	line segment AB	$C(n, r)$	combination of n objects taken r at a time

Algebraic Properties and Key Concepts

Identity	For any number a, $a + 0 = 0 + a = a$ and $a \cdot 1 = 1 \cdot a = a$.
Substitution (=)	If $a = b$, then a may be replaced by b.
Reflexive (=)	$a = a$
Symmetric (=)	If $a = b$, then $b = a$.
Transitive (=)	If $a = b$ and $b = c$, then $a = c$.
Commutative	For any numbers a and b, $a + b = b + a$ and $a \cdot b = b \cdot a$.
Associative	For any numbers a, b, and c, $(a + b) + c = a + (b + c)$ and $(a \cdot b) \cdot c = a \cdot (b \cdot c)$.
Distributive	For any numbers a, b, and c, $a(b + c) = ab + ac$ and $a(b - c) = ab - ac$.
Additive Inverse	For any number a, there is exactly one number $-a$ such that $a + (-a) = 0$.
Multiplicative Inverse	For any number $\frac{a}{b}$, where $a, b \neq 0$, there is exactly one number $\frac{b}{a}$ such that $\frac{a}{b} \cdot \frac{b}{a} = 1$.
Multiplicative (0)	For any number a, $a \cdot 0 = 0 \cdot a = 0$.
Addition (=)	For any numbers a, b, and c, if $a = b$, then $a + c = b + c$.
Subtraction (=)	For any numbers a, b, and c, if $a = b$, then $a - c = b - c$.
Multiplication and Division (=)	For any numbers a, b, and c, with $c \neq 0$, if $a = b$, then $ac = bc$ and $\frac{a}{c} = \frac{b}{c}$.
Addition (>)*	For any numbers a, b, and c, if $a > b$, then $a + c > b + c$.
Subtraction (>)*	For any numbers a, b, and c, if $a > b$, then $a - c > b - c$.
Multiplication and Division (>)*	For any numbers a, b, and c, 1. if $a > b$ and $c > 0$, then $ac > bc$ and $\frac{a}{c} > \frac{b}{c}$. 2. if $a > b$ and $c < 0$, then $ac < bc$ and $\frac{a}{c} < \frac{b}{c}$.
Zero Product	For any real numbers a and b, if $ab = 0$, then $a = 0$, $b = 0$, or both a and b equal 0.
Square of a Sum	$(a + b)^2 = (a + b)(a + b) = a^2 + 2ab + b^2$
Square of a Difference	$(a - b)^2 = (a - b)(a - b) = a^2 - 2ab + b^2$
Product of a Sum and a Difference	$(a + b)(a - b) = (a - b)(a + b) = a^2 - b^2$

* *These properties are also true for $<$, $\geq$, and $\leq$.*

Formulas

Slope	$m = \dfrac{y_2 - y_1}{x_2 - x_1}$
Distance on a coordinate plane	$d = \sqrt{(x_2 - x_1)^2 + (y_2 - y_1)^2}$
Midpoint on a coordinate plane	$M = \left(\dfrac{x_1 + x_2}{2}, \dfrac{y_1 + y_2}{2}\right)$
Pythagorean Theorem	$a^2 + b^2 = c^2$
Quadratic Formula	$x = \dfrac{-b \pm \sqrt{b^2 - 4ac}}{2a}$
Perimeter of a rectangle	$P = 2\ell + 2w$ or $P = 2(\ell + w)$
Circumference of a circle	$C = 2\pi r$ or $C = \pi d$

Area

rectangle	$A = \ell w$	**trapezoid**	$A = \frac{1}{2}h(b_1 + b_2)$
parallelogram	$A = bh$	**circle**	$A = \pi r^2$
triangle	$A = \frac{1}{2}bh$		

Surface Area

cube	$S = 6s^2$	**regular pyramid**	$S = \frac{1}{2}P\ell + B$
prism	$S = Ph + 2B$	**cone**	$S = \pi r\ell + \pi r^2$
cylinder	$S = 2\pi rh + 2\pi r^2$		

Volume

cube	$V = s^3$	**regular pyramid**	$V = \frac{1}{3}Bh$
prism	$V = Bh$	**cone**	$V = \frac{1}{3}\pi r^2 h$
cylinder	$V = \pi r^2 h$		

Measures

Metric	**Customary**
Length	
1 kilometer (km) = 1000 meters (m)	1 mile (mi) = 1760 yards (yd)
1 meter = 100 centimeters (cm)	1 mile = 5280 feet (ft)
1 centimeter = 10 millimeters (mm)	1 yard = 3 feet
	1 foot = 12 inches (in.)
	1 yard = 36 inches
Volume and Capacity	
1 liter (L) = 1000 milliliters (mL)	1 gallon (gal) = 4 quarts (qt)
1 kiloliter (kL) = 1000 liters	1 gallon = 128 fluid ounces (fl oz)
	1 quart = 2 pints (pt)
	1 pint = 2 cups (c)
	1 cup = 8 fluid ounces
Weight and Mass	
1 kilogram (kg) = 1000 grams (g)	1 ton (T) = 2000 pounds (lb)
1 gram = 1000 milligrams (mg)	1 pound = 16 ounces (oz)
1 metric ton (t) = 1000 kilograms	

Formulas

Coordinate Geometry

Slope	$m = \frac{y_2 - y_1}{x_2 - x_1}$
Distance on a number line:	$d = \lvert a - b \rvert$
Distance on a coordinate plane:	$d = \sqrt{(x_2 - x_1)^2 + (y_2 - y_1)^2}$
Distance in space:	$d = \sqrt{(x_2 - x_1)^2 + (y_2 - y_1)^2 + (z_2 - z_1)^2}$
Distance arc length:	$\ell = \frac{x}{360} \cdot 2\pi r$
Midpoint on a number line:	$M = \frac{a + b}{2}$
Midpoint on a coordinate plane:	$M = \left(\frac{x_1 + x_2}{2}, \frac{y_1 + y_2}{2}\right)$
Midpoint in space:	$M = \left(\frac{x_1 + x_2}{2}, \frac{y_1 + y_2}{2}, \frac{z_1 + z_2}{2}\right)$

Perimeter and Circumference

square	$P = 4s$	**rectangle**	$P = 2\ell + 2w$	**circle**	$C = 2\pi r$ or $C = \pi d$

Area

square	$A = s^2$	**triangle**	$A = \frac{1}{2}bh$
rectangle	$A = \ell w$ or $A = bh$	**regular polygon**	$A = \frac{1}{2}Pa$
parallelogram	$A = bh$	**circle**	$A = \pi r^2$
trapezoid	$A = \frac{1}{2}h(b_1 + b_2)$	**sector of a circle**	$A = \frac{x}{360} \cdot \pi r^2$
rhombus	$A = \frac{1}{2}d_1 d_2$ or $A = bh$		

Lateral Surface Area

prism	$L = Ph$	**pyramid**	$L = \frac{1}{2}P\ell$
cylinder	$L = 2\pi rh$	**cone**	$L = \pi r\ell$

Total Surface Area

prism	$S = Ph + 2B$	**cone**	$S = \pi r\ell + \pi r^2$
cylinder	$S = 2\pi rh + 2\pi r^2$	**sphere**	$S = 4\pi r^2$
pyramid	$S = \frac{1}{2}P\ell + B$		

Volume

cube	$V = s^3$	**pyramid**	$V = \frac{1}{3}Bh$
rectangular prism	$V = \ell wh$	**cone**	$V = \frac{1}{3}\pi r^2 h$
prism	$V = Bh$	**sphere**	$V = \frac{4}{3}\pi r^3$
cylinder	$V = \pi r^2 h$		

Equations for Figures on a Coordinate Plane

slope-intercept form of a line	$y = mx + b$	**circle**	$(x - h)^2 + (y - k)^2 = r^2$
point-slope form of a line	$y - y_1 = m(x - x_1)$		

Trigonometry

Law of Sines	$\frac{\sin A}{a} = \frac{\sin B}{b} = \frac{\sin C}{c}$	**Law of Cosines**	$a^2 = b^2 + c^2 - 2bc \cos A$ $b^2 = a^2 + c^2 - 2ac \cos B$ $c^2 = a^2 + b^2 - 2ab \cos C$
Pythagorean Theorem	$a^2 + b^2 = c^2$		

Symbols

$\neq$	is not equal to	$\parallel$	is parallel to	$\lvert\overrightarrow{AB}\rvert$	magnitude of the vector from *A* to *B*
$\approx$	is approximately equal to	$\nparallel$	is not parallel to	A'	the image of preimage *A*
$\cong$	is congruent to	$\perp$	is perpendicular to	$\rightarrow$	is mapped onto
$\sim$	is similar to	$\triangle$	triangle	$\odot A$	circle with center *A*
$\angle$, $\measuredangle$	angle, angles	$>, \geq$	is greater than, is greater than or equal to	π	pi
$m\angle A$	degree measure of $\angle A$	$<, \leq$	is less than, is less than or equal to	$\widehat{AB}$	minor arc with endpoints *A* and *B*
$^\circ$	degree	$\square$	parallelogram	$\widehat{ABC}$	major arc with endpoints *A* and *C*
$\overleftrightarrow{AB}$	line containing points *A* and *B*	*n*-gon	polygon with *n* sides	$m\widehat{AB}$	degree measure of arc *AB*
$\overline{AB}$	segment with endpoints *A* and *B*	$a:b$	ratio of *a* to *b*	$f(x)$	*f* of *x*, the value of *f* at *x*
$\overrightarrow{AB}$	ray with endpoint *A* containing *B*	(x, y)	ordered pair	!	factorial
AB	measure of $\overline{AB}$, distance between points *A* and *B*	(x, y, z)	ordered triple	${}_nP_r$	permutation of *n* objects taken *r* at a time
$\sim p$	negation of *p*, not *p*	$\sin x$	sine of *x*	${}_nC_r$	combination of *n* objects taken *r* at a time
$p \wedge q$	conjunction of *p* and *q*	$\cos x$	cosine of *x*	$P(A)$	probability of *A*
$p \vee q$	disjunction of *p* and *q*	$\tan x$	tangent of *x*	$P(A\mid B)$	the probability of *A* given that *B* has already occurred
$p \longrightarrow q$	conditional statement, if *p* then *q*	$\vec{a}$	vector *a*		
$p \longleftrightarrow q$	biconditional statement, *p* if and only if *q*	$\overrightarrow{AB}$	vector from *A* to *B*		

Measures

Metric	Customary
Length	
1 kilometer (km) = 1000 meters (m) 1 meter = 100 centimeters (cm) 1 centimeter = 10 millimeters (mm)	1 mile (mi) = 1760 yards (yd) 1 mile = 5280 feet (ft) 1 yard = 3 feet 1 yard = 36 inches (in.) 1 foot = 12 inches
Volume and Capacity	
1 liter (L) = 1000 milliliters (mL) 1 kiloliter (kL) = 1000 liters	1 gallon (gal) = 4 quarts (qt) 1 gallon = 128 fluid ounces (fl oz) 1 quart = 2 pints (pt) 1 pint = 2 cups (c) 1 cup = 8 fluid ounces
Weight and Mass	
1 kilogram (kg) = 1000 grams (g) 1 gram = 1000 milligrams (mg) 1 metric ton (t) = 1000 kilograms	1 ton (T) = 2000 pounds (lb) 1 pound = 16 ounces (oz)

Formulas

Coordinate Geometry

Midpoint	$M = \left(\frac{x_1 + x_2}{2}, \frac{y_1 + y_2}{2}\right)$	**Distance**	$d = \sqrt{(x_2 - x_1)^2 + (y_2 - y_1)^2}$
		Slope	$m = \frac{y_2 - y_1}{x_2 - x_1}, x_2 \neq x_1$

Matrices

Adding	$\begin{bmatrix} a & b \\ c & d \end{bmatrix} + \begin{bmatrix} e & f \\ g & h \end{bmatrix} = \begin{bmatrix} a+e & b+f \\ c+g & d+h \end{bmatrix}$	**Multiplying by a Scalar**	$k\begin{bmatrix} a & b \\ c & d \end{bmatrix} = \begin{bmatrix} ka & kb \\ kc & kd \end{bmatrix}$
Subtracting	$\begin{bmatrix} a & b \\ c & d \end{bmatrix} - \begin{bmatrix} e & f \\ g & h \end{bmatrix} = \begin{bmatrix} a-e & b-f \\ c-g & d-h \end{bmatrix}$	**Multiplying**	$\begin{bmatrix} a & b \\ c & d \end{bmatrix} \cdot \begin{bmatrix} e & f \\ g & h \end{bmatrix} = \begin{bmatrix} ab+bg & af-bh \\ ce+dg & cf-dh \end{bmatrix}$

Polynomials

Quadratic Formula	$x = \frac{-b \pm \sqrt{b^2 - 4ac}}{2a}, a \neq 0$	**Square of a Difference**	$(a - b)^2 = (a - b)(a - b)$ $= a^2 - 2ab + b^2$
Square of a Sum	$(a + b)^2 = (a + b)(a + b)$ $= a^2 + 2ab + b^2$	**Product of Sum and Difference**	$(a + b)(a - b) = (a - b)(a + b)$ $= a^2 - b^2$

Logarithms

Product Property	$\log_x ab = \log_x a + \log_x b$	**Power Property**	$\log_b m^p = p \log_b m$
Quotient Property	$\log_x \frac{a}{b} = \log_x a - \log_x b, b \neq 0$	**Change of Base**	$\log_a n = \frac{\log_b n}{\log_b a}$

Conic Sections

Parabola	$y = a(x - h)^2 + k$ or $x = a(y - k)^2 + h$	**Ellipse**	$\frac{x^2}{a^2} + \frac{y^2}{b^2} = 1$ or $\frac{y^2}{a^2} + \frac{x^2}{b^2} = 1, a, b \neq 0$
Circle	$x^2 + y^2 = r^2$ or $(x - h)^2 + (y - k)^2 = r^2$	**Hyperbola**	$\frac{x^2}{a^2} - \frac{y^2}{b^2} = 1$ or $\frac{y^2}{a^2} - \frac{x^2}{b^2} = 1, a, b \neq 0$

Sequences and Series

***n*th term, Arithmetic**	$a_n = a_1 + (n - 1)d$	***n*th term, Geometric**	$a_n = a_1 r^{n-1}$
Sum of Arithmetic Series	$S_n = n\left(\frac{a_1 + a_2}{2}\right)$ or $S_n = \frac{n}{2}[2a_1 + (n - 1)d]$	**Sum of Geometric Series**	$S_n = \frac{a_1 - a_1 r^n}{1 - r}$ or $S_n = \frac{a_1 - a_n r}{1 - r}, r \neq 1$

Trigonometry

Law of Sines	$\frac{\sin A}{a} = \frac{\sin B}{b} = \frac{\sin C}{c}, a, b, c \neq 0$		
Law of Cosines	$a^2 = b^2 + c^2 - 2bc \cos A$	$b^2 = a^2 + c^2 - 2ac \cos B$	$c^2 = a^2 + b^2 - 2ab \cos C$
Trigonometric Functions	$\sin \theta = \frac{\text{opp}}{\text{hyp}}$	$\cos \theta = \frac{\text{adj}}{\text{hyp}}$	$\tan \theta = \frac{\text{opp}}{\text{adj}} = \frac{\sin \theta}{\cos \theta}$
	$\csc \theta = \frac{\text{hyp}}{\text{opp}} = \frac{1}{\sin \theta}$	$\sec \theta = \frac{\text{hyp}}{\text{adj}} = \frac{1}{\cos \theta}$	$\cot \theta = \frac{\text{adj}}{\text{opp}} = \frac{\cos \theta}{\sin \theta}$
Pythagorean Identities	$\cos^2 \theta + \sin^2 \theta = 1$	$\tan^2 \theta + 1 = \sec^2 \theta$	$\cot^2 \theta + 1 = \csc^2 \theta$

Symbols

$f(x) = \{$	piecewise-defined function
$f(x) = \lvert x \rvert$	absolute value function
$f(x) = [\![x]\!]$	function of greatest integer not greater than a
$f(x, y)$	f of x and y, a function with two variables, x and y
$\overrightarrow{AB}$	vector AB
i	the imaginary unit
$[f \circ g](x)$	f of g of x, the composition of functions f and g
$f^{-1}(x)$	inverse of $f(x)$
$b^{\frac{1}{n}} = \sqrt[n]{b}$	nth root of b
$\log_b x$	logarithm base b of x
$\log x$	common logarithm of x
$\ln x$	natural logarithm of x

$\sum$	sigma, summation
$\bar{x}$	mean of a sample
μ	mean of a population
s	standard deviation of a sample
σ	standard deviation of a population
$P(B \mid A)$	the probability of B given that A has already occurred
nPr	permutation of n objects taken r at a time
nCr	combination of n objects taken r at a time
$\text{Sin}^{-1} x$	Arcsin x
$\text{Cos}^{-1} x$	Arccos x
$\text{Tan}^{-1} x$	Arctan x

Parent Functions

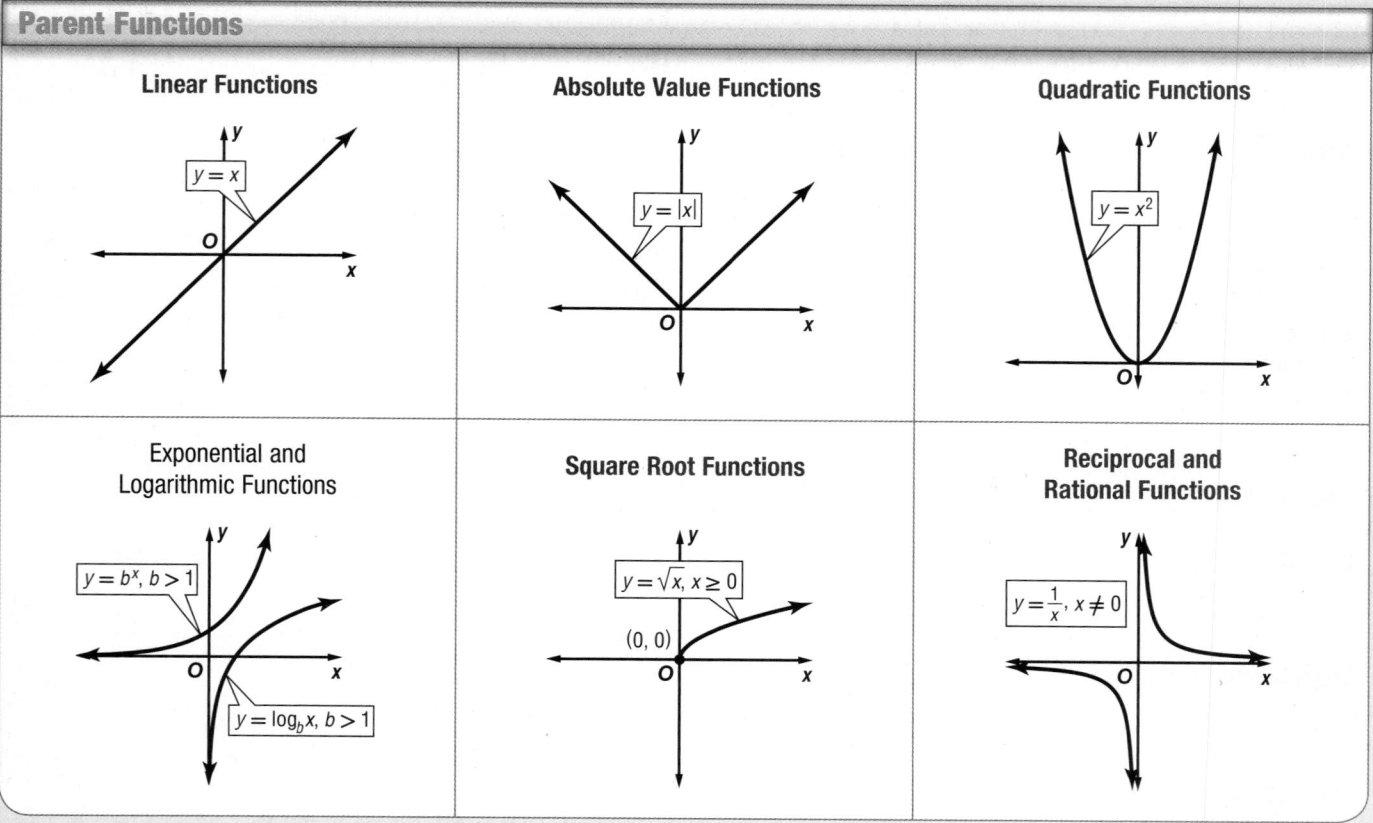

Linear Functions

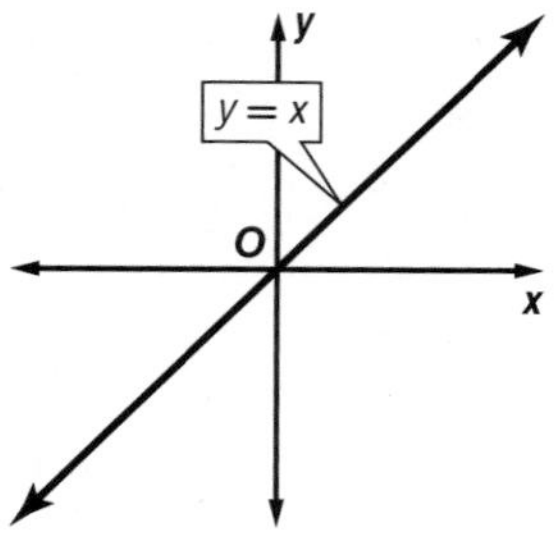

Absolute Value Functions

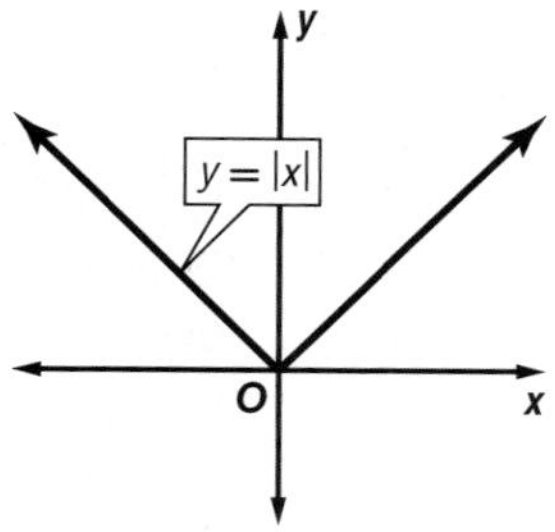

Quadratic Functions

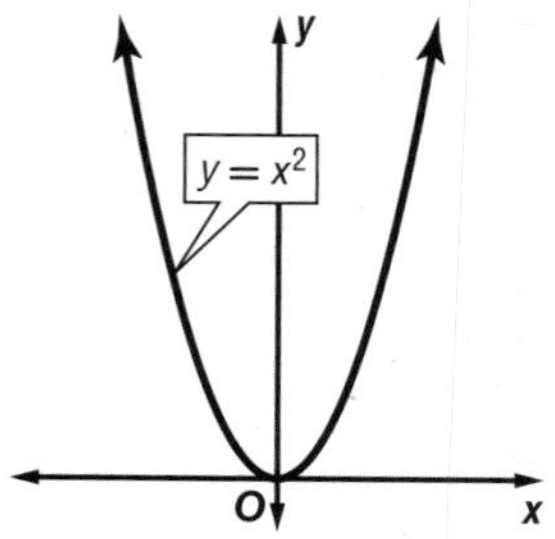

Exponential and Logarithmic Functions

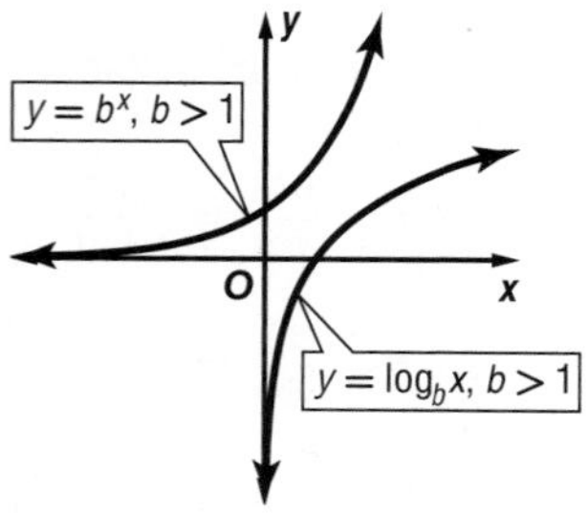

Square Root Functions

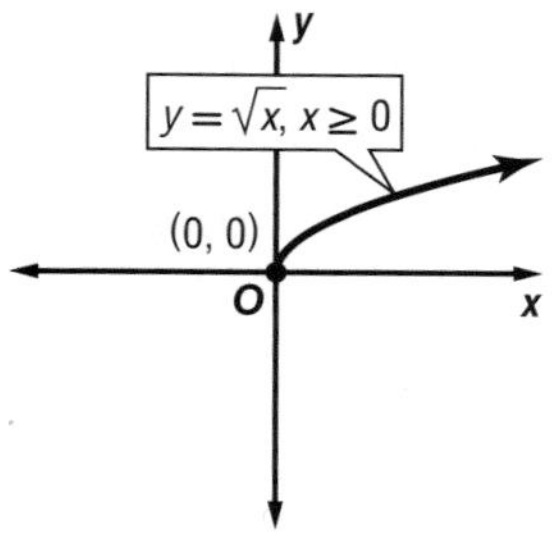

Reciprocal and Rational Functions

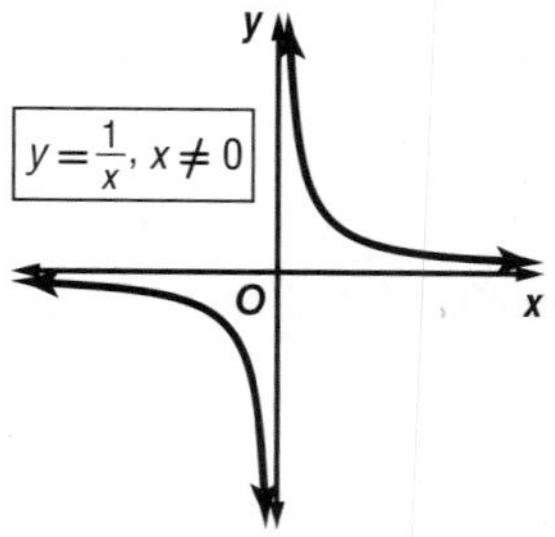